PERIMETER AND AREA FORMULAS

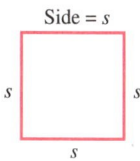
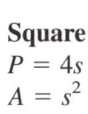

Square
$P = 4s$
$A = s^2$

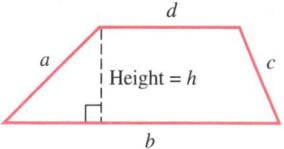

$r = a + b + c + d$
$A = \dfrac{1}{2}h(b + d)$

Rectangle
$P = 2l + 2w$
$A = lw$

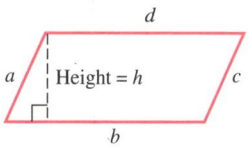

Parallelogram
$P = a + b + c + d$
$A = bh$

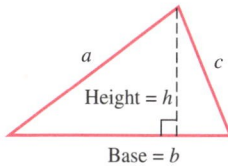

Triangle
$P = a + b + c$
$A = \dfrac{1}{2}bh$

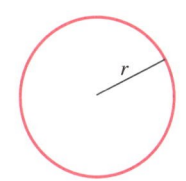

Circle
$C = 2\pi r \quad \text{or} \quad C = \pi D$
where $\pi = 3.14$
$A = \pi r^2$

VOLUME FORMULAS

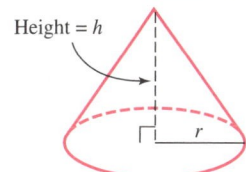

Cone
$V = \dfrac{1}{3}\pi r^2 h$

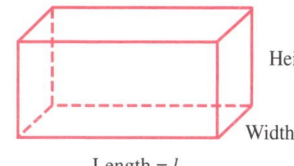

Rectangular solid
$V = lwh$

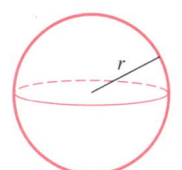

Sphere
$V = \dfrac{4}{3}\pi r^3$

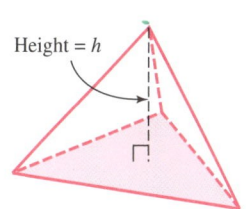

Pyramid
$V = \dfrac{1}{3}Bh*$

*B represents the area of the base.

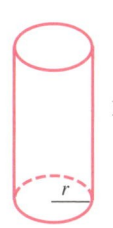

Height = h **Cylinder**
$V = \pi r^2 h$

WHAT DO YOU NEED TO LEARN *NOW*?

This powerful online learning companion helps you gauge your unique study needs and provides you with a *Personalized Learning Plan* so you can focus your study time where you need it most. Access to the program is included using the **1pass™** access code packaged with this text.

TOTALLY INTEGRATED WITH THIS TEXT!

As you read the text, watch for references like the one at right. They direct you to corresponding media-enhanced activities found in **Elementary & Intermediate AlgebraNow™**. This page-by-page integration lets you fully explore key course concepts through the interactive learning environment of **Elementary & Intermediate AlgebraNow**.

HOW DOES IT WORK?

Elementary & Intermediate AlgebraNow is made up of three powerful and easy-to-use components:

1) What Do I Know?

A diagnostic *Pre-Test* recognizes the key chapter concepts that you already know and highlights the areas that you need to study.

2) What Do I Need to Learn?

A *Personalized Learning Plan* outlines key elements for review and provides tutorial activities.

3) What Have I Learned?

A *Post-Test* assesses your mastery of core chapter concepts; results can be e-mailed to your instructor.

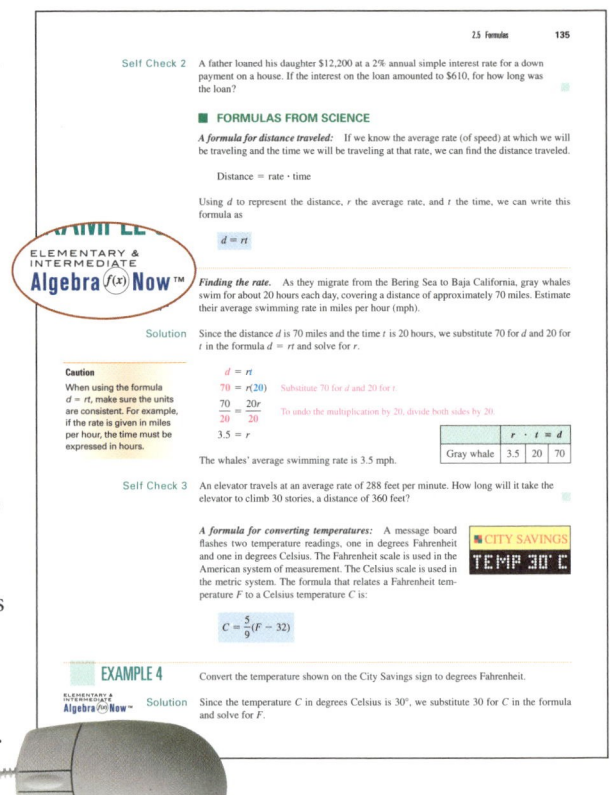

With a click of the mouse, the unique resources you'll access through **Elementary & Intermediate AlgebraNow** allow you to:

- ◆ Create a *Personalized Learning Plan* for each chapter that helps you focus on—and master—essential concepts
- ◆ Review for an exam by using the web quizzes and skillbuilder videos within the online tutorials
- ◆ Explore and reinforce mathematical concepts with **TLE Online Labs**
- ◆ Assess your understanding of core concepts by completing a *Post-Test* after you work through your *Personalized Learning Plan*

Study smarter—and make every minute count!

If an access card came with this book, you can start using **Elementary & Intermediate AlgebraNow** right away by following the directions on the card. If a card was not included, you may visit **http://www.thomsonedu.com** to purchase electronic access.

www.brookscole.com

www.brookscole.com is the World Wide Web site for Thomson Brooks/Cole and is your direct source to dozens of online resources.

At *www.brookscole.com* you can find out about supplements, demonstration software, and student resources. You can also send email to many of our authors and preview new publications and exciting new technologies.

www.brookscole.com
Changing the way the world learns®

Books in the Tussy and Gustafson Series

In hardcover:

Elementary Algebra, Third Edition
Intermediate Algebra, Third Edition
Elementary and Intermediate Algebra, Third Edition

In paperback:

Basic Mathematics for College Students, Second Edition
Basic Geometry for College Students
Prealgebra, Second Edition
Introductory Algebra, Second Edition
Intermediate Algebra, Second Edition
Developmental Mathematics

For more information, please visit www.brookscole.com

Edition

3

Elementary and Intermediate Algebra

Alan S. Tussy
Citrus College

R. David Gustafson
Rock Valley College

THOMSON

BROOKS/COLE

Australia • Canada • Mexico • Singapore • Spain
United Kingdom • United States

THOMSON

BROOKS/COLE

Executive Editor: Jennifer Huber
Executive Publisher: Curt Hinrichs
Development Editor: Kirsten Markson
Assistant Editor: Rebecca Subity
Editorial Assistant: Sarah Woicicki
Technology Project Manager: Rachael Sturgeon
Marketing Manager: Greta Kleinert
Marketing Assistant: Jessica Bothwell
Advertising Project Manager: Bryan Vann
Project Manager, Editorial Production: Hal Humphrey
Senior Art Director: Vernon T. Boes

Print/Media Buyer: Barbara Britton
Permissions Editor: Kiely Sisk
Production Service: Helen Walden
Text Designer: Kim Rokusek
Photo Researcher: Helen Walden
Illustrator: Lori Heckelman Illustration
Cover Designer: Patrick Devine
Cover Image: Kevin Tolman
Compositor: Graphic World, Inc.
Text and Cover Printer: CTPS

Printed in China
4 5 6 7 08 07 06

For more information about our products, contact us at:
Thomson Learning Academic Resource Center
1-800-423-0563
For permission to use material from this text or product,
submit a request online at http://www.thomson.com.
Any additional questions about permissions can be
submitted by e-mail to thomsonrights@thomson.com.

Library of Congress Control Number: 2004110437

Student Edition: ISBN 0-495-18874-3

Annotated Instructor's Edition: ISBN 0-534-41933-X

Thomson Higher Education
10 Davis Drive
Belmont, CA 94002
USA

Asia
Thomson Learning
5 Shenton Way #01-01
UIC Building
Singapore 068808

Australia/New Zealand
Thomson Learning
102 Dodds Street
Southbank, Victoria 3006
Australia

Canada
Nelson
1120 Birchmount Road
Toronto, Ontario M1K 5G4
Canada

Europe/Middle East/Africa
Thomson Learning
High Holborn House
50/51 Bedford Row
London WC1R 4LR
United Kingdom

Latin America
Thomson Learning
Seneca, 53
Colonia Polanco
11560 Mexico D.F.
Mexico

Spain/Portugal
Paraninfo
Calle Magallanes, 25
28015 Madrid, Spain

To three good friends

Jennifer,
Ellen,
and
Bob

Contents

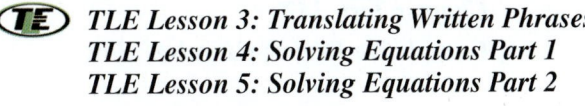

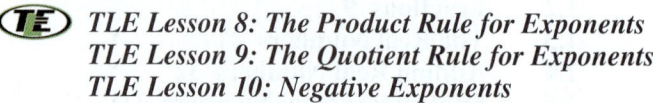
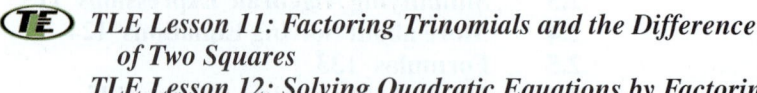

Chapter 6

Rational Expressions and Equations 418

Chapter 7

Solving Systems of Equations and Inequalities 506

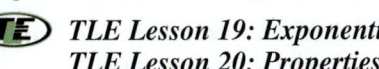

Preface

An increasing number of schools are offering the traditional elementary algebra and intermediate algebra courses in combination. There are several advantages in doing this:

- Much of the redundancy encountered by teaching the elementary/intermediate sequence as two separate courses is eliminated. As a result, the students have more time to master the material.
- A combined approach promotes a smooth transition from the elementary algebra topics to the intermediate algebra topics.
- For many students, the purchase of a single textbook saves money.

However, there are several concerns inherent in offering a combination course:

- The textbook used in such a course must include enough elementary algebra to ensure that students who complete the first half of the book, and then transfer, will have the prerequisite skills to enroll in an intermediate algebra course at another college.
- The elementary algebra material should not get too difficult too fast.
- Students entering the second half of the combination course must get some review of the basic topics so that they can compete with students continuing from the first half of the course.

Elementary and Intermediate Algebra has been written to address these concerns. The first seven chapter of this book provide a complete course in elementary algebra. The standard beginning algebra topics are introduced at a reasonable pace that allows the student to develop a strong conceptual foundation on which the second half of the course can build. Chapter 8 serves as the transitional chapter. It quickly reviews the topics taught in the first part of the course and extends those topics to the intermediate algebra level. Chapters 9–14 provide a complete course in intermediate algebra.

The purpose of this textbook is to teach students how to read, write, speak and think mathematically using the language of algebra. We have used a blend of the traditional and the reform instructional approaches to do this. In this book, you will find the vocabulary, practice, and well-defined pedagogy of a traditional approach. You will also find that we emphasize the reasoning, modeling, communicating, and technological skills that are such a big part of today's reform movement.

The third edition retains the basic philosophy of the second edition. However, we have made several improvements as a direct result of the comments and suggestions we received from instructors and students. Our goal has been to make the book more enjoyable to read, easier to understand, and more relevant.

■ NEW TO THIS MEDIA EDITION

- This new Media Edition is integrated with Elementary & Intermediate AlgebraNow™, a powerful online learning companion. Icons throughout the text indicate opportunities to reinforce concepts with online tutorials.

- New chapter openers reference the *TLE* computer lessons that accompany each chapter.
- The new Language of Algebra features, along with Success Tips, Notation, Calculator Boxes, and Cautions, are presented in the margins to promote understanding and increased clarity.
- Many additional applications involving real-life data have been added.
- Answers to the popular *Self Check* feature have been relocated to the end of each section, right before the *Study Set.*
- Several higher-level Challenge Problems have been added to each *Study Set.*
- The Accent on Teamwork feature has been redesigned to offer the instructor two or three collaborative activities per chapter that can be assigned as group work.
- More illustrations, diagrams, and color have been added for the visual learner.

■ REVISED TABLE OF CONTENTS

Chapter 1: *An Introduction to Algebra* Introductory concepts from Chapters 1 and 2 of the previous edition have been consolidated to create a less-repetitive, quicker-paced beginning for the course. More emphasis has been placed on simplifying and building fractions to ready students for Chapter 6, Rational Expressions and Equations. Addition, subtraction, multiplication, and division of signed numbers are now discussed in this chapter.

Chapter 2: *Equations, Inequalities, and Problem Solving* This chapter now contains two sections that are exclusively devoted to problem solving using the five-step problem-solving strategy. Circle graphs have been added to the study of percent.

Chapter 3: *Linear Equations and Inequalities in Two Variables* The nonlinear graphs previously in Section 3.2 were moved to Chapter 4. The slope section was rewritten so that slope of a line is introduced first, followed by a discussion of rates of change. Sections 3.5 and 3.6 have new titles: Slope–Intercept Form and Point–Slope Form. Graphing Linear Inequalities, previously a part of the Systems of Equations chapter, is now Section 3.7. The Introduction to functions has been moved to Chapter 8.

Chapter 4: *Exponents and Polynomials* Section 4.4, Polynomials, now includes graphs of nonlinear equations, such as $y = x^2$ and $y = x^3$. A separate section has been devoted to special products. Sections 4.7 and 4.8 from the second edition were combined so that division of polynomials is now covered in one section.

Chapter 5: *Factoring and Quadratic Equations* This chapter has been extensively rewritten. Added attention has been given to the grouping method (key number method) for factoring trinomials. The quadratic formula and graphing quadratic functions have been moved to Chapter 10.

Chapter 6: *Rational Expressions and Equations* Unit conversion is now introduced in Section 6.2. Addition and subtraction of rational expressions is now presented in two parts: In Section 6.3, the denominators are like, and in Section 6.4, the denominators are unlike. The concepts of unit price and best buy have been added to Section 6.8, Proportions and Similar Triangles.

Chapter 7: *Solving Systems of Equations and Inequalities* Many new real-life applications have been added.

Chapter 8: *Transition to Intermediate Algebra* This chapter has been extensively reorganized. It includes a thorough review of topics from the first part of the course.

Chapter 9: *Radical Expressions and Equations* The sections in this chapter have been reordered so that *Solving Radical Equations* follows the sections that discuss simplifying radical expressions. The topic of *Complex Numbers,* formerly in Chapter 10, is now the final section of Chapter 9.

Chapter 10: *Quadratic Equations, Functions, and Inequalities* The section, *The Discriminant and Equations That Can be Written in Quadratic Form,* was relocated so that it now follows Section 10.2, *The Quadratic Formula.* Section 10.5, *Quadratic and Other Nonlinear Inequalities,* was rewritten. It now includes a more detailed discussion of the interval testing method.

Chapter 11: *Exponential and Logarithmic Functions* Several sections have been edited to improve clarity. Section 11.8, *Exponential and Logarithmic Equations,* has been reorganized: Exponential equations whose sides can be written as a power of the same base appear first, followed by equations that can be solved by taking the logarithm of both sides.

Chapter 12: *More on Systems of Equations* In Section 12.3, *Solving Systems Using Matrices,* a more in-depth explanation of matrix solutions of systems of three equations is presented.

Chapter 13: *Conic Sections; More Graphing* More detailed drawings are included in Section 13.1, *The Circle and the Parabola,* where conic sections and their applications are introduced.

Chapter 14: *Miscellaneous Topics* In Section 14.3, *Geometric Sequences and Series,* more in-depth explanations of solution methods are presented.

■ ACKNOWLEDGMENTS

We are grateful to the following people who reviewed the manuscript at various stages of its development. They all had valuable suggestions that have been incorporated into the text.

The following people reviewed the first and second editions:

Julia Brown
Atlantic Community College

Lynn B. Cade
Pensacola Junior College

Tim Caldwell
Meridian Community College

John Coburn
Saint Louis Community College–Florissant Valley

Sally Copeland
Johnson County Community College

Ben Cornelius
Oregon Institute of Technology

James Edmondson
Santa Barbara Community College

Hector L. Esteban
Valencia Community College

Judith Jones
Valencia Community College

Therese Jones
Amarillo College

Mauricio Marroquin
Los Angeles Valley College

Marilyn Mays
North Lake College

Janice McFatter
Gulf Coast Community College

Elizabeth Morrison
Valencia Community College

Angelo Segalla
Orange Coast College

June Strohm
Pennsylvania State Community College–DuBois

Rita Sturgeon
San Bernardino Valley College

Jo Anne Temple
Texas Technical University

Sharon Testone
Onondaga Community College

Marilyn Treder
Rochester Community College

Betty Weissbecker
J. Sargeant Reynolds Community College

Cathleen Zucco
SUNY-New Paltz

The following people reviewed the third edition:

Mike Adams
Modesto Junior College

Ray Brinker
Western Illinois University

Cynthia J. Broughton
Arizona Western College

Don K. Brown
Macon State College

Light Bryant
Arizona Western College

Warren S. Butler
Daytona Beach Community College

John Scott Collins
Pima Community College

Lucy H. Edwards
Ohlone College

Hajrudin Fejzic
California State University, San Bernardino

Lee Gibbs
Arizona Western College

Barry T. Gibson
Daytona Beach Community College

Haile K. Haile
Minneapolis Community and Technical College

Suzanne Harris-Smith
Albuquerque Technical Vocational Institute

Kamal Hennayake
Chesapeake College

Doreen Kelly
Mesa Community College

Lynn Marecek
Santa Ana College

Michael Marzinske
Inver Hills Community College

Jamie McGill
East Tennessee State University

Margaret Michener
University of Nebraska, Kearney

Micheal Montano
Riverside Community College

Brian W. Moudry
Davis & Elkins College

William Peters
San Diego Mesa College

Bernard J. Pina
Dona Ana Branch Community College

Carol Purcell
Century Community College

Daniel Russow
Arizona Western College

Donald W. Solomon
University of Wisconsin, Milwaukee

John Thoo
Yuba College

Susan M. Twigg
Wor-Wic Community College

Gloria Upson
Winston-Salem State University

Gizelle Worley
California State University, Stanislaus

We want to express our gratitude to Karl Hunsicker, Cathy Gong, Dave Ryba, Terry Damron, Rob Everest, Marion Hammond, Lin Humphrey, Doug Keebaugh, Robin Carter, Tanja Rinkel, George Carlson, Jim Cope, Arnold Kondo, John McKeown, Kent Miller, Donna Neff, Eric Rabitoy, Chris Scott, Bill Tussy, Maryann Rachford, Bob Billups, Liz Tussy, Alexander Lee, Steve Odrich, and the Citrus College Library staff (including Barbara Rugeley) for their help with some of the application problems in the textbook.

Without the talents and dedication of the editorial, marketing, and production staff of Brooks/Cole, this revision of *Elementary and Intermediate Algebra* could not have been so well accomplished. We express our sincere appreciation for the hard work of Bob Pirtle, Jennifer Huber, Helen Walden, Lori Heckleman, Vernon Boes, Kim Rokusek, Sarah Woicicki, Greta Kleinert, Jessica Bothwell, Bryan Vann, Kirsten Markson, Rebecca Subity, Hal Humphrey, Tammy Fisher-Vasta, Christine Davis, Ellen Brownstein, Diane Koenig, and Graphic World for their help in creating the book.

Alan S. Tussy
R. David Gustafson

For the Student

■ SUCCESS IN ALGEBRA

To be successful in mathematics, you need to know how to study it. The following check-list will help you develop your own personal strategy to study and learn the material. The suggestions below require some time and self-discipline on your part, but it will be worth the effort. This will help you get the most out of the course.

As you read each of the following statements, place a check mark in the box if you can truthfully answer Yes. If you can't answer Yes, think of what you might do to make the suggestion part of your personal study plan. You should go over this checklist several times during the semester to be sure you are following it.

Preparing for the Class
❑ I have made a commitment to myself to give this course my best effort.
❑ I have the proper materials: a pencil with an eraser, paper, a notebook, a ruler, a calculator, and a calendar or day planner.
❑ I am willing to spend a minimum of two hours doing homework for every hour of class.
❑ I will try to work on this subject every day.
❑ I have a copy of the class syllabus. I understand the requirements of the course and how I will be graded.
❑ I have scheduled a free hour after the class to give me time to review my notes and begin the homework assignment.

Class Participation
❑ I know my instructor's name.
❑ I will regularly attend the class sessions and be on time.
❑ When I am absent, I will find out what the class studied, get a copy of any notes or handouts, and make up the work that was assigned when I was gone.
❑ I will sit where I can hear the instructor and see the board.
❑ I will pay attention in class and take careful notes.
❑ I will ask the instructor questions when I don't understand the material.
❑ When tests, quizzes, or homework papers are passed back and discussed in class, I will write down the correct solutions for the problems I missed so that I can learn from my mistakes.

Study Sessions
❑ I will find a comfortable and quiet place to study.
❑ I realize that reading a math book is different from reading a newspaper or a novel. Quite often, it will take more than one reading to understand the material.
❑ After studying an example in the textbook, I will work the accompanying Self Check.
❑ I will begin the homework assignment only after reading the assigned section.
❑ I will try to use the mathematical vocabulary mentioned in the book and used by my instructor when I am writing or talking about the topics studied in this course.
❑ I will look for opportunities to explain the material to others.
❑ I will check all my answers to the problems with those provided in the back of the book (or with the *Student Solutions Manual*) and resolve any differences.
❑ My homework will be organized and neat. My solutions will show all the necessary steps.
❑ I will work some review problems every day.

❑ After completing the homework assignment, I will read the next section to prepare for the coming class session.

❑ I will keep a notebook containing my class notes, homework papers, quizzes, tests, and any handouts—all in order by date.

Special Help

❑ I know my instructor's office hours and am willing to go in to ask for help.

❑ I have formed a study group with classmates that meets regularly to discuss the material and work on problems.

❑ When I need additional explanation of a topic, I use the tutorial videos and the interactive CD, as well as the website.

❑ I make use of extra tutorial assistance that my school offers for mathematics courses.

❑ I have purchased the *Students Solutions Manual* that accompanies this text, and I use it.

To follow each of these suggestions will take time. It takes a lot of practice to learn mathematics, just as with any other skill.

No doubt, you will sometimes become frustrated along the way. This is natural. When it occurs, take a break and come back to the material after you have had time to clear your thoughts. Keep in mind that the skills and discipline you learn in this course will help make for a brighter future. Good luck!

Elementary & Intermediate AlgebraNow Can Help You Succeed in Math

Elementary & Intermediate AlgebraNow is a powerful online learning companion that helps you gauge your unique study needs and provides you with a personalized learning plan. The resources in Elementary & Intermediate AlgebraNow give you all the learning tools you need to master core concepts. Elementary & Intermediate AlgebraNow and this new edition of Tussy and Gustafson's *Elementary and Intermediate Algebra* enhance each other, providing you with a seamless learning system. Completely integrated with the textbook, icons in the text direct you to the online tutorials and TLE Labs found in Elementary & Intermediate AlgebraNow.

The Elementary & Intermediate AlgebraNow system consists of three powerful, easy-to-use assessment components:

1. **What Do I Know?**
   Take a diagnostic pre-test to find out.
2. **What Do I Need to Learn?**
   Your personalized learning plan helps you focus on the areas you need to study.
3. **What Have I Learned?**
   Take a post-test to see how your understanding of key concepts has improved; results can be e-mailed to your instructor.

■ WHAT YOU'LL FIND IN YOUR PERSONALIZED LEARNING PLAN

Text-specific tutorials, live online tutoring, and TLE online labs provide you with multiple tools to help you get a better grade.

Text-Specific Tutorials

The view of a tutorial looks like this. To navigate between chapters and sections, use the drop-down menu below the top navigation bar. This will give you access to the study activities available for each section.

Math Toolbar

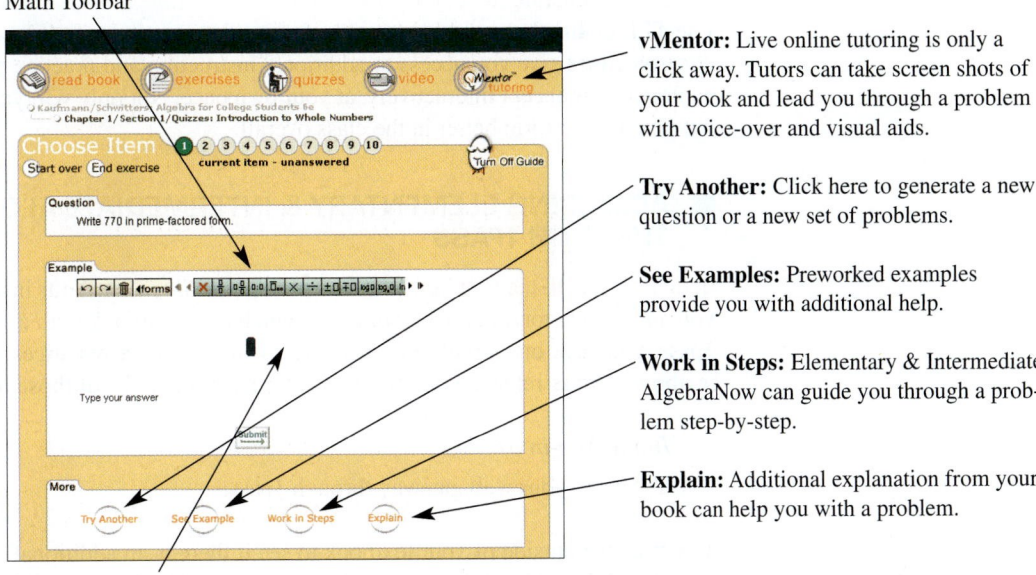

vMentor: Live online tutoring is only a click away. Tutors can take screen shots of your book and lead you through a problem with voice-over and visual aids.

Try Another: Click here to generate a new question or a new set of problems.

See Examples: Preworked examples provide you with additional help.

Work in Steps: Elementary & Intermediate AlgebraNow can guide you through a problem step-by-step.

Explain: Additional explanation from your book can help you with a problem.

Type your answer here.

Online Tutoring with vMentor

Access live online tutors and support through vMentor™. To access vMentor while you are working in the Exercises or "Tutorial" areas in iLrn, click on the **vMentor Tutoring** button at the top right of the navigation bar above the problem or exercise.

Next, click on the **vMentor** button; you will be taken to a Web page that lists the steps for entering a vMentor classroom. If you are a first-time user of vMentor, you might need to download Java software before entering the class for the first class. You can either take an Orientation Session or log in to a vClass from the links at the bottom of the opening screen.

All vMentor Tutoring is done through a vClass, an Internet-based virtual classroom that features two-way audio, a shared whiteboard, chat, messaging, and experienced tutors. You can access vMentor during the following times:

Monday through Thursday:
5 p.m. to 12 a.m. Pacific Time
6 p.m. to 1 a.m. Mountain Time
7 p.m. to 2 a.m. Central Time
8 p.m. to 3 a.m. Eastern Time

Sunday:
12 p.m. to 12 a.m. Pacific Time
1 p.m. to 1 a.m. Mountain Time
2 p.m. to 2 a.m. Central Time
3 p.m. to 3 a.m. Eastern Time

If you need additional help using vMentor, you can access the Participant Guide at this Website: **http://www.elluminate.com/support/guide/pdf.**

TLE Online Labs

Use TLE Online Labs to explore and reinforce key concepts introduced in this text. These electronic labs give you access to additional instruction and practice problems, so you can explore each concept interactively, at your own pace. Not only will you be better prepared, but you will perform better in the class overall.

■ ACCESSING ELEMENTARY & INTERMEDIATE ALGEBRANOW THROUGH 1PASS

Registering with the PIN Code on the 1pass Card *Situation:* Your instructor has not given you a PIN code for an online course, but you have a textbook with a 1pass PIN code. With 1pass, you have one simple PIN code access to all media resources associated with your textbook. Please refer to your 1pass card for a complete list of those resources.

Initial Log-in

To access your web gateway through 1pass:

1. Check the outside of your textbook to see if there is an additional 1pass card.
2. Take this card (and the additional 1pass card if appropriate) and go to http://1pass.thomson.com
3. Type in your 1pass acess code (or codes).
4. Follow the directions on the screen to set up your personal username and password.
5. Click through to launch your personal portal.
6. Access the media resources associated with your text . . . all the resources are just one click away.
7. Record your username and password for future visits and be sure to use the same username for all Thomson Learning resources.

For tech support, contact us at 1 (800) 423-0563.

> You will be asked to enter a valid e-mail address and password. Save your password in a safe place. You will need them to log in the next time you use 1pass. Only your e-mail address and password will allow you to reenter 1pass.

Subsequent Log-in

1. Go to **http://1pass.thomson.com.**
2. Type your e-mail address and password (see boxed information above) in the "Existing Users" box; then click on **Login.**

Applications Index

Examples that are applications are shown with boldface page numbers.
Exercises that are applications are shown with lightface page numbers.

1

An Introduction to Algebra

ELEMENTARY &
INTERMEDIATE
Algebra $f(x)$ **Now**™

Throughout the chapter, this icon introduces resources on the Elementary & Intermediate AlgebraNow Web site, accessed through **http://1pass. thomson.com**, that will

• Help you test your knowledge of the material with a pre-test and a post-test

• Provide a personalized learning plan targeting areas you should study

©Chuck Savage/CORBIS

Most banks are beehives of activity. Tellers and bank officials help customers make deposits and withdrawals, arrange loans, and set up credit card accounts. To describe these financial transactions, positive and negative numbers, fractions, and decimals are used. These numbers belong to a set that we call the *real numbers*.

To learn more about real numbers and how they are used in banking, visit *The Learning Equation* on the Internet at http://tle.brookscole.com. (The log-in instructions are in the Preface.) For Chapter 1, the online lessons are:

• *TLE* Lesson 1: The Real Numbers
• *TLE* Lesson 2: Order of Operations

Algebra is a mathematical language that can be used to solve many types of problems.

1.1 Introducing the Language of Algebra

- Tables and Graphs   • Vocabulary   • Notation
- Variables, Expressions, and Equations   • Constructing Tables

Algebra is the result of contributions from many cultures over thousands of years. The word *algebra* comes from the title of the book *Ihm Al-jabr wa'l muqābalah,* written by the Arabian mathematician al-Khwarizmi around A.D. 800. Using the vocabulary and notation of algebra, we can mathematically **model** many situations in the real world. In this section, we begin to explore the language of algebra by introducing some of its basic components.

■ TABLES AND GRAPHS

In algebra, we use tables to show relationships between quantities. For example, the following table lists the number of bicycle tires a production planner must order when a given number of bicycles are to be manufactured. For a production run of, say, 300 bikes, we locate 300 in the left-hand column and then scan across the table to see that the company must order 600 tires.

Bicycles to be manufactured	Tires to order
100	200
200	400
300	600
400	800

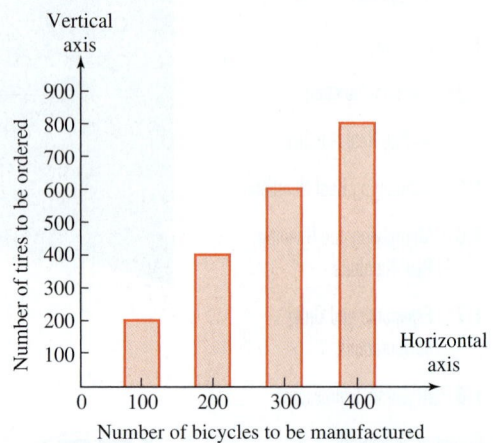

The Language of Algebra

Horizontal is a form of the word *horizon.* Think of the sun setting over the *horizon. Vertical* means in an upright position, like *vertical* blinds in a window.

This information can also be presented using a graph. The **bar graph** (shown above) has a **horizontal axis** labeled "Number of bicycles to be manufactured." The labels are in units of 100 bicycles. The **vertical axis,** labeled "Number of tires to be ordered," is scaled in units of 100 tires. The height of a bar indicates the number of tires to order. For example, if 200 bikes are to be manufactured, we see that the bar extends to 400, meaning 400 tires are needed.

Another way to present this information is with a **line graph.** Instead of using a bar to denote the number of tires to order, we use a dot drawn at the correct height. After drawing the data points for 100, 200, 300, and 400 bicycles, we connect them with line segments to create the following graph, on the right.

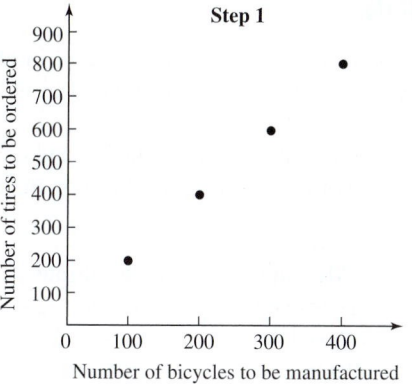

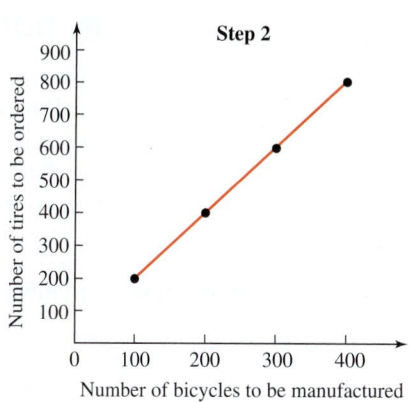

EXAMPLE 1

ELEMENTARY &
INTERMEDIATE
Algebra $f(x)$ **Now** ™

Use the following line graph to find the number of tires needed when 250 bicycles are to be manufactured.

Solution We locate 250 between 200 and 300 on the horizontal axis and draw a dashed line upward to intersect the graph. From the point of intersection, we draw a dashed horizontal line to the left that intersects the vertical axis at 500. This means that 500 tires should be ordered if 250 bicycles are to be manufactured.

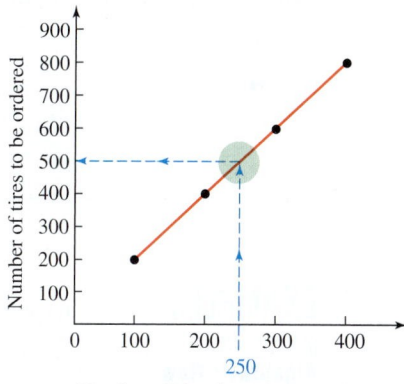

Success Tip

The video icons (see above) show which examples are taught on tutorial video tapes or disks.

Self Check 1 Use the graph to find the number of tires needed if 350 bicycles are to be manufactured.

■ VOCABULARY

From the table and graphs, it is clear that there is a relationship between the number of tires to order and the number of bicycles to be manufactured. Using words, we can express this relationship as a **verbal model:**

"The number of tires to order is two times the number of bicycles to be manufactured."

Since the word **product** indicates the result of a multiplication, we can write:

"The number of tires to order is the *product* of 2 and the number of bicycles to be manufactured."

To indicate other arithmetic operations, we will use the following words.

- A **sum** is the result of an addition: the sum of 5 and 6 is 11.
- A **difference** is the result of a subtraction: the difference of 3 and 2 is 1.
- A **quotient** is the result of a division: the quotient of 6 and 3 is 2.

■ NOTATION

Many symbols used in arithmetic are also used in algebra. For example, a + symbol is used to indicate addition, a − symbol is used to indicate subtraction, and an = symbol means *is equal to.*

Since the letter x is often used in algebra and could be confused with the multiplication symbol ×, we normally write multiplication in other ways.

Symbols for Multiplication	Symbol	Meaning	Example
	×	Times sign	$6 \times 4 = 24$
	·	Raised dot	$6 \cdot 4 = 24$
	( )	Parentheses	$(6)4 = 24$ or $6(4) = 24$ or $(6)(4) = 24$

In algebra, the symbol most often used to indicate division is the fraction bar.

Symbols for Division	Symbol	Meaning	Example
	÷	Division sign	$24 \div 4 = 6$
	$\overline{)}$	Long division	$4\overline{)24}$ with 6
	—	Fraction bar	$\dfrac{24}{4} = 6$

EXAMPLE 2

ELEMENTARY & INTERMEDIATE
Algebra $f(x)$ **Now**™

Express each statement in words, using one of the words *sum, product, difference,* or *quotient:*   **a.** $\frac{22}{11} = 2$ and   **b.** $22 + 11 = 33$.

Solution   **a.** We can represent the equal symbol = with the word *is.* Since the fraction bar indicates division, we have: the quotient of 22 and 11 is 2.

b. The + symbol indicates addition: the sum of 22 and 11 is 33.

Self Check 2   Express the following statement in words: $22 - 11 = 11$

■ VARIABLES, EXPRESSIONS, AND EQUATIONS

Another way to describe the tires–to–bicycles relationship uses *variables.* **Variables** are letters that stand for numbers. If we let the letter t stand for the number of tires to be ordered and b for the number of bicycles to be manufactured, we can translate the *verbal model* to mathematical symbols.

The Language of Algebra

The equal symbol = can be represented by words such as:

is   gives   yields   equals

The symbol ≠ is read as *"is not equal to."*

The number of tires to order	is	two	times	the number of bicycles to be manufactured.
t	$=$	2	$\cdot$	b

The statement $t = 2 \cdot b$ is called an *equation.* An **equation** is a mathematical sentence that contains an = symbol. Some examples are

$$3 + 5 = 8 \qquad x + 5 = 20 \qquad 17 - t = 14 - t \qquad p = 100 - d$$

When we multiply a variable by a number or multiply a variable by another variable, we can omit the symbol for multiplication.

$2b$ means $2 \cdot b$ xy means $x \cdot y$ abc means $a \cdot b \cdot c$

Using this form, we can write the equation $t = 2 \cdot b$ as $t = 2b$. The notation $2b$ on the right-hand side is called an **algebraic expression,** or more simply, an **expression.**

Algebraic Expressions	Variables and/or numbers can be combined with the operations of addition, subtraction, multiplication, and division to create **algebraic expressions.**

Here are some examples of algebraic expressions.

$4a + 7$ This expression is a combination of the numbers 4 and 7, the variable a, and the operations of multiplication and addition.

$\dfrac{10 - y}{3}$ This expression is a combination of the numbers 10 and 3, the variable y, and the operations of subtraction and division.

$15mn(2m)$ This expression is a combination of the numbers 15 and 2, the variables m and n, and the operation of multiplication.

In the bicycle manufacturing example, using the equation $t = 2b$ to describe the relationship has one major advantage over the other methods. It can be used to determine the number of tires needed for a production run of any size.

EXAMPLE 3

ELEMENTARY & INTERMEDIATE
Algebra *f(x)* Now™

Use the equation $t = 2b$ to find the number of tires needed for a production run of 178 bicycles.

Solution $t = 2\boldsymbol{b}$ This is the describing equation.

$t = 2(\mathbf{178})$ Replace b, which stands for the number of bicycles, with 178. Use parentheses to show the multiplication.

$t = 356$ Do the multiplication.

If 178 bicycles are manufactured, 356 tires will be needed.

Self Check 3   Use the equation $t = 2b$ to find the number of tires needed if 604 bicycles are to be manufactured.

■ CONSTRUCTING TABLES

Equations such as $t = 2b$, which express a relationship between two or more variables, are called **formulas.** Some applications require the repeated use of a formula.

EXAMPLE 4

ELEMENTARY & INTERMEDIATE
Algebra *f(x)* Now™

Find the number of tires to order for production runs of 233 and 852 bicycles. Present the results in a table.

Solution   **Step 1:** We construct a two-column table. Since *b* represents the number of bicycles to be manufactured, we use it as the heading of the first column. Since *t* represents the number of tires needed, we use it as the heading of the second column. Then we enter the size of each production run in the first column, as shown.

Bicycles to be manufactured *b*	Tires needed *t*
233	466
852	1,704

Step 2: In $t = 2b$, we replace *b* with 233 and with 852 and find each corresponding value of *t*.

$$t = 2\mathbf{b}$$
$$t = 2(\mathbf{233})$$
$$t = 466$$

$$t = 2\mathbf{b}$$
$$t = 2(\mathbf{852})$$
$$t = 1{,}704$$

Step 3: These results are entered in the second column.

Self Check 4   Find the number of tires needed for production runs of 87 and 487 bicycles. Present the results in a table.

Answers to Self Checks   **1.** 700   **2.** The difference of 22 and 11 is 11.   **3.** 1,208   **4.**

b	*t*
87	174
487	974

1.1   STUDY SET   ELEMENTARY & INTERMEDIATE Algebra *f(x)* Now™

VOCABULARY   Fill in the blanks.

1. The answer to an addition problem is called the _____. The answer to a subtraction problem is called the _____.

2. The answer to a multiplication problem is called the _____. The answer to a division problem is called the _____.

3. _____ are letters that stand for numbers.

4. Variables and numbers can be combined with the operations of addition, subtraction, multiplication, and division to create algebraic _____.

5. An _____ is a mathematical sentence that contains an = symbol.

6. An equation such as $t = 2b$, which expresses a relationship between two or more variables, is called a _____.

7. The _____ axis of a graph extends left and right and the _____ axis extends up and down.

8. The word _____ comes from the title of a book written by an Arabian mathematician around A.D. 800.

CONCEPTS   Classify each item as an algebraic expression or an equation.

9. $18 + m = 23$

10. $18 + m$

11. $y - 1$

12. $y - 1 = 2$

13. $30x$

14. $t = 16b$

15. $r = \dfrac{2}{3}$

16. $\dfrac{c - 7}{5}$

17. a. What operations does the expression $5x - 16$ contain?

   b. What variable does the expression contain?

18. a. What operations does the expression $\dfrac{12 + t}{25}$ contain?

  b. What variable does the expression contain?

19. a. What operations does the equation $4 + 1 = 20 - m$ contain?

  b. What variable does the equation contain?

20. a. What operations does the equation $y + 14 = 5(6)$ contain?

  b. What variable does the equation contain?

21. Construct a line graph using the data in the table.

Hours worked	Pay (dollars)
1	20
2	40
3	60
4	80
5	100

22. Use the data in the graph to complete the table.

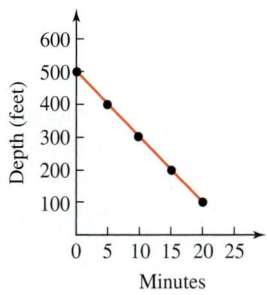

Minutes	Depth (feet)
0	
5	
10	
15	
20	

23. Explain what the dashed lines help us find in the graph.

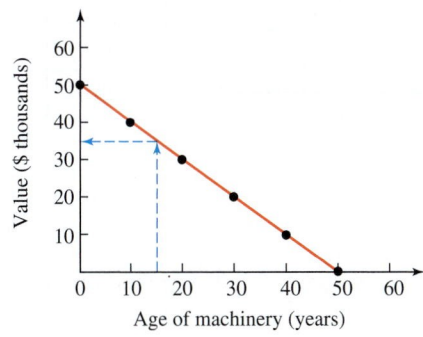

24. Use the line graph to find the income received from 30, 50, and 70 customers.

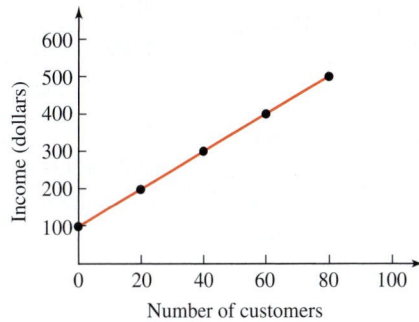

NOTATION Write each multiplication using a raised dot and then parentheses.

25. 5×6 **26.** 4×7

27. 34×75 **28.** 90×12

Write each expression without using a multiplication symbol or parentheses.

29. $4 \cdot x$ **30.** $5 \cdot y$

31. $3 \cdot r \cdot t$ **32.** $22 \cdot q \cdot s$

33. $l \cdot w$ **34.** $b \cdot h$

35. $P \cdot r \cdot t$ **36.** $l \cdot w \cdot h$

37. $2(w)$ **38.** $2(l)$

39. $(x)(y)$ **40.** $(r)(t)$

Write each division using a fraction bar.

41. $32 \div x$ **42.** $y \div 15$

43. $30\overline{)90}$ **44.** $20\overline{)80}$

PRACTICE Express each statement using one of the words *sum, difference, product,* or *quotient.*

45. $8(2) = 16$ **46.** $45 \cdot 12 = 540$

47. $11 - 9 = 2$ **48.** $65 + 89 = 154$

49. $2x = 10$ **50.** $16t = 4$

51. $\dfrac{66}{11} = 6$ **52.** $12 \div 3 = 4$

Translate each verbal model into an equation. (Answers may vary, depending on the variables chosen.)

53.

| The sale price | is | $100 | minus | the discount. |

54.

| The cost of dining out | equals | the cost of the meal | plus | $7 for parking. |

55.

| 7 | times | the age of a dog in years | gives | the dog's equivalent human age. |

56.

| The number of centuries | is | the number of years | divided by | 100. |

57. The amount of sand that should be used is the product of 3 and the amount of cement used.

58. The number of waiters needed is the quotient of the number of customers and 10.

59. The weight of the truck is the sum of the weight of the engine and 1,200.

60. The number of classes still open is the difference of 150 and the number of classes that are closed.

61. The profit is the difference of the revenue and 600.

62. The distance is the product of the rate and 3.

63. The quotient of the number of laps run and 4 is the number of miles run.

64. The sum of the tax and 35 is the total cost.

Use the formula to complete each table.

65. $d = 360 + L$

Lunch time (minutes) L	School day (minutes) d
30	
40	
45	

66. $b = 1,024k$

Kilobytes k	Bytes b
1	
5	
10	

67. $t = 1,500 - d$

Deductions d	Take-home pay t
200	
300	
400	

68. $w = \dfrac{s}{12}$

Inches of snow s	Inches of water w
12	
24	
72	

Use the data to find a formula that describes the relationship between the two quantities.

69.

Eggs e	Dozens d
24	2
36	3
48	4

70.

Couples c	Individuals I
20	40
100	200
200	400

APPLICATIONS

71. TRAFFIC SAFETY   As the railroad crossing guard drops, the measure of angle 1 ($\angle 1$) increases while the measure of $\angle 2$ decreases. At any instant the *sum* of the measures of the two angles is 90 degrees (denoted 90°). Complete the table. Then use the data to construct a line graph.

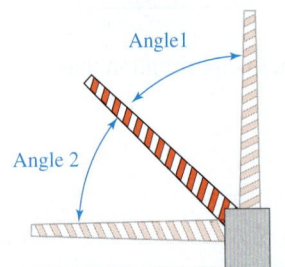

Angle 1 (degrees)	Angle 2 (degrees)
0	
30	
45	
60	
90	

72. U.S. CRIME STATISTICS   Property crimes include burglary, theft, and motor vehicle theft. Graph the following property crime rate data using a bar graph. Is an overall trend apparent?

Year	Crimes per 1,000 households
1991	354
1992	325
1993	319
1994	310
1995	291
1996	266

Year	Crimes per 1,000 households
1997	248
1998	217
1999	198
2000	178
2001	167

Source: Bureau of Justice Statistics

WRITING

73. Many people misuse the word *equation* when discussing mathematics. What is an equation? Give an example.

74. Explain the difference between an algebraic expression and an equation. Give an example of each.

75. In this section, four methods for describing numerical relationships were discussed: tables, words, graphs, and equations. Which method do you think is the most useful? Explain why.

76. In your own words, define *horizontal* and *vertical*.

CHALLENGE PROBLEMS

77. Complete the table and the formula.

s	t
10	
18	19
	34
47	48

$t = \boxed{}$

78. Suppose $h = 4n$ and $n = 2g$. Complete the following formula: $h = \boxed{}\, g$.

1.2   Fractions

- Factors and Prime Factorizations   • The Meaning of Fractions
- Multiplying and Dividing Fractions   • Building Equivalent Fractions
- Simplifying Fractions   • Adding and Subtracting Fractions
- Simplifying Answers   • Mixed Numbers

In arithmetic, you added, subtracted, multiplied, and divided **whole numbers:** 0, 1, 2, 3, 4, 5, and so on. Assuming that you have mastered those skills, we will now review the arithmetic of fractions.

■ FACTORS AND PRIME FACTORIZATIONS

The Language of Algebra

When we say "factor 8," we are using the word *factor* as a verb. When we say "2 is a *factor* of 8," we are using the word factor as a noun.

To compute with fractions, we need to know how to *factor* whole numbers. To **factor** a number means to express it as a product of two or more numbers. For example, some ways to factor 8 are

$$1 \cdot 8, \qquad 4 \cdot 2, \qquad \text{and} \qquad 2 \cdot 2 \cdot 2$$

The numbers 1, 2, 4, and 8 that were used to write the products are called **factors** of 8.

Sometimes a number has only two factors, itself and 1. We call these numbers *prime numbers.*

Prime Numbers and Composite Numbers	A **prime number** is a whole number greater than 1 that has only itself and 1 as factors. The first ten prime numbers are 2, 3, 5, 7, 11, 13, 17, 19, 23, and 29.
	A **composite number** is a whole number, greater than 1, that is not prime. The first ten composite numbers are 4, 6, 8, 9, 10, 12, 14, 15, 16, and 18.

Every composite number can be factored into the product of two or more prime numbers. This product of these prime numbers is called its **prime factorization.**

EXAMPLE 1

Find the prime factorization of 210.

Solution

First, write 210 as the product of two whole numbers other than 1.

$$210 = \mathbf{10 \cdot 21}$$ The resulting prime factorization will be the same no matter which two factors of 210 you begin with.

Neither 10 nor 21 are prime numbers, so we factor each of them.

$$210 = \mathbf{2 \cdot 5 \cdot 3 \cdot 7}$$ Factor 10 as 2 · 5 and factor 21 as 3 · 7.

Writing the factors in ascending order, the **prime-factored form** of 210 is 2 · 3 · 5 · 7. Two other methods for prime factoring 210 are shown as follows.

The Language of Algebra

In Example 1, the prime factors are written in *ascending* order. To *ascend* means to move upward.

Success Tip

When finding a prime factorization, remember that a whole number is divisible by

- 2 if it ends in 0, 2, 4, 6, or 8
- 3 if the sum of the digits is divisible by 3
- 5 if it ends in 0 or 5
- 10 if it ends in 0

Factor tree   *Division ladder*

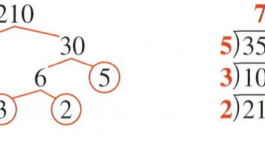

Work downward. Factor each number as a product of two numbers other than 1 and itself until all factors are prime.

Work upward. Perform repeated division by prime numbers until the final quotient is a prime number.

Either way, the factorization is 2 · 3 · 5 · 7.

Self Check 1   Find the prime factorization of 189.

■ THE MEANING OF FRACTIONS

In a fraction, the number above the **fraction bar** is called the **numerator,** and the number below is called the **denominator.**

$$\text{fraction bar} \longrightarrow \frac{1}{2} \begin{array}{l} \longleftarrow \text{numerator} \\ \longleftarrow \text{denominator} \end{array}$$

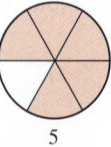

$$\frac{5}{6}$$

Fractions can describe the number of equal parts of a whole. To illustrate, we consider the circle with 5 of its 6 equal parts colored red. We say that $\frac{5}{6}$ (five-sixths) of the circle is shaded.

Fractions are also used to indicate division. For example, $\frac{8}{2}$ indicates that the numerator, 8, is to be divided by the denominator, 2:

$$\frac{8}{2} = 8 \div 2 = 4$$ We know that $\frac{8}{2} = 4$ because of its related multiplication statement: $4 \cdot 2 = 8$.

A numerator can be any number, including 0. If it is 0 and the denominator is not, the fraction is equal to 0. For example, $\frac{0}{6} = 0$ because $0 \cdot 6 = 0$.

A denominator can be any number except 0. To see why, consider $\frac{6}{0}$. If there were an answer to the division, then $\frac{6}{0} =$ answer. The related multiplication statement (answer $\cdot \, 0 = 6$) is impossible because no number, when multiplied by 0, gives 6. Fractions such as $\frac{6}{0}$, that indicate division of a nonzero number by 0, are undefined.

If the numerator and denominator of a fraction are the same nonzero number, the fraction indicates division of a number by itself, and the result is 1. For example, $\frac{9}{9} = 1$.

If a denominator is 1, the fraction indicates division by 1, and the result is simply the numerator. For example, $\frac{5}{1} = 5$.

The facts about special fraction forms are summarized as follows.

The Language of Algebra

The word *undefined* is used in mathematics to describe an expression that has no meaning.

Special Fraction Forms   For any nonzero number a,

$$\frac{a}{a} = 1 \qquad \frac{a}{1} = a \qquad \frac{0}{a} = 0 \qquad \frac{a}{0} \text{ is undefined}$$

■ MULTIPLYING AND DIVIDING FRACTIONS

We now discuss how to add, subtract, multiply, and divide fractions. We begin with multiplication.

Multiplying Fractions   To multiply two fractions, multiply the numerators and multiply the denominators.
Let a, b, c, and d represent numbers, where b and d are not 0,

$$\frac{a}{b} \cdot \frac{c}{d} = \frac{a \cdot c}{b \cdot d}$$

EXAMPLE 2

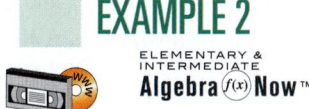

ELEMENTARY & INTERMEDIATE
Algebra *f(x)* **Now**™

Multiply: $\dfrac{7}{8} \cdot \dfrac{3}{5}$.

Solution $\dfrac{7}{8} \cdot \dfrac{3}{5} = \dfrac{7 \cdot 3}{8 \cdot 5}$ Multiply the numerators and multiply the denominators.

$= \dfrac{21}{40}$ Do the multiplications.

Self Check 2   Multiply: $\dfrac{5}{9} \cdot \dfrac{2}{3}$.

One number is called the **reciprocal** of another if their product is 1. To find the reciprocal of a fraction, we invert its numerator and denominator.

$\dfrac{3}{4}$ is the reciprocal of $\dfrac{4}{3}$, because $\dfrac{3}{4} \cdot \dfrac{4}{3} = \dfrac{12}{12} = 1$.

$\dfrac{1}{10}$ is the reciprocal of 10, because $\dfrac{1}{10} \cdot 10 = \dfrac{10}{10} = 1$.

We use reciprocals to divide fractions.

Dividing Fractions   To divide fractions, multiply the first fraction by the reciprocal of the second.
Let a, b, c, and d represent numbers, where b, c, and d are not 0,

$$\frac{a}{b} \div \frac{c}{d} = \frac{a}{b} \cdot \frac{d}{c}$$

EXAMPLE 3   Divide: $\dfrac{1}{3} \div \dfrac{4}{5}$.

ELEMENTARY &
INTERMEDIATE
Algebra $f(x)$ **Now**™

Solution $\dfrac{1}{3} \div \dfrac{4}{5} = \dfrac{1}{3} \cdot \dfrac{5}{4}$ Multiply the first fraction by the reciprocal of the second. The reciprocal of $\frac{4}{5}$ is $\frac{5}{4}$.

$= \dfrac{1 \cdot 5}{3 \cdot 4}$ Multiply the numerators and multiply the denominators.

$= \dfrac{5}{12}$ Do the multiplications.

Self Check 3   Divide: $\dfrac{6}{25} \div \dfrac{1}{2}$.

■ BUILDING EQUIVALENT FRACTIONS

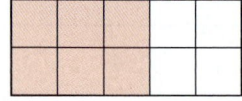

In the figure on the left, the rectangle is divided into 10 equal parts. Since 6 of those parts are red, $\dfrac{6}{10}$ of the figure is shaded.

Now consider the 5 equal parts of the rectangle created by the vertical lines. Since 3 of those parts are red, $\dfrac{3}{5}$ of the figure is shaded. We can conclude that $\dfrac{6}{10} = \dfrac{3}{5}$ because $\dfrac{6}{10}$ and $\dfrac{3}{5}$ represent the same shaded part of the rectangle. We say that $\dfrac{6}{10}$ and $\dfrac{3}{5}$ are *equivalent fractions*.

Equivalent Fractions   Two fractions are **equivalent** if they represent the same number.

Writing a fraction as an equivalent fraction with a larger denominator is called **building** the fraction. To build a fraction, we multiply it by a form of the number 1. Since any number multiplied by 1 remains the same (identical), 1 is called the **multiplicative identity**.

Multiplication Property of 1	The product of 1 and any number is that number.
	For any number a,
	$1 \cdot a = a$　　　and　　　$a \cdot 1 = a$

EXAMPLE 4

Write $\dfrac{3}{5}$ as an equivalent fraction with a denominator of 35.

Solution　We need to multiply the denominator of $\dfrac{3}{5}$ by 7 to obtain a denominator of 35. It follows that $\dfrac{7}{7}$ should be the form of 1 that is used to build $\dfrac{3}{5}$. Multiplying $\dfrac{3}{5}$ by $\dfrac{7}{7}$ does not change its value, because we are multiplying $\dfrac{3}{5}$ by 1.

$$\frac{3}{5} = \frac{3}{5} \cdot \frac{\mathbf{7}}{\mathbf{7}} \qquad \tfrac{7}{7} = 1$$

$$= \frac{3 \cdot 7}{5 \cdot 7} \qquad \text{Multiply the numerators and multiply the denominators.}$$

$$= \frac{21}{35}$$

Self Check 4　Write $\dfrac{5}{8}$ as an equivalent fraction with a denominator of 24.

■ SIMPLIFYING FRACTIONS

The Language of Algebra

The word *infinitely* is a form of the word *infinite*, which means endless.

Every fraction can be written in infinitely many equivalent forms. For example, some equivalent forms of $\dfrac{30}{36}$ are:

$$\frac{5}{6} = \frac{10}{12} = \frac{15}{18} = \frac{20}{24} = \frac{25}{30} = \frac{\mathbf{30}}{\mathbf{36}} = \frac{35}{42} = \frac{40}{48} = \frac{45}{54} = \frac{50}{60} = \cdots$$

Of all of the equivalent forms in which we can write a fraction, we often need to determine the one that is in *simplest form*.

Simplest Form of a Fraction	A fraction is in **simplest form**, or **lowest terms**, when the numerator and denominator have no common factors other than 1.

To **simplify a fraction,** we write it in simplest form by removing a factor equal to 1. For example, to simplify $\dfrac{30}{36}$, we note that the greatest factor common to the numerator and denominator is 6 and proceed as follows:

$$\frac{30}{36} = \frac{5 \cdot \mathbf{6}}{6 \cdot \mathbf{6}} \qquad \text{Factor 30 and 36, using their greatest common factor, 6.}$$

$$= \frac{5}{6} \cdot \frac{\mathbf{6}}{\mathbf{6}} \qquad \text{Use the rule for multiplying fractions in reverse: write } \tfrac{5 \cdot 6}{6 \cdot 6} \text{ as the product of two fractions, } \tfrac{5}{6} \text{ and } \tfrac{6}{6}.$$

$$= \frac{5}{6} \cdot \mathbf{1} \qquad \text{A number divided by itself is equal to 1: } \tfrac{6}{6} = 1.$$

$$= \frac{5}{6} \qquad \text{Use the multiplication property of 1: any number multiplied by 1 remains the same.}$$

To simplify $\frac{30}{36}$, we removed a factor equal to 1 in the form of $\frac{6}{6}$. The result, $\frac{5}{6}$, is equivalent to $\frac{30}{36}$.

We can easily identify the greatest common factor of the numerator and the denominator of a fraction if we write them in prime-factored form.

EXAMPLE 5

ELEMENTARY & INTERMEDIATE
Algebra *f(x)* **Now™**

Simplify each fraction, if possible: **a.** $\dfrac{63}{42}$ and   **b.** $\dfrac{33}{40}$.

Solution   **a.** After prime factoring 63 and 42, we see that the greatest common factor of the numerator and the denominator is $3 \cdot 7 = 21$.

Language of Algebra

What do Calvin Klein, Sheryl Swoopes, and Dustin Hoffman have in common? They all attended a community college. The word *common* means shared by two or more. In this section, we will work with *common* factors and *common* denominators.

$$\frac{63}{42} = \frac{3 \cdot \mathbf{3 \cdot 7}}{2 \cdot \mathbf{3 \cdot 7}} \qquad \text{Write 63 and 42 in prime-factored form.}$$

$$= \frac{3}{2} \cdot \frac{\mathbf{3 \cdot 7}}{\mathbf{3 \cdot 7}} \qquad \text{Write } \frac{3 \cdot 3 \cdot 7}{2 \cdot 3 \cdot 7} \text{ as the product of two fractions, } \frac{3}{2} \text{ and } \frac{3 \cdot 7}{3 \cdot 7}.$$

$$= \frac{3}{2} \cdot 1 \qquad \text{A nonzero number divided by itself is equal to 1: } \frac{3 \cdot 7}{3 \cdot 7} = 1.$$

$$= \frac{3}{2} \qquad \text{Any number multiplied by 1 remains the same.}$$

b. Prime factor 33 and 40.

$$\frac{33}{40} = \frac{3 \cdot 11}{2 \cdot 2 \cdot 2 \cdot 5}$$

Since the numerator and the denominator have no common factors other than 1, $\frac{33}{40}$ is in simplest form (lowest terms).

Self Check 5   Simplify each fraction, if possible: **a.** $\dfrac{24}{56}$ and   **b.** $\dfrac{16}{125}$.

To streamline the simplifying process, we can replace pairs of factors common to the numerator and denominator with the equivalent fraction $\frac{1}{1}$.

EXAMPLE 6

Simplify: $\dfrac{90}{105}$.

Solution
$$\frac{90}{105} = \frac{2 \cdot 3 \cdot 3 \cdot 5}{3 \cdot 5 \cdot 7} \qquad \text{Write 90 and 105 in prime-factored form.}$$

$$= \frac{2 \cdot \overset{1}{\cancel{3}} \cdot 3 \cdot \overset{1}{\cancel{5}}}{\underset{1}{\cancel{3}} \cdot \underset{1}{\cancel{5}} \cdot 7} \qquad \text{Slashes and 1's are used to show that } \frac{3}{3} \text{ and } \frac{5}{5} \text{ are replaced by the equivalent fraction } \frac{1}{1}.$$

$$= \frac{6}{7}$$

Multiply to find the numerator: $2 \cdot 1 \cdot 3 \cdot 1 = 6$. Multiply to find the denominator: $1 \cdot 1 \cdot 7 = 7$.

Self Check 6   Simplify: $\frac{126}{70}$.

Simplifying a Fraction	1. Factor (or prime factor) the numerator and denominator to determine all the factors common to both.
	2. Replace each pair of factors common to the numerator and denominator with the equivalent fraction $\frac{1}{1}$.
	3. Multiply the remaining factors in the numerator and in the denominator.

Caution   When all common factors of the numerator and the denominator of a fraction are removed, forgetting to write 1's above the slashes leads to a common mistake.

Correct

$$\frac{15}{45} = \frac{\overset{1}{\cancel{3}} \cdot \overset{1}{\cancel{5}}}{\underset{1}{\cancel{3}} \cdot 3 \cdot \underset{1}{\cancel{5}}} = \frac{1}{3}$$

Incorrect

$$\frac{15}{45} = \frac{\cancel{3} \cdot \cancel{5}}{\cancel{3} \cdot 3 \cdot \cancel{5}} = \frac{0}{3} = 0$$

■ ADDING AND SUBTRACTING FRACTIONS

To add or subtract fractions, they must have the same denominator.

Adding and Subtracting Fractions	To add (or subtract) two fractions with the same denominator, add (or subtract) their numerators and write the sum (or difference) over the common denominator.
	Let a, b, and d represent numbers, where d is not 0,
	$$\frac{a}{d} + \frac{b}{d} = \frac{a+b}{d} \qquad \frac{a}{d} - \frac{b}{d} = \frac{a-b}{d}$$

For example,

$$\frac{3}{7} + \frac{1}{7} = \frac{3+1}{7} = \frac{4}{7} \qquad \text{and} \qquad \frac{18}{25} - \frac{9}{25} = \frac{18-9}{25} = \frac{9}{25}$$

Caution   Only factors common to the numerator and the denominator of a fraction can be removed. It is incorrect to remove the 5's in $\frac{5+8}{5}$, because 5 is not used as a factor in the expression $5 + 8$. This error leads to an incorrect answer of 9.

Correct

$$\frac{5+8}{5} = \frac{13}{5}$$

Incorrect

$$\frac{5+8}{5} = \frac{\overset{1}{\cancel{5}} + 8}{\underset{1}{\cancel{5}}} = \frac{9}{1} = 9$$

To add (or subtract) fractions with different denominators, we express them as equivalent fractions that have a common denominator. The smallest common denominator, called the **least** or **lowest common denominator,** is usually the easiest common denominator to use.

Least Common Denominator (LCD)	The **least** or **lowest common denominator (LCD)** for a set of fractions is the smallest number each denominator will divide exactly (divide with no remainder).

The denominators of $\frac{2}{5}$ and $\frac{1}{3}$ are 5 and 3. The numbers 5 and 3 divide many numbers exactly (30, 45, and 60, to name a few), but the smallest number that they divide exactly is 15. Thus, 15 is the LCD for $\frac{2}{5}$ and $\frac{1}{3}$.

To find $\frac{2}{5} + \frac{1}{3}$, we find equivalent fractions that have denominators of 15 and we use the rule for adding fractions.

$$\frac{2}{5} + \frac{1}{3} = \frac{2}{5} \cdot \frac{3}{3} + \frac{1}{3} \cdot \frac{5}{5}$$ Multiply $\frac{2}{5}$ by 1 in the form of $\frac{3}{3}$. Multiply $\frac{1}{3}$ by 1 in the form of $\frac{5}{5}$.

$$= \frac{6}{15} + \frac{5}{15}$$ Multiply the numerators and multiply the denominators. Note that the denominators are now the same.

$$= \frac{6 + 5}{15}$$ Add the numerators. Write the sum over the common denominator.

$$= \frac{11}{15}$$ Do the addition.

When adding (or subtracting) fractions with unlike denominators, the least common denominator is not always obvious. Prime factorization is helpful in determining the LCD.

EXAMPLE 7   Subtract: $\dfrac{3}{10} - \dfrac{5}{28}$.

ELEMENTARY & INTERMEDIATE
Algebra *f(x)* **Now**™

Solution   To find the LCD, we find the prime factorization of both denominators and use each prime factor the *greatest* number of times it appears in any one factorization:

$$\left.\begin{array}{l} 10 = 2 \cdot 5 \\ 28 = 2 \cdot 2 \cdot 7 \end{array}\right\} \text{LCD} = 2 \cdot 2 \cdot 5 \cdot 7 = 140$$ 2 appears twice in the factorization of 28. 5 appears once in the factorization of 10. 7 appears once in the factorization of 28.

Since 140 is the smallest number that 10 and 28 divide exactly, we write both fractions as fractions with the LCD of 140.

$$\frac{3}{10} - \frac{5}{28} = \frac{3}{10} \cdot \frac{14}{14} - \frac{5}{28} \cdot \frac{5}{5}$$ We must multiply 10 by 14 to obtain 140. We must multiply 28 by 5 to obtain 140.

$$= \frac{42}{140} - \frac{25}{140}$$ Do the multiplications.

$$= \frac{42 - 25}{140}$$ Subtract the numerators. Write the difference over the common denominator.

$$= \frac{17}{140}$$ Do the subtraction.

Self Check 7   Subtract: $\dfrac{11}{48} - \dfrac{7}{40}$.

■ **SIMPLIFYING ANSWERS**

When adding, subtracting, multiplying, or dividing fractions, remember to express the answer in simplest form.

EXAMPLE 8

ELEMENTARY &
INTERMEDIATE
Algebra *f(x)* **Now**™

Perform each operation: **a.** $45\left(\dfrac{4}{9}\right)$ and   **b.** $\dfrac{5}{12} + \dfrac{3}{4}$.

Solution   **a.** $45\left(\dfrac{4}{9}\right) = \dfrac{45}{1}\left(\dfrac{4}{9}\right)$ Write 45 as a fraction: $45 = \dfrac{45}{1}$.

$$= \frac{45 \cdot 4}{1 \cdot 9}$$ Multiply the numerators. Multiply the denominators.

$$= \frac{\overset{1}{\cancel{3}} \cdot \overset{1}{\cancel{3}} \cdot 5 \cdot 2 \cdot 2}{1 \cdot \underset{1}{\cancel{3}} \cdot \underset{1}{\cancel{3}}}$$ To simplify the fraction, prime factor 45, 4, and 9. Then replace each $\dfrac{3}{3}$ with $\dfrac{1}{1}$.

$$= 20$$

b.   Since the smallest number that 12 and 4 divide exactly is 12, the LCD is 12.

$$\frac{5}{12} + \frac{3}{4} = \frac{5}{12} + \frac{3}{4} \cdot \frac{\mathbf{3}}{\mathbf{3}}$$ $\dfrac{5}{12}$ already has a denominator of 12. Build $\dfrac{3}{4}$ so that its denominator is 12.

$$= \frac{5}{12} + \frac{9}{12}$$ Multiply the numerators and denominators in the second term. The denominators are now the same.

$$= \frac{14}{12}$$ Add the numerators, 5 and 9, to get 14. Write that sum over the common denominator.

$$= \frac{\overset{1}{\cancel{2}} \cdot 7}{\underset{1}{\cancel{2}} \cdot 6}$$ To simplify $\dfrac{14}{12}$, factor 14 and 12, using their greatest common factor, 2. Then remove $\dfrac{2}{2} = 1$.

$$= \frac{7}{6}$$

Self Check 8   Perform each operation: **a.** $24\left(\dfrac{8}{6}\right)$,   **b.** $\dfrac{1}{15} + \dfrac{31}{30}$

■ **MIXED NUMBERS**

A **mixed number** represents the sum of a whole number and a fraction. For example, $5\frac{3}{4}$ means $5 + \frac{3}{4}$. To perform calculations involving mixed numbers, we often express them as improper fractions.

EXAMPLE 9

ELEMENTARY & INTERMEDIATE

Algebra $f(x)$ **Now**™

Divide: $5\frac{3}{4} \div 2$.

Solution

To write the mixed number $5\frac{3}{4}$ as a fraction, we use a two-step process.

The Language of Algebra

Fractions such as $\frac{23}{4}$, with a numerator greater than or equal to the denominator, are called **improper fractions**. This term is misleading. In algebra, such fractions are often preferable to their equivalent mixed number form.

$$5\frac{3}{4} = \frac{23}{4}$$

Step 1. Multiply the whole number by the denominator: $5 \cdot 4 = 20$. Then add that product to the numerator: $20 + 3 = 23$.

Step 2. Write the result from Step 1 over the denominator, 4.

Now we replace $5\frac{3}{4}$ with $\frac{23}{4}$ and divide.

$$5\frac{3}{4} \div 2 = \frac{23}{4} \div \frac{2}{1}$$ Write $5\frac{3}{4}$ as $\frac{23}{4}$. Write 2 as a fraction: $2 = \frac{2}{1}$.

$$= \frac{23}{4} \cdot \frac{1}{2}$$ Multiply by the reciprocal of $\frac{2}{1}$, which is $\frac{1}{2}$.

$$= \frac{23}{8}$$ Multiply the numerators. Multiply the denominators.

To write the answer, $\frac{23}{8}$, as a mixed number, we use a two-step process.

$$\frac{23}{8} = 2\frac{7}{8}$$

Step 1. Divide the numerator, 23, by the denominator, 8.

Step 2. The quotient, 2, is the whole-number part of the mixed number. Its fractional part is the remainder, 7, over the original denominator, 8.

$$\begin{array}{r} 2 \\ 8{\overline{\smash{\big)}\,23}} \\ \underline{16} \\ 7 \end{array}$$

Self Check 9

Multiply: $1\frac{1}{8} \cdot 9$.

Answers to Self Checks

1. $189 = 3 \cdot 3 \cdot 3 \cdot 7$ **2.** $\frac{10}{27}$ **3.** $\frac{12}{25}$ **4.** $\frac{15}{24}$ **5. a.** $\frac{3}{7}$,   **b.** in simplest form   **6.** $\frac{9}{5}$
7. $\frac{13}{240}$ **8. a.** 32,   **b.** $\frac{11}{10}$ **9.** $\frac{81}{8} = 10\frac{1}{8}$

1.2   STUDY SET

ELEMENTARY & INTERMEDIATE

Algebra $f(x)$ **Now**™

VOCABULARY   Fill in the blanks.

1. Numbers that have only 1 and themselves as factors, such as 23, 37, and 41, are called _____ numbers.

2. When we write 60 as $20 \cdot 3$, we say that we have _____ 60. When we write 60 as $5 \cdot 3 \cdot 2 \cdot 2$, we say that we have written 60 in _____ form.

3. The _____ of the fraction $\frac{3}{4}$ is 3, and the _____ is 4.

4. A fraction is in _____ form, or _____ terms, when the numerator and denominator have no common factors other than 1.

5. Two fractions that represent the same number, such as $\frac{1}{2}$ and $\frac{2}{4}$, are called _____ fractions.

6. The number $\frac{2}{3}$ is the _____ of the number $\frac{3}{2}$, because their product is 1.

7. The _____ common denominator for a set of fractions is the smallest number each denominator will divide exactly.

8. The _____ number $7\frac{1}{3}$ represents the sum of a whole number and a fraction: $7 + \frac{1}{3}$.

CONCEPTS

9. The prime factorization of a number is $2 \cdot 2 \cdot 3 \cdot 5$. What is the number?

10. Complete each fact about fractions. Assume $a \neq 0$.

$$\frac{a}{a} = \boxed{} \quad \frac{a}{1} = \boxed{} \quad \frac{0}{a} = \boxed{} \quad \frac{a}{0} \text{ is } \boxed{}$$

11. What equivalent fractions are shown in the illustration?

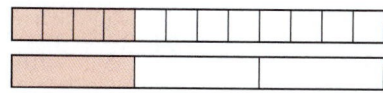

12. Complete each rule.

a. $\dfrac{a}{b} \cdot \dfrac{c}{d} = \boxed{}$ **b.** $\dfrac{a}{b} \div \dfrac{c}{d} = \boxed{}$

c. $\dfrac{a}{d} + \dfrac{b}{d} = \boxed{}$ **d.** $\dfrac{a}{d} - \dfrac{b}{d} = \boxed{}$

13. Simplify $\dfrac{2 \cdot 2 \cdot 3}{2 \cdot 3 \cdot 5}$.

14. To express $\frac{3}{8}$ as an equivalent fraction with a denominator of 40, by what number must we multiply the numerator and the denominator?

15. Complete each statement.

a. To build a fraction, we multiply it by $\boxed{}$ in the form of $\frac{2}{2}$, $\frac{3}{3}$, or $\frac{4}{4}$, and so on.

b. To simplify a fraction, we remove factors equal to $\boxed{}$ in the form of $\frac{2}{2}$, $\frac{3}{3}$, or $\frac{4}{4}$, and so on.

16. a. Give three numbers that 4 and 6 divide exactly.

b. What is the smallest number that 4 and 6 divide exactly?

17. The prime factorizations of 24 and 36 are

$$24 = 2 \cdot 2 \cdot 2 \cdot 3$$
$$36 = 2 \cdot 2 \cdot 3 \cdot 3$$

a. What is the greatest number of times 2 appears in any one factorization?

b. What is the greatest number of times 3 appears in any one factorization?

18. a. Write $2\frac{15}{16}$ as an improper fraction.

b. Write $\frac{49}{12}$ as a mixed number.

NOTATION

19. Consider $\dfrac{5}{16} = \dfrac{5}{16} \cdot \dfrac{\mathbf{3}}{\mathbf{3}}$.

a. What fraction is being built up?

b. Fill in the blank: $\dfrac{3}{3} = \boxed{}$.

c. What equivalent fraction is the result after building $\frac{5}{16}$?

20. Consider $\dfrac{70}{175} = \dfrac{\overset{1}{\cancel{7}} \cdot \overset{1}{\cancel{5}} \cdot 2}{\underset{1}{\cancel{7}} \cdot \underset{1}{\cancel{5}} \cdot 5}$.

a. What fraction is being simplified?

b. What are $\dfrac{7}{7}$ and $\dfrac{5}{5}$ replaced with?

c. What equivalent fraction is the result after simplifying?

PRACTICE   List the factors of each number.

21. 20   **22.** 50

23. 28   **24.** 36

Give the prime factorization of each number.

25. 75   **26.** 20

27. 28   **28.** 54

29. 117   **30.** 147

31. 220   **32.** 270

Build each fraction or whole number to an equivalent fraction having the indicated denominator.

33. $\dfrac{1}{3}$, denominator 9   **34.** $\dfrac{3}{8}$, denominator 24

35. $\dfrac{4}{9}$, denominator 54   **36.** $\dfrac{9}{16}$, denominator 64

37. 7, denominator 5   **38.** 12, denominator 3

Write each fraction in lowest terms. If the fraction is in lowest terms, so indicate.

39. $\dfrac{6}{12}$ **40.** $\dfrac{3}{9}$

41. $\dfrac{24}{18}$ **42.** $\dfrac{35}{14}$

43. $\dfrac{15}{20}$

44. $\dfrac{22}{77}$

45. $\dfrac{72}{64}$

46. $\dfrac{26}{21}$

47. $\dfrac{33}{56}$

48. $\dfrac{26}{39}$

49. $\dfrac{36}{225}$

50. $\dfrac{175}{490}$

Perform each operation and simplify the result when possible.

51. $\dfrac{1}{2} \cdot \dfrac{3}{5}$

52. $\dfrac{3}{4} \cdot \dfrac{5}{7}$

53. $\dfrac{4}{3}\left(\dfrac{6}{5}\right)$

54. $\dfrac{7}{8}\left(\dfrac{6}{15}\right)$

55. $\dfrac{5}{12} \cdot \dfrac{18}{5}$

56. $\dfrac{5}{4} \cdot \dfrac{12}{10}$

57. $21\left(\dfrac{10}{3}\right)$

58. $28\left(\dfrac{4}{7}\right)$

59. $7\dfrac{1}{2} \cdot 1\dfrac{2}{5}$

60. $3\dfrac{1}{4}\left(1\dfrac{1}{5}\right)$

61. $6 \cdot 2\dfrac{7}{24}$

62. $7 \cdot 1\dfrac{3}{28}$

63. $\dfrac{3}{5} \div \dfrac{2}{3}$

64. $\dfrac{4}{5} \div \dfrac{3}{7}$

65. $\dfrac{3}{4} \div \dfrac{6}{5}$

66. $\dfrac{3}{8} \div \dfrac{15}{28}$

67. $\dfrac{21}{35} \div \dfrac{3}{14}$

68. $\dfrac{23}{25} \div \dfrac{46}{5}$

69. $6 \div \dfrac{3}{14}$

70. $23 \div \dfrac{46}{5}$

71. $3\dfrac{1}{3} \div 1\dfrac{5}{6}$

72. $2\dfrac{1}{2} \div 1\dfrac{5}{8}$

73. $8 \div 3\dfrac{1}{5}$

74. $15 \div 3\dfrac{1}{3}$

75. $\dfrac{3}{5} + \dfrac{3}{5}$

76. $\dfrac{4}{13} - \dfrac{3}{13}$

77. $\dfrac{1}{6} + \dfrac{1}{24}$

78. $\dfrac{17}{25} - \dfrac{2}{5}$

79. $\dfrac{3}{5} + \dfrac{2}{3}$

80. $\dfrac{4}{3} + \dfrac{7}{2}$

81. $\dfrac{5}{12} + \dfrac{1}{3}$

82. $\dfrac{7}{15} + \dfrac{1}{5}$

83. $\dfrac{9}{4} - \dfrac{5}{6}$

84. $\dfrac{2}{15} + \dfrac{7}{9}$

85. $\dfrac{7}{10} - \dfrac{1}{14}$

86. $\dfrac{7}{25} + \dfrac{3}{10}$

87. $\dfrac{5}{14} - \dfrac{4}{21}$

88. $\dfrac{2}{33} + \dfrac{3}{22}$

89. $3 - \dfrac{3}{4}$

90. $\dfrac{17}{3} + 4$

91. $3\dfrac{3}{4} - 2\dfrac{1}{2}$

92. $15\dfrac{5}{6} + 11\dfrac{5}{8}$

93. $8\dfrac{2}{9} - 7\dfrac{2}{3}$

94. $3\dfrac{4}{5} - 3\dfrac{1}{10}$

APPLICATIONS

95. BOTANY To assess the effects of smog, botanists cut down a pine tree and measured the width of the growth rings for the last two years.
 a. What was the growth over this two-year period?

 b. What is the difference in the widths of the rings?

$\dfrac{5}{32}$ in. $\dfrac{1}{16}$ in.

96. HARDWARE To secure the bracket to the stock, a bolt and a nut are used. How long should the threaded part of the bolt be?

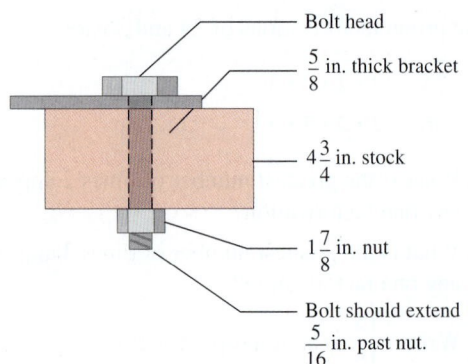

Bolt head

$\dfrac{5}{8}$ in. thick bracket

$4\dfrac{3}{4}$ in. stock

$1\dfrac{7}{8}$ in. nut

Bolt should extend $\dfrac{5}{16}$ in. past nut.

97. FRAMES   How much molding is needed to produce the square picture frame shown?

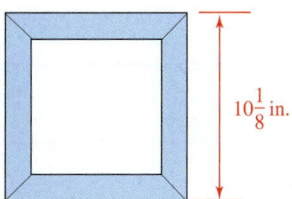

$10\frac{1}{8}$ in.

98. DECORATING   The materials used to make a pillow are shown here. Examine the inventory list to decide how many pillows can be manufactured in one production run with the materials in stock.

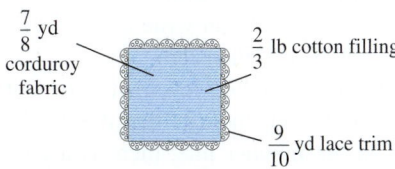

$\frac{7}{8}$ yd corduroy fabric

$\frac{2}{3}$ lb cotton filling

$\frac{9}{10}$ yd lace trim

Factory Inventory List

Materials	Amount in stock
Lace trim	135 yd
Corduroy fabric	154 yd
Cotton filling	98 lb

WRITING

99. Explain how to add two fractions having unlike denominators.

100. To multiply two fractions, must they have like denominators? Explain.

101. What are equivalent fractions?

102. Explain the error in the following work.

$$\text{Add: } \frac{4}{3} + \frac{3}{2} = \frac{4}{\cancel{3}} + \frac{\cancel{3}^{1}}{5}$$

$$= \frac{4}{1} + \frac{1}{5}$$

$$= 4 + \frac{1}{5}$$

$$= 4\frac{1}{5}$$

REVIEW   **Express each statement using one of the words sum, difference, product, or quotient.**

103. $7 - 5 = 2$

104. $5(6) = 30$

105. $30 \div 15 = 2$

106. $12 + 12 = 24$

Use the formula to complete each table.

107. $T = 15g$

Number of gears g	Number of teeth T
10	
12	

108. $p = r - 200$

Revenue r	Profit p
1,000	
5,000	

CHALLENGE PROBLEMS

109. Which is larger: $\frac{11}{12}$ or $\frac{8}{9}$?

110. If the circle represents a whole, find the missing value.

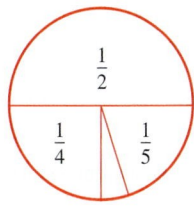

$\frac{1}{2}$ $\frac{1}{4}$ $\frac{1}{5}$

1.3   The Real Numbers

- The Integers   • Rational Numbers: Fractions and Mixed Numbers
- Rational Numbers: Decimals   • Irrational Numbers   • The Real Numbers
- Graphing on the Number Line   • Absolute Value

In this section, we will define many types of numbers. Then we will show that together, they form a collection or **set** of numbers called the *real numbers*.

■ THE INTEGERS

Natural numbers are the numbers that we use for counting. To write this set, we list its **members** (or **elements**) within **braces** {   }.

Natural Numbers	The set of **natural numbers** is {1, 2, 3, 4, 5, . . .}. Read as "the set containing one, two, three, four, five, and so on."

The three dots . . . in this definition mean that the list continues on forever.

The natural numbers, together with 0, form the set of **whole numbers.**

Whole Numbers	The set of **whole numbers** is {0, 1, 2, 3, 4, 5, . . .}.

Although whole numbers are used in a wide variety of settings, they are not adequate for describing many real-life situations. For example, if you write a check for more than what's in your account, the account balance will be less than zero.

We can use the **number line** below to visualize numbers less than zero. A number line is straight and has uniform markings. The arrowheads indicate that it extends forever in both directions. For each natural number on the number line, there is a corresponding number, called its *opposite,* to the left of 0. In the illustration, we see that 3 and −3 (negative three) are opposites, as are −5 (negative five) and 5. Note that 0 is its own opposite.

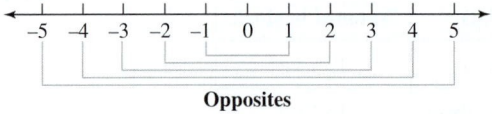

Opposites

Opposites	Two numbers that are the same distance from 0 on the number line, but on opposite sides of it, are called **opposites.**

The whole numbers, together with their opposites, form the set of **integers.**

Integers	The set of **integers** is {. . . , −4, −3, −2, −1, 0, 1, 2, 3, 4, . . .}.

The Language of Algebra

The *positive integers* are:
1, 2, 3, 4, 5, ...
The *negative integers* are:
−1, −2, −3, −4, −5, ...

On the number line, numbers greater than 0 are to the right of 0. They are called **positive numbers.** Positive numbers can be written with or without a **positive sign** +. For example, 2 = +2 (positive two). They can be used to describe such quantities as an elevation above sea level (+3,000 ft) or a stock market gain (25 points).

Numbers less than 0 are to the left of 0 on the number line. They are called **negative numbers.** Negative numbers are always written with a **negative sign** −. They can be used to describe such quantities as an overdrawn checking account (−$75) or a below-zero temperature (−12°).

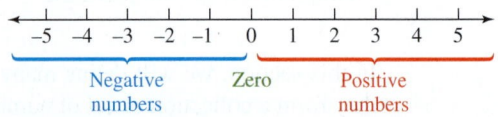

■ RATIONAL NUMBERS: FRACTIONS AND MIXED NUMBERS

Many situations cannot be described using integers. For example, a commute to work might take $\frac{1}{4}$ hour, or a hat might be size $7\frac{5}{8}$. To describe these situations, we need fractions, often called *rational numbers*.

Rational Numbers	A **rational number** is any number that can be expressed as a fraction with an integer numerator and a nonzero integer denominator.

Some examples of rational numbers are

$$\frac{1}{4}, \quad \frac{7}{8}, \quad \frac{25}{25}, \quad \text{and} \quad \frac{19}{12}$$

To show that negative fractions are rational numbers, we use the following fact.

Negative Fractions	Let a and b represent numbers, where b is not 0. $$-\frac{a}{b} = \frac{-a}{b} = \frac{a}{-b}$$

To illustrate this rule, we consider $-\frac{11}{16}$. It is a rational number because it can be written as $\frac{-11}{16}$ or as $\frac{11}{-16}$.

Positive and negative mixed numbers are also rational numbers because they can be expressed as fractions. For example,

The Language of Algebra

Rational numbers are so named because they can be expressed as the ratio (quotient) of two integers.

$$7\frac{5}{8} = \frac{61}{8} \qquad \text{and} \qquad -6\frac{1}{2} = -\frac{13}{2} = \frac{-13}{2}$$

Any natural number, whole number, or integer can be expressed as a fraction with a denominator of 1. For example, $5 = \frac{5}{1}$, $0 = \frac{0}{1}$, and $-3 = \frac{-3}{1}$. Therefore, every natural number, whole number, and integer is also a rational number.

■ RATIONAL NUMBERS: DECIMALS

Many numerical quantities are written in decimal notation. For instance, a candy bar might cost $0.89, a dragster might travel at 203.156 mph, or a business loss might be $-\$4.7$ million. These decimals are called **terminating decimals** because their representations terminate. As shown below, terminating decimals can be expressed as fractions. Therefore, terminating decimals are rational numbers.

The Language of Algebra

To *terminate* means to bring to an end. In the movie *The Terminator,* actor Arnold Schwarzenegger plays a heartless machine sent to Earth to bring an end to his enemies.

$$0.89 = \frac{89}{100} \qquad 203.156 = 203\frac{156}{1,000} = \frac{203,156}{1,000} \qquad -4.7 = -4\frac{7}{10} = \frac{-47}{10}$$

Decimals such as 0.3333 . . . and 2.8167167167 . . . , which have a digit (or block of digits) that repeats, are called **repeating decimals.** Since any repeating decimal can be expressed as a fraction, repeating decimals are rational numbers.

The set of rational numbers cannot be listed as we listed other sets in this section. Instead, we use **set-builder** notation.

Rational Numbers	The set of rational numbers is	
	$$\left\{\frac{a}{b} \,\middle|\, a \text{ and } b \text{ are integers, with } b \neq 0.\right\}$$	Read as "the set of all numbers of the form $\frac{a}{b}$, such that a and b are integers, with $b \neq 0$."

To find the decimal equivalent for a fraction, we divide its numerator by its denominator. For example, to write $\frac{1}{4}$ and $\frac{5}{22}$ as decimals, we proceed as follows:

$$\begin{array}{r} 0.25 \\ 4\overline{)1.00} \\ \underline{8} \\ 20 \\ \underline{20} \\ 0 \end{array}$$

Write a decimal point and additional zeros to the right of 1.

The remainder is 0.

$$\begin{array}{r} 0.22727\ldots \\ 22\overline{)5.00000} \\ \underline{4\ 4} \\ 60 \\ \underline{44} \\ 160 \\ \underline{154} \\ 60 \\ \underline{44} \\ 160 \end{array}$$

Write a decimal point and additional zeros to the right of 5.

60 and 160 continually appear as remainders. Therefore, 2 and 7 will continually appear in the quotient.

The decimal equivalent of $\frac{1}{4}$ is 0.25 and the decimal equivalent of $\frac{5}{22}$ is 0.2272727 . . . . We can use an **overbar** to write repeating decimals in more compact form: $0.2272727\ldots = 0.2\overline{27}$. Here are more fractions and their decimal equivalents.

Terminating decimals

$$\frac{1}{2} = 0.5$$

$$\frac{5}{8} = 0.625$$

$$\frac{3}{4} = 0.75$$

Repeating decimals

$$\frac{1}{6} = 0.16666\ldots \quad \text{or} \quad 0.1\overline{6}$$

$$\frac{1}{3} = 0.3333\ldots \quad \text{or} \quad 0.\overline{3}$$

$$\frac{5}{11} = 0.454545\ldots \quad \text{or} \quad 0.\overline{45}$$

■ IRRATIONAL NUMBERS

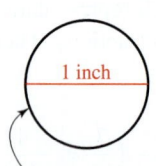

1 inch

$\sqrt{2}$ inches

1 inch

The distance around the circle is π inches.

Numbers that cannot be expressed as a fraction with an integer numerator and an integer denominator are called **irrational numbers.** One example is the square root of 2, denoted $\sqrt{2}$. It is the number that, when multiplied by itself, gives 2: $\sqrt{2} \cdot \sqrt{2} = 2$. It can be shown that a square with sides of length 1 inch has a diagonal that is $\sqrt{2}$ inches long.

The number represented by the Greek letter π (pi) is another example of an irrational number. A circle, with a 1-inch diameter, has a circumference of π inches.

Expressed in decimal form,

$$\sqrt{2} = 1.414213562\ldots \quad \text{and} \quad \pi = 3.141592654\ldots$$

These decimals neither terminate nor repeat.

Irrational Numbers	An **irrational number** is a nonterminating, nonrepeating decimal. An irrational number cannot be expressed as a fraction with an integer numerator and an integer denominator.

Other examples of irrational numbers are:

$$\sqrt{3} = 1.732050808\ldots \qquad -\sqrt{5} = -2.236067977\ldots$$

$$-\pi = -3.141592654\ldots \qquad 3\pi = 9.424777961\ldots \qquad 3\pi \text{ means } 3 \cdot \pi.$$

We can use a calculator to approximate the decimal value of an irrational number. To approximate $\sqrt{2}$ using a scientific calculator, we use the square root key $\sqrt{}$. To approximate π, we use the *pi* key π.

$$\sqrt{2} \approx 1.414213562 \qquad \text{and} \qquad \pi \approx 3.141592654 \qquad \text{Read } \approx \text{ as "is approximately equal to."}$$

Rounded to the nearest thousandth, $\sqrt{2} \approx 1.414$ and $\pi \approx 3.142$.

■ THE REAL NUMBERS

The set of **real numbers** is formed by combining the set of rational numbers and the set of irrational numbers. Every real number has a decimal representation. If it is rational, its corresponding decimal terminates or repeats. If it is irrational, its decimal representation is nonterminating and nonrepeating.

The Real Numbers	A **real number** is any number that is a rational number or an irrational number.

The following diagram shows how various sets of numbers are related. Note that a number can belong to more than one set. For example, -6 is an integer, a rational number, and a real number.

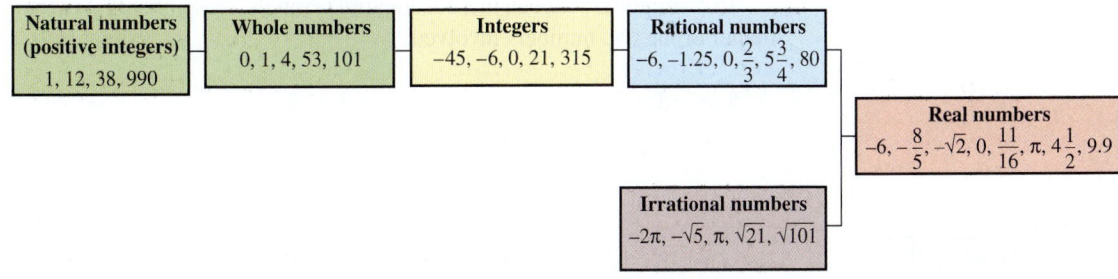

Natural numbers (positive integers)
1, 12, 38, 990

Whole numbers
0, 1, 4, 53, 101

Integers
$-45, -6, 0, 21, 315$

Rational numbers
$-6, -1.25, 0, \frac{2}{3}, 5\frac{3}{4}, 80$

Real numbers
$-6, -\frac{8}{5}, -\sqrt{2}, 0, \frac{11}{16}, \pi, 4\frac{1}{2}, 9.9$

Irrational numbers
$-2\pi, -\sqrt{5}, \pi, \sqrt{21}, \sqrt{101}$

EXAMPLE 1

ELEMENTARY & INTERMEDIATE
Algebra *f(x)* **Now**™

Solution

Which numbers in the following set are natural numbers, whole numbers, integers, rational numbers, irrational numbers, real numbers? $\left\{-3.4, \frac{2}{5}, 0, -6, 1\frac{3}{4}, \pi, 16\right\}$

Natural numbers: 16   16 is a member of $\{1, 2, 3, 4, 5, \ldots\}$.

Whole numbers: 0, 16   0 and 16 are members of $\{0, 1, 2, 3, 4, 5, \ldots\}$.

Integers: $0, -6, 16$ $0, -6,$ and 16 are members of $\{\ldots, -3, -2, -1, 0, 1, 2, 3, \ldots\}$.

Rational numbers:

$$-3.4, \frac{2}{5}, 0, -6, 1\frac{3}{4}, 16$$

A rational number can be expressed as a fraction: $-3.4 = \frac{-34}{10}, 0 = \frac{0}{1}, -6 = \frac{-6}{1}, 1\frac{3}{4} = \frac{7}{4},$ and $16 = \frac{16}{1}$.

Irrational numbers: π $\pi = 3.1415\ldots$ is a nonterminating, nonrepeating decimal.

Real numbers:

$$-3.4, \frac{2}{5}, 0, -6, 1\frac{3}{4}, \pi, 16$$

Every natural number, whole number, integer, rational number, and irrational number is a real number.

Self Check 1   Use the instructions for Example 1 with the set $\left\{0.1, \sqrt{2}, -\dfrac{2}{7}, 45, -2, \dfrac{13}{4}, -6\dfrac{7}{8}\right\}$.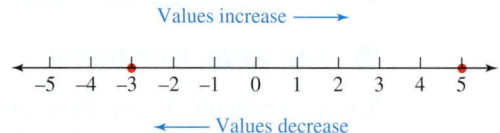

■ GRAPHING ON THE NUMBER LINE

Every real number corresponds to a point on the number line, and every point on the number line corresponds to exactly one real number. As we move right on the number line, the values of the numbers increase. As we move left, the values decrease. On the number line, we see that 5 is greater than -3, because 5 lies to the right of -3. Similarly, -3 is less than 5, because it lies to the left of 5.

Values increase ⟶

$$\begin{array}{ccccccccccc} \mid & \mid & \bullet & \mid & \mid & \mid & \mid & \mid & \mid & \bullet & \mid \\ -5 & -4 & -3 & -2 & -1 & 0 & 1 & 2 & 3 & 4 & 5 \end{array}$$

⟵ Values decrease

The Language of Algebra

The prefix *in* means *not*. For example:

inaccurate ↔ not accurate

inexpensive ↔ not expensive

inequality ↔ not equal

The **inequality symbol** $>$ means "is greater than." It is used to show that one number is greater than another. The inequality symbol $<$ means "is less than." It is used to show that one number is less than another. For example,

$5 > -3$ Read as "5 is greater than -3."

$-3 < 5$ Read as "-3 is less than 5."

To distinguish between these inequality symbols, remember that each one points to the smaller of the two numbers involved.

$5 > -3$ $-3 < 5$

└─── Points to the smaller number. ───┘

EXAMPLE 2

ELEMENTARY &
INTERMEDIATE
Algebra *f(x)* **Now**™

Use one of the symbols $>$ or $<$ to make each statement true: **a.** -4 ⬚ 4, **b.** -2 ⬚ -3,   **c.** 4.47 ⬚ 12.5,   and   **d.** $\dfrac{3}{4}$ ⬚ $\dfrac{5}{8}$.

Solution   **a.** Since -4 is to the left of 4 on the number line, we have $-4 < 4$.

b. Since -2 is to the right of -3 on the number line, we have $-2 > -3$.

c. Since 4.47 is to the left of 12.5 on the number line, we have $4.47 < 12.5$.

The Language of Algebra

To state that a number x is positive, we can write $x > 0$. To state that a number x is negative, we can write $x < 0$.

d. To compare fractions, express them in terms of the same denominator, preferably the LCD. If we write $\dfrac{3}{4}$ as an equivalent fraction with denominator 8, we see that $\dfrac{3}{4} = \dfrac{3 \cdot 2}{4 \cdot 2} = \dfrac{6}{8}$. Therefore, $\dfrac{3}{4} > \dfrac{5}{8}$.

To compare the fractions, we could also convert each to its decimal equivalent. Since $\dfrac{3}{4} = 0.75$ and $\dfrac{5}{8} = 0.625$, we know that $\dfrac{3}{4} > \dfrac{5}{8}$.

Self Check 2   Use one of the symbols $>$ or $<$ to make each statement true: **a.** 1 ⬚ -1, **b.** -5 ⬚ -4,   **c.** 6.7 ⬚ 4.999,   **d.** $\dfrac{3}{5}$ ⬚ $\dfrac{2}{3}$.

To **graph a number** means to mark its position on the number line.

EXAMPLE 3

ELEMENTARY & INTERMEDIATE
Algebra *f(x)* **Now**™

Graph each number in the following set: $\left\{ -2.43, \sqrt{2}, 1, -0.\overline{3}, 2\frac{5}{6}, -\frac{3}{2} \right\}$.

Solution   We locate the position of each number on the number line, draw a bold dot, and label it. It is often helpful to approximate the value of a number or to write the number in an equivalent form to determine its location on the number line.

- To locate -2.43, we round it to the nearest tenth: $-2.43 \approx -2.4$.
- Use a calculator to find that $\sqrt{2} \approx 1.4$.
- Recall that $0.\overline{3} = 0.333 \ldots = \frac{1}{3}$. Therefore, $-0.\overline{3} = -\frac{1}{3}$.
- In mixed-number form, $-\frac{3}{2} = -1\frac{1}{2}$. This is midway between -1 and -2.

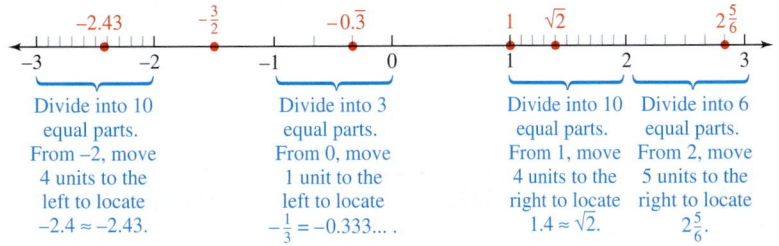

Self Check 3   Graph each number in the set: $\left\{ 1.7, \pi, -1\frac{3}{4}, 0.\overline{6}, \frac{5}{2}, -3 \right\}$.

■ ABSOLUTE VALUE

A number line can be used to measure the distance from one number to another. For example, in the following figure we see that the distance from 0 to -4 is 4 units and the distance from 0 to 3 is 3 units.

4 units   3 units

$$-5 \quad -4 \quad -3 \quad -2 \quad -1 \quad 0 \quad 1 \quad 2 \quad 3 \quad 4 \quad 5$$

To express the distance that a number is from 0 on a number line, we often use absolute values.

Absolute Value	The **absolute value** of a number is the distance from 0 to the number on the number line.

Success Tip

Since absolute value expresses distance, the absolute value of a number is always positive or zero.

To indicate the absolute value of a number, we write the number between two vertical bars. From the figure above, we see that $\left| -4 \right| = 4$. This is read as "the absolute value of negative 4 is 4" and it tells us that the distance from 0 to -4 is 4 units. It also follows from the figure that $\left| 3 \right| = 3$.

EXAMPLE 4

Find each absolute value: **a.** $|18|$, **b.** $\left|-\frac{7}{8}\right|$, **c.** $|98.6|$, and **d.** $|0|$.

ELEMENTARY & INTERMEDIATE
Algebra *(f(x))* **Now**™

Solution **a.** Since 18 is a distance of 18 from 0 on the number line, $|18| = 18$.

b. Since $-\frac{7}{8}$ is a distance of $\frac{7}{8}$ from 0 on the number line, $\left|-\frac{7}{8}\right| = \frac{7}{8}$.

c. Since 98.6 is a distance of 98.6 from 0 on the number line, $|98.6| = 98.6$.

d. Since 0 is a distance of 0 from 0 on the number line, $|0| = 0$.

Self Check 4 Find each absolute value: **a.** $|100|$, **b.** $|-4.7|$, **c.** $\left|-\sqrt{2}\right|$.

Answers to Self Checks
1. natural numbers: 45; whole numbers: 45; integers: 45, −2; rational numbers: 0.1, $-\frac{2}{7}$, 45, −2, $\frac{13}{4}$, $-6\frac{7}{8}$; irrational numbers: $\sqrt{2}$; real numbers: all   2. **a.** >,   **b.** <,   **c.** >,   **d.** <

3.

4. **a.** 100,   **b.** 4.7,   **c.** $\sqrt{2}$

1.3   STUDY SET   ELEMENTARY & INTERMEDIATE Algebra *(f(x))* **Now**™

VOCABULARY   Fill in the blanks.

1. The set of _____ numbers is {0, 1, 2, 3, 4, 5, . . .}.

2. The set of _____ numbers is {1, 2, 3, 4, 5, . . .}.

3. The set of _____ is {. . . , −2, −1, 0, 1, 2, . . .}.

4. Two numbers represented by points on the number line that are the same distance away from 0, but on opposite sides of it, are called _____.

5. Numbers less than zero are _____, and numbers greater than zero are _____.

6. The symbols < and > are _____ symbols.

7. A _____ number is any number that can be expressed as a fraction with an integer numerator and a nonzero integer denominator.

8. A decimal such as 0.25 is called a _____ decimal, and 0.333. . . is called a _____ decimal.

9. An _____ number is a nonterminating, nonrepeating decimal.

10. An _____ number cannot be expressed as a fraction.

11. Every point on the number line corresponds to exactly one _____ number.

12. The _____ of a number is the distance on the number line between the number and 0.

CONCEPTS

13. What concept is illustrated here?

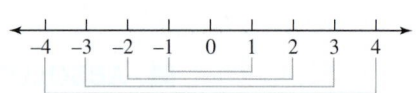

14. Fill in the blanks on the illustration.

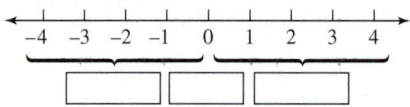

15. Show that each of the following numbers is a rational number by expressing it as a fraction with an integer numerator and a nonzero integer denominator: 6, −9, $-\frac{7}{8}$, $3\frac{1}{2}$, −0.3, 2.83.

16. Represent each situation using a signed number.
    a. A loss of $15 million
    b. A rainfall total 0.75 inch below average
    c. A score $12\frac{1}{2}$ points under the standard
    d. A building foundation $\frac{5}{16}$ inch above grade

17. What numbers are a distance of 8 away from 5 on the number line?

18. Suppose m stands for a negative number. Use m, an inequality symbol, and 0 to express this fact.

19. Refer to the graph. Use an inequality symbol, $<$ or $>$, to make each statement true.

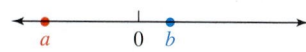

  a. a ▢ b **b.** b ▢ a
  c. b ▢ 0 and a ▢ 0   **d.** $|a|$ ▢ $|b|$

20. What is the length of the diagonal of the square shown below?

21. What is the circumference of the circle?

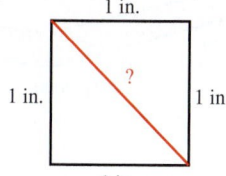
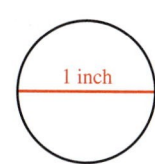

22. Place check marks in the table to show the set or sets to which each number belongs. For example, the checks show that $\sqrt{2}$ is an irrational and a real number.

	5	0	−3	$\frac{7}{8}$	0.17	$-9\frac{1}{4}$	$\sqrt{2}$	π
Real							✓	
Irrational							✓	
Rational								
Integer								
Whole								
Natural								

NOTATION   Fill in the blanks.

23. $\sqrt{2}$ is read "the _____ of 2."

24. $|-15|$ is read "the _____ of -15."

25. The symbol $\neq$ means _____.

26. The symbols $\{$ $\}$, called _____, are used when writing a set.

27. The symbol π is a letter from the _____ alphabet.

28. To find the decimal equivalent for the fraction $\frac{2}{3}$ we divide: ▢$\overline{)}$

29. Write $-\frac{4}{5}$ in two other equivalent fractional forms.

30. Write each repeating decimal using an overbar.
  a. 0.666. . .   **b.** 0.2444. . .
  c. 0.717171. . .   **d.** 0.456456456. . .

PRACTICE   **Write each fraction as a decimal. If the result is a repeating decimal, use an overbar.**

31. $\frac{5}{8}$ **32.** $\frac{3}{32}$

33. $\frac{1}{30}$ **34.** $\frac{7}{9}$

35. $\frac{21}{50}$ **36.** $\frac{2}{125}$

37. $\frac{5}{11}$ **38.** $\frac{1}{60}$

Insert one of the symbols $>$, $<$, or $=$ in the blank.

39. 5 ▢ 4   **40.** -5 ▢ -4
41. -2 ▢ -3 **42.** 0 ▢ 32
43. $|3.4|$ ▢ $\sqrt{2}$ **44.** 0.08 ▢ 0.079
45. $|-1.1|$ ▢ 1.2   **46.** -5.5 ▢ $-5\frac{1}{2}$
47. $-\frac{5}{8}$ ▢ $-\frac{3}{8}$ **48.** $-19\frac{2}{3}$ ▢ $-19\frac{1}{3}$
49. $|-\frac{15}{2}|$ ▢ 7.5   **50.** $\sqrt{2}$ ▢ π
51. $\frac{99}{100}$ ▢ 0.99   **52.** $|2|$ ▢ $|-2|$
53. 0.333. . . ▢ 0.3   **54.** $|-2\frac{2}{3}|$ ▢ $\frac{7}{3}$
55. 1 ▢ $|-\frac{15}{16}|$ **56.** -0.666. . . ▢ 0

Decide whether each statement is true or false.

57. a. Every whole number is an integer.
  b. Every integer is a natural number.
  c. Every integer is a whole number.
  d. Irrational numbers are nonterminating, nonrepeating decimals.

58. a. Irrational numbers are real numbers.
  b. Every whole number is a rational number.
  c. Every rational number can be written as a fraction.
  d. Every rational number is a whole number.

59. a. Write the statement $-6 < -5$ using an inequality symbol that points in the other direction.

b. Write the statement $16 > -25$ using an inequality symbol that points in the other direction.

60. If we begin with the number -4 and find its opposite, and then find the opposite of that result, what number do we obtain?

Graph each set of numbers on the number line.

61. $\left\{-\pi, 4.25, -1\frac{1}{2}, -0.333\ldots, \sqrt{2}, -\frac{35}{8}, 3\right\}$

$$\xleftarrow{\qquad} \overset{-5\ -4\ -3\ -2\ -1\ \ 0\ \ 1\ \ 2\ \ 3\ \ 4\ \ 5}{\vert\ \ \vert\ \ \vert\ \ \vert\ \ \vert\ \ \vert\ \ \vert\ \ \vert\ \ \vert\ \ \vert\ \ \vert} \xrightarrow{\qquad}$$

62. $\left\{\pi, -2\frac{1}{8}, 2.75, -\sqrt{2}, \frac{17}{4}, 0.666\ldots, -3\right\}$

$$\xleftarrow{\qquad} \overset{-5\ -4\ -3\ -2\ -1\ \ 0\ \ 1\ \ 2\ \ 3\ \ 4\ \ 5}{\vert\ \ \vert\ \ \vert\ \ \vert\ \ \vert\ \ \vert\ \ \vert\ \ \vert\ \ \vert\ \ \vert\ \ \vert} \xrightarrow{\qquad}$$

APPLICATIONS

63. DRAFTING   The drawing shows the dimensions of an aluminum bracket. Which numbers shown are natural numbers, whole numbers, integers, rational numbers, irrational numbers, and real numbers?

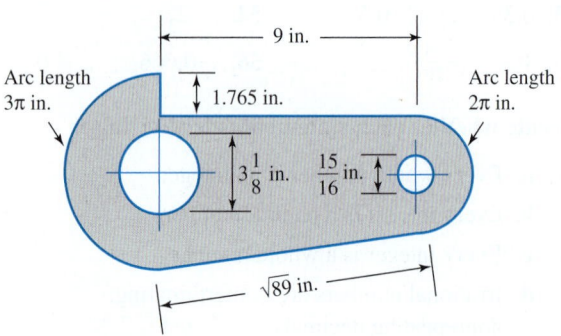

64. HISTORY   Refer to the time line.
a. What basic unit was used to scale the time line?

b. On the time line, what symbolism is used to represent zero?

c. On the time line, which numbers could be thought of as positive and which as negative?

d. Express the dates for the Maya civilization using positive and negative numbers.

MAYA CIVILIZATION

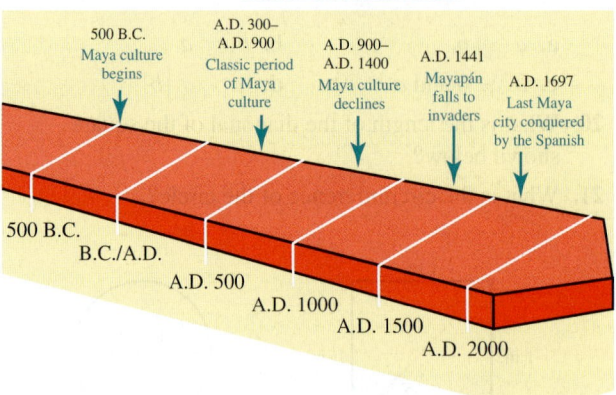

Based on data from *People in Time and Place, Western Hemisphere* (Silver Burdett & Ginn, 1991), p. 129

65. TARGET PRACTICE   Which artillery shell landed farther from the target? How can the concept of absolute value be applied to answer this question?

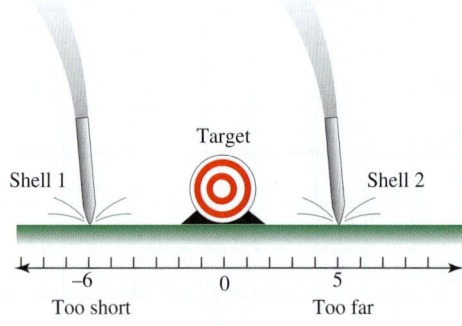

66. DRAFTING   An architect's scale has several measuring edges. The edge marked 16 divides each inch into 16 equal parts. Find the decimal form for each fractional part of one inch that is highlighted on the scale.

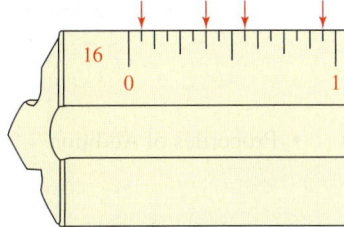

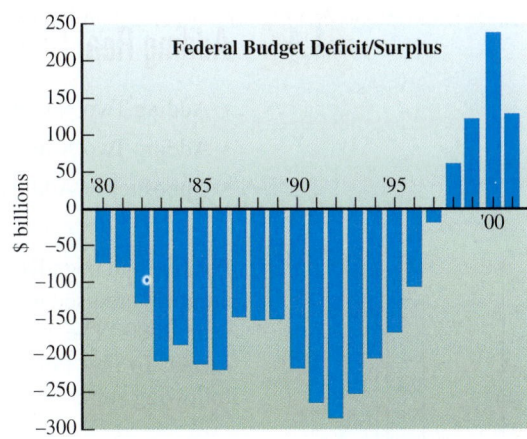

Source: U.S. Bureau of the Census

67. TRADE Each year from 1990 through 2002, the United States imported more goods and services from Japan than it exported to Japan. This caused trade deficits, which are represented by negative numbers on the graph.

a. In which three years was the deficit the worst? Estimate each of them.

b. In which year was the deficit the smallest? Estimate the deficit then.

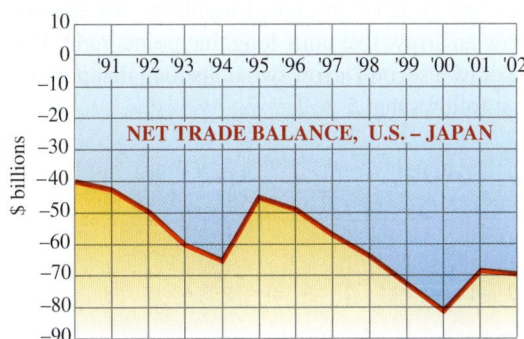

Source: U.S. Bureau of the Census

68. GOVERNMENT DEBT A budget deficit indicates that the government's expenditures were more than the revenue it took in that year. Deficits are represented by negative numbers on the graph.

a. For the years 1980–2001, when was the federal budget deficit the worst? Estimate the size of the deficit.

b. For the years 1980–2001, when did the first budget surplus occur? Estimate it. Explain what it means to have a budget surplus.

WRITING

69. Explain the difference between a rational and an irrational number.

70. Can two different numbers have the same absolute value? Explain.

71. Explain how to find the decimal equivalent of a fraction.

72. What is a real number?

REVIEW

73. Simplify: $\dfrac{24}{54}$.

74. Prime factor 60.

75. Find: $\dfrac{3}{4}\left(\dfrac{8}{5}\right)$.

76. Find: $5\dfrac{2}{3} \div 2\dfrac{5}{9}$.

77. Find: $\dfrac{3}{10} + \dfrac{2}{15}$.

78. Classify each of the following as an *expression* or an *equation*.
  a. $2x$
  b. $x = 2$

CHALLENGE PROBLEMS

79. Find the set of nonnegative integers.

80. Is 0.10100100010000 . . . a repeating decimal? Explain.

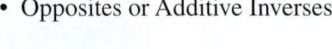

Adding Real Numbers

- Adding Two Numbers with the Same Sign
- Adding Two Numbers with Different Signs   • Properties of Addition
- Opposites or Additive Inverses

Net income ($ millions)

30
27
25
20   18
15   2nd 3rd
10
5
0   1st
      -1   Qtr. 2000   4th
-5   -6
-10   **Polaroid**
      Corporation
-15

Source: Hoover's Online

Positive and negative numbers are called **signed numbers.** In the graph on the left, signed numbers are used to denote the financial performance of the Polaroid Corporation for the year 2000. The positive numbers indicate *profits* and the negative numbers indicate *losses.* To find Polaroid's net earnings (in millions of dollars), we need to calculate the following sum:

$$\text{Net earnings} = -1 + 27 + 18 + (-6)$$

In this section, we discuss how to perform this addition and others involving signed numbers.

▪ ADDING TWO NUMBERS WITH THE SAME SIGN

A number line can be used to explain the addition of signed numbers. For example, to compute $5 + 2$, we begin at 0 and draw an arrow five units long that points right. It represents 5. From the tip of that arrow, we draw a second arrow two units long that points right. It represents 2. Since we end up at 7, it follows that $5 + 2 = 7$.

The Language of Algebra

The names of the parts of an addition fact are:

Addend Addend Sum
   ↘   ↓   ↙
      5 + 2 = 7

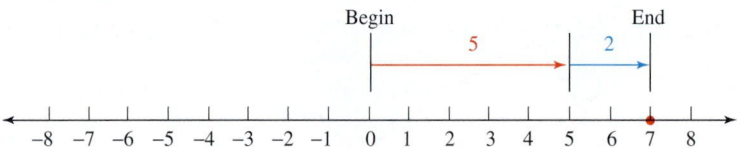

Notation

To avoid confusion, we write negative numbers within parentheses to separate the negative sign − from the addition symbol +.

$$-5 + (-2)$$

To compute $-5 + (-2)$, we begin at 0 and draw an arrow five units long that points left. It represents -5. From the tip of that arrow, we draw a second arrow two units long that points left. It represents -2. Since we end up at -7, it follows that $-5 + (-2) = -7$.

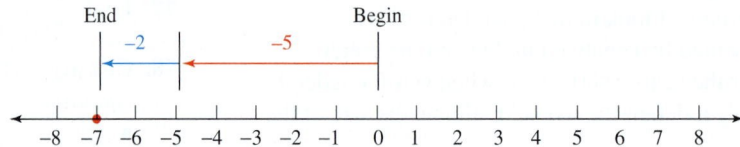

To check this result, think of the problem in terms of money. If you lost $5 ($-5$) and then lost another $2 ($-2$), you would have lost a total of $7 ($-7$).

When we use a number line to add numbers with the same sign, the arrows point in the same direction and they build upon each other. Furthermore, the answer has the same sign as the numbers that we added. We can conclude that *the sum of two positive numbers is positive* and *the sum of two negative numbers is negative.*

| **Adding Two Numbers with the Same Sign** | 1. To add two positive numbers, add them. The answer is positive. |
| | 2. To add two negative numbers, add their absolute values and make the answer negative. |

EXAMPLE 1   Find each sum: **a.** $-20 + (-15)$,   **b.** $-7.89 + (-0.6)$,   and   **c.** $-\dfrac{1}{3} + \left(-\dfrac{1}{2}\right)$.

Solution   **a.** $-20 + (-15) = -35$ Add their absolute values, 20 and 15, to get 35.
Make the answer negative.

b. Add their absolute values and make the answer negative: $-7.89 + (-0.6) = -8.49$.

c. Add their absolute values

$$\frac{1}{3} + \frac{1}{2} = \frac{2}{6} + \frac{3}{6}$$ The LCD is **6.** Build each fraction: $\dfrac{1}{3} \cdot \dfrac{2}{2} = \dfrac{2}{6}$ and $\dfrac{1}{2} \cdot \dfrac{3}{3} = \dfrac{3}{6}$.

$$= \frac{5}{6}$$ Add the numerators and write the sum over the LCD.

and make the answer negative: $-\dfrac{1}{3} + \left(-\dfrac{1}{2}\right) = -\dfrac{5}{6}$.

Self Check 1   Find the sum: **a.** $-51 + (-9)$,   **b.** $-12.3 + (-0.88)$,   **c.** $-\dfrac{1}{4} + \left(-\dfrac{2}{3}\right)$.

■ ADDING TWO NUMBERS WITH DIFFERENT SIGNS

To compute $5 + (-2)$, we begin at 0 and draw an arrow five units long that points right. From the tip of that arrow, we draw a second arrow two units long that points left. Since we end up at 3, it follows that $5 + (-2) = 3$. In terms of money, if you won $5 and then lost $2, you would have $3 left.

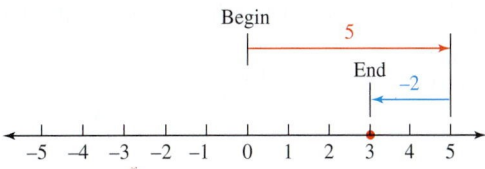

To compute $-5 + 2$, we begin at 0 and draw an arrow five units long that points left. From the tip of that arrow, we draw a second arrow two units long that points right. Since we end up at -3, it follows that $-5 + 2 = -3$. In terms of money, if you lost $5 and then won $2, you have lost $3.

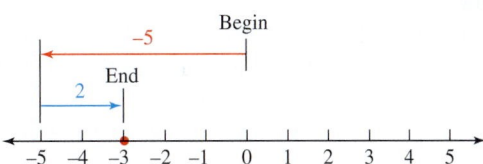

When we use a number line to add numbers with different signs, the arrows point in opposite directions and the longer arrow determines the sign of the answer. If the longer arrow represents a positive number, the sum is positive. If it represents a negative number, the sum is negative.

Adding Two Numbers with Different Signs	To add a positive number and a negative number, subtract the smaller absolute value from the larger.

1. If the positive number has the larger absolute value, the answer is positive.

2. If the negative number has the larger absolute value, make the answer negative.

EXAMPLE 2

ELEMENTARY & INTERMEDIATE
Algebra $f(x)$ **Now**™

Find each sum: **a.** $-20 + 32$,   **b.** $5.7 + (-7.4)$,   and   **c.** $-\dfrac{19}{25} + \dfrac{2}{5}$.

Solution   **a.** $-20 + 32 = 12$ Subtract the smaller absolute value from the larger: $32 - 20 = 12$. The positive number, 32, has the larger absolute value, so the answer is positive.

b. Subtract the smaller absolute value, 5.7, from the larger, 7.4, and determine the sign of the answer.

$$5.7 + (-7.4) = -1.7 \qquad \text{\color{red}The negative number, } -7.4, \text{ has the larger absolute value, so we make the answer negative.}$$

Calculators
Entering negative numbers

When using a calculator to add positive and negative numbers, we don't do anything special to enter positive numbers. To enter a negative number, say -1, on a scientific calculator, we press the *sign change* key $+/-$ after entering 1. If we use a graphing calculator, we press the *negation* key $(-)$ and then enter 1.

c. Since $\dfrac{2}{5} = \dfrac{10}{25}$, $-\dfrac{19}{25}$ has the larger absolute value. We subtract the smaller absolute value from the larger:

$$\frac{19}{25} - \frac{2}{5} = \frac{19}{25} - \frac{10}{25} \qquad \text{\color{red}Replace } \frac{2}{5} \text{ with the equivalent fraction } \frac{10}{25}.$$

$$= \frac{9}{25} \qquad \text{\color{red}Subtract the numerators and write the difference over the LCD.}$$

and determine the sign of the answer.

$$-\frac{19}{25} + \frac{10}{25} = -\frac{9}{25} \qquad \text{\color{red}Since } -\frac{19}{25} \text{ has the larger absolute value, make the answer negative.}$$

Self Check 2   Find each sum: **a.** $63 + (-87)$,   **b.** $-6.27 + 8$,   **c.** $-\dfrac{1}{10} + \dfrac{1}{2}$

EXAMPLE 3

ELEMENTARY & INTERMEDIATE
Algebra $f(x)$ **Now**™

Corporate earnings.   Find the net earnings of Polaroid Corporation for the year 2000 using the data in the graph on page 32.

Solution   To find the net earnings, we add the quarterly profits and losses (in millions of dollars), performing the additions as they occur from left to right.

$$-1 + 27 + 18 + (-6) = 26 + 18 + (-6) \qquad \text{\color{red}Add: } -1 + 27 = 26.$$

$$= 44 + (-6) \qquad \text{\color{red}Add: } 26 + 18 = 44.$$

$$= 38$$

In 2000, Polaroid's net earnings were $38 million.

Self Check 3   Add: $7 + (-13) + 8 + (-10)$.

■ PROPERTIES OF ADDITION

The addition of two numbers can be done in any order and the result is the same. For example, $8 + (-1) = 7$ and $-1 + 8 = 7$. This example illustrates that addition is **commutative.**

The Commutative Property of Addition	Changing the order when adding does not affect the answer. Let a and b represent real numbers, $$a + b = b + a$$

The Language of Algebra

Commutative is a form of the word *commute,* meaning to go back and forth.

In the following example, we add $-3 + 7 + 5$ in two ways. We will use grouping symbols ( ), called **parentheses,** to show this. Standard practice requires that the operation within the parentheses be performed first.

Method 1: Group −3 and 7

$(\mathbf{-3 + 7}) + 5 = \mathbf{4} + 5$
$\qquad\qquad\quad = 9$

Method 2: Group 7 and 5

$-3 + (\mathbf{7 + 5}) = -3 + \mathbf{12}$
$\qquad\qquad\quad = 9$

It doesn't matter how we group the numbers in this addition; the result is 9. This example illustrates that addition is **associative.**

The Associative Property of Addition	Changing the grouping when adding does not affect the answer. Let a, b, and c represent real numbers, $$(a + b) + c = a + (b + c)$$

Sometimes, an application of the associative property can simplify a computation.

EXAMPLE 4

Find the sum: $98 + (2 + 17)$.

Solution

If we use the associative property of addition to regroup, we have a convenient pair of numbers to add: $98 + 2 = 100$.

$$98 + (2 + 17) = (98 + 2) + 17$$
$$= 100 + 17$$
$$= 117$$

The Language of Algebra

Associative is a form of the word *associate,* meaning to join a group.

Self Check 4   Find the sum: $(39 + 25) + 75$.

EXAMPLE 5

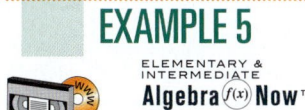

Game shows.   A contestant on *Jeopardy!* correctly answered the first question to win $100, missed the second to lose $200, correctly answered the third to win $300, and missed the fourth to lose $400. What is her score after answering four questions?

Solution We can represent money won by a positive number and money lost by a negative number. Her score is the sum of 100, −200, 300, and −400. Instead of doing the additions from left to right, we will use another approach. Applying the commutative and associative properties, we add the positives, add the negatives, and then add those results.

WHO'S WHO	WHODUNIT	WHO'S ON FIRST	HOUDINI	HOOVER	HOULIGANS
100	100	100	100	100	100
200	200	200	200	200	200
300	300	300	300	300	300
400	400	400	400	400	400
500	500	500	500	500	500

$$100 + (-200) + 300 + (-400)$$
$$= (100 + 300) + [(-200) + (-400)]$$ Reorder the numbers. Group the positives together. Group the negatives together using brackets [ ].
$$= 400 + (-600)$$ Add the positives. Add the negatives.
$$= -200$$ Add the results.

After four questions, her score was −$200, a loss of $200.

Self Check 5 Find $-6 + 1 + (-4) + (-5) + 9$.

The Language of Algebra

Identity is a form of the word *identical,* meaning the same. You have probably seen *identical* twins.

Whenever we add 0 to a number, the number remains the same:

$$8 + 0 = 8, \qquad 2.3 + 0 = 2.3, \qquad \text{and} \qquad 0 + (-16) = -16$$

These examples illustrate the **addition property of 0.** Since any number added to 0 remains the same, 0 is called the **identity element** for addition.

Addition Property of 0

When 0 is added to any real number, the result is the same real number.
For any real number a,

$$a + 0 = a \qquad \text{and} \qquad 0 + a = a$$

■ OPPOSITES OR ADDITIVE INVERSES

The Language of Algebra

Don't confuse the words *opposite* and *reciprocal.* The opposite of 4 is −4. The reciprocal of 4 is $\frac{1}{4}$.

Recall that two numbers that are the same distance from 0 on a number line, but on opposite sides of it, are called **opposites.** To develop a property of opposites, we will find $-4 + 4$ using a number line. We begin at 0 and draw an arrow four units long that points left, to represent −4. From the tip of that arrow, we draw a second arrow, four units long that points right, to represent 4. We end up at 0; therefore, $-4 + 4 = 0$.

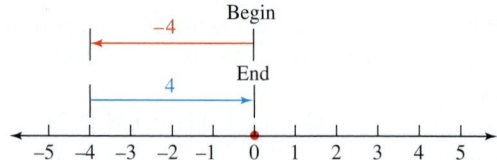

This example illustrates that when we add opposites, the result is 0. It also follows that whenever the sum of two numbers is 0, those numbers are opposites. For these reasons, opposites are also called **additive inverses.**

Addition Property of Opposites (Inverse Property of Addition)

The sum of a number and its opposite (additive inverse) is 0.
For any real number a and its opposite or additive inverse $-a$,

$$a + (-a) = 0 \qquad \text{Read } -a \text{ as "the opposite of } a.\text{"}$$

EXAMPLE 6

Find the sum: $12 + (-5) + 6 + 5 + (-12)$.

Solution   The commutative and associative properties of addition enable us to add pairs of opposites: $12 + (-12) = 0$ and $-5 + 5 = 0$.

opposites

$$12 + (-5) + 6 + 5 + (-12) = 0 + 0 + 6$$
$$= 6$$

opposites

Self Check 6   Find the sum: $8 + (-1) + 6 + 5 + (-8) + 1$.

Answers to Self Checks   **1. a.** -60,   **b.** -13.18,   **c.** $-\frac{11}{12}$ **2. a.** -24,   **b.** 1.73,   **c.** $\frac{2}{5}$ **3.** -8 **4.** 139
5. -5 **6.** 11

1.4   STUDY SET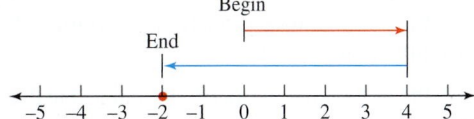

VOCABULARY   **Fill in the blanks.**

1. Positive and negative numbers are called _____ numbers.

2. In the addition statement $-2 + 5 = 3$, the result, 3, is called the _____.

3. Two numbers that are the same distance from 0 on a number line, but on opposite sides of it, are called _____.

4. The _____ property of addition states that changing the order when adding does not affect the answer. The _____ property of addition states that changing the grouping when adding does not affect the answer.

5. Since any number added to 0 remains the same (is identical), the number 0 is called the _____ element for addition.

6. The sum of a number and its opposite or additive _____ is 0.

CONCEPTS   **What addition fact is represented by each illustration?**

7.

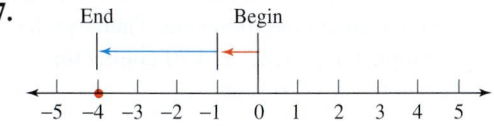

8.

Begin

End

```
  ┼──┼──┼──┼──┼──┼──┼──┼──┼──┼──┼─→
 -5 -4 -3 -2 -1  0  1  2  3  4  5
```

Add using a number line.

9. $-3 + (-2)$ **10.** $4 + (-6)$

11. $1 + (-4)$ **12.** $-3 + (-1)$

13. Complete each property of addition.
  a. $a + (-a) =$ ☐   **b.** $a + 0 =$ ☐
  c. $a + b = b +$ ☐
  d. $(a + b) + c = a +$ ☐

For each addition, determine the sign of the answer.

14. $59 + (-64)$

15. $-87 + 98$

16. Circle any opposites in the following expression.

$$12 + (-3) + (-6) + 3 + 1$$

17. Use the commutative property of addition to complete each statement.

a. $-5 + 1 =$ _____

b. $15 + (-80.5) =$ _____

c. $-20 + (4 + 20) = -20 +$ _____

18. Use the associative property of addition to complete each statement.

a. $(-6 + 2) + 8 =$ _____

b. $-7 + (7 + 3) =$ _____

19. Find each sum.

a. $5 + (-5)$ **b.** $-2.2 + 2.2$

c. $0 + (-6)$ **d.** $-\dfrac{15}{16} + 0$

e. $-\dfrac{3}{4} + \dfrac{3}{4}$ **f.** $19 + (-19)$

20. Consider $-3 + 6 + (-9) + 8 + (-4)$.

a. Add all the positives in the expression.

b. Add all of the negatives.

c. Add the results from parts **a** and **b**.

NOTATION

21. Express the commutative property of addition using the variables x and y.

22. Express the associative property of addition using the variables x, y, and z.

23. In $7 + (8 + 9)$, which addition should be done first?

24. Insert parentheses where they are needed.

$$6 + -8 + -10$$

PRACTICE   Add.

25. $6 + (-8)$ **26.** $4 + (-3)$

27. $-6 + 8$ **28.** $-21 + (-12)$

29. $-4 + (-4)$ **30.** $-5 + (-5)$

31. $9 + (-1)$ **32.** $11 + (-2)$

33. $-16 + 16$ **34.** $-25 + 25$

35. $-65 + (-12)$ **36.** $75 + (-13)$

37. $15 + (-11)$ **38.** $27 + (-30)$

39. $300 + (-335)$ **40.** $240 + (-340)$

41. $-10.5 + 2.3$ **42.** $-2.1 + 0.4$

43. $-9.1 + (-11)$ **44.** $-6.7 + (-7.1)$

45. $0.7 + (-0.5)$ **46.** $0.9 + (-0.2)$

47. $-\dfrac{9}{16} + \dfrac{7}{16}$ **48.** $-\dfrac{3}{4} + \dfrac{1}{4}$

49. $-\dfrac{1}{4} + \dfrac{2}{3}$ **50.** $\dfrac{3}{16} + \left(-\dfrac{1}{2}\right)$

51. $-\dfrac{4}{5} + \left(-\dfrac{1}{10}\right)$ **52.** $-\dfrac{3}{8} + \left(-\dfrac{1}{3}\right)$

53. $8 + (-5) + 13$ **54.** $17 + (-12) + (-23)$

55. $21 + (-27) + (-9)$ **56.** $-32 + 12 + 17$

57. $-27 + (-3) + (-13) + 22$

58. $53 + (-27) + (-32) + (-7)$

59. $-20 + (-16 + 10)$

60. $-13 + (-16 + 4)$

61. $19 + (-20 + 1)$

62. $33 + (-35 + 2)$

63. $(-7 + 8) + 2 + (-12 + 13)$

64. $(-9 + 5) + 10 + (-8 + 1)$

65. $-7 + 5 + (-10) + 7$

66. $-3 + 6 + (-9) + (-6)$

67. $-8 + 11 + (-11) + 8 + 1$

68. $2 + 15 + (-15) + 8 + (-2)$

69. $-2.1 + 6.5 + (-8.2) + 0.6$

70. $0.9 + 0.5 + (-0.2) + (-0.9)$

71. $-60 + 70 + (-10) + (-10) + 20$

72. $-100 + 200 + (-300) + (-100) + 200$

Apply the associative property of addition, and find the sum.

73. $-99 + (99 + 215)$

74. $67 + (-67 + 127)$

75. $(-112 + 56) + (-56)$

76. $(-67 + 5) + (-5)$

APPLICATIONS

77. MILITARY SCIENCE   During a battle, an army retreated 1,500 meters, regrouped, and advanced 2,400 meters. The next day, it advanced another 1,250 meters. Find the army's net gain.

78. HEALTH   Find the point total for the six risk factors (in blue) on the medical questionnaire. Then use the table to determine the patient's risk of contracting heart disease in the next 10 years.

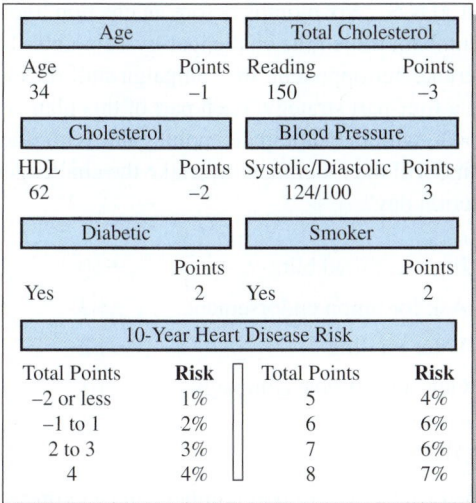

Age		Total Cholesterol	
Age	Points	Reading	Points
34	−1	150	−3
Cholesterol		**Blood Pressure**	
HDL	Points	Systolic/Diastolic	Points
62	−2	124/100	3
Diabetic		**Smoker**	
	Points		Points
Yes	2	Yes	2

10-Year Heart Disease Risk

Total Points	Risk	Total Points	Risk
−2 or less	1%	5	4%
−1 to 1	2%	6	6%
2 to 3	3%	7	6%
4	4%	8	7%

Previous Balance	New Purchases, Fees, Advances & Debts	Payments & Credits	New Balance
3,660.66	1,408.78	3,826.58	

04/21/03 Billing Date	05/16/03 Date Payment Due	9,100 Credit Line

Periodic rates may vary.
See reverse for explanation and important information.
Please allow sufficient time for mail to reach us.

79. GOLF The leaderboard shows the top finishers from the 1997 Masters Golf Tournament. Scores for each round are compared to *par,* the standard number of strokes necessary to complete the course. A score of −2, for example, indicates that the golfer used two strokes less than par to complete the course. A score of +5 indicates five strokes more than par. Determine the tournament total for each golfer.

Leaderboard

	Round				
	1	2	3	4	Total
Tiger Woods	−2	−6	−7	−3	
Tom Kite	+5	−3	−6	−2	
Tommy Tolles	0	0	0	−5	
Tom Watson	+3	−4	−3	0	

80. SUBMARINES A submarine was cruising at a depth of 1,250 feet. The captain gave the order to climb 550 feet. Relative to sea level, find the new depth of the sub.

81. CREDIT CARDS
   a. What amounts in the monthly credit card statement could be represented by negative numbers?
   b. Express the new balance as a negative number.

82. POLITICS The following proposal to limit campaign contributions was on the ballot in a state election, and it passed. What will be the net fiscal impact on the state government?

212 Campaign Spending Limits	YES ☐ NO ☐

Limits contributions to $200 in state campaigns. Fiscal impact: Costs of $4.5 million for implementation and enforcement. Increases state revenue by $6.7 million by eliminating tax deductions for lobbying.

83. MOVIE LOSSES According to the *Guinness Book of World Records 2000,* MGM's *Cutthroat Island* (1995), starring Geena Davis, cost about $100 million to produce, promote, and distribute. It has reportedly earned back just $11 million since being released. What dollar loss did the studio suffer on this film?

84. STOCK EXCHANGE Many newspapers publish daily summaries of the stock market's activity. The last entry on the line for June 12 indicates that one share of Walt Disney Co. stock lost $0.81 in value that day. How much did the value of a share of Disney stock rise or fall over the five-day period shown?

June 12	43.88	23.38	Disney	.21	0.5	87	−43	40.75	−.81
June 13	43.88	23.38	Disney	.21	0.5	86	−15	40.19	−.56
June 14	43.88	23.38	Disney	.21	0.5	87	−50	41.00	+.81
June 15	43.88	23.38	Disney	.21	0.5	89	−28	41.81	+.81
June 16	43.88	23.38	Disney				−15	41.19	−.63

Based on data from the *Los Angeles Times*

85. SAHARA DESERT   From 1980 to 1990, a satellite was used to trace the expansion and contraction of the southern boundary of the Sahara Desert in Africa. If movement southward is represented with a negative number and movement northward with a positive number, use the data in the table to determine the net movement of the Sahara Desert boundary over the 10-year period.

Years	Distance/Direction
1980–1984	240 km/South
1984–1985	110 km/North
1985–1986	30 km/North
1986–1987	55 km/South
1987–1988	100 km/North
1988–1990	77 km/South

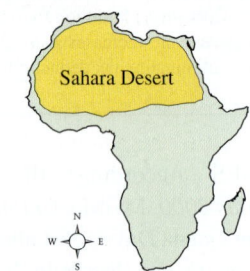

Sahara Desert

Based on data from A. Dolgoff, *Physical Geology* (D.C. Heath, 1996), p. 496

86. ELECTRONICS   A closed circuit contains two batteries and three resistors. The sum of the voltages in the loop must be 0. Is it?

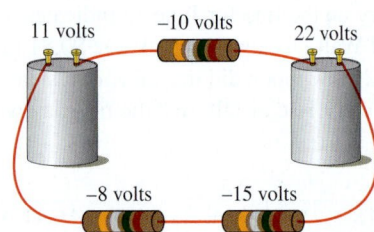

11 volts   −10 volts   22 volts
−8 volts   −15 volts

87. PROFITS AND LOSSES The 2001 quarterly profits and losses of Greyhound Bus Lines are shown in the table. Losses are denoted using parentheses. Use the data to construct a line graph. Then calculate the company's total net income for 2001.

Quarter	Net income ($ million)
1st	(10.7)
2nd	4.0
3rd	12.2
4th	(3.4)

Source: Edgar Online

88. POLITICS   Six months before an election, the incumbent trailed the challenger by 18 points. To overtake her opponent, the campaign staff decided to use a four-part strategy. Each part of this plan is shown, with the anticipated point gain. With these gains, will the incumbent overtake the challenger on election day?

   1. Intense TV ad blitz   +10
   2. Ask for union endorsement   +2
   3. Voter mailing   +3
   4. Get-out-the-vote campaign   +1

REVIEW

89. True or false: Every real number can be expressed as a decimal.

90. Multiply: $\dfrac{1}{3} \cdot \dfrac{1}{3}$.

91. What two numbers are a distance of 6 away from -3 on the number line?

92. Graph $\left\{ -2.5, \sqrt{2}, \dfrac{11}{3}, -0.333 \ldots, 0.75 \right\}$.

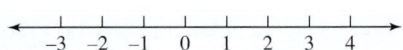

WRITING

93. Explain why the sum of two positive numbers is always positive and the sum of two negative numbers is always negative.

94. Explain why the sum of a negative number and a positive number is sometimes positive, sometimes negative, and sometimes zero.

CHALLENGE PROBLEMS

95. A set is said to be *closed under addition* if the sum of any two of its members is also a member of the set. Is the set $\{-1, 0, 1\}$ a closed set under addition? Explain.

96. Think of two numbers. First, add the absolute value of the two numbers, and write your answer. Second, add the two numbers, take the absolute value of that sum, and write that answer. Do the two answers agree? Can you find two numbers that produce different answers? When do you get answers that agree, and when don't you?

1.5 Subtracting Real Numbers

• Subtraction   • Applications Involving Subtraction

In this section, we discuss a rule to use when subtracting signed numbers.

■ SUBTRACTION

A minus symbol $-$ is used to indicate subtraction. However, this symbol is also used in two other ways, depending on where it appears in an expression.

$5 - 18$ This is read as "five minus eighteen."

-5 This is usually read as "negative five." It could also be read as "the additive inverse of five" or "the opposite of five."

$-(-5)$ This is usually read as "the opposite of negative five." It could also be read as "the additive inverse of negative five."

In $-(-5)$, parentheses are used to write the opposite of a negative number. When such expressions are encountered in computations, we simplify them by finding the opposite of the number within the parentheses.

$-(-5) = 5$ The opposite of negative five is five.

This observation illustrates the following rule.

Opposite of an Opposite   The opposite of the opposite of a number is that number.
For any real number a,

$$-(-a) = a \quad \text{Read as "the opposite of the opposite of } a \text{ is } a\text{."}$$

EXAMPLE 1

Simplify each expression: **a.** $-(-45)$,   **b.** $-(-h)$,   and   **c.** $-|-10|$.

ELEMENTARY & INTERMEDIATE
Algebra *f(x)* **Now™**

Solution

a. The number within the parentheses is -45. Its opposite is 45. Therefore, $-(-45) = 45$.

b. The opposite of the opposite of h is h. Therefore, $-(-h) = h$.

c. The notation $-|-10|$ means "the opposite of the absolute value of negative ten." Since $|-10| = 10$, we have:

$$-|-10| = -10$$

Self Check 1   Simplify each expression: **a.** $-(-1)$,   **b.** $-(-y)$,   and   **c.** $-|-500|$.

The subtraction $5 - 2$ can be thought of as taking away 2 from 5. To use a number line to illustrate this, we begin at 0 and draw an arrow 5 units long that points to the right. From the tip of that arrow, we move back two units to the left. Since we end up at 3, it follows that $5 - 2 = 3$.

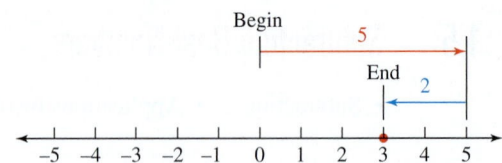

The names of the parts of a subtraction fact are:

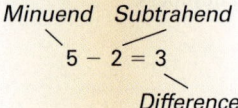

The previous illustration could also serve as a representation of the addition problem $5 + (-2)$. In the problem $5 - 2$, we subtracted 2 from 5. In the problem $5 + (-2)$, we added -2 to 5. In each case, the result is 3.

Subtracting 2.   Adding the opposite of 2.

$$5 - 2 = 3 \qquad\qquad 5 + (-2) = 3$$

The results are the same.

These observations suggest the following definition.

Subtraction of Real Numbers To subtract two real numbers, add the first number to the opposite (additive inverse) of the number to be subtracted.

Let a and b represent real numbers,

$$a - b = a + (-b)$$

EXAMPLE 2 Subtract: **a.** $-13 - 8$, **b.** $-7 - (-45)$, and **c.** $\dfrac{1}{4} - \left(-\dfrac{1}{8}\right)$.

ELEMENTARY &
INTERMEDIATE
Algebra *f(x)* **Now**™ **Solution** **a.** We read $-13 - 8$ as "negative thirteen *minus* eight."

Change the subtraction to addition.

$$-13 - 8 \quad = \quad -13 + (-8) \quad = \quad -21 \qquad \text{To subtract, add the opposite.}$$

Change the number being subtracted to its opposite.

To check, we add the difference, -21, and the subtrahend, 8, to obtain the minuend, -13.

$$-21 + 8 = -13$$

b. We read $-7 - (-45)$ as "negative seven *minus* negative forty-five."

Add

$$-7 - (-45) \quad = \quad -7 + 45 \quad = \quad 38 \qquad \text{To subtract, add the opposite.}$$

the opposite.

Check the result: $38 + (-45) = -7$.

c.
$$\frac{1}{4} - \left(-\frac{1}{8}\right) = \frac{2}{8} - \left(-\frac{1}{8}\right) \qquad \text{Build } \frac{1}{4}: \frac{1}{4} \cdot \frac{2}{2} = \frac{2}{8}.$$

$$= \frac{2}{8} + \frac{1}{8} \qquad \text{To subtract, add the opposite.}$$

$$= \frac{3}{8}$$

Check the result: $\dfrac{3}{8} + \left(-\dfrac{1}{8}\right) = \dfrac{2}{8} = \dfrac{1}{4}.$

Self Check 2   Subtract: **a.** $-32 - 25$,   **b.** $17 - (-12)$,   and   **c.** $-\dfrac{1}{3} - \left(-\dfrac{3}{4}\right).$

EXAMPLE 3

ELEMENTARY & INTERMEDIATE
Algebra *f(x)* **Now**™

Solution

Translate from words to symbols: **a.** Subtract 0.5 from 4.6,   and   **b.** subtract 4.6 from 0.5.

a. The number to be subtracted is 0.5. When we translate to mathematical symbols, we must reverse the order in which 0.5 and 4.6 appear in the sentence.

Subtract 0.5 from 4.6.

$$4.6 - 0.5 = 4.1$$

b. The number to be subtracted is 4.6. When we translate to mathematical symbols, we must reverse the order in which 4.6 and 0.5 appear in the sentence.

Subtract 4.6 from 0.5.

$$0.5 - 4.6 = 0.5 + (-4.6) \qquad \text{Add the opposite of 4.6.}$$
$$= -4.1$$

Self Check 3   **a.** Subtract 2.2 from 4.9,   and   **b.** subtract 4.9 from 2.2.

EXAMPLE 4

ELEMENTARY & INTERMEDIATE
Algebra *f(x)* **Now**™

Solution

Find: $-9 - 15 + 20 - (-6)$.

We write each subtraction as addition of the opposite and add.

$$-9 - 15 + 20 - (-6) = -9 + (-15) + 20 + 6$$
$$= -24 + 26$$
$$= 2$$

Self Check 4   Find: $-40 - (-10) + 7 - (-15)$.

■ APPLICATIONS INVOLVING SUBTRACTION

Subtraction finds the difference between two numbers. When we find the difference between the maximum value and the minimum value of a collection of measurements, we are finding the **range** of the values.

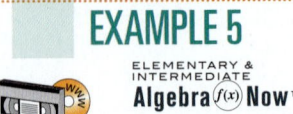

EXAMPLE 5

ELEMENTARY &
INTERMEDIATE

Algebra *f(x)* **Now**™

U.S. temperatures.   The record high temperature in the United States was 134°F in Death Valley, California, on July 10, 1913. The record low was −80°F at Prospeck Creek, Alaska, on January 23, 1971. Find the temperature range for these extremes.

Solution   To find the temperature range, we subtract the lowest temperature from the highest temperature.

$$134 - (-80) = 134 + 80$$
$$= 214$$

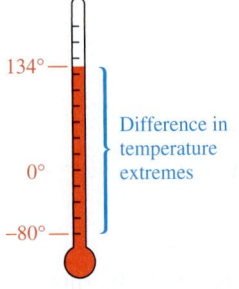

The temperature range for these extremes is 214°F.

Many things change in our lives. The price of a gallon of gasoline, the amount of money we have in the bank, and our ages are just a few examples. In general, to find the change in a quantity, we subtract the earlier value from the later value.

EXAMPLE 6

ELEMENTARY &
INTERMEDIATE

Algebra *f(x)* **Now**™

Water levels.   In one week, the water level in a storage tank went from 16 feet above normal to 14 feet below normal. Find the change in the water level.

Solution   We can represent a water level above normal using a positive number and a water level below normal using a negative number. To find the change in the water level, we subtract the previous measurement, 16, from the most recent measurement, −14.

$$-14 - 16 = -14 + (-16)$$
$$= -30$$

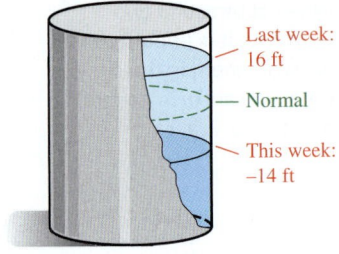

Last week:
16 ft

Normal

This week:
−14 ft

The negative result indicates that the water level fell 30 feet that week.

Answers to Self Checks   **1. a.** 1,   **b.** *y*,   **c.** −500   **2. a.** −57,   **b.** 29,   **c.** $\frac{5}{12}$ **3. a.** 2.7,   **b.** −2.7   **4.** −8

1.5   STUDY SET   ELEMENTARY & INTERMEDIATE   **Algebra** *f(x)* **Now**™

VOCABULARY   **Fill in the blanks.**

1. Two numbers that are the same distance from 0 on a number line, but on opposite sides of it, are called _____, or additive _____.

2. In the subtraction −2 − 5 = −7, the result of −7 is called the _____.

3. The difference between the maximum and the minimum value of a collection of measurements is called the _____ of the values.

4. To find the _____ in a quantity, subtract the earlier value from the later value.

CONCEPTS

5. Find the opposite, or additive inverse, of each number.

   a. 12

   b. $-\frac{1}{5}$

   c. 2.71

   d. 0

6. Complete each statement.

   a. $a - b = a +$ _____   **b.** $-(-a) =$ _____

   c. In general, $a - b$ _____ $b - a$.

7. Fill in the blanks.

   a. The opposite of the opposite of a number is that

   _____.

   b. To subtract two numbers, add the first number to the _____ of the number to be subtracted.

8. In each case, determine what number is being subtracted.

   a. $5 - 8$ **b.** $5 - (-8)$

   c. $-5 - 8$ **d.** $-5 - (-8)$

Apply the rule for subtraction and fill in the blanks.

9. $-1 - 9 = -1$ _____ _____

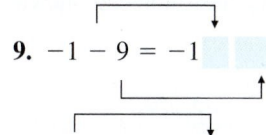

10. $1 - (-9) = 1$ _____ _____

11. Use addition to check this subtraction: $15 - (-8) = 7$. Is the result correct?

12. Which expression below represents the phrase *subtract 6 from 2*?

       $2 - 6$ or $6 - 2$

13. Write each subtraction in the following expression as addition of the opposite.

      $-10 - 8 + (-23) + 5 - (-34)$

14. Simplify each expression.

   a. $-(-1)$ **b.** $-(-0.5)$

15. Simplify each expression.

   a. $-|-500|$ **b.** $-(-y)$

NOTATION

16. Write each phrase using symbols.

   a. Negative four

   b. One minus negative seven

   c. The opposite of negative two

   d. The opposite of the absolute value of negative three

   e. The opposite of the opposite of m

PRACTICE   Subtract.

17. $4 - 7$ **18.** $1 - 6$

19. $2 - 15$ **20.** $3 - 14$

21. $8 - (-3)$ **22.** $17 - (-21)$

23. $-12 - 9$ **24.** $-25 - 17$

25. $0 - 6$ **26.** $0 - 9$

27. $0 - (-1)$ **28.** $0 - (-8)$

29. $10 - (-2)$ **30.** $11 - (-3)$

31. $-1 - (-3)$ **32.** $-1 - (-7)$

33. $20 - (-20)$ **34.** $30 - (-30)$

35. $-3 - (-3)$ **36.** $-6 - (-6)$

37. $-2 - (-7)$ **38.** $-9 - (-1)$

39. $-4 - 5$ **40.** $-3 - 4$

41. $-44 - 44$ **42.** $-33 - 33$

43. $0 - (-12)$ **44.** $0 - 12$

45. $-25 - (-25)$ **46.** $13 - (-13)$

47. $0 - 4$ **48.** $0 - (-3)$

49. $-19 - (-17)$ **50.** $-30 - (-11)$

51. $-\dfrac{1}{8} - \dfrac{3}{8}$ **52.** $-\dfrac{3}{4} - \dfrac{1}{4}$

53. $-\dfrac{9}{16} - \left(-\dfrac{1}{4}\right)$ **54.** $-\dfrac{1}{2} - \left(-\dfrac{1}{4}\right)$

55. $\dfrac{1}{3} - \dfrac{3}{4}$ **56.** $\dfrac{1}{6} - \dfrac{5}{8}$

57. $-0.9 - 0.2$ **58.** $-0.3 - 0.2$

59. $6.3 - 9.8$ **60.** $2.1 - 9.4$

61. $-1.5 - 0.8$ **62.** $-1.5 - (-0.8)$

63. $2.8 - (-1.8)$ **64.** $4.7 - (-1.9)$

65. Subtract -5 from 17.

66. Subtract 45 from -50.

67. Subtract 12 from -13.

68. Subtract -11 from -20.

Perform the operations.

69. $8 - 9 - 10$

70. $1 - 2 - 3$

71. $-25 - (-50) - 75$

72. $-33 - (-22) - 44$

73. $-6 + 8 - (-1) - 10$

74. $-4 + 5 - (-3) - 13$

75. $61 - (-62) + (-64) - 60$

76. $93 - (-92) + (-94) - 95$

77. $-6 - 7 - (-3) + 9$

78. $-1 - 3 - (-8) + 5$

79. $-20 - (-30) - 50 + 40$

80. $-24 - (-28) - 48 + 44$

APPLICATIONS

81. TEMPERATURE RECORDS   Find the difference between the record high temperature of 108°F set in 1926 and the record low of −52°F set in 1979 for New York State.

82. LIE DETECTOR TESTS   A burglar scored −18 on a lie detector test, a score that indicates deception. However, on a second test, he scored +3, a score that is inconclusive. Find the change in the scores.

83. LAND ELEVATIONS   The elevation of Death Valley, California, is 282 feet below sea level. The elevation of the Dead Sea in Israel is 1,312 feet below sea level. Find the change in their elevations.

84. CARD GAMES   Gonzalo won the second round of a card game and earned 50 points. Matt and Hydecki had to deduct the value of each of the cards left in their hands from their score on the first round. Use this information to update the score sheet. (Face cards are counted as 10 points and aces as 1 point.)

Matt	Hydecki

Running point total	Round 1	Round 2
Matt	+50	
Gonzalo	−15	
Hydecki	−2	

85. RACING   To improve handling, drivers often adjust the angle of the wheels of their car. When the wheel leans out, the degree measure is considered positive. When the wheel leans in, the degree measure is considered negative. Find the change in the position of the wheel shown.

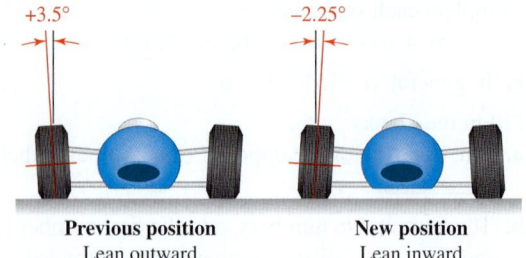

+3.5°	−2.25°
Previous position	**New position**
Lean outward	Lean inward

86. EYESIGHT   Nearsightedness, the condition where near objects are clear and far objects are blurry, is measured using negative numbers. Farsightedness, the condition where far objects are clear and near objects are blurry, is measured using positive numbers. Find the range in the measurements shown.

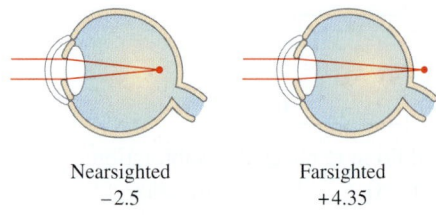

Nearsighted	Farsighted
−2.5	+4.35

87. HISTORY   Plato, a famous Greek philosopher, died in 347 B.C. (−347) at the age of 81. When was he born?

88. HISTORY   Julius Caesar, a famous Roman leader, died in 44 B.C. (−44) at the age of 56. When was he born?

89. GAUGES   Many automobiles have an ammeter like that shown. If the headlights, which draw a current of 7 amps, and the radio, which draws a current of 6 amps, are both turned on, which way will the arrow move? What will be the new reading?

90. U.S. JOBS   The graph shows the number of jobs gained or lost each month during the year 2002.
   a. In what month were the most jobs gained? Estimate the number.

b. In what month were the most jobs lost? Estimate the number.

c. Find the range for these two extremes.

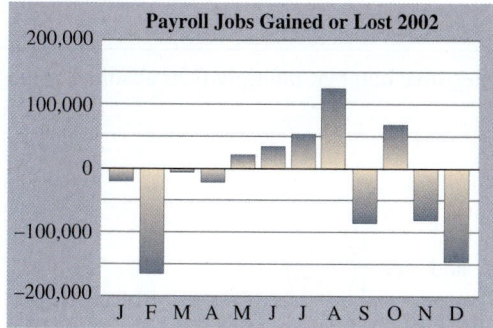

Source: Bureau of Labor Statistics

WRITING

91. Is subtracting 2 from 10 the same as subtracting 10 from 2? Explain.

92. How is $-1 - (-8)$ read?

93. Why is addition of signed numbers taught before subtraction of signed numbers?

94. Explain why we know that the answer to $4 - 10$ is negative without having to do any computation.

REVIEW

95. Find the prime factorization of 30.

96. What factor do the numerator and denominator of the fraction $\frac{15}{18}$ have in common? Simplify the fraction.

97. Build the fraction $\frac{3}{8}$ to an equivalent fraction with a denominator of 56.

98. Write the set of integers.

99. True or false: $-4 > -5$?

100. Use the associative property of addition to simplify the calculation: $-18 + (18 + 89)$.

CHALLENGE PROBLEMS

101. Suppose x is positive and y is negative. Determine whether each statement is true or false.
a. $x - y > 0$ **b.** $y - x < 0$
c. $-x < 0$ **d.** $-y < 0$

102. Find:

$$1 - 2 + 3 - 4 + 5 - 6 + \cdots + 99 - 100.$$

1.6 Multiplying and Dividing Real Numbers

- Multiplying Signed Numbers
- Dividing Signed Numbers
- Properties of Multiplication
- Properties of Division

In this section, we will develop rules for multiplying and dividing positive and negative numbers.

■ MULTIPLYING SIGNED NUMBERS

The Language of Algebra

The names of the parts of a multiplication fact are:

Factor Factor Product

$$4(3) = 12$$

Multiplication represents repeated addition. For example, 4(3) equals the sum of four 3's.

$$4(3) = 3 + 3 + 3 + 3$$
$$= 12$$

This example illustrates that *the product of two positive numbers is positive.*

To develop a rule for multiplying a positive number and a negative number, we will find $4(-3)$, which equals the sum of four -3's.

$$4(-3) = -3 + (-3) + (-3) + (-3)$$
$$= -12$$

We see that the result is negative. As a check, think of the problem in terms of money. If you lose $3 four times, you have lost a total of $12, which is denoted $-\$12$. This example illustrates that *the product of a positive number and a negative number is negative.*

Multiplying Two Numbers with Unlike Signs	To multiply a positive number and a negative number, multiply their absolute values. Then make the product negative.

EXAMPLE 1

ELEMENTARY & INTERMEDIATE
Algebra *f(x)* **Now**™

Multiply: **a.** $8(-12)$,   **b.** $-15 \cdot 5$,   and   **c.** $\dfrac{3}{4}\left(-\dfrac{4}{15}\right)$.

Solution   **a.** $8(-12) = -96$ Multiply the absolute values, 8 and 12, to get 96. The signs are unlike. Make the product negative.

b. $-15 \cdot 5 = -75$ Multiply the absolute values, 15 and 5, to get 75. The signs are unlike. Make the product negative.

The Language of Algebra

A positive number and a negative number are said to have *unlike* signs.

c. $\dfrac{3}{4}\left(-\dfrac{4}{15}\right) = -\dfrac{3 \cdot 4}{4 \cdot 15}$ Multiply the absolute values $\frac{3}{4}$ and $\frac{4}{15}$. The signs are unlike. Make the product negative.

$$= -\dfrac{\overset{1}{\cancel{3}} \cdot \overset{1}{\cancel{4}}}{\underset{1}{\cancel{4}} \cdot \underset{1}{\cancel{3}} \cdot 5}$$ To simplify the fraction, factor 15 as $3 \cdot 5$. Remove the common factors 3 and 4 in the numerator and denominator.

$$= -\dfrac{1}{5}$$

Self Check 1   Multiply: **a.** $20(-3)$,   **b.** $-3 \cdot 5$,   **c.** $-\dfrac{5}{8} \cdot \dfrac{16}{25}$

EXAMPLE 2

Medicine.   A doctor changes the setting on a heart monitor so that the screen display is magnified by a factor of 1.5. If the current low reading is -14, what will it be after the setting is changed?

Solution   To *magnify by a factor of 1.5* means to multiply by 1.5. Therefore, the new low will be $1.5(-14)$. To find this product, we multiply the absolute values, 1.5 and 14 and make the product negative.

$1.5(-14) = -21$ Multiply absolute values, 1.5 and 14, to get 21. Since 1.5 and -14 have unlike signs, the product is negative.

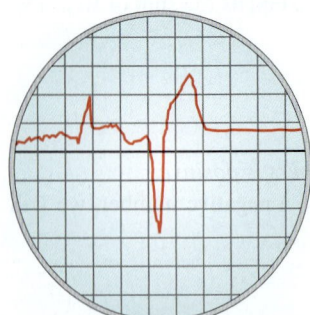

When the setting is changed, the low reading will be -21.

To develop a rule for multiplying negative numbers, we will find $-4(-1)$, $-4(-2)$, and $-4(-3)$. In the following list, we multiply -4 and a series of factors that decrease by 1. We know how to find the first four products. Graphing those results on a number line is helpful in determining the last three products.

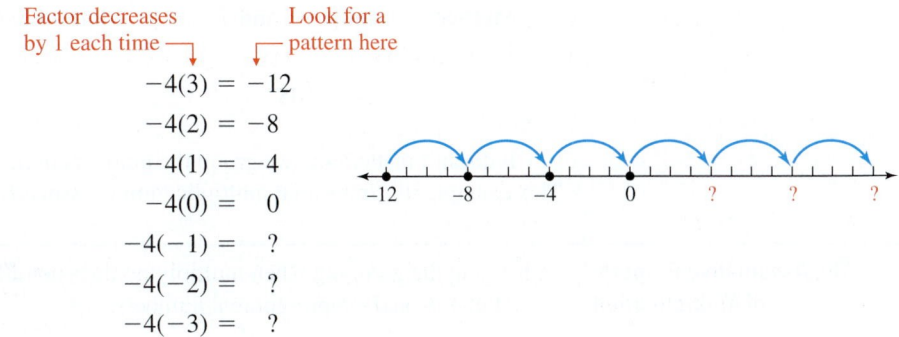

$$-4(3) = -12$$
$$-4(2) = -8$$
$$-4(1) = -4$$
$$-4(0) = 0$$
$$-4(-1) = ?$$
$$-4(-2) = ?$$
$$-4(-3) = ?$$

The Language of Algebra

Two negative numbers, as well as two positive numbers, are said to have *like* signs.

From the pattern, we see that the product increases by 4 each time. Thus,

$$-4(-1) = 4, \qquad -4(-2) = 8, \qquad \text{and} \qquad -4(-3) = 12$$

These results illustrate that *the product of two negative numbers is positive.* As a check, think of $-4(-3)$ as losing four debts of $3. This is equivalent to gaining $12. Therefore, $-4(-\$3) = \$12$.

Since the product of two positive numbers is positive, and the product of two negative numbers is also positive, we can summarize the multiplication rule as follows.

Multiplying Two Numbers with Like Signs	To multiply two real numbers with the same sign, multiply their absolute values. The product is positive.

EXAMPLE 3

ELEMENTARY & INTERMEDIATE
Algebra *f(x)* **Now**™

Multiply: **a.** $-5(-6)$ and **b.** $\left(-\dfrac{1}{2}\right)\left(-\dfrac{5}{8}\right)$.

Solution   **a.** $-5(-6) = 30$ Multiply the absolute values, 5 and 6, to get 30. Since both factors are negative, the product is positive.

b. $\left(-\dfrac{1}{2}\right)\left(-\dfrac{5}{8}\right) = \dfrac{5}{16}$ Multiply the absolute values, $\frac{1}{2}$ and $\frac{5}{8}$, to get $\frac{5}{16}$. Since the factors have like signs, the product is positive.

Self Check 3   Multiply: **a.** $-15(-8)$,   **b.** $-\dfrac{1}{4}\left(-\dfrac{1}{3}\right)$.

■ PROPERTIES OF MULTIPLICATION

The multiplication of two numbers can be done in any order; the result is the same. For example, $-6(5) = -30$ and $5(-6) = -30$. This shows that multiplication is **commutative.**

The Commutative Property of Multiplication	Changing the order when multiplying does not affect the answer. Let *a* and *b* represent real numbers, $$ab = ba$$

In the following example, we multiply $-3 \cdot 7 \cdot 5$ in two ways. Recall that the operation within the parentheses should be performed first.

Method 1: Group −3 and 7	**Method 2: Group 7 and 5**
$(-3 \cdot 7)5 = (-21)5$	$-3(7 \cdot 5) = -3(35)$
$= -105$	$= -105$

It doesn't matter how we group the numbers in this multiplication; the result is -105. This example illustrates that multiplication is **associative.**

The Associative Property of Multiplication	Changing the grouping when multiplying does not affect the answer. Let a, b, and c represent real numbers, $$(ab)c = a(bc)$$

EXAMPLE 4

ELEMENTARY &
INTERMEDIATE
Algebra $f(x)$ **Now**™

Multiply: **a.** $-5(-37)(-2)$ and   **b.** $-4(-3)(-2)(-1)$.

Solution   Using the commutative and associative properties of multiplication, we can reorder and regroup the factors to simplify computations.

a. Since it is easy to multiply by 10, we will find $-5(-2)$ first.

$$-5(-37)(-2) = 10(-37)$$
$$= -370$$

b. $-4(-3)(-2)(-1) = 12(2)$ Multiply the first two factors and multiply the last two factors.
$$= 24$$

Self Check 4   Multiply: **a.** $-25(-3)(-4)$,   **b.** $-1(-2)(-3)(-3)$.

In Example 4a, we multiplied three negative numbers. In Example 4b, we multiplied four negative numbers. The results illustrate the following fact.

Multiplying Negative Numbers	The product of an even number of negative numbers is positive. The product of an odd number of negative numbers is negative.

Success Tip

If 0 is a factor in a multiplication, the product is 0. For example,

$$25(-6)(0)(-17) = 0$$

Whenever we multiply 0 and a number, the product is 0:

$$0 \cdot 8 = 0, \qquad 6.5(0) = 0, \qquad \text{and} \qquad 0(-12) = 0$$

These examples illustrate the **multiplication property of 0.**

Multiplication Property of 0	The product of 0 and any real number is 0. For any real number a, $$0 \cdot a = 0 \qquad \text{and} \qquad a \cdot 0 = 0$$

Whenever we multiply a number by 1, the number remains the same:

$$1 \cdot 6 = 6, \qquad 4.57 \cdot 1 = 4.57, \qquad \text{and} \qquad 1(-9) = -9$$

These examples illustrate the **multiplication property of 1.** Since any number multiplied by 1 remains the same (is identical), the number 1 is called the **identity element** for multiplication.

Multiplication Property of 1 (Identity Property of Multiplication)	The product of 1 and any number is that number. For any real number a, $$1 \cdot a = a \qquad \text{and} \qquad a \cdot 1 = a$$

Two numbers whose product is 1 are **reciprocals** or **multiplicative inverses** of each other. For example, 8 is the multiplicative inverse of $\frac{1}{8}$, and $\frac{1}{8}$ is the multiplicative inverse of 8, because $8 \cdot \frac{1}{8} = 1$. Likewise, $-\frac{3}{4}$ and $-\frac{4}{3}$ are multiplicative inverses because $-\frac{3}{4}\left(-\frac{4}{3}\right) = 1$. All real numbers, except 0, have a multiplicative inverse.

Multiplicative Inverses (Inverse Property of Multiplication)	The product of any number and its multiplicative inverse (reciprocal) is 1. For any nonzero real number a, $$a\left(\frac{1}{a}\right) = 1$$

EXAMPLE 5

Find the reciprocal of each number.   **a.** $\frac{2}{3}$,   **b.** $-\frac{2}{3}$,   and   **c.** -11.

Solution   **a.** The reciprocal of $\frac{2}{3}$ is $\frac{3}{2}$ because $\frac{2}{3}\left(\frac{3}{2}\right) = 1$.

b. The reciprocal of $-\frac{2}{3}$ is $-\frac{3}{2}$ because $-\frac{2}{3}\left(-\frac{3}{2}\right) = 1$.

c. The reciprocal of -11 is $-\frac{1}{11}$ because $-11\left(-\frac{1}{11}\right) = 1$.

Self Check 5   Find the reciprocal of each number: **a.** $-\frac{15}{16}$,   **b.** $\frac{15}{16}$,   **c.** -27.

■ DIVIDING SIGNED NUMBERS

Every division fact can be written as an equivalent multiplication fact.

Division	Let a, b, and c represent real numbers, where $b \neq 0$, $$\frac{a}{b} = c \qquad \text{provided that} \qquad c \cdot b = a$$

We can use this relationship between multiplication and division to develop rules for dividing signed numbers. For example,

$$\frac{15}{5} = 3 \qquad \text{because} \qquad 3(5) = 15$$

From this example, we see that *the quotient of two positive numbers is positive.*

To determine the quotient of two negative numbers, we consider $\frac{-15}{-5}$.

$$\frac{-15}{-5} = 3 \qquad \text{because} \qquad 3(-5) = -15$$

The Language of Algebra

The names of the parts of a division fact are:

Dividend   Quotient

$$\frac{15}{5} = 3$$

Divisor

From this example, we see that the *quotient of two negative numbers is positive.*

To determine the quotient of a positive number and a negative number, we consider $\frac{15}{-5}$.

$$\frac{15}{-5} = -3 \qquad \text{because} \qquad -3(-5) = 15$$

From this example, we see that *the quotient of a positive number and a negative number is negative.*

To determine the quotient of a negative number and a positive number, we consider $\frac{-15}{5}$.

$$\frac{-15}{5} = -3 \qquad \text{because} \qquad -3(5) = -15$$

From this example, we see that *the quotient of a negative number and a positive number is negative.*

We summarize the rules from the previous examples and note that they are similar to the rules for multiplication.

Dividing Two Real Numbers   To divide two real numbers, divide their absolute values.

1. The quotient of two numbers with *like* signs is positive.

2. The quotient of two numbers with *unlike* signs is negative.

EXAMPLE 6

ELEMENTARY &
INTERMEDIATE
Algebra $f(x)$ **Now**™

Find each quotient: **a.** $\frac{-81}{-9}$,   **b.** $\frac{45}{-9}$,   and   **c.** $-2.87 \div 0.7$.

Solution   **a.** $\dfrac{-81}{-9} = 9$ Divide the absolute values, 81 by 9, to get 9. Since the signs are like, the quotient is positive.

Multiply to check the result: $9(-9) = -81$.

b. $\dfrac{45}{-9} = -5$ Divide the absolute values, 45 by 9, to get 5. Since the signs are unlike, make the quotient negative.

Multiply to check the result: $-5(-9) = 45$.

c. $-2.87 \div 0.7 = -4.1$ Since the signs are unlike, make the quotient negative.

Multiply to check the result: $-4.1(0.7) = -2.87$.

Self Check 6   Find each quotient: **a.** $\dfrac{-28}{-4}$,   **b.** $\dfrac{75}{-25}$,   **c.** $0.32 \div (-1.6)$.

EXAMPLE 7

ELEMENTARY &
INTERMEDIATE
Algebra $f(x)$ **Now**™

Divide: $-\dfrac{5}{16} \div \left(-\dfrac{1}{2}\right)$.

Solution
$$-\frac{5}{16} \div \left(-\frac{1}{2}\right) = -\frac{5}{16}\left(-\frac{2}{1}\right)$$

Multiply the first fraction by the reciprocal of the second fraction. The reciprocal of $-\frac{1}{2}$ is $-\frac{2}{1}$.

$$= \frac{5 \cdot 2}{16 \cdot 1}$$

Multiply the absolute values $\frac{5}{16}$ and $\frac{2}{1}$. Since the signs are like, the product is positive.

$$= \frac{5 \cdot \overset{1}{\cancel{2}}}{\underset{1}{\cancel{2}} \cdot 8 \cdot 1}$$

To simplify the fraction, factor 16 as $2 \cdot 8$. Replace $\frac{2}{2}$ with $\frac{1}{1}$.

$$= \frac{5}{8}$$

Self Check 7   Divide: $\dfrac{3}{4} \div \left(-\dfrac{5}{8}\right)$.

EXAMPLE 8

ELEMENTARY &
INTERMEDIATE
Algebra *f(x)* Now™

Depreciation.   Over an 8-year period, the value of a \$150,000 house fell at a uniform rate to \$110,000. Find the amount of depreciation per year.

Solution   First, we find the change in the value of the house.

$$110{,}000 - 150{,}000 = -40{,}000 \quad \text{Subtract the previous value from the current value.}$$

The result represents a drop in value of \$40,000. Since the depreciation occurred over 8 years, we divide $-40{,}000$ by 8.

$$\frac{-40{,}000}{8} = -5{,}000 \quad \text{Divide the absolute values, 40,000 by 8, to get 5,000, and make the quotient negative.}$$

The house depreciated \$5,000 per year.

The Language of Algebra

Depreciation is a form of the word *depreciate,* meaning to lose value. You've probably heard that the minute you drive a new car off the lot, it has depreciated.

■ PROPERTIES OF DIVISION

Whenever we divide a number by 1, the quotient is that number:

$$\frac{12}{1} = 12, \qquad \frac{-80}{1} = -80, \qquad \text{and} \qquad 7.75 \div 1 = 7.75$$

Furthermore, whenever we divide a nonzero number by itself, the quotient is 1:

$$\frac{35}{35} = 1, \qquad \frac{-4}{-4} = 1, \qquad \text{and} \qquad 0.9 \div 0.9 = 1$$

These observations suggest the following properties of division.

Division Properties   Any number divided by 1 is the number itself. Any number (except 0) divided by itself is 1.

For any real number a,

$$\frac{a}{1} = a \qquad \text{and} \qquad \frac{a}{a} = 1 \qquad (\text{where } a \neq 0)$$

We will now consider division that involves zero. First, we examine division of zero. As an example, let's consider $\frac{0}{2}$. We know that

$$\frac{0}{2} = 0 \qquad \text{because} \qquad 0(2) = 0$$

The Language of Algebra

When we say a division by 0, such as $\frac{2}{0}$, is *undefined,* we mean that $\frac{2}{0}$ does not represent a real number.

From this example, we see that *0 divided by a nonzero number is 0.*

Next, we consider division by zero by considering $\frac{2}{0}$. We know that $\frac{2}{0}$ has no answer because there is no number we can multiply 0 by to get 2. We say that such a division is **undefined.**

These results suggest the following division facts.

Division Involving 0   For any nonzero real number a,

$$\frac{0}{a} = 0 \qquad \text{and} \qquad \frac{a}{0} \text{ is undefined.}$$

EXAMPLE 9   Find each quotient, if possible: **a.** $\dfrac{0}{13}$ and   **b.** $\dfrac{-13}{0}$.

Solution   **a.** $\dfrac{0}{13} = 0$ because $0(13) = 0$.

b. Since $\dfrac{-13}{0}$ involves division by zero, the division is undefined.

Self Check 9   Find each quotient, if possible: **a.** $\dfrac{4}{0}$,   **b.** $\dfrac{0}{17}$.

Answers to Self Checks   **1. a.** -60,   **b.** -15,   **c.** $-\frac{2}{5}$ **3. a.** 120,   **b.** $\frac{1}{12}$ **4. a.** -300,   **b.** 18
5. a. $-\frac{16}{15}$,   **b.** $\frac{16}{15}$,   **c.** $-\frac{1}{27}$ **6. a.** 7,   **b.** -3,   **c.** -0.2 **7.** $-\frac{6}{5}$
9. a. undefined,   **b.** 0

1.6   STUDY SET   ELEMENTARY & INTERMEDIATE Algebra *f(x)* Now™

VOCABULARY   Fill in the blanks.

1. The answer to a multiplication problem is called a _____. The answer to a division problem is called a _____.

2. The numbers -4 and -6 are said to have _____ signs. The numbers -10 and 12 are said to have _____ signs.

3. The _____ property of multiplication states that changing the order when multiplying does not affect the answer.

4. The _____ property of multiplication states that changing the grouping when multiplying does not affect the answer.

5. Division of a nonzero number by zero is _____.

6. If the product of two numbers is 1, the numbers are called _____ or _____ inverses.

CONCEPTS   Fill in the blanks.

7. The expression $-5 + (-5) + (-5) + (-5)$ can be represented by the multiplication $4()$.

8. The quotient of two numbers with _____ signs is negative.

9. The product of two negative numbers is _____.

10. The product of zero and any number is ___.

11. The product of ___ and any number is that number.

12. We know that $\frac{20}{-2} = -10$ because $-10(\quad) = 20$.

13. **a.** If we multiply two different numbers and the answer is 0, what must be true about one of the numbers?

   b. If we multiply two different numbers and the answer is 1, what must be true about the numbers?

14. **a.** If we divide two numbers and the answer is 1, what must be true about the numbers?

   b. If we divide two numbers and the answer is 0, what must be true about the numbers?

Let POS stand for a positive number and NEG stand for a negative number. Determine the sign of each result, if possible.

15. **a.** POS · NEG   **b.** POS + NEG

   c. POS − NEG   **d.** $\dfrac{\text{POS}}{\text{NEG}}$

16. **a.** NEG · NEG   **b.** NEG + NEG

   c. NEG − NEG   **d.** $\dfrac{\text{NEG}}{\text{NEG}}$

17. Complete each property of multiplication.
   a. $a \cdot b = b \cdot \boxed{}$ **b.** $(ab)c = \boxed{}$
   c. $0 \cdot a = \boxed{}$ **d.** $1 \cdot a = \boxed{}$
   e. $a\left(\dfrac{1}{a}\right) = \boxed{}$ $(a \neq 0)$

18. Complete each property of division.
   a. $\dfrac{a}{1} = \boxed{}$ **b.** $\dfrac{a}{a} = \boxed{}$ $(a \neq 0)$
   c. $\dfrac{0}{a} = \boxed{}$ $(a \neq 0)$ **d.** $\dfrac{a}{0}$ is $\boxed{}$

19. Which property justifies each statement?
   a. $-5(2 \cdot 17) = (-5 \cdot 2)17$

   b. $-5\left(-\dfrac{1}{5}\right) = 1$

c. $-5 \cdot 2 = 2(-5)$

d. $-5(1) = -5$

20. Complete the table.

Number	Opposite (additive inverse)	Reciprocal (multiplicative inverse)
2		
$-\dfrac{4}{5}$		
-55		
1.75		

NOTATION

21. Write each sentence using symbols.
   a. The product of negative four and negative five is twenty.

   b. The quotient of sixteen and negative eight is negative two.

22. Write each expression without − signs.
   a. $\dfrac{-1}{-2}$ **b.** $\dfrac{-7}{-8}$

PRACTICE   Perform each operation.

23. $-2 \cdot 8$ 24. $-3 \cdot 4$

25. $(-6)(-9)$ 26. $(-8)(-7)$

27. $12(-5)$ 28. $(-9)(11)$

29. $-6 \cdot 4$ 30. $-8 \cdot 9$

31. $-20(40)$ 32. $-10(10)$

33. $(-6)(-6)$ 34. $(-1)(-1)$

35. $-0.6(-4)$ 36. $-0.7(-8)$

37. $1.2(-0.4)$ 38. $0(-0.2)$

39. $-1.1(-0.9)$ 40. $-2.3(-3.1)$

41. $7.2(-2.1)$ 42. $4.6(-5.4)$

43. $\dfrac{1}{2}\left(-\dfrac{3}{4}\right)$ 44. $\dfrac{1}{3}\left(-\dfrac{5}{16}\right)$

45. $\left(-\dfrac{7}{8}\right)\left(-\dfrac{2}{21}\right)$ 46. $\left(-\dfrac{5}{6}\right)\left(-\dfrac{2}{15}\right)$

47. $-\dfrac{16}{25} \cdot \dfrac{15}{64}$ 48. $-\dfrac{15}{16} \cdot \dfrac{8}{25}$

49. $-1\dfrac{1}{4}\left(-\dfrac{3}{4}\right)$ 50. $-1\dfrac{1}{8}\left(-\dfrac{3}{8}\right)$

51. $-5.2 \cdot 100$ 52. $-1.17 \cdot 1{,}000$

53. $0(-22)$

54. $-8 \cdot 0$

55. $-3(-4)(0)$

56. $15(0)(-22)$

57. $3(-4)(-5)$

58. $(-2)(-4)(-5)$

59. $(-4)(3)(-7)$

60. $5(-3)(-4)$

61. $(-2)(-3)(-4)(-5)$

62. $(-3)(-4)(5)(-6)$

63. $\dfrac{1}{2}\left(-\dfrac{1}{3}\right)\left(-\dfrac{1}{4}\right)$

64. $\dfrac{1}{3}\left(-\dfrac{1}{5}\right)\left(-\dfrac{1}{7}\right)$

65. $-2(-3)(-4)(-5)(-6)$

66. $-9(-7)(-5)(-3)(-1)$

67. $-30 \div (-3)$

68. $-12 \div (-2)$

69. $\dfrac{-6}{-2}$

70. $\dfrac{-36}{9}$

71. $\dfrac{4}{-2}$

72. $\dfrac{-9}{3}$

73. $\dfrac{80}{-20}$

74. $\dfrac{-66}{33}$

75. $\dfrac{17}{-17}$

76. $\dfrac{-24}{24}$

77. $\dfrac{-110}{-110}$

78. $\dfrac{-200}{-200}$

79. $\dfrac{-160}{40}$

80. $\dfrac{-250}{-50}$

81. $\dfrac{320}{-16}$

82. $\dfrac{-180}{36}$

83. $\dfrac{0.5}{-100}$

84. $\dfrac{-1.7}{10}$

85. $\dfrac{0}{150}$

86. $\dfrac{225}{0}$

87. $\dfrac{-17}{0}$

88. $\dfrac{0}{-12}$

89. $-\dfrac{1}{3} \div \dfrac{4}{5}$

90. $-\dfrac{2}{3} \div \dfrac{7}{8}$

91. $-\dfrac{9}{16} \div \left(-\dfrac{3}{20}\right)$

92. $-\dfrac{4}{5} \div \left(-\dfrac{8}{25}\right)$

93. $-3\dfrac{3}{8} \div \left(-2\dfrac{1}{4}\right)$

94. $-3\dfrac{4}{15} \div \left(-2\dfrac{1}{10}\right)$

95. $\dfrac{-23.5}{5}$

96. $\dfrac{-337.8}{6}$

97. $\dfrac{-24.24}{-0.8}$

98. $\dfrac{-55.02}{-0.7}$

Use the associative property of multiplication to find each product.

99. $-\dfrac{1}{2}(2 \cdot 67)$

100. $\left(-\dfrac{5}{16} \cdot \dfrac{1}{7}\right)7$

101. $-0.2(10 \cdot 3)$

102. $-1.5(100 \cdot 4)$

APPLICATIONS

103. TEMPERATURE CHANGE   In a lab, the temperature of a fluid was decreased 6° per hour for 12 hours. What signed number indicates the change in temperature?

104. BACTERIAL GROWTH   To slowly warm a bacterial culture, biologists programmed a heating pad under the culture to increase the temperature 4° every hour for 6 hours. What signed number indicates the change in the temperature of the pad?

105. GAMBLING   A gambler places a $40 bet and loses. He then decides to go "double or nothing" and loses again. Feeling that his luck has to change, he goes "double or nothing" once more and, for the third time, loses. What signed number indicates his gambling losses?

106. REAL ESTATE   A house has depreciated $1,250 each year for 8 years. What signed number indicates its change in value over that time period?

107. PLANETS   The temperature on Pluto gets as low as $-386°$ F. This is twice as low as the lowest temperature reached on Jupiter. What is the lowest temperature on Jupiter?

108. CAR RADIATORS   The instructions on the back of a container of antifreeze state, "A 50/50 mixture of antifreeze and water protects against freeze-ups down to $-34°$ F, while a 60/40 mix protects against freeze-ups down to one and one-half times that temperature." To what temperature does the 60/40 mixture protect?

109. ACCOUNTING   For 1999, the total net income for Converse, the sports shoe company, was about $-\$22.8$ million. The company's losses for 2000 were even worse, by a factor of about 1.9. What signed number indicates the company's total net income that year?

110. AIRLINES In the income statement for Trans World Airlines, numbers within parentheses represent a loss. Complete the statement given these facts. The second and fourth quarter losses were approximately the same and totaled $-\$1,000$ million. The third quarter loss was about $\frac{3}{5}$ of the first quarter loss.

TWA INCOME STATEMENT				2002
All amounts in millions of US dollars	1st Qtr (1,550)	2nd Qtr (?)	3rd Qtr (?)	4th Qtr (?)

111. THE QUEEN MARY The ocean liner Queen Mary was commissioned in 1936 and cost $22,500,000 to build. In 1967, the ship was purchased by the city of Long Beach, California for $3,450,000 and now serves as a hotel and convention center. What signed number indicates the annual average depreciation of the Queen Mary over the 31-year period from 1936 to 1967? Round to the nearest dollar.

112. COMPUTERS The formula $= A1*B1*C1$ in cell D1 of the following spreadsheet instructs the computer to multiply the values in cells A1, B1, and C1 and to print the result *in place of the formula* in cell D1. (The symbol $*$ represents multiplication.) What value will the computer print in the cell D1? What values will be printed in cells D2 and D3?

Microsoft Excel-Book 1			
File Edit View Insert Format Tools			
	A	B	C	D
1	4	–5	–17	= A1*B1*C1
2	22	–30	14	= A2*B2*C2
3	–60	–20	–34	= A3*B3*C3
4				

Sheet 1 / Sheet 2 / Sheet 3 / Sheet 4 / Sheet 5

113. PHYSICS An oscilloscope displays electrical signals which appear as wavy lines on a screen. By switching the magnification setting to $\times 2$, for example, the height of the "peak" and the depth of the "valley" of a graph will be doubled. Use signed numbers to indicate the height and depth of the display for each setting of the magnification dial.

a. normal   **b.** $\times 0.5$

c. $\times 1.5$ **d.** $\times 2$

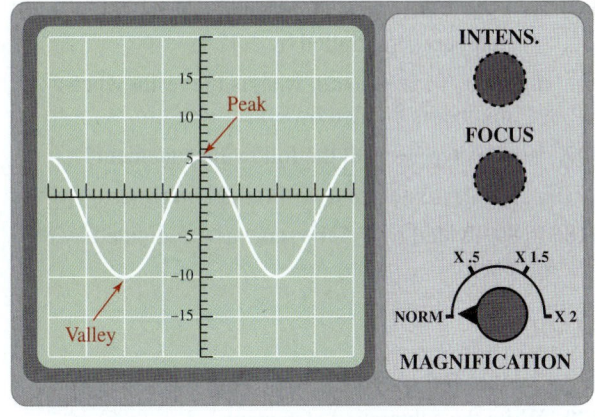

114. LIGHT Water acts as a selective filter of light. In the illustration, we see that red light waves penetrate water only to a depth of about 5 meters. How many times deeper does

a. yellow light penetrate than red light?

b. green light penetrate than orange light?

c. blue light penetrate than yellow light?

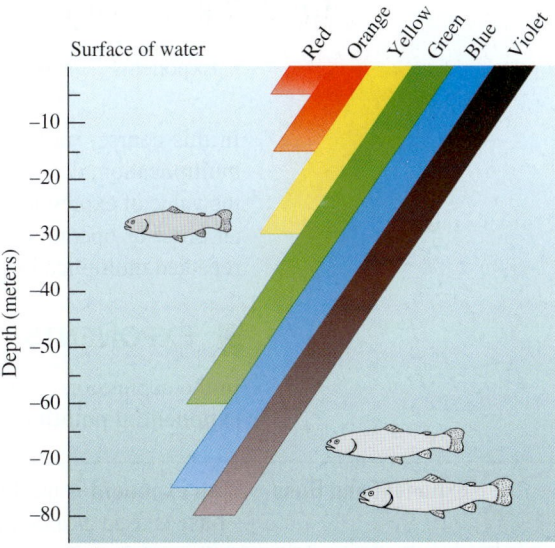

WRITING

115. When a calculator was used to compute $16 \div 0$, the message shown appeared on the display screen. Explain what the message means.

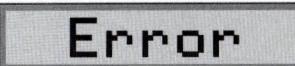

116. a. Find $-1(8)$. In general, what is the result when a number is multiplied by -1?

   b. Find $\frac{8}{-1}$. In general, what is the result when a number is divided by -1?

117. What is wrong with the following statement?

   A negative and a positive is a negative.

118. Is 80 divided by -5 the same as -5 divided by 80? Explain.

REVIEW

119. Add: $-3 + (-4) + (-5) + 4 + 3$.

120. Write the subtraction statement $-3 - (-5)$ as addition of the opposite.

121. Find $\frac{1}{2} + \frac{1}{4} + \frac{1}{3}$ and express the result as a decimal.

122. Describe the balance in a checking account that is overdrawn $65 using a signed number.

123. Remove the common factors of the numerator and denominator to simplify the fraction: $\frac{2 \cdot 3 \cdot 5 \cdot 5}{2 \cdot 5 \cdot 5 \cdot 7}$.

124. Find $|-2,345|$.

CHALLENGE PROBLEMS

125. If the product of five numbers is negative, how many of them could be negative? Explain.

126. Suppose a is a positive number and b is a negative number. Determine whether the given expression is positive or negative.

   a. $-a(-b)$

   b. $\frac{-a}{b}$

   c. $\frac{-a}{a}$

   d. $\frac{1}{b}$

1.7   Exponents and Order of Operations

• Exponents   • Order of Operations   • Grouping Symbols   • The Mean (Average)

In this course, we will perform six operations with real numbers: addition, subtraction, multiplication, division, raising to a power, and finding a root. Often, we will have to find the value of expressions that involve more than one operation. In this section, we introduce an order-of-operations rule to follow in such cases. But first, we discuss a way to write repeated multiplication using *exponents*.

■ EXPONENTS

In the expression $3 \cdot 3 \cdot 3 \cdot 3 \cdot 3$, the number 3 repeats as a factor five times. We can use **exponential notation** to write this product more concisely.

Exponent and Base	An **exponent** is used to indicate repeated multiplication. It tells how many times the **base** is used as a factor.

The Language of Algebra

5^2 represents the area of a square with sides 5 units long. 4^3 represents the volume of a cube with sides 4 units long.

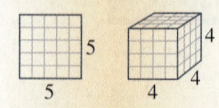

$$\underbrace{3 \cdot 3 \cdot 3 \cdot 3 \cdot 3}_{\text{Five repeated factors of 3.}} = 3^5 \quad \overset{\text{The exponent is 5.}}{\underset{\text{The base is 3.}}{}}$$

In the **exponential expression** 3^5, 3 is the base, and 5 is the exponent. The expression 3^5 is called a power of 3. Some other examples of exponential expressions are:

5^2 Read as "5 to the second power" or "5 squared."

4^3 Read as "4 to the third power" or "4 cubed."

$(-2)^5$ Read as "-2 to the fifth power."

EXAMPLE 1 Write each expression using exponents: **a.** $7 \cdot 7 \cdot 7$, **b.** $(-5)(-5)(-5)(-5)(-5)$, **c.** sixteen cubed, and **d.** $8 \cdot 8 \cdot 15 \cdot 15 \cdot 15 \cdot 15$.

Solution **a.** We can represent this repeated multiplication with an exponential expression having a base of 7 and an exponent of 3: $7 \cdot 7 \cdot 7 = 7^3$.

b. The factor -5 is repeated five times: $(-5)(-5)(-5)(-5)(-5) = (-5)^5$.

c. Sixteen cubed can be written as 16^3.

d. $8 \cdot 8 \cdot 15 \cdot 15 \cdot 15 \cdot 15 = 8^2 \cdot 15^4$

Self Check 1 Write each expression using exponents: **a.** $(12)(12)(12)(12)(12)(12)$, **b.** $2 \cdot 9 \cdot 9 \cdot 9$, **c.** fifty squared, **d.** $(-30)(-30)(-30)$.

EXAMPLE 2 Write each product using exponents: **a.** $a \cdot a \cdot a \cdot a \cdot a \cdot a$ and **b.** $4 \cdot \pi \cdot r \cdot r$.

Solution **a.** $a \cdot a \cdot a \cdot a \cdot a \cdot a = a^6$ *a* is repeated as a factor 6 times.

b. $4 \cdot \pi \cdot r \cdot r = 4\pi r^2$ *r* is repeated as a factor 2 times.

Self Check 2 Write each product using exponents: **a.** $y \cdot y \cdot y \cdot y$, **b.** $12 \cdot b \cdot b \cdot b \cdot c$.

EXAMPLE 3 Find the value of each expression: **a.** 5^3, **b.** 10^1, **c.** $(-3)^4$, and **d.** $(-3)^5$.

Solution **a.** $5^3 = 5 \cdot 5 \cdot 5 = 125$ The base is 5, the exponent is 3.

b. $10^1 = 10$ The base is 10. Since the exponent is 1, we write the base once.

c. $(-3)^4 = (-3)(-3)(-3)(-3)$ Write -3 as a factor four times.
$= 9(-3)(-3)$ Work from left to right.
$= -27(-3)$
$= 81$

d. $(-3)^5 = (-3)(-3)(-3)(-3)(-3)$ Write -3 as a factor five times.
$= 9(-3)(-3)(-3)$ Work from left to right.
$= -27(-3)(-3)$
$= 81(-3)$
$= -243$

Notation

If a number or a variable is written without an exponent, we assume the number or variable has an understood exponent of 1. For example,

$x = x^1$ and $8 = 8^1$

Self Check 3 Evaluate: **a.** 2^5, **b.** 9^1, **c.** $(-6)^2$, **d.** $(-5)^3$.

Caution   A common mistake when evaluating an exponential expression is to multiply the base and the exponent. For example, $5^3 \neq 15$. As we saw in Example 3a, $5^3 = 5 \cdot 5 \cdot 5 = 125$.

In part c of Example 3, we raised -3 to an even power, and the result was positive. In part d, we raised -3 to an odd power, and the result was negative. These results illustrate the following rule.

Even and Odd Powers of a Negative Number	When a negative number is raised to an even power, the result is positive.
	When a negative number is raised to an odd power, the result is negative.

Calculators
Finding a power

The squaring key x^2 can be used to find the square of a number. To raise a number to a power, we use the y^x key on a scientific calculator and the $\wedge$ key on a graphing calculator.

Although the expressions $(-4)^2$ and -4^2 look alike, they are not. When we find the value of each expression, it becomes clear that they are not equivalent.

$(-4)^2 = (-4)(-4)$ The base is -4, the exponent is 2.

$= 16$

$-4^2 = -(4 \cdot 4)$ The base is 4, the exponent is 2.

$= -16$

Different results

EXAMPLE 4   Evaluate: -2^4.

Solution

$-2^4 = -(2 \cdot 2 \cdot 2 \cdot 2)$ Since the base is 2, and the exponent is 4, write 2 as a factor four times within parentheses.

$= -16$ Do the multiplication within the parentheses.

Self Check 4   Evaluate: -5^4.

EXAMPLE 5   Evaluate: **a.** $\left(-\dfrac{2}{3}\right)^3$ and   **b.** $(0.6)^2$.

Solution   **a.** $\left(-\dfrac{2}{3}\right)^3 = \left(-\dfrac{2}{3}\right)\left(-\dfrac{2}{3}\right)\left(-\dfrac{2}{3}\right)$ Since $-\dfrac{2}{3}$ is the base and 3 is the exponent, we write $-\dfrac{2}{3}$ as a factor three times.

$= \dfrac{4}{9}\left(-\dfrac{2}{3}\right)$ Work from left to right. $\left(-\dfrac{2}{3}\right)\left(-\dfrac{2}{3}\right) = \dfrac{4}{9}$

$= -\dfrac{8}{27}$ We say $-\dfrac{8}{27}$ is the *cube* of $-\dfrac{2}{3}$.

b. $(0.6)^2 = (0.6)(0.6)$ Since 0.6 is the base and 2 is the exponent, we write 0.6 as a factor two times.

$= 0.36$ We say 0.36 is the *square* of 0.6.

Self Check 5   Evaluate: **a.** $\left(-\dfrac{3}{4}\right)^3$,   **b.** $(-0.3)^2$.

■ ORDER OF OPERATIONS

Suppose you have been asked to contact a friend if you see a Rolex watch for sale when you are traveling in Europe. While in Switzerland, you find the watch and send the e-mail message shown on the left. The next day, you get the response shown on the right.

E-Mail
FOUND WATCH. $5,000. SHOULD I BUY IT FOR YOU?

E-Mail
NO PRICE TOO HIGH! REPEAT...NO! PRICE TOO HIGH.

Something is wrong. The first part of the response (No price too high!) says to buy the watch at any price. The second part (No! Price too high.) says not to buy it, because it's too expensive. The placement of the exclamation point makes us read the two parts of the response differently, resulting in different meanings. When reading a mathematical statement, the same kind of confusion is possible. For example, consider the expression

$$2 + 3 \cdot 6$$

We can evaluate this expression in two ways. We can add first, and then multiply. Or we can multiply first, and then add. However, the results are different.

$$2 + \mathbf{3} \cdot 6 = \mathbf{5} \cdot 6 \quad \text{Add 2 and 3 first.} \qquad 2 + \mathbf{3} \cdot \mathbf{6} = 2 + \mathbf{18} \quad \text{Multiply 3 and 6 first.}$$
$$= 30 \quad \text{Multiply 5 and 6.} \qquad\qquad\qquad\quad = 20 \quad \text{Add 2 and 18.}$$

Different results

If we don't establish a uniform order of operations, the expression has two different values. To avoid this possibility, we will always use the following set of priority rules.

Order of Operations
1. Perform all calculations within parentheses and other grouping symbols following the order listed in Steps 2–4 below, working from the innermost pair of grouping symbols to the outermost pair.
2. Evaluate all exponential expressions.
3. Perform all multiplications and divisions as they occur from left to right.
4. Perform all additions and subtractions as they occur from left to right.

When grouping symbols have been removed, repeat Steps 2–4 to complete the calculation.

If a fraction is present, evaluate the expression above and the expression below the bar separately. Then do the division indicated by the fraction bar, if possible.

It isn't necessary to apply all of these steps in every problem. For example, the expression $2 + 3 \cdot 6$ does not contain any parentheses, and there are no exponential expressions. So we look for multiplications and divisions to perform and proceed as follows:

$$2 + \mathbf{3} \cdot \mathbf{6} = 2 + \mathbf{18} \quad \text{Do the multiplication first: } 3 \cdot 6 = 18.$$
$$= 20 \quad \text{Do the addition.}$$

EXAMPLE 6

Evaluate: $3 \cdot 2^3 - 4$.

Solution   Three operations need to be performed to find the value of this expression. By the rules for the order of operations, we evaluate 2^3 first.

$$3 \cdot \mathbf{2^3} - 4 = 3 \cdot \mathbf{8} - 4 \qquad \text{Evaluate the exponential expression: } 2^3 = 8.$$
$$= 24 - 4 \qquad \text{Do the multiplication: } 3 \cdot 8 = 24.$$
$$= 20 \qquad \text{Do the subtraction.}$$

Self Check 6   Evaluate: $2 \cdot 3^2 + 17$.

EXAMPLE 7

Evaluate: $-30 - 4 \cdot 5 + 9$.

Solution   This expression involves subtraction, multiplication, and addition. The rules for the order of operations tell us to multiply first.

The Language of Algebra

Sometimes, the word *simplify* is used in place of the word *evaluate*.

$$-30 - \mathbf{4 \cdot 5} + 9 = -30 - \mathbf{20} + 9 \qquad \text{Do the multiplication: } 4 \cdot 5 = 20.$$
$$= -50 + 9 \qquad \text{Working from left to right, do the subtraction:} \\ -30 - 20 = -30 + (-20) = -50.$$
$$= -41 \qquad \text{Do the addition.}$$

Self Check 7   Evaluate: $-40 - 9 \cdot 4 + 10$.

EXAMPLE 8

Evaluate: $160 \div (-4) - 6(-2)3$.

Solution   Although this expression contains parentheses, there are no operations to perform within them. Since there are no exponents, we do multiplications and divisions as they are encountered from left to right.

$$\mathbf{160 \div (-4)} - 6(-2)3 = \mathbf{-40} - 6(-2)3 \qquad \text{Do the division: } 160 \div (-4) = -40.$$
$$= -40 - (-12)3 \qquad \text{Do the multiplication: } 6(-2) = -12.$$
$$= -40 - (-36) \qquad \text{Do the multiplication: } (-12)3 = -36.$$
$$= -40 + 36 \qquad \text{Write the subtraction as addition of the opposite.}$$
$$= -4 \qquad \text{Do the addition.}$$

Self Check 8   Evaluate: $240 \div (-8) - 3(-2)4$.

■ GROUPING SYMBOLS

Grouping symbols serve as mathematical punctuation marks. They help determine the order in which an expression is to be evaluated. Examples of grouping symbols are parentheses ( ), brackets [ ], absolute value symbols | |, and the fraction bar —.

EXAMPLE 9

Evaluate: $(6 - 3)^2$.

Solution   This expression contains parentheses. By the rules for the order of operations, we must perform the operation within the parentheses first.

$$(\mathbf{6 - 3})^2 = \mathbf{3}^2 \qquad \text{Do the subtraction: } 6 - 3 = 3.$$
$$= 9 \qquad \text{Evaluate the exponential expression.}$$

Self Check 9   Evaluate: $(12 - 6)^3$.

EXAMPLE 10

Evaluate: $5^3 + 2(-8 - 3 \cdot 2)$.

ELEMENTARY &
INTERMEDIATE
Algebra $f(x)$ **Now**™

Solution   We begin by performing the operations within the parentheses in the proper order: multiplication first, and then subtraction.

Notation

Multiplication is indicated when a number is next to a grouping symbol.

$$\downarrow$$
$$5^3 + 2(-8 - 3 \cdot 2)$$

$$5^3 + 2(-8 - \mathbf{3 \cdot 2}) = 5^3 + 2(-8 - \mathbf{6}) \qquad \text{Do the multiplication: } 3 \cdot 2 = 6.$$
$$= 5^3 + 2(-14) \qquad \text{Do the subtraction: } -8 - 6 = -14.$$
$$= 125 + 2(-14) \qquad \text{Evaluate } 5^3.$$
$$= 125 + (-28) \qquad \text{Do the multiplication.}$$
$$= 97 \qquad \text{Do the addition.}$$

Self Check 10   Evaluate: $1^3 + 6(-6 - 3 \cdot 0)$.

Expressions can contain two or more pairs of grouping symbols. To evaluate the following expression, we begin by working within the innermost pair of grouping symbols. Then we work within the outermost pair.

Innermost pair
$$\downarrow \qquad \downarrow$$
$$-4[-2 - 3(4 - 8^2)] - 2$$
$$\uparrow \qquad\qquad \uparrow$$
Outermost pair

EXAMPLE 11

Evaluate: $-4[-2 - 3(4 - 8^2)] - 2$.

ELEMENTARY &
INTERMEDIATE
Algebra $f(x)$ **Now**™

Solution   We do the work within the innermost grouping symbols (the parentheses) first.

$$-4[-2 - 3(4 - \mathbf{8}^2)] - 2$$
$$= -4[-2 - 3(4 - \mathbf{64})] - 2 \qquad \text{Evaluate the exponential expression within the parentheses: } 8^2 = 64.$$
$$= -4[-2 - 3(-60)] - 2 \qquad \text{Do the subtraction within the parentheses: } 4 - 64 = 4 + (-64) = -60.$$
$$= -4[-2 - (-180)] - 2 \qquad \text{Do the multiplication within the brackets: } 3(-60) = -180.$$
$$= -4[178] - 2 \qquad \text{Do the subtraction within the brackets: } -2 - (-180) = -2 + 180 = 178.$$
$$= -712 - 2 \qquad \text{Do the multiplication: } -4[178] = -712.$$
$$= -714 \qquad \text{Do the subtraction.}$$

Self Check 11   Evaluate: $-5[2(5^2 - 15) + 4] - 10$

EXAMPLE 12

Evaluate: $\dfrac{-3(3 + 2) + 5}{17 - 3(-4)}$.

Solution   We simplify the numerator and the denominator separately.

Calculators
Order of operations

Calculators have the order of operations built in. A left parenthesis key **(** and a right parenthesis key **)** should be used when grouping symbols, including a fraction bar, are needed.

$$\frac{-3(\mathbf{3 + 2}) + 5}{17 - \mathbf{3(-4)}} = \frac{-3(\mathbf{5}) + 5}{17 - (\mathbf{-12})}$$

In the numerator, do the addition within the parentheses. In the denominator, do the multiplication.

$$= \frac{-15 + 5}{17 + 12}$$

In the numerator, do the multiplication. In the denominator, write the subtraction as addition of the opposite of -12, which is 12.

$$= \frac{-10}{29}$$

Do the additions.

$$= -\frac{10}{29}$$

Write the $-$ sign in front of the fraction: $\dfrac{-10}{29} = -\dfrac{10}{29}$.

Self Check 12   Evaluate: $\dfrac{-4(-2 + 8) + 6}{8 - 5(-2)}$.

EXAMPLE 13

Evaluate: $10|9 - 15| - 2^5$.

Solution   The absolute value bars are grouping symbols. We do the calculation within them first.

$$\begin{aligned}
10|\mathbf{9 - 15}| - 2^5 &= 10|\mathbf{-6}| - 2^5 & \text{Subtract: } 9 - 15 = 9 + (-15) = -6. \\
&= 10(6) - 2^5 & \text{Find the absolute value: } |-6| = 6. \\
&= 10(6) - 32 & \text{Evaluate the exponential expression: } 2^5 = 32. \\
&= 60 - 32 & \text{Do the multiplication: } 10(6) = 60. \\
&= 28 & \text{Do the subtraction.}
\end{aligned}$$

Self Check 13   Evaluate: $10^3 + 3|24 - 25|$.

■ THE MEAN (AVERAGE)

The **arithmetic mean** (or **average**) of a set of numbers is a value around which the values of the numbers are grouped.

Finding an Arithmetic Mean   To find the **mean** of a set of values, divide the sum of the values by the number of values.

EXAMPLE 14

Customer service.   To measure its effectiveness in serving customers, a store had the telephone company electronically record the number of times the telephone rang before an employee answered it. The results of the week-long survey are shown in the table. We see that for 11 calls, the phone was answered after it rang 1 time. For 46 calls, the phone was answered after it rang 2 times, and so on. Find the *average* number of times the phone rang before an employee answered it that week.

Number of rings	Number of calls
1	11
2	46
3	45
4	28
5	20

Solution To find the total number of rings, we multiply each *number of rings* (1, 2, 3, 4, and 5 rings) by the respective number of occurrences and add those subtotals.

$$\text{Total number of rings} = 11(1) + 46(2) + 45(3) + 28(4) + 20(5)$$

The total number of calls received was $11 + 46 + 45 + 28 + 20$. To find the average, we divide the total number of rings by the total number of calls.

$$\text{Average} = \frac{11(1) + 46(2) + 45(3) + 28(4) + 20(5)}{11 + 46 + 45 + 28 + 20}$$

In the numerator, do the multiplications. In the denominator, do the additions.

$$= \frac{11 + 92 + 135 + 112 + 100}{150}$$

Do the addition.

$$= \frac{450}{150}$$

$$= 3$$

The average number of times the phone rang before it was answered was 3.

Answers to Self Checks **1. a.** 12^6, **b.** $2 \cdot 9^3$, **c.** 50^2, **d.** $(-30)^3$ **2. a.** y^4, **b.** $12b^3c$ **3. a.** 32,

b. 9, **c.** 36, **d.** -125 **4.** -625 **5. a.** $-\frac{27}{64}$, **b.** 0.09 **6.** 35 **7.** -66

8. -6 **9.** 216 **10.** -35 **11.** -130 **12.** -1 **13.** 1,003

1.7   STUDY SET   ELEMENTARY & INTERMEDIATE Algebra *f(x)* Now ™

VOCABULARY   **Fill in the blanks.**

1. In the exponential expression 3^2, 3 is the _____, and 2 is the _____.

2. 10^2 can be read as ten _____, and 10^3 can be read as ten _____.

3. 7^5 is the fifth _____ of seven.

4. An _____ is used to represent repeated multiplication.

5. The rules for the _____ of operations guarantee that an evaluation of a numerical expression will result in a single answer.

6. The arithmetic _____ or _____ of a set of numbers is a value around which the values of the numbers are grouped.

CONCEPTS

7. Given: $4 + 5 \cdot 6$.
   a. Evaluate the expression in two different ways and state the two possible results.

   b. Which result from part a is correct, and why?

8. a. What repeated multiplication does 5^3 represent?

   b. Write a multiplication statement in which the factor x is repeated 4 times. Then write the expression in simpler form using an exponent.

9. In the expression $-8 + 2[15 - (-6 + 1)]$, which grouping symbols are innermost, and which are outermost?

10. a. What operations does the expression $12 + 5^2(-3)$ contain?

   b. In what order should they be performed?

11. a. What operations does the expression $20 - (-2)^2 + 3(-1)$ contain?

   b. In what order should they be performed?

12. Consider the expression $\frac{36 - 4(7)}{2(10 - 8)}$. In the numerator, what operation should be done first? In the denominator, what operation should be done first?

13. To evaluate each expression, what operation should be done first?

a. $24 - 4 + 2$ **b.** $24 \div 4 \cdot 2$

14. To evaluate each expression, what operation should be done first?

a. $-80 - 3 + 5 - 2^2$

b. $-80 - (3 + 5) - 2^2$

c. $-80 - 3 + (5 - 2)^2$

15. To evaluate each expression, what operation should be done first?

a. $(65 - 3)^3$

b. $65 - 3^3$

c. $6(5) - (3)^3$ **d.** $65 \cdot 3^3$

16. a. How is the mean (or average) of a set of scores found?

b. Find the average of 75, 81, 47, and 53.

NOTATION

17. Fill in the blanks.

a. $3^1 = \boxed{}$ **b.** $x^1 = \boxed{}$

c. $9 = 9\boxed{}$ **d.** $y = y\boxed{}$

18. Tell the name of each grouping symbol: ( ), [ ], | |, and —.

19. a. In the expression $(-5)^2$, what is the base?

b. In the expression -5^2, what is the base?

20. Write each expression using symbols.

a. Negative two squared.

b. The opposite of two squared.

Complete the evaluation of each expression.

21. $-19 - 2[(1 + 2) \cdot 3] = -19 - 2[\boxed{} \cdot 3]$
$$= -19 - 2[\boxed{}]$$
$$= -19 - \boxed{}$$
$$= -37$$

22. $\dfrac{46 - 2^3}{-3(5) - 4} = \dfrac{46 - \boxed{}}{\boxed{} - 4}$
$$= \dfrac{\boxed{}}{\boxed{}}$$
$$= -2$$

PRACTICE   **Write each product using exponents.**

23. $3 \cdot 3 \cdot 3 \cdot 3$ **24.** $m \cdot m \cdot m \cdot m \cdot m$

25. $10 \cdot 10 \cdot k \cdot k \cdot k$ **26.** $5(5)(5)(i)(i)$

27. $8 \cdot \pi \cdot r \cdot r \cdot r$ **28.** $4 \cdot \pi \cdot r \cdot r$

29. $6(x)(x)(y)(y)(y)$ **30.** $76 \cdot s \cdot s \cdot s \cdot s \cdot t$

Evaluate each expression.

31. $(-6)^2$ **32.** -6^2

33. -4^4 **34.** $(-4)^4$

35. $(-5)^3$ **36.** -5^3

37. $-(-6)^4$ **38.** $-(-7)^2$

39. $(-0.4)^2$ **40.** $(-0.5)^2$

41. $\left(-\dfrac{2}{5}\right)^3$ **42.** $\left(-\dfrac{1}{4}\right)^3$

43. $3 - 5 \cdot 4$ **44.** $-4 \cdot 6 + 5$

45. $3 \cdot 8^2$ **46.** $(3 \cdot 4)^2$

47. $8 \cdot 5^1 - 4 \div 2$ **48.** $9 \cdot 5^1 - 6 \div 3$

49. $100 - 8(10) + 60$ **50.** $50 - 2(5) - 7$

51. $-22 - (15 - 3)$ **52.** $-(33 - 8) - 10$

53. $-2(9) - 2(5)$ **54.** $-75 - 7^2$

55. $5^2 + 13^2$ **56.** $3^3 - 2^3$

57. $-4(6 + 5)$ **58.** $-3(5 - 4)$

59. $(9 - 3)(9 - 9)$ **60.** $-(-8 - 6)(6 - 6)$

61. $(-1 - 18)2$ **62.** $-5(128 - 5^3)^2$

63. $-2(-1)^2 + 3(-1) - 3$ **64.** $-4(-3)^2 + 3(-3) - 1$

65. $4^2 - (-2)^2$ **66.** $(-5 - 2)^2$

67. $12 + 2\left(-\dfrac{9}{3}\right) - (-2)$ **68.** $2 + 3\left(-\dfrac{25}{5}\right) - (-4)$

69. $\dfrac{-2 - 5}{-7 - (-7)}$ **70.** $\dfrac{-3 - (-1)}{-2 - (-2)}$

71. $200 - (-6 + 5)^3$ **72.** $19 - (-45 + 41)^3$

73. $|5 \cdot 2^2 \cdot 4| - 30$ **74.** $2 + |3 \cdot 2^2 \cdot 4|$

75. $[6(5) - 5(5)]4$ **76.** $175 - 2 \cdot 3^4$

77. $-6(130 - 4^3)$ **78.** $-5(150 - 3^3)$

79. $(17 - 5 \cdot 2)^3$ **80.** $(4 + 2 \cdot 3)^4$

81. $-5(-2)^3(3)^2$ **82.** $-3(-2)^5(2)^2$

83. $-2\left(\dfrac{15}{-5}\right) - \dfrac{6}{2} + 9$ **84.** $-6\left(\dfrac{25}{-5}\right) - \dfrac{36}{9} + 1$

85. $\dfrac{5 \cdot 50 - 160}{-9}$ **86.** $\dfrac{5(68 - 32)}{-9}$

87. $\dfrac{2(6 - 1)}{16 - (-4)^2}$ **88.** $\dfrac{6 - (-1)}{4 - 2^2}$

89. $5(10 + 2) - 1$ **90.** $14 + 3(7 - 5)$

91. $64 - 6[15 + (-3)3]$ **92.** $4 + 2[26 + 5(-3)]$

93. $(12 - 2)^3$ **94.** $(-2)^3\left(\dfrac{-6}{2}\right)(-1)$

95. $(-3)^3\left(\dfrac{-4}{2}\right)(-1)$ **96.** $\dfrac{-5 - 3^3}{2^3}$

97. $\dfrac{1}{2}\left(\dfrac{1}{8}\right) + \left(-\dfrac{1}{4}\right)^2$ **98.** $-\dfrac{1}{9}\left(\dfrac{1}{4}\right) + \left(-\dfrac{1}{6}\right)^2$

99. $-2|4 - 8|$ **100.** $-5|1 - 8|$

101. $|7 - 8(4 - 7)|$ **102.** $|9 - 5(1 - 8)|$

103. $3 + 2[-1 - 4(5)]$ **104.** $4 + 2[-7 - 3(9)]$

105. $-3[5^2 - (7 - 3)^2]$ **106.** $3 - [3^3 + (3 - 1)^3]$

107. $-(2 \cdot 3 - 4)^3$ **108.** $-(3 \cdot 5 - 2 \cdot 6)^2$

109. $\dfrac{(3 + 5)^2 + |-2|}{-2(5 - 8)}$ **110.** $\dfrac{|-25| - 8(-5)}{2^4 - 29}$

111. $\dfrac{2[-4 - 2(3 - 1)]}{3(-3)(-2)}$ **112.** $\dfrac{3[-9 + 2(7 - 3)]}{(5 - 8)(7 - 9)}$

113. $\dfrac{|6 - 4| + 2|-4|}{26 - 2^4}$ **114.** $\dfrac{4|9 - 7| + |-7|}{3^2 - 2^2}$

115. $\dfrac{(4^3 - 10) + (-4)}{5^2 - (-4)(-5)}$ **116.** $\dfrac{(6 - 5)^4 - (-21)}{(-9)(-3) - 4^2}$

117. $\dfrac{72 - (2 - 2 \cdot 1)}{10^2 - (90 + 2^2)}$ **118.** $\dfrac{13^2 - 5^2}{-3(5 - 9)}$

119. $-\left(\dfrac{40 - 1^3 - 2^4}{3(2 + 5) + 2}\right)$ **120.** $-\left(\dfrac{8^2 - 10}{2(3)(4) - 5(3)}\right)$

APPLICATIONS

121. LIGHT   The illustration shows that the light energy that passes through the first unit of area, 1 yard away from the bulb, spreads out as it travels away from the source. How much area does that light energy cover 2 yards, 3 yards, and 4 yards from the bulb? Express each answer using exponents.

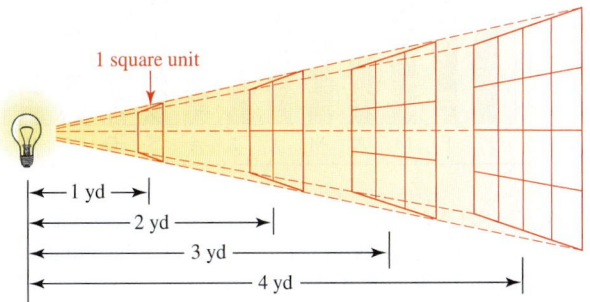

1 square unit

1 yd

2 yd

3 yd

4 yd

122. CHAIN LETTERS   A store owner sent two friends a letter advertising her store's low prices. The ad closed with the following request: "Please send a copy of this letter to two of your friends."

a. Assume that all those receiving letters respond and that everyone in the chain receives just one letter. Complete the table.

b. How many letters will be circulated in the 10th level of the mailing?

Level	Numbers of letters circulated
1st	$2 = 2^1$
2nd	$\ \ \ \ = 2$
3rd	$\ \ \ \ = 2$
4th	$\ \ \ \ = 2$

123. AUTO INSURANCE   See the following premium comparison. What is the average six-month insurance premium?

Allstate	**$2,672**	**Mercury**	**$1,370**
Auto Club	**$1,680**	**State Farm**	**$2,737**
Farmers	**$2,485**	**20th Century**	**$1,692**

Criteria: Six-month premium. Husband, 45, drives a 1995 Explorer, 12,000 annual miles. Wife, 43, drives a 1996 Dodge Caravan, 12,000 annual miles. Son, 17, is an occasional operator. All have clean driving records.

124. ENERGY USAGE Find the average number of therms of natural gas used per month.

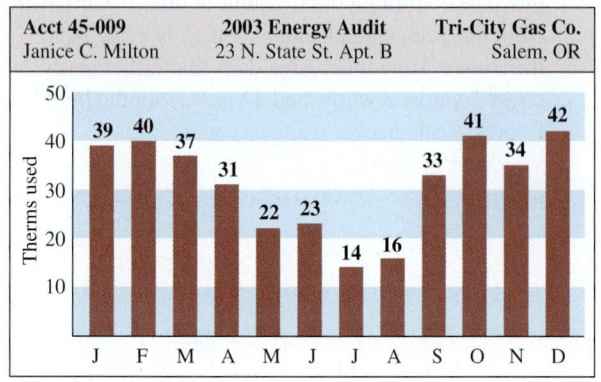

125. CASH AWARDS A contest is to be part of a promotional kickoff for a new children's cereal. The prizes to be awarded are shown.
  a. How much money will be awarded in the promotion?
  b. What is the average cash prize?

> ### Coloring Contest
> **Grand prize: Disney World vacation plus $2,500**
> Four 1st place prizes of $500
> Thirty-five 2nd place prizes of $150
> Eighty-five 3rd place prizes of $25

126. SURVEYS Some students were asked to rate the food at their college cafeteria on a scale from 1 to 5. The responses are shown on the tally sheet. Find the average rating.

Poor		Fair		Excellent
1	2	3	4	5
	|||	|||	ꓶꓶꓶ	ꓶꓶꓶ ||||

127. WRAPPING GIFTS How much ribbon is needed to wrap the package if 15 inches of ribbon are needed to make the bow?

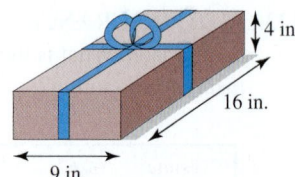

4 in.
16 in.
9 in.

128. SCRABBLE Illustration (a) shows a portion of the game board before and Illustration (b) shows it after the word *QUARTZY* is played. Determine the score. (The number on each tile gives the point value of the letter.)

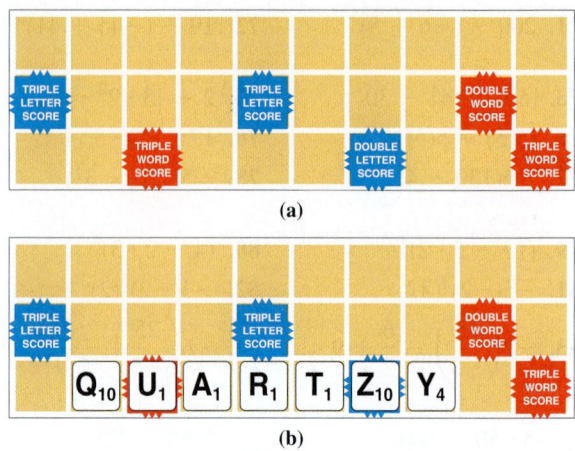

(a)

(b)

WRITING

129. Explain the difference between 2^3 and 3^2.

130. Explain why rules for the order of operations are necessary.

131. What does it mean when we say, do all additions and subtractions as they occur from left to right?

132. In what settings do you encounter or use the concept of arithmetic mean (average) in your everyday life?

REVIEW

133. Match each term with the proper operation.
  a. sum   **i.** division
  b. difference   **ii.** addition
  c. product   **iii.** subtraction
  d. quotient   **iv.** multiplication

134. a. What is the opposite of -8?
  b. What is the reciprocal of -8?

CHALLENGE PROBLEMS

135. Using each of the numbers 2, 3, and 4 only once, what is the greatest value that the following expression can have?

$$\left(\square^{\square}\right)^{\square}$$

136. Insert a pair of parentheses into the expression so that it has a value of 40.

$$4 \cdot 3^2 - 4 \cdot 2$$

1.8 Algebraic Expressions

- • Algebraic Expressions
- • Translating from Words to Symbols
- • Writing Algebraic Expressions
- • Analyzing Problems
- • Number and Value Problems
- • Evaluating Algebraic Expressions

Since problems in algebra are often presented in words, the ability to interpret what you read is important. In this section, we will introduce several strategies that will help you translate words into algebraic expressions.

■ ALGEBRAIC EXPRESSIONS

Recall that variables and/or numbers can be combined with the operations of arithmetic to create **algebraic expressions.** Addition symbols separate these expressions into parts called *terms.* For example, the expression $x + 8$ has two terms.

$$\underset{\text{First term}}{x} \quad + \quad \underset{\text{Second term}}{8}$$

The Language of Algebra

Note the difference between *terms* and *factors.* In the expression $x + 8$, x and 8 are *terms.* In the expression $8x$, x and 8 are *factors.*

Since subtraction can be written as addition of the opposite, the expression $a^2 - 3a - 9$ has three terms.

$$a^2 - 3a - 9 = \quad \underset{\text{First term}}{a^2} \quad + \quad \underset{\text{Second term}}{(-3a)} \quad + \quad \underset{\text{Third term}}{(-9)}$$

In general, a **term** is a product or quotient of numbers and/or variables. A single number or variable is also a term. Examples of terms include:

Notation

By the commutative property of multiplication, $r6 = 6r$ and $-15b^2a = -15ab^2$. However, we usually write the numerical factor first and the variable factors in alphabetical order.

$$8, \qquad y, \qquad 6r, \qquad -w^3, \qquad 3.7x^5, \qquad \frac{3}{n}, \qquad -15ab^2$$

The numerical factor of a term is called the **coefficient** of the term. For example, the term $6r$ has a coefficient of 6 because $6r = 6 \cdot r$. The coefficient of $-15ab^2$ is -15 because $-15ab^2 = -15 \cdot ab^2$. More examples are shown in the following table.

The Language of Algebra

Terms such as x and yz^3 have *implied* coefficients of 1. *Implied* means suggested without being precisely expressed.

Term	Coefficient	
$8y^2$	8	
$-0.9pq$	-0.9	
$\frac{3}{4}b$	$\frac{3}{4}$	This term could be written $\frac{3b}{4}$.
$-\frac{x}{6}$	$-\frac{1}{6}$	This term could be written $-\frac{1}{6}x$.
x	1	$x = 1x$.
$-t$	-1	$-t = -1t$.
15	15	

A term, such as 15, that consists of a single number is called a **constant term.**

EXAMPLE 1

ELEMENTARY & INTERMEDIATE
Algebra $f(x)$ **Now**™

Identify the coefficient of each term in the expression $7x^2 - x + 6$.

Solution If we write $7x^2 - x + 6$ as $7x^2 + (-x) + 6$, we see that it has three terms: $7x^2$, $-x$, and 6.

The coefficient of $7x^2$ is 7.
The coefficient of $-x$ is -1.
The coefficient of 6 is 6.

Self Check 1 Identify the coefficient of each term in the expression $p^3 - 12p^2 + 3p - 4$.

■ TRANSLATING FROM WORDS TO SYMBOLS

In the following tables, we list words and phrases and show how they can be translated into algebraic expressions.

Addition	
the sum of a and 8	$a + 8$
4 plus c	$4 + c$
16 added to m	$m + 16$
4 more than t	$t + 4$
20 greater than F	$F + 20$
T increased by r	$T + r$
exceeds y by 35	$y + 35$

Subtraction	
the difference of 23 and P	$23 - P$
550 minus h	$550 - h$
18 less than w	$w - 18$
7 decreased by j	$7 - j$
M reduced by x	$M - x$
12 subtracted from L	$L - 12$
5 less f	$5 - f$

The Language of Algebra

When a translation involves the phrase *less than,* note how the terms are reversed.

18 less than w
$w - 18$

Multiplication	
the product of 4 and x	$4x$
20 times B	$20B$
twice r	$2r$
triple the profit P	$3P$
$\frac{3}{4}$ of m	$\frac{3}{4}m$

Division	
the quotient of R and 19	$\frac{R}{19}$
s divided by d	$\frac{s}{d}$
the ratio of c to d	$\frac{c}{d}$
k split into 4 equal parts	$\frac{k}{4}$

EXAMPLE 2

Write each phrase as an algebraic expression: **a.** one-half of the profit P,   **b.** 5 less than the capacity c,   and   **c.** the product of the weight w and 2,000, increased by 300.

Solution **a. Key phrase:** *One-half of*   **Translation:** multiply by $\frac{1}{2}$
The translation is: $\frac{1}{2}P$.

b. Key phrase: *less than*   **Translation:** subtract

Sometimes, thinking in terms of specific numbers makes translating easier. Suppose the capacity was 100. Then 5 *less than* 100 would be $100 - 5$. If the capacity is c, then we need to make it 5 less. The translation is: $c - 5$.

c. We are given: The product of the weight w and 2,000, increased by 300.

> **Key word:** *product*   **Translation:** multiply
>
> **Key phrase:** *increased by*   **Translation:** add

The comma after 2,000 means w is first multiplied by 2,000 and then 300 is added to that product. The translation is: $2,000w + 300$.

Self Check 2 Write each phrase as an algebraic expression: **a.** 80 less than the total t, **b.** $\frac{2}{3}$ of the time T, **c.** the difference of twice a and 15.

■ WRITING ALGEBRAIC EXPRESSIONS

When solving problems, we usually begin by letting a variable stand for an unknown quantity.

EXAMPLE 3

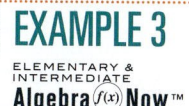
ELEMENTARY &
INTERMEDIATE
Algebra *f(x)* **Now**™

Swimming. The pool shown is x feet wide. If it is to be sectioned into 8 equally wide swimming lanes, write an algebraic expression that represents the width of each lane.

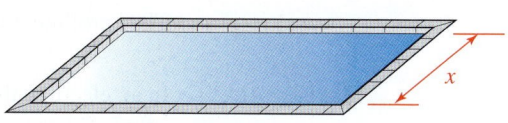

Solution We let x = the width of the swimming pool (in feet).

> **Key phrase:** *sectioned into 8 equally wide lanes*   **Translation:** divide

The width of each lane is $\dfrac{x}{8}$ feet.

Self Check 3 When a secretary rides the bus to work, it takes her m minutes. If she drives her own car, her travel time exceeds this by 15 minutes. How long does it take her to get to work by car?

EXAMPLE 4

Painting. A 10-inch-long paintbrush has two parts: a handle and bristles. Choose a variable to represent the length of one of the parts. Then write an expression to represent the length of the other part.

Solution There are two approaches. First, refer to the following drawing on the left. If we let h = the length of the handle (in inches), the length of the bristles is $10 - h$.

Now refer to the drawing on the right. If we let b = the length of the bristles (in inches), the length of the handle is $10 - b$.

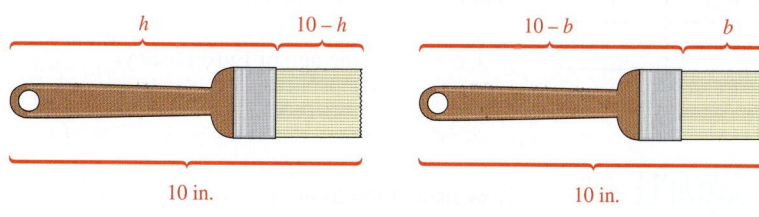

Self Check 4   Part of a $900 donation to a preschool was designated to go to the scholarship fund, the remainder to the building fund. Choose a variable to represent the amount donated to one of the funds. Write an expression for the amount donated to the other fund.

EXAMPLE 5   *College enrollments.*   In the second semester, enrollment in a retraining program at a college was 32 more than twice that of the first semester. Choose a variable to represent the enrollment in one of the semesters. Then write an expression for the enrollment in the other semester.

Solution   Since the second-semester enrollment is expressed in terms of the first-semester enrollment, we let x = the enrollment in the first semester.

> **Key phrase:** *more than*   **Translation:** add
> **Key word:** *twice*   **Translation:** multiply by 2

The enrollment for the second semester is $2x + 32$.

Self Check 5   In an election, the incumbent received 55 fewer votes than three times the challenger's votes. Choose a variable to represent the number of votes received by one candidate. Write an expression for the number of votes received by the other.

■ ANALYZING PROBLEMS

When solving problems, we aren't always given key words or key phrases to help establish what mathematical operation to use. Sometimes a careful reading of the problem is needed to determine any hidden operations.

EXAMPLE 6   *Vacationing.*   Disneyland, in California, was in operation 16 years before the opening of Disney World in Florida. Euro Disney, in France, was constructed 21 years after Disney World. Write an algebraic expression to represent the age (in years) of each Disney attraction.

Solution   The ages of Disneyland and Euro Disney are both related to the age of Disney World. Therefore, we will let x = the age of Disney World.
    In carefully reading the problem, we see that Disneyland was built 16 years *before* Disney World. That makes its age 16 years more than that of Disney World.

Attraction	Age
Disneyland	$x + 16$
Disney World	x
Euro Disney	$x - 21$

$x + 16$ = the age of Disneyland

Euro Disney was built 21 years *after* Disney World. That makes its age 21 years less than that of Disney World.

$x - 21$ = the age of Euro Disney

EXAMPLE 7   How many months are in x years?

Solution   Since there are no key words, we must carefully analyze the problem. It is often helpful to
consider some specific cases. For example,
let's calculate the number of months in 1 year,
2 years, and 3 years. When we write the results
in a table, a pattern is apparent.

Number of years	Number of months
1	12
2	24
3	36
x	$12x$

The number of months in x years is $12 \cdot x$ or $12x$.

We multiply the number of years
by 12 to find the number of months.

Self Check 7   How many days is h hours?

■ NUMBER AND VALUE PROBLEMS

In some problems, we must distinguish between *the number of* and *the value of* the
unknown quantity. For example, to find the value of 3 quarters, we multiply the number of
quarters by the value (in cents) of one quarter. Therefore, the value of 3 quarters is
$3 \cdot 25$ cents $= 75$ cents.

The same distinction must be made if the number is unknown. For example, the value
of n nickels is not n cents. The value of n nickels is $n \cdot 5$ cents $= 5n$ cents. For problems of
this type, we will use the relationship

Number $\cdot$ value $=$ total value

EXAMPLE 8

Find the total value of   **a.** five dimes,   **b.** q quarters,   and   **c.** $x + 1$ half-dollars.

Solution   To find the total value (in cents) of each collection of coins, we multiply the number of
coins by the value (in cents) of one coin, as shown in the table.

Type of coin	Number	· Value =	Total value
Dime	5	10	50
Quarter	q	25	$25q$
Half-dollar	$x + 1$	50	$50(x + 1)$

⟵ $q \cdot 25$ is written $25q$.

Self Check 8   Find the value of   **a.** six fifty-dollar savings bonds,   **b.** t one-hundred-dollar savings
bonds,   **c.** $x - 4$ one-thousand-dollar savings bonds.

■ EVALUATING ALGEBRAIC EXPRESSIONS

To evaluate an algebraic expression, we substitute given numbers for each variable and do
the necessary calculations.

EXAMPLE 9

ELEMENTARY &
INTERMEDIATE
Algebra $f(x)$ **Now**™

Evaluate each expression if $x = 3$ and $y = -4$:   **a.** $y^3 + y^2$,   **b.** $-y - x$,   **c.** $|5xy - 7|$,
and   **d.** $\dfrac{y - 0}{x - (-1)}$.

Solution   **a.** $y^3 + y^2 = (\mathbf{-4})^3 + (\mathbf{-4})^2$ Substitute -4 for each y. We must write -4 within parentheses so that it is the base of each exponential expression.

$= -64 + 16$ Evaluate each exponential expression.

$= -48$

Caution

When replacing a variable with its numerical value, we must often write the replacement number within parentheses to convey the proper meaning.

b. $-y - x = -(\mathbf{-4}) - \mathbf{3}$ Substitute 3 for x and -4 for y.

$= 4 - 3$ Simplify: $-(-4) = 4$.

$= 1$

c. $|5xy - 7| = |5(\mathbf{3})(\mathbf{-4}) - 7|$ Substitute 3 for x and -4 for y.

$= |-60 - 7|$ Do the multiplication within the absolute value symbols, working left to right: $5(3)(-4) = 15(-4) = -60$.

$= |-67|$ Do the subtraction: $-60 - 7 = -60 + (-7) = -67$.

$= 67$ Find the absolute value of -67.

d. $\dfrac{y - 0}{x - (-1)} = \dfrac{\mathbf{-4} - 0}{\mathbf{3} - (-1)}$ Substitute 3 for x and -4 for y.

$= \dfrac{-4}{4}$ In the denominator, do the subtraction: $3 - (-1) = 3 + 1 = 4$.

$= -1$

Self Check 9   Evaluate each expression if $a = -2$ and $b = 5$:   **a.** $|a^3 + b^2|$,   **b.** $-a + 2ab$,   and   **c.** $\dfrac{a + 2}{b - 3}$.

EXAMPLE 10

ELEMENTARY & INTERMEDIATE

Algebra $f(x)$ **Now**™

Diving.   Fins are used by divers to provide a larger area to push against the water. The fin shown on the right is in the shape of a trapezoid. The expression $\frac{1}{2}h(b + d)$ gives the area of a trapezoid, where h is the height and b and d are the lengths of the lower and upper bases, respectively. Find the area of the fin.

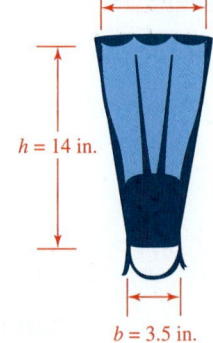

$d = 8.5$ in.

$h = 14$ in.

$b = 3.5$ in.

Solution $\dfrac{1}{2}\boldsymbol{h}(\boldsymbol{b} + \boldsymbol{d}) = \dfrac{1}{2}(\mathbf{14})(\mathbf{3.5} + \mathbf{8.5})$ Substitute 14 for h, 3.5 for b, and 8.5 for d.

$= \dfrac{1}{2}(14)(12)$ Do the addition within the parentheses.

$= 84$ $\frac{1}{2}(14) = 7$ and $7(12) = 84$.

The area of the fin is 84 square inches.

EXAMPLE 11

ELEMENTARY & INTERMEDIATE

Algebra $f(x)$ **Now**™

Rocketry.   If a toy rocket is shot into the air with an initial velocity of 80 feet per second, its height (in feet) after t seconds in flight is given by

$80t - 16t^2$

How many seconds after the launch will it hit the ground?

Solution   We can substitute positive values for t, the time in flight, until we find the one that gives a height of 0. At that time, the rocket will be on the ground. We begin by finding the height after the rocket has been in flight for 1 second ($t = 1$).

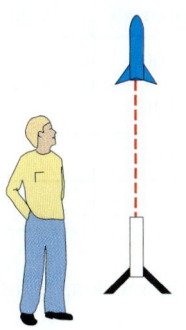

$$80t - 16t^2 = 80(\mathbf{1}) - 16(\mathbf{1})^2 \quad \text{Substitute 1 for } t.$$
$$= 64$$

As we evaluate $80t - 16t^2$ for several more values of t, we record each result in a table.

t	$80t - 16t^2$
1	64
2	96
3	96
4	64
5	0

Evaluate for $t = 2$: $80t - 16t^2 = 80(2) - 16(2)^2 = 96$

Evaluate for $t = 3$: $80t - 16t^2 = 80(3) - 16(3)^2 = 96$

Evaluate for $t = 4$: $80t - 16t^2 = 80(4) - 16(4)^2 = 64$

Evaluate for $t = 5$: $80t - 16t^2 = 80(5) - 16(5)^2 = 0$

The height of the rocket is 0 when $t = 5$. The rocket will hit the ground 5 seconds after being launched.

Self Check 11   In Example 11, suppose the height of the rocket is given by $112t - 16t^2$. In how many seconds after launch would the rocket hit the ground?

The two columns of the table in Example 11 can be headed with the terms **input** and **output.** The values of t are the inputs into the expression $80t - 16t^2$, and the resulting values are the outputs.

Input	Output
1	64
2	96
3	96
4	64
5	0

Answers to Self Checks   **1.** $1, -12, 3, -4$ **2. a.** $t - 80$,   **b.** $\frac{2}{3}T$,   **c.** $2a - 15$ **3.** $(m + 15)$ minutes
4. $s =$ amount donated to scholarship fund in dollars; $900 - s =$ amount donated to building fund
5. $x =$ the number of votes received by the challenger; $3x - 55 =$ the number of votes received
by the incumbent   **7.** $\frac{h}{24}$ **8. a.** \$300,   **b.** \$100t,   **c.** \$1,000$(x - 4)$
9. a. 17,   **b.** -18,   **c.** 0   **11.** 7 sec

1.8   STUDY SET   ELEMENTARY & INTERMEDIATE Algebra $f(x)$ Now™

VOCABULARY   Fill in the blanks.

1. Variables and/or numbers can be combined with the operations of arithmetic to create algebraic _____.

2. Addition symbols separate algebraic expressions into parts called _____.

3. A term, such as 27, that consists of a single number is called a _____ term.

4. The _____ of $10x$ is 10.

5. To _____ an algebraic expression, we substitute the values for the variables and simplify.

6. Consider the expression $2a + 8$. When we replace a with 4, we say we are _____ a value for the variable.

7. $2x + 5$ is an example of an algebraic _____, whereas $2x + 5 = 7$ is an example of an _____.

8. When we evaluate an expression for several values of x, we can keep track of the results in an input/output _____.

CONCEPTS

9. Consider the expression $11x^2 - 6x - 9$.
   a. How many terms does this expression have?
   b. What is the coefficient of the first term?
   c. What is the coefficient of the second term?
   d. What is the constant term?

10. Complete the table.

Term	Coefficient
$6m$	
$-75t$	
w	
$\frac{1}{2}bh$	
$\frac{x}{5}$	

11. In each expression, tell whether c is used as a *factor* or as a *term*.
   a. $c + 32$ **b.** $5c$
   c. $-18bc$ **d.** $a + b + c$
   e. $24c + 6$ **f.** $c - 9$

12. a. Write a term that has an implied coefficient of 1.

   b. Write a term that has an implied coefficient of -1.

13. a. Complete the table on the left to determine how many days are in w weeks.
   b. Complete the table on the right.

Number of weeks	Number of days		Number of seconds	Number of minutes
1			60	
2			120	
3			180	
w			s	

14. a. If the knife is 12 inches long, how long is the blade?

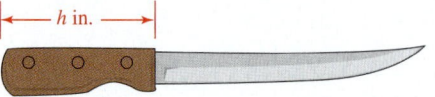
$\longleftarrow h$ in. $\longrightarrow$

   b. A student inherited $5,000 and deposits x dollars in American Savings. Write an expression to show how much she has left to deposit in a City Mutual account.

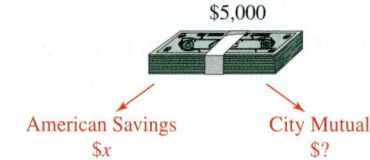
$5,000
American Savings   City Mutual
x $?

   c. Suppose solution 1 is poured into solution 2. Write an expression to show how many ounces of the mixture there will be.

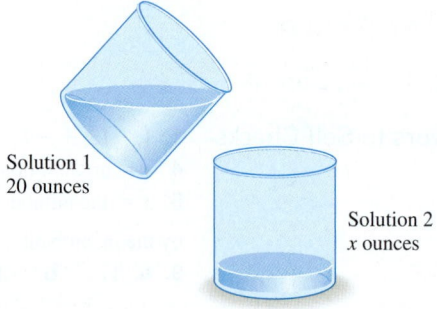
Solution 1
20 ounces
Solution 2
x ounces

15. a. The weight of the van in the illustration is 500 pounds less than twice the weight of the car. Express the weight of the van and the car using the variable x.

b. If the actual weight of the car is 2,000 pounds, what is the weight of the van?

16. a. If we let b represent the length of the beam, write an algebraic expression for the length of the pipe.

b. If we let p represent the length of the pipe, write an algebraic expression for the length of the beam.

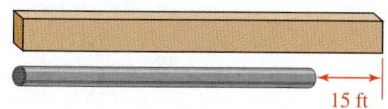

15 ft

17. Complete the table.

Type of coin	Number ·	Value (in cents) =	Total value (in cents)
Nickel	6		
Dime	d		
Half-dollar	$x + 5$		

18. If $x = -9$, find the value of $-x$.

NOTATION Complete each solution.

19. Evaluate the expression $9a - a^2$ for $a = 5$.

$$9a - a^2 = 9(\ ) - (5)^2$$
$$= 9(5) - \boxed{}$$
$$= \boxed{} - 25$$
$$= 20$$

20. Evaluate $-x + 6y$ for $x = -2$ and $y = 4$.

$$-x + 6y = -(\ ) + 6(\ )$$
$$= \boxed{} + 24$$
$$= 26$$

21. Write each term in standard form.
  a. $y8$ **b.** $c2d$
  c. $15xs$ **d.** $b^2(-9)a^3$

Fill in the blanks.

22. a. $\dfrac{2}{3}m = \dfrac{\boxed{}}{3}$ **b.** $\dfrac{t}{3} = \boxed{}\ t$

  c. $-\dfrac{w}{2} = \boxed{}\ w$ **d.** $-\dfrac{5}{3}s = -\dfrac{\boxed{}}{3}$

  e. $d = \boxed{}\ d$ **f.** $-h = \boxed{}\ h$

PRACTICE Translate each phrase to an algebraic expression. If no variable is given, use x as the variable.

23. The sum of the length l and 15

24. The difference of a number and 10

25. The product of a number and 50

26. Three-fourths of the population p

27. The ratio of the amount won w and lost l

28. The tax t added to c

29. P increased by p

30. 21 less than the total height h

31. The square of k minus 2,005

32. s subtracted from S

33. J reduced by 500

34. Twice the attendance a

35. 1,000 split n equal ways

36. Exceeds the cost c by 25,000

37. 90 more than the current price p

38. 64 divided by the cube of y

39. The total of 35, h, and 300

40. x decreased by 17

41. 680 fewer than the entire population p

42. Triple the number of expected participants

43. The product of d and 4, decreased by 15

44. Forty-five more than the quotient of y and 6

45. Twice the sum of 200 and t

46. The square of the quantity 14 less than x

47. The absolute value of the difference of a and 2

48. The absolute value of a, decreased by 2

49. How many minutes are there in   **a.** 5 hours   and **b.** h hours?

50. A woman watches television x hours a day. Express the number of hours she watches TV   **a.** in a week and   **b.** in a year.

51. a. How many feet are in y yards?

  b. How many yards are in f feet?

52. A sales clerk earns \$$x$ an hour. How much does he earn in **a.** an 8-hour day and **b.** a 40-hour week?

53. If a car rental agency charges 29¢ a mile, express the rental fee if a car is driven x miles.

54. A model's skirt is x inches long. The designer then lets the hem down 2 inches. How can we express the length (in inches) of the altered skirt?

55. A soft drink manufacturer produced c cans of cola during the morning shift. Write an expression for how many six-packs of cola can be assembled from the morning shift's production.

56. The tag on a new pair of 36-inch-long jeans warns that after washing, they will shrink x inches in length. Express the length (in inches) of the jeans after they are washed.

57. A caravan of b cars, each carrying 5 people, traveled to the state capital for a political rally. Express how many people were in the caravan.

58. A caterer always prepares food for 10 more people than the order specifies. If p people are to attend a reception, write an expression for the number of people she should prepare for.

59. Tickets to a circus cost \$5 each. Express how much tickets will cost for a family of x people if they also pay for two of their neighbors.

60. If each egg is worth e¢, express the value (in cents) of a dozen eggs.

Complete each table.

61.

x	$x^3 - 1$
0	
−1	
−3	

62.

g	$g^2 - 7g + 1$
0	
7	
−10	

63.

s	$\dfrac{5s + 36}{s}$
1	
6	
−12	

64.

a	$2{,}500a + a^3$
2	
4	
−5	

65.

Input x	Output $2x - \dfrac{x}{2}$
100	
−300	

66.

Input x	Output $\dfrac{x}{3} + \dfrac{x}{4}$
12	
−36	

67.

x	$(x + 1)(x + 5)$
−1	
−5	
−6	

68.

x	$\dfrac{1}{x + 8}$
−7	
−9	
−8	

Evaluate each expression, given that $x = 3$, $y = -2$, and $z = -4$.

69. $3y^2 - 6y - 4$

70. $-z^2 - z - 12$

71. $(3 + x)y$

72. $(4 + z)y$

73. $(x + y)^2 - |z + y|$

74. $[(z - 1)(z + 1)]^2$

75. $(4x)^2 + 3y^2$

76. $4x^2 + (3y)^2$

77. $-\dfrac{2x + y^3}{y + 2z}$

78. $-\dfrac{2z^2 - y}{2x - y^2}$

Evaluate each expression for the given values of the variables.

79. $b^2 - 4ac$ for $a = -1$, $b = 5$, and $c = -2$

80. $(x - a)^2 + (y - b)^2$ for $x = -2$, $y = 1$, $a = 5$, and $b = -3$

81. $a^2 + 2ab + b^2$ for $a = -5$ and $b = -1$

82. $\dfrac{x - a}{y - b}$ for $x = -2$, $y = 1$, $a = 5$, and $b = 2$

83. $\dfrac{n}{2}[2a + (n - 1)d]$ for $n = 10$, $a = -4$, and $d = 6$

84. $\dfrac{a(1 - r^n)}{1 - r}$ for $a = -5$, $r = 2$, and $n = 3$

85. $\dfrac{a^2 + b^2}{2}$ for $a = 0$ and $b = -10$

86. $(y^3 - 52y^2)^2$ for $y = 0$

APPLICATIONS

87. ROCKETRY The expression $64t - 16t^2$ gives the height of a toy rocket (in feet) t seconds after being launched. Find the height of the rocket for each of the times shown.

t	h
1	
2	
3	
4	

88. AMUSEMENT PARK RIDES   The distance in feet that an object will fall in t seconds is given by the expression $16t^2$. Find the distance that riders on "Drop Zone" will fall during the times listed in the table.

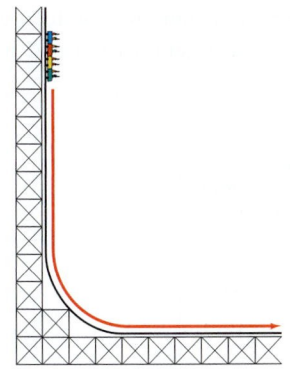

Time (seconds)	Distance (feet)
1	
2	
3	
4	

89. ANTIFREEZE   The expression $\frac{5(F - 32)}{9}$ converts a temperature in degrees Fahrenheit (given as F) to degrees Celsius. Convert the temperatures listed on the container of antifreeze shown to degrees Celsius. Round to the nearest degree.

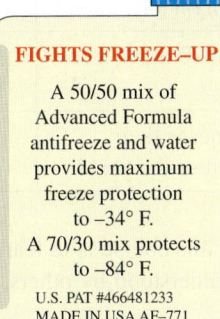

FIGHTS FREEZE–UP

A 50/50 mix of Advanced Formula antifreeze and water provides maximum freeze protection to –34° F.
A 70/30 mix protects to –84° F.

U.S. PAT #466481233
MADE IN USA AF–771

90. MARS   The expression $\frac{9C + 160}{5}$ converts a temperature in degrees Celsius (represented by C) to a temperature in degrees Fahrenheit. On Mars, daily temperatures average $-33°$ C. Convert this to degrees Fahrenheit. Round to the nearest degree.

91. TOOLS   The utility knife blade shown is in the shape of a trapezoid. Find the area of the front face of the blade.

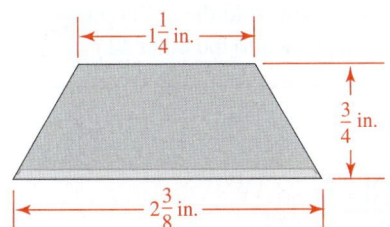

92. GROWING SOD   To determine the number of square feet of sod remaining in a field after filling an order, the manager of a sod farm uses the expression $20{,}000 - 3s$ (where s is the number of strips the customer has ordered). To sod a soccer field, a city orders 7,000 strips of sod. Evaluate the expression for this value of s and explain the result.

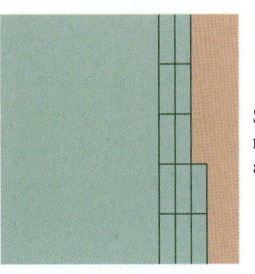

Strips of sod, cut and ready to be loaded on a truck for delivery

WRITING

93. What is an algebraic expression? Give some examples.

94. What is a variable? How are variables used in this section?

95. In this section, we substituted a number for a variable. List some other uses of the word *substitute* that you encounter in everyday life.

96. Explain why d dimes are not worth d¢.

REVIEW

97. Find the LCD for $\frac{5}{12}$ and $\frac{1}{15}$.

98. Remove the common factors of the numerator and denominator to simplify: $\frac{3 \cdot 5 \cdot 5}{3 \cdot 5 \cdot 5 \cdot 11}$.

99. Evaluate: $(-2 \cdot 3)^3 - 10 + 1$.

100. Find the result when $\frac{7}{8}$ is multiplied by its reciprocal.

CHALLENGE PROBLEMS

101. Evaluate:
$(8 - 1)(8 - 2)(8 - 3) \cdot \cdots \cdot (8 - 49)(8 - 50)$

102. If the values of x and y were doubled, what would happen to the value of $17xy$?

ACCENT ON TEAMWORK

WRITING FRACTIONS AS DECIMALS

Overview: This is a good activity to try at the beginning of the course. You can become acquainted with other students in your class while you review the process for finding decimal equivalents of fractions.

Instructions: Form groups of 6 students. Select one person from your group to record the group's responses on the questionnaire. Express the results in fraction form and in decimal form.

What fraction (decimal) of the students in your group . . .	Fraction	Decimal
have the letter *a* in their first names?		
have a birthday in January or February?		
work full-time or part-time?		
have ever been on television?		
live more than 10 miles from the campus?		
say that summer is their favorite season of the year?		

WRITING MATHEMATICAL SOLUTIONS

Overview: A major objective of this course is to learn how to put your thinking on paper in a form that can be read and understood by others. This activity will help you develop that ability.

Instructions: Form groups of 2 or 3 students. Have someone read the following problem out loud. Discuss it in your group. Then work together to present a written solution. Exchange your solution with another group and see whether you understand their explanation.

When three professors attending a convention in Las Vegas registered at the hotel, they were told that the room rate was $120. Each professor paid his $40 share. Later the desk clerk realized that the cost of the room should have been $115. To fix the mistake, she sent a bellhop to the room to refund the $5 overcharge. Realizing that $5 could not be evenly divided among the three professors and not wanting to start a quarrel, the bellhop refunded only $3 and kept the other $2. Since each professor received a $1 refund, each paid $39 for the room, and the bellhop kept $2. This gives $39 + $39 + $39 + $2, or $119. What happened to the other $1?

KEY CONCEPT: VARIABLES

One of the major objectives of this course is for you to become comfortable working with **variables.** In Chapter 1, we have used the concept of variables in several ways.

STATING MATHEMATICAL PROPERTIES

Variables have been used to state properties of mathematics. Match each statement in words with its proper description expressed with a variable (or variables). Assume that a, b, and c are real numbers and that there are no divisions by zero.

1. A nonzero number divided by itself is 1.

2. When we add opposites, the result is 0.

3. Two numbers can be multiplied in either order to get the same result.

4. It doesn't matter how we group numbers in multiplication.

5. When we multiply a number and its reciprocal, the result is 1.

6. When we multiply a number and 0, the result is 0.

7. It doesn't matter how we group numbers in addition.

8. Any number divided by 1 is the number itself.

9. Two numbers can be added in either order to get the same result.

10. When we add a number and 0, the number remains the same.

a. $(ab)c = a(bc)$

b. $a\left(\dfrac{1}{a}\right) = 1$

c. $(a + b) + c = a + (b + c)$

d. $a + b = b + a$

e. $a \cdot 0 = 0$

f. $\dfrac{a}{a} = 1$

g. $a + 0 = a$

h. $ab = ba$

i. $\dfrac{a}{1} = a$

j. $a + (-a) = 0$

STATING RELATIONSHIPS BETWEEN QUANTITIES

Variables are letters that stand for numbers. We have used **formulas** to express known relationships between two or more variables.

11. Translate the word model to an equation (formula) that mathematically describes the situation: The total cost is the sum of the purchase price of the item and the sales tax.

12. Use the data in the table to state the relationship between the quantities using a formula.

Picnic tables	Benches needed
2	4
3	6
4	8

WRITING ALGEBRAIC EXPRESSIONS

We have combined variables and numbers with the operations of arithmetic to create **algebraic expressions.**

13. One year, a cruise company did x million dollars' worth of business. After a celebrity was signed as a spokeswoman for the company, its business increased by \$4 million the next year. Write an expression that represents the amount of business the cruise company had in the year the celebrity was the spokeswoman.

14. Evaluate the expression for the given values of the variable, and enter the results in the table.

x	$3x^2 - 2x + 1$
0	
4	
6	

CHAPTER REVIEW

ELEMENTARY &
INTERMEDIATE
Algebra $f(x)$ **Now**™

SECTION 1.1	**Introducing the Language of Algebra**

CONCEPTS

Tables, bar graphs, and *line graphs* are used to describe numerical relationships.

REVIEW EXERCISES

The line graph shows the number of cars parked in a parking structure from 6 P.M. to 12 midnight on a Saturday.

1. What units are used to scale the horizontal and vertical axes?

2. How many cars were in the parking structure at 11 P.M.?

3. At what time did the parking structure have 500 cars in it?

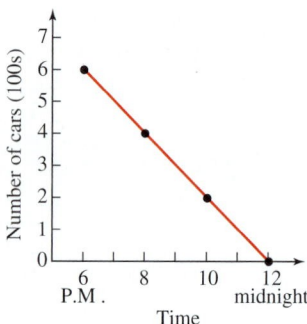

The result of an addition is called the *sum;* of a subtraction, the *difference;* of a multiplication, the *product;* and of a division, the *quotient.*

Express each statement in words.

4. $15 - 3 = 12$

5. $15 + 3 = 18$

6. $15 \div 3 = 5$

7. $15 \cdot 3 = 45$

8. Write the multiplication 4×9 with a raised dot and then with parentheses.

9. Write the division $9 \div 3$ using a fraction bar.

Variables are letters used to stand for numbers.

An *equation* is a mathematical sentence that contains an $=$ symbol. Variables and/or numbers can be combined with the operations of arithmetic to create *algebraic expressions.*

Equations that express a known relationship between two or more variables are called *formulas.*

Write each multiplication without a multiplication symbol.

10. $8 \cdot b$

11. $P \cdot r \cdot t$

Classify each item as either an expression or an equation.

12. $5 = 2x + 3$

13. $2x + 3$

14. Use the formula $n = b + 5$ to complete the table.

Number of brackets (*b*)	Number of nails (*n*)
5	
10	
20	

| SECTION 1.2 | **Fractions** |

Whole numbers can be written as the product of two or more whole-number *factors*.

A *prime number* is a whole number greater than 1 that has only 1 and itself as factors.

Division by 0 is *undefined*.

15. a. Write 24 as the product of two factors.

b. Write 24 as the product of three factors.

c. List the factors of 24.

Give the prime factorization of each number, if possible.

16. 54   **17.** 147   **18.** 385   **19.** 41

Perform each division, if possible.

20. $\dfrac{12}{12}$ **21.** $\dfrac{0}{10}$

Simplify each fraction.

22. $\dfrac{20}{35}$ **23.** $\dfrac{24}{18}$

Build each number to an equivalent fraction with the indicated denominator.

24. $\dfrac{5}{8}$, denominator 64   **25.** 12, denominator 3

To multiply two fractions, multiply their numerators and multiply their denominators.

To divide two fractions, multiply the first fraction by the reciprocal of the second fraction.

To add (or subtract) fractions with the same denominator, add (or subtract) the numerators and keep the common denominator.

The *least common denominator (LCD)* for a set of fractions is the smallest number each denominator will divide exactly.

Perform each operation.

26. $\dfrac{1}{8} \cdot \dfrac{7}{8}$ **27.** $\dfrac{16}{35} \cdot \dfrac{25}{48}$ **28.** $\dfrac{1}{3} \div \dfrac{15}{16}$ **29.** $16\dfrac{1}{4} \div 5$

30. $\dfrac{17}{25} - \dfrac{7}{25}$ **31.** $\dfrac{8}{11} - \dfrac{1}{2}$ **32.** $\dfrac{1}{4} + \dfrac{2}{3}$ **33.** $4\dfrac{1}{9} - 3\dfrac{5}{6}$

34. MACHINE SHOPS   How much must be milled off the $\frac{17}{24}$-inch-thick steel rod so that the collar will slip over it?

Steel rod

| SECTION 1.3 | The Real Numbers |

The *natural numbers:*
  {1, 2, 3, 4, 5, 6, . . .}
The *whole numbers:*
  {0, 1, 2, 3, 4, 5, 6, . . .}
The *integers:*
  {. . . , −2, −1, 0, 1,
  2, . . .}

Two *inequality symbols* are
  > "is greater than"
  < "is less than"

A *rational number* is any number that can be expressed as a fraction with an integer numerator and a nonzero integer denominator.

To write a fraction as a decimal, divide the numerator by the denominator.

An *irrational number* is a nonterminating, nonrepeating decimal. Irrational numbers cannot be written as the ratio of two integers.

A *real number* is any number that is either a rational or an irrational number.

The natural numbers are a *subset* of the whole numbers. The whole numbers are a subset of the integers. The integers are a subset of the rational numbers. The rational numbers are a subset of the real numbers.

35. Which number is a whole number but not a natural number?

Represent each of these situations with a signed number.

36. A budget deficit of $65 billion

37. 206 feet below sea level

Use one of the symbols > or < to make each statement true.

38. 0 ___ 5

39. −12 ___ −13

Show that each of the following numbers is a rational number by expressing it as a fraction.

40. 0.7

41. $4\frac{2}{3}$

Write each fraction as a decimal. Use an overbar if the result is a repeating decimal.

42. $\frac{1}{250}$

43. $\frac{17}{22}$

44. Graph each number on a number line: $\left\{\pi, 0.333. \ . \ . \ , 3.75, \sqrt{2}, -\frac{17}{4}, \frac{7}{8}, -2\right\}$.

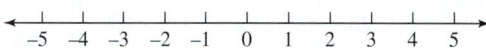

Decide whether each statement is true or false.

45. All integers are whole numbers.

46. π is a rational number.

47. The set of real numbers corresponds to all points on the number line.

48. A real number is either rational or irrational.

49. Tell which numbers in the given set are natural numbers, whole numbers, integers, rational numbers, irrational numbers, and real numbers.

$$\left\{-\frac{4}{5}, 99.99, 0, \sqrt{2}, -12, 4\frac{1}{2}, 0.666. \ . \ . \ , 8\right\}$$

The *absolute value* of a number is the distance on the number line between the number and 0.

Insert one of the symbols $>$, $<$, or $=$ in the blank to make each statement true.

50. $|-6|$ ____ $|5|$

51. -9 ____ $|-10|$

| **SECTION 1.4** | **Adding Real Numbers** |

To add two real numbers with *like signs,* add their absolute values and attach their common sign to the sum.

To add two real numbers with *unlike signs,* subtract their absolute values, the smaller from the larger. To that result, attach the sign of the number with the larger absolute value.

Add the numbers.

52. $-45 + (-37)$

53. $25 + (-13)$

54. $0 + (-7)$

55. $-7 + 7$

56. $12 + (-8) + (-15)$

57. $-9.9 + (-2.4)$

58. $\dfrac{5}{16} + \left(-\dfrac{1}{2}\right)$

59. $35 + (-13) + (-17) + 6$

The *commutative* and *associative* properties of addition:

  $a + b = b + a$

  $(a + b) + c$
    $= a + (b + c)$

Tell what property of addition guarantees that the quantities are equal.

60. $-2 + 5 = 5 + (-2)$

61. $(-2 + 5) + 1 = -2 + (5 + 1)$

62. $80 + (-80) = 0$

63. $-5.75 + 0 = -5.75$

64. ATOMS   An atom is composed of protons, neutrons, and electrons. A proton has a positive charge (represented by $+1$), a neutron has no charge, and an electron has a negative charge (-1). A simple model of an atom is shown here. What is its net charge?

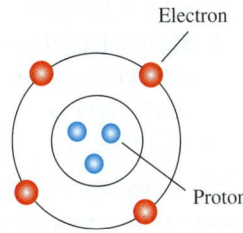

Electron

Proton

| SECTION 1.5 | **Subtracting Real Numbers** |

Two numbers represented by points on a number line that are the same distance away from 0, but on opposite sides of it, are called *opposites*.

To *subtract* real numbers, add the opposite:

$$a - b = a + (-b)$$

Write the expression in simpler form.

65. The opposite of 10

66. The opposite of -3

67. $-\left(-\dfrac{9}{16}\right)$

68. $-|-4|$

Perform the operations.

69. $45 - 64$

70. $-17 - 32$

71. $-27 - (-12)$

72. $3.6 - (-2.1)$

73. $0 - 10$

74. $-33 + 7 - 5 - (-2)$

75. GEOGRAPHY   The tallest peak on Earth is Mount Everest, at 29,028 feet, and the greatest ocean depth is the Mariana Trench, at $-36,205$ feet. Find the difference in the two elevations.

| SECTION 1.6 | **Multiplying and Dividing Real Numbers** |

To multiply two real numbers, multiply their absolute values.

1. The product of two real numbers with *like signs* is positive.
2. The product of two real numbers with *unlike signs* is negative.

The *commutative* and *associative* properties of multiplication:

$ab = ba$

$(ab)c = a(bc)$

Multiply.

76. $-8 \cdot 7$

77. $(-9)(-6)$

78. $2(-3)(-2)$

79. $(-4)(-1)(-3)(-3)$

80. $-1.2(-5.3)$

81. $0.002(-1,000)$

82. $-\dfrac{2}{3}\left(\dfrac{1}{5}\right)$

83. $-6(-3)(0)(-1)$

Tell what property of multiplication guarantees that the quantities are equal.

84. $(2 \cdot 3)5 = 2(3 \cdot 5)$

85. $(-5)(-6) = (-6)(-5)$

86. $-6 \cdot 1 = -6$

87. $\dfrac{1}{2}(2) = 1$

88. What is the additive inverse (opposite) of -3?

89. What is the multiplicative inverse (reciprocal) of -3?

To divide two real numbers, divide their absolute values.

1. The quotient of two real numbers with *like signs* is positive.

2. The quotient of two real numbers with *unlike signs* is negative.

Division *of zero* by a nonzero number is 0. Division *by zero* is undefined.

Perform each division.

90. $\dfrac{44}{-44}$

91. $\dfrac{-100}{25}$

92. $\dfrac{-81}{-27}$

93. $-\dfrac{3}{5} \div \dfrac{1}{2}$

94. $\dfrac{-60}{0}$

95. $\dfrac{-4.5}{1}$

96. Find the high and low reading that is displayed on the screen of the emissions-testing device.

97. The picture on the screen can be magnified by switching a setting on the monitor. What would be the new high and low if every value were to be doubled?

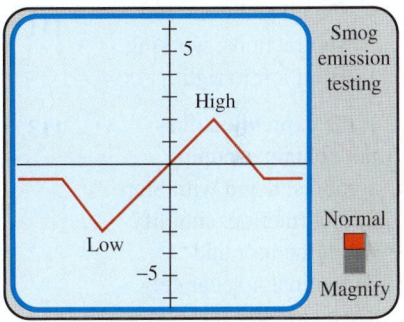

| SECTION 1.7 | **Exponents and Order of Operations** |

An *exponent* is used to represent repeated multiplication. In the *exponential expression* a^n, a is the *base*, and n is the *exponent*.

Write each expression using exponents.

98. $8 \cdot 8 \cdot 8 \cdot 8 \cdot 8$

99. $a(a)(a)(a)$

100. $9 \cdot \pi \cdot r \cdot r$

101. $x \cdot x \cdot x \cdot y \cdot y \cdot y \cdot y$

Order of operations:

1. Do all calculations within grouping symbols, working from the innermost pair to the outermost pair, in the following order:

2. Evaluate all exponential expressions.

3. Do all multiplications and divisions, working from left to right.

4. Do all additions and subtractions, working from left to right.

If the expression does not contain grouping symbols, begin with Step 2. In a fraction, simplify the numerator and denominator separately. Then simplify the fraction, if possible.

The *arithmetic mean* (or *average*) is a value around which number values are grouped.

$$\text{Mean} = \frac{\text{sum of values}}{\text{number of values}}$$

Evaluate each expression.

102. 9^2

103. $\left(-\dfrac{2}{3}\right)^3$

104. 2^5

105. 50^1

106. How many operations does the expression $5 \cdot 4 - 3^2 + 1$ contain, and in what order should they be performed?

Evaluate each expression.

107. $2 + 5 \cdot 3$

108. $24 - 3(6)(4)$

109. $-(6 - 3)^2$

110. $4^3 + 2(-6 - 2 \cdot 2)$

111. $10 - 5[-3 - 2(5 - 7^2)] - 5$

112. $\dfrac{-4(4 + 2) - 4}{|-18 - 4(5)|}$

113. $(-3)^3\left(\dfrac{-8}{2}\right) + 5$

114. $-9^2 + (-9)^2$

115. WALK-A-THONS   Use the data in the table to find the average (mean) donation to a charity walk-a-thon.

Donation	Number received
$5	20
$10	65
$20	25
$50	5
$100	10

SECTION 1.8

Algebraic Expressions

Addition signs separate algebraic expressions into terms.

In a term, the numerical factor is called the *coefficient*.

How many terms are in each expression?

116. $3x^2 + 2x - 5$

117. $-12xyz$

Identify the coefficient of each term.

118. $2x - 5$

119. $16x^2 - 5x + 25$

120. $\dfrac{x}{2} + y$

121. $9.6t^2 - t$

In order to describe numerical relationships, we need to translate the words of a problem into mathematical symbols.

Write each phrase as an algebraic expression.

122. 25 more than the height h

123. 15 less than the cutoff score s

124. $\dfrac{1}{2}$ of the time t

125. If we let n represent the length of the nail, write an algebraic expression for the length of the bolt (in inches).

4 in.

126. If we let b represent the length of the bolt, write an algebraic expression for the length of the nail (in inches).

Sometimes we must rely on common sense and insight to find *hidden operations.*

127. How many years are in d decades?

128. Five years after a house was constructed, a patio was added. How old, in years, is the patio if the house is x years old?

Number · value
          = total value

129. Complete the table.

Type of coin	Number	Value (¢)	Total value (¢)
Nickel	6	5	
Dime	d	10	

When we replace the variable, or variables, in an algebraic expression with specific numbers and then apply the rules for the order of operations, we are *evaluating* the algebraic expression.

130. Complete the table.

x	$20x - x^3$
0	
1	
-4	

Evaluate each algebraic expression for the given value(s) of the variable(s).

131. $b^2 - 4ac$ for $b = -10$, $a = 3$, and $c = 5$

132. $\dfrac{x + y}{-x - z}$ for $x = 19$, $y = 17$, and $z = -18$

CHAPTER 1 TEST   ELEMENTARY & INTERMEDIATE Algebra $f(x)$ Now™

The line graph shows the cost to hire a security guard. Use the graph to answer Problems 1 and 2.

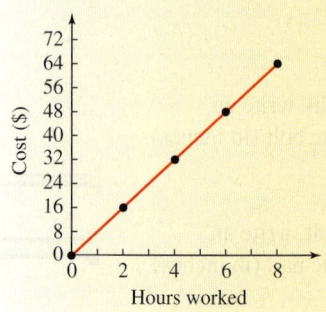

1. What will it cost to hire a security guard for 3 hours?

2. If a school was billed $40 for hiring a security guard for a dance, for how long did the guard work?

3. Use the formula $f = \dfrac{a}{5}$ to complete the table.

Area in square miles (a)	Number of fire stations (f)
15	
100	
350	

4. Give the prime factorization of 180.

5. Simplify: $\dfrac{42}{105}$.

6. Divide: $\dfrac{15}{16} \div \dfrac{5}{8}$.

7. Subtract: $\dfrac{11}{12} - \dfrac{2}{9}$.

8. Add: $1\dfrac{2}{3} + 8\dfrac{2}{5}$.

9. SHOPPING   Find the cost of the fruit on the scale.

Oranges
84 cents a pound

10. Write $\dfrac{5}{6}$ as a decimal.

11. Graph each member of the set on the number line.
$$\left\{-1\tfrac{1}{4},\ \sqrt{2},\ -3.75,\ \tfrac{7}{2},\ 0.5,\ -3\right\}$$

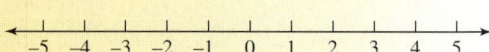

12. Decide whether each statement is true or false.

  a. Every integer is a rational number.
  b. Every rational number is an integer.
  c. π is an irrational number.
  d. 0 is a whole number.

13. Describe the set of real numbers.

14. Insert the proper symbol, $>$ or $<$, in the blank to make each statement true.

  a. $-2 \quad -3$ **b.** $-|-7| \quad 8$
  c. $|-4| \quad -(-5)$ **d.** $\left|-\dfrac{7}{8}\right| \quad 0.5$

15. SWEEPS WEEK   During "sweeps week," television networks make a special effort to gain viewers by showing unusually flashy programming. Use the information in the graph to determine the average daily gain (or loss) of ratings points by a network for the 7-day "sweeps period."

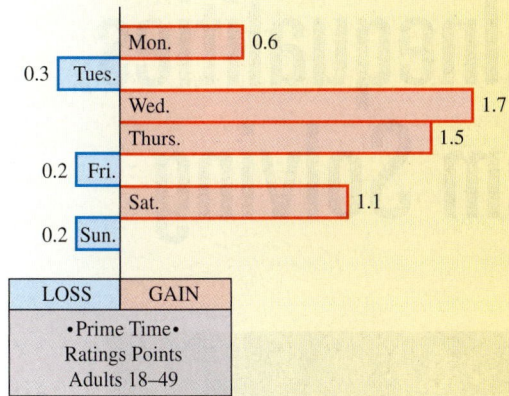

Perform the operations.

16. $(-6) + 8 + (-4)$

17. $-\dfrac{1}{2} + \dfrac{7}{8}$

18. $-10 - (-4)$

19. $(-2)(-3)(-5)$

20. $\dfrac{-22}{-11}$

21. $-6.1(0.4)$

22. $\dfrac{0}{-3}$

23. $0 - 3$

24. $3 + (-3)$

25. $-30 + 50 - 10 - (-40)$

26. $\left(-\dfrac{3}{5}\right)^3$

27. ASTRONOMY *Magnitude* is a term used in astronomy to designate the brightness of celestial objects as viewed from Earth. Smaller magnitudes are associated with brighter objects, and larger magnitudes refer to fainter objects. By how many magnitudes do a full moon and the sun differ?

Object	Magnitude
Sun	−26.5
Full moon	−12.5

28. What property of real numbers is illustrated below?
$(-12 + 97) + 3 = -12 + (97 + 3)$.

29. Write each product using exponents:
   a. $9(9)(9)(9)(9)$ **b.** $3 \cdot x \cdot x \cdot z \cdot z \cdot z.$

30. Evaluate: $8 + 2 \cdot 3^4$.

31. Evaluate: $9^2 - 3[45 - 3(6 + 4)]$.

32. Evaluate: $\dfrac{3(40 - 2^3)}{-2(6 - 4)^2}$.

33. Evaluate: -10^2.

34. Evaluate $3(x - y) - 5(x + y)$ for $x = 2$ and $y = -5$.

35. Complete the table.

x	$2x - \dfrac{30}{x}$
5	
10	
−30	

36. Translate to an algebraic expression: seven more than twice the width w.

37. A rock band recorded x songs for a CD. Technicians had to delete two songs from the album because of poor sound quality. Express the number of songs on the CD using an algebraic expression.

38. Find the value of q quarters in cents.

39. Explain the difference between an expression and an equation.

40. How many terms are in the expression $4x^2 + 5x - 7$? What is the coefficient of the second term?

Chapter

2

Equations, Inequalities, and Problem Solving

ELEMENTARY & INTERMEDIATE

Algebra *f(x)* **Now**™

Throughout the chapter, this icon introduces resources on the Elementary & Intermediate AlgebraNow Web site, accessed through **http://1pass. thomson.com**, that will

- Help you test your knowledge of the material with a pre-test and a post-test

- Provide a personalized learning plan targeting areas you should study

Comstock/Getty Images

When shopping for big-ticket items such as appliances, automobiles, or home furnishings, we often have to stay within a budget. Besides price, the quality and dependability of an item must also be considered. The problem-solving skills that we will discuss can help you make wise decisions when making large purchases such as these.

To learn more about the use of algebra in the marketplace, visit *The Learning Equation* on the Internet at http://tle.brookscole.com. (Log-in instructions are in the Preface.) For Chapter 2, the online lessons are:

- *TLE* Lesson 3: Translating Written Phrases
- *TLE* Lesson 4: Solving Equations Part 1
- *TLE* Lesson 5: Solving Equations Part 2

One of the most useful concepts in algebra is the equation. Writing and then solving an equation is a powerful problem-solving strategy.

2.1 Solving Equations

- Equations and Solutions   • The Addition Property of Equality
- The Subtraction Property of Equality   • The Multiplication Property of Equality
- The Division Property of Equality   • Geometry

In this section, we introduce basic types of equations and discuss four fundamental properties that are used to solve them.

■ EQUATIONS AND SOLUTIONS

An **equation** is a statement indicating that two expressions are equal. An example is $x + 5 = 15$. The equal symbol separates the equation into two parts: The expression $x + 5$ is the **left-hand side** and 15 is the **right-hand side.** The letter x is the **variable** (or the **unknown**). The sides of an equation can be reversed, so we can write $x + 5 = 15$ or $15 = x + 5$.

- An equation can be true: $6 + 3 = 9$.
- An equation can be false: $2 + 4 = 7$.
- An equation can be neither true nor false. For example, $x + 5 = 15$ is neither true nor false because we don't know what number x represents.

An equation that contains a variable is made true or false by substituting a number for the variable. If we substitute 10 for x in $x + 5 = 15$, the resulting equation is true: $10 + 5 = 15$. If we substitute 1 for x, the resulting equation is false: $1 + 5 = 15$. A number that makes an equation true is called a **solution** and it is said to *satisfy* the equation. Therefore, 10 is a solution of $x + 5 = 15$, and 1 is not.

EXAMPLE 1

Is 9 a solution of $3y - 1 = 2y + 7$?

ELEMENTARY & INTERMEDIATE
Algebra *f(x)* **Now**™   **Solution**

We begin by substituting 9 for each y in the equation. Then we evaluate each side separately. If 9 is a solution, we will obtain a true statement.

The Language of Algebra

Read $\stackrel{?}{=}$ as "is possibly equal to."

Evaluate the expression on the left-hand side.

$$3y - 1 = 2y + 7$$
$$3(9) - 1 \stackrel{?}{=} 2(9) + 7$$
$$27 - 1 \stackrel{?}{=} 18 + 7$$
$$26 = 25$$

Evaluate the expression on the right-hand side.

Since $26 = 25$ is false, 9 is not a solution.

Self Check 1   Is 25 a solution of $10 - x = 35 - 2x$?

■ THE ADDITION PROPERTY OF EQUALITY

To **solve an equation** means to find all values of the variable that make the equation true. We can develop an understanding of how to solve equations by referring to the scales shown on the right.

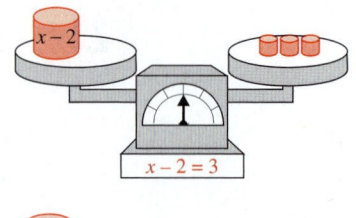

The first scale represents the equation $x - 2 = 3$. The scale is in balance because the weight on the left-hand side, $(x - 2)$ grams, and the weight on the right-hand side, 3 grams, are equal. To find x, we must add 2 grams to the left-hand side. To keep the scale in balance, we must also add 2 grams to the right-hand side. After doing this, we see in the second illustration that x grams is balanced by 5 grams. Therefore, x must be 5. We say that we have solved the equation $x - 2 = 3$ and that the solution is 5.

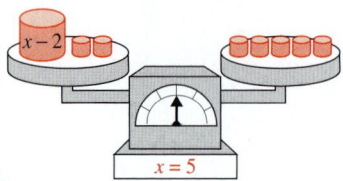

In this example, we solved $x - 2 = 3$ by transforming it to a simpler *equivalent equation,* $x = 5$.

Equivalent Equations	Equations with the same solutions are called **equivalent equations.**

The procedure that we used suggests the following property of equality.

Addition Property of Equality	Adding the same number to both sides of an equation does not change its solution.
	    For any real numbers a, b, and c,
	    if $a = b$, then $a + c = b + c$

When we use this property, the resulting equation is equivalent to the original one. We will now show how it is used to solve $x - 2 = 3$ algebraically.

EXAMPLE 2

Solve: $x - 2 = 3$.

Solution   To isolate x on the left-hand side of the equation, we use the addition property of equality. We can undo the subtraction of 2 by adding 2 to both sides.

The Language of Algebra

When solving an equation, we want to *isolate* the variable on one side of the equation. The word *isolate* means to be alone or by itself.

$$x - 2 = 3$$
$$x - 2 + 2 = 3 + 2 \quad \text{Add 2 to both sides.}$$
$$x + 0 = 5 \quad\quad\quad \text{Do the addition: } -2 + 2 = 0.$$
$$x = 5 \quad\quad\quad\quad \text{When 0 is added to a number, the result is the same number.}$$

To check, we substitute 5 for x in the original equation and simplify. If 5 is a solution, we will obtain a true statement.

$$x - 2 = 3 \quad \text{This is the original equation.}$$
$$5 - 2 \stackrel{?}{=} 3 \quad \text{Substitute 5 for } x.$$
$$3 = 3 \quad \text{True.}$$

Since the statement is true, 5 is the solution.

Self Check 2 Solve: $n - 16 = 33$.

EXAMPLE 3 Solve: $-19 = y - 7$.

Solution To isolate y on the right-hand side, we use the addition property of equality. We can undo the subtraction of 7 by adding 7 to both sides.

$$-19 = y - 7$$
$$-19 \mathbf{+ 7} = y - 7 \mathbf{+ 7} \qquad \text{Add 7 to both sides.}$$
$$-12 = y \qquad\qquad\;\; \text{Do the addition: } -7 + 7 = 0.$$

To check, we substitute -12 for y in the original equation and simplify.

$$-19 = \mathbf{y} - 7 \qquad \text{This is the original equation.}$$
$$-19 \overset{?}{=} \mathbf{-12} - 7 \qquad \text{Substitute } -12 \text{ for } y.$$
$$-19 = -19 \qquad\;\; \text{True.}$$

Since the statement is true, the solution is -12.

Self Check 3 Solve: $-5 = b - 38$.

Notation

We may solve an equation so that the variable is isolated on either side of the equation. In Example 3, note that $-12 = y$ is equivalent to $y = -12$.

EXAMPLE 4 Solve: $-27 + g = -3$.

Solution To isolate g, we use the addition property of equality. We can eliminate -27 on the left-hand side by adding its opposite (additive inverse) to both sides.

$$-27 + g = -3$$
$$-27 + g \mathbf{+ 27} = -3 \mathbf{+ 27} \qquad \text{Add 27 to both sides.}$$
$$g = 24 \qquad\qquad\;\; \text{Do the addition: } -27 + 27 = 0.$$

Check: $-27 + \mathbf{g} = -3$ $\qquad$ This is the original equation.
$\qquad\quad -27 + \mathbf{24} \overset{?}{=} -3$ $\qquad$ Substitute 24 for g.
$\qquad\qquad\qquad -3 = -3$ $\qquad$ True.

The solution is 24.

Self Check 4 Solve: $-20 + n = 29$.

ELEMENTARY & INTERMEDIATE: Algebra $f(x)$ Now™

Caution

After checking a result, be careful when stating your conclusion. For Example 4, it would be incorrect to say:

The solution is -3.

The number we were checking was 24, not -3.

■ THE SUBTRACTION PROPERTY OF EQUALITY

The first scale shown on the next page represents the equation $x + 2 = 5$. The scale is in balance because the weight on the left-hand side, $(x + 2)$ grams, and the weight on the right-hand side, 5 grams, are equal. To find x, we isolate it by subtracting 2 grams from the left-hand side. To keep the scale in balance, we must also subtract 2 grams from the right-hand side. After doing this, we see in the second illustration that x grams is balanced by 3 grams. Therefore, x must be 3. We say that we have solved the equation $x + 2 = 5$ and that the solution is 3.

In this example, we solved $x + 2 = 5$ by transforming it to a simpler equivalent equation, $x = 3$. The process that we used to isolate x on the left-hand side of the scale suggests the following property of equality.

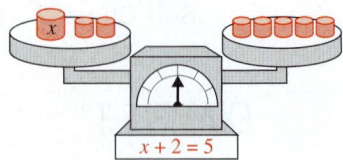

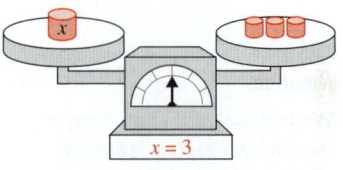

Subtraction Property of Equality	Subtracting the same number from both sides of an equation does not change its solution. For any real numbers a, b, and c,
	if $a = b$, then $a - c = b - c$

When we use this property, the resulting equation is equivalent to the original one. We will now show how it is used to solve $x + 2 = 5$ algebraically.

EXAMPLE 5

Solve: $x + 2 = 5$.

Solution   To isolate x on the left-hand side, we use the subtraction property of equality. We can undo the addition of 2 by subtracting 2 from both sides.

$$x + 2 = 5$$
$$x + 2 - \mathbf{2} = 5 - \mathbf{2} \qquad \text{Subtract 2 from both sides.}$$
$$x + 0 = 3$$
$$x = 3$$

Check: $\quad x + 2 = 5 \qquad$ This is the original equation.
$$\mathbf{3} + 2 \overset{?}{=} 5 \qquad \text{Substitute 3 for } x.$$
$$5 = 5 \qquad \text{True.}$$

The solution is 3.

Self Check 5   Solve: $x + 24 = 50$.

EXAMPLE 6

Solve: $54.9 + m = 45.2$.

ELEMENTARY &
INTERMEDIATE
Algebra $f(x)$ **Now**™   **Solution**   To isolate m, we can undo the addition of 54.9 by subtracting 54.9 from both sides.

$$54.9 + m = 45.2$$
$$54.9 + m - \mathbf{54.9} = 45.2 - \mathbf{54.9} \qquad \text{Subtract 54.9 from both sides.}$$
$$m = -9.7$$

Check: $54.9 + \boldsymbol{m} = 45.2$ This is the original equation.

$54.9 + (\boldsymbol{-9.7}) \overset{?}{=} 45.2$ Substitute -9.7 for m.

$45.2 = 45.2$ True.

The solution is -9.7.

Self Check 6   Solve: $0.7 + a = 0.2$

■ THE MULTIPLICATION PROPERTY OF EQUALITY

We can think of the first scale shown below as representing the equation $\frac{x}{3} = 25$. The weight on the left-hand side is $\frac{x}{3}$ grams, and the weight on the right-hand side is 25 grams. Because the weights are equal, the scale is in balance. To find x, we can triple (multiply by 3) the weight on each side. When we do this, the scale will remain in balance. We see that x grams will be balanced by 75 grams. Therefore, x is 75.

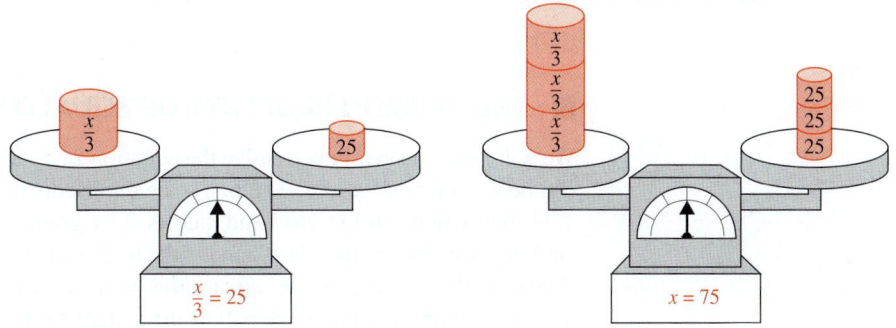

The process used to isolate x on the left-hand side of the scale suggests the following property of equality.

Multiplication Property of Equality	Multiplying both sides of an equation by the same nonzero number does not change its solution.
	
	  For any real numbers a, b, and c, where c is not 0,
	
	    if $a = b$, then $ca = cb$

When we use this property, the resulting equation is equivalent to the original one. We will now show how it is used to solve $\frac{x}{3} = 25$ algebraically.

EXAMPLE 7   Solve: $\dfrac{x}{3} = 25$.

ELEMENTARY & INTERMEDIATE
Algebra *f(x)* **Now**™

Solution   To isolate x on the left-hand side, we use the multiplication property of equality. We can undo the division by 3 by multiplying both sides by 3.

$$\frac{x}{3} = 25$$

$$\boldsymbol{3} \cdot \frac{x}{3} = \boldsymbol{3} \cdot 25 \quad \text{Multiply both sides by 3.}$$

$$\frac{3x}{3} = 75 \qquad \text{Do the multiplications.}$$

$$1x = 75 \qquad \text{Simplify the fraction by removing the common factor of 3 in the}$$
numerator and denominator: $\frac{3}{3} = 1$.

$$x = 75 \qquad \text{The product of 1 and any number is that number.}$$

Check: $\dfrac{x}{3} = 25$ This is the original equation.

$$\frac{75}{3} \stackrel{?}{=} 25 \qquad \text{Substitute 75 for } x.$$

$$25 = 25 \qquad \text{True.}$$

The solution is 75.

Self Check 7   Solve: $\dfrac{b}{24} = 3$.

■ THE DIVISION PROPERTY OF EQUALITY

We will now consider how to solve the equation $2x = 8$. Since $2x$ means $2 \cdot x$, the equation can be written as $2 \cdot x = 8$. The first scale represents this equation.

The weight on the left-hand side is $2 \cdot x$ grams and the weight on the right-hand side is 8 grams. Because these weights are equal, the scale is in balance. To find x, we remove half of the weight from each side. This is equivalent to dividing the weight on both sides by 2. When we do this, the scale remains in balance. We see that x grams is balanced by 4 grams. Therefore, x is 4. We say that we have solved the equation $2x = 8$ and that the solution is 4.

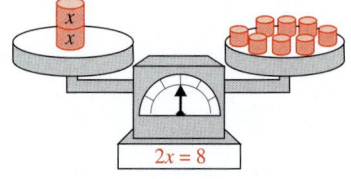

In this example, we solved $2x = 8$ by transforming it to a simpler equivalent equation, $x = 4$. The procedure that we used suggests the following property of equality.

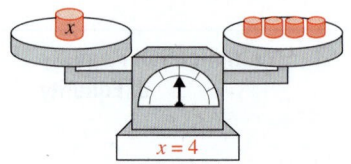

Division Property of Equality	Dividing both sides of an equation by the same nonzero number does not change its solution.
	For any real numbers a, b, and c, where c is not 0,
	$$\text{if } a = b, \text{ then } \frac{a}{c} = \frac{b}{c}$$

When we use this property, the resulting equation is equivalent to the original one. We will now show how it is used to solve $2x = 8$ algebraically.

EXAMPLE 8   Solve: $2x = 8$.

Solution   To isolate x on the left-hand side, we use the division property of equality to undo the multiplication by 2 by dividing both sides of the equation by 2.

$$2x = 8$$

$$\frac{2x}{2} = \frac{8}{2} \qquad \text{Divide both sides by 2.}$$

$$1x = 4 \qquad \text{Simplify the fraction by removing the common factor of 2 in the numerator and denominator: } \frac{2}{2} = 1.$$

$$x = 4 \qquad \text{The product of 1 and any number is that number: } 1x = x.$$

Check: $2x = 8$ This is the original equation.

$\qquad\quad 2(\mathbf{4}) \overset{?}{=} 8$ Substitute 4 for x.

$\qquad\qquad\quad 8 = 8$ True.

The solution is 4.

Self Check 8   Solve: $16x = 176$.

EXAMPLE 9

Solve: $-6.02 = -8.6t$.

Solution   To isolate t on the right-hand side, we can undo the multiplication by -8.6 by dividing both sides by -8.6.

Notation

In Example 9, if you prefer to isolate the variable on the left-hand side, you can solve $-6.02 = -8.6t$ by reversing both sides and solving $-8.6t = -6.02$.

$$-6.02 = -8.6t$$

$$\frac{-6.02}{-8.6} = \frac{-8.6t}{-8.6} \qquad \text{Divide both sides by } -8.6.$$

$$0.7 = t \qquad \text{Do the divisions.}$$

The check is left to the student. The solution is 0.7.

Self Check 9   Solve: $10.04 = -0.4r$.

ELEMENTARY &
INTERMEDIATE
Algebra *f(x)* **Now**™

■ **GEOMETRY**

The following three figures are called **angles.** Angles are measured in **degrees,** denoted by the symbol °. If an angle measures 90° (ninety degrees), it is called a **right angle.** If an angle measures 180°, it is called a **straight angle.**

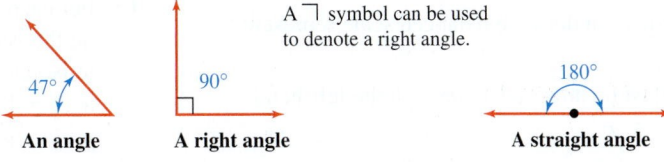

A ⌐ symbol can be used to denote a right angle.

47°

90°

180°

An angle   **A right angle**   **A straight angle**

EXAMPLE 10

ELEMENTARY &
INTERMEDIATE

Algebra *f(x)* **Now**™

Lawn mowers. Find x, the angle that the mower's handle makes with the ground.

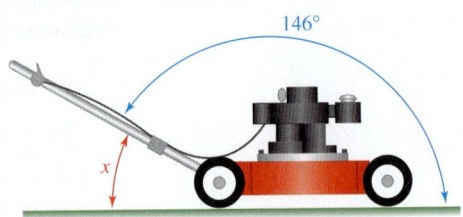

Solution   We can use algebra to find the unknown angle labeled x. Since the sum of the measures of the angles is 180°, we have

$$x + 146 = 180$$
$$x + 146 - \mathbf{146} = 180 - \mathbf{146} \qquad \text{Subtract 146 from both sides.}$$
$$x = 34$$

The angle that the handle makes with the ground is 34°.

Answers to Self Checks   **1.** yes   **2.** 49   **3.** 33   **4.** 49   **5.** 26   **6.** −0.5   **7.** 72   **8.** 11   **9.** −25.1

2.1   STUDY SET

ELEMENTARY &
INTERMEDIATE

Algebra *f(x)* **Now**™

VOCABULARY   Fill in the blanks.

1. An _____ is a statement indicating that two expressions are equal.

2. Any number that makes an equation true when substituted for its variable is said to _____ the equation. Such numbers are called _____.

3. To _____ the solution of an equation, we substitute the value for the variable in the original equation and see whether the result is a true statement.

4. In $30 = t - 12$, the _____ side of the equation is $t - 12$.

5. Equations with the same solutions are called _____ equations.

6. To _____ an equation means to find all values of the variable that make the equation true.

7. To solve an equation, we _____ the variable on one side of the equal symbol.

8. When solving an equation, the objective is to find all values of the _____ that will make the equation true.

CONCEPTS

9. a. What equation does the balanced scale represent?

   b. What must be done to isolate x on the left-hand side?

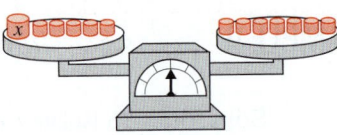

10. a. What equation does the balanced scale represent?

   b. What must be done to isolate x on the left-hand side?

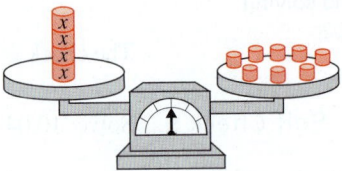

11. Given $x + 6 = 12$,
   a. What forms the left-hand side of the equation?

   b. Is this equation true or false?

   c. Is 5 a solution?

   d. Does 6 satisfy the equation?

12. For each equation, tell what operation is performed on the variable. Then tell how to undo that operation to isolate the variable.
   a. $x - 8 = 24$
   b. $x + 8 = 24$

c. $\dfrac{x}{8} = 24$

d. $8x = 24$

Complete each flow chart.

13.
Begin with the number 24.
↓
Add 6.
↓
Subtract 6.
↓
The result is [ ] .

14.
Begin with a number x.
↓
Multiply by 10.
↓
Divide by 10.
↓
The result is [ ] .

15.
Begin with a number n.
↓
Divide by 5.
↓
Multiply by 5.
↓
The result is [ ] .

16.
Begin with the number 45.
↓
Subtract 9.
↓
Add 9.
↓
The result is [ ] .

17. Complete the following properties of equality.
   a. If $x = y$, then $x + c = y +$ [ ] and
   $x - c = y -$ [ ] .

   b. If $x = y$, then $cx =$ [ ] y and $\dfrac{x}{c} = \dfrac{y}{ }$ ($c \neq 0$).

18. a. When solving $\dfrac{h}{10} = 20$, do we multiply both sides of the equation by 10 or 20?

   b. When solving $4k = 16$, do we subtract 4 from both sides of the equation or divide both sides by 4?

19. Simplify each expression.
   a. $x + 7 - 7$ **b.** $y - 2 + 2$
   c. $\dfrac{5t}{5}$ **d.** $6 \cdot \dfrac{h}{6}$

20. Complete each equation.
   a. $x + 20 =$ [ ]   **b.** $x + 20 =$ [ ]

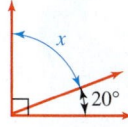

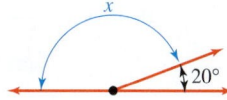

21. Solve: $x + 15 = 45$

$x + 15 -$ [ ] $= 45 -$ [ ]

$x = 30$

Check: $x + 15 = 45$

[ ] $+ 15 \overset{?}{=} 45$

$45 = 45$

[ ] is a solution.

22. Solve: $8x = 40$

$\dfrac{8x}{[ ]} = \dfrac{40}{[ ]}$

$x = 5$

Check: $8x = 40$

$8( ) \overset{?}{=} 40$

$40 = 40$

[ ] is a solution.

23. a. What does the symbol $\overset{?}{=}$ mean?

   b. Write twenty-seven degrees using symbols.

24. If you solve an equation and obtain $50 = x$, can you write $x = 50$?

PRACTICE   **Check to see whether the given number is a solution of the equation.**

25. $6, x + 12 = 18$

26. $110, x - 50 = 60$

27. $-8, 2b + 3 = -15$

28. $-2, 5t - 4 = -16$

29. $5, 0.5x = 2.9$

30. $3.5, 1.2 + x = 4.7$

31. $-6, 33 - \dfrac{x}{2} = 30$

32. $-8, \dfrac{x}{4} + 98 = 100$

33. $20, |c - 8| = 10$

34. $20, |30 - r| = 15$

35. $12, 3x - 2 = 4x - 5$

36. $5, 5y + 8 = 3y - 2$

37. $-3, x^2 - x - 6 = 0$

38. $-2, y^2 + 5y - 3 = 0$

39. $1, \dfrac{2}{a+1} + 5 = \dfrac{12}{a+1}$

40. $4, \dfrac{2t}{t-2} - \dfrac{4}{t-2} = 1$

41. $-3, (x-4)(x+3) = 0$

42. $5, (2x+1)(x-5) = 0$

Use a property of equality to solve each equation. Then check the result.

43. $x + 7 = 10$ **44.** $y + 15 = 24$

45. $a - 5 = 66$ **46.** $x - 34 = 19$

47. $0 = n - 9$ **48.** $3 = m - 20$

49. $9 + p = 9$ **50.** $88 + j = 88$

51. $x - 16 = -25$ **52.** $y - 12 = -13$

53. $a + 3 = 0$ **54.** $m + 1 = 0$

55. $f + 3.5 = 1.2$ **56.** $h + 9.4 = 8.1$

57. $-8 + p = -44$ **58.** $-2 + k = -41$

59. $8.9 = -4.1 + t$ **60.** $7.7 = -3.2 + s$

61. $d - \dfrac{1}{9} = \dfrac{7}{9}$ **62.** $\dfrac{7}{15} = b - \dfrac{1}{15}$

63. $s + \dfrac{4}{25} = \dfrac{11}{25}$ **64.** $\dfrac{8}{3} = h + \dfrac{1}{3}$

65. $4x = 16$ **66.** $5y = 45$

67. $369 = 9c$ **68.** $840 = 105t$

69. $4f = 0$ **70.** $0 = 60k$

71. $23b = 23$ **72.** $16 = 16h$

73. $-8h = 48$ **74.** $-9a = 72$

75. $-100 = -5g$ **76.** $-80 = -5w$

77. $-3.4y = -1.7$ **78.** $-2.1x = -1.26$

79. $\dfrac{x}{15} = 3$ **80.** $\dfrac{y}{7} = 12$

81. $0 = \dfrac{v}{11}$ **82.** $\dfrac{d}{49} = 0$

83. $\dfrac{w}{-7} = 15$ **84.** $\dfrac{h}{-2} = 3$

85. $\dfrac{d}{-7} = -3$ **86.** $\dfrac{c}{-2} = -11$

87. $\dfrac{y}{0.6} = -4.4$ **88.** $\dfrac{y}{0.8} = -2.9$

89. ▦ $a + 456{,}932 = 1{,}708{,}921$

90. ▦ $229{,}989 = x - 84{,}863$

91. ▦ $-1{,}563x = 43{,}764$

92. ▦ $999 = \dfrac{y}{-5{,}565}$

APPLICATIONS

93. SYNTHESIZERS Find the unknown angle measure.

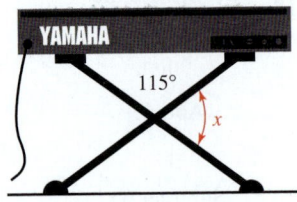

94. PHYSICS   A 15-pound block is suspended with two ropes, one of which is horizontal. Find the unknown angle measure.

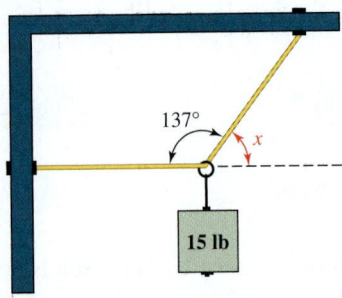

95. AVIATION   How many degrees from the horizontal position are the wings of the airplane?

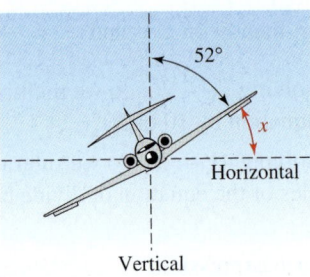

96. PLAYING A FLUTE   How many degrees from the horizontal is the position of the flute?

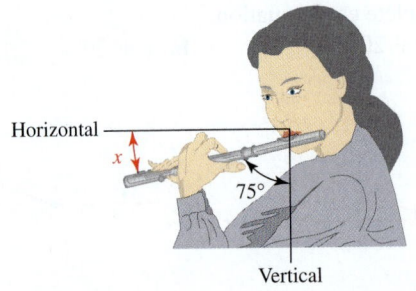

WRITING

97. What does it mean to solve an equation?

98. When solving an equation, we *isolate* the variable on one side of the equation. Write a sentence in which the word *isolate* is used in a different context.

99. Explain the error in the following work.

$$\text{Solve:} \quad x + 2 = 40$$
$$x + 2 - 2 = 40$$
$$x = 40$$

100. After solving an equation, how do we check the result?

REVIEW

101. Evaluate $-9 - 3x$ for $x = -3$.

102. Write a formula that would give the number of eggs in d dozen.

103. Translate to symbols: Subtract x from 45.

104. Evaluate: $\dfrac{2^3 + 3(5 - 3)}{15 - 4 \cdot 2}$.

CHALLENGE PROBLEMS

105. If $a + 80 = 50$, what is $a - 80$?

106. Find two solutions of $|x + 1| = 100$.

2.2 Problem Solving

- A Problem-Solving Strategy
- Drawing Diagrams
- Constructing Tables
- Solving Percent Problems

The Language of Algebra

A *strategy* is a plan for achieving a goal. Businesses often hire firms to develop an advertising *strategy* that will increase the sales of their products.

In this section, we combine the translating skills discussed in Chapter 1 and the equation-solving skills discussed in Section 2.1 to solve many applied problems.

■ A PROBLEM-SOLVING STRATEGY

To become a good problem solver, you need a plan to follow, such as the following five-step strategy.

Strategy for Problem Solving

1. **Analyze the problem** by reading it carefully to understand the given facts. What information is given? What are you asked to find? What vocabulary is given? Often, a diagram or table will help you visualize the facts of the problem.
2. **Form an equation** by picking a variable to represent the numerical value to be found. Then express all other unknown quantities as expressions involving that variable. Key words or phrases can be helpful. Finally, translate the words of the problem into an equation.
3. **Solve the equation.**
4. **State the conclusion.**
5. **Check the result** in the words of the problem.

EXAMPLE 1

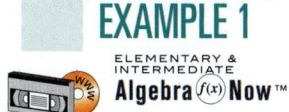

ELEMENTARY & INTERMEDIATE
Algebra $f(x)$ **Now**™

Systems analysis. A company's telephone use would have to increase by 350 calls per hour before the system would reach the maximum capacity of 1,500 calls per hour. Currently, how many calls are being made each hour on the system?

Analyze the Problem
- The maximum capacity of the system is 1,500 calls per hour.
- If the number of calls increases by 350, the system will reach capacity.
- We are to find the number of calls currently being made each hour.

Form an Equation   Let n = the number of calls currently being made each hour. To form an equation, we look for a key word or phrase in the problem.

Key phrase: *increase by 350*   **Translation:** addition

The key phrase tells us to add 350 to the current number of calls to obtain an expression for the maximum capacity of the system. Now we translate the words of the problem into an equation.

The current number of calls per hour	increased by	350	equals	the maximum capacity of the system.
n	+	350	=	1,500

Solve the Equation

$$n + 350 = 1,500$$

$$n + 350 - 350 = 1,500 - 350 \qquad \text{To undo the addition of 350, subtract 350 from both sides.}$$

$$n = 1,150 \qquad \text{Do the subtractions.}$$

State the Conclusion   Currently, 1,150 calls are being made per hour.

Check the Result   If 1,150 calls are currently being made each hour and an increase of 350 calls per hour occurs, then $1,150 + 350 = 1,500$ calls will be made each hour. This is the capacity of the system. The answer, 1,150, checks.

■ DRAWING DIAGRAMS

Diagrams are often helpful because they enable us to visualize the given facts of a problem.

EXAMPLE 2

ELEMENTARY & INTERMEDIATE

Airline travel.   On a book tour that took her from New York City to Chicago to Los Angeles and back to New York City, an author flew a total of 4,910 miles. The flight from New York to Chicago was 714 miles, and the flight from Chicago to L.A. was 1,745 miles. How long was the direct flight back to New York City?

Analyze the Problem
- The total miles flown on the tour was 4,910.
- The flight from New York to Chicago was 714 miles.
- The flight from Chicago to L.A. was 1,745 miles.
- We are to find the length of the flight from L.A. to New York City.

In the diagram, we see that the three parts of the tour form a triangle. We know the lengths of two of the sides of the triangle, 714 and 1,745, and the perimeter of the triangle, 4,910.

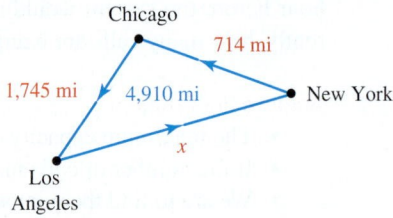

Form an Equation   We will let x = the length (in miles) of the flight from L.A. to New York and label the appropriate side of the triangle in the diagram. We then have:

The miles from New York to Chicago	plus	the miles from Chicago to L.A.	plus	the miles from L.A. to New York	is	4,910.
714	+	1,745	+	x	=	4,910

Notation

When forming an equation, we usually don't include the units.

714 ~~mi~~ + 1,745 ~~mi~~ + x ~~mi~~
        = 4,910 ~~mi~~

Solve the Equation

$$714 + 1,745 + x = 4,910$$

$$2,459 + x = 4,910 \qquad \text{Simplify the left-hand side of the equation:}$$
$$714 + 1,745 = 2,459.$$

$$2,459 + x - \mathbf{2,459} = 4,910 - \mathbf{2,459} \qquad \text{Subtract 2,459 from both sides to isolate } x.$$

$$x = 2,451 \qquad \text{Do the subtractions.}$$

State the Conclusion   The flight from L.A. to New York was 2,451 miles.

Check the Result   If we add the three flight lengths, we get $714 + 1,745 + 2,451 = 4,910$. This was the total number of miles flown on the book tour. The answer, 2,451, checks.

EXAMPLE 3

Eye surgery. A technique called **radial keratotomy** is sometimes used to correct nearsightedness. This procedure involves equally spaced incisions in the cornea, as shown in the figure. Find the angle between each incision. (Round to the nearest tenth of a degree.)

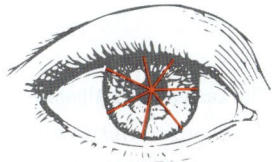

Analyze the Problem   A diagram labeled with the known and unknown information is shown.

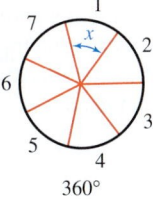

- There are 7 angles of equal measure.
- One complete revolution is 360°.
- Find the measure of one angle.

Form an Equation   We will let x = the measure of one of the angles. We then have:

7	times	the measure of one of the angles	is	360°
7	·	x	=	360

Solve the Equation

$$7x = 360$$

$$\frac{7x}{\mathbf{7}} = \frac{360}{\mathbf{7}} \qquad \text{Divide both sides by 7.}$$

$$x = 51.42 \ldots \qquad \text{Do the division.}$$

$$x \approx 51.4 \qquad \text{Round to the nearest tenth of a degree.}$$

State the Conclusion   The incisions are approximately 51.4° apart.

Check the Result   The angle measure is approximately 51°. Since $51 \cdot 7 = 357$, and this is close to 360, the result of 51.4° seems reasonable.

■ CONSTRUCTING TABLES

Sometimes it is helpful to organize the given facts of the problem in a table.

EXAMPLE 4

ELEMENTARY &
INTERMEDIATE
Algebra $f(x)$ **Now**™

Labor statistics.   The number of women (16 years or older) in the U.S. labor force has grown steadily over the past 40 years. From 1960 to 1970, the number grew by 8 million. By 1980, it had increased an additional 12 million. By 2000, the number rose another 21 million; by the end of that year, 63 million women were in the labor force. How many women were in the labor force in 1960?

Analyze the Problem
- The number grew by 8 million, increased by 12 million, and rose 21 million.
- The number of women in the labor force in 2000 was 63 million.
- We are to find the number of women in the U.S. labor force in 1960.

Form an Equation   We will let $x =$ the number of women (in millions) in the labor force in 1960. We can write algebraic expressions to represent the number of women in the work force in 1970, 1980, and 2000 by translating key words.

The Language of Algebra

The word *cumulative* means to increase by successive additions. You've probably heard of a *cumulative* final exam—one that covers all of the material that was studied in the course.

Year	Women in the labor force (millions)
1960	x
1970	$x + 8$
1980	$x + 8 + 12$
2000	$x + 8 + 12 + 21$

This table helps us keep track of the cumulative total as the number of women in the work force increased over the years.

Key word: *grew*   **Translation:** addition

Key word: *increased*   **Translation:** addition

Key word: *rose*   **Translation:** addition

There are two ways to represent the number of women (in millions) in the 2000 labor force: $x + 8 + 12 + 21$ and 63. Therefore,

$$x + 8 + 12 + 21 = 63$$

Solve the Equation

$$
\begin{aligned}
x + 8 + 12 + 21 &= 63 \\
x + 41 &= 63 && \text{Simplify: } 8 + 12 + 21 = 41. \\
x + 41 - 41 &= 63 - 41 && \text{To undo the addition of 41, subtract 41 from both sides.} \\
x &= 22 && \text{Do the subtractions.}
\end{aligned}
$$

State the Conclusion   There were 22 million women in the U.S. labor force in 1960.

Check the Result   Adding the number of women (in millions) in the labor force in 1960 and the increases, we get $22 + 8 + 12 + 21 = 63$. In 2000 there were 63 million, so the answer, 22, checks.

■ SOLVING PERCENT PROBLEMS

Percents are often used to present numeric information. Stores use them to advertise discounts, manufacturers use them to describe the contents of their products, and banks use them to list interest rates for loans and savings accounts.

Percent means parts per one hundred. For example, 93% means 93 out of 100 or $\frac{93}{100}$. There are three types of percent problems. Examples of these are as follows:

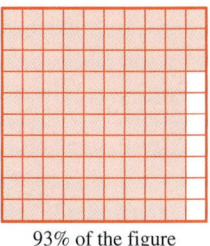

93% of the figure
is shaded.

- What number is 8% of 215?
- 14 is what percent of 52?
- 82 is 20.3% of what number?

EXAMPLE 5

What number is 8% of 215?

Solution

First, we translate the words into an equation. Here the word *of* indicates multiplication, and the word *is* means equals.

What number	is	8%	of	215?
↓	↓	↓	↓	↓
x	$=$	8%	·	215

Translate to mathematical symbols.

To do the multiplication on the right-hand side of the equation, we must change the percent to a decimal (or a fraction). To change 8% to a decimal, we proceed as follows.

$8\% = 8.0\%$ The number 8 has an understood decimal point to the right of the 8.

$= .08\,0$ Drop the % symbol and divide 8.0 by 100 by moving the decimal point 2 places to the left.

The Language of Algebra

The names of the parts of a percent sentence are:

17.2   is   8%   of   215.
amount   percent   base

They are related by the formula:

Amount = percent · base

To complete the solution, we replace 8% with its decimal equivalent, 0.08, and do the multiplication.

$x = 8\% \cdot 215$ This is the original equation.

$x = 0.08 \cdot 215$ $8\% = 0.08$

$x = 17.2$ Do the multiplication.

We have found that 17.2 is 8% of 215.

Self Check 5   What number is 5.6% of 40?

One method for solving applied percent problems is to read the problem carefully, and use the given facts to write a **percent sentence** of the form:

| | is | | % | of | | ? |

We enter the appropriate numbers in the first two blanks, and the word "what" in the remaining blank. Then we translate the sentence to mathematical symbols and solve the resulting equation.

EXAMPLE 6

Best-selling songs. In 1993, Whitney Houston's "I Will Always Love You" led the *Bill-board*'s music charts for 14 weeks. What percent of the year did she have the #1 song? (Round to the nearest one percent.)

Analyze the Problem

- For 14 out of 52 weeks in a year, she had the #1 song.
- We are to find what percent of the year she had the #1 song.

Form an Equation

Let x = the unknown percent and translate the words of the problem into an equation.

14	is	what percent	of	52?
14	=	x	·	52

14 is the amount, x is the percent, and 52 is the base.

Solve the Equation

$$14 = x \cdot 52$$

$$14 = 52x \qquad \text{Write } x \cdot 52 \text{ as } 52x.$$

$$\frac{14}{52} = \frac{52x}{52} \qquad \text{To isolate } x \text{, undo the multiplication by 52 by dividing both sides by 52.}$$

$$0.2692308 \approx x \qquad \text{Use a calculator to do the division.}$$

We were asked to find what *percent* of the year she had the #1 song. To change the decimal 0.2692307 to a percent, we proceed as follows.

$$0\,26.92308\% \approx x \qquad \text{Multiply 0.2692308 by 100 by moving the decimal point 2 places to the right, and then insert a \% symbol.}$$

$$26.92308\% \approx x$$

$$27\% \approx x \qquad \text{Round 26.92308\% to the nearest one percent.}$$

State the Conclusion

To the nearest one percent, Whitney Houston had the #1 song for 27% of the year.

Check the Result

We can check this result using estimation. Fourteen out of 52 weeks is approximately $\frac{14}{50}$ or $\frac{28}{100}$, which is 28%. The answer, 27%, seems reasonable.

EXAMPLE 7

ELEMENTARY & INTERMEDIATE
Algebra $f(x)$ **Now**™

Aging population. By the year 2050, the U.S. Bureau of the Census predicts that about 82 million residents will be 65 years of age or older. The **circle graph** (or **pie chart**) indicates that age group will make up 20.3% of the population. If the prediction is correct, what will the population of the United States be in 2050? (Round to the nearest million.)

Projection of the 2050 U.S. Population by Age

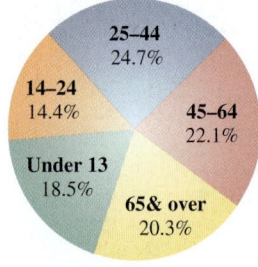

25–44
24.7%

14–24
14.4%

45–64
22.1%

Under 13
18.5%

65 & over
20.3%

Source: U.S. Bureau of the Census (2002).

Analyze the Problem

- 82 million people will be 65 years of age or older in 2050.
- 20.3% of the population will be 65 years of age or older in 2050.
- We are to find the predicted U.S. population for the year 2050.

Form an Equation   Let x = the predicted population in 2050. Translate to form an equation.

82	is	20.3%	of	what number?

$$82 \quad = \quad 20.3\% \quad \cdot \quad x$$

82 is the amount, 20.3% is the percent, and x is the base.

Solve the Equation

$$82 = 20.3\% \cdot x$$

$$82 = 0.203 \cdot x \qquad \text{Change 20.3\% to a decimal: } 20.3\% = 0.203.$$

$$82 = 0.203x \qquad \text{Write } 0.203 \cdot x \text{ as } 0.203x.$$

$$\frac{82}{\mathbf{0.203}} = \frac{0.203x}{\mathbf{0.203}} \qquad \text{To undo the multiplication by 0.203, divide both sides by 0.203.}$$

$$403.94 \approx x \qquad \text{Use a calculator to do the division.}$$

$$404 \approx x \qquad \text{Round to the nearest one million.}$$

State the Conclusion   The census bureau is predicting a population of about 404 million in the year 2050.

Check the Result   82 million out of a population of 404 million is about $\frac{80}{400} = \frac{40}{200} = \frac{20}{100}$, or 20%. The answer of 404 million seems reasonable.

Percents are often used to describe how a quantity has changed. For example, a health care provider might increase the cost of medical insurance by 3%, or a police department might decrease the number of officers assigned to street patrols by 10%. To describe such changes, we use **percent of increase** or **percent of decrease.**

EXAMPLE 8

ELEMENTARY &
INTERMEDIATE
Algebra $f(x)$ **Now**™

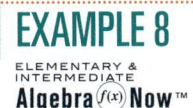

Identity theft.   The Federal Trade Commission receives complaints involving the theft of someone's identity information, such as a credit card/Social Security number or cell phone account. Refer to the data in the table. What was the percent of increase in the number of complaints from 2001 to 2002? (Round to the nearest percent.)

Number of Complaints

Year	2001	2002
	86,000	162,000

Caution

Always find the percent of increase (or decrease) with respect to the *original* amount.

Analyze the Problem   To find the *amount of increase,* we subtract the number of complaints in 2001 from the number of complaints in 2002.

$$162,000 - 86,000 = 76,000 \qquad \text{Subtract the earlier number from the later number.}$$

- The number of complaints increased by 76,000.
- Find what percent of the previous number of complaints a 76,000 increase is.

Form an Equation   Let x = the unknown percent and translate the words to an equation.

76,000	is	what percent	of	86,000?

$$76,000 \quad = \quad x \quad \cdot \quad 86,000$$

76,000 is the amount, x is the percent, and 86,000 is the base.

Solve the Equation

$$76,000 = x \cdot 86,000$$

$$76,000 = 86,000x \qquad \text{Write } x \cdot 86,000 \text{ as } 86,000x.$$

$$\frac{76,000}{86,000} = \frac{86,000x}{86,000} \qquad \text{To undo the multiplication by 86,000, divide both sides by 86,000.}$$

$$0.88372093 \approx x \qquad \text{Use a calculator to do the division.}$$

$$0\,88.372093\% \approx x \qquad \text{To write the decimal as a percent, multiply by 100 by moving the decimal point two places to the right and insert a \% symbol.}$$

$$88\% \approx x \qquad \text{Round to the nearest percent.}$$

State the Conclusion   In 2002, the number of complaints increased by about 88%.

Check the Result   A 100% increase in complaints would be 86,000 more complaints. Therefore, it seems reasonable that 76,000 complaints is an 88% increase.

Answer to Self Check   **5.** 2.24

2.2   STUDY SET   ELEMENTARY & INTERMEDIATE Algebra $f(x)$ Now™

VOCABULARY   Fill in the blanks.

1. A letter that is used to represent a number is called a _____.

2. An _____ is a mathematical statement that two quantities are equal. To _____ an equation means to find all the values of the variable that make the equation true.

3. To solve an applied problem, we let a _____ represent the unknown quantity. Then we write an _____ that models the situation. Finally, we _____ the equation for the variable to find the unknown.

4. _____ means parts per one hundred.

5. In the statement "10 is 50% of 20," 10 is called the _____, 50% is the _____, and 20 is the _____.

6. In mathematics, the word *of* often indicates _____, and __ means equals.

Write an equation to describe each situation.

7. A college choir's tour of three cities covers 1,240 miles.

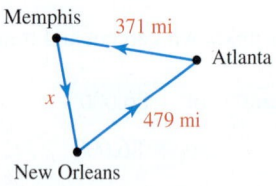

8. A woman participated in a triathlon.

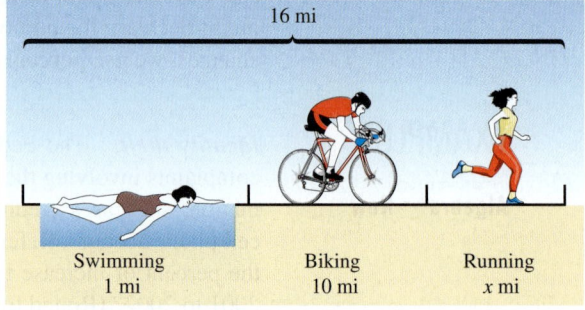

9. An airliner changed its altitude to avoid storm clouds.

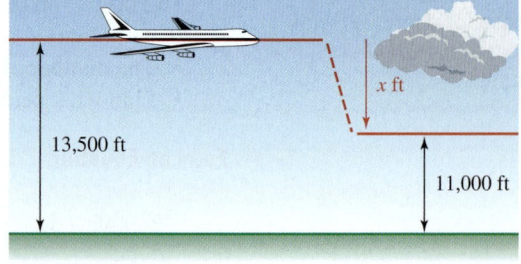

10. The sections of a 430-page book were assembled.

Section	Number of pages
Table of Contents	4
Preface	x
Text	400
Index	12

11. A hamburger chain sold a total of 31 million hamburgers in its first 4 years of business.

Year in business	Running total of hamburgers sold (millions)
1	x
2	$x + 5$
3	$x + 5 + 8$
4	$x + 5 + 8 + 16$

12. A pie was cut into equal-sized slices.

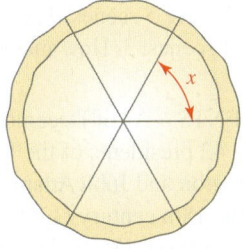

13. One method for solving percent problems is to read the problem carefully and use the given facts to write a percent sentence. What is the basic form of a percent sentence?

14.

High School Sports Programs Girl's Water Polo—Number of Participants	
1981	**2001**
282	14,792

Source: National Federation of State High School Associations

a. Find the *amount* of increase in participation.

b. Fill in blanks to find the percent of increase in participation: is   % of

NOTATION   **Translate each sentence into an equation.**

15. 12 is 40% of what number?

16. 99 is what percent of 200?

17. When computing with percents, the percent must be changed to a decimal or a fraction. Change each percent to a decimal.
  a. 35%   **b.** 3.5%
  c. 350%   **d.** $\frac{1}{2}$%

18. Change each decimal to a percent.
  a. 0.9   **b.** 0.09
  c. 9   **d.** 0.999

PRACTICE   **Translate each sentence into an equation, and then solve it.**

19. What number is 48% of 650?

20. What number is 60% of 200?

21. What percent of 300 is 78?

22. What percent of 325 is 143?

23. 75 is 25% of what number?

24. 78 is 6% of what number?

25. What number is 92.4% of 50?

26. What number is 2.8% of 220?

27. What percent of 16.8 is 0.42?

28. What percent of 2,352 is 199.92?

29. 128.1 is 8.75% of what number?

30. 1.12 is 140% of what number?

APPLICATIONS

31. GRAVITY   Suppose an astronaut, in full gear, was weighed on Earth. (See the illustration.) The weight of an object on Earth is 6 times greater than what it is on the moon. If we let x = the weight the scale would read on the moon, which equation is true?

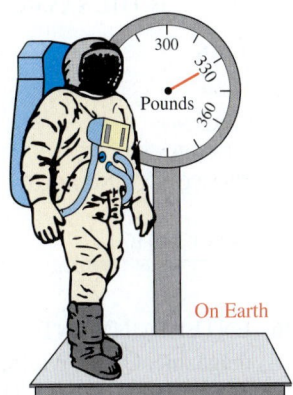

On Earth

$$330x = 6 \qquad x + 6 = 330 \qquad \frac{x}{6} = 330 \qquad 6x = 330$$

32. POWER OUTAGE   The electrical system in a building automatically shuts down when the meter shown reads 85. Suppose we let $x =$ the amount the reading must increase to cause the system to shut down. Which equation is true?

$$85 + x = 60 \quad 60 + x = 85 \quad 60x = 85 \quad 60 - 85 = x$$

You can probably solve Problems 33–38 without algebra. Nevertheless, use the methods discussed in this section so that you can gain experience with writing and then solving an equation to find the unknown.

33. MONARCHY   George III reigned as king of Great Britain for 59 years. This is 4 years less than the longest-reigning British monarch, Queen Victoria. For how many years did Queen Victoria rule?

34. TENNIS   Billie Jean King won 40 Grand Slam tennis titles in her career. This is 14 less than the all-time leader, Martina Navratilova. How many Grand Slam titles did Navratilova win?

35. ATM RECEIPT   Use the information on the automatic-teller receipt to find the balance in the account before the withdrawal.

HOME SAVINGS OF AMERICA			
TRAN.	DAT	TIM	T RM
0286.	1/16/03	11:46 AM	HSOA822
CARD NO.			61258
WITHDRAWAL OF			$35.00
FROM CHECKING ACCT.			3325256-612
CHECKING BAL.			$287.00

36. ENTERTAINMENT   According to *Forbes* magazine, Oprah Winfrey made an estimated $150 million in 2002. This was $92 million more than Mariah Carey's estimated earnings for that year. How much did Mariah Carey make in 2002?

37. TV NEWS   An interview with a world leader was edited into equally long segments and broadcast in parts over a 3-day period on a TV news program. If each daily segment of the interview lasted 9 minutes, how long was the original interview?

38. FLOODING   Torrential rains caused the width of a river to swell to 84 feet. If this was twice its normal size, how wide was the river before the flooding?

Use a table to help organize the facts of the problem, then find the solution.

39. STATEHOOD   From 1800 to 1850, 15 states joined the Union. From 1851 to 1900, an additional 14 states entered. Three states joined from 1901 to 1950. Since then, Alaska and Hawaii are the only others to enter the Union. How many states were part of the Union prior to 1800?

40. STUDIO TOUR   Over a 4-year span, improvements in a Hollywood movie studio tour caused it to take longer. The first year, 10 minutes were added to the tour length. In the second, third, and fourth years, 5 minutes were added each year. If the tour now lasts 135 minutes, how long was it originally?

41. THEATER   The play *Romeo and Juliet,* by William Shakespeare, has 5 acts and a total of 24 scenes. The second act has the most scenes, 6. The third and fourth acts each have 5 scenes. The last act has the least number of scenes, 3. How many scenes are in the first act?

42. U.S. PRESIDENTS   As of December 31, 1999, there had been 42 presidents of the United States. George Washington and John Adams were the only presidents in the 18th century (1700–1799). During the 19th century (1800–1899), there were 23 presidents. How many presidents were there during the 20th century (1900–1999)?

43. ORCHESTRAS   A 98-member orchestra is made up of a woodwind section with 19 musicians, a brass section with 23 players, a two-person percussion section, and a large string section. How many musicians make up the string section of the orchestra?

44. ANATOMY   A premed student has to know the names of all 206 bones that make up the human skeleton. So far, she has memorized the names of the 60 bones in the feet and legs, the 31 bones in the torso, and the 55 bones in the neck and head. How many more names does she have to memorize?

Draw a diagram to help organize the facts of the problem, and then find the solution.

45. BERMUDA TRIANGLE   The Bermuda Triangle is a triangular region in the Atlantic Ocean where many ships and airplanes have disappeared. The perimeter of the triangle is about 3,075 miles. It is formed by three imaginary lines. The first, 1,100 miles long, is from Melbourne, Florida, to Puerto Rico. The second, 1,000 miles long, stretches from Puerto Rico to Bermuda. The third extends from Bermuda back to Florida. Find its length.

46. FENCING   To cut down on vandalism, a lot on which a house was to be constructed was completely fenced. The north side of the lot was 205 feet in length. The west and east sides were 275 and 210 feet long, respectively. If 945 feet of fencing was used, how long is the south side of the lot?

47. SPACE TRAVEL   The 364-foot-tall *Saturn V* rocket carried the first astronauts to the moon. Its first, second, and third stages were 138, 98, and 46 feet tall, respectively. Atop the third stage was the lunar module, and from it extended a 28-foot escape tower. How tall was the lunar module?

48. PLANETS   Mercury, Venus, and Earth have approximately circular orbits around the sun. Earth is the farthest from the sun, at 93 million miles, and Mercury is the closest, at 36 million miles. The orbit of Venus is about 31 million miles from that of Mercury. How far is Earth's orbit from that of Venus?

49. STOP SIGNS   Find the measure of one angle of the octagonal stop sign. (*Hint:* The sum of the measures of the angles of an octagon is 1,080°.)

50. FERRIS WHEELS   What is the measure of the angle between each of the "spokes" of the Ferris wheel?

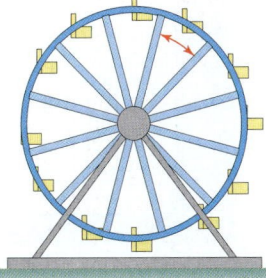

PERCENT PROBLEMS

51. ANTISEPTICS   Use the facts on the label to determine the amount of pure hydrogen peroxide in the bottle.

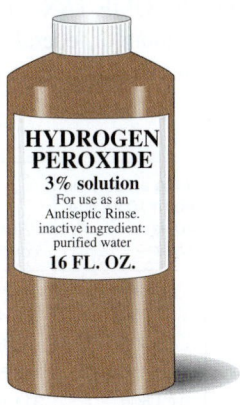

52. TIPPING   When paying with a Visa card, the user must fill in the amount of the gratuity (tip) and then compute the total. Complete the sales receipt if a 15% tip, rounded up to the nearest dollar, is to be left for the waiter.

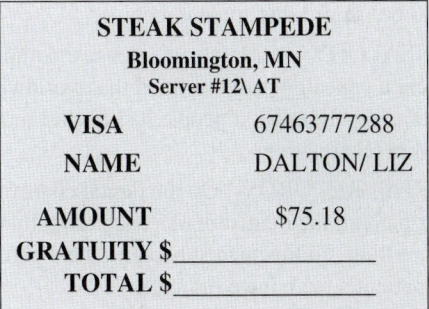

53. FEDERAL OUTLAYS   The **circle graph** shows the breakdown of the U.S. federal budget for fiscal year 2001. If total spending was approximately $1,900 billion, how much was paid for Social Security, Medicare, and other retirement programs?

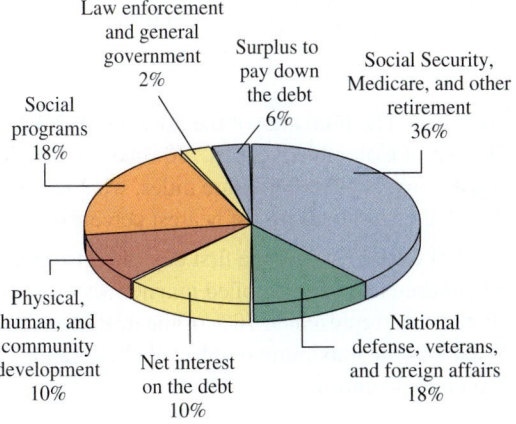

Based on 2002 Federal Income Tax Form 1040

54. INCOME TAX   Use the tax table to compute the amount of federal income tax if the amount of taxable income entered on Form 1040, line 40, is $39,909.

If the amount on Form 1040, line 40, is: Over—	But not over—	Enter on Form 1040, line 41	of the amount over—
$0	$7,000	.... 10%	$0
7,000	28,400	$700.00 + 15%	7,000
28,400	68,800	3,910.00 + 25%	28,400
68,800	143,500	14,010.00 + 28%	68,800
143,500	311,950	34,926.00 + 33%	143,500
311,950	....	90,514.50 + 35%	311,950

55. COLLEGE ENTRANCE EXAMS   On the Scholastic Aptitude Test, or SAT, a high school senior scored 550 on the mathematics portion and 700 on the verbal portion. What percent of the maximum 1,600 points did this student receive?

56. GENEALOGY   Through an extensive computer search, a genealogist determined that worldwide, 180 out of every 10 million people had his last name. What percent is this?

57. DENTAL RECORDS   On the dental chart for an adult patient, the dentist marks each tooth that has had a filling. To the nearest percent, what percent of this patient's teeth have fillings?

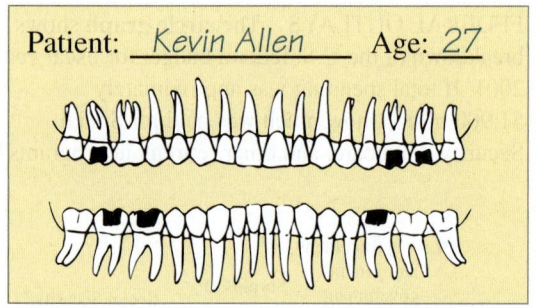

Patient: _Kevin Allen_   Age: _27_

58. AREAS   The total area of the 50 states and the District of Columbia is 3,618,770 square miles. If Alaska covers 591,004 square miles, what percent is this of the U.S. total (to the nearest percent)?

59. CHILD CARE   After the first day of registration, 84 children had been enrolled in a new day care center. That represented 70% of the available slots. What was the maximum number of children the center could enroll?

60. RACING PROGRAMS   One month before a stock car race, the sale of ads for the official race program was slow. Only 12 pages, or just 60% of the available pages, had been sold. What was the total number of pages devoted to advertising in the program?

61. NUTRITION   The Nutrition Facts label from a can of clam chowder is shown.
  a. Find the number of grams of saturated fat in one serving. What percent of a person's recommended daily intake is this?

Nutrition Facts	
Serving Size 1 cup (240mL) Servings Per Container about 2	
Amount per serving	
Calories 240 Calories from Fat 140	
	% Daily Value*
Total Fat 15 g	23%
Saturated Fat 5 g	25%
Cholesterol 10 mg	3%
Sodium 980 mg	41%
Total Carbohydrate 21 g	7%
Dietary Fiber 2 g	8%
Sugars 1 g	
Protein 7 g	

  b. Determine the recommended number of grams of saturated fat that a person should consume daily.

62. CUSTOMER GUARANTEES   To assure its customers of low prices, the Home Club offers a "10% Plus" guarantee. If the customer finds the same item selling for less somewhere else, he or she receives the difference in price plus 10% of the difference. A woman bought miniblinds at the Home Club for $120 but later saw the same blinds on sale for $98 at another store. How much can she expect to be reimbursed?

63. EXPORTS   The bar graph shows United States exports to Mexico for the years 1992 through 2001. Between what two years was there the greatest percent of decrease in exports?

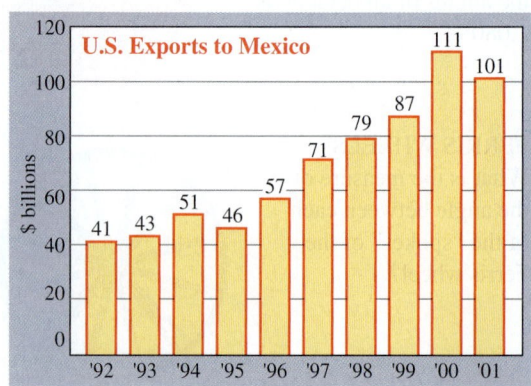

Based on data from www.census.gov/foreign-trade

64. AUCTIONS A pearl necklace of former First Lady Jacqueline Kennedy Onassis, originally valued at $700, was sold at auction in 1996 for $211,500. What was the percent of increase in the value of the necklace? (Round to the nearest percent.)

65. INSURANCE COSTS A college student's good grades earned her a student discount on her car insurance premium. What was the percent of decrease, to the nearest percent, if her annual premium was lowered from $1,050 to $925?

66. U.S. LIFE EXPECTANCY Use the following life expectancy data for 1900 and 2000 to determine the percent of increase for males and for females. Round to the nearest percent.

Years of life expected at birth		
	Male	**Female**
1900	46.3 yr	48.3 yr
2000	74.1 yr	79.5 yr

WRITING

67. Explain the relationship in a percent problem between the amount, the percent, and the base.

68. Write a real-life situation that could be described by "9 is what percent of 20?"

69. Explain why 150% of a number is more than the number.

70. Explain why "Find 9% of 100" is an easy problem to solve.

REVIEW

71. Divide: $-\dfrac{16}{25} \div \left(-\dfrac{4}{15}\right)$.

72. What two numbers are a distance of 8 away from 4 on the number line?

73. Is -34 a solution of $x + 15 = -49$?

74. Evaluate: $2 + 3[24 - 2(2 - 5)]$.

CHALLENGE PROBLEMS

75. SOAP A soap advertises itself as $99\frac{44}{100}\%$ pure. First, tell what percent of the soap is impurities. Then express your answer as a decimal.

76. Express $\frac{1}{20}$ of 1% as a percent using decimal notation.

2.3 Simplifying Algebraic Expressions

- Simplifying Products   • The Distributive Property   • Like Terms
- Combining Like Terms

In algebra, we frequently replace one algebraic expression with another that is equivalent and simpler in form. That process, called *simplifying an algebraic expression,* often involves the use of one or more properties of real numbers.

■ SIMPLIFYING PRODUCTS

Recall that the associative property of multiplication enables us to change the grouping of factors involved in a multiplication. The commutative property of multiplication enables us to change the order of the factors. These properties can be used to simplify certain products. For example, let's simplify $8(4x)$.

$$8(4x) = 8 \cdot (4 \cdot x) \quad \text{4x = 4 · x.}$$
$$= (8 \cdot 4) \cdot x \quad \text{Use the associative property of multiplication to group 4 with 8.}$$
$$= 32x \quad \text{Do the multiplication within the parentheses.}$$

We have found that $8(4x) = 32x$. We say that $8(4x)$ and $32x$ are **equivalent expressions** because for each value of x, they represent the same number.

If $x = 10$		*If* $x = -3$	
$8(4x) = 8[4(\mathbf{10})]$	$32x = 32(\mathbf{10})$	$8(4x) = 8[4(\mathbf{-3})]$	$32x = 32(\mathbf{-3})$
$= 8(40)$	$= 320$	$= 8(-12)$	$= -96$
$= 320$		$= -96$	

EXAMPLE 1

Simplify each expression: **a.** $-9(3b)$, **b.** $15a(6)$, **c.** $35\left(\dfrac{4}{5}x\right)$, **d.** $\dfrac{8}{3} \cdot \dfrac{3}{8}r$, and **e.** $3(7p)(-5)$.

Solution

a. $-9(3b) = -27b$

b. $15a(6) = 90a$ Use the commutative property of multiplication to reorder the factors.

c. $35\left(\dfrac{4}{5}x\right) = \left(35 \cdot \dfrac{4}{5}\right)x$ Use the associative property of multiplication to regroup the factors.

$\qquad\qquad = 28x$ Factor 35 as $5 \cdot 7$ to do the multiplication.

d. $\dfrac{8}{3} \cdot \dfrac{3}{8}r = \left(\dfrac{8}{3} \cdot \dfrac{3}{8}\right)r$ Use the associative property of multiplication to regroup the factors.

$\qquad\quad = 1r$ Multiply.

$\qquad\quad = r$ Simplify.

e. $3(7p)(-5) = -105p$ Use the commutative property of multiplication to reorder the factors.

> **The Language of Algebra**
>
> Two of the most often encountered instructions in algebra are *simplify* and *solve*. Remember that we *simplify* expressions and we *solve* equations.

Self Check 1   Multiply: **a.** $9 \cdot 6s$,   **b.** $36\left(\dfrac{2}{9}y\right)$,   **c.** $\dfrac{2}{3} \cdot \dfrac{3}{2}m$,   and   **d.** $-4(6u)(-2)$.

■ THE DISTRIBUTIVE PROPERTY

Another property that is used to simplify algebraic expressions is the **distributive property.** To introduce it, we will evaluate $4(5 + 3)$ in two ways.

Method 1   *Order of operations:*   Compute the sum within the parentheses first.

$\qquad 4(\mathbf{5 + 3}) = 4(\mathbf{8})$

$\qquad\qquad\quad\; = 32$ Do the multiplication.

Method 2   *The distributive property:*   Multiply 5 and 3 by 4 and then add the results.

$\qquad \mathbf{4}(5 + 3) = \mathbf{4}(5) + \mathbf{4}(3)$ Distribute the multiplication by 4.

$\qquad\qquad\quad\; = 20 + 12$ Do the multiplications.

$\qquad\qquad\quad\; = 32$

Each method gives a result of 32. This observation suggests the following property.

The Distributive Property	For any real numbers a, b, and c,

$$a(b + c) = ab + ac$$

The Language of Algebra

Formally, it is called the distributive property of multiplication over addition. When we use it to write a product, such as $5(x + 2)$ as a sum, $5x + 10$, we say that we have *removed* or *cleared* the parentheses.

To illustrate one use of the distributive property, let's consider the expression $5(x + 2)$. Since we are not given the value of x, we cannot add x and 2 within the parentheses. However, we can distribute the multiplication by the factor of 5 that is outside the parentheses to x and to 2 and add those products.

$$5(x + 2) = 5(x) + 5(2) \quad \text{Distribute the multiplication by 5.}$$
$$= 5x + 10$$

Since subtraction is the same as adding the opposite, the distributive property also holds for subtraction.

$$a(b - c) = ab - ac$$

EXAMPLE 2

ELEMENTARY & INTERMEDIATE
Algebra $f(x)$ **Now**™

Use the distributive property to remove parentheses: **a.** $6(a + 9)$, **b.** $3(3b - 8)$, **c.** $-12(a + 1)$, **d.** $-6(-3y - 8)$, and **e.** $15\left(\frac{x}{3} + \frac{2}{5}\right)$.

Solution **a.** $6(a + 9) = 6 \cdot a + 6 \cdot 9 \quad \text{Distribute the multiplication by 6.}$
$$= 6a + 54$$

The Language of Algebra

We read $6(a + 9)$ as "six times the *quantity* of a plus nine." The word *quantity* alerts us to the grouping symbols in the expression.

b. $3(3b - 8) = 3(3b) - 3(8) \quad \text{Distribute the multiplication by 3.}$
$$= 9b - 24 \quad \text{Do the multiplications.}$$

c. $-12(a + 1) = -12(a) + (-12)(1) \quad \text{Distribute the multiplication by } -12.$
$$= -12a - 12$$

d. $-6(-3y - 8) = -6(-3y) - (-6)(8) \quad \text{Distribute the multiplication by } -6.$
$$= 18y + 48$$

e. $15\left(\frac{x}{3} + \frac{2}{5}\right) = 15 \cdot \frac{x}{3} + 15 \cdot \frac{2}{5} \quad \text{Distribute the multiplication by 15.}$
$$= 5x + 6$$

Self Check 2 Use the distributive property to remove parentheses: **a.** $5(p + 2)$, **b.** $4(2x - 1)$, **c.** $-8(2x - 4)$, and **d.** $24\left(\frac{y}{6} + \frac{3}{8}\right)$.

Caution The distributive property does not apply to every expression that contains parentheses—only those where multiplication is distributed over addition (or subtraction). For example, to simplify $6(5x)$, we do not use the distributive property.

<table>
<tr><td align="center">*Correct*</td><td align="center">*Incorrect*</td></tr>
<tr><td align="center">$6(5x) = (6 \cdot 5)x = 30x$</td><td align="center">$6(5x) = 30 \cdot 6x = 180x$</td></tr>
</table>

The distributive property can be extended to several other useful forms. Since multiplication is commutative:

$$(b + c)a = ba + ca \qquad\qquad (b - c)a = ba - ca$$

For situations in which there are more than two terms within parentheses:

$$a(b + c + d) = ab + ac + ad \qquad a(b - c - d) = ab - ac - ad$$

EXAMPLE 3

ELEMENTARY &
INTERMEDIATE
Algebra *f(x)* **Now™**

Multiply: **a.** $(6x + 4y)\dfrac{1}{2}$,   **b.** $2(a - 3b)8$,   and   **c.** $-0.3(3a - 4b + 7)$.

Solution   **a.** $(6x + 4y)\dfrac{1}{2} = (6x)\dfrac{1}{2} + (4y)\dfrac{1}{2}$ Distribute the multiplication by $\dfrac{1}{2}$.

$\qquad\qquad\qquad = 3x + 2y$ Do the multiplications.

b. $2(a - 3b)\mathbf{8} = 2 \cdot \mathbf{8}(a - 3b)$ Use the commutative property of multiplication to reorder the factors.

$\qquad\qquad\quad = 16(a - 3b)$ Do the multiplication.

$\qquad\qquad\quad = 16a - 48b$ Distribute the multiplication by 16.

c. $-\mathbf{0.3}(3a - 4b + 7) = -\mathbf{0.3}(3a) - (-\mathbf{0.3})(4b) + (-\mathbf{0.3})(7)$ Distribute the multiplication by -0.3.

$\qquad\qquad\qquad\qquad = -0.9a + 1.2b - 2.1$ Do the three multiplications.

Self Check 3   Multiply: **a.** $(-6x - 24y)\dfrac{1}{3}$,   **b.** $6(c - 2d)9$,   **c.** $-0.7(2r + 5s - 8)$

Success Tip

Note that distributing the multiplication by -1 changes the sign of each term within the parentheses.

We can use the distributive property to find the opposite of a sum. For example, to find $-(x + 10)$, we interpret the $-$ symbol as a factor of -1, and proceed as follows:

$$-(x + 10) = -\mathbf{1}(x + 10)$$ Replace the $-$ symbol with -1.

$$\qquad\qquad = -\mathbf{1}(x) + (-\mathbf{1})(10)$$ Distribute the multiplication by -1.

$$\qquad\qquad = -x - 10$$ Multiply.

In general, we have the following property of real numbers.

The Opposite of a Sum   The opposite of a sum is the sum of the opposites. For any real numbers a and b,

$$-(a + b) = -a + (-b)$$

EXAMPLE 4

ELEMENTARY &
INTERMEDIATE
Algebra *f(x)* **Now™**

Simplify: $-(-9s - 3)$.

Solution $-(-9s - 3) = -\mathbf{1}(-9s - 3)$ Replace the $-$ symbol in front of the parentheses with -1.

$\qquad\qquad\qquad = -\mathbf{1}(-9s) - (-\mathbf{1})(3)$ Distribute the multiplication by -1.

$\qquad\qquad\qquad = 9s + 3$

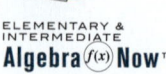

Self Check 4   Simplify: $-(-5x + 18)$.

■ LIKE TERMS

The distributive property can be used to simplify certain sums and differences. But before we can discuss this, we need to introduce some new vocabulary.

Like Terms	**Like terms** are terms with exactly the same variables raised to exactly the same powers. Any constant terms in an expression are considered to be like terms. Terms that are not like terms are called **unlike terms.**

Here are several examples.

Like terms	*Unlike terms*	
$4x$ and $7x$	$4x$ and $7y$	Different variables
$-10p^2$ and $25p^2$	$-10p$ and $25p^2$	Same variable, different powers
$\frac{1}{3}c^3d$ and c^3d	$\frac{1}{3}c^3d$ and c^3	Different variables

EXAMPLE 5

List the like terms in each expression: **a.** $7r + 5 + 3r$,   **b.** $6x^4 - 6x^2 - 6x$,   and   **c.** $-17m^3 + 3 - 2 + m^3$.

Solution   **a.** $7r + 5 + 3r$ contains the like terms $7r$ and $3r$.

b. $6x^4 - 6x^2 - 6x$ contains no like terms.

c. $-17m^3 + 3 - 2 + m^3$ contains two pairs of like terms: $-17m^3$ and m^3 are like terms, and the constant terms, 3 and -2, are like terms.

Self Check 5   List the like terms: **a.** $5x - 2y + 7y$ and   **b.** $-5p^2 - 12 + 17p^2 + 2$.

■ COMBINING LIKE TERMS

To add or subtract objects, they must have the same units. For example, we can add dollars to dollars and inches to inches, but we cannot add dollars to inches. When simplifying algebraic expressions, we can only add or subtract like terms.

Success Tip

When looking for like terms, don't look at the coefficients of the terms. Consider only the variable factors of each term.

This expression can be simplified, because it contains like terms.

$$3x + 4x$$

This expression cannot be simplified, because its terms are not like terms.

$$3x + 4y$$

Recall that the distributive property can be written in the following forms:

$$(b + c)a = ba + ca \qquad (b - c)a = ba - ca$$

We can use these forms of the distributive property in reverse to simplify a sum or difference of like terms. For example, we can simplify $3x + 4x$ as follows:

$$3x + 4x = (3 + 4)x$$
$$= 7x$$

We can simplify $15m^2 - 9m^2$ in a similar way:

$$15m^2 - 9m^2 = (15 - 9)m^2$$
$$= 6m^2$$

In each case, we say that we *combined like terms*. These examples suggest the following general rule.

Combining Like Terms   To add or subtract like terms, combine their coefficients and keep the same variables with the same exponents.

EXAMPLE 6   Simplify each expression, if possible: **a.** $-2x + 11x$,   **b.** $-8p + (-2p) + 4p$,   **c.** $0.5s^2 - 0.3s^2$,   and   **d.** $4w + 6$.

Solution   **a.** $-2x + 11x = 9x$ Think: $(-2 + 11)x = 9x$.

b. $-8p + (-2p) + 4p = -6p$ Think: $[-8 + (-2) + 4]p = -6p$.

c. $0.5s^2 - 0.3s^2 = 0.2s^2$ Think: $(0.5 - 0.3)s^2 = 0.2s^2$.

d. Since $4w$ and 6 are not like terms, they cannot be combined.

Self Check 6   Simplify, if possible: **a.** $-3x + 5x$,   **b.** $-6y + (-6y) + 9y$,   **c.** $4.4s^4 - 3.9s^4$,   and   **d.** $4a - 2$.

EXAMPLE 7   Simplify by combining like terms: **a.** $16t - 15t$,   **b.** $16t - t$,   **c.** $15t - 16t$,   and   **d.** $16t + t$.

ELEMENTARY &
INTERMEDIATE
Algebra $f(x)$ **Now**™

Solution   **a.** $16t - 15t = t$ Think: $(16 - 15)t = 1t = t$.

b. $16t - t = 15t$ Think: $16t - 1t = (16 - 1)t = 15t$.

c. $15t - 16t = -t$ Think: $(15 - 16)t = -1t = -t$.

d. $16t + t = 17t$ Think: $16t + 1t = (16 + 1)t = 17t$.

Self Check 7   Simplify: **a.** $9h - h$,   **b.** $9h + h$,   **c.** $9h - 8h$,   and   **d.** $8h - 9h$.

EXAMPLE 8   Simplify: $6t - 8 - 4t + 1$.

Solution   We combine the like terms that involve the variable t and we combine the constant terms.

ELEMENTARY &
INTERMEDIATE
Algebra $f(x)$ **Now**™

Think: $(6 - 4)t = 2t$

$$6t - 8 - 4t + 1 = 2t - 7$$

Think: $-8 + 1 = -7$

Self Check 8   Simplify: $50 + 70a - 60 - 10a$.

EXAMPLE 9

Simplify: $4(x + 5) - 5 - (2x - 4)$.

Solution   To simplify, we use the distributive property and combine like terms.

$$4(x + 5) - 5 - (2x - 4) = 4(x + 5) - 5 - 1(2x - 4)$$
$$= 4x + 20 - 5 - 2x + 4 \qquad \text{Distribute the multiplication by 4 and } -1.$$
$$= 2x + 19 \qquad \text{Simplify.}$$

Self Check 9   Simplify: $6(3y - 1) + 2 - (-3y + 4)$.

Answers to Self Checks   **1. a.** $54s$,   **b.** $8y$,   **c.** m,   **d.** $48u$ **2. a.** $5p + 10$,   **b.** $8x + 4$,   **c.** $-16x + 32$,   **d.** $4y + 9$ **3. a.** $-2x - 8y$,   **b.** $54c - 108d$,   **c.** $-1.4r - 3.5s + 5.6$ **4.** $5x - 18$ **5. a.** $-2y$ and $7y$ **b.** $-5p^2$ and $17p^2$; -12 and 2 **6. a.** $2x$,   **b.** $-3y$,   **c.** $0.5s^4$,   **d.** $4a - 2$ **7. a.** $8h$,   **b.** $10h$,   **c.** h,   **d.** $-h$ **8.** $60a - 10$ **9.** $21y - 8$

2.3   STUDY SET   ELEMENTARY & INTERMEDIATE Algebra $f(x)$ Now™

VOCABULARY   Fill in the blanks.

1. We can use the associative property of multiplication to _____ the expression $5(6x)$.

2. The _____ property of multiplication allows us to reorder factors. The _____ property of multiplication allows us to regroup factors.

3. We simplify _____ and we solve _____.

4. We can use the _____ property to remove or clear parentheses in the expression $2(x + 8)$.

5. We call $-(c + 9)$ the _____ of a sum.

6. Terms such as $7x^2$ and $5x^2$, which have the same variables raised to exactly the same exponents, are called _____ terms.

7. The _____ of the term $-23y$ is -23.

8. When we write $9x + x$ as $10x$, we say we have _____ like terms.

CONCEPTS

9. Fill in the blanks to simplify each product.
   a. $5 \cdot 6t = (\cdot)t = t$
   b. $-8(2x)(4) = (\cdot \cdot)x = x$

10. Fill in the blanks.
   a. $a(b + c) = ab + $ **b.** $a(b - c) = - ac$
   c. $(b + c)a = ba + $ **d.** $(b - c)a = - ca$
   e. $a(b + c + d) = + ac + $
   f. $-(a + b) = -a + $

11. Consider $3(x + 6)$. Why can't we add x and 6 within the parentheses?

12. Fill in the blanks.
   a. $2(x + 4) = 2x 8$
   b. $2(x - 4) = 2x 8$
   c. $-2(x + 4) = -2x 8$
   d. $-2(-x - 4) = 2x 8$

13. Fill in the blanks.

$$-(x + 10) = (x + 10)$$

Distributing the multiplication by -1 changes the _____ of each term within the parentheses.

14. Consider $33x - 8x^2 - 21x + 6$. Identify the terms of the expression and the coefficient of each term.

15. For each expression, identify any like terms.
   a. $3a + 8 + 2a$ **b.** $10 - 13h + 12$
   c. $3x^2 + 3x + 3$ **d.** $9y^2 - 9m - 8y^2$

16. Complete this statement: To add like terms, add their _____ and keep the same _____ and exponents.

17. Fill in the blanks to combine like terms.
   a. $4m + 6m = ()m = m$
   b. $30n - 50n = ()n = n$
   c. $12 - 32d + 15 = -32d + $

18. Simplify each expression, if possible.
    a. $5(2x)$ and $5 + 2x$

    b. $6(-7x)$ and $6 - 7x$

    c. $2(3x)(3)$ and $2 + 3x + 3$

NOTATION   **Complete each solution.**

19. Translate to symbols.
    a. Six times the quantity of h minus four.

    b. The opposite of the sum of z and sixteen.

20. Write an equivalent expression for the given expression using fewer symbols.
    a. $1x$ **b.** $-1d$ **c.** $0m$

    d. $5x - (-1)$ **e.** $16t + (-6)$

21. A student compared her answers to six homework problems to the answers in the back of the book. Are her answers equivalent? Write *yes* or *no*.

Student's answer	Book's answer	Equivalent?
$10x$	$10 + x$	
$3 + y$	$y + 3$	
$5 - 8a$	$8a - 5$	
$3x + 4$	$3(x + 4)$	
$3 - 2x$	$-2x + 3$	
$h^2 + (-16)$	$h^2 - 16$	

22. Draw arrows to illustrate how we distribute the factor outside the parentheses over the terms within the parentheses.
    a. $8(6g + 7)$ **b.** $(4x^2 - x + 6)2$

PRACTICE   **Simplify each expression.**

23. $9(7m)$ **24.** $12n(8)$

25. $5(-7q)$ **26.** $-7(5t)$

27. $5t \cdot 60$ **28.** $70a \cdot 10$

29. $(-5.6x)(-2)$ **30.** $(-4.4x)(-3)$

31. $\dfrac{5}{3} \cdot \dfrac{3}{5}g$ **32.** $\dfrac{9}{7} \cdot \dfrac{7}{9}k$

33. $12\left(\dfrac{5}{12}x\right)$ **34.** $15\left(\dfrac{4}{15}w\right)$

35. $8\left(\dfrac{3}{4}y\right)$ **36.** $27\left(\dfrac{2}{3}x\right)$

37. $-\dfrac{15}{4}\left(-\dfrac{4s}{15}\right)$ **38.** $-\dfrac{50}{3}\left(-\dfrac{3h}{50}\right)$

39. $24\left(-\dfrac{5}{6}r\right)$ **40.** $\dfrac{3}{4} \cdot \dfrac{1}{2}g$

41. $5(4c)(3)$ **42.** $9(2h)(2)$

43. $-4(-6)(-4m)$ **44.** $-5(-9)(-4n)$

Remove parentheses, and simplify.

45. $5(x + 3)$ **46.** $4(x + 2)$

47. $6(6c - 7)$ **48.** $9(9d - 3)$

49. $(3t + 2)8$ **50.** $(2q + 1)9$

51. $0.4(x - 4)$ **52.** $-2.2(2q + 1)$

53. $-5(-t - 1)$ **54.** $-8(-r - 1)$

55. $-4(3x + 5)$ **56.** $-4(6r + 4)$

57. $(13c - 3)(-6)$ **58.** $(10s - 11)(-2)$

59. $-\dfrac{2}{3}(3w - 6)$ **60.** $\dfrac{1}{2}(2y - 8)$

61. $45\left(\dfrac{x}{5} + \dfrac{2}{9}\right)$ **62.** $35\left(\dfrac{y}{5} + \dfrac{8}{7}\right)$

63. $60\left(\dfrac{3}{20}r - \dfrac{4}{15}\right)$ **64.** $72\left(\dfrac{7}{8}f - \dfrac{8}{9}\right)$

65. $-(x - 7)$ **66.** $-(y + 1)$

67. $-(-5.6y + 7)$ **68.** $-(-4.8a - 3)$

69. $2(4d + 5)5$ **70.** $4(2w + 3)5$

71. $-6(r + 5)2$ **72.** $-7(b + 3)3$

73. $-(-x - y + 5)$ **74.** $-(-14 + 3p - t)$

75. $5(1.2x - 4.2y - 3.2z)$

76. $5(2.4a + 5.4b - 6.4c)$

Simplify each expression.

77. $3x + 17x$ **78.** $12y - 15y$

79. $-4x + 4x$ **80.** $-16y + 16y$

81. $-7b^2 + 7b^2$ **82.** $-2c^3 + 2c^3$

83. $13r - 12r$ **84.** $25s + s$

85. $36y + y$

86. $32a - a$

87. $43s^3 - 44s^3$

88. $8j^3 - 9j^3$

89. $23w + 5 - 23w$

90. $19x + 3 - 19x$

91. $-4r - 7r + 2r - r$

92. $-v - 3v + 6v + 2v$

93. $a + a + a$

94. $t - t - t - t$

95. $0 - 3x$

96. $0 - 4a$

97. $3x + 5x - 7x$

98. $-5.7m + 4.3m$

99. $\dfrac{3}{5}t + \dfrac{1}{5}t$

100. $\dfrac{3}{16}x - \dfrac{5}{16}x$

101. $-0.2r - (-0.6r)$

102. $-1.1m - (-2.4m)$

103. $2z + 5(z - 3)$

104. $12(m + 11) - 11$

105. $3x + 4 - 5x + 1$

106. $4b + 9 - 9b + 9$

107. $10(2d - 7) + 4$

108. $5(3x - 2) + 5$

109. $-(c + 7) - 2(c - 3)$

110. $-(z + 2) + 5(3 - z)$

111. $2(s - 7) - (s - 2)$

112. $4(d - 3) - (d - 1)$

113. $6 - 4(-3c - 7)$

114. $10 - 5(-5g - 1)$

115. $36\left(\dfrac{2}{9}x - \dfrac{3}{4}\right) + 36\left(\dfrac{1}{2}\right)$

116. $40\left(\dfrac{3}{8}y - \dfrac{1}{4}\right) + 40\left(\dfrac{4}{5}\right)$

APPLICATIONS

117. THE AMERICAN RED CROSS
In 1891, Clara Barton founded the
Red Cross. Its symbol is a white
flag bearing a red cross. If each
side of the cross has length x, write
an algebraic expression for the
perimeter of the cross.

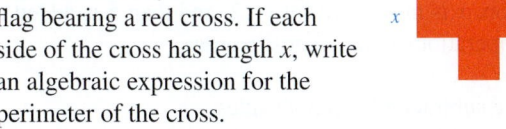

118. BILLIARDS   Billiard tables vary in size, but all
tables are twice as long as they are wide.
a. If the billiard table is x feet wide, write an
expression involving x that represents its length.

b. Write an expression for the perimeter of the
table.

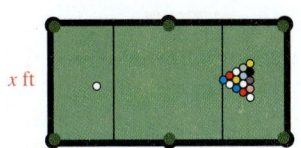

119. PING-PONG   Write an
expression for the
perimeter of the Ping-
Pong table.

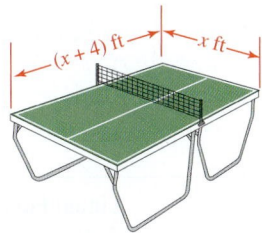

120. SEWING   Write an expression for the length of the
yellow trim needed to outline a pennant with the
given side lengths.

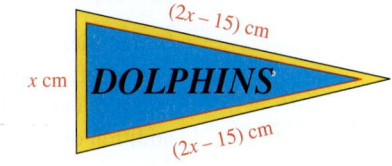

WRITING

121. Explain why the distributive property applies to
$2(3 + x)$ but not to $2(3x)$.

122. Tell how to combine like terms.

REVIEW   **Evaluate each expression for $x = -3$,
$y = -5$, and $z = 0$.**

123. $x^2z(y^3 - z)$

124. $|y^3 - z|$

125. $\dfrac{x - y^2}{2y - 1 + x}$

126. $\dfrac{2y + 1}{x} - x$

CHALLENGE PROBLEMS

127. The quantities below do not have the same units and
cannot be added as shown. Is there a way to find the
sum? If so, what is it?

2 inches + 2 feet + 2 yards

128. Fill in the blanks: $\left(\;\rule{1cm}{0.15mm}\; - \;\rule{1cm}{0.15mm}\;\right) = -75x + 40$.

Simplify.

129. $-2[x + 4(2x + 1)]$

130. $-5[y + 2(3y + 4)]$

2.4 More about Solving Equations

- Using More Than One Property of Equality
- Simplifying Expressions to Solve Equations
- Identities and Contradictions

We have solved simple equations by using properties of equality. We will now expand our equation-solving skills by considering more complicated equations. The objective is to develop a general strategy that we can use to solve any kind of *linear equation in one variable*.

Linear Equation in One Variable	A **linear equation in one variable** can be written in the form $$ax + b = c$$ where a, b, and c are real numbers and $a \neq 0$.

■ USING MORE THAN ONE PROPERTY OF EQUALITY

Sometimes we must use several properties of equality to solve an equation. For example, on the left-hand side of $2x + 6 = 10$, the variable is multiplied by 2, and then 6 is added to that product. To isolate x, we use the order of operations rules in reverse. First, we undo the addition of 6, and then we undo the multiplication by 2.

$$2x + 6 = 10$$
$$2x + 6 - 6 = 10 - 6 \qquad \text{To undo the addition of 6, subtract 6 from both sides.}$$
$$2x = 4 \qquad \text{Do the subtractions.}$$
$$\frac{2x}{2} = \frac{4}{2} \qquad \text{To undo the multiplication by 2, divide both sides by 2.}$$
$$x = 2$$

The solution is 2.

EXAMPLE 1

Solve: $-12x + 5 = 17$.

ELEMENTARY & INTERMEDIATE Algebra $f(x)$ Now™

Solution On the left-hand side of the equation, x is multiplied by -12, and then 5 is added to that product. To isolate x, we undo the operations in the opposite order.

- To undo the addition of 5, we subtract 5 from both sides.
- To undo the multiplication by -12, we divide both sides by -12.

$$-12x + 5 = 17$$
$$-12x + 5 - 5 = 17 - 5 \qquad \text{Subtract 5 from both sides.}$$
$$-12x = 12 \qquad \text{Do the subtractions: } 5 - 5 = 0 \text{ and } 17 - 5 = 12.$$
$$\frac{-12x}{-12} = \frac{12}{-12} \qquad \text{Divide both sides by } -12.$$
$$x = -1 \qquad \text{Do the divisions.}$$

Caution

When checking solutions, always use the original equation.

Check: $\quad -12x + 5 = 17$

$$-12(\mathbf{-1}) + 5 \stackrel{?}{=} 17 \qquad \text{Substitute } -1 \text{ for } x.$$

$$12 + 5 \stackrel{?}{=} 17 \qquad \text{Do the multiplication.}$$

$$17 = 17 \qquad \text{True.}$$

The solution is -1.

Self Check 1   Solve: $8x - 13 = 43$.

EXAMPLE 2   Solve: $\dfrac{2x}{3} = -6$.

Solution   On the left-hand side, x is multiplied by 2, and then that product is divided by 3. To solve this equation, we must undo these operations in the opposite order.

- To undo the division of 3, we multiply both sides by 3.
- To undo the multiplication by 2, we divide both sides by 2.

$$\frac{2x}{3} = -6$$

$$\mathbf{3}\left(\frac{2x}{3}\right) = \mathbf{3}(-6) \qquad \text{Multiply both sides by 3.}$$

$$2x = -18 \qquad \text{On the left-hand side: } 3\left(\tfrac{2x}{3}\right) = \frac{\overset{1}{\cancel{3}} \cdot 2 \cdot x}{\underset{1}{\cancel{3}}} = 2x.$$

$$\frac{2x}{\mathbf{2}} = \frac{-18}{\mathbf{2}} \qquad \text{Divide both sides by 2.}$$

$$x = -9 \qquad \text{Do the divisions.}$$

Check: $\quad \dfrac{2x}{3} = -6$

$$\frac{2(\mathbf{-9})}{3} \stackrel{?}{=} -6 \qquad \text{Substitute } -9 \text{ for } x.$$

$$\frac{-18}{3} \stackrel{?}{=} -6 \qquad \text{Do the multiplication.}$$

$$-6 = -6 \qquad \text{True.}$$

The solution is -9.

Self Check 2   Solve: $\dfrac{7h}{16} = -14$.

Recall that the product of a number and its **reciprocal,** or **multiplicative inverse,** is 1. We can use this fact to solve the equation from Example 2 in a different way. Since $\frac{2x}{3}$ is the same as $\frac{2}{3}x$, the equation can be written

$$\frac{2}{3}x = -6 \qquad \text{Note that the coefficient of } x \text{ is } \frac{2}{3}.$$

When the coefficient of the variable is a fraction, we can isolate the variable by multiplying both sides by the reciprocal.

$$\frac{3}{2}\left(\frac{2}{3}x\right) = \frac{3}{2}(-6)$$ Multiply both sides by the reciprocal of $\frac{2}{3}$, which is $\frac{3}{2}$.

$$\left(\frac{3}{2} \cdot \frac{2}{3}\right)x = \frac{3}{2}(-6)$$ On the left-hand side, regroup the factors.

$$1x = -9$$ Do the multiplications: $\frac{3}{2} \cdot \frac{2}{3} = 1$ and $\frac{3}{2}(-6) = -9$.

$$x = -9$$ Simplify.

EXAMPLE 3

ELEMENTARY &
INTERMEDIATE
Algebra $f(x)$ **Now**™

Solve: $-\dfrac{5}{8}m - 2 = -12$.

Solution   The coefficient of m is the fraction $-\dfrac{5}{8}$. We proceed as follows.

- To undo the subtraction of 2, we add 2 to both sides.
- To undo the multiplication by $-\frac{5}{8}$, we multiply both sides by its reciprocal, $-\frac{8}{5}$.

$$-\frac{5}{8}m - 2 = -12$$

$$-\frac{5}{8}m - 2 \mathbf{+ 2} = -12 \mathbf{+ 2}$$ Add 2 to both sides.

$$-\frac{5}{8}m = -10$$ Do the additions: $-2 + 2 = 0$ and $-12 + 2 = -10$.

$$\mathbf{-\frac{8}{5}}\left(-\frac{5}{8}m\right) = \mathbf{-\frac{8}{5}}(-10)$$ Multiply both sides by $-\frac{8}{5}$.

$$m = 16$$ On the left-hand side: $-\frac{8}{5}(-\frac{5}{8})m = 1m = m$.

On the right-hand side: $-\frac{8}{5}(-10) = \dfrac{8 \cdot 2 \cdot \overset{1}{\cancel{5}}}{\underset{1}{\cancel{5}}} = 16$.

Check that 16 is the solution.

Self Check 3   Solve: $\dfrac{7}{12}a - 6 = -27$.

EXAMPLE 4

Solve: $-0.2 = -0.8 - y$.

Solution   We begin by eliminating -0.8 from the right-hand side. We can do this by adding 0.8 to both sides.

$$-0.2 = -0.8 - y$$

$$-0.2 \mathbf{+ 0.8} = -0.8 - y \mathbf{+ 0.8}$$ Add 0.8 to both sides.

$$0.6 = -y$$ Do the additions.

Since the term $-y$ has an understood coefficient of -1, the equation can be rewritten as $0.6 = -1y$. To isolate y, either multiply both sides or divide both sides by -1.

Method 1	**Method 2**
$0.6 = -1y$	$0.6 = -1y$
$-\mathbf{1}(0.6) = -\mathbf{1}(-1y)$ Multiply both sides by -1.	$\dfrac{0.6}{-\mathbf{1}} = \dfrac{-1y}{-\mathbf{1}}$ To undo the multiplication by -1, divide both sides by -1.
$-0.6 = y$	$-0.6 = y$

Check that -0.6 is the solution.

Self Check 4   Solve: $-6.6 - m = -2.7$.

■ SIMPLIFYING EXPRESSIONS TO SOLVE EQUATIONS

When solving equations, we should simplify the expressions that make up the left- and right-hand sides before applying any properties of equality. Often, that involves removing parentheses and/or combining like terms.

EXAMPLE 5   Solve: $3(k + 1) - 5k = 0$.

Solution

$$3(k + 1) - 5k = 0$$
$$3k + 3 - 5k = 0 \qquad \text{Distribute the multiplication by 3.}$$
$$-2k + 3 = 0 \qquad \text{Combine like terms: } 3k - 5k = -2k.$$
$$-2k + 3 - \mathbf{3} = 0 - \mathbf{3} \qquad \text{To undo the addition of 3, subtract 3 from both sides.}$$
$$-2k = -3 \qquad \text{Do the subtractions: } 3 - 3 = 0 \text{ and } 0 - 3 = -3.$$
$$\frac{-2k}{-\mathbf{2}} = \frac{-3}{-\mathbf{2}} \qquad \text{To undo the multiplication by } -2, \text{ divide both sides by } -2.$$
$$k = \frac{3}{2} \qquad \text{Simplify: } \frac{-3}{-2} = \frac{3}{2}.$$

Check:
$$3(k + 1) - 5k = 0$$
$$3\left(\frac{\mathbf{3}}{\mathbf{2}} + 1\right) - 5\left(\frac{\mathbf{3}}{\mathbf{2}}\right) \stackrel{?}{=} 0 \qquad \text{Substitute } \frac{3}{2} \text{ for } k.$$
$$3\left(\frac{5}{2}\right) - 5\left(\frac{3}{2}\right) \stackrel{?}{=} 0 \qquad \text{Do the addition within the parentheses. Think of 1 as } \frac{2}{2} \text{ and then add: } \frac{3}{2} + \frac{2}{2} = \frac{5}{2}.$$
$$\frac{15}{2} - \frac{15}{2} \stackrel{?}{=} 0 \qquad \text{Do the multiplications.}$$
$$0 = 0 \qquad \text{True.}$$

The solution is $\frac{3}{2}$.

Self Check 5   Solve: $-5(x - 3) + 3x = 11$.

EXAMPLE 6

Solution

Success Tip

We could have eliminated $4x$ from the right-hand side by subtracting $4x$ from both sides:

$3x - 15 - 4x = 4x + 36 - 4x$
$-x - 15 = 36$

However, it is usually easier to isolate the variable term on the side that will result in a *positive* coefficient.

Solve: $3x - 15 = 4x + 36$.

To solve for x, all the terms containing x must be on the same side of the equation. We can eliminate $3x$ from the left-hand side by subtracting $3x$ from both sides.

$$3x - 15 = 4x + 36$$
$$3x - 15 - 3x = 4x + 36 - 3x \quad \text{Subtract } 3x \text{ from both sides.}$$
$$-15 = x + 36 \quad \text{Combine like terms: } 3x - 3x = 0 \text{ and } 4x - 3x = x.$$
$$-15 - 36 = x + 36 - 36 \quad \text{To undo the addition of 36, subtract 36 from both sides.}$$
$$-51 = x \quad \text{Do the subtractions.}$$

Check:
$$3x - 15 = 4x + 36$$
$$3(-51) - 15 \stackrel{?}{=} 4(-51) + 36 \quad \text{Substitute } -51 \text{ for } x.$$
$$-153 - 15 \stackrel{?}{=} -204 + 36 \quad \text{Do the multiplications.}$$
$$-168 = -168 \quad \text{True.}$$

The solution is -51.

Self Check 6   Solve: $30 + 6n = 4n - 2$.

Equations are usually easier to solve if they don't involve fractions. We can use the multiplication property of equality to *clear* an equation of fractions by multiplying both sides of the equation by the least common denominator.

EXAMPLE 7

Solution

Success Tip

Before multiplying both sides of an equation by the LCD, enclose the left-hand side and enclose the right-hand side with parentheses:

$$\left(\frac{x}{6} + \frac{5}{2}\right) = \left(\frac{1}{3}\right)$$

Solve: $\dfrac{x}{6} + \dfrac{5}{2} = \dfrac{1}{3}$.

To clear the equation of fractions, we multiply both sides by the LCD, 6.

$$\frac{x}{6} + \frac{5}{2} = \frac{1}{3}$$
$$6\left(\frac{x}{6} + \frac{5}{2}\right) = 6\left(\frac{1}{3}\right) \quad \text{Multiply both sides by the LCD of } \frac{x}{6}, \frac{5}{2}, \text{ and } \frac{1}{3}, \text{ which is 6. The parentheses are used to show that both } \frac{x}{6} \text{ and } \frac{5}{2} \text{ must be multiplied by 6.}$$
$$6\left(\frac{x}{6}\right) + 6\left(\frac{5}{2}\right) = 6\left(\frac{1}{3}\right) \quad \text{On the left-hand side, distribute the multiplication by 6.}$$
$$x + 15 = 2 \quad \text{Simplify.}$$
$$x + 15 - 15 = 2 - 15 \quad \text{To undo the addition of 15, subtract 15 from both sides.}$$
$$x = -13$$

Check that -13 is the solution.

Self Check 7   Solve: $\dfrac{x}{4} + \dfrac{1}{2} = -\dfrac{1}{8}$.

The preceding examples suggest the following strategy for solving equations.

Strategy for Solving Equations	**1.** Clear the equation of fractions.
	2. Use the distributive property to remove parentheses, if necessary.
	3. Combine like terms, if necessary.
	4. Undo the operations of addition and subtraction to get the variables on one side and the constant terms on the other.
	5. Undo the operations of multiplication and division to isolate the variable.
	6. Check the result.

The next example illustrates an important point: not all of these steps are necessary to solve every equation.

EXAMPLE 8

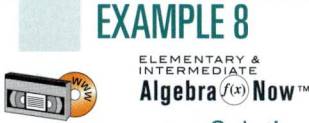

ELEMENTARY & INTERMEDIATE

Algebra $f(x)$ **Now™**

Solution

Solve: $\dfrac{7m + 5}{5} = -4m + 1$.

$$\frac{7m + 5}{5} = -4m + 1$$

$$\mathbf{5}\left(\frac{7m + 5}{5}\right) = \mathbf{5}(-4m + 1)$$ Clear the equation of the fraction by multiplying both sides by 5.

$$7m + 5 = -20m + 5$$ On the left-hand side, divide out the common factor of 5 in the numerator and denominator. On the right-hand side, distribute the multiplication by 5.

$$7m + 5 \mathbf{+ 20m} = -20m + 5 \mathbf{+ 20m}$$ To eliminate the term $-20m$ on the right-hand side, add $20m$ to both sides.

$$27m + 5 = 5$$ Combine like terms: $7m + 20m = 27m$ and $-20m + 20m = 0$.

$$27m + 5 \mathbf{- 5} = 5 \mathbf{- 5}$$ To undo the addition of 5 on the left-hand side, subtract 5 from both sides.

$$27m = 0$$ Do the subtractions.

$$\frac{27m}{\mathbf{27}} = \frac{0}{\mathbf{27}}$$ To undo the multiplication by 27, divide both sides by 27.

$$m = 0$$ 0 divided by any nonzero number is 0.

Substitute 0 for m in $\dfrac{7m + 5}{5} = -4m + 1$ to check that the solution is 0.

Caution

Remember that when you multiply one side of an equation by a nonzero number, you must multiply the other side of the equation by the same number.

Self Check 8   Solve: $6c + 2 = \dfrac{-c + 18}{9}$.

■ IDENTITIES AND CONTRADICTIONS

Each of the equations in Examples 1 through 8 had exactly one solution. However, some equations have no solutions while others have infinitely many solutions.

An equation that is true for all values of its variable is called an **identity.** An example is

$x + x = 2x$ If we substitute -10 for x, we get the true statement $-20 = -20$. If we substitute 0 for x, we get $0 = 0$. If we substitute 7 for x, we get $14 = 14$, and so on.

It is apparent that in an identity, we can replace x with any number and the equation will be true. We say that $x + x = 2x$ has infinitely many solutions.

An equation that is not true for any values of its variable is called a **contradiction.** An example is

$$x = x + 1 \qquad \text{No number is 1 greater than itself.}$$

We say that $x = x + 1$ has no solutions.

EXAMPLE 9

ELEMENTARY &
INTERMEDIATE
Algebra $f(x)$ **Now**™   **Solution**

Solve: $3(x + 8) + 5x = 2(12 + 4x)$.

$3(x + 8) + 5x = 2(12 + 4x)$	
$3x + 24 + 5x = 24 + 8x$	Distribute the multiplication by 3 and by 2.
$8x + 24 = 24 + 8x$	Combine like terms: $3x + 5x = 8x$. Note that the sides of the equation are identical.
$8x + 24 - \mathbf{8x} = 24 + 8x - \mathbf{8x}$	To eliminate the term $8x$ on the right-hand side, subtract $8x$ from both sides.
$24 = 24$	Combine like terms: $8x - 8x = 0$.

The terms involving x drop out and the result, $24 = 24$, is true. This means that any number substituted for x in the original equation will yield a true statement. Therefore, every real number is a solution and this equation is an identity.

Self Check 9   Solve: $3(x + 5) - 4(x + 4) = -x - 1$.

EXAMPLE 10

Solution

Solve: $3(d + 7) - d = 2(d + 10)$.

$3(d + 7) - d = 2(d + 10)$	
$3d + 21 - d = 2d + 20$	Distribute the multiplication by 3 and by 2.
$2d + 21 = 2d + 20$	Combine like terms: $3d - d = 2d$.
$2d + 21 - \mathbf{2d} = 2d + 20 - \mathbf{2d}$	To eliminate the term $2d$ on the right-hand side, subtract $2d$ from both sides.
$21 = 20$	Combine like terms: $2d - 2d = 0$.

The Language of Algebra

Contradiction is a form of the word *contradict,* meaning conflicting ideas. During a trial, evidence might be introduced that *contradicts* the testimony of a witness.

The terms involving d drop out and the result, $21 = 20$, is false. This means that any number that is substituted for x in the original equation will yield a false statement. This equation has no solution and it is a contradiction.

Self Check 10   Solve: $-4(c - 3) + 2c = 2(10 - c)$.

Answers to Self Checks   **1.** 7   **2.** -32 **3.** -36 **4.** -3.9 **5.** 2   **6.** -16 **7.** $-\frac{5}{2}$ **8.** 0   **9.** all real numbers; this equation is an identity   **10.** no solution; this equation is a contradiction

2.4   STUDY SET   ELEMENTARY & INTERMEDIATE Algebra *f(x)* Now™

VOCABULARY   Fill in the blanks.

1. An equation is a statement indicating that two expressions are _____.

2. To _____ an equation means to find all of the values of the variable that make the equation a true statement.

3. After solving an equation, we can check our result by substituting that value for the variable in the _____ equation.

4. The product of a number and its _____ is 1.

5. An equation that is true for all values of its variable is called an _____.

6. An equation that is not true for any values of its variable is called a _____.

CONCEPTS   Fill in the blanks.

7. To solve the equation $2x - 7 = 21$, we first undo the _____ of 7 by adding 7 to both sides. Then we undo the _____ by 2 by dividing both sides by 2.

8. To solve the equation $\frac{x}{2} + 3 = 5$, we first undo the _____ of 3 by subtracting 3 from both sides. Then we undo the _____ by 2 by multiplying both sides by 2.

9. To solve $\frac{s}{3} + \frac{1}{4} = -\frac{1}{2}$, we can clear the equation of the fractions by _____ both sides by 12.

10. To solve $15d = -2(3d + 7) + 2$, we begin by using the _____ property to remove parentheses.

11. One method of solving $-\frac{4}{5}x = 8$ is to multiply both sides of the equation by the reciprocal of $-\frac{4}{5}$. What is the reciprocal of $-\frac{4}{5}$?

12. **a.** Combine like terms on the left-hand side of $6x - 8 - 8x = -24$.

    b. Combine like terms on the right-hand side of $5a + 1 = 9a + 16 + a$.

    c. Combine like terms on both sides of $12 - 3r + 5r = -8 - r - 2$.

13. Find the LCD of the fractions in the equation $\frac{x}{3} - \frac{4}{5} = \frac{1}{2}$.

14. Simplify: $20\left(\frac{3}{5}x\right)$.

15. What must you multiply both sides of $\frac{2}{3} - \frac{b}{2} = -\frac{4}{3}$ by to clear the equation of fractions?

16. **a.** Simplify: $3x + 5 - x$.

    b. Solve: $3x + 5 - x = 9$.

    c. Evaluate $3x + 5 - x$ for $x = 9$.

    d. Check: Is -1 a solution of $3x + 5 - x = 9$?

NOTATION   Complete the solution.

17.
$$2x - 7 = 21$$
$$2x - 7 + \boxed{} = 21 + \boxed{}$$
$$2x = \boxed{}$$
$$\frac{2x}{\boxed{}} = \frac{28}{\boxed{}}$$
$$x = 14$$

18. Identify the like terms on the right-hand side of $5(9a + 1) = a + 6 - 7a$.

19. Fill in the blanks.

    a. $-x = \boxed{} x$.   **b.** $\frac{3x}{5} = \boxed{} x$.

20. What does the symbol $\stackrel{?}{=}$ mean?

PRACTICE   Solve each equation and check all solutions.

21. $2x + 5 = 17$

22. $3x - 5 = 13$

23. $5q - 2 = 23$

24. $4p + 3 = 43$

25. $-33 = 5t + 2$

26. $-55 = 3w + 5$

27. $20 = -x$

28. $10 = -a$

29. $-g = -4$

30. $-u = -20$

31. $1.2 - x = -1.7$

31. $0.6 = 4.1 - x$

33. $-3p + 7 = -3$

34. $-2r + 8 = -1$

35. $0 - 2y = 8$

36. $0 - 7x = -21$

37. $-8 - 3c = 0$

38. $-5 - 2d = 0$

39. $\frac{5}{6}k = 10$

40. $\frac{2c}{5} = 2$

41. $-\frac{7}{16}h = 21$

42. $-\frac{5}{8}h = 15$

43. $-\frac{t}{3} + 2 = 6$

44. $\frac{x}{5} - 5 = -12$

45. $2(-3) + 4y = 14$

46. $4(-1) + 3y = 8$

47. $7(0) - 4y = 17$

48. $3x - 4(0) = 2$

49. $10.08 = 4(0.5x + 2.5)$

50. $-3.28 = 8(1.5y - 0.5)$

51. $-(4 - m) = -10$

52. $-(6 - t) = -12$

53. $15s + 8 - s = 7 + 1$

54. $-7t - 9 + t = -10 + 1$

55. $-3(2y - 2) - y = 5$

56. $-(3a + 1) + a = 2$

57. $3x - 8 - 4x - 7x = -2 - 8$

58. $-6t - 7t - 5t - 1 = 12 - 3$

59. $4(5b) + 2(6b - 1) = -34$

60. $2(3x) + 5(3x - 1) = 58$

61. $9(x + 11) + 5(13 - x) = 0$

62. $-(19 - 3s) - (8s + 1) = 35$

63. $60r - 50 = 15r - 5$

64. $100f - 75 = 50f + 75$

65. $8y - 3 = 4y + 15$

66. $7 + 3w = 4 + 9w$

67. $5x + 7.2 = 4x$

68. $3x + 2.5 = 2x$

69. $8y + 328 = 4y$

70. $9y + 369 = 6y$

71. $15x = x$

72. $7y = 8y$

73. $3(a + 2) = 2(a - 7)$

74. $9(t - 1) = 6(t + 2) - t$

75. $2 - 3(x - 5) = 4(x - 1)$

76. $2 - (4x + 7) = 3 + 2(x + 2)$

77. $\dfrac{x + 5}{3} = 11$

78. $\dfrac{x + 2}{13} = 3$

79. $\dfrac{y}{6} + \dfrac{y}{4} = -1$

80. $\dfrac{x}{3} + \dfrac{x}{4} = -2$

81. $-\dfrac{2}{9} = \dfrac{5x}{6} - \dfrac{1}{3}$

82. $\dfrac{2}{3} = -\dfrac{2x}{3} + \dfrac{3}{4}$

83. $\dfrac{2}{3}y + 2 = \dfrac{1}{5} + y$

84. $\dfrac{2}{5}x + 1 = \dfrac{1}{3} + x$

85. $-\dfrac{3}{4}n + 2n = \dfrac{1}{2}n + \dfrac{13}{3}$

86. $-\dfrac{5}{6}n - 3n = \dfrac{1}{3}n + \dfrac{11}{9}$

87. $\dfrac{10 - 5s}{3} = s$

88. $\dfrac{40 - 8s}{5} = -2s$

89. $\dfrac{5(1 - x)}{6} = -x$

90. $\dfrac{3(14 - u)}{8} = -3u$

91. $\dfrac{3(d - 8)}{4} = \dfrac{2(d + 1)}{3}$

92. $\dfrac{3(c - 2)}{2} = \dfrac{2(2c + 3)}{5}$

93. $\dfrac{1}{2}(x + 3) + \dfrac{3}{4}(x - 2) = x + 1$

94. $\dfrac{3}{2}(t + 2) + \dfrac{1}{6}(t + 2) = 2 + t$

95. $8x + 3(2 - x) = 5(x + 2) - 4$

96. $5(x + 2) = 5x - 2$

97. $-3(s + 2) = -2(s + 4) - s$

98. $21(b - 1) + 3 = 3(7b - 6)$

99. $2(3z + 4) = 2(3z - 2) + 13$

100. $x + 7 = \dfrac{2x + 6}{2} + 4$

101. $4(y - 3) - y = 3(y - 4)$

102. $5(x + 3) - 3x = 2(x + 8)$

🖩 **Solve each equation.**

103. $\dfrac{h}{709} - 23{,}898 = -19{,}678$

104. $9.35 - 1.4y = 7.32 + 1.5y$

WRITING

105. To solve $3x - 4 = 5x + 1$, one student began by subtracting $3x$ from both sides. Another student solved the same equation by first subtracting $5x$ from both sides. Will the students get the same solution? Explain why or why not.

106. What does it mean to clear an equation such as $\frac{1}{4} + \frac{x}{2} = \frac{3}{8}$ of the fractions?

107. Explain the error in the following solution.

$$\text{Solve: } 2x + 4 = 30.$$
$$2x + 4 = 30$$
$$\dfrac{2x}{2} + 4 = \dfrac{30}{2}$$
$$x + 4 = 15$$
$$x + 4 - 4 = 15 - 4$$
$$x = 11$$

108. Write an equation that is an identity. Explain why every number is a solution.

REVIEW

109. Subtract: $-8 - (-8)$.

110. Add: $\dfrac{1}{8} + \dfrac{1}{8}$.

111. Multiply: $\dfrac{1}{8} \cdot \dfrac{1}{8}$.

112. Divide: $\dfrac{0.8}{8}$.

113. Simplify: $8x + 8 + 8x - 8$.

114. Evaluate: -1^8.

CHALLENGE PROBLEMS

115. In this section, we discussed equations that have no solution, one solution, and an infinite number of solutions. Do you think an equation could have exactly two solutions? If so, give an example.

116. The equation $4x - 3y = 5$ contains two different variables. Solve the equation by determining a value of x and a value for y that make the equation true.

2.5 Formulas

- Formulas from Business
- Formulas from Science
- Formulas from Geometry
- Solving for a Specified Variable

A **formula** is an equation that states a known relationship between two or more variables. Formulas are used in fields such as economics, physical education, biology, automotive repair, and nursing. In this section, we will consider formulas from business, science, and geometry.

■ FORMULAS FROM BUSINESS

A formula for retail price: To make a profit, a merchant must sell an item for more than he or she paid for it. The price at which the merchant sells the product, called the **retail price,** is the sum of what the item cost the merchant plus the **markup.**

Retail price = cost + markup

Using r to represent the retail price, c the cost, and m the markup, we can write this formula as

$$r = c + m$$

A formula for profit: The **profit** a business makes is the difference between the **revenue** (the money it takes in) and the costs.

Profit = revenue − costs

Using p to represent the profit, r the revenue, and c the costs, we can write this formula as

$$p = r - c$$

EXAMPLE 1

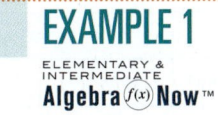
ELEMENTARY & INTERMEDIATE
Algebra *f(x)* Now™

Charitable giving.   In 2001, the Salvation Army received $2.31 billion in revenue. Of that amount, $1.92 billion went directly toward program services. Find the 2001 administrative costs of the organization.

Solution   The charity collected $2.31 billion. We can think of the $1.92 billion that was spent on programs as profit. We need to find the administrative costs, c.

$$p = r - c$$ This is the formula for profit.

$$1.92 = 2.31 - c$$ Substitute 1.92 for p and 2.31 for r.

$$1.92 - 2.31 = 2.31 - c - 2.31$$ To eliminate 2.31, subtract 2.31 from both sides.

$$-0.39 = -c$$ Do the subtractions.

$$\frac{-0.39}{-1} = \frac{-c}{-1}$$ Since $-c = -1c$, divide (or multiply) both sides by -1.

$$0.39 = c$$

In 2001, the Salvation Army had administrative costs of $0.39 billion.

Self Check 1   A PTA spaghetti dinner made a profit of $275.50. If the cost to host the dinner was $1,235, how much revenue did it generate?

The Language of Algebra

The word *annual* means occurring once a year. An *annual* interest rate is the interest rate paid per year.

A formula for simple interest: When money is borrowed, the lender expects to be paid back the amount of the loan plus an additional charge for the use of the money. The additional charge is called **interest.** When money is deposited in a bank, the depositor is paid for the use of the money. The money the deposit earns is also called interest.

Interest is computed in two ways: either as **simple interest** or as **compound interest.** Simple interest is the product of the principal (the amount of money that is invested, deposited, or borrowed), the annual interest rate, and the length of time in years.

Interest = principal · rate · time

Using I to represent the simple interest, P the principal, r the annual interest rate, and t the time in years, we can write this formula as

$$I = Prt$$

EXAMPLE 2

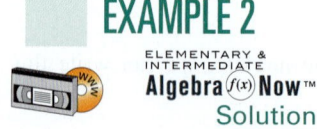

ELEMENTARY & INTERMEDIATE

Algebra $f(x)$ **Now**™

Solution

Retirement income.   One year after investing $15,000, a retired couple received a check for $1,125 in interest. Find the interest rate their money earned that year.

The couple invested $15,000 (the principal) for 1 year (the time) and made $1,125 (the interest). We need to find the annual interest rate, r.

Caution

When using the formula $I = Prt$, always write the interest rate r (which is given as a percent) as a decimal or fraction before performing any calculations.

$$I = Prt$$ This is the formula for simple interest.

$$1,125 = 15,000r(1)$$ Substitute 1,125 for I, 15,000 for P, and 1 for t.

$$1,125 = 15,000r$$ Simplify.

$$\frac{1,125}{15,000} = \frac{15,000r}{15,000}$$ To solve for r, undo the multiplication by 15,000 by dividing both sides by 15,000.

$$0.075 = r$$ Do the divisions.

$$7.5\% = r$$ To write 0.075 as a percent, multiply 0.075 by 100 by moving the decimal point two places to the right and inserting a % symbol.

The couple received an annual rate of 7.5% that year on their investment.

	P ·	r ·	$t =$	I
Investment	15,000	0.075	1	1,125

Self Check 2   A father loaned his daughter $12,200 at a 2% annual simple interest rate for a down payment on a house. If the interest on the loan amounted to $610, for how long was the loan?

■ FORMULAS FROM SCIENCE

A formula for distance traveled: If we know the average rate (of speed) at which we will be traveling and the time we will be traveling at that rate, we can find the distance traveled.

Distance = rate · time

Using d to represent the distance, r the average rate, and t the time, we can write this formula as

$$d = rt$$

EXAMPLE 3

ELEMENTARY &
INTERMEDIATE
Algebra *f(x)* **Now**™

Solution

Finding the rate.   As they migrate from the Bering Sea to Baja California, gray whales swim for about 20 hours each day, covering a distance of approximately 70 miles. Estimate their average swimming rate in miles per hour (mph).

Since the distance d is 70 miles and the time t is 20 hours, we substitute 70 for d and 20 for t in the formula $d = rt$ and solve for r.

Caution

When using the formula $d = rt$, make sure the units are consistent. For example, if the rate is given in miles per hour, the time must be expressed in hours.

$d = rt$

$70 = r(20)$ Substitute 70 for d and 20 for t.

$\dfrac{70}{20} = \dfrac{20r}{20}$ To undo the multiplication by 20, divide both sides by 20.

$3.5 = r$

The whales' average swimming rate is 3.5 mph.

	r	· t	= d
Gray whale	3.5	20	70

Self Check 3   An elevator travels at an average rate of 288 feet per minute. How long will it take the elevator to climb 30 stories, a distance of 360 feet?

A formula for converting temperatures: A message board flashes two temperature readings, one in degrees Fahrenheit and one in degrees Celsius. The Fahrenheit scale is used in the American system of measurement. The Celsius scale is used in the metric system. The formula that relates a Fahrenheit temperature F to a Celsius temperature C is:

$$C = \frac{5}{9}(F - 32)$$

EXAMPLE 4

ELEMENTARY &
INTERMEDIATE
Algebra *f(x)* **Now**™

Solution

Convert the temperature shown on the City Savings sign to degrees Fahrenheit.

Since the temperature C in degrees Celsius is 30°, we substitute 30 for C in the formula and solve for F.

$$C = \frac{5}{9}(F - 32)$$ This is the temperature conversion formula.

$$\mathbf{30} = \frac{5}{9}(F - 32)$$ Substitute 30 for C.

$$\frac{\mathbf{9}}{\mathbf{5}} \cdot 30 = \frac{\mathbf{9}}{\mathbf{5}} \cdot \frac{5}{9}(F - 32)$$ To undo the multiplication by $\frac{5}{9}$, multiply both sides by the reciprocal of $\frac{5}{9}$.

$$54 = F - 32$$ Do the multiplications.

$$54 + \mathbf{32} = F - 32 + \mathbf{32}$$ To undo the subtraction of 32, add 32 to both sides.

$$86 = F$$

30°C is equivalent to 86°F.

Self Check 4   Change -175°C, the temperature on Saturn, to degrees Fahrenheit.

■ FORMULAS FROM GEOMETRY

To find the **perimeter** of a geometric figure, we find the distance around the figure by computing the sum of the lengths of its sides. Perimeter is measured in linear units, such as inches, feet, yards, and meters. The **area** of a figure is the amount of surface that it encloses. Area is measured in square units, such as square inches, square feet, square yards, and square meters (denoted as in.^{2}, ft^2, yd^2, and m^2, respectively). Many formulas for perimeter and area are shown inside the front cover of the book.

EXAMPLE 5

ELEMENTARY &
INTERMEDIATE
Algebra $f^{(x)}$ **Now**™   Solution

The flag of Eritrea, a country in east Africa, is shown below.   **a.** Find the perimeter of the flag.   **b.** Find the area of the red triangular region of the flag.

a. The perimeter of the flag is given by the formula $P = 2l + 2w$, where l is the length and w is the width of the rectangle.

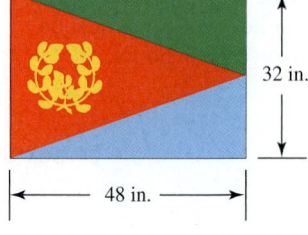

32 in.

48 in.

$$P = 2\mathbf{l} + 2\mathbf{w}$$
$$P = 2(\mathbf{48}) + 2(\mathbf{32})$$ Substitute 48 for l and 32 for w.
$$= 96 + 64$$
$$= 160$$

The perimeter of the flag is 160 inches.

b. The area of a triangle is given by the formula $A = \frac{1}{2}bh$, where b is the length of the base and h is the height. With the triangle positioned as it is, the base is 32 inches and the height is 48 inches.

$$A = \frac{1}{2}\mathbf{bh}$$

$$A = \frac{1}{2}(\mathbf{32})(\mathbf{48})$$ Substitute 32 for b and 48 for h.

$$= 16(48)$$ Multiply.

$$= 768$$

The area of the red triangular region of the flag is 768 in.^{2}.

Self Check 5   **a.** Find the perimeter of a square with sides 6 inches long.   **b.** Find the area of a triangle with base of 8 meters and height of 13 meters.

Formulas involving circles: A **circle** is the set of all points on a flat surface that are a fixed distance from a point called its **center.** A segment drawn from the center to a point on the circle is called a **radius.** Since a **diameter** of a circle is a segment passing through the center that joins two points on the circle, the diameter D of a circle is twice as long as its radius r. The perimeter of a circle is called its **circumference** C.

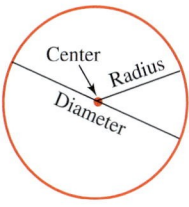

EXAMPLE 6

To the nearest tenth, find the area of a circle with a diameter of 14 feet.

Solution   The radius is one-half the diameter, or 7 feet. To find the area we substitute 7 for r in the formula for the area of a circle and proceed as follows.

Notation

We found the area of the circle to be *exactly* 49π ft². This form of the answer is convenient, but not very informative. To get a better feel for the area, we computed $49 \cdot \pi$ and rounded to the nearest tenth.

$$A = \pi r^2 \qquad \pi r^2 \text{ means } \pi \cdot r^2.$$
$$A = \pi (7)^2 \qquad \text{Substitute 7 for } r.$$
$$= 49\pi \qquad \text{Evaluate the exponential expression: } 7^2 = 49. \text{ The exact area is } 49\pi \text{ ft}^2.$$
$$\approx 153.93804 \qquad \text{Using a scientific calculator, enter these numbers and press these keys: } 49 \times \pi = . \text{ If you do not have a calculator, use 3.14 as an approximation of } \pi.$$

To the nearest tenth, the area is 153.9 ft².

Self Check 6   To the nearest hundredth, find the circumference of the circle.

The **volume** of a three-dimensional geometric solid is the amount of space it encloses. Volume is measured in cubic units, such as cubic inches, cubic feet, and cubic meters (denoted as in.³, ft³, and m³, respectively). Many formulas for volume are shown inside the front cover of the book.

EXAMPLE 7

Finding volumes.   To the nearest tenth, find the volume of the cylinder.

ELEMENTARY &
INTERMEDIATE
Algebra *f(x)* **Now**™

Solution   Since the radius of a circle is one-half its diameter, the radius of the cylinder is $\frac{1}{2}(6 \text{ cm}) = 3$ cm. The height of the cylinder is 12 cm. We substitute 3 for r and 12 for h in the formula for volume and proceed as follows.

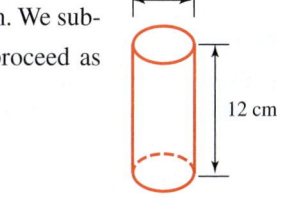

$$V = \pi r^2 h \qquad \pi r^2 h \text{ means } \pi \cdot r^2 \cdot h.$$
$$V = \pi (3)^2 (12) \qquad \text{Substitute 3 for } r \text{ and 12 for } h.$$
$$= \pi (9)(12) \qquad \text{Evaluate the exponential expression.}$$
$$= 108\pi \qquad \text{Multiply. The exact volume is } 108\pi \text{ cm}^3.$$
$$\approx 339.2920066 \qquad \text{Use a calculator.}$$

To the nearest tenth, the volume is 339.3 cubic centimeters. This can be written as 339.3 cm³.

Self Check 7   Find the volume of each figure: **a.** a rectangular solid with length 7 inches, width 12 inches, and height 15 inches;   and   **b.** a cone whose base has radius 12 meters and whose height is 9 meters. Give the answer to the nearest tenth.

■ SOLVING FOR A SPECIFIED VARIABLE

Suppose we wish to find the bases of several triangles whose areas and heights are known. It could be tedious to substitute values for A and h into the formula and then repeatedly solve the formula for b. A better way is to solve the formula $A = \frac{1}{2}bh$ for b first, and then substitute values for A and h and compute b directly.

To **solve an equation for a variable** means to isolate that variable on one side of the equation, with all other terms on the opposite side.

EXAMPLE 8   Solve $A = \frac{1}{2}bh$ for b.

Solution

$$A = \frac{1}{2}bh$$ We must isolate b on one side of the equation.

$$2 \cdot A = 2 \cdot \frac{1}{2}bh$$ To clear the equation of the fraction, multiply both sides by 2.

$$2A = bh$$ Simplify.

$$\frac{2A}{h} = \frac{bh}{h}$$ To undo the multiplication by h, divide both sides by h.

$$\frac{2A}{h} = b$$ On the right-hand side, remove the common factor of h: $\frac{b\overset{1}{\cancel{h}}}{\underset{1}{\cancel{h}}} = b$.

$$b = \frac{2A}{h}$$ Reverse the sides to write b on the left.

Self Check 8   Solve $V = lwh$ for w.

EXAMPLE 9   Solve $P = 2l + 2w$ for l.

Solution

$$P = 2l + 2w$$ We must isolate l on one side of the equation.

$$P - 2w = 2l + 2w - 2w$$ To undo the addition of $2w$, subtract $2w$ from both sides.

$$P - 2w = 2l$$ Combine like terms: $2w - 2w = 0$.

$$\frac{P - 2w}{2} = \frac{2l}{2}$$ To undo the multiplication by 2, divide both sides by 2.

$$\frac{P - 2w}{2} = l$$ Simplify the right-hand side.

Caution

In Example 9, do not try to simplify the result this way:

$$l = \frac{P - \overset{1}{\cancel{2}}w}{\underset{1}{\cancel{2}}}$$

This step is incorrect because 2 is not a factor of the entire numerator.

We can write the result as $l = \dfrac{P - 2w}{2}$.

Self Check 9   Solve $P = 2l + 2w$ for w.

EXAMPLE 10

In Chapter 3, we will work with equations that involve the variables x and y, such as $2y - 4 = 3x$. Solve this equation for y.

Solution

We must isolate y on one side of the equation.

$$2y - 4 = 3x$$

$2y - 4 + 4 = 3x + 4$ To undo the subtraction of 4, add 4 to both sides.

$2y = 3x + 4$ Do the addition.

$\dfrac{2y}{2} = \dfrac{3x + 4}{2}$ To undo the multiplication by 2, divide both sides by 2.

$y = \dfrac{3x}{2} + \dfrac{4}{2}$ On the right-hand side, write $\frac{3x+4}{2}$ as the sum of two fractions with like denominators, $\frac{3x}{2}$ and $\frac{4}{2}$.

$y = \dfrac{3}{2}x + 2$ Write $\frac{3x}{2}$ as $\frac{3}{2}x$. Simplify: $\frac{4}{2} = 2$.

Self Check 10   Solve $3y + 12 = x$ for y.

EXAMPLE 11

Solve $V = \pi r^2 h$ for r^2.

Solution

We must isolate r^2 on one side of the equation.

$$V = \pi r^2 h$$

$\dfrac{V}{\pi h} = \dfrac{\pi r^2 h}{\pi h}$ To undo the multiplication by π and h on the right-hand side, divide both sides by πh.

$\dfrac{V}{\pi h} = r^2$ Remove the common factors of π and h: $\dfrac{\overset{1}{\cancel{\pi}} r^2 \overset{1}{\cancel{h}}}{\underset{1}{\cancel{\pi}} \underset{1}{\cancel{h}}} = r^2$.

$r^2 = \dfrac{V}{\pi h}$ Reverse the sides of the equation so that r^2 is on the left.

> **Caution**
>
> When solving for a variable, that variable must be isolated on one side of the equation.

Self Check 11   Solve $a^2 + b^2 = c^2$ for b^2.

Answers to Self Checks
   1. $\$1,510.50$ **2.** 2.5 years   **3.** 1.25 minutes   **4.** $-283°F$ **5. a.** 24 in.,   **b.** $52\ \text{m}^2$
6. 43.98 ft   **7. a.** $1{,}260\ \text{in.}^3$,   **b.** $1{,}357.2\ \text{m}^3$ **8.** $w = \dfrac{V}{lh}$ **9.** $w = \dfrac{P - 2l}{2}$
10. $y = \frac{1}{3}x - 4$ **11.** $b^2 = c^2 - a^2$

2.5   STUDY SET ELEMENTARY & INTERMEDIATE Algebra *f(x)* Now ™

VOCABULARY   Fill in the blanks.

1. A _____ is an equation that is used to state a known relationship between two or more variables.

2. The _____ of a three-dimensional geometric solid is the amount of space it encloses.

3. The distance around a geometric figure is called its _____.

4. A _____ is the set of all points on a flat surface that are a fixed distance from a point called its center.

5. A line segment drawn from the center of a circle to a point on the circle is called a _____.

6. The amount of surface that is enclosed by a geometric figure is called its _____.

7. The perimeter of a circle is called its _____.

8. A line segment passing through the center of a circle and connecting two points on the circle is called a _____.

CONCEPTS

9. Use variables to write the formula relating the following:
   a. Time, distance, rate
   b. Markup, retail price, cost
   c. Costs, revenue, profit
   d. Interest rate, time, interest, principal
   e. Circumference, radius

10. Complete the table.

Principal ·	rate ·	time =	interest
$2,500	5%	2 yr	
$15,000	4.8%	1 yr	

11. Complete the table to find how far light and sound travel in 60 seconds. (*Hint:* mi/sec means miles per second.)

	Rate ·	time =	distance
Light	186,282 mi/sec	60 sec	
Sound	1,088 ft/sec	60 sec	

12. Give the name of each figure.

a.   **b.**

c.   **d.**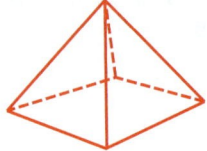

e.    **f.**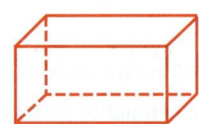

13. Tell which concept, perimeter, circumference, area, or volume, should be used to find the following:
   a. The amount of storage in a freezer
   b. How far a bicycle tire rolls in one revolution

c. The amount of land making up the Sahara Desert

d. The distance around a Monopoly game board

14. Tell which unit of measurement, ft, ft^2, or ft^3, would be appropriate when finding the following:
   a. The amount of storage inside a safe
   b. The ground covered by a sleeping bag lying on the floor
   c. The distance the tip of an airplane propeller travels in one revolution
   d. The size of the trunk of a car

15. a. Write an expression for the perimeter of the figure.

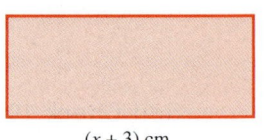

2 cm

$(x + 3)$ cm

   b. Write an expression for the area of the figure.

16. WHEELCHAIRS
   a. Find the diameter of the rear wheel.
   b. Find the radius of the front wheel.

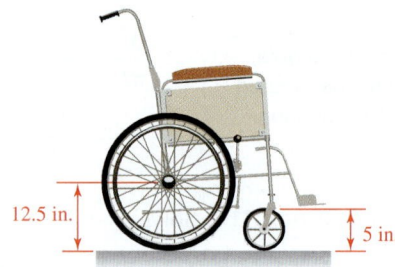

12.5 in.   5 in.

NOTATION   **Complete the solution.**

17. Solve $Ax + By = C$ for y.

$$Ax + By = C$$
$$Ax + By - \boxed{} = C - \boxed{}$$
$$\boxed{} = C - Ax$$
$$\frac{By}{\boxed{}} = \frac{C - Ax}{\boxed{}}$$
$$y = \frac{C - Ax}{B}$$

18. Enter the missing formulas in each table.
   a. The table contains information about an investment earning simple interest.

	?	·	?	·	?	=	?
Certificate of deposit	$3,500		0.04		1 yr		$140

b. The table contains information about a trip made by a cross-country skier.

	?	·	?	=	?
Skier	3 mph		2 hr		6 mi

19. a. Approximate π to the nearest hundredth.

b. What does 98π mean?

c. In the formula for the volume of a cylinder, $V = \pi r^2 h$, what does r represent? What does h represent?

20. a. What does ft^2 mean?

b. What does $in.^3$ mean?

PRACTICE   Use a formula to solve each problem.

21. SWIMMING   In 1930, a man swam down the Mississippi River from Minneapolis to New Orleans, a total of 1,826 miles. He was in the water for 742 hours. To the nearest tenth, what was his average swimming rate?

22. ROSE PARADE   Rose Parade floats travel down the 5.5-mile-long parade route at a rate of 2.5 mph. How long will it take a float to complete the route if there are no delays?

23. HOLLYWOOD   Figures for the summer of 1998 showed that the movie *Saving Private Ryan* had U.S. box-office receipts of $190 million. What were the production costs to make the movie if, at that time, the studio had made a $125 million profit?

24. SERVICE CLUBS   After expenses of $55.15 were paid, a Rotary Club donated $875.85 in proceeds from a pancake breakfast to a local health clinic. How much did the pancake breakfast gross?

25. ENTREPRENEURS   To start a mobile dog-grooming service, a woman borrowed $2,500. If the loan was for 2 years and the amount of interest was $175, what simple interest rate was she charged?

26. BANKING   Three years after opening an account that paid 6.45% annually, a depositor withdrew the $3,483 in interest earned. How much money was left in the account?

27. METALLURGY   Change 2,212°C, the temperature at which silver boils, to degrees Fahrenheit. Round to the nearest degree.

28. LOW TEMPERATURES   Cryobiologists freeze living matter to preserve it for future use. They can work with temperatures as low as −270°C. Change this to degrees Fahrenheit.

29. VALENTINE'S DAY   Find the markup on a dozen roses if a florist buys them wholesale for $12.95 and sells them for $37.50.

30. STICKER PRICES   The factory invoice for a minivan shows that the dealer paid $16,264.55 for the vehicle. If the sticker price of the van is $18,202, how much over factory invoice is the sticker price?

31. YO-YOS   How far does a yo-yo travel during one revolution of the "around the world" trick if the length of the string is 21 inches?

32. HORSES   A horse trots in a perfect circle around its trainer at the end of a 28-foot-long rope. How far does the horse travel as it circles the trainer once?

Solve each formula for the given variable.

33. $E = IR$; for R

34. $d = rt$; for t

35. $V = lwh$; for w

36. $I = Prt$; for r

37. $C = 2\pi r$; for r

38. $V = \pi r^2 h$; for h

39. $A = \dfrac{Bh}{3}$; for h

40. $C = \dfrac{Rt}{7}$; for R

41. $w = \dfrac{s}{f}$; for f

42. $P = \dfrac{ab}{c}$; for c

43. $P = a + b + c$; for b

44. $a + b + c = 180$; for a

45. $T = 2r + 2t$; for r

46. $y = mx + b$; for x

47. $Ax + By = C$; for x

48. $A = P + Prt$; for t

49. $K = \dfrac{1}{2}mv^2$; for m

50. $V = \dfrac{1}{3}\pi r^2 h$; for h

51. $A = \dfrac{a + b + c}{3}$; for c **52.** $x = \dfrac{a + b}{2}$; for b

53. $2E = \dfrac{T - t}{9}$; for t **54.** $D = \dfrac{C - s}{n}$; for s

55. $s = 4\pi r^2$; for r^2 **56.** $E = mc^2$; for c^2

57. $Kg = \dfrac{wv^2}{2}$; for v^2 **58.** $c^2 = a^2 + b^2$; for a^2

59. $V = \dfrac{4}{3}\pi r^3$; for r^3 **60.** $A = \dfrac{\pi r^2 S}{360}$; for r^2

61. $\dfrac{M}{2} - 9.9 = 2.1B$; for M

62. $\dfrac{G}{0.5} + 16r = -8t$; for G

63. $S = 2\pi rh + 2\pi r^2$; for h

64. $c = bn + 16t^2$; for t^2

65. $3x + y = 9$; for y **66.** $-5x + y = 4$

67. $3y - 9 = x$; for y **68.** $5y - 25 = x$; for y

69. $4y + 16 = -3x$; for y **70.** $6y + 12 = -5x$; for y

71. $A = \dfrac{1}{2}h(b + d)$; for b

72. $C = \dfrac{1}{4}s(t - d)$; for t

73. $\dfrac{7}{8}c + w = 9$; for c **74.** $\dfrac{3}{4}m - t = 5b$; for m

APPLICATIONS

75. PROPERTIES OF WATER   The boiling point and the freezing point of water are to be given in both degrees Celsius and degrees Fahrenheit on the thermometer. Find the missing degree measures.

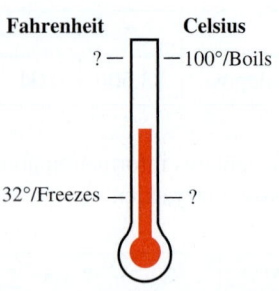

76. SPEED LIMITS   Several state speed limits for trucks are shown. At each of these speeds, how far would a truck travel in $2\frac{1}{2}$ hours?

77. AVON PRODUCTS   Complete the financial statement from the Hoover's Online business Web site.

Quarterly financials income statement (dollar amounts in millions except per share amounts)	Quarter ending Dec 02	Quarter ending Sep 02
Revenue	1,854.1	1,463.4
Cost of goods sold	679.5	506.5
Gross profit		

78. CREDIT CARDS   The finance charge section of a person's credit card statement says, "annual percentage rate (APR) is 19.8%." Determine how much finance charges (interest) the card owner would have to pay if the account's average balance for the year was $2,500.

79. CARPENTRY   Find the perimeter and area of the truss.

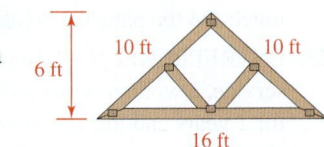

80. CAMPERS   Find the area of the window of the camper shell on the next page.

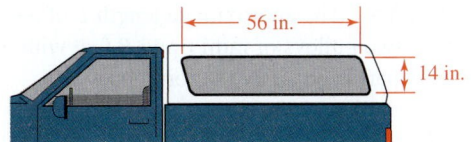

81. ARCHERY   To the nearest tenth, find the circumference and area of the target.

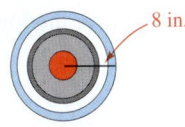

82. GEOGRAPHY   The circumference of the Earth is about 25,000 miles. Find its diameter to the nearest mile.

83. LANDSCAPING   Find the perimeter and the area of the redwood trellis.

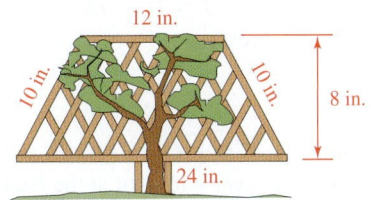

84. HAMSTER HABITATS   Find the amount of space in the tube.

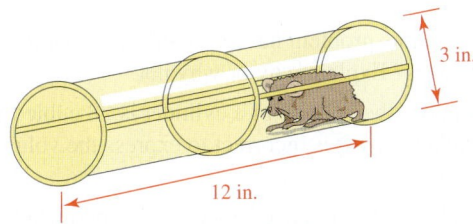

85. MEMORIALS   The Vietnam Veterans Memorial is a black granite wall recognizing the more than 58,000 Americans who lost their lives or remain missing. Find the total area of the two triangular-shaped surfaces on which the names are inscribed.

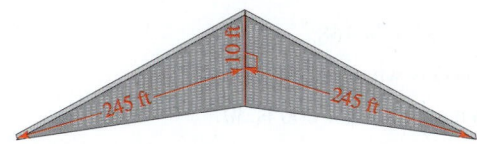

86. SIGNAGE   Find the perimeter and area of the service station sign.

87. RUBBER MEETS THE ROAD   A sport truck tire has the road surface footprint shown here. Estimate the perimeter and area of the tire's footprint.

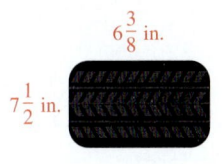

88. SOFTBALL   The strike zone in fast-pitch softball is between the batter's armpit and the top of her knees, as shown. Find the area of the strike zone.

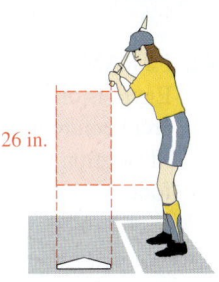

89. FIREWOOD   Find the area on which the wood is stacked and the volume the cord of firewood occupies.

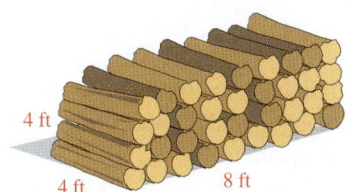

90. NATIVE AMERICAN DWELLINGS   The teepees constructed by the Blackfoot Indians were cone-shaped tents made of long poles and animal hide, about 10 feet high and about 15 feet across at the ground. Estimate the volume of a teepee with these dimensions, to the nearest cubic foot.

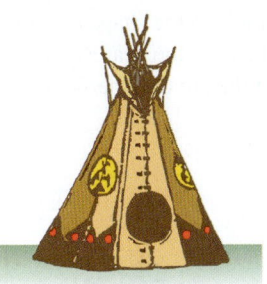

91. IGLOOS   During long journeys, some Canadian Inuit (Eskimos) built winter houses of snow blocks stacked in the dome shape shown on the next page. Estimate the volume of an igloo having an interior height of 5.5 feet to the nearest cubic foot.

92. PYRAMIDS The Great Pyramid at Giza in northern Egypt is one of the most famous works of architecture in the world. Use the information in the illustration to find the volume to the nearest cubic foot.

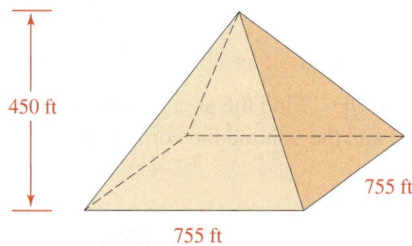

450 ft

755 ft
755 ft

93. BARBECUING Use the fact that the fish is 18 inches long to find the area of the barbecue grill to the nearest square inch.

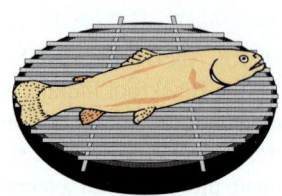

94. SKATEBOARDING A half-pipe ramp used for skateboarding is in the shape of a semicircle with a radius of 8 feet. To the nearest tenth of a foot, what is the length of the arc that the skateboarder travels on the ramp?

8 ft

Plywood

95. PULLEYS The approximate length L of a belt joining two pulleys of radii r and R feet with centers D feet apart is given by the formula

$$L = 2D + 3.25(r + R)$$

Solve the formula for D.

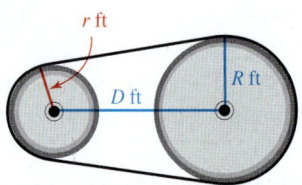

r ft
R ft
D ft

96. THERMODYNAMICS The Gibbs free-energy function is given by $G = U - TS + pV$. Solve this formula for the pressure p.

WRITING

97. After solving $A = B + C + D$ for B, a student compared her answer with that at the back of the textbook.

> Student's answer: $B = A - C - D$
> Book's answer: $B = A - D - C$

Could this problem have two different-looking answers? Explain why or why not.

98. Suppose the volume of a cylinder is 28 cubic feet. Explain why it is incorrect to express the volume as 28^3 ft.

99. Explain the difference between what perimeter measures and what area measures.

100. Explain the error made below.

$$y = \frac{3x + \overset{1}{\cancel{2}}}{\underset{1}{\cancel{2}}}$$

REVIEW

101. Find 82% of 168.

102. 29.05 is what percent of 415?

103. What percent of 200 is 30?

104. A woman bought a coat for $98.95 and some gloves for $7.95. If the sales tax was 6%, how much did the purchase cost her?

CHALLENGE PROBLEMS

105. In mathematics, letters from the Greek alphabet are often used as variables. Solve the following equation for α (read as "alpha"), the first letter of the Greek alphabet.

$$-7(\alpha - \beta) - (4\alpha - \theta) = \frac{\alpha}{2}$$

106. When a car of mass m collides with a wall, the energy of the collision is given by the formula $E = \frac{1}{2}mv^2$. Compare the energy of two collisions: a car striking a wall at 30 mph, and at 60 mph.

2.6   More about Problem Solving

- Finding More than One Unknown
- Solving Number–Value Problems
- Solving Uniform Motion Problems
- Solving Geometric Problems
- Solving Investment Problems
- Solving Mixture Problems

In this section, we will solve several types of problems using the five-step problem-solving strategy.

■ FINDING MORE THAN ONE UNKNOWN

EXAMPLE 1

California coastline. The first part of California's magnificent 17-Mile Drive begins at the Pacific Grove entrance and continues to Seal Rock. It is 1 mile longer than the second part of the drive, which extends from Seal Rock to the Lone Cypress. The final part of the tour winds through the Monterey Peninsula, eventually returning to the entrance. This part of the drive is 1 mile longer than four times the length of the second part. How long is each part of 17-Mile Drive?

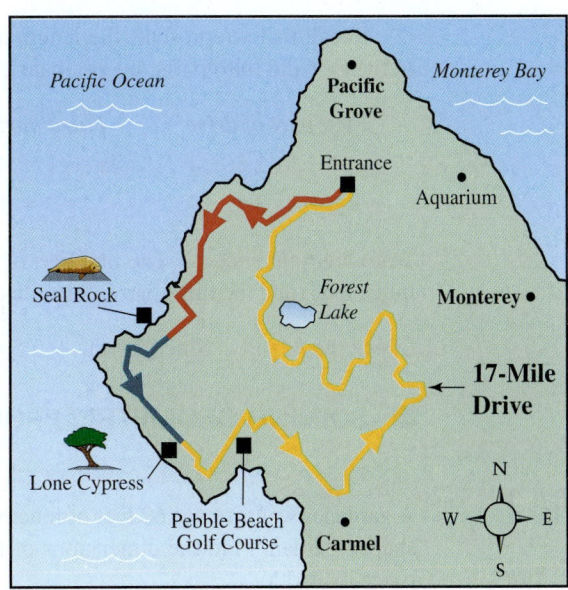

Analyze the Problem   The drive is composed of three parts. We need to find the length of each part. We can straighten out the winding 17-Mile Drive and model it with a line segment.

Form an Equation   Since the lengths of the first part and of the third part of the drive are related to the length of the second part, we will let x represent the length of that part. We then express the other lengths in terms of x. Let

$$x = \text{the length of the second part of the drive}$$
$$x + 1 = \text{the length of the first part of the drive}$$
$$4x + 1 = \text{the length of the third part of the drive}$$

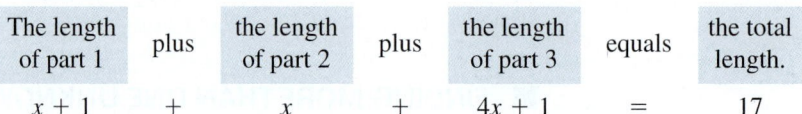

> **Caution**
>
> For this problem, one common mistake is to let
>
> x = the length of each part of the drive
>
> The three parts of the drive have different lengths; x cannot represent three different distances.

The sum of the lengths of the three parts must be 17 miles.

The length of part 1	plus	the length of part 2	plus	the length of part 3	equals	the total length.
$x + 1$	$+$	x	$+$	$4x + 1$	$=$	17

Solve the Equation

$$x + 1 + x + 4x + 1 = 17$$
$$6x + 2 = 17 \qquad \text{Combine like terms: } x + x + 4x = 6x \text{ and } 1 + 1 = 2.$$
$$6x = 15 \qquad \text{To undo the addition of 2, subtract 2 from both sides.}$$
$$\frac{6x}{6} = \frac{15}{6} \qquad \text{To undo the multiplication by 6, divide both sides by 6.}$$
$$x = 2.5$$

Recall that x represents the length of the second part of the drive. To find the lengths of the first and third parts, we evaluate $x + 1$ and $4x + 1$ for $x = 2.5$.

First part of drive	*Third part of drive*	
$x + 1 = \mathbf{2.5} + 1$	$4x + 1 = 4(\mathbf{2.5}) + 1$	Substitute 2.5 for x.
$= 3.5$	$= 11$	

State the Conclusion   The first part of the drive is 3.5 miles long, the second part is 2.5 miles long, and the third part is 11 miles long.

Check the Result   Since 3.5 mi + 2.5 mi + 11 mi = 17 mi, the answers check.

■ SOLVING GEOMETRIC PROBLEMS

EXAMPLE 2

ELEMENTARY & INTERMEDIATE
Algebra $f(x)$ **Now**™

A gardener wants to use 62 feet of fencing bought at a garage sale to enclose a rectangular-shaped garden. Find the dimensions of the garden if its length is to be 4 feet longer than twice its width.

Analyze the Problem   A sketch is often helpful when solving problems about geometric figures. We know that the length of the garden is to be 4 feet longer than twice the width. We also know that its perimeter is to be 62 feet. Recall that the perimeter of a rectangle is given by the formula $P = 2l + 2w$.

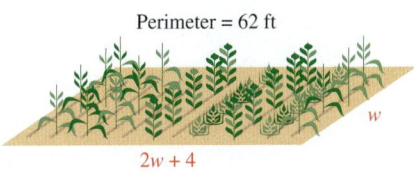

Perimeter = 62 ft

w

$2w + 4$

Form an Equation   Since the length of the garden is given in terms of the width, we will let $w =$ the width of the garden. Then the length $= 2w + 4$.

2	times	the length	plus	2	times	the width	is	the perimeter.
2	·	$(2w + 4)$	+	2	·	w	=	62

Solve the Equation

$2(2w + 4) + 2w = 62$	Be sure to write the parentheses so that the entire expression $2w + 4$ is multiplied by 2.
$4w + 8 + 2w = 62$	Use the distributive property to remove parentheses.
$6w + 8 = 62$	Combine like terms: $4w + 2w = 6w$.
$6w = 54$	To undo the addition of 8, subtract 8 from both sides.
$w = 9$	To undo the multiplication by 6, divide both sides by 6.

State the Conclusion   The width of the garden is 9 feet. Since $2w + 4 = 2(9) + 4 = 22$, the length is 22 feet.

Check the Result   If the garden has a width of 9 feet and a length of 22 feet, its length is 4 feet longer than twice the width $(2 \cdot 9 + 4 = 22)$. Since its perimeter is $2 \cdot 22$ ft $+ 2 \cdot 9$ ft $= 62$ ft, the answers check.

EXAMPLE 3

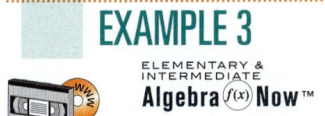

ELEMENTARY & INTERMEDIATE
Algebra $^{f(x)}$ **Now**™

Isosceles triangles.   If the vertex angle of an isosceles triangle is $56°$, find the measure of each base angle.

Analyze the Problem   An **isosceles triangle** has two sides of equal length, which meet to form the **vertex angle.** In this case, the measurement of the vertex angle is $56°$. We can sketch the triangle as shown. The **base angles** opposite the equal sides are also equal. We need to find their measure.

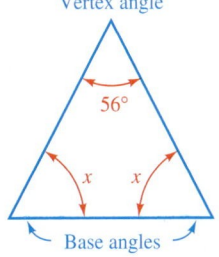

Vertex angle

$56°$

x x

Base angles

Form an Equation   If we let $x =$ the measure of one base angle, the measure of the other base angle is also x. Since the sum of the angles of any triangle is $180°$, the sum of the base angles and the vertex angle is $180°$. We can use this fact to form the equation.

One base angle	plus	the other base angle	plus	the vertex angle	is	$180°$.
x	+	x	+	56	=	180

Solve the Equation

$$x + x + 56 = 180$$

$$2x + 56 = 180 \qquad \text{Combine like terms: } x + x = 2x.$$

$$2x = 124 \qquad \text{To undo the addition of 56, subtract 56 from both sides.}$$

$$x = 62 \qquad \text{To undo the multiplication by 2, divide both sides by 2.}$$

State the Conclusion   The measure of each base angle is 62°.

Check the Result   Since 62° + 62° + 56° = 180°, the answer checks.

■ SOLVING NUMBER–VALUE PROBLEMS

Some problems deal with items that have monetary value. In these problems, we must distinguish between the *number of* and the *value of* the items. For problems of this type, we will use the fact that

Number · value = total value

EXAMPLE 4

Dining area improvements.   A restaurant owner needs to purchase some tables, chairs, and dinner plates for the dining area of her establishment. She plans to buy four chairs and four plates for each new table. She also plans to buy 20 additional plates in case of breakage. If a table costs $100, a chair $50, and a plate $5, how many of each can she buy if she takes out a loan for $6,500 to pay for the new items?

Analyze the Problem   We know the *value* of each item: Tables cost $100, chairs cost $50, and plates cost $5 each. We need to find the *number* of tables, chairs, and plates she can purchase for $6,500.

Form an Equation   The number of chairs and plates she needs depends on the number of tables she buys. So we let $t =$ the number of tables to be purchased. Since every table requires four chairs and four plates, she needs to order $4t$ chairs. Because 20 additional plates are needed, she should order $(4t + 20)$ plates. We can organize the facts of the problem in a table.

	Number ·	Value =	Total value
Tables	t	100	$100t$
Chairs	$4t$	50	$50(4t)$
Plates	$4t + 20$	5	$5(4t + 20)$

Enter this information first.   Multiply to get each of these entries.

We can use the information in the last column of the table to form an equation.

The value of the tables	plus	the value of the chairs	plus	the value of the plates	equals	the total value of the purchase.
$100t$	+	$50(4t)$	+	$5(4t + 20)$	=	6,500

Solve the Equation

$$100t + 50(4t) + 5(4t + 20) = 6,500$$
$$100t + 200t + 20t + 100 = 6,500 \qquad \text{Do the multiplications.}$$
$$320t + 100 = 6,500 \qquad \text{Combine like terms.}$$
$$320t = 6,400 \qquad \text{Subtract 100 from both sides.}$$
$$t = 20 \qquad \text{Divide both sides by 320.}$$

To find the number of chairs and plates to buy, we evaluate $4t$ and $4t + 20$ for $t = 20$.

$$\textit{Chairs: } 4t = 4(\mathbf{20}) \qquad \textit{Plates: } 4t + 20 = 4(\mathbf{20}) + 20 \qquad \text{Substitute 20 for } t.$$
$$= 80 \qquad\qquad\qquad\qquad\quad = 100$$

State the Conclusion   The owner needs to buy 20 tables, 80 chairs, and 100 plates.

Check the Result   The total value of 20 tables is $20(\$100) = \$2,000$, the total value of 80 chairs is $80(\$50) = \$4,000$, and the total value of 100 plates is $100(\$5) = \$500$. Because the total purchase is $\$2,000 + \$4,000 + \$500 = \$6,500$, the answers check.

■ SOLVING INVESTMENT PROBLEMS

To find the amount of simple interest I an investment earns, we use the formula

$$I = Prt$$

where P is the principal, r is the annual rate, and t is the time in years.

EXAMPLE 5

Paying tuition.   A college student invested the $12,000 inheritance he received and decided to use the annual interest earned to pay his tuition costs of $945. The highest rate offered by a bank at that time was 6% annual simple interest. At this rate, he could not earn the needed $945, so he invested some of the money in a riskier, but more profitable, investment offering a 9% return. How much did he invest at each rate?

Analyze the Problem   We know that $12,000 was invested for 1 year at two rates: 6% and 9%. We are asked to find the amount invested at each rate so that the total return would be $945.

Form an Equation   Let $x =$ the amount invested at 6%. Then $12,000 - x =$ the amount invested at 9%. To organize the facts of the problem, we enter the principal, rate, time, and interest earned in a table.

	P	$\cdot \; r \;$	$\cdot \; t =$	I
Bank	x	0.06	1	$0.06x$
Riskier investment	$12,000 - x$	0.09	1	$0.09(12,000 - x)$

Enter this information first.   Multiply to get each of these entries.

We can use the information in the last column of the table to form an equation.

The interest earned at 6%	plus	the interest earned at 9%	equals	the total interest.
$0.06x$	$+$	$0.09(12{,}000 - x)$	$=$	945

Solve the Equation

$$0.06x + 0.09(12{,}000 - x) = 945$$

$$\mathbf{100}[0.06x + 0.09(12{,}000 - x)] = \mathbf{100}(945)$$
Multiply both sides by 100 to clear the equation of decimals.

$$\mathbf{100}(0.06x) + \mathbf{100}(0.09)(12{,}000 - x) = 100(945)$$
Distribute the multiplication by 100.

$$6x + 9(12{,}000 - x) = 94{,}500$$
Do the multiplications by 100.

$$6x + 108{,}000 - 9x = 94{,}500$$
Use the distributive property.

$$-3x + 108{,}000 = 94{,}500$$
Combine like terms.

$$-3x = -13{,}500$$
Subtract 108,000 from both sides.

$$x = 4{,}500$$
Divide both sides by -3.

State the Conclusion   The student invested $4,500 at 6% and $12,000 − $4,500 = $7,500 at 9%.

Check the Result   The first investment earned 0.06($4,500), or $270. The second earned 0.09($7,500), or $675. The total return was $270 + $675 = $945. The answers check.

■ SOLVING UNIFORM MOTION PROBLEMS

If we know the rate r at which we will be traveling and the time t we will be traveling at that rate, we can find the distance d traveled by using the formula

$$d = rt$$

EXAMPLE 6

ELEMENTARY & INTERMEDIATE
Algebra *f(x)* **Now**™

Coast Guard rescues.   A cargo ship, heading into port, radios the Coast Guard that it is experiencing engine trouble and that its speed has dropped to 3 knots (3 nautical miles per hour). Immediately, a Coast Guard cutter leaves port and speeds at a rate of 25 knots directly toward the disabled ship, which is 21 nautical miles away. How long will it take the Coast Guard to reach the cargo ship?

Analyze the Problem   We know the *rate* of each ship (25 knots and 3 knots), and we know that they must close a *distance* of 21 nautical miles between them. We don't know the *time* it will take to do this.

Form an Equation   Let $t =$ the time it takes for the ships to meet. Using $d = rt$, we find that $25t$ represents the distance traveled by the Coast Guard cutter and $3t$ represents the distance traveled by the cargo ship. We can organize the facts of the problem in a table.

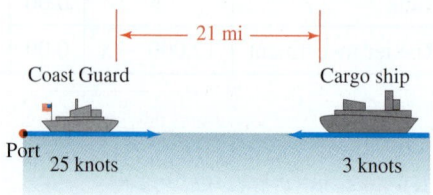

	r	$\cdot$ t	$=$ d
Coast Guard cutter	25	t	$25t$
Cargo ship	3	t	$3t$

Enter this information first.

We can use the information in the last column of the table to form an equation.

The distance the Coast Guard cutter travels	plus	the distance the cargo ship travels	equals	the initial distance between the two ships.
$25t$	$+$	$3t$	$=$	21

Solve the Equation

$$25t + 3t = 21$$
$$28t = 21 \quad \text{Combine like terms.}$$
$$t = \frac{21}{28} \quad \text{Divide both sides by 28.}$$
$$t = \frac{3}{4} \quad \text{Simplify the fraction.}$$

State the Conclusion   The ships will meet in $\frac{3}{4}$ hr, or 45 minutes.

Check the Result   In $\frac{3}{4}$ hr, the Coast Guard cutter travels $25 \cdot \frac{3}{4} = \frac{75}{4}$ nautical miles, and the cargo ship travels $3 \cdot \frac{3}{4} = \frac{9}{4}$ nautical miles. Together, they travel $\frac{75}{4} + \frac{9}{4} = \frac{84}{4} = 21$ nautical miles. This is the initial distance between the ships; the answer checks.

■ SOLVING MIXTURE PROBLEMS

We now discuss how to solve mixture problems. In the first type, a *liquid mixture* of a desired strength is made from two solutions with different concentrations. In the second type, a *dry mixture* of a specified value is created from two differently priced components.

EXAMPLE 7

ELEMENTARY & INTERMEDIATE
Algebra *f(x)* **Now**™

Mixing solutions.   A chemistry experiment calls for a 30% sulfuric acid solution. If the lab supply room has only 50% and 20% sulfuric acid solutions, how much of each should be mixed to obtain 12 liters of a 30% acid solution?

Analyze the Problem   The 50% solution is too strong and the 20% solution is too weak. We must find how much of each should be combined to obtain 12 liters of a 30% solution.

Form an Equation   If $x =$ the number of liters of the 50% solution used in the mixture, the remaining $(12 - x)$ liters must be the 20% solution.

The amount of pure acid in each solution is given by

Amount of solution · strength of solution = amount of pure acid

A table is helpful in organizing the facts of the problem.

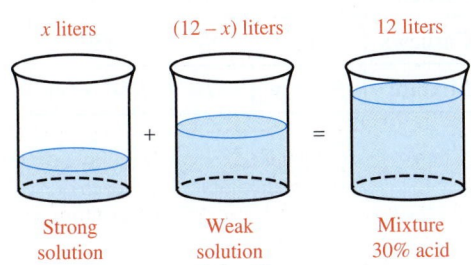

	Amount	· Strength =	Amount of acid
Strong	x	0.50	$0.50x$
Weak	$12 - x$	0.20	$0.20(12 - x)$
Mixture	12	0.30	$12(0.30)$

Enter this information first.   Multiply to get each of these entries.

x liters   (12 − *x*) liters   12 liters

Strong solution 50% acid   Weak solution 20% acid   Mixture 30% acid

We can use the information in the last column of the table to form an equation.

The acid in the 50% solution	plus	the acid in the 20% solution	equals	the acid in the final mixture.
$0.50x$	$+$	$0.20(12 - x)$	$=$	$12(0.30)$

Solve the Equation

$$0.50x + 0.20(12 - x) = 12(0.30) \qquad \text{50\% = 0.50, 20\% = 0.20, and 30\% = 0.30.}$$
$$0.5x + 2.4 - 0.2x = 3.6 \qquad \text{Distribute the multiplication by 0.20.}$$
$$0.3x + 2.4 = 3.6 \qquad \text{Combine like terms.}$$
$$0.3x = 1.2 \qquad \text{Subtract 2.4 from both sides.}$$
$$\frac{0.3x}{0.3} = \frac{1.2}{0.3} \qquad \text{To undo the multiplication by 0.3, divide both sides by 0.3.}$$
$$x = 4$$

Success Tip

We could begin by multiplying both sides of the equation by 10 to clear it of the decimals.

State the Conclusion   4 liters of 50% solution and $12 - 4 = 8$ liters of 20% solution should be used.

Check the Result   Acid in 4 liters of the 50% solution: $0.50(4) = 2.0$ liters.
Acid in 8 liters of the 20% solution: $0.20(8) = 1.6$ liters.   3.6 liters
Acid in 12 liters of the 30% mixture: $0.30(12) = 3.6$ liters. The answers check.

EXAMPLE 8

Snack foods.   Because cashews priced at $9 per pound were not selling, a produce clerk decided to combine them with less expensive peanuts and sell the mixture for $7 per pound. How many pounds of peanuts, selling at $6 per pound, should be mixed with 50 pounds of cashews to obtain such a mixture?

Analyze the Problem   We need to determine how many pounds of peanuts to mix with 50 pounds of cashews to obtain a mixture worth $7 per pound.

Form an Equation   Let $x =$ the number of pounds of peanuts to use in the mixture. Since 50 pounds of cashews will be combined with the peanuts, the mixture will weigh $50 + x$ pounds. The value of the mixture and of each of the components of the mixture is given by

Amount · price = total value

We can organize the facts of the problem in a table.

	Amount · Price = Total value		
Peanuts	x	6	$6x$
Cashews	50	9	450
Mixture	$50 + x$	7	$7(50 + x)$

Enter this information first.   Multiply to get each of these entries.

We can use the information in the last column of the table to form an equation.

The value of the peanuts	plus	the value of the cashews	equals	the value of the mixture.
$6x$	$+$	450	$=$	$7(50 + x)$

Solve the Equation

$6x + 450 = 7(50 + x)$

$6x + 450 = 350 + 7x$ Distribute the multiplication by 7.

$450 = 350 + x$ Subtract $6x$ from both sides.

$100 = x$ Subtract 350 from both sides.

State the Conclusion   100 pounds of peanuts should be used in the mixture.

Check the Result

Value of 100 pounds of peanuts, at \$6 per pound: $100(6) = \$600.$

Value of 50 pounds of cashews, at \$9 per pound: $50(9) = \$450.$

$\Big\rangle$ \$1,050

Value of 150 pounds of the mixture, at \$7 per pound: \$1,050. The answer checks.

2.6   STUDY SET   ELEMENTARY & INTERMEDIATE Algebra *f(x)* Now™

VOCABULARY   Fill in the blanks.

1. The _____ of a triangle or a rectangle is the distance around it.

2. An _____ triangle is a triangle with two sides of the same length.

3. The equal sides of an isosceles triangle meet to form the _____ angle. The angles opposite the equal sides are called _____ angles, and they have equal measures.

4. When asked to find the dimensions of a rectangle, we are to find its _____ and _____.

CONCEPTS

5. A plumber wants to cut a 17-foot pipe into three sections. The longest section is to be three times as long as the shortest, and the middle-sized section is to be 2 feet longer than the shortest.
   a. Complete the diagram.

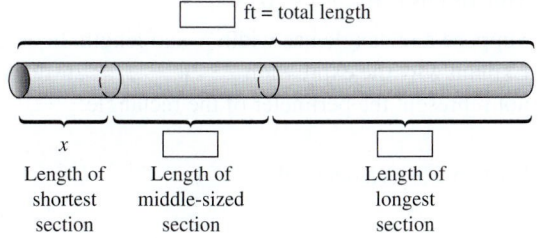

x / Length of shortest section / Length of middle-sized section / Length of longest section

b. To solve this problem, an equation is formed, it is solved, and it is found that $x = 3$. How long is each section of pipe?

6. Complete the expression, which represents the perimeter of the rectangle shown.

$2(\quad\quad) + \quad x$

$5x - 1$
x

7. What is the sum of the measures of the angles of any triangle?

8. a. Complete the table, which shows the inventory of nylon brushes that a paint store carries.

	Number ·	Value =	Total value
1 inch	$\frac{x}{2}$	4	
2 inch	x	5	
3 inch	$x + 10$	7	

b. Which type of brush does the store have the greatest number of?

c. What is the least expensive brush?

d. What is the total value of the inventory of nylon brushes?

9. In the advertisement, what are the principal, the rate, and the time for the investment opportunity shown?

> **Invest in Mini Malls!**
> Builder seeks daring people who want to earn big $$$$$$. In just 1 year, you will earn a gigantic 14% on an investment of only $30,000! Call now.

10. a. Complete the table, which gives the details about two investments made by a retired couple.

	P	$\cdot$ r	$\cdot$ $t =$	I
Certificate of deposit	x	0.04	1	
Brother-in-law's business	$2x$	0.06	1	

b. How much more money was invested in the brother-in-law's business than in the certificate of deposit?

c. What is the total amount of interest the couple will make from these investments?

11. When a husband and wife leave for work, they drive in opposite directions. Their average speeds are different; however, their drives last the same amount of time. Complete the table, which gives the details of each person's commute.

	r	$\cdot$ $t =$	d
Husband	35	t	
Wife	45		

12. a. How many gallons of acid are there in the second barrel?

b. Suppose the contents of the two barrels are poured into an empty third barrel. How many gallons of liquid will the third barrel contain?

c. What would be a *reasonable* estimate of the concentration of the solution in the third barrel: 15%, 35%, or 60% acid?

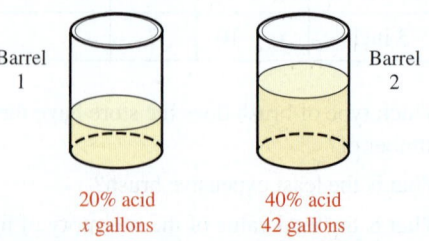

Barrel 1 — 20% acid, x gallons

Barrel 2 — 40% acid, 42 gallons

13. a. Two oil and vinegar salad dressings are combined to make a new mixture. Complete the table.

	Amount	$\cdot$ Strength =	Pure vinegar
Strong	x	0.06	
Weak		0.03	
Mixture	10	0.05	

b. Two antifreeze solutions are combined to form a mixture. Complete the table.

	Amount	$\cdot$ Strength =	Pure antifreeze
Strong	6	0.50	
Weak	x	0.25	
Mixture		0.30	

14. Use the information in the table to fill in the blanks. How many pounds of a _____ diet supplement worth $15.95 per pound should be mixed with _____ pounds of a vitamin diet supplement worth _____ per pound to obtain a _____ that is worth _____ per pound?

	Amount	$\cdot$ Value =	Total value
Protein	x	15.95	$15.95x$
Vitamin	20	7.99	$20(7.99)$
Mixture	$x + 20$	12.50	$12.50(x + 20)$

NOTATION

15. What concept about decimal multiplication is shown?

$$100(0.08) = 8$$

16. True or false: $x(0.09) = 0.09x$?

17. Suppose a rectangle has width w and length $2w - 3$. Explain why the expression $2 \cdot 2w - 3 + 2w$ does not represent the perimeter of the rectangle.

18. Write 5.5% as a decimal.

PRACTICE Solve each equation by first clearing it of decimals.

19. $0.08x + 0.07(15,000 - x) = 1,110$

20. $0.108x + 0.07(16,000 - x) = 1,500$

APPLICATIONS

21. CARPENTRY   A 12-foot board has been cut into two sections, one twice as long as the other. How long is each section?

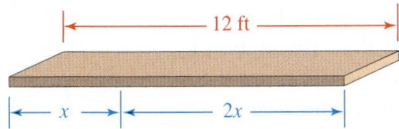

22. ROBOTICS   The robotic arm will extend a total distance of 18 feet. Find the length of each section.

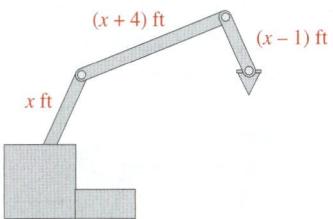

23. NATIONAL PARKS
The Natchez Trace
Parkway is a historical
444-mile route from
Natchez, Mississippi, to
Nashville, Tennessee. A
couple drove the Trace
in four days. Each day
they drove 6 miles more than the previous day. How many miles did they drive each day?

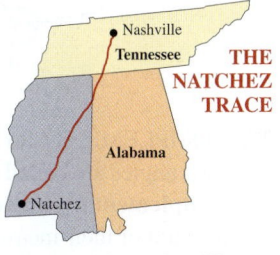

24. SOLAR HEATING   One solar panel is 3.4 feet wider than the other. Find the width of each panel.

25. TOURING   A rock group plans to travel for a total of 38 weeks, making three concert stops. They will be in Japan for 4 more weeks than they will be in Australia. Their stay in Sweden will be 2 weeks shorter than that in Australia. How many weeks will they be in each country?

26. LOCKS   The three numbers of the combination for a lock are **consecutive integers,** and their sum is 81. (Consecutive integers follow each other, like 7, 8, 9.) Find the combination.   (*Hint:* If x represents the smallest integer, $x + 1$ represents the next integer, and $x + 2$ represents the largest integer.)

27. COUNTING CALORIES   A slice of pie with a scoop of ice cream has 850 calories. The calories in the pie alone are 100 more than twice the calories in the ice cream alone. How many calories are in each food?

28. WASTE DISPOSAL   Two tanks hold a total of 45 gallons of a toxic solvent. One tank holds 6 gallons more than twice the amount in the other. How many gallons does each tank hold?

29. ACCOUNTING   Determine the 2002 income of Sears, Roebuck and Co. for each quarter from the data in the graph. (Source: Hoover's Online Internet service.)

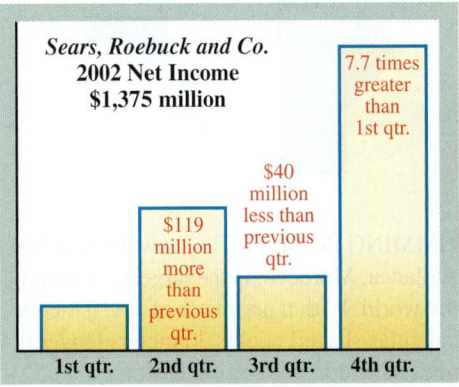

30. AMERICAN COLLEGE STUDENTS   A 1999–2000 survey found that the percent of college students that worked part time was 1% more than twice the percent that did not work. The percent that worked full time was 1% less than twice the percent that did not work. Find the missing percents in the graph.   (*Hint:* The sum of the percents is 100%.)

Most students work

Source: National Center for Education Statistics

31. ENGINEERING   A truss is in the form of an isosceles triangle. Each of the two equal sides is 4 feet shorter than the third side. If the perimeter is 25 feet, find the lengths of the sides.

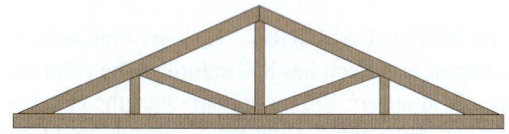

32. FIRST AID   A sling is in the shape of an isosceles triangle with a perimeter of 144 inches. The longest side of the sling is 18 inches longer than either of the other two sides. Find the lengths of each side.

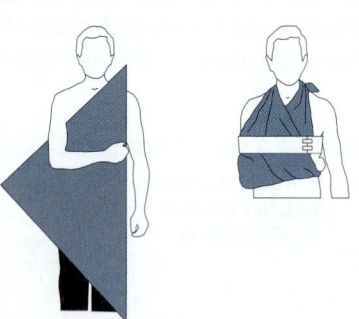

33. SWIMMING POOLS   The seawater Orthlieb Pool in Casablanca, Morocco, is the largest swimming pool in the world. With a perimeter of 1,110 meters, this rectangular-shaped pool is 30 meters longer than 6 times its width. Find its dimensions.

34. ART   The *Mona Lisa* was completed by Leonardo da Vinci in 1506. The length of the picture is 11.75 inches shorter than twice the width. If the perimeter of the picture is 102.5 inches, find its dimensions.

35. TV TOWERS   The two guy wires supporting a tower form an isosceles triangle with the ground. Each of the base angles of the triangle is 4 times the third angle (the vertex angle). Find the measure of the vertex angle.

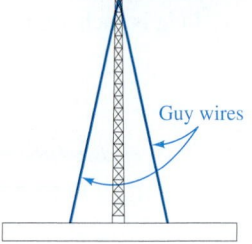

36. MOUNTAIN BICYCLES   For the bicycle frame shown, the angle that the horizontal crossbar makes with the seat support is 15° less than twice the angle at the steering column. The angle at the pedal gear is 25° more than the angle at the steering column. Find these three angle measures.

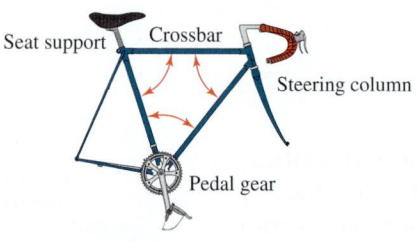

37. COMPLEMENTARY ANGLES   Two angles are called **complementary angles** when the sum of their measures is 90°. Find the measures of the complementary angles shown here.

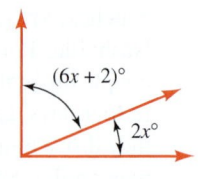

38. SUPPLEMENTARY ANGLES   Two angles are called **supplementary angles** when the sum of their measures is 180°. Find the measures of the supplementary angles shown here.

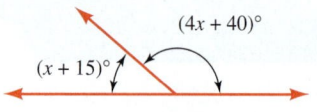

39. RENTALS The owners of an apartment building rent 1-, 2-, and 3-bedroom units. They rent equal numbers of each, with the monthly rents given in the table. If the total monthly income is $36,550, how many of each type of unit are there?

Unit	Rent
One-bedroom	$550
Two-bedroom	$700
Three-bedroom	$900

40. WAREHOUSING A store warehouses 40 more portables than big-screen TV sets, and 15 more consoles than big-screen sets. Storage costs for the different TV sets are shown in the table. If storage costs $276 per month, how many big-screen sets are in stock?

Type of TV	Monthly cost
Portable	$1.50
Console	$4.00
Big-screen	$7.50

41. SOFTWARE Three software applications are priced as shown. Spreadsheet and database programs sold in equal numbers, but 15 more word processing applications were sold than the other two combined. If the three applications generated sales of $72,000, how many spreadsheets were sold?

Software	Price
Spreadsheet	$150
Database	$195
Word processing	$210

42. INVENTORIES With summer approaching, the number of air conditioners sold is expected to be double that of stoves and refrigerators combined. Stoves sell for $350, refrigerators for $450, and air conditioners for $500, and sales of $56,000 are expected. If stoves and refrigerators sell in equal numbers, how many of each appliance should be stocked?

43. INVESTMENTS Equal amounts are invested in each of three accounts paying 7%, 8%, and 10.5% annually. If one year's combined interest income is $1,249.50, how much is invested in each account?

44. RETIREMENT A professor wants to supplement her pension with investment interest. If she invests $28,000 at 6% interest, how much more would she have to invest at 7% to achieve a goal of $3,500 per year in supplemental income?

45. INVESTMENT PLANS A financial planner recommends a plan for a client who has $65,000 to invest. (See the chart.) At the end of the presentation, the client asks, "How much will be invested at each rate?" Answer this question using the given information.

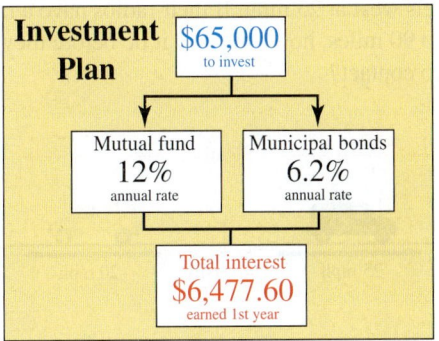

46. TAXES On January 2, 2003, Terrell Washington opened two savings accounts. At the end of the year his bank mailed him the form shown below for income tax purposes. If a total of $15,000 was initially deposited and if no further deposits or withdrawals were made, how much money was originally deposited in account number 721?

USA HOME SAVINGS	Copy B For Recipient Interest Income
This is important tax information and is being furnished to the Internal Revenue Service.	OMB No. 1545-0112 **2003** Form 1099–iNT
RECIPIENT'S name **TERRELL WASHINGTON**	

Acct. Number	Annual Percent Yield	Early Withdrawal Penalty
822	5%	.00
721	4.5%	.00

Total Interest Income 720.00

47. FINANCIAL PLANNING   A plumber has a choice of two investment plans:

- An insured fund that pays 11% interest
- A risky investment that pays a 13% return

If the same amount invested at the higher rate would generate an extra $150 per year, how much does the plumber have to invest?

48. INVESTMENTS   The amount of annual interest earned by $8,000 invested at a certain rate is $200 less than $12,000 would earn at a rate 1% lower. At what rate is the $8,000 invested?

49. TORNADOES   During a storm, two teams of scientists leave a university at the same time in specially designed vans to search for tornadoes. The first team travels east at 20 mph and the second travels west at 25 mph. If their radios have a range of up to 90 miles, how long will it be before they lose radio contact?

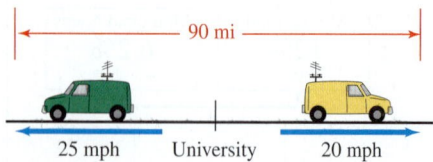

90 mi

25 mph   University   20 mph

50. SEARCH AND RESCUE   Two search-and-rescue teams leave base at the same time looking for a lost boy. The first team, on foot, heads north at 2 mph and the other, on horseback, south at 4 mph. How long will it take them to search a combined distance of 21 miles between them?

51. AIR TRAFFIC CONTROL   An airliner leaves Berlin, Germany, headed for Montreal, Canada, flying at an average speed of 450 mph. At the same time, an airliner leaves Montreal headed for Berlin, averaging 500 mph. If the airports are 3,800 miles apart, when will the air traffic controllers have to make the pilots aware that the planes are passing each other?

52. SPEED OF TRAINS   Two trains are 330 miles apart, and their speeds differ by 20 mph. Find the speed of each train if they are traveling toward each other and will meet in 3 hours.

53. ROAD TRIPS   A car averaged 40 mph for part of a trip and 50 mph for the remainder. If the 5-hour trip covered 210 miles, for how long did the car average 40 mph?

54. CYCLING   A cyclist leaves his training base for a morning workout, riding at the rate of 18 mph. One hour later, his support staff leaves the base in a car going 45 mph in the same direction. How long will it take the support staff to catch up with the cyclist?

55. PHOTOGRAPHIC CHEMICALS   A photographer wishes to mix 2 liters of a 5% acetic acid solution with a 10% solution to get a 7% solution. How many liters of 10% solution must be added?

56. SALT SOLUTIONS   How many gallons of a 3% salt solution must be mixed with 50 gallons of a 7% solution to obtain a 5% solution?

57. ANTISEPTIC SOLUTIONS   A nurse wants to add water to 30 ounces of a 10% solution of benzalkonium chloride to dilute it to an 8% solution. How much water must she add?   (*Hint:* Water is 0% benzalkonium chloride.)

58. MAKING CHEESE   To make low-fat cottage cheese, milk containing 4% butterfat is mixed with milk containing 1% butterfat to obtain 15 gallons of a mixture containing 2% butterfat. How many gallons of the richer milk must be used?

59. MIXING FUELS   How many gallons of fuel costing $1.15 per gallon must be mixed with 20 gallons of a fuel costing $0.85 per gallon to obtain a mixture costing $1 per gallon?

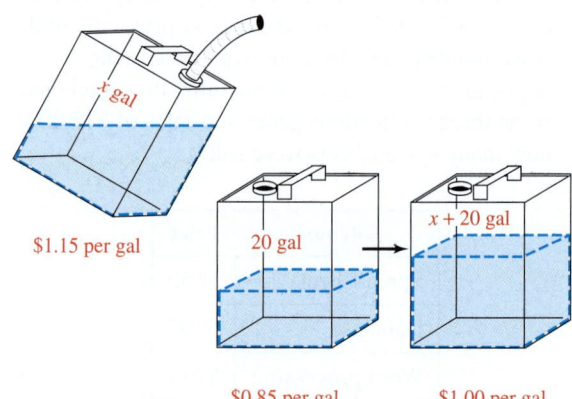

$1.15 per gal   20 gal $x + 20$ gal

$0.85 per gal $1.00 per gal

60. MIXING PAINT   Paint costing $19 per gallon is to be mixed with $3-per-gallon thinner to make 16 gallons of a paint that can be sold for $14 per gallon. How much paint thinner should be used?

61. BLENDING LAWN SEED   A store sells bluegrass seed for $6 per pound and ryegrass seed for $3 per pound. How much ryegrass must be mixed with 100 pounds of bluegrass to obtain a blend that will sell for $5 per pound?

62. BLENDING COFFEE   A store sells regular coffee for $4 a pound and gourmet coffee for $7 a pound. To get rid of 40 pounds of the gourmet coffee, a shopkeeper makes a blend to put on sale for $5 a pound. How many pounds of regular coffee should he use?

63. MIXING CANDY   Lemon drops worth $1.90 per pound are to be mixed with jelly beans that cost $1.20 per pound to make 100 pounds of a mixture worth $1.48 per pound. How many pounds of each candy should be used?

64. SNACK FOODS   A bag of peanuts is worth $.30 less than a bag of cashews. Equal amounts of peanuts and cashews are used to make 40 bags of a mixture that sells for $1.05 per bag. How much is a bag of cashews worth?

WRITING

65. Create a mixture problem of your own, and solve it.

66. Is it possible to mix a 10% sugar solution with a 20% sugar solution to get a 30% sugar solution?

67. A car travels at 60 mph for 15 minutes. Why can't we multiply the rate, 60, and the time, 15, to find the distance traveled by the car?

68. Create a geometry problem that could be answered by solving the equation $2w + 2(w + 5) = 26$.

REVIEW   Remove parentheses.

69. $-25(2x - 5)$

70. $-12(3a + 4b - 32)$

71. $-(-3x - 3)$

72. $\dfrac{1}{2}(4b - 8)$

73. $(4y - 4)4$

74. $3(5t + 1)2$

CHALLENGE PROBLEMS

75. EVAPORATION   How much water must be boiled away to increase the concentration of 300 milliliters of a 2% salt solution to a 3% salt solution?

76. TESTING   A teacher awarded 4 points for each correct answer and deducted 2 points for each incorrect answer when grading a 50-question true-false test. A student scored 56 points on the test, and did not leave any questions unanswered. How many questions did the student answer correctly?

2.7   Solving Inequalities

- Inequalities and Solutions
- Graphing Inequalities and Interval Notation
- Solving Inequalities
- Graphing Compound Inequalities
- Solving Compound Inequalities
- Applications

In our daily lives, we often speak of one value being greater than or less than another value.

- To melt ice, the temperature must be *greater than* 32° F.
- An airplane is rated to fly at altitudes that are *less than* 36,000 feet.
- To earn a B, a student needed a final exam score of *at least* 80%.

In mathematics, we use *inequalities* to show that one expression is greater than or less than another expression.

■ INEQUALITIES AND SOLUTIONS

An **inequality** is a statement that contains one of the following symbols.

Inequality Symbols	$<$ is less than	$>$ is greater than
	$\leq$ is less than or equal to	$\geq$ is greater than or equal to

Some examples of inequalities are:

$$2 < 3, \qquad 1.1 \geq -2.7, \qquad x + 1 > 5, \qquad \text{and} \qquad 5(x + 1) \leq 2(x - 3)$$

An inequality may be true, false, or neither true nor false.

- $9 \geq 9$ is true because $9 = 9$.
- $37 < 24$ is false.
- $x + 1 > 5$ is neither true nor false because we don't know what number x represents.

An inequality that contains a variable can be made true or false, depending on the number that we substitute for the variable. If we substitute 10 for x in $x + 1 > 5$, the resulting inequality is true: $10 + 1 > 5$. If we substitute 1 for x, the result is false: $1 + 1 > 5$. A number that makes an inequality true is called a **solution** of the inequality. Therefore, 10 is a solution of $x + 1 > 5$ and 1 is not.

EXAMPLE 1

ELEMENTARY & INTERMEDIATE
Algebra *f(x)* **Now**™

Is 9 a solution of $2x + 4 \leq 21$?

Solution   We substitute 9 for x in the inequality and evaluate the left-hand side. If 9 is a solution, we will obtain a true statement.

$$2x + 4 \leq 21$$
$$2(9) + 4 \stackrel{?}{\leq} 21$$
$$18 + 4 \stackrel{?}{\leq} 21$$
$$22 \leq 21$$

The statement $22 \leq 21$ is false because neither $22 < 21$ nor $22 = 21$ is true. Therefore, 9 is not a solution.

Self Check 1   Is 2 a solution of $3x - 1 \geq 0$?

■ GRAPHING INEQUALITIES AND INTERVAL NOTATION

The **solution set** of an inequality is the set of all of its solutions. Some solution sets are easy to determine. For example, if we replace the variable in $x > -3$ with a number greater than -3, the resulting inequality will be true. Because there are infinitely many real numbers greater than -3, it follows that $x > -3$ has infinitely many solutions. We can illustrate the solution set on a number line by **graphing the inequality.**

To graph $x > -3$, we shade all the points on the number line that are to the right of -3. We use a shaded arrowhead to show that the solutions continue indefinitely to the right. A **parenthesis** or an **open circle** is drawn at the endpoint -3 to indicate that -3 is not part of the graph.

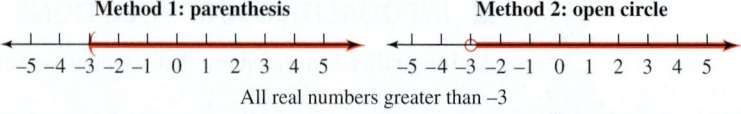

Method 1: parenthesis   Method 2: open circle

All real numbers greater than -3

Graphs of inequalities are **intervals** on the number line. The graph of $x > -3$ can be expressed in **interval notation** as $(-3, \infty)$. Again, the left parenthesis indicates that -3 is not included in the interval. The **infinity symbol** ∞ indicates that the interval continues without end to the right.

EXAMPLE 2   Graph: $x \leq 2$.

Solution   If we replace x with a number less than or equal to 2, the resulting inequality will be true. To graph the solution set, we shade the point 2 and all points to the left of 2 on the number line. A **bracket** or a **closed circle** is drawn at the endpoint 2 to indicate that 2 is part of the graph.

Notation

Since we use parentheses and brackets in interval notation, we will use them to graph inequalities. Note that parentheses, not brackets, are written next to infinity symbols:

$(-3, \infty)$ $(-\infty, 2]$

Method 1: bracket

$$-5 \; -4 \; -3 \; -2 \; -1 \;\; 0 \;\; 1 \;\; 2 \;\; 3 \;\; 4 \;\; 5$$

Method 2: closed circle

$$-5 \; -4 \; -3 \; -2 \; -1 \;\; 0 \;\; 1 \;\; 2 \;\; 3 \;\; 4 \;\; 5$$

All real numbers less than or equal to 2

The interval is written as $(-\infty, 2]$. The bracket indicates that 2 is included in the interval. The **negative infinity symbol** $-\infty$ shows that the interval continues indefinitely to the left.

Self Check 2   Graph: $x \geq 0$.

■ SOLVING INEQUALITIES

To **solve an inequality** means to find all values of the variable that make the inequality true. Inequalities are solved by isolating the variable on one side. We will use the following properties of inequality to do this.

Addition and Subtraction Properties of Inequality	Adding the same number to, or subtracting the same number from, both sides of an inequality does not change the solutions.
	
	For any real numbers a, b, and c,
	
	    If $a < b$, then $a + c < b + c$.
	    If $a < b$, then $a - c < b - c$.
	
	Similar statements can be made for the symbols $\leq$, $>$, or $\geq$.

After applying one of these properties, the resulting inequality is equivalent to the original one. Like equivalent equations, **equivalent inequalities** have the same solution set.

EXAMPLE 3   Solve: $x + 3 > 2$.

Solution   We can use the subtraction property of inequality to isolate x on the left-hand side.

$$x + 3 > 2$$
$$x + 3 - 3 > 2 - 3 \qquad \text{To undo the addition of 3, subtract 3 from both sides.}$$
$$x > -1$$

All real numbers greater than -1 are solutions of $x + 3 > 2$. The solution set can be written as $(-1, \infty)$. The graph of the solution set is shown below.

Since there are infinitely many solutions, we cannot check each of them. As an informal check, we can pick two numbers in the graph, say 1 and 30, substitute each for x in the original inequality, and see whether true statements result.

Check:
$x + 3 > 2$ $x + 3 > 2$

$1 + 3 \overset{?}{>} 2$ Substitute 1 for x. $30 + 3 \overset{?}{>} 2$ Substitute 30 for x.

$4 > 2$ This is a true inequality. $33 > 2$ This is a true inequality.

The solution set appears to be correct.

Self Check 3 Solve: $x - 3 < -2$. Write the solution set in interval notation and graph it.

As with equations, there are properties for multiplying and dividing both sides of an inequality by the same number. To develop what is called *the multiplication property of inequality,* consider the true statement $2 < 5$. If both sides are multiplied by a positive number, such as 3, another true inequality results.

$2 < 5$

$3 \cdot 2 < 3 \cdot 5$ Multiply both sides by 3.

$6 < 15$ This is a true inequality.

However, if we multiply both sides of $2 < 5$ by a negative number, such as -3, the direction of the inequality symbol is reversed to produce another true inequality.

$2 < 5$

$-3 \cdot 2 > -3 \cdot 5$ Multiply both sides by the negative number -3 and reverse the direction of the inequality.

$-6 > -15$ This is a true inequality.

The inequality $-6 > -15$ is true because -6 is to the right of -15 on the number line.

Dividing both sides of an inequality by the same negative number also requires that the direction of the inequality symbol be reversed.

$-4 < 6$ A true inequality.

$\dfrac{-4}{-2} > \dfrac{6}{-2}$ Divide both sides by -2 and change $<$ to $>$.

$2 > -3$ This is a true inequality.

These examples illustrate the multiplication and division properties of inequality.

Multiplication and Division Properties of Inequality

Multiplying or dividing both sides of an inequality by the same positive number does not change the solutions.

For any real numbers a, b, and c, where c is positive,

$$\text{If } a < b, \text{ then } ac < bc. \qquad \text{If } a < b, \text{ then } \frac{a}{c} < \frac{b}{c}.$$

If we multiply or divide both sides of an inequality by a negative number, the direction of the inequality symbol must be reversed for the inequalities to have the same solutions.

For any real numbers a, b, and c, where c is negative,

$$\text{If } a < b, \text{ then } ac > bc. \qquad \text{If } a < b, \text{ then } \frac{a}{c} > \frac{b}{c}.$$

Similar statements can be made for the symbols $\leq$, $>$, or $\geq$.

EXAMPLE 4

Solve: $-\dfrac{3}{2}t \geq -12$.

ELEMENTARY & INTERMEDIATE
Algebra *f(x)* **Now**™

Solution   To undo the multiplication by $-\frac{3}{2}$, we multiply both sides by the reciprocal, which is $-\frac{2}{3}$.

$$-\frac{3}{2}t \geq -12$$

$$-\frac{2}{3}\left(-\frac{3}{2}t\right) \leq -\frac{2}{3}(-12) \qquad \textcolor{red}{\text{Multiply both sides by } -\tfrac{2}{3}. \text{ Change } \geq \text{ to } \leq.}$$

$$t \leq 8 \qquad \textcolor{red}{\text{Do the multiplications.}}$$

The solution set is $(-\infty, 8]$ and it is graphed as shown.

Self Check 4   Solve: $-\dfrac{h}{20} \leq 10$. Write the solution set in interval notation and graph it.

EXAMPLE 5

Solve: $-5 > 3x + 7$.

ELEMENTARY & INTERMEDIATE
Algebra *f(x)* **Now**™

Solution

$$-5 > 3x + 7$$

$$-5 \textcolor{red}{- 7} > 3x + 7 \textcolor{red}{- 7} \qquad \textcolor{red}{\text{To undo the addition of 7, subtract 7 from both sides.}}$$

$$-12 > 3x \qquad \textcolor{red}{\text{Do the subtractions.}}$$

$$\frac{-12}{\textcolor{red}{3}} > \frac{3x}{\textcolor{red}{3}} \qquad \textcolor{red}{\text{To undo the multiplication by 3, divide both sides by 3.}}$$

$$-4 > x \qquad \textcolor{red}{\text{Do the divisions.}}$$

To find the solution set, it is useful to write $-4 > x$ with the variable on the left-hand side. If -4 is greater than x, then x must be less than -4.

$$x < -4$$

The solution set is $(-\infty, -4)$ whose graph is shown.

Caution

In Example 5, don't be confused by the negative number on the left-hand side. We didn't reverse the $>$ symbol because we divided both sides by *positive* 3.

$$\frac{-12}{3} > \frac{3x}{3}$$

Self Check 5   Solve: $-13 < 2r - 7$. Write the solution set in interval notation and graph it.

EXAMPLE 6

Solve: $5.1 - 3a < 19.5$.

ELEMENTARY &
INTERMEDIATE
Algebra $f(x)$ **Now**™   **Solution**

$$5.1 - 3a < 19.5$$

$5.1 - 3a \mathbf{- 5.1} < 19.5 \mathbf{- 5.1}$ To isolate $-3a$ on the left-hand side, subtract 5.1 from both sides.

$-3a < 14.4$ Do the subtractions.

$\dfrac{-3a}{\mathbf{-3}} > \dfrac{14.4}{\mathbf{-3}}$ To undo the multiplication by -3, divide both sides by -3. Since we are dividing by a negative number, we reverse the direction of the $<$ symbol.

$a > -4.8$ Do the divisions.

The solution set is $(-4.8, \infty)$ whose graph is shown.

Self Check 6   Solve: $-9n + 1.8 > -17.1$. Write the solution set in interval notation and graph it.

EXAMPLE 7

Solve: $8(y + 1) \geq 2(y - 4) + y$.

ELEMENTARY &
INTERMEDIATE
Algebra $f(x)$ **Now**™   **Solution**

To solve this inequality, we follow the same strategy used for solving equations.

$$8(y + 1) \geq 2(y - 4) + y$$

$8y + 8 \geq 2y - 8 + y$ Distribute the multiplication by 8 and by 2.

$8y + 8 \geq 3y - 8$ Combine like terms: $2y + y = 3y$.

$8y + 8 \mathbf{- 3y} \geq 3y - 8 \mathbf{- 3y}$ To eliminate $3y$ from the right-hand side, subtract $3y$ from both sides.

$5y + 8 \geq -8$ Combine like terms on both sides.

$5y + 8 \mathbf{- 8} \geq -8 \mathbf{- 8}$ To undo the addition of 8, subtract 8 from both sides.

$5y \geq -16$ Do the subtractions.

$\dfrac{5y}{\mathbf{5}} \geq \dfrac{-16}{\mathbf{5}}$ To undo the multiplication by 5, divide both sides by 5.

$y \geq -\dfrac{16}{5}$

The solution set is $\left[-\dfrac{16}{5}, \infty \right)$. To graph it, we note that

$-\dfrac{16}{5} = -3\dfrac{1}{5}$.

Self Check 7   Solve: $5(b - 2) \geq -(b - 3) + 2b$. Write the solution set in interval notation and graph it.

■ GRAPHING COMPOUND INEQUALITIES

The Language of Algebra

The word *compound* means made up of individual parts. For example, a *compound* inequality has three parts. In writing classes, students learn about *compound* sentences.

Two inequalities can be combined into a **compound inequality** to show that an expression lies between two fixed values. For example, $-2 < x < 3$ is a combination of

$$-2 < x \quad \text{and} \quad x < 3$$

It indicates that x is greater than -2 and that x is also less than 3. The solution set of $-2 < x < 3$ consists of all numbers that lie between -2 and 3, and we write it as $(-2, 3)$. The graph of the compound inequality is shown below.

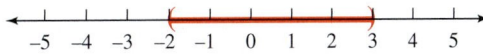

EXAMPLE 8

Graph: $-4 \leq x < 0$.

Solution

If we replace the variable in $-4 \leq x < 0$ with a number between -4 and 0, including -4, the resulting compound inequality will be true. Therefore, the solution set is $[-4, 0)$. To graph the interval, we draw a bracket at -4, a parenthesis at 0, and shade in between.

To check, we pick a number in the graph, such as -2, and see whether it satisfies the inequality. Since $-4 \leq -2 < 0$ is true, the answer appears to be correct.

Self Check 8

Graph $-2 \leq x < 1$ and write the solution set in interval notation.

■ SOLVING COMPOUND INEQUALITIES

To solve compound inequalities, we use the same methods used for solving equations. However, we will apply the properties of inequality to all *three* parts of the inequality.

EXAMPLE 9

Solve: $-4 < 2(x - 1) \leq 4$.

ELEMENTARY &
INTERMEDIATE
Algebra *f(x)* **Now**™

Solution

$$-4 < 2(x - 1) \leq 4$$

$$-4 < 2x - 2 \leq 4 \qquad \text{Distribute the multiplication by 2.}$$

$$-4 + 2 < 2x - 2 + 2 \leq 4 + 2 \qquad \text{To undo the subtraction of 2, add 2 to all three parts.}$$

$$-2 < 2x \leq 6 \qquad \text{Do the additions.}$$

$$\frac{-2}{2} < \frac{2x}{2} \leq \frac{6}{2} \qquad \text{To isolate } x \text{, we undo the multiplication by 2 by dividing all three parts by 2.}$$

$$-1 < x \leq 3$$

The solution set is $(-1, 3]$ and its graph is shown.

Self Check 9

Solve: $-6 \leq 3(t + 2) \leq 6$. Write the solution set in interval notation and graph it.

■ APPLICATIONS

When solving problems, phrases such as "not more than," or "should exceed" suggest that an *inequality* should be written instead of an *equation*.

EXAMPLE 10

Grades.   A student has scores of 72%, 74%, and 78% on three exams. What percent score does he need on the last exam to earn no less than a grade of B (80%)?

Analyze the Problem   We know three scores. We are to find what the student must score on the last exam to earn at least a B grade.

Form an Inequality   We can let x = the score on the fourth (and last) exam. To find the average grade, we add the four scores and divide by 4. To earn no less than a grade of B, the student's average must be greater than or equal to 80%.

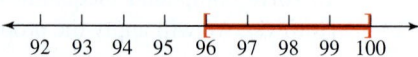

The average of the four grades	must be greater than or equal to	80.
$\dfrac{72 + 74 + 78 + x}{4}$	$\geq$	80

Solve the Inequality   We can solve this inequality for x.

$\dfrac{224 + x}{4} \geq 80$ Combine like terms in the numerator: $72 + 74 + 78 = 224$.

$224 + x \geq 320$ To clear the inequality of the fraction, multiply both sides by 4.

$x \geq 96$ To undo the addition of 224, subtract 224 from both sides.

State the Conclusion   To earn a B, the student must score 96% or better on the last exam. Assuming the student cannot score higher than 100% on the exam, the solution set is written as [96, 100]. The graph is shown below.

$$\xleftarrow{\quad}\underset{92\quad93\quad94\quad95\quad96\quad97\quad98\quad99\quad100}{\rule{6cm}{0.4pt}}\xrightarrow{\quad}$$

Check the Result   Pick some numbers in the interval, and verify that the average of the four scores will be 80% or greater.

Answers to Self Checks   1. yes   2. $[0, \infty)$ 3. $x < 1, (-\infty, 1)$

4. $h \geq -200, [-200, \infty)$

5. $r > -3, (-3, \infty)$ 6. $n < 2.1, (-\infty, 2.1)$

7. $b \geq \dfrac{13}{4}, \left[\dfrac{13}{4}, \infty\right)$ 8. $[-2, 1)$

9. $-4 \leq t \leq 0, [-4, 0]$

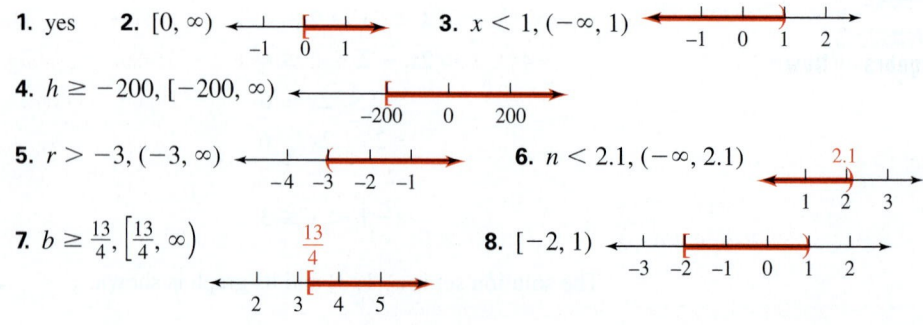

VOCABULARY   **Fill in the blanks.**

1. An _____ is a statement that contains one of the following symbols: $>$, $\geq$, $<$, or $\leq$.

2. A number that makes an inequality true is called a _____ of the inequality. The solution _____ of an inequality is the set of all solutions.

3. To _____ an inequality means to find all the values of the variable that make the inequality true.

4. Graphs of inequalities are _____ on the number line.

5. The solution set of $x > 2$ can be expressed in _____ notation as $(2, \infty)$.

6. The inequality $-4 < x \leq 10$ is an example of a _____ inequality.

CONCEPTS

7. Decide whether each statement is true or false.
   a. $35 \geq 34$
   b. $-16 \leq -17$
   c. $\dfrac{3}{4} \leq 0.75$
   d. $-0.6 \geq -0.5$

8. Decide whether each number is a solution of $3x + 7 < 4x - 2$.
   a. 12
   b. -6
   c. 0
   d. 9

9. Write each inequality so that the inequality symbol points in the opposite direction.
   a. $17 \geq -2$
   b. $32 < x$

10. The solution set of an inequality is graphed as shown.

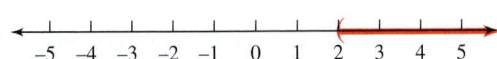

Which of the following numbers, when substituted for the variable in that inequality, would make it true?

$$3 \qquad -3 \qquad 2 \qquad 4.5$$

Fill in the blanks.

11. a. Adding the same number to, or subtracting the _____ number from, both sides of an inequality does not change the solutions.

    b. Multiplying or dividing both sides of an inequality by the same _____ number does not change the solutions.

12. If we multiply or divide both sides of an inequality by a _____ number, the direction of the inequality symbol must be reversed for the inequalities to have the same solutions.

13. Solve $x + 2 > 10$ and give the solution set:
    a. using a graph
    b. using interval notation
    c. in words

14. The solution set of a compound inequality is graphed as shown.

Which of the following numbers, when substituted for the variable in that compound inequality, would make it true?

$$3 \qquad -3 \qquad 2.75 \qquad -3.5$$

15. To solve compound inequalities, the properties of inequalities are applied to all _____ parts of the inequality.

16. Solve $-4 < 2x < 12$ and give the solution set:
    a. using a graph
    b. using interval notation
    c. in words

NOTATION   **Fill in the blanks.**

17. a. The symbol $<$ means "_____," and the symbol $>$ means "_____."

    b. The symbol $\geq$ means "_____ or equal to," and the symbol $\leq$ means "is less than _____."

18. In the interval $[4, 8)$, the endpoint 4 is _____, but the endpoint 8 is not included.

19. Give an example of each symbol: bracket, parenthesis, infinity, negative infinity.

20. Tell what is wrong with the notation that is used.
    a. The graph of the solution set for $x > 1$:

b. The solution set for $x > 5$ is $(5, \infty]$.

c. The solution set for $x < 7$ is $(7, -\infty)$.

d. $5 > 2 < 10$ is a true compound inequality.

Complete the solution to solve each inequality.

21. $4x - 5 \geq 7$

$4x - 5 + \boxed{} \geq 7 + \boxed{}$

$4x \geq \boxed{}$

$\dfrac{4x}{\boxed{}} \geq \dfrac{12}{\boxed{}}$

$x \geq 3$

Solution set: $\left[\,\boxed{}\,, \infty\right)$

22. $-6x > 12$

$\dfrac{-6x}{\boxed{}} \quad \dfrac{12}{-6}$

$x < \boxed{}$

Solution set: $\left(\boxed{}\,, -2\right)$

PRACTICE   Graph each inequality and describe the graph using interval notation.

23. $x < 5$ **24.** $x \geq -2$

25. $-3 < x \leq 1$ **26.** $-1 \leq x \leq 3$

Write the inequality that is represented by each graph. Then describe the graph using interval notation.

27.
-1

28.
2

29.
$-7 \quad 2$

30.
$-3 \quad 1$

Solve each inequality. Write the solution set in interval notation and graph it.

31. $x + 2 > 5$ **32.** $x + 5 \geq 2$

33. $3 + x < 2$ **34.** $5 + x > 3$

35. $g - 30 \geq -20$ **36.** $h - 18 \leq -3$

37. $\dfrac{2}{3}x \geq 2$ **38.** $\dfrac{3}{4}x < 3$

39. $\dfrac{y}{4} + 1 \leq -9$ **40.** $\dfrac{r}{8} - 7 \geq -8$

41. $7x - 1 > 5$ **42.** $3x + 10 \leq 5$

43. $0.5 \geq 2x - 0.3$ **44.** $0.8 > 7x - 0.04$

45. $-30y \leq -600$ **46.** $-6y \geq -600$

47. $-\dfrac{7}{8}x \leq 21$ **48.** $-\dfrac{3}{16}x \geq -9$

49. $-1 \leq -\dfrac{1}{2}n$ **50.** $-3 \geq -\dfrac{1}{3}t$

51. $\dfrac{m}{-42} - 1 > -1$ **52.** $\dfrac{a}{-25} + 3 < 3$

53. $-x - 3 \leq 7$ **54.** $-x - 9 > 3$

55. $-3x - 7 > -1$ **56.** $-5x + 7 \leq 12$

57. $-4x + 6 > 17$ **58.** $-3x - 0.5 < 0.4$

59. $2x + 9 \leq x + 8$ **60.** $3x + 7 \leq 4x - 2$

61. $9x + 13 \geq 8x$ **62.** $7x - 16 < 6x$

63. $8x + 4 > -(3x - 4)$ **64.** $7x + 6 \geq -(x - 6)$

65. $0.4x + 0.4 \leq 0.1x + 0.85$

66. $0.05 - 0.5x \leq -0.7 - 0.8x$

67. $7 < \dfrac{5}{3}a - 3$ **68.** $5 > \dfrac{7}{2}a - 9$

69. $7 - x \leq 3x - 2$ **70.** $9 - 3x \geq 6 + x$

71. $8(5 - x) \leq 10(8 - x)$ **72.** $17(3 - x) \geq 3 - 13x$

73. $\dfrac{1}{2} + \dfrac{n}{5} > \dfrac{3}{4}$ **74.** $\dfrac{1}{3} + \dfrac{c}{5} > -\dfrac{3}{2}$

75. $-\dfrac{2}{3} \geq \dfrac{2y}{3} - \dfrac{3}{4}$ **76.** $-\dfrac{2}{9} \geq \dfrac{5x}{6} - \dfrac{1}{3}$

77. $\dfrac{6x + 1}{4} \leq x + 1$ **78.** $\dfrac{3x - 10}{5} \leq x + 4$

79. $\dfrac{5}{2}(7x - 15) + x \geq \dfrac{13}{2}x - \dfrac{3}{2}$

80. $\dfrac{5}{3}(x + 1) \leq -x + \dfrac{2}{3}$

Solve each compound inequality. Write the solution set in interval notation and graph it.

81. $2 < x - 5 < 5$ **82.** $-8 < t - 8 < 8$

83. $0 \leq x + 10 \leq 10$
84. $-9 \leq x + 8 < 1$

85. $-3 \leq \dfrac{c}{2} \leq 5$ **86.** $-12 < \dfrac{b}{3} < 0$

87. $3 \leq 2x - 1 < 5$ **88.** $4 < 3x - 5 \leq 7$

89. $4 < -2x < 10$ **90.** $-4 \leq -4x < 12$

91. $0 < 10 - 5x \leq 15$ **92.** $-18 \leq 9(x - 5) < 27$

Solve each inequality. Write the solution set in interval notation and graph it.

93. $9(0.05 - 0.3x) + 0.162 \leq 0.081 + 15x$

94. $-1,630 \leq \dfrac{b + 312,451}{47} < 42,616$

APPLICATIONS

95. GRADES   A student has test scores of 68%, 75%, and 79% in a government class. What must she score on the last exam to earn a B (80% or better) in the course?

96. OCCUPATIONAL TESTING   Before taking on a client, an employment agency requires the applicant to average at least 70% on a battery of four job skills tests. If an applicant scored 70%, 74%, and 84% on the first three exams, what must he score on the fourth test to maintain a 70% or better average?

97. FLEET AVERAGES   A car manufacturer produces three models in equal quantities. One model has an economy rating of 17 miles per gallon, and the second model is rated for 19 mpg. If government regulations require the manufacturer to have a fleet average of at least 21 mpg, what economy rating is required for the third model?

98. SERVICE CHARGES   When the average daily balance of a customer's checking account falls below $500 in any week, the bank assesses a $5 service charge. The table shows the daily balances of one customer. What must Friday's balance be to avoid the service charge?

Day	Balance
Monday	$540.00
Tuesday	$435.50
Wednesday	$345.30
Thursday	$310.00

99. GEOMETRY   The perimeter of an equilateral triangle is at most 57 feet. What could the length of a side be?   (*Hint:* All three sides of an equilateral triangle are equal.)

100. GEOMETRY   The perimeter of a square is no less than 68 centimeters. How long can a side be?

101. COUNTER SPACE   In a large discount store, a rectangular counter is being built for the customer service department. If designers have determined that the outside perimeter of the counter (shown in red) needs to be at least 150 feet, determine the acceptable values for x.

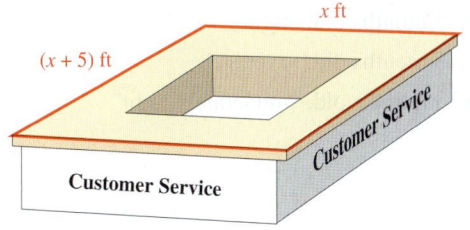

102. NUMBER PUZZLES What numbers satisfy the condition: Four more than three times the number is at most 10?

103. SAFETY CODES The illustration shows the acceptable and preferred angles of "pitch" or slope for ladders, stairs, and ramps. Use a compound inequality to describe each safe-angle range.
  a. ramps or inclines
  b. stairs
  c. preferred range for stairs
  d. ladders with cleats

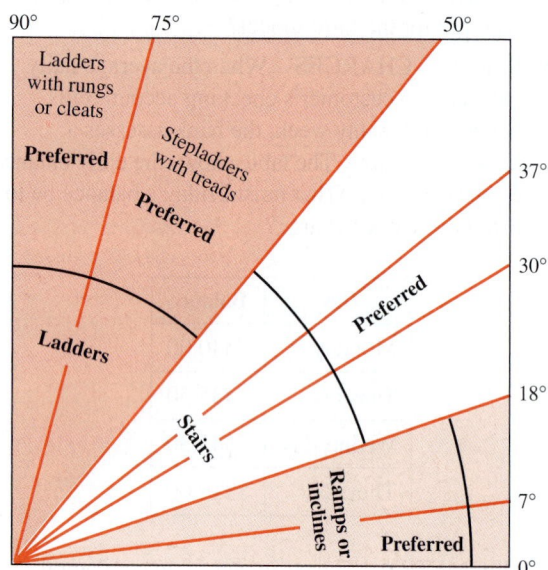

104. COMPARING TEMPERATURES To hold the temperature of a room between 19° and 22° Celsius, what Fahrenheit temperatures must be maintained? (*Hint:* Fahrenheit temperature (*F*) and Celsius temperature (*C*) are related by the formula $C = \frac{5}{9}(F - 32)$.)

105. WEIGHT CHART The graph is used to classify the weight of a baby boy from birth to 1 year. Estimate the weight range *w* for boys in the following classifications, using a compound inequality:
  a. 10 months old, "heavy"
  b. 5 months old, "light"
  c. 8 months old, "average"
  d. 3 months old, "moderately light"

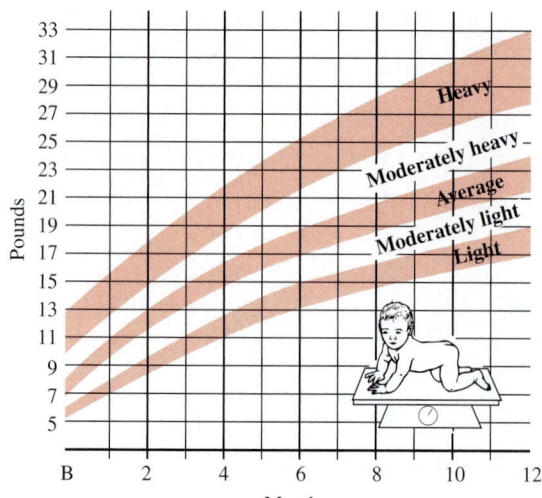

Based on data from *Better Homes and Gardens Baby Book* (Meredith Corp., 1969)

106. NUMBER PUZZLE What *whole* numbers satisfy the condition: Twice the number decreased by 1 is between 50 and 60?

WRITING

107. Explain why multiplying both sides of an inequality by a negative number reverses the direction of the inequality.

108. Explain the use of parentheses and brackets for graphing intervals.

REVIEW   Evaluate each expression.

109. -5^3 **110.** $(-3)^4$

Complete each table.

111.

x	$x^2 - 3$
-2	
0	
3	

112.

x	$\frac{x}{3} + 2$
-6	
0	
12	

CHALLENGE PROBLEMS

113. Solve the inequality. Write the solution set in interval notation and graph it.

$$3 - x < 5 < 7 - x$$

114. Use a guess-and-check approach to solve $\frac{1}{x} > 1$.

Write the solution set in interval notation and graph it.

ACCENT ON TEAMWORK

TRANSLATING KEY WORDS AND PHRASES

Overview: Students often say that the most challenging step of the five-step problem-solving strategy is forming an equation. This activity is designed to make that step easier by improving your translating skills.

Instructions: Form groups of 3 or 4 students. Select one person from your group to record the group's responses. Determine whether addition, subtraction, multiplication, or division is suggested by each of the following words or phrases. Then use the word or phrase in a sentence to illustrate its meaning.

deflate	recede	partition	evaporate	amplify
bisect	augment	hike	erode	boost
annexed	diminish	plummet	upsurge	wane
quadruple	corrode	taper off	trisect	broaden

GEOMETRIC SNACKS

Overview: This activity is designed to improve your ability to identify geometric figures and recall their perimeter, area, and volume formulas.

Instructions: Review the geometric figures and formulas inside the front cover. Then find some snack foods that have the shapes of the figures. For example, tortilla chips can be triangular in shape, and malted milk balls are spheres. If you are unable to find a particular shape already available, make a snack in that shape. Bring your collection of snacks to the next class. Form groups of 3 or 4 students. In your group, discuss the various shapes of the snacks, as well as their respective perimeter, area, and volume formulas.

COMPUTER SPREADSHEETS

Overview: In this activity, you will get some experience working with a spreadsheet.

Instructions: Form groups of 3 or 4 students. Examine the following spreadsheet, which consists of cells named by column and row. For example, 7 is entered in cell B3. In any cell you may enter data or a formula. For each formula in cells D1–D4 and E1–E4, the computer performs a calculation using values entered in other cells and prints the result in place of the formula. Find the value that will be printed in each formula cell. The symbol $*$ means multiply, / means divide, and $\wedge$ means raise to a power.

	A	B	C	D	E
1	-8	20	-6	$= 2*B1 - 3*C1 + 4$	$= B1 - 3*A1\wedge2$
2	39	2	-1	$= A2/(B2 - C2)$	$= B3*B2*C2*2$
3	50	7	3	$= A3/5 + C3\wedge3$	$= 65 - 2*(B3 - 5)\wedge5$
4	6.8	-2.8	-0.5	$= 100*A4 + B4*C4$	$= A4/10 + A3/2*5$

KEY CONCEPT: SIMPLIFY AND SOLVE

Two of the most often used instructions in this book are **simplify** and **solve**. In algebra, we *simplify expressions* and we *solve equations and inequalities.*

To simplify an expression, we write it in a less complicated form. To do so, we apply the rules of arithmetic as well as algebraic concepts such as combining like terms, the distributive property, and the properties of 0 and 1.

To solve an equation or an inequality means to find the numbers that make the equation or inequality true when substituted for its variable. We use the addition, subtraction, multiplication, and division properties of equality or inequality to solve equations and inequalities. Quite often, we must simplify expressions on the left- or right-hand sides of an equation or inequality when solving it.

Use the procedures and the properties that we have studied to simplify the expression in part a and to solve the equation or inequality in part b.

Simplify   **Solve**

1. a. $-3x + 2 + 5x - 10$ **b.** $-3x + 2 + 5x - 10 = 4$

2. a. $4(y + 2) - 3(y + 1)$ **b.** $4(y + 2) = 3(y + 1)$

3. a. $\frac{1}{3}a + \frac{1}{3}a$ **b.** $\frac{1}{3}a + \frac{1}{3} = \frac{1}{2}$

4. a. $-(2x + 10)$ **b.** $-2x \geq -10$

5. a. $\frac{2}{3}(x - 2) - \frac{1}{6}(4x - 8)$ **b.** $\frac{2}{3}(x - 2) - \frac{1}{6}(4x - 8) = 0$

6. In the student's work on the right, where was the mistake made? Explain what the student did wrong.

Simplify: $2(x + 3) - x - 12$.

$$2(x + 3) - x - 12 = 2x + 6 - x - 12$$
$$= x - 6$$
$$0 = x - 6$$
$$0 + 6 = x - 6 + 6$$
$$\boxed{6 = x}$$

CHAPTER REVIEW

ELEMENTARY &
INTERMEDIATE
Algebra $f(x)$ **Now**™

| **SECTION 2.1** | **Solving Equations** |

CONCEPTS

A number that makes an equation a true statement when substituted for the variable is called a *solution* of the equation.

REVIEW EXERCISES

Decide whether the given number is a solution of the equation.

1. $84, x - 34 = 50$

2. $3, 5y + 2 = 12$

3. $-30, \frac{x}{5} = 6$

4. $2, a^2 - a - 1 = 0$

5. $-3, 5b - 2 = 3b - 8$

6. $1, \frac{2}{y + 1} = \frac{12}{y + 1} - 5$

To *solve an equation,* isolate the variable on one side of the equation by undoing the operations performed on it.

Equations with the same solutions are called *equivalent equations.*

If the same number is added to, or subtracted from, both sides of an equation, an equivalent equation results.

If both sides of an equation are multiplied, or divided, by the same nonzero number, an equivalent equation results.

7. Fill in the blanks: To solve $x + 8 = 10$ means to find all the values of the _____ that make the equation a _____ statement.

Solve each equation. Check each result.

8. $x - 9 = 12$

9. $y + 15 = -32$

10. $a + 3.7 = 16.9$

11. $100 = -7 + r$

12. $120 = 15c$

13. $t - \dfrac{1}{2} = \dfrac{1}{2}$

14. $\dfrac{t}{8} = -12$

15. $3 = \dfrac{q}{2.6}$

16. $6b = 0$

17. $\dfrac{x}{14} = 0$

18. GEOMETRY   Find the unknown angle measure.

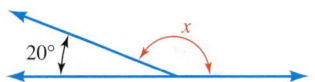

| SECTION 2.2 | **Problem Solving** |

To solve a problem, follow these steps:

1. Analyze the problem.
2. Form an equation.
3. Solve the equation.
4. State the conclusion.
5. Check the result.

Drawing a diagram or creating a table is often helpful in problem solving.

19. WOMEN'S SOCCER   The U.S. National Women's team won the 1999 World Cup. On the 20-player roster were 3 goalkeepers, 6 defenders, 6 midfielders, and a group of forwards. How many forwards were on the team?

20. HISTORIC TOURS   A driving tour of three historic cities is an 858-mile round trip. Beginning in Boston, the drive to Philadelphia is 296 miles. From Philadelphia to Washington, D.C., is another 133 miles. How long will the return trip to Boston be?

21. CHROME WHEELS   Find the measure of the angle between each of the spokes on the wheel shown.

To solve applied percent problems, use the facts of the problem to write a percent sentence of the form:

_____ is ____ % of _____ .

Translate the sentence to mathematical symbols: *is* translates to an = symbol and *of* means multiply. Then, solve the resulting equation.

22. ADVERTISING   In 2001, $231 billion was spent on advertising in the United States. The circle graph gives the individual expenditures in percents. Find the amount of money spent on television advertising. Round to the nearest billion dollars.

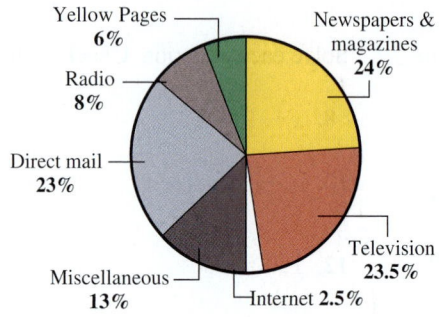

Based on data from *Advertising Age*

23. COST OF LIVING   A retired trucker receives a monthly Social Security check of $764. If she is to receive a 3.5% cost-of-living increase soon, how much larger will her check be?

24. 4.81 is 2.5% of what number?

25. FAMILY BUDGETS   It is recommended that a family pay no more than 30% of its monthly income (after taxes) on housing. If a family has an after-tax income of $1,890 per month and pays $625 in housing costs each month, are they within the recommended range?

To find *percent of increase or decrease,* find what percent the increase or decrease is of the original amount.

26. COLLECTIBLES   A collector of football trading cards paid $6 for a 1984 Dan Marino rookie card several years ago. If the card is now worth $100, what is the percent of increase in the card's value? (Round to the nearest one percent.)

| SECTION 2.3 | **Simplifying Algebraic Expressions** |

To *simplify* an algebraic expression means to write it in less complicated form.

The *distributive property:*
$a(b + c) = ab + ac$
$a(b - c) = ab - ac$
$a(b + c + d)$
$\quad = ab + ac + ad$

Simplify each expression.

27. $-4(7w)$

28. $-3r(-5)$

29. $3(-2x)(-4)$

30. $0.4(5.2f)$

31. $15\left(\dfrac{3}{5}a\right)$

32. $\dfrac{7}{2} \cdot \dfrac{2}{7}r$

Remove parentheses.

33. $5(x + 3)$

34. $-2(2x + 3 - y)$

35. $-(a - 4)$

36. $\dfrac{3}{4}(4c - 8)$

37. $40\left(\dfrac{x}{2} + \dfrac{4}{5}\right)$

38. $2(-3c - 7)(2.1)$

Like terms are terms with exactly the same variables raised to exactly the same powers.

To add or subtract like terms, combine their coefficients and keep the same variables with the same exponents.

Simplify each expression by combining like terms.

39. $8p + 5p - 4p$

40. $-5m + 2 - 2m - 2$

41. $n + n + n + n$

42. $5(p - 2) - 2(3p + 4)$

43. $55.7k - 55.6k$

44. $8a^3 + 4a^3 - 20a^3$

45. $\dfrac{3}{5}w - \left(-\dfrac{2}{5}w\right)$

46. $36\left(\dfrac{1}{9}h - \dfrac{3}{4}\right) + 36\left(\dfrac{1}{3}\right)$

47. Write an algebraic expression in simplified form for the perimeter of the triangle.

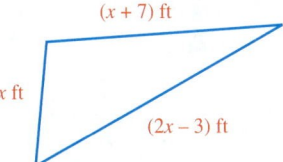

$(x + 7)$ ft

x ft

$(2x - 3)$ ft

SECTION 2.4	**More about Solving Equations**

To solve an equation means to find all the values of the variable that, when substituted for the variable, make a true statement.

An equation that is true for all values of its variable is called an *identity.*

An equation that is not true for any values of its variable is called a *contradiction.*

Solve each equation. Check the result.

48. $5x + 4 = 14$

49. $98.6 - t = 129.2$

50. $\dfrac{n}{5} - 2 = 4$

51. $\dfrac{b - 5}{4} = -6$

52. $5(2x - 4) - 5x = 0$

53. $-2(x - 5) = 5(-3x + 4) + 3$

54. $\dfrac{3}{4} = \dfrac{1}{2} + \dfrac{d}{5}$

55. $-\dfrac{2}{3}f = 4$

56. $\dfrac{3(2 - c)}{2} = \dfrac{-2(2c + 3)}{5}$

57. $\dfrac{b}{3} + \dfrac{11}{9} + 3b = -\dfrac{5}{6}b$

58. $3(a + 8) = 6(a + 4) - 3a$

59. $2(y + 10) + y = 3(y + 8)$

| SECTION 2.5 | **Formulas** |

A *formula* is an equation that is used to state a known relationship between two or more variables.

Retail price: $r = c + m$

Profit: $p = r - c$

Distance: $d = rt$

Temperature:
$C = \frac{5}{9}(F - 32)$

Formulas from geometry:
Square: $P = 4s, A = s^2$

Rectangle: $P = 2l + 2w$, $A = lw$

Triangle: $P = a + b + c$
$A = \frac{1}{2}bh$

Trapezoid:
$P = a + b + c + d$
$A = \frac{1}{2}h(b + d)$

Circle: $D = 2r$
$C = 2\pi r$
$A = \pi r^2$

Rectangular solid:
$V = lwh$

Cylinder: $V = \pi r^2 h$

Pyramid: $V = \frac{1}{3}Bh$
   (*B* is the area of the base)

Cone: $V = \frac{1}{3}\pi r^2 h$

Sphere: $V = \frac{4}{3}\pi r^3$

60. Find the markup on a CD player whose wholesale cost is $219 and whose retail price is $395.

61. One month, a restaurant had sales of $13,500 and made a profit of $1,700. Find the expenses for the month.

62. INDY 500   In 2002, the winner of the Indianapolis 500-mile automobile race averaged 166.499 mph. To the nearest hundredth of an hour, how long did it take him to complete the race?

63. JEWELRY MAKING   Gold melts at about 1,065°C. Change this to degrees Fahrenheit.

64. CAMPING   Find the perimeter of the air mattress.

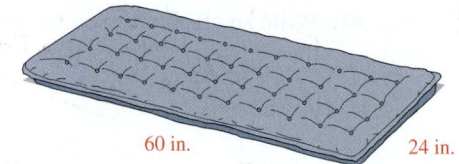

60 in.   24 in.

65. CAMPING   Find the amount of sleeping area on the top surface of the air mattress.

66. Find the area of a triangle with a base 17 meters long and a height of 9 meters.

67. Find the area of a trapezoid with bases 11 inches and 13 inches long and a height of 12 inches.

68. To the nearest hundredth, find the circumference of a circle with a radius of 8 centimeters.

69. To the nearest hundredth, find the area of the circle in Problem 68.

70. CAMPING   Find the approximate volume of the air mattress in Problem 64 if it is 3 inches thick.

71. Find the volume of a 12-foot cylinder whose circular base has a radius of 0.5 feet. Give the result to the nearest tenth.

72. Find the volume of a pyramid that has a square base, measuring 6 feet on a side, and a height of 10 feet.

73. HALLOWEEN   After being cleaned out, a spherical-shaped pumpkin has an inside diameter of 9 inches. To the nearest hundredth, what is its volume?

Solve each formula for the required variable.

74. $A = 2\pi rh$ for h

75. $A - BC = \dfrac{G - K}{3}$ for G

76. $a^2 + b^2 = c^2$ for b^2

77. $4y - 16 = 3x$ for y

<table>
<tr><td>**SECTION 2.6**</td><td>

More about Problem Solving
</td></tr>
</table>

To solve problems, use the five-step problem-solving strategy.

1. Analyze the problem.
2. Form an equation.
3. Solve the equation.
4. State the conclusion.
5. Check the result.

78. SOUND SYSTEMS   A 45-foot-long speaker wire is to be cut into three pieces. One piece is to be 15 feet long. Of the remaining pieces, one must be 2 feet less than 3 times the length of the other. Find the length of the shorter piece.

79. AUTOGRAPHS   Kesha collected the autographs of 8 more television celebrities than she has of movie stars. Each TV celebrity autograph is worth $75 and each movie star autograph is worth $250. If her collection is valued at $1,900, how many of each type of autograph does she have?

80. ART HISTORY   *American Gothic* was painted in 1930 by Grant Wood. The length of the rectangular painting is 5 inches more than the width. Find the dimensions of the painting if it has a perimeter of $109\frac{1}{2}$ inches.

The sum of the measures of the angles of a triangle is 180°.

81. Find the missing angle measures of the triangle.

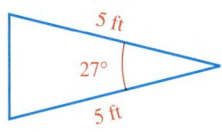

5 ft
27°
5 ft

Total value
= number · value
Interest = principal
· rate · time
$I = Prt$

82. Write an expression to represent the value of x video games each costing $45.

83. INVESTMENT INCOME   A woman has $27,000. Part is invested for 1 year in a certificate of deposit paying 7% interest, and the remaining amount in a cash management fund paying 9%. After 1 year, the total interest on the two investments is $2,110. How much is invested at each rate?

Distance = rate · time
$d = rt$

84. WALKING AND BICYCLING   A bicycle path is 5 miles long. A man walks from one end at the rate of 3 mph. At the same time, a friend bicycles from the other end, traveling at 12 mph. In how many minutes will they meet?

The value v of a commodity is its price per pound p times the number of pounds n:

$$v = pn$$

85. MIXTURES   A store manager mixes candy worth 90¢ per pound with gumdrops worth $1.50 per pound to make 20 pounds of a mixture worth $1.20 per pound. How many pounds of each kind of candy does he use?

86. SOLUTIONS   How much acetic acid is in x gallons of a solution that is 12% acetic acid?

| SECTION 2.7 | **Solving Inequalities** |

An *inequality* is a mathematical expression that contains a $>$, $<$, $\geq$, or $\leq$ symbol.

A *solution of an inequality* is any number that makes the inequality true.

A *parenthesis* indicates that a number is not on the graph. A *bracket* indicates that a number is included in the graph.

Interval notation can be used to describe a set of real numbers.

Solve each inequality. Write the solution set in interval notation and graph it.

87. $3x + 2 < 5$

88. $-\dfrac{3}{4}x \geq -9$

89. $\dfrac{3}{4} < \dfrac{d}{5} + \dfrac{1}{2}$

90. $5(3 - x) \leq 3(x - 3)$

91. $8 < x + 2 < 13$

92. $0 \leq 3 - 2x < 10$

93. SPORTS EQUIPMENT   The acceptable weight of Ping-Pong balls used in competition can range from 2.40 to 2.53 grams. Express this range using a compound inequality.

94. SIGNS   A large office complex has a strict policy about signs. Any sign to be posted in the building must meet three requirements:

- It must be rectangular in shape.
- Its width must be 18 inches.
- Its perimeter is not to exceed 132 inches.

What possible sign lengths meet these specifications?

CHAPTER 2 TEST

ELEMENTARY &
INTERMEDIATE
Algebra $f(x)$ Now™

1. Is 3 a solution of $2x + 3 = 4x - 6$?

2. EXERCISING
Find x.

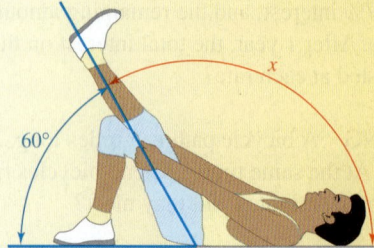

3. MULTIPLE BIRTHS IN THE UNITED STATES   In 2000, about 7,322 women gave birth to three or more babies at one time. This is about seven times the number of such births in 1971, 29 years earlier. How many multiple births occurred in 1971?

4. DOWN PAYMENTS   To buy a house, a woman was required to make a down payment of $11,400. What did the house sell for if this was 15% of the purchase price?

5. **SPORTS** In sports, percentages are most often expressed as three-place decimals instead of percents. For example, if a basketball player makes 75.8% of his free throws, the sports page will list this as .758. Use this format to complete the table.

All-time best regular-season winning percentages		
Team	Won–lost record	Winning percentage
1996 Chicago Bulls Basketball	72–10	
1972 Miami Dolphins Football	14–0	

6. **BODY TEMPERATURES** Suppose a person's body temperature rises from 98.6°F to 101.6°F. What is the percent increase? Round to the nearest one percent.

7. Find the expression that represents the perimeter of the rectangle.

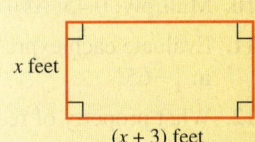

x feet

$(x + 3)$ feet

8. What property is illustrated below?

$$2(x + 7) = 2x + 2(7)$$

Simplify each expression.

9. $5(-4x)$

10. $-8(-7t)(4)$

11. $\dfrac{4}{5}(15a + 5) - 16a$

12. $-1.1d^2 - 3.8d^2 - d^2$

Solve each equation.

13. $5h + 8 - 3h + h = 8$

14. $\dfrac{4}{5}t = -4$

15. $\dfrac{11(b - 1)}{5} = 3b - 2$

16. $0.8x + 1.4 = 2.9 + 0.2x$

17. $\dfrac{m}{2} - \dfrac{1}{3} = \dfrac{1}{4}$

18. $23 - 5(x + 10) = -12$

19. Solve the equation for the variable indicated.
$A = P + Prt$; for r

20. On its first night of business, a pizza parlor brought in $445. The owner estimated his costs that night to be $295. What was the profit?

21. Find the Celsius temperature reading if the Fahrenheit reading is 14°.

22. **PETS** The spherical fishbowl is three-quarters full of water. To the nearest cubic inch, find the volume of water in the bowl.

10 in.

23. **TRAVEL TIMES** A car leaves Rockford, Illinois, at the rate of 65 mph, bound for Madison, Wisconsin. At the same time, a truck leaves Madison at the rate of 55 mph, bound for Rockford. If the cities are 72 miles apart, how long will it take for the car and the truck to meet?

24. **SALT SOLUTIONS** How many liters of a 2% brine solution must be added to 30 liters of a 10% brine solution to dilute it to an 8% solution?

25. **GEOMETRY** If the vertex angle of an isosceles triangle is 44°, find the measure of each base angle.

26. **INVESTMENT PROBLEM** Part of $13,750 is invested at 9% annual interest, and the rest is invested at 8%. After one year, the accounts paid $1,185 in interest. How much was invested at the lower rate?

Solve each inequality. Write the solution set in interval notation and graph it.

27. $-8x - 20 \le 4$ **28.** $-4 \le 2(x + 1) < 10$

29. DRAFTING   In the illustration, the $\pm$ (read "plus or minus") symbol means that the width of a plug a manufacturer produces can range from $1.497 - 0.001$ inches to $1.497 + 0.001$ inches. Write the range of acceptable widths w for the plug using a compound inequality.

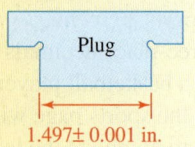

Plug

1.497± 0.001 in.

30. Solve: $2(y - 7) - 3y = -(y - 3) - 17$. Explain why the solution set is all real numbers.

CHAPTERS 1–2   CUMULATIVE REVIEW EXERCISES

1. Classify each of the following as an equation or an expression.
   a. $4m - 3 + 2m$ **b.** $4m = 3 + 2m$

2. Use the formula $t = \dfrac{w}{5}$ to complete the table.

Weight (lb)	Cooking time (hr)
15	
20	
25	

3. Give the prime factorization of 200.

4. Simplify: $\dfrac{24}{36}$.

5. Multiply: $\dfrac{11}{21}\left(-\dfrac{14}{33}\right)$.

6. COOKING   A recipe calls for $\frac{3}{4}$ cup of flour, and the only measuring container you have holds $\frac{1}{8}$ of a cup. How many $\frac{1}{8}$ cups of flour would you need to add to follow the recipe?

7. Add: $\dfrac{4}{5} + \dfrac{2}{3}$.

8. Subtract: $42\dfrac{1}{8} - 29\dfrac{2}{3}$.

9. Write $\dfrac{15}{16}$ as a decimal.

10. Multiply: $0.45(100)$.

11. Evaluate each expression.
   a. $|-65|$ **b.** $-|-12|$

12. What property of real numbers is illustrated below?

$x \cdot 5 = 5x$

Classify each number as a natural number, a whole number, an integer, a rational number, an irrational number, and a real number. Each number may have several classifications.

13. 3

14. -1.95

15. $\dfrac{17}{20}$

16. π

17. Write each product using exponents.
   a. $4 \cdot 4 \cdot 4$ **b.** $\pi \cdot r \cdot r \cdot h$

18. Perform each operation.
   a. $-6 + (-12) + 8$
   b. $-15 - (-1)$
   c. $2(-32)$
   d. $\dfrac{0}{35}$ **e.** $\dfrac{-11}{11}$

19. Write each phrase as an algebraic expression.
   a. The sum of the width w and 12.
   b. Four less than a number n.

20. SICK DAYS   Use the data in the table to find the average (mean) number of sick days used by this group of employees this year.

Name	Sick days	Name	Sick days
Chung	4	Ryba	0
Cruz	8	Nguyen	5
Damron	3	Tomaka	4
Hammond	2	Young	6

21. Complete the table.

x	$x^2 - 3$
-2	
0	
3	

22. Translate to mathematical symbols.

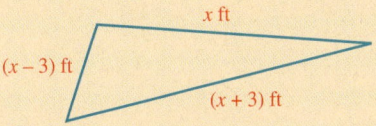

The loudness of a stereo speaker | is | 2,000 | divided by | the square of the distance of the listener from the speaker.

23. LAND OF THE RISING SUN   The flag of Japan is a red disc (representing sincerity and passion) on a white background (representing honesty and purity).
   a. What is the area of the rectangular-shaped flag?

   b. To the nearest tenth of a square foot, what is the area of the red disc?
   c. Use the results from parts a and b to find what percent of the area of the Japanese flag is occupied by the red disc.

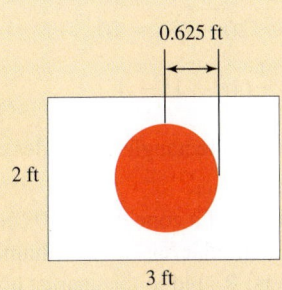

0.625 ft

2 ft

3 ft

24. 45 is 15% of what number?

Let $x = -5$, $y = 3$, and $z = 0$. Evaluate each expression.

25. $(3x - 2y)z$

26. $\dfrac{x - 3y + |z|}{2 - x}$

27. $x^2 - y^2 + z^2$

28. $\dfrac{x}{y} + \dfrac{y + 2}{3 - z}$

Simplify each expression.

29. $-8(4d)$

30. $5(2x - 3y + 1)$

31. $2x + 3x - x$

32. $3a^2 + 6a^2 - 17a^2$

33. $\dfrac{2}{3}(15t - 30) + t - 30$ **34.** $5(t - 4) + 3t$

35. What is the length of the longest side of the triangle shown below?

36. Write an algebraic expression in simplest form for the perimeter of the triangle.

x ft

$(x - 3)$ ft

$(x + 3)$ ft

Solve each equation.

37. $3x - 4 = 23$

38. $\dfrac{x}{5} + 3 = 7$

39. $-5p + 0.7 = 3.7$

40. $\dfrac{y - 4}{5} = 3 - y$

41. $-\dfrac{4}{5}x = 16$

42. $\dfrac{1}{2} + \dfrac{x}{5} = \dfrac{3}{4}$

43. $-9(n + 2) - 2(n - 3) = 10$

44. $\dfrac{2}{3}(r - 2) = \dfrac{1}{6}(4r - 1) + 1$

45. Find the area of a rectangle with sides of 5 meters and 13 meters.

46. Find the volume of a cone that is 10 centimeters tall and has a circular base whose diameter is 12 centimeters. Round to the nearest hundredth.

47. Solve $V = \dfrac{1}{3}\pi r^2 h$ for r^2.

48. WORK   Physicists say that work is done when an object is moved a distance d by a force F. To find the work done, we can use the formula $W = Fd$. Find the work done in lifting the bundle of newspapers onto the workbench.   (*Hint:* The force that must be applied to lift the newspapers is equal to the weight of the newspapers.)

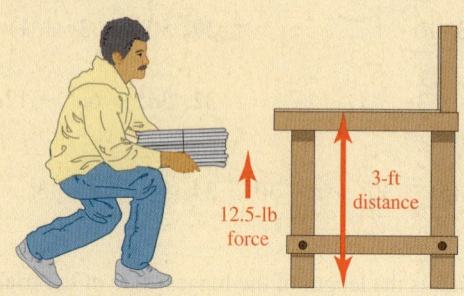

12.5-lb force

3-ft distance

49. WORK   See Exercise 48. Find the weight of a 1-gallon can of paint if the amount of work done to lift it onto the workbench is 28.35 foot-pounds.

50. Find the unknown angle measures.

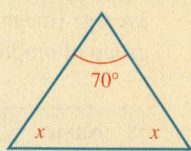

70°

x x

51. INVESTING   An investment club invested part of $10,000 at 9% annual interest and the rest at 8%. If the annual income from these investments was $860, how much was invested at 8%?

52. GOLDSMITH   How many ounces of a 40% gold alloy must be mixed with 10 ounces of a 10% gold alloy to obtain an alloy that is 25% gold?

Solve each inequality. Write the solution set in interval notation and graph it.

53. $x - 4 > -6$

54. $-6x \geq -12$

55. $8x + 4 \geq 5x + 1$

56. $-1 \leq 2x + 1 < 5$

3

Linear Equations and Inequalities in Two Variables

ELEMENTARY & INTERMEDIATE
Algebra *f(x)* **Now**™

Throughout the chapter, this icon introduces resources on the Elementary & Intermediate AlgebraNow Web site, accessed through **http://1pass. thomson.com**, that will

- Help you test your knowledge of the material with a pre-test and a post-test
- Provide a personalized learning plan targeting areas you should study

Getty/Stone/Glen Allison

Snow skiing is one of our country's most popular recreational activities. Whether on the gentle incline of a cross-country trip or racing down a near-vertical mountainside, a skier constantly adapts to the steepness of the course. In this chapter, we discuss lines and a means of measuring their steepness, called *slope*. The concept of slope has a wide variety of applications, including roofing, road design, and handicap accessible ramps.

To learn more about the slope of a line, visit *The Learning Equation* on the Internet at http://tle.brookscole.com. (The log-in instructions are in the Preface.) For Chapter 3, the online lessons are:

- *TLE* Lesson 6: Equations Containing Two Variables
- *TLE* Lesson 7: Rate of Change and the Slope of a Line

Relationships between two quantities can be described by a table, a graph, or an equation.

Graphing Using the Rectangular Coordinate System

- The Rectangular Coordinate System
- Reading Graphs
- Graphing Mathematical Relationships

It is often said, "A picture is worth a thousand words." That is certainly true when it comes to graphs. Graphs present data in an attractive and informative way. **Bar graphs** enable us to make quick comparisons, **line graphs** let us notice trends, and **circle graphs** show the relationship of the parts to the whole.

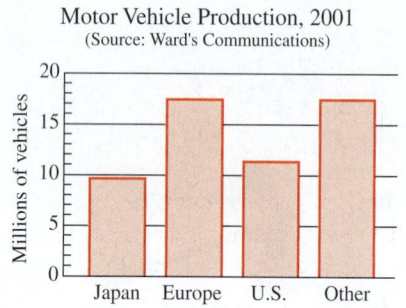

Motor Vehicle Production, 2001
(Source: Ward's Communications)

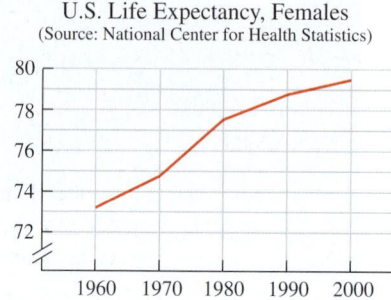

U.S. Life Expectancy, Females
(Source: National Center for Health Statistics)

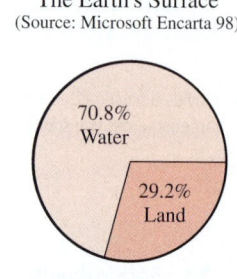

The Earth's Surface
(Source: Microsoft Encarta 98)

We will now introduce another type of graph that is widely used in mathematics called a *rectangular coordinate graph.*

■ THE RECTANGULAR COORDINATE SYSTEM

When designing the Gateway Arch in St. Louis, architects created a mathematical model called a **rectangular coordinate graph.** This graph, shown below on the right, is drawn on a grid called a **rectangular coordinate system.** This coordinate system is also called a **Cartesian coordinate system,** after the 17th-century French mathematician René Descartes.

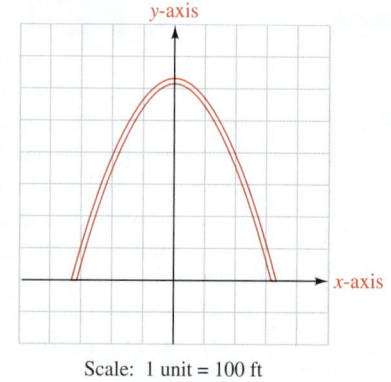

Scale: 1 unit = 100 ft

A rectangular coordinate system is formed by two perpendicular number lines. The horizontal number line is usually called the **x-axis,** and the vertical number line is usually called the **y-axis.** On the x-axis, the positive direction is to the right. On the y-axis, the positive direction is upward. Each axis should be scaled to fit the data. For example, the axes of the graph of the arch are scaled in units of 100 feet.

The point where the axes intersect is called the **origin.** This is the zero point on each axis. The axes form a **coordinate plane,** and they divide it into four regions called **quadrants,** which are numbered using Roman numerals.

The Language of Algebra

The word *axis* is used in mathematics and science. For example, Earth rotates on its *axis* once every 24 hours.

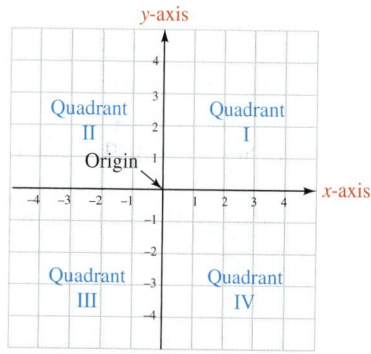

Each point in a coordinate plane can be identified by an **ordered pair** of real numbers x and y written in the form (x, y). The first number x in the pair is called the **x-coordinate,** and the second number y is called the **y-coordinate.** Some examples of such pairs are $(3, -4)$, $\left(-1, -\frac{3}{2}\right)$, and $(0, 2.5)$.

Notation

Don't be confused by this new use of parentheses. The notation $(3, -4)$ represents a point on the coordinate plane, whereas $3(-4)$ indicates multiplication. Also, don't confuse the ordered pair with interval notation.

$$(3, -4)$$
$$\uparrow \qquad \uparrow$$

The x-coordinate   The y-coordinate

The process of locating a point in the coordinate plane is called **graphing** or **plotting** the point. Below, we use blue arrows to show how to graph the point with coordinates $(3, -4)$. Since the x-coordinate, 3, is positive, we start at the origin and move 3 units to the *right* along the x-axis. Since the y-coordinate, -4, is negative, we then move *down* 4 units, and draw a dot. This locates the point $(3, -4)$.

In the figure, red arrows are used to show how to plot the point $(-4, 3)$. We start at the origin, move 4 units to the *left* along the x-axis, then move *up* 3 units and draw a dot. This locates the point $(-4, 3)$.

The Language of Algebra

Note that the point $(3, -4)$ has a different location than the point $(-4, 3)$. Since the order of the coordinates of a point is important, we call them *ordered pairs.*

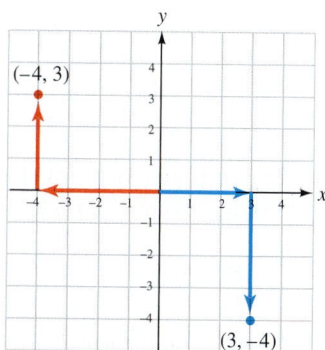

EXAMPLE 1

ELEMENTARY &
INTERMEDIATE

Algebra *f(x)* **Now**™

Plot each point and state the quadrant in which it lies.

a. (4, 4)   **b.** $\left(-1, -\frac{7}{2}\right)$ **c.** (0, 2.5)   **d.** (−3, 0)   **e.** (0, 0)

Solution

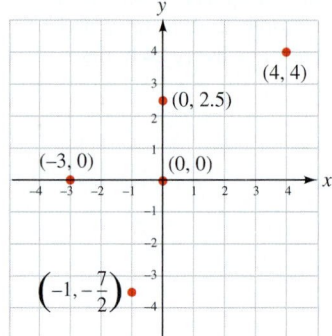

a. To plot the point (4, 4), we begin at the origin, move 4 units to the *right* on the *x*-axis, and then move 4 units *up*. The point lies in quadrant I.

b. To plot the point $\left(-1, -\frac{7}{2}\right)$, we begin at the origin, move 1 unit to the *left*, and then move $\frac{7}{2}$ units, or $3\frac{1}{2}$ units, *down*. The point lies in quadrant III.

Success Tip

Points with an *x*-coordinate that is 0 lie on the *y*-axis. Points with a *y*-coordinate that is 0 lie on the *x*-axis. Points that lie on an axis are not considered to be in any quadrant.

c. To plot the point (0, 2.5), we begin at the origin and do not move right or left, because the *x*-coordinate is 0. Since the *y*-coordinate is positive, we move 2.5 units *up*. The point lies on the *y*-axis.

d. To plot the point (−3, 0), we begin at the origin and move 3 units to the *left*. Since the *y*-coordinate is 0, we do not move up or down. The point lies on the *x*-axis.

e. To plot the point (0, 0), we begin at the origin, and we remain there because both coordinates are 0. The point with coordinates (0, 0) is the origin.

Self Check 1

Plot the points:

a. (2, −2)   **b.** (−4, 0)   **c.** $\left(1.5, \frac{5}{2}\right)$ **d.** (0, 5)

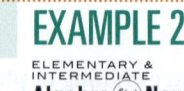

EXAMPLE 2

ELEMENTARY &
INTERMEDIATE

Algebra *f(x)* **Now**™

Find the coordinates of points *A*, *B*, *C*, *D*, *E*, and *F* plotted below.

Notation

Points are labeled with capital letters. The notation *A*(2, 3) indicates that point *A* has coordinates (2, 3).

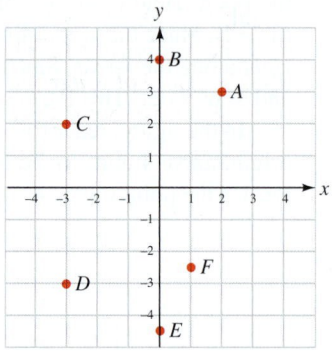

Solution

To locate point *A*, we start at the origin, move 2 units to the right, and then 3 units up. Its coordinates are (2, 3). The coordinates of the other points are found in the same manner.

$$B(0, 4) \qquad C(-3, 2) \qquad D(-3, -3) \qquad E(0, -4.5) \qquad F(1, -2.5)$$

Self Check 2

Find the coordinates of each point.

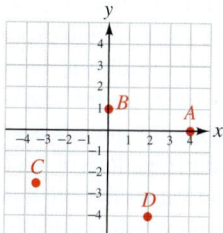

■ GRAPHING MATHEMATICAL RELATIONSHIPS

Every day, we deal with quantities that are related:

- The time it takes to cook a roast depends on the weight of the roast.
- The money we earn depends on the number of hours we work.
- The sales tax that we pay depends on the price of the item purchased.

We can use graphs to visualize such relationships. For example, suppose a tub is filling with water, as shown below. Obviously, the amount of water in the tub depends on how long the water has been running. To graph this relationship, we can use the measurements that were taken as the tub began to fill.

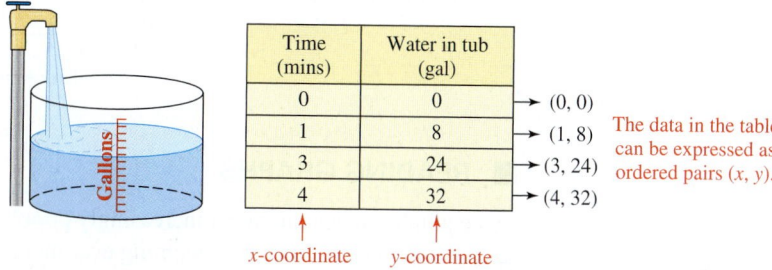

Time (mins)	Water in tub (gal)	
0	0	→ (0, 0)
1	8	→ (1, 8)
3	24	→ (3, 24)
4	32	→ (4, 32)

The data in the table can be expressed as ordered pairs (x, y).

x-coordinate   *y*-coordinate

The data in each row of the table can be written as an ordered pair and plotted on a rectangular coordinate system. Since the first coordinate of each ordered pair is a time, we label the *x*-axis *Time (min).* The second coordinate is an amount of water, so we label the *y*-axis *Amount of water (gal).* The *y*-axis is scaled in larger units (multiples of 4 gallons) because the size of the data ranges from 0 to 32 gallons.

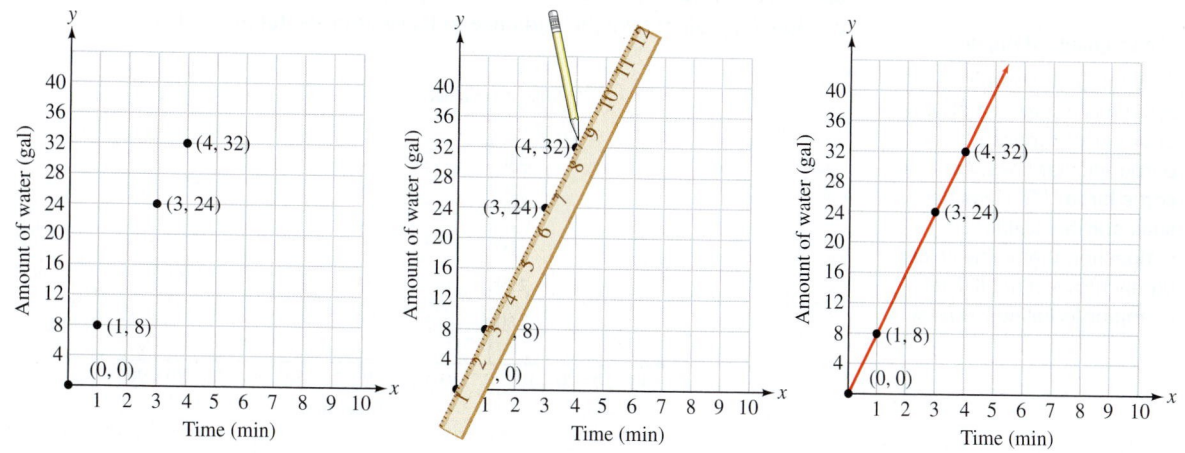

After plotting the ordered pairs, we use a straightedge to draw a line through the points. As we would expect, the completed graph shows that the amount of water in the tub increases steadily as the water is allowed to run.

We can use the graph to determine the amount of water in the tub at various times. For example, the green dashed line on the graph below shows that in 2 minutes, the tub will contain 16 gallons of water. This process, called **interpolation,** uses known information to predict values that are not known but are *within* the range of the data. The blue dashed line on the graph shows that in 5 minutes, the tub will contain 40 gallons of water. This process, called **extrapolation,** uses known information to predict values that are not known and are *outside* the range of the data.

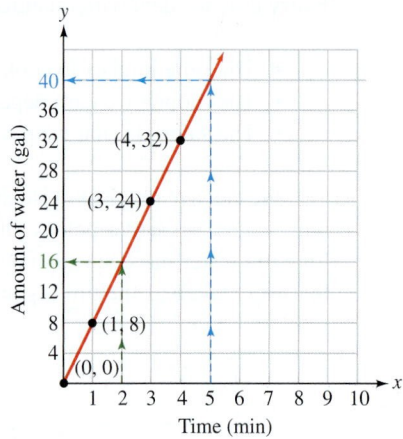

■ READING GRAPHS

Since graphs are becoming an increasingly popular way to present information, the ability to read and interpret them is becoming ever more important.

EXAMPLE 3

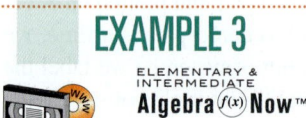

ELEMENTARY &
INTERMEDIATE
Algebra *f(x)* **Now**™

The Language of Algebra

A rectangular coordinate system is a *grid*—a network of uniformly spaced perpendicular lines. At times, some large U.S. cities have such horrible traffic congestion that vehicles can barely move, if at all. The condition is called *gridlock*.

TV taping. The graph below shows the number of people in an audience before, during, and after the taping of a television show. Use the graph to answer the following questions.

a. How many people were in the audience when the taping began?
b. At what times were there exactly 100 people in the audience?
c. How long did it take the audience to leave after the taping ended?

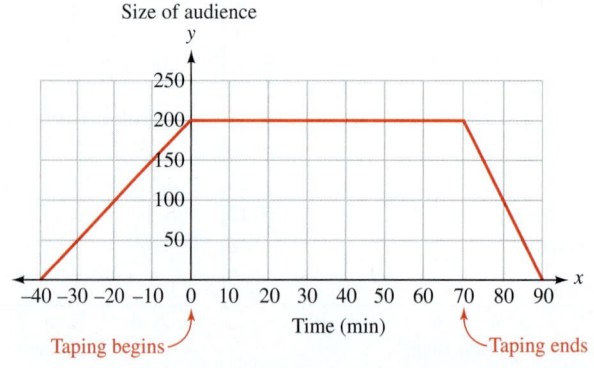

Solution   For each part of the solution, refer to the graph below. We can use the coordinates of specific points on the graph to answer these questions.

a. The time when the taping began is represented by 0 on the x-axis. The point on the graph directly above 0 is (0, **200**). The y-coordinate indicates that 200 people were in the audience when the taping began.

b. We can draw a horizontal line passing through 100 on the y-axis. Since the line intersects the graph twice, at (**−20**, 100) and at (**80**, 100), there are two times when 100 people were in the audience. The x-coordinates of the points tell us those times: 20 minutes before the taping began, and 80 minutes after.

c. The x-coordinate of the point (**70**, 200) tells us when the audience began to leave. The x-coordinate of (**90**, 0) tells when the exiting was completed. Subtracting the x-coordinates, we see that it took $90 - 70 = 20$ minutes for the audience to leave.

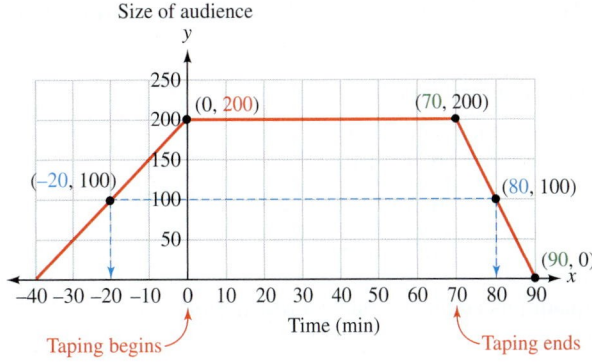

Self Check 3   Use the graph to answer the following questions.

a. At what times were there exactly 50 people in the audience?
b. How many people were in the audience when the taping took place?
c. When were the first audience members allowed into the taping session?

Answers to Self Checks   **1.** 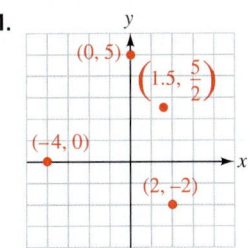   **2.** $A(4, 0)$; $B(0, 1)$; $C(-3.5, -2.5)$; $D(2, -4)$
3. a. 30 min before and 85 min after taping began,   **b.** 200,
c. 40 min before taping began

....

3.1   **STUDY SET**   ELEMENTARY &
INTERMEDIATE
Algebra $f(x)$ **Now**™

VOCABULARY   **Fill in the blanks.**

1. $(-1, -5)$ is called an _____ pair.
2. In the ordered pair $\left(-\frac{3}{2}, -5\right)$, the _____ is −5.

3. A rectangular coordinate system is formed by two perpendicular number lines called the _____ and the _____. The point where the axes cross is called the _____.

4. The *x*- and *y*-axes divide the coordinate plane into four regions called _____.

5. The point with coordinates (4, 2) can be graphed on a _____ coordinate system.

6. The process of locating the position of a point on a coordinate plane is called _____ the point.

CONCEPTS

7. Fill in the blanks.

   a. To plot the point with coordinates $(-5, 4)$, we start at the _____ and move 5 units to the _____ and then move 4 units _____.

   b. To plot the point with coordinates $\left(6, -\frac{3}{2}\right)$, we start at the _____ and move 6 units to the _____ and then move $\frac{3}{2}$ units _____.

8. In which quadrant is each point located?

   a. $(-2, 7)$ **b.** $(3, 16)$

   c. $(-1, -2.75)$ **d.** $(50, -16)$

   e. $\left(\frac{1}{2}, \frac{15}{16}\right)$ **f.** $(-6, \pi)$

9. a. In which quadrants are the second coordinates of points positive?

   b. In which quadrants are the first coordinates of points negative?

   c. In which quadrant do points with a negative *x*-coordinate and a positive *y*-coordinate lie?

   d. In which quadrant do points with a positive *x*-coordinate and a negative *y*-coordinate lie?

10. FARMING

   a. Write each row of data in the table as an ordered pair.

Rain (inches)	Bushels produced	
2	10	→
4	15	→
8	25	→

   b. Plot the ordered pairs on the following graph. Then draw a straight line through the points.

   c. Use the graph to determine how many bushels will be produced if 6 inches of rain fall.

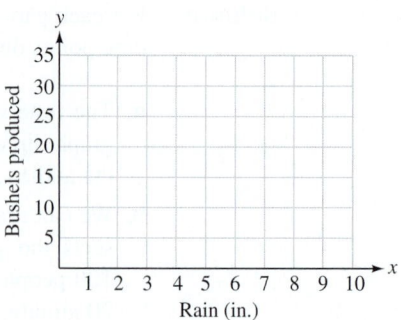

The graph that follows gives the heart rate of a woman before, during, and after an aerobic workout. In Problems 11–18, use the graph to answer the questions.

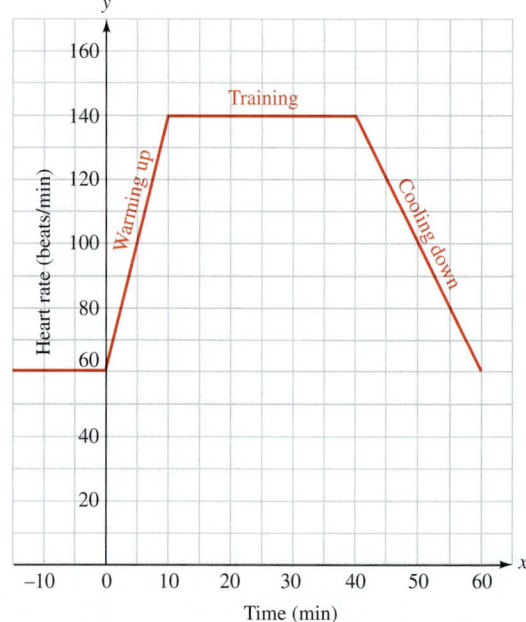

11. What was her heart rate before beginning the workout?

12. After beginning her workout, how long did it take the woman to reach her training-zone heart rate?

13. What was the woman's heart rate half an hour after beginning the workout?

14. For how long did the woman work out at her training zone?

15. At what time was her heart rate 100 beats per minute?

16. How long was her cool-down period?

17. What was the difference in the woman's heart rate before the workout and after the cool-down period?

18. What was her approximate heart rate 8 minutes after beginning?

19. BAR GRAPHS   Use the graph on page 184 to estimate the difference in the number of motor vehicles produced by Europe and the United States in 2001.

20. CIRCLE GRAPHS   The surface area of Earth is about 197,000,000 square miles. Use the graph on page 184 to find how many square miles are covered with water.

NOTATION

21. Explain the difference between (3, 5), 3(5), and 5(3 + 5).

22. In the table, which column contains values associated with the vertical axis of a graph?

x	y
2	0
5	3
−1	−3

23. Do these ordered pairs name the same point? $\left(2.5, -\frac{7}{2}\right), \left(2\frac{1}{2}, -3.5\right), \left(2.5, -3\frac{1}{2}\right)$

24. Do (3, 2) and (2, 3) represent the same point?

25. In the ordered pair (4, 5), is the number 4 associated with the horizontal or the vertical axis?

26. Fill in the blank: In the notation A(4, 5), the capital letter A is used to name a _____.

PRACTICE

27. Complete the coordinates for each point.

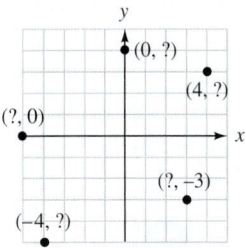

28. Find the coordinates of points $A, B, C, D, E,$ and F.

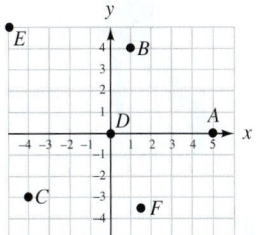

29. Graph each point: $(-3, 4), (4, 3.5), \left(-2, -\frac{5}{2}\right),$ $(0, -4), \left(\frac{3}{2}, 0\right), (3, -4).$

30. Graph each point: $(4, 4), (0.5, -3), (-4, -4),$ $(0, -1), (0, 0), (0, 3), (-2, 0).$

APPLICATIONS

31. CONSTRUCTION   The following graph shows a side view of a bridge design. Find the coordinates of each rivet, weld, and anchor.

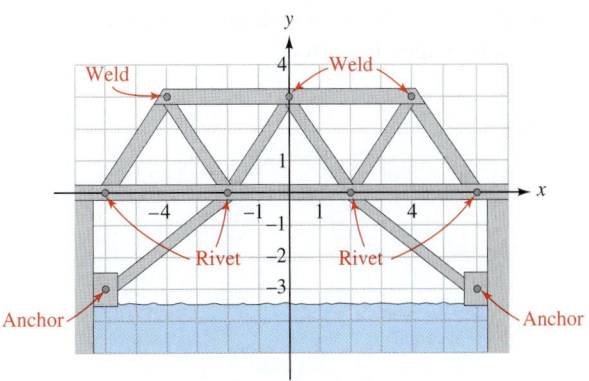

Scale: 1 unit = 8 ft

32. GOLF   To correct her swing, the golfer is videotaped and then has her image displayed on a computer monitor so that it can be analyzed by a golf pro. Give the coordinates of the points that are highlighted on the arc of her swing.

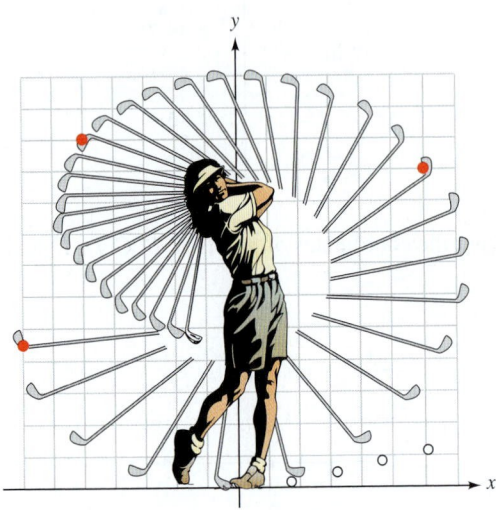

33. BATTLESHIP   In the game Battleship, the player uses coordinates to drop depth charges from a battleship to hit a hidden submarine. What coordinates should be used to make three hits on the exposed submarine shown? Express each answer in the form (letter, number).

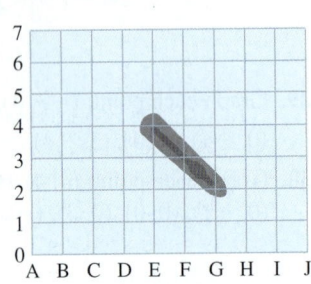

34. DENTISTRY   Orthodontists describe teeth as being located in one of four *quadrants* as shown below.
  a. How many teeth are in the *upper left quadrant?*

  b. Why would the upper left quadrant appear on the right in the illustration?

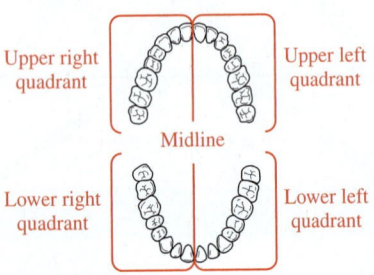

35. MAPS   Use coordinates that have the form (number, letter) to locate each of the following on the map: Rockford, Mount Carroll, Harvard, and the intersection of state Highway 251 and U.S. Highway 30.

36. WATER PRESSURE   The graphs show how the path of a stream of water changes when the hose is held at two different angles.
  a. At which angle does the stream of water shoot up higher? How much higher?

  b. At which angle does the stream of water shoot out farther? How much farther?

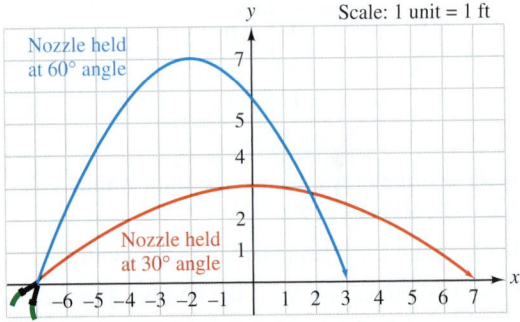

37. GEOMETRY   Three vertices (corners) of a rectangle are the points (2, 1), (6, 1), and (6, 4). Find the coordinates of the fourth vertex. Then find the area of the rectangle.

38. GEOMETRY   Three vertices (corners) of a right triangle are the points $(-1, -7)$, $(-5, -7)$, and $(-5, -2)$. Find the area of the triangle.

39. THE MILITARY   The table below shows the number of miles that a tank can be driven on a given number of gallons of diesel fuel. Write the data in the table as ordered pairs and plot them. Then draw a straight line through the points.
   a. How far can the tank go on 7 gallons of fuel?
   b. How many gallons of fuel are needed to travel a distance of 20 miles?
   c. How far can the tank go on 6.5 gallons of fuel?

Fuel (gal)	Distance (mi)
2	10
3	15
5	25

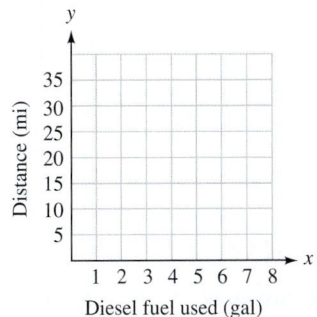

Diesel fuel used (gal)

40. BOATING   The table below shows the cost to rent a sailboat for a given number of hours. Write the data in the table as ordered pairs and plot them. Then draw a straight line through the points.
   a. What does it cost to rent the boat for 3 hours?
   b. For how long can the boat be rented for $60?

Rental time (hr)	Cost ($)
2	20
4	30
9	55

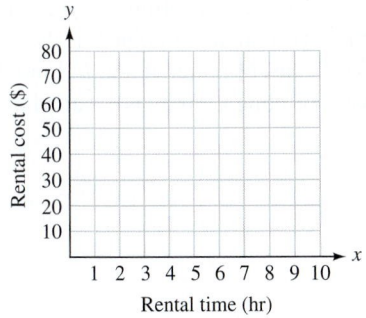

Rental time (hr)

41. DEPRECIATION   The table below shows the value (in thousands of dollars) of a car at various lengths of time after its purchase. Write the data in the table as ordered pairs and plot them. Then draw a straight line passing through the points.
   a. What does the point (3, 7) on the graph tell you?
   b. Find the value of the car when it is 7 years old.
   c. After how many years will the car be worth $2,500?

Age (yr)	Value ($1,000)
3	7
4	5.5
5	4

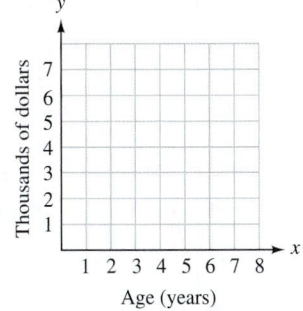

Age (years)

42. SWIMMING   The table below shows the number of people at a public swimming pool at various times during the day. Write the data in the table as ordered pairs and plot them. (On the x-axis, 0 represents noon, 1 represents 1 PM, and so on.) Then draw a straight line passing through the points.
   a. How many people will be at the pool at 6 PM?
   b. At what time will there be 250 people at the pool?
   c. When will the number of people at the pool be half of what it was at noon?

Time	Number of people
0	350
3	200
5	100

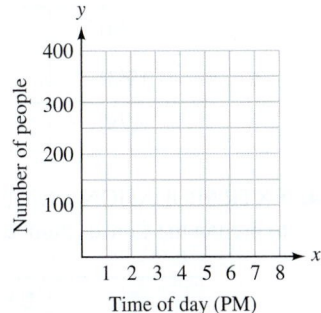

Time of day (PM)

In Problems 43–44, refer to the following information.
The approximate population of the United States for
the years 1950–2000 is given by the straight line graph
shown below.

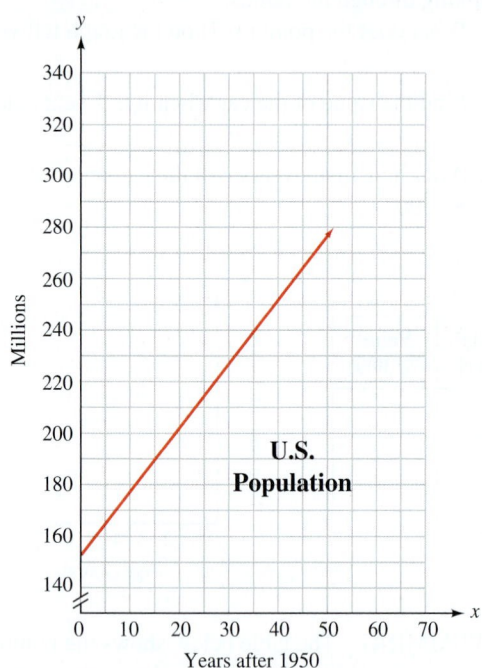

U.S.
Population

Millions

Years after 1950

43. CENSUS An official count of the population,
known as the Census, is conducted every 10 years in
the United States, as required by the Constitution.
Use the graph to estimate the Census numbers.

Census year	U.S. Population (millions)
1950	
1960	
1970	
1980	
1990	
2000	

44. EXTRAPOLATION On the graph, use a straightedge
to predict the Census numbers for 2010 and 2020.

Census year	U.S. Population (millions)
2010	
2020	

WRITING

45. Explain why the point $(-3, 3)$ is not the same as the
point $(3, -3)$.

46. Explain how to plot the point $(-2, 5)$.

47. Explain why the coordinates of the origin are $(0, 0)$.

48. Explain this diagram.

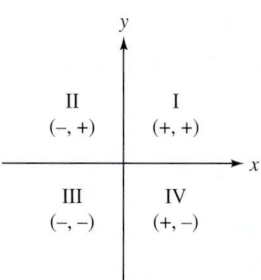

REVIEW

49. Solve $AC = \frac{2}{3}h - T$ for h.

50. Solve: $5(x + 1) \leq 2(x - 3)$. Write the solution set in
interval notation and graph it.

51. Evaluate: $\dfrac{-4(4 + 2) - 2^3}{|-12 - 4(5)|}$.

52. Simplify: $\dfrac{24}{54}$.

CHALLENGE PROBLEMS

53. In what quadrant does a point lie if the *sum* of its
coordinates is negative and the *product* of its
coordinates is positive?

54. Draw a line segment $\overline{AB}$ with endpoints $A(6, 5)$ and
$B(-4, 5)$. Suppose that the x-coordinate of a point C
is the average of the x-coordinates of points A and
B, and the y-coordinate of point C is the average of
the y-coordinates of points A and B. Find the
coordinates of point C. Why is C called the midpoint
of $\overline{AB}$?

3.2 Graphing Linear Equations

- Solutions of Equations in Two Variables   • Constructing Tables of Solutions
- Graphing Linear Equations   • Applications

In this section, we discuss equations that contain two variables. Such equations are often used to describe algebraic relationships between two quantities. To see a mathematical picture of these relationships, we will construct graphs of their equations.

■ SOLUTIONS OF EQUATIONS IN TWO VARIABLES

We have previously solved **equations in one variable.** For example, $x + 3 = 9$ is an equation in x. If we subtract 3 from both sides, we see that 6 is the solution. To verify this, we replace x with 6 and note that the result is a true statement: $9 = 9$.

In this chapter, we extend our equation-solving skills to find solutions of **equations in two variables.** To begin, let's consider $y = x - 1$, an equation in x and y.

A solution of $y = x - 1$ is a *pair* of values, one for x and one for y, that make the equation true. To illustrate, suppose x is 5 and y is 4. Then we have:

> **Notation**
>
> Equations in two variables often involve the variables x and y. However, other letters can be used. For example, $a - 3b = 5$ and $n = 4m + 6$ are equations in two variables.

$y = x - 1$

$4 \stackrel{?}{=} 5 - 1$ Substitute 5 for x and 4 for y.

$4 = 4$ True.

Since the result is a true statement, $x = 5$ and $y = 4$ is a solution of $y = x - 1$. We write the solution as the ordered pair $(5, 4)$, with the value of x listed first. We say that $(5, 4)$ *satisfies* the equation.

In general, a **solution of an equation in two variables** is an ordered pair of numbers that makes the equation a true statement.

EXAMPLE 1

Is $(-1, -3)$ a solution of $y = x - 1$?

Solution   We substitute -1 for x and -3 for y and see whether the resulting equation is true.

$y = x - 1$

$-3 \stackrel{?}{=} -1 - 1$ Substitute -1 for x and -3 for y.

$-3 = -2$

Since $-3 = -2$ is false, $(-1, -3)$ is *not* a solution.

Self Check 1   Is $(9, 8)$ a solution of $y = x - 1$?

We have seen that solutions of equations in two variables are written as ordered pairs. If only one of the values of an ordered-pair solution is known, we can substitute it into the equation to determine the other value.

EXAMPLE 2

Complete the solution $(-5, \;\;)$ of the equation $y = -2x + 3$.

Solution   In the ordered pair, the x-value is -5; the y-value is not known. To find y, we substitute -5 for x in the equation and evaluate the right-hand side.

$$y = -2x + 3$$
$$y = -2(-5) + 3 \qquad \text{Substitute } -5 \text{ for } x.$$
$$y = 10 + 3 \qquad\qquad \text{Do the multiplication.}$$
$$y = 13$$

The completed ordered pair is $(-5, 13)$.

Self Check 2   Complete the solution $(-2, \ )$ of the equation $y = 4x - 2$.

Solutions of equations in two variables are often listed in a **table of solutions** (or **table of values**).

EXAMPLE 3

ELEMENTARY &
INTERMEDIATE
Algebra *f(x)* **Now**™

Complete the table of solutions for $3x + 2y = 5$.

x	y	(x, y)
7		(7,)
	4	(, 4)

Solution   In the first row, we are given an x-value of 7. To find the corresponding y-value, we substitute 7 for x and solve for y.

$$3x + 2y = 5$$
$$3(7) + 2y = 5 \qquad \text{Substitute 7 for } x.$$
$$21 + 2y = 5 \qquad\quad \text{Do the multiplication.}$$
$$2y = -16 \qquad \text{Subtract 21 from both sides.}$$
$$y = -8 \qquad\quad \text{Divide both sides by 2.}$$

A solution of $3x + 2y = 5$ is $(7, -8)$.

In the second row, we are given a y-value of 4. To find the corresponding x-value, we substitute 4 for y and solve for x.

$$3x + 2y = 5$$
$$3x + 2(4) = 5 \qquad \text{Substitute 4 for } y.$$
$$3x + 8 = 5 \qquad\quad \text{Do the multiplication.}$$
$$3x = -3 \qquad \text{Subtract 8 from both sides.}$$
$$x = -1 \qquad \text{Divide both sides by 3.}$$

Another solution is $(-1, 4)$. The completed table is as follows:

x	y	(x, y)
7	-8	$(7, -8)$
-1	4	$(-1, 4)$

Self Check 3   Complete the table of solutions for $3x + 2y = 5$.

x	y	(x, y)
	-2	$(, -2)$
5		$(5,)$

■ CONSTRUCTING TABLES OF SOLUTIONS

To find a solution of an equation in two variables, we can select a number, substitute it for one of the variables, and find the corresponding value of the other variable. For example, to find a solution of $y = x - 1$, we can select a value for x, say, -4, substitute -4 for x in the equation, and find y.

x	y	(x, y)
-4	-5	$(-4, -5)$

$$y = \mathbf{x} - 1$$
$$y = \mathbf{-4} - 1 \qquad \text{Substitute } -4 \text{ for } x.$$
$$y = -5$$

The ordered pair $(-4, -5)$ is a solution. We list it in the table on the left.

To find another solution of $y = x - 1$, we select another value for x, say, -2, and find the corresponding y-value.

x	y	(x, y)
-4	-5	$(-4, -5)$
-2	-3	$(-2, -3)$

$$y = \mathbf{x} - 1$$
$$y = \mathbf{-2} - 1 \qquad \text{Substitute } -2 \text{ for } x.$$
$$y = -3$$

A second solution is $(-2, -3)$, and we list it in the table of solutions.

If we let $x = 0$, we can find a third ordered pair that satisfies $y = x - 1$.

x	y	(x, y)
-4	-5	$(-4, -5)$
-2	-3	$(-2, -3)$
0	-1	$(0, -1)$

$$y = \mathbf{x} - 1$$
$$y = \mathbf{0} - 1 \qquad \text{Substitute } 0 \text{ for } x.$$
$$y = -1$$

A third solution is $(0, -1)$, which we also add to the table of solutions.

We can find a fourth solution by letting $x = 2$, and a fifth solution by letting $x = 4$.

x	y	(x, y)
-4	-5	$(-4, -5)$
-2	-3	$(-2, -3)$
0	-1	$(0, -1)$
2	1	$(2, 1)$
4	3	$(4, 3)$

$$y = \mathbf{x} - 1 \qquad\qquad y = \mathbf{x} - 1$$
$$y = \mathbf{2} - 1 \quad \text{Substitute 2 for } x. \qquad y = \mathbf{4} - 1 \quad \text{Substitute 4 for } x.$$
$$y = 1 \qquad\qquad\qquad y = 3$$

A fourth solution is $(2, 1)$ and a fifth solution is $(4, 3)$. We add them to the table.

Since we can choose any real number for x, and since any choice of x will give a corresponding value of y, it is apparent that the equation $y = x - 1$ has *infinitely many solutions*. We have found five of them: $(-4, -5)$, $(-2, -3)$, $(0, -1)$, $(2, 1)$, and $(4, 3)$.

■ GRAPHING LINEAR EQUATIONS

It would be impossible to list the infinitely many solutions of the equation $y = x - 1$. To show all of its solutions, we draw a mathematical "picture" of them. We call this picture the *graph of the equation.*

To graph $y = x - 1$, we plot the ordered pairs shown in the table on a rectangular coordinate system. Then we draw a straight line through the points, because *the graph of any solution of $y = x - 1$ will lie on this line.* Furthermore, every point on this line represents a solution. We call the line the **graph of the equation;** it represents all of the solutions of $y = x - 1$.

$y = x - 1$

x	y	(x, y)
-4	-5	$(-4, -5)$
-2	-3	$(-2, -3)$
0	-1	$(0, -1)$
2	1	$(2, 1)$
4	3	$(4, 3)$

Construct a table of solutions.

Plot the ordered pairs.

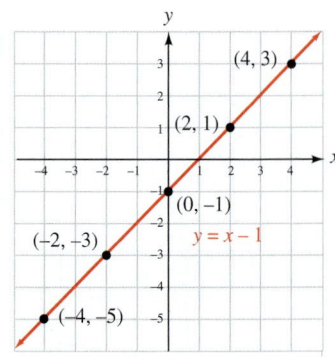

Draw a straight line through the points. This is the *graph of the equation.*

The equation $y = x - 1$ is said to be *linear* because its graph is a line. By definition, a linear equation in two variables is any equation that can be written in the following **general** (or **standard**) **form.**

Linear Equations

A **linear equation in two variables** is an equation that can be written in the form

$$Ax + By = C$$

where A, B, and C are real numbers and A and B are not both 0.

Success Tip

The exponent on each variable of a linear equation is an understood 1. For example, $y = 2x + 4$ can be thought of as $y^1 = 2x^1 + 4$.

Some more examples of linear equations are

$$5x - 6y = 10, \qquad 3y = -2x - 12, \qquad \text{and} \qquad y = 2x + 4$$

Linear equations can be graphed in several ways. Generally, the form in which an equation is written determines the method that we use to graph it. To graph linear equations solved for y, such as $y = 2x + 4$, we can use the following method.

Graphing Linear Equations Solved for y

1. Find three solutions of the equation by selecting three values for x and calculating the corresponding values of y.
2. Plot the solutions on a rectangular coordinate system.
3. Draw a straight line passing through the points. If the points do not lie on a line, check your computations.

EXAMPLE 4

Solution

Graph: $y = 2x + 4$.

To find three solutions of this linear equation, we select three values of x that will make the computations easy. Then we find each corresponding value of y.

If $x = -2$	If $x = 0$	If $x = 2$
$y = 2\boldsymbol{x} + 4$	$y = 2\boldsymbol{x} + 4$	$y = 2\boldsymbol{x} + 4$
$y = 2(\boldsymbol{-2}) + 4$	$y = 2(\boldsymbol{0}) + 4$	$y = 2(\boldsymbol{2}) + 4$
$y = -4 + 4$	$y = 0 + 4$	$y = 4 + 4$
$y = 0$	$y = 4$	$y = 8$
$(-2, 0)$ is a solution.	$(0, 4)$ is a solution.	$(2, 8)$ is a solution.

Success Tip

When selecting x-values for a table of solutions, a rule-of-thumb is to choose a negative number, a positive number, and 0. When $x = 0$, the computations to find y are usually quite simple.

We enter the results in a table of solutions and plot the points. Then we draw a straight line through the points and label it $y = 2x + 4$.

Success Tip

Since two points determine a line, only two points are needed to graph a linear equation. However, we should plot a third point as a check. If the three points do not lie on a straight line, then at least one of them is in error.

$y = 2x + 4$

x	y	(x, y)
-2	0	$(-2, 0)$
0	4	$(0, 4)$
2	8	$(2, 8)$

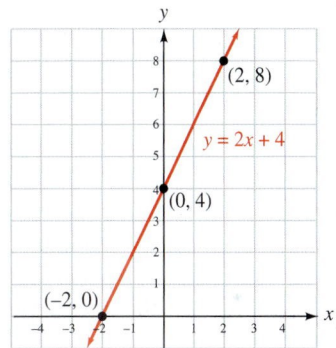

Self Check 4   Graph: $y = 2x - 3$.

EXAMPLE 5

Solution

Graph: $y = -3x$.

To find three solutions, we begin by choosing three x-values: -1, 0, and 1. Then we find the corresponding values of y. If $x = -1$, we have

$$y = -3\boldsymbol{x}$$
$$y = -3(\boldsymbol{-1}) \quad \text{Substitute } -1 \text{ for } x.$$
$$y = 3$$

$(-1, 3)$ is a solution.

In a similar manner, we find the y-values for x-values of 0 and 1, and record the results in a table of solutions. After plotting the ordered pairs, we draw a straight line through the points and label it $y = -3x$.

$$y = -3x$$

x	y	(x, y)
-1	3	$(-1, 3)$
0	0	$(0, 0)$
1	-3	$(1, -3)$

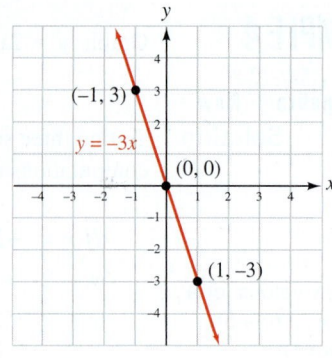

Self Check 5   Graph: $y = -4x$.

EXAMPLE 6   Graph: $3y = -2x - 12$.

Solution   The graph of $3y = -2x - 12$ can be found by solving for y and graphing the *equivalent equation* that results. To isolate y, we proceed as follows.

$$3y = -2x - 12$$

$$\frac{3y}{3} = \frac{-2x}{3} - \frac{12}{3} \qquad \text{To undo the multiplication by 3, divide both sides by 3.}$$

$$y = -\frac{2}{3}x - 4 \qquad \text{Write } \frac{-2x}{3} \text{ as } -\frac{2}{3}x. \text{ Simplify: } \frac{12}{3} = 4.$$

To find solutions of $y = -\frac{2}{3}x - 4$, each value of x must be multiplied by $-\frac{2}{3}$. This computation is made easier if we select x-values that are *multiples of 3,* such as $-3, 0$, and 6. For example, if $x = -3$, we have

$$y = -\frac{2}{3}x - 4$$

$$y = -\frac{2}{3}(-3) - 4 \qquad \text{Substitute } -3 \text{ for } x.$$

$$y = 2 - 4 \qquad \text{Do the multiplication: } -\frac{2}{3}(-3) = 2. \text{ This step is simpler if we choose } x\text{-values that are multiples of 3.}$$

$$y = -2$$

$(-3, -2)$ is a solution.

Two more solutions, one for $x = 0$ and one for $x = 6$, are found in a similar way, and entered in a table. We plot the ordered pairs, draw a straight line through the points, and label the line as $y = -\frac{2}{3}x - 4$ or as $3y = -2x - 12$.

$$y = -\frac{2}{3}x - 4$$

x	y	(x, y)
-3	-2	$(-3, -2)$
0	-4	$(0, -4)$
6	-8	$(6, -8)$

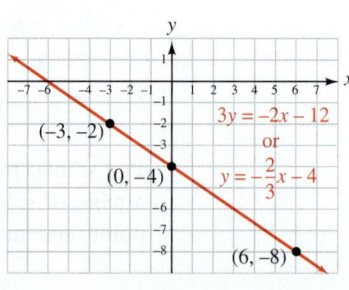

Self Check 6   Solve $2y = 5x + 2$ for y. Then graph the equation.

■ APPLICATIONS

When linear equations are used to model real-life situations, they are often written in variables other than x and y. In such cases, we must make the appropriate changes when labeling the table of solutions and the graph of the equation.

EXAMPLE 7

Cleaning windows.   The linear equation $A = -0.03n + 32$ estimates the amount A of glass cleaner (in ounces) that is left in the bottle after the sprayer trigger has been pulled a total of n times. Graph the equation and use the graph to estimate the amount of cleaner that is left after 500 sprays.

Solution   Since A depends on n in the equation $A = -0.03n + 32$, solutions will have the form (n, A). To find three solutions, we begin by selecting three values of n. Because the number of trials cannot be negative, and the computations to find A involve decimal multiplication, we select 0, 100, and 1,000. For example, if $n = 100$, we have

$$A = -0.03\mathbf{n} + 32$$
$$A = -0.03(\mathbf{100}) + 32$$
$$A = -3 + 32 \qquad \text{Do the multiplication: } -0.03(100) = -3.$$
$$A = 29$$

(100, 29) is a solution.   After 100 sprays, 29 ounces of cleaner are left in the bottle.

In a similar manner, solutions are found for $n = 0$ and $n = 1,000$, and listed in the following table. Then the ordered pairs are plotted and a straight line is drawn through the points.

To graphically estimate the amount of solution that is left after 500 sprays, we draw the dashed blue lines, as shown. Reading on the vertical A-axis, we see that after 500 sprays, about 17 ounces of glass cleaner would be left.

$$A = -0.03n + 32$$

n	A	(n, A)
0	32	(0, 32)
100	29	(100, 29)
1,000	2	(1,000, 2)

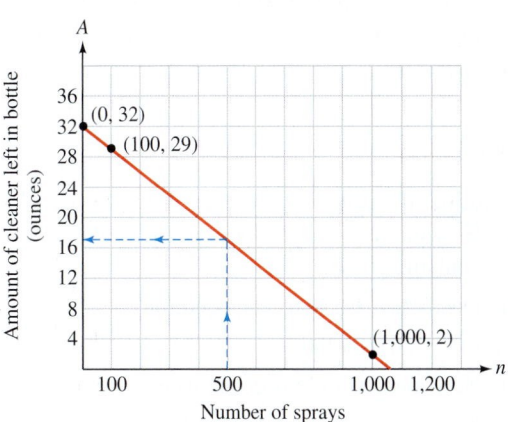

Answers to Self Checks   **1.** yes   **2.** $(-2, -10)$ **3.**

x	y	(x, y)
3	-2	$(3, -2)$
5	-5	$(5, -5)$

4.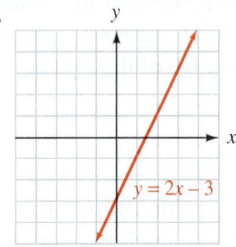

5.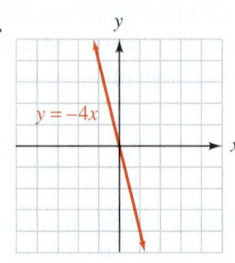

6.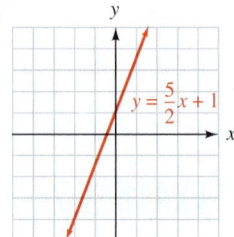

3.2   STUDY SET   ELEMENTARY & INTERMEDIATE   Algebra *f(x)* Now™

VOCABULARY   Fill in the blanks.

1. We say $y = 2x + 5$ is an equation in _____ variables, x and y.

2. A _____ of an equation in two variables is an ordered pair of numbers that makes the equation a true statement.

3. Solutions of equations in two variables are often listed in a _____ of solutions.

4. The line that represents all of the solutions of a linear equation is called the _____ of the equation.

5. The equation $y = 3x + 8$ is said to be _____ because its graph is a line.

6. The _____ form of a linear equation in two variables is $Ax + By = C$.

7. A linear equation in two variables has _____ many solutions.

8. Two points _____ a line.

CONCEPTS

9. Consider the equation $y = -2x + 6$.
   a. How many variables does the equation contain?
   b. Does $(4, -2)$ satisfy the equation?
   c. Is $(-3, 0)$ a solution?
   d. How many solutions does this equation have?

10. To graph a linear equation, three solutions were found, they were plotted (in black), and a straight line was drawn through them, as shown below.
   a. Construct the table of solutions for this graph.
   b. From the graph, determine three other solutions of the equation.

x	y	(x, y)

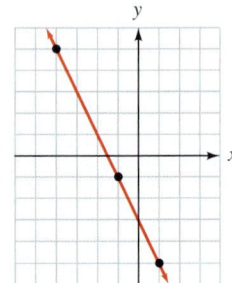

11. The graph of $y = -2x - 3$ is shown in Problem 10. Fill in the blanks: Every point on the graph represents an ordered-pair _____ of $y = -2x - 3$ and every ordered-pair solution is a _____ on the graph.

12. The graph of a linear equation is shown on the next page.
   a. If the coordinates of point M are substituted into the equation, will the result be true or false?
   b. If the coordinates of point N are substituted into the equation, will the result be true or false?

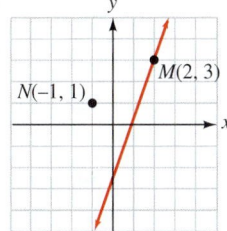

13. A student found three solutions of a linear equation and plotted them as shown in part (a) of the illustration.

   a. What conclusion can be made?

   b. What is wrong with the graph of $y = x - 3$, shown in illustration (b)?

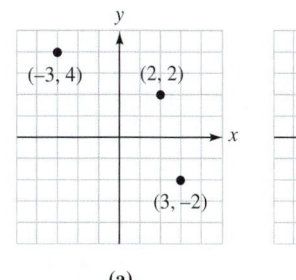

   (a)

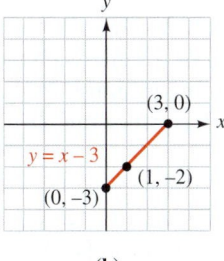

   (b)

14. **a.** Rewrite the linear equation $y = \frac{1}{2}x + 7$ showing the understood exponents on the variables.

   $$y^{} = \tfrac{1}{2}x^{} + 7$$

   b. Explain why $y = x^2 + 2$ and $y = x^3 - 4$ are not linear equations.

15. Solve each equation for y.

   a. $5y = 10x + 5$

   b. $3y = -5x - 6$

   c. $-7y = -x + 21$

16. **a.** Complete three lines of the table of solutions for $y = \frac{4}{5}x + 2$. Use the three values of x that make the computations the easiest.

x	y	(x, y)
-5		
-3		
0		
4		
10		

   b. Complete three lines of the table of solutions for $T = 0.6r + 560$. Use the three values of r that make the computations the easiest.

r	T	(r, T)
-12		
-10		
0		
8		
100		

NOTATION   **Complete the solution.**

17. Verify that $(-2, 6)$ is a solution of $y = -x + 4$.

   $$y = -x + 4$$
   $$\stackrel{?}{=} -(\quad) + 4$$
   $$6 \stackrel{?}{=} \quad + 4$$
   $$6 =$$

18. Determine whether each statement is true or false.

   a. $\dfrac{9x}{8} = \dfrac{9}{8}x$ **b.** $-\dfrac{1}{6}x = -\dfrac{x}{6}$

   c. When we divide both sides of $2y = 3x + 10$ by 2, we obtain $y = 3x + 5$.

19. Complete the labeling of the table of solutions and graph of $c = -a + 4$.

		$(\quad , \quad)$
-1	5	$(-1, 5)$
0	4	$(0, 4)$
2	2	$(2, 2)$

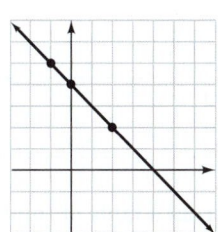

20. A table of solutions for a linear equation is shown below. When constructing the graph of the equation, how would you scale the x-axis and the y-axis?

x	y	(x, y)
-20	600	$(-20, 600)$
5	100	$(5, 100)$
35	-500	$(35, -500)$

PRACTICE   Determine whether each equation has the given ordered pair as a solution.

21. $y = 5x - 4$; $(1, 1)$

22. $y = -2x + 3$; $(2, -1)$

23. $y = -\frac{3}{4}x + 8$; $(-8, 12)$

24. $y = \frac{1}{6}x - 2$; $(-12, 4)$

25. $7x - 2y = 3$; $(2, 6)$

26. $10x - y = 10$; $(0, 0)$

27. $x + 12y = -12$; $(0, -1)$

28. $-2x + 3y = 0$; $(-3, -2)$

For each equation, complete the solution.

29. $y = -5x - 4$; $(-3, \quad)$

30. $y = 8x + 30$; $(-6, \quad)$

31. $4x - 5y = -4$; $(\quad, 4)$

32. $7x + y = -12$; $(\quad, 2)$

Complete each table of solutions.

33. $y = 2x - 4$

x	y	(x, y)
8		
	8	

34. $y = 3x + 1$

x	y	(x, y)
-3		
	-2	

35. $3x - y = -2$

x	y	(x, y)
-5		
	-1	

36. $5x - 2y = -15$

x	y	(x, y)
5		
	0	

Construct a table of solutions and then graph each equation.

37. $y = 2x - 3$

38. $y = 3x + 1$

39. $y = 5x - 4$

40. $y = 6x - 3$

41. $y = x$

42. $y = 4x$

43. $y = -3x + 2$

44. $y = -2x + 1$

45. $y = -x - 1$

46. $y = -x + 2$

47. $y = \frac{x}{3}$

48. $y = -\frac{x}{3} - 1$

49. $y = -\frac{1}{2}x$

50. $y = \frac{3}{4}x$

51. $y = \frac{3}{8}x - 6$

52. $y = \frac{5}{6}x - 5$

53. $y = \frac{2}{3}x - 2$

54. $y = -\frac{3}{2}x + 2$

Solve each equation for *y* and then graph it.

55. $7y = -2x$

56. $6y = -4x$

57. $3y = 12x + 15$

58. $5y = 20x - 30$

59. $5y = x + 20$

60. $4y = x - 16$

61. $y + 1 = 7x$

62. $y - 3 = 2x$

APPLICATIONS

63. BILLIARDS   The path traveled by the black eight ball on a game-winning shot is described by two linear equations. Complete the table of solutions for each equation and then graph the path of the ball.

$y = 2x - 4$

x	y	(x, y)
1		
2		
4		

$y = -2x + 12$

x	y	(x, y)
4		
6		
8		

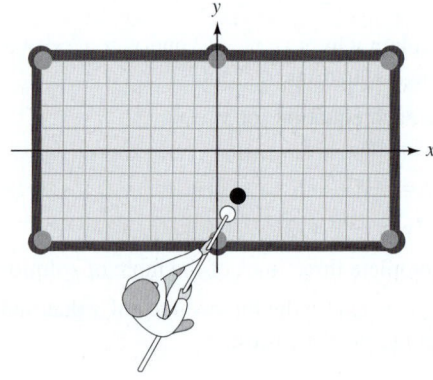

64. PING-PONG   The path traveled by a Ping-Pong ball is described by two linear equations. Complete the table of solutions for each equation and then graph the path of the ball.

$$y = \frac{1}{2}x + \frac{3}{2} \qquad\qquad y = -\frac{1}{2}x - \frac{3}{2}$$

x	y	(x, y)
7		
3		
-3		

x	y	(x, y)
-3		
-5		
-7		

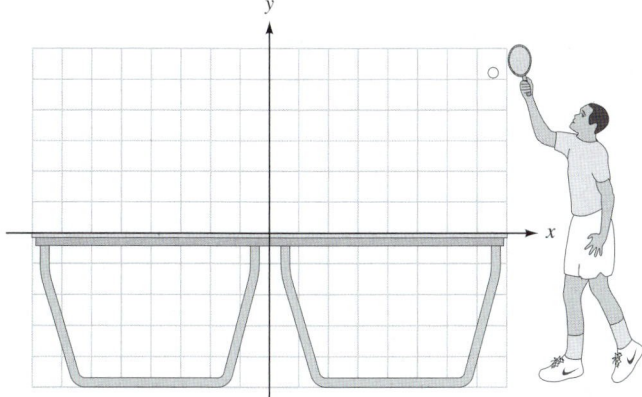

65. HOUSEKEEPING   The linear equation $A = -0.02n + 16$ estimates the amount A of furniture polish (in ounces) that is left in the bottle after the sprayer trigger has been pulled a total of n times. Graph the equation and use the graph to estimate the amount of polish that is left after 650 sprays.

66. SHARPENING PENCILS   The linear equation $L = -0.04t + 8$ estimates the length L (in inches) of a pencil after it has been inserted into a sharpener and the handle turned a total of t times. Graph the equation and use the graph to estimate the length of the pencil after 75 turns of the handle.

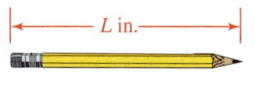

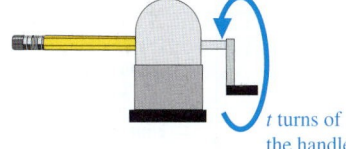

t turns of the handle

67. NFL TICKETS   The average ticket price p to a National Football League game during the years 1990–2002 is approximated by $p = \frac{9}{4}t + 23$, where t is the number of years after 1990. Graph this equation and use the graph to predict the average ticket price in 2010. (Source: Team Marketing Report, NFL.)

68. U.S. AUTOMOBILE ACCIDENTS   The number n of lives saved by seat belts during the years 1995–2001 is estimated by $n = 392t + 9{,}970$, where t is the number of years after 1995. Graph this equation and use the graph to predict the number of lives that will be saved by seat belts in 2020. (Source: Bureau of Transportation Statistics.)

69. RAFFLES   A private school is going to sell raffle tickets as a fund raiser. Suppose the number n of raffle tickets that will be sold is predicted by the equation $n = -20p + 300$, where p is the price of a raffle ticket in dollars. Graph the equation and use the graph to predict the number of raffle tickets that will be sold at a price of $6.

70. CATS   The number n of cat owners (in millions) in the U.S. during the years 1994–2002 is estimated by $n = \frac{3}{5}t + 31.5$, where t is the number of years after 1994. Graph this equation and use the graph to predict the number of cat owners in the U.S. in 2014. (Source: Pet Food Institute.)

WRITING

71. When we say that $(-2, -6)$ is a solution of $y = x - 4$, what do we mean?

72. What is a table of solutions?

73. What does it mean when we say that a linear equation in two variables has infinitely many solutions?

74. A linear equation and a graph are two ways of mathematically describing a relationship between two quantities. Which do you think is more informative and why?

75. From geometry, we know that two points determine a line. Explain why it is a good practice when graphing linear equations to find and plot three solutions instead of just two.

76. On a quiz, students were asked to graph $y = 3x - 1$. One student made the table of solutions on the left. Another student made the table on the right. The tables are completely different. Which table is incorrect? Or could they both be correct? Explain.

x	y	(x, y)
0	-1	$(0, -1)$
2	5	$(2, 5)$
3	8	$(3, 8)$

x	y	(x, y)
-2	-7	$(-2, -7)$
-1	-4	$(-1, -4)$
1	2	$(1, 2)$

REVIEW

77. Simplify: $-(-5 - 4c)$.

78. Denote the set of integers.

79. Evaluate: $-2^2 + 2^2$.

80. Find the volume, to the nearest tenth, of a sphere with radius 6 feet.

81. Evaluate: $1 + 2[-3 - 4(2 - 8^2)]$.

82. Solve: $-2(a + 3) = 3(a - 5)$.

CHALLENGE PROBLEMS   **Graph each of the following *nonlinear* equations in two variables by constructing a table of solutions consisting of seven ordered pairs. These equations are called nonlinear, because their graphs are not straight lines.**

83. $y = x^2 + 1$ **84.** $y = x^3 - 2$

85. $y = |x| - 2$ **86.** $y = (x + 2)^2$

3.3 More about Graphing Linear Equations

- Intercepts   • The Intercept Method   • Graphing Horizontal and Vertical Lines
- Information from Intercepts   • Graphing Calculators

In this section, we graph a linear equation by determining the points where the graph intersects the x-axis and the y-axis. These points are called the *intercepts* of the graph.

■ INTERCEPTS

The graph of $y = 2x - 4$ is shown below. We see that the graph crosses the y-axis at the point $(0, -4)$; this point is called the **y-intercept** of the graph. The graph crosses the x-axis at the point $(2, 0)$; this point is called the **x-intercept** of the graph.

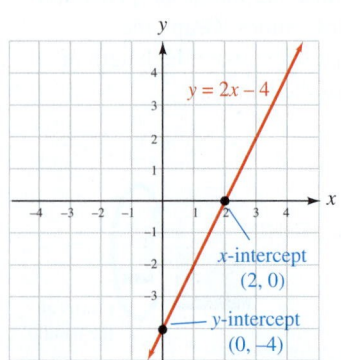

EXAMPLE 1

ELEMENTARY &
INTERMEDIATE
Algebra $f(x)$ **Now**™

For each graph, identify the *x*-intercept and the *y*-intercept.

a. 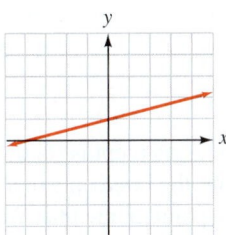   **b.** 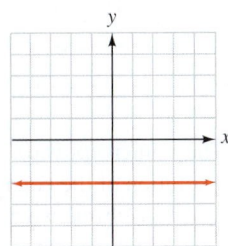   **c.**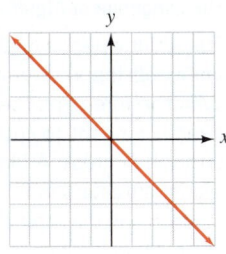

Solution   **a.** The graph crosses the *x*-axis at $(-4, 0)$. This is the *x*-intercept. The graph crosses the *y*-axis at $(0, 1)$. This is the *y*-intercept.

b. Since the horizontal line does not cross the *x*-axis, there is no *x*-intercept. The graph crosses the *y*-axis at $(0, -2)$. This is the *y*-intercept.

c. The graph crosses the *x*-axis and the *y*-axis at the same point, the origin. Therefore the *x*-intercept is $(0, 0)$ and the *y*-intercept is $(0, 0)$.

Self Check 1   Identify the *x*-intercept and the *y*-intercept of the graph.

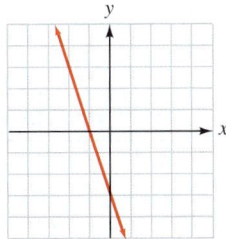

From the previous examples, we see that a *y*-intercept has an *x*-coordinate of 0, and an *x*-intercept has a *y*-coordinate of 0. These observations suggest the following procedures for finding the intercepts of a graph from its equation.

Finding Intercepts   To find the *y*-intercept, substitute 0 for *x* in the given equation and solve for *y*.

To find the *x*-intercept, substitute 0 for *y* in the given equation and solve for *x*.

■ THE INTERCEPT METHOD

Plotting the *x*- and *y*-intercepts of a graph and drawing a straight line through them is called the **intercept method of graphing a line.** This method is useful when graphing linear equations written in the general form $Ax + By = C$.

EXAMPLE 2

ELEMENTARY &
INTERMEDIATE
Algebra $f(x)$ **Now**™

Graph $x - 3y = 6$ by finding the intercepts.

Solution   To find the *y*-intercept, let $x = 0$ and solve for *y*. To find the *x*-intercept, let $y = 0$ and solve for *x*.

y-intercept		**x-intercept**	
$x - 3y = 6$		$x - 3y = 6$	
$\mathbf{0} - 3y = 6$	Substitute 0 for x.	$x - 3(\mathbf{0}) = 6$	Substitute 0 for y.
$-3y = 6$		$x - 0 = 6$	
$y = -2$	Divide both sides by -3.	$x = 6$	

The Language of Algebra

The point where a line *intersects* the x- or y-axis is called an intercept. The word *intersect* means to cut through or cross. A famous tourist attraction in Southern California is the *intersection* of Hollywood Blvd. & Vine St.

The y-intercept is $(0, -2)$ and the x-intercept is $(6, 0)$. Since each intercept of the graph is a solution of the equation, we enter the intercepts in a table of solutions.

As a check, we find one more point on the line. We select a convenient value for x, say, 3, and find the corresponding value of y.

$x - 3y = 6$	
$\mathbf{3} - 3y = 6$	Substitute 3 for x.
$-3y = 3$	Subtract 3 from both sides.
$y = -1$	Divide both sides by -3.

Therefore, $(3, -1)$ is a solution. It is also entered in the table.

The intercepts and the check point are plotted, a straight line is drawn through them, and the line is labeled $x - 3y = 6$.

$$x - 3y = 6$$

x	y	(x, y)	
0	-2	$(0, -2)$	← y-intercept
6	0	$(6, 0)$	← x-intercept
3	-1	$(3, -1)$	← check point

Self Check 2   Graph $x - 2y = 2$ by finding the intercepts.

The computations for finding intercepts can be simplified if we realize what logically follows when we substitute 0 for y or 0 for x in an equation written in the form $Ax + By = C$.

EXAMPLE 3

Graph $4x + 3y = -12$ by finding the intercepts.

Solution   When we substitute 0 for x, it follows that the term $4x$ will be equal to 0. Therefore, to find the y-intercept, we can cover $4x$ and solve the equation that remains for y.

$ + 3y = -12$	If $x = 0$, then $4x = 4(0) = 0$. Cover this term.
$y = -4$	To solve $3y = -12$, divide both sides by 3.

The y-intercept is $(0, -4)$.

When we substitute 0 for y, it follows that the term $3y$ will be equal to 0. Therefore, to find the x-intercept, we can cover $3y$ and solve the equation that remains for x.

$4x$ 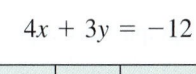 $= -12$ If $y = 0$, then $3y = 3(0) = 0$. Cover this term.

$x = -3$ To solve $4x = -12$, divide both sides by 4.

The x-intercept is $(-3, 0)$.

A third solution can be found by selecting a convenient value for x and finding the corresponding value for y. If we choose $x = -6$, we find that $y = 4$. The solution $(-6, 4)$ is entered in the table, and the equation is graphed as shown.

$$4x + 3y = -12$$

x	y	(x, y)
0	-4	$(0, -4)$
-3	0	$(-3, 0)$
-6	4	$(-6, 4)$

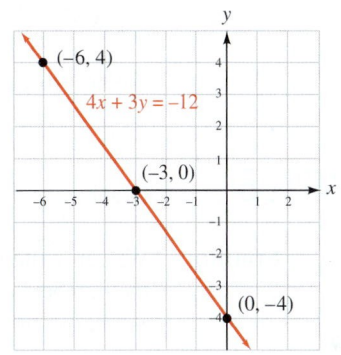

Self Check 3   Graph $2x + 5y = -10$ by finding the intercepts.

EXAMPLE 4

Graph $3x = -5y + 8$ by finding the intercepts.

Solution   We find the intercepts and select $x = 1$ to find a check point.

y-intercept: $x = 0$	*x-intercept: $y = 0$*	*Check point: $x = 1$*
$3x = -5y + 8$	$3x = -5y + 8$	$3x = -5y + 8$
$3(0) = -5y + 8$	$3x = -5(0) + 8$	$3(1) = -5y + 8$
$0 = -5y + 8$	$3x = 8$	$3 = -5y + 8$
$-8 = -5y$	$x = \dfrac{8}{3}$	$-5 = -5y$
$\dfrac{8}{5} = y$	$x = 2\dfrac{2}{3}$	$1 = y$
$1\dfrac{3}{5} = y$		

The y-intercept is $\left(0, 1\dfrac{3}{5}\right)$, the x-intercept is $\left(2\dfrac{2}{3}, 0\right)$, and the check point is $(1, 1)$. The ordered pairs are plotted as shown.

Success Tip

When graphing, it is often helpful to write any coordinates that are improper fractions as mixed numbers. For example:

$\left(\dfrac{8}{3}, 0\right) = \left(2\dfrac{2}{3}, 0\right)$.

$$3x = -5y + 8$$

x	y	(x, y)
0	$1\dfrac{3}{5}$	$\left(0, 1\dfrac{3}{5}\right)$
$\dfrac{8}{3} = 2\dfrac{2}{3}$	0	$\left(2\dfrac{2}{3}, 0\right)$
1	1	$(1, 1)$

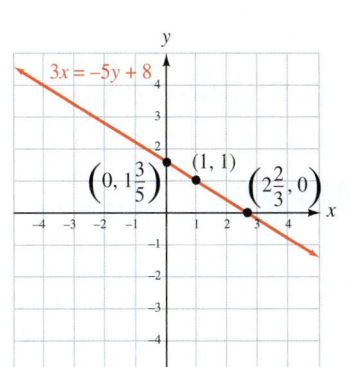

Self Check 4   Graph $8x = -4y + 15$ by finding the intercepts.

■ GRAPHING HORIZONTAL AND VERTICAL LINES

Equations such as $y = 4$ and $x = -3$ are linear equations, because they can be written in the general form $Ax + By = C$.

$y = 4$	is equivalent to	$0x + 1y = 4$
$x = -3$	is equivalent to	$1x + 0y = -3$

We now discuss how to graph these types of linear equations.

EXAMPLE 5

ELEMENTARY &
INTERMEDIATE
Algebra $f(x)$ **Now**™

Solution

Graph: $y = 4$.

We can write the equation in general form as $0x + y = 4$. Since the coefficient of x is 0, the numbers chosen for x have no effect on y. The value of y is always 4. For example, if $x = 2$, we have

$0x + y = 4$	This is the original equation written in general form.
$0(2) + y = 4$	Substitute 2 for x.
$y = 4$	Simplify the left-hand side.

One solution is $(2, 4)$. To find two more solutions, we choose $x = 0$ and $x = -3$. For any x-value, the y-value is always 4, so we enter $(0, 4)$ and $(-3, 4)$ in the table. If we plot the ordered pairs and draw a straight line through the points, the result is a horizontal line. The y-intercept is $(0, 4)$ and there is no x-intercept.

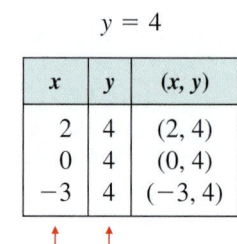

$y = 4$

x	y	(x, y)
2	4	$(2, 4)$
0	4	$(0, 4)$
-3	4	$(-3, 4)$

↑ Choose any number for x.   ↑ Each value of y must be 4.

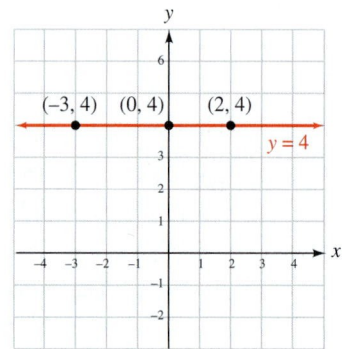

Self Check 5   Graph: $y = -2$.

EXAMPLE 6

ELEMENTARY &
INTERMEDIATE
Algebra $f(x)$ **Now**™

Solution

Graph: $x = -3$.

We can write the equation in general form as $x + 0y = -3$. Since the coefficient of y is 0, the numbers chosen for y have no effect on x. The value of x is always -3. For example, if $y = -2$, we have

$$x + 0y = -3 \quad \text{This is the original equation written in general form.}$$
$$x + 0(-2) = -3 \quad \text{Substitute } -2 \text{ for } y.$$
$$x = -3 \quad \text{Simplify the left-hand side.}$$

One solution is $(-3, -2)$. To find two more solutions, we choose $y = 0$ and $y = 3$. For any y-value, the x-value is always -3, so we enter $(-3, 0)$ and $(-3, 3)$ in the table. If we plot the ordered pairs and draw a straight line through the points, the result is a vertical line. The x-intercept is $(-3, 0)$ and there is no y-intercept.

$$x = -3$$

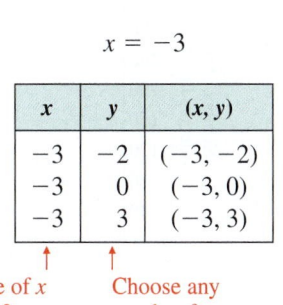

x	y	(x, y)
-3	-2	$(-3, -2)$
-3	0	$(-3, 0)$
-3	3	$(-3, 3)$

↑ Each value of x must be -3.

↑ Choose any number for y.

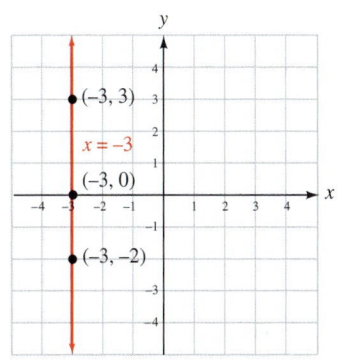

Self Check 6   Graph: $x = 4$.

From the results of Examples 5 and 6, we have the following facts.

Equations of Horizontal and Vertical Lines

The equation $y = b$ represents the horizontal line that intersects the y-axis at $(0, b)$.
The equation $x = a$ represents the vertical line that intersects the x-axis at $(a, 0)$.

The Language of Algebra

Two *parallel* lines are always the same distance apart. For example, think of the rails of train tracks. When graphing lines, remember that $y = b$ is *parallel* to the x-axis and $x = a$ is *parallel* to the y-axis.

The graph of the equation $y = 0$ has special significance; it is the x-axis. Similarly, the graph of the equation $x = 0$ is the y-axis.

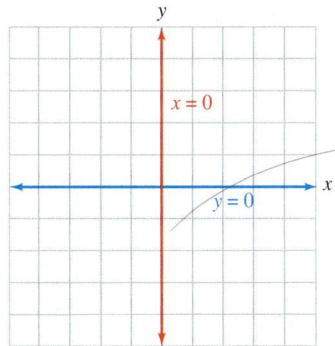

■ INFORMATION FROM INTERCEPTS

The ability to read and interpret graphs is a valuable skill. When analyzing a graph, we should locate and examine the intercepts. As the following example illustrates, the coordinates of the intercepts can yield useful information.

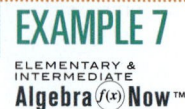

EXAMPLE 7

ELEMENTARY &
INTERMEDIATE
Algebra *f(x)* **Now**™

Camcorders. The number of feet of videotape that remain on a cassette depends on the number of minutes of videotaping that has already occurred. The following graph shows the relationship between these two quantities for a cassette in standard play mode. What information do the intercepts give about the cassette?

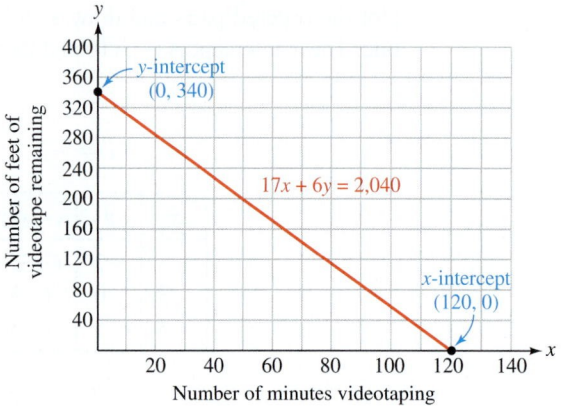

Solution The *y*-intercept of (0, 340) indicates that when 0 minutes of taping have occurred, 340 feet of videotape are available. That is, the cassette contains 340 feet of videotape.

The *x*-coordinate of (120, 0) indicates that when 120 minutes of videotaping have occurred, 0 feet of tape are available. In other words, the cassette can hold 120 minutes, or 2 hours, of videotaping.

Courtesy of Texas Instruments

■ GRAPHING CALCULATORS

So far, we have graphed linear equations by making tables of solutions and plotting points. A graphing calculator can make the task of graphing much easier.

However, a graphing calculator does not take the place of a working knowledge of the topics discussed in this chapter. It should serve as an aid to enhance your study of algebra.

The Viewing Window The screen on which a graph is displayed is called the **viewing window.** The **standard window** has settings of

$$\text{Xmin} = -10, \quad \text{Xmax} = 10, \quad \text{Ymin} = -10, \quad \text{and} \quad \text{Ymax} = 10$$

which indicate that the minimum *x*- and *y*-coordinates used in the graph will be -10, and that the maximum *x*- and *y*-coordinates will be 10.

Graphing an Equation To graph $y = x - 1$ using a graphing calculator, we press the Y = key and enter $x - 1$ after the symbol Y_1. Then we press the GRAPH key to see the graph.

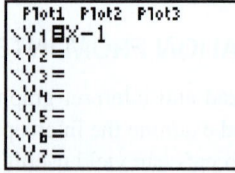

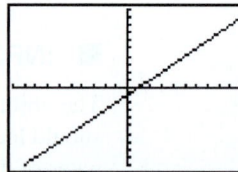

Changing the Viewing Window We can change the viewing window by pressing the WINDOW key and entering -4 for the minimum x- and y-coordinates and 4 for the maximum x- and y-coordinates. Then we press the GRAPH key to see the graph of $y = x - 1$ in more detail.

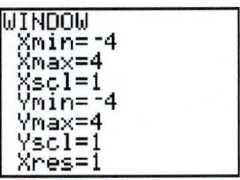   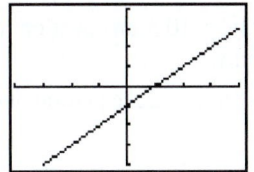

Solving an Equation for y To graph $3x + 2y = 12$, we must first solve the equation for y.

$$3x + 2y = 12$$
$$2y = -3x + 12 \qquad \text{Subtract } 3x \text{ from both sides.}$$
$$y = -\frac{3}{2}x + 6 \qquad \text{Divide both sides by 2.}$$

Next, we press the WINDOW key to reenter the standard window settings and press the Y = key to enter $y = -\frac{3}{2}x + 6$. Then we press the GRAPH key to see the graph.

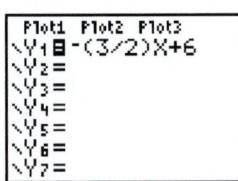   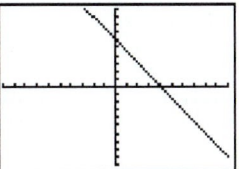

Answers to Self Checks   **1.** $(-1, 0); (0, -3)$

2.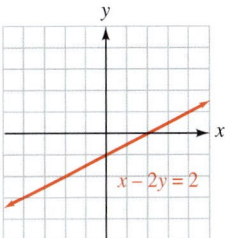

3.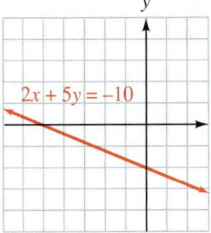

4.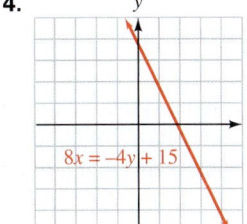

5.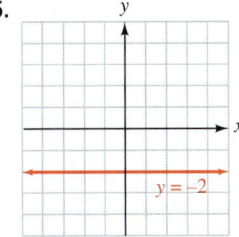

6.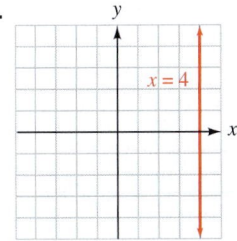

3.3   **STUDY SET**   ELEMENTARY & INTERMEDIATE Algebra $f(x)$ **Now**™

VOCABULARY   Fill in the blanks.

1. We say $5x + 3y = 10$ is an equation in _____ variables, x and y.

2. $2x - 3y = 6$ is a _____ equation; its graph is a line.

3. The equation $2x - 3y = 7$ is written in _____ form.

4. The _____ of a line is the point where the line intersects the x-axis.

5. The y-intercept of a line is the point where the line _____ the y-axis.

6. The graph of $y = 4$ is a _____ line, with y-intercept $(0, 4)$.

CONCEPTS   Identify the intercepts of each graph.

7. 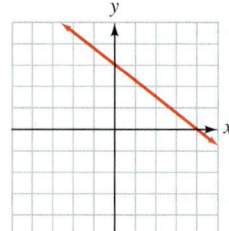   **8.**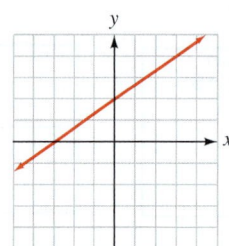

9. 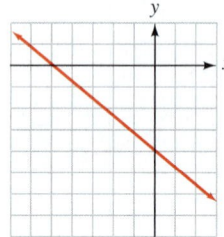   **10.**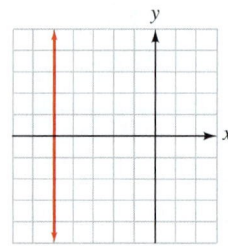

11. 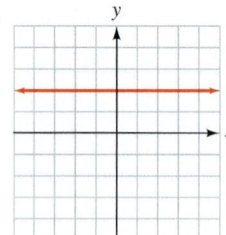   **12.**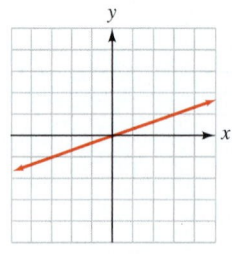

13. Estimate the intercepts of the line in the graph.

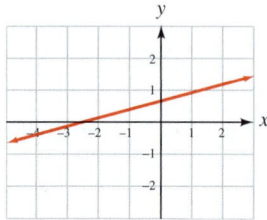

14. Fill in the blanks.
  a. To find the y-intercept of the graph of a line, substitute ▢ for x in the equation and solve for ▢.
  b. To find the x-intercept of the graph of a line, substitute ▢ for y in the equation and solve for ▢.

15. Consider the linear equation $3x + 2y = 6$.
  a. If we let $x = 0$, which term of the equation is equal to 0?
  b. Solve the equation that remains. What is the y-intercept of the graph of $3x + 2y = 6$?

16. In the table of solutions, which entry is the y-intercept of the graph and which entry is the x-intercept of the graph?

x	y	(x, y)
6	0	$(6, 0)$
0	-2	$(0, -2)$
-3	-3	$(-3, -3)$

17. What is the maximum number of intercepts that a line may have? What is the minimum number of intercepts that a line may have?

18. It is known that the value of a certain piece of farm machinery will steadily decrease after it is purchased.
  a. From the graph, which intercept tells the purchase price of the machinery? What was that price?

  b. Which intercept tells when the machinery will have lost all of its value? When is that?

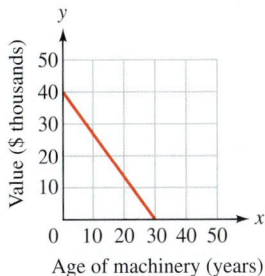

Age of machinery (years)

19. Match each graph with its equation
  a. $x = 2$ **b.** $y = 2$ **c.** $y = 2x$
  d. $2x - y = 2$ **e.** $y = 2x + 2$ **f.** $y = -2x$

i.
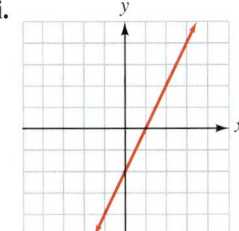

ii.
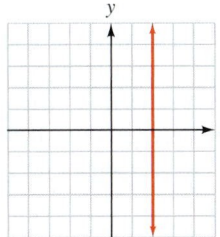

iii.
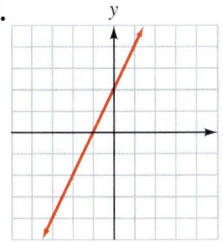

iv.
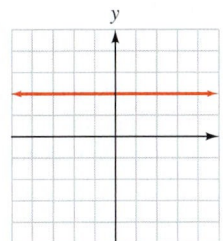

v.
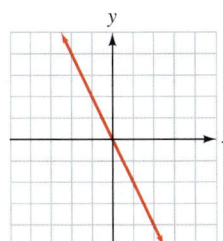

vi.
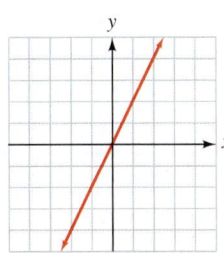

20. What linear equation could have the table of solutions shown below?

x	y	(x, y)
1	8	$(1, 8)$
0	8	$(0, 8)$
-1	8	$(-1, 8)$

NOTATION

21. a. Does the point $(0, 6)$ lie on the x-axis or the y-axis?

  b. Is it correct to say that the point $(0, 0)$ lies on the x-axis *and* on the y-axis?

22. True or false: $x = 5$ is equivalent to $1x + 0y = 5$.

23. What is the equation of the x-axis? What is the equation of the y-axis?

24. Write any coordinates that are improper fractions as mixed numbers.
  a. $\left(\frac{7}{2}, 0\right)$ **b.** $\left(0, -\frac{17}{3}\right)$

PRACTICE   Use the intercept method to graph each equation.

25. $4x + 5y = 20$ **26.** $3x + 4y = 12$
27. $x - y = -3$ **28.** $x - y = 3$
29. $5x + 15y = -15$ **30.** $8x + 4y = -24$
31. $x + 2y = -2$ **32.** $x + 2y = -4$
33. $4x - 3y = 12$ **34.** $5x - 10y = 20$
35. $3x + y = -3$ **36.** $2x - y = -2$
37. $9x - 4y = -9$ **38.** $5x - 4y = -15$
39. $8 = 3x + 4y$ **40.** $9 = 2x + 3y$
41. $4x - 2y = 6$ **42.** $6x - 3y = 3$
43. $3x - 4y = 11$ **44.** $5x - 4y = 13$
45. $9x + 3y = 10$ **46.** $4x + 4y = 5$
47. $3x = -15 - 5y$ **48.** $x = 5 - 5y$
49. $-4x = 8 - 2y$ **50.** $-5x = 10 + 5y$
51. $7x = 4y - 12$ **52.** $7x = 5y - 15$
53. $y - 3x = -\dfrac{4}{3}$ **54.** $y - 2x = -\dfrac{9}{8}$

Graph each equation.

55. $y = 4$

56. $y = -3$

57. $x = -2$

58. $x = 5$

59. $y = -\dfrac{1}{2}$

60. $y = \dfrac{5}{2}$

61. $x = \dfrac{4}{3}$

62. $x = -\dfrac{5}{3}$

63. $y - 2 = 0$

64. $x + 1 = 0$

65. $-2x + 3 = 11$

66. $-3y + 2 = 5$

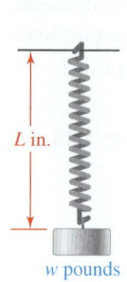

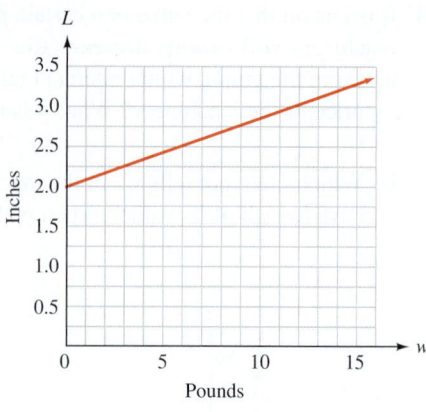

APPLICATIONS

67. CHEMISTRY   The relationship between the temperature and volume of a gas at a constant pressure is graphed below. The T-intercept of this graph is a very important scientific fact. It represents the lowest possible temperature, called **absolute zero.**
   a. Estimate absolute zero.

   b. What is the volume of the gas when the temperature is absolute zero?

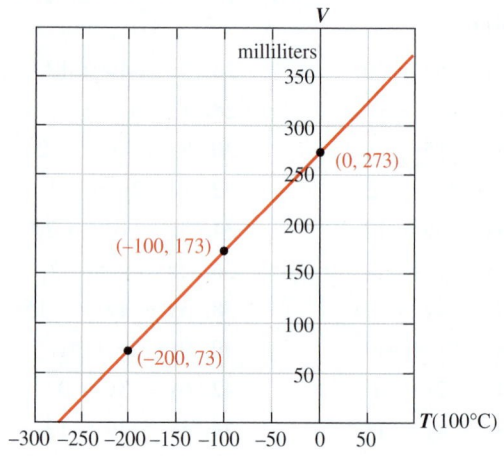

68. PHYSICS   The graph shows the length L of a stretched spring (in inches) as different weights w (in pounds) are attached to it. What information about the spring does the L-intercept give us?

69. LANDSCAPING   A developer is going to purchase x trees and y shrubs to landscape a new office complex. The trees cost $50 each and the shrubs cost $25 each. His budget is $5,000. This situation is modeled by the equation $50x + 25y = 5,000$. Use the intercept method to graph it.
   a. What information is given by the y-intercept?

   b. What information is given by the x-intercept?

70. THE MOTOR CITY   The linear equation $y = -192,000x + 1,850,000$ models the population y of Detroit, Michigan, where x is the number of decades since 1950. Without graphing, find the y-intercept of the graph. Then explain what it means.

WRITING

71. To graph $3x + 2y = 12$, a student found the intercepts and a check point, and graphed them, as shown on the left. Instead of drawing a crooked line through the points, what should he have done?

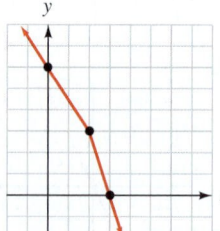

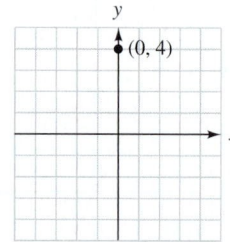

72. A student graphed the linear equation $y = 4$, as shown above on the right. Explain his error.

73. How do we find the intercepts of the graph of an equation without having to graph the equation?

74. In Section 3.2, we discussed a method to graph $y = 2x - 3$. In Section 3.3, we discussed a method to graph $2x + 3y = 6$. Briefly explain the steps involved in each method.

REVIEW

75. Simplify $\dfrac{3 \cdot 5 \cdot 5}{3 \cdot 5 \cdot 5 \cdot 5}$ by removing the common factors.

76. Simplify: $4\left(\dfrac{d}{2} - 3\right) - 5\left(\dfrac{2}{5}d - 1\right).$

77. Translate: Six less than twice x.

78. Is -5 a solution of $2(3x + 10) = 5x + 6$?

CHALLENGE PROBLEMS

79. Where will the line $y = b$ intersect the line $x = a$?

80. Write an equation of the line that has an x-intercept of $(4, 0)$ and a y-intercept of $(0, 3)$.

| 3.4 | # The Slope of a Line |

- Finding the Slope of a Line from Its Graph
- The Slope Formula
- Slopes of Horizontal and Vertical Lines
- Applications of Slope
- Rates of Change

In Sections 3.2 and 3.3, we graphed linear equations. All of the graphs were similar in one sense—they were lines. However, the lines slanted in different ways and had varying degrees of steepness. In this section, we introduce a means of measuring the steepness of a line. We call this measure the *slope of the line,* and it can be found in several ways.

■ FINDING THE SLOPE OF A LINE FROM ITS GRAPH

The **slope of a line** is a comparison of the vertical change to the corresponding horizontal change as we move along the line. The comparison is expressed as a **ratio** (a quotient of two numbers).

The Language of Algebra

Ratios are used in many settings. Mechanics speak of gear *ratios.* Colleges advertise their student-to-teacher *ratios.* Banks calculate debt-to-income *ratios* for loan applicants.

As an example, let's find the slope of the line graphed on the next page. To begin, we select two points on the line, $P(4, 2)$ and $Q(10, 7)$. As we move from point P to point Q, the y-coordinates change from 2 to 7. Therefore, the vertical change, called the **rise,** is $7 - 2$ or 5 units.

As we move from point P to point Q, the x-coordinates change from 4 to 10. Therefore, the horizontal change, called the **run,** is $10 - 4$ or 6 units.

The slope of a line is defined to be *the ratio of the vertical change to the horizontal change.* So we have

$$\text{slope} = \frac{\text{vertical change}}{\text{horizontal change}} = \frac{\text{change in } y}{\text{change in } x} = \frac{\text{rise}}{\text{run}} = \frac{5}{6}$$

The slope of the line is $\frac{5}{6}$. This indicates that there is a vertical change (rise) of 5 units for each horizontal change (run) of 6 units.

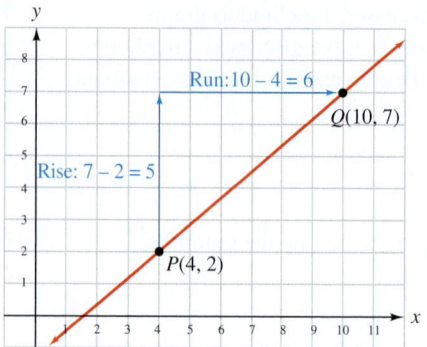

EXAMPLE 1

ELEMENTARY &
INTERMEDIATE
Algebra *f(x)* **Now**™

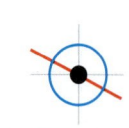

Find the slope of the line graphed below.

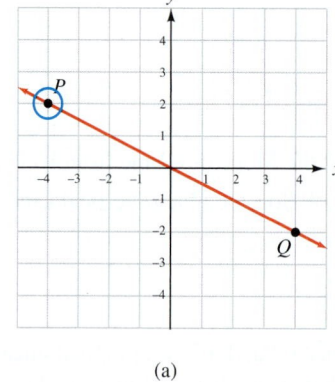

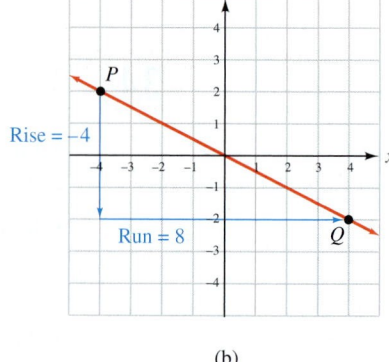

(a)   (b)

Solution   We begin by choosing two points on the line, *P* and *Q*, as shown in illustration (a). One way to move from *P* to *Q* is shown in illustration (b). Starting at *P*, we move downward, a rise of −4, and then to the right, a run of 8, to reach *Q*. These steps create a right triangle called a **slope triangle.**

To find the slope of the line, we write a ratio of the rise to the run. By tradition, the letter *m* is used to denote slope, so we have

The Language of Algebra

The symbol *m* is used to denote the slope of a line. Many historians credit this to the fact that it is the first letter of the French word *monter,* meaning to ascend or to climb.

$$m = \frac{\text{rise}}{\text{run}}$$ Slope is the ratio (quotient) of rise to run.

$$m = \frac{-4}{8}$$ Substitute −4 for the rise and 8 for the run.

$$m = -\frac{1}{2}$$ Simplify the fraction.

Success Tip

When drawing a slope triangle, remember that upward movements are positive, downward movements are negative, movements to the right are positive, and movements to the left are negative.

The slope of the line is $-\frac{1}{2}$.

The two-step process to move from *P* to *Q* can be reversed. Starting at *P*, we can move to the right, a run of 8; and then downward, a rise of −4, to reach *Q*. With this approach, the slope triangle is above the line. When we form the ratio to find the slope, we get the same result as before:

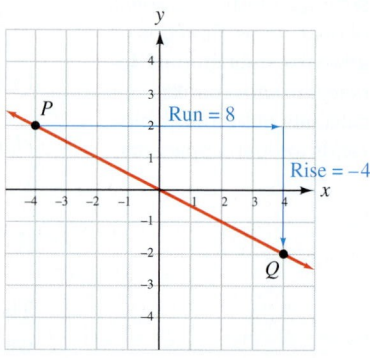

$$m = \frac{\text{rise}}{\text{run}} = \frac{-4}{8} = -\frac{1}{2}$$

Self Check 1   Find the slope of the line shown above using two points different from those used in the solution of Example 1.

The identical answers from Example 1 and its Self Check illustrate an important fact about slope: The same value will be obtained no matter which two points on a line are used to find the slope.

■ THE SLOPE FORMULA

We can generalize the graphic method for finding slope to develop a slope formula. To begin, we select two points on a line, as shown in the figure below. Call them P and Q. To distinguish between the coordinates of these points, we use **subscript notation.**

The Language of Algebra

The prefix *sub* means below or beneath, as in submarine or subway. In x_2, the *subscript* 2 is written lower than the variable.

- Point P is denoted as $P(x_1, y_1)$.   Read as "point P with coordinates of x sub 1 and y sub 1."

- Point Q is denoted as $Q(x_2, y_2)$.   Read as "point Q with coordinates of x sub 2 and y sub 2."

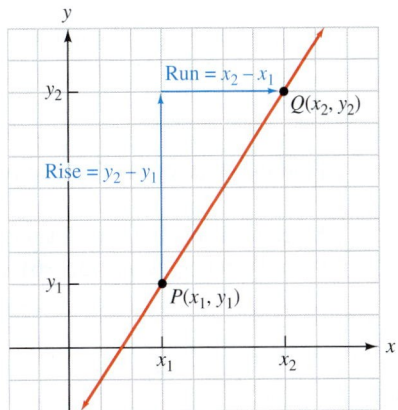

As we move from point P to point Q, the rise is the difference of the y-coordinates: $y_2 - y_1$. The run is the difference of the x-coordinates: $x_2 - x_1$. Since the slope is the ratio $\frac{\text{rise}}{\text{run}}$, we have the following formula for calculating slope.

Slope of a Line   The **slope** of a line passing through points (x_1, y_1) and (x_2, y_2) is

$$m = \frac{\text{vertical change}}{\text{horizontal change}} = \frac{\text{change in } y}{\text{change in } x} = \frac{\text{rise}}{\text{run}} = \frac{y_2 - y_1}{x_2 - x_1} \quad \text{if } x_2 \neq x_1.$$

EXAMPLE 2   Find the slope of the line passing through (1, 2) and (3, 8). Then graph the line.

Solution   When using the slope formula, it makes no difference which point you call (x_1, y_1) and which point you call (x_2, y_2). If we let (x_1, y_1) be (1, 2) and (x_2, y_2) be (3, 8), then

$$m = \frac{y_2 - y_1}{x_2 - x_1}$$ This is the slope formula.

$$m = \frac{8 - 2}{3 - 1}$$ Substitute 8 for y_2, 2 for y_1, 3 for x_2, and 1 for x_1.

$$m = \frac{6}{2}$$ Do the subtractions.

$$m = 3$$ Simplify. Think of this as a $\frac{3}{1}$ rise-to-run ratio.

Notation

We can write slopes that are integers in $\frac{\text{rise}}{\text{run}}$ form by writing them as fractions with a denominator of 1. For example, $m = 3 = \frac{3}{1}$ or $m = -5 = \frac{-5}{1}$.

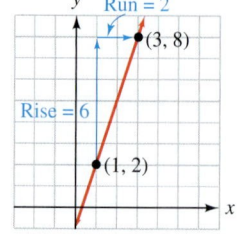

The slope of the line is 3. The graph of the line, including the slope triangle, is shown here. Note that we obtain the same value for the slope if we let $(x_1, y_1) = (3, 8)$ and $(x_2, y_2) = (1, 2)$.

$$m = \frac{y_2 - y_1}{x_2 - x_1} = \frac{2 - 8}{1 - 3} = \frac{-6}{-2} = 3$$

Self Check 2   Find the slope of the line passing through (2, 1) and (4, 11).

Caution   When using the slope formula, be sure to subtract the y-coordinates and the x-coordinates in the same order. Otherwise, your answer will have the wrong sign.

$$m \neq \frac{y_2 - y_1}{x_1 - x_2} \quad \text{and} \quad m \neq \frac{y_1 - y_2}{x_2 - x_1}$$

EXAMPLE 3

Find the slope of the line that passes through $(-2, 4)$ and $(5, -6)$ and graph the line.

ELEMENTARY &
INTERMEDIATE
Algebra $f(x)$ **Now**™   Solution   Since we know the coordinates of two points on the line, we can find its slope. If (x_1, y_1) is $(-2, 4)$ and (x_2, y_2) is $(5, -6)$, then

Notation

Slopes are normally written as fractions, sometimes as decimals, but never as mixed numbers.

As with any fractional answer, always express slope in lowest terms.

$$m = \frac{y_2 - y_1}{x_2 - x_1}$$ This is the slope formula.

$$m = \frac{-6 - 4}{5 - (-2)}$$ Substitute -6 for y_2, 4 for y_1, 5 for x_2, and -2 for x_1.

$$m = -\frac{10}{7}$$ Do the subtractions. We may write the result as $\frac{-10}{7}$ or $-\frac{10}{7}$.

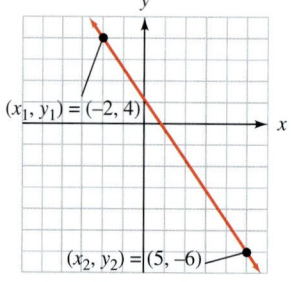

The slope of the line is $-\frac{10}{7}$. In the graph, we see that the line falls from left to right—a fact indicated by its negative slope.

Self Check 3   Find the slope of the line that passes through $(-1, -2)$ and $(1, -7)$.

From the previous examples, we see that the slope of the line in Example 2 was positive and the slopes of the lines in Examples 1 and 3 were negative. In general, lines that rise from left to right have a positive slope. Lines that fall from left to right have a negative slope.

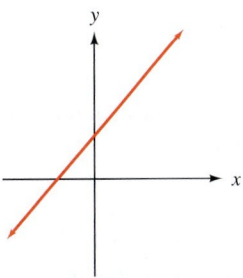

Positive slope

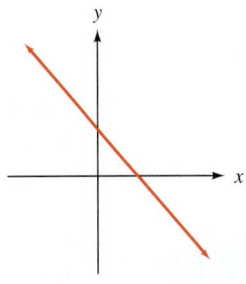

Negative slope

When we compare lines with positive slopes, we see that the larger the slope, the steeper the line. For example, a line with slope 3 is steeper than a line with slope $\frac{5}{6}$, and a line with slope $\frac{5}{6}$ is steeper than a line with slope $\frac{1}{4}$.

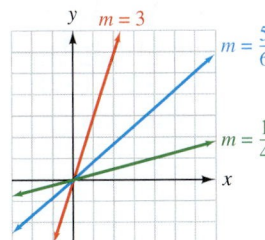

■ SLOPES OF HORIZONTAL AND VERTICAL LINES

In the next two examples, we calculate the slope of a horizontal line and we show that a vertical line has no defined slope.

EXAMPLE 4

Find the slope of the line $y = 3$.

Solution The graph of $y = 3$ is a horizontal line. To find its slope, we need to know two points on the line. From the graph, we select $(-2, 3)$ and $(3, 3)$. If (x_1, y_1) is $(-2, 3)$ and (x_2, y_2) is $(3, 3)$, we have

$$m = \frac{y_2 - y_1}{x_2 - x_1}$$ This is the slope formula.

$$m = \frac{3 - 3}{3 - (-2)}$$ Substitute 3 for y_2, 3 for y_1, 3 for x_2, and -2 for x_1.

$$m = \frac{0}{5}$$ Simplify the numerator and the denominator.

$$m = 0$$

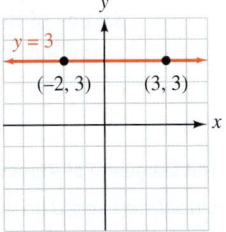

The slope of the line $y = 3$ is 0.

The y-coordinates of any two points on a horizontal line will be the same, and the x-coordinates will be different. Thus, the numerator of

$$\frac{y_2 - y_1}{x_2 - x_1}$$

will always be zero, and the denominator will always be nonzero. Therefore, the slope of a horizontal line is zero.

EXAMPLE 5

ELEMENTARY &
INTERMEDIATE
Algebra *f(x)* **Now**™

If possible, find the slope of the line $x = -2$.

Solution   The graph of $x = -2$ is a vertical line. To find its slope, we need to know two points on the line. From the graph, we select $(-2, 3)$ and $(-2, -1)$. If (x_2, y_2) is $(-2, 3)$ and (x_1, y_1) is $(-2, -1)$, we have

$$m = \frac{y_2 - y_1}{x_2 - x_1}$$ This is the slope formula.

$$m = \frac{3 - (-1)}{-2 - (-2)}$$ Substitute 3 for y_2, -1 for y_1, -2 for x_2, and -2 for x_1.

$$m = \frac{4}{0}$$ Simplify the numerator and the denominator.

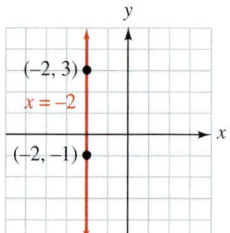

Since division by zero is undefined, $\frac{4}{0}$ has no meaning. The slope of the line $x = -2$ is undefined.

The y-coordinates of any two points on a vertical line will be different, and the x-coordinates will be the same. Thus, the numerator of

$$\frac{y_2 - y_1}{x_2 - x_1}$$

will always be nonzero, and the denominator will always be zero. Therefore, the slope of a vertical line is undefined.

We now summarize the results from Examples 4 and 5.

Slopes of Horizontal and Vertical Lines	Horizontal lines (lines with equations of the form $y = b$) have a slope of 0. Vertical lines (lines with equations of the form $x = a$) have undefined slope.

The Language of Algebra

Undefined and *0* do not mean the same thing. A horizontal line has a defined slope; it is 0. A vertical line does not have a defined slope; we say its slope is *undefined.*

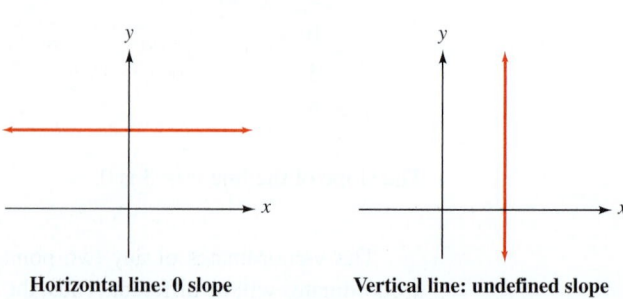

Horizontal line: 0 slope   Vertical line: undefined slope

■ APPLICATIONS OF SLOPE

The concept of slope has many applications. For example, architects use slope when designing ramps and roofs. Truckers must be aware of the slope, or *grade,* of the roads they travel. Mountain bikers ride up rocky trails and snow skiers speed down steep slopes.

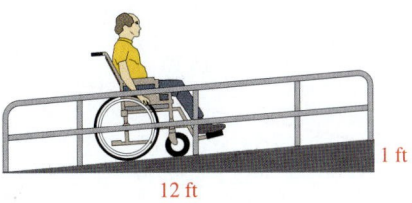

The Americans with Disabilities Act provides a guideline for the steepness of a ramp. The maximum slope for a wheelchair ramp is 1 foot of rise for every 12 feet of run: $m = \frac{1}{12}$.

The **grade** of an incline is its slope expressed as a percent: A 15% grade means a rise of 15 feet for every run of 100 feet: $m = \frac{15}{100}$.

EXAMPLE 6

ELEMENTARY & INTERMEDIATE
Algebra $f(x)$ **Now**™

Architecture. **Pitch** is the incline of a roof expressed as a ratio of the vertical rise to the horizontal run. Find the pitch of the roof shown in the illustration.

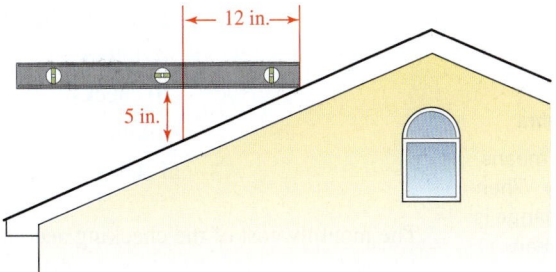

Solution   From the definition, we recognize that the pitch of a roof is simply its slope. In the illustration, a level is used to create a slope triangle. The rise is 5 and the run is 12.

$$m = \frac{\text{rise}}{\text{run}}$$

$$= \frac{5}{12}$$

The roof has a $\frac{5}{12}$ pitch.

■ RATES OF CHANGE

We have seen that the slope of a line is a *ratio* of two numbers. For many applications, however, we often attach units to a slope calculation. When we do so, we say that we have found a **rate of change.**

EXAMPLE 7

ELEMENTARY &
INTERMEDIATE
Algebra *f(x)* **Now**™

Checking accounts.   A checking plan at a bank charges customers a fixed monthly fee plus a small service charge for each check written. The relationship between the monthly cost y and the number x of checks written is graphed below. At what rate does the monthly cost change?

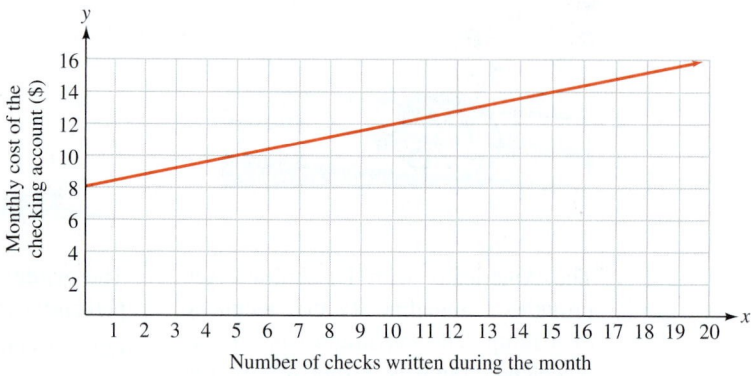

Number of checks written during the month

Solution   To find the rate of change, we find the slope of the line and attach the proper units. Two points on the line are $(5, 10)$ and $(15, 14)$. If we let $(x_1, y_1) = (5, 10)$ and $(x_2, y_2) = (15, 14)$, we have

$$\text{Rate of change} = \frac{(y_2 - y_1) \text{ dollars}}{(x_2 - x_1) \text{ checks}} \qquad \begin{array}{l}\text{This is the slope formula with the}\\\text{appropriate units attached.}\end{array}$$

$$= \frac{(14 - 10) \text{ dollars}}{(15 - 5) \text{ checks}} \qquad \begin{array}{l}\text{Substitute 14 for } y_2,\ 10 \text{ for } y_1,\ 15 \text{ for } x_2,\\\text{and 5 for } x_1.\end{array}$$

$$= \frac{4 \text{ dollars}}{10 \text{ checks}} \qquad \text{Do the subtractions.}$$

The Language of Algebra

The preposition *per* means for each, or for every. When we say the rate of change is 40¢ *per* check, we mean 40¢ for each check.

$$= \frac{2 \text{ dollars}}{5 \text{ checks}} \qquad \text{Simplify the fraction.}$$

The monthly cost of the checking account increases \$2 for every 5 checks written.

We can express $\frac{2}{5}$ in decimal form by dividing the numerator by the denominator. Then we can write the rate of change in two other ways, using the word *per*, which indicates division.

$$\text{Rate of change} = \$0.40 \text{ per check} \qquad \text{or} \qquad \text{Rate of change} = 40¢ \text{ per check}$$

Answers to Self Checks   **1.** $-\dfrac{1}{2}$ **2.** 5   **3.** $-\dfrac{5}{2}$

3.4   **STUDY SET**   ELEMENTARY &
INTERMEDIATE
Algebra *f(x)* **Now**™

VOCABULARY   **Fill in the blanks.**

1. A _____ is the quotient of two numbers.

2. The _____ of a line is a measure of the line's steepness.

3. The _____ of a line is defined to be the ratio of the change in y to the change in x.

4. $m = \dfrac{\text{change in } y}{\text{horizontal change}} = \dfrac{\text{rise}}{}$

5. The rate of _____ of a linear relationship can be found by finding the slope of the graph of the line and attaching the proper units.

6. _____ lines have a slope of 0. Vertical lines have _____ slope.

CONCEPTS

7. Fill in the blanks.
   a. A line with positive slope _____ from left to right.
   b. A line with negative slope _____ from left to right.

8. Which line graphed has
   a. a positive slope?
   b. a negative slope?
   c. zero slope?
   d. undefined slope?

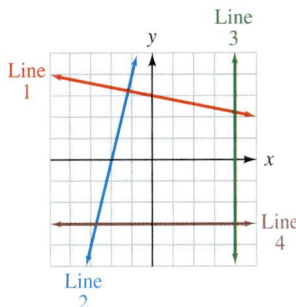

9. Suppose the rise of a line is 2 and the run is 15. Write the ratio of rise to run.

10. Consider the following graph of the line and the slope triangle.
   a. What is the rise?
   b. What is the run?
   c. What is the slope of the line?

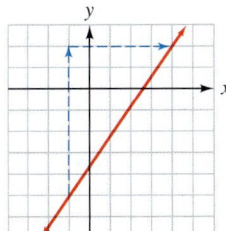

11. Consider the graph of the line and the slope triangle shown in the next column.
   a. What is the rise?
   b. What is the run?
   c. What is the slope of the line?

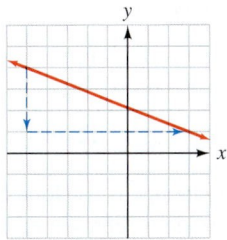

12. Which two labeled points should be used to find the slope of the line?

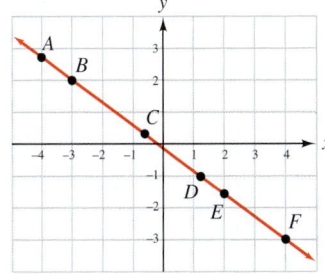

13. In this problem, you are to find the slope of the line graphed below by drawing a slope triangle.
   a. Find the slope using points A and B.
   b. Find the slope using points B and C.
   c. Find the slope using points A and C.
   d. What observation is suggested by your answers to parts a, b, and c?

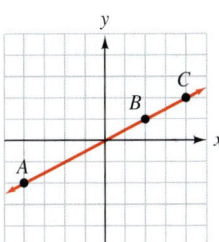

14. Evaluate each expression.
   a. $\dfrac{10-4}{6-5}$ **b.** $\dfrac{-1-1}{-2-(-7)}$

15. Express each of the following slope calculations in a better way.
   a. $m = \dfrac{0}{6}$ **b.** $m = \dfrac{8}{0}$

16. Simplify each slope.

  a. $m = \dfrac{3}{12}$ **b.** $m = -\dfrac{9}{6}$

  c. $m = \dfrac{-4}{4}$ **d.** $m = \dfrac{-10}{-5}$

17. The *grade* of an incline is its slope expressed as a percent. What grade is represented by each of the following slopes?

  a. $m = \dfrac{2}{5}$ **b.** $m = \dfrac{3}{20}$

18. GROWTH RATES   Refer to the graph. The slope of the line is 3. Fill in the correct units: The rate of change of the boy's height is   3 ___ per ___ .

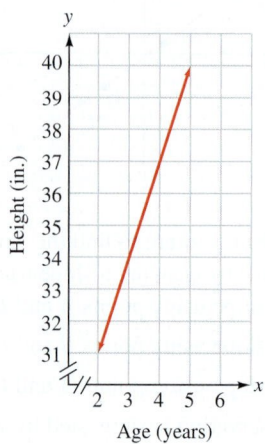

NOTATION

19. What is the formula used to find the slope of the line passing through (x_1, y_1) and (x_2, y_2)?

20. Fill in the blanks to state the slope formula in words:
  m equals y _____ two minus y _____ one _____ x sub _____ minus x sub _____ .

21. Explain the difference between y^2 and y_2.

22. Consider the points $(7, 2)$ and $(-4, 1)$. If we let $y_2 = 1$, then what is x_2?

PRACTICE   Find the slope of each line, if possible.

23.

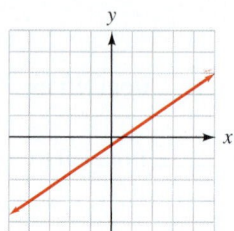

24.

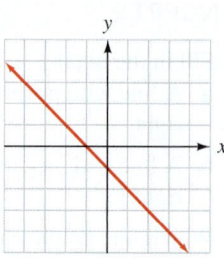

25.

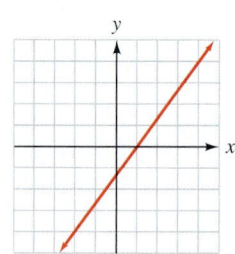

26.

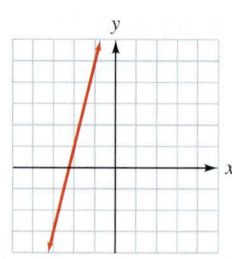

27.

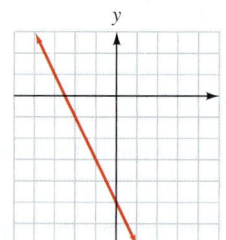

28.

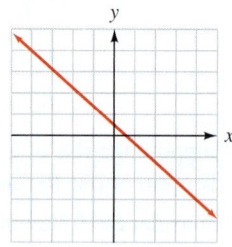

29.

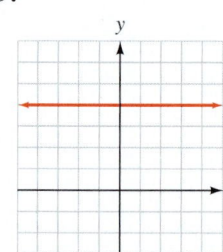

30.

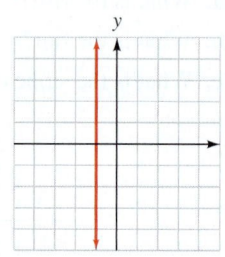

31.

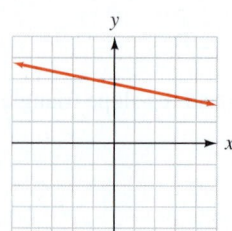

32.

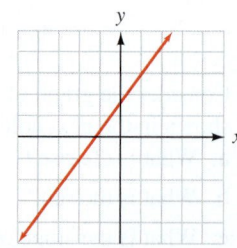

Find the slope of the line passing through the given points, when possible.

33. (2, 4) and (1, 3)

34. (1, 3) and (2, 5)

35. (3, 4) and (2, 7)

36. (3, 6) and (5, 2)

37. (0, 0) and (4, 5)

38. (4, 3) and (7, 8)

39. $(-3, 5)$ and $(-5, 6)$

40. $(6, -2)$ and $(-3, 2)$

41. $(-2, -2)$ and $(-12, -8)$

42. $(-1, -2)$ and $(-10, -5)$

43. (5, 7) and $(-4, 7)$

44. $(-1, -12)$ and $(6, -12)$

45. $(8, -4)$ and $(8, -3)$

46. $(-2, 8)$ and $(-2, 15)$

47. $(-6, 0)$ and $(0, -4)$

48. $(0, -9)$ and $(-6, 0)$

49. $(-2.5, 1.75)$ and $(-0.5, -7.75)$

50. $(6.4, -7.2)$ and $(-8.8, 4.2)$

Determine the slope of the graph of the line that has the given table of solutions.

51.

x	y	(x, y)
-3	-1	$(-3, -1)$
1	2	(1, 2)
5	5	(5, 5)

52.

x	y	(x, y)
-3	6	$(-3, 6)$
0	2	(0, 2)
3	-2	$(3, -2)$

53.

x	y	(x, y)
-3	6	$(-3, 6)$
0	6	(0, 6)
3	6	(3, 6)

54.

x	y	(x, y)
4	-5	$(4, -5)$
4	0	(4, 0)
4	3	(4, 3)

Find the slope of each line, if possible.

55. $x = 6$

56. $y = -2$

57. $y = 0$

58. $x = 0$

59. $x = -2$

60. $y = 8$

61. $y = -3$

62. $x = 6$

APPLICATIONS

63. POOL DESIGN   Find the slope of the bottom of the swimming pool as it drops off from the shallow end to the deep end.

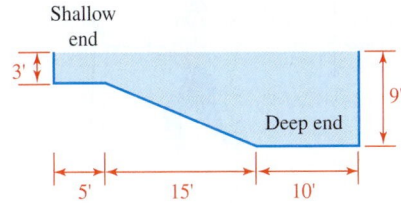

64. DRAINAGE   To measure the amount of fall (slope) of a concrete patio slab, a 10-foot-long 2-by-4, a 1-foot ruler, and a level were used. Find the amount of fall in the slab.

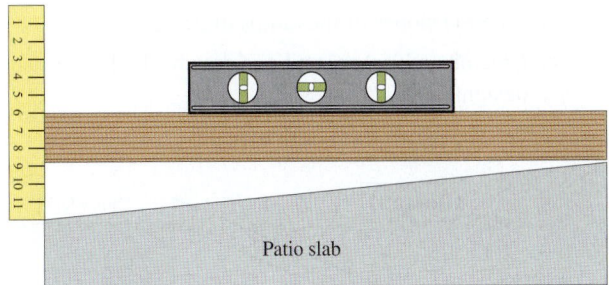

65. GRADE OF A ROAD   The vertical fall of the road shown in the illustration is 264 feet for a horizontal run of 1 mile. Find the slope of the decline and use that information to complete the roadside warning sign for truckers.

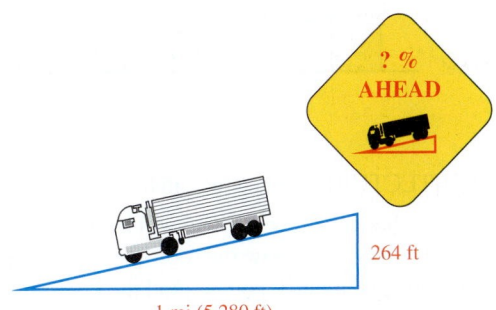

66. TREADMILLS   For each height setting listed in the table, find the resulting slope of the jogging surface of the treadmill. Express each incline as a percent.

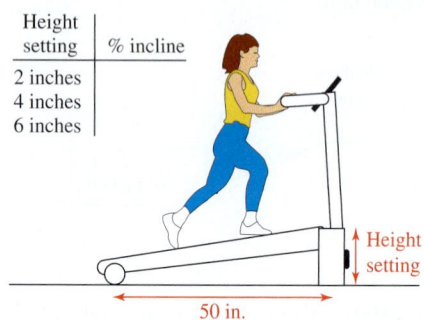

Height setting	% incline
2 inches	
4 inches	
6 inches	

67. ENGINEERING   The illustrations show two ramp designs.

   a. Find the slope of the ramp in design 1.

   b. Find the slopes of the ramps in design 2.

   c. Give one advantage and one drawback of each design.

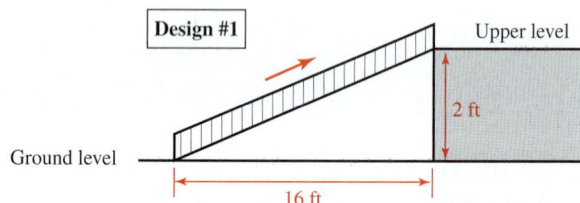

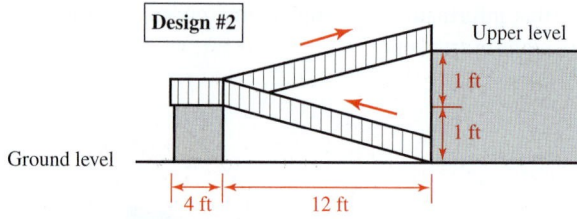

68. ARCHITECTURE   Since the pitch of the roof of the house shown is to be $\frac{2}{5}$, there will be a 2-foot rise for every 5-foot run. Draw the roof line if it is to pass through the given black points. Find the coordinates of the peak of the roof.

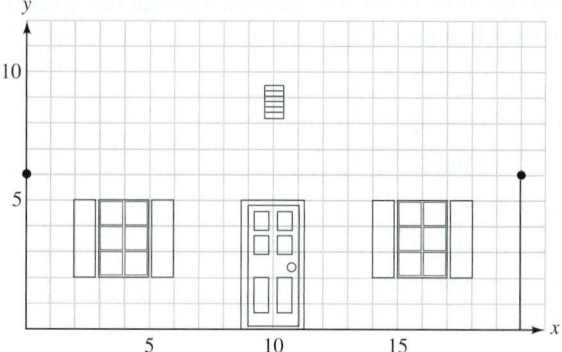

69. IRRIGATION   The graph shows the number of gallons of water remaining in a reservoir as water is discharged from it to irrigate a field. Find the rate of change in the number of gallons of water in the reservoir.

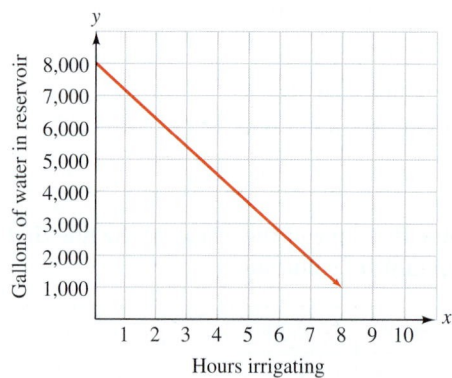

70. COMMERCIAL JETS   Examine the graph and consider trips of more than 7,000 miles by a Boeing 777. Use a rate of change to estimate how the maximum payload decreases as the distance traveled increases.

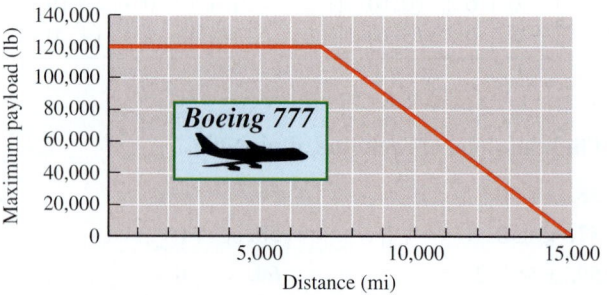

Based on data from Lawrence Livermore National Laboratory and *Los Angeles Times* (October 22, 1998).

71. MILK PRODUCTION   The following graph approximates the amount of milk produced per cow in the U.S. for the years 1993–2002. Find the rate of change.

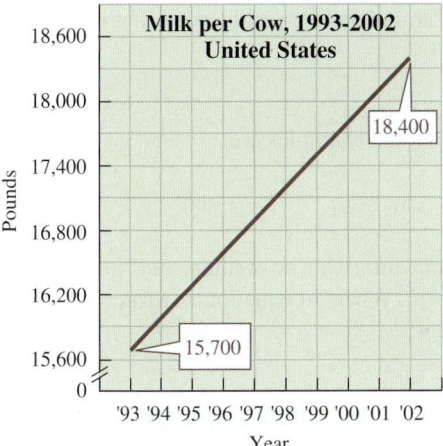

Source: United States Department of Agriculture

72. WAL-MART   The graph below approximates the net sales of Wal-Mart for the years 1991–2002.
   a. Find the rate of change in sales for the years 1991–1998.
   b. Find the rate of change in sales for the years 1998–2002.

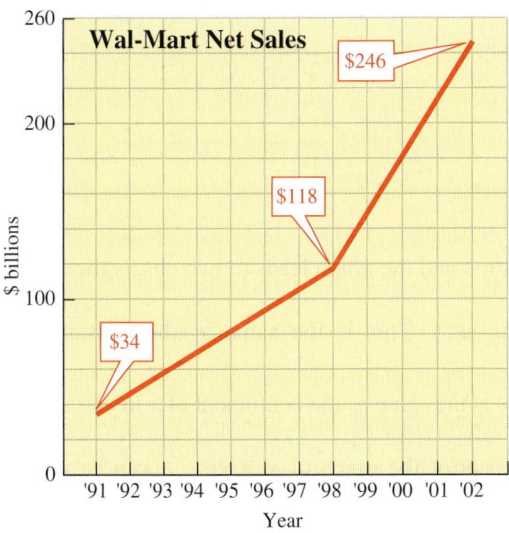

Based on data from Wal-Mart, *USA TODAY* (November 6, 1998), and Hoover's online.

WRITING

73. Explain why the slope of a vertical line is undefined.

74. How do we distinguish between a line with positive slope and a line with negative slope?

75. Give an example of a rate of change that government officials might be interested in knowing so they can plan for the future needs of our country.

76. Explain the difference between a rate of change that is positive and one that is negative. Give an example of each.

REVIEW

77. HALLOWEEN CANDY   A candy maker wants to make a 60-pound mixture of two candies to sell for $2 per pound. If black licorice bits sell for $1.90 per pound and orange gumdrops sell for $2.20 per pound, how many pounds of each should be used?

78. MEDICATIONS   A doctor prescribes an ointment that is 2% hydrocortisone. A pharmacist has 1% and 5% concentrations in stock. How many ounces of each should the pharmacist use to make a 1-ounce tube?

CHALLENGE PROBLEMS

79. Use the concept of slope to determine whether $A(-50, -10)$, $B(20, 0)$, and $C(34, 2)$ all lie on the same straight line.

80. A line having slope $\frac{2}{3}$ passes through the point $(10, -12)$. What is the y-coordinate of another point on the line whose x-coordinate is 16?

<table>
<tr><td>**3.5**</td><td></td></tr>
</table>

3.5 Slope–Intercept Form

- Slope–Intercept Form of the Equation of a Line
- Using the Slope and y-Intercept to Graph a Line
- Writing the Equation of a Line • Parallel and Perpendicular Lines
- Applications

Linear equations appear in many forms. Some examples are:

$$y = 2x + 1, \qquad 3x - 5y = 15, \qquad 8y = 6x + 7, \qquad \text{and} \qquad x = -4$$

Of all of the ways in which a linear equation can be written, one form, called *slope–intercept form,* is probably the most useful. When an equation is written in this form, two important features of its graph are evident.

■ SLOPE–INTERCEPT FORM OF THE EQUATION OF A LINE

To explore the relationship between a linear equation and its graph, let's consider $y = 2x + 1$ and its graph.

$$y = 2x + 1$$

x	y	(x, y)
-1	-1	$(-1, -1)$
0	1	$(0, 1)$
1	3	$(1, 3)$

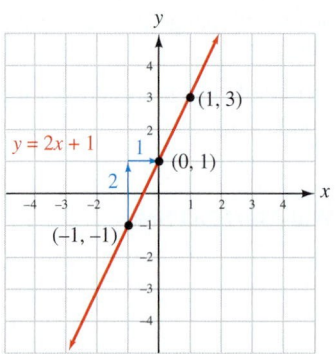

$$\text{Slope} = \frac{\text{rise}}{\text{run}} = \frac{2}{1} = 2$$

A close examination of the equation and the graph leads to two observations:

- The graph crosses the y-axis at 1. This is the same as the constant term in $y = 2x + \mathbf{1}$.
- The slope of the line is 2. This is the same as the coefficient of x in $y = \mathbf{2}x + 1$.

It appears that the slope and y-intercept of the graph of $y = 2x + 1$ can be determined from the equation.

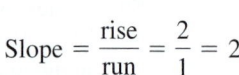

$$y = \mathbf{2}x + \mathbf{1}$$

The slope of   The y-intercept
the line is 2.   is $(0, 1)$.

These observations suggest the following form of an equation of a line.

Slope–Intercept Form of the Equation of a Line	If a linear equation is written in the form $$y = mx + b$$ the graph of the equation is a line with slope m and y-intercept $(0, b)$.

EXAMPLE 1

Find the slope and the y-intercept of the graph of each equation.

a. $y = 6x - 2$ **b.** $y = -\dfrac{5}{4}x$ **c.** $y = \dfrac{x}{2} + 3$ **d.** $y = 5 + 12x$

Solution

a. If we write the subtraction as the addition of the opposite, the equation will be in $y = mx + b$ form:

$$y = \underset{\uparrow}{6}x + \underset{\uparrow}{(-2)}$$
$$y = mx + \ b$$

> **Caution**
>
> For equations in $y = mx + b$ form, the slope of the line is the *coefficient* of x, not the x-term. For example, the graph of $y = 6x - 5$ has slope 6, *not* $6x$.

Since $m = 6$ and $b = -2$, the slope of the line is 6 and the y-intercept is $(0, -2)$.

b. Writing $y = -\frac{5}{4}x$ in slope–intercept form, we have

$$y = -\frac{5}{4}x + 0 \qquad \text{Add 0 to make the value of } b \text{ obvious.}$$

Since $m = -\frac{5}{4}$ and $b = 0$, the slope of the line is $-\frac{5}{4}$ and the y-intercept is $(0, 0)$.

c. Since $\frac{x}{2}$ means $\frac{1}{2}x$, we can rewrite $y = \frac{x}{2} + 3$ as

$$y = \frac{1}{2}x + 3$$

We see that $m = \frac{1}{2}$ and $b = 3$, so the slope of the line is $\frac{1}{2}$ and the y-intercept is $(0, 3)$.

d. We can use the commutative property of addition to reorder the terms on the right-hand side of the equation so that it is in $y = mx + b$ form.

$$y = 12x + 5$$

The slope is 12 and the y-intercept is $(0, 5)$.

Self Check 1

Find the slope and the y-intercept:

a. $y = -5x - 1$ **b.** $y = \dfrac{7}{8}x$ **c.** $y = 5 - \dfrac{x}{3}$

The equation of any nonvertical line can be written in slope–intercept form. To do so, we apply the properties of equality to solve for y.

EXAMPLE 2

ELEMENTARY &
INTERMEDIATE
Algebra $f(x)$ **Now**™   Solution

Find the slope and y-intercept of the line whose equation is $8x + y = 9$.

The slope and y-intercept are not immediately apparent because the equation is not in slope–intercept form. To write it in $y = mx + b$ form, we isolate y on the left-hand side.

$$8x + y = 9$$

$$8x + y - \mathbf{8x} = -\mathbf{8x} + 9 \qquad \text{To undo the addition of } 8x, \text{ subtract } 8x \text{ from both sides.}$$

$$y = -\mathbf{8}x + \mathbf{9} \qquad \text{On the left-hand side, combine like terms: } 8x - 8x = 0.$$

The slope is -8.   The y-intercept is $(0, 9)$.

Self Check 2   Find the slope and y-intercept of the line whose equation is $9x + y = -4$.

EXAMPLE 3

ELEMENTARY &
INTERMEDIATE
Algebra $f(x)$ **Now**™

Find the slope and y-intercept of the line with the given equation.

a. $x + 4y = 16$ **b.** $-9x - 3y = 11$

Solution   **a.** To write the equation in slope–intercept form, we solve for y.

$$x + 4y = 16$$

$$x + 4y - \mathbf{x} = -\mathbf{x} + 16 \qquad \text{To undo the addition of } x, \text{ subtract } x \text{ from both sides.}$$

$$4y = -x + 16 \qquad \text{Simplify the left-hand side.}$$

$$\frac{4y}{\mathbf{4}} = \frac{-x + 16}{\mathbf{4}} \qquad \text{To undo the multiplication by 4, divide both sides by 4.}$$

$$y = \frac{-x}{4} + \frac{16}{4} \qquad \text{On the right-hand side, rewrite } \tfrac{-x + 16}{4} \text{ as the sum of two fractions with like denominators, } \tfrac{-x}{4} \text{ and } \tfrac{16}{4}.$$

$$y = -\frac{1}{4}x + 4 \qquad \text{Write } \tfrac{-x}{4} \text{ as } -\tfrac{1}{4}x. \text{ Simplify: } \tfrac{16}{4} = 4.$$

Since $m = -\frac{1}{4}$ and $b = 4$, the slope is $-\frac{1}{4}$ and the y-intercept is $(0, 4)$.

b. To write the equation in $y = mx + b$ form, we isolate y on the left-hand side.

$$-9x - 3y = 11$$

$$-3y = 9x + 11 \qquad \text{To eliminate the term } -9x \text{ on the left-hand side, add } 9x \text{ to both sides: } -9x + 9x = 0.$$

$$\frac{-3y}{-\mathbf{3}} = \frac{9x}{-\mathbf{3}} + \frac{11}{-\mathbf{3}} \qquad \text{To undo the multiplication by } -3, \text{ divide both sides by } -3.$$

$$y = -3x - \frac{11}{3} \qquad \text{Simplify.}$$

Since $m = -3$ and $b = -\frac{11}{3}$, the slope is -3 and the y-intercept is $\left(0, -\frac{11}{3}\right)$.

Self Check 3   Find the slope and y-intercept of the line whose equation is $10x + 2y = 7$.

■ USING THE SLOPE AND *y*-INTERCEPT TO GRAPH A LINE

If we know the slope and *y*-intercept of a line, we can graph the line. To illustrate this, we graph $y = 5x - 4$, a line with slope 5 and *y*-intercept $(0, -4)$.

We begin by plotting the *y*-intercept. If we write the slope as $\frac{5}{1}$, we see that the rise is 5 and the run is 1. From $(0, -4)$, we move 5 units upward and then 1 unit to the right. This locates a second point on the line, $(1, 1)$. Then we draw a line through the two points. The result is a line with *y*-intercept $(0, -4)$ and slope 5.

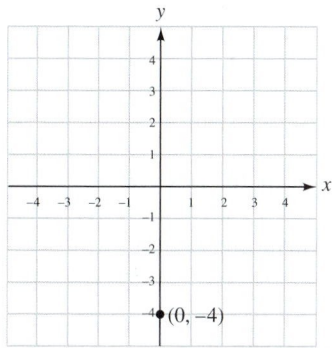

Plot the *y*-intercept, $(0, -4)$.

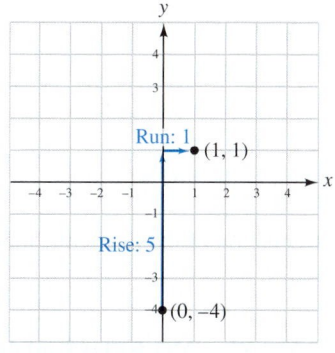

From the *y*-intercept, draw the rise and run components of the slope triangle.

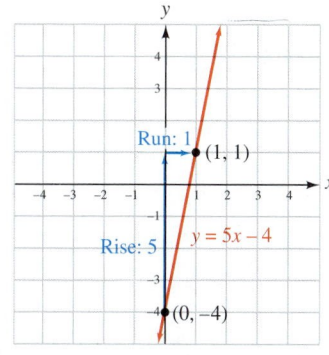

Use a straightedge to draw a line through the two points.

EXAMPLE 4

ELEMENTARY &
INTERMEDIATE
Algebra *f(x)* **Now**™

Graph the line whose equation is $y = -\frac{4}{3}x + 2$.

Solution The slope of the line is $-\frac{4}{3}$, which can be expressed as $\frac{-4}{3}$. After plotting the *y*-intercept, $(0, 2)$, we move 4 units downward and then 3 units to the right. This locates a second point on the line, $(3, -2)$. From this point, we can move another 4 units downward and 3 units to the right, to locate a *third point* on the line, $(6, -6)$. Then we draw a line through the three points to obtain a line with *y*-intercept $(0, 2)$ and slope $-\frac{4}{3}$.

Caution

When using the *y*-intercept and the slope to graph a line, remember to draw the slope triangle from the *y*-intercept, *not* from the origin.

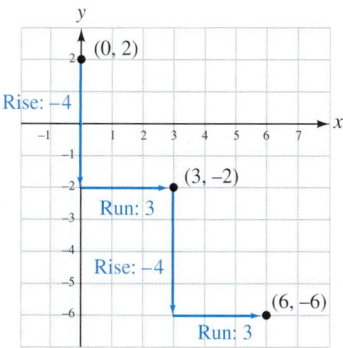

Plot the *y*-intercept. From $(0, 2)$, draw the rise and run components of a slope triangle. From $(3, -2)$, draw another slope triangle.

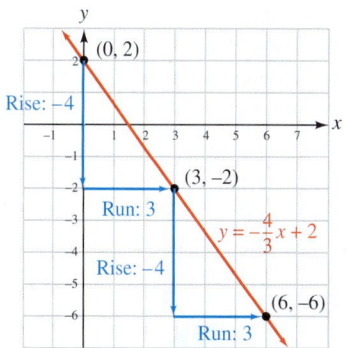

Use a straightedge to draw a line through the three points.

Self Check 4 Graph the line whose equation is $y = -\frac{5}{6}x + 2$.

■ **WRITING THE EQUATION OF A LINE**

If we are given the slope and y-intercept of a line, we can write an equation of the line by substituting for m and b in the slope–intercept form.

EXAMPLE 5   Write an equation of the line with slope -1 and y-intercept $(0, 9)$.

ELEMENTARY &
INTERMEDIATE
Algebra $f(x)$ **Now** ™

Solution   If the slope is -1 and the y-intercept is $(0, 9)$, then $m = -1$ and $b = 9$.

$y = mx + b$ This is the slope–intercept form.
$y = -1x + 9$ Substitute -1 for m and 9 for b.
$y = -x + 9$ Simplify: $-1x = -x$.

The equation of the line with slope -1 and y-intercept $(0, 9)$ is $y = -x + 9$.

Self Check 5   Write the equation of the line with slope 1 and y-intercept $(0, -12)$.

■ **PARALLEL AND PERPENDICULAR LINES**

Two lines that lie in the same plane are **parallel** if they do not intersect. When graphed, parallel lines have the same slope, but different y-intercepts.

Parallel Lines	Two different lines with the same slope are parallel.

EXAMPLE 6   Graph $y = -\dfrac{2}{3}x$ and $y = -\dfrac{2}{3}x + 3$ on the same coordinate system.

Solution   The graph of the first equation is a line with slope $-\dfrac{2}{3}$ and y-intercept $(0, 0)$. The graph of the second equation is a line with slope $-\dfrac{2}{3}$ and y-intercept $(0, 3)$. Since the lines have the same slope, they are parallel.

The Language of Algebra

The word *parallel* is used in many settings: drivers *parallel* park, and gymnasts perform on the *parallel* bars.

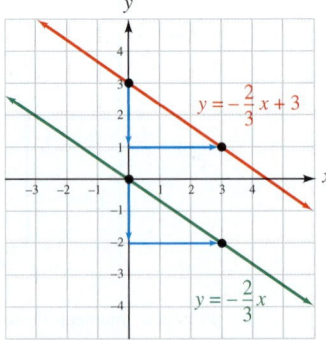

Self Check 6   Graph $y = \dfrac{5}{2}x - 2$ and $y = \dfrac{5}{2}x$ on the same coordinate system.

Unlike parallel lines, **perpendicular lines** intersect. And more important, they intersect to form four right angles (angles with measure 90°). The two lines graphed below are perpendicular. In the figure, the symbol ⌐ is used to denote a right angle.

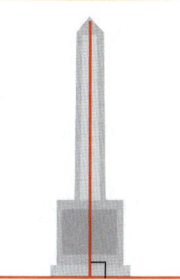

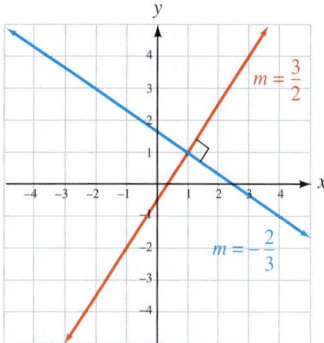

The product of the slopes of two (nonvertical) perpendicular lines is −1. To illustrate this, consider the lines in the illustration, with slopes $\frac{3}{2}$ and $-\frac{2}{3}$. If we find the product of their slopes, we have

$$\frac{3}{2}\left(-\frac{2}{3}\right) = -\frac{6}{6} = -1$$

Note that the slopes of the lines in this example, $\frac{3}{2}$ and $-\frac{2}{3}$, are **negative** (or **opposite**) **reciprocals.** We can use the term *negative reciprocal* to express the relationship between the slopes of perpendicular lines in an alternate way.

Slopes of Perpendicular Lines	**1.** Two nonvertical lines are perpendicular if the product of the slopes is −1, that is, if their slopes are negative reciprocals.

In symbols, two lines with slopes m_1 and m_2 are perpendicular if

$$m_1 \cdot m_2 = -1 \qquad \text{or} \qquad m_1 = -\frac{1}{m_2}$$

2. A horizontal line with 0 slope is perpendicular to a vertical line with undefined slope.

EXAMPLE 7

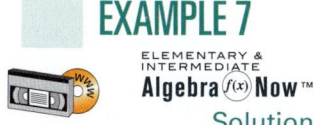

ELEMENTARY & INTERMEDIATE
Algebra *f(x)* **Now**™

Determine whether the graphs of $y = -5x + 6$ and $y = \frac{x}{5} - 2$ are parallel, perpendicular, or neither.

Solution   The slope of the line $y = -5x + 6$ is −5. The slope of the line $y = \frac{x}{5} - 2$ is $\frac{1}{5}$. $\left(\text{Recall that } \frac{x}{5} = \frac{1}{5}x.\right)$ Since the slopes are not equal, the lines are not parallel. If we find the product of their slopes, we have

$$-5\left(\frac{1}{5}\right) = -\frac{5}{5} = -1 \qquad \text{−5 and } \frac{1}{5} \text{ are negative reciprocals.}$$

Since the product of their slopes is −1, the lines are perpendicular.

Self Check 7   Determine whether the graphs of $y = 4x + 6$ and $y = \frac{1}{4}x$ are parallel, perpendicular, or neither.

■ APPLICATIONS

In the next example, as a means of making the equation more descriptive, we replace x and y in $y = mx + b$ with two other variables.

EXAMPLE 8

ELEMENTARY &
INTERMEDIATE
Algebra $f(x)$ **Now** ™

Group discounts.   To promote group sales for an Alaskan cruise, a travel agency reduces the regular cost of $4,500 by $5 for each person traveling in the group.

a. Write a linear equation in slope–intercept form that finds the cost c of the cruise, if a group of p people travel together.

b. Use the equation to determine the cost if a group of 55 retired teachers travel together.

Solution   **a.** Since the cost c of the cruise depends on the number p of people traveling in the group, the equation will have the form $c = mp + b$. We need to determine m and b.

   The cost of the cruise steadily *decreases* as the number of people in the group increases. This rate of change, -5 dollars per person, is the slope of the graph of the equation. Thus m is -5.

   If a group of 0 people take the cruise, there will be no discount; the cruise will cost $4,500. Written as an ordered pair of the form (p, c), we have $(0, 4,500)$. When graphed, this would be the c-intercept. Thus, b is 4,500.

   Substituting for m and b, we obtain the linear equation that models this situation.

$$c = -5p + 4,500$$

Cruise to Alaska

$4,500 per person

Group discounts available*

*For groups of up to 100

---

b. To find the cost of the cruise for a group of 55, we proceed as follows:

$$c = -5\boldsymbol{p} + 4,500$$
$$c = -5(\boldsymbol{55}) + 4,500 \qquad \text{Substitute 55 for } p, \text{ the number of people in the group.}$$
$$c = -275 + 4,500$$
$$= 4,225$$

If a group of 55 people travel together, the Alaskan cruise will cost each person $4,225.

Self Check 8   Write a linear equation in slope–intercept form that finds the cost c of the cruise if a $10-per-person discount is offered for groups.

Answers to Self Checks   **1. a.** $m = -5, (0, -1)$
   b. $m = \frac{7}{8}, (0, 0)$
   c. $m = -\frac{1}{3}, (0, 5)$
2. $m = -9; (0, -4)$
3. $m = -5; \left(0, \frac{7}{2}\right)$

4.
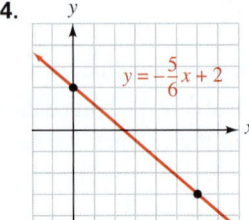

6.
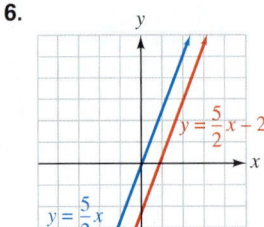

5. $y = x - 12$

7. neither
8. $c = -10p + 4,500$

3.5   STUDY SET   ELEMENTARY & INTERMEDIATE Algebra *f(x)* Now™

VOCABULARY   **Fill in the blanks.**

1. The equation $y = mx + b$ is called the _____ form of the equation of a line.

2. The graph of the linear equation $y = mx + b$ has a _____ of $(0, b)$ and a _____ of m.

3. _____ lines do not intersect.

4. The slope of a line is a _____ of change.

5. The numbers $\frac{5}{6}$ and $-\frac{6}{5}$ are called negative _____. Their product is -1.

6. The product of the slopes of _____ lines is -1.

CONCEPTS

7. Tell whether each equation is in slope–intercept form.
   a. $7x + 4y = 2$ **b.** $5y = 2x - 3$
   c. $y = 6x + 1$ **d.** $x = 4y - 8$
   e. $y = \frac{x}{5} - 3$ **f.** $y = 2x$

8. **a.** How do we solve $4x + y = 9$ for y?

   b. How do we solve $-2x + y = 9$ for y?

9. **a.** To solve $5y = 10x + 20$ for y, both sides of the equation were divided by 5. Complete the solution.

   $$\frac{5y}{5} = \frac{10x}{5} + \frac{20}{5}$$

   $$\boxed{} = \boxed{} + \boxed{}$$

   b. To solve $-2y = 6x - 12$ for y, both sides of the equation were divided by -2. Complete the solution.

   $$\frac{-2y}{-2} = \frac{6x}{-2} - \frac{12}{-2}$$

   $$\boxed{} = \boxed{} \boxed{} 6$$

10. Examine the work shown in the following graph. An equation in slope–intercept form is in the process of being graphed.
    a. What is the y-intercept of the line?
    b. What is the slope of the line?
    c. What equation is being graphed?

d. One more step needs to be completed. What is it?

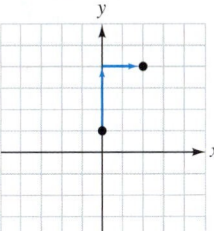

11. Use the graph of the line to determine m and b. Then write the equation of the line in slope–intercept form.

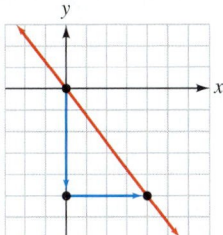

12. Find the negative reciprocal of each number.

Number	Negative reciprocal
6	
$\frac{7}{8}$	
$-\frac{1}{4}$	
1	

13. Fill in the blanks.
    a. Two different lines with the same slope are _____.
    b. If the slopes of two lines are negative reciprocals, the lines are _____.
    c. The product of the slopes of perpendicular lines is _____ .

14. a. What is the y-intercept of Line 1?

   b. What do Line 1 and Line 2 have in common? How are they different?

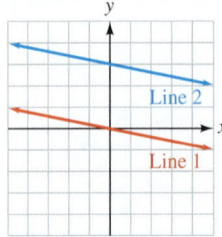

15. The slope of Line 1 shown in the following illustration is 2.

   a. What is the slope of Line 2?

   b. What is the slope of Line 3?

   c. What is the slope of Line 4?

   d. Which lines have the same y-intercept?

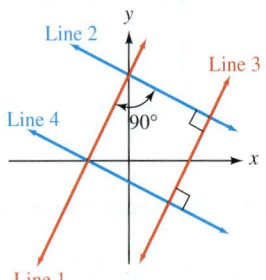

16. NAVIGATION   The graph shows the recommended speed at which a ship should proceed into head waves of various heights.

   a. What information does the y-intercept of the graph give?

   b. What is the rate of change in the recommended speed of the ship as the wave height increases?

   c. Write the equation of the graph.

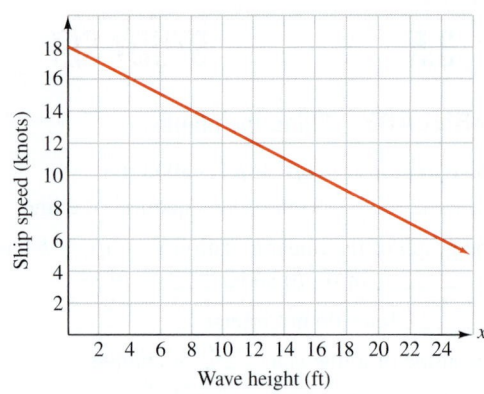

NOTATION   **Complete the solution by solving the equation for y. Then find the slope and the y-intercept of its graph.**

17.
$$2x + 5y = 15$$
$$2x + 5y - \boxed{} = \boxed{} + 15$$
$$\boxed{} = -2x + 15$$
$$\frac{5y}{\boxed{}} = \frac{-2x}{\boxed{}} + \frac{15}{\boxed{}}$$
$$y = -\frac{2}{5}x + \boxed{}$$

The slope is $\boxed{}$ and the y-intercept is $\boxed{}$.

18. What is the slope–intercept form of the equation of a line?

19. Simplify each expression.

   a. $\dfrac{8x}{2}$ **b.** $\dfrac{8x}{6}$

   c. $\dfrac{-8x}{-8}$ **d.** $\dfrac{-16}{8}$

20. Tell whether each statement is true or false.

   a. $\dfrac{x}{6} = \dfrac{1}{6}x$ **b.** $\dfrac{5}{3}x = \dfrac{5x}{3}$

21. What does the symbol $\sqsupset$ denote?

22. Write the phrase *ninety degrees* in symbols.

PRACTICE   **Find the slope and the y-intercept of the graph of each equation.**

23. $y = 4x + 2$ **24.** $y = -4x - 2$

25. $y = -5x - 8$ **26.** $y = 7x + 3$

27. $4x - 2 = y$ **28.** $6x - 1 = y$

29. $y = \dfrac{x}{4} - \dfrac{1}{2}$

30. $y = \dfrac{x}{15} - \dfrac{3}{4}$

31. $y = \dfrac{1}{2}x + 6$

32. $y = \dfrac{4}{5}x - 9$

33. $y = 6 - x$

34. $y = 12 + 4x$

35. $x + y = 8$

36. $x - y = -30$

37. $6y = x - 6$

38. $2y = x + 20$

39. $7y = -14x + 49$

40. $9y = -27x + 36$

41. $-4y = 6x - 4$

42. $-6y = 8x + 6$

43. $2x + 3y = 6$

44. $4x + 5y = 25$

45. $3x - 5y = 15$

46. $x - 6y = 6$

47. $-6x + 6y = -11$

48. $-4x + 4y = -9$

49. $y = x$

50. $y = -x$

51. $y = -5x$

52. $y = 14x$

53. $y = -2$

54. $y = 30$

55. $-5y - 2 = 0$

56. $3y - 13 = 0$

Write an equation of the line with the given slope and y-intercept. Then graph it.

57. Slope 5, y-intercept $(0, -3)$

58. Slope -2, y-intercept $(0, 1)$

59. Slope $\dfrac{1}{4}$, y-intercept $(0, -2)$

60. Slope $\dfrac{1}{3}$, y-intercept $(0, -5)$

61. Slope -3, y-intercept $(0, 6)$

62. Slope 4, y-intercept $(0, -1)$

63. Slope $-\dfrac{8}{3}$, y-intercept $(0, 5)$

64. Slope $-\dfrac{7}{6}$, y-intercept $(0, 2)$

Find the slope and the y-intercept of the graph of each equation. Then graph the equation.

65. $y = 3x + 3$

66. $y = -3x + 5$

67. $y = -\dfrac{x}{2} + 2$

68. $y = \dfrac{x}{3}$

69. $y = -3x$

70. $y = -4x$

71. $4x + y = -4$

72. $2x + y = -6$

73. $3x + 4y = 16$

74. $2x + 3y = 9$

75. $10x - 5y = 5$

76. $4x - 2y = 6$

For each pair of equations, determine whether their graphs are parallel, perpendicular, or neither.

77. $y = 6x + 8$
$y = 6x$

78. $y = 3x - 15$
$y = -\dfrac{1}{3}x + 4$

79. $y = x$
$y = -x$

80. $y = \dfrac{1}{2}x - \dfrac{4}{5}$
$y = 0.5x + 3$

81. $y = -2x - 9$
$y = 2x - 9$

82. $y = \dfrac{3}{4}x + 1$
$y = \dfrac{4}{3}x - 5$

83. $x - y = 12$
$-2x + 2y = -23$

84. $y = -3x + 1$
$3y = x - 5$

85. $x = 9$
$y = 8$

86. $-x + 4y = 10$
$2y + 16 = -8x$

APPLICATIONS

87. PRODUCTION COSTS   A television production company charges a basic fee of $5,000 and then $2,000 an hour when filming a commercial.
   a. Write a linear equation that describes the relationship between the total production costs c and the hours h of filming.
   b. Use your answer to part a to find the production costs if a commercial required 8 hours of filming.

88. COLLEGE FEES Each semester, students enrolling at a community college must pay tuition costs of $20 per unit as well as a $40 student services fee.

a. Write a linear equation that gives the total fees t to be paid by a student enrolling at the college and taking x units.

b. Use your answer to part a to find the enrollment cost for a student taking 12 units.

89. CHEMISTRY A portion of a student's chemistry lab manual is shown below. Use the information to write a linear equation relating the temperature F (in degrees Fahrenheit) of the compound to the time t (in minutes) elapsed during the lab procedure.

Chem. Lab #1   Aug. 13
Step 1: Removed compound from freezer @ –10°F.

Step 2: Used heating unit to raise temperature of compound 5° F every minute.

90. INCOME PROPERTY Use the information in the newspaper advertisement to write a linear equation that gives the amount of income A (in dollars) the apartment owner will receive when the unit is rented for m months.

APARTMENT FOR RENT
1 bedroom/1 bath, with garage
$500 per month +
$250 nonrefundable security fee.

91. EMPLOYMENT SERVICE A policy statement of LIZCO, Inc., is shown below. Suppose a secretary had to pay an employment service $500 to get placed in a new job at LIZCO. Write a linear equation that tells the secretary the actual cost c of the employment service to her m months after being hired.

Policy no. 23452–A new hire will be reimbursed by LIZCO for any employment service fees paid by the employee at the rate of $20 per month.

92. VIDEOTAPES A VHS videocassette contains 800 feet of tape. In the long play mode (LP), it plays 10 feet of tape every 3 minutes. Write a linear equation that relates the number of feet f of tape yet to be played and the number of minutes m the tape has been playing.

93. SEWING COSTS A tailor charges a basic fee of $20 plus $5 per letter to sew an athlete's name on the back of a jacket.

a. Write a linear equation that will find the cost c to have a name containing x letters sewn on the back of a jacket.

b. Graph the equation.

c. Suppose the tailor raises the basic fee to $30. On your graph from part b, draw the new graph showing the increased cost.

94. SALAD BARS For lunch, a delicatessen offers a "Salad and Soda" special where customers serve themselves at a well-stocked salad bar. The cost is $1.00 for the drink and 20¢ an ounce for the salad.

a. Write a linear equation that will find the cost c of a "Salad and Soda" lunch when a salad weighing x ounces is purchased.

b. Graph the equation.

c. How would the graph from part b change if the delicatessen began charging $2.00 for the drink?

d. How would the graph from part b change if the cost of the salad changed to 30¢ an ounce?

95. PROFESSIONAL HOCKEY Use the following facts to write a linear equation in slope–intercept form that approximates the average price of a National Hockey League ticket for the years 1995–2002.

- Let t represent the number of years since 1995 and c the average cost of a ticket.
- In 1995, the average ticket price was $33.50.
- From 1995 to 2002, the average ticket price increased $2 per year.

(Source: Team Marketing Report, NHL)

96. COMPUTER DRAFTING The illustration shows a computer-generated drawing of an automobile engine mount. When the designer clicks the mouse on a line of the drawing, the computer finds the equation of the line. Determine whether the two lines selected in the drawing are perpendicular.

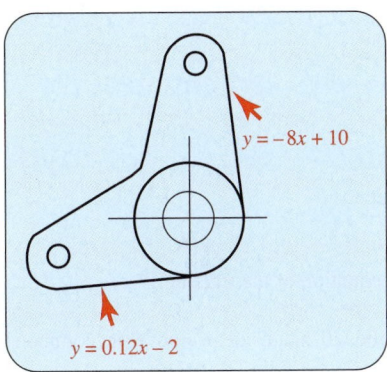

$y = -8x + 10$

$y = 0.12x - 2$

WRITING

97. Why is $y = mx + b$ called the slope–intercept form of the equation of a line?

98. On a quiz, a student was asked to find the slope of the graph of $y = 2x + 3$. She answered: $m = 2x$. Her instructor marked it wrong. Explain why the answer is incorrect.

REVIEW

99. CABLE TV   A 186-foot television cable is to be cut into four pieces. Find the length of each piece if each successive piece is 3 feet longer than the previous one.

100. INVESTMENTS   Joni received $25,000 as part of a settlement in a class action lawsuit. She invested some money at 10% and the rest at 9% simple interest rates. If her total annual income from these two investments was $2,430, how much did she invest at each rate?

CHALLENGE PROBLEMS

101. If the graph of $y = mx + b$ passes through quadrants I, II, and IV, what can be known about the constants m and b?

102. The equation $y = \frac{3}{4}x - 5$ is in slope–intercept form. Write it in general form, $Ax + By = C$, where $A > 0$.

3.6 Point–Slope Form

- Point–Slope Form of the Equation of a Line   • Writing the Equation of a Line
- Horizontal and Vertical Lines   • Using a Point and the Slope to Graph a Line
- Applications

If we know the slope of a line and its y-intercept, we can use the slope–intercept form to write the equation of the line. The question that now arises is, can *any* point on the line be used in combination with its slope to write its equation? In this section, we answer this question.

■ POINT–SLOPE FORM OF THE EQUATION OF A LINE

Refer to the line graphed on the left. Since the slope triangle has rise 3 and run 1, the slope of the line is $\frac{3}{1} = 3$. To develop a new form for the equation of a line, we will find the slope of this line in another way.

   If we pick another point on the line and call it $Q(x, y)$, we can find the slope of the line by substituting the coordinates of points P and Q into the slope formula.

$$\frac{y_2 - y_1}{x_2 - x_1} = m$$

$$\frac{y - 1}{x - 2} = m \qquad \text{Substitute } y \text{ for } y_2, 1 \text{ for } y_1, x \text{ for } x_2, \text{ and } 2 \text{ for } x_1.$$

Since the slope of the line is 3, we can substitute 3 for m in the previous equation.

$$\frac{y-1}{x-2} = m$$

$$\frac{y-1}{x-2} = 3$$

We then multiply both sides by $x - 2$ to get

$$\frac{y-1}{x-2}(x-2) = 3(x-2) \qquad \text{Clear the equation of the fraction.}$$

$$y - 1 = 3(x-2) \qquad \text{Simplify the left-hand side. Remove the common factor}$$
$$x - 2 \text{ in the numerator and denominator: } \frac{y-1}{x-2} \cdot \frac{x-2}{1}.$$

The resulting equation displays the slope of the line and the coordinates of one point on the line:

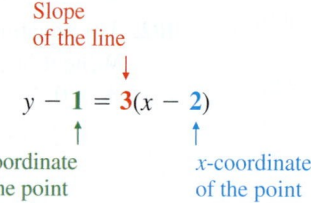

In general, suppose we know that the slope of a line is m and that the line passes through the point (x_1, y_1). Then if (x, y) is any other point on the line, we can use the definition of slope to write

$$\frac{y - y_1}{x - x_1} = m$$

If we multiply both sides by $x - x_1$, we have

$$y - y_1 = m(x - x_1)$$

This form of a linear equation is called **point–slope form.** It can be used to write the equation of a line when the slope and one point on the line are known.

Point–Slope Form of the Equation of a Line

If a line with slope m passes through the point (x_1, y_1), the equation of the line is

$$y - y_1 = m(x - x_1)$$

■ WRITING THE EQUATION OF A LINE

If we are given the slope of a line and a point on the line, we can write an equation of the line by substituting for m, x_1, and y_1 in the point–slope form.

EXAMPLE 1

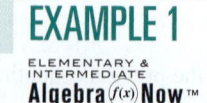

Write an equation of the line that has slope -8 and passes through $(-1, 5)$. Write the answer in slope–intercept form.

Solution   We begin by writing an equation in point–slope form.

$$y - y_1 = m(x - x_1)$$
$$y - 5 = -8[x - (-1)]$$ Substitute -8 for m, -1 for x_1, and 5 for y_1.
$$y - 5 = -8(x + 1)$$ Simplify within the brackets.

To write this equation in slope–intercept form, solve for y.

$$y - 5 = -8(x + 1)$$
$$y - 5 = -8x - 8$$ Distribute the multiplication by -8.
$$y = -8x - 3$$ To undo the subtraction of 5, add 5 to both sides.

In slope–intercept form, the equation is $y = -8x - 3$.

   To verify this result, we note that $m = -8$. Therefore, the slope of the line is -8, as required. To see whether the line passes through $(-1, 5)$, we substitute -1 for x and 5 for y in the equation. If this point is on the line, a true statement should result.

$$y = -8x - 3$$
$$5 \stackrel{?}{=} -8(-1) - 3$$
$$5 \stackrel{?}{=} 8 - 3$$
$$5 = 5$$ True.

Self Check 1   Write an equation of the line that has slope -2 and passes through $(4, -3)$. Write the answer in slope–intercept form.

   In the next example, we show that it is possible to write the equation of a line when we know the coordinates of two points on the line.

EXAMPLE 2

Write an equation of the line that passes through $(-2, 6)$ and $(4, 7)$. Write the answer in slope–intercept form.

Solution   We begin by writing an equation in point–slope form. To find the slope of the line, we use the slope formula.

$$m = \frac{y_2 - y_1}{x_2 - x_1} = \frac{7 - 6}{4 - (-2)} = \frac{1}{6}$$

Either point on the line can serve as (x_1, y_1). If we choose $(4, 7)$, we have

$$y - y_1 = m(x - x_1)$$ This is the point–slope form.
$$y - 7 = \frac{1}{6}(x - 4)$$ Substitute $\frac{1}{6}$ for m, 7 for y_1, and 4 for x_1.

To write this equation in slope–intercept form, solve for y.

Success Tip

In Example 2, either of the given points can be used as (x_1, y_1) when writing the point–slope equation.

   Looking ahead, we usually choose the point whose coordinates will make the computations the easiest.

$$y - 7 = \frac{1}{6}x - \frac{2}{3}$$ Distribute the multiplication by $\frac{1}{6}$.

$$y = \frac{1}{6}x - \frac{4}{6} + \frac{42}{6}$$ To undo the subtraction of 7, add 7 to both sides. On the right-hand side, add 7 in the form $\frac{42}{6}$ so that it can be added to $-\frac{4}{6}$.

$$y = \frac{1}{6}x + \frac{19}{3}$$ This is slope–intercept form.

An equation of the line that passes through $(-2, 6)$ and $(4, 7)$ is $y = \frac{1}{6}x + \frac{19}{3}$.

Self Check 2   Write an equation of the line that passes through $(-5, 4)$ and $(8, -6)$. Write the answer in slope–intercept form.

■ HORIZONTAL AND VERTICAL LINES

We have previously graphed horizontal and vertical lines. We will now discuss how to write their equations.

EXAMPLE 3

ELEMENTARY &
INTERMEDIATE
Algebra ƒ⁽ˣ⁾ **Now**™

Write an equation of each line and graph it: **a.** a horizontal line passing through $(-2, -4)$ and   **b.** a vertical line passing through $(1, 3)$.

Solution   **a.** The equation of a horizontal line can be written in the form $y = b$. Since the y-coordinate of $(-2, -4)$ is -4, the equation of the line is $y = -4$. The graph is shown in the figure.

b. The equation of a vertical line can be written in the form $x = a$. Since the x-coordinate of $(1, 3)$ is 1, the equation of the line is $x = 1$. The graph is shown in the figure.

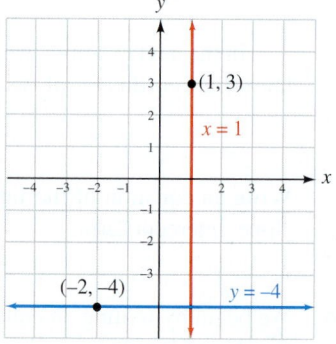

Self Check 3   Write an equation of each line: **a.** a horizontal line passing through $(3, 2)$
b. a vertical line passing through $(-1, -3)$

■ USING A POINT AND THE SLOPE TO GRAPH A LINE

If we know the coordinates of a point on a line, and if we know the slope of the line, we can use the slope to determine a second point on the line.

EXAMPLE 4

ELEMENTARY &
INTERMEDIATE
Algebra ƒ⁽ˣ⁾ **Now**™

Graph the line with slope $\frac{2}{5}$ that passes through $(-1, -3)$.

Solution   We begin by plotting the point $(-1, -3)$. From there, we move 2 units up and then 5 units to the right. This puts us at a second point on the line, $(4, -1)$. We use a straight-edge to draw a line through the two points.

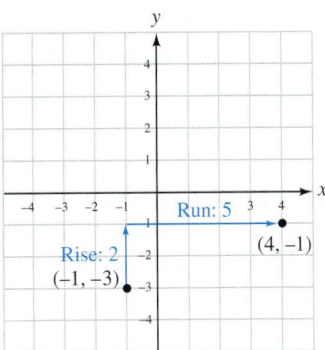 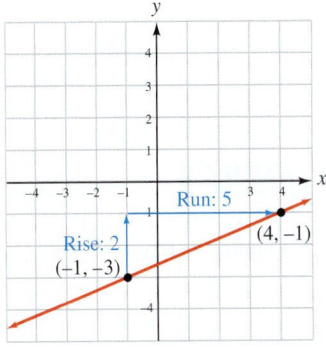

Self Check 4   Graph the line with slope -4 that passes through $(-4, 2)$.

■ APPLICATIONS

A variety of situations can be described by a linear equation. Quite often, these equations are written using variables other than x and y. In such cases, it is helpful to determine what an ordered-pair solution of the equation would look like.

EXAMPLE 5

ELEMENTARY & INTERMEDIATE
Algebra $f(x)$ **Now**™

Men's shoe sizes.   The length (in inches) of a man's foot is *not* his shoe size. For example, the smallest adult men's shoe size is 5, and it fits a 9-inch-long foot. There is, however, a linear relationship between the two. It can be stated this way: Shoe size increases by 3 sizes for each 1-inch increase in foot length.

a. Write a linear equation that relates shoe size s to foot length L.

b. Shaquille O'Neal's foot is about 14.6 inches long. What size shoe does he wear?

Solution   **a.** Since shoe size s depends on the length L of the foot, ordered pairs have the form (L, s). The relationship is linear, so the graph of the equation we are to write is a line.

- The line's slope is the rate of change: $\frac{3 \text{ sizes}}{1 \text{ inch}}$. Therefore, $m = 3$.
- A 9-inch-long foot wears size 5, so the line passes through $(9, 5)$.

We substitute these facts into the point–slope form and solve for s.

$s - s_1 = m(L - L_1)$	This is the point–slope form using the variables L and s.
$s - 5 = 3(L - 9)$	Substitute 3 for m, 9 for L_1, and 5 for s_1.
$s - 5 = 3L - 27$	Distribute the multiplication by 3.
$s = 3L - 22$	Add 5 to both sides.

The equation relating men's shoe size and foot length is $s = 3L - 22$.

b. To find Shaquille's shoe size, we substitute 14.6 inches for L in the equation.

$$s = 3L - 22$$
$$s = 3(14.6) - 22$$
$$s = 43.8 - 22$$
$$s = 21.8$$

Since men's shoes only come in full- and half-sizes, we round 21.8 up to 22. Shaquille O'Neal wears size 22 shoes.

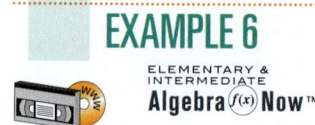

EXAMPLE 6

ELEMENTARY & INTERMEDIATE

Algebra $f(x)$ **Now** ™

Studying learning.   In a series of trials, a rat was released in a maze to search for food. Researchers recorded the time that it took the rat to complete the maze on a **scatter diagram.** After the 40th trial, they drew a straight line through the data to obtain a model of the rat's performance. Write an equation of the line in slope–intercept form.

The Language of Algebra

The term *scatter diagram* is somewhat misleading. Often, the data points are not scattered loosely about. In this case, they fall, more or less, along an imaginary straight line, indicating a linear relationship.

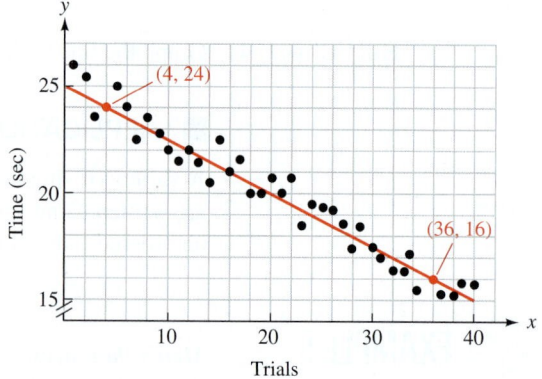

Solution   We begin by writing a point–slope equation. The line passes through several points; we will use (4, 24) and (36, 16) to find the slope.

$$m = \frac{y_2 - y_1}{x_2 - x_1} = \frac{16 - 24}{36 - 4} = \frac{-8}{32} = -\frac{1}{4}$$

Any point on the line can serve as (x_1, y_1). We will use (4, 24).

$$y - y_1 = m(x - x_1) \qquad \text{This is the point–slope form.}$$

$$y - 24 = -\frac{1}{4}(x - 4) \qquad \text{Substitute } -\frac{1}{4} \text{ for } m, 4 \text{ for } x_1, \text{ and } 24 \text{ for } y_1.$$

To write this equation in slope–intercept form, solve for y.

$$y - 24 = -\frac{1}{4}x + 1 \qquad \text{Distribute the multiplication by } -\frac{1}{4}: -\frac{1}{4}(-4) = 1.$$

$$y = -\frac{1}{4}x + 25 \qquad \text{Add 24 to both sides.}$$

A linear equation that models the rat's performance on the maze is $y = -\frac{1}{4}x + 25$, where x is the number of the trial and y is the time it took, in seconds.

Answers to Self Checks   **1.** $y = -2x + 5$

2. $y = -\dfrac{10}{13}x + \dfrac{2}{13}$

3. a. $y = 2$,   **b.** $x = -1$

4.

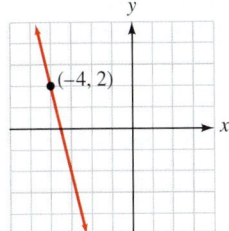

3.6   STUDY SET   ELEMENTARY & INTERMEDIATE Algebra $f(x)$ Now™

VOCABULARY   Fill in the blanks.

1. $y - y_1 = m(x - x_1)$ is called the _____ form of the equation of a line.

2. $y = mx + b$ is called _____ form of the equation of a line.

CONCEPTS

3. Tell what form each equation is written in.
  a. $y - 4 = 2(x - 5)$
  b. $y = 2x + 15$

4. The following equations are written in point–slope form. What point does the graph of the equation pass through, and what is the line's slope?
  a. $y - 2 = 6(x - 7)$
  b. $y + 3 = -8(x + 1)$

5. Simplify each expression.
  a. $x - (-6)$ **b.** $y - (-9)$

6. Use the distributive property to remove parentheses.
  a. $-3(x - 2)$ **b.** $\dfrac{4}{5}(x + 10)$

  c. $-\dfrac{1}{3}(x + 4)$ **d.** $4[x - (-12)]$

7. Find the slope of the line that passes through $(2, -3)$ and $(-4, 12)$.

8. Simplify: $\dfrac{1}{8}x - \dfrac{5}{8} + 2$.

9. On a quiz, a student was asked to write the slope–intercept equation of a line with slope 4 that passes through $(-1, 3)$. His answer was $y = 4x + 7$.
  a. Does $y = 4x + 7$ have the required slope?
  b. Does $y = 4x + 7$ pass through the point $(-1, 3)$?

10. Fill in the blanks.
  a. The equation of a horizontal line has the form $\boxed{} = b$.
  b. The equation of a vertical line has the form $\boxed{} = a$.

11. Examine the work shown in the following illustration, where a line is being graphed.
  a. What point does the line pass through?
  b. What is the slope of the line?
  c. What is a second point on the line?

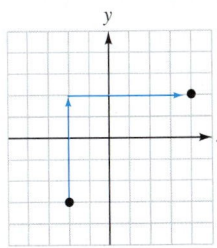

12. a. Find two points on the line graphed below whose coordinates are integers.
  b. What is the slope of the line?
  c. Use your answers to parts a and b to write the equation of the line. Give the answer in point–slope form.

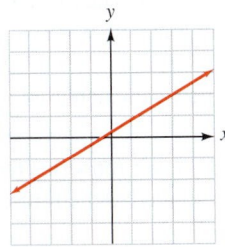

13. In each of the following cases, is the given information sufficient to write an equation of the line?

a. It passes through $(2, -7)$.

b. Its slope is $-\frac{3}{4}$.

c. Its y-intercept is $(0, -1)$ and its slope is 2.

d. It has the following table of solutions.

x	y
2	3
-3	-6

14. In each of the following cases, is the given information sufficient to write an equation of the line?

a. It is horizontal.

b. It is vertical and passes through $(-1, 1)$.

c. Its y-intercept is $(0, 5)$.

d. It has the following table of solutions.

x	y
4	5

15. Suppose you are asked to write an equation of the line in the scatter diagram below. What two points would you use to write the point–slope equation?

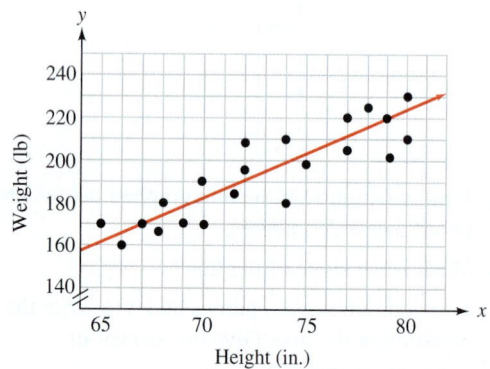

16. In each of the following cases, a linear relationship between two quantities is described. If the relationship were graphed, what would be the slope of the line?

a. The sales of new cars increased by 15 every 2 months.

b. There were 35 fewer robberies for each dozen police officers added to the force.

c. Withdrawals were occurring at the rate of $700 every 45 minutes.

d. One acre of forest is being destroyed every 30 seconds.

NOTATION

17. What is the point–slope form of the equation of a line?

18. Fill in the blanks to state the point–slope form in words.

> y minus y _____ one equals m _____ the quantity of x _____ x sub _____.

19. Consider these steps:

$$y - 3 = 2(x + 1)$$
$$y - 3 = 2x + 2$$
$$y = 2x + 5$$

Now fill in the blanks. The original equation was in _____ form. After solving for ▢, we obtain an equation in _____ form.

Complete the solution.

20. Write the equation of the line with slope -2 that passes through the point $(-1, 5)$. Write the answer in slope–intercept form.

$$y - y_1 = m(x - x_1)$$
$$y - \boxed{} = -2\big[x - \big(\,\boxed{}\,\big)\big]$$
$$y - 5 = -2x - \boxed{}$$
$$y = -2x + \boxed{}$$

PRACTICE Use the point–slope form to write an equation of the line with the given slope and point.

21. Slope 3, passes through $(2, 1)$

22. Slope 2, passes through $(4, 3)$

23. Slope $-\dfrac{4}{5}$, passes through $(-5, -1)$

24. Slope $-\dfrac{7}{8}$, passes through $(-2, -9)$

Use the point–slope form to write an equation of the line with the given slope and point. Then write your result in slope–intercept form.

25. Slope $\dfrac{1}{5}$, passes through $(10, 1)$

26. Slope $\dfrac{1}{4}$, passes through $(8, 1)$

27. Slope -5, passes through $(-9, 8)$

28. Slope -4, passes through $(-2, 10)$

29. Slope $-\dfrac{4}{3}$,

x	y
6	-4

30. Slope $-\dfrac{3}{2}$,

x	y
-2	1

31. Slope $-\dfrac{11}{6}$, passes through $(2, -6)$

32. Slope $-\dfrac{5}{4}$, passes through $(2, 0)$

33. Slope $-\dfrac{2}{3}$, passes through $(3, 0)$

34. Slope $-\dfrac{2}{5}$, passes through $(15, 0)$

35. Slope 8, passes through $(0, 4)$

36. Slope 6, passes through $(0, -4)$

37. Slope -3, passes through the origin

38. Slope -1, passes through the origin

Write an equation of the line that passes through the two given points. Write your result in slope–intercept form.

39. Passes through $(1, 7)$ and $(-2, 1)$

40. Passes through $(-2, 2)$ and $(2, -8)$

41.

x	y
-4	3
2	0

42.

x	y
-1	-4
1	-2

43. Passes through $(5, 5)$ and $(7, 5)$

44. Passes through $(-2, 1)$ and $(-2, 15)$

45. Passes through $(5, 1)$ and $(-5, 0)$

46. Passes through $(-3, 0)$ and $(3, 1)$

47. Passes through $(-8, 2)$ and $(-8, 17)$

48. Passes through $\left(\dfrac{2}{3}, 2\right)$ and $(0, 2)$

49. Passes through $(5, 0)$ and $(-11, -4)$

50. Passes through $(7, -3)$ and $(-5, 1)$

Write the equation of the line with the given characteristics.

51. Vertical, passes through $(4, 5)$

52. Vertical, passes through $(-2, -5)$

53. Horizontal, passes through $(4, 5)$

54. Horizontal, passes through $(-2, -5)$

Graph the line that passes through the given point and has the given slope.

55. $(1, -2)$, slope -1 **56.** $(-4, 1)$, slope -3

57. $(5, -3)$, $m = \dfrac{3}{4}$ **58.** $(2, -4)$, $m = \dfrac{2}{3}$

59. $(-2, -3)$, slope 2   **60.** $(-3, -3)$, slope 4

61. $(4, -3)$, slope $-\dfrac{7}{8}$ **62.** $(4, 2)$, slope $-\dfrac{1}{5}$

APPLICATIONS

63. ANATOMY   There is a linear relationship between a woman's height and the length of her radius bone. It can be stated this way: Height increases by 3.9 inches for each 1-inch increase in the length of the radius. Suppose a 64-inch-tall woman has a 9-inch-long radius bone. Use this information to write a linear equation that relates height h to the length r of the radius. Write your answer in slope–intercept form.

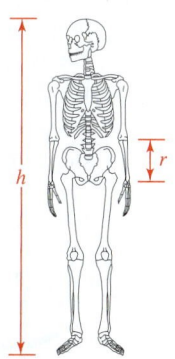

64. AUTOMATION   An automated production line uses distilled water at a rate of 300 gallons every 2 hours to make shampoo. After the line had run for 7 hours, planners noted that 2,500 gallons of distilled water remained in the storage tank. Write a linear equation relating the time t in hours since the production line began and the number g of gallons of distilled water in the storage tank. Write the answer in slope–intercept form.

65. POLE VAULTING   Write the equations of the lines that describe the positions of the pole for parts 1, 3, and 4 of the jump. Write the answers in slope–intercept form.

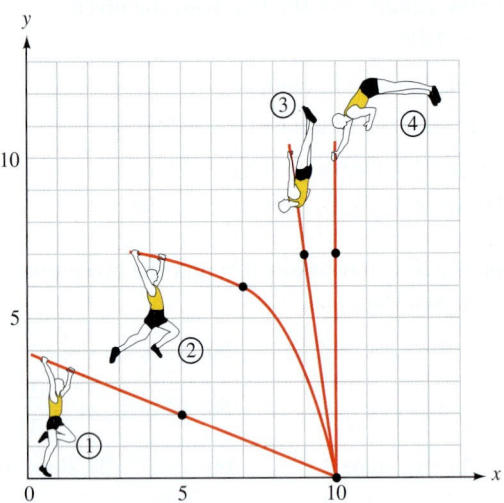

66. FREEWAY DESIGN   The graph below shows the route of a proposed freeway.

   a. Give the coordinates of the points where the proposed freeway will join Interstate 25 and Highway 40.

   b. Write the equation of the line that describes the route of the proposed freeway. Give the answer in slope–intercept form.

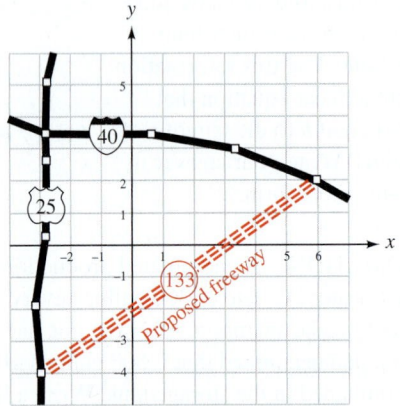

67. TOXIC CLEANUP   Three months after cleanup began at a dump site, 800 cubic yards of toxic waste had yet to be removed. Two months later, that number had been lowered to 720 cubic yards.

   a. Write an equation that describes the linear relationship between the length of time m (in months) the cleanup crew has been working and the number of cubic yards y of toxic waste remaining. Give the answer in slope–intercept form.

   b. Use your answer to part a to predict the number of cubic yards of waste that will still be on the site one year after the cleanup project began.

68. DEPRECIATION   To lower its corporate income tax, accountants of a large company depreciated a word processing system over several years using a linear model, as shown in the worksheet.

   a. Write a linear equation relating the years since the system was purchased, x, and its value, y, in dollars. Give the answer in slope–intercept form.

   b. Find the purchase price of the system.

Tax Worksheet

Method of depreciation: *Linear*

Property	Value	Years after purchase
Word processing system	$60,000	2
"	$30,000	4

69. TRAMPOLINES   There is a linear relationship between the length of the protective pad that wraps around a trampoline and the radius of the trampoline. Use the data in the table to write an equation that gives the length l of pad needed for any trampoline with radius r. Write the answer in slope–intercept form. Use units of feet for both l and r.

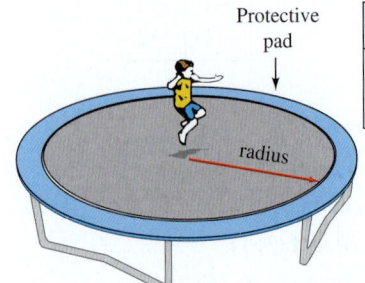

Protective pad

radius

Radius	Pad length
3 ft	19 ft
7 ft	44 ft

70. CONVERTING TEMPERATURES   The relationship between Fahrenheit temperature, F, and Celsius temperature, C, is linear.

  a. Use the data in the illustration to write two ordered pairs of the form (C, F).

  b. Use your answer to part a to write a linear equation relating the Fahrenheit and Celsius scales. Write the answer in slope–intercept form.

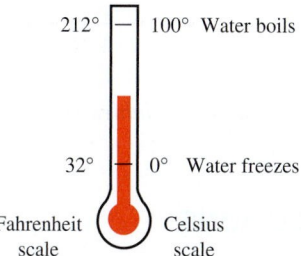

212° — 100° Water boils

32°   — 0°   Water freezes

Fahrenheit   Celsius
  scale   scale

71. GOT MILK?   The scatter diagram shows per capita milk consumption in the U.S. for the years 1975–2000. A straight line can be used to model the data.

  a. Use two points on the line to write its equation. Write the answer in slope–intercept form.

  b. Use your answer to part a to predict the per capita milk consumption in 2015.

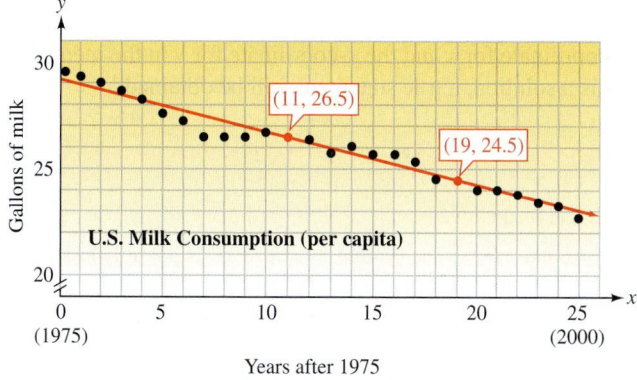

U.S. Milk Consumption (per capita)

(11, 26.5)

(19, 24.5)

0 (1975)   5   10   15   20   25 (2000)

Years after 1975

Source: United States Department of Agriculture

72. ENGINE OUTPUT   The horsepower produced by an automobile engine was recorded for various engine speeds in the range of 2,400–4,800 revolutions per minute (rpm). The data were recorded on the following scatter diagram. Write an equation of the line that models the relationship between engine speed s and horsepower h. Give the answer in slope–intercept form.

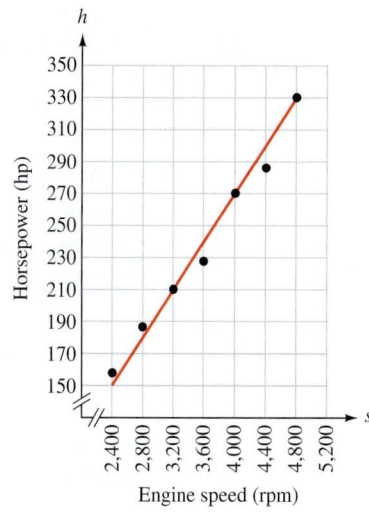

Engine speed (rpm)

WRITING

73. Why is $y - y_1 = m(x - x_1)$ called the point–slope form of the equation of a line?

74. If we know two points that a line passes through, we can write its equation. Explain how this is done.

75. Explain the steps involved in writing $y - 6 = 4(x - 1)$ in slope–intercept form.

76. Think of several points on the graph of the horizontal line $y = 4$. What do the points have in common? How do they differ?

REVIEW

77. FRAMING PICTURES   The length of a rectangular picture is 5 inches greater than twice the width. If the perimeter is 112 inches, find the dimensions of the frame.

78. SPEED OF AN AIRPLANE   Two planes are 6,000 miles apart, and their speeds differ by 200 mph. They travel toward each other and meet in 5 hours. Find the speed of the slower plane.

CHALLENGE PROBLEMS

79. Write an equation of the line that passes through $(2, 5)$ and is parallel to the line $y = 4x - 7$. Give the answer in slope–intercept form.

80. Write an equation of the line that passes through $(-6, 3)$ and is perpendicular to the line $y = -3x - 12$. Give the answer in slope–intercept form.

<table>
<tr><td>**3.7**</td></tr>
</table>

3.7 Graphing Linear Inequalities

• Linear Inequalities and Solutions   • Graphing Linear Inequalities   • Applications

Recall that an **inequality** is a statement that contains one of the symbols $<$, $\leq$, $>$, or $\geq$. Inequalities *in one variable,* such as $x + 6 < 8$ and $5x + 3 \geq 4x$, were solved in Section 2.7. Because they have an infinite number of solutions, we represented their solution sets graphically, by shading intervals on a number line.

We now extend that concept to linear inequalities *in two variables,* as we introduce a procedure that is used to graph their solution sets.

■ LINEAR INEQUALITIES AND SOLUTIONS

If the $=$ symbol in a linear equation in two variables is replaced with an inequality symbol, we have a **linear inequality in two variables.** Some examples are

$$x - y \leq 5, \qquad 4x + 3y < -6, \qquad \text{and} \qquad y > 2x$$

As with linear equations, a **solution of a linear inequality** is an ordered pair of numbers that makes the inequality true.

EXAMPLE 1

ELEMENTARY &
INTERMEDIATE
Algebra *f(x)* **Now**™

Determine whether each ordered pair is a solution of $x - y \leq 5$. Then graph each solution:
a. $(4, 2)$, **b.** $(0, -6)$, and **c.** $(1, -4)$.

Solution   In each case, we substitute the x-coordinate for x and the y-coordinate for y in the inequality $x - y \leq 5$. If the ordered pair is a solution, a true statement will be obtained.

a. For $(4, 2)$:

$$x - y \leq 5 \qquad \text{This is the original inequality.}$$
$$4 - 2 \overset{?}{\leq} 5 \qquad \text{Substitute 4 for } x \text{ and 2 for } y. \text{ The symbol } \overset{?}{\leq} \text{ is read as ``is possibly less than or equal to.''}$$
$$2 \leq 5 \qquad \text{True.}$$

Because $2 \leq 5$ is true, $(4, 2)$ is a solution of $x - y \leq 5$. We say that $(4, 2)$ *satisfies* the inequality. The solution is graphed as shown on the right.

b. For $(0, -6)$:

$$x - y \leq 5 \qquad \text{This is the original inequality.}$$
$$0 - (-6) \overset{?}{\leq} 5 \qquad \text{Substitute 0 for } x \text{ and } -6 \text{ for } y.$$
$$6 \leq 5 \qquad \text{False.}$$

Because $6 \leq 5$ is false, $(0, -6)$ is not a solution.

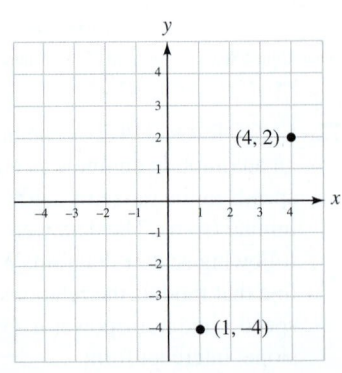

c. For $(1, -4)$:

$$x - y \leq 5 \qquad \text{This is the original inequality.}$$
$$1 - (-4) \stackrel{?}{\leq} 5 \qquad \text{Substitute 1 for } x \text{ and } -4 \text{ for } y.$$
$$5 \leq 5 \qquad \text{True.}$$

Because $5 \leq 5$ is true, $(1, -4)$ is a solution, and we graph it as shown.

Self Check 1   Using the inequality in Example 1, determine whether each ordered pair is a solution:
a. $(8, 2)$,   **b.** $(4, -1)$,   **c.** $(-2, 4)$,   and   **d.** $(-3, -5)$.

In Example 1, we graphed several of the solutions of $x - y \leq 5$. There are many more ordered pairs (x, y) that make the inequality true. It would not be reasonable to plot each of them. Fortunately, there is an easier way to show all of the solutions.

■ GRAPHING LINEAR INEQUALITIES

The graph of a linear inequality is a picture that represents the set of all points whose coordinates satisfy the inequality. In general, such graphs are areas bounded by a line. We call those areas **half-planes,** and we use a two-step procedure to find them.

EXAMPLE 2   Graph: $x - y \leq 5$.

ELEMENTARY &
INTERMEDIATE
Algebra *f(x)* **Now**™   **Solution**   Since the inequality symbol $\leq$ includes an equal symbol, the graph of $x - y \leq 5$ includes the graph of $x - y = 5$.

Step 1: To graph $x - y = 5$, we use the intercept method, as shown in part (a) of the illustration. The resulting line, called a **boundary line,** divides the coordinate plane into two half-planes. To show that the points on the boundary line are solutions of $x - y \leq 5$, we draw it as a solid line.

$x - y = 5$

x	y	(x, y)
0	-5	$(0, -5)$
5	0	$(5, 0)$
6	1	$(6, 1)$

Let $x = 0$, find y. Let $y = 0$, find x. As a check, let $x = 6$ and find y.

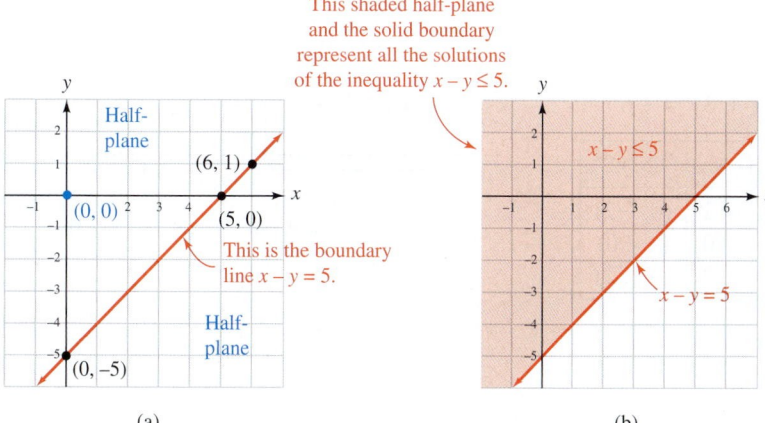

(a)   (b)

Step 2: Since the inequality $x - y \leq 5$ also allows $x - y$ to be less than 5, other ordered pairs, besides those on the boundary, satisfy the inequality. For example, consider the origin, with coordinates $(0, 0)$. If we substitute 0 for x and 0 for y in the given inequality, we have

$$x - y \leq 5$$
$$0 - 0 \stackrel{?}{\leq} 5$$
$$0 \leq 5 \quad \text{True.}$$

Because $0 \leq 5$, the coordinates of the origin satisfy $x - y \leq 5$. In fact, the coordinates of every point on the *same side* of the line as the origin satisfy the inequality. To indicate this, we shade the half-plane that contains $(0, 0)$, as shown in part (b). Every point in the shaded half-plane and every point on the boundary line satisfies $x - y \leq 5$.

As an *informal* check, we can pick an ordered pair that lies in the shaded region and one that does not lie in the shaded region. When we substitute their coordinates into the inequality, we should obtain a true statement and then a false statement.

For (3, 1), in the shaded region:

$$x - y \leq 5$$
$$3 - 1 \stackrel{?}{\leq} 5$$
$$2 \leq 5 \quad \text{True.}$$

For (5, −4), not in the shaded region:

$$x - y \leq 5$$
$$5 - (-4) \stackrel{?}{\leq} 5$$
$$9 \leq 5 \quad \text{False.}$$

Self Check 2   Graph: $x - y \leq 2$.

EXAMPLE 3

 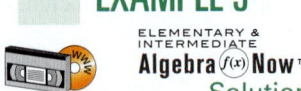

ELEMENTARY & INTERMEDIATE
Algebra $f(x)$ **Now**™

Solution

Graph: $4x + 3y < -6$.

To find the boundary line, we graph $4x + 3y = -6$. Since the inequality symbol $<$ does *not* include an equal symbol, the points on the graph of $4x + 3y = -6$ are not part of the graph of $4x + 3y < -6$. We draw the boundary line as a dashed line to show this. See part (a) of the illustration.

$4x + 3y = -6$

x	y	(x, y)
0	-2	$(0, -2)$
$-\frac{3}{2}$	0	$(-\frac{3}{2}, 0)$
-3	2	$(-3, 2)$

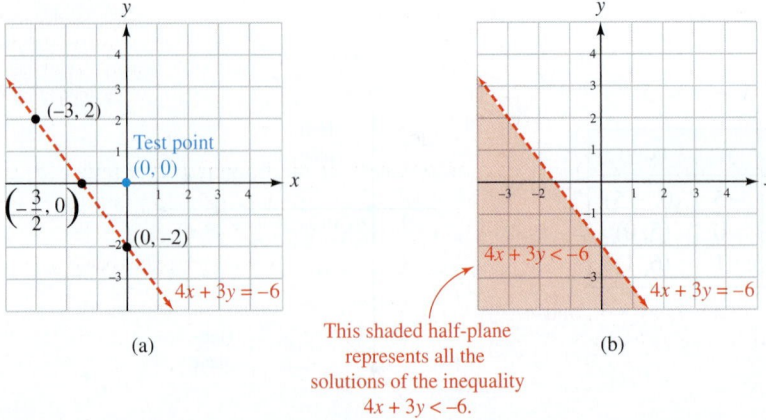

(a)

This shaded half-plane represents all the solutions of the inequality $4x + 3y < -6$.

(b)

To determine which half-plane to shade, we substitute the coordinates of a point that lies on one side of the boundary line into $4x + 3y < -6$. We choose the origin (0, 0) as the **test point** because the computations are usually easy when they involve 0. We substitute 0 for x and 0 for y in the inequality.

$$4x + 3y < -6$$
$$4(0) + 3(0) \overset{?}{<} -6 \qquad \text{The symbol } \overset{?}{<} \text{ is read as "is possibly less than."}$$
$$0 + 0 \overset{?}{<} -6$$
$$0 < -6 \qquad \text{False.}$$

Since $0 < -6$ is a false statement, the point (0, 0) does not satisfy the inequality. This indicates that it is *not* on the side of the dashed line we wish to shade. Instead, we shade the other side of the boundary line. The graph of the solution set of $4x + 3y < -6$ is the half-plane below the dashed line, as shown in part (b).

Self Check 3   Graph: $5x + 6y < -15$.

EXAMPLE 4

Solution

Graph: $y > 2x$.

To find the boundary line, we graph $y = 2x$. Since the symbol $>$ does *not* include an equal symbol, the points on the graph of $y = 2x$ are not part of the graph of $y > 2x$. Therefore, the boundary line should be dashed, as shown in part (a) of the illustration.

Success Tip

Draw a solid boundary line if the inequality has $\leq$ or $\geq$. Draw a dashed line if the inequality has $<$ or $>$.

$y = 2x$

x	y	(x, y)
0	0	$(0, 0)$
-1	-2	$(-1, -2)$
1	2	$(1, 2)$

↑
Choose three values for x and find the corresponding values of y.

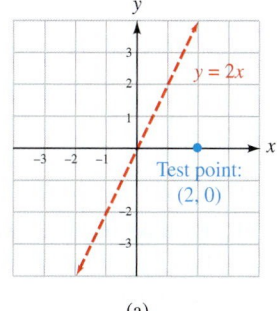

(a)

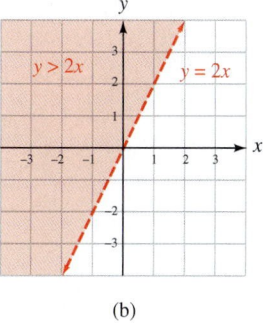

(b)

Success Tip

The origin (0, 0) is a smart choice for a test point because computations involving 0 are usually easy. If the origin is on the boundary, choose a test point not on the boundary that has one coordinate that is 0, such as (0, 1) or (2, 0).

To determine which half-plane to shade, we substitute the coordinates of a point that lies on one side of the boundary line into $y > 2x$. Since the origin is on the boundary, it cannot serve as a test point. One of the many possible choices for a test point is (2, 0), because it does not lie on the boundary line. To see whether it satisfies $y > 2x$, we substitute 2 for x and 0 for y in the inequality.

$$y > 2x$$
$$0 \overset{?}{>} 2(2) \qquad \text{The symbol } \overset{?}{>} \text{ is read as "is possibly greater than."}$$
$$0 > 4 \qquad \text{False.}$$

Since $0 > 4$ is a false statement, the point $(2, 0)$ does not satisfy the inequality. We shade the half-plane that does not contain $(2, 0)$, as shown in part (b).

Self Check 4   Graph: $y < 3x$.

EXAMPLE 5

Graph each linear inequality: **a.** $x < -3$ and   **b.** $y \geq 0$.

ELEMENTARY & INTERMEDIATE

Algebra $f(x)$ **Now**™   Solution

a. Because the inequality contains a $<$ symbol, we draw the boundary, $x = -3$, as a dashed vertical line. We can use $(0, 0)$ as the test point.

$$x < -3$$

$$0 < -3 \quad \text{Substitute 0 for } x. \text{ The } y\text{-coordinate of } (0, 0) \text{ is not used.}$$

Since the result is false, we shade the half-plane that does not contain $(0, 0)$, as shown in part (a) of the illustration below. Note that the solution consists of all points that have an x-coordinate that is less than -3.

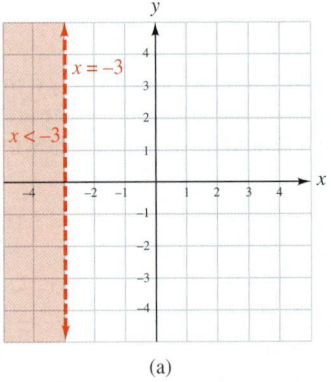
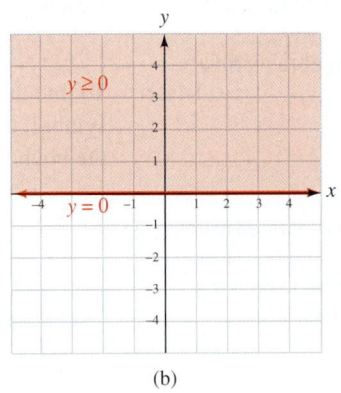

(a)   (b)

b. Because the inequality contains a $\geq$ symbol, we draw the boundary, $y = 0$, as a solid horizontal line. (Recall that the graph of $y = 0$ is the x-axis.) Next, we choose a test point not on the boundary. The point $(0, 1)$ is a convenient choice.

$$y \geq 0$$

$$1 \geq 0 \quad \text{Substitute 1 for } y. \text{ The } x\text{-coordinate of } (0, 1) \text{ is not used.}$$

Since the result is true, we shade the half-plane that contains $(0, 1)$, as shown in part (b) above. Note that the solution consists of all points that have a y-coordinate that is greater than or equal to 0.

Self Check 5   Graph each linear inequality: **a.** $x \geq 2$ and   **b.** $y < 4$.

The following is a summary of the procedure for graphing linear inequalities.

Graphing Linear Inequalities in Two Variables	**1.** Graph the boundary line of the region. If the inequality allows the possibility of equality (the symbol is either $\leq$ or $\geq$), draw the boundary line as a solid line. If equality is not allowed ($<$ or $>$), draw the boundary line as a dashed line.
	2. Pick a test point that is on one side of the boundary line. (Use the origin if possible.) Replace x and y in the inequality with the coordinates of that point. If the inequality is satisfied, shade the side that contains that point. If the inequality is not satisfied, shade the other side of the boundary.

■ APPLICATIONS

When solving applied problems, phrases such as *at least, at most,* and *should not exceed* indicate that an inequality should be used.

EXAMPLE 6

Working two jobs. Carlos has two part-time jobs, one paying $10 per hour and another paying $12 per hour. If x represents the number of hours he works on the first job, and y represents the number of hours he works on the second, the graph of $10x + 12y \geq 240$ shows the possible ways he can schedule his time to earn at least $240 per week to pay his college expenses. Find three possible combinations of hours he can work to achieve his financial goal.

Solution   The graph of the inequality is shown below in part (a) of the illustration. Any point in the shaded region represents a possible way Carlos can schedule his time and earn $240 or more per week. If each shift is a whole number of hours long, the highlighted points in part (b) represent the acceptable combinations. Three such combinations are

(6, 24):   6 hours on the first job, 24 hours on the second job
(12, 12): 12 hours on the first job, 12 hours on the second job
(22, 4):   22 hours on the first job, 4 hours on the second job

To verify one combination, suppose Carlos works 22 hours on the first job and 4 hours on the second job. He will earn

$$\$10(22) + \$12(4) = \$220 + \$48$$
$$= \$268$$

$10x + 12y = 240$

x	y	(x, y)
0	20	(0, 20)
24	0	(24, 0)

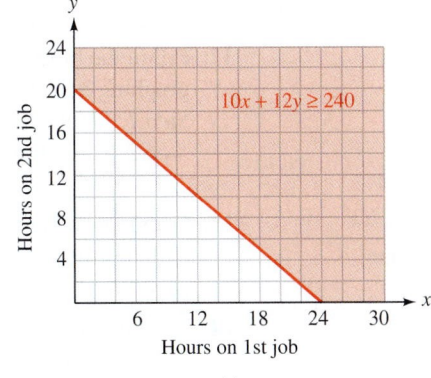

(a)

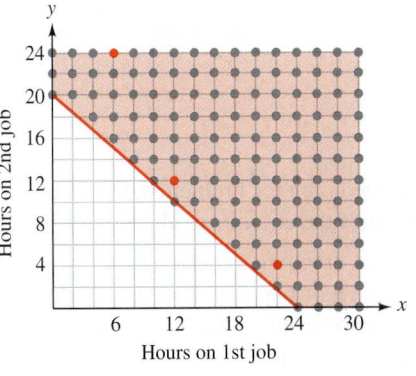

(b)

Answers to Self Checks   **1. a.** not a solution,   **b.** solution,   **c.** solution,   **d.** solution

2.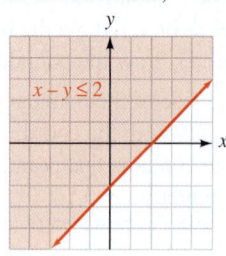

3.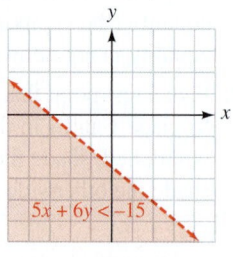

4.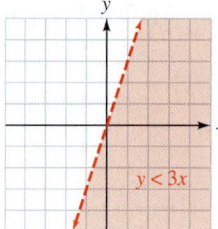

5.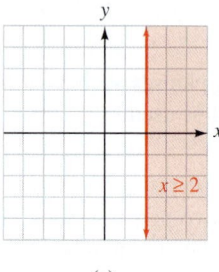

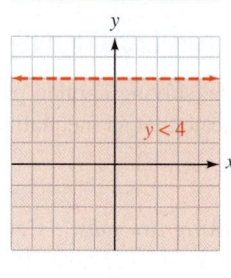

(a)   (b)

3.7   STUDY SET

ELEMENTARY &
INTERMEDIATE
Algebra *f(x)* Now™

VOCABULARY   Fill in the blanks.

1. An _____ is a statement that contains one of the symbols $<$, $\le$, $>$, or $\ge$.

2. $2x - y \le 4$ is a _____ inequality in two variables.

3. A _____ of a linear inequality is an ordered pair of numbers that makes the inequality true.

4. $(7, 2)$ is a solution of $x - y > 1$. We say that $(7, 2)$ _____ the inequality.

5. In the graph, the line $2x - y = 4$ is the _____.

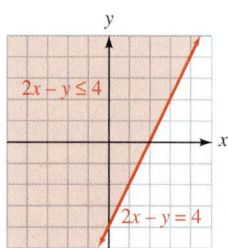

6. In the graph for Problem 5, the line $2x - y = 4$ divides the coordinate plane into two _____.

7. When graphing a linear inequality, we determine which half-plane to shade by substituting the coordinates of a test _____ into the inequality.

8. A _____ line indicates that points on the boundary are not solutions. A _____ line indicates that points on the boundary are solutions.

CONCEPTS

9. Decide whether each statement is true or false.
   a. $2 \le 0$ **b.** $-5 > 0$
   c. $0 \ge 0$ **d.** $0 < -6$

10. Decide whether each ordered pair is a solution of $5x - 3y \ge 0$.
   a. $(1, 1)$ **b.** $(-2, -3)$
   c. $(0, 0)$ **d.** $\left(\dfrac{1}{5}, \dfrac{4}{3}\right)$

11. Decide whether each ordered pair is a solution of $x + 4y < -1$.
   a. $(3, 1)$ **b.** $(-2, 0)$
   c. $(-0.5, 0.2)$ **d.** $\left(-2, \dfrac{1}{4}\right)$

12. Fill in the blanks to explain the procedure for graphing a linear inequality.
   Step 1: Graph the _____.
   Step 2: Use a test point to determine which side to _____.

13. Decide whether the graph of each linear inequality includes the boundary line.
   a. $y > -x$ **b.** $5x - 3y \le -2$

14. The boundary for the graph of a linear inequality follows. Why can't the origin be used as a test point to decide which side to shade?

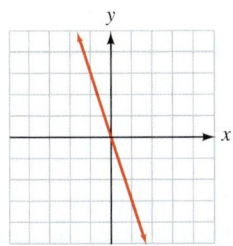

15. If a false statement results when the coordinates of a test point are substituted into a linear inequality, which half-plane should be shaded to represent the solution of the inequality?

16. A linear inequality has been graphed. Tell whether each point satisfies the inequality.
   a. $(1, -3)$
   b. $(-2, -1)$
   c. $(2, 3)$
   d. $(3, -4)$

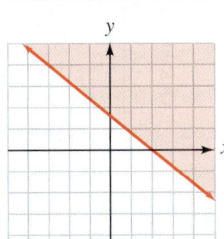

17. A linear inequality has been graphed. Tell whether each point satisfies the inequality.
   a. $(2, 1)$
   b. $(-2, -4)$
   c. $(4, -2)$
   d. $(-3, 4)$

18. To graph linear inequalities, we must be able to graph boundary lines. Complete the table of solutions for each given boundary line.
   a. $5x - 3y = 15$

x	y	(x, y)
0		
	0	
1		

   b. $y = 3x - 2$

x	y	(x, y)
-1		
0		
	4	

19. a. Is the graph of $y = 5$ a vertical or horizontal line?

   b. Is the graph of $x = -6$ a vertical or horizontal line?

20. To decide how many pallets x and barrels y a delivery truck can hold, a dispatcher refers to the loading sheet below. Can a truck make a delivery of 4 pallets and 10 barrels in one trip?

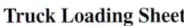
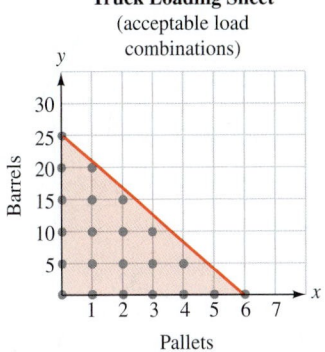

Truck Loading Sheet
(acceptable load combinations)

NOTATION

21. Write the meaning of each symbol in words.
   a. $<$ **b.** $>$

   c. $\leq$ **d.** $\geq$

22. When graphing linear inequalities, which inequality symbols are associated with a dashed boundary line?

23. When graphing linear inequalities, which inequality symbols are associated with a solid boundary line?

24. How do we read the symbol $\overset{?}{>}$?

PRACTICE   Complete the graph by shading the correct side of the boundary.

25. $x - y \geq -2$ **26.** $x - y < 3$

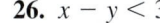
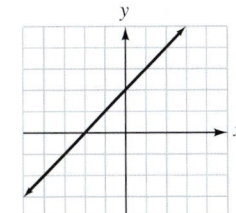
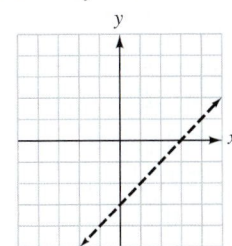

27. $y > 2x - 4$

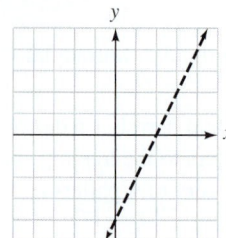

28. $y \leq -x + 1$

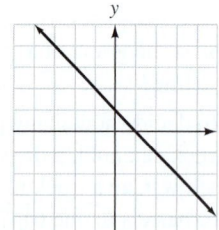

29. $x - 2y \geq 4$

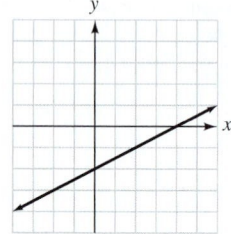

30. $3x + 2y > 12$

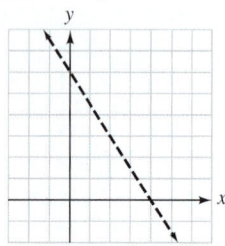

31. $y \leq 4x$

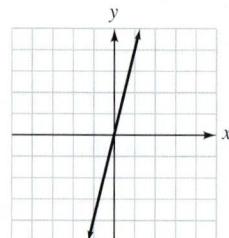

32. $y + 2x < 0$

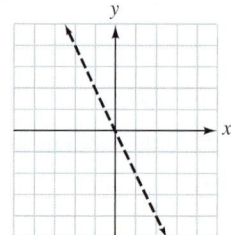

Graph each inequality.

33. $x + y \geq 3$ **34.** $x + y < 2$

35. $3x - 4y > 12$ **36.** $5x + 4y \geq 20$

37. $2x + 3y \leq -12$ **38.** $3x - 2y > 6$

39. $y < 2x - 1$ **40.** $y > x + 1$

41. $y < -3x + 2$ **42.** $y \geq -2x + 5$

43. $y \geq -\dfrac{3}{2}x + 1$ **44.** $y < \dfrac{x}{3} - 1$

45. $x - 2y \geq 4$ **46.** $4x + y \geq -4$

47. $2y - x < 8$ **48.** $y - x \geq 0$

49. $y + x < 0$ **50.** $y + 9x \geq 3$

51. $y \geq 2x$ **52.** $y < 3x$

53. $y < -\dfrac{x}{2}$ **54.** $y \geq x$

55. $x < 2$ **56.** $y > -3$

57. $y \leq 1$ **58.** $x \geq -4$

59. $x \leq 0$ **60.** $y < 0$

61. $7x - 2y < 21$ **62.** $3x - 3y \geq -10$

63. $2x - 3y \geq 4$ **64.** $4x + 3y < 6$

65. $5x + 3y < 0$ **66.** $2x + 5y > 0$

APPLICATIONS

67. ROLLING DICE The points on the graph represent all of the possible outcomes when two fair dice are rolled a single time. For example, (5, 2), shown in red, represents a 5 on the first die and a 2 on the second. Which of the following sentences best describes the outcomes that lie in the shaded area?

  (i) Their sum is at most 6.

 (ii) Their sum exceeds 6.

(iii) Their sum does not exceed 6.

 (iv) Their sum is at least 6.

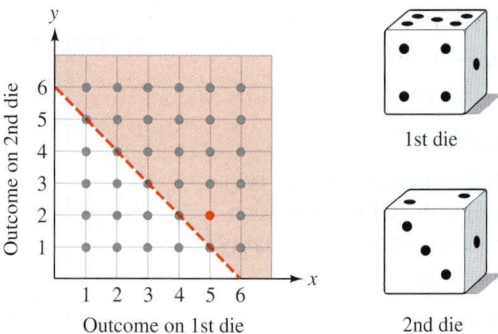

68. NATO In March 1999, NATO aircraft and cruise missiles targeted Serbian military forces that were south of the 44th parallel in Yugoslavia, Montenegro, and Kosovo. Shade the geographic area that NATO was trying to rid of Serbian forces.

Based on data from *Los Angeles Times* (March 24, 1999)

69. PRODUCTION PLANNING   It costs a bakery $3 to make a cake and $4 to make a pie. If x represents the number of cakes made, and y represents the number of pies made, the graph of $3x + 4y \leq 120$ shows the possible combinations of cakes and pies that can be produced so that costs do not exceed $120 per day. Graph the inequality. Then find three possible combinations of pies and cakes that can be made so that the daily costs are not exceeded.

70. HIRING BABYSITTERS   Mrs. Cansino has a choice of two babysitters. Sitter 1 charges $6 per hour, and Sitter 2 charges $7 per hour. If x represents the number of hours she uses Sitter 1 and y represents the number of hours she uses Sitter 2, the graph of $6x + 7y \leq 42$ shows the possible ways she can hire the sitters and not spend more than $42 per week. Graph the inequality. Then find three possible ways she can hire the babysitters so that her weekly budget for babysitting is not exceeded.

71. INVENTORIES   A clothing store advertises that it maintains an inventory of at least $4,400 worth of men's jackets at all times. At the store, leather jackets cost $100 and nylon jackets cost $88. If x represents the number of leather jackets in stock and y represents the number of nylon jackets in stock, the graph of $100x + 88y \geq 4,400$ shows the possible ways the jackets can be stocked. Graph the inequality. Then find three possible combinations of leather and nylon jackets so that the store lives up to its advertising claim.

72. MAKING SPORTING GOODS   A sporting goods manufacturer allocates at least 2,400 units of production time per day to make baseballs and footballs. It takes 20 units of time to make a baseball and 30 units of time to make a football. If x represents the number of baseballs made and y represents the number of footballs made, the graph of $20x + 30y \geq 2,400$ shows the possible ways to schedule the production time. Graph the inequality. Then find three possible combinations of production time for the company to make baseballs and footballs.

WRITING

73. Explain how to decide which side of the boundary line to shade when graphing a linear inequality in two variables.

74. Why is the origin usually a good test point to choose when graphing a linear inequality?

75. Why is (0, 0) not an acceptable choice for a test point when graphing a linear inequality whose boundary passes through the origin?

76. Explain the difference between the graph of the solution set of $x + 1 > 8$, an inequality *in one variable,* and the graph of $x + y > 8$, an inequality *in two variables.*

REVIEW

77. Solve $A = P + Prt$ for t.

78. What is the sum of the measures of the three angles of any triangle?

79. Simplify: $40\left(\dfrac{3}{8}x - \dfrac{1}{4}\right) + 40\left(\dfrac{4}{5}\right)$.

80. Evaluate: $-4 + 5 - (-3) - 13$.

CHALLENGE PROBLEMS

81. Find a linear inequality that has the following graph.

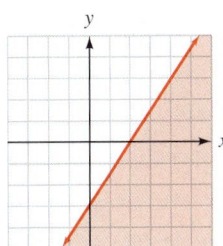

82. Graph the inequality: $4x - 3(x + 2y) \geq -6y$.

ACCENT ON TEAMWORK

MATCHING GAME

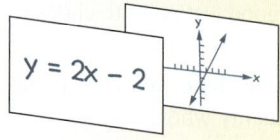

Overview: This activity is designed to improve your understanding of linear equations and their graphs.

Instructions: Form groups of 3 or 4 students. Then distribute ten 3 × 5 cards to every student in the class. Each student is to create a "deck" of equation–graph cards by following these steps.

1. On the first card, write a linear equation. (You may get ideas from the problems in the Study Sets in Sections 3.2, 3.3, and 3.5.) On the second card, draw the graph of that equation. Then, write "1" on the *backs* of those two cards to identify them as an equation–graph pair.
2. Create four more equation–graph pairs using the remaining cards. Label them pairs 2, 3, 4, and 5. When finished, shuffle your cards, and exchange decks with another member of your group.
3. Match each equation with its graph. To check your work, you can turn the cards over and compare the numbers on the back. If time allows, exchange decks of cards with another member of your group.

HEIGHT AND ARM SPAN

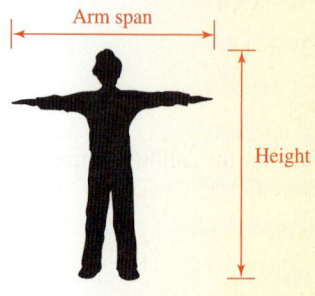

Overview: In this activity, you will explore the relationship between a person's height and arm span. Arm span is defined to be the distance between the tips of a person's fingers when their arms are held out to the side.

Instructions: Form groups of 5 or 6 students. Measure the height and arm span of each person in your group, and record the results in a table like the one shown below.

Name	Height (in.)	Arm span (in.)
1.		
2.		
3.		
4.		
5.		
6.		

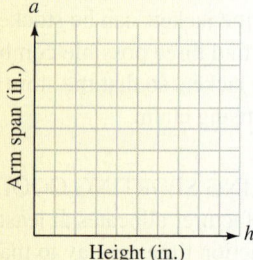

Plot the data in the table as ordered pairs of the form (height, arm span) on a graph like the one shown above. Then draw a straight-line model that best fits the data points.

Pick two convenient points on the line and find its slope. Use the point–slope form $a - a_1 = m(h - h_1)$ to find an equation of the line. Then, write the equation in slope–intercept form.

Ask a person from another group for his or her height measurement. Substitute that value into your linear model to predict that person's arm span. How close is your prediction to the person's actual arm span?

(From *Activities for Beginning and Intermediate Algebra* by Debbie Garrison, Judy Jones, and Jolene Rhodes)

KEY CONCEPT: DESCRIBING LINEAR RELATIONSHIPS

In Chapter 3, we discussed ways to mathematically describe linear relationships between two quantities using equations and graphs.

EQUATIONS IN TWO VARIABLES

The general form of the equation of a line is $Ax + By = C$. Two very useful forms of the equation of a line are the slope–intercept form and the point–slope form.

1. Write the equation of a line with a slope of -3 and a y-intercept of $(0, -4)$.

2. Write the equation of the line that passes through $(5, 2)$ and $(-5, 0)$. Give the answer in slope–intercept form.

3. CRICKETS   The equation $T = \frac{1}{4}c + 40$ predicts the outdoor temperature T in degrees Fahrenheit using the number c of cricket chirps per minute. Find the temperature if a cricket chirps 160 times in one minute.

4. U.S. HEALTH CARE   For the year 1990, the per capita health care expenditure was about \$2,660. Since then, the rate of increase has been about \$209 per year. Write a linear equation to model this. Let x represent the number of years since 1990 and let y represent the yearly per capita expenditure. Use your answer to predict the per capita expenditure in 2020. (Source: Centers for Medicare and Medicaid Services)

RECTANGULAR COORDINATE GRAPHS

The graph of an equation is a "picture" of all of its solutions. A thorough examination of a graph can yield a lot of useful information.

5. Complete the table of solutions for $2x - 4y = 8$. Then graph the equation.

$$2x - 4y = 8$$

x	y
0	
	0
-2	

6. a. What information does the y-intercept give?

b. What is the slope of the line and what does it tell?

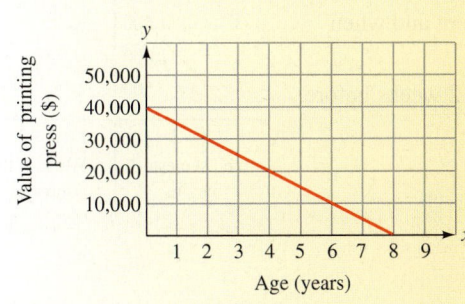

7. a. Find a point on the line.

b. Determine the slope of the line.

c. Write the equation of the line in slope–intercept form.

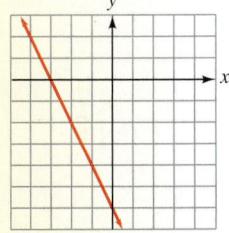

8. Write the equation of the line that passes through $(1, -1)$ and is parallel to the line graphed to the right.

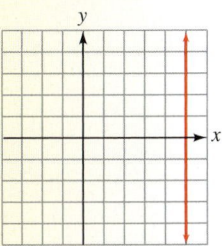

CHAPTER REVIEW   ELEMENTARY & INTERMEDIATE   **Algebra** *f(x)* **Now**™

Graphing Using the Rectangular Coordinate System

CONCEPTS

A *rectangular coordinate system* is composed of a horizontal number line called the *x*-axis and a vertical number line called the *y*-axis.

To *graph* ordered pairs means to locate their position on a coordinate system.

The two axes divide the coordinate plane into four regions called *quadrants*.

REVIEW EXERCISES

1. Graph the points with coordinates $(-1, 3)$, $(0, 1.5)$, $(-4, -4)$, $\left(2, \frac{7}{2}\right)$, and $(4, 0)$.

2. HAWAIIAN ISLANDS   Estimate the coordinates of Oahu using an ordered pair of the form (longitude, latitude).

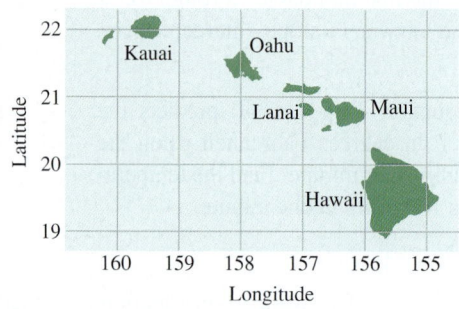

3. In what quadrant does the point $(-3, -4)$ lie?

4. What are the coordinates of the origin?

5. GEOMETRY   Three vertices (corners) of a square are the points $(-5, 4)$, $(-5, -2)$, and $(1, -2)$. Find the coordinates of the fourth vertex and find the area of the square.

6. COLLEGE ENROLLMENTS   The graph gives the number of students enrolled at a college for the period from 4 weeks before to 5 weeks after the semester began.

   a. What was the maximum enrollment and when did it occur?

   b. How many students had enrolled 2 weeks before the semester began?

   c. When was enrollment 2,250?

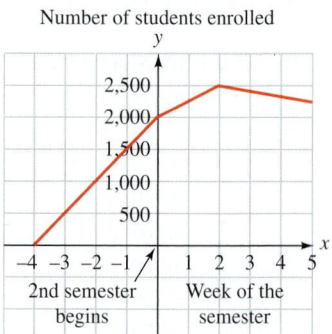

Graphing Linear Equations

A *solution* of an equation in two variables is an ordered pair of numbers that makes the equation a true statement.

An equation whose graph is a straight line and whose variables are raised to the first power is called a *linear equation*.

The *general* or *standard form* of a linear equation is $Ax + By = C$, where A, B, and C are real numbers and A and B are not both zero.

To graph a linear equation solved for y:

1. Find three solutions by selecting three x-values and finding the corresponding y-values.

2. Plot each ordered-pair solution.

3. Draw a straight line through the points.

7. Is $(-3, -2)$ a solution of $y = 2x + 4$?

8. Complete the table of solutions for $3x + 2y = -18$.

x	y	(x, y)
-2		
	3	

9. Which of the following equations are not linear equations?

$$8x - 2y = 6 \qquad y = x^2 + 1 \qquad y = x \qquad 3y = -x + 4 \qquad y - x^3 = 0$$

Graph each equation by constructing a table of solutions.

10. $y = 4x - 2$

11. $5y = -5x + 15$ (Solve for y first.)

12. The graph of a linear equation is shown here.

    a. When the coordinates of point A are substituted into the equation, will a true or false statement result?

    b. When the coordinates of point B are substituted into the equation, will a true or false statement result?

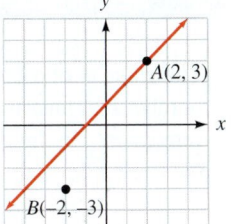

13. BIRTHDAY PARTIES   A restaurant offers a party package for children that includes everything: food, drinks, cake, and party favors. The cost c, in dollars, is given by the linear equation $c = 8n + 50$, where n is the number of children attending the party. Graph the equation and use the graph to estimate the cost of a party if 18 children attend.

More about Graphing Linear Equations

The point where a line intersects the x-axis is called the *x-intercept*. The point where a line intersects the y-axis is called the *y-intercept*.

14. Identify the x- and y-intercepts of the graph.

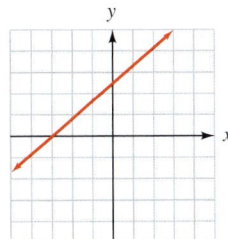

To find the *y*-intercept, substitute 0 for *x* in the given equation and solve for *y*. To find the *x*-intercept, substitute 0 for *y* in the given equation and solve for *x*.

The equation $y = b$ represents the horizontal line that intersects the *y*-axis at $(0, b)$. The equation $x = a$ represents the vertical line that intersects the *x*-axis at $(a, 0)$.

15. Graph $-4x + 2y = 8$ by finding its *x*- and *y*-intercepts.

16. Graph: $y = 4$.

17. Graph: $x = -1$.

18. DEPRECIATION   The graph shows how the value of some sound equipment decreased over the years. Find the intercepts of the graph. What information do the intercepts give about the equipment?

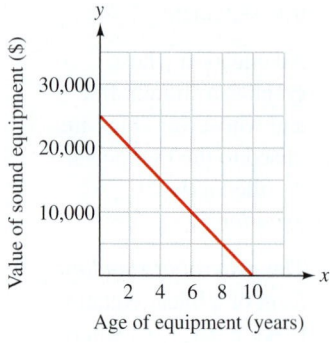

| SECTION 3.4 | **The Slope of a Line** |

The *slope m* of a line is a number that measures "steepness" by finding the ratio $\frac{\text{rise}}{\text{run}}$.

$$m = \frac{\text{change in the } y\text{-values}}{\text{change in the } x\text{-values}}$$

Lines that rise from left to right have a *positive slope,* and lines that fall from left to right have a *negative slope.* Horizontal lines have a slope of zero. Vertical lines have *undefined* slope.

If (x_1, y_1) and (x_2, y_2) are two points on a nonvertical line, the slope *m* of the line is

$$m = \frac{y_2 - y_1}{x_2 - x_1}$$

In each case, find the slope of the line.

19.

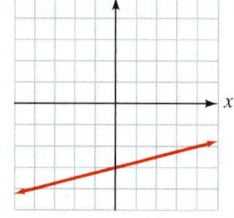

20.

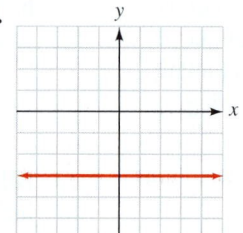

21. The line with the table of solutions shown below

x	y	(x, y)
2	−3	(2, −3)
4	−17	(4, −17)

22. The line passing through points $(1, -4)$ and $(3, -7)$

The *pitch* of a roof is its slope.

The *grade* of an incline or decline is its slope expressed as a percent.

When units are attached to a slope it becomes a *rate of change*.

23. CARPENTRY   Trusses like the one shown here will be used to construct the roof of a shed. Find the pitch of the roof.

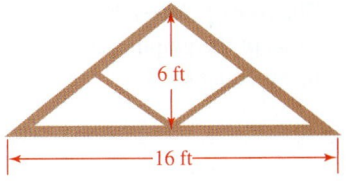

6 ft

16 ft

24. HANDICAP ACCESSIBILITY   Find the grade of the ramp. Round to the nearest tenth of a percent.

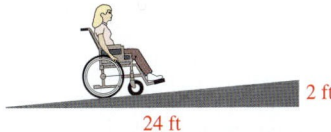

2 ft

24 ft

25. TOURISM   The graph shows the number of international travelers to the United States from 1986 to 2000, in two-year increments.

a. Between 1992 and 1994 the largest decline in the number of visitors occurred. What was the rate of change?

b. Between 1986 and 1988 the largest increase in the number of visitors occurred. What was the rate of change?

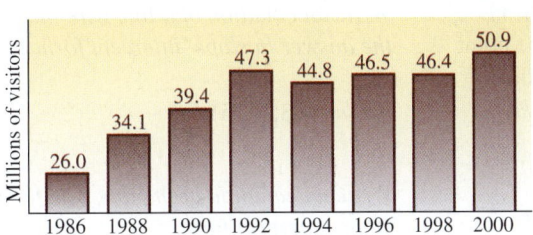

Based on data from *World Almanac* 2003.

| **SECTION 3.5** | **Slope–Intercept Form** |

If a linear equation is written in *slope–intercept* form,

$$y = mx + b$$

the graph of the equation is a line with slope m and y-intercept $(0, b)$.

Find the slope and the y-intercept of each line.

26. $y = \dfrac{3}{4}x - 2$

27. $y = -4x$

28. $y = \dfrac{x}{8} + 10$

29. $7x + 5y = -21$

30. Write an equation of the line with slope -6 and y-intercept $(0, 4)$.

31. Find the slope and the y-intercept of the line whose equation is $9x - 3y = 15$. Then graph it.

The *rate of change* is the slope of the graph of a linear equation.

32. COPIERS   A business buys a used copy machine that, when purchased, has already produced 75,000 copies.

    a. If the business plans to run 300 copies a week, write a linear equation that would find the number of copies c the machine has made in its lifetime after the business has used it for w weeks.

    b. Use your result in part a to predict the total number of copies that will have been made on the machine 1 year, or 52 weeks, after being purchased by the business.

Two lines with the same slope are *parallel.*

The product of the slopes of *perpendicular* lines is -1.

33. Without graphing, determine whether graphs of the given pairs of lines would be parallel, perpendicular, or neither.

    a. $y = -\dfrac{2}{3}x + 6$

         $y = -\dfrac{2}{3}x - 6$

    b. $x + 5y = -10$

         $y = 5x$

SECTION 3.6	**Point–Slope Form**

If a line with slope m passes through the point (x_1, y_1), the equation of the line in *point–slope* form is

$$y - y_1 = m(x - x_1)$$

Write an equation of a line with the given slope that passes through the given point. Give the answer in slope–intercept form and graph the equation.

34. $m = 3, (1, 5)$

35. $m = -\dfrac{1}{2}, (-4, -1)$

Write an equation of the line with the following characteristics. Give the answer in slope–intercept form.

36. passing through $(3, 7)$ and $(-6, 1)$

37. horizontal, passing through $(6, -8)$

38. CAR REGISTRATION   When it was 2 years old, the annual registration fee for a Dodge Caravan was \$380. When it was 4 years old, the registration fee dropped to \$310. If the relationship is linear, write an equation that gives the registration fee f in dollars for the van when it is x years old.

SECTION 3.7	**Graphing Linear Inequalities**

An ordered pair (x, y) is a *solution* of an inequality in x and y if a true statement results when the variables are replaced by the coordinates of the ordered pair.

39. Determine whether each ordered pair is a solution of $2x - y \le -4$.

    a. $(0, 5)$

    c. $(-3, -2)$

    b. $(2, 8)$

    d. $\left(\dfrac{1}{2}, -5\right)$

To graph a linear inequality:

1. Graph the *boundary line.* Draw a solid line if the inequality contains ≤ or ≥ and a dashed line if it contains < or >.

2. Pick a *test point* on one side of the boundary. Use the origin if possible. Replace x and y with the coordinates of that point. If the inequality is satisfied, shade the side that contains the point. If the inequality is not satisfied, shade the other side.

Graph each inequality.

40. $x - y < 5$

41. $2x - 3y \geq 6$

42. $y \leq -2x$

43. $y < -4$

44. The graph of a linear inequality is shown. Would a true or a false statement result if the coordinates of

   a. point A were substituted into the inequality?

   b. point B were substituted into the inequality?

   c. point C were substituted into the inequality?

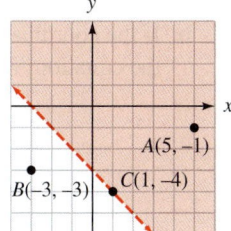

45. WORK SCHEDULES   A student told her employer that during the school year, she would be available for up to 30 hours a week, working either 3- or 5-hour shifts. If x represents the number of 3-hour shifts she works and y represents the number of 5-hour shifts she works, the inequality $3x + 5y \leq 30$ shows the possible combinations of shifts she can work. Graph the inequality, then find three possible combinations.

46. Explain the difference between an equation and an inequality.

CHAPTER 3 TEST

ELEMENTARY & INTERMEDIATE
Algebra *f(x)* Now™

The graph shows the number of dogs being boarded in a kennel over a 3-day holiday weekend.

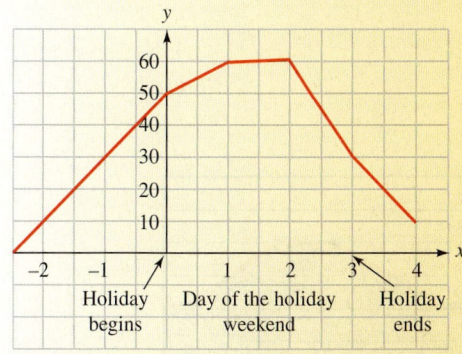

1. How many dogs were in the kennel 2 days before the holiday?

2. What is the maximum number of dogs that were boarded on the holiday weekend?

3. When were there 30 dogs in the kennel?

4. What information does the y-intercept of the graph give?

5. Draw a rectangular coordinate system and label each quadrant.

6. Complete the table of solutions for the linear equation.

$$x + 4y = 6$$

x	y	(x, y)
2		
	3	

7. Is $(-3, -4)$ a solution of $3x - 4y = 7$?

8. The graph of a linear equation is shown below.

   a. If the coordinates of point C are substituted into the equation, will the result be true or false?

   b. If the coordinates of point D are substituted into the equation, will the result be true or false?

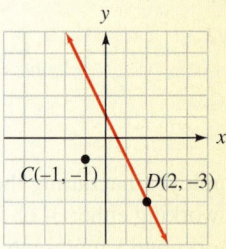

9. Graph: $y = \dfrac{x}{3}$.

10. Graph: $8x + 4y = -24$.

11. What are the x- and y-intercepts of the graph of $2x - 3y = 6$?

12. Find the slope of the line.

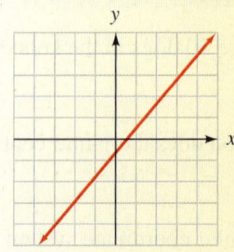

13. Find the slope of the line passing through $(-1, 3)$ and $(3, -1)$.

14. What is the slope of a vertical line?

15. What is the slope of a line that is perpendicular to a line with slope $-\dfrac{7}{8}$?

16. When graphed, are the lines $y = 2x + 6$ and $y = 2x$ parallel, perpendicular, or neither?

In Problems 17 and 18, refer to the illustration at the bottom of the page that shows the elevation changes in a 26-mile marathon course.

17. Find the rate of change of the decline on which the woman is running.

18. Find the rate of change of the incline on which the man is running.

19. Graph: $x = -4$.

20. Graph the line passing through $(-2, -4)$ having a slope of $\dfrac{2}{3}$.

21. Find the slope and the y-intercept of $x + 2y = 8$.

22. Write an equation of the line passing through $(-2, 5)$ with slope 7. Give the answer in slope–intercept form.

23. DEPRECIATION   After it is purchased, a $15,000 computer loses $1,500 in resale value every year. Write a linear equation that gives the resale value v of the computer x years after being purchased.

24. Determine whether $(6, 1)$ is a solution of $2x - 4y \geq 8$.

25. WATER HEATERS   The scatter diagram shows how excessively high temperatures affect the life of a water heater. Write an equation of the line that models the data for water temperatures between 140° and 180°. Let T represent the temperature of the water in degrees Fahrenheit and y represent the expected life of the heater in years. Give the answer in slope–intercept form.

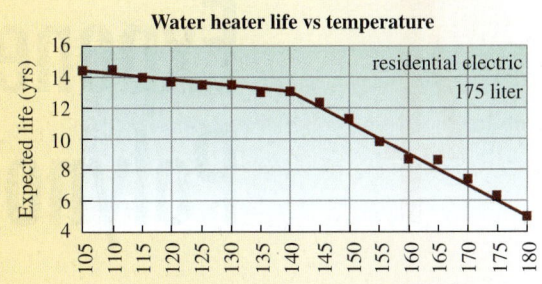

Water heater life vs temperature

Expected life (yrs)

residential electric
175 liter

Water: stored temperature (Fahrenheit)

Source: www.uniongas.com/WaterHeating

26. Graph the inequality $x - y > -2$.

4

Exponents and Polynomials

ELEMENTARY &
INTERMEDIATE
Algebra *f(x)* **Now**™

Throughout the chapter, this icon introduces resources on the Elementary & Intermediate AlgebraNow Web site, accessed through **http://1pass. thomson.com**, that will

- Help you test your knowledge of the material with a pre-test and a post-test
- Provide a personalized learning plan targeting areas you should study

©Pete Saloutos/CORBIS

Under certain conditions, bacteria can grow at an incredible rate. To model the growth, scientists use exponents. In this chapter, we will discuss several rules for simplifying expressions involving exponents. We will also see the role exponents play as part of a compact notation that is used to represent extremely large numbers, such as the distance from the Earth to the sun, and extremely small numbers, such as the mass of a proton.

To learn more about exponents visit *The Learning Equation* on the Internet at http://tle.brookscole.com. (The log-in instructions are in the Preface.) For Chapter 4, the online lessons are:

- *TLE* Lesson 8: The Product Rule for Exponents
- *TLE* Lesson 9: The Quotient Rule for Exponents
- *TLE* Lesson 10: Negative Exponents

In this chapter, we introduce the rules for exponents and use them to add, subtract, multiply, and divide polynomials.

4.1 Rules for Exponents

- Multiplying Exponential Expressions That Have Like Bases
- Dividing Exponential Expressions That Have Like Bases
- Raising an Exponential Expression to a Power   • Powers of Products and Quotients

Recall that an **exponent** indicates repeated multiplication. It tells how many times the **base** is used as a factor. For example, 3^5 represents the product of five 3's.

$$\underset{\text{base}}{\overset{\text{exponent}}{3^5}} = \overset{\text{5 factors of 3}}{3 \cdot 3 \cdot 3 \cdot 3 \cdot 3}$$

In general, we have the following definition for exponents that are natural numbers: 1, 2, 3, 4, 5, and so on.

Natural-Number Exponents	A natural-number exponent tells how many times its base is to be used as a factor. For any number x and any natural number n,
	$$x^n = \overset{n \text{ factors of } x}{\underbrace{x \cdot x \cdot x \cdot \cdots \cdot x}}$$

Expressions of the form x^n are called **exponential expressions.** The base of an exponential expression can be a number, a variable, or a combination of numbers and variables. Some examples are:

Notation

When an exponent is written outside parentheses, the expression within the parentheses is the base.

$$\underset{\text{base}}{(-2s)}^{\overset{\text{exponent}}{3}}$$

$$10^5 = 10 \cdot 10 \cdot 10 \cdot 10 \cdot 10 \qquad \text{Read } 10^5 \text{ as "10 to the fifth power."}$$
$$y^2 = y \cdot y \qquad \text{Read } y^2 \text{ as "}y \text{ to the second power" or "}y \text{ squared."}$$
$$(-2s)^3 = (-2s)(-2s)(-2s) \qquad \text{Read } (-2s)^3 \text{ as "the quantity of } -2s \text{ cubed."}$$
$$-8^4 = -(8 \cdot 8 \cdot 8 \cdot 8) \qquad \text{Read } -8^4 \text{ as "the opposite of 8 to the fourth power."}$$

In this section, we continue our study of exponents as we discuss how to simplify exponential expressions that are multiplied, divided, and raised to powers.

■ MULTIPLYING EXPONENTIAL EXPRESSIONS THAT HAVE LIKE BASES

To develop a rule for multiplying exponential expressions that have the same base, we consider the product $6^2 \cdot 6^3$. Since 6^2 means that 6 is to be used as a factor two times, and 6^3 means that 6 is to be used as a factor three times, we have

$$\overbrace{}^{\text{2 factors of 6}} \quad \overbrace{}^{\text{3 factors of 6}}$$
$$6^2 \cdot 6^3 = \quad 6 \cdot 6 \quad \cdot \quad 6 \cdot 6 \cdot 6$$
$$\overbrace{}^{\text{5 factors of 6}}$$
$$= 6 \cdot 6 \cdot 6 \cdot 6 \cdot 6$$
$$= 6^5$$

We can quickly find this result if we keep the common base 6 and add the exponents on 6^2 and 6^3.

$$6^2 \cdot 6^3 = 6^{2+3} = 6^5$$

This example suggests the following rule for exponents.

Product Rule for Exponents	To multiply exponential expressions with the same base, keep the common base and add the exponents. For any number x and any natural numbers m and n,
	$$x^m \cdot x^n = x^{m+n}$$

EXAMPLE 1

ELEMENTARY &
INTERMEDIATE
Algebra $f(x)$ **Now**™

Simplify: **a.** $9^5(9^6)$, **b.** $x^3 \cdot x^4$, **c.** $y^2 y^4 y$, **d.** $(x + 2)^8 (x + 2)^7$, and **e.** $(c^2 d^3)(c^4 d^5)$.

Solution **a.** To simplify $9^5(9^6)$ means to write it in an equivalent form using one base and one exponent.

$$9^5(9^6) = 9^{5+6} \quad \text{Keep the common base, which is 9, and add the exponents.}$$
$$= 9^{11}$$

Notation

An exponent of 1 means the base is to be used as a factor 1 time. For example, $y^1 = y$.

b. $x^3 \cdot x^4 = x^{3+4}$ Keep the common base x and add the exponents.
$$= x^7$$

c. $y^2 y^4 y = y^{2+4} y$ Working from left to right, keep the common base y and add the exponents.
$$= y^6 y \quad \text{Do the addition.}$$
$$= y^6 y^1 \quad \text{Write } y \text{ as } y^1.$$
$$= y^{6+1} \quad \text{Keep the common base } y \text{ and add the exponents.}$$
$$= y^7$$

d. $(x + 2)^8 (x + 2)^7 = (x + 2)^{8+7}$ Keep the common base $x + 2$ and add the exponents.
$$= (x + 2)^{15}$$

e. $(c^2 d^3)(c^4 d^5) = (c^2 c^4)(d^3 d^5)$ Group like bases together.
$$= (c^{2+4})(d^{3+5}) \quad \text{Keep the common base } c \text{ and add the exponents. Keep the common base } d \text{ and add the exponents.}$$
$$= c^6 d^8$$

Self Check 1   Simplify: **a.** $7^8(7^7)$, **b.** $z \cdot z^3$, **c.** $x^2 x^3 x^6$, **d.** $(y - 1)^5 (y - 1)^5$, **e.** $(s^4 t^3)(s^4 t^4)$.

■ DIVIDING EXPONENTIAL EXPRESSIONS THAT HAVE LIKE BASES

To develop a rule for dividing exponential expressions that have the same base, we consider the quotient

$$\frac{4^5}{4^2}$$

where the exponent in the numerator is greater than the exponent in the denominator. We can simplify this fraction as follows:

$$\frac{4^5}{4^2} = \frac{4 \cdot 4 \cdot 4 \cdot 4 \cdot 4}{4 \cdot 4}$$

$$= \frac{\overset{1}{\cancel{4}} \cdot \overset{1}{\cancel{4}} \cdot 4 \cdot 4 \cdot 4}{\underset{1}{\cancel{4}} \cdot \underset{1}{\cancel{4}}} \qquad \text{\textcolor{red}{Remove the common factors of 4 in the numerator and denominator.}}$$

$$= 4^3$$

We can quickly find this result if we keep the common base 4 and subtract the exponents on 4^5 and 4^2.

$$\frac{4^5}{4^2} = 4^{5-2} = 4^3$$

This example suggests another rule for exponents.

Quotient Rule for Exponents   To divide exponential expressions with the same base, keep the common base and subtract the exponents. For any nonzero number x and any natural numbers m and n, where $m > n$,

$$\frac{x^m}{x^n} = x^{m-n}$$

EXAMPLE 2

ELEMENTARY &
INTERMEDIATE

Algebra $f(x)$ **Now**™

Simplify each expression. Assume that there are no divisions by 0.   **a.** $\frac{20^{16}}{20^9}$,   **b.** $\frac{x^9}{x^3}$,   **c.** $\frac{(75n)^{12}}{(75n)^{11}}$,   and   **d.** $\frac{a^3 b^8}{ab^5}$.

Solution   **a.** To simplify $\frac{20^{16}}{20^9}$ means to write it in an equivalent form using one base and one exponent.

$$\frac{20^{16}}{20^9} = 20^{16-9} \qquad \text{\textcolor{red}{Keep the common base, which is 20, and subtract the exponents.}}$$

$$= 20^7$$

b. $\dfrac{x^9}{x^3} = x^{9-3} \qquad \text{\textcolor{red}{Keep the common base x and subtract the exponents.}}$

$$= x^6$$

c. $\dfrac{(75n)^{12}}{(75n)^{11}} = (75n)^{12-11} \qquad \text{\textcolor{red}{Keep the common base 75n and subtract the exponents.}}$

$$= (75n)^1$$

$$= 75n \qquad \text{\textcolor{red}{Any number raised to the first power is simply that number.}}$$

d. $\dfrac{a^3 b^8}{ab^5} = \dfrac{a^3}{a} \cdot \dfrac{b^8}{b^5} \qquad \text{\textcolor{red}{Group the common bases together.}}$

$$= a^{3-1} b^{8-5} \qquad \text{\textcolor{red}{Keep the common base a and subtract the exponents. Keep the common base b and subtract the exponents.}}$$

$$= a^2 b^3 \qquad \text{\textcolor{red}{Do the subtractions.}}$$

Self Check 2   Simplify: **a.** $\dfrac{55^{30}}{55^5}$,   **b.** $\dfrac{a^5}{a^3}$,   **c.** $\dfrac{(8t)^8}{(8t)^7}$,   **d.** $\dfrac{b^{15}c^4}{b^4c}$.

EXAMPLE 3

Simplify: $\dfrac{a^3a^5a^7}{a^4a}$.

ELEMENTARY & INTERMEDIATE
Algebra $f(x)$ **Now**™   **Solution**   We simplify the numerator and denominator separately and proceed as follows.

$$\frac{a^3a^5a^7}{a^4a} = \frac{a^{15}}{a^5} \qquad \text{In the numerator, keep the common base } a \text{ and add the exponents. In the denominator, keep the common base } a \text{ and add the exponents.}$$

$$= a^{15-5} \qquad \text{Keep the common base } a \text{ and subtract the exponents.}$$

$$= a^{10}$$

Self Check 3   Simplify: $\dfrac{b^2b^6b}{b^4b^4}$.

Caution   Recall that like terms are terms with exactly the same variables raised to exactly the same powers. To add or subtract exponential expressions, they must be like terms. To multiply or divide exponential expressions, only the bases need to be the same.

$$x^5 + x^2 \qquad \text{They are not like terms; the exponents are different. We cannot add.}$$

$$x^2 + x^2 = 2x^2 \qquad \text{They are like terms; we can add. Recall that } x^2 = 1x^2.$$

$$x^5 \cdot x^2 = x^7 \qquad \text{The bases are the same; we can multiply.}$$

$$\frac{x^5}{x^2} = x^3 \qquad \text{The bases are the same; we can divide.}$$

◼ RAISING AN EXPONENTIAL EXPRESSION TO A POWER

To develop another rule for exponents, we consider $(5^3)^4$. Here, an exponential expression, 5^3, is raised to a power. Since 5^3 is the base and 4 is the exponent, $(5^3)^4$ can be written as $5^3 \cdot 5^3 \cdot 5^3 \cdot 5^3$. Because each of the four factors of 5^3 contains three factors of 5, there are $4 \cdot 3$ or 12 factors of 5.

The Language of Algebra

An exponential expression raised to a power, such as $(5^3)^4$, is also called a *power of a power*.

$$(5^3)^4 = 5^3 \cdot 5^3 \cdot 5^3 \cdot 5^3$$

$$\overbrace{}^{\text{12 factors of 5}}$$

$$= \underbrace{5 \cdot 5 \cdot 5}_{5^3} \cdot \underbrace{5 \cdot 5 \cdot 5}_{5^3} \cdot \underbrace{5 \cdot 5 \cdot 5}_{5^3} \cdot \underbrace{5 \cdot 5 \cdot 5}_{5^3}$$

$$= 5^{12}$$

We can quickly find this result if we keep the common base 5 and multiply the exponents.

$$(5^3)^4 = 5^{3 \cdot 4} = 5^{12}$$

This example suggests the following rule for exponents.

Power Rule for Exponents   To raise an exponential expression to a power, keep the base and multiply the exponents. For any number x and any natural numbers m and n,

$$(x^m)^n = x^{mn}$$

EXAMPLE 4

Simplify: **a.** $(2^3)^7$ and   **b.** $(z^8)^8$.

Solution   **a.** To simplify $(2^3)^7$ means to write it in an equivalent form using one base and one exponent.

$$(2^3)^7 = 2^{3\cdot 7} \qquad \text{Keep the base and multiply the exponents.}$$
$$= 2^{21}$$

b. $(z^8)^8 = z^{8\cdot 8} \qquad \text{Keep the base and multiply the exponents.}$
$$= z^{64}$$

Self Check 4   Simplify: **a.** $(4^6)^5$,   **b.** $(y^5)^2$.

EXAMPLE 5

Simplify: **a.** $(x^2 x^5)^2$ and   **b.** $(z^2)^4(z^3)^3$.

Solution   **a.** We begin by using the product rule for exponents. Then we use the power rule.

$$(x^2 x^5)^2 = (x^7)^2 \qquad \text{Within the parentheses, keep the base and add the exponents.}$$
$$= x^{14} \qquad \text{Keep the base and multiply the exponents.}$$

b. We begin by using the power rule for exponents twice. Then we use the product rule.

$$(z^2)^4(z^3)^3 = z^8 z^9 \qquad \text{For each power of } z \text{ raised to a power, keep the base and multiply the exponents.}$$
$$= z^{17} \qquad \text{Keep the common base } z \text{ and add the exponents.}$$

Self Check 5   Simplify: **a.** $(a^4 a^3)^3$,   **b.** $(a^3)^3(a^4)^2$.

■ POWERS OF PRODUCTS AND QUOTIENTS

To develop more rules for exponents, we consider the expression $(2x)^3$, which is a *power of the product* of 2 and x, and the expression $\left(\frac{2}{x}\right)^3$, which is a *power of the quotient* of 2 and x.

$$(2x)^3 = (2x)(2x)(2x) \qquad\qquad \left(\frac{2}{x}\right)^3 = \left(\frac{2}{x}\right)\left(\frac{2}{x}\right)\left(\frac{2}{x}\right) \qquad \text{Assume } x \neq 0.$$

$$= (2\cdot 2\cdot 2)(x\cdot x\cdot x) \qquad\qquad = \frac{2\cdot 2\cdot 2}{x\cdot x\cdot x} \qquad \begin{array}{l}\text{Multiply the numerators.}\\ \text{Multiply the denominators.}\end{array}$$

$$= 2^3 x^3 \qquad\qquad\qquad\qquad = \frac{2^3}{x^3}$$

$$= 8x^3 \qquad\qquad\qquad\qquad = \frac{8}{x^3}$$

These examples suggest the following rules for exponents.

Powers of a Product and a Quotient	To raise a product to a power, raise each factor of the product to that power. To raise a quotient to a power, raise the numerator and the denominator to that power. For any numbers x and y, and any natural number n,

$$(xy)^n = x^n y^n \qquad \text{and} \qquad \left(\frac{x}{y}\right)^n = \frac{x^n}{y^n}, \qquad \text{where } y \neq 0$$

EXAMPLE 6

Simplify: **a.** $(3c)^3$, **b.** $(x^2y^3)^5$, and **c.** $(-2a^3b)^2$.

Solution   **a.** Since $3c$ is the product of 3 and c, the expression $(3c)^3$ is a power of a product.

$$(3c)^3 = 3^3c^3 \quad \text{Raise each factor of the product } 3c \text{ to the 3rd power.}$$
$$= 27c^3 \quad \text{Evaluate } 3^3.$$

ELEMENTARY &
INTERMEDIATE
Algebra $f(x)$ **Now**™

Caution

There is no rule for the power of a sum or power of a difference. To show why, consider this example:

$$(3 + 2)^2 \overset{?}{=} 3^2 + 2^2$$
$$5^2 \overset{?}{=} 9 + 4$$
$$25 \neq 13$$

b. $(x^2y^3)^5 = (x^2)^5(y^3)^5 \quad$ Raise each factor of the product x^2y^3 to the 5th power.
$$= x^{10}y^{15} \quad \text{For each power of a power, keep the base and multiply the exponents.}$$

c. $(-2a^3b)^2 = (-2)^2(a^3)^2b^2 \quad$ Raise each of the three factors of the product $-2a^3b$ to the 2nd power.
$$= 4a^6b^2 \quad \text{Evaluate } (-2)^2. \text{ Keep the base } a \text{ and multiply the exponents.}$$

Self Check 6   Simplify: **a.** $(2t)^4$, **b.** $(c^3d^4)^6$, **c.** $(-3ab^5)^3$.

EXAMPLE 7

Simplify: $\dfrac{(a^3b^4)^2}{ab^5}$.

ELEMENTARY &
INTERMEDIATE
Algebra $f(x)$ **Now**™

Solution

$$\frac{(a^3b^4)^2}{ab^5} = \frac{(a^3)^2(b^4)^2}{ab^5} \quad \text{In the numerator, raise each factor within the parentheses to the 2nd power.}$$

$$= \frac{a^6b^8}{ab^5} \quad \text{In the numerator, for each power of a power, keep the base and multiply the exponents.}$$

$$= a^{6-1}b^{8-5} \quad \text{Keep each of the bases, } a \text{ and } b, \text{ and subtract the exponents.}$$

$$= a^5b^3 \quad \text{Do the subtractions.}$$

Self Check 7   Simplify: $\dfrac{(c^4d^5)^3}{c^2d^3}$.

EXAMPLE 8

Simplify: **a.** $\left(\dfrac{4}{k}\right)^3$ and **b.** $\left(\dfrac{3x^2}{2y^3}\right)^5$.

Solution   **a.** Since $\frac{4}{k}$ is the quotient of 4 and k, the expression $\left(\frac{4}{k}\right)^3$ is a power of a quotient.

$$\left(\frac{4}{k}\right)^3 = \frac{4^3}{k^3} \quad \text{Raise the numerator and denominator to the 3rd power.}$$

$$= \frac{64}{k^3} \quad \text{Evaluate } 4^3.$$

b. $\left(\dfrac{3x^2}{2y^3}\right)^5 = \dfrac{(3x^2)^5}{(2y^3)^5} \quad$ Raise the numerator and the denominator to the 5th power.

$$= \frac{3^5(x^2)^5}{2^5(y^3)^5} \quad \text{In the numerator and denominator, raise each factor within the parentheses to the 5th power.}$$

$$= \frac{243x^{10}}{32y^{15}} \quad \text{Evaluate } 3^5 \text{ and } 2^5. \text{ For each power of a power, keep the base and multiply the exponents.}$$

Self Check 8   Simplify: **a.** $\left(\dfrac{x}{7}\right)^3$,   **b.** $\left(\dfrac{2x^3}{3y^2}\right)^4$.

EXAMPLE 9   Simplify: $\dfrac{(5b)^9}{(5b)^6}$.

Solution $\dfrac{(5b)^9}{(5b)^6} = (5b)^{9-6}$ Keep the common base $5b$, and subtract the exponents.

$= (5b)^3$ Do the subtraction.

$= 5^3b^3$ Raise each factor within the parentheses to the 3rd power.

$= 125b^3$ Evaluate 5^3.

Self Check 9   Simplify: $\dfrac{(-2h)^{20}}{(-2h)^{14}}$.

The rules for natural-number exponents are summarized as follows.

Rules for Exponents	If m and n represent natural numbers and there are no divisions by zero, then

1. $x^m x^n = x^{m+n}$ **2.** $\dfrac{x^m}{x^n} = x^{m-n}$ **3.** $(x^m)^n = x^{mn}$

4. $(xy)^n = x^n y^n$ **5.** $\left(\dfrac{x}{y}\right)^n = \dfrac{x^n}{y^n}$

Answers to Self Checks   **1. a.** 7^{15},   **b.** z^4,   **c.** x^{11},   **d.** $(y-1)^{10}$,   **e.** $s^8 t^7$ **2. a.** 55^{25},   **b.** a^2,   **c.** $8t$,   **d.** $b^{11}c^3$ **3.** b **4. a.** 4^{30},   **b.** y^{10} **5. a.** a^{21},   **b.** a^{17} **6. a.** $16t^4$,   **b.** $c^{18}d^{24}$,   **c.** $-27a^3b^{15}$ **7.** $c^{10}d^{12}$ **8. a.** $\frac{x^3}{343}$,   **b.** $\frac{16x^{12}}{81y^8}$ **9.** $64h^6$

4.1   **STUDY SET**   ELEMENTARY & INTERMEDIATE   Algebra $f(x)$ **Now**™

VOCABULARY   **Fill in the blanks.**

1. Expressions such as x^4, 10^3, and $(5t)^2$ are called _____ expressions.

2. The _____ of the expression 5^3 is 5. The _____ is 3.

3. The expression x^4 represents a repeated multiplication where x is to be written as a _____ four times.

4. $3^4 \cdot 3^8$ is a _____ of exponential expressions with the same base and $\frac{x^4}{x^2}$ is a _____ of exponential expressions with the same base.

5. $(h^3)^7$ is a _____ of an exponential expression.

6. The expression $(2xy)^3$ is a power of a _____ and $\left(\frac{x}{c}\right)^5$ is a power of a _____.

CONCEPTS   **Fill in the blanks.**

7. a. $(3x)^4$ means ⬜ $\cdot$ ⬜ $\cdot$ ⬜ $\cdot$ ⬜.

b. Using an exponent, $(-5y)(-5y)(-5y)$ can be written as ⬜.

8. a. $x^m x^n =$ ⬜   **b.** $(xy)^n =$ ⬜

c. $\left(\dfrac{a}{b}\right)^n =$ ⬜   **d.** $(a^b)^c =$ ⬜

e. $\dfrac{x^m}{x^n} =$ ⬜   **f.** $x = x$

9. a. Write a power of a product that has two factors.

  b. Write a power of a quotient.

10. a. To simplify $(2y^3z^2)^4$, how many factors within the parentheses must be raised to the fourth power?

  b. To simplify $\left(\dfrac{y^3}{z^2}\right)^4$ what two expressions must be raised to the fourth power?

Simplify each expression, if possible.

11. a. $x^2 + x^2$ **b.** $x^2 - x^2$

  c. $x^2 \cdot x^2$ **d.** $a^3 \cdot a^4$

12. a. $x^2 + x$ **b.** $x^2 - x$

  c. $x^2 \cdot x$ **d.** $\dfrac{x^2}{x}$

13. a. $x^3 + x^2$ **b.** $x^3 - x^2$

  c. $x^3 \cdot x^2$ **d.** $\dfrac{x^3}{x^2}$

14. a. $x^3 + y^3$ **b.** $x^3 - y^3$

  c. x^3y^3 **d.** $\dfrac{x^3}{y^3}$

Evaluate each expression.

15. a. $(-4)^2$ **b.** -4^2

16. a. $(-5)^2$ **b.** -5^2

NOTATION Complete each solution.

17. $(x^4x^2)^3 = \left(\ \rule{1.2em}{0.8em}\ \right)^3$

  $= x^{\rule{0.8em}{0.8em}}$

18. $\dfrac{a^3a^4}{a^2} = \dfrac{a^{\rule{0.8em}{0.8em}}}{a^2}$

  $= a^{\rule{0.8em}{0.8em} -2}$

  $= a^{\rule{0.8em}{0.8em}}$

Identify the base and the exponent in each expression.

19. 4^3 **20.** $(-8)^2$

21. x^5 **22.** $\left(\dfrac{5}{x}\right)^3$

23. $(-3x)^2$ **24.** $-x^4$

25. $-\dfrac{1}{3}y^6$ **26.** $3.14r^4$

27. $(y + 9)^4$ **28.** $(2xy)^{10}$

29. $(-3ab)^7$ **30.** $-(z - 2)^3$

Write each repeated multiplication without using exponents.

31. x^5 **32.** $(-7y)^4$

33. $\left(\dfrac{t}{2}\right)^3$ **34.** c^3d^2

35. $(x - 5)^2$ **36.** $(m + 4)^3$

Write each expression using an exponent.

37. $4t(4t)(4t)(4t)$ **38.** $-5u(-5u)$

39. $-4 \cdot t \cdot t \cdot t$ **40.** $-5 \cdot u \cdot u$

41. $(x - y)(x - y)(x - y)$ **42.** $\left(\dfrac{x}{c}\right)\left(\dfrac{x}{c}\right)\left(\dfrac{x}{c}\right)\left(\dfrac{x}{c}\right)$

PRACTICE Write each expression as an expression involving one base and one exponent.

43. $12^3 \cdot 12^4$ **44.** $3^4 \cdot 3^6$

45. $2(2^3)(2^2)$ **46.** $5(5^5)(5^3)$

47. $a^3 \cdot a^3$ **48.** $m^7 \cdot m^7$

49. x^4x^3 **50.** y^5y^2

51. a^3aa^5 **52.** b^2b^3b

53. $(-7)^2(-7)^3$ **54.** $(-10)^8(-10)^3$

55. $(8t)^{20}(8t)^{40}$ **56.** $(9p)^{80}(9p)^{10}$

57. $(n - 1)^2(n - 1)$ **58.** $(s - 14)(s - 14)^8$

59. $y^3(y^2y^4)$ **60.** $(y^4y)y^6$

61. $\dfrac{8^{12}}{8^4}$ **62.** $\dfrac{10^4}{10^2}$

63. $\dfrac{x^{15}}{x^3}$ **64.** $\dfrac{y^6}{y^3}$

65. $\dfrac{c^{10}}{c^9}$ **66.** $\dfrac{h^{20}}{h^{10}}$

67. $\dfrac{(k-2)^{15}}{(k-2)}$ **68.** $\dfrac{(m+8)^{20}}{(m+8)}$

69. $(3^2)^4$ **70.** $(4^3)^3$

71. $(y^5)^3$ **72.** $(b^3)^6$

73. $(m^{50})^{10}$ **74.** $(n^{25})^4$

Simplify. Assume there are no divisions by 0.

75. $(a^2b^3)(a^3b^3)$ **76.** $(u^3v^5)(u^4v^5)$

77. $(cd^4)(cd)$ **78.** $ab^3c^4 \cdot ab^4c^2$

79. $xy^2 \cdot x^2y$ **80.** $s^8t^2s^2t^7$

81. $\dfrac{y^3y^4}{yy^2}$ **82.** $\dfrac{b^4b^5}{b^2b^3}$

83. $\dfrac{c^3d^7}{cd}$ **84.** $\dfrac{r^8s^9}{rs}$

85. $(x^2x^3)^5$ **86.** $(y^3y^4)^4$

87. $(3zz^2z^3)^5$ **88.** $(4t^3t^6t^2)^2$

89. $(3n^8)^2$ **90.** $(y^3y)^2(y^2)^2$

91. $(uv)^4$ **92.** $(2m^4)^3$

93. $(a^3b^2)^3$ **94.** $(r^3s^2)^2$

95. $(-2r^2s^3)^3$ **96.** $(-3x^2y^4)^2$

97. $\left(\dfrac{a}{b}\right)^3$ **98.** $\left(\dfrac{r}{s}\right)^4$

99. $\left(\dfrac{x^2}{y^3}\right)^5$ **100.** $\left(\dfrac{u^4}{v^2}\right)^6$

101. $\left(\dfrac{-2a}{b}\right)^5$ **102.** $\left(\dfrac{-2t}{3}\right)^4$

103. $\dfrac{(6k)^7}{(6k)^4}$ **104.** $\dfrac{(-3a)^{12}}{(-3a)^{10}}$

105. $\dfrac{(a^2b)^{15}}{(a^2b)^9}$ **106.** $\dfrac{(s^2t^3)^4}{(s^2t^3)^2}$

107. $\dfrac{a^2a^3a^4}{(a^4)^2}$ **108.** $\dfrac{(aa^2)^3}{a^2a^3}$

109. $\dfrac{(ab^2)^3}{(ab)^2}$ **110.** $\dfrac{(m^3n^4)^3}{(mn^2)^3}$

111. $\dfrac{(r^4s^3)^4}{(rs^3)^3}$ **112.** $\dfrac{(x^2y^5)^5}{(x^3y)^2}$

113. $\left(\dfrac{y^3y}{2yy^2}\right)^3$ **114.** $\left(\dfrac{2y^3y}{yy^2}\right)^3$

115. $\left(\dfrac{3t^3t^4t^5}{4t^2t^6}\right)^3$ **116.** $\left(\dfrac{4t^3t^4t^5}{3t^2t^6}\right)^3$

APPLICATIONS   Write an expression for the area or volume of each figure. Leave π in your answer.

117.

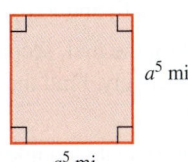

a^5 mi

a^5 mi

118.

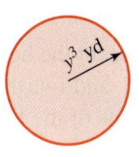

y^3 yd

119.

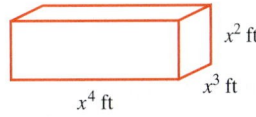

x^2 ft

x^4 ft

x^3 ft

120.

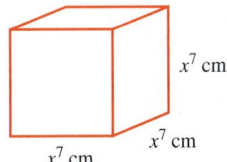

x^7 cm

x^7 cm

x^7 cm

121. ART HISTORY   Leonardo da Vinci's drawing relating a human figure to a square and a circle is shown.

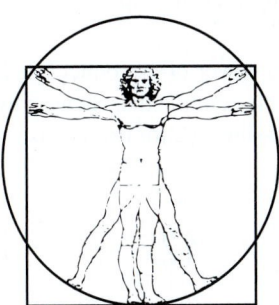

a. Find the area of the square if the man's height is $5x$ feet.

b. Find the area of the circle if the distance from his waist to his feet is $3a$ feet. Leave π in your answer.

122. PACKAGING   Find the volume of the bowling ball and the cardboard box it is packaged in. Leave π in your answer.

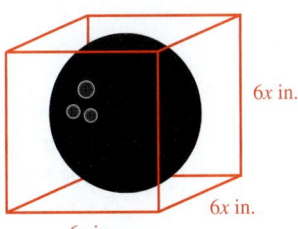

$6x$ in.

$6x$ in.

$6x$ in.

123. BOUNCING BALLS A ball is dropped from a height of 32 feet and always rebounds to one-half of its previous height. Draw a diagram of the path of the ball, showing four bounces. Then explain why the expressions $32\left(\frac{1}{2}\right)$, $32\left(\frac{1}{2}\right)^2$, $32\left(\frac{1}{2}\right)^3$, and $32\left(\frac{1}{2}\right)^4$ represent the height of the ball on the first, second, third, and fourth bounces, respectively. Find the heights of the first four bounces.

124. PROBABILITY The probability that a couple will have n baby boys in a row is given by the formula $\left(\frac{1}{2}\right)^n$. Find the probability that a couple will have four baby boys in a row.

WRITING

125. Explain the mistake in the following work.

$$2^3 \cdot 2^2 = 4^5$$
$$= 1,024$$

126. Are the expressions $2x^3$ and $(2x)^3$ equivalent? Explain.

127. Explain why we can simplify $x^4 \cdot x^5$, but cannot simplify $x^4 + x^5$.

128. Explain the power-of-a-product rule for exponents. Then give an example that shows why there is no power-of-a-sum rule for exponents.

REVIEW   **Match each equation with its graph below.**

129. $y = 2x - 1$

130. $y = 3x - 1$

131. $y = 3$

132. $x = 3$

a.

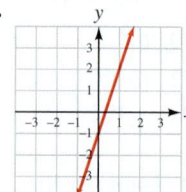

b.

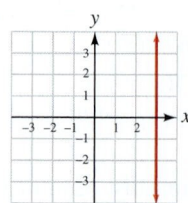

c.

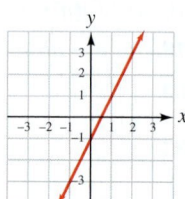

d.

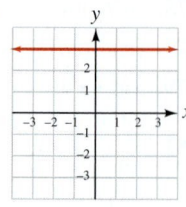

CHALLENGE PROBLEMS

133. Simplify each expression. The variables represent natural numbers.

a. $x^{2m}x^{3m}$

b. $(y^{5c})^4$

c. $\dfrac{m^{8x}}{m^{4x}}$

d. $(2a^{6y})^4$

134. Is the operation of raising to a power commutative? That is, is $a^b = b^a$? Explain.

4.2   Zero and Negative Exponents

- Zero Exponents   • Negative Integer Exponents
- Negative Exponents Appearing in Fractions   • Applications

We now extend the discussion of natural-number exponents to include exponents that are zero and exponents that are negative integers.

▪ ZERO EXPONENTS

To develop the definition of a zero exponent, we will simplify the expression

$$\frac{5^3}{5^3}$$

in two ways and compare the results.

First, we apply the quotient rule for exponents, where we subtract the equal exponents in the numerator and denominator. The result is 5^0. In the second approach, we write 5^3 as $5 \cdot 5 \cdot 5$ and remove the common factors of 5 in the numerator and denominator. The result is 1.

$$\frac{5^3}{5^3} = 5^{3-3} = \mathbf{5^0} \qquad \frac{5^3}{5^3} = \frac{\overset{1}{\cancel{5}} \cdot \overset{1}{\cancel{5}} \cdot \overset{1}{\cancel{5}}}{\underset{1}{\cancel{5}} \cdot \underset{1}{\cancel{5}} \cdot \underset{1}{\cancel{5}}} = \mathbf{1}$$

└─── These must be equal. ───┘

Since $\frac{5^3}{5^3} = 5^0$ and $\frac{5^3}{5^3} = 1$, we conclude that $5^0 = 1$. This observation suggests the following definition.

Zero Exponents   Any nonzero base raised to the 0 power is 1. For any nonzero real number x,

$$x^0 = 1$$

EXAMPLE 1

Simplify each expression, and assume $a \neq 0$:   **a.** $(-8)^0$,   **b.** $\left(\frac{14}{15}\right)^0$,   **c.** $(3a)^0$,   and
d. $3a^0$.

Solution   **a.** $(-8)^0 = 1$ Any nonzero number raised to the 0 power is 1.

The Language of Algebra

Note that the zero exponent definition doesn't define 0^0. This expression is said to be an *indeterminate form,* which is beyond the scope of this book.

b. $\left(\dfrac{14}{15}\right)^0 = 1$

c. $(3a)^0 = 1$ Because of the parentheses, the base is $3a$. Any nonzero base raised to the 0 power is 1.

d. $3a^0 = 3 \cdot \mathbf{a^0}$ Since there are no parentheses, the base is a.

   $= 3 \cdot \mathbf{1}$

   $= 3$

Self Check 1   Simplify each expression: **a.** $(0.75)^0$,   **b.** $-5c^0d$,   **c.** $(5c)^0$.

■ NEGATIVE INTEGER EXPONENTS

The Language of Algebra

The *negative integers* are: $-1, -2, -3, -4, -5$, and so on.

To develop the definition of a negative exponent, we will simplify

$$\frac{6^2}{6^5}$$

in two ways and compare the results.

   If we apply the quotient rule for exponents, where we subtract the greater exponent in the denominator from the lesser exponent in the numerator, we obtain 6^{-3}. In the second approach, we remove the two common factors of 6 to obtain $\frac{1}{6^3}$.

$$\frac{6^2}{6^5} = 6^{2-5} = \mathbf{6^{-3}} \qquad \frac{6^2}{6^5} = \frac{\overset{1}{\cancel{6}} \cdot \overset{1}{\cancel{6}}}{\underset{1}{\cancel{6}} \cdot \underset{1}{\cancel{6}} \cdot 6 \cdot 6 \cdot 6} = \frac{\mathbf{1}}{\mathbf{6^3}}$$

└─── These must be equal. ───┘

Since $\frac{6^2}{6^5} = 6^{-3}$ and $\frac{6^2}{6^5} = \frac{1}{6^3}$, we conclude that $6^{-3} = \frac{1}{6^3}$. Note that 6^{-3} is equal to the reciprocal of 6^3. This observation suggests the following definition.

| **Negative Exponents** | For any nonzero real number x and any integer n, |

$$x^{-n} = \frac{1}{x^n}$$

In words, x^{-n} is the reciprocal of x^n.

From the definition, we see that another way to write x^{-n} is to write its reciprocal and change the sign of the exponent. For example,

$$5^{-4} = \frac{1}{5^4}$$ Think of the reciprocal of 5^{-4}, which is $\dfrac{1}{5^{-4}}$. Then change the sign of the exponent.

EXAMPLE 2   Simplify: **a.** 3^{-2},   **b.** y^{-1},   and   **c.** $(-2)^{-3}$.

ELEMENTARY &
INTERMEDIATE
Algebra $f(x)$ **Now**™   Solution   **a.** $3^{-2} = \dfrac{1}{3^2}$ Write the reciprocal of 3^{-2} and change the sign of the exponent.

$$= \frac{1}{9}$$ $3^2 = 9$.

Caution

A negative exponent does not indicate a negative number. It indicates a reciprocal.

$$4^{-2} = \frac{1}{4^2} = \frac{1}{16}$$

b. $y^{-1} = \dfrac{1}{y^1}$ Write the reciprocal of y^{-1} and change the sign of the exponent.

$$= \frac{1}{y}$$

c. $(-2)^{-3} = \dfrac{1}{(-2)^3}$ Because of the parentheses, the base is -2. Write the reciprocal of $(-2)^{-3}$ and change the exponent from -3 to 3.

$$= -\frac{1}{8}$$ $(-2)^3 = -8$.

Self Check 2   Simplify: **a.** 8^{-2},   **b.** x^{-5},   **c.** $(-3)^{-3}$.

EXAMPLE 3   Simplify: **a.** $9m^{-3}$ and   **b.** -5^{-2}.

ELEMENTARY &
INTERMEDIATE
Algebra $f(x)$ **Now**™   Solution   **a.** $9m^{-3} = 9 \cdot m^{-3}$ The base is m.

$$= 9 \cdot \frac{1}{m^3}$$ Write the reciprocal of m^{-3} and change the sign of the exponent.

$$= \frac{9}{m^3}$$ Multiply.

Caution

Don't confuse *negative numbers* with *negative exponents*. For example, the expressions -2 and 2^{-1} are not the same.

$$2^{-1} = \frac{1}{2^1} = \frac{1}{2}$$

b. $-5^{-2} = -1 \cdot 5^{-2}$ The base is 5.

$$= -1 \cdot \frac{1}{5^2}$$ Write the reciprocal of 5^{-2} and change the sign of the exponent.

$$= -\frac{1}{25}$$ Evaluate 5^2 and multiply.

Self Check 3   Simplify: **a.** $12h^{-9}$ and   **b.** -2^{-4}.

■ NEGATIVE EXPONENTS APPEARING IN FRACTIONS

Negative exponents can appear in the numerator and/or the denominator of a fraction. To develop rules for such situations, we consider the following example.

$$\frac{a^{-4}}{b^{-3}} = \frac{\dfrac{1}{a^4}}{\dfrac{1}{b^3}} = \frac{1}{a^4} \cdot \frac{b^3}{1} = \frac{b^3}{a^4}$$

We can obtain this result in a simpler way. Beginning with $\frac{a^{-4}}{b^{-3}}$, move a^{-4} to the denominator and change the sign of the exponent. Then, move b^{-3} to the numerator and change the sign of the exponent.

$$\frac{a^{-4}}{b^{-3}} = \frac{b^3}{a^4}$$

This example suggests the following rules.

Changing from Negative to Positive Exponents

A factor can be moved from the denominator to the numerator or from the numerator to the denominator of a fraction if the sign of its exponent is changed. For any nonzero real numbers x and y, and any integers m and n,

$$\frac{1}{x^{-n}} = x^n \quad \text{and} \quad \frac{x^{-m}}{y^{-n}} = \frac{y^n}{x^m}$$

These rules streamline the process when simplifying expressions involving negative exponents.

EXAMPLE 4

ELEMENTARY & INTERMEDIATE
Algebra $f(x)$ **Now**™

Write each expression using positive exponents only: **a.** $\dfrac{1}{d^{-10}}$,   **b.** $\dfrac{2^{-3}}{3^{-4}}$,   and   **c.** $\dfrac{s^{-2}}{5t^{-9}}$.

Solution   **a.** $\dfrac{1}{d^{-10}} = d^{10}$ Move d^{-10} to the numerator and change the sign of the exponent.

Caution

This rule does not permit moving *terms* that have negative exponents. For example,

$$\frac{3^{-2} + 8}{5} \neq \frac{8}{3^2 \cdot 5}$$

b. $\dfrac{2^{-3}}{3^{-4}} = \dfrac{3^4}{2^3}$ Move 2^{-3} to the denominator and change the sign of the exponent. Move 3^{-4} to the numerator and change the sign of the exponent.

        $= \dfrac{81}{8}$ $3^4 = 81$ and $2^3 = 8$.

c. $\dfrac{s^{-2}}{5t^{-9}} = \dfrac{t^9}{5s^2}$ Move s^{-2} to the denominator and change the sign of the exponent. Since $5t^{-9}$ has no parentheses, t is the base. Move t^{-9} to the numerator and change the sign of the exponent.

Self Check 4   Write each expression using positive exponents only: **a.** $\dfrac{1}{w^{-5}}$,   **b.** $\dfrac{5^{-2}}{4^{-3}}$,   and   **c.** $\dfrac{h^{-6}}{8a^{-7}}$.

The rules for exponents involving products, powers, and quotients are also true for zero and negative exponents.

Summary of Exponent Rules   If m and n represent integers and there are no divisions by zero, then

Product rule	*Quotient rule*	*Power rule*
$x^m \cdot x^n = x^{m+n}$	$\dfrac{x^m}{x^n} = x^{m-n}$	$(x^m)^n = x^{mn}$

Power of a product	*Power of a quotient*	*Zero exponent*
$(xy)^n = x^n y^n$	$\left(\dfrac{x}{y}\right)^n = \dfrac{x^n}{y^n}$	$x^0 = 1$

Negative exponent	*Negative exponent*	*Negative exponent*
$x^{-n} = \dfrac{1}{x^n}$	$\dfrac{1}{x^{-n}} = x^n$	$\dfrac{x^{-m}}{y^{-n}} = \dfrac{y^n}{x^m}$

The rules for exponents are used to simplify expressions. In general, an expression involving exponents is simplified when

- Each base occurs only once
- There are no parentheses
- There are no negative or zero exponents

EXAMPLE 5   Simplify: **a.** $\left(\dfrac{5}{16}\right)^{-1}$,   **b.** $\dfrac{x^3}{x^7}$,   and   **c.** $(x^3)^{-2}$.

Solution   **a.** The expression is not simplified because it contains parentheses and a negative exponent. Since it is a power of a quotient, we proceed as follows.

$$\left(\frac{5}{16}\right)^{-1} = \frac{5^{-1}}{16^{-1}}$$ Raise the numerator and the denominator to the -1 power.

$$= \frac{16^1}{5^1}$$ Move 5^{-1} to the denominator. Change the sign of the exponent.
Move 16^{-1} to the numerator. Change the sign of the exponent.

$$= \frac{16}{5}$$

b. The expression is not simplified because the base x occurs more than once. Since it is a quotient of like bases, we proceed as follows.

$$\frac{x^3}{x^7} = x^{3-7}$$ Keep the base x and subtract the exponents.

$$= x^{-4}$$ Do the subtraction: $3 - 7 = -4$.

$$= \frac{1}{x^4}$$ Write the reciprocal of x^{-4} and change the sign of the exponent.

c. The expression is not simplified because it contains parentheses and a negative exponent. Since it is a power of a power, we proceed as follows.

$$(x^3)^{-2} = x^{-6}$$ Multiply exponents.

$$= \frac{1}{x^6}$$ Write the reciprocal of x^{-6} and change the sign of the exponent.

ELEMENTARY &
INTERMEDIATE
Algebra $f(x)$ Now™

Self Check 5   Simplify: **a.** $\left(\dfrac{3}{7}\right)^{-2}$,   **b.** $\dfrac{a^3}{a^8}$,   and   **c.** $(n^4)^{-5}$.

EXAMPLE 6

Simplify each expression: **a.** $\dfrac{y^{-4}y^{-3}}{y^{-20}}$,   **b.** $\dfrac{7^{-1}a^3b^4}{6^{-2}a^5b^2}$,   and   **c.** $\left(\dfrac{x^3y^2}{xy^{-3}}\right)^{-2}$.

Solution   **a.** $\dfrac{y^{-4}y^{-3}}{y^{-20}} = \dfrac{y^{-7}}{y^{-20}}$ In the numerator, add exponents: $-4 + (-3) = -7$.

$\phantom{\dfrac{y^{-4}y^{-3}}{y^{-20}}} = y^{-7-(-20)}$ Keep the common base y and subtract exponents.

$\phantom{\dfrac{y^{-4}y^{-3}}{y^{-20}}} = y^{13}$ $-7 - (-20) = -7 + 20 = 13$.

b. $\dfrac{7^{-1}a^3b^4}{6^{-2}a^5b^2} = \dfrac{6^2a^3b^4}{7^1a^5b^2}$ Move 7^{-1} to the denominator. Change the sign of the exponent. Move 6^{-2} to the numerator. Change the sign of the exponent.

$\phantom{\dfrac{7^{-1}a^3b^4}{6^{-2}a^5b^2}} = \dfrac{36a^{3-5}b^{4-2}}{7}$ Use the quotient rule twice and subtract exponents.

$\phantom{\dfrac{7^{-1}a^3b^4}{6^{-2}a^5b^2}} = \dfrac{36a^{-2}b^2}{7}$ Do the subtractions.

$\phantom{\dfrac{7^{-1}a^3b^4}{6^{-2}a^5b^2}} = \dfrac{36b^2}{7a^2}$ Move a^{-2} to the denominator and change the sign of the exponent.

c. $\left(\dfrac{x^3y^2}{xy^{-3}}\right)^{-2} = \left(x^{3-1}y^{2-(-3)}\right)^{-2}$ Use the quotient rule twice and subtract exponents.

$\phantom{\left(\dfrac{x^3y^2}{xy^{-3}}\right)^{-2}} = (x^2y^5)^{-2}$ Do the subtractions.

$\phantom{\left(\dfrac{x^3y^2}{xy^{-3}}\right)^{-2}} = \dfrac{1}{(x^2y^5)^2}$ Write the reciprocal of $(x^2y^5)^{-2}$ and change the sign of the exponent.

$\phantom{\left(\dfrac{x^3y^2}{xy^{-3}}\right)^{-2}} = \dfrac{1}{x^4y^{10}}$ Raise each factor within the parentheses to the 2nd power.

Self Check 6   Simplify: **a.** $\dfrac{a^{-4}a^{-5}}{a^{-3}}$,   **b.** $\dfrac{1^{-4}x^5y^3}{9^{-2}x^3y^6}$,   and   **c.** $\left(\dfrac{c^2d^2}{c^4d^{-3}}\right)^{-3}$.

■ APPLICATIONS

When we determine what sum of money must be invested today to be worth a given amount in the future, we are computing **present value.** The formula for present value P involves a negative exponent.

$$P = A(1 + i)^{-n}$$

where A is the amount of money that is needed in n years, and i is the annual interest rate (expressed as a decimal) that the investment earns.

EXAMPLE 7

ELEMENTARY &
INTERMEDIATE
Algebra $f(x)$ **Now**™

Saving for college.   How much money should the grandparents of a newborn baby girl invest at a 6% annual rate so that on her 18th birthday, she will have a college fund of $20,000?

Solution   We substitute 20,000 for A, 0.06 for i, and 18 for n in the formula to find P.

$$P = A(1 + i)^{-n}$$
$$P = 20,000(1 + 0.06)^{-18}$$
$$P = 20,000(1.06)^{-18}$$ Do the addition within the parentheses.

To evaluate $20,000(1.06)^{-18}$ with a scientific calculator, we use the exponential key y^x.

Enter 1.06 y^x 18 +/− × 20000 = .

Since the result is $P \approx 7,006.88$, the grandparents must invest \$7,006.88 to have \$20,000 in 18 years.

Answers to Self Checks   **1. a.** 1,   **b.** $-5d$,   **c.** 1   **2. a.** $\frac{1}{64}$,   **b.** $\frac{1}{x^5}$,   **c.** $-\frac{1}{27}$ **3. a.** $\frac{12}{h^9}$,   **b.** $-\frac{1}{16}$
4. a. w^5,   **b.** $\frac{64}{25}$,   **c.** $\frac{a^7}{8h^6}$ **5. a.** $\frac{49}{9}$,   **b.** $\frac{1}{a^5}$,   **c.** $\frac{1}{n^{20}}$ **6. a.** $\frac{1}{a^6}$,   **b.** $\frac{81x^2}{y^3}$,   **c.** $\frac{c^6}{d^{15}}$

4.2   STUDY SET   ELEMENTARY & INTERMEDIATE   Algebra $f(x)$ Now™

VOCABULARY   Fill in the blanks.

1. In the expression 8^{-3}, 8 is the _____ and -3 is the _____.

2. In the expression 5^{-1}, the exponent is a _____ integer.

3. Another way to write 2^{-3} is to write its _____ and change the sign of the exponent:

$$2^{-3} = \frac{1}{2^{}}$$

4. To _____ $(y^{-4})^{-3}$, keep the base y and multiply the exponents to get y^{12}.

CONCEPTS

5. In part a, fill in the blanks to simplify the fraction in two ways. Then complete the sentence in part b.

a. $\dfrac{6^4}{6^4} = 6^{}$ $\dfrac{6^4}{6^4} = \dfrac{\boxed{} \cdot \boxed{} \cdot \boxed{} \cdot \boxed{}}{6 \cdot 6 \cdot 6 \cdot 6}$

$= 6^{}$ $= \boxed{}$

└── These must be equal. ──┘

b. So we define 6^0 to be __, and in general, if x is any nonzero real number, then $x^0 = $ __.

6. In part a, fill in the blanks to simplify the fraction in two ways. Then complete the sentence in part b.

a. $\dfrac{8^3}{8^5} = 8^{}$ $\dfrac{8^3}{8^5} = \dfrac{\boxed{} \cdot \boxed{} \cdot \boxed{}}{8 \cdot 8 \cdot 8 \cdot 8 \cdot 8}$

$= 8^{}$ $= \dfrac{1}{8^{}}$

└── These must be equal. ──┘

b. So we define 8^{-2} to be __, and in general, if x is any nonzero real number, then $x^{-n} = $ __.

Complete each table.

7.

x	3^x
2	
1	
0	
-1	
-2	

8.

x	$(-9)^x$
2	
1	
0	
-1	
-2	

9. Complete each rule for exponents.

a. $x^m \cdot x^n = $ __   **b.** $\dfrac{x^m}{x^n} = $ __

c. $(x^m)^n = $ __   **d.** $(xy)^n = $ __

e. $\left(\dfrac{x}{y}\right)^n = $ __   **f.** $x^{-n} = $ __

g. $\dfrac{1}{x^{-n}} = $ __   **h.** $\dfrac{x^{-m}}{y^{-n}} = $ __

i. $x^0 = $ __

Fill in the blanks.

10. A factor can be moved from the denominator to the numerator or from the numerator to the denominator of a fraction if the sign of its exponent is _____.

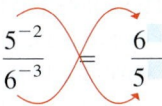

$$\frac{5^{-2}}{6^{-3}} = \frac{6}{5}$$

11. In general, an expression involving exponents is simplified when

- Each base occurs only _____
- There are _____ parentheses
- There are no negative or zero _____

12. Decide whether each statement is true or false.

a. $6^{-2} = -36$ **b.** $6^{-2} = -\dfrac{1}{36}$

c. $-6 = \dfrac{1}{6}$ **d.** $6^{-2} < 6^{-1}$

NOTATION   Complete each solution.

13. $(y^5 y^3)^{-5} = (\quad)^{-5}$

$\qquad = y^{}$

$\qquad = \dfrac{1}{y^{}}$

14. $\left(\dfrac{a^2 b^3}{a^{-3} b}\right)^{-3} = \left(a^{2-} b^{-1}\right)^{-3}$

$\qquad = (a^{} b^{})^{-3}$

$\qquad = \dfrac{1}{(a^5 b^2)^{}}$

$\qquad = \dfrac{1}{a^{15} b^{}}$

15. Complete the table.

Expression	Base	Exponent
4^{-2}		
$6x^{-5}$		
$\left(\dfrac{3}{y}\right)^{-8}$		
-7^{-1}		
$(-2)^{-3}$		
$10a^0 b$		

16. The formula for present value is

$$P = A(1 + i)^{-n}$$

On the right-hand side, what is the base and what is the exponent?

PRACTICE   **Simplify each expression. Write each answer without using parentheses or negative exponents.**

17. 7^0 **18.** 9^0

19. $\left(\dfrac{1}{4}\right)^0$ **20.** $\left(\dfrac{3}{8}\right)^0$

21. $2x^0$ **22.** $(2x)^0$

23. $(-x)^0$ **24.** $-x^0$

25. $\left(\dfrac{a^2 b^3}{ab^4}\right)^0$ **26.** $\dfrac{2}{3}\left(\dfrac{xyz}{x^2 y}\right)^0$

27. $\dfrac{5}{2x^0}$ **28.** $\dfrac{4}{3a^0}$

29. $-15x^0 y$ **30.** $24g^0 h^2$

31. 12^{-2} **32.** 11^{-2}

33. $(-4)^{-1}$ **34.** $(-8)^{-1}$

35. $44g^{-6}$ **36.** $55g^{-3}$

37. $(-10)^{-3}$ **38.** $(-20)^{-2}$

39. -4^{-3} **40.** -6^{-3}

41. $-(-4)^{-3}$ **42.** $-(-4)^{-2}$

43. x^{-2} **44.** y^{-3}

45. $-b^{-5}$ **46.** $-c^{-4}$

47. $\left(\dfrac{1}{6}\right)^{-2}$ **48.** $\left(\dfrac{1}{7}\right)^{-2}$

49. $\left(-\dfrac{1}{2}\right)^{-3}$ **50.** $\left(-\dfrac{1}{5}\right)^{-3}$

51. $\left(\dfrac{7}{8}\right)^{-1}$ **52.** $\left(\dfrac{16}{5}\right)^{-1}$

53. $\dfrac{1}{5^{-3}}$ **54.** $\dfrac{1}{3^{-3}}$

55. $\dfrac{2^{-4}}{3^{-1}}$ **56.** $\dfrac{7^{-2}}{2^{-3}}$

57. $\dfrac{a^{-5}}{b^{-2}}$ **58.** $\dfrac{r^{-6}}{s^{-1}}$

59. $\dfrac{r^{-50}}{r^{-70}}$ **60.** $\dfrac{m^{-30}}{m^{-40}}$

61. $-\dfrac{1}{p^{-10}}$ **62.** $-\dfrac{1}{n^{-30}}$

63. $\dfrac{h^{-5}}{h^2}$ **64.** $\dfrac{y^{-3}}{y^4}$

65. $\dfrac{8}{s^{-1}}$ **66.** $\dfrac{6}{k^{-2}}$

67. $\dfrac{h}{h^{-6}}$ **68.** $\dfrac{w}{w^{-9}}$

69. $(2y)^{-4}$ **70.** $(-3x)^{-1}$

71. $(ab^2)^{-3}$ **72.** $(m^2n^3)^{-2}$

73. $2^5 \cdot 2^{-2}$ **74.** $10^2 \cdot 10^{-4}$

75. $\left(\dfrac{y^4}{3}\right)^{-2}$ **76.** $\left(\dfrac{p^3}{2}\right)^{-3}$

77. $\dfrac{y^4}{y^5}$ **78.** $\dfrac{t^7}{t^{10}}$

79. $\dfrac{(r^2)^3}{(r^3)^4}$ **80.** $\dfrac{(b^3)^4}{(b^5)^4}$

81. $\dfrac{4s^{-5}}{t^{-2}}$ **82.** $\dfrac{9k^{-8}}{m^{-2}}$

83. $(5d^{-2})^3$ **84.** $(9s^{-6})^2$

85. $\dfrac{-2a^{-4}}{a^{-8}}$ **86.** $\dfrac{-3b^{-3}}{b^{-9}}$

87. $x^{-3}x^{-3}x^{-3}$ **88.** $y^{-2}y^{-2}y^{-2}$

89. $\dfrac{t(t^{-2})^{-2}}{t^{-5}}$ **90.** $\dfrac{d(d^{-3})^{-3}}{d^{-7}}$

91. $\dfrac{y^4y^3}{y^4y^{-2}}$ **92.** $\dfrac{x^{12}x^{-7}}{x^3x^4}$

93. $\dfrac{2a^4a^{-2}}{a^2a^0}$ **94.** $\dfrac{3b^0b^3}{b^{-3}b^4}$

95. $(ab^2)^{-2}$ **96.** $(c^2d^3)^{-2}$

97. $(x^2y)^{-3}$ **98.** $(-xy^2)^{-4}$

99. $(x^{-4}x^3)^3$ **100.** $(y^{-2}y)^3$

101. $(-2x^3y^{-2})^{-5}$ **102.** $(-3u^{-2}v^3)^{-3}$

103. $\left(\dfrac{a^3}{a^{-4}}\right)^2$ **104.** $\left(\dfrac{a^4}{a^{-3}}\right)^3$

105. $\left(\dfrac{4x^2}{3x^{-5}}\right)^4$ **106.** $\left(\dfrac{-3r^4r^{-3}}{r^{-3}r^7}\right)^3$

107. $\left(\dfrac{y^3z^{-2}}{3y^{-4}z^3}\right)^2$ **108.** $\left(\dfrac{6xy^3}{x^{-1}y}\right)^3$

109. $\dfrac{2^{-2}g^{-2}h^{-3}}{9^{-1}h^{-3}}$ **110.** $\dfrac{5^{-1}x^{-2}y^{-3}}{8^{-2}x^{-11}}$

Evaluate each expression.

111. $5^0 + (-7)^0$ **112.** $-4^0 + 5^0$

113. $2^{-2} + 4^{-1}$ **114.** $-9^{-1} + 9^{-2}$

115. $9^0 - 9^{-1}$ **116.** $7^{-1} - 7^0$

APPLICATIONS

117. THE DECIMAL NUMERATION SYSTEM
Decimal numbers are written by putting digits into place-value columns that are separated by a decimal point. Express the value of each of the columns using a power of 10.

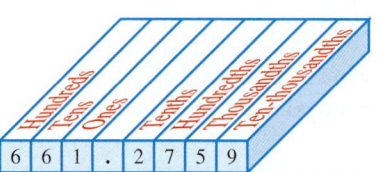

118. UNIT COMPARISONS Consider the relative sizes of the items listed in the table. In the column titled "measurement," write the most appropriate number from the following list. Each number is used only once.

10^0 meter 10^{-1} meter 10^{-2} meter
10^{-3} meter 10^{-4} meter 10^{-5} meter

Item	Measurement (m)
Thickness of a dime	
Height of a bathroom sink	
Length of a pencil eraser	
Thickness of soap bubble film	
Width of a video cassette	
Thickness of a piece of paper	

119. RETIREMENT How much money should a young married couple invest now at an 8% annual rate if they want to have $100,000 in the bank when they reach retirement age in 40 years?

120. SAVING FOR THE FUTURE How much money should a nursery school owner invest now at a 4% annual rate if he wants to have $20,000 in the bank in 5 years?

121. BIOLOGY During reproduction, the time required for a population to double is called the **generation time.** If b bacteria are introduced into a medium, then after the generation time has elapsed, there will be

$2b$ bacteria. After n generations, there will be $b \cdot 2^n$ bacteria. Explain what this expression represents when $n = 0$.

122. ELECTRONICS The total resistance R of a certain circuit is given by

$$R = \left(\frac{1}{R_1} + \frac{1}{R_2} \right)^{-1} + R_3$$

Find R if $R_1 = 4$, $R_2 = 2$, and $R_3 = 1$.

WRITING

123. Explain how you would help a friend understand that 2^{-3} is not equal to -8.

124. What does it mean to simplify $\left(\frac{xy^3}{2x^{-3}} \right)^{-2}$?

REVIEW

125. IQ TESTS An IQ (intelligence quotient) is a score derived from the formula

$$IQ = \frac{\text{mental age}}{\text{chronological age}} \cdot 100$$

Find the mental age of a 10-year-old girl if she has an IQ of 135.

126. DIVING When you are under water, the pressure in your ears is given by the formula

$$\text{Pressure} = \text{depth} \cdot \text{density of water}$$

Find the density of water (in lb/ft^3) if, at a depth of 9 feet, the pressure on your eardrum is 561.6 lb/ft^2.

127. Write the equation of the line having slope $\frac{3}{4}$ and y-intercept -5.

128. Write an equation of the line that passes through $(4, 4)$ and $(-6, -6)$. Express the answer in slope–intercept form.

CHALLENGE PROBLEMS

129. Simplify each expression. Assume there are no divisions by 0.

    a. $r^{5m}r^{-6m}$ **b.** $\dfrac{x^{3n}}{x^{6n}}$

130. If a positive number x is raised to a negative power, is the result greater than, equal to, or less than x? Explore the possibilities.

4.3 Scientific Notation

- Converting from Scientific to Standard Notation
- Writing Numbers in Scientific Notation   • Computations with Scientific Notation

Scientists often deal with extremely large and small numbers. Two examples are shown below.

The distance from the Earth to the sun is approximately 150,000,000 kilometers.

The influenza virus, which causes "flu" symptoms of cough, sore throat, headache, and congestion, has a diameter of 0.00000256 inch.

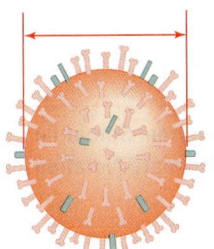

Because 150,000,000 and 0.00000256 contain many zeros, they are difficult to read and cumbersome to work with in calculations. In this section, we will discuss a more convenient form in which we can write such numbers.

■ CONVERTING FROM SCIENTIFIC TO STANDARD NOTATION

Scientific notation provides a compact way of writing very large or very small numbers.

Scientific Notation A positive number is written in **scientific notation** when it is written in the form $N \times 10^n$, where $1 \le N < 10$ and n is an integer.

Some examples of numbers written in scientific notation are shown below. Each is the product of a decimal number that is at least 1, but less than 10, and a power of 10.

Notation

A raised dot · is sometimes used when writing scientific notation.

$3.67 \times 10^2 = 3.67 \cdot 10^2$

An integer exponent
↓

$$3.67 \times 10^2 \qquad 2.15 \times 10^{-3} \qquad 9.875 \times 10^{22}$$

↑
A decimal that is at least 1, but less than 10

A number written in scientific notation can be written in **standard notation** by performing the indicated multiplication. For example, to write 3.67×10^2 in standard notation, we recall that multiplying a decimal by 100 moves the decimal point 2 places to the right.

$$3.67 \times \mathbf{10^2} = 3.67 \times \mathbf{100} = 3\,6\,7.$$

To write 2.15×10^{-3} in standard notation, we recall from arithmetic that dividing a decimal by 1,000 moves the decimal point 3 places to the left.

$$2.15 \times \mathbf{10^{-3}} = 2.15 \times \frac{1}{10^3} = 2.15 \times \frac{1}{\mathbf{1,000}} = \frac{2.15}{\mathbf{1,000}} = 0.0\,0\,2\,1\,5$$

In 3.67×10^2 and 2.15×10^{-3}, the exponent gives the number of decimal places that the decimal point moves, and the sign of the exponent indicates the direction in which it moves. Applying this observation to several other examples, we have

Success Tip

Since $10^0 = 1$, scientific notation involving 10^0 is easily simplified. For example,

$9.7 \times 10^0 = 9.7 \times 1 = 9.7$

$5.32 \times 10^6 = 5\,3\,2\,0\,0\,0\,0.$ Move the decimal point 6 places to the right.

$1.95 \times 10^{-5} = 0.0\,0\,0\,0\,1\,9\,5$ Move the decimal point 5 places to the left.

$9.7 \times 10^0 = 9.7$ There is no movement of the decimal point.

The following procedure summarizes our observations.

Converting from Scientific to Standard Notation

1. If the exponent is positive, move the decimal point the same number of places to the right as the exponent.
2. If the exponent is negative, move the decimal point the same number of places to the left as the absolute value of the exponent.

EXAMPLE 1

ELEMENTARY &
INTERMEDIATE
Algebra $f(x)$ **Now**™

Write each number in standard notation: **a.** 3.467×10^5 and   **b.** 8.9×10^{-4}.

Solution   **a.** Since the exponent is 5, the decimal point moves 5 places to the right.

$$3\,4\,6\,7\,0\,0.$$ To move 5 places to the right, two placeholder zeros must be written.

Thus, $3.467 \times 10^5 = 346{,}700$

b. Since the exponent is -4, the decimal point moves 4 places to the left.

$$0.0\,0\,0\,8\,9$$ To move 4 places to the left, three placeholder zeros must be written.

Thus, $8.9 \times 10^{-4} = 0.00089$

Self Check 1   Write each number in standard notation: **a.** 4.88×10^6 and   **b.** 9.8×10^{-3}.

■ WRITING NUMBERS IN SCIENTIFIC NOTATION

The next example shows how to write a number in scientific notation.

EXAMPLE 2

ELEMENTARY &
INTERMEDIATE
Algebra $f(x)$ **Now**™

Solution

Write each number in scientific notation: **a.** 150,000,000,   **b.** 0.00000256,   and
c. 432×10^5.

a. We must write 150,000,000 as the product of a number between 1 and 10 and a power of 10. We note that 1.5 lies between 1 and 10. To obtain 150,000,000, the decimal point in 1.5 must be moved 8 places to the right.

$$1\,5\,0\,0\,0\,0\,0\,0\,0.$$

This will happen if we multiply 1.5 by 10^8. Therefore,

$$150{,}000{,}000 = 1.5 \times 10^8$$

b. We must write 0.00000256 as the product of a number between 1 and 10 and a power of 10. We note that 2.56 lies between 1 and 10. To obtain 0.00000256, the decimal point in 2.56 must be moved 6 places to the left.

$$0\,0\,0\,0\,0\,0\,2.56$$

This will happen if we multiply 2.56 by 10^{-6}. Therefore,

$$0.00000256 = 2.56 \times 10^{-6}$$

c. The number 432×10^5 is not written in scientific notation because 432 is not a number between 1 and 10. To write this number in scientific notation, we proceed as follows:

$$432 \times 10^5 = \mathbf{4.32 \times 10^2} \times 10^5$$ Write 432 in scientific notation.
$$= 4.32 \times 10^7$$ Use the product rule to find $10^2 \times 10^5$. Keep the base of 10 and add the exponents.

Written in scientific notation, 432×10^5 is 4.32×10^7.

Notation

When writing numbers in scientific notation, keep the negative exponents. Don't apply the negative exponent rule.

2.56×10^{-6} ~~$2.56 \times \dfrac{1}{10^6}$~~

Calculators
Scientific notation

When displaying very large or very small answers, a scientific calculator uses scientific notation. The displays below show the results when two powers are computed.

$(453.46)^5 = 1.917321395 \times 10^{13}$

| 1.917321395 | 13 |

$(0.0005)^{12} = 2.44140625 \times 10^{-40}$

| 2.44140625 | -40 |

Self Check 2   Write each number in scientific notation: **a.** 93,000,000   **b.** 0.00009055,   and
c. 85×10^{-3}.

The results from Example 2 illustrate the following forms to use when converting numbers from standard to scientific notation.

For real numbers between 0 and 1: $\square \times 10^{\text{negative integer}}$
For real numbers at least 1, but less than 10: $\square \times 10^{0}$
For real numbers greater than or equal to 10: $\square \times 10^{\text{positive integer}}$

■ COMPUTATIONS WITH SCIENTIFIC NOTATION

Another advantage of scientific notation becomes apparent when we evaluate products or quotients that involve very large or small numbers. If we express those numbers in scientific notation, we can use rules for exponents to make the calculations easier.

EXAMPLE 3

ELEMENTARY &
INTERMEDIATE
Algebra $f(x)$ **Now**™

Astronomy.   Except for the sun, the nearest star visible to the naked eye from most parts of the United States is Sirius. Light from Sirius reaches Earth in about 70,000 hours. If light travels at approximately 670,000,000 mph, how far from Earth is Sirius?

Solution   We are given the rate at which light travels (670,000,000 mph) and the time it takes the light to travel from Sirius to Earth (70,000 hr). We can find the distance the light travels using the formula $d = rt$.

$d = \mathbf{rt}$
$d = \mathbf{670,000,000(70,000)}$ Substitute 670,000,000 for r and 70,000 for t.
$\quad = (6.7 \times 10^8)(7.0 \times 10^4)$ Write each number in scientific notation.
$\quad = (6.7 \cdot 7.0) \times (10^8 \times 10^4)$ Group the numbers together and the powers of 10 together.
$\quad = (6.7 \cdot 7.0) \times 10^{8+4}$ Use the product rule to find $10^8 \times 10^4$. Keep the base 10 and add exponents.
$\quad = 46.9 \times 10^{12}$ Do the multiplication. Do the addition.

We note that 46.9 is not between 0 and 1, so 46.9×10^{12} is not written in scientific notation. To answer in scientific notation, we proceed as follows.

$\quad = 4.69 \times 10^1 \times 10^{12}$ Write 46.9 in scientific notation as 4.69×10^1.
$\quad = 4.69 \times 10^{13}$ Keep the base of 10 and add the exponents.

Sirius is approximately 4.69×10^{13} or 46,900,000,000,000 miles from Earth.

EXAMPLE 4

Atoms.   Scientific notation is used in chemistry. As an example, we can approximate the weight (in grams) of one atom of the heaviest naturally occurring element, uranium, by evaluating the following expression.

$$\frac{2.4 \times 10^2}{6.0 \times 10^{23}}$$

Solution

$$\frac{2.4 \times 10^2}{6.0 \times 10^{23}} = \frac{2.4}{6.0} \times \frac{10^2}{10^{23}}$$ Divide the numbers and the powers of 10 separately.

$$= \frac{2.4}{6.0} \times 10^{2-23}$$ For the powers of 10, use the quotient rule. Keep the base 10 and subtract the exponents.

$$= 0.4 \times 10^{-21}$$ Do the division. Then subtract the exponents.

$$= 4.0 \times 10^{-1} \times 10^{-21}$$ Write 0.4 in scientific notation as 4.0×10^{-1}.

$$= 4.0 \times 10^{-22}$$ Keep the base of 10 and add the exponents.

Calculators
Entering scientific notation

We can evaluate the expression from Example 4 by entering the numbers in scientific notation using the EE key on a scientific calculator:

2.4 EE 2 ÷ 6 EE 23 =

One atom of uranium weighs 4.0×10^{-22} gram. Written in standard notation, this is 0.00000000000000000000004 g.

Self Check 4   Find the approximate weight (in grams) of one atom of gold by evaluating

$$\frac{1.98 \times 10^2}{6.0 \times 10^{23}}$$

Answers to Self Checks   **1. a.** 4,880,000,   **b.** 0.0098   **2. a.** 9.3×10^7,   **b.** 9.055×10^{-5},   **c.** 8.5×10^{-2}
4. 3.3×10^{-22} g

4.3   STUDY SET   ELEMENTARY & INTERMEDIATE
Algebra *f(x)* **Now**™

VOCABULARY   Fill in the blanks.

1. 4.84×10^9 and 1.05×10^{-2} are written in _____ notation.

2. The number 125,000 is written in _____ notation.

3. Scientific _____ provides a compact way of writing very large or very small numbers.

4. A number written in scientific notation is the product of a _____ that is at least 1, but less than 10, and a power of _____.

5. 10^{34}, 10^{50}, and 10^{-14} are _____ of 10.

6. The numbers 2, 5, 8, and 15 are examples of _____ integers. The numbers -1, -6, -30, and -45 are examples of negative _____.

CONCEPTS   Fill in the blanks.

7. When we multiply a decimal by 10^5, the decimal point moves 5 places to the _____.
When we multiply a decimal by 10^{-7}, the decimal point moves 7 places to the _____.

8. Multiplying a decimal by 10^0 does not move the decimal point, because $10^0 = $ ▮.

9. The arrows show the movement of a decimal point. By what power of 10 was each decimal multiplied?
a. 0.0 0 0 0 0 0 5 56
b. 8,0 4 1,0 0 0,0 0 0.

10. Fill in the blanks to describe the procedure for converting a number from scientific notation to standard form.
a. If the exponent is positive, move the decimal point the same number of places to the _____ as the exponent.
b. If the exponent is negative, move the decimal point the same number of places to the _____ as the absolute value of the exponent.

11. a. When a real number greater than 1 is written in scientific notation, the exponent on 10 is a _____ integer.
b. When a real number between 0 and 1 is written in scientific notation, the exponent on 10 is a _____ integer.

12. A portion of the scientific notation for a positive number is blocked out. Decide whether each statement is true or false.
a. ▮ $\times 10^{-8} > 0$ **b.** ▮ $\times 10^{-6} > 1$
c. ▮ $\times 10^3 > 0$ **d.** ▮ $\times 10^5 > 10$

Fill in the blanks to write each number in scientific notation.

13. a. $7,700 = $ ▮ $\times 10^3$
b. $500,000 = $ ▮ $\times 10^5$
c. $114,000,000 = 1.14 \times 10$ ▮

14. **a.** $0.0082 = \boxed{} \times 10^{-3}$

b. $0.0000001 = \boxed{} \times 10^{-7}$

c. $0.00003457 = 3.457 \times 10^{\boxed{}}$

15. Write each expression so that the decimal numbers are grouped together and the powers of ten are grouped together.

a. $(5.1 \times 10^9)(1.5 \times 10^{22})$

b. $\dfrac{8.8 \times 10^{30}}{2.2 \times 10^{19}}$

16. Simplify each expression.

a. $10^{24} \times 10^{33}$ **b.** $\dfrac{10^{50}}{10^{36}}$

c. $(10^{18})^3$ **d.** $\dfrac{10^{15} \times 10^{27}}{10^{40}}$

NOTATION

17. Fill in the blanks. A positive number is written in scientific notation when it is written in the form $N \times 10^n$, where $\boxed{} \leq N < \boxed{}$ and n is an _____

18. In the basic form for scientific notation, what type of number can be written as the exponent?

$$\blacksquare \times 10^{\blacksquare}$$

What type of number can be written as the coefficient?

PRACTICE   Write each number in standard notation.

19. 2.3×10^2 **20.** 3.75×10^4

21. 8.12×10^5 **22.** 1.2×10^3

23. 1.15×10^{-3} **24.** 4.9×10^{-2}

25. 9.76×10^{-4} **26.** 7.63×10^{-5}

27. 6.001×10^6 **28.** 9.998×10^5

29. 2.718×10^0 **30.** 3.14×10^0

31. 6.789×10^{-2} **32.** 4.321×10^{-1}

33. 2.0×10^{-5} **34.** 7.0×10^{-6}

35. 9.0×10^9 **36.** 8.0×10^8

Write each number in scientific notation.

37. 23,000   **38.** 4,750

39. 1,700,000   **40.** 290,000

41. 0.062   **42.** 0.00073

43. 0.0000051   **44.** 0.04

45. 5,000,000,000   **46.** 7,000,000

47. 0.0000003   **48.** 0.0001

49. 909,000,000   **50.** 7,007,000,000

51. 0.0345   **52.** 0.000000567

53. 9   **54.** 2

55. 1,718,000,000,000,000,000

56. 44,180,000,000,000,000,000

57. 0.0000000000000123

58. 0.0000000000000000555

Use scientific notation to perform the calculations. Give all answers in standard notation.

59. $(3.4 \times 10^2)(2.1 \times 10^3)$

60. $(4.1 \times 10^{-3})(3.4 \times 10^4)$

61. $(8.4 \times 10^{-13})(4.8 \times 10^9)$

62. $(5.5 \times 10^{-15})(2.2 \times 10^{13})$

63. $\dfrac{9.3 \times 10^2}{3.1 \times 10^{-2}}$ **64.** $\dfrac{7.2 \times 10^6}{1.2 \times 10^8}$

65. $\dfrac{0.00000129}{0.0003}$ **66.** $\dfrac{169,000,000,000}{26,000,000}$

67. $(0.0000000056)(5,500,000)$

68. $(0.000000061)(3,500,000,000)$

69. $\dfrac{96,000}{(12,000)(0.00004)}$ **70.** $\dfrac{(0.48)(14,400,000)}{96,000,000}$

⊞   Find each power.

71. $(456.4)^6$

72. $(0.009)^{-6}$

73. 225^{-5}

74. $\left(\dfrac{1}{3}\right)^{-55}$

APPLICATIONS

75. ASTRONOMY   The distance from Earth to Alpha Centauri (the nearest star outside our solar system) is about 25,700,000,000,000 miles. Express this number in scientific notation.

76. SPEED OF SOUND   The speed of sound in air is 33,100 centimeters per second. Express this number in scientific notation.

77. GEOGRAPHY   The largest ocean in the world is the Pacific Ocean, which covers 6.38×10^7 square miles. Express this number in standard notation.

78. ATOMS   The number of atoms in 1 gram of iron is approximately 1.08×10^{22}. Express this number in standard notation.

79. LENGTH OF A METER   One meter is approximately 0.00622 mile. Use scientific notation to express this number.

80. ANGSTROMS   One angstrom is 1.0×10^{-7} millimeter. Express this number in standard notation.

81. WAVELENGTHS   Transmitters, vacuum tubes, and lights emit energy that can be modeled as a wave, as shown below. Examples of the most common types of electromagnetic waves are given in the table. List the wavelengths in order from shortest to longest.

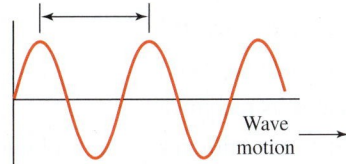

This distance between the two crests of the wave is called the wavelength.

Wave motion

Type	Use	Wavelength (m)
visible light	lighting	9.3×10^{-6}
infrared	photography	3.7×10^{-5}
x-ray	medical	2.3×10^{-11}
radio wave	communication	3.0×10^{2}
gamma ray	treating cancer	8.9×10^{-14}
microwave	cooking	1.1×10^{-2}
ultraviolet	sun lamp	6.1×10^{-8}

82. EXPLORATION   On July 4, 1997, the *Pathfinder*, carrying the rover vehicle called Sojourner, landed on Mars to perform a scientific investigation of the planet. The distance from Mars to Earth is approximately 3.5×10^7 miles. Use scientific notation to express this distance in feet.   (*Hint:* 5,280 feet = 1 mile.)

83. PROTONS   The mass of one proton is approximately 1.7×10^{-24} gram. Use scientific notation to express the mass of 1 million protons.

84. SPEED OF SOUND   The speed of sound in air is approximately 3.3×10^4 centimeters per second. Use scientific notation to express this speed in kilometers per second.   (*Hint:* 100 centimeters = 1 meter and 1,000 meters = 1 kilometer.)

85. LIGHT YEARS   One light year is about 5.87×10^{12} miles. Use scientific notation to express this distance in feet.   (*Hint:* 5,280 feet = 1 mile.)

86. OIL RESERVES   As of January 1, 2001, Saudi Arabia was believed to have crude oil reserves of about 2.617×10^{11} barrels. A barrel contains 42 gallons of oil. Use scientific notation to express its oil reserves in gallons. (Source: *The World Almanac and Book of Facts 2003.*)

87. INTEREST   As of December 2000, the Federal Deposit Insurance Corporation (FDIC) reported that the total insured deposits in U.S. banks and savings and loans was approximately 5.2×10^{12} dollars. If this money was invested at a rate of 4% simple annual interest, how much would it earn in 1 year? Use scientific notation to express the answer. (Source: *The World Almanac and Book of Facts 2003.*)

88. CURRENCY   As of June 30, 2002, the U.S. Treasury reported that the number of $20 bills in circulation was approximately 4.84×10^9. What was the total value of the currency? Use scientific notation to express the answer. (Source: *The World Almanac and Book of Facts 2003.*)

89. THE MILITARY   The graph shows the number of U.S. troops for 1983–2001. Estimate each of the following and express your answers in scientific and standard notation.
a. The number of troops in 1993
b. The largest number of troops during these years

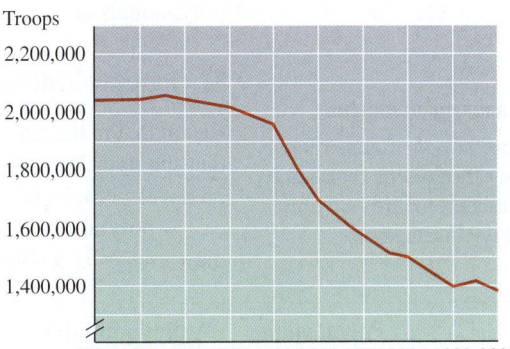

Source: The U.S. Department of Defense

90. THE NATIONAL DEBT   The graph shows the growth of the national debt for the fiscal years 1993–2002.

  a. Use scientific notation to express the debt as of 1994, 1995, and 2001.

  b. In 2002, the population of the United States was about 2.88×10^8. Estimate the share of the debt for each man, woman, and child in the United States. Answer in standard notation.

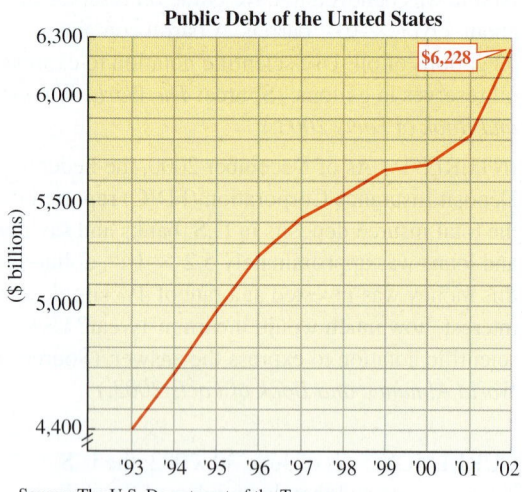

Public Debt of the United States

$6,228

($ billions)

'93 '94 '95 '96 '97 '98 '99 '00 '01 '02

Source: The U.S. Department of the Treasury

WRITING

91. In what situations would scientific notation be more convenient than standard notation?

92. To multiply a number by a power of 10, we move the decimal point. Which way, and how far? Explain.

93. 2.3×10^{-3} contains a negative sign but represents a positive number. Explain.

94. Explain why 237.8×10^8 is not written in scientific notation.

REVIEW

95. If $y = -1$, find the value of $-5y^{55}$.

96. What is the y-intercept of the graph of $y = -3x - 5$?

97. COUNSELING   In the first year of her practice, a family counselor saw 75 clients. In her second year, the number of clients grew to 105. If a linear trend continues, write an equation that gives the number of clients c the counselor will have t years after beginning her practice.

98. Is $(0, -5)$ a solution of $2x + 3y \geq -14$?

CHALLENGE PROBLEMS

99. Consider 2.5×10^{-4}.
  a. What is its opposite?
  b. What is its reciprocal?

100. a. Write one million and one millionth in scientific notation.
  b. By what number must we multiply one millionth to get one million?

	4.4	**Polynomials**

• Polynomials   • Evaluating Polynomials   • Graphing Nonlinear Equations

The simplest types of algebraic expressions are *polynomials*. In this section, we will define polynomials and introduce the vocabulary that is used to describe them.

■ POLYNOMIALS

The Language of Algebra

The prefix *poly* means many. Some other words that begin with this prefix are *poly*gon and *poly*unsaturated.

Recall that in the expression $x^4 + 6x^2 + 4x + 25$, x^4, $6x^2$, $4x$, and 25 are called **terms** and that the numerical factor of each term is called its **coefficient.** Therefore, x^4, $6x^2$, $4x$, and 25 have coefficients of 1, 6, 4, and 25, respectively.

  A **polynomial** is an expression that consists of one or more terms. The exponents on the variables in a polynomial must be whole numbers.

Polynomials	A **polynomial** is a single term or a sum of terms in which all variables have whole-number exponents. No variable appears in a denominator.

Here are some examples of polynomials:

$$3x + 2, \qquad 4y^2 - 2y - 3, \qquad -8xy^2, \qquad \text{and} \qquad a^3 + 3a^2b + 3ab^2 + b^3$$

The polynomial $3x + 2$ is the sum of two terms, $3x$ and 2, and we say it is a **polynomial in x.** A single number is called a **constant,** and so its last term, 2, is called the **constant term.**

Since $4y^2 - 2y - 3$ can be written as $4y^2 + (-2y) + (-3)$, it is the sum of three terms, $4y^2$, $-2y$, and -3. It is written in **descending powers of y,** because the exponents on y decrease from left to right. When a polynomial is written in descending powers, the first term, in this case $4y^2$, is called the **lead term.** The coefficient of the lead term, in this case 4, is called the **lead coefficient.**

$-8xy^2$ is a polynomial with just one term. We say that it is a **polynomial in x and y.** The four-term polynomial $a^3 + 3a^2b + 3ab^2 + b^3$ is written in descending powers of a and **ascending powers** of b.

Polynomials are classified according to the number of terms they have. A polynomial with exactly one term is called a **monomial;** exactly two terms, a **binomial;** and exactly three terms, a **trinomial.** Polynomials with four or more terms have no special names.

Polynomials

Monomials	Binomials	Trinomials
$-6x$	$3u^3 - 4u^2$	$-5t^2 + 4t + 3$
$5x^2y$	$18a^2b + 4ab$	$27x^3 - 6x - 2$
29	$-29z^{17} - 1$	$a^2 + 2ab + b^2$

Polynomials and their terms can be described according to the exponents on their variables.

Degree of a Term   The **degree of a term** of a polynomial in one variable is the value of the exponent on the variable. If a polynomial is in more than one variable, the **degree of a term** is the sum of the exponents on the variables. The **degree of a nonzero constant** is 0.

Here are some examples:

$7x^6$ has degree 6.
$-2a^4$ has degree 4.
$47x^2y$ has degree 3 because x^2y can be written x^2y^1 and $2 + 1 = 3$.
8 has degree 0 since it can be written as $8x^0$.

We can determine the *degree of a polynomial* by considering the degrees of each of its terms.

Degree of a Polynomial   The **degree of a polynomial** is the same as the highest degree of any term of the polynomial.

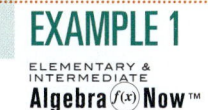

EXAMPLE 1

ELEMENTARY & INTERMEDIATE
Algebra $f(x)$ **Now**™

Describe each polynomial: **a.** $d^3 + 4d^2 - 16$, **b.** $\frac{1}{2}x^2 - x$, and
c. $-6y^{14} - 15y^9z^9 + 25y^8z^{10} + 4yz^{11}$.

Solution **a.** Since $d^3 + 4d^2 - 16$ has three terms, it is a trinomial. It is written in descending powers of d. The highest degree of any term is 3, so it is of degree 3.

Term	Coefficient	Degree
d^3	1	3
$4d^2$	4	2
-16	-16	0

The Language of Algebra

To *descend* means to move from higher to lower. A classic scene in the movie "Gone with the Wind" is Scarlett O'Hara *descending* the grand staircase. In $x^3 + x^2 + x$, the powers of x are written in *descending* order, as we read from left to right.

b. Since $\frac{1}{2}x^2 - x$ has two terms, it is a binomial. It is written in descending powers of x. The highest degree of any term is 2, so it is of degree 2.

Term	Coefficient	Degree
$\frac{1}{2}x^2$	$\frac{1}{2}$	2
$-x$	-1	1

c. $-6y^{14} - 15y^9z^9 + 25y^8z^{10} + 4yz^{11}$ is a polynomial with 4 terms. It is written in descending powers of y and ascending powers of z. The highest degree of any term is 18, so it is of degree 18.

Term	Coefficient	Degree
$-6y^{14}$	-6	14
$-15y^9z^9$	-15	18
$25y^8z^{10}$	25	18
$4yz^{11}$	4	12

Self Check 1   Describe each polynomial: **a.** $x^2 + 4x - 16$ and   **b.** $-14s^5t + s^4t^3$.

■ EVALUATING POLYNOMIALS

A polynomial can have different values depending on the number that is substituted for its variable.

EXAMPLE 2   Evaluate $3x^2 + 4x - 5$ for   **a.** $x = 0$ and   **b.** $x = -2$.

ELEMENTARY & INTERMEDIATE
Algebra *f(x)* **Now**™

Solution   We substitute the given value for each x and follow the rules for the order of operations.

a. $3x^2 + 4x - 5 = 3(\mathbf{0})^2 + 4(\mathbf{0}) - 5$ Substitute 0 for x.
$$= 3(0) + 4(0) - 5$$
$$= 0 + 0 - 5$$
$$= -5$$

b. $3x^2 + 4x - 5 = 3(\mathbf{-2})^2 + 4(\mathbf{-2}) - 5$ Substitute -2 for x.
$$= 3(4) + 4(-2) - 5$$
$$= 12 + (-8) - 5$$
$$= -1$$

Self Check 2   Evaluate $-x^3 + x - 2x + 3$ for   **a.** $x = 0$ and   **b.** $x = -3$.

EXAMPLE 3   ***Supermarket displays.***   The polynomial

$$\frac{1}{3}c^3 + \frac{1}{2}c^2 + \frac{1}{6}c$$

gives the number of cans used in a display shaped like a square pyramid, having a square base formed by c cans per side. Find the number of cans used in the display shown here.

Solution Since each side of the square base of the display is formed by 4 cans, c is 4.

$$\frac{1}{3}c^3 + \frac{1}{2}c^2 + \frac{1}{6}c = \frac{1}{3}(\mathbf{4})^3 + \frac{1}{2}(\mathbf{4})^2 + \frac{1}{6}(\mathbf{4}) \qquad \text{Substitute 4 for } c.$$

$$= \frac{1}{3}(64) + \frac{1}{2}(16) + \frac{1}{6}(4) \qquad \text{Find the powers.}$$

$$= \frac{64}{3} + 8 + \frac{2}{3} \qquad \text{Do the multiplication, and then simplify:} \quad \frac{4}{6} = \frac{2}{3}.$$

$$= 30 \qquad \text{Add the fractions: } \frac{64}{3} + \frac{2}{3} = \frac{66}{3} = 22.$$

30 cans of soup were used in the display.

■ GRAPHING NONLINEAR EQUATIONS

We have graphed equations such as $y = x$ and $y = 2x - 3$. These equations are said to be linear equations because their graphs are straight lines. Note that the right-hand side of each equation is a polynomial of degree 1.

$$y = \underset{\underset{\text{The degree of each polynomial is 1.}}{|}}{x} \qquad\qquad y = \underset{\underset{|}{}}{2x - 3}$$

The degree of each polynomial is 1.

We can also graph equations defined by polynomials with degrees greater than 1.

$$y = \underset{\underset{|}{}}{x^2} \qquad\qquad y = \underset{\underset{|}{}}{-x^2 + 2} \qquad\qquad y = \underset{\underset{|}{}}{x^3 + 1}$$

The degree of this polynomial is 2. The degree of this polynomial is 2. The degree of this polynomial is 3.

EXAMPLE 4

ELEMENTARY &
INTERMEDIATE
Algebra $f(x)$ **Now**™

Graph: $y = x^2$.

Solution To make a table of solutions, we choose values for x and find the corresponding values of y. If $x = -3$, we have

$$y = \mathbf{x^2}$$
$$y = (\mathbf{-3})^2 \qquad \text{Substitute } -3 \text{ for } x.$$
$$y = 9$$

Thus, $(-3, 9)$ is a solution. In a similar manner, we find the corresponding y-values for x-values of $-2, -1, 0, 1, 2,$ and 3. If we plot the ordered pairs listed in the table and

join the points with a smooth curve, we get the graph shown below, which is called a
parabola.

$$y = x^2$$

x	y	(x, y)
-3	9	$(-3, 9)$
-2	4	$(-2, 4)$
-1	1	$(-1, 1)$
0	0	$(0, 0)$
1	1	$(1, 1)$
2	4	$(2, 4)$
3	9	$(3, 9)$

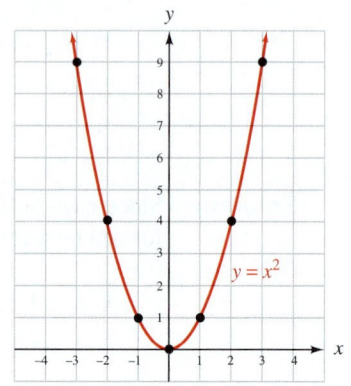

Self Check 4   Graph: $y = x^2 - 2$.

EXAMPLE 5   Graph: $y = -x^2 + 2$.

ELEMENTARY &
INTERMEDIATE
Algebra $f(x)$ **Now**™

Solution   To make a table of solutions, we select x-values of $-3, -2, -1, 0, 1, 2$, and 3 and find
each corresponding y-value. For example, if $x = -3$, we have

$$y = -x^2 + 2$$
$$y = -(-3)^2 + 2 \qquad \text{Substitute } -3 \text{ for } x.$$
$$y = -(9) + 2$$
$$y = -7$$

The ordered pair $(-3, -7)$ is a solution. Six other solutions appear in the table. After plotting each pair, we join the points with a smooth curve to obtain the graph, a parabola
opening downward.

x	y	(x, y)
-3	-7	$(-3, -7)$
-2	-2	$(-2, -2)$
-1	1	$(-1, 1)$
0	2	$(0, 2)$
1	1	$(1, 1)$
2	-2	$(2, -2)$
3	-7	$(3, -7)$

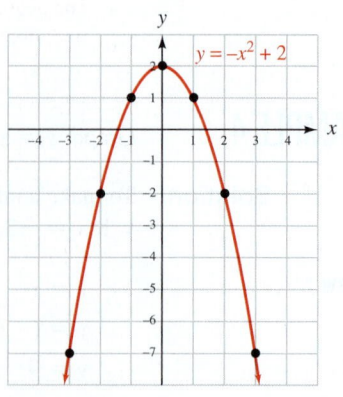

Self Check 5   Graph: $y = -x^2$.

EXAMPLE 6   Graph: $y = x^3 + 1$.

ELEMENTARY &
INTERMEDIATE
Algebra $f(x)$ **Now**™

Solution   If we let $x = -2$, we have

$$y = x^3 + 1$$
$$y = (-2)^3 + 1 \quad \text{Substitute } -2 \text{ for } x.$$
$$y = -8 + 1$$
$$y = -7$$

The ordered pair $(-2, -7)$ is a solution. This pair and others that satisfy the equation are listed in the table. Plotting the ordered pairs and joining the points with a smooth curve gives us the graph.

$$y = x^3 + 1$$

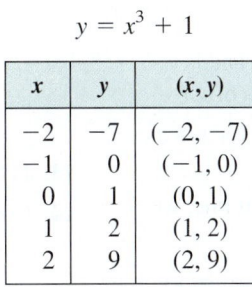

x	y	(x, y)
-2	-7	$(-2, -7)$
-1	0	$(-1, 0)$
0	1	$(0, 1)$
1	2	$(1, 2)$
2	9	$(2, 9)$

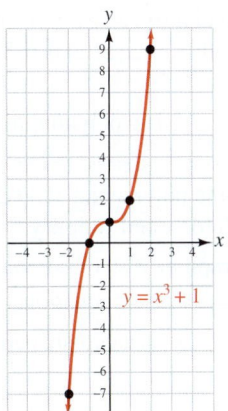

Self Check 6   Graph: $y = x^3 - 1$.

Since the graphs of $y = x^2$, $y = -x^2 + 2$, and $y = x^3 + 1$ are not lines, they are called **nonlinear equations.**

Answers to Self Checks   **1. a.** trinomial of degree 2; terms: x^2, $4x$, -16; coefficients: $1, 4, -16$; degree: 2, 1, 0.   **b.** binomial of degree 7; terms: $-14s^5t$, s^4t^3; coefficients: $-14, 1$; degree: 6, 7   **2. a.** 3,   **b.** 33   **4.** The graph has the same shape, but is 2 units lower.   **5.**   **6.**

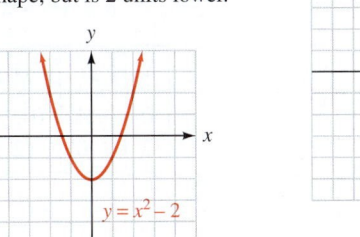

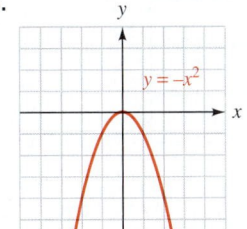

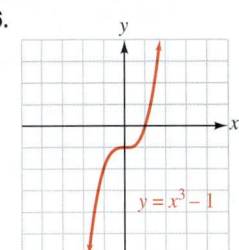

 **4.4**   **STUDY SET**   ELEMENTARY & INTERMEDIATE Algebra $f(x)$ Now™

VOCABULARY   Fill in the blanks.

1. A _____ is a term or a sum of terms in which all variables have whole-number exponents.

2. The numerical _____ of the term $-25x^2y^3$ is -25.

3. The degree of a polynomial is the same as the degree of its _____ with the highest degree.

4. A _____ is a polynomial with one term. A _____ is a polynomial with two terms. A _____ is a polynomial with three terms.

5. The _____ of the monomial $3x^7$ is 7.

6. For the polynomial $6x^2 + 3x - 1$, the _____ term is $6x^2$, and the lead _____ is 6. The _____ term is -1.

7. $x^3 - 6x^2 + 9x - 2$ is a polynomial in ____ and is written in _____ powers of x.

8. To _____ the polynomial $x^2 - 2x + 1$ for $x = 6$, we substitute 6 for x and simplify.

9. Because the graph of $y = x^3$ is not a straight line, we call $y = x^3$ a _____ equation.

10. The graph of $y = x^2$ is a cup-shaped curve called a _____.

CONCEPTS   Decide whether each expression is a polynomial.

11. **a.** $x^3 - 5x^2 - 2$ **b.** $x^{-4} - 5x$

    c. $x^2 - \dfrac{1}{2x} + 3$ **d.** $x^3 - 1$

    e. $x^2 - y^2$ **f.** $a^4 + a^3 + a^2 + a$

12. Complete the table for each polynomial.

    a. $8x^2 + x - 7$

Term	Coefficient	Degree

b. $y^4 - y^3 + 16y^2 + 3y$

Term	Coefficient	Degree

c. $8a^6b^3 - 27ab$

Term	Coefficient	Degree

13. Is $(-1, 2)$ a solution of $y = x^2 + 3$?

14. Complete the table of solutions.

$$y = x^3 + 5$$

x	y	(x, y)
-2		
-1		
0		
1		
2		

15. To graph $y = x^2 - 4$, a table of solutions is constructed and a graph is drawn. Explain the error.

$$y = x^2 - 4$$

x	y	(x, y)
0	-4	$(0, -4)$
2	0	$(2, 0)$

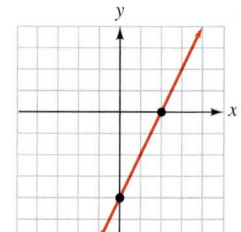

16. Explain the error in the graph of $y = x^2$.

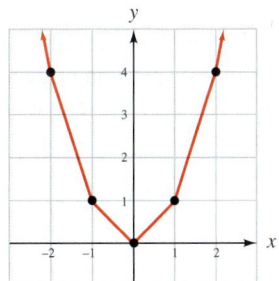

NOTATION   Complete the solution.

17. Evaluate $-2x^2 + 3x - 1$ for $x = -2$.

$$-2x^2 + 3x - 1 = -2\left(\boxed{}\right)^2 + 3\left(\boxed{}\right) - 1$$
$$= -2\left(\boxed{}\right) + 3\left(\boxed{}\right) - 1$$
$$= \boxed{} + (-6) - 1$$
$$= \boxed{} - 1$$
$$= \boxed{}$$

18. a. Write $x - 9 + 3x^2$ in descending powers of x.

   b. Write $-2xy + y^2 + x^2$ in ascending powers of y.

19. Which of the following equations are defined by a polynomial of degree 2 or higher?

$$y = x^2 - 1 \qquad y = x - 1 \qquad y = x^3 - 1$$

20. The expression $x + y$ is a binomial. Is xy also a binomial? Explain why or why not.

PRACTICE   Classify each polynomial as a monomial, a binomial, a trinomial, or none of these.

21. $3x + 7$ **22.** $3y - 5$

23. $y^2 + 4y + 3$ **24.** $3xy$

25. $3z^2$ **26.** $3x^4 - 2x^3 + 3x - 1$

27. $t - 32$ **28.** $9x^2y^3z^4$

29. $s^2 - 23s + 31$ **30.** $2x^3 - 5x^2 + 6x - 3$

31. $3x^5 - x^4 - 3x^3 + 7$ **32.** x^3

33. $2a^2 - 3ab + b^2$ **34.** $a^3 - b^3$

Find the degree of each polynomial.

35. $3x^4$ **36.** $3x^5$

37. $-2x^2 + 3x + 1$ **38.** $-5x^4 + 3x^2 - 3x$

39. $3x - 5$ **40.** $y^3 + 4y^2$

41. $-5r^2s^2 - r^3s + 3$ **42.** $4r^2s^3 - 5r^2s^8$

43. $x^{12} + 3x^2y^3$ **44.** $17ab^5 - 12a^3b$

45. 38 **46.** -24

Evaluate each expression.

47. $-9x + 1$ for
   a. $x = 5$ **b.** $x = -4$

48. $-10x + 6$ for
   a. $x = 8$ **b.** $x = -6$

49. $3x^2 - 2x + 8$ for
   a. $x = 1$ **b.** $x = 0$

50. $4x^2 + 2x - 8$ for
   a. $x = -1$ **b.** $x = -10$

51. $-x^2 - 6$ for
   a. $x = -4$ **b.** $x = 20$

52. $-x^3 + 2$ for
   a. $x = -3$ **b.** $x = 10$

53. $x^3 + 3x^2 + 2x + 4$ for
   a. $x = 2$ **b.** $x = -2$

54. $x^3 - 3x^2 - x + 9$ for
   a. $x = 3$ **b.** $x = -3$

55. $x^4 - x^3 + x^2 + 2x - 1$ for
   a. $x = 1$ **b.** $x = -1$

56. $-x^4 + x^3 + x^2 + x + 1$ for
   a. $x = 1$ **b.** $x = -1$

Complete the table of solutions and graph the equation.

57. $y = x^2$

x	y
-3	
-2	
-1	
0	
1	
2	
3	

58. $y = x^3$

x	y
-2	
-1	
0	
1	
2	

Construct a table of solutions. Then graph the equation.

59. $y = x^2 + 1$

60. $y = x^2 - 4$

61. $y = -x^2 - 2$

62. $y = -x^2 + 1$

63. $y = 2x^2 - 3$

64. $y = -2x^2 + 2$

65. $y = x^3 + 2$

66. $y = x^3 + 4$

67. $y = -x^3 - 1$

68. $y = -x^3$

APPLICATIONS

69. SUPERMARKETS A grocer plans to set up a pyramid-shaped display of cantaloupes like that shown in Example 3. If each side of the square base of the display is made of six cantaloupes, how many will be used in the display?

70. PACKAGING To make boxes, a manufacturer cuts equal-sized squares from each corner of a 10 in. × 12 in. piece of cardboard, and then folds up the sides. The polynomial $4x^3 - 44x^2 + 120x$ gives the volume (in cubic inches) of the resulting box when a square with sides x inches long is cut from each corner. Find the volume of a box if 3-inch squares are cut out.

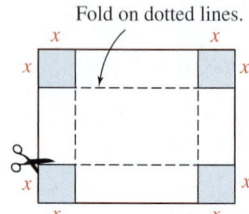

Fold on dotted lines.

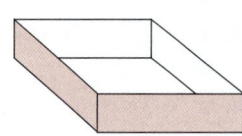

71. STOPPING DISTANCE The number of feet that a car travels before stopping depends on the driver's reaction time and the braking distance, as shown in the illustration. For one driver, the stopping distance is given by the polynomial

$$0.04v^2 + 0.9v$$

where v is the velocity of the car. Find the stopping distance when the driver is traveling at 30 mph.

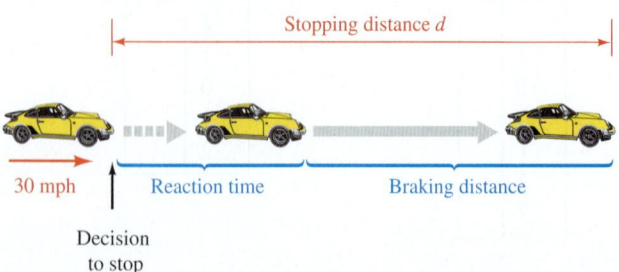

72.  SUSPENSION BRIDGES The polynomial

$$-0.0000001s^4 + 0.0066667s^2 + 400$$

approximates the length of the cable between the two vertical towers of a suspension bridge, where s is the sag in the cable. Estimate the length of the cable if the sag is 24.6 feet.

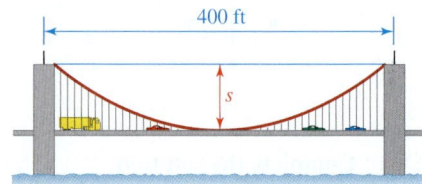

73. SCIENCE HISTORY The renowned Italian scientist Galileo Galilei (1564–1642) built an incline plane like that shown to study falling objects. As the ball rolled down, he measured the time it took the ball to travel different distances. Graph the data and then connect the points with a smooth curve.

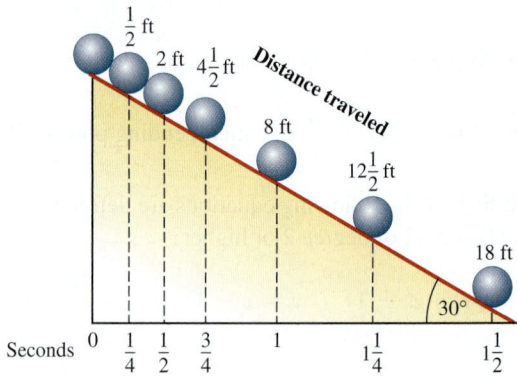

74. DOLPHINS At a marine park, three trained dolphins jump in unison over an arching stream of water whose path can be described by the equation

$$y = -0.05x^2 + 2x$$

Given the takeoff points for each dolphin, how high must each jump to clear the stream of water?

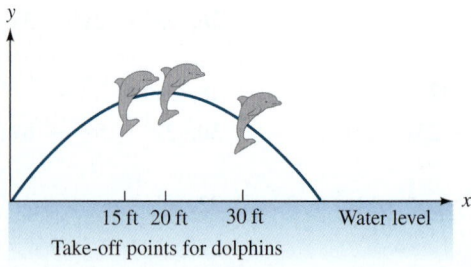

75. MANUFACTURING The graph shows the relationship between the length l (in inches) of a machine bolt and the cost C (in cents) to manufacture it.

a. What information does the point (2, 8) on the graph give us?

b. How much does it cost to make a 7-inch bolt?

c. What length bolt is the least expensive to make?

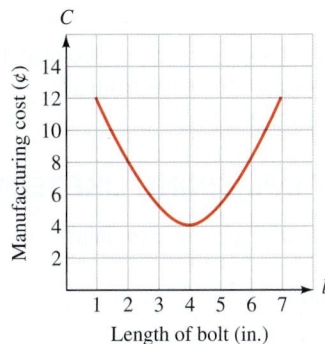

76. SOFTBALL The graph shows the relationship between the distance d (in feet) traveled by a batted softball and the height h (in feet) it attains.

a. What information does the point (40, 40) on the graph give us?

b. At what distance from home plate does the ball reach its maximum height?

c. Where will the ball land?

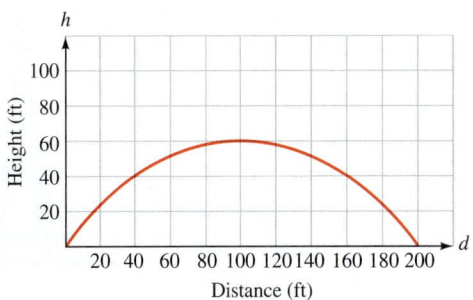

WRITING

77. Describe how to determine the degree of a polynomial.

78. List some words that contain the prefixes *mono, bi,* or *tri.*

REVIEW   Solve each inequality. Write the solution set in interval notation and graph it.

79. $-4(3y + 2) \le 28$ **80.** $-5 < 3t + 4 \le 13$

Simplify each expression.

81. $(x^2x^4)^3$ **82.** $(a^2)^3(a^3)^2$

83. $\left(\dfrac{y^2y^5}{y^4}\right)^3$ **84.** $\left(\dfrac{2t^3}{t}\right)^{-4}$

CHALLENGE PROBLEMS

85. Find a three-term polynomial of degree 2 whose value will be 1 when it is evaluated for $x = -2$.

86. Graph: $y = 2x^3 - 3x^2 - 11x + 6$

4.5 ## Adding and Subtracting Polynomials

- Simplifying Polynomials by Combining Like Terms   • Adding Polynomials
- Subtracting Polynomials   • Adding and Subtracting Multiples of Polynomials
- Applications

The heights of the Seattle Space Needle and the Eiffel Tower are shown in the illustration. We can find the difference in their heights by subtracting 607 from 984.

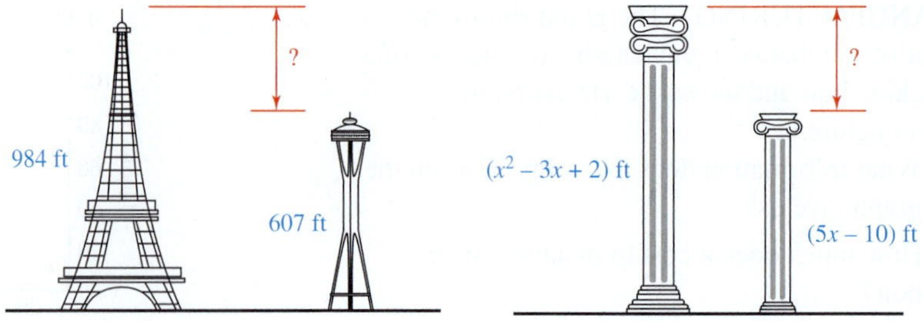

Arithmetic	Algebra
$984 - 607 = 377$	$(x^2 - 3x + 2) - (5x - 10) = ?$
The difference in height is 377 feet.	

The heights of the two Greek columns shown above are expressed as polynomials. To find the difference in their heights, we must subtract $5x - 10$ from $x^2 - 3x + 2$. In this section, we will learn how to perform this subtraction.

■ SIMPLIFYING POLYNOMIALS BY COMBINING LIKE TERMS

Recall that **like terms** have the same variables with the same exponents:

Like terms	*Unlike terms*
$-7x$ and $15x$	$-7x$ and $15a$
$4y^3$ and $16y^3$	$4y^3$ and $16y^2$
$\dfrac{1}{2}xy^2$ and $-\dfrac{1}{3}xy^2$	$\dfrac{1}{2}xy^2$ and $-\dfrac{1}{3}x^2y$

The Language of Algebra

Simplifying the sum or difference of like terms is called *combining like terms.*

Also recall that to **combine like terms,** we combine their coefficients and keep the same variables with the same exponents. For example,

$$4y + 5y = (4 + 5)y \qquad 8x^2 - x^2 = (8 - 1)x^2$$
$$= 9y \qquad\qquad\qquad = 7x^2$$

Polynomials with like terms can be simplified by combining like terms.

EXAMPLE 1

ELEMENTARY & INTERMEDIATE

Algebra *(f(x))* **Now**™

Solution

Simplify each polynomial: **a.** $4x^4 + 81x^4$, **b.** $17x^2y^2 + 2x^2y - 6x^2y^2$, **c.** $-3r - 4r + 6r$, and **d.** $ab + 8 - 15 + 4ab$.

a. $4x^4 + 81x^4 = 85x^4$ $(4 + 81)x^4 = 85x^4$.

b. The first and third terms are like terms.

$$17x^2y^2 + 2x^2y - 6x^2y^2 = 11x^2y^2 + 2x^2y \qquad (17 - 6)x^2y^2 = 11x^2y^2.$$

c. $-3r - 4r + 6r = -r$ $(-3 - 4 + 6)r = -1r = -r$.

d. The first and fourth terms are like terms, and the second and third terms are like terms.

$$ab + 8 - 15 + 4ab = 5ab - 7 \qquad (1 + 4)ab = 5ab \text{ and } 8 - 15 = -7.$$

Self Check 1

Simplify each polynomial: **a.** $6m^4 + 3m^4$, **b.** $17s^3t + 3s^2t - 6s^3t$, **c.** $-19x + 21x - x$, and **d.** $rs + 3r - 5rs + 4r$.

■ ADDING POLYNOMIALS

When adding polynomials horizontally, each polynomial is usually enclosed within parentheses. For example,

$$(3x^2 + 6x + 7) + (2x - 5)$$

is the sum of a trinomial and a binomial. To find the sum, we reorder and regroup the terms so that like terms are together.

$$(3x^2 + 6x + 7) + (2x - 5) = 3x^2 + (6x + 2x) + (7 - 5)$$
$$= 3x^2 + 8x + 2 \qquad \text{Combine like terms.}$$

This example suggests the following rule.

Adding Polynomials	To add polynomials, combine their like terms.

EXAMPLE 2

ELEMENTARY &
INTERMEDIATE

Algebra $f(x)$ **Now**™

Add the polynomials: **a.** $(-6a^3 + 5a^2 - 7a + 9) + (4a^3 - 5a^2 - a - 8)$ and
b. $(16g^2 - h^2) + (4g^2 + 2gh + 10h^2)$.

Solution   Reorder and regroup the terms to get like terms together and combine like terms.

a. $(-6a^3 + 5a^2 - 7a + 9) + (4a^3 - 5a^2 - a - 8)$
$$= (-6a^3 + 4a^3) + (5a^2 - 5a^2) + (-7a - a) + (9 - 8)$$
$$= -2a^3 + 0a^2 + (-8a) + 1 \qquad \text{Combine like terms.}$$
$$= -2a^3 - 8a + 1$$
b. $(16g^2 - h^2) + (4g^2 + 2gh + 10h^2)$
$$= (16g^2 + 4g^2) + 2gh + (-h^2 + 10h^2)$$
$$= 20g^2 + 2gh + 9h^2 \qquad \text{Combine like terms.}$$

Notation

When performing operations on polynomials, it is standard practice to write the terms of a result in descending powers of one variable.

Self Check 2   Add the polynomials: **a.** $(2a^2 - a + 4) + (5a^2 + 6a - 5)$ and
b. $(7x^2 - 2xy - y^2) + (4x^2 - y^2)$.

Polynomials can also be added vertically by aligning like terms in columns.

EXAMPLE 3

ELEMENTARY &
INTERMEDIATE

Algebra $f(x)$ **Now**™

Add $4x^2 - 3$ and $3x^2 - 8x + 8$.

Solution   We write one polynomial underneath the other. Since the first polynomial does not have an x-term, we leave a space so that the constant terms can be aligned. Next, we add like terms, column by column, writing each result under the horizontal bar.

Notation

In performing a vertical addition, any missing term may be written with a coefficient of 0:

$$4x^2 + 0x - 3$$
$$3x^2 - 8x + 8$$

$$
\begin{array}{r}
4x^2 \qquad - 3 \\
+\ 3x^2 - 8x + 8 \\
\hline
7x^2 - 8x + 5
\end{array}
$$

In the x^2-column, find $4x^2 + 3x^2$.
In the x-column, find $0x + (-8x)$.
In the constant column, find $-3 + 8$.

Self Check 3   Add $4q^2 - 7$ and $2q^2 - 8q + 9$ vertically.

■ SUBTRACTING POLYNOMIALS

Because of the distributive property, we can remove parentheses enclosing several terms when the sign preceding the parentheses is a − sign. We simply drop the − sign and the parentheses, and *change the sign of every term within the parentheses.*

$$-(3x^2 + 3x - 2) = \mathbf{-1}(3x^2 + 3x - 2)$$
$$= \mathbf{-1}(3x^2) + (\mathbf{-1})(3x) + (\mathbf{-1})(-2)$$
$$= -3x^2 + (-3x) + 2$$
$$= -3x^2 - 3x + 2$$

This suggests a way to subtract polynomials.

Subtracting Polynomials	To subtract two polynomials, change the signs of the terms of the polynomial being subtracted, drop the parentheses, and combine like terms.

EXAMPLE 4

ELEMENTARY & INTERMEDIATE

Algebra *(f(x))* **Now**™

Subtract the polynomials: **a.** $(3a^2 - 4a - 6) - (2a^2 - a + 9)$ and
b. $(-t^3u + 2t^2u - u + 1) - (-3t^2u - u + 8)$

Solution   **a.** $(3a^2 - 4a - 6) - (2a^2 - a + 9)$

$$= 3a^2 - 4a - 6 - 2a^2 + a - 9 \qquad \text{Change the sign of each term of } 2a^2 - a + 9 \text{ and drop the parentheses.}$$

$$= a^2 - 3a - 15 \qquad \text{Combine like terms.}$$

b. $(-t^3u + 2t^2u - u + 1) - (-3t^2u - u + 8)$

$$= -t^3u + 2t^2u - u + 1 + 3t^2u + u - 8 \qquad \text{Change the sign of each term of } -3t^2u - u + 8 \text{ and drop the parentheses.}$$

$$= -t^3u + 5t^2u - 7 \qquad \text{Combine like terms.}$$

Caution

When combining like terms, the exponents on the variables stay the same. Don't incorrectly apply a rule for exponents and add the exponents.

Self Check 4   Subtract the polynomials: **a.** $(8a^3 - 5a^2 + 5) - (a^3 - a^2 - 7)$ and
b. $(x^2y - 2x + y - 2) - (6x + 9y - 2)$.

EXAMPLE 5

Subtract $12a - 7$ from the sum of $6a + 5$ and $4a - 10$.

Solution   We will use brackets to show that $(12a - 7)$ is to be subtracted from the *sum* of $(6a + 5)$ and $(4a - 10)$.

$$[(6a + 5) + (4a - 10)] - (12a - 7)$$

Next, we remove the grouping symbols to obtain

$$= 6a + 5 + 4a - 10 - 12a + 7 \qquad \text{Change the sign of each term within } (12a - 7) \text{ and drop the parentheses.}$$

$$= -2a + 2 \qquad \text{Combine like terms.}$$

Self Check 5   Subtract $-2q^2 - 2q$ from the sum of $q^2 - 6q$ and $3q^2 + q$.

Polynomials can also be subtracted vertically by aligning like terms in columns.

EXAMPLE 6   Subtract $3x^2 - 2x + 3$ from $2x^2 + 4x - 1$ using vertical form.

Solution   Since $3x^2 - 2x + 3$ is to be subtracted from $2x^2 + 4x - 1$, we write $3x^2 - 2x + 3$ underneath $2x^2 + 4x - 1$. Then we add the opposite of $3x^2 - 2x + 3$ by changing the sign of each of its terms, and combining like terms, column-by-column.

$$\begin{array}{r} 2x^2 + 4x - 1 \\ -\ \underline{3x^2 - 2x + 3} \end{array} \longrightarrow \begin{array}{r} 2x^2 + 4x - 1 \\ +\ \underline{-3x^2 + 2x - 3} \\ -x^2 + 6x - 4 \end{array}$$

In the x^2-column, find $2x^2 + (-3x^2)$.
In the x-column, find $4x + 2x$.
In the constant-column, find $-1 + (-3)$.

Self Check 6   Subtract $2p^2 + 2p - 8$ from $5p^2 - 6p + 7$.

■ ADDING AND SUBTRACTING MULTIPLES OF POLYNOMIALS

Because of the distributive property, we can remove parentheses enclosing several terms when a monomial precedes the parentheses. We simply multiply every term within the parentheses by that monomial. For example, to add $3(2x + 5)$ and $2(4x - 3)$, we proceed as follows:

$$3(2x + 5) + 2(4x - 3) = 6x + 15 + 8x - 6$$ Use the distributive property to remove parentheses.

$$= 6x + 8x + 15 - 6$$ $15 + 8x = 8x + 15$.

$$= 14x + 9$$ Combine like terms.

EXAMPLE 7

ELEMENTARY & INTERMEDIATE
Algebra $f(x)$ **Now**™

Remove parentheses and simplify.

a. $3(x^2 + 4x) + 2(x^2 - 4) = 3x^2 + 12x + 2x^2 - 8$
$$= 5x^2 + 12x - 8$$

b. $-8(y^2 - 2y + 3) - 4(2y^2 + y - 6) = -8y^2 + 16y - 24 - 8y^2 - 4y + 24$
$$= -16y^2 + 12y$$

Self Check 7   Remove parentheses and simplify: $2(a^2 - 3a) + 5(a^2 + 2a)$

■ APPLICATIONS

EXAMPLE 8

ELEMENTARY & INTERMEDIATE
Algebra $f(x)$ **Now**™

Fireworks shows.   Two fireworks shells are simultaneously fired upward from different platforms. The height of the first shell is $(-16t^2 + 160t + 3)$ feet and the height of the higher-traveling second shell is $(-16t^2 + 200t + 1)$ feet, after t seconds.

a. Find a polynomial that represents the difference in the heights of the shells.

b. In 5 seconds, the first shell reaches its peak and explodes. How much higher is the second shell at that time?

Solution   **a.** To find the difference in heights, we subtract the height of the first shell from the height of the second.

$$(-16t^2 + 200t + 1) - (-16t^2 + 160t + 3)$$
$$= -16t^2 + 200t + 1 + 16t^2 - 160t - 3 \qquad \text{Change the sign of each term of } -16t^2 + 160t + 3 \text{ and remove parentheses.}$$

$$= 40t - 2 \qquad \text{Combine like terms.}$$

The difference in the heights of the shells t seconds after being fired is $(40t - 2)$ feet.

b. If we substitute 5 for t in the polynomial found in part a, we have

$$40t - 2 = 40(5) - 2 = 200 - 2 = 198$$

When the first shell explodes, the second shell will be 198 feet higher than the first shell.

Answers to Self Checks   **1. a.** $9m^4$,   **b.** $11s^3t + 3s^2t$,   **c.** x,   **d.** $-4rs + 7r$ **2. a.** $7a^2 + 5a - 1$,
b. $11x^2 - 2xy - 2y^2$ **3.** $6q^2 - 8q + 2$ **4. a.** $7a^3 - 4a^2 + 12$ **b.** $x^2y - 8x - 8y$
5. $6q^2 - 3q$ **6.** $3p^2 - 8p + 15$ **7.** $7a^2 + 4a$

4.5   STUDY SET   ELEMENTARY & INTERMEDIATE Algebra $f(x)$ Now™

VOCABULARY   **Fill in the blanks.**

1. The expression $(b^3 - b^2 - 9b + 1) +$ $(b^3 - b^2 - 9b + 1)$ is the sum of two _____.

2. The expression $(b^2 - 9b + 11) - (4b^2 - 14b)$ is the _____ of a trinomial and a binomial.

3. _____ terms have the same variables with the same exponents.

4. Simplifying the sum or difference of like terms is called _____ like terms.

5. The polynomial $2t^4 + 3t^3 - 4t^2 + 5t - 6$ is written in _____ powers of t.

6. When a $-$ symbol precedes a grouping symbol, as in $-(m^2 + 12m - 1)$, we interpret it as a factor of _____.

CONCEPTS   **Fill in the blanks.**

7. To add polynomials, _____ like terms.

8. To subtract polynomials, _____ the signs of the terms of the polynomial being subtracted, drop parentheses, and combine like terms.

9. Subtract using vertical form:

$$8x^2 - 7x - 1 \qquad 8x^2 - 7x - 1$$
$$-4x^2 + 6x - 9 \longrightarrow + \underline{\ - 6x\ 9}$$

10. $(y^2 - 6y + 7) - (2y^2 - 4) = y^2 - 6y + 7\ \boxed{}\ 2y^2\ \boxed{}\ 4$

11. Simplify.
   a. $2x^2 + 3x^2$ **b.** $15m^3 - m^3$
   c. $8a^3b - ab$ **d.** $6cd + 4c^2d$

12. Explain the error.

$$7x^2y + 6x^2y = \cancel{13x^4y^2}$$

13. Write without parentheses.
   a. $-(5x^2 - 8x + 23)$ **b.** $-(-5y^4 + 3y^2 - 7)$

14. What is the result when the addition is done in the x-column?

$$4x^2 + \ \ x - 12$$
$$+ \underline{5x^2 - 8x + 23}$$

NOTATION   **Complete the solution.**

15. $(6x^2 + 2x + 3) - (4x^2 - 7x + 1)$
   $= 6x^2 + 2x + 3\ \boxed{}\ 4x^2\ \boxed{}\ 7x\ \boxed{}\ 1$
   $= \boxed{} + 9x + \boxed{}$

16. Fill in the blank:
$-(7x^2 - 6x + 12) = -\ \boxed{\ }\ (7x^2 - 6x + 12)$.

17. True or false: $5x^2 - 120 = 5x^2 + 0x - 120$.

18. Write $3x^2 - 9 + 6x - 12x^3$ in descending powers of x.

PRACTICE   Simplify each polynomial.

19. $8t^2 + 4t^2$

20. $15x^2 + 10x^2$

21. $-32u^3 - 16u^3$

22. $-25x^3 - 7x^3$

23. $1.8x - 1.9x$

24. $1.7y - 2.2y$

25. $\dfrac{1}{2}st + \dfrac{3}{2}st$

26. $\dfrac{2}{5}at + \dfrac{1}{5}at$

27. $3r - 4r + 7r$

28. $-2b + 7b - 3b$

29. $-4ab + 4ab - ab$

30. $xy - 4xy - 2xy$

31. $10x^2 - 8x + 9x - 9x^2$ **32.** $-3y^2 - y - 6y^2 + 7y$

33. $6x^3 + 8x^4 + 7x^3 + (-8x^4)$

34. $-3rt - 7t^2 - 6rt + 7t^2 + (-6r^2)$

35. $4x^2y + 5 - 6x^3y - 3x^2y + 2x^3y$

36. $5b - 9ab^2 + 10a^3b - 8ab^2 - 9a^3b$

37. $\dfrac{2}{3}d^2 - \dfrac{1}{4}c^2 + \dfrac{5}{6}c^2 - \dfrac{1}{2}cd + \dfrac{1}{3}d^2$

38. $1 + \dfrac{3}{5}s^2 - \dfrac{2}{5}t^2 - \dfrac{1}{2}s^2 - \dfrac{7}{10}st - \dfrac{3}{10}st$

Perform the operations.

39. $(3x + 7) + (4x - 3)$

40. $(2y - 3) + (4y + 7)$

41. $(9a^2 + 3a) - (2a - 4a^2)$

42. $(4b^2 + 3b) - (7b - b^2)$

43. $(2x + 3y) + (5x - 10y)$

44. $(5x - 8y) - (-2x + 5y)$

45. $(-8x - 3y) - (-11x + y)$

46. $(-4a + b) + (5a - b)$

47. $(3x^2 - 3x - 2) + (3x^2 + 4x - 3)$

48. $(3a^2 - 2a + 4) - (a^2 - 3a + 7)$

49. $(2b^2 + 3b - 5) - (2b^2 - 4b - 9)$

50. $(4c^2 + 3c - 2) + (3c^2 + 4c + 2)$

51. $(2x^2 - 3x + 1) - (4x^2 - 3x + 2) + (2x^2 + 3x + 2)$

52. $(-3z^2 - 4z + 7) + (2z^2 + 2z - 1) - (2z^2 - 3z + 7)$

53. $(-4h^3 + 5h^2 + 15) - (h^3 - 15)$

54. $(-c^5 + 5c^4 - 12) - (2c^5 - c^4)$

55. $(1.04x^2 + 2.07x - 5.01) + (1.33x - 2.98x^2 + 5.02)$

56. $(0.03f^2 + 0.25g) - (0.17g - 0.23f^2)$

57. $\left(\dfrac{7}{8}r^4 + \dfrac{5}{3}r^2 - \dfrac{9}{4}\right) - \left(-\dfrac{3}{8}r^4 - \dfrac{2}{3}r^2 - \dfrac{1}{4}\right)$

58. $\left(\dfrac{1}{16}r^6 + \dfrac{1}{2}r^3 - \dfrac{11}{12}\right) + \left(\dfrac{9}{16}r^6 + \dfrac{9}{2}r^3 + \dfrac{1}{12}\right)$

59. $\begin{aligned}3x^2 + 4x + 5 \\ + \underline{\ 2x^2 - 3x + 6}\end{aligned}$

60. $\begin{aligned}-6x^3 - 4x^2 + 7 \\ + \underline{\ -7x^3 + 9x^2}\ \end{aligned}$

61. $\begin{aligned}3x^2 + 4x - 5 \\ - \underline{\ -2x^2 - 2x + 3}\end{aligned}$

62. $\begin{aligned}3y^2 - 4y + 7 \\ - \underline{\ 6y^2 - 6y - 13}\end{aligned}$

63. $\begin{aligned}4x^3 + 4x^2 - 3x + 10 \\ - \underline{\ 5x^3 - 2x^2 - 4x - \ 4}\end{aligned}$

64. $\begin{aligned}2x^3 + 2x^2 - 3x + 5 \\ + \underline{\ 3x^3 - 4x^2 - \ x - 7}\end{aligned}$

65. $\begin{aligned}-3x^3 + 4x^2 - 4x + 9 \\ + \underline{\ 2x^3 \qquad\quad + 9x - 3}\end{aligned}$

66. $\begin{aligned}-3x^2 + 4x + 25 \\ + \underline{\ 5x^2 \qquad - 12}\end{aligned}$

67. $\begin{aligned}3x^3 + 4x^2 + 7x + 12 \\ - \underline{\ -4x^3 + 6x^2 + 9x - \ 3}\end{aligned}$

68. $\begin{aligned}-2x^2y^2 \qquad\quad + 12y^2 \\ - \underline{\ 10x^2y^2 + 9xy - 24y^2}\end{aligned}$

69. $2(x + 3) + 4(x - 2)$

70. $3(y - 4) - 5(y + 3)$

71. $-2(x^2 + 7x - 1) - 3(x^2 - 2x + 2)$

72. $-5(y^2 - 2y - 6) + 6(2y^2 + 2y - 5)$

73. $2(2y^2 - 2y + 2) - 4(3y^2 - 4y - 1) + 4(y^3 - y^2 - y)$

74. $-4(z^2 - 5z) - 5(4z^2 - 1) + 6(2z - 3)$

75. Subtract $s^2 + 4s + 2$ from $5s^2 - s + 9$.

76. Subtract $4p^2 - 4p - 40$ from $10p^2 - p - 30$.

77. Subtract $-y^5 + 5y^4 - 1.2$ from $2y^5 - y^4$.

78. Subtract $-4w^3 + 5w^2 + 7.6$ from $w^3 - 15w^2$.

79. Find the difference when $t^3 - 2t^2 + 2$ is subtracted from the sum of $3t^3 + t^2$ and $-t^3 + 6t - 3$.

80. Find the difference when $-3z^3 - 4z + 7$ is subtracted from the sum of $2z^2 + 3z - 7$ and $-4z^3 - 2z - 3$.

81. Find the sum when $3x^2 + 4x - 7$ is added to the sum of $-2x^2 - 7x + 1$ and $-4x^2 + 8x - 1$.

82. Find the difference when $32x^2 - 17x + 45$ is subtracted from the sum of $23x^2 - 12x - 7$ and $-11x^2 + 12x + 7$.

APPLICATIONS

83. GREEK ARCHITECTURE
   a. Find a polynomial that represents the difference in the heights of the columns shown at the beginning of this section.

   b. If the columns were stacked one atop the other, to what height would they reach?

Find the polynomial that represents the perimeter of each figure.

84. a.

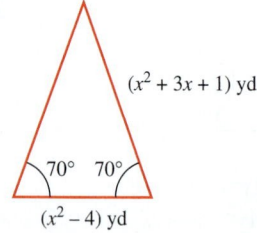

$(x^2 + 3x + 1)$ yd

70°   70°

$(x^2 - 4)$ yd

b.

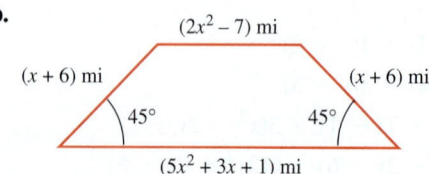

$(2x^2 - 7)$ mi

$(x + 6)$ mi $(x + 6)$ mi

45°   45°

$(5x^2 + 3x + 1)$ mi

85. JETS   Find the polynomial representing the length of the passenger jet.

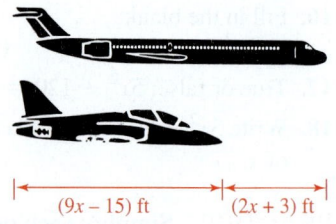

$(9x - 15)$ ft $(2x + 3)$ ft

86. PIÑATAS   Find the polynomial that represents the length of the rope used to hold up the piñata.

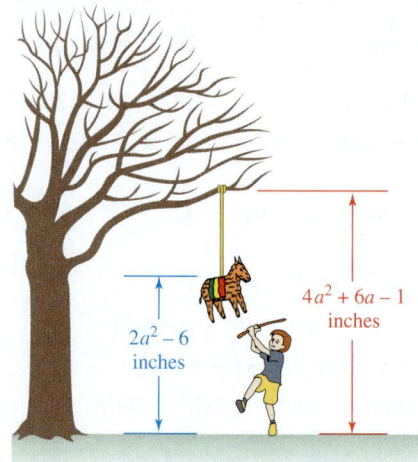

$4a^2 + 6a - 1$ inches

$2a^2 - 6$ inches

87. READING BLUEPRINTS
   a. Find a polynomial that represents the difference in the length and width of the one-bedroom apartment shown in the illustration.

   b. Find the perimeter of the apartment.

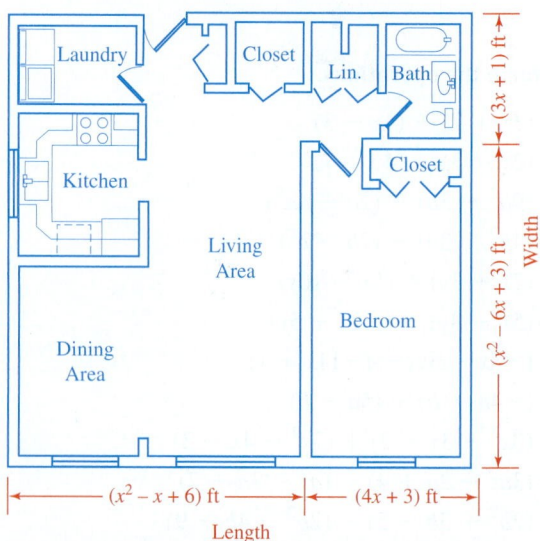

$(3x + 1)$ ft

Width

$(x^2 - 6x + 3)$ ft

$(x^2 - x + 6)$ ft $(4x + 3)$ ft

Length

88. AUTO MECHANICS Find the polynomial representing the length of the fan belt. The dimensions are in inches. Leave π in your answer.

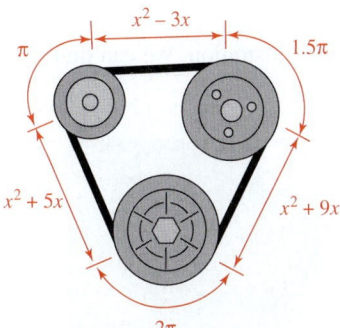

89. NAVAL OPERATIONS Two warning flares are simultaneously fired upward from different parts of a ship. The height of the first flare is $(-16t^2 + 128t + 20)$ feet and the height of the higher-traveling second flare is $(-16t^2 + 150t + 40)$ feet, after t seconds.

a. Find a polynomial that represents the difference in the heights of the flares.

b. In 4 seconds, the first flare reaches its peak, explodes, and lights up the sky. How much higher is the second flare at that time?

90. REAL ESTATE INVESTMENTS A house is purchased for $105,000 and is expected to appreciate $900 per year. Therefore, its value in x years is given by the polynomial $900x + 105,000$.

a. A second house is purchased at the same time for $120,000 and is expected to appreciate $1,000 per year. Find a polynomial that predicts the value of the second house in x years.

b. Find a polynomial that predicts the combined value of the houses x years after their purchase.

c. Use your answer to part b to find the predicted value of the two houses in 20 years.

WRITING

91. How do you recognize like terms?

92. How do you add like terms?

93. Explain the concept that is illustrated by the statement

$$-(x^2 + 3x - 1) = -1(x^2 + 3x - 1)$$

94. Explain the mistake made in the solution. Simplify: $(12x^2 - 4) - (3x^2 - 1)$

$$(12x^2 - 4) - (3x^2 - 1) = 12x^2 - 4 - 3x^2 - 1$$
$$= 9x^2 - 5$$

95. What can be concluded if
a. you add two polynomials and get 0?
b. you subtract two polynomials and get 0?

96. Explain the error.
Subtract $(2d^2 - d - 3)$ from $(d^2 - 9)$:

$$(2d^2 - d - 3) - (d^2 - 9)$$

REVIEW

97. What is the sum of the measures of the angles of a triangle?

98. Graph: $y = -\dfrac{x}{2} + 2$.   **99.** Graph: $2x + 3y = 9$.

100. CURLING IRONS A curling iron is plugged into a 110-volt electrical outlet and used for $\frac{1}{4}$ hour. If its resistance is 10 ohms, find the electrical power (in kilowatt hours, kwh) used by the curling iron by applying the formula

$$\text{kwh} = \frac{(\text{volts})^2}{1,000 \cdot \text{ohms}} \cdot \text{hours}$$

CHALLENGE PROBLEMS

101. What polynomial must be added to $2x^2 - x + 3$ so that the sum is $6x^2 - 7x - 8$?

102. Is the sum of two trinomials always a trinomial?

4.6 Multiplying Polynomials

- Multiplying Monomials
- Multiplying a Polynomial by a Monomial
- Multiplying Binomials
- The FOIL Method
- Multiplying Polynomials

The dimensions of a dollar bill are shown in the illustration. We can find its area by multiplying its length and width.

6.5 cm

$(3x - 1)$ cm

$\leftarrow$ 15.6 cm $\rightarrow$

$(2x + 1)$ cm

Arithmetic

$15.6(6.5) = 101.4$

The area is 101.4 cm².

Algebra

$(2x + 1)(3x - 1) = ?$

The length and width of the postage stamp shown above are expressed as polynomials. To find the area of the stamp, we need to multiply $2x + 1$ by $3x - 1$. In this section, we will discuss the rules for multiplying polynomials. To begin, we consider the product of two monomials.

■ MULTIPLYING MONOMIALS

Success Tip

In this section, you will see that every polynomial multiplication is a series of monomial multiplications.

We have seen that to multiply the monomials $8x^2$ and $3x^4$, we proceed as follows.

$(8x^2)(3x^4) = (8 \cdot 3)(x^2 \cdot x^4)$ Group the coefficients together and the variables together.

$= 24x^6$ $x^2 \cdot x^4 = x^{2+4} = x^6$.

This example suggests the following rule.

Multiplying Monomials   To multiply two monomials, multiply the numerical factors (the coefficients) and then multiply the variable factors.

EXAMPLE 1

ELEMENTARY & INTERMEDIATE

Algebra *f(x)* **Now**™

Multiply: **a.** $(6r)(r)$,   **b.** $3t^4(-2t^5)$,   **c.** $(-2a^2b^3)(-5ab^2)$,   and
d. $-4y^5z^2(2y^3z^3)(3yz)$.

Solution   **a.** $(6r)(r) = 6r^2$ Recall that $r = 1r$. $6 \cdot 1 = 6$ and $r \cdot r = r^2$.

b. $(3t^4)(-2t^5) = -6t^9$ $3(-2) = -6$ and $t^4 \cdot t^5 = t^9$.

c. $-2a^2b^3(-5ab^2) = 10a^3b^5$ $-2(-5) = 10$, $a^2 \cdot a = a^3$, and $b^3 \cdot b^2 = b^5$.

d. $-4y^5z^2(2y^3z^3)(3yz) = -24y^9z^6$ $-4(2)(3) = -24$, $y^5 \cdot y^3 \cdot y = y^9$, and $z^2 \cdot z^3 \cdot z = z^6$.

Self Check 1   Multiply: **a.** $18t(t)$,   **b.** $-10d^8(-6d^3)$,   and   **c.** $(5a^2b^3)(-6a^3b^4)(ab)$.

■ MULTIPLYING A POLYNOMIAL BY A MONOMIAL

To find the product of a polynomial and a monomial, we use the distributive property. To multiply $2x + 4$ by $5x$, for example, we proceed as follows.

$$5x(2x + 4) = 5x(2x) + 5x(4) \qquad \text{Distribute the multiplication by } 5x.$$
$$= 10x^2 + 20x \qquad \text{Multiply the monomials.}$$

This example suggests the following rule.

Multiplying Polynomials by Monomials	To multiply a monomial and a polynomial, multiply each term of the polynomial by the monomial.

EXAMPLE 2

ELEMENTARY & INTERMEDIATE
Algebra $f(x)$ **Now**™

Solution

Multiply: **a.** $3a^2(3a^2 - 5a + 2)$,   **b.** $-2xz^3(6x^3z + x^2z^2 - xz^3 + 7z^4)$,   and
c. $(-m^4 - 25)(4m^3)$.

a. Multiply each term of $3a^2 - 5a + 2$ by $3a^2$.

$$3a^2(3a^2 - 5a + 2)$$
$$= 3a^2(3a^2) + 3a^2(-5a) + 3a^2(2) \qquad \text{Distribute the multiplication by } 3a^2.$$
$$= 9a^4 - 15a^3 + 6a^2 \qquad \text{Multiply the monomials.}$$

b. Multiply each term of $6x^3z + x^2z^2 - xz^3 + 7z^4$ by $-2xz^3$.

$$-2xz^3(6x^3z + x^2z^2 - xz^3 + 7z^4)$$
$$= -2xz^3(6x^3z) - 2xz^3(x^2z^2) - 2xz^3(-xz^3) - 2xz^3(7z^4)$$
$$= -12x^4z^4 - 2x^3z^5 + 2x^2z^6 - 14xz^7 \qquad \text{Multiply the monomials.}$$

c. Multiply each term of $-m^4 - 25$ by $4m^3$.

$$(-m^4 - 25)(4m^3) = -m^4(4m^3) - 25(4m^3) \qquad \text{Distribute the multiplication by } 4m^3.$$
$$= -4m^7 - 100m^3 \qquad \text{Multiply the monomials.}$$

Success Tip

The rectangle can be used to visualize polynomial multiplication. The total area is $x(x + 2)$ and the sum of the two smaller areas is $x^2 + 2x$. Thus,

$$x(x + 2) = x^2 + 2x$$

x	x^2	$2x$
	x	2

$$x + 2$$

Self Check 2   Multiply: **a.** $5c^2(4c^3 - 9c - 8)$ and   **b.** $-s^2t^2(-s^4t^2 + s^3t^3 - st^4 + 7s)$.

■ MULTIPLYING BINOMIALS

The distributive property can also be used to multiply binomials. For example, to multiply $2a + 4$ and $3a + 5$, we think of $2a + 4$ as a single quantity and distribute it over each term of $3a + 5$.

$$(2a + 4)(3a + 5) = (2a + 4)3a + (2a + 4)5$$
$$= 3a(2a + 4) + 5(2a + 4) \qquad \text{Use the commutative property of multiplication.}$$

$$= 3a(2a) + 3a(4) + 5(2a) + 5(4)$$

Distribute the multiplication by $3a$. Distribute the multiplication by 5.

$$= 6a^2 + 12a + 10a + 20$$

Multiply the monomials.

$$= 6a^2 + 22a + 20$$

Combine like terms.

This example suggests the following rule.

Multiplying Binomials	To multiply two binomials, multiply each term of one binomial by each term of the other binomial, and then combine like terms.

EXAMPLE 3

Multiply: $(5x - 8)(x + 1)$.

Solution   To find the product, we multiply $x + 1$ by $5x$ and by -8. In that way, each term of $x + 1$ will be multiplied by each term of $5x - 8$.

$$(\mathbf{5x - 8})(x + 1) = \mathbf{5x}(x + 1) - \mathbf{8}(x + 1)$$
$$= 5x^2 + 5x - 8x - 8$$

Distribute the multiplication by $5x$. Distribute the multiplication by -8.

$$= 5x^2 - 3x - 8$$

Combine like terms.

Self Check 3   Multiply: $(9y + 3)(y - 4)$.

■ THE FOIL METHOD

We can use a shortcut method, called the **FOIL method,** to multiply binomials. FOIL is an acronym for **F**irst terms, **O**uter terms, **I**nner terms, **L**ast terms. To use the FOIL method to multiply $2a + 4$ by $3a + 5$, we

The Language of Algebra

The acronym FOIL helps us remember the order to follow when multiplying two binomials. Another popular acronym is PEMDAS. It represents the order of operations rules: **P**arentheses, **E**xponents, **M**ultiply, **D**ivide, **A**dd, **S**ubtract.

1. multiply the **F**irst terms $2a$ and $3a$ to obtain $6a^2$,
2. multiply the **O**uter terms $2a$ and 5 to obtain $10a$,
3. multiply the **I**nner terms 4 and $3a$ to obtain $12a$, and
4. multiply the **L**ast terms 4 and 5 to obtain 20.

Then we simplify the resulting polynomial, if possible.

First terms   Last terms

$$(2a + 4)(3a + 5) = \mathbf{2a(3a)} + \mathbf{2a(5)} + \mathbf{4(3a)} + \mathbf{4(5)}$$

Inner terms

$$= 6a^2 + 10a + 12a + 20$$

Do the multiplications.

$$= 6a^2 + 22a + 20$$

Combine like terms.

Outer terms

EXAMPLE 4

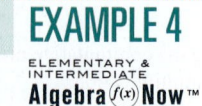

Find each product.

Solution   **a.** $(x + 5)(x + 7) = x(x) + x(7) + 5(x) + 5(7)$

$$= x^2 + 7x + 5x + 35$$

$$= x^2 + 12x + 35 \qquad \text{Combine like terms.}$$

Success Tip

The area of the large rectangle is given by $(x + 5)(x + 7)$. The sum of the areas of the smaller rectangles is $x^2 + 7x + 5x + 35$ or $x^2 + 12x + 35$. Thus,

$(x + 5)(x + 7) = x^2 + 12x + 35$.

5	$5x$	35
x	x^2	$7x$
	x	7

$x + 5$ { ... }

$x + 7$

b. $(3x + 4)(2x - 3) = 3x(2x) + 3x(-3) + 4(2x) + 4(-3)$

$$= 6x^2 - 9x + 8x - 12$$

$$= 6x^2 - x - 12 \qquad \text{Combine like terms.}$$

c. $(2r - 3s)(2r + t) = 2r(2r) + 2r(t) - 3s(2r) - 3s(t)$

$$= 4r^2 + 2rt - 6rs - 3st$$

d. $(3a^2 - 7b)(a^2 - b) = 3a^2(a^2) + 3a^2(-b) - 7b(a^2) - 7b(-b)$

$$= 3a^4 - 3a^2b - 7a^2b + 7b^2$$

$$= 3a^4 - 10a^2b + 7b^2 \qquad \text{Combine like terms.}$$

Self Check 4   Find each product: **a.** $(y + 3)(y + 1)$,   **b.** $(2a - 1)(3a + 2)$,
c. $(5y^3 - 2z)(2y^3 - z)$

■ MULTIPLYING POLYNOMIALS

To develop a general rule for multiplying any two polynomials, we will find the product of $2x + 3$ and $3x^2 + 3x + 5$. In the solution, the distributive property is used four times.

$$(2x + 3)(3x^2 + 3x + 5) = (2x + 3)3x^2 + (2x + 3)3x + (2x + 3)5$$

$$= 2x(3x^2) + 3(3x^2) + 2x(3x) + 3(3x) + 2x(5) + 3(5)$$

$$= 6x^3 + 9x^2 + 6x^2 + 9x + 10x + 15 \qquad \text{Multiply the monomials.}$$

$$= 6x^3 + 15x^2 + 19x + 15 \qquad \text{Combine like terms.}$$

In the second line of the solution, note that each term of $3x^2 + 3x - 5$ has been multiplied by each term of $2x + 3$. This example suggests the following rule.

Multiplying Polynomials   To multiply two polynomials, multiply each term of one polynomial by each term of the other polynomial, and then combine like terms.

EXAMPLE 5   Multiply: $(6y^3 + y^2 - 8y + 1)(7y + 3)$.

ELEMENTARY &
INTERMEDIATE
Algebra $_{f(x)}$ **Now**™   Solution   Multiply each term of $7y + 3$ by each term of $6y^3 + y^2 - 8y + 1$.

$$(6y^3 + y^2 - 8y + 1)(7y + 3)$$
$$= 6y^3(7y) + 6y^3(3) + y^2(7y) + y^2(3) - 8y(7y) - 8y(3) + 1(7y) + 1(3)$$
$$= 42y^4 + 18y^3 + 7y^3 + 3y^2 - 56y^2 - 24y + 7y + 3$$
$$= 42y^4 + 25y^3 - 53y^2 - 17y + 3$$

Self Check 5   Multiply: $(2a^4 - a^2 - a)(3a^2 - 1)$.

It is often convenient to multiply polynomials using a vertical format similar to that used to multiply whole numbers.

EXAMPLE 6

ELEMENTARY &
INTERMEDIATE
Algebra *f(x)* **Now**™

a. Multiply:

$$
\begin{array}{r}
3a^2 - 4a + 7 \\
2a + 5 \\
\hline
15a^2 - 20a + 35 \\
6a^3 - 8a^2 + 14a \\
\hline
6a^3 + 7a^2 - 6a + 35
\end{array}
$$

Multiply $3a^2 - 4a + 7$ by 5.

Multiply $3a^2 - 4a + 7$ by 2a.

In each column, combine like terms.

b. Multiply:

$$
\begin{array}{r}
6y^3 - 5y + 4 \\
-4y^2 - 3 \\
\hline
-18y^3 \qquad + 15y - 12 \\
-24y^5 + 20y^3 - 16y^2 \\
\hline
-24y^5 + 2y^3 - 16y^2 + 15y - 12
\end{array}
$$

Multiply $6y^3 - 5y + 4$ by -3.

Multiply $6y^3 - 5y + 4$ by $-4y^2$.

Leave a space for any missing powers of y. In each column, combine like terms.

Self Check 6   Multiply: **a.** $(3x + 2)(2x^2 - 4x + 5)$ and   **b.** $(-2x^2 + 3)(2x^2 - 4x - 1)$.

When finding the product of three polynomials, we begin by multiplying any two of them, and then we multiply that result by the third polynomial.

EXAMPLE 7

Find the product: $-3a(4a + 1)(a - 7)$.

Solution   We find the product of $4a + 1$ and $a - 7$ and then multiply that result by $-3a$.

$$
\begin{aligned}
-3a(4a + 1)(a - 7) &= -3a(4a^2 - 28a + a - 7) \\
&= -3a(4a^2 - 27a - 7) \\
&= -12a^3 + 81a^2 + 21a
\end{aligned}
$$

Multiply the two binomials.

Combine like terms.

Distribute the multiplication by $-3a$.

Self Check 7   Find the product: $-2y(y + 3)(3y - 2)$

Answers to Self Checks   **1. a.** $18t^2$,   **b.** $60d^{11}$,   **c.** $-30a^6b^8$ **2. a.** $20c^5 - 45c^3 - 40c^2$,
b. $s^6t^4 - s^5t^5 + s^3t^6 - 7s^3t^2$ **3.** $9y^2 - 33y - 12$ **4. a.** $y^2 + 4y + 3$,
b. $6a^2 + a - 2$,   **c.** $10y^6 - 9y^3z + 2z^2$ **5.** $6a^6 - 5a^4 - 3a^3 + a^2 + a$
6. a. $6x^3 - 8x^2 + 7x + 10$,   **b.** $-4x^4 + 8x^3 + 8x^2 - 12x - 3$ **7.** $-6y^3 - 14y^2 + 12y$

4.6   STUDY SET   ELEMENTARY & INTERMEDIATE Algebra *f(x)* Now™

VOCABULARY   **Fill in the blanks.**

1. The expression $(2x^3)(3x^4)$ is the product of two _____.

2. To find the product $5a^3(a^2 - 2a + 8)$, we _____ the multiplication by $5a^3$.

3. The expression $(2a - 4)(3a + 5)$ is the product of two _____.

4. In the acronym FOIL, F stands for _____ terms, O for _____ terms, I for _____ terms, and L for _____ terms.

5. To simplify $y^2 + 3y + 2y + 6$, we combine the _____ terms $3y$ and $2y$ to get $y^2 + 5y + 6$.

6. The expression $(2a - 4)(3a^2 + 5a - 1)$ is the product of a _____ and a _____.

CONCEPTS   **Fill in the blanks.**

7. a. To multiply two monomials, multiply the numerical factors (the _____) and then multiply the _____ factors.

  b. To multiply two polynomials, multiply _____ term of one polynomial by _____ term of the other polynomial, and then combine like terms.

  c. When multiplying three polynomials, we begin by multiplying _____ two of them, and then we multiply that result by the _____ polynomial.

  d. To find the area of a rectangle, multiply the _____ by the width.

8. To multiply two monomials, we use the commutative and associative properties of multiplication to reorder and regroup. Fill in the blanks.

$$(9n^3)(8n^2) = (9 \cdot \boxed{})(\boxed{} \cdot n^2) = \boxed{}$$

9. Label each arrow using one of the letters F, O, I, or L. Then fill in the blanks.

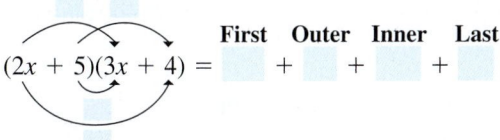

First Outer Inner Last

$$(2x + 5)(3x + 4) = \boxed{} + \boxed{} + \boxed{} + \boxed{}$$

10. Simplify each polynomial by combining like terms.
  a. $6x^2 - 8x + 9x - 12$
  b. $5x^4 + 3ax^2 + 5ax^2 + 3a^2$

11. a. Add: $(x - 4) + (x + 8)$.
  b. Subtract: $(x - 4) - (x + 8)$.
  c. Multiply: $(x - 4)(x + 8)$.

12. Fill in the blanks in the multiplication.

$$
\begin{array}{r}
3x^2 + \ 4x - 2 \\
2x + 3 \\
\hline
\boxed{} + 12x - 6 \\
6x^3 + \ 8x^2 - \ 4x \\
\hline
\boxed{} + 17x^2 + \boxed{} - 6
\end{array}
$$

NOTATION   **Complete each solution.**

13. $7x(3x^2 - 2x + 5) = \boxed{}(3x^2) - \boxed{}(2x) + \boxed{}(5)$
$$= \boxed{} - 14x^2 + 35x$$

14. $(2x + 5)(3x - 2) = 2x(3x) - \boxed{}(2) + \boxed{}(3x) - \boxed{}(2)$
$$= 6x^2 - \boxed{} + \boxed{} - 10$$
$$= 6x^2 + \boxed{} - 10$$

PRACTICE   **Find each product.**

15. $15m(m)$

16. $4s^9(s)$

17. $(3x^2)(4x^3)$

18. $(-2a^3)(3a^2)$

19. $(3b^2)(-2b)(4b^3)$

20. $(3y)(2y^2)(-y^4)$

21. $(2x^2y^3)(3x^3y^2)$

22. $(-5x^3y^6)(x^2y^2)$

23. $(8a^5)\left(-\dfrac{1}{4}a^6\right)$

24. $\left(-\dfrac{2}{3}x^6\right)(9x^3)$

25. $(1.2c^3)(5c^3)$

26. $(2.5h^4)(2h^4)$

27. $\left(\dfrac{1}{2}a\right)\left(\dfrac{1}{8}a^4\right)(a^5)$

28. $\left(\dfrac{1}{3}b\right)\left(\dfrac{7}{6}b\right)(b^4)$

29. $(x^2y^5)(x^2z^5)(-3z^3)$

30. $(-r^4st^2)(2r^2st)(rst)$

31. $3(x + 4)$

32. $-3(a - 2)$

33. $-4(t^2 + 7)$

34. $6(s^2 - 3)$

35. $3x(x - 2)$

36. $4y(y + 5)$

37. $-2x^2(3x^2 - x)$

38. $4b^3(2b^2 - 2b)$

39. $(x^2 - 12x)(6x^{12})$

40. $(w^9 - 11w)(2w^7)$

41. $\frac{5}{8}t^2(t^6 + 8t^2)$ **42.** $\frac{4}{9}a^2(9a^3 + a^2)$

43. $0.3p^5(0.4p^4 - 6p^2)$ **44.** $0.5u^5(0.4u^6 - 0.5u^3)$

45. $-4x^2z(3x^2 - z)$ **46.** $3xy(x + y)$

47. $2x^2(3x^2 + 4x - 7)$ **48.** $3y^3(2y^2 - 7y - 8)$

49. $3a(4a^2 + 3a - 4)$ **50.** $-2x(3x^2 - 3x + 2)$

51. $(-2a^2)(-3a^3)(3a - 2)$ **52.** $(3x)(-2x^2)(x + 4)$

53. $(y - 3)(y + 5)$ **54.** $(a + 4)(a + 5)$

55. $(t + 4)(2t - 3)$ **56.** $(3x - 2)(x + 4)$

57. $(2y - 5)(3y + 7)$ **58.** $(3x - 5)(2x + 1)$

59. $(2x + 3)(2x - 5)$ **60.** $(x + 3)(2x - 3)$

61. $(a + b)(a + b)$ **62.** $(m - n)(m - n)$

63. $(3a - 2b)(4a + b)$ **64.** $(2t + 3s)(3t - s)$

65. $(t^2 - 3)(t^2 + 4)$ **66.** $(s^3 + 6)(s^3 - 8)$

67. $(-3t + 2s)(2t - 3s)$ **68.** $(4t - u)(-3t + u)$

69. $\left(4a - \frac{5}{9}r\right)\left(2a + \frac{3}{4}r\right)$ **70.** $\left(5c - \frac{2}{3}t\right)\left(2c + \frac{1}{5}t\right)$

71. $4(2x + 1)(x - 2)$
72. $-5(3a - 2)(2a + 3)$
73. $3a(a + b)(a - b)$
74. $-2r(r + s)(r + s)$
75. $(x + 2)(x^2 - 2x + 3)$
76. $(x - 5)(x^2 + 2x - 3)$
77. $(4t + 3)(t^2 + 2t + 3)$
78. $(3x + 1)(2x^2 - 3x + 1)$
79. $(x^2 + 6x + 7)(2x - 5)$
80. $(y^2 - 2y + 1)(4y + 8)$
81. $(-3x + y)(x^2 - 8xy + 16y^2)$

82. $(3x - y)(x^2 + 3xy - y^2)$
83. $(r^2 - r + 3)(r^2 - 4r - 5)$

84. $(w^2 + w - 9)(w^2 - w + 3)$

Perform each multiplication.

85. $\begin{array}{r} x^2 - 2x + 1 \\ x + 2 \\ \hline \end{array}$ **86.** $\begin{array}{r} 5r^2 + r + 6 \\ 2r - 1 \\ \hline \end{array}$

87. $\begin{array}{r} 4x^2 + 3x - 4 \\ 3x + 2 \\ \hline \end{array}$ **88.** $\begin{array}{r} x^2 - x + 1 \\ x + 1 \\ \hline \end{array}$

89. $\begin{array}{r} 2a^2 + 3a + 1 \\ 3a^2 - 2a + 4 \\ \hline \end{array}$

90. $\begin{array}{r} 3y^2 + 2y - 4 \\ 2y^2 - 4y + 3 \\ \hline \end{array}$

APPLICATIONS

91. STAMPS   Find a polynomial that represents the area of the Louis Armstrong stamp shown at the beginning of the section.

92. PARKING   Find a polynomial to represent the total area of the van-accessible parking space and its access aisle.

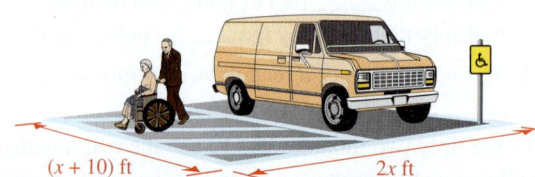

$(x + 10)$ ft $2x$ ft

Find the area of each figure. Leave π in your answer.

93.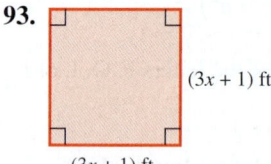

$(3x + 1)$ ft

$(3x + 1)$ ft

94.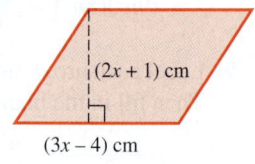

$(2x + 1)$ cm

$(3x - 4)$ cm

95.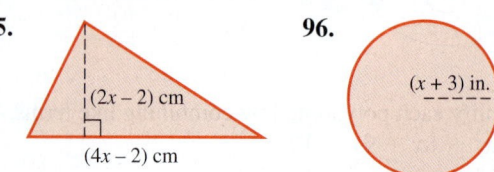

$(2x - 2)$ cm

$(4x - 2)$ cm

96.

$(x + 3)$ in.

97. SUNGLASSES An ellipse is an oval-shaped closed curve. The area of an ellipse is approximately 3.14*ab*, where *a* is its length and *b* is its width. Find a polynomial that approximates the total area of the elliptical-shaped lenses of the sunglasses.

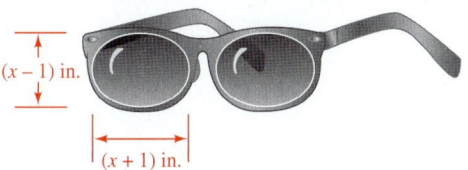

$(x - 1)$ in.

$(x + 1)$ in.

98. GARDENING

a. What is the area of the region planted with corn? tomatoes? beans? carrots? Use your answers to find the total area of the garden.

b. What is the length of the garden? What is its width? Use your answers to find its area.

c. How do the answers from parts a and b for the area of the garden compare?

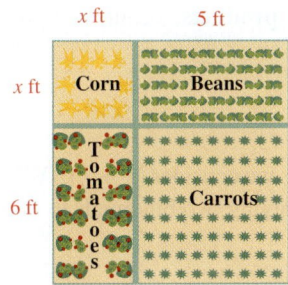

x ft   5 ft

x ft   Corn   Beans

6 ft   Tomatoes   Carrots

99. TOYS Find a polynomial that represents the area of the screen of the Etch A Sketch®

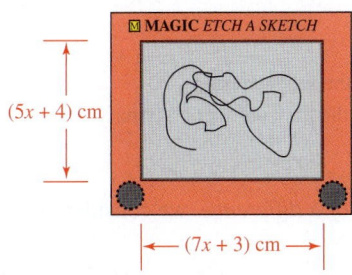

MAGIC *ETCH A SKETCH*

$(5x + 4)$ cm

$(7x + 3)$ cm

100. GRAPHIC ARTS A graduation announcement has a 1-inch-wide border around the written text. Find a polynomial that represents the area of the announcement.

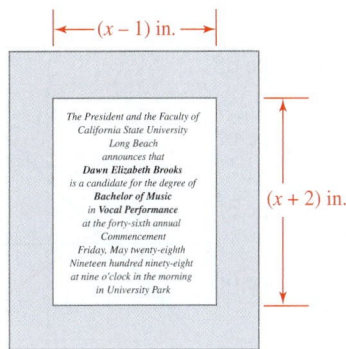

$(x - 1)$ in.

The President and the Faculty of
California State University
Long Beach
announces that
Dawn Elizabeth Brooks
is a candidate for the degree of
Bachelor of Music
in **Vocal Performance**
at the forty-sixth annual
Commencement
Friday, May twenty-eighth
Nineteen hundred ninety-eight
at nine o'clock in the morning
in University Park

$(x + 2)$ in.

101. LUGGAGE Find a polynomial that represents the volume of the garment bag.

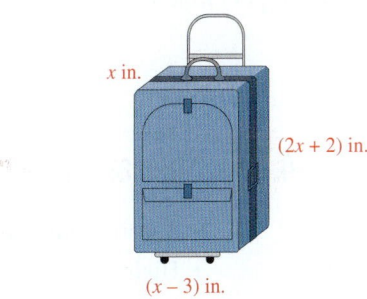

x in.

$(2x + 2)$ in.

$(x - 3)$ in.

102. BASEBALL Find a polynomial that represents the amount of space within the batting cage.

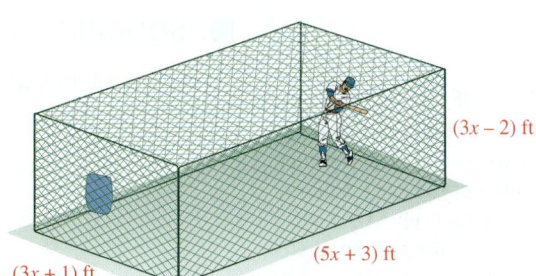

$(3x - 2)$ ft

$(3x + 1)$ ft

$(5x + 3)$ ft

WRITING

103. Is the product of a monomial and a monomial always a monomial? Explain why or why not.

104. Explain this diagram.

$$(5x + 6)(7x - 1)$$

105. Explain why the FOIL method cannot be used to find $(3x + 2)(4x^2 - x + 10)$.

106. Explain the error:

$$(x + 3)(x - 2) = x^2 - 6$$

REVIEW

107. What is the slope of Line 1?

108. What is the slope of Line 2?

109. What is the slope of Line 3?

110. What is the slope of the x-axis?

111. What is the y-intercept of Line 1?

112. What is the x-intercept of Line 1?

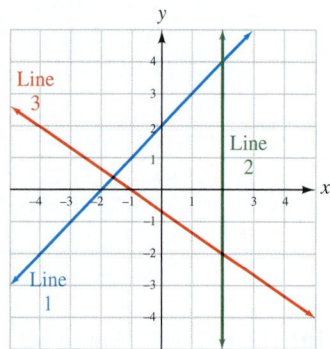

CHALLENGE PROBLEMS

113. Find each of the following.
  a. $(x - 1)(x + 1)$
  b. $(x - 1)(x^2 + x + 1)$
  c. $(x - 1)(x^3 + x^2 + x + 1)$

114. Solve: $(y - 1)(y + 6) = (y - 3)(y - 2) + 8$.

4.7 Special Products

- Squaring a Binomial
- The Product of a Sum and Difference
- Higher Powers of Binomials
- Simplifying Expressions Containing Polynomials

Certain products of binomials, called **special products,** occur so frequently that it is worthwhile to learn their forms.

■ SQUARING A BINOMIAL

Success Tip

The illustration can be used to visualize a special product. The area of the large square is $(x + y)(x + y) = (x + y)^2$. The sum of the four smaller areas is $x^2 + xy + xy + y^2$ or $x^2 + 2xy + y^2$. Thus,

$$(x + y)^2 = x^2 + 2xy + y^2$$

$$x + y \begin{cases} \begin{array}{|c|c|} \hline y & xy & y^2 \\ \hline x & x^2 & xy \\ \hline \end{array} \\ \qquad x \quad\ y \end{cases}$$
$$\underbrace{}_{x + y}$$

To develop a formula to find the *square of a sum,* we consider $(x + y)^2$. We can use the definition of exponent and the procedure for multiplying two binomials to find the product.

$$\begin{aligned} (x + y)^2 &= (x + y)(x + y) && \text{In } (x + y)^2, \text{ the base is } (x + y) \text{ and the exponent is } 2. \\ &= x^2 + xy + xy + y^2 && \text{Multiply the binomials.} \\ &= x^2 + 2xy + y^2 && \text{Combine like terms: } xy + xy = 1xy + 1xy = 2xy. \end{aligned}$$

Note that the terms of the result are related to the terms of the binomial that was squared.

$$(x + y)^2 = x^2 + 2xy + y^2$$

└ The square of the second term, y.
└ Twice the product of the first and second terms, x and y.
└ The square of the first term, x.

To develop a formula to find the *square of a difference,* we consider $(x - y)^2$.

$$\begin{aligned} (x - y)^2 &= (x - y)(x - y) && \text{In } (x - y)^2, \text{ the base is } (x - y) \text{ and the exponent is } 2. \\ &= x^2 - xy - xy + y^2 && \text{Multiply the binomials.} \\ &= x^2 - 2xy + y^2 && \text{Combine like terms: } -xy - xy = -2xy. \end{aligned}$$

Again, the terms of the result are related to the terms of the binomial that was squared.

$$(x - y)^2 = x^2 - 2xy + y^2$$

The square of the second term, $-y$.

Twice the product of the first and second terms, x and $-y$.

The square of the first term, x.

The results of these two examples suggest the following rules.

Squaring a Binomial	The **square of a binomial** is the square of its first term, plus twice the product of both of its terms, plus the square of its second term.
	$$(x + y)^2 = x^2 + 2xy + y^2 \qquad (x - y)^2 = x^2 - 2xy + y^2$$

EXAMPLE 1

ELEMENTARY & INTERMEDIATE
Algebra $f(x)$ **Now™**

Solution

Find each square: **a.** $(t + 9)^2$,   **b.** $(8a - 5)^2$,   **c.** $(-d + 0.5)^2$,   and   **d.** $(c^3 - 7d)^2$.

a. $(t + 9)^2$ is the square of a sum. The first term is t and the second term is 9.

$$(t + 9)^2 = \underbrace{t^2}_{\substack{\text{The square} \\ \text{of the first} \\ \text{term, } t.}} + \underbrace{2(t)(9)}_{\substack{\text{Twice the} \\ \text{product of} \\ \text{both terms.}}} + \underbrace{9^2}_{\substack{\text{The square} \\ \text{of the second} \\ \text{term, 9.}}}$$

$$= t^2 + 18t + 81$$

The Language of Algebra

When squaring a binomial, the result is called a *perfect square trinomial*. For example,

$$(t + 9)^2 = t^2 + 18t + 81$$

Perfect square trinomial

b. $(8a - 5)^2$ is the square of a difference. The first term is $8a$ and the second term is -5.

$$(8a - 5)^2 = \underbrace{(8a)^2}_{\substack{\text{The square} \\ \text{of the first} \\ \text{term, } 8a.}} - \underbrace{2(8a)(5)}_{\substack{\text{Twice the product} \\ \text{of both terms.}}} + \underbrace{(-5)^2}_{\substack{\text{The square} \\ \text{of the second} \\ \text{term, } -5.}}$$

$$= 64a^2 - 80a + 25 \qquad \begin{array}{l}\text{Use the power rule for products:} \\ (8a)^2 = 8^2a^2 = 64a^2.\end{array}$$

Caution

Remember that the square of a binomial is a *trinomial*. A common error when squaring a binomial is to forget the middle term of the product. For example,

$$(x + 3)^2 \neq x^2 + 9$$

Missing $6x$

c. $(-d + 0.5)^2$ is the square of a sum. The first term is $-d$ and the second term is 0.5.

$$(-d + 0.5)^2 = (-d)^2 + 2(-d)(0.5) + (0.5)^2$$
$$= d^2 - d + 0.25$$

d. $(c^3 - 7d)^2$ is the square of a difference. The first term is c^3 and the second term is $-7d$.

$$(c^3 - 7d)^2 = (c^3)^2 - 2(c^3)(7d) + (-7d)^2$$
$$= c^6 - 14c^3d + 49d^2 \qquad \begin{array}{l}\text{Use rules for exponents to find } (c^3)^2 \\ \text{and } (-7d)^2.\end{array}$$

Self Check 1   Find each square: **a.** $(r + 6)^2$,   **b.** $(7g - 2)^2$,   **c.** $(-v + 0.8)^2$,   and **d.** $(w^4 - 3y)^2$

■ THE PRODUCT OF A SUM AND DIFFERENCE

The final special product is the product of two binomials that differ only in the signs of the second terms. To develop a rule to find the product of a *sum and a difference*, we consider $(x + y)(x - y)$.

$$(x + y)(x - y) = x^2 - xy + xy - y^2 \qquad \text{Multiply the binomials.}$$
$$= x^2 - y^2 \qquad\qquad\qquad \text{Combine like terms: } -xy + xy = 0.$$

Success Tip

We can use the FOIL method to find each of the special products discussed in this section. However, these forms occur so often, it is worthwhile to learn the special product rules.

Note that when we combined like terms, we added opposites. This will always be the case for products of this type; the sum of the outer and inner products will be 0.

$$(x + y)(x - y) = x^2 - y^2$$

The square of the second term, y.

The square of the first term, x.

These observations suggest the following rule.

Multiplying the Sum and Difference of Two Terms

The product of the sum and difference of the two terms x and y is the square of x minus the square of y.

$$(x + y)(x - y) = x^2 - y^2$$

EXAMPLE 2

ELEMENTARY &
INTERMEDIATE
Algebra $f(x)$ **Now**™

Solution

Multiply: **a.** $(m + 2)(m - 2)$, **b.** $(3y + 4)(3y - 4)$, **c.** $\left(b - \frac{2}{3}\right)\left(b + \frac{2}{3}\right)$, and **d.** $(t^4 - 6u)(t^4 + 6u)$.

a. In $m + 2$, the first term is m and the second term is 2.

$$(m + 2)(m - 2) = \underbrace{m^2}_{\substack{\text{The square} \\ \text{of the first} \\ \text{term, } m.}} - \underbrace{2^2}_{\substack{\text{The square} \\ \text{of the second} \\ \text{term, 2.}}}$$

$$= m^2 - 4$$

The Language of Algebra

When multiplying the sum and difference of two terms, the result is called a *difference of two squares.* For example:

$$(m + 2)(m - 2) = \underline{m^2 - 4}$$

Difference
of two
squares

b. In $3y + 4$, the first term is $3y$ and the second term is 4.

$$(3y + 4)(3y - 4) = (3y)^2 - 4^2$$
$$= 9y^2 - 16 \qquad (3y)^2 = 3^2 y^2 = 9y^2.$$

c. By the commutative property of multiplication, the special product rule can be written with the factor containing the $-$ symbol first. That is, $(x - y)(x + y) = x^2 - y^2$. So we have

$$\left(b - \frac{2}{3}\right)\left(b + \frac{2}{3}\right) = b^2 - \left(\frac{2}{3}\right)^2$$

$$= b^2 - \frac{4}{9}$$

d. Apply the sum and difference special product rule and simplify.

$$(t^4 - 6u)(t^4 + 6u) = (t^4)^2 - (6u)^2$$
$$= t^8 - 36u^2 \qquad (t^4)^2 = t^{4 \cdot 2} = t^8 \text{ and } (6u)^2 = 6^2 u^2 = 36u^2.$$

Self Check 2 Multiply: **a.** $(b + 4)(b - 4)$, **b.** $(5m + 9)(5m - 9)$, **c.** $\left(s - \frac{3}{4}\right)\left(s + \frac{3}{4}\right)$, and **d.** $(c^3 + 2d)(c^3 - 2d)$.

EXAMPLE 3

Paper towels.   The amount of space (volume) occupied by the paper on a roll of paper towels is given by the expression $\pi h(R + r)(R - r)$, where R is the outer radius and r is the inner radius. Perform the indicated multiplication.

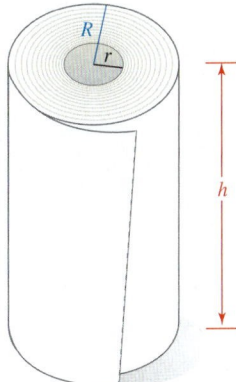

Solution   We can multiply the polynomials πh, $(R + r)$, and $(R - r)$ in any order. Since $(R + r)$ and $(R - r)$ form a special product, we will do that multiplication first. Then we multiply the result by the monomial πh.

$$\pi h(R + r)(R - r) = \pi h(R^2 - r^2)$$

To find $(R + r)(R - r)$, use a special product rule. The first term squared is R^2. The second term squared is r^2.

$$= \pi h R^2 - \pi h r^2$$

Distribute the multiplication by πh.

■ HIGHER POWERS OF BINOMIALS

To find the third, fourth, or even higher powers of a binomial, we can use the special product rules.

EXAMPLE 4

Expand: $(x + 1)^3$.

Solution   Write $(x + 1)^3$ as $(x + 1)^2(x + 1)$ so that a special product rule can be used.

The Language of Algebra

To *expand* means to increase in size. When we *expand* a power of a binomial, the result is an expression that has more terms than the original binomial.

$$(x + 1)^3 = (x + 1)^2(x + 1)$$
$$= (x^2 + 2x + 1)(x + 1)$$

Find $(x + 1)^2$ using the square of a sum rule.

$$= x^2(x) + x^2(1) + 2x(x) + 2x(1) + 1(x) + 1(1)$$

Multiply each term of $x + 1$ by each term of $x^2 + 2x + 1$.

$$= x^3 + x^2 + 2x^2 + 2x + x + 1$$

Multiply the monomials.

$$= x^3 + 3x^2 + 3x + 1$$

Combine like terms.

Self Check 4   Expand: $(n - 3)^3$.

■ SIMPLIFYING EXPRESSIONS CONTAINING POLYNOMIALS

We can use the methods discussed in Sections 4.5, 4.6, and 4.7 to simplify expressions that involve addition, subtraction, and multiplication of polynomials.

EXAMPLE 5

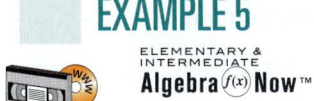

Simplify each expression: **a.** $(x + 1)(x - 2) + 3x(x + 3)$ and
b. $(3y - 2)^2 - (y - 5)(y + 5)$.

Solution **a.** $(x + 1)(x - 2) + 3x(x + 3)$

$= x^2 - x - 2 + 3x^2 + 9x$ Use the FOIL method to find $(x + 1)(x - 2)$. Distribute the multiplication by $3x$.

$= 4x^2 + 8x - 2$ Combine like terms.

b. $(3y - 2)^2 - (y - 5)(y + 5)$

$= 9y^2 - 12y + 4 - (y^2 - 25)$ Find the special products: $(3y - 2)^2$ and $(y - 5)(y + 5)$. Write $y^2 - 25$ within parentheses so that both terms get subtracted.

$= 9y^2 - 12y + 4 - 1(y^2 - 25)$ Interpret the $-$ symbol preceding the parenthesis as -1.

$= 9y^2 - 12y + 4 - y^2 + 25$ Distribute the multiplication by -1.

$= 8y^2 - 12y + 29$ Combine like terms.

Self Check 5 Simplify each expression: **a.** $(x - 4)(x + 6) + 5x(2x - 1)$ and
b. $(a + 9)(a - 9) - (2a - 4)^2$.

Answers to Self Checks
1. a. $r^2 + 12r + 36$, **b.** $49g^2 - 28g + 4$, **c.** $v^2 - 1.6v + 0.64$, **d.** $w^8 - 6w^4y + 9y^2$
2. a. $b^2 - 16$, **b.** $25m^2 - 81$, **c.** $s^2 - \frac{9}{16}$, **d.** $c^6 - 4d^2$ **4.** $n^3 - 9n^2 + 27n - 27$
5. a. $11x^2 - 3x - 24$, **b.** $-3a^2 + 16a - 97$

4.7   STUDY SET   ELEMENTARY & INTERMEDIATE Algebra $f(x)$ Now™

VOCABULARY   Fill in the blanks.

1. The first _____ of $3x + 6$ is $3x$ and the second _____ is 6.

2. $(x + 4)^2$ is the _____ of a sum and $(m - 9)^2$ is the square of a _____.

3. $(b + 1)(b - 1)$ is the product of the _____ and difference of two terms.

4. Since $x^2 + 16x + 64$ is the square of $x + 8$, it is called a _____ square trinomial.

5. An expression of the form $x^2 - y^2$ is called a _____ of two squares.

6. Expressions of the form $(x + y)^2$, $(x - y)^2$, and $(x + y)(x - y)$ occur so frequently in algebra that they are called special _____.

CONCEPTS

7. Complete the rules for exponents.
  a. $(x^m)^n =$
  b. $(xy)^n =$

8. Simplify each expression.
  a. $(5x)^2$
  b. $2(x)(3)$
  c. $(d^3)^2$
  d. $-2(6y)(5)$

9. Fill in the blanks.
  a. $(a + 5)(a + 5) = (a + 5)$
  b. $(n - 12)(n - 12) = (n - 12)$

10. Complete the special product.

$$(x + y)^2 = x^2 + 2xy + y^2$$

  └ The _____ of the second term
  _____ the product of the first and second terms
  └ The square of the _____ term

11. Complete the special product.

$$(x + y)(x - y) = x^2 - y^2$$

  └ The square of the _____ term
  └ The _____ of the first term

12. Fill in the blanks.
  a. $(x + 2)^3 = (x + 2)\ (x + 2)$
  b. $(c - 1)^4 = (c - 1)^2(c - 1)$

NOTATION   Complete each solution.

13. $(x + 4)^2 = \quad^2 + 2(x)(\quad) + \quad^2$

$= x^2 + \quad + 16$

14. $(6r - 1)^2 = (\quad)^2 \quad 2(6r)(1) + (-1)^2$

$\quad = \quad - \quad + 1$

15. $(s + 5)(s - 5) = \quad^2 \quad^2$

$\quad = s^2 - \quad$

16. $(h - 3)(h + 3) = \quad^2 \quad^2$

$\quad = \quad - 9$

PRACTICE Find each product.

17. $(x + 1)^2$

18. $(y + 7)^2$

19. $(r + 2)^2$

20. $(n + 10)^2$

21. $(m - 6)^2$

22. $(b - 1)^2$

23. $(f - 8)^2$

24. $(w - 9)^2$

25. $(d + 7)(d - 7)$

26. $(t + 2)(t - 2)$

27. $(n + 6)(n - 6)$

28. $(a + 12)(a - 12)$

29. $(4x + 5)^2$

30. $(6y + 3)^2$

31. $(7m - 2)^2$

32. $(9b - 2)^2$

33. $(y^2 + 9)^2$

34. $(d^2 + 2)^2$

35. $(2v^3 - 8)^2$

36. $(8x^4 - 3)^2$

37. $(4f + 0.4)(4f - 0.4)$

38. $(4t + 0.6)(4t - 0.6)$

39. $(3n + 1)(3n - 1)$

40. $(5a + 4)(5a - 4)$

41. $(1 - 3y)^2$

42. $(1 - 4a)^2$

43. $(x - 2y)^2$

44. $(3a + 2b)^2$

45. $(2a - 3b)^2$

46. $(2x + 5y)^2$

47. $\left(s + \dfrac{3}{4}\right)^2$

48. $\left(y - \dfrac{5}{3}\right)^2$

49. $(a + b)^2$

50. $(c + d)^2$

51. $(r - s)^2$

52. $(t - u)^2$

53. $\left(6b + \dfrac{1}{2}\right)\left(6b - \dfrac{1}{2}\right)$

54. $\left(4h + \dfrac{2}{3}\right)\left(4h - \dfrac{2}{3}\right)$

55. $(r + 10s)^2$

56. $(m + 8n)^2$

57. $(6 - 2d^3)^2$

58. $(6 - 5p^2)^2$

59. $-(8x + 3)^2$

60. $-(4b - 8)^2$

61. $-(5 - 6g)(5 + 6g)$

62. $-(6 - c^2)(6 + c^2)$

63. $3x(2x + 3)^2$

64. $4y(3y + 4)^2$

65. $-5d(4d - 1)^2$

66. $-2h(7h - 2)^2$

67. $4d(d^2 + g^3)(d^2 - g^3)$

68. $8y(x^2 + y^2)(x^2 - y^2)$

Expand each binomial.

69. $(x + 4)^3$

70. $(y + 2)^3$

71. $(n - 6)^3$

72. $(m - 5)^3$

73. $(2g - 3)^3$

74. $(3x - 2)^3$

75. $(n - 2)^4$

76. $(c + d)^4$

Perform the operations.

77. $2t(t + 2) + (t - 1)(t + 9)$

78. $3y(y + 2) + (y + 1)(y - 1)$

79. $(x + y)(x - y) + x(x + y)$

80. $(3x + 4)(2x - 2) - (2x + 1)(x + 3)$

81. $(3x - 2)^2 + (2x + 1)^2$

82. $(4a - 3)^2 + (a + 6)^2$

83. $(m + 10)^2 - (m - 8)^2$

84. $(5y - 1)^2 - (y + 7)(y - 7)$

APPLICATIONS

85. DINNER PLATES   The expression $\dfrac{\pi}{4}(D - d)(D + d)$ estimates the difference in area of two plates, the larger with diameter D and smaller with diameter d, as shown on the next page. Perform the indicated multiplication.

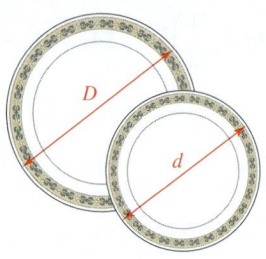

86. RETAIL STORES Refer to the illustration, which shows the floor space of a department store.

   a. What area do the men's items occupy? the women's items? the children's items? the household items? Use these answers to find the total floor space of the department store.

   b. What is the length of the floor space? the width? Use these answers and a special product rule to find the floor space area.

   c. How do your final answers from parts a and b compare?

87. PLAYPENS Find a polynomial that represent the area of the floor of the playpen.

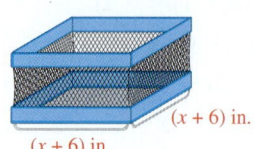

88. STORAGE Find a polynomial that represents the volume of the cubicle.

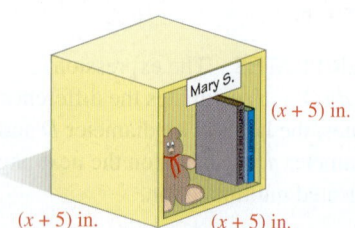

89. PAINTING To purchase the correct amount of enamel to paint the two garage doors, a painter must find their areas. Find a polynomial that gives the number of square feet to be painted. All dimensions are in feet, and the windows are squares with sides of x feet.

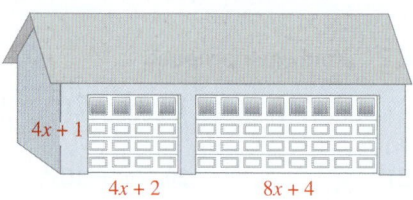

90. SIGNAL FLAGS Find a polynomial that represents the area in blue of the maritime signal flag for the letter p. The dimensions are given in centimeters.

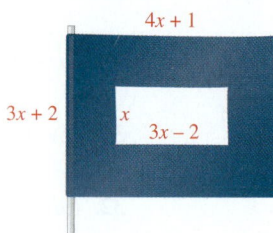

WRITING

91. What is a binomial? Explain how to square it.

92. Writing $(x + y)^2$ as $x^2 + y^2$ illustrates a common error. Explain.

93. We can find $(2x + 3)^2$ and $(5y - 6)^2$ using the FOIL method or using special product rules. Explain why the special product rules are faster.

94. Explain why $(x + 2)^3 \neq x^3 + 8$.

REVIEW

95. Find the prime factorization of 189.

96. Complete each statement. For any nonzero number a,

   a. $\dfrac{a}{a} =$

   b. $\dfrac{a}{1} =$

   c. $\dfrac{0}{a} =$

   d. $\dfrac{a}{0}$ is

97. Simplify: $\dfrac{30}{36}$.

98. Add: $\dfrac{5}{12} + \dfrac{1}{4}$.

99. Multiply: $\dfrac{7}{8} \cdot \dfrac{3}{5}$.

100. Divide: $\dfrac{1}{3} \div \dfrac{4}{5}$.

CHALLENGE PROBLEMS

101. a. Find two binomials whose product is a binomial.

b. Find two binomials whose product is a trinomial.

c. Find two binomials whose product is a four-term polynomial.

102. A special product rule can be used to find $31 \cdot 29$.

$$31 \cdot 29 = (30 + 1)(30 - 1)$$
$$= 30^2 - 1^2 = 900 - 1 = 899$$

Use this method to find $52 \cdot 48$.

4.8 Division of Polynomials

- Dividing a Monomial by a Monomial
- Dividing a Polynomial by a Monomial
- Dividing a Polynomial by a Polynomial

In this section, we will conclude our study of operations with polynomials by discussing division of polynomials. To begin, we consider the simplest case, the quotient of two monomials.

■ DIVIDING A MONOMIAL BY A MONOMIAL

To divide monomials, we can use the method for simplifying fractions or the quotient rule for exponents.

EXAMPLE 1

Simplify: **a.** $\dfrac{21x^5}{7x^2}$ and **b.** $\dfrac{10r^6 s}{6rs^3}$.

ELEMENTARY & INTERMEDIATE
Algebra *f(x)* Now™

Solution

By simplifying fractions

a. $\dfrac{21x^5}{7x^2} = \dfrac{3 \cdot \overset{1}{\cancel{7}} \cdot \overset{1}{\cancel{x}} \cdot \overset{1}{\cancel{x}} \cdot x \cdot x \cdot x}{\underset{1}{\cancel{7}} \cdot \underset{1}{\cancel{x}} \cdot \underset{1}{\cancel{x}}}$

$= 3x^3$

b. $\dfrac{10r^6 s}{6rs^3} = \dfrac{\overset{1}{\cancel{2}} \cdot 5 \cdot \overset{1}{\cancel{r}} \cdot r \cdot r \cdot r \cdot r \cdot r \cdot \overset{1}{\cancel{s}}}{\underset{1}{\cancel{2}} \cdot 3 \cdot \underset{1}{\cancel{r}} \cdot \underset{1}{\cancel{s}} \cdot s \cdot s}$

$= \dfrac{5r^5}{3s^2}$

Using the rules for exponents

$\dfrac{21x^5}{7x^2} = 3x^{5-2}$ Divide the coefficients.

$= 3x^3$

$\dfrac{10r^6 s}{6rs^3} = \dfrac{5}{3} r^{6-1} s^{1-3}$ Simplify $\frac{10}{6}$.

$= \dfrac{5}{3} r^5 s^{-2}$

$= \dfrac{5r^5}{3s^2}$ Move s^{-2} to the denominator and change the sign of the exponent.

Success Tip

In this section, you will see that regardless of the number of terms involved, every polynomial division is a series of monomial divisions.

Self Check 1 Simplify: **a.** $\dfrac{30y^4}{5y^2}$ and **b.** $\dfrac{8c^2 d^6}{32c^5 d^2}$.

■ DIVIDING A POLYNOMIAL BY A MONOMIAL

Recall that to add two fractions with the same denominator, we add their numerators and keep the common denominator.

$$\frac{a}{d} + \frac{b}{d} = \frac{a + b}{d}$$

We can use this rule in reverse to divide polynomials by monomials.

Dividing a Polynomial by a Monomial   To divide a polynomial by a monomial, divide each term of the polynomial by the monomial. Let a, b, and d represent monomials, where d is not 0

$$\frac{a + b}{d} = \frac{a}{d} + \frac{b}{d}$$

EXAMPLE 2   Divide   **a.** $\dfrac{9x^2 + 6x}{3x}$ and   **b.** $\dfrac{12a^4b^3 - 18a^3b^2 + 2a^2}{6a^2b^2}$.

ELEMENTARY &
INTERMEDIATE
Algebra $f(x)$**Now**™   **Solution**   **a.** Here, we have a binomial divided by a monomial.

$$\frac{9x^2 + 6x}{3x} = \frac{9x^2}{3x} + \frac{6x}{3x}$$ Divide each term of the numerator by the denominator, $3x$.

The Language of Algebra

The names of the parts of a division statement are

Dividend

$$\overset{\overbrace{\qquad}}{\underset{\underset{\textit{Divisor}}{\underbrace{\qquad}}}{\frac{9x^2 + 6x}{3x}}} = \underset{\textit{Quotient}}{\underbrace{3x + 2}}$$

$$= 3x^{2-1} + 2x^{1-1}$$ Do each monomial division. Divide the coefficients. Subtract the exponents.

$$= 3x + 2$$ Recall that $x^0 = 1$.

Check: We multiply the divisor, $3x$, and the quotient, $3x + 2$. The result should be the dividend, $9x^2 + 6x$.

$$3x(3x + 2) = 9x^2 + 6x$$ The answer checks.

b. Here, we have a trinomial divided by a monomial.

$$\frac{12a^4b^3 - 18a^3b^2 + 2a^2}{6a^2b^2} = \frac{12a^4b^3}{6a^2b^2} - \frac{18a^3b^2}{6a^2b^2} + \frac{2a^2}{6a^2b^2}$$ Divide each term of the numerator by the denominator, $6a^2b^2$.

$$= 2a^{4-2}b^{3-2} - 3a^{3-2}b^{2-2} + \frac{a^{2-2}}{3b^2}$$ Do each monomial division. Simplify: $\frac{2}{6} = \frac{1}{3}$.

$$= 2a^2b - 3a + \frac{1}{3b^2}$$

Success Tip

The sum, difference, and product of two polynomials are always polynomials. However, as seen in Example 2b, the quotient of two polynomials is not always a polynomial.

Recall that the variables in a polynomial must have whole-number exponents. Therefore, the result, $2a^2b - 3a + \frac{1}{3b^2}$, is not a polynomial because the last term can be written $\frac{1}{3}b^{-2}$.

Check: $6a^2b^2\left(2a^2b - 3a + \dfrac{1}{3b^2}\right) = 12a^4b^3 - 18a^3b^2 + 2a^2$ The answer checks.

Self Check 2   Divide: **a.** $\dfrac{50h^3 + 15h^2}{5h^2}$ and   **b.** $\dfrac{22s^5t^2 - s^4t^3 + 44s^2t^4}{11s^2t^2}$.

EXAMPLE 3

Formulas. The area of a trapezoid is given by the formula $A = \frac{1}{2}h(B + b)$, where B and b are its bases and h is its height. Solve the formula for b.

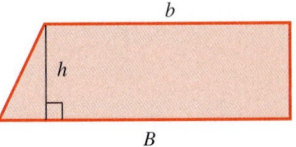

$$A = \frac{1}{2}h(B + b)$$

$$2A = h(B + b) \qquad \text{Multiply both sides by 2 to clear the equation of the fraction.}$$

$$2A = hB + hb \qquad \text{Distribute the multiplication by } h.$$

$$2A - \mathbf{hB} = hB + hb - \mathbf{hB} \qquad \text{Subtract } hB \text{ from both sides.}$$

$$2A - hB = hb \qquad \text{Combine like terms: } hB - hB = 0.$$

$$\frac{2A - hB}{\mathbf{h}} = \frac{hb}{\mathbf{h}} \qquad \text{To undo the multiplication by } h, \text{ divide both sides by } h.$$

$$\frac{2A - hB}{h} = b$$

We can stop at this point, or, as shown below, we can perform the division on the left-hand side of the equation.

$$\frac{2A}{h} - \frac{hB}{h} = b \qquad \text{Divide each term of the numerator by the denominator, } h.$$

$$\frac{2A}{h} - B = b \qquad \text{Simplify: } \frac{\overset{1}{\cancel{h}}B}{\underset{1}{\cancel{h}}} = B.$$

■ DIVIDING A POLYNOMIAL BY A POLYNOMIAL

To divide one polynomial by another, we use a method similar to long division in arithmetic.

EXAMPLE 4

Divide $x^2 + 5x + 6$ by $x + 2$.

Solution

Here the divisor is $x + 2$, and the dividend is $x^2 + 5x + 6$.

Success Tip

The long-division process is a series of four steps that are repeated: Divide, multiply, subtract, and bring down.

Step 1: $\begin{array}{r} x \\ x + 2)\overline{x^2 + 5x + 6} \end{array}$

Divide x^2 by x: $\frac{x^2}{x} = x$. Write the result, x, above the division symbol.

Step 2: $\begin{array}{r} x \\ x + 2)\overline{x^2 + 5x + 6} \\ x^2 + 2x \end{array}$

Multiply each term of the divisor by x. Write the result, $x^2 + 2x$, under $x^2 + 5x$, and draw a line.

Step 3: $\begin{array}{r} x \\ x + 2)\overline{x^2 + 5x + 6} \\ x^2 + 2x \\ \hline 3x + 6 \end{array}$

Subtract $x^2 + 2x$ from $x^2 + 5x$. Work vertically, column by column: $x^2 - x^2 = 0$ and $5x - 2x = 3x$.

Bring down the 6.

Step 4: $\begin{array}{r} x + 3 \\ x + 2)\overline{x^2 + 5x + 6} \\ x^2 + 2x \\ \hline 3x + 6 \end{array}$

Divide $3x$ by x: $\frac{3x}{x} = 3$. Write $+3$ above the division symbol to form the second term of the quotient.

$$
\begin{array}{r}
x + 3 \\
x + 2 \overline{\smash{)}x^2 + 5x + 6} \\
\underline{x^2 + 2x} \\
3x + 6 \\
\underline{3x + 6}
\end{array}
$$

Step 5: Multiply each term of the divisor by 3. Write the result, $3x + 6$, under $3x + 6$, and draw a line.

$$
\begin{array}{r}
x + 3 \\
x + 2 \overline{\smash{)}x^2 + 5x + 6} \\
\underline{x^2 + 2x} \\
3x + 6 \\
\underline{3x + 6} \\
0
\end{array}
$$

Step 6: Subtract $3x + 6$ from $3x + 6$. Work vertically: $3x - 3x = 0$ and $6 - 6 = 0$.

$0 \leftarrow$ This is the remainder.

The quotient is $x + 3$ and the remainder is 0.

Step 7: Check the work by verifying that $(x + 2)(x + 3)$ is $x^2 + 5x + 6$.

$$
\begin{aligned}
(x + 2)(x + 3) &= x^2 + 3x + 2x + 6 \\
&= x^2 + 5x + 6
\end{aligned}
$$

The answer checks.

Self Check 4   Divide $x^2 + 7x + 12$ by $x + 3$.

EXAMPLE 5

Divide: $\dfrac{6x^2 - 7x - 2}{2x - 1}$.

ELEMENTARY &
INTERMEDIATE
Algebra $f(x)$ **Now**™   **Solution**   Here the divisor is $2x - 1$ and the dividend is $6x^2 - 7x - 2$.

$$
\begin{array}{r}
3x \\
2x - 1 \overline{\smash{)}6x^2 - 7x - 2}
\end{array}
$$

Step 1: Divide $6x^2$ by $2x$: $\frac{6x^2}{2x} = 3x$. Write the result, $3x$, above the division symbol.

$$
\begin{array}{r}
3x \\
2x - 1 \overline{\smash{)}6x^2 - 7x - 2} \\
6x^2 - 3x
\end{array}
$$

Step 2: Multiply each term of the divisor by $3x$. Write the result, $6x^2 - 3x$, under $6x^2 - 7x$, and draw a line.

$$
\begin{array}{r}
3x \\
2x - 1 \overline{\smash{)}6x^2 - 7x - 2} \\
\underline{6x^2 - 3x} \\
-4x - 2
\end{array}
$$

Step 3: Subtract $6x^2 - 3x$ from $6x^2 - 7x$. Work vertically: $6x^2 - 6x^2 = 0$ and $-7x - (-3x) = -7x + 3x = -4x$.

Bring down the -2.

$$
\begin{array}{r}
3x - 2 \\
2x - 1 \overline{\smash{)}6x^2 - 7x - 2} \\
\underline{6x^2 - 3x} \\
-4x - 2
\end{array}
$$

Step 4: Divide $-4x$ by $2x$: $\frac{-4x}{2x} = -2$. Write -2 above the division symbol to form the second term of the quotient.

$$
\begin{array}{r}
3x - 2 \\
2x - 1 \overline{\smash{)}6x^2 - 7x - 2} \\
\underline{6x^2 - 3x} \\
-4x - 2 \\
-4x + 2
\end{array}
$$

Step 5: Multiply each term of the divisor by -2. Write the result, $-4x + 2$, under $-4x - 2$, and draw a line.

$$
\begin{array}{r}
3x \;-\; 2 \\[2pt]
2x - 1\overline{)6x^2 - 7x - 2}
\end{array}
$$

Step 6:

$$
\begin{array}{r}
6x^2 - 3x \\ \hline
-4x - 2 \\
-4x + 2 \\ \hline
-4
\end{array}
$$

Subtract $-4x + 2$ from $-4x - 2$. Work vertically:
$-4x - (-4x) = -4x + 4x = 0$ and $-2 - 2 = -4$.

Success Tip

The division process for polynomials continues until the degree of the remainder is less than the degree of the divisor. In Example 5, the remainder, -4, has degree 0. The divisor, $2x - 1$, has degree 1. Therefore, the division ends.

Here the quotient is $3x - 2$ and the remainder is -4. It is common to write the answer as either

$$
3x - 2 + \frac{-4}{2x - 1} \qquad \text{or} \qquad 3x - 2 - \frac{4}{2x - 1} \qquad \text{Quotient} + \frac{\text{remainder}}{\text{divisor}}.
$$

Step 7: We can check the answer using the fact that for any division:

$$
\text{divisor} \cdot \text{quotient} + \text{remainder} = \text{dividend}
$$

$$
\begin{aligned}
(2x - 1)(3x - 2) + (-4) &= 6x^2 - 4x - 3x + 2 + (-4) \\
&= 6x^2 - 7x - 2 \qquad \text{The answer checks.}
\end{aligned}
$$

Self Check 5   Divide: $\dfrac{8x^2 + 6x - 3}{2x + 3}$.

The division method works best when the terms of the divisor and the dividend are written in descending powers of the variable. If the powers in the dividend or divisor are not in descending order, we use the commutative property of addition to write them that way.

EXAMPLE 6

Divide $4x^2 + 2x^3 + 12 - 2x$ by $x + 3$.

ELEMENTARY & INTERMEDIATE
Algebra $f(x)$ Now™

Solution   If we write the dividend in descending powers of x, the division is routine.

$$
\begin{array}{r}
2x^2 - 2x \;+\; 4 \\[2pt]
x + 3\overline{)2x^3 + 4x^2 - 2x + 12} \\
2x^3 + 6x^2 \\ \hline
-2x^2 - 2x \\
-2x^2 - 6x \\ \hline
4x + 12 \\
4x + 12 \\ \hline
0
\end{array}
$$

$\dfrac{2x^3}{x} = 2x^2$.

$\dfrac{-2x^2}{x} = -2x$.

$\dfrac{4x}{x} = 4$.

Check:
$$
\begin{aligned}
(x + 3)(2x^2 - 2x + 4) &= 2x^3 - 2x^2 + 4x + 6x^2 - 6x + 12 \\
&= 2x^3 + 4x^2 - 2x + 12
\end{aligned}
$$

Self Check 6   Divide $x^2 - 10x + 6x^3 + 4$ by $2x - 1$.

When we write the terms of a dividend in descending powers, we must determine whether some powers of the variable are missing. If any are missing, we should write such terms with a coefficient of 0 or leave blank spaces for them.

EXAMPLE 7

Divide: $\dfrac{27x^3 + 1}{3x + 1}$.

Solution   The dividend, $27x^3 + 1$, does not have an x^2-term or an x-term. We can either insert a $0x^2$ term and a $0x$ term as placeholders, or leave spaces for them.

$$
\begin{array}{r}
9x^2 - 3x + 1 \\
3x + 1\overline{\smash{)}27x^3 + 0x^2 + 0x + 1} \\
\underline{27x^3 + 9x^2} \\
-9x^2 + 0x \\
\underline{-9x^2 - 3x} \\
3x + 1 \\
\underline{3x + 1} \\
0
\end{array}
$$

$\dfrac{27x^3}{3x} = 9x^2.$

$\dfrac{-9x^2}{3x} = -3x.$

$\dfrac{3x}{3x} = 1.$

Check: $(3x + 1)(9x^2 - 3x + 1) = 27x^3 - 9x^2 + 3x + 9x^2 - 3x + 1$
$$= 27x^3 + 1$$

Self Check 7   Divide: $\dfrac{x^2 - 9}{x - 3}$.

Answers to Self Checks   **1. a.** $6y^2$,   **b.** $\dfrac{d^4}{4c^3}$ **2. a.** $10h + 3$,   **b.** $2s^3 - \dfrac{s^2 t}{11} + 4t^2$ **4.** $x + 4$
5. $4x - 3 + \dfrac{6}{2x + 3}$ **6.** $3x^2 + 2x - 4$ **7.** $x + 3$

4.8   STUDY SET   ELEMENTARY & INTERMEDIATE Algebra *f(x)* Now™

VOCABULARY   **Fill in the blanks.**

1. The _____ of $\dfrac{15x^2 - 25x}{5x}$ is $15x^2 - 25x$ and the _____ is $5x$.

2. The expression $\dfrac{18x^7}{9x^4}$ is a monomial divided by a _____.

3. The expression $\dfrac{6x^3 y - 4x^2 y^2 + 8xy^3 - 2y^4}{2x^4}$ is a _____ divided by a monomial.

4. The expression $\dfrac{x^2 - 8x + 12}{x - 6}$ is a trinomial divided by a _____.

5. The powers of x in $2x^4 + 3x^3 + 4x^2 - 7x - 8$ are written in _____ order.

6.
$$
\begin{array}{r}
x - 2 \\
x - 6\overline{\smash{)}x^2 - 8x - 4} \\
\underline{x^2 - 6x} \\
-2x - 4 \\
\underline{-2x + 12} \\
-16
\end{array}
$$

7. The expression $5x^2 + 6$ is missing an x-term. We can insert a _____ $0x$ term and write it as $5x^2 + 0x + 6$.

CONCEPTS   **Fill in the blanks.**

8. a. To divide a polynomial by a monomial, divide each _____ of the polynomial by the monomial.

b. $\dfrac{18x + 9}{9} = \dfrac{18x}{} + \dfrac{9}{}$

c. $\dfrac{30x^2 + 12x - 24}{6} = \dfrac{30x^2}{} + \dfrac{12x}{} - \dfrac{24}{}$

9. Complete each rule of exponents.

a. $\dfrac{x^m}{x^n} = $ **b.** $x^{-n} = $

Simplify.

10. a. $\dfrac{x^8}{x^3}$ **b.** $\dfrac{y^5}{y^7}$

c. $\dfrac{a^4}{a^4}$ **d.** h^{1-5}

e. s^{10-4} **f.** t^{5-5}

11. Write each polynomial with the powers in descending order.

a. $5x^2 + 7x^3 - 3x - 9$

b. $9x + 2x^2 - x^3 + 6x^4$

12. The long division process is a series of four steps that are repeated. Put them in the correct order:

subtract   multiply   bring down   divide

13. In the long division below, a subtraction must be performed. What is the answer to the subtraction?

$$
\begin{array}{r}
x \\
x - 7 \overline{)\, x^2 - 9x - 6} \\
\underline{x^2 - 7x}
\end{array}
$$

14. Fill in the blanks: To check an answer of a long division, we use the fact that

divisor · [   ] + remainder = [   ]

15. Using long division, a student found that

$$\frac{3x^2 + 8x + 4}{3x + 2} = x + 2$$

Check to see whether the result is correct.

16. Using long division, a student found that

$$\frac{x^2 + 4x - 20}{x - 3} = x + 7 + \frac{1}{x - 3}$$

Check to see whether the result is correct.

NOTATION   Complete each solution.

17. $\dfrac{28x^5 - x^3 + 7x^2}{7x^2} = \dfrac{28x^5}{\boxed{}} - \dfrac{\boxed{}}{7x^2} + \dfrac{7x^2}{\boxed{}}$

$= 4x^{5-2} - \dfrac{x^{3-2}}{\boxed{}} + x^{\boxed{}}$

$= \boxed{} - \dfrac{x}{7} + \boxed{}$

18.

$$
\begin{array}{r}
\boxed{} + 2 \\
x + 2 \overline{)\, x^2 + 4x + 5} \\
\underline{x^2 + \boxed{}} \\
\boxed{} + 5 \\
\underline{2x + 4} \\
\boxed{}
\end{array}
$$

19. Insert placeholders for each missing term in the polynomial.

a. $5x^4 + 2x^2 - 1$

b. $-3x^5 - 2x^3 + 4x - 6$

20. True or false: $6x + 4 + \dfrac{-3}{x + 2} = 6x + 4 - \dfrac{3}{x + 2}$.

PRACTICE   Perform each division by simplifying the fraction.

21. $\dfrac{8}{6}$ **22.** $\dfrac{15}{9}$

23. $\dfrac{x^5}{x^2}$ **24.** $\dfrac{a^{12}}{a^8}$

25. $\dfrac{45m^{10}}{9m^5}$ **26.** $\dfrac{24n^{12}}{8n^4}$

27. $\dfrac{12h^8}{9h^6}$ **28.** $\dfrac{22b^9}{6b^6}$

29. $\dfrac{-3d^4}{15d^8}$ **30.** $\dfrac{-4x^3}{16x^5}$

31. $\dfrac{r^3s^2}{rs^3}$ **32.** $\dfrac{y^4z^3}{y^2z^2}$

33. $\dfrac{8x^3y^2}{4xy^3}$ **34.** $\dfrac{-3y^3z}{6yz^2}$

35. $\dfrac{-16r^3y^2}{-4r^2y^4}$ **36.** $\dfrac{-35xyz^2}{-7x^2yz}$

37. $\dfrac{-65rs^2t}{15r^2s^3t}$ **38.** $\dfrac{112u^3z^6}{-42u^3z^6}$

Perform each division.

39. $\dfrac{6x + 9}{3}$ **40.** $\dfrac{8x + 12}{4}$

41. $\dfrac{8x^9 - 32x^6}{4x^4}$ **42.** $\dfrac{30y^8 + 40y^7}{10y^6}$

43. $\dfrac{6h^{12} + 48h^9}{24h^{10}}$ **44.** $\dfrac{4x^{14} - 36x^8}{36x^{12}}$

45. $\dfrac{-18w^6 - 9}{9w^4}$

46. $\dfrac{-40f^4 + 16}{8f^3}$

47. $\dfrac{9s^8 - 18s^5 + 12s^4}{3s^3}$

48. $\dfrac{16b^{10} + 4b^6 - 20b^4}{4b^2}$

49. $\dfrac{7c^5 + 21c^4 - 14c^3 - 35c}{7c^2}$

50. $\dfrac{12r^{15} - 48r^{12} + r^{10} - 18r^8}{6r^{10}}$

51. $\dfrac{5x - 10y}{25xy}$

52. $\dfrac{2x - 32}{16x}$

53. $\dfrac{15a^3b^2 - 10a^2b^3}{5a^2b^2}$

54. $\dfrac{9a^4b^3 - 16a^3b^4}{12a^2b}$

55. $\dfrac{12x^3y^2 - 8x^2y - 4x}{4xy}$

56. $\dfrac{12a^2b^2 - 8a^2b - 4ab}{4ab}$

57. $\dfrac{-25x^2y + 30xy^2 - 5xy}{-5xy}$

58. $\dfrac{-30a^2b^2 - 15a^2b - 10ab^2}{-10ab}$

59. Divide $x^2 + 8x + 12$ by $x + 2$.

60. Divide $x^2 + 5x + 6$ by $x + 2$.

61. Divide $y^2 + 13y + 12$ by $y + 1$.

62. Divide $z^2 + 7z + 12$ by $z + 3$.

63. $\dfrac{6a^2 + 5a - 6}{2a + 3}$

64. $\dfrac{3b^2 - 5b + 2}{3b - 2}$

65. $\dfrac{3b^2 + 11b + 6}{3b + 2}$

66. $\dfrac{8a^2 + 2a - 3}{2a - 1}$

67. $5x + 3 \overline{)\, 11x + 10x^2 + 3}$

68. $2x - 7 \overline{)\, -x - 21 + 2x^2}$

69. $4 + 2x \overline{)\, -10x - 28 + 2x^2}$

70. $1 + 3x \overline{)\, 9x^2 + 1 + 6x}$

71. $2x - 1 \overline{)\, x - 2 + 6x^2}$

72. $2 + x \overline{)\, 3x + 2x^2 - 2}$

73. $2x + 3 \overline{)\, 2x^3 + 7x^2 + 4x - 3}$

74. $2x - 1 \overline{)\, 2x^3 - 3x^2 + 5x - 2}$

75. $3x + 2 \overline{)\, 6x^3 + 10x^2 + 7x + 2}$

76. $4x + 3 \overline{)\, 4x^3 - 5x^2 - 2x + 3}$

77. $2x + 1 \overline{)\, 2x^3 + 3x^2 + 3x + 1}$

78. $3x - 2 \overline{)\, 6x^3 - x^2 + 4x - 4}$

79. $\dfrac{x^2 - 1}{x - 1}$

80. $\dfrac{x^2 - 9}{x + 3}$

81. $\dfrac{4x^2 - 9}{2x + 3}$

82. $\dfrac{25x^2 - 16}{5x - 4}$

83. $\dfrac{x^3 + 1}{x + 1}$

84. $\dfrac{x^3 - 8}{x - 2}$

85. $\dfrac{a^3 + a}{a + 3}$

86. $\dfrac{y^3 - 50}{y - 5}$

87. $\dfrac{2x^2 + 5x + 2}{2x + 3}$

88. $\dfrac{3x^2 - 8x + 3}{3x - 2}$

89. $\dfrac{4x^2 + 6x - 1}{2x + 1}$

90. $\dfrac{6x^2 - 11x + 2}{3x - 1}$

91. $\dfrac{2x^3 + 7x^2 + 4x + 3}{2x + 3}$

92. $\dfrac{2x^3 + 4x^2 - 2x + 3}{x - 2}$

93. $\dfrac{6x^3 + x^2 + 2x + 1}{3x - 1}$

94. $\dfrac{3y^3 - 4y^2 + 2y + 3}{y + 3}$

APPLICATIONS

95. POOL   The rack shown in the illustration is used to set up the balls for a game of pool. If the perimeter of the rack, in inches, is given by the polynomial $6x^2 - 3x + 9$, what is the length of one side?

96. CHECKERBOARD   If the perimeter (in inches) of a checkerboard is $12x^2 - 8x + 32$, what is the length of one side?

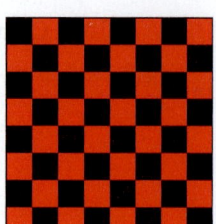

97. a. Solve the formula $d = rt$ for t.

b. Use your answer from part a to complete the table.

	r	$\cdot$	t	$=$	d
Motorcycle	$2x$				$6x^3$

c. Use your answer to part a to complete the table.

	r	$\cdot$	t	$=$	d
Subway	$x - 3$				$x^2 + x - 12$

98. FURNACE FILTERS The area of the furnace filter is $(x^2 - 2x - 24)$ square inches. Find its length.

$(x + 4)$ in.

99. AIR CONDITIONING If the volume occupied by the air conditioning unit is $(36x^3 - 24x^2)$ cubic feet, find its height.

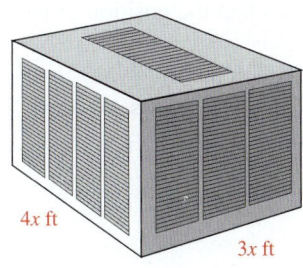

4x ft

3x ft

100. MINI-BLINDS The area covered by the mini-blinds is $(3x^3 - 6x)$ square feet. How long are the blinds?

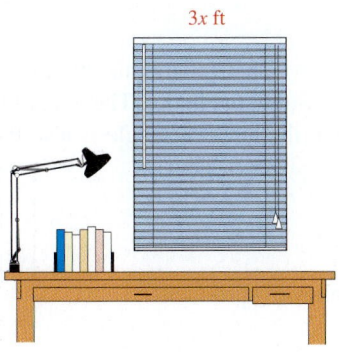

3x ft

101. COMMUNICATIONS Telephone poles were installed every $(2x - 3)$ feet along a stretch of railroad track $(8x^3 - 6x^2 + 5x - 21)$ feet long. How many poles were used?

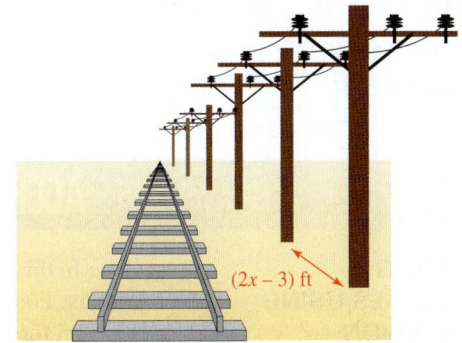

$(2x - 3)$ ft

102. ELECTRIC BILLS On an electric bill, the formula

$$A = \frac{0.08x + 5}{x}$$

is used to compute the average cost of x kilowatt hours of electricity. Could the following formula be used instead?

$$A = 0.08 + \frac{5}{x}$$

WRITING

103. Explain how to check the following long division.

$$
\begin{array}{r}
x + 5 \\
3x + 5 \overline{\smash{)}3x^2 + 20x - 5} \\
\underline{3x^2 + 5x} \\
15x - 5 \\
\underline{15x + 25} \\
-30
\end{array}
$$

104. Explain the difference in the methods used to divide $\dfrac{x^2 - 3x + 2}{x}$ as compared to $\dfrac{x^2 - 3x + 2}{x - 2}$.

105. How do you know when to stop the long division process when dividing polynomials?

106. When dividing $x^3 + 1$ by $x + 1$, why is it helpful to write $x^3 + 1$ as $x^3 + 0x^2 + 0x + 1$?

REVIEW

107. Write an equation of the line with slope $-\frac{11}{6}$ that passes through $(2, -6)$. Write the answer in slope–intercept form.

108. Solve $S = 2\pi rh + 2\pi r^2$ for h.

109. Evaluate: $-10(18 - 4^2)^3$.

110. Evaluate: -5^2.

CHALLENGE PROBLEMS

111. Divide: $\dfrac{6x^{6m}y^{6n} + 15x^{4m}y^{7n} - 24x^{2m}y^{8n}}{3x^{2m}y^n}$.

112. Divide: $\dfrac{6a^3 - 17a^2b + 14ab^2 - 3b^3}{2a - 3b}$.

ACCENT ON TEAMWORK

EVALUATING POLYNOMIALS USING LONG DIVISION

Overview: In this activity, you will learn an alternate way to evaluate polynomials.

Instructions: Form groups of 2 or 3 students. Have each student evaluate the polynomial $2x^2 - 3x - 5$ for $x = 1$ and for $x = 3$.

Next, divide the polynomial by $x - 1$ and by $x - 3$. What do you notice about the remainders of these divisions when compared to your answers to the evaluations?

Continue by evaluating the polynomial for $x = 2$ and dividing it by $x - 2$. Does the pattern still hold?

Finally, evaluate the polynomial for $x = -2$. By what must you divide to get a remainder that matches the evaluation?

Does the pattern hold for other polynomials? Try some polynomials of your own, experiment, and report your conclusions.

SHIFTING THE GRAPH OF $y = x^2$

Overview: In this activity, some minor changes are made to the equation $y = x^2$. You are to determine how these changes affect the position of the graph.

Instructions: Form groups of 2 or 3 students. Graph the following equations on the *same* coordinate system using a table of solutions as shown. Compare the graphs. How are they similar? How do they differ?

$$y = x^2$$
$$y = (x - 1)^2$$
$$y = (x + 1)^2$$

x	y
-4	
-3	
-2	
-1	
0	
1	
2	
3	
4	

On another coordinate system, graph $y = x^2$, $y = (x - 2)^2$, and $y = (x + 2)^2$. Compare these graphs. How are they similar? How do they differ?

From these examples, what conclusions can be made about shifting the graph of $y = x^2$ to the left and to the right?

BINOMIAL MULTIPLICATION AND THE AREA OF RECTANGLES

Overview: In this activity, rectangles are used to visualize binomial multiplication.

Instructions: Form groups of 2 or 3 students. Study the figure. The area of the large rectangle is given by $(x + 2)(x + 3)$. The area of the large rectangle is also the sum of the areas of the four smaller rectangles: $x^2 + 3x + 2x + 6$. Thus,

$$(x + 2)(x + 3) = x^2 + 3x + 2x + 6 = x^2 + 5x + 6.$$

	x	2
3	$3x$	6
x	x^2	$2x$

$x + 3$ (left), $x + 2$ (bottom)

Draw three similar models to represent the following products.

1. $(x + 4)(x + 5)$ **2.** $x(x + 6)$ **3.** $(x + 2)^2$

KEY CONCEPT: POLYNOMIALS

A **polynomial** is a single term or a sum of terms in which all variables have whole-number exponents. Some examples are

$$-16a^4b, \qquad y + 8, \qquad x^2 + 2xy - y^2, \qquad \text{and} \qquad x^3 - 2x^2 + 6x - 8$$

THE VOCABULARY OF POLYNOMIALS

1. Consider $x^3 - 2x^2 + 6x - 8$.
  a. Fill in: This is a polynomial in ☐. It is written in _____ powers of ☐.
  b. How many terms does the polynomial have?
  c. Give the degree of each term.
  d. What is the degree of the polynomial?
  e. Give the coefficient of each term.

2. Classify each polynomial as a monomial, binomial, trinomial, or none of these.
  a. $x^2 - y^2$
  b. $s^2t + st^2 - st + 1$
  c. $4y^2 - 10y + 16$
  d. $15h$

OPERATIONS WITH POLYNOMIALS

Just like numbers, polynomials can be added, subtracted, multiplied, divided, and raised to powers. The key to performing these operations with polynomials is knowing how to perform these operations with monomials.

Perform the operations.

3. $4x^3 + 3x^3$

4. $7m^{10} + (-6m^{10})$

5. $7a^2b - 9a^2b$

6. $(6y^5)(-7y^8)$

7. $\dfrac{16c^4d^5}{8c^2d^6}$

8. $(5f^3)^2$

Fill in the blanks.

9. To add polynomials, _____ like terms.

10. To subtract two polynomials, change the _____ of the terms of the polynomial being subtracted, drop parentheses, and combine like terms.

11. To multiply two polynomials, multiply _____ term of one polynomial by _____ term of the other polynomial and combine like terms.

12. To divide a polynomial by a monomial, divide each _____ of the polynomial by the monomial.

13. To divide two polynomials, use the _____ division method.

Perform the operations.

14. $(8x^3 + 4x^2 - 8x + 1) + (6x^3 - 5x^2 - 2x + 3)$

15. $(20s^3t + s^2t^2 - 6st^3) - (8s^3t - 9s^2t^2 + 12st^3)$

16. $(2x + 3)(x - 8)$

17. $(2x^2 + 3)^2$

18. $(4h^5 + 8t)(4h^5 - 8t)$

19. $(y^2 + y - 6)(y + 3)$

20. $\dfrac{9x^6 + 27x^7 - 18x^5}{3x^2}$

21. $\dfrac{x^3 + 3x^2 + 5x + 3}{x + 1}$

SECTION 4.1	**Rules for Exponents**

CONCEPTS

If n represents a natural number, then

$$\overbrace{x^n = x \cdot x \cdot x \cdot \cdots \cdot x}^{n \text{ factors of } x}$$

where x is called the *base* and n is called the *exponent*.

REVIEW EXERCISES

1. Write the repeated multiplication that is indicated.

    a. $-3x^4$ **b.** $\left(\dfrac{1}{2}pq\right)^3$

2. Identify the base and the exponent in each expression.

    a. $2x^6$ **b.** $(2x)^6$

Evaluate each expression.

3. 5^3 **4.** $(-8)^2$

5. -8^2 **6.** $(5 - 3)^2$

Rules for exponents:
If m and n represent integers, and there are no divisions by 0, then

$$x^m x^n = x^{m+n}$$
$$\frac{x^m}{x^n} = x^{m-n}$$
$$(x^m)^n = x^{mn}$$
$$(xy)^n = x^n y^n$$
$$\left(\frac{x}{y}\right)^n = \frac{x^n}{y^n}$$

Simplify each expression. Assume there are no divisions by 0.

7. $7^4 \cdot 7^8$ **8.** $mmnn$

9. $(y^7)^3$ **10.** $(3x)^4$

11. $b^3 b^4 b^5$ **12.** $-z^2(z^3 y^2)$

13. $(-16s)^2 s$ **14.** $(2x^2 y)^2$

15. $(x^2 x^3)^3$ **16.** $\left(\dfrac{x^2 y}{xy^2}\right)^2$

17. $\dfrac{(m - 25)^{16}}{(m - 25)^4}$ **18.** $\dfrac{(5y^2 z^3)^3}{(yz)^5}$

Find the area or the volume of each figure, whichever is appropriate.

19.

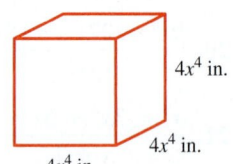

$4x^4$ in.
$4x^4$ in.
$4x^4$ in.

20.

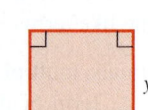

y^2 m
y^2 m

| SECTION 4.2 | **Zero and Negative Exponents** |

For any nonzero real numbers x and y and any integers m and n,

$$x^0 = 1$$

$$x^{-n} = \frac{1}{x^n}$$

$$\frac{1}{x^{-n}} = x^n$$

$$\frac{x^{-m}}{y^{-n}} = \frac{y^n}{x^m}$$

Simplify each expression. Write each answer without using negative exponents or parentheses.

21. x^0

22. $(3x^2y^2)^0$

23. $(3x^0)^2$

24. 10^{-3}

25. $\left(\dfrac{3}{4}\right)^{-1}$

26. -5^{-2}

27. x^{-5}

28. $-6y^4y^{-5}$

29. $\dfrac{7^{-2}}{2^{-3}}$

30. $(x^{-3}x^{-4})^{-2}$

31. $\left(\dfrac{-3r^4r^{-3}}{r^{-3}r^7}\right)^3$

32. $\left(\dfrac{4z^4}{z^3}\right)^{-2}$

| SECTION 4.3 | **Scientific Notation** |

A positive number is written in *scientific notation* when it is written in the form $N \times 10^n$, where $1 \le N < 10$ and n is an integer.

Write each number in scientific notation.

33. 728

34. 9,370,000,000,000,000

35. 0.0136

36. 0.00942

37. 0.018×10^{-2}

38. 753×10^3

Write each number in standard notation.

39. 7.26×10^5

40. 3.91×10^{-8}

41. 2.68×10^0

42. 5.76×10^1

Scientific notation provides an easier way to perform computations involving very large or very small numbers.

Evaluate each expression by first writing each number in scientific notation. Then do the arithmetic. Express the result in standard notation.

43. $\dfrac{(0.00012)(0.00004)}{0.00000016}$

44. $\dfrac{(4,800)(20,000)}{600,000}$

45. WORLD POPULATION   As of 2003, the world's population was estimated to be 6.31 billion. Write this number in standard notation and in scientific notation.

46. ATOMS   The illustration shows a cross section of an atom. How many nuclei, placed end to end, would it take to stretch across the atom?

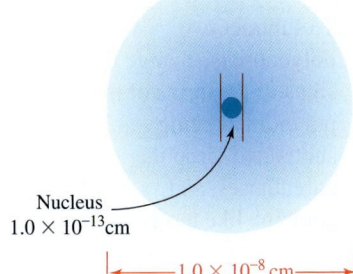

Nucleus
1.0×10^{-13}cm

$\leftarrow$ 1.0×10^{-8} cm $\rightarrow$

| SECTION 4.4 | **Polynomials** |

A *polynomial* is a single term or a sum of terms in which all variables have whole-number exponents.

47. Consider the polynomial $3x^3 - x^2 + x + 10$.

    a. How many terms does the polynomial have?

    b. What is the lead term?

    c. What is the coefficient of each term?

    d. What is the constant term?

The *degree of a term* of a polynomial in one variable is the value of the exponent on the variable. If a polynomial is in more than one variable, the *degree of a term* is the sum of the exponents on the variables. The *degree of a nonzero constant* is 0.

48. Find the degree of each polynomial and classify it as a monomial, binomial, trinomial, or none of these.

    a. $13x^7$ **b.** $-16a^2b$

    c. $5^3x + x^2$ **d.** $-3x^5 + x - 1$

    e. $9xy^2 + 21x^3y^3$ **f.** $4s^4 - 3s^2 + 5s + 4$

The *degree of a polynomial* is the highest degree of any term of the polynomial.

49. Evaluate $-x^5 - 3x^4 + 3$ for $x = 0$ and $x = -2$.

The graph of a *nonlinear equation* is not a straight line.

A *parabola* is a cup-shaped curve.

50.   DIVING   See the illustration. The number of inches that the woman deflects the diving board is given by the polynomial

$$0.1875x^2 - 0.0078125x^3$$

where x is the number of feet that she stands from the front anchor point of the board. Find the amount of deflection if she stands on the end of the diving board, 8 feet from the anchor point.

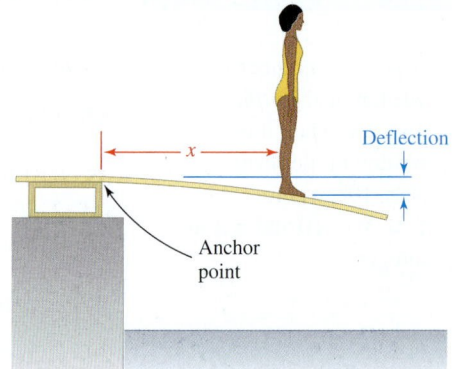

Construct a table of solutions and then graph the equation.

51. $y = x^2$ **52.** $y = x^3 + 1$

| SECTION 4.5 | **Adding and Subtracting Polynomials** |

To add polynomials, combine like terms.

To subtract two polynomials, change the signs of the terms of the polynomial being subtracted, drop the parentheses, and combine like terms.

Simplify each polynomial.

53. $6y^3 + 8y^4 + 7y^3 + (-8y^4)$ **54.** $4a^2b + 5 - 6a^3b - 3a^2b + 2a^3b$

55. $\frac{5}{6}x^2 + \frac{1}{3}y^2 - \frac{1}{4}x^2 - \frac{3}{4}xy + \frac{2}{3}y^2$ **56.** $-(c^5 + 5c^4 - 12)$

Perform the operations.

57. $(2r^6 + 14r^3) + (23r^6 - 5r^3 + 5r)$

58. $(7a^2 + 2a - 5) - (3a^2 - 2a + 1)$

59. $(3r^3s + r^2s^2 - 3rs^3 - 3s^4) + (r^3s - 8r^2s^2 - 4rs^3 + s^4)$

60. $3(9x^2 + 3x + 7) - 2(11x^2 - 5x + 9)$

61. Find the difference when $-3z^3 - 4z + 7$ is subtracted from the sum of $2z^2 + 3z - 7$ and $-4z^3 - 2z - 3$.

62. $\quad 3x^2 + 5x + 2$
$\quad + \; \underline{x^2 - 3x + 6}$

63. $\quad 20x^3 \qquad\quad + 12x$
$\quad - \; \underline{12x^3 + 7x^2 - \;\; 7x}$

64. GARDENING   Find a polynomial that represents the length of the wooden handle of the shovel.

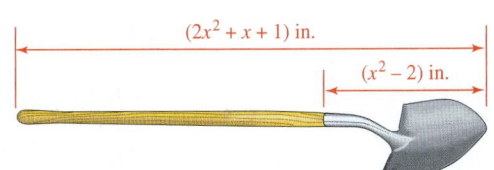

$(2x^2 + x + 1)$ in.

$(x^2 - 2)$ in.

SECTION 4.6

Multiplying Polynomials

To multiply two monomials, multiply the numerical factors (the coefficients) and then multiply the variable factors.

Find each product.

65. $(2x^2)(5x)$

66. $(-6x^4z^3)(x^6z^2)$

67. $5b^3 \cdot 6b^2 \cdot 4b^6$

68. $\frac{2}{3}h^5(3h^9 + 12h^6)$

To multiply a monomial and a polynomial, multiply each term of the polynomial by the monomial.

69. $x^2y(y^2 - xy)$

70. $3n^2(3n^2 - 5n + 2)$

71. $2x(3x^4)(x + 2)$

72. $-a^2b^2(-a^4b^2 + a^3b^3 - ab^4 + 7a)$

To multiply two binomials, use the *FOIL method:*
  F: First
  O: Outer
  I: Inner
  L: Last

73. $(x + 3)(x + 2)$

74. $(2x + 1)(x - 1)$

75. $(3a - 3)(2a + 2)$

76. $6(a - 1)(a + 1)$

77. $(a - b)(2a + b)$

78. $(3n^4 - 5n^2)(2n^4 - n^2)$

To multiply two polynomials, multiply each term of one polynomial by each term of the other polynomial, and then combine like terms.

79. $(2a - 3)(4a^2 + 6a + 9)$

80. $(8x^2 + x - 2)(7x^2 + x - 1)$

81. Multiply: $4x^2 - 2x + 1$
$\qquad\qquad\quad \underline{2x + 1}$

82. APPLIANCES   Find the perimeter of the base, the area of the base, and the volume occupied by the dishwasher.

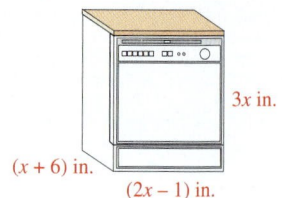

3x in.

(x + 6) in.

(2x − 1) in.

Special Products

Special products:

$(x + y)^2 = x^2 + 2xy + y^2$

$(x - y)^2 = x^2 - 2xy + y^2$

$(x + y)(x - y) = x^2 - y^2$

Find each product.

83. $(x + 3)(x + 3)$

84. $(2x - 0.9)(2x + 0.9)$

85. $(a - 3)^2$

86. $(x + 4)^2$

87. $(-2y + 1)^2$

88. $(y^2 + 1)(y^2 - 1)$

89. $(6r^2 + 10s)^2$

90. $-(8a - 3)^2$

91. $80s(r^2 + s^2)(r^2 - s^2)$

92. $4b(3b - 4)^2$

93. $\left(t - \dfrac{3}{4}\right)^2$

94. $(m + 2)^3$

95. Perform the operations.

$$(5c - 1)^2 - (c + 6)(c - 6)$$

96. GRAPHIC ARTS   A Dr. Martin Luther King poster has his picture with a $\frac{1}{2}$-inch wide border around it. The length of the poster is $(x + 3)$ inches and the width is $(x - 1)$ inches. Find a polynomial that represents the area of the picture of Dr. King.

Division of Polynomials

To divide monomials, use the method for simplifying fractions or use the rules for exponents.

Perform each division.

97. $\dfrac{16n^8}{8n^5}$

98. $\dfrac{-14x^2y}{21xy^3}$

99. $\dfrac{a^{15} - 24a^8}{6a^{12}}$

100. $\dfrac{15a^2b + 20ab^2 - 25ab}{5ab}$

To divide a polynomial by a monomial, divide each term of the numerator by the denominator.

Long division is used to divide one polynomial by another. When a division has a remainder, write the answer in the form

$$\text{Quotient} + \frac{\text{remainder}}{\text{divisor}}$$

The division method works best when the exponents of the terms of the divisor and the dividend are written in descending order.

When the dividend is missing a term, write it with a coefficient of zero or leave a blank space.

101. $x - 1\overline{)x^2 - 6x + 5}$

102. $\dfrac{2x^2 + 3 + 7x}{x + 3}$

103. $2x - 1\overline{)6x^3 + x^2 + 1}$

104. $3x + 1\overline{)-13x - 4 + 9x^3}$

105. Use multiplication to show that the answer when dividing $3y^2 + 11y + 6$ by $y + 3$ is $3y + 2$.

106. SAVINGS BONDS   How many \$50 savings bonds would have a total value of $(50x + 250)$ dollars?

CHAPTER 4 TEST   ELEMENTARY & INTERMEDIATE Algebra $f(x)$ Now™

1. Use exponents to rewrite $2xxxyyyy$.

2. Evaluate: -6^2.

Simplify each expression. Write answers without using parentheses or negative exponents.

3. $y^2(yy^3)$

4. $(2x^3)^5(x^2)^3$

5. $3x^0$

6. $2y^{-5}y^2$

7. 5^{-3}

8. $\dfrac{(x + 1)^{15}}{(x + 1)^6}$

9. $\dfrac{y^2}{yy^{-2}}$

10. $\left(\dfrac{a^2b^{-1}}{4a^3b^{-2}}\right)^{-3}$

11. Find the volume of a cube that has sides of length $10y^4$ inches.

12. ELECTRICITY   One ampere (amp) corresponds to the flow of 6,250,000,000,000,000,000 electrons per second past any point in a direct current (DC) circuit. Write this number in scientific notation.

13. Write 9.3×10^{-5} in standard notation.

14. Evaluate: $(2.3 \times 10^{18})(4.0 \times 10^{-15})$. Write the answer in standard notation.

15. Complete the table for the polynomial $x^4 + 8x^2 - 12$.

Term	Coefficient	Degree
Degree of the polynomial		

16. Identify $3x^2 + 2$ as a monomial, binomial, or trinomial.

17. Find the degree of the polynomial $3x^2y^3 + 2x^3y - 5x^2y$.

18. Graph: $y = x^2 + 2$.

19. FREE FALL   A visitor standing on the rim of the Grand Canyon drops a rock over the side. The distance (in feet) that the rock is from the canyon floor t seconds after being dropped is given by the polynomial $-16t^2 + 5{,}184$. Find the position of the rock 18 seconds after being dropped. Explain your answer.

20. Simplify: $4a^2b + 5 - 6a^3b - 3a^2b + 2a^3b$.

Perform the operations:

21. $(3a^2 - 4a - 6) + (2a^2 - a + 9)$

22. Subtract $(b^3c - 3bc + 12)$ from $(7b^3c - 5bc)$.

23.
$$\begin{array}{r} -5y^3 + 4y^2 - 11y + 3 \\ -\underline{-2y^3 - 14y^2 + 17y - 32} \end{array}$$

24. Simplify: $-6(x - y) + 2(x + y)$.

Find each product.

25. $(-2x^3)(2x^2y)$

26. $3y^2(y^2 - 2y + 3)$

27. $(2x - 5)(3x + 4)$

28. $(2x - 3)(x^2 - 2x + 4)$

29. $(1 + 10c)(1 - 10c)$

30. $(7b^3 - 3)^2$

31. Perform the operations: $(x + y)(x - y) + x(x + y)$

Perform each division.

32. $\dfrac{6a^2 - 12b^2}{24ab}$

33. $\dfrac{2x^2 - x - 6}{2x + 3}$

34. $2x - 1 \overline{)6x^3 + x^2 + 1}$

35. Find the width of the rectangle.

> Area: $(x^2 - 6x + 5)$ ft^2
>
> Length: $(x - 1)$ ft

CHAPTERS 1–4   CUMULATIVE REVIEW EXERCISES

1. SPORTS CARS   The graph shows the Porsche vehicle sales in the United States for the years 1986–2001.
   a. In what year were sales the lowest?
   b. In what year were sales the greatest?
   c. Between what two years was there the greatest increase in sales?

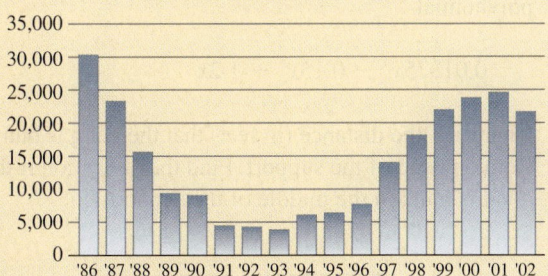

Porsche vehicle sales in U.S.

Source: Porsche Cars North America

2. Divide: $\dfrac{3}{4} \div \dfrac{6}{5}$.

3. Subtract: $\dfrac{7}{10} - \dfrac{1}{14}$.

4. Is π a rational or irrational number?

5. RACING   Suppose a driver has completed x laps of a 250-lap race. Write an expression for how many more laps he must make to finish the race.

6. CLINICAL TRIALS   In a clinical test of Aricept, a drug to treat Alzheimer's disease, one group of patients took a placebo (a sugar pill) while another group took the actual medication. Use the data in the table to determine the number of patients in each group who experienced nausea.

Comparison of rates of adverse events in patients		
Adverse event	**Group 1—Placebo** (number = 300)	**Group 2—Aricept** (number = 320)
Nausea	6%	5%

7. Give the prime factorization of 100.

8. Graph each member of the set on the number line.

$$\left\{ -2\tfrac{1}{4},\ \sqrt{2},\ -1.75,\ \tfrac{7}{2},\ 0.5 \right\}$$

9. Write $\dfrac{2}{3}$ as a decimal.

10. What property of real numbers is illustrated?

$$3(2x) = (3 \cdot 2)x$$

11. What is the value of d dimes?

Simplify each expression.

12. $13r - 12r$

13. $27\left(\dfrac{2}{3}x\right)$

14. $4(d - 3) - (d - 1)$

15. $(13c - 3)(-6)$

Evaluate each expression.

16. $-3^2 + |4^2 - 5^2|$

17. $(4 - 5)^{20}$

18. $\dfrac{-3 - (-7)}{2^2 - 3}$

19. $12 - 2[1 - (-8 + 2)]$

Solve each equation.

20. $3(x - 5) + 2 = 2x$

21. $\dfrac{x - 5}{3} - 5 = 7$

22. $\dfrac{2}{5}x + 1 = \dfrac{1}{3} + x$

23. $-\dfrac{5}{8}h = 15$

24. Solve: $8(4 + x) > 10(6 + x)$. Write the solution set in interval notation and graph it.

25. Solve: $A = \dfrac{1}{2}h(b + B)$ for h.

26. CANDY   The owner of a candy store wants to make a 30-pound mixture of two candies to sell for $2 per pound. If red licorice bits sell for $1.90 per pound and lemon gumdrops sell for $2.20 per pound, how many pounds of each should be used?

Graph each equation.

27. $4x - 3y = 12$

28. $x = 4$

Find the slope of the line with the given properties.

29. Passing through $(-2, 4)$ and $(6, 8)$

30. A line that is horizontal

31. An equation of $2x - 3y = 12$

Write an equation of the line with the following properties.

32. Slope $= \dfrac{2}{3}$, y-intercept $= (0, 5)$

33. Passing through $(-2, 4)$ and $(6, 10)$

34. A horizontal line passing through $(2, 4)$

35. Are the graphs of the lines parallel or perpendicular?

$$y = -\frac{3}{4}x + \frac{15}{4}$$

$$4x - 3y = 25$$

36. Subtract: $\begin{array}{r} 17x^4 - 3x^2 - 65x - 12 \\ -\ 23x^4 + 14x^2 + 3x - 23 \end{array}$

Simplify each expression. Write each answer without using parentheses or negative exponents.

37. $(-3x^2y^4)^2$

38. $(2y)^{-4}$

39. $(x^3x^4)^2$

40. $ab^3c^4 \cdot ab^4c^2$

41. $\dfrac{a^4b^0}{a^{-3}}$

42. $\left(\dfrac{4t^3t^4t^5}{3t^2t^6}\right)^3$

Perform the operations.

43. $(4c^2 + 3c - 2) + (3c^2 + 4c + 2)$

44. $3x(2x + 3)^2$

45. $(2t + 3s)(3t - s)$

46. $5x + 3\overline{)11x + 10x^2 + 3}$

47. Graph: $y = x^2$.

Write each number in scientific notation.

48. 615,000

49. 0.0000013

50. 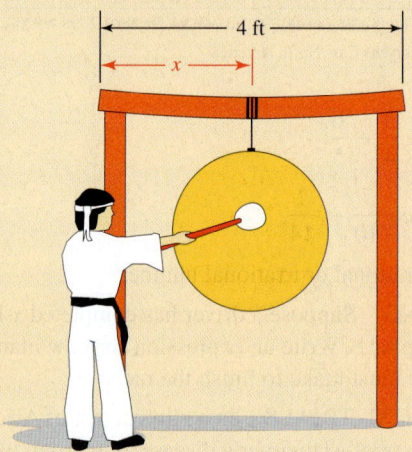 MUSICAL INSTRUMENTS   The gong shown in the illustration is a percussion instrument used throughout Southeast Asia. The amount of deflection of the horizontal support (in inches) is given by the polynomial

$$0.01875x^4 - 0.15x^3 + 1.2x$$

where x is the distance (in feet) that the gong is hung from one end of the support. Find the deflection if the gong is hung in the middle of the support.

5

Factoring and Quadratic Equations

ELEMENTARY & INTERMEDIATE
Algebra $f(x)$ Now™

Throughout the chapter, this icon introduces resources on the Elementary & Intermediate AlgebraNow Web site, accessed through **http://1pass.thomson.com**, that will

- Help you test your knowledge of the material with a pre-test and a post-test

- Provide a personalized learning plan targeting areas you should study

Getty Images

Construction projects come in all shapes and sizes. Whether remodeling a kitchen or erecting a giant skyscraper, algebra and geometry play an important role from start to finish. In this chapter, we explore other applications that involve these branches of mathematics. Using an algebraic process called *factoring,* we will solve applied problems that deal with area, one of the fundamental concepts of geometry.

To learn more about the use of algebra in construction projects, visit *The Learning Equation* on the Internet at http://tle.brookscole.com. (The log-in instructions are in the Preface.) For Chapter 5, the online lessons are:

- *TLE* Lesson 11: Factoring Trinomials and the Difference of Two Squares
- *TLE* Lesson 12: Solving Quadratic Equations by Factoring

Recall that whole numbers can be factored into products of prime numbers. We will now extend that concept and discuss methods for factoring polynomials.

5.1 The Greatest Common Factor; Factoring by Grouping

- The Greatest Common Factor (GCF)
- Factoring Out the GCF
- Factoring by Grouping

In Chapter 4, we learned how to multiply polynomials. For example, to multiply $3x + 5$ by $4x$, we use the distributive property.

$$4x(3x + 5) = 4x \cdot 3x + 4x \cdot 5$$
$$= 12x^2 + 20x$$

Success Tip

On the game show Jeopardy!, answers are revealed and contestants respond with the appropriate questions. Factoring is similar. Answers to multiplications are given. You are to respond by telling what factors were multiplied.

To factor the polynomial $12x^2 + 20x$, we reverse the previous steps and determine what factors were multiplied to obtain this result. This process is called *factoring the polynomial.*

$$\textit{Multiplication} \longrightarrow$$
$$4x(3x + 5) = 12x^2 + 20x$$
$$\longleftarrow \textit{Factoring}$$

When factoring a polynomial, the first step is to determine whether its terms have any common factors.

■ THE GREATEST COMMON FACTOR (GCF)

To determine whether integers have common factors, it is helpful to write them as products of prime numbers. For example, the prime factorizations of 42 and 90 are given below.

$$42 = 2 \cdot 3 \cdot 7 \qquad\qquad 90 = 2 \cdot 3 \cdot 3 \cdot 5$$

The highlighting shows that 42 and 90 have one factor of 2 and one factor of 3 in common. To find their *greatest common factor (GCF)*, we multiply the common factors: $2 \cdot 3 = 6$. Thus, the GCF of 42 and 90 is 6.

The Greatest Common Factor (GCF)	The **greatest common factor (GCF)** of a list of integers is the largest common factor of those integers.

Recall from arithmetic that the factors of a number divide the number exactly, leaving no remainder. Therefore, the greatest common factor of two or more integers is the *largest natural number that divides each of the integers exactly.*

EXAMPLE 1

ELEMENTARY & INTERMEDIATE
Algebra *f(x)* **Now**™

Solution

Find the GCF of each list of numbers: **a.** 21 and 140,   **b.** 24, 60, and 96,   and **c.** 9, 10, and 30.

a. We write each number as a product of primes and determine the factors they have in common.

$$21 = 3 \cdot 7$$
$$140 = 2 \cdot 2 \cdot 5 \cdot 7$$

The Language of Algebra

Recall that the *prime numbers* are: 2, 3, 5, 7, 11, 13, 17, 19, ...

Since the only prime factor common to 21 and 140 is 7, the GCF is 7.

b. To find the GCF of three numbers, we proceed in a similar way.

$$24 = 2 \cdot 2 \cdot 2 \cdot 3$$
$$60 = 2 \cdot 2 \cdot 3 \cdot 5$$
$$96 = 2 \cdot 2 \cdot 2 \cdot 2 \cdot 2 \cdot 3$$

Since 24, 60, and 96 have two factors of 2 and one factor of 3 in common, we have

$$\text{GCF} = 2 \cdot 2 \cdot 3 = 12$$

c. Since there are no prime factors common to all three numbers, the GCF is 1.

$$9 = 3 \cdot 3$$
$$10 = 2 \cdot 5$$
$$30 = 2 \cdot 3 \cdot 5$$

Self Check 1 Find the GCF of each list of numbers: **a.** 24 and 70, **b.** 22, 25, and 98, and **c.** 45, 60, and 75.

To find the greatest common factor of a list of terms, we can use the following approach.

Strategy for Finding the GCF

1. Write each coefficient as a product of prime factors.
2. Identify the numerical and variable factors common to each term.
3. Multiply the common factors identified in Step 2 to obtain the GCF. If there are no common factors, the GCF is 1.

EXAMPLE 2   Find the GCF of each list of terms: **a.** $12x^2$ and $20x$ and **b.** $9a^5b^2$, $15a^4b^2$, and $90a^3b^3$.

Solution   **a.** *Step 1:* We write each coefficient, 12 and 20, as a product of prime factors.

Success Tip

One way to identify common factors is to circle them:

$12x^2 = \boxed{2} \cdot \boxed{2} \cdot 3 \cdot \boxed{x} \cdot x$
$20x = \boxed{2} \cdot \boxed{2} \cdot 5 \cdot \boxed{x}$

$\text{GCF} = 2 \cdot 2 \cdot x = 4x$

$$12x^2 = 2 \cdot 2 \cdot 3 \cdot x \cdot x$$
$$20x = 2 \cdot 2 \cdot 5 \cdot x$$

Step 2: There are two common factors of 2 and one common factor of x.

Step 3: We multiply the common factors, 2, 2, and x, to obtain the GCF.

$$\text{GCF} = 2 \cdot 2 \cdot x = 4x$$

Success Tip

The exponent on any variable in a GCF is the *smallest* exponent that appears on that variable in all of the terms under consideration.

b. *Step 1:* The coefficients, 9, 15, and 90, are written as products of primes.

$$9a^5b^2 = 3 \cdot 3 \cdot a \cdot a \cdot a \cdot a \cdot a \cdot b \cdot b$$
$$15a^4b^2 = 3 \cdot 5 \cdot a \cdot a \cdot a \cdot a \cdot b \cdot b$$
$$90a^3b^3 = 2 \cdot 3 \cdot 3 \cdot 5 \cdot a \cdot a \cdot a \cdot b \cdot b \cdot b$$

Step 2: The highlighting identifies one common factor of 3, three common factors of a, and two common factors of b.

Step 3: GCF $= \mathbf{3} \cdot \mathbf{a} \cdot \mathbf{a} \cdot \mathbf{a} \cdot \mathbf{b} \cdot \mathbf{b} = 3a^3b^2$

Self Check 2   Find the GCF of each list of terms: **a.** $33c$ and $22c^4$,   **b.** $42s^3t^2$, $63s^2t^4$, and $21st^3$.

■ FACTORING OUT THE GCF

The concept of greatest common factor is used to factor polynomials. For example, to factor $12x^2 + 20x$, we note that there are two terms, $12x^2$ and $20x$. We previously determined that the GCF of $12x^2$ and $20x$ is $4x$. With this in mind, we write each term of $12x^2 + 20x$ as a product of the GCF and one other factor. Then we apply the distributive property: $ab + ac = a(b + c)$.

$$12x^2 + 20x = \mathbf{4x} \cdot 3x + \mathbf{4x} \cdot 5 \qquad \text{Write } 12x^2 \text{ and write } 20x \text{ as the product of the GCF, } 4x, \text{ and one other factor.}$$

$$= \mathbf{4x}(3x + 5) \qquad \text{Write an expression so that the multiplication by } 4x \text{ distributes over the terms } 3x \text{ and } 5.$$

We have found that the factored form of $12x^2 + 20x$ is $4x(3x + 5)$. This process is called **factoring out the greatest common factor.**

EXAMPLE 3

ELEMENTARY &
INTERMEDIATE
Algebra *f(x)* **Now**™

Success Tip

Always verify a factorization by doing the indicated multiplication. The result should be the original polynomial.

Factor: **a.** $8m - 48$ and   **b.** $35a^3b^2 + 14a^2b^3$.

Solution   **a.** The greatest common factor of $8m$ and 48 is 8.

$$8m - 48 = \mathbf{8} \cdot m - \mathbf{8} \cdot 6$$
$$= \mathbf{8}(m - 6) \qquad \text{Factor out the GCF, which is } 8.$$

To check, we multiply: $8(m - 6) = 8 \cdot m - 8 \cdot 6 = 8m - 48$. Since we obtain the original polynomial, $8m - 48$, the factorization is correct.

b. First, find the GCF of $35a^3b^2$ and $14a^2b^3$.

$$\left. \begin{array}{l} 35a^3b^2 = 5 \cdot \mathbf{7} \cdot \mathbf{a} \cdot \mathbf{a} \cdot a \cdot \mathbf{b} \cdot \mathbf{b} \\ 14a^2b^3 = 2 \cdot \mathbf{7} \cdot \mathbf{a} \cdot \mathbf{a} \cdot \mathbf{b} \cdot \mathbf{b} \cdot b \end{array} \right\} \quad \text{The GCF is } 7a^2b^2.$$

Now, we write $35a^3b^2$ and $14a^2b^3$ as the product of the GCF, $7a^2b^2$, and one other factor.

$$35a^3b^2 + 14a^2b^3 = \mathbf{7a^2b^2} \cdot 5a + \mathbf{7a^2b^2} \cdot 2b$$
$$= \mathbf{7a^2b^2}(5a + 2b) \qquad \text{Factor out the GCF, } 7a^2b^2.$$

We check by multiplying: $7a^2b^2(5a + 2b) = 35a^3b^2 + 14a^2b^3$.

Self Check 3   Factor: **a.** $6f - 36$ and   **b.** $48s^2t^2 + 84s^3t$.

EXAMPLE 4

ELEMENTARY &
INTERMEDIATE
Algebra *f(x)* **Now**™

Factor: $3x^4 - 5x^3 + x^2$.

Solution   The polynomial has three terms. We factor out the GCF, which is x^2.

$$3x^4 - 5x^3 + x^2 = \mathbf{x^2}(3x^2) - \mathbf{x^2}(5x) + \mathbf{x^2}(\mathbf{1})$$

The last term has an implied coefficient of 1.

$$= \mathbf{x^2}(3x^2 - 5x + \mathbf{1})$$

Factor out the GCF.

We check by multiplying: $x^2(3x^2 - 5x + 1) = 3x^4 - 5x^3 + x^2$.

Caution

Sometimes, the GCF of the terms of a polynomial is the same as one of the terms.

Self Check 4   Factor: $y^6 - 10y^4 - y^3$.

EXAMPLE 5

Crayons.   The amount of wax used to make the crayon shown in the illustration can be found by computing its volume using the formula

$$V = \pi r^2 H + \frac{1}{3}\pi r^2 h$$

Factor the expression on the right-hand side.

Solution   Each term on the right-hand side of the formula contains a factor of π and r^2.

$$V = \boldsymbol{\pi r^2}H + \frac{1}{3}\boldsymbol{\pi r^2}h$$

$$= \pi r^2\left(H + \frac{1}{3}h\right)$$

Factor out πr^2.

The formula can be expressed as $V = \pi r^2\left(H + \frac{1}{3}h\right)$.

It is often useful to factor out a common factor having a negative coefficient.

EXAMPLE 6

Factor -1 out of $-a^3 + 2a^2 - 4$.

Solution   First, we write each term of the polynomial as the product of -1 and another factor. Then we factor out the common factor, -1.

Success Tip

The result of Example 6 suggests a quick way to factor out -1. Simply change the sign of each term of $-a^3 + 2a^2 - 4$ and write a $-$ symbol in front of the parentheses.

$$-a^3 + 2a^2 - 4 = (\mathbf{-1})a^3 + (\mathbf{-1})(-2a^2) + (\mathbf{-1})4$$

$$= \mathbf{-1}(a^3 - 2a^2 + 4)$$

Factor out -1.

$$= -(a^3 - 2a^2 + 4)$$

The coefficient of 1 need not be written.

We check by multiplying: $-(a^3 - 2a^2 + 4) = -a^3 + 2a^2 - 4$.

Self Check 6   Factor -1 out of $-b^4 - 3b^2 + 2$.

EXAMPLE 7

Factor out the opposite of the GCF in $-20m + 30$.

ELEMENTARY & INTERMEDIATE
Algebra $f(x)$ **Now**™

Solution   The GCF is 10. To factor out its opposite, we write each term of the polynomial as the product of -10 and another factor. Then we factor out -10.

$$-20m + 30 = (\mathbf{-10})(2m) + (\mathbf{-10})(-3)$$
$$= \mathbf{-10}(2m - 3)$$

We check by multiplying: $-10(2m - 3) = -20m + 30$.

Self Check 7   Factor out the opposite of the GCF in $-44c + 55$.

EXAMPLE 8   Factor: $x(x + 4) + 3(x + 4)$.

Solution   The polynomial has two terms: $\underline{x(x + 4)} + \underline{3(x + 4)}$.

The first term   The second term

The GCF of the terms is the binomial $x + 4$, which can be factored out.

$$x(x + 4) + 3(x + 4) = (\mathbf{x + 4})x + (\mathbf{x + 4})3 \qquad \text{Use the commutative property of multiplication twice to reorder the factors.}$$

$$= (\mathbf{x + 4})(x + 3) \qquad \text{Factor out the common factor, which is } (x + 4).$$

Self Check 8   Factor: $2y(y - 1) + 7(y - 1)$.

■ FACTORING BY GROUPING

To factor the polynomial

$$2x^3 + x^2 + 12x + 6$$

We note that no factor, other than 1, is common to all four terms. However, there is a common factor of x^2 in $2x^3 + x^2$ and a common factor of 6 in $12x + 6$.

$$2x^3 + x^2 = x^2(2x + 1) \qquad \text{and} \qquad 12x + 6 = 6(2x + 1)$$

Caution

Factoring by grouping can be attempted on any polynomial with four or more terms. However, not every such polynomial can be factored in this way.

We now see that $2x^3 + x^2$ and $12x + 6$ have a common factor of $2x + 1$, which can be factored out.

$$2x^3 + x^2 + 12x + 6 = x^2(\mathbf{2x + 1}) + 6(\mathbf{2x + 1}) \qquad \text{Factor } 2x^3 + x^2 \text{ and } 12x + 6.$$

$$= (\mathbf{2x + 1})(x^2 + 6) \qquad \text{Factor out } 2x + 1.$$

This type of factoring is called **factoring by grouping.**

Factoring by Grouping   1. Group the terms of the polynomial so that the first two terms have a common factor and the last two terms have a common factor.
2. Factor out the common factor from each group.
3. Factor out the resulting common binomial factor. If there is no common binomial factor, regroup the terms of the polynomial and repeat steps 2 and 3.

EXAMPLE 9

ELEMENTARY &
INTERMEDIATE
Algebra *f(x)* **Now**™

Solution

Factor: **a.** $2c - 2d + cd - d^2$ and   **b.** $x^2 - ax - x + a$.

a. The first two terms have a common factor of 2 and the last two terms have a common factor of d.

$$2c - 2d + cd - d^2 = 2(c - d) + d(c - d)$$ Factor out 2 from $2c - 2d$ and d from $cd - d^2$.

$$= (c - d)(2 + d)$$ Factor out the common binomial factor, $c - d$.

Caution

When factoring polynomials such as the one in Example 9a, don't think that $2(c - d) + d(c - d)$ is in factored form. It is a sum of two terms. To be in factored form, the result must be a product.

We check by multiplying:

$$(c - d)(2 + d) = 2c + cd - 2d - d^2$$
$$= 2c - 2d + cd - d^2$$ Rearrange the terms to get the original polynomial.

b. Since x is a common factor of the first two terms, we can factor it out and proceed as follows.

$$x^2 - ax - x + a = x(x - a) - x + a$$ Factor out x from $x^2 - ax$.

If we factor -1 from $-x + a$, a common binomial factor $(x - a)$ appears, which we can factor out.

$$x^2 - ax - x + a = x(x - a) - 1(x - a)$$
$$= (x - a)(x - 1)$$ Factor out the common factor of $x - a$.

Check by multiplying.

Self Check 9   Factor: **a.** $7x - 7y + xy - y^2$ and   **b.** $b^2 - bc - b + c$.

The next example illustrates that when factoring a polynomial, we should always look for a common factor first.

EXAMPLE 10

ELEMENTARY &
INTERMEDIATE
Algebra *f(x)* **Now**™

Solution

Factor: $10k + 10m - 2km - 2m^2$.

Since the four terms have a common factor of 2, we factor it out first. Then we use factoring by grouping to factor the polynomial within the parentheses. The first two terms have a common factor of 5. The last two terms have a common factor of $-m$.

$$10k + 10m - 2km - 2m^2 = 2(5k + 5m - km - m^2)$$ Factor out the GCF, 2.
$$= 2[5(k + m) - m(k + m)]$$
$$= 2[(k + m)(5 - m)]$$ Factor out $k + m$.
$$= 2(k + m)(5 - m)$$

Check by multiplying.

Self Check 10   Factor: $-4t - 4s - 4tz - 4sz$.

5.1   STUDY SET   ELEMENTARY & INTERMEDIATE Algebra $f(x)$ Now™

VOCABULARY   Fill in the blanks.

1. The letters GCF stand for _____ _____
_____.

2. In the multiplication $2x(x + 7) = 2x^2 + 14x$, the
_____ are $2x$ and $x + 7$ and the product is
$2x^2 + 14x$.

3. When we write 24 as $2 \cdot 2 \cdot 2 \cdot 3$, we say that 24 has
been written as a product of _____.

4. When we write $2x + 4$ as $2(x + 2)$, we say that we
have _____ _____ the GCF, 2.

5. To factor $m^3 + 3m^2 + 4m + 12$ by _____, we
begin by writing $m^2(m + 3) + 4(m + 3)$.

6. The terms $x(x - 1)$ and $4(x - 1)$ have a common
_____ factor, $x - 1$.

CONCEPTS

7. Complete each prime factorization.
   a. $15 = \boxed{} \cdot 5$ **b.** $8 = 2 \cdot 2 \cdot \boxed{}$
   c. $36 = 2 \cdot 2 \cdot \boxed{} \cdot 3$ **d.** $98 = 2 \cdot 7 \cdot \boxed{}$

8. Find the GCF of $30x^2$ and $105x^3$.

$$30x^2 = 2 \cdot 3 \cdot 5 \cdot x \cdot x$$
$$105x^3 = 3 \cdot 5 \cdot 7 \cdot x \cdot x \cdot x$$

9. Find the GCF of $12a^2b^2$, $15a^3b$, and $75a^4b^2$.

$$12a^2b^2 = 2 \cdot 2 \cdot 3 \cdot a \cdot a \cdot b \cdot b$$
$$15a^3b = 3 \cdot 5 \cdot a \cdot a \cdot a \cdot b$$
$$75a^4b^2 = 3 \cdot 5 \cdot 5 \cdot a \cdot a \cdot a \cdot a \cdot b \cdot b$$

10. Write a binomial such that the GCF of its terms is $2x^2$.

11. a. What property is illustrated here?

$$2x(x - 3) = 2x^2 - 6x$$

b. Fill in the blank: $2x^2 - 6x = 2x \cdot x - 2x \cdot 3$
$$= \boxed{} (x - 3)$$

12. Is $3y(3y^2 + 2y - 5)$ the factorization of
$9y^3 + 5y^2 - 15y$? Explain.

13. Explain the error in each solution.
   a. Factor out the GCF: $30a^3 - 12a^2 = 6a(5a^2 - 2a)$.

   b. Factor: $6a + 9b + 3 = 3(2a + 3b + 0)$
   $$= 3(2a + 3b)$$

14. Consider the polynomial $2k - 8 + hk - 4h$.
   a. How many terms does the polynomial have?
   b. Is there a common factor of all the terms?
   c. What is the common factor of the first two
   terms?
   d. What is the common factor of the last two
   terms?

15. Is $5(c - d) + r(c - d)$ in factored form?
Explain.

16. What is the first step in factoring
$8y^2 - 16yz - 6y + 12z$?

NOTATION   Complete each factorization.

17. $40m^4 - 8m^3 + 32m = \boxed{} (5m^3 - m^2 + \boxed{})$

18. $b^3 - 6b^2 + 2b - 12 = \boxed{} (b - 6) + \boxed{} (b - 6)$
$$= (\boxed{})(b^2 + 2)$$

PRACTICE   Find the prime factorization of each number.

19. 12	**20.** 24
21. 40	**22.** 62
23. 98	**24.** 112
25. 225	**26.** 288

Find the GCF of each list.

27. $18, 24$	**28.** $60, 72$
29. m^4, m^3	**30.** c^2, c^7
31. $20c^2, 12c$	**32.** $18r, 27r^3$
33. $6m^4n, 12m^3n^2, 9m^3n^3$	**34.** $15cd^4, 10cd, 40cd^3$

Fill in the blanks.

35. $6x = 3(\quad)$

36. $9y = 3(\quad)$

37. $24y^2 = 8y \cdot \quad$

38. $35h^2 = 7h \cdot \quad$

39. $30t^3 = 15t(\quad)$

40. $54h^3 = 6h^2(\quad)$

41. $32x^3z^4 = 4xz^2(\quad)$

42. $48t^2w^5 = 8tw^2(\quad)$

Factor out the GCF.

43. $3x + 6$

44. $2y - 10$

45. $18m - 9$

46. $24s + 8$

47. $18x + 24$

48. $15s - 35$

49. $d^2 - 7d$

50. $a^2 + 9a$

51. $14c^3 + 63$

52. $33h^4 - 22$

53. $12x^2 - 6x - 24$

54. $27a^2 - 9a + 45$

55. $t^3 + 2t^2$

56. $b^3 - 3b^2$

57. $ab + ac - ad$

58. $rs - rt + ru$

59. $a^3 - a^2$

60. $r^3 + r^2$

61. $24x^2y^3 + 8xy^2$

62. $3x^2y^3 - 9x^4y^3$

63. $12uvw^3 - 18uv^2w^2$

64. $14xyz - 16x^2y^2z$

65. $12r^2 - 3rs + 9r^2s^2$

66. $6a^2 - 12a^3b + 36ab$

67. $\pi R^2 - \pi ab$

68. $\dfrac{1}{3}\pi R^2 h - \dfrac{1}{3}\pi rh$

69. $3(x + 2) - x(x + 2)$

70. $t(5 - s) + 4(5 - s)$

71. $h^2(14 + r) + 14 + r$

72. $k^2(14 + v) - 7(14 + v)$

Factor out −1 from each polynomial.

73. $-a - b$

74. $-x - 2y$

75. $-2x + 5y$

76. $-3x + 8z$

77. $-3r + 2s - 3$

78. $-6yz + 12xz - 5xy$

Factor each polynomial by factoring out the opposite of the GCF.

79. $-3x^2 - 6x$

80. $-4a^2 + 6a$

81. $-4a^2b^3 + 12a^3b^2$

82. $-25x^4y^3 + 30x^2y^3$

83. $-4a^2b^2c^2 + 14a^2b^2c - 10ab^2c^2$

84. $-10x^4y^3z^2 + 8x^3y^2z - 20x^2y$

Factor by grouping.

85. $2x + 2y + ax + ay$

86. $bx + bz + 5x + 5z$

87. $9p - 9q + mp - mq$

88. $xr + xs + yr + ys$

89. $pm - pn + qm - qn$

90. $2ax + 2bx + 3a + 3b$

91. $2xy - 3y^2 + 2x - 3y$

92. $2ab + 2ac + b + c$

93. $ax + bx - a - b$

94. $2xy + y^2 - 2x - y$

Factor by grouping. Remember to factor out the GCF first.

95. $ax^3 + bx^3 + 2ax^2y + 2bx^2y$

96. $x^3y^2 - 2x^2y^2 + 3xy^2 - 6y^2$

97. $2x^3z - 4x^2z + 32xz - 64z$

98. $4a^2b + 12a^2 - 8ab - 24a$

Factor each polynomial.

99. $x(y + 9) - 11(y + 9)$

100. $2\pi R - 2\pi r$

101. $9mp + 3mq - 3np - nq$

102. $-4abc - 4ac^2 + 2bc + 2c^2$

103. $25x^5y^7z^3 - 45x^3y^2z^6$

104. $(t^3 + 7)x + (t^3 + 7)y$

105. $24m - 12n + 16$

106. $6x^2 - 2x - 15x + 5$

107. $-60P^2 - 80P$

108. $\dfrac{1}{3}a^2 - \dfrac{2}{3}a$

APPLICATIONS

109. REARVIEW MIRRORS   The dimensions of the three rearview mirrors on an automobile are given in the illustration on the next page. Write an algebraic expression that gives

   a. the area of the rearview mirror mounted on the windshield.

   b. the total area of the two side mirrors.

c. the total area of all three mirrors. Express the result in factored form.

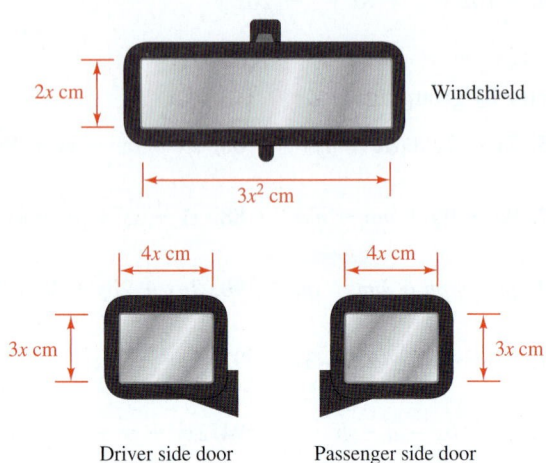

Windshield

$2x$ cm

$3x^2$ cm

$4x$ cm $4x$ cm

$3x$ cm $3x$ cm

Driver side door   Passenger side door

110. COOKING

a. What is the length of a side of the square griddle, in terms of r? What is the area of the cooking surface of the griddle, in terms of r?

b. How many square inches of the cooking surface do the pancakes cover, in terms of r?

c. Find the amount of cooking surface that is not covered by the pancakes. Express the result in factored form.

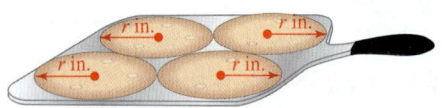

r in. r in.

r in. r in.

111. AIRCRAFT CARRIERS   The rectangular-shaped landing area of $(x^3 + 4x^2 + 5x + 20)$ ft^2 is shaded. The dimensions of the landing area can be found by factoring. What are the length and width of the landing area?

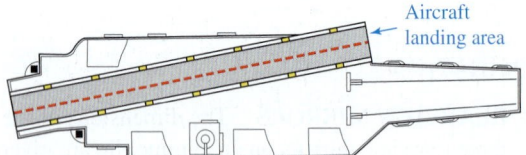

Aircraft landing area

112. INTERIOR DECORATING   The expression $\pi rs + \pi Rs$ can be used to find the amount of material needed to make the lamp shade shown. Factor the expression.

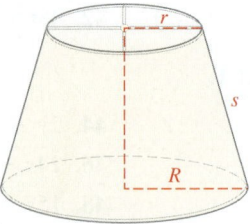

r

s

R

WRITING

113. Explain how to find the GCF of $32a^3$, $16a^2$, and $24a^3$.

114. Explain this diagram.

Multiplication $\longrightarrow$
$$3x^2(5x^2 - 6x + 4) = 15x^4 - 18x^3 + 12x^2$$
$\longleftarrow$ *Factoring*

115. Explain how factorizations of polynomials are checked.

116. What is a factor? What does it mean to factor?

REVIEW

117. INSURANCE COSTS   A college student's good grades earned her a student discount on her car insurance premium. What was the percent of decrease, to the nearest percent, if her annual premium was lowered from $1,050 to $925?

118. CALCULATING GRADES   A student has test scores of 68%, 75%, and 79% in a government class. What must she score on the last exam to earn a B (80% or better) in the course?

CHALLENGE PROBLEMS

119. Factor: $6x^{4m}y^n + 21x^{3m}y^{2n} - 15x^{2m}y^{3n}$.

120. Factor $ax + ay + bx + by$ by grouping the first two terms and the last two terms. Then rearrange two of the terms of $ax + ay + bx + by$ and factor that polynomial by grouping. Do the results agree?

5.2 # Factoring Trinomials of the Form $x^2 + bx + c$

- Factoring Trinomials with a Lead Coefficient of 1   • Factoring Out the GCF
- Prime Polynomials   • The Grouping Method

We have learned how to multiply binomials. For example, to multiply $x + 1$ and $x + 2$, we proceed as follows.

$$(x + 1)(x + 2) = x^2 + 2x + x + 2$$
$$= x^2 + 3x + 2$$

To factor the trinomial $x^2 + 3x + 2$, we will reverse the multiplication process and determine what factors were multiplied to obtain this result. This process is called *factoring the trinomial.* Since the product of two binomials is often a trinomial, many trinomials factor into two binomials.

Multiplication ⎯⎯⎯⎯⎯⎯⎯⎯⎯⎯→

$$(x + 1)(x + 2) = x^2 + 3x + 2$$

←⎯⎯⎯⎯⎯⎯⎯⎯⎯⎯ *Factoring*

To begin the discussion of trinomial factoring, we consider trinomials of the form $x^2 + bx + c$, such as

$$x^2 + 8x + 15, \qquad y^2 - 13y + 12, \qquad a^2 + a - 20, \qquad \text{and} \qquad z^2 - 20z - 21$$

In each case, the **lead coefficient**—the coefficient of the squared variable—is an implied 1.

■ FACTORING TRINOMIALS WITH A LEAD COEFFICIENT OF 1

To develop a method for factoring trinomials, we will find the product of $x + 6$ and $x + 4$ and make some observations about the result.

$$(x + 6)(x + 4) = x \cdot x + 4x + 6x + 6 \cdot 4 \qquad \text{Use the FOIL method to multiply.}$$
$$= x^2 + 10x + 24$$

First term ⎤   Middle term   ⎣ Last term

The Language of Algebra

If a term of a trinomial is a number only, it is called a *constant term.*

$$x^2 + 10x + 24$$

Constant term

The result is a trinomial, where

- the first term, x^2, is the product of x and x
- the last term, 24, is the product of 6 and 4
- the coefficient of the middle term, 10, is the sum of 6 and 4

These observations suggest a strategy to use to factor trinomials with a lead coefficient of 1.

EXAMPLE 1

Factor: $x^2 + 8x + 15$.

Solution   We assume that $x^2 + 8x + 15$ factors as the product of two binomials, and we represent the binomials using two sets of parentheses. Since the first term of the trinomial is x^2, we enter x and x as the first terms of the binomial factors.

$$x^2 + 8x + 15 = \left(x \;\boxed{}\;\right)\left(x \;\boxed{}\;\right) \qquad \text{Because } x \cdot x \text{ will give } x^2$$

The second terms of the binomials must be two integers whose product is 15 and whose sum is 8. Since the integers must have a positive product and a positive sum, we consider only pairs of positive integer factors of 15. The only such pairs, $1 \cdot 15$ and $3 \cdot 5$, are listed in the table. Then we find the sum of each pair and enter each result in the table.

Positive factors of 15	Sum of the factors of 15
$1 \cdot 15 = 15$	$1 + 15 = 16$
$\mathbf{3 \cdot 5 = 15}$	$\mathbf{3 + 5 = 8}$

List all of the pairs of positive integers that multiply to give 15. ⟶   Add each pair of factors.

Notation

By the commutative property of multiplication, the binomial factors of a trinomial can be written in either order. For Example 1, we can write:

$x^2 + 8x + 15 = (x + 5)(x + 3)$

The second row of the table contains the correct pair of integers 3 and 5, whose product is 15 and whose sum is 8. To complete the factorization, we enter 3 and 5 as the second terms of the binomial factors.

$$x^2 + 8x + 15 = (x + 3)(x + 5)$$

We check by multiplying: $(x + 3)(x + 5) = x^2 + 5x + 3x + 15$
$$= x^2 + 8x + 15$$

Self Check 1   Factor: $y^2 + 7y + 10$.

EXAMPLE 2

Factor: $y^2 - 13y + 12$.

ELEMENTARY &
INTERMEDIATE
Algebra $f(x)$ **Now**™

Solution   Again, we assume that the trinomial factors as the product of two binomials. Since the first term of the trinomial is y^2, the first term of each binomial factor must be y.

$$y^2 - 13y + 12 = \left(y \;\boxed{}\;\right)\left(y \;\boxed{}\;\right) \qquad \text{Because } y \cdot y \text{ will give } y^2$$

The second terms of the binomials must be two integers whose product is 12 and whose sum is -13. Since the integers must have a positive product and a negative sum, we only consider pairs of negative integer factors of 12. The possible pairs are listed in the table.

Negative factors of 12	Sum of the factors of 12
$\mathbf{-1(-12) = 12}$	$\mathbf{-1 + (-12) = -13}$
$-2(-6) = 12$	$-2 + (-6) = -8$
$-3(-4) = 12$	$-3 + (-4) = -7$

The first row of the table contains the correct pair of integers -1 and -12, whose product is 12 and whose sum is -13. To complete the factorization, we enter -1 and -12 as the second terms of the binomial factors.

$$y^2 - 13y + 12 = (y - \mathbf{1})(y - \mathbf{12})$$

We check by multiplying: $(y - 1)(y - 12) = y^2 - y - 12y + 12$
$$= y^2 - 13y + 12$$

Self Check 2   Factor: $p^2 - 6p + 8$.

EXAMPLE 3   Factor: $a^2 + a - 20$.

Solution   Since the first term of the trinomial is a^2, the first term of each binomial factor must be a.

$$a^2 + a - 20 = \left(a \;\boxed{}\right)\left(a \;\boxed{}\right) \qquad \text{Because } a \cdot a \text{ will give } a^2$$

To determine the second terms of the binomials, we must find two integers whose product is -20 and whose sum is 1. Because the integers must have a negative product, their signs must be different. The possible pairs are listed in the table.

Factors of -20	Sum of the factors of -20
$1(-20) = -20$	$1 + (-20) = -19$
$2(-10) = -20$	$2 + (-10) = -8$
$4(-5) = -20$	$4 + (-5) = -1$
$\mathbf{5(-4) = -20}$	$\mathbf{5 + (-4) =}$ 1
$10(-2) = -20$	$10 + (-2) =$ 8
$20(-1) = -20$	$20 + (-1) =$ 19

The Language of Algebra

Make sure you understand the following vocabulary: *Many trinomials factor as the product of two binomials.*

Trinomial   Product of two binomials

$$a^2 + a - 20 = (a + 5)(a - 4)$$

The fourth row of the table contains the correct pair of integers 5 and -4, whose product is -20 and whose sum is 1. To complete the factorization, we enter 5 and -4 as the second terms of the binomial factors.

$$a^2 + a - 20 = (a + \mathbf{5})(a - \mathbf{4})$$

Check by multiplying.

Self Check 3   Factor: $m^2 + m - 42$.

EXAMPLE 4   Factor: $z^2 - 4z - 21$.

Solution   Since the first term of the trinomial is z^2, the first term of each binomial factor must be z.

$$z^2 - 4z - 21 = \left(z \;\boxed{}\right)\left(z \;\boxed{}\right) \qquad \text{Because } z \cdot z \text{ will give } z^2$$

To determine the second terms of the binomials, we must find two integers whose product is -21 and whose sum is -4. Because the integers must have a negative product, their signs must be different. The possible pairs are listed in the table.

Factors of -21	Sum of the factors of -21
$1(-21) = -21$	$1 + (-21) = -20$
$3(-7) = -21$	$3 + (-7) = -4$
$7(-3) = -21$	$7 + (-3) = 4$
$21(-1) = -21$	$21 + (-1) = 20$

The second row of the table contains the correct pair of integers 3 and -7, whose product is -21 and whose sum if -4. To complete the factorization, enter 3 and -7 as the second terms of the binomial factors.

$$z^2 - 4z - 21 = (z + 3)(z - 7)$$

Check by multiplying.

Self Check 4 Factor: $q^2 - 2q - 24$.

The following guidelines are helpful when factoring trinomials.

Factoring Trinomials with a Lead Coefficient of 1

To factor a trinomial of the form $x^2 + bx + c$, find two numbers whose product is c and whose sum is b.

1. If c is positive, the numbers have the same sign.
2. If c is negative, the numbers have different signs.

Then write the trinomial as a product of two binomials. You can check by multiplying.

The product of these numbers must be c.

$$x^2 + bx + c = \left(x \;\boxed{}\right)\left(x \;\boxed{}\right)$$

The sum of these numbers must be b.

EXAMPLE 5

Factor: $-h^2 + 2h + 63$.

Solution Because it is easier to factor trinomials that have a positive lead coefficient, we factor out -1. Then we factor the resulting trinomial.

$$\begin{aligned}
-h^2 + 2h + 63 &= -1(h^2 - 2h - 63) &&\text{Factor out } -1. \\
&= -(h^2 - 2h - 63) &&\text{The coefficient of 1 need not be written.} \\
&= -(h + 7)(h - 9) &&\text{Factor } h^2 - 2h - 63.
\end{aligned}$$

We check by multiplying:

$$\begin{aligned}
-(h + 7)(h - 9) &= -(h^2 - 9h + 7h - 63) &&\text{Multiply the binomials first.} \\
&= -(h^2 - 2h - 63) &&\text{Combine like terms.} \\
&= -h^2 + 2h + 63 &&\text{Remove parentheses.}
\end{aligned}$$

Self Check 5 Factor: $-x^2 + 11x - 28$.

EXAMPLE 6

ELEMENTARY & INTERMEDIATE
Algebra (f(x)) **Now**™

Factor: $x^2 - 4xy - 5y^2$.

Solution   Since the first term of the trinomial is x^2, the first term of each binomial factor must be x. Since the third term contains y^2, the last term of each binomial factor must contain y. To complete the factorization, we need to determine the coefficient of each y-term.

$$x^2 - 4xy - 5y^2 = \left(x\ \boxed{}\,y\right)\left(x\ \boxed{}\,y\right)$$

The coefficients of y must be two integers whose product is -5 and whose sum is -4. Such a pair is 1 and -5. Instead of writing the first factor as $(x + 1y)$, we write it as $(x + y)$, because $1y = y$.

$$x^2 - 4xy - 5y^2 = (x + y)(x - 5y)$$

Check: $(x + y)(x - 5y) = x^2 - 5xy + xy - 5y^2$
$$= x^2 - 4xy - 5y^2$$

Self Check 6   Factor: $s^2 + 6st - 7t^2$.

■ FACTORING OUT THE GCF

If the terms of a trinomial have a common factor, it should be factored out first. A trinomial is **factored completely** when no factor can be factored further.

EXAMPLE 7

ELEMENTARY & INTERMEDIATE
Algebra (f(x)) **Now**™

Factor completely: $2x^4 + 26x^3 + 80x^2$.

Solution   We begin by factoring out the GCF $2x^2$.

Caution

When prime factoring 30, you wouldn't stop here because 6 can be factored:

```
      30
     /  \
    5    6
```

When factoring polynomials, make sure that no factor can be factored further.

$$2x^4 + 26x^3 + 80x^2 = 2x^2(x^2 + 13x + 40)$$

Next, we factor $x^2 + 13x + 40$. The integers 8 and 5 have a product of 40 and a sum of 13, so the completely factored form of the given trinomial is

$$2x^4 + 26x^3 + 80x^2 = 2x^2(x + 8)(x + 5)$$

Check by multiplying.

Self Check 7   Factor completely: $4m^5 + 8m^4 - 32m^3$.

EXAMPLE 8

Factor completely: $-13g^2 + 36g + g^3$.

Solution   Before factoring the trinomial, we write its terms in descending powers of g.

Caution

For multistep factorizations, don't forget to write the GCF in the final factored form.

$$\begin{aligned}
-13g^2 + 36g + g^3 &= g^3 - 13g^2 + 36g && \text{Rearrange the terms.} \\
&= g(g^2 - 13g + 36) && \text{Factor out } g. \\
&= g(g - 9)(g - 4) && \text{Factor the trinomial.}
\end{aligned}$$

Check by multiplying.

Self Check 8   Factor completely: $-12t + t^3 + 4t^2$.

■ PRIME POLYNOMIALS

If a trinomial with integer coefficients cannot be factored using only integers, it is called a **prime trinomial.**

EXAMPLE 9

ELEMENTARY &
INTERMEDIATE
Algebra *f(x)* **Now**™

Factor $x^2 + 2x + 3$, if possible.

Solution   To factor the trinomial, we must find two integers whose product is 3 and whose sum is 2. The possible factorizations are shown in the table.

Factors of 3	Sum of the factors of 3
$1(3) = 3$	$1 + 3 = \quad 4$
$-1(-3) = 3$	$-1 + (-3) = -4$

The Language of Algebra

When a trinomial is not factorable using only integers, we say it is *prime* and that it does not factor *over* the integers.

Since there are no two integers whose product is 3 and whose sum is 2, $x^2 + 2x + 3$ cannot be factored. It is a prime trinomial.

Self Check 9   Factor $x^2 - 4x + 6$, if possible.

■ THE GROUPING METHOD

Another way to factor trinomials of the form $x^2 + bx + c$ is to write them as equivalent four-termed polynomials and factor by grouping. To factor $x^2 + 8x + 15$ using this method, we proceed as follows.

1. First, identify b as the coefficient of the x-term, and c as the last term. For trinomials of the form $x^2 + bx + c$, we call c the **key number.**

$$\left.\begin{array}{c} x^2 + bx + c \\ \downarrow \quad\quad \downarrow \\ x^2 + 8x + 15 \end{array}\right\} b = 8 \text{ and } c = 15$$

2. Now find two integers whose product is the key number, 15, and whose sum is $b = 8$. Since the integers must have a positive product and a positive sum, we consider only positive factors of 15.

Key number = 15

Positive factors of 15	Sum of the factors of 15
$1 \cdot 15 = 15$	$1 + 15 = 16$
$3 \cdot 5 = 15$	$3 + 5 = \quad 8$

The second row of the table contains the correct pair of integers 3 and 5, whose product is 15 and whose sum is 8.

3. Use the integers 3 and 5 as coefficients of two terms to be placed between x^2 and 15.

$$x^2 + \mathbf{8x} + 15 = x^2 + \mathbf{3x + 5x} + 15 \qquad \text{Express } 8x \text{ as } 3x + 5x.$$

4. Factor the four-termed polynomial by grouping:

$$x^2 + 3x + 5x + 15 = x(x + 3) + 5(x + 3) \qquad \begin{array}{l}\text{Factor } x \text{ out of } x^2 + 3x \text{ and } 5 \text{ out of} \\ 5x + 15.\end{array}$$

$$= (x + 3)(x + 5) \qquad \text{Factor out } x + 3.$$

Check by multiplying.

The grouping method is an alternative to the method for factoring trinomials discussed earlier in this section. It is especially useful when the constant term, c, has many factors.

Factoring Trinomials of the Form $x^2 + bx + c$ Using Grouping

To factor a trinomial that has a lead coefficient of 1:

1. Identify b and the key number c.
2. Find two numbers whose product is the key number and whose sum is b.
3. Enter the two numbers as coefficients of x in the form shown below. Then factor the polynomial by grouping.

The product of these numbers must be c.

$$x^2 + \boxed{}x + \boxed{}x + c$$

The sum of these numbers must be b.

4. Check using multiplication.

EXAMPLE 10

ELEMENTARY & INTERMEDIATE
Algebra $\widehat{f(x)}$ **Now** ™

Solution

Factor: $a^2 + a - 20$.

Since $a^2 + a - 20 = a^2 + 1a - 20$, we identify b as 1 and the key number c as -20. We must find two integers whose product is -20 and whose sum is 1. Since the integers must have a negative product, their signs must be different.

Key number $= -20$

Factors of -20	Sum of the factors of -20
$1(-20) = -20$	$1 + (-20) = -19$
$2(-10) = -20$	$2 + (-10) = -8$
$4(-5) = -20$	$4 + (-5) = -1$
$\mathbf{5(-4)} = -20$	$\mathbf{5 + (-4)} = 1$
$10(-2) = -20$	$10 + (-2) = 8$
$20(-1) = -20$	$20 + (-1) = 19$

The fourth row of the table contains the correct pair of integers 5 and -4, whose product is -20 and whose sum is 1. They serve as the coefficients of $5a$ and $-4a$ that we place between a^2 and -20.

$$a^2 + a - 20 = a^2 + 5a - 4a - 20 \qquad \text{Express } a \text{ as } 5a - 4a.$$
$$= a(a + 5) - 4(a + 5) \qquad \text{Factor } a \text{ out of } a^2 + 5a \text{ and } -4 \text{ out of } -4a - 20.$$
$$= (a + 5)(a - 4) \qquad \text{Factor out } a + 5.$$

Check by multiplying.

Self Check 10   Factor: $m^2 + m - 42$.

EXAMPLE 11

ELEMENTARY &
INTERMEDIATE
Algebra $f(x)$ **Now**™

Factor: $2x^3 - 20x^2 + 18x$.

Solution   First, we factor out $2x$.

$$2x^3 - 20x^2 + 18x = 2x(x^2 - 10x + 9)$$

To factor $x^2 - 10x + 9$ by grouping, we must find two integers whose product is the key number 9 and whose sum is $b = -10$. Such a pair is -9 and -1.

$$x^2 - 10x + 9 = x^2 - 9x - 1x + 9 \qquad \text{Express } -10x \text{ as } -9x - 1x.$$
$$= x(x - 9) - 1(x - 9) \qquad \text{Factor } x \text{ out of } x^2 - 9x \text{ and } -1 \text{ out of } -1x + 9.$$
$$= (x - 9)(x - 1) \qquad \text{Factor out } x - 9.$$

The complete factorization of the original trinomial is

$$2x^3 - 20x^2 + 18x = 2x(x - 9)(x - 1)$$

Check by multiplying.

Self Check 11   Factor: $3m^3 - 27m^2 + 24m$.

EXAMPLE 12

Factor: $x^2 - 4xy - 5y^2$.

Solution   We identify b as -4 and the key number c as -5. We must find two integers whose product is -5 and whose sum is -4. Such a pair is -5 and 1. They serve as the coefficients of $-5xy$ and $1xy$ that we place between x^2 and $-5y^2$.

$$x^2 - 4xy - 5y^2 = x^2 - 5xy + 1xy - 5y^2 \qquad \text{Express } -4xy \text{ as } -5xy + 1xy.$$
$$= x(x - 5y) + y(x - 5y) \qquad \text{Factor } x \text{ out of } x^2 - 5xy \text{ and } y \text{ out of } 1xy - 5y^2.$$
$$= (x - 5y)(x + y) \qquad \text{Factor out } x - 5y.$$

Check by multiplying.

Self Check 12   Factor: $q^2 - 2qt - 24t^2$.

Answers to Self Checks   **1.** $(y + 2)(y + 5)$ **2.** $(p - 2)(p - 4)$ **3.** $(m + 7)(m - 6)$ **4.** $(q + 4)(q - 6)$ **5.** $-(x - 4)(x - 7)$ **6.** $(s + 7t)(s - t)$ **7.** $4m^3(m + 4)(m - 2)$ **8.** $t(t - 2)(t + 6)$ **9.** prime trinomial   **10.** $(m + 7)(m - 6)$ **11.** $3m(m - 8)(m - 1)$ **12.** $(q + 4t)(q - 6t)$

5.2   STUDY SET   ELEMENTARY & INTERMEDIATE Algebra $f(x)$ Now™

VOCABULARY   **Fill in the blanks.**

1. A polynomial that has three terms is called a
   _____. A polynomial that has two terms is called
   a _____.

2. The statement $x^2 - x - 12 = (x - 4)(x + 3)$ shows
   that $x^2 - x - 12$ _____ into the product of two
   binomials.

3. Since $10 = (-5)(-2)$, we say -5 and -2 are
   _____ of 10.

4. A _____ polynomial cannot be factored by using
   only integers.

5. The _____ coefficient of the trinomial $x^2 - 3x + 2$
   is 1, the _____ of the middle term is -3, and
   the last _____ is 2.

6. A trinomial is factored _____ when no factor
   can be factored further.

CONCEPTS   **Fill in the blanks.**

7. Two factorizations of 4 that involve only positive
   numbers are _____ and _____. Two factorizations of 4
   that involve only negative numbers are _____ and
   _____ .

8. **a.** Before attempting to factor a trinomial, be sure
   that it is written in _____ powers of a
   variable.

   b. Before attempting to factor a trinomial into two
   binomials, always factor out any _____
   factors first.

9. To factor $x^2 + x - 56$, we must find two integers
   whose _____ is -56 and whose _____ is 1.

10. Two factors of 18 whose sum is -9 are _____ and _____ .

11. $x^2 + 5x + 3$ cannot be factored because we cannot
    find two integers whose product is _____ and whose
    sum is _____ .

12. Complete the table.

Factors of 8	Sum of the factors of 8
$1(8) = 8$	
$2(4) = 8$	
$-1(-8) = 8$	
$-2(-4) = 8$	

13. If we use the FOIL method to do the multiplication
    $(x + 5)(x + 4)$, we obtain $x^2 + 9x + 20$.
    a. What step of the FOIL process produced 20?

    b. What steps of the FOIL process produced $9x$?

14. Find two integers whose
    a. product is 10 and whose sum is 7.

    b. product is 8 and whose sum is -6.

    c. product is -6 and whose sum is 1.

    d. product is -9 and whose sum is -8.

15. Given $x^2 + 8x + 15$:
    a. What is the coefficient of the x^2-term?

    b. What is the last term? What is the coefficient of
    the middle term?

    c. What two integers have a product of 15 and a
    sum of 8?

16. Complete the factorization table.

Factors of -9	Sum of the factors of -9
$1(-9) = -9$	
$3(-3) = -9$	
$-1(9) = -9$	

17. Consider factoring a trinomial of the form
    $x^2 + bx + c$.
    a. If c is positive, what can be said about the two
    integers that should be chosen for the
    factorization?

    b. If c is negative, what can be said about the two
    integers that should be chosen for the
    factorization?

18. What trinomial has the factorization $(x + 8)(x - 2)$?

19. Factor out the GCF: $x(x + 1) + 5(x + 1)$.

20. Factor each polynomial using factoring by grouping.
    a. $x^2 + 2x + 5x + 10$

    b. $y^2 - 4y + 3y - 12$

21. Complete the table.

Key number = 8

Negative factors of 8	Sum of factors of 8
$-1(-8) = 8$	
	$-2 + (-4) = -6$

22. Complete each sentence to explain how to factor $x^2 + 7x + 10$ by grouping.

The product of these numbers must be ▢.

$x^2 + 7x + 10 = x^2 + \Box x + \Box x + 10$

The sum of these numbers must be ▢.

NOTATION

23. To factor a trinomial, a student made a table and circled the correct pair of integers, as shown. Complete the factorization of the trinomial.

$(x \quad)(x \quad)$

Factors	Sum
$1(-6)$	-5
$2(-3)$	-1
$\boxed{3(-2)}$	$\boxed{1}$
$6(-1)$	5

24. To factor a trinomial by grouping, a student made a table and circled the correct pair of integers, as shown. Enter the correct coefficients.

$x^2 + \ x + \ x + 16$

Key number = 16

Factors	Sum
$1 \cdot 16$	17
$\boxed{2 \cdot 8}$	$\boxed{10}$
$4 \cdot 4$	8

PRACTICE Complete each factorization.

25. $x^2 + 3x + 2 = (x \quad 2)(x \quad 1)$
26. $y^2 + 4y + 3 = (y \quad 3)(y \quad 1)$
27. $t^2 - 9t + 14 = (t \quad 7)(t \quad 2)$
28. $c^2 - 9c + 8 = (c \quad 8)(c \quad 1)$
29. $a^2 + 6a - 16 = (a \quad 8)(a \quad 2)$
30. $x^2 - 3x - 40 = (x \quad 8)(x \quad 5)$

Factor each trinomial. If it can't be factored, write "prime."

31. $z^2 + 12z + 11$ **32.** $x^2 + 7x + 10$

33. $m^2 - 5m + 6$ **34.** $n^2 - 7n + 10$
35. $a^2 - 4a - 5$ **36.** $b^2 + 6b - 7$
37. $x^2 + 5x - 24$ **38.** $t^2 - 5t - 50$
39. $a^2 - 10a - 39$ **40.** $r^2 - 9r - 12$
41. $u^2 + 10u + 15$ **42.** $v^2 + 9v + 15$
43. $r^2 - 2r + 4$ **44.** $y^2 - 17y + 72$
45. $r^2 - 9r + 18$ **46.** $u^2 + u - 42$
47. $x^2 + 4xy + 4y^2$ **48.** $m^2 - mn - 12n^2$
49. $a^2 - 4ab - 12b^2$ **50.** $p^2 + pq - 6q^2$
51. $r^2 - 2rs + 4s^2$ **52.** $m^2 + 3mn - 20n^2$

Factor each trinomial. Factor out -1 first.

53. $-x^2 - 7x - 10$ **54.** $-x^2 + 9x - 20$
55. $-t^2 - 15t + 34$ **56.** $-t^2 - t + 30$
57. $-a^2 - 4ab - 3b^2$ **58.** $-a^2 - 6ab - 5b^2$
59. $-x^2 + 6xy + 7y^2$ **60.** $-x^2 - 10xy + 11y^2$

Write each trinomial in descending powers of one variable and factor.

61. $4 - 5x + x^2$ **62.** $y^2 + 5 + 6y$
63. $10y + 9 + y^2$ **64.** $x^2 - 13 - 12x$
65. $r^2 - 2 - r$ **66.** $u^2 - 3 + 2u$
67. $4rx + r^2 + 3x^2$ **68.** $a^2 + 5b^2 + 6ab$

Completely factor each trinomial. Factor out any common monomials first (including -1 if necessary).

69. $2x^2 + 10x + 12$ **70.** $3y^2 - 21y + 18$

71. $-5a^2 + 25a - 30$ **72.** $-2b^2 + 20b - 18$

73. $z^3 - 29z^2 + 100z$ **74.** $m^3 - m^2 - 56m$

75. $12xy + 4x^2y - 72y$ **76.** $48xy + 6xy^2 + 96x$

77. $-r^2 + 14r - 40$ **78.** $-y^2 - 2y + 99$

79. $-13yz + y^2 - 14z^2$ **80.** $2x^2 - 12x + 16$

81. $s^2 + 11s - 26$ **82.** $x^2 + 14x + 45$

83. $a^2 + 10ab + 9b^2$ **84.** $-r^2 + 14r - 45$

85. $-x^2 + 21x + 22$ **86.** $-3ab + a^2 + 2b^2$

87. $d^3 - 11d^2 - 26d$ **88.** $m^2 + 3mn - 10n^2$

APPLICATIONS

89. PETS   The cage shown in the illustration is used for transporting dogs. Its volume is $(x^3 + 12x^2 + 27x)$ in.^{3}. The dimensions of the cage can be found by factoring. If the cage is longer than it is tall and taller than it is wide, determine its length, width, and height.

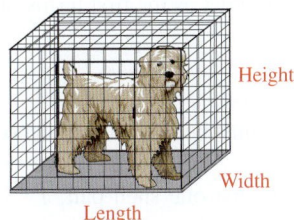
Height
Width
Length

90. PHOTOGRAPHY   A picture cube is a clever way to display 6 photographs in a small amount of space. Suppose the surface area of the cube is given by the polynomial $(6s^2 + 12s + 6)$ in.^{2}. Find the expression that represents the length of an edge of the cube.

WRITING

91. Explain what it means when we say that a trinomial is the product of two binomials. Give an example.

92. Are $2x^2 - 12x + 16$ and $x^2 - 6x + 8$ factored in the same way? Explain.

93. When factoring $x^2 - 2x - 3$, one student got $(x - 3)(x + 1)$, and another got $(x + 1)(x - 3)$. Are both answers acceptable? Explain.

94. In the partial solution shown below, a student began to factor the trinomial. Write a note to the student explaining his initial mistake.

Factor: $x^2 - 2x - 63$.
$(x - \quad)(x - \quad)$
???

95. Explain the error in the following factorization.

$$x^3 + 8x^2 + 15x = x(x^2 + 8x + 15)$$
$$= (x + 3)(x + 5)$$

96. Explain why the factorization is not complete.

$$2y^2 - 12y + 16 = 2(y^2 - 6y + 8)$$

REVIEW   **Simplify each expression. Write each answer without using parentheses or negative exponents.**

97. $\dfrac{x^{12}x^{-7}}{x^3x^4}$ **98.** $\dfrac{a^4a^{-2}}{a^2a^0}$

99. $(x^{-3}x^{-2})^2$ **100.** $\left(\dfrac{18a^2b^3c^{-4}}{3a^{-1}b^2c}\right)^{-3}$

CHALLENGE PROBLEMS   **Factor completely.**

101. $x^2 - \dfrac{6}{5}x + \dfrac{9}{25}$

102. $x^2 - 0.5x + 0.06$

103. $x^{2m} - 12x^m - 45$

104. $x^2(y + 1) - 3x(y + 1) - 70(y + 1)$

105. Find all positive integer values of c that make $n^2 + 6n + c$ factorable.

106. Find all integer values of b that make $x^2 + bx - 44$ factorable.

5.3 Factoring Trinomials of the Form $ax^2 + bx + c$

• The Trial-and-Check Method   • Factoring Out the GCF   • The Grouping Method

In this section we will factor trinomials with lead coefficients other than 1, such as

$$2x^2 + 5x + 3, \qquad 6a^2 - 17a + 5, \qquad \text{and} \qquad 4b^2 + 8bc - 45c^2$$

We can use two methods to factor these trinomials. With the first method, we make educated guesses and then check them with multiplication. The correct factorization is determined through a process of elimination. The second method is an extension of factoring by grouping.

■ THE TRIAL-AND-CHECK METHOD

EXAMPLE 1

ELEMENTARY &
INTERMEDIATE
Algebra $f(x)$ **Now**™

Factor: $2x^2 + 5x + 3$.

Solution   We assume that $2x^2 + 5x + 3$ factors as the product of two binomials. Since the first term of the trinomial is $2x^2$, we enter $2x$ and x as the first terms of the binomial factors.

$$\left(2x \ \boxed{}\right)\left(x \ \boxed{}\right) \qquad \textcolor{red}{\text{Because } 2x \cdot x \text{ will give } 2x^2}$$

The Language of Algebra

To *interchange* means to put each in the place of the other. We create all of the possible factorizations by *interchanging* the second terms of the binomials.

$(2x + 1)(x + 3)$

$(2x + 3)(x + 1)$

The second terms of the binomials must be two integers whose product is 3. Since the coefficients of the terms of $2x^2 + 5x + 3$ are positive, we only consider pairs of positive integer factors of 3. Since there is just one such pair, $1 \cdot 3$, we can enter 1 and 3 as the second terms of the binomials, or we can reverse the order and enter 3 and 1.

$$(2x + 1)(x + 3) \qquad \text{or} \qquad (2x + 3)(x + 1)$$

The first possibility is incorrect, because when we find the outer and inner products and combine like terms, we obtain an incorrect middle term of $7x$.

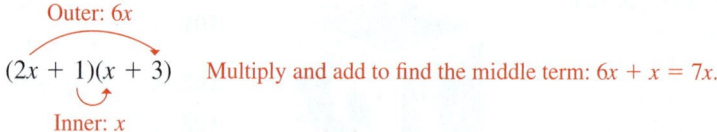

$$\textcolor{red}{\text{Outer: } 6x}$$
$$(2x + 1)(x + 3) \qquad \textcolor{red}{\text{Multiply and add to find the middle term: } 6x + x = 7x.}$$
$$\textcolor{red}{\text{Inner: } x}$$

The second possibility is correct, because it gives a middle term of $5x$.

$$\textcolor{red}{\text{Outer: } 2x}$$
$$(2x + 3)(x + 1) \qquad \textcolor{red}{\text{Multiply and add to find the middle term: } 2x + 3x = 5x.}$$
$$\textcolor{red}{\text{Inner: } 3x}$$

Thus,

$$2x^2 + 5x + 3 = (2x + 3)(x + 1)$$

We check by multiplying: $(2x + 3)(x + 1) = 2x^2 + 2x + 3x + 3$
$$= 2x^2 + 5x + 3$$

Self Check 1   Factor: $2x^2 + 5x + 2$.

EXAMPLE 2

Factor: $6a^2 - 17a + 5$.

Solution   Since the first term is $6a^2$, the first terms of the factors must be $6a$ and a or $3a$ and $2a$.

$$\left(6a \;\boxed{}\right)\left(a \;\boxed{}\right) \text{ or } \left(3a \;\boxed{}\right)\left(2a \;\boxed{}\right)$$ Because $6a \cdot a$ or $3a \cdot 2a$ will give $6a^2$

The second terms of the binomials must be two integers whose product is 5. Since the last term of $6a^2 - 17a + 5$ is positive and the coefficient of the middle term is negative, we only consider negative integer factors of the last term. Since there is just one such pair, $-1(-5)$, we can enter -1 and -5, or we can reverse the order and enter -5 and -1 as second terms of the binomials.

$$\overset{-30a}{(6a - 1)(a - 5)} \underset{-a}{\qquad} -30a - a = -31a.$$

$$\overset{-6a}{(6a - 5)(a - 1)} \underset{-5a}{\qquad} -6a - 5a = -11a.$$

$$\overset{-15a}{\mathbf{(3a - 1)(2a - 5)}} \underset{-2a}{\qquad} -15a - 2a = -17a.$$

$$\overset{-3a}{(3a - 5)(2a - 1)} \underset{-10a}{\qquad} -3a - 10a = -13a.$$

Only the possibility shown in blue gives the correct middle term of $-17a$. Thus,

$$6a^2 - 17a + 5 = (3a - 1)(2a - 5)$$

We check by multiplying: $(3a - 1)(2a - 5) = 6a^2 - 17a + 5$.

Self Check 2   Factor: $6b^2 - 19b + 3$.

EXAMPLE 3

Factor: $3y^2 - 7y - 6$.

Solution   Since the first term is $3y^2$, the first terms of the binomial factors must be $3y$ and y.

$$\left(3y \;\boxed{}\right)\left(y \;\boxed{}\right)$$ Because $3y \cdot y$ will give $3y^2$

The second terms of the binomials must be two integers whose product is -6. There are four such pairs: $1(-6)$, $-1(6)$, $2(-3)$, and $-2(3)$. When these pairs are entered, and then reversed, as second terms of the binomials, there are eight possibilities to consider. Four of them can be discarded because they include a binomial whose terms have a common factor. If $3y^2 - 7y - 6$ does not have a common factor, neither can any of its binomial factors.

For 1 and −6:

$$-18y$$

$$(3y + 1)(y - 6) \qquad (3y - 6)(y + 1)$$

$$y$$

$$-18y + y = -17y \qquad 3y - 6 \text{ has a common factor of 3.}$$

For −1 and 6:

$$18y$$

$$(3y - 1)(y + 6) \qquad (3y + 6)(y - 1)$$

$$-y$$

$$18y - y = 17y \qquad 3y + 6 \text{ has a common factor of 3.}$$

For 2 and −3:

$$-9y$$

$$(3y + 2)(y - 3) \qquad (3y - 3)(y + 2)$$

$$2y$$

$$-9y + 2y = -7y \qquad 3y - 3 \text{ has a common factor of 3.}$$

For −2 and 3:

$$9y$$

$$(3y - 2)(y + 3) \qquad (3y + 3)(y - 2)$$

$$-2y$$

$$9y - 2y = 7y \qquad 3y + 3 \text{ has a common factor of 3.}$$

Only the possibility shown in blue gives the correct middle term of $-7y$. Thus,

$$3y^2 - 7y - 6 = (3y + 2)(y - 3)$$

Check by multiplying.

Self Check 3 Factor: $5t^2 - 23t - 10$.

EXAMPLE 4

ELEMENTARY & INTERMEDIATE Algebra $f(x)$ Now™

Factor: $4b^2 + 8bc - 45c^2$.

Solution Since the first term is $4b^2$, the first terms of the binomial factors must be $4b$ and b or $2b$ and $2b$. Since the last term contains c^2, the second terms of the binomial factors must contain c.

$$\left(4b \;\boxed{}\; c\right)\!\left(b \;\boxed{}\; c\right) \text{ or } \left(2b \;\boxed{}\; c\right)\!\left(2b \;\boxed{}\; c\right)$$

Because $4b \cdot b$ or $2b \cdot 2b$ gives $4b^2$, and because $c \cdot c$ gives c^2

The coefficients of c must be two integers whose product is -45. Since the coefficient of the last term is negative, the signs of the integers must be different. If we pick factors of $4b$ and b for the first terms, and -1 and 45 for the coefficients of c, the multiplication gives an incorrect middle term of $179bc$.

$$180bc$$

$$(4b - c)(b + 45c) \qquad 180bc - bc = 179bc.$$

$$-bc$$

If we pick factors of $4b$ and b for the first terms, and 15 and -3 for the coefficients of c, the multiplication gives an incorrect middle term of $3bc$.

$$\overset{\overset{\displaystyle -12bc}{\frown}}{(4b + 15c)(b - 3c)} \qquad -12bc + 15bc = 3bc.$$
$$\underset{15bc}{}$$

If we pick factors of $2b$ and $2b$ for the first terms, and -5 and 9 for the coefficients of c, we have

$$\overset{\overset{\displaystyle 18bc}{\frown}}{(2b - 5c)(2b + 9c)} \qquad 18bc - 10bc = 8bc.$$
$$\underset{-10bc}{}$$

which gives the correct middle term of $8bc$. Thus,

$$4b^2 + 8bc - 45c^2 = (2b - 5c)(2b + 9c)$$

Check by multiplying.

Self Check 4   Factor: $4x^2 + 4xy - 3y^2$.

Because guesswork is often necessary, it is difficult to give specific rules for factoring trinomials with lead coefficients other than 1. However, the following hints are helpful.

Factoring Trinomials with Lead Coefficients Other Than 1

To factor trinomials with lead coefficients other than 1:

1. Factor out any GCF (including -1 if that is necessary to make $a > 0$ in a trinomial of the form $ax^2 + bx + c$).
2. Write the trinomial as a product of two binomials. The coefficients of the first terms of each binomial factor must be factors of a, and the last terms must be factors of c.

The product of these numbers must be a.

$$ax^2 + bx + c = (\boxed{}x \ \boxed{})(\boxed{}x \ \boxed{})$$

The product of these numbers must be c.

3. If c is positive, the signs within the binomial factors match the sign of b. If c is negative, the signs within the binomial factors are opposites.
4. Try combinations of first terms and second terms until you find the one that gives the proper middle term. If no combination works, the trinomial is prime.
5. Check by multiplying.

■ FACTORING OUT THE GCF

EXAMPLE 5   Factor: $2x^2 - 8x^3 + 3x$.

Solution   We write the trinomial in descending powers of x,

$$-8x^3 + 2x^2 + 3x$$

and we factor out the opposite of the GCF, which is $-x$.

$$-8x^3 + 2x^2 + 3x = -x(8x^2 - 2x - 3)$$

We now factor $8x^2 - 2x - 3$. Its factorization has the form

$$\left(8x \ \boxed{}\right)\left(x \ \boxed{}\right) \text{ or } \left(2x \ \boxed{}\right)\left(4x \ \boxed{}\right) \qquad \textcolor{red}{\text{Because } 8x \cdot x \text{ or } 4x \cdot 2x \text{ gives } 8x^2}$$

The second terms of the binomials must be two integers whose product is -3. There are two such pairs: $1(-3)$ and $-1(3)$. Since the coefficient of middle term $-2x$ is small, we pick the smaller factors of $8x^2$, which are $2x$ and $4x$, for the first terms and 1 and -3 for the second terms.

$$\overset{\textcolor{red}{-6x}}{(2x + 1)(4x - 3)} \qquad \textcolor{red}{-6x + 4x = -2x}$$
$$\textcolor{red}{4x}$$

This factorization gives the correct middle term of $-2x$. Thus,

$$8x^2 - 2x - 3 = (2x + 1)(4x - 3)$$

We can now give the complete factorization of the original trinomial.

$$-8x^3 + 2x^2 + 3x = -x(8x^2 - 2x - 3)$$
$$= -x(2x + 1)(4x - 3)$$

Check by multiplying.

Self Check 5   Factor: $12y - 2y^3 - 2y^2$.

■ THE GROUPING METHOD

Another way to factor a trinomial of the form $ax^2 + bx + c$ is to write it as an equivalent four-termed polynomial and factor it by grouping. For example, to factor $2x^2 + 5x + 3$, we proceed as follows.

1. Identify the values of a, b, and c.

$$\left.\begin{array}{ccc} ax^2 & + bx & + c \\ \downarrow & \downarrow & \downarrow \\ 2x^2 & + 5x & + 3 \end{array}\right\} a = 2, b = 5, \text{ and } c = 3$$

Then, find the product ac, called the **key number:** $ac = 2(3) = 6$.

2. Next, find two numbers whose product is $ac = 6$ and whose sum is $b = 5$. Since the numbers must have a positive product and a positive sum, we consider only positive factors of 6.

Key number $= 6$

Positive factors of 6	Sum of the factors of 6
$1 \cdot 6 = 6$	$1 + 6 = 7$
$2 \cdot 3 = 6$	$2 + 3 = 5$

The second row of the table contains the correct pair of integers 2 and 3, whose product is 6 and whose sum is 5.

3. Use the factors 2 and 3 as coefficients of two terms to be placed between $2x^2$ and 3.

$$2x^2 + \mathbf{5x} + 3 = 2x^2 + \mathbf{2x + 3x} + 3 \qquad \text{Express } 5x \text{ as } 2x + 3x.$$

4. Factor the four-termed polynomial by grouping:

$$2x^2 + 2x + 3x + 3 = 2x(\mathbf{x + 1}) + 3(\mathbf{x + 1}) \qquad \text{Factor } 2x \text{ out of } 2x^2 + 2x \text{ and } 3 \text{ out of } 3x + 3.$$

$$= (\mathbf{x + 1})(2x + 3) \qquad \text{Factor out } x + 1.$$

Check by multiplying.

Factoring by grouping is especially useful when the lead coefficient, a, and the constant term, c, have many factors.

Factoring Trinomials by Grouping

To factor a trinomial by grouping:

1. Factor out any GCF (including -1 if that is necessary to make $a > 0$ in a trinomial of the form $ax^2 + bx + c$).
2. Identify a, b, and c, and find the key number ac.
3. Find two numbers whose product is the key number and whose sum is b.
4. Enter the two numbers as coefficients of x between the first and last terms and factor the polynomial by grouping.

The product of these numbers must be ac.

$$ax^2 + \boxed{}x + \boxed{}x + c$$

The sum of these numbers must be b.

5. Check by multiplying.

EXAMPLE 6

ELEMENTARY & INTERMEDIATE
Algebra $\widehat{f(x)}$ Now™

Factor by grouping: $10x^2 + 13x - 3$.

Solution   In $10x^2 + 13x - 3$, $a = 10$, $b = 13$, and $c = -3$. The key number is $ac = 10(-3) = -30$. We must find a factorization of -30 in which the sum of the factors is $b = 13$. Since the factors must have a negative product, their signs must be different. The possible factor pairs are listed in the table.

$$\text{Key number} = -30$$

Factors of -30	Sum of the factors of -30
$1(-30) = -30$	$1 + (-30) = -29$
$2(-15) = -30$	$2 + (-15) = -13$
$3(-10) = -30$	$3 + (-10) = -7$
$5(-6) = -30$	$5 + (-6) = -1$
$6(-5) = -30$	$6 + (-5) = 1$
$10(-3) = -30$	$10 + (-3) = 7$
$\mathbf{15(-2)} = -30$	$\mathbf{15 + (-2)} = 13$
$30(-1) = -30$	$30 + (-1) = 29$

The seventh row contains the correct pair of numbers 15 and -2, whose product is -30 and whose sum is 13. They serve as the coefficients of two terms, $15x$ and $-2x$, that we place between $10x^2$ and -3.

$$10x^2 + 13x - 3 = 10x^2 + 15x - 2x - 3 \quad \text{\textcolor{red}{Express $13x$ as $15x - 2x$.}}$$

Notation

In Example 6, the middle term, $13x$, may be expressed as $15x - 2x$ or as $-2x + 15x$ when using factoring by grouping. The resulting factorizations will be equivalent.

Finally, we factor by grouping.

$$10x^2 + 15x - 2x - 3 = 5x(2x + 3) - 1(2x + 3) \quad \text{\textcolor{red}{Factor out $5x$ from $10x^2 + 15x$.}}$$
$$\text{\textcolor{red}{Factor out -1 from $-2x - 3$.}}$$
$$= (2x + 3)(5x - 1) \quad \text{\textcolor{red}{Factor out $2x + 3$.}}$$

So $10x^2 + 13x - 3 = (2x + 3)(5x - 1)$. Check by multiplying.

Self Check 6   Factor: $15a^2 + 17a - 4$.

EXAMPLE 7

ELEMENTARY &
INTERMEDIATE
Algebra $f(x)$ **Now**™

Factor by grouping: $12x^5 - 17x^4 + 6x^3$.

Solution   First, we factor out the GCF, which is x^3.

$$12x^5 - 17x^4 + 6x^3 = x^3(12x^2 - 17x + 6)$$

To factor $12x^2 - 17x + 6$, we must find two integers whose product is $12(6) = 72$ and whose sum is -17. Two such numbers are -8 and -9.

$$12x^2 - 17x + 6 = 12x^2 - 8x - 9x + 6 \quad \text{\textcolor{red}{Express $-17x$ as $-8x - 9x$.}}$$
$$= 4x(3x - 2) - 3(3x - 2) \quad \text{\textcolor{red}{Factor out $4x$ and factor out -3.}}$$
$$= (3x - 2)(4x - 3) \quad \text{\textcolor{red}{Factor out $3x - 2$.}}$$

The complete factorization of the original trinomial is

$$12x^5 - 17x^4 + 6x^3 = x^3(3x - 2)(4x - 3)$$

Check by multiplying.

Self Check 7   Factor: $21a^4 - 13a^3 + 2a^2$.

Answers to Self Checks   **1.** $(2x + 1)(x + 2)$ **2.** $(6b - 1)(b - 3)$ **3.** $(5t + 2)(t - 5)$ **4.** $(2x + 3y)(2x - y)$
   5. $-2y(y + 3)(y - 2)$ **6.** $(3a + 4)(5a - 1)$ **7.** $a^2(7a - 2)(3a - 1)$

5.3   STUDY SET   ELEMENTARY & INTERMEDIATE Algebra $f(x)$ Now™

VOCABULARY   Fill in the blanks.

1. The trinomial $3x^2 - x - 12$ has a _____ coefficient of 3. The middle term is $-x$ and the last _____ is -12.

2. Given: $5y^2 - 16y + 3 = (5y - 1)(y - 3)$. We say that $5y^2 - 16y + 3$ factors as the product of two _____.

3. The numbers 6 and -2 are two integers whose _____ is -12 and whose _____ is 4.

4. To factor $2m^2 + 11m + 12$ by _____, we write it as $2m^2 + 8m + 3m + 12$.

5. To factor a trinomial by grouping, we begin by finding ac, the key _____.

6. A _____ trinomial cannot be factored using only integers.

CONCEPTS

7. **a.** Give an example of a pair of positive integer factors of 9.

   b. Give an example of a pair of negative integer factors of 16.

   c. Give an example of a pair of integer factors of -10.

8. Consider the factorization $(7m - 2)(5m + 1)$.
   a. What are the first terms of the binomial factors?

   b. What are the second terms of the binomial factors?

9. Consider $(4y - 8)(3y + 2)$.
   a. Find the outer product.

   b. Find the inner product.

   c. Combine the inner and outer products.

10. Check to see whether $(3t - 1)(5t - 6)$ is the correct factorization of $15t^2 - 19t + 6$.

11. If $10x^2 - 27x + 5$ is to be factored, what are the possible first terms of the binomial factors.

12. $(3a - 4)(4a - 3)$ is an incomplete factorization of $12a^2 - 25ab + 12b^2$. What is missing?

13. **a.** Fill in the blanks. When factoring a trinomial, we write it in _____ powers of the variable. Then we factor out any _____ (including -1 if that is necessary to make the lead coefficient _____).

   b. What is the GCF of the terms of $6s^4 + 33s^3 + 36s^2$?

   c. Factor out -1 from $-2d^2 + 19d - 8$.

14. Complete each sentence.

The product of these numbers must be ▨.

$$5x^2 + 6x - 8 = (\square x \;\square)(\square x \;\square)$$

The product of these numbers must be ▨.

A trinomial has been partially factored. Complete each statement that describes the type of integers we should consider for the blanks.

15. $5y^2 - 13y + 6 = \left(5y \;\boxed{}\right)\left(y \;\boxed{}\right)$
Since the last term of the trinomial is _____ and the middle term is _____, the integers must be _____ factors of 6.

16. $5y^2 + 13y + 6 = \left(5y \;\boxed{}\right)\left(y \;\boxed{}\right)$
Since the last term of the trinomial is _____ and the middle term is _____, the integers must be _____ factors of 6.

17. $5y^2 - 7y - 6 = \left(5y \;\boxed{}\right)\left(y \;\boxed{}\right)$
Since the last term of the trinomial is _____, the signs of the integers will be _____.

18. $5y^2 + 7y - 6 = \left(5y \;\boxed{}\right)\left(y \;\boxed{}\right)$
Since the last term of the trinomial is _____, the signs of the integers will be _____.

19. Factor out the GCF: $3x(5x - 2) + 2(5x - 2)$.

20. Factor each polynomial using factoring by grouping.

   a. $5x^2 - 10x + 3x - 6$

   b. $8h^2 + 32h + h + 4$

21. Complete the key number table.

Key number = 12

Negative factors of 12	Sum of factors of 12
$-1(-12) = 12$	
$-3(-4) = 12$	

22. Complete each sentence to explain how to factor $3x^2 + 16x + 5$ by grouping.

The product of these numbers must be   .

$$3x^2 + 16x + 5 = 3x^2 + \boxed{}x + \boxed{}x + 5$$

The sum of these numbers must be   .

NOTATION

23. a. Give an example of a trinomial that has a lead coefficient of 1.

   b. Give an example of a trinomial that has a lead coefficient that is not 1.

24. Write the terms of the trinomial $40 - t - 4t^2$ in descending powers of the variable.

25. a. Suppose we wish to factor $12b^2 + 20b - 9$ by grouping. Identify a, b, and c.

   b. What is the key number, ac?

26. To factor a trinomial by grouping, a student made a table and circled the correct pair of integers, as shown. Enter the correct coefficients.

$6x^2 + \boxed{}x + \boxed{}x + 6$

Key number = 36

Factors	Sum
$1 \cdot 36$	37
$2 \cdot 18$	20
$3 \cdot 12$	15
$\boxed{4 \cdot 9}$	$\boxed{13}$

PRACTICE   Complete each factorization.

27. $3a^2 + 13a + 4 = (3a\ \boxed{}\ 1)(a\ \boxed{}\ 4)$

28. $2b^2 + 7b + 6 = (2b\ \boxed{}\ 3)(b\ \boxed{}\ 2)$

29. $4z^2 - 13z + 3 = (z\ \boxed{}\ 3)(4z\ \boxed{}\ 1)$

30. $4t^2 - 4t + 1 = (2t\ \boxed{}\ 1)(2t\ \boxed{}\ 1)$

31. $2m^2 + 5m - 12 = (2m\ \boxed{}\ 3)(m\ \boxed{}\ 4)$

32. $10u^2 - 13u - 3 = (2u\ \boxed{}\ 3)(5u\ \boxed{}\ 1)$

Complete each step of the factorization of the trinomial by grouping.

33. $12t^2 + 17t + 6 = 12t^2 + \boxed{}t + \boxed{}t + 6$

$$= \boxed{}(4t + 3) + \boxed{}(4t + 3)$$

$$= (\boxed{})(3t + 2)$$

34. $35t^2 - 11t - 6 = 35t^2 + \boxed{}t - 21t - 6$

$$= 5t(7t + 2)\ \boxed{}\ 3(7t + 2)$$

$$= (\boxed{})(5t - 3)$$

Factor each trinomial, if possible.

35. $3a^2 + 13a + 4$

36. $2b^2 + 7b + 6$

37. $5x^2 + 11x + 2$

38. $7t^2 + 10t + 3$

39. $4x^2 + 8x + 3$

40. $4z^2 + 13z + 3$

41. $6x^2 + 25x + 21$

42. $6y^2 + 7y + 2$

43. $2x^2 - 3x + 1$

44. $2y^2 - 7y + 3$

45. $4t^2 - 4t + 1$

46. $9x^2 - 32x + 15$

47. $15t^2 - 34t + 8$

48. $7x^2 - 9x + 2$

49. $2x^2 - 3x - 2$

50. $3a^2 - 4a - 4$

51. $12y^2 - y - 1$

52. $8u^2 - 2u - 15$

53. $10y^2 - 3y - 1$

54. $6m^2 + 19m + 3$

55. $12y^2 - 5y - 2$

56. $10x^2 + 21x - 10$

57. $2m^2 + 5m - 10$

58. $10u^2 - 13u - 6$

59. $-13x + 3x^2 - 10$

60. $-14 + 3a^2 - a$

61. $6r^2 + rs - 2s^2$

62. $3m^2 + 5mn + 2n^2$

63. $8n^2 + 91n + 33$

64. $2m^2 + 27m + 70$

65. $4a^2 - 15ab + 9b^2$

66. $12x^2 + 5xy - 3y^2$

67. $130r^2 + 20r - 110$

68. $170h^2 - 210h - 260$

69. $-5t^2 - 13t - 6$

70. $-16y^2 - 10y - 1$

71. $36y^2 - 88y + 32$

72. $70a^2 - 95a + 30$

73. $4x^2 + 8xy + 3y^2$

74. $4b^2 + 15bc - 4c^2$

75. $12y^2 + 12 - 25y$

76. $12t^2 - 1 - 4t$

77. $18x^2 + 31x - 10$

78. $20y^2 - 93y - 35$

79. $-y^3 - 13y^2 - 12y$

80. $-2xy^2 - 8xy + 24x$

81. $3x^2 + 6 + x$

82. $25 + 2u^2 + 3u$

83. $30r^5 + 63r^4 - 30r^3$

84. $6s^5 - 26s^4 - 20s^3$

85. $2a^2 + 3b^2 + 5ab$

86. $11uv + 3u^2 + 6v^2$

87. $pq + 6p^2 - q^2$

88. $-11mn + 12m^2 + 2n^2$

89. $6x^3 - 15x^2 - 9x$

90. $9y^3 + 3y^2 - 6y$

91. $15 + 8a^2 - 26a$

92. $16 - 40a + 25a^2$

93. $16m^3n + 20m^2n^2 + 6mn^3$

94. $-28u^3v^3 + 26u^2v^4 - 6uv^5$

APPLICATIONS

95. FURNITURE   The area of a desktop is given by the trinomial $(4x^2 + 20x - 11)$ in.^{2}. Factor it to find the expressions that represent its length and width. Then determine the difference in the length and width of the desktop.

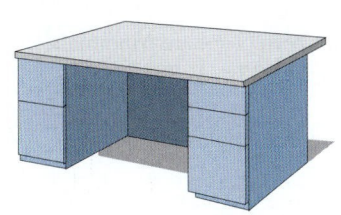

96. STORAGE   The volume of an 8-foot-wide portable storage container is given by the trinomial $(72x^2 + 120x - 400)$ ft^3. Its dimensions can be determined by factoring the trinomial. Find the height and the length of the container.

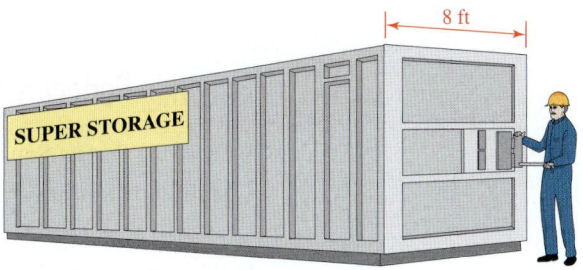

WRITING

97. In the work below, a student began to factor the trinomial. Explain his mistake.

$$\text{Factor: } 3x^2 - 5x + 2.$$
$$(3x - \quad)(x + \quad)$$
$$???$$

98. Two students factor $2x^2 + 20x + 42$ and get two different answers:

$$(2x + 6)(x + 7) \qquad \text{and} \qquad (x + 3)(2x + 14)$$

Do both answers check? Why don't they agree? Is either answer completely correct? Explain.

99. Why is the process of factoring $6x^2 - 5x - 6$ more complicated than the process of factoring $x^2 - 5x - 6$?

100. How can the factorization shown below be checked?

$$6x^2 - 5x - 6 = (3x + 2)(2x - 3)$$

101. Suppose a factorization check of $(3x - 9)(5x + 7)$ gives a middle term $-24x$, but a middle term of $24x$ is actually needed. Explain how to quickly obtain the correct factorization.

102. Suppose we want to factor $2x^2 + 7x - 72$. Explain why $(2x - 1)(x + 72)$ is not a wise choice to try first.

REVIEW   Evaluate each expression.

103. -7^2

104. $(-7)^2$

105. 7^0

106. 7^{-2}

107. $\dfrac{1}{7^{-2}}$

108. $2 \cdot 7^2$

109. $6a^{10} + 5a^5 - 21$

110. $3x^4y^2 - 29x^2y + 56$

111. $8x^2(c^2 + c - 2) - 2x(c^2 + c - 2) - 1(c^2 + c - 2)$

112. Find all integer values of b that make $2x^2 + bx - 5$ factorable.

5.4   Factoring Perfect Square Trinomials and the Difference of Two Squares

- Factoring Perfect Square Trinomials
- Multistep Factoring
- Factoring the Difference of Two Squares

In this section, we will discuss a method that can be used to factor two types of trinomials, called *perfect square trinomials*. We also develop techniques for factoring a type of binomial called the *difference of two squares*.

■ FACTORING PERFECT SQUARE TRINOMIALS

We have seen that the square of a binomial is a trinomial. We have also seen that the special product rules shown below can be used to quickly find the square of a sum and the square of a difference.

$$(x + y)^2 = x^2 + 2xy + y^2$$

This is the square of the first term of the binomial.   This is twice the product of the two terms of the binomial.   This is the square of the last term of the binomial.

$$(x - y)^2 = x^2 - 2xy + y^2$$

Notation

In the expression $(y + 3)^2$, the exponent 2 indicates repeated multiplication. It tells how many times $y + 3$ is to be used as a factor:

$$(y + 3)^2 = (y + 3)(y + 3)$$

Trinomials that are squares of a binomial are called **perfect square trinomials.** Some examples of perfect square trinomials are

$y^2 + 6y + 9$ — Because it is the square of $(y + 3)$: $(y + 3)^2 = y^2 + 6y + 9$

$t^2 - 14t + 49$ — Because it is the square of $(t - 7)$: $(t - 7)^2 = t^2 - 14t + 49$

$4m^2 - 20m + 25$ — Because it is the square of $(2m - 5)$: $(2m - 5)^2 = 4m^2 - 20m + 25$

EXAMPLE 1

Determine whether the following trinomials are perfect square trinomials:
a. $x^2 + 10x + 25$, **b.** $c^2 - 12c - 36$, and **c.** $25y^2 - 30y + 9$.

Solution   **a.** To determine whether this is a perfect square trinomial, we note that

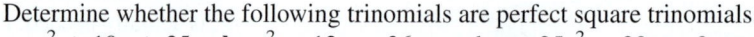

$$x^2 + 10x + 25$$

The first term is the square of x.   The middle term is twice the product of x and 5: $2 \cdot x \cdot 5 = 10x$.   The last term is the square of 5.

Thus, $x^2 + 10x + 25$ is a perfect square trinomial.

b. To determine whether this is a perfect square trinomial, we note that

$$c^2 - 12c - 36$$

The last term, -36, is not
the square of a real number.

Since the last term is negative, $c^2 - 12c - 36$ is not a perfect square trinomial.

c. To determine whether this is a perfect square trinomial, we note that

$$25y^2 - 30y + 9$$

The first term is   The middle term is   The last term is
the square of $5y$.   twice the product of $5y$ the square of -3.
     and -3: $2(5y)(-3) = -30y$.

Thus, $25y^2 - 30y + 9$ is a perfect square trinomial.

Self Check 1   Determine whether the following are perfect square trinomials: **a.** $y^2 + 4y + 4$,
b. $b^2 - 6b - 9$,   **c.** $4z^2 + 4z + 4$.

Although we can factor perfect square trinomials using methods discussed earlier, we can also factor them by inspecting their terms and applying the special product formulas in reverse.

Factoring Perfect Square Trinomials	$x^2 + 2xy + y^2 = (x + y)^2$ $x^2 - 2xy + y^2 = (x - y)^2$

EXAMPLE 2

Factor: $N^2 + 20N + 100$.

Solution $N^2 + 20N + 100$ is a perfect square trinomial, because:

Success Tip

Note that the sign of the second term of a perfect square trinomial is the same as the sign of the second term of the squared binomial.

$$x^2 + 2xy + y^2 = (x + y)^2$$
$$x^2 - 2xy + y^2 = (x - y)^2$$

- The first term N^2 is the square of **N**.
- The last term 100 is the square of **10**: $10^2 = 100$.
- The middle term is twice the product of **N** and **10**: $2(N)(10) = 20N$.

To find the factorization, we match the given trinomial to the proper special product formula.

$$x^2 + 2 \cdot x \cdot y + y^2 = (x + y)^2$$
$$N^2 + 20N + 10^2 = N^2 + 2 \cdot N \cdot 10 + 10^2 = (N + 10)^2$$

Therefore, $N^2 + 20N + 10^2 = (N + 10)^2$. Check by finding $(N + 10)^2$.

Self Check 2   Factor: $x^2 + 18x + 81$.

EXAMPLE 3

Factor: $9x^2 - 30xy + 25y^2$.

Solution $9x^2 - 30xy + 25y^2$ is a perfect square trinomial, because:

- The first term $9x^2$ is the square of $\mathbf{3x}$: $(3x)^2 = 9x^2$.
- The last term $25y^2$ is the square of $\mathbf{-5y}$: $(-5y)^2 = 25y^2$.
- The middle term is twice the product of $\mathbf{3x}$ and $\mathbf{-5y}$: $2(3x)(-5y) = -30xy$.

By inspection, we match the given trinomial to the special product formula to find the factorization.

$$9x^2 - 30xy + 25y^2 = (\mathbf{3x})^2 - 2(\mathbf{3x})(\mathbf{5y}) + (\mathbf{-5y})^2 \qquad 2(3x)(-5y) = -2(3x)(5y).$$
$$= (\mathbf{3x - 5y})^2$$

Therefore, $9x^2 - 30xy + 25y^2 = (3x - 5y)^2$. Check by finding $(3x - 5y)^2$.

Self Check 3 Factor: $16x^2 - 8xy + y^2$.

■ FACTORING THE DIFFERENCE OF TWO SQUARES

Recall the special product formula for multiplying the sum and difference of two terms:

$$(x + y)(x - y) = x^2 - y^2$$

The binomial $x^2 - y^2$ is called a **difference of two squares,** because x^2 is the square of x and y^2 is the square of y. If we reverse this formula, we obtain a method for factoring a difference of two squares.

The Language of Algebra

The expression $x^2 - y^2$ is a *difference of two squares,* whereas $(x - y)^2$ is the *square of a difference.* They are not equivalent because $(x - y)^2 \neq x^2 - y^2$.

Factoring
$$x^2 - y^2 = (x + y)(x - y)$$

This pattern is easy to remember if we think of a difference of two squares as the square of a **F**irst quantity minus the square of a **L**ast quantity.

Factoring a Difference of Two Squares	To factor the square of a First quantity minus the square of a Last quantity, multiply the First plus the Last by the First minus the Last. $$F^2 - L^2 = (F + L)(F - L)$$

When factoring a difference of two squares, it is helpful to know the integers that are perfect squares. The number 400, for example, is a perfect square, because $400 = 20^2$. The **perfect integer squares** through 400 are

1, 4, 9, 16, 25, 36, 49, 64, 81, 100, 121, 144, 169, 196, 225, 256, 289, 324, 361, 400

EXAMPLE 4 Factor each of the following, if possible: **a.** $x^2 - 9$, **b.** $1 - b^2$, and **c.** $n^2 - 45$.

Solution **a.** $x^2 - 9$ is the difference of two squares because it can be written as $x^2 - 3^2$. We can match it to the formula for factoring a difference of two squares to find the factorization.

$$\mathbf{F}^2 - \mathbf{L}^2 = (\mathbf{F} + \mathbf{L})(\mathbf{F} - \mathbf{L})$$
$$x^2 - 3^2 = (x + 3)(x - 3)$$

Notation

By the commutative property of multiplication, the factors of a difference of two squares can be written in either order. For example, we can write:

$$x^2 - 9 = (x - 3)(x + 3)$$

Therefore, $x^2 - 9 = (x + 3)(x - 3)$.

Check by multiplying: $(x + 3)(x - 3) = x^2 - 9$.

b. $1 - b^2$ is the difference of two squares because $1 - b^2 = 1^2 - b^2$. Therefore,

$$1 - b^2 = (1 + b)(1 - b) \qquad \text{1 is a perfect integer square: } 1 = 1^2.$$

Check by multiplying.

c. Since 45 is not a perfect integer square, $n^2 - 45$ cannot be factored using integers. It is prime.

Self Check 4   Factor each of the following, if possible: **a.** $c^2 - 4$,   **b.** $121 - t^2$,   and   **c.** $x^2 - 24$.

Terms containing variables such as $25x^2$ are also perfect squares, because they can be written as the square of a quantity:

$$25x^2 = (5x)^2$$

EXAMPLE 5

Factor: $25x^2 - 49$.

Solution   We can write $25x^2 - 49$ in the form $(5x)^2 - 7^2$ and match it to the formula for factoring the difference of two squares:

$$
\begin{array}{ccccccccc}
\mathbf{F}^2 & - & \mathbf{L}^2 & = & (\mathbf{F} & + & \mathbf{L}) & (\mathbf{F} & - & \mathbf{L}) \\
\downarrow & & \downarrow & & \downarrow & & \downarrow & \downarrow & & \downarrow \\
(\mathbf{5}x)^2 & - & \mathbf{7}^2 & = & (\mathbf{5}x & + & \mathbf{7}) & (\mathbf{5}x & - & \mathbf{7})
\end{array}
$$

Therefore, $25x^2 - 49 = (5x + 7)(5x - 7)$. Check by multiplying.

Self Check 5   Factor: $16y^2 - 9$.

EXAMPLE 6

Factor: $4y^4 - 121z^2$.

Solution   We can write $4y^4 - 121z^2$ in the form $(2y^2)^2 - (11z)^2$ and match it to the formula for factoring the difference of two squares:

Success Tip

Remember that a *difference of two squares* is a binomial. Each term is a square and the terms have different signs. The powers of the variables in the terms must be even.

$$
\begin{array}{ccccccccc}
\mathbf{F}^2 & - & \mathbf{L}^2 & = & (\mathbf{F} & + & \mathbf{L}) & (\mathbf{F} & - & \mathbf{L}) \\
\downarrow & & \downarrow & & \downarrow & & \downarrow & \downarrow & & \downarrow \\
(\mathbf{2}y^2)^2 & - & (\mathbf{11}z)^2 & = & (\mathbf{2}y^2 & + & \mathbf{11}z) & (\mathbf{2}y^2 & - & \mathbf{11}z)
\end{array}
$$

Therefore, $4y^4 - 121z^2 = (2y^2 + 11z)(2y^2 - 11z)$. Check by multiplying.

Self Check 6   Factor: $9m^2 - 64n^4$.

■ MULTISTEP FACTORING

When factoring a polynomial, always factor out the greatest common factor first.

EXAMPLE 7

Factor: $8x^2 - 8$.

Solution   We factor out the GCF of 8, and then factor the resulting difference of two squares.

$$8x^2 - 8 = 8(x^2 - 1) \qquad \text{The GCF is 8.}$$
$$= 8(x + 1)(x - 1) \qquad \text{Think of } x^2 - 1 \text{ as } x^2 - 1^2 \text{ and factor the}$$
$$\text{difference of two squares.}$$

Check by multiplying.

$$8(x + 1)(x - 1) = 8(x^2 - 1) \qquad \text{Multiply the binomials first.}$$
$$= 8x^2 - 8 \qquad \text{Distribute the multiplication by 8.}$$

Self Check 7   Factor: $2p^2 - 200$.

Sometimes we must factor a difference of two squares more than once to completely factor a polynomial.

EXAMPLE 8

Factor: $x^4 - 16$.

Solution
$$x^4 - 16 = (x^2 + 4)(x^2 - 4) \qquad \text{Factor the difference of two squares.}$$
$$= (x^2 + 4)(x + 2)(x - 2) \qquad \text{Factor another difference of two squares: } x^2 - 4.$$

Caution   The binomial $x^2 + 4$ is the **sum of two squares.** In general, after any common factor is removed, a sum of two squares cannot be factored using real numbers.

Self Check 8   Factor: $a^4 - 81$.

Answers to Self Checks   **1. a.** yes,   **b.** no,   **c.** no   **2.** $(x + 9)^2$ **3.** $(4x - y)^2$ **4. a.** $(c + 2)(c - 2)$,
b. $(11 + t)(11 - t)$,   **c.** prime   **5.** $(4y + 3)(4y - 3)$ **6.** $(3m + 8n^2)(3m - 8n^2)$
7. $2(p + 10)(p - 10)$ **8.** $(a^2 + 9)(a + 3)(a - 3)$

5.4   STUDY SET

VOCABULARY   Fill in the blanks.

1. $x^2 + 6x + 9$ is a _____ square trinomial because it is the square of the binomial $(x + 3)$.

2. The binomial $x^2 - 25$ is called a _____ of two squares and it factors as $(x + 5)(x - 5)$. The binomial $x^2 + 25$ is a _____ of two squares and it does not factor using integers.

CONCEPTS   Fill in the blanks.

3. Consider $25x^2 + 30x + 9$.
   a. The first term is the square of ▮.
   b. The last term is the square of ▮.
   c. The middle term is twice the product of ▮ and ▮.

4. Consider $49x^2 - 28xy + 4y^2$.
   a. The first term is the square of ▮.
   b. The last term is the square of ▮.
   c. The middle term is twice the product of ▮ and ▮.

5. **a.** $x^2 + 2xy + y^2 = (\ ▮\ + \ ▮\ )^2$
   b. $x^2 - 2xy + y^2 = (x\ ▮\ )^2$
   c. $x^2 - y^2 = (x\ ▮\ )(\ ▮ - ▮\ )$

6. **a.** $36x^2 = (\ ▮\ )^2$ **b.** $100x^4 = (\ ▮\ )^2$
   c. $4x^2 - 9 = (\ ▮\ )^2 - (\ ▮\ )^2$

7. List the first ten perfect integer squares.

8. Explain why each trinomial is not a perfect square trinomial.
  a. $9h^2 - 6h + 7$
  b. $j^2 - 8j - 16$

  c. $25r^2 + 20r + 16$

9. a. Three incorrect factorizations of $x^2 + 36$ are given below. Use the FOIL method to show why each is wrong.

$$(x + 6)(x - 6)$$
$$(x + 6)(x + 6)$$
$$(x - 6)(x - 6)$$

  b. Can $x^2 + 36$ be factored using only integers?

10. $(2b - 7)^2$ is the factorization of a trinomial. What is the trinomial?

11. $(3x + 4y)(3x - 4y)$ is the factorization of a binomial. What is the binomial?

12. $4b(a + 3)(a - 3)$ is the factorization of a binomial. What is the binomial?

NOTATION

13. Give an example of each type of expression.
  a. difference of two squares
  b. square of a difference
  c. perfect square trinomial
  d. sum of two squares

14. a. Write $(2s)^2 + 2(2s)(9t) + (9t)^2$ as a perfect square trinomial in simplest form.
  b. Write $(6x)^2 - (5y)^2$ as a difference of two squares in simplest form.

15. a. What is the base and what is the exponent of the expression $(x + 3)^2$?
  b. What repeated multiplication is represented by $(x - 8)^2$?

16. What is F and what is L?

$$\begin{array}{cc} F^2 & - & L^2 \\ \downarrow & & \downarrow \\ x^2 & - & 9^2 \end{array}$$

PRACTICE Complete each factorization.

17. $a^2 - 6a + 9 = (a - \boxed{})^2$
18. $t^2 + 2t + 1 = (t \boxed{} 1)^2$

19. $4x^2 + 4x + 1 = (2x \boxed{} 1)^2$
20. $9y^2 - 12y + 4 = (3y - \boxed{})^2$

Factor each polynomial.

21. $x^2 + 6x + 9$ **22.** $x^2 + 10x + 25$

23. $b^2 + 2b + 1$ **24.** $m^2 + 12m + 36$

25. $c^2 - 12c + 36$ **26.** $d^2 - 10d + 25$

27. $y^2 - 8y + 16$ **28.** $z^2 - 2z + 1$

29. $t^2 + 20t + 100$ **30.** $r^2 + 24r + 144$

31. $2u^2 - 36u + 162$ **32.** $3v^2 - 42v + 147$

33. $36x^3 + 12x^2 + x$ **34.** $4x^4 - 20x^3 + 25x^2$

35. $9 + 4x^2 + 12x$ **36.** $1 + 4x^2 - 4x$

37. $a^2 + 2ab + b^2$ **38.** $a^2 - 2ab + b^2$

39. $25m^2 + 70mn + 49n^2$ **40.** $25x^2 + 20xy + 4y^2$

41. $9x^2y^2 + 30xy + 25$ **42.** $s^2t^2 - 20st + 100$

43. $t^2 - \dfrac{2}{3}t + \dfrac{1}{9}$ **44.** $p^2 + p + \dfrac{1}{4}$

45. $s^2 - 1.2s + 0.36$ **46.** $c^2 + 1.6c + 0.64$

Complete each factorization.

47. $y^2 - 49 = (y + \boxed{})(y - \boxed{})$
48. $p^4 - q^2 = (p^2 + q)(\boxed{} - \boxed{})$
49. $t^2 - w^2 = (\boxed{} + \boxed{})(t - w)$
50. $49u^2 - 64v^2 = (\boxed{} + 8v)(7u - 8v)$

Factor each polynomial, if possible.

51. $x^2 - 16$ **52.** $x^2 - 25$

53. $4y^2 - 1$ **54.** $9z^2 - 1$

55. $9x^2 - y^2$ **56.** $4x^2 - z^2$

57. $16a^2 - 25b^2$ **58.** $36a^2 - 121b^2$

59. $36 - y^2$ **60.** $49 - w^2$

61. $a^2 + b^2$ **62.** $121a^2 + 144b^2$

63. $a^4 - 144b^2$ **64.** $81y^4 - 100z^2$

65. $t^2z^2 - 64$ **66.** $900 - B^2C^2$

67. $y^2 - 63$ **68.** $x^2 - 27$

69. $8x^2 - 32y^2$ **70.** $2a^2 - 200b^2$

71. $7a^2 - 7$ **72.** $20x^2 - 5$

73. $-25 + v^2$ **74.** $-144 + h^2$

75. $6x^4 - 6x^2y^2$ **76.** $4b^2y - 16c^2y$

77. $x^4 - 81$ **78.** $y^4 - 625$

79. $a^4 - 16$ **80.** $b^4 - 256$

81. $c^2 - \dfrac{1}{16}$ **82.** $t^2 - \dfrac{9}{25}$

APPLICATIONS

83. GENETICS   The Hardy–Weinberg equation, one of the fundamental concepts in population genetics, is

$$p^2 + 2pq + q^2 = 1$$

where p represents the frequency of a certain dominant gene and q represents the frequency of a certain recessive gene. Factor the left-hand side of the equation.

84. SIGNAL FLAGS   The maritime signal flag for the letter X is shown. Find the polynomial that represents the area of the shaded region and express it in factored form.

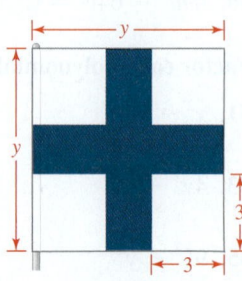

85. PHYSICS   The illustration shows a time-sequence picture of a falling apple. Factor the expression, which gives the difference in the distance fallen by the apple during the time interval from t_1 to t_2 seconds.

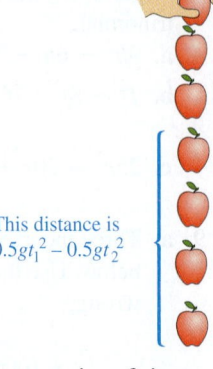

This distance is $0.5gt_1^2 - 0.5gt_2^2$

86. DARTS   A circular dart board has a series of rings around a solid center, called the bullseye. To find the area of the outer white ring, we can use the formula

$$A = \pi R^2 - \pi r^2$$

Factor the expression on the right-hand side of the equation.

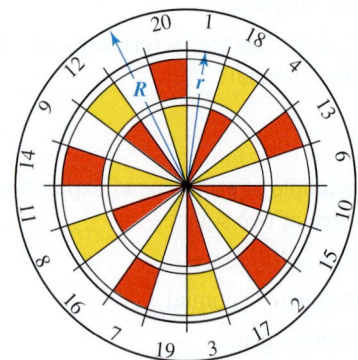

WRITING

87. When asked to factor $x^2 - 25$, one student wrote $(x + 5)(x - 5)$, and another student wrote $(x - 5)(x + 5)$. Are both answers correct? Explain.

88. Explain the error.

$$x^2 - 100 = (x + 50)(x - 50)$$

89. Explain why the following factorization isn't complete.

$$x^4 - 625 = (x^2 + 25)(x^2 - 25)$$

90. Explain why $a^2 + 2a + 1$ is a perfect square trinomial and why $a^2 + 4a + 1$ isn't a perfect square trinomial.

91. Explain the following diagram.

$$x^2 + 2xy + y^2 = (x + y)^2$$
$$x^2 - 2xy + y^2 = (x - y)^2$$

92. Write a comment explaining the error that was made in the following factorization.

Factor: $4x^2 - 16y^2$.

$(2x + 4y)(2x - 4y)$

REVIEW Perform each division.

93. $\dfrac{5x^2 + 10y^2 - 15xy}{5xy}$

94. $\dfrac{-30c^2d^2 - 15c^2d - 10cd^2}{-10cd}$

95. $2a - 1\overline{)a - 2 + 6a^2}$

96. $4b + 3\overline{)4b^3 - 5b^2 - 2b + 3}$

CHALLENGE PROBLEMS

97. For what value of c does $80x^2 - c$ factor as $5(4x + 3)(4x - 3)$?

98. Find all values of b so that the trinomial is a perfect square trinomial.

$$0.16x^2 + bxy + 0.25y^2$$

Factor completely.

99. $81x^6 + 36x^3y^2 + 4y^4$

100. $x^{2n} - y^{4n}$

101. $(x + 5)^2 - y^2$

102. $\dfrac{1}{2} - 2a^2$

5.5 Factoring the Sum and Difference of Two Cubes

- Factoring the Sum and Difference of Two Cubes   • Factoring Out the GCF

In this section we will discuss how to factor two types of binomials, called the *sum* and the *difference of two cubes*.

■ FACTORING THE SUM AND DIFFERENCE OF TWO CUBES

We have seen that the sum of two squares, such as $x^2 + 4$ or $25a^2 + 9b^2$, cannot be factored. However, the sum of two cubes and the difference of two cubes can be factored.

The sum of two cubes

$$x^3 + 8$$

This is x cubed.   This is 2 cubed: $2^3 = 8$.

The difference of two cubes

$$a^3 - 64b^3$$

This is a cubed.   This is $4b$ cubed: $(4b)^3 = 64b^3$.

To find formulas for factoring the sum of two cubes and the difference of two cubes, we need to find the products shown below.

The Language of Algebra

The expression $x^3 + y^3$ is a *sum of two cubes,* whereas, $(x + y)^3$ is the *cube of a sum.* If you expand $(x + y)^3$, you will see that they are not equivalent.

$$(x + y)(x^2 - xy + y^2) = x^3 - x^2y + xy^2 + x^2y - xy^2 + y^3$$
$$= x^3 + y^3 \qquad \text{Combine like terms.}$$

$$(x - y)(x^2 + xy + y^2) = x^3 + x^2y + xy^2 - x^2y - xy^2 - y^3 \qquad \text{Multiply each term of the trinomial by each term of the binomial.}$$

$$= x^3 - y^3 \qquad \text{Combine like terms.}$$

These results justify the formulas for factoring the **sum and difference of two cubes.** They are easier to remember if we think of a sum (or a difference) of two cubes as the cube of a **F**irst quantity plus (or minus) the cube of the **L**ast quantity.

Factoring the Sum and Difference of Two Cubes	To factor the cube of a First quantity plus the cube of a Last quantity, multiply the First plus the Last by the First squared, minus the First times the Last, plus the Last squared. $$F^3 + L^3 = (F + L)(F^2 - FL + L^2)$$ To factor the cube of a First quantity minus the cube of a Last quantity, multiply the First minus the Last by the First squared, plus the First times the Last, plus the Last squared. $$F^3 - L^3 = (F - L)(F^2 + FL + L^2)$$

To factor the sum or difference of two cubes, it's helpful to know the cubes of the numbers from 1 to 10:

1, 8, 27, 64, 125, 216, 343, 512, 729, 1,000

EXAMPLE 1

ELEMENTARY & INTERMEDIATE
Algebra $f(x)$ **Now**™

Solution   Factor: $x^3 + 8$.

$x^3 + 8$ is the sum of two cubes because it can be written as $x^3 + 2^3$. We can match it to the formula for factoring the sum of two cubes to find its factorization.

$$\mathbf{F}^3 + \mathbf{L}^3 = (\mathbf{F} + \mathbf{L})(\mathbf{F}^2 - \mathbf{FL} + \mathbf{L}^2)$$

$$x^3 + 2^3 = (x + 2)(x^2 - x\,2 + 2^2)$$
$$= (x + 2)(x^2 - 2x + 4) \qquad x^2 - 2x + 4 \text{ does not factor.}$$

Caution

In Example 1, a common error is to try to factor $x^2 - 2x + 4$. It is not a perfect square trinomial, because the middle term needs to be $-4x$. Furthermore, it cannot be factored by the methods of Section 5.2. It is prime.

Therefore, $x^3 + 8 = (x + 2)(x^2 - 2x + 4)$. We can check by multiplying.

$$(x + 2)(x^2 - 2x + 4) = x^3 + 2x^2 - 2x^2 - 4x + 4x + 8$$
$$= x^3 + 8$$

Self Check 1   Factor: $h^3 + 27$.

Expressions containing variables such as $64b^3$ are also perfect cubes, because they can be written as the cube of a quantity:

$$64b^3 = (4b)^3$$

EXAMPLE 2

ELEMENTARY & INTERMEDIATE
Algebra $f(x)$ **Now**™

Solution   Factor: $a^3 - 64b^3$.

$a^3 - 64b^3$ is the difference of two cubes because it can be written as $a^3 - (4b)^3$. We can match it to the formula for factoring the difference of two cubes to find its factorization.

$$\begin{array}{ccccccc} \mathbf{F}^3 & - & \mathbf{L}^3 & = (\mathbf{F} & - & \mathbf{L})(\mathbf{F}^2 & + & \mathbf{F}\ \mathbf{L} & + & \mathbf{L}^2) \\ \downarrow & & \downarrow & \downarrow & & \downarrow & & \downarrow\ \downarrow & & \downarrow \end{array}$$

$$a^3 - (\mathbf{4b})^3 = (a - \mathbf{4b})[a^2 + a(\mathbf{4b}) + (\mathbf{4b})^2]$$
$$= (a - 4b)(a^2 + 4ab + 16b^2) \qquad a^2 + 4ab + 16b^2 \text{ does not factor.}$$

Therefore, $a^3 - 64b^3 = (a - 4b)(a^2 + 4ab + 16b^2)$. Check by multiplying.

Self Check 2   Factor: $8c^3 - 1$.

You should memorize the formulas for factoring the sum and the difference of two cubes. Note that each has the form

(a binomial)(a trinomial)

and that there is a relationship between the signs that appear in these forms.

same

$$F^3 + L^3 = (F + L)(F^2 - FL + L^2)$$

opposite   positive

same

$$F^3 - L^3 = (F - L)(F^2 + FL + L^2)$$

opposite   positive

■ FACTORING OUT THE GCF

If the terms of a binomial have a common factor, the GCF (or opposite of the GCF) should always be factored out first.

EXAMPLE 3   Factor: $-2t^5 + 250t^2$.

ELEMENTARY &
INTERMEDIATE
Algebra $f(x)$ **Now**™

Solution   Each term contains the factor $-2t^2$.

$$\begin{aligned} -2t^5 + 250t^2 &= -2t^2(t^3 - 125) & \text{Factor out } -2t^2. \\ &= -2t^2(t - 5)(t^2 + 5t + 25) & \text{Factor } t^3 - 125. \end{aligned}$$

Therefore, $-2t^5 + 250t^2 = -2t^2(t - 5)(t^2 + 5t + 25)$. Check by multiplying.

Self Check 3   Factor: $4c^3 + 4d^3$.

Answers to Self Checks   **1.** $(h + 3)(h^2 - 3h + 9)$ **2.** $(2c - 1)(4c^2 + 2c + 1)$ **3.** $4(c + d)(c^2 - cd + d^2)$

 **5.5**   **STUDY SET**   ELEMENTARY & INTERMEDIATE
Algebra $f(x)$ **Now**™

VOCABULARY   Fill in the blanks.

1. $x^3 + 27$ is the _____ of two cubes.

2. $a^3 - 125$ is the difference of two _____.

3. The factorization of $x^3 + 8$ is $(x + 2)(x^2 - 2x + 4)$. The first factor is a _____ and the second is a trinomial.

4. To factor $2x^3 - 16$, we begin by factoring out the _____, which is 2.

CONCEPTS   Fill in the blanks.

5. **a.** $F^3 + L^3 = (\boxed{} + \boxed{})(F^2 - FL + L^2)$
   b. $F^3 - L^3 = (F \boxed{} L)(\boxed{} + FL + \boxed{})$

6. **a.** $125 = (\boxed{})^3$ **b.** $27m^3 = (\boxed{})^3$
   c. $8x^3 - 27 = (\boxed{})^3 - (\boxed{})^3$
   d. $x^3 + 64y^3 = (\boxed{})^3 + (\boxed{})^3$

7. $m^3 + 64$
       ↑   ↑
   This is   This is
   ▨ cubed.   ▨ cubed.

8. $216n^3 - 125$
       ↑   ↑
   This is   This is
   ▨ cubed.   ▨ cubed.

9. List the first six positive integer cubes.

10. $(x - 2)(x^2 + 2x + 4)$ is the factorization of what binomial?

11. The factorization of $y^3 + 27$ is $(y + 3)(y^2 - 3y + 9)$. Is this factored completely, or does $y^2 - 3y + 9$ factor further?

12. Complete each factorization.
    a. $(x - 8)(x^2 + \boxed{} x + \boxed{})$
    b. $(r - \boxed{})(r^2 + 9r + \boxed{})$

NOTATION   Give an example of each type of expression.

13. **a.** sum of two cubes
    b. cube of a sum

14. **a.** difference of two cubes
    b. cube of a difference

PRACTICE   Complete each factorization.

15. $a^3 + 8 = (a + 2)(a^2 - \boxed{} + 4)$

16. $x^3 - 1 = (x - 1)(x^2 + \boxed{} + 1)$

17. $b^3 + 27 = (\boxed{})(b^2 - 3b + 9)$

18. $z^3 - 125 = (\boxed{})(z^2 + 5z + 25)$

Factor completely.

19. $y^3 + 1$ 20. $x^3 - 8$

21. $a^3 - 27$ 22. $b^3 + 125$

23. $8 + x^3$ 24. $27 - y^3$

25. $s^3 - t^3$ 26. $8u^3 + w^3$

27. $a^3 + 8b^3$ 28. $27a^3 - b^3$

29. $64x^3 - 27$ 30. $27x^3 + 125$

31. $a^6 - b^3$ 32. $a^3 + b^6$

33. $2x^3 + 54$ 34. $2x^3 - 2$

35. $-x^3 + 216$ 36. $-x^3 - 125$

37. $64m^3x - 8n^3x$
38. $16r^4 + 128rs^3$

APPLICATIONS

39. MAILING BREAKABLES   Write an expression that describes the amount of space in the larger box that must be filled with styrofoam chips if the smaller box containing a glass tea cup is to be placed within the larger box for mailing. Then factor the expression.

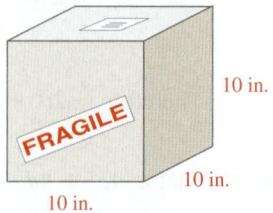

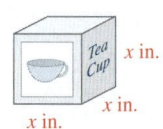

40. MELTING ICE   In one hour, the block of ice shown below had melted to the size shown on the right. Write an expression that describes the volume of ice that melted away. Then factor the expression.

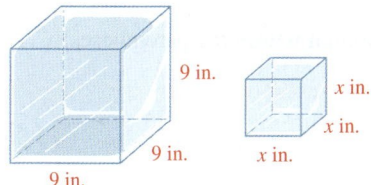

9 in.

9 in.

9 in.

x in.

x in.

x in.

WRITING

41. Explain why $x^3 - 25$ is not a difference of two cubes.

42. Explain this diagram. Then draw a similar diagram for the difference of two cubes.

$$\overbrace{}^{\text{same}}$$
$$F^3 + L^3 = (F + L)(F^2 - FL + L^2)$$
$$\underbrace{}_{\text{opposite}} \quad \uparrow_{\text{positive}}$$

REVIEW

43. When expressed as a decimal, is $\frac{7}{9}$ a terminating or a repeating decimal?

44. Solve: $x + 20 = 4x - 1 + 2x$.

45. List the integers.

46. Solve: $2x + 2 = \frac{2}{3}x - 2$.

47. Evaluate $2x^2 + 5x - 3$ for $x = -3$.

48. Solve $T - R = ma$ for R.

CHALLENGE PROBLEMS

49. Consider $x^6 - 1$.

   a. Write the binomial as a difference of two squares. Then factor.

   b. Write the binomial as a difference of two cubes. Then factor.

   c. Why do the factorizations in parts (a) and (b) appear to be different?

50. What binomial multiplied by $(a^2b^2 + 7ab + 49)$ gives a difference of two cubes?

Factor completely using rational numbers.

51. $x^{3m} - y^{3n}$

52. $\frac{125}{8}s^3 + \frac{1}{27}t^3$

5.6 A Factoring Strategy

The factoring methods introduced so far will be used in the remaining chapters to simplify expressions and to solve equations. In such cases, we must determine the factoring method to use ourselves—it will not be specified. This section will give you practice in selecting the appropriate factoring method to use given a randomly chosen polynomial.

The following strategy is helpful when factoring polynomials.

Steps for Factoring a Polynomial	**1.** Is there a common factor? If so, factor out the GCF.
	2. How many terms does the polynomial have?
	If it has *two terms,* look for the following problem types:
	a. The difference of two squares
	b. The sum of two cubes
	c. The difference of two cubes
	If it has *three terms,* look for the following problem types:
	a. A perfect square trinomial
	b. If the trinomial is not a perfect square, use the trial-and-check-method or the grouping method.
	If it has *four or more terms,* try to factor by grouping.
	3. Can any factors be factored further? If so, factor them completely.
	4. Does the factorization check? Check by multiplying.

EXAMPLE 1

Factor: $2x^4 - 162$.

Solution

**ELEMENTARY &
INTERMEDIATE
Algebra** $f(x)$ **Now**™

The Language of Algebra

Remember that the instruction to *factor* means to *factor completely*. A polynomial is *factored completely* when no factor can be factored further.

Is there a common factor? Yes. Factor out the GCF, which is 2.

$$2x^4 - 162 = 2(x^4 - 81)$$

How many terms does it have? The polynomial within the parentheses, $x^4 - 81$, has two terms. It is a difference of two squares.

$$2x^4 - 162 = 2(\mathbf{x^4 - 81})$$
$$= 2(\mathbf{x^2 + 9})(\mathbf{x^2 - 9})$$

Is it factored completely? No. There is another difference of two squares, $x^2 - 9$.

$$2x^4 - 162 = 2(x^4 - 81)$$
$$= 2(x^2 + 9)(\mathbf{x^2 - 9})$$
$$= 2(x^2 + 9)(\mathbf{x + 3})(\mathbf{x - 3}) \quad x^2 + 9 \text{ is a sum of two squares and does not factor.}$$

Therefore, $2x^4 - 162 = 2(x^2 + 9)(x + 3)(x - 3)$.

Does it check? Yes.

$$2(x^2 + 9)(x + 3)(x - 3) = 2(x^2 + 9)(x^2 - 9) \quad \text{Multiply } (x + 3)(x - 3) \text{ first.}$$
$$= 2(x^4 - 81) \quad \text{Multiply } (x^2 + 9)(x^2 - 9).$$
$$= 2x^4 - 162$$

Self Check 1   Factor: $11a^6 - 11a^2$.

EXAMPLE 2

Factor: $-4c^5d^2 - 12c^4d^3 - 9c^3d^4$.

Solution

**ELEMENTARY &
INTERMEDIATE
Algebra** $f(x)$ **Now**™

Is there a common factor? Yes. Factor out the opposite of the GCF, $-c^3d^2$, so that the lead coefficient is positive.

$$-4c^5d^2 - 12c^4d^3 - 9c^3d^4 = -c^3d^2(4c^2 + 12cd + 9d^2)$$

How many terms does it have? The polynomial within the parentheses, $4c^2 + 12cd + 9d^2$, has three terms. It is a perfect square trinomial because $4c^2 = (2c)^2$, $9d^2 = (3d)^2$, and $12cd = 2 \cdot 2c \cdot 3d$.

$$-4c^5d^2 - 12c^4d^3 - 9c^3d^4 = -c^3d^2(\mathbf{4c^2 + 12cd + 9d^2})$$
$$= -c^3d^2(\mathbf{2c + 3d})^2$$

Is it factored completely? Yes. The binomial $2c + 3d$ does not factor further.

Therefore, $-4c^5d^2 - 12c^4d^3 - 9c^3d^4 = -c^3d^2(2c + 3d)^2$.

Does it check? Yes.

$$-c^3d^2(2c + 3d)^2 = -c^3d^2(4c^2 + 12cd + 9d^2) \quad \text{Use a special product formula.}$$
$$= -4c^5d^2 - 12c^4d^3 - 9c^3d^4$$

Self Check 2   Factor: $-32h^4 - 80h^3 - 50h^2$.

EXAMPLE 3

ELEMENTARY &
INTERMEDIATE
Algebra $f(x)$ **Now**™

Factor: $y^4 - 3y^3 + y - 3$.

Solution   *Is there a common factor?* No. There is no common factor (other than 1).

How many terms does it have? The polynomial has four terms. Try factoring by grouping.

$$y^4 - 3y^3 + y - 3 = y^3(y - 3) + 1(y - 3) \quad \text{Factor } y^3 \text{ from } y^4 - 3y^3.$$
$$= (y - 3)(y^3 + 1)$$

Is it factored completely? No. We can factor $y^3 + 1$ as a sum of two cubes.

$$y^4 - 3y^3 + y - 3 = y^3(y - 3) + 1(y - 3)$$
$$= (y - 3)(y^3 + 1)$$
$$= (y - 3)(y + 1)(y^2 - y + 1) \quad y^2 - y + 1 \text{ does not factor further.}$$

Therefore, $y^4 - 3y^3 + y - 3 = (y - 3)(y + 1)(y^2 - y + 1)$.

Does it check? Yes.

$$(y - 3)(y + 1)(y^2 - y + 1) = (y - 3)(y^3 + 1) \quad \text{Multiply the last two factors.}$$
$$= y^4 + y - 3y^3 - 3 \quad \text{Use the FOIL method.}$$
$$= y^4 - 3y^3 + y - 3$$

Self Check 3   Factor: $b^4 + b^3 + 8b + 8$

EXAMPLE 4

ELEMENTARY &
INTERMEDIATE
Algebra $f(x)$ **Now**™

Factor: $32n - 4n^2 + 4n^3$.

Solution   *Is there a common factor?* Yes. When we write the terms in descending powers of n, we see that the GCF is $4n$.

$$4n^3 - 4n^2 + 32n = 4n(n^2 - n + 8)$$

How many terms does it have? The polynomial within the parentheses has three terms. It is not a perfect square trinomial because the last term, 8, is not a perfect integer square.

To factor the trinomial $n^2 - n + 8$, we must find two integers whose product is 8 and whose sum is -1. As we see in the table, there are no such integers. Thus, $n^2 - n + 8$ is prime.

Negative factors of 8	Sum of the factors of 8
$-1(-8) = 8$	$-1 + (-8) = -9$
$-2(-4) = 8$	$-2 + (-4) = -6$

Is it factored completely? Yes.

Therefore, $4n^3 - 4n^2 + 32n = 4n(n^2 - n + 8)$.

Does it check? Yes.

$$4n(n^2 - n + 8) = 4n^3 - 4n^2 + 32n$$

Self Check 4   Factor: $6m^2 - 54m + 6m^3$.

EXAMPLE 5

Factor: $3y^3 - 4y^2 - 4y$.

ELEMENTARY &
INTERMEDIATE
Algebra $f(x)$ **Now™**

Solution   *Is there a common factor?* Yes. The GCF is y.

$$3y^3 - 4y^2 - 4y = y(3y^2 - 4y - 4)$$

How many terms does it have? The polynomial within the parentheses has three terms. It is not a perfect square trinomial because the first term, $3y^2$, is not a perfect square.
     If we use grouping to factor $3y^2 - 4y - 4$, the key number is $ac = 3(-4) = -12$. We must find two integers whose product is -12 and whose sum is $b = -4$.

Key number $= -12$

Factors of -12	Sum of the factors of -12
$2(-6) = -12$	$2 + (-6) = -4$

From the table, the correct pair is 2 and -6. These numbers serve as the coefficients of two terms, $2y$ and $-6y$, that we place between $3y^2$ and -4.

$$
\begin{aligned}
3y^2 - \mathbf{4y} - 4 &= 3y^2 + \mathbf{2y} - \mathbf{6y} - 4 \\
&= y(3y + 2) - 2(3y + 2) \\
&= (3y + 2)(y - 2)
\end{aligned}
$$

The trinomial $3y^2 - 4y - 4$ factors as $(3y + 2)(y - 2)$.

Is it factored completely? Yes. Because $3y + 2$ and $y - 2$ do not factor.

Therefore, $3y^3 - 4y^2 - 4y = y(3y + 2)(y - 2)$.   Remember to write the GCF y from the first step.

Does it check? Yes.

$$
\begin{aligned}
y(3y + 2)(y - 2) &= y(3y^2 - 4y - 4) \quad \text{Multiply the binomials} \\
&= 3y^3 - 4y^2 - 4y
\end{aligned}
$$

Self Check 5   Factor: $6y^3 + 21y^2 - 12y$.

Answers to Self Checks   **1.** $11a^2(a^2 + 1)(a + 1)(a - 1)$ **2.** $-2h^2(4h + 5)^2$ **3.** $(b + 1)(b + 2)(b^2 - 2b + 4)$
4. $6m(m^2 + m - 9)$ **5.** $3y(2y - 1)(y + 4)$

5.6   STUDY SET   ELEMENTARY & INTERMEDIATE Algebra *f(x)* Now™

VOCABULARY   **Fill in the blanks.**

1. A polynomial is factored _____ when no factor can be factored further.

2. The polynomial $20 + 5x + 12y + 3xy$ has four _____. It can be factored by _____.

3. A polynomial that does not factor using integers is called a _____ polynomial.

4. The binomial $x^2 - 25$ is a _____ of two squares, whereas $x^3 + 125$ is a sum of two _____.

5. $4x^2 - 12x + 9$ is a perfect square _____ because it is the square of $(2x - 3)$.

6. The lead _____ of $x^2 + 3x - 40$ is 1.

7. When factoring a polynomial, always factor out the _____ first.

8. To _____ a factorization, use multiplication.

CONCEPTS   **For each of the following polynomials, which factoring method would you use first?**

9. $2x^5y - 4x^3y$ 10. $9b^2 + 12y - 5$

11. $x^2 + 18x + 81$ 12. $ax + ay - x - y$

13. $x^3 + 27$ 14. $y^3 - 64$

15. $m^2 + 3mn + 2n^2$ 16. $16 - 25z^2$

NOTATION

17. Find the GCF in the following factorization.

$$8m^5 - 8m^4 - 48m^3 = 8m^3(m + 2)(m - 3)$$

18. Give an example of a product of two binomials.

PRACTICE   **Factor each polynomial completely. If a polynomial is not factorable, write "prime."**

19. $2ab^2 + 8ab - 24a$ 20. $32 - 2t^4$

21. $-8p^3q^7 - 4p^2q^3$ 22. $8m^2n^3 - 24mn^4$

23. $20m^2 + 100m + 125$ 24. $3rs + 6r^2 - 18s^2$

25. $x^2 + 7x + 1$

26. $3a^3 + 24b^3$

27. $-2x^5 + 128x^2$

28. $16 - 40z + 25z^2$

29. $a^2c + a^2d^2 + bc + bd^2$

30. $14t^3 - 40t^2 + 6t^4$

31. $-9x^2y^2 + 6xy - 1$

32. $x^2y^2 - 2x^2 - y^2 + 2$

33. $5x^3y^3z^4 + 25x^2y^3z^2 - 35x^3y^2z^5$

34. $2 + 24y + 40y^2$

35. $2c^2 - 5cd - 3d^2$

36. $125p^3 - 64y^3$

37. $8a^2x^3y - 2b^2xy$

38. $a^2 + 8a + 3$

39. $a^2(x - a) - b^2(x - a)$

40. $70p^4q^3 - 35p^4q^2 + 49p^5q^2$

41. $a^2b^2 - 144$

42. $-16x^4y^2z + 24x^5y^3z^4 - 15x^2y^3z^7$

43. $2ac + 4ad + bc + 2bd$

44. $u^2 - 18u + 81$

45. $v^2 - 14v + 49$

46. $28 - 3m - m^2$

47. $39 + 10n - n^2$

48. $81r^4 - 256s^4$

49. $16y^8 - 81z^4$

50. $12x^2 + 14x - 6$

51. $6x^2 - 14x + 8$

52. $12x^2 - 12$

53. $4x^2y^2 + 4xy^2 + y^2$

54. $81r^4s^2 - 24rs^5$

55. $4m^5n + 500m^2n^4$

56. $ae + bf + af + be$

57. $a^2x^2 + b^2y^2 + b^2x^2 + a^2y^2$

58. $6x^2 - x - 16$

59. $4x^2 + 9y^2$

60. $x^4y + 216xy^4$

61. $16a^5 - 54a^2b^3$

62. $25x^2 - 16y^2$

63. $27x - 27y - 27z$

64. $12x^2 + 52x + 35$

65. $xy - ty + xs - ts$

66. $bc + b + cd + d$

67. $35x^8 - 2x^7 - x^6$

68. $x^3 - 25$

69. $5(x - 2) + 10y(x - 2)$

70. $16x^2 - 40x^3 + 25x^4$

71. $49p^2 + 28pq + 4q^2$

72. $x^2y^2 - 6xy - 16$

73. $4t^2 + 36$

74. $m(5 - y) - 5 + y$

75. $p^3 - 2p^2 + 3p - 6$

76. $z^2 + 6yz^2 + 9y^2z^2$

WRITING

77. Which factoring method do you find the most difficult? Why?

78. What four questions make up the factoring strategy for polynomials discussed in this section?

79. What does it mean to factor a polynomial?

80. How is a factorization checked?

REVIEW

81. Graph the real numbers $-3, 0, 2,$ and $-\frac{3}{2}$ on a number line.

82. Graph the interval $(-2, 3]$ on a number line.

83. Graph: $y = \frac{x}{2} + 1$.   **84.** Graph: $y < 2 - 3x$.

CHALLENGE PROBLEMS   Factor completely using rational numbers.

85. $x^4 - 2x^2 - 8$

86. $x(x - y) - y(y - x)$

87. $24 - x^3 + 8x^2 - 3x$

88. $25b^2 + 14b + \frac{49}{25}$

89. $x^9 + y^6$

90. $\frac{1}{4} - \frac{u^2}{81}$

5.7   Solving Quadratic Equations by Factoring

• Quadratic Equations   • Solving Quadratic Equations   • Applications

The Language of Algebra

Quadratic equations involve the square of a variable, not the fourth power as *quad* might suggest. Why the inconsistency? A closer look at the origin of the word *quadratic* reveals that it comes from the Latin word *quadratus,* meaning square.

In Chapter 2, we solved mixture, investment, and uniform motion problems. To model those situations, we used *linear equations in one variable.*

We will now consider situations that are modeled by *quadratic equations.* Unlike linear equations, quadratic equations contain a variable squared term, such as x^2 or t^2. For example, if a pitcher throws a ball upward with an initial velocity of 63 feet per second (about 45 mph), we can find how long the ball will be in the air by solving the quadratic equation $-16t^2 + 63t + 4 = 0$.

■ QUADRATIC EQUATIONS

In a linear, or first degree equation, such as $2x + 3 = 8$, the exponent on the variable is 1. A quadratic, or second degree equation, has a term in which the exponent on the variable is 2, and has no other terms of higher degree.

Quadratic Equations   A **quadratic equation** is an equation that can be written in the **standard form**

$$ax^2 + bx + c = 0$$

where a, b, and c represent real numbers, and a is not 0.

Some examples of quadratic equations are

$$x^2 - 2x - 63 = 0, \qquad x^2 - 25 = 0, \qquad \text{and} \qquad 2x^2 + 3x = 2$$

The first two equations are in standard form. To write the third equation in standard form, we subtract 2 from both sides to get $2x^2 + 3x - 2 = 0$.

To solve a quadratic equation, we find all values of the variable that make the equation true. Some quadratic equations can be solved using factoring methods in combination with the following property of real numbers.

The Zero-Factor Property	When the product of two real numbers is 0, at least one of them is 0. If a and b represent real numbers, and
	if $ab = 0$, then $a = 0$ or $b = 0$

EXAMPLE 1

Solution

Solve: $(4x - 1)(x + 6) = 0$.

By the zero-factor property, if the product of $4x - 1$ and $x + 6$ is 0, then $4x - 1$ must be 0, or $x + 6$ must be 0. In symbols, we write this as

$$4x - 1 = 0 \qquad \text{or} \qquad x + 6 = 0$$

Now we solve each equation.

$$\begin{array}{ccc} 4x - 1 = 0 & \text{or} & x + 6 = 0 \\ 4x = 1 & & x = -6 \\ x = \dfrac{1}{4} & & \end{array}$$

The Language of Algebra

In the zero-factor property, the word *or* means one or the other or both. If the product of two numbers is 0, then one factor is 0, or the other factor is 0, or both factors can be 0.

The results must be checked separately to see whether each of them produces a true statement. We substitute $\frac{1}{4}$ and -6 for x in the original equation and simplify.

Check for $x = \dfrac{1}{4}$	*Check for $x = -6$*
$(4x - 1)(x + 6) = 0$	$(4x - 1)(x + 6) = 0$
$\left[4\left(\dfrac{1}{4}\right) - 1\right]\left(\dfrac{1}{4} + 6\right) \overset{?}{=} 0$	$[4(-6) - 1](-6 + 6) \overset{?}{=} 0$
$(1 - 1)\left(\dfrac{25}{4}\right) \overset{?}{=} 0$	$(-24 - 1)(0) \overset{?}{=} 0$
$0\left(\dfrac{25}{4}\right) \overset{?}{=} 0$	$-25(0) \overset{?}{=} 0$
$0 = 0$	$0 = 0$

The resulting true statements indicate that the equation has two solutions: $\frac{1}{4}$ and -6.

Self Check 1   Solve: $(x - 12)(5x + 6) = 0$.

■ SOLVING QUADRATIC EQUATIONS

In Example 1, the left-hand side of $(4x - 1)(x + 6) = 0$ is in factored form, so we can use the zero-factor property. However, to solve many quadratic equations, we must factor before using the zero-factor property.

EXAMPLE 2

Solve: $x^2 - 2x - 63 = 0$.

Solution   We begin by factoring the trinomial on the left-hand side.

$$x^2 - 2x - 63 = 0$$
$$(x + 7)(x - 9) = 0$$

$x + 7 = 0$ or	$x - 9 = 0$	Set each factor equal to 0.
$x = -7$	$x = 9$	Solve each equation.

Success Tip

When you see the word *solve* in Example 2, you probably think of steps such as combining like terms, distributing, or doing something to both sides. However, to solve this quadratic equation, we begin by factoring $x^2 - 2x - 63$.

To check the results, we substitute -7 and 9 for x in the original equation and simplify.

Check for x = -7	*Check for x = 9*
$x^2 - 2x - 63 = 0$	$x^2 - 2x - 63 = 0$
$(-7)^2 - 2(-7) - 63 \overset{?}{=} 0$	$(9)^2 - 2(9) - 63 \overset{?}{=} 0$
$49 - (-14) - 63 \overset{?}{=} 0$	$81 - 18 - 63 \overset{?}{=} 0$
$63 - 63 \overset{?}{=} 0$	$63 - 63 \overset{?}{=} 0$
$0 = 0$	$0 = 0$

The solutions of $x^2 - 2x - 63 = 0$ are -7 and 9.

Self Check 2   Solve: $x^2 + 5x + 6 = 0$.

The observations made in Example 2 suggest a strategy for solving quadratic equations by factoring.

The Factoring Method for Solving a Quadratic Equation

1. Write the equation in standard form: $ax^2 + bx + c = 0$.
2. Factor the left-hand side.
3. Use the zero-factor property.
4. Solve each resulting equation.
5. Check the results in the original equation.

EXAMPLE 3

Solve: $x^2 - 25 = 0$.

ELEMENTARY & INTERMEDIATE
Algebra $f(x)$ **Now™**   **Solution**   Although $x^2 - 25 = 0$ is missing an x-term, it is a quadratic equation. To solve the equation, we factor the left-hand side and proceed as follows.

Success Tip

In Chapter 9, we discuss methods for solving quadratic equations that cannot be solved by factoring.

$$x^2 - 25 = 0$$

$(x + 5)(x - 5) = 0$		Factor the difference of two squares.
$x + 5 = 0$ or	$x - 5 = 0$	Set each factor equal to 0.
$x = -5$	$x = 5$	Solve each equation.

Check each solution by substituting it into the original equation.

Check for $x = -5$	**Check for $x = 5$**
$x^2 - 25 = 0$	$x^2 - 25 = 0$
$(-5)^2 - 25 \stackrel{?}{=} 0$	$5^2 - 25 \stackrel{?}{=} 0$
$25 - 25 \stackrel{?}{=} 0$	$25 - 25 \stackrel{?}{=} 0$
$0 = 0$	$0 = 0$

The solutions are -5 and 5.

Self Check 3   Solve: $x^2 - 49 = 0$.

EXAMPLE 4

ELEMENTARY &
INTERMEDIATE
Algebra $f(x)$ **Now**™

Solution

Solve: $6x^2 = 12x$.

To write the equation in standard form, we subtract $12x$ from both sides.

$$6x^2 = 12x$$
$$6x^2 - 12x = 12x - 12x$$
$$6x^2 - 12x = 0$$

To solve this equation, we factor the left-hand side.

$6x(x - 2) = 0$		Factor out $6x$.
$6x = 0$ or	$x - 2 = 0$	Set each factor equal to 0.
$x = \dfrac{0}{6}$	$x = 2$	Solve each equation.
$x = 0$		

The solutions are 0 and 2. Check each solution in the original equation.

Caution

In Example 4, a creative, but incorrect, approach is to divide both sides of $6x^2 = 12x$ by $6x$.

$$\frac{6x^2}{6x} = \frac{12x}{6x}$$

You will obtain $x = 2$; however, you will lose the second solution, 0.

Self Check 4   Solve: $5x^2 = 25x$.

EXAMPLE 5

ELEMENTARY &
INTERMEDIATE
Algebra $f(x)$ **Now**™

Solution

Solve: $2x^2 - 2 = -3x$.

We begin by writing the equation in standard form and factoring the left-hand side.

$2x^2 - 2 = -3x$		
$2x^2 + 3x - 2 = -3x + 3x$		To get 0 on the right-hand side, add $3x$ to both sides.
$2x^2 + 3x - 2 = 0$		Combine like terms: $-3x + 3x = 0$.
$(2x - 1)(x + 2) = 0$		Factor $2x^2 + 3x - 2$.
$2x - 1 = 0$ or	$x + 2 = 0$	Set each factor equal to 0.
$2x = 1$	$x = -2$	Solve each equation.
$x = \dfrac{1}{2}$		

The solutions are $\frac{1}{2}$ and -2. Check each solution.

Self Check 5   Solve: $3x^2 - 8 = -10x$.

EXAMPLE 6   Solve: $x(18x - 24) = -8$.

Solution   We begin by writing the equation in standard form

$$x(18x - 24) = -8$$
$$18x^2 - 24x = -8$$ Distribute the multiplication by x.

$$18x^2 - 24x + 8 = 0$$ Add 8 to both sides to make the right-hand side 0.

$$2(9x^2 - 12x + 4) = 0$$ Factor out 2.

$$2(3x - 2)(3x - 2) = 0$$ Factor $9x^2 - 12x + 4$.

$2 = 0$ or $3x - 2 = 0$ or $3x - 2 = 0$ Set each factor equal to 0.

$3x = 2$ $3x = 2$ Solve each equation.

$x = \dfrac{2}{3}$ $x = \dfrac{2}{3}$

> **Caution**
>
> To use the zero-factor property, one side of the equation must be 0. In Example 6, it would be incorrect to set each factor equal to -8.
>
> $x = -8$ or $18x - 24 = -8$
>
> If the product of two numbers is -8, one of them does not have to be -8. For example, $2(-4) = -8$.

Since 2 is a constant, it cannot equal 0. Therefore, we can discard that possibility. After solving the other equations, we see the two solutions are the same. We call $\frac{2}{3}$ a *repeated solution*. Check by substituting it into the original equation.

Self Check 6   Solve: $x(12x + 36) = -27$.

Some equations involving polynomials with degrees higher than 2 can also be solved using the factoring method.

EXAMPLE 7   Solve: $6x^3 + 12x = 17x^2$.

Solution   We can solve this equation as follows:

$$6x^3 + 12x = 17x^2$$
$$6x^3 - 17x^2 + 12x = 0$$ Subtract $17x^2$ from both sides to get 0 on the right-hand side.

$$x(6x^2 - 17x + 12) = 0$$ Factor out x.

$$x(2x - 3)(3x - 4) = 0$$ Factor $6x^2 - 17x + 12$.

$x = 0$ or $2x - 3 = 0$ or $3x - 4 = 0$ Set each factor equal to 0.

$2x = 3$ $3x = 4$ Solve each equation.

$x = \dfrac{3}{2}$ $x = \dfrac{4}{3}$

The solutions are $0, \frac{3}{2}$, and $\frac{4}{3}$. Check each one.

Self Check 7   Solve: $10x^3 + x^2 = 2x$.

■ APPLICATIONS

Each of the following problems is modeled by a quadratic equation.

EXAMPLE 8

Softball.   A pitcher can throw a fastball underhand at 63 feet per second (about 45 mph). If she throws a ball into the air with that velocity, its height h in feet, t seconds after being released, is given by the formula

$$h = -16t^2 + 63t + 4$$

After the ball is thrown, in how many seconds will it hit the ground?

Solution   When the ball hits the ground, its height will be 0 feet. To find the time that it will take for the ball to hit the ground, we set h equal to 0, and solve the quadratic equation for t.

$$h = -16t^2 + 63t + 4$$

$$0 = -16t^2 + 63t + 4 \qquad \text{Substitute 0 for the height } h.$$

$$0 = -1(16t^2 - 63t - 4) \qquad \text{Factor out } -1.$$

$$0 = -1(16t + 1)(t - 4) \qquad \text{Factor } 16t^2 - 63t - 4.$$

$16t + 1 = 0 \qquad \text{or} \qquad t - 4 = 0 \qquad$ Set each factor that contains a variable equal to 0.

$16t = -1 \qquad\qquad\qquad t = 4 \qquad$ Solve each equation.

$$t = -\frac{1}{16}$$

The Language of Algebra

Consecutive means following one another in uninterrupted order. Elton John holds the record for the most *consecutive* years with a song on the Top 50 music chart: 31 years (from 1970 to 2000).

Since time cannot be negative, we discard the solution $t = -\frac{1}{16}$. The second solution, $t = 4$, indicates that the ball hits the ground 4 seconds after being released. Check this answer by substituting 4 for t in $h = -16t^2 + 63t + 4$. You should get $h = 0$.

Consecutive integers are integers that follow one another, such as 15 and 16. They are 1 unit apart. Two consecutive integers can be represented as x and $x + 1$. **Consecutive even integers** are even integers that differ by 2 units, such as 12 and 14. Similarly, **consecutive odd integers** differ by 2 units, such as 9 and 11. We can represent two consecutive even or two consecutive odd integers as x and $x + 2$.

EXAMPLE 9

Women's Tennis.   In the 1998 Australian Open, sisters Venus and Serena Williams played against each other for the first time as professionals. Venus was victorious over her younger sister. At that time, their ages were consecutive integers whose product was 272. How old were Venus and Serena when they met in this match?

Analyze the Problem
- Venus is older than Serena.
- Their ages were consecutive integers.
- The product of their ages was 272.
- Find Venus' and Serena's age when they played this match.

Form an Equation   Let x = Serena's age when she played in the 1998 Australian Open. Since their ages were consecutive integers, Venus' age was $x + 1$. The word *product* indicates multiplication.

Serena's age	times	Venus' age	was	272.
x	$\cdot$	$(x + 1)$	$=$	272

Solve the Equation

$$x(x + 1) = 272$$
$$x^2 + x = 272$$ Distribute the multiplication by x.
$$x^2 + x - 272 = 0$$ Subtract 272 from both sides to make the right-hand side 0.
$$(x + 17)(x - 16) = 0$$ Factor $x^2 + x - 272$. Two numbers whose product is -272 and whose sum is 1 are 17 and -16.

$x + 17 = 0$	or	$x - 16 = 0$	Set each factor equal to 0.
$x = -17$		$x = 16$	Solve each equation.

State the Conclusion   We discard the solution -17, because an age cannot be negative. Thus, Serena Williams was 16 years old and Venus Williams was $16 + 1 = 17$ years old when they played their first professional match against each other.

Check the Result   Since 16 and 17 are consecutive integers, and since $16 \cdot 17 = 272$, the answers check.

EXAMPLE 10

X-rays.   A rectangular-shaped x-ray film has an area of 80 square inches. The length is 2 inches more than the width. Find its length and width.

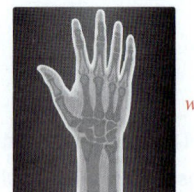

$w + 2$

w

Analyze the Problem
- The area of the film is 80 square inches.
- Its length is 2 inches greater than its width.
- Find its length and width.

Form an Equation   Since the width is related to the length, let w = the width of the film. Then $w + 2$ represents the length of the film. To form an equation, we use the formula for the area of a rectangle: $A = lw$.

The length	$\cdot$	the width	equals	the area of the rectangle.
$(w + 2)$	$\cdot$	w	$=$	80

Solve the Equation

$$(w + 2)w = 80$$
$$w^2 + 2w = 80$$ Distribute the multiplication by w.
$$w^2 + 2w - 80 = 0$$ Subtract 80 from both sides to make the right-hand side 0.
$$(w + 10)(w - 8) = 0$$ Factor the trinomial.

$w + 10 = 0$	or	$w - 8 = 0$	Set each factor equal to 0.
$w = -10$		$w = 8$	Solve each equation.

State the Conclusion   Since the width cannot be negative, we discard the result $w = -10$. Thus, the width of the x-ray film is 8 inches, and the length is $8 + 2 = 10$ inches.

Check the Result   A rectangle with dimensions of 8 inches by 10 inches does have an area of 80 square inches, and the length is 2 inches more than the width. The answers check.

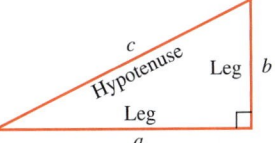

The Language of Algebra

A *theorem* is a mathematical statement that can be proved. The *Pythagorean theorem* is named after *Pythagoras,* a Greek mathematician who lived about 2,500 years ago. He is thought to have been the first to prove the theorem.

The next example involves a right triangle. A **right triangle** is a triangle that has a 90° (right) angle. The longest side of a right triangle is the **hypotenuse,** which is the side opposite the right angle. The remaining two sides are the **legs** of the triangle. The **Pythagorean theorem** provides a formula relating the lengths of the three sides of a right triangle.

The Pythagorean Theorem   If a and b are the lengths of the legs of a right triangle and c is the length of the hypotenuse, then

$$a^2 + b^2 = c^2$$

Right triangles.   The longer leg of a right triangle is 3 units longer than the shorter leg. If the hypotenuse is 6 units longer than the shorter leg, find the lengths of the sides of the triangle.

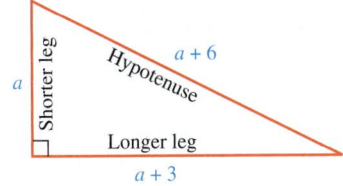

Analyze the Problem   We begin by drawing a right triangle and labeling the legs and the hypotenuse.

Form an Equation   We let a = the length of the shorter leg. Then the length of the hypotenuse is $a + 6$ and the length of the longer leg is $a + 3$. By the Pythagorean theorem, we have

$\left(\begin{matrix}\text{The length of}\\\text{the shorter leg}\end{matrix}\right)^2$	plus	$\left(\begin{matrix}\text{the length of}\\\text{the longer leg}\end{matrix}\right)^2$	equals	$\left(\begin{matrix}\text{the length of the}\\\text{hypotenuse.}\end{matrix}\right)^2$
a^2	$+$	$(a + 3)^2$	$=$	$(a + 6)^2$

Solve the Equation

$$a^2 + (a + 3)^2 = (a + 6)^2$$
$$a^2 + a^2 + 6a + 9 = a^2 + 12a + 36 \qquad \text{Find } (a + 3)^2 \text{ and } (a + 6)^2.$$
$$2a^2 + 6a + 9 = a^2 + 12a + 36 \qquad \text{Combine like terms on the left-hand side.}$$
$$a^2 - 6a - 27 = 0 \qquad \begin{matrix}\text{Subtract } a^2, 12a, \text{ and } 36 \text{ from both sides to make}\\\text{the right-hand side 0.}\end{matrix}$$

Now solve the quadratic equation for a.

$$a^2 - 6a - 27 = 0$$
$$(a - 9)(a + 3) = 0 \qquad\qquad\qquad \text{Factor.}$$
$$a - 9 = 0 \quad \text{or} \quad a + 3 = 0 \qquad \text{Set each factor equal to 0.}$$
$$a = 9 \quad \bigg| \quad\quad a = -3 \qquad \text{Solve each equation.}$$

State the Conclusion   Since a side cannot have a negative length, we discard the result $a = -3$. Thus, the shorter leg is 9 units long, the hypotenuse is $9 + 6 = 15$ units long, and the longer leg is $9 + 3 = 12$ units long.

Check the Result   The longer leg, 12, is 3 units longer than the shorter leg, 9. The hypotenuse, 15, is 6 units longer than the shorter leg, 9, and the lengths satisfy the Pythagorean theorem. So the results check.

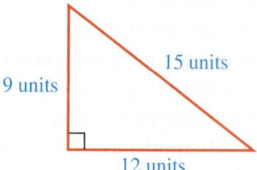

9 units

15 units

12 units

$$9^2 + 12^2 \overset{?}{=} 15^2$$
$$81 + 144 \overset{?}{=} 225$$
$$225 = 225$$

Answers to Self Checks   **1.** $12, -\dfrac{6}{5}$ **2.** $-2, -3$ **3.** $-7, 7$ **4.** $0, 5$ **5.** $\dfrac{2}{3}, -4$ **6.** $-\dfrac{3}{2}$ **7.** $0, \dfrac{2}{5}, -\dfrac{1}{2}$

5.7   STUDY SET

ELEMENTARY &
INTERMEDIATE
Algebra *f(x)* Now™

VOCABULARY   Fill in the blanks.

1. Any equation that can be written in the form $ax^2 + bx + c = 0 \ (a \neq 0)$ is called a _____ equation.

2. To _____ a quadratic equation, we find all values of the variable that make the equation true.

3. Integers that are 1 unit apart, such as 8 and 9, are called _____ integers.

4. A _____ triangle is a triangle that has a 90° angle.

5. The longest side of a right triangle is the _____. The remaining two sides are the _____ of the triangle.

6. The _____ theorem is a formula that relates the lengths of the three sides of a right triangle.

CONCEPTS

7. Which of the following are quadratic equations?
  a. $x^2 + 2x - 10 = 0$ **b.** $2x - 10 = 0$
  c. $x^2 = 15x$ **d.** $x^3 + x^2 + 2x = 0$

8. Write each equation in the form $ax^2 + bx + c = 0$.
  a. $x^2 + 2x = 6$ **b.** $x^2 = 5x$
  c. $3x(x - 8) = -9$ **d.** $4x^2 = 25$

9. Set $5x + 4$ equal to 0 and solve for x.

10. To solve $3(x + 2)(x - 8) = 0$, which factors should be set equal to 0?

11. What step (or steps) should be performed first before factoring is used to solve each equation?
  a. $x^2 + 7x = -6$
  b. $x(x + 7) = 3$

12. Check to see whether the given number is a solution of the given quadratic equation.
  a. $x^2 - 4x = 0; x = 4$
  b. $x^2 + 2x - 4 = 0; x = -2$
  c. $4x^2 - x + 3 = 0; x = 1$

13. a. Factor: $x^2 + 6x - 16$.
  b. Solve: $x^2 + 6x - 16 = 0$.

14. a. Give an example of two consecutive positive integers.
  b. Fill in the blank. Two consecutive integers can be represented by x and _____.

15. A ball is thrown into the air. Its height h in feet, t seconds after being released, is given by the formula

$$h = -16t^2 + 24t + 6$$

When the ball hits the ground, what is the value of h?

16. In a right triangle, the sum of the _____ of the lengths of the two legs is equal to the square of the length of the _____.

17. Fill in the blank: If the length of the hypotenuse of a right triangle is c and the lengths of the other two legs are a and b, then [____] $= c^2$.

18. a. What kind of triangle is shown here?

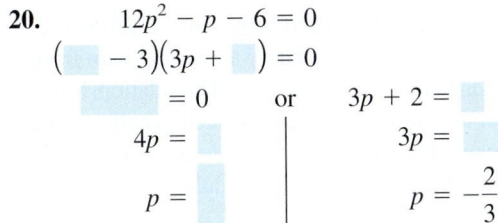
$(x + 9)$ ft
x ft
$(x + 1)$ ft

b. What are the lengths of the legs of the triangle?

c. How much longer is the hypotenuse than the shorter leg?

NOTATION Complete each solution.

19. $7y^2 + 14y = 0$

   [____]$(y + 2) = 0$

$7y = 0$ or   [____] $= 0$

$y =$ [____] | $y = -2$

20. $12p^2 - p - 6 = 0$

$($ [____] $- 3)(3p +$ [____]$) = 0$

   [____] $= 0$ or $3p + 2 =$ [____]

   $4p =$ [____] | $3p =$ [____]

   $p =$ [____] | $p = -\dfrac{2}{3}$

PRACTICE Solve each equation.

21. $(x - 2)(x + 3) = 0$

22. $(x - 3)(x - 2) = 0$

23. $(2s - 5)(s + 6) = 0$

24. $(3h - 4)(h + 1) = 0$

25. $2(t - 7)(t + 8) = 0$

26. $-(n + 3)(n - 6) = 0$

27. $(x - 1)(x + 2)(x - 3) = 0$

28. $(x + 2)(x + 3)(x - 4) = 0$

29. $x(x - 3) = 0$ **30.** $x(x + 5) = 0$

31. $6x(2x - 5) = 0$ **32.** $5x(5x + 7) = 0$

33. $w^2 - 7w = 0$ **34.** $p^2 + 5p = 0$

35. $3x^2 + 8x = 0$ **36.** $5x^2 - x = 0$

37. $8s^2 - 16s = 0$ **38.** $15s^2 - 20s = 0$

39. $x^2 - 25 = 0$ **40.** $x^2 - 36 = 0$

41. $4x^2 - 1 = 0$ **42.** $9y^2 - 1 = 0$

43. $9y^2 - 4 = 0$ **44.** $16z^2 - 25 = 0$

45. $x^2 = 100$ **46.** $z^2 = 25$

47. $4x^2 = 81$ **48.** $9y^2 = 64$

49. $x^2 - 13x + 12 = 0$ **50.** $x^2 + 7x + 6 = 0$

51. $x^2 - 4x - 21 = 0$ **52.** $x^2 + 2x - 15 = 0$

53. $x^2 - 9x + 8 = 0$ **54.** $x^2 - 14x + 45 = 0$

55. $a^2 + 8a = -15$ **56.** $a^2 - 16a = -64$

57. $4y + 4 = -y^2$ **58.** $-3y + 18 = y^2$

59. $0 = x^2 - 16x + 64$ **60.** $0 = 3h^2 + h - 2$

61. $2x^2 - 5x + 2 = 0$ **62.** $2x^2 + x - 3 = 0$

63. $5x^2 - 6x + 1 = 0$ **64.** $6x^2 - 5x + 1 = 0$

65. $4r^2 + 4r = -1$ **66.** $9m^2 + 6m = -1$

67. $12b^2 + 26b + 12 = 0$ **68.** $25f^2 - 80f + 15 = 0$

69. $-15x^2 + 2 = -7x$ **70.** $-8x^2 - 10x = -3$

71. $x(2x - 3) = 20$ **72.** $x(2x - 3) = 14$

73. $(n + 8)(n - 3) = -30$ **74.** $(2s + 5)(s + 1) = -1$

75. $3b^2 - 30b = 6b - 60$ **76.** $2m^2 - 8m = 2m - 12$

77. $(d + 1)(8d + 1) = 18d$ **78.** $4h(3h + 2) = h + 12$

79. $(x - 2)(x^2 - 8x + 7) = 0$

80. $(x - 1)(x^2 + 5x + 6) = 0$

81. $x^3 + 3x^2 + 2x = 0$ **82.** $x^3 - 7x^2 + 10x = 0$

83. $k^3 - 27k - 6k^2 = 0$ **84.** $j^3 - 22j - 9j^2 = 0$

85. $2x^3 = 2x(x + 2)$ **86.** $x^3 + 7x^2 = x^2 - 9x$

APPLICATIONS

87. OFFICIATING Before a football game, a coin toss is used to determine which team will kick off. The height h (in feet) of a coin above the ground t seconds after being flipped up into the air is given by $h = -16t^2 + 22t + 3$. How long does a team captain have to call heads or tails if it must be done while the coin is in the air (see next page)?

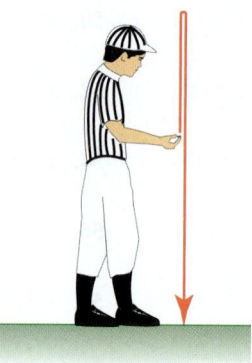

88. DOLPHINS   The height h in feet reached by a dolphin t seconds after breaking the surface of the water is given by

$$h = -16t^2 + 32t$$

How long will it take the dolphin to jump out of the water and touch the trainer's hand?

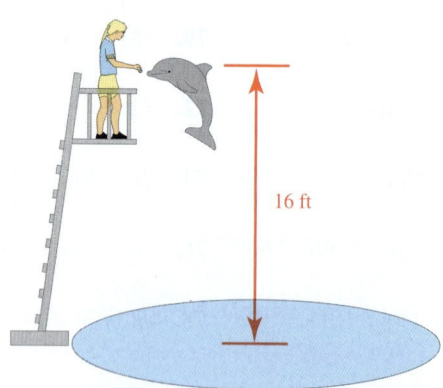

16 ft

89. EXHIBITION DIVING   In Acapulco, Mexico, men diving from a cliff to the water 64 feet below are quite a tourist attraction. A diver's height h above the water t seconds after diving is given by $h = -16t^2 + 64$. How long does a dive last?

90. TIME OF FLIGHT   The formula $h = -16t^2 + vt$ gives the height h in feet of an object after t seconds, when it is shot upward into the air with an initial velocity v in feet per second. After how many seconds will the object hit the ground if it is shot with a velocity of 144 feet per second?

91. CHOREOGRAPHY   For the finale of a musical, 36 dancers are to assemble in a triangular-shaped series of rows, where each successive row has one more

dancer than the previous row. The illustration shows the beginning of such a formation. The relationship between the number of rows r and the number of dancers d is given by

$$d = \frac{1}{2}r(r + 1)$$

Determine the number of rows in the formation. (*Hint:* Multiply both sides of the equation by 2.)

92. CRAFTS   The illustration shows how a geometric wall hanging can be created by stretching yarn from peg to peg across a wooden ring. The relationship between the number of pegs p placed evenly around the ring and the number of yarn segments s that criss-cross the ring is given by the formula

$$s = \frac{p(p - 3)}{2}$$

How many pegs are needed if the designer wants 27 segments to criss-cross the ring?   (*Hint:* Multiply both sides of the equation by 2.)

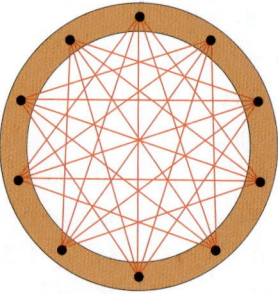

93. CUSTOMER SERVICE   At a pharmacy, customers take a ticket to reserve their turn for service. If the product of the ticket number now being served and the next ticket number to be served is 156, what number is now being served?

94. HISTORY Delaware was the first state to enter the Union and Hawaii was the 50th. If we order the positions of entry for the rest of the states, we find that Tennessee entered the Union right after Kentucky, and the product of their order-of-entry numbers is 240. Use the given information to complete these statements:

Kentucky was the ▮ th state to enter the Union.
Tennessee was the ▮ th state to enter the Union.

95. PLOTTING POINTS The *x*-coordinate and *y*-coordinate of a point in quadrant I are consecutive odd integers whose product is 143. Find the coordinates of the point.

96. PRESIDENTS George Washington was born on 2-22-1732 (February 22, 1732). He died in 1799 at the age of 67. The month in which he died and the day of the month on which he died are consecutive even integers whose product is 168. When did Washington die?

97. INSULATION The area of the rectangular slab of foam insulation in the illustration is 36 square meters. Find the dimensions of the slab.

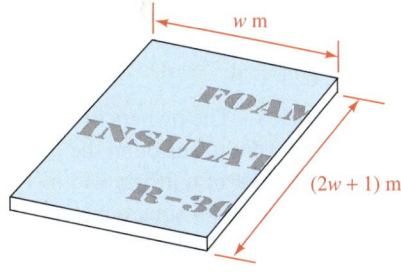

98. SHIPPING PALLETS The length of a rectangular shipping pallet is 2 feet less than 3 times its width. Its area is 21 square feet. Find the dimensions of the pallet.

99. DESIGNING A TENT The length of the base of the triangular sheet of canvas above the door of a tent is 2 feet more than twice its height. The area is 30 square feet. Find the height and the length of the base of the triangle. (*Hint:* Substitute into the formula for the area of a triangle, $A = \frac{1}{2}bh$, then multiply both sides by 2 to clear the equation of the fraction.)

100. TUBING A piece of cardboard in the shape of a parallelogram is twisted to form the tube for a roll of paper towels. The parallelogram has an area of 60 square inches. If its height *h* is 7 inches more than the length of the base *b*, what is the length of the base? (*Hint:* The formula for the area of a parallelogram is $A = bh$.)

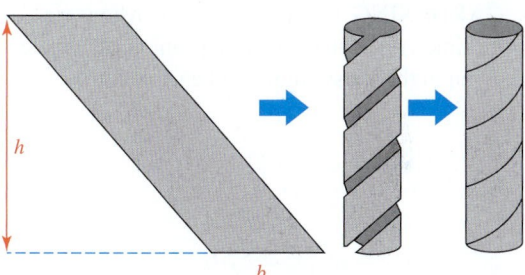

101. WIND DAMAGE A tree was blown over in a wind storm. Find *x*. Then find the height of the tree when it was standing upright.

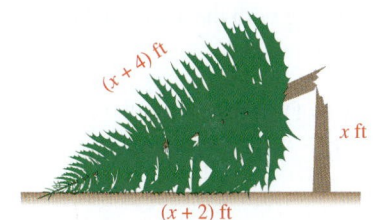

102. BOATING   The inclined ramp of the boat launch is 8 meters longer than the rise of the ramp. The run is 7 meters longer than the rise. How long are the three sides of the ramp?

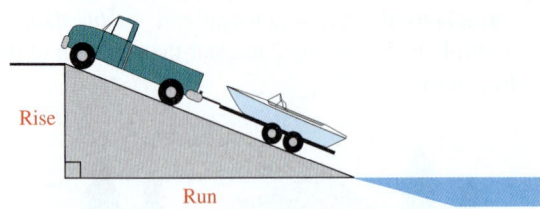

103. CAR REPAIRS   To create some space to work under the front end of a car, a mechanic drives it up steel ramps. A ramp is 1 foot longer than the back, and the base is 2 feet longer than the back of the ramp. Find the length of each side of the ramp.

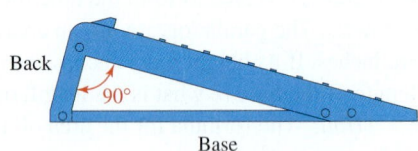

104. GARDENING TOOLS   The dimensions (in millimeters) of the teeth of a pruning saw blade are given in the illustration. Find each length.

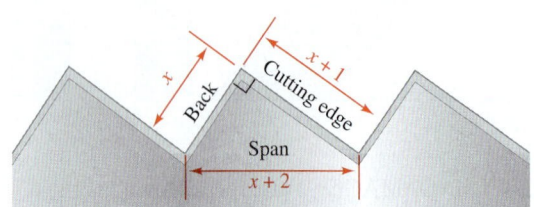

WRITING

105. What is wrong with the logic used to solve $x^2 + x = 6$?

$$x(x + 1) = 6$$

$$x = 6 \quad \text{or} \quad x + 1 = 6$$
$$x = 5$$

So the solutions are 6 and 5.

106. A student solved $x^2 - 5x + 6 = 0$ and obtained two solutions: 2 and 3. Explain the error in his check.

Check: $x^2 - 5x + 6 = 0$
$$2^2 - 5(3) + 6 \overset{?}{=} 0$$
$$4 - 15 + 6 \overset{?}{=} 0$$
$$-5 \neq 0 \quad \text{False.}$$
2 is not a solution.   3 is not a solution.

107. Suppose that to find the length of the base of a triangle, you write a quadratic equation and solve it to find $b = 6$ or $b = -8$. Explain why one solution should be discarded.

108. What error is apparent in the following illustration?

REVIEW

109. EXERCISE   A doctor advises a patient to exercise at least 15 minutes but less than 30 minutes per day. Use a compound inequality to express the range of these times t in minutes.

110. SNACKS   A bag of peanuts is worth $0.30 less than a bag of cashews. Equal amounts of peanuts and cashews are used to make 40 bags of a mixture that is worth $1.05 per bag. How much is a bag of cashews worth?

CHALLENGE PROBLEMS

111. Solve: $x^4 - 625 = 0$.

112. POOL BORDERS   The owners of a rectangular swimming pool want to surround the pool with a crushed-stone border of uniform width. They have enough stone to cover 74 square meters. How wide should they make the border?

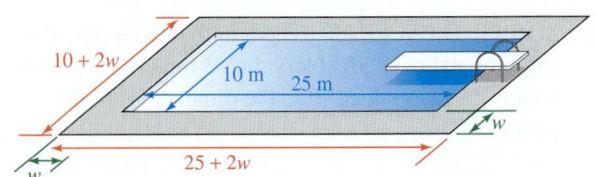

ACCENT ON TEAMWORK

FACTORING MODELS

Overview: In this activity, you will construct geometric models to find factorizations of several trinomials.

Instructions: Form groups of 2 or 3 students.

1. Copy and cut out each of the following figures. On each figure, write its area.

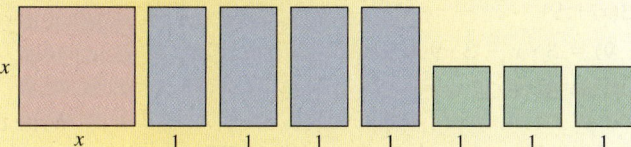

Write a trinomial that represents the *sum* of the areas of the eight figures by combining any like terms: _____ + _____ + _____

2. Now assemble the eight figures to form the large rectangle shown below.

Write an expression that represents the *length* of the rectangle: _____ + _____

Write an expression that represents the *width* of the rectangle: _____ + _____

Express the area of the rectangle as the product of its length and width:

(_____)(_____)

3. The set of figures used in step 1 and the set of figures used in step 2 are the same. Therefore, the expressions for the areas must be equal. Set your answers from steps 1 and 2 equal to find the factorization of the trinomial $x^2 + 4x + 3$.

$$\frac{}{\text{Answer from step 1}} = \frac{}{\text{Answer from step 2}}$$

4. Make a new model to find the factorization of $x^2 + 5x + 4$. (*Hint:* You will need to make one more 1-by-x figure and one more 1-by-1 figure.)

$$\underline{} = \underline{}$$

5. Make a new model to find the factorization of $2x^2 + 5x + 2$. (*Hint:* You will need to make one more x-by-x figure.)

$$\underline{} = \underline{}$$

KEY CONCEPT: FACTORING

Factoring polynomials is the reverse of the process of multiplying polynomials. When we factor a polynomial, we write it as a product of two or more factors.

1. In the following problem, the distributive property is used to multiply a monomial and a binomial.

$$\text{Find } 3(x + 9).$$
$$3(x + 9) = 3 \cdot x + 3 \cdot 9$$
$$= 3x + 27$$

Write this so that it becomes a factoring problem. What would you start with? What would the answer be?

2. In the following problem, we multiply two binomials.

$$\text{Find } (x + 3)(x + 9).$$
$$(x + 3)(x + 9) = x^2 + 9x + 3x + 27$$
$$= x^2 + 12x + 27$$

Write this so that it becomes a factoring problem. What would you start with? What would the answer be?

A FACTORING STRATEGY

The following flowchart leads you through the correct steps to identify the type(s) of factoring necessary to factor any given polynomial having two or more terms.

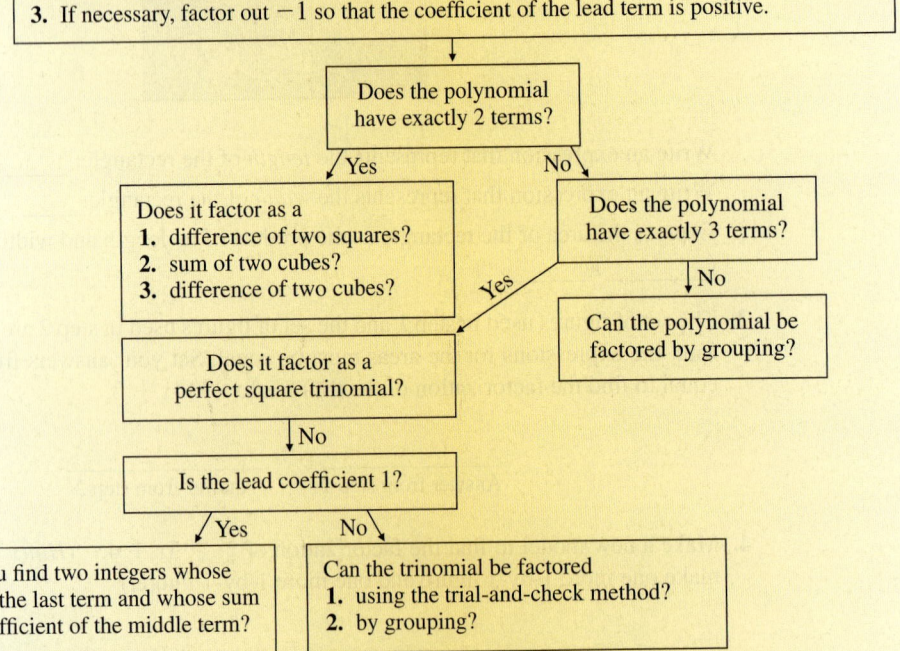

Factor each polynomial completely.

3. $-3a^2 + 21a - 36$

4. $x^2 - 121y^2$

5. $rt + 2r + st + 2s$

6. $v^3 - 8$

7. $6t^2 - 19t + 15$

8. $25y^2 - 20y + 4$

9. $2r^3 - 50r$

10. $46w - 6 + 16w^2$

CHAPTER REVIEW   ELEMENTARY & INTERMEDIATE Algebra $f(x)$ Now™

SECTION 5.1

The Greatest Common Factor; Factoring by Grouping

CONCEPTS

A natural number is in *prime-factored form* when it is written as the product of prime numbers.

To find the *greatest common factor* of a list of integers, prime-factor each to identify the common prime factors.

The first step of factoring a polynomial is to see whether the terms of the polynomial have a common factor. If they do, factor out the GCF.

If a polynomial has four or more terms, consider *factoring it by grouping*.

REVIEW EXERCISES

Find the prime factorization of each number.

1. 35

2. 96

Find the GCF of each list.

3. 28 and 35

4. $36a^4$, $54a^3$, and $126a^6$

Factor out the GCF.

5. $3x + 9y$

6. $5ax^2 + 15a$

7. $7s^5 + 14s^3$

8. $\pi ab - \pi ac$

9. $2x^3 + 4x^2 - 8x$

10. $x^2y^2z + xy^3z^2 - xy^2z$

11. $-5ab^2 + 10a^2b - 15ab$

12. $4(x - 2) - x(x - 2)$

Factor out -1.

13. $-a - 7$

14. $-4t^2 + 3t - 1$

Factor.

15. $2c + 2d + ac + ad$

16. $3xy + 9x - 2y - 6$

17. $2a^3 - a + 2a^2 - 1$

18. $4m^2n + 12m^2 - 8mn - 24m$

SECTION 5.2

Factoring Trinomials of the Form $x^2 + bx + c$

To *factor a trinomial* of the form $x^2 + bx + c$ means to write it as the product of two binomials.

To factor $x^2 + bx + c$, find two integers whose product is c and whose sum is b.

19. What is the lead coefficient of $x^2 + 8x - 9$?

20. Complete the table.

Factors of 6	Sum of the Factors of 6
1(6)	
2(3)	
$-1(-6)$	
$-2(-3)$	

Before factoring a trinomial, write it in *descending powers* of the variable. Also, factor out -1 if that is necessary to make the lead coefficient positive.

If a trinomial cannot be factored using only integers, it is called a *prime polynomial*.

The *GCF* should always be factored out first. A trinomial is *factored completely* when it is expressed as a product of prime polynomials.

Factor each trinomial, if possible.

21. $x^2 + 2x - 24$

22. $x^2 - 4x - 12$

23. $x^2 - 7x + 10$

24. $t^2 + 10t + 15$

25. $-y^2 + 9y - 20$

26. $10y + 9 + y^2$

27. $c^2 + 3cd - 10d^2$

28. $-3mn + m^2 + 2n^2$

29. Explain how we can check to see if $(x - 4)(x + 5)$ is the factorization of $x^2 + x - 20$.

30. Explain why $x^2 + 7x + 11$ is prime.

Completely factor each trinomial.

31. $5a^5 + 45a^4 - 50a^3$

32. $-4x^2y - 4x^3 + 24xy^2$

| SECTION 5.3 | **Factoring Trinomials of the Form $ax^2 + bx + c$** |

To factor $ax^2 + bx + c$ using the *trial-and-check* factoring method, we must determine four integers. Use the FOIL method to check your work.

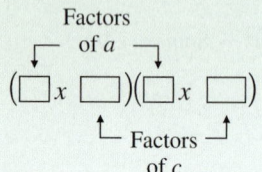

Factors
of a

$(\square x \ \square)(\square x \ \square)$

Factors
of c

To factor $ax^2 + bx + c$ using the *grouping* method, we write it as

$ax^2 + \square x + \square x + c$

The product of these numbers is ac, and their sum is b.

Factor each trinomial completely, if possible.

33. $2x^2 - 5x - 3$

34. $10y^2 + 21y - 10$

35. $-3x^2 + 14x + 5$

36. $-9p^2 - 6p + 6p^3$

37. $4b^2 - 17bc + 4c^2$

38. $7y^2 + 7y - 18$

39. ENTERTAINING   The rectangular-shaped area occupied by a table setting is $(12x^2 - x - 1)$ square inches. Factor the expression to find the binomials that represent the length and width of the table setting.

40. In the following work, a student began to factor $5x^2 - 8x + 3$. Explain his mistake.

$(5x - \ \ )(x + \ \ )$

SECTION 5.4	**Factoring Perfect Square Trinomials and the Difference of Two Squares**

Special product formulas are used to factor *perfect square trinomials*.

$$x^2 + 2xy + y^2$$
$$= (x + y)^2$$
$$x^2 - 2xy + y^2$$
$$= (x - y)^2$$

To factor the *difference of two squares*, use the formula

$$F^2 - L^2$$
$$= (F + L)(F - L)$$

Factor each polynomial completely.

41. $x^2 + 10x + 25$

42. $9y^2 + 16 - 24y$

43. $-z^2 + 2z - 1$

44. $25a^2 + 20ab + 4b^2$

Factor each polynomial completely, if possible.

45. $x^2 - 9$

46. $49t^2 - 25y^2$

47. $x^2y^2 - 400$

48. $8at^2 - 32a$

49. $c^4 - 256$

50. $h^2 + 36$

SECTION 5.5	**Factoring the Sum and Difference of Two Cubes**

To factor the *sum* and *difference of two cubes,* use the formulas

$$F^3 + L^3 =$$
$$(F + L)(F^2 - FL + L^2)$$

$$F^3 - L^3 =$$
$$(F - L)(F^2 + FL + L^2)$$

Factor each polynomial completely.

51. $h^3 + 1$

52. $125p^3 + q^3$

53. $x^3 - 27$

54. $16x^5 - 54x^2y^3$

SECTION 5.6	**A Factoring Strategy**

To factor a random polynomial use the *factoring strategy* discussed in Section 5.6.

Factor each polynomial completely, if possible.

55. $14y^3 + 6y^4 - 40y^2$

56. $s^2t + s^2u^2 + tv + u^2v$

57. $j^4 - 16$

58. $-3j^3 - 24k^3$

59. $12w^2 - 36w + 27$

60. $121p^2 + 36q^2$

61. $2t^3 + 10$

62. $400 - m^2$

63. $x^2 + 64y^2 + 16xy$

64. $18c^3d^2 - 12c^3d - 24c^2d$

| SECTION 5.7 | # Solving Quadratic Equations by Factoring |

A *quadratic equation* is an equation of the form

$$ax^2 + bx + c = 0$$

where a, b, and c represent real numbers, and $a \neq 0$.

To use the *factoring method* to solve a quadratic equation:

1. Write the equation in $ax^2 + bx + c = 0$ form.
2. Factor the left-hand side.
3. Use the *zero-factor property:* Set each factor equal to 0.
4. Solve each resulting equation.
5. Check the results in the original equation.

The Pythagorean theorem: If the length of the hypotenuse of a right triangle is c and the lengths of the two legs are a and b, then $c^2 = a^2 + b^2$.

Solve each quadratic equation by factoring.

65. $x^2 + 2x = 0$

66. $x(x - 6) = 0$

67. $x^2 - 9 = 0$

68. $a^2 - 7a + 12 = 0$

69. $2t^2 + 8t + 8 = 0$

70. $2x - x^2 = -24$

71. $5a^2 - 6a + 1 = 0$

72. $2p^3 = 2p(p + 2)$

73. CONSTRUCTION   The face of the triangular preformed concrete panel has an area of 45 square meters, and its base is 3 meters longer than twice its height. How long is its base?

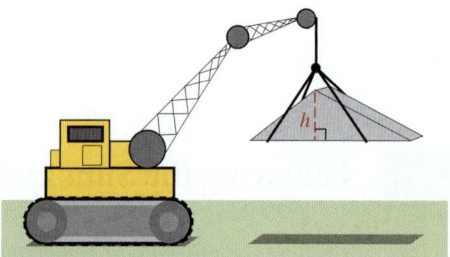

74. ACADEMY AWARDS   In 2003, Meryl Streep surpassed Katherine Hepburn as most-nominated actress. The number of times Streep has been nominated and the number of times Hepburn was nominated are consecutive integers whose product is 156. How many times was each nominated?

75. TIGHTROPE WALKERS   A circus performer intends to walk up a taut cable to a platform atop a pole, as shown in the illustration. How high above the ground is the platform?

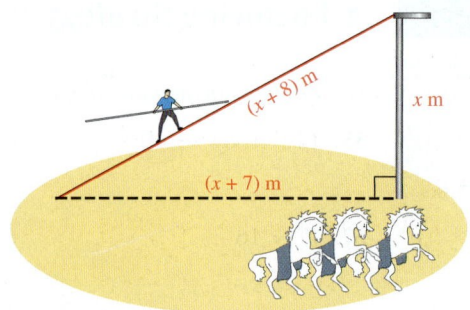

76. BALLOONING   A hot-air balloonist accidentally dropped his camera overboard while traveling at a height of 1,600 feet. The height h, in feet, of the camera t seconds after being dropped is given by $h = -16t^2 + 1,600$. In how many seconds will the camera hit the ground?

CHAPTER 5 TEST

ELEMENTARY & INTERMEDIATE
Algebra *f(x)* **Now**™

Find the prime factorization of each number.

1. 196

2. 111

3. Find the greatest common factor of $45x^4y^6$ and $30x^3y^8$.

Factor each polynomial completely. If a polynomial cannot be factored, write "prime."

4. $4x + 16$

5. $30a^2b^3 - 20a^3b^2 + 5abc$

6. $q^2 - 81$

7. $x^2 + 9$

8. $16x^4 - 81$

9. $x^2 + 4x + 3$

10. $-x^2 + 9x + 22$

11. $9a - 9b + ax - bx$

12. $2a^2 + 5a - 12$

13. $18x^2 - 60xy + 50y^2$

14. $x^3 + 8$

15. $20m^8 - 15m^6$

16. $3a^3 - 81$

17. LANDSCAPING
The combined area of the portions of the square lot that the sprinkler doesn't reach is given by $4r^2 - \pi r^2$, where r is the radius of the circular spray. Factor this expression.

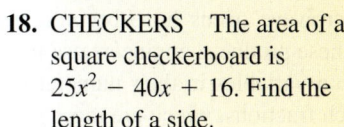

18. CHECKERS   The area of a square checkerboard is $25x^2 - 40x + 16$. Find the length of a side.

19. Factor $x^2 - 3x - 54$. Show a check of your answer.

Solve each equation.

20. $(x + 3)(x - 2) = 0$

21. $x^2 - 25 = 0$

22. $36x^2 - 6x = 0$

23. $x(x + 6) = -9$

24. $6x^2 + x - 1 = 0$

25. $(a - 2)(a - 5) = 28$

26. $x^3 + 7x^2 = -6x$

27. DRIVING SAFETY
Virtually all cars have a "blind spot" where it is difficult for the driver to see a car behind and to the right. The area of the rectangular blind spot shown is 54 square feet. Its length is 3 feet longer than its width. Find its dimensions.

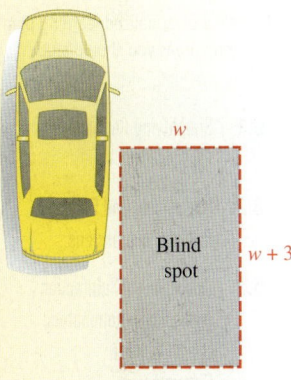

w

Blind spot

$w + 3$

28. What is a quadratic equation? Give an example.

29. Find the length of the hypotenuse of the right triangle shown.

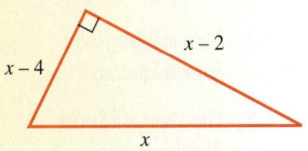

$x - 4$

$x - 2$

x

30. If the product of two numbers is 0, what conclusion can be drawn about the numbers?

6 Rational Expressions and Equations

ELEMENTARY & INTERMEDIATE
Algebra *f(x)* **Now**™

Throughout the chapter, this icon introduces resources on the Elementary & Intermediate AlgebraNow Web site, accessed through **http://1pass. thomson.com**, that will

- Help you test your knowledge of the material with a pre-test and a post-test

- Provide a personalized learning plan targeting areas you should study

Norbert Schaeffer/CORBIS

In today's fast-paced world, where would we be without technology? Cell phones, computers, CD players, and microwave ovens have become an important part of our daily lives. To design these products, engineers use formulas that involve fractions. Often, these fractions contain variables in their numerators and/or denominators. In this chapter, we will consider such fractions, which are more formally referred to as *rational expressions*.

To learn more about the use of algebra in the field of electronics, visit *The Learning Equation* on the Internet at http://tle.brookscole.com. (The log-in instructions are in the Preface.) For Chapter 6, the online lesson is

- *TLE* Lesson 13: Adding and Subtracting Rational Expressions

In Chapter 1, we discussed methods for adding, subtracting, multiplying, and dividing fractions. We will now apply these skills to a new type of fraction, called a rational expression. The numerators and denominators of rational expressions usually contain variables.

6.1 Simplifying Rational Expressions

- Evaluating Rational Expressions
- Factors That Are Opposites
- Simplifying Rational Expressions

Fractions that are the quotient of two integers are *rational numbers*. Examples are $\frac{1}{2}$ and $\frac{3}{4}$. Fractions such as

$$\frac{3}{2y}, \qquad \frac{x}{x+2}, \qquad \text{and} \qquad \frac{2a^2 - 8a}{a^2 - 6a + 8}$$

that are the quotient of two polynomials are called *rational expressions*.

Rational Expressions	A **rational expression** is an expression of the form $\frac{P}{Q}$, where P and Q are polynomials and Q does not equal 0.

■ EVALUATING RATIONAL EXPRESSIONS

Rational expressions can have different values depending on the number that is substituted for the variable.

EXAMPLE 1   Find the value of $\dfrac{2x-1}{x^2+1}$ for $x = -3$ and for $x = 0$.

ELEMENTARY &
INTERMEDIATE
Algebra $f(x)$ **Now**™

Solution   We replace each x in the expression with the given value, evaluate the numerator and denominator separately, and simplify, if possible.

For $x = -3$	*For $x = 0$*
$\dfrac{2x-1}{x^2+1} = \dfrac{2(-3)-1}{(-3)^2+1}$	$\dfrac{2x-1}{x^2+1} = \dfrac{2(0)-1}{0^2+1}$
$= \dfrac{-6-1}{9+1}$	$= \dfrac{0-1}{0+1}$
$= -\dfrac{7}{10}$	$= -1$

Self Check 1   Find the value of $\dfrac{2x-1}{x^2+1}$ for $x = 7$.

Since division by 0 is undefined, we must make sure that the denominator of a rational expression is not equal to 0.

EXAMPLE 2   Find all values of x for which the rational expression is undefined: **a.** $\dfrac{7x}{x-5}$, **b.** $\dfrac{3x-2}{x^2-x-6}$,   and   **c.** $\dfrac{8}{x^2+1}$.

Solution   **a.** The numerator of a rational expression can be any value, including 0. A denominator equal to 0 is what makes a rational expression undefined. The denominator of $\dfrac{7x}{x-5}$ will be 0 if we replace x with 5.

$$\frac{7\mathbf{x}}{\mathbf{x}-5} = \frac{7(\mathbf{5})}{\mathbf{5}-5} = \frac{35}{0}$$

Since $\dfrac{35}{0}$ is undefined, the expression is undefined for $x = 5$.

b. The expression $\dfrac{3x-2}{x^2-x-6}$ will be undefined for values of x that make the denominator 0. To find these values, we solve $x^2 - x - 6 = 0$.

$x^2 - x - 6 = 0$	Set the denominator equal to 0.
$(x-3)(x+2) = 0$	Factor the trinomial.
$x - 3 = 0 \quad$ or $\quad x + 2 = 0$	Set each factor equal to 0.
$x = 3 \qquad\qquad x = -2$	Solve each equation.

Since 3 and -2 make the denominator 0, the expression is undefined for $x = 3$ and $x = -2$.

> **The Language of Algebra**
>
> Another way that Example 2 could be phrased is: State the *restrictions* on the variable. For $\dfrac{7x}{x-5}$, we can state the *restriction* by writing $x \neq 5$. For $\dfrac{3x-2}{x^2-x-6}$, we can write $x \neq 3$ and $x \neq -2$.

For $x = 3$

$$\frac{3x-2}{x^2-x-6} = \frac{3(\mathbf{3})-2}{\mathbf{3}^2-\mathbf{3}-6}$$

$$= \frac{9-2}{9-3-6}$$

$$= \frac{7}{0} \quad \text{This expression is undefined.}$$

For $x = -2$

$$\frac{3x-2}{x^2-x-6} = \frac{3(\mathbf{-2})-2}{(\mathbf{-2})^2-(\mathbf{-2})-6}$$

$$= \frac{-6-2}{4+2-6}$$

$$= \frac{-8}{0} \quad \text{This expression is undefined.}$$

c. No matter what real number is substituted for x, $x^2 + 1$ will not be 0. Thus, no real numbers make $\dfrac{8}{x^2+1}$ undefined.

Self Check 2   Find any values of x for which the rational expression is undefined: **a.** $\dfrac{x}{x+9}$, **b.** $\dfrac{9x+7}{x^2-25}$,   and   **c.** $\dfrac{4-x}{x^2+64}$.

■ SIMPLIFYING RATIONAL EXPRESSIONS

In Section 1.2, we simplified fractions by removing a factor equal to 1. For example, to simplify $\dfrac{6}{15}$, we factor 6 and 15, and then remove the factor $\dfrac{3}{3}$.

$$\frac{6}{15} = \frac{2 \cdot 3}{5 \cdot 3} = \frac{2}{5} \cdot \frac{\mathbf{3}}{\mathbf{3}} = \frac{2}{5} \cdot \mathbf{1} = \frac{2}{5}$$

To streamline this process, we can replace $\dfrac{3}{3}$ in $\dfrac{2 \cdot 3}{5 \cdot 3}$ with the equivalent fraction $\dfrac{1}{1}$.

$$\frac{6}{15} = \frac{2 \cdot 3}{5 \cdot 3} = \frac{2 \cdot \overset{1}{\cancel{3}}}{5 \cdot \underset{1}{\cancel{3}}} = \frac{2}{5} \qquad \text{We are removing } \frac{3}{3} = 1.$$

To **simplify a rational expression** means to write it so that the numerator and denominator have no common factors other than 1.

Simplifying Rational Expressions	**1.** Factor the numerator and denominator completely to determine their common factors.
	2. Remove factors equal to 1 by replacing each pair of factors common to the numerator and denominator with the equivalent fraction $\frac{1}{1}$.

EXAMPLE 3

Simplify: $\dfrac{21x^3}{14x^2}$.

ELEMENTARY & INTERMEDIATE
Algebra $f(x)$ Now™   **Solution**

We write the numerator and denominator in factored form and remove a factor equal to 1.

$$\frac{21x^3}{14x^2} = \frac{3 \cdot 7 \cdot x \cdot x \cdot x}{2 \cdot 7 \cdot x \cdot x} \qquad \text{Factor the numerator and the denominator.}$$

The Language of Algebra

When a rational expression is simplified, the result is an *equivalent expression*. In Example 3, this means that $\frac{21x^3}{14x^2}$ has the same value as $\frac{3x}{2}$ for all values of x, except those that make either denominator 0.

$$= \frac{3 \cdot \overset{1}{\cancel{7}} \cdot \overset{1}{\cancel{x}} \cdot \overset{1}{\cancel{x}} \cdot x}{2 \cdot \underset{1}{\cancel{7}} \cdot \underset{1}{\cancel{x}} \cdot \underset{1}{\cancel{x}}} \qquad \text{Replace } \frac{7}{7} \text{ and each } \frac{x}{x} \text{ with the equivalent fraction } \frac{1}{1}. \text{ This removes the factor } \frac{7 \cdot x \cdot x}{7 \cdot x \cdot x}, \text{ which is equal to 1.}$$

$$= \frac{3x}{2} \qquad \text{Do the multiplications in the numerator and in the denominator.}$$

We say that $\dfrac{21x^3}{14x^2}$ simplifies to $\dfrac{3x}{2}$.

Self Check 3   Simplify: $\dfrac{32a^3}{24a}$.

To simplify rational expressions, we often make use of the factoring methods discussed in the preceding chapter.

EXAMPLE 4

Simplify: $\dfrac{30t - 6}{36}$.

ELEMENTARY & INTERMEDIATE
Algebra $f(x)$ Now™

Solution

$$\frac{30t - 6}{36} = \frac{6(5t - 1)}{6 \cdot 6} \qquad \text{Factor the numerator and denominator.}$$

$$= \frac{\overset{1}{\cancel{6}}(5t - 1)}{\underset{1}{\cancel{6}} \cdot 6} \qquad \text{Remove a factor equal to 1 by replacing } \frac{6}{6} \text{ with } \frac{1}{1}.$$

$$= \frac{5t - 1}{6}$$

Self Check 4   Simplify: $\dfrac{4t - 20}{12}$.

EXAMPLE 5

ELEMENTARY &
INTERMEDIATE
Algebra $f(x)$ **Now**™

Simplify: $\dfrac{x^2 + 13x + 12}{x^2 + 12x}$.

Solution

$$\dfrac{x^2 + 13x + 12}{x^2 + 12x} = \dfrac{(x+1)(x+12)}{x(x+12)}$$
Factor the numerator.
Factor the denominator.

$$= \dfrac{(x+1)\cancel{(x+12)}^{1}}{x\cancel{(x+12)}_{1}}$$
Replace $\frac{x+12}{x+12}$ with the equivalent fraction $\frac{1}{1}$. This removes the factor $\frac{x+12}{x+12} = 1$.

$$= \dfrac{x+1}{x}$$
This rational expression cannot be simplified further.

Caution

Unless otherwise stated, we will now assume that the variables in rational expressions represent real numbers for which the denominator is not zero.

Self Check 5   Simplify: $\dfrac{x^2 - x - 6}{x^2 - 3x}$.

Caution   When simplifying rational expressions, we can only remove factors common to the entire numerator and denominator. It is incorrect to remove terms common to the numerator and denominator.

$$\dfrac{\cancel{x}^{1} + 1}{\cancel{x}_{1}} \qquad\qquad \dfrac{a^2 - 3a + \cancel{2}^{1}}{a + \cancel{2}_{1}} \qquad\qquad \dfrac{\cancel{y^2}^{1} - 36}{\cancel{y^2}_{1} - y - 7}$$

x is a term of $x + 1$.   2 is a term of $a^2 - 3a + 2$ y^2 is a term of $y^2 - 36$
and a term of $a + 2$.   and a term of $y^2 - y - 7$.

EXAMPLE 6

ELEMENTARY &
INTERMEDIATE
Algebra $f(x)$ **Now**™

Simplify: **a.** $\dfrac{3x^2 - 8x - 3}{x^2 - 9}$ and **b.** $\dfrac{x^2 - 2xy + y^2}{(x-y)^4}$.

Solution

a. $\dfrac{3x^2 - 8x - 3}{x^2 - 9} = \dfrac{(3x+1)(x-3)}{(x+3)(x-3)}$
Factor the numerator.
Factor the denominator.

$$= \dfrac{(3x+1)\cancel{(x-3)}^{1}}{(x+3)\cancel{(x-3)}_{1}}$$
Replace $\frac{x-3}{x-3}$ with the equivalent fraction $\frac{1}{1}$. This removes the factor $\frac{x-3}{x-3} = 1$.

$$= \dfrac{3x+1}{x+3}$$

Notation

For Example 6a, we do not need to write the parentheses in the numerator or denominator of the answer.

b. $\dfrac{x^2 - 2xy + y^2}{(x-y)^4} = \dfrac{(x-y)^2}{(x-y)^4}$
Factor $x^2 - 2xy + y^2$.

$$= \dfrac{(x-y)(x-y)}{(x-y)(x-y)(x-y)(x-y)}$$
Write the repeated multiplication indicated by each exponent.

$$= \dfrac{\cancel{(x-y)}^{1}\cancel{(x-y)}^{1}}{\cancel{(x-y)}_{1}\cancel{(x-y)}_{1}(x-y)(x-y)}$$
Replace each $\frac{x-y}{x-y}$ with $\frac{1}{1}$.

$$= \dfrac{1}{(x-y)^2}$$
Use an exponent to write the repeated multiplication in the denominator.

Self Check 6 Simplify: **a.** $\dfrac{4x^2 - 8x - 21}{4x^2 - 49}$ and **b.** $\dfrac{(a + 3b)^5}{a^2 + 6ab + 9b^2}$.

EXAMPLE 7

ELEMENTARY &
INTERMEDIATE
Algebra _f(x)_ **Now™**

Simplify: $\dfrac{5(x + 3) - 5}{7(x + 3) - 7}$.

Solution   Since $x + 3$ is not a factor of the entire numerator and the entire denominator, it cannot be removed. Instead, we simplify the numerator and denominator separately, factor them, and remove any common factors.

$$\frac{5(x + 3) - 5}{7(x + 3) - 7} = \frac{5x + 15 - 5}{7x + 21 - 7} \qquad \text{Use the distributive property twice.}$$

$$= \frac{5x + 10}{7x + 14} \qquad \text{Combine like terms.}$$

$$= \frac{5(x + 2)}{7(x + 2)} \qquad \text{Factor the numerator and the denominator.}$$

$$= \frac{\overset{1}{5(\cancel{x + 2})}}{\underset{1}{7(\cancel{x + 2})}} \qquad \text{Replace } \tfrac{x + 2}{x + 2} \text{ with the equivalent fraction } \tfrac{1}{1}.$$
$$\qquad \qquad \qquad \text{This removes the factor } \tfrac{x + 2}{x + 2} = 1.$$

$$= \frac{5}{7}$$

The Language of Algebra

Some rational expressions cannot be simplified. For example, to attempt to simplify the following rational expression, we factor its numerator and denominator:

$$\frac{x^2 + x - 2}{x^2 + x} = \frac{(x + 2)(x - 1)}{x(x + 1)}$$

Since there are no common factors, we say it is in _simplest form_ or _lowest terms_.

Self Check 7 Simplify: $\dfrac{4(x - 2) + 4}{3(x - 2) + 3}$.

■ FACTORS THAT ARE OPPOSITES

If the terms of two polynomials are the same, except that they are opposite in sign, the polynomials are **opposites.** For example, the following pairs of polynomials are opposites.

$$\begin{array}{cc} 2a - 1 & \text{and} & 1 - 2a \\ \text{Compare terms: } 2a \text{ and } -2a; -1 \text{ and } 1. \end{array} \qquad \begin{array}{cc} -3x^2 - x + 5 & \text{and} & 3x^2 + x - 5 \\ \text{Compare terms: } -3x^2, 3x^2; -x \text{ and } x; 5 \text{ and } -5. \end{array}$$

Success Tip

When a difference is reversed, the original binomial and the resulting binomial are opposites. Here are some pairs of opposites:

$b - 11$	and	$11 - b$
$x - y$	and	$y - x$
$x^2 - 4$	and	$4 - x^2$

We have seen that the quotient of two real numbers that are opposites is always -1:

$$\frac{2}{-2} = -1 \qquad \frac{-78}{78} = -1 \qquad \frac{3.5}{-3.5} = -1$$

Likewise, the quotient of two binomials that are opposites is always -1.

EXAMPLE 8

Simplify: $\dfrac{2a - 1}{1 - 2a}$.

Solution   We can rearrange the terms of the numerator and factor out -1.

$$\frac{2a - 1}{1 - 2a} = \frac{-1 + 2a}{1 - 2a}$$

In the numerator, think of $2a - 1$ as $2a + (-1)$. Then change the order of the terms: $2a + (-1) = -1 + 2a$.

$$= \frac{-1(1 - 2a)}{1 - 2a}$$

In the numerator, factor out -1.

$$= \frac{-1(\overset{1}{\cancel{1 - 2a}})}{\underset{1}{\cancel{1 - 2a}}}$$

Replace $\frac{1 - 2a}{1 - 2a}$ with the equivalent fraction $\frac{1}{1}$. This removes the factor $\frac{1 - 2a}{1 - 2a}$, which is equal to 1.

$$= \frac{-1}{1}$$

$$= -1$$

Any number divided by 1 is itself.

Self Check 8   Simplify: $\dfrac{3p - 2}{2 - 3p}$.

In general, we have this fact.

The Quotient of Opposites   The quotient of any nonzero polynomial and its opposite is -1.

For each of the following rational expressions, the numerator and denominator are opposites. Thus, each expression is equal to -1.

$$\frac{x - 6}{6 - x} = -1 \qquad \frac{2a - 9b}{9b - 2a} = -1 \qquad \frac{-3x^2 - x + 5}{3x^2 + x - 5} = -1$$

This fact can be used to simplify certain rational expressions by removing a factor equal to -1. If a factor of the numerator is the opposite of a factor of the denominator, we can replace them with the equivalent fraction $\frac{-1}{1}$, as shown below.

Caution

It would be incorrect to apply the rule for opposites to a rational expression such as $\frac{x + 1}{1 + x}$. By the commutative property of addition, this is the quotient of a number and itself. The result is 1, not -1.

$$\frac{x + 1}{1 + x} = \frac{\overset{1}{\cancel{x + 1}}}{\underset{1}{\cancel{x + 1}}} = 1$$

$$\frac{\overset{-1}{\cancel{x - 6}}}{\underset{1}{\cancel{6 - x}}}$$ Use slashes / to show that $\dfrac{x - 6}{6 - x}$ is replaced with the equivalent fraction $\dfrac{-1}{1}$.

EXAMPLE 9   Simplify: $\dfrac{y^2 - 1}{3 - 3y}$.

Solution

$$\frac{y^2 - 1}{3 - 3y} = \frac{(y + 1)(y - 1)}{3(1 - y)}$$

Factor the numerator.
Factor the denominator.

$$= \frac{(y + 1)(\overset{-1}{\cancel{y - 1}})}{3(\underset{1}{\cancel{1 - y}})}$$

Since $y - 1$ and $1 - y$ are opposites, simplify by replacing $\frac{y - 1}{1 - y}$ with the equivalent fraction $\frac{-1}{1}$.

$$= \frac{-(y + 1)}{3}$$

This result may be written in other equivalent forms.

Caution

A − symbol preceding a fraction may be applied to the numerator or to the denominator, but not to both:

$$-\frac{y+1}{3} \neq \frac{-(y+1)}{-3}$$

$$\frac{-(y+1)}{3} = -\frac{y+1}{3}$$ The − symbol in $-(y+1)$ can be written in front of the fraction, and the parentheses can be dropped.

$$\frac{-(y+1)}{3} = \frac{-y-1}{3}$$ The − symbol in $-(y+1)$ represents a factor of -1. Distribute the multiplication by -1 in the numerator.

$$\frac{-(y+1)}{3} = \frac{y+1}{-3}$$ The − symbol in $-(y+1)$ can be applied to the denominator. However, we don't usually use this form.

Self Check 9 Simplify: $\dfrac{m^2 - 100}{10m - m^2}$.

Answers to Self Checks 1. $\frac{13}{50}$ 2. a. -9, b. $-5, 5$, c. none   3. $\frac{4a^2}{3}$ 4. $\frac{t-5}{3}$ 5. $\frac{x+2}{x}$
6. a. $\frac{2x+3}{2x+7}$, b. $(a+3b)^3$ 7. $\frac{4}{3}$ 8. -1 9. $-\frac{m+10}{m}$

6.1   STUDY SET

ELEMENTARY & INTERMEDIATE
Algebra $f(x)$ Now™

VOCABULARY   Fill in the blanks.

1. A quotient of two polynomials, such as $\frac{x^2+x}{x^2-3x}$, is called a _____ expression.

2. In a rational expression, the polynomial above the fraction bar is called the _____ and the polynomial below the fraction bar is called the _____.

3. Because of the division by 0, the expression $\frac{8}{0}$ is _____.

4. To _____ the rational expression $\frac{x^2-1}{x-3}$ for $x = -2$, we substitute -2 for each x, and simplify.

5. To _____ a rational expression, we remove factors common to the numerator and denominator.

6. The binomials $x - 15$ and $15 - x$ are called _____, because their terms are the same, except that they are opposite in sign.

CONCEPTS

7. What value of x makes each rational expression undefined?
   a. $\dfrac{x+2}{x}$ b. $\dfrac{x+2}{x-6}$ c. $\dfrac{x+2}{x+6}$

8. When we simplify $\frac{x^2+5x}{4x+20}$, the result is $\frac{x}{4}$. They have the same value for all real numbers, except $x = -5$. Show that they have the same value for $x = 1$.

9. a. For the following rational expression, what factor is common to the numerator and denominator?

$$\frac{x^2+2x+1}{x^2+4x+3} = \frac{\overset{1}{\cancel{(x+1)}}(x+1)}{(x+3)\cancel{(x+1)}} = \frac{x+1}{x+3}$$

   b. Fill in the blanks: To simplify the rational expression, we replace $\frac{x+1}{x+1}$ with the equivalent fraction . This removes the factor $\frac{x+1}{x+1}$, which is equal to ___.

10. Tell whether each pair of polynomials are opposites.
    a. $y + 7$ and $y - 7$
    b. $b - 20$ and $20 - b$
    c. $x^2 - 9x$ and $9x - x^2$
    d. $x^2 + 2x - 1$ and $-x^2 - 2x - 1$

11. a. For the given rational expression, what factor of the numerator and what factor of the denominator are opposites?

$$\frac{x^2-2x}{20-10x} = \frac{x(x-2)}{10(2-x)} = -\frac{x}{10}$$

   b. Fill in the blanks: To simplify the rational expression, we replace $\frac{x-2}{2-x}$ with the equivalent fraction ___. This removes the factor $\frac{x-2}{2-x}$, which is equal to ___.

12. Simplify each expression.
    a. $\dfrac{x-8}{x-8}$ b. $\dfrac{x-8}{8-x}$
    c. $\dfrac{x+8}{8+x}$ d. $\dfrac{x-8}{-x+8}$

13. Write each expression in simplest form.

a. $\dfrac{(x + 2)(x - 2)}{(x + 1)(x + 2)}$ **b.** $\dfrac{y(y - 2)}{9(2 - y)}$

c. $\dfrac{(2m + 7)(m - 5)}{(2m + 7)}$ **d.** $\dfrac{x \cdot x}{x \cdot x(x - 30)}$

14. What is the first step in the process of simplifying
$\dfrac{3(x + 1) + 2x + 2}{x + 1}$

NOTATION

15. In the following table, the answers to three homework problems are compared to the answers in the back of the book. Are the answers equivalent?

Answer	Book's answer	Equivalent?
$\dfrac{-3}{x + 3}$	$-\dfrac{3}{x + 3}$	
$\dfrac{-x + 4}{6x + 1}$	$\dfrac{-(x - 4)}{6x + 1}$	
$\dfrac{x + 7}{(x - 4)(x + 2)}$	$\dfrac{x + 7}{(x + 2)(x - 4)}$	

16. a. In $\dfrac{(x + 5)(x \cancel{- 5})^{\,1}}{x(x \cancel{- 5})_{\,1}}$, what do the slashes show?

b. In $\dfrac{(x - 3)(x \cancel{- 7})^{\,-1}}{(x + 3)(7 \cancel{- x})_{\,1}}$, what do the slashes show?

PRACTICE   Evaluate each expression for $x = 6$.

17. $\dfrac{x - 2}{x - 5}$ **18.** $\dfrac{3x - 2}{x - 2}$

19. $\dfrac{x^2 - 4x - 12}{x^2 + x - 2}$ **20.** $\dfrac{x^2 - 1}{x^3 - 1}$

Evaluate each expression for $y = -3$.

21. $\dfrac{y + 5}{3y - 2}$ **22.** $\dfrac{2y + 9}{y^2 + 25}$

23. $\dfrac{y^3}{3y^2 + 1}$ **24.** $\dfrac{-y}{y^2 - y + 6}$

Find all values of x for which the rational expression is undefined.

25. $\dfrac{x + 5}{8x}$ **26.** $\dfrac{4x - 1}{6x}$

27. $\dfrac{15}{x - 2}$ **28.** $\dfrac{5x}{x + 5}$

29. $\dfrac{15x + 2}{x^2 + 6}$ **30.** $\dfrac{x^2 - 4x}{x^2 + 4}$

31. $\dfrac{x + 1}{2x - 1}$ **32.** $\dfrac{-6x}{3x - 1}$

33. $\dfrac{30x}{x^2 - 36}$ **34.** $\dfrac{2x - 15}{x^2 - 49}$

35. $\dfrac{15}{x^2 + x - 2}$ **36.** $\dfrac{x - 20}{x^2 + 2x - 8}$

Simplify each expression, if possible.

37. $\dfrac{45}{9a}$ **38.** $\dfrac{48}{16y}$

39. $\dfrac{6x^2}{4x^2}$ **40.** $\dfrac{9xy}{6xy}$

41. $\dfrac{2x^2}{x + 2}$ **42.** $\dfrac{5y^2}{y + 5}$

43. $\dfrac{15x^2y}{5xy^2}$ **44.** $\dfrac{12xz}{4xz^2}$

45. $\dfrac{6x + 3}{3y}$ **46.** $\dfrac{4x + 12}{2y}$

47. $\dfrac{x + 3}{3x + 9}$ **48.** $\dfrac{2x - 14}{x - 7}$

49. $\dfrac{a^3 - a^2}{a^4 - a^3}$ **50.** $\dfrac{2c^4 + 2c^3}{4c^5 + 4c^4}$

51. $\dfrac{4x + 16}{5x + 20}$ **52.** $\dfrac{14x - 7}{12x - 6}$

53. $\dfrac{4c + 4d}{d + c}$ **54.** $\dfrac{a + b}{5b + 5a}$

55. $\dfrac{x - 7}{7 - x}$ **56.** $\dfrac{18 - d}{d - 18}$

57. $\dfrac{6x - 30}{5 - x}$ **58.** $\dfrac{6t - 42}{7 - t}$

59. $\dfrac{x^2 - 4}{x^2 - x - 2}$ **60.** $\dfrac{y^2 - 81}{y^2 + 10y + 9}$

61. $\dfrac{x^2 + 3x + 2}{x^2 + x - 2}$

62. $\dfrac{x^2 + x - 6}{x^2 - x - 2}$

63. $\dfrac{2x^2 - 8x}{x^2 - 6x + 8}$

64. $\dfrac{3y^2 - 15y}{y^2 - 3y - 10}$

65. $\dfrac{2 - a}{a^2 - a - 2}$

66. $\dfrac{4 - b}{b^2 - 5b + 4}$

67. $\dfrac{b^2 + 2b + 1}{(b + 1)^3}$

68. $\dfrac{y^2 - 4y + 4}{(y - 2)^3}$

69. $\dfrac{6x^2 - 13x + 6}{3x^2 + x - 2}$

70. $\dfrac{x^2 + 3x + 2}{x^3 + x^2}$

71. $\dfrac{10(c - 3) + 10}{3(c - 3) + 3}$

72. $\dfrac{6(d + 3) - 6}{7(d + 3) - 7}$

73. $\dfrac{6a + 3(a + 2) + 12}{a + 2}$

74. $\dfrac{2y + 4(y - 1) - 2}{y - 1}$

75. $\dfrac{3x^2 - 27}{2x^2 - 5x - 3}$

76. $\dfrac{2x^2 - 8}{3x^2 - 5x - 2}$

77. $\dfrac{-x^2 - 4x + 77}{x^2 - 4x - 21}$

78. $\dfrac{-5x^2 - 2x + 3}{x^2 + 2x - 15}$

79. $\dfrac{x(x - 8) + 16}{16 - x^2}$

80. $\dfrac{x^2 - 3(2x - 3)}{9 - x^2}$

81. $\dfrac{m^2 - 2mn + n^2}{2m^2 - 2n^2}$

82. $\dfrac{c^2 - d^2}{c^2 - 2cd + d^2}$

83. $\dfrac{16a^2 - 1}{4a + 4}$

84. $\dfrac{25m^2 - 1}{5m + 5}$

85. $\dfrac{8u^2 - 2u - 15}{4u^4 + 5u^3}$

86. $\dfrac{6n^2 - 7n + 2}{3n^3 - 2n^2}$

87. $\dfrac{y - xy}{xy - x}$

88. $\dfrac{x^2 + y^2}{x + y}$

89. $\dfrac{6a - 6b + 6c}{9a - 9b + 9c}$

90. $\dfrac{3a - 3b - 6}{2a - 2b - 4}$

91. $\dfrac{15x - 3x^2}{25y - 5xy}$

92. $\dfrac{a + b - c}{c - a - b}$

APPLICATIONS

93. ROOFING The *pitch* of a roof is a measure of how steep the roof is. If pitch $= \frac{\text{rise}}{\text{run}}$, find the pitch of the roof of the cabin shown. Express the result in lowest terms.

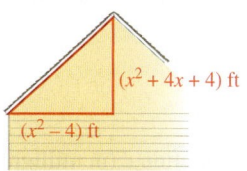

$(x^2 + 4x + 4)$ ft

$(x^2 - 4)$ ft

94. GRAPHIC DESIGN A chart of the basic food groups, in the shape of an equilateral triangle, is to be enlarged and distributed to schools for display in their health classes. What is the length of a side of the original design divided by the length of a side of the enlargement? Express the result in lowest terms.

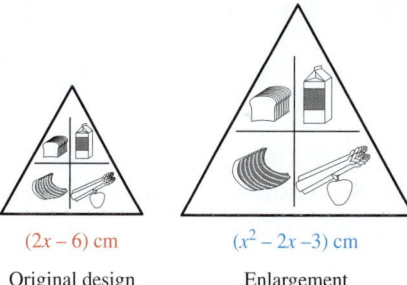

$(2x - 6)$ cm $(x^2 - 2x - 3)$ cm

Original design   Enlargement

95. ORGAN PIPES The number of vibrations n per second of an organ pipe is given by the formula

$$n = \frac{512}{L}$$

where L is the length of the pipe in feet. How many times per second will a 6-foot pipe vibrate?

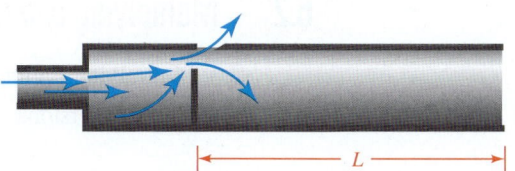

L

96. RAISING TURKEYS   The formula

$$T = \frac{2{,}000m}{m + 1}$$

gives the number T of turkeys on a poultry farm m months after the beginning of the year. How many turkeys will there be on the farm by the end of July?

97. MEDICAL DOSAGES   The formula

$$c = \frac{4t}{t^2 + 1}$$

gives the concentration c (in milligrams per liter) of a certain dosage of medication in a patient's blood stream t hours after the medication is administered. Suppose the patient received the medication at noon. Find the concentration of medication in his blood at the following times later that afternoon.

98. MANUFACTURING   The average cost c (in dollars) for a company to produce x child car seats is given by the formula

$$c = \frac{50x + 50{,}000}{x}$$

Find the company's average cost to produce a car seat if 1,000 are produced.

WRITING

99. Explain why $\dfrac{x - 7}{7 - x} = -1$.

100. Explain the difference between a factor and a term. Give several examples.

101. Explain the error in the following work.

$$\frac{x}{x + 2} = \frac{\overset{1}{\cancel{x}}}{\cancel{x} + 2} = \frac{1}{3}$$
$$\phantom{\frac{x}{x + 2}} \scriptstyle 1$$

102. Explain why there are no values for x for which $\dfrac{x - 7}{x^2 + 49}$ is undefined.

REVIEW

103. State the associative property of addition using the variables a, b, and c.

104. State the distributive property using the variables x, y, and z.

105. If $ab = 0$, what must be true about a or b?

106. What is the product of a number and 1?

CHALLENGE PROBLEMS   Simplify each expression.

107. $\dfrac{(x^2 + 2x + 1)(x^2 - 2x + 1)}{(x^2 - 1)^2}$

108. $\dfrac{2x^2 + 2x - 12}{x^3 + 3x^2 - 4x - 12}$

109. $\dfrac{x^3 - 27}{x^3 - 9x}$

110. $\dfrac{b^3 + a^3}{a^2 - ab + b^2}$

6.2   Multiplying and Dividing Rational Expressions

- Multiplying Rational Expressions
- Unit Conversions
- Dividing Rational Expressions

In this section, we will extend the rules for multiplying and dividing fractions to problems involving multiplication and division of rational expressions.

■ MULTIPLYING RATIONAL EXPRESSIONS

Recall that to multiply fractions, we multiply their numerators and multiply their denominators. For example,

$$\frac{4}{7} \cdot \frac{3}{5} = \frac{4 \cdot 3}{7 \cdot 5} \qquad \text{Multiply the numerators and multiply the denominators.}$$

$$= \frac{12}{35} \qquad \begin{array}{l} \text{Do the multiplication in the numerator.} \\ \text{Do the multiplication in the denominator.} \end{array}$$

We use the same procedure to multiply rational expressions.

Multiplying Rational Expressions	Let A, B, C, and D represent polynomials, where B and D are not 0, $$\frac{A}{B} \cdot \frac{C}{D} = \frac{AC}{BD}$$ Then simplify, if possible.

EXAMPLE 1

ELEMENTARY & INTERMEDIATE
Algebra $f(x)$ **Now**™

Multiply: **a.** $\dfrac{x + 1}{3} \cdot \dfrac{x + 2}{x}$ and   **b.** $\dfrac{35x^3}{17y} \cdot \dfrac{y}{5x}$.

Solution   **a.** $\dfrac{x + 1}{3} \cdot \dfrac{x + 2}{x} = \dfrac{(x + 1)(x + 2)}{3x}$ Multiply the numerators and multiply the denominators.

Since the numerator and denominator do not share any common factors, $\dfrac{(x + 1)(x + 2)}{3x}$ cannot be simplified. We will leave the numerator in factored form.

Caution

When multiplying rational expressions, always write the result in simplest form by removing any factors common to the numerator and denominator.

b. $\dfrac{35x^3}{17y} \cdot \dfrac{y}{5x} = \dfrac{35x^3 \cdot y}{17y \cdot 5x}$ Multiply the numerators and multiply the denominators.

Since $\dfrac{35x^3 \cdot y}{17y \cdot 5x}$ is not in simplest form, we simplify it as follows.

$$\frac{35x^3 \cdot y}{17y \cdot 5x} = \frac{5 \cdot 7 \cdot x \cdot x \cdot x \cdot y}{17 \cdot y \cdot 5 \cdot x} \qquad \text{Factor } 35x^3.$$

$$= \frac{\overset{1}{\cancel{5}} \cdot 7 \cdot \overset{1}{\cancel{x}} \cdot x \cdot x \cdot \overset{1}{\cancel{y}}}{17 \cdot \underset{1}{\cancel{y}} \cdot \underset{1}{\cancel{5}} \cdot \underset{1}{\cancel{x}}} \qquad \begin{array}{l} \text{Replace } \frac{5}{5}, \frac{x}{x}, \text{ and } \frac{y}{y} \text{ with the equivalent fraction } \frac{1}{1}. \text{ This} \\ \text{removes the factor } \frac{5 \cdot x \cdot y}{5 \cdot x \cdot y}, \text{ which is equal to 1.} \end{array}$$

$$= \frac{7x^2}{17}$$

Self Check 1   Multiply and simplify the result, if possible: **a.** $\dfrac{y}{y + 6} \cdot \dfrac{12}{y - 4}$ and   **b.** $\dfrac{a^4}{8b} \cdot \dfrac{24b}{11a^3}$.

EXAMPLE 2

ELEMENTARY & INTERMEDIATE
Algebra $f(x)$ **Now**™

Multiply: **a.** $\dfrac{x + 3}{2x + 4} \cdot \dfrac{6}{x^2 - 9}$ and   **b.** $\dfrac{8x^2 - 8x}{x^2 + x - 56} \cdot \dfrac{3x^2 - 22x + 7}{x - x^2}$.

Solution   **a.** $\dfrac{x + 3}{2x + 4} \cdot \dfrac{6}{x^2 - 9} = \dfrac{(x + 3)6}{(2x + 4)(x^2 - 9)}$ Multiply the numerators and multiply the denominators.

$$= \frac{(x + 3) \cdot 3 \cdot 2}{2(x + 2)(x + 3)(x - 3)}$$

Factor 6. Factor out the GCF from $2x + 4$.
Factor the difference of two squares, $x^2 - 9$.

$$= \frac{\overset{1}{(x + 3)} \cdot 3 \cdot \overset{1}{2}}{\underset{1}{2}(x + 2)\underset{1}{(x + 3)}(x - 3)}$$

Simplify by replacing $\frac{x+3}{x+3}$ and $\frac{2}{2}$ with $\frac{1}{1}$.
This removes the factor $\frac{2 \cdot (x+3)}{2 \cdot (x+3)}$, which
is equal to 1.

$$= \frac{3}{(x + 2)(x - 3)}$$

b. $\dfrac{8x^2 - 8x}{x^2 + x - 56} \cdot \dfrac{3x^2 - 22x + 7}{x - x^2}$

$$= \frac{(8x^2 - 8x)(3x^2 - 22x + 7)}{(x^2 + x - 56)(x - x^2)}$$

Multiply the numerators and multiply the
denominators.

$$= \frac{8x(x - 1)(3x - 1)(x - 7)}{(x + 8)(x - 7)x(1 - x)}$$

Factor all four polynomials.

$$= \frac{\overset{1}{8x}\overset{-1}{(x - 1)}(3x - 1)\overset{1}{(x - 7)}}{(x + 8)\underset{1}{(x - 7)}\underset{1}{x}\underset{1}{(1 - x)}}$$

Since $x - 1$ and $1 - x$ are opposites, simplify
by replacing $\frac{x-1}{1-x}$ with $\frac{-1}{1}$.

$$= \frac{-8(3x - 1)}{x + 8}$$

The result can also be written as $-\dfrac{8(3x - 1)}{x + 8}$.

Self Check 2   Multiply and simplify the result, if possible: **a.** $\dfrac{3n - 9}{3n + 2} \cdot \dfrac{9n^2 - 4}{6}$ and
b. $\dfrac{m^2 - 4m - 5}{2m - m^2} \cdot \dfrac{2m^2 - 4m}{3m^2 - 14m - 5}$.

EXAMPLE 3

Multiply: **a.** $63x\left(\dfrac{1}{7x}\right)$ and   **b.** $5a\left(\dfrac{3a - 1}{a}\right)$.

Solution   **a.** Since any number divided by 1 remains unchanged, we can write any polynomial as a
fraction by inserting a denominator of 1.

$$63x\left(\frac{1}{7x}\right) = \frac{63x}{1}\left(\frac{1}{7x}\right)$$

Write $63x$ as a fraction: $63x = \dfrac{63x}{1}$.

$$= \frac{63x \cdot 1}{1 \cdot 7 \cdot x}$$

Multiply the numerators and the denominators.

$$= \frac{9 \cdot \overset{1}{7} \cdot \overset{1}{x} \cdot 1}{1 \cdot \underset{1}{7} \cdot \underset{1}{x}}$$

Write $63x$ in factored form as $9 \cdot 7 \cdot x$. Then
simplify by removing a factor equal to 1: $\frac{7x}{7x}$.

$$= 9$$

b. $5a\left(\dfrac{3a - 1}{a}\right) = \dfrac{5a}{1}\left(\dfrac{3a - 1}{a}\right)$

Write $5a$ as a fraction: $5a = \dfrac{5a}{1}$.

$$= \frac{5\overset{1}{a}(3a - 1)}{1 \cdot \underset{1}{a}}$$

Multiply the numerators and the denominators.
Then simplify by removing a factor equal to 1: $\frac{a}{a}$.

$$= 5(3a - 1)$$

$$= 15a - 5 \qquad \text{Distribute the multiplication by 5.}$$

Self Check 3   Multiply and simplify the result, if possible: **a.** $36b\left(\dfrac{1}{6b}\right)$ and   **b.** $4x\left(\dfrac{x+3}{x}\right)$.

■ DIVIDING RATIONAL EXPRESSIONS

Recall that one number is called the **reciprocal** of another if their product is 1. To find the reciprocal of a fraction, we invert its numerator and denominator. We have seen that to divide fractions, we multiply the first fraction by the reciprocal of the second fraction.

$$\frac{4}{7} \div \frac{3}{5} = \frac{4}{7} \cdot \frac{5}{3} \qquad \text{Invert } \frac{3}{5} \text{ and change the division to a multiplication.}$$

$$= \frac{20}{21} \qquad \text{Multiply the numerators and multiply the denominators.}$$

We use the same procedure to divide rational expressions.

Dividing Rational Expressions	Let A, B, C, and D represent polynomials, where B, C, and D are not 0, $$\frac{A}{B} \div \frac{C}{D} = \frac{A}{B} \cdot \frac{D}{C} = \frac{AD}{BC}$$ Then simplify, if possible.

EXAMPLE 4

ELEMENTARY & INTERMEDIATE
Algebra *f(x)* **Now**™

Solution

Divide: **a.** $\dfrac{a}{13} \div \dfrac{17}{26}$ and   **b.** $\dfrac{9x}{35y} \div \dfrac{15x^2}{14}$.

a. $\dfrac{a}{13} \div \dfrac{17}{26} = \dfrac{a}{13} \cdot \dfrac{26}{17} \qquad$ Multiply by the reciprocal of $\dfrac{17}{26}$.

$$= \frac{a \cdot 2 \cdot 13}{13 \cdot 17} \qquad \begin{array}{l}\text{Multiply the numerators and denominators.}\\ \text{Then factor 26 as } 2 \cdot 13.\end{array}$$

$$= \frac{a \cdot 2 \cdot \overset{1}{\cancel{13}}}{\underset{1}{\cancel{13}} \cdot 17} \qquad \text{Simplify by removing a factor equal to 1.}$$

$$= \frac{2a}{17}$$

Caution

When dividing rational expressions, always write the result in simplest form (lowest terms), by removing any factors common to the numerator and denominator.

b. $\dfrac{9x}{35y} \div \dfrac{15x^2}{14} = \dfrac{9x}{35y} \cdot \dfrac{14}{15x^2} \qquad$ Multiply by the reciprocal of $\dfrac{15x^2}{14}$.

$$= \frac{3 \cdot 3 \cdot x \cdot 2 \cdot 7}{5 \cdot 7 \cdot y \cdot 3 \cdot 5 \cdot x \cdot x} \qquad \begin{array}{l}\text{Multiply the numerators and denominators.}\\ \text{Then factor 9, 35, 14, and } 15x^2.\end{array}$$

$$= \frac{3 \cdot \overset{1}{\cancel{3}} \cdot \overset{1}{\cancel{x}} \cdot 2 \cdot \overset{1}{\cancel{7}}}{5 \cdot \underset{1}{\cancel{7}} \cdot y \cdot \underset{1}{\cancel{3}} \cdot 5 \cdot \underset{1}{\cancel{x}} \cdot x} \qquad \text{Simplify by removing factors equal to 1.}$$

$$= \frac{6}{25xy}$$

Self Check 4   Divide and simplify the result, if possible: $\dfrac{8a}{3b} \div \dfrac{16a^2}{9b^2}$.

EXAMPLE 5

ELEMENTARY &
INTERMEDIATE

Algebra *f(x)* **Now**™

Divide: $\dfrac{x^2 + x}{3x - 15} \div \dfrac{x^2 + 2x + 1}{6x - 30}$.

Solution

$$\dfrac{x^2 + x}{3x - 15} \div \dfrac{x^2 + 2x + 1}{6x - 30}$$

The Language of Algebra

To find the reciprocal of $\dfrac{x^2 + 2x + 1}{6x - 30}$, we invert it. To *invert* means to turn upside down: $\dfrac{6x - 30}{x^2 + 2x + 1}$. Some amusement park thrill rides have giant loops where the riders become *inverted*.

$$= \dfrac{x^2 + x}{3x - 15} \cdot \dfrac{6x - 30}{x^2 + 2x + 1} \qquad \text{Multiply by the reciprocal of } \dfrac{x^2 + 2x + 1}{6x - 30}.$$

$$= \dfrac{x(x + 1) \cdot 2 \cdot 3(x - 5)}{3(x - 5)(x + 1)(x + 1)} \qquad \begin{array}{l}\text{Multiply the numerators and denominators.} \\ \text{Then factor the polynomials.}\end{array}$$

$$= \dfrac{x(\cancel{x + 1}) \cdot 2 \cdot \cancel{3}(\cancel{x - 5})}{\cancel{3}(\cancel{x - 5})(\cancel{x + 1})(x + 1)} \qquad \text{Simplify by removing factors equal to 1.}$$

$$= \dfrac{2x}{x + 1}$$

Self Check 5   Divide and simplify the result, if possible: $\dfrac{z^2 - 1}{z^2 + 4z + 3} \div \dfrac{z - 1}{z^2 + 2z - 3}$.

EXAMPLE 6

ELEMENTARY &
INTERMEDIATE

Algebra *f(x)* **Now**™

Divide: $\dfrac{2x^2 - 3x - 2}{2x + 1} \div (4 - x^2)$.

Solution   To divide a rational expression by a polynomial, we write the polynomial as a fraction by inserting a denominator of 1, and then we divide the fractions.

$$\dfrac{2x^2 - 3x - 2}{2x + 1} \div (4 - x^2)$$

$$= \dfrac{2x^2 - 3x - 2}{2x + 1} \div \dfrac{4 - x^2}{1} \qquad \text{Write } 4 - x^2 \text{ as a fraction with a denominator of 1.}$$

$$= \dfrac{2x^2 - 3x - 2}{2x + 1} \cdot \dfrac{1}{4 - x^2} \qquad \text{Multiply by the reciprocal of } \dfrac{4 - x^2}{1}.$$

$$= \dfrac{(2x + 1)(x - 2) \cdot 1}{(2x + 1)(2 + x)(2 - x)} \qquad \begin{array}{l}\text{Multiply the numerators and denominators.} \\ \text{Then factor } 2x^2 - 3x - 2 \text{ and } 4 - x^2.\end{array}$$

$$= \dfrac{(\cancel{2x + 1})(\overset{-1}{\cancel{x - 2}}) \cdot 1}{(\cancel{2x + 1})(2 + x)(\cancel{2 - x})} \qquad \begin{array}{l}\text{Since } x - 2 \text{ and } 2 - x \text{ are opposites, simplify} \\ \text{by replacing } \dfrac{x - 2}{2 - x} \text{ with } \dfrac{-1}{1}.\end{array}$$

$$= \dfrac{-1}{2 + x}$$

$$= -\dfrac{1}{2 + x} \qquad \text{Write the } - \text{ symbol in front of the rational expression.}$$

Self Check 6   Divide and simplify the result, if possible: $(b - a) \div \dfrac{a^2 - b^2}{a^2 + ab}$.

■ UNIT CONVERSIONS

We can use the concepts discussed in this section to make conversions from one unit of measure to another. *Unit conversion factors* play an important role in this process. A **unit conversion factor** is a fraction that has value 1. For example, we can use the fact that 1 square yard = 9 square feet to form two unit conversion factors:

$$\frac{1 \text{ yd}^2}{9 \text{ ft}^2} = 1 \quad \text{Read as "1 square yard per 9 square feet."} \qquad \frac{9 \text{ ft}^2}{1 \text{ yd}^2} = 1 \quad \text{Read as "9 square feet per 1 square yard."}$$

Since a unit conversion factor is equal to 1, multiplying a measurement by a unit conversion factor does not change the measurement, it only changes the units of measure.

EXAMPLE 7

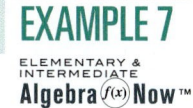

Carpeting. A roll of carpeting is 12 feet wide and 150 feet long. Find the number of square yards of carpeting on the roll.

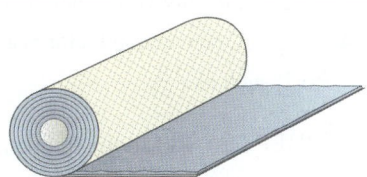

Solution When unrolled, the carpeting forms a rectangular shape with an area of $12 \cdot 150 = 1{,}800$ square feet.

We multiply 1,800 ft^2 by a unit conversion factor such that the units of ft^2 are removed and the units of yd^2 are introduced. Then we can remove units common to the numerator and denominator.

$$1{,}800 \text{ ft}^2 = \frac{1{,}800 \text{ ft}^2}{1} \cdot \frac{1 \text{ yd}^2}{9 \text{ ft}^2} \qquad \text{Multiply by a unit conversion factor that relates yd}^2 \text{ to ft}^2.$$

$$= \frac{1{,}800 \text{ ft}^2}{1} \cdot \frac{1 \text{ yd}^2}{9 \text{ ft}^2} \qquad \text{Remove the units of ft}^2 \text{ that are common to the numerator and denominator.}$$

$$= 200 \text{ yd}^2 \qquad \text{Divide 1,800 by 9.}$$

There are 200 yd^2 of carpeting on the roll.

EXAMPLE 8

The speed of light. The speed with which light moves through space is about 186,000 miles per second. Express this speed in miles per minute.

Solution The speed of light can be expressed as $\frac{186{,}000 \text{ miles}}{1 \text{ sec}}$. We multiply this fraction by a unit conversion factor such that the units of seconds are removed and the units of minutes are introduced. Since 60 seconds = 1 minute, we will use $\frac{60 \text{ sec}}{1 \text{ min}}$.

$$\frac{186{,}000 \text{ miles}}{1 \text{ sec}} = \frac{186{,}000 \text{ miles}}{1 \text{ sec}} \cdot \frac{60 \text{ sec}}{1 \text{ min}} \qquad \text{Multiply by a unit conversion factor that relates seconds to minutes.}$$

$$= \frac{186{,}000 \text{ miles}}{1 \text{ sec}} \cdot \frac{60 \text{ sec}}{1 \text{ min}} \qquad \text{Remove the units of seconds that are common to the numerator and denominator.}$$

$$= \frac{11{,}160{,}000 \text{ miles}}{1 \text{ min}} \qquad \text{Multiply 186,000 and 60.}$$

The speed of light is about 11,160,000 miles per minute.

Answers to Self Checks   **1. a.** $\dfrac{12y}{(y + 6)(y - 4)}$,   **b.** $\dfrac{3a}{11}$ **2. a.** $\dfrac{(n - 3)(3n - 2)}{2}$,   **b.** $-\dfrac{2(m + 1)}{3m + 1}$ **3. a.** 6,
b. $4x + 12$ **4.** $\dfrac{3b}{2a}$ **5.** $z - 1$ **6.** $-a$

6.2   STUDY SET   ELEMENTARY & INTERMEDIATE Algebra *f(x)* Now™

VOCABULARY   Fill in the blanks.

1. The quotient of _____ is -1. For example,
   $\dfrac{x - 8}{8 - x} = -1$.

2. The _____ of $\dfrac{x^2 + 6x + 1}{10x}$ is $\dfrac{10x}{x^2 + 6x + 1}$.

3. To find the reciprocal of a rational expression, we
   _____ its numerator and denominator.

4. A _____ conversion factor is a fraction that is equal
   to 1, such as $\dfrac{3 \text{ ft}}{1 \text{ yd}}$.

CONCEPTS   Fill in the blanks.

5. To multiply rational expressions, multiply their
   _____ and multiply their _____. In
   symbols,

   $$\frac{A}{B} \cdot \frac{C}{D} = \quad$$

6. To divide two rational expressions, multiply the first
   by the _____ of the second. In symbols,

   $$\frac{A}{B} \div \frac{C}{D} = \frac{A}{B} \cdot \quad$$

Simplify each fraction.

7. $\dfrac{(x + 7) \cdot 2 \cdot 5}{5(x + 1)(x + 7)(x - 9)}$

8. $\dfrac{y \cdot y \cdot y(15 - y)}{y(y - 15)(y + 1)}$

Write each polynomial in fractional form.

9. $6n$

10. $3x + 5$

Find the reciprocal of each polynomial.

11. $18x$

12. $\dfrac{7m + 11}{m - 2}$

13. $\dfrac{x^2 + 1}{x^2 + 2x + 1}$

14. Find the product of the rational expression and its
    reciprocal.

   $$\frac{3}{x + 2} \cdot \frac{x + 2}{3}$$

15. What units are common to the numerator and
    denominator?

   $$\frac{45 \text{ ft}}{1} \cdot \frac{1 \text{ yd}}{3 \text{ ft}}$$

16. Use the following fact to write two unit conversion
    factors.

   1 square mile $=$ 640 acres

NOTATION

17. The abbreviation for meter is m.
    a. How do we write square meters in symbols?
    b. How do we write cubic meters in symbols?

18. **a.** What fact is indicated by the unit conversion
        factor $\dfrac{1 \text{ day}}{24 \text{ hours}}$?
    b. Fill in the blank: $\dfrac{1 \text{ day}}{24 \text{ hours}} = \quad$.

PRACTICE   Multiply, and then simplify, if possible.

19. $\dfrac{3}{y} \cdot \dfrac{y}{2}$

20. $\dfrac{2}{z} \cdot \dfrac{z}{3}$

21. $\dfrac{35n}{12} \cdot \dfrac{16}{7n^2}$

22. $\dfrac{11m}{21} \cdot \dfrac{14}{55m^3}$

23. $\dfrac{2x^2y}{3xy} \cdot \dfrac{3xy^2}{2}$

24. $\dfrac{2x^2z}{z} \cdot \dfrac{5x}{z}$

25. $\dfrac{10r^2st^3}{6rs^2} \cdot \dfrac{3r^3t}{2rst}$

26. $\dfrac{3a^3b}{25cd^3} \cdot \dfrac{5cd^2}{6ab}$

27. $\dfrac{z + 7}{7} \cdot \dfrac{z + 2}{z}$

28. $\dfrac{a - 3}{a} \cdot \dfrac{a + 3}{5}$

29. $\dfrac{x+5}{5} \cdot \dfrac{x}{x+5}$

30. $\dfrac{a-9}{9} \cdot \dfrac{8a}{a-9}$

31. $\dfrac{x-2}{2} \cdot \dfrac{2x}{2-x}$

32. $\dfrac{y-3}{y} \cdot \dfrac{3y}{3-y}$

33. $\dfrac{5}{m} \cdot m$

34. $p \cdot \dfrac{10}{p}$

35. $15x\left(\dfrac{x+1}{15x}\right)$

36. $30t\left(\dfrac{t-7}{30t}\right)$

37. $12y\left(\dfrac{y+8}{6y}\right)$

38. $16x\left(\dfrac{3x+8}{4x}\right)$

39. $(x+8)\dfrac{x+5}{x+8}$

40. $(y-2)\dfrac{y+3}{y-2}$

41. $10(h+9)\dfrac{h-3}{h+9}$

42. $r(r-25)\dfrac{r+4}{r-25}$

43. $\dfrac{(x+1)^2}{x+1} \cdot \dfrac{x+2}{x+1}$

44. $\dfrac{(y-3)^2}{y-3} \cdot \dfrac{y-3}{y-3}$

45. $\dfrac{2x+6}{x+3} \cdot \dfrac{3}{4x}$

46. $\dfrac{3y-9}{y-3} \cdot \dfrac{y}{3y^2}$

47. $\dfrac{x^2-x}{x} \cdot \dfrac{3x-6}{3-3x}$

48. $\dfrac{5z-10}{z+2} \cdot \dfrac{3}{6-3z}$

49. $\dfrac{x^2+x-6}{5x} \cdot \dfrac{5x-10}{x+3}$

50. $\dfrac{z^2+4z-5}{5z-5} \cdot \dfrac{5z}{z+5}$

51. $\dfrac{m^2-2m-3}{2m+4} \cdot \dfrac{m^2-4}{m^2+3m+2}$

52. $\dfrac{p^2-p-6}{3p-9} \cdot \dfrac{p^2-9}{p^2+6p+9}$

53. $\dfrac{2x^2+17x+21}{x^2+2x-35} \cdot \dfrac{x^2-25}{2x^2-7x-15}$

54. $\dfrac{2x^2-9x+9}{x^2-1} \cdot \dfrac{2x^2-5x+3}{x^2-3x}$

55. $\dfrac{4x^2-12xy+9y^2}{x^3y^2} \cdot \dfrac{x^2y}{9y^2-4x^2}$

56. $\dfrac{ab^4}{16b^2-25a^2} \cdot \dfrac{25a^2-40ab+16b^2}{a^2b^5}$

57. $\dfrac{3x^2+5x+2}{x^2-9} \cdot \dfrac{x-3}{x^2-4} \cdot \dfrac{x^2+5x+6}{6x+4}$

58. $\dfrac{x^2-25}{3x+6} \cdot \dfrac{x^2+x-2}{2x+10} \cdot \dfrac{6x}{3x^2-18x+15}$

Divide, then simplify, if possible.

59. $\dfrac{2}{y} \div \dfrac{4}{3}$

60. $\dfrac{3}{a} \div \dfrac{a}{9}$

61. $\dfrac{3x}{y} \div \dfrac{2x}{4}$

62. $\dfrac{3y}{8} \div \dfrac{2y}{4y}$

63. $\dfrac{x^2y}{3xy} \div \dfrac{xy^2}{6y}$

64. $\dfrac{2xz}{z} \div \dfrac{4x^2}{z^2}$

65. $24n^2 \div \dfrac{18n^3}{n-1}$

66. $12m \div \dfrac{16m^2}{m+4}$

67. $\dfrac{x+2}{3x} \div \dfrac{x+2}{2}$

68. $\dfrac{z-3}{3z} \div \dfrac{z+3}{z}$

69. $\dfrac{(z-2)^2}{3z^2} \div \dfrac{z-2}{6z}$

70. $\dfrac{(x-3)^4}{x+7} \div \dfrac{(x-3)^2}{x+7}$

71. $\dfrac{9a-18}{28} \div \dfrac{9a^3}{35}$

72. $\dfrac{3x+6}{40} \div \dfrac{3x^2}{24}$

73. $\dfrac{x^2-4}{3x+6} \div \dfrac{2-x}{x+2}$

74. $\dfrac{x^2-9}{5x+15} \div \dfrac{3-x}{x+3}$

75. $\dfrac{x^2-1}{3x-3} \div (x+1)$

76. $\dfrac{x^2-16}{x-4} \div (3x+12)$

77. $\dfrac{x^2-2x-35}{3x^2+27x} \div \dfrac{x^2+7x+10}{6x^2+12x}$

78. $\dfrac{x^2-x-6}{2x^2+9x+10} \div \dfrac{x^2-25}{2x^2+15x+25}$

79. $\dfrac{36c^2-49d^2}{3d^3} \div \dfrac{12c+14d}{d^4}$

80. $\dfrac{25y^2-16z^2}{2yz} \div \dfrac{10y-8z}{y^2}$

81. $\dfrac{2d^2+8d-42}{3-d} \div \dfrac{2d^2+14d}{d^2+5d}$

82. $\dfrac{5x^2 + 13x - 6}{x + 3} \div \dfrac{5x^2 - 17x + 6}{x - 2}$

83. $\dfrac{2r - 3s}{12} \div (4r^2 - 12rs + 9s^2)$

84. $\dfrac{3m + n}{18} \div (9m^2 + 6mn + n^2)$

Complete each unit conversion.

85. $\dfrac{150 \text{ yd}}{1} \cdot \dfrac{3 \text{ ft}}{1 \text{ yd}} = ?$ 86. $\dfrac{60 \text{ in.}}{1} \cdot \dfrac{1 \text{ ft}}{12 \text{ in.}} = ?$

87. $\dfrac{30 \text{ meters}}{1 \text{ sec}} \cdot \dfrac{60 \text{ sec}}{1 \text{ min}} = ?$ 88. $\dfrac{288 \text{ in.}^2}{1 \text{ year}} \cdot \dfrac{1 \text{ ft}^2}{144 \text{ in.}^2} = ?$

APPLICATIONS

89. INTERNATIONAL ALPHABET   The symbols representing the letters A, B, C, D, E, and F in an international code used at sea are printed six to a sheet and then cut into separate cards. If each card is a square, find the area of the large printed sheet shown.

$\dfrac{2x + 1}{2}$ in.

90. PHYSICS EXPERIMENTS   The table contains algebraic expressions for the rate an object travels, and the time traveled at that rate in terms of a constant k. Complete the table.

Rate (mph)	Time (hr)	Distance (mi)
$\dfrac{k^2 + k - 6}{k - 3}$	$\dfrac{k^2 - 9}{k^2 - 4}$	

91. TALKING   According to the *Sacramento Bee* newspaper, the number of words an average man speaks a day is about 12,000. How many words does an average man speak in 1 year?   (*Hint:* 365 days = 1 year.)

92. CLASSROOM SPACE   The recommended size of an elementary school classroom in the United States is approximately 900 square feet. Convert this to square yards.

93. NATURAL LIGHT   According to the University of Georgia School Design and Planning Laboratory, the basic classroom should have at least 72 square feet of windows for natural light. Convert this to square yards.

94. TRUCKING   A cement truck holds 9 cubic yards of concrete. How many cubic feet of concrete does it hold?   (*Hint:* 27 cubic feet = 1 cubic yard.)

95. BEARS   The maximum speed a grizzly bear can run is about 30 miles per hour. What is its average speed in miles per minute?

96. FUEL ECONOMY   Use the information that follows to determine the miles per fluid ounce of gasoline for city and for highway driving for the Chevy Astro Van.   (*Hint:* 1 gallon = 128 fluid ounces.)

2003 Chevrolet Astro 2WD

Fuel Economy

Fuel Type	Regular
MPG (city)	16
MPG (highway)	20

97. TV TRIVIA   On the comedy television series *Green Acres* (1965–1971), New York socialites Oliver Wendell Douglas (played by Eddie Albert) and his wife, Lisa Douglas (played by Eva Gabor), move from New York to purchase a 160-acre farm in Hooterville. Convert this to square miles.   (*Hint:* 1 square mile = 640 acres.)

98. CAMPING   The capacity of backpacks is usually given in cubic inches. Convert a backpack capacity of 5,400 cubic inches to cubic feet. (*Hint:* 1 cubic foot = 1,728 cubic inches.)

WRITING

99. Explain how to multiply rational expressions.

100. To divide rational expressions, you must first know how to multiply rational expressions. Explain why.

101. Explain how to find the reciprocal of a rational expression.

102. Explain how to write $x^2 + 2x - 1$ as a fraction.

103. Explain why 60 miles per hour and 1 mile per minute are the same speed.

104. Explain why $\dfrac{1 \text{ ft}}{12 \text{ in.}}$ is equal to 1.

REVIEW

105. HARDWARE   An aluminum brace used to support a wooden shelf has a length that is 2 inches less than twice the width of the shelf. The brace is anchored to the wall 8 inches below the shelf, as shown. Find the width of the shelf and the length of the brace.

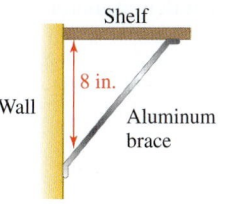

Shelf

Wall

8 in.

Aluminum brace

106. FORENSIC MEDICINE   The kinetic energy E of a moving object is given by $E = \frac{1}{2}mv^2$, where m is the mass of the object (in kilograms) and v is the object's velocity (in meters per second). Kinetic energy is measured in joules. Examining the damage done to a victim, a police pathologist determines that the energy of a 3-kilogram mass at impact was 54 joules. Find the velocity at impact.

CHALLENGE PROBLEMS   **Perform the operations. Simplify, if possible.**

107. $\dfrac{c^3 - 2c^2 + 5c - 10}{c^2 - c - 2} \cdot \dfrac{c^3 + c^2 - 5c - 5}{c^4 - 25}$

108. $\dfrac{x^3 - y^3}{x^3 + y^3} \div \dfrac{x^3 + x^2y + xy^2}{x^2y - xy^2 + y^3}$

109. $\dfrac{y^2}{x + 1} \cdot \dfrac{x^2 + 2x + 1}{x^2 - 1} \div \dfrac{3y}{xy - y}$

110. $\dfrac{x^2 - y^2}{x^4 - x^3} \div \dfrac{x - y}{x^2} \div \dfrac{x^2 + 2xy + y^2}{x + y}$

6.3   # Addition and Subtraction with Like Denominators; Least Common Denominators

- Adding and Subtracting Rational Expressions Having Like Denominators
- Finding the Least Common Denominator   • Building Rational Expressions

In this section, we extend the rules for adding and subtracting fractions to problems involving addition and subtraction of rational expressions.

■ ADDING AND SUBTRACTING RATIONAL EXPRESSIONS HAVING LIKE DENOMINATORS

To add or subtract fractions that have a common denominator, we add or subtract their numerators and write the sum or difference over the common denominator. For example,

$$\frac{3}{7} + \frac{2}{7} = \frac{3 + 2}{7} \qquad\qquad \frac{3}{7} - \frac{2}{7} = \frac{3 - 2}{7}$$

$$= \frac{5}{7} \qquad\qquad\qquad = \frac{1}{7}$$

We use the same procedure to add and subtract rational expressions with like denominators.

Adding and Subtracting Rational Expressions	If $\dfrac{A}{D}$ and $\dfrac{B}{D}$ are rational expressions,

$$\frac{A}{D} + \frac{B}{D} = \frac{A + B}{D} \qquad\qquad \frac{A}{D} - \frac{B}{D} = \frac{A - B}{D}$$

Then simplify, if possible.

EXAMPLE 1 Add: **a.** $\dfrac{x}{8} + \dfrac{3x}{8}$ and **b.** $\dfrac{4s - 9}{9t} + \dfrac{7}{9t}$.

ELEMENTARY &
INTERMEDIATE
Algebra$f(x)$ **Now**™

Solution **a.** The rational expressions have the same denominator, 8. We add their numerators and write the sum over the common denominator.

Caution

When adding or subtracting rational expressions, always write the result in simplest form by removing any factors common to the numerator and denominator.

$$\frac{x}{8} + \frac{3x}{8} = \frac{x + 3x}{8}$$

$$= \frac{4x}{8} \qquad \text{Combine like terms in the numerator.}$$

This result can be simplified as follows:

$$= \frac{\overset{1}{\cancel{4}} \cdot x}{\underset{1}{\cancel{4}} \cdot 2} \qquad \text{Factor 8 as } 4 \cdot 2. \text{ Then simplify by removing a factor equal to 1.}$$

$$= \frac{x}{2}$$

Notation

In Example 1b, the numerator of the result may be written two ways:

Not factored   Factored

$$\dfrac{4s - 2}{9t} \quad \text{or} \quad \dfrac{2(2s - 1)}{9t}$$

Check with your instructor to see which form he or she prefers.

---

b. $\dfrac{4s - 9}{9t} + \dfrac{7}{9t} = \dfrac{4s - 9 + 7}{9t}$ Add the numerators. Write the sum over the common denominator, $9t$.

$$= \frac{4s - 2}{9t} \qquad \text{Combine like terms in the numerator.}$$

To attempt to simplify the result, we factor the numerator to get $\dfrac{2(2s - 1)}{9t}$. Since the numerator and denominator do not have any common factors, $\dfrac{4s - 2}{9t}$ cannot be simplified. Thus,

$$\frac{4s - 9}{9t} + \frac{7}{9t} = \frac{4s - 2}{9t} \quad \text{or} \quad \frac{2(2s - 1)}{9t}$$

Self Check 1 Add and simplify the result, if possible: **a.** $\dfrac{2x}{15} + \dfrac{4x}{15}$ and **b.** $\dfrac{3m - 8}{23n} + \dfrac{2}{23n}$. ◼

EXAMPLE 2 Add: **a.** $\dfrac{3x + 21}{5x + 10} + \dfrac{8x + 1}{5x + 10}$ and **b.** $\dfrac{x^2 + 9x - 7}{2x(x - 6)} + \dfrac{x^2 - 9x}{(x - 6)2x}$.

ELEMENTARY &
INTERMEDIATE
Algebra$f(x)$ **Now**™

Solution **a.** $\dfrac{3x + 21}{5x + 10} + \dfrac{8x + 1}{5x + 10} = \dfrac{3x + 21 + 8x + 1}{5x + 10}$ Add the numerators. Write the sum over the common denominator, $5x + 10$.

$$= \frac{11x + 22}{5x + 10} \qquad \text{Combine like terms in the numerator.}$$

$$= \frac{11(\overset{1}{\cancel{x + 2}})}{5(\underset{1}{\cancel{x + 2}})} \qquad \begin{array}{l}\text{Factor the numerator and the denominator.} \\ \text{Then simplify by removing a factor equal} \\ \text{to 1.}\end{array}$$

$$= \frac{11}{5}$$

b. By the commutative property of multiplication, $2x(x - 6) = (x - 6)2x$. Therefore, the denominators are the same. We add the numerators and write the sum over the common denominator.

$$\frac{x^2 + 9x - 7}{2x(x - 6)} + \frac{x^2 - 9x}{(x - 6)2x} = \frac{x^2 + 9x - 7 + x^2 - 9x}{2x(x - 6)}$$

$$= \frac{2x^2 - 7}{2x(x - 6)} \qquad \text{Combine like terms in the numerator.}$$

Since the numerator, $2x^2 - 7$, does not factor, $\dfrac{2x^2 - 7}{2x(x - 6)}$ is in simplest form.

Self Check 2   Add and simplify the result, if possible: **a.** $\frac{m + 3}{3m - 9} + \frac{m - 9}{3m - 9}$ and

b. $\frac{c^2 - c}{(c - 1)(c + 2)} + \frac{c^2 - 10c}{(c + 2)(c - 1)}$.

EXAMPLE 3

ELEMENTARY & INTERMEDIATE
Algebra $f(x)$ Now™

Subtract: $\dfrac{x + 6}{x^2 + 4x - 5} - \dfrac{1}{x^2 + 4x - 5}$.

Solution   Because the rational expressions have the same denominator, we subtract their numerators and keep the common denominator.

$$\frac{x + 6}{x^2 + 4x - 5} - \frac{1}{x^2 + 4x - 5} = \frac{x + 6 - 1}{x^2 + 4x - 5} \qquad \begin{array}{l}\text{Subtract the numerators. Write the}\\ \text{difference over the common}\\ \text{denominator, } x^2 + 4x - 5.\end{array}$$

$$= \frac{x + 5}{x^2 + 4x - 5} \qquad \begin{array}{l}\text{Combine like terms in the}\\ \text{numerator.}\end{array}$$

$$= \frac{x + 5}{(x + 5)(x - 1)} \qquad \text{Factor the denominator.}$$

$$= \frac{\overset{1}{\cancel{x + 5}}}{\underset{1}{\cancel{(x + 5)}}(x - 1)} \qquad \begin{array}{l}\text{Simplify by removing a factor equal}\\ \text{to 1.}\end{array}$$

$$= \frac{1}{x - 1}$$

Self Check 3   Subtract and simplify the result, if possible: $\dfrac{n - 3}{n^2 - 16} - \dfrac{1}{n^2 - 16}$.

EXAMPLE 4

ELEMENTARY & INTERMEDIATE
Algebra $f(x)$ Now™

Subtract: **a.** $\dfrac{x^2 + 10x}{x + 3} - \dfrac{4x - 9}{x + 3}$ and   **b.** $\dfrac{x^2}{(x + 7)(x - 8)} - \dfrac{x^2 + 14x}{(x + 7)(x - 8)}$.

Solution   **a.** To subtract the numerators, each term of $4x - 9$ must be subtracted from $x^2 + 10x$.

This sign applies to   This numerator is written within parentheses
the entire numerator $4x - 9$.   to make sure that we subtract both of its terms.

$$\frac{x^2 + 10x}{x + 3} - \frac{4x - 9}{x + 3} = \frac{x^2 + 10x - (4x - 9)}{x + 3} \qquad \begin{array}{l}\text{Subtract the numerators. Write the}\\ \text{difference over the common}\\ \text{denominator.}\end{array}$$

Notation

A fraction bar groups the terms of the numerator together and the terms of the denominator together. It is helpful to think of an implied pair of parentheses being around each.

$$\frac{4x - 9}{x + 3} = \frac{(4x - 9)}{(x + 3)}$$

$$= \frac{x^2 + 10x - 4x + 9}{x + 3}$$

In the numerator, use the distributive property: $-(4x - 9) = -1(4x - 9) = -4x + 9$.

$$= \frac{x^2 + 6x + 9}{x + 3}$$

Combine like terms in the numerator.

$$= \frac{(x + 3)(x + 3)}{x + 3}$$

Factor the numerator.

$$= \frac{\overset{1}{\cancel{(x + 3)}}(x + 3)}{\underset{1}{\cancel{x + 3}}}$$

Simplify by removing a factor equal to 1.

$$= x + 3$$

b. We subtract the numerators and write the difference over the common denominator.

Success Tip

In Example 4b, $-(x^2 + 14x)$ means $-1(x^2 + 14x)$. The multiplication by -1 changes the signs of x^2 and $14x$:

$$-(x^2 + 14x) = -x^2 - 14x$$

$$\frac{x^2}{(x + 7)(x - 8)} - \frac{x^2 + 14x}{(x + 7)(x - 8)} = \frac{x^2 - (x^2 + 14x)}{(x + 7)(x - 8)}$$

Write the second numerator within parentheses.

$$= \frac{x^2 - x^2 - 14x}{(x + 7)(x - 8)}$$

Use the distributive property: $-(x^2 + 14x) = -x^2 - 14x$.

$$= \frac{-14x}{(x + 7)(x - 8)}$$

In the numerator, combine like terms.

The result can also be written as $-\dfrac{14x}{(x + 7)(x - 8)}$.

Self Check 4   Subtract and simplify the result, if possible: **a.** $\dfrac{x^2 + 3x}{x - 1} - \dfrac{5x - 1}{x - 1}$ and
b. $\dfrac{3y^2}{(y + 3)(y - 3)} - \dfrac{3y^2 + y}{(y + 3)(y - 3)}$.

■ FINDING THE LEAST COMMON DENOMINATOR

We will now discuss two skills that are needed for adding and subtracting rational expressions that have unlike denominators. To begin, let's consider

$$\frac{11}{8x} + \frac{7}{18x^2}$$

To add these expressions, we must express them as equivalent expressions with a common denominator. The **least common denominator (LCD)** is usually the easiest one to use. The least common denominator of several rational expressions can be found as follows.

Finding the LCD   1. Factor each denominator completely.
2. The LCD is a product that uses each different factor obtained in step 1 the greatest number of times it appears in any one factorization.

EXAMPLE 5

ELEMENTARY & INTERMEDIATE
Algebra $f(x)$ **Now**™

Find the LCD of each pair of rational expressions: **a.** $\dfrac{11}{8x}$ and $\dfrac{7}{18x^2}$ and
b. $\dfrac{20}{x}$ and $\dfrac{4x}{x - 9}$.

Solution **a.** We begin by factoring each denominator completely.

$$8x = 2 \cdot 2 \cdot 2 \cdot x \qquad \text{Prime factor 8.}$$
$$18x^2 = 2 \cdot 3 \cdot 3 \cdot x \cdot x \qquad \text{Prime factor 18.}$$

The factorizations of $8x$ and $18x^2$ contain the factors 2, 3, and x. We form a product using each factor the greatest number of times it appears in any one factorization.

Success Tip

For Example 5a, the factorizations can be written:

$$8x = 2^3 \cdot x$$
$$18x^2 = 2 \cdot 3^2 \cdot x^2$$

Note that the highest power of each factor is used to form the LCD.

$$\text{LCD} = 2^3 \cdot 3^2 \cdot x^2 = 72x^2$$

┌ The greatest number of times the factor 2 appears is three times.
       ┌ The greatest number of times the factor 3 appears is twice.
             ┌ The greatest number of times the factor x appears is twice.

$$\text{LCD} = \mathbf{2 \cdot 2 \cdot 2 \cdot 3 \cdot 3 \cdot x \cdot x}$$
$$= 72x^2$$

The LCD for $\dfrac{11}{8x}$ and $\dfrac{7}{18x^2}$ is $72x^2$.

b. Since the denominators of $\dfrac{20}{x}$ and $\dfrac{4x}{x - 9}$ are completely factored, the factor x appears once and the factor $x - 9$ appears once. Thus, the LCD is $x(x - 9)$.

Self Check 5 Find the LCD of each pair of rational expressions: **a.** $\dfrac{y + 7}{6y^3}$ and $\dfrac{7}{75y}$ and **b.** $\dfrac{a - 3}{a + 3}$ and $\dfrac{21}{a}$.

EXAMPLE 6 Find the LCD of each pair of rational expressions: **a.** $\dfrac{x}{7x + 7}$ and $\dfrac{x - 2}{5x + 5}$ and **b.** $\dfrac{6 - x}{x^2 + 8x + 16}$ and $\dfrac{15x}{x^2 - 16}$.

Solution **a.** Factor each denominator completely.

$$7x + 7 = 7(x + 1)$$
$$5x + 5 = 5(x + 1)$$

The factorizations of $7x + 7$ and $5x + 5$ contain the factors 7, 5, and $x + 1$. We form a product using each factor the greatest number of times it appears in any one factorization.

┌ The greatest number of times the factor 7 appears is once.
       ┌ The greatest number of times the factor 5 appears is once.
             ┌ The greatest number of times the factor $x + 1$ appears is once.

$$\text{LCD} = \mathbf{7 \cdot 5 \cdot (x + 1)} = 35(x + 1)$$

b. Factor each denominator completely.

$$x^2 + 8x + 16 = (x + 4)(x + 4)$$
$$x^2 - 16 = (x + 4)(x - 4)$$

The factorizations of $x^2 + 8x + 16$ and $x^2 - 16$ contain the factors $x + 4$ and $x - 4$.

Notation

Rather than doing a multiplication, it is often better to leave an LCD in factored form. Therefore, in Example 6, we have:

$$\text{LCD} = 35(x + 1)$$
$$\text{LCD} = (x + 4)^2(x - 4)$$

┌ The greatest number of times the factor $x + 4$ appears is twice.
       ┌ The greatest number of times the factor $x - 4$ appears is once.

$$\text{LCD} = \mathbf{(x + 4)(x + 4)(x - 4)} = (x + 4)^2(x - 4)$$

Self Check 6   Find the LCD of each pair of rational expressions: **a.** $\frac{x^3}{x^2 - 6x}$ and $\frac{25x}{2x - 12}$ and
b. $\frac{m + 1}{m^2 - 9}$ and $\frac{6m^2}{m^2 - 6m + 9}$.

■ BUILDING RATIONAL EXPRESSIONS

Recall that writing a fraction as an equivalent fraction with a larger denominator is called building the fraction. For example, to write $\frac{3}{5}$ as an equivalent fraction with a denominator of 35, we multiply it by 1 in the form of $\frac{7}{7}$:

$$\frac{3}{5} = \frac{3}{5} \cdot \frac{\mathbf{7}}{\mathbf{7}} = \frac{21}{35}$$

To add and subtract rational expressions with different denominators, we must write them as equivalent expressions having a common denominator. To do so, we build rational expressions.

Building Rational Expressions	To build a rational expression, multiply it by 1 in the form of $\frac{c}{c}$, where c is any nonzero number or expression.

EXAMPLE 7   Write $\frac{7}{15n}$ as an equivalent expression with a denominator of $30n^3$.

Solution   We need to multiply the denominator of $\frac{7}{15n}$ by $2n^2$ to obtain a denominator of $30n^3$. It follows that $\frac{2n^2}{2n^2}$ is the form of 1 that should be used to build an equivalent expression.

The Language of Algebra

When we build the rational expression in Example 7, $\frac{7}{15n}$ has the same value as $\frac{14n^2}{30n^3}$ for all values of n, except those that make either denominator 0.

$$\frac{7}{15n} = \frac{7}{15n} \cdot \frac{\mathbf{2n^2}}{\mathbf{2n^2}} \qquad \text{Multiply the given rational expression by 1, in the form of } \frac{2n^2}{2n^2}.$$

$$= \frac{14n^2}{30n^3} \qquad \begin{array}{l}\text{Multiply the numerators.}\\ \text{Multiply the denominators.}\end{array}$$

Self Check 7   Write $\frac{7}{20m^2}$ as an equivalent expression with a denominator of $60m^3$.

EXAMPLE 8   Write $\frac{x + 1}{x^2 + 6x}$ as an equivalent expression with a denominator of $x(x + 6)(x + 2)$.

Solution   We factor the denominator to determine what factors are missing.

$$\frac{x + 1}{x^2 + 6x} = \frac{x + 1}{x(x + 6)} \qquad \text{Factor out } x \text{ from } x^2 + 6x.$$

It is now apparent that we need to multiply the denominator by $x + 2$ to obtain a denominator of $x(x + 6)(x + 2)$. It follows that $\frac{x + 2}{x + 2}$ is the form of 1 that should be used to build an equivalent expression.

$$\frac{x + 1}{x^2 + 6x} = \frac{x + 1}{x(x + 6)} \cdot \frac{\mathbf{x + 2}}{\mathbf{x + 2}} \qquad \begin{array}{l}\text{Multiply the given rational expression by 1, in the}\\ \text{form of } \frac{x + 2}{x + 2}.\end{array}$$

$$= \frac{(x + 1)(x + 2)}{x(x + 6)(x + 2)} \qquad \text{Multiply the numerators. Multiply the denominators.}$$

Self Check 8   Write $\dfrac{x - 3}{x^2 - 4x}$ as an equivalent expression with a denominator of $x(x - 4)(x + 8)$.

Answers to Self Checks   **1. a.** $\frac{2x}{5}$,   **b.** $\frac{3m - 6}{23n}$ or $\frac{3(m - 2)}{23n}$ **2. a.** $\frac{2}{3}$,   **b.** $\frac{2c^2 - 11c}{(c - 1)(c + 2)}$ or $\frac{c(2c - 11)}{(c - 1)(c + 2)}$ **3.** $\frac{1}{n + 4}$
4. a. $x - 1$,   **b.** $-\frac{y}{(y + 3)(y - 3)}$ **5. a.** $150y^3$,   **b.** $a(a + 3)$ **6. a.** $2x(x - 6)$,
b. $(m + 3)(m - 3)^2$ **7.** $\frac{21m}{60m^3}$ **8.** $\frac{(x - 3)(x + 8)}{x(x - 4)(x + 8)}$

6.3   STUDY SET   ELEMENTARY & INTERMEDIATE   Algebra ƒ(x) Now™

VOCABULARY   Fill in the blanks.

1. The rational expressions $\frac{7}{6n}$ and $\frac{n + 1}{6n}$ have a _____ denominator of $6n$.

2. The _____ of the numerators of $\frac{9m}{m + 1}$ and $\frac{6m}{m + 1}$ is $15m$, and the _____ of their numerators is $3m$.

3. The _____ _____ _____ of $\frac{x - 8}{x + 6}$ and $\frac{6 - 5x}{x}$ is $x(x + 6)$.

4. To _____ a rational expression, we multiply it by a form of 1. For example, $\frac{2}{n^2} \cdot \frac{8}{8} = \frac{16}{8n^2}$.

5. To simplify $5y + 8 - y + 4$, we combine _____ terms.

6. Since $\frac{7x}{x^2 + 1}$ has the same value as $\frac{14x}{2(x^2 + 1)}$ for all values of x, they are called _____ expressions.

CONCEPTS   Fill in the blanks.

7. To add or subtract rational expressions that have the same denominator, add or subtract the _____, and write the sum or difference over the common _____.

   In symbols, if $\frac{A}{D}$ and $\frac{B}{D}$ are rational expressions,

   $$\frac{A}{D} + \frac{B}{D} = \frac{\quad}{D} \qquad \text{or} \qquad \frac{A}{D} - \frac{B}{D} = \frac{\quad}{D}$$

8. When a $-$ symbol precedes an expression within parentheses, such as in $-(15x - 2)$, we drop the parentheses and change the _____ of each term of the expression. For example,

   $$-(15x - 2) = \boxed{}$$

9. Simplify each expression.
   a. $-(6x + 9)$ **b.** $x^2 + 3x - 1 + x^2 - 5x$

   c. $7x - 1 - (5x - 6)$ **d.** $4x^2 - 2x - (x^2 + x)$

10. The sum of two rational expressions is $\frac{4x + 4}{5(x + 1)}$. Simplify the result.

11. The difference of two rational expressions is $\frac{x^2 - 14x + 49}{x^2 - 49}$. Simplify the result.

12. Fill in the blanks: To find the least common denominator of several rational expressions, _____ each denominator completely. The LCD is a product that uses each different factor the _____ number of times it appears in any one factorization.

13. Factor each denominator completely.

   a. $\dfrac{17}{40x^2}$

   b. $\dfrac{x + 25}{2x^2 - 6x}$

   c. $\dfrac{n^2 + 3n - 4}{n^2 - 64}$

14. Consider the following factorizations.

   $$18x - 36 = 2 \cdot 3 \cdot 3 \cdot (x - 2)$$
   $$3x^2 - 3x - 6 = 3(x - 2)(x + 1)$$

   a. What is the greatest number of times the factor 3 appears in any one factorization?

   b. What is the greatest number of times the factor $x - 2$ appears in any one factorization?

15. Consider $\dfrac{5}{21a} = \dfrac{5}{21a} \cdot \dfrac{a+4}{a+4}$.

   a. What rational expression is being built?

   b. To build a rational expression, we multiply it by a form of 1. What form of 1 is used here?

16. Fill in the blanks. We need to multiply the denominator of $\dfrac{x}{x-9}$ by ▮ to obtain a denominator of $3x(x-9)$. It follows that ▮ is the form of 1 that should be used to build $\dfrac{x}{x-9}$.

NOTATION

17. Do these rational expressions have the same denominator?

$$\dfrac{7-x}{(x+2)(x-3)} \qquad \dfrac{15x+4}{(x-3)(x+2)}$$

18. Write each LCD using exponents.

   a. LCD $= 2 \cdot 3 \cdot 5 \cdot x \cdot x \cdot x$

   b. LCD $= (y+3)(y+3)(y-9)$

PRACTICE   Add or subtract. Simplify the result, if possible.

19. $\dfrac{9}{x} + \dfrac{2}{x}$

20. $\dfrac{4}{s} + \dfrac{4}{s}$

21. $\dfrac{x}{18} + \dfrac{5}{18}$

22. $\dfrac{7}{10} + \dfrac{3y}{10}$

23. $\dfrac{m-3}{m^3} - \dfrac{5}{m^3}$

24. $\dfrac{c+7}{c^4} - \dfrac{3}{c^4}$

25. $\dfrac{2x}{y} - \dfrac{x}{y}$

26. $\dfrac{16c}{d} - \dfrac{4c}{d}$

27. $\dfrac{13t}{99} - \dfrac{35t}{99}$

28. $\dfrac{35y}{72} - \dfrac{44y}{72}$

29. $\dfrac{x}{9} + \dfrac{2x}{9}$

30. $\dfrac{5x}{7} + \dfrac{9x}{7}$

31. $\dfrac{50}{r^3-25} + \dfrac{r}{r^3-25}$

32. $\dfrac{75}{h^2-27} + \dfrac{h}{h^2-27}$

33. $\dfrac{6a}{a+2} - \dfrac{4a}{a+2}$

34. $\dfrac{10b}{b-6} - \dfrac{4b}{b-6}$

35. $\dfrac{7}{t+5} - \dfrac{9}{t+5}$

36. $\dfrac{3}{r-11} - \dfrac{6}{r-11}$

37. $\dfrac{3x-5}{x-2} + \dfrac{6x-13}{x-2}$

38. $\dfrac{8x-7}{x+3} + \dfrac{2x+37}{x+3}$

39. $\dfrac{6x-5}{3xy} - \dfrac{3x-5}{3xy}$

40. $\dfrac{7x+7}{5y} - \dfrac{2x+7}{5y}$

41. $\dfrac{3y-2}{2y+6} - \dfrac{2y-5}{2y+6}$

42. $\dfrac{5x+8}{3x+15} - \dfrac{3x-2}{3x+15}$

43. $\dfrac{x+3}{2y} + \dfrac{x+5}{2y}$

44. $\dfrac{y+2}{10z} + \dfrac{y+4}{10z}$

45. $\dfrac{6x^2-11x}{3x+2} - \dfrac{10}{3x+2}$

46. $\dfrac{8a^2}{2a+5} - \dfrac{4a^2+25}{2a+5}$

47. $\dfrac{2-p}{p^2-p} - \dfrac{-p+2}{p^2-p}$

48. $\dfrac{7n+2}{n^2+5} - \dfrac{2+7n}{n^2+5}$

49. $\dfrac{3x^2}{x+1} + \dfrac{x-2}{x+1}$

50. $\dfrac{10b^2}{3b-2} - \dfrac{b^2+4}{3b-2}$

51. $\dfrac{11w+1}{3w(w-9)} - \dfrac{11w}{3w(w-9)}$

52. $\dfrac{y+2}{y^2(y-14)} - \dfrac{y}{y^2(y-14)}$

53. $\dfrac{a}{a^2+5a+6} + \dfrac{3}{a^2+5a+6}$

54. $\dfrac{b}{b^2-4} + \dfrac{2}{b^2-4}$

55. $\dfrac{2c}{c^2-d^2} - \dfrac{2d}{c^2-d^2}$

56. $\dfrac{3t}{t^2-8t+7} - \dfrac{3}{t^2-8t+7}$

57. $\dfrac{11n-1}{(n+4)(n-2)} - \dfrac{4n}{(n-2)(n+4)}$

58. $\dfrac{r-32}{(r-4)(r-2)} + \dfrac{4r}{(r-2)(r-4)}$

59. $\dfrac{1}{t^2-2t+1} - \dfrac{6-t}{t^2-2t+1}$

60. $\dfrac{2}{d^2-10d+25} - \dfrac{10-d}{d^2-10d+25}$

Find the LCD of each pair of rational expressions.

61. $\dfrac{1}{2x}, \dfrac{9}{6x}$

62. $\dfrac{4}{9y}, \dfrac{11}{3y}$

63. $\dfrac{35}{3a^2 b}, \dfrac{23}{a^2 b^3}$

64. $\dfrac{27}{c^2 d}, \dfrac{17}{2c^2 d^3}$

65. $\dfrac{8}{c}, \dfrac{8 - c}{c + 2}$

66. $\dfrac{d^2 - 5}{d + 9}, \dfrac{d - 3}{d}$

67. $\dfrac{33}{15a^3}, \dfrac{9}{10a}$

68. $\dfrac{m - 21}{12m^4}, \dfrac{m + 1}{18m}$

69. $\dfrac{b - 9}{4b + 8}, \dfrac{b}{6}$

70. $\dfrac{b^2 - b}{10b - 15}, \dfrac{11b}{10}$

71. $\dfrac{-2x}{x^2 - 1}, \dfrac{5x}{x + 1}$

72. $\dfrac{7 - y^2}{y^2 - 4}, \dfrac{y - 49}{y + 2}$

73. $\dfrac{3x + 1}{3x - 1}, \dfrac{3x}{3x + 1}$

74. $\dfrac{b + 1}{b - 1}, \dfrac{b}{b + 1}$

75. $\dfrac{4x - 5}{x^2 - 4x - 5}, \dfrac{3x + 1}{x^2 - 25}$

76. $\dfrac{44}{s^2 - 9}, \dfrac{s + 9}{s^2 - s - 6}$

77. $\dfrac{5n^2 - 16}{2n^2 + 13n + 20}, \dfrac{3n^2}{n^2 + 8n + 16}$

78. $\dfrac{4y + 25}{y^2 + 10y + 25}, \dfrac{y^2 - 7}{2y^2 + 17y + 35}$

Build each rational expression into an equivalent expression with the given denominator.

79. $\dfrac{25}{4}; 20x$

80. $\dfrac{5}{y}; y^2$

81. $\dfrac{8}{x}; x^2 y$

82. $\dfrac{7}{y}; xy^2$

83. $\dfrac{3x}{x + 1}; (x + 1)^2$

84. $\dfrac{5y}{y - 2}; (y - 2)^2$

85. $\dfrac{2y}{x}; x(x + 3)$

86. $\dfrac{3x}{y}; y(y - 9)$

87. $\dfrac{10}{b - 1}; 3(b - 1)$

88. $\dfrac{8}{c + 9}; 6(c + 9)$

89. $\dfrac{t + 5}{4t + 8}; 20(t + 2)$

90. $\dfrac{x + 7}{3x - 15}; 6(x - 5)$

91. $\dfrac{y + 3}{y^2 - 5y + 6}; 4y(y - 2)(y - 3)$

92. $\dfrac{3x - 4}{x^2 + 3x + 2}; 8x(x + 1)(x + 2)$

93. $\dfrac{12 - h}{h^2 - 81}; 3(h + 9)(h - 9)$

94. $\dfrac{m^2}{m^2 - 100}; 9(m + 10)(m - 10)$

95. $\dfrac{6t}{t^2 + 4t + 3}; (t + 1)(t - 2)(t + 3)$

96. $\dfrac{9s}{s^2 + 6s + 5}; (s + 1)(s - 4)(s + 5)$

APPLICATIONS

97. DOING LAUNDRY   At a laundromat, it takes 30 minutes to wash a load of clothes and 45 minutes to dry a load. Suppose a customer starts a washer and dryer at the same time. If she unloads, refills, and restarts each appliance when a cycle is complete, in how many minutes will the washing and drying cycles end simultaneously?

98. GEOMETRY
 a. What is the difference of the length and width of the rectangle?
 b. What is the perimeter?

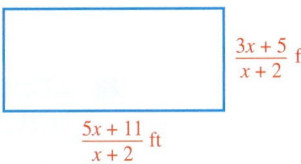

$\dfrac{3x + 5}{x + 2}$ ft

$\dfrac{5x + 11}{x + 2}$ ft

WRITING

99. Explain how to add fractions with the same denominator.

100. Explain how to find a least common denominator.

101. Explain the error in the following solution:

$$\frac{2x + 3}{x + 5} - \frac{x + 2}{x + 5} = \frac{2x + 3 - x + 2}{x + 5}$$

$$= \frac{x + 5}{x + 5}$$

$$= 1$$

102. Explain the error in the following solution:

$$\frac{5x - 4}{y} + \frac{x}{y} = \frac{5x - 4 + x}{y + y}$$

$$= \frac{6x - 4}{2y}$$

$$= \frac{2(3x - 2)}{2y}$$

$$= \frac{3x - 2}{y}$$

103. Explain why the LCD of $\frac{5}{h^2}$ and $\frac{3}{h}$ is h^2 and not h^3.

104. Explain how multiplication by 1 is used to build a rational expression.

REVIEW

105. Give the formula for simple interest.

106. Give the formula that finds the slope of a line.

107. Give the formula for the area of a triangle.

108. Give the slope–intercept form for the equation of a line.

109. Give the formula that is used to solve uniform motion problems.

110. Give the point–slope form for the equation of a line.

CHALLENGE PROBLEMS

111. Perform the operations. Simplify the result, if possible.

$$\frac{3xy}{x - y} - \frac{x(3y - x)}{x - y} - \frac{x(x - y)}{x - y}$$

112. Find the LCD of these rational expressions.

$$\frac{2}{a^3 + 8}, \quad \frac{a}{a^2 - 4}, \quad \frac{2a + 5}{a^3 - 8}$$

6.4 Addition and Subtraction with Unlike Denominators

- Adding and Subtracting Rational Expressions With Unlike Denominators
- Denominators That Are Opposites

We have discussed a method for finding the least common denominator (LCD) of two rational expressions. We have also built rational expressions into equivalent expressions having specified denominators. We will now use these skills to add and subtract rational expressions with unlike denominators.

■ ADDING AND SUBTRACTING RATIONAL EXPRESSIONS WITH UNLIKE DENOMINATORS

The following steps summarize how to add or subtract rational expressions with different denominators.

Adding and Subtracting Rational Expressions With Unlike Denominators	1. Find the LCD. 2. Write each rational expression as an equivalent expression whose denominator is the LCD. 3. Add or subtract the numerators and write the sum or difference over the LCD. 4. Simplify the resulting rational expression, if possible.

EXAMPLE 1

Add: $\dfrac{9x}{7} + \dfrac{3x}{5}$.

ELEMENTARY &
INTERMEDIATE
Algebra *f(x)* **Now**™

Solution

Step 1: The LCD is $7 \cdot 5 = 35$.

Step 2: We need to multiply the denominator of $\frac{9x}{7}$ by 5 and we need to multiply the denominator of $\frac{3x}{5}$ by 7 to obtain the LCD, 35. It follows that $\frac{5}{5}$ and $\frac{7}{7}$ are the forms of 1 that should be used to write the equivalent rational expressions.

$$\frac{9x}{7} + \frac{3x}{5} = \frac{9x}{7} \cdot \frac{5}{5} + \frac{3x}{5} \cdot \frac{7}{7}$$ Build the rational expressions so that each has a denominator of 35.

$$= \frac{45x}{35} + \frac{21x}{35}$$ Multiply the numerators.
Multiply the denominators.

Step 3:
$$= \frac{45x + 21x}{35}$$ Add the numerators. Write the sum over the common denominator.

$$= \frac{66x}{35}$$ Combine like terms in the numerator.

Step 4: Since 66 and 35 have no common factor other than 1, the result cannot be simplified.

Self Check 1   Add: $\dfrac{y}{2} + \dfrac{6y}{7}$.

EXAMPLE 2

Subtract: $\dfrac{13}{18b^2} - \dfrac{1}{24b}$.

Solution

To find the LCD, we form a product that uses each different factor of $18b^2$ and $24b$ the greatest number of times it appears in any one factorization.

$$\left.\begin{array}{l} 18b^2 = 2 \cdot 3 \cdot 3 \cdot b \cdot b \\ 24b = 2 \cdot 2 \cdot 2 \cdot 3 \cdot b \end{array}\right\} \text{LCD} = 2 \cdot 2 \cdot 2 \cdot 3 \cdot 3 \cdot b \cdot b = 72b^2$$

Notation

When checking your answers with those in the back of the book, remember that the result of an addition or subtraction of rational expressions can often be presented in several equivalent forms. For instance, the result of Example 2 may also be expressed as

$$\frac{-3b + 52}{72b^2} \text{ or } -\frac{3b - 52}{72b^2}$$

We need to multiply $18b^2$ by 4 to obtain $72b^2$, and $24b$ by $3b$ to obtain $72b^2$. It follows that we should use $\frac{4}{4}$ and $\frac{3b}{3b}$ to build the equivalent rational expressions.

$$\frac{13}{18b^2} - \frac{1}{24b} = \frac{13}{18b^2} \cdot \frac{4}{4} - \frac{1}{24b} \cdot \frac{3b}{3b}$$ Build the rational expressions so that each has a denominator of $72b^2$.

$$= \frac{52}{72b^2} - \frac{3b}{72b^2}$$ Multiply the numerators.
Multiply the denominators.

$$= \frac{52 - 3b}{72b^2}$$ Subtract the numerators. Write the difference over the common denominator.

The result cannot be simplified.

Self Check 2   Subtract: $\dfrac{5}{21z^2} - \dfrac{3}{28z}$.

EXAMPLE 3

ELEMENTARY &
INTERMEDIATE
Algebra $f(x)$ **Now**™

Success Tip

In Example 3, we use the distributive property to multiply the numerators of $\frac{3}{2(x+9)}$ and $\frac{x-9}{x-9}$. Note that we don't multiply out the denominators.

$$\frac{3}{2(x+9)} \cdot \frac{x-9}{x-9}$$

The result is $\frac{3x-27}{2(x+9)(x-9)}$.

Add: $\dfrac{3}{2x+18} + \dfrac{27}{x^2-81}$.

Solution   After factoring the denominators, we see that the greatest number of times each of the factors 2, $x+9$, and $x-9$ appear in any one of the factorizations is once.

$$\left.\begin{array}{l} 2x+18 = 2(x+9) \\ x^2-81 = (x+9)(x-9) \end{array}\right\} \text{LCD} = 2(x+9)(x-9)$$

Since we need to multiply $2(x+9)$ by $x-9$ to obtain the LCD and $(x+9)(x-9)$ by 2 to obtain the LCD, $\frac{x-9}{x-9}$ and $\frac{2}{2}$ are the forms of 1 to use to build the equivalent rational expressions.

$$\frac{3}{2x+18} + \frac{27}{x^2-81} = \frac{3}{2(x+9)} + \frac{27}{(x+9)(x-9)}$$

Write each denominator in factored form.

$$= \frac{3}{2(x+9)} \cdot \frac{x-9}{x-9} + \frac{27}{(x+9)(x-9)} \cdot \frac{2}{2}$$

Build the expressions so that each has a denominator of $2(x+9)(x-9)$.

$$= \frac{3x-27}{2(x+9)(x-9)} + \frac{54}{2(x+9)(x-9)}$$

Multiply the numerators. Multiply the denominators.

Although it is not mandatory, the factors of each denominator are written in the same order.

$$= \frac{3x-27+54}{2(x+9)(x-9)}$$

Add the numerators. Write the sum over the common denominator.

$$= \frac{3x+27}{2(x+9)(x-9)}$$

Combine like terms in the numerator.

$$= \frac{3\overset{1}{\cancel{(x+9)}}}{2\underset{1}{\cancel{(x+9)}}(x-9)}$$

Factor the numerator. Then simplify the expression by removing a factor equal to 1.

$$= \frac{3}{2(x-9)}$$

Self Check 3   Add: $\dfrac{2}{5x+25} + \dfrac{4}{x^2-25}$.

EXAMPLE 4

ELEMENTARY &
INTERMEDIATE
Algebra $f(x)$ **Now**™

Subtract: $\dfrac{x}{x-1} - \dfrac{x-6}{x-4}$.

Solution   The denominators of $\frac{x}{x-1}$ and $\frac{x-6}{x-4}$ are completely factored. The factor $x-1$ appears once and the factor $x-4$ appears once. Thus, the LCD $= (x-1)(x-4)$.

We need to multiply the first denominator by $x-4$ to obtain the LCD and the second denominator by $x-1$ to obtain the LCD. It follows that $\frac{x-4}{x-4}$ and $\frac{x-1}{x-1}$ are the forms of 1 to use to build the equivalent rational expressions.

$$\frac{x}{x - 1} - \frac{x - 6}{x - 4} = \frac{x}{x - 1} \cdot \frac{x - 4}{x - 4} - \frac{x - 6}{x - 4} \cdot \frac{x - 1}{x - 1}$$

Build the expressions so that each has a denominator of $(x - 1)(x - 4)$.

Success Tip

In Example 4, we use the FOIL method to multiply the numerators of $\frac{x - 6}{x - 4}$ and $\frac{x - 1}{x - 1}$. Note that we do not multiply out the denominators.

$$= \frac{x^2 - 4x}{(x - 1)(x - 4)} - \frac{x^2 - 7x + 6}{(x - 4)(x - 1)}$$

Multiply the numerators. Multiply the denominators.

By the commutative property of multiplication, these are like denominators.

$$= \frac{x^2 - 4x - (x^2 - 7x + 6)}{(x - 1)(x - 4)}$$

Subtract the numerators. Write the difference over the common denominator.

$$= \frac{x^2 - 4x - x^2 + 7x - 6}{(x - 1)(x - 4)}$$

Use the distributive property.

$$= \frac{3x - 6}{(x - 1)(x - 4)}$$

Combine like terms in the numerator.

The numerator factors as $3(x - 2)$. Since the numerator and denominator have no common factor, the result is in simplest form.

Self Check 4   Subtact: $\dfrac{x}{x + 9} - \dfrac{x - 7}{x + 8}$.

EXAMPLE 5

Subtract: $\dfrac{m}{m^2 + 5m + 6} - \dfrac{2}{m^2 + 3m + 2}$.

ELEMENTARY & INTERMEDIATE
Algebra $f(x)$ **Now**™

Solution   Factor each denominator and form the LCD.

$$\left.\begin{array}{l} m^2 + 5m + 6 = (m + 2)(m + 3) \\ m^2 + 3m + 2 = (m + 2)(m + 1) \end{array}\right\} \text{LCD} = (m + 2)(m + 3)(m + 1)$$

Examining the factored forms, we see that the first denominator must be multiplied by $m + 1$, and the second must be multiplied by $m + 3$ to obtain the LCD. To build the expressions, we will use $\frac{m + 1}{m + 1}$ and $\frac{m + 3}{m + 3}$.

$$\frac{m}{m^2 + 5m + 6} - \frac{2}{m^2 + 3m + 2}$$

$$= \frac{m}{(m + 2)(m + 3)} - \frac{2}{(m + 2)(m + 1)}$$

Write each denominator in factored form.

$$= \frac{m}{(m + 2)(m + 3)} \cdot \frac{m + 1}{m + 1} - \frac{2}{(m + 2)(m + 1)} \cdot \frac{m + 3}{m + 3}$$

Build each expression.

$$= \frac{m^2 + m}{(m + 2)(m + 3)(m + 1)} - \frac{2m + 6}{(m + 2)(m + 1)(m + 3)}$$

Multiply numerators. Multiply denominators.

By the commutative property of multiplication, these are like denominators.

$$= \frac{m^2 + m - (2m + 6)}{(m + 2)(m + 3)(m + 1)}$$

Subtract the numerators. Write the difference over the common denominator. Remember the parentheses.

$$= \frac{m^2 + m - 2m - 6}{(m + 2)(m + 3)(m + 1)}$$

Use the distributive property:
$-(2m + 6) = -1(2m + 6) = -2m - 6$.

$$= \frac{m^2 - m - 6}{(m + 2)(m + 3)(m + 1)}$$

Combine like terms in the numerator.

$$= \frac{\overset{1}{(m - 3)(\cancel{m + 2})}}{\underset{1}{(\cancel{m + 2})(m + 3)(m + 1)}}$$

Factor the numerator and simplify the expression by removing a factor equal to 1.

$$= \frac{m - 3}{(m + 3)(m + 1)}$$

Self Check 5 Subtract: $\dfrac{b}{b^2 - 2b - 8} - \dfrac{6}{b^2 + b - 20}$.

EXAMPLE 6

ELEMENTARY &
INTERMEDIATE
Algebra $f(x)$ **Now**™

Add: $\dfrac{4b}{a - 5} + b$.

Solution   We can write b as $\frac{b}{1}$. The LCD of $\frac{4b}{a - 5}$ and $\frac{b}{1}$ is $1(a - 5)$, or simply $a - 5$. Since we must multiply the denominator of $\frac{b}{1}$ by $a - 5$ to obtain the LCD, we will use $\frac{a - 5}{a - 5}$ to write an equivalent rational expression.

$$\frac{4b}{a - 5} + b = \frac{4b}{a - 5} + \frac{b}{1} \cdot \frac{\boldsymbol{a - 5}}{\boldsymbol{a - 5}}$$

Build $\frac{b}{1}$ so that it has a denominator of $a - 5$.

$$= \frac{4b}{a - 5} + \frac{ab - 5b}{a - 5}$$

Multiply numerators: $b(a - 5) = ab - 5b$.
Multiply denominators: $1(a - 5) = a - 5$.

$$= \frac{4b + ab - 5b}{a - 5}$$

Add the numerators. Write the sum over the common denominator.

$$= \frac{ab - b}{a - 5}$$

Combine like terms in the numerator:
$4b - 5b = -b$.

Although the numerator factors as $b(a - 1)$, the numerator and denominator do not have a common factor. Therefore, the result does not simplify any further.

Self Check 6 Add: $\dfrac{10y}{n + 4} + y$.

■ DENOMINATORS THAT ARE OPPOSITES

Recall that two polynomials are **opposites** if their terms are the same but they are opposite in sign. For example, $x - 4$ and $4 - x$ are opposites. If we multiply one of these binomials by -1, the result is the other binomial.

$$-1(x - 4) = -x + 4$$
$$= 4 - x \quad \text{Write the expression with 4 first.}$$

$$-1(4 - x) = -4 + x$$
$$= x - 4 \quad \text{Write the expression with } x \text{ first.}$$

These results suggest a general fact.

| Multiplying by −1 | When a polynomial is multiplied by −1, the result is its opposite. |

This fact can be used when adding or subtracting rational expressions whose denominators are opposites.

EXAMPLE 7

ELEMENTARY & INTERMEDIATE
Algebra $f(x)$ Now™

Success Tip

Either denominator in Example 7 can serve as the LCD. However, it is common to have a result whose denominator is written in descending powers of the variable. Therefore, we chose $x - 7$, as opposed to $7 - x$, as the LCD.

Subtract: $\dfrac{x}{x - 7} - \dfrac{1}{7 - x}$.

Solution   We note that the denominators are opposites. Either can serve as the LCD; we will choose $x - 7$.

We must multiply the denominator of $\dfrac{1}{7 - x}$ by -1 to obtain the LCD. It follows that $\dfrac{-1}{-1}$ should be the form of 1 that is used to write an equivalent rational expression.

$$\frac{x}{x - 7} - \frac{1}{7 - x} = \frac{x}{x - 7} - \frac{1}{7 - x} \cdot \frac{-1}{-1}$$ Build $\dfrac{1}{7 - x}$ so that it has a denominator of $x - 7$.

$$= \frac{x}{x - 7} - \frac{-1}{-7 + x}$$ Multiply the numerators. Multiply the denominators.

$$= \frac{x}{x - 7} - \frac{-1}{x - 7}$$ Rewrite the second denominator: $-7 + x = x - 7$.

$$= \frac{x - (-1)}{x - 7}$$ Subtract the numerators. Write the difference over the common denominator.

$$= \frac{x + 1}{x - 7}$$ Simplify the numerator.

The result does not simplify any further.

Self Check 7   Add: $\dfrac{n}{n - 8} + \dfrac{12}{8 - n}$.

Answers to Self Checks   1. $\dfrac{19y}{14}$ 2. $\dfrac{20 - 9z}{84z^2}$ 3. $\dfrac{2}{5(x - 5)}$ 4. $\dfrac{6x + 63}{(x + 9)(x + 8)}$ 5. $\dfrac{b + 3}{(b + 2)(b + 5)}$ 6. $\dfrac{ny + 14y}{n + 4}$
7. $\dfrac{n - 12}{n - 8}$

6.4   STUDY SET   ELEMENTARY & INTERMEDIATE Algebra $f(x)$ Now™

VOCABULARY   Fill in the blanks.

1. The rational expressions $\dfrac{x}{x - 7}$ and $\dfrac{1}{x - 7}$ have like denominators. The rational expressions $\dfrac{x + 5}{x - 7}$ and $\dfrac{4x}{x + 7}$ have _____ denominators.

2. Two polynomials are _____ if their terms are the same, but are opposite in sign.

CONCEPTS

3. Write each denominator in factored form.

   a. $\dfrac{x + 1}{20x^2}$

   b. $\dfrac{3x^2 - 4}{x^2 + 4x - 12}$

4. The factorizations of the denominators of two rational expressions follow. Find their LCD.

$$12a = 2 \cdot 2 \cdot 3 \cdot a$$
$$18a^2 = 2 \cdot 3 \cdot 3 \cdot a \cdot a$$

5. What is the LCD for $\dfrac{x-1}{x+6}$ and $\dfrac{1}{x+3}$?

6. The factorizations of the denominators of two rational expressions follow. Find their LCD.

$$x^2 - 36 = (x+6)(x-6)$$
$$3x - 18 = 3(x-6)$$

7. To build $\dfrac{x}{x+2}$ so that it has a denominator of $5(x+2)$, we multiply it by 1 in the form of ☐.

8. To build $\dfrac{x-3}{x-4}$ so that it has a denominator of $(x+5)(x-4)$, we multiply it by 1 in the form of ☐.

9. The LCD for $\dfrac{1}{9n^2}$ and $\dfrac{37}{15n^3}$ is

$$LCD = 3 \cdot 3 \cdot 5 \cdot n \cdot n \cdot n$$

If we want to add these rational expressions, what form of 1 should be used

a. to build $\dfrac{1}{9n^2}$?   **b.** to build $\dfrac{37}{15n^3}$?

10. The LCD for $\dfrac{2x+1}{x^2+5x+6}$ and $\dfrac{3x}{x^2-4}$ is

$$LCD = (x+2)(x+3)(x-2)$$

If we want to subtract these rational expressions, what form of 1 should be used

a. to build $\dfrac{2x+1}{x^2+5x+6}$?

b. to build $\dfrac{3x}{x^2-4}$?

Fill in the blanks. Write each numerator without using parentheses.

11. $\dfrac{x-3}{x-4} \cdot \dfrac{8}{8} = \dfrac{}{8(x-4)}$

12. $\dfrac{x-3}{x-4} \cdot \dfrac{x+9}{x+2} = \dfrac{}{(x-4)(x+2)}$

13. $\dfrac{7x}{3x^2(x-5)} + \dfrac{10}{3x^2(x-5)} = \dfrac{}{3x^2(x-5)}$

14. $\dfrac{x^2 - 3x + 9 - (x^2 - 5x)}{(x-9)(x+3)} = \dfrac{}{(x-9)(x+3)}$

15. Multiply: $-1(x-10)$.

16. What is the opposite of $x - 25$?

17. By what must $y - 4$ be multiplied to obtain $4 - y$?

18. To write $\dfrac{8x}{2-x}$ so that it has a denominator of $x - 2$, we multiply it by 1 in the form of ☐.

19. Write x in fraction form.

20. Fill in the blank. To write $\dfrac{7}{1}$ so that it has a denominator of $2c + 1$, we multiply it by 1 in the form of ☐.

NOTATION

21. In the following table, a student's answers to three homework problems are compared to the answers in the back of the book. Are the answers equivalent?

Student's answer	Book's answer	Equivalent?
$\dfrac{m^2 + 2m}{(m-1)(m-4)}$	$\dfrac{m^2 + 2m}{(m-4)(m-1)}$	
$\dfrac{-5x^2 - 7}{4x(x+3)}$	$-\dfrac{5x^2 - 7}{4x(x+3)}$	
$\dfrac{-2x}{x-y}$	$\dfrac{2x}{y-x}$	

22. Simplify each expression.

a. $\dfrac{6}{6}$ **b.** $\dfrac{3x}{3x}$

c. $\dfrac{x+7}{x+7}$ **d.** $\dfrac{-1}{-1}$

PRACTICE **Perform the operations. Simplify, if possible.**

23. $\dfrac{x}{3} + \dfrac{2x}{7}$ **24.** $\dfrac{y}{4} + \dfrac{3y}{5}$

25. $\dfrac{2y}{9} + \dfrac{y}{3}$ **26.** $\dfrac{4b}{3} - \dfrac{5b}{12}$

27. $\dfrac{11}{5x} - \dfrac{5}{6x}$ **28.** $\dfrac{5}{9y} - \dfrac{1}{4y}$

29. $\dfrac{7}{m^2} - \dfrac{2}{m}$ **30.** $\dfrac{6}{n^2} + \dfrac{2}{n}$

31. $\dfrac{4x}{3} + \dfrac{2x}{y}$ **32.** $\dfrac{7b}{10} + \dfrac{4b}{t}$

33. $\dfrac{1}{6c^4} + \dfrac{8}{9c^2}$

34. $\dfrac{7}{8b^2} + \dfrac{5}{6b^3}$

55. $\dfrac{9}{t+3} + \dfrac{8}{t+2}$

56. $\dfrac{2}{m-3} + \dfrac{7}{m-4}$

35. $\dfrac{y}{8} + \dfrac{y-3}{16}$

36. $\dfrac{m}{4} + \dfrac{m-4}{8}$

57. $\dfrac{3x}{2x-1} - \dfrac{2x}{2x+3}$

58. $\dfrac{2y}{5y-1} - \dfrac{2y}{3y+2}$

37. $\dfrac{n}{5} - \dfrac{n-2}{15}$

38. $\dfrac{m}{9} - \dfrac{m+1}{27}$

59. $\dfrac{4}{a+2} - \dfrac{7}{(a+2)^2}$

60. $\dfrac{9}{(b-1)^2} - \dfrac{2}{b-1}$

39. $\dfrac{2-b}{6} - \dfrac{b-7}{21}$

40. $\dfrac{7-x}{4} - \dfrac{x-3}{14}$

61. $\dfrac{3m}{m-2} + \dfrac{m-3}{m+5}$

62. $\dfrac{2x}{x+2} + \dfrac{x+1}{x-3}$

41. $\dfrac{x-1}{9x} - \dfrac{x-2}{x^3}$

42. $\dfrac{a-3}{7a} - \dfrac{a-4}{a^3}$

63. $\dfrac{s+7}{s+3} - \dfrac{s-3}{s+7}$

64. $\dfrac{t+5}{t-5} - \dfrac{t-5}{t+5}$

43. $\dfrac{y+2}{5y^2} + \dfrac{y+4}{15y}$

44. $\dfrac{x+3}{x^2} + \dfrac{x+5}{2x}$

65. $\dfrac{7}{(a+1)(a+3)} + \dfrac{5}{(a+3)^2}$

66. $\dfrac{1}{(b-1)(b-4)} + \dfrac{100}{(b-4)^2}$

45. $\dfrac{x+5}{xy} - \dfrac{x-1}{x^2y}$

46. $\dfrac{y-7}{y^2} - \dfrac{y+7}{2y}$

67. $\dfrac{2c-3}{(2c+1)(c-3)} - \dfrac{3}{c-3}$

68. $\dfrac{3y-7}{(3y+5)(y-2)} - \dfrac{5}{y-2}$

47. $\dfrac{x-3}{6x} + \dfrac{x+4}{8x}$

48. $\dfrac{3y-2}{12y} + \dfrac{3-y}{18y}$

69. $\dfrac{b}{b+1} - \dfrac{b+1}{2b+2}$

70. $\dfrac{4x+1}{8x-12} + \dfrac{x-3}{2x-3}$

49. $\dfrac{a+2}{b} + \dfrac{b-2}{a}$

50. $\dfrac{x-1}{x} + \dfrac{y+1}{y}$

71. $\dfrac{2}{a^2+4a+3} + \dfrac{1}{a+3}$

72. $\dfrac{1}{c+6} - \dfrac{-4}{c^2+8c+12}$

51. $\dfrac{x}{x+1} + \dfrac{x-1}{x}$

52. $\dfrac{t-2}{t} + \dfrac{t}{t+3}$

73. $\dfrac{7s}{s^2+s-12} - \dfrac{4}{s+4}$

74. $\dfrac{2y}{y^2-5y+6} + \dfrac{4}{y-2}$

53. $\dfrac{1}{5x} + \dfrac{7x}{x+5}$

54. $\dfrac{10h}{h-3} + \dfrac{7}{9h}$

75. $\dfrac{x}{x-2} + \dfrac{4+2x}{x^2-4}$

76. $\dfrac{y}{y+3} - \dfrac{2y-6}{y^2-9}$

77. $\dfrac{x+1}{x+2} - \dfrac{x^2+1}{x^2-x-6}$ **78.** $\dfrac{6w}{w^2-4} - \dfrac{3}{w-2}$

79. $\dfrac{2}{3h-6} + \dfrac{3}{4h+8}$ **80.** $\dfrac{4}{3d-9} - \dfrac{3}{2d+4}$

81. $\dfrac{8}{y^2-16} - \dfrac{7}{y^2-y-12}$

82. $\dfrac{6}{s^2-9} - \dfrac{5}{s^2-s-6}$

83. $\dfrac{4}{s^2+5s+4} + \dfrac{s}{s^2+2s+1}$

84. $\dfrac{d}{d^2+6d+5} - \dfrac{3}{d^2+5d+4}$

85. $\dfrac{5}{x^2-9x+8} - \dfrac{3}{x^2-6x-16}$

86. $\dfrac{3}{t^2+t-6} + \dfrac{1}{t^2+3t-10}$

87. $\dfrac{x+1}{2x+4} - \dfrac{x^2}{2x^2-8}$ **88.** $\dfrac{-x}{3x^2-27} + \dfrac{1}{3x+9}$

89. $\dfrac{8}{x} + 6$ **90.** $\dfrac{2}{y} + 7$

91. $b - \dfrac{3}{a^2}$ **92.** $c - \dfrac{5}{3b}$

93. $\dfrac{9}{x-4} + x$ **94.** $\dfrac{9}{m+4} + 9$

95. $\dfrac{x+2}{x+1} - 5$ **96.** $\dfrac{y+8}{y-8} - 4$

97. $\dfrac{5}{a-4} + \dfrac{7}{4-a}$ **98.** $\dfrac{4}{b-6} - \dfrac{b}{6-b}$

99. $\dfrac{r+2}{r^2-4} + \dfrac{4}{4-r^2}$ **100.** $\dfrac{2x+2}{x-2} - \dfrac{2x}{2-x}$

101. $\dfrac{y+3}{y-1} - \dfrac{y+4}{1-y}$ **102.** $\dfrac{t+1}{t-7} - \dfrac{t+1}{7-t}$

APPLICATIONS

103. Find the total height of the funnel.

104. What is the difference between the diameter of the opening at the top of the funnel and the diameter of its spout?

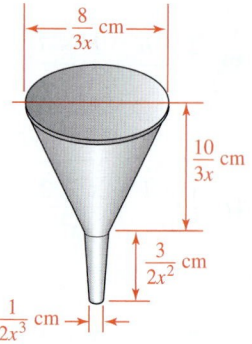

WRITING

105. Explain the error:

$$\frac{3}{x} + \frac{3}{y} = \frac{3+3}{x+y} = \frac{6}{x+y}$$

106. Explain how to add two rational expressions with unlike denominators.

107. When will the LCD of two rational expressions be the product of the denominators of those rational expressions? Give an example.

108. In general, what is the opposite of a polynomial? Explain how to find the opposite of a polynomial. Give examples.

REVIEW

109. Find the slope and y-intercept of the graph of $y = 8x + 2$.

110. Find the slope and y-intercept of the graph of $3x + 4y = -36$.

111. What is the slope of the graph of $y = -2$?

112. Is the graph of the equation $x = 0$ the x-axis or the y-axis?

CHALLENGE PROBLEMS   **Perform the operations and simplify the result, if possible.**

113. $\dfrac{a}{a-1} - \dfrac{2}{a+2} + \dfrac{3(a-2)}{a^2+a-2}$

114. $\dfrac{2x}{x^2-3x+2} + \dfrac{2x}{x-1} - \dfrac{x}{x-2}$

115. $\dfrac{1}{a+1} + \dfrac{a^2-7a+10}{2a^2-2a-4} \cdot \dfrac{2a^2-50}{a^2+10a+25}$

116. $1 - \dfrac{(x-2)^2}{(x+2)^2}$

6.5   Simplifying Complex Fractions

- Simplifying Complex Fractions Using Division
- Simplifying Complex Fractions Using the LCD

A rational expression whose numerator and/or denominator contain rational expressions is called a **complex rational expression** or, more simply, a **complex fraction.** The expression above the main fraction bar of a complex fraction is the numerator, and the expression below the main fraction bar is the denominator. Two examples are:

$$\underbrace{\dfrac{5x}{3}}_{\substack{\text{Denominator}}} \quad \begin{array}{l}\leftarrow \text{Numerator} \rightarrow\\ \leftarrow \text{Main fraction bar} \rightarrow\\ \leftarrow \text{Denominator} \rightarrow\end{array} \quad \dfrac{\dfrac{1}{2}-\dfrac{1}{x}}{\dfrac{x}{3}+\dfrac{1}{5}}$$

In this section, we will simplify complex fractions.

■ SIMPLIFYING COMPLEX FRACTIONS USING DIVISION

One method for simplifying complex fractions uses the fact that fractions indicate division.

Simplifying Complex Fractions	1. Write the numerator and the denominator of the complex fraction as single rational expressions. 2. Perform the division by multiplying the numerator of the complex fraction by the reciprocal of the denominator. 3. Simplify the result, if possible.

EXAMPLE 1

ELEMENTARY & INTERMEDIATE
Algebra $f(x)$ **Now** ™

Simplify: $\dfrac{\dfrac{5x}{3}}{\dfrac{2x}{9}}$.

Solution   Since the numerator and the denominator of the complex fraction are already single rational expressions, we can perform the division.

$$\dfrac{\dfrac{5x}{3}}{\dfrac{2x}{9}} = \dfrac{5x}{3} \div \dfrac{2x}{9} \qquad \text{\color{red}Write the division indicated by the main fraction bar using a ÷ symbol.}$$

$$= \frac{5x}{3} \cdot \frac{9}{2x}$$

To divide rational expressions, multiply the first by the reciprocal of the second.

$$= \frac{5x \cdot 9}{3 \cdot 2x}$$

Multiply the numerators.
Multiply the denominators.

$$= \frac{\overset{1}{5x} \cdot \overset{1}{3} \cdot 3}{\underset{1}{3} \cdot \underset{1}{2x}}$$

Factor 9 as 3 · 3. Then simplify the result by removing factors equal to 1.

$$= \frac{15}{2}$$

The Language of Algebra

The second step of this method could also be phrased: Perform the division by *inverting the denominator of the complex fraction and multiplying.*

Self Check 1   Simplify: $\dfrac{\frac{7y}{8}}{\frac{21y}{20}}$.

EXAMPLE 2

ELEMENTARY & INTERMEDIATE
Algebra *f(x)* **Now**™

Simplify: $\dfrac{\frac{1}{2} - \frac{1}{x}}{\frac{x}{3} + \frac{1}{5}}$.

Solution   We consider the numerator and the denominator of the complex fraction separately. To write the numerator as a single rational expression, we build $\frac{1}{2}$ and $\frac{1}{x}$ to have an LCD of $2x$, and then subtract. To write the denominator as a single rational expression, we build $\frac{x}{3}$ and $\frac{1}{5}$ to have an LCD of 15, and then add.

$$\frac{\frac{1}{2} - \frac{1}{x}}{\frac{x}{3} + \frac{1}{5}} = \frac{\frac{1}{2} \cdot \frac{x}{x} - \frac{1}{x} \cdot \frac{2}{2}}{\frac{x}{3} \cdot \frac{5}{5} + \frac{1}{5} \cdot \frac{3}{3}}$$

The LCD for the numerator is $2x$.
The LCD for the denominator is 15.

$$= \frac{\frac{x}{2x} - \frac{2}{2x}}{\frac{5x}{15} + \frac{3}{15}}$$

Multiply the numerators.
Multiply the denominators.

$$= \frac{\frac{x - 2}{2x}}{\frac{5x + 3}{15}}$$

Subtract in the numerator and add in the denominator.

Now that the numerator and the denominator of the complex fraction are single rational expressions, we perform the division.

$$\frac{\frac{x - 2}{2x}}{\frac{5x + 3}{15}} = \frac{x - 2}{2x} \div \frac{5x + 3}{15}$$

Write the division indicated by the main fraction bar using a ÷ symbol.

Notation

The result after simplifying a complex fraction can often have several equivalent forms. The result for Example 2 could be written:

$$\frac{15x - 30}{2x(5x + 3)}$$

$$= \frac{x - 2}{2x} \cdot \frac{15}{5x + 3} \qquad \text{Multiply by the reciprocal of } \frac{5x + 3}{15}.$$

$$= \frac{15(x - 2)}{2x(5x + 3)} \qquad \begin{array}{l}\text{Multiply the numerators.}\\ \text{Multiply the denominators.}\end{array}$$

Since the numerator and denominator have no common factor, the result does not simplify.

Self Check 2   Simplify: $\dfrac{\dfrac{1}{3} + \dfrac{1}{x}}{\dfrac{x}{5} - \dfrac{1}{2}}.$

EXAMPLE 3

ELEMENTARY &
INTERMEDIATE
Algebra ƒ(x) **Now**™

Simplify: $\dfrac{\dfrac{6}{x} + y}{\dfrac{6}{y} + x}.$

Solution   To write $\frac{6}{x} + y$ as a single rational expression, we build y to a fraction with a denominator of x and add. To write $\frac{6}{y} + x$ as a single rational expression, we build x to a fraction with a denominator of y and add.

$$\frac{\dfrac{6}{x} + y}{\dfrac{6}{y} + x} = \frac{\dfrac{6}{x} + \dfrac{y}{1} \cdot \dfrac{x}{x}}{\dfrac{6}{y} + \dfrac{x}{1} \cdot \dfrac{y}{y}} \qquad \begin{array}{l}\text{The LCD for the numerator is } x.\\ \text{The LCD for the denominator is } y.\end{array}$$

$$= \frac{\dfrac{6}{x} + \dfrac{xy}{x}}{\dfrac{6}{y} + \dfrac{xy}{y}} \qquad \begin{array}{l}\text{Multiply the numerators.}\\ \text{Multiply the denominators.}\end{array}$$

$$= \frac{\dfrac{6 + xy}{x}}{\dfrac{6 + xy}{y}} \qquad \text{Add in the numerator and denominator.}$$

Now that the numerator and the denominator of the complex fraction are single expressions, we can perform the division.

Success Tip

Simplifying using division works well when a complex fraction is written, or can be easily written, as a quotient of two single rational expressions.

$$\frac{\dfrac{6 + xy}{x}}{\dfrac{6 + xy}{y}} = \frac{6 + xy}{x} \div \frac{6 + xy}{y} \qquad \text{Write the division using a } \div \text{ symbol.}$$

$$= \frac{6 + xy}{x} \cdot \frac{y}{6 + xy} \qquad \text{Multiply by the reciprocal of } \frac{6 + xy}{y}.$$

$$= \frac{y(6 + xy)}{x(6 + xy)} \qquad \begin{array}{l}\text{Multiply the numerators.}\\ \text{Multiply the denominators.}\end{array}$$

$$= \frac{y(6 + \cancel{xy})}{x(6 + \cancel{xy})}$$ Simplify the result by removing a factor equal to 1.

$$= \frac{y}{x}$$

Self Check 3   Simplify: $\dfrac{\dfrac{2}{a} - b}{\dfrac{2}{b} - a}$.

■ SIMPLIFYING COMPLEX FRACTIONS USING THE LCD

A second method for simplifying complex fractions uses the concepts of LCD and multiplication by a form of 1.

Simplifying Complex Fractions	1. Find the LCD of all rational expressions in the complex fraction.
	2. Multiply the complex fraction by 1 in the form $\frac{\text{LCD}}{\text{LCD}}$.
	3. Perform the operations in the numerator and denominator. No fractional expressions should remain within the complex fraction.
	4. Simplify the result, if possible.

EXAMPLE 4

ELEMENTARY &
INTERMEDIATE
Algebra *f(x)* **Now**™

Simplify: $\dfrac{\dfrac{1}{2} - \dfrac{1}{x}}{\dfrac{x}{3} + \dfrac{1}{5}}$.

Solution   The denominators of the rational expressions that appear in the complex fraction are 2, x, 3, and 5. Thus, their LCD is $2 \cdot x \cdot 3 \cdot 5 = 30x$.

We now multiply the complex fraction by a factor equal to 1, using the LCD: $\frac{30x}{30x} = 1$.

Success Tip

With method 2, each term of the numerator and each term of the denominator of the complex fraction is multiplied by the LCD. Arrows can be helpful in showing this. For Example 4, we can write

$$\frac{\left(\dfrac{1}{2} - \dfrac{1}{x}\right)30x}{\left(\dfrac{x}{3} + \dfrac{1}{5}\right)30x}$$

$$\frac{\dfrac{1}{2} - \dfrac{1}{x}}{\dfrac{x}{3} + \dfrac{1}{5}} = \frac{\dfrac{1}{2} - \dfrac{1}{x}}{\dfrac{x}{3} + \dfrac{1}{5}} \cdot \frac{\mathbf{30x}}{\mathbf{30x}}$$

$$= \frac{\left(\dfrac{1}{2} - \dfrac{1}{x}\right)\mathbf{30x}}{\left(\dfrac{x}{3} + \dfrac{1}{5}\right)\mathbf{30x}}$$ Multiply the numerators.
Multiply the denominators.

$$= \frac{\dfrac{1}{2}(30x) - \dfrac{1}{x}(30x)}{\dfrac{x}{3}(30x) + \dfrac{1}{5}(30x)}$$ In the numerator and the denominator, distribute the multiplication by 30x.

$$= \frac{15x - 30}{10x^2 + 6x}$$ Perform the multiplications by 30x.

To attempt to simplify the result, factor the numerator and denominator. Since they do not have a common factor, the result is in simplest form.

$$\frac{15x - 30}{10x^2 + 6x} = \frac{15(x - 2)}{2x(5x + 3)}$$

Self Check 4 Use method 2 to simplify: $\dfrac{\dfrac{1}{4} - \dfrac{1}{x}}{\dfrac{x}{5} + \dfrac{1}{3}}$.

EXAMPLE 5

ELEMENTARY &
INTERMEDIATE
Algebra *(f(x))* **Now**™

Simplify: $\dfrac{\dfrac{1}{8} - \dfrac{1}{y}}{\dfrac{8 - y}{8}}$.

Solution The denominators of the rational expressions that appear in the complex fraction are 8, y, and 8. Therefore, the LCD is $8y$ and we multiply the complex fraction by a factor equal to 1, using the LCD: $\dfrac{8y}{8y} = 1$.

$$\frac{\dfrac{1}{8} - \dfrac{1}{y}}{\dfrac{8 - y}{8}} = \frac{\dfrac{1}{8} - \dfrac{1}{y}}{\dfrac{8 - y}{8}} \cdot \frac{\mathbf{8y}}{\mathbf{8y}}$$

$$= \frac{\left(\dfrac{1}{8} - \dfrac{1}{y}\right)\mathbf{8y}}{\left(\dfrac{8 - y}{8}\right)\mathbf{8y}}$$

Multiply the numerators.
Multiply the denominators.

$$= \frac{\dfrac{1}{8}(\mathbf{8y}) - \dfrac{1}{y}(\mathbf{8y})}{\left(\dfrac{8 - y}{8}\right)\mathbf{8y}}$$

In the numerator, distribute the multiplication by $8y$.

$$= \frac{y - 8}{(8 - y)y}$$

Perform each multiplication.

$$= \frac{\overset{-1}{\cancel{y - 8}}}{\underset{1}{\cancel{(8 - y)}}y}$$

Since $y - 8$ and $8 - y$ are opposites, replace $\dfrac{y - 8}{8 - y}$ with $\dfrac{-1}{1}$.

$$= -\frac{1}{y}$$

Self Check 5 Simplify: $\dfrac{\dfrac{10 - n}{10}}{\dfrac{1}{10} - \dfrac{1}{n}}$.

EXAMPLE 6

Simplify: $\dfrac{1}{1+\dfrac{1}{x+1}}$.

Solution   The only rational expression in the complex fraction has the denominator $x + 1$. Therefore, the LCD is $x + 1$. So we multiply the complex fraction by a factor equal to 1, using the LCD: $\dfrac{x+1}{x+1} = 1$.

$$\dfrac{1}{1+\dfrac{1}{x+1}} = \dfrac{1}{1+\dfrac{1}{x+1}} \cdot \dfrac{x+1}{x+1}$$

$$= \dfrac{1(x+1)}{\left(1+\dfrac{1}{x+1}\right)(x+1)}$$ Multiply the numerators.
Multiply the denominators.

$$= \dfrac{1(x+1)}{1(x+1)+\dfrac{1}{x+1}(x+1)}$$ In the denominator, distribute the multiplication by $x + 1$.

$$= \dfrac{x+1}{x+1+1}$$ Perform each multiplication.

$$= \dfrac{x+1}{x+2}$$ Combine like terms in the denominator.

The result does not simplify.

Success Tip

Simplifying using the LCD works well when the complex fraction has sums and/or differences in the numerator or denominator.

Self Check 6   Simplify: $\dfrac{2}{\dfrac{1}{x+2}+2}$.

Answers to Self Checks   **1.** $\dfrac{5}{6}$ **2.** $\dfrac{10(x+3)}{3x(2x-5)}$ **3.** $\dfrac{b}{a}$ **4.** $\dfrac{15(x-4)}{4x(3x+5)}$ **5.** $-n$ **6.** $\dfrac{2(x+2)}{2x+5}$

6.5   STUDY SET

VOCABULARY   Fill in the blanks.

1. The expression $\dfrac{\dfrac{2}{3}-\dfrac{1}{x}}{\dfrac{x-3}{4}}$ is called a _____ rational

   expression or, more simply, a _____ fraction.

2. In a complex fraction, the _____ is above the main fraction bar and the _____ is below it.

3. To find the _____ of $\dfrac{x+8}{x+7}$, we invert it.

4. The _____ common denominator of all the rational

   expressions in $\dfrac{\dfrac{1}{3}-\dfrac{1}{x}}{\dfrac{3}{x}}$ is $3x$.

CONCEPTS   Fill in the blanks.

5. Method 1: To simplify a complex fraction, write its numerator and denominator as _____ rational expressions. Then perform the division by multiplying the numerator of the complex fraction by the _____ of the denominator of the complex fraction.

6. Method 2: To simplify a complex fraction, find the _____ of all rational expressions in the complex fraction. Multiply the complex fraction by ____ in the form $\dfrac{\text{LCD}}{\text{LCD}}$. Then perform the operations.

7. Consider $\dfrac{\dfrac{x-3}{4}}{\dfrac{1}{12}-\dfrac{x}{6}}$.

   a. What is the numerator of the complex fraction?

   b. Is the numerator a single rational expression?

   c. What is the denominator of the complex fraction?

   d. Is the denominator a single rational expression?

8. Fill in the blank.

$$\dfrac{\dfrac{12}{y^2}}{\dfrac{4}{y^3}}=\dfrac{12}{y^2}\;\boxed{}\;\dfrac{4}{y^3}$$

9. Consider the complex fraction $\dfrac{\dfrac{1}{y}-\dfrac{1}{3}}{\dfrac{5}{6}+\dfrac{1}{y}}$.

   a. What are the denominators of the rational expressions in the complex fraction?

   b. What is the LCD of all rational expressions in the complex fraction?

   c. To simplify the complex fraction using method 2, it should be multiplied by what form of 1?

10. Fill in the blanks.

$$\dfrac{\left(\dfrac{1}{5}-\dfrac{1}{a}\right)20a}{\left(\dfrac{a}{4}+\dfrac{1}{a}\right)20a}=\dfrac{\dfrac{1}{5}\left(\boxed{}\right)-\dfrac{1}{a}\left(\boxed{}\right)}{\dfrac{a}{4}\left(\boxed{}\right)+\dfrac{1}{a}\left(\boxed{}\right)}$$

Find each product.

11. $\dfrac{x}{12}(24)$

12. $\dfrac{1}{2}(20x)$

13. $\dfrac{2}{3}(6y^2)$

14. $\dfrac{9}{x-8}(x-8)$

NOTATION

15. Write the following expression as a complex fraction.

$$\dfrac{4x^2}{15}\div\dfrac{16x}{25}$$

16. Draw arrows to show how the distributive property should be applied in the denominator.

$$\dfrac{\dfrac{1}{x-6}(x-6)}{\left(\dfrac{1}{x-6}+2\right)(x-6)}$$

PRACTICE   **Simplify each complex fraction.**

17. $\dfrac{\dfrac{x}{2}}{\dfrac{6}{5}}$

18. $\dfrac{\dfrac{9}{4}}{\dfrac{7}{x}}$

19. $\dfrac{\dfrac{2}{3}}{\dfrac{3}{4}}$

20. $\dfrac{\dfrac{3}{5}}{\dfrac{2}{7}}$

21. $\dfrac{\dfrac{x}{y}}{\dfrac{1}{x}}$

22. $\dfrac{\dfrac{y}{x}}{\dfrac{x}{xy}}$

23. $\dfrac{\dfrac{n}{8}}{\dfrac{1}{n^2}}$

24. $\dfrac{\dfrac{1}{m}}{\dfrac{m^3}{15}}$

25. $\dfrac{\dfrac{1}{x}-3}{\dfrac{5}{x}+2}$

26. $\dfrac{\dfrac{1}{y}+3}{\dfrac{3}{y}-2}$

27. $\dfrac{\dfrac{2}{3}+1}{\dfrac{1}{3}+1}$

28. $\dfrac{\dfrac{3}{5}-2}{\dfrac{2}{5}-2}$

29. $\dfrac{\dfrac{4a}{11}}{\dfrac{6a}{55}}$

30. $\dfrac{\dfrac{14}{15m}}{\dfrac{21}{25m}}$

31. $\dfrac{\dfrac{40x^2}{20x}}{9}$

32. $\dfrac{\dfrac{18n^2}{6n}}{13}$

33. $\dfrac{\dfrac{1}{y}-\dfrac{5}{2}}{\dfrac{3}{y}}$

34. $\dfrac{\dfrac{1}{6}-\dfrac{5}{s}}{\dfrac{2}{s}}$

35. $\dfrac{\dfrac{d+2}{2}}{\dfrac{d}{3}-\dfrac{d}{4}}$

36. $\dfrac{\dfrac{d^2}{4}+\dfrac{4d}{5}}{\dfrac{d+1}{2}}$

57. $\dfrac{\dfrac{3y}{x}-y}{y-\dfrac{y}{x}}$

58. $\dfrac{\dfrac{y}{x}+3y}{y+\dfrac{2y}{x}}$

37. $\dfrac{\dfrac{s+15}{16}}{\dfrac{s+15}{8}}$

38. $\dfrac{\dfrac{t-5}{12}}{\dfrac{t-5}{4}}$

59. $\dfrac{\dfrac{b^2-81}{18a^2}}{\dfrac{4b-36}{9a}}$

60. $\dfrac{\dfrac{8x-64}{y}}{\dfrac{x^2-64}{y^2}}$

39. $\dfrac{\dfrac{1}{2}+\dfrac{3}{4}}{\dfrac{3}{2}+\dfrac{1}{4}}$

40. $\dfrac{\dfrac{2}{3}-\dfrac{5}{2}}{\dfrac{2}{3}-\dfrac{3}{2}}$

61. $\dfrac{4-\dfrac{1}{8h}}{12+\dfrac{3}{4h}}$

62. $\dfrac{12+\dfrac{1}{3b}}{12-\dfrac{1}{b^2}}$

41. $\dfrac{\dfrac{x^4}{30}}{\dfrac{7x}{15}}$

42. $\dfrac{\dfrac{5x^2}{24}}{\dfrac{x^5}{56}}$

63. $\dfrac{1}{\dfrac{1}{x}+\dfrac{1}{y}}$

64. $\dfrac{1}{\dfrac{b}{a}-\dfrac{a}{b}}$

43. $\dfrac{\dfrac{2}{s}-\dfrac{2}{s^2}}{\dfrac{4}{s^3}+\dfrac{4}{s^2}}$

44. $\dfrac{\dfrac{2}{x^3}-\dfrac{2}{x}}{\dfrac{4}{x}+\dfrac{8}{x^2}}$

65. $\dfrac{\dfrac{1}{x+1}}{1+\dfrac{1}{x+1}}$

66. $\dfrac{\dfrac{1}{x-1}}{1-\dfrac{1}{x-1}}$

45. $\dfrac{-\dfrac{3}{x^3}}{\dfrac{6}{x^5}}$

46. $\dfrac{\dfrac{18}{x^6}}{-\dfrac{2}{x^8}}$

67. $\dfrac{\dfrac{x}{x+2}}{\dfrac{x}{x+2}+x}$

68. $\dfrac{\dfrac{2}{x-2}}{\dfrac{2}{x-2}-1}$

47. $\dfrac{\dfrac{2}{x}+2}{\dfrac{4}{x}+2}$

48. $\dfrac{\dfrac{3}{x}-3}{\dfrac{9}{x}-3}$

69. $\dfrac{\dfrac{5t^2}{9x^2}}{\dfrac{3t}{x^2t}}$

70. $\dfrac{\dfrac{5w^2}{4tz}}{\dfrac{15wt}{z^2}}$

49. $\dfrac{\dfrac{2x-8}{15}}{\dfrac{3x-12}{35x}}$

50. $\dfrac{\dfrac{3x-9}{8x}}{\dfrac{5x-15}{32}}$

71. $\dfrac{\dfrac{m^2-4}{3m+3}}{\dfrac{2m+4}{m+1}}$

72. $\dfrac{\dfrac{d^2-16}{5d+10}}{\dfrac{2d-8}{25}}$

51. $\dfrac{\dfrac{t-6}{16}}{\dfrac{12-2t}{t}}$

52. $\dfrac{\dfrac{15-30c}{28}}{\dfrac{2c-1}{c}}$

73. $\dfrac{\dfrac{2}{x}}{\dfrac{2}{y}-\dfrac{4}{x}}$

74. $\dfrac{\dfrac{2y}{3}}{\dfrac{2y}{3}-\dfrac{8}{y}}$

53. $\dfrac{\dfrac{2}{c^2}}{\dfrac{1}{c}+\dfrac{5}{4}}$

54. $\dfrac{\dfrac{7}{s^2}}{\dfrac{1}{s}+\dfrac{10}{3}}$

75. $\dfrac{\dfrac{m}{n}+\dfrac{n}{m}}{\dfrac{m}{n}-\dfrac{n}{m}}$

76. $\dfrac{\dfrac{2a}{b}-\dfrac{b}{a}}{\dfrac{2a}{b}+\dfrac{b}{a}}$

55. $\dfrac{\dfrac{1}{a^2b}-\dfrac{5}{ab}}{\dfrac{3}{ab}-\dfrac{7}{ab^2}}$

56. $\dfrac{\dfrac{3}{ab^2}+\dfrac{6}{a^2b}}{\dfrac{6}{a}-\dfrac{9}{b^2}}$

77. $\dfrac{3+\dfrac{3}{x-1}}{3-\dfrac{3}{x-1}}$

78. $\dfrac{2-\dfrac{2}{x+1}}{2+\dfrac{2}{x+1}}$

APPLICATIONS

79. GARDENING TOOLS What is the result when the opening of the cutting blades is divided by the opening of the handles? Express the result in simplest form.

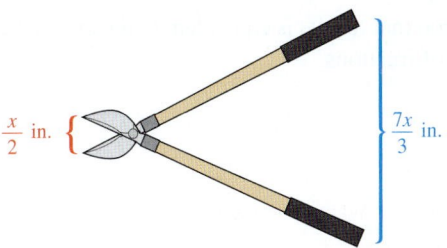

$\frac{x}{2}$ in.

$\frac{7x}{3}$ in.

80. EARNED RUN AVERAGE The earned run average (ERA) is a statistic that gives the average number of earned runs a pitcher allows. For a softball pitcher, this is based on a six-inning game. The formula for ERA is

$$\text{ERA} = \frac{\text{earned runs}}{\dfrac{\text{innings pitched}}{6}}$$

Simplify the complex fraction on the right-hand side of the equation.

81. ELECTRONICS In electronic circuits, resistors oppose the flow of an electric current. To find the total resistance of a parallel combination of two resistors, we can use the formula

$$\text{Total resistance} = \frac{1}{\dfrac{1}{R_1} + \dfrac{1}{R_2}}$$

where R_1 is the resistance of the first resistor and R_2 is the resistance of the second. Simplify the complex fraction on the right-hand side of the formula.

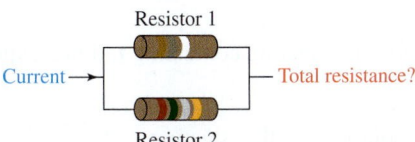

Resistor 1
Current → | Total resistance?
Resistor 2

82. DATA ANALYSIS Use the data in the table to find the average measurement for the three-trial experiment.

	Trial 1	Trial 2	Trial 3
Measurement	$\dfrac{k}{2}$	$\dfrac{k}{3}$	$\dfrac{k}{2}$

WRITING

83. What is a complex fraction? Give several examples.

84. Explain how to use method 1 to simplify $\dfrac{1 + \dfrac{1}{x}}{3 - \dfrac{1}{x}}$.

85. Explain how to use method 2 to simplify the expression in Problem 84.

86. a. List an advantage and a disadvantage of using method 1 to simplify a complex fraction.

b. List an advantage and a disadvantage of using method 2 to simplify a complex fraction.

REVIEW **Simplify each expression. Write each answer without using parentheses or negative exponents.**

87. $(8x)^0$

88. $\dfrac{8x^0}{4}$

89. $\left(\dfrac{3r}{4r^3}\right)^4$

90. $\left(\dfrac{12y^{-3}}{3y^2}\right)^{-2}$

91. $\left(\dfrac{6r^{-2}}{2r^3}\right)^{-2}$

92. $\left(\dfrac{4x^3}{5x^{-3}}\right)^{-2}$

CHALLENGE PROBLEMS Simplify.

93. $\dfrac{\dfrac{h}{h^2 + 3h + 2}}{\dfrac{4}{h + 2} - \dfrac{4}{h + 1}}$

94. $\dfrac{\dfrac{2}{b^2 - 1} - \dfrac{3}{ab - a}}{\dfrac{3}{ab - a} - \dfrac{2}{b^2 - 1}}$

95. $a + \dfrac{a}{1 + \dfrac{a}{a + 1}}$

96. $\dfrac{y^{-2} + 1}{y^{-2} - 1}$

6.6   Solving Rational Equations

- Solving Rational Equations
- Solving Formulas for a Specified Variable

We have solved equations such as $\frac{2x}{3} = \frac{x}{6} + \frac{3}{2}$ by multiplying both sides of the equation by the LCD. With this approach, the equation that results is equivalent to the original equation but easier to solve because it is cleared of fractions.

$$\frac{2x}{3} = \frac{x}{6} + \frac{3}{2}$$

$$6\left(\frac{2x}{3}\right) = 6\left(\frac{x}{6} + \frac{3}{2}\right) \qquad \text{Multiply both sides of the equation by the LCD of } \frac{2x}{3}, \frac{x}{6},$$
$$\text{and } \frac{3}{2}, \text{ which is } 6.$$

$$6\left(\frac{2x}{3}\right) = 6\left(\frac{x}{6}\right) + 6\left(\frac{3}{2}\right) \qquad \text{Distribute the multiplication by } 6.$$

$$4x = x + 9 \qquad \text{Perform each multiplication.}$$

$$3x = 9 \qquad \text{Subtract } x \text{ from both sides.}$$

$$x = 3 \qquad \text{Divide both sides by } 3.$$

Equations such as $\dfrac{2x}{3} = \dfrac{x}{6} + \dfrac{3}{2}$ are called *rational equations*.

Success Tip

Before multiplying both sides of an equation by the LCD, enclose the left-hand and right-hand sides with parentheses:

$$\left(\frac{2x}{3}\right) = \left(\frac{x}{6} + \frac{3}{2}\right)$$

Rational Equations   A **rational equation** is an equation that contains one or more rational expressions.

Rational equations often have a variable in the denominator. Some examples are:

$$\frac{2}{x} + \frac{1}{2} = \frac{5}{2x}, \qquad y - \frac{12}{y} = 4, \qquad \text{and} \qquad \frac{11x}{x-5} = 6 + \frac{55}{x-5}$$

In this section, we will extend the fraction-clearing strategy to solve rational equations like those listed above.

■ SOLVING RATIONAL EQUATIONS

To solve a rational equation, we must find all values of the variable that make the equation true. However, before solving a rational equation, we must identify any values of the variable that make an expression in the equation undefined. A number cannot be a solution of the equation if, when substituted for the variable, it makes a denominator equal to 0.

The following steps can be used to solve an equation containing rational expressions.

Solving Rational Equations   1. Determine which numbers cannot be solutions of the equation.
2. Multiply both sides of the equation by the LCD of all rational expressions in the equation. This clears the equation of fractions.
3. Solve the resulting equation.
4. Check all possible solutions in the original equation.

EXAMPLE 1

Solution

Solve: $\dfrac{2}{x} + \dfrac{1}{2} = \dfrac{5}{2x}$.

If x is 0, the denominators of $\frac{2}{x}$ and $\frac{5}{2x}$ are 0 and the expressions would be undefined. Therefore, 0 cannot be a solution.

Since the denominators are x, 2, and $2x$, we multiply both sides of the equation by the LCD, $2x$, to clear the fractions.

$$\frac{2}{x} + \frac{1}{2} = \frac{5}{2x}$$

$$\mathbf{2x}\left(\frac{2}{x} + \frac{1}{2}\right) = \mathbf{2x}\left(\frac{5}{2x}\right) \qquad \text{Multiply both sides by } 2x.$$

$$\mathbf{2x}\left(\frac{2}{x}\right) + \mathbf{2x}\left(\frac{1}{2}\right) = \mathbf{2x}\left(\frac{5}{2x}\right) \qquad \text{Distribute the multiplication by } 2x.$$

$$4 + x = 5 \qquad \text{Simplify.}$$

To solve the resulting equation, subtract 4 from both sides.

$$4 + x - \mathbf{4} = 5 - \mathbf{4}$$

$$x = 1$$

To check, we substitute 1 for x in the original equation.

$$\frac{2}{\mathbf{x}} + \frac{1}{2} = \frac{5}{2\mathbf{x}}$$

$$\frac{2}{\mathbf{1}} + \frac{1}{2} \overset{?}{=} \frac{5}{2(\mathbf{1})} \qquad \text{Substitute 1 for } x.$$

$$\frac{5}{2} = \frac{5}{2} \qquad \text{Evaluate the left-hand side: } \frac{2}{1} + \frac{1}{2} = \frac{4}{2} + \frac{1}{2} = \frac{5}{2}.$$

Since we obtain a true statement, 1 is the solution.

Self Check 1   Solve: $\dfrac{1}{3} + \dfrac{4}{3x} = \dfrac{5}{x}$.

Notation

To streamline multiplication, we can remove factors equal to 1 in the following way:

$$2x\left(\dfrac{2}{\overset{1}{\cancel{x}}}\right)^{\!\!1}$$

Here, we simplify by replacing $\frac{x}{x}$ with $\frac{1}{1}$.

Success Tip

Don't confuse procedures. To *simplify the expression* $\frac{2}{x} + \frac{1}{2}$, we build each fraction to have the LCD $2x$, add the numerators, and write the sum over the LCD. To *solve the equation* $\frac{2}{x} + \frac{1}{2} = \frac{5}{2x}$, we multiply both sides by the LCD $2x$ to eliminate the denominators.

EXAMPLE 2

Solution

Solve: $y - \dfrac{12}{y} = 4$.

If y is 0, the denominator of $\frac{12}{y}$ is 0, and the expression would be undefined. Therefore, 0 cannot be a solution.

Since the only denominator is y, we multiply both sides of the equation by y to clear the fractions.

$$y - \frac{12}{y} = 4$$

$$\mathbf{y}\left(y - \frac{12}{y}\right) = \mathbf{y}(4) \qquad \text{Multiply both sides of the equation by } y.$$

$$y(y) - y\left(\frac{12}{y}\right) = y(4) \qquad \text{Distribute the multiplication by } y.$$

$$y^2 - 12 = 4y \qquad \text{Simplify.}$$

We can solve the resulting quadratic equation using the factoring method.

$$y^2 - 4y - 12 = 0 \qquad\qquad\qquad \text{Subtract } 4y \text{ from both sides.}$$

$$(y - 6)(y + 2) = 0 \qquad\qquad\qquad \text{Factor the trinomial.}$$

$$y - 6 = 0 \qquad \text{or} \qquad y + 2 = 0 \qquad \text{Set each factor equal to 0.}$$

$$y = 6 \qquad\qquad\qquad y = -2 \qquad \text{Solve each equation.}$$

There are two possible solutions, 6 and -2.

Check: $y = 6$	*Check: $y = -2$*
$y - \dfrac{12}{y} = 4$	$y - \dfrac{12}{y} = 4$
$6 - \dfrac{12}{6} \overset{?}{=} 4$	$-2 - \dfrac{12}{-2} \overset{?}{=} 4$
$6 - 2 \overset{?}{=} 4$	$-2 - (-6) \overset{?}{=} 4$
$4 = 4$	$4 = 4$

The solutions of $y - \dfrac{12}{y} = 4$ are 6 and -2.

Self Check 2   Solve: $x - \dfrac{24}{x} = -5$.

EXAMPLE 3

Solve: $\dfrac{11x}{x - 5} = 6 + \dfrac{55}{x - 5}$.

ELEMENTARY &
INTERMEDIATE
Algebra ⨍⁽ˣ⁾ **Now**™

Solution   If x is 5, the denominators of $\dfrac{11x}{x - 5}$ and $\dfrac{55}{x - 5}$ are 0, and the expressions are undefined. Therefore, 5 cannot be a solution of the equation.

Since both denominators are $x - 5$, we multiply both sides by the LCD, $x - 5$, to clear the fractions.

Caution

When solving rational equations, each term on both sides must be multiplied by the LCD.

$$\frac{11x}{x - 5} = 6 + \frac{55}{x - 5}$$

$$(x - 5)\left(\frac{11x}{x - 5}\right) = (x - 5)\left(6 + \frac{55}{x - 5}\right) \qquad \begin{array}{l}\text{Multiply both sides of the equation} \\ \text{by } x - 5.\end{array}$$

$$(x - 5)\left(\frac{11x}{x - 5}\right) = (x - 5)6 + (x - 5)\left(\frac{55}{x - 5}\right) \qquad \begin{array}{l}\text{Distribute the multiplication by} \\ x - 5.\end{array}$$

$$11x = (x - 5)6 + 55 \qquad \text{Simplify: } \tfrac{x - 5}{x - 5} = 1.$$

$$11x = 6x - 30 + 55 \qquad \text{Distribute the multiplication by 6.}$$

To solve the resulting equation, we proceed as follows:

The Language of Algebra

Extraneous means not a vital part. Mathematicians speak of *extraneous* solutions. Rock groups don't want any *extraneous* sounds (like humming or feedback) coming from their amplifiers. Artists erase any *extraneous* marks on their sketches.

$11x = 6x + 25$ Simplify: $-30 + 55 = 25$.

$5x = 25$ Subtract $6x$ from both sides.

$x = 5$ Divide both sides by 5.

We have determined that 5 makes both denominators in the original equation 0. Therefore, 5 is not a solution. Since 5 is the only possible solution, and it must be rejected, it follows that $\frac{11x}{x-5} = 6 + \frac{55}{x-5}$ has no solution.

When solving an equation, a possible solution that does not satisfy the original equation is called an **extraneous solution.** In this example, 5 is an extraneous solution.

Self Check 3   Solve: $\dfrac{9x}{x-6} = 3 + \dfrac{54}{x-6}$, if possible.

EXAMPLE 4   Solve: $\dfrac{x+5}{x+3} + \dfrac{1}{x^2+2x-3} = 1$.

ELEMENTARY & INTERMEDIATE
Algebra $f(x)$ Now™   Solution   To determine the restrictions on the variable, we factor the second denominator.

$$\frac{x+5}{x+3} + \frac{1}{(x+3)(x-1)} = 1$$ If x is -3, the first denominator is 0. If x is -3 or 1, the second denominator is 0.

We see that -3 and 1 cannot be solutions of the equation, because they make rational expressions in the equation undefined.

Since the denominators are $x + 3$ and $(x + 3)(x - 1)$, we multiply both sides of the equation by the LCD, $(x + 3)(x - 1)$, to clear the fractions.

$$(x+3)(x-1)\left[\frac{x+5}{x+3} + \frac{1}{(x+3)(x-1)}\right] = (x+3)(x-1)[1]$$

$$(x+3)(x-1)\frac{x+5}{x+3} + (x+3)(x-1)\frac{1}{(x+3)(x-1)} = (x+3)(x-1)1$$ Distribute the multiplication by $(x+3)(x-1)$.

$$(x-1)(x+5) + 1 = (x+3)(x-1)$$ Simplify: $\frac{x+3}{x+3} = 1$ and $\frac{x-1}{x-1} = 1$.

To solve the resulting equation, we multiply the binomials on the left-hand side and the right-hand side, and proceed as follows.

$x^2 + 4x - 5 + 1 = x^2 + 2x - 3$ Find $(x - 1)(x + 5)$ and $(x + 3)(x - 1)$.

$x^2 + 4x - 4 = x^2 + 2x - 3$ Combine like terms: $-5 + 1 = -4$.

$4x - 4 = 2x - 3$ Subtract x^2 from both sides.

$2x - 4 = -3$ Subtract $2x$ from both sides.

$2x = 1$ Add 4 to both sides.

$x = \dfrac{1}{2}$ Divide both sides by 2.

A check will show that $\dfrac{1}{2}$ is the solution of the original equation.

Self Check 4 Solve: $\dfrac{1}{x+3} + \dfrac{1}{x-3} = \dfrac{5}{x^2-9}$.

■ SOLVING FORMULAS FOR A SPECIFIED VARIABLE

Recall that a **formula** is an equation that states a known relationship between two or more variables. Many formulas are expressed as rational equations.

EXAMPLE 5

ELEMENTARY &
INTERMEDIATE
Algebra *f(x)* **Now**™

Determining a child's dosage. The formula

$$C = \frac{AD}{A+12}$$

is called Young's rule. It is a way to find the approximate child's dose C of a prescribed medication, where A is the age of the child in years and D is the recommended dosage for an adult. Solve the formula for D.

Solution To clear the equation of the fraction, we multiply both sides by $A + 12$.

$$C = \frac{AD}{A+12}$$

$$(A+12)(C) = (A+12)\left(\frac{AD}{A+12}\right)$$

$$(A+12)C = AD \qquad \text{Simplify: } \tfrac{A+12}{A+12} = 1.$$

$$AC + 12C = AD \qquad \text{Distribute the multiplication by } C.$$

$$\frac{AC + 12C}{A} = D \qquad \text{To isolate } D, \text{ divide both sides by } A.$$

Solving Young's formula for D, we have $D = \dfrac{AC + 12C}{A}$.

EXAMPLE 6

Photography. The design of a camera lens uses the formula

$$\frac{1}{f} = \frac{1}{p} + \frac{1}{q}$$

where f is the focal length of the lens, p is the distance from the lens to the object, and q is the distance from the lens to the image. Solve the formula for q.

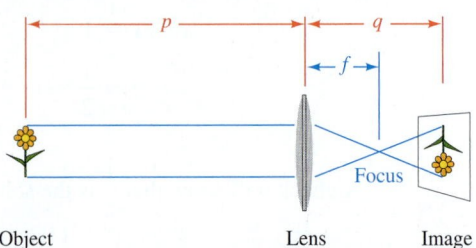

Object   Lens   Image

Solution   The denominators are f, p, and q. To clear the equation of fractions, we multiply both sides by the LCD, fpq.

$$\frac{1}{f} = \frac{1}{p} + \frac{1}{q}$$

$$\boldsymbol{fpq}\left(\frac{1}{f}\right) = \boldsymbol{fpq}\left(\frac{1}{p} + \frac{1}{q}\right) \qquad \text{Multiply both sides of the equation by the LCD.}$$

$$fpq\left(\frac{1}{f}\right) = \boldsymbol{fpq}\left(\frac{1}{p}\right) + \boldsymbol{fpq}\left(\frac{1}{q}\right) \qquad \text{Distribute the multiplication by } fpq.$$

$$pq = fq + fp \qquad \text{Simplify: } \frac{f}{f} = 1, \frac{p}{p} = 1, \text{ and } \frac{q}{q} = 1.$$

If we subtract fq from both sides, all terms that contain q will be on the left-hand side.

$$pq - fq = fp \qquad \text{Subtract } fq \text{ from both sides.}$$

$$q(p - f) = fp \qquad \text{Factor out } q \text{ from the terms on the left-hand side.}$$

$$\frac{q(p - f)}{\boldsymbol{p - f}} = \frac{fp}{\boldsymbol{p - f}} \qquad \text{To isolate } q, \text{ divide both sides by } p - f.$$

$$q = \frac{fp}{p - f} \qquad \text{Simplify: } \frac{p - f}{p - f} = 1.$$

Solving the formula for q, we have $q = \dfrac{fp}{p - f}$.

Self Check 6   Solve the formula in Example 6 for p.

Answers to Self Checks   **1.** 11   **2.** 3, -8 **3.** no solution   **4.** $\dfrac{5}{2}$ **6.** $p = \dfrac{fq}{q - f}$

6.6   STUDY SET   ELEMENTARY & INTERMEDIATE Algebra $f(x)$ Now™

VOCABULARY   Fill in the blanks.

1. Equations that contain one or more rational expressions, such as $\frac{x}{x + 2} = 4 + \frac{10}{x + 2}$, are called _____ equations.

2. To _____ a rational equation we find all the values of the variable that make the equation true.

3. To _____ a rational equation of fractions, multiply both sides by the LCD of all rational expressions in the equation.

4. If x is 3, the denominator of rational expression $\frac{4x}{x - 3}$ is 0, and the expression is _____.

5. $x^2 - x + 2 = 0$ is a _____ equation.

6. When solving a rational equation, if we obtain a number that does not satisfy the original equation, the number is called an _____ solution.

CONCEPTS

7. Is 5 a solution of the following equations?

  a. $\dfrac{1}{x - 1} = 1 - \dfrac{3}{x - 1}$

  b. $\dfrac{x}{x - 5} = 3 + \dfrac{5}{x - 5}$

8. A student was asked to solve a rational equation. The first step of his solution is as follows:

$$\boldsymbol{12x}\left(\frac{5}{x} + \frac{2}{3}\right) = \boldsymbol{12x}\left(\frac{7}{4x}\right)$$

  a. What equation was he asked to solve?

  b. What LCD is used to clear the equation of fractions?

9. Consider the rational equation $\dfrac{x}{x-3} = \dfrac{1}{x} + \dfrac{2}{x-3}$.

   a. What values of x make a denominator 0?

   b. What values of x make a rational expression undefined?

   c. What numbers can't be solutions of the equation?

By what should both sides of the equation be multiplied to clear it of fractions?

10. $x + \dfrac{11}{x} = 3$

11. $\dfrac{1}{y} = 20 - \dfrac{5}{y}$

12. $\dfrac{2x}{x-6} = 4 + \dfrac{1}{x-6}$

13. $\dfrac{x}{x+8} = \dfrac{4}{x-8}$

14. $\dfrac{x}{5} = \dfrac{3x}{10} + \dfrac{7}{2x}$

15. $\dfrac{x}{x^2-4} = \dfrac{4}{x-2}$

16. Perform each multiplication.

   a. $4x\left(\dfrac{3}{4x}\right)$ **b.** $(x+6)(x-2)\left(\dfrac{3}{x-2}\right)$

17. Fill in the blanks.

$$8x\left(\dfrac{3}{2x}\right) = 8x\left(\dfrac{1}{8x}\right) + 8x\left(\dfrac{5}{4}\right)$$

$$\underline{} = \underline{} + \underline{}$$

18. Factor out a from the terms on the left-hand side of $ab - ca = cb$.

NOTATION

19. Complete the solution.

$$\dfrac{2}{a} + \dfrac{1}{2} = \dfrac{7}{2a}$$

$$\underline{}\left(\dfrac{2}{a} + \dfrac{1}{2}\right) = \underline{}\left(\dfrac{7}{2a}\right)$$

$$\underline{}\left(\dfrac{2}{a}\right) + \underline{}\left(\dfrac{1}{2}\right) = \underline{}\left(\dfrac{7}{2a}\right)$$

$$\underline{} + a = \underline{}$$

$$4 + a - 4 = 7 - 4$$

$$a = \underline{}$$

20. A student solved a rational equation and found 8 to be a possible solution. When she checked 8, she obtained $\dfrac{3}{0} = \dfrac{1}{0} + \dfrac{2}{3}$. What conclusion can be drawn?

21. What operation is indicated when we write fpq?

22. Can $5x\left(\dfrac{2}{x} + \dfrac{4}{5}\right)$ be written as $5x \cdot \dfrac{2}{x} + \dfrac{4}{5}$? Explain.

PRACTICE   Solve each equation and check the result. If an equation has no solution, so indicate.

23. $\dfrac{2}{3} = \dfrac{1}{2} + \dfrac{x}{6}$

24. $\dfrac{7}{4} = \dfrac{x}{8} + \dfrac{5}{2}$

25. $\dfrac{x}{18} = \dfrac{1}{3} - \dfrac{x}{2}$

26. $\dfrac{x}{4} = \dfrac{1}{2} - \dfrac{3x}{20}$

27. $\dfrac{a-1}{7} - \dfrac{a-2}{14} = \dfrac{1}{2}$

28. $\dfrac{3x-1}{6} - \dfrac{x+3}{2} = \dfrac{3x+4}{3}$

29. $\dfrac{3}{x} + 2 = 3$

30. $\dfrac{2}{x} + 9 = 11$

31. $\dfrac{x}{x-5} - \dfrac{5}{x-5} = 3$

32. $\dfrac{3}{y-2} + 1 = \dfrac{3}{y-2}$

33. $\dfrac{a}{4} - \dfrac{4}{a} = 0$

34. $0 = \dfrac{t}{3} - \dfrac{12}{t}$

35. $\dfrac{2}{y+1} + 5 = \dfrac{12}{y+1}$

36. $\dfrac{3}{p+6} - 2 = \dfrac{7}{p+6}$

37. $-\dfrac{1}{b} = \dfrac{5}{b} - 9 - \dfrac{6}{b}$

38. $\dfrac{3}{a} = \dfrac{4}{a} + 8 - \dfrac{1}{a}$

39. $\dfrac{1}{t+2} = \dfrac{t-1}{t+7}$

40. $\dfrac{d-7}{d-20} = \dfrac{1}{d+8}$

41. $\dfrac{5}{n} + \dfrac{5}{12} = 0$

42. $\dfrac{2}{m} + \dfrac{2}{33} = 0$

43. $\dfrac{1}{8} + \dfrac{2}{y} = \dfrac{1}{y} + \dfrac{1}{10}$

44. $\dfrac{7}{10} + \dfrac{4}{c} = \dfrac{1}{c} + \dfrac{11}{15}$

45. $\dfrac{1}{8} + \dfrac{2}{b} - \dfrac{1}{12} = 0$

46. $\dfrac{1}{14} + \dfrac{2}{n} - \dfrac{2}{21} = 0$

47. $\dfrac{4}{5x} = \dfrac{8}{x-5}$

48. $\dfrac{1}{6x} = \dfrac{2}{x-6}$

49. $\dfrac{1}{4} - \dfrac{5}{6} = \dfrac{1}{a}$

50. $\dfrac{5}{9} - \dfrac{1}{3} = \dfrac{1}{b}$

51. $\dfrac{3}{4h} + \dfrac{2}{h} = 1$

52. $\dfrac{5}{3k} + \dfrac{1}{k} = -2$

53. $\dfrac{3r}{2} - \dfrac{3}{r} = \dfrac{3r}{2} + 3$

54. $\dfrac{2p}{3} - \dfrac{1}{p} = \dfrac{2p - 1}{3}$

55. $\dfrac{1}{3} + \dfrac{2}{x - 3} = 1$

56. $\dfrac{3}{5} + \dfrac{7}{x + 2} = 2$

57. $\dfrac{z - 4}{z - 3} = \dfrac{z + 2}{z + 1}$

58. $\dfrac{a + 2}{a + 8} = \dfrac{a - 3}{a - 2}$

59. $\dfrac{v}{v + 2} + \dfrac{1}{v - 1} = 1$

60. $\dfrac{x}{x - 2} = 1 + \dfrac{1}{x - 3}$

61. $\dfrac{a^2}{a + 2} - \dfrac{4}{a + 2} = a$

62. $\dfrac{z^2}{z + 1} + 2 = \dfrac{1}{z + 1}$

63. $\dfrac{5}{x + 4} + \dfrac{1}{x + 4} = x - 1$

64. $\dfrac{7}{x - 3} + \dfrac{1}{x - 3} = x - 5$

65. $\dfrac{3}{x + 1} - \dfrac{x - 2}{2} = \dfrac{x - 2}{x + 1}$

66. $\dfrac{2}{x - 1} + \dfrac{x - 2}{3} = \dfrac{4}{x - 1}$

67. $\dfrac{b + 2}{b + 3} + 1 = \dfrac{-7}{b - 5}$

68. $\dfrac{x - 4}{x - 3} + \dfrac{x - 2}{x - 3} = x - 3$

69. $\dfrac{u}{u - 1} + \dfrac{1}{u} = \dfrac{u^2 + 1}{u^2 - u}$

70. $\dfrac{3}{x - 2} + \dfrac{1}{x} = \dfrac{2(3x + 2)}{x^2 - 2x}$

71. $\dfrac{n}{n^2 - 9} + \dfrac{n + 8}{n + 3} = \dfrac{n - 8}{n - 3}$

72. $\dfrac{7}{x - 5} - \dfrac{3}{x + 5} = \dfrac{40}{x^2 - 25}$

73. $\dfrac{x}{x - 1} - \dfrac{12}{x^2 - x} = \dfrac{-1}{x - 1}$

74. $y + \dfrac{2}{3} = \dfrac{2y - 12}{3y - 9}$

75. $1 - \dfrac{3}{b} = \dfrac{-8b}{b^2 + 3b}$

76. $\dfrac{7}{q^2 - q - 2} + \dfrac{1}{q + 1} = \dfrac{3}{q - 2}$

77. $\dfrac{3}{x - 1} - \dfrac{1}{x + 9} = \dfrac{18}{x^2 + 8x - 9}$

78. $\dfrac{5}{4y + 12} - \dfrac{3}{4} = \dfrac{5}{4y + 12} - \dfrac{y}{4}$

Solve each formula for the indicated variable.

79. $\dfrac{P}{n} = rt$ for P

80. $\dfrac{F}{m} = a$ for F

81. $\dfrac{a}{b} = \dfrac{c}{d}$ for d

82. $\dfrac{pc}{s} = \dfrac{t}{r}$ for c

83. $h = \dfrac{2A}{b + d}$ for A

84. $T = \dfrac{3R}{M - n}$ for R

85. $\dfrac{1}{a} + \dfrac{1}{b} = 1$ for a

86. $\dfrac{1}{a} - \dfrac{1}{b} = 1$ for b

87. $I = \dfrac{E}{R + r}$ for r

88. $\dfrac{S}{k + h} = E$ for k

89. $\dfrac{1}{r} + \dfrac{1}{s} = \dfrac{1}{t}$ for r

90. $\dfrac{5}{x} - \dfrac{4}{y} = \dfrac{5}{z}$ for x

91. $F = \dfrac{L^2}{6d} + \dfrac{d}{2}$ for L^2

92. $H = \dfrac{J^3}{cd} - \dfrac{K^3}{d}$ for J^3

APPLICATIONS

93. MEDICINE Radioactive tracers are used for diagnostic work in nuclear medicine. The **effective half-life** H of a radioactive material in an organism is given by the formula

$$H = \dfrac{RB}{R + B}$$

where R is the radioactive half-life and B is the biological half-life of the tracer. Solve the formula for R.

94. CHEMISTRY Charles's law describes the relationship between the volume and temperature of a gas that is kept at a constant pressure. It can be expressed as

$$\frac{V_1}{V_2} = \frac{T_1}{T_2}$$

where V_1 and V_2 are variables representing two different volumes, and T_1 and T_2 are variables representing two different temperatures. (Recall that the notation V_1 is read as *V sub one*.) Solve for V_2.

95. ELECTRONICS Most electronic circuits require resistors to make them work properly. Resistors are components that limit current. An important formula about resistors in a circuit is

$$\frac{1}{r} = \frac{1}{r_1} + \frac{1}{r_2}$$

Solve for r.

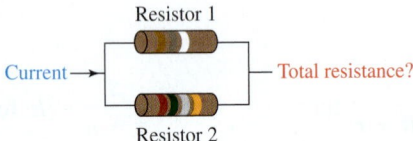

96. MATHEMATICAL FORMULAS To quickly find the sum of a list of fractions, such as

$$\frac{1}{2} + \frac{1}{4} + \frac{1}{8} + \frac{1}{16} + \frac{1}{32} + \frac{1}{64} + \frac{1}{128}$$

mathematicians use the formula

$$S = \frac{a(1 - r^n)}{1 - r}$$

Solve the formula for a.

WRITING

97. Explain how the multiplication property of equality is used to solve rational equations. Give an example.

98. When solving rational equations, how do you know whether a solution is extraneous?

99. What is meant by clearing a rational equation of fractions? Give an example.

100. Explain the difference between the procedure used to simplify

$$\frac{1}{x} + \frac{1}{3}$$

and the procedure used to solve

$$\frac{1}{x} + \frac{1}{3} = \frac{1}{2}$$

REVIEW   Factor completely.

101. $x^2 + 4x$
102. $x^2 - 16y^2$
103. $2x^2 + x - 3$
104. $6a^2 - 5a - 6$
105. $x^4 - 81$
106. $4x^2 + 10x - 6$

CHALLENGE PROBLEMS

107. Solve: $x^{-2} + 2x^{-1} + 1 = 0$.

108. ENGINES A formula that is used in the design and testing of diesel engines is

$$E = 1 - \frac{T_4 - T_1}{a(T_3 - T_2)}$$

Solve the formula for T_1.

6.7   Problem Solving Using Rational Equations

• Number Problems   • Uniform Motion Problems   • Shared-Work Problems
• Investment Problems

We will now use the five-step problem-solving strategy to solve application problems from a variety of areas, including banking, petroleum engineering, sports, and travel. In each case, we will use a rational equation to model the situation. Then we will apply the skills we have learned to solve the equation. We begin with an example in which we find an unknown number.

■ NUMBER PROBLEMS

EXAMPLE 1

ELEMENTARY &
INTERMEDIATE
Algebra $f(x)$ **Now**™

Number problem.   If the same number is added to both the numerator and the denominator of the fraction $\frac{3}{5}$, the result is $\frac{4}{5}$. Find the number.

Analyze the Problem
- Begin with the fraction $\frac{3}{5}$.
- Add the same number to the numerator and to the denominator.
- The result is $\frac{4}{5}$.
- Find the number.

Form an Equation   Let $n =$ the unknown number. To form an equation, add the unknown number to the numerator and to the denominator of $\frac{3}{5}$. Then set the result equal to $\frac{4}{5}$.

$$\frac{3 + n}{5 + n} = \frac{4}{5}$$

Solve the Equation   To solve this equation, we begin by clearing it of fractions.

$$\frac{3 + n}{5 + n} = \frac{4}{5}$$

$$\mathbf{5(5 + n)}\left(\frac{3 + n}{5 + n}\right) = \mathbf{5(5 + n)}\left(\frac{4}{5}\right) \qquad \text{Multiply both sides by } 5(5 + n), \text{ which is the LCD}$$
of the rational expressions appearing in the equation.

$$5(3 + n) = (5 + n)4 \qquad \text{Simplify: } \tfrac{5 + n}{5 + n} = 1 \text{ and } \tfrac{5}{5} = 1.$$

$$15 + 5n = 20 + 4n \qquad \text{Distribute the multiplication by 5 and by 4.}$$

$$15 + n = 20 \qquad \text{Subtract } 4n \text{ from both sides.}$$

$$n = 5 \qquad \text{Subtract 15 from both sides.}$$

State the Conclusion   The number is 5.

Check the Result   When we add 5 to both the numerator and denominator of $\frac{3}{5}$, we get

$$\frac{3 + 5}{5 + 5} = \frac{8}{10} = \frac{4}{5}$$

The result checks.

■ UNIFORM MOTION PROBLEMS

Recall that we use the distance formula $d = rt$ to solve motion problems. The relationship between distance, rate, and time can be expressed in another way, by solving for t.

$$d = rt \qquad \text{Distance} = \text{rate} \cdot \text{time.}$$

$$\frac{d}{r} = \frac{rt}{r} \qquad \text{Divide both sides by } r.$$

$$\frac{d}{r} = t \qquad \text{Simplify: } \frac{r}{r} = 1.$$

$$t = \frac{d}{r}$$

The Language of Algebra

In uniform motion problems, the word *speed* is often used in place of the word *rate*. For example, we can say a car travels at a *rate* of 50 mph or its *speed* is 50 mph.

This alternate form of the distance formula is used to solve the next example.

EXAMPLE 2

ELEMENTARY & INTERMEDIATE

Distance runners. A coach can run 10 miles in the same amount of time as his best student-athlete can run 12 miles. If the student runs 1 mile per hour faster than the coach, find the running speeds of the coach and the student.

Analyze the Problem
- The coach runs 10 miles in the same time that the student runs 12 miles.
- The student runs 1 mph faster than the coach.
- Find the speed that each runs.

Form an Equation
Since the student's speed is 1 mph faster than the coach's, let $r =$ the speed that the coach runs. Then, $r + 1 =$ the speed that the student runs. The expressions for the rates are entered in the Rate column of the table. The distances run by the coach and by the student are entered in the Distance column of the table.

Using $t = \frac{d}{r}$, we find that the time it takes the coach to run 10 miles, at a rate of r mph, is $\frac{10}{r}$ hours. Similarly, we find that the time it takes the student to run 12 miles, at a rate of $(r + 1)$ mph, is $\frac{12}{r + 1}$ hours. These expressions are entered in the Time column of the table.

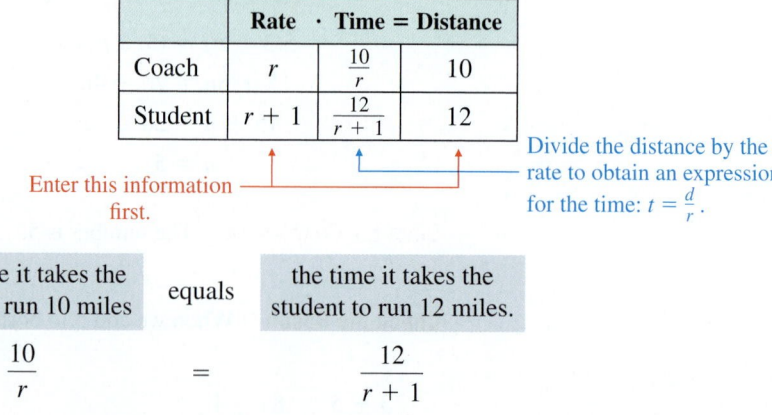

	Rate	·	Time	=	Distance
Coach	r		$\frac{10}{r}$		10
Student	$r + 1$		$\frac{12}{r + 1}$		12

Enter this information first.

Divide the distance by the rate to obtain an expression for the time: $t = \frac{d}{r}$.

The time it takes the coach to run 10 miles	equals	the time it takes the student to run 12 miles.
$\dfrac{10}{r}$	$=$	$\dfrac{12}{r + 1}$

Solve the Equation
To solve this rational equation, we begin by clearing it of fractions.

$$\frac{10}{r} = \frac{12}{r + 1}$$

$$r(r + 1)\left(\frac{10}{r}\right) = r(r + 1)\left(\frac{12}{r + 1}\right) \qquad \text{Multiply both sides by the LCD, } r(r + 1).$$

$$(r + 1)10 = 12r \qquad \text{Simplify.}$$

$$10r + 10 = 12r \qquad \text{Distribute the multiplication by 10.}$$
$$10 = 2r \qquad \text{Subtract } 10r \text{ from both sides.}$$
$$5 = r \qquad \text{Divide both sides by 2.}$$

If $r = 5$, then $r + 1 = 6$.

State the Conclusion The coach's running speed is 5 mph and the student's running speed is 6 mph.

Check the Result The coach will run 10 miles in $\frac{10 \text{ miles}}{5 \text{ mph}} = 2$ hours. The student will run 12 miles in $\frac{12 \text{ miles}}{6 \text{ mph}} = 2$ hours. The times are the same; the results check.

■ SHARED-WORK PROBLEMS

Problems in which two or more people (or machines) work together to complete a job are called *shared-work problems*. To solve such problems, we must determine the **rate of work** for each person or machine involved. For example, suppose it takes you 4 hours to clean your house. Your rate of work can be expressed as $\frac{1}{4}$ of the job is completed per hour. If someone else takes 5 hours to clean the same house, they complete $\frac{1}{5}$ of the job per hour. In general, a rate of work can be determined in the following way.

Rate of Work	If a job can be completed in x hours, the rate of work can be expressed as: $$\frac{1}{x} \text{ of the job is completed per hour.}$$ If a job is completed in some other unit of time, such as x minutes or x days, then the rate of work is expressed in that unit.

To solve shared-work problems, we must also determine the *amount of work* completed. To do this, we use a formula similar to the distance formula $d = rt$ used for motion problems.

$$\text{Work completed} = \text{rate of work} \cdot \text{time worked} \qquad \text{or} \qquad W = rt$$

EXAMPLE 3

ELEMENTARY &
INTERMEDIATE
Algebra $f(x)$ **Now**™

Payroll checks. At the end of a pay period, it takes the president of a company 15 minutes to sign all of her employees' payroll checks. What fractional part of the job is accomplished if the president signs checks for 10 minutes?

Solution If all of the checks can be signed in 15 minutes, the president's work rate can be expressed as $\frac{1}{15}$ of the job completed per minute. To find the work that has been completed after 10 minutes, use the work formula.

$$W = rt$$
$$= \frac{1}{15} \cdot 10 \qquad \text{Substitute } \frac{1}{15} \text{ for } r, \text{ the work rate, and 10 for } t, \text{ the time worked.}$$

$$= \frac{10}{15} \qquad \text{Multiply.}$$

$$= \frac{2}{3} \qquad \text{Simplify the fraction.}$$

In 10 minutes, the president will complete $\frac{2}{3}$ of the job of signing the payroll checks.

Self Check 3   It takes a farmer 8 days to harvest a wheat crop. What part of the job is completed in 6 days?

EXAMPLE 4

ELEMENTARY & INTERMEDIATE

Algebra *f(x)* **Now**™

Filling a tank.   An inlet pipe can fill an oil storage tank in 7 days, and a second inlet pipe can fill the same tank in 9 days. If both pipes are used, how long will it take to fill the tank?

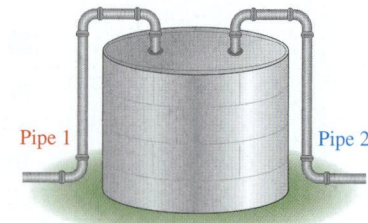

Pipe 1   Pipe 2

Analyze the Problem
• The first pipe can fill the tank in 7 days.
• The second pipe can fill the tank in 9 days.
• How long will it take the two pipes, working together, to fill the tank?

Form an Equation   Let $x =$ the number of days it will take to fill the tank if both pipes are used. It is helpful to organize the facts of the problem in a table. Since the pipes will be open for the same amount of time as they fill the tank, enter x as the time worked for each pipe.

The first pipe can fill the tank in 7 days; its rate working alone is $\frac{1}{7}$ of the job per day. The second pipe can fill the tank in 9 days; its rate working alone is $\frac{1}{9}$ of the job per day. To determine the work completed by each pipe, multiply the rate by the time.

	Rate	· Time	= Work completed
1st pipe	$\frac{1}{7}$	x	$\frac{x}{7}$
2nd pipe	$\frac{1}{9}$	x	$\frac{x}{9}$

Enter this information first.   Multiply to get each of these entries: $W = rt$.

In shared-work problems, the number 1 represents one whole job completed. So we have

The part of job done by 1st pipe	plus	part of job done by 2nd pipe	equals	1 job completed.
$\frac{x}{7}$	$+$	$\frac{x}{9}$	$=$	1

Solve the Equation

$$\frac{x}{7} + \frac{x}{9} = 1$$

$$63\left(\frac{x}{7} + \frac{x}{9}\right) = 63(1) \qquad \text{Clear the equation of fractions by multiplying both sides by the LCD, 63.}$$

$$63\left(\frac{x}{7}\right) + 63\left(\frac{x}{9}\right) = 63 \qquad \text{Distribute the multiplication by 63.}$$

$$9x + 7x = 63 \qquad \text{On the left-hand side, do the multiplications.}$$

$$16x = 63 \qquad \text{Combine like terms.}$$

$$x = \frac{63}{16} \qquad \text{Divide both sides by 16.}$$

State the Conclusion   If both pipes are used, it will take $\frac{63}{16}$ or $3\frac{15}{16}$ days to fill the tank.

Check the Result   In $\frac{63}{16}$ days, the first pipe fills $\frac{1}{7} \cdot \frac{63}{16} = \frac{9}{16}$ of the tank and the second pipe fills $\frac{1}{9} \cdot \frac{63}{16} = \frac{7}{16}$ of the tank. The sum of these efforts, $\frac{9}{16} + \frac{7}{16}$, is $\frac{16}{16}$ or 1 full tank. The result checks.

Example 4 can be solved in a different way by considering *the amount of work done by each pipe in 1 day.* As before, if we let $x =$ the number of days it will take to fill the tank if both inlet pipes are used, then together, in 1 day, they will complete $\frac{1}{x}$ of the job. If we add what the first pipe can do in 1 day to what the second pipe can do in 1 day, the sum is what they can do together in 1 day.

What the first inlet pipe can do in 1 day	plus	what the second inlet pipe can do in 1 day	equals	what they can do together in 1 day.
$\frac{1}{7}$	$+$	$\frac{1}{9}$	$=$	$\frac{1}{x}$

To solve the equation, begin by clearing it of fractions.

$$\frac{1}{7} + \frac{1}{9} = \frac{1}{x}$$

$$63x\left(\frac{1}{7} + \frac{1}{9}\right) = 63x\left(\frac{1}{x}\right) \qquad \text{Multiply both sides by the LCD, } 63x.$$

$$9x + 7x = 63 \qquad \text{Distribute the multiplication by } 63x \text{ and simplify.}$$

$$16x = 63 \qquad \text{Combine like terms.}$$

$$x = \frac{63}{16} \qquad \text{Divide both sides by 16.}$$

This is the same as the solution obtained in Example 4.

■ INVESTMENT PROBLEMS

We have used the interest formula $I = Prt$ to solve investment problems. The relationship between interest, principal, rate, and time can be expressed in another way, by solving for P.

$$I = Prt \qquad \text{Interest} = \text{principal} \cdot \text{rate} \cdot \text{time.}$$

$$\frac{I}{rt} = \frac{Prt}{rt} \qquad \text{Divide both sides by } rt.$$

$$P = \frac{I}{rt} \qquad \text{Simplify: } \frac{rt}{rt} = 1.$$

This alternate form of the interest formula is used to solve the next example.

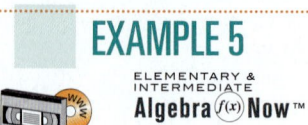

EXAMPLE 5

ELEMENTARY &
INTERMEDIATE
Algebra *(x)* **Now**™

Comparing investments.   An amount of money invested for one year in bonds will earn $120. At a bank, that same amount of money will only earn $75 interest, because the interest paid by the bank is 3% less than that paid by the bonds. Find the rate of interest paid by each investment.

Analyze the Problem

- The investment in bonds earns $120 in one year.
- The same amount of money, invested in a bank, earns $75 in one year.
- The interest rate paid by the bank is 3% less than that paid by the bonds.
- Find the bond's rate of interest and the bank's rate of interest.

Form an Equation

Since the interest rate paid by the bank is 3% less than that paid by the bonds, let r = the bond's rate of interest, and $r - 0.03$ = the bank's interest rate. (Recall that 3% = 0.03.)

If an investment earns $120 interest in 1 year at some rate r, we can use $P = \frac{I}{rt}$ to find that the principal invested was $\frac{120}{r}$ dollars. Similarly, if another investment earns $75 interest in 1 year at some rate $r - 0.03$, the principal invested was $\frac{75}{r - 0.03}$ dollars. We can organize the facts of the problem in a table.

	Principal ·	Rate	· Time =	Interest
Bonds	$\frac{120}{r}$	r	1	120
Bank	$\frac{75}{r - 0.03}$	$r - 0.03$	1	75

Divide to get each of these entries: $P = \frac{I}{rt}$.   Enter this information first.

The amount invested in the bonds	equals	the amount invested in the bank.
$\dfrac{120}{r}$	$=$	$\dfrac{75}{r - 0.03}$

Solve the Equation

$$\frac{120}{r} = \frac{75}{r - 0.03}$$

$$r(r - 0.03)\left(\frac{120}{r}\right) = \left(\frac{75}{r - 0.03}\right)r(r - 0.03) \qquad \text{Multiply both sides by the LCD, } r(r - 0.03).$$

$$(r - 0.03)120 = 75r \qquad \text{Simplify.}$$

$$120r - 3.6 = 75r \qquad \text{Distribute the multiplication by 120.}$$

$$45r - 3.6 = 0 \qquad \text{Subtract } 75r \text{ from both sides.}$$

$$45r = 3.6 \qquad \text{Add 3.6 to both sides.}$$

$$r = 0.08 \qquad \text{Divide both sides by 45.}$$

If $r = 0.08$, then $r - 0.03 = 0.05$.

State the Conclusion

The bonds pay 0.08, or 8%, interest. The bank's interest rate is 5%.

Check the Result   The amount invested at 8% that will earn $120 interest in 1 year is $\frac{120}{(0.08)1} = \$1,500$. The amount invested at 5% that will earn $75 interest in 1 year is $\frac{75}{(0.05)1} = \$1,500$. The amounts invested in the bonds and the bank are the same. The results check.

Answer to Self Check   **3.** $\frac{3}{4}$

6.7 STUDY SET

 ELEMENTARY & INTERMEDIATE Algebra *f(x)* Now™

VOCABULARY   Fill in the blanks.

1. In the formula $d = rt$, the variable d stands for the _____ traveled, r is the _____, and t is the _____.

2. In the formula $W = rt$, the variable W stands for the _____ completed, r is the _____, and t is the _____.

3. In the formula $I = Prt$, the variable I stands for the amount of _____ earned, P is the _____, r stands for the annual interest _____, and t is the _____.

4. If a job can be completed in x hours, then the rate of work can be expressed as $\frac{1}{x}$ of the job is completed _____ hour.

CONCEPTS

5. Choose the equation that can be used to solve the following: *If the same number is added to the numerator and the denominator of the fraction $\frac{5}{8}$, the result is $\frac{2}{3}$. Find the number.*

(i) $\frac{5}{8} + x = \frac{2}{3}$ (ii) $\frac{5 + x}{8} = \frac{2}{3}$

(iii) $\frac{5 + x}{8 + x} = \frac{2}{3}$ (iv) $\frac{5}{8} = \frac{2 + x}{3 + x}$

6. Choose the equation that can be used to solve the following: *The sum of a number and its reciprocal is $\frac{7}{2}$. Find the number.*

(i) $x + (-x) = \frac{7}{2}$ (ii) $x + \frac{1}{x} = \frac{7}{2}$

(iii) $x + \frac{1}{x} = x + \frac{7}{2}$ (iv) $x + \frac{1}{x} = \frac{2}{7}$

7. Solve $d = rt$ for t.

8. Solve $I = Prt$ for P.

9. a. Write 9% as a decimal.

   b. Write 0.035 as a percent.

10. It takes a night security officer 45 minutes to check each of the doors in an office building to make sure they are locked. What is the officer's rate of work?

11. If a custodian can sweep a cafeteria floor in 12 minutes, what fractional part of the job does he accomplish in 4 minutes?

12. If the exits at the front of a theater are opened, a full theater can be emptied of all occupants in 6 minutes. How much of the theater is emptied in 1 minute?

13. It takes an elementary school teacher 4 hours to make out the semester report cards. What part of the job does she complete in x hours?

14. Complete the table.

	r	$\cdot$	t	$= d$
Snowmobile	r			4
4 × 4 truck	$r - 5$			3

15. Complete the table.

	P	$\cdot$	r	$\cdot t = I$	
City savings bank			r	1	50
Credit union			$r - 0.02$	1	75

16. Complete the table.

	Rate	$\cdot$ Time	= Work completed
1st printer	$\frac{1}{15}$	x	
2nd printer	$\frac{1}{8}$	x	

NOTATION

17. Write $\frac{55}{9}$ days using a mixed number.

18. How much greater is the rate paid by the mutual funds compared to the rate paid by the certificate of deposit?

	P	$\cdot$	r	$\cdot t =$	I
Certificate of deposit	$\frac{600}{r}$		r	1	600
Mutual funds	$\frac{1,400}{r + 0.04}$		$r + 0.04$	1	1,400

PRACTICE

19. NUMBER PROBLEM   If the same number is added to both the numerator and the denominator of $\frac{2}{5}$, the result is $\frac{2}{3}$. Find the number.

20. NUMBER PROBLEM   If the same number is subtracted from both the numerator and the denominator of $\frac{11}{13}$, the result is $\frac{3}{4}$. Find the number.

21. NUMBER PROBLEM   If the denominator of $\frac{3}{4}$ is increased by a number and the numerator is doubled, the result is 1. Find the number.

22. NUMBER PROBLEM   If a number is added to the numerator of $\frac{7}{8}$ and the same number is subtracted from the denominator, the result is 2. Find the number.

23. NUMBER PROBLEM   If a number is added to the numerator of $\frac{3}{4}$ and twice as much is added to the denominator, the result is $\frac{4}{7}$. Find the number.

24. NUMBER PROBLEM   If a number is added to the numerator of $\frac{5}{7}$ and twice as much is subtracted from the denominator, the result is 8. Find the number.

25. NUMBER PROBLEM   The sum of a number and its reciprocal is $\frac{13}{6}$. Find the numbers.

26. NUMBER PROBLEM   The sum of the reciprocals of two consecutive even integers is $\frac{7}{24}$. Find the integers.

APPLICATIONS

27. COOKING   If the same number is added to both the numerator and the denominator of the amount of butter used in the following recipe for toffee, the result is the amount of brown sugar to be used. Find the number.

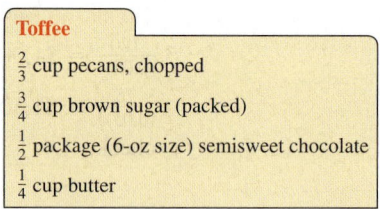

Toffee

$\frac{2}{3}$ cup pecans, chopped

$\frac{3}{4}$ cup brown sugar (packed)

$\frac{1}{2}$ package (6-oz size) semisweet chocolate

$\frac{1}{4}$ cup butter

28. TAPE MEASURES   If the same number is added to both the numerator and the denominator of the first measurement, the result is the second measurement. Find the number.

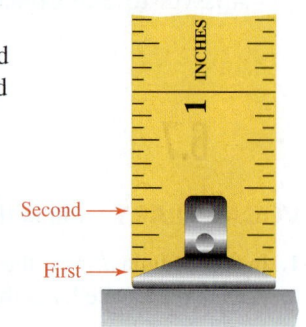

29. FILLING A POOL   An inlet pipe can fill an empty swimming pool in 5 hours, and another inlet pipe can fill the pool in 4 hours. How long will it take both pipes to fill the pool?

30. ROOFING HOUSES   A homeowner estimates that it will take her 7 days to roof her house. A professional roofer estimates that he could roof the house in 4 days. How long will it take if the homeowner helps the roofer?

31. HOLIDAY DECORATING   One crew can put up holiday decorations in the mall in 8 hours. A second crew can put up the decorations in 10 hours. How long will it take if both crews work together to decorate the mall?

32. GROUNDS KEEPING   It takes a grounds keeper 45 minutes to prepare a softball field for a game. It takes his assistant 55 minutes to prepare the same field. How long will it take if they work together to prepare the field?

33. FILLING A POOL   One inlet pipe can fill an empty pool in 4 hours, and a drain can empty the pool in 8 hours. How long will it take the pipe to fill the pool if the drain is left open?

34. SEWAGE TREATMENT   A sludge pool is filled by two inlet pipes. One pipe can fill the pool in 15 days, and the other can fill it in 21 days. However, if no sewage is added, continuous waste removal will empty the pool in 36 days. How long will it take the two inlet pipes to fill an empty sludge pool?

35. GRADING PAPERS   On average, it takes a teacher 30 minutes to grade a set of quizzes. It takes her teacher's aide twice as long to do the same grading. How long will it take if they work together to grade a set of quizzes?

36. DOG KENNELS   It takes the owner/operator of a dog kennel 6 hours to clean all of the cages. It takes his assistant 2 hours more than that to clean the same cages. How long will it take if they work together?

37. PRINTERS   It takes a printer 6 hours to print the class schedules for all of the students enrolled in a community college. A faster printer can print the schedules in 4 hours. How long will it take the two printers working together to print $\frac{3}{4}$ of the class schedules?

38. OFFICE WORK   In 5 hours, a secretary can address 100 envelopes. Another secretary can address 100 envelopes in 6 hours. How long would it take the secretaries, working together, to address 300 envelopes.   (*Hint:* Think of addressing 300 envelopes as three 100-envelope jobs.)

39. PHYSICAL FITNESS   A woman can bicycle 28 miles in the same time as it takes her to walk 8 miles. She can ride 10 mph faster than she can walk. How fast can she walk?

40. COMPARING TRAVEL   A plane can fly 300 miles in the same time as it takes a car to go 120 miles. If the car travels 90 mph slower than the plane, find the speed of the plane.

41. PACKAGING FRUIT   The diagram shows how apples are processed for market. Although the second conveyor belt is shorter, an apple spends the same amount of time on each belt because the second conveyor moves 1 foot per second slower than the first. Determine the speed of each conveyor belt.

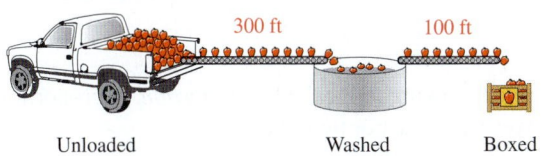

| 300 ft | 100 ft |
| Unloaded | Washed | Boxed |

42. BIRDS IN FLIGHT   Although flight speed is dependent upon the weather and the wind, in general, a Canada goose can fly about 10 mph faster than a Great Blue heron. In the same time that a Canada goose travels 120 miles, a Great Blue heron travels 80 miles. Find their flying speeds.

43. WIND SPEED   When a plane flies downwind, the wind pushes the plane so that its speed is the *sum* of the speed of the plane in still air and the speed of the wind. Traveling upwind, the wind pushes against the plane so that its speed is the *difference* of the speed of the plane in still air and the speed of the wind. Suppose a plane that travels 255 mph in still air can travel 300 miles downwind in the same time as it takes to travel 210 miles upwind. Complete the table and find the speed of the wind.

	Rate	· Time	= Distance
Downwind	$255 + x$		300
Upwind	$255 - x$		210

44. BOATING   A boat that travels 18 mph in still water can travel 22 miles downstream in the same time as it takes to travel 14 miles upstream. Find the speed of the current in the river.

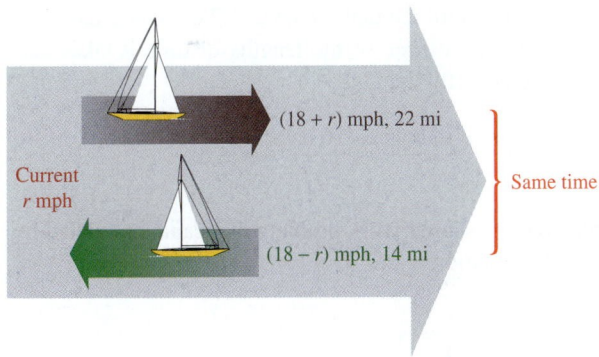

45. COMPARING INVESTMENTS   An amount of money invested for 1 year in tax-free bonds will earn $300. In a certain credit union account, that same amount of money will only earn $200 interest in a year, because the interest paid is 2% less than that paid by the bonds. Find the rate of interest paid by each investment.

46. COMPARING INVESTMENTS   An amount of money invested for 1 year in a savings account will earn $1,500. That same amount of money, invested in a mini-mall development will earn $6,500 interest in a year, because the interest paid is 10% more than that paid by the savings account. Find the rate of interest paid by each investment.

47. COMPARING INVESTMENTS Two certificates of deposit (CDs) pay interest at rates that differ by 1%. Money invested for 1 year in the first CD earns $175 interest. The same principal invested in the second CD earns $200. Find the two rates of interest.

48. COMPARING INTEREST RATES Two bond funds pay interest at rates that differ by 2%. Money invested for 1 year in the first fund earns $315 interest. The same amount invested in the second fund earns $385. Find the lower rate of interest.

WRITING

49. In Example 4, one inlet pipe could fill an oil tank in 7 days, and another could fill the same tank in 9 days. We were asked to find how long it would take if both pipes were used. Explain why each of the following approaches is incorrect.

The time it would take to fill the tank

  • is the *sum* of the lengths of time it takes each pipe to fill the tank: 7 days + 9 days = 16 days.
  • is the *difference* in the lengths of time it takes each pipe to fill the tank: 9 days − 7 days = 2 days.
  • is the *average* of the lengths of time it takes each pipe to fill the tank:

$$\frac{7 \text{ days} + 9 \text{ days}}{2} = \frac{16 \text{ days}}{2} = 8 \text{ days}$$

50. Write a shared-work problem that can be modeled by the equation

$$\frac{x}{3} + \frac{x}{4} = 1$$

REVIEW

51. When expressed as a decimal, is $\frac{7}{9}$ a terminating or a repeating decimal?

52. Solve: $x + 20 = 4x - 1 + 2x$.

53. List the set of integers.

54. Solve: $4x^2 + 8x = 0$.

55. Evaluate $2x^2 + 5x - 3$ for $x = -3$.

56. Solve $T - R = ma$ for R.

CHALLENGE PROBLEMS

57. RIVER TOURS A river boat tour begins by going 60 miles upstream against a 5-mph current. There, the boat turns around and returns with the current. What still-water speed should the captain use to complete the tour in 5 hours?

58. TRAVEL TIME A company president flew 680 miles one way in the corporate jet but returned in a smaller plane that could fly only half as fast. If the total travel time was 6 hours, find the speeds of the planes.

59. SALES A dealer bought some radios for a total of $1,200. She gave away 6 radios as gifts, sold the rest for $10 more than she paid for each radio, and broke even. How many radios did she buy?

60. FURNACE REPAIRS A repairman purchased several furnace-blower motors for a total cost of $210. If his cost per motor had been $5 less, he could have purchased one additional motor. How many motors did he buy at the regular rate?

6.8 Proportions and Similar Triangles

  • Ratios, Rates, and Proportions   • Solving Proportions   • Problem Solving
  • Similar Triangles

In this section, we will discuss a problem-solving tool called a *proportion*. A proportion is a type of rational equation that involves two *ratios* or two *rates*.

■ RATIOS, RATES, AND PROPORTIONS

Ratios enable us to compare numerical quantities. Here are some examples.

  • To prepare fuel for a lawnmower, gasoline is mixed with oil in a 50 to 1 ratio.
  • In the stock market, winning stocks might outnumber losers by a ratio of 7 to 4.
  • Gold is combined with other metals in the ratio of 14 to 10 to make 14-karat jewelry.

Ratios	A **ratio** is the quotient of two numbers or the quotient of two quantities that have the same units.

There are three common ways to write a ratio: as a fraction, using the word *to,* or with a colon. For example, the comparison of the number of winning stocks to the number of losing stocks mentioned earlier can be written as

$$\frac{7}{4}, \qquad 7 \text{ to } 4, \qquad \text{or} \qquad 7\!:\!4$$

Each of these forms can be read as "the ratio of 7 to 4."

EXAMPLE 1

Translate each phrase into a ratio written in fractional form: **a.** The ratio of 5 to 9   and **b.** 12 ounces to 2 pounds.

Solution   **a.** The ratio of 5 to 9 is written $\dfrac{5}{9}$.

Notation

A ratio that is the quotient of two quantities having the same units should be simplified so that no units appear in the final answer.

b. To write a ratio of two quantities with the same units, we must express 2 pounds in terms of ounces. Since 1 pound = 16 ounces, 2 pounds = 32 ounces. The ratio of 12 ounces to 32 ounces can be simplified so that no units appear in the final form.

$$\frac{12 \text{ ounces}}{32 \text{ ounces}} = \frac{3 \cdot \overset{1}{\cancel{4}} \ \overset{1}{\cancel{\text{ounces}}}}{\underset{1}{\cancel{4}} \cdot 8 \ \underset{1}{\cancel{\text{ounces}}}} = \frac{3}{8}$$

Self Check 1   Translate each phrase into a ratio written in fractional form: **a.** The ratio of 15 to 2 and   **b.** 12 hours to 2 days.

A quotient that compares quantities with different units is called a **rate.** For example, if the 495-mile drive from New Orleans to Dallas takes 9 hours, the average rate of speed is the quotient of the miles driven and the length of time the trip takes.

$$\text{Average rate of speed} = \frac{495 \text{ miles}}{9 \text{ hours}} = \frac{\overset{1}{\cancel{9}} \cdot 55 \text{ miles}}{\underset{1}{\cancel{9}} \cdot 1 \text{ hours}} = \frac{55 \text{ miles}}{1 \text{ hour}}$$

Rates	A **rate** is a quotient of two quantities that have different units.

If two ratios or two rates are equal, we say that they are *in proportion.*

Proportion	A **proportion** is a mathematical statement that two ratios or two rates are equal.

Some examples of proportions are

$$\frac{1}{2} = \frac{3}{6}, \qquad \frac{3 \text{ waiters}}{7 \text{ tables}} = \frac{9 \text{ waiters}}{21 \text{ tables}}, \qquad \text{and} \qquad \frac{a}{b} = \frac{c}{d}$$

- The proportion $\frac{1}{2} = \frac{3}{6}$ can be read as "1 is to 2 as 3 is to 6."

- The proportion $\frac{3 \text{ waiters}}{7 \text{ tables}} = \frac{9 \text{ waiters}}{21 \text{ tables}}$ can be read as "3 waiters is to 7 tables as 9 waiters is to 21 tables."

- The proportion $\frac{a}{b} = \frac{c}{d}$ can be read as "a is to b as c is to d."

Each of the four numbers in a proportion is called a **term.** The first and fourth terms are called the **extremes,** and the second and third terms are called the **means.**

$$\begin{array}{c} \text{First term} \longrightarrow a \\ \text{Second term} \longrightarrow b \end{array} = \begin{array}{c} c \longleftarrow \text{Third term} \\ d \longleftarrow \text{Fourth term} \end{array} \qquad a \text{ and } d \text{ are the extremes. } b \text{ and } c \text{ are the means.}$$

For the proportion $\frac{a}{b} = \frac{c}{d}$, we can show that the product of the extremes, ad, is equal to the product of the means, bc, by multiplying both sides of the proportion by bd, and observing that $ad = bc$.

$$\frac{a}{b} = \frac{c}{d}$$

$$bd \cdot \frac{a}{b} = bd \cdot \frac{c}{d} \qquad \text{To clear the fractions, multiply both sides by the LCD, } bd.$$

$$ad = bc \qquad \text{Simplify: } \frac{b}{b} = 1 \text{ and } \frac{d}{d} = 1.$$

Since $ad = bc$, the product of the extremes equals the product of the means.

The same products ad and bc can be found by multiplying diagonally in the proportion $\frac{a}{b} = \frac{c}{d}$. We call ad and bc **cross products.**

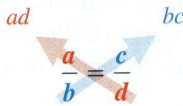

$$\begin{array}{cc} ad & bc \\ \frac{a}{b} & \frac{c}{d} \end{array}$$

The Fundamental Property of Proportions

In a proportion, the product of the extremes is equal to the product of the means.

$$\text{If } \frac{a}{b} = \frac{c}{d}, \text{ then } ad = bc \qquad \text{and} \qquad \text{if } ad = bc, \text{ then } \frac{a}{b} = \frac{c}{d}.$$

EXAMPLE 2

ELEMENTARY & INTERMEDIATE

Algebra $f(x)$ **Now**™

Solution

Determine whether each equation is a proportion: **a.** $\frac{3}{7} = \frac{9}{21}$ and **b.** $\frac{8}{3} = \frac{13}{5}$.

In each case, we check to see whether the product of the extremes is equal to the product of the means.

a. The product of the extremes is $3 \cdot 21 = 63$. The product of the means is $7 \cdot 9 = 63$. Since the cross products are equal, $\frac{3}{7} = \frac{9}{21}$ is a proportion.

$$3 \cdot 21 = 63 \qquad 7 \cdot 9 = 63$$

$$\frac{3}{7} = \frac{9}{21}$$

b. The product of the extremes is $8 \cdot 5 = 40$. The product of the means is $3 \cdot 13 = 39$. Since the cross products are not equal, the equation is not a proportion: $\frac{8}{3} \neq \frac{13}{5}$.

$$8 \cdot 5 = 40 \qquad 3 \cdot 13 = 39$$

$$\frac{8}{3} = \frac{13}{5}$$

Self Check 2   Determine whether the equation is a proportion: $\dfrac{6}{13} = \dfrac{24}{53}$.

■ SOLVING PROPORTIONS

The fundamental property of proportions provides us with a way to solve proportions.

EXAMPLE 3   Solve: $\dfrac{12}{18} = \dfrac{4}{x}$.

Solution   To solve for x, we set the cross products equal.

$$\frac{12}{18} = \frac{4}{x}$$

Caution

Remember that a cross product is the product of the means or the extremes of a *proportion*. For example, it would be incorrect to try to compute cross products to solve $\frac{12}{18} = \frac{4}{x} + \frac{1}{2}$. It is not a proportion. The right-hand side is not a ratio.

$12 \cdot x = 18 \cdot 4$ In a proportion, the product of the extremes equals the product of the means.

$12x = 72$ Do the multiplications.

$\dfrac{12x}{12} = \dfrac{72}{12}$ To isolate x, divide both sides by 12.

$x = 6$

Check: To check the result, we substitute 6 for x in $\dfrac{12}{18} = \dfrac{4}{x}$ and find the cross products.

$$12 \cdot 6 = 72 \qquad 18 \cdot 4 = 72$$

$$\frac{12}{18} \overset{?}{=} \frac{4}{6}$$

Since the cross products are equal, the solution of $\dfrac{12}{18} = \dfrac{4}{x}$ is 6.

Self Check 3   Solve: $\dfrac{15}{x} = \dfrac{25}{40}$.

EXAMPLE 4

ELEMENTARY &
INTERMEDIATE
Algebra $f(x)$ **Now**™

Solution

Solve: $\dfrac{2a + 1}{4} = \dfrac{10}{8}$.

$$\frac{2a + 1}{4} = \frac{10}{8}$$

$8(2a + 1) = 40$ In a proportion, the product of the extremes equals the product of the means.

$16a + 8 = 40$ Distribute the multiplication by 8.

$16a + 8 - 8 = 40 - 8$ Subtract 8 from both sides.

$16a = 32$ Combine like terms.

$\dfrac{16a}{16} = \dfrac{32}{16}$ Divide both sides by 16.

$a = 2$

The solution is 2.

Success Tip

Since proportions are rational equations, they can also be solved by multiplying both sides by the LCD. For Example 4, an alternate approach is to multiply both sides by 8:

$$8\left(\frac{2a + 1}{4}\right) = 8\left(\frac{10}{8}\right)$$

Self Check 4   Solve: $\dfrac{3x - 1}{2} = \dfrac{12.5}{5}$.

■ PROBLEM SOLVING

We can use proportions to solve many problems. If we are given a ratio (or rate) comparing two quantities, the words of the problem can be translated to a proportion, and we can solve it to find the unknown.

EXAMPLE 5

ELEMENTARY &
INTERMEDIATE
Algebra $f(x)$ **Now**™

Solution

Grocery shopping.   If 6 apples cost $1.38, how much will 16 apples cost?

Analyze the Problem   We know the cost of 6 apples; we are to find the cost of 16 apples.

Form a Proportion   Let c = the cost of 16 apples. If we compare the number of apples to their cost, we know that the two rates are equal.

6 apples is to $1.38 as 16 apples is to $c.

6 apples ⟶ $\dfrac{6}{1.38} = \dfrac{16}{c}$ ⟵ 16 apples
Cost of 6 apples ⟶   ⟵ Cost of 16 apples

Solve the Proportion

$6 \cdot c = 1.38(16)$ In a proportion, the product of the extremes equals the product of the means.

$6c = 22.08$ Multiply: $1.38(16) = 22.08$.

$\dfrac{6c}{6} = \dfrac{22.08}{6}$ Divide both sides by 6.

$c = 3.68$

State the Conclusion   Sixteen apples will cost $3.68.

Check the Result   16 apples are about 3 times as many as 6 apples, which cost $1.38. If we multiply $1.38 by 3, we get an estimate of the cost of 16 apples: $1.38 \cdot 3 = \$4.14$. The result, $3.68, seems reasonable.

Self Check 5 If 9 tickets to a concert cost $112.50, how much will 15 tickets cost?

Caution When solving problems using proportions, we must make sure that the units of both numerators are the same and the units of both denominators are the same. In Example 5, it would be incorrect to write

Cost of 6 apples ⟶ $\dfrac{1.38}{6} = \dfrac{16}{c}$ ⟵ 16 apples

6 apples ⟶ $\dfrac{}{6}$ = $\dfrac{16}{c}$ ⟵ Cost of 16 apples

---

EXAMPLE 6

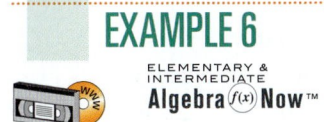

ELEMENTARY & INTERMEDIATE
Algebra $f(x)$ **Now**™

The Language of Algebra

Architects, interior decorators, landscapers, and automotive engineers are a few of the professionals who construct *scale* drawings or *scale* models of the projects they are designing.

Miniatures. A **scale** is a ratio (or rate) that compares the size of a model, drawing, or map to the size of an actual object. The scale indicates that 1 inch on the model carousel is equivalent to 160 inches on the actual carousel. How wide should the model be if the actual carousel is 35 feet wide?

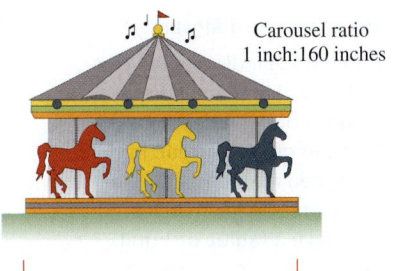

Carousel ratio
1 inch:160 inches

Analyze the Problem We are asked to determine the width of the miniature carousel, if a ratio of 1 inch to 160 inches is used. We would like the width of the model to be given in inches, not feet, so we will express the 35-foot width of the actual carousel as $35 \cdot 12 = 420$ inches.

Form a Proportion Let w = the width of the model. The ratios of the dimensions of the model to the corresponding dimensions of the actual carousel are equal.

1 inch is to 160 inches as w inches is to 420 inches.

model ⟶ $\dfrac{1}{160} = \dfrac{w}{420}$ ⟵ model
actual ⟶ $\dfrac{1}{160}$ = $\dfrac{w}{420}$ ⟵ actual

Solve the Proportion

$420 = 160w$ In a proportion, the product of the extremes is equal to the product of the means.

$\dfrac{420}{\mathbf{160}} = \dfrac{160w}{\mathbf{160}}$ Divide both sides by 160.

$2.625 = w$ Simplify.

State the Conclusion The width of the miniature carousel should be 2.625 in., or $2\dfrac{5}{8}$ in.

Check the Result A width of $2\dfrac{5}{8}$ in. is approximately 3 in. When we write the ratio of the model's approximate width to the width of the actual carousel, we get $\dfrac{3}{420} = \dfrac{1}{140}$, which is about $\dfrac{1}{160}$. The answer seems reasonable.

When shopping, *unit prices* can be used to compare costs of different sizes of the same brand to determine the best buy. The **unit price** gives the cost per unit, such as cost per ounce, cost per pound, or cost per sheet. We can find the unit price of an item using a proportion.

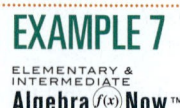

EXAMPLE 7

ELEMENTARY &
INTERMEDIATE
Algebra *f(x)* Now™

Comparison shopping.   Which size of toothpaste is the better buy?

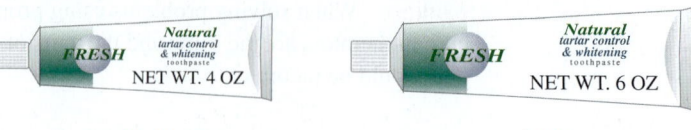

$2.19 $2.79

Solution   To find the unit price for each tube, we let x = the price of 1 ounce of toothpaste. Then we set up and solve the following proportions.

The Language of Algebra

A unit price indicates the cost of 1 unit of an item, such as 1 ounce of bottled water or 1 pound of hamburger. In advanced mathematics, we study unit circles—circles that have a radius of 1 unit.

For the 4-ounce tube:

Ounces ⟶ $\dfrac{4}{2.19} = \dfrac{1}{x}$ ⟵ Ounce
Price ⟶ $\phantom{\dfrac{4}{2.19}}$ ⟵ Price

$$4x = 2.19$$

$$x = \frac{2.19}{4}$$

$$x \approx 0.55$$ The unit price is approximately $0.55.

For the 6-ounce tube:

Ounces ⟶ $\dfrac{6}{2.79} = \dfrac{1}{x}$ ⟵ Ounce
Price ⟶ $\phantom{\dfrac{6}{2.79}}$ ⟵ Price

$$6x = 2.79$$

$$x = \frac{2.79}{6}$$

$$x \approx 0.47$$ The unit price is approximately $0.47.

The price of 1 ounce of toothpaste from the 4-ounce tube is about 55¢. The price for 1 ounce of toothpaste from the 6-ounce tube is about 47¢. Since the 6-ounce tube has the lower unit price, it is the better buy.

Self Check 7   Which is the better buy: 3 pounds of hamburger for $6.89 or 5 pounds for $12.49?

■ SIMILAR TRIANGLES

If two angles of one triangle have the same measures as two angles of a second triangle, the triangles have the same shape. Triangles with the same shape, but not necessarily the same size, are called **similar triangles.** In the following figure, $\triangle ABC \sim \triangle DEF$. (Read the symbol $\sim$ as "is similar to.")

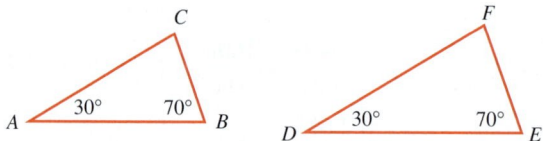

Property of Similar Triangles   If two triangles are **similar,** all pairs of corresponding sides are in proportion.

For the similar triangles shown above, the following proportions are true.

$$\frac{AB}{DE} = \frac{BC}{EF}, \qquad \frac{BC}{EF} = \frac{CA}{FD}, \qquad \text{and} \qquad \frac{CA}{FD} = \frac{AB}{DE}$$ Read *AB* as "the length of segment *AB*."

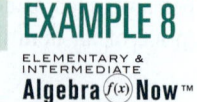

EXAMPLE 8

ELEMENTARY &
INTERMEDIATE
Algebra *f(x)* Now™

Finding the height of a tree.   A tree casts a shadow 18 feet long at the same time as a woman 5 feet tall casts a shadow 1.5 feet long. Find the height of the tree.

Solution   **Analyze the Problem**   The figure shows the similar triangles determined by the tree and its shadow and the woman and her shadow. Since the triangles are similar, the lengths of their corresponding sides are in proportion. We can use this fact to find the height of the tree.

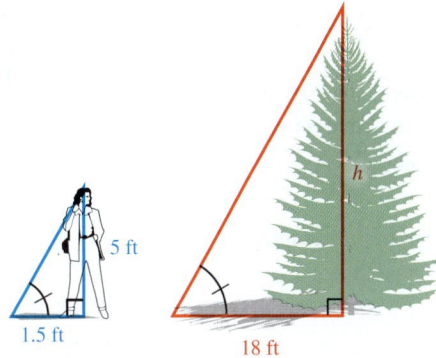

Each triangle has a right angle. Since the sun's rays strike the ground at the same angle, the angles highlighted with a tick mark have the same measure. Therefore, two angles of the smaller triangle have the same measures as two angles of the larger triangle; the triangles are similar.

Form a Proportion   If we let h = the height of the tree, we can find h by solving the following proportion.

$$\frac{h}{5} = \frac{18}{1.5}$$ $\dfrac{\text{Height of the tree}}{\text{Height of the woman}} = \dfrac{\text{Length of shadow of the tree}}{\text{Length of shadow of the woman}}$

Solve the Proportion

$1.5h = 5(18)$ In a proportion, the product of the extremes equals the product of the means.

$1.5h = 90$ Multiply.

$\dfrac{1.5h}{\mathbf{1.5}} = \dfrac{90}{\mathbf{1.5}}$ Divide both sides by 1.5.

$h = 60$ $\frac{90}{1.5} = 60$.

State the Conclusion   The tree is 60 feet tall.

Check the Result $\dfrac{18}{1.5} = 12$ and $\dfrac{60}{5} = 12$. The ratios are the same. The result checks.

Self Check 8   Find the height of the tree in Example 8 if the woman is 5 feet 6 inches tall.

Answers to Self Checks   **1. a.** $\frac{15}{2}$,   **b.** $\frac{1}{4}$ **2.** no   **3.** 24   **4.** 2   **5.** $187.50   **7.** 3 pounds for $6.89

8. 66 ft

6.8   **STUDY SET**   ELEMENTARY & INTERMEDIATE Algebra $f(x)$ Now™

VOCABULARY   **Fill in the blanks.**

1. A _____ is the quotient of two numbers or the quotient of two quantities with the same units. A _____ is a quotient of two quantities that have different units.

2. A _____ is a mathematical statement that two ratios or two rates are equal.

3. The _____ of the proportion $\frac{2}{x} = \frac{16}{40}$ are 2, x, 16, and 40.

4. In $\frac{50}{3} = \frac{x}{9}$, the terms 50 and 9 are called the

_____ and the terms 3 and x are called the

_____ of the proportion.

5. The product of the extremes and the product of the means of a proportion are also known as _____ products.

6. A _____ is a ratio (or rate) that compares the size of a model, drawing, or map to the size of an actual object.

7. Examples of _____ prices are $1.65 per gallon, 17¢ per day, and $50 per foot.

8. Two triangles with the same shape, but not necessarily the same size, are called _____ triangles.

CONCEPTS

9. WEST AFRICA   Write the ratio (in fractional form) of the number of red stripes to the number of white stripes on the flag of Liberia.

10. Fill in the blanks: In a proportion, the product of the extremes is _____ to the product of the means.

If $\frac{a}{b} = \frac{c}{d}$, then [ ] = [ ].

11. What are the cross products for each proportion?

  a. $\frac{6}{5} = \frac{12}{10}$ **b.** $\frac{15}{2} = \frac{45}{x}$

12. a. Is 45 a solution of $\frac{5}{3} = \frac{75}{x}$?

  b. Is -2 a solution of $\frac{a+4}{8} = \frac{3}{16}$?

13. SNACK FOODS   In a sample of 25 bags of potato chips, 2 were found to be underweight. Complete the following proportion that could be used to find the number of underweight bags that would be expected in a shipment of 1,000 bags of potato chips.

number of bags → [ ]   [ ] ← number of bags
_____ = _____
number underweight → [ ]   [ ] ← number underweight

14. MINIATURES   A model of a "high wheeler" bicycle is to be made using a scale of 2 inches to 15 inches. The following proportion was set up to determine the height of the front wheel of the model. Explain the error.

$$\frac{2}{15} = \frac{48}{h}$$

48 in.

Actual size

15. GROCERY SHOPPING   Examine the following pricing stickers. Which item is the better buy?

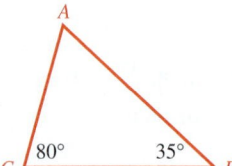   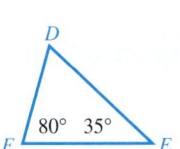

AQUACLEAR WATER   1.79
12- 8 OZ BOTTLES

AQUACLEAR WATER   4.49
24- 12 OZ BOTTLES

16. Complete the following proportion that can be used to find the unit price of facial tissue if a box of 85 tissues sells for $1.19.

price → $\dfrac{1.19}{85}$ = $\dfrac{x}{[\ ]}$ ← price
number of sheets →   ← number of sheets

17. Two similar triangles are shown. Fill in the blanks to make the proportions true.

$\dfrac{AB}{DE} = \dfrac{[\ ]}{EF}$ $\dfrac{BC}{[\ ]} = \dfrac{CA}{FD}$ $\dfrac{CA}{FD} = \dfrac{AB}{[\ ]}$

18. The two triangles shown in the following illustration are similar. Complete the proportion.

$\dfrac{x}{[\ ]} = \dfrac{[\ ]}{[\ ]}$

Solution   **Analyze the Problem**   The figure shows the similar triangles determined by the tree and its shadow and the woman and her shadow. Since the triangles are similar, the lengths of their corresponding sides are in proportion. We can use this fact to find the height of the tree.

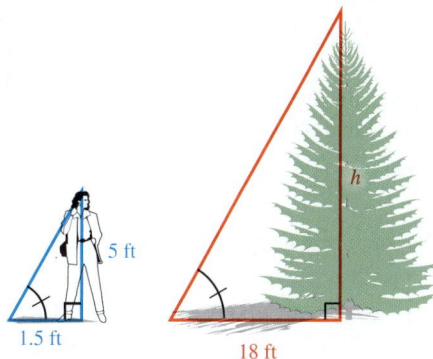

Each triangle has a right angle. Since the sun's rays strike the ground at the same angle, the angles highlighted with a tick mark have the same measure. Therefore, two angles of the smaller triangle have the same measures as two angles of the larger triangle; the triangles are similar.

Form a Proportion   If we let h = the height of the tree, we can find h by solving the following proportion.

$$\frac{h}{5} = \frac{18}{1.5} \qquad \frac{\text{Height of the tree}}{\text{Height of the woman}} = \frac{\text{Length of shadow of the tree}}{\text{Length of shadow of the woman}}$$

Solve the Proportion

$1.5h = 5(18)$ In a proportion, the product of the extremes equals the product of the means.

$1.5h = 90$ Multiply.

$\dfrac{1.5h}{\mathbf{1.5}} = \dfrac{90}{\mathbf{1.5}}$ Divide both sides by 1.5.

$h = 60$ $\frac{90}{1.5} = 60.$

State the Conclusion   The tree is 60 feet tall.

Check the Result $\dfrac{18}{1.5} = 12$ and $\dfrac{60}{5} = 12$. The ratios are the same. The result checks.

Self Check 8   Find the height of the tree in Example 8 if the woman is 5 feet 6 inches tall.

Answers to Self Checks   **1. a.** $\frac{15}{2}$,   **b.** $\frac{1}{4}$ **2.** no   **3.** 24   **4.** 2   **5.** $187.50   **7.** 3 pounds for $6.89   **8.** 66 ft

---

6.8   **STUDY SET**   ELEMENTARY & INTERMEDIATE   **Algebra** $f(x)$ **Now**™

VOCABULARY   **Fill in the blanks.**

1. A _____ is the quotient of two numbers or the quotient of two quantities with the same units. A _____ is a quotient of two quantities that have different units.

2. A _____ is a mathematical statement that two ratios or two rates are equal.

3. The _____ of the proportion $\frac{2}{x} = \frac{16}{40}$ are 2, x, 16, and 40.

4. In $\frac{50}{3} = \frac{x}{9}$, the terms 50 and 9 are called the _____ and the terms 3 and x are called the _____ of the proportion.

5. The product of the extremes and the product of the means of a proportion are also known as _____ products.

6. A _____ is a ratio (or rate) that compares the size of a model, drawing, or map to the size of an actual object.

7. Examples of _____ prices are $1.65 per gallon, 17¢ per day, and $50 per foot.

8. Two triangles with the same shape, but not necessarily the same size, are called _____ triangles.

CONCEPTS

9. **WEST AFRICA** Write the ratio (in fractional form) of the number of red stripes to the number of white stripes on the flag of Liberia.

10. Fill in the blanks: In a proportion, the product of the extremes is _____ to the product of the means.

$$\text{If } \frac{a}{b} = \frac{c}{d}, \text{ then } \boxed{} = \boxed{}.$$

11. What are the cross products for each proportion?

   a. $\dfrac{6}{5} = \dfrac{12}{10}$ **b.** $\dfrac{15}{2} = \dfrac{45}{x}$

12. **a.** Is 45 a solution of $\dfrac{5}{3} = \dfrac{75}{x}$?

   b. Is -2 a solution of $\dfrac{a+4}{8} = \dfrac{3}{16}$?

13. **SNACK FOODS** In a sample of 25 bags of potato chips, 2 were found to be underweight. Complete the following proportion that could be used to find the number of underweight bags that would be expected in a shipment of 1,000 bags of potato chips.

number of bags → $\dfrac{\boxed{}}{\boxed{}} = \dfrac{\boxed{}}{\boxed{}}$ ← number of bags

number underweight →   ← number underweight

14. **MINIATURES** A model of a "high wheeler" bicycle is to be made using a scale of 2 inches to 15 inches. The following proportion was set up to determine the height of the front wheel of the model. Explain the error.

$$\frac{2}{15} = \frac{48}{h}$$

48 in.

Actual size

15. **GROCERY SHOPPING** Examine the following pricing stickers. Which item is the better buy?

AQUACLEAR WATER
1.79
12- 8 OZ BOTTLES

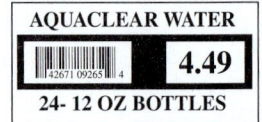

AQUACLEAR WATER
4.49
24- 12 OZ BOTTLES

16. Complete the following proportion that can be used to find the unit price of facial tissue if a box of 85 tissues sells for $1.19.

price → $\dfrac{1.19}{85} = \dfrac{x}{\boxed{}}$ ← price
number of sheets →   ← number of sheets

17. Two similar triangles are shown. Fill in the blanks to make the proportions true.

$$\frac{AB}{DE} = \frac{\boxed{}}{EF} \qquad \frac{BC}{\boxed{}} = \frac{CA}{FD} \qquad \frac{CA}{FD} = \frac{AB}{\boxed{}}$$

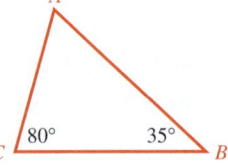
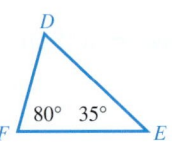

18. The two triangles shown in the following illustration are similar. Complete the proportion.

$$\frac{x}{\boxed{}} = \frac{\boxed{}}{\boxed{}}$$

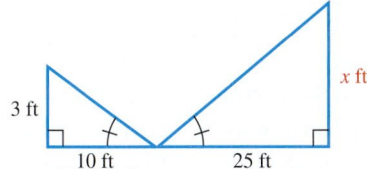

NOTATION   Complete the solution.

19. Solve for x: $\dfrac{12}{18} = \dfrac{x}{24}$.

$$12 \cdot 24 = 18 \cdot \square$$
$$\square = 18x$$
$$\dfrac{288}{\square} = \dfrac{18x}{\square}$$
$$\square = x$$

20. Write the ratio of 25 to 4 in two forms.

21. Write each ratio in simplified form.

　a. $\dfrac{12}{15}$　　　　　　**b.** $\dfrac{9 \text{ crates}}{7 \text{ crates}}$

22. Fill in the blanks: The proportion $\dfrac{20}{1.6} = \dfrac{100}{8}$ can be read: 20 is to 1.6 ____ 100 is ____ 8.

23. Fill in the blank: We read $\triangle XYZ \sim \triangle MNO$ as: triangle XYZ is _____ to triangle MNO.

24. Write the statement *x is approximately 0.75* using symbols.

PRACTICE   Write each ratio as a fraction in simplest form.

25. 4 boxes to 15 boxes   **26.** 2 miles to 9 miles

27. 30 days to 24 days   **28.** 45 people to 30 people

29. 90 minutes to 3 hours   **30.** 20 inches to 2 feet

31. 13 quarts to 2 gallons   **32.** 11 dimes to 1 dollar

Tell whether each statement is a proportion.

33. $\dfrac{9}{7} = \dfrac{81}{70}$ **34.** $\dfrac{5}{2} = \dfrac{20}{8}$

35. $\dfrac{7}{3} = \dfrac{14}{6}$ **36.** $\dfrac{13}{19} = \dfrac{65}{95}$

37. $\dfrac{9}{19} = \dfrac{38}{80}$ **38.** $\dfrac{40}{29} = \dfrac{29}{22}$

Solve each proportion.

39. $\dfrac{2}{3} = \dfrac{x}{6}$ **40.** $\dfrac{3}{6} = \dfrac{x}{8}$

41. $\dfrac{5}{10} = \dfrac{3}{c}$ **42.** $\dfrac{7}{14} = \dfrac{2}{x}$

43. $\dfrac{6}{x} = \dfrac{8}{4}$ **44.** $\dfrac{4}{x} = \dfrac{2}{8}$

45. $\dfrac{x+1}{5} = \dfrac{3}{15}$ **46.** $\dfrac{x-1}{7} = \dfrac{2}{21}$

47. $\dfrac{x+7}{-4} = \dfrac{1}{4}$ **48.** $\dfrac{x+3}{12} = \dfrac{-7}{6}$

49. $\dfrac{5-x}{17} = \dfrac{13}{34}$ **50.** $\dfrac{4-x}{13} = \dfrac{11}{26}$

51. $\dfrac{2x-1}{18} = \dfrac{9}{54}$ **52.** $\dfrac{2x+1}{18} = \dfrac{14}{3}$

53. $\dfrac{x-1}{9} = \dfrac{2x}{3}$ **54.** $\dfrac{x+1}{4} = \dfrac{3x}{8}$

55. $\dfrac{8x}{3} = \dfrac{11x+9}{4}$ **56.** $\dfrac{3x}{16} = \dfrac{x+2}{5}$

57. $\dfrac{2}{3x} = \dfrac{x}{6}$ **58.** $\dfrac{y}{4} = \dfrac{4}{y}$

59. $\dfrac{b-5}{3} = \dfrac{2}{b}$ **60.** $\dfrac{2}{c} = \dfrac{c-3}{2}$

61. $\dfrac{x-1}{x+1} = \dfrac{2}{3x}$ **62.** $\dfrac{2}{x+6} = \dfrac{-2x}{5}$

Each pair of triangles is similar. Find the missing side length.

63.   **64.**

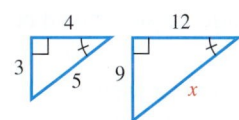   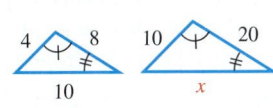

65.   **66.**

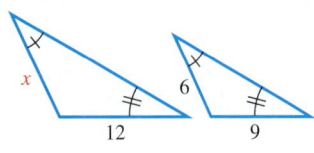   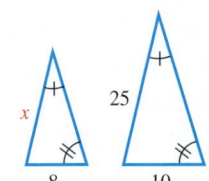

APPLICATIONS

67. GEAR RATIOS   Write each ratio in two ways: as a fraction in simplest form and using a colon.
    a. The number of teeth of the larger gear to the number of teeth of the smaller gear
    b. The number of teeth of the smaller gear to the number of teeth of the larger gear

68. FACULTY–STUDENT RATIOS   At a college, there are 300 faculty members and 2,850 students. Find the rate of faculty to students. (This is often referred to as the faculty to student ratio, even though the units are different.)

69. SHOPPING FOR CLOTHES   If shirts are on sale at two for $25, how much do five shirts cost?

70. COMPUTING A PAYCHECK   Billie earns $412 for a 40-hour week. If she missed 10 hours of work last week, how much did she get paid?

71. COOKING   A recipe for spaghetti sauce requires four 16-ounce bottles of ketchup to make 2 gallons of sauce. How many bottles of ketchup are needed to make 10 gallons of sauce?

72. MIXING PERFUME   A perfume is to be mixed in the ratio of 3 drops of pure essence to 7 drops of alcohol. How many drops of pure essence should be mixed with 56 drops of alcohol?

73. CPR   A first aid handbook states that when performing cardiopulmonary resuscitation on an adult, the ratio of chest compressions to breaths should be $5 : 2$. If 210 compressions were administered to an adult patient, how many breaths should have been given?

74. COOKING   A recipe for wild rice soup follows. Find the amounts of chicken broth, rice, and flour needed to make 15 servings.

> **Wild Rice Soup**
>
> *A sumptuous side dish with a nutty flavor*
>
> | | |
> |---|---|
> | 3 cups chicken broth | 1 cup light cream |
> | $\frac{2}{3}$ cup uncooked rice | 2 tablespoons flour |
> | $\frac{1}{4}$ cup sliced onions | $\frac{1}{8}$ teaspoon pepper |
> | $\frac{1}{2}$ cup shredded carrots | |
> | | Serves: 6 |

75. NUTRITION   The table shows the nutritional facts about a 10-oz chocolate milkshake sold by a fast-food restaurant. Use the information to complete the table for the 16-oz shake. Round to the nearest unit when an answer is not exact.

	Calories	Fat (gm)	Protein (gm)
10-oz chocolate milkshake	355	8	9
16-oz chocolate milkshake			

76. STRUCTURAL ENGINEERING   A portion of a bridge is shown. Use the fact that $\frac{AB}{BC}$ is in proportion to $\frac{FE}{ED}$ to find FE.

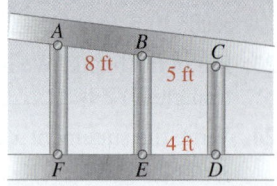

77. QUALITY CONTROL   Out of a sample of 500 men's shirts, 17 were rejected because of crooked collars. How many crooked collars would you expect to find in a run of 15,000 shirts?

78. PHOTO ENLARGEMENTS   The 3-by-5 photo is to be blown up to the larger size. Find x.

79. MIXING FUEL   The instructions on a can of oil intended to be added to lawnmower gasoline are shown on the next page. Are these instructions correct?   (*Hint:* There are 128 ounces in 1 gallon.)

Recommended	Gasoline	Oil
50 to 1	6 gal	16 oz

80. DRIVER'S LICENSES Of the 50 states, Alabama has one of the highest ratios of licensed drivers to residents. If the ratio is 800 : 1,000 and the population of Alabama is 4,500,000, how many residents of that state have a driver's license?

81. CROP DAMAGE To estimate the ground squirrel population on his acreage, a farmer trapped, tagged, and then released a dozen squirrels. Two weeks later, the farmer trapped 35 squirrels and noted that 3 were tagged. Use this information to estimate the number of ground squirrels on his acreage.

82. CONCRETE A 2 : 3 concrete mix means that for every two parts of sand, three parts of gravel are used. How much sand should be used in a mix composed of 25 cubic feet of gravel?

83. MODEL RAILROADS A model railroad engine is 9 inches long. If the scale is 87 feet to 1 foot, how long is a real engine?

84. MODEL RAILROADS A model railroad caboose is 3.5 inches long. If the scale is 169 feet to 1 foot, how long is a real caboose?

85. BLUEPRINTS The scale for the drawing shown means that a $\frac{1}{4}$-inch length $\left(\frac{1}{4}''\right)$ on the drawing corresponds to an actual size of 1 foot ($1'0''$). Suppose the length of the kitchen is $2\frac{1}{2}$ inches on the drawing. How long is the actual kitchen?

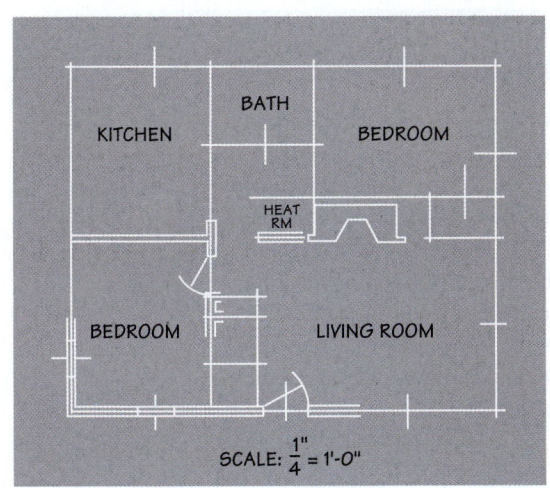

SCALE: $\frac{1''}{4}$ = 1'-0"

86. THE *TITANIC* A 1 : 144 scale model of the *Titanic* is to be built. If the ship was 882 feet long, find the length of the model?

For each of the following purchases, determine the better buy.

87. Trumpet lessons: 45 minutes for $25 or 60 minutes for $35

88. Memory for a computer: 128 megabytes for $26 or 512 megabytes for $110

89. Business cards: 100 for $9.99 or 150 for $12.99

90. Dog food: 20 pounds for $7.49 or 44 pounds for $14.99

91. Soft drinks: 6-pack for $1.50 or a case (24 cans) for $6.25

92. Donuts: A dozen for $6.24 or a baker's dozen (13) for $6.65

93. HEIGHT OF A TREE A tree casts a shadow of 26 feet at the same time as a 6-foot man casts a shadow of 4 feet. Find the height of the tree.

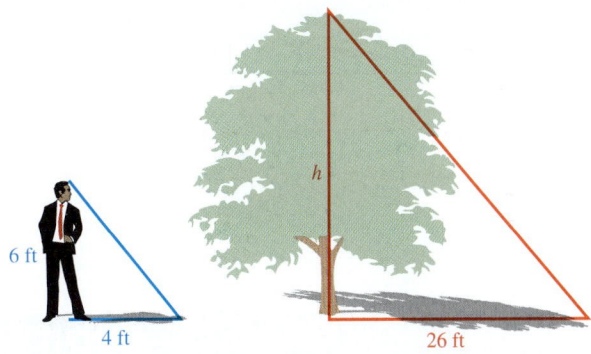

6 ft

4 ft

h

26 ft

94. HEIGHT OF A BUILDING A man places a mirror on the ground and sees the reflection of the top of a building, as shown. The two triangles in the illustration are similar. Find the height, h, of the building.

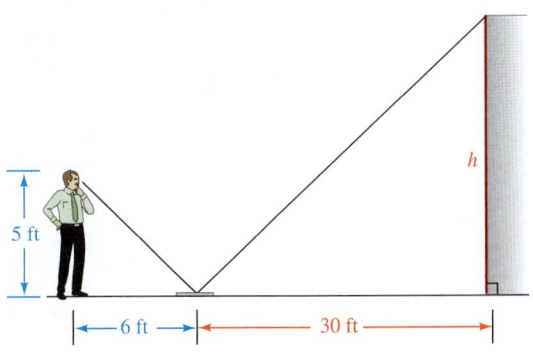

5 ft

h

6 ft

30 ft

95. SURVEYING   To determine the width of a river, a surveyor laid out the similar triangles as shown. Find w.

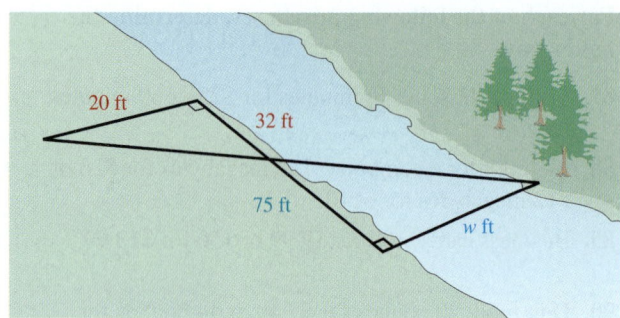

96. FLIGHT PATHS   An airplane ascends 100 feet as it flies a horizontal distance of 1,000 feet. How much altitude will it gain as it flies a horizontal distance of 1 mile?   (*Hint:* 5,280 feet = 1 mile.)

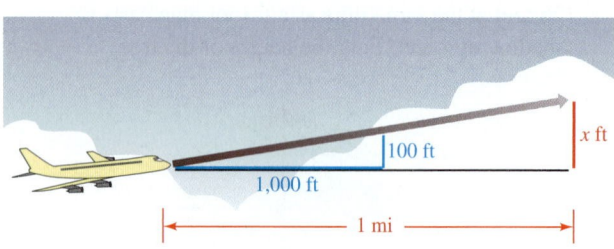

WRITING

97. Explain the difference between a ratio and a proportion.

98. Explain why the concept of cross products cannot be used to solve the equation

$$\frac{x}{3} - \frac{3x}{4} = \frac{1}{12}$$

99. What are similar triangles?

100. What is a unit price? Give an example.

REVIEW

101. Change $\frac{9}{10}$ to a percent.

102. Change $33\frac{1}{3}\%$ to a fraction.

103. Find 30% of 1,600.

104. SHOPPING   Maria bought a dress for 25% off the original price of $98. How much did the dress cost?

CHALLENGE PROBLEMS

105. Suppose $\frac{a}{b} = \frac{c}{d}$. Write three other proportions using $a, b, c,$ and d.

106. Verify that $\frac{3}{5} = \frac{12}{20} = \frac{3 + 12}{5 + 20}$. Is the following rule always true? Explain.

$$\frac{a}{b} = \frac{c}{d} = \frac{a + c}{b + d}$$

ACCENT ON TEAMWORK

WHAT IS π?

Overview: In this activity, you will discover an important fact about the ratio of the circumference to the diameter of a circle.

Instructions: Form groups of 2 or 3 students. With a piece of string or a cloth tape measure, find the circumference and the diameter of objects that are circular in shape. You can measure anything that is round: for example, a coin, the top of a can, a tire, or a waste paper basket. Enter your results in a table, as shown below. Convert each measurement to a decimal, and then use a calculator to determine a decimal approximation of the ratio of the circumference C to diameter d.

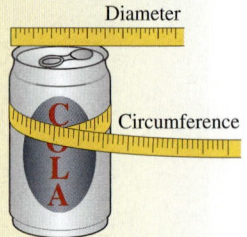

Object	Circumference C	Diameter d	$\frac{C}{d}$ (approx.)
A quarter	$2\frac{15}{16}$ in. = 2.9375 in.	$\frac{15}{16}$ in. = 0.9375 in.	3.13333

Since early history, mathematicians have known that the ratio of the circumference to the diameter of a circle is the same for any size circle, approximately 3. Today, following centuries of study, we know that this ratio is exactly 3.141592653589 . . . .

$$\frac{C}{d} = 3.141592653589 \ldots$$

The Greek letter π (pi) is used to represent the ratio of circumference to diameter:

$$\pi = \frac{C}{d}, \qquad \text{where } \pi = 3.141592653589 \ldots$$

Are the ratios in your table numerically close to π? Give some reasons why they aren't exactly 3.141592653589 . . . in each case.

USING PROPORTIONS WHEN COOKING

Overview: In this activity, you will use proportions to adjust the ingredients in your favorite recipe so that it would serve everyone in your class.

Instructions: Each student should bring his/her favorite recipe to class. Working in pairs, begin with the first recipe and determine the amount of each ingredient that is needed to make enough of it to serve the exact number of people in the class. For example, if the recipe serves 8 and there are 25 students and an instructor in the class, write and solve proportions to determine the amount of flour, sugar, milk, and so on, to make enough of the recipe to serve 26 people. Then do the same for the second recipe.

When finished, share with the class how you made the calculations, as well as any difficulties that you encountered adjusting the ingredients of the recipes.

KEY CONCEPT: EXPRESSIONS AND EQUATIONS

In this chapter, we have discussed procedures for simplifying rational expressions and procedures for solving rational equations.

SIMPLIFYING RATIONAL EXPRESSIONS

To simplify a rational expression:

1. Factor the numerator and denominator completely to determine all the factors common to both.

2. Remove factors equal to 1 by replacing each pair of factors common to the numerator and denominator with the equivalent fraction $\frac{1}{1}$.

This procedure is also used when multiplying and dividing rational expressions.

1. a. Simplify: $\dfrac{2x^2 - 8x}{x^2 - 6x + 8}$.

b. What common factor was removed?

2. a. Multiply: $\dfrac{x^2 + 2x + 1}{x} \cdot \dfrac{x^2 - x}{x^2 - 1}$.

b. What common factors were removed?

BUILDING RATIONAL EXPRESSIONS

To build a rational expression, we multiply it by a form of 1. We use this concept to add and subtract rational expressions having unlike denominators and when simplifying complex fractions.

3. a. Add: $\dfrac{x}{x+1} + \dfrac{x-1}{x}$.

b. By what did you multiply the first fraction to rewrite it in terms of the LCD? The second fraction?

4. a. Simplify: $\dfrac{n - 1 - \dfrac{2}{n}}{\dfrac{n}{3}}$.

b. By what did you multiply the numerator and denominator to simplify the complex fraction?

SOLVING RATIONAL EQUATIONS

The multiplication property of equality states that *multiplying both sides of an equation by the same nonzero number does not change its solution.* We use this property when solving rational equations. If we multiply both sides of the equation by the LCD of the rational expressions in the equation, we can clear it of fractions.

5. a. Solve: $\dfrac{11}{b} + \dfrac{13}{b} = 12$.

b. By what did you multiply both sides to clear the equation of fractions?

6. a. Solve: $\dfrac{3}{s+2} - \dfrac{5}{s^2+s-2} = \dfrac{1}{s-1}$.

b. By what did you multiply both sides to clear the equation of fractions?

7. a. Solve: $y + \dfrac{3}{4} = \dfrac{3y-50}{4y-24}$.

b. By what did you multiply both sides to clear the equation of fractions?

8. a. Solve: $\dfrac{1}{a} - \dfrac{1}{b} = 1$ for b.

b. By what did you multiply both sides to clear the equation of fractions?

CHAPTER REVIEW

SECTION 6.1	**Simplifying Rational Expressions**

CONCEPTS

Since division by 0 is undefined, we must make sure that the denominator of a rational expression is not 0.

REVIEW EXERCISES

1. Find the values of x for which the rational expression $\dfrac{x-1}{x^2-16}$ is undefined.

2. Evaluate $\dfrac{x^2-1}{x-5}$ for $x = -2$.

Simplify each rational expression, if possible. Assume that no denominators are zero.

3. $\dfrac{3x^2}{6x^3}$

4. $\dfrac{5xy^2}{2x^2y^2}$

5. $\dfrac{x^2}{x^2+x}$

6. $\dfrac{a^2-4}{a+2}$

To *simplify* a rational expression:

1. Factor the numerator and denominator completely.

2. Remove factors equal to 1 by replacing each pair of factors common to the numerator and denominator with the equivalent fraction $\frac{1}{1}$.

The quotient of any nonzero expression and its opposite is -1.

7. $\dfrac{3p - 2}{2 - 3p}$

8. $\dfrac{8 - x}{x^2 - 5x - 24}$

9. $\dfrac{2x^2 - 16x}{2x^2 - 18x + 16}$

10. $\dfrac{x^2 + x - 2}{x^2 - x - 2}$

11. $\dfrac{4(t + 3) + 8}{3(t + 3) + 6}$

12. Explain the error in the following work: $\dfrac{x + 1}{x} = \dfrac{\overset{1}{\cancel{x}} + 1}{\underset{1}{\cancel{x}}} = \dfrac{2}{1} = 2.$

13. DOSAGES   Cowling's rule is a formula that can be used to determine the dosage of a prescription medication for children. If C is the proper child's dosage, D is an adult dosage, and A is the child's age in years, then

$$C = \frac{D(A + 1)}{24}$$

Find the daily dosage of an antibiotic for an 11-year-old child if the adult daily dosage is 300 milligrams.

SECTION 6.2

Multiplying and Dividing Rational Expressions

To *multiply* rational expressions, multiply the numerators and multiply the denominators.

$$\frac{A}{B} \cdot \frac{C}{D} = \frac{AC}{BD}$$

Then simplify, if possible.

To divide rational expressions, multiply the first by the reciprocal of the second.

$$\frac{A}{B} \div \frac{C}{D} = \frac{A}{B} \cdot \frac{D}{C} = \frac{AD}{BC}$$

Then simplify, if possible.

Multiply and simplify, if possible.

14. $\dfrac{3xy}{2x} \cdot \dfrac{4x}{2y^2}$

15. $56x\left(\dfrac{12}{7x}\right)$

16. $\dfrac{x^2 - 1}{x^2 + 2x} \cdot \dfrac{x}{x + 1}$

17. $\dfrac{x^2 + x}{3x - 15} \cdot \dfrac{6x - 30}{x^2 + 2x + 1}$

Divide and simplify, if possible.

18. $\dfrac{3x^2}{5x^2 y} \div \dfrac{6x}{15xy^2}$

19. $\dfrac{x^2 + 5x}{x^2 + 4x - 5} \div x^2$

20. $\dfrac{x^2 - x - 6}{1 - 2x} \div \dfrac{x^2 - 2x - 3}{2x^2 + x - 1}$

21. Tell whether the given fraction is a unit conversion factor.

a. $\dfrac{1 \text{ ft}}{12 \text{ in.}}$

b. $\dfrac{60 \text{ min}}{1 \text{ day}}$

c. $\dfrac{2,000 \text{ lbs.}}{1 \text{ ton}}$

d. $\dfrac{1 \text{ gal}}{4 \text{ qt}}$

To find the *reciprocal* of a rational expression, we invert its numerator and denominator.

A *unit conversion factor* is a fraction that has value 1.

22. TRAFFIC SIGNS   Convert the speed limit on the sign from miles per hour to miles per minute.

SPEED LIMIT
20

SECTION 6.3

Addition and Subtraction with Like Denominators; Least Common Denominators

To *add* or *subtract* rational expressions that have the same denominator, add or subtract their numerators and write the sum or difference over the common denominator.

$$\frac{A}{D} + \frac{B}{D} = \frac{A + B}{D}$$

$$\frac{A}{D} - \frac{B}{D} = \frac{A - B}{D}$$

Then simplify, if possible.

To find the *LCD*, factor each denominator completely. Form a product using each different factor the greatest number of times it appears in any one factorization.

To *build* an equivalent rational expression, multiply the given expression by 1 in the form of $\frac{c}{c}$, where c is any nonzero number or expression.

Add or subtract and simplify, if possible.

23. $\dfrac{13}{15d} - \dfrac{8}{15d}$

24. $\dfrac{x}{x + y} + \dfrac{y}{x + y}$

25. $\dfrac{3x}{x - 7} - \dfrac{x - 2}{x - 7}$

26. $\dfrac{a}{a^2 - 2a - 8} + \dfrac{2}{a^2 - 2a - 8}$

Find the LCD of each pair of rational expressions.

27. $\dfrac{12}{x}, \dfrac{1}{9}$

28. $\dfrac{1}{2x^3}, \dfrac{5}{8x}$

29. $\dfrac{7}{m}, \dfrac{m + 2}{m - 8}$

30. $\dfrac{x}{5x + 1}, \dfrac{5x}{5x - 1}$

31. $\dfrac{6 - a}{a^2 - 25}, \dfrac{a^2}{a - 5}$

32. $\dfrac{4t + 25}{t^2 + 10t + 25}, \dfrac{t^2 - 7}{2t^2 + 17t + 35}$

Build each rational expression into an equivalent fraction having the denominator shown in red.

33. $\dfrac{9}{a}; \; 7a$

34. $\dfrac{2y + 1}{x - 9}; \; x(x - 9)$

35. $\dfrac{b + 7}{3b - 15}; \; 6(b - 5)$

36. $\dfrac{9r}{r^2 + 6r + 5}; \; (r + 1)(r - 4)(r + 5)$

| SECTION 6.4 | **Addition and Subtraction with Unlike Denominators** |

To *add* or *subtract* rational expressions with unlike denominators:

1. Find the LCD.

2. Write each rational expression as an equivalent expression whose denominator is the LCD.

3. Add or subtract the numerators and write the sum or difference over the LCD.

4. Simplify the resulting rational expression, if possible.

When a polynomial is multiplied by -1, the result is its opposite.

Add or subtract and simplify, if possible.

37. $\dfrac{1}{7} - \dfrac{1}{a}$

38. $\dfrac{x}{x-1} + \dfrac{1}{x}$

39. $\dfrac{2t+2}{t^2+2t+1} - \dfrac{1}{t+1}$

40. $\dfrac{x+2}{2x} - \dfrac{2-x}{x^2}$

41. $\dfrac{6}{b-1} - \dfrac{b}{1-b}$

42. $\dfrac{8}{c} + 6$

43. $\dfrac{n+7}{n+3} - \dfrac{n-3}{n+7}$

44. $\dfrac{4}{t+2} - \dfrac{7}{(t+2)^2}$

45. $\dfrac{6}{a^2-9} - \dfrac{5}{a^2-a-6}$

46. $\dfrac{2}{3y-6} + \dfrac{3}{4y+8}$

47. Working on a homework assignment, a student added two rational expressions and obtained $\dfrac{-5n^3-7}{3n(n+6)}$. The answer given in the back of the book was $-\dfrac{5n^3+7}{3n(n+6)}$. Are the answers equivalent?

48. VIDEO CAMERAS   Find the perimeter and the area of the LED screen of the camera.

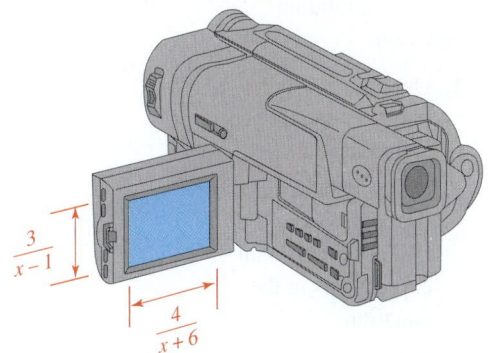

$\dfrac{3}{x-1}$ $\dfrac{4}{x+6}$

| SECTION 6.5 | **Simplifying Complex Fractions** |

Complex fractions contain fractions in their numerators and/or their denominators.

Simplify each complex fraction.

49. $\dfrac{\dfrac{n^4}{30}}{\dfrac{7n}{15}}$

50. $\dfrac{\dfrac{r^2-81}{18s^2}}{\dfrac{4r-36}{9s}}$

To *simplify* a complex fraction:

Method 1
Write the numerator and the denominator as single rational expressions. Perform the indicated division (multiply by the reciprocal).

Method 2
Determine the LCD for all rational expressions in the complex fraction. Multiply the complex fraction by 1 in the form of $\frac{\text{LCD}}{\text{LCD}}$.

51. $\dfrac{\dfrac{1}{y} + 1}{\dfrac{1}{y} - 1}$

52. $\dfrac{\dfrac{7}{a^2}}{\dfrac{1}{a} + \dfrac{10}{3}}$

53. $\dfrac{\dfrac{2}{x-1} + \dfrac{x-1}{x+1}}{\dfrac{1}{x^2 - 1}}$

54. $\dfrac{\dfrac{1}{x^2 y} - \dfrac{5}{xy}}{\dfrac{3}{xy} - \dfrac{7}{xy^2}}$

SECTION 6.6

Solving Rational Equations

To *solve* a rational equation:

1. Determine which numbers cannot be solutions.

2. Multiply both sides of the equation by the LCD of all rational expressions in the equation.

3. Solve the resulting equation.

4. Check all possible solutions in the original equation.

A possible solution that does not satisfy the original equation is called an *extraneous* solution.

Solve each equation and check the result. If an equation has no solution, so indicate.

55. $\dfrac{3}{x} = \dfrac{2}{x-1}$

56. $\dfrac{a}{a-5} = 3 + \dfrac{5}{a-5}$

57. $\dfrac{2}{3t} + \dfrac{1}{t} = \dfrac{5}{9}$

58. $a = \dfrac{3a-50}{4a-24} - \dfrac{3}{4}$

59. $\dfrac{4}{x+2} - \dfrac{3}{x+3} = \dfrac{6}{x^2 + 5x + 6}$

60. $\dfrac{3}{x+1} - \dfrac{x-2}{2} = \dfrac{x-2}{x+1}$

61. ENGINEERING   The efficiency E of a Carnot engine is given by the following formula. Solve it for T_1.

$$E = 1 - \frac{T_2}{T_1}$$

62. Solve for y: $\dfrac{1}{x} = \dfrac{1}{y} + \dfrac{1}{z}$.

SECTION 6.7	**Problem Solving Using Rational Equations**

To solve a problem follow these steps:

1. Analyze the problem.
2. Form an equation.
3. Solve the equation.
4. State the conclusion.
5. Check the result.

Distance = rate · time

Work completed = rate of work · time

Interest = Principal · rate · time

63. NUMBER PROBLEM   If a number is subtracted from the denominator of $\frac{4}{5}$ and twice as much is added to the numerator, the result is 5. Find the number.

64. EXERCISE   A jogger can bicycle 30 miles in the same time that it takes her to jog 10 miles. If she can ride 10 mph faster than she can jog, how fast can she jog?

65. HOUSE CLEANING   A maid can clean a house in 4 hours. What is her rate of work?

66. HOUSE PAINTING   If a homeowner can paint a house in 14 days and a professional painter can paint it in 10 days, how long will it take if they work together?

67. INVESTMENTS   In 1 year, a student earned $100 interest on money she deposited at a savings and loan. She later learned that the money would have earned $120 if she had deposited it at a credit union, because the credit union paid 1% more interest at the time. Find the rate she received from the savings and loan.

68. WIND SPEED   A plane flies 400 miles downwind in the same amount of time as it takes to travel 320 miles upwind. If the plane can fly at 360 mph in still air, find the velocity of the wind.

SECTION 6.8	**Proportions and Similar Triangles**

A *proportion* is a statement that two ratios or two rates are equal.

In the proportion $\frac{a}{b} = \frac{c}{d}$, *a* and *d* are the *extremes*, and *b* and *c* are the *means*.

In any proportion, the product of the extremes is equal to the product of the means. That is, the *cross products* are equal.

Determine whether each equation is a proportion.

69. $\dfrac{4}{7} = \dfrac{20}{34}$

70. $\dfrac{5}{7} = \dfrac{30}{42}$

Solve each proportion.

71. $\dfrac{3}{x} = \dfrac{6}{9}$

72. $\dfrac{x}{3} = \dfrac{x}{5}$

73. $\dfrac{x-2}{5} = \dfrac{x}{7}$

74. $\dfrac{2x}{x+4} = \dfrac{3}{x-1}$

75. DENTISTRY   The diagram on the right was displayed in a dentist's office. According to the diagram, if the dentist has 340 adult patients, how many will develop gum disease?

3 out of 4 adults will develop gum disease.

The lengths of corresponding sides of *similar triangles* are in proportion.

A *scale* is a ratio (or rate) that compares the size of a model to the size of an actual object.

Unit prices can be used to compare costs of different sizes of the same brand to determine the best buy.

76. A telephone pole casts a shadow 12 feet long at the same time that a man 6 feet tall casts a shadow of 3.6 feet. How tall is the pole?

77. PORCELAIN FIGURINES   A model of a flutist, standing and playing at a music stand, was made using a 1/12th scale. If the scale model is 5.5 inches tall, how tall is the flutist?

78. COMPARISON SHOPPING   Which is the better buy for recordable compact discs: 150 for $60 or 250 for $98?

CHAPTER 6 TEST   ELEMENTARY & INTERMEDIATE Algebra $f(x)$ Now ™

For what numbers is each rational expression undefined?

1. $\dfrac{6x - 9}{5x}$

2. $\dfrac{x}{x^2 + x - 6}$

3. MEMORY   The formula $n = \dfrac{35 + 5d}{d}$ approximates the number of words n that a certain person can recall d days after memorizing a list of 50 words. How many words will the person remember in 1 week?

4. U.S. SCHOOL CONSTRUCTION   The table shows the average school size for projects finished in 2002. Convert the size of a middle school to square yards.

	Square feet
Elementary school	70,174
Middle school	109,512
High school	125,304

(Source: American School & University's 29th Annual Official Education Construction Report)

Simplify each rational expression.

5. $\dfrac{48x^2 y}{54xy^2}$

6. $\dfrac{7m - 49}{7 - m}$

7. $\dfrac{2x^2 - x - 3}{4x^2 - 9}$

8. $\dfrac{3(x + 2) - 3}{6x + 5 - (3x + 2)}$

Find the LCD of each pair of rational expressions.

9. $\dfrac{19}{3c^2 d}, \dfrac{6}{c^2 d^3}$

10. $\dfrac{4n + 25}{n^2 - 4n - 5}, \dfrac{6n}{n^2 - 25}$

Perform the operations. Simplify, if possible.

11. $\dfrac{12x^2 y}{15xy} \cdot \dfrac{25y^2}{16x}$

12. $\dfrac{x^2 + 3x + 2}{3x + 9} \cdot \dfrac{x + 3}{x^2 - 4}$

13. $\dfrac{x - x^2}{3x^2 + 6x} \div \dfrac{3x - 3}{3x^3 + 6x^2}$

14. $\dfrac{a^2 - 16}{a - 4} \div (6a + 24)$

15. $\dfrac{3y + 7}{2y + 3} - \dfrac{3y - 6}{2y + 3}$

16. $\dfrac{2n}{5m} - \dfrac{n}{2}$

17. $\dfrac{x + 1}{x} + \dfrac{x - 1}{x + 1}$

18. $\dfrac{a + 3}{a - 1} - \dfrac{a + 4}{1 - a}$

19. $\dfrac{9}{c - 4} + c$

20. $\dfrac{6}{t^2 - 9} - \dfrac{5}{t^2 - t - 6}$

Simplify each complex fraction.

21. $\dfrac{\dfrac{3m-9}{8m}}{\dfrac{5m-15}{32}}$

22. $\dfrac{\dfrac{3}{as^2}+\dfrac{6}{a^2s}}{\dfrac{6}{a}-\dfrac{9}{s^2}}$

Solve each equation. If an equation has no solution, so indicate.

23. $\dfrac{1}{3}+\dfrac{4}{3y}=\dfrac{5}{y}$

24. $\dfrac{9n}{n-6}=3+\dfrac{54}{n-6}$

25. $\dfrac{7}{q^2-q-2}+\dfrac{1}{q+1}=\dfrac{3}{q-2}$

26. $\dfrac{2}{3}=\dfrac{2c-12}{3c-9}-c$

27. $\dfrac{y}{y-1}=\dfrac{y-2}{y}$

28. Solve for B: $H=\dfrac{RB}{R+B}$.

29. HEALTH RISKS A medical newsletter states that a "healthy" waist-to-hip ratio for men is 19:20 or less. Does the patient shown in the illustration fall within the "healthy" range?

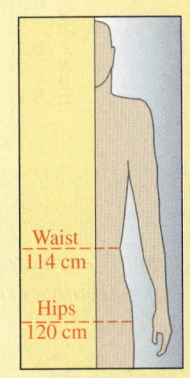

Waist
114 cm

Hips
120 cm

30. CURRENCY EXCHANGE RATES Preparing for a visit to London, a New York resident exchanged 1,500 U.S. dollars for British pounds. (A pound is the basic monetary unit of Great Britain.) If the exchange rate was 5 U.S. dollars for 3 British pounds, how many British pounds did the traveler receive?

31. TV TOWERS A television tower casts a shadow 114 feet long at the same time that a 6-foot-tall television reporter casts a shadow of 4 feet. Find the height of the tower.

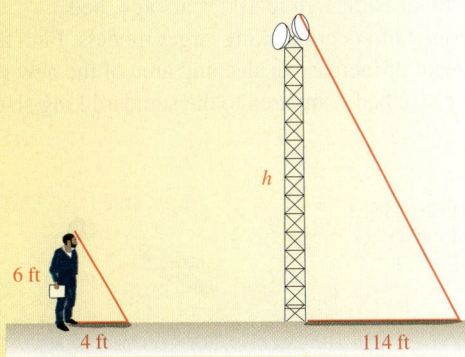

h

6 ft

4 ft   114 ft

32. COMPARISON SHOPPING Which is the better buy for fabric softener: 80 sheets for \$3.89 or 120 sheets for \$6.19?

33. CLEANING HIGHWAYS One highway worker can pick up all the trash on a strip of highway in 7 hours, and his helper can pick up the trash in 9 hours. How long will it take them if they work together?

34. PHYSICAL FITNESS A man roller-blades at a rate 6 miles per hour faster than he jogs. In the same time it takes him to roller-blade 5 miles he can jog 2 miles. How fast does he jog?

35. Explain the error: $\dfrac{x+5}{5}=\dfrac{x+\overset{1}{\cancel{5}}}{\underset{1}{\cancel{5}}}=x+1$.

36. Explain what it means to clear the following equation of fractions.

$$\dfrac{u}{u-1}+\dfrac{1}{u}=\dfrac{u^2+1}{u^2-u}$$

Why is this a helpful first step in solving the equation?

CHAPTERS 1–6   CUMULATIVE REVIEW EXERCISES

1. Evaluate: $9^2 - 3[45 - 3(6 + 4)]$.

2. GRAND KING SIZE BEDS   Because Americans are taller compared to 100 years ago, bed manufacturers are making larger models. Find the percent of increase in sleeping area of the new grand king size bed compared to the standard king size.

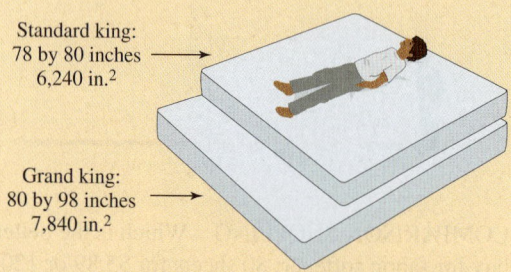

Standard king:
78 by 80 inches
6,240 in.²

Grand king:
80 by 98 inches
7,840 in.²

3. Insert the proper symbol, $<$ or $>$, in the blank to make a true statement.

$$|2 - 4| \quad\rule{1cm}{0.4pt}\quad -(-6)$$

4. Find the average (mean) test score of a student in a history class with scores of 80, 73, 61, 73, and 98.

5. Change $40°$ C to degrees Fahrenheit.

6. Find the volume of a pyramid that has a square base, measuring 6 feet on a side, and whose height is 20 feet.

7. Tell whether each statement is true or false.
   a. Every integer is a whole number.
   b. 0 is not a rational number.
   c. π is an irrational number.
   d. The set of integers is the set of whole numbers and their opposites.

8. Solve: $2 - 3(x - 5) = 4(x - 1)$.

9. Simplify: $8(c + 7) - 2(c - 3)$.

10. Solve $A - c = 2B + r$ for B.

11. Solve: $7x + 2 \geq 4x - 1$. Write the solution set in interval notation and graph it.

12. Solve: $\dfrac{4}{5}d = -4$.

13. BLENDING TEA   One grade of tea (worth $3.20 per pound) is to be mixed with another grade (worth $2 per pound) to make 20 pounds of a mixture that will be worth $2.72 per pound. How much of each grade of tea must be used?

14. SPEED OF A PLANE   Two planes are 6,000 miles apart and their speeds differ by 200 mph. If they travel toward each other and meet in 5 hours, find the speed of the slower plane.

15. Graph: $y = 2x - 3$.

16. Graph: $3x - 2y \leq 6$.

17. What is the slope of a line perpendicular to the line $y = -\dfrac{7}{8}x - 6$?

18. Find the slope of the line passing through $(-1, 3)$ and $(3, -1)$.

19. Write the equation of a line that has slope 3 and passes through the point $(1, 5)$.

20. CUTTING STEEL   The graph shows the amount of wear (in mm) on a cutting blade for a given length of a cut (in m). Find the rate of change in the length of the cutting blade.

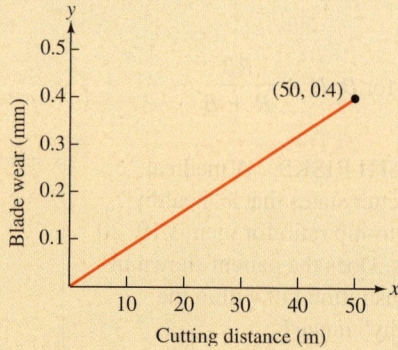

Simplify each expression. Write each answer without using negative exponents.

21. $x^4 x^3$

22. $(x^2 x^3)^5$

23. $\left(\dfrac{y^3 y}{2yy^2}\right)^3$

24. $\left(\dfrac{-2a}{b}\right)^5$

25. $(a^{-2}b^3)^{-4}$

26. $\dfrac{9b^0 b^3}{3b^{-3}b^4}$

27. Write 290,000 in scientific notation.

28. What is the degree of the polynomial $5x^3 - 4x + 16$?

29. CONCENTRIC CIRCLES
The area of the ring between
the two concentric circles of
radius r and R is given by
the formula

$$A = \pi(R + r)(R - r)$$

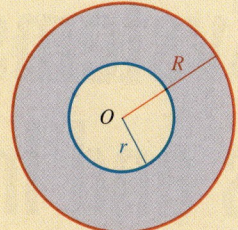

Do the multiplication on the right-hand side of the
equation.

30. Graph: $y = -x^3$.

Perform the operations.

31. $(3x^2 - 3x - 2) + (3x^2 + 4x - 3)$

32. $(2x^2y^3)(3x^3y^2)$

33. $(2y - 5)(3y + 7)$

34. $-4x^2z(3x^2 - z)$

35. $\dfrac{6x + 9}{3}$

36. $2x + 3\overline{)2x^3 + 7x^2 + 4x - 3}$

Factor each polynomial completely, if possible.

37. $k^3t - 3k^2t$

38. $2ab + 2ac + 3b + 3c$

39. $2a^2 - 200b^2$

40. $b^3 + 125$

41. $u^2 - 18u + 81$

42. $6x^2 - 63 - 13x$

43. $-r^2 + 2 + r$

44. $u^2 + 10u + 15$

Solve each equation by factoring.

45. $5x^2 + x = 0$

46. $6x^2 - 5x = -1$

47. COOKING   The electric griddle shown has a
cooking surface of 160 square inches. Find the length
and the width of the griddle.

$w + 6$

w

48. For what values of x is the rational expression $\dfrac{3x^2}{x^2 - 25}$
undefined?

Perform the operations. Simplify, if possible.

49. $\dfrac{x^2 - 16}{x - 4} \div \dfrac{3x + 12}{x}$

50. $\dfrac{4}{x - 3} + \dfrac{5}{3 - x}$

51. $\dfrac{m}{m^2 + 5m + 6} - \dfrac{2}{m^2 + 3m + 2}$

52. Simplify: $\dfrac{2 - \dfrac{2}{x + 1}}{2 + \dfrac{2}{x}}$.

Solve each equation.

53. $\dfrac{7}{5x} - \dfrac{1}{2} = \dfrac{5}{6x} + \dfrac{1}{3}$

54. $\dfrac{u}{u - 1} + \dfrac{1}{u} = \dfrac{u^2 + 1}{u^2 - u}$

55. DRAINING A TANK   If one outlet pipe can drain a
tank in 24 hours, and another pipe can drain the tank
in 36 hours, how long will it take for both pipes to
drain the tank?

56. HEIGHT OF A TREE   A tree casts a shadow of
29 feet at the same time as a vertical yardstick casts a
shadow of 2.5 feet. Find the height of the tree.

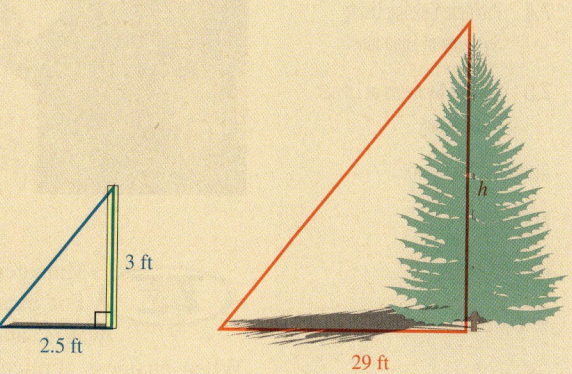

3 ft

2.5 ft

h

29 ft

7

Solving Systems of Equations and Inequalities

ELEMENTARY &
INTERMEDIATE
Algebra $f(x)$ Now ™

Throughout the chapter, this icon introduces resources on the Elementary & Intermediate AlgebraNow Web site, accessed through **http://1pass. thomson.com**, that will

- Help you test your knowledge of the material with a pre-test and a post-test

- Provide a personalized learning plan targeting areas you should study

Getty Images

Managers make many decisions in the day-to-day operation of a business. For example, a theater owner must determine ticket prices, employees' wages, and show times. Algebra can be helpful in making these decisions. When business applications involve two variable quantities, often we can model them using a pair of equations called a *system of equations*.

To learn more about the use of systems of equations, visit *The Learning Equation* on the Internet at http://tle.brookscole.com. (The log-in instructions are in the Preface.) For Chapter 7, the online lesson is

- *TLE* Lesson 14: Solving Systems of Equations by Graphing

Some problems are more easily solved using a pair of equations in two variables rather than a single equation in one variable. We call a pair of equations with common variables a system of equations.

7.1   Solving Systems of Equations by Graphing

- Systems of Equations and Their Solutions   • The Graphing Method
- Inconsistent Systems and Dependent Equations   • Graphing Calculators

The illustration shows the per capita consumption of chicken and beef in the United States for the years 1985–2002. Plotting both graphs on the same coordinate system makes it easy to compare recent trends. The point of intersection of the graphs indicates that Americans consumed equal amounts of chicken and beef in 1992—about 66 pounds of each, per person.

In this section, we will use a similar graphical approach to solve *systems of equations*.

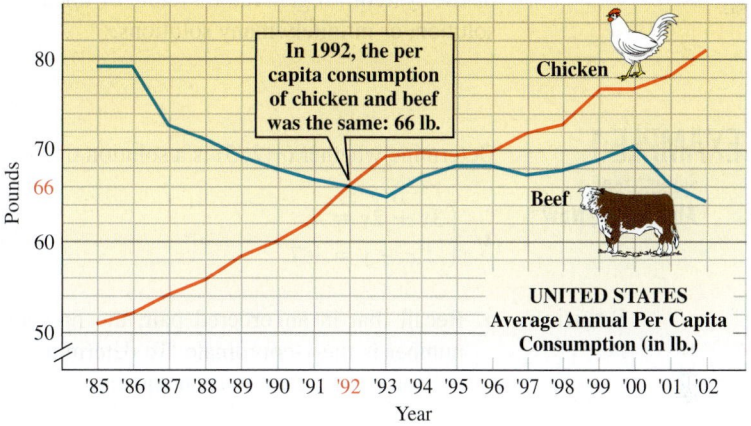

Source: American Meat Institute

■ SYSTEMS OF EQUATIONS AND THEIR SOLUTIONS

We have previously discussed equations in two variables, such as $x + y = 3$. Because there are infinitely many pairs of numbers whose sum is 3, there are infinitely many pairs (x, y) that satisfy this equation. Some of these pairs are listed in the following table.

The Language of Algebra

We say that (2, 1) *satisfies* $x + y = 3$, because the x-coordinate, 2, and the y-coordinate, 1, make the equation true when substituted for x and y: $2 + 1 = 3$. To *satisfy* means to make content, as in *satisfy* your thirst or a *satisfied* customer.

$$x + y = 3$$

x	y	(x, y)
0	3	(0, 3)
1	2	(1, 2)
2	1	(2, 1)
3	0	(3, 0)

Now consider the equation $x - y = 1$. Because there are infinitely many pairs of numbers whose difference is 1, there are infinitely many pairs (x, y) that satisfy $x - y = 1$. Some of these pairs are listed in the following table.

$$x - y = 1$$

x	y	(x, y)
0	-1	$(0, -1)$
1	0	$(1, 0)$
2	**1**	**(2, 1)**
3	2	$(3, 2)$

The Language of Algebra

A system of equations is two (or more) equations that we consider *simultaneously*—at the same time. Some professional sports teams *simulcast* their games. That is, the announcer's play-by-play description is broadcast on radio and television at the same time.

From the two tables, we see that (2, 1) satisfies both equations.

When two equations are considered simultaneously, we say that they form a **system of equations.** Using a left brace $\{$, we can write the equations from the previous example as a system:

$$\begin{cases} x + y = 3 \\ x - y = 1 \end{cases}$$ Read as "the system of equations $x + y = 3$ and $x - y = 1$."

Because the ordered pair (2, 1) satisfies both of these equations, it is called a **solution of the system.** In general, a system of linear equations can have exactly one solution, no solution, or infinitely many solutions.

EXAMPLE 1

Determine whether $(-2, 5)$ is a solution of each system of equations.

a. $\begin{cases} 3x + 2y = 4 \\ x - y = -7 \end{cases}$

b. $\begin{cases} 4y = 18 - x \\ y = 2x \end{cases}$

Solution **a.** Recall that in an ordered pair, the first number is the x-coordinate and the second number is the y-coordinate. To determine whether $(-2, 5)$ is a solution, we substitute -2 for x and 5 for y in each equation.

Check: $\begin{aligned} 3x + 2y &= 4 \\ 3(-2) + 2(5) &\stackrel{?}{=} 4 \\ -6 + 10 &\stackrel{?}{=} 4 \\ 4 &= 4 \quad \text{True} \end{aligned}$ $\begin{aligned} x - y &= -7 \\ -2 - 5 &\stackrel{?}{=} -7 \\ -7 &= -7 \quad \text{True} \end{aligned}$

Since $(-2, 5)$ satisfies both equations, it is a solution of the system.

b. Again, we substitute -2 for x and 5 for y in each equation.

Check: $\begin{aligned} 4y &= 18 - x \\ 4(5) &\stackrel{?}{=} 18 - (-2) \\ 20 &\stackrel{?}{=} 18 + 2 \\ 20 &= 20 \quad \text{True} \end{aligned}$ $\begin{aligned} y &= 2x \\ 5 &\stackrel{?}{=} 2(-2) \\ 5 &= -4 \quad \text{False} \end{aligned}$

Although $(-2, 5)$ satisfies the first equation, it does not satisfy the second. Thus, $(-2, 5)$ is not a solution of the system.

Self Check 1   Determine whether $(4, -1)$ is a solution of $\begin{cases} x - 2y = 6 \\ y = 3x - 11 \end{cases}$.

■ THE GRAPHING METHOD

To **solve a system** of equations means to find all of the solutions of the system. One way to solve a system of linear equations is to graph the equations on the same set of axes.

EXAMPLE 2   Solve the following system by graphing: $\begin{cases} 2x + 3y = 2 \\ 3x - 2y = 16 \end{cases}$.

ELEMENTARY &
INTERMEDIATE
Algebra $f(x)$ **Now**™

Solution   We graph both equations on the same coordinate system.

Success Tip

Accuracy is crucial when using the graphing method to solve a system. Here are some suggestions for improving your accuracy:

• Use graph paper.
• Use a sharp pencil.
• Use a ruler or straightedge.

$2x + 3y = 2$

x	y	(x, y)
0	$\frac{2}{3}$	$\left(0, \frac{2}{3}\right)$
1	0	$(1, 0)$
-2	2	$(-2, 2)$

$3x - 2y = 16$

x	y	(x, y)
0	-8	$(0, -8)$
$\frac{16}{3}$	0	$\left(\frac{16}{3}, 0\right)$
2	-5	$(2, -5)$

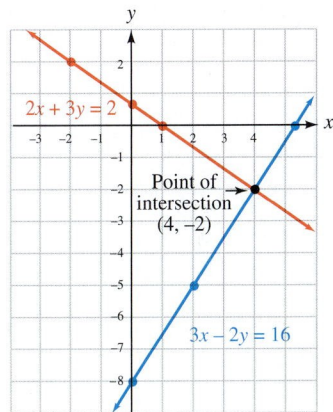

The coordinates of each point on the line graphed in red satisfy $2x + 3y = 2$ and the coordinates of each point on the line graphed in blue satisfy $3x - 2y = 16$. Because the point of intersection is on both graphs, its coordinates satisfy both equations.

It appears that the graphs intersect at the point $(4, -2)$. To verify that it is the solution of the system, we substitute 4 for x and -2 for y in each equation.

Check:

$$2x + 3y = 2 \qquad\qquad 3x - 2y = 16$$
$$2(4) + 3(-2) \stackrel{?}{=} 2 \qquad 3(4) - 2(-2) \stackrel{?}{=} 16$$
$$8 + (-6) \stackrel{?}{=} 2 \qquad 12 - (-4) \stackrel{?}{=} 16$$
$$2 = 2 \qquad\qquad\qquad 16 = 16$$

Since $(4, -2)$ makes both equations true, it is the solution of the system.

Self Check 2   Solve the following system by graphing: $\begin{cases} 2x - y = -5 \\ x + y = -1 \end{cases}$.

To solve a system of equations in two variables by graphing, follow these steps.

The Graphing Method	**1.** Carefully graph each equation on the same rectangular coordinate system.
	2. If the lines intersect, determine the coordinates of the point of intersection of the graphs. That ordered pair is the solution of the system.
	3. Check the proposed solution in each equation of the original system.

■ **INCONSISTENT SYSTEMS AND DEPENDENT EQUATIONS**

A system of equations that has at least one solution, like that in Example 2, is called a **consistent system.** A system with no solution is called an **inconsistent system.**

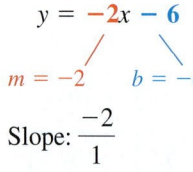

EXAMPLE 3

Solve the following system by graphing: $\begin{cases} y = -2x - 6 \\ 4x + 2y = 8 \end{cases}$.

Solution   Since $y = -2x - 6$ is written in slope–intercept form, we can graph it by plotting the y-intercept $(0, -6)$ and then drawing a slope of -2. (The rise is -2 and the run is 1.) We graph $4x + 2y = 8$ using the intercept method.

$$y = \mathbf{-2}x - \mathbf{6}$$

$$m = -2 \qquad b = -6$$

Slope: $\dfrac{-2}{1}$

y-intercept: $(0, -6)$

$4x + 2y = 8$

x	y	(x, y)
0	4	$(0, 4)$
2	0	$(2, 0)$
1	2	$(1, 2)$

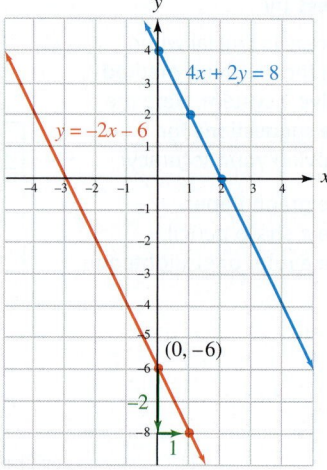

The lines appear to be parallel. We can verify this by writing the second equation in slope–intercept form and observing that the lines have the same slope, -2, and different y-intercepts, $(0, -6)$ and $(0, 4)$.

$$y = -2x - 6 \qquad 4x + 2y = 8$$
$$2y = -4x + 8 \qquad \text{Subtract } 4x \text{ from both sides.}$$
$$y = -2x + 4 \qquad \text{Divide both sides by 2.}$$

Different y-intercepts
Same slope

Because the lines are parallel, there is no point of intersection. Such a system has no solution.

Self Check 3   Solve the following system by graphing: $\begin{cases} y = \dfrac{3}{2}x \\ 3x - 2y = 6 \end{cases}$.

Some systems of equations have infinitely many solutions.

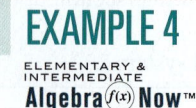

EXAMPLE 4

Solve the following system by graphing: $\begin{cases} 4x + 8 = 2y \\ y = 2x + 4 \end{cases}$.

Solution   To graph $4x + 8 = 2y$, we use the intercept method and to graph $y = 2x + 4$, we use the slope and y-intercept.

$4x + 8 = 2y$

x	y	(x, y)
0	4	$(0, 4)$
-2	0	$(-2, 0)$
-3	-2	$(-3, -2)$

$y = \mathbf{2}x + \mathbf{4}$

$m = 2 \qquad b = 4$

Slope: $\dfrac{2}{1}$

y-intercept: $(0, 4)$

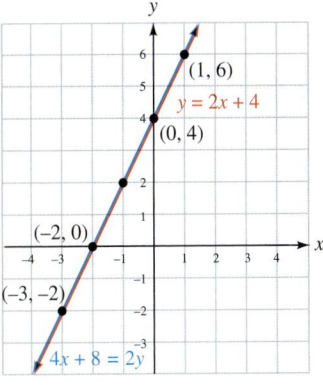

The graphs appear to be identical. We can verify this by writing the first equation in slope–intercept form and observing that it is the same as the second equation.

Since the graphs are the same line, they have infinitely many points in common. For such a system, there are infinitely many solutions.

From the graph, we see that four of the solutions are $(-3, -2)$, $(-2, 0)$, $(0, 4)$, and $(1, 6)$. Checks for two ordered pairs follow.

The Language of Algebra

In Example 4, the graphs of the lines *coincide.* That is, they occupy the same location. To illustrate this concept, think of a clock. At noon and midnight, the hands of the clock *coincide.*

For $(-3, -2)$:

$$4x + 8 = 2y \qquad y = 2x + 4$$
$$4(\mathbf{-3}) + 8 \stackrel{?}{=} 2(\mathbf{-2}) \qquad \mathbf{-2} \stackrel{?}{=} 2(\mathbf{-3}) + 4$$
$$-12 + 8 \stackrel{?}{=} -4 \qquad -2 \stackrel{?}{=} -6 + 4$$
$$-4 = -4 \qquad -2 = -2$$

For $(0, 4)$:

$$4x + 8 = 2y \qquad y = 2x + 4$$
$$4(\mathbf{0}) + 8 \stackrel{?}{=} 2(\mathbf{4}) \qquad \mathbf{4} \stackrel{?}{=} 2(\mathbf{0}) + 4$$
$$0 + 8 \stackrel{?}{=} 8 \qquad 4 \stackrel{?}{=} 0 + 4$$
$$8 = 8 \qquad 4 = 4$$

Self Check 4   Solve the following system by graphing: $\begin{cases} 6x - 4 = 2y \\ y = 3x - 2 \end{cases}$.

In Examples 2 and 3, the graphs of the equations of the system were different lines. We call equations with different graphs **independent equations.** In Example 4, the equations have the same graph and are called **dependent equations.**

There are three possible outcomes when we solve a system of two linear equations using the graphing method:

The two lines intersect at one point.

The two lines are parallel.

The two lines are identical.

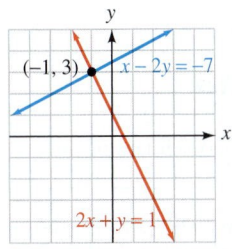

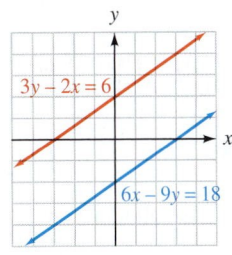

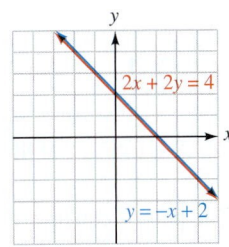

Exactly one solution
(the point of intersection)

No solution

Infinitely many solutions
(any point on the line is a solution)

Consistent system
Independent equations

Inconsistent system
Independent equations

Consistent system
Dependent equations

■ GRAPHING CALCULATORS

A graphing calculator can be used to solve systems of equations, such as

$$\begin{cases} 2x + y = 12 \\ 2x - y = -2 \end{cases}$$

Before we can enter the equations into the calculator, we must solve them for y.

$$2x + y = 12 \qquad\qquad 2x - y = -2$$
$$y = -2x + 12 \qquad\qquad -y = -2x - 2$$
$$\qquad\qquad\qquad\qquad y = 2x + 2$$

We enter the resulting equations as Y_1 and Y_2 and graph them on the same axes. If we use the standard window setting, their graphs will look like figure (a).

To find the solution of the system, we can use the INTERSECT feature found on most graphing calculators. With this option, the cursor automatically moves to the point of intersection of the graphs and displays the coordinates of that point. In figure (b), we see that the solution is $(2.5, 7)$.

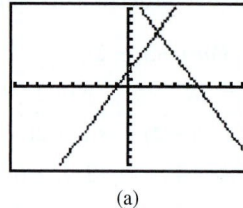   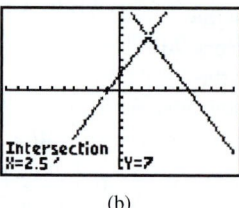

(a)   (b)

Answers to Self Checks   **1.** no   **2.** $(-2, 1)$ **3.** no solution   **4.** infinitely many solutions

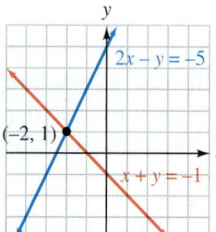   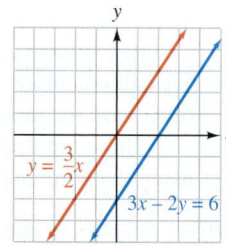   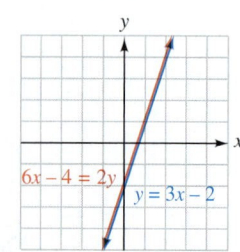

7.1   STUDY SET

ELEMENTARY & INTERMEDIATE
Algebra $f(x)$ Now™

VOCABULARY   **Fill in the blanks.**

1. The pair of equations $\begin{cases} x - y = -1 \\ 2x - y = 1 \end{cases}$ is called a

_____ of equations.

2. Because the ordered pair $(2, 3)$ satisfies both equations in Problem 1, it is called a _____ of the system of equations.

3. The x-coordinate of the ordered pair $(-4, 7)$ is -4 and the _____ is 7.

4. We say that $(1, 4)$ _____ $x + y = 5$, because the x-coordinate, 1, and the y-coordinate, 4, make the equation true when substituted for x and y.

5. The point of _____ of the lines graphed in part (a) of the following figure is $(1, 2)$.

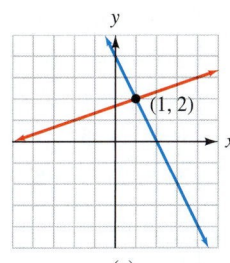
(a)

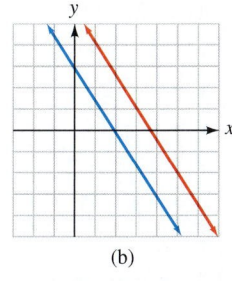
(b)

6. The lines graphed in part (b) of the figure above do not intersect. They are _____ lines.

7. A system of equations that has at least one solution is called a _____ system. A system with no solution is called an _____ system.

8. We call equations with different graphs _____ equations. Because _____ equations are different forms of the same equation, they have the same graph.

CONCEPTS

9. The following tables were created to graph the two linear equations in a system. What is the solution of the system?

Equation 1	
x	y
0	-5
-5	0
-4	-1
1	-6
2	-7

Equation 2	
x	y
0	3
-3	0
-2	1
-4	-1
1	4

Refer to the illustration. Determine whether a true or false statement will result if

10. The coordinates of point A are substituted into the equation for Line 1.

11. The coordinates of point C are substituted into the equation for Line 1.

12. The coordinates of point C are substituted into the equation for Line 2.

13. The coordinates of point B are substituted into the equation for Line 1.

14. To graph $5x - 2y = 10$, we can use the intercept method. Complete the table.

x	y
0	
	0

15. To graph $y = 3x - 2$, we can use the slope and y-intercept. Fill in the blanks.

slope: ▨ $= \dfrac{▨}{1}$ y-intercept: ▨▨

16. What is the solution of the system graphed on the right? Is it consistent or inconsistent?

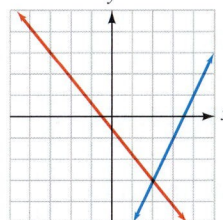

17. How many solutions does the system graphed on the right have? Are the equations dependent or independent?

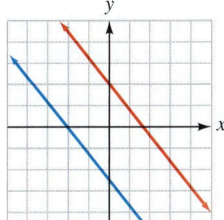

18. How many solutions does the system graphed on the right have? Give three of the solutions. Is the system consistent or inconsistent?

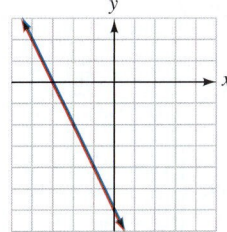

NOTATION

19. The symbol $\stackrel{?}{=}$ is used when checking a solution. What does it mean?

20. Draw the graphs of two linear equations so that the system has one solution, $(-3, -2)$.

PRACTICE **Decide whether the ordered pair is a solution of the given system.**

21. $(1, 1)$, $\begin{cases} x + y = 2 \\ 2x - y = 1 \end{cases}$

22. $(1, 3)$, $\begin{cases} 2x + y = 5 \\ 3x - y = 0 \end{cases}$

23. $(3, -2)$, $\begin{cases} 2x + y = 4 \\ y = 1 - x \end{cases}$

24. $(-2, 4)$, $\begin{cases} 2x + 2y = 4 \\ 3y = 10 - x \end{cases}$

25. $(-2, -4)$, $\begin{cases} 4x + 5y = -23 \\ -3x + 2y = 0 \end{cases}$

26. $(-5, 2)$, $\begin{cases} -2x + 7y = 17 \\ 3x - 4y = -19 \end{cases}$

27. $\left(\dfrac{1}{2}, 3\right)$, $\begin{cases} 2x + y = 4 \\ 4x - 11 = 3y \end{cases}$

28. $\left(2, \dfrac{1}{3}\right)$, $\begin{cases} x - 3y = 1 \\ -2x + 6 = -6y \end{cases}$

29. $(2.5, 3.5)$, $\begin{cases} 4x - 3 = 2y \\ 4y + 1 = 6x \end{cases}$

30. $(0.2, 0.3)$, $\begin{cases} 20x + 10y = 7 \\ 20y = 15x + 3 \end{cases}$

Solve each system by graphing. If a system has no solution or infinitely many, so state.

31. $\begin{cases} 2x + 3y = 12 \\ 2x - y = 4 \end{cases}$

32. $\begin{cases} 5x + y = 5 \\ 5x + 3y = 15 \end{cases}$

33. $\begin{cases} x + y = 4 \\ x - y = -6 \end{cases}$

34. $\begin{cases} x + y = 4 \\ x - y = -2 \end{cases}$

35. $\begin{cases} y = 3x + 6 \\ y = -2x - 4 \end{cases}$

36. $\begin{cases} y = x + 3 \\ y = -2x - 3 \end{cases}$

37. $\begin{cases} y = x - 1 \\ 3x - 3y = 3 \end{cases}$

38. $\begin{cases} y = -x + 1 \\ 4x + 4y = 4 \end{cases}$

39. $\begin{cases} y = -\dfrac{1}{3}x - 4 \\ x + 3y = 6 \end{cases}$

40. $\begin{cases} y = -\dfrac{1}{2}x - 3 \\ x + 2y = 2 \end{cases}$

41. $\begin{cases} y = -x - 2 \\ y = -3x + 6 \end{cases}$

42. $\begin{cases} y = 2x - 4 \\ y = -5x + 3 \end{cases}$

43. $\begin{cases} -x + 3y = -11 \\ 3x - y = 17 \end{cases}$

44. $\begin{cases} 2x - 3y = -18 \\ 3x + 2y = -1 \end{cases}$

45. $\begin{cases} x + y = 2 \\ y = x \end{cases}$

46. $\begin{cases} x + y = 4 \\ y = x \end{cases}$

47. $\begin{cases} y = \dfrac{3}{4}x + 3 \\ y = -\dfrac{x}{4} - 1 \end{cases}$

48. $\begin{cases} y = \dfrac{2}{3}x + 4 \\ y = -\dfrac{x}{3} + 7 \end{cases}$

49. $\begin{cases} 2y = 3x + 2 \\ 3x - 2y = 6 \end{cases}$

50. $\begin{cases} 3x - 6y = 18 \\ x = 2y + 3 \end{cases}$

51. $\begin{cases} 4x - 2y = 8 \\ y = 2x - 4 \end{cases}$

52. $\begin{cases} 2y = -6x - 12 \\ 3x + y = -6 \end{cases}$

53. $\begin{cases} x + y = 2 \\ y = x - 4 \end{cases}$

54. $\begin{cases} x + y = 1 \\ y = x + 5 \end{cases}$

55. $\begin{cases} x + 4y = -2 \\ y = -x - 5 \end{cases}$

56. $\begin{cases} 3x + 2y = -8 \\ 2x - 3y = -1 \end{cases}$

57. $\begin{cases} x = 3 \\ 3y = 6 - 2x \end{cases}$

58. $\begin{cases} x = 4 \\ 2y = 12 - 4x \end{cases}$

59. $\begin{cases} y = -3 \\ -x + 2y = -4 \end{cases}$

60. $\begin{cases} y = -4 \\ -2x - y = 8 \end{cases}$

61. $\begin{cases} x + 2y = -4 \\ x - \dfrac{1}{2}y = 6 \end{cases}$

62. $\begin{cases} \dfrac{2}{3}x - y = -3 \\ 3x + y = 3 \end{cases}$

Determine the slope and the y-intercept of the graph of each line in the system. Then, use that information to determine the number of solutions of the system.

63. $\begin{cases} y = 6x - 7 \\ y = -2x + 1 \end{cases}$

64. $\begin{cases} y = \dfrac{1}{2}x + 8 \\ y = 4x - 10 \end{cases}$

65. $\begin{cases} 3x - y = -3 \\ y - 3x = 3 \end{cases}$

66. $\begin{cases} x + 4y = 4 \\ 12y = 12 - 3x \end{cases}$

67. $\begin{cases} y = -x + 6 \\ x + y = 8 \end{cases}$

68. $\begin{cases} 5x + y = 0 \\ y = -5x + 6 \end{cases}$

69. $\begin{cases} 6x + y = 0 \\ 2x + 2y = 0 \end{cases}$

70. $\begin{cases} x + y = 1 \\ 2x - 2y = 5 \end{cases}$

Use a graphing calculator to solve each system, if possible.

71. $\begin{cases} y = 4 - x \\ y = 2 + x \end{cases}$

72. $\begin{cases} 3x - 6y = 4 \\ 2x + y = 1 \end{cases}$

73. $\begin{cases} 6x - 2y = 5 \\ 3x = y + 10 \end{cases}$

74. $\begin{cases} x - 3y = -2 \\ 5x + y = 10 \end{cases}$

APPLICATIONS

75. TRANSPLANTS   Refer to the graph. In what year were the number of donors and the number waiting for a liver transplant the same? Estimate the number.

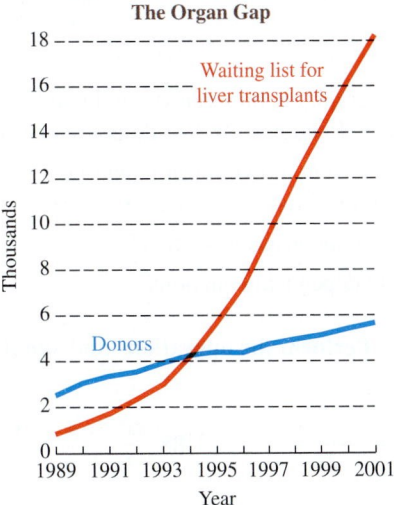

The Organ Gap

Waiting list for liver transplants

Donors

Thousands

1989 1991 1993 1995 1997 1999 2001

Year

Source: Organ Procurement and Transportation Network

76. COLLEGE BOARDS   Many colleges and universities use the SAT test as one indicator of a high school student's readiness to do college level work. Refer to the graph. For any given year, were the average math score and the average verbal score ever the same?

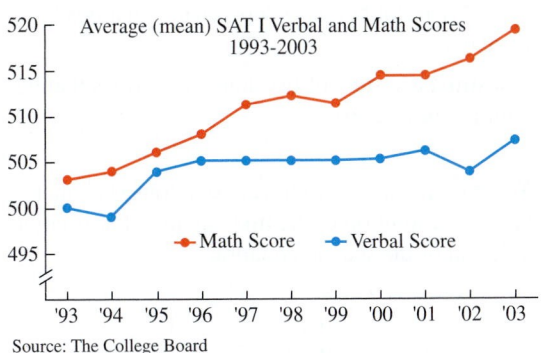

Average (mean) SAT I Verbal and Math Scores
1993-2003

520
515
510
505
500
495

Math Score   Verbal Score

'93 '94 '95 '96 '97 '98 '99 '00 '01 '02 '03

Source: The College Board

77. LATITUDE AND LONGITUDE   Refer to the following map.

a. Name three American cities that lie on a latitude line of 30° north.

b. Name three American cities that lie on a longitude line of 90° west.

c. What city lies on both lines?

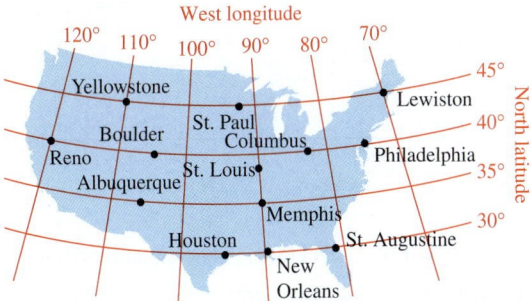

West longitude

120° 110° 100° 90° 80° 70°

Yellowstone   Lewiston   45°

Boulder   St. Paul   40°
                 Columbus
Reno   Philadelphia
         Albuquerque St. Louis   35°

                          Memphis   30°
           Houston   St. Augustine

New
Orleans

North latitude

78. ECONOMICS   The graph illustrates the law of supply and demand.

a. Complete this sentence: As the price of an item increases, the *supply* of the item _____.

b. Complete this sentence: As the price of an item increases, the *demand* for the item _____.

c. For what price will the supply equal the demand? How many items will be supplied for this price?

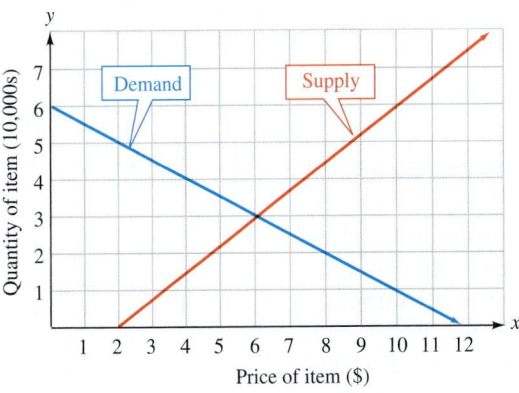

y

Demand   Supply

7
6
5
4
3
2
1

Quantity of item (10,000s)

1 2 3 4 5 6 7 8 9 10 11 12   *x*

Price of item ($)

79. DAILY TRACKING POLLS   Use the graph on the next page to answer the following.

a. Which political candidate was ahead on October 28 and by how much?

b. On what day did the challenger pull even with the incumbent?

c. If the election was held November 4, who did the poll predict would win, and by how many percentage points?

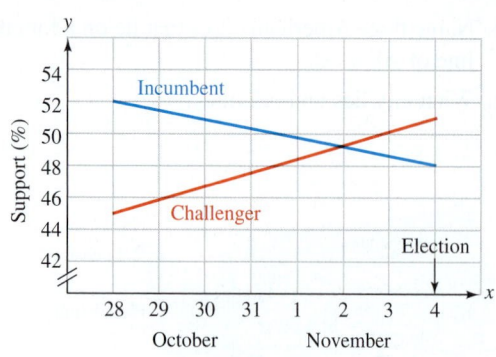

80. AIR TRAFFIC CONTROL The equations describing the paths of two airplanes are $y = -\frac{1}{2}x + 3$ and $3y = 2x + 2$. Graph each equation on the radar screen shown. Is there a possibility of a midair collision? If so, where?

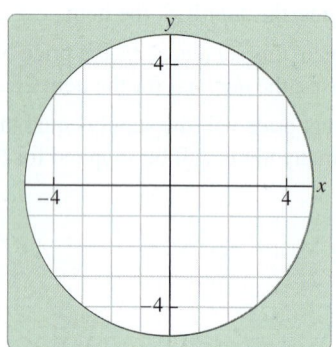

81. TV COVERAGE
A television camera is located at $(-2, 0)$ and will follow the launch of a space shuttle, as shown in the graph. (Each unit in the illustration is 1 mile.) As the shuttle rises vertically on a path described by $x = 2$, the farthest the camera can tilt back is a line of sight given by $y = \frac{5}{2}x + 5$. For how many miles of the shuttle's flight will it be in view of the camera?

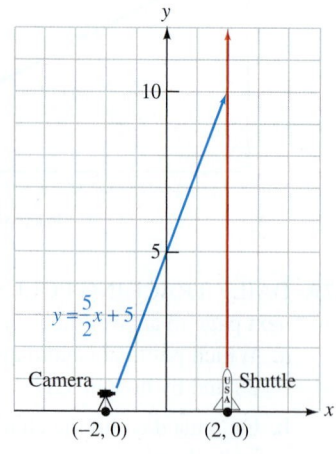

WRITING

82. Explain why it is difficult to determine the solution of the system in the graph.

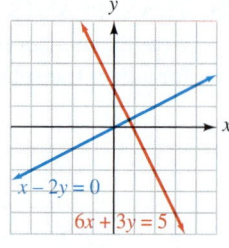

83. Without graphing, how can you tell that the graphs of $y = 2x + 1$ and $y = 2x + 2$ do not intersect?

84. Could a system of two linear equations have exactly two solutions? Explain why or why not.

85. What is an inconsistent system?

86. What are dependent equations?

REVIEW Perform the operations and simplify, if possible.

87. $\dfrac{x + 3}{x^2} + \dfrac{x + 5}{2x}$

88. $\dfrac{x^2 - 4}{3x + 6} \div \dfrac{x - 2}{x + 2}$

89. $\dfrac{z^2 + 4z - 5}{5z - 5} \cdot \dfrac{5z}{z + 5}$

90. $\dfrac{6x - 5}{3xy} - \dfrac{3x - 5}{3xy}$

CHALLENGE PROBLEMS

91. Can a system of two linear equations in two variables be inconsistent but have dependent equations? Explain.

92. Construct a system of two linear equations that has a solution of $(-2, 6)$.

93. Write a system of two linear equations such that $(2, 3)$ is a solution of the first equation but is not a solution of the second equation.

94. Solve by graphing: $\begin{cases} \dfrac{1}{3}x - \dfrac{1}{2}y = \dfrac{1}{6} \\ \dfrac{2x}{5} + \dfrac{y}{2} = \dfrac{13}{10} \end{cases}$.

7.2 Solving Systems of Equations by Substitution

- The Substitution Method   • Finding a Substitution Equation
- Inconsistent Systems and Dependent Equations

When solving a system of equations by graphing, it is often difficult to determine the coordinates of the intersection point. For example, a solution $\left(\frac{7}{8}, \frac{3}{5}\right)$ is virtually impossible to identify. In this section, we will discuss a second, more precise method for solving systems.

■ THE SUBSTITUTION METHOD

EXAMPLE 1

ELEMENTARY &
INTERMEDIATE
Algebra $f(x)$ **Now**™

Solve the system: $\begin{cases} y = 3x - 2 \\ 2x + y = 8 \end{cases}$.

Solution   Note that the first equation, $y = 3x - 2$, is solved for y. Because y and $3x - 2$ represent the same value, we can substitute $3x - 2$ for y in the second equation. We call $y = 3x - 2$ the *substitution equation*.

$$\begin{cases} y = \boxed{3x - 2} \\ 2x + y = 8 \end{cases}$$

To find the solution of the system, we proceed as follows:

$2x + y = 8$ This is the second equation of the system.

$2x + 3x - 2 = 8$ Substitute $3x - 2$ for y.

The resulting equation has only one variable and can be solved for x.

$2x + 3x - 2 = 8$

$5x - 2 = 8$ Combine like terms.

$5x = 10$ Add 2 to both sides.

$x = 2$ Divide both sides by 5.

After finding x, we can find y by substituting 2 for x in either equation of the original system. Because the substitution equation, $y = 3x - 2$, is already solved for y, it is easy to substitute into this equation.

$y = 3x - 2$ Use the substitution equation to find y.

$y = 3(2) - 2$ Substitute 2 for x.

$y = 6 - 2$

$y = 4$ This is the y-value of the solution.

The ordered pair $(2, 4)$ appears to be the solution of the system. To check, we substitute 2 for x and 4 for y in each equation.

Caution

When using the substitution method, a common error is to find the value of one of the variables, say x, and forget to find the value of the other. Remember that a solution of a system is an ordered pair (x, y).

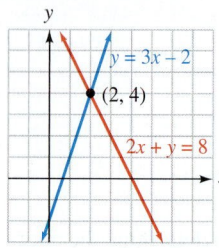

$y = 3x - 2$

(2, 4)

$2x + y = 8$

Check: $y = 3x - 2$ $2x + y = 8$

$4 \stackrel{?}{=} 3(2) - 2$ $2(2) + 4 \stackrel{?}{=} 8$

$4 \stackrel{?}{=} 6 - 2$ $4 + 4 \stackrel{?}{=} 8$

$4 = 4$ $8 = 8$

Since (2, 4) satisfies both equations, it is the solution. A graph of the equations of the system shows an intersection point of (2, 4). This illustrates an important fact: *The solution found using the substitution method will be the same as the solution found using the graphing method.*

Self Check 1   Solve the system: $\begin{cases} x + 4y = 7 \\ x = 6y - 3 \end{cases}$.

To solve a system of equations in x and y by the substitution method, follow these steps.

The Substitution Method

1. Solve one of the equations for either x or y. If this is already done, go to step 2.
2. Substitute the expression for x or for y obtained in step 1 into the other equation and solve that equation.
3. Substitute the value of the variable found in step 2 into the substitution equation to find the value of the remaining variable.
4. Check the proposed solution in the equations of the original system. Write the solution as an ordered pair.

EXAMPLE 2   Solve the system: $\begin{cases} 4x + 27 = 7y \\ x = -5y \end{cases}$.

Solution   ***Step 1:*** Note that the second equation, $x = -5y$, is solved for x. Because x and $-5y$ represent the same value, we can substitute $-5y$ for x in the first equation.

Success Tip

With this method, the basic objective is to use an appropriate substitution to obtain one equation in one variable.

$\begin{cases} 4x + 27 = 7y \\ x = \boxed{-5y} \end{cases}$

Step 2: When we substitute for x in the first equation, the resulting equation contains only one variable and can be solved for y.

$4x + 27 = 7y$ This is the first equation of the system.

$4(-5y) + 27 = 7y$ Substitute $-5y$ for x.

$-20y + 27 = 7y$ Do the multiplication.

$27 = 27y$ Add $20y$ to both sides.

$1 = y$ Divide both sides by 27.

The Language of Algebra

The phrase *back-substitute* can also be used to describe step 3 of the substitution method. In Example 2, to find x, we *back-substitute* 1 for y in the equation $x = -5y$.

Step 3: To find x, substitute 1 for y in the equation $x = -5y$.

$x = -5y$

$x = -5(1)$ Substitute 1 for y.

$x = -5$

Step 4: The check below verifies that the solution is $(-5, 1)$.

Check:

$$4x + 27 = 7y \qquad\qquad x = -5y$$
$$4(-5) + 27 \overset{?}{=} 7(1) \qquad -5 \overset{?}{=} -5(1)$$
$$-20 + 27 \overset{?}{=} 7 \qquad\qquad -5 = -5$$
$$7 = 7$$

Self Check 2 Solve the system: $\begin{cases} 3x + 40 = 8y \\ x = -4y \end{cases}$.

■ FINDING A SUBSTITUTION EQUATION

Sometimes neither equation of a system is solved for a variable. In such cases, we can find a substitution equation by solving one of the equations for one of its variables.

EXAMPLE 3

ELEMENTARY &
INTERMEDIATE
Algebra *f(x)* **Now**™

Solve the system: $\begin{cases} 4x + y = 3 \\ 3x + 5y = 15 \end{cases}$.

Solution

Step 1: Since the system does not contain an equation solved for x or y, we must choose an equation and solve it for x or y. It is easiest to solve for y in the first equation, because it has a coefficient of 1.

$$4x + y = 3$$
$$y = 3 - 4x \qquad \text{Subtract } 4x \text{ from both sides to isolate } y.$$

Success Tip

To find a substitution equation, solve one of the equations of the system for one of its variables. If possible, solve for a variable whose coefficient is 1 or −1 to avoid working with fractions.

Because y and $3 - 4x$ represent the same value, we can substitute $3 - 4x$ for y in the second equation of the system.

$$y = \boxed{3 - 4x} \qquad 3x + 5y = 15$$

Step 2: When we substitute for y in the second equation, the resulting equation contains only one variable and can be solved for x.

$$3x + 5y = 15$$
$$3x + 5(3 - 4x) = 15 \qquad \text{Substitute } 3 - 4x \text{ for } y.$$
$$3x + 15 - 20x = 15 \qquad \text{Distribute the multiplication by 5.}$$
$$15 - 17x = 15 \qquad \text{Combine like terms.}$$
$$-17x = 0 \qquad \text{Subtract 15 from both sides.}$$
$$x = 0 \qquad \text{Divide both sides by } -17.$$

Caution

In Example 3, use parentheses when substituting $3 - 4x$ for y so that the multiplication by 5 is distributed over both terms of $3 - 4x$.

$$3x + 5(3 - 4x) = 15$$

Step 3: To find y, substitute 0 for x in the equation $y = 3 - 4x$.

$$y = 3 - 4x$$
$$y = 3 - 4(0) \qquad \text{Substitute 0 for } x.$$
$$y = 3 - 0$$
$$y = 3$$

Step 4: The solution appears to be (0, 3). Check it in the original equations.

Check:

$$4x + y = 3 \qquad\qquad 3x + 5y = 15$$
$$4(0) + 3 \stackrel{?}{=} 3 \qquad\qquad 3(0) + 5(3) \stackrel{?}{=} 15$$
$$0 + 3 \stackrel{?}{=} 3 \qquad\qquad 0 + 15 \stackrel{?}{=} 15$$
$$3 = 3 \qquad\qquad\qquad 15 = 15$$

Self Check 3 Solve the system: $\begin{cases} 2x - 3y = 10 \\ 3x + y = 15 \end{cases}$.

EXAMPLE 4

ELEMENTARY &
INTERMEDIATE
Algebra (f(x)) Now™

Solve the system: $\begin{cases} 3a - 3b = 5 \\ 3 - a = -2b \end{cases}$.

Solution

Step 1: Since the coefficient of a in the second equation is -1, solve that equation for a.

$$3 - a = -2b$$
$$-a = -2b - 3 \qquad \text{Subtract 3 from both sides.}$$

To obtain a on the left-hand side, multiply both sides of the equation by -1.

$$-1(-a) = -1(-2b - 3) \qquad \text{Multiply both sides by } -1.$$
$$a = 2b + 3 \qquad\qquad \text{Do the multiplications.}$$

Because a and $2b + 3$ represent the same value, we can substitute $2b + 3$ for a in the first equation.

Caution

In Example 4, do not substitute $2b + 3$ into the equation from which it came, $3 - a = -2b$. Substitute it into the other equation of the system. This common error leads to an identity.

$$3 - a = -2b$$
$$3 - (2b + 3) = -2b$$
$$3 - 2b - 3 = -2b$$
$$0 = 0$$

$$3a - 3b = 5 \qquad a = \boxed{2b + 3}$$

Step 2: Substitute $2b + 3$ for a in the first equation and solve for b.

$$3a - 3b = 5$$
$$3(2b + 3) - 3b = 5 \qquad \text{Substitute } 2b + 3 \text{ for } a.$$
$$6b + 9 - 3b = 5 \qquad \text{Distribute the multiplication by 3.}$$
$$3b + 9 = 5 \qquad \text{Combine like terms.}$$
$$3b = -4 \qquad \text{Subtract 9 from both sides.}$$
$$b = -\frac{4}{3} \qquad \text{Divide both sides by 3.}$$

Step 3: To find a, substitute $-\frac{4}{3}$ for b in the equation $a = 2b + 3$.

Notation

Unless told otherwise, list the values of the variables of a solution in *alphabetical* order. For example, if the equations of a system involve the variables a and b, write the solution in the form (a, b).

$$a = 2b + 3$$
$$a = 2\left(-\frac{4}{3}\right) + 3 \qquad \text{Substitute } -\frac{4}{3} \text{ for } b.$$
$$a = -\frac{8}{3} + \frac{9}{3} \qquad \text{Do the multiplication.}$$
$$a = \frac{1}{3} \qquad \text{Add.}$$

Step 4: The solution is $\left(\frac{1}{3}, -\frac{4}{3}\right)$. Check it in the original equations.

Self Check 4   Solve the system: $\begin{cases} 2s - t = 4 \\ 3s - 5t = 2 \end{cases}$.

It is often helpful to clear any equations of fractions and combine any like terms before performing a substitution.

EXAMPLE 5

ELEMENTARY &
INTERMEDIATE

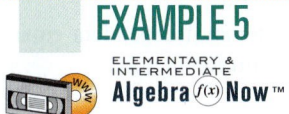
Algebra $f(x)$ **Now**™

Solve the system: $\begin{cases} \dfrac{y}{4} = -\dfrac{x}{2} - \dfrac{3}{4} \\ 2x - y = -1 + y - x \end{cases}$.

Solution   We can clear the first equation of fractions by multiplying both sides by the LCD.

$$\frac{y}{4} = -\frac{x}{2} - \frac{3}{4}$$

$$\mathbf{4}\left(\frac{y}{4}\right) = \mathbf{4}\left(-\frac{x}{2} - \frac{3}{4}\right) \qquad \text{Multiply both sides by 4.}$$

(1) $\qquad\qquad\quad y = -2x - 3 \qquad\qquad \text{Distribute the multiplication by 4.}$

We can write the second equation of the system in general form ($Ax + By = C$) by adding x and subtracting y from both sides.

$$2x - y = -1 + y - x$$

$$2x - y \mathbf{+ x - y} = -1 + y - x \mathbf{+ x - y}$$

(2) $\qquad\qquad\quad 3x - 2y = -1 \qquad\qquad\qquad \text{Combine like terms.}$

Step 1: Equations 1 and 2 form a system, which has the same solution as the original one. To solve this system, we proceed as follows:

(1)
(2) $\qquad \begin{cases} y = \boxed{-2x - 3} \\ 3x - 2y = -1 \end{cases}$

Step 2: To find x, substitute $-2x - 3$ for y in equation 2 and proceed as follows:

$$\begin{aligned} 3x - 2y &= -1 & \text{This is equation 2.} \\ 3x - 2(\mathbf{-2x - 3}) &= -1 & \text{Substitute } -2x - 3 \text{ for } y. \\ 3x + 4x + 6 &= -1 & \text{Distribute the multiplication by } -2. \\ 7x + 6 &= -1 & \text{Combine like terms.} \\ 7x &= -7 & \text{Subtract 6 from both sides.} \\ x &= -1 & \text{Divide both sides by 7.} \end{aligned}$$

Caution

Always use the original equations when checking a solution. Do not use a substitution equation or an equivalent equation that you found algebraically. If an error was made, a proposed solution that would not satisfy the original system might appear to be correct.

Step 3: To find y, we substitute -1 for x in equation 1.

$$\begin{aligned} y &= -2\mathbf{x} - 3 & \text{This is equation 1.} \\ y &= -2(\mathbf{-1}) - 3 & \text{Substitute } -1 \text{ for } x. \\ y &= 2 - 3 & \text{Do the multiplication.} \\ y &= -1 \end{aligned}$$

Step 4: The solution is $(-1, -1)$. Check it in the original system.

Self Check 5 Solve the system: $\begin{cases} \dfrac{y}{6} = \dfrac{x}{3} + \dfrac{1}{2} \\ 2x - y = -3 + y - x \end{cases}$.

■ INCONSISTENT SYSTEMS AND DEPENDENT EQUATIONS

In the previous section, we solved inconsistent systems and systems of dependent equations graphically. We can also solve these systems using the substitution method.

EXAMPLE 6

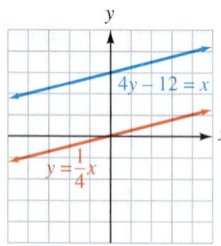

Solution

Solve the system: $\begin{cases} 4y - 12 = x \\ y = \dfrac{1}{4}x \end{cases}$.

To solve this system, substitute $\dfrac{1}{4}x$ for y in the first equation and solve for x.

$$4\mathbf{y} - 12 = x$$
$$4\left(\mathbf{\dfrac{1}{4}x}\right) - 12 = x \qquad \text{Substitute } \dfrac{1}{4}x \text{ for } y.$$
$$x - 12 = x \qquad \text{Do the multiplication.}$$
$$x - 12 \mathbf{- x} = x \mathbf{- x} \qquad \text{Subtract } x \text{ from both sides.}$$
$$-12 = 0$$

Here, the terms involving x drop out, and we get $-12 = 0$. This false statement indicates that the system has no solution, and is inconsistent. The graphs of the equations of the system verify this; they are parallel lines.

Self Check 6 Solve the system: $\begin{cases} x - 4 = y \\ -2y = 4 - 2x \end{cases}$.

EXAMPLE 7

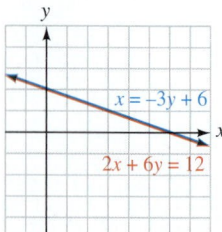

Solution

Solve the system: $\begin{cases} x = -3y + 6 \\ 2x + 6y = 12 \end{cases}$.

To solve this system, substitute $-3y + 6$ for x in the second equation and solve for y.

$$2\mathbf{x} + 6y = 12$$
$$2(\mathbf{-3y + 6}) + 6y = 12 \qquad \text{Substitute } -3y + 6 \text{ for } x.$$
$$-6y + 12 + 6y = 12 \qquad \text{Distribute the multiplication by 2.}$$
$$12 = 12$$

Here, the terms involving y drop out, and we get $12 = 12$. This true statement indicates that the two equations of the system are equivalent. Therefore, they are dependent equations and the system has infinitely many solutions. The graphs of the equations verify this; they are the same line.

Any ordered pair that satisfies one equation of this system also satisfies the other. To find some solutions, we can substitute some values of x, say 0, 3, and 6, in either equation and solve for y. The results are: $(0, 2)$, $(3, 1)$, and $(6, 0)$.

Self Check 7 Solve the system: $\begin{cases} y = 2 - x \\ 3x + 3y = 6 \end{cases}$.

Answers to Self Checks **1.** $(3, 1)$ **2.** $(-8, 2)$ **3.** $(5, 0)$ **4.** $\left(\frac{18}{7}, \frac{8}{7}\right)$ **5.** $(-3, -3)$ **6.** no solution
7. infinitely many solutions

7.2   STUDY SET   ELEMENTARY & INTERMEDIATE Algebra *f(x)* Now™

VOCABULARY   **Fill in the blanks.**

1. $\begin{cases} y = x + 3 \\ 3x - y = -1 \end{cases}$ is called a _____ of equations.

2. The ordered pair $(1, 4)$ is a _____ of the system in Problem 1 because it satisfies both equations.

3. When checking a proposed solution of a system of equations, always use the _____ equations.

4. We say that the equation $y = 2x + 4$ is solved for ___.

5. In mathematics, "to _____" means to replace an expression with one that has the same value.

6. In $x - y = 4$, the _____ of x is 1 and the coefficient of y is ___.

7. With the substitution method, the basic objective is to use an appropriate substitution to obtain one equation in ___ variable.

8. To _____ $\frac{x}{2} + \frac{y}{3} = 8$ of fractions, multiply both sides by the LCD, which is 6.

9. When the graphs of the equations of a system are identical lines, the equations are called dependent and the system has _____ many solutions.

10. A system that has no solution is called an _____ system.

CONCEPTS

11. Suppose the substitution method will be used to solve each system. Which equation should be used as the substitution equation?

a. $\begin{cases} 5x + y = 2 \\ y = -3x \end{cases}$ **b.** $\begin{cases} x = 2y + 1 \\ 7y - 3x = 1 \end{cases}$

12. Multiply both sides of the equation $-x = 4y - 15$ by -1.

13. Suppose the substitution method will be used to solve each system. Find a substitution equation by solving one of the equations for one of the variables.

a. $\begin{cases} x - 2y = 2 \\ 2x + 3y = 11 \end{cases}$ **b.** $\begin{cases} 2x - 3y = 2 \\ 2x - y = 11 \end{cases}$

14. Is $(6, 2)$ a solution of the system $\begin{cases} x = 3y \\ x - y = 9 \end{cases}$.

15. Suppose $x - 4$ is substituted for y in the equation $x + 3y = 8$. In the following equation, insert parentheses where they are needed.

$$x + 3x - 4 = 8$$

16. A student uses the substitution method to solve the system $\begin{cases} 4a + 5b = 2 \\ b = 3a - 11 \end{cases}$. She finds that a is 3. What is the easiest way for her to determine the value of b?

17. a. What is the LCD of the fractions in $\frac{x}{5} + \frac{2y}{3} = 1$?

b. Clear the equation of fractions.

18. Write $2x + y = x - 5y + 3$ in the form $Ax + By = C$.

19. Suppose the equation $-2 = 1$ is obtained when a system is solved by the substitution method.

a. Does the system have a solution?

b. Which of the following is a possible graph of the system?

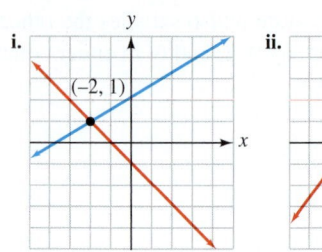

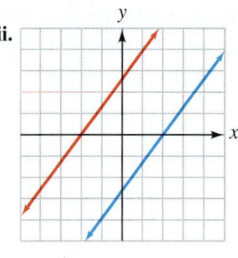

20. Suppose the equation $2 = 2$ is obtained when a system is solved by the substitution method.

 a. Does the system have a solution?

 b. Which graph is a possible graph of the system?

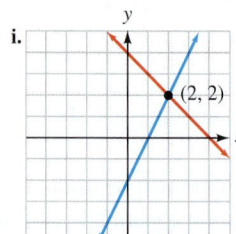

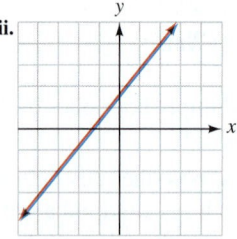

NOTATION   Complete the solution of the system.

21. Solve: $\begin{cases} y = 3x \\ x - y = 4 \end{cases}$

$$x - y = 4 \qquad \text{This is the second equation.}$$
$$x - \left(\right) = 4$$
$$-2x = $$
$$x = $$

$$y = 3x \qquad \text{This is the first equation.}$$
$$y = 3\left(\right)$$
$$y = $$

The solution is $\left(, \right)$.

22. The system $\begin{cases} a = 3b + 2 \\ a + 3b = 8 \end{cases}$ was solved and it was found that b is 1 and a is 5. Write the solution as an ordered pair.

PRACTICE   Use the substitution method to solve each system. If a system has no solution or infinitely many, so state.

23. $\begin{cases} y = 2x \\ x + y = 6 \end{cases}$

24. $\begin{cases} y = 3x \\ x + y = 4 \end{cases}$

25. $\begin{cases} y = 2x - 6 \\ 2x + y = 6 \end{cases}$

26. $\begin{cases} y = 2x - 9 \\ x + 3y = 8 \end{cases}$

27. $\begin{cases} y = 2x + 5 \\ x + 2y = -5 \end{cases}$

28. $\begin{cases} y = -2x \\ 3x + 2y = -1 \end{cases}$

29. $\begin{cases} 3x + y = -4 \\ x = y \end{cases}$

30. $\begin{cases} x + 2y = -6 \\ x = y \end{cases}$

31. $\begin{cases} 2a + 4b = -24 \\ a = 20 - 2b \end{cases}$

32. $\begin{cases} 3a + 6b = -15 \\ a = -2b - 5 \end{cases}$

33. $\begin{cases} 2a - 3b = -13 \\ -b = -2a - 7 \end{cases}$

34. $\begin{cases} a - 3b = -1 \\ -b = -2a - 2 \end{cases}$

35. $\begin{cases} -y = 11 - 3x \\ 2x + 5y = -4 \end{cases}$

36. $\begin{cases} -x = 10 - 3y \\ 2x + 8y = -6 \end{cases}$

37. $\begin{cases} 2x + 3 = -4y \\ x - 6 = -8y \end{cases}$

38. $\begin{cases} 5y + 2 = -4x \\ x + 2y = -2 \end{cases}$

39. $\begin{cases} r + 3s = 9 \\ 3r + 2s = 13 \end{cases}$

40. $\begin{cases} x - 2y = 2 \\ 2x + 3y = 11 \end{cases}$

41. $\begin{cases} 6x - 3y = 5 \\ 2y + x = 0 \end{cases}$

42. $\begin{cases} 5s + 10t = 3 \\ 2s + t = 0 \end{cases}$

43. $\begin{cases} y - 3x = -5 \\ 21x = 7y + 35 \end{cases}$

44. $\begin{cases} 8y = 15 - 4x \\ x + 2y = 4 \end{cases}$

45. $\begin{cases} y = 3x + 6 \\ y = -2x - 4 \end{cases}$

46. $\begin{cases} y = x + 3 \\ y = -2x - 3 \end{cases}$

47. $\begin{cases} x = \dfrac{1}{3}y - 1 \\ x = y + 1 \end{cases}$

48. $\begin{cases} x = \dfrac{1}{2}y + 2 \\ x = y + 1 \end{cases}$

49. $\begin{cases} 4x + 5y = 2 \\ 3x - y = 11 \end{cases}$

50. $\begin{cases} 5u + 3v = 5 \\ 4u - v = 4 \end{cases}$

51. $\begin{cases} 3x + 4y = -7 \\ 2y - x = -1 \end{cases}$

52. $\begin{cases} 5x - 2y = -7 \\ 5 - y = -3x \end{cases}$

53. $\begin{cases} 6 - y = 4x \\ 2y = -8x - 20 \end{cases}$

54. $\begin{cases} 9x = 3y + 12 \\ 4 = 3x - y \end{cases}$

55. $\begin{cases} b = \dfrac{2}{3}a \\ 8a - 3b = 3 \end{cases}$

56. $\begin{cases} a = \dfrac{2}{3}b \\ 9a + 4b = 5 \end{cases}$

57. $\begin{cases} 2x + 5y = -2 \\ -\dfrac{x}{2} = y \end{cases}$

58. $\begin{cases} y = -\dfrac{x}{2} \\ 2x - 3y = -7 \end{cases}$

59. $\begin{cases} \dfrac{x}{2} + \dfrac{y}{2} = -1 \\ \dfrac{x}{3} - \dfrac{y}{2} = -4 \end{cases}$

60. $\begin{cases} \dfrac{2a}{3} + \dfrac{b}{5} = 1 \\ \dfrac{a}{3} - \dfrac{2b}{3} = \dfrac{13}{3} \end{cases}$

61. $\begin{cases} 5x = \dfrac{1}{2}y - 1 \\ \dfrac{1}{4}y = 10x - 1 \end{cases}$

62. $\begin{cases} \dfrac{x}{4} + y = \dfrac{1}{4} \\ \dfrac{y}{2} + \dfrac{11}{20} = \dfrac{x}{10} \end{cases}$

63. $\begin{cases} x - \dfrac{4y}{5} = 4 \\ \dfrac{y}{3} = \dfrac{x}{2} - \dfrac{5}{2} \end{cases}$

64. $\begin{cases} 3x - 2y = \dfrac{9}{2} \\ \dfrac{x}{2} - \dfrac{3}{4} = 2y \end{cases}$

65. $\begin{cases} y + x = 2x + 2 \\ 6x - 4y = 21 - y \end{cases}$

66. $\begin{cases} y - x = 3x \\ 2x + 2y = 14 - y \end{cases}$

67. $\begin{cases} x = -3y + 6 \\ 2x + 4y = 6 + x + y \end{cases}$

68. $\begin{cases} 2x - y = x + y \\ -2x + 4y = 6 \end{cases}$

69. $\begin{cases} 4x + 5y + 1 = -12 + 2x \\ x - 3y + 2 = -3 - x \end{cases}$

70. $\begin{cases} 6x + y = -8 + 3x - y \\ 3x - y = 2y + x - 1 \end{cases}$

71. $\begin{cases} 3(x - 1) + 3 = 8 + 2y \\ 2(x + 1) = 8 + y \end{cases}$

72. $\begin{cases} 4(x - 2) = 19 - 5y \\ 3(x - 2) - 2y = -y \end{cases}$

APPLICATIONS

73. DINING   When a customer orders the $5.95 Country Breakfast from the menu, he wants to substitute something from the a la carte menu in place

of the hash browns. What items can he pick so that the breakfast doesn't increase in price? Why?

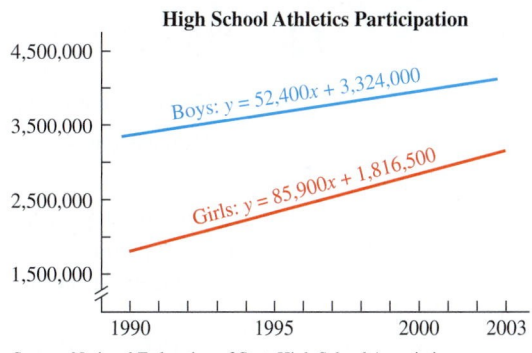

Village Vault Restaurant			
Country Breakfast			**$5.95**
2 eggs, 3 pancakes, bacon, sausage, hash browns, coffee			
A la Carte Menu–Single Servings			
Strawberries	$1.25	Melon	$0.95
Croissant	$1.70	Orange juice	$1.65
Hash browns	$0.95	Oatmeal	$1.95
Muffin	$1.30	Ham	$1.80

74. HIGH SCHOOL SPORTS   The equations shown model the number of boys and girls taking part in high school sports. In both models, x is the number of years since 1990, and y is the number of participants. If the trends continue, the graphs will intersect. Use the substitution method to predict the year when the number of boys and girls participating in high school sports will be the same.

High School Athletics Participation

Boys: $y = 52,400x + 3,324,000$

Girls: $y = 85,900x + 1,816,500$

Source: National Federation of State High School Associations

WRITING

75. What concept does this diagram illustrate?

$$\begin{cases} 6x + 5y = 11 \\ y = 3x - 2 \end{cases}$$

76. When using the substitution method, how can you tell whether

a. a system of linear equations has no solution?

b. a system of linear equations has infinitely many solutions?

77. When solving a system, what advantages are there with the substitution method compared to the graphing method?

78. Consider the equation $5x + y = 12$. Explain why it is easier to solve for y than it is for x.

REVIEW **Use a check to decide whether 3 is a solution of the given equation.**

79. $3x - 8 = 1$

80. $9y^2 - 81 = 0$

81. $3(x + 8) + 5x = 2(12 + 4x)$

82. $\dfrac{7}{x + 4} - \dfrac{1}{2} = \dfrac{3}{x + 4}$

83. $x^3 + 7x^2 = x^2 - 9x$

84. $\dfrac{11(x - 12)}{2} = 9 - 2x$

CHALLENGE PROBLEMS **Use the substitution method to solve each system.**

85. $\begin{cases} \dfrac{6x - 1}{3} - \dfrac{5}{3} = \dfrac{3y + 1}{2} \\ \dfrac{1 + 5y}{4} + \dfrac{x + 3}{4} = \dfrac{17}{2} \end{cases}$

86. $\begin{cases} 0.5x + 0.5y = 6 \\ 0.001x - 0.001y = -0.004 \end{cases}$

87. The system $\begin{cases} \dfrac{1}{2}x = y + 3 \\ x - 2y = 6 \end{cases}$ has infinitely many solutions. Find three of them.

88. Could the substitution method be used to solve the following system?

$$\begin{cases} y = -2 \\ x = 5 \end{cases}$$

How would you solve it?

....

7.3 **Solving Systems of Equations by Elimination**

- The Elimination Method
- Using Multiplication to Eliminate a Variable
- Inconsistent Systems and Dependent Equations

In the first step of the substitution method for solving a system of equations, we solve one of the equations for one of the variables. At times, this can be difficult, especially if neither variable has a coefficient of 1 or -1. This is the case for the system

$$\begin{cases} 2x + 5y = 11 \\ 7x - 5y = 16 \end{cases}$$

Solving either equation for x or y involves working with cumbersome fractions. Fortunately, we can solve systems like this one using a simpler algebraic method called the *elimination* or the *addition method*.

■ **THE ELIMINATION METHOD**

The elimination method for solving a system is based on the **addition property of equality:** *When equal quantities are added to both sides of an equation, the results are equal.* In symbols, if $A = B$ and $C = D$, then adding the left-hand sides and the right-hand sides of these equations, we have $A + C = B + D$. This procedure is called *adding the equations*.

The Language of Algebra

To *eliminate* means to remove. People allergic to peanuts need to *eliminate* any foods containing peanuts from their diets.

Add the terms on the left-hand sides.

$$\begin{array}{c} A = B \\ C = D \\ \hline A + C = B + D \end{array}$$

Add the terms on the right-hand sides.

EXAMPLE 1   Solve the system: $\begin{cases} 2x + 5y = 11 \\ 6x - 5y = 13 \end{cases}$.

ELEMENTARY &
INTERMEDIATE
Algebra $f(x)$ **Now**™   **Solution**   Since $6x - 5y$ and 13 are equal quantities, we can add $6x - 5y$ to the left-hand side and 13 to the right-hand side of the first equation, $2x + 5y = 11$.

$$\begin{array}{l} 2x + 5y = 11 \\ \underline{6x - 5y = 13} \qquad \text{To add the equations, add the like terms, column by column:} \\ 8x \qquad\quad = 24 \end{array}$$

$$11 + 13 = 24$$
$$5y + (-5y) = 0$$
$$2x + 6x = 8x$$

Because the sum of the terms $5y$ and $-5y$ is 0, we say that the variable y has been eliminated. Since the resulting equation has only one variable, we can solve it for x.

$$8x = 24$$
$$x = 3 \qquad \text{Divide both sides by 8.}$$

To find the y-value of the solution, substitute 3 for x in either equation of the system.

$$\begin{array}{ll} 2x + 5y = 11 & \text{This is the first equation of the system.} \\ 2(3) + 5y = 11 & \text{Substitute 3 for } x. \\ 6 + 5y = 11 & \text{Multiply.} \\ 5y = 5 & \text{Subtract 6 from both sides.} \\ y = 1 & \text{Divide both sides by 5.} \end{array}$$

Now we check the proposed solution $(3, 1)$ in the equations of the original system.

Check: $\begin{array}{ll} 2x + 5y = 11 & 6x - 5y = 13 \\ 2(3) + 5(1) \stackrel{?}{=} 11 & 6(3) - 5(1) \stackrel{?}{=} 13 \\ 6 + 5 \stackrel{?}{=} 11 & 18 - 5 \stackrel{?}{=} 13 \\ 11 = 11 & 13 = 13 \end{array}$

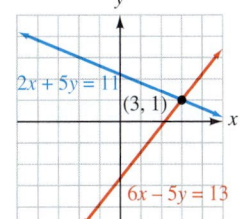

Since $(3, 1)$ satisfies both equations, it is the solution. The graph on the left verifies this.

Self Check 1   Solve the system: $\begin{cases} -4x + 3y = 4 \\ 4x + 5y = 28 \end{cases}$.

To solve a system of equations in x and y by the elimination method, follow these steps.

The Elimination Method   **1.** Write both equations in general form: $Ax + By = C$.
2. If necessary, multiply one or both of the equations by nonzero quantities to make the coefficients of x (or the coefficients of y) opposites.
3. Add the equations to eliminate the terms involving x (or y).
4. Solve the equation resulting from step 3.
5. Find the value of the other variable by substituting the solution found in step 4 into any equation containing both variables.
6. Check the proposed solution in the equations of the original system. Write the solution as an ordered pair.

■ USING MULTIPLICATION TO ELIMINATE A VARIABLE

The system in Example 1 was easy to solve using elimination because the coefficients of the terms $5y$ in the first equation and $-5y$ in the second equation were opposites. For many systems, however, we are not able to immediately eliminate a variable by adding. In such cases, we use the multiplication property of equality to create coefficients of x or y that are opposites.

EXAMPLE 2

ELEMENTARY &
INTERMEDIATE
Algebra $f(x)$ **Now**™

Solve the system: $\begin{cases} 2x + 7y = -18 \\ 2x + 3y = -10 \end{cases}$.

Solution

Step 1: We see that neither the coefficients of x nor the coefficients of y are opposites. Adding these equations as they are does not eliminate a variable.

Step 2: To eliminate x, we can multiply both sides of the second equation by -1. This creates the term $-2x$, whose coefficient is opposite that of the $2x$ term in the first equation.

$$\begin{cases} 2x + 7y = -18 \xrightarrow{\text{Unchanged}} \\ 2x + 3y = -10 \xrightarrow[\text{Multiply by } -1]{} \end{cases} \quad \begin{aligned} 2x + 7y &= -18 \\ -1(2x + 3y) &= -1(-10) \end{aligned} \xrightarrow[\text{Simplify}]{\text{Unchanged}} \begin{cases} 2x + 7y = -18 \\ -2x - 3y = 10 \end{cases}$$

Step 3: When the equations are added, x is eliminated.

$$\begin{array}{r} 2x + 7y = -18 \\ \underline{-2x - 3y = 10} \\ 4y = -8 \end{array} \quad \text{In the left column: } 2x + (-2x) = 0.$$

Step 4: Solve the resulting equation to find y.

$$\begin{aligned} 4y &= -8 \\ y &= -2 \quad \text{Divide both sides by 4.} \end{aligned}$$

Step 5: To find x, we can substitute -2 for y in either of the equations of the original system, or in $-2x - 3y = 10$. It appears the computations will be simplest if we use $2x + 3y = -10$.

$$\begin{aligned} 2x + 3y &= -10 \\ 2x + 3(-2) &= -10 \quad \text{Substitute } -2 \text{ for } y. \\ 2x - 6 &= -10 \quad \text{Multiply.} \\ 2x &= -4 \quad \text{Add 6 to both sides.} \\ x &= -2 \quad \text{Divide both sides by 2.} \end{aligned}$$

Step 6: The solution is $(-2, -2)$. Check this result in the original equations.

Self Check 2 Solve the system: $\begin{cases} x + 7y = -24 \\ 3x + 7y = -30 \end{cases}$.

EXAMPLE 3

ELEMENTARY &
INTERMEDIATE
Algebra $f(x)$ **Now**™

Solve the system: $\begin{cases} 7x + 2y - 14 = 0 \\ 9x = 4y - 28 \end{cases}$.

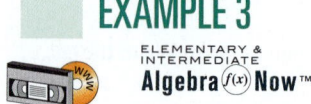

Solution **Step 1:** To compare coefficients, write each equation in the form $Ax + By = C$. Since each of the original equations will be written in an equivalent form, the resulting system will have the same solution as the original system.

$$\begin{cases} 7x + 2y = 14 \\ 9x - 4y = -28 \end{cases} \quad \begin{array}{l} \text{Add 14 to both sides of } 7x + 2y - 14 = 0. \\ \text{Subtract } 4y \text{ from both sides of } 9x = 4y - 28. \end{array}$$

Step 2: Neither the coefficients of x nor the coefficients of y are opposites. To eliminate y, we can multiply both sides of the first equation by 2. This creates the term $4y$, whose coefficient is opposite that of the $-4y$ term in the second equation.

$$\begin{cases} 7x + 2y = 14 \\ 9x - 4y = -28 \end{cases} \xrightarrow[\text{Unchanged}]{\text{Multiply by 2}} \begin{array}{l} 2(7x + 2y) = 2(14) \\ 9x - 4y = -28 \end{array} \xrightarrow[\text{Unchanged}]{\text{Simplify}} \begin{cases} 14x + 4y = 28 \\ 9x - 4y = -28 \end{cases}$$

Caution

When using the elimination method, don't forget to multiply both sides of an equation by the appropriate number. For instance, in Example 3:

Multiply both sides by 2

$$2(7x + 2y) = 2(14)$$

Step 3: When the equations are added, y is eliminated.

$$\begin{array}{r} 14x + 4y = 28 \\ \underline{9x - 4y = -28} \\ 23x \qquad = 0 \end{array} \quad \text{In the middle column, } 4y + (-4y) = 0.$$

Step 4: Since the result of the addition is an equation in one variable, we can solve for x.

$$23x = 0$$
$$x = 0 \quad \text{Divide both sides by 23.}$$

Step 5: To find y, we can substitute 0 for x in any equation that contains both variables. It appears the computations will be simplest if we use $7x + 2y = 14$.

$$\begin{aligned} 7x + 2y &= 14 \\ 7(0) + 2y &= 14 \quad \text{Substitute 0 for } x. \\ 0 + 2y &= 14 \quad \text{Multiply.} \\ 2y &= 14 \\ y &= 7 \quad \text{Divide both sides by 2.} \end{aligned}$$

Step 6: The solution is $(0, 7)$. Check this result in the original equations.

Self Check 3 Solve the system: $\begin{cases} 3x = 10 - 2y \\ 5x - 6y + 30 = 0 \end{cases}$.

EXAMPLE 4

Solve the system: $\begin{cases} 4a + 7b = -8 \\ 5a + 6b = 1 \end{cases}$.

Solution **Step 1:** Both equations are written in general form.

Step 2: In this example, we must rewrite both equations to obtain like terms that are opposites. To eliminate a, we can multiply the first equation by 5 to create the term $20a$, and we can multiply the second equation by -4 to create the term $-20a$.

$$\begin{cases} 4a + 7b = -8 \\ 5a + 6b = 1 \end{cases} \begin{array}{l} \xrightarrow{\text{Multiply by 5}} \\ \xrightarrow[\text{Multiply by } -4]{} \end{array} \begin{array}{l} 5(4a + 7b) = 5(-8) \\ -4(5a + 6b) = -4(1) \end{array} \begin{array}{l} \xrightarrow{\text{Simplify}} \\ \xrightarrow[\text{Simplify}]{} \end{array} \begin{cases} 20a + 35b = -40 \\ -20a - 24b = -4 \end{cases}$$

Step 3: When we add the resulting equations, a is eliminated.

$$\begin{array}{l} 20a + 35b = -40 \\ \underline{-20a - 24b = -4} \qquad \text{In the left column, } 20a + (-20a) = 0. \\ 11b = -44 \end{array}$$

Step 4: Solve the resulting equation for b.

$$11b = -44$$
$$b = -4 \qquad \text{Divide both sides by 11.}$$

Step 5: To find a, we can substitute -4 for b in any equation that contains both variables. It appears the computations will be simplest if we use $5a + 6b = 1$.

$$\begin{array}{ll} 5a + 6\boldsymbol{b} = 1 & \\ 5a + 6(\boldsymbol{-4}) = 1 & \text{Substitute } -4 \text{ for } b. \\ 5a - 24 = 1 & \text{Multiply.} \\ 5a = 25 & \text{Add 24 to both sides.} \\ a = 5 & \text{Divide both sides by 5. This is the } a\text{-value of the solution.} \end{array}$$

Step 6: Written in (a, b) form, the solution is $(5, -4)$. Check it in the original equations.

Self Check 4 Solve the system: $\begin{cases} 5a + 3b = -7 \\ 3a + 4b = 9 \end{cases}$.

EXAMPLE 5

Solve the system: $\begin{cases} \dfrac{1}{6}x + \dfrac{1}{2}y = \dfrac{1}{3} \\ -\dfrac{x}{9} + y = \dfrac{5}{9} \end{cases}$.

Solution *Step 1:* To clear the equations of the fractions, multiply both sides of the first equation by 6 and both sides of the second equation by 9.

$$\begin{cases} \dfrac{1}{6}x + \dfrac{1}{2}y = \dfrac{1}{3} \xrightarrow{\text{Multiply by 6}} 6\left(\dfrac{1}{6}x + \dfrac{1}{2}y\right) = 6\left(\dfrac{1}{3}\right) \xrightarrow{\text{Simplify}} \\ -\dfrac{x}{9} + y = \dfrac{5}{9} \xrightarrow{\text{Multiply by 9}} 9\left(-\dfrac{x}{9} + y\right) = 9\left(\dfrac{5}{9}\right) \xrightarrow{\text{Simplify}} \end{cases} \begin{cases} x + 3y = 2 \\ -x + 9y = 5 \end{cases}$$

Step 2: The coefficients of x are opposites.

Step 3: The variable x is eliminated when we add the resulting equations.

$$\begin{array}{l} x + 3y = 2 \\ \underline{-x + 9y = 5} \qquad \text{In the left column: } x + (-x) = 0. \\ 12y = 7 \end{array}$$

Step 4: Now solve to find y.

$$12y = 7$$
$$y = \dfrac{7}{12} \qquad \text{Divide both sides by 12.}$$

Step 5: We can find x by substituting $\frac{7}{12}$ for y in any equation containing both variables. However, that computation could be complicated, because $\frac{7}{12}$ is a fraction. Instead, we can begin again with the system that is cleared of fractions, but this time, eliminate y. If we multiply both sides of the first equation by -3, this creates the term $-9y$, whose coefficient is opposite that of the $9y$ term in the second equation.

$$\begin{cases} x + 3y = 2 \\ -x + 9y = 5 \end{cases} \xrightarrow[\text{Unchanged}]{\text{Multiply by } -3} \begin{matrix} -3(x + 3y) = -3(2) \\ -x + 9y = 5 \end{matrix} \xrightarrow[\text{Unchanged}]{\text{Simplify}} \begin{cases} -3x - 9y = -6 \\ -x + 9y = 5 \end{cases}$$

When we add the resulting equations, y is eliminated.

$$\begin{array}{r} -3x - 9y = -6 \\ \underline{-x + 9y = 5} \\ -4x = -1 \end{array}$$ In the middle column: $9y + (-9y) = 0$.

Now we solve the previous equation to find x.

$$-4x = -1$$
$$x = \frac{1}{4} \qquad \text{Divide both sides by } -4.$$

Step 6: The solution is $\left(\dfrac{1}{4}, \dfrac{7}{12}\right)$. To verify this, check it in the original equations.

Self Check 5 Solve the system: $\begin{cases} -\dfrac{1}{5}x + y = \dfrac{8}{5} \\ \dfrac{x}{8} + \dfrac{y}{2} = \dfrac{1}{4} \end{cases}$.

■ INCONSISTENT SYSTEMS AND DEPENDENT EQUATIONS

We have solved inconsistent systems and systems of dependent equations graphically and by substitution. We can also solve these systems using the elimination method.

EXAMPLE 6 Solve the system: $\begin{cases} 3x - 2y = 2 \\ -3x + 2y = -12 \end{cases}$.

Solution If we add the equations as they are, x is eliminated.

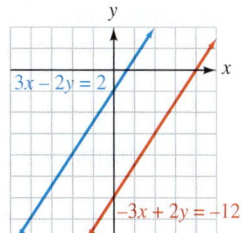

$$\begin{array}{r} 3x - 2y = 2 \\ \underline{-3x + 2y = -12} \\ 0 = -10 \end{array}$$ In the left column: $3x + (-3x) = 0$. In the middle column: $-2y + 2y = 0$.

In eliminating x, the variable y is eliminated as well. The resulting false statement, $0 = -10$, indicates that the system has no solution and is inconsistent. The graphs of the equations verify this; they are parallel lines.

Self Check 6 Solve the system: $\begin{cases} 2x - 7y = 5 \\ -2x + 7y = 3 \end{cases}$.

EXAMPLE 7

ELEMENTARY &
INTERMEDIATE

Algebra $f(x)$ **Now**™

Solve the system: $\begin{cases} \dfrac{2x - 5y}{15} = \dfrac{8}{15} \\ -0.2x + 0.5y = -0.8 \end{cases}$.

Solution   We can multiply both sides of the first equation by 15 to clear it of fractions and both sides of the second equation by 10 to clear it of decimals.

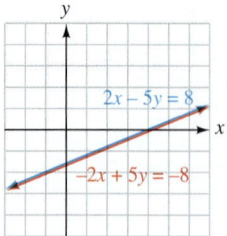

$$\begin{cases} \dfrac{2x - 5y}{15} = \dfrac{8}{15} \\ -0.2x + 0.5y = -0.8 \end{cases} \longrightarrow \begin{array}{c} 15\left(\dfrac{2x - 5y}{15}\right) = 15\left(\dfrac{8}{15}\right) \\ 10(-0.2x + 0.5y) = 10(-0.8) \end{array} \longrightarrow \begin{cases} 2x - 5y = 8 \\ -2x + 5y = -8 \end{cases}$$

We add the resulting equations to get

$$\begin{array}{rl} 2x - 5y = & 8 \\ \underline{-2x + 5y = -8} \\ 0 = & 0 \end{array}$$

In the left column: $2x + (-2x) = 0$.
In the middle column: $-5y + 5y = 0$.
In the right column: $8 + (-8) = 0$.

As in Example 6, both variables are eliminated. However, this time a true statement, $0 = 0$, is obtained. It indicates that the equations are dependent and that the system has infinitely many solutions. The graphs of the equations verify this; they are identical.

To find some solutions, we can substitute some values of x, say -1, 4, and 9, in either equation and solve for y. The results are: $(-1, -2)$, $(4, 0)$, and $(9, 2)$.

Self Check 7   Solve the system: $\begin{cases} \dfrac{3x + y}{6} = \dfrac{1}{3} \\ -0.3x - 0.1y = -0.2 \end{cases}$.

Answers to Self Checks   **1.** $(2, 4)$ **2.** $(-3, -3)$ **3.** $(0, 5)$ **4.** $(-5, 6)$ **5.** $\left(-\dfrac{22}{9}, \dfrac{10}{9}\right)$ **6.** no solution
7. infinitely many solutions

7.3   STUDY SET   ELEMENTARY & INTERMEDIATE   Algebra $f(x)$ Now™

VOCABULARY   Fill in the blanks.

1. The coefficients of $3x$ and $-3x$ are _____.

2. The general form of the equation of a line is

$$\boxed{} + By = \boxed{} .$$

3. When the following equations are added, the variable y will be _____.

$$\begin{array}{r} 5x - 6y = 10 \\ -3x + 6y = 24 \end{array}$$

4. By the _____ property of equality, we know that if we multiply both sides of $2x - 3y = 4$ by 6, we will obtain an equivalent equation.

5. The elimination method for solving a system is based on the _____ property of equality: *When equal quantities are added to both sides of an equation, the results are* _____.

6. The objective of the elimination method is to obtain two equations whose sum will be one equation in _____ variable.

CONCEPTS

7. In the following system, which terms have coefficients that are opposites?

$$\begin{cases} 3x + 7y = -25 \\ 4x - 7y = 12 \end{cases}$$

8. Add each pair of equations.

a. $8x + 5y = 11$
$\underline{-4x + y = -2}$

b. $2a + 2b = -6$
$\underline{3a - 2b = 2}$

c. $x - 3y = 15$
$\underline{-x - y = -14}$

d. $5x - y = 7$
$\underline{-5x + y = -7}$

9. a. Fill in the blank:

$$\overset{\text{Multiply by 3}}{4x + y = 2} \to \overset{\text{Simplify}}{\mathbf{3}(4x + y) = \mathbf{3}(2)} \to \rule{2cm}{0.4pt}$$

b. Fill in the blank:

$$\overset{\text{Multiply by } -2}{x - 3y = 4} \to \overset{\text{Simplify}}{\mathbf{-2}(x - 3y) = \mathbf{-2}(4)} \to \rule{2cm}{0.4pt}$$

10. If the elimination method is used to solve

$$\begin{cases} 3x + 12y = 4 \\ 6x - 4y = 7 \end{cases}$$

a. By what would we multiply the first equation to eliminate x?

b. By what would we multiply the second equation to eliminate y?

11. Suppose the following system is solved using the elimination method and it is found that x is 2. Find the value of y.

$$\begin{cases} 4x + 3y = 11 \\ 3x - 2y = 4 \end{cases}$$

12. What algebraic step should be performed to

a. Clear $\dfrac{2}{3}x + 4y = -\dfrac{4}{5}$ of fractions?

b. Clear $0.2x - 0.9y = 6.4$ of decimals?

13. Is $(1, -1)$ a solution of the system $\begin{cases} 2x + 3y = -1 \\ 3x + 5y = -2 \end{cases}$?

14. a. Suppose $0 = 0$ is obtained when a system is solved by the elimination method. Does the system have a solution? Which of the following is a possible graph of the system?

b. Suppose $0 = 2$ is obtained when a system is solved by the elimination method. Does the system have a solution? Which of the following is a possible graph of the system?

i 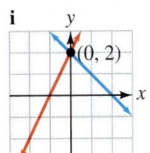   ii 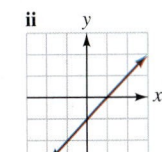   iii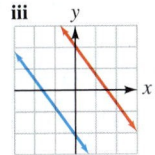

NOTATION   **Complete the solution.**

15. Solve: $\begin{cases} x + y = 5 \\ x - y = -3 \end{cases}$.

$$\begin{aligned} x + y &= 5 \\ \underline{x - y} &= \underline{-3} \\ \rule{1cm}{0.4pt} &= 2 \\ x &= \rule{1cm}{0.4pt} \end{aligned}$$

$x + y = 5$ This is the first equation.

$\rule{1cm}{0.4pt} + y = 5$

$y = 4$

The solution is $\rule{1cm}{0.4pt}$.

16. Write each equation of each system in general form: $Ax + By = C$.

a. $\begin{cases} 7x + y + 3 = 0 \\ 8x + 4 = -y \end{cases} \to \begin{cases} \rule{1.5cm}{0.4pt} \\ \rule{1.5cm}{0.4pt} \end{cases}$

b. $\begin{cases} x = 2y \\ 7 - 5y = 2x \end{cases} \to \begin{cases} \rule{1.5cm}{0.4pt} \\ \rule{1.5cm}{0.4pt} \end{cases}$

PRACTICE   **Use the elimination method to solve each system. If a system has no solution or infinitely many solutions, so indicate.**

17. $\begin{cases} x + y = 5 \\ x - y = 1 \end{cases}$

18. $\begin{cases} x - y = 4 \\ x + y = 8 \end{cases}$

19. $\begin{cases} x + y = 1 \\ x - y = 5 \end{cases}$

20. $\begin{cases} x - y = -5 \\ x + y = 1 \end{cases}$

21. $\begin{cases} x + y = -5 \\ -x + y = -1 \end{cases}$

22. $\begin{cases} -x + y = -3 \\ x + y = 1 \end{cases}$

23. $\begin{cases} 4x + 3y = 24 \\ 4x - 3y = -24 \end{cases}$

24. $\begin{cases} -9x + 5y = -9 \\ -9x - 5y = -9 \end{cases}$

25. $\begin{cases} 2s + t = -2 \\ -2s - 3t = -6 \end{cases}$

26. $\begin{cases} -2x + 4y = 12 \\ 2x + 4y = 28 \end{cases}$

27. $\begin{cases} 5x - 4y = 8 \\ -5x - 4y = 8 \end{cases}$

28. $\begin{cases} 2r + s = -8 \\ -2r + 4s = 28 \end{cases}$

29. $\begin{cases} 4x - 7y = -19 \\ -4x + 7y = 19 \end{cases}$

30. $\begin{cases} x + 20y = -2 \\ -x - 20y = 2 \end{cases}$

31. $\begin{cases} x + 3y = -9 \\ x + 8y = -4 \end{cases}$

32. $\begin{cases} x + 7y = -22 \\ x + 9y = -24 \end{cases}$

33. $\begin{cases} 5c + d = -15 \\ 6c + d = -20 \end{cases}$

34. $\begin{cases} 11c + d = -65 \\ 10c + d = -60 \end{cases}$

35. $\begin{cases} 7x - y = 10 \\ 8x - y = 13 \end{cases}$

36. $\begin{cases} 6x - y = 4 \\ 9x - y = 10 \end{cases}$

37. $\begin{cases} 3x - 5y = -29 \\ 3x - 5y = 15 \end{cases}$

38. $\begin{cases} 2a - 3b = -6 \\ 2a - 3b = 8 \end{cases}$

39. $\begin{cases} 6x - 3y = -7 \\ 9x + y = 6 \end{cases}$

40. $\begin{cases} 9x + 4y = 31 \\ 6x - y = -5 \end{cases}$

41. $\begin{cases} 9x + 4y = 31 \\ 6x - y = -5 \end{cases}$

42. $\begin{cases} 5x - 14y = -12 \\ -x - 6y = -2 \end{cases}$

43. $\begin{cases} 8x + 8y = -16 \\ 3x + y = -4 \end{cases}$

44. $\begin{cases} 4x + 11y = -11 \\ 5x + y = -1 \end{cases}$

45. $\begin{cases} 7x - 50y = -43 \\ x + 3y = 4 \end{cases}$

46. $\begin{cases} x - 2y = -1 \\ 12x + 11y = 23 \end{cases}$

47. $\begin{cases} 8x - 4y = 18 \\ 3x - 2y = 8 \end{cases}$

48. $\begin{cases} 4x + 6y = 5 \\ 8x - 9y = 3 \end{cases}$

49. $\begin{cases} 4x + 3y = 7 \\ 3x - 2y = -16 \end{cases}$

50. $\begin{cases} 3x - 2y = 20 \\ 2x + 7y = 5 \end{cases}$

51. $\begin{cases} 3x + 4y = 12 \\ 4x + 5y = 17 \end{cases}$

52. $\begin{cases} 2x + 11y = -10 \\ 5x + 4y = 22 \end{cases}$

53. $\begin{cases} -3x + 6y = -9 \\ -5x + 4y = -15 \end{cases}$

54. $\begin{cases} -4x + 3y = -13 \\ -6x + 8y = -16 \end{cases}$

55. $\begin{cases} 4a + 7b = 2 \\ 9a - 3b = 1 \end{cases}$

56. $\begin{cases} 5a - 7b = 6 \\ 7a - 6b = 8 \end{cases}$

57. $\begin{cases} 9x = 10y \\ 3x - 2y = 12 \end{cases}$

58. $\begin{cases} 8x = 9y \\ 2x - 3y = -6 \end{cases}$

59. $\begin{cases} 2x + 5y + 13 = 0 \\ 2x + 5 = 3y \end{cases}$

60. $\begin{cases} 3x - 16 = 5y \\ 4x + 5y - 33 = 0 \end{cases}$

61. $\begin{cases} 0 = 4x - 3y \\ 5x = 4y - 2 \end{cases}$

62. $\begin{cases} 6x + 3y = 0 \\ 5y = 2x + 12 \end{cases}$

63. $\begin{cases} 3x - 16 = 5y \\ -3x + 5y - 33 = 0 \end{cases}$

64. $\begin{cases} 2x + 5y - 13 = 0 \\ -2x + 13 = 5y \end{cases}$

65. $\begin{cases} \dfrac{3}{5}s + \dfrac{4}{5}t = 1 \\ -\dfrac{1}{4}s + \dfrac{3}{8}t = 1 \end{cases}$

66. $\begin{cases} \dfrac{1}{2}x + \dfrac{4}{7}y = -1 \\ 5x - \dfrac{4}{5}y = -10 \end{cases}$

67. $\begin{cases} \dfrac{1}{2}s - \dfrac{1}{4}t = 1 \\ \dfrac{1}{3}s + t = 3 \end{cases}$

68. $\begin{cases} \dfrac{3}{5}x + y = 1 \\ \dfrac{4}{5}x - y = -1 \end{cases}$

69. $\begin{cases} -\dfrac{m}{4} - \dfrac{n}{3} = \dfrac{1}{12} \\ \dfrac{m}{2} - \dfrac{5n}{4} = \dfrac{7}{4} \end{cases}$

70. $\begin{cases} -\dfrac{x}{2} + \dfrac{y}{3} = 2 \\ \dfrac{x}{3} + \dfrac{2y}{3} = \dfrac{4}{3} \end{cases}$

71. $\begin{cases} \dfrac{x}{2} - 3y = 7 \\ -x + 6y = -14 \end{cases}$

72. $\begin{cases} -9x + \dfrac{y}{2} = 7 \\ 18x - y = 0 \end{cases}$

73. $\begin{cases} 2x + y = 10 \\ 0.1x + 0.2y = 1.0 \end{cases}$

74. $\begin{cases} 0.3x + 0.2y = 0 \\ 2x - 3y = -13 \end{cases}$

75. $\begin{cases} 2x - y = 16 \\ 0.03x + 0.02y = 0.03 \end{cases}$

76. $\begin{cases} -5y + 2x = 4 \\ -0.02y + 0.03x = 0.04 \end{cases}$

APPLICATIONS

77. EDUCATION The graph shows educational trends during the years 1980–2001 for persons 25 years or older. The equation $9x + 11y = 352$ approximates the percent y who had less than 12 years of schooling. The equation $5x - 11y = -198$ approximates the percent y who had 4 or more years of college. In each case, x is the number of years since 1980. Use the elimination method to determine in what year the percents were equal.

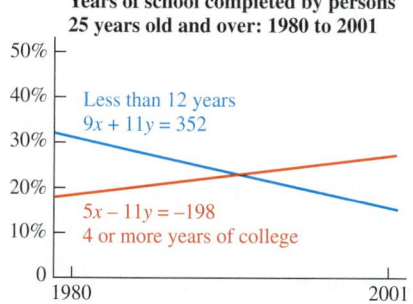

Years of school completed by persons 25 years old and over: 1980 to 2001

Less than 12 years
$9x + 11y = 352$

$5x - 11y = -198$
4 or more years of college

Source: U.S. Department of Commerce, Bureau of the Census

78. EDUCATION Answer Problem 77 by solving the system of equations using the substitution method. (*Hint:* Solve one equation for $11y$, then substitute for $11y$ in the other equation.)

WRITING

79. Why is the method for solving systems that is discussed in this section called the *elimination method?*

80. If the elimination method is to be used to solve this system, what is wrong with the form in which it is written?

$$\begin{cases} 2x - 5y = -3 \\ -2y + 3x = 10 \end{cases}$$

81. Can the system $\begin{cases} 2x + 5y = -13 \\ -2x - 3y = -5 \end{cases}$ be solved more easily using the elimination method or the substitution method? Explain.

82. Explain the error in the following work.

Solve: $\begin{cases} x + y = 1 \\ x - y = 5 \end{cases}$.

$$\begin{aligned} x + y &= 1 \\ +x - y &= 5 \\ \hline 2x &= 6 \end{aligned}$$

$$\frac{2x}{2} = \frac{6}{2}$$

$$\boxed{x = 3} \quad \text{The solution is 3.}$$

REVIEW

83. Translate into symbols: 10 less than x.

84. Solve: $3y + \dfrac{y + 2}{2} = \dfrac{2(y + 3)}{3} + 16$.

85. Simplify: $x - x$.

86. Simplify: $3.2m - 4.4 + 2.1m + 16$.

87. Find the area of a triangular-shaped sign with a base of 4 feet and a height of 3.75 feet.

88. Factor: $6x^2 + 7x - 20$.

CHALLENGE PROBLEMS   Use the elimination method to solve each system.

89. $\begin{cases} \dfrac{x - 3}{2} + \dfrac{y + 5}{3} = \dfrac{11}{6} \\ \dfrac{x + 3}{3} - \dfrac{5}{12} = \dfrac{y + 3}{4} \end{cases}$

90. $\begin{cases} 4(x + 1) = 17 - 3(y - 1) \\ 2(x + 2) + 3(y - 1) = 9 \end{cases}$

7.4 Problem Solving Using Systems of Equations

- Assigning Variables to Two Unknowns   • Geometry Problems
- Number-Value Problems   • The Break Point
- Interest, Uniform Motion, and Mixture Problems

In previous chapters, many applied problems were modeled and solved with an equation in one variable. In this section, the application problems involve two unknowns. It is often easier to solve such problems using a two-variable approach.

■ ASSIGNING VARIABLES TO TWO UNKNOWNS

The following steps are helpful when solving problems involving two unknown quantities.

Problem-Solving Strategy	1. **Analyze the problem** by reading it carefully to understand the given facts. Often a diagram or table will help you visualize the facts of the problem. 2. Pick different variables to represent two unknown quantities. Translate the words of the problem to **form two equations** involving each of the two variables. 3. **Solve the system** of equations using graphing, substitution, or elimination. 4. **State the conclusion.** 5. **Check the results** in the words of the problem.

EXAMPLE 1

Motion pictures. Each year, Academy Award winners are presented with Oscars. The 13.5-inch statuette has a base on which a gold plated figure stands. The figure itself is 7.5 inches taller than the base. Find the height of the figure and the height of the base.

Analyze the Problem
* The statuette is a total of 13.5 inches tall.
* The figure is 7.5 inches taller than the base.
* Find the height of the figure and the height of the base.

Caution

If two variables are used to represent two unknown quantities, we must form a system of equations to find the unknown.

Form Two Equations   Let x = the height of the figure and y = the height of the base. We can translate the words of the problem into two equations, each involving x and y.

The height of the figure	plus	the height of the base	is	13.5 inches.
x	$+$	y	$=$	13.5

The height of the figure	is	the height of the base	plus	7.5 inches.
x	$=$	y	$+$	7.5

The resulting system is $\begin{cases} x + y = 13.5 \\ x = y + 7.5 \end{cases}$.

Solve the System   Since the second equation is solved for x, we will use substitution to solve the system.

$$x + y = 13.5 \qquad \text{This is the first equation of the system.}$$
$$y + 7.5 + y = 13.5 \qquad \text{Substitute } y + 7.5 \text{ for } x.$$
$$2y + 7.5 = 13.5 \qquad \text{Combine like terms.}$$

$$2y = 6 \qquad \text{Subtract 7.5 from both sides.}$$
$$y = 3 \qquad \text{Divide both sides by 2. This is the height of the base.}$$

To find x, substitute 3 for y in the second equation of the system.

$$x = \mathbf{y} + 7.5$$
$$x = \mathbf{3} + 7.5 \qquad \text{Substitute for } y.$$
$$x = 10.5 \qquad \text{This is the height of the figure.}$$

State the Conclusion   The height of the figure is 10.5 inches and the height of the base is 3 inches.

Check the Results   The sum of 10.5 inches and 3 inches is 13.5 inches, and the 10.5-inch figure is 7.5 inches taller than the 3-inch base. The results check.

■ GEOMETRY PROBLEMS

Two angles are said to be **complementary** if the sum of their measures is 90°. Two angles are said to be **supplementary** if the sum of their measures is 180°.

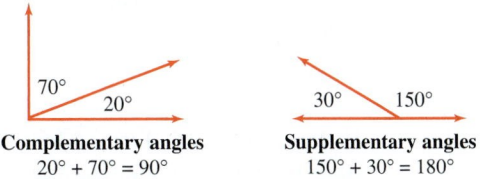

Complementary angles
$20° + 70° = 90°$

Supplementary angles
$150° + 30° = 180°$

EXAMPLE 2

Angles.   The difference of the measures of two complementary angles is 6°. Find the measure of each angle.

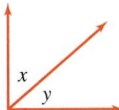

Analyze the Problem
- Since the angles are complementary, the sum of their measures is 90°.
- The word *difference* indicates subtraction. If the measure of the smaller angle is subtracted from the measure of the larger angle, the result will be 6°.
- Find the measure of the larger angle and the measure of the smaller angle.

Form Two Equations   Let $x =$ the measure of the larger angle and $y =$ the measure of the smaller angle. We can translate the words of the problem into two equations, each involving x and y.

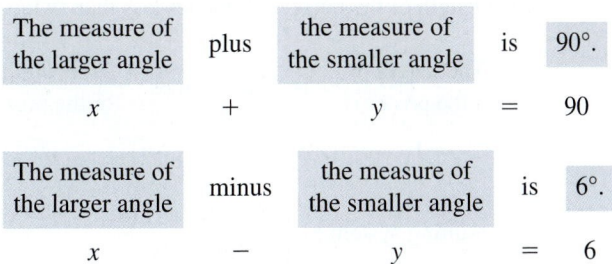

The resulting system is $\begin{cases} x + y = 90 \\ x - y = 6 \end{cases}$.

Solve the System   Since the coefficients of y are opposites, we will use elimination to solve the system.

$$\begin{array}{rl} x + y = 90 & \\ \underline{x - y = 6} & \\ 2x = 96 & \text{Add the equations to eliminate } y. \\ x = 48 & \text{Divide both sides by 2. This is the measure of the larger angle.} \end{array}$$

To find y, substitute 48 for x in the first equation of the system.

$$\begin{array}{rl} x + y = 90 & \\ 48 + y = 90 & \text{Substitute 48 for } x. \\ y = 42 & \text{Subtract 48 from both sides. This is the measure of the smaller angle.} \end{array}$$

State the Conclusion   The measure of the larger angle is 48° and the measure of the smaller angle is 42°.

Check the Results   The sum of 48° and 42° is 90°, and the difference is 6°. The results check.

EXAMPLE 3

History.   In 1917, James Montgomery Flagg created the classic *I Want You* poster to help recruiting for World War I. The perimeter of the poster is 114 inches, and its length is 9 inches less than twice its width. Find the length and the width of the poster.

Analyze the Problem
• The perimeter of the rectangular poster is 114 inches.
• The length is 9 inches less than twice the width.
• Find the length and the width of the poster.

Form Two Equations   Let l = the length of the poster and w = the width of the poster. The perimeter of a rectangle is the sum of two lengths and two widths, so we have

2 times	the length of the poster	plus	2 times	the width of the poster	is	114 inches.
2 ·	l	+	2 ·	w	=	114

If the length of the poster is 9 inches less than twice the width, we have

The length of the poster	is	2 times	the width of the poster	minus	9 inches.
l	=	2 ·	w	−	9

The resulting system is $\begin{cases} 2l + 2w = 114 \\ l = 2w - 9 \end{cases}$.

Solve the System   Since the second equation is solved for *l*, we will use substitution to solve the system.

$$2l + 2w = 114$$ This is the first equation of the system.

$$2(2w - 9) + 2w = 114$$ Substitute $2w - 9$ for *l*.

$$4w - 18 + 2w = 114$$ Distribute the multiplication by 2.

$$6w - 18 = 114$$ Combine like terms.

$$6w = 132$$ Add 18 to both sides.

$$w = 22$$ Divide both sides by 6. This is the width of the poster.

To find *l*, substitute 22 for *w* in the second equation of the system.

$$l = 2w - 9$$

$$l = 2(22) - 9$$

$$l = 44 - 9$$

$$l = 35$$ This is the length of the poster.

State the Conclusion   The length of the poster is 35 inches and the width is 22 inches.

Check the Results   The perimeter is $2(35) + 2(22) = 70 + 44 = 114$ inches, and 35 inches is 9 inches less than twice 22 inches. The results check.

■ NUMBER-VALUE PROBLEMS

EXAMPLE 4

ELEMENTARY &
INTERMEDIATE
Algebra $f(x)$ **Now**™

Photography.   At a school, two picture packages are available, as shown in the illustration. Find the cost of a class picture and the cost of an individual wallet-size picture.

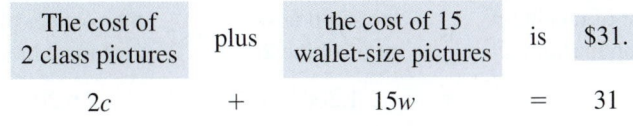

Package 1
1 class picture
10 wallet-size
Only $19

FOOTHILL ELEMENTARY

Package 2
2 class pictures
15 wallet-size
Only $31

Solution   **Analyze the Problem**
- Package 1 contains 1 class picture and 10 wallet-size pictures.
- Package 2 contains 2 class pictures and 15 wallet-size pictures.
- Find the cost of a class picture and the cost of a wallet-size picture.

Form Two Equations   Let $c =$ the cost of 1 class picture and $w =$ the cost of 1 wallet-size picture. To write an equation that models the first package, we note that (in dollars) the cost of 1 class picture is c and the cost of 10 wallet-size pictures is $10 \cdot w = 10w$.

The cost of 1 class picture	plus	the cost of 10 wallet-size pictures	is	$19.
c	$+$	$10w$	$=$	19

To write an equation that models the second package, we note that (in dollars) the cost of 2 class pictures is $2 \cdot c = 2c$, and the cost of 15 wallet-size pictures is $15 \cdot w = 15w$.

The cost of 2 class pictures	plus	the cost of 15 wallet-size pictures	is	$31.
$2c$	$+$	$15w$	$=$	31

The resulting system is $\begin{cases} c + 10w = 19 \\ 2c + 15w = 31 \end{cases}$.

Solve the System   We can use elimination to solve this system. To eliminate c, we proceed as follows.

$$-2c - 20w = -38 \qquad \text{Multiply both sides of } c + 10w = 19 \text{ by } -2.$$
$$\underline{2c + 15w = 31}$$
$$-5w = -7 \qquad \text{Add the equations to eliminate } c.$$
$$w = 1.4 \qquad \text{Divide both sides by } -5. \text{ This is the cost of a wallet-size picture.}$$

To find c, substitute 1.4 for w in the first equation of the original system.

$$c + 10w = 19$$
$$c + 10(\mathbf{1.4}) = 19 \qquad \text{Substitute 1.4 for } w.$$
$$c + 14 = 19 \qquad \text{Multiply.}$$
$$c = 5 \qquad \text{Subtract 14 from both sides. This is the cost of a class picture.}$$

State the Conclusion   A class picture costs \$5 and a wallet-size picture costs \$1.40.

Check the Results   Package 1 has 1 class picture and 10 wallets: \$5 + 10(\$1.40) = \$5 + \$14 = \$19. Package 2 has 2 class pictures and 15 wallets: 2(\$5) + 15(\$1.40) = \$10 + \$21 = \$31. The results check.

■ THE BREAK POINT

EXAMPLE 5

ELEMENTARY &
INTERMEDIATE
Algebra *f(x)* **Now**™

Manufacturing.   The setup cost of a machine that mills brass plates is \$750. After setup, it costs \$0.25 to mill each plate. Management is considering the purchase of a larger machine that can produce the same plate at a cost of \$0.20 per plate. If the setup cost of the larger machine is \$1,200, how many plates would the company have to produce to make the purchase worthwhile?

Analyze the Problem   We need to find the number of plates (called the **break point**) that will cost equal amounts to produce on either machine.

Form Two Equations   We can let c represent the cost of milling p plates. If we call the machine currently being used machine 1, and the new, larger one machine 2, we can form the two equations.

The cost of making p plates on machine 1	is	the setup cost of machine 1	plus	the cost per plate of machine 1	times	the number of plates p to be made.
c	=	750	+	0.25	·	p

The cost of making p plates on machine 2	is	the setup cost of machine 2	plus	the cost per plate of machine 2	times	the number of plates p to be made.
c	=	1,200	+	0.20	·	p

Solve the System   Since the costs are equal, we can use the substitution method to solve the system

$$\begin{cases} c = \mathbf{750 + 0.25p} \\ c = 1{,}200 + 0.20p \end{cases}$$

$\mathbf{750 + 0.25p} = 1{,}200 + 0.20p$	Substitute $750 + 0.25p$ for c in the second equation.
$0.25p = 450 + 0.20p$	Subtract 750 from both sides.
$0.05p = 450$	Subtract $0.20p$ from both sides.
$p = 9{,}000$	Divide both sides by 0.05.

State the Conclusion   If 9,000 plates are milled, the cost will be the same on either machine. If more than 9,000 plates are milled, the cost will be cheaper on the larger machine, because it mills the plates less expensively than the smaller machine.

Check the Result   We check the solution by substituting 9,000 for p in each equation of the system and verifying that 3,000 is the value of c in both cases.
    If we graph the two equations, we can illustrate the break point.

Machine 1
$c = 750 + 0.25p$

p	c
0	750
1,000	1,000
5,000	2,000

Machine 2
$c = 1{,}200 + 0.20p$

p	c
0	1,200
4,000	2,000
12,000	3,600

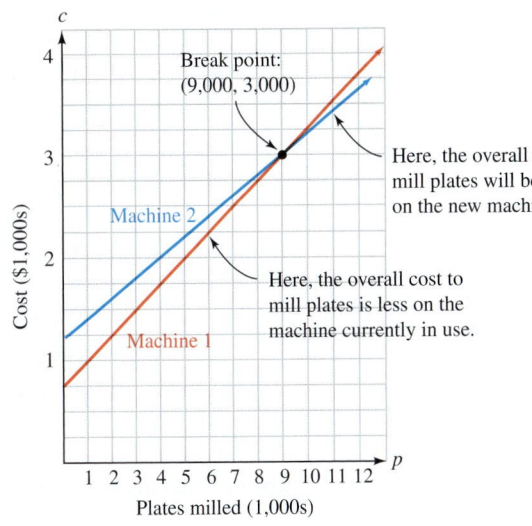

■ INTEREST, UNIFORM MOTION, AND MIXTURE PROBLEMS

EXAMPLE 6

ELEMENTARY &
INTERMEDIATE
Algebra $f(x)$ **Now**™

White-collar crime.   Investigators discovered that a company secretly moved $150,000 out of the country to avoid paying income tax. Some of the money was invested in a Swiss bank account that paid 8% interest annually. The remainder was deposited in a Cayman Islands account, paying 7% annual interest. The investigation also revealed that the combined interest earned the first year was $11,500. How much money was invested in each account?

Analyze the Problem   We are told that an unknown part of the $150,000 was invested at an annual rate of 8% and the rest at 7%. Together, the accounts earned $11,500 in interest.

Form Two Equations   Let x = the amount invested in the Swiss account and y = the amount invested in the Cayman Islands account. Because the total investment was $150,000, we have

The amount invested in the Swiss account	plus	the amount invested in the Cayman Is. account	is	$150,000.
x	+	y	=	150,000

We can use the formula $I = Prt$ to determine that x dollars invested for 1 year at 8% earns $x \cdot 0.08 \cdot 1 = 0.08x$ dollars. Similarly, y dollars invested for 1 year at 7% earns $y \cdot 0.07 \cdot 1 = 0.07y$ dollars. If the total combined interest earned was $11,500, we have

The income on the 8% investment	plus	the income on the 7% investment	is	$11,500.
$0.08x$	+	$0.07y$	=	11,500

The resulting system is $\begin{cases} x + y = 150,000 \\ 0.08x + 0.07y = 11,500 \end{cases}$.

Caution

In Example 6, it is incorrect to let

~~x = the amount invested in each account~~

This implies that *equal amounts* were invested in the Swiss and Cayman Island accounts. We do not know that.

Solve the System   To solve the system, clear the second equation of decimals. Then eliminate x.

$$
\begin{aligned}
-8x - 8y &= -1,200,000 \quad &\text{Multiply both sides of the first equation by } -8. \\
\underline{8x + 7y =\ \ 1,150,000} \quad &\text{Multiply both sides of the second equation by 100.} \\
-y &= \ \ \ -50,000 \\
y &= 50,000 \quad &\text{Multiply both sides by } -1.
\end{aligned}
$$

To find x, substitute 50,000 for y in the first equation of the original system.

$$
\begin{aligned}
x + y &= 150,000 \\
x + \mathbf{50,000} &= 150,000 \quad &\text{Substitute.} \\
x &= 100,000 \quad &\text{Subtract 50,000 from both sides.}
\end{aligned}
$$

State the Conclusion $100,000 was invested in the Swiss bank account, and $50,000 was invested in the Cayman Islands account.

Check the Results

$$
\begin{aligned}
\$100,000 + \$50,000 &= \$150,000 \quad &\text{The two investments total \$150,000.} \\
0.08(\$100,000) &= \$8,000 \quad &\text{The Swiss bank account earned \$8,000.} \\
0.07(\$50,000) &= \$3,500 \quad &\text{The Cayman Islands account earned \$3,500.}
\end{aligned}
$$

The combined interest is $8,000 + $3,500 = $11,500. The results check.

EXAMPLE 7

ELEMENTARY & INTERMEDIATE
Algebra *f(x)* Now™

Boating.   A boat traveled 30 miles downstream in 3 hours and made the return trip in 5 hours. Find the speed of the boat in still water and the speed of the current.

Analyze the Problem   Traveling downstream, the speed of the boat will be *faster* than it would be in still water. Traveling upstream, the speed of the boat will be *slower* than it would be in still water.

Form Two Equations   Let s = the speed of the boat in still water and c = the speed of the current. Then the speed of the boat going downstream is $s + c$ and the speed of the boat going upstream is $s - c$. We can organize the facts of the problem in a table.

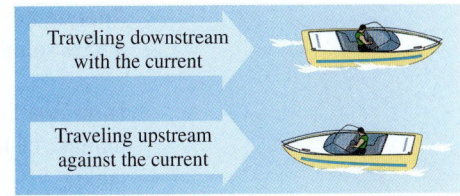

Traveling downstream with the current

Traveling upstream against the current

	Rate	· Time =	Distance
Downstream	$s + c$	3	$3(s + c)$
Upstream	$s - c$	5	$5(s - c)$

Enter this information first.

Set each of these expressions for distance traveled equal to 30.

Since each trip is 30 miles long, the Distance column of the table gives two equations in two variables. To write each equation in general form, use the distributive property.

$$\begin{cases} 3(s + c) = 30 \\ 5(s - c) = 30 \end{cases} \xrightarrow[\text{Distribute}]{\text{Distribute}} \begin{cases} 3s + 3c = 30 \\ 5s - 5c = 30 \end{cases}$$

Solve the System   To eliminate c, we proceed as follows.

$$\begin{array}{ll} 15s + 15c = 150 & \text{Multiply both sides of } 3s + 3c = 30 \text{ by 5.} \\ \underline{15s - 15c = 90} & \text{Multiply both sides of } 5s - 5c = 30 \text{ by 3.} \\ 30s = 240 \\ s = 8 & \text{Divide both sides by 30. This is the speed of the boat in still water.} \end{array}$$

To find c, it appears that the computations will be easiest if we use $3s + 3c = 30$.

$$\begin{array}{ll} 3s + 3c = 30 \\ 3(8) + 3c = 30 & \text{Substitute 8 for } s. \\ 24 + 3c = 30 & \text{Multiply.} \\ 3c = 6 & \text{Subtract 24 from both sides.} \\ c = 2 & \text{Divide both sides by 3. This is the speed of the current.} \end{array}$$

State the Conclusion   The speed of the boat in still water is 8 mph and the speed of the current is 2 mph.

Check the Results   With a 2-mph current, the boat's downstream speed will be $8 + 2 = 10$ mph. In 3 hours, it will travel $10 \cdot 3 = 30$ miles. With a 2-mph current, the boat's upstream speed will be $8 - 2 = 6$ mph. In 5 hours, it will cover $6 \cdot 5 = 30$ miles. The results check.

EXAMPLE 8

ELEMENTARY & INTERMEDIATE

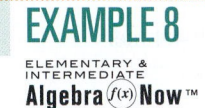

Medical technology.   A laboratory technician has one batch of antiseptic that is 40% alcohol and a second batch that is 60% alcohol. She would like to make 8 liters of solution that is 55% alcohol. How many liters of each batch should she use?

Analyze the Problem   Some 60% solution must be added to some 40% solution to make a 55% solution.

Form Two Equations   Let x = the number of liters to be used from batch 1 and y = the number of liters to be used from batch 2. We can organize the facts of the problem in a table.

	Amount ·	Strength =	Amount of alcohol
Batch 1 (too weak)	x	0.40	$0.40x$
Batch 2 (too strong)	y	0.60	$0.60y$
Mixture	8	0.55	0.55(8)

One equation comes from information in this column.

40%, 60%, and 55% have been expressed as decimals.

Another equation comes from information in this column.

The information in the table provides two equations.

$$\begin{cases} x + y = 8 \\ 0.40x + 0.60y = 0.55(8) \end{cases}$$

The number of liters of batch 1 plus the number of liters of batch 2 equals the total number of liters in the mixture.

The amount of alcohol in batch 1 plus the amount of alcohol in batch 2 equals the amount of alcohol in the mixture.

Solve the System   We can solve this system by elimination. To eliminate x, we proceed as follows.

$$\begin{aligned} -40x - 40y &= -320 \\ \underline{40x + 60y} &= \underline{440} \\ 20y &= 120 \\ y &= 6 \end{aligned}$$

Multiply both sides of the first equation by -40.

Multiply both sides of the second equation by 100.

Divide both sides by 20. This is the number of liters of batch 2 needed.

To find x, we substitute 6 for y in the first equation of the original system.

$$x + y = 8$$
$$x + 6 = 8 \qquad \text{Substitute.}$$
$$x = 2 \qquad \text{Subtract 6 from both sides. This is the number of liters of batch 1 needed.}$$

State the Conclusion   The technician should use 2 liters of the 40% solution and 6 liters of the 60% solution.

Check the Results   Note that 2 liters + 6 liters = 8 liters, the required number. Also, the amount of alcohol in the two solutions is equal to the amount of alcohol in the mixture. The results check.

Alcohol in batch 1: $0.40x = 0.40(2) = 0.8$ liters

Alcohol in batch 2: $0.60y = 0.60(6) = 3.6$ liters

Total: 4.4 liters

Alcohol in the mixture: $0.55(8) = 4.4$ liters

VOCABULARY   Fill in the blanks.

1. A _____ is a letter that stands for a number.

2. An _____ is a statement indicating that two quantities are equal.

3. $\begin{cases} a + b = 20 \\ a = 2b + 4 \end{cases}$ is a _____ of equations.

4. A _____ of a system of equations satisfies both equations.

5. Two angles are said to be _____ if the sum of their measures is 90°.

6. Two angles are said to be _____ if the sum of their measures is 180°.

CONCEPTS

7. A length of pipe is to be cut into two pieces. The longer piece is to be 1 foot less than twice the shorter piece. Write two equations that model the situation.

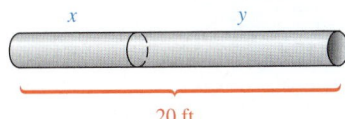

20 ft

8. Two angles are complementary. The measure of the larger angle is four times the measure of the smaller angle. Write two equations that model the situation.

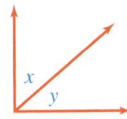

9. Two angles are supplementary. The measure of the smaller angle is 25° less than the measure of the larger angle. Write two equations that model the situation.

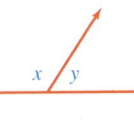

10. The perimeter of a ping pong table is 28 feet. The length is 4 feet more than the width. Write two equations that model the situation.

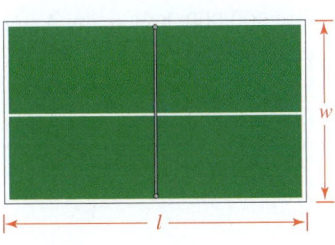

11. Let $x =$ the cost of a chicken taco and $y =$ the cost of a beef taco. Write an equation that models the offer shown in the advertisement.

TUESDAY TACO SPECIAL

5 CHICKEN TACOS   2 BEEF TACOS

only $10

12. **a.** Complete the following table.

	Principal · Rate · Time = Interest			
City Bank	x	5%	1 yr	
USA Savings	y	11%	1 yr	

b. A total of $50,000 was deposited in the two accounts. Use that information to write an equation about the principal.

c. A total of $4,300 was earned by the two accounts. Use that information to write an equation about the interest.

13. In still water, a man can paddle a canoe so that it travels x mph. What will the speed of the canoe be for each situation shown?

a. Downstream

Current c mph

b. Upstream

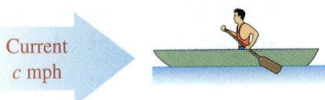

Current c mph

14. Complete the table, which contains information about an airplane flying in windy conditions.

	Rate · Time = Distance		
With wind	$x + y$	3	
Against wind	$x - y$	5	

15. a. If the contents of the two test tubes are poured into a third tube, how much solution will the third tube contain?

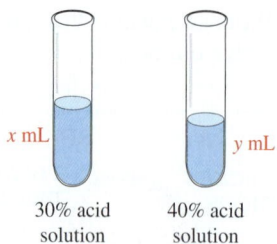

30% acid solution   40% acid solution

b. Which of the following strengths could the mixture possibly be: 27%, 33%, or 44% acid solution?

16. a. Complete the table, which contains information about mixing two salt solutions to get 12 gallons of a 3% salt solution.

	Amount	· Strength =	Amount of salt
Weak	x	0.01	
Strong	y	0.06	
Mix			

b. Use the information from the Amount column to write an equation.

c. Use the information from the Amount of salt column to write an equation.

PRACTICE

17. COMPLEMENTARY ANGLES   Two angles are complementary. The measure of one angle is 10° more than three times the measure of the other. Find the measure of each angle.

18. SUPPLEMENTARY ANGLES   Two angles are supplementary. The measure of one angle is 20° less than 19 times the measure of the other. Find the measure of each angle.

19. SUPPLEMENTARY ANGLES   The difference of the measures of two supplementary angles is 80°. Find the measure of each angle.

20. COMPLEMENTARY ANGLES   Two angles are complementary. The measure of one angle is 15° more than one-half of the measure of the other. Find the measure of each angle.

APPLICATIONS   Write a system of two equations in two variables to solve each problem.

21. TREE TRIMMING When fully extended, the arm on a tree service truck is 51 feet long. If the upper part of the arm is 7 feet shorter than the lower part, how long is each part of the arm?

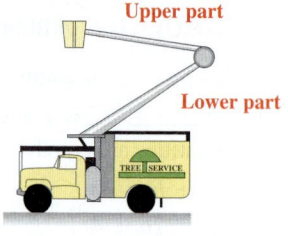

Upper part

Lower part

22. ALASKA   Most of the 1,422-mile-long Alaskan Highway is actually in Canada. Find the length of the highway that is in Alaska and the length of the highway that is in Canada, if it is known that the difference in the lengths is 1,020 miles.

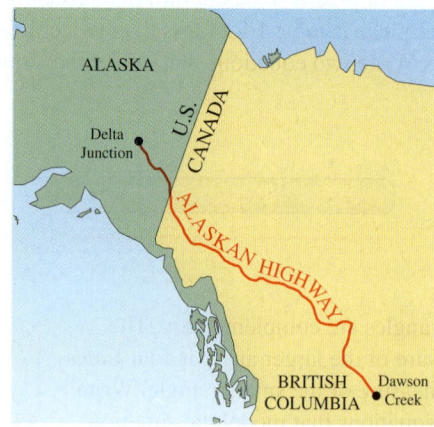

23. EXECUTIVE BRANCH   The salaries of the president and vice president of the United States total $592,600 a year. If the president makes $207,400 more than the vice president, find each of their salaries.

24. CAUSES OF DEATH   According to the *National Vital Statistics Reports,* in 2001, the number of Americans who died from heart disease was about 7 times the number who died from accidents. If the number of deaths from these two causes totaled approximately 800,000, how many Americans died from each cause in 2001?

25. MARINE CORPS   The Marine Corps War Memorial in Arlington, Virginia, portrays the raising of the U.S. flag on Iwo Jima during World War II. Find the measures of the two angles shown if the measure of $\angle 2$ is 15° less than twice the measure of $\angle 1$.

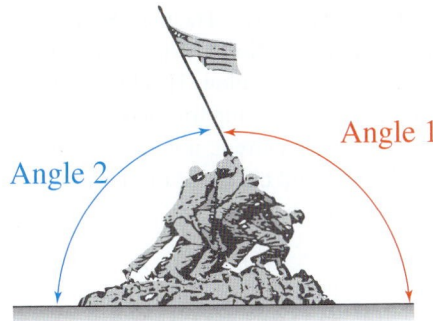

26. PHYSICAL THERAPY   To rehabilitate her knee, an athlete does leg extensions. Her goal is to regain a full 90° range of motion in this exercise. Use the information in the illustration to determine her current range of motion in degrees and the number of degrees of improvement she still needs to make.

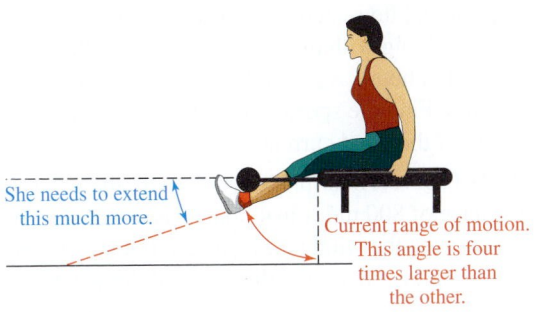

27. THEATER SCREENS   At an IMAX theater, the giant rectangular movie screen has a width 26 feet less than its length. If its perimeter is 332 feet, find the length and the width of the screen.

28. ENGLISH ARTISTS   In 1770, Thomas Gainsborough painted *The Blue Boy*. The sum of the length and width of the painting is 118 inches. The difference of the length and width is 22 inches. Find the length and width.

29. GEOMETRY   A 50-meter path surrounds a rectangular garden. The width of the garden is two-thirds its length. Find the length and width.

30. BALLROOM DANCING   A rectangular-shaped dance floor has a perimeter of 200 feet. If the floor were 20 feet wider, its width would equal its length. Find the length and width of the dance floor.

31. BUYING PAINTING SUPPLIES   Two partial receipts for paint supplies are shown. (Assume no sales tax was charged.) Find the cost of one gallon of paint and the cost of one paint brush.

32. WEDDING PICTURES   A photographer sells the two wedding picture packages shown. How much does a 10 × 14 photo cost? How much does an 8 × 10 photo cost?

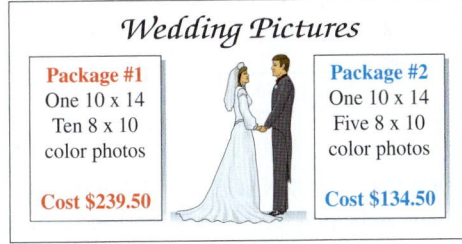

33. SELLING ICE CREAM   At a store, ice cream cones cost $0.90 and sundaes cost $1.65. One day the receipts for a total of 148 cones and sundaes were $180.45. How many cones were sold? How many sundaes?

34. BUYING TICKETS   The ticket prices for a movie are shown on the next page. Receipts for one showing were $1,440 for an audience of 190 people. How many general admission tickets and how many senior citizen tickets were sold?

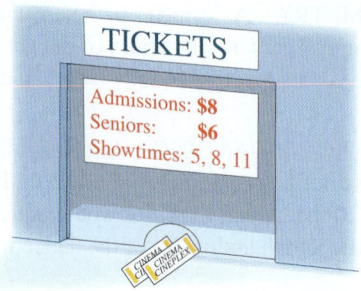

35. MAKING TIRES A company has two molds to form tires. One mold has a setup cost of $1,000 and the other has a setup cost of $3,000. The cost to make each tire with the first mold is $15, and the cost to make each tire with the second mold is $10.

a. Find the break point.

b. Check your result by graphing both equations on the coordinate system below.

c. If a production run of 500 tires is planned, determine which mold should be used.

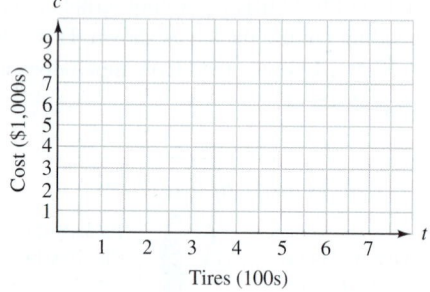

36. CHOOSING A FURNACE A high-efficiency 90+ furnace can be purchased for $2,250 and costs an average of $412 per year to operate in Rockford, Illinois. An 80+ furnace can be purchased for only $1,710, but it costs $466 per year to operate.

a. Find the break point.

b. If you intended to live in a Rockford house for 7 years, which furnace would you choose?

37. TELEPHONE RATES A long distance provider offers two plans. Plan 1 has a monthly fee of $10 plus 8¢ per minute. Plan 2 has a monthly fee of $15 plus 4¢ a minute. For what number of minutes will the costs of the plans be the same?

38. SUPPLY AND DEMAND Suppose that the number c of cheesecakes that a bakery will supply a day is given by $c = 60p - 200$, and the number c of cheesecakes that are purchased a day is given by $c = -40p + 1,000$, where p is the price (in dollars) of a cheesecake. For what price will supply equal demand?

39. THE GULF STREAM The Gulf Stream is a warm ocean current of the North Atlantic Ocean that flows northward, as shown below. Heading north with the Gulf Stream, a cruise ship traveled 300 miles in 10 hours. Against the current, it took 15 hours to make the return trip. Find the speed of the ship in still water and the speed of the current.

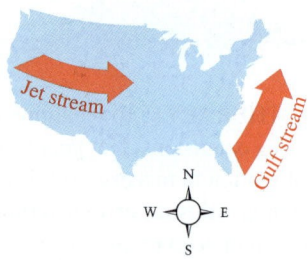

40. THE JET STREAM The jet stream is a strong wind current that flows across the United States, as shown above. Flying with the jet stream, a plane flew 3,000 miles in 5 hours. Against the same wind, the trip took 6 hours. Find the speed of the plane in still air and the speed of the wind current.

41. AVIATION An airplane can fly with the wind a distance of 800 miles in 4 hours. However, the return trip against the wind takes 5 hours. Find the speed of the plane in still air and the speed of the wind.

42. BOATING A boat can travel 24 miles downstream in 2 hours and can make the return trip in 3 hours. Find the speed of the boat in still water and the speed of the current.

43. STUDENT LOANS A college used a $5,000 gift from an alumnus to make two student loans. The first was at 5% annual interest to a nursing student. The second was at 7% to a business major. If the college collected $310 in interest the first year, how much was loaned to each student?

44. FINANCIAL PLANNING   In investing $6,000 of a couple's money, a financial planner put some of it into a savings account paying 6% annual interest. The rest was invested in a riskier mini-mall development plan paying 12% annually. The combined interest earned for the first year was $540. How much money was invested at each rate?

45. MARINE BIOLOGY   A marine biologist wants to set up an aquarium containing 3% salt water. He has two tanks on hand that contain 6% and 2% salt water. How much water from each tank must he use to fill a 16-liter aquarium with a 3% saltwater mixture?

46. COMMEMORATIVE COINS   A foundry has been commissioned to make souvenir coins. The coins are to be made from an alloy that is 40% silver. The foundry has on hand two alloys, one with 50% silver content and one with a 25% silver content. How many kilograms of each alloy should be used to make 20 kilograms of the 40% silver alloy?

47. COFFEE SALES   A coffee supply store waits until the orders for its special blend reach 100 pounds before making up a batch. Columbian coffee selling for $8.75 a pound is blended with Brazilian coffee selling for $3.75 a pound to make a product that sells for $6.35 a pound. How much of each type of coffee should be used to make the blend that will fill the orders?

	Amount ·	Price =	Total value
Columbian	x	8.75	8.75x
Brazilian	y	3.75	3.75y
Mixture	100	6.35	100(6.35)

48. MIXING NUTS   A merchant wants to mix peanuts with cashews, as shown in the illustration, to get 48 pounds of mixed nuts that will be sold at $4 per pound. How many pounds of each should the merchant use?

WRITING

49. What is a *break point?* Give an example.

50. A man paid $89 for two shirts and four pairs of socks. If we let $x =$ the cost of a shirt and $y =$ the cost of a pair of socks, an equation modeling the purchase is $2x + 4y = 89$. Explain why there is not enough information to determine the cost of a shirt or the cost of a pair of socks.

REVIEW   Graph each inequality. Then describe the graph using interval notation.

51. $x < 4$

52. $x \geq -3$

53. $-1 < x \leq 2$

54. $-2 \leq x \leq 0$

CHALLENGE PROBLEMS

55. On the last scale, how many nails will it take to balance 1 nut?

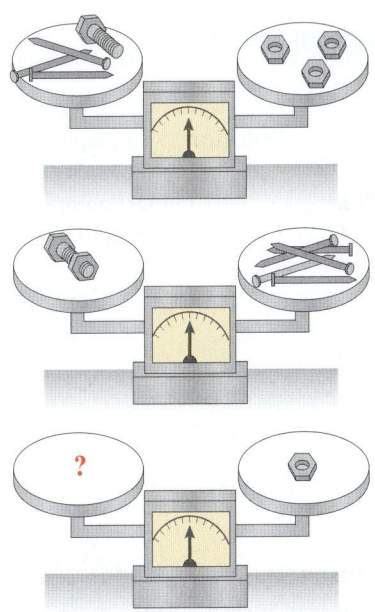

56. FARMING   In a pen of goats and chickens, there are 40 heads and 130 feet. How many goats and chickens are in the pen?

7.5   Solving Systems of Linear Inequalities

• Graphing Systems of Linear Inequalities   • An Application

In Section 7.1, we solved systems of linear equations graphically by finding the point of intersection of two lines. Now we consider systems of linear inequalities, such as

$$\begin{cases} x + y \geq -1 \\ x - y \geq 1 \end{cases}$$

To solve systems of linear inequalities, we again find the points of intersection of graphs. In this case, however, we are not looking for an intersection of two lines, but an intersection of two half-planes.

■ GRAPHING SYSTEMS OF LINEAR INEQUALITIES

A solution of a **system of linear inequalities** is an ordered pair that satisfies each inequality. *To solve a system of linear inequalities* means to find all of its solutions. This can be done by graphing each inequality on the same set of axes and finding the points that are common to every graph in the system.

EXAMPLE 1

ELEMENTARY &
INTERMEDIATE
Algebra $f(x)$ **Now**™

Graph the solutions of the system: $\begin{cases} x + y \geq -1 \\ x - y \geq 1 \end{cases}$.

Solution   To graph $x + y \geq -1$, we begin by graphing the boundary line $x + y = -1$. Since the inequality contains a $\geq$ symbol, the boundary is a solid line. Because the coordinates of the test point $(0, 0)$ satisfy $x + y \geq -1$, we shade (in red) the side of the boundary that contains $(0, 0)$. See part (a) of the figure.

Graph the boundary: The intercept method

$$x + y = -1$$

x	y	(x, y)
0	-1	$(0, -1)$
-1	0	$(-1, 0)$

Shading: Check the test point $(0, 0)$

$x + y \geq -1$

$0 + 0 \overset{?}{\geq} -1$ Substitute.

$0 \geq -1$

$(0, 0)$ is a solution of $x + y \geq -1$.

The Language of Algebra

To solve a system of linear inequalities, we *superimpose* the graphs of the inequalities. That is, we place one graph over the other. Most video camcorders can *superimpose* the date and time over the picture being recorded.

In part (b) of the figure, we superimpose the graph of $x - y \geq 1$ on the graph of $x + y \geq -1$ so that we can determine the points that the graphs have in common. To graph $x - y \geq 1$, we graph the boundary $x - y = 1$ as a solid line. Since the test point $(0, 0)$ does not satisfy $x - y \geq 1$, we shade (in blue) the half-plane that does not contain $(0, 0)$.

Graph the boundary: The intercept method

$$x - y = 1$$

x	y	(x, y)
0	-1	$(0, -1)$
1	0	$(1, 0)$

Shading: Check the test point $(0, 0)$

$x - y \geq 1$

$0 - 0 \overset{?}{\geq} 1$ Substitute.

$0 \geq 1$

$(0, 0)$ is not a solution of $x - y \geq 1$.

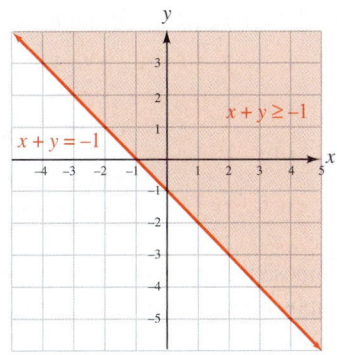

The graph of $x + y \geq -1$ is shaded in red.

(a)

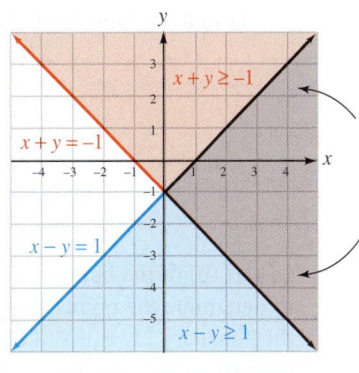

The solutions of the system are shaded in purple. The purple region is the intersection or overlap of the red and blue shaded regions. It includes portions of each boundary.

The graph of $x - y \geq 1$ is shaded in blue. It is drawn over the graph of $x + y \geq -1$.

(b)

In part (b) of the figure, the area that is shaded twice represents the solutions of the given system. Any point in the doubly shaded region in purple (including the purple portions of each boundary) has coordinates that satisfy both inequalities.

Since there are infinitely many solutions, we cannot check each of them. However, as an informal check, we can select one ordered pair, say (4, 1), that lies in the doubly shaded region and show that its coordinates satisfy both inequalities of the system.

Success Tip

Colored pencils are often used to graph systems of inequalities. A standard pencil can also be used. Just draw different patterns of lines instead of shading.

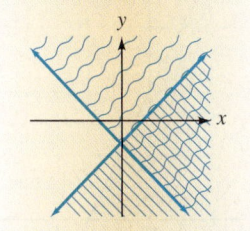

Check:

$$x + y \geq -1 \qquad x - y \geq 1$$
$$4 + 1 \overset{?}{\geq} -1 \qquad 4 - 1 \overset{?}{\geq} 1$$
$$5 \geq -1 \qquad 3 \geq 1$$

The resulting true statements verify that (4, 1) is a solution of the system. If we pick a point that is not in the doubly shaded region, such as (1, 3), (−2, −2), or (0, −4), the coordinates of that point will fail to satisfy one or both of the inequalities.

Self Check 1 Graph the solutions of the system: $\begin{cases} x - y \leq 2 \\ x + y \geq -1 \end{cases}$.

In general, to solve systems of linear inequalities, we will follow these steps.

Solving Systems of Linear Inequalities

1. Graph each inequality on the same rectangular coordinate system.
2. Use shading to highlight the intersection of the graphs (the region where the graphs overlap). The points in this region are the solutions of the system.
3. As an informal check, pick a point from the region and verify that its coordinates satisfy each inequality of the original system.

EXAMPLE 2

ELEMENTARY & INTERMEDIATE
Algebra (f(x)) **Now™**

Graph the solutions of the system: $\begin{cases} y > 3x \\ 2x + y < 4 \end{cases}$.

Solution To graph $y > 3x$, we begin by graphing the boundary line $y = 3x$. Since the inequality contains a $>$ symbol, the boundary is a dashed line. Because the boundary passes through (0, 0), we use (2, 0) as the test point instead. Since (2, 0) does not satisfy $y > 3x$, we shade (in red) the half-plane that does not contain (2, 0). See part (a) of the following figure.

Graph the boundary: Slope and y-intercept

$$y = 3x + 0$$

$$m = 3 \qquad b = 0$$

Slope: $\dfrac{3}{1}$ y-intercept: $(0, 0)$

Shading: Check the test point (2, 0)

$$y > 3x$$

$$0 > 3(2) \qquad \text{Substitute.}$$

$$0 > 6$$

Since $0 > 6$ is false, $(2, 0)$ is not a solution of $y > 3x$.

In part (b) of the figure, we superimpose the graph of $2x + y < 4$ on the graph of $y > 3x$ to determine the points that the graphs have in common. To graph $2x + y < 4$, we graph the boundary $2x + y = 4$ as a dashed line. Then we shade (in blue) the half-plane that contains $(0, 0)$, because the coordinates of the test point $(0, 0)$ satisfy $2x + y < 4$.

Graph the boundary: The intercept method

$$2x + y = 4$$

x	y	(x, y)
0	4	$(0, 4)$
2	0	$(2, 0)$

Shading: Use the test point (0, 0)

$$2x + y < 4$$

$$2(0) + 0 \overset{?}{<} 4 \qquad \text{Substitute.}$$

$$0 < 4$$

Since $0 < 4$ is true, $(0, 0)$ is a solution of $2x + y < 4$.

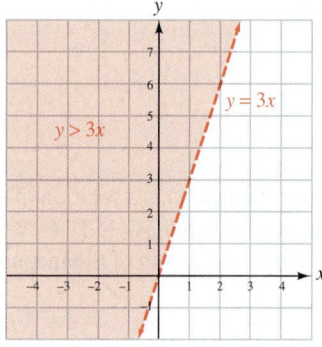

The graph of $y > 3x$ is shaded in red.

(a)

The solutions of the system are shaded in purple. Points on the boundaries are not solutions.

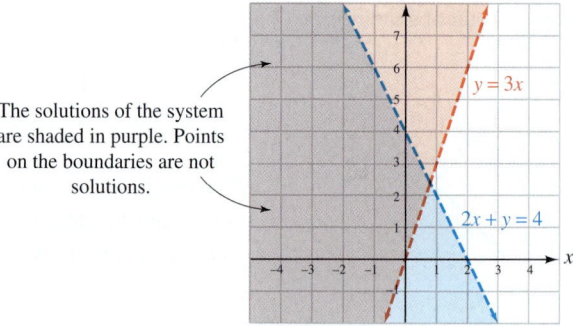

The graph of $2x + y < 4$ is shaded in blue. It is drawn over the graph of $y > 3x$.

(b)

In part (b) of the figure, the area that is shaded twice represents the solutions of the given system. Any point in the doubly shaded region in purple has coordinates that satisfy both inequalities. Pick a point in the region and show that this is true. Note that the region does not include either boundary; points on the boundaries are not solutions of the system.

Self Check 2   Graph the solutions of the system: $\begin{cases} x + 3y < 3 \\ y > \dfrac{1}{3}x \end{cases}$.

EXAMPLE 3

Graph the solutions of the system: $\begin{cases} x \le 2 \\ y > 3 \end{cases}$.

ELEMENTARY &
INTERMEDIATE
Algebra *f(x)* **Now**™   **Solution**   The boundary of the graph of $x \le 2$ is the line $x = 2$. Since the inequality contains the symbol $\le$, we draw the boundary as a solid line. The test point $(0, 0)$ makes $x \le 2$ true, so we shade the side of the boundary that contains $(0, 0)$. See part (a) of the figure.

Graph the boundary: A table of solutions

$$x = 2$$

x	y	(x, y)
2	0	$(2, 0)$
2	2	$(2, 2)$
2	4	$(2, 4)$

Shading: Check the test point $(0, 0)$

$$x \leq 2$$
$$0 \leq 2$$

Since $0 \leq 2$ is true, $(0, 0)$ is a solution of $x \leq 2$.

In part (b) of the figure, the graph of $y > 3$ is superimposed over the graph of $x \leq 2$. The boundary of the graph of $y > 3$ is the line $y = 3$. Since the inequality contains the symbol $>$, we draw the boundary as a dashed line. The test point $(0, 0)$ makes $y > 3$ false, so we shade the side of the boundary that does not contain $(0, 0)$.

Graph the boundary: A table of solutions

$$y = 3$$

x	y	(x, y)
0	3	$(0, 3)$
1	3	$(1, 3)$
4	3	$(4, 3)$

Shading: Check the test point $(0, 0)$

$$y > 3$$
$$0 > 3$$

Since $0 > 3$ is false, $(0, 0)$ is not a solution of $y > 3$.

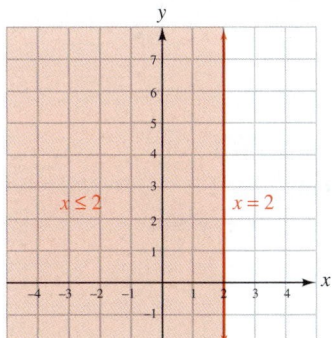

The graph of $x \leq 2$ is shaded in red.

(a)

The solutions of the system are shaded in purple. Points on the purple portion of $x = 2$ are solutions. Points on the dashed boundary line are not.

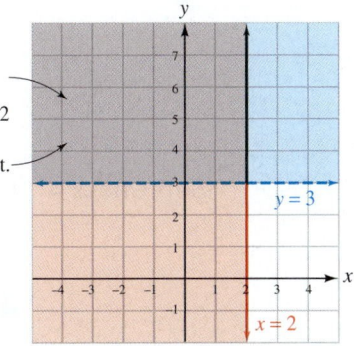

The graph of $y > 3$ is shaded in blue. It is drawn over the graph of $x \leq 2$.

(b)

The area that is shaded twice represents the solutions of the system of inequalities. Any point in the doubly shaded region in purple has coordinates that satisfy both inequalities, including the purple portion of the $x = 2$ boundary. Pick a point in the region and show that this is true.

Self Check 3   Graph the solutions of the system: $\begin{cases} y \leq 1 \\ x > 2 \end{cases}$.

EXAMPLE 4

ELEMENTARY & INTERMEDIATE
Algebra *f(x)* **Now**™

Graph the solutions of the system: $\begin{cases} x \geq 0 \\ y \geq 0 \\ x + 2y \leq 6 \end{cases}$.

Solution   This is a system of three linear inequalities. If shading is used to graph them on the same set of axes, it can become difficult to interpret the results. Instead, we can draw directional arrows attached to each boundary line in place of the shading.

- The graph of $x \geq 0$ has the boundary $x = 0$ and includes all points on the y-axis and to the right.
- The graph of $y \geq 0$ has the boundary $y = 0$ and includes all points on the x-axis and above.
- The graph of $x + 2y \leq 6$ has the boundary $x + 2y = 6$. Because the coordinates of the origin satisfy $x + 2y \leq 6$, the graph includes all points on and below the boundary.

The solutions of the system are the points that lie on triangle OPQ and the shaded triangular region that it encloses.

$x + 2y = 6$

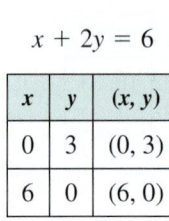

x	y	(x, y)
0	3	$(0, 3)$
6	0	$(6, 0)$

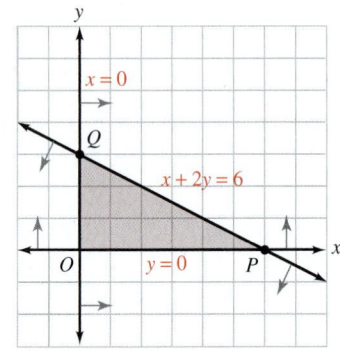

Self Check 4   Graph the solutions of the system: $\begin{cases} x \leq 1 \\ y \leq 2 \\ 2x - y \leq 4 \end{cases}$.

■ AN APPLICATION

EXAMPLE 5

ELEMENTARY &
INTERMEDIATE
Algebra $f(x)$ **Now**™

Landscaping.   A homeowner budgets from \$300 to \$600 for trees and bushes to landscape his yard. After shopping around, he finds that good trees cost \$150 and mature bushes cost \$75. What combinations of trees and bushes can he afford to buy?

Analyze the Problem
- At least \$300 but not more than \$600 is to be spent for trees and bushes.
- Trees cost \$150 and bushes cost \$75.
- What combination of trees and bushes can he buy?

Form Two Inequalities   Let $x =$ the number of trees purchased and $y =$ the number of bushes purchased. We then form the following system of inequalities:

The cost of a tree	times	the number of trees purchased	plus	the cost of a bush	times	the number of bushes purchased	should at least be	\$300.
\$150	·	x	+	\$75	·	y	≥	\$300

The cost of a tree	times	the number of trees purchased	plus	the cost of a bush	times	the number of bushes purchased	should not be more than	$600.
$150	·	x	+	$75	·	y	≤	$600

Solve the System   To solve this system of linear inequalities

$$\begin{cases} 150x + 75y \geq 300 \\ 150x + 75y \leq 600 \end{cases}$$

we use the graphing methods discussed in this section. Neither a negative number of trees nor a negative number of bushes can be purchased, so we restrict the graph to Quadrant I.

State the Conclusion   The coordinates of each point highlighted in the graph give a possible combination of the number of trees, x, and the number of bushes, y, that can be purchased. Written as ordered pairs, these possibilities are

$(0, 4), (0, 5), (0, 6), (0, 7), (0, 8),$
$(1, 2), (1, 3), (1, 4), (1, 5), (1, 6),$
$(2, 0), (2, 1), (2, 2), (2, 3), (2, 4),$
$(3, 0), (3, 1), (3, 2), (4, 0)$

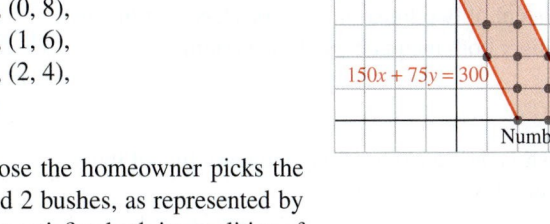

Check the Result   Suppose the homeowner picks the combination of 3 trees and 2 bushes, as represented by $(3, 2)$. Show that this point satisfies both inequalities of the system.

Answers to Self Checks

1.
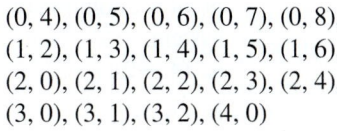

2.
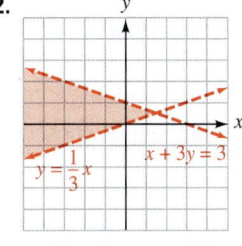

3.
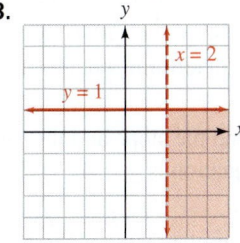

4.
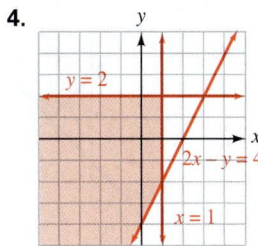

7.5   STUDY SET   ELEMENTARY & INTERMEDIATE Algebra $f(x)$ Now™

VOCABULARY   Fill in the blanks.

1. $\begin{cases} x + y > 2 \\ x + y < 4 \end{cases}$ is a system of linear _____.

2. A _____ of a system of linear inequalities is an ordered pair that satisfies each inequality.

3. To _____ a system of linear inequalities means to find all of its solutions.

4. To graph the linear inequality $x + y > 2$, first graph the _____ $x + y = 2$. Then pick the test _____ $(0, 0)$ to determine which half-plane to shade.

5. To find the solutions of a system of two linear inequalities graphically, look for the _____, or overlap, of the two shaded regions.

6. Any point in the doubly _____ region of the graph of a system of two linear inequalities has coordinates that satisfy both inequalities of the system.

CONCEPTS

7. **a.** What is the equation of the boundary line of the graph of $3x - y < 5$?

   b. Is the boundary a solid or dashed line?

8. **a.** What is the equation of the boundary line of the graph of $y \geq 4x$?

   b. Is the boundary a solid or dashed line?

   c. Why can't $(0, 0)$ be used as a test point to determine what to shade?

9. Find the slope and the y-intercept of the line whose equation is $y = 4x - 3$.

10. Complete the table to find the x- and y-intercepts of the line whose equation is $3x - 2y = 6$.

x	y
0	
	0

11. The boundary of the graph of $2x + y > 4$ is shown.

    a. Does the point $(0, 0)$ make the inequality true?

    b. Should the region above or below the boundary be shaded?

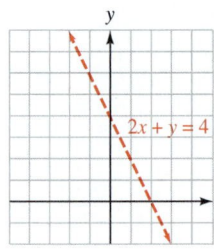

12. Linear inequality 1 is graphed in red and linear inequality 2 is graphed in blue. Decide whether a true or false statement results when the coordinates of the given point are substituted into the given inequality.

    a. A, inequality 1

    b. A, inequality 2

    c. B, inequality 1

    d. B, inequality 2

    e. C, inequality 1

    f. C, inequality 2

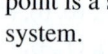

Inequality 1 solutions in red

Inequality 2 solutions in blue

13. The graph of a system of two linear inequalities is shown. Tell whether each point is a solution of the system.

    a. $(4, -2)$

    b. $(1, 3)$

    c. the origin

14. Use a check to determine whether each ordered pair is a solution of the system.

$$\begin{cases} x + 2y \geq -1 \\ x - y < 2 \end{cases}$$

    a. $(1, 4)$ **b.** $(-2, 0)$

15. Match each equation, inequality, or system with the graph of its solution.

    a. $x + y = 2$ **b.** $x + y \geq 2$

    c. $\begin{cases} x + y = 2 \\ x - y = 2 \end{cases}$ **d.** $\begin{cases} x + y \geq 2 \\ x - y \leq 2 \end{cases}$

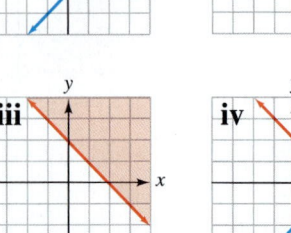
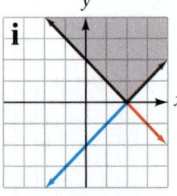
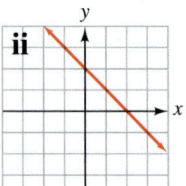
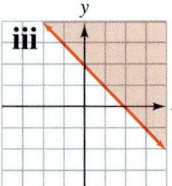
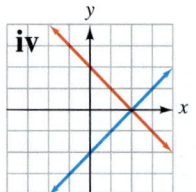

16. Match the system of inequalities with the correct graph.

a. $\begin{cases} x \geq 2 \\ y < 1 \end{cases}$ **b.** $\begin{cases} x > 2 \\ y \leq 1 \end{cases}$

c. $\begin{cases} x \geq 2 \\ y \geq 1 \end{cases}$ **d.** $\begin{cases} x > 2 \\ y > -1 \end{cases}$

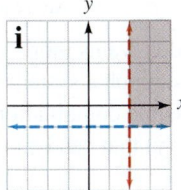
i

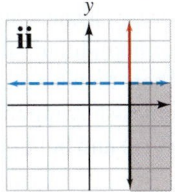
ii

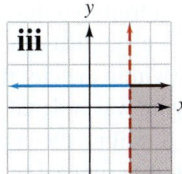
iii

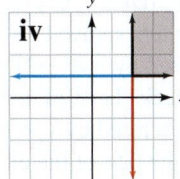
iv

NOTATION

17. Fill in the blank: This graph of the solutions of a system of linear inequalities can be described as the triangle _____ and the triangular region it encloses.

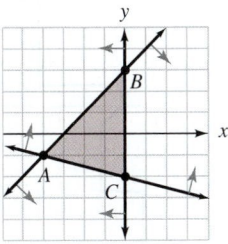

18. Represent each phrase using either $>$, $<$, $\geq$, or $\leq$.
  a. is not more than
  b. must be at least
  c. should not surpass
  d. cannot go below

PRACTICE   Graph the solutions of each system.

19. $\begin{cases} x + 2y \leq 3 \\ 2x - y \geq 1 \end{cases}$ **20.** $\begin{cases} 2x + y \geq 3 \\ x - 2y \leq -1 \end{cases}$

21. $\begin{cases} x + y < -1 \\ x - y > -1 \end{cases}$ **22.** $\begin{cases} x + y > 2 \\ x - y < -2 \end{cases}$

23. $\begin{cases} 2x - 3y \leq 0 \\ y \geq x - 1 \end{cases}$ **24.** $\begin{cases} y > 2x - 4 \\ y \geq -x - 1 \end{cases}$

25. $\begin{cases} x + y < 2 \\ x + y \leq 1 \end{cases}$ **26.** $\begin{cases} y > -x + 2 \\ y < -x + 4 \end{cases}$

27. $\begin{cases} x \geq 2 \\ y \leq 3 \end{cases}$ **28.** $\begin{cases} x \geq -1 \\ y > -2 \end{cases}$

29. $\begin{cases} x > 0 \\ y > 0 \end{cases}$ **30.** $\begin{cases} x \leq 0 \\ y < 0 \end{cases}$

31. $\begin{cases} 3x + 4y \geq -7 \\ 2x - 3y \geq 1 \end{cases}$ **32.** $\begin{cases} 3x + y \leq 1 \\ 4x - y \geq -8 \end{cases}$

33. $\begin{cases} 2x + y < 7 \\ y > 2 - 2x \end{cases}$ **34.** $\begin{cases} 2x + y \geq 6 \\ y \leq 4x - 6 \end{cases}$

35. $\begin{cases} 2x - 4y > -6 \\ 3x + y \geq 5 \end{cases}$ **36.** $\begin{cases} 2x - 3y < 0 \\ 2x + 3y \geq 12 \end{cases}$

37. $\begin{cases} 3x - y + 4 \leq 0 \\ 3y > -2x - 10 \end{cases}$ **38.** $\begin{cases} 3x + 2y - 12 \geq 0 \\ x < -2 + y \end{cases}$

39. $\begin{cases} y \geq x \\ y \leq \dfrac{1}{3}x + 1 \end{cases}$ **40.** $\begin{cases} y > 3x \\ y \leq -x - 1 \end{cases}$

41. $\begin{cases} x + y > 0 \\ y - x < -2 \end{cases}$ **42.** $\begin{cases} y + 2x \leq 0 \\ y \leq \dfrac{1}{2}x + 2 \end{cases}$

43. $\begin{cases} x \geq 0 \\ y \geq 0 \\ x + y \leq 3 \end{cases}$ **44.** $\begin{cases} x - y \leq 6 \\ x + 2y \leq 6 \\ x \geq 0 \end{cases}$

45. $\begin{cases} x - y < 4 \\ y \leq 0 \\ x \geq 0 \end{cases}$ **46.** $\begin{cases} 2x + y \leq 2 \\ y > x \\ x \geq 0 \end{cases}$

APPLICATIONS

47. BIRDS OF PREY   Parts (a) and (b) of the illustration show the individual fields of vision for each eye of an owl. In part (c), shade the area where the fields of vision overlap—that is, the area that is seen by both eyes.

(a)

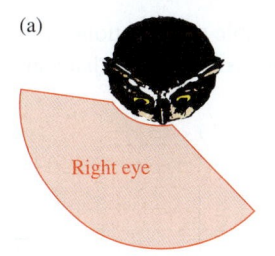

Right eye

(b)

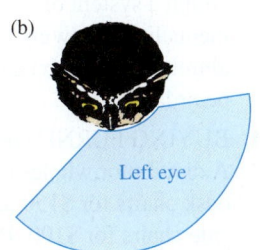

Left eye

(c)

Right   Left

48. EARTH SCIENCE
Shade the area of the earth's surface that is north of the Tropic of Capricorn and south of the Tropic of Cancer.

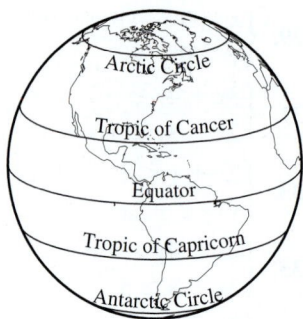

In Problems 49–52, graph each system of inequalities and give two possible solutions.

49. BUYING COMPACT DISCS   Melodic Music has compact discs on sale for either $10 or $15. If a customer wants to spend at least $30 but no more than $60 on CDs, graph a system of inequalities showing the possible combinations of $10 CDs ($x$) and $15 CDs ($y$) that the customer can buy.

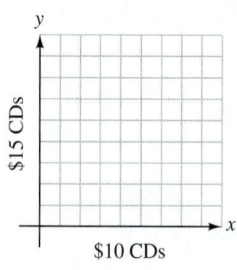

50. BUYING BOATS   Dry Boatworks wholesales aluminum boats for $800 and fiberglass boats for $600. Northland Marina wants to make a purchase totaling at least $2,400 but no more than $4,800. Graph a system of inequalities showing the possible combinations of aluminum boats (x) and fiberglass boats (y) that can be ordered.

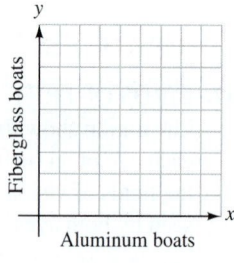

51. BUYING FURNITURE   A distributor wholesales desk chairs for $150 and side chairs for $100. Best Furniture wants its order to total no more than $900; Best also wants to order more side chairs than desk chairs. Graph a system of inequalities showing the possible combinations of desk chairs (x) and side chairs (y) that can be ordered.

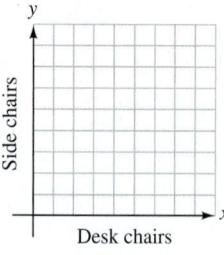

52. ORDERING FURNACE EQUIPMENT   J. Bolden Heating Company wants to order no more than $2,000 worth of electronic air cleaners and humidifiers from a wholesaler that charges $500 for air cleaners and $200 for humidifiers. If Bolden wants more humidifiers than air cleaners, graph a system of inequalities showing the possible combinations of air cleaners (x) and humidifiers (y) that can be ordered.

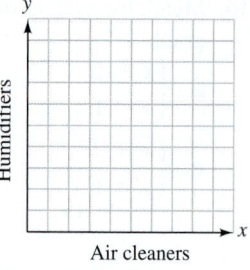

53. PESTICIDES   To eradicate a fruit fly infestation, helicopters sprayed an area of a city that can be described by $y \geq -2x + 1$ (within the city limits). Two weeks later, more spraying was ordered over the area described by $y \geq \frac{1}{4}x - 4$ (within the city limits). Show the part of the city that was sprayed twice.

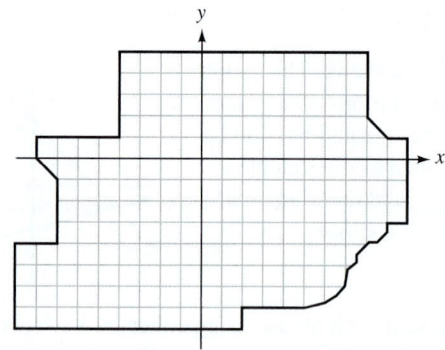

54. REDEVELOPMENT   Refer to the following diagram. A government agency has declared an area of a city east of First Street, north of Second Avenue, south of Sixth Avenue, and west of Fifth Street as eligible for federal redevelopment funds. Describe this area of the city mathematically using a system of four inequalities, if the corner of Central Avenue and Main Street is considered the origin.

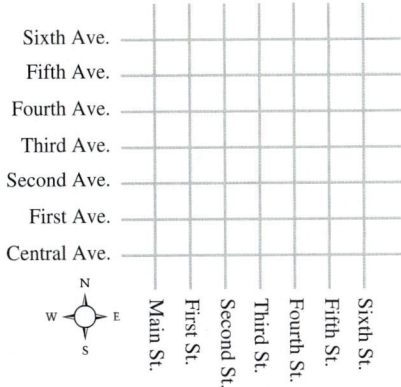

58. Explain when a system of inequalities will have no solutions.

REVIEW   Write each number in standard notation.

59. 2.3×10^2

60. 3.75×10^4

61. 9.76×10^{-4}

62. 7.63×10^{-5}

Write each number in scientific notation.

63. 290,000

64. 1,700,000

65. 0.0000051

66. 0.00073

WRITING

55. Explain how to use graphing to solve a system of inequalities.

56. When a solution of a system of linear inequalities is graphed, what does the shading represent?

57. Describe how the graphs of the solutions of these systems are similar and how they differ.

$$\begin{cases} x + y = 4 \\ x - y = 4 \end{cases} \quad \text{and} \quad \begin{cases} x + y \geq 4 \\ x - y \geq 4 \end{cases}$$

CHALLENGE PROBLEMS   Graph the solutions of each system.

67. $\begin{cases} \dfrac{x}{3} - \dfrac{y}{2} < -3 \\ \dfrac{x}{3} + \dfrac{y}{2} > -1 \end{cases}$

68. $\begin{cases} 3x + y < -2 \\ y > 3(1 - x) \end{cases}$

69. $\begin{cases} 2x + 3y \leq 6 \\ 3x + y \leq 1 \\ x \leq 0 \end{cases}$

70. $\begin{cases} x \geq 0 \\ y \geq 0 \\ 9x + 3y \leq 18 \\ 3x + 6y \leq 18 \end{cases}$

ACCENT ON TEAMWORK

WRITING APPLICATION PROBLEMS

Overview: In Section 7.4, you solved application problems by translating the words of the problem into a system of two equations. In this activity, you will reverse these steps.
Instructions: Form groups of 2 or 3 students. For each type of application, write a problem that could be solved using the given equations. If you need help getting started, refer to the specific problem types in the text. When finished writing the five applications, pick one problem and solve it completely.

A rectangle problem:
$$\begin{cases} 2l + 2w = 320 \\ l = w + 40 \end{cases}$$

A number-value problem:
$$\begin{cases} 5x + 2y = 23 \\ 3x + 7y = 37 \end{cases}$$

An interest problem:
$$\begin{cases} x + y = 75{,}000 \\ 0.03x + 0.05y = 2{,}750 \end{cases}$$

A with–against the wind problem:
$$\begin{cases} 2(x + y) = 600 \\ 3(x - y) = 600 \end{cases}$$

A liquid mixture problem:
$$\begin{cases} x + y = 36 \\ 0.50x + 0.20y = 0.30(36) \end{cases}$$

INTERPRETING GRAPHS OF LINEAR INEQUALITIES

Overview: This activity will strengthen your understanding of shading and boundary lines, which are used when graphing the solutions of systems of linear inequalities.

Instructions: Form groups of 2 or 3 students. Find the solutions of the system of linear inequalities by graphing the inequalities on the given coordinate system.

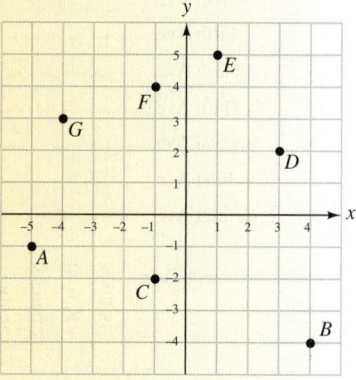

$$\begin{cases} 3x + 2y < 4 \\ y \le \dfrac{3}{5}x + 2 \end{cases}$$

For each point $A, B, C, D, E, F,$ and G on the graph, determine whether its coordinates make the first inequality true or false. Then determine whether each point's coordinates make the second inequality true or false. Write your answers in a table like that shown below.

Point	Coordinates	1st inequality	2nd inequality
A			

KEY CONCEPT: THREE METHODS FOR SOLVING SYSTEMS OF EQUATIONS

Each method that we used to solve systems of linear equations has advantages and disadvantages. Here are some guidelines to follow when deciding whether to use graphing, substitution, or elimination.

Method	Advantages	Disadvantages
Graphing	• You see the solutions. • The graphs allow you to observe trends.	• Inaccurate when the solutions are not integers, or are large numbers off the graph.
Substitution	• Always gives the exact solutions. • Works well if one of the equations is solved for one of the variables, or if it is easy to solve for one of the variables.	• You do not see the solution. • If no variable has a coefficient of 1 or -1, solving for one of the variables often involves fractions.
Elimination	• Always gives the exact solutions. • Works well if no variable has a coefficient of 1 or -1.	• You do not see the solution. • Equations must be written in the form $Ax + By = C$.

SOLVING SYSTEMS OF EQUATIONS BY GRAPHING

A system of linear equations can be solved by graphing both equations and locating the point of intersection of the two lines.

1. **FOOD SERVICE** The equations in the table give the fees two catering companies charge a Hollywood studio for on-location meal service, where y = the total cost (in dollars) for x meals. Graph the system to find the break point. That is, find the number of meals and the corresponding fee for which the two caterers will charge the same amount.

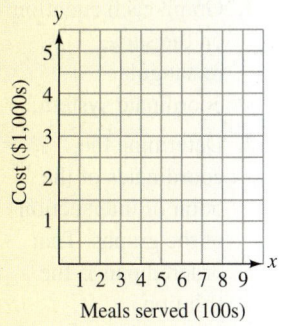

Meals served (100s)

Caterer	Setup fee	Cost per meal	Equation
Sunshine	$1,000	$4	$y = 4x + 1,000$
Lucy's	$500	$5	$y = 5x + 500$

SOLVING SYSTEMS OF EQUATIONS BY SUBSTITUTION

When using the substitution method, the objective is to substitute for one of the variables to obtain one equation containing only one unknown.

2. Solve by substitution: $\begin{cases} y = 2x - 9 \\ x + 3y = 8 \end{cases}$

3. Solve by substitution: $\begin{cases} 3x + 4y = -7 \\ 2y - x = -1 \end{cases}$

SOLVING SYSTEMS OF EQUATIONS BY ELIMINATION

When using the elimination method, the objective is to obtain two equations whose sum is an equation containing only one unknown.

4. Solve by elimination: $\begin{cases} x + y = 1 \\ x - y = 5 \end{cases}$

5. Solve by elimination: $\begin{cases} 2x - 3y = -18 \\ 3x + 2y = -1 \end{cases}$

CHAPTER REVIEW   ELEMENTARY & INTERMEDIATE Algebra $f(x)$ Now™

SECTION 7.1	**Solving Systems of Equations by Graphing**

CONCEPTS

A *solution* of a system of equations is an ordered pair that satisfies both equations of the system.

REVIEW EXERCISES

Determine whether the ordered pair is a solution of the system.

1. $(2, -3)$, $\begin{cases} 3x - 2y = 12 \\ 2x + 3y = -5 \end{cases}$

2. $\left(\dfrac{7}{2}, -\dfrac{2}{3}\right)$, $\begin{cases} 3y = 2x - 9 \\ 2x + 3y = 5 \end{cases}$

To solve a system *graphically*:

1. Graph each equation on the same rectangular coordinate system.

2. Determine the coordinates of the point of intersection of the graphs. That ordered pair is the solution.

3. Check the solution in each equation of the original system.

A system of equations that has at least one solution is called a *consistent system*. If the graphs are parallel lines, the system has no solution, and it is called an *inconsistent system*.

Equations with different graphs are called *independent equations*. If the graphs are the same line, the system has infinitely many solutions. The equations are called *dependent equations*.

3. COLLEGE ENROLLMENT Estimate the point of intersection of the graphs. Explain its significance.

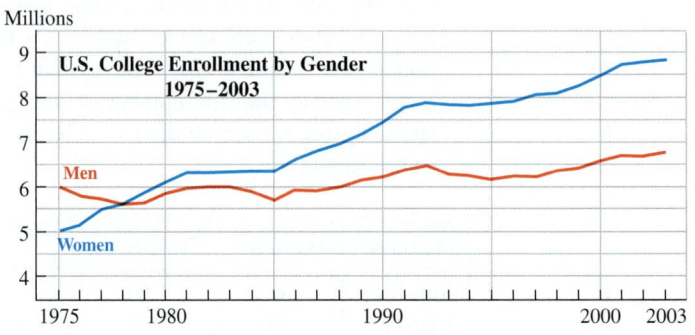

Source: *Digest of Education Statistics*

Use the graphing method to solve each system.

4. $\begin{cases} x + y = 7 \\ 2x - y = 5 \end{cases}$

5. $\begin{cases} 2x + y = 5 \\ y = -\dfrac{x}{3} \end{cases}$

6. $\begin{cases} 3x + 6y = 6 \\ x + 2y - 2 = 0 \end{cases}$

7. $\begin{cases} 6x + 3y = 12 \\ y = -2x + 2 \end{cases}$

SECTION 7.2   **Solving Systems of Equations by Substitution**

To solve a system of equations in *x* and *y* by substitution:

1. Solve one of the equations for either *x* or *y*.

2. Substitute the expression for *x* or for *y* obtained in step 1 into the other equation and solve that equation.

Use the substitution method to solve each system.

8. $\begin{cases} y = 15 - 3x \\ 7y + 3x = 15 \end{cases}$

9. $\begin{cases} x = y \\ 5x - 4y = 3 \end{cases}$

10. $\begin{cases} 6x + 2y = 8 - y + x \\ 3x = 2 - y \end{cases}$

11. $\begin{cases} r = 3s + 7 \\ r = 2s + 5 \end{cases}$

12. $\begin{cases} 9x + 3y - 5 = 0 \\ 3x + y = \dfrac{5}{3} \end{cases}$

13. $\begin{cases} \dfrac{x}{2} + \dfrac{y}{2} = 11 \\ \dfrac{5x}{16} - \dfrac{3y}{16} = \dfrac{15}{8} \end{cases}$

3. Substitute the value of the variable found in step 2 into the substitution equation to find the value of the remaining variable.

4. Check the proposed solution in the equations of the original system.

14. In solving a system using the substitution method, suppose you obtain the result of $8 = 9$.

    a. How many solutions does the system have?

    b. Describe the graph of the system.

    c. What term is used to describe the system?

Solving Systems of Equations by Elimination

To solve a system of equations in x and y using elimination:

1. Write each equation in the form $Ax + By = C$.

2. Multiply one or both equations by nonzero quantities to make the coefficients of x (or y) opposites.

3. Add the equations to eliminate the terms involving x (or y).

4. Solve the equation resulting from step 3.

5. Find the value of the other variable by substituting the value of the variable found in step 4 into any equation containing both variables.

6. Check the solution in the original equations.

Solve each system using the elimination method.

15. $\begin{cases} 2x + y = 1 \\ 5x - y = 20 \end{cases}$

16. $\begin{cases} x + 8y = 7 \\ x - 4y = 1 \end{cases}$

17. $\begin{cases} 5a + b = 2 \\ 3a + 2b = 11 \end{cases}$

18. $\begin{cases} 11x + 3y = 27 \\ 8x + 4y = 36 \end{cases}$

19. $\begin{cases} 9x + 3y = 15 \\ 3x = 5 - y \end{cases}$

20. $\begin{cases} -\dfrac{a}{4} - \dfrac{b}{3} = \dfrac{1}{12} \\ \dfrac{a}{2} - \dfrac{5b}{4} = \dfrac{7}{4} \end{cases}$

21. $\begin{cases} 0.02x + 0.05y = 0 \\ 0.3x - 0.2y = -1.9 \end{cases}$

22. $\begin{cases} -\dfrac{1}{4}x = 1 - \dfrac{2}{3}y \\ 6x - 18y = 5 - 2y \end{cases}$

For each system, tell which method, substitution or elimination, would be easier to use to solve the system and why.

23. $\begin{cases} 6x + 2y = 5 \\ 3x - 3y = -4 \end{cases}$

24. $\begin{cases} x = 5 - 7y \\ 3x - 3y = -4 \end{cases}$

Problem Solving Using Systems of Equations

In this section, we considered ways to solve problems by using *two* variables.

Write a system of two equations in two variables to solve each problem.

25. ELEVATIONS   The elevation of Las Vegas, Nevada, is 20 times greater than that of Baltimore, Maryland. The sum of their elevations is 2,100 feet. Find the elevation of each city.

To solve problems involving two unknown quantities:

1. *Analyze* the facts of the problem. Make a table or diagram if necessary.

2. Pick different variables to represent two unknown quantities. *Form* two equations involving the variables.

3. *Solve* the system of equations.

4. *State* the conclusion.

5. *Check* the results.

The *break point* of a linear system is the point of intersection of the graphs.

26. PAINTING EQUIPMENT   When fully extended, a ladder is 35 feet in length. If the extension is 7 feet shorter than the base, how long is each part of the ladder?

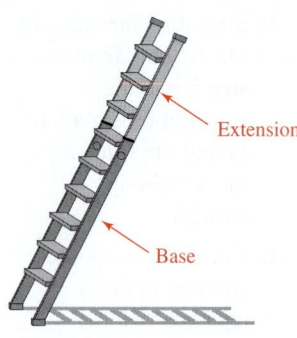

Extension

Base

27. CRASH INVESTIGATION   In an effort to protect evidence, investigators used 420 yards of yellow "Police Line—Do Not Cross" tape to seal off a large rectangular-shaped area around an airplane crash site. How much area will the investigators have to search if the width of the rectangle is three-fourths of the length?

28. CELEBRITY ENDORSEMENT   A company selling a home juicing machine is contemplating hiring either an athlete or an actor to serve as a spokesperson for the product. The terms of each contract would be as follows:

Celebrity	Base pay	Commission per item sold
Athlete	$30,000	$5
Actor	$20,000	$10

a. For each celebrity, write an equation giving the money y in dollars the celebrity would earn if x juicers were sold.

b. For what number of juicers would the athlete and the actor earn the same amount?

c. Graph the equations from part (a). The company expects to sell over 3,000 juicers. Which celebrity would cost the company the least money to serve as a spokesperson?

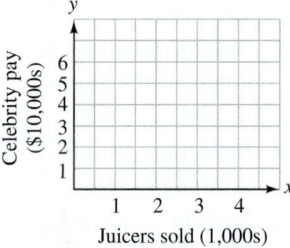

29. CANDY OUTLET STORE   A merchant wants to mix gummy worms worth $3 per pound and gummy bears worth $1.50 per pound to make 30 pounds of a mixture worth $2.10 per pound. How many pounds of each type of candy should he use?

30. BOATING   It takes a motorboat 4 hours to travel 56 miles down a river, and 3 hours longer to make the return trip. Find the speed of the current.

31. SHOPPING   Packages containing two bottles of contact lens cleaner and three bottles of soaking solution cost $63.40, and packages containing three bottles of cleaner and two bottles of soaking solution cost $69.60. Find the cost of a bottle of cleaner and a bottle of soaking solution.

32. INVESTING   Carlos invested part of $3,000 in a 10% certificate account and the rest in a 6% passbook account. The total annual interest from both accounts is $270. How much did he invest at 6%?

33. ANTIFREEZE   How much of a 40% antifreeze solution must a mechanic mix with a 70% antifreeze solution if he needs 20 gallons of a 50% antifreeze solution?

SECTION 7.5	**Solving Systems of Linear Inequalities**

A *solution* of a system of linear inequalities is an ordered pair that satisfies each inequality.

Solve each system of inequalities.

34. $\begin{cases} 5x + 3y < 15 \\ 3x - y > 3 \end{cases}$

35. $\begin{cases} 3y \leq x \\ y > 3x \end{cases}$

To solve a system of linear inequalities:

1. Graph each inequality on the same rectangular coordinate system.

2. Use shading to highlight the intersection of the graphs. The points in this region are the solutions of the system.

3. As an informal check, pick a point from the region and verify that its coordinates satisfy each inequality of the original system.

36. GIFT SHOPPING   A grandmother wants to spend at least $40 but no more than $60 on school clothes for her grandson. If T-shirts sell for $10 and pants sell for $20, write a system of inequalities that describes the possible numbers of T-shirts x and pairs of pants y that she can buy. Graph the system and give two possible solutions.

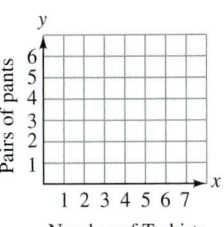

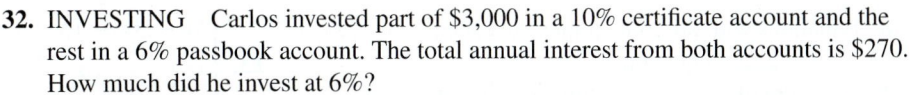

CHAPTER 7 TEST

ELEMENTARY &
INTERMEDIATE
Algebra $f(x)$ **Now**™

Determine whether the ordered pair is a solution of the system.

1. $(5, 3)$, $\begin{cases} 3x + 2y = 21 \\ x + y = 8 \end{cases}$

2. $(-2, -1)$, $\begin{cases} 4x + y = -9 \\ 2x - 3y = -7 \end{cases}$

3. Fill in the blanks.
   a. A _____ of a system of linear equations is an ordered pair that satisfies each equation.
   b. A system of equations that has at least one solution is called a _____ system.
   c. A system of equations that has no solution is called an _____ system.

d. Equations with different graphs are called
_____ equations.

e. A system of _____ equations has an infinite
number of solutions.

Solve each system by graphing.

4. $\begin{cases} y = 2x - 1 \\ x - 2y = -4 \end{cases}$

5. $\begin{cases} x + y = 5 \\ y = -x \end{cases}$

*The following graph shows two different ways in which a
salesperson can be paid according to the number of items
she sells each month.*

6. What is the point of intersection of the graphs? Explain
its significance.

7. Which plan is better for the salesperson if she feels
that selling 30 items per month is virtually
impossible?

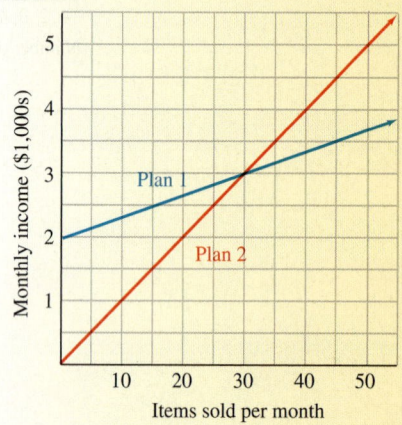

Solve each system by substitution.

8. $\begin{cases} y = x - 1 \\ 2x + y = -7 \end{cases}$

9. $\begin{cases} 3x + 6y = -15 \\ x + 2y = -5 \end{cases}$

10. $\begin{cases} 3a + 4b = -7 \\ 2b - a = -1 \end{cases}$

Solve each system using elimination.

11. $\begin{cases} 3x - y = 2 \\ 2x + y = 8 \end{cases}$

12. $\begin{cases} 4x + 3y = -3 \\ -3x = -4y + 21 \end{cases}$

13. $\begin{cases} 3x - 5y - 16 = 0 \\ \dfrac{x}{2} - \dfrac{5}{6}y = \dfrac{1}{3} \end{cases}$

14. Which method would be most efficient to solve the
following system? Explain your answer. (**You do not
need to solve the system.**)

$$\begin{cases} 5x - 3y = 5 \\ 3x + 3y = 3 \end{cases}$$

*Write a system of two equations in two variables to solve
each problem.*

15. **CHILD CARE** On a mother's 22-mile commute to
work, she drops her daughter off at a child care
facility. The first part of the trip is 6 miles less than
the second part. How long is each part of her morning
commute?

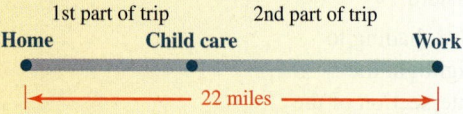

16. **VACATIONING** It cost a family of 7 a total of $119
for general admission tickets to the San Diego Zoo.
How many adult tickets and how many child tickets
were purchased?

SAN DIEGO ZOO

General Admission	
ADULT TICKETS	$21
CHILD TICKETS	$14
(ages 3-11)	

17. **FINANCIAL PLANNING** A woman invested some
money at 8% and some at 9%. The interest for 1 year
on the combined investment of $10,000 was $840.
How much was invested at 8% and how much was
invested at 9%?

18. **KAYAKING** A man can paddle his kayak *x* miles
per hour in still water. How fast will his kayak travel
if he paddles against a river current of *c* miles per
hour?

19. TETHER BALL   Translate this fact into an equation in two variables: the angles shown in the illustration are complementary.

20. Solve the system by graphing.

$$\begin{cases} 2x + 3y \le 6 \\ x > 2 \end{cases}$$

21. CLOTHES SHOPPING   This system of inequalities describes the number of $20 shirts, x, and $40 pairs of pants, y, a person can buy if he or she plans to spend not less than $80 but not more than $120. Graph the system. Then give three solutions.

$$\begin{cases} 20x + 40y \ge 80 \\ 20x + 40y \le 120 \end{cases}$$

8

Transition to Intermediate Algebra

ELEMENTARY & INTERMEDIATE
Algebra *f(x)* Now™

Throughout the chapter, this icon introduces resources on the Elementary & Intermediate AlgebraNow Web site, accessed through **http://1pass. thomson.com**, that will

- Help you test your knowledge of the material with a pre-test and a post-test

- Provide a personalized learning plan targeting areas you should study

There are many ways to measure distance. For example, on long trips, motorists use the car's odometer to measure the distance traveled. When building a staircase, carpenters use a tape measure to make sure the distances between the vertical posts are the same. Scientists use a beam of light to measure the distances from Earth to the planets. In mathematics, we use absolute value to measure distance. Recall that the absolute value of a real number is its distance from zero on the number line. In this chapter, we will define absolute value more formally and we will solve equations and inequalities that contain the absolute value of a variable expression.

To learn more about absolute value, visit *The Learning Equation* on the Internet at http://tle.brookscole.com. (The log-in instructions are in the Preface.) For Chapter 8, the online lesson is:

- *TLE* Lesson 15: Absolute Value Equations

In this chapter, we will review many of the ideas covered in the first seven chapters and then extend them to the intermediate algebra level.

8.1 Review of Equations and Inequalities

- Solving equations
- Inequalities
- Identities and contradictions
- Compound inequalities
- Solving formulas
- Problem solving

In this section, we will review how to solve equations and formulas. We will then extend our review to include inequalities.

■ SOLVING EQUATIONS

Recall that an **equation** is a statement indicating that two expressions are equal. The set of numbers that satisfy an equation is called its **solution set,** and the elements in the solution set are called **solutions** of the equation. Finding all of the solutions of an equation is called **solving the equation.**

To solve an equation, we use the following properties of equality to replace the equation with simpler **equivalent equations** that have the same solution set. We continue this process until the variable is isolated on one side of the $=$ symbol.

1. Adding the same number to, or subtracting the same number from, both sides of an equation does not change its solution.
2. Multiplying or dividing both sides of an equation by the same nonzero number does not change its solution.

Success Tip

You may want to review the *properties of equality* on pages 94, 96, 97, and 98.

EXAMPLE 1

Solve: $3(2x - 1) = 2x + 9$.

Solution   We use the distributive property to remove parentheses and then isolate x on the left-hand side of the equation.

$$3(2x - 1) = 2x + 9$$
$$6x - 3 = 2x + 9 \qquad \text{Distribute the multiplication by 3.}$$
$$6x - 3 + 3 = 2x + 9 + 3 \qquad \text{To undo the subtraction by 3, add 3 to both sides.}$$
$$6x = 2x + 12 \qquad \text{Combine like terms.}$$
$$6x - 2x = 2x + 12 - 2x \qquad \text{To eliminate } 2x \text{ from the right-hand side, subtract } 2x \text{ from both sides.}$$
$$4x = 12 \qquad \text{Combine like terms.}$$
$$x = 3 \qquad \text{To undo the multiplication by 4, divide both sides by 4.}$$

Check: We substitute 3 for x in the original equation to see whether it satisfies the equation.

Evaluate the expression on the left-hand side.

$$3(2x - 1) = 2x + 9$$
$$3(2 \cdot 3 - 1) \overset{?}{=} 2 \cdot 3 + 9$$
$$3(5) \overset{?}{=} 6 + 9$$
$$15 = 15$$

Evaluate the expression on the right-hand side.

Since 3 satisfies the original equation, it is a solution. The solution set is {3}.

Self Check 1 Solve: $2(3x - 2) = 3x - 13$.

To solve more complicated linear equations we follow these steps.

Strategy for Solving Equations

1. Clear the equation of fractions.
2. Use the distributive property to remove parentheses, if necessary.
3. Combine like terms, if necessary.
4. Undo the operations of addition and subtraction to get the variable terms on one side and the constant terms on the other.
5. Undo the operations of multiplication and division to isolate the variable.
6. Check the result.

EXAMPLE 2

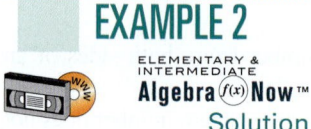

ELEMENTARY & INTERMEDIATE
Algebra *f(x)* **Now**™

Solution

Solve: $\dfrac{1}{3}(6x + 15) = \dfrac{3}{2}(x + 2) - 2$.

Step 1: We can clear the equation of fractions by multiplying both sides by the least common denominator (LCD) of $\frac{1}{3}$ and $\frac{3}{2}$. The LCD of these fractions is the smallest number that can be divided by both 2 and 3 exactly. That number is 6.

$$\frac{1}{3}(6x + 15) = \frac{3}{2}(x + 2) - 2$$

Success Tip

Before multiplying both sides of an equation by the LCD, frame the left-hand side and frame the right-hand side with parentheses or brackets.

$$\mathbf{6}\left[\frac{1}{3}(6x + 15)\right] = \mathbf{6}\left[\frac{3}{2}(x + 2) - 2\right]$$

To eliminate the fractions, multiply both sides by the LCD, 6.

$$2(6x + 15) = 6 \cdot \frac{3}{2}(x + 2) - 6 \cdot 2$$

On the left-hand side, perform the multiplication: $6 \cdot \frac{1}{3} = 2$. On the right-hand side, distribute the multiplication by 6.

$$2(6x + 15) = 9(x + 2) - 12$$

Multiply on the right-hand side.

Step 2: Use the distributive property to remove parentheses.

$$12x + 30 = 9x + 18 - 12$$

Distribute the multiplication by 2 and the multiplication by 9.

Step 3: Combine like terms on the right-hand side.

$$12x + 30 = 9x + 6$$

Step 4: To get the variable term on the left-hand side and the constant on the right-hand side, subtract $9x$ and 30 from both sides.

$$12x + 30 - 9x - 30 = 9x + 6 - 9x - 30$$
$$3x = -24 \qquad \text{On each side, combine like terms.}$$

Step 5: The coefficient of the variable x is 3. To undo the multiplication by 3, we divide both sides by 3.

$$\frac{3x}{3} = \frac{-24}{3}$$
$$x = -8$$

Step 6: We check by substituting -8 for x in the original equation and simplifying:

$$\frac{1}{3}(6x + 15) = \frac{3}{2}(x + 2) - 2$$
$$\frac{1}{3}[6(-8) + 15] \stackrel{?}{=} \frac{3}{2}(-8 + 2) - 2$$
$$\frac{1}{3}(-48 + 15) \stackrel{?}{=} \frac{3}{2}(-6) - 2$$
$$\frac{1}{3}(-33) \stackrel{?}{=} -9 - 2$$
$$-11 = -11 \qquad \text{True.}$$

The solution is -8.

Self Check 2 Solve: $\frac{1}{3}(2x - 2) = \frac{1}{4}(5x + 1) + 2$

■ IDENTITIES AND CONTRADICTIONS

The equations discussed so far are called **conditional equations.** For these equations, some numbers satisfy the equation and others do not. An **identity** is an equation that is satisfied by every number for which both sides of the equation are defined.

EXAMPLE 3

ELEMENTARY & INTERMEDIATE
Algebra *f(x)* **Now**™

Solution

Solve: $-2(x - 1) - 4 = -4(1 + x) + 2x + 2$.

We can use the distributve property to remove the parentheses on the left- and right-hand sides of the equation.

$$-2(x - 1) - 4 = -4(1 + x) + 2x + 2$$
$$-2x + 2 - 4 = -4 - 4x + 2x + 2 \qquad \text{Use the distributive property.}$$
$$-2x - 2 = -2x - 2 \qquad \text{On each side, combine like terms.}$$
$$-2 = -2 \qquad \text{True.}$$

The terms involving x drop out. The resulting true statement indicates that the original equation is true for every value of x. The solution set is the set of real numbers denoted $\mathbb{R}$. The equation is an identity.

Self Check 3 Solve: $3(a + 1) - (20 + a) = 5(a - 1) - 3(a + 4)$.

A **contradiction** is an equation that is never true.

EXAMPLE 4 Solve: $-6.2(-x - 1) - 4 = 4.2x - (-2x)$.

Solution

The Language of Algebra

Contradiction is a form of the word *contradict,* meaning conflicting ideas. During a trial, evidence might be introduced that *contradicts* the testimony of a witness.

$$-6.2(-x - 1) - 4 = 4.2x - (-2x)$$
$$6.2x + 6.2 - 4 = 4.2x + 2x$$

On the left-hand side, remove parentheses. On the right-hand side, write the subtraction as addition of the opposite.

$$6.2x + 2.2 = 6.2x$$

On each side, combine like terms.

$$6.2x + 2.2 - \mathbf{6.2x} = 6.2x - \mathbf{6.2x}$$

Subtract 6.2x from both sides.

$$2.2 = 0$$

False.

The terms involving x drop out. The resulting false statement indicates that no value for x makes the original equation true. The solution set contains no elements and can be denoted as the **empty set** { } or the **null set** $\varnothing$. The equation is a contradiction.

Self Check 4 Solve: $3(a + 4) + 2 = 2(a - 1) + a + 19$.

■ SOLVING FORMULAS

To solve a formula for a variable means to isolate that variable on one side of the equation and write all other quantities on the other side.

EXAMPLE 5 For simple interest, the formula $A = P + Prt$ gives the amount of money in an account at the end of a specific time. A represents the amount, P the principal, r the rate of interest, and t the time. Solve the formula for t.

We must isolate t on one side of the equation.

Solution
$$A = P + Prt$$
$$A - P = Prt$$

To isolate the term involving t, subtract P from both sides.

$$\frac{A - P}{Pr} = t$$

To isolate t, multiply both sides by $\frac{1}{Pr}$ or divide both sides by Pr.

$$t = \frac{A - P}{Pr}$$

Write the equation with t on the left-hand side.

Self Check 5 Solve $A = P + Prt$ for r.

EXAMPLE 6

ELEMENTARY &
INTERMEDIATE
Algebra *(f(x))* Now™

The formula $F = \dfrac{9}{5}C + 32$ converts degrees Celsius to degrees Fahrenheit. Solve it for C.

We most isolate C on one side of the equation.

Solution

$$F = \frac{9}{5}C + 32$$

$$F - 32 = \frac{9}{5}C \qquad \text{To isolate the term involving } C, \text{ subtract 32 from both sides.}$$

$$\frac{5}{9}(F - 32) = \frac{5}{9}\left(\frac{9}{5}C\right) \qquad \text{To isolate } C, \text{ multiply both sides by } \frac{5}{9}.$$

$$\frac{5}{9}(F - 32) = C \qquad \qquad \frac{5}{9} \cdot \frac{9}{5} = 1.$$

$$C = \frac{5}{9}(F - 32)$$

To convert degrees Fahrenheit to degrees Celsius, we can use the formula $C = \dfrac{5}{9}(F - 32)$.

Self Check 6   Solve $S = \dfrac{180(t - 2)}{7}$ for t.

■ INEQUALITIES

Inequalities are statements indicating that two quantities are unequal. Inequalities contain one or more of the following symbols.

Inequality Symbols			
	$<$ is less than	$>$ is greater than	$\neq$ is not equal to
	$\leq$ is less than or equal to	$\geq$ is greater than or equal to	

In this section, we will work with **linear inequalities** in one variable.

Linear Inequalities	A **linear inequality** in one variable (say, x) is any inequality that can be expressed in one of the following forms, where a, b, and c represent real numbers and $a \neq 0$.

$$ax + b < c \qquad ax + b \leq c \qquad ax + b > c \qquad \text{or} \qquad ax + b \geq c$$

Some examples of linear inequalities are

$$3x \leq 0 \qquad 3(2x - 9) < 9 \qquad \text{and} \qquad -12x - 8 \geq 16$$

To **solve a linear inequality** means to find all the values that, when substituted for the variable, make the inequality true. The set of all solutions of an inequality is called its **solution set.** The solution set of an inequality can be graphed as an **interval** on the number line. We can also write a solution set using **interval notation** and **set-builder notation.**

We use the following properties to solve inequalities.

Success Tip

You may want to review the *properties of inequality* on pages 161 and 163.

1. Adding the same number to, or subtracting the same number from, both sides of an inequality does not change the solutions.
2. Multiplying or dividing both sides of an inequality by the same positive number does not change the solutions.
3. If we multiply or divide both sides of an inequality by a negative number, the direction of the inequality symbol must be *reversed* for the inequalities to have the same solutions.

After applying one of the properties of inequality, the resulting inequality is equivalent to the original one. Like equivalent equations, **equivalent inequalities** have the same solution set.

EXAMPLE 7

Solve: $3(2x - 9) < 9$. Write the solution set in interval notation and graph it.

Solution   We want to isolate x on one side of the inequality symbol. To do that, we use the same strategy as we used to solve equations.

$$3(2x - 9) < 9$$
$$6x - 27 < 9 \qquad \text{Distribute the multiplication by 3.}$$
$$6x < 36 \qquad \text{To undo the subtraction of 27, add 27 to both sides.}$$
$$x < 6 \qquad \text{To undo the multiplication by 6, divide both sides by 6.}$$

The solution set is the interval $(-\infty, 6)$, whose graph is shown. We can also write the solution set using set-builder notation: $\{x \mid x < 6\}$.

The solution set contains infinitely many real numbers. We cannot check to see if all of them satisfy the original inequality. As an informal check, we pick one number on the graph, such as 4, and see whether it satisfies the inequality.

Check:
$$3(2x - 9) < 9 \qquad \text{This is the original inequality.}$$
$$3[2(4) - 9] \stackrel{?}{<} 9 \qquad \text{Substitute 4 for } x. \text{ Read } \stackrel{?}{<} \text{ as "is possibly less than."}$$
$$3(8 - 9) \stackrel{?}{<} 9 \qquad \text{Perform the multiplication: } 2(4) = 8.$$
$$3(-1) \stackrel{?}{<} 9 \qquad \text{Perform the subtraction: } 8 - 9 = -1.$$
$$-3 < 9 \qquad \text{This is a true statement.}$$

Since $-3 < 9$, the number 4 satisfies the inequality. The solution appears to be correct.

Self Check 7   Solve: $2(3x + 2) > -44$.

EXAMPLE 8

Solve: $-12x - 8 \leq 16$. Write the solution set in interval notation and graph it.

Solution   To solve this inequality, we isolate x.

$$-12x - 8 \leq 16$$
$$-12x \leq 24 \qquad \text{To undo the subtraction of 8, add 8 to both sides.}$$
$$x \geq -2 \qquad \text{To undo the multiplication by } -12, \text{ divide both sides by } -12. \text{ Because we are dividing by a negative number, we reverse the } \leq \text{ symbol.}$$

Success Tip

We must remember to reverse the inequality symbol every time we multiply or divide both sides by a negative number.

The solution set is $\{x \mid x \geq -2\}$ or the interval $[-2, \infty)$, whose graph is shown.

Self Check 8   Solve: $-6x + 6 \leq 0$.

EXAMPLE 9

Solve: $3a - 4 < 3(a + 5)$. Write the solution set in interval notation and graph it.

Solution

$$3a - 4 < 3(a + 5)$$
$$3a - 4 < 3a + 15 \qquad \text{Distribute the multiplication by 3.}$$
$$3a - 4 - 3a < 3a + 15 - 3a \qquad \text{Subtract } 3a \text{ from both sides.}$$
$$-4 < 15 \qquad \text{True.}$$

Success Tip

When solving an inequality, if the variables drop out and the result is a false statement, the solution set contains no elements and is denoted $\varnothing$. The graph is an unshaded number line.

The terms involving a drop out. The resulting true statement indicates that the original inequality is true for all values of a. Therefore, the solution set is the set of real numbers, denoted $(-\infty, \infty)$ or $\mathbb{R}$, and its graph is as shown.

(number line marked -1, 0, 1)

(number line marked -1, 0, 1)

Self Check 9

Solve: $-8n + 10 \geq 1 - 2(4n - 2)$.

■ COMPOUND INEQUALITIES

The Language of Algebra

Compound means composed of two or more parts, as in *compound* inequalities, chemical *compounds,* and *compound* sentences.

When two inequalities are joined with the word *and* or the word *or*, we call the statement a **compound inequality.** Some examples are

$$x \geq -3 \qquad \text{and} \qquad x \leq 6$$

$$\frac{x}{2} + 1 > 0 \qquad \text{and} \qquad 2x - 3 < 5$$

$$\frac{x}{3} > \frac{2}{3} \qquad \text{or} \qquad -(x - 2) > 3$$

The solution set of a compound inequality containing the word *and* includes all numbers that make *both* of the inequalities true. When solving a compound inequality containing *and*, the solution set is the *intersection* of the solutions sets of the two inequalities. In general, the **intersection** of two sets is the set of elements that are common to both sets. We can denote the intersection of two sets using the symbol $\cap$, which is read as "intersection."

EXAMPLE 10

Solve: $\frac{x}{2} + 1 > 0$ and $2x - 3 < 5$. Graph the solution set.

Solution

We solve each linear inequality separately.

The Language of Algebra

The *intersection* of two sets is the collection of elements that they have in common. When two streets cross, we call the area of pavement that they have in common an *intersection.*

$$\frac{x}{2} + 1 > 0 \qquad \text{and} \qquad 2x - 3 < 5$$

$$\frac{x}{2} > -1 \qquad\qquad\qquad 2x < 8$$

$$x > -2 \qquad\qquad\qquad x < 4$$

Next, we graph the solutions of each inequality on the same number line and determine their intersection.

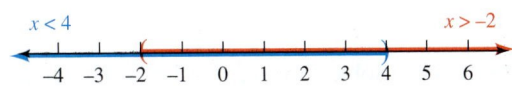

Notation

When graphing on the number line, $(-2, 4)$ represents an *interval*. When graphing on a rectangular coordinate system, $(-2, 4)$ is an *ordered pair* that gives the coordinates of a point.

The intersection of the graphs is the set of all real numbers between -2 and 4. The solution set of the compound inequality is the interval $(-2, 4)$, whose graph is shown below. This bounded interval, which does not include either endpoint, is called an **open interval.**

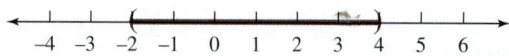

Self Check 10 Solve: $3x > -18$ and $\dfrac{x}{5} - 1 \leq 1$. Graph the solution set.

Inequalities containing two inequality symbols are called **double inequalities.** An example is

$$-3 \leq 2x + 5 < 7 \qquad \text{Read as "}-3 \text{ is less than or equal to } 2x + 5 \text{ and } 2x + 5 \text{ is less than 7."}$$

Any double linear inequality can be written as a compound inequality containing the word *and*. In general, the following is true.

Double Linear Inequalities	The double inequality $c < x < d$ is equivalent to $c < x$ and $x < d$.

EXAMPLE 11

ELEMENTARY & INTERMEDIATE
Algebra $f(x)$ **Now** ™

Solve: $-3 \leq 2x + 5 < 7$. Graph the solution set.

Solution This double inequality $-3 \leq 2x + 5 < 7$ means that

$$-3 \leq 2x + 5 \quad \text{and} \quad 2x + 5 < 7$$

We could solve each linear inequality separately, but we note that each solution would involve the same steps: subtracting 5 from both sides and dividing both sides by 2. We can solve the double inequality more efficiently by leaving it in its original form and applying these steps to each of its three parts to isolate x in the middle.

$$-3 \leq 2x + 5 < 7$$
$$-3 - 5 \leq 2x + 5 - 5 < 7 - 5 \qquad \text{To undo the addition of 5, subtract 5 from all three parts.}$$
$$-8 \leq 2x < 2 \qquad \text{Perform the subtractions.}$$
$$\frac{-8}{2} \leq \frac{2x}{2} < \frac{2}{2} \qquad \text{To undo the multiplication by 2, divide all three parts by 2.}$$
$$-4 \leq x < 1 \qquad \text{Perform the divisions.}$$

The solution set of the double linear inequality is the half-open interval $[-4, 1)$, whose graph is shown below.

Self Check 11 Solve: $-5 \leq 3x - 8 \leq 7$. Graph the solution set.

Caution   When multiplying or dividing all three parts of a double inequality by a negative number, don't forget to reverse the direction of *both* inequalities. As an example, we solve $-15 < -5x \le 25$.

$$-15 < -5x \le 25$$

$$\frac{-15}{-5} > \frac{-5x}{-5} \ge \frac{25}{-5}$$ Divide all three parts by -5 to isolate x in the middle. Reverse both inequality signs.

$$3 > x \ge -5$$ Perform the divisions.

$$-5 \le x < 3$$ Write an equivalent double inequality with the smaller number, -5, on the left.

The solution set of a compound inequality containing the word *or* includes all numbers that make *one or the other or both* inequalities true. When solving a compound inequality containing *or,* the solution set is the *union* of the solutions sets of the two inequalities. In general, the **union** of two sets is the set of elements that are in either of the sets or both. We can denote the union of two sets using the symbol $\cup$, which is read as "union."

The Language of Algebra

The *union* of two sets is the collection of elements that belong to either set. The concept is similar to that of a family re*union,* which brings together the members of several families.

EXAMPLE 12

ELEMENTARY & INTERMEDIATE
Algebra (f(x)) **Now**™   Solution

Solve: $\dfrac{x}{3} > \dfrac{2}{3}$ or $-(x - 2) > 3$. Graph the solution set.

We solve each inequality separately.

$$\frac{x}{3} > \frac{2}{3} \qquad \text{or} \qquad -(x - 2) > 3$$
$$x > 2 \qquad\qquad\quad -x + 2 > 3$$
$$\qquad\qquad\qquad\qquad -x > 1$$
$$\qquad\qquad\qquad\qquad\quad x < -1$$

Next, we graph the solutions of each inequality on the same number line and determine their union.

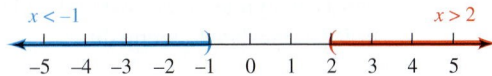

The union of the two solution sets consists of all real numbers less than -1 or greater than 2. The solution set of the compound inequality is the interval $(-\infty, -1) \cup (2, \infty)$. Its graph appears below.

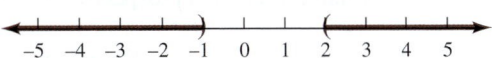

The Language of Algebra

The meaning of the word *or* in a compound inequality differs from our everyday use of the word. For example, when we say, "I will go shopping today *or* tomorrow," we mean that we will go one day or the other, but *not* both. With compound inequalities, *or* includes one possibility, or the other, or both.

Self Check 12   Solve: $\dfrac{x}{2} > 2$ or $-3(x - 2) > 0$. Graph the solution set.

■ **PROBLEM SOLVING**

We can use equations and inequalities to solve applied problems.

EXAMPLE 13

Travel promotions.   The price of a 7-day Alaskan cruise, normally $2,752 per person, is reduced by $1.75 per person for large groups traveling together. How large a group is needed for the price to be $2,500 per person?

Analyze the problem   For each member of the group, the cost is reduced by $1.75. For a group of 20 people, the $2,752 price is reduced by 20($1.75) = $35.

The per-person price of the cruise = $2,752 − 20($1.75)

For a group of 30 people, the $2,752 cost is reduced by 30($1.75) = $52.50.

The per-person price of the cruise = $2,752 − 30($1.75)

Form an equation   If we let x = the group size necessary for the price of the cruise to be $2,500 per person, we can form the following equation:

The price of the cruise	is	$2,752	minus	the number of people in the group	times	$1.75.
2,500	=	2,752	−	x	·	1.75

Solve the equation

$$2,500 = 2,752 - 1.75x$$
$$2,500 - \mathbf{2,752} = 2,752 - 1.75x - \mathbf{2,752} \qquad \text{Subtract 2,752 from both sides.}$$
$$-252 = -1.75x \qquad \text{Simplify each side.}$$
$$144 = x \qquad \text{Divide both sides by } -1.75.$$

State the conclusion   If 144 people travel together, the price will be $2,500 per person.

Check the result   For 144 people, the cruise cost of $2,752 will be reduced by 144($1.75) = $252. If we subtract, $2,752 − $252 = $2,500. The answer checks.

To decide whether to use an equation or an inequality to solve a problem, look for key words and phrases. For example, phrases such as *does not exceed, is no more than,* and *is at least* translate to inequalities.

EXAMPLE 14

Political contributions.   Some volunteers are making long-distance telephone calls to solicit contributions for their candidate. The calls are billed at the rate of 25¢ for the first three minutes and 7¢ for each additional minute or part thereof. If the campaign chairperson has ordered that the cost of each call is not to exceed $1.00, how many minutes can a volunteer talk to a prospective donor on the phone?

Analyze the problem   We are given the rate at which a call is billed. Since the cost of a call is not to exceed $1.00, the cost must be *less than or equal to* $1.00. This phrase indicates that we should write an inequality to find how long a volunteer can talk to a prospective donor.

Form an inequality   We will let x = the total number of minutes that a call can last. Then the cost of a call will be 25¢ for the first three minutes plus 7¢ times the number of additional minutes, where the number of *additional* minutes is $x - 3$ (the total number of minutes minus the first 3 minutes). With this information, we can form an inequality.

The cost of the first three minutes	plus	the cost of the additional minutes	is not to exceed	$1.00.
0.25	+	$0.07(x - 3)$	$\leq$	1

Solve the inequality   To simplify the computations, we first clear the inequality of decimals.

$$0.25 + 0.07(x - 3) \leq 1$$
$$25 + 7(x - 3) \leq 100 \qquad \text{To eliminate the decimals, multiply both sides by 100.}$$
$$25 + 7x - 21 \leq 100 \qquad \text{Distribute the multiplication by 7.}$$
$$7x + 4 \leq 100 \qquad \text{Combine like terms.}$$
$$7x \leq 96 \qquad \text{Subtract 4 from both sides.}$$
$$x \leq 13.\overline{714285} \qquad \text{Divide both sides by 7.}$$

State the conclusion   Since the phone company doesn't bill for part of a minute, the longest time a call can last is 13 minutes. If a call lasts for $13.\overline{714285}$ minutes, it will be charged as a 14-minute call, and the cost will be $0.25 + $0.07(11) = $1.02.

Check the result   If the call lasts 13 minutes, the cost will be $0.25 + $0.07(10) = $0.95. This is less than $1.00. The result checks.

Answers to Self Checks   **1.** -3 **2.** -5 **3.** all real numbers   **4.** $\varnothing$ **5.** $r = \dfrac{A - P}{Pt}$ **6.** $t = \dfrac{7S + 360}{180}$

7. $(-8, \infty)$ 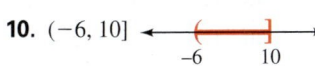 **8.** $[1, \infty)$ 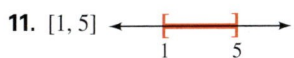 **9.** $(-\infty, \infty)$

10. $(-6, 10]$![number line from -6 to 10] **11.** $[1, 5]$![number line from 1 to 5]

12. $(-\infty, 2) \cup (4, \infty)$

8.1   STUDY SET   ◉   ELEMENTARY & INTERMEDIATE Algebra *f(x)* Now™

VOCABULARY   Fill in the blanks.

1. An _____ is a statement indicating that two expressions are equal.

2. Any number that makes an equation true when substituted for its variable is said to _____ the equation. Such numbers are called _____.

3. An equation that is true for all values of its variable is called an _____.

4. An equation that is not true for any values of its variable is called a _____.

5. $<, >, \leq,$ and $\geq$ are _____ symbols.

6. To _____ an inequality means to find all values of the variable that make the inequality true.

7. $x \geq 3$ and $x < 4$ is a _____ inequality and $-6 < x + 1 \leq 1$ is a _____ linear inequality.

8. The _____ of two sets is the set of elements that are common to both sets.

9. The _____ of two sets is the set of elements that are in one set, or the other, or both.

10. The parenthesis on the right of the _____ notation $(-\infty, 5)$ indicates that 5 is not included in the interval.

CONCEPTS   Fill in the blanks.

11. If any quantity is _____ to (or _____ from) both sides of an equation, a new equation is formed that is equivalent to the original one.

12. If both sides of an equation are _____ (or _____) by the same nonzero number, a new equation is formed that is equivalent to the original one.

13. If both sides of an inequality are multiplied by a _____ number, a new inequality is formed that has the same direction as the original one.

14. If both sides of an inequality are multiplied by a _____ number, a new inequality is formed that has the opposite direction from the original one.

15. The solution set of a compound inequality containing the word *and* includes all numbers that make _____ inequalities true.

16. The solution set of a compound inequality containing the word *or* includes all numbers that make _____, or the other, or _____ inequalities true.

17. The double inequality $4 < 3x + 5 \le 15$ is equivalent to $4 < 3x + 5$ _____ $3x + 5 \le 15$.

18. When solving a compound inequality containing the word *and,* the solution set is the _____ of the solution sets of the inequalities.

19. When solving a compound inequality containing the word *or,* the solution set is the _____ of the solution sets of the inequalities.

20. When multiplying or dividing all three parts of a double inequality by a negative number, the direction of both inequality symbols must be _____.

21. In each case, decide whether -3 is a solution of the compound inequality.

   a. $\dfrac{x}{3} + 1 \ge 0$ and $2x - 3 < -10$

   b. $2x \le 0$ or $-3x < -5$

22. In each case, decide whether -3 is a solution of the double linear inequality.
   a. $-1 < -3x + 4 < 12$
   b. $-1 < -3x + 4 < 14$

23. Use interval notation to describe the intersection of the graphs.

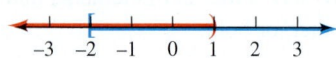

24. Use interval notation to describe the union of the graphs.

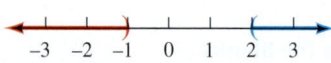

NOTATION

25. Fill in the blanks: We read $\cup$ as _____ and $\cap$ as _____.

26. Match each interval with its corresponding graph.

   a. $[2, 3)$ **i.**

   b. $(2, 3)$ **ii.**

   c. $[2, 3]$ **iii.**

PRACTICE Solve each equation. If an equation is an identity or a contradiction, so indicate.

27. $2x + 1 = 13$ **28.** $2x - 4 = 16$

29. $3x + 1 = 3$ **30.** $8k - 2 = 13$

31. $3(x + 1) = 15$ **32.** $-2(x + 5) = 30$

33. $2r - 5 = 1 - r$ **34.** $5s - 13 = s - 1$

35. $3(2y - 4) - 6 = 3y$ **36.** $2x + (2x - 3) = 5$

37. $5(5 - a) = 37 - 2a$ **38.** $4a + 17 = 7(a + 2)$

39. $2(a - 5) - (3a + 1) = 0$

40. $8(3a - 5) - 4(2a + 3) = 12$

41. $2x - 6 = -2x + 4(x - 2)$

42. $4(2 - 3t) + 6t = -6t + 8$

43. $9(x + 2) = -6(4 - x) + 18$

44. $3(x + 2) - 2 = -(5 + x) + x$

45. $12 + 3(x - 4) - 21 = 5[5 - 4(4 - x)]$

46. $1 + 3[-2 + 6(4 - 2x)] = -(x + 3)$

47. $\dfrac{1}{2}x - 4 = -1 + 2x$ **48.** $2x + 3 = \dfrac{2}{3}x - 1$

49. $2y + 1 = 5(0.2y + 1) - (4 - y)$

50. $-3x = -2x + 1 - (5 + x)$

51. $\dfrac{x}{2} - \dfrac{x}{3} = 4$ **52.** $\dfrac{x}{2} + \dfrac{x}{3} = 10$

53. $\dfrac{x}{6} + 1 = \dfrac{x}{3}$ **54.** $\dfrac{3}{2}(y + 4) = \dfrac{20 - y}{2}$

55. $\dfrac{3 + p}{3} - 4p = 1 - \dfrac{p + 7}{2}$

56. $\dfrac{4 - t}{2} - \dfrac{3t}{5} = 2 + \dfrac{t + 1}{3}$

57. $3(x - 4) + 6 = -2(x + 4) + 5x$

58. $2(x - 3) = \dfrac{3}{2}(x - 4) + \dfrac{x}{2}$

59. $\dfrac{4}{5}(x + 5) = \dfrac{7}{8}(3x + 23) - 7$

60. $\dfrac{2}{3}(2x + 2) + 4 = \dfrac{1}{6}(5x + 29)$

Solve each formula for the indicated variable.

61. $V = \dfrac{1}{3}Bh$ for B

62. $A = \dfrac{1}{2}bh$ for b

63. $p = 2l + 2w$ for w

64. $p = 2l + 2w$ for l

65. $z = \dfrac{x - \mu}{\sigma}$ for x

66. $P = L + \dfrac{s}{f}i$ for s

67. $z = \dfrac{x - \mu}{\sigma}$ for μ

68. $T - W = ma$ for W

69. $S = \dfrac{n(a + l)}{2}$ for l

70. $\ell = a + (n - 1)d$ for d

Solve each inequality. Write the solution set in interval notation and then graph it.

71. $5x - 3 > 7$

72. $7x - 9 < 5$

73. $-3x - 1 \le 5$

74. $-2x + 6 \ge 16$

75. $8 - 9y \ge -y$

76. $4 - 3x \le x$

77. $7 < \dfrac{5}{3}a - 3$

78. $5 > \dfrac{7}{2}a - 9$

79. $-3(a + 2) > 2(a + 1)$

80. $-4(y - 1) < y + 8$

81. $\dfrac{1}{2}y + 2 \ge \dfrac{1}{3}y - 4$

82. $\dfrac{1}{4}x - \dfrac{1}{3} \le x + 2$

83. $\dfrac{x - 7}{2} - \dfrac{x - 1}{5} \le -\dfrac{x}{4}$

84. $\dfrac{3a + 1}{3} - \dfrac{4 - 3a}{5} \le -\dfrac{1}{15}$

85. $x > -2$ and $x \le 5$

86. $x \le -4$ and $x \ge -7$

87. $x + 3 < 3x - 1$ and $4x - 3 \le 3x$

88. $4x \ge -x + 5$ and $6 \ge 4x - 3$

89. $x - 1 \le 2(x + 2)$ and $x \le 2x - 5$

90. $5(x + 1) \le 4(x + 3)$ and $x + 12 < -3$

91. $4 \le x + 3 \le 7$

92. $-5.3 < x - 2.3 < -1.3$

93. $-6 \le \dfrac{1}{3}a + 1 < 0$

94. $10 < \dfrac{1}{2}c - 4 \le 12$

95. $-2 < -b + 3 < 5$

96. $4 < -t - 2 < 9$

97. $0 \le \dfrac{4 - x}{3} \le 2$

98. $-2 \le \dfrac{5 - 3x}{2} \le 2$

99. $x \le -2$ or $x > 6$

100. $x \ge -1$ or $x \le -3$

101. $x - 3 < -4$ or $x - 2 > 0$

102. $4x < -12$ or $\dfrac{x}{2} > 4$

103. $3x + 2 < 8$ or $2x - 3 > 11$

104. $3x + 4 < -2$ or $3x + 4 > 10$

APPLICATIONS

105. SPRING TOURS   A group of junior high students will be touring Washington, D.C. Their chaperons will have the $1,810 cost of the tour reduced by $15.50 for each student they personally supervise. How many students will a chaperon have to supervise so that his or her cost to take the tour will be $1,500?

106. DELAYS   An engineering company was to be paid $1,050,000 for the construction of a building. However, the company was unable to complete the project on time, and was fined $1,500 for each day past the scheduled completion date. If the company was paid $999,000, how many days late was the building completed?

107. FENCING PENS   A man has 150 feet of fencing to build the pen shown in the illustration. If one end is a square, find the outside dimensions.

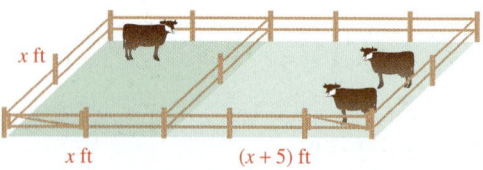

108. FENCING PASTURES   A farmer has 624 feet of fencing to enclose a pasture. Because a river runs along one side, fencing will be needed on only three sides. Find the dimensions of the pasture if its length is double its width.

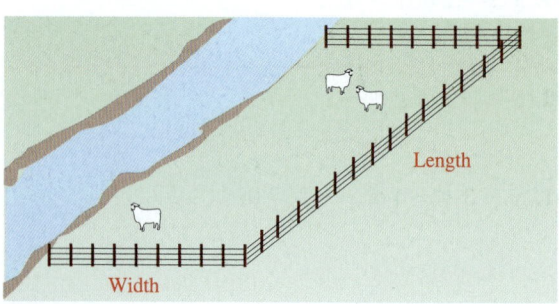

109. PROFITS   The wholesale cost of a clock is $27. A store owner knows that in order to sell, the clock must be priced under $42. If p is the profit, express the possible profit as an inequality.

110. INVESTING MONEY   If a woman invests $10,000 at 8% annual interest, how much more must she invest at 9% so that her annual income will exceed $1,250?

111. BUYING COMPACT DISCS   A student can afford to spend up to $330 on a stereo system and some compact discs. If the stereo costs $175 and the discs cost $8.50 each, find the greatest number of discs the student can buy.

112. GRADE AVERAGES   A student has scores of 70, 77, and 85 on three exams. What score is needed on a fourth exam to make the average 80 or better?

113. BABY FURNITURE   A company makes various square playpens having perimeters between 128 and 192 inches, inclusive.
   a. Complete the double inequality that describes the range of the perimeters of the playpens.

   $$? \le 4s \le ?$$

   b. Solve the double inequality to find the range of the side lengths of the playpens.

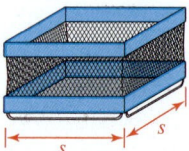

114. THERMOSTATS   The *Temp range* control on the thermostat shown on the next page directs the heater to come on when the room temperature gets 5° below the *Temp setting,* and it directs the air conditioner to come on when the room temperature gets 5° above the *Temp setting.* Use interval notation to describe
   a. the temperature range for the room when neither the heater nor the air conditioner will be on.

   b. the temperature range for the room when either the heater or the air conditioner will be on. (*Note:* The lowest theoretical temperature possible is −460° F, called **absolute zero.**)

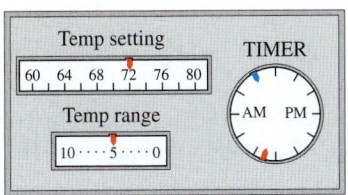

WRITING

115. Explain how to find the union and how to find the intersection of $(-\infty, 5)$ and $(-2, \infty)$ graphically.

116. Explain why the double inequality

$$2 < x < 8$$

can be written in the equivalent form

$$2 < x \text{ and } x < 8$$

REVIEW   Simplify each expression.

117. $\left(\dfrac{t^3 t^5 t^{-6}}{t^2 t^{-4}}\right)^{-3}$

118. $\left(\dfrac{a^{-2} b^3 a^5 b^{-2}}{a^6 b^{-5}}\right)^{-4}$

CHALLENGE PROBLEMS   Solve each inequality. Write the solution set (if one exists) in interval notation and graph it.

119. $2(-2) \le 3x - 1$ and $3x - 1 \le -1 - 3$

120. $4.5x - 2 > 2.5$ or $\dfrac{1}{2}x \le 1$

121. $x - 1 \le 2(x + 2)$ and $x \le 2x - 5$

122. $-x < -2x$ and $3x > 2x$

8.2   Solving Absolute Value Equations and Inequalities

- Equations of the form $|X| = k$
- Inequalities of the form $|X| < k$
- Equations with two absolute values
- Inequalities of the form $|X| > k$

Many quantities studied in mathematics, science, and engineering are expressed as positive numbers. To guarantee that a quantity is positive, we often use absolute value. In this section, we will consider equations and inequalities involving the absolute value of an algebraic expression. Some examples are

$$|3x - 2| = 5, \qquad |2x - 3| < 9, \qquad \text{and} \qquad \left|\frac{3 - x}{5}\right| \ge 6$$

To solve these *absolute value equations* and *inequalities,* we write and then solve equivalent compound equations and inequalities.

■ EQUATIONS OF THE FORM $|X| = k$

Recall that the absolute value of a real number is its distance from 0 on a number line. To solve the **absolute value equation** $|x| = 5$, we must find all real numbers x whose distance from 0 on the number line is 5. There are two such numbers: 5 and -5. We say that the solutions of $|x| = 5$ are 5 and -5 and the solution set is $\{5, -5\}$.

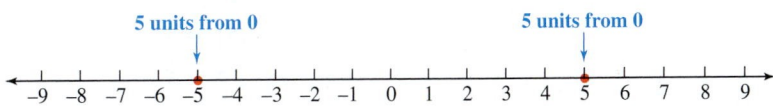

EXAMPLE 1   Solve: **a.** $|x| = 8$,   **b.** $|s| = 0.003$,   and   **c.** $|c| = -15$.

Solution   **a.** To solve $|x| = 8$, we must find all real numbers x whose distance from 0 on the number line is 8. Therefore, the solutions are 8 and -8 and the solutions set is $\{8, -8\}$.

b. To solve $|s| = 0.003$, we must find all real numbers s whose distance from 0 on the number line is 0.003. Therefore, the solutions are 0.003 and -0.003.

c. Recall that the absolute value of a number is either positive or zero, but never negative. Therefore, there is no value for c for which $|c| = -15$. The equation has no solution and the solution set is $\varnothing$.

Self Check 1   Solve: **a.** $|y| = 24$,   **b.** $|x| = \frac{1}{2}$,   and   **c.** $|a| = -1.1$.

The results from Example 1 suggest the following approach for solving absolute value equations.

Solving Absolute Value Equations	For any positive number k and any algebraic expression X:
	To solve $|X| = k$, solve the equivalent compound equation $X = k$ or $X = -k$.

EXAMPLE 2   Solve: **a.** $|3x - 2| = 5$ and   **b.** $|10 - x| = -40$.

Solution   **a.** To solve $|3x - 2| = 5$, we write and then solve an equivalent compound equation.

The Language of Algebra

When two equations are joined with the word *or*, we call the statement a *compound equation*.

$$|3x - 2| = 5$$

means

$$3x - 2 = 5 \quad \text{or} \quad 3x - 2 = -5$$

Now we solve each equation for x:

$$
\begin{array}{lll}
3x - 2 = 5 & \text{or} & 3x - 2 = -5 \\
3x = 7 & & 3x = -3 \\
x = \dfrac{7}{3} & & x = -1
\end{array}
$$

The Language of Algebra

When we say that the absolute value equation and a compound equation are *equivalent*, we mean that they have the same solution(s).

The results must be checked separately to see whether each of them produces a true statement. We substitute $\frac{7}{3}$ for x and then -1 for x in the original equation.

Check: *For* $x = \dfrac{7}{3}$ *For* $x = -1$

$$
\begin{array}{ll}
|3x - 2| = 5 & |3x - 2| = 5 \\
\left| 3\left(\dfrac{7}{3}\right) - 2 \right| \overset{?}{=} 5 & |3(-1) - 2| \overset{?}{=} 5 \\
|7 - 2| \overset{?}{=} 5 & |-3 - 2| \overset{?}{=} 5 \\
|5| \overset{?}{=} 5 & |-5| \overset{?}{=} 5 \\
5 = 5 & 5 = 5
\end{array}
$$

The resulting true statements indicate that the equation has two solutions: $\dfrac{7}{3}$ and -1.

b. Since an absolute value can never be negative, there are no real numbers x that make $|10 - x| = -40$ true. The equation has no solution. The solution set is $\varnothing$.

Self Check 2   Solve: **a.** $|2x - 3| = 7$ and   **b.** $\left| \dfrac{x}{4} - 1 \right| = -3$.

Caution   When solving absolute value equations (or inequalities), isolate the absolute value expression on one side *before* writing the equivalent compound statement.

EXAMPLE 3

ELEMENTARY &
INTERMEDIATE
Algebra $f(x)$ **Now**™

Solve: $\left| \dfrac{2}{3}x + 3 \right| + 4 = 10$.

Solution   We can isolate $\left| \dfrac{2}{3}x + 3 \right|$ on the left-hand side by subtracting 4 from both sides.

$$\left| \frac{2}{3}x + 3 \right| + 4 = 10$$

$$\left| \frac{2}{3}x + 3 \right| = 6 \qquad \text{Subtract 4 from both sides. The equation is in the form } |X| = k.$$

With the absolute value now isolated, we can solve $\left| \dfrac{2}{3}x + 3 \right| = 6$ by writing and solving an equivalent compound equation.

$$\left| \frac{2}{3}x + 3 \right| = 6$$

means

$$\frac{2}{3}x + 3 = 6 \qquad \text{or} \qquad \frac{2}{3}x + 3 = -6$$

Now we solve each equation for x:

$$\frac{2}{3}x + 3 = 6 \qquad \text{or} \qquad \frac{2}{3}x + 3 = -6$$

$$\frac{2}{3}x = 3 \qquad\qquad\qquad \frac{2}{3}x = -9$$

$$2x = 9 \qquad\qquad\qquad 2x = -27$$

$$x = \frac{9}{2} \qquad\qquad\qquad x = -\frac{27}{2}$$

Verify that both solutions check.

Caution

A common error when solving absolute value equations is to forget to isolate the absolute value expression first. Note:

$$\left| \frac{2}{3}x + 3 \right| + 4 = 10$$

does not mean

$$\frac{2}{3}x + 3 + 4 = 10$$

or

$$\frac{2}{3}x + 3 + 4 = -10$$

Self Check 3   Solve: $|0.4x - 2| - 0.6 = 0.4$.

EXAMPLE 4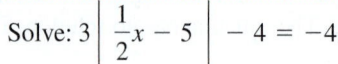

Solve: $3\left|\dfrac{1}{2}x - 5\right| - 4 = -4$.

Solution We first isolate $\left|\dfrac{1}{2}x - 5\right|$ on the left-hand side.

$$3\left|\dfrac{1}{2}x - 5\right| - 4 = -4$$

$$3\left|\dfrac{1}{2}x - 5\right| = 0 \qquad \text{Add 4 to both sides.}$$

$$\left|\dfrac{1}{2}x - 5\right| = 0 \qquad \text{Divide both sides by 3. The equation is in the form } |X| = k.$$

Since 0 is the only number whose absolute value is 0, the expression $\frac{1}{2}x - 5$ must be 0, and we have

$$\dfrac{1}{2}x - 5 = 0$$

$$\dfrac{1}{2}x = 5 \qquad \text{Add 5 to both sides.}$$

$$x = 10 \qquad \text{Multiply both sides by 2.}$$

Verify that 10 satisfies the original equation.

Self Check 4 Solve: $-5\left|\dfrac{2x}{3} + 4\right| + 1 = 1$.

Success Tip

To solve most absolute value equations, we must consider two cases. However, if an absolute value is equal to 0, we need only consider one: the case when the expression within the absolute value bars is equal to 0.

■ EQUATIONS WITH TWO ABSOLUTE VALUES

Equations can contain two absolute value expressions. To develop a strategy to solve them, consider the following example.

$$|3| = |3| \qquad \text{or} \qquad |-3| = |-3| \qquad \text{or} \qquad |3| = |-3| \qquad \text{or} \qquad |-3| = |3|$$

The same number. The same number. These numbers are opposites. These numbers are opposites.

Look closely to see that these four possible cases are really just two cases: *For two expressions to have the same absolute value, they must either be equal or be opposites of each other.* This observation suggests the following approach for solving equations having two absolute value expressions.

Solving Equations with Two Absolute Values

For any algebraic expressions X and Y:

To solve $|X| = |Y|$, solve the equivalent compound equation $X = Y$ or $X = -Y$.

EXAMPLE 5

Solve: $|5x + 3| = |3x + 25|$.

Solution   To solve $|5x + 3| = |3x + 25|$, we write and then solve an equivalent compound equation.

$$|5x + 3| = |3x + 25|$$

means

The expressions within the absolute value symbols are equal.		The expressions within the absolute value symbols are opposites.
$5x + 3 = 3x + 25$	or	$5x + 3 = -(3x + 25)$
$2x = 22$		$5x + 3 = -3x - 25$
$x = 11$		$8x = -28$
		$x = -\dfrac{28}{8}$
		$x = -\dfrac{7}{2}$

Verify that both solutions check.

Self Check 5   Solve: $|2x - 3| = |4x + 9|$.

■ INEQUALITIES OF THE FORM $|X| < k$

To solve the **absolute value inequality** $|x| < 5$, we must find all real numbers x whose distance from 0 on the number line is less than 5. From the graph, we see that there are many such numbers. For example, $-4.999, -3, -2.4, -1\frac{7}{8}, -\frac{3}{4}, 0, 1, 2.8, 3.001,$ and 4.999 all meet this requirement. We conclude that the solution set is all numbers between -5 and 5, which can be written $(-5, 5)$.

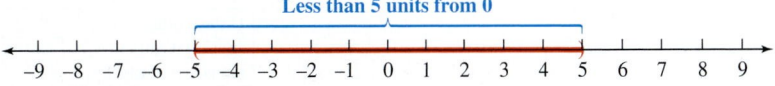

Less than 5 units from 0

Since x is between -5 and 5, it follows that $|x| < 5$ is equivalent to $-5 < x < 5$. The observation suggests the following approach for solving absolute value inequalities of the form $|X| < k$ and $|X| \le k$.

Solving $|X| < k$ and $|X| \le k$

For any positive number k and any algebraic expression X:

To solve $|X| < k$, solve the equivalent double inequality $-k < X < k$.

To solve $|X| \le k$, solve the equivalent double inequality $-k \le X \le k$.

MPLE 6

Solve $|2x - 3| < 9$ and graph the solution set.

Solution   To solve $|2x - 3| < 9$, we write and then solve an equivalent double inequality.

$$|2x - 3| < 9 \qquad \text{means} \qquad -9 < 2x - 3 < 9$$

Now we solve for x:

$$-9 < 2x - 3 < 9$$
$$-6 < 2x < 12 \qquad \text{Add 3 to all three parts.}$$
$$-3 < x < 6 \qquad \text{Divide all parts by 2.}$$

Any number between -3 and 6 is in the solution set. This is the interval $(-3, 6)$; its graph is shown on the right.

Self Check 6   Solve $|3x + 2| < 4$ and graph the solution set.

EXAMPLE 7

ELEMENTARY &
INTERMEDIATE
Algebra $f(x)$ **Now**™

Tolerances.   When manufactured parts are inspected by a quality control engineer, they are classified as acceptable if each dimension falls within a given *tolerance range* of the dimensions listed on the blueprint. For the bracket shown in the figure, the distance between the two drilled holes is given as 2.900 inches. Because the tolerance is ± 0.015 inch, this distance can be as much as 0.015 inch longer or 0.015 inch shorter, and the part will be considered acceptable. The acceptable distance d between holes can be represented by the absolute value inequality $|d - 2.900| \leq 0.015$. Solve the inequality and explain the result.

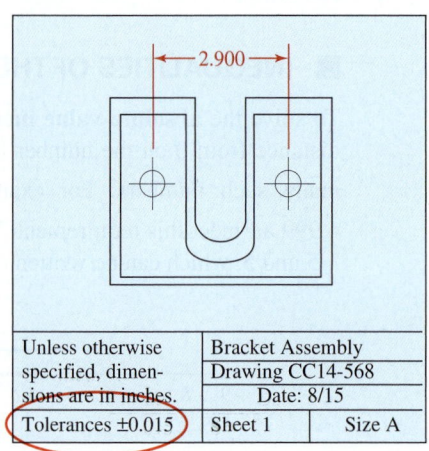

Solution   To solve the absolute value inequality, we write and then solve an equivalent double inequality.

$$|d - 2.900| \leq 0.015 \qquad \text{means} \qquad -0.015 \leq d - 2.900 \leq 0.015$$

Now we solve for d:

$$-0.015 \leq d - 2.900 \leq 0.015$$
$$2.885 \leq d \leq 2.915 \qquad \text{Add 2.900 to all three parts.}$$

The solution set is the interval $[2.885, 2.915]$. This means that the distance between the two holes should be between 2.885 and 2.915 inches, inclusive. If the distance is less than 2.885 inches or more than 2.915 inches, the part should be rejected.

EXAMPLE 8

Solve: $|4x - 5| < -2$.

Solution Since $|4x - 5|$ is always greater than or equal to 0 for any real number x, this absolute value inequality has no solution. The solution set is $\varnothing$.

Self Check 8 Solve: $|6x + 24| < -51$

■ INEQUALITIES OF THE FORM $|X| > k$

To solve the absolute value inequality $|x| > 5$, we must find all real numbers x whose distance from 0 on the number line is greater than 5. From the graph, we see that there are many such numbers. For example, -5.001, -6, -7.5, $-8\frac{3}{8}$, 5.001, 6.2, 7, 8, and $9\frac{1}{2}$ all meet this requirement. We conclude that the solution set is all numbers less than -5 or greater than 5, which can be written $(-\infty, -5) \cup (5, \infty)$.

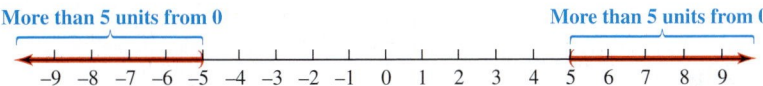

Since x is less than -5 or greater than 5, it follows that $|x| > 5$ is equivalent to $x < -5$ or $x > 5$. The observation suggests the following approach for solving absolute value inequalities of the form $|X| > k$ and $|X| \geq k$.

Solving $|X| > k$ and $|X| \geq k$

For any positive number k and any algebraic expression X:

To solve $|X| > k$, solve the equivalent compound inequality $X < -k$ or $X > k$.

To solve $|X| \geq k$, solve the equivalent compound inequality $X \leq -k$ or $X \geq k$.

EXAMPLE 9

Solve $\left| \dfrac{3 - x}{5} \right| \geq 6$ and graph the solution set.

Solution To solve $\left| \dfrac{3 - x}{5} \right| \geq 6$, we write and then solve an equivalent compound inequality.

$$\left| \frac{3 - x}{5} \right| \geq 6$$

means

$$\frac{3 - x}{5} \leq -6 \qquad \text{or} \qquad \frac{3 - x}{5} \geq 6$$

Then we solve each inequality for x:

$\dfrac{3 - x}{5} \leq -6$ or	$\dfrac{3 - x}{5} \geq 6$	
$3 - x \leq -30$	$3 - x \geq 30$	Multiply both sides by 5.
$-x \leq -33$	$-x \geq 27$	Subtract 3 from both sides.
$x \geq 33$	$x \leq -27$	Divide both sides by -1 and reverse the direction of the inequality symbol.

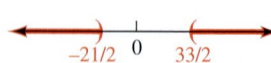

The solution set is the interval $(-\infty, -27] \cup [33, \infty)$. Its graph appears on the left.

Self Check 9   Solve $\left| \dfrac{2 - x}{4} \right| \geq 1$ and graph the solution set.

EXAMPLE 10

ELEMENTARY &
INTERMEDIATE
Algebra *f(x)* **Now**™

Solve $\left| \dfrac{2}{3}x - 2 \right| - 3 > 6$ and graph the solution set.

Solution   We add 3 to both sides to isolate the absolute value on the left-hand side.

$$\left| \frac{2}{3}x - 2 \right| - 3 > 6$$

$$\left| \frac{2}{3}x - 2 \right| > 9 \qquad \text{Add 3 to both sides to isolate the absolute value.}$$

To solve this absolute value inequality, we write and then solve an equivalent compound inequality.

$$\frac{2}{3}x - 2 < -9 \qquad \text{or} \qquad \frac{2}{3}x - 2 > 9$$

$$\frac{2}{3}x < -7 \qquad\qquad \frac{2}{3}x > 11 \qquad \text{Add 2 to both sides.}$$

$$2x < -21 \qquad\qquad 2x > 33 \qquad \text{Multiply both sides by 3.}$$

$$x < -\frac{21}{2} \qquad\qquad x > \frac{33}{2} \qquad \text{Divide both sides by 2.}$$

The solution set is $\left(-\infty, -\dfrac{21}{2}\right) \cup \left(\dfrac{33}{2}, \infty\right)$. Its graph appears on the left.

Self Check 10   Solve $\left| \dfrac{3}{4}x + 2 \right| - 1 > 3$ and graph the solution set.

EXAMPLE 11

Solve: $\left| \dfrac{x}{8} - 1 \right| \geq -4$.

Solution   Since $\left| \dfrac{x}{8} - 1 \right|$ is always greater than or equal to 0 for any real number x, this absolute value inequality is true for all real numbers. The solution set is $(-\infty, \infty)$ or $\mathbb{R}$.

Self Check 11   Solve: $|-x - 9| > -0.5$.

The following summary shows how we can interpret absolute value in three ways. Assume $k > 0$.

Geometric description	Graphic description	Algebraic description
1. $\lvert x \rvert = k$ means that x is k units from 0 on the number line.		$\lvert x \rvert = k$ is equivalent to $x = k$ or $x = -k$.
2. $\lvert x \rvert < k$ means that x is less than k units from 0 on the number line.		$\lvert x \rvert < k$ is equivalent to $-k < x < k$.
3. $\lvert x \rvert > k$ means that x is more than k units from 0 on the number line.		$\lvert x \rvert > k$ is equivalent to $x > k$ or $x < -k$.

ACCENT ON TECHNOLOGY: SOLVING ABSOLUTE VALUE EQUATIONS AND INEQUALITIES

We can solve absolute value equations and inequalities with a graphing calculator. For example, to solve $\lvert 2x - 3 \rvert = 9$, we graph the equations $y = \lvert 2x - 3 \rvert$ and $y = 9$ on the same coordinate system, as shown in the figure. The equation $\lvert 2x - 3 \rvert = 9$ will be true for all x-coordinates of points that lie on *both* graphs. Using the TRACE or INTERSECT feature, we can see that the graphs intersect at the points 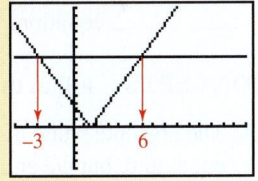 $(-3, 9)$ and $(6, 9)$. Thus, the solutions of the absolute value equation are -3 and 6.

The inequality $\lvert 2x - 3 \rvert < 9$ will be true for all x-coordinates of points that lie on the graph of $y = \lvert 2x - 3 \rvert$ and *below* the graph of $y = 9$. We see that these values of x are between -3 and 6. Thus, the solution set is the interval $(-3, 6)$.

The inequality $\lvert 2x - 3 \rvert > 9$ will be true for all x-coordinates of points that lie on the graph of $y = \lvert 2x - 3 \rvert$ and *above* the graph of $y = 9$. We see that these values of x are less than -3 or greater than 6. Thus, the solution set is the interval $(-\infty, -3) \cup (6, \infty)$.

Answers to Self Checks   **1. a.** $24, -24,$ **b.** $\frac{1}{2}, -\frac{1}{2}$ **2. a.** $5, -2,$ **b.** no solution

3. $7.5, 2.5$ **4.** -6 **5.** $-1, -6$ **6.** $\left(-2, \frac{2}{3}\right)$ **8.** no solution

9. $(-\infty, -2] \cup [6, \infty)$

10. $(-\infty, -8) \cup \left(\frac{8}{3}, \infty\right)$ **11.** $(-\infty, \infty)$

8.2 STUDY SET

ELEMENTARY &
INTERMEDIATE
Algebra *f(x)* **Now**™

VOCABULARY   Fill in the blanks.

1. The _____ _____ of a number is its distance from 0 on a number line.

2. $|2x - 1| = 10$ is an absolute value _____.

3. $|2x - 1| > 10$ is an absolute value _____.

4. To _____ an absolute value inequality means to find all the values of the variable that make the inequality true.

5. To _____ the absolute value in $|3 - x| - 4 = 5$, we add 4 to both sides.

6. $-(2x + 9)$ is the _____ of $2x + 9$.

7. When we say that the absolute value equation and a compound equation are equivalent, we mean that they have the same _____.

8. When two equations are joined by the word *or*, such as $x + 1 = 5$ or $x + 1 = -5$, we call the statement a _____ equation.

CONCEPTS   Fill in the blanks.

9. The absolute value of a real number is greater than or equal to 0, but never _____.

10. For two expressions to have the same absolute value, they must either be equal or _____ of each other.

11. To solve $|x| > 5$, we must find the coordinates of all points on a number line that are _____ 5 units from the origin.

12. To solve $|x| < 5$, we must find the coordinates of all points on a number line that are _____ 5 units from the origin.

13. To solve $|x| = 5$, we must find the coordinates of all points on a number line that are _____ units from the origin.

14. To solve absolute value equations and inequalities, we write and then solve equivalent _____ equations and inequalities.

15. Consider the following real numbers: $-4, -3, -2.01, -2, -1.99, -1, 0, 1, 1.99, 2, 2.01, 3, 4$
   a. Which of them make $|x| = 2$ true?
   b. Which of them make $|x| < 2$ true?

   c. Which of them make $|x| > 2$ true?

16. Decide whether -3 is a solution of the given equation or inequality.
   a. $|x - 1| = 4$ **b.** $|x - 1| > 4$
   c. $|x - 1| \leq 4$ **d.** $|5 - x| = |x + 12|$

For each absolute value equation, write an equivalent compound equation.

17. a. $|x - 7| = 8$ means

   _____ = _____   or   _____ = _____

   b. $|x + 10| = |x - 3|$ means

   _____ = _____   or   _____ = _____

For each absolute value inequality, write an equivalent compound inequality.

18. a. $|x + 5| < 1$ means

   _____ < _____ < _____

   b. $|x - 6| \geq 3$ means

   _____ $\geq$ _____   or   _____ $\leq$ _____

19. For each absolute value equation or inequality, write an equivalent compound equation or inequality.
   a. $|x| = 8$ **b.** $|x| \geq 8$

   c. $|x| \leq 8$ **d.** $|5x - 1| = |x + 3|$

20. Perform the necessary steps to isolate the absolute value expression on one side of the equation. **Do not solve.**
   a. $|3x + 2| - 7 = -5$
   b. $6 + |5x - 19| \leq 40$

NOTATION

21. Match each equation or inequality with its graph.
   a. $|x| = 1$ **i.**
   b. $|x| > 1$ **ii.**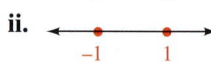
   c. $|x| < 1$ **iii.** (graph)

22. Match each graph with its corresponding equation or inequality.
   a.    **i.** $|x| \geq 2$
   b.    **ii.** $|x| \leq 2$
   c. 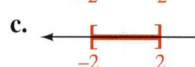   **iii.** $|x| = 2$

23. Describe the set graphed below using interval notation.

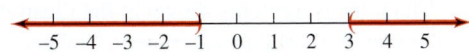

24. a. If an absolute value inequality has no solution, what symbol is used to represent the solution set?

b. If the solution set of an absolute value inequality is all real numbers, what notation is used to represent the solution set?

PRACTICE Solve each equation, if possible.

25. $|x| = 23$

26. $|x| = 90$

27. $|5x| = 20$

28. $|6x| = 12$

29. $|x - 3.1| = 6$

30. $|x + 4.3| = 8.9$

31. $|3x + 2| = 16$

32. $|5x - 3| = 22$

33. $|x| - 3 = 9$

34. $|x| + 6 = 11$

35. $\left|\dfrac{7}{2}x + 3\right| = -5$

36. $5|x - 21| = -8$

37. $|3 - 4x| + 1 = 6$

38. $|8 - 5x| - 8 = 10$

39. $2|3x + 24| = 0$

40. $\left|\dfrac{2x}{3} + 10\right| = 0$

41. $\left|\dfrac{3x + 48}{3}\right| = 12$

42. $\left|\dfrac{4x - 64}{4}\right| = 32$

43. $-7 = 2 - |0.3x - 3|$

44. $-1 = 1 - |0.1x + 8|$

45. $\dfrac{6}{5} = \left|\dfrac{3x}{5} + \dfrac{x}{2}\right|$

46. $\dfrac{11}{12} = \left|\dfrac{x}{3} - \dfrac{3x}{4}\right|$

47. $|2x + 1| = |3(x + 1)|$

48. $|5x - 7| = |4(x + 1)|$

49. $|2 - x| = |3x + 2|$

50. $|4x + 3| = |9 - 2x|$

51. $\left|\dfrac{x}{2} + 2\right| = \left|\dfrac{x}{2} - 2\right|$

52. $|7x + 12| = |x - 6|$

53. $\left|x + \dfrac{1}{3}\right| = |x - 3|$

54. $\left|x - \dfrac{1}{4}\right| = |x + 4|$

Solve each inequality. Write the solution set in interval notation (if possible) and then graph it.

55. $|x| < 4$

56. $|x| < 9$

57. $|x + 9| \leq 12$

58. $|x - 8| \leq 12$

59. $|3x - 2| < 10$

60. $|4 - 3x| \leq 13$

61. $|3x + 2| \leq -3$

62. $|5x - 12| < -5$

63. $|x| > 3$

64. $|x| > 7$

65. $|x - 12| > 24$

66. $|x + 5| \geq 7$

67. $|3x + 2| > 14$

68. $|2x - 5| > 25$

69. $|4x + 3| > -5$

70. $|7x + 2| > -8$

71. $|2 - 3x| \geq 8$

72. $|-1 - 2x| > 5$

73. $-|2x - 3| < -7$

74. $-|3x + 1| < -8$

75. $\left|\dfrac{x - 2}{3}\right| \leq 4$

76. $\left|\dfrac{x - 2}{3}\right| > 4$

77. $|3x + 1| + 2 < 6$

78. $1 + \left|\dfrac{1}{7}x + 1\right| \leq 1$

79. $\left|\dfrac{1}{3}x + 7\right| + 5 > 6$

80. $-2|3x - 4| < 16$

81. $|0.5x + 1| + 2 \leq 0$

82. $15 \geq 7 - |1.4x + 9|$

APPLICATIONS

83. TEMPERATURE RANGES   The temperatures on a sunny summer day satisfied the inequality $|t - 78°| \le 8°$, where t is a temperature in degrees Fahrenheit. Solve this inequality and express the range of temperatures as a double inequality.

84. OPERATING TEMPERATURES   A car CD player has an operating temperature of $|t - 40°| < 80°$, where t is a temperature in degrees Fahrenheit. Solve the inequality and express this range of temperatures as an interval.

85. AUTO MECHANICS   On most cars, the bottoms of the front wheels are closer together than the tops, creating a *camber angle*. This lessens road shock to the steering system. (See below.) The specifications for a certain car state that the camber angle c of its wheels should be $0.6° \pm 0.5°$.
  a. Express the range with an inequality containing absolute value symbols.
  b. Solve the inequality and express this range of camber angles as an interval.

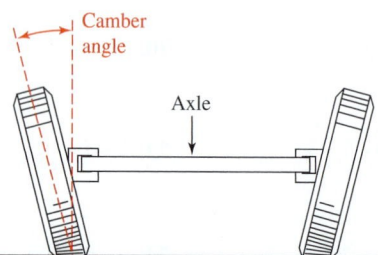

86. STEEL PRODUCTION   A sheet of steel is to be 0.250 inch thick with a tolerance of 0.025 inch.
  a. Express this specification with an inequality containing absolute value symbols, using x to represent the thickness of a sheet of steel.
  b. Solve the inequality and express the range of thickness as an interval.

87. ERROR ANALYSIS
In a lab, students measured the percent of copper p in a sample of copper sulfate. The students know that copper sulfate is actually 25.46% copper by mass. They are to compare their

Lab 4	Section A
Title: "Percent copper (Cu) in copper sulfate ($CuSO_4 \cdot 5H_2O$)"	
Results	
	% Copper
Trial #1:	22.91%
Trial #2:	26.45%
Trial #3:	26.49%
Trial #4:	24.76%

results to the actual value and find the amount of *experimental error*.
  a. Which measurements shown in the illustration in the previous column satisfy the absolute value inequality $|p - 25.46| \le 1.00$?
  b. What can be said about the amount of error for each of the trials listed in part a?

88. ERROR ANALYSIS   See Exercise 87.
  a. Which measurements satisfy the absolute value inequality $|p - 25.46| > 1.00$?
  b. What can be said about the amount of error for each of the trials listed in part a?

WRITING

89. Explain the error.

$$\text{Solve: } |x| + 2 = 6$$
$$x + 2 = 6 \quad \text{or} \quad x + 2 = -6$$
$$x = 4 \quad | \quad x = -8$$

90. Explain why the equation $|x - 4| = -5$ has no solutions.

91. Explain the differences between the solution sets of $|x| < 8$ and $|x| > 8$.

92. Explain how to use the graph in the illustration to solve the following.
  a. $|x - 2| = 3$
  b. $|x - 2| \le 3$
  c. $|x - 2| \ge 3$

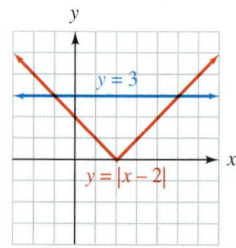

REVIEW

93. RAILROAD CROSSINGS   The warning sign in the illustration is to be painted on the street in front of a railroad crossing. If y is 30° more than twice x, find x and y.

94. GEOMETRY   Refer to the illustration. What is $2x + 2y$?

CHALLENGE PROBLEMS

95. For what values of k does $|x| + k = 0$ have exactly two solutions?

96. For what values of k does $|x| + k = 0$ have exactly one solution?

97. Under what conditions is $|x| + |y| > |x + y|$?

98. Under what conditions is $|x| + |y| = |x + y|$? (Assume that x and y are nonzero.)

8.3 Review of Factoring

- Factoring out the greatest common factor   • Factoring by grouping   • Formulas
- Factoring trinomials   • Using substitution to factor trinomials
- Using grouping to factor trinomials

In Chapter 5, we discussed how to factor polynomials. In this section, we will review that material. Recall that *when we factor a polynomial, we write a sum of terms as a product of factors*. To perform the most basic type of factoring, we determine whether the terms of the given polynomial have any common factors.

■ FACTORING OUT THE GREATEST COMMON FACTOR

Factoring out a common monomial factor is based on the distributive property.

EXAMPLE 1

Factor: $3xy^2z^3 + 6xz^2 - 9xyz^4$.

ELEMENTARY & INTERMEDIATE
Algebra $f(x)$ **Now** ™

Solution   We begin by factoring each term:

Success Tip

Always verify a factorization by performing the indicated multiplication. The result should be the original polynomial.

$$\left.\begin{array}{l} 3xy^2z^3 = \mathbf{3} \cdot \mathbf{x} \cdot y \cdot y \cdot \mathbf{z} \cdot \mathbf{z} \cdot z \\ 6xz^2 = \mathbf{3} \cdot 2 \cdot \mathbf{x} \cdot \mathbf{z} \cdot \mathbf{z} \\ 9xyz^4 = \mathbf{3} \cdot 3 \cdot \mathbf{x} \cdot y \cdot \mathbf{z} \cdot \mathbf{z} \cdot z \cdot z \end{array}\right\} \text{GCF} = \mathbf{3} \cdot \mathbf{x} \cdot \mathbf{z} \cdot \mathbf{z} = 3xz^2$$

Since each term has one factor of 3, one factor of x, and two factors of z, and there are no other common factors, $3xz^2$ is the greatest common factor of the three terms. We can use the distributive property to factor out $3xz^2$.

$$3xy^2z^3 + 6xz^2 - 9xyz^4 = \mathbf{3xz^2} \cdot y^2z + \mathbf{3xz^2} \cdot 2 - \mathbf{3xz^2} \cdot 3yz^2$$
$$= \mathbf{3xz^2}(y^2z + 2 - 3yz^2)$$

Check: $3xz^2(y^2z + 2 - 3yz^2) = 3xy^2z^3 + 6xz^2 - 9xyz^4$.

Self Check 1   Factor: $6a^2b^2 - 4ab^3 + 2ab^2$.

EXAMPLE 2

Factor the opposite of the greatest common factor from $-6u^2v^3 + 8u^3v^2$.

Solution   Because the greatest common factor of the two terms is $2u^2v^2$, the opposite of the greatest common factor is $-2u^2v^2$. To factor out $-2u^2v^2$, we proceed as follows:

$$-6u^2v^3 + 8u^3v^2 = -2u^2v^2 \cdot 3v + 2u^2v^2 \cdot 4u$$
$$= -2u^2v^2 \cdot 3v - (-2u^2v^2)4u$$
$$= -2u^2v^2(3v - 4u)$$

Self Check 2 Factor the opposite of the greatest common factor from $-3p^3q + 6p^2q^2$.

A polynomial that cannot be factored is called a **prime polynomial** or an **irreducible polynomial.**

EXAMPLE 3 Factor $3x^2 + 4y + 7$, if possible.

Solution We factor each term:

$$3x^2 = 3 \cdot x \cdot x \qquad 4y = 2 \cdot 2 \cdot y \qquad 7 = 7$$

Since there are no common factors other than 1, this polynomial cannot be factored. It is a prime polynomial.

Self Check 3 Factor: $6a^3 + 7b^2 + 5$.

A common factor can have more than one term.

EXAMPLE 4 Factor: $a(x - y + z) - b(x - y + z) + 3(x - y + z)$.

Solution Since the trinomial $x - y + z$ is a factor of each term of the given polynomial, we factor it out.

$$a(x - y + z) - b(x - y + z) + 3(x - y + z) = (x - y + z)(a - b + 3)$$

Self Check 4 Factor: $c^2(y^2 + 1) + d^2(y^2 + 1)$.

■ FACTORING BY GROUPING

Sometimes polynomials having four or more terms can be factored by removing common factors from groups of terms. This process is called *factoring by grouping*.

EXAMPLE 5 Factor: $2c - 2d + cd - d^2$.

ELEMENTARY &
INTERMEDIATE
Algebra *f(x)* **Now**™

Solution The first two terms have a common factor, 2, and the last two terms have a common factor, d.

Success Tip

You may want to review the steps in the process of *factoring by grouping* on page 356.

$$2c - 2d + cd - d^2 = 2(c - d) + d(c - d) \qquad \text{Factor out 2 from } 2c - 2d \text{ and } d \text{ from } cd - d^2.$$

$$= (c - d)(2 + d) \qquad \text{Factor out the common binomial factor, } c - d.$$

We check by multiplying:

$$(c - d)(2 + d) = 2c + cd - 2d - d^2$$
$$= 2c - 2d + cd - d^2$$

Rearrange the terms to get the original polynomial.

Self Check 5   Factor: $7m - 7n + mn - n^2$.

To factor a polynomial, it is often necessary to factor more than once. When factoring a polynomial, *always look for a common factor first.*

EXAMPLE 6

Factor: $3x^3y - 4x^2y^2 - 6x^2y + 8xy^2$.

Solution   We begin by factoring out the common factor of xy.

$$3x^3y - 4x^2y^2 - 6x^2y + 8xy^2 = xy(3x^2 - 4xy - 6x + 8y)$$

We can now factor $3x^2 - 4xy - 6x + 8y$ by grouping:

$$3x^3y - 4x^2y^2 - 6x^2y + 8xy^2$$
$$= xy(3x^2 - 4xy - 6x + 8y)$$
$$= xy[x(3x - 4y) - 2(3x - 4y)] \quad \text{Factor } x \text{ from } 3x^2 - 4xy \text{ and } -2 \text{ from } -6x + 8y.$$
$$= xy(3x - 4y)(x - 2) \quad \text{Factor out } 3x - 4y.$$

Caution

The instruction "Factor" means to factor the given expression *completely.* Each factor of a completely factored expression will be prime.

Because no more factoring can be done, the factorization is complete.

Self Check 6   Factor: $3a^3b + 3a^2b - 2a^2b^2 - 2ab^2$.

■ FORMULAS

Factoring is often required to solve a formula for one of its variables.

EXAMPLE 7

The formula $r_1r_2 = rr_2 + rr_1$ is used in electronics. Solve for r_2.

Solution   To isolate r_2 on one side of the equation, we get all terms involving r_2 on the left-hand side and all terms not involving r_2 on the right-hand side. We then proceed as follows:

$$r_1r_2 = rr_2 + rr_1$$
$$r_1r_2 - rr_2 = rr_1 \quad \text{Subtract } rr_2 \text{ from both sides.}$$
$$r_2(r_1 - r) = rr_1 \quad \text{Factor out } r_2 \text{ on the left-hand side.}$$
$$r_2 = \frac{rr_1}{r_1 - r} \quad \text{Divide both sides by } r_1 - r.$$

Self Check 7   Solve $f_1f_2 = ff_1 + ff_2$ for f_1.

■ FACTORING TRINOMIALS

Recall that many trinomials factor as the product of two binomials.

EXAMPLE 8

**ELEMENTARY &
INTERMEDIATE
Algebra** $f(x)$ **Now**™

Solution

Success Tip

You may want to review the procedure for *factoring trinomials with a lead coefficient of 1* on page 364.

The Language of Algebra

Make sure you understand the following vocabulary:
Many trinomials factor as the product of two binomials.

$$\underbrace{\text{Trinomial}}_{} \quad \overset{\text{Product of}}{\underbrace{\text{two binomials}}}$$

$$x^2 - 6x + 8 = (x - 2)(x - 4)$$

Factor: $x^2 - 6x + 8$.

We attempt to factor $x^2 - 6x + 8$ as the product of two binomials. Since the first term of the trinomial is x^2, we enter x and x as the first terms of the binomial factors.

$$x^2 - 6x + 8 = (x \;\boxed{})(x \;\boxed{}) \qquad \text{\color{red}Because } x \cdot x \text{ will give } x^2$$

The second terms of the binomials must be two integers whose product is 8 and whose sum is -6. All possible integer-pair factors of 8 are listed in the table.

Factors of 8	Sum of factors
1(8)	$1 + 8 = 9$
2(4)	$2 + 4 = 6$
$-1(-8)$	$-1 + (-8) = -9$
$-2(-4)$	**$-2 + (-4) = -6$**	← This is the one to choose.

The fourth row of the table contains the correct pair of integers -2 and -4, whose product is 8 and whose sum is -6. To complete the factorization, we enter -2 and -4 as the second terms of the binomial factors.

$$x^2 - 6x + 8 = (x - 2)(x - 4)$$

Check: We can verify this result by multiplication:

$$(x - 2)(x - 4) = x^2 - 4x - 2x + 8 \qquad \text{\color{red}Use the FOIL method.}$$
$$= x^2 - 6x + 8$$

Self Check 8 Factor: $a^2 - 7a + 12$.

EXAMPLE 9

Solution

Success Tip

Always write the terms of a trinomial in descending powers of one variable before attempting to factor it.

Factor: $30x - 4xy - 2xy^2$.

We begin by writing the trinomial in descending powers of y:

$$30x - 4xy - 2xy^2 = -2xy^2 - 4xy + 30x$$

Each term in this trinomial has a common factor of $-2x$, which can be factored out.

$$30x - 4xy - 2xy^2 = -2x(y^2 + 2y - 15)$$

To factor $y^2 + 2y - 15$, we list the factors of -15 and find the pair whose sum is 2.

This is the one to choose.
$\downarrow$

$$15(-1) \qquad 5(-3) \qquad 1(-15) \qquad 3(-5)$$

The only factorization where the sum of the factors is 2 (the coefficient of the middle term of $y^2 + 2y - 15$) is $5(-3)$. Thus,

Caution

Be sure to include all factors in the final answer. Here, a common error is to forget to write the $-2x$.

$$30x - 4xy - 2xy^2 = -2x(y^2 + 2y - 15)$$
$$= -2x(y + 5)(y - 3)$$

Verify this result by multiplication.

Self Check 9   Factor: $16a - 2ap^2 - 4ap$.

There are more combinations of coefficients to consider when factoring trinomials with lead coefficients other than 1. It is not easy to give specific rules for factoring such trinomials. However, the following hints are helpful. This approach is called the **trial-and-check method.**

EXAMPLE 10

Factor: $3p^2 - 4p - 4$.

Solution   To factor the trinomial, we note that the first terms of the binomial factors must be $3p$ and p to give the first term of $3p^2$.

$$3p^2 - 4p - 4 = \left(3p \;\boxed{}\;\right)\left(p \;\boxed{}\;\right)$$

Success Tip

You may want to review the procedure for factoring trinomials using the *trial-and-check method* on page 375.

The product of the last terms must be -4, and the sum of the products of the outer terms and the inner terms must be $-4p$.

$$3p^2 - 4p - 4 = \left(3p \;\boxed{}\;\right)\left(p \;\boxed{}\;\right)$$

$$O + I = -4p$$

Notation

By the commutative property of multiplication, the factors of a trinomial can be written in either order. Thus, we could also write:

$$3p^2 - 4p - 4 = (p - 2)(3p + 2)$$

Because $1(-4)$, $-1(4)$, and $-2(2)$ all give a product of -4, there are six possible combinations to consider:

$(3p + 1)(p - 4)$	$(3p - 4)(p + 1)$
$(3p - 1)(p + 4)$	$(3p + 4)(p - 1)$
$(3p - 2)(p + 2)$	$(3p + 2)(p - 2)$

Of these possibilities, only the one in blue gives the required middle term of $-4p$.

$$3p^2 - 4p - 4 = (3p + 2)(p - 2)$$

Self Check 10   Factor: $4q^2 - 9q - 9$.

EXAMPLE 11

Factor: $6y^3 + 13x^2y^3 + 6x^4y^3$.

ELEMENTARY &
INTERMEDIATE
Algebra $f(x)$ **Now**™   **Solution**

We write the expression in descending powers of x and then factor out the common factor y^3.

$$6y^3 + 13x^2y^3 + 6x^4y^3 = 6x^4y^3 + 13x^2y^3 + 6y^3$$
$$= y^3(6x^4 + 13x^2 + 6)$$

A test for factorability will show that $6x^4 + 13x^2 + 6$ will factor.
To factor $6x^4 + 13x^2 + 6$, we examine its terms.

- Since the first term is $6x^4$, the first terms of the binomial factors must be either $2x^2$ and $3x^2$ or x^2 and $6x^2$.

$$6x^4 + 13x^2 + 6 = \left(2x^2 \; \boxed{}\right)\left(3x^2 \; \boxed{}\right) \quad \text{or} \quad \left(x^2 \; \boxed{}\right)\left(6x^2 \; \boxed{}\right)$$

- Since the signs of the middle term and the last term of the trinomial are positive, the signs within each binomial factor will be positive.
- Since the product of the last terms of the binomial factors must be 6, we must find two numbers whose product is 6 that will lead to a middle term of $13x^2$.

After trying some combinations, we find the one that works.

$$6x^4y^3 + 13x^2y^3 + 6y^3 = y^3(6x^4 + 13x^2 + 6)$$
$$= y^3(2x^2 + 3)(3x^2 + 2)$$

Self Check 11   Factor: $4b + 11a^2b + 6a^4b$.

■ USING SUBSTITUTION TO FACTOR TRINOMIALS

For more complicated expressions, a substitution sometimes helps to simplify the factoring process.

EXAMPLE 12

Factor: $(x + y)^2 + 7(x + y) + 12$.

ELEMENTARY &
INTERMEDIATE
Algebra $f(x)$ **Now**™

Solution

We rewrite the trinomial $(x + y)^2 + 7(x + y) + 12$ as $z^2 + 7z + 12$, where $z = x + y$. The trinomial $z^2 + 7z + 12$ factors as $(z + 4)(z + 3)$.
To find the factorization of $(x + y)^2 + 7(x + y) + 12$, we substitute $x + y$ for z in the expression $(z + 4)(z + 3)$ to obtain

$$z^2 + 7z + 12 = (z + 4)(z + 3)$$
$$(x + y)^2 + 7(x + y) + 12 = (x + y + 4)(x + y + 3) \qquad \text{Replace each } z \text{ with } x + y.$$

Self Check 12   Factor: $(a + b)^2 - 3(a + b) - 10$.

■ USING GROUPING TO FACTOR TRINOMIALS

Another way to factor trinomials is to write them as equivalent four-termed polynomials and factor by grouping.

EXAMPLE 13

ELEMENTARY & INTERMEDIATE

Algebra *f(x)* **Now**™

Success Tip

You may want to review the procedure for factoring trinomials using *factoring by grouping* (the *key number method*) on page 377.

Factor by grouping: **a.** $x^2 + 8x + 15$ and   **b.** $10x^2 + 13x - 3$.

Solution   **a.** Since $x^2 + 8x + 15 = 1x^2 + 8x + 15$, we identify a as 1, b as 8, and c as 15. The key number is $ac = 1(15) = 15$. We must find two integers whose product is 15 and whose sum is $b = 8$. Since the integers must have a positive product and a positive sum, we consider only positive factors of 15.

Key number = 15

Positive factors of 15	Sum of the factors
$1 \cdot 15 = 15$	$1 + 15 = 16$
$3 \cdot 5 = 15$	**$3 + 5 = 8$**

The second row of the table contains the correct pair of integers 3 and 5, whose product is 15 and whose sum is 8. They serve as the coefficients of $3x$ and $5x$ that we place between x^2 and 15.

$$x^2 + 8x + 15 = x^2 + 3x + 5x + 15 \qquad \text{Express } 8x \text{ as } 3x + 5x.$$
$$= x(x + 3) + 5(x + 3) \qquad \text{Factor } x \text{ out of } x^2 + 3x \text{ and 5 out of } 5x + 15.$$
$$= (x + 3)(x + 5) \qquad \text{Factor out } x + 3.$$

The factorization is $(x + 3)(x + 5)$. Check by multiplying.

b. In $10x^2 + 13x - 3$, $a = 10$, $b = 13$, and $c = -3$. The key number is $ac = 10(-3) = -30$. We must find a factorization of -30 in which the sum of the factors is $b = 13$. Since the factors must have a negative product, their signs must be different. The possible factor pairs are listed in the table.

Key number = -30

Factors of -30	Sum of the factors
$1(-30) = -30$	$1 + (-30) = -29$
$2(-15) = -30$	$2 + (-15) = -13$
$3(-10) = -30$	$3 + (-10) = -7$
$5(-6) = -30$	$5 + (-6) = -1$
$6(-5) = -30$	$6 + (-5) = 1$
$10(-3) = -30$	$10 + (-3) = 7$
$15(-2) = -30$	**$15 + (-2) = 13$**
$30(-1) = -30$	$30 + (-1) = 29$

The seventh row contains the correct pair of numbers 15 and -2, whose product is -30 and whose sum is 13. They serve as the coefficients of two terms, $15x$ and $-2x$, that we place between $10x^2$ and -3.

$$10x^2 + \mathbf{13x} - 3 = 10x^2 + \mathbf{15x - 2x} - 3 \qquad \text{Express } 13x \text{ as } 15x - 2x.$$

Notation

The middle term, $13x$, may be expressed as $15x - 2x$ or as $-2x + 15x$ when using factoring by grouping. The resulting factorizations will be equivalent.

Finally, we factor by grouping.

$$10x^2 + 15x - 2x - 3 = 5x(2x + 3) - 1(2x + 3)$$
  Factor out $5x$ from $10x^2 + 15x$.
Factor out -1 from $-2x - 3$.

$$= (2x + 3)(5x - 1)$$
  Factor out $2x + 3$.

So $10x^2 + 13x - 3 = (2x + 3)(5x - 1)$. Check by multiplying.

Self Check 13 Factor by grouping: **a.** $m^2 + 13m + 42$ and **b.** $15a^2 + 17a - 4$.

Answers to Self Checks
1. $2ab^2(3a - 2b + 1)$ **2.** $-3p^2q(p - 2q)$ **3.** a prime polynomial   **4.** $(y^2 + 1)(c^2 + d^2)$
5. $(m - n)(7 + n)$ **6.** $ab(3a - 2b)(a + 1)$ **7.** $f_1 = \dfrac{ff_2}{f_2 - f}$ **8.** $(a - 4)(a - 3)$
9. $-2a(p + 4)(p - 2)$ **10.** $(4q + 3)(q - 3)$ **11.** $b(2a^2 + 1)(3a^2 + 4)$
12. $(a + b + 2)(a + b - 5)$ **13. a.** $(m + 7)(m + 6)$,   **b.** $(3a + 4)(5a - 1)$

8.3 STUDY SET

ELEMENTARY &
INTERMEDIATE
Algebra *f(x)* Now™

VOCABULARY Fill in the blanks.

1. When we write $2x + 4$ as $2(x + 2)$, we say that we have _____ $2x + 4$.

2. When we _____ a polynomial, we write a sum of terms as a product of factors.

3. The abbreviation GCF stands for _____ _____ _____.

4. If a polynomial cannot be factored, it is called a _____ polynomial or an irreducible polynomial.

5. To factor means to factor _____. Each factor of a completely factored expression will be _____.

6. To factor $ab + 6a + 2b + 12$ by _____, we begin by factoring out a from the first two terms and 2 from the last two terms.

7. A polynomial with three terms, such as $3x^2 - 2x + 4$, is called a _____.

8. The trinomial $4a^2 - 5a - 6$ is written in _____ powers of a.

9. The _____ coefficient of the trinomial $x^2 - 3x + 2$ is 1, the _____ of the middle term is -3, and the last term is ▢.

10. The statement $x^2 - x - 12 = (x - 4)(x + 3)$ shows that $x^2 - x - 12$ factors into the _____ of two binomials.

CONCEPTS

11. The prime factorizations of three terms are shown here. Find their GCF.

$$2 \cdot 2 \cdot 3 \cdot x \cdot x \cdot y \cdot y \cdot y$$
$$2 \cdot 3 \cdot 3 \cdot x \cdot y \cdot y \cdot y \cdot y$$
$$2 \cdot 3 \cdot 3 \cdot 7 \cdot x \cdot x \cdot x \cdot y \cdot y$$

12. Check to see whether $(3t - 1)(5t - 6)$ is the correct factorization of $15t^2 - 19t + 6$.

13. Complete the table.

Factors of 8	Sum of the factors of 8
$1(8) = 8$	
$2(4) = 8$	
$-1(-8) = 8$	
$-2(-4) = 8$	

14. Find two integers whose
 a. product is 10 and whose sum is 7.
 b. product is 8 and whose sum is -6.
 c. product is -6 and whose sum is 1.
 d. product is -9 and whose sum is -8.

15. Complete the key number table.

Key number $= 12$

Negative factors of 12	Sum of factors of 12
$-1(-12) = 12$	
$-3(-4) = 12$	

16. Use the substitution $x = a + b$ to rewrite the trinomial $6(a + b)^2 - 17(a + b) - 3$.

NOTATION   Complete each factorization.

17. $15c^3d^4 - 25c^2d^4 + 5c^3d^6 = \boxed{}(3c - 5 + cd^2)$

18. $x^3 - x^2 + 2x - 2 = \boxed{}(x - 1) + \boxed{}(x - 1)$
$$= \left(\boxed{}\right)(x^2 + 2)$$

19. $6m^2 + 7m - 3 = \left(\boxed{} - 1\right)\left(2m + \boxed{}\right)$

20. $2y^2 + 10y + 12 = \boxed{}(y^2 + 5y + 6)$
$$= 2\left(y + \boxed{}\right)\left(\boxed{} + 2\right)$$

PRACTICE   Factor each expression. Factor out all common factors first (including -1 if the first term is negative). If an expression is prime, so indicate.

21. $2x^2 - 6x$

22. $3y^3 + 3y^2$

23. $15x^2y - 10x^2y^2$

24. $63x^3y^2 + 81x^2y^4$

25. $27z^3 + 12z^2 + 3z$

26. $25t^6 - 10t^3 + 5t^2$

27. $11m^3n^2 - 12x^2y$

28. $14r^2s^3 + 15t^6$

29. $-5xy + y - 4$

30. $-7m - 12n + 16$

31. $24s^3 - 12s^2t + 6st^2$

32. $18y^2z^2 + 12y^2z^3 - 24y^4z^3$

33. $(x + y)u + (x + y)v$

34. $4(x + y) + t(x + y)$

35. $-18a^2b - 12ab^2$

36. $-21t^5 + 28t^3$

37. $\dfrac{3}{5}ax^4 + \dfrac{1}{5}bx^2 - \dfrac{4}{5}ax^3$

38. $\dfrac{3}{2}t^2y^4 - \dfrac{1}{2}ty^4 - \dfrac{5}{2}ry^3$

39. $5(a - b) - t(a - b)$

40. $(a - b)r - (a - b)s$

41. $3(m + n + p) + x(m + n + p)$

42. $x(x - y - z) + y(x - y - z)$

43. $-63u^3v^6z^9 + 28u^2v^7z^2 - 21u^3v^3z^4$

44. $-56x^4y^3z^2 - 72x^3y^4z^5 + 80xy^2z^3$

45. $4(x^2 + 1)^2 + 2(x^2 + 1)^3$

46. $6(x^3 - 7x + 1)^2 - 3(x^3 - 7x + 1)^3$

Solve for the indicated variable.

47. $r_1r_2 = rr_2 + rr_1$ for r_1

48. $r_1r_2 = rr_2 + rr_1$ for r

49. $S(1 - r) = a - \ell r$ for r

50. $Sn = (n - 2)180°$ for n

51. $b^2x^2 + a^2y^2 = a^2b^2$ for a^2

52. $b^2x^2 + a^2y^2 = a^2b^2$ for b^2

Factor by grouping. Factor out the GCF first.

53. $ax + bx + ay + by$

54. $ar - br + as - bs$

55. $x^2 + yx + x + y$

56. $c + d + cd + d^2$

57. $a^2 - 4b + ab - 4a$

58. $3c - cd + 3d - c^2$

59. $x^2 + 4y - xy - 4x$

60. $7u + v^2 - 7v - uv$

61. $a^2x + bx - a^2 - b$

62. $x^2y - ax - xy + a$

63. $x^2 + xy + xz + xy + y^2 + zy$

64. $ab - b^2 - bc + ac - bc - c^2$

65. $1 - m + mn - n$

66. $a^2x^2 - 10 - 2x^2 + 5a^2$

67. $2ax^2 - 4 + a - 8x^2$

68. $a^3b^2 - 3 + a^3 - 3b^2$

69. $mpx + mqx + npx + nqx$

70. $abd - abe + acd - ace$

71. $x^2y + xy^2 + 2xyz + xy^2 + y^3 + 2y^2z$

72. $a^3 - 2a^2b + a^2c - a^2b + 2ab^2 - abc$

Factor each trinomial. Factor out all common factors first (including −1 if the first term is negative). If a trinomial is prime, so indicate.

73. $x^2 - 5x + 6$ **74.** $y^2 + 7y + 6$

75. $x^2 - 7x + 10$ **76.** $c^2 - 7c + 12$

77. $b^2 + 8b + 18$ **78.** $x^2 + 4x - 28$

79. $-x + x^2 - 30$ **80.** $a^2 - 45 + 4a$

81. $a^2 - 18a + 81$ **82.** $b^2 + 12b + 36$

83. $x^2 - 4xy - 21y^2$ **84.** $a^2 + 4ab - 5b^2$

85. $s^2 - 10st + 16t^2$ **86.** $h^2 - 8hk + 15k^2$

87. $3x^2 + 12x - 63$ **88.** $2y^2 + 4y - 48$

89. $32 - a^2 + 4a$ **90.** $15 - x^2 - 2x$

91. $-3a^2x^2 + 15a^2x - 18a^2$
92. $-2bcy^2 - 16bcy + 40bc$
93. $y^4 - 13y^2 + 30$ **94.** $y^4 - 13y^2 + 42$

95. $b^4x^2 - 12b^2x^2 + 35x^2$
96. $c^3x^4 + 11c^3x^2 - 42c^3$
97. $6y^2 + 7y + 2$ **98.** $6x^2 - 11x + 3$

99. $8a^2 + 6a - 9$ **100.** $15b^2 + 4b - 4$

101. $6x^2 - 5xy - 4y^2$ **102.** $18y^2 - 3yz - 10z^2$

103. $5x^2 + 4x + 1$ **104.** $3 + 4a^2 + 20a$

105. $6z^2 + 17z + 12$ **106.** $3 - 10x + 8x^2$

107. $4y^2 + 4y + 1$ **108.** $9x^2 + 6x + 1$

109. $-3a^2 + ab + 2b^2$ **110.** $-2x^2 + 3xy + 5y^2$

111. $20a^2 + 60b^2 + 45ab$ **112.** $-4x^2 - 9 + 12x$

113. $64h^6 + 24h^5 - 4h^4$
114. $27x^2yz + 90xyz - 72yz$
115. $6a^2(m + n) + 13a(m + n) - 15(m + n)$

116. $15n^2(q - r) - 17n(q - r) - 18(q - r)$

Use substitution to help factor each expression.

117. $(x + a)^2 + 2(x + a) + 1$
118. $(a + b)^2 - 2(a + b) + 1$
119. $(a + b)^2 - 2(a + b) - 24$

120. $(x - y)^2 + 3(x - y) - 10$
121. $14(q - r)^2 - 17(q - r) - 6$

122. $8(h + s)^2 + 34(h + s) + 35$

APPLICATIONS

123. CRAYONS   The amount of colored wax used to make the crayon shown in the illustration can be found by computing its volume using the formula

$$V = \pi r^2 h_1 + \frac{1}{3}\pi r^2 h_2$$

Factor the expression on the right-hand side of this equation.

124. PACKAGING   The amount of cardboard needed to make the following cereal box can be found by finding the area A, which is given by the formula

$$A = 2wh + 4wl + 2lh$$

where w is the width, h the height, and l the length. Solve the equation for the width.

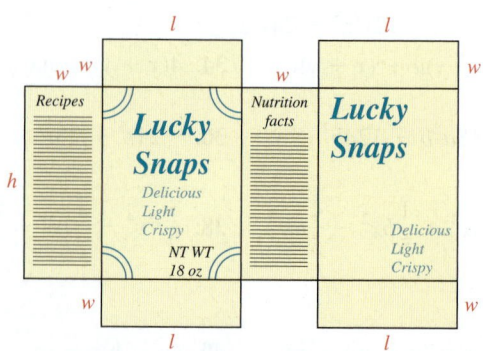

125. ICE The surface area of the ice cube is $6x^2 + 36x + 54$. Find the length of an edge of the cube.

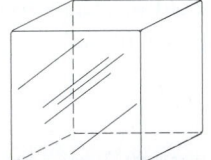

126. CHECKERS The area of the checkerboard is $25x^2 - 40x + 16$. Find the length of each side.

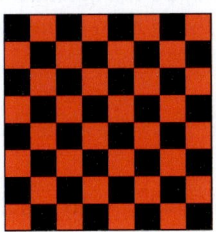

WRITING

127. Explain the error in the following solution.

Solve for r_1:
$$r_1 r_2 = rr_2 + rr_1$$
$$\frac{r_1 r_2}{r_2} = \frac{rr_2 + rr_1}{r_2}$$
$$\boxed{r_1 = \frac{rr_2 + rr_1}{r_2}}$$

128. Explain the error.

Factor: $2x^2 - 4x - 6$.

$$2x^2 - 4x - 6 = (2x + 2)(x - 3)$$

REVIEW

129. INVESTMENTS Equal amounts are invested in each of three accounts paying 7%, 8%, and 10.5% annually. If one year's combined interest income is $1,249.50, how much is invested in each account?

130. SEARCH AND RESCUE Two search-and-rescue teams leave base at the same time looking for a lost boy. The first team, on foot, heads north at 2 mph and the other, on horseback, south at 4 mph. How long will it take them to search a distance of 21 miles between them?

CHALLENGE PROBLEMS   Factor out the designated factor.

131. x^2 from $x^{n+2} + x^{n+3}$

132. y^n from $2y^{n+2} - 3y^{n+3}$

Factor. Assume that n is a natural number.

133. $x^{2n} + 2x^n + 1$ **134.** $2a^{6n} - 3a^{3n} - 2$

135. $x^{4n} + 2x^{2n}y^{2n} + y^{4n}$ **136.** $6x^{2n} + 7x^n - 3$

8.4   The Difference of Two Squares; the Sum and Difference of Two Cubes

- Factoring the difference of two squares
- Factoring the sum and difference of two cubes

We will now review some special rules of factoring. These rules are applied to polynomials that can be written as the difference of two squares or the sum or difference of two cubes.

■ FACTORING THE DIFFERENCE OF TWO SQUARES

Recall that the difference of the squares of two quantities factors into the product of two binomials.

Difference of Two Squares	$x^2 - y^2 = (x + y)(x - y)$

If we think of the difference of two squares as the square of a **F**irst quantity minus the square of a **L**ast quantity, we have the formula

$$F^2 - L^2 = (F + L)(F - L)$$

and we say: *To factor the square of a **F**irst quantity minus the square of a **L**ast quantity, we multiply the **F**irst plus the **L**ast by the **F**irst minus the **L**ast.*

EXAMPLE 1

ELEMENTARY & INTERMEDIATE
Algebra $f(x)$ **Now**™

Solution

Factor: $49x^2 - 16$.

We begin by rewriting the binomial $49x^2 - 16$ as a difference of two squares: $(7x)^2 - (4)^2$. Then we use the formula for factoring the difference of two squares:

$$\begin{array}{ccccccc} \mathbf{F}^2 & - & \mathbf{L}^2 & = & (\mathbf{F} & + & \mathbf{L})(\mathbf{F} & - & \mathbf{L}) \\ \downarrow & & \downarrow & & \downarrow & & \downarrow \quad \downarrow & & \downarrow \\ (\mathbf{7x})^2 & - & \mathbf{4}^2 & = & (\mathbf{7x} & + & \mathbf{4})(\mathbf{7x} & - & \mathbf{4}) \end{array}$$

We can verify this result using the FOIL method to do the multiplication.

$$(7x + 4)(7x - 4) = 49x^2 - 28x + 28x - 16$$
$$= 49x^2 - 16$$

Self Check 1   Factor: $81p^2 - 25$.

EXAMPLE 2

Solution

Factor: $x^4 - 1$.

Because the binomial is the difference of the squares of x^2 and 1, it factors into the sum of x^2 and 1 and the difference of x^2 and 1.

$$x^4 - 1 = (\mathbf{x^2})^2 - (\mathbf{1})^2$$
$$= (\mathbf{x^2} + \mathbf{1})(\mathbf{x^2} - \mathbf{1})$$

The factor $x^2 + 1$ is the sum of two quantities and is prime. However, the factor $x^2 - 1$ is the difference of two squares and can be factored as $(x + 1)(x - 1)$. Thus,

$$x^4 - 1 = (x^2 + 1)(x^2 - 1)$$
$$= (x^2 + 1)(x + 1)(x - 1)$$

Self Check 2   Factor: $a^4 - 81$.

EXAMPLE 3

ELEMENTARY & INTERMEDIATE
Algebra $f(x)$ **Now**™

Solution

Factor: $(x + y)^4 - z^4$.

This expression is the difference of two squares and can be factored:

$$(x + y)^4 - z^4 = [(x + y)^2]^2 - (z^2)^2$$
$$= [(x + y)^2 + z^2][(x + y)^2 - z^2]$$

Caution

When factoring a polynomial, be sure to factor it completely. Always check to see whether any of the factors of your result can be factored further.

The factor $(x + y)^2 + z^2$ is the sum of two squares and is prime. However, the factor $(x + y)^2 - z^2$ is the difference of two squares and can be factored as $(x + y + z)(x + y - z)$. Thus,

$$(x + y)^4 - z^4 = [(x + y)^2 + z^2][(x + y)^2 - z^2]$$
$$= [(x + y)^2 + z^2](x + y + z)(x + y - z)$$

Self Check 3   Factor: $(a - b)^4 - c^4$.

When possible, we always factor out a common factor before factoring the difference of two squares. The factoring process is easier when all common factors are factored out first.

EXAMPLE 4

Factor: $2x^4y - 32y$.

Solution

$$
\begin{aligned}
2x^4y - 32y &= 2y(x^4 - 16) && \text{\color{red}Factor out the GCF, which is } 2y. \\
&= 2y(x^2 + 4)(x^2 - 4) && \text{\color{red}Factor } x^4 - 16. \\
&= 2y(x^2 + 4)(x + 2)(x - 2) && \text{\color{red}Factor } x^2 - 4.
\end{aligned}
$$

Self Check 4   Factor: $3a^4 - 3$.

EXAMPLE 5

Factor: $x^2 - y^2 + x - y$.

Solution

If we group the first two terms and factor the difference of two squares, we have

$$
\begin{aligned}
x^2 - y^2 + x - y &= (x + y)(x - y) + (x - y) && \text{\color{red}Factor } x^2 - y^2. \\
&= (x - y)(x + y + 1) && \text{\color{red}Factor out } x - y.
\end{aligned}
$$

Self Check 5   Factor: $a^2 - b^2 + a + b$.

EXAMPLE 6

Factor: $x^2 + 6x + 9 - z^2$.

Solution

We group the first three terms together and factor the trinomial to get

$$
\begin{aligned}
x^2 + 6x + 9 - z^2 &= (x + 3)(x + 3) - z^2 \\
&= (x + 3)^2 - z^2
\end{aligned}
$$

We can now factor the difference of two squares to get

$$x^2 + 6x + 9 - z^2 = (x + 3 + z)(x + 3 - z)$$

Self Check 6   Factor: $a^2 + 4a + 4 - b^2$.

■ FACTORING THE SUM AND DIFFERENCE OF TWO CUBES

Recall that the sum and difference of two cubes factor as the product of a binomial and a trinomial.

Sum and Difference of Two Cubes	$x^3 + y^3 = (x + y)(x^2 - xy + y^2)$ $x^3 - y^3 = (x - y)(x^2 + xy + y^2)$

Success Tip

The formulas for factoring the *sum and difference of two cubes* were developed on page 390.

If we think of the sum of two cubes as the sum of the cube of a **F**irst quantity plus the cube of a **L**ast quantity, we have the formula

$$F^3 + L^3 = (F + L)(F^2 - FL + L^2)$$

*To factor the cube of a **F**irst quantity plus the cube of a **L**ast quantity, we multiply the sum of the **F**irst and **L**ast by*

- *the **F**irst squared*
- *minus the **F**irst times the **L**ast*
- *plus the **L**ast squared.*

The formula for the difference of two cubes is

$$F^3 - L^3 = (F - L)(F^2 + FL + L^2)$$

*To factor the cube of a **F**irst quantity minus the cube of a **L**ast quantity, we multiply the difference of the **F**irst and **L**ast by*

- *the **F**irst squared*
- *plus the **F**irst times the **L**ast*
- *plus the **L**ast squared.*

EXAMPLE 7

Factor: $a^3 + 8$.

Algebra *f(x)* **Now**™   **Solution**

Since $a^3 + 8$ can be written as $a^3 + 2^3$, we have the sum of two cubes, which factors as follows:

$$F^3 + L^3 = (F + L)(F^2 - FL + L^2)$$
$$\downarrow \quad \downarrow \qquad \downarrow \quad \downarrow \quad \downarrow \qquad \downarrow\downarrow \quad \downarrow$$
$$a^3 + 2^3 = (a + 2)(a^2 - a2 + 2^2)$$
$$= (a + 2)(a^2 - 2a + 4) \qquad a^2 - 2a + 4 \text{ does not factor.}$$

Caution

In Example 7, a common error is to try to factor $a^2 - 2a + 4$. It is not a perfect square trinomial, because the middle term needs to be $-4a$. Furthermore, it cannot be factored. It is prime.

Therefore, $a^3 + 8 = (a + 2)(a^2 - 2a + 4)$. We can check by multiplying.

$$(a + 2)(a^2 - 2a + 4) = a^3 - 2a^2 + 4a + 2a^2 - 4a + 8$$
$$= a^3 + 8$$

Self Check 7   Factor: $p^3 + 27$.

You should memorize the formulas for factoring the sum and the difference of two cubes. Note that each has the form

(a binomial)(a trinomial)

and that there is a relationship between the signs that appear in these forms.

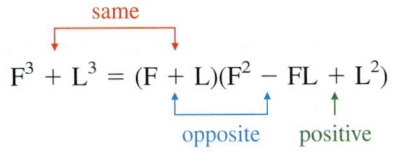

$$F^3 + L^3 = (F + L)(F^2 - FL + L^2) \qquad F^3 - L^3 = (F - L)(F^2 + FL + L^2)$$

EXAMPLE 8

Factor: $27a^3 - 64b^3$.

Solution

Since $27a^3 - 64b^3$ can be written as $(3a)^3 - (4b)^3$, we have the difference of two cubes, which factors as follows:

$$\mathbf{F^3 \;-\; L^3 = (F \;-\; L)\,(F^2 \;+\; F\;\;L \;+\; L^2)}$$
$$(\mathbf{3a})^3 - (\mathbf{4b})^3 = (\mathbf{3a - 4b})[(\mathbf{3a})^2 + (\mathbf{3a})(\mathbf{4b}) + (\mathbf{4b})^2]$$
$$= (3a - 4b)(9a^2 + 12ab + 16b^2)$$

Thus, $27a^3 - 64b^3 = (3a - 4b)(9a^2 + 12ab + 16b^2)$.

Success Tip

To factor sums or differences of cubes, it is helpful to know these **perfect integer cubes**: 1, 8, 27, 64, 125, 216, 343, 512, 729, 1,000.

Self Check 8   Factor: $8c^3 - 125d^3$.

EXAMPLE 9

Factor: $a^3 - (c + d)^3$.

Solution

$$\mathbf{a^3 - (c + d)^3 = [a - (c + d)][a^2 + a(c + d) + (c + d)^2]}$$

Now we simplify the expressions inside both sets of brackets.

$$a^3 - (c + d)^3 = (a - c - d)(a^2 + ac + ad + c^2 + 2cd + d^2)$$

Self Check 9   Factor: $(p + q)^3 - r^3$.

EXAMPLE 10

Factor: $x^6 - 64$.

ELEMENTARY &
INTERMEDIATE
Algebra *f(x)* **Now**™

Solution

This expression is both the difference of two squares and the difference of two cubes. It is easier to factor it as the difference of two squares first.

$$x^6 - 64 = (x^3)^2 - 8^2$$
$$= (x^3 + 8)(x^3 - 8)$$

Each of these factors can be factored further. One is the sum of two cubes and the other is the difference of two cubes:

$$x^6 - 64 = (x + 2)(x^2 - 2x + 4)(x - 2)(x^2 + 2x + 4)$$

Self Check 10   Factor: $x^6 - 1$.

EXAMPLE 11 Factor: $2a^5 + 250a^2$.

Solution We first factor out the common monomial factor of $2a^2$ to obtain

$$2a^5 + 250a^2 = 2a^2(a^3 + 125)$$

Then we factor $a^3 + 125$ as the sum of two cubes to obtain

$$2a^5 + 250a^2 = 2a^2(a + 5)(a^2 - 5a + 25)$$

Self Check 11 Factor: $3x^5 + 24x^2$.

Answers to Self Checks **1.** $(9p + 5)(9p - 5)$ **2.** $(a^2 + 9)(a + 3)(a - 3)$
3. $[(a - b)^2 + c^2](a - b + c)(a - b - c)$ **4.** $3(a^2 + 1)(a + 1)(a - 1)$
5. $(a + b)(a - b + 1)$ **6.** $(a + 2 + b)(a + 2 - b)$ **7.** $(p + 3)(p^2 - 3p + 9)$
8. $(2c - 5d)(4c^2 + 10cd + 25d^2)$ **9.** $(p + q - r)(p^2 + 2pq + q^2 + pr + qr + r^2)$
10. $(x + 1)(x^2 - x + 1)(x - 1)(x^2 + x + 1)$ **11.** $3x^2(x + 2)(x^2 - 2x + 4)$

8.4 STUDY SET   ELEMENTARY & INTERMEDIATE Algebra *f(x)* Now™

VOCABULARY Fill in the blanks.

1. When the polynomial $4x^2 - 25$ is written as $(2x)^2 - (5)^2$, we see that it is the difference of two _____.

2. When the polynomial $8x^3 + 125$ is written as $(2x)^3 + (5)^3$, we see that it is the sum of two _____.

CONCEPTS

3. Write the first ten perfect integer squares.

4. Write the first ten perfect integer cubes.

5. a. Use multiplication to verify that the sum of two squares $x^2 + 25$ does not factor as $(x + 5)(x + 5)$.

b. Use multiplication to verify that the difference of two squares $x^2 - 25$ factors as $(x + 5)(x - 5)$.

6. Explain the error.
a. Factor: $4g^2 - 16 = (2g + 4)(2g - 4)$

b. Factor: $1 - t^8 = (1 + t^4)(1 - t^4)$

7. When asked to factor $81t^2 - 16$, one student answered $(9t - 4)(9t + 4)$, and another answered $(9t + 4)(9t - 4)$. Explain why both students are correct.

8. Factor each polynomial.
a. $5p^2 + 20$
b. $5p^2 - 20$
c. $5p^3 + 20$
d. $5p^3 + 40$

Complete each factorization.

9. $p^2 - q^2 = (p + q)$ _____

10. $36y^2 - 49m^2 = (\quad)^2 - (7m)^2$
$$= (6y \quad 7m)(6y - \quad)$$

11. $p^2q + pq^2 = \quad (p + q)$

12. $p^3 + q^3 = (p + q)$ _____

13. $p^3 - q^3 = (p - q)$ _____

14. $h^3 - 27k^3 = (h)^3 - (\quad)^3$
$$= (h \quad 3k)(h^2 + \quad + 9k^2)$$

NOTATION

15. Give an example of each.

    a. A difference of 2 squares.

    b. A square of a difference.

    c. A sum of two squares.

    d. A sum of two cubes.

    e. A cube of a sum.

16. Fill in the blanks.

    a. $x^2 - y^2 = (x \quad y)(x \quad y)$

    b. $x^3 + y^3 = (x \quad y)(x^2 \quad xy \quad y^2)$

    c. $x^3 - y^3 = (x \quad y)(x^2 \quad xy \quad y^2)$

PRACTICE Factor, if possible.

17. $x^2 - 4$

18. $y^2 - 9$

19. $9y^2 - 64$

20. $16x^4 - 81y^2$

21. $x^2 + 25$

22. $144a^2 - b^4$

23. $400 - c^2$

24. $900 - t^2$

25. $625a^2 - 169b^4$

26. $4y^2 + 9z^4$

27. $81a^4 - 49b^2$

28. $64r^6 - 121s^2$

29. $36x^4y^2 - 49z^4$

30. $4a^2b^4c^6 - 9d^8$

31. $(x + y)^2 - z^2$

32. $a^2 - (b - c)^2$

33. $(a - b)^2 - c^2$

34. $(m + n)^2 - p^4$

35. $x^4 - y^4$

36. $16a^4 - 81b^4$

37. $256x^4y^4 - z^8$

38. $225a^4 - 16b^8c^{12}$

39. $\dfrac{1}{36} - y^4$

40. $\dfrac{4}{81} - m^4$

41. $2x^2 - 288$

42. $8x^2 - 72$

43. $2x^3 - 32x$

44. $3x^3 - 243x$

45. $5x^3 - 125x$

46. $6x^4 - 216x^2$

47. $r^2s^2t^2 - t^2x^4y^2$

48. $16a^4b^3c^4 - 64a^2bc^6$

49. $a^2 - b^2 + a + b$

50. $x^2 - y^2 - x - y$

51. $a^2 - b^2 + 2a - 2b$

52. $m^2 - n^2 + 3m + 3n$

53. $2x + y + 4x^2 - y^2$

54. $m - 2n + m^2 - 4n^2$

55. $x^3 - xy^2 - 4x^2 + 4y^2$

56. $m^2n - 9n + 9m^2 - 81$

57. $x^2 + 4x + 4 - y^2$

58. $x^2 - 6x + 9 - 4y^2$

59. $x^2 + 2x + 1 - 9z^2$

60. $x^2 + 10x + 25 - 16z^2$

61. $c^2 - 4a^2 + 4ab - b^2$

62. $4c^2 - a^2 - 6ab - 9b^2$

63. $r^3 + s^3$

64. $t^3 - v^3$

65. $x^3 - 8y^3$

66. $27a^3 + b^3$

67. $64a^3 - 125b^6$

68. $8x^6 + 125y^3$

69. $125x^3y^6 + 216z^9$

70. $1,000a^6 - 343b^3c^6$

71. $x^6 + y^6$

72. $x^9 + y^9$

73. $5x^3 + 625$

74. $2x^3 - 128$

75. $4x^5 - 256x^2$

76. $2x^6 + 54x^3$

77. $128u^2v^3 - 2t^3u^2$

78. $56rs^2t^3 + 7rs^2v^6$

79. $(a + b)x^3 + 27(a + b)$

80. $(c - d)r^3 - (c - d)s^3$

81. $x^9 - y^{12}z^{15}$

82. $r^{12} + s^{18}t^{24}$

83. $(a + b)^3 + 27$

84. $(b - c)^3 - 1,000$

85. $y^3(y^2 - 1) - 27(y^2 - 1)$

86. $z^3(y^2 - 4) + 8(y^2 - 4)$

Factor each expression completely. Factor a difference of two squares first.

87. $x^6 - 1$

88. $x^6 - y^6$

89. $x^{12} - y^6$

90. $a^{12} - 64$

Factor each trinomial.

91. $a^4 - 13a^2 + 36$

92. $b^4 - 17b^2 + 16$

APPLICATIONS

93. CANDY   To find the amount of chocolate used in the outer coating of the malted-milk ball shown in the illustration, we can find the volume V of the chocolate shell using the formula

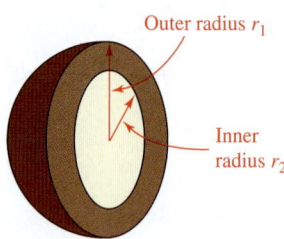

Outer radius r_1

Inner radius r_2

$$V = \frac{4}{3}\pi r_1^3 - \frac{4}{3}\pi r_2^3$$

Factor the expression on the right-hand side of the formula.

94. MOVIE STUNTS   The function that gives the distance a stuntwoman is above the ground t seconds after she falls over the side of a 144-foot tall building is $h(t) = 144 - 16t^2$. Factor the right-hand side.

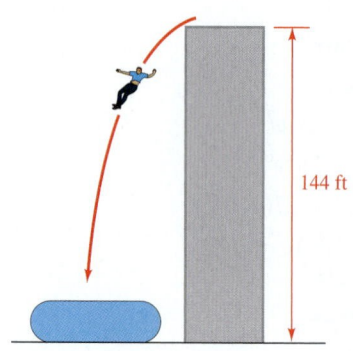

144 ft

WRITING

95. Describe the pattern used to factor the difference of two squares.

96. Describe the patterns used to factor the sum and the difference of two cubes.

REVIEW   **For each of the following purchases, determine the better buy.**

97. Flute lessons: 45 minutes for \$25 or 1 hour for \$35.

98. Tissue paper: 15 sheets for \$1.39 or a dozen sheets for \$1.10.

CHALLENGE PROBLEMS   **Factor. Assume all variables are natural numbers.**

99. $4x^{2n} - 9y^{2n}$

100. $25 - x^{6n}$

101. $a^{3b} - c^{3b}$

102. $8 - x^{3n}$

103. $27x^{3n} + y^{3n}$

104. $a^{3b} + b^{3c}$

105. Factor: $x^{32} - y^{32}$.

106. Find the error in this proof that $2 = 1$.

$$x = y$$
$$x^2 = xy$$
$$x^2 - y^2 = xy - y^2$$
$$(x + y)(x - y) = y(x - y)$$
$$\frac{(x + y)(x - y)}{(x - y)} = \frac{y(x - y)}{x - y}$$
$$x + y = y$$
$$y + y = y$$
$$2y = y$$
$$\frac{2y}{y} = \frac{y}{y}$$
$$2 = 1$$

8.5 Review of Rational Expressions

- Simplifying rational expressions
- Multiplying and dividing rational expressions
- Adding and subtracting rational expressions
- Complex fractions

Recall that rational expressions are algebraic fractions with polynomial numerators and denominators.

Rational Expressions	A **rational expression** is an expression of the form $\frac{P}{Q}$, where P and Q are polynomials and Q does not equal 0.

Some examples of rational expressions are

$$\frac{8y^3z^5}{6y^4z^3}, \qquad \frac{3x}{x-7}, \qquad \frac{5m+n}{8m+16}, \qquad \text{and} \qquad \frac{6a^2-13a+6}{3a^2+a-2}$$

Caution   Since division by 0 is undefined, the value of a polynomial in the denominator of a rational expression cannot be 0. For example, x cannot be 7 in the rational expression $\frac{3x}{x-7}$, because the value of the denominator would be 0. In the rational expression $\frac{5m+n}{8m+16}$, m cannot be -2, because the value of the denominator would be 0.

To manipulate rational expressions, we use the same rules as we use to simplify, multiply, divide, add, and subtract arithmetic fractions.

■ SIMPLIFYING RATIONAL EXPRESSIONS

To **simplify a rational expression** we use the following procedure to write it so that the numerator and denominator have no common factors other than 1.

Simplifying Rational Expressions	1. Factor the numerator and denominator completely to determine their common factors. 2. Remove factors equal to 1 by replacing each pair of factors common to the numerator and denominator with the equivalent fraction $\frac{1}{1}$.

EXAMPLE 1   Simplify: $\frac{8y^3z^5}{6y^4z^3}$.

Solution   We factor the numerator and denominator and remove factors common to the numerator and denominator.

$$\frac{8y^3z^5}{6y^4z^3} = \frac{\overset{1}{2} \cdot 4 \cdot \overset{1}{y} \cdot \overset{1}{y} \cdot \overset{1}{y} \cdot \overset{1}{z} \cdot \overset{1}{z} \cdot \overset{1}{z} \cdot z \cdot z}{\underset{1}{2} \cdot 3 \cdot \underset{1}{y} \cdot \underset{1}{y} \cdot \underset{1}{y} \cdot y \cdot \underset{1}{z} \cdot \underset{1}{z} \cdot z} \qquad \text{Simplify: } \frac{2}{2}=1, \frac{y}{y}=1, \frac{z}{z}=1.$$

$$= \frac{4z^2}{3y}$$

Self Check 1   Simplify: $\frac{10k}{25k^2}$.

The fractions in Example 1 and the Self Check can also be simplified by using the rules of exponents:

$$\frac{8y^3z^5}{6y^4z^3} = \frac{\overset{1}{\cancel{2}} \cdot 4 \cdot y^{3-4} \cdot z^{5-3}}{\underset{1}{\cancel{2}} \cdot 3}$$

$$= \frac{4 \cdot y^{-1} \cdot z^2}{3}$$

$$= \frac{4z^2}{3y}$$

$$\frac{10k}{25k^2} = \frac{\overset{1}{\cancel{5}} \cdot 2 \cdot k^{1-2}}{\underset{1}{\cancel{5}} \cdot 5}$$

$$= \frac{2 \cdot k^{-1}}{5}$$

$$= \frac{2}{5k}$$

EXAMPLE 2   Simplify: $\dfrac{2x^2 + 11x + 12}{3x^2 + 11x - 4}$.

ELEMENTARY &
INTERMEDIATE
Algebra $f(x)$ **Now**™

Solution   We factor the numerator and denominator and remove factors common to the numerator and denominator.

$$\frac{2x^2 + 11x + 12}{3x^2 + 11x - 4} = \frac{(2x + 3)\cancel{(x + 4)}^{\,1}}{(3x - 1)\underset{1}{\cancel{(x + 4)}}} \qquad \text{Simplify: } \frac{x + 4}{x + 4} = 1.$$

$$= \frac{2x + 3}{3x - 1} \qquad\qquad \text{This expression does not simplify further.}$$

Self Check 2   Simplify: $\dfrac{2x^2 + 5x + 2}{3x^2 + 5x - 2}$.

Caution   In Example 2, do not remove the x's in the fraction $\frac{2x + 3}{3x - 1}$. The x in the numerator is a factor of the first term only. It is not a factor of the entire numerator. Likewise, the x in the denominator is not a factor of the entire denominator. This error leads to the incorrect answer $\frac{5}{2}$.

$$\frac{\overset{1}{\cancel{2x}} + 3}{\underset{1}{\cancel{3x}} - 1} = \frac{2 + 3}{3 - 1} = \frac{5}{2}$$

Recall that if the terms of two polynomials are the same except that they are opposite in sign, the polynomials are **opposites.**

The Quotient of Opposites   The quotient of any nonzero polynomial and its opposite is -1.

EXAMPLE 3   Simplify: $\dfrac{3x^2 - 10xy - 8y^2}{4y^2 - xy}$.

Solution   We factor the numerator and denominator. Because $x - 4y$ and $4y - x$ are opposites, their quotient is -1.

Caution

A − symbol preceding a fraction may be applied to the numerator or to the denominator, but not to both. For example,

$$-\frac{3x + 2y}{y} \neq \frac{-3x - 2y}{-y}$$

$$\frac{3x^2 - 10xy - 8y^2}{4y^2 - xy} = \frac{(3x + 2y)\overset{-1}{\cancel{(x - 4y)}}}{\underset{1}{y\cancel{(4y - x)}}}$$

Since $x - 4y$ and $4y - x$ are opposites, simplify by replacing $\frac{x - 4y}{4y - x}$ with the equivalent fraction $\frac{-1}{1} = -1$.

$$= \frac{-(3x + 2y)}{y}$$

$$= \frac{-3x - 2y}{y}$$

This result may also be written as $\dfrac{-(3x + 2y)}{y}$ or $-\dfrac{3x + 2y}{y}$.

Self Check 3 Simplify: $\dfrac{2a^2 - 3ab - 9b^2}{3b^2 - ab}$.

■ MULTIPLYING AND DIVIDING RATIONAL EXPRESSIONS

Recall that to multiply two fractions, we multiply the numerators and multiply the denominators. We use the same procedure to multiply rational expressions.

Multiplying Rational Expressions	If A, B, C, and D represent polynomials, where B and D are not 0, $$\frac{A}{B} \cdot \frac{C}{D} = \frac{AC}{BD}$$ Then simplify, if possible.

EXAMPLE 4

ELEMENTARY & INTERMEDIATE
Algebra *f(x)* **Now**™

Multiply: $\dfrac{x^2 - 6x + 9}{20x} \cdot \dfrac{5x^2}{x^2 - 9}$.

Solution We multiply the numerators and multiply the denominators and then factor to simplify the resulting fraction.

$$\frac{x^2 - 6x + 9}{20x} \cdot \frac{5x^2}{x^2 - 9} = \frac{(x^2 - 6x + 9)5x^2}{20x(x^2 - 9)}$$

Multiply the numerators and multiply the denominators.

$$= \frac{(x - 3)(x - 3)5xx}{4 \cdot 5 \cdot x(x + 3)(x - 3)}$$

Factor in the numerator and denominator. Write 20 as $4 \cdot 5$.

$$= \frac{\overset{1}{\cancel{(x - 3)}}(x - 3)\overset{1}{\cancel{5}}\overset{1}{\cancel{x}}x}{4 \cdot \cancel{5} \cdot \cancel{x}(x + 3)\cancel{(x - 3)}}$$

Simplify.

$$= \frac{x(x - 3)}{4(x + 3)}$$

Caution

When multiplying rational expressions, always write the result in simplest form by removing any factors common to the numerator and denominator.

Self Check 4 Multiply: $\dfrac{a^2 + 6a + 9}{18a} \cdot \dfrac{3a^3}{a + 3}$.

EXAMPLE 5

Multiply: $(2x - x^2) \cdot \dfrac{x}{x^2 - xb - 2x + 2b}$.

Solution

$(2x - x^2) \cdot \dfrac{x}{x^2 - xb - 2x + 2b}$

$= \dfrac{2x - x^2}{1} \cdot \dfrac{x}{x^2 - xb - 2x + 2b}$ Write $2x - x^2$ as the fraction $\dfrac{2x - x^2}{1}$.

$= \dfrac{(2x - x^2)x}{1(x^2 - xb - 2x + 2b)}$ Multiply the numerators and multiply the denominators.

$= \dfrac{x \overset{-1}{\cancel{(2 - x)}} x}{1(x - b)\underset{1}{\cancel{(x - 2)}}}$ Factor out x in the numerator. In the denominator, factor by grouping. Recall that the quotient of any nonzero quantity and its opposite is -1: $\dfrac{2 - x}{x - 2} = -1$.

$= \dfrac{-x^2}{x - b}$

Success Tip

We would obtain the same answer if we had factored the numerators and denominators first and simplified before we multiplied.

Since the $-$ symbol can be written in front of the fraction, this result can also be written as

$$-\dfrac{x^2}{x - b}$$

Self Check 5   Multiply: $\dfrac{x^2 + 5x + 6}{4x + 8 - x^2 - 2x} (x^2 - 4x)$.

Recall that to divide fractions, we multiply the first fraction by the reciprocal of the second fraction. We use the same procedure to divide rational expressions.

Dividing Rational Expressions

If A, B, C, and D represent polynomials, where B, C, and D are not 0,

$$\dfrac{A}{B} \div \dfrac{C}{D} = \dfrac{A}{B} \cdot \dfrac{D}{C} = \dfrac{AD}{BC}$$

Then simplify, if possible.

EXAMPLE 6

Divide: $\dfrac{x^3 + 8}{4x + 4} \div \dfrac{x^2 - 2x + 4}{2x^2 - 2}$.

Solution

$\dfrac{x^3 + 8}{4x + 4} \div \dfrac{x^2 - 2x + 4}{2x^2 - 2}$

$= \dfrac{x^3 + 8}{4x + 4} \cdot \dfrac{2x^2 - 2}{x^2 - 2x + 4}$ Multiply the first rational expression by the reciprocal of the second.

$= \dfrac{(x^3 + 8)(2x^2 - 2)}{(4x + 4)(x^2 - 2x + 4)}$ Multiply the numerators and multiply the denominators.

Caution
When dividing rational expressions, always write the result in simplest form by removing any factors common to the numerator and denominator.

$$= \frac{(x + 2)(\overset{1}{\cancel{x^2 - 2x + 4}})\overset{1}{\cancel{2}}\overset{1}{\cancel{(x + 1)}}(x - 1)}{2 \cdot \underset{1}{\cancel{2}}\underset{1}{\cancel{(x + 1)}}\underset{1}{\cancel{(x^2 - 2x + 4)}}}$$

Factor $x^3 + 8$, $2x^2 - 2$, and $4x + 4$.
The polynomial $x^2 - 2x + 4$ does not factor. Write 4 as $2 \cdot 2$. Then simplify.

$$= \frac{(x + 2)(x - 1)}{2}$$

Self Check 6   Divide: $\dfrac{x^3 - 8}{9x - 9} \div \dfrac{x^2 + 2x + 4}{3x^2 - 3x}$.

EXAMPLE 7

ELEMENTARY &
INTERMEDIATE
Algebra $f(x)$ **Now**™

Simplify: $\dfrac{x^2 + 2x - 3}{6x^2 + 5x + 1} \div \dfrac{2x^2 - 2}{2x^2 - 5x - 3} \cdot \dfrac{6x^2 + 4x - 2}{x^2 - 2x - 3}$.

Solution   Since multiplications and divisions are done in order from left to right, we begin by focusing on the division. We introduce grouping symbols to emphasize this. To divide the expressions within the parentheses, we invert $\dfrac{2x^2 - 2}{2x^2 - 5x - 3}$ and multiply.

$$\left(\frac{x^2 + 2x - 3}{6x^2 + 5x + 1} \div \frac{2x^2 - 2}{2x^2 - 5x - 3}\right)\frac{6x^2 + 4x - 2}{x^2 - 2x - 3} = \left(\frac{x^2 + 2x - 3}{6x^2 + 5x + 1} \cdot \frac{2x^2 - 5x - 3}{2x^2 - 2}\right)\frac{6x^2 + 4x - 2}{x^2 - 2x - 3}$$

Next, we multiply the three fractions and simplify the result.

$$= \frac{(x^2 + 2x - 3)(2x^2 - 5x - 3)(6x^2 + 4x - 2)}{(6x^2 + 5x + 1)(2x^2 - 2)(x^2 - 2x - 3)}$$

$$= \frac{(x + 3)\overset{1}{\cancel{(x - 1)}}\overset{1}{\cancel{(2x + 1)}}\overset{1}{\cancel{(x - 3)}}\overset{1}{\cancel{2}}(3x - 1)\overset{1}{\cancel{(x + 1)}}}{(3x + 1)\underset{1}{\cancel{(2x + 1)}}\underset{1}{\cancel{2}}(x + 1)\underset{1}{\cancel{(x - 1)}}\underset{1}{\cancel{(x - 3)}}\underset{1}{\cancel{(x + 1)}}}$$

$$= \frac{(x + 3)(3x - 1)}{(3x + 1)(x + 1)}$$

Self Check 7   Simplify: $\dfrac{x^2 - 25}{4x^2 + 12x + 9} \div \dfrac{x^2 - 5x}{3x - 1} \cdot \dfrac{2x + 3}{3x^2 + 14x - 5}$.

■ ADDING AND SUBTRACTING RATIONAL EXPRESSIONS

To add or subtract fractions with like denominators, we add or subtract the numerators and keep the same denominator. We use the same procedure to add and subtract rational expressions with like denominators.

Adding and Subtracting Rational Expressions

If $\dfrac{A}{D}$ and $\dfrac{B}{D}$ are rational expressions,

$$\frac{A}{D} + \frac{B}{D} = \frac{A + B}{D} \qquad\qquad \frac{A}{D} - \frac{B}{D} = \frac{A - B}{D}$$

Then simplify, if possible.

EXAMPLE 8   Add: $\dfrac{a^2}{a^2 - 36} + \dfrac{6a}{a^2 - 36}$. Simplify the result.

Solution

$$\dfrac{a^2}{a^2 - 36} + \dfrac{6a}{a^2 - 36} = \dfrac{a^2 + 6a}{a^2 - 36} \qquad \text{\color{red}{Add the numerators. Write the sum over the common denominator } } a^2 - 36.$$

We can factor the binomials in the numerator and the denominator.

$$\dfrac{a^2}{a^2 - 36} + \dfrac{6a}{a^2 - 36} = \dfrac{a(a + 6)}{(a + 6)(a - 6)}$$

$$= \dfrac{\overset{1}{a\cancel{(a + 6)}}}{\underset{1}{\cancel{(a + 6)}(a - 6)}} \qquad \text{\color{red}{Simplify.}}$$

$$= \dfrac{a}{a - 6}$$

Caution

When adding or subtracting rational expressions, always write the result in simplest form by removing any factors common to the numerator and denominator.

Self Check 8   Add: $\dfrac{2b}{b^2 - 4} + \dfrac{b^2}{b^2 - 4}$.

To add or subtract rational expressions with unlike denominators, we build them to rational expressions with the same denominator.

Building Rational Expressions	To build a rational expression, multiply it by 1 in the form of $\dfrac{c}{c}$, where c is any nonzero number or expression.

When adding or subtracting rational expressions with unlike denominators, it is easiest if we write the rational expressions in terms of the smallest common denominator possible, called the **least** (or lowest) **common denominator (LCD).** To find the least common denominator of several rational expressions, we follow these steps.

Finding the LCD	1. Factor each denominator completely. 2. The LCD is a product that uses each different factor obtained in step 1 the greatest number of times it appears in any one factorization.

EXAMPLE 9   Add: $\dfrac{5a}{24b} + \dfrac{11a}{18b^2}$.

Solution   We write each denominator as the product of prime numbers and variables.

$$24b = 2 \cdot 2 \cdot 2 \cdot 3 \cdot b = 2^3 \cdot 3 \cdot b$$
$$18b^2 = 2 \cdot 3 \cdot 3 \cdot b \cdot b = 2 \cdot 3^2 \cdot b^2$$

To find the LCD, we form a product using each of these factors the greatest number of times it appears in any one factorization.

Success Tip

Note that the highest power of each factor is used to form the LCD:

$24b = 2^{\textcircled{3}} \cdot 3 \cdot b$

$18b^2 = 2 \cdot 3^{\textcircled{2}} \cdot b^{\textcircled{2}}$

$\text{LCD} = 2^3 \cdot 3^2 \cdot b^2 = 72b^2$

The greatest number of times the factor 2 appears is three times.
The greatest number of times the factor 3 appears is twice.
The greatest number of times the factor b appears is twice.

$$\text{LCD} = \mathbf{2 \cdot 2 \cdot 2 \cdot 3 \cdot 3 \cdot b \cdot b}$$
$$= 72b^2$$

Now we multiply each numerator and denominator by whatever it takes to build the denominator to $72b^2$.

$$\frac{5a}{24b} + \frac{11a}{18b^2} = \frac{5a}{24b} \cdot \frac{\mathbf{3b}}{\mathbf{3b}} + \frac{11a}{18b^2} \cdot \frac{\mathbf{4}}{\mathbf{4}}$$ Build each rational expression.

$$= \frac{15ab}{72b^2} + \frac{44a}{72b^2}$$ Multiply the numerators. Multiply the denominators.

$$= \frac{15ab + 44a}{72b^2}$$ Add the numerators. Write the sum over the common denominator. The result does not simplify.

Self Check 9   Add: $\dfrac{3y}{28z^3} + \dfrac{5x}{21z}$.

- -

EXAMPLE 10

ELEMENTARY & INTERMEDIATE
Algebra $f(x)$ **Now**™

Subtract: $\dfrac{x+1}{x^2 - 2x + 1} - \dfrac{x-4}{x^2 - 1}$.

Solution   We factor each denominator to find the LCD:

$$x^2 - 2x + 1 = (x-1)(x-1) = (x-1)^2$$ The greatest number of times the factor $x - 1$ appears is twice.

$$x^2 - 1 = (x+1)(x-1)$$ The greatest number of times the factor $x + 1$ appears is once.

The LCD is $(x - 1)^2(x + 1)$ or $(x - 1)(x - 1)(x + 1)$.

We now write each rational expression with its denominator in factored form. Then we multiply each numerator and denominator by the missing factor, so that each rational expression has a denominator of $(x - 1)(x - 1)(x + 1)$.

$$\frac{x+1}{x^2 - 2x + 1} - \frac{x-4}{x^2 - 1}$$

$$= \frac{x+1}{(x-1)(x-1)} - \frac{x-4}{(x+1)(x-1)}$$ Write each denominator in factored form.

$$= \frac{x+1}{(x-1)(x-1)} \cdot \frac{\mathbf{x+1}}{\mathbf{x+1}} - \frac{x-4}{(x+1)(x-1)} \cdot \frac{\mathbf{x-1}}{\mathbf{x-1}}$$ Build each rational expression.

$$= \frac{x^2 + 2x + 1}{(x-1)(x-1)(x+1)} - \frac{x^2 - 5x + 4}{(x+1)(x-1)(x-1)}$$ Multiply the numerators using the FOIL method. Multiply the denominators.

$$= \frac{x^2 + 2x + 1 - (x^2 - 5x + 4)}{(x - 1)(x - 1)(x + 1)}$$ This numerator is written within parentheses to make sure we subtract all three of its terms.

Subtract the numerators. Write the difference over the common denominator.

$$= \frac{x^2 + 2x + 1 - x^2 + 5x - 4}{(x - 1)(x - 1)(x + 1)}$$ In the numerator, subtract the polynomials.

$$= \frac{7x - 3}{(x - 1)(x - 1)(x + 1)}$$ Combine like terms. The result does not simplify.

$$= \frac{7x - 3}{(x - 1)^2(x + 1)}$$

Self Check 10   Subtract: $\dfrac{a + 2}{a^2 - 4a + 4} - \dfrac{a - 3}{a^2 - 4}$.

We can use the following fact to add or subtract rational expressions whose denominators are opposites.

Multiplying by −1   When a polynomial is multiplied by -1, the result is its opposite.

EXAMPLE 11   Add: $\dfrac{x}{x - y} + \dfrac{y}{y - x}$.

Solution   We note that the denominators are opposites. Either can serve as the LCD; we will choose $x - y$.

We must multiply the denominator of $\dfrac{y}{y - x}$ by -1 to obtain the LCD, $x - y$. It follows that $\dfrac{-1}{-1}$ should be the form of 1 that is used to build an equivalent rational expression.

$$\frac{x}{x - y} + \frac{y}{y - x} = \frac{x}{x - y} + \frac{y}{y - x} \cdot \frac{-1}{-1}$$ Build $\dfrac{y}{y - x}$ so that it has a denominator of $x - y$.

$$= \frac{x}{x - y} + \frac{-y}{-y + x}$$ Multiply the numerators. Multiply the denominators.

$$= \frac{x}{x - y} + \frac{-y}{x - y}$$ Rewrite the second denominator, $-y + x$, as $x - y$. The fractions now have a common denominator.

$$= \frac{x - y}{x - y}$$ Add the numerators. Write the difference over the common denominator $x - y$.

$$= 1$$ Simplify.

Self Check 11   Add: $\dfrac{2a}{a - b} + \dfrac{b}{b - a}$.

■ COMPLEX FRACTIONS

A rational expression whose numerator and/or denominator contain rational expressions is called a **complex rational expression** or, more simply, a **complex fraction.** The expression above the main fraction bar of a complex fraction is the numerator, and the expression below the main fraction bar is the denominator. Two examples are:

$$\frac{\dfrac{3a}{b}}{\dfrac{6ac}{b^2}}, \qquad \frac{\dfrac{1}{a^2 - 3a + 2}}{\dfrac{3}{a - 2} - \dfrac{2}{a - 1}}$$

Numerator ← → Numerator

← Main fraction bar →

← Denominator →

Success Tip

You may want to review Method 1 and Method 2 for *simplifying complex fractions* on pages 455 and 458.

The first method to **simplify complex fractions** uses the fact that the main fraction bar indicates division. With the second method, we multiply the numerator and denominator of the complex fraction by the LCD of all fractions in the complex fraction.

EXAMPLE 12

ELEMENTARY & INTERMEDIATE

Algebra $f(x)$ **Now**™

Simplify: $\dfrac{\dfrac{3a}{b}}{\dfrac{6ac}{b^2}}$.

Solution   We will use Method 1 to do the simplification. We write the complex fraction as a division and proceed as follows:

Success Tip

Simplifying using division works well when a complex fraction is written, or can be easily written, as a quotient of two single rational expressions.

$$\frac{\dfrac{3a}{b}}{\dfrac{6ac}{b^2}} = \frac{3a}{b} \div \frac{6ac}{b^2} \qquad \text{The main fraction bar of the complex fraction indicates division.}$$

$$= \frac{3a}{b} \cdot \frac{b^2}{6ac} \qquad \text{Multiply by the reciprocal of } \dfrac{6ac}{b^2}.$$

$$= \frac{b}{2c} \qquad \text{Multiply the rational expressions and then simplify.}$$

Self Check 12   Simplify: $\dfrac{\dfrac{2x}{y^2}}{\dfrac{6xz}{y}}$.

EXAMPLE 13

Simplify: $\dfrac{\dfrac{1}{a^2 - 3a + 2}}{\dfrac{3}{a - 2} - \dfrac{2}{a - 1}}$.

Solution   We will use Method 2 to do the simplification. To determine the LCD for all the fractions appearing in the complex fraction, we must factor $a^2 - 3a + 2$.

$$\frac{\dfrac{1}{a^2 - 3a + 2}}{\dfrac{3}{a - 2} - \dfrac{2}{a - 1}} = \frac{\dfrac{1}{(a - 2)(a - 1)}}{\dfrac{3}{a - 2} - \dfrac{2}{a - 1}}$$

The LCD of the fractions in the numerator and denominator of the complex fraction is $(a - 2)(a - 1)$. We multiply the numerator and the denominator by the LCD.

Success Tip

Simplifying using the LCD works well when the complex fraction has sums and/or differences in the numerator or denominator.

$$= \frac{\dfrac{1}{(a-2)(a-1)}}{\dfrac{3}{a-2} - \dfrac{2}{a-1}} \cdot \frac{(a-2)(a-1)}{(a-2)(a-1)}$$

The Language of Algebra

After multiplying a complex fraction by $\frac{LCD}{LCD}$ and performing the multiplications, the numerator and denominator of the complex fraction will be *cleared* of fractions.

$$= \frac{\left(\dfrac{1}{(a-2)(a-1)}\right)(a-2)(a-1)}{\left(\dfrac{3}{a-2} - \dfrac{2}{a-1}\right)(a-2)(a-1)}$$

Multiply the numerators.
Multiply the denominators.

$$= \frac{\dfrac{(a-2)(a-1)}{(a-2)(a-1)}}{\dfrac{3(a-2)(a-1)}{a-2} - \dfrac{2(a-2)(a-1)}{a-1}}$$

Perform the multiplication in the numerator. In the denominator, distribute the LCD.

$$= \frac{1}{3(a-1) - 2(a-2)}$$

Simplify each of the three rational expressions.

$$= \frac{1}{3a - 3 - 2a + 4}$$

In the denominator, remove parentheses.

$$= \frac{1}{a+1}$$

Combine like terms.

Self Check 13 Simplify: $\dfrac{\dfrac{b}{b+4} + \dfrac{3}{b+3}}{\dfrac{b}{b^2 + 7b + 12}}$.

Answers to Self Checks
1. $\frac{2}{5k}$
2. $\frac{2x+1}{3x-1}$
3. $-\frac{2a+3b}{b}$ or $\frac{-2a-3b}{b}$
4. $\frac{a^2(a+3)}{6}$
5. $-x(x+3)$
6. $\frac{x(x-2)}{3}$
7. $\frac{1}{x(2x+3)}$
8. $\frac{b}{b-2}$
9. $\frac{9y+20xz^2}{84z^3}$
10. $\frac{9a-2}{(a-2)^2(a+2)}$
11. $\frac{2a-b}{a-b}$
12. $\frac{1}{3yz}$
13. $\frac{b^2+6b+12}{b}$

8.5 STUDY SET ELEMENTARY & INTERMEDIATE Algebra *f(x)* Now™

VOCABULARY Fill in the blanks.

1. A quotient of two polynomials, such as $\frac{x^2+x}{x^2-3x}$, is called a _____ expression.

2. To _____ a rational expression, we remove factors common to the numerator and denominator.

3. The quotient of _____ is -1. For example, $\frac{x-8}{8-x} = -1$.

4. In the rational expression $\frac{(x+2)(3x-1)}{(x+2)(4x+2)}$, $x + 2$ is a common _____ of the numerator and the denominator.

5. The _____ of $\frac{a+3}{a+7}$ is $\frac{a+7}{a+3}$.

6. The _____ _____ of $\frac{x-8}{x+6}$ and $\frac{6-5x}{x}$ is $x(x+6)$.

7. To _____ a rational expression, we multiply it by a form of 1. For example, $\frac{2}{n^2} \cdot \frac{8}{8} = \frac{16}{8n^2}$.

8. $\dfrac{\frac{x}{y} + \frac{1}{x}}{\frac{1}{y} + \frac{2}{x}}$ and $\dfrac{\frac{5a^2}{b}}{\frac{b}{2a^3}}$ are examples of _____ rational expressions or, more simply, complex _____.

CONCEPTS **Fill in the blanks.**

9. To multiply rational expressions, multiply their _____ and multiply their _____. To divide two rational expressions, multiply the first by the _____ of the second.

$$\frac{A}{B} \cdot \frac{C}{D} = \boxed{} \qquad \frac{A}{B} \div \frac{C}{D} = \boxed{}$$

10. To add or subtract rational expressions that have the same denominator, add or subtract the _____, and write the sum or difference over the common _____.

$$\frac{A}{D} + \frac{B}{D} = \boxed{} \qquad \frac{A}{D} - \frac{B}{D} = \boxed{}$$

11. To find the least common denominator of several rational expressions, _____ each denominator completely. The LCD is a product that uses each different factor the _____ number of times it appears in any one factorization.

12. The expression $4 - y$ must be multiplied by $\boxed{}$ to obtain $y - 4$.

13. Consider the following two procedures.

i. $\dfrac{x^2 - 2x}{x^2 + 4x - 12} = \dfrac{x(x - 2)}{(x + 6)(x - 2)} = \dfrac{x}{x + 6}$

ii. $\dfrac{x}{x + 6} = \dfrac{x}{x + 6} \cdot \dfrac{x - 2}{x - 2} = \dfrac{x^2 - 2x}{x^2 + 4x - 12}$

   a. In which of these procedures are we *building* a rational expression?

   b. For what type of problem is this procedure often necessary?

   c. What name is used to describe the other procedure?

14. Simplify each rational expression, if possible.

   a. $\dfrac{x + 8}{x}$ **b.** $\dfrac{x + 8}{8}$

   c. $\dfrac{a^3 + 8}{2}$ **d.** $\dfrac{x^2 + 5x + 6}{x^2 + x - 12}$

15. Consider the following factorizations.

$$18x - 36 = 2 \cdot 3 \cdot 3 \cdot (x - 2)$$
$$3x^2 - 3x - 6 = 3(x - 2)(x + 1)$$

   a. What is the greatest number of times the factor 3 appears in any one factorization?

   b. What is the greatest number of times the factor $x - 2$ appears in any one factorization?

16. The LCD for $\dfrac{2x + 1}{x^2 + 5x + 6}$ and $\dfrac{3x}{x^2 - 4}$ is

$$\text{LCD} = (x + 2)(x + 3)(x - 2)$$

If we want to subtract these rational expressions, what form of 1 should be used

   a. to build $\dfrac{2x + 1}{x^2 + 5x + 6}$?

   b. to build $\dfrac{3x}{x^2 - 4}$?

17. Determine the LCD of the rational expressions appearing in each complex fraction.

   a. $\dfrac{1 + \frac{4}{c}}{\frac{2}{c} + c}$ **b.** $\dfrac{\frac{p}{p + 2} + \frac{12}{p + 3}}{\frac{p - 1}{p^2 + 5p + 6}}$

18. To simplify the complex fraction shown below, it is multiplied by a form of 1. What form of 1 is used?

$$\dfrac{\frac{4}{t^2} + \frac{b}{t}}{\frac{3b}{t}} = \dfrac{\frac{4}{t^2} + \frac{b}{t}}{\frac{3b}{t}} \cdot \frac{t^2}{t^2}$$

NOTATION   Fill in the blanks.

19. $\dfrac{x^2 + 3x}{x - 1} - \dfrac{2x - 1}{x - 1} = \dfrac{x^2 + 3x - ()}{x - 1}$

20. The fraction $\dfrac{\frac{a}{b}}{\frac{c}{d}}$ is equivalent to $\dfrac{a}{b} \dfrac{c}{d}$.

PRACTICE   Simplify each rational expression.

21. $\dfrac{24x^3y^4}{18x^4y^3}$

22. $\dfrac{15a^5b^4}{21b^3c^2}$

23. $\dfrac{9y^2(y - z)}{21y(y - z)^2}$

24. $\dfrac{3ab^2(a - b)}{9ab(a - b)^3}$

25. $\dfrac{(a - b)(b - c)(c - d)}{(c - d)(b - c)(a - b)}$

26. $\dfrac{(p + q)(p - r)(r + s)}{(r - p)(r + s)(p + q)}$

27. $\dfrac{3m - 6n}{3n - 6m}$

28. $\dfrac{4c - 8d}{4d + 8c}$

29. $\dfrac{x^2 + 2x + 1}{x^2 + 4x + 3}$

30. $\dfrac{x^2 + 2x - 15}{x^2 - 25}$

31. $\dfrac{6x^2 - 7x - 5}{2x^2 + 5x + 2}$

32. $\dfrac{6x^2 + x - 2}{8x^2 + 2x - 3}$

33. $\dfrac{ax + by + ay + bx}{a^2 - b^2}$

34. $\dfrac{3x^2 - 3y^2}{x^2 + 2y + 2x + yx}$

35. $\dfrac{12 - 3x^2}{x^2 - x - 2}$

36. $\dfrac{2x^2 - 4x - 30}{25 - x^2}$

37. $\dfrac{a^3 + 27}{4a^2 - 36}$

38. $\dfrac{a^2 - 4}{a^3 - 8}$

39. $\dfrac{2x^2 + 2x - 12}{x^3 + 3x^2 - 4x - 12}$

40. $\dfrac{x - y}{x^3 - y^3 - x + y}$

41. $\dfrac{p^3 + p^2q - 2pq^2}{pq^2 + p^2q - 2p^3}$

42. $\dfrac{m^3 - mn^2}{mn^2 + m^2n - 2m^3}$

43. $\dfrac{(2x^2 + 3xy + y^2)(3a + b)}{(x + y)(2xy + 2bx + y^2 + by)}$

44. $\dfrac{(x^2 + 2x + 1)(x^2 - 2x + 1)}{(x^2 - 1)^2}$

Perform the operations and simplify, if possible.

45. $\dfrac{10a^2}{3b^4} \cdot \dfrac{12b^3}{5a^2}$

46. $\dfrac{16c^3}{5d^2} \cdot \dfrac{25d}{12c}$

47. $\dfrac{m^2n}{4} \div \dfrac{mn^3}{6}$

48. $\dfrac{a^4b}{14} \div \dfrac{a^3b^2}{21}$

49. $12y\left(\dfrac{y + 8}{6y}\right)$

50. $16x\left(\dfrac{3x + 8}{4x}\right)$

51. $\dfrac{x^2 - 16}{x^2 - 25} \div \dfrac{x + 4}{x - 5}$

52. $\dfrac{a^2 - 9}{a^2 - 49} \div \dfrac{a + 3}{a + 7}$

53. $\dfrac{x^2 + 2x + 1}{9x} \cdot \dfrac{2x^2 - 2x}{2x^2 - 2}$

54. $\dfrac{a + 6}{16 - a^2} \cdot \dfrac{3a - 12}{3a + 18}$

55. $\dfrac{2x^2 - x - 3}{x^2 - 1} \cdot \dfrac{x^2 + x - 2}{2x^2 + x - 6}$

56. $\dfrac{2p^2 - 5p - 3}{p^2 - 9} \cdot \dfrac{2p^2 + 5p - 3}{2p^2 + 5p + 2}$

57. $(2x^2 - 15x + 25) \div \dfrac{2x^2 - 3x - 5}{x + 1}$

58. $(x^2 - 6x + 9) \div \dfrac{x^2 - 9}{x + 3}$

59. $\dfrac{3n^2 + 5n - 2}{12n^2 - 13n + 3} \div \dfrac{n^2 + 3n + 2}{4n^2 + 5n - 6}$

60. $\dfrac{8y^2 - 14y - 15}{6y^2 - 11y - 10} \div \dfrac{4y^2 - 9y - 9}{3y^2 - 7y - 6}$

61. $\dfrac{2x^2 + 5xy + 3y^2}{3x^2 - 5xy + 2y^2} \div \dfrac{2x^2 + xy - 3y^2}{-3x^2 + 5xy - 2y^2}$

62. $\dfrac{2p^2 - 5pq - 3q^2}{p^2 - 9q^2} \div \dfrac{2p^2 + 5pq + 2q^2}{2p^2 + 5pq - 3q^2}$

63. $\dfrac{p^3 - q^3}{q^2 - p^2} \cdot \dfrac{q^2 + pq}{p^3 + p^2q + pq^2}$

64. $\dfrac{x^3 + y^3}{x^3 - y^3} \div \dfrac{x^2 - xy + y^2}{x^2 + xy + y^2}$

65. $\dfrac{y^3 - x^3}{2x^2 + 2xy + x + y} \cdot \dfrac{2x^2 - 5x - 3}{yx - 3y - x^2 + 3x}$

66. $\dfrac{ax + ay + bx + by}{x^3 - 27} \cdot \dfrac{x^2 + 3x + 9}{xc + xd + yc + yd}$

67. $(x^2 + x - 2cx - 2c) \cdot \dfrac{x^2 + 3x + 2}{x^2 - 4c^2}$

68. $(2ax - 10x + a - 5) \cdot \dfrac{x}{2x^2 + x}$

69. $\dfrac{15c^2d^3}{8x} \div \dfrac{25cd^4x}{16} \cdot \dfrac{5x^2}{4d}$

70. $\dfrac{9ab^3}{7xy} \cdot \dfrac{14xy^2}{27z^3} \div \dfrac{18a^2b^2x}{3z^2}$

71. $\dfrac{4x^2 - 10x + 6}{x^4 - 3x^3} \div \dfrac{2x - 3}{2x^3} \cdot \dfrac{x - 3}{2 - 2x}$

72. $\dfrac{2x^2 - 2x - 4}{x^2 + 2x - 8} \cdot \dfrac{3x^2 + 15x}{x + 1} \div \dfrac{100 - 4x^2}{x^2 - x - 20}$

73. $\dfrac{2x^2 + x - 1}{x^2 - 1} \div \left(\dfrac{x^2 + 2x - 35}{x^2 - 6x + 5} \div \dfrac{x^2 - 9x + 14}{2x^2 - 5x + 2} \right)$

74. $\dfrac{x^2 - 4}{x^2 - x - 6} \div \left(\dfrac{x^2 - x - 2}{x^2 - 8x + 15} \cdot \dfrac{x^2 - 3x - 10}{x^2 + 3x + 2} \right)$

Perform the operations and simplify, if possible.

75. $\dfrac{3}{a + b} - \dfrac{a}{a + b}$

76. $\dfrac{x}{x + 4} + \dfrac{5}{x + 4}$

77. $\dfrac{3x}{2x + 2} + \dfrac{x + 4}{2x + 2}$

78. $\dfrac{4y}{y - 4} - \dfrac{16}{y - 4}$

79. $\dfrac{5x}{x + 1} + \dfrac{3}{x + 1} - \dfrac{2x}{x + 1}$

80. $\dfrac{4}{a + 4} - \dfrac{2a}{a + 4} + \dfrac{3a}{a + 4}$

81. $\dfrac{3y - 2}{2y + 6} - \dfrac{2y - 5}{2y + 6}$

82. $\dfrac{5x + 8}{3x + 15} - \dfrac{3x - 2}{3x + 15}$

83. $\dfrac{2x + 1}{x^4 - 81} + \dfrac{2 - x}{x^4 - 81}$

84. $\dfrac{2m^2 - 7}{m^4 - 9} + \dfrac{4 - m^2}{m^4 - 9}$

85. $\dfrac{3}{4x} + \dfrac{2}{3x}$

86. $\dfrac{2}{5a} + \dfrac{3}{2b}$

87. $\dfrac{3}{x + 2} + \dfrac{5}{x - 4}$

88. $\dfrac{2}{a + 4} - \dfrac{6}{a + 3}$

89. $\dfrac{x + 2}{x + 5} - \dfrac{x - 3}{x + 7}$

90. $\dfrac{7}{x + 3} + \dfrac{4x}{x + 6}$

91. $\dfrac{x + 8}{x - 3} - \dfrac{x - 14}{3 - x}$

92. $\dfrac{3 - x}{2 - x} + \dfrac{x - 1}{x - 2}$

93. $\dfrac{a^2 + ab}{a^3 - b^3} - \dfrac{b^2}{b^3 - a^3}$

94. $\dfrac{y^2 - 3xy}{x^3 - y^3} - \dfrac{x^2 + 4xy}{y^3 - x^3}$

95. $\dfrac{x + 5}{xy} - \dfrac{x - 1}{x^2y}$

96. $\dfrac{y - 7}{y^2} - \dfrac{y + 7}{2y}$

97. $\dfrac{x}{x^2 + 5x + 6} + \dfrac{x}{x^2 - 4}$

98. $\dfrac{x}{3x^2 - 2x - 1} + \dfrac{4}{3x^2 + 10x + 3}$

99. $\dfrac{4}{x^2 - 2x - 3} - \dfrac{x}{3x^2 - 7x - 6}$

100. $\dfrac{2a}{a^2 - 2a - 8} + \dfrac{3}{a^2 - 5a + 4}$

101. $\dfrac{x}{x^2 - 4} - \dfrac{x}{x + 2} + \dfrac{2}{x}$

102. $\dfrac{8}{x^2 - 9} + \dfrac{2}{x - 3} - \dfrac{6}{x}$

103. $1 + x - \dfrac{x}{x - 5}$

104. $2 - x + \dfrac{3}{x - 9}$

105. $\dfrac{2}{x - 1} - \dfrac{2x}{x^2 - 1} - \dfrac{x}{x^2 + 2x + 1}$

106. $\dfrac{x - 2}{x^2 - 3x} + \dfrac{2x - 1}{x^2 + 3x} - \dfrac{2}{x^2 - 9}$

Simplify each complex fraction.

107. $\dfrac{\dfrac{4x}{y}}{\dfrac{6xz}{y^2}}$

108. $\dfrac{\dfrac{5t^4}{9x}}{\dfrac{2t}{18x}}$

109. $\dfrac{\dfrac{x - y}{xy}}{\dfrac{y - x}{x}}$

110. $\dfrac{\dfrac{x^2 + 5x + 6}{3xy}}{\dfrac{x^2 - 9}{6xy}}$

111. $\dfrac{\dfrac{ac - ad - c + d}{a^3 - 1}}{\dfrac{c^2 - 2cd + d^2}{a^2 + a + 1}}$

112. $\dfrac{\dfrac{2x - tx + 2y - ty}{x^2 + 2xy + y^2}}{\dfrac{t^3 - 8}{15x + 15y}}$

113. $\dfrac{\dfrac{1}{a} + \dfrac{1}{b}}{\dfrac{1}{a}}$

114. $\dfrac{\dfrac{1}{b}}{\dfrac{1}{a} - \dfrac{1}{b}}$

115. $\dfrac{\dfrac{y}{x} - \dfrac{x}{y}}{\dfrac{1}{x} + \dfrac{1}{y}}$

116. $\dfrac{\dfrac{y}{x} - \dfrac{x}{y}}{\dfrac{1}{y} - \dfrac{1}{x}}$

117. $\dfrac{1 + \dfrac{6}{x} + \dfrac{8}{x^2}}{1 + \dfrac{1}{x} - \dfrac{12}{x^2}}$

118. $\dfrac{1 - x - \dfrac{2}{x}}{\dfrac{6}{x^2} + \dfrac{1}{x} - 1}$

119. $\dfrac{\dfrac{1}{a + 1} + 1}{\dfrac{3}{a - 1} + 1}$

120. $\dfrac{2 + \dfrac{3}{x + 1}}{\dfrac{1}{x} + x + x^2}$

121. $\dfrac{x^{-1} + y^{-1}}{x^{-1} - y^{-1}}$

122. $\dfrac{x + y}{x^{-1} + y^{-1}}$

123. $\dfrac{\dfrac{2}{x + 3} - \dfrac{1}{x - 3}}{\dfrac{3}{x^2 - 9}}$

124. $\dfrac{2 + \dfrac{1}{x^2 - 1}}{1 + \dfrac{1}{x - 1}}$

125. $\dfrac{\dfrac{h}{h^2 + 3h + 2}}{\dfrac{4}{h + 2} - \dfrac{4}{h + 1}}$

126. $\dfrac{\dfrac{1}{r^2 + 4r + 4}}{\dfrac{r}{r + 2} + \dfrac{r}{r + 2}}$

127. $\dfrac{1 + \dfrac{a}{b}}{1 - \dfrac{a}{1 - \dfrac{a}{b}}}$

128. $\dfrac{1 + \dfrac{2}{1 + \dfrac{a}{b}}}{1 - \dfrac{a}{b}}$

APPLICATIONS

129. PHYSICS The following table contains data from a physics experiment. k_1 and k_2 are constants. Complete the table.

Trial	Rate (m/sec)	Time (sec)	Distance (m)
1	$\dfrac{k_1^2 + 3k_1 + 2}{k_1 - 3}$	$\dfrac{k_1^2 - 3k_1}{k_1 + 1}$	
2	$\dfrac{k_2^2 + 6k_2 + 5}{k_2 + 1}$		$k_2^2 + 11k_2 + 30$

130. DRAFTING Among the tools used in drafting are 45°–45°–90° and 30°–60°–90° triangles. Find the perimeter of each triangle and express each result as a rational expression.

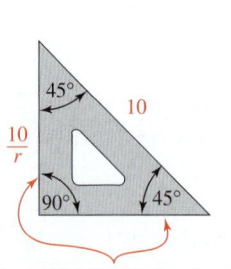

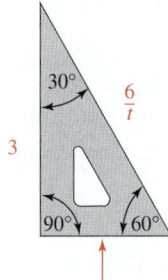

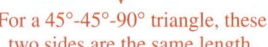

For a 45°-45°-90° triangle, these two sides are the same length.

For a 30°-60°-90° triangle, this side is half as long as the hypotenuse.

131. THE AMAZON   The Amazon River flows in a general eastern direction to the Atlantic Ocean. In Brazil, when the river is at low stage, the rate of flow is about 5 mph. Suppose that a river guide can canoe in still water at a rate of r mph.

   a. Complete the table to find rational expressions that represent the time it would take the guide to canoe 3 miles downriver and to canoe 3 miles upriver.

   b. Find the difference in the times for the trips upriver and downriver. Express the result as a rational expression.

	Rate (mph)	Time (hr)	Distance (mi)
Downriver	$r + 5$		3
Upriver	$r - 5$		3

132. ENGINEERING   The stiffness k of the shaft shown below is given by the formula

$$k = \cfrac{1}{\dfrac{1}{k_1} + \dfrac{1}{k_2}}$$

Section 1   Section 2

where k_1 and k_2 are the individual stiffnesses of each section. Simplify the complex fraction.

WRITING

133. A student compared his answer, $\frac{a - 3b}{2b - a}$, with the answer, $\frac{3b - a}{a - 2b}$, in the back of the text. Is the student's work correct? Explain.

134. Explain the error that is made in the following work:

$$\frac{3x^2 + 1}{3y} = \frac{\cancel{3}x^2 + 1}{\cancel{3}y} = \frac{x^2 + 1}{y}$$

135. Write some comments to the student who wrote the following solution, explaining his misunderstanding.

$$\text{Multiply:} \quad \frac{1}{x} \cdot \frac{3}{2} = \frac{1 \cdot 2}{x \cdot 2} \cdot \frac{3 \cdot x}{2 \cdot x}$$

$$= \frac{2}{2x} \cdot \frac{3x}{2x}$$

$$= \frac{6x}{2x}$$

136. Explain how to find the least common denominator of a set of rational expressions.

REVIEW   Write a system of two equations in two variables to solve each problem.

137. INTEGER PROBLEMS   The sum of two integers is 38, and their difference is 12. Find the integers.

138. INTEGER PROBLEMS   Twice one integer plus another integer is 21. If the first integer plus 3 times the second is 33, find the integers.

CHALLENGE PROBLEMS

139. Simplify: $\dfrac{a^6 - 64}{(a^2 + 2a + 4)(a^2 - 2a + 4)}$.

140. Add: $x^{-1} + x^{-2} + x^{-3} + x^{-4} + x^{-5}$.

141. Simplify: $[(x^{-1} + 1)^{-1} + 1]^{-1}$.

142. Find two rational expressions, each with denominator $x^2 + 5x + 6$, such that their sum is $\dfrac{1}{x + 2}$.

8.6 # Review of Linear Equations in Two Variables

- Graphing linear equations
- The slope of a line
- Slopes of parallel and perpendicular lines
- Writing equations of lines
- The midpoint formula

Recall the definition of a *linear equation in two variables.*

General (Standard) Form of a Linear Equation	A **linear equation in two variables** is an equation that can be written in the form $$Ax + By = C$$ where A, B, and C are real numbers and A and B are not both 0.

Some examples of linear equations are

$$y = -\frac{1}{2}x + 3, \qquad 2x - 5y = 10, \qquad y = 3, \qquad \text{and} \qquad x = 2$$

Linear equations can be graphed in several ways. Generally, the form in which an equation is written determines the method that we use to graph it.

■ GRAPHING LINEAR EQUATIONS

To graph linear equations solved for y, we can select values of x and calculate the corresponding values of y.

EXAMPLE 1

ELEMENTARY & INTERMEDIATE
Algebra *f(x)* **Now™**   Solution

Graph: $y = -\frac{1}{2}x + 3$.

To graph $y = -\frac{1}{2}x + 3$, we construct a **table of solutions** by choosing several values of x and finding the corresponding values of y. For example, if x is -2, we have

$$y = -\frac{1}{2}x + 3$$

$$y = -\frac{1}{2}(-2) + 3 \qquad \text{Substitute } -2 \text{ for } x.$$

$$y = 1 + 3$$

$$y = 4$$

Success Tip

Choose x-values that are multiples of 2 to make the computations easier when multiplying them by $-\frac{1}{2}$.

The Language of Algebra

The *graph of a linear equation* is a mathematical "picture" of the infinitely many solutions of the equation.

Thus, $(-2, 4)$ is a solution. In a similar manner, we find corresponding y-values for x-values of 0, 2, and 4, and enter the solutions in a table.

When we plot the ordered-pair solutions on a rectangular coordinate system, we see that they lie in a straight line. Using a straight edge or ruler, we then draw a line through the points because the graph of any solution of $y = -\frac{1}{2}x + 3$ will lie on this line. Furthermore, every point on this line represents a solution. We call the line the **graph of the equation.** It represents all of the solutions of $y = -\frac{1}{2}x + 3$.

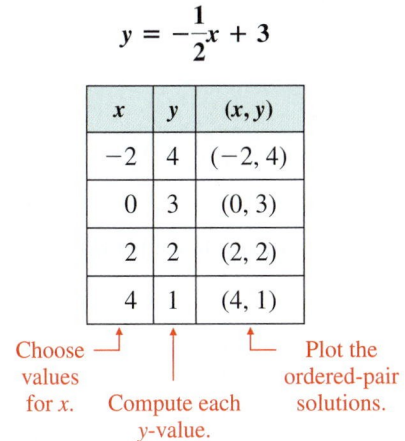

$$y = -\frac{1}{2}x + 3$$

x	y	(x, y)
-2	4	$(-2, 4)$
0	3	$(0, 3)$
2	2	$(2, 2)$
4	1	$(4, 1)$

Choose values for x. Compute each y-value. Plot the ordered-pair solutions.

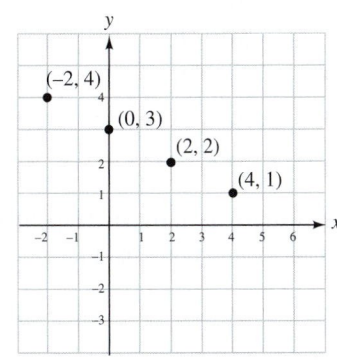

Plot the ordered pairs.

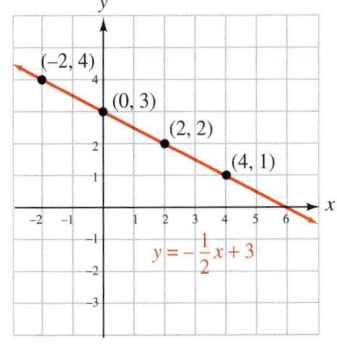

Draw a straight line through the points. This is the *graph of the equation.*

Self Check 1 Graph: $y = \frac{1}{3}x + 1$.

In Example 1, the graph intersected the y-axis at the point $(0, 3)$ (called the **y-intercept**) and intersected the x-axis at the point $(6, 0)$ (called the **x-intercept**). In general, we have the following definitions.

Intercepts of a Line	The **y-intercept** of a line is the point $(0, b)$, where the line intersects the y-axis. To find b, substitute 0 for x in the equation of the line and solve for y.
	The **x-intercept** of a line is the point $(a, 0)$, where the line intersects the x-axis. To find a, substitute 0 for y in the equation of the line and solve for x.

EXAMPLE 2

ELEMENTARY & INTERMEDIATE
Algebra *f(x)* **Now**™

Use the x- and y-intercepts to graph $2x - 5y = 10$.

Solution To find the y-intercept of the graph, we substitute 0 for x and solve for y. To find the x-intercept of the graph, we substitute 0 for y and solve for x:

Success Tip

The exponent on each variable of a linear equation is an understood 1. For example, $2x - 5y = 10$ can be thought of as $2x^1 - 5y^1 = 10$.

y-intercept:

$2x - 5y = 10$

$2(0) - 5y = 10$ Substitute 0 for x.

$-5y = 10$

$y = -2$

The y-intercept is $(0, -2)$.

x-intercept:

$2x - 5y = 10$

$2x - 5(0) = 10$ Substitute 0 for y.

$2x = 10$

$x = 5$

The x-intercept is $(5, 0)$.

Although two points are enough to draw the line, it is a good idea to find and plot a third point as a check. To find the coordinates of a third point, we can substitute any convenient number (such as -5) for x and solve for y:

$$2x - 5y = 10$$
$$2(-5) - 5y = 10 \qquad \text{Substitute } -5 \text{ for } x.$$
$$-10 - 5y = 10$$
$$-5y = 20 \qquad \text{Add 10 to both sides.}$$
$$y = -4 \qquad \text{Divide both sides by } -5.$$

The Language of Algebra

For any two points, exactly one line passes through them. We say two points *determine* a line.

The line also passes through the point $(-5, -4)$.

A table of solutions and the graph of $2x - 5y = 10$ are shown below.

2x − 5y = 10

x	y	(x, y)
0	-2	$(0, -2)$
5	0	$(5, 0)$
-5	-4	$(-5, -4)$

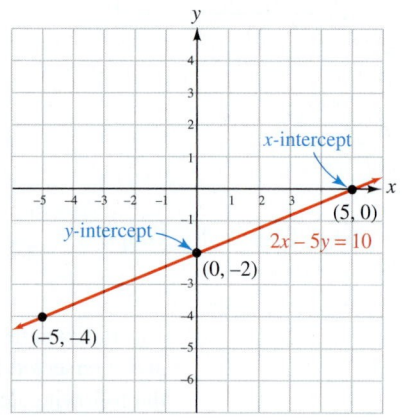

Self Check 2   Find the x- and y-intercepts and graph $5x + 15y = -15$.

Equations such as $y = 3$ and $x = -2$ are linear equations, because they can be written in the form $Ax + By = C$.

$$y = 3 \qquad \text{is equivalent to} \qquad 0x + 1y = 3$$
$$x = -2 \qquad \text{is equivalent to} \qquad 1x + 0y = -2$$

....

EXAMPLE 3   Graph: **a.** $y = 3$ and   **b.** $x = -2$.

ELEMENTARY &
INTERMEDIATE
Algebra $f(x)$ **Now**™

Solution   **a.** Since the equation $y = 3$ does not contain x, the numbers chosen for x have no effect on y. The value of y is always 3.

After plotting the ordered pairs shown in the table on the next page, we see that the graph is a horizontal line, parallel to the x-axis, with a y-intercept of $(0, 3)$. The line has no x-intercept.

b. Since the equation $x = -2$ does not contain y, the value of y can be any number. The value of x is always -2.

After plotting the ordered pairs shown in the table on the next page, we see that the graph is a vertical line, parallel to the y-axis, with an x-intercept of $(-2, 0)$. The line has no y-intercept.

Success Tip

The graph of $y = 0$ is the x-axis and the graph of $x = 0$ is the y-axis.

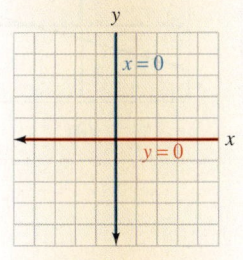

$y = 3$

x	y	(x, y)
-3	3	$(-3, 3)$
0	3	$(0, 3)$
2	3	$(2, 3)$
4	3	$(4, 3)$

The value of x can be any number.

$x = -2$

x	y	(x, y)
-2	-2	$(-2, -2)$
-2	0	$(-2, 0)$
-2	2	$(-2, 2)$
-2	6	$(-2, 6)$

The value of y can be any number.

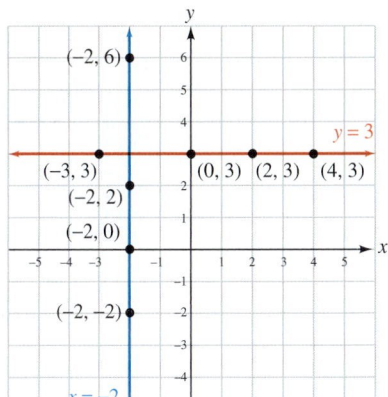

Self Check 3   Graph: $x = 4$ and $y = -3$.

The results of Example 3 suggest the following facts.

Equations of Horizontal and Vertical Lines	The equation $y = b$ represents the horizontal line that intersects the y-axis at $(0, b)$.
	The equation $x = a$ represents the vertical line that intersects the x-axis at $(a, 0)$.

ACCENT ON TECHNOLOGY: GENERATING TABLES OF SOLUTIONS

Courtesy of Texas Instruments

If an equation in x and y is solved for y, we can use a graphing calculator to generate a table of solutions. The instructions in this discussion are for a TI-83 or a TI-83 Plus graphing calculator. For specific details about other brands, please consult the owner's manual.

To construct a table of solutions for $2x - 5y = 10$, we first solve for y.

$$2x - 5y = 10$$
$$-5y = -2x + 10 \qquad \text{Subtract } 2x \text{ from both sides.}$$
$$y = \frac{2}{5}x - 2 \qquad \text{Divide both sides by } -5 \text{ and simplify.}$$

To enter $y = \frac{2}{5}x - 2$, we press $\boxed{Y =}$ and enter $(2/5)x - 2$, as shown in figure (a) on the next page. (Ignore the subscript 1 on y; it is not relevant at this time.)

To enter the x-values that are to appear in the table, we press $\boxed{2nd}$ $\boxed{TBLSET}$ and enter the first value for x on the line labeled TblStart =. In figure (b), -5 has been entered on this line. Other values for x that are to appear in the table are determined by setting an **increment value** on the line labeled $\triangle$Tbl =. Figure (b) shows that an increment of 1 was entered. This means that each x-value in the table will be 1 unit larger than the previous x-value.

The final step is to press the keys $\boxed{2nd}$ $\boxed{TABLE}$. This displays a table of solutions, as shown in figure (c).

To see the graph of $y = \frac{2}{5}x - 2$, we press $\boxed{\text{GRAPH}}$, as shown in figure (d).

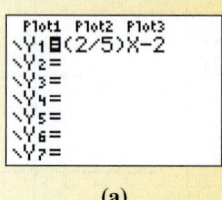
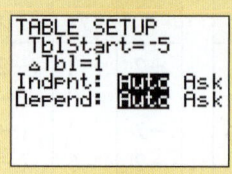
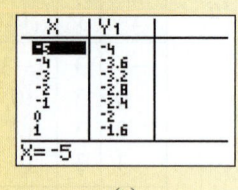
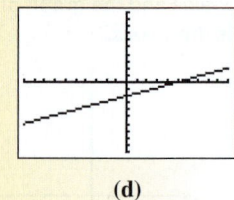

(a)   (b)   (c)   (d)

■ THE SLOPE OF A LINE

Recall that the *slope* of a line is a measure of the steepness of the line.

Slope of a Line   The **slope** of a line passing through points (x_1, y_1) and (x_2, y_2) is

$$m = \frac{\text{change in } y}{\text{change in } x} = \frac{y_2 - y_1}{x_2 - x_1} \qquad \text{where} \qquad x_2 \neq x_1$$

The Language of Algebra

The symbol Δ is the letter *delta* from the Greek alphabet.

The change in y (denoted as Δy and read as "delta y") is the **rise** of the line between two points on the line. The change in x (denoted as Δx and read as "delta x") is the **run.** Using this terminology, we can define slope as the ratio of the rise to the run:

$$m = \frac{\Delta y}{\Delta x} = \frac{\text{rise}}{\text{run}} \qquad \text{where} \qquad \Delta x \neq 0$$

EXAMPLE 4

ELEMENTARY & INTERMEDIATE
Algebra $f^{(x)}$ **Now**™

Success Tip

When calculating slope, it doesn't matter which point we call (x_1, y_1) and which point we call (x_2, y_2). We will obtain the same result in Example 4 if we let $(x_1, y_1) = (3, -4)$ and $(x_2, y_2) = (-2, 4)$.

Find the slope of the line passing through $(-2, 4)$ and $(3, -4)$.

Solution   We can let $(x_1, y_1) = (-2, 4)$ and $(x_2, y_2) = (3, -4)$. Then we have

$$m = \frac{y_2 - y_1}{x_2 - x_1} \qquad \text{This is the slope formula.}$$

$$= \frac{-4 - 4}{3 - (-2)} \qquad \text{Substitute } -4 \text{ for } y_2, 4 \text{ for } y_1, 3 \text{ for } x_2, \text{ and } -2 \text{ for } x_1.$$

$$= \frac{-8}{5}$$

$$= -\frac{8}{5}$$

The slope of the line is $-\frac{8}{5}$.

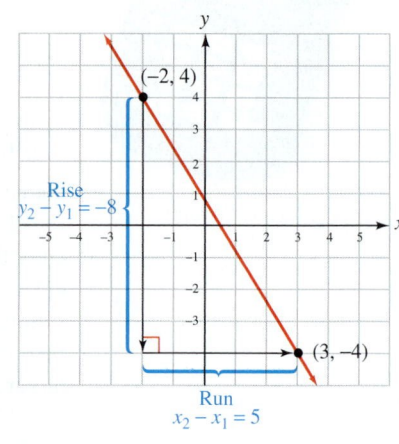

Self Check 4   Find the slope of the line passing through $(-3, 6)$ and $(4, -8)$.

Caution

Note that 0 slope and undefined slope do not mean the same thing.

If a line rises as we follow it from left to right, its slope is positive. If a line drops as we follow it from left to right, its slope is negative. If a line is horizontal, its slope is 0. If a line is vertical, it has undefined slope.

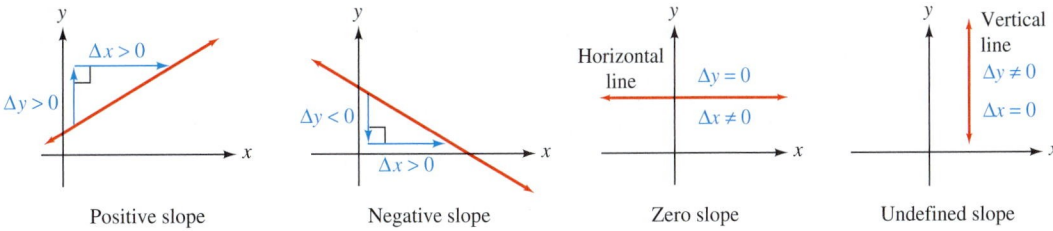

| Positive slope | Negative slope | Zero slope | Undefined slope |

■ SLOPES OF PARALLEL AND PERPENDICULAR LINES

Recall the following facts about parallel and perpendicular lines.

1. If two nonvertical lines are parallel, they have the same slope.
2. If two lines have the same slope, they are parallel.
3. If two nonvertical lines are perpendicular, their slopes are negative reciprocals.
4. If the slopes of two lines are negative reciprocals, the lines are perpendicular.

EXAMPLE 5

Are the lines l_1 and l_2 shown in the following figure perpendicular?

Solution

To find the slope of a line, we can pick two points on the line and determine the rise and run by drawing a **slope triangle**.

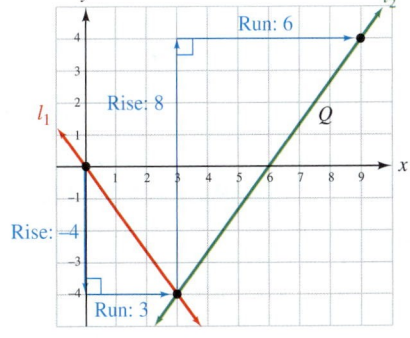

$$\text{Slope } l_1: m = \frac{\text{rise}}{\text{run}} = \frac{-4}{3}$$

Success Tip

Two lines with slopes m_1 and m_2 are perpendicular if

$$m_1 \cdot m_2 = -1$$

$$\text{Slope } l_2: m = \frac{\text{rise}}{\text{run}} = \frac{8}{6} = \frac{4}{3}$$

Since their slopes are not negative reciprocals, the lines are not perpendicular.

■ WRITING EQUATIONS OF LINES

Recall the following two forms of the equation of a line.

Equations of Lines

Point–slope form: The equation of the line passing through (x_1, y_1) and with slope m is

$$y - y_1 = m(x - x_1)$$

Slope–intercept form: The equation of the line with slope m and y-intercept $(0, b)$ is

$$y = mx + b$$

EXAMPLE 6

ELEMENTARY & INTERMEDIATE

Algebra $f(x)$ **Now**™

Find an equation of the line passing through $(-5, 4)$ and $(8, -6)$. Write the result in slope–intercept form.

Solution   First we find the slope of the line.

$$m = \frac{y_2 - y_1}{x_2 - x_1} \qquad \text{This is the slope formula.}$$

$$= \frac{-6 - 4}{8 - (-5)} \qquad \text{Substitute } -6 \text{ for } y_2, 4 \text{ for } y_1, 8 \text{ for } x_2, \text{ and } -5 \text{ for } x_1.$$

$$= -\frac{10}{13}$$

Since the line passes through $(-5, 4)$ and $(8, -6)$, we can choose either point and substitute its coordinates into the point–slope form. If we select $(-5, 4)$, we substitute -5 for x_1, 4 for y_1, and $-\frac{10}{13}$ for m and proceed as follows.

$$y - y_1 = m(x - x_1) \qquad \text{This is point–slope form.}$$

$$y - 4 = -\frac{10}{13}[x - (-5)] \qquad \text{Substitute } -\frac{10}{13} \text{ for } m, -5 \text{ for } x_1, \text{ and } 4 \text{ for } y_1.$$

$$y - 4 = -\frac{10}{13}(x + 5) \qquad \text{Simplify the expression within the brackets.}$$

$$y - 4 = -\frac{10}{13}x - \frac{50}{13} \qquad \text{Remove parentheses: distribute } -\frac{10}{13}.$$

$$y = -\frac{10}{13}x + \frac{2}{13} \qquad \text{To solve for } y, \text{ add 4 in the form of } \frac{52}{13} \text{ to both sides and simplify.}$$

The equation of the line is $y = -\frac{10}{13}x + \frac{2}{13}$.

Success Tip

In Example 6, either of the given points can be used as (x_1, y_1) when writing the point–slope equation.

Looking ahead, we usually choose the point whose coordinates will make the computations the easiest.

Self Check 6   Find an equation of the line passing through $(-2, 5)$ and $(4, -3)$. Write the result in slope–intercept form.

EXAMPLE 7

ELEMENTARY & INTERMEDIATE

Algebra $f(x)$ **Now**™

Find an equation of the line that passes through $(-2, 5)$ and is parallel to the line $y = 8x - 3$. Write the result in general form.

Solution   Since the slope of the line given by $y = 8x - 3$ is the coefficient of x, the slope is 8. Since the desired equation is to have a graph that is parallel to the graph of $y = 8x - 3$, its slope must also be 8.

We substitute -2 for x_1, 5 for y_1, and 8 for m in the point–slope form and simplify.

$$y - y_1 = m(x - x_1)$$

$$y - 5 = 8[x - (-2)] \qquad \text{Substitute 5 for } y_1, 8 \text{ for } m, \text{ and } -2 \text{ for } x_1.$$

$$y - 5 = 8(x + 2) \qquad \text{Simplify the expression inside the brackets.}$$

$$y - 5 = 8x + 16 \qquad \text{Use the distributive property to remove parentheses.}$$

$$y = 8x + 21 \qquad \text{To solve for } y, \text{ add 5 to both sides.}$$

To write the equation in general form, $Ax + By + C$, we proceed as follows:

$$y = 8x + 21$$
$$8x + 21 = y$$
$$8x - y = -21 \qquad \text{Subtract } y \text{ from both sides. Subtract 21 from both sides.}$$

The equation is $8x - y = -21$.

Self Check 7   Find an equation of the line that is parallel to the line $y = -4x - 6$ and passes through $(-2, 10)$. Write the result in general form.

Linear models can be used to describe certain types of financial gain or loss. For example, **straight-line depreciation** is used when aging equipment declines in value and **straight-line appreciation** is used when property or collectibles increase in value.

EXAMPLE 8

Accounting.   After purchasing a new drill press, a machine shop owner had his accountant prepare a depreciation worksheet for tax purposes. See the illustration on the right.

> **Depreciation Worksheet**
>
> Drill press $1,970
> (new)
>
> Salvage value $270
> (in 10 years)

a. Assuming straight-line depreciation, write an equation that gives the value v of the drill press after x years of use.

b. Find the value of the drill press after $2\frac{1}{2}$ years of use.

c. What is the economic meaning of the v-intercept of the line?

d. What is the economic meaning of the slope of the line?

Solution   **a.** The facts presented in the worksheet can be expressed as ordered pairs of the form

$$(x, v)$$

number of years of use ⎯↑ ↑⎯ value of the drill press

- When purchased, the new $1,970 drill press had been used 0 years: $(0, 1{,}970)$.
- After 10 years of use, the value of the drill press will be $270: (10, 270)$.

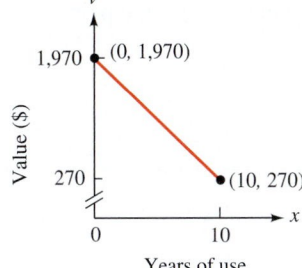

A simple sketch showing these ordered pairs on the line of depreciation is helpful in visualizing the situation.

Since we know two points that lie on the line, we can write its equation using the point–slope form. First, we find the slope of the line.

$$m = \frac{v_2 - v_1}{x_2 - x_1} \qquad \text{This is the slope formula written in terms of } x \text{ and } v.$$

$$= \frac{270 - 1{,}970}{10 - 0} \qquad (x_1, v_1) = (0, 1{,}970) \text{ and } (x_2, v_2) = (10, 270).$$

$$= \frac{-1{,}700}{10}$$

$$= -170$$

To find the equation of the line, we substitute -170 for m, 0 for x_1, and 1,970 for v_1 in the point–slope form and simplify.

$$v - v_1 = m(x - x_1)$$ This is the point–slope form written in terms x and v.

$$v - 1{,}970 = -170(x - 0)$$

$$v = -170x + 1{,}970$$ This is the straight-line depreciation equation.

The value v of drill press after x years of use is given by the linear model $v = -170x + 1{,}970$.

b. To find the value of the drill press after $2\frac{1}{2}$ years of use, we substitute 2.5 for x in the depreciation equation and find v.

$$v = -170x + 1{,}970$$

$$= -170(2.5) + 1{,}970$$

$$= -425 + 1{,}970$$

$$= 1{,}545$$

In $2\frac{1}{2}$ years, the drill press will be worth \$1,545.

c. From the sketch on the previous page, we see that the v-intercept of the graph of the depreciation line is $(0, 1{,}970)$. This gives the original cost of the drill press, \$1,970.

d. Each year, the value of the drill press decreases by \$170, because the slope of the line is -170. The slope of the line is the *annual depreciation rate*.

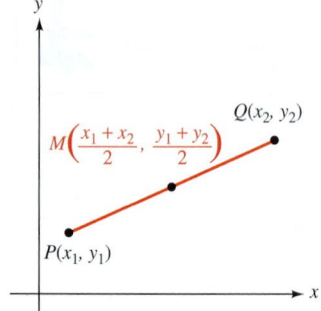

■ THE MIDPOINT FORMULA

If point M in the figure on the left lies midway between points $P(x_1, y_1)$ and $Q(x_2, y_2)$, point M is called the **midpoint** of segment PQ. To find the coordinates of M, we average the x-coordinates and average the y-coordinates of P and Q.

The Midpoint Formula   The **midpoint** of the line segment with endpoints at (x_1, y_1) and (x_2, y_2) is the point with coordinates of

$$\left(\frac{x_1 + x_2}{2}, \frac{y_1 + y_2}{2} \right)$$

EXAMPLE 9

Find the midpoint of the line segment joining $(-2, 3)$ and $(3, -5)$.

Solution   To find the midpoint, we average the x-coordinates and average the y-coordinates to get

$$\frac{x_1 + x_2}{2} = \frac{-2 + 3}{2} \quad \text{and} \quad \frac{y_1 + y_2}{2} = \frac{3 + (-5)}{2}$$

$$= \frac{1}{2} \qquad\qquad\qquad = -1$$

The midpoint of the segment is the point $\left(\frac{1}{2}, -1 \right)$.

Self Check 9   Find the midpoint of the segment joining $(5, -4)$ and $(-3, 5)$.

Answers to Self Checks

1.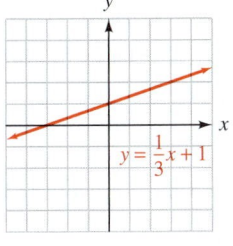

$y = \frac{1}{3}x + 1$

2.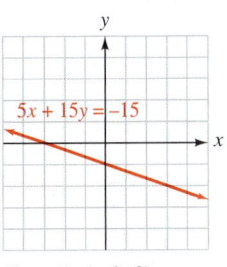

$5x + 15y = -15$

$(0, -1), (-3, 0)$

3.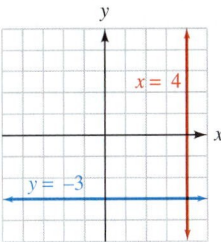

$x = 4$

$y = -3$

4. -2 **6.** $y = -\frac{4}{3}x + \frac{7}{3}$ **7.** $4x + y = 2$ **9.** $\left(1, \frac{1}{2}\right)$

8.6   STUDY SET   Elementary & Intermediate Algebra *f(x)* Now™

VOCABULARY   Fill in the blanks.

1. $y = 3x - 1$ is an equation in _____ variables, x and y.

2. A _____ of an equation in two variables is an ordered pair of numbers that makes the equation a true statement.

3. Solutions of equations in two variables are often listed in a _____ of solutions.

4. The line that represents all of the solutions of a linear equation is called the _____ of the equation.

5. The equation $y = 3x + 8$ is said to be _____ because its graph is a line.

6. A linear equation in two variables has _____ many solutions.

7. The point where a graph intersects the y-axis is called the _____ and the point where it intersects the x-axis is called the _____.

8. _____ is defined as the change in y divided by the change in x.

CONCEPTS   Fill in the blanks.

9. The graph of any equation of the form $x = a$ is a _____ line.

10. The graph of any equation of the form $y = b$ is a _____ line.

11. The formula to compute slope is $m =$ _____.

12. The slope of a _____ line is 0. A _____ line has no defined slope.

13. The graph of the equation $y = 0$ is the _____. The graph of the equation $x = 0$ is the _____.

14. If a line rises as x increases, its slope is _____.

15. _____ lines have the same slope.

16. The slopes of _____ lines are negative reciprocals.

17. The point–slope form of the equation of a line is _____.

18. The slope–intercept form of the equation of a line is _____.

19. The general form of the equation of a line is _____.

20. The midpoint of a line segment joining (x_1, y_1) and (x_2, y_2) is given by the formula _____.

21. Refer to the graph.

a. What is the x-intercept and what is the y-intercept of the line?

b. If the coordinates of point P are substituted into the equation of the line that is graphed here, will a true or a false statement result?

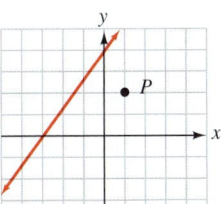

22. Use the graph to determine three solutions of $2x + 3y = 9$.

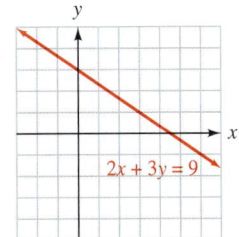

$2x + 3y = 9$

NOTATION Fill in the blanks.

23. The symbol x_1 is read as x _____.

24. The symbol Δ is the letter _____ from the Greek alphabet. The change in x is denoted _____.

PRACTICE Graph each equation.

25. $y = x - 2$　　　　　**26.** $y = -x + 4$

27. $y = x$　　　　　　**28.** $y = -2x$

29. $y = 2x - 3$　　　　**30.** $y = -3x + 2$

31. $y = -\dfrac{1}{3}x - 1$　　　**32.** $y = -\dfrac{1}{2}x + \dfrac{5}{2}$

33. $x = 3$　　　　　　**34.** $y = -4$

Write each equation in $y = b$ or $x = a$ form. Then graph it.

35. $y - 2 = 0$　　　　**36.** $x + 1 = 0$

37. $-2x + 3 = 11$　　　**38.** $-3y + 2 = 5$

Graph each equation using the intercept method. Label the intercepts on each graph.

39. $3x + 4y = 12$　　　**40.** $4x - 3y = 12$

41. $3y = 6x - 9$　　　　**42.** $2x = 4y - 10$

43. $2y + x = -2$　　　　**44.** $4y + 2x = -8$

Find the slope of each line.

45.　　　　　　　　　**46.**

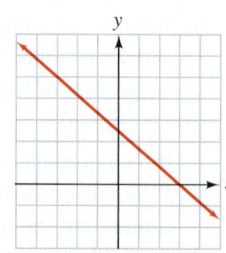

47.　　　　　　　　　**48.**

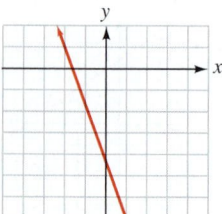　　　　　　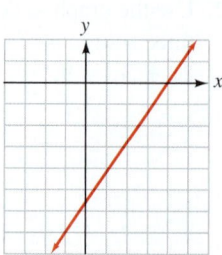

Find the slope of the line that passes through the given points, if possible.

49. $(0, 0), (3, 9)$　　　　**50.** $(9, 6), (0, 0)$

51. $(-1, 8), (6, 1)$　　　**52.** $(-5, -8), (3, 8)$

53. $(3, -1), (-6, 2)$　　**54.** $(0, -8), (-5, 0)$

55. $(-7, 5), (-7, 2)$　　**56.** $(3, -5), (3, 8)$

Find the slope of the line determined by each equation.

57. $3x + 2y = 12$　　　**58.** $2x - y = 6$

59. $3x = 4y - 2$　　　　**60.** $x = y$

The equations of two lines are given. Tell whether the lines are parallel, perpendicular, or neither.

61. $3x - y = 4$,　$3x - y = 7$

62. $4x - y = 13$,　$x - 4y = 13$

63. $x + y = 2$,　$y = x + 5$

64. $5x + 6y = 30$,　$6x + 5y = 24$

65. $9x - 12y = 17$,　$3x - 4y = 17$

66. $y = -3$,　$y = -7$

Find an equation of the line with the given properties. Write the equation in general form.

67. $m = 5$, passing through $(0, 7)$

68. $m = -8$, passing through $(0, -2)$

69. Passing through $(0, 0)$ and $(4, 4)$

70. Passing through $(-5, -5)$ and $(0, 0)$

71. Passing through $(3, 4)$ and $(0, -3)$

72. Passing through $(4, 0)$ and $(6, -8)$

Find an equation of the line with the given properties. Write the equation in slope–intercept form.

73. $m = 3, b = 17$

74. $m = -2, b = 11$

75. $m = -7$, passing through $(7, 5)$

76. $m = 3$, passing through $(-2, -5)$

77. Passing through $(6, 8)$ and $(2, 10)$

78. Passing through $(-4, 5)$ and $(2, -6)$

Find an equation of the line that passes through the given point and its parallel to the given line. Write the equation in slope–intercept form.

79. $(0, 0), y = 4x - 7$

80. $(0, 0), x = -3y - 12$

81. $(2, 5), 4x - y = 7$

82. $(-6, 3), y + 3x = -12$

Find an equation of the line that passes through the given point and is perpendicular to the given line. Write the equation in slope–intercept form.

83. $(0, 0), y = 4x - 7$

84. $(0, 0), x = -3y - 12$

85. $(2, 5), 4x - y = 7$

86. $(-6, 3), y + 3x = -12$

Find the midpoint of segment PQ.

87. $P(0, 0), Q(6, 8)$

88. $P(10, 12), Q(0, 0)$

89. $P(6, 8), Q(12, 16)$

90. $P(10, 4), Q(2, -2)$

91. $P(-2, -8), Q(3, 4)$

92. $P(-5, -2), Q(7, 3)$

93. FINDING THE ENDPOINT OF A SEGMENT
If $(-2, 3)$ is the midpoint of segment PQ and the coordinates of P are $(-8, 5)$, find the coordinates of Q.

94. FINDING THE ENDPOINT OF A SEGMENT
If $(6, -5)$ is the midpoint of segment PQ and the coordinates of Q are $(-5, -8)$, find the coordinates of P.

APPLICATIONS

95. HIGHWAY GRADES   Find the **grade** of the road shown below by expressing the slope as a percent.

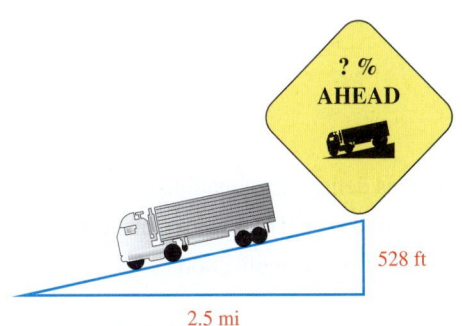

96. DECK DESIGNS   See the illustration. Find the slopes of the cross-brace and the supports. Is the cross-brace perpendicular to either support?

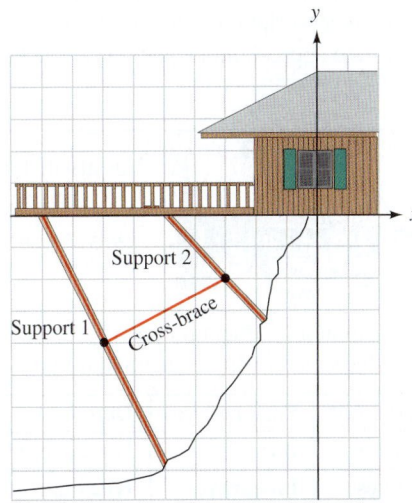

97. RATES OF GROWTH   When a college started an aviation program, the administration agreed to predict enrollments using a straight-line method. If the enrollment during the first year was 80 and the enrollment during the fifth year was 200, find the rate of growth per year (the slope of the line).

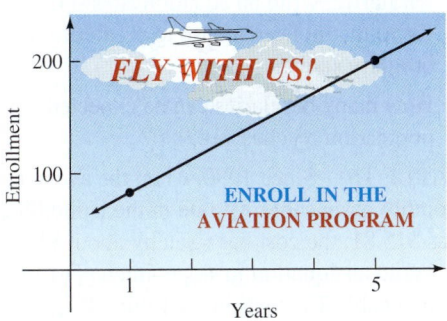

98. BIG-SCREEN TVS   Find the straight-line depreciation equation for the TV in the want ad shown below.

For Sale: 3-year-old 65-inch TV, with matrix surround sound & picture within picture, remote. $1,750 new. Asking $800. Call 875-5555. Ask for Mike.

99. SALVAGE VALUES A truck was purchased for $19,984. Its salvage value at the end of 8 years is expected to be $1,600. Find the straight-line depreciation equation.

100. REAL ESTATE LISTINGS Use the information given in the description of the property to write a straight-line appreciation equation for the house.

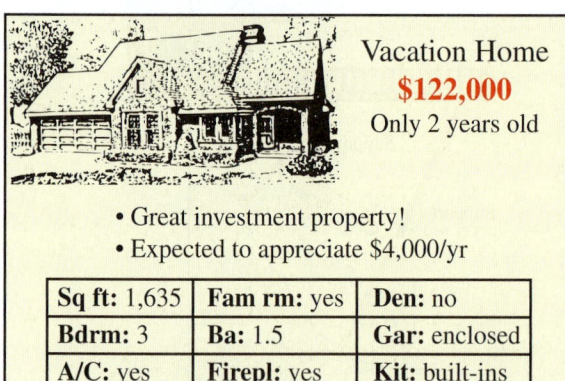

Vacation Home
$122,000
Only 2 years old

• Great investment property!
• Expected to appreciate $4,000/yr

Sq ft: 1,635	Fam rm: yes	Den: no
Bdrm: 3	Ba: 1.5	Gar: enclosed
A/C: yes	Firepl: yes	Kit: built-ins

101. CRIMINOLOGY City growth and the number of burglaries are related by a linear equation. Records show that 575 burglaries were reported in a year when the local population was 77,000 and that the rate of increase in the number of burglaries was 1 for every 100 new residents.

a. Using the variables p for population and B for burglaries, write an equation (in slope–intercept form) that police can use to predict future burglary statistics.

b. How many burglaries can be expected when the population reaches 110,000?

102. CABLE TV Since 1990, when the average monthly basic cable TV rate in the United States was $15.81, the cost has risen by about $1.52 a year.

a. Write an equation in slope–intercept form to predict cable TV costs in the future. Use t to represent time in years after 1990 and C to represent the average basic monthly cost. (Source: Kagan World Media)

b. If the equation in part a were graphed, what would be the meaning of the C-intercept and the slope of the line?

WRITING

103. Explain how to graph a line using the intercept method.

104. When graphing a line by plotting points, why is it a good practice to find three solutions instead of two?

105. A student was asked to determine the slope of the graph of the line $y = 6x - 4$. His answer was $m = 6x$. Explain his error.

106. Explain why a vertical line has no defined slope.

REVIEW

107. INVESTMENTS Equal amounts are invested at 6%, 7%, and 8% annual interest. The three investments yield a total of $2,037 annual interest. Find the total amount of money invested.

108. MEDICATIONS A doctor prescribes an ointment that is 2% hydrocortisone. A pharmacist has 1% and 5% concentrations in stock. How many ounces of each should the pharmacist use to make a 1-ounce tube?

CHALLENGE PROBLEMS

109. The line passing through $(1, 3)$ and $(-2, 7)$ is perpendicular to the line passing through points $(4, b)$ and $(8, -1)$. Without graphing, find b.

110. a. Solve $Ax + By = C$ for y, and thereby show that the slope of its graph is $-\frac{A}{B}$ and its y-intercept is $\left(0, \frac{C}{B}\right)$.

b. Show that the x-intercept of the graph of $Ax + By = C$ is $\left(\frac{C}{A}, 0\right)$.

8.7 An Introduction to Functions

• Functions; domain and range   • Functions defined by equations
• Function notation   • The graph of a function   • The vertical line test
• Finding the domain and range of a function   • An application

The concept of a *function* is one of the most important ideas in all of mathematics. To introduce this topic, let's look at a table that one might see on television or printed in a newspaper.

■ FUNCTIONS; DOMAIN AND RANGE

The following table shows the number of women serving in the House of Representatives during the most recent sessions of Congress.

Women in the U.S. House of Representatives						
Session of Congress	103rd	104th	105th	106th	107th	108th
Women members	47	48	54	56	59	59

For each session of Congress, there corresponds exactly one number of women representatives. Such a correspondence is an example of a *function*.

Functions	A **function** is a rule (or correspondence) that assigns to each value of one variable (called the **independent variable**) exactly one value of another variable (called the **dependent variable**).
Domain and Range	The set of all possible values that can be used for the independent variable is called the **domain.** The set of all values of the dependent variable is called the **range.**

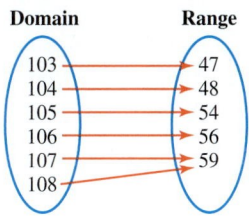

An **arrow** or **mapping diagram** can be used to show how a function assigns to each member of the domain exactly one member of the range. For the House of Representatives example, we have the diagram shown on the right.

We can restate the definition of a function using the variables *x* and *y*.

y* Is a Function of *x	If to each value of *x* in the domain there is assigned exactly one value of *y* in the range, then *y* is said to be a function of *x*.

EXAMPLE 1

Determine whether the arrow diagram and the tables define *y* as a function of *x*:

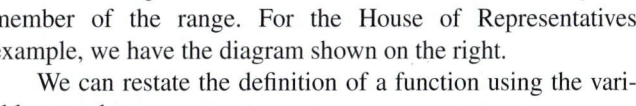

a.

b.
x	y
8	2
1	4
8	3
9	9

c.
x	y
−2	3
−1	3
0	3
1	3

Solution

a. The arrow diagram defines a function because each *x*-value is assigned exactly one *y*-value: $5 \rightarrow 4$, $7 \rightarrow 6$, and $11 \rightarrow 10$.

b. This table does not define a function, because to the *x*-value 8 there is assigned more than one *y*-value. In the first row, 2 is assigned to 8, and in the third row, 3 is also assigned to 8.

c. Since the table assigns to each *x*-value exactly one *y*-value, it defines a function. It also illustrates an important fact about functions: *Different values of x may be assigned the same value of y.* In this case, each *x*-value is assigned the *y*-value 3.

Self Check 1   Determine whether the arrow diagram and the table define *y* as a function of *x*.

a.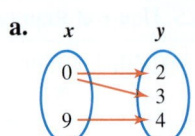

b.

x	y
−1	−60
0	55
3	0

■ FUNCTIONS DEFINED BY EQUATIONS

A function can also be defined by an equation. For example, $y = \frac{1}{2}x + 3$ is a rule that assigns to each value of *x* exactly one value of *y*. To find the *y*-value (called an **output**) that is assigned to the *x*-value 4 (called an **input**), we substitute 4 for *x* and evaluate the right-hand side of the equation.

$$y = \frac{1}{2}x + 3$$

$$= \frac{1}{2}(4) + 3 \quad \text{Substitute 4 for } x.$$

$$= 2 + 3$$

$$= 5 \qquad \text{The output is 5.}$$

The function $y = \frac{1}{2}x + 3$ assigns the *y*-value 5 to the *x*-value 4.

Not all equations define functions, as we see in the following example.

EXAMPLE 2
ELEMENTARY &
INTERMEDIATE
Algebra *f(x)* **Now**™

Determine whether each equation defines *y* to be a function of *x*:   **a.** $y = 2x - 5$ and   **b.** $y^2 = x$.

Solution   **a.** To find the output value *y* that is assigned to an input value *x*, we *multiply x by 2 and then subtract 5.* Since this arithmetic gives one result, to each value of *x* there is assigned exactly one *y*-value. Thus, $y = 2x - 5$ defines *y* to be a function of *x*.

b. The equation $y^2 = x$ does not define *y* to be a function of *x*, because we can find an input value *x* that is assigned more than one output value *y*. For example, consider $x = 16$. It is assigned two values of *y*, 4 and −4, because $4^2 = 16$ and $(-4)^2 = 16$.

x	y
16	4
16	−4

Self Check 2   Determine whether each equation defines *y* to be a function of *x*:   **a.** $y = -2x + 5$ and   **b.** $|y| = x$.

■ FUNCTION NOTATION

A special notation is used to name functions that are defined by equations.

Function Notation	The notation $y = f(x)$ denotes that the variable y is a function of x.

Caution

The symbol $f(x)$ denotes a function. It does not mean $f \cdot x$ (f times x).

In Example 2a, we saw that $y = 2x - 5$ defines y to be a function of x. To write this equation using function notation, we replace y with $f(x)$, to get $f(x) = 2x - 5$. This is read as "f of x is equal to $2x$ minus 5."

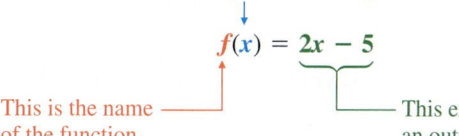

This variable represents the input.

$$f(x) = 2x - 5$$

This is the name of the function.

This expression shows how to obtain an output from a given input.

Function notation provides a compact way of denoting the output value that is assigned to some input value x. For example, if $f(x) = 2x - 5$, the value that is assigned to an x-value 6 is represented by $f(6)$.

The Language of Algebra

Another way to read $f(6) = 7$ is to say "the value of the function at 6 is 7."

$$f(x) = 2x - 5$$
$$f(6) = 2(6) - 5 \qquad \text{Substitute 6 for each } x. \text{ (The input is 6.)}$$
$$= 12 - 5 \qquad \text{Evaluate the right-hand side.}$$
$$= 7$$

Thus, $f(6) = 7$. The output 7 is called a **function value.**

To see why function notation is helpful, consider these equivalent sentences:

1. If $y = 2x - 5$, find the value of y when x is 6.
2. If $f(x) = 2x - 5$, find $f(6)$.

Statement 2, which uses $f(x)$ notation, is much more concise.

EXAMPLE 3

ELEMENTARY & INTERMEDIATE
Algebra *f(x)* **Now**™

Let $f(x) = 4x + 3$. Find **a.** $f(3)$, **b.** $f(-1)$, **c.** $f(0)$, and **d.** $f(r + 1)$.

Solution **a.** To find $f(3)$, we replace x with 3:

$$f(x) = 4x + 3$$
$$f(3) = 4(3) + 3$$
$$= 12 + 3$$
$$= 15$$

b. To find $f(-1)$, we replace x with -1:

$$f(x) = 4x + 3$$
$$f(-1) = 4(-1) + 3$$
$$= -4 + 3$$
$$= -1$$

c. To find $f(0)$, we replace x with 0:

$$f(x) = 4x + 3$$
$$f(0) = 4(0) + 3$$
$$= 3$$

d. To find $f(r + 1)$, we replace x with $r + 1$:

$$f(x) = 4x + 3$$
$$f(r + 1) = 4(r + 1) + 3$$
$$= 4r + 4 + 3$$
$$= 4r + 7$$

Self Check 3   If $f(x) = -2x - 1$, find **a.** $f(2)$, **b.** $f(-3)$, and **c.** $f(-t)$.

The letter f used in the notation $y = f(x)$ represents the word *function*. However, other letters can be used to represent functions. For example, the notations $y = g(x)$ and $y = h(x)$ are often used to denote functions involving the independent variable x.

EXAMPLE 4   Let $g(x) = x^2 - 2x$. Find   **a.** $g\left(\dfrac{2}{5}\right)$ and   **b.** $g(-2.4)$.

Solution   **a.** To find $g\left(\dfrac{2}{5}\right)$, we replace x with $\dfrac{2}{5}$:   **b.** To find $g(-2.4)$, we replace x with -2.4:

$$g(x) = x^2 - 2x$$

$$g\left(\frac{2}{5}\right) = \left(\frac{2}{5}\right)^2 - 2\left(\frac{2}{5}\right)$$

$$= \frac{4}{25} - \frac{4}{5}$$

$$= -\frac{16}{25}$$

$$g(x) = x^2 - 2x$$

$$g(-2.4) = (-2.4)^2 - 2(-2.4)$$

$$= 5.76 + 4.8$$

$$= 10.56$$

Self Check 4   Let $h(x) = -\dfrac{x^2 + 2}{2}$. Find   **a.** $h(4)$ and   **b.** $h(-0.6)$.

EXAMPLE 5   *Archery.*   The area of a circle with a diameter of length d is given by the function $A(d) = \pi\left(\frac{d}{2}\right)^2$. Find the area of the archery target.

Solution   Since the diameter of the circular target is 48 inches, $A(48)$ gives the area of the target. To find $A(48)$, we replace d with 48.

The Language of Algebra

The function was named A because it finds the *area* of a circle. Since the area of a circle is a function of its *diameter, d* was used for the independent variable.

$$A(d) = \pi\left(\frac{d}{2}\right)^2$$

$$A(48) = \pi\left(\frac{48}{2}\right)^2 \qquad \text{Substitute 48 for } d.$$

$$= \pi(24)^2$$

$$= 576\pi$$

$$\approx 1{,}809.557368 \qquad \text{Use a calculator.}$$

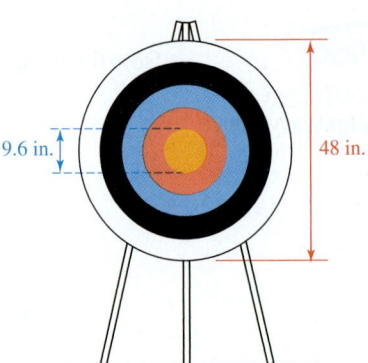

To the nearest tenth, the area of the target is $1{,}809.6$ in.^{2}.

Self Check 5   Find the area of the "bull's eye" to the nearest tenth.

If we are given an output of a function, we can work in reverse to find the corresponding input(s).

EXAMPLE 6

Let: $f(x) = \dfrac{1}{3}x + 4$. For what value(s) of x is $f(x) = 2$?

Solution   To find the value(s) where $f(x) = 2$, we substitute 2 for $f(x)$ and solve for x.

$$\boldsymbol{f(x)} = \frac{1}{3}x + 4$$

$$\boldsymbol{2} = \frac{1}{3}x + 4 \qquad \text{Substitute 2 for } f(x).$$

$$-2 = \frac{1}{3}x \qquad \text{Subtract 4 from both sides.}$$

$$-6 = x \qquad \text{Multiply both sides by 3.}$$

To check, we can substitute -6 for x and verify that $f(-6) = 2$.

$$f(\boldsymbol{x}) = \frac{1}{3}x + 4$$

$$f(\boldsymbol{-6}) = \frac{1}{3}(\boldsymbol{-6}) + 4$$

$$= -2 + 4$$

$$= 2$$

Self Check 6   For what value(s) of x is $f(x) = -5$?

■ THE GRAPH OF A FUNCTION

We have seen that a function assigns to each value of x a single value $f(x)$. The "input-output" pairs that a function generates can be plotted on a rectangular coordinate system to get the graph of the function.

EXAMPLE 7

Graph: $f(x) = \dfrac{1}{2}x + 3$.

Solution   We begin by constructing a table of function values. To make a table, we choose several values for x and find the corresponding values of $f(x)$. If x is -2, we have

$$f(\boldsymbol{x}) = \frac{1}{2}\boldsymbol{x} + 3 \qquad \text{This is the function to graph.}$$

$$f(\boldsymbol{-2}) = \frac{1}{2}(\boldsymbol{-2}) + 3 \qquad \text{Substitute } -2 \text{ for each } x.$$

$$= -1 + 3$$

$$= 2$$

Thus, $f(-2) = 2$. This means that, when x is -2, $f(x)$ is 2, and it indicates that the ordered pair $(-2, 2)$ lies on the graph of f.

In a similar manner, we find the corresponding values of $f(x)$ for x-values of 0, 2, and 4 and record them in the table. Then we plot the ordered pairs and draw a straight line through the points to get the graph of $f(x) = \frac{1}{2}x + 3$.

$$f(x) = \frac{1}{2}x + 3$$

x	$f(x)$
-2	2	→ $(-2, 2)$
0	3	→ $(0, 3)$
2	4	→ $(2, 4)$
4	5	→ $(4, 5)$

Since $y = f(x)$, this column may be labeled $f(x)$ or y.

This axis can be labeled y or $f(x)$.

Self Check 7   Graph: $f(x) = -3x - 2$.

We call $f(x) = \frac{1}{2}x + 3$ from Example 7 a **linear function** because its graph is a nonvertical straight line. Any linear equation, except those of the form $x = a$, can be written in function notation by writing it in slope–intercept form ($y = mx + b$) and then replacing y with $f(x)$.

■ THE VERTICAL LINE TEST

Some graphs define functions and some do not. If any vertical line intersects a graph more than once, the graph cannot represent a function, because to one value of x there would be assigned more than one value of y.

The Vertical Line Test	If a vertical line intersects a graph in more than one point, the graph is not the graph of a function.

The Language of Algebra

Graphs that do not represent functions are called *relations*. A *relation* is simply a set of ordered pairs.

The graph shown in red in figure (a) is not the graph of a function because the vertical line intersects the graph at more than one point. The points of intersection indicate that the x-value 3 is assigned two y-values, 2.5 and -2.5.

The graph shown in red in figure (b) represents a function, because no vertical line intersects the graph at more than one point. Several vertical lines are drawn to illustrate this.

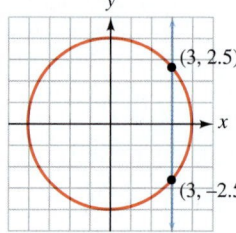

x	y
3	2.5
3	-2.5

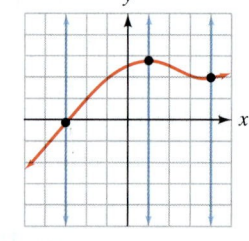

(a)   (b)

■ FINDING THE DOMAIN AND RANGE OF A FUNCTION

We can think of a function as a machine that takes some input x and turns it into some output $f(x)$, as shown in figure (a). The machine shown in figure (b) turns the input -6 into the output -11. The set of numbers that we put into the machine is the domain of the function, and the set of numbers that comes out is the range.

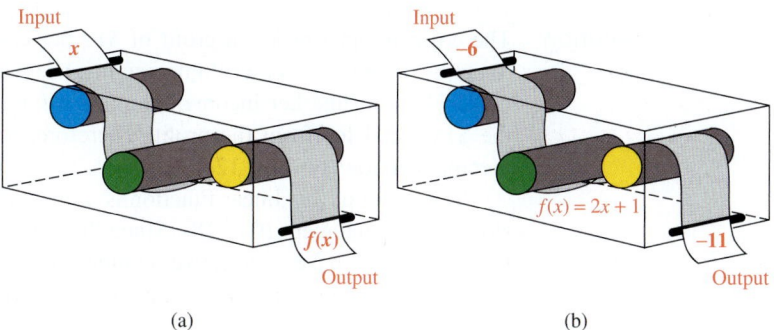

(a)   (b)

EXAMPLE 8   Find the domain and range of each function: **a.** $\{(-2, 4), (0, 6), (2, 8)\}$,
b. $f(x) = 3x + 1$,   and   **c.** $f(x) = \dfrac{1}{x - 2}$.

Solution   **a.** This function consists of only three ordered pairs. The ordered pairs set up a correspondence between x (the input) and y (the output), where a single value of y is assigned to each x.

   • The domain is the set of first coordinates in the set of ordered pairs: $\{-2, 0, 2\}$.
   • The range is the set of second coordinates in the set of ordered pairs: $\{4, 6, 8\}$.

b. We will be able to evaluate $3x + 1$ for any real-number input x. So the domain of the function is the set of real numbers. Since the output y can be any real number, the range is the set of real numbers.

c. To find the domain of $f(x) = \dfrac{1}{x - 2}$, we exclude any real-number x inputs for which we would be unable to compute $\dfrac{1}{x - 2}$. The number 2 cannot be substituted for x, because that would make the denominator equal to zero. Since any real number except 2 can be substituted for x in the equation $f(x) = \dfrac{1}{x - 2}$, the domain is the set of all real numbers except 2.

Since a fraction with a numerator of 1 cannot be 0, the range is the set of all real numbers except 0.

Self Check 8   Find the domain and range of each function: **a.** $\{(-3, 5), (-2, 7), (1, 11)\}$ and
b. $f(x) = \dfrac{2}{x + 3}$.

■ AN APPLICATION

Functions are used to mathematically describe certain relationships where one quantity depends upon another. Letters other than f and x are often chosen to more clearly describe these situations.

EXAMPLE 9

Cosmetology. A cosmetologist rents a station from the owner of a beauty salon for $18 a day. She expects to make $12 profit from each customer she serves. Write a linear function describing her daily income if she serves c customers per day. Then graph the function.

Solution The cosmetologist makes a profit of $12 per customer, so if she serves c customers a day, she will make $12c$. To find her income, we must *subtract* the $18 rental fee from the profit. Therefore, the income function is $I(c) = 12c - 18$.

The graph of this linear function is a line with slope 12 and intercept $(0, -18)$. Since the cosmetologist cannot have a negative number of customers, we do not extend the line into quadrant III.

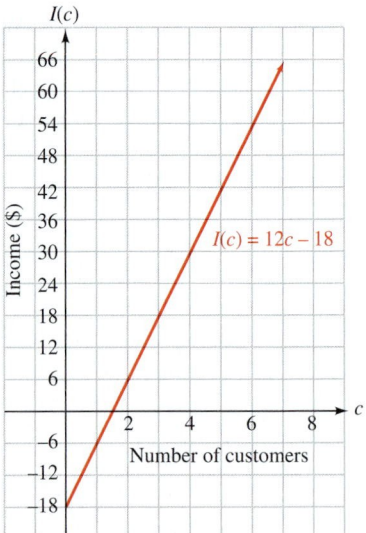

ACCENT ON TECHNOLOGY: EVALUATING FUNCTIONS

We can use a graphing calculator to find function values.

For example, to find the income earned by the cosmetologist in Example 9 for different numbers of customers, we first graph the income function $I(c) = 12c - 18$ as $y = 12x - 18$, using window settings of [0, 10] for x and [0, 100] for y to obtain figure (a). To find her income when she serves seven customers, we trace and move the cursor until the x-coordinate on the screen is nearly 7, as in figure (b). From the screen, we see that her income is about $66.25.

To find her income when she serves nine customers, we trace and move the cursor until the x-coordinate is nearly 9, as in figure (c). From the screen, we see that her income is about $90.51.

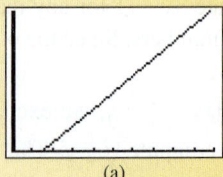
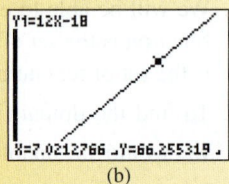
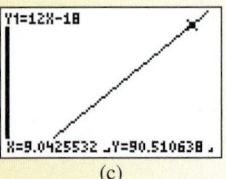

(a)   (b)   (c)

With some graphing calculator models, we can evaluate a function by entering function notation. To find $I(15)$, the income earned by the cosmetologist of Example 9 if she serves 15 customers, we use the following steps on a TI-83 Plus calculator.

With $I(c) = 12c - 18$ entered as $Y_1 = 12x - 18$, we call up the home screen by pressing $\boxed{2nd}$ $\boxed{QUIT}$. Then we enter $\boxed{VARS}$ $\boxed{\blacktriangleright}$ $\boxed{1}$ $\boxed{ENTER}$. The symbolism Y_1 will be displayed. See figure (a). Next, we enter the input value 15, as shown in figure (b), and press $\boxed{ENTER}$. In figure (c) we see that $Y_1(15) = 162$. That is, $I(15) = 162$. The cosmetologist will earn $162 if she serves 15 customers in one day.

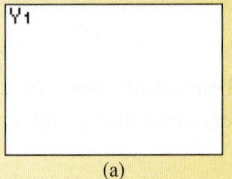
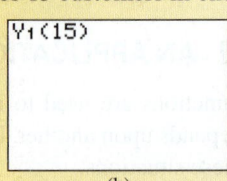
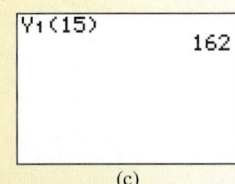

(a)   (b)   (c)

Answers to Self Checks   **1. a.** no,   **b.** yes   **2. a.** yes,   **b.** no; $x = 3$ is assigned two y-values, 3 and -3.
3. a. -5,   **b.** 5,   **c.** $2t - 1$ **4. a.** -9,   **b.** -1.18 **5.** 72.4 in.²   **6.** -27
7.   **8. a.** $\{-3, -2, 1\}, \{5, 7, 11\}$,   **b.** D: the set of all real numbers except -3; R: the set of all real numbers except 0

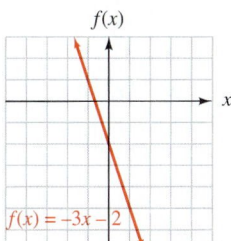

8.7   STUDY SET   ELEMENTARY & INTERMEDIATE Algebra $f(x)$ Now™

VOCABULARY   Fill in the blanks.

1. A _____ is a rule (or correspondence) that assigns to each value of one variable (called the independent variable) exactly _____ value of another variable (called the dependent variable).

2. The set of all possible values that can be used for the independent variable is called the _____. The set of all values of the dependent variable is called the _____.

3. We can think of a function as a machine that takes some _____ x and turns it into some _____ $f(x)$.

4. The notation $y = f(x)$ denotes that the variable y is a _____ of x.

5. If $f(2) = -1$, we call -1 a function _____.

6. We call $f(x) = 2x + 1$ a _____ function because its graph is a straight line.

CONCEPTS

7. RECYCLING   The following table gives the annual average price (in cents) paid for one pound of aluminum cans. Use an arrow diagram to show how members of the domain are assigned members of the range.

Year	1996	1997	1998	1999	2000	2001
Cents per lb	33	35	30	30	32	35

Source: *Midwest Assistance Program*

8. The arrow diagram describes a function.
   a. What is the domain of the function?
   b. What is the range of the function?

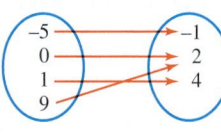

9. Fill in the blank so that the statements are equivalent:
   • If $y = 5x + 1$, find the value of y when $x = 8$.
   • If $f(x) = 5x + 1$, find _____.

10. For the given input, what value will the function machine output?

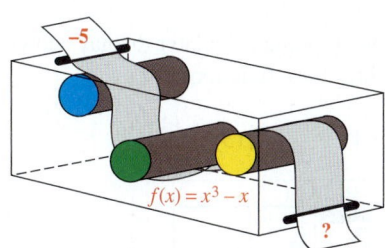

11. Complete the table of function values. Then give the corresponding ordered pairs.

$f(x) = 2x^2 - 1$

x	y
-3	→
0	→
2	→

12. Fill in the blank: If a _____ line intersects a graph in more than one point, the graph is not the graph of a function.

13. a. Give the coordinates of the points where the given vertical line intersects the graph.

   b. Is this the graph of a function? Explain your answer.

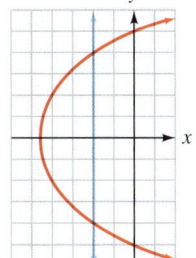

14. Explain why -4 isn't in the domain of $f(x) = \dfrac{1}{x+4}$.

NOTATION   Fill in the blanks.

15. We read $f(x) = 5x - 6$ as "f ___ x is $5x$ minus 6."

16. This variable represents the ___ .

$$f(x) = 2x - 5$$

This is the ___   Use this expression
of the function.   to find the ___ .

17. Since $y = $ ___ , the equations $y = 3x + 2$ and $f(x) = 3x + 2$ are equivalent.

18. The notation $f(2) = 7$ indicates that when the x-value ___ is input into a function rule, the output is ___ . This fact can be shown graphically by plotting the ordered pair (___ , ___).

19. When graphing the function $f(x) = -x + 5$, the vertical axis of the rectangular coordinate system can be labeled ___ or ___ .

20. The graphing calculator display shows a table of values for a function f.

$$f(-1) = \quad\quad f(3) = $$

PRACTICE   **Determine whether each arrow diagram and table defines y as a function of x. If it does not, indicate a value of x that is assigned more than one value of y.**

21.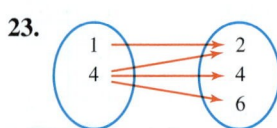

22.

23.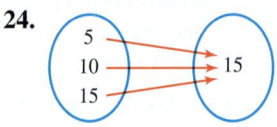

24.

25.

x	y
1	7
2	15
3	23
4	16
5	8

26.

x	y
30	2
30	4
30	6
30	8
30	10

27.

x	y
-4	6
-1	0
0	-3
2	4
-1	2

28.

x	y
1	1
2	2
3	3
4	4

29.

x	y
3	4
3	-4
4	3
4	-3

30.

x	y
-1	1
-3	1
-5	1
-7	1
-9	1

Decide whether the equation defines y as a function of x.

31. $y = 2x + 3$ **32.** $y = 4x - 1$

33. $y = 2x^2$ **34.** $y^2 = x + 1$

35. $y^2 = 3 - 2x$ **36.** $y = 3 + 7x^2$

37. $x = |y|$ **38.** $y = |x|$

Find $f(3)$ and $f(-1)$.

39. $f(x) = 3x$ **40.** $f(x) = -4x$

41. $f(x) = 2x - 3$ **42.** $f(x) = 3x - 5$

43. $f(x) = 7 + 5x$ **44.** $f(x) = 3 + 3x$

45. $f(x) = 9 - 2x$ **46.** $f(x) = 12 + 3x$

Find $g(2)$ and $g(3)$.

47. $g(x) = x^2$ **48.** $g(x) = x^2 - 2$

49. $g(x) = x^3 - 1$ **50.** $g(x) = x^3$

51. $g(x) = (x + 1)^2$ **52.** $g(x) = (x - 3)^2$

53. $g(x) = 2x^2 - x$ **54.** $g(x) = 5x^2 + 2x$

Find $h(2)$ and $h(-2)$.

55. $h(x) = |x| + 2$

56. $h(x) = |x| - 5$

57. $h(x) = x^2 - 2$

58. $h(x) = x^2 + 3$

59. $h(x) = \dfrac{1}{x + 3}$

60. $h(x) = \dfrac{3}{x - 4}$

61. $h(x) = \dfrac{x}{x - 3}$

62. $h(x) = \dfrac{x}{x^2 + 2}$

Complete each table.

63. $f(t) = |t - 2|$

t	$f(t)$
-1.7	
0.9	
5.4	

64. $f(r) = -2r^2 + 1$

Input	Output
-1.7	
0.9	
5.4	

65. $g(x) = x^3$

Input	Output
$-\dfrac{3}{4}$	
$\dfrac{1}{6}$	
$\dfrac{5}{2}$	

66. $g(x) = 2\left(-x - \dfrac{1}{4}\right)$

x	$g(x)$
$-\dfrac{3}{4}$	
$\dfrac{1}{8}$	
$\dfrac{5}{2}$	

Find $g(w)$ and $g(w + 1)$.

67. $g(x) = 2x$

68. $g(x) = -3x$

69. $g(x) = 3x - 5$

70. $g(x) = 2x - 7$

Let $f(x) = -2x + 5$. For what value(s) of x is

71. $f(x) = 5$?

72. $f(x) = -7$?

Let $f(x) = \dfrac{3}{2}x - 2$. For what value(s) of x is

73. $f(x) = -\dfrac{1}{2}$?

74. $f(x) = \dfrac{2}{3}$?

Find the domain and range of each function.

75. $\{(-2, 3), (4, 5), (6, 7)\}$

76. $\{(0, 2), (1, 2), (3, 4)\}$

77. $s(x) = 3x + 6$

78. $h(x) = \dfrac{4}{5}x - 8$

79. $f(x) = x^2$

80. $g(x) = x^3$

81. $s(x) = |x - 7|$

82. $t(x) = \left|\dfrac{2x}{3} + 1\right|$

83. $f(x) = \dfrac{1}{x - 4}$

84. $f(x) = \dfrac{5}{x + 1}$

Use the vertical line test to decide whether the given graph represents a function.

85.
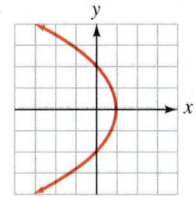

86.
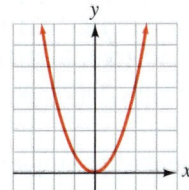

87.
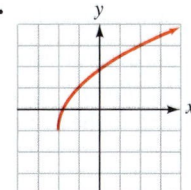

88.
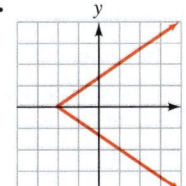

89.
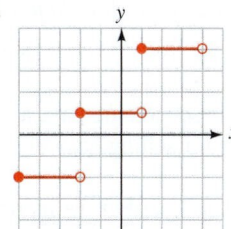

90.
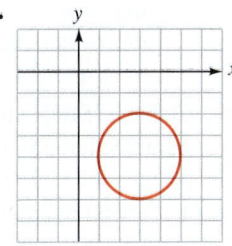

91.
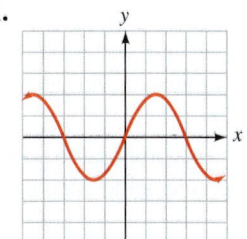

92.
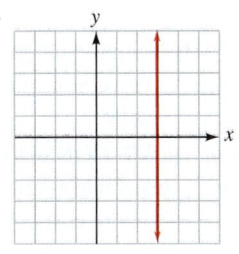

Graph each function.

93. $f(x) = 2x - 1$

94. $f(x) = -x + 2$

95. $f(x) = \dfrac{2}{3}x - 2$

96. $f(x) = -\dfrac{3}{2}x - 3$

APPLICATIONS

97. DECONGESTANTS   The temperature in degrees Celsius that is equivalent to a temperature in degrees Fahrenheit is given by the linear function $C(F) = \dfrac{5}{9}(F - 32)$. Use this function to find the temperature range, in degrees Celsius, at which a bottle of Dimetapp should be stored. The label directions follow.

> **DIRECTIONS:** Adults and children 12 years of age and over: Two teaspoons every 4 hours. DO NOT EXCEED 6 DOSES IN A 24-HOUR PERIOD. Store at a controlled room temperature between 68°F and 77°F.

98. BODY TEMPERATURES   The temperature in degrees Fahrenheit that is equivalent to a temperature in degrees Celsius is given by the linear function $F(C) = \dfrac{9}{5}C + 32$. Convert each of the temperatures in the following excerpt from *The Good Housekeeping Family Health and Medical Guide* to degrees Fahrenheit. (Round to the nearest degree.)

> *In disease, the temperature of the human body may vary from about 32.2°C to 43.3°C for a time, but there is grave danger to life should it drop and remain below 35°C or rise and remain at or above 41°C.*

99. CONCESSIONAIRES   A baseball club pays a peanut vendor $50 per game for selling bags of peanuts for $1.75 each.

a. Write a linear function that describes the income the vendor makes for the baseball club during a game if she sells b bags of peanuts.

b. Find the income the baseball club will make if the vendor sells 110 bags of peanuts during a game.

100. HOME CONSTRUCTION   In a proposal to some prospective clients, a housing contractor listed the following costs.

Fees, permits, miscellaneous	$12,000
Construction, per square foot	$75

a. Write a linear function that the clients could use to determine the cost of building a home having f square feet.

b. Find the cost to build a home having 1,950 square feet.

101. EARTH'S ATMOSPHERE   The illustration shows a graph of the temperatures of the atmosphere at various altitudes above the Earth's surface. The temperature is expressed in degrees Kelvin, a scale widely used in scientific work.

a. Estimate the coordinates of three points on the graph that have an x-coordinate of 200.

b. Explain why this is not the graph of a function.

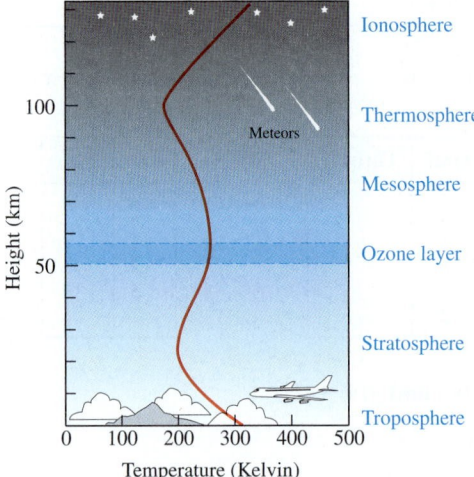

102. CHEMICAL REACTIONS   When students in a chemistry laboratory mixed solutions of acetone and chloroform, they found that heat was immediately generated. As time went by, the mixture cooled down. The illustration on the next page shows a graph of data points of the form (time, temperature) taken by the students.

a. The linear function $T(t) = -\dfrac{t}{240} + 30$ models the relationship between the elapsed time t since the solutions were combined and the temperature $T(t)$ of the mixture. Graph the function.

b. Predict the temperature of the mixture immediately after the two solutions are combined.

c. Is $T(180)$ more or less than the temperature recorded by the students for $t = 300$?

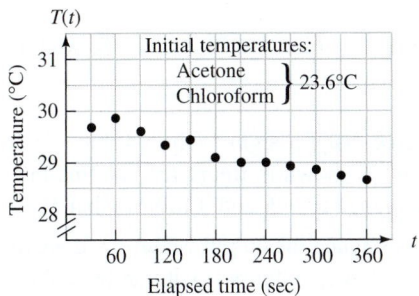

103. TAXES The function

$$T(a) = 700 + 0.15(a - 7,000)$$

(where a is adjusted gross income) is a model of the instructions given on the first line of the following tax rate Schedule X.

a. Find $T(25,000)$ and interpret the result.

b. Write a function that models the second line on Schedule X.

Schedule X–Use if your filing status is **Single**			2003
If your adjusted gross income is: *Over —*	*But not over —*	Your tax is	*of the amount over —*
$ 7,000	$28,400	$ 700 + 15%	$ 7,000
$28,400	$68,800	$3,910 + 25%	$28,400

104. COST FUNCTIONS An electronics firm manufactures DVD recorders, receiving $120 for each recorder it makes. If x represents the number of recorders produced, the income received is determined by the *revenue function* $R(x) = 120x$. The manufacturer has fixed costs of $12,000 per month and variable costs of $57.50 for each recorder manufactured. Thus, the *cost function* is $C(x) = 57.50x + 12,000$. How many recorders must the company sell for revenue to equal cost? (*Hint:* Set $R(x) = C(x)$.)

WRITING

105. What is a function?

106. Explain why we can think of a function as a machine.

REVIEW   **Show that each number is a rational number by expressing it as a ratio of two integers.**

107. $-3\frac{3}{4}$

108. 4.7

109. 0.333. . .

110. $-0.\overline{6}$

CHALLENGE PROBLEMS   **Let $f(x) = 2x + 1$ and $g(x) = x^2$.**

111. Is $f(x) + g(x)$ equal to $g(x) + f(x)$?

112. Is $f(x) - g(x)$ equal to $g(x) - f(x)$?

8.8 Graphs of Functions

- Finding function values graphically
- Finding domain and range graphically
- Graphs of nonlinear functions
- Translations of graphs
- Reflections of graphs

Since a graph is often the best way to describe a function, we need to know how to construct and interpret graphs of functions.

■ FINDING FUNCTION VALUES GRAPHICALLY

Recall that the graph of a function is a picture of the ordered pairs $(x, f(x))$ that define the function. From the graph of a function, we can determine function values.

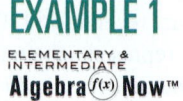

EXAMPLE 1

ELEMENTARY &
INTERMEDIATE
Algebra *(f(x))* **Now**™

Refer to the graph of function f in figure (a). **a.** Find $f(-3)$, and **b.** find the value of x for which $f(x) = -2$.

Solution   **a.** To find $f(-3)$, we need to find the y-value that f assigns to the x-value -3. If we draw a vertical line through -3 on the x-axis, as shown in figure (b), the line intersects the graph of f at $(-3, 5)$. Therefore, 5 is assigned to -3, and it follows that $f(-3) = 5$.

b. We need to find the input value x that is assigned the output value -2. If we draw a horizontal line through -2 on the y-axis, as shown in figure (c), it intersects the graph of f at $(4, -2)$. Therefore, the function assigns -2 to 4, and it follows that $f(4) = -2$.

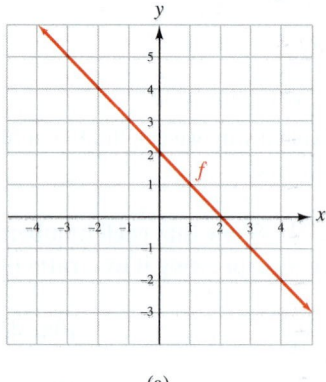

(a)

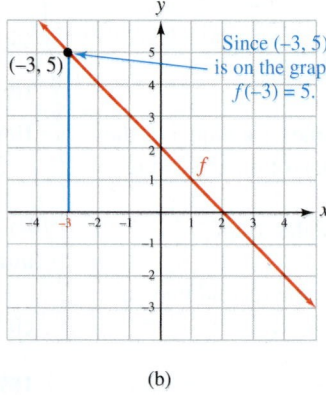

(b)

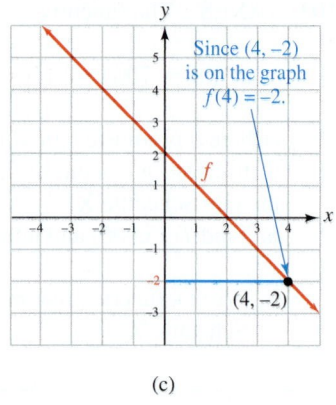

(c)

Self Check 1   From the graph of function g: **a.** find $g(-3)$, and **b.** find the x-value for which $g(x) = 4$.

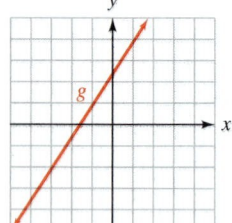

■ FINDING DOMAIN AND RANGE GRAPHICALLY

We can determine the domain and range of a function from its graph. For example, to find the domain of the linear function graphed in figure (a), we *project* the graph onto the x-axis. Because the graph of the function extends indefinitely to the left and to the right, the projection includes all the real numbers. Therefore, the domain of the function is the set of real numbers.

To determine the range of the same linear function, we project the graph onto the y-axis, as shown in figure (b). Because the graph of the function extends indefinitely upward and downward, the projection includes all the real numbers. Therefore, the range of the function is the set of real numbers.

The Language of Algebra

Think of the *projection* of a graph on an axis as the "shadow" that the graph makes on the axis.

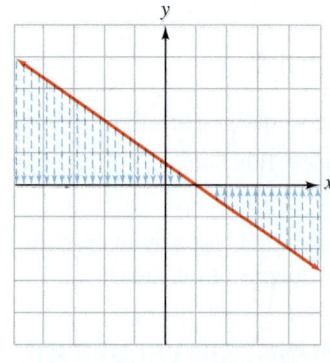

(a)

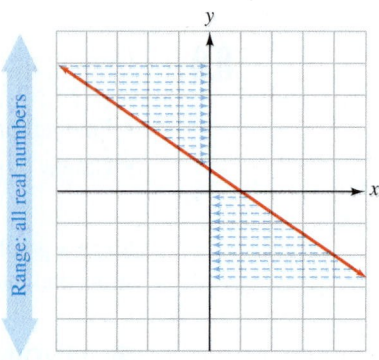

(b)

■ **GRAPHS OF NONLINEAR FUNCTIONS**

We have seen that the graph of a linear function is a straight line. We will now consider several examples of **nonlinear functions.** Their graphs are not straight lines. We will begin with $f(x) = x^2$, called the **squaring function.**

EXAMPLE 2

ELEMENTARY &
INTERMEDIATE
Algebra $f(x)$ **Now**™   Solution

Graph $f(x) = x^2$ and find its domain and range.

To graph the function, we select numbers for x and find the corresponding values for $f(x)$. For example, if we choose -3 for x, we have

$$f(x) = x^2$$
$$f(-3) = (-3)^2 \quad \text{Substitute } -3 \text{ for } x.$$
$$= 9$$

Since $f(-3) = 9$, the ordered pair $(-3, 9)$ lies on the graph of f. In a similar manner, we find the corresponding values of $f(x)$ for other x-values and list the ordered pairs in the table of values. Then we plot the points and draw a smooth curve through them to get the graph, called a **parabola.**

The Language of Algebra

The cup-like shape of a *parabola* has many real-life applications. For example, a satellite TV dish is often called a *parabolic* dish.

$$f(x) = x^2$$

x	$f(x)$	
-3	9	→ $(-3, 9)$
-2	4	→ $(-2, 4)$
-1	1	→ $(-1, 1)$
0	0	→ $(0, 0)$
1	1	→ $(1, 1)$
2	4	→ $(2, 4)$
3	9	→ $(3, 9)$

↑ Choose values for x.   ↑ Compute each $f(x)$.   ↑ Plot these points.

The Language of Algebra

The set of **nonnegative real numbers** is the set of real numbers greater than or equal to 0.

Because the graph extends indefinitely to the left and to the right, the projection of the graph onto the x-axis includes all the real numbers. This means that the domain of the squaring function is the set of real numbers.

Because the graph extends upward indefinitely from the point $(0, 0)$, the projection of the graph on the y-axis includes only positive real numbers and zero. This means that the range of the squaring function is the set of nonnegative real numbers.

Self Check 2   Graph $g(x) = x^2 - 2$ and find its domain and range. Compare the graph to the graph of $f(x) = x^2$.

Another important nonlinear function is $f(x) = x^3$, called the **cubing function.**

EXAMPLE 3   Graph $f(x) = x^3$ and find its domain and range.

ELEMENTARY &
INTERMEDIATE
Algebra $f(x)$ **Now**™   Solution   To graph the function, we select numbers for x and find the corresponding values for $f(x)$. For example, if we choose -2 for x, we have

$$f(x) = x^3$$
$$f(-2) = (-2)^3 \quad \text{Substitute } -2 \text{ for } x.$$
$$= -8$$

Since $f(-2) = -8$, the ordered pair $(-2, -8)$ lies on the graph of f. In a similar manner, we find the corresponding values of $f(x)$ for other x-values and list the ordered pairs in the table. Then we plot the points and draw a smooth curve through them to get the graph.

$f(x) = x^3$

x	$f(x)$	
-2	-8	$\to (-2, -8)$
-1	-1	$\to (-1, -1)$
0	0	$\to (0, 0)$
1	1	$\to (1, 1)$
2	8	$\to (2, 8)$

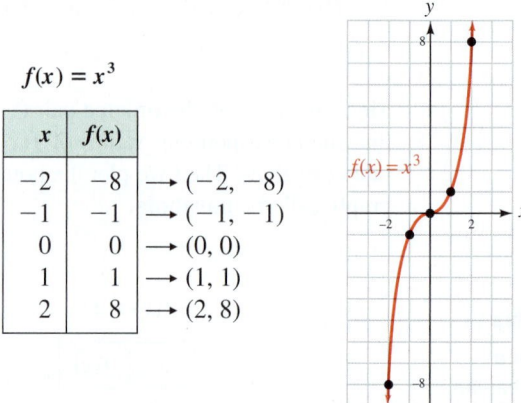

Because the graph of the function extends indefinitely to the left and to the right, the projection includes all the real numbers. Therefore, the domain of the cubing function is the set of real numbers.

Because the graph of the function extends indefinitely upward and downward, the projection includes all the real numbers. Therefore, the range of the cubing function is the set of real numbers.

Self Check 3   Graph $g(x) = x^3 + 1$ and find its domain and range. Compare the graph to the graph of $f(x) = x^3$.

A third nonlinear function is $f(x) = |x|$, called the **absolute value function.**

EXAMPLE 4   Graph $f(x) = |x|$ and find its domain and range.

Solution   To graph the function, we select numbers for x and find the corresponding values for $f(x)$. For example, if we choose -3 for x, we have

$$f(x) = |x|$$
$$f(-3) = |-3| \quad \text{Substitute } -3 \text{ for } x.$$
$$= 3$$

Since $f(-3) = 3$, the ordered pair $(-3, 3)$ lies on the graph of f. In a similar manner, we find the corresponding values of $f(x)$ for other x-values and list the ordered pairs in the table. Then we plot the points and connect them to get the following V-shaped graph.

$f(x) = |x|$

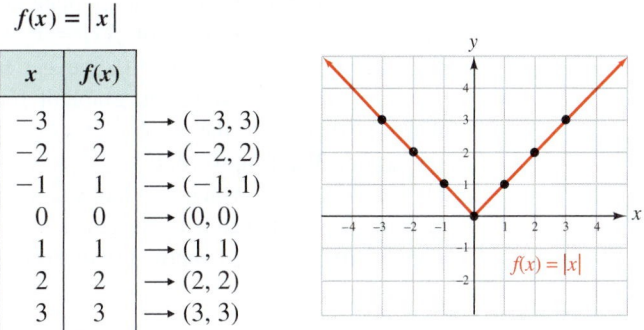

x	$f(x)$	
-3	3	$\longrightarrow (-3, 3)$
-2	2	$\longrightarrow (-2, 2)$
-1	1	$\longrightarrow (-1, 1)$
0	0	$\longrightarrow (0, 0)$
1	1	$\longrightarrow (1, 1)$
2	2	$\longrightarrow (2, 2)$
3	3	$\longrightarrow (3, 3)$

Because the graph extends indefinitely to the left and to the right, the projection of the graph onto the x-axis includes all the real numbers. The domain of the absolute value function is the set of real numbers.

Because the graph extends upward indefinitely from the point $(0, 0)$, the projection of the graph on the y-axis includes only positive real numbers and zero. The range of the absolute value function is the set of nonnegative real numbers.

Self Check 4   Graph $g(x) = |x - 2|$ and find its domain and range. Compare the graph to the graph of $f(x) = |x|$.

ACCENT ON TECHNOLOGY: GRAPHING FUNCTIONS

We can graph nonlinear functions with a graphing calculator. For example, to graph $f(x) = x^2$ in a standard window of $[-10, 10]$ for x and $[-10, 10]$ for y, we press $\boxed{Y =}$, enter the function by typing $\boxed{x^2}$, and press the $\boxed{\text{GRAPH}}$ key. We will obtain the graph shown in figure (a).

To graph $f(x) = x^3$, we enter the function by typing $x \wedge 3$ and then press the $\boxed{\text{GRAPH}}$ key to obtain the graph in figure (b). To graph $f(x) = |x|$, we enter the function by selecting abs from the NUM option within the MATH menu, typing x, and pressing the $\boxed{\text{GRAPH}}$ key to obtain the graph in figure (c).

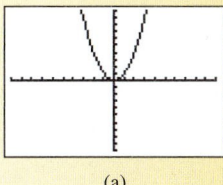 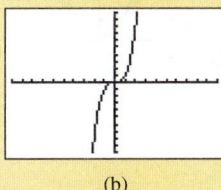 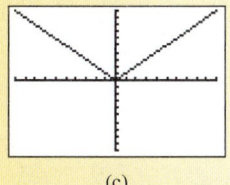 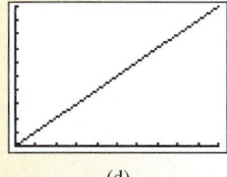

   (a)   (b)   (c)   (d)

When using a graphing calculator, we must be sure that the viewing window does not show a misleading graph. For example, if we graph $f(x) = |x|$ in the window $[0, 10]$ for x and $[0, 10]$ for y, we will obtain a misleading graph that looks like a line. See figure (d). This is not correct. The proper graph is the V-shaped graph shown in figure (c). One of the challenges of using graphing calculators is finding an appropriate viewing window.

■ TRANSLATIONS OF GRAPHS

Examples 2, 3, and 4 and their Self Checks suggest that the graphs of different functions may be identical except for their positions in the xy-plane. For example, the figure shows the graph of $f(x) = x^2 + k$ for three different values of k. If $k = 0$, we get the graph of $f(x) = x^2$. If $k = 3$, we get the graph of $f(x) = x^2 + 3$, which is identical to the graph of $f(x) = x^2$ except that it is shifted 3 units upward. If $k = -4$, we get the graph of $f(x) = x^2 - 4$, which is identical to the graph of $f(x) = x^2$ except that it is shifted 4 units downward. These shifts are called **vertical translations.**

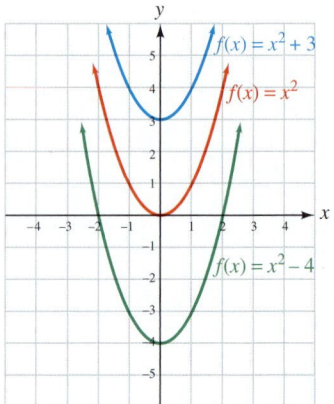

In general, we can make these observations.

Vertical Translations	If f is a function and k represents a positive number, then

- The graph of $y = f(x) + k$ is identical to the graph of $y = f(x)$ except that it is translated k units upward.
- The graph of $y = f(x) - k$ is identical to the graph of $y = f(x)$ except that it is translated k units downward.

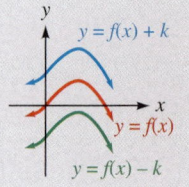

EXAMPLE 5

ELEMENTARY & INTERMEDIATE
Algebra *f(x)* Now™

Graph: $g(x) = |x| + 2$.

Solution   The graph of $g(x) = |x| + 2$ will be the same V-shaped graph as $f(x) = |x|$, except that it is shifted 2 units upward.

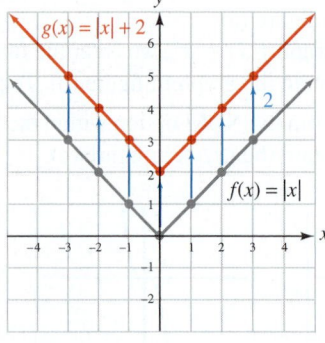

To graph $g(x) = |x| + 2$, translate each point on the graph of $f(x) = |x|$ up 2 units.

Self Check 5   Graph: $g(x) = |x| - 3$.

The figure on the next page shows the graph of $f(x) = (x + h)^2$ for three different values of h. If $h = 0$, we get the graph of $f(x) = x^2$. The graph of $f(x) = (x - 3)^2$ is identical to the graph of $f(x) = x^2$ except that it is shifted 3 units to the right. The graph of $f(x) = (x + 2)^2$ is identical to the graph of $f(x) = x^2$ except that it is shifted 2 units to the left. These shifts are called **horizontal translations.**

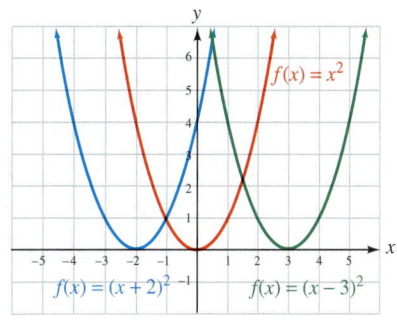

In general, we can make these observations.

Horizontal Translations	If f is a function and h is a positive number, then

- The graph of $y = f(x - h)$ is identical to the graph of $y = f(x)$ except that it is translated h units to the right.
- The graph of $y = f(x + h)$ is identical to the graph of $y = f(x)$ except that it is translated h units to the left.

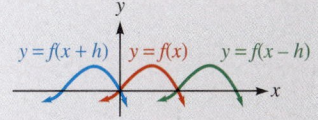

EXAMPLE 6   Graph: $g(x) = (x + 3)^3$.

Solution   The graph of $g(x) = (x + 3)^3$ will be the same shape as the graph of $f(x) = x^3$ except that it is shifted 3 units to the left.

Success Tip

To determine the direction of the horizontal translation, find the value of x that makes the expression within the parentheses, $x + 3$, equal to 0. Since -3 makes $x + 3 = 0$, the translation is 3 units to the *left*.

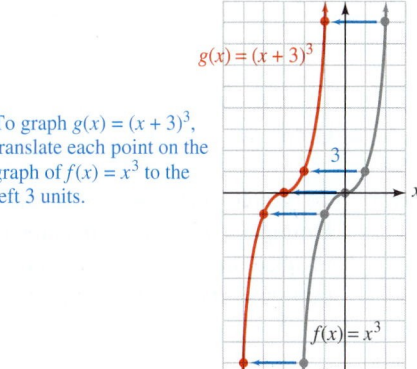

To graph $g(x) = (x + 3)^3$, translate each point on the graph of $f(x) = x^3$ to the left 3 units.

Self Check 6   Graph $g(x) = (x - 2)^2$.

EXAMPLE 7   Graph: $g(x) = (x - 3)^2 + 2$.

ELEMENTARY &
INTERMEDIATE
Algebra $f(x)$ **Now**™

Solution   Two translations are made to a basic graph. We can graph this function by translating the graph of $f(x) = x^2$ to the right 3 units and then 2 units up, as follows.

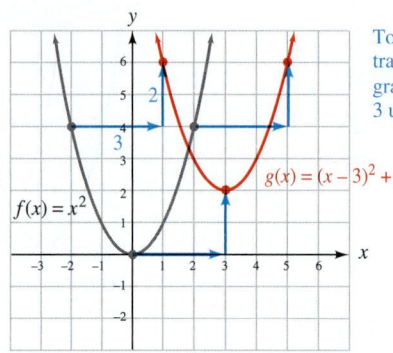

To graph $g(x) = (x - 3)^2 + 2$, translate each point on the graph of $f(x) = x^2$ to the right 3 units and then 2 units up.

$g(x) = (x - 3)^2 + 2$

$f(x) = x^2$

Self Check 7   Graph: $g(x) = |x + 2| - 3$.

■ REFLECTIONS OF GRAPHS

The following figure shows a table of values for $f(x) = x^2$ and for $g(x) = -x^2$. We note that for a given value of x, the corresponding y-values in the tables are opposites. When graphed, we see that the $-$ in $g(x) = -x^2$ has the effect of flipping the graph of $f(x) = x^2$ over the x-axis so that the parabola opens downward. We say that the graph of $g(x) = -x^2$ is a **reflection** of the graph of $f(x) = x^2$ about the x-axis.

$f(x) = x^2$

x	$f(x)$	
-2	4	$\rightarrow (-2, 4)$
-1	1	$\rightarrow (-1, 1)$
0	0	$\rightarrow (0, 0)$
1	1	$\rightarrow (1, 1)$
2	4	$\rightarrow (2, 4)$

$g(x) = -x^2$

x	$g(x)$	
-2	-4	$\rightarrow (-2, -4)$
-1	-1	$\rightarrow (-1, -1)$
0	0	$\rightarrow (0, 0)$
1	-1	$\rightarrow (1, -1)$
2	-4	$\rightarrow (2, -4)$

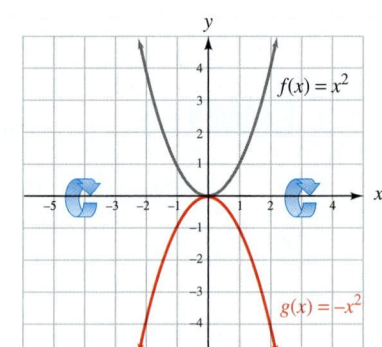

$f(x) = x^2$

$g(x) = -x^2$

EXAMPLE 8

Graph: $g(x) = -x^3$.

Solution   To graph $g(x) = -x^3$, we use the graph of $f(x) = x^3$ from Example 3. First, we reflect the portion of the graph of $f(x) = x^3$ in quadrant I to quadrant IV, as shown. Then we reflect the portion of the graph of $f(x) = x^3$ in quadrant III to quadrant II.

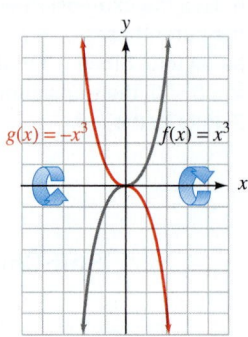

$g(x) = -x^3$ $f(x) = x^3$

Self Check 8   Graph: $g(x) = -|x|$.

Reflection of a Graph   The graph of $y = -f(x)$ is the graph of $y = f(x)$ reflected about the x-axis.

Answers to Self Checks

1. a. -2, **b.** 1

2. D: the set of real numbers, R: the set of all real numbers greater than or equal to -2; the graph has the same shape but is 2 units lower.

3. D: the set of real numbers, R: the set of real numbers; the graph has the same shape but is 1 unit higher.

4. D: the set of real numbers, R: the set of nonnegative real numbers; the graph has the same shape but is 2 units to the right.

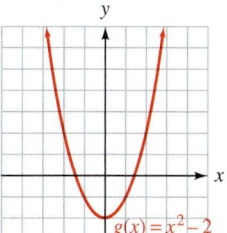

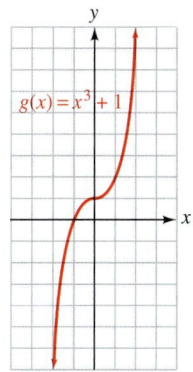

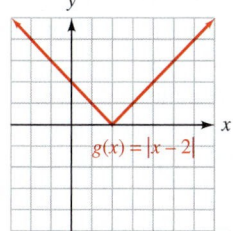

5.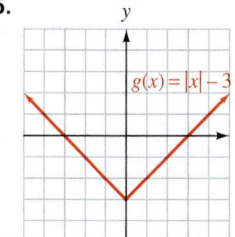

6.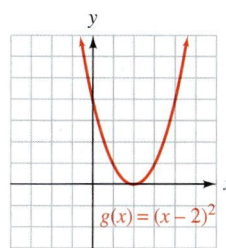

7.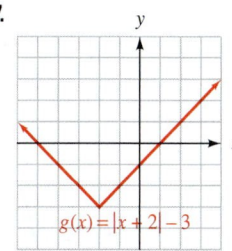

8.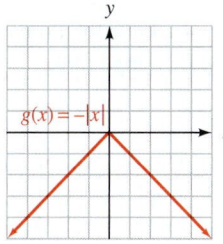

8.8 **STUDY SET** ELEMENTARY & INTERMEDIATE Algebra ⓕ Now™

VOCABULARY Fill in the blanks.

1. Functions whose graphs are not straight lines are called _____ functions.

2. The function $f(x) = x^2$ is called the _____ function.

3. The graph of $f(x) = x^2$ is a cup-like shape called a _____.

4. The set of _____ real numbers is the set of real numbers greater than or equal to 0.

5. The function $f(x) = x^3$ is called the _____ function.

6. The function $f(x) = |x|$ is called the _____ function.

CONCEPTS

7. Use the graph of function f to find each of the following.
   a. $f(-2)$ **b.** $f(0)$
   c. The value of x for which $f(x) = 4$.
   d. The value of x for which $f(x) = -2$.

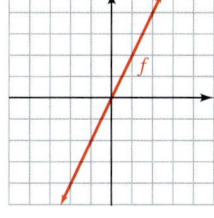

8. Use the graph of function g to find each of the following.
   a. $g(-2)$ **b.** $g(0)$
   c. The value of x for which $g(x) = 3$.
   d. The values of x for which $g(x) = -1$.

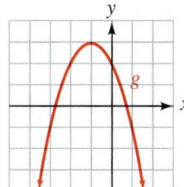

9. Use the graph of function h to find each of the following.
  a. $h(-3)$ **b.** $h(4)$
  c. The value(s) of x for which $h(x) = 1$.
  d. The value(s) of x for which $h(x) = 0$.

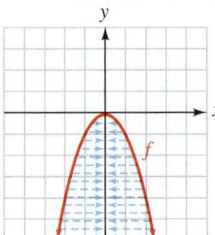

10. Fill in the blanks. The illustration shows the projection of the graph of function f on the _____. We see that the _____ of f is the set of real numbers less than or equal to 0.

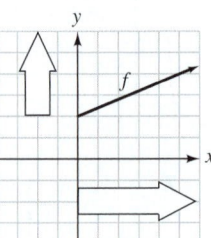

11. Consider the graph of the function f.
  a. Label each arrow in the illustration with the appropriate term: *domain* or *range*.
  b. Give the domain and range of f.

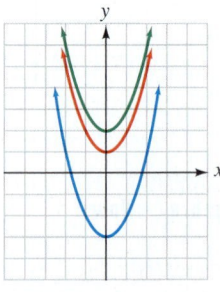

12. The illustration shows the graph of $f(x) = x^2 + k$ for three values of k. What are the three values?

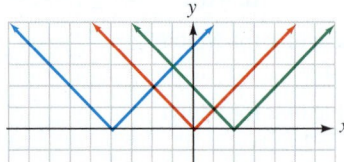

13. The illustration shows the graph of $f(x) = |x + h|$ for three values of h. What are the three values?

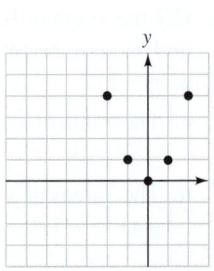

14. Translate each point plotted on the graph to the left 5 units and then up 1 unit.

15. Translate each point plotted on the graph to the right 4 units and then down 3 units.

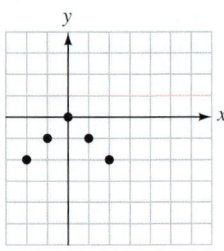

16. Use a graphing calculator to sketch the reflection of the following graph.

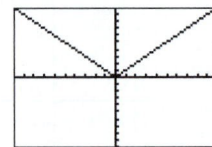

NOTATION **Fill in the blanks.**

17. The graph of $f(x) = (x + 4)^3$ is the same as the graph of $f(x) = x^3$ except that it is shifted ____ units to the _____.

18. The graph of $f(x) = x^3 - 2$ is the same as the graph of $f(x) = x^3$ except that it is shifted ____ units _____.

19. The graph of $f(x) = x^2 + 5$ is the same as the graph of $f(x) = x^2$ except that it is shifted ____ units _____.

20. The graph of $f(x) = |x - 5|$ is the same as the graph of $f(x) = |x|$ except that it is shifted ____ units to the _____.

PRACTICE **Graph each function by plotting points. Give the domain and range.**

21. $f(x) = x^2 - 3$

22. $f(x) = x^2 + 2$

23. $f(x) = (x - 1)^3$

24. $f(x) = (x + 1)^3$

25. $f(x) = |x| - 2$

26. $f(x) = |x| + 1$

27. $f(x) = |x - 1|$

28. $f(x) = |x + 2|$

29. $f(x) = -3x$

30. $f(x) = \dfrac{1}{4}x + 4$

Graph each function using window settings of [−4, 4] for *x* and [−4, 4] for *y*. The graph is not what it appears to be. Pick a better viewing window and find a better representation of the true graph.

31. $f(x) = x^2 + 8$ **32.** $f(x) = x^3 - 8$

33. $f(x) = |x + 5|$ **34.** $f(x) = |x - 5|$

35. $f(x) = (x - 6)^2$ **36.** $f(x) = (x + 9)^2$

37. $f(x) = x^3 + 8$ **38.** $f(x) = x^3 - 12$

For each function, first sketch the graph of its associated function, $f(x) = x^2, f(x) = x^3,$ or $f(x) = |x|$. Then draw each graph using a translation or a reflection.

39. $f(x) = x^2 - 5$ **40.** $f(x) = x^3 + 4$

41. $f(x) = (x - 1)^3$ **42.** $f(x) = (x + 4)^2$

43. $f(x) = |x - 2| - 1$ **44.** $f(x) = (x + 2)^2 - 1$

45. $f(x) = (x + 1)^3 - 2$ **46.** $f(x) = |x + 4| + 3$

47. $f(x) = -x^3$ **48.** $f(x) = -|x|$

49. $f(x) = -x^2$ **50.** $f(x) = -(x + 1)^2$

APPLICATIONS

51. OPTICS   See the illustration. The law of reflection states that the angle of reflection is equal to the angle of incidence. What function studied in this section models the path of the reflected light beam with an angle of incidence measuring 45°?

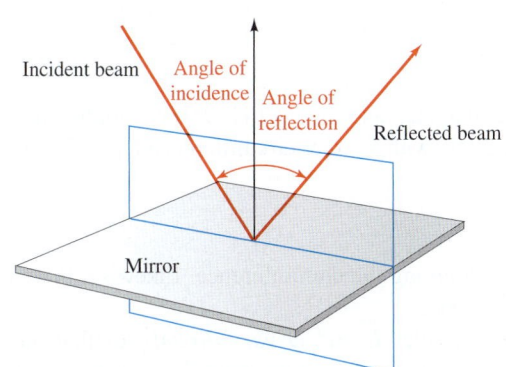

52. BILLIARDS   In the illustration, a rectangular coordinate system has been superimposed over a billiard table. Write a function that models the path of the ball that is shown banking off of the far cushion.

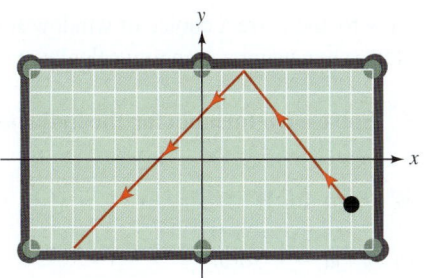

53. CENTER OF GRAVITY See the illustration. As a diver performs a $1\frac{1}{2}$-somersault in the tuck position, her center of gravity follows a path that can be described by a graph shape studied in this section. What graph shape is that?

54. FLASHLIGHTS   Light beams coming from a bulb are reflected outward by a parabolic mirror as parallel rays.

a. The cross-sectional view of a parabolic mirror is given by the function $f(x) = x^2$ for the following values of *x*: −0.7, −0.6, −0.5, −0.4, −0.3, −0.2, −0.1, 0, 0.1, 0.2, 0.3, 0.4, 0.5, 0.6, 0.7. Sketch the parabolic mirror using the following graph.

b. From the lightbulb filament at $(0, 0.25)$, draw a line segment representing a beam of light that strikes the mirror at $(-0.4, 0.16)$ and then reflects outward, parallel to the *y*-axis.

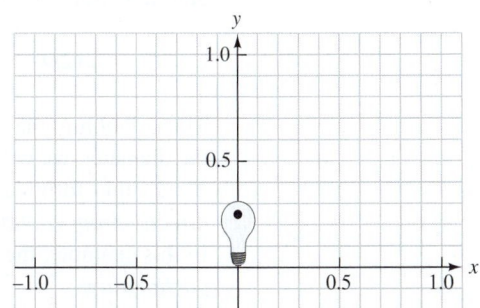

WRITING

55. Explain how to graph a function by plotting points.

56. What does it mean when we say that the domain of a function is the set of all real numbers?

57. What does it mean to vertically translate a graph?

58. Explain why the correct choice of window settings is important when using a graphing calculator.

REVIEW   Solve each formula for the indicated variable.

59. $T - W = ma$ for W

60. $a + (n - 1)d = \ell$ for n

61. $s = \dfrac{1}{2}gt^2 + vt$ for g

62. $e = mc^2$ for m

63. BUDGETING   Last year, Rock Valley College had an operating budget of \$4.5 million. Due to salary increases and a new robotics program, the budget was increased by 20%. Find the operating budget for this year.

64. In the illustration, the line passing through points R, C, and S is parallel to line segment AB. Find the measure of $\angle ACB$. (Read $\angle ACB$ as "angle ACB." *Hint:* Recall from geometry that alternate interior angles have the same measure.)

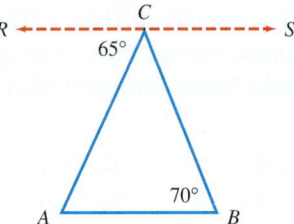

CHALLENGE PROBLEMS   Graph each function.

65. $f(x) = \begin{cases} |x| & \text{for } x \geq 0 \\ x^3 & \text{for } x < 0 \end{cases}$

66. $f(x) = \begin{cases} x^2 & \text{for } x \geq 0 \\ |x| & \text{for } x < 0 \end{cases}$

8.9   Variation

- Direct variation   • Inverse variation   • Joint variation   • Combined variation

In this section, we introduce four types of *variation models,* each of which expresses a special relationship between two or more quantities. We will use these models to solve problems involving travel, lighting, geometry, and highway construction.

■ DIRECT VARIATION

To introduce direct variation, we consider the formula for the circumference of a circle

$$C = \pi D$$

where C is the circumference, D is the diameter, and $\pi \approx 3.14159$. If we double the diameter of a circle, we determine another circle with a larger circumference C_1 such that

$$C_1 = \pi(2D) = 2\pi D = 2C$$

Thus, doubling the diameter results in doubling the circumference. Likewise, if we triple the diameter, we will triple the circumference.

In this formula, we say that the variables C and D *vary directly,* or that they are *directly proportional.* This is because C is always found by multiplying D by a constant. In this example, the constant π is called the *constant of variation* or the *constant of proportionality.*

Direct Variation	The words "*y* varies directly with *x*" or "*y* is directly proportional to *x*" mean that $y = kx$ for some nonzero constant k. The constant k is called the **constant of variation** or the **constant of proportionality.**

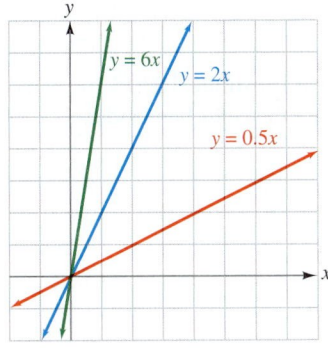

Since the formula for direct variation ($y = kx$) defines a linear function, its graph is always a line with a *y*-intercept at the origin. The graph of $y = kx$ appears in the figure for three positive values of k.

One example of direct variation is Hooke's law from physics. Hooke's law states that the distance a spring will stretch varies directly with the force that is applied to it.

If *d* represents a distance and *f* represents a force, this verbal model of Hooke's law can be expressed as

$$d = kf \qquad \text{This direct variation model can also be read as "d is directly proportional to f."}$$

where *k* is the constant of variation. Suppose we know that a certain spring stretches 10 inches when a weight of 6 pounds is attached (see the figure). We can find *k* as follows:

$$d = k\mathbf{f}$$
$$\mathbf{10} = k(\mathbf{6}) \qquad \text{Substitute 10 for } d \text{ and 6 for } f.$$
$$\frac{5}{3} = k$$

To find the force required to stretch the spring a distance of 35 inches, we can solve the equation $d = kf$ for *f*, with $d = 35$ and $k = \frac{5}{3}$.

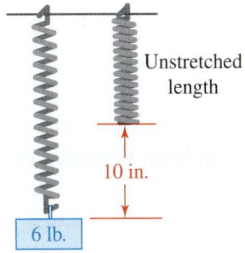

$$d = k\mathbf{f}$$
$$\mathbf{35} = \frac{5}{3}f \qquad \text{Substitute 35 for } d \text{ and } \frac{5}{3} \text{ for } k.$$
$$105 = 5f \qquad \text{Multiply both sides by 3.}$$
$$21 = f \qquad \text{Divide both sides by 5.}$$

Thus, the force required to stretch the spring a distance of 35 inches is 21 pounds.

EXAMPLE 1

ELEMENTARY &
INTERMEDIATE

Algebra *f(x)* **Now**™

Currency exchange. The currency calculator shown on the right converts from U.S. dollars to Japanese yen. When exchanging these currencies, the number of yen received is directly proportional to the number of dollars to be exchanged. How many yen will an exchange of $1,200 bring?

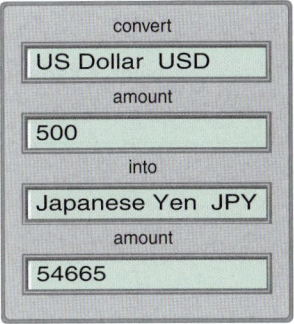

Solution   The verbal model *the number of yen is directly proportional to the number of dollars* can be expressed by the equation

$$y = kd \qquad \text{This is a direct variation model.}$$

where y is the number of yen, k is the constant of variation, and d is the number of dollars. From the illustration, we see that an exchange of \$500 brings 54,665 yen. To find k, we substitute 500 for d and 54,665 for y, and then we solve for k.

$$y = kd$$
$$54{,}665 = k(500)$$
$$109.33 = k \qquad \text{Divide both sides by 500.}$$

To find how many yen an exchange of \$1,200 will bring, we substitute 109.33 for k and 1,200 for d in the direct variation model, and then we evaluate the right-hand side.

$$y = kd$$
$$y = 109.33(1{,}200)$$
$$y = 131{,}196$$

An exchange of \$1,200 will bring 131,196 yen.

Self Check 1   When exchanging currencies, the number of British pounds received is directly proportional to the number of U.S. dollars to be exchanged. If \$800 converts to 440 pounds, how many pounds will be received if \$1,500 is exchanged?

Solving Variation Problems

To solve a variation problem:

1. Translate the verbal model into an equation.
2. Substitute the first set of values into the equation from step 1 to determine the value of k.
3. Substitute the value of k into the equation from step 1.
4. Substitute the remaining set of values into the equation from step 3 and solve for the unknown.

■ INVERSE VARIATION

In the formula $w = \frac{12}{l}$, w gets smaller as l gets larger, and w gets larger as l gets smaller. Since these variables vary in opposite directions in a predictable way, we say that the variables *vary inversely,* or that they are *inversely proportional.* The constant 12 is the constant of variation.

Inverse Variation

The words "y varies inversely with x" or "y is inversely proportional to x" mean that $y = \frac{k}{x}$ for some nonzero constant k. The constant k is called the **constant of variation.**

The formula for inverse variation, $y = \frac{k}{x}$, defines a rational function whose graph will have the x- and y-axes as asymptotes. The graph of $y = \frac{k}{x}$ appears in the figure for three positive values of k.

In an elevator, the amount of floor space per person varies inversely with the number of people in the elevator. If f represents the amount of floor space per person and n the

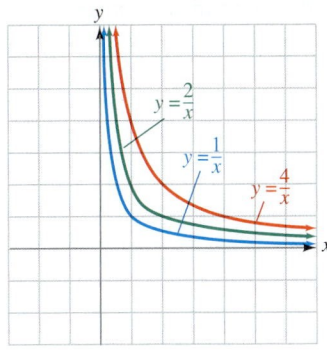

number of people in the elevator, the relationship between f and n can be expressed by the equation

$$f = \frac{k}{n}$$ This inverse variation model can also be read as "f is inversely proportional to n."

The figure shows 6 people in an elevator; each has 8.25 square feet of floor space. To determine how much floor space each person would have if 15 people were in the elevator, we begin by determining k.

$$\mathbf{f} = \frac{\mathbf{k}}{\mathbf{n}}$$

$$\mathbf{8.25} = \frac{k}{\mathbf{6}}$$ Substitute 8.25 for f and 6 for n.

$$k = 49.5$$ Multiply both sides by 6 to solve for k.

To find the amount of floor space per person if 15 people are in the elevator, we proceed as follows:

$$f = \frac{\mathbf{k}}{\mathbf{n}}$$

$$f = \frac{\mathbf{49.5}}{\mathbf{15}}$$ Substitute 49.5 for k and 15 for n.

$$f = 3.3$$ Do the division.

If 15 people were in the elevator, each would have 3.3 square feet of floor space.

EXAMPLE 2

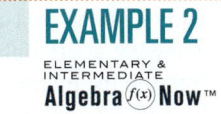

ELEMENTARY &
INTERMEDIATE
Algebra $f(x)$ **Now**™

Photography.   The intensity I of light received from a light source varies inversely with the square of the distance from the light source. If a photographer, 16 feet away from his subject, has a light meter reading of 4 foot-candles of illuminance, what will the meter read if the photographer moves in for a close-up 4 feet away from the subject?

Solution   The words *intensity varies inversely with the square of the distance d* can be expressed by the equation

$$I = \frac{k}{d^2}$$ This inverse variation model can also be read as "I is inversely proportional to d^2."

Success Tip

The constant of variation is usually positive, because most real-life applications involve only positive quantities. However, the definition of direct, inverse, joint, and combined variation allow for a negative constant of variation.

To find k, we substitute 4 for I and 16 for d and solve for k.

$$\mathbf{I} = \frac{\mathbf{k}}{\mathbf{d^2}}$$

$$\mathbf{4} = \frac{k}{\mathbf{16^2}}$$

$$4 = \frac{k}{256}$$

$$1{,}024 = k$$

To find the intensity when the photographer is 4 feet away from the subject, we substitute 4 for d and 1,024 for k and simplify.

$$I = \frac{k}{d^2}$$

$$I = \frac{1{,}024}{4^2}$$

$$= 64$$

The intensity at 4 feet is 64 foot-candles.

Self Check 2   Find the intensity when the photographer is 8 feet away from the subject.

■ JOINT VARIATION

There are times when one variable varies with the product of several variables. For example, the area of a triangle varies directly with the product of its base and height:

$$A = \frac{1}{2}bh$$

Such variation is called *joint variation.*

Joint Variation   If one variable varies directly with the product of two or more variables, the relationship is called **joint variation.** If y varies jointly with x and z, then $y = kxz$. The nonzero constant k is called the **constant of variation.**

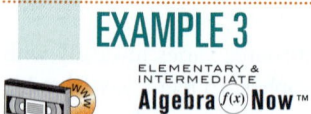

EXAMPLE 3

ELEMENTARY & INTERMEDIATE
Algebra $f(x)$ **Now**™

Force of the wind.   The force of the wind on a billboard varies jointly with the area of the billboard and the square of the wind velocity. When the wind is blowing at 20 mph, the force on a billboard 30 feet wide and 18 feet high is 972 pounds. (Refer to the figure.) Find the force on a billboard having an area of 300 square feet caused by a 40-mph wind.

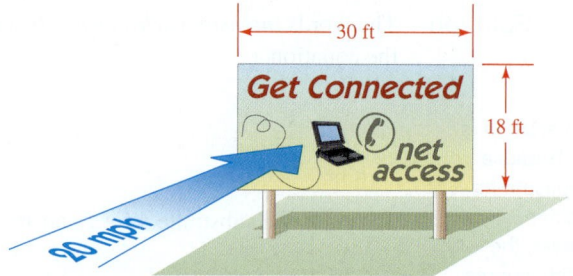

Solution   We will let f represent the force of the wind, A the area of the billboard, and v the velocity of the wind. The words *the force of the wind on a billboard varies jointly with the area of the billboard and the square of the wind velocity* mean that f *varies directly as the product of A and v^2.* Thus,

$$f = kAv^2 \qquad \text{The joint variation model can also be read as "f is directly proportional to the product of A and v^2."}$$

Since the billboard is 30 feet wide and 18 feet high, it has an area of $30 \cdot 18 = 540$ square feet. We can find k by substituting 972 for f, 540 for A, and 20 for v.

$$f = kAv^2$$
$$972 = k(540)(20)^2$$
$$972 = k(216,000) \qquad \text{First, find the power: } (20)^2 = 400. \text{ Then do the multiplication.}$$
$$0.0045 = k \qquad\qquad\quad \text{Divide both sides by 216,000 to solve for } k.$$

To find the force exerted on a 300-square-foot billboard by a 40-mph wind, we use the formula $f = 0.0045Av^2$ and substitute 300 for A and 40 for v.

$$f = 0.0045Av^2$$
$$= 0.0045(300)(40)^2$$
$$= 2,160$$

The 40-mph wind exerts a force of 2,160 pounds on the billboard.

■ COMBINED VARIATION

Many applied problems involve a combination of direct and inverse variation. Such variation is called **combined variation.**

EXAMPLE 4

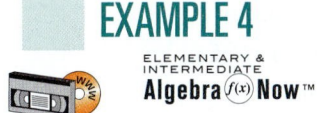

ELEMENTARY &
INTERMEDIATE
Algebra $f(x)$ **Now**™

Highway construction. The time it takes to build a highway varies directly with the length of the road, but inversely with the number of workers. If it takes 100 workers 4 weeks to build 2 miles of highway, how long will it take 80 workers to build 10 miles of highway?

Solution

We can let t represent the time in weeks to build a highway, l represent the length of the highway in miles, and w represent the number of workers. The relationship between these variables can be expressed by the equation

$$t = \frac{k\ell}{w} \qquad \text{This is a combined variation model.}$$

We substitute 4 for t, 100 for w, and 2 for ℓ to find k:

$$4 = \frac{k(2)}{100}$$
$$400 = 2k \qquad \text{Multiply both sides by 100.}$$
$$200 = k \qquad \text{Divide both sides by 2.}$$

We now substitute 80 for w, 10 for ℓ, and 200 for k in the equation $t = \frac{k\ell}{w}$ and simplify:

$$t = \frac{k\ell}{w}$$
$$t = \frac{200(10)}{80}$$
$$= 25$$

It will take 25 weeks for 80 workers to build 10 miles of highway.

Self Check 4   How long will it take 60 workers to build 6 miles of highway?

Answers to Self Checks   **1.** 825 British pounds   **2.** 16 foot-candles   **4.** 20 weeks

---

 8.9   **STUDY SET**   ELEMENTARY & INTERMEDIATE **Algebra** f(x) **Now**™

VOCABULARY   Fill in the blanks.

1. The equation $y = kx$ defines _____ variation.

2. The equation $y = \dfrac{k}{x}$ defines _____ variation.

3. In $y = kx$, the _____ of variation is k.

4. A constant is a _____.

5. The equation $y = kxz$ represents _____ variation.

6. The equation $y = \dfrac{kx}{z}$ means that y varies _____ with x and _____ with z.

CONCEPTS   Decide whether direct or inverse variation applies and sketch a possible graph for the situation.

7.
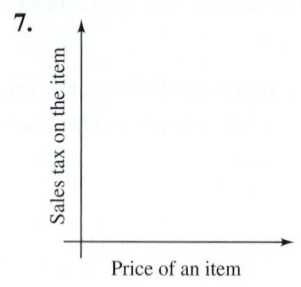

8.
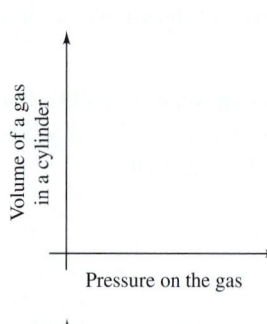

9.
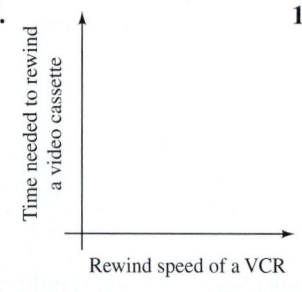

10.
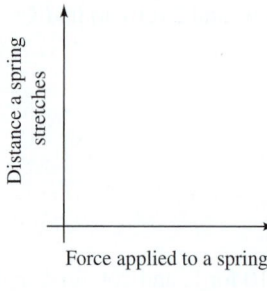

NOTATION

11. Decide whether the equation defines direct variation.

   a. $y = kx$ **b.** $y = k + x$

   c. $y = \dfrac{k}{x}$ **d.** $m = kc$

12. Decide whether each equation defines inverse variation.

   a. $y = kx$ **b.** $y = \dfrac{k}{x}$

   c. $y = \dfrac{x}{k}$ **d.** $d = \dfrac{k}{g}$

PRACTICE   Express each verbal model as a formula.

13. A varies directly with the square of p.

14. z varies inversely with the cube of t.

15. v varies inversely with the cube of r.

16. r varies directly with the square of s.

17. B varies jointly with m and n.

18. C varies jointly with x, y, and z.

19. P varies directly with the square of a, and inversely with the cube of j.

20. M varies inversely with the cube of n, and jointly with x and the square of z.

Express each variation model in words. In each formula, k is the constant of variation.

21. $L = kmn$ **22.** $P = \dfrac{km}{n}$

23. $E = kab^2$ **24.** $U = krs^2t$

25. $X = \dfrac{kx^2}{y^2}$ **26.** $Z = \dfrac{kw}{xy}$

27. $R = \dfrac{kL}{d^2}$ **28.** $e = \dfrac{kPL}{A}$

APPLICATIONS

29. **FREE FALL**   An object in free fall travels a distance *s* that is directly proportional to the square of the time *t*. If an object falls 1,024 feet in 8 seconds, how far will it fall in 10 seconds?

30. **FINDING DISTANCE**   The distance that a car can go is directly proportional to the number of gallons of gasoline it consumes. If a car can go 288 miles on 12 gallons of gasoline, how far can it go on a full tank of 18 gallons?

31. **FARMING**   A farmer's harvest in bushels varies directly with the number of acres planted. If 8 acres can produce 144 bushels, how many acres are required to produce 1,152 bushels?

32. **REAL ESTATE**   The following table shows the listing price for three homes in the same general locality. Write the variation model (direct or inverse) that describes the relationship between the listing price and the number of square feet of a house in this area.

Number of square feet	Listing price
1,720	$180,600
1,205	$126,525
1,080	$113,400

33. **FARMING**   The length of time that a given number of bushels of corn will last when feeding cattle varies inversely with the number of animals. If *x* bushels will feed 25 cows for 10 days, how long will the feed last for 10 cows?

34. **GEOMETRY**   For a fixed area, the length of a rectangle is inversely proportional to its width. A rectangle has a width of 18 feet and a length of 12 feet. If the length is increased to 16 feet, find the width.

35. **GAS PRESSURE**   Under constant temperature, the volume occupied by a gas is inversely proportional to the pressure applied. If the gas occupies a volume of 20 cubic inches under a pressure of 6 pounds per square inch, find the volume when the gas is subjected to a pressure of 10 pounds per square inch.

36. **VALUE OF A CAR**   The value of a car usually varies inversely with its age. If a car is worth $7,000 when it is 3 years old, how much will it be worth when it is 7 years old?

37. **ORGAN PIPES**   The frequency of vibration of air in an organ pipe is inversely proportional to the length of the pipe. If a pipe 2 feet long vibrates 256 times per second, how many times per second will a 6-foot pipe vibrate?

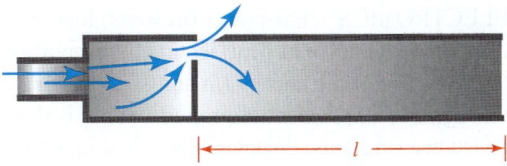

38. **GEOMETRY**   The area of a rectangle varies jointly with its length and width. If both the length and the width are tripled, by what factor is the area multiplied?

39. **GEOMETRY**   The volume of a rectangular solid varies jointly with its length, width, and height. If the length is doubled, the width is tripled, and the height is doubled, by what factor is the volume multiplied?

40. **COSTS OF A TRUCKING COMPANY**   The costs incurred by a trucking company vary jointly with the number of trucks in service and the number of hours they are used. When 4 trucks are used for 6 hours each, the costs are $1,800. Find the costs of using 10 trucks, each for 12 hours.

41. **STORING OIL**   The number of gallons of oil that can be stored in a cylindrical tank varies jointly with the height of the tank and the square of the radius of its base. The constant of proportionality is 23.5. Find the number of gallons that can be stored in the cylindrical tank shown below.

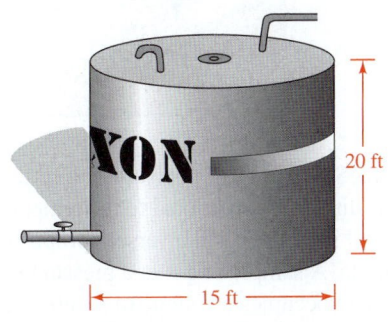

42. **FINDING THE CONSTANT OF VARIATION** A quantity *l* varies jointly with *x* and *y* and inversely with *z*. If the value of *l* is 30 when $x = 15$, $y = 5$, and $z = 10$, find *k*.

43. ELECTRONICS The voltage (in volts) measured across a resistor is directly proportional to the current (in amperes) flowing through the resistor. The constant of variation is the **resistance** (in ohms). If 6 volts is measured across a resistor carrying a current of 2 amperes, find the resistance.

44. ELECTRONICS The power (in watts) lost in a resistor (in the form of heat) is directly proportional to the square of the current (in amperes) passing through it. The constant of proportionality is the resistance (in ohms). What power is lost in a 5-ohm resistor carrying a 3-ampere current?

45. STRUCTURAL ENGINEERING The deflection of a beam is inversely proportional to its width and the cube of its depth. If the deflection of a 4-inch-by-4-inch beam is 1.1 inches, find the deflection of a 2-inch-by-8-inch beam positioned as shown below.

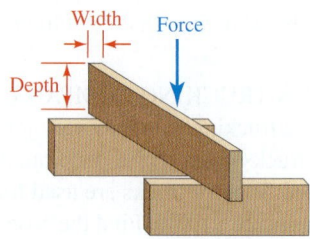

46. STRUCTURAL ENGINEERING Find the deflection of the beam in Exercise 45 when the beam is positioned as shown below.

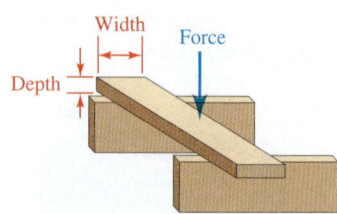

47. GAS PRESSURE The pressure of a certain amount of gas is directly proportional to the temperature (measured in Kelvin) and inversely proportional to the volume. A sample of gas at a pressure of 1 atmosphere occupies a volume of 1 cubic meter at a temperature of 273 Kelvin. When heated, the gas expands to twice its volume, but the pressure remains constant. To what temperature is it heated?

48. TENSION IN A STRING When playing with a Skip It toy, a child swings a weighted ball on the end of a string in a circular motion around one leg while jumping over the revolving string with the other leg. The tension T in the string is directly proportional to the square of the speed s of the ball and inversely proportional to the radius r of the circle. If the tension in the string is 6 pounds when the speed of the ball is 6 feet per second and the radius is 3 feet, find the tension when the speed is 8 feet per second and the radius is 2.5 feet.

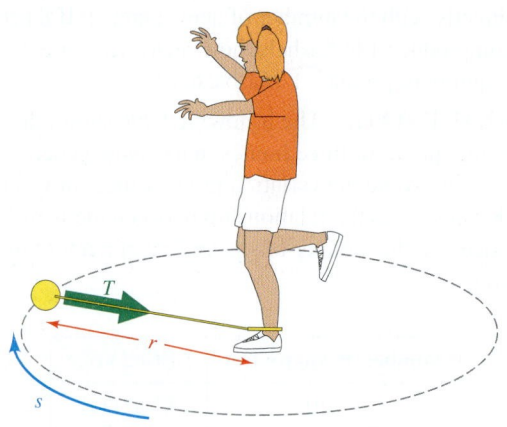

WRITING

49. Explain the term *direct variation*.

50. Explain the term *inverse variation*.

51. Explain the term *joint variation*.

52. From everyday life, give examples of two quantities that vary directly and two quantities that vary inversely.

REVIEW

53. Write 35,000 in scientific notation.

54. Write 0.00035 in scientific notation.

55. Write 2.5×10^{-3} in standard notation.

56. Write 2.5×10^4 in standard notation.

CHALLENGE PROBLEMS

57. As the cost of a purchase that is less than $5 increases, the amount of change received from a five-dollar bill decreases. Is this inverse variation? Explain.

58. You've probably heard of Murphy's First Law: *If anything can go wrong, it will.* Another of Murphy's Laws is: *The chances of a piece of bread falling with the grape jelly side down varies directly with the cost of the carpet.* Write one of your own witty sayings using the phrase *varies directly*.

ACCENT ON TEAMWORK

MEASURING SLOPE

Overview: This hands-on activity will give you a better understanding of slope.

Instructions: Form groups of 2 or 3 students. Use a ruler and a level to find the slopes of five ramps or inclines by measuring $\frac{\text{rise}}{\text{run}}$, as shown below. Record your results in a table, listing the slopes in order from smallest to largest.

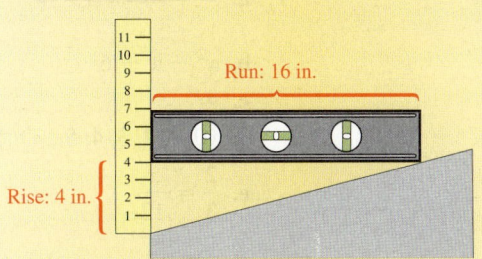

Object/location	Slope		
Ramp outside the cafeteria	$\dfrac{\text{Rise}}{\text{Run}}$	$= \dfrac{4 \; in.}{16 \; in.}$	$= \dfrac{1}{4}$

CONTINUED FRACTIONS

Overview: In this activity, as you gain experience simplifying complex fractions, you will make an interesting discovery about continued fractions.

Instructions: Form groups of 2 or 3 students. Working as a group, simplify each expression in the following list. Note that the third, fourth, fifth, and all the subsequent fractions in the list have a complex fraction in their denominator. These expressions are called **continued fractions.**

$$1 + \frac{1}{2}, \qquad 1 + \cfrac{1}{1 + \cfrac{1}{2}}, \qquad 1 + \cfrac{1}{1 + \cfrac{1}{1 + \cfrac{1}{2}}},$$

$$1 + \cfrac{1}{1 + \cfrac{1}{1 + \cfrac{1}{1 + \cfrac{1}{2}}}}, \qquad 1 + \cfrac{1}{1 + \cfrac{1}{1 + \cfrac{1}{1 + \cfrac{1}{1 + \cfrac{1}{2}}}}}, \; \dots$$

Each of these expressions can be simplified by using the value of the expression preceding it. For example, to simplify the second expression in the list, replace $1 + \frac{1}{2}$ with $\frac{3}{2}$. Show that the expressions in the list simplify to $\frac{3}{2}, \frac{5}{3}, \frac{8}{5}, \frac{13}{8}, \frac{21}{13}, \dots$. Then write the next 3 continued fractions in the list. From what you have learned, *predict* the answers if each of them was simplified.

KEY CONCEPT: THE SYMBOLS OF ALGEBRA

One of the keys to becoming a good algebra student is to know the symbols of algebra. Match each item in column I with the most appropriate item in column II. Each letter in column II is used only once.

COLUMN I

_____ **1.** Parentheses

_____ **2.** Scientific notation

_____ **3.** Function notation

_____ **4.** Rational expression

_____ **5.** System of linear equations

_____ **6.** Exponential expression

_____ **7.** Absolute value

_____ **8.** Subscript notation

_____ **9.** Ordered pair

_____ **10.** Difference of two squares

_____ **11.** Equation

_____ **12.** Numerical expression

_____ **13.** Brackets

_____ **14.** Positive infinity

_____ **15.** Trinomial

_____ **16.** Inequality

_____ **17.** Set

_____ **18.** Number line

_____ **19.** Rectangular coordinate system

_____ **20.** The graph of the real numbers between -1 and 3

_____ **21.** The graph of the interval $[-1, 3]$

_____ **22.** Is approximately equal to

_____ **23.** Polynomial in a

_____ **24.** Complex fraction

_____ **25.** Quadrant

_____ **26.** Proportion

COLUMN II

a.
-1 3

b. $x^2 - x - 6$

c. ∞

d. $\{0, 1, 2, 3, 4, 5, \ldots\}$

e. $\dfrac{2}{3} = \dfrac{x}{12}$

f. x_2

g.
-1 3

h. $\dfrac{\frac{x}{y}}{\frac{2x}{3y}}$

i.
$-3 \ -2 \ -1 \ \ 0 \ \ 1 \ \ 2 \ \ 3$

j. 5.78×10^{-6}

k. $x^2 - 36$

l. II

m. $6c - 8 = 15(c + 1)$

n. $-4x + 6 \geq 20$

o. $(-7)(-3) + 3^3$

p. $\dfrac{12x^2 - 4x + 1}{3x^2 + 2x - 5}$

q. $\begin{cases} 4x - 2y = 3 \\ x + 3y = -12 \end{cases}$

r. x^4

s. $(-5, 6)$

t. $a^3 - 3a^2 + 2a + 1$

u. $(\quad)$

v. $|-6|$

w. $f(x) = 3x^2 - 4x$

x.

y. $\approx$

z. $[\ \ ]$

CHAPTER REVIEW ELEMENTARY & INTERMEDIATE Algebra $f(x)$ Now™

| SECTION 8.1 | **Review of Equations and Inequalities** |

CONCEPTS

To solve a linear equation:

1. Clear the equation of fractions.

2. Remove all parentheses.

3. Combine like terms.

4. Get all variables on one side and all constants on the other.

5. Isolate the variable.

6. Check the result.

An equation that is true for all values of its variable is called an *identity*.

An equation that is not true for any values of its variable is called a *contradiction*.

If both sides of an inequality are multiplied (or divided) by a negative number, another inequality results, but with the opposite direction from the original inequality.

The solution set of a compound inequality containing *and* is the *intersection* of the two solution sets.

The solution set of a compound inequality containing *or* is the *union* of the two solution sets.

REVIEW EXERCISES

Solve each equation. If an equation is an identity or a contradiction, so indicate.

1. $5x + 12 = 0$

2. $-3x - 7 + x = 6x + 20 - 5x$

3. $4(y - 1) = 28$

4. $2 - 13(x - 1) = 4 - 6x$

5. $\dfrac{8}{3}(x - 5) = \dfrac{2}{5}(x - 4)$

6. $\dfrac{3y}{4} - 14 = -\dfrac{y}{3} - 1$

7. $2x + 4 = 2(x + 3) - 2$

8. $3x - 2 - x = 2(x - 4)$

9. $-\dfrac{5}{4}p = 10$

10. $\dfrac{4t + 1}{3} - \dfrac{t + 5}{6} = \dfrac{t - 3}{6}$

Solve each formula for the indicated variable.

11. $V = \pi r^2 h$ for h

12. $v = \dfrac{1}{6}ab(x + y)$ for x

Solve each inequality. Give each solution set in interval notation and graph it.

13. $0.3x - 0.4 \geq 1.2 - 0.1x$

14. $\dfrac{7}{4}(x + 3) < \dfrac{3}{8}(x - 3)$

15. $-16 < -\dfrac{4}{5}x$

16. $3 < 3x + 4 < 10$

17. $-2x > 8$ and $x + 4 \geq -6$

18. $x + 1 < -4$ or $x - 4 > 0$

19. CARPENTRY   A carpenter wants to cut a 20-foot rafter so that one piece is 3 times as long as the other. Where should he cut the board?

20. GEOMETRY   A rectangle is 4 meters longer than it is wide. If the perimeter of the rectangle is 28 meters, find its area.

SECTION 8.2

Solving Absolute Value Equations and Inequalities

Absolute value equations:
For $k > 0$ and any algebraic expressions X and Y:

$|X| = k$ is equivalent to

$X = k$ or $X = -k$

$|X| = |Y|$ is equivalent to

$X = Y$ or $X = -Y$

Absolute value inequalities:
For $k > 0$ and any algebraic expression X:

$|X| < k$ is equivalent to

$-k < X < k$

$|X| > k$ is equivalent to

$X < -k$ or $X > k$

Solve each absolute value equation.

21. $|4x| = 8$

22. $2|3x + 1| - 1 = 19$

23. $\left| \dfrac{3}{2}x - 4 \right| - 10 = -1$

24. $\left| \dfrac{2 - x}{3} \right| = -4$

25. $|3x + 2| = |2x - 3|$

26. $\left| \dfrac{2(1 - x) + 1}{2} \right| = \left| \dfrac{3x - 2}{3} \right|$

Solve each absolute value inequality. Give the solution in interval notation and graph it.

27. $|x| \leq 3$

28. $|2x + 7| < 3$

29. $2|5 - 3x| \leq 28$

30. $\left| \dfrac{2}{3}x + 14 \right| + 6 < 6$

31. $|x| > 1$

32. $|3x - 8| - 4 \geq 0$

SECTION 8.3

Review of Factoring

Always factor out all *common factors* as the first step in a factoring problem.

To factor trinomials with a *lead coefficient of 1,* list the factorizations of the third term.

To factor trinomials with a *lead coefficient other than 1,* factor by trial-and-check or by the key number/grouping method.

If an expression has four or more terms, try to factor the expression by *grouping.*

Factor, if possible.

33. $z^2 - 11z + 30$

34. $x^4 + 4y + 4x^2 + x^2y$

35. $4a^2 - 5a + 1$

36. $27x^3y^3z^3 + 81x^4y^5z^2 - 90x^2y^3z^7$

37. $y^2 + 3y + 2 + 2x + xy$

38. $-x^2 - 3x + 28$

39. $15x^2 - 57xy - 12y^2$

40. $w^8 - w^4 - 90$

41. $r^2y - ar - ry + a + r - 1$

42. $49a^6 + 84a^3b^2 + 36b^4$

43. $3b^2 + 2b + 1$

44. $2a^4 + 4a^3 - 6a^2$

45. Use a substitution to factor $(s + t)^2 - 2(s + t) + 1$.

46. Solve $m_1m_2 = mm_2 + mm_1$ for m_1.

SECTION 8.4	**The Difference of Two Squares; the Sum and Difference of Two Cubes**

Factoring the *difference of two squares:*

$$x^2 - y^2 = (x + y)(x - y)$$

Factoring the *sum of two cubes:*

$$\begin{aligned} x^3 + y^3 \\ = (x + y)(x^2 - xy + y^2) \end{aligned}$$

the *difference of two cubes:*

$$\begin{aligned} x^3 - y^3 \\ = (x - y)(x^2 + xy + y^2) \end{aligned}$$

Factor, if possible.

47. $z^2 - 16$

48. $x^2y^4 - 64z^6$

49. $a^2b^2 + c^2$

50. $c^2 - (a + b)^2$

51. $32a^4c - 162b^4c$

52. $k^2 + 2k + 1 - 9m^2$

53. $m^2 - n^2 - m - n$

54. $t^3 + 64$

55. $8a^3 - 125b^9$

56. SPANISH ROOF TILE   The amount of clay used to make a roof tile is given by

$$V = \frac{\pi}{2}r_1^2h - \frac{\pi}{2}r_2^2h$$

Factor the right-hand side of the formula completely.

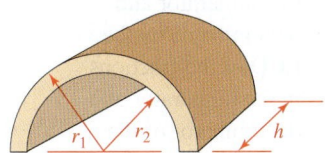

SECTION 8.5	**Review of Rational Expressions**

To *simplify* rational expressions, remove common factors that are equal to 1.

To *multiply* rational expressions, multiply the numerators and multiply the denominators.

To *divide* rational expressions, multiply the first by the reciprocal of the second.

Simplify each rational expression.

57. $\dfrac{248x^2y}{576xy^2}$

58. $\dfrac{2m - 2n}{n - m}$

Perform the operations and simplify, if possible.

59. $\dfrac{3x^3y^4}{c^2d} \cdot \dfrac{c^3d^2}{21x^5y^4}$

60. $\dfrac{2a^2 - 5a - 3}{a^2 - 9} \div \dfrac{2a^2 + 5a + 2}{2a^2 + 5a - 3}$

61. $\dfrac{m^2 + 3m + 9}{m^2 + mp + mr + pr} \div \dfrac{m^3 - 27}{am + ar + bm + br}$

62. $\dfrac{x^3 + 3x^2 + 2x}{2x^2 - 2x - 12} \cdot \dfrac{3x^2 - 3x}{x^3 - 3x^2 - 4x} \div \dfrac{x^2 + 3x + 2}{2x^2 - 4x - 16}$

To add or subtract rational expressions with *unlike denominators*, find the LCD and express each rational expression with a denominator that is the LCD. Add (or subtract) the resulting fractions and simplify the result if possible.

63. $\dfrac{d^2}{c^3 - d^3} + \dfrac{c^2 + cd}{c^3 - d^3}$

64. $\dfrac{4}{t - 3} + \dfrac{6}{3 - t}$

65. $\dfrac{5x}{14z^2} + \dfrac{y^2}{16z}$

66. $\dfrac{4}{3xy - 6y} - \dfrac{4}{10 - 5x}$

67. $\dfrac{y + 7}{y + 3} - \dfrac{y - 3}{y + 7}$

68. $\dfrac{2x}{x + 1} + \dfrac{3x}{x + 2} + \dfrac{4x}{x^2 + 3x + 2}$

To *simplify a complex fraction:*

Method 1: Write the numerator and denominator as single fractions. Then divide the fractions and simplify.

Method 2: Multiply the numerator and denominator by the LCD of the fractions in the numerator and denominator of the complex fraction. Then simplify the results.

Simplify each complex fraction.

69. $\dfrac{\dfrac{4a^3b^2}{9c}}{\dfrac{14a^3b}{9c^4}}$

70. $\dfrac{\dfrac{p^2 - 9}{6pt}}{\dfrac{p^2 + 5p + 6}{3pt}}$

71. $\dfrac{1 - \dfrac{1}{x} - \dfrac{2}{x^2}}{1 + \dfrac{4}{x} + \dfrac{3}{x^2}}$

72. $\dfrac{\dfrac{2}{b^2 - 1} - \dfrac{3}{ab - a}}{\dfrac{3}{ab - a} - \dfrac{2}{b^2 - 1}}$

| **SECTION 8.6** | **Review of Linear Equations in Two Variables** |

The *graph of an equation* is the graph of all points on the rectangular coordinate system whose coordinates satisfy the equation.

To find the *y-intercept* of a line, substitute 0 for *x* in the equation and solve for *y*. To find the *x-intercept* of a line, substitute 0 for *y* in the equation and solve for *x*.

Graph each equation.

73. $y = -\dfrac{1}{3}x - 1$

74. $x = -2$

Graph each equation using the intercept method.

75. $2x + y = 4$

76. $3x - 4y - 8 = 0$

Slope of a nonvertical line:

$$m = \frac{\Delta y}{\Delta x} = \frac{y_2 - y_1}{x_2 - x_1} = \frac{\text{rise}}{\text{run}}$$

Horizontal lines have a slope of 0. Vertical lines have no defined slope.

77. Find the slope of the graph of $2x - 3y = 18$.

78. Find the slope of lines l_1 and l_2 in the illustration.

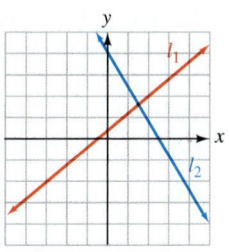

Parallel lines have the same slope. The slopes of two nonvertical perpendicular lines are negative reciprocals.

Find the slope of the line passing through the given points

79. $(2, 5)$ and $(5, 8)$ **80.** $(3, -2)$ and $(-6, 12)$

81. $(-2, 4)$ and $(8, 4)$ **82.** $(-5, -4)$ and $(-5, 8)$

Determine whether the lines with the given slopes are parallel, perpendicular, or neither.

83. $m_1 = 4, m_2 = -\dfrac{1}{4}$ **84.** $m_1 = 0.5, m_2 = \dfrac{1}{2}$

Equations of a line:

Point–slope form:

$$y - y_1 = m(x - x_1)$$

Slope–intercept form:

$$y = mx + b$$

General form:

$$Ax + By = C$$

Find an equation of the line with the given properties. Write the equation in general form.

85. Slope of 3; passing through $(-8, 5)$

86. Passing through $(-2, 4)$ and $(6, -9)$

Find an equation of the line with the given properties. Write the equation in slope–intercept form.

87. Passing through $(-3, -5)$; parallel to the graph of $3x - 2y = 7$

88. Passing through $(-3, -5)$; perpendicular to the graph of $3x - 2y = 7$

Midpoint formula:

For $P(x_1, y_1)$ and $Q(x_2, y_2)$, the midpoint of PQ is

$$\left(\dfrac{x_1 + x_2}{2}, \dfrac{y_1 + y_2}{2}\right)$$

89. A business purchases a copy machine for \$8,700 and will depreciate it on a straight-line basis over the next 5 years. At the end of its useful life, it will be sold as scrap for \$100. Find its depreciation equation.

90. Find the midpoint of the line segment joining $(-3, 5)$ and $(6, 11)$.

SECTION 8.7	**An Introduction to Functions**

A *function* is a rule (or correspondence) that assigns to each value of one variable (called the independent variable) exactly one value of another variable (called the dependent variable).

Determine whether the arrow diagram and the table define y as a function of x.

91.

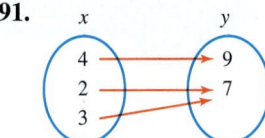

92.

x	y
-1	8
0	5
4	1
-1	9

The notation $y = f(x)$ denotes that the variable y (the *dependent* variable) is a function of x (the *independent* variable).

Determine whether each equation determines y to be a function of x.

93. $y = 6x - 4$

94. $|y| = x$

Let $f(x) = 3x + 2$ and $g(x) = \dfrac{x^2 - 4x + 4}{2}$. Find each function value.

95. $f(-3)$

96. $g(8)$

97. $g(-2)$

98. $f(t)$

99. Let $f(x) = -5x + 7$. For what value of x is $f(x) = -8$?

100. Let $g(x) = \dfrac{3}{4}x - 1$. For what value of x is $g(x) = 0$?

The *domain* of a function is the set of input values. The *range* is the set of output values.

Find the domain and range of each function.

101. $f(x) = 4x - 1$

102. $f(x) = \dfrac{4}{2 - x}$

The *vertical line test* can be used to determine whether a graph represents a function.

Determine whether each graph represents a function.

103.

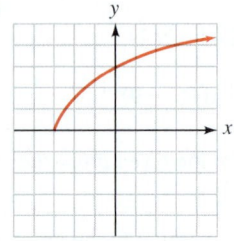

104.

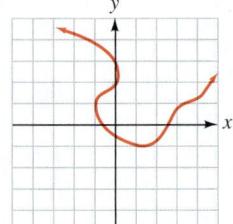

105. U.S. VEHICLE SALES On the graph in the illustration, draw a line through the points (0, 21.2) and (20, 48.6). Write an equation of the line. Express your result using function notation. Then use the function to predict the market share for the year 2004 if the trend continues.

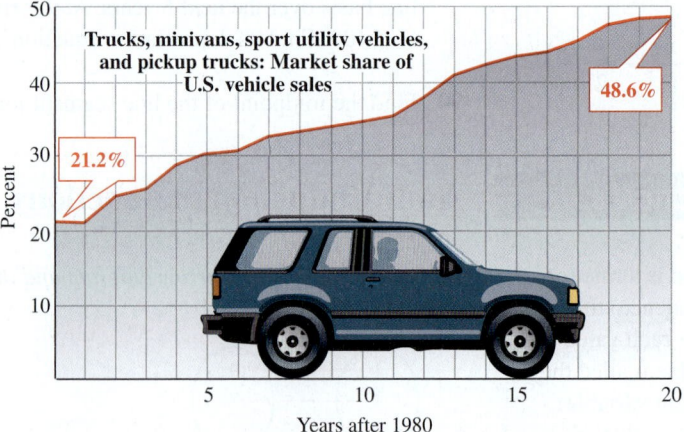

Source: American Automotive Association and U.S. Bureau of Economic Analysis

106. Graph: $f(x) = \dfrac{2}{3}x - 2$.

Graphs of Functions

Graphs of *nonlinear functions* are not lines.

The *squaring function:*

$$f(x) = x^2$$

The *cubing function:*

$$f(x) = x^3$$

The *absolute value function:*

$$f(x) = |x|$$

A *horizontal translation* shifts a graph left or right. A *vertical translation* shifts a graph upward or downward. A *reflection* "flips" a graph on the *x*-axis.

The domain of a function is the *projection* of its graph onto the *x*-axis. The range of a function is the *projection* of its graph onto the *y*-axis.

107. Use the graph in the illustration to find each value.

   a. $f(-2)$ **b.** $f(3)$

   c. The value(s) for which $f(x) = -1$

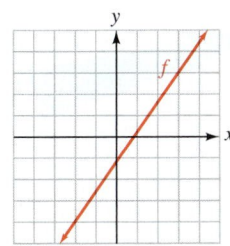

Graph each function.

109. $f(x) = x^2 - 3$

110. $f(x) = (x - 2)^3 + 1$

Give the domain and range of each function.

111.

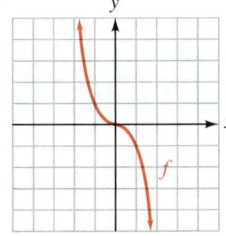

112.

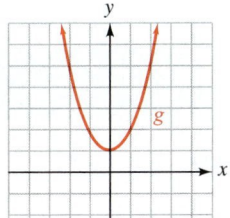

108. Graph $f(x) = |x + 2|$, $g(x) = |x| - 3$, and $h(x) = -|x|$ on the same coordinate system.

Variation

Direct variation: As one variable gets larger, the other gets larger as described by the equation $y = kx$, where k is the *constant of proportionality.*

113. PROPERTY TAX   The property tax in a certain county varies directly as assessed valuation. If a tax of $1,575 is levied on a single-family home assessed at $90,000, determine the property tax on an apartment complex assessed at $312,000.

114. ELECTRICITY   For a fixed voltage, the current in an electrical circuit varies inversely as the resistance in the circuit. If a certain circuit has a current of $2\frac{1}{2}$ amps when the resistance is 150 ohms, find the current in the circuit when the resistance is doubled.

115. Assume that y varies jointly with x and z. Find the constant of variation if $x = 24$ when $y = 3$ and $z = 4$.

Inverse variation: As one variable gets larger, the other gets smaller as described by the equation

$$y = \frac{k}{x} \quad (k \text{ is a constant})$$

Joint variation: One variable varies with the product of several variables. For example, $y = kxz$ (k is a constant).

Combined variation: a combination of direct and inverse variation. For example,

$$y = \frac{kx}{z} \quad (k \text{ is a constant})$$

116. HURRICANE WINDS The wind force on a vertical surface varies jointly as the area of the surface and the square of the wind's velocity. If a 10-mph wind exerts a force of 1.98 pounds on the sign shown below, find the force on the sign when the wind is blowing at 80 mph.

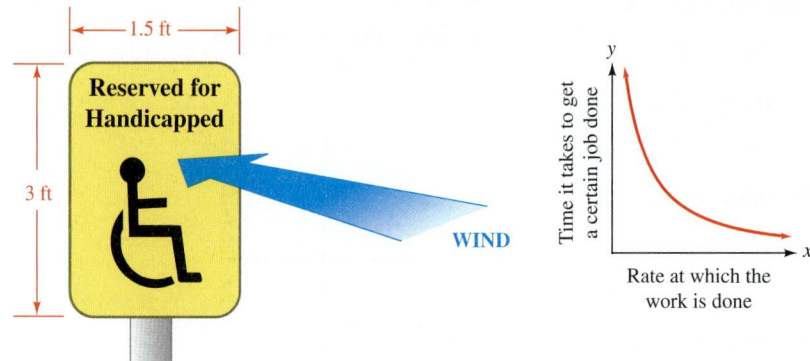

117. Does the graph given above show direct or inverse variation?

118. Assume that x_1 varies directly with the third power of t and inversely with x_2. Find the constant of variation if $x_1 = 1.6$ when $t = 8$ and $x_2 = 64$.

CHAPTER 8 TEST

ELEMENTARY & INTERMEDIATE
Algebra $f(x)$ **Now** ™

Solve each equation.

1. $t + 18 = 5t - 3 + t$

2. $\frac{2}{3}(2s + 2) = \frac{1}{6}(5s + 29) - 4$

3. $6 - (x - 3) - 5x = 3[1 - 2(x + 2)]$

4. Solve $n = \dfrac{360}{180 - a}$ for a.

5. CALCULATORS The viewing window of a calculator has a perimeter of 26 centimeters and is 5 centimeters longer than it is wide. Find the dimensions of the window.

6. AVERAGING GRADES Use the information from the gradebook to determine what score Karen Nelson-Sims needs on the fifth exam so that her exam average exceeds 80.

Sociology 101 8:00-10:00 pm MW	Exam 1	Exam 2	Exam 3	Exam 4	Exam 5
Nelson-Sims, Karen	70	79	85	88	

Solve each inequality. Write the solution set in interval notation and graph it.

7. $-2(2x + 3) \geq 14$

8. $-2 < \dfrac{x - 4}{3} < 4$

9. $3x \geq -2x + 5$ and $7 \geq 4x - 2$

10. $3x < -9$ or $-\dfrac{x}{4} < -2$

11. $|2x - 4| > 22$

12. $2|3(x - 2)| \leq 4$

Solve each equation.

13. $|2x + 3| - 19 = 0$ **14.** $|3x + 4| = |x + 12|$

Factor, if possible.

15. $12a^3b^2c - 3a^2b^2c^2 + 6abc^3$

16. $4y^4 - 64$

17. $b^3 + 125$ **18.** $6u^2 + 9u - 6$

19. $ax - xy + ay - y^2$ **20.** $25m^8 - 60m^4n + 36n^2$

21. $144b^2 + 25$ **22.** $x^2 + 6x + 9 - y^2$

23. $64a^3 - 125b^6$

24. $(x - y)^2 + 3(x - y) - 10$

Simplify each rational expression.

25. $\dfrac{3y - 6z}{2z - y}$ **26.** $\dfrac{2x^2 + 7xy + 3y^2}{4xy + 12y^2}$

Perform the operations.

27. $\dfrac{x^3 + y^3}{4} \div \dfrac{x^2 - xy + y^2}{2x + 2y}$

28. $\dfrac{xu + 2u + 3x + 6}{u^2 - 9} \cdot \dfrac{13u - 39}{x^2 + 3x + 2}$

29. $\dfrac{-3t + 4}{t^2 + t - 20} + \dfrac{6 + 5t}{t^2 + t - 20}$

30. $\dfrac{a + 3}{a^2 - a - 2} - \dfrac{a - 4}{a^2 - 2a - 3}$

Simplify each complex fraction.

31. $\dfrac{\dfrac{2u^2w^3}{v^2}}{\dfrac{4uw^4}{uv}}$ **32.** $\dfrac{\dfrac{4}{3k} + \dfrac{k}{k + 1}}{\dfrac{k}{k + 1} - \dfrac{3}{k}}$

33. Find an equation of the line that passes through $(-2, 1)$ and is parallel to the graph of $y = -\dfrac{3}{2}x - 7$. Write the equation in general form.

34. Find the slope and the y-intercept of the graph of $2x + 9 = -6y$.

35. Find the slope of the line passing through $(-3, 5)$ and $(4, -6)$.

36. ACCOUNTING   After purchasing a new color copier, a business owner had his accountant prepare a depreciation worksheet for tax purposes. (See the illustration.)

   a. Assuming straight-line depreciation, write an equation that gives the value v of the copier after x years of use.

   b. If the depreciation equation is graphed, explain the significance of its v-intercept.

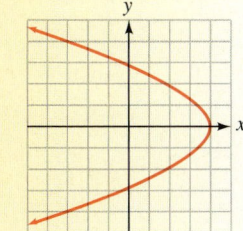

Depreciation Worksheet

Color copier $\cdots\cdots$ $4,000
(new)

Salvage value $\cdots\cdots$ $400
(in 6 years)

37. Find the x- and y-intercepts of the graph of $2x - 5y = 10$. Then graph the equation.

38. Graph: $y = -2$.

39. Does the table define y as a function of x?

x	y
-3	4
4	-3
1	4
2	5

40. Determine whether the graph represents a function.

41. Let $g(x) = x^2 - 2x - 1$. Find $g(0)$.

42. Let $f(x) = -\frac{4}{5}x - 12$. For what value of x is $f(x) = 4$?

43. Graph:
$g(x) = -|x + 2|$.

44. Give the domain and range of function f graphed below.

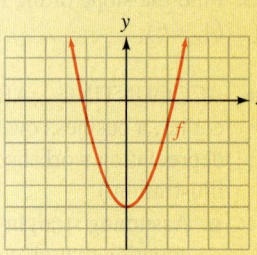

Refer to the graph of function f.

45. Find $f(-2)$.

46. Find the value of x for which $f(x) = 3$.

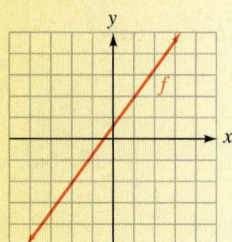

47. Assume that y varies directly with x. If $x = 30$ when $y = 4$, find y when $x = 9$.

48. SOUND   Sound intensity (loudness) varies inversely as the square of the distance from the source. If a rock band has a sound intensity of 100 decibels 30 feet away from the amplifier, find the sound intensity 60 feet away from the amplifier.

CHAPTERS 1–8   CUMULATIVE REVIEW EXERCISES

1. **CANDY SALES**   The circle graph shows how $6,300,000,000 in seasonal candy sales for 2002 was spent. Find the candy sales for Halloween.

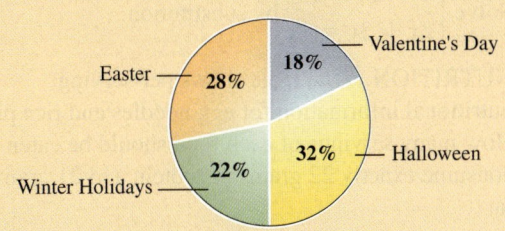

Source: National Confectioners Association

2. List the set of integers.

Evaluate each expression.

3. $3 - 4[-10 - 4(-5)]$

4. $\dfrac{|-45| - 2(-5) + 1^5}{2 \cdot 9 - 2^4}$

5. **AIR CONDITIONING**   Find the volume of air contained in the duct. Round to the nearest tenth of a cubic foot.

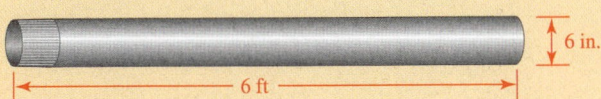

6. The diagram shows the sets that compose the set of real numbers. Which of the indicated sets make up the *rational numbers* and the *irrational numbers?*

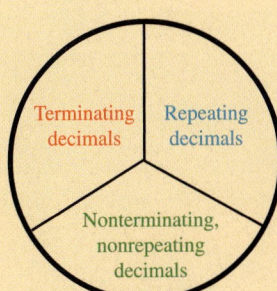

Evaluate each expression when $x = 2$ and $y = -4$.

7. $|x| - xy$

8. $\dfrac{x^2 - y^2}{3x + y}$

Simplify each expression.

9. $3p^2 - 6(5p^2 + p) + p^2$

10. $-(a + 2) - (a - b)$

11. **ANGLE OF ELEVATION**   Find x.

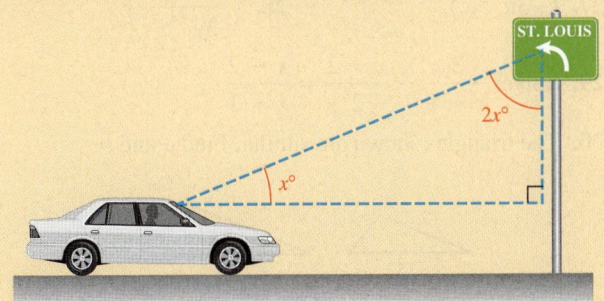

12. **THE STOCK MARKET**   An investment club invested part of $45,000 in a high-yield mutual fund that earned 12% annual simple interest. The remainder of the money was invested in Treasury bonds that earned 6.5% simple annual interest. The two investments earned $4,300 in one year. How much was invested in each account?

13. **PLASTIC WRAP**   Estimate the number of *square feet* of plastic wrap on a roll if the dimensions printed on the box describe the roll as 205 feet long by $11\frac{3}{4}$ inches wide.

14. Give the formula for
   a. the perimeter of a rectangle
   b. the area of a rectangle
   c. the area of a circle
   d. the distance traveled

Perform the operation(s).

15. $(5x - 8y) - (-2x + 5y)$

16. $2x^2(3x^2 + 4x - 7)$

17. $(6p - 5q)^2$

18. $(x + 3)(2x - 3)$

19. $\dfrac{2x - 32}{16x}$

20. $3x + 1 \overline{)9x^2 + 6x + 1}$

21. $(-3x + y)(x^2 - 8xy + 16y^2)$

22. $(x + y)(x - y) + x(x + y)$

Simplify each expression. Write each answer without using negative exponents.

23. $(x^5)^2(x^7)^3$

24. $\dfrac{16(aa^2)^3}{2a^2a^3}$

25. $\dfrac{2^{-4}}{3^{-1}}$

26. $(2x)^0$

27. $(-3u^{-2}v^3)^{-3}$

28. $\left(\dfrac{12y^3z^{-2}}{3y^{-4}z^3}\right)^2$

29. Solve: $\dfrac{3}{x+1} - \dfrac{x-2}{2} = \dfrac{x-2}{x+1}$

30. The triangles shown are similar. Find a and b.

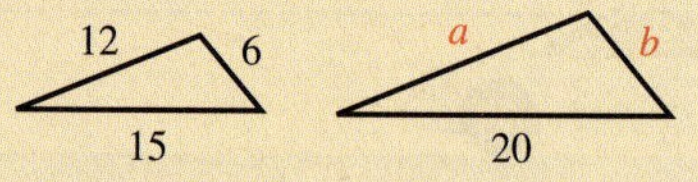

Solve each equation.

31. $8s^2 - 16s = 0$

32. $x^2 + 2x - 15 = 0$

Solve each system by graphing.

33. $\begin{cases} x + 4y = -2 \\ y = -x - 5 \end{cases}$

34. $\begin{cases} 2x - 3y < 0 \\ y > x - 1 \end{cases}$

35. Solve $\begin{cases} x - 2y = 2 \\ 2x + 3y = 11 \end{cases}$ by substitution.

36. NUTRITION   The table shows per-serving nutritional information for egg noodles and rice pilaf. How many servings of each food should be eaten to consume exactly 22 grams of protein and 21 grams of fat?

	Protein (g)	Fat (g)
Egg noodles	5	3
Rice pilaf	4	5

Radical Expressions and Equations

CORBIS

When investigating an automobile accident, police and insurance professionals use clues from the scene to reconstruct the events that led to the collision. They estimate the speed of a vehicle prior to braking using a formula that involves the length of the skid marks and the condition of the road surface. This formula, $s = \sqrt{30Df}$, contains a square root. Square roots are more formally known as *radical expressions*.

To learn more about radical expressions and radical equations, visit *The Learning Equation* on the Internet at http://tle.brookscole.com. (The log-in instructions are in the Preface.) For Chapter 9, the online lessons are:

• *TLE* Lesson 16: Simplifying Radical Expressions
• *TLE* Lesson 17: Radical Equations

Radical expressions have the form $\sqrt[n]{a}$. In this chapter, we will see how they are used to model many real-world situations.

9.1 Radical Expressions and Radical Functions

- Square roots • Square roots of expressions containing variables
- The square root function • Cube roots • The cube root function • *n*th roots

In this section, we will reverse the squaring process and learn how to find *square roots* of numbers. Then we will generalize the concept of root and consider cube roots, fourth roots, and so on. We will also discuss a new family of functions, called *radical functions*.

■ SQUARE ROOTS

When we raise a number to the second power, we are squaring it, or finding its **square.**

- The square of 5 is 25 because $5^2 = 25$.
- The square of -5 is 25, because $(-5)^2 = 25$.

We can reverse the squaring process to find **square roots** of numbers. For example, to find the square roots of 25, we ask ourselves "What number, when squared, is equal to 25?" There are two possible answers.

- 5 is a square root of 25, because $5^2 = 25$.
- -5 is a square root of 25, because $(-5)^2 = 25$.

In general, we have the following definition.

Square Root of *a*   The number b is a **square root** of the number a if $b^2 = a$.

Every positive number has two square roots, one positive and one negative. For example, the two square roots of 9 are 3 and -3, and the two square roots of 144 are 12 and -12. The number 0 is the only real number with exactly one square root. In fact, it is its own square root, because $0^2 = 0$.

A **radical symbol** $\sqrt{}$ represents the **positive** or **principal square root** of a number. Since 3 is the positive square root of 9, we can write

The Language of Algebra

We can read $\sqrt{9}$ as "the square root of 9" or as "radical 9."

$$\sqrt{9} = 3$$

The symbol $-\sqrt{}$ represents the **negative square root** of a number. It is the opposite of the principal square root. Since -12 is the negative square root of 144, we can write

$$-\sqrt{144} = -12 \quad \text{Read as "the negative square root of 144 is } -12\text{" or "the opposite of the square root of 144 is } -12\text{."}$$

Square Root Notation   If a is a positive real number,

1. $\sqrt{a}$ represents the **positive** or **principal square root** of a. It is the positive number we square to get a.
2. $-\sqrt{a}$ represents the **negative square root** of a. It is the opposite of the principal square root of a: $-\sqrt{a} = -1 \cdot \sqrt{a}$.
3. The principal square root of 0 is 0: $\sqrt{0} = 0$.

The number or variable expression under a radical symbol is called the **radicand,** and the radical symbol and radicand are called a **radical.** An algebraic expression containing a radical is called a **radical expression.**

Radical symbol

$$\sqrt{81} \leftarrow \text{Radicand}$$

Radical

EXAMPLE 1

Find each square root: **a.** $\sqrt{81}$,   **b.** $-\sqrt{225}$,   **c.** $\sqrt{\dfrac{49}{4}}$,   and   **d.** $\sqrt{0.36}$.

Solution **a.** $\sqrt{81} = 9$ Because $9^2 = 81$.   **b.** $-\sqrt{225} = -15$ Because $(15)^2 = 225$.

c. $\sqrt{\dfrac{49}{4}} = \dfrac{7}{2}$ Because $\left(\dfrac{7}{2}\right)^2 = \dfrac{49}{4}$.   **d.** $\sqrt{0.36} = 0.6$ Because $(0.6)^2 = 0.36$.

Self Check 1 Find each square root: **a.** $\sqrt{64}$,   **b.** $-\sqrt{1}$,   **c.** $\sqrt{\dfrac{1}{16}}$,   and   **d.** $\sqrt{0.09}$.

A table of square roots

n	$\sqrt{n}$
1	1.000
2	1.414
3	1.732
4	2.000
5	**2.236**
6	2.449
7	2.646
8	2.828
9	3.000
10	3.162

A number such as 81, 225, $\frac{1}{4}$, and 0.36, that is the square of some rational number, is called a **perfect square.** In Example 1, we saw that the square root of a perfect square is a rational number.

If a positive number is not a perfect square, its square root is irrational. For example, $\sqrt{5}$ is an irrational number because 5 is not a perfect square. Since $\sqrt{5}$ is irrational, its decimal representation is nonterminating and nonrepeating. We can find an approximate value of $\sqrt{5}$ using the square root key $\sqrt{}$ on a calculator or from the table of square roots found in Appendix II.

$$\sqrt{5} \approx 2.236067978$$

Caution Square roots of negative numbers are not real numbers. For example, $\sqrt{-9}$ is not a real number, because no real number squared equals -9. Square roots of negative numbers come from a set called the **imaginary numbers,** which we will discuss later in this chapter. If we attempt to evaluate $\sqrt{-9}$ using a calculator, we will get an error message.

Error

Scientific calculator

ERR:NONREAL ANS
1▪Quit
2:Goto

Graphing calculator

Caution

Although they look similar, these radical expressions have very different meanings.

$-\sqrt{9} = -3$

$\sqrt{-9}$ is not a real number.

■ **SQUARE ROOTS OF EXPRESSIONS CONTAINING VARIABLES**

If $x \neq 0$, the positive number x^2 has x and $-x$ for its two square roots. To denote the positive square root of $\sqrt{x^2}$, we must know whether x is positive or negative.

If x is positive, we can write

$$\sqrt{x^2} = x \qquad \sqrt{x^2} \text{ represents the positive square root of } x^2, \text{ which is } x.$$

If x is negative, then $-x > 0$, and we can write

$$\sqrt{x^2} = -x \qquad \sqrt{x^2} \text{ represents the positive square root of } x^2, \text{ which is } -x.$$

If we don't know whether x is positive or negative, we can use absolute value symbols to guarantee that $\sqrt{x^2}$ is positive.

Definition of $\sqrt{x^2}$	For any real number x, $$\sqrt{x^2} = |x|$$

We use this definition to *simplify* square root radical expressions.

EXAMPLE 2

Simplify: **a.** $\sqrt{16x^2}$, **b.** $\sqrt{x^2 + 2x + 1}$, and **c.** $\sqrt{m^4}$.

ELEMENTARY &
INTERMEDIATE
Algebra *f(x)* **Now**™   Solution

If x can be any real number, we have

a. $\sqrt{16x^2} = \sqrt{(4x)^2}$ Write the radicand $16x^2$ as $(4x)^2$.

$= |4x|$ Because $(|4x|)^2 = 16x^2$. Since x could be negative, absolute value symbols are needed.

$= 4|x|$ Since 4 is a positive constant in the product $4x$, we can write it outside the absolute value symbols.

b. $\sqrt{x^2 + 2x + 1} = \sqrt{(x + 1)^2}$ Factor the radicand: $x^2 + 2x + 1 = (x + 1)^2$.

$= |x + 1|$ Since $x + 1$ can be negative (for example, when $x = -5$, $x + 1$ is -4), absolute value symbols are needed.

c. $\sqrt{m^4} = m^2$ Because $(m^2)^2 = m^4$. Since $m^2 \geq 0$, no absolute value symbols are needed.

Self Check 2   Simplify: **a.** $\sqrt{25a^2}$ and **b.** $\sqrt{16a^4}$.

If we know that x is positive in parts a and b of Example 2, we don't need to use absolute value symbols. For example, if $x > 0$, then

$\sqrt{16x^2} = 4x$ If x is positive, $4x$ is positive.

$\sqrt{x^2 + 2x + 1} = x + 1$ If x is positive, $x + 1$ is positive.

■ THE SQUARE ROOT FUNCTION

Since there is one principal square root for every nonnegative real number x, the equation $f(x) = \sqrt{x}$ determines a function, called a **square root function.** Square root functions belong to a larger family of functions known as **radical functions.**

EXAMPLE 3

Graph $f(x) = \sqrt{x}$ and find its domain and range.

Solution

To graph this square root function, we will evaluate it for several values of x. We begin with $x = 0$, since 0 is the smallest input for which $\sqrt{x}$ is defined.

$f(x) = \sqrt{x}$
$f(0) = \sqrt{0}$ Substitute 0 for x.
$f(0) = 0$

We enter 0 for x and 0 for $f(x)$ in the table. Then we let $x = 1, 4, 9,$ and 16, and list each corresponding function value in the table. After plotting the ordered pairs, we draw a smooth curve through the points to get the graph shown in figure (a). Since the equation defines a function, its graph passes the vertical line test.

We can use a graphing calculator to get the graph shown in figure (b). From either graph, we can see that the domain and the range are the set of nonnegative real numbers. Expressed in interval notation, the domain is $[0, \infty)$, and the range is $[0, \infty)$.

$f(x) = \sqrt{x}$

x	$f(x)$	
0	**0**	→ $(0, 0)$
1	1	→ $(1, 1)$
4	2	→ $(4, 2)$
9	3	→ $(9, 3)$
16	4	→ $(16, 4)$

↑
Select values
of x that are
perfect squares.

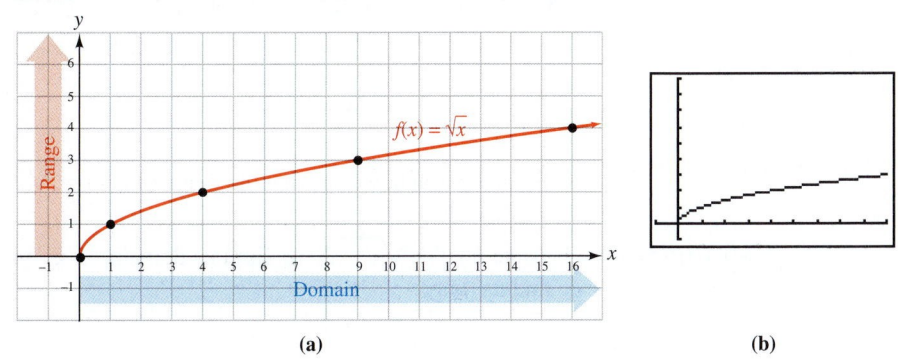

(a)　　　　　　　　　　(b)

Self Check 3　Graph: $g(x) = \sqrt{x} + 2$. Then give its domain and range and compare it to the graph of $f(x) = \sqrt{x}$.

EXAMPLE 4
ELEMENTARY &
INTERMEDIATE
Algebra *f(x)* **Now**™

Consider the function $g(x) = \sqrt{x + 3}$. **a.** Find its domain, **b.** graph the function, and **c.** find its range.

Solution　**a.** We can determine the domain algebraically. Since the expression $\sqrt{x + 3}$ is not a real number when $x + 3$ is negative, we must require that

$$x + 3 \geq 0$$

To solve for x, we subtract 3 from both sides,

$$x \geq -3 \qquad \text{The x-inputs must be real numbers greater than or equal to } -3.$$

Thus, the domain of $g(x) = \sqrt{x + 3}$ is the interval $[-3, \infty)$.

b. To graph the function, we construct a table of function values. We begin by selecting $x = -3$, since -3 is the smallest input for which $\sqrt{x + 3}$ is defined.

$$g(\textbf{\textit{x}}) = \sqrt{\textbf{\textit{x}} + 3}$$
$$g(\textbf{−3}) = \sqrt{\textbf{−3} + 3}$$
$$g(-3) = \sqrt{0} = 0$$

We enter -3 for x and 0 for $g(x)$ in the table. Then we let $x = -2, 1,$ and 6 and list each corresponding function value in the table. After plotting the ordered pairs, we draw a smooth curve through the points to get the graph shown in figure (a) on the next page. In figure (b), we see that the graph of $g(x) = \sqrt{x + 3}$ is the graph of $f(x) = \sqrt{x}$, translated 3 units to the left.

$g(x) = \sqrt{x + 3}$

x	$g(x)$
−3	**0**
−2	1
1	2
6	3

$\rightarrow$ (−3, 0)
$\rightarrow$ (−2, 1)
$\rightarrow$ (1, 2)
$\rightarrow$ (6, 3)

↑
Select values of
x that make $x + 3$
a perfect square.

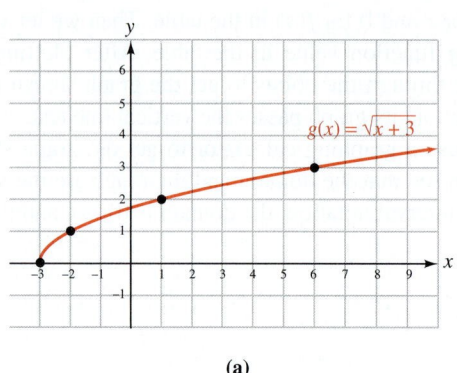

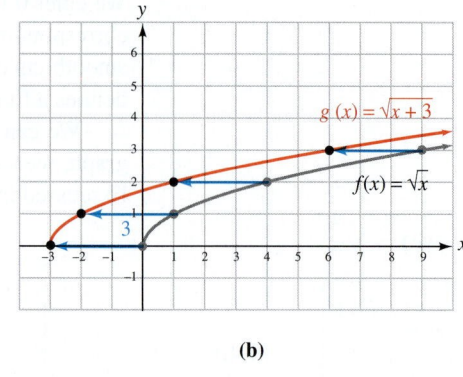

(a)

(b)

c. From the graph, we see that the range of $g(x) = \sqrt{x + 3}$ is $[0, \infty)$.

Self Check 4   Consider $h(x) = \sqrt{x - 2}$.   **a.** Find its domain,   **b.** graph the function,   and   **c.** find its range.

EXAMPLE 5

ELEMENTARY &
INTERMEDIATE
Algebra *f(x)* **Now**™

Pendulums.   The **period of a pendulum** is the time required for the pendulum to swing back and forth to complete one cycle. The period (in seconds) is a function of the pendulum's length L (in feet) and is given by

$$f(L) = 2\pi \sqrt{\frac{L}{32}}$$

Find the period of the 5-foot-long pendulum of a clock.

Solution   To determine the period, we substitute 5 for L.

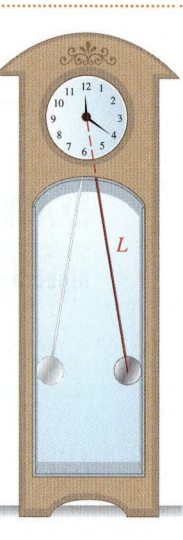

Notation

$2\pi \sqrt{\dfrac{5}{32}}$ means $2 \cdot \pi \cdot \sqrt{\dfrac{5}{32}}$.

$$f(\mathbf{L}) = 2\pi \sqrt{\frac{\mathbf{L}}{32}}$$

$$f(\mathbf{5}) = 2\pi \sqrt{\frac{\mathbf{5}}{32}}$$

≈ 2.483647066 Use a calculator to find an approximation.

The period is approximately 2.5 seconds.

Self Check 5   To the nearest hundredth, find the period of a pendulum that is 3 feet long.

ACCENT ON TECHNOLOGY: EVALUATING A SQUARE ROOT FUNCTION

To solve Example 5 with a graphing calculator, we graph the function $f(x) = 2\pi \sqrt{\frac{x}{32}}$, as in figure (a). We then trace and move the cursor toward an x-value of 5 until we see the coordinates shown in figure (b). The pendulum's period is given by the y-value shown on the screen. By zooming in, we can get better results.

After entering $Y_1 = 2\pi\sqrt{\dfrac{x}{32}}$, we can also use the TABLE mode to find $f(5)$. See figure (c).

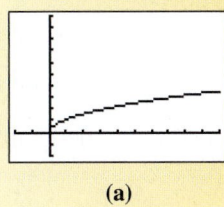

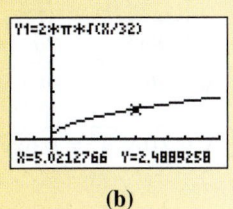

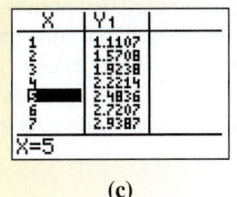

(a)   (b)   (c)

■ CUBE ROOTS

When we raise a number to the third power, we are cubing it, or finding its **cube.** We can reverse the cubing process to find **cube roots** of numbers. To find the cube root of 8, we ask "What number, when cubed, is equal to 8?" It follows that 2 is a cube root of 8, because $2^3 = 8$.

In general, we have this definition.

Cube Root of a	The number b is a **cube root** of the number a if $b^3 = a$.

All real numbers have one real cube root. A positive number has a positive cube root, a negative number has a negative cube root, and the cube root of 0 is 0.

Cube Root Notation	The **cube root of a** is denoted by $\sqrt[3]{a}$. By definition, $$\sqrt[3]{a} = b \quad \text{if} \quad b^3 = a$$

Earlier, we determined that the cube root of 8 is 2. In symbols, we can write: $\sqrt[3]{8} = 2$. The number 3 is called the **index.**

Notation

For the square root symbol $\sqrt{}$, the unwritten index is understood to be 2.

$$\sqrt{a} = \sqrt[2]{a}$$

Index
$$\overset{\text{Index}}{\searrow}\atop \sqrt[3]{8}$$

A number such as 125, $\frac{1}{64}$, -27, and -8, that is the cube of some rational number, is called a **perfect cube.** To simplify cube root radical expressions, we look for perfect cubes and apply the following definition.

Definition of $\sqrt[3]{x^3}$	For any real number x, $$\sqrt[3]{x^3} = x$$

EXAMPLE 6

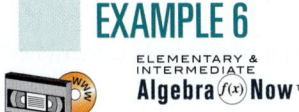

ELEMENTARY & INTERMEDIATE
Algebra *f(x)* Now ™

Simplify: **a.** $\sqrt[3]{125}$,   **b.** $\sqrt[3]{\dfrac{1}{64}}$,   **c.** $\sqrt[3]{-27x^3}$,   **d.** $\sqrt[3]{-\dfrac{8a^3}{b^3}}$,   and   **e.** $\sqrt[3]{0.216x^3y^6}$.

Solution   **a.** $\sqrt[3]{125} = 5$ Because $5^3 = 5 \cdot 5 \cdot 5 = 125$.

b. $\sqrt[3]{\dfrac{1}{64}} = \dfrac{1}{4}$ Because $\left(\dfrac{1}{4}\right)^3 = \dfrac{1}{4} \cdot \dfrac{1}{4} \cdot \dfrac{1}{4} = \dfrac{1}{64}$.

c. $\sqrt[3]{-27x^3} = -3x$ Because $(-3x)^3 = (-3x)(-3x)(-3x) = -27x^3$.

d. $\sqrt[3]{-\dfrac{8a^3}{b^3}} = -\dfrac{2a}{b}$ Because $\left(-\dfrac{2a}{b}\right)^3 = \left(-\dfrac{2a}{b}\right)\left(-\dfrac{2a}{b}\right)\left(-\dfrac{2a}{b}\right) = -\dfrac{8a^3}{b^3}$.

e. $\sqrt[3]{0.216x^3y^6} = 0.6xy^2$ Because $(0.6xy^2)^3 = (0.6xy^2)(0.6xy^2)(0.6xy^2) = 0.216x^3y^6$.

Success Tip

Since every real number has exactly one real cube root, it is unnecessary to use absolute value symbols when simplifying cube roots.

Self Check 6   Simplify: **a.** $\sqrt[3]{1,000}$,   **b.** $\sqrt[3]{\dfrac{1}{27}}$,   and   **c.** $\sqrt[3]{125a^3}$.

■ THE CUBE ROOT FUNCTION

Since there is one cube root for every real number x, the equation $f(x) = \sqrt[3]{x}$ defines a function, called the **cube root function.** Like square root functions, cube root functions belong to the family of radical functions.

EXAMPLE 7

ELEMENTARY & INTERMEDIATE
Algebra *f(x)* **Now**™

Consider $f(x) = \sqrt[3]{x}$. **a.** Graph the function,   **b.** find its domain and range,   and
c. graph $g(x) = \sqrt[3]{x} - 2$.

Solution   **a.** To graph this cube root function, we will evaluate it for several values of x. We begin
with $x = -8$.

$$f(x) = \sqrt[3]{x}$$
$$f(-8) = \sqrt[3]{-8}$$ Substitute -8 for x.
$$f(-8) = -2$$

We enter -8 for x and -2 for $f(x)$ in the table. Then we let $x = -1, 0, 1,$ and 8, and list each corresponding function value in the table. After plotting the ordered pairs, we draw a smooth curve through the points to get the graph shown in figure (a).

$f(x) = \sqrt[3]{x}$

x	$f(x)$	
-8	-2	→ $(-8, -2)$
-1	-1	→ $(-1, -1)$
0	0	→ $(0, 0)$
1	1	→ $(1, 1)$
8	2	→ $(8, 2)$

↑
Select values
of x that are
perfect cubes.

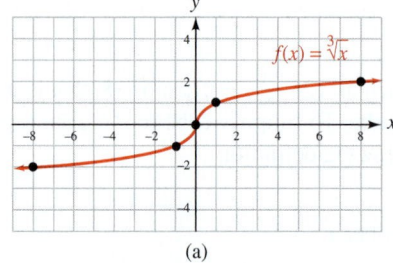

(a)

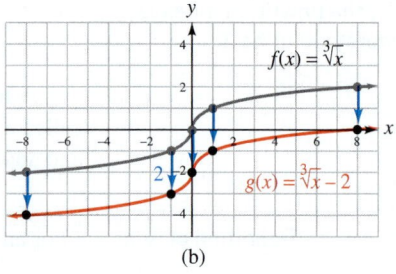

(b)

b. From the graph in figure (a), we see that the domain and the range of function f are the set of real numbers. Thus, the domain is $(-\infty, \infty)$ and the range is $(-\infty, \infty)$.

c. Refer to figure (b). The graph of $g(x) = \sqrt[3]{x} - 2$ is the graph of $f(x) = \sqrt[3]{x}$, translated 2 units downward.

Self Check 7   Consider $f(x) = \sqrt[3]{x} + 1$. **a.** Graph the function and **b.** find its domain and range.

■ nTH ROOTS

Just as there are square roots and cube roots, there are fourth roots, fifth roots, sixth roots, and so on. In general, we have the following definition.

nth Roots of a	The **nth root of a** is denoted by $\sqrt[n]{a}$, and

$$\sqrt[n]{a} = b \quad \text{if} \quad b^n = a$$

The number n is called the **index** (or **order**) of the radical. If n is an even natural number, a must be positive or zero, and b must be positive.

When n is an odd natural number, the expression $\sqrt[n]{x}$ ($n > 1$) represents an **odd root.** Since every real number has just one real nth root when n is odd, we don't need absolute value symbols when finding odd roots. For example,

$$\sqrt[5]{243} = \sqrt[5]{3^5} = 3 \qquad \text{Because } 3^5 = 243.$$

$$\sqrt[7]{-128x^7} = \sqrt[7]{(-2x)^7} = -2x \qquad \text{Because } (-2x)^7 = -128x^7.$$

When n is an even natural number, the expression $\sqrt[n]{x}$, where $n > 1$, $x > 0$, represents an **even root.** In this case, there will be one positive and one negative real nth root. For example, the real sixth roots of 729 are 3 and -3, because $3^6 = 729$ and $(-3)^6 = 729$. When finding even roots, we can use absolute value symbols to guarantee that the nth root is positive.

$$\sqrt[4]{(-3)^4} = |-3| = 3 \qquad \begin{array}{l}\text{We could also simplify this as follows:}\\ \sqrt[4]{(-3)^4} = \sqrt[4]{81} = 3.\end{array}$$

$$\sqrt[6]{729x^6} = \sqrt[6]{(3x)^6} = |3x| = 3|x| \qquad \begin{array}{l}\text{The absolute value symbols guarantee that}\\ \text{the sixth root is positive.}\end{array}$$

In general, we have the following rules.

Rules for $\sqrt[n]{x^n}$	If x is a real number and $n > 1$, then

If n is an odd natural number, $\sqrt[n]{x^n} = x$.

If n is an even natural number, $\sqrt[n]{x^n} = |x|$.

EXAMPLE 8

Simplify each radical expression, if possible: **a.** $\sqrt[4]{625}$, **b.** $\sqrt[4]{-1}$, **c.** $\sqrt[5]{-32}$,

d. $\sqrt[6]{\dfrac{1}{64}}$, and **e.** $\sqrt[7]{10^7}$.

Solution   **a.** $\sqrt[4]{625} = 5$, because $5^4 = 625$.   Read $\sqrt[4]{625}$ as "the fourth root of 625."

b. $\sqrt[4]{-1}$ is not a real number.   This is an even root of a negative number.

c. $\sqrt[5]{-32} = -2$, because $(-2)^5 = -32$.   Read $\sqrt[5]{-32}$ as "the fifth root of -32."

d. $\sqrt[6]{\dfrac{1}{64}} = \dfrac{1}{2}$, because $\left(\dfrac{1}{2}\right)^6 = \dfrac{1}{64}$.   Read $\sqrt[6]{\dfrac{1}{64}}$ as "the sixth root of $\dfrac{1}{64}$."

e. $\sqrt[7]{10^7} = 10$, because $10^7 = 10^7$.   Read $\sqrt[7]{10^7}$ as "the seventh root of 10^7."

Caution

When n is even ($n > 1$) and $x < 0$, $\sqrt[n]{x}$ is not a real number. For example, $\sqrt[4]{-81}$ is not a real number, because no real number raised to the fourth power is -81.

Self Check 8   Simplify: **a.** $\sqrt[4]{\dfrac{1}{81}}$, **b.** $\sqrt[5]{10^5}$, and **c.** $\sqrt[6]{-64}$.

ACCENT ON TECHNOLOGY: FINDING ROOTS

The square root key $\boxed{\sqrt{}}$ on a scientific calculator can be used to evaluate square roots. To evaluate roots with an index greater than 2, we can use the root key $\boxed{\sqrt[x]{y}}$. For example, the function

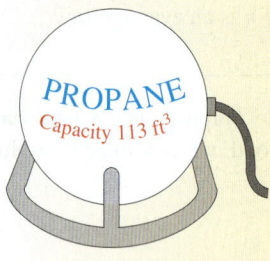

PROPANE
Capacity 113 ft³

$$r(V) = \sqrt[3]{\frac{3V}{4\pi}}$$

gives the radius of a sphere with volume V. To find the radius of the spherical propane tank shown on the left, we substitute 113 for V to get

$$r(\textcolor{red}{V}) = \sqrt[3]{\frac{3\textcolor{red}{V}}{4\pi}}$$

$$r(\textcolor{red}{113}) = \sqrt[3]{\frac{3(\textcolor{red}{113})}{4\pi}}$$

To evaluate a root, we enter the radicand and press the root key $\boxed{\sqrt[x]{y}}$ followed by the index of the radical, which in this case is 3.

$$3\boxed{\times}113\boxed{\div}\boxed{(}\boxed{4}\boxed{\times}\boxed{\pi}\boxed{)}\boxed{=}\boxed{2\text{nd}}\boxed{\sqrt[x]{y}}3\boxed{=}$$

$$\boxed{2.999139118}$$

To evaluate the cube root of $\frac{3(113)}{4\pi}$ with a graphing calculator, we enter

$$\boxed{\text{MATH}}4\boxed{(}3\boxed{\times}113\boxed{)}\boxed{\div}\boxed{(}4\boxed{\times}\boxed{2\text{nd}}\boxed{\pi}\boxed{)}\boxed{)}\boxed{\text{ENTER}}$$

$$\sqrt[3]{}((3*113)/(4*\pi)$$
$$)$$
$$2.999139118$$

The radius of the propane tank is about 3 feet.

EXAMPLE 9

ELEMENTARY &
INTERMEDIATE
Algebra *f(x)* **Now**™

Simplify each radical expression. Assume that x can be any real number. **a.** $\sqrt[5]{x^5}$, **b.** $\sqrt[4]{16x^4}$,   **c.** $\sqrt[6]{(x+4)^6}$,   **d.** $\sqrt[3]{(x+1)^3}$,   and **e.** $\sqrt{9x^4}$.

Solution

a. $\sqrt[5]{x^5} = x$ Since n is odd, absolute value symbols aren't needed.

b. $\sqrt[4]{16x^4} = |2x| = 2|x|$ Since n is even and x can be negative, absolute value symbols are needed to guarantee that the result is positive.

c. $\sqrt[6]{(x+4)^6} = |x+4|$ Absolute value symbols are needed to guarantee that the result is positive.

d. $\sqrt[3]{(x+1)^3} = x+1$ Since n is odd, absolute value symbols aren't needed.

e. $\sqrt{9x^4} = 3x^2$ Since x^2 is always nonnegative, we don't need absolute value symbols.

The Language of Algebra

Another way to say that x can be any real number is to say that the variable is *unrestricted*.

Self Check 9   Simplify: **a.** $\sqrt[6]{x^6}$,   **b.** $\sqrt[5]{(a+5)^5}$,   **c.** $\sqrt{(x^2+4x+4)^2}$,   and **d.** $\sqrt[4]{16a^8}$.

If we know that x is positive in parts b and c of Example 9, we don't need to use absolute value symbols. For example, if $x > 0$, then

$$\sqrt[4]{16x^4} = 2x \qquad \text{If } x \text{ is positive, } 2x \text{ is positive.}$$

$$\sqrt[6]{(x + 4)^6} = x + 4 \qquad \text{If } x \text{ is positive, } x + 4 \text{ is positive.}$$

We summarize the definitions concerning $\sqrt[n]{x}$ as follows.

Summary of the Definitions of $\sqrt[n]{x}$	If n is a natural number greater than 1 and x is a real number, then If $x > 0$, then $\sqrt[n]{x}$ is the positive number such that $\left(\sqrt[n]{x}\right)^n = x$. If $x = 0$, then $\sqrt[n]{x} = 0$. If $x < 0$ $\begin{cases} \text{and } n \text{ is odd, then } \sqrt[n]{x} \text{ is the negative number such that } \left(\sqrt[n]{x}\right)^n = x. \\ \text{and } n \text{ is even, then } \sqrt[n]{x} \text{ is not a real number.} \end{cases}$

Answers to Self Checks   **1. a.** 8,   **b.** -1,   **c.** $\dfrac{1}{4}$,   **d.** 0.3   **2. a.** $5|a|$,   **b.** $4a^2$

3. 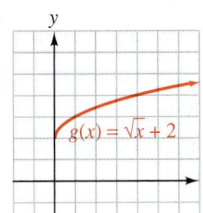   D: $[0, \infty)$,   R: $[2, \infty)$;   the graph is 2 units higher   **4. a.** $[2, \infty)$ **b.** 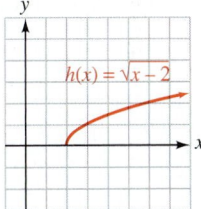   **c.** $[0, \infty)$

5. 1.92 sec   **6. a.** 10,   **b.** $\dfrac{1}{3}$,   **c.** $5a$

7. a.   **8. a.** $\dfrac{1}{3}$,   **b.** 10,   **c.** not a real number

9. a. $|x|$,   **b.** $a + 5$,   **c.** $(x + 2)^2$,   **d.** $2a^2$

b. D: $(-\infty, \infty)$; R: $(-\infty, \infty)$

9.1   STUDY SET   ELEMENTARY & INTERMEDIATE Algebra f(x) Now™

VOCABULARY   Fill in the blanks.

1. $5x^2$ is the _____ root of $25x^4$, because $(5x^2)^2 = 25x^4$. The _____ root of 216 is 6 because $6^3 = 216$.

2. The symbol $\sqrt{}$ is called a _____ symbol.

3. In the expression $\sqrt[3]{27x^6}$, the _____ is 3 and $27x^6$ is the _____.

4. When we write $\sqrt{b^4} = b^2$, we say that we have _____ the radical expression.

5. When n is an odd number, $\sqrt[n]{x}$ represents an _____ root. When n is an _____ number, $\sqrt[n]{x}$ represents an even root.

6. $f(x) = \sqrt{x}$ and $g(x) = \sqrt[3]{x}$ are _____ functions.

CONCEPTS   Fill in the blanks.

7. b is a square root of a if $b^2 = \boxed{}$.

8. $\sqrt{0} = \boxed{}$ and $\sqrt[3]{0} = \boxed{}$.

9. The number 25 has _____ square roots. The principal square root of 25 is [ ].

10. $\sqrt{-4}$ is not a real number, because no real number _____ equals -4.

11. $\sqrt[3]{x} = y$ if $y^3 =$ [ ].

12. $\sqrt{x^2} =$ [ ] and $\sqrt[3]{x^3} =$ [ ].

13. The graph of $g(x) = \sqrt{x} + 3$ is the graph of $f(x) = \sqrt{x}$ translated [ ] units _____.

14. The graph of $g(x) = \sqrt{x + 5}$ is the graph of $f(x) = \sqrt{x}$ translated [ ] units to the _____.

15. The graph of a square root function f is shown below. Find each of the following.

   a. $f(11)$ **b.** $f(2)$

   c. The value(s) of x for which $f(x) = 2$

   d. The domain and range of f

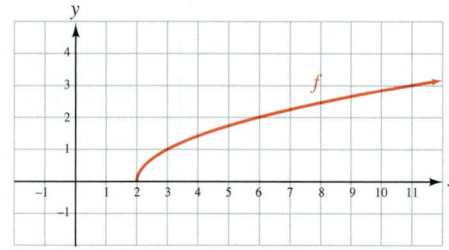

16. The graph of a cube root function f is shown below. Find each of the following.

   a. $f(-8)$ **b.** $f(0)$

   c. The value(s) of x for which $f(x) = -2$

   d. The domain and range of f

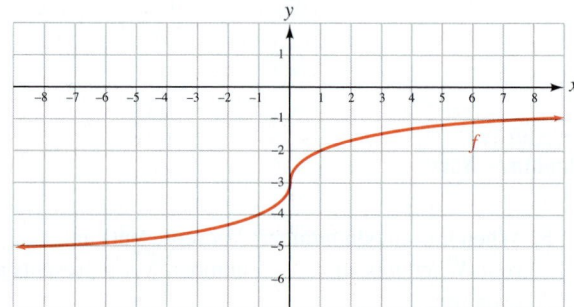

Complete each table of values and graph the function. Then find the domain and range.

17. $f(x) = \sqrt{x}$

x	y
0	
1	
4	
9	
16	

18. $f(x) = \sqrt[3]{x}$

x	y
-8	
-1	
0	
1	
8	

NOTATION   Translate each sentence into mathematical symbols.

19. The square root of x squared is the absolute value of x.

20. The cube root of x cubed is x.

21. f of x equals the square root of the quantity x minus five.

22. The fifth root of negative thirty-two is negative two.

PRACTICE   Find each square root, if possible.

23. $\sqrt{121}$ **24.** $\sqrt{144}$

25. $-\sqrt{64}$ **26.** $-\sqrt{1}$

27. $\sqrt{\dfrac{1}{9}}$ **28.** $-\sqrt{\dfrac{4}{25}}$

29. $\sqrt{0.25}$ **30.** $\sqrt{0.16}$

31. $\sqrt{-25}$ **32.** $-\sqrt{-49}$

33. $\sqrt{(-4)^2}$ **34.** $\sqrt{(-9)^2}$

Use a calculator to find each square root. Give each answer to four decimal places.

35. $\sqrt{12}$ **36.** $\sqrt{340}$

37. $\sqrt{679.25}$ **38.** $\sqrt{0.0063}$

Simplify each expression. Assume that all variables are unrestricted and use absolute value symbols when necessary.

39. $\sqrt{4x^2}$ **40.** $\sqrt{16y^4}$

41. $\sqrt{(t + 5)^2}$ **42.** $\sqrt{(a + 6)^2}$

43. $\sqrt{(-5b)^2}$ **44.** $\sqrt{(-8c)^2}$

45. $\sqrt{a^2 + 6a + 9}$ **46.** $\sqrt{x^2 + 10x + 25}$

Find each value given that $f(x) = \sqrt{x-4}$ and $g(x) = \sqrt[3]{x-4}$.

47. $f(4)$

48. $f(8)$

49. $f(20)$

50. $f(29)$

51. $g(12)$

52. $g(-4)$

53. $g(-996)$

54. $g(1,004)$

Find each value given that $f(x) = \sqrt{x^2+1}$ and $g(x) = \sqrt[3]{x^2+1}$. Give each answer to four decimal places.

55. $f(4)$

56. $f(2.35)$

57. $g(6)$

58. $g(21.57)$

Complete each table and graph the function. Find the domain and range.

59. $f(x) = -\sqrt{x}$

x	y
0	
1	
4	
9	
16	

60. $f(x) = -\sqrt[3]{x}$

x	y
-8	
-1	
0	
1	
8	

Graph each function and find its domain and range.

61. $f(x) = \sqrt{x+4}$

62. $f(x) = \sqrt{x-1}$

63. $f(x) = \sqrt[3]{x} + 3$

64. $f(x) = \sqrt[3]{x-3}$

Simplify each cube root.

65. $\sqrt[3]{1}$

66. $\sqrt[3]{-8}$

67. $\sqrt[3]{-125}$

68. $\sqrt[3]{512}$

69. $\sqrt[3]{-\dfrac{8}{27}}$

70. $\sqrt[3]{\dfrac{125}{216}}$

71. $\sqrt[3]{0.064}$

72. $\sqrt[3]{0.001}$

73. $\sqrt[3]{8a^3}$

74. $\sqrt[3]{-27x^6}$

75. $\sqrt[3]{-1,000p^3q^3}$

76. $\sqrt[3]{343a^6b^3}$

77. $\sqrt[3]{-\dfrac{1}{8}m^6n^3}$

78. $\sqrt[3]{0.008z^9}$

79. $\sqrt[3]{-0.064s^9t^6}$

80. $\sqrt[3]{\dfrac{27}{1,000}a^6b^6}$

Simplify each radical, if possible. Assume that all variables represent positive real numbers.

81. $\sqrt[4]{81}$

82. $\sqrt[6]{64}$

83. $-\sqrt[5]{243}$

84. $-\sqrt[4]{625}$

85. $\sqrt[4]{-256}$

86. $\sqrt[6]{-729}$

87. $\sqrt[4]{\dfrac{16}{625}}$

88. $\sqrt[5]{-\dfrac{243}{32}}$

89. $-\sqrt[5]{-\dfrac{1}{32}}$

90. $-\sqrt[4]{\dfrac{81}{256}}$

91. $\sqrt[5]{32a^5}$

92. $\sqrt[5]{-32x^5}$

93. $\sqrt[4]{16a^4}$

94. $\sqrt[8]{x^{24}}$

95. $\sqrt[4]{k^{12}}$

96. $\sqrt[6]{64b^6}$

97. $\sqrt[4]{\dfrac{1}{16}m^4}$

98. $\sqrt[4]{\dfrac{1}{81}x^8}$

99. $\sqrt[25]{(x+2)^{25}}$

100. $\sqrt[44]{(x+4)^{44}}$

APPLICATIONS Use a calculator to solve each problem. Round answers to the nearest tenth.

101. EMBROIDERY   The radius r of a circle is given by the formula

$$r = \sqrt{\dfrac{A}{\pi}}$$

where A is its area. Find the diameter of the embroidery hoop if there are 38.5 in.^{2} of stretched fabric on which to embroider.

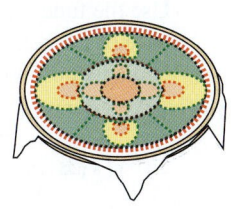

102. BASEBALL   The length of a diagonal of a square is given by the function $d(s) = \sqrt{2s^2}$, where s is the length of a side of the square. Find the distance from home plate to second base on a softball diamond and on a baseball diamond. The illustration gives the dimensions of each type of infield.

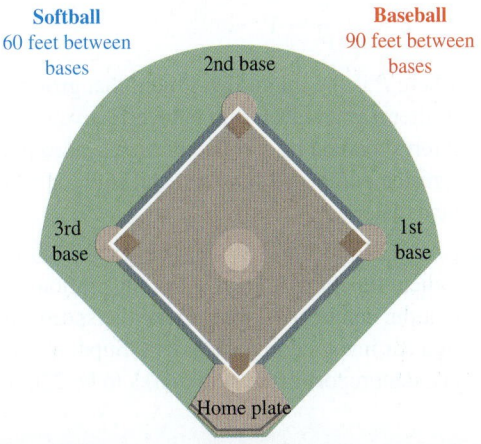

Softball
60 feet between bases

Baseball
90 feet between bases

2nd base

3rd base

1st base

Home plate

103. PULSE RATES The approximate pulse rate (in beats per minute) of an adult who is t inches tall is given by the function

$$p(t) = \frac{590}{\sqrt{t}}$$

The *Guinness Book of World Records 1998* lists Ri Myong-hun of North Korea as the tallest living man, at 7 ft $8\frac{1}{2}$ in. Find his approximate pulse rate as predicted by the function.

104. THE GRAND CANYON The time t (in seconds) that it takes for an object to fall a distance of s feet is given by the formula

$$t = \frac{\sqrt{s}}{4}$$

In some places, the Grand Canyon is one mile (5,280 feet) deep. How long would it take a stone dropped over the edge of the canyon to hit bottom?

105. BIOLOGY Scientists will place five rats inside a controlled environment to study the rats' behavior. The function

$$d(V) = \sqrt[3]{12\left(\frac{V}{\pi}\right)}$$

gives the diameter of a hemisphere with volume V. Use the function to determine the diameter of the base of the hemisphere, if each rat requires 125 cubic feet of living space.

106. AQUARIUMS The function

$$s(g) = \sqrt[3]{\frac{g}{7.5}}$$

determines how long (in feet) an edge of a cube-shaped tank must be if it is to hold g gallons of water. What dimensions should a cube-shaped aquarium have if it is to hold 1,250 gallons of water?

107. COLLECTIBLES The *effective rate of interest r* earned by an investment is given by the formula

$$r = \sqrt[n]{\frac{A}{P}} - 1$$

where P is the initial investment that grows to value A after n years. Determine the effective rate of interest earned by a collector on a Lladró porcelain figurine purchased for $800 and sold for $950 five years later.

108. LAW ENFORCEMENT The graphs of the two radical functions shown in the illustration in the next column can be used to estimate the speed (in mph) of a car involved in an accident. Suppose a police accident report listed skid marks to be 220 feet long but failed to give the road conditions. Estimate the possible speeds the car was traveling prior to the brakes being applied.

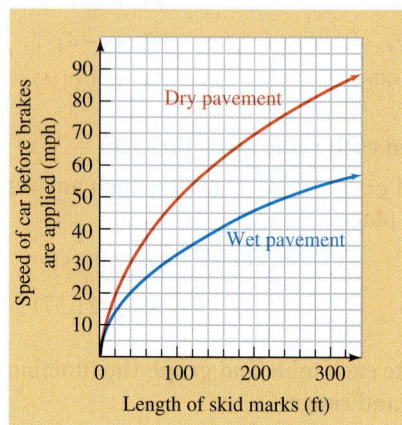

WRITING

109. Explain why 36 has two square roots, but $\sqrt{36}$ is just 6, and not -6.

110. If x is any real number, then $\sqrt{x^2} = x$ is not correct. Explain.

111. Explain what is wrong with the graph in the illustration if it is supposed to be the graph of $f(x) = \sqrt{x}$.

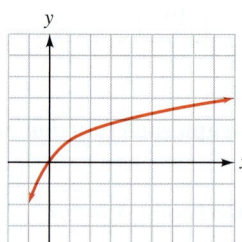

112. Explain how to estimate the domain and range of the radical function shown below.

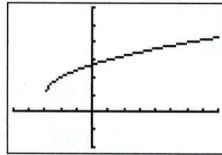

REVIEW Perform the operations.

113. $\dfrac{x^2 - x - 6}{x^2 - 2x - 3} \cdot \dfrac{x^2 - 1}{x^2 + x - 2}$

114. $\dfrac{x^2 - 3xy - 4y^2}{x^2 + cx - 2yx - 2cy} \div \dfrac{x^2 - 2xy - 3y^2}{x^2 + cx - 4yx - 4cy}$

115. $\dfrac{3}{m+1} + \dfrac{3m}{m-1}$

116. $\dfrac{2x+3}{3x-1} - \dfrac{x-4}{2x+1}$

CHALLENGE PROBLEMS

117. Graph: $f(x) = -\sqrt{x-2} + 3$. Find the domain and range.

118. Simplify: $\sqrt{9a^{16} + 12a^8 b^{25} + 4b^{50}}$. Assume that $a > 0$ and $b > 0$.

9.2 Rational Exponents

- Rational exponents
- Exponential expressions with variables in their bases
- Rational exponents with numerators other than 1
- Negative rational exponents
- Applying the rules for exponents
- Simplifying radical expressions

In this section, we will extend the definition of exponent to include rational (fractional) exponents. We will see how expressions such as $9^{1/2}$, $\left(\frac{1}{16}\right)^{3/4}$, and $(-32x^5)^{-2/5}$ can be simplified by writing them in an equivalent radical form using two new rules for exponents.

■ RATIONAL EXPONENTS

The Language of Algebra

Rational exponents are also called *fractional exponents.*

It is possible to raise numbers to fractional powers. To give meaning to rational exponents, we first consider $\sqrt{7}$. Because $\sqrt{7}$ is the positive number whose square is 7, we have

$$\left(\sqrt{7}\right)^2 = 7$$

We now consider the notation $7^{1/2}$. If rational exponents are to follow the same rules as integer exponents, the square of $7^{1/2}$ must be 7, because

$$(7^{1/2})^2 = 7^{1/2 \cdot 2} \qquad \text{Keep the base and multiply the exponents.}$$
$$= 7^1 \qquad \text{Do the multiplication: } \tfrac{1}{2} \cdot 2 = 1$$
$$= 7$$

Since the square of $7^{1/2}$ and the square of $\sqrt{7}$ are both equal to 7, we define $7^{1/2}$ to be $\sqrt{7}$. Similarly,

$$7^{1/3} = \sqrt[3]{7}, \qquad 7^{1/4} = \sqrt[4]{7}, \qquad \text{and} \qquad 7^{1/5} = \sqrt[5]{7}$$

In general, we have the following definition.

The Definition of $x^{1/n}$

A **rational exponent** of $\frac{1}{n}$ indicates the nth root of its base.

If n represents a positive integer greater than 1 and $\sqrt[n]{x}$ represents a real number,

$$x^{1/n} = \sqrt[n]{x}$$

We can use this definition to simplify exponential expressions that have rational exponents with a numerator of 1. For example, to simplify $8^{1/3}$, we write it as an equivalent expression in radical form and proceed as follows:

Index

$$8^{1/3} = \sqrt[3]{8} = 2$$

Radicand

The base of the exponential expression, 8, is the radicand of the radical expression. The denominator of the fractional exponent, 3, is the index of the radical.

Thus, $8^{1/3} = 2$.

EXAMPLE 1

ELEMENTARY &
INTERMEDIATE
Algebra *f(x)* **Now**™

Write each expression in radical form and simplify, if possible: **a.** $9^{1/2}$, **b.** $(-64)^{1/3}$,   **c.** $-\left(\dfrac{1}{32}\right)^{1/5}$,   **d.** $16^{1/4}$,   and   **e.** $(2x^5)^{1/6}$.

Solution

a. $9^{1/2} = \sqrt{9} = 3$

Because the denominator of the exponent is 2, find the square root of the base, 9.

b. $(-64)^{1/3} = \sqrt[3]{-64} = -4$

Because the denominator of the exponent is 3, find the cube root of the base, -64.

c. $-\left(\dfrac{1}{32}\right)^{1/5} = -\sqrt[5]{\dfrac{1}{32}} = -\dfrac{1}{2}$

Because the denominator of the exponent is 5, find the fifth root of the base, $\frac{1}{32}$.

d. $16^{1/4} = \sqrt[4]{16} = 2$

Because the denominator of the exponent is 4, find the fourth root of the base, 16.

e. $(2x^5)^{1/6} = \sqrt[6]{2x^5}$

Because the denominator of the exponent is 6, find the sixth root of the base, $2x^5$. This expression does not simplify further.

Self Check 1

Write each expression in radical form and simplify, if possible: **a.** $16^{1/2}$, **b.** $\left(-\dfrac{27}{8}\right)^{1/3}$,   and   **c.** $-(6x^3)^{1/4}$.

EXAMPLE 2

Write $\sqrt{5xyz}$ as an exponential expression with a rational exponent.

Solution

The radicand is $5xyz$, so the base of the exponential expression is $5xyz$. The index of the radical is an understood 2, so the denominator of the fractional exponent is 2.

$$\sqrt{5xyz} = (5xyz)^{1/2} \qquad \text{Recall: } \sqrt{5xyz} = \sqrt[2]{5xyz}.$$

Self Check 2

Write the radical with a fractional exponent: $\sqrt[6]{7ab}$.

Rational exponents appear in formulas used in many disciplines, such as science and engineering.

EXAMPLE 3

Satellites. The formula

$$r = \left(\frac{GMP^2}{4\pi^2} \right)^{1/3}$$

gives the orbital radius (in meters) of a satellite circling the Earth, where G and M are constants and P is the time in seconds for the satellite to make one complete revolution. Write the formula using a radical.

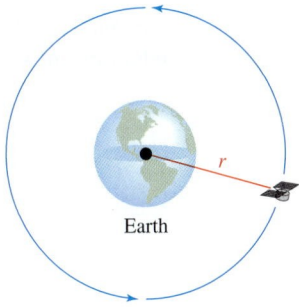

Earth

Solution   The fractional exponent $\frac{1}{3}$ has a denominator of 3, which indicates that we are to find the cube root of the base of the exponential expression. So we have

$$r = \sqrt[3]{\frac{GMP^2}{4\pi^2}}$$

■ EXPONENTIAL EXPRESSIONS WITH VARIABLES IN THEIR BASES

As with radicals, when n is an *odd natural number* in the expression $x^{1/n}$ ($n > 1$), there is exactly one real nth root, and we don't need to use absolute value symbols.

When n is an *even natural number,* there are two nth roots. Since we want the expression $x^{1/n}$ to represent the positive nth root, we must often use absolute value symbols to guarantee that the simplified result is positive. Thus, if n is even,

$$(x^n)^{1/n} = |x|$$

When n is even and x is negative, the expression $x^{1/n}$ is not a real number.

EXAMPLE 4

ELEMENTARY & INTERMEDIATE
Algebra *f(x)* **Now**™

Simplify each expression. Assume that the variables can be any real number.
a. $(-27x^3)^{1/3}$,   **b.** $(256a^8)^{1/8}$,   **c.** $[(y + 4)^2]^{1/2}$,   **d.** $(25b^4)^{1/2}$,   and   **e.** $(-256x^4)^{1/4}$.

Solution   **a.** $(-27x^3)^{1/3} = -3x$ Because $(-3x)^3 = -27x^3$. Since n is odd, no absolute value symbols are needed.

b. $(256a^8)^{1/8} = 2|a|$ Because $\left(2|a|\right)^8 = 256a^8$. Since n is even and a can be any real number, $2a$ can be negative. Thus, absolute value symbols are needed.

c. $[(y + 4)^2]^{1/2} = |y + 4|$ Because $|y + 4|^2 = (y + 4)^2$. Since n is even and y can be any real number, $y + 4$ can be negative. Thus, absolute value symbols are needed.

d. $(25b^4)^{1/2} = 5b^2$ Because $(5b^2)^2 = 25b^4$. Since $b^2 \geq 0$, no absolute value symbols are needed.

e. $(-256x^4)^{1/4}$ is not a real number   Because no real number raised to the 4th power is $-256x^4$.

Self Check 4   Simplify each expression: **a.** $(625a^4)^{1/4}$ and   **b.** $(b^4)^{1/2}$.

If we are told that the variables represent positive real numbers in parts b and c of Example 4, the absolute value symbols in the answers are not needed.

$(256a^8)^{1/8} = 2a$ If a represents a positive number, then $2a$ is positive.
$[(y + 4)^2]^{1/2} = y + 4$ If y represents a positive number, then $y + 4$ is positive.

We summarize the cases as follows.

Summary of the Definitions of $x^{1/n}$

If n is a natural number greater than 1 and x is a real number,

If $x > 0$, then $x^{1/n}$ is the positive number such that $(x^{1/n})^n = x$.

If $x = 0$, then $x^{1/n} = 0$.

If $x < 0$ $\begin{cases} \text{and } n \text{ is odd, then } x^{1/n} \text{ is the negative number such that } (x^{1/n})^n = x. \\ \text{and } n \text{ is even, then } x^{1/n} \text{ is not a real number.} \end{cases}$

■ RATIONAL EXPONENTS WITH NUMERATORS OTHER THAN 1

We can extend the definition of $x^{1/n}$ to include fractional exponents with numerators other than 1. For example, since $8^{2/3}$ can be written as $(8^{1/3})^2$, we have

$8^{2/3} = (\mathbf{8^{1/3}})^2$

$\qquad = \left(\sqrt[3]{8}\right)^2$ Write $8^{1/3}$ in radical form.

$\qquad = 2^2$ Find the cube root first: $\sqrt[3]{8} = 2$.

$\qquad = 4$ Then find the power.

Thus, we can simplify $8^{2/3}$ by finding the second power of the cube root of 8.

The numerator of the rational exponent is the power.

$$\mathbf{8^{2/3}} \;=\; \left(\sqrt[3]{\mathbf{8}}\right)^{\mathbf{2}}$$ The base of the exponential expression is the radicand.

The denominator of the exponent is the index of the radical.

We can also simplify $8^{2/3}$ by taking the cube root of 8 squared.

$8^{2/3} = (\mathbf{8^2})^{1/3}$

$\qquad = \mathbf{64}^{1/3}$ Find the power first: $8^2 = 64$.

$\qquad = \sqrt[3]{64}$ Write $64^{1/3}$ in radical form.

$\qquad = 4$ Now find the cube root.

In general, we have the following definition.

The Definition of $x^{m/n}$

If m and n represent positive integers ($n \neq 1$) and $\sqrt[n]{x}$ represents a real number,

$$x^{m/n} = \left(\sqrt[n]{x}\right)^m \quad \text{and} \quad x^{m/n} = \sqrt[n]{x^m}$$

Because of the previous definition, we can interpret $x^{m/n}$ in two ways:

1. $x^{m/n}$ means the nth root of the mth power of x.

2. $x^{m/n}$ means the mth power of the nth root of x.

We can use this definition to evaluate exponential expressions that have rational exponents with a numerator that is not 1. To avoid large numbers, we usually find the root of the base first and then calculate the power using the relationship $x^{m/n} = \left(\sqrt[n]{x}\right)^m$.

EXAMPLE 5

Evaluate: **a.** $32^{2/5}$, **b.** $81^{3/4}$, and **c.** $-25^{3/2}$.

ELEMENTARY & INTERMEDIATE
Algebra (f(x)) Now™

Solution **a.** To evaluate $32^{2/5}$, we write it in an equivalent radical form. The denominator of the rational exponent is the same as the index of the corresponding radical. The numerator of the rational exponent indicates the power to which the radical base is raised.

Power
Root

$$32^{2/5} = \left(\sqrt[5]{32}\right)^2 = (2)^2 = 4$$

Because the exponent is 2/5, find the fifth root of the base, 32, to get 2. Then find the second power of 2.

Caution

We can also evaluate $x^{m/n}$ using $\sqrt[n]{x^m}$, however the resulting radicand is often extremely large. For example,

$$81^{3/4} = \sqrt[4]{81^3}$$
$$= \sqrt[4]{531,441}$$
$$= 27$$

b. $81^{3/4} = \left(\sqrt[4]{81}\right)^3 = (3)^3 = 27$

Because the exponent is 3/4, find the fourth root of the base, 81, to get 3. Then find the third power of 3.

c. For $-25^{3/2}$, the base is 25, not -25.

$$-25^{3/2} = -(25^{3/2}) = -\left(\sqrt{25}\right)^3 = -(5)^3 = -125$$

Because the exponent is 3/2, find the square root of the base, 25, to get 5. Then find the third power of 5.

Self Check 5 Simplify: **a.** $16^{3/2}$, **b.** $125^{4/3}$, and **c.** $-32^{4/5}$.

EXAMPLE 6

Simplify each expression. Assume that the variables can represent any real number.
a. $(36m^4)^{3/2}$, **b.** $(-8x^3)^{4/3}$, and **c.** $(x^5y^5)^{2/5}$.

Power
Root

Solution **a.** $(36m^4)^{3/2} = \left(\sqrt{36m^4}\right)^3 = (6m^2)^3 = 216m^6$

Because the exponent is $\frac{3}{2}$, find the square root of the base, $36m^4$, to get $6m^2$. Then find the third power of $6m^2$.

b. $(-8x^3)^{4/3} = \left(\sqrt[3]{-8x^3}\right)^4 = (-2x)^4 = 16x^4$

Because the exponent is $\frac{4}{3}$, find the cube root of the base, $-8x^3$, to get $-2x$. Then find the fourth power of $-2x$.

c. $(x^5y^5)^{2/5} = \left(\sqrt[5]{x^5y^5}\right)^2 = (xy)^2 = x^2y^2$

Because the exponent is $\frac{2}{5}$, find the fifth root of the base, x^5y^5, to get xy. Then find the second power of xy.

Self Check 6 Simplify: **a.** $(4c^4)^{3/2}$ and **b.** $(-27m^3n^3)^{2/3}$.

ACCENT ON TECHNOLOGY: RATIONAL EXPONENTS

We can evaluate expressions containing rational exponents using the exponential key $\boxed{y^x}$ or $\boxed{x^y}$ on a scientific calculator. For example, to evaluate $10^{2/3}$, we enter

$$10\; \boxed{y^x}\; \boxed{(}\; \boxed{2}\; \boxed{\div}\; \boxed{3}\; \boxed{)}\; \boxed{=}$$

$\boxed{4.641588834}$

Note that parentheses were used when entering the power. Without them, the calculator would interpret the entry as $10^2 \div 3$.

To evaluate the exponential expression using a graphing calculator, we use the $\boxed{\wedge}$ key, which raises a base to a power. Again, we use parentheses when entering the power.

$10 \boxed{\wedge} \boxed{(} 2 \boxed{\div} 3 \boxed{)} \boxed{\text{ENTER}}$

```
10 ^ (2/3)
          4.641588834
```

To the nearest hundredth, $10^{2/3} \approx 4.64$.

■ NEGATIVE RATIONAL EXPONENTS

To be consistent with the definition of negative-integer exponents, we define $x^{-m/n}$ as follows.

Definition of $x^{-m/n}$	If m and n are positive integers, $\frac{m}{n}$ is in simplified form, and $x^{1/n}$ is a real number, then
	$$x^{-m/n} = \frac{1}{x^{m/n}} \quad \text{and} \quad \frac{1}{x^{-m/n}} = x^{m/n} \quad (x \neq 0)$$

From the definition, we see that another way to write $x^{-m/n}$ is to write its reciprocal and change the sign of the exponent.

EXAMPLE 7

ELEMENTARY & INTERMEDIATE
Algebra *f(x)* Now™

Simplify each expression. Assume that x can represent any nonzero real number.

a. $64^{-1/2}$,   **b.** $(-16)^{-5/4}$,   **c.** $-625^{-3/4}$,   **d.** $(-32x^5)^{-2/5}$,   and   **e.** $\frac{1}{25^{-3/2}}$.

Solution   **a.**

Reciprocal

$$64^{-1/2} = \frac{1}{64^{1/2}} = \frac{1}{\sqrt{64}} = \frac{1}{8}$$

Change sign

Because the exponent is negative, write the reciprocal of $64^{-1/2}$, and change the sign of the exponent.

Caution

A negative exponent does not indicate a negative number. For example,

$$64^{-1/2} = \frac{1}{8}$$

b. $(-16)^{-5/4}$ is not a real number because $(-16)^{5/4}$ is not a real number.

c. In $-625^{-3/4}$, the base is 625.

$$-625^{-3/4} = -\frac{1}{625^{3/4}} = -\frac{1}{} = -\frac{1}{(5)^3} = -\frac{1}{125}$$

d. $(-32x^5)^{-2/5} = \frac{1}{(-32x^5)^{2/5}} = \frac{1}{} = \frac{1}{(-2x)^2} = \frac{1}{4x^2}$

Reciprocal

Caution

A base of 0 raised to a negative power is undefined. For example, $0^{-2} = \frac{1}{0^2}$ is undefined because we cannot divide by 0.

e. $\frac{1}{25^{-3/2}} = 25^{3/2} = \left(\sqrt{25}\right)^3 = (5)^3 = 125$

Change sign

Because the exponent is negative, write the reciprocal of $\frac{1}{25^{-3/2}}$, and change the sign of the exponent.

Self Check 7   Simplify: **a.** $(36)^{-3/2}$ and   **b.** $(-27a^3)^{-2/3}$.

■ APPLYING THE RULES FOR EXPONENTS

We can use the rules for exponents to simplify many expressions with fractional exponents. If all variables represent positive numbers, absolute value symbols are not needed.

EXAMPLE 8

ELEMENTARY &
INTERMEDIATE
Algebra *f(x)* **Now**™

Simplify each expression. Assume that all variables represent positive numbers. Write all answers using positive exponents only.

a. $5^{2/7}5^{3/7}$,   **b.** $(5^{2/7})^3$,   **c.** $(a^{2/3}b^{1/2})^6$,   and   **d.** $\dfrac{a^{8/3}a^{1/3}}{a^2}$.

Solution   **a.** $5^{2/7}5^{3/7} = 5^{2/7+3/7}$ Use the rule $x^m x^n = x^{m+n}$.

$\qquad\qquad = 5^{5/7}$ Add: $\frac{2}{7} + \frac{3}{7} = \frac{5}{7}$.

b. $(5^{2/7})^3 = 5^{(2/7)(3)}$ Use the rule $(x^m)^n = x^{mn}$.

$\qquad\qquad = 5^{6/7}$ Multiply: $\frac{2}{7}(3) = \frac{6}{7}$.

c. $(a^{2/3}b^{1/2})^6 = (a^{2/3})^6(b^{1/2})^6$ Use the rule $(xy)^n = x^n y^n$.

$\qquad\qquad = a^{12/3}b^{6/2}$ Use the rule $(x^m)^n = x^{mn}$ twice.

$\qquad\qquad = a^4 b^3$ Simplify the exponents.

d. $\dfrac{a^{8/3}a^{1/3}}{a^2} = a^{8/3+1/3-2}$ Use the rules $x^m x^n = x^{m+n}$ and $\dfrac{x^m}{x^n} = x^{m-n}$.

$\qquad\qquad = a^{8/3+1/3-6/3}$ $2 = \frac{6}{3}$.

$\qquad\qquad = a^{3/3}$ $\frac{8}{3} + \frac{1}{3} - \frac{6}{3} = \frac{3}{3}$.

$\qquad\qquad = a$ $\frac{3}{3} = 1$.

Self Check 8   Simplify: **a.** $(x^{1/3}y^{3/2})^6$ and   **b.** $\dfrac{x^{5/3}x^{2/3}}{x^{1/3}}$.

EXAMPLE 9

Perform each multiplication and then simplify if possible. Assume all variables represent positive numbers. Write all answers using positive exponents only. **a.** $a^{4/5}(a^{1/5} + a^{3/5})$ and   **b.** $x^{1/2}(x^{-1/2} + x^{1/2})$.

Solution   **a.** $a^{4/5}(a^{1/5} + a^{3/5}) = a^{4/5}a^{1/5} + a^{4/5}a^{3/5}$ Use the distributive property.

$\qquad\qquad = a^{4/5+1/5} + a^{4/5+3/5}$ Use the rule $x^m x^n = x^{m+n}$.

$\qquad\qquad = a^{5/5} + a^{7/5}$ Simplify the exponents.

$\qquad\qquad = a + a^{7/5}$ We cannot add these terms because they are not like terms.

b. $x^{1/2}(x^{-1/2} + x^{1/2}) = x^{1/2}x^{-1/2} + x^{1/2}x^{1/2}$ Use the distributive property.

$\qquad\qquad = x^{1/2+(-1/2)} + x^{1/2+1/2}$ Use the rule $x^m x^n = x^{m+n}$.

$\qquad\qquad = x^0 + x^1$ Simplify each exponent.

$\qquad\qquad = 1 + x$ $x^0 = 1$.

Self Check 9   Simplify: $t^{5/8}(t^{3/8} + t^{-5/8})$.

■ SIMPLIFYING RADICAL EXPRESSIONS

We can simplify many radical expressions by using the following steps.

Using Rational Exponents to Simplify Radicals	**1.** Change the radical expression into an exponential expression. **2.** Simplify the rational exponents. **3.** Change the exponential expression back into a radical.

EXAMPLE 10

ELEMENTARY &
INTERMEDIATE
Algebra $f(x)$ Now™

Simplify: **a.** $\sqrt[4]{3^2}$, **b.** $\sqrt[8]{x^6}$, **c.** $\sqrt[9]{27x^6y^3}$, and **d.** $\sqrt[5]{\sqrt[3]{t}}$.

Solution

a. $\sqrt[4]{3^2} = (3^2)^{1/4}$ Change the radical to an exponential expression.

$= 3^{2/4}$ Use the rule $(x^m)^n = x^{mn}$.

$= 3^{1/2}$ $\frac{2}{4} = \frac{1}{2}$.

$= \sqrt{3}$ Change back to radical form.

b. $\sqrt[8]{x^6} = (x^6)^{1/8}$ Change the radical to an exponential expression.

$= x^{6/8}$ Use the rule $(x^m)^n = x^{mn}$.

$= x^{3/4}$ $\frac{6}{8} = \frac{3}{4}$.

$= (x^3)^{1/4}$ $\frac{3}{4} = 3\left(\frac{1}{4}\right)$.

$= \sqrt[4]{x^3}$ Change back to radical form.

c. $\sqrt[9]{27x^6y^3} = (3^3x^6y^3)^{1/9}$ Write 27 as 3^3 and change the radical to an exponential expression.

$= 3^{3/9}x^{6/9}y^{3/9}$ Raise each factor to the $\frac{1}{9}$ power by multiplying the fractional exponents.

$= 3^{1/3}x^{2/3}y^{1/3}$ Simplify each fractional exponent.

$= (3x^2y)^{1/3}$ Use the rule $(xy)^n = x^ny^n$.

$= \sqrt[3]{3x^2y}$ Change back to radical form.

d. $\sqrt[5]{\sqrt[3]{t}} = \sqrt[5]{t^{1/3}}$ Change the radical $\sqrt[3]{t}$ to exponential notation.

$= (t^{1/3})^{1/5}$ Change the radical $\sqrt[5]{t^{1/3}}$ to exponential notation.

$= t^{1/15}$ Use the rule $(x^m)^n = x^{mn}$. Multiply: $\frac{1}{3} \cdot \frac{1}{5} = \frac{1}{15}$.

$= \sqrt[15]{t}$ Change back to radical form.

Self Check 10 Simplify: **a.** $\sqrt[6]{3^3}$, **b.** $\sqrt[4]{49x^2y^2}$, and **c.** $\sqrt[3]{\sqrt[4]{m}}$.

Answers to Self Checks

1. a. 4, **b.** $-\dfrac{3}{2}$, **c.** $-\sqrt[4]{6x^3}$ **2.** $(7ab)^{1/6}$ **4. a.** $5|a|$, **b.** b^2

5. a. 64, **b.** 625, **c.** -16 **6. a.** $8c^6$, **b.** $9m^2n^2$ **7. a.** $\dfrac{1}{216}$, **b.** $\dfrac{1}{9a^2}$

8. a. x^2y^9, **b.** x^2 **9.** $t+1$ **10. a.** $\sqrt{3}$, **b.** $\sqrt{7xy}$ **c.** $\sqrt[12]{m}$

9.2 STUDY SET ELEMENTARY & INTERMEDIATE Algebra *f(x)* Now™

VOCABULARY **Fill in the blanks.**

1. The expressions $4^{1/2}$ and $(-8)^{-2/3}$ have
_____ exponents.

2. In the exponential expression $27^{4/3}$, 27 is the _____,
and 4/3 is the _____.

3. In the radical expression $\sqrt[3]{4{,}096x^{12}}$, 3 is the _____,
and $4{,}096x^{12}$ is the _____.

4. $32^{4/5}$ means the fourth _____ of the fifth _____ of 32.

CONCEPTS

5. Complete the table by writing the given expression in
the alternate form.

Radical form	Exponential form
$\sqrt[5]{25}$	
	$(-27)^{2/3}$
$\left(\sqrt[4]{16}\right)^{-3}$	
	$81^{3/2}$
$-\sqrt{\dfrac{9}{64}}$	

6. In your own words, explain the three rules for rational
exponents illustrated in the diagrams below.

a. $(-32)^{1/5} = \sqrt[5]{-32}$

b. $125^{4/3} = \left(\sqrt[3]{125}\right)^{4}$

c. $8^{-1/3} = \dfrac{1}{8^{1/3}}$

7. Graph each number on the number line.

$$\left\{ 8^{2/3},\ (-125)^{1/3},\ -16^{-1/4},\ 4^{3/2},\ -\left(\frac{9}{100}\right)^{-1/2} \right\}$$

8. Evaluate $25^{3/2}$ in two ways. Which way is easier?

Complete each rule for exponents.

9. $x^{1/n} = $

10. $x^{m/n} = $ $= \sqrt[n]{x^m}$

11. $x^{-m/n} = $

12. $\dfrac{1}{x^{-m/n}} = $

NOTATION **Complete each solution.**

13. Simplify: $(100a^4)^{3/2}$.

$$(100a^4)^{3/2} = \left(\sqrt{}\ \right)^3$$
$$= \left(\right)^3$$
$$= 1{,}000a^6$$

14. Simplify: $(m^{1/3}n^{1/2})^6$.

$$(m^{1/3}n^{1/2})^6 = \left(\right)^6 (n^{1/2})^6$$
$$= m^{}n^{6/2}$$
$$= m^2 n^3$$

PRACTICE **Write each expression in radical form.**

15. $x^{1/3}$

16. $b^{1/2}$

17. $(3x)^{1/4}$

18. $(4ab)^{1/6}$

19. $(6x^3y)^{1/4}$

20. $(7a^2b^2)^{1/5}$

21. $(x^2 + y^2)^{1/2}$

22. $(x^3 + y^3)^{1/3}$

Change each radical to an exponential expression.

23. $\sqrt{m}$

24. $\sqrt[3]{r}$

25. $\sqrt[4]{3a}$

26. $3\sqrt[5]{a}$

27. $\sqrt[6]{8abc}$

28. $\sqrt[7]{7p^2q}$

29. $\sqrt[3]{a^2 - b^2}$

30. $\sqrt{x^2 + y^2}$

Simplify each expression, if possible.

31. $4^{1/2}$

32. $25^{1/2}$

33. $125^{1/3}$

34. $8^{1/3}$

35. $81^{1/4}$

36. $625^{1/4}$

37. $32^{1/5}$

38. $0^{1/5}$

39. $\left(\dfrac{1}{4}\right)^{1/2}$

40. $\left(\dfrac{1}{16}\right)^{1/2}$

41. $-16^{1/4}$

42. $-125^{1/3}$

43. $(-64)^{1/2}$

44. $(-216)^{1/2}$

45. $(-216)^{1/3}$

46. $(-1,000)^{1/3}$

Simplify each expression, if possible. Assume that all variables are unrestricted and use absolute value symbols when necessary.

47. $(x^2)^{1/2}$

48. $(x^3)^{1/3}$

49. $(m^4)^{1/2}$

50. $(a^4)^{1/4}$

51. $(n^9)^{1/3}$

52. $(t^{10})^{1/5}$

53. $(25y^2)^{1/2}$

54. $(-27x^3)^{1/3}$

55. $(16x^4)^{1/4}$

56. $(-16x^4)^{1/2}$

57. $(-64x^8)^{1/4}$

58. $(243x^{10})^{1/5}$

59. $[(x+1)^4]^{1/4}$

60. $[(x+5)^3]^{1/3}$

Simplify each expression. Assume that all variables represent positive numbers.

61. $36^{3/2}$

62. $27^{2/3}$

63. $81^{3/4}$

64. $100^{3/2}$

65. $144^{3/2}$

66. $1,000^{2/3}$

67. $\left(\dfrac{1}{8}\right)^{2/3}$

68. $\left(\dfrac{4}{9}\right)^{3/2}$

69. $(25x^4)^{3/2}$

70. $(27a^3b^3)^{2/3}$

71. $\left(\dfrac{8x^3}{27}\right)^{2/3}$

72. $\left(\dfrac{27}{64y^6}\right)^{2/3}$

Write each expression without using negative exponents. Assume that all variables represent positive numbers.

73. $4^{-1/2}$

74. $8^{-1/3}$

75. $25^{-5/2}$

76. $4^{-3/2}$

77. $(16x^2)^{-3/2}$

78. $(81c^4)^{-3/2}$

79. $(-27y^3)^{-2/3}$

80. $(-8z^9)^{-2/3}$

81. $\left(\dfrac{27}{8}\right)^{-4/3}$

82. $\left(\dfrac{25}{49}\right)^{-3/2}$

83. $\left(-\dfrac{8x^3}{27}\right)^{-1/3}$

84. $\left(\dfrac{16}{81y^4}\right)^{-3/4}$

Perform the operations. Write the answers without negative exponents. Assume that all variables represent positive numbers.

85. $9^{3/7}9^{2/7}$

86. $4^{2/5}4^{2/5}$

87. $6^{-2/3}6^{-4/3}$

88. $5^{1/3}5^{-5/3}$

89. $\dfrac{3^{4/3}3^{1/3}}{3^{2/3}}$

90. $\dfrac{2^{5/6}2^{1/3}}{2^{1/2}}$

91. $a^{2/3}a^{1/3}$

92. $b^{3/5}b^{1/5}$

93. $(a^{2/3})^{1/3}$

94. $(t^{4/5})^{10}$

95. $(a^{1/2}b^{1/3})^{3/2}$

96. $(mn^{-2/3})^{-3/5}$

97. $(27x^{-3})^{-1/3}$

98. $(16a^{-2})^{-1/2}$

Perform the multiplications. Assume that all variables are positive.

99. $y^{1/3}(y^{2/3} + y^{5/3})$

100. $y^{2/5}(y^{-2/5} + y^{3/5})$

101. $x^{3/5}(x^{7/5} - x^{2/5} + 1)$

102. $x^{4/3}(x^{2/3} + 3x^{5/3} - 4)$

Use rational exponents to simplify each radical. Assume that all variables represent positive numbers.

103. $\sqrt[6]{p^3}$

104. $\sqrt[8]{q^2}$

105. $\sqrt[4]{25b^2}$

106. $\sqrt[9]{-8x^6}$

107. $\sqrt[10]{x^2y^2}$

108. $\sqrt[6]{x^2y^2}$

109. $\sqrt[9]{\sqrt{c}}$

110. $\sqrt[4]{\sqrt{x}}$

111. $\sqrt[5]{\sqrt[3]{7m}}$

112. $\sqrt[3]{\sqrt[4]{21x}}$

Use a calculator to evaluate each expression. Round to the nearest hundredth.

113. $\sqrt[3]{15}$

114. $\sqrt[4]{50.5}$

115. $\sqrt[5]{1.045}$

116. $\sqrt[5]{-1,000}$

APPLICATIONS

117. BALLISTIC PENDULUMS   The formula

$$v = \frac{m+M}{m}(2gh)^{1/2}$$

gives the velocity (in ft/sec) of a bullet with weight m fired into a block with weight M, that raises the height of the block h feet after the collision. The letter g represents the constant, 32. Find the velocity of the bullet to the nearest ft/sec.

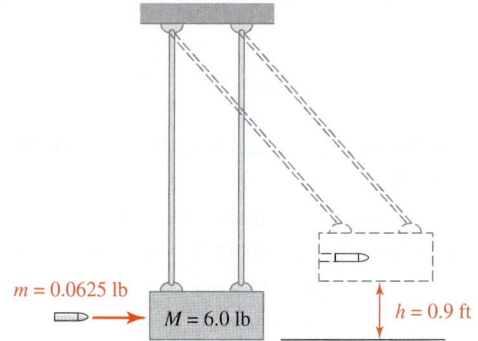

$m = 0.0625$ lb

$M = 6.0$ lb

$h = 0.9$ ft

118. GEOGRAPHY   The formula

$$A = [s(s - a)(s - b)(s - c)]^{1/2}$$

gives the area of a triangle with sides of length a, b, and c, where s is one-half of the perimeter. Estimate the area of Virginia (to the nearest square mile) using the data given in the illustration.

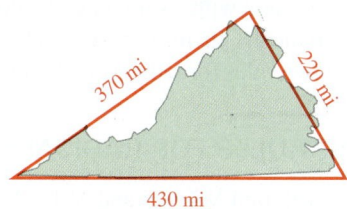

370 mi

220 mi

430 mi

119. RELATIVITY   One concept of relativity theory is that an object moving past an observer at a speed near the speed of light appears to have a larger mass because of its motion. If the mass of the object is m_0 when the object is at rest relative to the observer, its mass m will be given by the formula

$$m = m_0\left(1 - \frac{v^2}{c^2}\right)^{-1/2}$$

when it is moving with speed v (in miles per second) past the observer. The variable c is the speed of light, 186,000 mi/sec. If a proton with a rest mass of 1 unit is accelerated by a nuclear accelerator to a speed of 160,000 mi/sec, what mass will the technicians observe it to have? Round to the nearest hundredth.

120. LOGGING   The width w and height h of the strongest rectangular beam that can be cut from a cylindrical log of radius a are given by

$$w = \frac{2a}{3}(3^{1/2}) \qquad h = a\left(\frac{8}{3}\right)^{1/2}$$

Find the width, height, and cross-sectional area of the strongest beam that can be cut from a log with *diameter* 4 feet. Round to the nearest hundredth.

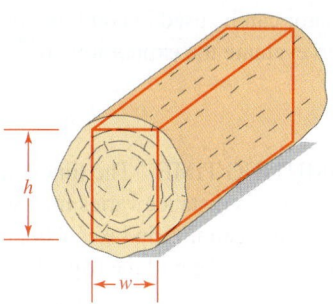

h

w

121. CUBICLES   The area of the base of a cube is given by the function $A(V) = V^{2/3}$, where V is the volume of the cube. In a preschool room, 18 children's cubicles like the one shown are placed on the floor around the room. Estimate how much floor space is lost to the cubicles. Give your answer in square inches and in square feet.

Mary S.

Storage capacity 4,096 in.^{3}

122. CARPENTRY   The length L of the longest board that can be carried horizontally around the right-angle corner of two intersecting hallways is given by the formula

$$L = (a^{2/3} + b^{2/3})^{3/2}$$

where a and b represent the widths of the hallways. Find the longest shelf that a carpenter can carry around the corner if $a = 40$ in. and $b = 64$ in. Give your result in inches and in feet. In each case, round to the nearest tenth.

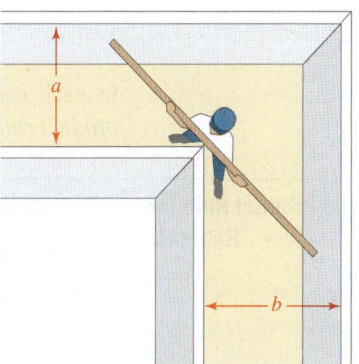

a

b

WRITING

123. What is a rational exponent? Give some examples.

124. Explain how the root key $\sqrt[x]{y}$ on a scientific calculator can be used in combination with other keys to evaluate the expression $16^{3/4}$.

REVIEW

125. COMMUTING TIME   The time it takes a car to travel a certain distance varies inversely with its rate of speed. If a certain trip takes 3 hours at 50 miles per hour, how long will the trip take at 60 miles per hour?

126. BANKRUPTCY   After filing for bankruptcy, a company was able to pay its creditors only 15 cents on the dollar. If the company owed a lumberyard $9,712, how much could the lumberyard expect to be paid?

CHALLENGE PROBLEMS

127. The fraction $\frac{2}{4}$ is equal to $\frac{1}{2}$. Is $16^{2/4}$ equal to $16^{1/2}$? Explain.

128. How would you evaluate an expression with a mixed-number exponent? For example, what is $8^{1\frac{1}{3}}$? What is $25^{2\frac{1}{2}}$? Discuss.

9.3   Simplifying and Combining Radical Expressions

- The product and quotient rules for radicals
- Simplifying radical expressions
- Adding and subtracting radical expressions

In algebra, it is often helpful to replace an expression with a simpler equivalent expression. This is certainly true when working with radicals. In most cases, radical expressions should be written in simplified form. We use two rules for radicals to do this.

■ THE PRODUCT AND QUOTIENT RULES FOR RADICALS

To introduce the product rule for radicals, we will find $\sqrt{4 \cdot 25}$ and $\sqrt{4}\sqrt{25}$, and compare the results.

Square root of a product	*Product of square roots*
$\sqrt{4 \cdot 25} = \sqrt{100}$	$\sqrt{4}\sqrt{25} = 2 \cdot 5$
$= 10$	$= 10$

In each case, the answer is 10. Thus, $\sqrt{4 \cdot 25} = \sqrt{4}\sqrt{25}$.

Similarly, we will find $\sqrt[3]{8 \cdot 27}$ and $\sqrt[3]{8}\sqrt[3]{27}$, and compare the results.

Cube root of a product	*Product of cube roots*
$\sqrt[3]{8 \cdot 27} = \sqrt[3]{216}$	$\sqrt[3]{8}\sqrt[3]{27} = 2 \cdot 3$
$= 6$	$= 6$

In each case, the answer is 6. Thus, $\sqrt[3]{8 \cdot 27} = \sqrt[3]{8}\sqrt[3]{27}$. These results illustrate the *product rule for radicals.*

Notation

The products $\sqrt{4}\sqrt{25}$ and $\sqrt[3]{8}\sqrt[3]{27}$ can also be written using a raised dot:

$\sqrt{4} \cdot \sqrt{25}$ $\sqrt[3]{8} \cdot \sqrt[3]{27}$

The Product Rule for Radicals   The nth root of the product of two numbers is equal to the product of their nth roots.

If $\sqrt[n]{a}$ and $\sqrt[n]{b}$ are real numbers,

$$\sqrt[n]{ab} = \sqrt[n]{a}\sqrt[n]{b}$$

Caution The product rule for radicals applies to the nth root of a product. There is no such property for sums or differences. For example,

$$\sqrt{9+4} \neq \sqrt{9} + \sqrt{4} \qquad \sqrt{9-4} \neq \sqrt{9} - \sqrt{4}$$
$$\sqrt{13} \neq 3 + 2 \qquad\qquad \sqrt{5} \neq 3 - 2$$
$$\sqrt{13} \neq 5 \qquad\qquad\quad \sqrt{5} \neq 1$$

Thus, $\sqrt{a+b} \neq \sqrt{a} + \sqrt{b}$ and $\sqrt{a-b} \neq \sqrt{a} - \sqrt{b}$.

To introduce the quotient rule for radicals, we will find $\sqrt{\dfrac{100}{4}}$ and $\dfrac{\sqrt{100}}{\sqrt{4}}$, and compare the results.

Square root of a quotient
Quotient of square roots

$$\sqrt{\frac{100}{4}} = \sqrt{25} \qquad\qquad \frac{\sqrt{100}}{\sqrt{4}} = \frac{10}{2}$$
$$= 5 \qquad\qquad\qquad\qquad = 5$$

Since the answer is 5 in each case, $\sqrt{\dfrac{100}{4}} = \dfrac{\sqrt{100}}{\sqrt{4}}$.

Similarly, we will find $\sqrt[3]{\dfrac{64}{8}}$ and $\dfrac{\sqrt[3]{64}}{\sqrt[3]{8}}$, and compare the results.

Cube root of a quotient
Quotient of cube roots

$$\sqrt[3]{\frac{64}{8}} = \sqrt[3]{8} \qquad\qquad \frac{\sqrt[3]{64}}{\sqrt[3]{8}} = \frac{4}{2}$$
$$= 2 \qquad\qquad\qquad\qquad = 2$$

Since the answer is 2 in each case, $\sqrt[3]{\dfrac{64}{8}} = \dfrac{\sqrt[3]{64}}{\sqrt[3]{8}}$. These results illustrate the *quotient rule for radicals*.

The Quotient Rule for Radicals	The nth root of the quotient of two numbers is equal to the quotient of their nth roots. If $\sqrt[n]{a}$ and $\sqrt[n]{b}$ are real numbers, and b is not 0, $$\sqrt[n]{\frac{a}{b}} = \frac{\sqrt[n]{a}}{\sqrt[n]{b}}$$

■ SIMPLIFYING RADICAL EXPRESSIONS

When a radical expression is written in **simplified form,** each of the following is true.

Simplified Form of a Radical Expression	1. Each factor in the radicand is to a power that is less than the index of the radical. 2. The radicand contains no fractions or negative numbers. 3. No radicals appear in the denominator of a fraction.

To simplify radical expressions, we must often factor the radicand using two natural-number factors. To simplify square root and cube root radicals, it is helpful to have the following lists memorized.

Perfect squares: **1, 4, 9, 16, 25, 36, 49, 64, 81, 100, 121, 144, 169, 196, 225, . . .**
Perfect cubes: **1, 8, 27, 64, 125, 216, 343, 512, 729, 1,000, . . .**

EXAMPLE 1

Simplify: **a.** $\sqrt{12}$,   **b.** $\sqrt{98}$,   **c.** $\sqrt[3]{54}$,   and   **d.** $-\sqrt[4]{48}$.

ELEMENTARY &
INTERMEDIATE
Algebra *f(x)* **Now**™   Solution

a. To simplify $\sqrt{12}$, we first factor 12 so that one factor is the largest perfect square that divides 12. Since 4 is the largest perfect square factor of 12, we write 12 as $4 \cdot 3$, use the multiplication property of radicals, and simplify.

$$\sqrt{12} = \sqrt{4 \cdot 3}$$ Write 12 as $12 = 4 \cdot 3$.

↑————————Write the perfect square factor first.

$$= \sqrt{4}\sqrt{3}$$ The square root of a product is equal to the product of the square roots.

$$= 2\sqrt{3}$$ Find the square root of the perfect square factor: $\sqrt{4} = 2$. Read as "2 times the square root of 3" or as "2 radical 3."

We say that $2\sqrt{3}$ is the simplified form of $\sqrt{12}$.

b. The largest perfect square factor of 98 is 49. Thus,

$$\sqrt{98} = \sqrt{49 \cdot 2}$$ Write 98 in factored form: $98 = 49 \cdot 2$.

$$= \sqrt{49}\sqrt{2}$$ The square root of a product is equal to the product of the square roots: $\sqrt{49 \cdot 2} = \sqrt{49}\sqrt{2}$.

$$= 7\sqrt{2}$$ Simplify: $\sqrt{49} = 7$.

c. Since the largest perfect cube factor of 54 is 27, we have

$$\sqrt[3]{54} = \sqrt[3]{27 \cdot 2}$$ Write 54 as $27 \cdot 2$.

$$= \sqrt[3]{27}\sqrt[3]{2}$$ The cube root of a product is equal to the product of the cube roots: $\sqrt[3]{27 \cdot 2} = \sqrt[3]{27}\sqrt[3]{2}$.

$$= 3\sqrt[3]{2}$$ Simplify: $\sqrt[3]{27} = 3$.

d. The largest perfect fourth-power factor of 48 is 16. Thus,

$$-\sqrt[4]{48} = -\sqrt[4]{16 \cdot 3}$$ Write 48 as $16 \cdot 3$.

$$= -\sqrt[4]{16}\sqrt[4]{3}$$ The fourth root of a product is equal to the product of the fourth roots: $\sqrt[4]{16 \cdot 3} = \sqrt[4]{16} \cdot \sqrt[4]{3}$.

$$= -2\sqrt[4]{3}$$ Simplify: $\sqrt[4]{16} = 2$.

The Language of Algebra

Perfect fourth-powers are

1, 16, 81, 256, 625, . . .

Self Check 1   Simplify: **a.** $\sqrt{20}$,   **b.** $\sqrt[3]{24}$,   and   **c.** $\sqrt[5]{-128}$.

Variable expressions can also be perfect squares, perfect cubes, perfect fourth-powers, and so on. For example,

Perfect squares: $x^2, x^4, x^6, x^8, x^{10}, \ldots$ Perfect cubes: $x^3, x^6, x^9, x^{12}, x^{15}, \ldots$

EXAMPLE 2

Simplify: **a.** $\sqrt{m^9}$,   **b.** $\sqrt{128a^5}$,   **c.** $\sqrt[3]{24x^5}$,   and   **d.** $\sqrt[5]{a^9b^5}$. Assume that all variables represent positive real numbers.

Solution   **a.** The largest perfect square factor of m^9 is m^8.

$$\sqrt{m^9} = \sqrt{m^8 \cdot m} \qquad \text{Write } m^9 \text{ in factored form as } m^8 \cdot m.$$
$$= \sqrt{m^8}\sqrt{m} \qquad \text{Use the product rule for radicals.}$$
$$= m^4\sqrt{m} \qquad \text{Simplify: } \sqrt{m^8} = m^4.$$

b. Since the largest perfect square factor of 128 is 64 and the largest perfect square factor of a^5 is a^4, the largest perfect square factor of $128a^5$ is $64a^4$. We write $128a^5$ as $64a^4 \cdot 2a$ and proceed as follows:

$$\sqrt{128a^5} = \sqrt{64a^4 \cdot 2a} \qquad \text{Write } 128a^5 \text{ in factored form as } 64a^4 \cdot 2a.$$
$$= \sqrt{64a^4}\sqrt{2a} \qquad \text{Use the product rule for radicals.}$$
$$= 8a^2\sqrt{2a} \qquad \text{Simplify: } \sqrt{64a^4} = 8a^2.$$

c. We write $24x^5$ as $8x^3 \cdot 3x^2$ and proceed as follows:

$$\sqrt[3]{24x^5} = \sqrt[3]{8x^3 \cdot 3x^2} \qquad 8x^3 \text{ is the largest perfect cube factor of } 24x^5.$$
$$= \sqrt[3]{8x^3}\sqrt[3]{3x^2} \qquad \text{Use the product rule for radicals.}$$
$$= 2x\sqrt[3]{3x^2} \qquad \text{Simplify: } \sqrt[3]{8x^3} = 2x.$$

d. The largest perfect fifth-power factor of a^9 is a^5, and b^5 is a perfect fifth power.

$$\sqrt[5]{a^9b^5} = \sqrt[5]{a^5b^5 \cdot a^4} \qquad a^5b^5 \text{ is the largest perfect fifth-power factor of } a^9b^5.$$
$$= \sqrt[5]{a^5b^5}\sqrt[5]{a^4} \qquad \text{Use the product rule for radicals.}$$
$$= ab\sqrt[5]{a^4} \qquad \text{Simplify: } \sqrt[5]{a^5b^5} = ab.$$

The Language of Algebra

Perfect fifth-powers of a are

$$a^5, a^{10}, a^{15}, a^{20}, a^{25}, \dots$$

Self Check 2   Simplify: **a.** $\sqrt{98b^3}$,   **b.** $\sqrt[3]{54y^5}$,   and   **c.** $\sqrt[4]{t^8u^{15}}$.

EXAMPLE 3

Simplify each expression:

a. $\sqrt{\dfrac{7}{64}}$,   **b.** $\sqrt{\dfrac{15}{49x^2}}$,   and   **c.** $\sqrt[3]{\dfrac{10x^2}{27y^6}}$. Assume that the variables represent positive real numbers.

Solution   **a.** We can use the quotient rule for radicals to simplify each expression.

$$\sqrt{\frac{7}{64}} = \frac{\sqrt{7}}{\sqrt{64}} \qquad \text{The square root of a quotient is equal to the quotient of the square roots.}$$
$$= \frac{\sqrt{7}}{8} \qquad \text{Simplify the denominator: } \sqrt{64} = 8.$$

Success Tip

In Example 3, a radical of a quotient is written as a quotient of radicals.

$$\sqrt[n]{\frac{a}{b}} = \frac{\sqrt[n]{a}}{\sqrt[n]{b}}$$

b. $$\sqrt{\frac{15}{49x^2}} = \frac{\sqrt{15}}{\sqrt{49x^2}} \qquad \text{The square root of a quotient is equal to the quotient of the square roots.}$$
$$= \frac{\sqrt{15}}{7x} \qquad \text{Simplify the denominator: } \sqrt{49x^2} = 7x.$$

c. $\sqrt[3]{\dfrac{10x^2}{27y^6}} = \dfrac{\sqrt[3]{10x^2}}{\sqrt[3]{27y^6}}$ The cube root of a quotient is equal to the quotient of the cube roots.

$= \dfrac{\sqrt[3]{10x^2}}{3y^2}$ Simplify the denominator.

Self Check 3   Simplify: **a.** $\sqrt{\dfrac{11}{36a^2}}$ $(a > 0)$ and   **b.** $\sqrt[4]{\dfrac{a^3}{625y^{12}}}$ $(a > 0, y > 0)$.

EXAMPLE 4   Simplify each expression. Assume that all variables represent positive numbers.

a. $\dfrac{\sqrt{45xy^2}}{\sqrt{5x}}$ and   **b.** $\dfrac{\sqrt[3]{-432x^5}}{\sqrt[3]{8x}}$.

Solution   **a.** We can write the quotient of the square roots as the square root of a quotient.

$\dfrac{\sqrt{45xy^2}}{\sqrt{5x}} = \sqrt{\dfrac{45xy^2}{5x}}$ Use the quotient rule for radicals.

$= \sqrt{9y^2}$ Simplify the radicand: $\dfrac{45xy^2}{5x} = \dfrac{\overset{1}{\cancel{5}} \cdot 9 \cdot \overset{1}{\cancel{x}} \cdot y^2}{\underset{1}{\cancel{5}} \cdot \underset{1}{\cancel{x}}} = 9y^2$.

$= 3y$ Simplify the radical.

> **Success Tip**
>
> In Example 4, a quotient of radicals is written as a radical of a quotient.
>
> $\dfrac{\sqrt[n]{a}}{\sqrt[n]{b}} = \sqrt[n]{\dfrac{a}{b}}$

b. We can write the quotient of the cube roots as the cube root of a quotient.

$\dfrac{\sqrt[3]{-432x^5}}{\sqrt[3]{8x}} = \sqrt[3]{\dfrac{-432x^5}{8x}}$ Use the quotient rule for radicals.

$= \sqrt[3]{-54x^4}$ Simplify the radicand: $\dfrac{-432x^5}{8x} = -54x^4$.

$= \sqrt[3]{-27x^3 \cdot 2x}$ $-27x^3$ is the largest perfect cube that divides $-54x^4$.

$= \sqrt[3]{-27x^3}\sqrt[3]{2x}$ Use the product rule for radicals.

$= -3x\sqrt[3]{2x}$ Simplify: $\sqrt[3]{-27x^3} = -3x$.

Self Check 4   Simplify each expression (assume that all variables represent positive numbers):

a. $\dfrac{\sqrt{50ab^2}}{\sqrt{2a}}$ and   **b.** $\dfrac{\sqrt[3]{-2,000x^5v^3}}{\sqrt[3]{2x}}$.

■ ADDING AND SUBTRACTING RADICAL EXPRESSIONS

Radical expressions with the same index and the same radicand are called **like** or **similar radicals**. For example, $3\sqrt{2}$ and $2\sqrt{2}$ are like radicals. However,

- $3\sqrt{5}$ and $4\sqrt{2}$ are not like radicals, because the radicands are different.
- $3\sqrt[4]{5}$ and $2\sqrt[3]{5}$ are not like radicals, because the indices are different.

For an expression with two or more radical terms, we should attempt to combine like radicals, if possible. For example, to simplify the expression $3\sqrt{2} + 2\sqrt{2}$, we use the distributive property to factor out $\sqrt{2}$ and simplify.

> **Success Tip**
>
> Combining like radicals is similar to combining like terms.
>
> $3\sqrt{2} + 2\sqrt{2} = 5\sqrt{2}$
>
> $3x + 2x = 5x$

$3\sqrt{2} + 2\sqrt{2} = (3 + 2)\sqrt{2}$

$= 5\sqrt{2}$

Radicals with the same index but different radicands can often be written as like radicals. For example, to simplify the expression $\sqrt{27} - \sqrt{12}$, we simplify both radicals first and then combine the like radicals.

$$\begin{aligned}
\sqrt{27} - \sqrt{12} &= \sqrt{9 \cdot 3} - \sqrt{4 \cdot 3} && \text{Write 27 and 12 in factored form.} \\
&= \sqrt{9}\sqrt{3} - \sqrt{4}\sqrt{3} && \text{Use the product rule for radicals.} \\
&= \mathbf{3}\sqrt{3} - \mathbf{2}\sqrt{3} && \text{Simplify } \sqrt{9} \text{ and } \sqrt{4}. \\
&= (\mathbf{3} - \mathbf{2})\sqrt{3} && \text{Factor out } \sqrt{3}. \\
&= \sqrt{3} && 1\sqrt{3} = \sqrt{3}.
\end{aligned}$$

As the previous examples suggest, we can add or subtract radicals as follows.

Adding and Subtracting Radicals	1. To add or subtract radicals, simplify each radical expression and combine all like radicals. 2. To add or subtract like radicals, combine the coefficients and keep the common radical.

EXAMPLE 5

Simplify: $2\sqrt{12} - 3\sqrt{48} + 3\sqrt{3}$.

ELEMENTARY & INTERMEDIATE
Algebra $f(x)$ **Now**™

Solution   We simplify $2\sqrt{12}$ and $3\sqrt{48}$ and then combine like radicals.

$$\begin{aligned}
2\sqrt{12} - 3\sqrt{48} + 3\sqrt{3} &= 2\sqrt{4 \cdot 3} - 3\sqrt{16 \cdot 3} + 3\sqrt{3} \\
&= 2\sqrt{4}\sqrt{3} - 3\sqrt{16}\sqrt{3} + 3\sqrt{3} \\
&= 2(2)\sqrt{3} - 3(4)\sqrt{3} + 3\sqrt{3} \\
&= \mathbf{4}\sqrt{3} - \mathbf{12}\sqrt{3} + \mathbf{3}\sqrt{3} && \text{All three expressions have the same index and radicand.} \\
&= (\mathbf{4} - \mathbf{12} + \mathbf{3})\sqrt{3} && \text{Combine the coefficients of these like radicals and keep } \sqrt{3}. \\
&= -5\sqrt{3}
\end{aligned}$$

Self Check 5   Simplify: $3\sqrt{75} - 2\sqrt{12} + 2\sqrt{48}$.

EXAMPLE 6

Simplify: $\sqrt[3]{16} - \sqrt[3]{54} + \sqrt[3]{24}$.

ELEMENTARY & INTERMEDIATE
Algebra $f(x)$ **Now**™

Solution   We begin by simplifying each radical expression:

$$\begin{aligned}
\sqrt[3]{16} - \sqrt[3]{54} + \sqrt[3]{24} &= \sqrt[3]{8 \cdot 2} - \sqrt[3]{27 \cdot 2} + \sqrt[3]{8 \cdot 3} \\
&= \sqrt[3]{8}\sqrt[3]{2} - \sqrt[3]{27}\sqrt[3]{2} + \sqrt[3]{8}\sqrt[3]{3} \\
&= \mathbf{2}\sqrt[3]{\mathbf{2}} - \mathbf{3}\sqrt[3]{\mathbf{2}} + 2\sqrt[3]{3}
\end{aligned}$$

Notation

Just as $-1x = -x$,

$$-1\sqrt[n]{x} = -\sqrt[n]{x}$$

And just as $1x = x$,

$$1\sqrt[n]{x} = \sqrt[n]{x}$$

Now we combine the two radical expressions that have the same index and radicand.

$$\sqrt[3]{16} - \sqrt[3]{54} + \sqrt[3]{24} = -\sqrt[3]{2} + 2\sqrt[3]{3} \qquad 2\sqrt[3]{2} - 3\sqrt[3]{2} = -1\sqrt[3]{2} = -\sqrt[3]{2}.$$

Self Check 6   Simplify: $\sqrt[3]{24} - \sqrt[3]{16} + \sqrt[3]{54}$.

Caution Even though the expressions $-\sqrt[3]{2}$ and $2\sqrt[3]{3}$ in the last line of Example 6 have the same index, we cannot combine them, because their radicands are different. Neither can we combine radical expressions having the same radicand but a different index. For example, the expression $\sqrt[3]{2} + \sqrt[4]{2}$ cannot be simplified.

EXAMPLE 7   Simplify: $\sqrt[3]{16x^4} + \sqrt[3]{54x^4} - \sqrt[3]{-128x^4}$.

ELEMENTARY &
INTERMEDIATE
Algebra *f(x)* **Now**™

Solution   We simplify each expression and then combine like radicals.

$$\sqrt[3]{16x^4} + \sqrt[3]{54x^4} - \sqrt[3]{-128x^4}$$
$$= \sqrt[3]{8x^3 \cdot 2x} + \sqrt[3]{27x^3 \cdot 2x} - \sqrt[3]{-64x^3 \cdot 2x}$$
$$= \sqrt[3]{8x^3}\sqrt[3]{2x} + \sqrt[3]{27x^3}\sqrt[3]{2x} - \sqrt[3]{-64x^3}\sqrt[3]{2x}$$
$$= 2x\sqrt[3]{2x} + 3x\sqrt[3]{2x} + 4x\sqrt[3]{2x} \qquad \text{All three radicals have the same index and radicand.}$$
$$= (2x + 3x + 4x)\sqrt[3]{2x}$$
$$= 9x\sqrt[3]{2x} \qquad \text{Within the parentheses, combine like terms.}$$

Self Check 7   Simplify: $\sqrt{32x^3} + \sqrt{50x^3} - \sqrt{18x^3}$.

Answers to Self Checks   **1. a.** $2\sqrt{5}$,   **b.** $2\sqrt[3]{3}$,   **c.** $-2\sqrt[5]{4}$ **2. a.** $7b\sqrt{2b}$,   **b.** $3y\sqrt[3]{2y^2}$,   **c.** $t^2u^3\sqrt[4]{u^3}$

3. a. $\dfrac{\sqrt{11}}{6a}$,   **b.** $\dfrac{\sqrt[4]{a^3}}{5y^3}$ **4. a.** $5b$,   **b.** $-10xv\sqrt[3]{x}$ **5.** $19\sqrt{3}$

6. $2\sqrt[3]{3} + \sqrt[3]{2}$ **7.** $6x\sqrt{2x}$

9.3   STUDY SET   ⊙   ELEMENTARY & INTERMEDIATE   **Algebra** *f(x)* **Now**™

VOCABULARY   **Fill in the blanks.**

1. Radical expressions such as $\sqrt[3]{4}$ and $6\sqrt[3]{4}$ with the same index and the same radicand are called _____ radicals.

2. Numbers such as 1, 4, 9, 16, 25, and 36 are called perfect _____. Numbers such as 1, 8, 27, 64, and 125 are called perfect _____.

3. The largest perfect square _____ of 27 is 9.

4. "To _____ $\sqrt{24}$" means to write it as $2\sqrt{6}$.

CONCEPTS   **Fill in the blanks.**

5. $\sqrt[n]{ab} =$

In words, the nth root of the _____ of two numbers is equal to the product of their nth _____.

6. $\sqrt[n]{\dfrac{a}{b}} =$

In words, the nth root of the _____ of two numbers is equal to the quotient of their nth _____.

7. Consider the expressions
$$\sqrt{4 \cdot 5} \quad \text{and} \quad \sqrt{4}\sqrt{5}$$
Which expression is

a. the square root of a product?

b. the product of square roots?

c. How are these two expressions related?

8. Consider the expressions
$$\frac{\sqrt[3]{a}}{\sqrt[3]{x^2}} \quad \text{and} \quad \sqrt[3]{\frac{a}{x^2}}$$
Which expression is

a. the cube root of a quotient?

b. the quotient of cube roots?

c. How are these two expressions related?

9. a. Write two radical expressions that have the same radicand but a different index. Can the expressions be added?

   b. Write two radical expressions that have the same index but a different radicand. Can the expressions be added?

10. Explain the mistake in the student's solution shown below.

Simplify: $\sqrt[3]{54}$.

$$\sqrt[3]{54} = \sqrt[3]{27 + 27}$$
$$= \sqrt[3]{27} + \sqrt[3]{27}$$
$$= 3 + 3$$
$$= 6$$

NOTATION   Complete each solution.

11. Simplify:

$$\sqrt[3]{32k^4} = \sqrt[3]{ \cdot 4k}$$
$$= \sqrt[3]{} \sqrt[3]{4k}$$
$$= 2k\sqrt[3]{}$$

12. Simplify:

$$\frac{\sqrt{80s^2t^4}}{\sqrt{5s^2}} = \sqrt{\frac{80s^2t^4}{5s^2}}$$
$$= \sqrt{}$$
$$= 4t^2$$

PRACTICE   Simplify each expression.

13. $\sqrt{20}$ **14.** $\sqrt{8}$

15. $\sqrt{200}$ **16.** $\sqrt{250}$

17. $\sqrt[3]{80}$ **18.** $\sqrt[3]{270}$

19. $\sqrt[3]{-81}$ **20.** $\sqrt[3]{-72}$

21. $\sqrt[4]{32}$ **22.** $\sqrt[4]{48}$

23. $-\sqrt[5]{96}$ **24.** $-\sqrt[7]{256}$

25. $\sqrt[6]{320}$ **26.** $\sqrt[6]{192}$

27. $\sqrt{\dfrac{7}{9}}$ **28.** $\sqrt{\dfrac{3}{4}}$

29. $\sqrt[3]{\dfrac{7}{64}}$ **30.** $\sqrt[3]{\dfrac{4}{125}}$

31. $\sqrt[4]{\dfrac{3}{10,000}}$ **32.** $\sqrt[5]{\dfrac{4}{243}}$

33. $\sqrt[5]{\dfrac{3}{32}}$ **34.** $\sqrt[6]{\dfrac{5}{64}}$

35. $\dfrac{\sqrt{500}}{\sqrt{5}}$ **36.** $\dfrac{\sqrt{128}}{\sqrt{2}}$

37. $\dfrac{\sqrt[3]{48}}{\sqrt[3]{6}}$ **38.** $\dfrac{\sqrt[3]{64}}{\sqrt[3]{8}}$

Simplify each radical expression. All variables represent positive numbers.

39. $\sqrt{50x^2}$ **40.** $\sqrt{75a^2}$

41. $\sqrt{32b}$ **42.** $\sqrt{80c}$

43. $-\sqrt{112a^3}$ **44.** $\sqrt{147a^5}$

45. $\sqrt{175a^2b^3}$ **46.** $\sqrt{128a^3b^5}$

47. $-\sqrt{300xy}$ **48.** $\sqrt{200x^2y}$

49. $\sqrt[3]{-54x^6}$ **50.** $-\sqrt[3]{-81a^3}$

51. $\sqrt[3]{16x^{12}y^3}$ **52.** $\sqrt[3]{40a^3b^6}$

53. $\sqrt[4]{32x^{12}y^4}$ **54.** $\sqrt[5]{64x^{10}y^5}$

55. $\sqrt[5]{a^7}$ **56.** $\sqrt[5]{b^8}$

57. $\sqrt[6]{m^{11}}$ **58.** $\sqrt[6]{n^{13}}$

59. $\sqrt[5]{32t^{11}}$ **60.** $\sqrt[5]{243r^{22}}$

61. $\dfrac{\sqrt[3]{189a^4}}{\sqrt[3]{7a}}$ **62.** $\dfrac{\sqrt[3]{243x^7}}{\sqrt[3]{9x}}$

63. $\dfrac{\sqrt{98x^3}}{\sqrt{2x}}$ **64.** $\dfrac{\sqrt{75y^5}}{\sqrt{3y}}$

65. $\sqrt{\dfrac{z^2}{16x^2}}$ **66.** $\sqrt{\dfrac{b^4}{64a^8}}$

67. $\sqrt[4]{\dfrac{5x}{16z^4}}$ **68.** $\sqrt[3]{\dfrac{11a^2}{125b^6}}$

Simplify and combine like radicals. All variables represent positive numbers.

69. $4\sqrt{2x} + 6\sqrt{2x}$ **70.** $6\sqrt[3]{5y} + 3\sqrt[3]{5y}$

71. $8\sqrt[5]{7a^2} - 7\sqrt[5]{7a^2}$ **72.** $10\sqrt[6]{12xyz} - \sqrt[6]{12xyz}$

73. $\sqrt{2} - \sqrt{8}$ **74.** $\sqrt{20} - \sqrt{125}$

75. $\sqrt{98} - \sqrt{50}$ **76.** $\sqrt{72} - \sqrt{200}$

77. $3\sqrt{24} + \sqrt{54}$ **78.** $\sqrt{18} + 2\sqrt{50}$

79. $\sqrt[3]{24x} + \sqrt[3]{3x}$ **80.** $\sqrt[3]{16y} + \sqrt[3]{128y}$

81. $\sqrt[3]{32} - \sqrt[3]{108}$ **82.** $\sqrt[3]{80} - \sqrt[3]{10,000}$

83. $2\sqrt[3]{125} - 5\sqrt[3]{64}$

84. $3\sqrt[3]{27} + 12\sqrt[3]{216}$

85. $14\sqrt[4]{32} - 15\sqrt[4]{162}$

86. $23\sqrt[4]{768} + \sqrt[4]{48}$

87. $3\sqrt[4]{512} + 2\sqrt[4]{32}$

88. $4\sqrt[4]{243} - \sqrt[4]{48}$

89. $\sqrt{98} - \sqrt{50} - \sqrt{72}$

90. $\sqrt{20} + \sqrt{125} - \sqrt{80}$

91. $\sqrt{18t} + \sqrt{300t} - \sqrt{243t}$

92. $\sqrt{80m} - \sqrt{128m} + \sqrt{288m}$

93. $2\sqrt[3]{16} - \sqrt[3]{54} - 3\sqrt[3]{128}$

94. $\sqrt[4]{48} - \sqrt[4]{243} - \sqrt[4]{768}$

95. $\sqrt{25y^2z} - \sqrt{16y^2z}$

96. $\sqrt{25yz^2} + \sqrt{9yz^2}$

97. $\sqrt{36xy^2} + \sqrt{49xy^2}$

98. $3\sqrt{2x} - \sqrt{8x}$

99. $2\sqrt[3]{64a} + 2\sqrt[3]{8a}$

100. $3\sqrt[4]{x^4y} - 2\sqrt[4]{x^4y}$

101. $\sqrt{y^5} - \sqrt{9y^5} - \sqrt{25y^5}$

102. $\sqrt{8y^7} + \sqrt{32y^7} - \sqrt{2y^7}$

103. $\sqrt[5]{x^6y^2} + \sqrt[5]{32x^6y^2} + \sqrt[5]{x^6y^2}$

104. $\sqrt[3]{xy^4} + \sqrt[3]{8xy^4} - \sqrt[3]{27xy^4}$

APPLICATIONS   First give the exact answer, expressed as a simplified radical expression. Then give an approximation, rounded to the nearest tenth.

105. UMBRELLAS   The surface area of a cone is given by the formula $S = \pi r\sqrt{r^2 + h^2}$, where r is the radius of the base and h is its height. Use this formula to find the number of square feet of waterproof cloth used to make the umbrella shown below.

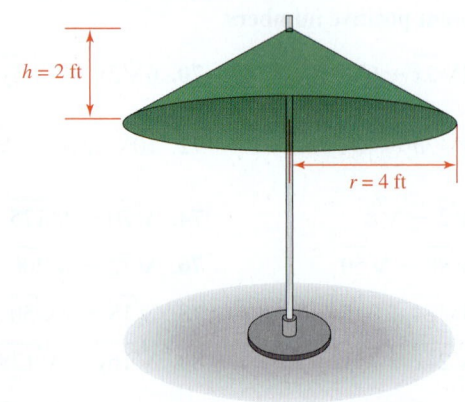

$h = 2$ ft

$r = 4$ ft

106. STRUCTURAL ENGINEERING   Engineers have determined that two additional supports need to be added to strengthen the truss shown. Find the length L of each new support using the formula

$$L = \sqrt{\frac{b^2}{2} + \frac{c^2}{2} - \frac{a^2}{4}}$$

$b = 16$ ft

$c = 14$ ft

L L

$a = 20$ ft

Joins at the midpoint of this segment

107. BLOW DRYERS   The current I (in amps), the power P (in watts), and the resistance R (in ohms) are related by the formula $I = \sqrt{\dfrac{P}{R}}$. What current is needed for a 1,200-watt hair dryer if the resistance is 16 ohms?

108. COMMUNICATIONS SATELLITES   Engineers have determined that a spherical communications satellite needs to have a capacity of 565.2 cubic feet to house all of its operating systems. The volume V of a sphere is related to its radius r by the formula $r = \sqrt[3]{\dfrac{3V}{4\pi}}$. What radius must the satellite have to meet the engineer's specification? Use 3.14 for π.

109. DUCTWORK   The following pattern is laid out on a sheet of galvanized tin. Then it is cut out with snips and bent to make an air conditioning duct connection. Find the total length of the cut that must be made with the tin snips. (All measurements are in inches.)

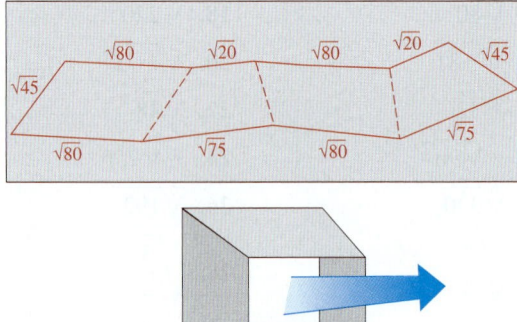

110. OUTDOOR COOKING   The diameter of a circle is given by the function $d(A) = 2\sqrt{\dfrac{A}{\pi}}$, where A is the area of the circle. Find the difference between the diameters of the barbecue grills on the next page.

Cooking area
147π in.^{3}

Cooking area
48π in.^{3}

WRITING

111. Explain why $\sqrt[3]{9x^4}$ is not in simplified form.

112. How are the procedures used to simplify $3x + 4x$ and $3\sqrt{x} + 4\sqrt{x}$ similar?

113. Explain how the graphs of $Y_1 = 3\sqrt{24x} + \sqrt{54x}$ (on the left) and $Y_2 = 9\sqrt{6x}$ (on the right) can be used to verify the simplification $3\sqrt{24x} + \sqrt{54x} = 9\sqrt{6x}$. In each graph, settings of $[-5, 20]$ for x and $[-5, 100]$ for y were used.

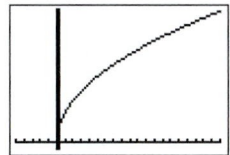
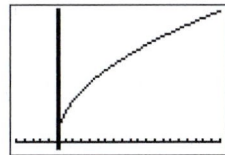

114. Explain how to verify algebraically that
$$\sqrt{200x^3y^5} = 10xy^2\sqrt{2xy}$$

REVIEW   **Perform each operation.**

115. $3x^2y^3(-5x^3y^{-4})$

116. $(2x^2 - 9x - 5) \cdot \dfrac{x}{2x^2 + x}$

117. $2p - 5\overline{)6p^2 - 7p - 25}$

118. $\dfrac{xy}{\dfrac{1}{x} - \dfrac{1}{y}}$

CHALLENGE PROBLEMS

119. Can you find any numbers a and b such that $\sqrt{a + b} = \sqrt{a} + \sqrt{b}$?

120. Find the sum:
$$\sqrt{3} + \sqrt{3^2} + \sqrt{3^3} + \sqrt{3^4} + \sqrt{3^5}$$

| **9.4** | **Multiplying and Dividing Radical Expressions** |

- Multiplying radical expressions
- Powers of radical expressions
- Rationalizing denominators
- Rationalizing two-term denominators
- Rationalizing numerators

In this section, we will discuss the methods used to multiply and divide radical expressions.

■ MULTIPLYING RADICAL EXPRESSIONS

We have used the *product rule for radicals* to write radical expressions in simplified form. We can also use this rule to multiply radical expressions that have the same index.

The Product Rule for Radicals   The product of the nth roots of two nonnegative numbers is equal to the nth root of the product of those numbers.

If $\sqrt[n]{a}$ and $\sqrt[n]{b}$ are real numbers,

$$\sqrt[n]{a} \cdot \sqrt[n]{b} = \sqrt[n]{a \cdot b}$$

EXAMPLE 1

ELEMENTARY &
INTERMEDIATE
Algebra *f(x)* **Now** ™

Multiply and then simplify, if possible: **a.** $\sqrt{5}\sqrt{10}$,   **b.** $3\sqrt{6}(2\sqrt{3})$,   and

c. $-2\sqrt[3]{7x} \cdot 6\sqrt[3]{49x^2}$.

Solution   **a.** $\sqrt{5}\sqrt{10} = \sqrt{5 \cdot 10}$ Use the product rule for radicals.

$= \sqrt{50}$ Multiply under the radical. Note that $\sqrt{50}$ can be simplified.

$= \sqrt{25 \cdot 2}$ Begin the process of simplifying $\sqrt{50}$ by factoring 50.

$= 5\sqrt{2}$ $\sqrt{25 \cdot 2} = \sqrt{25}\sqrt{2} = 5\sqrt{2}$.

Caution

Note that to multiply radical expressions, they must have the same index.

$$\sqrt[n]{a} \cdot \sqrt[n]{b} = \sqrt[n]{a \cdot b}$$

b. We use the commutative and associative properties of multiplication to multiply the coefficients and the radicals separately. Then we simplify any radicals in the product, if possible.

$3\sqrt{6}(2\sqrt{3}) = 3(2)\sqrt{6}\sqrt{3}$ Multiply the coefficients and multiply the radicals.

$= 6\sqrt{18}$ Use the product rule for radicals.

$= 6\sqrt{9}\sqrt{2}$ Simplify: $\sqrt{18} = \sqrt{9 \cdot 2} = \sqrt{9}\sqrt{2}$.

$= 6(3)\sqrt{2}$ $\sqrt{9} = 3$.

$= 18\sqrt{2}$ Multiply.

c. $-2\sqrt[3]{7x} \cdot 6\sqrt[3]{49x^2} = -2(6)\sqrt[3]{7x}\sqrt[3]{49x^2}$ Write the coefficients together and the radicals together.

$= -12\sqrt[3]{7x \cdot 49x^2}$ Multiply the coefficients and multiply the radicals.

$= -12\sqrt[3]{7x \cdot 7^2x^2}$ Write 49 as 7^2.

$= -12\sqrt[3]{7^3x^3}$ Write $7x \cdot 7^2x^2$ as 7^3x^3.

$= -12(7x)$ Simplify: $\sqrt[3]{7^3x^3} = 7x$.

$= -84x$ Multiply.

Self Check 1   Multiply: **a.** $\sqrt{7}\sqrt{14}$,   **b.** $-2\sqrt{7}(5\sqrt{2})$,   and   **c.** $\sqrt[4]{4x^3} \cdot 9\sqrt[4]{8x^2}$.

Recall that to multiply a polynomial by a monomial, we use the distributive property. We use the same technique to multiply a radical expression that has two or more terms by a radical expression that has only one term.

EXAMPLE 2

Multiply and then simplify, if possible: $3\sqrt{3}(4\sqrt{8} - 5\sqrt{10})$.

Solution $3\sqrt{3}(4\sqrt{8} - 5\sqrt{10})$

$= 3\sqrt{3} \cdot 4\sqrt{8} - 3\sqrt{3} \cdot 5\sqrt{10}$ Distribute the multiplication by $3\sqrt{3}$.

$= 12\sqrt{24} - 15\sqrt{30}$ Multiply the coefficients and multiply the radicals.

$= 12\sqrt{4}\sqrt{6} - 15\sqrt{30}$ Simplify: $\sqrt{24} = \sqrt{4 \cdot 6} = \sqrt{4}\sqrt{6}$.

$= 12(2)\sqrt{6} - 15\sqrt{30}$ $\sqrt{4} = 2$.

$= 24\sqrt{6} - 15\sqrt{30}$

Self Check 2   Simplify: $4\sqrt{2}(3\sqrt{5} - 2\sqrt{8})$.

Recall that to multiply two binomials, we multiply each term of one binomial by each term of the other binomial and simplify. We multiply two radical expressions, each having two terms, in the same way.

EXAMPLE 3

ELEMENTARY &
INTERMEDIATE
Algebra $f(x)$ **Now**™

Multiply and then simplify, if possible: **a.** $\left(\sqrt{7} + \sqrt{2}\right)\left(\sqrt{7} - 3\sqrt{2}\right)$ and
b. $\left(\sqrt[3]{x^2} - 4\sqrt[3]{5}\right)\left(\sqrt[3]{x} + \sqrt[3]{2}\right)$.

Solution **a.** $\left(\sqrt{7} + \sqrt{2}\right)\left(\sqrt{7} - 3\sqrt{2}\right)$

$$= \sqrt{7}\sqrt{7} - 3\sqrt{7}\sqrt{2} + \sqrt{2}\sqrt{7} - 3\sqrt{2}\sqrt{2} \qquad \text{Use the FOIL method.}$$
$$= 7 - 3\sqrt{14} + \sqrt{14} - 3(2) \qquad \text{Perform each multiplication.}$$
$$= 7 - 2\sqrt{14} - 6 \qquad \text{Combine like radicals.}$$
$$= 1 - 2\sqrt{14} \qquad \text{Combine like terms.}$$

b. $\left(\sqrt[3]{x^2} - 4\sqrt[3]{5}\right)\left(\sqrt[3]{x} + \sqrt[3]{2}\right)$

$$= \sqrt[3]{x^2}\sqrt[3]{x} + \sqrt[3]{x^2}\sqrt[3]{2} - 4\sqrt[3]{5}\sqrt[3]{x} - 4\sqrt[3]{5}\sqrt[3]{2} \qquad \text{Use the FOIL method.}$$
$$= \sqrt[3]{x^3} + \sqrt[3]{2x^2} - 4\sqrt[3]{5x} - 4\sqrt[3]{10} \qquad \text{Perform each multiplication.}$$
$$= x + \sqrt[3]{2x^2} - 4\sqrt[3]{5x} - 4\sqrt[3]{10} \qquad \text{Simplify the first term.}$$

Self Check 3 Multiply: **a.** $\left(\sqrt{5} + 2\sqrt{3}\right)\left(\sqrt{5} - \sqrt{3}\right)$ and   **b.** $\left(\sqrt[3]{a} + 9\sqrt[3]{2}\right)\left(\sqrt[3]{a^2} - \sqrt[3]{3}\right)$.

■ POWERS OF RADICAL EXPRESSIONS

To find the power of a radical expression, such as $\left(\sqrt{5}\right)^2$ or $\left(\sqrt[3]{2}\right)^3$, we can use the definition of exponent and the product rule for radicals.

$$\left(\sqrt{5}\right)^2 = \sqrt{5} \cdot \sqrt{5} \qquad\qquad \left(\sqrt[3]{2}\right)^3 = \sqrt[3]{2} \cdot \sqrt[3]{2} \cdot \sqrt[3]{2}$$
$$= \sqrt{25} \qquad\qquad\qquad\qquad = \sqrt[3]{8}$$
$$= 5 \qquad\qquad\qquad\qquad\quad = 2$$

These results illustrate the following property of radicals.

The *n*th Power of the *n*th Root	If $\sqrt[n]{a}$ is a real number, $$\left(\sqrt[n]{a}\right)^n = a$$

EXAMPLE 4

ELEMENTARY &
INTERMEDIATE
Algebra $f(x)$ **Now**™

Find: **a.** $\left(\sqrt{5}\right)^2$,   **b.** $\left(2\sqrt[3]{7x^2}\right)^3$,   and   **c.** $\left(\sqrt{m+1} + 2\right)^2$, where $m > 0$.

Solution **a.** $\left(\sqrt{5}\right)^2 = 5$ Because the square of the square root of a positive number is that number.

b. We can use the power of a product rule for exponents to find $\left(2\sqrt[3]{7x^2}\right)^3$.

$$\left(2\sqrt[3]{7x^2}\right)^3 = 2^3\left(\sqrt[3]{7x^2}\right)^3 \qquad \text{Raise each factor of } 2\sqrt[3]{7x^2} \text{ to the 3rd power.}$$
$$= 8(7x^2) \qquad \text{Evaluate: } 2^3 = 8. \text{ Use } \left(\sqrt[n]{a}\right)^n = a.$$
$$= 56x^2$$

c. We can use the FOIL method to find the product.

$$\left(\sqrt{m+1}+2\right)^2 = \left(\sqrt{m+1}+2\right)\left(\sqrt{m+1}+2\right)$$

$$= \left(\sqrt{m+1}\right)^2 + 2\sqrt{m+1} + 2\sqrt{m+1} + 2\cdot 2$$

$$= m+1 + 2\sqrt{m+1} + 2\sqrt{m+1} + 4 \qquad \text{Use } \left(\sqrt[n]{a}\right)^n = a.$$

$$= m + 4\sqrt{m+1} + 5 \qquad\qquad\qquad \text{Combine like terms.}$$

Self Check 4   Find: **a.** $\left(\sqrt{11}\right)^2$,   **b.** $\left(3\sqrt[3]{4y}\right)^3$,   and   **c.** $\left(\sqrt{x-8}-5\right)^2$.

■ RATIONALIZING DENOMINATORS

We have seen that when a radical expression is written in simplified form, each of the following statements is true.

1. Each factor in the radicand is to a power that is less than the index of the radical.
2. The radicand contains no fractions or negative numbers.
3. No radicals appear in the denominator of a fraction.

We now consider radical expressions that do not satisfy requirement 2 and those that do not satisfy requirement 3. We will introduce an algebraic technique, called *rationalizing the denominator,* that is used to write such expressions in an equivalent simplified form.

The Language of Algebra

Since $\sqrt{3}$ is an irrational number, the fraction $\dfrac{\sqrt{5}}{\sqrt{3}}$ has an *irrational* denominator.

To divide radical expressions, we **rationalize the denominator** of a fraction to replace the denominator with a rational number. For example, to divide $\sqrt{5}$ by $\sqrt{3}$, we write the division as the fraction

$$\frac{\sqrt{5}}{\sqrt{3}} \qquad \text{This radical expression is not in simplified form, because a radical appears in the denominator.}$$

We want to find a fraction equivalent to $\dfrac{\sqrt{5}}{\sqrt{3}}$ that does not have a radical in its denominator. If we multiply $\dfrac{\sqrt{5}}{\sqrt{3}}$ by $\dfrac{\sqrt{3}}{\sqrt{3}}$, the denominator becomes $\sqrt{3}\cdot\sqrt{3}=3$, a rational number.

Success Tip

As an informal check, we can use a calculator to evaluate each expression.

$$\frac{\sqrt{5}}{\sqrt{3}} \approx 1.290994449$$

$$\frac{\sqrt{15}}{3} \approx 1.290994449$$

$$\frac{\sqrt{5}}{\sqrt{3}} = \frac{\sqrt{5}}{\sqrt{3}} \cdot \frac{\sqrt{3}}{\sqrt{3}} \qquad \text{To build an equivalent fraction, multiply by } \frac{\sqrt{3}}{\sqrt{3}} = 1.$$

$$= \frac{\sqrt{15}}{3} \qquad\qquad \text{Multiply the numerators: } \sqrt{5}\cdot\sqrt{3}=\sqrt{15}.$$
$$\text{Multiply the denominators: } \sqrt{3}\cdot\sqrt{3}=\left(\sqrt{3}\right)^2=3.$$

Thus, $\dfrac{\sqrt{5}}{\sqrt{3}} = \dfrac{\sqrt{15}}{3}$. These equivalent fractions represent the same number, but have different forms. Since there is no radical in the denominator, and $\sqrt{15}$ is in simplest form, the division is complete.

EXAMPLE 5

ELEMENTARY & INTERMEDIATE
Algebra *f(x)* **Now**™

Simplify by rationalizing the denominator:

a. $\sqrt{\dfrac{20}{7}}$ and   **b.** $\dfrac{4}{\sqrt[3]{2}}$.

Solution   **a.** This radical expression is not in simplified form, because the radicand contains a fraction. We begin by writing the square root of the quotient as the quotient of two square roots:

$$\sqrt{\frac{20}{7}} = \frac{\sqrt{20}}{\sqrt{7}} \qquad \text{Use the division property of radicals: } \sqrt[n]{\frac{a}{b}} = \frac{\sqrt[n]{a}}{\sqrt[n]{b}}.$$

To rationalize the denominator, we proceed as follows:

Caution

Do not attempt to remove a common factor of 7 from the numerator and denominator of $\frac{2\sqrt{35}}{7}$. The numerator, $2\sqrt{35}$, does not have a factor of 7.

$$\frac{2\sqrt{35}}{7} = \frac{2 \cdot \sqrt{5 \cdot 7}}{7}$$

$$\frac{\sqrt{20}}{\sqrt{7}} = \frac{\sqrt{20}}{\sqrt{7}} \cdot \frac{\sqrt{7}}{\sqrt{7}} \qquad \text{To build an equivalent fraction, multiply by } \frac{\sqrt{7}}{\sqrt{7}} = 1.$$

$$= \frac{\sqrt{140}}{7} \qquad \begin{array}{l}\text{Multiply the numerators.}\\ \text{Multiply the denominators: } \left(\sqrt{7}\right)^2 = 7.\end{array}$$

$$= \frac{2\sqrt{35}}{7} \qquad \text{Simplify: } \sqrt{140} = \sqrt{4 \cdot 35} = \sqrt{4}\sqrt{35} = 2\sqrt{35}.$$

b. This expression is not in simplified form because a radical appears in the denominator of a fraction. Here, we must rationalize a denominator that is a cube root. We multiply the numerator and the denominator by a number that will give a perfect cube under the radical. Since $2 \cdot 4 = 8$ is a perfect cube, $\sqrt[3]{4}$ is such a number.

Caution

Note that multiplying $\frac{4}{\sqrt[3]{2}}$ by $\frac{\sqrt[3]{2}}{\sqrt[3]{2}}$ does not rationalize the denominator.

$$\frac{4}{\sqrt[3]{2}} \cdot \frac{\sqrt[3]{2}}{\sqrt[3]{2}} = \frac{4\sqrt[3]{2}}{\sqrt[3]{4}}$$

Since 4 is not a perfect cube, this radical does not simplify.

$$\frac{4}{\sqrt[3]{2}} = \frac{4}{\sqrt[3]{2}} \cdot \frac{\sqrt[3]{4}}{\sqrt[3]{4}} \qquad \text{To build an equivalent fraction, multiply by } \frac{\sqrt[3]{4}}{\sqrt[3]{4}} = 1.$$

$$= \frac{4\sqrt[3]{4}}{\sqrt[3]{8}} \qquad \begin{array}{l}\text{Multiply the numerators. Multiply the denominators.}\\ \text{This radicand is now a perfect cube.}\end{array}$$

$$= \frac{4\sqrt[3]{4}}{2} \qquad \text{Simplify: } \sqrt[3]{8} = 2.$$

$$= 2\sqrt[3]{4} \qquad \text{Simplify: } \frac{4\sqrt[3]{4}}{2} = \frac{\overset{2}{\cancel{4}} \cdot 2\sqrt[3]{4}}{\underset{1}{\cancel{2}}} = 2\sqrt[3]{4}.$$

Self Check 5   Rationalize the denominator: **a.** $\sqrt{\dfrac{24}{5}}$ and   **b.** $\dfrac{5}{\sqrt[4]{3}}$.

EXAMPLE 6   Rationalize the denominator:

$$\frac{\sqrt{5xy^2}}{\sqrt{xy^3}}$$

Solution   Two possible methods for rationalizing the denominator are shown on the next page. In each case, we simplify the expression first.

Caution

We will assume that all of the variables appearing in the following examples represent positive numbers.

Method 1

$$\frac{\sqrt{5xy^2}}{\sqrt{xy^3}} = \sqrt{\frac{5xy^2}{xy^3}}$$

$$= \sqrt{\frac{5}{y}}$$

$$= \frac{\sqrt{5}}{\sqrt{y}}$$

$$= \frac{\sqrt{5}}{\sqrt{y}} \cdot \frac{\sqrt{y}}{\sqrt{y}} \qquad \text{Multiply outside the radical.}$$

$$= \frac{\sqrt{5y}}{y}$$

Method 2

$$\frac{\sqrt{5xy^2}}{\sqrt{xy^3}} = \sqrt{\frac{5xy^2}{xy^3}}$$

$$= \sqrt{\frac{5}{y}}$$

$$= \sqrt{\frac{5}{y} \cdot \frac{y}{y}} \qquad \text{Multiply within the radical.}$$

$$= \frac{\sqrt{5y}}{\sqrt{y^2}}$$

$$= \frac{\sqrt{5y}}{y}$$

Self Check 6   Rationalize the denominator: $\dfrac{\sqrt{4ab^3}}{\sqrt{2a^2b^2}}$.

EXAMPLE 7   Rationalize the denominator: $\dfrac{11}{\sqrt{20q^5}}$.

Solution   We could begin by multiplying $\dfrac{11}{\sqrt{20q^5}}$ by $\dfrac{\sqrt{20q^5}}{\sqrt{20q^5}}$. However, to work with smaller

numbers, it is easier if we simplify $\sqrt{20q^5}$ first, and then rationalize the denominator.

Success Tip

We usually simplify a radical expression before rationalizing the denominator.

$$\frac{11}{\sqrt{20q^5}} = \frac{11}{\sqrt{4q^4 \cdot 5q}} \qquad \text{To simplify } \sqrt{20q^5}\text{, factor } 20q^5 \text{ as } 4q^4 \cdot 5q.$$

$$= \frac{11}{2q^2\sqrt{5q}} \qquad \sqrt{4q^4 \cdot 5q} = \sqrt{4q^4}\sqrt{5q} = 2q^2\sqrt{5q}.$$

$$= \frac{11}{2q^2\sqrt{5q}} \cdot \frac{\sqrt{5q}}{\sqrt{5q}} \qquad \text{To rationalize the denominator, multiply by } \dfrac{\sqrt{5q}}{\sqrt{5q}} = 1.$$

$$= \frac{11\sqrt{5q}}{2q^2(5q)} \qquad \begin{array}{l}\text{Multiply the numerators.}\\ \text{Multiply the denominators: } \left(\sqrt{5q}\right)^2 = 5q.\end{array}$$

$$= \frac{11\sqrt{5q}}{10q^3} \qquad \text{Multiply in the denominator.}$$

Self Check 7   Rationalize the denominator: $\sqrt[3]{\dfrac{1}{16h^4}}$.

EXAMPLE 8   Rationalize each denominator:

a. $\dfrac{5}{\sqrt[3]{6mn^2}}$ and   **b.** $\dfrac{\sqrt[4]{b}}{\sqrt[4]{9a}}$.

Solution   **a.** To rationalize the denominator $\sqrt[3]{6mn^2}$, we need the radicand to be a perfect cube. Since $6mn^2 = 6 \cdot m \cdot n \cdot n$, the radicand needs two more factors of 6, two more factors of m, and one more factor of n. It follows that we should multiply the given expression by $\dfrac{\sqrt[3]{36m^2n}}{\sqrt[3]{36m^2n}}$.

$$\dfrac{5}{\sqrt[3]{6mn^2}} = \dfrac{5}{\sqrt[3]{6mn^2}} \cdot \dfrac{\sqrt[3]{36m^2n}}{\sqrt[3]{36m^2n}}$$
Multiply by a form of 1 to rationalize the denominator.

$$= \dfrac{5\sqrt[3]{36m^2n}}{\sqrt[3]{216m^3n^3}}$$
Multiply the numerators. Multiply the denominators. This radicand is now a perfect cube.

$$= \dfrac{5\sqrt[3]{36m^2n}}{6mn}$$
Simplify: $\sqrt[3]{216m^3n^3} = 6mn$.

b. To rationalize the denominator $\sqrt[4]{9a}$, we need the radicand to be a perfect fourth power. Since $9a = 3 \cdot 3 \cdot a$, the radicand needs two more factors of 3 and three more factors of a. It follows that we should multiply the given expression by $\dfrac{\sqrt[4]{9a^3}}{\sqrt[4]{9a^3}}$.

$$\dfrac{\sqrt[4]{b}}{\sqrt[4]{9a}} = \dfrac{\sqrt[4]{b}}{\sqrt[4]{9a}} \cdot \dfrac{\sqrt[4]{9a^3}}{\sqrt[4]{9a^3}}$$
Multiply by a form of 1 to rationalize the denominator.

$$= \dfrac{\sqrt[4]{9a^3b}}{\sqrt[4]{81a^4}}$$
Multiply the numerators. Multiply the denominators. This radicand is now a perfect fourth power.

$$= \dfrac{\sqrt[4]{9a^3b}}{3a}$$
Simplify: $\sqrt[4]{81a^4} = 3a$.

Self Check 8 Rationalize each denominator: **a.** $\dfrac{11}{\sqrt[3]{100ab^2}}$ and **b.** $\dfrac{\sqrt[4]{s}}{\sqrt[4]{x^3y^2}}$.

■ RATIONALIZING TWO-TERM DENOMINATORS

So far, we have rationalized denominators that had only one term. We will now discuss a method to rationalize denominators that have two terms.

One-term denominators   *Two-term denominators*

$$\dfrac{\sqrt{5}}{\sqrt{3}}, \quad \dfrac{11}{\sqrt{20q^5}}, \quad \dfrac{4}{\sqrt[3]{2}} \qquad \dfrac{1}{\sqrt{2}+1}, \quad \dfrac{\sqrt{x}+\sqrt{2}}{\sqrt{x}-\sqrt{2}}$$

Success Tip

Recall the special product formula for finding the product of the sum and difference of two terms:

$$(x+y)(x-y) = x^2 - y^2$$

To rationalize the denominator of $\dfrac{1}{\sqrt{2}+1}$, for example, we multiply the numerator and denominator by $\sqrt{2}-1$, because the product $(\sqrt{2}+1)(\sqrt{2}-1)$ contains no radicals.

$$(\sqrt{2}+1)(\sqrt{2}-1) = (\sqrt{2})^2 - (1)^2$$
Use a special product formula.

$$= 2 - 1$$

$$= 1$$

Radical expressions that involve the sum and difference of the same two terms, such as $\sqrt{2}+1$ and $\sqrt{2}-1$, are called **conjugates**.

EXAMPLE 9

ELEMENTARY &
INTERMEDIATE
Algebra *f(x)* **Now**™

Rationalize the denominator:

a. $\dfrac{1}{\sqrt{2} + 1}$ and   **b.** $\dfrac{\sqrt{x} + \sqrt{2}}{\sqrt{x} - \sqrt{2}}$.

Solution **a.** To find a fraction equivalent to $\dfrac{1}{\sqrt{2} + 1}$ that does not have a radical in its denominator, we multiply $\dfrac{1}{\sqrt{2} + 1}$ by a form of 1 that uses the conjugate of $\sqrt{2} + 1$.

$$\frac{1}{\sqrt{2} + 1} = \frac{1}{\sqrt{2} + 1} \cdot \frac{\sqrt{2} - 1}{\sqrt{2} - 1}$$

$$= \frac{\sqrt{2} - 1}{} \qquad \text{\textcolor{red}{Multiply the numerators.}}$$
$$\text{\textcolor{red}{Multiply the denominators.}}$$

$$= \frac{\sqrt{2} - 1}{2 - 1}$$

$$= \frac{\sqrt{2} - 1}{1}$$

$$= \sqrt{2} - 1$$

b. We multiply the numerator and denominator by $\sqrt{x} + \sqrt{2}$, which is the conjugate of $\sqrt{x} - \sqrt{2}$, and simplify.

$$\frac{\sqrt{x} + \sqrt{2}}{\sqrt{x} - \sqrt{2}} = \frac{\sqrt{x} + \sqrt{2}}{\sqrt{x} - \sqrt{2}} \cdot \frac{\sqrt{x} + \sqrt{2}}{\sqrt{x} + \sqrt{2}}$$

$$= \frac{x + \sqrt{2x} + \sqrt{2x} + 2}{} \qquad \text{\textcolor{red}{Multiply the numerators.}}$$
$$\text{\textcolor{red}{Multiply the denominators.}}$$

$$= \frac{x + \sqrt{2x} + \sqrt{2x} + 2}{x - 2}$$

$$= \frac{x + 2\sqrt{2x} + 2}{x - 2} \qquad \text{\textcolor{red}{In the numerator, combine like radicals.}}$$

Self Check 9 Rationalize the denominator: $\dfrac{\sqrt{x} - \sqrt{2}}{\sqrt{x} + \sqrt{2}}$.

■ RATIONALIZING NUMERATORS

In calculus, we sometimes have to rationalize a numerator by multiplying the numerator and denominator of the fraction by the conjugate of the numerator.

EXAMPLE 10

ELEMENTARY &
INTERMEDIATE
Algebra *f(x)* **Now**™

Rationalize the numerator:

$$\frac{\sqrt{x} - 3}{\sqrt{x}}$$

Solution   We multiply the numerator and denominator by $\sqrt{x} + 3$, which is the conjugate of the numerator.

$$\frac{\sqrt{x} - 3}{\sqrt{x}} = \frac{\sqrt{x} - 3}{\sqrt{x}} \cdot \frac{\sqrt{x} + 3}{\sqrt{x} + 3}$$ Multiply by a form of 1 to rationalize the numerator.

$$= \frac{\left(\sqrt{x}\right)^2 - (3)^2}{x + 3\sqrt{x}}$$ Multiply the numerators.
Multiply the denominators.

$$= \frac{x - 9}{x + 3\sqrt{x}}$$

The final expression is not in simplified form. However, this nonsimplified form is often desirable in calculus.

Self Check 10   Rationalize the numerator: $\dfrac{\sqrt{x} + 3}{\sqrt{x}}$.

Answers to Self Checks   **1. a.** $7\sqrt{2}$,   **b.** $-10\sqrt{14}$,   **c.** $18x\sqrt[4]{2x}$ **2.** $12\sqrt{10} - 32$

3. a. $-1 + \sqrt{15}$,   **b.** $a - \sqrt[3]{3a} + 9\sqrt[3]{2a^2} - 9\sqrt[3]{6}$ **4. a.** 11,   **b.** $108y$,

c. $x - 10\sqrt{x - 8} + 17$ **5. a.** $\dfrac{2\sqrt{30}}{5}$,   **b.** $\dfrac{5\sqrt[4]{27}}{3}$ **6.** $\dfrac{\sqrt{2ab}}{a}$ **7.** $\dfrac{\sqrt[3]{4h^2}}{4h^2}$

8. a. $\dfrac{11\sqrt[3]{10a^2b}}{10ab}$,   **b.** $\dfrac{\sqrt[4]{sxy^2}}{xy}$ **9.** $\dfrac{x - 2\sqrt{2x} + 2}{}$ **10.** $\dfrac{x - 9}{}$

9.4   STUDY SET ELEMENTARY & INTERMEDIATE Algebra *f(x)* Now™

VOCABULARY   **Fill in the blanks.**

1. To multiply $(\sqrt{3} + \sqrt{2})(\sqrt{3} - 2\sqrt{2})$, we can use the _____ method.

2. To multiply $2\sqrt{5}(3\sqrt{8} + \sqrt{3})$, use the _____ property to remove parentheses.

3. The denominator of the fraction $\dfrac{4}{\sqrt{5}}$ is an _____ number.

4. The _____ of $\sqrt{x} + 1$ is $\sqrt{x} - 1$.

5. To obtain a _____ cube radicand in the denominator of $\dfrac{\sqrt[3]{7}}{\sqrt[3]{5n}}$, we multiply the fraction by $\dfrac{\sqrt[3]{25n^2}}{\sqrt[3]{25n^2}}$.

6. To _____ the denominator of $\dfrac{4}{\sqrt{5}}$, we multiply the fraction by $\dfrac{\sqrt{5}}{\sqrt{5}}$.

CONCEPTS

7. Perform each operation, if possible.

   a. $4\sqrt{6} + 2\sqrt{6}$ **b.** $4\sqrt{6}(2\sqrt{6})$

   c. $3\sqrt{2} - 2\sqrt{3}$ **d.** $3\sqrt{2}(-2\sqrt{3})$

8. Perform each operation, if possible.

   a. $5 + 6\sqrt[3]{6}$ **b.** $5(6\sqrt[3]{6})$

   c. $\dfrac{30\sqrt[3]{15}}{5}$ **d.** $\dfrac{\sqrt[3]{15}}{5}$

9. Consider $\dfrac{\sqrt{3}}{\sqrt{7}} = \dfrac{\sqrt{3}}{\sqrt{7}} \cdot \dfrac{\sqrt{7}}{\sqrt{7}}$. Explain why the expressions on the left-hand side and the right-hand side of the equation are equal.

10. To rationalize the denominator of $\dfrac{\sqrt[4]{2}}{\sqrt[4]{3}}$, why wouldn't we multiply the numerator and denominator by $\dfrac{\sqrt[4]{3}}{\sqrt[4]{3}}$?

11. Explain why $\dfrac{\sqrt[3]{12}}{\sqrt[3]{5}}$ is not in simplified form.

12. Explain why $\sqrt{\dfrac{3a}{11k}}$ is not in simplified form.

NOTATION   Fill in the blanks.

13. Multiply: $5\sqrt{8} \cdot 7\sqrt{6}$.

$$5\sqrt{8} \cdot 7\sqrt{6} = 5(7)\sqrt{8\ \square}$$
$$= 35\sqrt{\square}$$
$$= 35\sqrt{\square \cdot 3}$$
$$= 35(\ \square\ )\sqrt{3}$$
$$= 140\sqrt{3}$$

14. Rationalize the denominator: $\dfrac{9}{\sqrt[3]{4a^2}}$.

$$\frac{9}{\sqrt[3]{4a^2}} = \frac{9}{\sqrt[3]{4a^2}} \cdot \frac{\sqrt[3]{2a}}{\square}$$
$$= \frac{9\sqrt[3]{2a}}{\sqrt[3]{\square}}$$
$$= \frac{9\sqrt[3]{2a}}{\square}$$

PRACTICE   Perform each multiplication and simplify, if possible. All variables represent positive real numbers.

15. $\sqrt{11}\sqrt{11}$ **16.** $\sqrt{35}\sqrt{35}$

17. $\left(\sqrt{7}\right)^2$ **18.** $\left(\sqrt{23}\right)^2$

19. $\sqrt{2}\sqrt{8}$ **20.** $\sqrt{3}\sqrt{27}$

21. $\sqrt{5} \cdot \sqrt{10}$ **22.** $\sqrt{7} \cdot \sqrt{35}$

23. $2\sqrt{3}\sqrt{6}$ **24.** $-3\sqrt{11}\sqrt{33}$

25. $\sqrt[3]{5}\sqrt[3]{25}$ **26.** $-\sqrt[3]{7}\sqrt[3]{49}$

27. $\left(3\sqrt{2}\right)^2$ **28.** $\left(2\sqrt{5}\right)^2$

29. $\left(-2\sqrt{2}\right)^2$ **30.** $\left(-3\sqrt{10}\right)^2$

31. $\left(3\sqrt[3]{9}\right)\left(2\sqrt[3]{3}\right)$ **32.** $\left(2\sqrt[3]{16}\right)\left(-\sqrt[3]{4}\right)$

33. $\sqrt[3]{2} \cdot \sqrt[3]{12}$ **34.** $\sqrt[3]{3} \cdot \sqrt[3]{18}$

35. $\sqrt{ab^3}\left(\sqrt{ab}\right)$ **36.** $\sqrt{8x}\left(\sqrt{2x^3 y}\right)$

37. $\sqrt{5ab}\sqrt{5a}$ **38.** $\sqrt{15rs^2}\sqrt{10r}$

39. $-4\sqrt[3]{5r^2 s}\left(5\sqrt[3]{2r}\right)$ **40.** $-\sqrt[3]{3xy^2}\left(-\sqrt[3]{9x^3}\right)$

41. $\sqrt{x(x+3)}\sqrt{x^3(x+3)}$ **42.** $\sqrt{y^2(x+y)}\sqrt{(x+y)^3}$

43. $\left(\sqrt[3]{9b}\right)^3$ **44.** $\left(\sqrt[3]{5c^2}\right)^3$

45. $\sqrt[4]{2a^3} \cdot \sqrt[4]{3a^2 b}$ **46.** $\sqrt[4]{9m^3 n} \cdot \sqrt[4]{2mn^5}$

47. $\sqrt[5]{2t}\left(\sqrt[5]{16t}\right)$ **48.** $\sqrt[5]{27y^3}\left(\sqrt[5]{9y^4}\right)$

Perform each multiplication and simplify. All variables represent positive real numbers.

49. $3\sqrt{5}\left(4 - \sqrt{5}\right)$ **50.** $2\sqrt{7}\left(3\sqrt{7} - 1\right)$

51. $3\sqrt{2}\left(4\sqrt{6} + 2\sqrt{7}\right)$ **52.** $-\sqrt{3}\left(\sqrt{7} - \sqrt{15}\right)$

53. $-2\sqrt{5x}\left(4\sqrt{2x} - 3\sqrt{3}\right)$

54. $3\sqrt{7t}\left(2\sqrt{7t} + 3\sqrt{3t^2}\right)$

55. $\left(\sqrt{2} + 1\right)\left(\sqrt{2} - 3\right)$

56. $\left(2\sqrt{3} + 1\right)\left(\sqrt{3} - 1\right)$

57. $\left(\sqrt[3]{5z} + \sqrt[3]{3}\right)\left(\sqrt[3]{5z} + 2\sqrt[3]{3}\right)$

58. $\left(\sqrt[3]{3p} - 2\sqrt[3]{2}\right)\left(\sqrt[3]{3p} + \sqrt[3]{2}\right)$

59. $\left(\sqrt{3x} - \sqrt{2y}\right)\left(\sqrt{3x} + \sqrt{2y}\right)$

60. $\left(\sqrt{3m} + \sqrt{2n}\right)\left(\sqrt{3m} - \sqrt{2n}\right)$

61. $\left(2\sqrt{3a} - \sqrt{b}\right)\left(\sqrt{3a} + 3\sqrt{b}\right)$

62. $\left(5\sqrt{p} - \sqrt{3q}\right)\left(\sqrt{p} + 2\sqrt{3q}\right)$

63. $\left(3\sqrt{2r} - 2\right)^2$ **64.** $\left(2\sqrt{3t} + 5\right)^2$

65. $-2\left(\sqrt{3x} + \sqrt{3}\right)^2$ **66.** $3\left(\sqrt{5x} - \sqrt{3}\right)^2$

67. $\left(2\sqrt[3]{4a^2} + 1\right)\left(\sqrt[3]{4a^2} - 3\right)$ **68.** $\left(\sqrt[3]{9b^2} - 5\right)\left(\sqrt[3]{9b^2} - 2\right)$

Simplify each radical expression by rationalizing the denominator. All variables represent positive real numbers.

69. $\sqrt{\dfrac{1}{7}}$ **70.** $\sqrt{\dfrac{5}{3}}$

71. $\dfrac{6}{\sqrt{30}}$ **72.** $\dfrac{8}{\sqrt{10}}$

73. $\dfrac{\sqrt{5}}{\sqrt{8}}$ **74.** $\dfrac{\sqrt{3}}{\sqrt{50}}$

75. $\dfrac{1}{\sqrt[3]{2}}$ **76.** $\dfrac{2}{\sqrt[3]{6}}$

77. $\dfrac{3}{\sqrt[3]{9}}$ **78.** $\dfrac{2}{\sqrt[3]{a}}$

79. $\dfrac{\sqrt[3]{2}}{\sqrt[3]{9}}$ **80.** $\dfrac{\sqrt[3]{9}}{\sqrt[3]{54}}$

81. $\dfrac{\sqrt{8}}{\sqrt{xy}}$ **82.** $\dfrac{\sqrt{9xy}}{\sqrt{3x^2 y}}$

83. $\dfrac{\sqrt{10xy^2}}{\sqrt{2xy^3}}$

84. $\dfrac{\sqrt{5ab^2c}}{\sqrt{10abc}}$

85. $\dfrac{\sqrt[3]{4a^2}}{\sqrt[3]{2ab}}$

86. $\dfrac{\sqrt[3]{9x}}{\sqrt[3]{3xy}}$

87. $\dfrac{1}{\sqrt[4]{4}}$

88. $\dfrac{1}{\sqrt[5]{2}}$

89. $\dfrac{1}{\sqrt[5]{16}}$

90. $\dfrac{4}{\sqrt[4]{32}}$

91. $\dfrac{\sqrt[4]{s}}{\sqrt[4]{3t^2}}$

92. $\dfrac{\sqrt[4]{c^2}}{\sqrt[4]{5b^3}}$

93. $\dfrac{t}{\sqrt[5]{27a}}$

94. $\dfrac{n}{\sqrt[5]{8m^2}}$

Rationalize each denominator. All variables represent positive real numbers.

95. $\dfrac{\sqrt{2}}{\sqrt{5}+3}$

96. $\dfrac{\sqrt{3}}{\sqrt{3}-2}$

97. $\dfrac{\sqrt{7}-\sqrt{2}}{\sqrt{2}+\sqrt{7}}$

98. $\dfrac{\sqrt{3}+\sqrt{2}}{\sqrt{3}-\sqrt{2}}$

99. $\dfrac{3\sqrt{2}-5\sqrt{3}}{2\sqrt{3}-3\sqrt{2}}$

100. $\dfrac{3\sqrt{6}+5\sqrt{5}}{2\sqrt{5}-3\sqrt{6}}$

101. $\dfrac{2}{\sqrt{x}+1}$

102. $\dfrac{3}{\sqrt{x}-2}$

103. $\dfrac{2z-1}{\sqrt{2z}-1}$

104. $\dfrac{3t-1}{\sqrt{3t}+1}$

105. $\dfrac{\sqrt{x}-\sqrt{y}}{\sqrt{x}+\sqrt{y}}$

106. $\dfrac{\sqrt{x}+\sqrt{y}}{\sqrt{x}-\sqrt{y}}$

107. $\dfrac{\sqrt{a}+3\sqrt{b}}{\sqrt{a}-3\sqrt{b}}$

108. $\dfrac{4\sqrt{m}-2\sqrt{n}}{\sqrt{n}+4\sqrt{m}}$

Rationalize each numerator. All variables represent positive numbers.

109. $\dfrac{\sqrt{x}+3}{x}$

110. $\dfrac{2+\sqrt{x}}{5x}$

111. $\dfrac{\sqrt{x}+\sqrt{y}}{\sqrt{x}}$

112. $\dfrac{\sqrt{x}-\sqrt{y}}{\sqrt{x}+\sqrt{y}}$

APPLICATIONS

113. STATISTICS   An example of a normal distribution curve, or *bell-shaped* curve, is shown below. A fraction that is part of the equation that models this curve is

$$\dfrac{1}{\sigma\sqrt{2\pi}}$$

where σ is a letter from the Greek alphabet. Rationalize the denominator of the fraction.

114. ANALYTIC GEOMETRY   The length of the perpendicular segment drawn from $(-2, 2)$ to the line with equation $2x - 4y = 4$ is given by

$$L = \dfrac{\left| 2(-2) + (-4)(2) + (-4) \right|}{\sqrt{(2)^2 + (-4)^2}}$$

Find L. Express the result in simplified radical form. Then give an approximation to the nearest tenth.

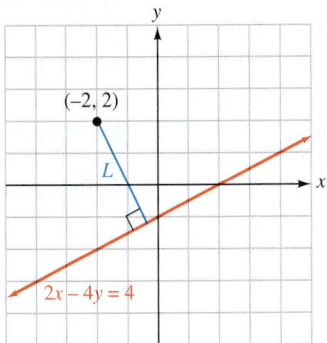

115. TRIGONOMETRY   In trigonometry, we must often find the ratio of the lengths of two sides of right triangles. Use the information in the illustration to find the ratio

$$\dfrac{\text{length of side } AC}{\text{length of side } AB}$$

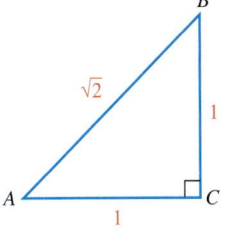

Write the result in simplified radical form.

116. ENGINEERING   A measure of how fast the block shown below will oscillate when the system is set in motion is given by the formula

$$\omega = \sqrt{\dfrac{k_1 + k_2}{m}}$$

where k_1 and k_2 indicate the stiffness of the springs and m is the mass of the block. Rationalize the right-hand side and restate the formula.

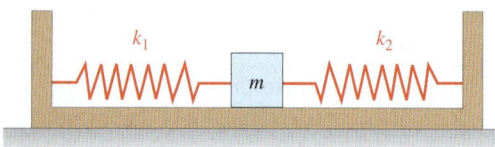

WRITING

117. Explain why $\sqrt{m} \cdot \sqrt{m} = m$ but $\sqrt[3]{m} \cdot \sqrt[3]{m} \neq m$. (Assume that $m > 0$.)

118. Explain why the product of $\sqrt{m} + 3$ and $\sqrt{m} - 3$ does not contain a radical.

REVIEW   **Solve each equation.**

119. $\dfrac{8}{b-2} + \dfrac{3}{2-b} = -\dfrac{1}{b}$

120. $\dfrac{2}{x-2} + \dfrac{1}{x+1} = \dfrac{1}{(x+1)(x-2)}$

CHALLENGE PROBLEMS

121. Multiply: $\sqrt{2} \cdot \sqrt[3]{2}$. (*Hint:* Keep in mind two things. The indices (plural for index) must be the same to use the product rule for radicals, and radical expressions can be written using rational exponents.)

122. Show that $\dfrac{\sqrt[3]{a^2} + \sqrt[3]{a}\sqrt[3]{b} + \sqrt[3]{b^2}}{\sqrt[3]{a^2} + \sqrt[3]{a}\sqrt[3]{b} + \sqrt[3]{b^2}}$ can be used to rationalize the denominator of $\dfrac{1}{\sqrt[3]{a} - \sqrt[3]{b}}$.

9.5   Solving Radical Equations

- The power rule   • Equations containing one radical
- Equations containing two radicals   • Solving formulas containing radicals

When we solve equations containing fractions, we clear them of the fractions by multiplying both sides by the LCD. To solve equations containing radical expressions, we take a similar approach. The first step is to clear them of the radicals. To do this, we raise both sides to a power.

■ **THE POWER RULE**

Radical equations contain a radical expression with a variable radicand. Some examples are

$$\sqrt{x+3} = 4 \qquad \sqrt[3]{x^3+7} = x+1 \qquad \sqrt{x} + \sqrt{x+2} = 2$$

To solve equations containing radicals, we will use the *power rule*.

The Power Rule	If we raise two equal quantities to the same power, the results are equal quantities. If x, y, and n are real numbers and $x = y$, then $$x^n = y^n$$

If both sides of an equation are raised to the same power, all solutions of the original equation are also solutions of the new equation. However, the resulting equation might not be equivalent to the original equation. For example, if we square both sides of the equation

(1) $x = 3$

with a solution set of $\{3\}$

we obtain the equation

(2) $x^2 = 9$

with a solution set of $\{3, -3\}$.

Equations 1 and 2 are not equivalent, because they have different solution sets, and the solution -3 of Equation 2 does not satisfy Equation 1. Since raising both sides of an equation to the same power can produce an equation with apparent solutions that don't satisfy the original equation, we must always check each apparent solution in the original equation and discard any *extraneous solutions*.

◼ EQUATIONS CONTAINING ONE RADICAL

To develop a method for solving any radical equation, let's see how the power rule can be used to solve an equation that contains a square root.

EXAMPLE 1

ELEMENTARY & INTERMEDIATE
Algebra *f(x)* **Now**™

Solve: $\sqrt{x + 3} = 4$.

Solution   To eliminate the radical, we use the power rule by squaring both sides of the equation and proceeding as follows:

The Language of Algebra

When we square both sides of an equation, we are *raising both sides* to the second power.

$$\sqrt{x + 3} = 4$$
$$\left(\sqrt{x + 3}\right)^2 = (4)^2 \qquad \text{Square both sides.}$$
$$x + 3 = 16$$
$$x = 13 \qquad \text{Subtract 3 from both sides.}$$

We must check the apparent solution 13 to see whether it satisfies the original equation.

Check: $\sqrt{x + 3} = 4$
$$\sqrt{13 + 3} \stackrel{?}{=} 4 \qquad \text{Substitute 13 for } x.$$
$$\sqrt{16} \stackrel{?}{=} 4$$
$$4 = 4$$

Since 13 satisfies the original equation, it is the solution. The solution set is $\{13\}$.

Self Check 1   Solve: $\sqrt{a - 2} = 3$.

The method used in Example 1 to solve a radical equation containing a square root can be generalized, as follows.

Solving an Equation Containing Radicals
1. Isolate one radical expression on one side of the equation.
2. Raise both sides of the equation to the power that is the same as the index of the radical.
3. Solve the resulting equation. If it still contains a radical, go back to step 1.
4. Check the results to eliminate extraneous solutions.

EXAMPLE 2

ELEMENTARY &
INTERMEDIATE
Algebra $f(x)$ **Now**™

Amusement park rides. The distance d in feet that an object will fall in t seconds is given by the formula

$$t = \sqrt{\frac{d}{16}}$$

If the designers of the amusement park attraction want the riders to experience 3 seconds of vertical free fall, what length of vertical drop is needed?

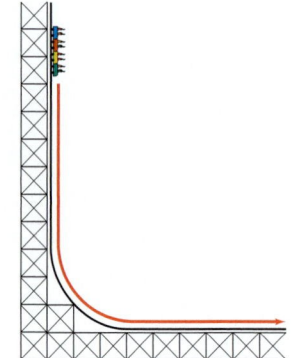

Solution   We substitute 3 for t in the formula and solve for d.

Caution

When using the power rule, don't forget to raise both sides to the same power. For this example, a common error would be to write

$$3 = \left(\sqrt{\frac{d}{16}}\right)^2$$

$$t = \sqrt{\frac{d}{16}}$$

$$\mathbf{3} = \sqrt{\frac{d}{16}} \qquad \text{Here the radical is isolated on the right-hand side.}$$

$$(3)^2 = \left(\sqrt{\frac{d}{16}}\right)^2 \qquad \begin{array}{l}\text{Use the power rule:} \\ \text{Raise both sides to the second power.}\end{array}$$

$$9 = \frac{d}{16} \qquad \text{Simplify.}$$

$$144 = d \qquad \text{Solve the resulting equation by multiplying both sides by 16.}$$

The amount of vertical drop needs to be 144 feet.

Self Check 2   How long a vertical drop is needed if the riders are to free fall for 3.5 seconds?

EXAMPLE 3

ELEMENTARY &
INTERMEDIATE
Algebra $f(x)$ **Now**™

Solve: $\sqrt{3x + 1} + 1 = x$.

Solution   We first subtract 1 from both sides to isolate the radical. Then, to eliminate the radical, we square both sides of the equation and proceed as follows:

Notation

We have seen that expressions with rational exponents can be written as radical expressions. The equation in this example,

$$\sqrt{3x + 1} + 1 = x$$

could also be written as

$$(3x + 1)^{1/2} + 1 = x$$

$$\sqrt{3x + 1} + 1 = x$$

$$\sqrt{3x + 1} = x - 1 \qquad \text{Subtract 1 from both sides.}$$

$$\left(\sqrt{3x + 1}\right)^2 = (x - 1)^2 \qquad \text{Square both sides to eliminate the square root.}$$

$$3x + 1 = x^2 - 2x + 1 \qquad \begin{array}{l}\text{On the right-hand side, use the FOIL method:} \\ (x - 1)^2 = (x - 1)(x - 1) = x^2 - x - x + 1 = x^2 - 2x + 1.\end{array}$$

$$0 = x^2 - 5x \qquad \begin{array}{l}\text{Subtract } 3x \text{ and 1 from both sides. This is a quadratic} \\ \text{equation. Use factoring to solve it.}\end{array}$$

$$0 = x(x - 5) \qquad \text{Factor } x^2 - 5x.$$

$$x = 0 \quad \text{or} \quad x - 5 = 0 \qquad \text{Set each factor equal to 0.}$$

$$x = 0 \qquad\qquad\quad x = 5$$

We must check each apparent solution to see whether it satisfies the original equation.

Check:

$$\sqrt{3x+1} + 1 = x$$
$$\sqrt{3(0)+1} + 1 \overset{?}{=} 0$$
$$\sqrt{1} + 1 \overset{?}{=} 0$$
$$2 \neq 0$$

$$\sqrt{3x+1} + 1 = x$$
$$\sqrt{3(5)+1} + 1 \overset{?}{=} 5$$
$$\sqrt{16} + 1 \overset{?}{=} 5$$
$$5 = 5$$

Since 0 does not check, it must be discarded. The only solution is 5.

Self Check 3 Solve: $\sqrt{4x+1} + 1 = x$.

ACCENT ON TECHNOLOGY: SOLVING RADICAL EQUATIONS

To find solutions for $\sqrt{3x+1} + 1 = x$ with a graphing calculator, we graph the functions $f(x) = \sqrt{3x+1} + 1$ and $g(x) = x$, as in figure (a). We then trace to find the approximate x-coordinate of their intersection point, as in figure (b). After repeated zooms, we will see that x is 5.

We can also use the INTERSECT feature to approximate the point of intersection of the graphs. See figure (c). The intersection point of (5, 5) implies that $x = 5$ is a solution of the radical equation.

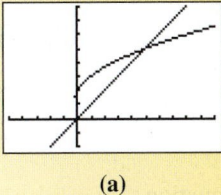
(a)

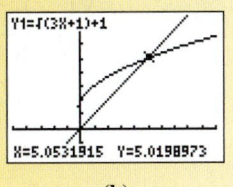
(b)

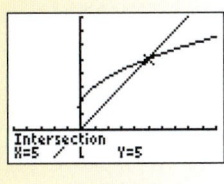
(c)

EXAMPLE 4

Solve: $\sqrt{3x} + 6 = 0$.

Solution To isolate the radical, we subtract 6 from both sides and proceed as follows:

$$\sqrt{3x} + 6 = 0$$
$$\sqrt{3x} = -6$$
$$\left(\sqrt{3x}\right)^2 = (-6)^2 \qquad \text{Square both sides to eliminate the square root.}$$
$$3x = 36$$
$$x = 12$$

We check the proposed solution 12 in the original equation.

$$\sqrt{3x} + 6 = 0$$
$$\sqrt{3(12)} + 6 \overset{?}{=} 0 \qquad \text{Substitute 12 for } x.$$
$$\sqrt{36} + 6 \overset{?}{=} 0$$
$$12 \neq 0$$

Success Tip

After isolating the radical, we obtained the equation $\sqrt{3x} = -6$. Since $\sqrt{3x}$ cannot be negative, we immediately know that $\sqrt{3x} + 6 = 0$ has no solution.

Since 12 does not satisfy the original equation, it is extraneous. The equation has no solution. The solution set is $\varnothing$.

Self Check 4 Solve: $\sqrt{4x+1} + 5 = 0$.

EXAMPLE 5

Solve: $\sqrt[3]{x^3 + 7} = x + 1$.

Solution   To eliminate the radical, we cube both sides of the equation and proceed as follows:

$$\sqrt[3]{x^3 + 7} = x + 1$$

$$\left(\sqrt[3]{x^3 + 7}\right)^3 = (x + 1)^3 \qquad \text{Cube both sides to eliminate the cube root.}$$

$$x^3 + 7 = x^3 + 3x^2 + 3x + 1 \qquad (x + 1)^3 = (x + 1)(x + 1)(x + 1).$$

$$0 = 3x^2 + 3x - 6 \qquad \text{Subtract } x^3 \text{ and 7 from both sides.}$$

$$0 = x^2 + x - 2 \qquad \text{Divide both sides by 3. To solve this quadratic equation, use factoring.}$$

$$0 = (x + 2)(x - 1) \qquad \text{Factor the trinomial.}$$

$$x + 2 = 0 \quad \text{or} \quad x - 1 = 0$$

$$x = -2 \quad | \quad x = 1$$

Success Tip

After raising both sides of a radical equation to a power, we use $\left(\sqrt[n]{a}\right)^n = a$ to simplify one side. For example:

$$\left(\sqrt[3]{x^3 + 7}\right)^3 = x^3 + 7$$

We check each apparent solution to see whether it satisfies the original equation.

Check:

$$\sqrt[3]{x^3 + 7} = x + 1 \qquad\qquad \sqrt[3]{x^3 + 7} = x + 1$$

$$\sqrt[3]{(-2)^3 + 7} \stackrel{?}{=} -2 + 1 \qquad\qquad \sqrt[3]{1^3 + 7} \stackrel{?}{=} 1 + 1$$

$$\sqrt[3]{-8 + 7} \stackrel{?}{=} -1 \qquad\qquad \sqrt[3]{1 + 7} \stackrel{?}{=} 2$$

$$\sqrt[3]{-1} \stackrel{?}{=} -1 \qquad\qquad \sqrt[3]{8} \stackrel{?}{=} 2$$

$$-1 = -1 \qquad\qquad 2 = 2$$

Both -2 and 1 satisfy the original equation.

Self Check 5   Solve: $\sqrt[3]{x^3 + 8} = x + 2$.

EXAMPLE 6

Let $f(x) = \sqrt[4]{2x + 1}$. For what value(s) of x is $f(x) = 5$?

ELEMENTARY & INTERMEDIATE

Algebra $f(x)$ **Now**™   **Solution**   To find the value(s) where $f(x) = 5$, we substitute 5 for $f(x)$ and solve for x.

$$f(x) = \sqrt[4]{2x + 1}$$

$$5 = \sqrt[4]{2x + 1}$$

Since the equation contains a fourth root, we raise both sides to the fourth power to solve for x.

$$(5)^4 = \left(\sqrt[4]{2x + 1}\right)^4 \qquad \text{Use the power rule to eliminate the radical.}$$

$$625 = 2x + 1$$

$$624 = 2x$$

$$312 = x$$

If x is 312, then $f(x) = 5$. Verify this by evaluating $f(312)$.

Self Check 6   Let $g(x) = \sqrt[5]{10x + 1}$. For what value(s) of x is $g(x) = 1$?

■ EQUATIONS CONTAINING TWO RADICALS

EXAMPLE 7

ELEMENTARY &
INTERMEDIATE
Algebra $f(x)$ **Now**™   Solution

Solve: $\sqrt{5x + 9} = 2\sqrt{3x + 4}$.

Each radical is isolated on one side of the equation, so we square both sides to eliminate them.

$$\sqrt{5x + 9} = 2\sqrt{3x + 4}$$
$$\left(\sqrt{5x + 9}\right)^2 = \left(2\sqrt{3x + 4}\right)^2 \qquad \text{Square both sides.}$$
$$5x + 9 = 4(3x + 4) \qquad \begin{array}{l}\text{On the right-hand side:}\\ \left(2\sqrt{3x + 4}\right)^2 = 2^2\left(\sqrt{3x + 4}\right)^2 = 4(3x + 4).\end{array}$$
$$5x + 9 = 12x + 16 \qquad \text{Remove parentheses.}$$
$$-7 = 7x \qquad \text{Subtract } 5x \text{ and } 16 \text{ from both sides.}$$
$$-1 = x \qquad \text{Divide both sides by 7.}$$

> **Caution**
> When finding $\left(2\sqrt{3x + 4}\right)^2$, remember to square both 2 and $\sqrt{3x + 4}$ to get:
> $$2^2\left(\sqrt{3x + 4}\right)^2$$

We check the solution by substituting -1 for x in the original equation.

$$\sqrt{5x + 9} = 2\sqrt{3x + 4}$$
$$\sqrt{5(-1) + 9} \stackrel{?}{=} 2\sqrt{3(-1) + 4} \qquad \text{Substitute } -1 \text{ for } x.$$
$$\sqrt{4} \stackrel{?}{=} 2\sqrt{1}$$
$$2 = 2$$

The solution checks.

Self Check 7   Solve: $\sqrt{x - 4} = 2\sqrt{x - 16}$.

When more than one radical appears in an equation, we can use the power rule more than once.

EXAMPLE 8

Solution

Solve: $\sqrt{x} + \sqrt{x + 2} = 2$.

To remove the radicals, we square both sides of the equation. Since this is easier to do if one radical is on each side of the equation, we subtract $\sqrt{x}$ from both sides to isolate $\sqrt{x + 2}$ on the left-hand side.

$$\sqrt{x} + \sqrt{x + 2} = 2$$
$$\sqrt{x + 2} = 2 - \sqrt{x} \qquad \text{Subtract } \sqrt{x} \text{ from both sides.}$$
$$\left(\sqrt{x + 2}\right)^2 = \left(2 - \sqrt{x}\right)^2 \qquad \text{Square both sides to eliminate the square root.}$$
$$x + 2 = 4 - 4\sqrt{x} + x \qquad \begin{array}{l}\text{Use FOIL: } \left(2 - \sqrt{x}\right)^2 = \left(2 - \sqrt{x}\right)\left(2 - \sqrt{x}\right) =\\ 4 - 2\sqrt{x} - 2\sqrt{x} + x = 4 - 4\sqrt{x} + x.\end{array}$$
$$2 = 4 - 4\sqrt{x} \qquad \text{Subtract } x \text{ from both sides.}$$
$$-2 = -4\sqrt{x} \qquad \text{Subtract 4 from both sides.}$$
$$\frac{1}{2} = \sqrt{x} \qquad \text{Divide both sides by } -4 \text{ and simplify.}$$
$$\frac{1}{4} = x \qquad \text{Square both sides.}$$

> **Caution**
> When finding $\left(2 - \sqrt{x}\right)^2$, remember to use FOIL or a special product formula:
> $$\left(2 - \sqrt{x}\right)^2 \neq 2^2 - \left(\sqrt{x}\right)^2$$

Check: $\sqrt{x} + \sqrt{x+2} = 2$

$$\sqrt{\frac{1}{4}} + \sqrt{\frac{1}{4} + 2} \overset{?}{=} 2$$

$$\frac{1}{2} + \sqrt{\frac{9}{4}} \overset{?}{=} 2$$

$$\frac{1}{2} + \frac{3}{2} \overset{?}{=} 2$$

$$2 = 2$$

The result $\frac{1}{4}$ checks.

Self Check 8   Solve: $\sqrt{a} + \sqrt{a+3} = 3$.

ACCENT ON TECHNOLOGY: SOLVING RADICAL EQUATIONS

To find solutions for $\sqrt{x} + \sqrt{x+2} = 4$ (an equation similar to Example 8) with a graphing calculator, we graph the functions $f(x) = \sqrt{x} + \sqrt{x+2}$ and $g(x) = 4$. We then trace to find an approximation of the x-coordinate of their intersection point, as in figure (a). From the figure, we can see that $x \approx 2.98$. We can zoom to get better results.

Figure (b) shows that the INTERSECT feature gives the approximate coordinates of the point of intersection of the two graphs as (3.06, 4). Therefore, an approximate solution of the radical equation is 3.06. Check its reasonableness.

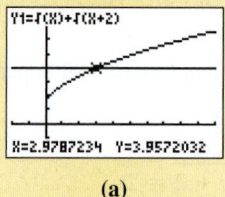

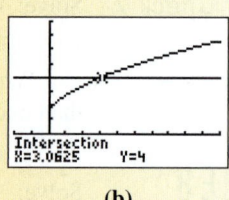

(a)   (b)

■ SOLVING FORMULAS CONTAINING RADICALS

To *solve a formula for a variable* means to isolate that variable on one side of the equation, with all other quantities on the other side.

EXAMPLE 9

Depreciation rates.   Some office equipment that is now worth V dollars originally cost C dollars 3 years ago. The rate r at which it has depreciated is given by

$$r = 1 - \sqrt[3]{\frac{V}{C}}$$

Solve the formula for C.

Solution   We begin by isolating the cube root on the right-hand side of the equation.

$$r = 1 - \sqrt[3]{\dfrac{V}{C}}$$

$$r - 1 = -\sqrt[3]{\dfrac{V}{C}} \qquad \text{Subtract 1 from both sides.}$$

$$(r - 1)^3 = \left[-\sqrt[3]{\dfrac{V}{C}} \right]^3 \qquad \text{To eliminate the radical, cube both sides.}$$

$$(r - 1)^3 = -\dfrac{V}{C} \qquad \text{Simplify the right-hand side.}$$

$$C(r - 1)^3 = -V \qquad \text{Multiply both sides by } C.$$

$$C = -\dfrac{V}{(r - 1)^3} \qquad \text{Divide both sides by } (r - 1)^3.$$

Self Check 9   A formula used in statistics to determine the size of a sample to obtain a desired degree of accuracy is

$$E = z_0 \sqrt{\dfrac{pq}{n}}$$

Solve the formula for n.

Answers to Self Checks   **1.** 11   **2.** 196 ft   **3.** 6, 0 is extraneous   **4.** 6 is extraneous, no solution   **5.** 0, -2 **6.** 0

7. 20   **8.** 1   **9.** $n = \dfrac{z_0^2 pq}{E^2}$

9.5   STUDY SET   ELEMENTARY & INTERMEDIATE Algebra $f(x)$ Now™

VOCABULARY   Fill in the blanks.

1. Equations such as $\sqrt{x + 4} - 4 = 5$ and $\sqrt[3]{x + 1} = 12$ are called _____ equations.

2. When solving equations containing radicals, try to _____ one radical expression on one side of the equation.

3. Squaring both sides of an equation can introduce _____ solutions.

4. To _____ an apparent solution means to substitute it into the original equation and see whether a true statement results.

CONCEPTS

5. Fill in the blanks: The power rule states that if x, y, and n are real numbers and $x = y$, then
$$x^{\blacksquare} = y^{\blacksquare}$$

6. Determine whether 6 is a solution of each radical equation.

a. $\sqrt{x + 3} = x - 3$

b. $\sqrt{4x + 1} = \sqrt{6x - 1}$

c. $\sqrt[3]{5x - 3} = x - 9$

7. What is the first step in solving each equation?

   a. $\sqrt{x + 11} = 5$

   b. $2 = \sqrt[3]{x - 2}$

   c. $\sqrt[3]{5x + 4} + 3 = 30$

8. What is the first step in solving each equation?

   a. $\sqrt{5x + 4} + \sqrt{x + 4} = 0$

   b. $\sqrt{x + 8} - \sqrt{2x + 9} = 1$

9. Simplify each expression.

   a. $\left(\sqrt{x}\right)^2$ **b.** $\left(\sqrt{x - 5}\right)^2$

   c. $\left(4\sqrt{2x}\right)^2$ **d.** $\left(-\sqrt{x + 3}\right)^2$

10. Simplify each expression.

   a. $\left(\sqrt[3]{x}\right)^3$ **b.** $\left(\sqrt[4]{x}\right)^4$

   c. $\left(-\sqrt[3]{2x}\right)^3$ **d.** $\left(2\sqrt[3]{x + 3}\right)^3$

11. Multiply.

   a. $\left(\sqrt{x} - 3\right)^2$

   b. $\left(\sqrt{2y + 1} + 5\right)^2$

12. Perform the necessary steps to isolate the radical on the right-hand side of the equation.

$$3x - 6 = 2x + \sqrt{2y + 1} - 9$$

13. Explain why it is immediately apparent that $\sqrt{8x - 7} = -2$ has no solution.

14. Solve $\sqrt{x - 2} + 2 = 4$ graphically, using the graphs in the illustration.

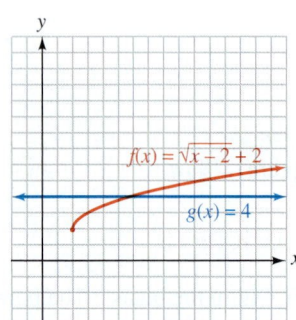

NOTATION   Complete each solution.

15. Solve: $\sqrt{3x} - 1 = 5$

$$\sqrt{3x} = \boxed{}$$
$$\left(\sqrt{3x}\right)^{\boxed{}} = (6)^{\boxed{}}$$
$$\boxed{} = 36$$
$$x = 12$$

Does 12 check?

16. Solve: $\sqrt{1 - 2x} = \sqrt{x + 10}$.

$$\left(\boxed{}\right)^2 = \left(\sqrt{x + 10}\right)^2$$
$$\boxed{} = x + 10$$
$$\boxed{} = 9$$
$$x = -3$$

Does -3 check?

PRACTICE   Solve each equation. Write all apparent solutions. Cross out those that are extraneous.

17. $\sqrt{5x - 6} = 2$ **18.** $\sqrt{7x - 10} = 12$

19. $\sqrt{6x + 1} + 2 = 7$ **20.** $\sqrt{6x + 13} - 2 = 5$

21. $2\sqrt{4x + 1} = \sqrt{x + 4}$

22. $\sqrt{3(x + 4)} = \sqrt{5x - 12}$

23. $\sqrt[3]{7n - 1} = 3$ **24.** $\sqrt[3]{12m + 4} = 4$

25. $\sqrt[4]{10p + 1} = \sqrt[4]{11p - 7}$

26. $\sqrt[4]{10y + 6} = 2\sqrt[4]{y}$

27. $(6x + 2)^{1/2} = (5x + 3)^{1/2}$

28. $(5x + 3)^{1/2} = (x + 11)^{1/2}$

29. $(x + 8)^{1/3} = -2$ **30.** $(x + 4)^{1/3} = -1$

31. $\sqrt{5 - x} + 10 = 9$

32. $1 = 2 + \sqrt{4x + 75}$

33. $x = \dfrac{\sqrt{12x - 5}}{2}$ **34.** $x = \dfrac{\sqrt{16x - 12}}{2}$

35. $\sqrt{x + 2} - \sqrt{4 - x} = 0$

36. $\sqrt{6 - x} - \sqrt{2x + 3} = 0$

37. $2\sqrt{x} = \sqrt{5x - 16}$ **38.** $3\sqrt{x} = \sqrt{3x + 54}$

39. $r - 9 = \sqrt{2r - 3}$ **40.** $-s - 3 = 2\sqrt{5 - s}$

41. $(m^4 + m^2 - 25)^{1/4} = m$

42. $n = (n^3 + n^2 - 1)^{1/3}$

43. $\sqrt{-5x + 24} = 6 - x$ **44.** $\sqrt{-x + 2} = x - 2$

45. $\sqrt{y + 2} = 4 - y$ **46.** $\sqrt{22y + 86} = y + 9$

47. $\sqrt[3]{x^3 - 7} = x - 1$ **48.** $\sqrt[3]{x^3 + 56} - 2 = x$

49. $\sqrt[4]{x^4 + 4x^2 - 4} = -x$ **50.** $u = \sqrt[4]{u^4 - 6u^2 + 24}$

51. $\sqrt[4]{12t + 4} + 2 = 0$ **52.** $\sqrt[4]{8x - 8} + 2 = 0$

53. $\sqrt{2y + 1} = 1 - 2\sqrt{y}$ **54.** $\sqrt{u + 3} = \sqrt{u - 3}$

55. $\sqrt{n^2 + 6n + 3} = \sqrt{n^2 - 6n - 3}$

56. $\sqrt{m^2 - 12m - 3} = \sqrt{m^2 + 12m + 3}$

57. $\sqrt{7t^2 + 4} = \sqrt{17t - 8t^2}$

58. $\sqrt{b^2 + b} = \sqrt{3 - b^2}$

59. $\sqrt{y + 7} + 3 = \sqrt{y + 4}$

60. $1 + \sqrt{z} = \sqrt{z + 3}$

61. $2 + \sqrt{u} = \sqrt{2u + 7}$

62. $5r + 4 = \sqrt{5r + 20} + 4r$

63. $\sqrt{6t + 1} - 3\sqrt{t} = -1$

64. $\sqrt{4s + 1} - \sqrt{6s} = -1$

65. $\sqrt{2x + 5} + \sqrt{x + 2} = 5$

66. $\sqrt{2x + 5} + \sqrt{2x + 1} + 4 = 0$

67. $\sqrt{x - 5} - \sqrt{x + 3} = 4$

68. $\sqrt{x + 8} - \sqrt{x - 4} = -2$

69. Let $f(x) = \sqrt{3x - 6}$. For what value(s) of x is $f(x) = 3$?

70. Let $f(x) = \sqrt{2x^2 - 7x}$. For what value(s) of x is $f(x) = 2$?

71. Let $f(x) = \sqrt{x + 8} - \sqrt{x}$. For what value(s) of x is $f(x) = 2$?

72. Let $f(x) = \sqrt{x} - \sqrt{x + 5}$. For what value(s) of x is $f(x) = -1$?

Solve each equation for the indicated variable.

73. $v = \sqrt{2gh}$ for h

74. $d = 1.4\sqrt{h}$ for h

75. $T = 2\pi\sqrt{\dfrac{\ell}{32}}$ for ℓ

76. $d = \sqrt[3]{\dfrac{12V}{\pi}}$ for V

77. $r = \sqrt[3]{\dfrac{A}{P}} - 1$ for A

78. $r = \sqrt[3]{\dfrac{A}{P}} - 1$ for P

79. $L_A = L_B\sqrt{1 - \dfrac{v^2}{c^2}}$ for v^2

80. $R_1 = \sqrt{\dfrac{A}{\pi} - R_2^2}$ for A

APPLICATIONS

81. HIGHWAY DESIGN   A curved road will accommodate traffic traveling s mph if the radius of the curve is r feet, according to the formula $s = 3\sqrt{r}$. If engineers expect 40-mph traffic, what radius should they specify? Give the result to the nearest foot.

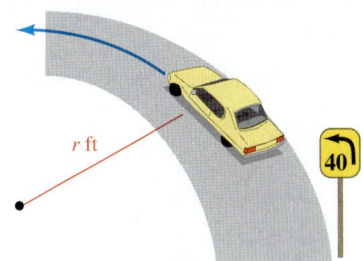

82. FORESTRY   The taller a lookout tower, the farther an observer can see. That distance d (called the *horizon distance,* measured in miles) is related to the height h of the observer (measured in feet) by the formula $d = 1.22\sqrt{h}$. How tall must a lookout tower be to see the edge of the forest, 25 miles away? (Round to the nearest foot.)

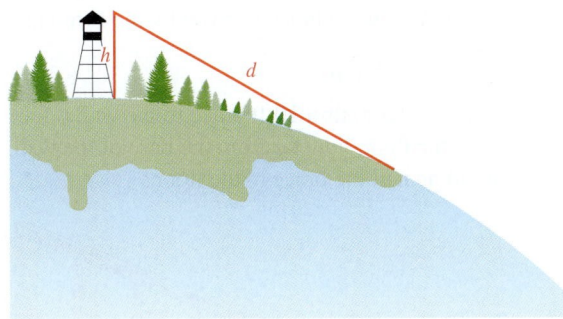

83. WIND POWER   The power generated by a windmill is related to the velocity of the wind by the formula

$$v = \sqrt[3]{\dfrac{P}{0.02}}$$

where P is the power (in watts) and v is the velocity of the wind (in mph). Find how much power the windmill is generating when the wind is 29 mph.

84. DIAMONDS   The *effective rate of interest r* earned by an investment is given by the formula

$$r = \sqrt[n]{\dfrac{A}{P}} - 1$$

where P is the initial investment that grows to value A after n years. If a diamond buyer got \$4,000 for a 1.73-carat diamond that he had purchased 4 years earlier, and earned an annual rate of return of 6.5% on the investment, what did he originally pay for the diamond?

85. THEATER PRODUCTIONS   The ropes, pulleys, and sandbags shown in the illustration are part of a mechanical system used to raise and lower scenery for a stage play. For the scenery to be in the proper position, the following formula must apply:

$$w_2 = \sqrt{w_1^2 + w_3^2}$$

If $w_2 = 12.5$ lb and $w_3 = 7.5$ lb, find w_1.

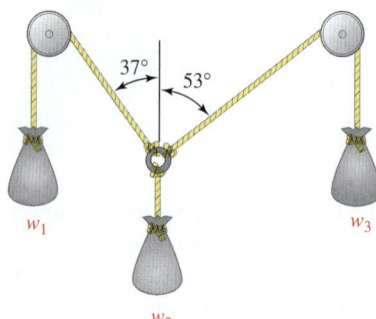

86. CARPENTRY   During construction, carpenters often brace walls as shown in the illustration, where the length ℓ of the brace is given by the formula

$$\ell = \sqrt{f^2 + h^2}$$

If a carpenter nails a 10-ft brace to the wall 6 feet above the floor, how far from the base of the wall should he nail the brace to the floor?

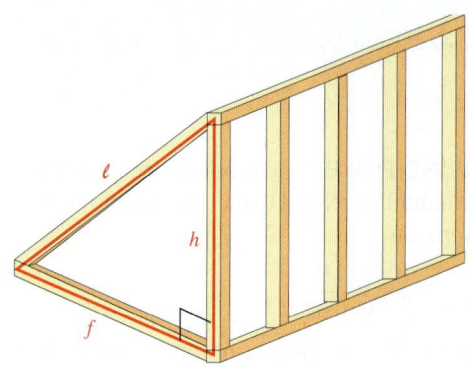

87. SUPPLY AND DEMAND   The number of wrenches that will be produced at a given price can be predicted by the formula $s = \sqrt{5x}$, where s is the supply (in thousands) and x is the price (in dollars). The demand d for wrenches can be predicted by the formula $d = \sqrt{100 - 3x^2}$. Find the equilibrium price—that is, find the price at which supply will equal demand.

88. SUPPLY AND DEMAND   The number of mirrors that will be produced at a given price can be predicted by the formula $s = \sqrt{23x}$, where s is the supply (in thousands) and x is the price (in dollars). The demand d for mirrors can be predicted by the formula $d = \sqrt{312 - 2x^2}$. Find the equilibrium price—that is, find the price at which supply will equal demand.

WRITING

89. If both sides of an equation are raised to the same power, the resulting equation might not be equivalent to the original equation. Explain.

90. What is wrong with the student's work shown below?

Solve: $\sqrt{x + 1} - 3 = 8$.
$$\sqrt{x + 1} = 11$$
$$\left(\sqrt{x + 1}\right)^2 = 11$$
$$x + 1 = 11$$
$$x = 10$$

91. The first step of a student's solution is shown below. What is a better way to begin the solution?

Solve: $\sqrt{x} + \sqrt{x + 22} = 12$.
$$\left(\sqrt{x} + \sqrt{x + 22}\right)^2 = 10^2$$

92. Explain how $\sqrt{2x - 1} = x$ can be solved graphically.

93. Explain how the table can be used to solve

$$\sqrt{4x - 3} - 2 = \sqrt{2x - 5}$$

if $Y_1 = \sqrt{4x - 3} - 2$ and $Y_2 = \sqrt{2x - 5}$.

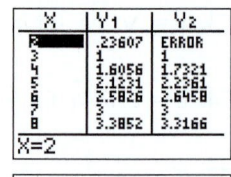

94. Explain how to use the graph of $f(x) = \sqrt[3]{x - 0.5} - 1$, shown in the illustration, to approximate the solution of

$$\sqrt[3]{x - 0.5} = 1.$$

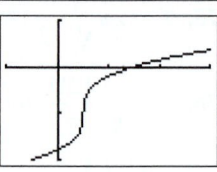

REVIEW

95. LIGHTING   The intensity of light from a light bulb varies inversely as the square of the distance from the bulb. If you are 5 feet away from a bulb and the intensity is 40 foot-candles, what will the intensity be if you move 20 feet away from the bulb?

96. COMMITTEES   What type of variation is shown in the illustration? As the number of people on this committee increased, what happened to its effectiveness?

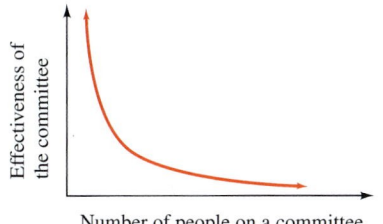

Number of people on a committee

97. TYPESETTING   If 12-point type is 0.166044 inch tall, how tall is 30-point type?

98. GUITAR STRINGS   The frequency of vibration of a string varies directly as the square root of the tension and inversely as the length of the string. Suppose a string 2.5 feet long, under a tension of 16 pounds, vibrates 25 times per second. Find k, the constant of proportionality.

CHALLENGE PROBLEMS   Solve each equation.

99. $\sqrt[3]{2x} = \sqrt{x}$ (*Hint:* Square and then cube both sides.)

100. $\sqrt[4]{x} = \sqrt{\dfrac{x}{4}}$

101. $\sqrt{x + 2} + \sqrt{2x} = \sqrt{18 - x}$

102. $\sqrt{8 - x} - \sqrt{3x - 8} = \sqrt{x - 4}$

9.6   Geometric Applications of Radicals

- The Pythagorean theorem
- $45°$–$45°$–$90°$ triangles
- $30°$–$60°$–$90°$ triangles
- The distance formula

We will now consider applications of square roots in geometry. Then we will find the distance between two points on a rectangular coordinate system, using a formula that contains a square root. We begin by considering an important theorem about right triangles.

The Language of Algebra

A *theorem* is a mathematical statement that can be proved. The *Pythagorean theorem* is named after *Pythagoras,* a Greek mathematician who lived about 2,500 years ago. He is thought to have been the first to prove the theorem.

■ THE PYTHAGOREAN THEOREM

If we know the lengths of two legs of a right triangle, we can find the length of the **hypotenuse** (the side opposite the $90°$ angle) by using the **Pythagorean theorem.**

The Pythagorean Theorem   If a and b are the lengths of the legs of a right triangle and c is the length of the hypotenuse.

$$a^2 + b^2 = c^2$$

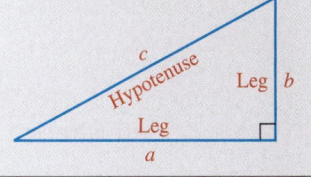

In words, the Pythagorean theorem is expressed as follows:

In any right triangle, the square of the hypotenuse is equal to the sum of the squares of the two legs.

Suppose the right triangle shown in the figure has legs of length 3 and 4 units. To find the length of the hypotenuse, we use the Pythagorean theorem.

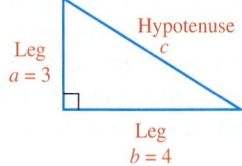

Leg
$a = 3$

Hypotenuse
c

Leg
$b = 4$

$$a^2 + b^2 = c^2$$
$$3^2 + 4^2 = c^2 \quad \text{Substitute 3 for } a \text{ and 4 for } b.$$
$$9 + 16 = c^2$$
$$25 = c^2$$

To find c, we ask "What number, when squared, is equal to 25?" There are two such numbers: the positive square root of 25 and the negative square root of 25. Since c represents the length of the hypotenuse, and it cannot be negative, it follows that c is the positive square root of 25.

$$\sqrt{25} = c \quad \text{Recall that a radical symbol } \sqrt{} \text{ is used to represent the positive, or principal square root of a number.}$$
$$5 = c$$

The length of the hypotenuse is 5 units.

EXAMPLE 1

ELEMENTARY & INTERMEDIATE
Algebra $f(x)$ **Now**™

Firefighting.   To fight a fire, the forestry department plans to clear a rectangular fire break around the fire, as shown in the illustration. Crews are equipped with mobile communications that have a 3,000-yard range. Can crews at points A and B remain in radio contact?

Solution   Points A, B, and C form a right triangle. To find the distance c from point A to point B, we can use the Pythagorean theorem, substituting 2,400 for a and 1,000 for b and solving for c.

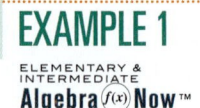

A

1,000 yd

c yd

C

2,400 yd

B

> **Caution**
>
> When using the Pythagorean theorem $a^2 + b^2 = c^2$, we can let a represent the length of either leg of the right triangle. We then let b represent the length of the other leg. The variable c must always represent the length of the hypotenuse.

$$a^2 + b^2 = c^2$$
$$2{,}400^2 + 1{,}000^2 = c^2$$
$$5{,}760{,}000 + 1{,}000{,}000 = c^2$$
$$6{,}760{,}000 = c^2$$
$$\sqrt{6{,}760{,}000} = c \quad \text{If } c^2 = 6{,}760{,}000, \text{ then } c \text{ must be a square root of } 6{,}760{,}000. \text{ Because } c \text{ represents a length, it must be the positive square root of } 6{,}760{,}000.$$
$$2{,}600 = c \quad \text{Use a calculator to find the square root.}$$

The two crews are 2,600 yards apart. Because this distance is less than the range of the radios, they can communicate by radio.

Self Check 1   Can the crews communicate if $b = 1{,}500$ yards?

◼ 45°–45°–90° TRIANGLES

An **isosceles right triangle** is a right triangle with two legs of equal length. Isosceles right triangles have angle measures of 45°, 45°, and 90°. If we know the length of one leg of an isosceles right triangle, we can use the Pythagorean theorem to find the length of the hypotenuse. Since the triangle shown in the figure is a right triangle, we have

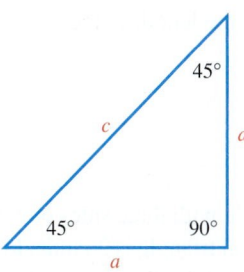

$$c^2 = a^2 + b^2$$

$$c^2 = a^2 + a^2$$ Both legs are a units long, so replace b with a.

$$c^2 = 2a^2$$ Combine like terms.

$$c = \sqrt{2a^2}$$ If $c^2 = 2a^2$, then c must be a square root of $2a^2$. Because c represents a length, it must be the positive square root of $2a^2$.

$$c = a\sqrt{2}$$ Simplify the radical: $\sqrt{2a^2} = \sqrt{2}\sqrt{a^2} = \sqrt{2}a = a\sqrt{2}$.

Thus, *in an isosceles right triangle, the length of the hypotenuse is the length of one leg times* $\sqrt{2}$.

EXAMPLE 2

ELEMENTARY &
INTERMEDIATE
Algebra $f(x)$ **Now**™

If one leg of the isosceles right triangle shown above is 10 feet long, find the length of the hypotenuse.

Solution Since the length of the hypotenuse is the length of a leg times $\sqrt{2}$, we have

$$c = 10\sqrt{2}$$

The length of the hypotenuse is $10\sqrt{2}$ feet. To two decimal places, the length is 14.14 feet.

Self Check 2 Find the length of the hypotenuse of an isosceles right triangle if one leg is 12 meters long.

If the length of the hypotenuse of an isosceles right triangle is known, we can use the Pythagorean theorem to find the length of each leg.

EXAMPLE 3

ELEMENTARY &
INTERMEDIATE
Algebra $f(x)$ **Now**™

Find the exact length of each leg of the following isosceles right triangle.

Solution We use the Pythagorean theorem.

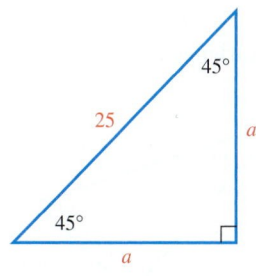

$$c^2 = a^2 + b^2$$

$$25^2 = a^2 + a^2$$ Since both legs are a units long, substitute a for b. The hypotenuse is 25 units long. Substitute 25 for c.

$$25^2 = 2a^2$$ Combine like terms.

$$\frac{625}{2} = a^2$$ Square 25 and divide both sides by 2.

$$\sqrt{\frac{625}{2}} = a$$ If $a^2 = \frac{625}{2}$, then a must be the positive square root of $\frac{625}{2}$.

$$\frac{\sqrt{625}}{\sqrt{2}} \cdot \frac{\sqrt{2}}{\sqrt{2}} = a$$ Write $\sqrt{\frac{625}{2}}$ as $\frac{\sqrt{625}}{\sqrt{2}}$. Then rationalize the denominator.

$$\frac{25\sqrt{2}}{2} = a$$ In the numerator, simplify the radical: $\sqrt{625} = 25$. In the denominator, do the multiplication: $\sqrt{2} \cdot \sqrt{2} = 2$.

The exact length of each leg is $\dfrac{25\sqrt{2}}{2}$ units. To two decimal places, the length is 17.68 units.

Self Check 3   Find the exact length of each leg of an isosceles right triangle if the length of the hypotenuse is 9 inches.

■ 30°–60°–90° TRIANGLES

From geometry, we know that an **equilateral triangle** is a triangle with three sides of equal length and three 60° angles. Each side of the following equilateral triangle is $2a$ units long. If an **altitude** is drawn to its base the altitude bisects the base and divides the equilateral triangle, into two 30°–60°–90° triangles. We can see that the shorter leg of each 30°–60°–90° triangle (the side *opposite* the 30° angle) is a units long. Thus,

> *The length of the shorter leg of a 30°–60°–90° right triangle is half as long as the hypotenuse.*

We can discover another important relationship between the legs of a 30°–60°–90° triangle if we find the length of the altitude h in the figure. We begin by applying the Pythagorean theorem to one of the 30°–60°–90° triangles.

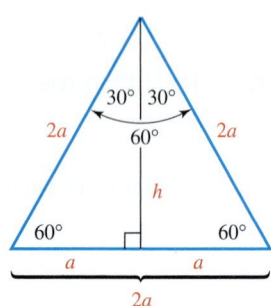

$$a^2 + \mathbf{b}^2 = \mathbf{c}^2$$
$$a^2 + \mathbf{h}^2 = (\mathbf{2a})^2 \qquad \text{One leg is } h \text{ units long, so replace } b \text{ with } h. \text{ The hypotenuse is } 2a \text{ units long, so replace } c \text{ with } 2a.$$
$$a^2 + h^2 = 4a^2 \qquad (2a)^2 = (2a)(2a) = 4a^2.$$
$$h^2 = 3a^2 \qquad \text{Subtract } a^2 \text{ from both sides.}$$
$$h = \sqrt{3a^2} \qquad \text{If } h^2 = 3a^2, \text{ then } h \text{ must be the positive square root of } 3a^2.$$
$$h = a\sqrt{3} \qquad \text{Simplify the radical: } \sqrt{3a^2} = \sqrt{3}\sqrt{a^2} = a\sqrt{3}.$$

We see that the altitude—the longer leg of the 30°–60°–90° triangle—is $\sqrt{3}$ times as long as the shorter leg. Thus,

> *The length of the longer leg of a 30°–60°–90° triangle is the length of the shorter leg times $\sqrt{3}$.*

··

EXAMPLE 4   Find the length of the hypotenuse and the longer leg of the right triangle.

Solution   Since the shorter leg of a 30°–60°–90° triangle is half as long as the hypotenuse, the hypotenuse is 12 centimeters long.

Since the length of the longer leg is the length of the shorter leg times $\sqrt{3}$, the longer leg is $6\sqrt{3}$ (about 10.39) centimeters long.

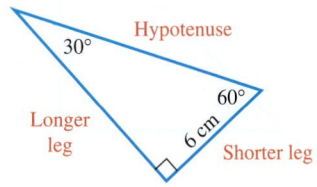

Self Check 4   Find the length of the hypotenuse and the longer leg of a 30°–60°–90° triangle if the shorter leg is 8 centimeters long.

EXAMPLE 5

ELEMENTARY &
INTERMEDIATE
Algebra *f(x)* **Now**™

Stretching exercises. A doctor prescribed the exercise shown in figure (a). The patient was instructed to raise his leg to an angle of 60° and hold the position for 10 seconds. If the patient's leg is 36 inches long, how high off the floor will his foot be when his leg is held at the proper angle?

Solution   In figure (b), we see that a 30°–60°–90° triangle, which we will call triangle *ABC*, models the situation. Since the side opposite the 30° angle of a 30°–60°–90° triangle is half as long as the hypotenuse, side *AC* is 18 inches long.

   Since the length of the side opposite the 60° angle is the length of the side opposite the 30° angle times $\sqrt{3}$, side *BC* is $18\sqrt{3}$, or about 31 inches long. So the patient's foot will be about 31 inches from the floor when his leg is in the proper position.

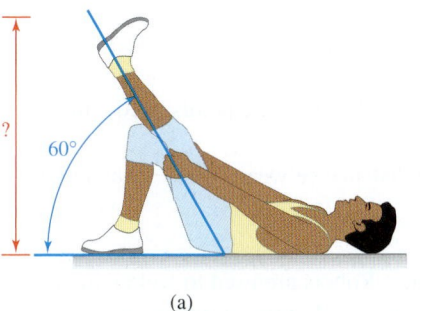

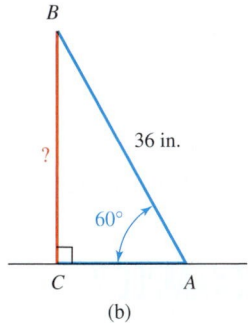

(a)   (b)

■ THE DISTANCE FORMULA

With the *distance formula,* we can find the distance between any two points graphed on a rectangular coordinate system.

   To find the distance *d* between points $P(x_1, y_1)$ and $Q(x_2, y_2)$ shown in the figure, we construct the right triangle *PRQ*. The distance between *P* and *R* is $|x_2 - x_1|$, and the distance between *R* and *Q* is $|y_2 - y_1|$. We apply the Pythagorean theorem to the right triangle *PRQ* to get

$$d^2 = |x_2 - x_1|^2 + |y_2 - y_1|^2$$
$$= (x_2 - x_1)^2 + (y_2 - y_1)^2 \quad \text{Because } |x_2 - x_1|^2 = (x_2 - x_1)^2 \text{ and } |y_2 - y_1|^2 = (y_2 - y_1)^2.$$

Because *d* represents the distance between two points, it must be equal to the positive square root of $(x_2 - x_1)^2 + (y_2 - y_1)^2$.

$$d = \sqrt{(x_2 - x_1)^2 + (y_2 - y_1)^2}$$

We call this result the *distance formula.*

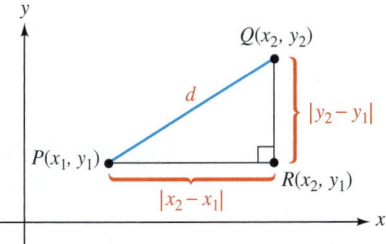

Distance Formula	The distance *d* between two points with coordinates (x_1, y_1) and (x_2, y_2) is given by

$$d = \sqrt{(x_2 - x_1)^2 + (y_2 - y_1)^2}$$

EXAMPLE 6

ELEMENTARY &
INTERMEDIATE
Algebra $f(x)$ **Now**™

Find the distance between the points $(-2, 3)$ and $(4, -5)$.

Solution   To find the distance, we can use the distance formula by substituting 4 for x_2, -2 for x_1, -5 for y_2, and 3 for y_1.

$$d = \sqrt{(x_2 - x_1)^2 + (y_2 - y_1)^2}$$
$$= \sqrt{[4 - (-2)]^2 + (-5 - 3)^2}$$
$$= \sqrt{(4 + 2)^2 + (-5 - 3)^2}$$
$$= \sqrt{6^2 + (-8)^2}$$
$$= \sqrt{36 + 64}$$
$$= \sqrt{100}$$
$$= 10$$

The distance between the points is 10 units.

Self Check 6   Find the distance between $(-2, -2)$ and $(3, 10)$.

EXAMPLE 7

Robotics.   Robots are used to weld parts of an automobile chassis on an automated production line. To do this, an imaginary coordinate system is superimposed on the side of the vehicle, and the robot is programmed to move to specific positions to make each weld. See the figure, which is scaled in inches. If the welder unit moves from point to point at an average rate of speed of 48 in./sec, how long will it take it to move from position 1 to position 2?

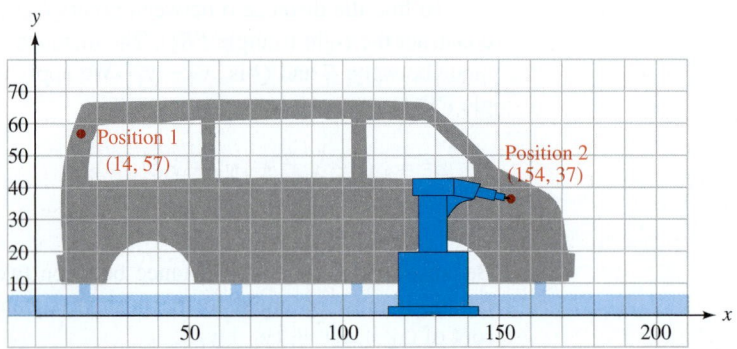

Solution   This is a uniform motion problem. We can use the formula $t = \frac{d}{r}$ to find the time it takes for the welder to move from position 1 at $(14, 57)$ to position 2 at $(154, 37)$.
    We can use the distance formula to find the distance d that the welder unit moves.

$$d = \sqrt{(x_2 - x_1)^2 + (y_2 - y_1)^2}$$
$$d = \sqrt{(154 - 14)^2 + (37 - 57)^2} \qquad \text{Substitute 154 for } x_2, \text{ 14 for } x_1, \text{ 37 for } y_2, \text{ and 57 for } y_1.$$
$$= \sqrt{140^2 + (-20)^2}$$
$$= \sqrt{20,000} \qquad\qquad\qquad 140^2 + (-20)^2 = 19,600 + 400 = 20,000.$$
$$= 100\sqrt{2} \qquad\qquad\qquad \text{Simplify: } \sqrt{20,000} = \sqrt{100 \cdot 100 \cdot 2} = 100\sqrt{2}.$$

The welder travels $100\sqrt{2}$ inches as it moves from position 1 to position 2. To find the time this will take, we divide the distance by the average rate of speed, 48 in./sec.

$$t = \frac{d}{r}$$

$$t = \frac{100\sqrt{2}}{48}$$ Substitute $100\sqrt{2}$ for d and 48 for r.

$$t \approx 2.9$$ Use a calculator to find an approximation to the nearest tenth.

It will take the welder about 2.9 seconds to travel from position 1 to position 2.

Answers to Self Checks   **1.** yes   **2.** $12\sqrt{2}$ m   **3.** $\dfrac{9\sqrt{2}}{2}$ in.   **4.** 16 cm, $8\sqrt{3}$ cm   **6.** 13

9.6   STUDY SET   ELEMENTARY & INTERMEDIATE   Algebra $f(x)$ Now™

VOCABULARY   Fill in the blanks.

1. In a right triangle, the side opposite the 90° angle is called the _____.

2. An _____ right triangle is a right triangle with two legs of equal length.

3. The _____ theorem states that in any right triangle, the square of the hypotenuse is equal to the sum of the squares of the lengths of the two legs.

4. An _____ triangle has three sides of equal length and three 60° angles.

CONCEPTS   Fill in the blanks.

5. If a and b are the lengths of the legs of a right triangle and c is the length of the hypotenuse, then _____ .

6. In any right triangle, the square of the hypotenuse is equal to the _____ of the squares of the two _____.

7. In an isosceles right triangle, the length of the hypotenuse is the length of one leg times ___.

8. The shorter leg of a 30°–60°–90° triangle is _____ as long as the hypotenuse.

9. The length of the longer leg of a 30°–60°–90° triangle is the length of the shorter leg times ___.

10. The formula to find the distance between two points (x_1, y_1) and (x_2, y_2) is $d =$ _____ .

11. In a right triangle, the shorter leg is opposite the _____ angle, and the longer leg is opposite the _____ angle.

12. An isosceles triangle has _____ sides of equal length.

13. Solve for c, where c represents the length of the hypotenuse of a right triangle.

   a. $c^2 = 64$

   b. $c^2 = 15$

   c. $c^2 = 24$

14. When the lengths of the sides of a certain triangle are substituted into the equation $a^2 + b^2 = c^2$, the result is a false statement. Explain why.

$$a^2 + b^2 = c^2$$
$$2^2 + 4^2 = 5^2$$
$$4 + 16 = 25$$
$$20 = 25$$

NOTATION   Complete each solution.

15. Evaluate:

$$\sqrt{(-1-3)^2 + [2-(-4)]^2} = \sqrt{(-4)^2 + [\quad]^2}$$
$$= \sqrt{\boxed{}}$$
$$= \sqrt{\boxed{} \cdot 13}$$
$$= \boxed{}\sqrt{13}$$
$$\approx 7.21$$

16. Solve: $8^2 + 4^2 = c^2$. Assume $c > 0$.

$$\boxed{} + 16 = c^2$$
$$\boxed{} = c^2$$
$$\sqrt{\boxed{}} = \boxed{}$$
$$\sqrt{\boxed{}} \cdot 5 = c$$
$$\boxed{}\sqrt{5} = c$$
$$c \approx 8.94$$

PRACTICE   The lengths of two sides of the right triangle *ABC* are given. Find the length of the missing side.

17. $a = 6$ ft and $b = 8$ ft

18. $a = 10$ cm and $c = 26$ cm

19. $b = 18$ m and $c = 82$ m

20. $a = 14$ in. and $c = 50$ in.

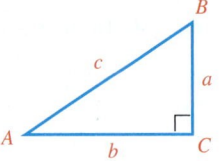

 Find the missing lengths in each triangle. Give the exact answer and then an approximation to two decimal places, when applicable.

21.

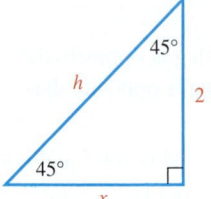

22.

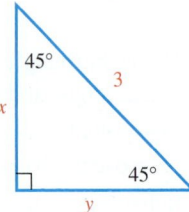

23.

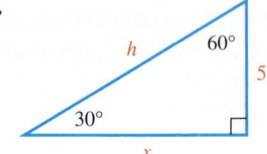

24.

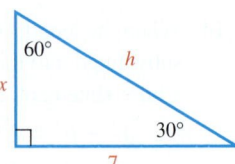

 Find the missing lengths in each triangle. Give the answer to two decimal places.

25.

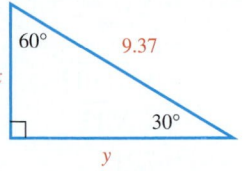

26.

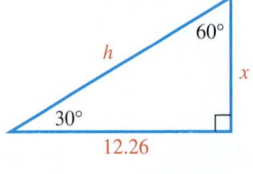

27.

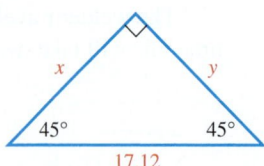

28.

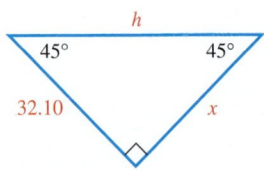

29. GEOMETRY   Find the exact length of the diagonal (in blue) of one of the *faces* of the cube shown below.

30. GEOMETRY   Find the exact length of the diagonal (in green) of the cube shown below.

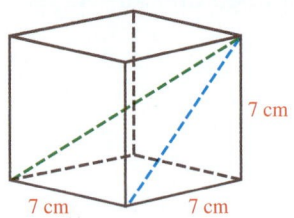

Find the distance between each pair of points.

31. $(0, 0), (3, -4)$

32. $(0, 0), (-12, 16)$

33. $(-2, -8), (3, 4)$

34. $(-5, -2), (7, 3)$

35. $(6, 8), (12, 16)$

36. $(10, 4), (2, -2)$

37. $(-3, 5), (-5, -5)$

38. $(2, -3), (4, -8)$

39. ISOSCELES TRIANGLES   Use the distance formula to show that a triangle with vertices $(-2, 4)$, $(2, 8)$, and $(6, 4)$ is isosceles.

40. RIGHT TRIANGLES   Use the distance formula and the Pythagorean theorem to show that a triangle with vertices $(2, 3)$, $(-3, 4)$, and $(1, -2)$ is a right triangle.

APPLICATIONS   Find the exact answer. Then give an approximation to two decimal places.

41. WASHINGTON, D.C.   The square in the map shows the 100-square-mile site selected by George Washington in 1790 to serve as a permanent capital for the United States. In 1847, the part of the district lying on the west bank of the Potomac was returned to Virginia. Find the coordinates of each corner of the original square that outlined the District of Columbia.

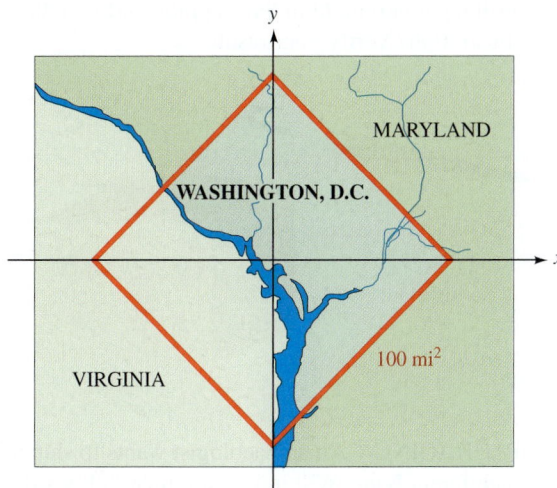

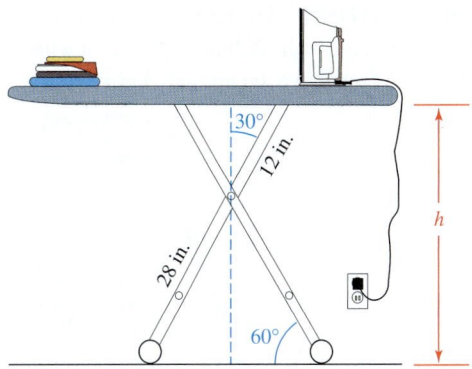

42. PAPER AIRPLANES The illustration gives the directions for making a paper airplane from a square piece of paper with sides 8 inches long. Find the length ℓ of the plane when it is completed.

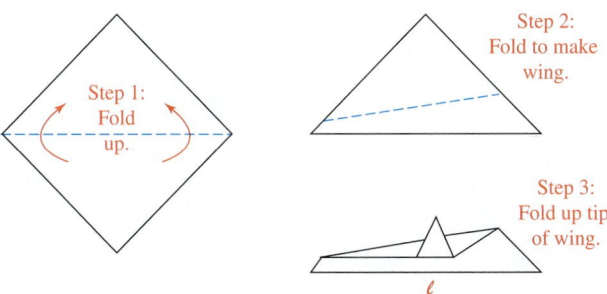

43. HARDWARE The sides of a regular hexagonal nut are 10 millimeters long. Find the height h of the nut.

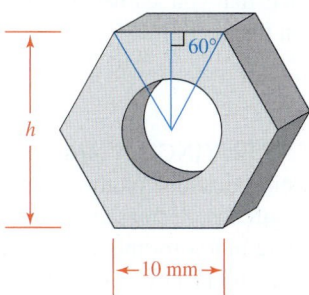

44. IRONING BOARDS Find the height h of the ironing board shown in the illustration in the next column.

45. BASEBALL A baseball diamond is a square, 90 feet on a side. If the third baseman fields a ground ball 10 feet directly behind third base, how far must he throw the ball to throw a runner out at first base?

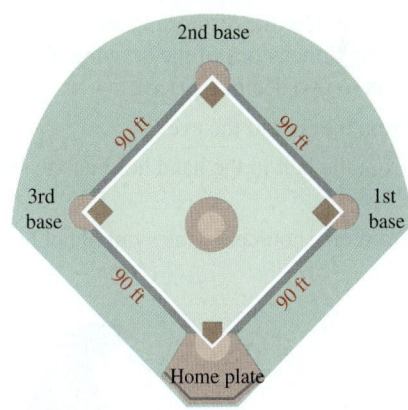

46. BASEBALL A shortstop fields a grounder at a point one-third of the way from second base to third base. How far will he have to throw the ball to make an out at first base?

47. CLOTHESLINES A pair of damp jeans are hung on a clothesline to dry. They pull the center down 1 foot. By how much is the line stretched?

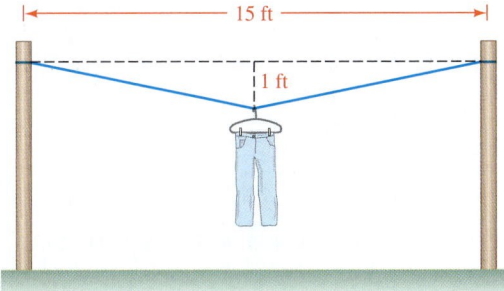

48. FIREFIGHTING   The base of the 37-foot ladder is 9 feet from the wall. Will the top reach a window ledge that is 35 feet above the ground? Verify your result.

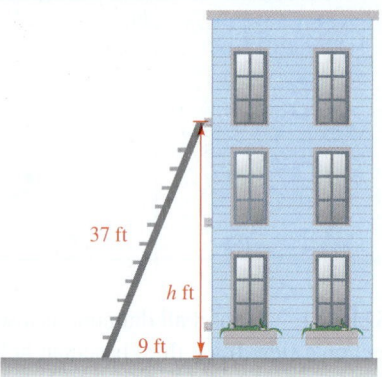

37 ft

h ft

9 ft

49. ART HISTORY   A figure displaying some of the characteristics of Egyptian art is shown in the illustration. Use the distance formula to find the following dimensions of the drawing. Round your answers to two decimal places.

a. From the foot to the eye

b. From the belt to the hand holding the staff

c. From the shoulder to the symbol held in the hand

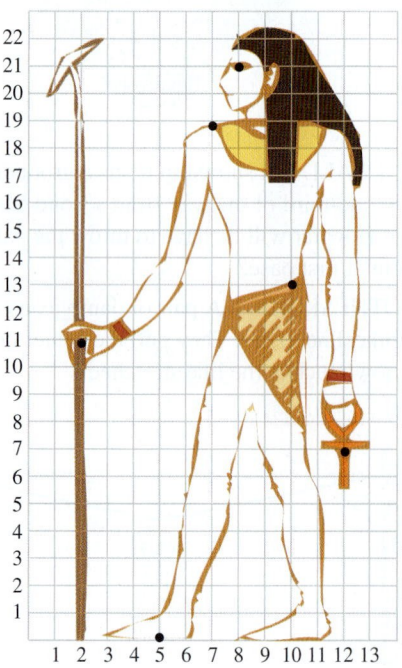

50. PACKAGING   The diagonal d of a rectangular box with dimensions $a \times b \times c$ is given by

$$d = \sqrt{a^2 + b^2 + c^2}$$

Will the umbrella fit in the shipping carton in the illustration? Verify your result.

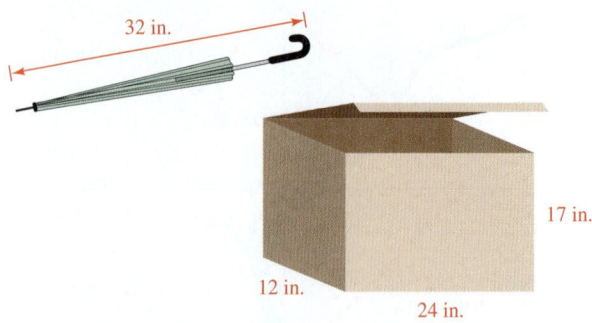

32 in.

17 in.

12 in.

24 in.

51. PACKAGING   An archaeologist wants to ship a 34-inch femur bone. Will it fit in a 4-inch-tall box that has a 24-inch-square base? (See Exercise 50.) Verify your result.

52. TELEPHONE SERVICE   The telephone cable in the illustration runs from A to B to C to D. How much cable is required to run from A to D directly?

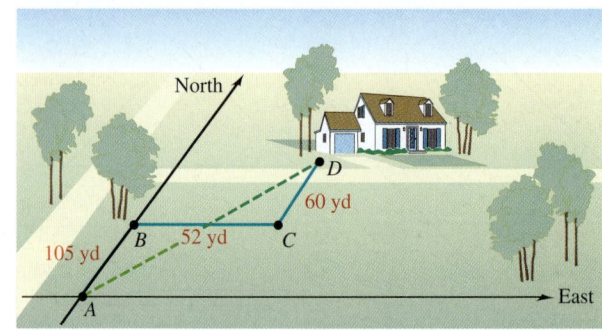

North

D

60 yd

105 yd

B   52 yd   C

A   East

WRITING

53. State the Pythagorean theorem in words.

54. List the facts that you learned about special right triangles in this section.

REVIEW

55. DISCOUNT BUYING   A repairman purchased some washing-machine motors for a total of $224. When the unit cost decreased by $4, he was able to buy one extra motor for the same total price. How many motors did he buy originally?

56. AVIATION   An airplane can fly 650 miles with the wind in the same amount of time as it can fly 475 miles against the wind. If the wind speed is 40 mph, find the speed of the plane in still air.

57. Find the mean of 16, 6, 10, 4, 5, 13

58. Find the median of 16, 6, 10, 4, 5

CHALLENGE PROBLEMS

59. Find the length of the diagonal of the cube.

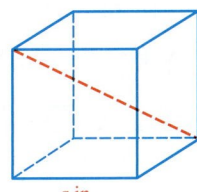

a in.

60. Show that the length of the diagonal of the rectangular solid shown is $\sqrt{a^2 + b^2 + c^2}$ cm.

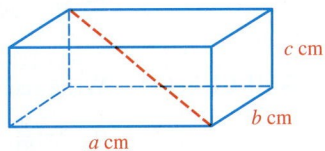

c cm

b cm

a cm

9.7 Complex Numbers

- The imaginary number i • Simplifying square roots of negative numbers
- Complex numbers   • Arithmetic of complex numbers   • Complex conjugates
- Division of complex numbers   • Powers of i

The Language of Algebra

For years, mathematicians thought numbers like $\sqrt{-9}$ and $\sqrt{-25}$ were useless. In the 17th century, René Descartes (1596–1650) called them *imaginary numbers*. Today they have important uses such as describing alternating electric current.

Recall that the square root of a negative number is not a real number. However, an expanded number system, called the *complex number system*, has been devised to give meaning to $\sqrt{-9}$, $\sqrt{-25}$, and the like. To define complex numbers, we use a new type of number that is denoted by the letter i.

■ THE IMAGINARY NUMBER i

Some equations do not have real-number solutions. For example, $x^2 = -1$ has no real-number solutions because the square of a real number is never negative. To provide a solution to this equation, mathematicians have defined the number i in such a way that $i^2 = -1$.

The Number i	The **imaginary number i** is defined as
	$i = \sqrt{-1}$
	From the definition, it follows that $i^2 = -1$.

This definition enables us to write the square root of any negative number in terms of i.

■ SIMPLIFYING SQUARE ROOTS OF NEGATIVE NUMBERS

We can use extensions of the product and quotient rules for radicals to write the square root of a negative number as the product of a real number and i.

EXAMPLE 1

Write each expression in terms of i: **a.** $\sqrt{-9}$, **b.** $\sqrt{-7}$, **c.** $-\sqrt{-18}$, and **d.** $\sqrt{-\dfrac{24}{49}}$.

Solution   We write each negative radicand as the product of -1 and a positive number and use the product rule for radicals. (The product rule, $\sqrt{ab} = \sqrt{a}\sqrt{b}$, holds when a is a negative real number and b is a positive real number.) Then we replace $\sqrt{-1}$ with i.

a. $\sqrt{-9} = \sqrt{-1 \cdot 9} = \sqrt{-1}\sqrt{9} = i \cdot 3 = 3i$

b. $\sqrt{-7} = \sqrt{-1 \cdot 7} = \sqrt{-1}\sqrt{7} = i\sqrt{7}$ or $\sqrt{7}i$

c. $-\sqrt{-18} = -\sqrt{-1 \cdot 9 \cdot 2} = -\sqrt{-1}\sqrt{9}\sqrt{2} = -i \cdot 3 \cdot \sqrt{2} = -3i\sqrt{2}$ or $-3\sqrt{2}i$

d. $\sqrt{-\dfrac{24}{49}} = \sqrt{-1 \cdot \dfrac{24}{49}} = \dfrac{\sqrt{-1 \cdot 24}}{\sqrt{49}} = \dfrac{\sqrt{-1}\sqrt{4}\sqrt{6}}{\sqrt{49}} = \dfrac{2i\sqrt{6}}{7}$ or $\dfrac{2\sqrt{6}}{7}i$.

Self Check 1   Write each expression in terms of i: **a.** $\sqrt{-25}$,   **b.** $-\sqrt{-19}$,   **c.** $\sqrt{-45}$,   and

d. $\sqrt{-\dfrac{50}{81}}$.

The results from Example 1 illustrate a rule for simplifying square roots of negative numbers.

Square Root of a Negative Number

For any positive real number b,

$$\sqrt{-b} = i\sqrt{b}$$

To justify this rule, we use the fact that $\sqrt{-1} = i$.

$$\sqrt{-b} = \sqrt{-1 \cdot b}$$
$$= \sqrt{-1}\sqrt{b}$$
$$= i\sqrt{b}$$

■ COMPLEX NUMBERS

The imaginary number i is used to define *complex numbers.*

Complex Numbers

A **complex number** is any number that can be written in the form $a + bi$, where a and b are real numbers and $i = \sqrt{-1}$.

Complex numbers of the form $a + bi$, where $b \neq 0$, are also called **imaginary numbers.***

For a complex number written in the **standard form** $a + bi$, we call a the **real part** and b the **imaginary part.** Some examples of complex numbers written in standard form are

$$2 + 11i \qquad\qquad 6 - 9i \qquad\qquad -\dfrac{1}{2} + 0i \qquad\qquad 0 + i\sqrt{3}$$

Two complex numbers are equal if and only if their real parts are equal and their imaginary parts are equal. Thus, $0.5 + 0.9i = \dfrac{1}{2} + \dfrac{9}{10}i$ because $0.5 = \dfrac{1}{2}$ and $0.9 = \dfrac{9}{10}$.

EXAMPLE 2   Write each number in the form $a + bi$: **a.** 6,   **b.** $\sqrt{-64}$,   and   **c.** $-2 + \sqrt{-63}$.

Algebra *f(x)* **Now**™   **Solution**   **a.** $6 = 6 + 0i$ The imaginary part is 0.

                       b. $\sqrt{-64} = 0 + 8i$ The real part is 0. $\sqrt{-64} = \sqrt{-1}\sqrt{64} = 8i$.

                       c. $-2 + \sqrt{-63} = -2 + 3i\sqrt{7}$ $\sqrt{-63} = \sqrt{-1}\sqrt{63} = \sqrt{-1}\sqrt{9}\sqrt{7} = 3i\sqrt{7}$.

*Some textbooks define imaginary numbers as complex numbers with $a = 0$ and $b \neq 0$.

Self Check 2 Write each number in the form $a + bi$: **a.** -18, **b.** $\sqrt{-36}$, and **c.** $1 + \sqrt{-24}$.

The following illustration shows the relationship between the real numbers, the imaginary numbers, and the complex numbers.

Success Tip

Just as real numbers are either rational or irrational, but not both, complex numbers are either real or imaginary, but not both.

Complex numbers

Real numbers				Imaginary numbers		
-6	$\frac{5}{16}$	-1.75	π	$9 + 7i$	$-2i$	$\frac{1}{4} - \frac{3}{4}i$
$48 + 0i$	0	$-\sqrt{10}$	$-\frac{7}{2}$	$0.56i$	$\sqrt{-10}$	$6 + i\sqrt{3}$

ARITHMETIC OF COMPLEX NUMBERS

We now consider how to add, subtract, multiply, and divide complex numbers.

Addition and Subtraction of Complex Numbers To add (or subtract) two complex numbers, add (or subtract) their real parts and add (or subtract) their imaginary parts.

EXAMPLE 3

ELEMENTARY & INTERMEDIATE Algebra f(x) Now™

Perform each operation: **a.** $(8 + 4i) + (12 + 8i)$, **b.** $(-6 + i) - (3 + 2i)$, and **c.** $\left(7 - \sqrt{-16}\right) + \left(9 + \sqrt{-4}\right)$.

Solution

a. $(8 + 4i) + (12 + 8i) = (8 + 12) + (4 + 8)i$ — Add the real parts. Add the imaginary parts.

$= 20 + 12i$

b. $(-6 + i) - (3 + 2i) = (-6 - 3) + (1 - 2)i$ — Subtract the real parts. Subtract the imaginary parts.

$= -9 - i$

c. $\left(7 - \sqrt{-16}\right) + \left(9 + \sqrt{-4}\right)$

$= (7 - 4i) + (9 + 2i)$ — Write $\sqrt{-16}$ and $\sqrt{-4}$ in terms of i.

$= (7 + 9) + (-4 + 2)i$ — Add the real parts. Add the imaginary parts.

$= 16 - 2i$ — Write $16 + (-2i)$ in the form $16 - 2i$.

Success Tip

Always change complex numbers to $a + bi$ form before performing any arithmetic.

Self Check 3 Perform the operations: **a.** $(3 - 5i) + (-2 + 7i)$ and **b.** $\left(3 - \sqrt{-25}\right) - \left(-2 + \sqrt{-49}\right)$.

Imaginary numbers are not real numbers; some properties of real numbers do not apply to imaginary numbers. For example, we cannot use the product rule for radicals to multiply two imaginary numbers.

Caution If a and b are both negative, then $\sqrt{a}\sqrt{b} \neq \sqrt{ab}$. For example, if $a = -4$ and $b = -9$,

$$\cancel{\sqrt{-4}\sqrt{-9} = \sqrt{-4(-9)} = \sqrt{36} = 6} \qquad \sqrt{-4}\sqrt{-9} = 2i(3i) = 6i^2 = 6(-1) = -6$$

EXAMPLE 4

Multiply: $\sqrt{-2}\sqrt{-20}$.

Solution We first express $\sqrt{-2}$ and $\sqrt{-20}$ in terms of i. Then, we can multiply the radical expressions as usual because the radicands are positive numbers.

$$\sqrt{-2}\sqrt{-20} = \left(i\sqrt{2}\right)\left(2i\sqrt{5}\right) \qquad \text{Simplify: } \sqrt{-20} = i\sqrt{20} = 2i\sqrt{5}.$$
$$= 2i^2\sqrt{10} \qquad\qquad i \cdot 2i = 2i^2 \text{ and } \sqrt{2}\sqrt{5} = \sqrt{10}.$$
$$= -2\sqrt{10} \qquad\qquad \text{Simplify: } i^2 = -1.$$

Self Check 4 Multiply: $\sqrt{-3}\sqrt{-32}$.

EXAMPLE 5

Multiply: **a.** $6(2 + 9i)$ and **b.** $-5i(4 - 8i)$.

ELEMENTARY &
INTERMEDIATE
Algebra *f(x)* **Now**™

Solution **a.** To multiply a complex number by a real number, we use the distributive property to remove parentheses and then simplify. For example,

$$6(2 + 9i) = 6(2) + 6(9i) \qquad \text{Use the distributive property.}$$
$$= 12 + 54i \qquad\qquad \text{Simplify.}$$

> **Caution**
>
> A common mistake is to replace i with -1. Remember, $i \neq -1$. By definition, $i = \sqrt{-1}$ and $i^2 = -1$.

b. To multiply a complex number by an imaginary number, we use the distributive property to remove parentheses and then simplify. For example,

$$-5i(4 - 8i) = -5i(4) - (-5i)8i \qquad \text{Use the distributive property.}$$
$$= -20i + 40i^2 \qquad\qquad \text{Simplify.}$$
$$= -40 - 20i \qquad\qquad \text{Since } i^2 = -1, 40i^2 = 40(-1) = -40.$$

Self Check 5 Multiply: **a.** $-2(-9 - i)$ and **b.** $10i(7 + 4i)$.

To multiply two complex numbers, we can use the FOIL method.

EXAMPLE 5

Multiply: **a.** $(2 + 3i)(3 - 2i)$ and **b.** $(-4 + 2i)(2 + i)$.

ELEMENTARY &
INTERMEDIATE
Algebra *f(x)* **Now**™

Solution **a.** $(2 + 3i)(3 - 2i) = 6 - 4i + 9i - 6i^2 \qquad$ Use the FOIL method.
$$= 6 + 5i - (-6) \qquad \begin{array}{l}\text{Combine the imaginary terms: } -4i + 9i = 5i.\\ \text{Simplify: } i^2 = -1, \text{ so } 6i^2 = -6.\end{array}$$
$$= 6 + 5i + 6$$
$$= 12 + 5i \qquad\qquad \text{Combine like terms.}$$

Success Tip

i is not a variable, but you can think of it as one when adding, subtracting, and multiplying. For example:

$-4i + 9i = 5i$
$6i - 2i = 4i$
$i \cdot i = i^2$

b. $(-4 + 2i)(2 + i) = -8 - 4i + 4i + 2i^2$ Use the FOIL method.

$\qquad\qquad\qquad\quad = -8 + 0i - 2$ $-4i + 4i = 0i$. Since $i^2 = -1, 2i^2 = -2$.

$\qquad\qquad\qquad\quad = -10 + 0i$

Self Check 6 Multiply: $(-2 + 3i)(3 - 2i)$.

■ COMPLEX CONJUGATES

Before we can discuss division of complex numbers, we must introduce an important fact about *complex conjugates*.

| **Complex Conjugates** | The complex numbers $a + bi$ and $a - bi$ are called **complex conjugates.** |

For example,

- $7 + 4i$ and $7 - 4i$ are complex conjugates.
- $5 - i$ and $5 + i$ are complex conjugates.
- $-6i$ and $6i$ are complex conjugates, because $-6i = 0 - 6i$ and $6i = 0 + 6i$.

The Language of Algebra

Recall that the word *conjugate* was used earlier when we rationalized the denominators of radical expressions such as

$$\frac{5}{\sqrt{6} - 1}.$$

In general, the product of the complex number $a + bi$ and its complex conjugate $a - bi$ is the real number $a^2 + b^2$, as the following work shows:

$(a + bi)(a - bi) = a^2 - abi + abi - b^2i^2$ Use the FOIL method.

$\qquad\qquad\qquad = a^2 - b^2(-1)$ $-abi + abi = 0$. Replace i^2 with -1.

$\qquad\qquad\qquad = a^2 + b^2$

EXAMPLE 7

Find the product of $3 + i$ and its complex conjugate.

Solution The complex conjugate of $3 + i$ is $3 - i$. We can find the product as follows:

$(3 + i)(3 - i) = 9 - 3i + 3i - i^2$ Use the FOIL method.

$\qquad\qquad\quad = 9 - i^2$ Combine like terms: $-3i + 3i = 0$.

$\qquad\qquad\quad = 9 - (-1)$ $i^2 = -1$.

$\qquad\qquad\quad = 10$

The product of the complex numbers $3 + i$ and $3 - i$ is the real number 10.

Self Check 7 Multiply: $(2 + 3i)(2 - 3i)$.

■ DIVISION OF COMPLEX NUMBERS

Recall that to divide *radical expressions,* we rationalize the denominator. We use a similar approach to divide complex numbers.

Division of Complex Numbers	To divide complex numbers, multiply the numerator and denominator by the complex conjugate of the denominator.

EXAMPLE 8   Find the quotient: **a.** $\dfrac{1}{3 + i}$ and   **b.** $\dfrac{3 - i}{2 + i}$.

Solution   **a.** We multiply the numerator and the denominator of the fraction by the complex conjugate of the denominator, which is $3 - i$.

$$\frac{1}{3 + i} = \frac{1}{3 + i} \cdot \frac{3 - i}{3 - i}$$ Multiply $\dfrac{1}{3 + i}$ by a form of 1: $\dfrac{3 - i}{3 - i} = 1$.

$$= \frac{3 - i}{9 - 3i + 3i - i^2}$$ Multiply the numerators and multiply the denominators.

$$= \frac{3 - i}{9 - (-1)}$$ $i^2 = -1$. Note that the denominator no longer contains i.

$$= \frac{3 - i}{10}$$ Simplify in the denominator.

$$= \frac{3}{10} - \frac{1}{10}i$$ Write the complex number in $a + bi$ form.

b. $\dfrac{3 - i}{2 + i} = \dfrac{3 - i}{2 + i} \cdot \dfrac{2 - i}{2 - i}$ The complex conjugate of the denominator of the fraction is $2 - i$. Multiply by $\dfrac{2 - i}{2 - i} = 1$.

$$= \frac{6 - 3i - 2i + i^2}{4 - 2i + 2i - i^2}$$ Multiply the numerators and multiply the denominators.

$$= \frac{5 - 5i}{4 - (-1)}$$ $i^2 = -1$. The denominator is now a real number.

$$= \frac{\overset{1}{\cancel{5}}(1 - i)}{\underset{1}{\cancel{5}}}$$ Factor out 5 in the numerator and remove the common factor of 5.

$$= 1 - i$$

Self Check 8   Find the quotient: **a.** $\dfrac{1}{5 - i}$ and   **b.** $\dfrac{5 + 4i}{3 + 2i}$.

EXAMPLE 9   Find the quotient: $\dfrac{4 + \sqrt{-16}}{2 + \sqrt{-4}}$. Write the result in $a + bi$ form.

Solution $\dfrac{4 + \sqrt{-16}}{2 + \sqrt{-4}} = \dfrac{4 + 4i}{2 + 2i}$ Write the numerator and denominator in $a + bi$ form.

$$= \frac{2(\overset{1}{\cancel{2 + 2i}})}{\underset{1}{\cancel{2 + 2i}}}$$ Factor out 2 in the numerator and remove the common factor of $2 + 2i$.

$$= 2 + 0i$$ Write 2 in the form $a + bi$.

Self Check 9   Find the quotient: $\dfrac{3 + \sqrt{-9}}{4 + \sqrt{-16}}$.

EXAMPLE 10   Find the quotient: $\dfrac{7}{2i}$. Write the result in $a + bi$ form.

Solution   The denominator can be expressed as $0 + 2i$. Its conjugate is $0 - 2i$, or just $-2i$.

Success Tip

In this example, the denominator of $\dfrac{7}{2i}$ is of the form bi. In such cases, we can eliminate i in the denominator by simply multiplying by $\dfrac{i}{i}$.

$$\dfrac{7}{2i} = \dfrac{7}{2i} \cdot \dfrac{i}{i} = \dfrac{7i}{2i^2} = -\dfrac{7}{2}i$$

$$\begin{aligned}
\dfrac{7}{2i} &= \dfrac{7}{2i} \cdot \dfrac{-2i}{-2i} && \text{Multiply } \dfrac{7}{2i} \text{ by a form of 1 using the complex conjugate of the} \\
&&& \text{denominator: } \dfrac{-2i}{-2i} = 1. \\
&= \dfrac{-14i}{-4i^2} \\
&= \dfrac{-14i}{4} && i^2 = -1. \text{ The denominator is now a real number.} \\
&= \dfrac{-7i}{2} && \text{Simplify.} \\
&= 0 - \dfrac{7}{2}i && \text{Write in } a + bi \text{ form.}
\end{aligned}$$

Self Check 10   Divide: $\dfrac{5}{-i}$.

■ POWERS OF i

The powers of i produce an interesting pattern:

Success Tip

Note that the powers of i cycle through four possible outcomes:

$$\begin{aligned}
i &= \sqrt{-1} = i \\
i^2 &= \left(\sqrt{-1}\right)^2 = -1 \\
i^3 &= i^2 i = -1i = -i \\
i^4 &= i^2 i^2 = (-1)(-1) = 1
\end{aligned}
\qquad
\begin{aligned}
i^5 &= i^4 i = 1i = i \\
i^6 &= i^4 i^2 = 1(-1) = -1 \\
i^7 &= i^4 i^3 = 1(-i) = -i \\
i^8 &= i^4 i^4 = (1)(1) = 1
\end{aligned}$$

The pattern continues: $i, -1, -i, 1, \ldots$.

Larger powers of i can be simplified by using the fact that $i^4 = 1$. For example, to simplify i^{29}, we note that 29 divided by 4 gives a quotient of 7 and a remainder of 1. Thus, $29 = 4 \cdot 7 + 1$ and

$$\begin{aligned}
i^{29} &= i^{4 \cdot 7 + 1} && 4 \cdot 7 = 28. \\
&= (i^4)^7 \cdot i^1 \\
&= 1^7 \cdot i && i^4 = 1. \\
&= i && 1 \cdot i = i.
\end{aligned}$$

The result of this example illustrates the following fact.

Powers of i   If n is a natural number that has a remainder of r when divided by 4, then

$$i^n = i^r$$

EXAMPLE 11

Simplify: i^{55}.

Solution   We divide 55 by 4 and get a remainder of 3. Therefore,

$$i^{55} = i^3 = -i$$

Self Check 11   Simplify: i^{62}.

Answers to Self Checks   1. a. $5i$,   b. $-i\sqrt{19}$,   c. $3i\sqrt{5}$,   d. $\dfrac{5\sqrt{2}}{9}i$ 2.   a. $-18 + 0i$,   b. $0 + 6i$,

c. $1 + 2i\sqrt{6}$ 3. a. $1 + 2i$,   b. $5 - 12i$ 4. $-4\sqrt{6}$ 5. a. $18 + 2i$,   b. $-40 + 70i$

6. $0 + 13i$ 7. 13 8. a. $\dfrac{5}{26} + \dfrac{1}{26}i$,   b. $\dfrac{23}{13} + \dfrac{2}{13}i$ 9. $\dfrac{3}{4} + 0i$

10. $0 + 5i$ 11. -1

9.7   STUDY SET   ELEMENTARY & INTERMEDIATE   Algebra $f(x)$ Now™

VOCABULARY   Fill in the blanks.

1. The _____ number i is defined as $i = \sqrt{-1}$.
2. A _____ number is any number that can be written in the form $a + bi$, where a and b are real numbers and $i = \sqrt{-1}$.
3. For the complex number $2 + 5i$, we call 2 the _____ part and 5 the _____ part.
4. Complex numbers such as $5i$ and $-10i$ are called _____ imaginary numbers.
5. $6 + 3i$ and $6 - 3i$ are called complex _____.
6. i^{25} is called a _____ of i.

CONCEPTS   Fill in the blanks.

7. a. $i =$ b. $i^2 =$
   c. $i^3 =$ d. $i^4 =$
8. Simplify:

   $$\sqrt{-36} = \sqrt{ \cdot 36} = \sqrt{}\sqrt{36} = 6$$

9. To add (or subtract) complex numbers, add (or subtract) their _____ parts and add (or subtract) their _____ parts.
10. We _____ two complex numbers by using the FOIL method.
11. To divide complex numbers, multiply the numerator and denominator by the complex conjugate of the _____.

12. The powers of i cycle through _____ possible outcomes.
13. Explain the error. Then find the correct result.
    a. Add: $\sqrt{-16} + \sqrt{-9} = \sqrt{-25}$.
    b. Multiply: $\sqrt{-2}\sqrt{-3} = \sqrt{-2(-3)} = \sqrt{6}$.
14. Give the complex conjugate of each number.
    a. $2 - 3i$ b. 2   c. $-3i$
15. Complete the illustration. Label the real numbers, the imaginary numbers, the complex numbers, the rational numbers, and the irrational numbers.

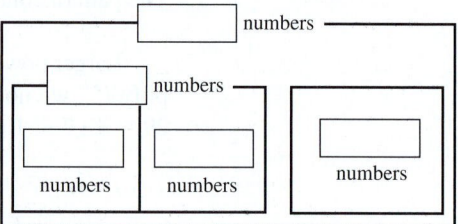

16. Decide whether each statement is true or false.
    a. Every complex number is a real number.
    b. Every real number is a complex number.
    c. i is a real number.
    d. The square root of a negative number is an imaginary number.

17. Decide whether the complex numbers are equal.

   a. $4 - \dfrac{2}{5}i,\ \dfrac{8}{2} - 0.4i$

   b. $0.25 + 0.7i,\ \dfrac{1}{4} + \dfrac{7}{10}i$

18. To divide $6 + 7i$ by $1 - 8i$, we multiply $\dfrac{6 + 7i}{1 - 8i}$ by a form of 1. What form of 1 do we use?

NOTATION Complete each operation.

19. $(3 + 2i)(3 - i) = \quad - 3i + \quad - 2i^2$

$= 9 + 3i + \quad$

$= \quad + 3i$

20. $\dfrac{3}{2 - i} = \dfrac{3}{2 - i} \cdot \dfrac{3}{\quad}$

$= \dfrac{6 + \quad}{4 + \quad - 2i - i^2}$

$= \dfrac{6 + 3i}{\quad}$

$= \quad + \dfrac{3}{5}i$

21. Decide whether each statement is true or false.

   a. $\sqrt{6}i = i\sqrt{6}$ **b.** $\sqrt{8}i = \sqrt{8i}$

   c. $\sqrt{-25} = -\sqrt{25}$ **d.** $-i = i$

22. Write each number in the form $a + bi$.

   a. $\dfrac{9 + 11i}{4}$ **b.** $\dfrac{1 - i}{18}$

PRACTICE Express each number in terms of i.

23. $\sqrt{-9}$ **24.** $\sqrt{-4}$

25. $\sqrt{-7}$ **26.** $\sqrt{-11}$

27. $\sqrt{-24}$ **28.** $\sqrt{-28}$

29. $-\sqrt{-24}$ **30.** $-\sqrt{-72}$

31. $5\sqrt{-81}$ **32.** $6\sqrt{-49}$

33. $\sqrt{-\dfrac{25}{9}}$ **34.** $-\sqrt{-\dfrac{121}{144}}$

Simplify each expression.

35. $\sqrt{-1}\sqrt{-36}$ **36.** $\sqrt{-9}\sqrt{-100}$

37. $\sqrt{-2}\sqrt{-6}$ **38.** $\sqrt{-3}\sqrt{-6}$

39. $\dfrac{\sqrt{-25}}{\sqrt{-64}}$ **40.** $\dfrac{\sqrt{-4}}{\sqrt{-1}}$

41. $-\dfrac{\sqrt{-400}}{\sqrt{-1}}$ **42.** $-\dfrac{\sqrt{-225}}{\sqrt{-16}}$

Perform the operations. Write all answers in $a + bi$ form.

43. $(3 + 4i) + (5 - 6i)$ **44.** $(5 + 3i) - (6 - 9i)$

45. $(7 - 3i) - (4 + 2i)$ **46.** $(8 + 3i) + (-7 - 2i)$

47. $(6 - i) + (9 + 3i)$ **48.** $(5 - 4i) - (3 + 2i)$

49. $\left(8 + \sqrt{-25}\right) + \left(7 + \sqrt{-4}\right)$

50. $\left(-7 + \sqrt{-81}\right) - \left(-2 - \sqrt{-64}\right)$

51. $3(2 - i)$ **52.** $-4(3 + 4i)$

53. $-5i(5 - 5i)$ **54.** $2i(7 + 2i)$

55. $(2 + i)(3 - i)$ **56.** $(4 - i)(2 + i)$

57. $(3 - 2i)(2 + 3i)$ **58.** $(3 - i)(2 + 3i)$

59. $(4 + i)(3 - i)$ **60.** $(1 - 5i)(1 - 4i)$

61. $\left(2 - \sqrt{-16}\right)\left(3 + \sqrt{-4}\right)$

62. $\left(3 - \sqrt{-4}\right)\left(4 - \sqrt{-9}\right)$

63. $\left(2 + i\sqrt{2}\right)\left(3 - i\sqrt{2}\right)$ **64.** $\left(5 + i\sqrt{3}\right)\left(2 - i\sqrt{3}\right)$

65. $(2 + i)^2$ **66.** $(3 - 2i)^2$

67. $(3i)^2$ **68.** $(5i)^2$

69. $\left(i\sqrt{6}\right)^2$ **70.** $\left(i\sqrt{2}\right)^2$

71. $\dfrac{1}{i}$ **72.** $\dfrac{1}{i^3}$

73. $\dfrac{4}{5i^3}$ **74.** $\dfrac{3}{2i}$

75. $\dfrac{3i}{8\sqrt{-9}}$ **76.** $\dfrac{5i^3}{2\sqrt{-4}}$

77. $\dfrac{-3}{5i^5}$ **78.** $\dfrac{-4}{6i^7}$

79. $\dfrac{5}{2 - i}$ **80.** $\dfrac{3}{5 + i}$

81. $\dfrac{-12}{7 - \sqrt{-1}}$ **82.** $\dfrac{-4}{3 + \sqrt{-1}}$

83. $\dfrac{5i}{6 + 2i}$ **84.** $\dfrac{3i}{6 - i}$

85. $\dfrac{-2i}{3+2i}$

86. $\dfrac{-4i}{2-6i}$

87. $\dfrac{3-2i}{3+2i}$

88. $\dfrac{2+3i}{2-3i}$

89. $\dfrac{3+2i}{3+i}$

90. $\dfrac{2-5i}{2+5i}$

91. $\dfrac{\sqrt{5}-i\sqrt{3}}{\sqrt{5}+i\sqrt{3}}$

92. $\dfrac{\sqrt{3}+i\sqrt{2}}{\sqrt{3}-i\sqrt{2}}$

Simplify each expression.

93. i^{21}

94. i^{19}

95. i^{27}

96. i^{22}

97. i^{100}

98. i^{42}

99. i^{97}

100. i^{200}

APPLICATIONS

101. FRACTALS   Complex numbers are fundamental in the creation of the intricate geometric shape shown below, called a *fractal*. The process of creating this image is based on the following sequence of steps, which begins by picking any complex number, which we will call *z*.

   1. Square *z*, and then add that result to *z*.

   2. Square the result from step 1, and then add it to *z*.

   3. Square the result from step 2, and then add it to *z*.

If we begin with the complex number *i*, what is the result after performing steps 1, 2, and 3?

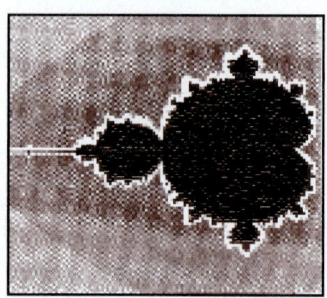

102. ELECTRONICS   The impedance *Z* in an ac (alternating current) circuit is a measure of how much the circuit impedes (hinders) the flow of current through it. The impedance is related to the voltage *V* and the current *I* by the formula

$$V = IZ$$

If a circuit has a current of $(0.5 + 2.0i)$ amps and an impedance of $(0.4 - 3.0i)$ ohms, find the voltage.

WRITING

103. What is an imaginary number? What is a complex number?

104. The method used to divide complex numbers is similar to the method used to divide radical expressions. Explain why. Give an example.

REVIEW

105. WIND SPEEDS   A plane that can fly 200 mph in still air makes a 330-mile flight with a tail wind and returns, flying into the same wind. Find the speed of the wind if the total flying time is $3\frac{1}{3}$ hours.

106. FINDING RATES   A student drove a distance of 135 miles at an average speed of 50 mph. How much faster would she have to drive on the return trip to save 30 minutes of driving time?

CHALLENGE PROBLEMS

107. Rationalize the numerator of $\dfrac{2+3i}{2-3i}$.

108. Simplify: $(2+3i)^{-2}$. Write the result in the form $a + bi$.

ACCENT ON TEAMWORK

A SPIRAL OF ROOTS

Overview: In this activity, you will create a visual representation of a collection of square roots.

Instructions: Form groups of 2 or 3 students. You will need a piece of unlined paper, a protractor, a ruler, and a pencil. Begin by drawing an isosceles right triangle with legs of length 1 inch in the middle of the paper. (See the illustration.) Use the Pythagorean theorem to determine the length of the hypotenuse. Draw a second right triangle using the hypotenuse of the first right triangle as one leg. Draw its second leg with length 1 inch. Find the length of the hypotenuse of the second triangle.

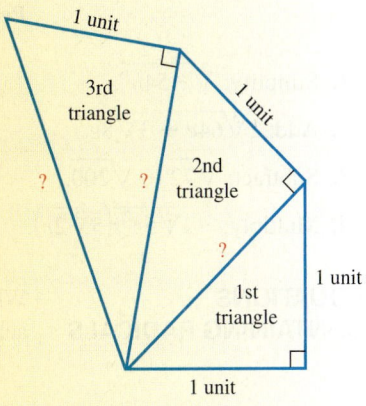

Continue creating right triangles, using the previous hypotenuse as one leg and drawing a new second leg of length 1 inch each time. Calculate the length of each resulting hypotenuse. When the figure begins to spiral onto itself, you may stop the process. Make a list of the lengths of each hypotenuse. What pattern do you see?

GRAPHING IN THREE DIMENSIONS

Overview: In this activity, you will find the distance between two points that lie in three-dimensional space.

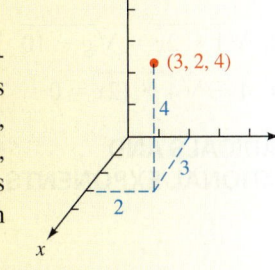

Instructions: Form groups of 2 or 3 students. In a three-dimensional Cartesian coordinate system, the positive x-axis is horizontal and pointing toward the viewer (out of the page), the positive y-axis is also horizontal and pointing to the right, and the positive z-axis is vertical, pointing up. A point is located by plotting an ordered triple of numbers (x, y, z). In the illustration, the point $(3, 2, 4)$ is plotted.

In three dimensions, the distance formula is

$$d = \sqrt{(x_2 - x_1)^2 + (y_2 - y_1)^2 + (z_2 - z_1)^2}$$

1. Copy the illustration shown above. Then plot the point $(1, 4, 3)$. Use the distance formula to find the distance between these two points.

2. Draw another three-dimensional coordinate system and plot the points $(-3, 3, -4)$ and $(2, -3, 2)$. Use the distance formula to find the distance between these two points.

KEY CONCEPT: RADICALS

The expression $\sqrt[n]{a}$ is called a **radical expression.** In this chapter, we have discussed the properties and procedures used when simplifying radical expressions, solving radical equations, and writing radical expressions using rational exponents.

EXPRESSIONS CONTAINING RADICALS

When working with expressions containing radicals, we must often apply the product rule and/or the quotient rule for radicals to simplify the expression. Recall that if a and b are both nonnegative,

$$\sqrt[n]{ab} = \sqrt[n]{a}\sqrt[n]{b} \qquad \sqrt[n]{\frac{a}{b}} = \frac{\sqrt[n]{a}}{\sqrt[n]{b}}, \qquad \text{where } b \neq 0$$

Perform each operation and simplify the expression.

1. Simplify: $\sqrt[3]{-54h^6}$.

2. Add: $2\sqrt[3]{64e} + 3\sqrt[3]{8e}$.

3. Subtract: $\sqrt{72} - \sqrt{200}$.

4. Multiply: $-4\sqrt[3]{5r^2s}\left(5\sqrt[3]{2r}\right)$.

5. Multiply: $\left(\sqrt{3s} - \sqrt{2t}\right)\left(\sqrt{3s} + \sqrt{2t}\right)$.

6. Multiply: $-\sqrt{3}\left(\sqrt{7} - \sqrt{5}\right)$.

7. Find the power: $\left(3\sqrt{2n} - 2\right)^2$.

8. Rationalize the denominator: $\dfrac{\sqrt[3]{9j}}{\sqrt[3]{3jk}}$.

EQUATIONS CONTAINING RADICALS

When solving radical equations, our objective is to rid the equation of the radical. This is achieved by using the *power rule:*

> If x, y, and n are real numbers and $x = y$, then $x^n = y^n$.

If we raise both sides of an equation to the same power, the resulting equation might not be equivalent to the original equation. We must always check for extraneous solutions.

Solve each equation, if possible.

9. $\sqrt{1 - 2g} = \sqrt{g + 10}$

10. $4 - \sqrt[3]{4 + 12x} = 0$

11. $\sqrt{y + 2} - 4 = -y$

12. $\sqrt[4]{12t + 4} + 2 = 0$

RADICALS AND RATIONAL EXPONENTS

Radicals can be written using rational (fractional) exponents, and exponential expressions having fractional exponents can be written in radical form. To do this, we use two rules for exponents introduced in this chapter.

$$x^{1/n} = \sqrt[n]{x} \qquad x^{m/n} = \sqrt[n]{x^m} = \left(\sqrt[n]{x}\right)^m$$

13. Express using a rational exponent: $\sqrt[3]{3}$.

14. Express in radical form: $5a^{2/5}$.

CHAPTER REVIEW

Radical Expressions and Radical Functions

SECTION 9.1

CONCEPTS

The number b is a *square root* of a if $b^2 = a$.

If $x > 0$, the *principal square root of x* is the positive square root of x, denoted $\sqrt{x}$. If x can be any real number, then $\sqrt{x^2} = |x|$.

The *cube root of x* is denoted as $\sqrt[3]{x}$ and is defined by

$\sqrt[3]{x} = y$ if $y^3 = x$

REVIEW EXERCISES

Simplify each expression, if possible. Assume that x and y can be any real number.

1. $\sqrt{49}$

2. $-\sqrt{121}$

3. $\sqrt{\dfrac{225}{49}}$

4. $\sqrt{-4}$

5. $\sqrt{0.01}$

6. $\sqrt{25x^2}$

7. $\sqrt{x^8}$

8. $\sqrt{x^2 + 4x + 4}$

9. $\sqrt[3]{-27}$

10. $-\sqrt[3]{216}$

11. $\sqrt[3]{64x^6y^3}$

12. $\sqrt[3]{\dfrac{x^9}{125}}$

13. $\sqrt[4]{625}$

14. $\sqrt[5]{-32}$

If n is an even natural number,
$$\sqrt[n]{a^n} = |a|$$

If n is an odd natural number,
$$\sqrt[n]{a^n} = a$$

If n is a natural number greater than 1 and x is a real number, then
- If $x > 0$, then $\sqrt[n]{x}$ is the positive number such that $\left(\sqrt[n]{x}\right)^n = x$.
- If $x = 0$, then $\sqrt[n]{x} = 0$.
- If $x < 0$, and n is odd, $\sqrt[n]{x}$ is the negative number such that $\left(\sqrt[n]{x}\right)^n = x$.
- If $x < 0$, and n is even, $\sqrt[n]{x}$ is not a real number.

15. $\sqrt[4]{256x^8y^4}$

16. $\sqrt[15]{(x+1)^{15}}$

17. $-\sqrt[4]{\dfrac{1}{16}}$

18. $\sqrt[6]{-1}$

19. $\sqrt{0}$

20. $\sqrt[3]{0}$

21. GEOMETRY   The side of a square with area A square feet is given by the function $s(A) = \sqrt{A}$. Find the *perimeter* of a square with an area of 144 square feet.

22. VOLUME OF A CUBE   The total surface area of a cube is related to its volume V by the function $A(V) = 6\sqrt[3]{V^2}$. Find the surface area of a cube with a volume of 8 cm³.

Graph each function. Find the domain and range.

23. $f(x) = \sqrt{x+2}$

24. $f(x) = -\sqrt[3]{x} + 3$

SECTION 9.2   **Rational Exponents**

If n is a natural number greater than 1 and $\sqrt[n]{x}$ is a real number, then
$$x^{1/n} = \sqrt[n]{x}$$

If n is a natural number greater than 1 and x is a real number,
- If $x > 0$, then $x^{1/n}$ is the positive number such that $(x^{1/n})^n = x$.
- If $x = 0$, then $x^{1/n} = 0$.
- If $x < 0$, and n is odd, then $x^{1/n}$ is the negative number such that $(x^{1/n})^n = x$.
- If $x < 0$ and n is even, then $x^{1/n}$ is not a real number.

If m and n are positive integers, $x > 0$, and $\frac{m}{n}$ is in simplest form,
$$x^{m/n} = \sqrt[n]{x^m} = \left(\sqrt[n]{x}\right)^m$$

$$x^{-m/n} = \frac{1}{x^{m/n}}$$

$$\frac{1}{x^{-m/n}} = x^{m/n}$$

Write each expression in radical form.

25. $t^{1/2}$

26. $(5xy^3)^{1/4}$

Simplify each expression, if possible. Assume that all variables represent positive real numbers.

27. $25^{1/2}$

28. $-36^{1/2}$

29. $(-36)^{1/2}$

30. $1^{1/5}$

31. $\left(\dfrac{9}{x^2}\right)^{1/2}$

32. $(-8)^{1/3}$

33. $625^{1/4}$

34. $(81c^4d^4)^{1/4}$

35. $9^{3/2}$

36. $8^{-2/3}$

37. $-49^{5/2}$

38. $\dfrac{1}{100^{-1/2}}$

39. $\left(\dfrac{4}{9}\right)^{-3/2}$

40. $\dfrac{1}{25^{5/2}}$

41. $(25x^2y^4)^{3/2}$

42. $(8u^6v^3)^{-2/3}$

The *rules for exponents* can be used to simplify expressions with fractional exponents.

Perform the operations. Write answers without negative exponents. Assume that all variables represent positive real numbers.

43. $5^{1/4}5^{1/2}$

44. $a^{3/7}a^{-2/7}$

45. $(k^{4/5})^{10}$

46. $\dfrac{3^{5/6}3^{1/3}}{3^{1/2}}$

Perform the multiplications. Assume all variables represent positive real numbers.

47. $u^{1/2}(u^{1/2} - u^{-1/2})$

48. $v^{2/3}(v^{1/3} + v^{4/3})$

Use rational exponents to simplify each radical. All variables represent positive real numbers.

49. $\sqrt[4]{a^2}$

50. $\sqrt[3]{\sqrt{c}}$

51. VISIBILITY   The distance d in miles a person in an airplane can see to the horizon on a clear day is given by the formula $d = 1.22a^{1/2}$, where a is the altitude of the plane in feet. Find d.

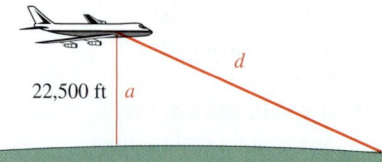

22,500 ft $\quad a$

d

52. Substitute the x- and y-coordinates of each point labeled in the graph into the equation

$$x^{2/3} + y^{2/3} = 32$$

Show that each one satisfies the equation.

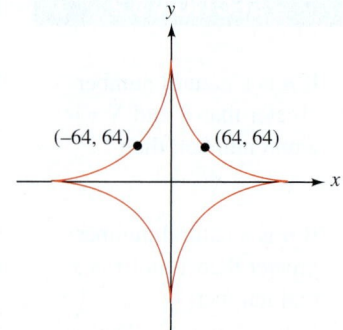

$(-64, 64)$ $(64, 64)$

Simplifying and Combining Radical Expressions

A radical is in *simplest form* when:

1. Each factor in the radicand appears to a power less than the index.
2. The radicand contains no fractions or negative numbers.
3. No radicals appear in a denominator.

Simplify each expression. Assume that all variables represent positive real numbers.

53. $\sqrt{240}$

54. $\sqrt[3]{54}$

55. $\sqrt[4]{32}$

56. $-2\sqrt[5]{-96}$

57. $\sqrt{8x^5}$

58. $\sqrt[3]{r^{17}}$

59. $\sqrt[3]{16x^5y^4}$

60. $3\sqrt[3]{27j^7k}$

61. $\dfrac{\sqrt{32x^3}}{\sqrt{2x}}$

62. $\sqrt{\dfrac{17xy}{64a^4}}$

Rules for radicals:

Multiplication:

$$\sqrt[n]{ab} = \sqrt[n]{a}\sqrt[n]{b}$$

Division:

$$\sqrt[n]{\frac{a}{b}} = \frac{\sqrt[n]{a}}{\sqrt[n]{b}} \quad (b \neq 0)$$

Like radicals can be combined by addition and subtraction.

Radicals that are not like can often be converted to radicals that are and then combined.

Simplify and combine like radicals. Assume that all variables represent positive real numbers.

63. $\sqrt{2} + 2\sqrt{2}$

64. $6\sqrt{20} - \sqrt{5}$

65. $2\sqrt[3]{3} - \sqrt[3]{24}$

66. $-\sqrt[4]{32a^5} - 2\sqrt[4]{162a^5}$

67. $2x\sqrt{8} + 2\sqrt{200x^2} + \sqrt{50x^2}$

68. $\sqrt[3]{54x^3} - 3\sqrt[3]{16x^3} + 4\sqrt[3]{128x^3}$

69. Explain the error that was made in each simplification.

 a. $2\sqrt{5x} + 3\sqrt{5x} = 5\sqrt{10x}$

 b. $30 + 30\sqrt[4]{2} = 60\sqrt[4]{2}$

 c. $7\sqrt[3]{y^2} - 5\sqrt[3]{y^2} = 2$

 d. $6\sqrt{11ab} - 3\sqrt{5ab} = 3\sqrt{6ab}$

70. SEWING   A corner of fabric is folded over to form a collar and stitched down as shown. From the dimensions given in the figure, determine the exact number of inches of stitching that must be made. Then give an approximation to one decimal place. (All measurements are in inches.)

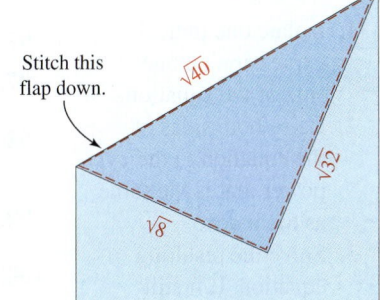

Stitch this flap down.

$\sqrt{40}$ $\sqrt{32}$ $\sqrt{8}$

SECTION 9.4

Multiplying and Dividing Radical Expressions

If two radicals have the same index, they can be multiplied:

$$\sqrt[n]{a}\sqrt[n]{b} = \sqrt[n]{ab}$$

If $\sqrt[n]{a}$ is a real number,

$$\left(\sqrt[n]{a}\right)^n = a$$

If a radical appears in a denominator of a fraction, or if a radicand contains a fraction, we can write the radical in simplest form by *rationalizing the denominator.*

To *rationalize a two-term denominator* of a fraction, multiply the numerator and the denominator by the conjugate of the denominator.

Simplify each expression. Assume that all variables represent positive real numbers.

71. $\sqrt{7}\sqrt{7}$

72. $\left(2\sqrt{5}\right)\left(3\sqrt{2}\right)$

73. $\left(-2\sqrt{8}\right)^2$

74. $2\sqrt{6}\sqrt{216}$

75. $\sqrt{9x}\sqrt{x}$

76. $\left(\sqrt[3]{x} + 1\right)^3$

77. $-\sqrt[3]{2x^2}\sqrt[3]{4x^8}$

78. $\sqrt[5]{9} \cdot \sqrt[5]{27}$

79. $3\sqrt{7t}\left(2\sqrt{7t} + 3\sqrt{3t^2}\right)$

80. $-\sqrt[4]{256x^5y^{11}}\sqrt[4]{625x^9y^3}$

81. $\left(\sqrt{3b} + \sqrt{3}\right)^2$

82. $\left(\sqrt[3]{3p} - 2\sqrt[3]{2}\right)\left(\sqrt[3]{3p} + \sqrt[3]{2}\right)$

Rationalize each denominator.

83. $\dfrac{10}{\sqrt{3}}$

84. $\sqrt{\dfrac{3}{5xy}}$

85. $\dfrac{\sqrt[3]{uv}}{\sqrt[3]{u^5v^7}}$

86. $\dfrac{\sqrt[4]{a}}{\sqrt[4]{3b^2}}$

87. $\dfrac{2}{\sqrt{2} - 1}$

88. $\dfrac{4\sqrt{x} - 2\sqrt{z}}{\sqrt{z} + 4\sqrt{x}}$

89. Rationalize the numerator: $\dfrac{\sqrt{a} - \sqrt{b}}{\sqrt{a}}$.

90. VOLUME   The formula relating the radius r of a sphere and its volume V is $r = \sqrt[3]{\dfrac{3V}{4\pi}}$. Write the radical in simplest form.

Solving Radical Equations

The power rule:

   If $x = y$, then $x^n = y^n$.

Solving equations containing radicals:

1. Isolate one radical expression on one side of the equation.
2. Raise both sides of the equation to the power that is the same as the index.
3. Solve the resulting equation. If it still contains a radical, go back to step 1.
4. Check the solutions to eliminate *extraneous* solutions.

Solve each equation. Write all solutions. Cross out those that are extraneous.

91. $\sqrt{7x - 10} - 1 = 11$

92. $u = \sqrt{25u - 144}$

93. $2\sqrt{y - 3} = \sqrt{2y + 1}$

94. $\sqrt{z + 1} + \sqrt{z} = 2$

95. $\sqrt[3]{x^3 + 56} - 2 = x$

96. $\sqrt[4]{8x - 8} + 2 = 0$

97. $(x + 2)^{1/2} - (4 - x)^{1/2} = 0$

98. $\sqrt{b^2 + b} = \sqrt{3 - b^2}$

99. Let $f(x) = \sqrt{2x^2 - 7x}$. For what value(s) of x is $f(x) = 2$?

100. Using the graphs of $f(x) = \sqrt{2x - 3}$ and $g(x) = -2x + 5$, estimate the solution of
$$\sqrt{2x - 3} = -2x + 5$$
Check the result.

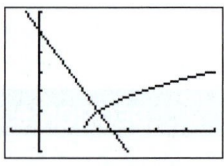

Solve each equation for the indicated variable.

101. $r = \sqrt{\dfrac{A}{P}} - 1$ for P

102. $h = \sqrt[3]{\dfrac{12I}{b}}$ for I

Geometric Applications of Radicals

The Pythagorean theorem:

If a and b are the lengths of the *legs* of a right triangle and c is the length of the *hypotenuse,* then $a^2 + b^2 = c^2$.

103. CARPENTRY   The gable end of the roof shown below is divided in half by a vertical brace, 8 feet in height. Find the length of the roof line.

104. SAILING   A technique called *tacking* allows a sailboat to make progress into the wind. A sailboat follows the course shown below. Find d, the distance the boat advances into the wind after tacking.

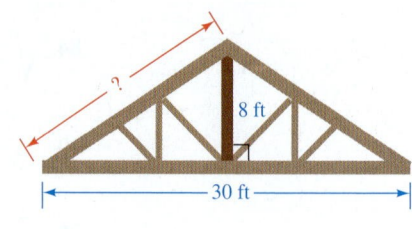

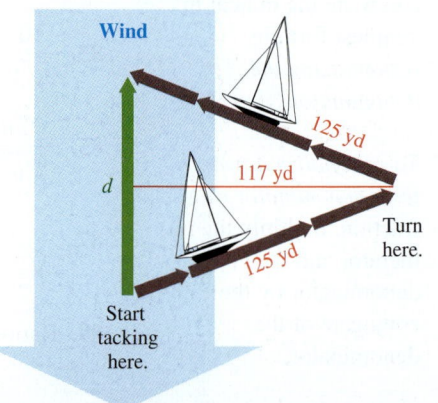

In an *isosceles right triangle,* the length of the hypotenuse is the length of one leg times $\sqrt{2}$.

The shorter leg of a *30°–60°–90° triangle* (the side opposite the 30° angle) is half as long as the hypotenuse. The longer leg (the side opposite the 60° angle) is the length of the shorter leg times $\sqrt{3}$.

105. Find the length of the hypotenuse of an isosceles right triangle whose legs measure 7 meters.

106. The hypotenuse of a 30°–60°–90° triangle measures $12\sqrt{3}$ centimeters. Find the length of each leg.

Find x to two decimal places.

107.

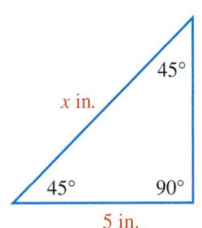

108.

Find the distance between the points.

The distance formula:

$$d = \sqrt{(x_2 - x_1)^2 + (y_2 - y_1)^2}$$

109. $(0, 0)$ and $(5, -12)$

110. $(-4, 6)$ and $(-2, 8)$

SECTION 9.7   # Complex Numbers

The *imaginary number i* is defined as
$$i = \sqrt{-1}$$

From the definition, it follows that $i^2 = -1$.

A *complex number* is any number that can be written in the form $a + bi$, where a and b are real numbers and $i = \sqrt{-1}$. We call a the *real part* and b the *imaginary part.*

Complex numbers of the form $a + bi$, where $b \neq 0$, are also called *imaginary numbers.*

The complex numbers $a + bi$ and $a - bi$ are called *complex conjugates.*

To add complex numbers, add their real parts and add their imaginary parts.

Write each expression in terms of i.

111. $\sqrt{-25}$

112. $4\sqrt{-18}$

113. $-\sqrt{-6}$

114. $\sqrt{-\dfrac{9}{64}}$

115. Complete the diagram.

Complex numbers

| _____ numbers | _____ numbers |

116. Determine whether each statement is true or false.

a. Every real number is a complex number.

b. $3 - 4i$ is a complex number.

c. $\sqrt{-4}$ is a real number.

d. i is a real number.

Give the complex conjugate of each number.

117. $3 + 6i$

118. $-1 - 7i$

119. $19i$

120. $-i$

Perform the operations. Write all answers in the form $a + bi$.

121. $(3 + 4i) + (5 - 6i)$

122. $\left(7 - \sqrt{-9}\right) - \left(4 + \sqrt{-4}\right)$

123. $3i(2 - i)$

124. $(2 - 7i)(-3 + 4i)$

To subtract complex numbers, add the opposite of the complex number being subtracted.

125. $\sqrt{-3} \cdot \sqrt{-9}$

126. $(9i)^2$

127. $\dfrac{3}{4i}$

128. $\dfrac{2 + 3i}{2 - 3i}$

Multiplying complex numbers is similar to multiplying polynomials.

To divide complex numbers, multiply the numerator and denominator by the complex conjugate of the denominator.

The powers of i cycle through four possible outcomes: i, -1, $-i$, and 1.

Simplify each expression.

129. i^{42}

130. i^{97}

CHAPTER 9 TEST   ELEMENTARY & INTERMEDIATE Algebra $f(x)$ Now™

1. Graph: $f(x) = \sqrt{x - 1}$. Find the domain and range.

2. DIVING   The velocity v of an object in feet per second after it has fallen a distance of d feet is approximated by the function $v(d) = \sqrt{64.4d}$. Olympic diving platforms are 10 meters (32.8 feet) tall. Estimate the velocity at which a diver hits the water from this height. Round to the nearest foot per second.

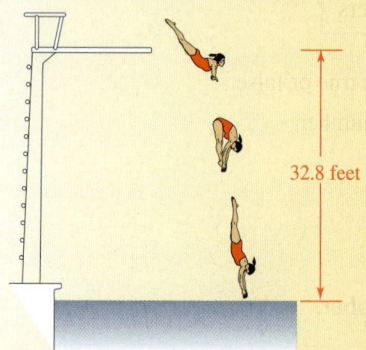

32.8 feet

3. Use the graph in the next column to find each of the following.

a $f(-1)$ **b.** $f(8)$

c. The value(s) of x for which $f(x) = 1$

d. The domain and range of f

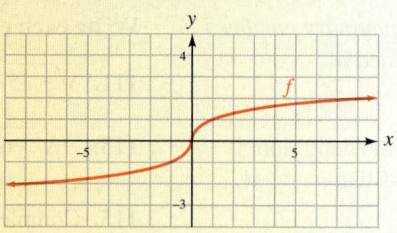

4. Explain why $\sqrt[4]{-16}$ is not a real number.

Simplify each expression. Assume that all variables represent positive real numbers and write answers without using negative exponents.

5. $(49x^4)^{1/2}$

6. $-27^{2/3}$

7. $36^{-3/2}$

8. $\left(-\dfrac{8}{125n^6}\right)^{-2/3}$

9. $\dfrac{2^{5/3}2^{1/6}}{2^{1/2}}$

10. $(a^{2/3})^{1/6}$

Simplify each expression. Assume that the variables are unrestricted.

11. $\sqrt{x^2}$

12. $\sqrt{y^2 + 10y + 25}$

Simplify each expression. Assume that all variables represent positive real numbers.

13. $\sqrt[3]{-64x^3y^6}$

14. $\sqrt{\dfrac{4a^2}{9}}$

15. $\sqrt[5]{(t+8)^5}$

16. $\sqrt{250x^3y^5}$

17. $\dfrac{\sqrt[3]{24x^{15}y^4}}{\sqrt[3]{y}}$

18. $\sqrt[7]{256}$

Perform the operations and simplify. Assume that all variables represent positive real numbers.

19. $2\sqrt{48y^5} - 3y\sqrt{12y^3}$

20. $2\sqrt[3]{40} - \sqrt[3]{5,000} + 4\sqrt[3]{625}$

21. $\sqrt[4]{243z^{13}} + z\sqrt[4]{48z^9}$

22. $-2\sqrt{xy}\left(3\sqrt{x} + \sqrt{xy^3}\right)$

23. $\left(3\sqrt{2} + \sqrt{3}\right)\left(2\sqrt{2} - 3\sqrt{3}\right)$

24. $\left(\sqrt[3]{2a} + 9\right)^2$

25. $\dfrac{8}{\sqrt{10}}$

26. $\dfrac{3t-1}{\sqrt{3t}-1}$

27. $\sqrt[3]{\dfrac{9}{4a}}$

28. Rationalize the numerator: $\dfrac{\sqrt{5}+3}{-4\sqrt{2}}$.

Solve each equation and check each result.

29. $2\sqrt{x} = \sqrt{x+1}$

30. $\sqrt[3]{6n+4} - 4 = 0$

31. $1 - \sqrt{u} = \sqrt{u-3}$

32. $(2m^2 - 9)^{1/2} = m$

33. Explain why, without having to perform any algebraic steps, it is obvious that the equation

$$\sqrt{x-8} = -10$$

has no solutions.

34. Solve $r = \sqrt[3]{\dfrac{GMt^2}{4\pi^2}}$ for G.

Find x to two decimal places.

35.

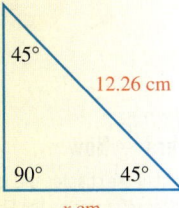

36.

37. Find the distance between $(-2, 5)$ and $(22, 12)$.

38. SHIPPING CRATES   The diagonal brace on the shipping crate in the illustration is 53 inches. Find the height h of the crate.

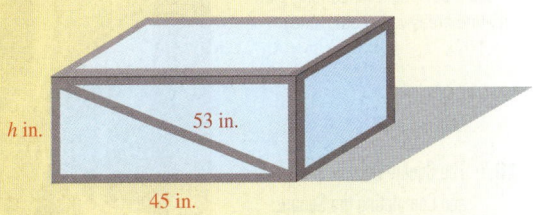

39. Express $\sqrt{-5}$ in terms of i.

40. Simplify: i^{22}.

Perform the operations. Give answers in $a + bi$ form.

41. $(2 + 4i) + (-3 + 7i)$

42. $\left(3 - \sqrt{-9}\right) - \left(-1 + \sqrt{-16}\right)$

43. $2i(3 - 4i)$

44. $(3 + 2i)(-4 - i)$

45. $\dfrac{1}{i\sqrt{2}}$

46. $\dfrac{2+i}{3-i}$

Chapter

10

Quadratic Equations, Functions, and Inequalities

ELEMENTARY & INTERMEDIATE
Algebra $f(x)$ Now™

Throughout the chapter, this icon introduces resources on the Elementary & Intermediate AlgebraNow Web site, accessed through **http://1pass. thomson.com**, that will

- Help you test your knowledge of the material with a pre-test and a post-test

- Provide a personalized learning plan targeting areas you should study

©Bob Krist, CORBIS

In a watercolor class, students learn that light, shadow, color, and perspective are fundamental components of an attractive painting. They also learn that the appropriate matting and frame can enhance their work. In this section, we will use mathematics to determine the dimensions of a uniform matting that is to have the same area as the picture it frames. To do this, we will write a quadratic equation and then solve it by *completing the square*. The technique of completing the square can be used to derive *the quadratic formula*. This formula is a valuable algebraic tool for solving any quadratic equation.

To learn more about quadratic equations, visit *The Learning Equation* on the Internet at http://tle.brookscole.com. (The log-in instructions are in the Preface.) For Chapter 10, the online lesson is:

- *TLE* Lesson 18: The Quadratic Formula

We have previously solved quadratic equations by factoring. In this chapter, we will discuss more general methods for solving quadratic equations, and we will consider the graphs of quadratic functions.

10.1 The Square Root Property and Completing the Square

- The square root property
- Completing the square
- Solving equations by completing the square
- Problem solving

Recall that a *quadratic equation* is an equation of the form $ax^2 + bx + c = 0$ where a, b, and c are real numbers and $a \neq 0$. We have solved quadratic equations using factoring and the zero-factor property. For example, to solve $6x^2 - 7x - 3 = 0$, we proceed as follows:

$$6x^2 - 7x - 3 = 0$$
$$(2x - 3)(3x + 1) = 0 \qquad \text{Factor.}$$
$$2x - 3 = 0 \quad \text{or} \quad 3x + 1 = 0 \qquad \text{Set each factor equal to 0.}$$
$$x = \frac{3}{2} \qquad\qquad x = -\frac{1}{3} \qquad \text{Solve each linear equation.}$$

Many expressions do not factor as easily as $6x^2 - 7x - 3$. For example, it would be difficult to solve $2x^2 + 4x + 1 = 0$ by factoring, because $2x^2 + 4x + 1$ cannot be factored by using only integers. With this in mind, we will now develop a more general method that enables us to solve *any* quadratic equation. It is based on the *square root property*.

■ THE SQUARE ROOT PROPERTY

To develop general methods for solving all quadratic equations, we first consider the equation $x^2 = c$. If $c \geq 0$, we can find the real solutions of $x^2 = c$ as follows:

$$x^2 = c$$
$$x^2 - c = 0 \qquad \text{Subtract } c \text{ from both sides.}$$
$$x^2 - \left(\sqrt{c}\right)^2 = 0 \qquad \text{Replace } c \text{ with } \left(\sqrt{c}\right)^2, \text{ since } c = \left(\sqrt{c}\right)^2.$$
$$\left(x + \sqrt{c}\right)\left(x - \sqrt{c}\right) = 0 \qquad \text{Factor the difference of two squares.}$$
$$x + \sqrt{c} = 0 \quad \text{or} \quad x - \sqrt{c} = 0 \qquad \text{Set each factor equal to 0.}$$
$$x = -\sqrt{c} \qquad\qquad x = \sqrt{c} \qquad \text{Solve each linear equation.}$$

The solutions of $x^2 = c$ are $\sqrt{c}$ and $-\sqrt{c}$.

The Square Root Property	For any nonnegative real number c, if $x^2 = c$, then $$x = \sqrt{c} \quad \text{or} \quad x = -\sqrt{c}$$

EXAMPLE 1

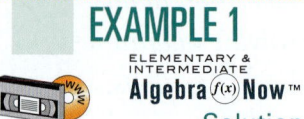

ELEMENTARY &
INTERMEDIATE
Algebra *f(x)* Now™

Solve: $x^2 - 12 = 0$.

Solution   We isolate x^2 on the left-hand side and use the square root property.

Notation

We can use **double-sign notation** $\pm$ to write the solutions in more compact form as $\pm 2\sqrt{3}$. Read $\pm$ as "positive or negative."

$$x^2 - 12 = 0$$

$$x^2 = 12 \qquad \text{Add 12 to both sides.}$$

$$x = \sqrt{12} \quad \text{or} \quad x = -\sqrt{12} \qquad \text{Use the square root property.}$$

$$x = 2\sqrt{3} \quad \Big| \quad x = -2\sqrt{3} \qquad \text{Simplify: } \sqrt{12} = \sqrt{4}\sqrt{3} = 2\sqrt{3}.$$

Check:

$x^2 - 12 = 0$	$x^2 - 12 = 0$
$\left(2\sqrt{3}\right)^2 - 12 \overset{?}{=} 0$	$\left(-2\sqrt{3}\right)^2 - 12 \overset{?}{=} 0$
$12 - 12 \overset{?}{=} 0$	$12 - 12 \overset{?}{=} 0$
$0 = 0$	$0 = 0$

The solutions are $2\sqrt{3}$ and $-2\sqrt{3}$ and the solution set is $\left\{2\sqrt{3}, -2\sqrt{3}\right\}$.

Self Check 1   Solve: $x^2 - 18 = 0$.

EXAMPLE 2

Phonograph records. Before compact discs, one way of recording music was by engraving grooves on thin vinyl discs called records. The vinyl discs used for long-playing records had a surface area of about 111 square inches per side and were played at $33\frac{1}{3}$ revolutions per minute on a turntable. What is the radius of a long-playing record?

CD

Solution   The relationship between the area of a circle and its radius is given by the formula $A = \pi r^2$. We can find the radius of a record by substituting 111 for A and solving for r.

$$A = \pi r^2 \qquad \text{This is the formula for the area of a circle.}$$

$$111 = \pi r^2 \qquad \text{Substitute 111 for } A.$$

$$\frac{111}{\pi} = r^2 \qquad \begin{array}{l}\text{To undo the multiplication by } \pi, \\ \text{divide both sides by } \pi.\end{array}$$

$$r = \sqrt{\frac{111}{\pi}} \quad \text{or} \quad r = -\sqrt{\frac{111}{\pi}} \qquad \begin{array}{l}\text{Use the square root property. Since the} \\ \text{radius of the record cannot be negative,} \\ \text{discard the second solution.}\end{array}$$

The radius of a record is $\sqrt{\dfrac{111}{\pi}}$ inches—to the nearest tenth, 5.9 inches.

EXAMPLE 3   Solve: $(x - 1)^2 = 16$.

ELEMENTARY & INTERMEDIATE
Algebra $f(x)$ **Now™**

Solution

$$(x - 1)^2 = 16$$
$$x - 1 = \pm\sqrt{16} \qquad \text{Use the square root property.}$$
$$x - 1 = \pm4 \qquad \text{Simplify: } \sqrt{16} = 4.$$
$$x = 1 \pm 4 \qquad \text{Add 1 to both sides.}$$

$x = 1 + 4$ or $x = 1 - 4$ To find one solution use $+$. To find the other use $-$.

$x = 5$ | $x = -3$ Add (subtract).

Verify that 5 and -3 satisfy the original equation.

Notation

We read 1 ± 4 as "one plus or minus four."

Self Check 3   Solve: $(x + 2)^2 = 9$.

Some quadratic equations have solutions that are not real numbers.

EXAMPLE 4   Solve: $4x^2 + 25 = 0$.

Solution $4x^2 + 25 = 0$

$$x^2 = -\frac{25}{4} \qquad \text{To isolate } x^2, \text{ subtract 25 from both sides and divide both sides by 4.}$$

$$x = \pm\sqrt{-\frac{25}{4}} \qquad \text{Use the square root property.}$$

Since

$$\sqrt{-\frac{25}{4}} = \sqrt{-1 \cdot \frac{25}{4}} = \sqrt{-1}\,\frac{\sqrt{25}}{\sqrt{4}} = \frac{5}{2}i$$

we have

$$x = \pm\frac{5}{2}i$$

The solutions are $\frac{5}{2}i$ and $-\frac{5}{2}i$.

The Language of Algebra

The $\pm$ symbol is often seen in surveys and polls. Suppose a poll predicts a candidate will receive 48% of the vote, $\pm4\%$. That means the candidate's support could be anywhere between $48 - 4 = 44\%$ and $48 + 4 = 52\%$.

Check:

$$4x^2 + 25 = 0 \qquad\qquad\qquad 4x^2 + 25 = 0$$
$$4\left(\frac{5}{2}i\right)^2 + 25 \stackrel{?}{=} 0 \qquad\qquad 4\left(-\frac{5}{2}i\right)^2 + 25 \stackrel{?}{=} 0$$
$$4\left(\frac{25}{4}\right)i^2 + 25 \stackrel{?}{=} 0 \qquad\qquad 4\left(\frac{25}{4}\right)i^2 + 25 \stackrel{?}{=} 0$$
$$25(-1) + 25 \stackrel{?}{=} 0 \qquad\qquad 25(-1) + 25 \stackrel{?}{=} 0$$
$$0 = 0 \qquad\qquad\qquad\qquad 0 = 0$$

Self Check 4   Solve: $16x^2 + 49 = 0$.

■ COMPLETING THE SQUARE

When the polynomial in a quadratic equation doesn't factor easily, we can solve the equation by *completing the square*. This method is based on the following special products:

$$x^2 + 2bx + b^2 = (x + b)^2 \quad \text{and} \quad x^2 - 2bx + b^2 = (x - b)^2$$

In each of these perfect square trinomials, the third term is the square of one-half of the coefficient of x.

The Language of Algebra

Recall that trinomials that are the square of a binomial are called **perfect square trinomials.**

- In $x^2 + 2bx + b^2$, the coefficient of x is $2b$. If we find $\frac{1}{2} \cdot 2b$, which is b, and square it, we get the third term, b^2.
- In $x^2 - 2bx + b^2$, the coefficient of x is $-2b$. If we find $\frac{1}{2}(-2b)$, which is $-b$, and square it, we get the third term: $(-b)^2 = b^2$.

We can use these observations to change certain binomials into perfect square trinomials. For example, to change $x^2 + 12x$ into a perfect square trinomial, we find one-half of the coefficient of x, square the result, and add the square to $x^2 + 12x$.

$$x^2 + 12x + \boxed{}$$

Find one-half of the coefficient of x.

Add the square to the binomial.

$$\frac{1}{2} \cdot 12 = 6 \qquad 6^2 = 36$$

Square the result.

We obtain the perfect square trinomial $x^2 + 12x + 36$ that factors as $(x + 6)^2$. By adding 36 to $x^2 + 12x$, we **completed the square** on $x^2 + 12x$.

Completing the Square   To complete the square on $x^2 + bx$, add the square of one-half of the coefficient of x:

$$x^2 + bx + \left(\frac{1}{2}b\right)^2$$

EXAMPLE 5

ELEMENTARY & INTERMEDIATE
Algebra *f(x)* **Now** ™

Complete the square and factor the resulting perfect square trinomial: **a.** $x^2 + 10x$ and **b.** $x^2 - 11x$.

Solution   **a.** To make $x^2 + 10x$ a perfect square trinomial, we find one-half of 10, square it, and add that result to $x^2 + 10x$.

$$x^2 + 10x + 25 \qquad \frac{1}{2} \cdot 10 = 5 \text{ and } 5^2 = 25. \text{ Add 25 to the binomial.}$$

This trinomial factors as $(x + 5)^2$.

b. To make $x^2 - 11x$ a perfect square trinomial, we find one-half of -11, square it, and add that result to $x^2 - 11x$.

$$x^2 - 11x + \frac{121}{4} \qquad \frac{1}{2}(-11) = \frac{-11}{2} \text{ and } \left(-\frac{11}{2}\right)^2 = \frac{121}{4}. \text{ Add } \frac{121}{4} \text{ to the binomial.}$$

This trinomial factors as $\left(x - \frac{11}{2}\right)^2$.

Self Check 5   Complete the square on $a^2 - 5a$ and factor the resulting trinomial.

■ SOLVING EQUATIONS BY COMPLETING THE SQUARE

To solve an equation of the form $ax^2 + bx + c = 0$ by completing the square, we use the following steps.

Completing the Square to Solve a Quadratic Equation	1. If the coefficient of x^2 is 1, go to step 2. If it is not 1, make it 1 by dividing both sides of the equation by the coefficient of x^2.
	2. Get all variable terms on one side of the equation and constants on the other side.
	3. Complete the square by finding one-half of the coefficient of x, squaring the result, and adding the square to both sides of the equation.
	4. Factor the perfect square trinomial as the square of a binomial.
	5. Solve the resulting equation using the square root property.
	6. Check the answers in the original equation.

EXAMPLE 6

Use completing the square to solve $x^2 + 8x + 7 = 0$.

Solution

Step 1:   In this example, the coefficient of x^2 is an understood 1.

Step 2:   We subtract 7 from both sides so that the variable terms are on one side and the constant is on the other side of the equation.

The Language of Algebra

In $x^2 + 8x + 7 = 0$, x^2 and $8x$ are called *variable terms* and 7 is called the *constant term*.

$$x^2 + 8x + 7 = 0$$
$$x^2 + 8x \quad\quad = -7$$

Step 3:   The coefficient of x is 8, one-half of 8 is 4, and $4^2 = 16$. To complete the square, we add 16 to both sides.

$$x^2 + 8x + \mathbf{16} = \mathbf{16} - 7$$
(1) $\quad x^2 + 8x + 16 = 9$ Simplify: $16 - 7 = 9$.

Step 4:   Since the left-hand side of Equation 1 is a perfect square trinomial, we can factor it to get $(x + 4)^2$.

$$x^2 + 8x + 16 = 9$$
(2) $\quad\quad (x + 4)^2 = 9$

Step 5:   We solve Equation 2 by using the square root property.

$$x + 4 = \pm\sqrt{9}$$
$$x + 4 = \pm 3 \quad\quad \text{Simplify: } \sqrt{9} = 3.$$
$$x = -4 \pm 3 \quad\quad \text{Subtract 4 from both sides.}$$

This result represents two solutions. To find the first, we add 3, and to find the second, we subtract 3.

$$x = -4 + 3 \quad \text{or} \quad x = -4 - 3 \qquad \pm \text{ represents } + \text{ or } -.$$
$$x = -1 \quad \Big| \quad x = -7 \qquad \text{Add and subtract.}$$

Step 6: The solutions are -1 and -7. Verify that they satisfy the original equation.

Self Check 6 Solve: $x^2 + 12x + 11 = 0$.

EXAMPLE 7

Solve: $6x^2 + 5x - 6 = 0$.

Solution **Step 1:** To make the coefficient of x^2 equal to 1, we divide both sides of the equation by 6.

$$6x^2 + 5x - 6 = 0$$

$$\frac{6x^2}{6} + \frac{5}{6}x - \frac{6}{6} = \frac{0}{6} \qquad \text{Divide both sides by 6.}$$

$$x^2 + \frac{5}{6}x - 1 = 0 \qquad \text{Simplify.}$$

Step 2: To have the constant term on one side of the equation and the variable terms on the other, add 1 to both sides.

$$x^2 + \frac{5}{6}x = 1$$

Step 3: The coefficient of x is $\frac{5}{6}$, one-half of $\frac{5}{6}$ is $\frac{5}{12}$, and $\left(\frac{5}{12}\right)^2 = \frac{25}{144}$. To complete the square, we add $\frac{25}{144}$ to both sides.

$$x^2 + \frac{5}{6}x + \frac{25}{144} = 1 + \frac{25}{144}$$

(3) $$x^2 + \frac{5}{6}x + \frac{25}{144} = \frac{169}{144} \qquad \text{Simplify: } 1 + \frac{25}{144} = \frac{144}{144} + \frac{25}{144} = \frac{169}{144}.$$

Step 4: Factor the left-hand side of Equation 3.

(4) $$\left(x + \frac{5}{12}\right)^2 = \frac{169}{144} \qquad x^2 + \frac{5}{6}x + \frac{25}{144} \text{ is a perfect square trinomial.}$$

Step 5: We can solve Equation 4 by using the square root property.

$$x + \frac{5}{12} = \pm\sqrt{\frac{169}{144}}$$

$$x + \frac{5}{12} = \pm\frac{13}{12} \qquad \text{Simplify: } \sqrt{\frac{169}{144}} = \frac{13}{12}.$$

$$x = -\frac{5}{12} \pm \frac{13}{12} \qquad \text{To isolate } x, \text{ subtract } \frac{5}{12} \text{ from both sides.}$$

$$x = -\frac{5}{12} \mathbin{\color{red}+} \frac{13}{12} \quad \text{or} \quad x = -\frac{5}{12} - \frac{13}{12}$$

$$x = \frac{8}{12} \qquad\qquad\qquad x = -\frac{18}{12} \qquad \text{Add (subtract) the fractions.}$$

$$x = \frac{2}{3} \qquad\qquad\qquad x = -\frac{3}{2} \qquad \text{Simplify each fraction.}$$

Step 6: Verify that $\dfrac{2}{3}$ and $-\dfrac{3}{2}$ satisfy the original equation.

Self Check 7   Solve: $3x^2 + 2x - 8 = 0$.

EXAMPLE 8

Solve: $2x^2 + 4x + 1 = 0$.

ELEMENTARY &
INTERMEDIATE
Algebra *f(x)* **Now**™   Solution

$$2x^2 + 4x + 1 = 0$$

$$x^2 + 2x + \frac{1}{2} = 0 \qquad \text{Divide both sides by 2 to make the coefficient of } x^2 \text{ equal to 1.}$$

$$x^2 + 2x \qquad = -\frac{1}{2} \qquad \text{Subtract } \frac{1}{2} \text{ from both sides.}$$

$$x^2 + 2x \mathbin{\color{red}+ 1} = \mathbin{\color{red}1} - \frac{1}{2} \qquad \text{Square one-half of the coefficient of } x \text{ and add it to both sides.}$$

Caution

A common error is to add a constant to one side of an equation to complete the square and forget to add it to the other side.

$$(x + 1)^2 = \frac{1}{2} \qquad \text{Factor and combine like terms.}$$

$$x + 1 = \pm\sqrt{\frac{1}{2}} \qquad \text{Use the square root property.}$$

$$x = -1 \pm \sqrt{\frac{1}{2}} \qquad \text{To isolate } x, \text{ subtract 1 from both sides.}$$

To write $\sqrt{\dfrac{1}{2}}$ in simplified radical form, we write it as a quotient of square roots and then rationalize the denominator.

$$x = -1 + \frac{\sqrt{2}}{2} \quad \text{or} \quad x = -1 - \frac{\sqrt{2}}{2} \qquad \sqrt{\frac{1}{2}} = \frac{\sqrt{1}}{\sqrt{2}} = \frac{1 \cdot \sqrt{2}}{\sqrt{2}\sqrt{2}} = \frac{\sqrt{2}}{2}.$$

We can express each solution in an alternate form if we write -1 as a fraction with a denominator of 2.

Caution

Recall that to simplify a fraction, we divide out common *factors* of the numerator and denominator. Since -2 is a *term* of the numerator of $\dfrac{-2 + \sqrt{2}}{2}$, no further simplification of this expression can be made.

$$x = -\frac{2}{2} + \frac{\sqrt{2}}{2} \quad \text{or} \quad x = -\frac{2}{2} - \frac{\sqrt{2}}{2} \qquad \text{Write } -1 \text{ as } -\frac{2}{2}.$$

$$x = \frac{-2 + \sqrt{2}}{2} \qquad\qquad x = \frac{-2 - \sqrt{2}}{2} \qquad \text{Add (subtract) the numerators and keep the common denominator of 2.}$$

The exact solutions are $\dfrac{-2 + \sqrt{2}}{2}$ and $\dfrac{-2 - \sqrt{2}}{2}$, or simply, $\dfrac{-2 \pm \sqrt{2}}{2}$. We can use a calculator to approximate them. To the nearest hundredth, they are -0.29 and -1.71.

Self Check 8   Solve: $3x^2 + 6x + 1 = 0$.

ACCENT ON TECHNOLOGY: CHECKING SOLUTIONS OF QUADRATIC EQUATIONS

We can use a graphing calculator to check the solutions of the quadratic equation $2x^2 + 4x + 1 = 0$ found in Example 8. After entering $Y_1 = 2x^2 + 4x + 1$, we call up the home screen by pressing $\boxed{\text{2nd}}$ QUIT. Then we press the $\boxed{\text{VARS}}$ key, arrow $\boxed{\blacktriangleright}$ to Y-VARS, and enter 1 to get the display shown in figure (a). We evaluate $2x^2 + 4x + 1$ for $x = \dfrac{-2 + \sqrt{2}}{2}$ by entering the solution using function notation, as shown in figure (b).

When $\boxed{\text{ENTER}}$ is pressed, the result of 0 is confirmation that $x = \dfrac{-2 + \sqrt{2}}{2}$ is a solution of the equation.

```
Y1
```
```
Y1((-2+√(2))/2)
                    0
■
```

(a)   (b)

In the next example, the solutions of the equation are two complex numbers that contain i.

EXAMPLE 9

Solve: $3x^2 + 2x + 2 = 0$.

Solution

$$3x^2 + 2x + 2 = 0$$

$$x^2 + \frac{2}{3}x + \frac{2}{3} = \frac{0}{3}$$ Divide both sides by 3 to make the coefficient of x^2 equal to 1.

$$x^2 + \frac{2}{3}x \qquad = -\frac{2}{3}$$ Subtract $\dfrac{2}{3}$ from both sides.

$$x^2 + \frac{2}{3}x + \frac{1}{9} = \frac{1}{9} - \frac{2}{3}$$ $\dfrac{1}{2} \cdot \dfrac{2}{3} = \dfrac{1}{3}$ and $\left(\dfrac{1}{3}\right)^2 = \dfrac{1}{9}$. Add $\dfrac{1}{9}$ to both sides.

$$\left(x + \frac{1}{3}\right)^2 = -\frac{5}{9}$$ Factor and combine terms: $\dfrac{1}{9} - \dfrac{2}{3} = \dfrac{1}{9} - \dfrac{6}{9} = -\dfrac{5}{9}$.

$$x + \frac{1}{3} = \pm\sqrt{-\frac{5}{9}}$$ Use the square root property.

$$x = -\frac{1}{3} \pm \sqrt{-\frac{5}{9}}$$ Subtract $\dfrac{1}{3}$ from both sides.

Since

$$\sqrt{-\frac{5}{9}} = \sqrt{-1 \cdot \frac{5}{9}} = \sqrt{-1}\,\frac{\sqrt{5}}{\sqrt{9}} = \frac{\sqrt{5}}{3}i$$

we have

$$x = -\frac{1}{3} \pm \frac{\sqrt{5}}{3}i$$

The solutions are $-\dfrac{1}{3} + \dfrac{\sqrt{5}}{3}i$ and $-\dfrac{1}{3} - \dfrac{\sqrt{5}}{3}i$.

Self Check 9   Solve: $x^2 + 4x + 6 = 0$.

Success Tip

Perhaps you noticed that Examples 6 and 7 can be solved by factoring. However, that is not the case for Examples 8 and 9. These observations illustrate that completing the square can be used to solve *any* quadratic equation.

Notation

The solutions are written in complex number form $a + bi$. They could also be written as

$$\frac{-1 \pm i\sqrt{5}}{3}$$

■ PROBLEM SOLVING

EXAMPLE 10

ELEMENTARY &
INTERMEDIATE
Algebra $f(x)$ **Now**™

Graduation announcements. To create the announcement shown to the right, a graphic artist must follow two design requirements:

- A border of uniform width should surround the text.
- Equal areas should be devoted to the text and to the border.

To meet these requirements, how wide should the border be?

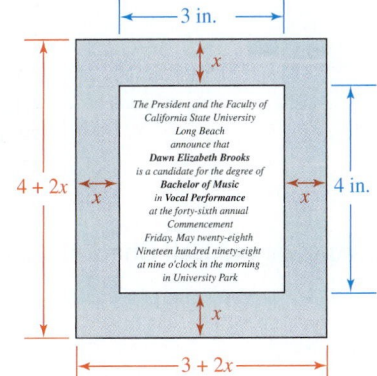

Analyze the problem The text occupies $4 \cdot 3 = 12$ in.^{2} of space. The border must also have an area of 12 in.^{2}.

Form an Equation If we let $x =$ the width of the border, the length of the announcement is $(4 + 2x)$ inches and the width is $(3 + 2x)$ inches. We can now form the equation.

The area of the announcement	minus	the area of the text	equals	the area of the border.
$(4 + 2x)(3 + 2x)$	$-$	12	$=$	12

Solve the Equation

$$(4 + 2x)(3 + 2x) - 12 = 12$$

$$12 + 8x + 6x + 4x^2 - 12 = 12 \qquad \text{On the left-hand side, use the FOIL method.}$$

$$4x^2 + 14x = 12 \qquad \text{Combine like terms.}$$

$$2x^2 + 7x - 6 = 0 \qquad \text{Subtract 12 from both sides. Then divide both sides of } 4x^2 + 14x - 12 = 0 \text{ by 2.}$$

We note that the trinomial on the left-hand side does not factor. We will solve the equation by completing the square.

$$x^2 + \frac{7}{2}x - 3 = 0 \qquad \text{Divide both sides by 2 so that the coefficient of } x^2 \text{ is 1.}$$

$$x^2 + \frac{7}{2}x = 3 \qquad \text{Add 3 to both sides.}$$

$$x^2 + \frac{7}{2}x + \frac{49}{16} = 3 + \frac{49}{16} \qquad \text{One-half of } \frac{7}{2} \text{ is } \frac{7}{4}. \text{ Square } \frac{7}{4}, \text{ which is } \frac{49}{16}, \text{ and add it to both sides.}$$

$$\left(x + \frac{7}{4}\right)^2 = \frac{97}{16} \qquad \text{On the left-hand side, factor the trinomial. On the right-hand side, } 3 = \frac{3 \cdot 16}{1 \cdot 16} = \frac{48}{16} \text{ and } \frac{48}{16} + \frac{49}{16} = \frac{97}{16}.$$

$$x + \frac{7}{4} = \pm\sqrt{\frac{97}{16}} \qquad \text{Apply the square root property.}$$

$$x = -\frac{7}{4} \pm \frac{\sqrt{97}}{4} \qquad \text{Subtract } \frac{7}{4} \text{ from both sides and simplify: } \sqrt{\frac{97}{16}} = \frac{\sqrt{97}}{\sqrt{16}} = \frac{\sqrt{97}}{4}.$$

$$x = \frac{-7 + \sqrt{97}}{4} \qquad \text{or} \qquad x = \frac{-7 - \sqrt{97}}{4} \qquad \text{Write each expression as a single fraction.}$$

State the Conclusion   The width of the border should be $\dfrac{-7 + \sqrt{97}}{4} \approx 0.71$ inch. (We discard the solution

$\dfrac{-7 - \sqrt{97}}{4}$, since it is negative.)

Check the Result   If the border is 0.71 inch wide, the announcement has an area of about $5.42 \cdot 4.42 \approx 23.96$ in.^{2}. If we subtract the area of the text from the area of the announcement, we get $23.96 - 12 = 11.96$ in.^{2}. This represents the area of the border, which was to be 12 in.^{2}. The answer seems reasonable.

Answers to Self Checks   **1.** $\pm 3\sqrt{2}$ **3.** $1, -5$ **4.** $\dfrac{7}{4}i, -\dfrac{7}{4}i$ **5.** $\left(a - \dfrac{5}{2}\right)^2$ **6.** $-1, -11$

7. $\dfrac{4}{3}, -2$ **8.** $\dfrac{-3 \pm \sqrt{6}}{3}$ **9.** $-2 \pm i\sqrt{2}$

10.1   STUDY SET   ELEMENTARY & INTERMEDIATE Algebra $f(x)$ Now™

VOCABULARY   Fill in the blanks.

1. An equation of the form $ax^2 + bx + c = 0$, where $a \neq 0$, is called a _____ equation.

2. We read $8 \pm \sqrt{3}$ as "eight _____ the square root of 3."

3. $x^2 + 6x + 9$ is called a _____ square trinomial because it factors as $(x + 3)^2$.

4. The _____ of x^2 in $x^2 - 12x + 36 = 0$ is 1, and the _____ term is 36.

CONCEPTS   Fill in the blanks.

5. For any nonnegative number c, if $x^2 = c$, then
$x = \boxed{}$ or $x = \boxed{}$.

6. To complete the square on x in $x^2 + 6x$, find one-half of $\boxed{}$, square it to get $\boxed{}$, and add $\boxed{}$ to get $\boxed{}$.

7. $\dfrac{1}{5} \pm \dfrac{\sqrt{2}}{5} = \dfrac{1 \pm \sqrt{2}}{\boxed{}}$

8. Suppose $x = 5 \pm 9$. Then $x = \boxed{}$ or $x = \boxed{}$.

9. Check to see whether $-3\sqrt{2}$ is a solution of $x^2 - 18 = 0$.

10. Check to see whether $-2 + \sqrt{2}$ is a solution of $x^2 + 4x + 2 = 0$.

11. Find one-half of the coefficient of x and then square it.

   a. $x^2 + 12x$ **b.** $x^2 - 5x$

   c. $x^2 - \dfrac{x}{2}$ **d.** $x^2 + \dfrac{3}{4}x$

12. Add a number to make each binomial a perfect square trinomial. Then factor the result.

   a. $x^2 + 8x$

   b. $x^2 - 4x$

   c. $x^2 - x$

13. What is the first step in solving the equation $x^2 + 12x = 35$

   a. by the factoring method?

   b. by completing the square?

14. Solve $x^2 = 16$

   a. by the factoring method.

   b. by the square root method.

15. Solve for x: $x + 7 = \pm\sqrt{6}$.

16. **a.** In the expression $8 \pm \dfrac{\sqrt{15}}{2}$, replace 8 with an equivalent fraction that has a denominator of 2.

   b. Write your answer to part a as a single fraction with denominator 2.

17. Explain the error in the work shown below.

   a. $\dfrac{4 \pm \sqrt{3}}{8} = \dfrac{\overset{1}{\cancel{4}} \pm \sqrt{3}}{\underset{1}{\cancel{4}} \cdot 2}$

      $= \dfrac{1 \pm \sqrt{3}}{2}$

b. $\dfrac{1 \pm \sqrt{5}}{5} = \dfrac{1 \pm \overset{1}{\cancel{\sqrt{5}}}}{\underset{1}{\cancel{5}}}$

$= \dfrac{1 \pm 1}{1}$

18. a. Write an expression that represents the width of the larger rectangle shown below.

b. Write an expression that represents the length of the larger rectangle shown below.

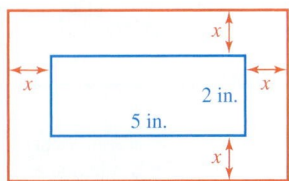

NOTATION

19. When solving a quadratic equation, a student obtains $x = \pm 2\sqrt{5}$.

a. How many solutions are represented by this notation? List them.

b. Approximate the solutions to the nearest hundredth.

20. When solving a quadratic equation, a student obtains $x = \dfrac{-5 \pm \sqrt{7}}{3}$.

a. How many solutions are represented by this notation? List them.

b. Approximate the solutions to the nearest hundredth.

PRACTICE   Use factoring to solve each equation.

21. $6x^2 + 12x = 0$

22. $5x^2 + 11x = 0$

23. $y^2 - 25 = 0$

24. $y^2 - 16 = 0$

25. $r^2 + 6r + 8 = 0$

26. $x^2 + 9x + 20 = 0$

27. $2z^2 = -2 + 5z$

28. $3x^2 = 8 - 10x$

Use the square root property to solve each equation.

29. $x^2 = 36$

30. $x^2 = 144$

31. $z^2 = 5$

32. $u^2 = 24$

33. $3x^2 - 16 = 0$

34. $5x^2 - 49 = 0$

35. $(x + 1)^2 = 1$

36. $(x - 1)^2 = 4$

37. $(s - 7)^2 = 9$

38. $(t + 4)^2 = 16$

39. $(x + 5)^2 - 3 = 0$

40. $(x + 3)^2 - 7 = 0$

41. $(a + 2)^2 = 8$

42. $(c + 2)^2 = 12$

43. $(3x - 1)^2 = 25$

44. $(5x - 2)^2 = 64$

45. $p^2 = -16$

46. $q^2 = -25$

47. $4m^2 + 81 = 0$

48. $9n^2 + 121 = 0$

49. $(x - 3)^2 = -5$

50. $(x + 2)^2 = -3$

Use the square root property to solve for the indicated variable. Assume that all variables represent positive numbers. Express all radicals in simplified form.

51. $2d^2 = 3h$ for d

52. $2x^2 = d^2$ for d

53. $E = mc^2$ for c

54. $A = \pi r^2$ for r

Use completing the square to solve each equation.

55. $x^2 + 2x - 8 = 0$

56. $x^2 + 6x + 5 = 0$

57. $k^2 - 8k + 12 = 0$

58. $p^2 - 4p + 3 = 0$

59. $g^2 + 5g - 6 = 0$

60. $s^2 + 5s - 14 = 0$

61. $x^2 - 3x - 4 = 0$

62. $x^2 - 7x + 12 = 0$

63. $x^2 + 8x + 6 = 0$

64. $x^2 + 6x + 4 = 0$

65. $x^2 - 2x = 17$

66. $x^2 + 10x = 7$

67. $m^2 - 7m + 3 = 0$

68. $m^2 - 5m + 3 = 0$

69. $a^2 - a = 3$

70. $b^2 - 3b = 5$

71. $2x^2 - x - 1 = 0$

72. $2x^2 - 5x + 2 = 0$

73. $3x^2 - 6x = 1$

74. $2x^2 - 6x = -3$

75. $4x^2 - 4x = 7$

76. $4x^2 - 4x = 1$

77. $2x^2 + 5x - 2 = 0$

78. $2x^2 - 8x + 5 = 0$

79. $\dfrac{7x + 1}{5} = -x^2$

80. $\dfrac{3x^2}{8} = \dfrac{1}{8} - x$

81. $p^2 + 2p + 2 = 0$

82. $x^2 - 6x + 10 = 0$

83. $y^2 + 8y + 18 = 0$

84. $t^2 + t + 3 = 0$

85. $3m^2 - 2m + 3 = 0$

86. $4p^2 + 2p + 3 = 0$

APPLICATIONS

87. FLAGS   In 1912, an order by President Taft fixed the width and length of the U.S. flag in the ratio 1 to 1.9. If 100 square feet of cloth are to be used to make a U.S. flag, estimate its dimensions to the nearest $\frac{1}{4}$ foot.

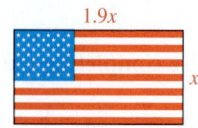

88. MOVIE STUNTS   According to the *Guinness Book of World Records, 1998,* stuntman Dan Koko fell a distance of 312 feet into an airbag after jumping from the Vegas World Hotel and Casino. The distance d in feet traveled by a free-falling object in t seconds is given by the formula $d = 16t^2$. To the nearest tenth of a second, how long did the fall last?

89. ACCIDENTS   The height h (in feet) of an object that is dropped from a height of s feet is given by the formula $h = s - 16t^2$, where t is the time the object has been falling. A 5-foot-tall woman on a sidewalk looks directly overhead and sees a window washer drop a bottle from 4 stories up. How long does she have to get out of the way? Round to the nearest tenth. (A story is 12 feet.)

90. GEOGRAPHY   The surface area S of a sphere is given by the formula $S = 4\pi r^2$, where r is the radius of the sphere. An almanac lists the surface area of the Earth as 196,938,800 square miles. Assuming the Earth to be spherical, what is its radius to the nearest mile?

91. AUTOMOBILE ENGINES   As the piston shown moves upward, it pushes a cylinder of a gasoline/air mixture that is ignited by the spark plug. The formula that gives the volume of a cylinder is $V = \pi r^2 h$, where r is the radius and h the height. Find the radius of the piston (to the nearest hundredth of an inch) if it displaces 47.75 cubic inches of gasoline/air mixture as it moves from its lowest to its highest point.

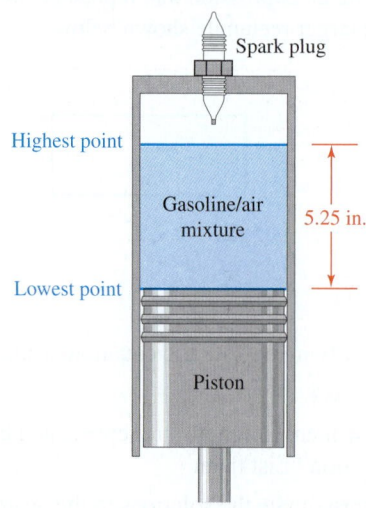

92. INVESTMENTS   If P dollars are deposited in an account that pays an annual rate of interest r, then in n years, the amount of money A in the account is given by the formula $A = P(1 + r)^n$. A savings account was opened on January 3, 1996, with a deposit of $10,000 and closed on January 2, 1998, with an ending balance of $11,772.25. Find the rate of interest.

93. PICTURE FRAMING   The matting around the picture has a uniform width. How wide is the matting if its area equals the area of the picture? Round to the nearest hundredth of an inch.

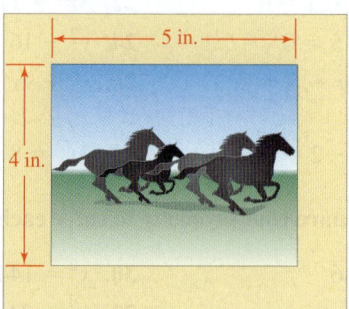

94. SWIMMING POOLS   In the advertisement shown, how wide will the free concrete decking be if a uniform width is constructed around the perimeter of the pool? Round to the nearest hundredth of a yard. (*Hint:* Note the difference in units.)

SAHARA POOL & SPA

SUMMER
SPECIAL

This 18 ft x 30 ft pool: only $18,500

Buy now and receive 28 square yards of concrete decking FREE!

95. DIMENSIONS OF A RECTANGLE   A rectangle is 4 feet longer than it is wide, and its area is 20 square feet. Find its dimensions to the nearest tenth of a foot.

96. DIMENSIONS OF A TRIANGLE   The height of a triangle is 4 meters longer than twice its base. Find the base and height if the area of the triangle is 10 square meters. Round to the nearest hundredth of a meter.

WRITING

97. Give an example of a perfect square trinomial. Why do you think the word "perfect" is used to describe it?

98. Explain why completing the square on $x^2 + 5x$ is more difficult than completing the square on $x^2 + 4x$.

REVIEW   **Simplify each expression. All variables represent positive real numbers.**

99. $\sqrt[3]{40a^3b^6}$

100. $\sqrt[3]{-27x^6}$

101. $\sqrt[8]{x^{24}}$

102. $\sqrt[4]{\dfrac{16}{625}}$

103. $\sqrt{175a^2b^3}$

104. $\sqrt{\dfrac{z^2}{16x^2}}$

CHALLENGE PROBLEMS

105. What number must be added to $x^2 + \sqrt{3}x$ to make a perfect square trinomial?

106. Solve $x^2 + \sqrt{3}x - \dfrac{1}{4} = 0$ by completing the square.

10.2   The Quadratic Formula

- The quadratic formula   • Solving quadratic equations using the quadratic formula
- Solving an equivalent equation   • Problem solving

We can solve any quadratic equation by the method of completing the square, but the work is often tedious. In this section, we will develop a formula, called the *quadratic formula*, that lets us solve quadratic equations with less effort.

■ THE QUADRATIC FORMULA

To develop a formula that will produce the solutions of any given quadratic equation, we start with the **general quadratic equation** $ax^2 + bx + c = 0$, with $a > 0$, and solve it for x by completing the square.

$$ax^2 + bx + c = 0$$

$$\frac{ax^2}{a} + \frac{bx}{a} + \frac{c}{a} = \frac{0}{a}$$ Divide both sides by a so that the coefficient of x^2 is 1.

$$x^2 + \frac{b}{a}x + \frac{c}{a} = 0$$ Simplify: $\frac{ax^2}{a} = x^2$. Write $\frac{bx}{a}$ as $\frac{b}{a}x$.

$$x^2 + \frac{b}{a}x \phantom{+ \frac{c}{a}} = -\frac{c}{a}$$ Subtract $\frac{c}{a}$ from both sides so that only the terms involving x are on the left-hand side of the equation.

We can complete the square on $x^2 + \frac{b}{a}x$ by adding the square of one-half of the coefficient of x. Since the coefficient of x is $\frac{b}{a}$, we have $\frac{1}{2} \cdot \frac{b}{a} = \frac{b}{2a}$ and $\left(\frac{b}{2a}\right)^2 = \frac{b^2}{4a^2}$.

$$x^2 + \frac{b}{a}x + \frac{b^2}{4a^2} = -\frac{c}{a} + \frac{b^2}{4a^2}$$

To complete the square, add $\frac{b^2}{4a^2}$ to both sides.

$$x^2 + \frac{b}{a}x + \frac{b^2}{4a^2} = -\frac{4ac}{4aa} + \frac{b^2}{4a^2}$$

Multiply $-\frac{c}{a}$ by $\frac{4a}{4a}$. Now the fractions on the right side have the common denominator $4a^2$.

$$\left(x + \frac{b}{2a}\right)^2 = \frac{b^2 - 4ac}{4a^2}$$

On the left-hand side, factor. On the right-hand side, add the fractions.

$$x + \frac{b}{2a} = \pm\sqrt{\frac{b^2 - 4ac}{4a^2}}$$

Use the square root property.

$$x + \frac{b}{2a} = \pm\frac{\sqrt{b^2 - 4ac}}{\sqrt{4a^2}}$$

The square root of a quotient is the quotient of square roots.

$$x + \frac{b}{2a} = \pm\frac{\sqrt{b^2 - 4ac}}{2a}$$

Since $a > 0$, $\sqrt{4a^2} = 2a$.

$$x = -\frac{b}{2a} \pm \frac{\sqrt{b^2 - 4ac}}{2a}$$

To isolate x, subtract $\frac{b}{2a}$ from both sides.

$$x = \frac{-b \pm \sqrt{b^2 - 4ac}}{2a}$$

Combine the fractions.

The Language of Algebra

Your instructor may ask you to *derive* the quadratic formula. That means to solve $ax^2 + bx + c = 0$ for x, using the series of steps shown here, to obtain

$$x = \frac{-b \pm \sqrt{b^2 - 4ac}}{2a}$$

This result is called the *quadratic formula*. To develop this formula, we assumed that a was positive. If a is negative, similar steps are used, and we obtain the same result.

Quadratic Formula   The solutions of $ax^2 + bx + c = 0$, with $a \neq 0$, are

$$x = \frac{-b \pm \sqrt{b^2 - 4ac}}{2a}$$

■ SOLVING QUADRATIC EQUATIONS USING THE QUADRATIC FORMULA

EXAMPLE 1   Solve: $2x^2 - 5x - 3 = 0$.

ELEMENTARY & INTERMEDIATE
Algebra $f(x)$ **Now**™   Solution   To use the quadratic formula, we need to identify a, b, and c. We do this by comparing the given equation to the general quadratic equation $ax^2 + bx + c = 0$.

$$2x^2 - 5x - 3 = 0$$
$$\uparrow \qquad \uparrow \qquad \uparrow$$
$$ax^2 + bx + c = 0$$

We see that $a = 2$, $b = -5$, and $c = -3$. To find the solutions of the equation, we substitute these values into the formula and evaluate the right-hand side.

$$x = \frac{-b \pm \sqrt{b^2 - 4ac}}{2a}$$ This is the quadratic formula.

$$x = \frac{-(-5) \pm \sqrt{(-5)^2 - 4(2)(-3)}}{2(2)}$$ Substitute 2 for a, -5 for b, and -3 for c.

$$x = \frac{5 \pm \sqrt{25 - (-24)}}{4}$$ Simplify: $-(-5) = 5$. Evaluate the power and multiply within the radical. Multiply in the denominator.

$$x = \frac{5 \pm \sqrt{49}}{4}$$ Simplify within the radical.

$$x = \frac{5 \pm 7}{4}$$ Simplify: $\sqrt{49} = 7$.

To find the first solution, evaluate the expression using the $+$ symbol. To find the second solution, evaluate the expression using the $-$ symbol.

$$x = \frac{5 + 7}{4} \quad \text{or} \quad x = \frac{5 - 7}{4}$$

$$x = \frac{12}{4} \qquad\qquad x = \frac{-2}{4}$$

$$x = 3 \qquad\qquad\quad x = -\frac{1}{2}$$

The solutions are 3 and $-\dfrac{1}{2}$. Verify that both satisfy the original equation.

Self Check 1 Solve: $4x^2 - 7x - 2 = 0$.

When using the quadratic formula, we should write the equation in $ax^2 + bx + c = 0$ form (called **quadratic form**) so that a, b, and c can be determined.

EXAMPLE 2

Solve: $2x^2 = -4x - 1$.

Solution We begin by writing the equation in quadratic form.

$$2x^2 = -4x - 1$$
$$2x^2 + 4x + 1 = 0 \quad \text{Add } 4x \text{ and 1 to both sides.}$$

In this equation, $a = 2$, $b = 4$, and $c = 1$.

$$x = \frac{-b \pm \sqrt{b^2 - 4ac}}{2a}$$ This is the quadratic formula.

$$x = \frac{-4 \pm \sqrt{4^2 - 4(2)(1)}}{2(2)}$$ Substitute 2 for a, 4 for b, and 1 for c.

$$x = \frac{-4 \pm \sqrt{16 - 8}}{4}$$ Evaluate the expression within the radical. Multiply in the denominator.

$$x = \frac{-4 \pm \sqrt{8}}{4}$$

$$x = \frac{-4 \pm 2\sqrt{2}}{4}$$ Simplify: $\sqrt{8} = \sqrt{4 \cdot 2} = 2\sqrt{2}$.

ELEMENTARY &
INTERMEDIATE
Algebra $f(x)$ **Now**™

We can write the solutions in simpler form by factoring out 2 from the terms in the numerator and then removing the common factor of 2 in the numerator and denominator.

$$x = \frac{-4 \pm 2\sqrt{2}}{4} = \frac{2(-2 \pm \sqrt{2})}{4} = \frac{\overset{1}{\cancel{2}}(-2 \pm \sqrt{2})}{\underset{1}{\cancel{2}} \cdot 2} = \frac{-2 \pm \sqrt{2}}{2}$$

Notation

The solutions can also be written as

$$-\frac{2}{2} \pm \frac{\sqrt{2}}{2} = -1 \pm \frac{\sqrt{2}}{2}$$

The solutions are $\dfrac{-2 + \sqrt{2}}{2}$ and $\dfrac{-2 - \sqrt{2}}{2}$. We can approximate the solutions using a calculator. To two decimal places, they are -0.29 and -1.71.

Self Check 2   Solve $3x^2 = 2x + 3$. Approximate the solutions to two decimal places.

The solutions to the next example are imaginary numbers.

EXAMPLE 3

ELEMENTARY & INTERMEDIATE
Algebra *f(x)* **Now**™   **Solution**

Solve: $x^2 + x = -1$.

We begin by writing the equation in quadratic form before identifying a, b, and c.

$$x^2 + x + 1 = 0$$

In this equation, $a = 1$, $b = 1$, and $c = 1$:

$$x = \frac{-b \pm \sqrt{b^2 - 4ac}}{2a}$$

$$x = \frac{-1 \pm \sqrt{1^2 - 4(1)(1)}}{2(1)} \qquad \text{Substitute 1 for } a, \text{1 for } b, \text{ and 1 for } c.$$

$$x = \frac{-1 \pm \sqrt{1 - 4}}{2} \qquad \text{Evaluate the expression within the radical.}$$

$$x = \frac{-1 \pm \sqrt{-3}}{2}$$

$$x = \frac{-1 \pm i\sqrt{3}}{2} \qquad \sqrt{-3} = \sqrt{-1 \cdot 3} = \sqrt{-1}\sqrt{3} = i\sqrt{3}.$$

Notation

The solutions are written in complex number form $a + bi$. They could also be written as

$$\frac{-1 \pm i\sqrt{3}}{2}$$

The solutions are $-\dfrac{1}{2} + \dfrac{\sqrt{3}}{2}i$ and $-\dfrac{1}{2} - \dfrac{\sqrt{3}}{2}i$.

Self Check 3   Solve: $a^2 + 3a + 5 = 0$.

■ SOLVING AN EQUIVALENT EQUATION

When solving a quadratic equation by the quadratic formula, we can often simplify the computations by solving an equivalent equation.

EXAMPLE 4

ELEMENTARY & INTERMEDIATE
Algebra *f(x)* **Now**™

For each equation, write an equivalent equation so that the quadratic formula computations will be easier:

a. $-2x^2 + 4x - 1 = 0$,   **b.** $x^2 + \frac{4}{5}x - \frac{1}{3} = 0$,   and   **c.** $20x^2 - 60x - 40 = 0$.

Solution   **a.** It is often easier to solve a quadratic equation using the quadratic formula if a is positive. If we multiply (or divide) both sides of $-2x^2 + 4x - 1 = 0$ by -1, we obtain an equivalent equation with $a > 0$.

$$-2x^2 + 4x - 1 = 0 \qquad \text{Here, } a = -2.$$
$$(-1)(-2x^2 + 4x - 1) = (-1)(0)$$
$$2x^2 - 4x + 1 = 0 \qquad \text{Now } a = 2.$$

b. For $x^2 + \dfrac{4}{5}x - \dfrac{1}{3} = 0$, two coefficients are fractions: $b = \dfrac{4}{5}$ and $c = -\dfrac{1}{3}$. We can multiply both sides of the equation by their least common denominator, 15, to obtain an equivalent equation having coefficients that are integers.

$$x^2 + \frac{4}{5}x - \frac{1}{3} = 0 \qquad \text{Here, } a = 1, b = \frac{4}{5}, \text{ and } c = -\frac{1}{3}.$$
$$15\left(x^2 + \frac{4}{5}x - \frac{1}{3}\right) = 15(0)$$
$$15x^2 + 12x - 5 = 0 \qquad \text{Now } a = 15, b = 12, \text{ and } c = -5.$$

c. For $20x^2 - 60x - 40 = 0$, the coefficients 20, -60, and -40 have a common factor of 20. If we divide both sides of the equation by their GCF, we obtain an equivalent equation having smaller coefficients.

$$20x^2 - 60x - 40 = 0 \qquad \text{Here, } a = 20, b = -60, \text{ and } c = -40.$$
$$\frac{20x^2}{20} - \frac{60x}{20} - \frac{40}{20} = \frac{0}{20}$$
$$x^2 - 3x - 2 = 0 \qquad \text{Now } a = 1, b = -3, \text{ and } c = -2.$$

Self Check 4   For each equation, write an equivalent equation so that the quadratic formula computations will be simpler: **a.** $-6x^2 + 7x - 9 = 0$,   **b.** $\dfrac{1}{3}x^2 - \dfrac{2}{3}x - \dfrac{5}{6} = 0$,   and   **c.** $44x^2 + 66x - 55 = 0$.

■ **PROBLEM SOLVING**

EXAMPLE 5

ELEMENTARY &
INTERMEDIATE
Algebra $f(x)$ **Now**™

Shortcuts.   Instead of using the hallways, students are wearing a path through a planted quad area to walk 195 feet directly from the classrooms to the cafeteria. If the length of the hallway from the office to the cafeteria is 105 feet longer than the hallway from the office to the classrooms, how much walking are the students saving by taking the shortcut?

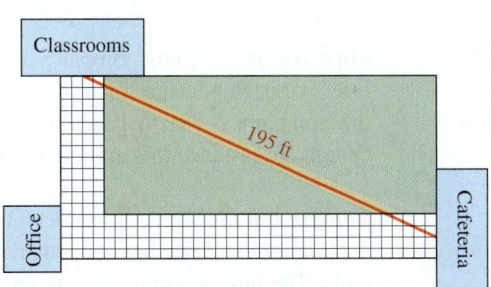

Analyze the Problem   The two hallways and the shortcut form a right triangle with a hypotenuse 195 feet long. We will use the Pythagorean theorem to solve this problem.

Form an Equation   If we let x = the length (in feet) of the hallway from the classrooms to the office, then the length of the hallway from the office to the cafeteria is $(x + 105)$ feet. Substituting these lengths into the Pythagorean theorem, we have

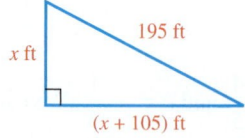

195 ft

x ft

$(x + 105)$ ft

$$a^2 + b^2 = c^2$$ This is the Pythagorean theorem.

$$x^2 + (x + 105)^2 = 195^2$$ Substitute x for a, $(x + 105)$ for b, and 195 for c.

$$x^2 + x^2 + 105x + 105x + 11{,}025 = 38{,}025$$ Find $(x + 105)^2$.

$$2x^2 + 210x + 11{,}025 = 38{,}025$$ Combine like terms.

$$2x^2 + 210x - 27{,}000 = 0$$ Subtract 38,025 from both sides.

$$x^2 + 105x - 13{,}500 = 0$$ Divide both sides by 2.

Solve the Equation   To solve $x^2 + 105x - 13{,}500 = 0$, we will use the quadratic formula with $a = 1$, $b = 105$, and $c = -13{,}500$.

$$x = \frac{-b \pm \sqrt{b^2 - 4ac}}{2a}$$

$$x = \frac{-105 \pm \sqrt{(105)^2 - 4(1)(-13{,}500)}}{2(1)}$$

$$x = \frac{-105 \pm \sqrt{65{,}025}}{2}$$ Simplify: $(105)^2 - 4(1)(-13{,}500) = 11{,}025 + 54{,}000 = 65{,}025$.

$$x = \frac{-105 \pm 255}{2}$$ Use a calculator: $\sqrt{65{,}025} = 255$.

$$x = \frac{150}{2} \quad \text{or} \quad x = \frac{-360}{2}$$

$$x = 75 \quad \quad \quad x = -180$$ Since the length of the hallway can't be negative, discard the solution $x = -180$.

State the Conclusion   The length of the hallway from the classrooms to the office is 75 feet. The length of the hallway from the office to the cafeteria is $75 + 105 = 180$ feet. Instead of using the hallways, a distance of $75 + 180 = 255$ feet, the students are taking the 195-foot shortcut to the cafeteria, a savings of $(255 - 195)$, or 60 feet.

Check the Result   The length of the 180-foot hallway is 105 feet longer than the length of the 75-foot hallway. The sum of the squares of the lengths of the hallways is $75^2 + 180^2 = 38{,}025$. This equals the square of the length of the 195-foot shortcut. The result checks. ▪

EXAMPLE 6

ELEMENTARY &
INTERMEDIATE
Algebra $f(x)$ **Now**™

Mass transit.   A bus company has 4,000 passengers daily, each currently paying a 75¢ fare. For each 15¢ fare increase, the company estimates that it will lose 50 passengers. If the company needs to bring in $6,570 per day to stay in business, what fare must be charged to produce this amount of revenue?

Analyze the Problem   To understand how a fare increase affects the number of passengers, let's consider what happens if there are two fare increases. We organize the data in a table. The fares are expressed in terms of dollars.

Number of increases	New fare	Number of passengers
One $0.15 increase	$0.75 + $0.15(1) = $0.90	4,000 − 50(1) = 3,950
Two $0.15 increases	$0.75 + $0.15(2) = $1.05	4,000 − 50(2) = 3,900

In general, the new fare will be the old fare ($0.75) plus the number of fare increases times $0.15. The number of passengers who will pay the new fare is 4,000 minus 50 times the number of $0.15 fare increases.

Form an Equation   If we let $x =$ the number of $0.15 fare increases necessary to bring in $6,570 daily, then $(0.75 + 0.15x)$ is the fare that must be charged. The number of passengers who will pay this fare is $4,000 − 50x$. We can now form the equation.

The bus fare	times	the number of passengers who will pay that fare	equals	$6,570.
$(0.75 + 0.15x)$	$\cdot$	$(4,000 − 50x)$	=	6,570

Solve the Equation

$(0.75 + 0.15x)(4,000 − 50x) = 6,570$

$3,000 − 37.5x + 600x − 7.5x^2 = 6,570$ Multiply the binomials.

$-7.5x^2 + 562.5x + 3,000 = 6,570$ Combine like terms: $-37.5x + 600x = 562.5x$.

$-7.5x^2 + 562.5x − 3,570 = 0$ Subtract 6,570 from both sides.

$7.5x^2 − 562.5x + 3,570 = 0$ Multiply both sides by $−1$ so that a, 7.5, is positive.

To solve this equation, we will use the quadratic formula.

$$x = \frac{-b \pm \sqrt{b^2 - 4ac}}{2a}$$

$$x = \frac{-(-562.5) \pm \sqrt{(-562.5)^2 - 4(7.5)(3,570)}}{2(7.5)}$$ Substitute 7.5 for a, $−562.5$ for b, and 3,570 for c.

$$x = \frac{562.5 \pm \sqrt{209,306.25}}{15}$$ Simplify: $(-562.5)^2 - 4(7.5)(3,570) = 316,406.25 - 107,100 = 209,306.25$.

$$x = \frac{562.5 \pm 457.5}{15}$$ Use a calculator: $\sqrt{209,306.25} = 457.5$.

$$x = \frac{1,020}{15} \quad \text{or} \quad x = \frac{105}{15}$$

$$x = 68 \quad \bigg| \quad x = 7$$

State the Conclusion   If there are 7 fifteen-cent increases in the fare, the new fare will be $0.75 + $0.15(7) = $1.80. If there are 68 fifteen-cent increases in the fare, the new fare will be $0.75 + $0.15(68) = $10.95. Although this fare would bring in the necessary revenue, a $10.95 bus fare is unreasonable, so we discard it.

Check the Result   A fare of $1.80 will be paid by $[4,000 − 50(7)] = 3,650$ bus riders. The amount of revenue brought in would be $1.80(3,650) = $6,570. The result checks.

EXAMPLE 7

Lawyers.   The number of lawyers N in the United States each year from 1980 to 2002 is approximated by $N = 222x^2 + 17{,}630x + 571{,}178$, where $x = 0$ corresponds to the year 1980, $x = 1$ corresponds to 1981, $x = 2$ corresponds to 1982, and so on. (Thus, $0 \le x \le 22$). In what year does this model indicate that the United States had one million lawyers?

Solution   We will substitute 1,000,000 for N in the equation. Then we can solve for x, which will give the number of years after 1980 that the United States had approximately 1,000,000 lawyers.

$$N = 222x^2 + 17{,}630x + 571{,}178$$

$$1{,}000{,}000 = 222x^2 + 17{,}630x + 571{,}178 \qquad \text{Replace } N \text{ with } 1{,}000{,}000.$$

$$0 = 222x^2 + 17{,}630x - 428{,}822 \qquad \begin{array}{l}\text{Subtract 1,000,000 from both sides so that}\\ \text{the equation is in quadratic form.}\end{array}$$

We can simplify the computations by dividing both sides of the equation by 2, which is the greatest common factor of 222, 17,630, and 428,822.

$$111x^2 + 8{,}815x - 214{,}411 = 0 \qquad \text{Divide both sides by 2.}$$

We solve this equation using the quadratic formula.

$$x = \frac{-b \pm \sqrt{b^2 - 4ac}}{2a}$$

$$x = \frac{-8{,}815 \pm \sqrt{(8{,}815)^2 - 4(111)(-214{,}411)}}{2(111)} \qquad \begin{array}{l}\text{Substitute 111 for } a, 8{,}815 \text{ for } b,\\ \text{and } -214{,}411 \text{ for } c.\end{array}$$

$$x = \frac{-8{,}815 \pm \sqrt{172{,}902{,}709}}{222} \qquad \begin{array}{l}\text{Evaluate the expression within the}\\ \text{radical.}\end{array}$$

$$x \approx \frac{4{,}334}{222} \quad \text{or} \quad x \approx \frac{-21{,}964}{222} \qquad \text{Use a calculator.}$$

$$x \approx 19.5 \quad \Big| \quad x \approx -98.9 \qquad \begin{array}{l}\text{Since the model is defined only}\\ \text{for } 0 \le x \le 22, \text{ we discard the}\\ \text{second solution.}\end{array}$$

In 19.5 years after 1980, or midway through 1999, the United States had approximately 1,000,000 lawyers.

Answers to Self Checks   **1.** $2, -\dfrac{1}{4}$ **2.** $\dfrac{1 \pm \sqrt{10}}{3}$; $-0.72, 1.39$ **3.** $-\dfrac{3}{2} \pm \dfrac{\sqrt{11}}{2}i$

4. a. $6x^2 - 7x + 9 = 0$,   **b.** $2x^2 - 4x - 5 = 0$,   **c.** $4x^2 + 6x - 5 = 0$

10.2   STUDY SET   ELEMENTARY & INTERMEDIATE Algebra $f(x)$ Now™

VOCABULARY   **Fill in the blanks.**

1. An equation of the form $ax^2 + bx + c = 0$, with $a \ne 0$, is a _____ equation.

2. The formula

$$x = \boxed{}$$

is called the quadratic formula.

CONCEPTS

3. Write each equation in quadratic form.

    a. $x^2 + 2x = -5$ **b.** $3x^2 = -2x + 1$

4. For each quadratic equation, find a, b, and c.

    a. $x^2 + 5x + 6 = 0$ **b.** $8x^2 - x = 10$

5. Decide whether each statement is true or false.

    a. Any quadratic equation can be solved by using the quadratic formula.

    b. Any quadratic equation can be solved by completing the square.

6. What is wrong with the beginning of the solution shown below?

    Solve: $x^2 - 3x = 2$.

    $a = 1$ $b = -3$ $c = 2$

Evaluate each expression.

7. a. $\dfrac{-2 \pm \sqrt{2^2 - 4(1)(-8)}}{2(1)}$

    b. $\dfrac{-(-1) \pm \sqrt{(-1)^2 - 4(2)(-4)}}{2(2)}$

8. A student used the quadratic formula to solve a quadratic equation and obtained $x = \dfrac{-2 \pm \sqrt{3}}{2}$.

    a. How many solutions does the equation have? What are they exactly?

    b. Graph the solutions on a number line.

9. Simplify each of the following.

    a. $\dfrac{3 \pm 6\sqrt{2}}{3}$

    b. $\dfrac{-12 \pm 4\sqrt{7}}{8}$

10. For each of the following, write an equivalent equation so that the quadratic formula computations will be easier to perform.

    a. $-5x^2 + 9x - 2 = 0$

    b. $\dfrac{1}{8}x^2 + \dfrac{1}{2}x - \dfrac{3}{4} = 0$

    c. $45x^2 + 30x - 15 = 0$

NOTATION

11. On a quiz, students were asked to write the quadratic formula. What is wrong with each answer shown below?

    a. $x = -b \pm \dfrac{\sqrt{b^2 - 4ac}}{2a}$

    b. $x = \dfrac{-b\sqrt{b^2 - 4ac}}{2a}$

12. In reading $\dfrac{-b \pm \sqrt{b^2 - 4ac}}{2a}$, we say, "The _____ of b, plus or _____ the square _____ of b _____ minus   times a times c, all _____ $2a$."

PRACTICE   Use the quadratic formula to solve each equation.

13. $x^2 + 3x + 2 = 0$ **14.** $x^2 - 3x + 2 = 0$

15. $x^2 + 12x = -36$ **16.** $y^2 - 18y = -81$

17. $2x^2 + 5x - 3 = 0$ **18.** $6x^2 - x - 1 = 0$

19. $5x^2 + 5x + 1 = 0$ **20.** $4w^2 + 6w + 1 = 0$

21. $8u = -4u^2 - 3$ **22.** $4t + 3 = 4t^2$

23. $-16y^2 - 8y + 3 = 0$ **24.** $-16x^2 - 16x - 3 = 0$

25. $x^2 - \dfrac{14}{15}x = \dfrac{8}{15}$ **26.** $x^2 = -\dfrac{5}{4}x + \dfrac{3}{2}$

27. $\dfrac{x^2}{2} + \dfrac{5}{2}x = -1$ **28.** $\dfrac{x^2}{8} - \dfrac{x}{4} = \dfrac{1}{2}$

29. $2x^2 - 1 = 3x$ **30.** $-9x = 2 - 3x^2$

31. $-x^2 + 10x = 18$ **32.** $-3x = \dfrac{x^2}{2} + 2$

33. $x^2 - 6x = 391$ **34.** $-x^2 + 27x = -280$

35. $x^2 - \dfrac{5}{3} = -\dfrac{11}{6}x$ **36.** $x^2 - \dfrac{1}{2} = \dfrac{2}{3}x$

37. $x^2 + 2x + 2 = 0$ **38.** $x^2 + 3x + 3 = 0$

39. $2x^2 + x + 1 = 0$ **40.** $3x^2 + 2x + 1 = 0$

41. $3x^2 - 4x = -2$ **42.** $2x^2 + 3x = -3$

43. $3x^2 - 2x = -3$ **44.** $5x^2 = 2x - 1$

45. $\dfrac{x^2}{8} - \dfrac{x}{2} + 1 = 0$ **46.** $\dfrac{x^2}{2} + 3x + \dfrac{13}{2} = 0$

47. $\dfrac{a^2}{10} - \dfrac{3a}{5} + \dfrac{7}{5} = 0$ **48.** $\dfrac{c^2}{4} + c + \dfrac{11}{4} = 0$

49. $50x^2 + 30x - 10 = 0$

50. $120b^2 + 120b - 40 = 0$

51. $900x^2 - 8,100x = 1,800$

52. $-14x^2 + 21x = -49$

53. $-0.6x^2 - 0.03 = -0.4x$

54. $2x^2 + 0.1x = 0.04$

🖩 **Use the quadratic formula and a scientific calculator to solve each equation. Give all answers to the nearest hundredth.**

55. $x^2 + 8x + 5 = 0$
56. $2x^2 - x - 9 = 0$
57. $3x^2 - 2x - 2 = 0$
58. $81x^2 + 12x - 80 = 0$
59. $0.7x^2 - 3.5x - 25 = 0$
60. $-4.5x^2 + 0.2x + 3.75 = 0$

APPLICATIONS

61. IMAX SCREENS   The largest permanent movie screen is in the Panasonic Imax theater at Darling Harbor, Sydney, Australia. The rectangular screen has an area of 11,349 square feet. Find the dimensions of the screen if it is 20 feet longer than it is wide.

62. ROCK CONCERTS   During a 1997 tour, the rock group U2 used an LED (light emitting diode) electronic screen as part of the stage backdrop. The rectangular screen, with an area of 9,520 square feet, had a length that was 2 feet more than three times its width. Find the dimensions of the LED screen.

63. PARKS   Central Park is one of New York's best-known landmarks. Rectangular in shape, its length is 5 times its width. When measured in miles, its perimeter numerically exceeds its area by 4.75. Find the dimensions of Central Park if we know that its width is less than 1 mile.

64. HISTORY   One of the important cities of the ancient world was Babylon. Greek historians wrote that the city was square-shaped. Measured in miles, its area numerically exceeded its perimeter by about 124. Find its dimensions. (Round to the nearest tenth.)

65. BADMINTON   The person who wrote the instructions for setting up the badminton net shown below forgot to give the specific dimensions for securing the pole. How long is the support string?

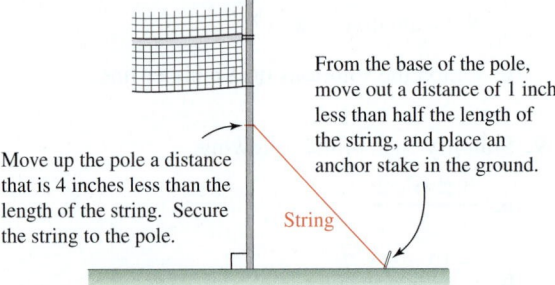

From the base of the pole, move out a distance of 1 inch less than half the length of the string, and place an anchor stake in the ground.

Move up the pole a distance that is 4 inches less than the length of the string. Secure the string to the pole.

String

66. RIGHT TRIANGLES   The hypotenuse of a right triangle is 2.5 units long. The longer leg is 1.7 units longer than the shorter leg. Find the lengths of the sides of the triangle.

67. DANCES   Tickets to a school dance cost $4, and the projected attendance is 300 persons. It is further projected that for every 10¢ increase in ticket price, the average attendance will decrease by 5. At what ticket price will the receipts from the dance be $1,248?

68. TICKET SALES   A carnival usually sells three thousand 75¢ ride tickets on a Saturday. For each 15¢ increase in price, management estimates that 80 fewer tickets will be sold. What increase in ticket price will produce $2,982 of revenue on Saturday?

69. MAGAZINE SALES   The *Gazette's* profit is $20 per year for each of its 3,000 subscribers. Management estimates that the profit per subscriber will increase by 1¢ for each additional subscriber over the current 3,000. How many subscribers will bring a total profit of $120,000?

70. POLYGONS   A five-sided polygon, called a *pentagon,* has 5 diagonals. The number of diagonals *d* of a polygon of *n* sides is given by the formula

$$d = \frac{n(n-3)}{2}$$

Find the number of sides of a polygon if it has 275 diagonals.

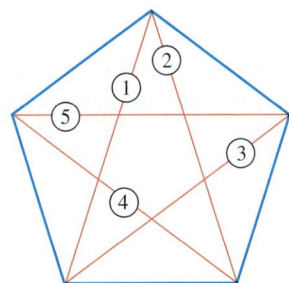

71. INVESTMENT RATES   A woman invests $1,000 in a fund for which interest is compounded annually at a rate *r*. After one year, she deposits an additional $2,000. After two years, the balance in the account is

$$\$1,000(1 + r)^2 + \$2,000(1 + r)$$

If this amount is $3,368.10, find *r*.

72. METAL FABRICATION   A box with no top is to be made by cutting a 2-inch square from each corner of the square sheet of metal. After bending up the sides, the volume of the box is to be 220 cubic inches. How large should the piece of metal be? Round to the nearest hundredth.

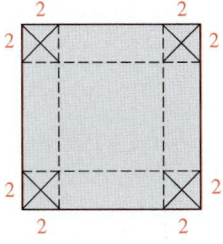

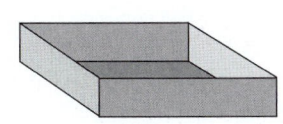

73. RETIREMENT   The labor force participation rate *P* (in percent) for men ages 55–64 from 1970 to 2000 is approximated by the quadratic equation

$$P = 0.03x^2 - 1.37x + 82.51$$

where $x = 0$ corresponds to the year 1970, $x = 1$ corresponds to 1971, $x = 2$ corresponds to 1972, and so on. (Thus, $0 \le x \le 30$.) When does the model indicate that 75% of the men ages 55–64 were part of the workforce?

74. SPACE PROGRAM   The yearly budget *B* (in billions of dollars) for the National Aeronautics and Space Administration (NASA) is approximated by the quadratic equation

$$B = 0.0596x^2 - 0.3811x + 14.2709$$

where *x* is the number of years since 1995 and $0 \le x \le 9$. In what year does the model indicate that NASA's budget was about $15 billion?

WRITING

75. Explain why the quadratic formula, in most cases, is less tedious to use in solving a quadratic equation than is the method of completing the square.

76. On an exam, a student was asked to solve the equation $-4w^2 - 6w - 1 = 0$. Her first step was to multiply both sides of the equation by -1. She then used the quadratic formula to solve $4w^2 + 6w + 1 = 0$ instead. Is this a valid approach? Explain.

REVIEW   Change each radical to an exponential expression.

77. $\sqrt{n}$

78. $\sqrt[7]{\frac{3}{8}r^2s}$

79. $\sqrt[4]{3b}$

80. $3\sqrt[3]{c^2 - d^2}$

Write each expression in radical form.

81. $t^{1/3}$

82. $\left(\frac{3}{4}m^2n^2\right)^{1/5}$

83. $(3t)^{1/4}$

84. $(c^2 + d^2)^{1/2}$

CHALLENGE PROBLEMS   All of the equations we have solved so far have had rational-number coefficients. However, the quadratic formula can be used to solve quadratic equations with irrational or even imaginary coefficients. Solve each equation.

85. $x^2 + 2\sqrt{2}x - 6 = 0$

86. $\sqrt{2}x^2 + x - \sqrt{2} = 0$

87. $x^2 - 3ix - 2 = 0$

88. $100ix^2 + 300x - 200i = 0$

10.3   The Discriminant and Equations That Can Be Written in Quadratic Form

• The discriminant   • Equations that are quadratic in form   • Problem solving

In this section, we will discuss how to predict what type of solutions a quadratic equation will have without solving the equation. We will then solve some special equations that can be written in quadratic form. Finally, we will use the equation-solving methods of this chapter to solve a shared-work problem.

■ THE DISCRIMINANT

We can predict what type of solutions a particular quadratic equation will have without solving it. To see how, we suppose that the coefficients a, b, and c in the equation $ax^2 + bx + c = 0$ represent real numbers and $a \neq 0$. Then the solutions of the equation are given by the quadratic formula

$$x = \frac{-b \pm \sqrt{b^2 - 4ac}}{2a}$$

If $b^2 - 4ac \geq 0$, the solutions are real numbers. If $b^2 - 4ac < 0$, the solutions are not real numbers. Thus, the value of $b^2 - 4ac$, called the **discriminant,** determines the type of solutions for a particular quadratic equation.

The Discriminant   If a, b, and c represent real numbers and

if $b^2 - 4ac$ is . . .	the solutions are . . .
positive,	two different real numbers.
0,	two real numbers that are equal.
negative,	two different complex numbers containing i that are complex conjugates.

If a, b, and c represent rational numbers and

if $b^2 - 4ac$ is . . .	the solutions are . . .
a perfect square,	two different rational numbers.
positive and not a perfect square,	two different irrational numbers.

EXAMPLE 1

ELEMENTARY &
INTERMEDIATE
Algebra $f(x)$ **Now**™

Determine the type of solutions for each equation: **a.** $x^2 + x + 1 = 0$ and
b. $3x^2 + 5x + 2 = 0$.

Solution   **a.** We calculate the discriminant for $x^2 + x + 1 = 0$:

$$b^2 - 4ac = 1^2 - 4(1)(1) \qquad a = 1, b = 1, \text{ and } c = 1.$$
$$= -3 \qquad\qquad \text{The result is a negative number.}$$

Since $b^2 - 4ac < 0$, the solutions of $x^2 + x + 1 = 0$ are two complex numbers containing i that are complex conjugates.

b. For $3x^2 + 5x + 2 = 0$,

$$b^2 - 4ac = 5^2 - 4(3)(2) \qquad a = 3, b = 5, \text{ and } c = 2.$$
$$= 25 - 24$$
$$= 1 \qquad \text{The result is a positive number.}$$

Since $b^2 - 4ac > 0$ and $b^2 - 4ac$ is a perfect square, the two solutions of $3x^2 + 5x + 2 = 0$ are rational and unequal.

Self Check 1   Determine the type of solutions for **a.** $x^2 + x - 1 = 0$ and **b.** $4x^2 - 10x + 25 = 0$.

■ EQUATIONS THAT ARE QUADRATIC IN FORM

Many equations that are not quadratic can be written in quadratic form ($ax^2 + bx + c = 0$) and then solved using the techniques discussed in previous sections. For example, a careful inspection of the equation $x^4 - 5x^2 + 4 = 0$ leads to the following observations:

The lead term, x^4, is the square of the expression x^2, here in the middle term: $x^4 = (x^2)^2$.

$$x^4 - 5x^2 + 4 = 0$$

The last term is a constant.

Equations that contain an expression, the same expression squared, and a constant term are said to be *quadratic in form*. One method used to solve such equations is to make a substitution.

EXAMPLE 2   Solve: $x^4 - 3x^2 - 4 = 0$.

Solution   If we write x^4 as $(x^2)^2$, then the equation takes the form

$$(x^2)^2 - 3x^2 - 4 = 0$$

Notation

The choice of the letter y for the substitution is arbitrary. We could just as well let $b = x^2$.

and it is said to be *quadratic in x^2*. We can solve this equation by letting $y = x^2$.

$$y^2 - 3y - 4 = 0 \qquad \text{Replace each } x^2 \text{ with } y.$$

We can solve this quadratic equation by factoring.

$$(y - 4)(y + 1) = 0 \qquad\qquad \text{Factor } y^2 - 3y - 4.$$
$$y - 4 = 0 \quad \text{or} \quad y + 1 = 0 \qquad \text{Set each factor equal to } 0.$$
$$y = 4 \qquad\qquad y = -1$$

Caution

If you are solving an equation in x, you can't answer with values of y. Remember to undo any substitutions, and solve for the variable in the original equation.

These are *not* the solutions for x. To find x, we now undo the earlier substitutions by replacing each y with x^2. Then we solve for x.

$$x^2 = 4 \quad \text{or} \quad x^2 = -1 \qquad \text{Substitute } x^2 \text{ for } y.$$
$$x = \pm\sqrt{4} \qquad\qquad x = \pm\sqrt{-1} \qquad \text{Use the square root property.}$$
$$x = \pm 2 \qquad\qquad x = \pm i$$

This equation has four solutions: 2, -2, i, and $-i$. Verify that each of them satisfies the original equation.

Self Check 2   Solve: $x^4 - 5x^2 - 36 = 0$.

EXAMPLE 3

ELEMENTARY &
INTERMEDIATE
Algebra $f(x)$ **Now**™

Solve: $x - 7\sqrt{x} + 12 = 0$.

Solution   We examine the lead term and middle term.

The lead term, x,
is the square of the
expression $\sqrt{x}$, here in
the middle term:
$x = \left(\sqrt{x}\right)^2$.

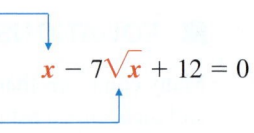

$$x - 7\sqrt{x} + 12 = 0$$

If we write x as $\left(\sqrt{x}\right)^2$, then the equation takes the form

$$\left(\sqrt{x}\right)^2 - 7\sqrt{x} + 12 = 0$$

The Language of Algebra

Equations such as
$x - 7\sqrt{x} + 12 = 0$ that
are quadratic in form are
also said to be *reducible
to a quadratic.*

and it is said to be *quadratic in* $\sqrt{x}$. We can solve this equation by letting $y = \sqrt{x}$ and factoring.

$$y^2 - 7y + 12 = 0 \qquad \text{Replace each } \sqrt{x} \text{ with } y.$$
$$(y - 3)(y - 4) = 0 \qquad \text{Factor } y^2 - 7y + 12.$$
$$y - 3 = 0 \quad \text{or} \quad y - 4 = 0 \qquad \text{Set each factor equal to 0.}$$
$$y = 3 \quad | \quad y = 4$$

To find x, we undo the substitutions by replacing each y with $\sqrt{x}$. Then we solve the radical equations by squaring both sides.

$$\sqrt{x} = 3 \quad \text{or} \quad \sqrt{x} = 4$$
$$x = 9 \quad | \quad x = 16$$

The solutions are 9 and 16. Verify that both satisfy the original equation.

Self Check 3   Solve: $x + \sqrt{x} - 6 = 0$.

EXAMPLE 4

ELEMENTARY &
INTERMEDIATE
Algebra $f(x)$ **Now**™

Solve: $2m^{2/3} - 2 = 3m^{1/3}$.

Solution   After writing the equation in descending powers of m, we see that

$$2m^{2/3} - 3m^{1/3} - 2 = 0$$

is *quadratic in* $m^{1/3}$, because $m^{2/3} = (m^{1/3})^2$. We will use the substitution $y = m^{1/3}$ to write this equation in quadratic form.

$$2m^{2/3} - 3m^{1/3} - 2 = 0$$

$$2(m^{1/3})^2 - 3m^{1/3} - 2 = 0 \qquad \text{Write } m^{2/3} \text{ as } (m^{1/3})^2.$$

$$2y^2 - 3y - 2 = 0 \qquad \text{Replace each } m^{1/3} \text{ with } y.$$

$$(2y + 1)(y - 2) = 0 \qquad \text{Factor } 2y^2 - 3y - 2.$$

$$2y + 1 = 0 \qquad \text{or} \qquad y - 2 = 0 \qquad \text{Set each factor equal to 0.}$$

$$y = -\frac{1}{2} \qquad\qquad y = 2$$

To find m, we undo the substitutions by replacing each y with $m^{1/3}$. Then we solve the equations by cubing both sides.

$$m^{1/3} = -\frac{1}{2} \qquad \text{or} \qquad m^{1/3} = 2$$

$$(m^{1/3})^3 = \left(-\frac{1}{2}\right)^3 \qquad (m^{1/3})^3 = (2)^3 \qquad \text{Recall that } m^{1/3} = \sqrt[3]{m}. \text{ To solve for } m, \text{ cube both sides.}$$

$$m = -\frac{1}{8} \qquad\qquad m = 8$$

The solutions are $-\dfrac{1}{8}$ and 8. Verify that both satisfy the original equation.

Self Check 4 Solve: $a^{2/3} = -3a^{1/3} + 10$.

EXAMPLE 5

Solve: $(4t + 2)^2 - 30(4t + 2) + 224 = 0$.

Solution This equation is *quadratic in* $4t + 2$. If we make the substitution $y = 4t + 2$, we have

$$y^2 - 30y + 224 = 0$$

which can be solved by using the quadratic formula.

$$y = \frac{-b \pm \sqrt{b^2 - 4ac}}{2a}$$

$$y = \frac{-(-30) \pm \sqrt{(-30)^2 - 4(1)(224)}}{2(1)} \qquad \text{Substitute 1 for } a, -30 \text{ for } b, \text{ and 224 for } c.$$

$$y = \frac{30 \pm \sqrt{900 - 896}}{2} \qquad \text{Simplify within the radical.}$$

$$y = \frac{30 \pm 2}{2} \qquad \sqrt{900 - 896} = \sqrt{4} = 2.$$

$$y = 16 \quad \text{or} \quad y = 14$$

To find t, we replace y with $4t + 2$ and solve for t.

$$4t + 2 = 16 \qquad \text{or} \qquad 4t + 2 = 14$$

$$4t = 14 \qquad\qquad 4t = 12$$

$$t = 3.5 \qquad\qquad t = 3$$

Verify that 3.5 and 3 satisfy the original equation.

Self Check 5   Solve: $(n + 3)^2 - 6(n + 3) = -8$.

EXAMPLE 6

Solve: $15a^{-2} - 8a^{-1} + 1 = 0$.

Solution   When we write the terms $15a^{-2}$ and $-8a^{-1}$ using positive exponents, we see that this equation is *quadratic in $\frac{1}{a}$.*

$$\frac{15}{a^2} - \frac{8}{a} + 1 = 0 \qquad \textcolor{red}{\text{Think of this equation as } 15 \cdot \left(\frac{1}{a}\right)^2 - 8 \cdot \frac{1}{a} + 1 = 0.}$$

If we let $y = \frac{1}{a}$, the resulting quadratic equation can be solved by factoring.

$$15y^2 - 8y + 1 = 0 \qquad \textcolor{red}{\text{Substitute } \frac{1}{a} \text{ for } y \text{ and } \frac{1}{a^2} \text{ for } y^2.}$$
$$(5y - 1)(3y - 1) = 0 \qquad \textcolor{red}{\text{Factor } 15y^2 - 8y + 1 = 0.}$$
$$5y - 1 = 0 \quad \text{or} \quad 3y - 1 = 0$$
$$y = \frac{1}{5} \qquad \qquad y = \frac{1}{3}$$

To find y, we undo the substitution by replacing each y with $\frac{1}{a}$.

$$\frac{1}{a} = \frac{1}{5} \quad \text{or} \quad \frac{1}{a} = \frac{1}{3}$$
$$5 = a \qquad \qquad 3 = a \qquad \textcolor{red}{\text{Solve the proportions.}}$$

The solutions are 5 and 3. Verify that they satisfy the original equation.

Self Check 6   Solve: $28c^{-2} - 3c^{-1} - 1 = 0$.

■ PROBLEM SOLVING

EXAMPLE 7

Household appliances.   The illustration shows a water temperature control on a washing machine. When the "warm" setting is selected, both the hot and cold water inlets open to fill the tub in 2 minutes 15 seconds. When the "cold" temperature setting is chosen, the cold water inlet fills the tub 45 seconds faster than when the "hot" setting is used. How long does it take to fill the washing machine with hot water?

Electronic Temperature Control

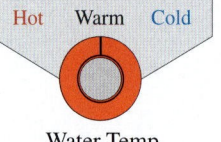

Analyze the Problem   It is helpful to organize the facts of this shared-work problem in a table.

Form an Equation   Let $x =$ the number of seconds it takes to fill the tub with hot water. Since the cold water inlet fills the tub in 45 seconds less time, $x - 45 =$ the number of seconds it takes to fill the tub with cold water. The hot and cold water inlets will be open for the same time: 2 minutes 15 seconds, or 135 seconds.

To determine the work completed by each inlet, multiply the rate by the time.

	Rate ·	Time	=	Work completed
Hot water	$\dfrac{1}{x}$	135		$\dfrac{135}{x}$
Cold water	$\dfrac{1}{x-45}$	135		$\dfrac{135}{x-45}$

Enter this information first.   Multiply to get each of these entries: $W = rt$.

In shared-work problems, the number 1 represents one whole job completed. So we have,

The fraction of tub filled with hot water	plus	the fraction of the tub filled with cold water	equals	1 tub filled.
$\dfrac{135}{x}$	$+$	$\dfrac{135}{x-45}$	$=$	1

Success Tip

An alternate way to form an equation is to note that what the hot water inlet can do in 1 second plus what the cold water inlet can do in 1 second equals what they can do together in 1 second:

$$\frac{1}{x} + \frac{1}{x-45} = \frac{1}{135}$$

Solve the Equation

$$\frac{135}{x} + \frac{135}{x-45} = 1$$

$$x(x-45)\left(\frac{135}{x} + \frac{135}{x-45}\right) = x(x-45)(1)$$

Multiply both sides by the LCD $x(x-45)$ to clear the equation of fractions.

$$135(x-45) + 135x = x(x-45)$$

$$135x - 6{,}075 + 135x = x^2 - 45x$$

Distribute the multiplication by 135 and by x.

$$270x - 6{,}075 = x^2 - 45x$$

Combine like terms.

$$0 = x^2 - 315x + 6{,}075$$

Subtract $270x$ from both sides. Add 6,075 to both sides.

To solve this equation, we will use the quadratic formula, with $a = 1$, $b = -315$, and $c = 6{,}075$.

$$x = \frac{-b \pm \sqrt{b^2 - 4ac}}{2a}$$

$$x = \frac{-(-315) \pm \sqrt{(-315)^2 - 4(1)(6{,}075)}}{2(1)}$$

Substitute 1 for a, -315 for b, and 6,075 for c.

$$x = \frac{315 \pm \sqrt{99{,}225 - 24{,}300}}{2}$$

Simplify within the radical.

$$x = \frac{315 \pm \sqrt{74{,}925}}{2}$$

$$x \approx \frac{589}{2} \quad \text{or} \quad x \approx \frac{41}{2}$$

$$x \approx 294 \qquad\quad x \approx 21$$

State the Conclusion We can discard the solution of 21 seconds, because this would imply that the cold water inlet fills the tub in a negative number of seconds ($21 - 45 = -24$). Therefore, the hot water inlet fills the washing machine tub in about 294 seconds, which is 4 minutes 54 seconds.

Check the Result Use estimation to check the result.

Answers to Self Checks   **1. a.** real numbers that are irrational and unequal,   **b.** two complex numbers containing i that are complex conjugates   **2.** $3, -3, 2i, -2i$ **3.** 4   **4.** $-125, 8$ **5.** $-1, 1$ **6.** $-7, 4$

10.3 STUDY SET

ELEMENTARY & INTERMEDIATE
Algebra *(f(x))* Now™

VOCABULARY   **Fill in the blanks.**

1. For the quadratic equation $ax^2 + bx + c = 0$, the _____ is $b^2 - 4ac$.

2. We can solve $x - 2\sqrt{x} - 8 = 0$ by making a _____: Let $y = \sqrt{x}$.

CONCEPTS   **Consider the quadratic equation $ax^2 + bx + c = 0$, where a, b, and c represent rational numbers, and fill in the blanks.**

3. If $b^2 - 4ac < 0$, the solutions of the equation are two complex numbers containing i that are complex _____.

4. If $b^2 - 4ac = $ ____, the solutions of the equation are equal real numbers.

5. If $b^2 - 4ac$ is a perfect square, the solutions are _____ numbers and _____.

6. If $b^2 - 4ac$ is positive and not a perfect square, the solutions are _____ numbers and _____.

7. For each equation, determine the substitution that should be made to write the equation in quadratic form.

a. $x^4 - 12x^2 + 27 = 0$ Let $y = $ ▢
b. $x - 13\sqrt{x} + 40 = 0$ Let $y = $ ▢
c. $x^{2/3} + 2x^{1/3} - 3 = 0$ Let $y = $ ▢
d. $x^{-2} - x^{-1} - 30 = 0$ Let $y = $ ▢
e. $(x+1)^2 - (x+1) - 6 = 0$ Let $y = $ ▢

8. Fill in the blanks.

a. $x^4 = ($ ▢ $)^2$ **b.** $x = ($ ▢ $)^2$
c. $x^{2/3} = ($ ▢ $)^2$ **d.** $\dfrac{1}{x^2} = ($ ▢ $)^2$

NOTATION   **Complete each solution.**

9. To find the type of solutions for the equation $x^2 + 5x + 6 = 0$, we compute the discriminant.

$$b^2 - \boxed{} = \boxed{}^2 - 4(1)(\boxed{})$$
$$= 25 - \boxed{}$$
$$= 1$$

Since a, b, and c are rational numbers and the value of the discriminant is a perfect square, the solutions are _____ numbers and unequal.

10. Change $\dfrac{3}{4} + x = \dfrac{3x - 50}{4(x - 6)}$ to quadratic form.

$$\boxed{}\left(\dfrac{3}{4} + x\right) = \boxed{}\dfrac{3x - 50}{4(x - 6)}$$
$$3(x - 6) + 4x(\boxed{}) = 3x - 50$$
$$3x - \boxed{} + 4x^2 - \boxed{} = 3x - 50$$
$$4x^2 - 24x + \boxed{} = 0$$
$$\boxed{} - 6x + 8 = 0$$

PRACTICE   **Use the discriminant to determine what type of solutions exist for each equation. Do not solve the equation.**

11. $4x^2 - 4x + 1 = 0$ **12.** $6x^2 - 5x - 6 = 0$

13. $5x^2 + x + 2 = 0$ **14.** $3x^2 + 10x - 2 = 0$

15. $2x^2 = 4x - 1$ **16.** $9x^2 = 12x - 4$

17. $x(2x - 3) = 20$ **18.** $x(x - 3) = -10$

19. Use the discriminant to determine whether the solutions of $1{,}492x^2 + 1{,}776x - 2{,}000 = 0$ are real numbers.

20. Use the discriminant to determine whether the solutions of $1{,}776x^2 - 1{,}492x + 2{,}000 = 0$ are real numbers.

Solve each equation.

21. $x^4 - 17x^2 + 16 = 0$ **22.** $x^4 - 10x^2 + 9 = 0$

23. $x^4 = 6x^2 - 5$ **24.** $2x^4 + 24 = 26x^2$

25. $t^4 + 3t^2 = 28$ **26.** $3h^4 + h^2 - 2 = 0$

27. $x^4 + 19x^2 + 18 = 0$ **28.** $t^4 + 4t^2 - 5 = 0$

29. $2x + \sqrt{x} - 3 = 0$ **30.** $2x - \sqrt{x} - 1 = 0$

31. $3x + 5\sqrt{x} + 2 = 0$ **32.** $3x - 4\sqrt{x} + 1 = 0$

33. $x - 6x^{1/2} = -8$ **34.** $x - 5x^{1/2} + 4 = 0$

35. $2x - \sqrt{x} = 3$ **36.** $3x + 4\sqrt{x} = 4$

37. $x^{2/3} + 5x^{1/3} + 6 = 0$ **38.** $x^{2/3} - 7x^{1/3} + 12 = 0$

39. $a^{2/3} - 2a^{1/3} - 3 = 0$ **40.** $r^{2/3} + 4r^{1/3} - 5 = 0$

41. $2x^{2/5} - 5x^{1/5} = -3$ **42.** $2x^{2/5} + 3x^{1/5} = -1$

43. $2(2x + 1)^2 - 7(2x + 1) + 6 = 0$

44. $3(2 - x)^2 + 10(2 - x) - 8 = 0$

45. $(c + 1)^2 - 4(c + 1) + 8 = 0$

46. $(k - 7)^2 + 6(k - 7) + 10 = 0$

47. $(a^2 - 4)^2 - 4(a^2 - 4) - 32 = 0$

48. $(y^2 - 9)^2 + 2(y^2 - 9) - 99 = 0$

49. $9\left(\dfrac{3m + 2}{m}\right)^2 - 30\left(\dfrac{3m + 2}{m}\right) + 25 = 0$

50. $4\left(\dfrac{c - 7}{c}\right)^2 - 12\left(\dfrac{c - 7}{c}\right) + 9 = 0$

51. $\left(8 - \sqrt{a}\right)^2 + 6\left(8 - \sqrt{a}\right) - 7 = 0$

52. $\left(10 - \sqrt{t}\right)^2 - 4\left(10 - \sqrt{t}\right) - 45 = 0$

53. $8x^{-2} - 10x^{-1} - 3 = 0$

54. $2x^{-2} - 5x^{-1} - 3 = 0$

55. $8(t + 1)^{-2} - 30(t + 1)^{-1} + 7 = 0$

56. $2(s - 2)^{-2} + 3(s - 2)^{-1} - 5 = 0$

57. $x^{-4} - 2x^{-2} + 1 = 0$ **58.** $4x^{-4} + 1 = 5x^{-2}$

59. $x + \dfrac{2}{x - 2} = 0$ **60.** $x + \dfrac{x + 5}{x - 3} = 0$

61. $x + 5 + \dfrac{4}{x} = 0$ **62.** $x - 4 + \dfrac{3}{x} = 0$

63. $\dfrac{1}{x + 2} + \dfrac{24}{x + 3} = 13$ **64.** $\dfrac{3}{x} + \dfrac{4}{x + 1} = 2$

65. $\dfrac{2}{x - 1} + \dfrac{1}{x + 1} = 3$ **66.** $\dfrac{3}{x - 2} - \dfrac{1}{x + 2} = 5$

APPLICATIONS

67. FLOWER ARRANGING   A florist needs to determine the height h of the flowers shown in the illustration. The radius r, the width w, and the height h of the circular-shaped arrangement are related by the formula

$$r = \frac{4h^2 + w^2}{8h}$$

If w is to be 34 inches and r is to be 18 inches, find h to the nearest tenth of an inch.

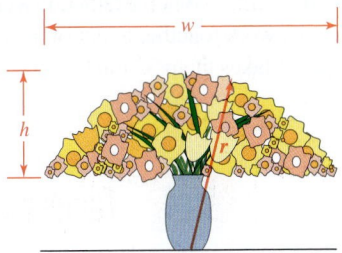

68. ARCHITECTURE   A **golden rectangle** is one of the most visually appealing of all geometric forms. The Parthenon, built by the Greeks in the 5th century B.C., fits into a golden rectangle if its ruined triangular pediment is included. See the illustration on the next page.

In a golden rectangle, the length ℓ and width w must satisfy the equation

$$\frac{\ell}{w} = \frac{w}{\ell - w}$$

If a rectangular billboard is to have a width of 20 feet, what should its length be so that it is a golden rectangle? Round to the nearest tenth.

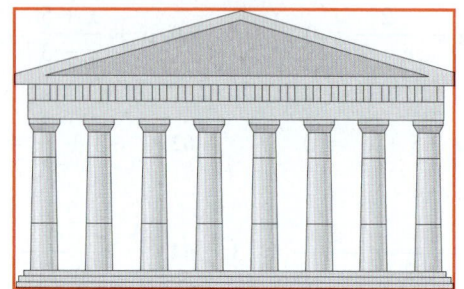

69. SNOWMOBILES   A woman drives her snowmobile 150 miles at a rate of r mph. She could have gone the same distance in 2 hours less time if she had increased her speed by 20 mph. Find r.

70. BICYCLING   Tina bicycles 160 miles at the rate of r mph. The same trip would have taken 2 hours longer if she had decreased her speed by 4 mph. Find r.

71. CROWD CONTROL   After a performance at a county fair, security guards have found that the grandstand area can be emptied in 6 minutes if both the east and west exits are opened. If just the east exit is used, it takes 4 minutes longer to clear the grandstand than it does if just the west exit is opened. How long does it take to clear the grandstand if everyone must file through the east exit?

72. PAPER ROUTES   When a father, in a car, and his son, on a bicycle, work together to distribute the morning newspaper, it takes them 35 minutes to complete the route. Working alone, it takes the son 25 minutes longer than the father. To the nearest minute, how long does it take the son to cover the route on his bicycle?

WRITING

73. Describe how to predict what type of solutions the equation $3x^2 - 4x + 5 = 0$ will have.

74. What error is made in the following solution?

Solve: $x^4 - 12x^2 + 27 = 0$

Let $y = x^2$

$y^2 - 12y + 27 = 0$

$(y - 9)(y - 3) = 0$

$y - 9 = 0$ or $y - 3 = 0$

$y = 9$ | $y = 3$

The solutions are 9 and 3.

REVIEW

75. Write an equation of the vertical line that passes through $(3, 4)$.

76. Write an equation of the line that passes through $(-1, -6)$ and $(-2, -1)$. Express the result in slope–intercept form.

77. Write an equation of the line with slope $\frac{2}{3}$ that passes through the origin.

78. Write an equation of the line that passes through $(2, -3)$ and is perpendicular to the line whose equation is $y = \frac{x}{5} + 6$. Express the result in slope–intercept form.

CHALLENGE PROBLEMS

79. Solve: $x^6 + 17x^3 + 16 = 0$.

80. Find the real-number solutions of $x^4 - 3x^2 - 2 = 0$. Rationalize the denominators of the solutions.

10.4   Quadratic Functions and Their Graphs

- Graphing $f(x) = ax^2$ • Graphing $f(x) = ax^2 + k$ • Graphing $f(x) = a(x - h)^2$
- Graphing $f(x) = a(x - h)^2 + k$ • Graphing $f(x) = ax^2 + bx + c$ by completing the square
- A formula to find the vertex   • Determining minimum and maximum values
- Solving quadratic equations graphically

In this section, we will discuss methods for graphing *quadratic functions*.

Quadratic Functions

A **quadratic function** is a second-degree polynomial function that can be written in the form

$$f(x) = ax^2 + bx + c$$

where a, b, and c are real numbers and $a \neq 0$.

Quadratic functions are often written in an alternate form, called **standard form,**

$$f(x) = a(x - h)^2 + k$$

Notation

Since $y = f(x)$, quadratic functions can also be written as $y = a(x - h)^2 + k$ and $y = ax^2 + bx + c$.

where a, h, and k are real numbers and $a \neq 0$. This form is useful because a, h, and k give us important information about the graph of the function. To develop a strategy for graphing quadratic functions written in standard form, we will begin by considering the simplest case, $f(x) = ax^2$.

■ GRAPHING $f(x) = ax^2$

One way to graph quadratic functions is to plot points.

EXAMPLE 1

ELEMENTARY &
INTERMEDIATE
Algebra $f(x)$ **Now**™

Graph: **a.** $f(x) = x^2$,   **b.** $g(x) = 3x^2$,   and   **c.** $h(x) = \dfrac{1}{3}x^2$.

Solution   We can make a table of values for each function, plot each point, and join them with a smooth curve. We note that the graph of $g(x) = 3x^2$ is narrower than the graph of $f(x) = x^2$, and the graph of $h(x) = \frac{1}{3}x^2$ is wider than the graph of $f(x) = x^2$. For $f(x) = ax^2$, the smaller the value of $|a|$, the wider the graph.

$f(x) = x^2$

x	$f(x)$
-2	4
-1	1
0	0
1	1
2	4

$g(x) = 3x^2$

x	$g(x)$
-2	12
-1	3
0	0
1	3
2	12

The values of $g(x)$ increase faster than the values of $f(x)$, making its graph steeper.

$h(x) = \frac{1}{3}x^2$

x	$h(x)$
-2	$\dfrac{4}{3}$
-1	$\dfrac{1}{3}$
0	0
1	$\dfrac{1}{3}$
2	$\dfrac{4}{3}$

The values of $h(x)$ increase more slowly than the values of $f(x)$, making its graph flatter.

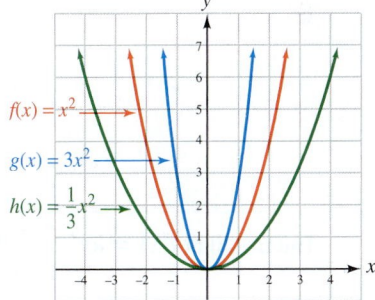

EXAMPLE 2

Graph: $f(x) = -3x^2$.

Solution   We make a table of values for the function, plot each point, and join them with a smooth curve. We see that the parabola opens downward and has the same shape as the graph of $g(x) = 3x^2$ that was graphed in Example 1.

$$f(x) = -3x^2$$

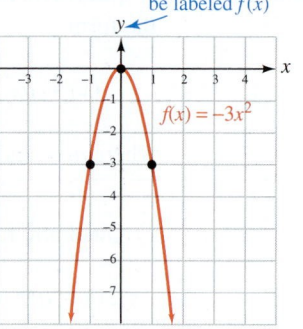

This axis could also be labeled $f(x)$

x	$f(x)$	
-2	-12	→ $(-2, -12)$
-1	-3	→ $(-1, -3)$
0	0	→ $(0, 0)$
1	-3	→ $(1, -3)$
2	-12	→ $(2, -12)$

Self Check 2   Graph: $f(x) = -\dfrac{1}{3}x^2$.

The graphs of functions of the form $f(x) = ax^2$ are **parabolas.** The lowest point on a parabola that opens upward, or the highest point on a parabola that opens downward, is called the **vertex** of the parabola. The vertical line, called an **axis of symmetry,** that passes through the vertex divides the parabola into two congruent halves. If we fold the paper along the axis of symmetry, the two sides of the parabola will match.

The Language of Algebra

An axis of symmetry divides a parabola into two matching sides. The sides are said to be *mirror images* of each other.

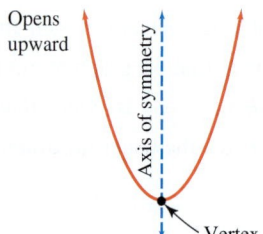

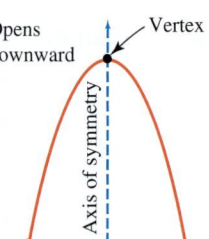

The results from Examples 1 and 2 confirm the following facts.

The Graph of $f(x) = ax^2$ The graph of $f(x) = ax^2$ is a parabola opening upward when $a > 0$ and downward when $a < 0$, with vertex at the point $(0, 0)$ and axis of symmetry the line $x = 0$.

■ GRAPHING $f(x) = ax^2 + k$

EXAMPLE 3   Graph: **a.** $f(x) = 2x^2$,   **b.** $g(x) = 2x^2 + 3$,   and   **c.** $h(x) = 2x^2 - 3$.

ELEMENTARY & INTERMEDIATE
Algebra $f(x)$ **Now™**   **Solution**   We make a table of values for each function, plot each point, and join them with a smooth curve. We note that the graph of $g(x) = 2x^2 + 3$ is identical to the graph of $f(x) = 2x^2$, except that it has been translated 3 units upward. The graph of $h(x) = 2x^2 - 3$ is identical to the graph of $f(x) = 2x^2$, except that it has been translated 3 units downward. In each case, the axis of symmetry is the line $x = 0$.

$f(x) = 2x^2$

x	$f(x)$
-2	8
-1	2
0	0
1	2
2	8

$g(x) = 2x^2 + 3$

x	$g(x)$
-2	11
-1	5
0	3
1	5
2	11

$h(x) = 2x^2 - 3$

x	$h(x)$
-2	5
-1	-1
0	-3
1	-1
2	5

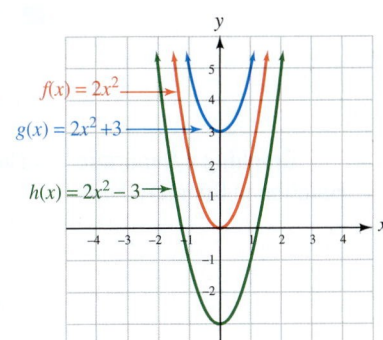

For each x-value, $g(x)$ is 3 more than $f(x)$.

For each x-value, $h(x)$ is 3 less than $f(x)$.

The results of Example 3 confirm the following facts.

The Graph of $f(x) = ax^2 + k$	The graph of $f(x) = ax^2 + k$ is a parabola having the same shape as $f(x) = ax^2$ but translated upward k units if k is positive and downward $\lvert k \rvert$ units if k is negative. The vertex is at the point $(0, k)$, and the axis of symmetry is the line $x = 0$.

■ GRAPHING $f(x) = a(x - h)^2$

EXAMPLE 4

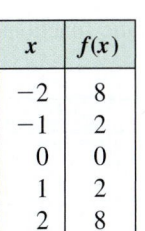

ELEMENTARY & INTERMEDIATE
Algebra *(f(x))* **Now**™

Graph: **a.** $f(x) = 2x^2$, **b.** $g(x) = 2(x - 3)^2$, and **c.** $h(x) = 2(x + 3)^2$.

Solution   We make a table of values for each function, plot each point, and join them with a smooth curve. We note that the graph of $g(x) = 2(x - 3)^2$ is identical to the graph of $f(x) = 2x^2$, except that it has been translated 3 units to the right. The graph of $h(x) = 2(x + 3)^2$ is identical to the graph of $f(x) = 2x^2$, except that it has been translated 3 units to the left.

$f(x) = 2x^2$

x	$f(x)$
-2	8
-1	2
0	0
1	2
2	8

$g(x) = 2(x - 3)^2$

x	$g(x)$
1	8
2	2
3	0
4	2
5	8

$h(x) = 2(x + 3)^2$

x	$h(x)$
-5	8
-4	2
-3	0
-2	2
-1	8

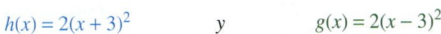
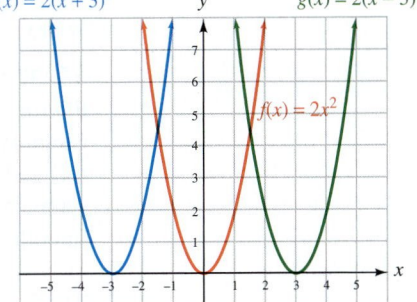

When an x-value is increased by 3, the function's outputs are the same.

When an x-value is decreased by 3, the function's outputs are the same.

The results of Example 4 confirm the following facts.

The Graph of $f(x) = a(x - h)^2$	The graph of $f(x) = a(x - h)^2$ is a parabola having the same shape as $f(x) = ax^2$ but translated h units to the right if h is positive and $\lvert h \rvert$ units to the left if h is negative. The vertex is at the point $(h, 0)$, and the axis of symmetry is the line $x = h$.

■ GRAPHING $f(x) = a(x - h)^2 + k$

The results of Examples 1–4 suggest a general strategy for graphing quadratic functions that are written in the form $f(x) = a(x - h)^2 + k$.

Graphing a Quadratic Function in Standard Form

The graph of the quadratic function

$$f(x) = a(x - h)^2 + k \quad \text{where } a \neq 0$$

is a parabola with vertex at (h, k). The axis of symmetry is the line $x = h$. The parabola opens upward when $a > 0$ and downward when $a < 0$.

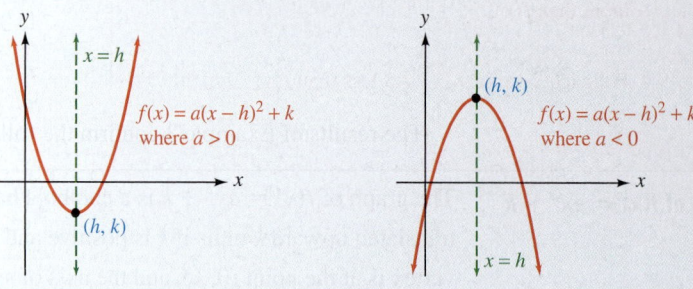

EXAMPLE 5

Graph: $f(x) = 2(x - 3)^2 - 4$. Label the vertex and draw the axis of symmetry.

Solution   The graph of $f(x) = 2(x - 3)^2 - 4$ is identical to the graph of $g(x) = 2(x - 3)^2$, except that it has been translated 4 units downward. The graph of $g(x) = 2(x - 3)^2$ is identical to the graph of $h(x) = 2x^2$, except that it has been translated 3 units to the right.

We can learn more about the graph of $f(x) = 2(x - 3)^2 - 4$ by determining a, h, and k.

$$\left.\begin{array}{l} f(x) = \mathbf{2}(x - \mathbf{3})^2 - \mathbf{4} \\ f(x) = \mathbf{a}(x - \mathbf{h})^2 + \mathbf{k} \end{array}\right\} \quad a = 2, h = 3, \text{ and } k = -4$$

Upward/downward: Since $a = 2$ and $2 > 0$, the parabola opens upward.

Vertex: The vertex of the parabola is $(h, k) = (3, -4)$, as shown below.

Axis of symmetry: Since $h = 3$, the axis of symmetry is the line $x = 3$, as shown below.

Plotting points: We can construct a table of values to determine several points on the parabola. Since the x-coordinate of the vertex is 3, we choose the x-values of 4 and 5, find $f(4)$ and $f(5)$, and record the results in a table. Then we plot $(4, -2)$ and $(5, 4)$, and use symmetry to locate two other points on the parabola: $(2, -2)$ and $(1, 4)$. Finally, we draw a smooth curve through the points to get the graph.

$$f(x) = 2(x - 3)^2 - 4$$

x	$f(x)$	
4	-2	→ $(4, -2)$
5	4	→ $(5, 4)$

↑
The x-coordinate of the vertex is 3. Choose values for x close to 3 and on the same side of the axis of symmetry.

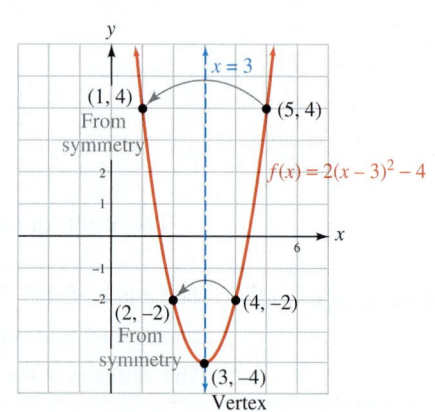

Self Check 5   Graph: $f(x) = 2(x - 1)^2 - 2$. Label the vertex and draw the axis of symmetry.

■ GRAPHING $f(x) = ax^2 + bx + c$ BY COMPLETING THE SQUARE

To graph functions of the form $f(x) = ax^2 + bx + c$, we can complete the square to write the function in standard form $f(x) = a(x - h)^2 + k$.

EXAMPLE 6   Determine the vertex and the axis of symmetry of the graph of $f(x) = x^2 + 8x + 21$. Will the graph open upward or downward?

Solution   To determine the vertex and the axis of symmetry of the graph, we complete the square on the right-hand side so that we can write the function in $f(x) = a(x - h)^2 + k$ form.

$$f(x) = x^2 + 8x + 21$$
$$f(x) = (x^2 + 8x \quad) + 21 \qquad \text{Prepare to complete the square on } x \text{ by writing parentheses around } x^2 + 8x.$$

To complete the square on $x^2 + 8x$, we note that one-half of the coefficient of x is $\frac{1}{2} \cdot 8 = 4$, and $4^2 = 16$. If we add 16 to $x^2 + 8x$, we obtain a perfect square trinomial within the parentheses. Since this step adds 16 to the right-hand side, we must also subtract 16 from the right-hand side so that it remains in an equivalent form.

Success Tip

When a number is added to and that same number is subtracted from one side of an equation, the value of that side of the equation remains the same.

Add 16 to the right-hand side. ⟶   Subtract 16 from ⟵ the right-hand side.

$$f(x) = (x^2 + 8x \mathbf{+ 16}) + 21 \mathbf{- 16}$$
$$f(x) = (x + 4)^2 + 5 \qquad \text{Factor } x^2 + 8x + 16 \text{ and combine like terms.}$$

The function is now written in standard form and we can determine a, h, and k.

The standard form requires a minus symbol here.

$$f(x) = \underset{a}{\uparrow} \left[(x - \underset{h}{(\mathbf{-4})}\right]^2 + \underset{k}{\mathbf{5}} \qquad \begin{array}{l}\text{Write } x + 4 \text{ as } x - (-4) \text{ to determine } h. \\ a = 1, h = -4, \text{ and } k = 5.\end{array}$$

The vertex is $(h, k) = (-4, 5)$ and the axis of symmetry is the line $x = -4$. Since $a = 1$ and $1 > 0$, the parabola opens upward.

Self Check 6   Determine the vertex and the axis of symmetry of the graph of $f(x) = x^2 + 4x + 10$. Will the graph open upward or downward?

EXAMPLE 7   Graph: $f(x) = 2x^2 - 4x - 1$.

ELEMENTARY &
INTERMEDIATE
Algebra $f(x)$ **Now**™   **Solution**   Recall that to complete the square on $2x^2 - 4x$, the coefficient of x^2 must be equal to 1. Therefore, we factor 2 from $2x^2 - 4x$.

$$f(x) = 2x^2 - 4x - 1$$
$$f(x) = 2(x^2 - 2x \quad) - 1$$

To complete the square on $x^2 - 2x$, we note that one-half of the coefficient of x is $\frac{1}{2}(-2) = -1$, and $(-1)^2 = 1$. If we add 1 to $x^2 - 2x$, we obtain a perfect square trinomial within the parentheses. Since this step adds 2 to the right-hand side, we must also subtract 2 from the right-hand side so that it remains in an equivalent form.

By the distributive property, when 1 is added to the expression within the parentheses, $2 \cdot 1 = 2$ is added to the right-hand side.

Subtract 2 to counteract the addition of 2.

$$f(x) = \mathbf{2}(x^2 - 2x + \mathbf{1}) - 1 \,\mathbf{-2}$$
$$f(x) = 2(x - 1)^2 - 3 \qquad \text{Factor } x^2 - 2x + 1 \text{ and combine like terms.}$$

We see that $a = 2$, $h = 1$, and $k = -3$. Thus, the vertex is at the point $(1, -3)$, and the axis of symmetry is $x = 1$. Since $a = 2$ and $2 > 0$, the parabola opens upward. We plot the vertex and axis of symmetry as shown below.

Finally, we construct a table of values, plot the points, use symmetry to plot the corresponding points, and then draw the graph.

$$f(x) = 2x^2 - 4x - 1$$
$$\text{or}$$
$$f(x) = 2(x - 1)^2 - 3$$

x	$f(x)$
2	−1
3	5

→ $(2, -1)$
→ $(3, 5)$

The x-coordinate of the vertex is 1. Choose values for x close to 1 and on the same side of the axis of symmetry.

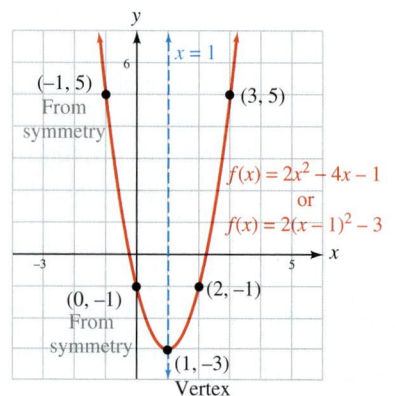

$f(x) = 2x^2 - 4x - 1$
or
$f(x) = 2(x - 1)^2 - 3$

Self Check 7   Graph: $f(x) = 3x^2 - 12x + 8$.

■ A FORMULA TO FIND THE VERTEX

We can derive a formula for the vertex of the graph of $f(x) = ax^2 + bx + c$ by completing the square in the same manner as we did in Example 7. After using similar steps, the result is

$$f(x) = a\left[x - \left(-\frac{b}{2a}\right)\right]^2 + \frac{4ac - b^2}{4a}$$

$$\underset{h}{\uparrow} \qquad\qquad \underset{k}{\uparrow}$$

The x-coordinate of the vertex is $-\dfrac{b}{2a}$. The y-coordinate of the vertex is $\dfrac{4ac - b^2}{4a}$.

However, we can also find the y-coordinate of the vertex by substituting the x-coordinate, $-\dfrac{b}{2a}$, for x in the quadratic function.

Formula for the Vertex of a Parabola   The vertex of the graph of the quadratic function $f(x) = ax^2 + bx + c$ is

$$\left(-\frac{b}{2a}, \, f\left(-\frac{b}{2a}\right)\right)$$

and the axis of symmetry of the parabola is the line $x = -\dfrac{b}{2a}$.

EXAMPLE 8

Find the vertex of the graph of $f(x) = 2x^2 - 4x - 1$.

ELEMENTARY &
INTERMEDIATE
Algebra $f(x)$ **Now**™

Solution

The function is written in $f(x) = ax^2 + bx + c$ form, where $a = 2$, $b = -4$, and $c = -1$. To find the vertex of its graph, we compute

Success Tip

We can find the vertex of the graph of a quadratic function by completing the square or by using the formula.

$$-\frac{b}{2a} = -\frac{-4}{2(2)} \qquad\qquad f\left(-\frac{b}{2a}\right) = f(1)$$

$$= -\frac{-4}{4} \qquad\qquad\qquad = 2(1)^2 - 4(1) - 1$$

$$= 1 \qquad\qquad\qquad\qquad = -3$$

The vertex is the point $(1, -3)$. This agrees with the result we obtained in Example 7 by completing the square.

Self Check 8   Find the vertex of the graph of $f(x) = 3x^2 - 12x + 8$.

ACCENT ON TECHNOLOGY: FINDING THE VERTEX

We can use a graphing calculator to graph the function $f(x) = 2x^2 + 6x - 3$ and find the coordinates of the vertex and the axis of symmetry of the parabola. If we enter the function, we will obtain the graph shown in figure (a).

We then trace to move the cursor to the lowest point on the graph, as shown in figure (b). By zooming in, we can see that the vertex is the point $(-1.5, -7.5)$, or $\left(-\frac{3}{2}, -\frac{15}{2}\right)$, and that the line $x = -\frac{3}{2}$ is the axis of symmetry.

Some calculators have an fmin or fmax feature that can also be used to find the vertex.

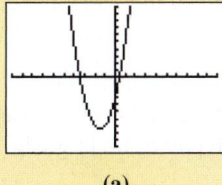   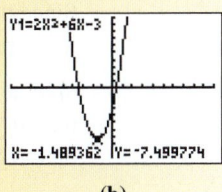

(a)   (b)

Much can be determined about the graph of $f(x) = ax^2 + bx + c$ from the coefficients a, b, and c. This information is summarized as follows:

Graphing a
Quadratic Function
$f(x) = ax^2 + bx + c$

Determine whether the parabola opens upward or downward by examining a.

The x-coordinate of the vertex of the parabola is $x = -\frac{b}{2a}$.

To find the y-coordinate of the vertex, substitute $-\frac{b}{2a}$ for x and find $f\left(-\frac{b}{2a}\right)$.

The axis of symmetry is the vertical line passing through the vertex.

The y-intercept is determined by the value of $f(x)$ when $x = 0$: the y-intercept is $(0, c)$.

The x-intercepts (if any) are determined by the values of x that make $f(x) = 0$. To find them, solve the quadratic equation $ax^2 + bx + c = 0$.

EXAMPLE 9

Graph: $f(x) = -2x^2 - 8x - 8$.

Solution

Step 1: *Determine whether the parabola opens upward or downward.* The function is in the form $f(x) = ax^2 + bx + c$, with $a = -2$, $b = -8$, and $c = -8$. Since $a < 0$, the parabola opens downward.

Step 2: *Find the vertex and draw the axis of symmetry.* To find the coordinates of the vertex, we compute

$$x = -\frac{b}{2a}$$

$$x = -\frac{-8}{2(-2)}$$

Substitute -2 for a and -8 for b.

$$= -2$$

$$f\left(-\frac{b}{2a}\right) = f(-2)$$

$$= -2(-2)^2 - 8(-2) - 8$$

$$= -8 + 16 - 8$$

$$= 0$$

The vertex of the parabola is the point $(-2, 0)$. This point is in blue on the graph. The axis of symmetry is the line $x = -2$.

Step 3: *Find the x- and y-intercepts.* Since $c = -8$, the y-intercept of the parabola is $(0, -8)$. The point $(-4, -8)$, two units to the left of the axis of symmetry, must also be on the graph. We plot both points in black on the graph.

To find the x-intercepts, we set $f(x)$ equal to 0 and solve the resulting quadratic equation.

$$f(x) = -2x^2 - 8x - 8$$

$$0 = -2x^2 - 8x - 8 \qquad \text{Set } f(x) = 0.$$

$$0 = x^2 + 4x + 4 \qquad \text{Divide both sides by } -2.$$

$$0 = (x + 2)(x + 2) \qquad \text{Factor the trinomial.}$$

$$x + 2 = 0 \quad \text{or} \quad x + 2 = 0 \qquad \text{Set each factor equal to 0.}$$

$$x = -2 \qquad\qquad x = -2$$

Since the solutions are the same, the graph has only one x-intercept: $(-2, 0)$. This point is the vertex of the parabola and has already been plotted.

Step 4: *Plot another point.* Finally, we find another point on the parabola. If $x = -3$, then $f(-3) = -2$. We plot $(-3, -2)$ and use symmetry to determine that $(-1, -2)$ is also on the graph. Both points are in green.

Step 5: Draw a smooth curve through the points, as shown.

$$f(x) = -2x^2 - 8x - 8$$

x	$f(x)$
-3	-2

$\rightarrow (-3, -2)$

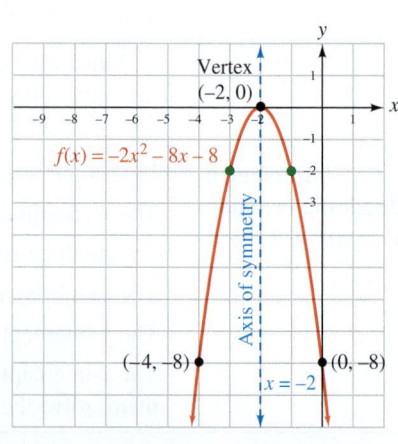

■ DETERMINING MINIMUM AND MAXIMUM VALUES

It is often useful to know the smallest or largest possible value a quantity can assume. For example, companies try to minimize their costs and maximize their profits. If the quantity is expressed by a quadratic function, the y-coordinate of the vertex of the graph of the function gives its minimum or maximum value.

EXAMPLE 10

Minimizing costs.   A glassworks that makes lead crystal vases has daily production costs given by the function $C(x) = 0.2x^2 - 10x + 650$, where x is the number of vases made each day. How many vases should be produced to minimize the per-day costs? What will the costs be?

Solution   The graph of $C(x) = 0.2x^2 - 10x + 650$ is a parabola opening upward. The vertex is the lowest point on the graph. To find the vertex, we compute

$$-\frac{b}{2a} = -\frac{-10}{2(0.2)} \quad \begin{array}{l} b = -10 \text{ and} \\ a = 0.2. \end{array} \qquad f\left(-\frac{b}{2a}\right) = f(25)$$

$$= -\frac{-10}{0.4} \qquad\qquad\qquad\qquad = 0.2(25)^2 - 10(25) + 650$$

$$= 25 \qquad\qquad\qquad\qquad\qquad = 525$$

The Language of Algebra

We say that 25 is the value for which the function $C(x) = 0.2x^2 - 10x + 650$ is a minimum.

The vertex is (25, 525), and it indicates that the costs are a minimum of $525 when 25 vases are made daily.

To solve this problem with a graphing calculator, we graph the function $C(x) = 0.2x^2 - 10x + 650$. By using TRACE and ZOOM, we can locate the vertex of the graph. The coordinates of the vertex indicate that the minimum cost is $525 when the number of vases produced is 25.

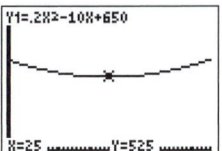

■ SOLVING QUADRATIC EQUATIONS GRAPHICALLY

When solving quadratic equations graphically, we must consider three possibilities. If the graph of the associated quadratic function has two x-intercepts, the quadratic equation has two real-number solutions. Figure (a) shows an example of this. If the graph has one x-intercept, as shown in figure (b), the equation has one real-number solution. Finally, if the graph does not have an x-intercept, as shown in figure (c), the equation does not have any real-number solutions.

Success Tip

Note that the solutions of each quadratic equation are given by the x-coordinate of the x-intercept of each respective graph.

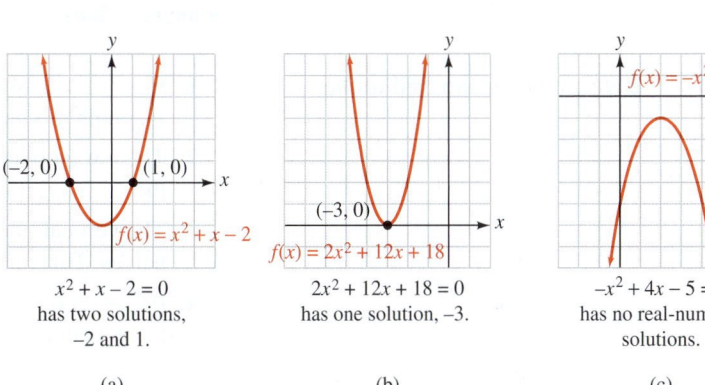

| (a) | (b) | (c) |

ACCENT ON TECHNOLOGY: SOLVING QUADRATIC EQUATIONS GRAPHICALLY

We can use a graphing calculator to find approximate solutions of quadratic equations. For example, the solutions of $0.7x^2 + 2x - 3.5 = 0$ are the numbers x that will make $y = 0$ in the quadratic function $f(x) = 0.7x^2 + 2x - 3.5$. To approximate these numbers, we graph the quadratic function and read the x-intercepts from the graph using the ZERO feature. In the figure, we see that the x-coordinate of the left-most x-intercept of the graph is given as -4.082025. This means that an approximate solution of the equation is -4.08. To find the positive x-intercept, we use similar steps.

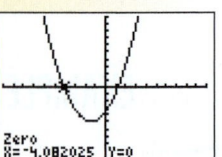

Answers to Self Checks

2.

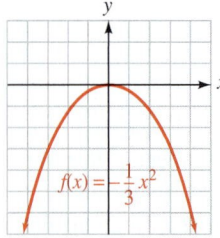

$f(x) = -\dfrac{1}{3}x^2$

5.

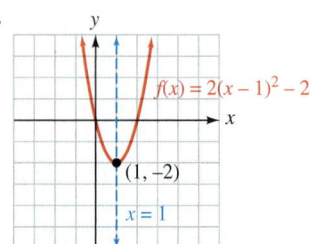

$f(x) = 2(x - 1)^2 - 2$

$(1, -2)$

$x = 1$

6. $(-2, 6)$; $x = -2$; opens upward

7.

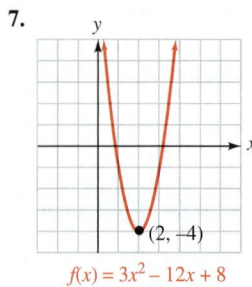

$(2, -4)$

$f(x) = 3x^2 - 12x + 8$

8. $(2, -4)$

10.4 STUDY SET

ELEMENTARY & INTERMEDIATE
Algebra $f(x)$ Now™

VOCABULARY **Refer to the graph. Fill in the blanks.**

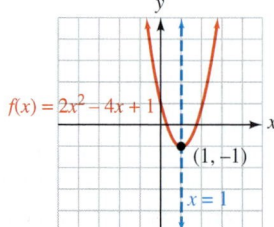

$f(x) = 2x^2 - 4x + 1$

$(1, -1)$

$x = 1$

1. $f(x) = 2x^2 - 4x + 1$ is called a _____ function. Its graph is a cup-shaped figure called a _____.

2. The lowest point on the graph is $(1, -1)$. This is called the _____ of the parabola.

3. The vertical line $x = 1$ divides the parabola into two halves. This line is called the _____.

4. $f(x) = a(x - h)^2 + k$ is called the _____ form of the equation of a quadratic function.

CONCEPTS

5. Refer to the graph below.

  a. What do we call the curve shown there?

  b. What are the x-intercepts of the graph?

  c. What is the y-intercept of the graph?

  d. What is the vertex?

  e. What is the axis of symmetry?

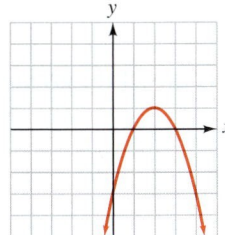

6. The vertex of a parabola is at $(1, -3)$, its y-intercept is $(0, -2)$, and it passes through the point $(3, 1)$, as shown in the illustration. Draw the axis of symmetry and use it to help determine two other points on the parabola.

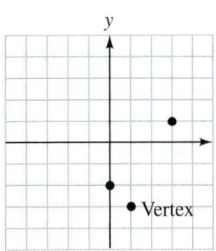

7. Draw the graph of a quadratic function using the given facts about its graph.

  • Opens upward   • y-intercept: $(0, -3)$

  • Vertex: $(-1, -4)$ • x-intercepts: $(-3, 0), (1, 0)$

  •

x	$f(x)$
2	5

8. For $f(x) = -x^2 + 6x - 7$, the value of $-\frac{b}{2a}$ is 3. Find the y-coordinate of the vertex of the graph of this function.

9. a. To complete the square on the right-hand side of $f(x) = 2x^2 + 12x + 11$, what should be factored from the first two terms?

$$f(x) = \boxed{}(x^2 + 6x) + 11$$

  b. To complete the square on $x^2 + 6x$ shown below, what should be added within the parentheses and what should be subtracted outside the parentheses?

$$f(x) = 2(x^2 + 6x + \boxed{}) + 11 - \boxed{}$$

10. To complete the square on $x^2 + 4x$ shown below, what should be added within the parentheses and what should be added outside the parentheses?

$$f(x) = -5(x^2 + 4x + \boxed{}) + 7 + \boxed{}$$

11. Use the graph of $f(x) = \dfrac{x^2}{10} - \dfrac{x}{5} - \dfrac{3}{2}$, shown below, to estimate the solutions of the equation $\dfrac{x^2}{10} - \dfrac{x}{5} - \dfrac{3}{2} = 0$.

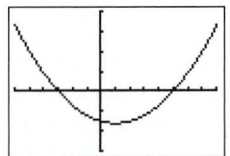

12. [graphing calculator icon] Three quadratic equations are to be solved graphically. The graphs of their associated quadratic functions are shown below. Decide which graph indicates that the equation has

  a. two real solutions.

  b. one real solution.

  c. no real solutions.

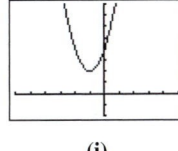 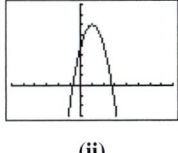 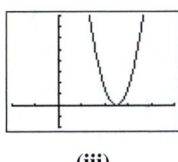

   (i)   **(ii)**   **(iii)**

NOTATION

13. The function $f(x) = 2(x + 1)^2 + 6$ is written in the form $f(x) = a(x - h)^2 + k$. Is $h = -1$ or is $h = 1$? Explain.

14. Consider the function $f(x) = 2x^2 + 4x - 8$.

  a. What are a, b, and c?

  b. Find $-\dfrac{b}{2a}$.

Make a table of values for each function. Then graph them on the same coordinate system.

15. $f(x) = x^2$, $g(x) = 2x^2$, $h(x) = \dfrac{1}{2}x^2$

16. $f(x) = -x^2$, $g(x) = -\dfrac{1}{4}x^2$, $h(x) = -4x^2$

Make a table of values to graph function f. Then use a translation to graph the other two functions on the same coordinate system.

17. $f(x) = 4x^2$, $g(x) = 4x^2 + 3$, $h(x) = 4x^2 - 2$

18. $f(x) = 3x^2$, $g(x) = 3(x + 2)^2$, $h(x) = 3(x - 3)^2$

Make a table of values to graph function f. Then use a series of translations to graph function g on the same coordinate system.

19. $f(x) = -3x^2$, $g(x) = -3(x - 2)^2 - 1$

20. $f(x) = -\dfrac{1}{2}x^2$, $g(x) = -\dfrac{1}{2}(x + 1)^2 + 2$

Find the vertex and the axis of symmetry of the graph of each function. If necessary, complete the square on x to write the equation in the form $f(x) = a(x - h)^2 + k$. Do not graph the equation, but tell whether the graph will open upward or downward.

21. $f(x) = (x - 1)^2 + 2$

22. $f(x) = 2(x - 2)^2 - 1$

23. $f(x) = 2(x + 3)^2 - 4$

24. $f(x) = -3(x + 1)^2 + 3$

25. $f(x) = 0.5(x - 7.5)^2 + 8.5$

26. $f(x) = -\dfrac{3}{2}\left(x + \dfrac{1}{4}\right)^2 + \dfrac{7}{8}$

27. $f(x) = 2x^2 - 4x$

28. $f(x) = 3x^2 - 3$

29. $f(x) = -4x^2 + 16x + 5$

30. $f(x) = 5x^2 + 20x + 25$

31. $f(x) = 3x^2 + 4x + 2$

32. $f(x) = -6x^2 + 5x - 7$

First determine the vertex and the axis of symmetry of the graph of the function. Then plot several points and complete the graph. (See Example 5.)

33. $f(x) = (x - 3)^2 + 2$ **34.** $f(x) = (x + 1)^2 - 2$

35. $f(x) = -(x - 2)^2$ **36.** $f(x) = -(x + 2)^2$

37. $f(x) = -2(x + 3)^2 + 4$ **38.** $f(x) = 2(x - 2)^2 - 4$

39. $f(x) = \dfrac{1}{2}(x + 1)^2 - 3$ **40.** $f(x) = \dfrac{1}{3}(x - 1)^2 + 2$

Complete the square to write each function in $f(x) = a(x - h)^2 + k$ form. Determine the vertex and the axis of symmetry of the graph of the function. Then plot several points and complete the graph. (See Examples 6 and 7.)

41. $f(x) = x^2 + 2x - 3$ **42.** $f(x) = x^2 + 6x + 5$

43. $f(x) = 3x^2 - 12x + 10$ **44.** $f(x) = 4x^2 + 24x + 37$

45. $f(x) = 2x^2 + 8x + 6$ **46.** $f(x) = 3x^2 - 12x + 9$

47. $f(x) = x^2 + x - 6$ **48.** $f(x) = x^2 - x - 6$

49. $f(x) = -x^2 - 8x - 17$ **50.** $f(x) = -x^2 + 6x - 8$

51. $f(x) = -4x^2 + 16x - 10$ **52.** $f(x) = -2x^2 + 4x + 3$

Find the x- and y-intercepts of the graph of the quadratic function.

53. $f(x) = x^2 - 2x + 1$

54. $f(x) = 2x^2 - 4x$

55. $f(x) = -x^2 - 10x - 21$

56. $f(x) = 3x^2 + 6x - 9$

First determine the coordinates of the vertex and the axis of symmetry of the graph of the function using the vertex formula. Then determine the x- and y-intercepts of the graph. Finally, plot several points and complete the graph. (See Example 9.)

57. $f(x) = x^2 + 4x + 4$ **58.** $f(x) = x^2 - 6x + 9$

59. $f(x) = -x^2 + 2x - 1$ **60.** $f(x) = -x^2 - 2x - 1$

61. $f(x) = x^2 - 2x$ **62.** $f(x) = x^2 + x$

63. $f(x) = 4x^2 - 12x + 9$ **64.** $f(x) = 3x^2 - 12x + 12$

65. $f(x) = 2x^2 - 8x + 6$ **66.** $f(x) = 4x^2 + 4x - 3$

67. $f(x) = -6x^2 - 12x - 8$ **68.** $f(x) = -2x^2 + 8x - 10$

Use a graphing calculator to find the coordinates of the vertex of the graph of each quadratic function. Round to the nearest hundredth.

69. $f(x) = 2x^2 - x + 1$ **70.** $f(x) = x^2 + 5x - 6$

71. $f(x) = 7 + x - x^2$ **72.** $f(x) = 2x^2 - 3x + 2$

Use a graphing calculator to solve each equation. If an answer is not exact, round to the nearest hundredth.

73. $x^2 + x - 6 = 0$

74. $2x^2 - 5x - 3 = 0$

75. $0.5x^2 - 0.7x - 3 = 0$

76. $2x^2 - 0.5x - 2 = 0$

APPLICATIONS

77. CROSSWORD PUZZLES Darken the appropriate squares to the right of the dashed red line so that the puzzle has symmetry with respect to that line.

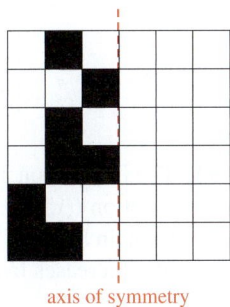

axis of symmetry

78. GRAPHIC ARTS Draw an axis of symmetry over the letter shown.

79. FIREWORKS A fireworks shell is shot straight up with an initial velocity of 120 feet per second. Its height s after t seconds is given by the equation $s = 120t - 16t^2$. If the shell is designed to explode when it reaches its maximum height, how long after being fired, and at what height, will the fireworks appear in the sky?

80. BALLISTICS From the top of the building in the illustration, a ball is thrown straight up with an initial velocity of 32 feet per second. The equation

$$s = -16t^2 + 32t + 48$$

gives the height s of the ball t seconds after it is thrown. Find the maximum height reached by the ball and the time it takes for the ball to hit the ground.

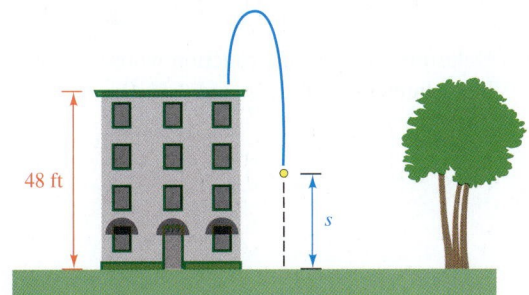

48 ft

s

81. FENCING A FIELD See the illustration in the next column. A farmer wants to fence in three sides of a rectangular field with 1,000 feet of fencing. The other side of the rectangle will be a river. If the enclosed area is to be maximum, find the dimensions of the field.

1,000 ft

82. POLICE INVESTIGATIONS A police officer seals off the scene of a car collision using a roll of yellow police tape that is 300 feet long, as shown in the illustration. What dimensions should be used to seal off the maximum rectangular area around the collision? What is the maximum area?

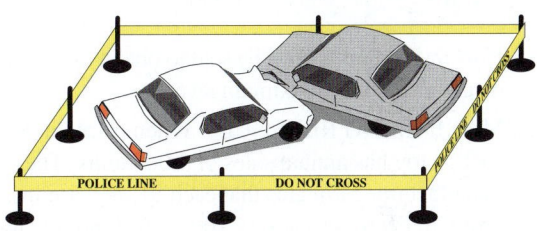

POLICE LINE   DO NOT CROSS

83. OPERATING COSTS The cost C in dollars of operating a certain concrete-cutting machine is related to the number of minutes n the machine is run by the function

$$C(n) = 2.2n^2 - 66n + 655$$

For what number of minutes is the cost of running the machine a minimum? What is the minimum cost?

84. WATER USAGE The height (in feet) of the water level in a reservoir over a 1-year period is modeled by the function

$$H(t) = 3.3(t - 9)^2 + 14$$

where $t = 1$ represents January, $t = 2$ represents February, and so on. How low did the water level get that year, and when did it reach the low mark?

85. U.S. ARMY The function

$$N(x) = -0.0534x^2 + 0.337x + 0.97$$

gives the number of active-duty military personnel in the United States Army (in millions) for the years 1965–1972, where $x = 0$ corresponds to 1965, $x = 1$ corresponds to 1966, and so on. For this period, when was the army's personnel strength level at its highest, and what was it? Historically, can you explain why?

86. SCHOOL ENROLLMENT The total annual enrollment (in millions) in U.S. elementary and secondary schools for the years 1975–1996 is given by the model

$$E(x) = 0.058x^2 - 1.162x + 50.604$$

where $x = 0$ corresponds to 1975, $x = 1$ corresponds to 1976, and so on. For this period, when was enrollment the lowest? What was the enrollment?

87. MAXIMIZING REVENUE The revenue R received for selling x stereos is given by the formula

$$R = -\frac{x^2}{5} + 80x - 1,000$$

How many stereos must be sold to obtain the maximum revenue? Find the maximum revenue.

88. MAXIMIZING REVENUE When priced at $30 each, a toy has annual sales of 4,000 units. The manufacturer estimates that each $1 increase in cost will decrease sales by 100 units. Find the unit price that will maximize total revenue. (*Hint:* Total revenue = price · the number of units sold.)

WRITING

89. Use the example of a stream of water from a drinking fountain to explain the concepts of the vertex and the axis of symmetry of a parabola.

90. What are some quantities that are good to maximize? What are some quantities that are good to minimize?

91. A mirror is held against the y-axis of the graph of a quadratic function. What fact about parabolas does this illustrate?

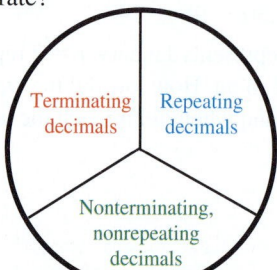

92. The vertex of a quadratic function $f(x) = ax^2 + bx + c$ is given by the formula $\left(-\dfrac{b}{2a}, f\left(-\dfrac{b}{2a}\right)\right)$. Explain what is meant by the notation $f\left(-\dfrac{b}{2a}\right)$.

93. A table of values for $f(x) = 2x^2 - 4x + 3$ is shown. Explain why it appears that the vertex of the graph of f is the point $(1, 1)$.

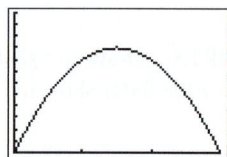

94. The illustration shows the graph of the quadratic function $f(x) = -4x^2 + 12x$ with domain $[0, 3]$. Explain how the value of $f(x)$ changes as the value of x increases from 0 to 3.

REVIEW   Simplify each expression. Assume all variables represent positive numbers.

95. $\sqrt{8a}\sqrt{2a^3b}$

96. $\left(\sqrt{23}\right)^2$

97. $\dfrac{\sqrt{3}}{\sqrt{50}}$

98. $\dfrac{3}{\sqrt[3]{9}}$

99. $3\left(\sqrt{5b} - \sqrt{3}\right)^2$

100. $-2\sqrt{5b}\left(4\sqrt{2b} - 3\sqrt{3}\right)$

CHALLENGE PROBLEMS

101. Use completing the square to show that the vertex of the graph of the quadratic function

$$f(x) = ax^2 + bx + c \text{ is } \left(-\frac{b}{2a}, \frac{4ac - b^2}{4a}\right).$$

102. Determine a quadratic function whose graph has x-intercepts of $(2, 0)$ and $(-4, 0)$.

10.5 Quadratic and Other Nonlinear Inequalities

- Solving quadratic inequalities
- Solving rational inequalities
- Graphs of nonlinear inequalities in two variables

We have previously solved *linear* inequalities in one variable such as $2x + 3 > 8$ and $6x - 7 < 4x - 9$. To find their solution sets, we used properties of inequalities to isolate the variable on one side of the inequality.

In this section, we will solve *quadratic* inequalities in one variable such as $x^2 + x - 6 < 0$ and $x^2 + 4x \geq 5$. We will use an interval testing method on the number line to determine their solution sets.

■ SOLVING QUADRATIC INEQUALITIES

Recall that a quadratic equation can be written in the form $ax^2 + bx + c = 0$. If we replace the = symbol with an inequality symbol, we have a quadratic inequality.

Quadratic Inequalities

A **quadratic inequality** can be written in one of the standard forms

$$ax^2 + bx + c < 0 \quad ax^2 + bx + c > 0 \quad ax^2 + bx + c \leq 0 \quad ax^2 + bx + c \geq 0$$

where a, b, and c are real numbers and $a \neq 0$.

To solve a quadratic inequality in one variable, we find all values of the variable that make the inequality true using the following steps.

Solving Quadratic Inequalities

1. Write the inequality in standard form and then solve its related quadratic equation.

2. Locate the solutions (called **critical numbers**) of the related quadratic equation on the number line.

3. Test each interval created in step 2 by choosing a test value from the interval and determining whether it satisfies the inequality. The solution set includes the interval(s) whose test value makes the inequality true.

4. Determine whether the endpoints of the intervals are included in the solution set.

EXAMPLE 1

ELEMENTARY &
INTERMEDIATE
Algebra *f(x)* **Now**™

Solution

Solve: $x^2 + x - 6 < 0$.

The expression $x^2 + x - 6$ can be positive, or negative, or 0, depending on what value is substituted for x. Solutions of the inequality are x-values that make $x^2 + x - 6$ less than 0. To find them, we first find the values of x that make $x^2 + x - 6$ equal to 0.

Step 1: *Solve the related quadratic equation.* For the quadratic inequality $x^2 + x - 6 < 0$, the related quadratic equation is $x^2 + x - 6 = 0$.

$$x^2 + x - 6 = 0$$

$$(x + 3)(x - 2) = 0 \qquad \text{Factor the trinomial.}$$

$$x + 3 = 0 \quad \text{or} \quad x - 2 = 0 \qquad \text{Set each factor equal to 0.}$$

$$x = -3 \qquad \qquad x = 2$$

The solutions of $x^2 + x - 6 = 0$ are -3 and 2. These solutions are the critical numbers.

The Language of Algebra

We say that the critical numbers *partition* the real-number line into test intervals. Interior decorators use freestanding screens to *partition* off parts of a room.

Step 2: *Locate the critical numbers.* When we highlight -3 and 2 on a number line, we see that they separate it into three intervals:

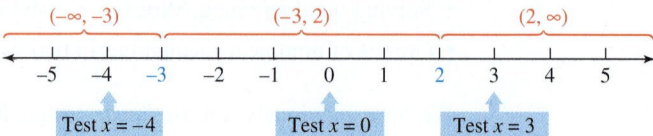

Step 3: *Test each interval.* To determine whether the numbers in $(-\infty, -3)$ are solutions of the inequality, we choose any number from that interval, substitute it for x, and see whether it satisfies $x^2 + x - 6 < 0$. *If one number in that interval satisfies the inequality, all numbers in that interval will satisfy the inequality.*

If we choose -4 from $(-\infty, -3)$, we have:

$$x^2 + x - 6 < 0 \qquad \text{This is the original inequality.}$$
$$(-4)^2 + (-4) - 6 \overset{?}{<} 0 \qquad \text{Substitute } -4 \text{ for } x.$$
$$16 + (-4) - 6 \overset{?}{<} 0$$
$$6 < 0 \qquad \text{False.}$$

Since -4 does not satisfy the inequality, none of the numbers in $(-\infty, -3)$ are solutions.

To test the second interval, $(-3, 2)$, we choose $x = 0$.

$$x^2 + x - 6 < 0 \qquad \text{This is the original inequality.}$$
$$0^2 + 0 - 6 \overset{?}{<} 0 \qquad \text{Substitute 0 for } x.$$
$$-6 < 0 \qquad \text{True.}$$

Since 0 satisfies the inequality, all of the numbers in $(-3, 2)$ are solutions.

To test the third interval, $(2, \infty)$, we choose $x = 3$.

$$x^2 + x - 6 < 0 \qquad \text{This is the original inequality.}$$
$$3^2 + 3 - 6 \overset{?}{<} 0 \qquad \text{Substitute 3 for } x.$$
$$9 + 3 - 6 \overset{?}{<} 0$$
$$6 < 0 \qquad \text{False.}$$

Since 3 does not satisfy the inequality, none of the numbers in $(2, \infty)$ are solutions.

Success Tip

If a quadratic inequality contains $\leq$ or $\geq$, the endpoints of the intervals are included in the solution set. If the inequality contains $<$ or $>$, they are not.

Step 4: *Are the endpoints included?* From the interval testing, we see that only numbers from $(-3, 2)$ satisfy $x^2 + x - 6 < 0$. The endpoints -3 and 2 are not included in the solution set because they do not satisfy the inequality. (Recall that -3 and 2 make $x^2 + x - 6$ equal to 0.) The solution set is $(-3, 2)$ as graphed below.

Self Check 1   Solve: $x^2 + x - 12 < 0$.

EXAMPLE 2
ELEMENTARY &
INTERMEDIATE
Algebra *f(x)* **Now**™
            Solution

Solve: $x^2 + 4x \geq 5$.

The inequality is not in standard form. To get 0 on the right-hand side, we subtract 5 from both sides.

$$x^2 + 4x \geq 5$$
$$x^2 + 4x - 5 \geq 0 \qquad \text{Write the inequality in the form } ax^2 + bx + c \geq 0.$$

Now we solve the related quadratic equation $x^2 + 4x - 5 = 0$.

$$x^2 + 4x - 5 = 0$$
$$(x + 5)(x - 1) = 0 \qquad \text{Factor the trinomial.}$$
$$x + 5 = 0 \quad \text{or} \quad x - 1 = 0 \qquad \text{Set each factor equal to 0.}$$
$$x = -5 \qquad \qquad x = 1$$

The critical numbers -5 and 1 separate the number line into three intervals. We pick a test value from each interval to see whether it satisfies $x^2 + 4x - 5 \geq 0$.

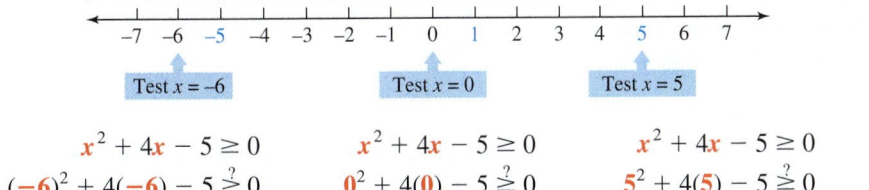

$$x^2 + 4x - 5 \geq 0 \qquad\qquad x^2 + 4x - 5 \geq 0 \qquad\qquad x^2 + 4x - 5 \geq 0$$
$$(-6)^2 + 4(-6) - 5 \overset{?}{\geq} 0 \qquad 0^2 + 4(0) - 5 \overset{?}{\geq} 0 \qquad 5^2 + 4(5) - 5 \overset{?}{\geq} 0$$
$$7 \geq 0 \quad \text{True.} \qquad\qquad -5 \geq 0 \quad \text{False.} \qquad\qquad 40 \geq 0 \quad \text{True.}$$

The numbers in the intervals $(-\infty, -5)$ and $(1, \infty)$ satisfy the inequality. Since the endpoints -5 and 1 also satisfy $x^2 + 4x - 5 \geq 0$, they are included in the solution set. (Recall that -5 and 1 make $x^2 + 4x - 5$ equal to 0.) Thus, the solution set is the union of two intervals: $(-\infty, -5] \cup [1, \infty)$. The graph of the solution set is shown below.

Self Check 2   Solve: $x^2 + 3x \geq 40$.

■ SOLVING RATIONAL INEQUALITIES

Rational inequalities in one variable such as $\dfrac{9}{x} < 8$ and $\dfrac{x^2 + x - 2}{x - 4} \geq 0$ can also be solved using the interval testing method.

Solving Rational Inequalities

1. Write the inequality in standard form and then solve its related rational equation.

2. Set the denominator equal to zero and solve that equation.

3. Locate the solutions (called critical numbers) found in steps 1 and 2 on the number line.

4. Test each interval on the number line created in step 3 by choosing a test value from the interval and determining whether it satisfies the inequality. The solution set includes the interval(s) whose test point makes the inequality true.

5. Determine whether the endpoints of the intervals are included in the solution set. Exclude any values that make the denominator 0.

EXAMPLE 3

ELEMENTARY &
INTERMEDIATE
Algebra *f(x)* **Now**™   **Solution**

Solve: $\dfrac{9}{x} < 8$.

The inequality is not in standard form. To get 0 on the right-hand side, we subtract 8 from both sides. We then find a common denominator to simplify the left-hand side.

$$\frac{9}{x} < 8$$

$$\frac{9}{x} - 8 < 0 \qquad \text{Subtract 8 from both sides.}$$

$$\frac{9}{x} - 8 \cdot \frac{x}{x} < 0 \qquad \text{Build 8 to a fraction with denominator } x.$$

$$\frac{9}{x} - \frac{8x}{x} < 0$$

$$\frac{9 - 8x}{x} < 0 \qquad \text{Subtract the numerators and keep the common denominator, } x.$$

Caution

When solving rational inequalities such as $\frac{9}{x} < 8$, a common error is to multiply both sides by x to clear it of the fraction. However, we don't know whether x is positive or negative, so we don't know whether or not to reverse the inequality symbol.

Now we solve the related rational equation.

$$\frac{9 - 8x}{x} = 0$$

$$9 - 8x = 0 \qquad \text{To clear the equation of the fraction, multiply both sides by } x.$$

$$-8x = -9 \qquad \text{Subtract 9 from both sides.}$$

$$x = \frac{9}{8} \qquad \text{This is a critical number.}$$

If we set the denominator of $\frac{9 - 8x}{x}$ equal to 0, we obtain a second critical number, $x = 0$. When graphed, the critical numbers 0 and $\frac{9}{8}$ separate the number line into three intervals. We pick a test value from each interval to see whether it satisfies $\frac{9 - 8x}{x} < 0$.

$$\frac{9 - 8x}{x} < 0 \qquad\qquad \frac{9 - 8x}{x} < 0 \qquad\qquad \frac{9 - 8x}{x} < 0$$

$$\frac{9 - 8(-1)}{-1} \overset{?}{<} 0 \qquad \frac{9 - 8(1)}{1} \overset{?}{<} 0 \qquad \frac{9 - 8(2)}{2} \overset{?}{<} 0$$

$$-17 < 0 \quad \text{True.} \qquad 1 < 0 \quad \text{False.} \qquad -\frac{7}{2} < 0 \quad \text{True.}$$

The numbers in the intervals $(-\infty, 0)$ and $\left(\frac{9}{8}, \infty\right)$ satisfy the inequality. We do not include the endpoint 0 in the solution set, because it makes the denominator of the original inequality 0. Neither do we include $\frac{9}{8}$, because it does not satisfy $\frac{9 - 8x}{x} < 0$. (Recall that $\frac{9}{8}$ makes

$\frac{9 - 8x}{x}$ equal to 0.) Thus, the solution set is the union of two intervals: $(-\infty, 0) \cup (\frac{9}{8}, \infty)$. Its graph is shown below.

Self Check 3   Solve: $\frac{3}{x} < 5$.

EXAMPLE 4

ELEMENTARY & INTERMEDIATE
Algebra *f(x)* **Now**™

Solve: $\frac{x^2 + x - 2}{x - 4} \geq 0$.

Solution   We first solve the related rational equation to find the critical numbers.

$$\frac{x^2 + x - 2}{x - 4} = 0$$

$x^2 + x - 2 = 0$ To clear the equation of the fraction, multiply both sides by $x - 4$.

$(x + 2)(x - 1) = 0$ Factor the trinomial.

$x + 2 = 0$ or $x - 1 = 0$ Set each factor equal to 0.

$\qquad x = -2$ | $x = 1$ These are critical numbers.

If we set the denominator of $\frac{x^2 + x - 2}{x - 4}$ equal to 0, we see that $x = 4$ is also a critical number. When graphed, the critical numbers, -2, 1, and 4, separate the number line into four intervals. We pick a test value from each interval to see whether it satisfies $\frac{x^2 + x - 2}{x - 4} \geq 0$.

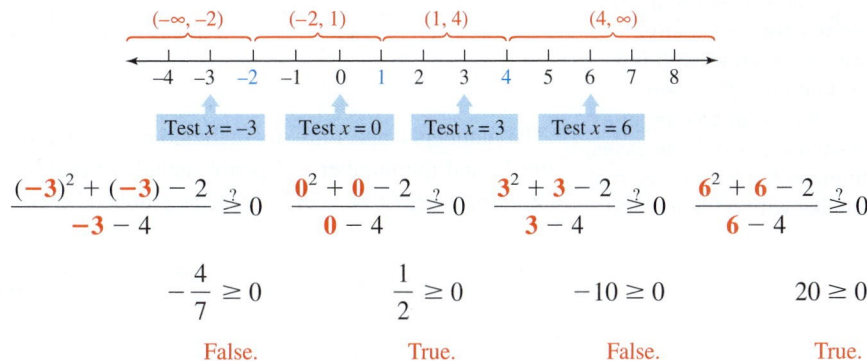

The numbers in the intervals $(-2, 1)$ and $(4, \infty)$ satisfy the inequality. We include the endpoints -2 and 1 in the solution set, because they satisfy the inequality. We do not include 4, because it makes the denominator of the inequality 0. Thus, the solution set is $[-2, 1] \cup (4, \infty)$ as graphed below.

Self Check 4   Solve: $\frac{x + 2}{x^2 - 2x - 3} \geq 0$.

EXAMPLE 5   Solve: $\dfrac{3}{x-1} < \dfrac{2}{x}$.

Solution   We subtract $\dfrac{2}{x}$ from both sides to get 0 on the right-hand side and proceed as follows:

$$\frac{3}{x-1} < \frac{2}{x}$$

$$\frac{3}{x-1} - \frac{2}{x} < 0 \qquad \text{Subtract } \frac{2}{x} \text{ from both sides.}$$

$$\frac{3}{x-1} \cdot \frac{x}{x} - \frac{2}{x} \cdot \frac{x-1}{x-1} < 0 \qquad \text{Build each rational expression to have the common denominator } x(x-1).$$

$$\frac{3x - 2x + 2}{x(x-1)} < 0 \qquad \text{Subtract the numerators and keep the common denominator.}$$

$$\frac{x+2}{x(x-1)} < 0 \qquad \text{Combine like terms.}$$

The only solution of the related rational equation $\dfrac{x+2}{x(x-1)}$ is -2. Thus, -2 is a critical number. When we set the denominator equal to 0 and solve $x(x-1) = 0$, we find two more critical numbers, 0 and 1. These three critical numbers create four intervals to test.

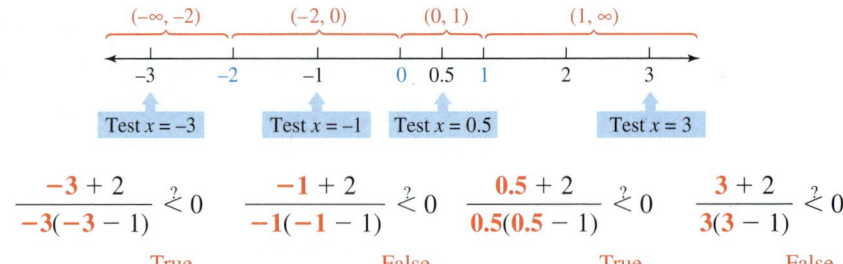

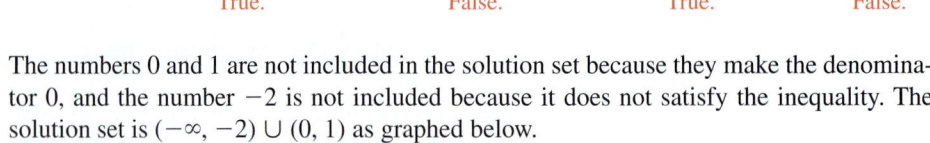

$\dfrac{-3+2}{-3(-3-1)} \overset{?}{<} 0$	$\dfrac{-1+2}{-1(-1-1)} \overset{?}{<} 0$	$\dfrac{0.5+2}{0.5(0.5-1)} \overset{?}{<} 0$	$\dfrac{3+2}{3(3-1)} \overset{?}{<} 0$
True.	False.	True.	False.

Success Tip

When the endpoints of an interval are consecutive integers, such as with the third interval (0, 1), we cannot choose an integer as a test value. For these cases, choose a fraction or decimal that lies within the interval.

The numbers 0 and 1 are not included in the solution set because they make the denominator 0, and the number -2 is not included because it does not satisfy the inequality. The solution set is $(-\infty, -2) \cup (0, 1)$ as graphed below.

Self Check 5   Solve: $\dfrac{2}{x+1} > \dfrac{1}{x}$.

ACCENT ON TECHNOLOGY: SOLVING INEQUALITIES GRAPHICALLY

We can solve $x^2 + 4x \geq 5$ (Example 2) graphically by first writing the inequality as $x^2 + 4x - 5 \geq 0$, and then graphing the quadratic function $f(x) = x^2 + 4x - 5$, as shown in figure (a). The solution set of the inequality will be those values of x for which the graph lies on or above the x-axis. We can trace to determine that this interval is $(-\infty, -5] \cup [1, \infty)$.

To solve $\dfrac{3}{x-1} < \dfrac{2}{x}$ (Example 5) graphically, we first write the inequality in the form $\dfrac{x+2}{x(x-1)} < 0$, and then we graph the rational function $f(x) = \dfrac{x+2}{x(x-1)}$, as shown in figure (b). The solution of the inequality will be those values of x for which the graph lies below the axis.

We can trace to see that the graph is below the x-axis when x is less than -2. Since we cannot see the graph in the interval $0 < x < 1$, we redraw the graph using window settings of $[-1, 2]$ for x and $[-25, 10]$ for y, as shown in figure (c).

Now we see that the graph is below the x-axis in the interval $(0, 1)$. Thus, the solution set of the inequality is the union of the two intervals: $(-\infty, -2) \cup (0, 1)$.

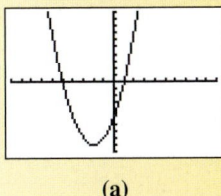
(a)

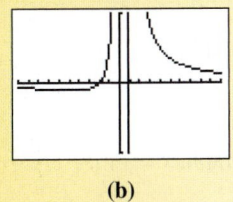
(b)

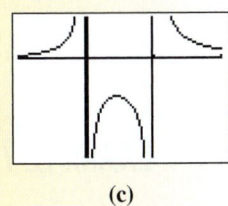
(c)

■ GRAPHS OF NONLINEAR INEQUALITIES IN TWO VARIABLES

We have previously graphed linear inequalities in two variables such as $y > 3x + 2$ and $2x - 3y \le 6$ using the following steps.

Graphing Inequalities in Two Variables	1. Graph the boundary line of the region. If the inequality allows equality (the symbol is either $\le$ or $\ge$), draw the boundary line as a solid line. If equality is not allowed ($<$ or $>$), draw the boundary line as a dashed line. 2. Pick a test point that is on one side of the boundary line. (Use the origin if possible.) Replace x and y in the original inequality with the coordinates of that point. If the inequality is satisfied, shade the side that contains that point. If the inequality is not satisfied, shade the other side of the boundary.

We use the same procedure to graph *nonlinear* inequalities in two variables.

EXAMPLE 6

Graph: $y < -x^2 + 4$.

Solution   The graph of the boundary $y = -x^2 + 4$ is a parabola opening downward, with vertex at $(0, 4)$ and axis of symmetry $x = 0$ (the y-axis). Since the inequality contains a $<$ symbol, and equality is not allowed, we draw the parabola using a dashed line.

To determine which region to shade, we pick the test point $(0, 0)$ and substitute its coordinates into the inequality. We shade the region containing $(0, 0)$ because its coordinates satisfy $y < -x^2 + 4$.

Graph the boundary

$y = -x^2 + 4$

Compare to $y = a(x - h)^2 + k$

$a = -1$: Opens downward

$h = 0$ and $k = 4$: Vertex $(0, 4)$

Axis of symmetry $x = 0$

x	y
1	3
2	0

Shading: Use the test point $(0, 0)$

$y < -x^2 + 4$

$0 \overset{?}{<} -0^2 + 4$

$0 < 4$ True.

Since $0 < 4$ is true, $(0, 0)$ is

a solution of $y < -x^2 + 4$.

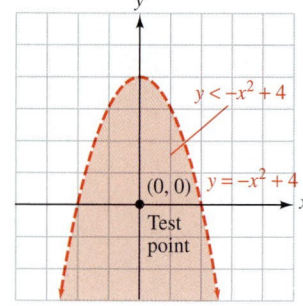

Self Check 6   Graph: $y \geq -x^2 + 4$.

─────────────────────────────────────

EXAMPLE 7

Graph: $x \leq |y|$.

Solution   To graph the boundary, $x = |y|$, we construct a table of solutions, as shown in figure (a). In figure (b), the boundary is graphed using a solid line because the inequality contains a $\leq$ symbol and equality is permitted. Since the origin is on the graph, we cannot use it as a test point. However, any other point, such as $(1, 0)$, will do. We substitute 1 for x and 0 for y into the inequality to get

$$x \leq |y|$$
$$1 \overset{?}{\leq} |0|$$
$$1 \leq 0 \qquad \text{False.}$$

Since $1 \leq 0$ is a false statement, the point $(1, 0)$ does not satisfy the inequality and is not part of the graph. Thus, the graph of $x \leq |y|$ is to the left of the boundary.

The complete graph is shown in figure (c).

$x = |y|$

x	y
0	0
1	1
1	−1
2	2
2	−2

(a)

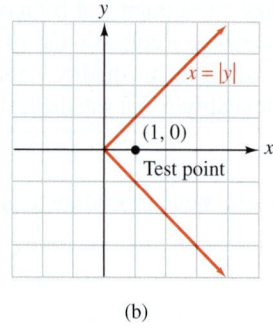

(b)

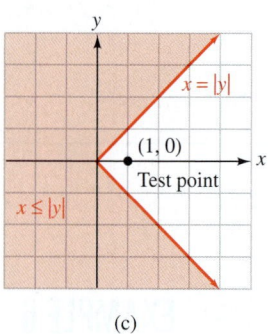

(c)

Self Check 7   Graph: $x \geq -|y|$.

Answers to Self Checks   **1.** $(-4, 3)$ **2.** $(-\infty, -8] \cup [5, \infty)$

3. $(-\infty, 0) \cup \left(\frac{3}{5}, \infty\right)$ **4.** $[-2, -1) \cup (3, \infty)$

5. $(-1, 0) \cup (1, \infty)$

6.

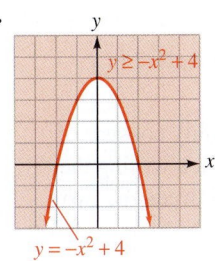

$y \geq -x^2 + 4$

$y = -x^2 + 4$

7.

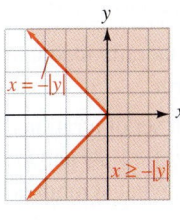

$x = -|y|$

$x \geq -|y|$

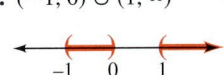 **10.5**   **STUDY SET**   Algebra *(x)* Now™
ELEMENTARY & INTERMEDIATE

VOCABULARY   **Fill in the blanks.**

1. $x^2 + 3x - 18 < 0$ is an example of a _____ inequality in one variable.

2. $\frac{x-1}{x^2 - x - 20} \leq 0$ is an example of a _____ inequality in one variable.

3. $y \leq x^2 - 4x + 3$ is an example of a nonlinear inequality in _____ variables.

4. The set of real numbers greater than 3 can be represented using the _____ notation $(3, \infty)$.

CONCEPTS

5. The critical numbers of a quadratic inequality are highlighted in red on the number line shown below. Use interval notation to represent each interval that must be tested to solve the inequality.

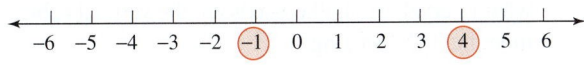

6. The graph of the solution set of a rational inequality in one variable is shown below. Determine whether each of the following numbers is a solution of the inequality.

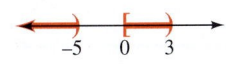

a. -10 **b.** -5

c. 0 **d.** 4

7. Graph each of the following solution sets.

a. $(-2, 4)$

b. $(-\infty, -2) \cup (3, 5]$

8. What are the critical numbers for each inequality?

a. $x^2 - 2x - 48 \geq 0$

b. $\frac{x-3}{x(x+4)} > 0$

9. a. The results after interval testing for a quadratic inequality containing a $>$ symbol are shown below. (The critical numbers are highlighted in red.) What is the solution set?

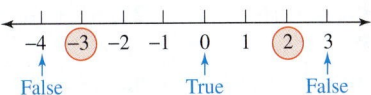

False   True   False

b. The results after interval testing for a quadratic inequality containing a $\leq$ symbol are shown below. (The critical numbers are highlighted in red.) What is the solution set?

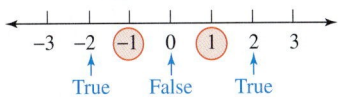

True   False   True

10. Fill in the blank to complete this important fact about the interval testing method discussed in this section: *If one number in an interval satisfies the inequality, _____ numbers in that interval will satisfy the inequality.*

11. a. When graphing the solution of $y \leq x^2 + 2x + 1$, should the boundary be solid or dashed?

b. Does the test point $(0, 0)$ satisfy the inequality?

12. **a.** Estimate the solution of $x^2 - x - 6 > 0$ using the graph of $y = x^2 - x - 6$ shown in figure (a) below.

b. Estimate the solution of $\frac{x-3}{x} \leq 0$ using the graph of $y = \frac{x-3}{x}$ shown in figure (b) below.

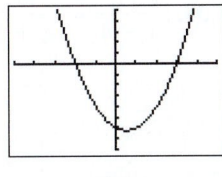   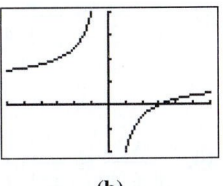

(a)   (b)

NOTATION

13. Write the quadratic inequality $x^2 - 6x \geq 7$ in standard form.

14. The solution set of a rational inequality consists of the intervals, $(-1, 4]$ and $(7, \infty)$. When writing the solution set, what symbol is used between the two intervals?

PRACTICE   Solve each inequality. Write the solution set in interval notation and graph it.

15. $x^2 - 5x + 4 < 0$

16. $x^2 - 3x - 4 > 0$

17. $x^2 - 8x + 15 > 0$

18. $x^2 + 2x - 8 < 0$

19. $x^2 + x - 12 \leq 0$

20. $x^2 - 8x \leq -15$

21. $x^2 + 8x < -16$

22. $x^2 + 6x \geq -9$

23. $x^2 \geq 9$

24. $x^2 \geq 16$

25. $2x^2 - 50 < 0$

26. $3x^2 - 243 < 0$

27. $\dfrac{1}{x} < 2$

28. $\dfrac{1}{x} > 3$

29. $-\dfrac{5}{x} < 3$

30. $\dfrac{4}{x} \geq 8$

31. $\dfrac{x^2 - x - 12}{x - 1} < 0$

32. $\dfrac{x^2 + x - 6}{x - 4} \geq 0$

33. $\dfrac{6x^2 - 5x + 1}{2x + 1} > 0$

34. $\dfrac{6x^2 + 11x + 3}{3x - 1} < 0$

35. $\dfrac{3}{x - 2} < \dfrac{4}{x}$

36. $\dfrac{-6}{x + 1} \geq \dfrac{1}{x}$

37. $\dfrac{7}{x - 3} \geq \dfrac{2}{x + 4}$

38. $\dfrac{-5}{x - 4} < \dfrac{3}{x + 1}$

39. $\dfrac{x}{x + 4} \leq \dfrac{1}{x + 1}$

40. $\dfrac{x}{x + 9} \geq \dfrac{1}{x + 1}$

41. $(x + 2)^2 > 0$

42. $(x - 3)^2 < 0$

Use a graphing calculator to solve each inequality. Write the solution set in interval notation.

43. $x^2 - 2x - 3 < 0$

44. $x^2 + x - 6 > 0$

45. $\dfrac{x + 3}{x - 2} > 0$

46. $\dfrac{3}{x} < 2$

Graph each inequality.

47. $y < x^2 + 1$

48. $y > x^2 - 3$

49. $y \leq x^2 + 5x + 6$

50. $y \geq x^2 + 5x + 4$

51. $y < |x + 4|$

52. $y \geq |x - 3|$

53. $y \leq -|x| + 2$

54. $y > |x| - 2$

APPLICATIONS

55. BRIDGES   If an x-axis is superimposed over the roadway of the Golden Gate Bridge, with the origin at the center of the bridge, the length L in feet of a vertical support cable can be approximated by the formula

$$L = \dfrac{1}{9,000}x^2 + 5$$

For the Golden Gate Bridge, $-2,100 < x < 2,100$. For what intervals along the x-axis are the vertical cables more than 95 feet long?

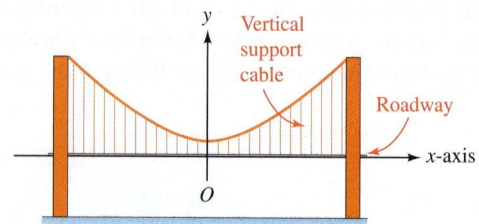

56. MALLS   The number of people n in a mall is modeled by the formula

$$n = -100x^2 + 1,200x$$

where x is the number of hours since the mall opened. If the mall opened at 9 A.M., when were there 2,000 or more people in it?

WRITING

57. How are critical numbers used when solving a quadratic inequality in one variable?

58. Explain how to find the graph of $y \geq x^2$.

59. The graph of $f(x) = x^2 - 3x + 4$ is shown below. Explain why the quadratic inequality $x^2 - 3x + 4 < 0$ has no solution.

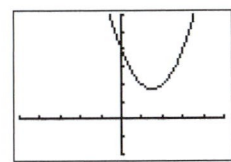

60. Describe the following solution set of a rational inequality in words: $(-\infty, 4] \cup (6, 7)$.

REVIEW   Translate each statement into an equation.

61. x varies directly with y.

62. y varies inversely with t.

63. t varies jointly with x and y.

64. d varies directly with t and inversely with u^2.

CHALLENGE PROBLEMS

65. a. Solve: $x^2 - x - 12 > 0$.

    b. Find a rational inequality in one variable that has the same solution set as the quadratic inequality in part a.

66. a. Solve: $\dfrac{1}{x} < 1$.

    b. Now incorrectly "solve" $\dfrac{1}{x} < 1$ by multiplying both sides by x to clear it of the fraction. What part of the solution set is not obtained with this faulty approach?

ACCENT ON TEAMWORK

PICTURE FRAMING

Overview: When framing pictures, mats are often used to enhance the images and give them a sense of depth. In this activity, you will use the quadratic formula to design the matting for several pictures.

Instructions: Form groups of 3 students. Each person in your group is to bring a picture to class. You can use a picture from a magazine or newspaper, a picture postcard, or a photograph that is no larger than 5 in. × 7 in. You will also need a pair of scissors, a ruler, glue, and three pieces of construction paper (12 in. × 18 in.).

Select one of the pictures and find its area. A mat of *uniform* width is to be placed around the picture. The area of the mat should equal the area of the picture. To determine the proper width of the matting, follow the steps of Example 10 in Section 8.1. However, use the quadratic formula, instead of completing the square, to solve the equation. Once you have determined the proper width, cut out the mat from the construction paper and glue it to the picture.

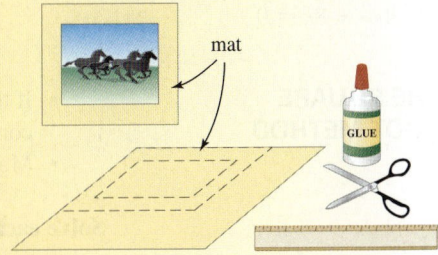

Then, choose another picture and find its area. Determine the uniform width that a matting should have so that its area is double that of the picture. Cut out the proper size matting from the construction paper and glue it to the second picture.

Finally, find the area of the third picture and determine the uniform width that a matting should have so that its area is one-half that of the picture. Cut out the proper-size matting from the construction paper and glue it to the third picture.

Is one size matting more visually appealing than another? Discuss this among the members of your group.

SOLUTIONS OF QUADRATIC EQUATIONS

Overview: In this activity, you will learn of an interesting fact about the solutions of a quadratic equation.

Instructions: Form groups of 2 or 3 students. Consider the following property about the sum and product of the solutions of a quadratic equation:

$$r_1 + r_2 = -\frac{b}{a} \quad \text{and} \quad r_1 \cdot r_2 = \frac{c}{a}, \quad \text{then } r_1 \text{ and } r_2 \text{ are the solutions of the}$$

quadratic equation $ax^2 + bx + c = 0$, where $a \neq 0$.

Use this property to show that

1. $\dfrac{3}{2}$ and $-\dfrac{1}{3}$ are solutions of $6x^2 - 7x - 3 = 0$.

2. $1 + 3\sqrt{2}$ and $1 - 3\sqrt{2}$ are solutions of $x^2 - 2x - 17 = 0$.

3. $i\sqrt{51}$ and $-i\sqrt{51}$ are solutions of $x^2 + 51 = 0$.

KEY CONCEPT: SOLVING QUADRATIC EQUATIONS

We have discussed five methods for solving **quadratic equations.** Let's review each of them and list an advantage and a drawback of each method.

FACTORING

- Factoring can be very quick and simple if the factoring pattern is evident.
- Much of the time, $ax^2 + bx + c$ cannot be factored or is not easily factored.

Solve each equation by factoring.

1. $4k^2 + 8k = 0$
2. $z^2 + 8z + 15 = 0$
3. $2r^2 + 5r = -3$

THE SQUARE ROOT METHOD

- If the equation can be written in the form $x^2 = a$ or $(x + d)^2 = a$, where a is a constant, the square root method is a fast method, requiring few computations.
- Most quadratic equations that we must solve are not written in either of these forms.

Solve each equation by the square root method.

4. $u^2 = 24$
5. $(s - 7)^2 - 9 = 0$
6. $3x^2 + 16 = 0$

COMPLETING THE SQUARE

- Completing the square can be used to solve any quadratic equation.
- Most often, it involves more steps than the other methods.

Solve each equation by completing the square.

7. $x^2 + 10x - 7 = 0$
8. $4x^2 - 4x - 1 = 0$
9. $x^2 + 2x + 2 = 0$

THE QUADRATIC FORMULA

- The quadratic formula simply involves an evaluation of the expression $\dfrac{-b \pm \sqrt{b^2 - 4ac}}{2a}$.

- If applicable, the factoring method and the square root method are usually faster.

Solve each equation by using the quadratic formula.

10. $2x^2 - 1 = 3x$

11. $x^2 - 6x - 391 = 0$

12. $3x^2 + 2x + 1 = 0$

THE GRAPHING METHOD

- We can solve the equation using a graphing calculator. It doesn't require any computations.
- It usually gives only approximations of the solutions.

13. Use the graph of $y = x^2 + x - 2$ to solve $x^2 + x - 2 = 0$.

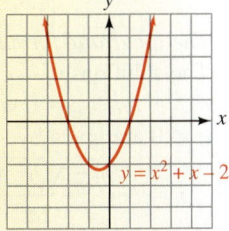

CHAPTER REVIEW ELEMENTARY & INTERMEDIATE **Algebra** $f(x)$ **Now**™

SECTION 10.1	**The Square Root Property and Completing the Square**

CONCEPTS

The square root property: If $c > 0$, the equation $x^2 = c$ has two real solutions:

$x = \sqrt{c}$ and $x = -\sqrt{c}$

To *complete the square:*

1. Make sure the coefficient of x^2 is 1.
2. Make sure the constant term is on the right-hand side of the equation.
3. Add the square of one-half of the coefficient of x to both sides.
4. Factor the trinomial.
5. Use the square root property.
6. Check the answers.

REVIEW EXERCISES

Solve each equation by factoring or using the square root property.

1. $x^2 + 9x + 20 = 0$

2. $6x^2 + 17x + 5 = 0$

3. $x^2 = 28$

4. $(t + 2)^2 = 36$

5. $5a^2 + 11a = 0$

6. $5x^2 - 49 = 0$

7. $a^2 + 25 = 0$

8. What number must be added to $x^2 - x$ to make a perfect square trinomial?

Solve each equation by completing the square.

9. $x^2 + 6x + 8 = 0$

10. $2x^2 - 6x + 3 = 0$

11. $x^2 - 2x + 13 = 0$

12. Solve $A = \pi r^2$ for r. Assume that all variables represent positive numbers. Express any radical in simplified form.

13. Explain the error:

$$\frac{2 \pm \sqrt{7}}{2} = \frac{\overset{1}{\cancel{2}} \pm \sqrt{7}}{\underset{1}{\cancel{2}}} = 1 \pm \sqrt{7}$$

14. HAPPY NEW YEAR As part of a New Year's Eve celebration, a huge ball is to be dropped from the top of a 605-foot-tall building at the proper moment so that it strikes the ground at exactly 12:00 midnight. The distance d in feet traveled by a free-falling object in t seconds is given by the formula $d = 16t^2$. To the nearest second, when should the ball be dropped from the building?

| SECTION 10.2 | **The Quadratic Formula** |

The quadratic formula:
The solutions of
$ax^2 + bx + c = 0$ where
$a \neq 0$ are given by
$$x = \frac{-b \pm \sqrt{b^2 - 4ac}}{2a}$$

When solving a quadratic equation by the quadratic formula, we can often simplify the computations by solving an equivalent equation that does not involve fractions or decimals, and whose lead coefficient is positive.

Solve each equation using the quadratic formula.

15. $x^2 - 10x = 0$

16. $-x^2 + 10x - 18 = 0$

17. $2x^2 + 13x = 7$

18. $26y - 3y^2 = 2$

19. $\dfrac{p^2}{3} + \dfrac{p}{2} + \dfrac{1}{2} = 0$

20. $3{,}000t^2 - 4{,}000t = -2{,}000$

21. $0.5x^2 + 0.3x - 0.1 = 0$

22. TUTORING A private tutoring company charges $20 for a 1-hour session. Currently, 300 students are tutored each week. Since the company is losing money, the owner has decided to increase the price. For each 50¢ increase, she estimates that 5 fewer students will participate. If the center needs to bring in $6,240 per week to stay in business, what price must be charged for a 1-hour tutoring session to produce this amount of revenue?

23. POSTERS The specifications for a poster of Cesar Chavez call for a 615-square-inch photograph to be surrounded by a blue border. (See below.) The borders on the sides of the poster are to be half as wide as those at the top and bottom. Find the width of each border.

35 in.
23 in.

24. ACROBATS To begin his routine on a trapeze, an acrobat is catapulted upward as shown in the illustration above. His distance d (in feet) from the arena floor during this maneuver is given by the formula $d = -16t^2 + 40t + 5$, where t is the time in seconds since being launched. If the trapeze bar is 25 feet in the air, at what two times will he be able to grab it? Round to the nearest tenth.

| **SECTION 10.3** | **The Discriminant and Equations That Can Be Written in Quadratic Form** |

The *discriminant* predicts the type of solutions of $ax^2 + bx + c = 0$:

1. If $b^2 - 4ac > 0$, the solutions are unequal real numbers.
2. If $b^2 - 4ac = 0$, the solutions are equal real numbers.
3. If $b^2 - 4ac < 0$, the solutions are complex conjugates.

Equations that contain an expression to a power, the same expression to that power squared, and a constant term are said to be *quadratic in form*. One method used to solve such equations is to make a substitution.

Use the discriminant to determine the type of solutions for each equation.

25. $3x^2 + 4x - 3 = 0$

26. $4x^2 - 5x + 7 = 0$

27. $9x^2 - 12x + 4 = 0$

28. $m(2m - 3) = 20$

Solve each equation.

29. $x - 13\sqrt{x} + 12 = 0$

30. $a^{2/3} + a^{1/3} - 6 = 0$

31. $3x^4 + x^2 - 2 = 0$

32. $\dfrac{6}{x + 2} + \dfrac{6}{x + 1} = 5$

33. $(x - 7)^2 + 6(x - 7) + 10 = 0$

34. $m^{-4} - 2m^{-2} + 1 = 0$

35. $4\left(\dfrac{x + 1}{x}\right)^2 + 12\left(\dfrac{x + 1}{x}\right) + 9 = 0$

36. WEEKLY CHORES Working together, two sisters can do the yard work at their house in 45 minutes. When the older girl does it all herself, she can complete the job in 20 minutes less time than it takes the younger girl working alone. How long does it take the older girl to do the yard work?

| **SECTION 10.4** | **Quadratic Functions and Their Graphs** |

A *quadratic function* is a second-degree polynomial function of the form $f(x) = ax^2 + bx + c$.

The graph of $f(x) = ax^2$ is a *parabola* opening upward when $a > 0$ and downward when $a < 0$, with *vertex* at the point $(0, 0)$ and *axis of symmetry* the line $x = 0$.

Each of the following functions has a graph that is the same shape as $f(x) = ax^2$ but involves a vertical or horizontal translation.

1. $f(x) = ax^2 + k$: translated upward if $k > 0$, downward if $k < 0$.

37. AEROSPACE INDUSTRY The annual sales of the Boeing Company in billions of dollars for the years 1993–1998 can be modeled by the quadratic function

$$S(x) = 2.2x^2 - 7.7x + 39.9$$

where x is the number of years since 1993. Use the function to determine the annual sales for 1997.

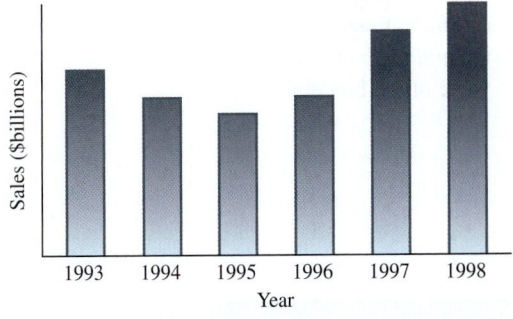

Source: The Boeing Company Annual Report, 1999 and 2002

Make a table of values to graph function f. Then use a series of translations to graph function g on the same coordinate system.

38. $f(x) = 2x^2$ $g(x) = 2x^2 - 3$

39. $f(x) = -4x^2$ $g(x) = -4(x - 2)^2 + 1$

40. Find the vertex and the axis of symmetry of the graph of $f(x) = -2(x - 1)^2 + 4$. Then plot several points and complete the graph.

2. $f(x) = a(x - h)^2$: translated right if $h > 0$, and left if $h < 0$.

The graph of $f(x) = a(x - h)^2 + k$ is a parabola with vertex at (h, k). It opens upward when $a > 0$ and downward when $a < 0$. The axis of symmetry is the line $x = h$.

The vertex of the graph of $f(x) = ax^2 + bx + c$ is

$$\left(-\frac{b}{2a}, f\left(-\frac{b}{2a}\right)\right)$$

and the axis of symmetry is the line

$$x = -\frac{b}{2a}$$

The *y-intercept* is determined by the value of $f(x)$ when $x = 0$: the *y*-intercept is $(0, c)$. To find the *x-intercepts*, let $f(x) = 0$ and solve $ax^2 + bx + c = 0$.

The *y*-coordinate of the vertex of the graph of a quadratic function gives the *minimum* or *maximum* value of the function.

41. Complete the square to write $f(x) = 4x^2 + 16x + 9$ in the form $f(x) = a(x - h)^2 + k$. Determine the vertex and the axis of symmetry of the graph. Then plot several points and complete the graph.

42. First determine the coordinates of the vertex and the axis of symmetry of the graph of $f(x) = x^2 + x - 2$ using the vertex formula. Then determine the *x*- and *y*-intercepts of the graph. Finally, plot several points and complete the graph.

43. FARMING   The number of farms in the United States for the years 1870–1970 is approximated by

$$N(x) = -1,526x^2 + 155,652x + 2,500,200$$

where $x = 0$ represents 1870, $x = 1$ represents 1871, and so on. For this period, when was the number of U.S. farms a maximum? How many farms were there?

44. Estimate the solutions of $-3x^2 - 5x + 2 = 0$ from the graph of $f(x) = -3x^2 - 5x + 2$, shown on the right.

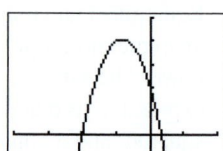

Quadratic and Other Nonlinear Inequalities

To solve a *quadratic inequality,* get 0 on one side and solve the related quadratic equation. Locate the *critical numbers* on a number line and test each interval. Finally, check the endpoints.

Solve each inequality. Write the solution set in interval notation and graph it.

45. $x^2 + 2x - 35 > 0$

46. $x^2 \leq 81$

47. $\dfrac{3}{x} \leq 5$

48. $\dfrac{2x^2 - x - 28}{x - 1} > 0$

To solve a *rational inequality,* get 0 on one side and solve the related rational equation. Locate the *critical numbers* (including any values that make the denominator 0) on a number line and test each interval. Finally, check the endpoints.

To graph a *nonlinear inequality in two variables,* first graph the boundary. Then use a test point to determine which half-plane to shade.

49. Estimate the solution set of $3x^2 + 10x - 8 \le 0$ from the graph of $f(x) = 3x^2 + 10x - 8$ shown in figure (a) below.

50. Estimate the solution set of $\dfrac{x-1}{x} > 0$ from the graph of $f(x) = \dfrac{x-1}{x}$ shown in figure (b).

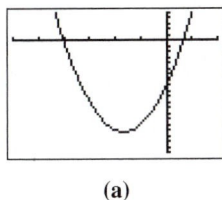   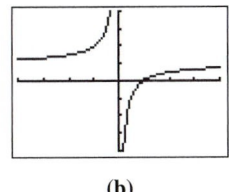

(a)   (b)

Graph each inequality.

51. $y < \dfrac{1}{2}x^2 - 1$

52. $y \ge -|x|$

CHAPTER 10 TEST   ELEMENTARY & INTERMEDIATE Algebra *f(x)* Now™

Solve each equation by factoring or using the square root property.

1. $3x^2 + 18x = 0$

2. $m^2 + 4 = 0$

3. $(a + 7)^2 = 50$

4. $x(6x + 19) = -15$

5. Determine what number must be added to $x^2 + 24x$ to make it a perfect square trinomial.

6. Solve $3x^2 + x - 24 = 0$ by completing the square.

Use the quadratic formula to solve each equation.

7. $2x^2 - 8x + 5 = 0$

8. $\dfrac{t^2}{8} - \dfrac{t}{4} = \dfrac{1}{2}$

9. $-t^2 + 4t - 13 = 0$

10. $0.01x^2 = -0.08x - 0.15$

Solve by any method.

11. $2y - 3\sqrt{y} + 1 = 0$

12. $m^{-2} + m^{-1} = -1$

13. $x^4 - x^2 - 12 = 0$

14. $4\left(\dfrac{x+2}{3x}\right)^2 - 4\left(\dfrac{x+2}{3x}\right) - 3 = 0$

15. Solve $E = mc^2$ for c. Assume that all variables represent positive numbers. Express any radical in simplified form.

16. Use the discriminant to determine the type of solutions for each equation.
   a. $3x^2 + 5x + 17 = 0$
   b. $9m^2 - 12m = -4$

17. TABLECLOTHS   In 1990, Sportex of Highland, Illinois, made what was at the time the world's longest tablecloth. Find the dimensions of the rectangular tablecloth if it covered an area of 6,759 square feet and its length was 8 feet more than 332 times its width.

18. COOKING   Working together, a chef and his assistant can make a pastry dessert in 25 minutes. When the chef makes it himself, it takes him 8 minutes less time than it takes his assistant working alone. How long does it take the chef to make the dessert?

19. DRAWING   An artist uses four equal-sized right triangles to block out a perspective drawing of an old hotel. For each triangle, one leg is 14 inches longer than the other, and the hypotenuse is 26 inches. On the centerline of the drawing, what is the length of the segment extending from the ground to the top of the building?

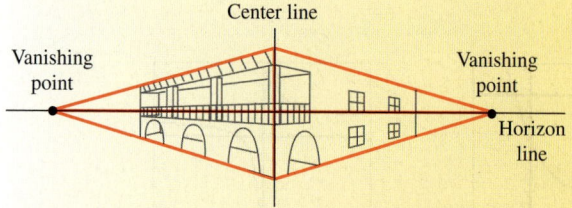

Center line

Vanishing point

Vanishing point

Horizon line

20. ANTHROPOLOGY   Anthropologists refer to the shape of the human jaw as a *parabolic dental arcade.* Which function is the best mathematical model of the parabola shown in the illustration?

   i. $f(x) = -\dfrac{3}{8}(x - 4)^2 + 6$

   ii. $f(x) = -\dfrac{3}{8}(x - 6)^2 + 4$

   iii. $f(x) = -\dfrac{3}{8}x^2 + 6$

   iv. $f(x) = \dfrac{3}{8}x^2 + 6$

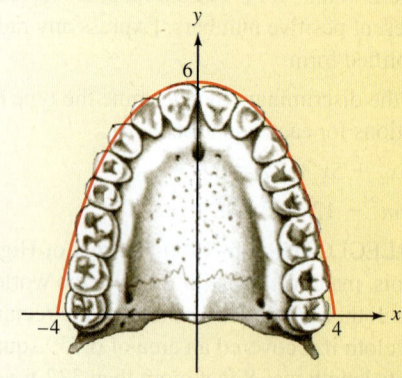

21. Find the vertex and the axis of symmetry of the graph of $f(x) = -3(x - 1)^2 - 2$. Then plot several points and complete the graph.

22. Complete the square to write the function $f(x) = 5x^2 + 10x - 1$ in the form $f(x) = a(x - h)^2 + k$. Determine the vertex and the axis of symmetry of the graph. Then plot several points and complete the graph.

23. First determine the coordinates of the vertex and the axis of symmetry of the graph of $f(x) = 2x^2 + x - 1$ using the vertex formula. Then determine the x- and y-intercepts of the graph. Finally, plot several points and complete the graph.

24. DISTRESS SIGNALS   A flare is fired directly upward into the air from a boat that is experiencing engine problems. The height of the flare (in feet) above the water, t seconds after being fired, is given by the formula $h = -16t^2 + 112t + 15$. If the flare is designed to explode when it reaches its highest point, at what height will this occur?

Solve each inequality. Write the solution set in interval notation and then graph it.

25. $x^2 - 2x > 8$

26. $\dfrac{x - 2}{x + 3} \le 0$

27. Explain the error.

   Solve: $\dfrac{12}{x} > 6$

$$12 > 6x$$
$$2 > x$$
$$x < 2$$
$$(-\infty, 2)$$

28. Graph: $y \le -x^2 + 3$.

29. The graph of a quadratic function of the form $f(x) = ax^2 + bx + c$ is shown. Estimate the solutions of the corresponding quadratic equation $ax^2 + bx + c = 0$.

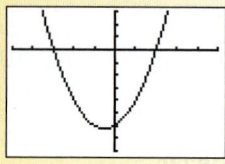

30. See Problem 29. Estimate the solution of the quadratic inequality $ax^2 + bx + c \le 0$.

CHAPTERS 1–10   CUMULATIVE REVIEW EXERCISES

Write an equation of the line with the given properties.

1. $m = 3$, passing through $(-2, -4)$

2. Parallel to the graph of $2x + 3y = 6$ and passing through $(0, -2)$

3. AIRPORT TRAFFIC   From the graph in the illustration, determine the projected average rate of change in the number of takeoffs and landings at Los Angeles International Airport for the years 2000–2015.

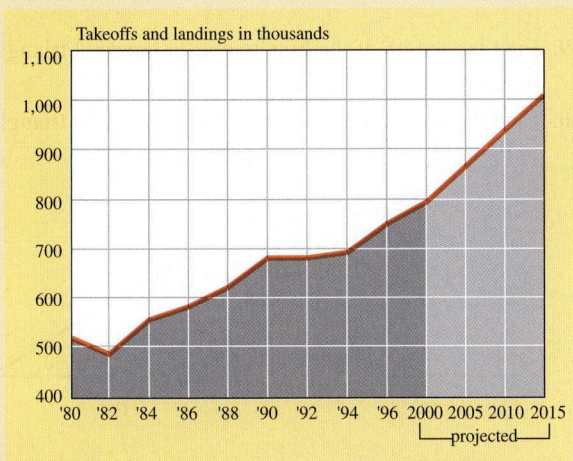

Based on data from *Los Angeles Times* (July 6, 1998) p. 83

4. Solve the system by graphing.

$$\begin{cases} y = -\dfrac{5}{2}x + \dfrac{1}{2} \\ 2x - \dfrac{3}{2}y = 5 \end{cases}$$

5. Solve the system using Cramer's rule.

$$\begin{cases} x - y + z = 4 \\ x + 2y - z = -1 \\ x + y - 3z = -2 \end{cases}$$

6. Graph the solution set of the system

$$\begin{cases} 3x + 2y > 6 \\ x + 3y \le 2 \end{cases}$$

Solve each inequality. Write the solution set in interval notation and then graph it.

7. $5(-2x + 2) > 20 - x$

8. $|2x - 5| \ge 25$

9. Simplify: $\left(\dfrac{-3x^4y^2}{-9x^5y^{-2}}\right)^2$.

10. TIDES   The illustration shows the graph of a function f, which gives the height of the tide for a 24-hour period in Seattle, Washington. (Note that military time is used on the x-axis: 3 A.M. = 3, noon = 12, 3 P.M. = 15, 9 P.M. = 21, and so on.)

  a. Find the domain of the function.

  b. Find $f(3)$.

  c. Find $f(6)$.

  d. Find $f(15)$.

  e. What information does $f(12)$ give?

  f. Estimate the values of x for which $f(x) = 0$.

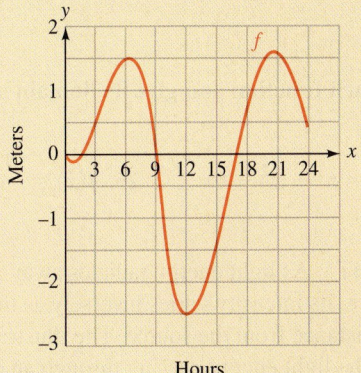

Perform the operations.

11. $(a + 2)(3a^2 + 4a - 2)$

12. $\dfrac{x^3 + y^3}{x^3 - y^3} \div \dfrac{x^2 - xy + y^2}{x^2 + xy + y^2}$

13. $\dfrac{1}{x + y} - \dfrac{1}{x - y} - \dfrac{2y}{y^2 - x^2}$

14. Simplify: $\dfrac{\dfrac{1}{r^2 + 4r + 4}}{\dfrac{r}{r + 2} + \dfrac{r}{r + 2}}$.

Factor each expression.

15. $x^4 - 16y^4$

16. $30a^4 - 4a^3 - 16a^2$

17. $x^2 + 4y - xy - 4x$

18. $8x^6 + 125y^3$

19. $x^2 + 10x + 25 - y^8$

20. $49s^6 - 84s^3n^2 + 36n^4$

Solve each equation.

21. $(m + 4)(2m + 3) - 22 = 10m$

22. $6a^3 - 2a = a^2$

23. $\dfrac{x - 4}{x - 3} + \dfrac{x - 2}{x - 3} = x - 3$

24. $P + \dfrac{a}{V^2} = \dfrac{RT}{V - b}$ solve for b

Graph each function and give its domain and range.

25. $f(x) = x^3 + x^2 - 6x$ **26.** $f(x) = \dfrac{4}{x}$ for $x > 0$

27. LIGHT   As light energy radiates away from its source, its intensity varies inversely as the square of the distance from the source. The illustration shows that the light energy passing through an area 1 foot from the source spreads out over 4 units of area 2 feet from the source. That energy is therefore less intense 2 feet from the source than it was 1 foot from the source. Over how many units of area will the light energy spread out 3 feet from the source?

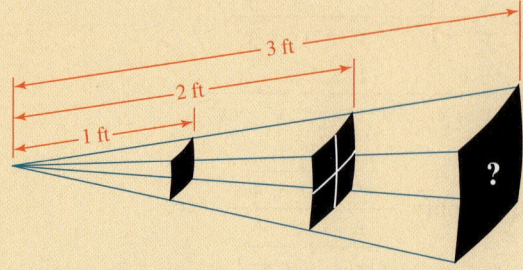

28. Graph the function $f(x) = \sqrt{x - 2}$ and give its domain and range.

Simplify each expression.

29. $\sqrt[3]{-27x^3}$ **30.** $\sqrt{48t^3}$

31. $64^{-2/3}$ **32.** $\dfrac{x^{5/3}x^{1/2}}{x^{3/4}}$

33. $-3\sqrt[4]{32} - 2\sqrt[4]{162} + 5\sqrt[4]{48}$

34. $3\sqrt{2}\left(2\sqrt{3} - 4\sqrt{12}\right)$

35. $\dfrac{\sqrt{x} + 2}{\sqrt{x} - 1}$ **36.** $\dfrac{5}{\sqrt[3]{x}}$

Solve each equation.

37. $5\sqrt{x + 2} = x + 8$ **38.** $\sqrt{x} + \sqrt{x + 2} = 2$

39. Find the length of the hypotenuse of the right triangle in figure (a).

40. Find the length of the hypotenuse of the right triangle in figure (b).

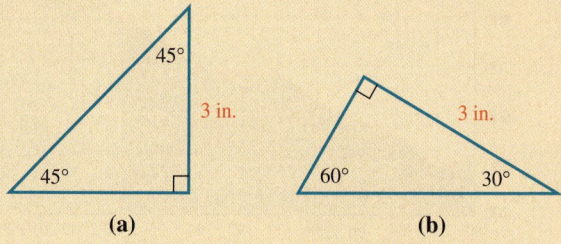

(a)   (b)

41. Find the distance between $(-2, 6)$ and $(4, 14)$.

42. Simplify: i^{43}.

Perform the operations. Write each result in $a + bi$ form.

43. $\left(-7 + \sqrt{-81}\right) - \left(-2 - \sqrt{-64}\right)$

44. $\dfrac{5}{3 - i}$

45. $(2 + i)^2$

46. $\dfrac{-4}{6i^7}$

47. What number must be added to $x^2 + 6x$ to make a perfect square trinomial?

48. Use the method of completing the square to solve $2x^2 + x - 3 = 0$.

49. Use the quadratic formula to solve $\dfrac{a^2}{8} - \dfrac{a}{2} = -1$.

50. COMMUNITY GARDENS   Residents of a community can work their own 16 ft × 24 ft plot of city-owned land if they agree to the following stipulations:

- The area of the garden cannot exceed 180 square feet.
- A path of uniform width must be maintained around the garden.

Find the dimensions of the largest possible garden.

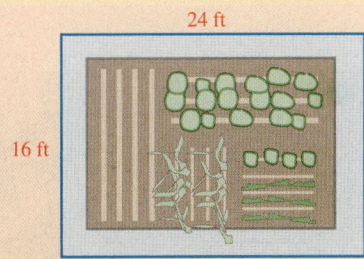

24 ft

16 ft

51. INSTALLING A SIDEWALK   A 170-meter-long sidewalk from the mathematics building M to the student center C is shown in red in the illustration. However, students prefer to walk directly from M to C. How long are the two segments of the existing sidewalk?

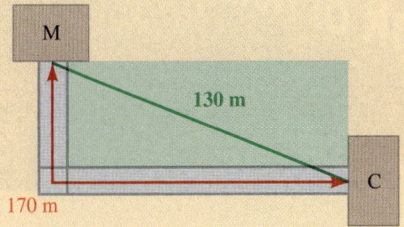

M

130 m

C

170 m

52. First determine the vertex and the axis of symmetry of the graph of $f(x) = -x^2 - 4x$ using the vertex formula. Then determine the x- and y-intercepts of the graph. Finally, plot several points and complete the graph.

Solve each equation.

53. $a - 7a^{1/2} + 12 = 0$

54. $x^{-4} - 2x^{-2} + 1 = 0$

55. The graph of $f(x) = 16x^2 + 24x + 9$ is shown below. Estimate the solution(s) of $16x^2 + 24x + 9 = 0$.

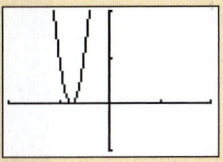

56. Use the graph above to determine the solution of $16x^2 + 24x + 9 < 0$.

Chapter

11

Exponential and Logarithmic Functions

ELEMENTARY &
INTERMEDIATE
Algebra ⓕ⁽ˣ⁾ **Now**™

Throughout the chapter, this icon introduces resources on the Elementary & Intermediate AlgebraNow Web site, accessed through **http://1pass.thomson.com**, that will

- Help you test your knowledge of the material with a pre-test and a post-test

- Provide a personalized learning plan targeting areas you should study

840

CORBIS

Financial planners advise us that we should begin saving for our retirement at a young age. However, it's difficult to make saving a budgeting priority with today's high cost of living. Fortunately, we don't have to choose between paying our current financial obligations and saving for retirement. Thanks to the power of compounding, a modest amount of money invested wisely can turn into a large sum over time. That is because compounding calculates interest on the principal and on the prior period's interest.

To learn more about compound interest, visit *The Learning Equation* on the Internet at http://tle.brookscole.com. (The log-in instructions are in the Preface.) For Chapter 11, the online lessons are:

- *TLE* Lesson 19: Exponential Functions
- *TLE* Lesson 20: Properties of Logarithms

In this chapter, we discuss the concept of function in more depth. We also introduce two new families of functions—exponential and logarithmic functions—which have applications in many areas.

11.1 Algebra and Composition of Functions

- Algebra of functions
- Composition of functions
- The identity function
- Writing composite functions

Just as it is possible to perform arithmetic operations on real numbers, it is also possible to perform those operations on functions. We call the process of adding, subtracting, multiplying, and dividing functions the *algebra of functions.*

■ ALGEBRA OF FUNCTIONS

The sum, difference, product, and quotient of two functions are themselves functions.

Operations on Functions

If the domains and ranges of functions f and g are subsets of the real numbers, then

The **sum** of f and g, denoted as $\boldsymbol{f + g}$, *is defined by*

$$(f + g)(x) = f(x) + g(x)$$

The **difference** of f and g, denoted as $\boldsymbol{f - g}$, is defined by

$$(f - g)(x) = f(x) - g(x)$$

The **product** of f and g, denoted as $\boldsymbol{f \cdot g}$, is defined by

$$(f \cdot g)(x) = f(x)g(x)$$

The **quotient** of f and g, denoted as $\boldsymbol{f/g}$, is defined by

$$(f/g)(x) = \frac{f(x)}{g(x)} \text{ where } g(x) \neq 0$$

The domain of each of these functions is the set of real numbers x that are in the domain of both f and g. In the case of the quotient, there is the further restriction that $g(x) \neq 0$.

EXAMPLE 1

ELEMENTARY & INTERMEDIATE

Let $f(x) = 2x^2 + 1$ and $g(x) = 5x - 3$. Find each function and its domain: **a.** $f + g$ and **b.** $f - g$.

Solution

a. $(f + g)(x) = \boldsymbol{f(x)} + \boldsymbol{g(x)}$

$\qquad\qquad = (\boldsymbol{2x^2 + 1}) + (\boldsymbol{5x - 3})$ Replace $f(x)$ with $2x^2 + 1$ and $g(x)$ with $5x - 3$.

$\qquad\qquad = 2x^2 + 5x - 2$ Combine like terms.

The domain of $f + g$ is the set of real numbers that are in the domain of both f and g. Since the domain of both f and g is the interval $(-\infty, \infty)$, the domain of $f + g$ is also the interval $(-\infty, \infty)$.

b. $\quad (f - g)(x) = f(x) - g(x)$
$$= (2x^2 + 1) - (5x - 3)$$
$$= 2x^2 + 1 - 5x + 3 \qquad \text{Remove parentheses.}$$
$$= 2x^2 - 5x + 4 \qquad \text{Combine like terms.}$$

Since the domain of both f and g is $(-\infty, \infty)$, the domain of $f - g$ is also the interval $(-\infty, \infty)$.

Self Check 1   Let $f(x) = 3x - 2$ and $g(x) = 2x^2 + 3x$. Find **a.** $f + g$ and   **b.** $f - g$.

EXAMPLE 2

ELEMENTARY &
INTERMEDIATE
Algebra (f(x)) **Now**™

Let $f(x) = 2x^2 + 1$ and $g(x) = 5x - 3$. Find each function and its domain: **a.** $f \cdot g$ and   **b.** f/g.

Solution   **a.** $\quad (f \cdot g)(x) = f(x)g(x) \qquad$ Replace $f(x)$ with $2x^2 + 1$ and $g(x)$ with $5x - 3$.
$$= (2x^2 + 1)(5x - 3)$$
$$= 10x^3 - 6x^2 + 5x - 3 \qquad \text{Use the FOIL method.}$$

The domain of $f \cdot g$ is the set of real numbers that are in the domain of both f and g. Since the domain of both f and g is the interval $(-\infty, \infty)$, the domain of $f \cdot g$ is also the interval $(-\infty, \infty)$.

b. $\quad (f/g)(x) = \dfrac{f(x)}{g(x)}$

$$= \dfrac{2x^2 + 1}{5x - 3}$$

Since the denominator of the fraction cannot be 0, $x \neq \dfrac{3}{5}$. Thus, the domain of f/g is the interval $\left(-\infty, \frac{3}{5}\right) \cup \left(\frac{3}{5}, \infty\right)$.

Self Check 2   Let $f(x) = 2x^2 - 3$ and $g(x) = x^2 + 1$. Find **a.** $f \cdot g$ and   **b.** f/g.

■ COMPOSITION OF FUNCTIONS

We have seen that a function can be represented by a machine: We put in a number from the domain, and a number from the range comes out. For example, if we put the number 2 into the machine shown in figure (a) on the next page, the number $f(2) = 8$ comes out. In general, if we put x into the machine shown in figure (b), the value $f(x)$ comes out.

Often one quantity is a function of a second quantity that depends, in turn, on a third quantity. For example, the cost of a car trip is a function of the gasoline consumed. The amount of gasoline consumed, in turn, is a function of the number of miles driven. Such chains of dependence can be analyzed mathematically as **compositions of functions.**

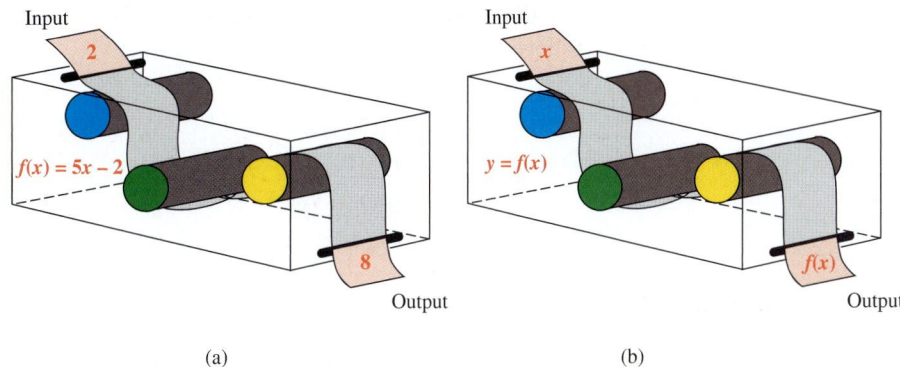

(a)   (b)

Suppose that $y = f(x)$ and $y = g(x)$ define two functions. Any number x in the domain of g will produce the corresponding value $g(x)$ in the range of g. If $g(x)$ is in the domain of function f, then $g(x)$ can be substituted into f, and a corresponding value $f(g(x))$ will be determined. This two-step process defines a new function, called a **composite function,** denoted by $f \circ g$. (This is read as "f composed with g.")

The function machines shown below illustrate the composition $f \circ g$. When we put a number x into the function g, $g(x)$ comes out. The value $g(x)$ goes into function f, which transforms $g(x)$ into $f(g(x))$. (This is read as "f of g of x.") If the function machines for g and f were connected to make a single machine, that machine would be named $f \circ g$.

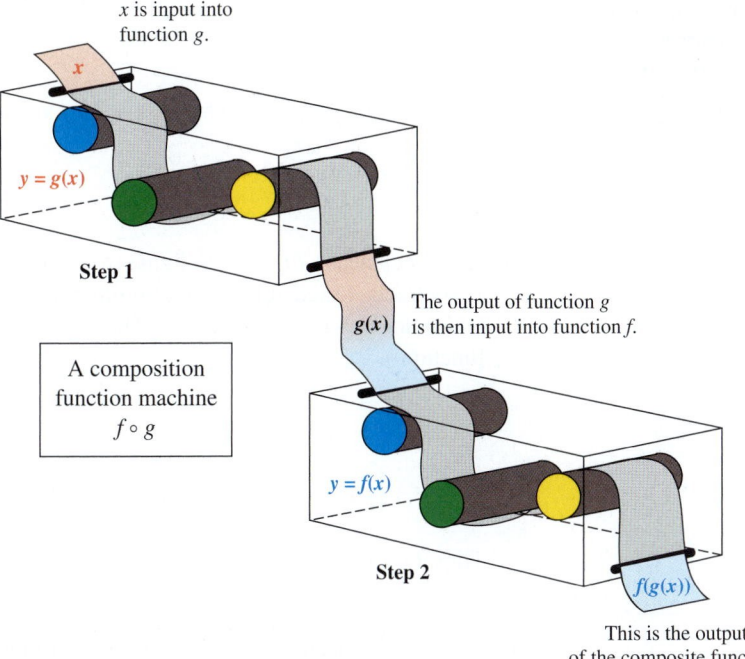

To be in the domain of the composite function $f \circ g$, a number x has to be in the domain of g and the output of g must be in the domain of f. Thus, the domain of $f \circ g$ consists of those numbers x that are in the domain of g, and for which $g(x)$ is in the domain of f.

Composite Functions   The **composite function** $f \circ g$ is defined by

$$(f \circ g)(x) = f(g(x))$$

For example, if $f(x) = 4x$ and $g(x) = 3x + 2$, to find $f \circ g$ and $g \circ f$, we proceed as follows.

$$(f \circ g)(x) = f(g(x)) \qquad\qquad (g \circ f)(x) = g(f(x))$$
$$= f(3x + 2) \qquad\qquad\qquad = g(4x)$$
$$= 4(3x + 2) \qquad\qquad\qquad = 3(4x) + 2$$
$$= 12x + 8 \qquad\qquad\qquad = 12x + 2$$

——————— Different results ———————

The different results illustrate that the composition of functions is not commutative: $(f \circ g)(x) \neq (g \circ f)(x)$.

EXAMPLE 3

ELEMENTARY &
INTERMEDIATE
Algebra $f(x)$ **Now**™

Let $f(x) = 2x + 1$ and $g(x) = x - 4$. Find **a.** $(f \circ g)(9)$, **b.** $(f \circ g)(x)$, and **c.** $(g \circ f)(-2)$.

Solution

a. $(f \circ g)(9)$ means $f(g(9))$. In figure (a) on the next page, function g receives the number 9, subtracts 4, and releases the number $g(9) = 5$. Then 5 goes into the f function, which doubles 5 and adds 1. The final result, 11, is the output of the composite function $f \circ g$:

Read as "f of g of 9."
↓
$$(f \circ g)(9) = f(g(9)) = f(5) = 2(5) + 1 = 11$$

Thus, $(f \circ g)(9) = 11$.

Notation

The notation $f \circ g$ can also be read as "f circle g." Remember, it means that the function g is applied first and function f is applied second.

b. $(f \circ g)(x)$ means $f(g(x))$. In figure (a) on the next page, function g receives the number x, subtracts 4, and releases the number $x - 4$. Then $x - 4$ goes into the f function, which doubles $x - 4$ and adds 1. The final result, $2x - 7$, is the output of the composite function $f \circ g$.

Read as "f of g of x."
↓
$$(f \circ g)(x) = f(g(x)) = f(x - 4) = 2(x - 4) + 1 = 2x - 7$$

Thus, $(f \circ g)(x) = 2x - 7$.

c. $(g \circ f)(-2)$ means $g(f(-2))$. In figure (b) on the next page, function f receives the number -2, doubles it and adds 1, and releases -3 into the g function. Function g subtracts 4 from -3 and outputs a final result of -7. Thus,

Read as "g of f of -2."
↓
$$(g \circ f)(-2) = g(f(-2)) = g(-3) = -3 - 4 = -7$$

Thus, $(g \circ f)(-2) = -7$.

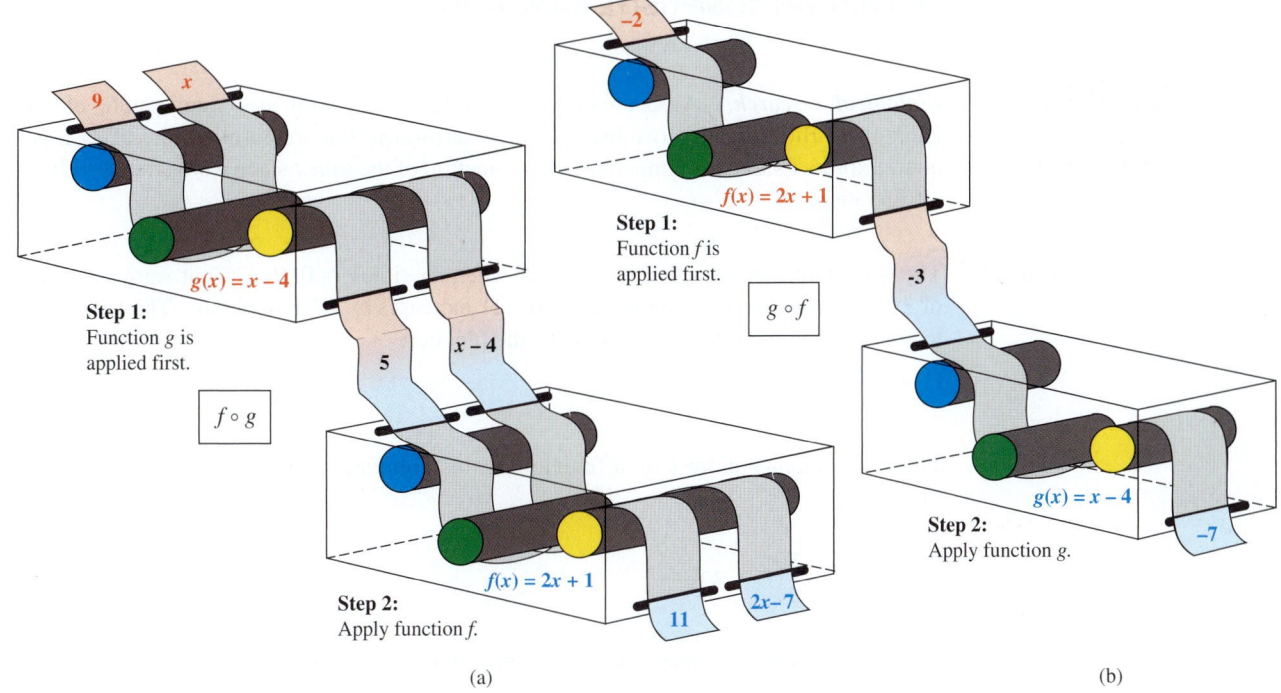

(a)   (b)

Self Check 3   Let $f(x) = x^3$ and $g(x) = 6 - x$. Find **a.** $(f \circ g)(8)$,   **b.** $(g \circ f)(1)$,   and
c. $(g \circ f)(x)$.

■ THE IDENTITY FUNCTION

The **identity function** is defined by the equation $I(x) = x$. Under this function, the value
that is assigned to any real number x is x itself. For example $I(2) = 2$, $I(-3) = -3$, and
$I(7.5) = 7.5$. If f is any function, the composition of f with the identity function is just the
function f:

$$(f \circ I)(x) = (I \circ f)(x) = f(x)$$

EXAMPLE 4

Let f be any function and let I be the identity function, $I(x) = x$. Show that
a. $(f \circ I)(x) = f(x)$ and   **b.** $(I \circ f)(x) = f(x)$.

Solution   **a.** $(f \circ I)(x)$ means $f(I(x))$. Because $I(x) = x$, we have

$$(f \circ I)(x) = f(I(x)) = f(x)$$

The Language of Algebra

The identity function pairs
each real number with itself
such that each output is
identical to its corresponding
input.

b. $(I \circ f)(x)$ means $I(f(x))$. Because I passes any number through unchanged, we have

$$(I \circ f)(x) = I(f(x)) = f(x)$$

■ WRITING COMPOSITE FUNCTIONS

EXAMPLE 5

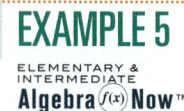
ELEMENTARY &
INTERMEDIATE
Algebra (f(x)) **Now**™

Biological research. A specimen is stored in refrigeration at a temperature of 15° Fahrenheit. Biologists remove the specimen and warm it at a controlled rate of 3° F per hour. Express its Celsius temperature as a function of the time *t* since it was removed from refrigeration.

Solution   The temperature of the specimen is 15° F when the time $t = 0$. Because it warms at a rate of 3° F per hour, its initial temperature of 15° increases by $3t°$ F in *t* hours. The Fahrenheit temperature of the specimen is given by the function

$$F(t) = 3t + 15$$

The Celsius temperature *C* is a function of this Fahrenheit temperature *F*, given by the function

$$C(F) = \frac{5}{9}(F - 32)$$

To express the specimen's Celsius temperature as a function of *time,* we find the composite function

$$
\begin{aligned}
(C \circ F)(t) &= C(F(t)) \\
&= C(3t + 15) \qquad &\text{Substitute } 3t + 15 \text{ for } F(t). \\
&= \frac{5}{9}[(3t + 15) - 32] \qquad &\text{Substitute } 3t + 15 \text{ for } F \text{ in } \frac{5}{9}(F - 32). \\
&= \frac{5}{9}(3t - 17) \qquad &\text{Simplify.} \\
&= \frac{15}{9}t + \frac{85}{9} \\
&= \frac{5}{3}t + \frac{85}{9}
\end{aligned}
$$

The composite function, $C(t) = \frac{5}{3}t + \frac{85}{9}$, gives the temperature of the specimen in degrees Celsius *t* hours after it is removed from refrigeration.

Answers to Self Checks   **1. a.** $(f + g)(x) = 2x^2 + 6x - 2$,   **b.** $(f - g)(x) = -2x^2 - 2$ **2. a.** $(f \cdot g)(x) = 2x^4 - x^2 - 3$,

b. $(f/g)(x) = \dfrac{2x^2 - 3}{}$ **3. a.** -8,   **b.** 5,   **c.** $(g \circ f)(x) = 6 - x^3$

11.1   STUDY SET   ELEMENTARY & INTERMEDIATE **Algebra** (f(x)) **Now**™

VOCABULARY   **Fill in the blanks.**

1. The _____ of *f* and *g*, denoted as $f + g$, is defined by $(f + g)(x) = $ _____ and the _____ of *f* and *g*, denoted as $f - g$, is defined by $(f - g)(x) = $ _____.

2. The _____ of *f* and *g*, denoted as $f \cdot g$, is defined by $(f \cdot g)(x) = $ _____ and the _____ of *f* and *g*, denoted as f/g, is defined by $(f/g)(x) = $ _____.

3. The _____ of the function $f + g$ is the set of real numbers x that are in the domain of both f and g.

4. The _____ function $f \circ g$ is defined by $(f \circ g)(x) = $ ▢ .

5. Under the _____ function, the value that is assigned to any real number x is x itself.

6. When reading the notation $f(g(x))$, we say "f ___ g ___ x."

CONCEPTS

7. Fill in the blanks to make the statements true.

a. $(f \circ g)(3) = f($ ▢ $)$

b. To find $f(g(3))$, we first find ▢ and then substitute that value for x in $f(x)$.

8. a. If $f(x) = 3x + 1$ and $g(x) = 1 - 2x$, find $f(g(3))$ and $g(f(3))$.

b. Is the composition of functions commutative?

9. Fill in the blanks in the drawing of the function machines that show how to compute $g(f(-2))$.

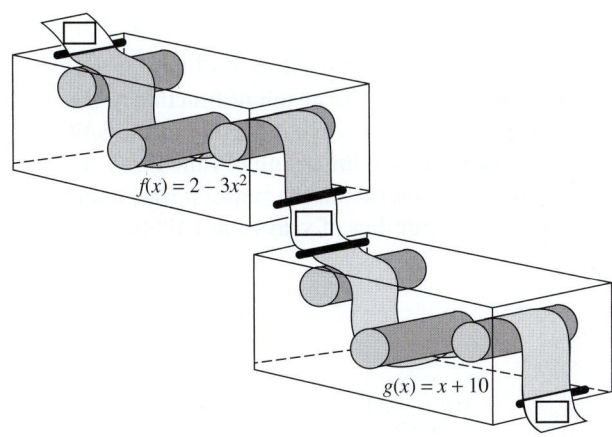

$f(x) = 2 - 3x^2$

$g(x) = x + 10$

10. Complete the table of values for the identity function, $I(x) = x$. Then graph it.

x	$I(x)$
-3	
-2	
-1	
0	
1	
2	
3	

11. Use the table of values for functions f and g to find each of the following.

a. $(f + g)(1)$ **b.** $(f - g)(5)$

c. $(f \cdot g)(1)$ **d.** $(g/f)(5)$

x	$f(x)$
1	3
5	8

x	$g(x)$
1	4
5	0

12. Use the table of values for functions f and g to find each of the following.

a. $(f \circ g)(1)$ **b.** $(g \circ f)(2)$

x	$f(x)$
2	5
4	7

x	$g(x)$
1	2
5	-3

NOTATION   Complete each solution.

13. Let $f(x) = 3x - 1$ and $g(x) = 2x + 3$. Find $f \cdot g$.

$$(f \cdot g)(x) = f(x) \cdot \text{▢}$$
$$= \text{▢} (2x + 3)$$
$$= 6x^2 + \text{▢} - \text{▢} - 3$$
$$(f \cdot g)(x) = 6x^2 + 7x - 3$$

14. Let $f(x) = 3x - 1$ and $g(x) = 2x + 3$. Find $f \circ g$.

$$(f \circ g)(x) = f(\text{▢})$$
$$= f(\text{▢})$$
$$= 3(\text{▢}) - 1$$
$$= \text{▢} + \text{▢} - 1$$
$$(f \circ g)(x) = 6x + 8$$

PRACTICE   Let $f(x) = 3x$ and $g(x) = 4x$. Find each function and its domain.

15. $f + g$ **16.** $f - g$

17. $f \cdot g$ **18.** f/g

19. $g - f$ **20.** $g + f$

21. g/f **22.** $g \cdot f$

Let $f(x) = 2x + 1$ and $g(x) = x - 3$. Find each function and its domain.

23. $f + g$ **24.** $f - g$

25. $f \cdot g$ **26.** f/g

27. $g - f$ **28.** $g + f$

29. g/f

30. $g \cdot f$

Let $f(x) = 3x - 2$ and $g(x) = 2x^2 + 1$. Find each function and its domain.

31. $f - g$ **32.** $f + g$

33. f/g **34.** $f \cdot g$

Let $f(x) = x^2 - 1$ and $g(x) = x^2 - 4$. Find each function and its domain.

35. $f - g$ **36.** $f + g$

37. g/f

38. $g \cdot f$

Let $f(x) = 2x + 1$ and $g(x) = x^2 - 1$. Find each composition.

39. $(f \circ g)(2)$ **40.** $(g \circ f)(2)$

41. $(g \circ f)(-3)$ **42.** $(f \circ g)(-3)$

43. $(f \circ g)(0)$ **44.** $(g \circ f)(0)$

45. $(f \circ g)\left(\dfrac{1}{2}\right)$ **46.** $(g \circ f)\left(\dfrac{1}{3}\right)$

47. $(f \circ g)(x)$ **48.** $(g \circ f)(x)$

49. $(g \circ f)(2x)$ **50.** $(f \circ g)(2x)$

Let $f(x) = 3x - 2$ and $g(x) = x^2 + x$. Find each composition.

51. $(f \circ g)(4)$ **52.** $(g \circ f)(4)$
53. $(g \circ f)(-3)$ **54.** $(f \circ g)(-3)$
55. $(g \circ f)(0)$ **56.** $(f \circ g)(0)$
57. $(g \circ f)(x)$ **58.** $(f \circ g)(x)$

Let $f(x) = \dfrac{1}{x}$ and $g(x) = \dfrac{1}{x^2}$. Find each composition.

59. $(f \circ g)(4)$ **60.** $(f \circ g)(6)$

61. $(g \circ f)\left(\dfrac{1}{3}\right)$ **62.** $(g \circ f)\left(\dfrac{1}{10}\right)$

63. $(g \circ f)(8x)$ **64.** $(f \circ g)(5x)$

65. If $f(x) = x + 1$ and $g(x) = 2x - 5$, show that $(f \circ g)(x) \neq (g \circ f)(x)$.

66. If $f(x) = x^2 + 1$ and $g(x) = 3x^2 - 2$, show that $(f \circ g)(x) \neq (g \circ f)(x)$.

APPLICATIONS   In Exercises 67–70, refer to the following illustration. The graph of function f gives the average score on the verbal portion of the SAT college entrance exam, the graph of function g gives the average score on the mathematics portion, and x represents the number of years since 1990.

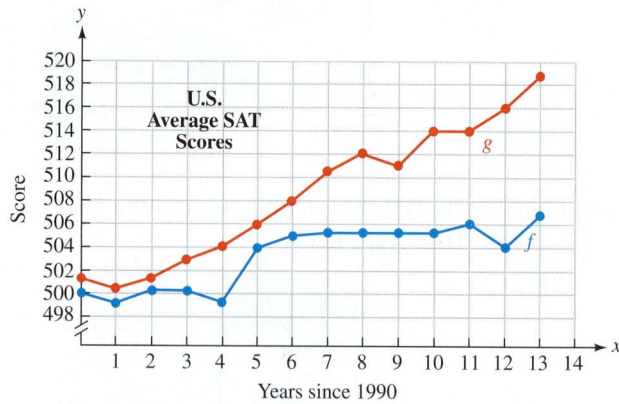

Source: *The World Almanac*, 1999 and 2004

67. From the graph, determine $f(3)$ and $g(3)$. Then use those results to find $(f + g)(3)$.

68. From the graph, determine $f(6)$ and $g(6)$. Then use those results to find $(g - f)(6)$.

69. Find $(f + g)(10)$ and explain what information about SAT scores it gives.

70. Find $(g - f)(12)$ and explain what information about SAT scores it gives.

71. METALLURGY   A molten alloy must be cooled slowly to control crystallization. When removed from the furnace, its temperature is 2,700° F, and it will be cooled at 200° per hour. Express the Celsius temperature as a function of the number of hours t since cooling began.

72. WEATHER FORECASTING   A high-pressure area promises increasingly warmer weather for the next 48 hours. The temperature is now 34° Celsius and is expected to rise 1° every 6 hours. Express the Fahrenheit temperature as a function of the number of hours from now. $\left(Hint: F = \frac{9}{5}C + 32.\right)$

73. VACATION MILEAGE COSTS

   a. Use the following graphs to determine the cost of the gasoline consumed if a family drove 500 miles on a summer vacation.

   b. Write a composition function that expresses the cost of the gasoline consumed on the vacation as a function of the miles driven.

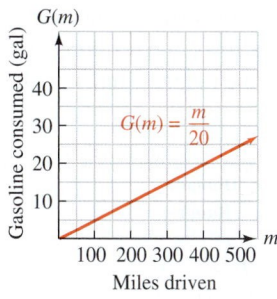

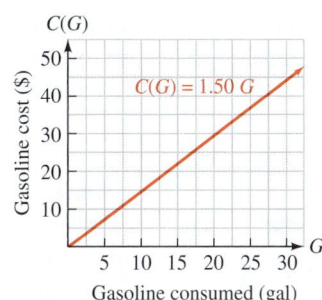

74. HALLOWEEN COSTUMES   The tables on the back of a pattern package (see the next column) can be used to determine the number of yards of material needed to make a rabbit costume for a child.

   a. How many yards of material are needed if the child's chest measures 29 inches?

   b. In this exercise, one quantity is a function of a second quantity that depends, in turn, on a third quantity. Explain this dependence.

PATTERN 9810   **Simplicity**

Costumes have front zipper, long raglan sleeves, elastic sleeve and leg casings, hood. Fabrics: Fleece or suede

BODY MEASUREMENTS

Chest (in.)	21	22	23	25	26	27	$28\frac{1}{2}$	29	30
Pattern Size	2	3	4	6	7	8	10	11	12

YARDAGE NEEDED

Pattern Size	2-4	6-8	10-12
Yards	$2\frac{5}{8}$	$3\frac{3}{8}$	$3\frac{3}{4}$

WRITING

75. Exercise 73 illustrates a chain of dependence between the cost of the gasoline, the gasoline consumed, and the miles driven. Describe another chain of dependence that could be represented by a composition function.

76. In this section, what operations are performed on functions? Give an example of each.

77. Write out in words how to say each of the following:
   $(f \circ g)(2)$ $g(f(-8))$

78. If $Y_1 = f(x)$ and $Y_2 = g(x)$, explain how to use the following tables to find $g(f(2))$.

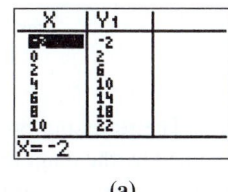

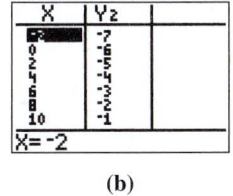

   (a)   (b)

REVIEW   **Simplify each complex fraction.**

79. $\dfrac{\dfrac{ac - ad - c + d}{a^3 - 1}}{\dfrac{c^2 - 2cd + d^2}{a^2 + a + 1}}$

80. $\dfrac{2 + \dfrac{1}{x^2 - 1}}{1 + \dfrac{1}{x - 1}}$

CHALLENGE PROBLEMS   **Fill in the blanks.**

81. If $f(x) = x^2$ and $g(x) = $ _____ , then $(f \circ g)(x) = 4x^2 + 20x + 25$.

82. If $f(x) = \sqrt{3x}$ and $g(x) = $ _____ , then $(g \circ f)(x) = 9x^2 + 7$.

11.2   Inverse Functions

- The inverse of a function
- One-to-one functions
- The horizontal line test
- Finding the inverse of a function
- The composition of a function and its inverse
- Graphing a function and its inverse

In the previous section, we used the operations of arithmetic and composition to create new functions from given functions. Another way that a new function can be created from a given function is to find its *inverse.*

■ THE INVERSE OF A FUNCTION

In figure (a) below, the arrow diagram defines a function f. If we reverse the arrows, as shown in figure (b), the domain of f becomes the range, and the range of f becomes the domain, of a new correspondence. The new correspondence is a function, because it assigns to each member of the domain exactly one member of the range. We call this new correspondence the **inverse** of f, or f inverse.

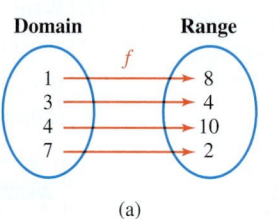
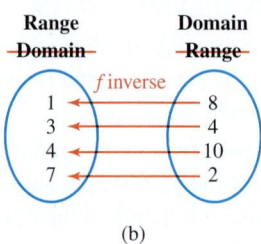

The reversing process does not always produce a function. Consider the function g defined by the diagram in figure (a) below. When we reverse the arrows, the resulting correspondence is not a function, because it assigns two members of the range, 8 and 4, to the number 2.

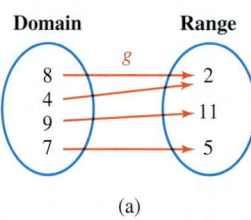
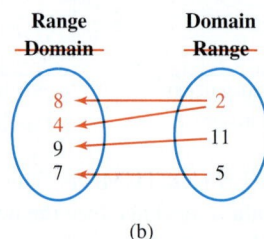

The question that arises is, "What must be true of the original function to guarantee that the reversing process produces a function?" The answer to that question is: *the original function must be one-to-one.*

■ ONE-TO-ONE FUNCTIONS

We have seen that a function assigns to each input exactly one output. For some functions, different inputs are assigned different outputs, as shown in figure (a) on the next page. For other functions, different inputs are assigned the *same* output, as in figure (b). When each output corresponds to exactly one input, as in figure (a), we say the function is *one-to-one*.

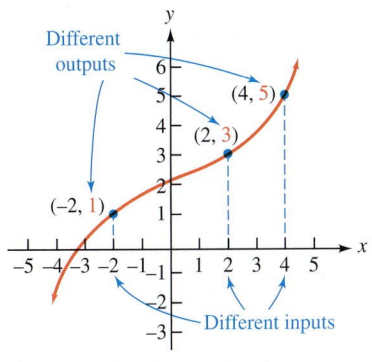

A one-to-one function

(a)

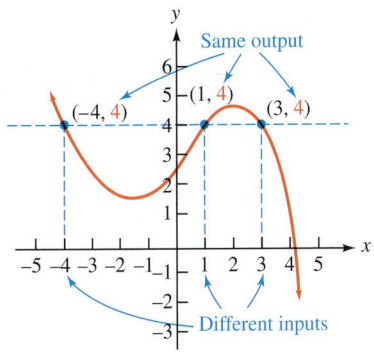

Not a one-to-one function

(b)

One-to-One Functions	For a **one-to-one function,** each input is assigned exactly one output, and each output corresponds to exactly one input.

EXAMPLE 1

Solution

Determine whether each function is one-to-one: **a.** $f(x) = x^2$ and **b.** $f(x) = x^3$.

a. Since the output 9 corresponds to two different inputs, -3 and 3, $f(x) = x^2$ is not one-to-one.

$$f(-3) = (-3)^2 = 9 \quad \text{and} \quad f(3) = 3^2 = 9$$

x	$f(x)$
-3	9
3	9

The output 9 does not correspond to exactly one input.

Success Tip

Example 1 illustrates that not every function is one-to-one.

b. Since different numbers have different cubes, each output of $f(x) = x^3$ corresponds to exactly one input. The function is one-to-one.

Self Check 1 Determine whether each function is one-to-one. If it is not, find an output that corresponds to more than one input: **a.** $f(x) = 2x + 3$ and **b.** $f(x) = x^4$.

■ THE HORIZONTAL LINE TEST

It is often easier to examine the graph of a function, rather than its defining equation, to determine whether the function is one-to-one. If two (or more) points on the graph of a function have the same y-coordinate, the function is not one-to-one. This observation suggests the following *horizontal line test*.

The Horizontal Line Test	A function is one-to-one if every horizontal line intersects the graph of the function at most once.

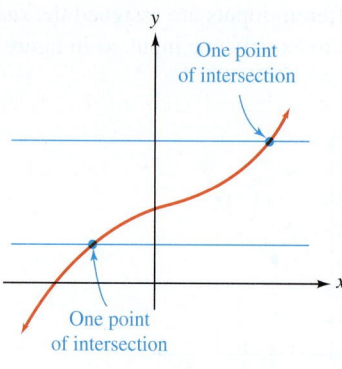

One point of intersection

One point of intersection

A one-to-one function

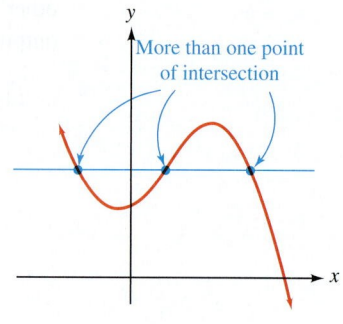

More than one point of intersection

Not a one-to-one function

EXAMPLE 2

ELEMENTARY & INTERMEDIATE

Algebra $f(x)$ Now™

Use the horizontal line test to decide whether the following graphs represent one-to-one functions.

Solution

a. Because we can draw a horizontal line that intersects the graph shown in figure (a) twice, the graph does not represent a one-to-one function.

b. Because every horizontal line that intersects the graph in figure (b) does so exactly once, the graph represents a one-to-one function.

> **Success Tip**
>
> Recall that we use the *vertical line test* to determine whether a graph is the graph of a function. We use the *horizontal line test* to determine whether the function that is graphed is one-to-one.

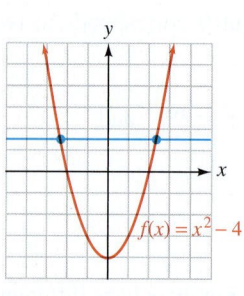

$f(x) = x^2 - 4$

(a)

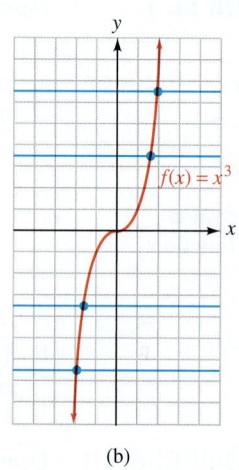

$f(x) = x^3$

(b)

Self Check 2 Determine whether the following graphs represent one-to-one functions.

a.

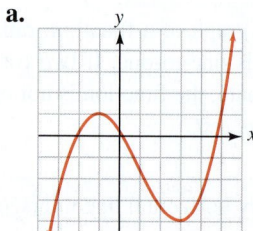

b.

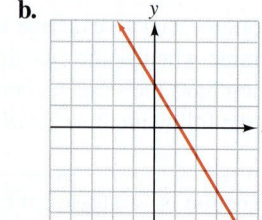

■ FINDING THE INVERSE OF A FUNCTION

If f is the one-to-one function defined by the arrow diagram in figure (a), it turns the number 1 into 10, 2 into 20, and 3 into 30. The ordered pairs that determine f can be listed in a table. Since the inverse of f must turn 10 back into 1, 20 back into 2, and 30 back into 3, it consists of the ordered pairs shown in the table in figure (b).

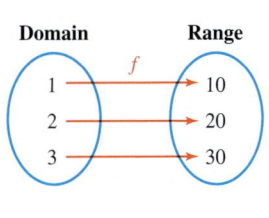

Function f

x	y	(x, y)
1	10	(1, 10)
2	20	(2, 20)
3	30	(3, 30)

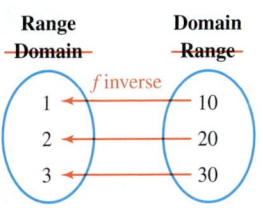

f inverse

x	y	(x, y)
10	1	(10, 1)
20	2	(20, 2)
30	3	(30, 3)

The x- and y-coordinates are interchanged.

(a)   (b)

Caution

The -1 in the notation $f^{-1}(x)$ is not an exponent:

$$f^{-1}(x) \neq \frac{1}{f(x)}$$

We note that the domain of f and the range of its inverse is $\{1, 2, 3\}$. The range of f and the domain of its inverse is $\{10, 20, 30\}$.

This example suggests that to form the inverse of a function f, we simply interchange the coordinates of each ordered pair that determines f. When the inverse of a function is also a function, we call it f **inverse** and denote it with the symbol f^{-1}. The symbol $f^{-1}(x)$ is read as "the inverse of $f(x)$" or "f inverse of x."

The Inverse of a Function   If f is a one-to-one function consisting of ordered pairs of the form (x, y), the **inverse** of f, denoted f^{-1}, is the one-to-one function consisting of all ordered pairs of the form (y, x).

When a one-to-one function is defined by an equation, we use the following method to find its inverse.

Finding the Inverse of a Function   If a function is one-to-one, we find its inverse as follows:

1. If the function is written using function notation, replace $f(x)$ with y.

2. Interchange the variables x and y.

3. Solve the resulting equation for y.

4. We can substitute $f^{-1}(x)$ for y.

EXAMPLE 3

ELEMENTARY & INTERMEDIATE
Algebra *f(x)* **Now**™

Determine whether each function is one-to-one. If it is, find the equation of its inverse:
a. $f(x) = 4x + 2$ and   **b.** $f(x) = x^3$.

Solution   **a.** We recognize $f(x) = 4x + 2$ as a linear function whose graph is a straight line with slope 4 and y-intercept $(0, 2)$. Since such a graph would pass the horizontal line test, we conclude that f is one-to-one.

To find the inverse, we proceed as follows:

Success Tip

Every linear function,
except those of the form
$f(x)$ = constant, is one-to-one.

$$f(x) = 4x + 2$$

$$y = 4x + 2 \qquad \text{Replace } f(x) \text{ with } y.$$

$$x = 4y + 2 \qquad \text{Interchange the variables } x \text{ and } y.$$

$$x - 2 = 4y \qquad \text{Subtract 2 from both sides.}$$

$$\frac{x - 2}{4} = y \qquad \text{Divide both sides by 4.}$$

$$y = \frac{x - 2}{4} \qquad \text{Write the equation with } y \text{ on the left-hand side.}$$

To denote that this equation is the inverse of function f, we replace y with $f^{-1}(x)$.

$$f^{-1}(x) = \frac{x - 2}{4}$$

b. In Example 2, we used the horizontal line test to determine that $f(x) = x^3$ is a one-to-one function.

To find the inverse, we proceed as follows:

Caution

Only one-to-one functions
have inverse functions.

$$y = x^3 \qquad \text{Replace } f(x) \text{ with } y.$$

$$x = y^3 \qquad \text{Interchange the variables } x \text{ and } y.$$

$$\sqrt[3]{x} = y \qquad \text{Take the cube root of both sides.}$$

$$y = \sqrt[3]{x}$$

Replacing y with $f^{-1}(x)$, we have

$$f^{-1}(x) = \sqrt[3]{x}$$

Self Check 3   Determine whether each function is one-to-one. If it is, find the equation of its inverse:
a. $f(x) = -5x - 3$ and   **b.** $f(x) = x^5$.

■ THE COMPOSITION OF A FUNCTION AND ITS INVERSE

To emphasize an important relationship between a function and its inverse, we substitute some number x, such as $x = 3$, into the function $f(x) = 4x + 2$ of Example 3. The corresponding value of y that is produced is

$$f(3) = 4(3) + 2 = 14 \qquad f \text{ determines the ordered pair (3, 14).}$$

If we substitute 14 into the inverse function, $f^{-1}(x) = \frac{x - 2}{4}$, the corresponding value of y that is produced is

$$f^{-1}(14) = \frac{14 - 2}{4} = 3 \qquad f^{-1} \text{ determines the ordered pair (14, 3).}$$

Thus, the function f turns 3 into 14, and the inverse function f^{-1} turns 14 back into 3.

In general, the composition of a function and its inverse function is the identity function such that any input x is assigned the output x. This fact can be stated symbolically as follows.

The Composition of Inverse Functions	For any one-to-one function f and its inverse, f^{-1}, $$(f \circ f^{-1})(x) = x \quad \text{and} \quad (f^{-1} \circ f)(x) = x$$

We can use this property to determine whether two functions are inverses.

EXAMPLE 4

ELEMENTARY & INTERMEDIATE
Algebra ⨍(x) Now™

Prove that $f(x) = 4x + 2$ and $f^{-1}(x) = \dfrac{x - 2}{4}$ are inverses.

Solution   To prove that $f(x) = 4x + 2$ and $f^{-1}(x) = \dfrac{x - 2}{4}$ are inverses, we must show that for each composition, an input x is assigned an output of x.

$$(f \circ f^{-1})(x) = f(f^{-1}(x)) \qquad\qquad (f^{-1} \circ f)(x) = f^{-1}(f(x))$$

$$= f\!\left(\frac{x - 2}{4}\right) \qquad\qquad\qquad = f^{-1}(4x + 2)$$

$$= 4\!\left(\frac{x - 2}{4}\right) + 2 \qquad\qquad = \frac{4x + 2 - 2}{4}$$

$$= x - 2 + 2 \qquad\qquad\qquad = \frac{4x}{4}$$

$$= x \qquad\qquad\qquad\qquad = x$$

Because $(f \circ f^{-1})(x) = x$ and $(f^{-1} \circ f)(x) = x$, the functions are inverses.

Self Check 4   Use composition to determine whether $f(x) = x - 4$ and $g(x) = x + 4$ are inverses.

■ GRAPHING A FUNCTION AND ITS INVERSE

Success Tip

Recall that the line $y = x$ passes through points whose x- and y-coordinates are equal: $(-1, -1)$, $(0, 0)$, $(1, 1)$, $(2, 2)$, and so on.

If a point (a, b) is on the graph of function f, it follows that the point (b, a) is on the graph of f^{-1}, and vice versa. There is a geometric relationship between a pair of points whose coordinates are interchanged. For example, in the graph, we see that the line segment between $(1, 3)$ and $(3, 1)$ is perpendicular to and cut in half by the line $y = x$. We say that $(1, 3)$ and $(3, 1)$ are mirror images of each other with respect to $y = x$.

Since each point on the graph of f^{-1} is a mirror image of a point on the graph of f, and vice versa, the graphs of f and f^{-1} must be mirror images of each other with respect to $y = x$.

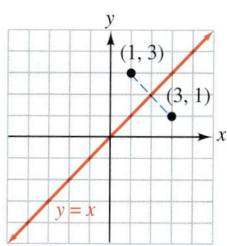

EXAMPLE 5

ELEMENTARY & INTERMEDIATE
Algebra ⨍(x) Now™

Find the inverse of $f(x) = -\dfrac{3}{2}x + 3$. Then graph f and its inverse on one coordinate system.

Solution   Since $f(x) = -\dfrac{3}{2}x + 3$ is a linear function, it is one-to-one and has an inverse. To find the inverse function, we replace $f(x)$ with y, and interchange x and y to obtain

$$x = -\frac{3}{2}y + 3$$

Then we solve for y to get

$$x - 3 = -\frac{3}{2}y \qquad \text{Subtract 3 from both sides.}$$

$$-\frac{2}{3}x + 2 = y \qquad \text{Multiply both sides by } -\frac{2}{3}.$$

Success Tip

To graph f^{-1}, we don't need to construct a table of values. We can simply interchange the coordinates of the ordered pairs in the table for f and use them to graph f^{-1}.

When we replace y with $f^{-1}(x)$, we have $f^{-1}(x) = -\frac{2}{3}x + 2.$

To graph f and f^{-1}, we construct tables of values and plot points. Because the functions are inverses of each other, their graphs are mirror images about the line $y = x$.

$f(x) = -\dfrac{3}{2}x + 3$

x	$f(x)$	
0	3	→ (0, 3)
2	0	→ (2, 0)
4	−3	→ (4, −3)

$f^{-1}(x) = -\dfrac{2}{3}x + 2$

x	$f^{-1}(x)$	
3	0	→ (3, 0)
0	2	→ (0, 2)
−3	4	→ (−3, 4)

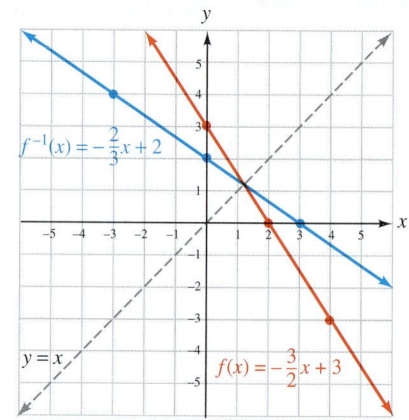

Self Check 5   Find the inverse of $f(x) = \dfrac{2}{3}x - 2$. Then graph the function and its inverse on one coordinate system.

ACCENT ON TECHNOLOGY: GRAPHING THE INVERSE OF A FUNCTION

We can use a graphing calculator to check the result found in Example 5. First, we enter $f(x) = -\frac{3}{2}x + 3$. Then we enter what we believe to be the inverse function, $f^{-1}(x) = -\frac{2}{3}x + 2$, as well as the equation $y = x$. See figure (a). Before graphing, we adjust the display so that the graphing grid will be composed of squares. The axis of symmetry is then at a 45° angle to the positive x-axis.

In figure (b), it appears that the two graphs are symmetric about the line $y = x$. Although it is not definitive, this visual check does help to validate the result of Example 5.

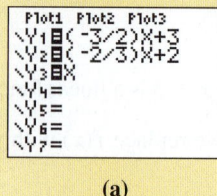

(a)

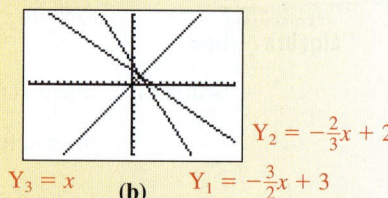

$Y_3 = x$ (b) $Y_1 = -\frac{3}{2}x + 3$

$Y_2 = -\frac{2}{3}x + 2$

EXAMPLE 6

Graph the inverse of function f shown in figure (a).

ELEMENTARY & INTERMEDIATE Algebra ⨍(x) Now™   Solution

To graph the inverse, we determine the coordinates of several points on the graph of f, interchange their coordinates, and plot them, as shown in figure (b). Then we draw a smooth curve with those points to get the graph of f^{-1}. We also graph the line $y = x$ to emphasize the symmetry.

The Language of Algebra

We can also say that the graphs of f and f^{-1} are *reflections* of each other about the line $y = x$, or they are *symmetric* about $y = x$.

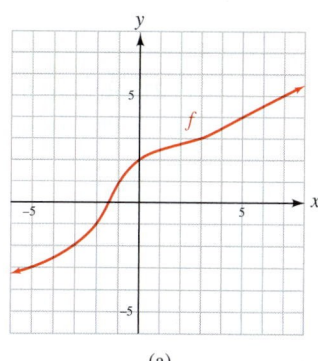

(a)

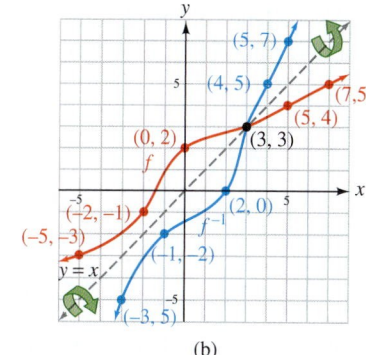

(b)

Answers to Self Checks   **1. a.** yes,   **b.** no, $(-1, 1), (1, 1)$ **2. a.** no,   **b.** yes   **3. a.** $f^{-1}(x) = \dfrac{-x - 3}{5}$,

b. $f^{-1}(x) = \sqrt[5]{x}$ **4.** They are inverses.

5.

11.2   STYLE SET   ⦿   ELEMENTARY & INTERMEDIATE Algebra ⨍(x) Now™

VOCABULARY   Fill in the blanks.

1. For a _____ function, each input is assigned exactly one output, and each output corresponds to exactly one input.

2. The _____ line test can be used to decide whether the graph of a function represents a one-to-one function.

3. The functions f and f^{-1} are _____.

4. When we _____ the coordinates of the point $(2, 5)$ we get the point $(5, 2)$.

5. The graphs of a function and its inverse are mirror _____ of each other with respect to $y = x$. We also say that their graphs are _____ with respect to the line $y = x$.

6. $(f \circ f^{-1})(x)$ is the _____ of a function f and its inverse f^{-1}.

CONCEPTS   Fill in the blanks.

7. a. If every horizontal line that intersects the graph of a function does so only _____, the function is one-to-one.

b. If any horizontal line that intersects the graph of a function does so more than once, the function is not _____.

8. If a function turns an input of 2 into an output of 5, the inverse function will turn an input of 5 into an output of ____.

9. The graphs of a function and its inverse are symmetrical about the line $y = \boxed{}$.

10. $(f \circ f^{-1})(x) = \boxed{}$ and $(f^{-1} \circ f)(x) = \boxed{}$.

11. To find the inverse of the function $f(x) = 2x - 3$, we begin by replacing $f(x)$ with $\boxed{}$, and then we _____ x and y.

12. If f is a one-to-one function, the domain of f is the _____ f^{-1}, and the range of f is the _____ f^{-1}.

13. Is the correspondence defined by the following arrow diagram a function? If it is, is the function one-to-one?

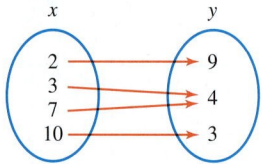

14. How can we tell that function f is not one-to-one from the table of values?

x	$f(x)$
-2	4
-1	1
0	0
2	4
3	9

15. Is the inverse of a one-to-one function always a function?

16. Name four points that the line $y = x$ passes through.

17. If f is a one-to-one function, and if $f(2) = 6$, then what is $f^{-1}(6)$?

18. If the point $(2, -4)$ is on the graph of the one-to-one function f, then what point is on the graph of f^{-1}?

19. Two functions are graphed on the following square grid along with the line $y = x$. Explain why we know that the functions are not inverses of each other.

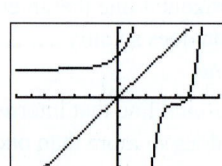

20. Use the table of values of the one-to-one function f to complete a table of values for f^{-1}.

x	$f(x)$
-6	-3
-4	-2
0	0
2	1
8	4

x	$f^{-1}(x)$
-3	
-2	
0	
1	
4	

21. Redraw the graph of function f. Then graph f^{-1} and the axis of symmetry on the same coordinate system.

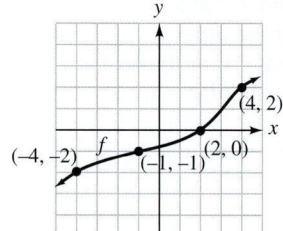

22. A table of values for a function f is shown in figure (a). A table of values for f^{-1} is shown in figure (b). Use the tables to find $f^{-1}(f(4))$ and $f(f^{-1}(2))$.

X	Y₁
-2	-2
0	2
2	6
4	10
6	14
8	18
10	22
X= -2	

X	Y₁
-2	-2
0	-1
2	0
4	1
6	2
8	3
10	4
X= -2	

(a)   (b)

NOTATION   **Complete each solution.**

23. Find the inverse of $f(x) = 2x - 3$.

$$\boxed{} = 2x - 3$$
$$x = \boxed{} - 3$$
$$x + \boxed{} = 2y$$
$$\frac{x + 3}{2} = \boxed{}$$

The inverse of $f(x) = 2x - 3$ is $\boxed{} = \dfrac{x + 3}{2}$.

24. Find the inverse of $f(x) = \sqrt[3]{x} + 2$.

$$\boxed{} = \sqrt[3]{x} + 2$$
$$x = \sqrt[3]{\boxed{}} + 2$$
$$x - \boxed{} = \sqrt[3]{y}$$
$$(x - 2)^3 = \boxed{}$$

The inverse of $f(x) = \sqrt[3]{x} + 2$ is $\boxed{} = (x - 2)^3$.

25. The symbol f^{-1} is read as "the _____ f" or "f _____."

26. Explain the difference in the meaning of the -1 in the notation $f^{-1}(x)$ as compared to x^{-1}.

PRACTICE Determine whether each function is one-to-one.

27. $f(x) = 2x$ **28.** $f(x) = |x|$

29. $f(x) = x^4$ **30.** $f(x) = x^3 + 1$

31. $f(x) = -x^2 + 3x$ **32.** $f(x) = \dfrac{2}{3}x + 8$

33. $\{(1, 1), (2, 1), (3, 1), (4, 1)\}$

34. $\{(3, 2), (2, 1), (1, 0)\}$

Each graph represents a function. Use the horizontal line test to decide whether the function is one-to-one.

35. 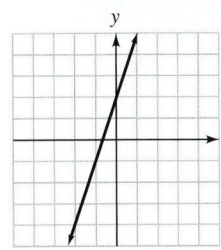 **36.**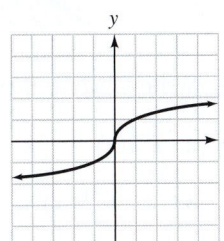

37. 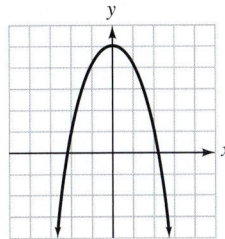 **38.**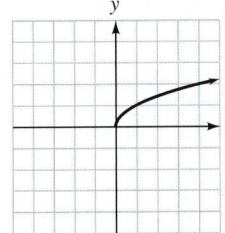

39. 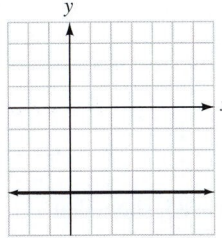 **40.**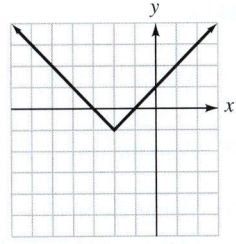

Find the inverse of the function and express it using $f^{-1}(x)$ notation.

41. $f(x) = 2x + 4$ **42.** $f(x) = 5x - 1$

43. $f(x) = \dfrac{x}{5} + \dfrac{4}{5}$ **44.** $f(x) = \dfrac{x}{3} - \dfrac{1}{3}$

45. $f(x) = \dfrac{x - 4}{5}$ **46.** $f(x) = \dfrac{2x + 6}{3}$

47. $f(x) = \dfrac{2}{x - 3}$ **48.** $f(x) = \dfrac{3}{x + 1}$

49. $f(x) = \dfrac{4}{x}$ **50.** $f(x) = \dfrac{1}{x}$

51. $f(x) = x^3 + 8$ **52.** $f(x) = x^3 - 4$

53. $f(x) = \sqrt[3]{x}$ **54.** $f(x) = \sqrt[3]{x - 5}$

55. $f(x) = (x + 10)^3$ **56.** $f(x) = (x - 9)^3$

57. $f(x) = 2x^3 - 3$ **58.** $f(x) = \dfrac{3}{x^3} - 1$

Use composition to show that each pair of functions are inverses. (See Example 4.)

59. $f(x) = 2x + 9$, $f^{-1}(x) = \dfrac{x - 9}{2}$

60. $f(x) = 5x - 1$, $f^{-1}(x) = \dfrac{x + 1}{5}$

61. $f(x) = \dfrac{2}{x - 3}$, $f^{-1}(x) = \dfrac{2}{x} + 3$

62. $f(x) = \sqrt[3]{x - 6}$, $f^{-1}(x) = x^3 + 6$

Find the inverse of each function. Then graph the function and its inverse on one coordinate system. Show the line of symmetry on the graph.

63. $f(x) = 2x$ **64.** $f(x) = -3x$

65. $f(x) = 4x + 3$ **66.** $f(x) = \dfrac{x}{3} + \dfrac{1}{3}$

67. $f(x) = -\dfrac{2}{3}x + 3$ **68.** $f(x) = -\dfrac{1}{3}x + \dfrac{4}{3}$

69. $f(x) = x^3$ **70.** $f(x) = x^3 + 1$

71. $f(x) = x^2 - 1$ $(x \geq 0)$

72. $f(x) = x^2 + 1$ $(x \geq 0)$

APPLICATIONS

73. INTERPERSONAL RELATIONSHIPS   Feelings of anxiety in a relationship can increase or decrease, depending on what is going on in the relationship. The graph shows how a person's anxiety might vary as a relationship develops over time.

a. Is this the graph of a function? Is its inverse a function?

b. Does each anxiety level correspond to exactly one point in time? Use the dashed lined labeled *Maximum threshold* to explain.

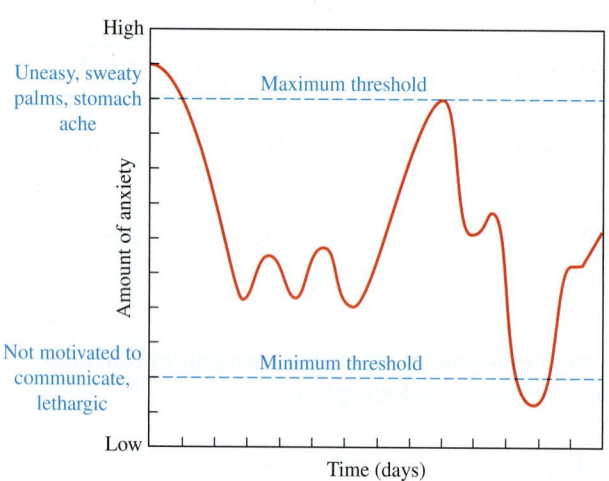

Source: Gudykunst, Building Bridges: Interpersonal Skills for a Changing World (Houghton Mifflin, 1994)

74. LIGHTING LEVELS The ability of the eye to see detail increases as the level of illumination increases. This relationship can be modeled by a function E, whose graph is shown here.

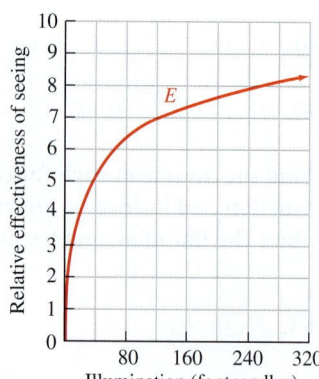

a. From the graph, determine $E(240)$.

b. Is function E one-to-one? Does E have an inverse?

c. If the effectiveness of seeing in an office is 7, what is the illumination in the office? How can this question be asked using inverse function notation?

WRITING

75. In your own words, what is a one-to-one function?

76. Explain the purpose of the horizontal line test.

77. In the illustration, a function f and its inverse f^{-1} have been graphed on the same coordinate system. Explain what concept can be demonstrated by folding the graph paper on the dashed line.

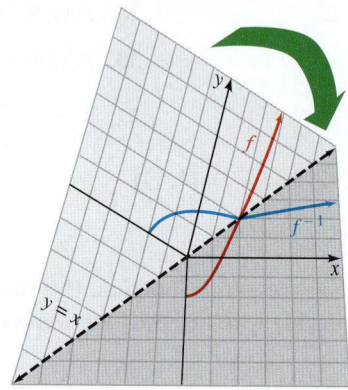

78. Write in words how to read the notation.

a. $f^{-1}(x) = \dfrac{1}{2}x - 3$

b. $(f \circ f^{-1})(x) = x$

REVIEW   Simplify. Write the result in $a + bi$ form.

79. $3 - \sqrt{-64}$

80. $(2 - 3i) + (4 + 5i)$

81. $(3 + 4i)(2 - 3i)$

82. $\dfrac{6 + 7i}{3 - 4i}$

83. $(6 - 8i)^2$

84. i^{100}

CHALLENGE PROBLEMS

85. Find the inverse of $f(x) = \dfrac{x + 1}{x - 1}$.

86. Using the functions of Exercise 85, show that $(f \circ f^{-1})(x) = x$ and $(f^{-1} \circ f)(x) = x$.

11.3 Exponential Functions

- Irrational exponents • Exponential functions • Graphing exponential functions
- Vertical and horizontal translations • Compound interest
- Exponential functions as models

The graph below shows the balance in a bank account in which \$10,000 was invested in 1998 at 9%, compounded monthly. The graph shows that in the year 2008, the value of the account will be approximately \$25,000, and in the year 2028, the value will be approximately \$147,000. The red curve is the graph of a function called an *exponential function.*

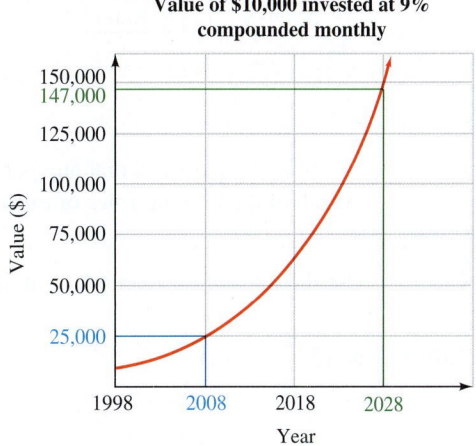

Exponential functions are also used to model many other situations, such as population growth, the spread of an epidemic, the temperature of a heated object as it cools, and radioactive decay. Before we can discuss exponential functions in more detail, we must define irrational exponents.

■ IRRATIONAL EXPONENTS

We have discussed expressions of the form b^x, where x is a rational number.

$8^{1/2}$ means "the square root of 8."

$5^{1/3}$ means "the cube root of 5."

$3^{-2/5} = \dfrac{1}{3^{2/5}}$ means "the reciprocal of the fifth root of 3^2."

To give meaning to b^x when x is an irrational number, we consider the expression

$5^{\sqrt{2}}$ where $\sqrt{2}$ is the irrational number 1.414213562 . . .

Each number in the following list is defined, because each exponent is a rational number.

$5^{1.4}, \quad 5^{1.41}, \quad 5^{1.414}, \quad 5^{1.4142}, \quad 5^{1.41421}, \quad . . .$

Since the exponents are getting closer to $\sqrt{2}$, the numbers in this list are successively better approximations of $5^{\sqrt{2}}$. We can use a calculator to obtain a very good approximation.

ACCENT ON TECHNOLOGY: EVALUATING EXPONENTIAL EXPRESSIONS

To find the value of $5^{\sqrt{2}}$ with a scientific calculator, we enter these numbers and press these keys:

$5 \boxed{y^x} 2 \boxed{\sqrt{\;}} \boxed{=}$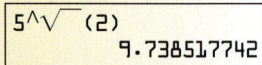

With a graphing calculator, we enter these numbers and press these keys:

$5 \boxed{\wedge} \boxed{\sqrt{\;}} 2 \boxed{)} \boxed{\text{Enter}}$

```
5^√ (2)
         9.738517742
```

If $b > 0$ and x is a real number, b^x represents a positive number. It can be shown that all of the familiar rules of exponents are also true for irrational exponents.

EXAMPLE 1   Use the rules of exponents to simplify **a.** $\left(5^{\sqrt{2}}\right)^{\sqrt{2}}$ and   **b.** $b^{\sqrt{3}} \cdot b^{\sqrt{12}}$.

ELEMENTARY & INTERMEDIATE
Algebra *f(x)* Now™   **Solution**   **a.** $\left(5^{\sqrt{2}}\right)^{\sqrt{2}} = 5^{\sqrt{2}\sqrt{2}}$ Keep the base and multiply the exponents.

$\qquad\qquad = 5^2$ Multiply: $\sqrt{2}\sqrt{2} = \sqrt{4} = 2$.

$\qquad\qquad = 25$

b. $b^{\sqrt{3}} \cdot b^{\sqrt{12}} = b^{\sqrt{3}+\sqrt{12}}$ Keep the base and add the exponents.

$\qquad\qquad = b^{\sqrt{3}+2\sqrt{3}}$ Simplify: $\sqrt{12} = \sqrt{4}\sqrt{3} = 2\sqrt{3}$.

$\qquad\qquad = b^{3\sqrt{3}}$ Combine like radicals. $\sqrt{3} + 2\sqrt{3} = 3\sqrt{3}$.

Self Check 1   Simplify: **a.** $\left(3^{\sqrt{2}}\right)^{\sqrt{8}}$ and   **b.** $b^{\sqrt{2}} \cdot b^{\sqrt{18}}$.

■ EXPONENTIAL FUNCTIONS

If $b > 0$ and $b \neq 1$, the function $f(x) = b^x$ is called an **exponential function.** Since x can be any real number, its domain is the set of real numbers. This is the interval $(-\infty, \infty)$.

Because b is positive, the value of $f(x)$ is positive, and the range is the set of positive numbers. This is the interval $(0, \infty)$.

Since $b \neq 1$, an exponential function *cannot* be the constant function $f(x) = 1^x$, in which $f(x) = 1$ for every real number x.

Exponential Functions   An **exponential function with base b** is defined by the equation

$$f(x) = b^x \quad \text{or} \quad y = b^x \quad \text{where } b > 0, b \neq 1, \text{ and } x \text{ is a real number}$$

The domain of $f(x) = b^x$ is the interval $(-\infty, \infty)$, and the range is the interval $(0, \infty)$.

■ GRAPHING EXPONENTIAL FUNCTIONS

Since the domain and range of $f(x) = b^x$ are sets of real numbers, we can graph exponential functions on a rectangular coordinate system.

EXAMPLE 2

ELEMENTARY &
INTERMEDIATE
Algebra *(f(x))* **Now**™

Graph: $f(x) = 2^x$.

Solution   To graph $f(x) = 2^x$, we construct a table of function values by choosing several values for x and finding the corresponding values of $f(x)$. If x is -1, we have

$$f(x) = 2^x$$
$$f(-1) = 2^{-1} \qquad \text{Substitute } -1 \text{ for } x.$$
$$= \frac{1}{2}$$

Notation

We have previously graphed the linear function $f(x) = 2x$ and the squaring function $f(x) = x^2$. For the exponential function $f(x) = 2^x$, note that the variable is in the exponent.

The point $\left(-1, \frac{1}{2}\right)$ is on the graph of $f(x) = 2^x$. In a similar manner, we find the corresponding values of $f(x)$ for x values of 0, 1, 2, 3, and 4 and record them in a table. Then we plot the ordered pairs and draw a smooth curve through them.

$f(x) = 2^x$

x	$f(x)$	
-1	$\frac{1}{2}$	$\rightarrow \left(-1, \frac{1}{2}\right)$
0	1	$\rightarrow (0, 1)$
1	2	$\rightarrow (1, 2)$
2	4	$\rightarrow (2, 4)$
3	8	$\rightarrow (3, 8)$
4	16	$\rightarrow (4, 16)$

The graph steadily approaches the x-axis, but never touches or crosses it.

The Language of Algebra

We encountered the word *asymptote* earlier, when we graphed rational functions. Recall that an asymptote is not part of the graph. It is a line that the graph approaches and, in this case, never touches.

From the graph, we can verify that the domain of $f(x) = 2^x$ is the interval $(-\infty, \infty)$ and the range is the interval $(0, \infty)$. Since the graph passes the horizontal line test, the function is one-to-one.

Note that as x decreases, the values of $f(x)$ decrease and approach 0. Thus, the x-axis is a horizontal asymptote of the graph. The graph does not have an x-intercept, the y-intercept is $(0, 1)$, and the graph passes through the point $(1, 2)$.

Self Check 2   Graph: $f(x) = 4^x$.

EXAMPLE 3   Graph: $f(x) = \left(\dfrac{1}{2}\right)^x$.

Solution   We make a table of values for the function. For example, if $x = -4$, we have

$$f(x) = \left(\frac{1}{2}\right)^x$$

$$f(-4) = \left(\frac{1}{2}\right)^{-4}$$

$$= \left(\frac{2}{1}\right)^4 \qquad \text{Recall: } \left(\frac{x}{y}\right)^{-n} = \left(\frac{y}{x}\right)^n.$$

$$= 16$$

The point $(-4, 16)$ is on the graph of $f(x) = \left(\frac{1}{2}\right)^x$. In a similar manner, we find the corresponding values of $f(x)$ for other x-values and record them in a table.

$$f(x) = \left(\frac{1}{2}\right)^x$$

x	$f(x)$	
-4	16	$\rightarrow (-4, 16)$
-3	8	$\rightarrow (-3, 8)$
-2	4	$\rightarrow (-2, 4)$
-1	2	$\rightarrow (-1, 2)$
0	1	$\rightarrow (0, 1)$
1	$\frac{1}{2}$	$\rightarrow \left(1, \frac{1}{2}\right)$

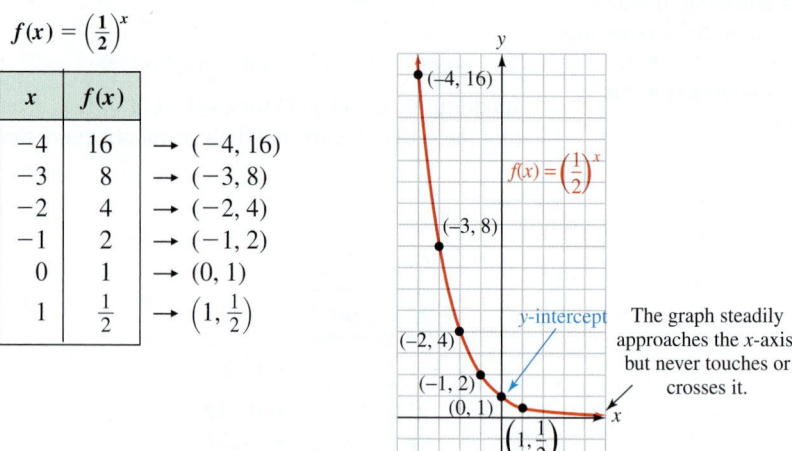

The graph steadily approaches the x-axis, but never touches or crosses it.

From the graph, we can verify that the domain of $f(x) = \left(\frac{1}{2}\right)^x$ is the interval $(-\infty, \infty)$ and the range is the interval $(0, \infty)$. Since the graph passes the horizontal line test, the function is one-to-one.

Note that as x increases, the values of $f(x)$ decrease and approach 0. Thus, the x-axis is a horizontal asymptote of the graph. The graph does not have an x-intercept, the y-intercept is $(0, 1)$, and the graph passes through the point $\left(1, \frac{1}{2}\right)$.

Self Check 3   Graph: $g(x) = \left(\dfrac{1}{3}\right)^x$.

Examples 2 and 3 illustrate the following properties of exponential functions.

Properties of Exponential Functions

The domain of the exponential function $f(x) = b^x$ is the interval $(-\infty, \infty)$.

The range is the interval $(0, \infty)$.

The graph has a y-intercept of $(0, 1)$.

The x-axis is an asymptote of the graph.

The graph of $f(x) = b^x$ passes through the point $(1, b)$.

In Example 2 (where $b = 2$), the values of y increase as the values of x increase. Since the graph rises as we move to the right, we call the function an *increasing function*. When $b > 1$, the larger the value of b, the steeper the curve.

In Example 3 $\left(\text{where } b = \frac{1}{2}\right)$, the values of y decrease as the values of x increase. Since the graph drops as we move to the right, we call the function a *decreasing function*. When $0 < b < 1$, the smaller the value of b, the steeper the curve.

In general, the following is true.

Increasing and Decreasing Functions	If $b > 1$, then $f(x) = b^x$ is an **increasing function.** If $0 < b < 1$, then $f(x) = b^x$ is a **decreasing function.**

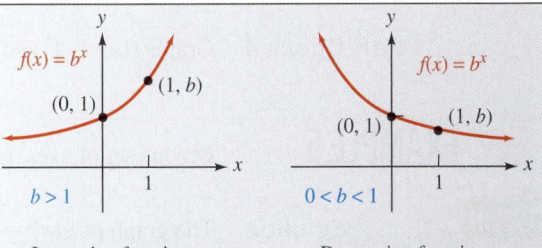

Increasing function   Decreasing function

An exponential function with base b is either increasing (for $b > 1$) or decreasing ($0 < b < 1$). Since different real numbers x determine different values of b^x, exponential functions are one-to-one.

ACCENT ON TECHNOLOGY: GRAPHING EXPONENTIAL FUNCTIONS

To use a graphing calculator to graph $f(x) = \left(\frac{2}{3}\right)^x$ and $g(x) = \left(\frac{3}{2}\right)^x$, we enter the right-hand sides of the equations after the symbols $Y_1 =$ and $Y_2 =$. The screen will show the following equations.

$Y_1 = (2/3)\wedge X$
$Y_2 = (3/2)\wedge X$

If we press the $\boxed{\text{GRAPH}}$ key, we will obtain the display shown.

We note that the graph of $f(x) = \left(\frac{2}{3}\right)^x$ passes through the points $(0, 1)$ and $\left(1, \frac{2}{3}\right)$. Since $\frac{2}{3} < 1$, the function is decreasing. The graph of $g(x) = \left(\frac{3}{2}\right)^x$ passes through the points $(0, 1)$ and $\left(1, \frac{3}{2}\right)$. Since $\frac{3}{2} > 1$, the function is increasing.

Since both graphs pass the horizontal line test, each function is one-to-one.

■ VERTICAL AND HORIZONTAL TRANSLATIONS

EXAMPLE 4   On one set of axes, graph $f(x) = 2^x$ and $g(x) = 2^x + 3$, and describe the translation.

Solution   The graph of $g(x) = 2^x + 3$ is identical to the graph of $f(x) = 2^x$, except that it is translated 3 units upward.

$f(x) = 2^x$

x	$f(x)$	
-4	$\frac{1}{16}$	$\rightarrow \left(-4, \frac{1}{16}\right)$
0	1	$\rightarrow (0, 1)$
2	4	$\rightarrow (2, 4)$

$g(x) = 2^x + 3$

x	$g(x)$	
-4	$3\frac{1}{16}$	$\rightarrow \left(-4, 3\frac{1}{16}\right)$
0	4	$\rightarrow (0, 4)$
2	7	$\rightarrow (2, 7)$

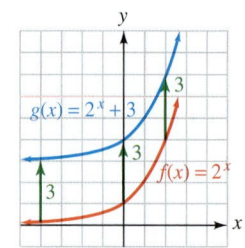

Self Check 4   Graph $f(x) = 4^x$ and $g(x) = 4^x - 3$, and describe the translation.

EXAMPLE 5

ELEMENTARY &
INTERMEDIATE
Algebra *f(x)* **Now**™

On one set of axes, graph $f(x) = 2^x$ and $g(x) = 2^{x+3}$, and describe the translation.

Solution   The graph of $g(x) = 2^{x+3}$ is identical to the graph of $f(x) = 2^x$, except that it is translated 3 units to the left.

$f(x) = 2^x$

x	$f(x)$	
-1	$\frac{1}{2}$	$\rightarrow \left(-1, \frac{1}{2}\right)$
0	1	$\rightarrow (0, 1)$
1	2	$\rightarrow (1, 2)$

$g(x) = 2^{x+3}$

x	$g(x)$	
-1	4	$\rightarrow (-1, 4)$
0	8	$\rightarrow (0, 8)$
1	16	$\rightarrow (1, 16)$

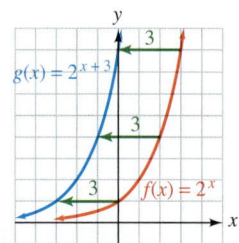

Self Check 5   On one set of axes, graph $f(x) = 4^x$ and $g(x) = 4^{x-3}$, and describe the translation.

The graphs of $f(x) = kb^x$ and $f(x) = b^{kx}$ are vertical and horizontal stretchings of the graph of $f(x) = b^x$. To graph these functions, we can plot several points and join them with a smooth curve, or we can use a graphing calculator.

ACCENT ON TECHNOLOGY: GRAPHING EXPONENTIAL FUNCTIONS

To use a graphing calculator to graph the exponential function $f(x) = 3(2^{x/3})$, we enter the right-hand side of the equation after the symbol $Y_1 = $. The display will show the equation

$$Y_1 = 3(2 \wedge (X/3))$$

If we press the graph key, we will obtain the display shown.

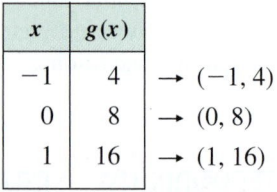

■ COMPOUND INTEREST

If we deposit $\$P$ in an account paying an annual interest rate r, we can find the amount A in the account at the end of t years by using the formula $A = P + Prt$, or $A = P(1 + rt)$.

Suppose that we deposit $500 in such an account that pays interest every 6 months. Then $P = 500$, and after 6 months $\left(\frac{1}{2} \text{ year}\right)$, the amount in the account will be

$$A = 500(1 + rt)$$

$$= 500\left(1 + r \cdot \frac{1}{2}\right) \qquad \text{Substitute } \frac{1}{2} \text{ for } t.$$

$$= 500\left(1 + \frac{r}{2}\right)$$

The account will begin the second 6-month period with a value of $\$500\left(1 + \frac{r}{2}\right)$. After the second 6-month period, the amount in the account will be

$$A = P(1 + rt)$$

$$A = \left[500\left(1 + \frac{r}{2}\right)\right]\left(1 + r \cdot \frac{1}{2}\right) \qquad \text{Substitute } 500\left(1 + \frac{r}{2}\right) \text{ for } P \text{ and } \frac{1}{2} \text{ for } t.$$

$$= 500\left(1 + \frac{r}{2}\right)\left(1 + \frac{r}{2}\right)$$

$$= 500\left(1 + \frac{r}{2}\right)^2$$

At the end of a third 6-month period, the amount in the account will be

$$A = 500\left(1 + \frac{r}{2}\right)^3$$

In this discussion, the earned interest is deposited back in the account and also earns interest, and we say that the account is earning **compound interest.** The preceding discussion suggests the following formula for compound interest.

Formula for Compound Interest	If $\$P$ is deposited in an account and interest is paid k times a year at an annual rate r, the amount A in the account after t years is given by $$A = P\left(1 + \frac{r}{k}\right)^{kt}$$

EXAMPLE 6

ELEMENTARY & INTERMEDIATE
Algebra $f(x)$ **Now**™

Saving for college. To save for college, parents invest $12,000 for their newborn child in a mutual fund that should average a 10% annual return. If the quarterly dividends are reinvested, how much will be available in 18 years?

Solution   We substitute 12,000 for P, 0.10 for r, and 18 for t in the formula for compound interest and find A. Since interest is paid quarterly, $k = 4$.

$$A = P\left(1 + \frac{r}{k}\right)^{kt}$$

$$A = 12{,}000\left(1 + \frac{0.10}{4}\right)^{4(18)} \qquad \text{Express } r = 10\% \text{ as a decimal.}$$

$$= 12{,}000(1 + 0.025)^{72}$$

$$= 12{,}000(1.025)^{72}$$

$$= 71{,}006.74 \qquad \begin{array}{l}\text{Use a scientific calculator and press these keys:} \\ 1.025 \;\boxed{y^x}\; 72 \;\boxed{=}\; \boxed{\times}\; 12{,}000 \;\boxed{=}\; .\end{array}$$

In 18 years, the account will be worth $71,006.74.

Self Check 6 How much would be available if the parents invested $20,000?

In business applications, the initial amount of money deposited is often called the **present value** (*PV*). The amount to which the money will grow is called the **future value** (*FV*). The interest rate used for each compounding period is the **periodic interest rate** (*i*), and the number of times interest is compounded is the number of **compounding periods** (*n*). Using these definitions, we have an alternate formula for compound interest.

Formula for Compound Interest	$FV = PV(1 + i)^n$

This alternate formula appears on business calculators. To use this formula to solve Example 6, we proceed as follows:

$$FV = \textbf{PV}(1 + \textbf{i})^n$$

$$FV = \textbf{12,000}(1 + \textbf{0.025})^{72} \qquad i = \frac{0.10}{4} = 0.025 \text{ and } n = 4(18) = 72.$$

$$\approx 71{,}006.74 \qquad \text{Use a calculator to evaluate the expression.}$$

ACCENT ON TECHNOLOGY: SOLVING INVESTMENT PROBLEMS

Suppose $1 is deposited in an account earning 6% annual interest, compounded monthly. To use a graphing calculator to estimate how much will be in the account in 100 years, we can substitute 1 for *P*, 0.06 for *r*, and 12 for *k* in the formula

$$A = P\left(1 + \frac{r}{k}\right)^{kt}$$

$$A = \textbf{1}\left(1 + \frac{\textbf{0.06}}{\textbf{12}}\right)^{12t}$$

and simplify to get

$$A = (1.005)^{12t}$$

We now graph $A = (1.005)^{12t}$ using window settings of $[0, 120]$ for *t* and $[0, 400]$ for *A* to obtain the graph shown. We can then trace and zoom to estimate that $1 grows to be approximately $397 in 100 years. From the graph, we can see that the money grows slowly in the early years and rapidly in the later years.

■ EXPONENTIAL FUNCTIONS AS MODELS

EXAMPLE 7

ELEMENTARY & INTERMEDIATE
Algebra *f(x)* **Now**™

Cellular phones. For the years 1990–1997, the U.S. cellular telephone industry experienced exponential growth. The exponential function $S(n) = 5.74(1.39)^n$ approximates the number of cellular telephone subscribers in millions, where *n* is the number of years since 1990 and $0 \le n \le 7$.

Use the function to answer the following.

a. How many subscribers were there in 1990?

b. How many subscribers were there in 1997?

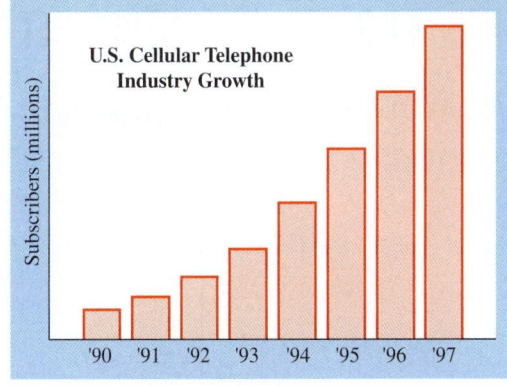

U.S. Cellular Telephone Industry Growth

Source: *New York Times Almanac*, 1998

Solution

a. The year 1990 is 0 years after 1990. To find the number of subscribers in 1990, we substitute 0 for n in the function and find $S(0)$.

$$S(n) = 5.74(1.39)^n$$
$$S(0) = 5.74(1.39)^0$$
$$= 5.74 \cdot 1 \quad (1.39)^0 = 1.$$
$$= 5.74$$

In 1990, there were approximately 5.74 million cellular telephone subscribers.

b. The year 1997 is 7 years after 1990. We need to find $S(7)$.

$$S(n) = 5.74(1.39)^n$$
$$S(7) = 5.74(1.39)^7 \quad \text{Substitute 7 for } n.$$
$$\approx 57.54604675 \quad \text{Use a calculator to find an approximation.}$$

In 1997, there were approximately 57.55 million cellular telephone subscribers.

Answers to Self Checks   **1. a.** 81,   **b.** $b^{4\sqrt{2}}$ **2.**

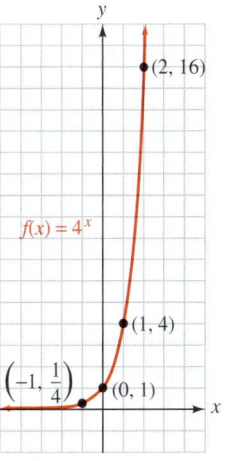

3.

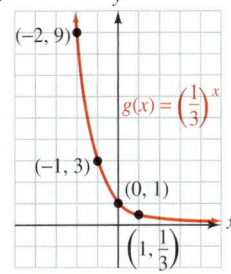

4. The graph of $f(x) = 4^x$ is translated 3 units downward.

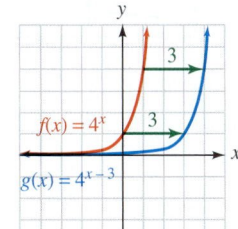

5. The graph of $f(x) = 4^x$ is translated 3 units to the right.

6. $118,344.56

11.3   **STUDY SET**   ELEMENTARY & INTERMEDIATE Algebra *f(x)* **Now**™

VOCABULARY   **Refer to the graph of $f(x) = 3^x$.**

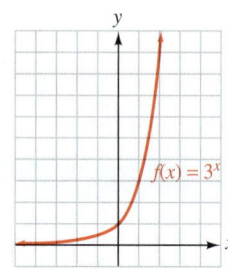

$f(x) = 3^x$

1. What type of function is $f(x) = 3^x$?

2. What is the domain of the function?

3. What is the range of the function?

4. a. What is the y-intercept of the graph?

   b. What is the x-intercept of the graph?

5. Is the function one-to-one?

6. What is an asymptote of the graph?

7. Is f an increasing or a decreasing function?

8. The graph passes through the point $(1, y)$. What is y?

CONCEPTS

9. Graph $f(x) = x^2$ and $g(x) = 2^x$ on the same set of coordinate axes.

10. Graph $y = x^{1/2}$ and $y = \left(\frac{1}{2}\right)^x$ on the same set of coordinate axes.

11. What are the two formulas that are used to determine the amount of money in a savings account that is earning compound interest?

12. Explain the order in which the expression $20{,}000(1.036)^{72}$ should be evaluated.

13. The illustration shows the graph of $f(x) = 2^x$, as well as two vertical translations of that graph. Using the notation $g(x)$ for one translation and $h(x)$ for the other, write the defining equation for each function.

14. The illustration shows the graph of $f(x) = 2^x$, as well as a horizontal translation of that graph. Using the notation $g(x)$ for the translation, write its defining equation.

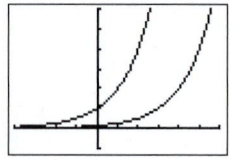

NOTATION

15. In $A = P\left(1 + \frac{r}{k}\right)^{kt}$, what is the base, and what is the exponent?

16. For an exponential function of the form $f(x) = b^x$, what are the restrictions on b?

PRACTICE   **Find each value to four decimal places.**

17. $2^{\sqrt{2}}$

18. $7^{\sqrt{2}}$

19. $5^{\sqrt{5}}$

20. $6^{\sqrt{3}}$

Simplify each expression.

21. $\left(2^{\sqrt{3}}\right)^{\sqrt{3}}$

22. $3^{\sqrt{2}}3^{\sqrt{18}}$

23. $7^{\sqrt{3}}7^{\sqrt{12}}$

24. $\left(3^{\sqrt{5}}\right)^{\sqrt{5}}$

Graph each function.

25. $f(x) = 5^x$

26. $f(x) = 6^x$

27. $y = \left(\frac{1}{4}\right)^x$

28. $y = \left(\frac{1}{5}\right)^x$

29. $f(x) = 3^x - 2$

30. $y = 2^x + 1$

31. $f(x) = 3^{x-1}$

32. $f(x) = 2^{x+1}$

Use a graphing calculator to graph each function. Determine whether the function is an increasing or a decreasing function.

33. $f(x) = \frac{1}{2}(3^{x/2})$

34. $f(x) = -3(2^{x/3})$

35. $y = 2(3^{-x/2})$

36. $y = -\frac{1}{4}(2^{-x/2})$

APPLICATIONS   **In Exercises 37–42, assume that there are no deposits or withdrawals.**

37. COMPOUND INTEREST   An initial deposit of $10,000 earns 8% interest, compounded quarterly. How much will be in the account after 10 years?

38. COMPOUND INTEREST   An initial deposit of $10,000 earns 8% interest, compounded monthly. How much will be in the account after 10 years?

39. COMPARING INTEREST RATES   How much more interest could $1,000 earn in 5 years, compounded quarterly, if the annual interest rate were $5\frac{1}{2}\%$ instead of 5%?

40. COMPARING SAVINGS PLANS   Which institution in the ads provides the better investment?

Fidelity Savings & Loan
Earn 5.25%
compounded monthly

Union Trust
Money Market Account
paying 5.35%
compounded annually

---

41. COMPOUND INTEREST   If $1 had been invested on July 4, 1776, at 5% interest, compounded annually, what would it be worth on July 4, 2076?

42. FREQUENCY OF COMPOUNDING $10,000 is invested in each of two accounts, both paying 6% annual interest. In the first account, interest compounds quarterly, and in the second account, interest compounds daily. Find the difference between the accounts after 20 years.

43. WORLD POPULATION   See the following graph.

a. Estimate when the world's population reached $\frac{1}{2}$ billion and when it reached 1 billion.

b. Estimate the world's population in the year 2000.

c. What type of function does it appear could be used to model the population growth?

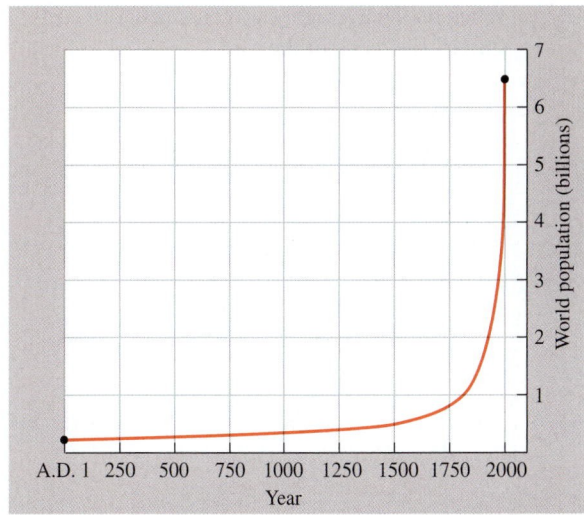

Source: *The Blue Planet* (Wiley, 1995)

44. THE STOCK MARKET   The Dow Jones Industrial Average is a measure of how well the stock market is doing. Graph the following Dow milestones as ordered pairs of the form (year, average). What type of function could be used to model the growth of the stock market between 1906 and 1999?

Dow Jones Milestones			
Year	**Average**	**Year**	**Average**
Jan. 1906	100	Nov. 1995	5,000
Mar. 1956	500	Oct. 1996	6,000
Nov. 1972	1,000	Feb. 1997	7,000
Jan. 1987	2,000	July 1997	8,000
Apr. 1991	3,000	April 1998	9,000
Feb. 1995	4,000	March 1999	10,000
		May 1999	11,000

Source: finfacts.com

45. VALUE OF A CAR   The graph on the next page shows how the value of the average car depreciates as a percent of its original value over a 10-year period. It also shows the yearly maintenance costs as a percent of the car's value.

a. When is the car worth half of its purchase price?

b. When is the car worth a quarter of its purchase price?

c. When do the average yearly maintenance costs surpass the value of the car?

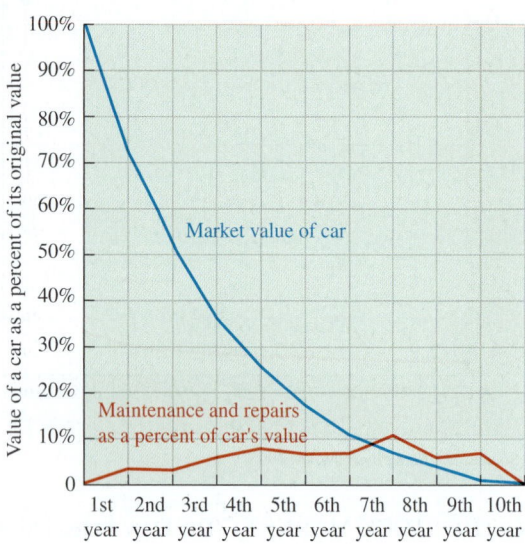

Source: U.S. Department of Transportation

46. DIVING *Bottom time* is the time a scuba diver spends descending plus the actual time spent at a certain depth. Graph the bottom time limits given in the table.

Bottom time limits			
Depth (ft)	**Bottom time (min)**	**Depth (ft)**	**Bottom time (min)**
30	no limit	80	40
35	310	90	30
40	200	100	25
50	100	110	20
60	60	120	15
70	50	130	10

47. BACTERIAL CULTURES A colony of 6 million bacteria is growing in a culture medium. The population P after t hours is given by the formula $P = (6 \times 10^6)(2.3)^t$. Find the population after 4 hours.

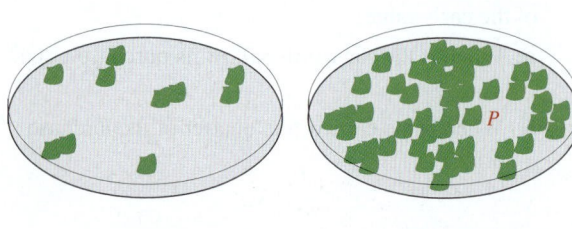

12:00 P.M.   4:00 P.M.

48. RADIOACTIVE DECAY A radioactive material decays according to the formula $A = A_0\left(\frac{2}{3}\right)^t$, where A_0 is the initial amount present and t is measured in years. Find the amount present in 5 years.

49. DISCHARGING A BATTERY The charge remaining in a battery decreases as the battery discharges. The charge C (in coulombs) after t days is given by the formula $C = (3 \times 10^{-4})(0.7)^t$. Find the charge after 5 days.

50. POPULATION GROWTH The population of North Rivers is decreasing exponentially according to the formula $P = 3,745(0.93)^t$, where t is measured in years from the present date. Find the population in 6 years, 9 months.

51. SALVAGE VALUES A small business purchased a computer for $4,700. It is expected that its value each year will be 75% of its value in the preceding year. If the business disposes of the computer after 5 years, find its salvage value (the value after 5 years).

52. THE LOUISIANA PURCHASE In 1803, the United States negotiated the Louisiana Purchase with France. The country doubled its territory by adding 827,000 square miles of land for $15 million. If the land appreciated at the rate of 6% each year, what would one square mile of land be worth in 2005?

WRITING

53. If world population is increasing exponentially, why is there cause for concern?

54. How do the graphs of $f(x) = 3^x$ and $g(x) = \left(\frac{1}{3}\right)^x$ differ? How are they similar?

55. A snowball rolling downhill grows *exponentially* with time. Explain what this means. Sketch a simple graph that models the situation.

56. Explain why the graph of $f(x) = 3^x$ gets closer and closer to the x-axis as the values of x decrease. Does the graph ever cross the x-axis? Explain why or why not.

REVIEW In Exercises 57–60, refer to the illustration below in which lines r and s are parallel.

57. Find x.

58. Find the measure of $\angle 1$.

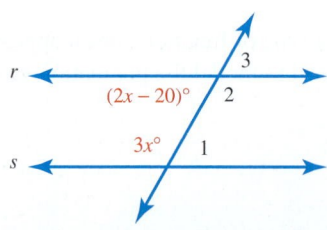

CHALLENGE PROBLEMS

59. Find the measure of $\angle 2$.

60. Find the measure of $\angle 3$.

61. In the definition of the exponential function, b could not be negative. Why?

62. Graph $f(x) = 3^x$. Then use the graph to estimate the value of $3^{1.5}$.

11.4 Base-*e* Exponential Functions

- Continuous compound interest
- The natural exponential function
- Graphing the natural exponential function
- Vertical and horizontal translations
- Malthusian population growth
- Base-*e* exponential function models

An exponential function that has many real-life applications is the base-*e* exponential function. In this section, we will show how to evaluate *e*, graph the base-*e* exponential function, and discuss one of its applications in analyzing population growth.

■ CONTINUOUS COMPOUND INTEREST

If a bank pays interest twice a year, we say that interest is compounded semiannually. If it pays interest four times a year, we say that interest is compounded quarterly. If it pays interest continuously (infinitely many times in a year), we say that interest is compounded continuously.

To develop the formula for continuous compound interest, we start with the formula

$$A = P\left(1 + \frac{r}{k}\right)^{kt}$$ This is the formula for compound interest: r is the annual rate and k is the number of times per year interest is paid.

and let $rn = k$. Since r and k are positive numbers, so is n.

$$A = P\left(1 + \frac{r}{rn}\right)^{rnt}$$

We can then simplify the fraction $\frac{r}{rn}$ and use the commutative property of multiplication to change the order of the exponents.

$$A = P\left(1 + \frac{1}{n}\right)^{nrt}$$

Finally, we can use a property of exponents to write the formula as

(1) $$A = P\left[\left(1 + \frac{1}{n}\right)^{n}\right]^{rt}$$ Use the property $a^{mn} = (a^m)^n$.

To find the value of $\left(1 + \frac{1}{n}\right)^n$, we evaluate it for several values of n, as shown in the table.

n	$\left(1 + \frac{1}{n}\right)^n$
1	2
2	2.25
4	2.44140625...
12	2.61303529...
365	2.71456748...
1,000	2.71692393...
100,000	2.71826830...
1,000,000	2.71828137...

The results suggest that as n gets larger, the value of $\left(1 + \frac{1}{n}\right)^n$ approaches the number $2.71828\ldots$. This number is called e, which has the following value.

$$e = 2.718281828459\ldots$$

Like π, the number e is irrational. Its decimal representation is nonterminating and nonrepeating. Rounded to four decimal places, $e \approx 2.7183$.

In continuous compound interest, k (the number of compoundings) is infinitely large. Since k, r, and n are all positive, r is a fixed rate, and $k = rn$, as k gets very large (approaches infinity), so does n. Therefore, we can replace $\left(1 + \frac{1}{n}\right)^n$ in Equation 1 with e to get

$$A = Pe^{rt}$$

Formula for Exponential Growth	If a quantity P increases or decreases at an annual rate r, compounded continuously, the amount A after t years is given by $$A = Pe^{rt}$$

If time is measured in years, then r is called the **annual growth rate.** If r is negative, the growth represents a decrease.

■ THE NATURAL EXPONENTIAL FUNCTION

Of all possible bases for an exponential function, e is the most convenient for applied problems involving growth or decay. Since these situations occur often in natural settings, we call $f(x) = e^x$ the *natural exponential function*.

The Natural Exponential Function	The function defined by $f(x) = e^x$ is the **natural exponential function** (or the **base-e exponential function**) where $e = 2.71828\ldots$. The domain of $f(x) = e^x$ is the interval $(-\infty, \infty)$. The range is the interval $(0, \infty)$.

The $\boxed{e^x}$ key on a calculator is used to evaluate the natural exponential function.

ACCENT ON TECHNOLOGY: THE NATURAL EXPONENTIAL FUNCTION KEY

To compute the amount to which $12,000 will grow if invested for 18 years at 10% annual interest, compounded continuously, we substitute 12,000 for P, 0.10 for r, and 18 for t in the formula for continuous compound interest.

$$A = Pe^{rt}$$
$$A = 12{,}000e^{0.10(18)} \qquad \text{Write 10% as 0.10.}$$
$$= 12{,}000e^{1.8}$$

To evaluate this expression using a scientific calculator, we enter

$$1.8 \boxed{e^x} \boxed{\times} 12000 \boxed{=} \qquad\qquad \boxed{72595.76957}$$

Using a graphing calculator, we enter

$$12000 \boxed{\times} \boxed{\text{2nd}} \boxed{e^x} 1.8 \boxed{)} \boxed{\text{ENTER}} \qquad \boxed{\begin{array}{l} 12000*e^{\wedge}(1.8) \\ \qquad 72595.76957 \end{array}}$$

After 18 years, the account will contain $72,595.77. This is $1,589.03 more than the result in Example 6 in the previous section, where interest was compounded quarterly.

EXAMPLE 1

ELEMENTARY & INTERMEDIATE
Algebra *f(x)* **Now**™

Solution

Investing. If $25,000 accumulates interest at an annual rate of 8%, compounded continuously, find the balance in the account in 50 years.

$$A = Pe^{rt} \qquad\qquad \text{This is the formula for continuous compound interest.}$$
$$A = 25{,}000e^{0.08(50)} \qquad \text{Write 8% as 0.08.}$$
$$= 25{,}000e^{4}$$
$$\approx 1{,}364{,}953.75 \qquad \text{Use a calculator.}$$

In 50 years, the balance will be $1,364,953.75—more than a million dollars.

Self Check 1 Find the balance in 60 years.

■ GRAPHING THE NATURAL EXPONENTIAL FUNCTION

To graph $f(x) = e^x$, we construct a table of function values by choosing several values for x and finding the corresponding values of $f(x)$. For example, if $x = -2$, we have

$$f(x) = e^x$$
$$f(-2) = e^{-2}$$
$$= 0.135335283\ldots \qquad \text{Use a calculator.}$$
$$\approx 0.1 \qquad\qquad \text{Round to the nearest tenth.}$$

We enter $(-2, 0.1)$ in the table. Similarly, we find $f(-1)$, $f(0)$, $f(1)$ and $f(2)$, enter each result in the table, and plot the ordered pairs. We draw a smooth curve through the points to get the graph on the next page.

From the graph, we can verify that the domain of $f(x) = e^x$ is the interval $(-\infty, \infty)$ and the range is the interval $(0, \infty)$. Since the graph passes the horizontal line test, the function is one-to-one.

Note that as x decreases, the values of $f(x)$ decrease and approach 0. Thus, the x-axis is a horizontal asymptote of the graph. The graph does not have an x-intercept, the y-intercept is $(0, 1)$, and the graph passes through the point $(1, e)$.

$f(x) = e^x$

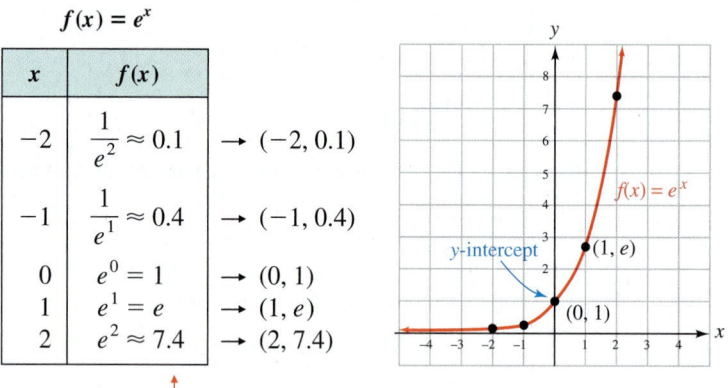

x	$f(x)$	
-2	$\dfrac{1}{e^2} \approx 0.1$	$\rightarrow (-2, 0.1)$
-1	$\dfrac{1}{e^1} \approx 0.4$	$\rightarrow (-1, 0.4)$
0	$e^0 = 1$	$\rightarrow (0, 1)$
1	$e^1 = e$	$\rightarrow (1, e)$
2	$e^2 \approx 7.4$	$\rightarrow (2, 7.4)$

The outputs can be found using the $\boxed{e^x}$ key on a calculator. Some are rounded to the nearest tenth to make point-plotting easier.

To graph more complicated natural exponential functions, point-plotting can be tedious. In such cases, a graphing calculator is a useful tool.

ACCENT ON TECHNOLOGY: GRAPHING EXPONENTIAL FUNCTIONS

The figure shows the calculator graph of $f(x) = 3e^{-x/2}$. To graph this function, we enter the right-hand side of the equation after the symbol $Y_1 =$. The display will show the equation

$$Y_1 = 3(e \wedge (-X/2))$$

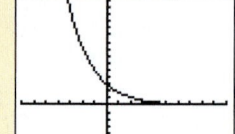

■ VERTICAL AND HORIZONTAL TRANSLATIONS

We can illustrate the effects of vertical and horizontal translations of the natural exponential function by using a graphing calculator.

ACCENT ON TECHNOLOGY: TRANSLATIONS OF THE NATURAL EXPONENTIAL FUNCTION

Figure (a) on the next page shows the calculator graphs of $f(x) = e^x$, $g(x) = e^x + 5$, and $h(x) = e^x - 3$. To graph these functions, we enter the right-hand sides of the equations after the symbols $Y_1 =$, $Y_2 =$, and $Y_3 =$. The display will show

$$Y_1 = e \wedge (X) \qquad Y_2 = e \wedge (X) + 5 \qquad Y_3 = e \wedge (X) - 3$$

After graphing these functions, we can see that the graph of $g(x) = e^x + 5$ is 5 units above the graph of $f(x) = e^x$, and that the graph of $h(x) = e^{x-3}$ is 3 units to the right of the graph of $f(x) = e^x$.

Figure (b) shows the calculator graphs of $f(x) = e^x$, $g(x) = e^{x+5}$ and $h(x) = e^{x-3}$. To graph these functions, we enter the right-hand sides of the equations after the symbols $Y_1 =$, $Y_2 =$, and $Y_3 =$. The display will show

$$Y_1 = e \wedge (X) \qquad Y_2 = e \wedge (X + 5) \qquad Y_3 = e \wedge (X - 3)$$

After graphing these functions, we can see that the graph of $g(x) = e^{x+5}$ is 5 units to the left of the graph of $f(x) = e^x$, and that the graph of $h(x) = e^{x-3}$ is 3 units to the right of the graph of $f(x) = e^x$.

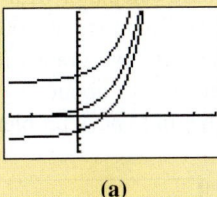   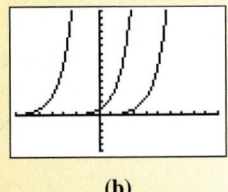

(a)   **(b)**

■ MALTHUSIAN POPULATION GROWTH

An equation based on the natural exponential function provides a model for **population growth.** In the **Malthusian model for population growth,** the future population of a colony is related to the present population by the formula $A = Pe^{rt}$.

EXAMPLE 2

City planning.   The population of a city is currently 15,000, but economic conditions are causing the population to decrease 3% each year. If this trend continues, find the population in 30 years.

Solution   Since the population is decreasing 3% each year, the annual growth rate is -3%, or -0.03. We can substitute -0.03 for r, 30 for t, and 15,000 for P in the formula for exponential growth and find A.

$$A = Pe^{rt}$$
$$A = 15{,}000e^{-0.03(30)}$$
$$= 15{,}000e^{-0.9}$$
$$\approx 6{,}099$$

In 30 years, the expected population will be 6,099.

Success Tip

For quantities that are decreasing, remember to enter a negative value for r, the annual rate, in the formula $A = Pe^{rt}$.

Self Check 2   Find the population in 50 years.

The English economist Thomas Robert Malthus (1766–1834) was a pioneer in studying population. He believed that poverty and starvation were unavoidable, because the human population tends to grow exponentially but the food supply tends to grow linearly.

EXAMPLE 3

ELEMENTARY &
INTERMEDIATE
Algebra *f(x)* **Now**™

The Malthusian model. Suppose that a country with a population of 1,000 people is growing exponentially according to the formula

$$P = 1,000e^{0.02t} \qquad \text{The annual growth rate is 2% = 0.02.}$$

where t is in years. Furthermore, assume that the food supply F, measured in adequate food per day per person, is growing linearly according to the formula

$$F = 30.625t + 2,000 \qquad (t \text{ is time in years})$$

In how many years will the population outstrip the food supply?

Solution   We can use a graphing calculator with window settings of $[0, 100]$ for x and $[0, 10,000]$ for y. After graphing the functions, we obtain figure (a). If we trace, as in figure (b), we can find the point where the two graphs intersect. From the graph, we can see that the food supply will be adequate for about 71 years. At that time, the population of approximately 4,200 people will begin to have problems.

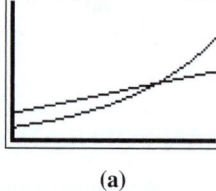

(a)

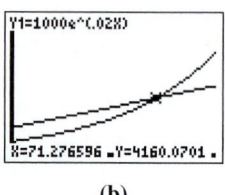

(b)

Self Check 3   Suppose that the population grows at a 2.5% rate. Use a graphing calculator to determine for how many years the food supply will be adequate.

■ BASE-*e* EXPONENTIAL FUNCTION MODELS

EXAMPLE 4

ELEMENTARY &
INTERMEDIATE
Algebra *f(x)* **Now**™

Baking.   A mother takes a cake out of the oven and sets it on a rack to cool. The function $T(t) = 68 + 220e^{-0.18t}$ gives the cake's temperature in degrees Fahrenheit after it has cooled for t minutes. If her children will be home from school in 20 minutes, will the cake have cooled enough for the children to eat it?

Solution   When the children arrive home, the cake will have cooled for 20 minutes. To find the temperature of the cake at that time, we need to find $T(20)$.

$$T(t) = 68 + 220e^{-0.18t}$$
$$T(20) = 68 + 220e^{-0.18(20)} \qquad \text{Substitute 20 for } t.$$
$$= 68 + 220e^{-3.6}$$
$$\approx 74.0 \qquad \text{Use a calculator.}$$

When the children return home, the temperature of the cake will be about 74°, and it can be eaten.

Answers to Self Checks   **1.** $3,037,760.44   **2.** 3,347   **3.** about 51 years

11.4   STUDY SET   ELEMENTARY & INTERMEDIATE Algebra *f(x)* Now™

VOCABULARY   **Refer to the graph of $f(x) = e^x$.**

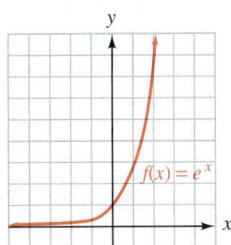

$f(x) = e^x$

1. What is the name of the function $f(x) = e^x$?

2. What is the domain of the function?

3. What is the range of the function?

4. a. What is the y-intercept of the graph?

   b. What is the x-intercept of the graph?

5. Is the function one-to-one?

6. What is an asymptote of the graph?

7. Is f an increasing or a decreasing function?

8. The graph passes through the point $(1, y)$. What is y?

CONCEPTS   **Fill in the blanks.**

9. In _____ compound interest, the number of compoundings is infinitely large.

10. The formula for continuous compound interest is $A = $ ▢ .

11. To two decimal places, the value of e is ▢ .

12. If n gets larger and larger, the value of $\left(1 + \frac{1}{n}\right)^n$ approaches the value of ▢ .

13. Graph each irrational number on the number line. $\left\{\pi, e, \sqrt{2}\right\}$

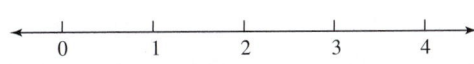

14. Complete the table of values. Round to the nearest hundredth.

x	-2	-1	0	1	2
e^x					

15. POPULATION OF THE UNITED STATES
Graph the U.S. census population figures shown in the table (in millions). What type of function does it appear could be used to model the population?

Year	Population	Year	Population
1790	3.9	1900	76.0
1800	5.3	1910	92.2
1810	7.2	1920	106.0
1820	9.6	1930	123.2
1830	12.9	1940	132.1
1840	17.0	1950	151.3
1850	23.1	1960	179.3
1860	31.4	1970	203.3
1870	38.5	1980	226.5
1880	50.1	1990	248.7
1890	62.9	2000	281.4

16. What is the Malthusian population growth formula?

17. The function $f(x) = e^x$ is graphed to the right and the TRACE feature is used. What is the y-coordinate of the point on the graph having an x-coordinate of 1? What is the name given this number?

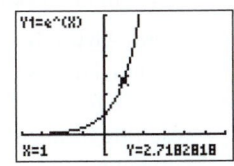

18. The illustration shows a table of values for $f(x) = e^x$. As x decreases, what happens to the values of $f(x)$ listed in the Y_1 column? Will the value of $f(x)$ ever be 0 or negative?

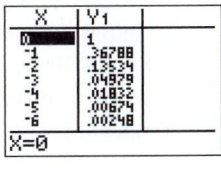

NOTATION   **Evaluate A in the formula $A = Pe^{rt}$ for the following values of P, r, and t.**

19. $P = 1,000$, $r = 0.09$ and $t = 10$

$$A = \boxed{} \, e^{(0.09)()}$$

$$= 1,000e^{\boxed{}}$$

$$\approx \boxed{} \, (2.459603111)$$

$$\approx 2,459.603111$$

20. $P = 1,000$, $r = 0.12$ and $t = 50$

$$A = 1,000e^{()(50)}$$
$$= \boxed{}\, e^6$$
$$\approx 1,000(\boxed{})$$
$$\approx 403,428.7935$$

PRACTICE   Graph each function.

21. $f(x) = e^x + 1$ **22.** $f(x) = e^x - 2$

23. $y = e^{x+3}$ **24.** $y = e^{x-5}$

25. $f(x) = -e^x$ **26.** $f(x) = -e^x + 1$

27. $f(x) = 2e^x$ **28.** $f(x) = \dfrac{1}{2}e^x$

APPLICATIONS   In Exercises 29–34, assume that there are no deposits or withdrawals.

29. CONTINUOUS COMPOUND INTEREST   An initial investment of $5,000 earns 8.2% interest, compounded continuously. What will the investment be worth in 12 years?

30. CONTINUOUS COMPOUND INTEREST   An initial investment of $2,000 earns 8% interest, compounded continuously. What will the investment be worth in 15 years?

31. COMPARISON OF COMPOUNDING METHODS   An initial deposit of $5,000 grows at an annual rate of 8.5% for 5 years. Compare the final balances resulting from annual compounding and continuous compounding.

32. COMPARISON OF COMPOUNDING METHODS   An initial deposit of $30,000 grows at an annual rate of 8% for 20 years. Compare the final balances resulting from annual compounding and continuous compounding.

33. DETERMINING THE INITIAL DEPOSIT   An account now contains $11,180 and has been accumulating interest at 7% annual interest, compounded continuously, for 7 years. Find the initial deposit.

34. DETERMINING THE PREVIOUS BALANCE   An account now contains $3,610 and has been accumulating interest at 8% annual interest, compounded continuously. How much was in the account 4 years ago?

35. WORLD POPULATION GROWTH   The population of the Earth is approximately 6.1 billion people and is growing at an annual rate of 1.4%. Assuming a Malthusian growth model, find the world population in 30 years.

36. HIGHS AND LOWS   Somalia, in eastern Africa, has one of the greatest population growth rates in the world. Bulgaria, in southeastern Europe, has one of the smallest. Assuming a Malthusian growth model, complete the table.

Country	Population 2003	Annual growth rate	Estimated population 2015
Somalia	8,025,190	3.43%	
Bulgaria	7,537,929	−1.09%	

Source: nationmaster.com

37. POPULATION GROWTH   The growth of a population is modeled by

$$P = 173e^{0.03t}$$

How large will the population be when $t = 20$?

38. POPULATION DECLINE   The decline of a population is modeled by

$$P = (1.2 \times 10^6)e^{-0.008t}$$

How large will the population be when $t = 30$?

39. EPIDEMICS   The spread of hoof and mouth disease through a herd of cattle can be modeled by the formula

$$P = P_0 e^{0.27t} \qquad (t \text{ is in days})$$

If a rancher does not act quickly to treat two cases, how many cattle will have the disease in 12 days?

40. OCEANOGRAPHY   The width w (in millimeters) of successive growth spirals of the sea shell *Catapulus voluto,* shown below, is given by the exponential function

$$w = 1.54e^{0.503n}$$

where n is the spiral number. Find the width, to the nearest tenth of a millimeter, of the sixth spiral.

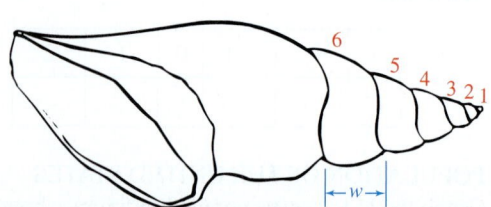

41. HALF-LIFE OF A DRUG The quantity of a prescription drug in the bloodstream of a patient t hours after it is administered can be modeled by an exponential function. (See the graph.) Determine the time it takes to eliminate half of the initial dose from the body.

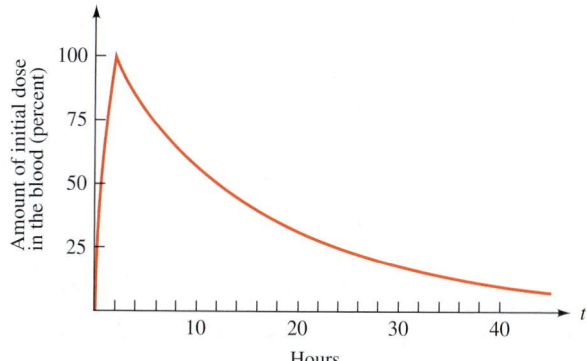

Hours

42. MEDICINE The concentration x of a certain prescription drug in an organ after t minutes is given by

$$x = 0.08\left(1 - e^{-0.1t}\right)$$

Find the concentration of the drug at 30 minutes.

43. SKYDIVING Before the parachute opens, a skydiver's velocity v in meters per second is given by

$$v = 50\left(1 - e^{-0.2t}\right)$$

Find the velocity after 20 seconds of free fall.

44. FREE FALL After t seconds a certain falling object has a velocity v given by

$$v = 50\left(1 - e^{-0.3t}\right)$$

Which is falling faster after 2 seconds—the object or the skydiver in Exercise 43?

Use a graphing calculator to solve each problem.

45. THE MALTHUSIAN MODEL In Example 3, suppose that better farming methods changed the formula for food growth to $F = 31t + 2,000$. How long would the food supply be adequate?

46. THE MALTHUSIAN MODEL In Example 3, suppose that a birth-control program changed the formula for population growth to $P = 1,000e^{0.01t}$. How long would the food supply be adequate?

WRITING

47. Explain why the graph of $y = e^x - 5$ is five units below the graph of $y = e^x$.

48. A feature article in a newspaper stated that the sport of snowboarding was growing *exponentially*. Explain what the author of the article meant by that.

49. As of 2003, the population growth rate for Russia was -0.3% annually. What are some of the consequences for a country that has a negative population growth?

50. What is e?

REVIEW Simplify each expression. Assume that all variables represent positive numbers.

51. $\sqrt{240x^5}$

52. $\sqrt[3]{-125x^5y^4}$

53. $4\sqrt{48y^3} - 3y\sqrt{12y}$

54. $\sqrt[4]{48z^5} + \sqrt[4]{768z^5}$

CHALLENGE PROBLEMS

55. Is the statement $e^e > e^3$ true or false? Explain your reasoning.

56. Graph the function defined by the equation

$$f(x) = \frac{e^x + e^{-x}}{2}$$

from $x = -2$ to $x = 2$. The graph will look like a parabola, but it is not. The graph, called a **catenary,** is important in the design of power distribution networks, because it represents the shape of a uniform flexible cable whose ends are suspended from the same height. The function is called the **hyperbolic cosine function.**

57. If $e^{t+5} = ke^t$, find k.

58. If $e^{5t} = k^t$, find k.

11.5 Logarithmic Functions

- The definition of logarithm
- Base-10 logarithms
- Vertical and horizontal translations
- Exponential and logarithmic form
- Logarithmic functions and their graphs
- Applications of logarithms

In this section, we will study inverses of exponential functions. These functions are called *logarithmic* functions and they can be used to solve applied problems from fields such as electronics, seismology (the study of earthquakes), and business.

■ THE DEFINITION OF LOGARITHM

The graph of the exponential function $f(x) = 2^x$ is shown in red below. Since it passes the horizontal line test, it is a one-to-one function, and therefore, has an inverse. To graph f^{-1}, we interchange the coordinates of the ordered pairs in the table, plot those points, and draw a smooth curve through them, as shown below in blue. As expected, the graphs of f and f^{-1} are symmetric with respect to the line $y = x$.

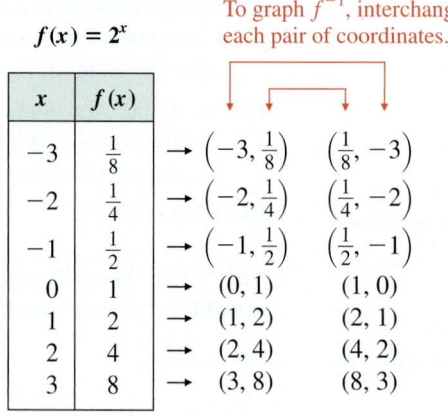

$$f(x) = 2^x$$

To graph f^{-1}, interchange each pair of coordinates.

x	$f(x)$		
-3	$\frac{1}{8}$	→ $\left(-3, \frac{1}{8}\right)$	$\left(\frac{1}{8}, -3\right)$
-2	$\frac{1}{4}$	→ $\left(-2, \frac{1}{4}\right)$	$\left(\frac{1}{4}, -2\right)$
-1	$\frac{1}{2}$	→ $\left(-1, \frac{1}{2}\right)$	$\left(\frac{1}{2}, -1\right)$
0	1	→ $(0, 1)$	$(1, 0)$
1	2	→ $(1, 2)$	$(2, 1)$
2	4	→ $(2, 4)$	$(4, 2)$
3	8	→ $(3, 8)$	$(8, 3)$

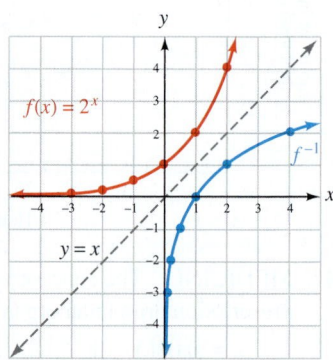

To write an equation for the inverse of $f(x) = 2^x$, we proceed as follows:

$$f(x) = 2^x$$
$$y = 2^x \quad \text{Replace } f(x) \text{ with } y.$$
$$x = 2^y \quad \text{Interchange the variables } x \text{ and } y.$$

Now we must solve for y. However, we have not studied methods for solving for a variable located in an exponent. Instead, we translate the relationship $x = 2^y$ into words:

$$y = \text{the power to which we raise 2 to get } x$$

Finally, we substitute $f^{-1}(x)$ for y.

$$f^{-1}(x) = \text{the power to which we raise 2 to get } x$$

A new notation, called **logarithmic notation,** enables us to write the inverse in simpler form. If we define the symbol $\log_2 x$ to mean *the power to which we raise 2 to get x,* we can write the equation for the inverse as

$$f^{-1}(x) = \log_2 x \quad \text{Read } \log_2 x \text{ as "the logarithm, base 2, of } x\text{" or "log, base 2, of } x.\text{"}$$

We have found that the inverse of the exponential function $f(x) = 2^x$ is $f^{-1}(x) = \log_2 x$. To find the inverse of exponential functions with other bases, such as $f(x) = 3^x$ and $f(x) = 10^x$, we define logarithm in the following way.

Definition of Logarithm	For all positive numbers b, where $b \neq 1$, and all positive numbers x,
	$y = \log_b x$ is equivalent to $x = b^y$

This definition guarantees that any pair (x, y) that satisfies the logarithmic equation $y = \log_b x$ also satisfies the exponential equation $x = b^y$. Because of this relationship, a statement written in logarithmic form can be written in an equivalent exponential form, and vice versa. The following diagram will help you remember the respective positions of the exponent and base in each form.

$$y = \log_b x \qquad x = b^y$$

Exponent

Base

■ EXPONENTIAL AND LOGARITHMIC FORM

Success Tip

Study the relationship between exponential and logarithmic statements carefully. Your success with the material in the rest of this chapter depends greatly on your understanding of this definition.

The following table shows several pairs of equivalent statements.

Logarithmic form	*Exponential form*
$\log_2 8 = 3$	$2^3 = 8$
$\log_3 81 = 4$	$3^4 = 81$
$\log_4 4 = 1$	$4^1 = 4$
$\log_5 \dfrac{1}{125} = -3$	$5^{-3} = \dfrac{1}{125}$

EXAMPLE 1   Write as an exponential equation: **a.** $\log_4 64 = 3$,   **b.** $\log_7 \sqrt{7} = \dfrac{1}{2}$,   and   **c.** $\log_6 \dfrac{1}{36} = -2$.

Solution   **a.** $\log_4 64 = 3$ is equivalent to $4^3 = 64$.

b. $\log_7 \sqrt{7} = \dfrac{1}{2}$ is equivalent to $7^{1/2} = \sqrt{7}$.

c. $\log_6 \dfrac{1}{36} = -2$ is equivalent to $6^{-2} = \dfrac{1}{36}$.

Self Check 1   Write $\log_2 128 = 7$ as an exponential equation.

EXAMPLE 2   Write as a logarithmic equation: **a.** $8^0 = 1$,   **b.** $6^{1/3} = \sqrt[3]{6}$,   and   **c.** $\left(\dfrac{1}{4}\right)^2 = \dfrac{1}{16}$.

Solution   **a.** $8^0 = 1$ is equivalent to $\log_8 1 = 0$

b. $6^{1/3} = \sqrt[3]{6}$ is equivalent to $\log_6 \sqrt[3]{6} = \dfrac{1}{3}$

c. $\left(\dfrac{1}{4}\right)^2 = \dfrac{1}{16}$ is equivalent to $\log_{1/4} \dfrac{1}{16} = 2$

Self Check 2   Write $9^{-1} = \dfrac{1}{9}$ as a logarithmic equation.

Certain logarithmic equations can be solved by writing them as exponential equations.

EXAMPLE 3

ELEMENTARY &
INTERMEDIATE
Algebra *f(x)* **Now**™

Solve each equation for x: **a.** $\log_x 25 = 2$,   **b.** $\log_3 x = -3$,   and   **c.** $\log_{1/2} \dfrac{1}{16} = x$.

Solution   **a.** Since $\log_x 25 = 2$ is equivalent to $x^2 = 25$, we can solve $x^2 = 25$ to find x.

$$x^2 = 25$$
$$x = \pm\sqrt{25} \qquad \text{Use the square root property.}$$
$$x = \pm 5$$

In the expression $\log_x 25$, the base of the logarithm is x. Because the base must be positive, we discard -5 and we have

$$x = 5$$

The solution is 5. To check, verify that $\log_5 25 = 2$.

b. Since $\log_3 x = -3$ is equivalent to $3^{-3} = x$, we can solve $3^{-3} = x$ to find x.

$$3^{-3} = x$$
$$\dfrac{1}{3^3} = x$$
$$x = \dfrac{1}{27}$$

The solution is $\dfrac{1}{27}$. To check, verify that $\log_3 \dfrac{1}{27} = -3$.

Success Tip

To solve this equation, we note that if the bases are equal, the exponents must be equal.

$$\left(\dfrac{1}{2}\right)^x = \left(\dfrac{1}{2}\right)^4$$

c. Since $\log_{1/2} \dfrac{1}{16} = x$ is equivalent to $\left(\dfrac{1}{2}\right)^x = \dfrac{1}{16}$, we can solve $\left(\dfrac{1}{2}\right)^x = \dfrac{1}{16}$ to find x.

$$\left(\dfrac{1}{2}\right)^x = \dfrac{1}{16}$$
$$\left(\dfrac{1}{2}\right)^x = \left(\dfrac{1}{2}\right)^4 \qquad \text{Write } \dfrac{1}{16} \text{ as a power of } \dfrac{1}{2} \text{ to match the bases: } \dfrac{1}{2} \cdot \dfrac{1}{2} \cdot \dfrac{1}{2} \cdot \dfrac{1}{2} = \dfrac{1}{16}.$$
$$x = 4 \qquad \text{Since the bases are the same, and since exponential functions are one-to-one, the exponents must be equal.}$$

The solution is 4. To check, verify that $\log_{1/2} \dfrac{1}{16} = 4$.

Self Check 3 Solve each equation for x: **a.** $\log_x 49 = 2$, **b.** $\log_{1/3} x = 2$, and **c.** $\log_6 216 = x$.

In the previous examples, we have seen that the logarithm of a number is an exponent. In fact,

$\log_b x$ is the exponent to which b is raised to get x.

Translating this statement into symbols, we have

$$b^{\log_b x} = x$$

EXAMPLE 4

Evaluate each logarithmic expression: **a.** $\log_8 64$, **b.** $\log_3 \dfrac{1}{3}$, and **c.** $\log_4 2$.

Solution **a.** $\log_8 64 = 2$ Ask: "To what power must we raise 8 to get 64?" Since $8^2 = 64$, the answer is: the 2nd power.

b. $\log_3 \dfrac{1}{3} = -1$ Ask: "To what power must we raise 3 to get $\frac{1}{3}$?" Since $3^{-1} = \frac{1}{3}$, the answer is: the -1 power.

c. $\log_4 2 = \dfrac{1}{2}$ Ask: "To what power must we raise 4 to get 2?" Since $\sqrt{4} = 4^{1/2} = 2$, the answer is: the $\frac{1}{2}$ power.

Self Check 4 Evaluate each expression: **a.** $\log_9 81$, **b.** $\log_4 \dfrac{1}{16}$, and **c.** $\log_9 3$.

■ BASE-10 LOGARITHMS

The Language of Algebra

London professor Henry Briggs (1561–1630) and Scottish lord John Napier (1550–1617) are credited with developing the concept of *common logarithms*. Their tables of logarithms were useful tools at that time for those performing large calculations.

For computational purposes and in many applications, we will use base-10 logarithms (also called **common logarithms**). When the base b is not indicated in the notation $\log x$, we assume that $b = 10$:

$$\log x \quad \text{means} \quad \log_{10} x$$

The table below shows several pairs of equivalent statements involving base-10 logarithms.

Logarithmic form	*Exponential form*	
$\log 100 = 2$	$10^2 = 100$	Read log 100 as "log of 100."
$\log \dfrac{1}{10} = -1$	$10^{-1} = \dfrac{1}{10}$	
$\log 1 = 0$	$10^0 = 1$	

In general, we have

$$\log_{10} 10^x = x$$

EXAMPLE 5   Evaluate each logarithmic expression, if possible: **a.** $\log 1,000$,   **b.** $\log \dfrac{1}{100}$,
c. $\log 10$,   and   **d.** $\log(-10)$.

Solution   **a.** $\log 1,000 = 3$ Ask: "To what power must we raise 10 to get 1,000?" Since $10^3 = 1,000$, the answer is: the 3rd power.

b. $\log \dfrac{1}{100} = -2$ Ask: "To what power must we raise 10 to get $\frac{1}{100}$?" Since $10^{-2} = \frac{1}{100}$, the answer is: the -2 power.

c. $\log 10 = 1$ Ask: "To what power must we raise 10 to get 10?" Since $10^1 = 10$, the answer is: the 1st power.

d. To find $\log(-10)$, we must find a power of 10 such that $10^? = -10$. There is no such number. Thus,

$$\log(-10) \text{ is undefined}$$

Self Check 5   Evaluate each expression: **a.** $\log 10,000$,   **b.** $\log \dfrac{1}{1,000}$,   and   **c.** $\log 0$.

Many logarithmic expressions are not as easy to evaluate as those in the previous example. For instance, to find $\log 2.34$, we ask, "To what power must we raise 10 to get 2.34?" The answer certainly isn't obvious. In such cases, we use a calculator.

ACCENT ON TECHNOLOGY: EVALUATING LOGARITHMS

To find $\log 2.34$ with a scientific calculator we enter

2.34 $\boxed{\text{LOG}}$ $\boxed{.369215857}$

On some calculators, the $\boxed{10^x}$ key also serves as the $\boxed{\log}$ key when $\boxed{\text{2nd}}$ or $\boxed{\text{SHIFT}}$ is pressed. This is because $f(x) = 10^x$ and $f(x) = \log x$ are inverses.
To use a graphing calculator, we enter

$\boxed{\text{LOG}}$ 2.34 $\boxed{)}$ $\boxed{\text{ENTER}}$

```
log(2.34)
        .3692158574
```

To four decimal places, $\log 2.34 = 0.3692$. This means, $10^{0.3692} \approx 2.34$.
If we attempt to evaluate logarithmic expressions such as $\log 0$, or the logarithm of a negative number, such as $\log(-5)$, the following error statements are displayed.

$\boxed{\text{Error}}$

```
ERR:DOMAIN
1:QUIT
2:Goto
```

```
ERR:NONREAL ANS
1:QUIT
2:Goto
```

EXAMPLE 6

Solve: $\log x = 0.3568$. Round to four decimal places.

Solution   The equation $\log x = 0.3568$ is equivalent to $10^{0.3568} = x$. To find x with a scientific calculator, we enter

$$10 \quad \boxed{y^x} \quad .3568 \quad \boxed{=}$$

The display will read $\boxed{2.274049951}$. To four decimal places,

$$x = 2.2740$$

If your calculator has a $\boxed{10^x}$ key, enter .3568 and press it to get the same result. The solution is 2.2740. To check, use your calculator to verify that $\log 2.2740 \approx 0.3568$.

Self Check 6   Solve: $\log x = 1.87737$. Round to four decimal places.

■ LOGARITHMIC FUNCTIONS AND THEIR GRAPHS

Because an exponential function defined by $f(x) = b^x$ is one-to-one, it has an inverse function that is defined by $x = b^y$. When we write $x = b^y$ in the equivalent form $y = \log_b x$, the result is called a *logarithmic function*.

Logarithmic Functions   If $b > 0$ and $b \neq 1$, the **logarithmic function with base b** is defined by

$$f(x) = \log_b x \quad \text{or} \quad y = \log_b x$$

The domain of $f(x) = \log_b x$ is the interval $(0, \infty)$ and the range is the interval $(-\infty, \infty)$.

Since every logarithmic function is the inverse of a one-to-one exponential function, logarithmic functions are one-to-one.

We can use the point-plotting method to graph logarithmic functions. For example, to graph $f(x) = \log_2 x$, we construct a table of function values, plot the resulting ordered pairs, and then draw a smooth curve through the points to get the graph, as shown in figure (a). To graph $f(x) = \log_{1/2} x$, we use the same procedure, as shown in figure (b).

$f(x) = \log_2 x$

x	$f(x)$	
$\frac{1}{4}$	-2	$\rightarrow \left(\frac{1}{4}, -2\right)$
$\frac{1}{2}$	-1	$\rightarrow \left(\frac{1}{2}, -1\right)$
1	0	$\rightarrow (1, 0)$
2	1	$\rightarrow (2, 1)$
4	2	$\rightarrow (4, 2)$
8	3	$\rightarrow (8, 3)$

Because the base of the function is 2, choose values for x that are integer powers of 2.

(a)

$f(x) = \log_{1/2} x$

x	$f(x)$	
$\frac{1}{4}$	2	$\rightarrow \left(\frac{1}{4}, 2\right)$
$\frac{1}{2}$	1	$\rightarrow \left(\frac{1}{2}, 1\right)$
1	0	$\rightarrow (1, 0)$
2	-1	$\rightarrow (2, -1)$
4	-2	$\rightarrow (4, -2)$
8	-3	$\rightarrow (8, -3)$

Because the base of the function is $\frac{1}{2}$, choose values for x that are integer powers of $\frac{1}{2}$.

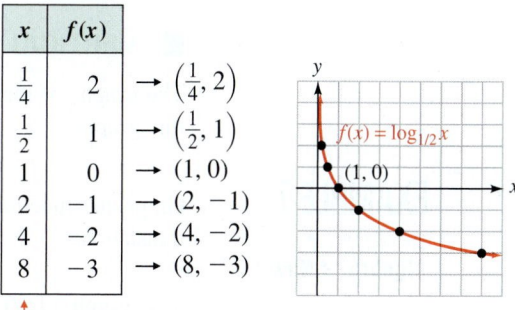

(b)

The graphs of all logarithmic functions are similar to those shown below. If $b > 1$, the logarithmic function is increasing, as in figure (a). If $0 < b < 1$, the logarithmic function is decreasing, as in figure (b).

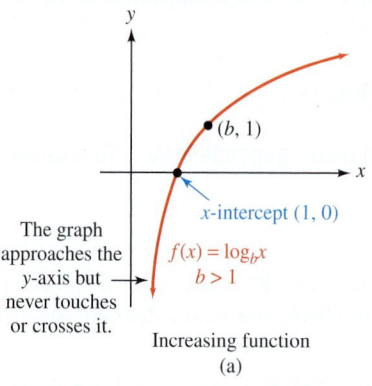

The graph approaches the y-axis but never touches or crosses it.

Increasing function
(a)

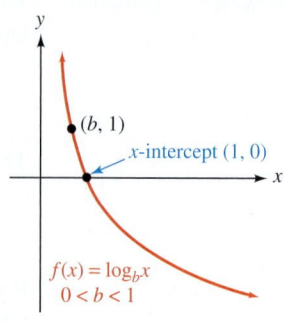

Decreasing function
(b)

Properties of Logarithmic Functions

The graph of $f(x) = \log_b x$ (or $y = \log_b x$) has the following properties.

1. It passes through the point $(1, 0)$.
2. It passes through the point $(b, 1)$.
3. The y-axis (the line $x = 0$) is an asymptote.
4. The domain is $(0, \infty)$ and the range is $(-\infty, \infty)$.

Caution

Since the domain of the logarithmic function is the set of positive real numbers, it is impossible to find the logarithm of 0 or the logarithm of a negative number. For example,

$$\log_2(-4) \quad \text{and} \quad \log_2 0$$

are undefined.

The exponential and logarithmic functions are inverses of each other, so their graphs have symmetry about the line $y = x$. The graphs of $f(x) = \log_b x$ and $g(x) = b^x$ are shown in figure (a) when $b > 1$ and in figure (b) when $0 < b < 1$.

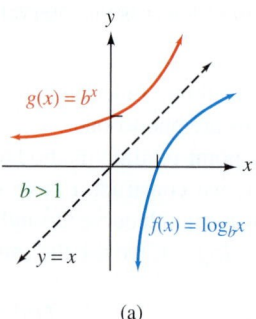

(a)

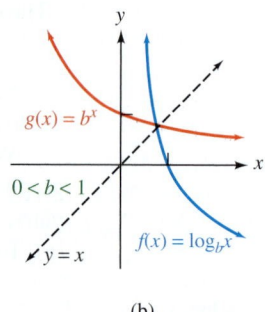

(b)

■ VERTICAL AND HORIZONTAL TRANSLATIONS

The graphs of many functions involving logarithms are translations of the basic logarithmic graphs.

EXAMPLE 7

ELEMENTARY & INTERMEDIATE

Algebra $f(x)$ **Now**™

Notation

Since $y = f(x)$, we can write $f(x) = 3 + \log_2 x$ as $y = 3 + \log_2 x$.

Graph the function $f(x) = 3 + \log_2 x$ and describe the translation.

The graph of $f(x) = 3 + \log_2 x$ is identical to the graph of $g(x) = \log_2 x$, except that it is translated 3 units upward.

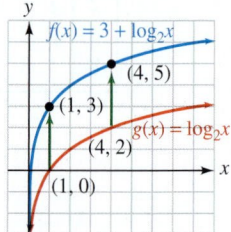

Self Check 7 Graph $y = (\log_3 x) - 2$ and describe the translation.

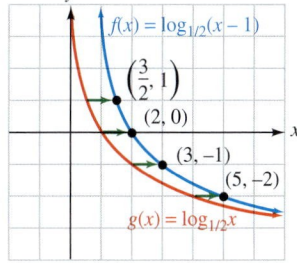

EXAMPLE 8

ELEMENTARY &
INTERMEDIATE
Algebra *f(x)* **Now**™ **Solution**

Graph $f(x) = \log_{1/2}(x - 1)$ and describe the translation.

The graph of $f(x) = \log_{1/2}(x - 1)$ is identical to the graph of $g(x) = \log_{1/2} x$, except that it is translated 1 unit to the right.

Notation

Parentheses are used to write the function in Example 8, because $f(x) = \log_{1/2} x - 1$ could be interpreted as $f(x) = \log_{1/2}(x - 1)$ or as $f(x) = (\log_{1/2} x) - 1$.

Self Check 8 Graph $f(x) = \log_{1/3}(x + 2)$ and describe the translation.

ACCENT ON TECHNOLOGY: GRAPHING LOGARITHMIC FUNCTIONS

To use a calculator to graph the logarithmic function $f(x) = -2 + \log_{10} \frac{x}{2}$, we enter the right-hand side of the equation after the symbol $Y_1 =$. The display will show the equation

$$Y_1 = -2 + \log(X/2)$$

If we use window settings of $[-1, 5]$ for x and $[-4, 1]$ for y and press the $\boxed{\text{GRAPH}}$ key, we will obtain the graph shown.

■ APPLICATIONS OF LOGARITHMS

Common logarithms are used in electrical engineering to express the voltage gain (or loss) of an electronic device such as an amplifier. The unit of gain (or loss), called the **decibel,** is defined by a logarithmic relation.

Decibel Voltage Gain If E_O is the output voltage of a device and E_I is the input voltage, the decibel voltage gain of the device (db gain) is given by

$$\text{db gain} = 20 \log \frac{E_O}{E_I}$$

EXAMPLE 9

ELEMENTARY &
INTERMEDIATE
Algebra $f(x)$ Now™

db gain. If the input to an amplifier is 0.5 volt and the output is 40 volts, find the decibel voltage gain of the amplifier.

Solution   We can find the decibel voltage gain by substituting 0.5 for E_I and 40 for E_O into the formula for db gain:

$$\text{db gain} = 20 \log \frac{E_O}{E_I}$$

$$\text{db gain} = 20 \log \frac{40}{0.5}$$

$$= 20 \log 80 \qquad \text{Divide: } \frac{40}{0.5} = 80.$$

$$\approx 38 \qquad \text{Use a calculator: } 20 \log 80 \text{ means } 20 \cdot \log 80.$$

The amplifier provides a 38-decibel voltage gain.

In seismology, common logarithms are used to measure the intensity of earthquakes on the **Richter scale.** The intensity of an earthquake is given by the following logarithmic function.

Richter Scale   If R is the intensity of an earthquake, A is the amplitude (measured in micrometers), and P is the period (the time of one oscillation of the Earth's surface measured in seconds), then

$$R = \log \frac{A}{P}$$

EXAMPLE 10

Earthquakes. Find the measure on the Richter scale of an earthquake with an amplitude of 5,000 micrometers (0.5 centimeter) and a period of 0.1 second.

Solution   We substitute 5,000 for A and 0.1 for P in the Richter scale formula and simplify:

The Language of Algebra

The *Richter* scale was developed in 1935 by Charles F. Richter of the California Institute of Technology.

$$R = \log \frac{A}{P}$$

$$R = \log \frac{5,000}{0.1}$$

$$= \log 50,000 \qquad \text{Divide: } \frac{5,000}{0.1} = 50,000.$$

$$\approx 4.698970004 \qquad \text{Use a calculator.}$$

The earthquake measures about 4.7 on the Richter scale.

Answers to Self Checks   **1.** $2^7 = 128$ **2.** $\log_9 \dfrac{1}{9} = -1$ **3. a.** 7,   **b.** $\dfrac{1}{9}$,   **c.** 3   **4. a.** 2,   **b.** -2,   **c.** $\dfrac{1}{2}$

5. a. 4,   **b.** -3,   **c.** undefined   **6.** 75.3998

7. The graph of $y = \log_3 x$ is translated 2 units downward.

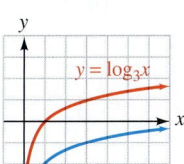

8. The graph of $g(x) = \log_{1/3} x$ is translated 2 units to the left.

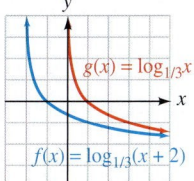

11.5   STUDY SET   ELEMENTARY & INTERMEDIATE Algebra *f(x)* Now™

VOCABULARY   Refer to the graph of $f(x) = \log_4 x$.

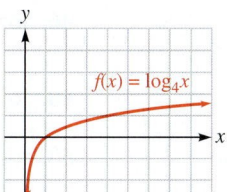

1. What type of function is $f(x) = \log_4 x$?
2. What is the domain of the function?
3. What is the range of the function?
4. **a.** What is the y-intercept of the graph?
   b. What is the x-intercept of the graph?
5. Is f a one-to-one function?
6. What is an asymptote of the graph?
7. Is f an increasing or a decreasing function?
8. The graph passes through the point $(4, y)$. What is y?

CONCEPTS   Fill in the blanks.

9. The equation $y = \log_b x$ is equivalent to the exponential equation .
10. $\log_b x$ is the _____ to which b is raised to get x.
11. The functions $f(x) = \log_{10} x$ and $f(x) = 10^x$ are _____ functions.
12. The inverse of an exponential function is called a _____ function.

Complete the table of values, where possible.

13. $y = \log x$

x	y
$\dfrac{1}{100}$	
$\dfrac{1}{10}$	
1	
10	
100	

14. $f(x) = \log_5 x$

x	$f(x)$
$\dfrac{1}{25}$	
$\dfrac{1}{5}$	
1	
5	
25	

15. $f(x) = \log_6 x$

Input	Output
-6	
0	
$\dfrac{1}{216}$	
$\sqrt{6}$	
6^8	

16. $f(x) = \log_8 x$

x	$f(x)$
-8	
0	
$\dfrac{1}{8}$	
$\sqrt{8}$	
64	

17. Use a calculator to complete the table of values for $f(x) = \log x$. Round to the nearest hundredth.

x	$f(x)$
0.5	
1	
2	
4	
6	
8	
10	

18. Graph $f(x) = \log x$ in the illustration. (See Exercise 17.) Note that the units on the x- and y-axes are different.

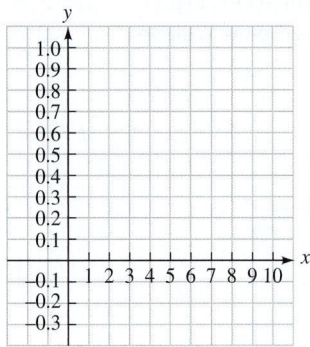

19. A table of solutions for $f(x) = \log x$ is shown below. As x decreases and gets close to 0, what happens to the values of $f(x)$?

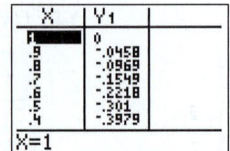

20. For each function, determine $f^{-1}(x)$.

    a. $f(x) = 10^x$ **b.** $f(x) = 3^x$

    c. $f(x) = \log x$ **d.** $f(x) = \log_2 x$

NOTATION   Fill in the blanks.

21. a. $\log x = \log \quad x$ **b.** $\log_{10} 10^x = \quad$

22. a. Read $\log_5 25$ as "log,   5,   25."

    b. Read $\log x$ as "   of x."

PRACTICE   Write each equation in exponential form.

23. $\log_3 81 = 4$ **24.** $\log_7 7 = 1$

25. $\log 10 = 1$ **26.** $\log 100 = 2$

27. $\log_4 \dfrac{1}{64} = -3$ **28.** $\log_6 \dfrac{1}{36} = -2$

29. $\log_5 \sqrt{5} = \dfrac{1}{2}$ **30.** $\log_8 \sqrt[3]{8} = \dfrac{1}{3}$

Write each equation in logarithmic form.

31. $8^2 = 64$ **32.** $10^3 = 1{,}000$

33. $4^{-2} = \dfrac{1}{16}$ **34.** $3^{-4} = \dfrac{1}{81}$

35. $\left(\dfrac{1}{2}\right)^{-5} = 32$ **36.** $\left(\dfrac{1}{3}\right)^{-3} = 27$

37. $x^y = z$ **38.** $m^n = p$

Evaluate each expression.

39. $\log_2 8$ **40.** $\log_3 9$

41. $\log_4 16$ **42.** $\log_6 216$

43. $\log_{1/2} \dfrac{1}{32}$ **44.** $\log_{1/3} \dfrac{1}{81}$

45. $\log_9 3$ **46.** $\log_{125} 5$

47. $\log \dfrac{1}{10}$ **48.** $\log \dfrac{1}{100}$

49. $\log 1{,}000{,}000$ **50.** $\log 100{,}000$

Solve for x.

51. $\log_8 x = 2$ **52.** $\log_7 x = 0$

53. $\log_{25} x = \dfrac{1}{2}$ **54.** $\log_4 x = \dfrac{1}{2}$

55. $\log_5 x = -2$ **56.** $\log_3 x = -4$

57. $\log_{36} x = -\dfrac{1}{2}$ **58.** $\log_{27} x = -\dfrac{1}{3}$

59. $\log_x 0.01 = -2$ **60.** $\log_x 0.001 = -3$

61. $\log_{27} 9 = x$ **62.** $\log_{12} x = 0$

63. $\log_x 5^3 = 3$ **64.** $\log_x 5 = 1$

65. $\log_x \dfrac{\sqrt{3}}{3} = \dfrac{1}{2}$ **66.** $\log_x \dfrac{9}{4} = 2$

67. $\log_{100} x = \dfrac{3}{2}$ **68.** $\log_x \dfrac{1}{1{,}000} = -\dfrac{3}{2}$

69. $\log_x \dfrac{1}{64} = -3$ **70.** $\log_x \dfrac{1}{100} = -2$

71. $\log_8 x = 0$ **72.** $\log_4 8 = x$

Use a calculator to find each value. Give answers to four decimal places.

73. $\log 3.25$

74. $\log 0.57$

75. $\log 0.00467$

76. $\log 375.876$

Use a calculator to find each value of x. Give answers to two decimal places.

77. $\log x = 1.4023$

78. $\log x = 0.926$

79. $\log x = -1.71$

80. $\log x = -0.5$

Graph each function. Decide whether each function is an increasing or a decreasing function.

81. $f(x) = \log_3 x$

82. $f(x) = \log_{1/3} x$

83. $y = \log_{1/2} x$

84. $y = \log_4 x$

Graph each function.

85. $f(x) = 3 + \log_3 x$

86. $f(x) = \log_{1/3} x - 1$

87. $y = \log_{1/2}(x - 2)$

88. $y = \log_4(x + 2)$

Graph each pair of inverse functions on a single coordinate system. Draw the axis of symmetry.

89. $f(x) = 6^x$
    $f^{-1}(x) = \log_6 x$

90. $f(x) = 3^x$
    $f^{-1}(x) = \log_3 x$

91. $f(x) = 5^x$
    $f^{-1}(x) = \log_5 x$

92. $f(x) = 8^x$
    $f^{-1}(x) = \log_8 x$

APPLICATIONS

93. INPUT VOLTAGE   Find the db gain of an amplifier if the input voltage is 0.71 volt when the output voltage is 20 volts.

94. OUTPUT VOLTAGE   Find the db gain of an amplifier if the output voltage is 2.8 volts when the input voltage is 0.05 volt.

95. db GAIN   Find the db gain of the amplifier shown below.

96. db GAIN   An amplifier produces an output of 80 volts when driven by an input of 0.12 volt. Find the amplifier's db gain.

97. THE RICHTER SCALE   An earthquake has amplitude of 5,000 micrometers and a period of 0.2 second. Find its measure on the Richter scale.

98. EARTHQUAKES   Find the period of an earthquake with amplitude of 80,000 micrometers that measures 6 on the Richter scale.

99. EARTHQUAKES   An earthquake with a period of $\frac{1}{4}$ second measures 4 on the Richter scale. Find its amplitude.

100. EARTHQUAKES   In 1985, Mexico City experienced an earthquake of magnitude 8.1 on the Richter scale. In 1989, the San Francisco Bay area was rocked by an earthquake measuring 7.1. By what factor must the amplitude of an earthquake change to increase its severity by 1 point on the Richter scale? (Assume that the period remains constant.)

101. DEPRECIATION   In business, equipment is often depreciated using the double declining-balance method. In this method, a piece of equipment with a life expectancy of N years, costing $\$C$, will depreciate to a value of $\$V$ in n years, where n is given by the formula

$$n = \frac{\log V - \log C}{}$$

A computer that cost $\$37,000$ has a life expectancy of 5 years. If it has depreciated to a value of $\$8,000$, how old is it?

102. DEPRECIATION   A printer worth $\$470$ when new had a life expectancy of 12 years. If it is now worth $\$189$, how old is it? (See Exercise 101.)

103. INVESTING   If $\$P$ is invested at the end of each year in an annuity earning annual interest at a rate r, the amount in the account will be $\$A$ after n years, where

$$n = \frac{\log\left[\dfrac{Ar}{P} + 1\right]}{\log(1 + r)}$$

If $\$1,000$ is invested each year in an annuity earning 12% annual interest, how long will it take for the account to be worth $\$20,000$?

104. GROWTH OF MONEY   If $\$5,000$ is invested each year in an annuity earning 8% annual interest, how long will it take for the account to be worth $\$50,000$? (See Exercise 103.)

WRITING

105. Explain the mathematical relationship between $f(x) = \log x$ and $g(x) = 10^x$.

106. Explain why it is impossible to find the logarithm of a negative number.

REVIEW Solve each equation.

107. $\sqrt[3]{6x + 4} = 4$

108. $\sqrt{3x + 4} = \sqrt{7x + 2}$

109. $\sqrt{a + 1} - 1 = 3a$

110. $3 - \sqrt{t - 3} = \sqrt{t}$

CHALLENGE PROBLEMS

111. Without graphing, determine the domain of the function $f(x) = \log_5 (x^2 - 1)$. Express the result in interval notation.

112. Evaluate: $\log_6 (\log_5 (\log_4 1,024))$.

11.6   Base-*e* Logarithms

- Base-*e* logarithms
- The natural logarithmic function and its graph
- An application of base-*e* logarithms

We have seen the importance of e in mathematically modeling the growth and decay of natural events. Just as $f(x) = e^x$ is called the natural exponential function, its inverse, the base-*e* logarithmic function, is called the *natural logarithmic function*. Natural logarithmic functions have many applications. They play a very important role in advanced mathematics courses, such as calculus.

■ BASE-*e* LOGARITHMS

Base-*e* logarithms are called **natural logarithms** or **Napierian logarithms** after John Napier (1550–1617). They are usually written as $\ln x$ rather than $\log_e x$:

> **ln *x*   means   log_e *x***   Read ln *x* letter-by-letter as "*ℓ*-n of *x*."

In general, the logarithm of a number is an exponent. For natural logarithms,

ln x is the exponent to which e is raised to get x.

Translating this statement into symbols, we have

$$e^{\ln x} = x$$

Caution

Because of the font used to print the natural log of *x*, some students initially misread the notation as In *x*. In handwriting, In *x* should look like

$\ell n\, x$

EXAMPLE 1

ELEMENTARY & INTERMEDIATE
Algebra $f(x)$ **Now**™

Evaluate each natural logarithmic expression: **a.** $\ln e$,   **b.** $\ln \dfrac{1}{e^2}$,   **c.** $\ln 1$,   and **d.** $\ln \sqrt{e}$.

Solution   **a.** $\ln e = 1$ Ask: "To what power must we raise e to get e?" Since $e^1 = e$, the answer is: the 1st power.

b. $\ln \dfrac{1}{e^2} = -2$ Ask: "To what power must we raise e to get $\frac{1}{e^2}$?" Since $e^{-2} = \frac{1}{e^2}$, the answer is: the -2 power.

c. $\ln 1 = 0$ Ask: "To what power must we raise e to get 1?" Since $e^0 = 1$, the answer is: the 0 power.

d. $\ln \sqrt{e} = \dfrac{1}{2}$ Ask: "To what power must we raise *e* to get $\sqrt{e}$?" Since $e^{1/2} = \sqrt{e}$, the answer is: the $\frac{1}{2}$ power.

Self Check 1   Evaluate each expression: **a.** $\ln e^3$,   **b.** $\ln \dfrac{1}{e}$,   and   **c.** $\ln \sqrt[3]{e}$.

Many natural logarithmic expressions are not as easy to evaluate as those in the previous example. For instance, to find $\ln 2.34$, we ask, "To what power must we raise *e* to get 2.34?" The answer certainly isn't obvious. In such cases, we use a calculator.

ACCENT ON TECHNOLOGY: EVALUATING BASE-*e* (NATURAL) LOGARITHMS

To find $\ln 2.34$ with a scientific calculator, we enter

2.34 $\boxed{\text{LN}}$ $\boxed{.850150929}$

On some calculators, the $\boxed{e^x}$ key also serves as the $\boxed{\text{LN}}$ key when $\boxed{\text{2nd}}$ or $\boxed{\text{SHIFT}}$ is pressed. This is because $f(x) = e^x$ and $g(x) = \ln x$ are inverses.
To use a graphing calculator, we enter

$\boxed{\text{LN}}$ 2.34 $\boxed{)}$ $\boxed{\text{ENTER}}$ $\boxed{\begin{array}{l}\text{ln(2.34)}\\ \qquad .8501509294\end{array}}$

To four decimal places, $\ln 2.34 = 0.8502$. This means $e^{0.8502} \approx 2.34$.
If we attempt to evaluate logarithmic expressions such as $\ln 0$, or the logarithm of a negative number, such as $\ln(-5)$, the following error statements are displayed.

$\boxed{\text{Error}}$ $\boxed{\begin{array}{l}\text{ERR:DOMAIN}\\ \text{1:QUIT}\\ \text{2:Goto}\end{array}}$ $\boxed{\begin{array}{l}\text{ERR:NONREAL ANS}\\ \text{1:QUIT}\\ \text{2:Goto}\end{array}}$

Certain natural logarithmic equations can be solved by writing them as natural exponential equations.

EXAMPLE 2

ELEMENTARY &
INTERMEDIATE
Algebra *f(x)* Now™

Solve each equation: **a.** $\ln x = 1.335$ and   **b.** $\ln x = -5.5$. Give each result to four decimal places.

Solution   **a.** Since the base of the natural logarithmic function is *e*, the equation $\ln x = 1.335$ is equivalent to $e^{1.335} = x$. To use a scientific calculator to find *x*, enter:

1.335 $\boxed{e^x}$

The display will read 3.799995946. To four decimal places,

$x = 3.8000$

The solution is 3.8000. To check, use your calculator to verify that $\ln 3.8000 \approx 1.335$.

b. The equation $\ln x = -5.5$ is equivalent to $e^{-5.5} = x$. To use a scientific calculator to find x, enter:

$$5.5 \quad \boxed{+/-} \quad \boxed{e^x}$$

The display will read 0.004086771. To four decimal places,

$$x = 0.0041$$

The solution is 0.0041. To check, use your calculator to verify that $\ln 0.0041 \approx -5.5$.

Self Check 2   Solve: **a.** $\ln x = 1.9344$ and   **b.** $-3 = \ln x$. Give each result to four decimal places.

■ THE NATURAL LOGARITHMIC FUNCTION AND ITS GRAPH

Because the natural exponential function defined by $f(x) = e^x$ is one-to-one, it has an inverse function that is defined by $x = e^y$. When we write $x = e^y$ in the equivalent form $y = \ln x$, the result is called the *natural logarithmic function.*

The Natural Logarithmic Function	The **natural logarithmic function with base e** is defined by $f(x) = \ln x \quad$ or $\quad y = \ln x, \quad$ where $\ln x = \log_e x.$ The domain of $f(x) = \ln x$ is the interval $(0, \infty)$ and the range is the interval $(-\infty, \infty)$.

Since the natural logarithmic function is the inverse of the one-to-one natural exponential function, the natural logarithmic function is one-to-one.

We can use the point-plotting method to graph $f(x) = \ln x$. We construct a table of function values, plot the resulting ordered pairs, and then draw a smooth curve through the points to get the graph, as shown in figure (a). Figure (b) shows the calculator graph of $f(x) = \ln x$.

$f(x) = \ln x$

To plot these ordered pairs, use a calculator to approximate some x-coordinates.

x	$f(x)$
$\dfrac{1}{e}$	-1
1	0
e	1
e^2	2

$\rightarrow \left(\dfrac{1}{e}, -1\right) \rightarrow (0.4, -1)$
$\rightarrow (1, 0) \quad \rightarrow (1, 0)$
$\rightarrow (e, 1) \quad \rightarrow (2.7, 1)$
$\rightarrow (e^2, 2) \quad \rightarrow (7.4, 2)$

Since the base of the natural logarithmic function is e, choose x-values that are integer powers of e.

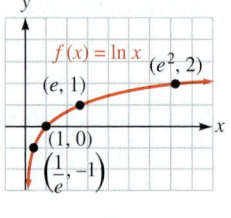

(a)

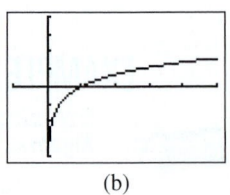

(b)

The exponential function and the natural logarithm function are inverse functions. The figure shows that their graphs are symmetric to the line $y = x$.

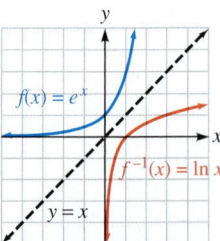

ACCENT ON TECHNOLOGY: GRAPHING BASE-*e* LOGARITHMIC FUNCTIONS

Many graphs of logarithmic functions involve translations of the graph of $f(x) = \ln x$. For example, the figure below shows calculator graphs of the functions $f(x) = \ln x$, $g(x) = (\ln x) + 2$, and $h(x) = (\ln x) - 3$.

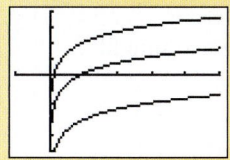

The graph of $g(x) = (\ln x) + 2$ is 2 units above the graph of $f(x) = \ln x$.

The graph of $h(x) = (\ln x) - 3$ is 3 units below the graph of $f(x) = \ln x$.

The next figure shows the calculator graph of the functions $f(x) = \ln x$, $g(x) = \ln(x - 2)$, and $h(x) = \ln(x + 3)$.

The graph of $h(x) = \ln(x + 3)$ is 3 units to the left of the graph of $f(x) = \ln x$.

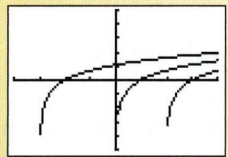

The graph of $g(x) = \ln(x - 2)$ is 2 units to the right of the graph of $f(x) = \ln x$.

■ AN APPLICATION OF BASE-*e* LOGARITHMS

Base-*e* logarithms have many applications. If a population grows exponentially at a certain annual rate, the time required for the population to double is called the **doubling time.** It is given by the following formula.

Formula for Doubling Time	If r is the annual rate, compounded continuously, and t is time required for a population to double, then $$t = \frac{\ln 2}{r}$$

EXAMPLE 3

ELEMENTARY &
INTERMEDIATE
Algebra *f(x)* **Now**™

Doubling time. The population of the Earth is growing at the approximate rate of 2% per year. If this rate continues, how long will it take for the population to double?

Solution   Because the population is growing at the rate of 2% per year, we substitute 0.02 for r in the formula for doubling time and simplify.

$$t = \frac{\ln 2}{r}$$

$$t = \frac{\ln 2}{0.02}$$

≈ 34.65735903 Use a calculator. Find ln 2 first, then divide the result by 0.02.

The population will double in about 35 years.

Self Check 3   If the population's annual growth rate could be reduced to 1.5% per year, what would be the doubling time?

EXAMPLE 4

ELEMENTARY &
INTERMEDIATE
Algebra *f(x)* **Now**™

Doubling time. How long will it take $1,000 to double at an annual rate of 8%, compounded continuously?

Solution   We substitute 0.08 for r and simplify:

$$t = \frac{\ln 2}{r}$$

$$t = \frac{\ln 2}{0.08}$$

$$\approx 8.664339757 \quad \text{Use a calculator. Find } \ln 2 \text{ first, then divide the result by 0.08.}$$

It will take about $8\frac{2}{3}$ years for the money to double.

Self Check 4   How long will it take at 9%, compounded continuously?

Answers to Self Checks   **1. a.** 3,   **b.** -1,   **c.** $\frac{1}{3}$ **2. a.** 6.9199,   **b.** 0.0498   **3.** about 46 years   **4.** about 7.7 years

11.6   STESTUDY SET   ELEMENTARY & INTERMEDIATE Algebra *f(x)* Now™

VOCABULARY   Fill in the blanks.

1. $f(x) = \ln x$ is called the _____ logarithmic function. The base is .

2. $f(x) = \ln x$ and $g(x) = e^x$ are _____ functions.

CONCEPTS

3. Use a calculator to complete the table of values for $f(x) = \ln x$. Round to the nearest hundredth.

x	$f(x)$
0.5	
1	
2	
4	
6	
8	
10	

4. Graph $f(x) = \ln x$ in the illustration. (See Exercise 3.) Note that the units on the x- and y-axes are different.

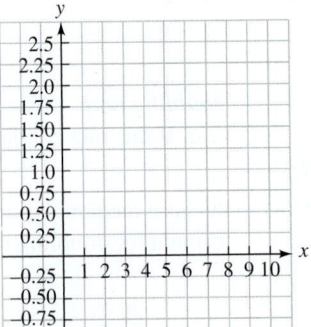

Fill in the blanks.

5. The graph of $f(x) = \ln x$ has the _____ as an asymptote.

6. The domain of the function $f(x) = \ln x$ is the interval _____ .

7. The range of the function $f(x) = \ln x$ is the interval
[blank] .

8. To find $\ln e^2$, we ask, "To what power must we raise
[blank] to get e^2?" Since the answer is the 2nd power,
$\ln e^2 =$ [blank] .

9. The graph of $f(x) = \ln x$ has the x-intercept ([blank] , 0).
The y-axis is an _____ of the graph.

10. The logarithmic statement $\ln x = 1.5318$ is equivalent
to the exponential statement [blank] .

11. A table of values for $f(x) = \ln x$ is shown below.
Explain why ERROR appears in the Y_1 column
for the first three entries.

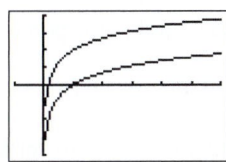

12. The illustration shows the graph of $f(x) = \ln x$, as
well as a vertical translation of that graph. Using the
notation $g(x)$ for the translation, write the defining
equation for the function.

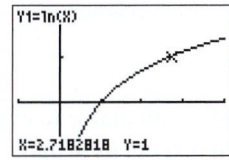

13. In the illustration, $f(x) = \ln x$ was graphed, and the
TRACE feature was used. What is the x-coordinate of
the point on the graph having a y-coordinate of 1? What
is the name given this number?

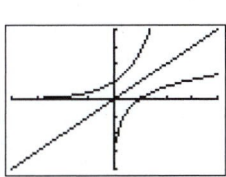

14. The graphs of $f(x) = \ln x$, $g(x) = e^x$, and $y = x$ are
shown below. What phrase is used to describe the
relationship between the graphs?

NOTATION   **Fill in the blanks.**

15. We read $\ln x$ letter-by-letter as " [blank] - [blank] of x."

16. a. $\ln 2$ means $\log_{[blank]} 2$.

  b. $\log 2$ means $\log_{[blank]} 2$.

17. If a population grows exponentially at a rate r, the
time it will take the population to double is given by

the formula $t =$ [blank] .

18. To evaluate a base-10 logarithm with a calculator, use
the [blank] key. To evaluate the base-e logarithm, use
the [blank] key.

Evaluate each expression without using a calculator.

19. $\ln e$ **20.** $\ln e^2$

21. $\ln e^6$ **22.** $\ln e^4$

23. $\ln \dfrac{1}{e}$ **24.** $\ln \dfrac{1}{e^3}$

25. $\ln \sqrt[4]{e}$ **26.** $\ln \sqrt[5]{e}$

PRACTICE   **Use a calculator to find each value,
if possible. Express all answers to four decimal
places.**

27. $\ln 35.15$ **28.** $\ln 0.675$

29. $\ln 0.00465$ **30.** $\ln 378.96$

31. $\ln 1.72$ **32.** $\ln 2.7$

33. $\ln(-0.1)$ **34.** $\ln(-10)$

**Solve each equation. Express all answers to four
decimal places.**

35. $\ln x = 1.4023$ **36.** $\ln x = 2.6490$

37. $\ln x = 4.24$ **38.** $\ln x = 0.926$

39. $\ln x = -3.71$ **40.** $\ln x = -0.28$

41. $1.001 = \ln x$ **42.** $\ln x = -0.001$

Use a graphing calculator to graph each function.

43. $y = \ln\left(\dfrac{1}{2}x\right)$ **44.** $y = \ln x^2$

45. $f(x) = \ln(-x)$ **46.** $f(x) = \ln(3x)$

APPLICATIONS   Use a calculator to solve each problem.

47. POPULATION GROWTH   How long will it take the population of River City to double?

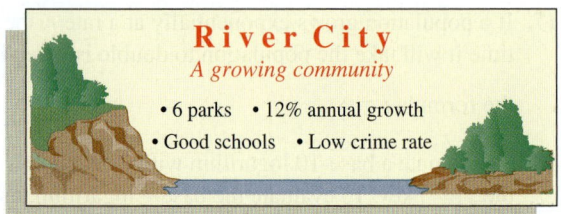

River City
A growing community

• 6 parks   • 12% annual growth
• Good schools   • Low crime rate

48. DOUBLING MONEY   How long will it take $1,000 to double if it is invested at an annual rate of 5% compounded continuously?

49. POPULATION GROWTH   A population growing continuously at an annual rate r will triple in a time t given by the formula

$$t = \frac{\ln 3}{r}$$

How long will it take the population of a town to triple if it is growing at the rate of 12% per year?

50. TRIPLING MONEY   Find the length of time for $25,000 to triple when it is invested at 6% annual interest, compounded continuously. See Exercise 49.

51. FORENSIC MEDICINE   To estimate the number of hours t that a murder victim had been dead, a coroner used the formula

$$t = \frac{1}{0.25} \ln \frac{98.6 - T_s}{82 - T_s}$$

where T_s is the temperature of the surroundings where the body was found. If the crime took place in an apartment where the thermostat was set at 70° F, approximately how long ago did the murder occur?

52. MAKING JELL-O®   After the contents of a package of JELL-O® are combined with boiling water, the mixture is placed in a refrigerator whose temperature remains a constant 42° F. Estimate the number of hours t that it will take for the JELL-O® to cool to 50° F using the formula

$$t = -\frac{1}{0.9} \ln \frac{50 - T_r}{200 - T_r}$$

where T_r is the temperature of the refrigerator.

WRITING

53. Explain the difference between the functions $f(x) = \log x$ and $g(x) = \ln x$.

54. How are the functions $f(x) = \ln x$ and $g(x) = e^x$ related?

55. Explain why $\ln e = 1$.

56. Why is $f(x) = \ln x$ called the natural logarithmic function?

REVIEW   Write the equation of the required line.

57. Parallel to $y = 5x - 8$ and passing through the origin

58. Having a slope of 7 and a y-intercept of 3

59. Passing through the point $(3, 2)$ and perpendicular to the line $y = \frac{2}{3}x - 12$

60. Parallel to the line $3x + 2y = 9$ and passing through the point $(-3, 5)$

61. Vertical line through the point $(2, 3)$

62. Horizontal line through the point $(2, 3)$

CHALLENGE PROBLEMS

63. Use the formula $P = P_0 e^{rt}$ to verify that P will be twice P_0 when $t = \frac{\ln 2}{r}$.

64. Use the formula $P = P_0 e^{rt}$ to verify that P will be three times as large as P_0 when $t = \frac{\ln 3}{r}$.

65. Find a formula to find how long it will take money to quadruple.

Use a graphing calculator to graph the following function and discuss its graph.

66. $f(x) = \dfrac{1}{1 + e^{-2x}}$

11.7 Properties of Logarithms

- Basic properties for logarithms
- The quotient rule for logarithms
- Writing expressions as a single logarithm
- An application from chemistry
- The product rule for logarithms
- The power rule for logarithms
- The change-of-base formula

Since a logarithm is an exponent, we would expect there to be properties of logarithms just as there are properties of exponents. In this section, we will introduce seven properties of logarithms and use them to simplify logarithmic expressions.

■ BASIC PROPERTIES FOR LOGARITHMS

The first four properties of logarithms follow directly from the definition of logarithm.

Properties of Logarithms	For all positive numbers b, where $b \neq 1$,
	1. $\log_b 1 = 0$ **2.** $\log_b b = 1$ **3.** $\log_b b^x = x$ **4.** $b^{\log_b x} = x$ $(x > 0)$

We can use properties of exponents to prove that these properties of logarithms are true.

1. $\log_b 1 = 0$, because $b^0 = 1$.
2. $\log_b b = 1$, because $b^1 = b$.
3. $\log_b b^x = x$, because $b^x = b^x$.
4. $b^{\log_b x} = x$, because $\log_b x$ is the exponent to which b is raised to get x.

Properties 3 and 4 also indicate that the composition of the exponential and logarithmic functions (in both directions) is the identity function. This is expected, because the exponential and logarithmic functions are inverse functions.

EXAMPLE 1

Simplify each expression: **a.** $\log_5 1$,   **b.** $\log_3 3$,   **c.** $\ln e^3$,   and   **d.** $b^{\log_b 7}$.

Solution

a. By property 1, $\log_5 1 = 0$, because $5^0 = 1$.
b. By property 2, $\log_3 3 = 1$, because $3^1 = 3$.
c. By property 3, $\ln e^3 = 3$, because $e^3 = e^3$.
d. By property 4, $b^{\log_b 7} = 7$, because $\log_b 7$ is the power to which b is raised to get 7.

Self Check 1

Simplify: **a.** $\log_4 1$,   **b.** $\log_4 4$,   **c.** $\log_2 2^4$,   and   **d.** $5^{\log_5 2}$.

■ THE PRODUCT RULE FOR LOGARITHMS

The next property of logarithms is related to the product rule for exponents: $x^m \cdot x^n = x^{m+n}$.

| **The Product Rule for Logarithms** | The logarithm of a product is equal to the sum of the logarithms. For all positive numbers b, where $b \neq 1$, $$\log_b MN = \log_b M + \log_b N$$ |

EXAMPLE 2

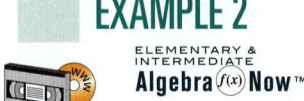

ELEMENTARY & INTERMEDIATE
Algebra *f(x)* Now™

Use the product rule for logarithms to write each expression as a sum of logarithms. Then simplify, if possible. **a.** $\log_2(2 \cdot 7)$,   **b.** $\log 100x$,   and   **c.** $\log_5 125yz$.

Solution   **a.** $\log_2(2 \cdot 7) = \log_2 2 + \log_2 7$ The log of a product is the sum of the logs.

$\qquad\qquad\qquad = 1 + \log_2 7$ By property 2, $\log_2 2 = 1$.

b. $\log 100x = \log 100 + \log x$ The log of a product is the sum of the logs.

$\qquad\qquad = 2 + \log x$ By property 3, $\log 100 = \log 10^2 = 2$.

Caution

As we apply properties of logarithms to rewrite expressions, assume that all variables represent positive numbers.

c. $\log_5 125yz = \log_5 (125y)z$ Group the first two factors together.

$\qquad\qquad = \log_5 (125y) + \log_5 z$ The log of a product is the sum of the logs.

$\qquad\qquad = \log_5 125 + \log_5 y + \log_5 z$ The log of a product is the sum of the logs.

$\qquad\qquad = 3 + \log_5 y + \log_5 z$ By property 3, $\log_5 125 = \log_5 5^3 = 3$.

Self Check 2   Write each expression as the sum of logarithms. Then simplify, if possible. **a.** $\log_3(4 \cdot 3)$, **b.** $\log 1{,}000y$,   and   **c.** $\log_5 25cd$.

PROOF

To prove the product rule for logarithms, we let $x = \log_b M$ and $y = \log_b N$. We use the definition of logarithm to write each equation in exponential form.

$$M = b^x \quad \text{and} \quad N = b^y$$

Then $MN = b^x b^y$, and a property of exponents gives

$$MN = b^{x+y} \qquad \text{Keep the base and add the exponents: } b^x b^y = b^{x+y}.$$

We write this exponential equation in logarithmic form as

$$\log_b MN = x + y$$

Substituting the values of x and y completes the proof.

$$\log_b MN = \log_b M + \log_b N$$

☐

Caution

The log of a sum *does not* equal the sum of the logs. The log of a difference *does not* equal the difference of the logs.

By the product rule, the logarithm of a *product* is equal to the *sum* of the logarithms. The logarithm of a sum or a difference usually does not simplify. In general,

$$\log_b(M + N) \neq \log_b M + \log_b N \quad \text{and} \quad \log_b(M - N) \neq \log_b M - \log_b N$$

For example,

$$\log_2(2 + 7) \neq \log_2 2 + \log_2 7 \quad \text{and} \quad \log(100 - y) \neq \log 100 - \log y$$

ACCENT ON TECHNOLOGY: VERIFYING PROPERTIES OF LOGARITHMS

We can use a calculator to illustrate the product rule for logarithms by showing that

$$\ln[(3.7)(15.9)] = \ln 3.7 + \ln 15.9$$

We calculate the left- and right-hand sides of the equation separately and compare the results. To use a scientific calculator to find $\ln[(3.7)(15.9)]$, we enter

3.7 $\boxed{\times}$ 15.9 $\boxed{=}$ $\boxed{\text{LN}}$ $\boxed{\text{4.074651929}}$

To find $\ln 3.7 + \ln 15.9$, we enter

3.7 $\boxed{\text{LN}}$ $\boxed{+}$ 15.9 $\boxed{\text{LN}}$ $\boxed{=}$ $\boxed{\text{4.074651929}}$

Since the left- and right-hand sides are equal, the equation $\ln[(3.7)(15.9)] = \ln 3.7 + \ln 15.9$ is true.

■ THE QUOTIENT RULE FOR LOGARITHMS

The next property of logarithms is related to the quotient rule for exponents: $\dfrac{x^m}{x^n} = x^{m-n}$.

The Quotient Rule for Logarithms	The logarithm of a quotient is equal to the difference of the logarithms. For all positive numbers b, where $b \neq 1$, $$\log_b \frac{M}{N} = \log_b M - \log_b N$$

The proof of the quotient rule for logarithms is similar to the proof for the product rule for logarithms.

EXAMPLE 3

ELEMENTARY & INTERMEDIATE
Algebra $f(x)$ **Now**™

Use the quotient rule for logarithms to write each expression as a difference of logarithms. Then simplify, if possible. **a.** $\ln \dfrac{10}{7}$ and   **b.** $\log_4 \dfrac{x}{64}$.

Solution **a.** $\ln \dfrac{10}{7} = \ln 10 - \ln 7$ The log of a quotient is the difference of the logs.

b. $\log_4 \dfrac{x}{64} = \log_4 x - \log_4 64$ The log of a quotient is the difference of the logs.

$= \log_4 x - 3$ $\log_4 64 = \log_4 4^3 = 3$.

Self Check 3 Write each expression as a difference of logarithms. Then simplify, if possible. **a.** $\log_6 \dfrac{6}{5}$

and   **b.** $\ln \dfrac{y}{100}$.

By the quotient rule, the logarithm of a *quotient* is equal to the *difference* of the logarithms. The logarithm of a quotient is not the quotient of the logarithms:

$$\log_b \frac{M}{N} \neq \frac{\log_b M}{\log_b N}$$

For example,

$$\ln \frac{10}{7} \neq \frac{\ln 10}{\ln 7} \quad \text{and} \quad \log_4 \frac{x}{64} \neq \frac{\log_4 x}{\log_4 64}$$

In the next example, the product and quotient rules for logarithms are used in combination to rewrite an expression.

EXAMPLE 4 Use properties of logarithms to rewrite $\log \dfrac{xy}{10z}$.

Solution We begin by recognizing that the expression is the logarithm of the quotient $\dfrac{xy}{10z}$.

$$\log \frac{xy}{10z} = \log xy - \log 10z \qquad \text{The log of a quotient is the difference of the logs.}$$

$$= \log x + \log y - (\log 10 + \log z) \qquad \text{The log of a product is the sum of the logs.}$$

Write parentheses here so that the sum, $\log 10 + \log z$, is subtracted.

$$= \log x + \log y - \log 10 - \log z \qquad \text{Remove parentheses.}$$

$$= \log x + \log y - 1 - \log z \qquad \text{Simplify: } \log 10 = 1.$$

Self Check 4 Use properties of logarithms to rewrite $\log_b \dfrac{x}{yz}$.

■ THE POWER RULE FOR LOGARITHMS

The next property of logarithms is related to the power rule for exponents: $(x^m)^n = x^{mn}$.

The Power Rule for Logarithms	The logarithm of a power is equal to the power times the logarithm. For all positive numbers b, where $b \neq 1$, and any real number p, $$\log_b M^p = p \log_b M$$

EXAMPLE 5 Use the power rule for logarithms to rewrite each of the following: **a.** $\log_5 6^2$ and **b.** $\log \sqrt{10}$.

Solution **a.** $\log_5 6^2 = 2 \log_5 6$ \qquad The log of a power is equal to the power times the log.

b. $\log \sqrt{10} = \log (10)^{1/2}$ \qquad Write $\sqrt{10}$ using a fractional exponent: $\sqrt{10} = (10)^{1/2}$.

$$= \frac{1}{2} \log 10 \qquad \text{The log of a power is equal to the power times the log.}$$

$$= \frac{1}{2} \qquad \text{Simplify: } \log 10 = 1.$$

Self Check 5 Use the power rule for logarithms to rewrite **a.** $\ln x^4$ and **b.** $\log_2 \sqrt[3]{3}$.

PROOF

To prove the power rule, we let $x = \log_b M$, write the expression in exponential form, and raise both sides to the pth power:

$$M = b^x$$
$$(M)^p = (b^x)^p \qquad \text{Raise both sides to the } p\text{th power.}$$
$$M^p = b^{px} \qquad \text{Keep the base and multiply the exponents.}$$

Using the definition of logarithms gives

$$\log_b M^p = px$$

Substituting the value for x completes the proof.

$$\log_b M^p = p \log_b M$$

□

EXAMPLE 6

ELEMENTARY &
INTERMEDIATE
Algebra $f(x)$ **Now**™

Use logarithm properties to rewrite each expression: **a.** $\log_b x^2 y^3 z$ and
b. $\ln \dfrac{y^3 \sqrt{x}}{z}$.

Solution

a. We begin by recognizing that $\log_b x^2 y^3 z$ is the logarithm of a product.

$$\log_b x^2 y^3 z = \log_b x^2 + \log_b y^3 + \log_b z \qquad \text{The log of a product is the sum of the logs.}$$
$$= 2 \log_b x + 3 \log_b y + \log_b z \qquad \text{The log of a power is the power times the log.}$$

The Language of Algebra

In Examples 2, 3, 4, and 6, we use properties of logarithms to *expand* logarithmic expressions.

b. The expression $\ln \dfrac{y^3 \sqrt{x}}{z}$ is the logarithm of a quotient.

$$\ln \frac{y^3 \sqrt{x}}{z} = \ln (y^3 \sqrt{x}) - \ln z \qquad \text{The log of a quotient is the difference of the logs.}$$
$$= \ln y^3 + \ln \sqrt{x} - \ln z \qquad \text{The log of a product is the sum of the logs.}$$
$$= \ln y^3 + \ln x^{1/2} - \ln z \qquad \text{Write } \sqrt{x} \text{ as } x^{1/2}.$$
$$= 3 \ln y + \frac{1}{2} \ln x - \ln z \qquad \text{The log of a power is the power times the log.}$$

Self Check 6 Expand: $\log \sqrt[4]{\dfrac{x^3 y}{z}}$.

■ WRITING EXPRESSIONS AS A SINGLE LOGARITHM

EXAMPLE 7

ELEMENTARY &
INTERMEDIATE
Algebra $f(x)$ **Now**™

Write each of the given expressions as one logarithm: **a.** $3 \log_5 x + \frac{1}{2} \log_5 y$ and
b. $\frac{1}{2} \log_b (x - 2) - \log_b y + 3 \log_b z$.

Solution

a. We begin by applying the power rule to each term of the expression.

$$3 \log_5 x + \frac{1}{2} \log_5 y = \log_5 x^3 + \log_5 y^{1/2} \qquad \text{A power times a log is the log of the power.}$$
$$= \log_5 x^3 y^{1/2} \qquad \text{The sum of two logs is the log of the product.}$$

b. The first and third terms of this expression can be rewritten using the power rule of logarithms.

$$\frac{1}{2} \log_b (x - 2) - \log_b y + 3 \log_b z$$

$$= \log_b (x - 2)^{1/2} - \log_b y + \log_b z^3 \qquad \text{A power times a log is the log of the power.}$$

$$= \log_b \frac{(x - 2)^{1/2}}{y} + \log_b z^3 \qquad \text{The difference of two logs is the log of the quotient.}$$

$$= \log_b \frac{z^3 \sqrt{x - 2}}{y} \qquad \text{The sum of two logs is the log of the product. Write } (x - 2)^{1/2} \text{ as } \sqrt{x - 2}.$$

Self Check 7   Write the expression as one logarithm: $2 \log_a x + \dfrac{1}{2} \log_a y - 2 \log_a (x - y)$.

The properties of logarithms can be used when working with numerical values.

EXAMPLE 8   Given that $\log 2 \approx 0.3010$ and $\log 3 \approx 0.4771$, find approximations for   **a.** $\log 6$ and   **b.** $\log 18$.

Solution   **a.** $\log 6 = \log (2 \cdot 3)$ Write 6 using the factors 2 and 3.

$\qquad\qquad = \log 2 + \log 3$ The log of a product is the sum of the logs.

$\qquad\qquad \approx 0.3010 + 0.4771$ Substitute the value of each logarithm.

$\qquad\qquad \approx 0.7781$

b. $\log 18 = \log (2 \cdot 3^2)$ Write 18 using the factors 2 and 3.

$\qquad\qquad = \log 2 + \log 3^2$ The log of a product is the sum of the logs.

$\qquad\qquad = \log 2 + 2 \log 3$ The log of a power is the power times the log.

$\qquad\qquad \approx 0.3010 + 2(0.4771)$ Substitute the value of each logarithm.

$\qquad\qquad \approx 1.2552$

Self Check 8   Find: **a.** $\log 1.5$ and   **b.** $\log 0.75$.

We summarize the properties of logarithms as follows.

Properties of Logarithms	If b, M, and N are positive numbers, $b \neq 1$, and p is any real number,

1. $\log_b 1 = 0$ $\qquad\qquad\qquad\qquad$ **2.** $\log_b b = 1$

3. $\log_b b^x = x$ $\qquad\qquad\qquad\quad$ **4.** $b^{\log_b x} = x$

5. $\log_b MN = \log_b M + \log_b N$ $\qquad$ **6.** $\log_b \dfrac{M}{N} = \log_b M - \log_b N$

7. $\log_b M^p = p \log_b M$

■ THE CHANGE-OF-BASE FORMULA

Most calculators can find common logarithms and natural logarithms. If we need to find a logarithm with some other base, we use a conversion formula.

If we know the base-a logarithm of a number, we can find its logarithm to some other base b by using a formula called the **change-of-base formula.**

Change-of-Base Formula	If a, b, and x are positive, $a \neq 1$, and $b \neq 1$,

$$\log_b x = \frac{\log_a x}{\log_a b}$$

If we know logarithms to base a (for example, $a = 10$), we can find the logarithm of x to a new base b. We simply divide the base-a logarithm of x by the base-a logarithm of b.

EXAMPLE 9

Find $\log_3 5$.

ELEMENTARY & INTERMEDIATE
Algebra $f(x)$ **Now**™   **Solution**

Caution

Don't misapply the quotient rule:

$\dfrac{\log_{10} 5}{\log_{10} 3}$ means $\log_{10} 5 \div \log_{10} 3$.

It is the expression $\log_{10} \dfrac{5}{3}$ that means $\log_{10} 5 - \log_{10} 3$.

We can use base-10 logarithms to find a base-3 logarithm. To do this, we substitute 3 for b, 10 for a, and 5 for x in the change-of-base formula:

$$\log_b \boldsymbol{x} = \frac{\log_a \boldsymbol{x}}{\log_a \boldsymbol{b}}$$

$$\log_3 \boldsymbol{5} = \frac{\log_{10} \boldsymbol{5}}{\log_{10} \boldsymbol{3}} \qquad b = 3, x = 5, \text{ and } a = 10.$$

$$\approx 1.464973521 \qquad \text{Use a scientific calculator and enter 5 } \boxed{\log} \div 3 \boxed{\log} \boxed{=} .$$

To four decimal places, $\log_3 5 = 1.4650$.

We can also use the natural logarithm function (base e) in the change-of-base formula to find a base-3 logarithm.

Caution

Wait until the final calculation has been made to round. Don't round any values when performing intermediate calculations. That could make the final result incorrect because of a build-up of rounding error.

$$\log_b \boldsymbol{x} = \frac{\log_a \boldsymbol{x}}{\log_a \boldsymbol{b}}$$

$$\log_3 \boldsymbol{5} = \frac{\ln \boldsymbol{5}}{\ln \boldsymbol{3}} \qquad \begin{array}{l} b = 3, x = 5, \text{ and } a = e. \\ \log_e 5 = \ln 5 \text{ and } \log_e 3 = \ln 3. \end{array}$$

$$\approx 1.464973521 \qquad \text{Use a calculator.}$$

We obtain the same result.

Self Check 9   Find $\log_5 3$ to four decimal places.

PROOF

To prove the change-of-base formula, we begin with the equation $\log_b x = y$.

$$y = \log_b x$$
$$x = b^y \qquad \text{Change the equation from logarithmic to exponential form.}$$
$$\boldsymbol{\log_a x = \log_a b^y} \qquad \text{Take the base-}a \text{ logarithm of both sides.}$$
$$\log_a x = y \log_a b \qquad \text{The log of a power is the power times the log.}$$
$$y = \frac{\log_a x}{\log_a b} \qquad \text{Divide both sides by } \log_a b.$$
$$\log_b x = \frac{\log_a x}{\log_a b} \qquad \text{Refer to the first equation and substitute } \log_b x \text{ for } y.$$

■ AN APPLICATION FROM CHEMISTRY

In chemistry, common logarithms are used to express the acidity of solutions. The more acidic a solution, the greater the concentration of hydrogen ions. This concentration is indicated indirectly by the *pH scale,* or *hydrogen ion index.* The pH of a solution is defined as follows.

pH of a Solution	If $[\text{H}^+]$ is the hydrogen ion concentration in gram-ions per liter, then $$\text{pH} = -\log[\text{H}^+]$$

EXAMPLE 10

pH meters. One of the most accurate ways to measure pH is with a probe and meter. What reading should the meter give for pure water if water has a hydrogen ion concentration $[\text{H}^+]$ of approximately 10^{-7} gram-ions per liter?

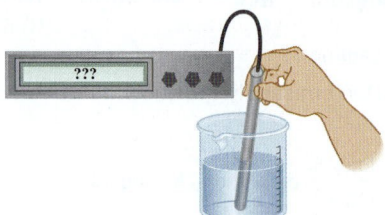

Solution   Since pure water has approximately 10^{-7} gram-ions per liter, its pH is

$$\text{pH} = -\log[\mathbf{H^+}] \qquad \text{This is the formula for pH.}$$
$$\text{pH} = -\log \mathbf{10^{-7}}$$
$$= -(-7)\log 10 \qquad \text{The log of a power is the power times the log.}$$
$$= -(-7)\cdot 1 \qquad \text{Simplify: } \log 10 = 1.$$
$$= 7$$

The meter should give a reading of 7.

EXAMPLE 11

ELEMENTARY &
INTERMEDIATE
Algebra *f(x)* **Now**™

Hydrogen ion concentration. Find the hydrogen ion concentration of seawater if its pH is 8.5.

Solution   To find its hydrogen ion concentration, we solve the following equation for $[\text{H}^+]$.

$$\mathbf{pH} = -\log[\text{H}^+] \qquad \text{This is the formula for pH.}$$
$$\mathbf{8.5} = -\log[\text{H}^+] \qquad \text{Substitute 8.5 for pH.}$$
$$-8.5 = \log[\text{H}^+] \qquad \text{Multiply both sides by } -1.$$
$$[\text{H}^+] = 10^{-8.5} \qquad \text{Write the equation in the equivalent exponential form.}$$

We can use a calculator to find that

$$[\text{H}^+] \approx 3.2 \times 10^{-9} \text{ gram-ions per liter}$$

Answers to Self Checks   **1. a.** 0,   **b.** 1,   **c.** 4,   **d.** 2   **2. a.** $\log_3 4 + 1$,   **b.** $3 + \log y$,   **c.** $2 + \log_5 c + \log_5 d$

3. a. $1 - \log_6 5$,   **b.** $\ln y - \ln 100$ **4.** $\log_b x - \log_b y - \log_b z$ **5. a.** $4\ln x$,   **b.** $\dfrac{1}{3}\log_2 3$

6. $\dfrac{1}{4}(3\log x + \log y - \log z)$ **7.** $\log_a \dfrac{x^2\sqrt{y}}{(x-y)^2}$ **8. a.** 0.1761,   **b.** −0.1249   **9.** 0.6826

11.7   STUDY SET

VOCABULARY   Fill in the blanks.

1. The expression $\log_3(4x)$ is the logarithm of a _____.

2. The expression $\log_2 \dfrac{5}{x}$ is the logarithm of a _____.

3. The expression $\log 4^x$ is the logarithm of a _____.

4. In the expression $\log_5 4$, the number 5 is the _____ of the logarithm.

CONCEPTS   Fill in the blanks.

5. $\log_b 1 = \boxed{}$

6. $\log_b b = \boxed{}$

7. $\log_b MN = \log_b \boxed{} + \log_b \boxed{}$

8. $b^{\log_b x} = \boxed{}$

9. $\log_b \dfrac{M}{N} = \log_b M \boxed{} \log_b N$

10. $\log_b M^p = p \log_b \boxed{}$

11. $\log_b b^x = \boxed{}$

12. $\log_b(A+B) \boxed{} \log_b A + \log_b B$

13. $\log_b \dfrac{M}{N} \boxed{} \dfrac{\log_b M}{\log_b N}$

14. $\log_b AB \boxed{} \log_b A + \log_b B$

15. $\log_b x = \dfrac{\log_a x}{\boxed{}}$

16. $\text{pH} = \boxed{}$

17. Three logarithmic expressions have been evaluated, and the results are shown on the calculator display. Show that each result is correct by writing the equivalent base-10 exponential statement.

log(1)	
log(10)	0
log(10²)	1
	2

18. Three logarithmic expressions have been evaluated, and the results are shown on the calculator display. Show that each result is correct by writing the equivalent base-*e* exponential statement. (The notation ln(*e* ∧ (2)) means $\ln e^2$.)

ln(e^(2))	
ln(e^(3))	2
ln(e^(4))	3
	4

Evaluate each expression.

19. $\log_4 1$

20. $\log_4 4$

21. $\log_4 4^7$

22. $\ln e^8$

23. $5^{\log_5 10}$

24. $8^{\log_8 10}$

25. $\log_5 5^2$

26. $\log_4 4^2$

27. $\ln e$

28. $\log_7 1$

29. $\log_3 3^7$

30. $5^{\log_5 8}$

NOTATION   Complete each solution.

31. $\log_b rst = \log_b\left(\boxed{}\right)t$

$= \log_b(rs) + \log_b \boxed{}$

$= \log_b \boxed{} + \log_b \boxed{} + \log_b t$

32. $\log \dfrac{r}{st} = \log r - \log\left(\boxed{}\right)$

$= \log r - \left(\log \boxed{} + \log t\right)$

$= \log r - \log s \boxed{} \log t$

PRACTICE   Use a calculator to verify that each equation is true.

33. $\log[(2.5)(3.7)] = \log 2.5 + \log 3.7$

34. $\log 45.37 = \dfrac{\ln 45.37}{\ln 10}$

35. $\ln(2.25)^4 = 4 \ln 2.25$

36. $\ln \dfrac{11.3}{6.1} = \ln 11.3 - \ln 6.1$

37. $\log \sqrt{24.3} = \dfrac{1}{2} \log 24.3$

38. $\ln 8.75 = \dfrac{\log 8.75}{\log e}$

Use the properties of logarithms to rewrite each expression. Assume that *x*, *y*, and *z* are positive numbers.

39. $\log_2(4 \cdot 5)$

40. $\log_3(27 \cdot 5)$

41. $\log_6 \dfrac{x}{36}$

42. $\log_8 \dfrac{y}{8}$

43. $\ln y^4$

44. $\ln z^9$

45. $\log \sqrt{5}$

46. $\log \sqrt[3]{7}$

Assume that $x, y, z,$ and b are positive numbers and $b \neq 1$. Use the properties of logarithms to write each expression in terms of the logarithms of $x, y,$ and z.

47. $\log xyz$

48. $\log 4xz$

49. $\log_2 \dfrac{2x}{y}$

50. $\log_3 \dfrac{x}{yz}$

51. $\log x^3 y^2$

52. $\log xy^2 z^3$

53. $\log_b (xy)^{1/2}$

54. $\log_b x^3 y^{1/2}$

55. $\log_a \dfrac{\sqrt[3]{x}}{\sqrt[4]{yz}}$

56. $\log_b \sqrt[4]{\dfrac{x^3 y^2}{z^4}}$

57. $\ln x\sqrt{z}$

58. $\ln \sqrt{xy}$

Assume that $x, y, z,$ and b are positive numbers and $b \neq 1$. Use the properties of logarithms to write each expression as the logarithm of a single quantity.

59. $\log_2 (x + 1) - \log_2 x$

60. $\log_3 x + \log_3 (x + 2) - \log_3 8$

61. $2 \log x + \dfrac{1}{2} \log y$

62. $-2 \log x - 3 \log y + \log z$

63. $-3 \log_b x - 2 \log_b y + \dfrac{1}{2} \log_b z$

64. $3 \log_b (x + 1) - 2 \log_b (x + 2) + \log_b x$

65. $\ln \left(\dfrac{x}{z} + x \right) - \ln \left(\dfrac{y}{z} + y \right)$

66. $\ln (xy + y^2) - \ln (xz + yz) + \ln z$

Determine whether the given statement is true. If a statement is false, explain why.

67. $\log xy = (\log x)(\log y)$

68. $\log ab = \log a + 1$

69. $\log_b (A - B) = \dfrac{\log_b A}{\log_b B}$

70. $\dfrac{\log_b A}{\log_b B} = \log_b A - \log_b B$

71. $\log_b \dfrac{A}{B} = \log_b A - \log_b B$

72. $\log_b 2 = \log_2 b$

Assume that $\log_b 4 = 0.6021$, $\log_b 7 = 0.8451$, and $\log_b 9 = 0.9542$. Use these values and the properties of logarithms to find each value.

73. $\log_b 28$

74. $\log_b \dfrac{7}{4}$

75. $\log_b \dfrac{4}{63}$

76. $\log_b 36$

77. $\log_b \dfrac{63}{4}$

78. $\log_b 2.25$

79. $\log_b 64$

80. $\log_b 49$

Use the change-of-base formula to find each logarithm to four decimal places.

81. $\log_3 7$

82. $\log_7 3$

83. $\log_{1/3} 3$

84. $\log_{1/2} 6$

85. $\log_3 8$

86. $\log_5 10$

87. $\log_{\sqrt{2}} \sqrt{5}$

88. $\log_\pi e$

APPLICATIONS

89. pH OF A SOLUTION   Find the pH of a solution with a hydrogen ion concentration of 1.7×10^{-5} gram-ions per liter.

90. HYDROGEN ION CONCENTRATION   Find the hydrogen ion concentration of a saturated solution of calcium hydroxide whose pH is 13.2.

91. AQUARIUMS   To test for safe pH levels in a freshwater aquarium, a test strip is compared with the scale shown below. Find the corresponding range in the hydrogen ion concentration.

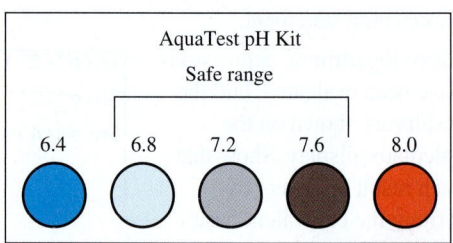

92. pH OF PICKLES   The hydrogen ion concentration of sour pickles is 6.31×10^{-4}. Find the pH.

WRITING

93. Explain the difference between a logarithm of a product and the product of logarithms.

94. How can the LOG key on a calculator be used to find $\log_2 7$?

REVIEW   **Consider the line that passes through $P(-2, 3)$ and $Q(4, -4)$.**

95. Find the slope of line PQ.

96. Find the distance PQ.

97. Find the midpoint of line segment PQ.

98. Write the equation of line PQ.

CHALLENGE PROBLEMS

99. Explain why $e^{\ln x} = x$.

100. If $\log_b 3x = 1 + \log_b x$, find b.

101. Show that $\log_{b^2} x = \dfrac{1}{2} \log_b x$.

102. Show that $e^{x \ln a} = a^x$.

11.8   # Exponential and Logarithmic Equations

- Solving exponential equations   • Solving logarithmic equations
- Radioactive decay   • Population growth

An **exponential equation** is an equation that contains a variable in one of its exponents. Some examples of exponential equations are

$$3^{x+1} = 81, \qquad 6^{x-3} = 2^x, \qquad \text{and} \qquad e^{0.9t} = 8$$

A **logarithmic equation** is an equation with a logarithmic expression that contains a variable. Some examples of logarithmic equations are

$$\log 5x = 3, \qquad \log(3x + 2) = \log(2x - 3), \qquad \text{and} \qquad \frac{\log_2(5x - 6)}{\log_2 x} = 2$$

In this section, we will learn how to solve exponential and logarithmic equations.

■ SOLVING EXPONENTIAL EQUATIONS

If both sides of an exponential equation can be expressed as a power of the same base, we can use the following property to solve the equation.

Exponent Property of Equality   If two exponential expressions with the same base are equal, their exponents are equal. For any real number b, where $b \neq -1, 0,$ or 1,

$$b^x = b^y \text{ is equivalent to } x = y$$

EXAMPLE 1   Solve: $3^{x+1} = 81$.

ELEMENTARY &
INTERMEDIATE
Algebra *f(x)* **Now**™   Solution   Since $81 = 3^4$, each side of the equation can be expressed as a power of 3.

$$3^{x+1} = 81$$
$$3^{x+1} = 3^4 \qquad \text{Write 81 as } 3^4.$$
$$x + 1 = 4 \qquad \text{If two exponential expressions with the same base are equal, their exponents are equal.}$$
$$x = 3$$

The solution is 3. To check this result, we substitute 3 for x in the original equation.

Check: $3^{x+1} = 81$

$3^{3+1} \stackrel{?}{=} 81$

$3^4 \stackrel{?}{=} 81$

$81 = 81$

Self Check 1 Solve: $5^{3x-4} = 25$.

EXAMPLE 2 Solve: $2^{x^2+2x} = \dfrac{1}{2}$.

Solution Since $\dfrac{1}{2} = 2^{-1}$, each side of the equation can be expressed as a power of 2.

$$2^{x^2+2x} = \frac{1}{2}$$

$$\mathbf{2^{x^2+2x} = 2^{-1}}$$ Write $\frac{1}{2}$ as 2^{-1}.

$$x^2 + 2x = -1$$ If two exponential expressions with the same base are equal, their exponents are equal.

$$x^2 + 2x + 1 = 0$$ Add 1 to both sides.

$$(x + 1)(x + 1) = 0$$ Factor the trinomial.

$$x + 1 = 0 \quad \text{or} \quad x + 1 = 0$$ Set each factor equal to 0.

$$x = -1 \quad \bigg| \quad x = -1$$

The solution is -1. Verify that it satisfies the original equation.

Self Check 2 Solve: $3^{x^2-2x} = \dfrac{1}{3}$.

ACCENT ON TECHNOLOGY: SOLVING EXPONENTIAL EQUATIONS GRAPHICALLY

To use a graphing calculator to approximate the solutions of $2^{x^2+2x} = \frac{1}{2}$ (see Example 2), we can subtract $\frac{1}{2}$ from both sides of the equation to get

$$2^{x^2+2x} - \frac{1}{2} = 0$$

and graph the corresponding function

$$y = 2^{x^2+2x} - \frac{1}{2}$$

as shown in figure (a) on the next page.

The solutions of $2^{x^2+2x} - \frac{1}{2} = 0$ are the x-coordinates of the x-intercepts of the graph of $y = 2^{x^2+2x} - \frac{1}{2}$. Using the ZERO feature, we see in figure (a) that the graph has only one x-intercept, $(-1, 0)$. Therefore, $x = -1$ is the only solution of $2^{x^2+2x} - \frac{1}{2} = 0$.

We can also solve $2^{x^2+2x} = \frac{1}{2}$ using the INTERSECT feature found on most graphing calculators. After graphing $Y_1 = 2^{x^2+2x}$ and $Y_2 = \frac{1}{2}$, we select INTERSECT, which approximates the coordinates of the point of intersection of the two graphs. From the display shown in figure (b), we can conclude that the solution is $x = -1$. Verify this by checking.

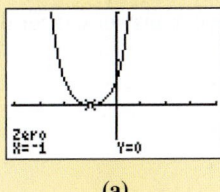

(a)

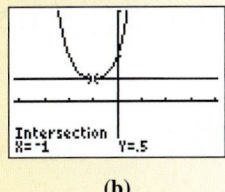

(b)

When it is difficult or impossible to write each side of an exponential equation as a power of the same base, we can often use the following property of logarithms to solve the equation.

The Logarithm Property of Equality

If two positive numbers are equal, then the logarithms base-b of the numbers are equal. For any positive number b, where $b \neq 1$, and positive numbers x and y,

$$\log_b x = \log_b y \quad \text{is equivalent to} \quad x = y$$

EXAMPLE 3

Solve: $3^x = 5$.

Solution

Unlike Example 1, $3^{x+1} = 81$, it is not possible to write each side of $3^x = 5$ as a power of the same base 3. Instead, we use the logarithm property of equality and *take the logarithm of both sides* to solve the equation. Any base logarithm can be chosen, however, the computations with a calculator are usually simplest if the common or the natural logarithm is used.

$$3^x = 5$$
$$\mathbf{log}\, 3^x = \mathbf{log}\, 5 \qquad \text{Take the common logarithm of each side.}$$
$$x \log 3 = \log 5 \qquad \text{The log of a power is the power times the log: } \log 3^x = x \log 3.$$
$$x = \frac{\log 5}{\log 3} \qquad \text{Divide both sides by } \log 3. \text{ This is the exact solution.}$$
$$x \approx 1.464973521 \qquad \text{Use a calculator.}$$

The exact solution is $\dfrac{\log 5}{\log 3}$. To four decimal places, the solution is 1.4650.

We can also take the natural logarithm of each side of the equation to solve for x.

$$3^x = 5$$
$$\mathbf{ln}\, 3^x = \mathbf{ln}\, 5 \qquad \text{Take the natural logarithm of each side.}$$
$$x \ln 3 = \ln 5 \qquad \text{Use the power rule of logarithms: } \ln 3^x = x \ln 3.$$
$$x = \frac{\ln 5}{\ln 3} \qquad \text{Divide both sides by } \ln 3.$$
$$x \approx 1.464973521 \qquad \text{Use a calculator.}$$

Success Tip

The power rule of logarithms provides a way of moving the variable x from its position in the exponent to a position as a factor of $x \log 3$.

$$\log 3^x = \log 5$$
$$x \log 3 = \log 5$$

Caution

Don't misapply the quotient rule:

$$\frac{\ln 5}{\ln 3} \text{ means } \ln 5 \div \ln 3.$$

It is the expression $\ln \dfrac{5}{3}$ that means $\ln 5 - \ln 3$.

The result is the same using the natural logarithm. To check, we substitute 1.4650 for x in 3^x and see if $3^{1.4650}$ is close to 5.

Check:
$$3^x = 5$$
$$3^{1.4650} \overset{?}{=} 5$$
$$5.000145454 \approx 5 \qquad \text{Use a calculator: Enter } 3 \boxed{y^x} \ 1.4650 \ \boxed{=}.$$

Self Check 3   Solve $5^x = 4$. Give the answer to four decimal places.

EXAMPLE 4

Solve: $6^{x-3} = 2^x$.

Solution

$$6^{x-3} = 2^x$$
$$\log 6^{x-3} = \log 2^x \qquad \text{Take the common logarithm of each side.}$$
$$(x-3)\log 6 = x \log 2 \qquad \text{The log of a power is the power times the log.}$$
$$x \log 6 - 3 \log 6 = x \log 2 \qquad \text{Use the distributive property.}$$
$$x \log 6 - x \log 2 = 3 \log 6 \qquad \text{On both sides, add 3 log 6 and subtract } x \log 2.$$
$$x (\log 6 - \log 2) = 3 \log 6 \qquad \text{Factor out } x \text{ on the left-hand side.}$$
$$x = \frac{3 \log 6}{\log 6 - \log 2} \qquad \text{Divide both sides by } \log 6 - \log 2.$$
$$x \approx 4.892789261 \qquad \text{Use a calculator.}$$

To four decimal places, the solution is 4.8928.

The Language of Algebra

$\dfrac{3 \log 6}{\log 6 - \log 2}$ is the *exact* solution of $6^{x-3} = 2^x$. An *approximate* solution is 4.8928.

ELEMENTARY &
INTERMEDIATE
Algebra *f(x)* **Now**™

Self Check 4   Solve: $5^{x-2} = 3^x$.

EXAMPLE 5

ELEMENTARY &
INTERMEDIATE
Algebra *f(x)* **Now**™

Solve: $e^{0.9t} = 10$.

Solution   The exponential expression on the left-hand side has base e. In such cases, the computations are somewhat simpler if we take the natural logarithm of each side.

$$e^{0.9t} = 10$$
$$\ln e^{0.9t} = \ln 10 \qquad \text{Take the natural logarithm of each side.}$$
$$0.9t \ln e = \ln 10 \qquad \text{Use the power rule of logarithms: } \ln e^{0.9t} = 0.9t \ln e.$$
$$0.9t \cdot 1 = \ln 10 \qquad \text{Simplify: } \ln e = 1.$$
$$0.9t = \ln 10$$
$$t = \frac{\ln 10}{0.9} \qquad \text{Divide both sides by } 0.9.$$
$$t \approx 2.558427881 \qquad \text{Use a calculator.}$$

To four decimal places, the solution is 2.5584

Self Check 5   Solve: $e^{2.1t} = 35$.

■ SOLVING LOGARITHMIC EQUATIONS

We can solve many logarithmic equations using properties of logarithms.

EXAMPLE 6

Solve: $\log 5x = 3$.

Solution

Recall that $\log 5x = \log_{10} 5x$. We can write the equation $\log 5x = 3$ as an equivalent base-10 exponential equation $10^3 = 5x$ and solve for x.

$$\log 5x = 3$$
$$10^3 = 5x \qquad \text{Write using exponential form.}$$
$$1{,}000 = 5x \qquad \text{Simplify: } 10^3 = 1{,}000.$$
$$200 = x \qquad \text{Divide both sides by 5.}$$

The solution is 200.

Caution

Always check your solutions to a logarithmic equation.

Check:
$$\log 5x = 3$$
$$\log 5(200) \overset{?}{=} 3 \qquad \text{Substitute 200 for } x.$$
$$\log 1{,}000 \overset{?}{=} 3$$
$$3 = 3 \qquad \log 1{,}000 = \log 10^3 = 3.$$

Self Check 6

Solve: $\log_2 (x - 3) = -1$.

EXAMPLE 7

Solve: $\log (3x + 2) = \log (2x - 3)$.

Solution

We can use the logarithm property of equality to solve the equation.

$$\log (3x + 2) = \log (2x - 3)$$
$$(3x + 2) = (2x - 3) \qquad \text{If the logarithms of two numbers are equal, the numbers are equal.}$$
$$x + 2 = -3 \qquad \text{Subtract } 2x \text{ from both sides.}$$
$$x = -5$$

Caution

Don't make this error of trying to "distribute" log:

$\log (3x + 2)$

The notation *log* is not a number, it is the name of a function and cannot be distributed.

Check:
$$\log (3x + 2) = \log (2x - 3)$$
$$\log [3(-5) + 2] \overset{?}{=} \log [2(-5) - 3] \qquad \text{Substitute } -5 \text{ for } x.$$
$$\log (-13) \overset{?}{=} \log (-13)$$

Since the logarithm of a negative number does not exist, the proposed solution of -5 must be discarded. This equation has no solutions.

Self Check 7

Solve: $\log (5x + 2) = \log (7x - 2)$.

EXAMPLE 8

Solve: $\log x + \log (x - 3) = 1$.

Solution

$$\log x + \log (x - 3) = 1$$
$$\log x(x - 3) = 1 \qquad \text{Use the product rule of logarithms.}$$
$$\log_{10} x(x-3) = 1 \qquad \text{The base is 10.}$$
$$x(x-3) = 10^1 \qquad \text{Write the equation in exponential form.}$$
$$x^2 - 3x - 10 = 0 \qquad \text{Remove parentheses. Subtract 10 from both sides.}$$
$$(x + 2)(x - 5) = 0 \qquad \text{Factor the trinomial.}$$
$$x + 2 = 0 \quad \text{or} \quad x - 5 = 0 \qquad \text{Set each factor equal to 0.}$$
$$x = -2 \quad \bigg| \quad x = 5$$

Caution

Proposed solutions of a logarithmic equation must be checked to see whether they produce undefined logarithms in the original equation.

Check: The number -2 is not a solution, because it does not satisfy the equation (a negative number does not have a logarithm). We will check the other result, 5.

$$\log x + \log (x - 3) = 1$$
$$\log 5 + \log (5 - 3) \stackrel{?}{=} 1 \qquad \text{Substitute 5 for } x.$$
$$\log 5 + \log 2 \stackrel{?}{=} 1$$
$$\log 10 \stackrel{?}{=} 1 \qquad \text{Use the product rule of logarithms:}$$
$$\qquad\qquad\qquad \log 5 + \log 2 = \log (5 \cdot 2) = \log 10.$$
$$1 = 1 \qquad \text{Simplify: } \log 10 = 1.$$

Since 5 satisfies the equation, it is the solution.

Self Check 8 Solve: $\log x + \log (x + 3) = 1$.

ACCENT ON TECHNOLOGY: SOLVING LOGARITHMIC EQUATIONS GRAPHICALLY

To use a graphing calculator to approximate the solutions of the logarithmic equation $\log x + \log (x - 3) = 1$ (see Example 8), we can subtract 1 from both sides of the equation to get

$$\log x + \log (x - 3) - 1 = 0$$

and graph the corresponding function

$$y = \log x + \log (x - 3) - 1$$

as shown in figure (a). Since the solution of the equation is the x-coordinate of the x-intercept, we can find the solutions using the ZERO feature.

We can also solve $\log x + \log (x - 3) = 1$ using the INTERSECT feature. After graphing $Y_1 = \log x + \log (x - 3)$ and $Y_2 = 1$, we select INTERSECT, which approximates the coordinates of the point of intersection of the two graphs. From the display shown in figure (b), we can conclude that the solution is $x = 5$.

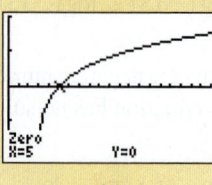

(a)   (b)

EXAMPLE 9

ELEMENTARY &
INTERMEDIATE
Algebra $f(x)$ **Now**™

Solve: $\dfrac{\log_2(5x - 6)}{\log_2 x} = 2$.

Solution   We can multiply both sides of the equation by $\log_2 x$ to get

$$\log_2(5x - 6) = 2 \log_2 x$$

and apply the power rule of logarithms to get

$$\log_2(5x - 6) = \log_2 x^2$$

By the logarithm property of equality, $5x - 6 = x^2$, because they have equal logarithms. Thus,

$$5x - 6 = x^2$$
$$0 = x^2 - 5x + 6$$
$$0 = (x - 3)(x - 2)$$
$$x - 3 = 0 \quad \text{or} \quad x - 2 = 0$$
$$x = 3 \quad \mid \quad x = 2$$

Verify that both 2 and 3 satisfy the equation.

Self Check 9   Solve: $\dfrac{\log_3(5x + 6)}{\log_3 x} = 2$.

■ RADIOACTIVE DECAY

Experiments have determined the time it takes for half of a sample of a given radioactive material to decompose. This time is a constant, called the material's **half-life.**

   When living organisms die, the oxygen/carbon dioxide cycle common to all living things ceases, and carbon-14, a radioactive isotope with a half-life of 5,700 years, is no longer absorbed. By measuring the amount of carbon-14 present in an ancient object, archaeologists can estimate the object's age by using the radioactive decay formula.

Radioactive Decay Formula   If A is the amount of radioactive material present at time t, A_0 was the amount present at $t = 0$, and h is the material's half-life, then

$$A = A_0 2^{-t/h}$$

EXAMPLE 10

Carbon-14 dating.   How old is a wooden statue that retains only one-third of its original carbon-14 content?

Solution   To find the time t when $A = \frac{1}{3}A_0$, we substitute $\frac{A_0}{3}$ for A and 5,700 for h in the radioactive decay formula and solve for t:

$$A = A_0 2^{-t/h}$$

$$\frac{A_0}{3} = A_0 2^{-t/5,700} \qquad \text{The half-life of carbon-14 is 5,700 years.}$$

$$1 = 3\left(2^{-t/5,700}\right) \qquad \text{Divide both sides by } A_0 \text{ and multiply both sides by 3.}$$

$$\log 1 = \log 3\left(2^{-t/5,700}\right) \qquad \text{Take the common logarithm of each side.}$$

$$0 = \log 3 + \log 2^{-t/5,700} \qquad \log 1 = 0, \text{ and use the product rule of logarithms.}$$

$$-\log 3 = -\frac{t}{5,700}\log 2 \qquad \text{Subtract log 3 from both sides and use the power rule of logarithms.}$$

$$5,700\left(\frac{\log 3}{\log 2}\right) = t \qquad \text{Multiply both sides by } -\dfrac{5,700}{\log 2}.$$

$$t \approx 9,034.286254 \qquad \text{Use a calculator.}$$

The statue is approximately 9,000 years old.

Self Check 10   How old is a statue that retains 25% of its original carbon-14 content?

Notation

The initial amount of radioactive material is represented by A_0, and it is read as "A sub 0."

■ POPULATION GROWTH

Recall that when there is sufficient food and space, populations of living organisms tend to increase exponentially according to the Malthusian growth model.

Malthusian Growth Model   If P is the population at some time t, P_0 is the initial population at $t = 0$, and k depends on the rate of growth, then

$$P = P_0 e^{kt}$$

EXAMPLE 11

ELEMENTARY & INTERMEDIATE
Algebra $f(x)$ **Now**™

Population growth.   The bacteria in a laboratory culture increased from an initial population of 500 to 1,500 in 3 hours. How long will it take for the population to reach 10,000?

Solution   We substitute 500 for P_0, 1,500 for P, and 3 for t and simplify to find k:

Notation

The initial population of bacteria is represented by P_0, and it is read as "P sub 0."

$$P = P_0 e^{kt} \qquad \text{This is the population growth formula.}$$

$$1{,}500 = 500(e^{k3}) \qquad \text{Substitute 1,500 for } P, \text{ 500 for } P_0, \text{ and 3 for } t.$$

$$3 = e^{3k} \qquad \text{Divide both sides by 500.}$$

$$3k = \ln 3 \qquad \text{Change the equation from exponential to logarithmic form.}$$

$$k = \frac{\ln 3}{3} \qquad \text{Divide both sides by 3.}$$

To find when the population will reach 10,000, we substitute 10,000 for P, 500 for P_0, and $\frac{\ln 3}{3}$ for k in the equation $P = P_0 e^{kt}$ and solve for t:

$$P = P_0 e^{kt}$$

$$10{,}000 = 500e^{[(\ln 3)/3]t}$$

$$20 = e^{[(\ln 3)/3]t} \qquad \text{Divide both sides by 500.}$$

$$\left(\frac{\ln 3}{3}\right)t = \ln 20 \qquad \text{Change the equation to logarithmic form.}$$

$$t = \frac{3\ln 20}{\ln 3} \qquad \text{Multiply both sides by } \frac{3}{\ln 3}.$$

$$\approx 8.180499084 \qquad \text{Use a calculator.}$$

The culture will reach 10,000 bacteria in about 8 hours.

Self Check 11   How long will it take the population to reach 20,000?

EXAMPLE 12

Generation time.   If a medium is inoculated with a bacterial culture that contains 1,000 cells per milliliter, how many generations will pass by the time the culture has grown to a population of 1 million cells per milliliter?

Solution   During bacterial reproduction, the time required for a population to double is called the *generation time.* If b bacteria are introduced into a medium, then after the generation time of the organism has elapsed, there are $2b$ cells. After another generation, there are $2(2b)$ or $4b$ cells, and so on. After n generations, the number of cells present will be

(1) $B = b \cdot 2^n$

To find the number of generations that have passed while the population grows from b bacteria to B bacteria, we solve Equation 1 for n. To do this, we can take the common logarithm or the natural logarithm of each side.

$$\ln B = \ln (b \cdot 2^n) \qquad \text{Take the natural logarithm of each side.}$$

$$\ln B = \ln b + n \ln 2 \qquad \text{Apply the product and power rules of logarithms.}$$

$$\ln B - \ln b = n \ln 2 \qquad \text{Subtract } \ln b \text{ from both sides.}$$

$$n = \frac{1}{\ln 2}(\ln B - \ln b) \qquad \text{Multiply both sides by } \frac{1}{\ln 2}.$$

(2) $n = \dfrac{1}{\ln 2}\left(\ln \dfrac{B}{b}\right) \qquad \text{Use the quotient rule of logarithms.}$

Equation 2 is a formula that gives the number of generations that will pass as the population grows from b bacteria to B bacteria.

To find the number of generations that have passed while a population of 1,000 cells per milliliter has grown to a population of 1 million cells per milliliter, we substitute 1,000 for b and 1,000,000 for B in Equation 2 and solve for n.

$$n = \frac{1}{\ln 2} \ln \frac{1{,}000{,}000}{1{,}000}$$

$$= \frac{1}{\ln 2} \ln 1{,}000 \qquad \text{Simplify.}$$

$$n \approx 9.965784285 \qquad \text{Use a calculator:}$$

Approximately 10 generations will have passed.

11.8   STUDY SET   ELEMENTARY & INTERMEDIATE Algebra $f(x)$ Now™

VOCABULARY   Fill in the blanks.

1. An equation with a variable in its exponent, such as $3^{2x} = 8$, is called a(n) _____ equation.

2. An equation with a logarithmic expression that contains a variable, such as $\log_5 (2x - 3) = \log_5 (x + 4)$, is a(n) _____ equation.

CONCEPTS

3. Perform a check to determine whether -2 is a solution of $5^{2x+3} = \dfrac{1}{5}$.

4. Perform a check to determine whether -4 is a solution of $\log_5 (x + 3) = \dfrac{1}{5}$.

5. Use a calculator to determine whether 2.5646 is an approximate solution of $2^{2x+1} = 70$.

6. a. The exponential property of equality: If two exponential expressions with the same base are equal, their exponents are _____.

$b^x = b^y$ is equivalent to

b. The logarithm property of equality: If the logarithms base-b of two numbers are equal, the numbers are _____

$\log_b x = \log_b y$ is equivalent to

7. Both sides of the exponential equation $5^{x-3} = 125$ can be written as a power of   .

8. If $6^{4x} = 6^{-2}$, then $4x =$.

9. a. Write the equivalent base-10 exponential equation for $\log (x + 1) = 2$.

b. Write the equivalent base-e exponential equation for $\ln (x + 1) = 2$.

10. Fill in the blanks: To solve $5^x = 21$, we can take the _____ of both sides of the equation to get

$\log 5^x = \log 21$

The power rule for logarithms then provides a way of moving the variable x from its position as an _____ to a position as a factor of $x \log 5$.

11. If $2^x = 9$, then $\log 2^x =$.

12. If $e^{x+2} = 4$, then $\ln e^{x+2} =$.

13. Apply the power rule for logarithms to the left-hand side of the equation

$\log 7^x = 12$

14. How do we solve $x \ln 3 = \ln 5$ for x?

15.   **a.** Find $\dfrac{\log 8}{\log 5}$. Round to four decimal places.

b. Find $\dfrac{2 \ln 12}{\ln 9}$. Round to four decimal places.

16. Does $\dfrac{\log 7}{\log 3} = \log 7 - \log 3$?

17. Complete each formula.

a. Radioactive decay: $A =$.

b. Population growth: $P =$.

18. Use the graphs to estimate the solution of each equation.

a. $2^x = 3^{-x+3}$ **b.** $3 \log (x - 1) = 2 \log x$

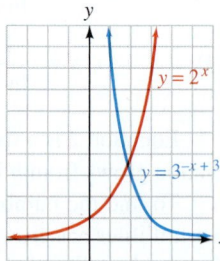

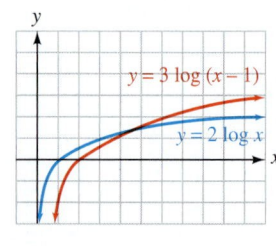

NOTATION   Complete each solution.

19. Solve: $2^x = 7$.

$$2^x = 7$$
$$2^x = \log 7$$
$$x \quad = \log 7$$
$$x = \dfrac{\log 7}{\log 2}$$

20. Solve: $\log_2(2x - 3) = \log_2(x + 4)$.

$$\log_2(2x - 3) = \log_2(x + 4)$$
$$ = x + 4$$
$$x = 7$$

PRACTICE Solve each equation. Give answers to four decimal places when necessary.

21. $2^{x-2} = 64$

22. $3^{-3x+1} = 243$

23. $5^{4x} = \dfrac{1}{125}$

24. $8^{-x+1} = \dfrac{1}{64}$

25. $2^{x^2-2x} = 8$

26. $3^{x^2-3x} = 81$

27. $3^{x^2+4x} = \dfrac{1}{81}$

28. $7^{x^2+3x} = \dfrac{1}{49}$

29. $4^x = 5$

30. $7^x = 12$

31. $13^{x-1} = 2$

32. $5^{x+1} = 3$

33. $2^{x+1} = 3^x$

34. $5^{x-3} = 3^{2x}$

35. $2^x = 3^x$

36. $3^{2x} = 4^x$

37. $7^{x^2} = 10$

38. $8^{x^2} = 11$

39. $8^{x^2} = 9^x$

40. $5^{x^2} = 2^{5x}$

41. $e^{3x} = 9$

42. $e^{4x} = 60$

43. $e^{-0.2t} = 14.2$

44. $e^{0.3t} = 9.1$

Use a graphing calculator to solve each equation. Give all answers to the nearest tenth.

45. $2^{x+1} = 7$

46. $3^{x-1} = 2^x$

47. $3^x - 10 = 3^{-x}$

48. $4(2^{x^2}) = 8^{3x}$

Solve each equation.

49. $\log(x + 2) = 4$

50. $\log 5x = 4$

51. $\log(7 - x) = 2$

52. $\log(2 - x) = 3$

53. $\ln x = 1$

54. $\ln x = 5$

55. $\ln(x + 1) = 3$

56. $\ln 2x = 5$

57. $\log 2x = \log 4$

58. $\log 3x = \log 9$

59. $\ln(3x + 1) = \ln(x + 7)$

60. $\ln(x^2 + 4x) = \ln(x^2 + 16)$

61. $\log(3 - 2x) - \log(x + 24) = 0$

62. $\log(3x + 5) - \log(2x + 6) = 0$

63. $\log \dfrac{4x + 1}{2x + 9} = 0$

64. $\log \dfrac{2 - 5x}{2(x + 8)} = 0$

65. $\log x^2 = 2$

66. $\log x^3 = 3$

67. $\log x + \log(x - 48) = 2$

68. $\log x + \log(x + 9) = 1$

69. $\log x + \log(x - 15) = 2$

70. $\log x + \log(x + 21) = 2$

71. $\log(x + 90) = 3 - \log x$

72. $\log(x - 90) = 3 - \log x$

73. $\log(x - 6) - \log(x - 2) = \log \dfrac{5}{x}$

74. $\log(3 - 2x) - \log(x + 9) = 0$

75. $\dfrac{\log(3x - 4)}{\log x} = 2$

76. $\dfrac{\log(8x - 7)}{\log x} = 2$

77. $\dfrac{\log(5x + 6)}{2} = \log x$

78. $\dfrac{1}{2}\log(4x + 5) = \log x$

79. $\log_3 x = \log_3\left(\dfrac{1}{x}\right) + 4$

80. $\log_5(7 + x) + \log_5(8 - x) - \log_5 2 = 2$

81. $2\log_2 x = 3 + \log_2(x - 2)$

82. $2\log_3 x - \log_3(x - 4) = 2 + \log_3 2$

83. $\log(7y + 1) = 2\log(y + 3) - \log 2$

84. $2\log(y + 2) = \log(y + 2) - \log 12$

Use a graphing calculator to solve each equation. If an answer is not exact, round to the nearest tenth.

85. $\log x + \log(x - 15) = 2$

86. $\log x + \log(x + 3) = 1$

87. $\ln(2x + 5) - \ln 3 = \ln(x - 1)$

88. $2\log(x^2 + 4x) = 1$

APPLICATIONS

89. TRITIUM DECAY The half-life of tritium is 12.4 years. How long will it take for 25% of a sample of tritium to decompose?

90. RADIOACTIVE DECAY In two years, 20% of a radioactive element decays. Find its half-life.

91. THORIUM DECAY An isotope of thorium, ^{227}Th, has a half-life of 18.4 days. How long will it take for 80% of the sample to decompose?

92. LEAD DECAY An isotope of lead, ^{201}Pb, has a half-life of 8.4 hours. How many hours ago was there 30% more of the substance?

93. CARBON-14 DATING A bone fragment analyzed by archaeologists contains 60% of the carbon-14 that it is assumed to have had initially. How old is it?

94. CARBON-14 DATING   Only 10% of the carbon-14 in a small wooden bowl remains. How old is the bowl?

95. COMPOUND INTEREST   If $500 is deposited in an account paying 8.5% annual interest, compounded semiannually, how long will it take for the account to increase to $800?

96. CONTINUOUS COMPOUND INTEREST   In Exercise 95, how long will it take if the interest is compounded continuously?

97. COMPOUND INTEREST   If $1,300 is deposited in a savings account paying 9% interest, compounded quarterly, how long will it take the account to increase to $2,100?

98. COMPOUND INTEREST   A sum of $5,000 deposited in an account grows to $7,000 in 5 years. Assuming annual compounding, what interest rate is being paid?

99. RULE OF SEVENTY   A rule of thumb for finding how long it takes an investment to double is called the **rule of seventy.** To apply the rule, divide 70 by the interest rate written as a percent. At 5%, an investment takes $\frac{70}{5} = 14$ years to double. At 7%, it takes $\frac{70}{7} = 10$ years. Explain why this formula works.

100. BACTERIAL GROWTH   A bacterial culture grows according to the formula

$$P = P_0 a^t$$

If it takes 5 days for the culture to triple in size, how long will it take to double in size?

101. RODENT CONTROL   The rodent population in a city is currently estimated at 30,000. If it is expected to double every 5 years, when will the population reach 1 million?

102. POPULATION GROWTH   The population of a city is expected to triple every 15 years. When can the city planners expect the present population of 140 persons to double?

103. BACTERIA CULTURE   A bacteria culture doubles in size every 24 hours. By how much will it have increased in 36 hours?

104. OCEANOGRAPHY   The intensity I of a light a distance x meters beneath the surface of a lake decreases exponentially. Use the illustration in the next column to find the depth at which the intensity will be 20%.

105. MEDICINE   If a medium is inoculated with a bacteria culture containing 500 cells per milliliter, how many generations will have passed by the time the culture contains 5×10^6 cells per milliliter?

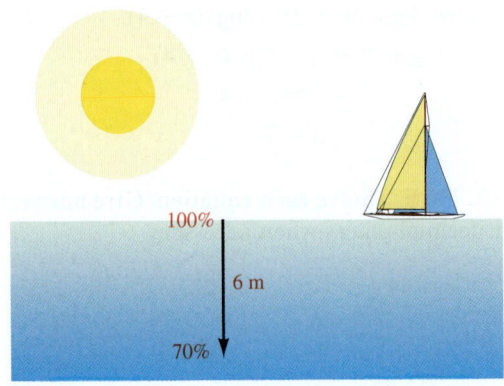

106. MEDICINE   If a medium is inoculated with a bacteria culture containing 800 cells per milliliter, how many generations will have passed by the time the culture contains 6×10^7 cells per milliliter?

107. NEWTON'S LAW OF COOLING   Water initially at 100°C is left to cool in a room at temperature 60°C. After 3 minutes, the water temperature is 90°. The water temperature T is a function of time t given by

$$T = 60 + 40e^{kt}$$

Find k.

108. NEWTON'S LAW OF COOLING   Refer to Exercise 107 and find the time for the water temperature to reach 70°C.

WRITING

109. Explain how to solve the equation $2^{x+1} = 31$.

110. Explain how to solve the equation $2^{x+1} = 32$.

111. Write a justification for each step of the solution.

$$15^x = 9 \qquad \text{This is the equation to solve.}$$

$$\log 15^x = \log 9 \qquad \text{\underline{\hspace{3cm}}.}$$

$$x \log 15 = \log 9 \qquad \text{\underline{\hspace{3cm}}.}$$

$$x = \frac{\log 9}{\log 15} \qquad \text{\underline{\hspace{3cm}}.}$$

112. What is meant by the term *half-life*?

REVIEW

113. Find the length of leg AC.

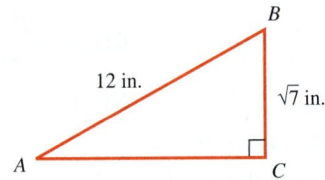

114. The amount of medicine a patient should take is often proportional to his or her weight. If a patient weighing 83 kilograms needs 150 milligrams of medicine, how much will be needed by a person weighing 99.6 kilograms?

CHALLENGE PROBLEMS

115. Without solving the following equation, find the values of x that cannot be a solution:

$$\log(x - 3) - \log(x^2 + 2) = 0$$

116. Solve the equation: $x^{\log x} = 10,000$.

ACCENT ON TEAMWORK

THE NUMBER e

Overview: In this activity, you will use a calculator to find progressively more accurate approximations of e.

Instructions: Form groups of 2 students. Each student will need a scientific calculator.

Begin by finding an approximation of e using the $\boxed{e^x}$ key on your calculator. Copy the table shown below, and write the number displayed on the calculator screen at the top of the table.

The value of e can be calculated to any degree of accuracy by adding the terms of the following pattern:

$$e = 1 + 1 + \frac{1}{2} + \frac{1}{2 \cdot 3} + \frac{1}{2 \cdot 3 \cdot 4} + \frac{1}{2 \cdot 3 \cdot 4 \cdot 5} + \cdots$$

The more terms that are added, the closer the sum will be to e.

You are to add as many terms as necessary until you obtain a sum that matches the value of e given by the $\boxed{e^x}$ key. Work together as a team. One member of the group should compute the fractional form of the term to be added. (See the middle column of the table.) The other member should take that information and calculate the cumulative sum. (See the right-hand column of the table.)

How many terms must be added so that the cumulative sum approximation and the $\boxed{e^x}$ key approximation match in each decimal place?

Approximation of e found using the $\boxed{e^x}$ key: $e \approx$ _____

Number of terms in the sum	Term (Expressed as a fraction)	Cumulative sum (An approximation of e)
1	1	1
2	1	2
3	$\dfrac{1}{2}$	2.5
4	$\dfrac{1}{2 \cdot 3} = \dfrac{1}{6}$	2.666666667
⋮	⋮	⋮

KEY CONCEPT: INVERSE FUNCTIONS

ONE-TO-ONE FUNCTIONS

For a one-to-one function, each input is assigned exactly one output, and each output corresponds to exactly one input. Determine whether each function is one-to-one.

1. $f(x) = x^2$

2. $f(x) = |x|$

3.

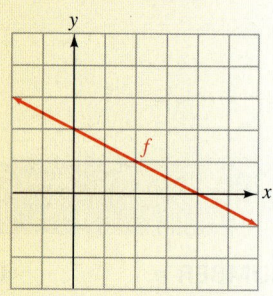

THE INVERSE OF A FUNCTION

If a function is one-to-one, its inverse is a function. To find the inverse of a function, replace $f^{-1}(x)$ with y, interchange x and y, and solve for y.

4. Find $f^{-1}(x)$ if $f(x) = -2x - 1$.

5. Given the table of values for a one-to-one function f, complete the table of values for f^{-1}.

x	-2	1	3
$f(x)$	4	-2	-6

x	4	-2	-6
$f^{-1}(x)$			

EXPONENTIAL AND LOGARITHMIC FUNCTIONS

The exponential function $f(x) = b^x$ and logarithmic function $g(x) = \log_b x$ (where $b > 0$ and $b \neq 1$) are inverse functions.

6. Write the exponential statement $10^3 = 1{,}000$ in logarithmic form.

7. Write the logarithmic statement $\log_2 \frac{1}{8} = -3$ in exponential form.

8. If $\log_4 x = \frac{1}{2}$, what is x?

9. If $\log_x \frac{9}{4} = 2$, what is x?

THE NATURAL EXPONENTIAL AND NATURAL LOGARITHMIC FUNCTIONS

A special exponential function that is used in many real-life applications involving growth and decay is the base-e exponential function, $f(x) = e^x$. Its inverse is the natural logarithm function $g(x) = \ln x$.

10. What is an approximate value of e?

11. What is the base of the logarithmic function $f(x) = \ln x$?

12. Use a calculator to find x: $\ln x = -0.28$.

13. Graph $f(x) = e^x$ and $g(x) = \ln x$ on the coordinate system shown. Label the axis of symmetry.

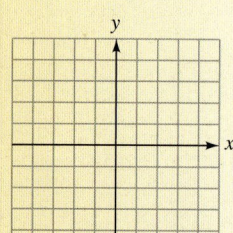

CHAPTER REVIEW
ELEMENTARY &
INTERMEDIATE
Algebra *f(x)* **Now**™

SECTION 11.1

Algebra and Composition of Functions

CONCEPTS

Sum, difference, product, and quotient functions are defined as:

$(f + g)(x) = f(x) + g(x)$

$(f - g)(x) = f(x) - g(x)$

$(f \cdot g)(x) = f(x)g(x)$

$(f/g)(x) = \dfrac{f(x)}{g(x)}$, with

$g(x) \neq 0$

Composition of functions:

$(f \circ g)(x) = f(g(x))$

REVIEW EXERCISES

Let $f(x) = 2x$ and $g(x) = x + 1$. Find each function and its domain.

1. $f + g$ **2.** $f - g$

3. $f \cdot g$ **4.** f/g

Let $f(x) = x^2 + 2$ and $g(x) = 2x + 1$. Find each composition.

5. $(f \circ g)(-1)$ **6.** $(g \circ f)(0)$

7. $(f \circ g)(x)$ **8.** $(g \circ f)(x)$

9. Use the table of values for functions f and g to find each of the following.

    a. $(f + g)(2)$ **b.** $(f \cdot g)(2)$ **c.** $(f \circ g)(2)$ **d.** $(g \circ f)(2)$

x	$f(x)$
2	3
9	7

x	$g(x)$
2	9
3	0

10. MILEAGE COSTS   The function $f(m) = \dfrac{m}{8}$ gives the number of gallons of fuel consumed if a bus travels m miles. The function $C(f) = 1.85f$ gives the cost (in dollars) of f gallons of fuel. Write a composition function that expresses the cost of the fuel consumed as a function of the number of miles driven.

SECTION 11.2

Inverse Functions

For a *one-to-one function,* each input is assigned exactly one output, and each output corresponds to exactly one input.

Horizontal line test: A function is one-to-one if every horizontal line intersects the graph of the function at most once.

In Exercises 11–16, determine whether the function is one-to-one.

11. $f(x) = x^2 + 3$ **12.** $f(x) = \dfrac{1}{3}x - 8$

13. $\{(3, 4), (5, 10), (10, -1), (6, 6)\}$ **14.**

x	$f(x)$
0	-5
2	10
4	-5
6	15

15.

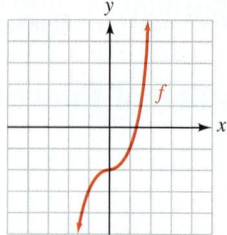

16.

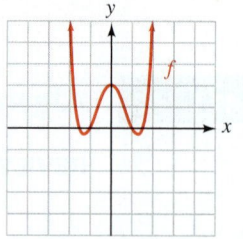

The graph of a function and its inverse are symmetric about the line $y = x$.

17. Use the table of values of the one-to-one function f to complete a table of values for f^{-1}.

x	$f(x)$
-6	-6
-1	-3
7	12
20	3

x	$f^{-1}(x)$
-6	
-3	
12	
3	

18. Given the graph of function f, graph f^{-1} on the same coordinate axes. Label the axis of symmetry.

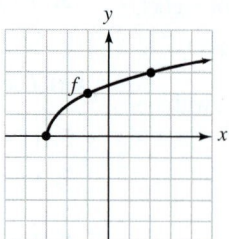

To *find the inverse of a function*, replace $f(x)$ with y, interchange the variables x and y, solve for y, and replace y with $f^{-1}(x)$.

Find the inverse of each function.

19. $f(x) = 6x - 3$

20. $f(x) = \dfrac{4}{x - 1}$

21. $f(x) = (x + 2)^3$

22. $f(x) = \dfrac{x}{6} - \dfrac{1}{6}$

For any one-to-one function f and its inverse, f^{-1},

$(f \circ f^{-1})(x) = x$
$(f^{-1} \circ f)(x) = x$

23. Find the inverse of $f(x) = \sqrt[3]{x - 1}$. Then graph the function and its inverse on one coordinate system. Show the line of symmetry on the graph.

24. Use composition to show that $f(x) = 5 - 4x$ and $f^{-1}(x) = -\dfrac{x - 5}{4}$ are inverse functions.

SECTION 11.3   Exponential Functions

An *exponential function* with base b is defined by the equation

$f(x) = b^x$, with $b > 0$, $b \neq 1$

Use properties of exponents to simplify each expression.

25. $5^{\sqrt{6}} \cdot 5^{3\sqrt{6}}$

26. $\left(2^{\sqrt{14}}\right)^{\sqrt{2}}$

Graph each function and give the domain and the range. Label the y-intercept.

If $b > 1$, then $f(x) = b^x$ is an *increasing function*.

If $0 < b < 1$, then $f(x) = b^x$ is a *decreasing function*.

27. $f(x) = 3^x$

28. $f(x) = \left(\dfrac{1}{3}\right)^x$

29. $f(x) = \left(\dfrac{1}{2}\right)^x - 2$

30. $f(x) = 3^{x-1}$

31. In Exercise 30, what is the asymptote of the graph of $f(x) = 3^{x-1}$?

Exponential functions
are suitable models for
describing m~~...~~

32. COAL PRODUCTION   The table gives the number of tons of coal produced in
~~...~~ ited States for the years 1800–1920. Graph the data. What type of function
~~...~~ appear could be used to model coal production over this period?

Year	Tons	Year	Tons
1800	108,000	1870	40,429,000
1810	178,000	1880	79,407,000
1820	881,000	1890	157,771,000
1830	1,334,000	1900	269,684,000
1840	2,474,000	1910	501,596,000
1850	8,356,000	1920	658,265,000
1860	20,041,000		

Source: *World Book Encyclopedia*

~~...~~ INTEREST   How much will $10,500 become if it earns 9% annual
~~...~~ ounded quarterly, for 60 years?

~~...~~ ON   The value (in dollars) of a certain model car is given by the
~~...~~ $= 12,000\left(10^{-0.155t}\right)$, where t is the number of years from the present.
~~...~~ alue of the car be in 5 years?

~~...~~ntial Functions

~~...~~ ES   The average annual interest rate on a 30-year fixed-rate
~~...~~ the years 1980–1996 can be approximated by the function
~~...~~ where t is the number of years since 1980. To the nearest
~~...~~ nt, what does this model predict was the 30-year fixed rate

~~...~~ UNDED CONTINUOUSLY   If $10,500 accumulates interest
~~...~~ %, compounded continuously, how much will be in the account

~~...~~ **give the domain and the range.**

38. $f(x) = e^{x-3}$

Chapter 11A Review

1-8, 11-14   p. 925
15-20, 24-28   p. 926
33, 36   p. 927
Example 2, se❦check   p. 877

If
tity

Malthusian population growth is modeled by the formula

$$A = Pe^{rt}$$

39. POPULATION GROWTH   As of July 2003, the population of the United States was estimated to be 290,340,000, with an annual growth rate of 0.92%. If the growth rate remains the same, how large will the population be in 50 years?

40. MEDICAL TESTS   A radioactive dye is injected into a patient as part of a test to detect heart disease. The amount of dye remaining in his bloodstream t hours after the injection is given by the function $f(t) = 10e^{-0.27t}$. How can you tell from the function that the amount of dye in the bloodstream is decreasing?

SECTION 11.5	**Logarithmic Functions**

If $b > 0$ and $b \neq 1$, then the *logarithmic function with base b* is defined by $f(x) = \log_b x$.

$y = \log_b x$ means $x = b^y$.

41. Give the domain and range of $f(x) = \log x$.

42. When a student used a calculator to evaluate $\log 0$, she obtained the following message:

$\boxed{\text{Error}}$

Explain why.

43. Write the statement $\log_4 64 = 3$ in exponential form.

44. Write the statement $7^{-1} = \dfrac{1}{7}$ in logarithmic form.

$\log_b x$ is the exponent to which b is raised to get x.

$$b^{\log_b x} = x$$

Find each value, if possible.

45. $\log_3 9$

46. $\log_9 \dfrac{1}{81}$

47. $\log_{1/2} 1$

48. $\log_5 (-25)$

49. $\log_6 \sqrt{6}$

50. $\log 1{,}000$

For computational purposes and in many applications, we use base-10 logarithms called *common logarithms*.

$\log x$ means $\log_{10} x$

Solve each equation.

51. $\log_2 x = 5$

52. $\log_3 x = -4$

53. $\log_x 16 = 2$

54. $\log_x \dfrac{1}{100} = -2$

55. $\log_9 3 = x$

56. $\log_{27} x = \dfrac{2}{3}$

Use a calculator to find the value of x to four decimal places.

57. $\log 4.51 = x$

58. $\log x = 1.43$

If $b > 1$, then $f(x) = \log_b x$ is an *increasing function*. If $0 < b < 1$, then $f(x) = \log_b x$ is a *decreasing function*.

Graph each pair of equations on one set of coordinate axes. Draw the axis of symmetry.

59. $f(x) = \log_4 x$ and $g(x) = 4^x$

60. $f(x) = \log_{1/3} x$ and $g(x) = \left(\dfrac{1}{3}\right)^x$

The exponential function $f(x) = b^x$ and the logarithmic function $f(x) = \log_b x$ are inverses of each other.

Graph each function. Label the x-intercept.

61. $f(x) = \log(x - 2)$

62. $f(x) = 3 + \log x$

Decibel voltage gain:

$$\text{db gain} = 20 \log \frac{E_O}{E_I}$$

63. ELECTRICAL ENGINEERING   An amplifier has an output of 18 volts when the input is 0.04 volt. Find the db gain.

The Richter scale:

$$R = \log \frac{A}{P}$$

64. EARTHQUAKES   An earthquake had a period of 0.3 second and an amplitude of 7,500 micrometers. Find its measure on the Richter scale.

| SECTION 11.6 | **Base-*e* Logarithms** |

Natural logarithms:

$\ln x$ means $\log_e x$

$\ln x$ is the exponent to which e is raised to get x.

$e^{\ln x} = x$

The *natural logarithmic function* with base e is defined by

$f(x) = \ln x$

The domain of $f(x) = \ln x$ is the interval $(0, \infty)$ and the range is the interval $(-\infty, \infty)$.

Evaluate each expression, if possible. Do not use a calculator.

65. $\ln e$

66. $\ln e^2$

67. $\ln \dfrac{1}{e^5}$

68. $\ln \sqrt{e}$

69. $\ln(-e)$

70. $\ln 0$

71. $\ln 1$

72. $\ln e^{-7}$

Use a calculator to find each value to four decimal places.

73. $\ln 452$

74. $\ln 0.85$

Use a calculator to find the value of x to four decimal places.

75. $\ln x = 2.336$

76. $\ln x = -8.8$

77. Explain the difference between the functions $f(x) = \log x$ and $g(x) = \ln x$.

78. What function is the inverse of $f(x) = \ln x$?

Graph each function.

79. $f(x) = 1 + \ln x$

80. $f(x) = \ln(x + 1)$

Population doubling time:

$$t = \frac{\ln 2}{r}$$

81. POPULATION GROWTH   How long will it take the population of Mexico to double if the growth rate is currently about 1.43%?

82. BOTANY   The height (in inches) of a certain plant is approximated by the function $H(a) = 13 + 20.03 \ln a$, where a is its age in years. How tall will it be when it is 19 years old?

| SECTION 11.7 | **Properties of Logarithms** |

Properties of logarithms:
If b is a positive number and $b \neq 1$,

1. $\log_b 1 = 0$
2. $\log_b b = 1$
3. $\log_b b^x = x$
4. $b^{\log_b x} = x$
5. $\log_b MN =$
   $\log_b M + \log_b N$
6. $\log_b \dfrac{M}{N} =$
   $\log_b M - \log_b N$
7. $\log_b M^P = p \log_b M$

Simplify each expression.

83. $\log_2 1$
84. $\log_9 9$
85. $\log 10^3$
86. $7^{\log_7 4}$

Use the properties of logarithms to rewrite each expression.

87. $\log_3 27x$
88. $\log \dfrac{100}{x}$
89. $\log_5 \sqrt{27}$
90. $\log_b 10ab$

Write each expression in terms of the logarithms of x, y and z.

91. $\log_b \dfrac{x^2 y^3}{z}$
92. $\ln \sqrt{\dfrac{x}{yz^2}}$

Properties of logarithms can be used to condense expressions.

Write each expression as the logarithm of one quantity.

93. $3 \log_2 x - 5 \log_2 y + 7 \log_2 z$

94. $-3 \log_b y - 7 \log_b z + \dfrac{1}{2} \log_b (x + 2)$

Assume that $\log_b 5 = 1.1609$ and $\log_b 8 = 1.5000$. Use these values and the properties of logarithms to find each value to four decimal places.

95. $\log_b 40$
96. $\log_b 64$

Change-of-base formula:

$\log_b x = \dfrac{\log_a x}{\log_a b}$

pH scale:
$\mathrm{pH} = -\log [\mathrm{H}^+]$

97. Find $\log_5 17$ to four decimal places.
98. pH OF GRAPEFRUIT   The pH of grapefruit juice is about 3.1. Find its hydrogen ion concentration.

| SECTION 11.8 | **Exponential and Logarithmic Equations** |

If two exponential expressions with the same base are equal, their exponents are equal.
$b^x = b^y$
is equivalent to $x = y$

Solve each equation for x. Give answers to four decimal places when necessary.

99. $5^{x+2} = 625$
100. $2^{x^2+4x} = \dfrac{1}{8}$
101. $3^x = 7$
102. $2^x = 3^{x-1}$
103. $e^x = 7$
104. $e^{-0.4t} = 25$

When it is difficult to write each side of an exponential equation as a power of the same base, take the logarithm of each side.

If two positive numbers are equal, then the logarithms base-b of the numbers are equal.

$$\log_b x = \log_b y$$

is equivalent to $x = y$

Carbon dating:

$$A = A_0 2^{-t/h}$$

Population growth

$$P = P_0 e^{kt}$$

Solve each equation for x.

105. $\log(x - 4) = 2$

106. $\ln(2x - 3) = \ln 15$

107. $\log x + \log(29 - x) = 2$

108. $\log_2 x + \log_2(x - 2) = 3$

109. $\dfrac{\log(7x - 12)}{\log x} = 2$

110. $\log_2(x + 2) + \log_2(x - 1) = 2$

111. $\log x + \log(x - 5) = \log 6$

112. $\log 3 - \log(x - 1) = -1$

113. Evaluate both sides of the statement $\dfrac{\log 8}{\log 15} \neq \log 8 - \log 15$ to show that the sides are indeed not equal.

114. CARBON-14 DATING   A wooden statue found in Egypt has a carbon-14 content that is two-thirds of that found in living wood. If the half-life of carbon-14 is 5,700 years, how old is the statue?

115. ANTS   The number of ants in a colony is estimated to be 800. If the ant population is expected to triple every 14 days, how long will it take for the population to reach one million?

116. The approximate coordinates of the points of intersection of the graphs of $f(x) = \log x$ and $g(x) = 1 - \log(7 - x)$ are shown in parts (a) and (b) of the illustration. Use the graphs to estimate the solutions of the logarithmic equation $\log x = 1 - \log(7 - x)$. Then check your answers.

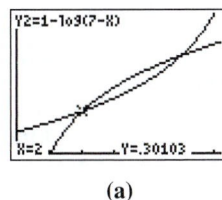   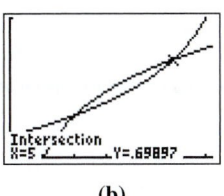

(a)   (b)

CHAPTER 11 TEST   ELEMENTARY & INTERMEDIATE Algebra *f(x)* Now™

Let $f(x) = x + 9$ and $g(x) = 4x^2 - 3x + 2$. Find each function and give its domain.

1. $f + g$

2. g/f

Let $f(x) = 2x^2 + 3$ and $g(x) = 4x - 8$. Find each composition.

3. $(g \circ f)(-3)$

4. $(f \circ g)(x)$

Use the tables of values for functions f and g to find each of the following.

5. $(f \cdot g)(9)$

6. $(f \circ g)(-3)$

x	$f(x)$
9	-1
10	17

x	$g(x)$
-3	10
9	16

Determine whether each function is one-to-one.

7. $f(x) = x$ **8.**

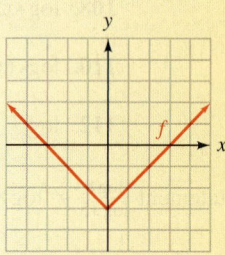

9. Find the inverse of $f(x) = -\frac{1}{3}x$ and then graph f and its inverse on the same coordinate axes.

10. Find the inverse of $f(x) = (x - 15)^3$.

11. Use composition to show that $f(x) = 4x + 4$ and $f^{-1}(x) = \dfrac{x - 4}{4}$ are inverse functions.

12. Consider the following graph of the function f.

   a. Is f a one-to-one function?

   b. Is its inverse a function?

   c. What is $f^{-1}(260)$? What information does it give?

Relationship between car speed and tire temperature

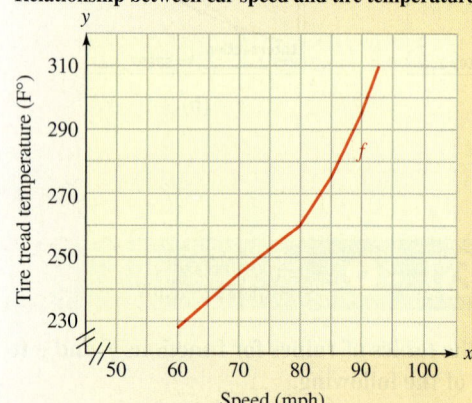

Graph each function and give the domain and the range.

13. $f(x) = 2^x + 1$ **14.** $f(x) = 3^{-x}$

15. RADIOACTIVE DECAY   A radioactive material decays according to the formula $A = A_0(2)^{-t}$. How much of a 3-gram sample will be left in 6 years?

16. COMPOUND INTEREST   An initial deposit of $1,000 earns 6% interest, compounded twice a year. How much will be in the account in one year?

17. Graph: $f(x) = e^x$. Label the y-intercept and the asymptote of the graph.

18. POPULATION GROWTH   As of July 2003, the population of India was estimated to be 1,050,000,000, with an annual growth rate of 1.47%. If the growth rate remains the same, how large will the population be in 30 years?

19. Write the statement $\log_6 \frac{1}{36} = -2$ in exponential form.

20. **a.** What are the domain and range of the function $f(x) = \log x$?

   b. What is the inverse of $f(x) = \log x$?

Evaluate each logarithmic expression, if possible.

21. $\log_5 25$ **22.** $\log_9 \dfrac{1}{81}$

23. $\log(-100)$ **24.** $\ln \dfrac{1}{e^6}$

25. $\log_4 2$ **26.** $\log_{1/3} 1$

Find x.

27. $\log_x 32 = 5$ **28.** $\log_8 x = \dfrac{4}{3}$

29. $\log_3 x = -3$ **30.** $\ln x = 1$

Graph each function.

31. $f(x) = -\log_3 x$ **32.** $f(x) = \ln x$

33. pH   Find the pH of a solution with a hydrogen ion concentration of 3.7×10^{-7}. (*Hint:* pH $= -\log[\text{H}^+]$.)

34. ELECTRONICS   Find the db gain of an amplifier when $E_O = 60$ volts and $E_I = 0.3$ volt.

   Hint: db gain $= 20 \log\left(\dfrac{E_O}{E_I}\right)$.

35. Use a calculator to find x to four decimal places.

$$\log x = -1.06$$

36. Use the change-of-base formula to find $\log_7 3$ to four decimal places.

37. Write the expression $\log_b a^2bc^3$ in terms of the logarithms of a, b, and c.

38. Write the expression $\frac{1}{2}\ln(a+2) + \ln b - 3\ln c$ as a logarithm of a single quantity.

Solve each equation. Round to four decimal places when necessary.

39. $5^x = 3$ **40.** $3^{x-1} = 27$

41. $\ln(5x+2) = \ln(2x+5)$

42. $\log x + \log(x-9) = 1$

43. The illustration shows the graphs of $y = \frac{1}{2}\ln(x-1)$ and $y = \ln 2$ and the approximate coordinates of their point of intersection. Estimate the solution of the logarithmic equation $\frac{1}{2}\ln(x-1) = \ln 2$. Then check the result.

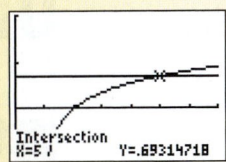

44. INSECTS   The number of insects attracted to a bright light is currently 5. If the number is expected to quadruple every 6 minutes, how long will it take for the number to reach 500?

More on Systems of Equations

ELEMENTARY & INTERMEDIATE
Algebra *f(x)* **Now**™

Throughout the chapter, this icon introduces resources on the Elementary & Intermediate AlgebraNow Web site, accessed through **http://1pass. thomson.com**, that will

- Help you test your knowledge of the material with a pre-test and a post-test

- Provide a personalized learning plan targeting areas you should study

Nutritionists plan food programs for hospitals, nursing care facilities, and schools. They often work with food service managers to supervise the preparation and serving of meals. In this chapter, we will use algebra to determine the proper serving size so that a meal contains the necessary amounts of carbohydrates, fats, and proteins. To answer such a question, we first write a system of three linear equations. To solve the system, we can use a process called Cramer's rule, in which we evaluate square arrays of numbers called *determinants*.

To learn more about determinants, visit *The Learning Equation* on the Internet at http://tle.brookscole.com. (The log-in instructions are in the Preface.) For Chapter 12, the online lesson is:

- *TLE* Lesson 21: Solving Linear Systems by Determinants

To solve many problems, we must use two and sometimes three variables. This requires that we solve a system of equations.

12.1 Systems with Two Variables

- The graphing method
- The substitution method
- The elimination method
- Inconsistent systems; dependent equations
- Problem solving
- Systems of inequalities

In this section, we will discuss graphical and algebraic methods for solving systems of two linear equations in two variables.

■ THE GRAPHING METHOD

To solve a system of two equations in two variables by graphing, we use the following steps.

The Graphing Method	1. Carefully graph each equation on the same rectangular coordinate system. 2. If the lines intersect, determine the coordinates of the point of intersection of the graphs. That ordered pair is the solution of the system. 3. Check the proposed solution in each equation of the original system.

When a system of equations (as in Example 1) has at least one solution, the system is called a **consistent system.**

EXAMPLE 1

Solve the system: $\begin{cases} x + 2y = 4 \\ 2x - y = 3 \end{cases}$.

 ELEMENTARY & INTERMEDIATE Algebra $f(x)$ Now™

Solution We graph both equations on one set of coordinate axes, as shown.

x + 2y = 4

x	y	(x, y)
4	0	(4, 0)
0	2	(0, 2)
−2	3	(−2, 3)

2x − y = 3

x	y	(x, y)
$\frac{3}{2}$	0	$\left(\frac{3}{2}, 0\right)$
0	−3	(0, −3)
−1	−5	(−1, −5)

Use the intercept method to graph each line.

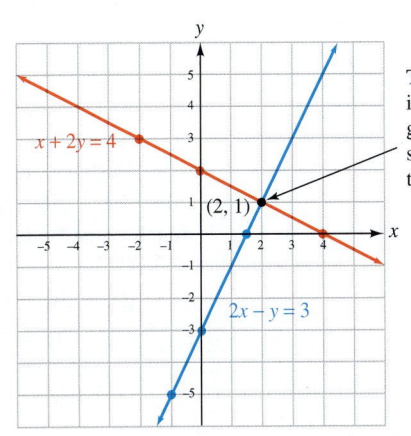

The point of intersection gives the solution of the system.

Although infinitely many ordered pairs (x, y) satisfy $x + 2y = 4$, and infinitely many ordered pairs (x, y) satisfy $2x - y = 3$, only the coordinates of the point where the graphs intersect satisfy both equations. From the graph, it appears that the intersection point has coordinates $(2, 1)$. To verify that it is the solution, we substitute 2 for x and 1 for y in both equations and verify that $(2, 1)$ satisfies each one.

Check:

The first equation	**The second equation**
$x + 2y = 4$	$2x - y = 3$
$2 + 2(1) \overset{?}{=} 4$	$2(2) - 1 \overset{?}{=} 3$
$2 + 2 \overset{?}{=} 4$	$4 - 1 \overset{?}{=} 3$
$4 = 4$ True.	$3 = 3$ True.

Since $(2, 1)$ makes both equations true, it is the solution of the system.

Self Check 1   Solve: $\begin{cases} x - 3y = -5 \\ 2x + y = 4 \end{cases}$.

When a system has no solution (as in Example 2), it is called an **inconsistent system.**

EXAMPLE 2

Solve $\begin{cases} 2x + 3y = 6 \\ 4x + 6y = 24 \end{cases}$, if possible.

Solution   Using the intercept method, we graph both equations on one set of coordinate axes, as shown in the figure.

Success Tip

Since accuracy is crucial when using the graphing method to solve a system:
- Use graph paper.
- Use a sharp pencil.
- Use a straight-edge.

$2x + 3y = 6$

x	y	(x, y)
3	0	$(3, 0)$
0	2	$(0, 2)$
-3	4	$(-3, 4)$

$4x + 6y = 24$

x	y	(x, y)
6	0	$(6, 0)$
0	4	$(0, 4)$
-3	6	$(-3, 6)$

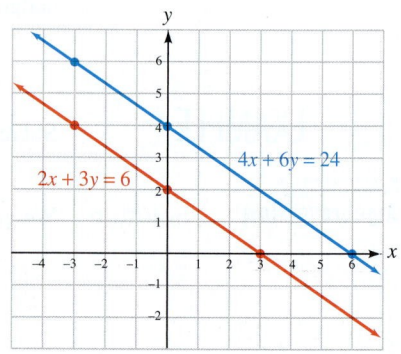

In this example, the graphs are parallel, because the slopes of the lines are equal and they have different y-intercepts. We can see that the slope of each line is $-\frac{2}{3}$ by writing each equation in slope–intercept form.

$$2x + 3y = 6 \qquad\qquad 4x + 6y = 24$$

$$3y = -2x + 6 \qquad\qquad 6y = -4x + 24$$

$$y = -\frac{2}{3}x + 2 \qquad\qquad y = -\frac{2}{3}x + 4$$

Since the graphs are parallel lines, the lines do not intersect, and the system does not have a solution. It is an inconsistent system.

Self Check 2 Solve: $\begin{cases} 3y - 2x = 6 \\ 2x - 3y = 6 \end{cases}$.

When the equations of a system have different graphs (as in Examples 1 and 2), the equations are called **independent equations.** Two equations with the same graph are called **dependent equations.**

EXAMPLE 3

Solve: $\begin{cases} y = \dfrac{1}{2}x + 2 \\ 2x + 8 = 4y \end{cases}$.

Solution We graph each equation on one set of coordinate axes, as shown in the figure. Since the graphs coincide, the system has an infinite number of solutions. Any ordered pair (x, y) that satisfies one equation also satisfies the other. From the graph, we see that $(-4, 0)$, $(0, 2)$, and $(2, 3)$ are three of infinitely many solutions. Because the two equations have the same graph, they are dependent equations.

Graph by using the slope and y-intercept.

$$y = \frac{1}{2}x + 2$$

$$m = \frac{1}{2} \qquad b = 2$$

$$\text{Slope} = \frac{1}{2} \qquad y\text{-intercept: } (0, 2)$$

Graph by using the intercept method.

$$2x + 8 = 4y$$

x	y	(x, y)
-4	0	$(-4, 0)$
0	2	$(0, 2)$
2	3	$(2, 3)$

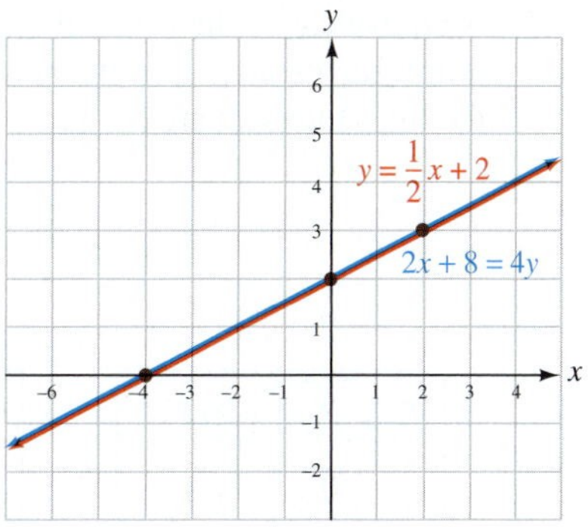

The Language of Algebra

Here the graphs of the lines *coincide.* That is, they occupy the same location. To illustrate this concept, think of a clock. At noon and midnight, the hands of the clock *coincide.*

Self Check 3 Solve: $\begin{cases} 2x - y = 4 \\ y = 2x - 4 \end{cases}$.

We now summarize the possibilities that can occur when two linear equations, each in two variables, are graphed.

Solving a System of Equations by the Graphing Method

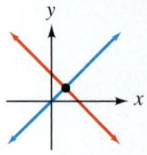

If the lines are different and intersect, the equations are independent, and the system is consistent. **One solution exists.** It is the point of intersection.

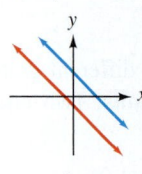

If the lines are different and parallel, the equations are independent, and the system is inconsistent. **No solution exists.**

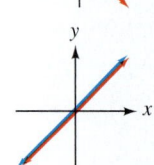

If the lines coincide, the equations are dependent, and the system is consistent. **Infinitely many solutions exist.** Any point on the line is a solution.

If each equation in one system is equivalent to a corresponding equation in another system, the systems are called **equivalent.**

EXAMPLE 4

ELEMENTARY & INTERMEDIATE
Algebra $f(x)$ **Now**™

Solve:
$$
\begin{cases}
\dfrac{3}{2}x - y = \dfrac{5}{2} \\
x + \dfrac{1}{2}y = 4
\end{cases}.
$$

Solution   We multiply both sides of $\frac{3}{2}x - y = \frac{5}{2}$ by 2 to eliminate the fractions and obtain the equation $3x - 2y = 5$. We multiply both sides of $x + \frac{1}{2}y = 4$ by 2 to eliminate the fraction and obtain the equation $2x + y = 8$.

The new system

$$
\begin{cases}
3x - 2y = 5 \\
2x + y = 8
\end{cases}
$$

is equivalent to the original system and is easier to solve, since it has no fractions. If we graph each equation in the new system, it appears that the coordinates of the point where the lines intersect are (3, 2).

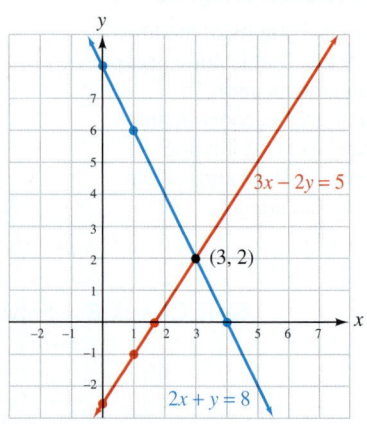

$3x - 2y = 5$

x	y	(x, y)
$\frac{5}{3}$	0	$\left(\frac{5}{3}, 0\right)$
0	$-\frac{5}{2}$	$\left(0, -\frac{5}{2}\right)$
1	-1	$(1, -1)$

$2x + y = 8$

x	y	(x, y)
4	0	$(4, 0)$
0	8	$(0, 8)$
1	6	$(1, 6)$

To verify that (3, 2) is the solution, we substitute 3 for x and 2 for y in each equation of the original system.

Check:
$$\frac{3}{2}x - y = \frac{5}{2} \qquad\qquad x + \frac{1}{2}y = 4$$

$$\frac{3}{2}(3) - 2 \overset{?}{=} \frac{5}{2} \qquad\qquad 3 + \frac{1}{2}(2) \overset{?}{=} 4$$

$$\frac{9}{2} - 2 \overset{?}{=} \frac{5}{2} \qquad\qquad 3 + 1 \overset{?}{=} 4$$

$$\qquad\qquad\qquad\qquad\qquad 4 = 4 \quad \text{True.}$$

$$\frac{5}{2} = \frac{5}{2} \quad \text{True.}$$

Caution

When checking the solution of a system of equations, always substitute the values of the variables into the original equations.

Self Check 4   Solve $\begin{cases} \dfrac{5}{2}x - y = 2 \\ x + \dfrac{1}{3}y = 3 \end{cases}$ by the graphing method.

ACCENT ON TECHNOLOGY: SOLVING SYSTEMS BY GRAPHING

The graphing method is limited to equations with two variables. Systems with three or more variables cannot be solved graphically. Also, it is often difficult to find exact solutions graphically. However, the TRACE and ZOOM capabilities of graphing calculators enable us to get very good approximations of such solutions.

To solve the system $\begin{cases} 3x + 2y = 12 \\ 2x - 3y = 12 \end{cases}$

with a graphing calculator, we must first solve each equation for y so that we can enter the equations into the calculator. After solving for y, we obtain the following equivalent system:

$$\begin{cases} y = -\dfrac{3}{2}x + 6 \\ y = \dfrac{2}{3}x - 4 \end{cases}$$

If we use window settings of $[-10, 10]$ for x and for y, the graphs of the equations will look like those in figure (a). If we zoom in on the intersection point of the two lines and trace, we will get an approximate solution like the one shown in figure (b). To get better results, we can do more zooms. We would then find that, to the nearest hundredth, the solution is $(4.63, -0.94)$. Verify that this is reasonable.

We can also find the intersection of two lines by using the INTERSECT feature found on most graphing calculators. After graphing the lines and using INTERSECT, we obtain a graph similar to figure (c). The display shows the approximate coordinates of the point of intersection.

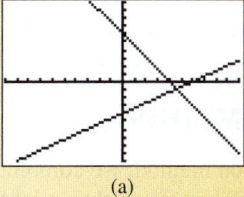

(a)

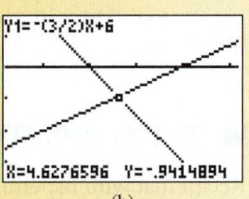

(b)

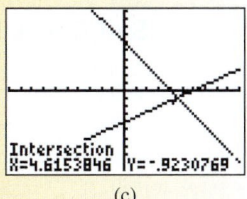

(c)

■ THE SUBSTITUTION METHOD

Recall that we use the following steps to solve a system of two equations in two variables by substitution.

The Substitution Method	
	1. Solve one of the equations for either x or y. If this is already done, go to step 2.
	2. Substitute the expression for x or for y obtained in step 1 into the other equation and solve that equation.
	3. Substitute the value of the variable found in step 2 into the substitution equation to find the value of the remaining variable.
	4. Check the proposed solution in the equations of the original system. Write the solution as an ordered pair.

EXAMPLE 5

ELEMENTARY & INTERMEDIATE

Algebra $f(x)$ **Now**™

Solution

Solve: $\begin{cases} 4x + y = 13 \\ -2x + 3y = -17 \end{cases}$.

We solve the first equation for y, because y has a coefficient of 1.

$$4x + y = 13$$
$$y = -4x + 13 \qquad \text{To isolate } y, \text{ subtract } 4x \text{ from both sides. This is the substitution equation.}$$

We then substitute $-4x + 13$ for y in the second equation to eliminate the variable y from that equation. The result will be an equation containing only one variable, x.

$$-2x + 3y = -17 \qquad \text{This is the second equation.}$$
$$-2x + 3(\mathbf{-4x + 13}) = -17 \qquad \text{Substitute } -4x + 13 \text{ for } y. \text{ The variable } y \text{ is eliminated from the equation.}$$
$$-2x - 12x + 39 = -17 \qquad \text{Distribute the multiplication by 3.}$$
$$-14x = -56 \qquad \text{To solve for } x, \text{ first combine like terms and then subtract 39 from both sides.}$$
$$x = 4 \qquad \text{Divide both sides by } -14.$$

Success Tip

With this method, the objective is to use an appropriate substitution to obtain *one* equation in *one* variable.

To find y, we substitute 4 for x in the substitution equation and simplify:

$$y = -4\mathbf{x} + 13$$
$$= -4(\mathbf{4}) + 13 \qquad \text{Substitute 4 for } x.$$
$$= -3$$

The Language of Algebra

The phrase *back-substitute* can also be used to describe Step 3 of the substitution method.

To find y, we *back-substitute* 4 for x in the equation $y = -4x + 13$.

The solution is $(4, -3)$. If graphed, the equations of the given system would intersect at the point $(4, -3)$. The check is left to the reader.

Self Check 5   Solve: $\begin{cases} x + 3y = 9 \\ 2x - y = -10 \end{cases}$.

■ THE ELIMINATION METHOD

Recall that with the elimination method, we combine the equations of the system in a way that will eliminate terms involving one of the variables.

The Elimination Method

1. Write both equations in general form: $Ax + By = C$.
2. If necessary, multiply one or both of the equations by nonzero quantities to make the coefficients of x (or the coefficients of y) opposites.
3. Add the equations to eliminate the terms involving x (or y).
4. Solve the equation resulting from step 3.
5. Find the value of the other variable by substituting the solution found in step 4 into any equation containing both variables.
6. Check the proposed solution in the equations of the original system. Write the solution as an ordered pair.

EXAMPLE 6

ELEMENTARY &
INTERMEDIATE
Algebra $f(x)$ **Now**™

Solve: $\begin{cases} \dfrac{4}{3}x + \dfrac{1}{2}y = -\dfrac{2}{3} \\ \dfrac{1}{2}x + \dfrac{2}{3}y = \dfrac{5}{3} \end{cases}$.

Solution
To solve this system by elimination, we find an equivalent system with no fractions by multiplying both sides of each equation by 6 to obtain

$$\begin{cases} 8x + 3y = -4 \\ 3x + 4y = 10 \end{cases}$$

To make the y-terms drop out when we add the equations, we multiply both sides of $8x + 3y = -4$ by 4 and both sides of $3x + 4y = 10$ by -3 to get

$$\begin{cases} 8x + 3y = -4 \\ 3x + 4y = 10 \end{cases} \xrightarrow[\text{Multiply by } -3]{\text{Multiply by } 4} \begin{cases} 32x + 12y = -16 \\ -9x - 12y = -30 \end{cases}$$

Success Tip

The basic objective of the elimination method is to obtain two equations whose sum will be *one* equation in *one* variable.

When these equations are added, the y-terms drop out.

$$\begin{array}{r} 32x + 12y = -16 \\ -9x - 12y = -30 \\ \hline 23x = -46 \end{array}$$

We solve the resulting equation to find x.

$$23x = -46$$
$$x = -2 \qquad \textcolor{red}{\text{Divide both sides by 23.}}$$

To find y, we can substitute -2 for x in either of the equations of the original system or either of the equations of the equivalent system. It appears the computations will be the simplest if we use $3x + 4y = 10$.

$$\begin{aligned} 3\boldsymbol{x} + 4y &= 10 \\ 3(\boldsymbol{-2}) + 4y &= 10 \qquad \textcolor{red}{\text{Substitute } -2 \text{ for } x.} \\ -6 + 4y &= 10 \qquad \textcolor{red}{\text{Simplify.}} \\ 4y &= 16 \qquad \textcolor{red}{\text{Add 6 to both sides.}} \\ y &= 4 \qquad \textcolor{red}{\text{Divide both sides by 4.}} \end{aligned}$$

The solution is $(-2, 4)$. The check is left to the reader.

Self Check 6   Solve: $\begin{cases} \dfrac{2}{3}x - \dfrac{2}{5}y = 10 \\ \dfrac{1}{2}x + \dfrac{2}{3}y = -7 \end{cases}$.

■ INCONSISTENT SYSTEMS; DEPENDENT EQUATIONS

EXAMPLE 7

Solve $\begin{cases} y = 2x + 4 \\ 8x - 4y = 7 \end{cases}$, if possible.

Solution   Because the first equation is already solved for y, we use the substitution method.

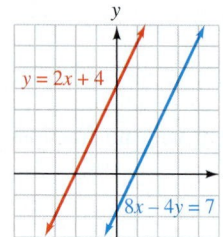

$$8x - 4y = 7 \qquad \text{This is the second equation.}$$
$$8x - 4(\mathbf{2x + 4}) = 7 \qquad \text{Substitute } 2x + 4 \text{ for } y.$$

We then solve this equation for x:

$$8x - 8x - 16 = 7 \qquad \text{Use the distributive property to remove parentheses.}$$
$$-16 \neq 7 \qquad \text{Combine like terms.}$$

Here, the terms involving x drop out, and we get $-16 = 7$. This false statement indicates that the system has *no solution,* and is, therefore, inconsistent. The graphs of the equations of the system verify this; they are parallel lines.

Self Check 7   Solve: $\begin{cases} 4x - 8y = 10 \\ y = \dfrac{1}{2}x - \dfrac{9}{8} \end{cases}$.

EXAMPLE 8

Solve: $\begin{cases} 4x + 6y = 12 \\ -2x - 3y = -6 \end{cases}$.

Solution   Since the equations are written in general form, we use the elimination method. We copy the first equation and multiply both sides of the second equation by 2 to get

$$\begin{array}{r} 4x + 6y = 12 \\ \underline{-4x - 6y = -12} \end{array}$$

After adding the left-hand sides and the right-hand sides, we get

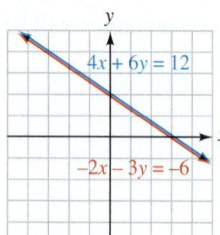

$$0x + 0y = 0$$
$$0 = 0$$

Here, both the x- and y-terms drop out. The resulting true statement $0 = 0$ indicates that the equations are dependent and that the system has infinitely many solutions.

Note that the equations of the system are equivalent; when the second equation is multiplied by -2, it becomes the first equation. The graphs of these equations coincide. Any ordered pair that satisfies one of the equations also satisfies the other. To find some solutions, we can substitute 0, 3, and -3 for x in either equation to obtain $(0, 2)$, $(3, 0)$, and $(-3, 4)$.

Self Check 8   Solve: $\begin{cases} 2(x + y) - y = 12 \\ y = -2x + 12 \end{cases}$.

■ **PROBLEM SOLVING**

EXAMPLE 9

ELEMENTARY &
INTERMEDIATE

Algebra $f(x)$ **Now**™

Retail sales.   Hi-Fi Electronics advertises two types of car radios, one selling for $67 and the other for $100. If the receipts from the sale of 36 radios totaled $2,940, how many of each type were sold?

Analyze the Problem   We can let x = the number of radios sold for $67 and let y = the number of radios sold for $100. Then the receipts from the sale of the less expensive radios are $67x$, and the receipts from the sale of the more expensive radios are $100y$.

Form Two Equations   The information in the problem gives the following two equations:

The number of less expensive radios sold	plus	the number of more expensive radios sold	is	the total number of radios sold.
x	$+$	y	$=$	36

The receipts from the sale of the less expensive radios sold	plus	the receipts from the sale of the more expensive radios sold	is	the total receipts.
$67x$	$+$	$100y$	$=$	2,940

Solve the System of Equations   We can solve the following system for x and y to find out how many of each type were sold:

$$\begin{cases} x + y = 36 \\ 67x + 100y = 2,940 \end{cases}$$

We multiply both sides of $x + y = 36$ by -100, add the resulting equation to $67x + 100y = 2,940$, and solve for x:

$$\begin{array}{rcl} -100x - 100y &=& -3,600 \\ \underline{67x + 100y} &=& \underline{2,940} \\ -33x &=& -660 \\ x &=& 20 \end{array}$$ Divide both sides by -33.

To find y, we substitute 20 for x in the first equation and solve for y:

$$\begin{array}{rl} \boldsymbol{x} + y =& 36 \\ \boldsymbol{20} + y =& 36 \qquad \text{Substitute 20 for } x. \\ y =& 16 \qquad \text{Subtract 20 from both sides.} \end{array}$$

State the Conclusion   The store sold 20 of the less expensive radios and 16 of the more expensive radios.

Check the Result   If 20 of one type were sold and 16 of the other type were sold, a total of 36 radios were sold. Since the value of the less expensive radios is 20($67) = $1,340, and the value of the more expensive radios is 16($100) = $1,600, the total receipts are $2,940.

■ **SYSTEMS OF INEQUALITIES**

We now review the graphing method of solving systems of two linear inequalities in two variables. Recall that the solutions are usually the intersection of half-planes.

EXAMPLE 10   Graph the solution set of the system: $\begin{cases} x + y \leq 1 \\ 2x - y > 2 \end{cases}$.

Solution   On the same set of coordinate axes, we graph each inequality as in the following figure.

The graph of the inequality $x + y \leq 1$ includes the line graph of $x + y = 1$ and all points below it. Since the boundary line is included, it is drawn as a solid line.

The graph of the inequality $2x - y > 2$ contains only those points below the graph of $2x - y = 2$. Since the boundary line is not included, it is drawn as a broken line.

The area where the half-planes intersect represents the solutions of the given system of inequalities, because any point in that region has coordinates that will satisfy both inequalities.

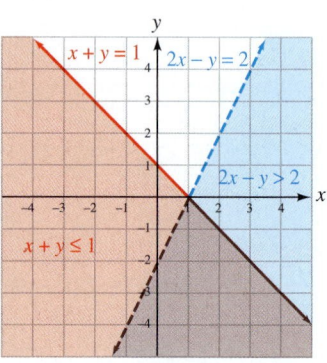

Graph each boundary using the intercept method.

$x + y = 1$

x	y	(x, y)
0	1	$(0, 1)$
1	0	$(1, 0)$

$2x - y = 2$

x	y	(x, y)
0	-2	$(0, -2)$
1	0	$(1, 0)$

Shading: Check the test point (0, 0)

$x + y \leq 1$

$0 + 0 \overset{?}{\leq} 1$

$0 \leq 1$ True.

$(0, 0)$ is a solution of $x + y \leq 1$.

Shading: Check the test point (0, 0)

$2x - y > 2$

$2(0) - 0 \overset{?}{>} 2$

$0 > 2$ False.

$(0, 0)$ is not a solution of $2x - y > 2$.

Answers to Self Checks   **1.** $(1, 2)$ **2.** no solution   **3.** There are infinitely many solutions; three of them are $(0, -4)$, $(2, 0)$, and $(4, 4)$.

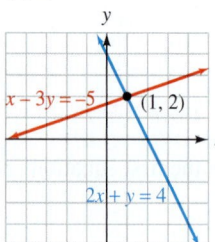

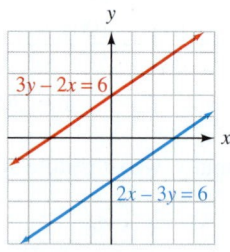

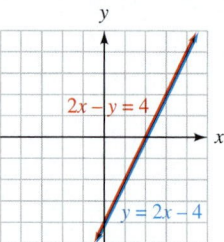

4. $(2, 3)$

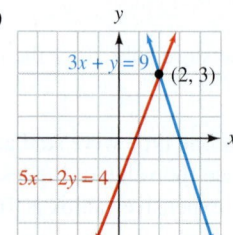

5. $(-3, 4)$ **6.** $(6, -15)$ **7.** no solution

8. There are infinitely many solutions. Any ordered pair that satisfies one equation also satisfies the other.

12.1 STUDY SET

ELEMENTARY & INTERMEDIATE Algebra $f(x)$ Now™

VOCABULARY Fill in the blanks.

1. The pair of equations $\begin{cases} x - y = -1 \\ 2x - y = 1 \end{cases}$ is called a _____ of equations.

2. Because the ordered pair (2, 3) satisfies both equations in Exercise 1, it is called a _____ of the system of equations.

3. When the graphs of the equations of a system are identical lines, the equations are called dependent and the system has _____ many solutions.

4. A system that has no solution is called an _____ system.

CONCEPTS

5. Refer to the illustration. Decide whether a true or a false statement would be obtained when the coordinates of
   a. point A are substituted into the equation for line l_1.
   b. point B are substituted into the equation for line l_1.
   c. point C are substituted into the equation for line l_1.
   d. point C are substituted into the equation for line l_2.

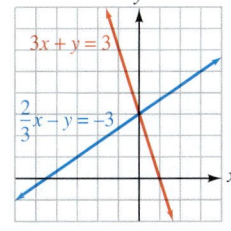

6. Refer to the illustration.
   a. How many ordered pairs satisfy the equation $3x + y = 3$? Name three.
   b. How many ordered pairs satisfy the equation $\frac{2}{3}x - y = -3$? Name three.

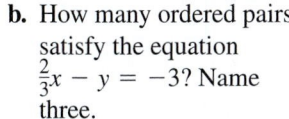

   c. How many ordered pairs satisfy both equations? Name it or them.

7. If the system $\begin{cases} 4x - 3y = 7 \\ 3x - 2y = 6 \end{cases}$ is to be solved using the elimination method, by what constant should each equation be multiplied if
   a. the x-terms are to drop out?
   b. the y-terms are to drop out?

8. Consider the system $\begin{cases} \frac{2}{3}x - \frac{y}{6} = \frac{16}{9} \\ 0.03x + 0.02y = 0.03 \end{cases}$.
   a. What step should be performed to clear the first equation of fractions?
   b. What step should be performed to clear the second equation of decimals?

NOTATION Complete each solution.

9. Solve: $\begin{cases} y = 3x - 7 \\ x + y = 5 \end{cases}$.

$x + (\quad) = 5$
$x + 3x - 7 = \quad$
$\quad x - 7 = 5$
$4x = \quad$
$x = 3$

$y = 3x - 7$
$y = 3(\ ) - 7$
$y = \quad$

The solution is (3, 2).

10. Solve: $\begin{cases} 3x + y = 0 \\ x - 2y = 7 \end{cases}$.

$6x + 2y = 0$
$\underline{x - 2y = 7}$
$7x \quad = \quad$
$x = \quad$

$x - 2y = 7$
$\quad - 2y = 7$
$-2y = \quad$
$y = \quad$

The solution is (1, −3).

PRACTICE Solve each system by graphing, if possible. Check your graphs with a graphing calculator. If a system is inconsistent or if the equations are dependent, so indicate.

11. $\begin{cases} y = x - 4 \\ 2x + y = 5 \end{cases}$

12. $\begin{cases} y = -2x + 1 \\ x - 2y = -7 \end{cases}$

13. $\begin{cases} x = 13 - 4y \\ 3x = 4 + 2y \end{cases}$ 14. $\begin{cases} 3x = 7 - 2y \\ 2x = 2 + 4y \end{cases}$

15. $\begin{cases} x = 3 - 2y \\ 2x + 4y = 6 \end{cases}$ 16. $\begin{cases} 3x = 5 - 2y \\ 3x + 2y = 7 \end{cases}$

17. $\begin{cases} y = 3 \\ x = 2 \end{cases}$ 18. $\begin{cases} 2x + 3y = -15 \\ 2x + y = -9 \end{cases}$

19. $\begin{cases} x = \dfrac{11 - 2y}{3} \\ y = \dfrac{11 - 6x}{4} \end{cases}$ 20. $\begin{cases} x = \dfrac{1 - 3y}{4} \\ y = \dfrac{12 + 3x}{2} \end{cases}$

21. $\begin{cases} y = -\dfrac{5}{2}x + \dfrac{1}{2} \\ 2x - \dfrac{3}{2}y = 5 \end{cases}$ 22. $\begin{cases} \dfrac{5}{2}x + 3y = 6 \\ y = -\dfrac{5}{6}x + 2 \end{cases}$

Use a graphing calculator to solve each system. Give all answers to the nearest hundredth.

23. $\begin{cases} y = 3.2x - 1.5 \\ y = -2.7x - 3.7 \end{cases}$

24. $\begin{cases} y = -0.45x + 5 \\ y = 5.55x - 13.7 \end{cases}$

25. $\begin{cases} 1.7x + 2.3y = 3.2 \\ y = 0.25x + 8.95 \end{cases}$

26. $\begin{cases} 2.75x = 12.9y - 3.79 \\ 7.1x - y = 35.76 \end{cases}$

Solve each system by substitution, if possible. If a system is inconsistent or if the equations are dependent, so indicate.

27. $\begin{cases} y = x \\ x + y = 4 \end{cases}$ 28. $\begin{cases} y = x + 2 \\ x + 2y = 16 \end{cases}$

29. $\begin{cases} x = 2 + y \\ 2x + y = 13 \end{cases}$ 30. $\begin{cases} x = -4 + y \\ 3x - 2y = -5 \end{cases}$

31. $\begin{cases} x + 2y = 6 \\ 3x - y = -10 \end{cases}$ 32. $\begin{cases} 2x - y = -21 \\ 4x + 5y = 7 \end{cases}$

33. $\begin{cases} y - 3x = -5 \\ 21x = 7y + 35 \end{cases}$ 34. $\begin{cases} 8y = 15 - 4x \\ x + 2y = 4 \end{cases}$

35. $\begin{cases} 3x - 4y = 9 \\ x + 2y = 8 \end{cases}$ 36. $\begin{cases} 3x - 2y = -10 \\ 6x + 5y = 25 \end{cases}$

37. $\begin{cases} x - \dfrac{4y}{5} = 4 \\ \dfrac{y}{3} = \dfrac{x}{2} - \dfrac{5}{2} \end{cases}$ 38. $\begin{cases} 3x - 2y = \dfrac{9}{2} \\ \dfrac{x}{2} - \dfrac{3}{4} = 2y \end{cases}$

Solve each system by elimination if possible. If a system is inconsistent or if the equations are dependent, so indicate.

39. $\begin{cases} x - y = 3 \\ x + y = 7 \end{cases}$ 40. $\begin{cases} x + y = 1 \\ x - y = 7 \end{cases}$

41. $\begin{cases} 2x + y = -10 \\ 2x - y = -6 \end{cases}$ 42. $\begin{cases} x + 2y = -9 \\ x - 2y = -1 \end{cases}$

43. $\begin{cases} 2x + 3y = 8 \\ 3x - 2y = -1 \end{cases}$ 44. $\begin{cases} 5x - 2y = 19 \\ 3x + 4y = 1 \end{cases}$

45. $\begin{cases} 4(x - 2) = -9y \\ 2(x - 3y) = -3 \end{cases}$ 46. $\begin{cases} 2(2x + 3y) = 5 \\ 8x = 3(1 + 3y) \end{cases}$

47. $\begin{cases} 8x - 4y = 16 \\ 2x - 4 = y \end{cases}$ 48. $\begin{cases} 2y - 3x = -13 \\ 3x - 17 = 4y \end{cases}$

49. $\begin{cases} x = \dfrac{3}{2}y + 5 \\ 2x - 3y = 8 \end{cases}$ 50. $\begin{cases} x = \dfrac{2}{3}y \\ y = 4x + 5 \end{cases}$

51. $\begin{cases} \dfrac{x}{2} + \dfrac{y}{2} = 6 \\ \dfrac{x}{2} - \dfrac{y}{2} = -2 \end{cases}$ 52. $\begin{cases} \dfrac{x}{2} - \dfrac{y}{3} = -4 \\ \dfrac{x}{2} + \dfrac{y}{9} = 0 \end{cases}$

53. $\begin{cases} \dfrac{2}{3}x - \dfrac{1}{4}y = -8 \\ 0.5x - 0.375y = -9 \end{cases}$ 54. $\begin{cases} 0.5x + 0.5y = 6 \\ \dfrac{x}{2} - \dfrac{y}{2} = -2 \end{cases}$

Solve each system. When writing the solution as an ordered pair, write the values for the variables in alphabetical order.

55. $\begin{cases} \dfrac{3}{2}p + \dfrac{1}{3}q = 2 \\ \dfrac{2}{3}p + \dfrac{1}{9}q = 1 \end{cases}$

56. $\begin{cases} a + \dfrac{b}{3} = \dfrac{5}{3} \\ \dfrac{a + b}{3} = 3 - a \end{cases}$

57. $\begin{cases} \dfrac{m - n}{5} + \dfrac{m + n}{2} = 6 \\ \dfrac{m - n}{2} - \dfrac{m + n}{4} = 3 \end{cases}$

58. $\begin{cases} \dfrac{r - 2}{5} + \dfrac{s + 3}{2} = 5 \\ \dfrac{r + 3}{2} + \dfrac{s - 2}{3} = 6 \end{cases}$

Solve each system. To do so, substitute a for $\dfrac{1}{x}$ and b for $\dfrac{1}{y}$ and solve for a and b. Then find x and y using the fact that $a = \dfrac{1}{x}$ and $b = \dfrac{1}{y}$.

59. $\begin{cases} \dfrac{1}{x} + \dfrac{1}{y} = \dfrac{5}{6} \\ \dfrac{1}{x} - \dfrac{1}{y} = \dfrac{1}{6} \end{cases}$

60. $\begin{cases} \dfrac{1}{x} + \dfrac{1}{y} = \dfrac{9}{20} \\ \dfrac{1}{x} - \dfrac{1}{y} = \dfrac{1}{20} \end{cases}$

61. $\begin{cases} \dfrac{1}{x} + \dfrac{2}{y} = -1 \\ \dfrac{2}{x} - \dfrac{1}{y} = -7 \end{cases}$

62. $\begin{cases} \dfrac{3}{x} - \dfrac{2}{y} = -30 \\ \dfrac{2}{x} - \dfrac{3}{y} = -30 \end{cases}$

Graph the solution set of each system of inequalities.

63. $\begin{cases} y < 3x + 2 \\ y < -2x + 3 \end{cases}$

64. $\begin{cases} y \leq x - 2 \\ y \geq 2x + 1 \end{cases}$

65. $\begin{cases} 3x + 2y > 6 \\ x + 3y \leq 2 \end{cases}$

66. $\begin{cases} 3x + y \leq 1 \\ -x + 2y \geq 6 \end{cases}$

APPLICATIONS

67. HEARING TESTS See the illustration. At what frequency and decibel level were the hearing test results the same for the left and right ear? Write your answer as an ordered pair.

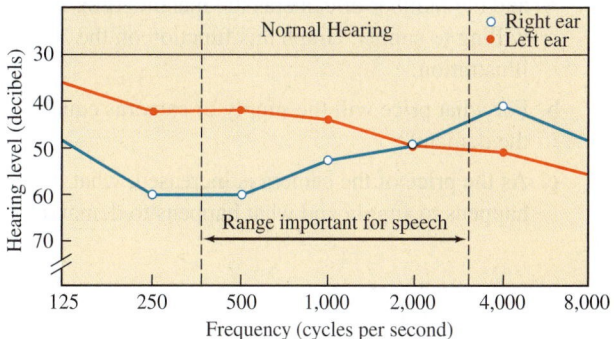

68. MAPS On the following map, what New Mexico city lies on the intersection of Interstate 25 and Interstate 40?

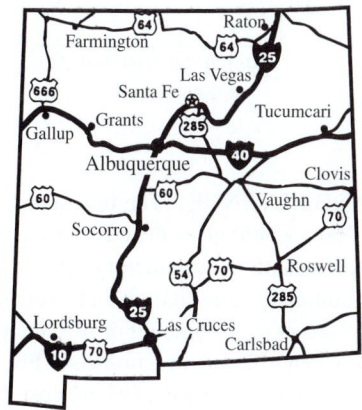

69. SUPPLY AND DEMAND The demand function, graphed in the illustration, describes the relationship between the price x of a certain camera and the demand for the camera.

a. The supply function, $S(x) = \frac{25}{4}x - 525$, describes the relationship between the price x of the camera and the number of cameras the manufacturer is willing to supply. Graph this function on the illustration.

b. For what price will the supply of cameras equal the demand?

c. As the price of the camera is increased, what happens to supply and what happens to demand?

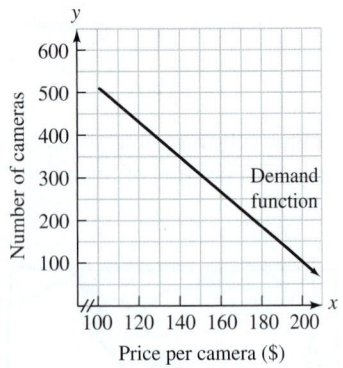

Price per camera ($)

70. COST AND REVENUE The function $C(x) = 200x + 400$ gives the cost for a college to offer x sections of an introductory class in CPR (cardiopulmonary resuscitation). The function $R(x) = 280x$ gives the amount of revenue the college brings in when offering x sections of CPR.

a. Find the *break-even point* (where cost = revenue) by graphing each function on the same coordinate system.

b. How many sections does the college need to offer to make a profit on the CPR training course?

71. BUSINESS Estimate the break-even point (where cost = revenue) on the graph in the illustration. Then describe why it is called the break-even point.

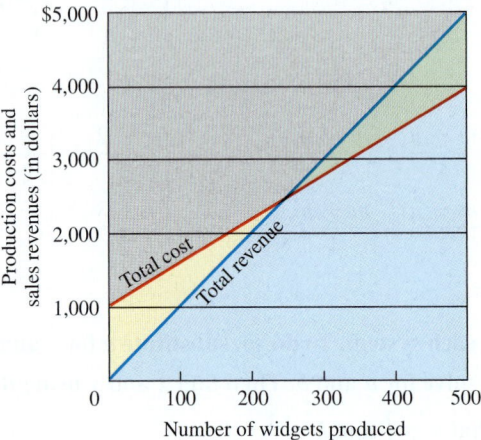

Number of widgets produced

72. NAVIGATION The paths of two ships are tracked on the same coordinate system. One ship is following a path described by the equation $2x + 3y = 6$, and the other is following a path described by the equation $y = \frac{2}{3}x - 3$.

a. Is there a possibility of a collision?

b. What are the coordinates of the danger point?

c. Is a collision a certainty?

In Exercises 73–88, use two variables to solve each problem.

73. ADVERTISING   Use the information in the ad to find the cost of a 15-second and a 30-second radio commercial on radio station KLIZ.

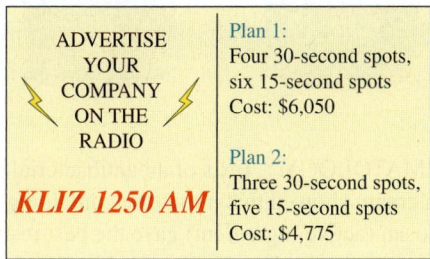

ADVERTISE
YOUR
COMPANY
ON THE
RADIO

KLIZ 1250 AM

Plan 1:
Four 30-second spots,
six 15-second spots
Cost: $6,050

Plan 2:
Three 30-second spots,
five 15-second spots
Cost: $4,775

74. ELECTRONICS   In the illustration, two resistors in the voltage divider circuit have a total resistance of 1,375 ohms. To provide the required voltage, R_1 must be 125 ohms greater than R_2. Find both resistances.

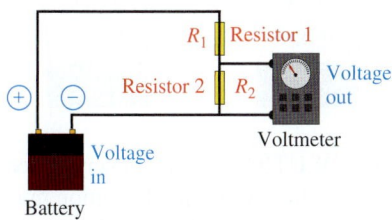

75. FENCING A FIELD   The rectangular field is surrounded by 72 meters of fencing. If the field is partitioned as shown, a total of 88 meters of fencing are required. Find the dimensions of the field.

76. GEOMETRY   In a right triangle, one acute angle is 15° greater than two times the other acute angle. Find the difference between the measures of the angles.

77. TRAFFIC SIGNALS   In the illustration, brace A and brace B are perpendicular. Find the values of x and y.

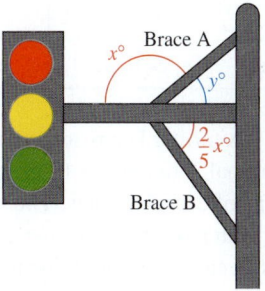

78. INVESTMENT CLUBS   Part of $8,000 was invested by an investment club at 10% interest and the rest at 12%. If the annual income from these investments is $900, how much was invested at each rate?

79. RETIREMENT INCOME   A retired couple invested part of $12,000 at 6% interest and the rest at 7.5%. If their annual income from these investments is $810, how much was invested at each rate?

80. TV NEWS   A news van and a helicopter left a TV station parking lot at the same time, headed in opposite directions to cover breaking news stories that were 145 miles apart. If the helicopter had to travel 55 miles farther than the van, how far did the van have to travel to reach the location of the news story?

81. DELIVERY SERVICES   A delivery truck travels 50 miles in the same time that a cargo plane travels 180 miles. The speed of the plane is 143 mph faster than the speed of the truck. Find the speed of the delivery truck.

82. PRODUCTION PLANNING   A bicycle manufacturer builds racing bikes and mountain bikes, with the per-unit manufacturing costs shown in the table. The company has budgeted $15,900 for labor and $13,075 for materials. How many bicycles of each type can be built?

Model	Cost of materials	Cost of labor
Racing	$55	$60
Mountain	$70	$90

83. FARMING   A farmer keeps some animals on a strict diet. Each animal is to receive 15 grams of protein and 7.5 grams of carbohydrates. The farmer uses two food mixes, with nutrients as shown in the table. How many grams of each mix should be used to provide the correct nutrients for each animal?

Mix	Protein	Carbohydrates
Mix A	12%	9%
Mix B	15%	5%

84. RECORDING COMPANIES   Three people invest a total of $105,000 to start a recording company that will produce reissues of classic jazz. Each release will be a set of 3 CDs that will retail for $45 per set. If each set can be produced for $18.95, how many sets must be sold for the investors to make a profit?

85. COSMETOLOGY   A beauty shop specializing in permanents has fixed costs of $2,101.20 per month. The owner estimates that the cost for each permanent is $23.60, which covers labor, chemicals, and electricity. If her shop can give as many permanents as she wants at a price of $44 each, how many must be given each month for her to break even?

86. MIXING CANDY   How many pounds of each candy shown in the illustration must be mixed to obtain 60 pounds of candy that would be worth $4 per pound?

Gummy Bears $3.50/lb   Jelly Beans $5.50/lb

87. DERMATOLOGY   Tests of an antibacterial face-wash cream showed that a mixture containing 0.3% Triclosan (active ingredient) gave the best results. How many grams of cream from each tube should be used to make an equal-size tube of the 0.3% cream?

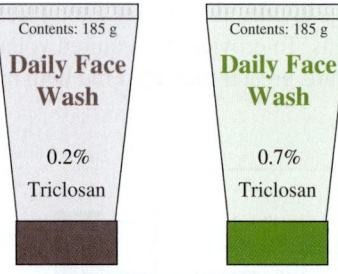

Contents: 185 g   **Daily Face Wash**   0.2% Triclosan

Contents: 185 g   **Daily Face Wash**   0.7% Triclosan

88. MIXING SOLUTIONS   How many ounces of the two alcohol solutions in the illustration must be mixed to obtain 100 ounces of a 12.2% solution?

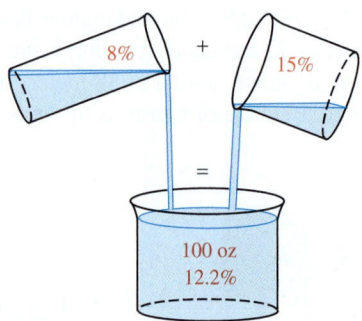

8%   +   15%

=

100 oz 12.2%

WRITING

89. Which method would you use to solve the system

$$\begin{cases} y - 1 = 3x \\ 3x + 2y = 12 \end{cases}? \text{ Explain.}$$

90. Which method would you use to solve the system

$$\begin{cases} 2x + 4y = 9 \\ 3x - 5y = 20 \end{cases}? \text{ Explain.}$$

91. When solving a system, what advantages are there with the substitution and elimination methods compared to the graphing method?

92. When graphing a system of linear inequalities, explain how to decide which region to shade.

REVIEW **Solve each formula for the given variable.**

93. $A = p + prt$ for r

94. $A = p + prt$ for p

95. $\dfrac{V_2}{V_1} = \dfrac{P_1}{P_2}$ for P_1

96. $\dfrac{1}{r} = \dfrac{1}{r_1} + \dfrac{1}{r_2}$ for r

CHALLENGE PROBLEMS

97. If the solution of the system $\begin{cases} Ax + By = -2 \\ Bx - Ay = -26 \end{cases}$ is $(-3, 5)$, what are the values of the constants A and B?

98. Solve: $\begin{cases} 2ab - 3cd = 1 \\ 3ab - 2cd = 1 \end{cases}$. Assume that b and d are constants.

....

12.2   Systems with Three Variables

- Solving three equations with three variables   • Consistent systems
- An inconsistent system   • Systems with dependent equations   • Problem solving
- Curve fitting

In the preceding sections, we solved systems of two linear equations with two variables. In this section, we will solve systems of linear equations with three variables by using a combination of the elimination method and the substitution method. We will then use that procedure to solve problems involving three variables.

■ SOLVING THREE EQUATIONS WITH THREE VARIABLES

We now extend the definition of a linear equation to include equations of the form $Ax + By + Cz = D$. The solution of a system of three linear equations with three variables is an **ordered triple** of numbers. For example, the solution of the system

$$\begin{cases} 2x + 3y + 4z = 20 \\ 3x + 4y + 2z = 17 \\ 3x + 2y + 3z = 16 \end{cases}$$

is the triple $(1, 2, 3)$, since each equation is satisfied if $x = 1$, $y = 2$, and $z = 3$.

The Language of Algebra

A system of equations is two (or more) equations that we consider *simultaneously*—at the same time. Some professional sports teams *simulcast* their games. That is, the announcer's play-by-play description is broadcast on radio and television at the same time.

$$2x + 3y + 4z = 20$$
$$2(1) + 3(2) + 4(3) \stackrel{?}{=} 20$$
$$2 + 6 + 12 \stackrel{?}{=} 20$$
$$20 = 20$$

$$3x + 4y + 2z = 17$$
$$3(1) + 4(2) + 2(3) \stackrel{?}{=} 17$$
$$3 + 8 + 6 \stackrel{?}{=} 17$$
$$17 = 17$$

$$3x + 2y + 3z = 16$$
$$3(1) + 2(2) + 3(3) \stackrel{?}{=} 16$$
$$3 + 4 + 9 \stackrel{?}{=} 16$$
$$16 = 16$$

The graph of an equation of the form $Ax + By + Cz = D$ is a flat surface called a *plane*. A system of three linear equations with three variables is consistent or inconsistent, depending on how the three planes corresponding to the three equations intersect. The following illustration shows some of the possibilities.

Consistent system

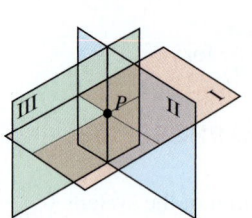

The three planes intersect at a
single point P: one solution

(a)

Consistent system

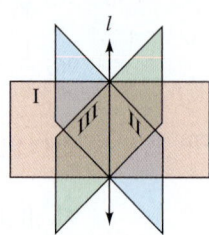

The three planes have a line l
in common: infinitely many
solutions

(b)

Inconsistent systems

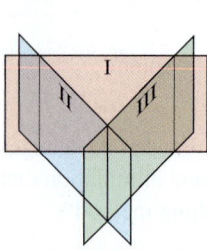

The three planes have no point
in common: no solutions

(c)

To solve a system of three linear equations with three variables, we follow these steps.

Solving Three Equations with Three Variables	1. Pick any two equations and eliminate a variable.
	2. Pick a different pair of equations and eliminate the same variable as in step 1.
	3. Solve the resulting pair of two equations in two variables.
	4. Use the values of the two variables found in step 3 to find the value of the third variable by substituting into an appropriate equation.
	5. Check the proposed solution in all three of the original equations. Write the solution as an ordered triple.

■ CONSISTENT SYSTEMS

Recall that when a system has a solution, it is called a *consistent* system.

EXAMPLE 1

Solve: $\begin{cases} 2x + y + 4z = 12 \\ x + 2y + 2z = 9 \\ 3x - 3y - 2z = 1 \end{cases}$.

Solution   **Step 1:** We are given the system

Notation

To clarify the solution process, we number the equations.

(1) $\quad \begin{cases} 2x + y + 4z = 12 \\ x + 2y + 2z = 9 \quad\quad \text{To clarify the solution process, we number each equation.} \\ 3x - 3y - 2z = 1 \end{cases}$

(2)

(3)

If we pick Equations 2 and 3 and add them, the variable z is eliminated.

(2) $\quad x + 2y + 2z = 9$

(3) $\quad \underline{3x - 3y - 2z = 1}$

(4) $\quad 4x - y = 10$ This equation does not contain z.

Step 2: We now pick a different pair of equations (Equations 1 and 3) and eliminate z again. If each side of Equation 3 is multiplied by 2 and the resulting equation is added to Equation 1, z is eliminated.

(1) $\quad 2x + \quad y + 4z = 12$
$\quad\quad \underline{6x - 6y - 4z = \quad 2} \quad$ Multiply both sides of Equation 3 by 2.
(5) $\quad 8x - 5y \quad\quad\quad = 14 \quad$ This equation does not contain z.

Step 3: Equations 4 and 5 form a system of two equations with two variables, x and y.

(4) $\quad \begin{cases} 4x - y = 10 \\ 8x - 5y = 14 \end{cases}$
(5)

To solve this system, we multiply Equation 4 by -5 and add the resulting equation to Equation 5 to eliminate y:

$\quad\quad -20x + 5y = -50 \quad$ Multiply both sides of Equation 4 by -5.
(5) $\quad \underline{\quad 8x - 5y = \quad 14}$
$\quad\quad -12x \quad\quad\quad = -36$
$\quad\quad\quad\quad\quad x = 3 \quad$ To find x, divide both sides by -12.

To find y, we substitute 3 for x in any equation containing x and y (such as Equation 5) and solve for y:

(5) $\quad 8x - 5y = 14$
$\quad\quad 8(3) - 5y = 14 \quad$ Substitute 3 for x.
$\quad\quad 24 - 5y = 14 \quad$ Simplify.
$\quad\quad -5y = -10 \quad$ Subtract 24 from both sides.
$\quad\quad y = 2 \quad$ Divide both sides by -5.

Step 4: To find z, we substitute 3 for x and 2 for y in any equation containing x, y, and z (such as Equation 1) and solve for z:

(1) $\quad 2x + y + 4z = 12$
$\quad\quad 2(3) + 2 + 4z = 12 \quad$ Substitute 3 for x and 2 for y.
$\quad\quad 8 + 4z = 12 \quad$ Simplify.
$\quad\quad 4z = 4 \quad$ Subtract 8 from both sides.
$\quad\quad z = 1 \quad$ Divide both sides by 4.

The solution of the system is (3, 2, 1). Because this system has a solution, it is a consistent system.

Step 5: Verify that these values satisfy each equation in the original system.

Self Check 1 $\quad$ Solve: $\begin{cases} 2x + y + 4z = 16 \\ x + 2y + 2z = 11 \\ 3x - 3y - 2z = -9 \end{cases}$.

When one or more of the equations of a system is missing a term, the elimination of a variable that is normally performed in step 1 of the solution process can be skipped.

EXAMPLE 2

Solve: $\begin{cases} 3x = 6 - 2y + z \\ -y - 2z = -8 - x. \\ x = 1 - 2z \end{cases}$

Solution   *Step 1:* First, we write each equation in the form $Ax + By + Cz = D$.

(1) $\begin{cases} 3x + 2y - z = 6 \\ (2) \quad x - y - 2z = -8 \\ (3) \quad x + 2z = 1 \end{cases}$

Since Equation 3 does not have a *y*-term, we can proceed to step 2, where we will find another equation that does not contain a *y*-term.

Step 2: If each side of Equation 2 is multiplied by 2 and the resulting equation is added to Equation 1, *y* is eliminated.

(1) $\quad 3x + 2y - z = 6$

$\underline{\quad 2x - 2y - 4z = -16} \quad$ Multiply both sides of Equation 2 by 2.

(4) $\quad 5x - 5z = -10$

Step 3: Equations 3 and 4 form a system of two equations with two variables, *x* and *z*:

(3) $\begin{cases} x + 2z = 1 \\ (4) \quad 5x - 5z = -10 \end{cases}$

To solve this system, we multiply Equation 3 by -5 and add the resulting equation to Equation 4 to eliminate *x*:

$\quad -5x - 10z = -5 \quad$ Multiply both sides of Equation 3 by -5.

(4) $\quad \underline{5x - 5z = -10}$

$\quad\quad\quad -15z = -15$

$\quad\quad\quad\quad\quad z = 1 \quad$ To find *z*, divide both sides by -15.

To find *x*, we substitute 1 for *z* in Equation 3.

(3) $\quad x + 2z = 1$

$\quad x + 2(\mathbf{1}) = 1 \quad$ Substitute 1 for *z*.

$\quad\quad x + 2 = 1 \quad$ Multiply.

$\quad\quad\quad\quad x = -1 \quad$ Subtract 2 from both sides.

Step 4: To find *y*, we substitute -1 for *x* and 1 for *z* in Equation 1:

(1) $\quad\quad 3x + 2y - z = 6$

$\quad 3(\mathbf{-1}) + 2y - \mathbf{1} = 6 \quad$ Substitute -1 for *x* and 1 for *z*.

$\quad\quad -3 + 2y - 1 = 6 \quad$ Multiply.

$\quad\quad\quad\quad\quad 2y = 10 \quad$ Add 4 to both sides.

$\quad\quad\quad\quad\quad y = 5 \quad$ Divide both sides by 2.

The solution of the system is $(-1, 5, 1)$.

Step 5: Check the proposed solution in all three of the original equations.

Self Check 2 Solve: $\begin{cases} x + 2y - z = 1 \\ 2x - y + z = 3. \\ x + z = 3 \end{cases}$

■ AN INCONSISTENT SYSTEM

Recall that when a system has no solution, it is called an *inconsistent system.*

EXAMPLE 3

ELEMENTARY &
INTERMEDIATE
Algebra *f(x)* **Now**™

Solve: $\begin{cases} 2a + b - 3c = -3 \\ 3a - 2b + 4c = 2 \\ 4a + 2b - 6c = -7 \end{cases}$.

Solution We can multiply the first equation of the system by 2 and add the resulting equation to the second equation to eliminate b:

$$4a + 2b - 6c = -6 \qquad \text{Multiply both sides of the first equation by 2.}$$
$$\underline{3a - 2b + 4c = 2}$$
(1) $\quad 7a \qquad\;\; - 2c = -4$

Now add the second and third equations of the system to eliminate b again:

$$3a - 2b + 4c = 2$$
$$\underline{4a + 2b - 6c = -7}$$
(2) $\quad 7a \qquad\;\; - 2c = -5$

Equations 1 and 2 form the system

(1) $\begin{cases} 7a - 2c = -4 \\ 7a - 2c = -5 \end{cases}$
(2)

Since $7a - 2c$ cannot equal both -4 and -5, the system is inconsistent and has no solution.

Self Check 3 Solve: $\begin{cases} 2a + b - 3c = 8 \\ 3a - 2b + 4c = 10 \\ 4a + 2b - 6c = -5 \end{cases}$.

■ SYSTEMS WITH DEPENDENT EQUATIONS

When the equations in a system of two equations with two variables are dependent, the system has infinitely many solutions. This is not always true for systems of three equations with three variables. In fact, a system can have dependent equations and still be inconsistent. The following illustration shows the different possibilities.

Consistent system

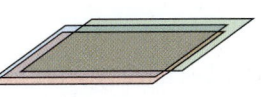

When three planes coincide, the equations are dependent, and there are infinitely many solutions.

(a)

Consistent system

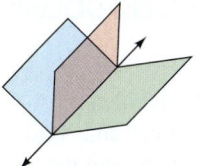

When three planes intersect in a common line, the equations are dependent, and there are infinitely many solutions.

(b)

Inconsistent system

When two planes coincide and are parallel to a third plane, the system is inconsistent, and there are no solutions.

(c)

EXAMPLE 4

ELEMENTARY &
INTERMEDIATE
Algebra $f(x)$ **Now**™

Solve: $\begin{cases} 3x - 2y + z = -1 \\ 2x + y - z = 5 \\ 5x - y = 4 \end{cases}$.

Solution   We can add the first two equations to get

$$\begin{array}{l} 3x - 2y + z = -1 \\ \underline{2x + y - z = 5} \\ \textbf{(1)} \quad 5x - y = 4 \end{array}$$

Since Equation 1 is the same as the third equation of the system, the equations of the system are dependent, and there are infinitely many solutions. From a graphical perspective, the equations represent three planes that intersect in a common line.

To write the general solution of this system, we can solve Equation 1 for y to get

$$5x - y = 4$$
$$-y = -5x + 4 \qquad \text{Subtract } 5x \text{ from both sides.}$$
$$y = 5x - 4 \qquad \text{Multiply both sides by } -1.$$

We can then substitute $5x - 4$ for y in the first equation of the system and solve for z to get

$$3x - 2y + z = -1$$
$$3x - 2(\mathbf{5x - 4}) + z = -1 \qquad \text{Substitute } 5x - 4 \text{ for } y.$$
$$3x - 10x + 8 + z = -1 \qquad \text{Use the distributive property to remove parentheses.}$$
$$-7x + 8 + z = -1 \qquad \text{Combine like terms.}$$
$$z = 7x - 9 \qquad \text{Add } 7x \text{ and } -8 \text{ to both sides.}$$

Since we have found the values of y and z in terms of x, every solution of the system has the form $(x, 5x - 4, 7x - 9)$, where x can be any real number. For example,

If $x = 1$, a solution is $(1, 1, -2)$. $5(1) - 4 = 1$, and $7(1) - 9 = -2$.
If $x = 2$, a solution is $(2, 6, 5)$. $5(2) - 4 = 6$, and $7(2) - 9 = 5$.
If $x = 3$, a solution is $(3, 11, 12)$. $5(3) - 4 = 11$, and $7(3) - 9 = 12$.

Self Check 4   Solve: $\begin{cases} 3x + 2y + z = -1 \\ 2x - y - z = 5 \\ 5x + y = 4 \end{cases}$.

■ PROBLEM SOLVING

EXAMPLE 5

ELEMENTARY &
INTERMEDIATE
Algebra $f(x)$ **Now**™

Tool manufacturing.   A company makes three types of hammers, which are marketed as "good," "better," and "best." The cost of manufacturing each type of hammer is $4, $6, and $7, respectively, and the hammers sell for $6, $9, and $12. Each day, the cost of manufacturing 100 hammers is $520, and the daily revenue from their sale is $810. How many hammers of each type are manufactured?

Analyze the Problem   We need to find how many of each type of hammer are manufactured daily. We must write three equations to find three unknowns.

Form Three Equations   If we let x represent the number of good hammers, y represent the number of better hammers, and z represent the number of best hammers, we know that

The total number of hammers is $x + y + z$.
The cost of manufacturing the good hammers is $4x$ ($4 times x hammers).
The cost of manufacturing the better hammers is $6y$ ($6 times y hammers).
The cost of manufacturing the best hammers is $7z$ ($7 times z hammers).
The revenue received by selling the good hammers is $6x$ ($6 times x hammers).
The revenue received by selling the better hammers is $9y$ ($9 times y hammers).
The revenue received by selling the best hammers is $12z$ ($12 times z hammers).

We can assemble the facts of the problem to write three equations.

The number of good hammers	plus	the number of better hammers	plus	the number of best hammers	is	the total number of hammers.
x	$+$	y	$+$	z	$=$	100

The cost of good hammers	plus	the cost of better hammers	plus	the cost of best hammers	is	the total cost.
$4x$	$+$	$6y$	$+$	$7z$	$=$	520

The revenue from good hammers	plus	the revenue from better hammers	plus	the revenue from best hammers	is	the total revenue.
$6x$	$+$	$9y$	$+$	$12z$	$=$	810

Solve the System   We must now solve the system

$$\begin{aligned}
\textbf{(1)} \quad & x + y + z = 100 \\
\textbf{(2)} \quad & 4x + 6y + 7z = 520 \\
\textbf{(3)} \quad & 6x + 9y + 12z = 810
\end{aligned}$$

If we multiply Equation 1 by -7 and add the result to Equation 2, we get

$$\begin{aligned}
& -7x - 7y - 7z = -700 \\
& \underline{4x + 6y + 7z = 520} \\
\textbf{(4)} \quad & -3x - y = -180
\end{aligned}$$

If we multiply Equation 1 by -12 and add the result to Equation 3, we get

$$\begin{aligned}
& -12x - 12y - 12z = -1{,}200 \\
& \underline{6x + 9y + 12z = 810} \\
\textbf{(5)} \quad & -6x - 3y = -390
\end{aligned}$$

If we multiply Equation 4 by -3 and add it to Equation 5, we get

$$\begin{aligned}
& 9x + 3y = 540 \\
& \underline{-6x - 3y = -390} \\
& 3x = 150 \\
& x = 50 \qquad \text{\color{red}To find x, divide both sides by 3.}
\end{aligned}$$

To find y, we substitute 50 for x in Equation 4:

$$-3x - y = -180$$
$$-3(50) - y = -180 \qquad \text{Substitute 50 for } x.$$
$$-150 - y = -180 \qquad -3(50) = -150.$$
$$-y = -30 \qquad \text{Add 150 to both sides.}$$
$$y = 30 \qquad \text{Divide both sides by } -1.$$

To find z, we substitute 50 for x and 30 for y in Equation 1:

$$x + y + z = 100$$
$$50 + 30 + z = 100$$
$$z = 20 \qquad \text{Subtract 80 from both sides.}$$

State the Conclusion   The company manufactures 50 good hammers, 30 better hammers, and 20 best hammers each day.

Check the Result   Check the proposed solution in each equation in the original system.

■ CURVE FITTING

EXAMPLE 6

ELEMENTARY &
INTERMEDIATE
Algebra *f(x)* **Now**™

The equation of a parabola opening upward or downward is of the form $y = ax^2 + bx + c$. Find the equation of the parabola shown to the right by determining the values of a, b, and c.

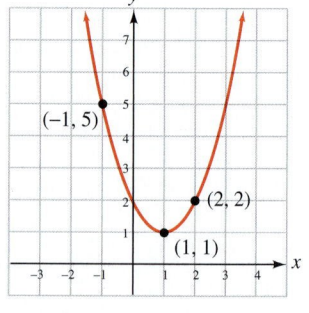

Solution

Since the parabola passes through the points $(-1, 5)$, $(1, 1)$, and $(2, 2)$, each pair of coordinates must satisfy the equation $y = ax^2 + bx + c$. If we substitute the x- and y-coordinates of each point into the equation and simplify, we obtain the following system of three equations with three variables.

(1) $\begin{cases} a - b + c = 5 \end{cases}$ Substitute the coordinates of $(-1, 5)$ into $y = ax^2 + bx + c$ and simplify.
(2) $\begin{cases} a + b + c = 1 \end{cases}$ Substitute the coordinates of $(1, 1)$ into $y = ax^2 + bx + c$ and simplify.
(3) $\begin{cases} 4a + 2b + c = 2 \end{cases}$ Substitute the coordinates of $(2, 2)$ into $y = ax^2 + bx + c$ and simplify.

If we add Equations 1 and 2, we obtain

$$a - b + c = 5$$
$$\underline{a + b + c = 1}$$
(4) $\quad 2a \qquad + 2c = 6$

If we multiply Equation 1 by 2 and add the result to Equation 3, we get

$$2a - 2b + 2c = 10$$
$$\underline{4a + 2b + c = 2}$$
(5) $\quad 6a \qquad + 3c = 12$

We can then divide both sides of Equation 4 by 2 to get Equation 6 and divide both sides of Equation 5 by 3 to get Equation 7. We now have the system

(6) $\quad a + c = 3$
(7) $\quad 2a + c = 4$

To eliminate c, we multiply Equation 6 by -1 and add the result to Equation 7. We get

$$
\begin{array}{r}
-a - c = -3 \\
2a + c = 4 \\
\hline
a = 1
\end{array}
$$

To find c, we can substitute 1 for a in Equation 6 and find that $c = 2$. To find b, we can substitute 1 for a and 2 for c in Equation 2 and find that $b = -2$.

After we substitute these values of a, b, and c into the equation $y = ax^2 + bx + c$, we have the equation of the parabola.

$$
\begin{aligned}
y &= \mathbf{a}x^2 + \mathbf{b}x + \mathbf{c} \\
y &= \mathbf{1}x^2 - \mathbf{2}x + \mathbf{2} \\
y &= x^2 - 2x + 2
\end{aligned}
$$

Answers to Self Checks **1.** $(1, 2, 3)$ **2.** $(1, 1, 2)$ **3.** no solution **4.** There are infinitely many solutions. A general solution is $(x, 4 - 5x, -9 + 7x)$. Three solutions are $(1, -1, -2)$, $(2, -6, 5)$, and $(3, -11, 12)$.

12.2 STUDY SET   ELEMENTARY & INTERMEDIATE Algebra $f(x)$ Now™

VOCABULARY Fill in the blanks.

1. $\begin{cases} 2x + y - 3z = 0 \\ 3x - y + 4z = 5 \\ 4x + 2y - 6z = 0 \end{cases}$ is called a _____ of three linear equations.

2. If the first two equations of the system in Exercise 1 are added, the variable y is _____.

3. The equation $2x + 3y + 4z = 5$ is a linear equation with _____ variables.

4. The graph of the equation $2x + 3y + 4z = 5$ is a flat surface called a _____.

5. When three planes coincide, the equations of the system are _____, and there are infinitely many solutions.

6. When three planes intersect in a line, the system will have _____ many solutions.

CONCEPTS

7. For each graph of a system of three equations, determine whether the solution set contains one solution, infinitely many solutions, or no solution.

a. 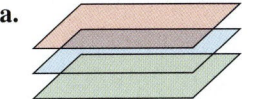   **b.**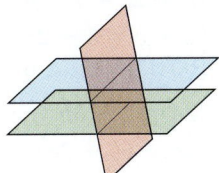

8. Consider the system $\begin{cases} -2x + y + 4z = 3 \\ x - y + 2z = 1 \\ x + y - 3z = 2 \end{cases}$.

a. What is the result if Equation 1 and Equation 2 are added?

b. What is the result if Equation 2 and Equation 3 are added?

c. What variable was eliminated in the steps performed in parts a and b?

NOTATION

9. Write the equation $3z - 2y = x + 6$ in $Ax + By + Cz = D$ form.

10. Fill in the blank: Solutions of a system of three equations in three variables, x, y, and z, are written in the form (x, y, z) and are called ordered _____.

PRACTICE   **Determine whether the given ordered triple is a solution of the given system.**

11. $(2, 1, 1)$, $\begin{cases} x - y + z = 2 \\ 2x + y - z = 4 \\ 2x - 3y + z = 2 \end{cases}$

12. $(-3, 2, -1)$, $\begin{cases} 2x + 2y + 3z = -1 \\ 3x + y - z = -6 \\ x + y + 2z = 1 \end{cases}$

Solve each system. If a system is inconsistent or if the equations are dependent, so indicate.

13. $\begin{cases} x + y + z = 4 \\ 2x + y - z = 1 \\ 2x - 3y + z = 1 \end{cases}$

14. $\begin{cases} x + y + z = 4 \\ x - y + z = 2 \\ x - y - z = 0 \end{cases}$

15. $\begin{cases} 2x + 2y + 3z = 10 \\ 3x + y - z = 0 \\ x + y + 2z = 6 \end{cases}$

16. $\begin{cases} x - y + z = 4 \\ x + 2y - z = -1 \\ x + y - 3z = -2 \end{cases}$

17. $\begin{cases} b + 2c = 7 - a \\ a + c = 8 - 2b \\ 2a + b + c = 9 \end{cases}$

18. $\begin{cases} 2a = 2 - 3b - c \\ 4a + 6b + 2c - 5 = 0 \\ a + c = 3 + 2b \end{cases}$

19. $\begin{cases} 2x + y - z = 1 \\ x + 2y + 2z = 2 \\ 4x + 5y + 3z = 3 \end{cases}$

20. $\begin{cases} 4x + 3z = 4 \\ 2y - 6z = -1 \\ 8x + 4y + 3z = 9 \end{cases}$

21. $\begin{cases} a + b + c = 180 \\ \dfrac{a}{4} + \dfrac{b}{2} + \dfrac{c}{3} = 60 \\ 2b + 3c - 330 = 0 \end{cases}$

22. $\begin{cases} 2a + 3b - 2c = 18 \\ 5a - 6b + c = 21 \\ 4b - 2c - 6 = 0 \end{cases}$

23. $\begin{cases} 0.5a + 0.3b = 2.2 \\ 1.2c - 8.5b = -24.4 \\ 3.3c + 1.3a = 29 \end{cases}$

24. $\begin{cases} 4a - 3b = 1 \\ 6a - 8c = 1 \\ 2b - 4c = 0 \end{cases}$

25. $\begin{cases} 2x + 3y + 4z = 6 \\ 2x - 3y - 4z = -4 \\ 4x + 6y + 8z = 12 \end{cases}$

26. $\begin{cases} x - 3y + 4z = 2 \\ 2x + y + 2z = 3 \\ 4x - 5y + 10z = 7 \end{cases}$

27. $\begin{cases} x + \dfrac{1}{3}y + z = 13 \\ \dfrac{1}{2}x - y + \dfrac{1}{3}z = -2 \\ x + \dfrac{1}{2}y - \dfrac{1}{3}z = 2 \end{cases}$

28. $\begin{cases} x - \dfrac{1}{5}y - z = 9 \\ \dfrac{1}{4}x + \dfrac{1}{5}y - \dfrac{1}{2}z = 5 \\ 2x + y + \dfrac{1}{6}z = 12 \end{cases}$

APPLICATIONS

29. MAKING STATUES   An artist makes three types of ceramic statues at a monthly cost of $650 for 180 statues. The manufacturing costs for the three types are $5, $4, and $3. If the statues sell for $20, $12, and $9, respectively, how many of each type should be made to produce $2,100 in monthly revenue?

30. POTPOURRI   The owner of a home decorating shop wants to mix dried rose petals selling for $6 per pound, dried lavender selling for $5 per pound, and buckwheat hulls selling for $4 per pound to get 10 pounds of a mixture that would sell for $5.50 per pound. She wants to use twice as many pounds of rose petals as lavender. How many pounds of each should she use?

31. NUTRITION   A dietitian is to design a meal that will provide a patient with exactly 14 grams (g) of fat, 9 g of carbohydrates, and 9 g of protein. She is to use a combination of the three foods listed in the table. If one ounce of each of the foods has the nutrient content shown in the table, how many ounces of each food should be used?

Food	Fat	Carbohydrates	Protein
A	2 g	1 g	2 g
B	3 g	2 g	1 g
C	1 g	1 g	2 g

32. NUTRITIONAL PLANNING One ounce of each of three foods has the vitamin and mineral content shown in the table. How many ounces of each must be used to provide exactly 22 milligrams (mg) of niacin, 12 mg of zinc, and 20 mg of vitamin C?

Food	Niacin	Zinc	Vitamin C
A	1 mg	1 mg	2 mg
B	2 mg	1 mg	1 mg
C	2 mg	1 mg	2 mg

33. CHAINSAW SCULPTING A wood sculptor carves three types of statues with a chainsaw. The number of hours required for carving, sanding, and painting a totem pole, a bear, and a deer are shown in the table. How many of each should be produced to use all available labor hours?

	Totem pole	Bear	Deer	Time available
Carving	2 hr	2 hr	1 hr	14 hr
Sanding	1 hr	2 hr	2 hr	15 hr
Painting	3 hr	2 hr	2 hr	21 hr

34. MAKING CLOTHES A clothing manufacturer makes coats, shirts, and slacks. The time required for cutting, sewing, and packaging each item is shown in the table. How many of each should be made to use all available labor hours?

	Coats	Shirts	Slacks	Time available
Cutting	20 min	15 min	10 min	115 hr
Sewing	60 min	30 min	24 min	280 hr
Packaging	5 min	12 min	6 min	65 hr

35. EARTH'S ATMOSPHERE Use the information in the circle graph to determine what percent of Earth's atmosphere is nitrogen, is oxygen, and is other gases.

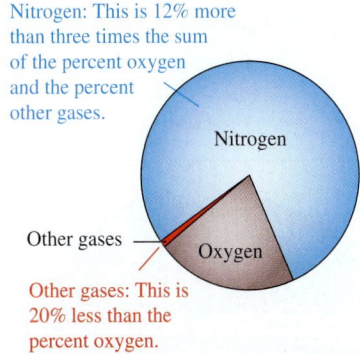

Nitrogen: This is 12% more than three times the sum of the percent oxygen and the percent other gases.

Other gases: This is 20% less than the percent oxygen.

36. NFL RECORDS Jerry Rice, who played with the San Francisco 49ers and the Oakland Raiders, holds the all-time record for touchdown passes caught. Here are some interesting facts about this feat.

- He caught 30 more TD passes from Steve Young than he did from Joe Montana.
- He caught 39 more TD passes from Joe Montana than he did from Rich Gannon.
- He caught a total of 156 TD passes from Young, Montana, and Gannon.

Determine the number of touchdown passes Rice has caught from Young, from Montana, and from Gannon as of 2003.

37. GRAPHS OF SYSTEMS Explain how each of the following pictures could be thought of as an example of the graph of a system of three equations. Then describe the solution, if there is any.

a.

b.
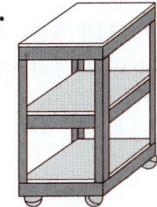

c.

d.
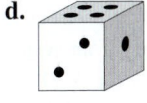

38. ZOOLOGY An X-ray of a mouse revealed a cancerous tumor located at the intersection of the coronal, sagittal, and transverse planes. From this description, would you expect the tumor to be at the base of the tail, on the back, in the stomach, on the tip of the right ear, or in the mouth of the mouse?

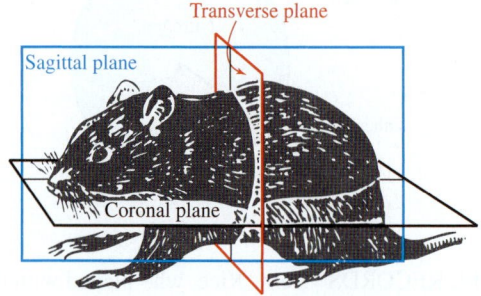

39. ASTRONOMY Comets have elliptical orbits, but the orbits of some comets are so vast that they are indistinguishable from parabolas. Find the equation of the parabola that closely describes the orbit of the comet shown in the illustration.

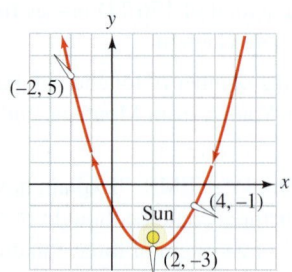

40. CURVE FITTING Find the equation of the parabola shown in the illustration.

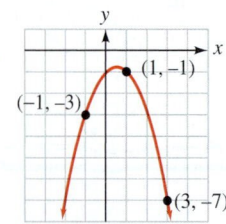

41. WALKWAYS A circular sidewalk is to be constructed in a city park. The walk is to pass by three particular areas of the park, as shown in the illustration. If an equation of a circle is of the form $x^2 + y^2 + Cx + Dy + E = 0$, find the equation that describes the path of the sidewalk by determining C, D, and E.

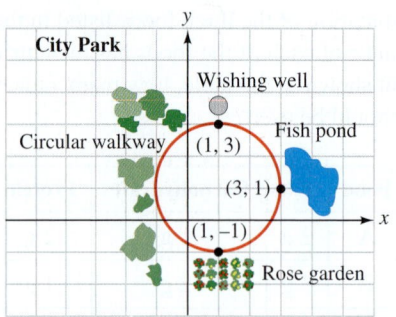

42. CURVE FITTING The equation of a circle is of the form $x^2 + y^2 + Cx + Dy + E = 0$. Find the equation of the circle shown in the illustration by determining C, D, and E.

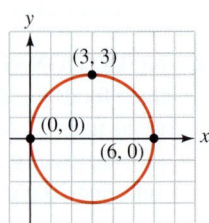

43. TRIANGLES The sum of the measures of the angles of any triangle is 180°. In $\triangle ABC$, $\angle A$ measures 100° less than the sum of the measures of $\angle B$ and $\angle C$, and the measure of $\angle C$ is 40° less than twice the measure of $\angle B$. Find the measure of each angle of the triangle.

44. QUADRILATERALS The quadrilateral is a four-sided polygon. The sum of the measures of the angles of any quadrilateral is 360°. In the illustration below, the measures of $\angle A$ and $\angle B$ are the same. The measure of $\angle C$ is 20° greater than the measure of $\angle A$, and $\angle D$ measures 40°. Find the measure of $\angle A$, $\angle B$, and $\angle C$.

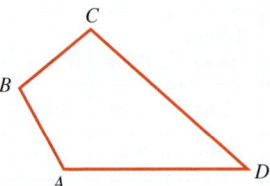

45. INTEGER PROBLEM The sum of three integers is 48. If the first integer is doubled, the sum is 60. If the second integer is doubled, the sum is 63. Find the integers.

46. INTEGER PROBLEM The sum of three integers is 18. The third integer is four times the second, and the second integer is 6 more than the first. Find the integers.

WRITING

47. Explain how a system of three equations with three variables can be reduced to a system of two equations with two variables.

48. What makes a system of three equations with three variables inconsistent?

REVIEW   Graph each function.

49. $f(x) = |x|$

50. $g(x) = x^2$

51. $h(x) = x^3$

52. $S(x) = x$

CHALLENGE PROBLEMS   Solve each system.

53.
$$\begin{cases} w + x + y + z = 3 \\ w - x + y + z = 1 \\ w + x - y + z = 1 \\ w + x + y - z = 3 \end{cases}$$

54.
$$\begin{cases} w + 2x + y + z = 3 \\ w + x - 2y - z = -3 \\ w - x + y + 2z = 3 \\ 2w + x + y - z = 4 \end{cases}$$

12.3 Solving Systems Using Matrices

- Matrices
- Augmented matrices
- Gaussian elimination
- Solving a system of three equations
- Inconsistent systems and dependent equations

In this section, we will discuss another method for solving systems of linear equations. This technique uses a mathematical tool called a *matrix* in a series of steps that are based on the elimination method.

■ MATRICES

Another method of solving systems of equations involves rectangular arrays of numbers called *matrices* (plural for *matrix*).

Matrices	A **matrix** is any rectangular array of numbers arranged in rows and columns, written within brackets.

The Language of Algebra

An *array* is an orderly arrangement. For example, a jewelry store might display an impressive *array* of gemstones.

Some examples of matrices are

$$A = \begin{bmatrix} 1 & -3 & 8 \\ 2 & 5 & -1 \end{bmatrix} \begin{array}{l} \leftarrow \text{Row 1} \\ \leftarrow \text{Row 2} \end{array}$$

$$\begin{array}{ccc} \uparrow & \uparrow & \uparrow \\ \text{Column} & \text{Column} & \text{Column} \\ 1 & 2 & 3 \end{array}$$

$$B = \begin{bmatrix} 1 & 4 & -2 & -4 \\ 6 & -2 & 6 & 1 \\ 3 & 8 & -3 & 12 \end{bmatrix} \begin{array}{l} \leftarrow \text{Row 1} \\ \leftarrow \text{Row 2} \\ \leftarrow \text{Row 3} \end{array}$$

$$\begin{array}{cccc} \uparrow & \uparrow & \uparrow & \uparrow \\ \text{Column} & \text{Column} & \text{Column} & \text{Column} \\ 1 & 2 & 3 & 4 \end{array}$$

The numbers in each matrix are called **elements.** Because matrix A has two rows and three columns, it is called a 2×3 matrix (read "2 by 3" matrix). Matrix B is a 3×4 matrix (three rows and four columns).

■ AUGMENTED MATRICES

To show how to use matrices to solve systems of linear equations, we consider the system

$$\begin{cases} x - y = 4 \\ 2x + y = 5 \end{cases}$$

which can be represented by the following matrix, called an **augmented matrix:**

$$\begin{bmatrix} 1 & -1 & \vdots & 4 \\ 2 & 1 & \vdots & 5 \end{bmatrix}$$

Each row of the augmented matrix represents one equation of the system. The first two columns of the augmented matrix are determined by the coefficients of x and y in the equations of the system. The last column is determined by the constants in the equations.

$$\begin{bmatrix} 1 & -1 & \vdots & 4 \\ 2 & 1 & \vdots & 5 \end{bmatrix}$$ ← This row represents the equation $x - y = 4$.
← This row represents the equation $2x + y = 5$.

Coefficients   Coefficients   Constants
of x of y

EXAMPLE 1

ELEMENTARY &
INTERMEDIATE
Algebra *f(x)* **Now**™

Represent each system using an augmented matrix:

a. $\begin{cases} 3x + y = 11 \\ x - 8y = 0 \end{cases}$ and   **b.** $\begin{cases} 2a + b - 3c = -3 \\ 9a + 4c = 2 \\ a - b - 6c = -7 \end{cases}$.

Solution   **a.** $\begin{cases} 3x + y = 11 \\ x - 8y = 0 \end{cases}$ $\leftrightarrow$ $\begin{bmatrix} 3 & 1 & \vdots & 11 \\ 1 & -8 & \vdots & 0 \end{bmatrix}$

b. $\begin{cases} 2a + b - 3c = -3 \\ 9a + 4c = 2 \\ a - b - 6c = -7 \end{cases}$ $\begin{matrix} \leftrightarrow \\ \leftrightarrow \\ \leftrightarrow \end{matrix}$ $\begin{bmatrix} 2 & 1 & -3 & \vdots & -3 \\ 9 & 0 & 4 & \vdots & 2 \\ 1 & -1 & -6 & \vdots & -7 \end{bmatrix}$

Self Check 1   Represent each system using an augmented matrix:

a. $\begin{cases} 2x - 4y = 9 \\ 5x - y = -2 \end{cases}$ and   **b.** $\begin{cases} a + b - c = -4 \\ -2b + 7c = 0 \\ 10a + 8b - 4c = 5 \end{cases}$.

■ GAUSSIAN ELIMINATION

To solve a 2×2 system of equations by **Gaussian elimination,** we transform the augmented matrix into a matrix that has 1's down its main diagonal and a 0 below the 1 in the first column.

$$\begin{bmatrix} 1 & a & \vdots & b \\ 0 & 1 & \vdots & c \end{bmatrix}$$ $a, b,$ and c represent real numbers.

Main diagonal

To write the augmented matrix in this form, we use three operations called **elementary row operations.**

Elementary Row Operations   **Type 1:**   Any two rows of a matrix can be interchanged.
Type 2:   Any row of a matrix can be multiplied by a nonzero constant.
Type 3:   Any row of a matrix can be changed by adding a nonzero constant multiple of another row to it.

- A type 1 row operation corresponds to interchanging two equations of the system.
- A type 2 row operation corresponds to multiplying both sides of an equation by a nonzero constant.
- A type 3 row operation corresponds to adding a nonzero multiple of one equation to another.

None of these row operations will change the solution of the given system of equations.

EXAMPLE 2

ELEMENTARY &
INTERMEDIATE
Algebra $f(x)$ **Now**™

Consider the augmented matrices

$$A = \begin{bmatrix} 2 & 4 & \vdots & -3 \\ 1 & -8 & \vdots & 0 \end{bmatrix} \qquad B = \begin{bmatrix} 1 & -1 & \vdots & 2 \\ 4 & -8 & \vdots & 0 \end{bmatrix} \qquad C = \begin{bmatrix} 2 & 1 & -8 & \vdots & 4 \\ 0 & 1 & 4 & \vdots & -2 \\ 0 & 0 & -6 & \vdots & 24 \end{bmatrix}$$

a. Interchange rows 1 and 2 of matrix A.

b. Multiply row 3 of matrix C by $-\frac{1}{6}$.

c. To the numbers in row 2 of matrix B, add the results of multiplying each number in row 1 by -4.

Solution

a. Interchanging the rows of matrix A, we obtain $\begin{bmatrix} 1 & -8 & \vdots & 0 \\ 2 & 4 & \vdots & -3 \end{bmatrix}$.

b. We multiply each number in row 3 by $-\frac{1}{6}$. Rows 1 and 2 remain unchanged.

$$\begin{bmatrix} 2 & 1 & -8 & \vdots & 4 \\ 0 & 1 & 4 & \vdots & -2 \\ 0 & 0 & 1 & \vdots & -4 \end{bmatrix}$$ We can represent the instruction to multiply the third row by $-\frac{1}{6}$ with the symbolism $-\frac{1}{6}R_3$.

c. If we multiply each number in row 1 of matrix B by -4, we get

$$-4 \quad 4 \quad -8$$

We then add these numbers to row 2. (Note that row 1 remains unchanged.)

$$\begin{bmatrix} 1 & -1 & \vdots & 2 \\ 4 + (-4) & -8 + 4 & \vdots & 0 + (-8) \end{bmatrix}$$ We can abbreviate this procedure using the notation $-4R_1 + R_2$, which means "Multiply row 1 by -4 and add the result to row 2."

After simplifying, we have the matrix

$$\begin{bmatrix} 1 & -1 & \vdots & 2 \\ 0 & -4 & \vdots & -8 \end{bmatrix}$$

Self Check 2

Refer to Example 2.

a. Interchange the rows of matrix B.

b. To the numbers in row 1 of matrix A, add the results of multiplying each number in row 2 by -2.

c. Interchange rows 2 and 3 of matrix C.

We now solve a system of two linear equations using the **Gaussian elimination** process, which involves a series of elementary row operations.

EXAMPLE 3

ELEMENTARY &
INTERMEDIATE

Algebra *(fx)* **Now**™

Solve the system: $\begin{cases} 2x + y = 5 \\ x - y = 4 \end{cases}$.

Solution   We can represent the system with the following augmented matrix:

$$\begin{bmatrix} \mathbf{2} & 1 & \vdots & 5 \\ 1 & -1 & \vdots & 4 \end{bmatrix}$$

First, we want to get a 1 in the top row of the first column where the red 2 is. This can be achieved by applying a type 1 row operation: Interchange rows 1 and 2.

$$\begin{bmatrix} 1 & -1 & \vdots & 4 \\ \mathbf{2} & 1 & \vdots & 5 \end{bmatrix}$$ Interchanging row 1 and row 2 can be abbreviated as $R_1 \leftrightarrow R_2$.

To get a 0 under the 1 in the first column where the red 2 is, we use a type 3 row operation. To row 2, we add the results of multiplying each number in row 1 by -2.

$$\begin{bmatrix} 1 & -1 & \vdots & 4 \\ 0 & \mathbf{3} & \vdots & -3 \end{bmatrix}$$ $-2R_1 + R_2$

To get a 1 in the bottom row of the second column where the red 3 is, we use a type 2 row operation: Multiply row 2 by $\frac{1}{3}$.

$$\begin{bmatrix} 1 & -1 & \vdots & 4 \\ 0 & 1 & \vdots & -1 \end{bmatrix}$$ $\frac{1}{3}R_2$

This augmented matrix represents the equations

$$1x - 1y = 4$$
$$0x + 1y = -1$$

Writing the equations without the coefficients of 1 and -1, we have

(1) $x - y = 4$
(2) $y = -1$

From Equation 2, we see that $y = -1$. We can *back-substitute* -1 for y in Equation 1 to find x.

$$\begin{aligned} x - \mathbf{y} &= 4 \\ x - (\mathbf{-1}) &= 4 &&\text{Substitute } -1 \text{ for } y. \\ x + 1 &= 4 &&-(-1) = 1. \\ x &= 3 &&\text{Subtract 1 from both sides.} \end{aligned}$$

The solution of the system is $(3, -1)$. Verify that this ordered pair satisfies the original system.

Self Check 3 Solve: $\begin{cases} 3x - 2y = -5 \\ x - y = -4 \end{cases}$.

In general, if a system of linear equations has a single solution, we can use the following steps to solve the system using matrices.

Solving Systems of Linear Equations Using Matrices	**1.** Write an augmented matrix for the system.
	2. Use elementary row operations to transform the augmented matrix into a matrix with 1's down its main diagonal and 0's under the 1's.
	3. When step 2 is complete, write the resulting system. Then use back substitution to find the solution.
	4. Check the proposed solution in the equations of the original system.

■ SOLVING A SYSTEM OF THREE EQUATIONS

To show how to use matrices to solve systems of three linear equations containing three variables, we consider the system

$$\begin{cases} x - 2y - z = 6 \\ 2x + 2y - z = 1 \\ -x - y + 2z = 1 \end{cases}$$

which can be represented by the augmented matrix

$$\left[\begin{array}{ccc|c} 1 & -2 & -1 & 6 \\ 2 & 2 & -1 & 1 \\ -1 & -1 & 2 & 1 \end{array} \right]$$

To solve a 3×3 system of equations by Gaussian elimination, we transform the augmented matrix into a matrix with 1's down its main diagonal and 0's below its main diagonal.

$$\left[\begin{array}{ccc|c} 1 & a & b & c \\ 0 & 1 & d & e \\ 0 & 0 & 1 & f \end{array} \right] \qquad a, b, c, \ldots, f \text{ represent real numbers.}$$

Main diagonal

EXAMPLE 4

ELEMENTARY & INTERMEDIATE
Algebra $f(x)$ **Now**™

Solve the system:

$$\begin{cases} 3x + y + 5z = 8 \\ 2x + 3y - z = 6 \\ x + 2y + 2z = 10 \end{cases}$$

Solution This system can be represented by the augmented matrix

$$\left[\begin{array}{ccc|c} 3 & 1 & 5 & 8 \\ 2 & 3 & -1 & 6 \\ 1 & 2 & 2 & 10 \end{array} \right]$$

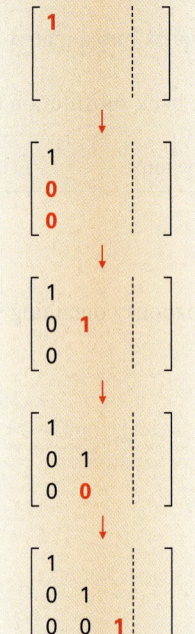

To get a 1 in the first column where the red 3 is, we perform a type 1 row operation: Interchange rows 1 and 3.

$$\begin{bmatrix} 1 & 2 & 2 & | & 10 \\ \mathbf{2} & 3 & -1 & | & 6 \\ \mathbf{3} & 1 & 5 & | & 8 \end{bmatrix} \quad R_1 \leftrightarrow R_3$$

To get a 0 under the 1 in the first column where the red 2 is, we perform a type 3 row operation: Multiply row 1 by -2 and add the results to row 2. Note that row 1 remains the same.

$$\begin{bmatrix} 1 & 2 & 2 & | & 10 \\ 0 & -1 & -5 & | & -14 \\ \mathbf{3} & 1 & 5 & | & 8 \end{bmatrix} \quad -2R_1 + R_2$$

To get a 0 under the 0 in the first column where the red 3 is, we perform another type 3 row operation: Multiply row 1 by -3 and add the results to row 3. Again, row 1 remains the same.

$$\begin{bmatrix} 1 & 2 & 2 & | & 10 \\ 0 & \mathbf{-1} & -5 & | & -14 \\ 0 & -5 & -1 & | & -22 \end{bmatrix} \quad -3R_1 + R_3$$

To get a 1 under the 2 in the second column where the red -1 is, we perform a type 2 row operation: Multiply row 2 by -1.

$$\begin{bmatrix} 1 & 2 & 2 & | & 10 \\ 0 & 1 & 5 & | & 14 \\ 0 & \mathbf{-5} & -1 & | & -22 \end{bmatrix} \quad -1R_2$$

To get a 0 under the 1 in the second column where the red -5 is, we perform a type 3 row operation: Multiply row 2 by 5 and add the results to row 3. Row 2 remains the same.

$$\begin{bmatrix} 1 & 2 & 2 & | & 10 \\ 0 & 1 & 5 & | & 14 \\ 0 & 0 & \mathbf{24} & | & 48 \end{bmatrix} \quad 5R_2 + R_3$$

To get a 1 under the 5 in the third column where the red 24 is, we perform a type 2 row operation: Multiply row 3 by $\frac{1}{24}$.

$$\begin{bmatrix} 1 & 2 & 2 & | & 10 \\ 0 & 1 & 5 & | & 14 \\ 0 & 0 & 1 & | & 2 \end{bmatrix} \quad \frac{1}{24}R_3$$

The final matrix represents the system

$$\begin{cases} 1x + 2y + 2z = 10 \\ 0x + 1y + 5z = 14 \\ 0x + 0y + 1z = 2 \end{cases}$$

which can be written without the coefficients of 0 and 1 as

$$\begin{cases} x + 2y + 2z = 10 & \textbf{(1)} \\ y + 5z = 14 & \textbf{(2)} \\ z = 2 & \textbf{(3)} \end{cases}$$

From Equation 3, we can read that z is 2. To find y, we back substitute 2 for z in Equation 2 and solve for y:

$$y + 5z = 14 \qquad \text{This is Equation 2.}$$
$$y + 5(\mathbf{2}) = 14 \qquad \text{Substitute 2 for } z.$$
$$y + 10 = 14$$
$$y = 4 \qquad \text{Subtract 10 from both sides.}$$

Thus, y is 4. To find x, we back substitute 2 for z and 4 for y in Equation 1 and solve for x:

$$x + 2y + 2z = 10 \qquad \text{This is Equation 1.}$$
$$x + 2(\mathbf{4}) + 2(\mathbf{2}) = 10 \qquad \text{Substitute 2 for } z \text{ and 4 for } y.$$
$$x + 8 + 4 = 10$$
$$x + 12 = 10$$
$$x = -2 \qquad \text{Subtract 12 from both sides.}$$

Thus, x is -2. The solution of the given system is $(-2, 4, 2)$. Verify that this ordered triple satisfies each equation of the original system.

Self Check 4   Solve: $\begin{cases} 2x - y + z = 5 \\ x + y - z = -2 \\ -x + 2y + 2z = 1 \end{cases}$.

■ INCONSISTENT SYSTEMS AND DEPENDENT EQUATIONS

In the next example, we consider a system with no solution.

EXAMPLE 5

ELEMENTARY &
INTERMEDIATE
Algebra $f(x)$ **Now**™

Using matrices, solve the system: $\begin{cases} x + y = -1 \\ -3x - 3y = -5 \end{cases}$.

Solution   This system can be represented by the augmented matrix

$$\begin{bmatrix} 1 & 1 & \vdots & -1 \\ -3 & -3 & \vdots & -5 \end{bmatrix}$$

Since the matrix has a 1 in the top row of the first column, we proceed to get a 0 under it by multiplying row 1 by 3 and adding the results to row 2.

$$\begin{bmatrix} 1 & 1 & \vdots & -1 \\ 0 & 0 & \vdots & -8 \end{bmatrix} \quad 3R_1 + R_2$$

This matrix represents the system

$$\begin{cases} x + y = -1 \\ 0 + 0 = -8 \end{cases}$$

This system has no solution, because the second equation is never true. Therefore, the system is inconsistent. It has no solutions.

Self Check 5 Solve: $\begin{cases} 4x - 8y = 9 \\ x - 2y = -5 \end{cases}$.

In the next example, we consider a system with infinitely many solutions.

EXAMPLE 6

ELEMENTARY &
INTERMEDIATE
Algebra $f(x)$**Now**™

Using matrices, solve the system:

$$\begin{cases} 2x + 3y - 4z = 6 \\ 4x + 6y - 8z = 12 \\ -6x - 9y + 12z = -18 \end{cases}$$

Solution This system can be represented by the augmented matrix

$$\left[\begin{array}{ccc|c} 2 & 3 & -4 & 6 \\ 4 & 6 & -8 & 12 \\ -6 & -9 & 12 & -18 \end{array} \right]$$

To get a 1 in the top row of the first column, we multiply row 1 by $\dfrac{1}{2}$.

$$\left[\begin{array}{ccc|c} 1 & \frac{3}{2} & -2 & 3 \\ 4 & 6 & -8 & 12 \\ -6 & -9 & 12 & -18 \end{array} \right] \quad \frac{1}{2}R_1$$

Next, we want to get 0's under the 1 in the first column. This can be achieved by multiplying row 1 by -4 and adding the results to row 2, and multiplying row 1 by 6 and adding the results to row 3.

$$\left[\begin{array}{ccc|c} 1 & \frac{3}{2} & -2 & 3 \\ 0 & 0 & 0 & 0 \\ 0 & 0 & 0 & 0 \end{array} \right] \quad \begin{array}{l} -4R_1 + R_2 \\ 6R_1 + R_3 \end{array}$$

The last matrix represents the system

$$\begin{cases} x + \frac{3}{2}y - 2z = 3 \\ 0x + 0y + 0z = 0 \\ 0x + 0y + 0z = 0 \end{cases}$$

If we clear the first equation of fractions, we have the system

$$\begin{cases} 2x + 3y - 4z = 6 \\ 0 = 0 \\ 0 = 0 \end{cases}$$

This system has dependent equations and infinitely many solutions. Solutions of this system would be any triple (x, y, z) that satisfies the equation $2x + 3y - 4z = 6$. Two such solutions would be $(0, 2, 0)$ and $(1, 0, -1)$.

Self Check 6   Solve: $\begin{cases} 5x - 10y + 15z = 35 \\ -3x + 6y - 9z = -21. \\ 2x - 4y + 6z = 14 \end{cases}$

Answers to Self Checks   **1. a.** $\begin{bmatrix} 2 & -4 & \vdots & 9 \\ 5 & -1 & \vdots & -2 \end{bmatrix}$, **b.** $\begin{bmatrix} 1 & 1 & -1 & \vdots & -4 \\ 0 & -2 & 7 & \vdots & 0 \\ 10 & 8 & -4 & \vdots & 5 \end{bmatrix}$

2. a. $\begin{bmatrix} 4 & -8 & \vdots & 0 \\ 1 & -1 & \vdots & 2 \end{bmatrix}$, **b.** $\begin{bmatrix} 0 & 20 & \vdots & -3 \\ 1 & -8 & \vdots & 0 \end{bmatrix}$, **c.** $\begin{bmatrix} 2 & 1 & -8 & \vdots & 4 \\ 0 & 0 & -6 & \vdots & 24 \\ 0 & 1 & 4 & \vdots & -2 \end{bmatrix}$

3. $(3, 7)$ **4.** $(1, -1, 2)$ **5.** no solution

6. There are infinitely many solutions—any triple satisfying the equation $x - 2y + 3z = 7$.

12.3   STUDY SET   ELEMENTARY & INTERMEDIATE
Algebra *(x)* Now™

VOCABULARY   Fill in the blanks.

1. A _____ is a rectangular array of numbers.

2. The numbers in a matrix are called its _____.

3. A 3 × 4 matrix has 3 _____ and 4 _____.

4. Elementary _____ operations are used to produce new matrices that lead to the solution of a system.

5. A matrix that represents the equations of a system is called an _____ matrix.

6. The augmented matrix $\begin{bmatrix} 1 & 3 & \vdots & -2 \\ 0 & 1 & \vdots & 4 \end{bmatrix}$ has 1's down its main _____.

CONCEPTS

7. For each matrix, determine the number of rows and the number of columns.

a. $\begin{bmatrix} 4 & 6 & \vdots & -1 \\ \frac{1}{2} & 9 & \vdots & -3 \end{bmatrix}$

b. $\begin{bmatrix} 1 & -2 & 3 & \vdots & 1 \\ 0 & 1 & 6 & \vdots & 4 \\ 0 & 0 & 1 & \vdots & \frac{1}{3} \end{bmatrix}$

8. For each augmented matrix, give the system of equations it represents.

a. $\begin{bmatrix} 1 & 6 & \vdots & 7 \\ 0 & 1 & \vdots & 4 \end{bmatrix}$

b. $\begin{bmatrix} 2 & -2 & 9 & \vdots & 1 \\ 3 & 1 & 1 & \vdots & 0 \\ 2 & -6 & 8 & \vdots & -7 \end{bmatrix}$

9. Write the system of equations represented by the augmented matrix and use back substitution to find the solution.

$\begin{bmatrix} 1 & -1 & \vdots & -10 \\ 0 & 1 & \vdots & 6 \end{bmatrix}$

10. Write the system of equations represented by the augmented matrix and use back substitution to find the solution.

$\begin{bmatrix} 1 & -2 & 1 & \vdots & -16 \\ 0 & 1 & 2 & \vdots & 8 \\ 0 & 0 & 1 & \vdots & 4 \end{bmatrix}$

11. Matrices were used to solve a system of two linear equations. The final matrix is shown here. Explain what the result tells about the system.

$\begin{bmatrix} 1 & 2 & \vdots & -4 \\ 0 & 0 & \vdots & 2 \end{bmatrix}$

12. Matrices were used to solve a system of two linear equations. The final matrix is shown here. Explain what the result tells about the equations.

$\begin{bmatrix} 1 & 2 & \vdots & -4 \\ 0 & 0 & \vdots & 0 \end{bmatrix}$

NOTATION

13. Consider the matrix

$$A = \begin{bmatrix} 3 & 6 & -9 & \vdots & 0 \\ 1 & 5 & -2 & \vdots & 1 \\ -2 & 2 & -2 & \vdots & 5 \end{bmatrix}.$$

a. Explain what is meant by $\frac{1}{3}R_1$. Then perform the operation on matrix A.

b. Explain what is meant by $-R_1 + R_2$. Then perform the operation on the answer to part a.

14. Consider the matrix $B = \begin{bmatrix} -3 & 1 & \vdots & -6 \\ 1 & -4 & \vdots & 4 \end{bmatrix}.$

a. Explain what is meant by $R_1 \leftrightarrow R_2$. Then perform the operation on matrix B.

b. Explain what is meant by $3R_1 + R_2$. Then perform the operation on the answer to part a.

Complete each solution.

15. Solve: $\begin{cases} 4x - y = 14 \\ x + y = 6 \end{cases}.$

$$\begin{bmatrix} 4 & \blacksquare & \vdots & 14 \\ 1 & 1 & \vdots & 6 \end{bmatrix}$$

$$\begin{bmatrix} \blacksquare & 1 & \vdots & 6 \\ 4 & -1 & \vdots & 14 \end{bmatrix} \quad R_1 \leftrightarrow R_2$$

$$\begin{bmatrix} 1 & 1 & \vdots & 6 \\ 0 & \blacksquare & \vdots & -10 \end{bmatrix} \quad -4R_1 + R_2$$

$$\begin{bmatrix} 1 & 1 & \vdots & 6 \\ 0 & 1 & \vdots & \blacksquare \end{bmatrix} \quad -\frac{1}{5}R_2$$

This matrix represents the system

$$\begin{cases} x + y = 6 \\ \blacksquare = 2 \end{cases}$$

The solution is $\left(\blacksquare, 2 \right)$.

16. Solve: $\begin{cases} 2x + 2y = 18 \\ x - y = 5 \end{cases}.$

$$\begin{bmatrix} 2 & 2 & \vdots & 18 \\ \blacksquare & -1 & \vdots & 5 \end{bmatrix}$$

$$\begin{bmatrix} 1 & 1 & \vdots & 9 \\ \blacksquare & -1 & \vdots & 5 \end{bmatrix} \quad \frac{1}{2}R_1$$

$$\begin{bmatrix} 1 & 1 & \vdots & 9 \\ 0 & \blacksquare & \vdots & -4 \end{bmatrix} \quad -R_1 + R_2$$

$$\begin{bmatrix} 1 & 1 & \vdots & 9 \\ 0 & 1 & \vdots & \blacksquare \end{bmatrix} \quad -\frac{1}{2}R_2$$

This matrix represents the system

$$\begin{cases} x + y = \blacksquare \\ y = 2 \end{cases}$$

The solution is $\left(\blacksquare, 2 \right)$.

PRACTICE   Use matrices to solve each system of equations. If the equations of a system are dependent or if a system is inconsistent, so indicate.

17. $\begin{cases} x + y = 2 \\ x - y = 0 \end{cases}$

18. $\begin{cases} x + y = 3 \\ x - y = -1 \end{cases}$

19. $\begin{cases} 2x + y = 1 \\ x + 2y = -4 \end{cases}$

20. $\begin{cases} 5x - 4y = 10 \\ x - 7y = 2 \end{cases}$

21. $\begin{cases} 2x - y = -1 \\ x - 2y = 1 \end{cases}$

22. $\begin{cases} 2x - y = 0 \\ x + y = 3 \end{cases}$

23. $\begin{cases} 3x + 4y = -12 \\ 9x - 2y = 6 \end{cases}$

24. $\begin{cases} 2x - 3y = 16 \\ -4x + y = -22 \end{cases}$

25. $\begin{cases} x + y + z = 6 \\ x + 2y + z = 8 \\ x + y + 2z = 9 \end{cases}$

26. $\begin{cases} x - y + z = 2 \\ x + 2y - z = 6 \\ 2x - y - z = 3 \end{cases}$

27. $\begin{cases} 3x + y - 3z = 5 \\ x - 2y + 4z = 10 \\ x + y + z = 13 \end{cases}$

28. $\begin{cases} 2x + y - 3z = -1 \\ 3x - 2y - z = -5 \\ x - 3y - 2z = -12 \end{cases}$

29. $\begin{cases} 3x - 2y + 4z = 4 \\ x + y + z = 3 \\ 6x - 2y - 3z = 10 \end{cases}$

30. $\begin{cases} 2x + 3y - z = -8 \\ x - y - z = -2 \\ -4x + 3y + z = 6 \end{cases}$

31. $\begin{cases} 2a + b + 3c = 3 \\ -2a - b + c = 5 \\ 4a - 2b + 2c = 2 \end{cases}$

32. $\begin{cases} 3a + 2b + c = 8 \\ 6a - b + 2c = 16 \\ -9a + b - c = -20 \end{cases}$

33. $\begin{cases} 2x + y - 3z = -7 \\ 3x - y + 2z = -9 \\ -2x - y - z = 3 \end{cases}$

34. $\begin{cases} -2x + 3y + z = -12 \\ 3x + y - z = 12 \\ 3x - y - z = 14 \end{cases}$

35. $\begin{cases} 2x + y - 2z = 6 \\ 4x - y + z = -1 \\ 6x - 2y + 3z = -5 \end{cases}$
36. $\begin{cases} 2x - 3y + 3z = 14 \\ 3x + 3y - z = 2 \\ -2x + 6y + 5z = 9 \end{cases}$

37. $\begin{cases} x - 3y = 9 \\ -2x + 6y = 18 \end{cases}$
38. $\begin{cases} -6x + 12y = 10 \\ 2x - 4y = 8 \end{cases}$

39. $\begin{cases} 4x + 4y = 12 \\ -x - y = -3 \end{cases}$
40. $\begin{cases} 5x - 15y = 10 \\ 2x - 6y = 4 \end{cases}$

41. $\begin{cases} 6x + y - z = -2 \\ x + 2y + z = 5 \\ 5y - z = 2 \end{cases}$
42. $\begin{cases} 2x + 3y - 2z = 18 \\ 5x - 6y + z = 21 \\ 4y - 2z = 6 \end{cases}$

43. $\begin{cases} 2x + y - z = 1 \\ x + 2y + 2z = 2 \\ 4x + 5y + 3z = 3 \end{cases}$
44. $\begin{cases} x - 3y + 4z = 2 \\ 2x + y + 2z = 3 \\ 4x - 5y + 10z = 7 \end{cases}$

45. $\begin{cases} 5x + 3y = 4 \\ 3y - 4z = 4 \\ x + z = 1 \end{cases}$
46. $\begin{cases} y + 2z = -2 \\ x + y = 1 \\ 2x - z = 0 \end{cases}$

47. $\begin{cases} x - y = 1 \\ 2x - z = 0 \\ 2y - z = -2 \end{cases}$
48. $\begin{cases} x + y - 3z = 4 \\ 2x + 2y - 6z = 5 \\ -3x + y - z = 2 \end{cases}$

Remember these facts from geometry. The sum of the measures of complementary angles is 90°, and the sum of the measures of supplementary angles is 180°.

49. One angle measures 46° more than the measure of its complement. Find the measure of each angle.

50. One angle measures 14° more than the measure of its supplement. Find the measure of each angle.

51. In the illustration, $\angle B$ measures 25° more than the measure of $\angle A$, and the measure of $\angle C$ is 5° less than twice the measure of $\angle A$. Find the measure of each angle of the triangle.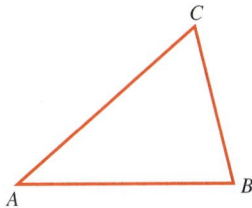

52. In the illustration, $\angle A$ measures 10° less than the measure of $\angle B$, and the measure of $\angle B$ is 10° less than the measure of $\angle C$. Find the measure of each angle of the triangle.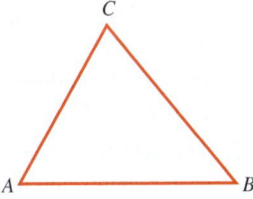

APPLICATIONS

53. DIGITAL PHOTOGRAPHY   A digital camera stores the black and white photograph shown below as a 512×512 matrix. Each element of the matrix corresponds to a small dot of grey scale shading, called a *pixel,* in the picture. How many elements does a 512×512 matrix have?

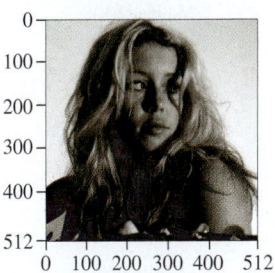

54. DIGITAL IMAGING   A scanner stores a black and white photograph as a matrix that has a total of 307,200 elements. If the matrix has 480 rows, how many columns does it have?

Write a system of equations to solve each problem. Use matrices to solve the system.

55. PHYSICAL THERAPY After an elbow injury, a volleyball player has restricted movement of her arm. Her range of motion (the measure of $\angle 1$) is 28° less than the measure of $\angle 2$. Find the measure of each angle.

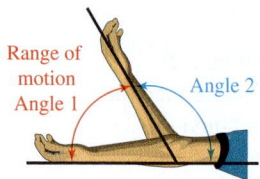

56. PIGGY BANKS When a child breaks open her piggy bank, she finds a total of 64 coins, consisting of nickels, dimes, and quarters. The total value of the coins is $6. If the nickels were dimes, and dimes were nickels, the value of the coins would be $5. How many nickels, dimes, and quarters were in the piggy bank?

57. THEATER SEATING The illustration shows the cash receipts and the ticket prices from two sold-out performances of a play. Find the number of seats in each of the three sections of the 800-seat theater.

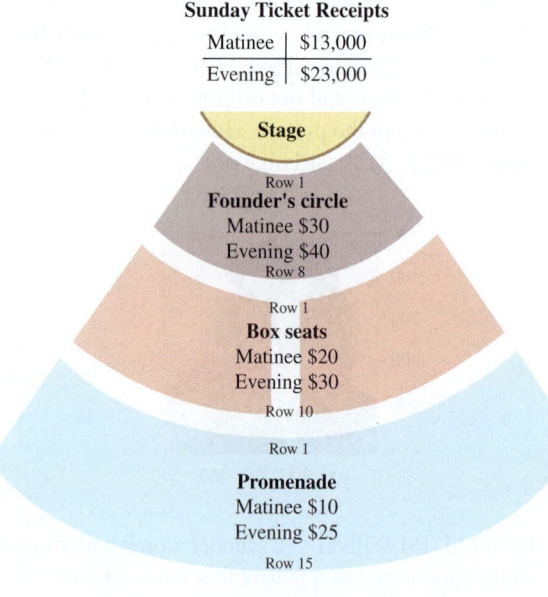

Sunday Ticket Receipts

Matinee	$13,000
Evening	$23,000

58. ICE SKATING Three circles are traced out by a figure skater during her performance. If the centers of the circles are the given distances apart, find the radius of each circle.

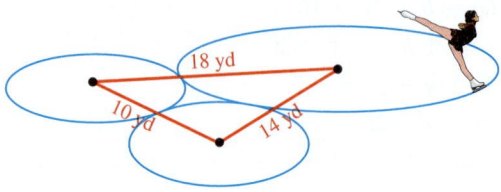

WRITING

59. Explain what is meant by the phrase *back substitution*.

60. Explain how a type 3 row operation is similar to the elimination method of solving a system of equations.

REVIEW

61. What is the formula used to find the slope of a line, given two points on the line?

62. What is the form of the equation of a horizontal line? Of a vertical line?

63. What is the point-slope form of the equation of a line?

64. What is the slope-intercept form of the equation of a line?

CHALLENGE PROBLEMS

65. If the system represented by

$$\begin{bmatrix} 1 & 1 & 0 & | & 1 \\ 0 & 0 & 1 & | & 2 \\ 0 & 0 & 0 & | & k \end{bmatrix}$$

has no solution, what do you know about k?

Use matrices to solve the system.

66. $\begin{cases} w + x + y + z = 0 \\ w - 2x + y - 3z = -3 \\ 2w + 3x + y - 2z = -1 \\ 2w - 2x - 2y + z = -12 \end{cases}$

12.4 Solving Systems Using Determinants

- Determinants   • Evaluating a determinant
- Using Cramer's rule to solve a system of two equations
- Using Cramer's rule to solve a system of three equations

In this section, we will discuss another method for solving systems of linear equations. With this method, called *Cramer's rule,* we work with combinations of the coefficients and the constants of the equations written as *determinants.*

▪ DETERMINANTS

An idea related to the concept of matrix is the **determinant.** A determinant is a number that is associated with a **square matrix,** a matrix that has the same number of rows and columns. For any square matrix A, the symbol $|A|$ represents the determinant of A. To write a determinant, we put the elements of a square matrix between two vertical lines.

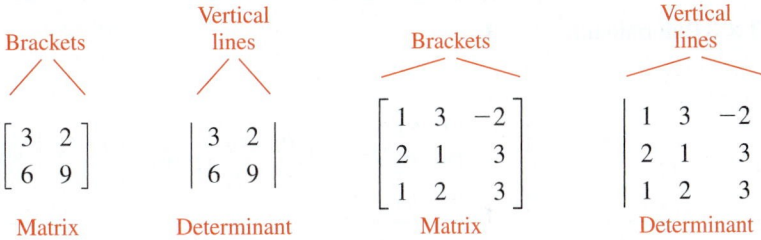

Like matrices, determinants are classified according to the number of rows and columns they contain. The determinant on the left is a 2×2 determinant. The other is a 3×3 determinant.

▪ EVALUATING A DETERMINANT

The determinant of a 2×2 matrix is the number that is equal to the product of the numbers on the main diagonal minus the product of the numbers on the other diagonal.

$$\begin{vmatrix} a & b \\ c & d \end{vmatrix}$$
Main diagonal

$$\begin{vmatrix} a & b \\ c & d \end{vmatrix}$$
Other diagonal

Value of a 2 × 2 Determinant

If a, b, c, and d are numbers, the **determinant** of the matrix $\begin{bmatrix} a & b \\ c & d \end{bmatrix}$ is

$$\begin{vmatrix} a & b \\ c & d \end{vmatrix} = ad - bc$$

EXAMPLE 1   Find each value: **a.** $\begin{vmatrix} 3 & 2 \\ 6 & 9 \end{vmatrix}$ and   **b.** $\begin{vmatrix} -5 & \frac{1}{2} \\ -1 & 0 \end{vmatrix}$.

ELEMENTARY &
INTERMEDIATE

Algebra $f(x)$ **Now™**   Solution   From the product of the numbers along the main diagonal, we subtract the product of the numbers along the other diagonal.

a. $\begin{vmatrix} 3 & 2 \\ 6 & 9 \end{vmatrix} = 3(9) - 2(6)$ **b.** $\begin{vmatrix} -5 & \frac{1}{2} \\ -1 & 0 \end{vmatrix} = -5(0) - \frac{1}{2}(-1)$

$$= 27 - 12$$

$$= 15$$

$$= 0 + \frac{1}{2}$$

$$= \frac{1}{2}$$

Self Check 1   Evaluate: $\begin{vmatrix} 4 & -3 \\ 2 & 1 \end{vmatrix}$.

A 3×3 determinant is evaluated by **expanding by minors.**

Value of a 3 × 3 Determinant

| | Minor of a_1 | Minor of b_1 | Minor of c_1 |

$$\begin{vmatrix} a_1 & b_1 & c_1 \\ a_2 & b_2 & c_2 \\ a_3 & b_3 & c_3 \end{vmatrix} = a_1 \begin{vmatrix} b_2 & c_2 \\ b_3 & c_3 \end{vmatrix} - b_1 \begin{vmatrix} a_2 & c_2 \\ a_3 & c_3 \end{vmatrix} + c_1 \begin{vmatrix} a_2 & b_2 \\ a_3 & b_3 \end{vmatrix}$$

To find the minor of a_1, we cross out the elements of the determinant that are in the same row and column as a_1:

$$\begin{vmatrix} a_1 & b_1 & c_1 \\ a_2 & b_2 & c_2 \\ a_3 & b_3 & c_3 \end{vmatrix}$$ The minor of a_1 is $\begin{vmatrix} b_2 & c_2 \\ b_3 & c_3 \end{vmatrix}$.

To find the minor of b_1, we cross out the elements of the determinant that are in the same row and column as b_1:

$$\begin{vmatrix} a_1 & b_1 & c_1 \\ a_2 & b_2 & c_2 \\ a_3 & b_3 & c_3 \end{vmatrix}$$ The minor of b_1 is $\begin{vmatrix} a_2 & c_2 \\ a_3 & c_3 \end{vmatrix}$.

To find the minor of c_1, we cross out the elements of the determinant that are in the same row and column as c_1:

$$\begin{vmatrix} a_1 & b_1 & c_1 \\ a_2 & b_2 & c_2 \\ a_3 & b_3 & c_3 \end{vmatrix}$$ The minor of c_1 is $\begin{vmatrix} a_2 & b_2 \\ a_3 & b_3 \end{vmatrix}$.

EXAMPLE 2

ELEMENTARY &
INTERMEDIATE
Algebra $f(x)$ **Now** ™

Find the value of $\begin{vmatrix} 1 & 3 & -2 \\ 2 & 0 & 3 \\ 1 & 2 & 3 \end{vmatrix}$.

Solution   We evaluate this determinant by expanding by minors along the first row of the determinant.

$$\begin{array}{ccc} \text{Minor} & \text{Minor} & \text{Minor} \\ \text{of 1} & \text{of 3} & \text{of } -2 \\ \downarrow & \downarrow & \downarrow \end{array}$$

$$\begin{vmatrix} \mathbf{1} & \mathbf{3} & \mathbf{-2} \\ 2 & 0 & 3 \\ 1 & 2 & 3 \end{vmatrix} = \mathbf{1}\begin{vmatrix} 0 & 3 \\ 2 & 3 \end{vmatrix} - \mathbf{3}\begin{vmatrix} 2 & 3 \\ 1 & 3 \end{vmatrix} + (\mathbf{-2})\begin{vmatrix} 2 & 0 \\ 1 & 2 \end{vmatrix}$$

$$= 1(0 - 6) - 3(6 - 3) - 2(4 - 0) \qquad \text{Evaluate each } 2 \times 2 \text{ determinant.}$$

$$= 1(-6) - 3(3) - 2(4)$$

$$= -6 - 9 - 8$$

$$= -23$$

Self Check 2   Evaluate: $\begin{vmatrix} 2 & 3 & -1 \\ 0 & 2 & 4 \\ -2 & 5 & 6 \end{vmatrix}$.

We can evaluate a 3×3 determinant by expanding it along any row or column. To determine the signs between the terms of the expansion of a 3×3 determinant, we use the following array of signs.

Array of Signs for a 3×3 Determinant	$\begin{array}{ccc} + & - & + \\ - & + & - \\ + & - & + \end{array}$	This array of signs is commonly referred to as the **checkerboard pattern.**

EXAMPLE 3

ELEMENTARY &
INTERMEDIATE
Algebra $f(x)$ **Now** ™

Evaluate the determinant $\begin{vmatrix} 1 & 3 & -2 \\ 2 & 0 & 3 \\ 1 & 2 & 3 \end{vmatrix}$ by expanding on the middle column.

Solution   This is the determinant of Example 2. To expand it along the middle column, we use the signs of the middle column of the array of signs:

$$\begin{array}{ccc} \text{Minor} & \text{Minor} & \text{Minor} \\ \text{of 3} & \text{of 0} & \text{of 2} \\ \downarrow & \downarrow & \downarrow \end{array}$$

$$\begin{vmatrix} 1 & \mathbf{3} & -2 \\ 2 & \mathbf{0} & 3 \\ 1 & \mathbf{2} & 3 \end{vmatrix} = \mathbf{-3}\begin{vmatrix} 2 & 3 \\ 1 & 3 \end{vmatrix} + \mathbf{0}\begin{vmatrix} 1 & -2 \\ 1 & 3 \end{vmatrix} - \mathbf{2}\begin{vmatrix} 1 & -2 \\ 2 & 3 \end{vmatrix}$$

Use the middle column of the checkerboard pattern:

$$\begin{array}{ccc} + & - & + \\ - & + & - \\ + & - & + \end{array}$$

Success Tip

When evaluating a determinant, expanding along a row or column that contains 0's can simplify the computations.

$$= -3(6 - 3) + 0 - 2[3 - (-4)]$$

$$= -3(3) + 0 - 2(7)$$

$$= -9 + 0 - 14$$

$$= -23$$

Evaluate each 2×2 determinant.

As expected, we get the same value as in Example 2.

Self Check 3 Evaluate: $\begin{vmatrix} 2 & 3 & -1 \\ 0 & 2 & 4 \\ -2 & 5 & 6 \end{vmatrix}$ by expanding along the first column.

ACCENT ON TECHNOLOGY: EVALUATING DETERMINANTS

It is possible to use a graphing calculator to evaluate determinants. For example, to evaluate the determinant in Example 3, we first enter the matrix by pressing the [MATRIX] key, selecting EDIT, and pressing the [ENTER] key. Next, we enter the dimensions and the elements of the matrix to get figure (a). We then press [2nd] [QUIT] to clear the screen, press [MATRIX], select MATH, and press 1 to get figure (b). We then press [MATRIX], select NAMES, press 1, and press [)] and [ENTER] to get the value of the determinant. Figure (c) shows that the value of the determinant is -23.

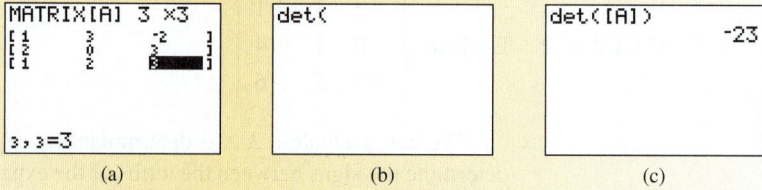

(a)   (b)   (c)

■ USING CRAMER'S RULE TO SOLVE A SYSTEM OF TWO EQUATIONS

The method of using determinants to solve systems of linear equations is called **Cramer's rule,** named after the 18th-century mathematician Gabriel Cramer. To develop Cramer's rule, we consider the system

$$\begin{cases} ax + by = e \\ cx + dy = f \end{cases}$$

where x and y are variables and a, b, c, d, e, and f are constants.

If we multiply both sides of the first equation by d and multiply both sides of the second equation by $-b$, we can add the equations and eliminate y:

$$adx + bdy = ed$$
$$\underline{-bcx - bdy = -bf}$$
$$adx - bcx \quad\quad = ed - bf$$

To solve for x, we use the distributive property to write $adx - bcx$ as $(ad - bc)x$ on the left-hand side and divide each side by $ad - bc$:

$$(ad - bc)x = ed - bf$$

$$x = \frac{ed - bf}{ad - bc} \quad \text{where } ad - bc \neq 0$$

We can find y in a similar manner. After eliminating the variable x, we get

$$y = \frac{af - ec}{ad - bc} \quad \text{where } ad - bc \neq 0$$

Determinants provide an easy way of remembering these formulas. Note that the denominator for both x and y is

$$\begin{vmatrix} a & b \\ c & d \end{vmatrix} = ad - bc$$

The numerators can be expressed as determinants also:

$$x = \frac{ed - bf}{ad - bc} = \frac{\begin{vmatrix} e & b \\ f & d \end{vmatrix}}{\begin{vmatrix} a & b \\ c & d \end{vmatrix}} \quad \text{and} \quad y = \frac{af - ec}{ad - bc} = \frac{\begin{vmatrix} a & e \\ c & f \end{vmatrix}}{\begin{vmatrix} a & b \\ c & d \end{vmatrix}}$$

If we compare these formulas with the original system

$$\begin{cases} ax + by = e \\ cx + dy = f \end{cases}$$

we note that in the expressions for x and y above, the denominator determinant is formed by using the coefficients a, b, c, and d of the variables in the equations. The numerator determinants are the same as the denominator determinant, except that the column of coefficients of the variable for which we are solving is replaced with the column of constants e and f.

Cramer's Rule for Two Equations in Two Variables

The solution of the system $\begin{cases} ax + by = e \\ cx + dy = f \end{cases}$ is given by

$$x = \frac{D_x}{D} = \frac{\begin{vmatrix} e & b \\ f & d \end{vmatrix}}{\begin{vmatrix} a & b \\ c & d \end{vmatrix}} \quad \text{and} \quad y = \frac{D_y}{D} = \frac{\begin{vmatrix} a & e \\ c & f \end{vmatrix}}{\begin{vmatrix} a & b \\ c & d \end{vmatrix}}$$

If every determinant is 0, the system is consistent, but the equations are dependent.

If $D = 0$ and D_x or D_y is nonzero, the system is inconsistent. If $D \neq 0$, the system is consistent, and the equations are independent.

EXAMPLE 4

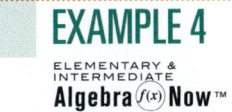

ELEMENTARY & INTERMEDIATE
Algebra $f(x)$ **Now**™

Use Cramer's rule to solve $\begin{cases} 4x - 3y = 6 \\ -2x + 5y = 4 \end{cases}$.

Solution   The value of x is the quotient of two determinants, D_x and D. The denominator determinant D is made up of the coefficients of x and y:

$$D = \begin{vmatrix} 4 & -3 \\ -2 & 5 \end{vmatrix}$$

To solve for x, we form the numerator determinant D_x from D by replacing its first column (the coefficients of x) with the column of constants (6 and 4).

To solve for y, we form the numerator determinant D_y from D by replacing the second column (the coefficients of y) with the column of constants (6 and 4).

To find the values of x and y, we evaluate each determinant:

$$x = \frac{D_x}{D} = \frac{\begin{vmatrix} 6 & -3 \\ 4 & 5 \end{vmatrix}}{\begin{vmatrix} 4 & -3 \\ -2 & 5 \end{vmatrix}} = \frac{6(5) - (-3)(4)}{4(5) - (-3)(-2)} = \frac{30 + 12}{20 - 6} = \frac{42}{14} = 3$$

$$y = \frac{D_y}{D} = \frac{\begin{vmatrix} 4 & 6 \\ -2 & 4 \end{vmatrix}}{\begin{vmatrix} 4 & -3 \\ -2 & 5 \end{vmatrix}} = \frac{4(4) - 6(-2)}{14} = \frac{16 + 12}{14} = \frac{28}{14} = 2$$

The solution of this system is (3, 2). Verify that it satisfies both equations.

Self Check 4   Use Cramer's rule to solve $\begin{cases} 2x - 3y = -16 \\ 3x + 5y = 14 \end{cases}$.

EXAMPLE 5

ELEMENTARY &
INTERMEDIATE
Algebra $f(x)$ **Now**™

Use Cramer's rule to solve $\begin{cases} 7x = 8 - 4y \\ 2y = 3 - \dfrac{7}{2}x \end{cases}$.

Solution   We multiply both sides of the second equation by 2 to eliminate the fraction and write the system in the form

$$\begin{cases} 7x + 4y = 8 \\ 7x + 4y = 6 \end{cases}$$

When we attempt to use Cramer's rule to solve this system for x, we obtain

Success Tip

If any two rows or any two columns of a determinant are identical, the value of the determinant is 0.

$$x = \frac{D_x}{D} = \frac{\begin{vmatrix} 8 & 4 \\ 6 & 4 \end{vmatrix}}{\begin{vmatrix} 7 & 4 \\ 7 & 4 \end{vmatrix}} = \frac{8}{0}, \quad \text{which is undefined}$$

Since the denominator determinant D is 0 and the numerator determinant D_x is not 0, the system is inconsistent. It has no solution.

We can see directly from the system that it is inconsistent. For any values of x and y, it is impossible that 7 times x plus 4 times y could be both 8 and 6.

Self Check 5   Use Cramer's rule to solve $\begin{cases} 3x = 8 - 4y \\ y = \dfrac{5}{2} - \dfrac{3}{4}x \end{cases}$.

■ USING CRAMER'S RULE TO SOLVE A SYSTEM OF THREE EQUATIONS

Cramer's rule can be extended to solve systems of three linear equations with three variables.

Cramer's Rule for Three Equations with Three Variables

The solution of the system $\begin{cases} ax + by + cz = j \\ dx + ey + fz = k \\ gx + hy + iz = l \end{cases}$ is given by

$$x = \frac{D_x}{D}, \qquad y = \frac{D_y}{D}, \qquad \text{and} \qquad z = \frac{D_z}{D}$$

where

$$D = \begin{vmatrix} a & b & c \\ d & e & f \\ g & h & i \end{vmatrix} \quad D_x = \begin{vmatrix} j & b & c \\ k & e & f \\ l & h & i \end{vmatrix} \quad D_y = \begin{vmatrix} a & j & c \\ d & k & f \\ g & l & i \end{vmatrix} \quad D_z = \begin{vmatrix} a & b & j \\ d & e & k \\ g & h & l \end{vmatrix}$$

If every determinant is 0, the system is consistent, but the equations are dependent.

If $D = 0$ and D_x or D_y or D_z is nonzero, the system is inconsistent. If $D \neq 0$, the system is consistent, and the equations are independent.

EXAMPLE 6

Use Cramer's rule to solve

$$\begin{cases} 2x + y + 4z = 12 \\ x + 2y + 2z = 9 \\ 3x - 3y - 2z = 1 \end{cases}.$$

Solution   The denominator determinant D is the determinant formed by the coefficients of the variables. The numerator determinants, D_x, D_y, and D_z, are formed by replacing the coefficients of the variable being solved for by the column of constants. We form the quotients for x, y, and z and evaluate each determinant by expanding by minors about the first row:

$$x = \frac{D_x}{D} = \frac{\begin{vmatrix} 12 & 1 & 4 \\ 9 & 2 & 2 \\ 1 & -3 & -2 \end{vmatrix}}{\begin{vmatrix} 2 & 1 & 4 \\ 1 & 2 & 2 \\ 3 & -3 & -2 \end{vmatrix}} = \frac{12\begin{vmatrix} 2 & 2 \\ -3 & -2 \end{vmatrix} - 1\begin{vmatrix} 9 & 2 \\ 1 & -2 \end{vmatrix} + 4\begin{vmatrix} 9 & 2 \\ 1 & -3 \end{vmatrix}}{2\begin{vmatrix} 2 & 2 \\ -3 & -2 \end{vmatrix} - 1\begin{vmatrix} 1 & 2 \\ 3 & -2 \end{vmatrix} + 4\begin{vmatrix} 1 & 2 \\ 3 & -3 \end{vmatrix}}$$

$$= \frac{12(2) - 1(-20) + 4(-29)}{2(2) - 1(-8) + 4(-9)} = \frac{-72}{-24} = 3$$

$$y = \frac{D_y}{D} = \frac{\begin{vmatrix} 2 & 12 & 4 \\ 1 & 9 & 2 \\ 3 & 1 & -2 \end{vmatrix}}{\begin{vmatrix} 2 & 1 & 4 \\ 1 & 2 & 2 \\ 3 & -3 & -2 \end{vmatrix}} = \frac{2\begin{vmatrix} 9 & 2 \\ 1 & -2 \end{vmatrix} - 12\begin{vmatrix} 1 & 2 \\ 3 & -2 \end{vmatrix} + 4\begin{vmatrix} 1 & 9 \\ 3 & 1 \end{vmatrix}}{-24}$$

$$= \frac{2(-20) - 12(-8) + 4(-26)}{-24} = \frac{-48}{-24} = 2$$

$$z = \frac{D_z}{D} = \frac{\begin{vmatrix} 2 & 1 & 12 \\ 1 & 2 & 9 \\ 3 & -3 & 1 \end{vmatrix}}{\begin{vmatrix} 2 & 1 & 4 \\ 1 & 2 & 2 \\ 3 & -3 & -2 \end{vmatrix}} = \frac{2\begin{vmatrix} 2 & 9 \\ -3 & 1 \end{vmatrix} - 1\begin{vmatrix} 1 & 9 \\ 3 & 1 \end{vmatrix} + 12\begin{vmatrix} 1 & 2 \\ 3 & -3 \end{vmatrix}}{-24}$$

$$= \frac{2(29) - 1(-26) + 12(-9)}{-24} = \frac{-24}{-24} = 1$$

The solution of this system is (3, 2, 1). Verify that it satisfies the three original equations.

Self Check 6   Use Cramer's rule to solve $\begin{cases} x + y + 2z = 6 \\ 2x - y + z = 9 \\ x + y - 2z = -6 \end{cases}$.

Answers to Self Checks   **1.** 10   **2.** -44 **3.** -44 **4.** $(-2, 4)$ **5.** no solution   **6.** $(2, -2, 3)$

12.4   **STUDY SET**   ELEMENTARY & INTERMEDIATE **Algebra** $f(x)$ **Now**™

VOCABULARY   Fill in the blanks.

1. A determinant is a _____ that is associated with a square matrix.

2. $\begin{vmatrix} 2 & 1 \\ -6 & 1 \end{vmatrix}$ is a 2 × 2 _____.

3. The _____ of b_1 in $\begin{vmatrix} a_1 & b_1 & c_1 \\ a_2 & b_2 & c_2 \\ a_3 & b_3 & c_3 \end{vmatrix}$ is $\begin{vmatrix} a_2 & c_2 \\ a_3 & c_3 \end{vmatrix}$.

4. In $\begin{vmatrix} 7 & -3 \\ 1 & 2 \end{vmatrix}$, 7 and 2 lie along the main _____.

5. A 3 × 3 determinant has 3 _____ and 3 _____.

6. _____ rule uses determinants to solve systems of linear equations.

CONCEPTS   Fill in the blanks.

7. If the denominator determinant D for a system of equations is zero, the equations of the system are _____ or the system is _____.

8. To find the minor of 5, we _____ the elements of the determinant that are in the same row and column as 5.

$$\begin{vmatrix} 3 & 5 & 1 \\ 6 & 2 & 2 \\ 8 & 1 & 4 \end{vmatrix}$$

9. $\begin{vmatrix} a & b \\ c & d \end{vmatrix} = $ $-$ ▨

10. $\begin{vmatrix} 5 & 1 & -1 \\ 8 & 7 & 4 \\ 9 & 7 & 6 \end{vmatrix} = -1 \begin{vmatrix} 8 & 7 \\ 9 & 7 \end{vmatrix} - 4 \begin{vmatrix} 5 & 1 \\ 9 & 7 \end{vmatrix} + 6 \begin{vmatrix} 5 & 1 \\ 8 & 7 \end{vmatrix}$

In evaluating this determinant, about what row or column was it expanded?

11. What is the denominator determinant D for the system $\begin{cases} 3x + 4y = 7 \\ 2x - 3y = 5 \end{cases}$?

12. What is the denominator determinant D for the system $\begin{cases} x + 2y = -8 \\ 3x + y - z = -2 \\ 8x + 4y - z = 6 \end{cases}$?

13. For the system $\begin{cases} 3x + 2y = 1 \\ 4x - y = 3 \end{cases}$, $D_x = -7, D_y = 5$, and $D = -11$. What is the solution of the system?

14. For the system $\begin{cases} 2x + 3y - z = -8 \\ x - y - z = -2 \\ -4x + 3y + z = 6 \end{cases}$, $D_x = -28$, $D_y = -14, D_z = 14$, and $D = 14$. What is the solution?

NOTATION Complete the evaluation of each determinant.

15. $\begin{vmatrix} 5 & -2 \\ -2 & 6 \end{vmatrix} = 5(\quad) - (-2)(-2)$

$\qquad = \boxed{} - 4$

$\qquad = 26$

16. $\begin{vmatrix} 2 & 1 & 3 \\ 3 & 4 & 2 \\ 1 & 5 & 3 \end{vmatrix}$

$= 2 \begin{vmatrix} 4 & \\ 5 & 3 \end{vmatrix} - 1 \begin{vmatrix} 3 & 2 \\ & 3 \end{vmatrix} + 3 \begin{vmatrix} 3 & 4 \\ 1 & \end{vmatrix}$

$= 2(\boxed{} - 10) - 1(9 - \boxed{}) + 3(15 - \boxed{})$

$= 2(2) - 1(\boxed{}) + \boxed{}(11)$

$= 4 - 7 + \boxed{}$

$= 30$

PRACTICE Evaluate each determinant.

17. $\begin{vmatrix} 2 & 3 \\ -2 & 1 \end{vmatrix}$

18. $\begin{vmatrix} 3 & -2 \\ -2 & 4 \end{vmatrix}$

19. $\begin{vmatrix} -1 & 2 \\ 3 & -4 \end{vmatrix}$

20. $\begin{vmatrix} -1 & -2 \\ -3 & -4 \end{vmatrix}$

21. $\begin{vmatrix} 10 & 0 \\ 1 & 20 \end{vmatrix}$

22. $\begin{vmatrix} 1 & 15 \\ 15 & 0 \end{vmatrix}$

23. $\begin{vmatrix} -6 & -2 \\ 15 & 4 \end{vmatrix}$

24. $\begin{vmatrix} 3 & -2 \\ 12 & -8 \end{vmatrix}$

25. $\begin{vmatrix} 1 & 2 & 0 \\ 0 & 1 & 2 \\ 0 & 0 & 1 \end{vmatrix}$

26. $\begin{vmatrix} -1 & 2 & 1 \\ 2 & 1 & -3 \\ 1 & 1 & 1 \end{vmatrix}$

27. $\begin{vmatrix} 1 & -2 & 3 \\ -2 & 1 & 1 \\ -3 & -2 & 1 \end{vmatrix}$

28. $\begin{vmatrix} 1 & 1 & 2 \\ 2 & 1 & -2 \\ 3 & 1 & 3 \end{vmatrix}$

29. $\begin{vmatrix} 1 & 0 & 1 \\ 0 & 1 & 0 \\ 1 & 1 & 1 \end{vmatrix}$

30. $\begin{vmatrix} 3 & 5 & 1 \\ 6 & -2 & 2 \\ 8 & -1 & 4 \end{vmatrix}$

31. $\begin{vmatrix} 1 & 2 & 1 \\ -3 & 7 & 3 \\ -4 & 3 & -5 \end{vmatrix}$

32. $\begin{vmatrix} 1 & 4 & 7 \\ 2 & 5 & 8 \\ 3 & 6 & 9 \end{vmatrix}$

Use Cramer's rule to solve each system of equations, if possible. If a system is inconsistent or if the equations are dependent, so indicate.

33. $\begin{cases} x + y = 6 \\ x - y = 2 \end{cases}$

34. $\begin{cases} x - y = 4 \\ 2x + y = 5 \end{cases}$

35. $\begin{cases} 2x + 3y = 0 \\ 4x - 6y = -4 \end{cases}$

36. $\begin{cases} 4x - 3y = -1 \\ 8x + 3y = 4 \end{cases}$

37. $\begin{cases} 3x + 2y = 11 \\ 6x + 4y = 11 \end{cases}$

38. $\begin{cases} 5x + 6y = 12 \\ 10x + 12y = 24 \end{cases}$

39. $\begin{cases} y = \dfrac{-2x + 1}{3} \\ 3x - 2y = 8 \end{cases}$

40. $\begin{cases} 2x + 3y = -1 \\ x = \dfrac{y - 9}{4} \end{cases}$

41. $\begin{cases} x + y + z = 4 \\ x + y - z = 0 \\ x - y + z = 2 \end{cases}$

42. $\begin{cases} x + y + z = 4 \\ x - y + z = 2 \\ x - y - z = 0 \end{cases}$

43. $\begin{cases} x + y + 2z = 7 \\ x + 2y + z = 8 \\ 2x + y + z = 9 \end{cases}$

44. $\begin{cases} x + 2y + 2z = 10 \\ 2x + y + 2z = 9 \\ 2x + 2y + z = 1 \end{cases}$

ACCENT ON TEAMWORK

INTERSECTION POINTS ON GRAPHS

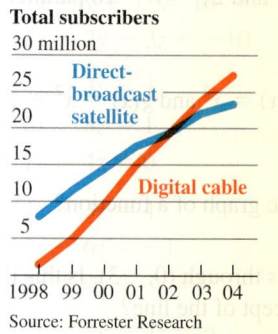

Total subscribers

30 million

Direct-broadcast satellite

Digital cable

1998 99 00 01 02 03 04

Source: Forrester Research

Overview: This activity will improve your ability to read and interpret graphs.

Instructions: Each student in the class should find a graph that involves intersecting lines. (See the example shown here.) Your school library is a good resource to find such graphs. Ask to look through the collection of recent magazines and newspapers, or scan encyclopedias and books from other disciplines such as nursing and science. You might also use the Internet to find a graph.

Form groups of 5 or 6 students. Have each student show his or her graph to the group and explain the information that is given by the point (or points) of intersection of the lines in the graph. After everyone has taken their turn, vote to determine which graph is the most interesting. The winner from each group should then present his or her graph to the entire class.

BREAK-POINT ANALYSIS

Overview: In this activity, you are to interpret a graph that contains a break point and submit your observations in writing in the form of a financial report.

Instructions: Form groups of 2 or 3 students. Suppose you are a financial analyst for a coathanger company. The setup cost of a machine that makes wooden coathangers is $400. After setup, it costs $1.50 to make each hanger (the unit cost). Management is considering the purchase of a new machine that can manufacture the same type of coathanger at a cost of $1.25 per hanger. If the setup cost of the new machine is $500, find the number of coathangers that the company would need to manufacture to make the cost the same using either machine. This is called the **break point.**

Then write a brief report that could be given to company managers, explaining their options concerning the purchase of the new machine. Under what conditions should they keep the machine currently in use? Under what conditions should they buy the new machine?

METHODS OF SOLUTION

Overview: In this activity, you will explore the advantages and disadvantages of several methods for solving a system of linear equations.

Instructions: Form groups of 5 students. Have each member of your group solve the system

$$\begin{cases} x - y = 4 \\ 2x + y = 5 \end{cases}$$

in a different way. The methods to use are graphing, substitution, elimination, matrices, and Cramer's rule. Have each person briefly explain his or her method of solution to the group. After everyone has presented a solution, discuss the advantages and drawbacks of each method. Then rank the five methods, from most desirable to least desirable.

KEY CONCEPT: SYSTEMS OF EQUATIONS

In Chapter 12, we have solved problems involving two and three variables by writing and solving a **system of equations.**

SOLUTIONS OF A SYSTEM OF EQUATIONS

A solution of a system of equations involving two or three variables is an ordered pair or an ordered triple whose coordinates satisfy each equation of the system. In Exercises 1 and 2, decide whether the given ordered pair or ordered triple is a solution of the system.

1. $\begin{cases} 2x - y = 1 \\ 4x + 2y = 0 \end{cases}, \left(\dfrac{1}{4}, -\dfrac{1}{2}\right)$

2. $\begin{cases} 2x - y + z = 9 \\ 3x + y - 4z = 8, \ (4, 0, 1) \\ 2x - 7z = -1 \end{cases}$

METHODS OF SOLVING SYSTEMS OF LINEAR EQUATIONS

There are several methods for solving systems of two and three linear equations.

3. Solve $\begin{cases} 2x + 5y = 8 \\ y = 3x + 5 \end{cases}$ using the *graphing method.*

4. Solve $\begin{cases} 9x - 8y = 1 \\ 6x + 12y = 5 \end{cases}$ using the *elimination method.*

5. Solve $\begin{cases} 4x - y - 10 = 0 \\ 3x + 5y = 19 \end{cases}$ using the *substitution* method.

6. Solve $\begin{cases} -x + 3y + 2z = 5 \\ 3x + 2y + z = -1 \\ 2x - y + 3z = 4 \end{cases}$ using the *elimination* method.

7. Solve $\begin{cases} x - 6y = 3 \\ x + 3y = 21 \end{cases}$ using *matrices.*

8. Solve $\begin{cases} x + 2z = 7 \\ 2x - y + 3z = 9 \\ y - z = 1 \end{cases}$ using *Cramer's rule.*

DEPENDENT EQUATIONS AND INCONSISTENT SYSTEMS

If the equations in a system of two linear equations are dependent, the system has infinitely many solutions. An inconsistent system has no solutions.

9. Suppose you are solving a system of two equations by the elimination method, and you obtain the following.

$$2x - 3y = 4$$
$$\underline{-2x + 3y = -4}$$
$$0 = 0$$

What can you conclude?

10. Suppose you are solving a system of two equations by the substitution method, and you obtain

$$-2(x - 3) + 2x = 7$$
$$\underline{-2x + 6 + 2x = 7}$$
$$6 = 7$$

What can you conclude?

CHAPTER REVIEW

ELEMENTARY & INTERMEDIATE
Algebra $f(x)$ **Now**™

Systems with Two Variables

CONCEPTS

The graph of a linear equation is the graph of all points (x, y) on the rectangular coordinate system whose coordinates satisfy the equation.

A *solution* of a system of equations is an ordered pair that satisfies both equations of the system.

To solve a system *graphically:*

1. Graph each equation on the same rectangular coordinate system.
2. Determine the coordinates of the point of intersection of the graphs. That ordered pair is the solution.
3. Check the proposed solution in each equation of the original system.

A system of equations that has at least one solution is called a *consistent system.* If the graphs are parallel lines, the system has no solution, and it is called an *inconsistent system.*

Equations with different graphs are called *independent equations.* If the graphs are the same line, the system has an infinite number of solutions. The equations are called *dependent equations.*

REVIEW EXERCISES

1. See the illustration.
   a. Give three points that satisfy the equation $2x + y = 5$.

   b. Give three points that satisfy the equation $x - y = 4$.

   c. What is the solution of $\begin{cases} 2x + y = 5 \\ x - y = 4 \end{cases}$?

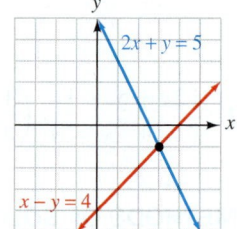

2. POLITICS   Explain the importance of the points of intersection of the graphs shown below.

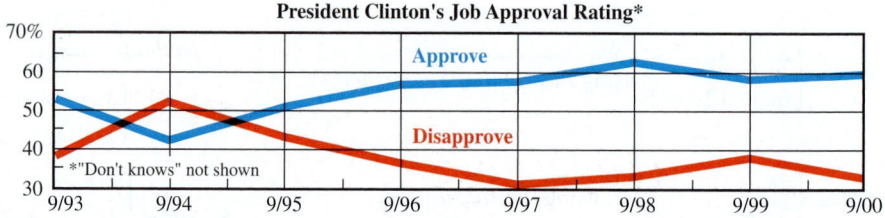

Solve each system by the graphing method, if possible. If a system is inconsistent or if the equations are dependent, so indicate.

3. $\begin{cases} 2x + y = 11 \\ -x + 2y = 7 \end{cases}$

4. $\begin{cases} y = -\dfrac{3}{2}x \\ 2x - 3y + 13 = 0 \end{cases}$

5. $\begin{cases} \dfrac{1}{2}x + \dfrac{1}{3}y = 2 \\ y = 6 - \dfrac{3}{2}x \end{cases}$

6. $\begin{cases} \dfrac{x}{3} - \dfrac{y}{2} = 1 \\ 6x - 9y = 3 \end{cases}$

Use the graphs in the illustration to solve each equation. Check each answer.

7. $2(2 - x) + x = x$

8. $2(2 - x) + x = 5$

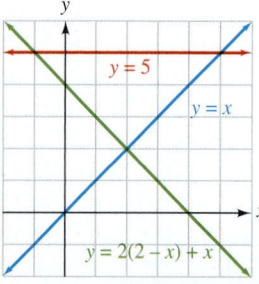

To solve a system by the *substitution method*:

1. Solve one equation for one of its variables.
2. Substitute the resulting expression for that variable into the other equation and solve that equation.
3. Find the value of the other variable by substituting the value of the variable found in step 2 into the equation from step 1.

To solve a system by the *elimination method*:

1. Write both equations in general form: $Ax + By = C$.
2. Multiply the terms of one or both equations by constants so that the coefficients of one variable differ only in sign.
3. Add the equations from step 2 and solve the resulting equation.
4. Substitute the value obtained in step 3 into either original equation and solve for the remaining variable.

Solve each system using the substitution method, if possible. If a system is inconsistent or if the equations are dependent, so indicate.

9. $\begin{cases} x = y - 4 \\ 2x + 3y = 7 \end{cases}$

10. $\begin{cases} y = 2x + 5 \\ 3x - 5y = -4 \end{cases}$

11. $\begin{cases} 0.1x + 0.2y = 1.1 \\ 2x - y = 2 \end{cases}$

12. $\begin{cases} x = -2 - 3y \\ -2x - 6y = 4 \end{cases}$

Solve each system using the elimination method, if possible.

13. $\begin{cases} x + y = -2 \\ 2x + 3y = -3 \end{cases}$

14. $\begin{cases} 2x - 3y = 5 \\ 2x - 3y = 8 \end{cases}$

15. $\begin{cases} x + \dfrac{1}{2}y = 7 \\ -2x = 3y - 6 \end{cases}$

16. $\begin{cases} y = \dfrac{x - 3}{2} \\ x = \dfrac{2y + 7}{2} \end{cases}$

17. To solve $\begin{cases} 5x - 2y = 19 \\ 3x + 4y = 1 \end{cases}$, which method, addition or substitution, would you use?
    Explain why. Using the addition method, the computations are easier.

18. Estimate the solution of the system $\begin{cases} y = -\dfrac{2}{3}x \\ 2x - 3y = -4 \end{cases}$
    from the graphs in the illustration. Then solve the system algebraically.

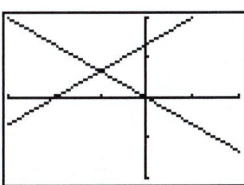

Use two equations to solve each problem.

19. MAPS   See the illustration. The distance between Austin and Houston is 4 miles less than twice the distance between Austin and San Antonio. The round trip from Houston to Austin to San Antonio and back to Houston is 442 miles. Determine the mileages between Austin and Houston and between Austin and San Antonio.

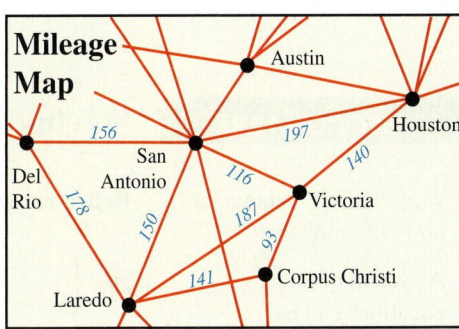

20. RIVERBOATS   A Mississippi riverboat travels 30 miles downstream in three hours and then makes the return trip upstream in five hours. Find the speed of the riverboat in still water and the speed of the current.

21. BREAK POINTS   A bottling company is considering purchasing a new piece of equipment for their production line. The machine they currently use has a setup cost of $250 and a cost of $0.04 per bottle. The new machine has a setup cost of $600 and a cost of $0.02 per bottle. Find the break point.

<table>
<tr><td>

The solution of a system of three linear equations is an *ordered triple*.

To solve a system of linear equations with three variables:

1. Pick any two equations and eliminate a variable.

2. Pick a different pair of equations and eliminate the same variable.

3. Solve the resulting pair of equations.

4. Use substitution to find the value of the third variable.

</td></tr>
</table>

Systems with Three Variables

22. Determine whether $(2, -1, 1)$ is a solution of the system $\begin{cases} x - y + z = 4 \\ x + 2y - z = -1. \\ x + y - 3z = -1 \end{cases}$

Solve each system, if possible.

23. $\begin{cases} x + y + z = 6 \\ x - y - z = -4 \\ -x + y - z = -2 \end{cases}$

24. $\begin{cases} 2x + 3y + z = -5 \\ -x + 2y - z = -6 \\ 3x + y + 2z = 4 \end{cases}$

25. $\begin{cases} x + y - z = -3 \\ x + z = 2 \\ 2x - y + 2z = 3 \end{cases}$

26. $\begin{cases} 3x + 3y + 6z = -6 \\ -x - y - 2z = 2 \\ 2x + 2y + 4z = -4 \end{cases}$

27. A system of three linear equations in three variables is graphed on the right. Does the system have a solution? If so, how many solutions does it have?

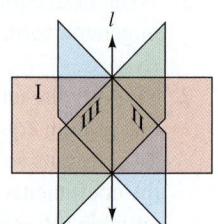

28. MIXING NUTS   The owner of a produce store wanted to mix peanuts selling for $3 per pound, cashews selling for $9 per pound, and Brazil nuts selling for $9 per pound to get 50 pounds of a mixture that would sell for $6 per pound. She used 15 fewer pounds of cashews than peanuts. How many pounds of each did she use?

Solving Systems Using Matrices

A *matrix* is a rectangular array of numbers.

A system of linear equations can be represented by an *augmented matrix*.

Represent each system of equations using an augmented matrix.

29. $\begin{cases} 5x + 4y = 3 \\ x - y = -3 \end{cases}$

30. $\begin{cases} x + 2y + 3z = 6 \\ x - 3y - z = 4 \\ 6x + y - 2z = -1 \end{cases}$

Solve each system using matrices, if possible.

31. $\begin{cases} x - y = 4 \\ 3x + 7y = -18 \end{cases}$

32. $\begin{cases} x + 2y - 3z = 5 \\ x + y + z = 0 \\ 3x + 4y + 2z = -1 \end{cases}$

Systems of linear equations can be solved using *Gaussian elimination* and *elementary row operations:*

1. Any two rows can be interchanged.

2. Any row can be multiplied by a nonzero constant.

3. Any row can be changed by adding a nonzero constant multiple of another row to it.

33. $\begin{cases} 16x - 8y = 32 \\ -2x + y = -4 \end{cases}$

34. $\begin{cases} x + 2y + 2z = 2 \\ 4x + 5y + 3z = 3 \\ 2x + y - z = 1 \end{cases}$

35. INVESTING   One year, a couple invested a total of $10,000 in two projects. The first investment, a mini-mall, made a 6% profit. The other investment, a skateboard park, made a 12% profit. If their investments made $960, how much was invested at each rate? To answer this question, write a system of two equations and solve it using matrices.

| **SECTION 12.4** | **Solving Systems Using Determinants** |

A *determinant* of a *square matrix* is a number.

To evaluate a 2×2 determinant:

$$\begin{vmatrix} a & b \\ c & d \end{vmatrix} = ad - bc$$

To evaluate a 3×3 determinant, we expand it by *minors* along any row or column using the *array of signs.*

Cramer's rule can be used to solve systems of linear equations.

Evaluate each determinant.

36. $\begin{vmatrix} 2 & 3 \\ -4 & 3 \end{vmatrix}$

37. $\begin{vmatrix} -3 & -4 \\ 5 & -6 \end{vmatrix}$

38. $\begin{vmatrix} -1 & 2 & -1 \\ 2 & -1 & 3 \\ 1 & -2 & 2 \end{vmatrix}$

39. $\begin{vmatrix} 3 & -2 & 2 \\ 1 & -2 & -2 \\ 2 & 1 & -1 \end{vmatrix}$

Use Cramer's rule to solve each system, if possible.

40. $\begin{cases} 3x + 4y = 10 \\ 2x - 3y = 1 \end{cases}$

41. $\begin{cases} -6x - 4y = -6 \\ 3x + 2y = 5 \end{cases}$

42. $\begin{cases} x + 2y + z = 0 \\ 2x + y + z = 3 \\ x + y + 2z = 5 \end{cases}$

43. $\begin{cases} 2x + 3y + z = 2 \\ x + 3y + 2z = 7 \\ x - y - z = -7 \end{cases}$

44. VETERINARY MEDICINE   The daily requirements of a balanced diet for an animal are shown in the following nutritional pyramid. The number of grams per cup of nutrients in three food mixes are shown in the table. How many cups of each mix should be used to meet the daily requirements for protein, carbohydrates, and essential fatty acids in the animal's diet? To answer this problem, write a system of three equations and solve it using Cramer's rule.

Vitamins

Minerals

Essential fatty acids: 5 grams

Carbohydrates: 10 grams

Quality protein: 24 grams

	Grams per cup		
	Protein	**Carbohydrates**	**Fatty acids**
Mix A	5	2	1
Mix B	6	3	2
Mix C	8	3	1

CHAPTER 12 TEST

ELEMENTARY &
INTERMEDIATE
Algebra $f(x)$ **Now**™

1. Solve $\begin{cases} 2x + y = 5 \\ y = 2x - 3 \end{cases}$ by graphing.

2. Use the graphs in the illustration to solve
$3(x - 2) - 2(-2 + x) = 1$.

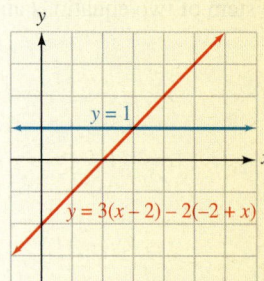

3. Use substitution to solve $\begin{cases} 2x - 4y = 14 \\ x + 2y = 7 \end{cases}$.

4. Use elimination to solve $\begin{cases} 2x + 3y = -5 \\ 3x - 2y = 12 \end{cases}$.

5. Are the equations of the system

$$\begin{cases} 3(x + y) = x - 3 \\ -y = \dfrac{2x + 3}{3} \end{cases}$$

dependent or independent?

6. Is $\left(-1, -\dfrac{1}{2}, 5\right)$ a solution of $\begin{cases} x - 2y + z = 4 \\ 2x + 4y = -4 \\ -6y + 4z = 22 \end{cases}$?

7. Solve the system $\begin{cases} x + y + z = 4 \\ x + y - z = 6 \\ 2x - 3y + z = -1 \end{cases}$
using elimination.

Write a system of equations to solve each problem.

8. In the sign, find x and y, if y is 15 more than x.

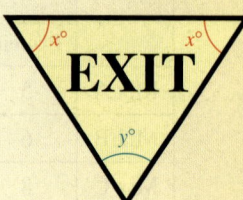

9. **ANTIFREEZE** How much of a 40% antifreeze solution must a mechanic mix with an 80% antifreeze solution if 20 gallons of a 50% antifreeze solution are needed?

10. **BREAK POINTS** A metal stamping plant is considering purchasing a new piece of equipment. The machine they currently use has a setup cost of $1,775 and a cost of $5.75 per impression. The new machine has a setup cost of $3,975 and a cost of $4.15 per impression. Find the break point.

Use matrices to solve each system, if possible.

11. $\begin{cases} x + y = 4 \\ 2x - y = 2 \end{cases}$

12. $\begin{cases} 2x + y - z = 1 \\ x + 2y + 2z = 2 \\ 4x + 5y + 3z = 3 \end{cases}$

Evaluate each determinant.

13. $\begin{vmatrix} 2 & -3 \\ 4 & 5 \end{vmatrix}$

14. $\begin{vmatrix} 1 & 2 & 0 \\ 2 & 0 & 3 \\ 1 & -2 & 2 \end{vmatrix}$

Consider the system $\begin{cases} x - y = -6 \\ 3x + y = -6 \end{cases}$, **which is to be solved using Cramer's rule.**

15. **a.** When solving for x, what is the numerator determinant D_x? (**Don't evaluate it.**)

     b. When solving for y, what is the denominator determinant D? (**Don't evaluate it.**)

16. Solve the system for x:

17. Solve the system for y:

18. Solve the following system for z only, using Cramer's rule.

$$\begin{cases} x + y + z = 4 \\ x + y - z = 6 \\ 2x - 3y + z = -1 \end{cases}$$

19. MOVIE TICKETS   The receipts for one showing of a movie were $410 for an audience of 100 people. The ticket prices are given in the table. If twice as many children's tickets as general admission tickets were purchased, how many of each type of ticket were sold?

Ticket prices	
Children	$3
General Admission	$6
Seniors	$5

20. Which method, substitution or elimination, would you use to solve the following system? Explain your reasoning.

$$\begin{cases} \dfrac{x}{2} - \dfrac{y}{3} = -4 \\ y = -2 - x \end{cases}$$

21. What does it mean to say that a system of two linear equations in two variables is an *inconsistent* system?

22. POPULATION PROJECTIONS   See the illustration. If the population trends for the years 2005–2015 continue as projected, estimate the point of intersection of the graphs. Interpret your answer.

Children under age 18 and adults 65 and older as a percent of the U.S. population

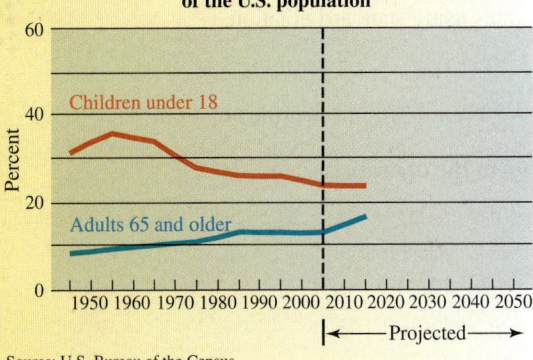

Source: U.S. Bureau of the Census

CHAPTERS 1–12   CUMULATIVE REVIEW EXERCISES

1. True or false:
  a. Every rational number can be written as a ratio of two integers.
  b. The set of real numbers corresponds to all points on the number line.
  c. The whole numbers and their opposites form the set of integers.

2. Evaluate: $-4 + 2[-7 - 3(-9)]$.

3. EMPLOYMENT   The following newspaper headline appeared in early 2000. How many employees did Xerox have at that time?

> **Xerox to Cut 5,200 Jobs, or 5.3% of Workforce, on Falling Profits**

4. Solve: $\dfrac{5}{6}k = 10$.

5. Solve: $-(3a + 1) + a = 2$.

6. Solve: $\dfrac{2z + 3}{3} + \dfrac{3z - 4}{6} = \dfrac{z - 2}{2}$.

7. Solve $5x + 7 < 2x + 1$ and graph the solution set. Then use interval notation to describe the solution.

8. Graph the line $y = -3$.

9. Graph the line passing through $(-2, -1)$ and having slope $\dfrac{4}{3}$.

10. Write the equation of the line whose graph has slope $m = \dfrac{1}{4}$ and passes through the point $(8, 1)$. Answer in slope–intercept form.

11. INSURANCE   The graph in red approximates the average annual expenditure on homeowner's insurance in the U.S. for the years 1995–2002. Find the rate of increase over this period of time.

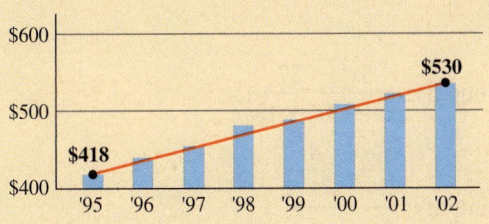

Source: Insurance Information Institute

Simplify each expression. Write each answer without using parentheses or negative exponents.

12. $y^3(y^2y^4)^3$

13. $\left(\dfrac{21x^{-2}y^2z^{-2}}{7x^3y^{-1}}\right)^{-2}$

14. FIVE-CARD POKER   The odds against being dealt the hand shown are about 2.6×10^6 to 1. Express the odds using standard notation.

15. Write 0.00073 in scientific notation.

Perform the operations.

16. $4(4x^3 + 2x^2 - 3x - 8) - 5(2x^3 - 3x + 8)$

17. $(-2a^3)(3a^2)$

18. $(2b - 1)(3b + 4)$

19. $(2x + 5y)^2$

20. $(3x + y)(2x^2 - 3xy + y^2)$

21. $x - 3\overline{)2x^2 - 3 - 5x}$

Factor each expression completely.

22. $6a^2 - 12a^3b + 36ab$ **23.** $2x + 2y + ax + ay$

24. $b^3 + 125$ **25.** $t^4 - 16$

Solve each equation.

26. $3x^2 + 8x = 0$ **27.** $15x^2 - 2 = 7x$

Perform the operations.

28. $\dfrac{x^2 - x - 6}{2x^2 + 9x + 10} \div \dfrac{x^2 - 25}{2x^2 + 15x + 25}$

29. $\dfrac{x + 5}{xy} - \dfrac{x - 1}{x^2y}$

30. Simplify: $\dfrac{3x^2 - 27}{x^2 + 3x - 18}$.

31. Simplify: $\dfrac{\dfrac{5}{y} + \dfrac{4}{y + 1}}{\dfrac{4}{y} - \dfrac{5}{y + 1}}$

32. Solve: $\dfrac{7}{q^2 - q - 2} + \dfrac{1}{q + 1} = \dfrac{3}{q - 2}$

33. ROOFING   A homeowner estimates that it will take him 7 days to roof his house. A professional roofer estimates that he can roof the house in 4 days. How long will it take if the homeowner helps the roofer?

34. LOSING WEIGHT   If a person cuts his or her daily calorie intake by 100, it will take 350 days for that person to lose 10 pounds. How long will it take for the person to lose 25 pounds?

Solve each system using the method indicated.

35. Graphing
$\begin{cases} x + y = 1 \\ y = x + 5 \end{cases}$

36. Substitution
$\begin{cases} y = 2x + 5 \\ x + 2y = -5 \end{cases}$

37. Elimination
$\begin{cases} \dfrac{3}{5}s + \dfrac{4}{5}t = 1 \\ -\dfrac{1}{4}s + \dfrac{3}{8}t = 1 \end{cases}$

38. MIXING CANDY   How many pounds of each candy must be mixed to obtain 48 pounds of candy that would be worth $4 per pound? Use two variables to solve this problem.

Solve each inequality if possible. Write the solution set in interval notation and graph it.

39. $\left|\dfrac{x - 2}{3}\right| - 4 \le 0$

40. $3x + 2 < 8$ or $2x - 3 > 11$

Factor completely.

41. $x^2 + 4x + 4 - y^2$

42. $b^4 - 17b^2 + 16$

43. Solve the system of linear inequalities.

$$\begin{cases} 3x + 4y \ge -7 \\ 2x - 3y \ge 1 \end{cases}$$

44. If $f(x) = 3x^2 + 3x - 8$, find $f(-1)$.

45. GEARS   The speed of a gear varies inversely with the number of teeth. If a gear with 10 teeth makes 3 revolutions per second, how many revolutions per second will a gear with 25 teeth make?

46. Graph $f(x) = (x - 1)^3$ and determine the domain and range of f.

47. Use synthetic division to find
$(3x^3 - 10x^2 + 5x - 6) \div (x - 3)$.

Simplify each expression. All variables represent positive numbers.

48. $\sqrt{50x^2}$

49. $\sqrt{100a^6b^4}$

50. $3\sqrt{24} + \sqrt{54}$

51. $\sqrt{\dfrac{72x^3}{y^2}}$

52. $\sqrt[5]{x^6y^2} + \sqrt[5]{32x^6y^2} + \sqrt[5]{x^6y^2}$

53. $\sqrt[3]{\dfrac{27m^3}{8n^6}}$

54. $(-8)^{-4/3}$

Rationalize the denominator.

55. $\dfrac{2}{\sqrt[3]{a}}$

56. $\dfrac{\sqrt{x} - \sqrt{y}}{\sqrt{x} + \sqrt{y}}$

57. Solve: $2 + \sqrt{u} = \sqrt{2u + 7}$.

58. Solve $x^2 + 8x + 12 = 0$ by completing the square.

59. STORAGE CUBES   The diagonal distance across the face of each of the stacking cubes is 15 inches. What is the height of the entire storage arrangement? Round to the nearest tenth of an inch.

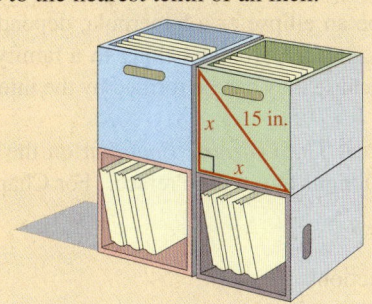

60. Solve $4x^2 - x - 2 = 0$ using the quadratic formula. Give the exact solutions, and then approximate each to the nearest hundredth.

Write each expression in terms of i.

61. $\sqrt{-49}$

62. $\sqrt{-54}$

Perform the operations. Express each answer in the form $a + bi$.

63. $(2 + 3i) - (1 - 2i)$

64. $(7 - 4i) + (9 + 2i)$

65. $(3 - 2i)(4 - 3i)$

66. $\dfrac{3 - i}{2 + i}$

Solve each equation. Express the solutions in the form $a + bi$.

67. $x^2 + 16 = 0$

68. $x^2 - 4x = -5$

69. Graph the quadratic equation $y = 2x^2 + 8x + 6$. Find the vertex, the x- and y-intercepts, and the axis of symmetry of the graph.

70. Solve: $a^{2/3} + a^{1/3} - 6 = 0$.

71. Find the inverse of $f(x) = -\dfrac{3}{2}x + 3$.

72. Let $f(x) = 3x - 2$ and $g(x) = x^2 + x$, find $(f \circ g)(-3)$.

Graph each function. Determine the domain and range.

73. $f(x) = 5^x$

74. $f(x) = \ln x$

Find each value of x.

75. $\log_x 5 = 1$

76. $\log_8 x = 2$

77. $\log_9 \dfrac{1}{81} = x$

Use logarithm properties to rewrite the expression.

78. $\ln \dfrac{y^3\sqrt{x}}{z}$.

Solve each equation. Give answers to four decimal places when necessary.

79. $5^{x-3} = 3^{2x}$

80. $\log(x + 90) = 3 - \log x$

13

Conic Sections; More Graphing

ELEMENTARY &
INTERMEDIATE
Algebra $f(x)$ **Now**™

Throughout the chapter, this icon introduces resources on the Elementary & Intermediate AlgebraNow Web site, accessed through **http://1pass. thomson.com**, that will

- Help you test your knowledge of the material with a pre-test and a post-test

- Provide a personalized learning plan targeting areas you should study

Getty Images/Image Bank

The solar system consists of the sun, the nine planets, more than 130 satellites of the planets, and a large number of comets and asteroids. The orbits of the planets are ellipses, though all except Mercury and Pluto are very nearly circular. The orbit of a comet can be an ellipse or a hyperbola, depending on the total energy of the comet. Circles, ellipses, and hyperbolas belong to a family of curves called *conic sections*. They are so named because they can be formed by the intersection of a plane and a right-circular cone.

To learn more about conic sections, visit *The Learning Equation* on the Internet at http://tle.brookscole.com. (The log-in instructions are in the Preface.) For Chapter 13, the online lesson is:

- *TLE* Lesson 22: Introduction to Conic Sections

In this chapter, we will discuss graphs that do not represent functions. Since these curves are cross sections of cones, they are called conic sections.

13.1 The Circle and the Parabola

- Conic sections   • The circle   • Problem solving using circles   • The parabola

We have previously graphed first-degree equations in two variables such as $y = 3x + 8$ and $4x - 3y = 12$. Their graphs are lines. In this section, we will graph second-degree equations in two variables such as $x^2 + y^2 = 25$ and $x = -3y^2 - 12y - 13$. The graphs of these equations are *conic sections.*

◼ CONIC SECTIONS

The curves formed by the intersection of a plane with an infinite right-circular cone are called **conic sections.** Those curves have four basic shapes, called **circles, parabolas, ellipses,** and **hyperbolas,** as shown below.

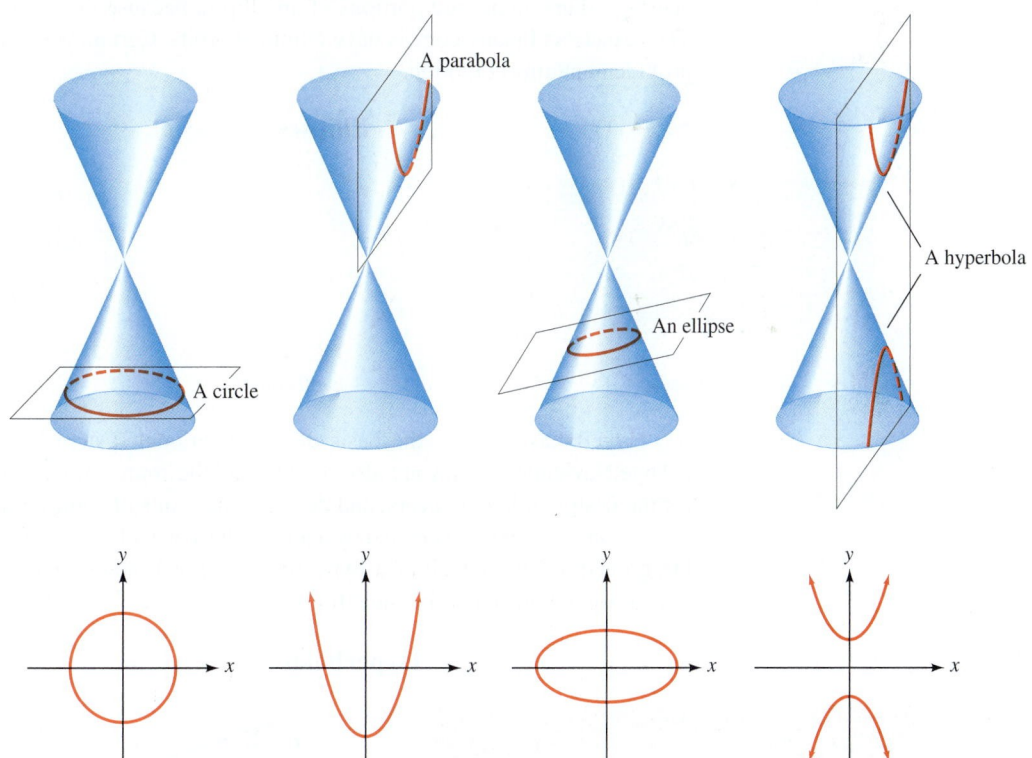

Conic sections have many applications. For example, everyone is familiar with circular wheels and gears, pizza cutters, and hula hoops.

Parabolas can be rotated to generate dish-shaped surfaces called **paraboloids.** Any light or sound placed at the **focus** of a paraboloid is reflected outward in parallel paths. This property makes parabolic surfaces ideal for flashlight and headlight reflectors. It also makes parabolic surfaces good antennas, because signals captured by such antennas are concentrated at the focus. Parabolic mirrors are capable of concentrating the rays of the sun at a single point, thereby generating tremendous heat. This property is used in the design of solar furnaces.

The Language of Algebra

Conic sections are often simply called *conics.*

Any object thrown upward and outward travels in a parabolic path. An example of this is a stream of water flowing from a drinking fountain. In architecture, many arches are parabolic in shape, because this gives them strength. Cables that support suspension bridges hang in the shape of a parabola.

Parabolas

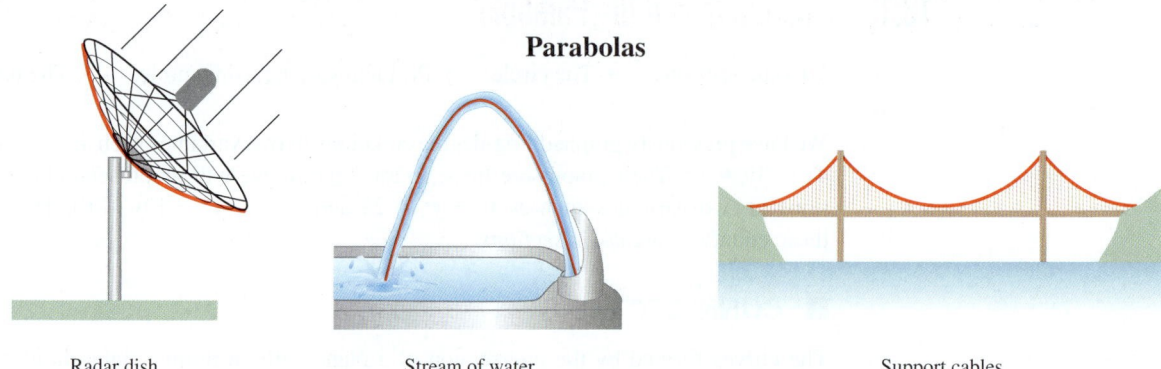

Radar dish   Stream of water   Support cables

Ellipses have optical and acoustical properties that are useful in architecture and engineering. Many arches are portions of an ellipse, because the shape is pleasing to the eye. The planets and many comets have elliptical orbits. Certain gears have elliptical shapes to provide nonuniform motion.

Ellipses

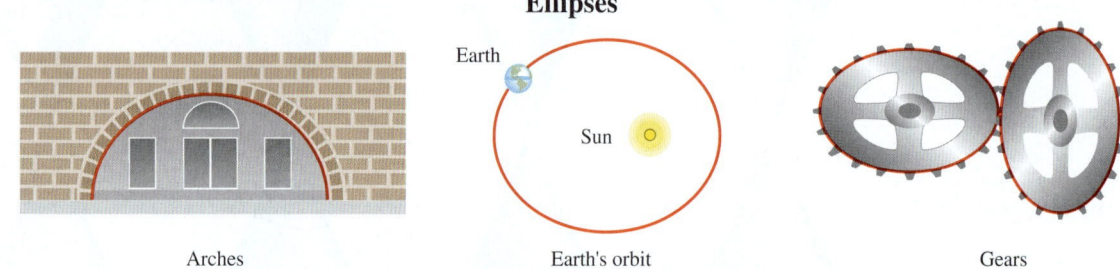

Arches   Earth's orbit   Gears

Hyperbolas serve as the basis of a navigational system known as LORAN (LOng RAnge Navigation). They are also used to find the source of a distress signal, are the basis for the design of hypoid gears, and describe the orbits of some comets.

A sonic shock wave created by a jet has the shape of a cone. The sound wave intersects the ground as one branch of a hyperbola, as shown below. The sonic boom is heard by anyone on the branch at the same time.

Hyperbolas

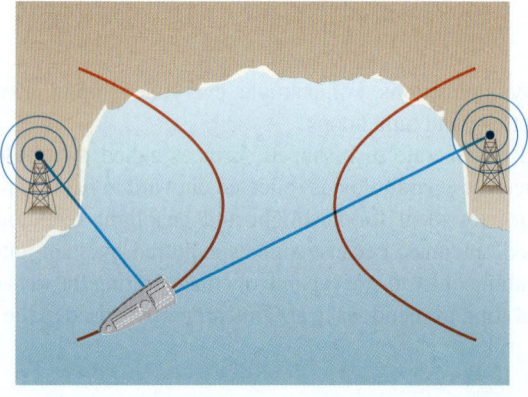
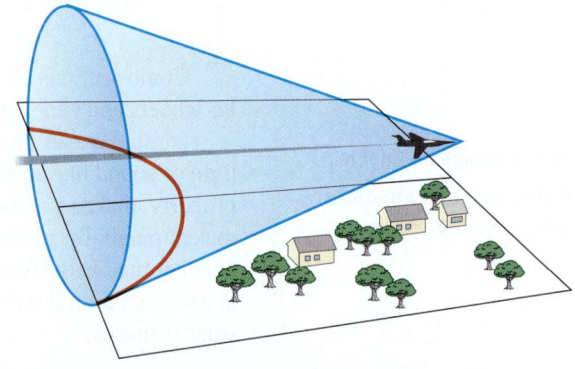

■ THE CIRCLE

Every conic section can be represented by a second-degree equation in x and y. To find the form of an equation of a circle, we use the following definition.

Definition of a Circle	A **circle** is the set of all points in a plane that are a fixed distance from a fixed point called its **center.** The fixed distance is called the **radius** of the circle.

On a rectangular coordinate system, if we let (h, k) be the center of a circle, and (x, y) be some point on the circle, then the distance from (h, k) to (x, y) is the radius of the circle, r units. We can use the distance formula to find r.

$$r = \sqrt{(x - h)^2 + (y - k)^2}$$

We then square both sides to eliminate the radical and obtain

$$r^2 = (x - h)^2 + (y - k)^2$$

This result is called the *standard form of the equation of a circle* with radius r and center at (h, k).

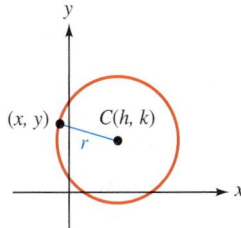

Equation of a Circle	The **standard form of the equation of a circle** with radius r and center at (h, k) is $$(x - h)^2 + (y - k)^2 = r^2$$

EXAMPLE 1

Find the center and the radius of each circle and then graph it: **a.** $(x - 4)^2 + (y - 1)^2 = 9$, **b.** $x^2 + y^2 = 25$,   and   **c.** $(x + 3)^2 + y^2 = 12$.

Solution   **a.** It is easy to determine the center and the radius of a circle when its equation is written in standard form.

Success Tip

A compass can be used to draw a circle. The distance between the compass point and pencil should be set to equal the radius.

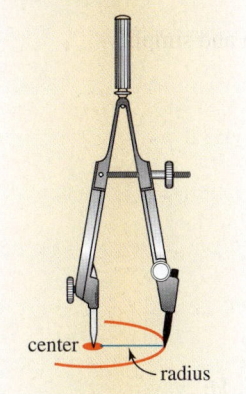

center
radius

$$(x - \mathbf{4})^2 + (y - \mathbf{1})^2 = \mathbf{9}$$

$(x - \mathbf{h})^2 + (y - \mathbf{k})^2 = \mathbf{r}^2$ $h = 4$, $k = 1$, and $r^2 = 9$. Since the radius of a circle must be positive, $r = 3$.

The center of the circle is $(h, k) = (4, 1)$ and the radius is 3.

To plot four points on the circle, we move up, down, left, and right 3 units from the center, as shown in figure (a). Then we draw a circle through the points to get the graph of $(x - 4)^2 + (y - 1)^2 = 9$, as shown in figure (b).

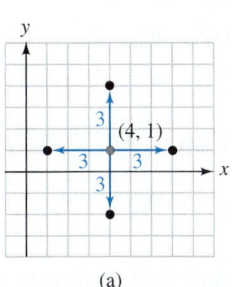

(a)

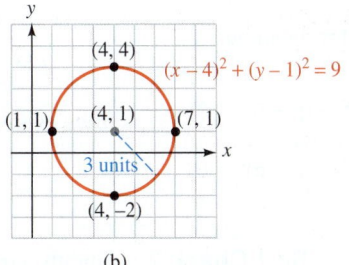

(b)

b. To determine h and k, it is helpful to write $x^2 + y^2 = 25$ in the following way:

$$(x - \mathbf{0})^2 + (y - \mathbf{0})^2 = \mathbf{25}$$
$$\uparrow \qquad\qquad \uparrow \qquad \uparrow$$
$$\mathbf{\textit{h}} \qquad\qquad \mathbf{\textit{k}} \qquad \mathbf{\textit{r}^2}$$

$h = 0, k = 0,$ and $r^2 = 25.$ Since the radius must be positive, $r = 5.$

The center of the circle is at (0, 0) and the radius is 5.

To plot four points on the circle, we move up, down, left, and right 5 units from the center. Then we draw a circle through the points to get the graph of $x^2 + y^2 = 25$, as shown.

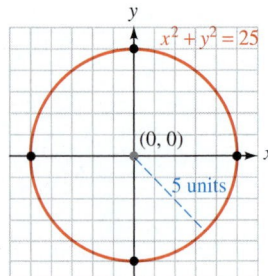

c. To determine h, it is helpful to write $x + 3$ as $x - (-3)$.

Standard form requires a minus symbol here.
$$\downarrow$$
$$[x - (\mathbf{-3})]^2 + (y - \mathbf{0})^2 = \mathbf{12}$$
$$\uparrow \qquad\qquad\quad \uparrow \qquad \uparrow$$
$$\mathbf{\textit{h}} \qquad\qquad\quad \mathbf{\textit{k}} \qquad \mathbf{\textit{r}^2}$$

$h = -3, k = 0,$ and $r^2 = 12.$

If $r^2 = 12$, by the square root property

$$r = \pm\sqrt{12} = \pm 2\sqrt{3}$$

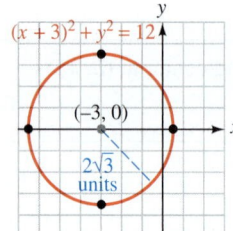

Since the radius can't be negative, we get $r = 2\sqrt{3}$. The center of the circle is at $(-3, 0)$ and the radius is $2\sqrt{3}$.

To plot four points on the circle, we move up, down, left, and right $2\sqrt{3} \approx 3.5$ units from the center. Then we draw a circle through the points to get the graph of $(x + 3)^2 + y^2 = 12$, as shown.

Self Check 1   Find the center and the radius of each circle and then graph it: **a.** $(x - 3)^2 + (y + 4)^2 = 4$ and   **b.** $x^2 + y^2 = 8$.

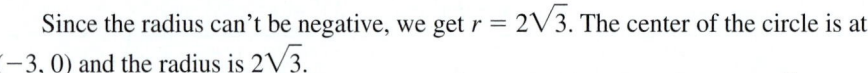

EXAMPLE 2   Find the equation of the circle with radius 9 and center at $(6, -5)$.

Solution   We substitute 9 for r, 6 for h, and -5 for k in the standard form and simplify.

$$(x - \mathbf{\textit{h}})^2 + (y - \mathbf{\textit{k}})^2 = \mathbf{\textit{r}}^2$$
$$(x - \mathbf{6})^2 + [y - (\mathbf{-5})]^2 = \mathbf{9}^2$$
$$(x - 6)^2 + (y + 5)^2 = 9^2 \quad \text{Write } y - (-5) \text{ as } y + 5.$$

If we express 9^2 as 81, we have

$$(x - 6)^2 + (y + 5)^2 = 81$$

Self Check 2   Find the equation of the circle with radius 10 and center at $(-7, 1)$.

In Example 2, the result was written in standard form: $(x - 6)^2 + (y + 5)^2 = 81$. If we square $x - 6$ and $y + 5$, we obtain a different form for the circle's equation.

$$(x - 6)^2 + (y + 5)^2 = 9^2$$
$$x^2 - 12x + 36 + y^2 + 10y + 25 = 81$$
$$x^2 - 12x + y^2 + 10y - 20 = 0 \qquad \text{Subtract 81 from both sides. Combine like terms.}$$
$$x^2 + y^2 - 12x + 10y - 20 = 0 \qquad \text{Rearrange the terms, writing the squared terms first.}$$

This result is written in the *general form of the equation of a circle*.

Equation of a Circle	The **general form of the equation of a circle** is $$x^2 + y^2 + Dx + Ey + F = 0$$

We can convert from the general form to the standard form of the equation of a circle by completing the square.

EXAMPLE 3

Write in standard form: $x^2 + y^2 - 4x + 2y - 11 = 0$.

ELEMENTARY & INTERMEDIATE
Algebra *f(x)* Now™   **Solution**   To write the equation in standard form, we complete the square twice.

$$x^2 + y^2 - 4x + 2y - 11 = 0$$
$$x^2 - 4x + y^2 + 2y - 11 = 0 \qquad \text{Write the } x\text{-terms together and the } y\text{-terms together.}$$
$$x^2 - 4x \quad + y^2 + 2y \quad = 11 \qquad \text{Add 11 to both sides.}$$

To complete the square on $x^2 - 4x$, we note that $\frac{1}{2}(-4) = -2$ and $(-2)^2 = $ **4**. To complete the square on $y^2 + 2y$, we note that $\frac{1}{2}(2) = 1$ and $1^2 = $ **1**. We add **4** and **1** to both sides of the equation.

$$x^2 - 4x + \mathbf{4} + y^2 + 2y + \mathbf{1} = 11 + \mathbf{4} + \mathbf{1}$$
$$(x - 2)^2 + (y + 1)^2 = 16 \qquad \text{Factor } x^2 - 4x + 4 \text{ and } y^2 + 2y + 1.$$

The equation could also be written as $(x - 2)^2 + (y + 1)^2 = 4^2$.

Self Check 3   Write in standard form: $x^2 + y^2 + 12x - 6y - 4 = 0$.

ACCENT ON TECHNOLOGY: GRAPHING CIRCLES

Since the graphs of circles fail the vertical line test, their equations do not represent functions. It is somewhat more difficult to use a graphing calculator to graph equations that are not functions. For example, to graph the circle described by $(x - 1)^2 + (y - 2)^2 = 4$, we must split the equation into two functions and graph each one separately. We begin by solving the equation for y.

$$(x - 1)^2 + (y - 2)^2 = 4$$
$$(y - 2)^2 = 4 - (x - 1)^2 \qquad \text{Subtract } (x - 1)^2 \text{ from both sides.}$$
$$y - 2 = \pm\sqrt{4 - (x - 1)^2} \qquad \text{Use the square root property.}$$
$$y = 2 \pm \sqrt{4 - (x - 1)^2} \qquad \text{Add 2 to both sides.}$$

This equation defines two functions. If we graph

$$y = 2 + \sqrt{4 - (x - 1)^2} \quad \text{and} \quad y = 2 - \sqrt{4 - (x - 1)^2}$$

we get the distorted circle shown in figure (a). To get a better circle, we can use the graphing calculator's square window feature, which gives an equal unit distance on both the x- and y-axes. Using this feature, we get the circle shown in figure (b). Sometimes the two arcs will not join because of approximations made by the calculator at each endpoint.

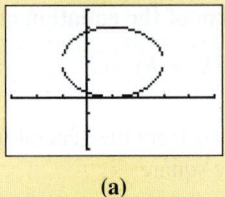

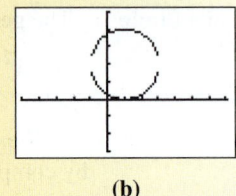

(a)   (b)

■ PROBLEM SOLVING USING CIRCLES

EXAMPLE 4

ELEMENTARY &
INTERMEDIATE
Algebra $f(x)$ **Now**™

Radio translators. The broadcast area of a television station is bounded by the circle $x^2 + y^2 = 3{,}600$, where x and y are measured in miles. A translator station picks up the signal and retransmits it from the center of a circular area bounded by

$$(x + 30)^2 + (y - 40)^2 = 1{,}600$$

Find the location of the translator and the greatest distance from the main transmitter that the signal can be received.

Solution The coverage of the TV station is bounded by $x^2 + y^2 = 60^2$, a circle centered at the origin with a radius of 60 miles, as shown in yellow in the figure. Because the translator is at the center of the circle $(x + 30)^2 + (y - 40)^2 = 1{,}600$, it is located at $(-30, 40)$, a point 30 miles west and 40 miles north of the TV station. The radius of the translator's coverage is $\sqrt{1{,}600}$, or 40 miles.

As shown in the figure, the greatest distance of reception is the sum of d, the distance from the translator to the television station, and 40 miles, the radius of the translator's coverage.

To find d, we use the distance formula to find the distance between $(x_1, y_1) = (-30, 40)$ and the origin, $(x_2, y_2) = (0, 0)$.

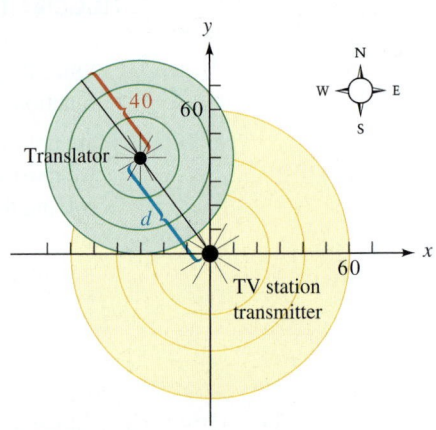

$$d = \sqrt{(\mathbf{x_1} - \mathbf{x_2})^2 + (\mathbf{y_1} - \mathbf{y_2})^2}$$

$$d = \sqrt{(\mathbf{-30} - \mathbf{0})^2 + (\mathbf{40} - \mathbf{0})^2}$$

$$d = \sqrt{(-30)^2 + 40^2}$$

$$= \sqrt{900 + 1{,}600}$$

$$= \sqrt{2{,}500}$$

$$= 50$$

The translator is located 50 miles from the television station, and it broadcasts the signal 40 miles. The greatest reception distance from the main transmitter signal is, therefore, $50 + 40$, or 90 miles.

■ THE PARABOLA

Another type of conic section is the parabola. We have previously discussed the equations of parabolas whose graphs open upward or downward. Parabolas can also open to the right and to the left, but the equations for these parabolas are not functions because their graphs fail the vertical line test.

The two *general forms of the equation of a parabola* are similar.

Equation of a Parabola

The **general forms of the equation of a parabola** are:

1. $y = ax^2 + bx + c$ The graph opens upward if $a > 0$ and downward if $a < 0$.

2. $x = ay^2 + by + c$ The graph opens to the right if $a > 0$ and to the left if $a < 0$.

Recall from Chapter 8 that equations written in the **standard form** $y = a(x - h)^2 + k$ represent parabolas with vertex at (h, k) and axis of symmetry $x = h$. They open upward when $a > 0$ and downward when $a < 0$.

EXAMPLE 5

ELEMENTARY &
INTERMEDIATE
Algebra $f(x)$ Now™

Graph: $y = -2x^2 + 12x - 15$.

Solution   Because the equation is not in standard form, the coordinates of the vertex are not obvious. To write the equation in standard form, we complete the square.

$$y = -2x^2 + 12x - 15$$
$$y = -2(x^2 - 6x \qquad) - 15 \qquad \text{Factor out } -2 \text{ from } -2x^2 + 12x.$$

This step adds $-2 \cdot 9$ Add 18 to counteract
or -18 to this side.   the addition of -18.

$$y = -2(x^2 - 6x + 9) - 15 + 18 \qquad \text{Complete the square on } x^2 - 6x.$$
$$y = -2(x - 3)^2 + 3 \qquad \text{Factor } x^2 - 6x + 9 \text{ and combine like terms.}$$

This equation is written in the form $y = a(x - h)^2 + k$, where $a = -2$, $h = 3$, and $k = 3$. Thus, the graph of the equation is a parabola that opens downward with vertex at $(3, 3)$ and an axis of symmetry $x = 3$. We can construct a table of solutions and use symmetry to plot several points on the parabola. Then we draw a smooth curve through the points to get the graph of $y = -2x^2 + 12x - 15$, as shown below.

Success Tip

Recall that we can find the x-coordinate of the vertex using

$$x = -\frac{b}{2a} = -\frac{12}{2(-2)} = 3$$

To find the y-coordinate, substitute:

$$y = -2(\mathbf{3})^2 + 12(\mathbf{3}) - 15$$
$$= 3$$

The vertex is at $(3, 3)$.

$$y = -2x^2 + 12x - 15$$

x	y
1	-5
2	1

→ $(1, -5)$
→ $(2, 1)$

↑

The x-coordinate of the vertex is 3. Choose values for x close to 3 on the same side of the axis of symmetry.

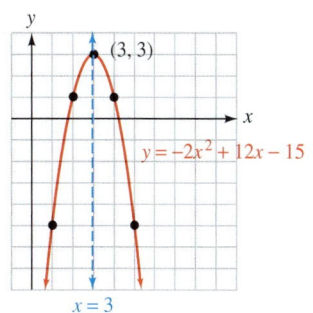

Self Check 5   Graph: $y = 2x^2 + 4x + 5$.

The *standard form* for the equation of a parabola that opens to the right or left is similar to $y = a(x - h)^2 + k$, except that the variables, x and y, exchange positions as do the constants, h and k.

Standard Form of the Equation of a Parabola

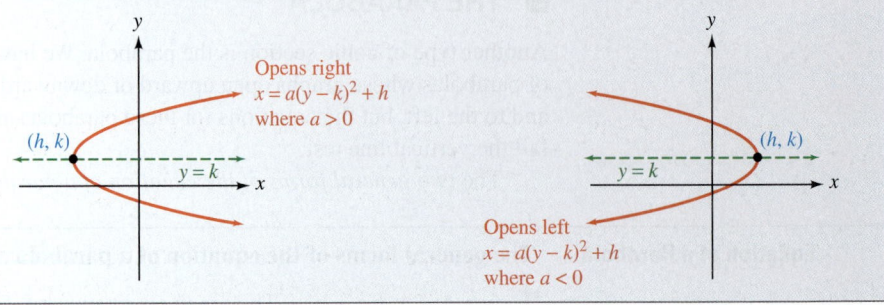

Opens right
$x = a(y - k)^2 + h$
where $a > 0$

(h, k) $y = k$

Opens left
$x = a(y - k)^2 + h$
where $a < 0$

(h, k) $y = k$

EXAMPLE 6

Graph: $x = \dfrac{1}{2}y^2$.

ELEMENTARY & INTERMEDIATE
Algebra *f(x)* Now™   **Solution**   This equation is written in the form $x = a(y - k)^2 + h$, where $a = \frac{1}{2}$, $k = 0$, and $h = 0$. The graph of the equation is a parabola that opens to the right with vertex at $(0, 0)$ and an axis of symmetry $y = 0$.

To construct a table of solutions, we choose *values of y* and find their corresponding values of x. For example, if $y = 1$, we have

$$x = \frac{1}{2}y^2$$

$$x = \frac{1}{2}(\mathbf{1})^2 \qquad \text{Substitute 1 for } y.$$

$$x = \frac{1}{2}$$

The point $\left(\frac{1}{2}, 1\right)$ is on the parabola.

We plot the ordered pairs from the table and use symmetry to plot three more points on the parabola. Then we draw a smooth curve through the points to get the graph of $x = \frac{1}{2}y^2$, as shown below.

$$x = \frac{1}{2}y^2$$

x	y	
$\frac{1}{2}$	1	$\rightarrow \left(\frac{1}{2}, 1\right)$
2	2	$\rightarrow (2, 2)$
8	4	$\rightarrow (8, 4)$

↑
The y-coordinate of the vertex is 0.
Choose values for y close to 0 on the same side of the axis of symmetry.

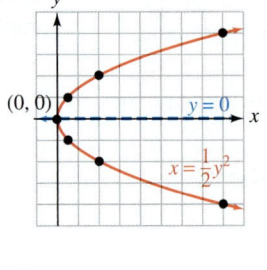

$(0, 0)$ $y = 0$

$x = \frac{1}{2}y^2$

Self Check 6   Graph: $x = -\dfrac{2}{3}y^2$.

EXAMPLE 7

Graph: $x = -3y^2 - 12y - 13$.

ELEMENTARY &
INTERMEDIATE
Algebra $f(x)$ **Now**™   **Solution**

To write the equation in standard form, we complete the square.

$$x = -3y^2 - 12y - 13$$
$$x = -3(y^2 + 4y \quad) - 13 \qquad \text{Factor out } -3 \text{ from } -3y^2 - 12y.$$
$$x = \mathbf{-3}(y^2 + 4y \mathbf{+ 4}) - 13 \mathbf{+ 12} \qquad \text{Complete the square on } y^2 + 4y. \text{ Then add 12 to the}$$
$$\text{right-hand side to counteract } -3 \cdot 4 = -12.$$
$$x = -3(y + 2)^2 - 1 \qquad \text{Factor } y^2 + 4y + 4 \text{ and combine like terms.}$$

Success Tip

When an equation is of the form $x = ay^2\ by + c$, we can find the *y*-coordinate of the vertex using

$$y = -\frac{b}{2a} = -\frac{-12}{2(-3)} = \mathbf{-2}$$

To find the *x*-coordinate, substitute:

$$x = -3(\mathbf{-2})^2 - 12(\mathbf{-2}) - 13$$
$$= -1$$

The vertex is at $(-1, -2)$.

This equation is in the standard form $x = a(y - k)^2 + h$, where $a = -3$, $k = -2$, and $h = -1$. The graph of the equation is a parabola that opens to the left with vertex at $(-1, -2)$ and an axis of symmetry $y = -2$.

We can construct a table of solutions and use symmetry to plot several points on the parabola. Then we draw a smooth curve through the points to get the graph of $x = -3y^2 - 12y - 13$, as shown below.

$$x = -3y^2 - 12y - 13$$
$$\text{or}$$
$$x = -3(y + 2)^2 - 1$$

x	y
-4	-1
-13	0

Choose values for *y*, and find the corresponding *x*-values.

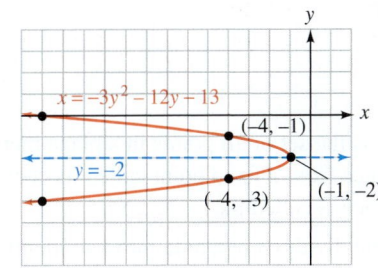

Success Tip

The equation of a circle contains an x^2 and a y^2 term. The equation of a parabola has either an x^2 term or a y^2 term, but not both.

Self Check 7   Graph: $x = 3y^2 - 6y - 1$.

Answers to Self Checks

1. a.
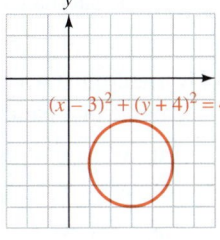

b.
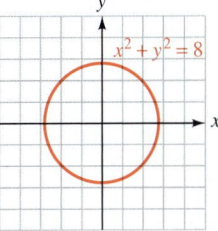

2. $(x + 7)^2 + (y - 1)^2 = 100$

3. $(x + 6)^2 + (y - 3)^2 = 49$

5.
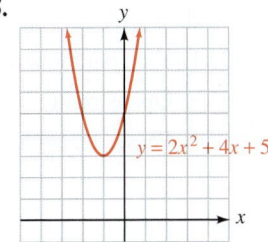

6.
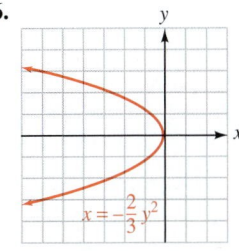

7.
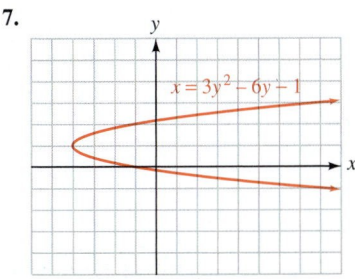

13.1 **STUDY SET** ELEMENTARY & INTERMEDIATE Algebra $f(x)$ Now™

VOCABULARY Fill in the blanks.

1. The curves formed by the intersection of a plane with an infinite right-circular cone are called _____ _____.

2. Give the name of each curve shown below.

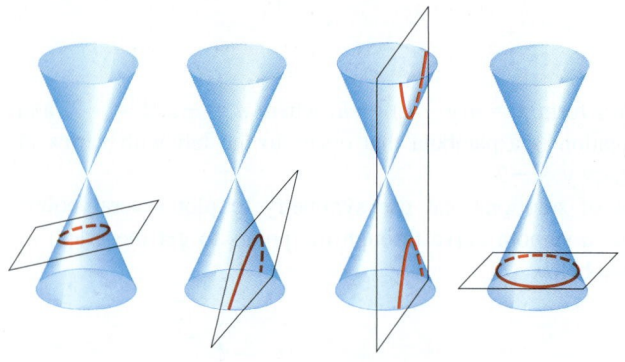

3. A _____ is the set of all points in a plane that are a fixed distance from a fixed point called its center. The fixed distance is called the _____.

4. The line that divides a parabola into two identical halves is called the axis of _____.

CONCEPTS

5. a. What is the standard form for the equation of a circle?

b. What is the standard form for the equation of a circle with the center at the origin?

6. a. What is the center and the radius of the circle graphed below?

b. What is the equation of the circle?

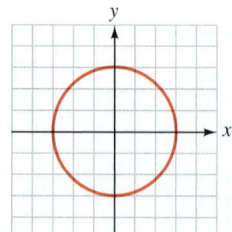

7. a. What are the center and the radius of the circle graphed below?

b. What is the equation of the circle?

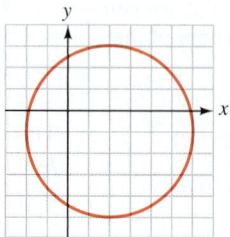

8. Fill in the blanks. To complete the square on $x^2 + 2x$ and on $y^2 - 6y$, what numbers must be added to each side of the equation?

$$x^2 + 2x + \boxed{} + y^2 - 6y + \boxed{} = 2 + \boxed{} + \boxed{}$$

9. a. What is the standard form for the equation of a parabola opening upward or downward?

b. What is the standard form for the equation of a parabola opening to the right or left?

10. Fill in the blanks.

a. To complete the square on the right-hand side, what should be factored from the first two terms?

$$x = 4y^2 + 16y + 9$$
$$x = \boxed{}(y^2 + 4y) + 9$$

b. To complete the square on $y^2 + 4y$, what should be added within the parentheses, and what should be subtracted outside the parentheses?

$$x = 4\left(y^2 + 4y + \boxed{}\right) + 9 - \boxed{}$$

11. Determine whether the graph of each equation is a circle or a parabola.

a. $x^2 + y^2 - 6x + 8y - 10 = 0$

b. $y^2 - 2x + 3y - 9 = 0$

c. $x^2 + 5x - y = 0$

d. $x^2 + 12x + y^2 = 0$

12. Draw a parabola using the given facts.

- Opens right
- Passes through $(-2, 1)$
- Vertex $(-3, 2)$
- x-intercept $(1, 0)$

NOTATION

13. Find h, k, and r: $(x - 6)^2 + (y + 2)^2 = 9$.

14. **a.** Find a, h, and k: $y = 6(x - 5)^2 - 9$.

  b. Find a, h, and k: $x = -3(y + 2)^2 + 1$.

PRACTICE   Graph each equation.

15. $x^2 + y^2 = 9$ **16.** $x^2 + y^2 = 16$

17. $(x - 2)^2 + y^2 = 9$ **18.** $x^2 + (y - 3)^2 = 4$

19. $(x - 2)^2 + (y - 4)^2 = 4$ **20.** $(x - 3)^2 + (y - 2)^2 = 4$

21. $(x + 3)^2 + (y - 1)^2 = 16$ **22.** $(x - 1)^2 + (y + 4)^2 = 9$

23. $x^2 + (y + 3)^2 = 1$ **24.** $(x + 4)^2 + y^2 = 1$

25. $x^2 + y^2 = 6$ **26.** $x^2 + y^2 = 10$

27. $(x - 1)^2 + (y - 3)^2 = 15$ **28.** $(x + 1)^2 + (y + 1)^2 = 8$

Use a graphing calculator to graph each equation.

29. $x^2 + y^2 = 7$ **30.** $x^2 + y^2 = 5$

31. $(x + 1)^2 + y^2 = 16$ **32.** $x^2 + (y - 2)^2 = 4$

Write the equation of the circle with the following properties.

33. Center at the origin; radius 1

34. Center at the origin; radius 4

35. Center at (6, 8); radius 5

36. Center at (5, 3); radius 2

37. Center at $(-2, 6)$; radius 12

38. Center at $(5, -4)$; radius 6

39. Center (0, 0); radius $\dfrac{1}{4}$

40. Center (0, 0); radius $\dfrac{1}{3}$

41. Center $\left(\dfrac{2}{3}, -\dfrac{7}{8}\right)$; radius $\sqrt{2}$

42. Center $(-0.7, -0.2)$; radius $\sqrt{11}$

43. Center at the origin; diameter $4\sqrt{2}$

44. Center at the origin; diameter $8\sqrt{3}$

Graph each circle and give the coordinates of the center and the radius.

45. $x^2 + y^2 - 2x + 4y = -1$

46. $x^2 + y^2 + 6x - 4y = -12$

47. $x^2 + y^2 + 4x + 2y = 4$

48. $x^2 + y^2 + 8x + 2y = -13$

49. $x^2 + y^2 + 2x - 8 = 0$

50. $x^2 + y^2 - 4y = 12$

51. $x^2 + y^2 - 6x + 8y + 18 = 0$

52. $x^2 + y^2 - 4x + 4y - 3 = 0$

Graph each parabola and give the coordinates of the vertex.

53. $x = y^2$ **54.** $x = -y^2 + 1$

55. $x = -\dfrac{1}{4}y^2$ **56.** $x = 4y^2$

57. $x = 2(y + 1)^2 + 3$ **58.** $x = 3(y - 2)^2 - 1$

59. $x = -3y^2 + 18y - 25$ **60.** $x = -2y^2 + 4y + 1$

61. $x = \dfrac{1}{2}y^2 + 2y$ **62.** $x = -\dfrac{1}{3}y^2 - 2y$

63. $y = 2x^2 - 4x + 5$ **64.** $y = -2x^2 - 4x$

65. $y = -x^2 - 2x + 3$ **66.** $y = x^2 + 4x + 5$

67. $y^2 + 4x - 6y = -1$ **68.** $x^2 - 2y - 2x = -7$

Use a graphing calculator to graph each equation. (*Hint:* Solve for y and graph two functions when necessary.)

69. $x = 2y^2$ **70.** $x = y^2 - 4$

71. $x^2 - 2x + y = 6$ **72.** $x = -2(y - 1)^2 + 2$

APPLICATIONS

73. MESHING GEARS   For design purposes, the large gear is described by the circle $x^2 + y^2 = 16$. The smaller gear is a circle centered at (7, 0) and tangent to the larger circle. Find the equation of the smaller gear.

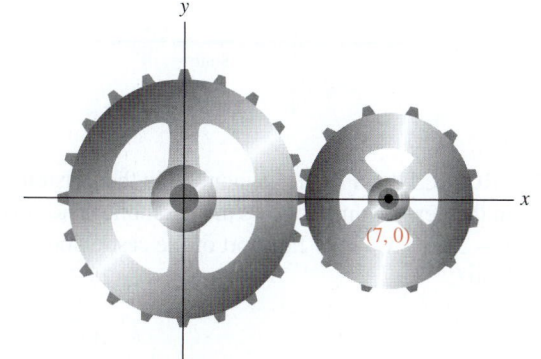

74. WALKWAYS The walkway shown is bounded by the two circles $x^2 + y^2 = 2,500$ and $(x - 10)^2 + y^2 = 900$, measured in feet. Find the largest and the smallest width of the walkway.

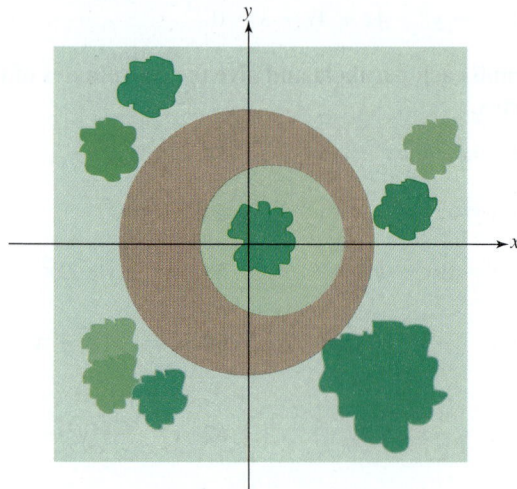

75. BROADCAST RANGES Radio stations applying for licensing may not use the same frequency if their broadcast areas overlap. One station's coverage is bounded by $x^2 + y^2 - 8x - 20y + 16 = 0$, and the other's by $x^2 + y^2 + 2x + 4y - 11 = 0$. May they be licensed for the same frequency?

76. HIGHWAY DESIGN Engineers want to join two sections of highway with a curve that is one-quarter of a circle, as shown. The equation of the circle is $x^2 + y^2 - 16x - 20y + 155 = 0$, where distances are measured in kilometers. Find the locations (relative to the center of town) of the intersections of the highway with State and with Main.

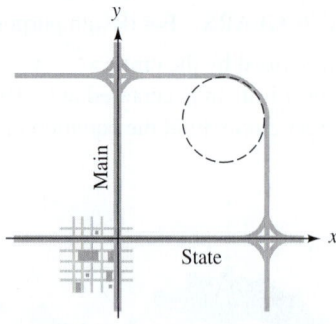

77. PROJECTILES The cannonball in the illustration in the next column follows the parabolic trajectory $y = 30x - x^2$. How far short of the castle does it land?

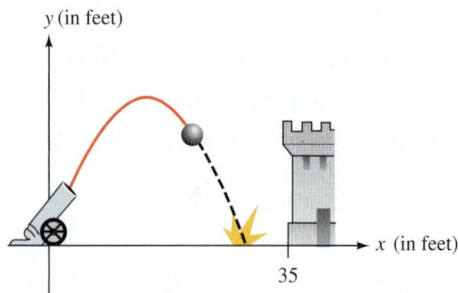

78. PROJECTILES In Exercise 77, how high does the cannonball get?

79. COMETS If the orbit of the comet is approximated by the equation $2y^2 - 9x = 18$, how far is it from the sun at the vertex of the orbit? Distances are measured in astronomical units (AU).

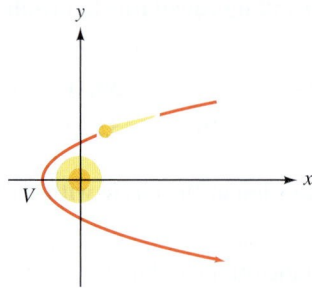

80. SATELLITE ANTENNAS The cross section of the satellite antenna in the illustration is a parabola given by the equation $y = \frac{1}{16}x^2$, with distances measured in feet. If the dish is 8 feet wide, how deep is it?

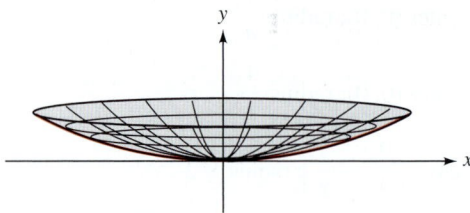

WRITING

81. Explain how to decide from its equation whether the graph of a parabola opens up, down, right, or left.

82. From the equation of a circle, explain how to determine the radius and the coordinates of the center.

83. On the day of an election, the following warning was posted in front of a school. Explain what it means.

> *No electioneering within a 1,000-foot radius of this polling place.*

84. What is meant by the *turning radius* of a truck?

REVIEW **Solve each equation.**

85. $|3x - 4| = 11$

86. $\left| \dfrac{4 - 3x}{5} \right| = 12$

87. $|3x + 4| = |5x - 2|$

88. $|6 - 4x| = |x + 2|$

CHALLENGE PROBLEMS

89. Could the intersection of a plane with a pair of right-circular cones be a single point? If so, draw a picture that illustrates this.

90. Under what conditions will the graph of $x = a(y - k)^2 + h$ have no y-intercepts?

91. Write the equation of a circle with a diameter whose endpoints are at $(-2, -6)$ and $(8, 10)$.

92. Write the equation of a circle with a diameter whose endpoints are at $(-5, 4)$ and $(7, -3)$.

13.2 The Ellipse

- The definition of an ellipse
- Graphing ellipses centered at the origin
- Graphing ellipses centered at (h, k)
- Problem solving using ellipses

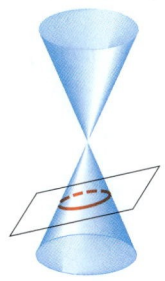

A third conic section is an oval-shaped curve called an *ellipse*. Ellipses can be nearly round and look almost like a circle, or they can be long and narrow. In this section, we will learn how to construct ellipses and how to graph equations that represent ellipses.

■ THE DEFINITION OF AN ELLIPSE

To define a circle, we considered a fixed distance from a fixed point. The definition of an ellipse involves *two* distances from *two* fixed points.

Definition of an Ellipse	An **ellipse** is the set of all points in a plane for which the sum of the distances from two fixed points is a constant.

The figure below illustrates that any point on an ellipse is a constant distance $d_1 + d_2$ from two fixed points, each of which is called a **focus.** Midway between the **foci** is the **center** of the ellipse.

The Language of Algebra

The word *foci* (pronounced *foe-sigh*) is the plural form of the word *focus.* In the illustration on the right, the foci are labeled using subscript notation. One focus is F_1 and the other is F_2.

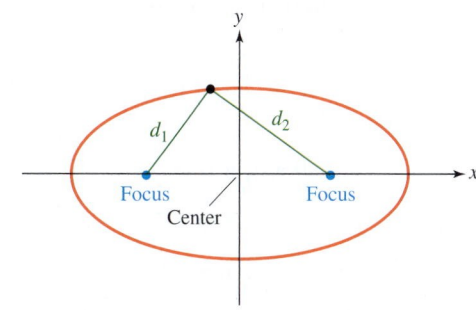

We can construct an ellipse by placing two thumbtacks fairly close together to serve as foci. We then tie each end of a piece of string to a thumbtack, catch the loop with the point of a pencil, and (keeping the string taut) draw the ellipse.

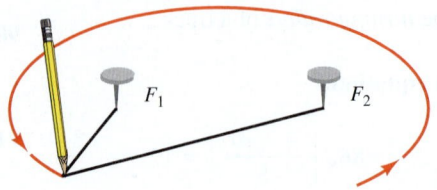

■ GRAPHING ELLIPSES CENTERED AT THE ORIGIN

The definition of an ellipse can be used to develop the standard form for the equation of an ellipse. To learn more about the derivation, see Problem 60 in the Challenge Problem section of the Study Set.

Equation of an Ellipse Centered at the Origin	The **standard form of the equation of an ellipse** that is symmetric with respect to both axes and centered at $(0, 0)$ is $$\frac{x^2}{a^2} + \frac{y^2}{b^2} = 1 \quad \text{where } a > 0 \text{ and } b > 0$$

To graph an ellipse centered at the origin, it is helpful to know the intercepts of the graph. To find the x-intercepts of the graph of

$$\frac{x^2}{a^2} + \frac{y^2}{b^2} = 1$$

we let $y = 0$ and solve for x.

$$\frac{x^2}{a^2} + \frac{\mathbf{0}^2}{b^2} = 1 \qquad \text{Substitute 0 for } y.$$

$$\frac{x^2}{a^2} + 0 = 1 \qquad \frac{0^2}{b^2} = 0.$$

$$x^2 = a^2 \qquad \text{Simplify and multiply both sides by } a^2.$$

$$x = \pm a \qquad \text{Use the square root property.}$$

The x-intercepts are $(a, 0)$ and $(-a, 0)$.

To find the y-intercepts of the graph, we can let $x = 0$ and solve for y.

$$\frac{\mathbf{0}^2}{a^2} + \frac{y^2}{b^2} = 1$$

$$0 + \frac{y^2}{b^2} = 1$$

$$y^2 = b^2 \qquad \text{Simplify and multiply both sides by } b^2.$$

$$y = \pm b$$

The y-intercepts are $(0, b)$ and $(0, -b)$.

In general, we have the following results.

The Intercepts of an Ellipse	The graph of $\dfrac{x^2}{a^2} + \dfrac{y^2}{b^2} = 1$ is an ellipse, centered at the origin, with x-intercepts $(a, 0)$ and $(-a, 0)$ and y-intercepts $(0, b)$ and $(0, -b)$.

For $\dfrac{x^2}{a^2} + \dfrac{y^2}{b^2} = 1$, if $a > b$, the ellipse is horizontal, as shown in figure (a). If $b > a$, the ellipse is vertical, as shown in figure (b). The points V_1 and V_2 are the **vertices** of the ellipse. The line segment joining the vertices is called the **major axis,** and its midpoint is the **center** of the ellipse. The line segment perpendicular to the major axis at the center is the **minor axis** of the ellipse.

Horizontal ellipse
(a)

Vertical ellipse
(b)

EXAMPLE 1

ELEMENTARY &
INTERMEDIATE
Algebra ⒻⓍ Now™   Solution

Graph: $\dfrac{x^2}{36} + \dfrac{y^2}{9} = 1$.

This is the equation of an ellipse centered at the origin. To determine the intercepts of the graph, we find a and b.

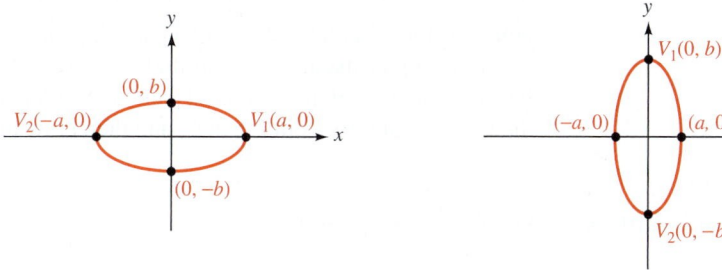

$$\dfrac{x^2}{36} + \dfrac{y^2}{9} = 1 \qquad \dfrac{x^2}{a^2} + \dfrac{y^2}{b^2} = 1$$

Since $a^2 = 36$, it follows that $a = 6$.   Since $b^2 = 9$, it follows that $b = 3$.

The x-intercepts are $(a, 0)$ and $(-a, 0)$, or $(6, 0)$ and $(-6, 0)$. The y-intercepts are $(0, b)$ and $(0, -b)$, or $(0, 3)$ and $(0, -3)$. Using these four points as a guide, we draw an oval-shaped curve through them, as shown in figure (a). The result is a horizontal ellipse.

The Language of Algebra

The word *vertices* is the plural form of the word *vertex.* From the graph, we see that one vertex of this horizontal ellipse is the point $(6, 0)$ and the other vertex is the point $(-6, 0)$.

$$\dfrac{x^2}{36} + \dfrac{y^2}{9} = 1$$

x	y	
2	$\pm 2\sqrt{2}$	$\rightarrow \left(2, \pm 2\sqrt{2}\right)$
4	$\pm\sqrt{5}$	$\rightarrow \left(4, \pm\sqrt{5}\right)$

Approximate the
radicals to graph.

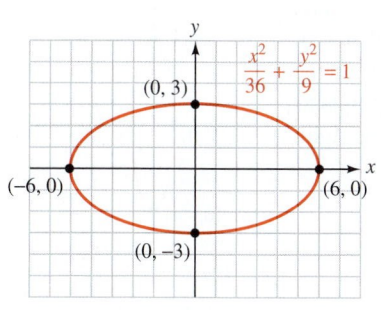

(a)

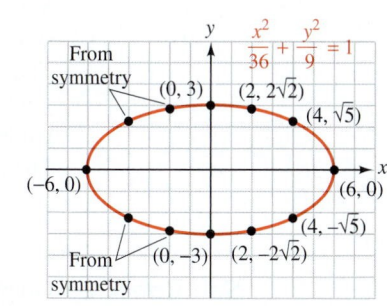

(b)

To increase the accuracy of the graph, we can find additional ordered pairs that satisfy the equation and plot them. For example, if $x = 2$, we have

$$\frac{2^2}{36} + \frac{y^2}{9} = 1$$ Substitute 2 for x in the equation of the ellipse.

$$36\left(\frac{4}{36} + \frac{y^2}{9}\right) = 36(1)$$ To clear the equation of fractions, multiply both sides by the LCD, 36.

$$4 + 4y^2 = 36$$ Distribute the multiplication by 36.

$$y^2 = 8$$ Subtract 4 from both sides, then divide both sides by 4.

$$y = \pm\sqrt{8}$$ Use the square root property.

$$y = \pm 2\sqrt{2}$$ Simplify the radical.

Since two values of y, $2\sqrt{2}$ and $-2\sqrt{2}$, correspond to the x-value 2, we have found two points on the ellipse: $\left(2, 2\sqrt{2}\right)$ and $\left(2, -2\sqrt{2}\right)$.

In a similar manner, we can find the corresponding values of y for the x-value 4. In figure (b) we record these ordered pairs in a table, plot them, use symmetry with respect to the y-axis to plot four other points, and then draw the graph of the ellipse.

Self Check 1   Graph: $\dfrac{x^2}{49} + \dfrac{y^2}{25} = 1$.

EXAMPLE 2

Graph: $16x^2 + y^2 = 16$.

ELEMENTARY &
INTERMEDIATE
Algebra *f(x)* **Now**™   **Solution**

This equation is not in standard form. To write it in standard form with 1 on the right-hand side, we divide both sides by 16.

$$16x^2 + y^2 = 16$$

$$\frac{16x^2}{16} + \frac{y^2}{16} = \frac{16}{16}$$ Divide both sides by 16.

$$\frac{x^2}{1} + \frac{y^2}{16} = 1$$ Simplify: $\dfrac{16x^2}{16} = x^2 = \dfrac{x^2}{1}$, and $\dfrac{16}{16} = 1$.

Success Tip

Although the term $\frac{16x^2}{16}$ simplifies to x^2, we write it as the fraction $\frac{x^2}{1}$ so that it has the form $\frac{x^2}{a^2}$.

To determine a and b, we can write the equation in the form

$$\frac{x^2}{1^2} + \frac{y^2}{4^2} = 1$$ To find a, write 1 as 1^2. To find b, write 16 as 4^2.

Since a^2 (the denominator of x^2) is 1^2, it follows that $a = 1$, and since b^2 (the denominator of y^2) is 4^2, it follows that $b = 4$. Thus, the x-intercepts of the graph are $(1, 0)$ and $(-1, 0)$ and the y-intercepts are $(0, 4)$ and $(0, -4)$. We use these four points as guides to sketch the graph of the ellipse, as shown on the right. The result is a vertical ellipse.

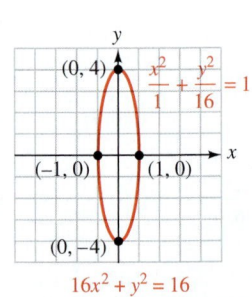

$16x^2 + y^2 = 16$

Self Check 2   Graph: $9x^2 + y^2 = 9$.

■ **GRAPHING ELLIPSES CENTERED AT (*h, k*)**

Not all ellipses are centered at the origin. As with the graphs of circles and parabolas, the graph of an ellipse can be translated horizontally and vertically.

The Equation of an Ellipse Centered at (*h, k*)	The **standard form of the equation of a horizontal or vertical ellipse** centered at (*h, k*) is $$\frac{(x-h)^2}{a^2} + \frac{(y-k)^2}{b^2} = 1 \quad \text{where } a > 0 \text{ and } b > 0$$ For a horizontal ellipse, *a* is the distance from the center to a vertex. For a vertical ellipse, *b* is the distance from the center to a vertex.

EXAMPLE 3

ELEMENTARY & INTERMEDIATE
Algebra *f(x)* **Now**™

Graph: $\dfrac{(x-2)^2}{16} + \dfrac{(y+3)^2}{25} = 1$.

Solution To determine *h*, *k*, *a*, and *b*, we write the equation in the form

$$\frac{(x-\mathbf{2})^2}{\mathbf{4}^2} + \frac{[y-(\mathbf{-3})]^2}{\mathbf{5}^2} = 1$$

To find *k*, write *y* + 3 as *y* − (−3). To find *a*, write 16 as 4^2. To find *b*, write 25 as 5^2.

We find the center of the ellipse in the same way we would find the center of a circle, by examining $(x - 2)^2$ and $(y + 3)^2$. Since *h* = 2 and *k* = −3, this is the equation of an ellipse centered at (*h, k*) = (2, −3). From the denominators, 4^2 and 5^2, we find that *a* = 4 and *b* = 5. Because *b* > *a*, it is a vertical ellipse.

We first plot the center, as shown below. Since *b* is the distance from the center to a vertex for a vertical ellipse, we can locate the vertices by counting 5 units above and 5 units below the center. The vertices are the points (2, 2) and (2, −8).

To locate two more points on the ellipse, we use the fact that *a* is 4 and count 4 units to the left and to the right of the center. We see that the points (−2, −3) and (6, −3) are also on the graph.

Using these four points as guides, we draw the graph shown below.

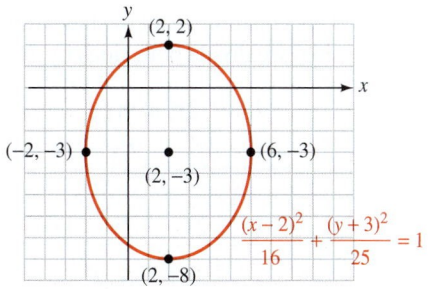

Self Check 3 Graph: $\dfrac{(x-1)^2}{9} + \dfrac{(y+2)^2}{16} = 1$.

ACCENT ON TECHNOLOGY: GRAPHING ELLIPSES

To use a graphing calculator to graph the equation from Example 3,

$$\frac{(x-2)^2}{16} + \frac{(y+3)^2}{25} = 1$$

we first clear the equation of fractions and then we solve for y.

$$25(x-2)^2 + 16(y+3)^2 = 400 \qquad \text{Multiply both sides by 400.}$$

$$16(y+3)^2 = 400 - 25(x-2)^2 \qquad \text{Subtract } 25(x-2)^2 \text{ from both sides.}$$

$$(y+3)^2 = \frac{400 - 25(x-2)^2}{16} \qquad \text{Divide both sides by 16.}$$

$$y+3 = \pm\frac{\sqrt{400 - 25(x-2)^2}}{4} \qquad \text{Use the square root property.}$$

$$y = -3 \pm \frac{\sqrt{400 - 25(x-2)^2}}{4} \qquad \text{Subtract 3 from both sides.}$$

If we graph the two functions that $y = -3 \pm \dfrac{\sqrt{400 - 25(x-2)^2}}{4}$ represents in a square window, we get the graph of the ellipse shown below.

$$y = -3 + \frac{\sqrt{400 - 25(x-2)^2}}{4} \quad \text{and} \quad y = -3 - \frac{\sqrt{400 - 25(x-2)^2}}{4}$$

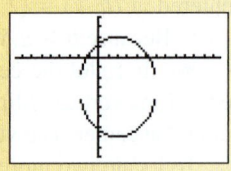

As we saw with circles, the two portions of the ellipse do not quite connect. This is because the graphs are nearly vertical there.

EXAMPLE 4

Graph: $4(x-2)^2 + 9(y-1)^2 = 36$.

ELEMENTARY & INTERMEDIATE
Algebra *f(x)* **Now**™

Solution   This equation is not in standard form. To write it in standard form with 1 on the right-hand side, we divide both sides by 36.

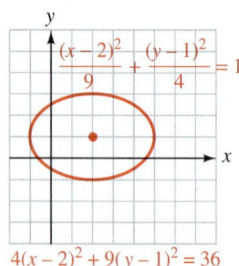

$$4(x-2)^2 + 9(y-1)^2 = 36$$

$$\frac{4(x-2)^2}{36} + \frac{9(y-1)^2}{36} = \frac{36}{36} \qquad \text{Divide both sides by 36.}$$

$$\frac{(x-2)^2}{9} + \frac{(y-1)^2}{4} = 1 \qquad \text{Simplify: } \frac{4}{36} = \frac{1}{9}, \frac{9}{36} = \frac{1}{4}, \text{ and } \frac{36}{36} = 1.$$

This is the standard form of the equation of a horizontal ellipse, centered at (2, 1), with $a = 3$ and $b = 2$. The graph of the ellipse is shown in the margin.

Self Check 4   Graph: $12(x-1)^2 + 3(y+1)^2 = 48$.

■ PROBLEM SOLVING USING ELLIPSES

EXAMPLE 5

ELEMENTARY &
INTERMEDIATE
Algebra *(x)* **Now**™

Landscape design.   A landscape architect is designing an elliptical pool that will fit in the center of a 20-by-30-foot rectangular garden, leaving at least 5 feet of space on all sides. Find the equation of the ellipse.

Solution   We place the rectangular garden in the coordinate system shown below.

To maintain 5 feet of clearance at the ends of the ellipse, the *x*-intercepts must be the points $(10, 0)$ and $(-10, 0)$. Similarly, the *y*-intercepts are the points $(0, 5)$ and $(0, -5)$.

Since the ellipse is centered at the origin, its equation has the form

$$\frac{x^2}{a^2} + \frac{y^2}{b^2} = 1$$

with $a = 10$ and $b = 5$. Thus, the equation of the boundary of the pool is

$$\frac{x^2}{100} + \frac{y^2}{25} = 1$$

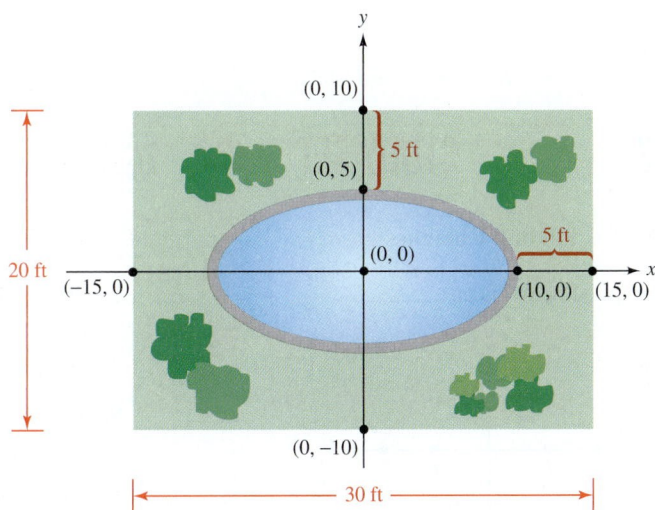

Ellipses, like parabolas, have reflective properties that are used in many practical applications. For example, any light or sound originating at one focus of an ellipse is reflected by the interior of the figure to the other focus.

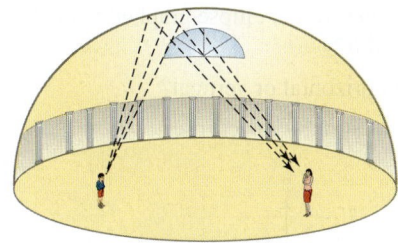

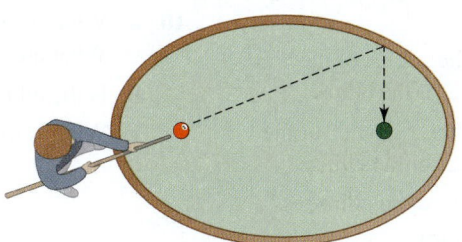

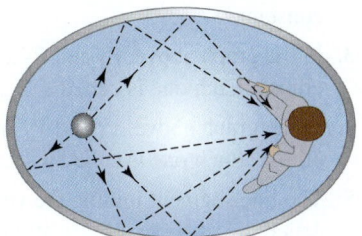

Whispering galleries

In an elliptical dome, even the slightest whisper made by a person standing at one focus can be heard by a person standing at the other focus.

Elliptical billiards tables

When a ball is shot from one focus, it will rebound off the side of the table into a pocket located at the other focus.

Treatment for kidney stones

The patient is positioned in an elliptical tank of water so that the kidney stone is at one focus. High-intensity sound waves generated at another focus are reflected to the stone to shatter it.

Answers to Self Checks   **1.**

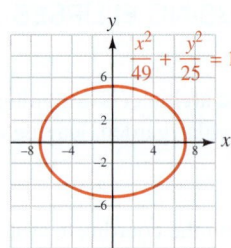

2.

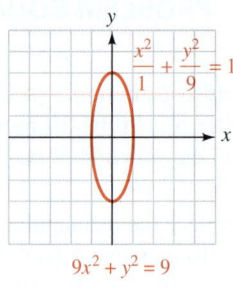

$9x^2 + y^2 = 9$

3.

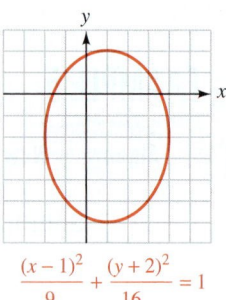

$$\frac{(x-1)^2}{9} + \frac{(y+2)^2}{16} = 1$$

4.

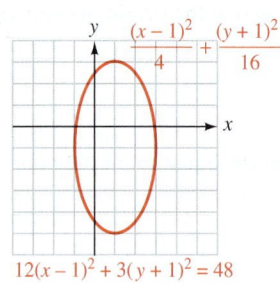

$12(x-1)^2 + 3(y+1)^2 = 48$

13.2   STUDY SET   ELEMENTARY & INTERMEDIATE Algebra *f(x)* Now™

VOCABULARY   Fill in the blanks.

1. The curve graphed below is an _____.

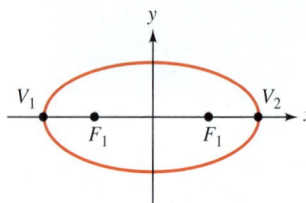

2. An _____ is the set of all points in a plane for which the sum of the distances from two fixed points is a constant.

3. In the graph above, F_1 and F_2 are the _____ of the ellipse. Each one is called a _____ of the ellipse.

4. In the graph above, V_1 and V_2 are the _____ of the ellipse. Each one is called a _____ of the ellipse.

5. The line segment joining the vertices of an ellipse is called the _____ _____ of the ellipse.

6. The midpoint of the major axis of an ellipse is the _____ of the ellipse.

CONCEPTS

7. What is the standard form for the equation of an ellipse centered at the origin and symmetric to both axes?

8. What is the standard form for the equation of a horizontal or vertical ellipse centered at (h, k)?

9. What are the x- and the y-intercepts of the graph of $\frac{x^2}{a^2} + \frac{y^2}{b^2} = 1$?

10. a. What is the center of the ellipse graphed below? What are a and b?

  b. Is the ellipse horizontal or vertical?

  c. What is the equation of the ellipse?

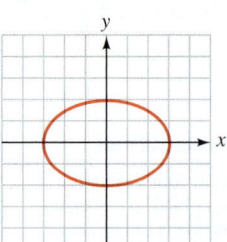

11. a. What is the center of the ellipse graphed below? What are a and b?

   b. Is the ellipse horizontal or vertical?

   c. What is the equation of the ellipse?

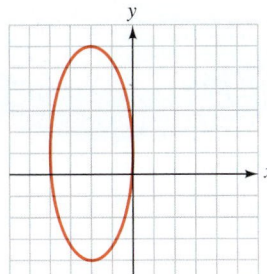

12. Find two points on the graph of $\frac{x^2}{16} + \frac{y^2}{4} = 1$ by letting $x = 2$ and finding the corresponding values of y.

13. Divide both sides of the equation by 64 and write the equation in standard form:

$$4(x - 1)^2 + 64(y + 5)^2 = 64$$

14. Determine whether the equation, when graphed, will be a circle, a parabola, or an ellipse.

   a. $x = y^2 - 2y + 10$ **b.** $\frac{x^2}{49} + \frac{y^2}{64} = 1$

   c. $(x - 3)^2 + (y + 4)^2 = 25$

   d. $2(x - 1)^2 + 8(y + 5)^2 = 32$

NOTATION

15. Find h, k, a, and b: $\dfrac{(x + 8)^2}{100} + \dfrac{(y - 6)^2}{144} = 1$.

16. Write each denominator in the equation $\dfrac{x^2}{81} + \dfrac{y^2}{49} = 1$ as the square of a number.

PRACTICE   Graph each equation.

17. $\dfrac{x^2}{25} + \dfrac{y^2}{4} = 1$ **18.** $\dfrac{x^2}{16} + \dfrac{y^2}{9} = 1$

19. $\dfrac{x^2}{4} + \dfrac{y^2}{9} = 1$ **20.** $\dfrac{x^2}{16} + \dfrac{y^2}{25} = 1$

21. $\dfrac{x^2}{16} + \dfrac{y^2}{1} = 1$ **22.** $\dfrac{x^2}{1} + \dfrac{y^2}{9} = 1$

23. $x^2 + 9y^2 = 9$ **24.** $25x^2 + 9y^2 = 225$

25. $16x^2 + 4y^2 = 64$ **26.** $4x^2 + 9y^2 = 36$

27. $x^2 = 100 - 4y^2$ **28.** $x^2 = 36 - 4y^2$

29. $\dfrac{(x - 2)^2}{9} + \dfrac{(y - 1)^2}{4} = 1$

30. $\dfrac{(x - 1)^2}{9} + \dfrac{(y - 3)^2}{4} = 1$

31. $\dfrac{(x + 2)^2}{64} + \dfrac{(y - 2)^2}{100} = 1$

32. $\dfrac{(x - 6)^2}{36} + \dfrac{(y + 6)^2}{144} = 1$

33. $(x + 1)^2 + 4(y + 2)^2 = 4$

34. $25(x + 1)^2 + 9y^2 = 225$

35. $(x - 2)^2 + 4(y + 1)^2 = 4$

36. $(x - 1)^2 + 4(y - 2)^2 = 4$

37. $9(x - 1)^2 = 36 - 4(y + 2)^2$

38. $16(x - 5)^2 = 400 - 25(y - 4)^2$

  Use a graphing calculator to graph each equation.

39. $\dfrac{x^2}{9} + \dfrac{y^2}{4} = 1$ **40.** $x^2 + 16y^2 = 16$

41. $\dfrac{x^2}{4} + \dfrac{(y - 1)^2}{9} = 1$

42. $\dfrac{(x + 1)^2}{9} + \dfrac{(y - 2)^2}{4} = 1$

APPLICATIONS

43. FITNESS EQUIPMENT   With elliptical cross-training equipment, the feet move through the natural elliptical pattern that one experiences when walking, jogging, or running. Write the equation of the elliptical pattern shown below.

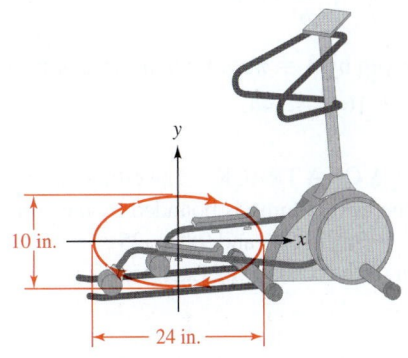

44. DESIGNING AN UNDERPASS   The arch of an underpass is a part of an ellipse. Find the equation of the ellipse.

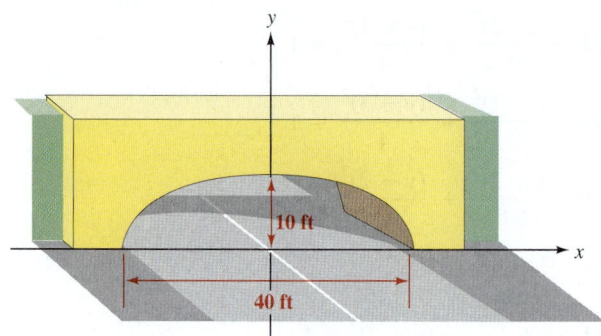

45. CALCULATING CLEARANCE   Find the height of the elliptical arch in Exercise 44 at a point 10 feet from the center of the roadway.

46. POOL TABLES   Find the equation of the outer edge of the elliptical pool table shown below.

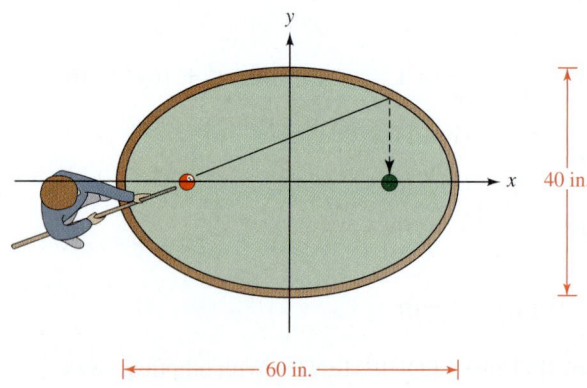

47. AREA OF AN ELLIPSE   The area A of the ellipse

$$\frac{x^2}{a^2} + \frac{y^2}{b^2} = 1$$

is given by $A = \pi ab$. Find the area of the ellipse $9x^2 + 16y^2 = 144$.

48. AREA OF A TRACK   The elliptical track shown in the next column is bounded by the ellipses $4x^2 + 9y^2 = 576$ and $9x^2 + 25y^2 = 900$. Find the area of the track. (See Exercise 47.)

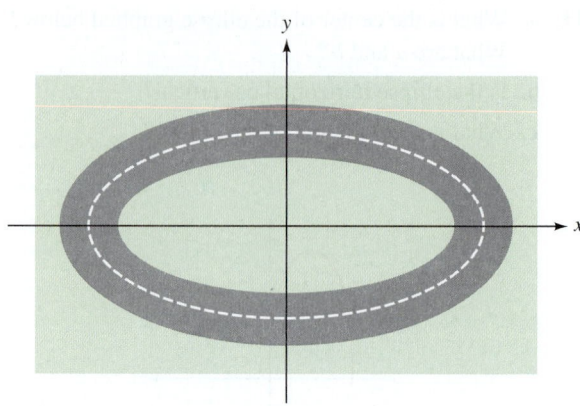

WRITING

49. What is an ellipse?

50. Explain the difference between the focus of an ellipse and the vertex of an ellipse.

51. Compare the graphs of $\frac{x^2}{81} + \frac{y^2}{64} = 1$ and $\frac{x^2}{64} + \frac{y^2}{81} = 1$. Do they have any similarities?

52. What are the reflective properties of an ellipse?

REVIEW   Find each product.

53. $3x^{-2}y^2(4x^2 + 3y^{-2})$

54. $(2a^{-2} - b^{-2})(2a^{-2} + b^{-2})$

Simplify each complex fraction.

55. $\dfrac{x^{-2} + y^{-2}}{x^{-2} - y^{-2}}$

56. $\dfrac{2x^{-3} - 2y^{-3}}{4x^{-3} + 4y^{-3}}$

CHALLENGE PROBLEMS

57. What happens to the graph of

$$\frac{x^2}{a^2} + \frac{y^2}{b^2} = 1$$

when $a = b$?

58. Graph: $9x^2 + 4y^2 = 1$.

59. Write the equation $9x^2 + 4y^2 - 18x + 16y = 11$ in the standard form of the equation of an ellipse.

60. Let the foci of an ellipse be $(c, 0)$ and $(-c, 0)$. Suppose that the sum of the distances from any point (x, y) on the ellipse to the two foci is the constant $2a$. Show that the equation for the ellipse is

$$\frac{x^2}{a^2} + \frac{y^2}{a^2 - c^2} = 1$$

Then let $b^2 = a^2 - c^2$ to obtain the standard form of the equation of an ellipse.

13.3   The Hyperbola

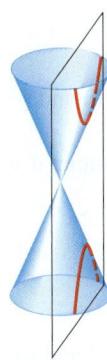

- The definition of a hyperbola
- Graphing hyperbolas centered at the origin
- Graphing hyperbolas centered at (h, k)
- Problem solving using hyperbolas

The final conic section that we will discuss, the *hyperbola*, is a curve that has two branches. In this section, we will learn how to graph equations that represent hyperbolas.

■ THE DEFINITION OF A HYPERBOLA

Ellipses and hyperbolas have completely different shapes, but their definitions are similar. Instead of the *sum* of distances, the definition of a hyperbola involves a *difference* of distances.

Definition of a Hyperbola   A **hyperbola** is the set of all points in a plane for which the difference of the distances from two fixed points is a constant.

The figure below illustrates that any point on the hyperbola is a constant distance $d_1 - d_2$ from two fixed points, each of which is called a **focus.** Midway between the **foci** is the **center** of the hyperbola.

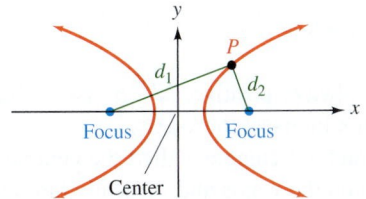

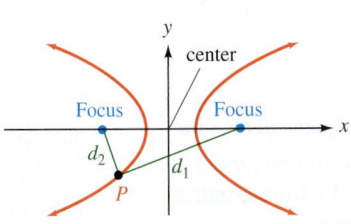

■ GRAPHING HYPERBOLAS CENTERED AT THE ORIGIN

The graph of the equation

$$\frac{x^2}{25} - \frac{y^2}{9} = 1$$

is a hyperbola. To graph the equation, we make a table of solutions that satisfy the equation, plot each point, and join them with a smooth curve.

$$\frac{x^2}{25} - \frac{y^2}{9} = 1$$

x	y	
-7	± 2.9	$\rightarrow$ $(-7, \pm 2.9)$
-6	± 2.0	$\rightarrow$ $(-6, \pm 2.0)$
-5	0	$\rightarrow$ $(-5, 0)$
5	0	$\rightarrow$ $(5, 0)$
6	± 2.0	$\rightarrow$ $(6, \pm 2.0)$
7	± 2.9	$\rightarrow$ $(7, \pm 2.9)$

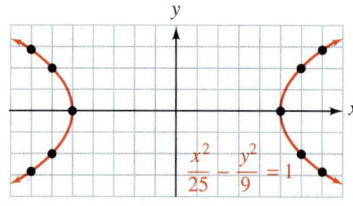

> **Caution**
>
> Although the two branches of a hyperbola look like parabolas, they are not parabolas.

This graph is centered at the origin and intersects the x-axis at $(5, 0)$ and $(-5, 0)$. We also note that the graph does not intersect the y-axis.

It is possible to draw a hyperbola without plotting points. For example, if we want to graph the hyperbola with an equation of

$$\frac{x^2}{a^2} - \frac{y^2}{b^2} = 1$$

we first find the x- and y-intercepts. To find the x-intercepts, we let $y = 0$ and solve for x:

$$\frac{x^2}{a^2} - \frac{\mathbf{0}^2}{b^2} = 1$$

$$x^2 = a^2$$

$$x = \pm a \qquad \text{Use the square root property.}$$

The hyperbola crosses the x-axis at the points $V_1(a, 0)$ and $V_2(-a, 0)$, called the **vertices** of the hyperbola.

To attempt to find the y-intercepts, we let $x = 0$ and solve for y:

$$\frac{\mathbf{0}^2}{a^2} - \frac{y^2}{b^2} = 1$$

$$y^2 = -b^2$$

$$y = \pm\sqrt{-b^2}$$

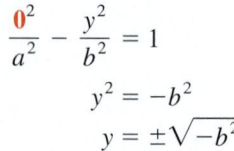
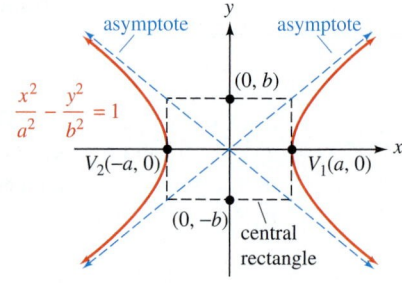

> **The Language of Algebra**
>
> The central rectangle is also called the **fundamental rectangle**.

Since b^2 is always positive, $\sqrt{-b^2}$ is an imaginary number. This means that the hyperbola does not intersect the y-axis.

If we construct a rectangle, called the **central rectangle,** whose sides pass horizontally through $\pm b$ on the y-axis and vertically through $\pm a$ on the x-axis, the extended diagonals of the rectangle will be **asymptotes** of the hyperbola.

Standard Form for the Equation of a Hyperbola Centered at the Origin

Any equation that can be written in the form

$$\frac{x^2}{a^2} - \frac{y^2}{b^2} = 1$$

has a graph that is a hyperbola centered at the origin. The x-intercepts are the vertices $V_1(a, 0)$ and $V_2(-a, 0)$. There are no y-intercepts.

The asymptotes of the hyperbola are the extended diagonals of the central rectangle.

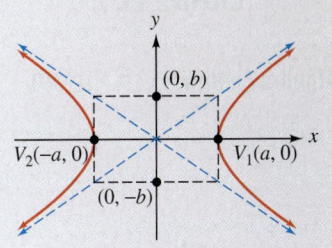

The branches of the hyperbola in previous discussions open to the left and to the right. It is possible for hyperbolas to have different orientations with respect to the x- and y-axes. For example, the branches of a hyperbola can open upward and downward. In that case, the following equation applies.

Standard Form for the Equation of a Hyperbola Centered at the Origin

Any equation that can be written in the form

$$\frac{y^2}{a^2} - \frac{x^2}{b^2} = 1$$

has a graph that is a hyperbola centered at the origin. The y-intercepts are the vertices $V_1(0, a)$ and $V_2(0, -a)$. There are no x-intercepts.

The asymptotes of the hyperbola are the extended diagonals of the central rectangle.

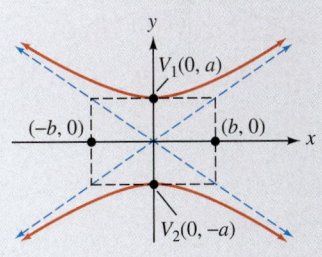

EXAMPLE 1

ELEMENTARY & INTERMEDIATE
Algebra *f(x)* **Now**™

Graph: $\dfrac{x^2}{9} - \dfrac{y^2}{16} = 1$.

Solution

This is the standard form of the equation of a hyperbola, centered at the origin, that opens left and right. To determine the vertices and the central rectangle of the graph, we find a and b.

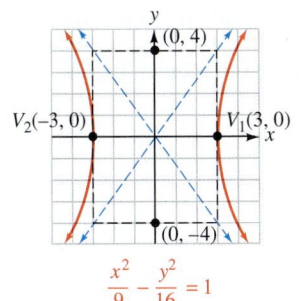

$$\frac{x^2}{9} - \frac{y^2}{16} = 1 \qquad \frac{x^2}{a^2} - \frac{y^2}{b^2} = 1$$

Since $a^2 = 9$, it follows that $a = 3$

Since $b^2 = 16$, it follows that $b = 4$.

The x-intercepts are $(a, 0)$ and $(-a, 0)$, or $(3, 0)$ and $(-3, 0)$. They are also the vertices of the hyperbola.

To construct the central rectangle, we use the values of $a = 3$ and $b = 4$. The rectangle passes through $(3, 0)$ and $(-3, 0)$ on the x-axis, and $(0, 4)$ and $(0, -4)$ on the y-axis. We draw extended diagonal dashed lines through the rectangle to obtain the asymptotes. Then we draw a smooth curve through each vertex that gets close to the asymptotes.

Self Check 1   Graph: $\dfrac{x^2}{25} - \dfrac{y^2}{4} = 1$.

EXAMPLE 2

ELEMENTARY &
INTERMEDIATE
Algebra *f(x)* **Now**™   Solution

Graph: $9y^2 - 4x^2 = 36$.

To write the equation in standard from, we divide both sides by 36 to obtain

$$\frac{9y^2}{36} - \frac{4x^2}{36} = 1$$

$$\frac{y^2}{4} - \frac{x^2}{9} = 1 \qquad \text{Simplify each fraction.}$$

This is the standard form of the equation of a hyperbola, centered at the origin, that opens up and down. To determine the vertices and the central rectangle of the graph, we find a and b.

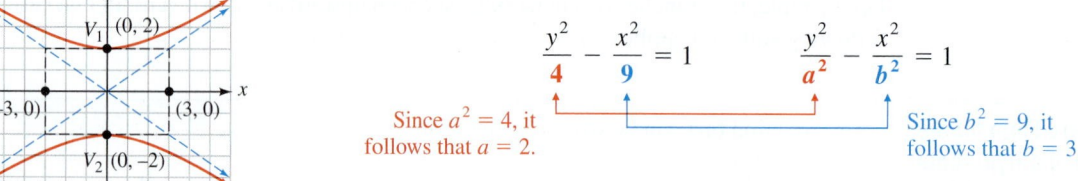

Since $a^2 = 4$, it follows that $a = 2$.   Since $b^2 = 9$, it follows that $b = 3$.

The y-intercepts are $(0, a)$ and $(0, -a)$, or $(0, 2)$ and $(0, -2)$. They are also the vertices of the hyperbola.

Since $a = 2$ and $b = 3$, the central rectangle passes through $(0, 2)$ and $(0, -2)$, and $(3, 0)$ and $(-3, 0)$. We draw its extended diagonals and sketch the hyperbola.

(left margin graph labels)
V_1 $(0, 2)$
$(-3, 0)$
$(3, 0)$
V_2 $(0, -2)$
$9y^2 - 4x^2 = 36$
or
$\dfrac{y^2}{4} - \dfrac{x^2}{9} = 1$

Self Check 2   Graph: $16y^2 - x^2 = 16$.

ACCENT ON TECHNOLOGY: GRAPHING HYPERBOLAS

To graph $\dfrac{x^2}{9} - \dfrac{y^2}{16} = 1$ from Example 1 using a graphing calculator, we follow the same procedure that we used for circles and ellipses. To write the equation as two functions, we solve for y to get $y = \pm \dfrac{\sqrt{16x^2 - 144}}{3}$. Then we graph the following two functions in a square window setting to get the graph of the hyperbola shown below.

$$y = \frac{\sqrt{16x^2 - 144}}{3} \qquad \text{and} \qquad y = -\frac{\sqrt{16x^2 - 144}}{3}$$

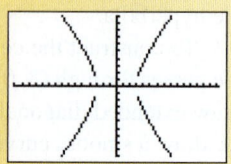

■ GRAPHING HYPERBOLAS CENTERED AT (*h*, *k*)

If a hyperbola is centered at a point with coordinates (h, k), the following equations apply.

Standard Forms for the Equations of Hyperbolas Centered at (*h*, *k*)

Any equation that can be written in the form

$$\frac{(x-h)^2}{a^2} - \frac{(y-k)^2}{b^2} = 1$$

is a hyperbola that has its center at (h, k) and opens left and right.
    Any equation of the form

$$\frac{(y-k)^2}{a^2} - \frac{(x-h)^2}{b^2} = 1$$

is a hyperbola that has its center at (h, k) and opens up and down.

EXAMPLE 3

ELEMENTARY & INTERMEDIATE
Algebra $f(x)$ **Now**™

Graph: $\dfrac{(x-3)^2}{16} - \dfrac{(y+1)^2}{4} = 1$.

Solution   We write the equation in the form

$$\frac{(x-3)^2}{4^2} - \frac{[y-(-1)]^2}{2^2} = 1$$

to see that its graph will be a hyperbola centered at the point $(h, k) = (3, -1)$. Its vertices are located at $a = 4$ units to the right and left of the center, at $(7, -1)$ and $(-1, -1)$. Since $b = 2$, we can count 2 units above and below the center to locate points $(3, 1)$ and $(3, -3)$. With these four points, we can draw the central rectangle along with its extended diagonals. We can then sketch the hyperbola, as shown.

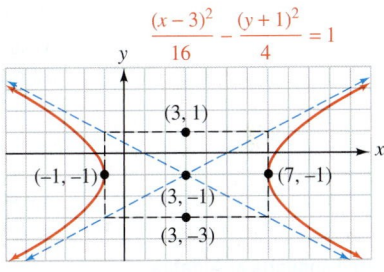

Self Check 3   Graph: $\dfrac{(x+2)^2}{9} - \dfrac{(y-1)^2}{4} = 1$.

EXAMPLE 4

ELEMENTARY & INTERMEDIATE
Algebra $f(x)$ **Now**™

Write the equation $x^2 - y^2 - 2x + 4y = 12$ in standard form to show that the equation represents a hyperbola. Then graph it.

Solution   We proceed as follows.

$$x^2 - y^2 - 2x + 4y = 12$$

$$x^2 - 2x - y^2 + 4y = 12 \qquad \text{Use the commutative property to rearrange terms.}$$

$$x^2 - 2x - (y^2 - 4y) = 12 \qquad \text{Factor } -1 \text{ from } -y^2 + 4y.$$

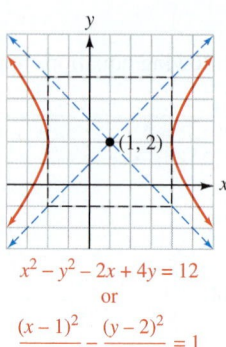

$x^2 - y^2 - 2x + 4y = 12$
or
$$\frac{(x-1)^2}{9} - \frac{(y-2)^2}{9} = 1$$

We then complete the square on x and y to make $x^2 - 2x$ and $y^2 - 4y$ perfect square trinomials.

$$x^2 - 2x + 1 - (y^2 - 4y + 4) = 12 + 1 - 4$$ Add 1 to both sides and subtract 4 from both sides.

We then factor $x^2 - 2x + 1$ and $y^2 - 4y + 4$ to get

$$(x-1)^2 - (y-2)^2 = 9$$
$$\frac{(x-1)^2}{9} - \frac{(y-2)^2}{9} = 1$$ Divide both sides by 9.

This is the equation of a hyperbola with center at $(1, 2)$. Its graph is shown in the figure.

Self Check 4   Graph: $x^2 - 4y^2 + 2x - 8y = 7$.

There is a special type of hyperbola (also centered at the origin) that does not intersect either the x- or the y-axis. These hyperbolas have equations of the form $xy = k$, where $k \neq 0$.

EXAMPLE 5

**ELEMENTARY &
INTERMEDIATE
Algebra $f(x)$ Now™**

Graph: $xy = -8$.

Solution   To make a table of solutions, we can solve the equation $xy = -8$ for y:

$$y = \frac{-8}{x}$$

Then we choose several values for x, find the corresponding values of y, and record the results in the table below. We plot the ordered pairs and join them with a smooth curve to obtain the graph of the hyperbola.

The Language of Algebra

The asymptotes of this hyperbola are the x- and y-axes. A hyperbola for which the asymptotes are perpendicular is called a **rectangular hyperbola**.

$$xy = -8 \quad \text{or} \quad y = \frac{-8}{x}$$

x	y	
1	-8	→ $(1, -8)$
2	-4	→ $(2, -4)$
4	-2	→ $(4, -2)$
8	-1	→ $(8, -1)$
-1	8	→ $(-1, 8)$
-2	4	→ $(-2, 4)$
-4	2	→ $(-4, 2)$
-8	1	→ $(-8, 1)$

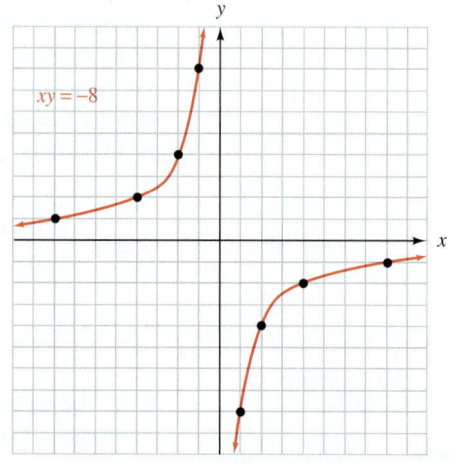

Self Check 5   Graph: $xy = 6$.

The result in Example 5 illustrates the following general equation.

Equations of Hyperbolas of the Form $xy = k$ Any equation of the form $xy = k$, where $k \neq 0$, has a graph that is a **hyperbola**, which does not intersect either the x- or the y-axis.

■ PROBLEM SOLVING USING HYPERBOLAS

EXAMPLE 6

ELEMENTARY &
INTERMEDIATE
Algebra *f(x)* **Now**™

Atomic structure. In an experiment that led to the discovery of the atomic structure of matter, Lord Rutherford (1871–1937) shot high-energy alpha particles toward a thin sheet of gold. Many of them were reflected, and Rutherford showed the existence of the nucleus of a gold atom. An alpha particle is repelled by the nucleus at the origin; it travels along the hyperbolic path given by $4x^2 - y^2 = 16$. How close does the particle come to the nucleus?

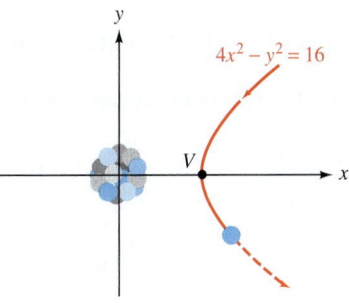

Solution

To find the distance from the nucleus at the origin, we must find the coordinates of the vertex *V*. To do so, we write the equation of the particle's path in standard form:

$$4x^2 - y^2 = 16$$

$$\frac{4x^2}{16} - \frac{y^2}{16} = \frac{16}{16} \qquad \text{Divide both sides by 16.}$$

$$\frac{x^2}{4} - \frac{y^2}{16} = 1 \qquad \text{Simplify.}$$

$$\frac{x^2}{2^2} - \frac{y^2}{4^2} = 1 \qquad \text{To determine } a \text{ and } b, \text{ write 4 as } 2^2 \text{ and 16 as } 4^2.$$

This equation is in the form

$$\frac{x^2}{a^2} - \frac{y^2}{b^2} = 1$$

with $a = 2$. Thus, the vertex of the path is (2, 0). The particle is never closer than 2 units from the nucleus

Answers to Self Checks

1.

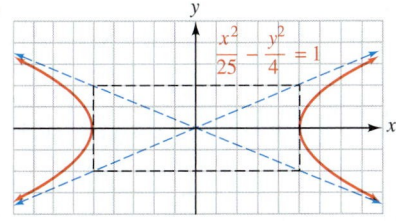

2.

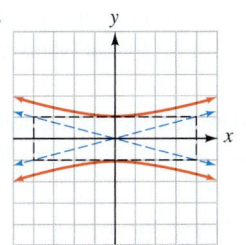

$16y^2 - x^2 = 16$
or
$\dfrac{y^2}{1} - \dfrac{x^2}{16} = 1$

3.

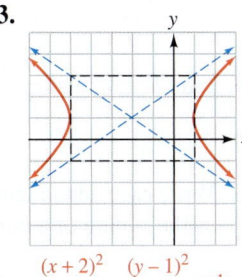

$\dfrac{(x+2)^2}{9} - \dfrac{(y-1)^2}{4} = 1$

4.

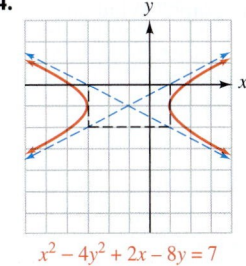

$x^2 - 4y^2 + 2x - 8y = 7$
or
$\dfrac{(x+1)^2}{4} - \dfrac{(y+1)^2}{1} = 1$

5.

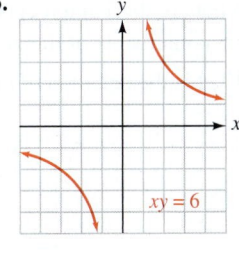

$xy = 6$

13.3   **STUDY SET**   ELEMENTARY &
INTERMEDIATE
Algebra *f(x)* **Now**™

VOCABULARY   Fill in the blanks.

1. The two-branch curve graphed below is a _____.

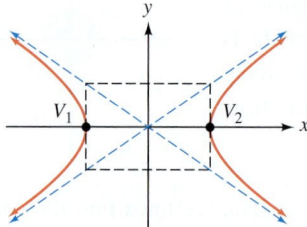

2. A _____ is the set of all points in a plane for which the difference of the distances from two fixed points is a constant.

3. In the graph above, V_1 and V_2 are the _____ of the hyperbola.

4. In the graph above, the figure drawn using dashed black lines is called the _____ _____.

5. The extended _____ of the central rectangle are asymptotes of the hyperbola.

6. To write $9x^2 - 4y^2 = 36$ in _____ form, we divide both sides by 36.

CONCEPTS

7. What is the standard form for the equation of a hyperbola centered at the origin that opens left and right?

8. What is the standard form for the equation of a hyperbola centered at (h, k) that opens up and down?

9. What is the standard form for the equation of a hyperbola centered at (h, k) that opens left and right?

10. a. What is the center of the hyperbola graphed below? What are a and b?

   b. What are the x-intercepts of the graph? What are the y-intercepts of the graph?

   c. What is the equation of the hyperbola?

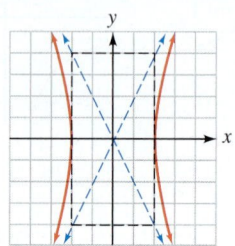

11. a. What is the center of the hyperbola graphed below? What are a and b?

   b. What is the equation of the hyperbola?

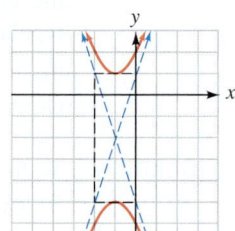

12. a. Fill in the blank: An equation of the form $xy = k$, where $k \neq 0$, has a graph that is a _____ that does not intersect either the x-axis or the y-axis.

   b. Complete the table of solutions for $xy = 10$.

x	y
-2	
5	

13. Divide both sides of the equation by 100 and write the equation in standard form:
$$100(x + 1)^2 - 25(y - 5)^2 = 100$$

14. Determine whether the equation, when graphed, will be a circle, a parabola, an ellipse, or a hyperbola.

   a. $(x - 8)^2 + y^2 = 10$ **b.** $\dfrac{y^2}{16} - \dfrac{x^2}{9} = 1$

   c. $x = y^2 - 3y + 6$

   d. $\dfrac{(x - 4)^2}{1} + \dfrac{(y + 12)^2}{49} = 1$

NOTATION

15. Find h, k, a, and b: $\dfrac{(x - 5)^2}{25} - \dfrac{(y + 11)^2}{36} = 1$.

16. Write each denominator in the equation $\dfrac{x^2}{36} - \dfrac{y^2}{81} = 1$ as the square of a number.

PRACTICE   Graph each hyperbola.

17. $\dfrac{x^2}{9} - \dfrac{y^2}{4} = 1$

18. $\dfrac{x^2}{4} - \dfrac{y^2}{4} = 1$

19. $\dfrac{y^2}{4} - \dfrac{x^2}{9} = 1$

20. $\dfrac{y^2}{4} - \dfrac{x^2}{64} = 1$

21. $25x^2 - y^2 = 25$

22. $9x^2 - 4y^2 = 36$

23. $\dfrac{(x-2)^2}{9} - \dfrac{y^2}{16} = 1$

24. $\dfrac{(x+2)^2}{16} - \dfrac{(y-3)^2}{25} = 1$

25. $\dfrac{(y+1)^2}{1} - \dfrac{(x-2)^2}{4} = 1$

26. $\dfrac{(y-2)^2}{4} - \dfrac{(x+1)^2}{1} = 1$

27. $4(x+3)^2 - (y-1)^2 = 4$

28. $(x+5)^2 - 16y^2 = 16$

29. $\dfrac{y^2}{25} - \dfrac{(x-2)^2}{4} = 1$

30. $\dfrac{y^2}{36} - \dfrac{(x+2)^2}{4} = 1$

Write each equation in standard form and graph it.

31. $x^2 + 2x - y^2 - 2y = 9$
32. $x^2 - 4x - y^2 + 2y = 13$
33. $x^2 - y^2 - 6y = 34$
34. $y^2 - 2y - x^2 - 8 = 0$
35. $xy = 8$ **36.** $xy = -10$

Use a graphing calculator to graph each equation.

37. $\dfrac{x^2}{9} - \dfrac{y^2}{4} = 1$ **38.** $y^2 - 16x^2 = 16$

39. $\dfrac{x^2}{4} - \dfrac{(y-1)^2}{9} = 1$

40. $\dfrac{(y+1)^2}{9} - \dfrac{(x-2)^2}{4} = 1$

APPLICATIONS

41. ALPHA PARTICLES   The particle in the illustration approaches the nucleus at the origin along the path $9y^2 - x^2 = 81$ in the coordinate system shown. How close does the particle come to the nucleus?

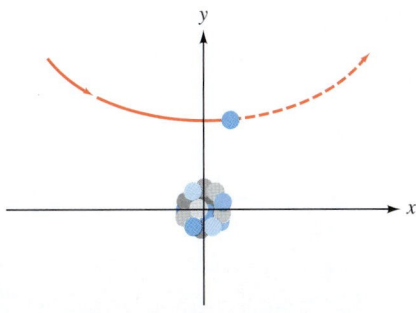

42. LORAN   By determining the difference of the distances between the ship in the illustration and two radio transmitters, the LORAN navigation system places the ship on the hyperbola $x^2 - 4y^2 = 576$ in the coordinate system shown. If the ship is 5 miles out to sea, find its coordinates.

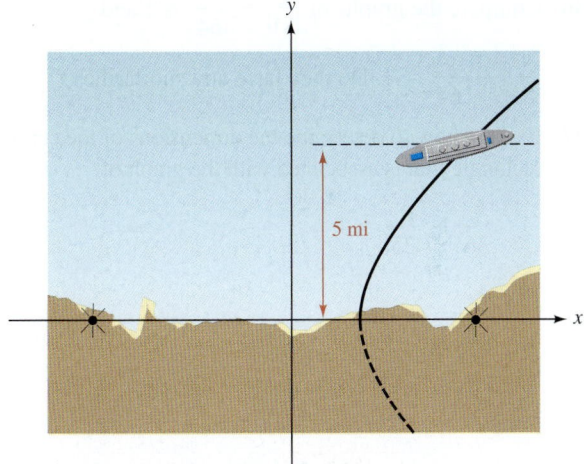

5 mi

43. SONIC BOOM   The position of a sonic boom caused by the faster-than-sound aircraft is one branch of the hyperbola $y^2 - x^2 = 25$ in the coordinate system shown. How wide is the hyperbola 5 miles from its vertex?

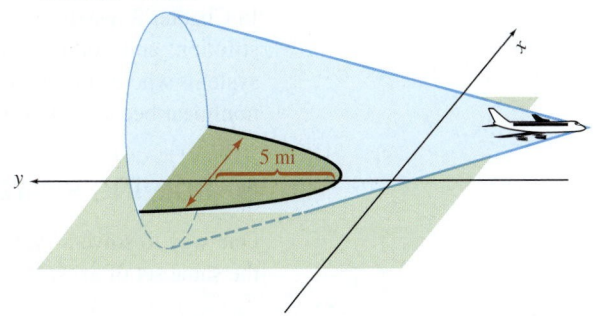

5 mi

44. FLUIDS   See the illustration below. Two glass plates in contact at the left, and separated by about 5 millimeters on the right, are dipped in beet juice, which rises by capillary action to form a hyperbola. The hyperbola is modeled by an equation of the form $xy = k$. If the curve passes through the point $(12, 2)$, what is k?

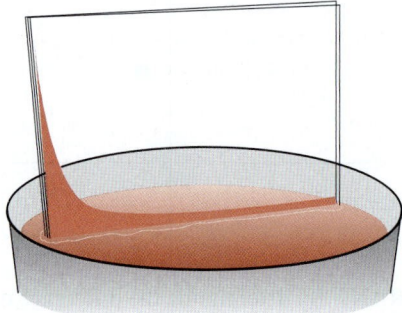

WRITING

45. What is a hyperbola?

46. Compare the graphs of $\dfrac{x^2}{81} - \dfrac{y^2}{64} = 1$ and

$\dfrac{y^2}{81} - \dfrac{x^2}{64} = 1$. Do they have any similarities?

47. Explain how to determine the dimensions of the central rectangle that is associated with the graph of

$$\frac{x^2}{36} - \frac{y^2}{25} = 1$$

48. Explain why the graph of the hyperbola

$$\frac{x^2}{a^2} - \frac{y^2}{b^2} = 1$$

has no y-intercept.

REVIEW   **Find each value of x.**

49. $\log_8 x = 2$

50. $\log_{25} x = \dfrac{1}{2}$

51. $\log_{1/2} \dfrac{1}{8} = x$

52. $\log_{12} x = 0$

53. $\log_x \dfrac{9}{4} = 2$

54. $\log_6 216 = x$

CHALLENGE PROBLEMS

55. Write $36x^2 - 25y^2 - 72x - 100y = 964$ in the standard form of the equation of a hyperbola.

56. Explain how a plane could intersect two right-circular cones to form two intersecting lines. Draw a picture to illustrate this.

57. Write an equation of a hyperbola whose graph has the following characteristics:

- vertices $(\pm 1, 0)$
- equations of asymptotes: $y = \pm 5x$

58. Graph: $16x^2 - 25y^2 = 1$.

13.4   Solving Nonlinear Systems of Equations

- Solving systems by graphing   • Solving systems by substitution
- Solving systems by elimination

In Chapter 3, we discussed how to solve systems of linear equations by the graphing, substitution, and elimination methods. In this section, we will use these methods to solve systems where at least one of the equations is nonlinear. Recall that equations are classified nonlinear because their graphs are not straight lines.

■ **SOLVING SYSTEMS BY GRAPHING**

One way to solve a system of two equations in two variables is to graph the equations on the same set of axes.

EXAMPLE 1

ELEMENTARY &
INTERMEDIATE
Algebra *f(x)* **Now**™   Solution

Success Tip

It is helpful to sketch the possibilities before solving the system:

2 points of intersection:
(2 real solutions)

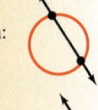

1 point of intersection:
(1 real solution)

1 point of intersection:
(1 real solution)

No points of intersection:
(0 real solutions)

Solve $\begin{cases} x^2 + y^2 = 25 \\ 2x + y = 10 \end{cases}$ by graphing.

The graph of $x^2 + y^2 = 25$ is a circle with center at the origin and radius of 5. The graph of $2x + y = 10$ is a line. Depending on whether the line is a **secant** (intersecting the circle at two points) or a **tangent** (intersecting the circle at one point) or does not intersect the circle at all, there are two, one, or no solutions to the system, respectively.

After graphing the circle and the line, it appears that the points of intersection are $(5, 0)$ and $(3, 4)$. To verify that they are solutions of the system, we need to check each one.

Check:

For (5, 0)

$$2x + y = 10 \qquad x^2 + y^2 = 25$$
$$2(5) + 0 \overset{?}{=} 10 \qquad 5^2 + 0^2 \overset{?}{=} 25$$
$$10 = 10 \qquad 25 = 25$$

For (3, 4)

$$2x + y = 10 \qquad x^2 + y^2 = 25$$
$$2(3) + 4 \overset{?}{=} 10 \qquad 3^2 + 4^2 \overset{?}{=} 25$$
$$10 = 10 \qquad 25 = 25$$

The ordered pair $(5, 0)$ satisfies both equations of the system, and so does $(3, 4)$. Thus, there are two solutions, $(5, 0)$ and $(3, 4)$, and the solution set is $\{(5, 0), (3, 4)\}$.

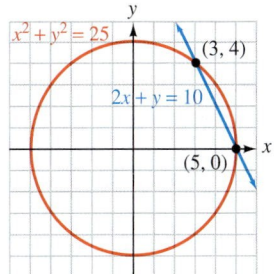

Self Check 1   Solve: $\begin{cases} x^2 + y^2 = 13 \\ y = -\dfrac{1}{5}x + \dfrac{13}{5} \end{cases}$.

ACCENT ON TECHNOLOGY: SOLVING SYSTEMS OF EQUATIONS

To solve Example 1 with a graphing calculator, we graph the circle and the line on one set of coordinate axes. (See figure (a).) We then trace to find the coordinates of the intersection points of the graphs. (See figures (b) and (c).)

We can zoom for better results.

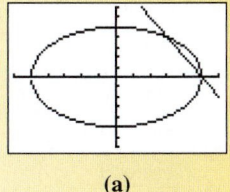

(a)

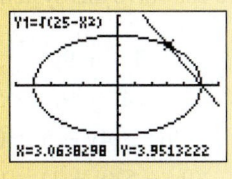

(b)

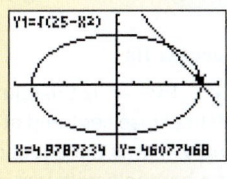

(c)

■ SOLVING SYSTEMS BY SUBSTITUTION

When solving a system by graphing, it is often difficult to determine the coordinates of the intersection points. A more precise algebraic method called the **substitution method** can be used to solve certain systems involving nonlinear equations.

EXAMPLE 2

Solution

Solve $\begin{cases} x^2 + y^2 = 2 \\ 2x - y = 1 \end{cases}$ by substitution.

This system has one second-degree equation and one first-degree equation. We can solve this type of system by substitution. Solving the linear equation for y gives

$$2x - y = 1$$
$$-y = -2x + 1 \qquad \text{Subtract } 2x \text{ from both sides.}$$
$$y = 2x - 1 \qquad \text{Multiply both sides by } -1. \text{ We call this the substitution equation.}$$

We can substitute $2x - 1$ for y in the second-degree equation and solve the resulting quadratic equation for x:

$$x^2 + y^2 = 2$$
$$x^2 + (2x - 1)^2 = 2$$
$$x^2 + 4x^2 - 4x + 1 = 2 \qquad \text{Find } (2x - 1)^2.$$
$$5x^2 - 4x - 1 = 0 \qquad \text{Combine like terms and subtract 2 from both sides.}$$
$$(5x + 1)(x - 1) = 0 \qquad \text{Factor.}$$
$$5x + 1 = 0 \qquad \text{or} \qquad x - 1 = 0 \qquad \text{Set each factor equal to 0.}$$
$$x = -\frac{1}{5} \qquad\qquad x = 1$$

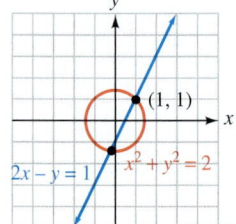

If we substitute $-\frac{1}{5}$ for x in the equation $y = 2x - 1$, we get $y = -\frac{7}{5}$. If we substitute 1 for x in $y = 2x - 1$, we get $y = 1$. Thus, the system has two solutions, $\left(-\frac{1}{5}, -\frac{7}{5}\right)$ and $(1, 1)$. Verify that each ordered pair satisfies both equations of the original system.

The graph in the margin confirms that the system has two solutions, and that one of them is $(1, 1)$. However, it would be virtually impossible to determine that the coordinates of the second point of intersection are $\left(-\frac{1}{5}, -\frac{7}{5}\right)$ from the graph.

Self Check 2 Solve $\begin{cases} x^2 + y^2 = 10 \\ y = x + 2 \end{cases}$ by substitution.

EXAMPLE 3

Solution

Solve: $\begin{cases} 4x^2 + 9y^2 = 5 \\ y = x^2 \end{cases}$.

We can solve this system by substitution.

$$4x^2 + 9y^2 = 5$$
$$4y + 9y^2 = 5 \qquad \text{Substitute } y \text{ for } x^2.$$
$$9y^2 + 4y - 5 = 0 \qquad \text{Subtract 5 from both sides.}$$
$$(9y - 5)(y + 1) = 0 \qquad \text{Factor } 9y^2 + 4y - 5.$$
$$9y - 5 = 0 \qquad \text{or} \qquad y + 1 = 0 \qquad \text{Set each factor equal to 0.}$$
$$y = \frac{5}{9} \qquad\qquad y = -1$$

Success Tip

$4x^2 + 9y^2 = 5$ is the equation of an ellipse centered at $(0, 0)$, and $y = x^2$ is the equation of a parabola with vertex at $(0, 0)$, opening upward. We would expect two solutions.

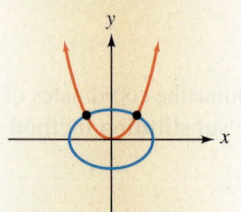

Since $y = x^2$, the values of x are found by solving the equations

$$x^2 = \frac{5}{9} \qquad \text{and} \qquad \bcancel{x^2 = -1}$$

Because $x^2 = -1$ has no real solutions, this possibility is discarded. The solutions of $x^2 = \frac{5}{9}$ are

$$x = \frac{\sqrt{5}}{3} \quad \text{or} \quad x = -\frac{\sqrt{5}}{3}$$

Thus, the solutions of the system are

$$\left(\frac{\sqrt{5}}{3}, \frac{5}{9} \right) \quad \text{and} \quad \left(-\frac{\sqrt{5}}{3}, \frac{5}{9} \right)$$

Self Check 3   Solve: $\begin{cases} x^2 + y^2 = 20 \\ y = x^2 \end{cases}$.

■ SOLVING SYSTEMS BY ELIMINATION

EXAMPLE 4

Solve: $\begin{cases} 3x^2 + 2y^2 = 36 \\ 4x^2 - y^2 = 4 \end{cases}$.

Solution   To solve this system of two second-degree equations, we can use either the substitution or the elimination method. We will use the elimination method because the y^2-terms can be eliminated by multiplying the second equation by 2 and adding it to the first equation.

$$\begin{cases} 3x^2 + 2y^2 = 36 \\ 4x^2 - y^2 = 4 \end{cases} \xrightarrow[\text{Multiply by 2}]{\text{Unchanged}} \begin{array}{l} 3x^2 + 2y^2 = 36 \\ 8x^2 - 2y^2 = 8 \end{array}$$

We add the two equations on the right to eliminate y^2 and solve the resulting equation for x:

$$11x^2 = 44$$
$$x^2 = 4$$
$$x = 2 \quad \text{or} \quad x = -2$$

To find y, we can substitute 2 for x and then -2 for x into any equation containing both variables. It appears that the computations will be simplest if we use $3x^2 + 2y^2 = 36$.

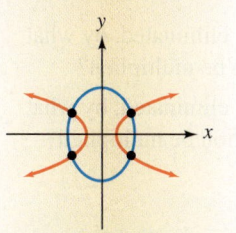

For x = 2	*For x = -2*
$3x^2 + 2y^2 = 36$	$3x^2 + 2y^2 = 36$
$3(2)^2 + 2y^2 = 36$	$3(-2)^2 + 2y^2 = 36$
$12 + 2y^2 = 36$	$12 + 2y^2 = 36$
$2y^2 = 24$	$2y^2 = 24$
$y^2 = 12$	$y^2 = 12$

$$y = \sqrt{12} \quad \text{or} \quad y = -\sqrt{12} \qquad\qquad y = \sqrt{12} \quad \text{or} \quad y = -\sqrt{12}$$
$$y = 2\sqrt{3} \quad\;\;\; \mid \quad\;\; y = -2\sqrt{3} \qquad\qquad y = 2\sqrt{3} \quad\;\;\; \mid \quad\;\; y = -2\sqrt{3}$$

The four solutions of this system are

$$\left(2, 2\sqrt{3}\right), \quad \left(2, -2\sqrt{3}\right), \quad \left(-2, 2\sqrt{3}\right), \quad \text{and} \quad \left(-2, -2\sqrt{3}\right)$$

Self Check 4   Solve: $\begin{cases} x^2 + 4y^2 = 16 \\ x^2 - y^2 = 1 \end{cases}$.

Answers to Self Checks   **1.** $(3, 2), (-2, 3)$

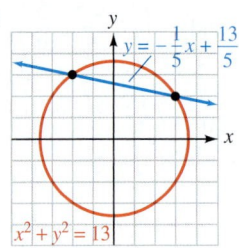

2. $(1, 3), (-3, -1)$ **3.** $(2, 4), (-2, 4)$

4. $\left(2, \sqrt{3}\right), \left(2, -\sqrt{3}\right), \left(-2, \sqrt{3}\right), \left(-2, -\sqrt{3}\right)$

13.4   STUDY SET   ELEMENTARY & INTERMEDIATE   Algebra $f(x)$ Now™

VOCABULARY   Fill in the blanks.

1. $\begin{cases} 4x^2 + 6y^2 = 24 \\ 9x^2 - y^2 = 9 \end{cases}$ is a _____ of two nonlinear equations.

2. _____ equations have graphs that are not straight lines.

3. When solving a system by _____, it is often difficult to determine the coordinates of the intersection points.

4. Two algebraic methods for solving systems of nonlinear equations are the _____ method and the _____ method.

5. A _____ is a line that intersects a circle at two points.

6. A _____ is a line that intersects a circle at one point.

CONCEPTS

7. a. At most, a line can intersect an ellipse at _____ points.

b. At most, an ellipse can intersect a parabola at _____ points.

c. At most, an ellipse can intersect a circle at _____ points.

d. At most, a hyperbola can intersect a circle at _____ points.

8. Check to determine whether $(1, -1)$ is a solution of the system $\begin{cases} 2x + y - 1 = 0 \\ x^2 + y^2 = 3 \end{cases}$.

9. Determine the solutions of the system $\begin{cases} x^2 + 4y^2 = 25 \\ x^2 - 2y^2 = 1 \end{cases}$ that is graphed below.

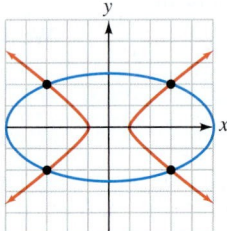

10. What should be used as the substitution equation to solve the system $\begin{cases} 2x^2 - y^2 = 6 \\ x^2 - y = 3 \end{cases}$?

11. Consider the system $\begin{cases} 6x^2 + y^2 = 9 \\ 3x^2 + 4y^2 = 36 \end{cases}$.

a. If the y^2-terms are to be eliminated, by what should the first equation be multiplied?

b. If the x^2-terms are to be eliminated, by what should the second equation be multiplied?

12. Suppose you begin to solve the system $\begin{cases} x^2 + y^2 = 10 \\ 4x^2 + y^2 = 13 \end{cases}$ and find that x is ± 1. Use the first equation to find the corresponding y-values for $x = 1$ and $x = -1$. State the solutions as ordered pairs.

NOTATION Complete each solution.

13. Solve: $\begin{cases} x^2 + y^2 = 5 \\ y = 2x \end{cases}$.

$$x^2 + y^2 = 5$$
$$x^2 + (\;\;\;)^2 = 5$$
$$x^2 + 4x^2 = \;\;\;$$
$$\;\;\;x^2 = 5$$
$$x^2 = \;\;\;$$
$$x = 1 \quad \text{or} \quad x = -1$$
If $x = 1$, then $y = 2(\;\;\;) = 2$.
If $x = -1$, then $y = 2(\;\;\;) = -2$.
The solutions are $(1, 2)$ and $\left(-1, \;\;\;\right)$.

14. Solve: $\begin{cases} x^2 + y^2 = 13 \\ x^2 - y^2 = 5 \end{cases}$

$$2x^2 = \;\;\; \qquad \text{Add the equations.}$$
$$x^2 = \;\;\;$$
$$x = \;\;\; \quad \text{or} \quad x = \;\;\;$$

$$2y^2 = \;\;\; \qquad \text{Subtract the equations.}$$
$$y^2 = \;\;\;$$
$$y = \;\;\; \quad \text{or} \quad y = \;\;\;$$
The solutions are
$$\left(3, \;\;\;\right), \left(3, \;\;\;\right), (-3, 2), (-3, -2)$$

PRACTICE Solve each system of equations by graphing.

15. $\begin{cases} 8x^2 + 32y^2 = 256 \\ x = 2y \end{cases}$
16. $\begin{cases} x^2 + y^2 = 2 \\ x + y = 2 \end{cases}$

17. $\begin{cases} x^2 + y^2 = 10 \\ y = 3x^2 \end{cases}$
18. $\begin{cases} x^2 + y^2 = 5 \\ x + y = 3 \end{cases}$

19. $\begin{cases} x^2 + y^2 = 25 \\ 12x^2 + 64y^2 = 768 \end{cases}$
20. $\begin{cases} x^2 + y^2 = 13 \\ y = x^2 - 1 \end{cases}$

21. $\begin{cases} x^2 - 13 = -y^2 \\ y = \frac{2}{3}x \end{cases}$
22. $\begin{cases} x^2 + y^2 = 20 \\ y = x^2 \end{cases}$

Use a graphing calculator to solve each system.

23. $\begin{cases} x^2 - 6x - y = -5 \\ x^2 - 6x + y = -5 \end{cases}$

24. $\begin{cases} x^2 - y^2 = -5 \\ 3x^2 + 2y^2 = 30 \end{cases}$

Solve each system of equations algebraically for real values of x and y.

25. $\begin{cases} 25x^2 + 9y^2 = 225 \\ 5x + 3y = 15 \end{cases}$
26. $\begin{cases} x^2 + y^2 = 20 \\ y = x^2 \end{cases}$

27. $\begin{cases} x^2 + y^2 = 2 \\ x + y = 2 \end{cases}$
28. $\begin{cases} x^2 + y^2 = 36 \\ 49x^2 + 36y^2 = 1{,}764 \end{cases}$

29. $\begin{cases} x^2 + y^2 = 5 \\ x + y = 3 \end{cases}$
30. $\begin{cases} x^2 - x - y = 2 \\ 4x - 3y = 0 \end{cases}$

31. $\begin{cases} x^2 + y^2 = 13 \\ y = x^2 - 1 \end{cases}$
32. $\begin{cases} x^2 + y^2 = 25 \\ 2x^2 - 3y^2 = 5 \end{cases}$

33. $\begin{cases} x^2 + y^2 = 30 \\ y = x^2 \end{cases}$
34. $\begin{cases} 9x^2 - 7y^2 = 81 \\ x^2 + y^2 = 9 \end{cases}$

35. $\begin{cases} x^2 + y^2 = 13 \\ x^2 - y^2 = 5 \end{cases}$
36. $\begin{cases} 2x^2 + y^2 = 6 \\ x^2 - y^2 = 3 \end{cases}$

37. $\begin{cases} x^2 + y^2 = 20 \\ x^2 - y^2 = -12 \end{cases}$
38. $\begin{cases} xy = -\dfrac{9}{2} \\ 3x + 2y = 6 \end{cases}$

39. $\begin{cases} y^2 = 40 - x^2 \\ y = x^2 - 10 \end{cases}$
40. $\begin{cases} x^2 - 6x - y = -5 \\ x^2 - 6x + y = -5 \end{cases}$

41. $\begin{cases} y = x^2 - 4 \\ x^2 - y^2 = -16 \end{cases}$
42. $\begin{cases} 6x^2 + 8y^2 = 182 \\ 8x^2 - 3y^2 = 24 \end{cases}$

43. $\begin{cases} x^2 - y^2 = -5 \\ 3x^2 + 2y^2 = 30 \end{cases}$ **44.** $\begin{cases} \dfrac{1}{x} + \dfrac{1}{y} = 5 \\ \dfrac{1}{x} - \dfrac{1}{y} = -3 \end{cases}$

45. $\begin{cases} \dfrac{1}{x} + \dfrac{2}{y} = 1 \\ \dfrac{2}{x} - \dfrac{1}{y} = \dfrac{1}{3} \end{cases}$ **46.** $\begin{cases} \dfrac{1}{x} + \dfrac{3}{y} = 4 \\ \dfrac{2}{x} - \dfrac{1}{y} = 7 \end{cases}$

47. $\begin{cases} 3y^2 = xy \\ 2x^2 + xy - 84 = 0 \end{cases}$ **48.** $\begin{cases} x^2 + y^2 = 10 \\ 2x^2 - 3y^2 = 5 \end{cases}$

49. $\begin{cases} xy = \dfrac{1}{6} \\ y + x = 5xy \end{cases}$ **50.** $\begin{cases} xy = \dfrac{1}{12} \\ y + x = 7xy \end{cases}$

51. INTEGER PROBLEM   The product of two integers is 32, and their sum is 12. Find the integers.

52. NUMBER PROBLEM   The sum of the squares of two numbers is 221, and the sum of the numbers is 9. Find the numbers.

APPLICATIONS

53. GEOMETRY   The area of a rectangle is 63 square centimeters, and its perimeter is 32 centimeters. Find the dimensions of the rectangle.

54. INVESTING   Grant receives $225 annual income from one investment. Jeff invested $500 more than Grant, but at an annual rate of 1% less. Jeff's annual income is $240. What are the amount and rate of Grant's investment?

55. INVESTING   Carol receives $67.50 annual income from one investment. John invested $150 more than Carol at an annual rate of $1\frac{1}{2}$% more. John's annual income is $94.50. What are the amount and rate of Carol's investment? (*Hint:* There are two answers.)

56. ARTILLERY   See the illustration in the next column. A shell fired from the base of a hill follows the parabolic path $y = -\frac{1}{6}x^2 + 2x$, with distances measured in miles. The hill has a slope of $\frac{1}{3}$. How far from the cannon is the point of impact? (*Hint:* Find the coordinates of the point and then the distance.)

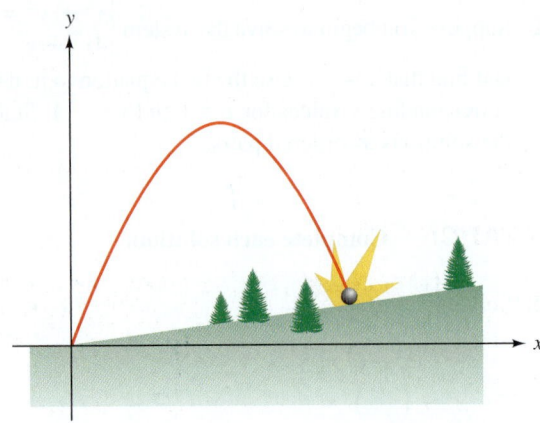

57. DRIVING RATES   Jim drove 306 miles. Jim's brother made the same trip at a speed 17 mph slower than Jim did and required an extra $1\frac{1}{2}$ hours. What was Jim's rate and time?

58. FENCING PASTURES   The rectangular pasture shown below is to be fenced in along a riverbank. If 260 feet of fencing is to enclose an area of 8,000 square feet, find the dimensions of the pasture.

WRITING

59. a. Describe the benefits of the graphical method for solving a system of equations.

   b. Describe the drawbacks of the graphical method.

60. Explain why the elimination method, not the substitution method, is the better method to solve the system

$$\begin{cases} 4x^2 + 9y^2 = 52 \\ 9x^2 + 4y^2 = 52 \end{cases}$$

REVIEW   Solve each equation.

61. $\log 5x = 4$ **62.** $\log 3x = \log 9$

63. $\dfrac{\log(8x - 7)}{\log x} = 2$ **64.** $\log x + \log(x + 9) = 1$

CHALLENGE PROBLEMS

65. a. The graphs of the two independent equations of a system are parabolas. How many solutions might the system have?

b. The graphs of the two independent equations of a system are hyperbolas. How many solutions might the system have?

66. Solve the system $\begin{cases} x^2 - y^2 = 16 \\ x^2 + y^2 = 9 \end{cases}$ over the complex numbers.

ACCENT ON TEAMWORK

CONIC SECTIONS

Overview: In this activity, you will construct the four basic conic sections from clay.

Instructions: Form groups of 3 students. Mold some clay into the shape of a right-circular cone, as shown. With both hands, one student should pull a thin wire through the clay to slice it in such a way that a circular shape results. Then he or she should slice it to get an elliptical shape, a parabolic shape, and one branch of a hyperbolic shape.

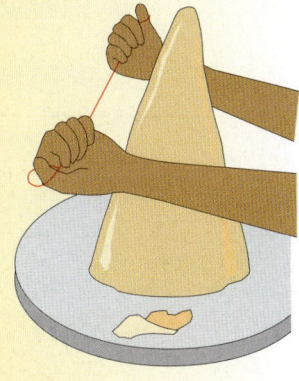

The second student should then mold the pieces back together into a right-circular cone whose base is wider and whose height is shorter than the first model, and slice it to create the four conics.

Finally, the third student should mold the pieces back together into a right-circular cone whose base is narrower and whose height is taller than the first model and create the four conics.

PARABOLAS

Overview: In this activity, you will construct several models of parabolas.

Instructions: Form groups of 2 or 3 students. You will need a T-square, string, paper, pencil, and thumbtack. To construct a parabola, secure one end of a piece of string that is as long as the T-square to a large piece of paper using a brad or thumbtack, as shown below. Attach the other end of the string to the upper end of the T-square. Hold the string taut against the T-square with a pencil and slide the T-square along the edge of the table. As the T-square moves, the pencil will trace a parabola.

Each point on the parabola is the same distance away from a point as it is from a given line. With this model, what is the given point, and what is the given line?

Make other models by moving the fixed point closer and further away from the edge of the table. How is the shape of the parabola affected?

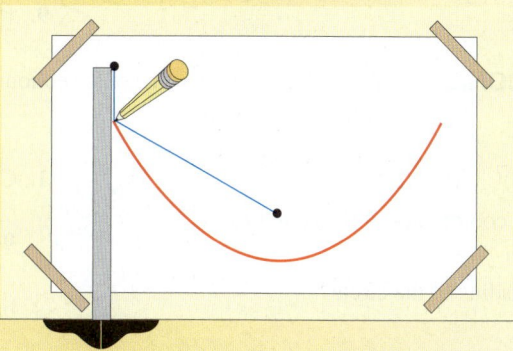

ELLIPSES

Overview: In this activity, you will construct several ellipses.

Instructions: Form groups of 2 or 3 students. You will need two thumbtacks, a pencil, and a length of string with a loop tied at one end. To construct an ellipse, place two thumbtacks (or brads) fairly close together, as shown in the illustration. Catch the loop of the string with the point of the pencil and, keeping the string taut, draw the ellipse.

Make several models by moving one of the thumbtacks farther away and then closer to the other thumbtack. How does the shape of the ellipse change?

For each point on the ellipse, the sum of the distances of the point from two given points is a constant. With this method of construction, what are the two points? What is the constant distance?

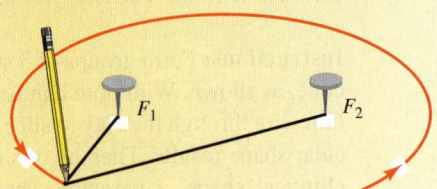

KEY CONCEPT: CONIC SECTIONS

In this chapter, we have studied four conic sections: the circle, the parabola, the ellipse, and the hyperbola. They are called conic sections because they are formed by the intersection of a plane and a pair of right-circular cones.

CLASSIFYING CONICS

In Exercises 1–9, classify the graph of each equation as a circle, a parabola, an ellipse, or a hyperbola.

1. $\dfrac{(x-2)^2}{9} + \dfrac{(y-1)^2}{4} = 1$

2. $(x-3)^2 + (y-2)^2 = 4$

3. $y = 4x^2 - 2x + 3$

4. $\dfrac{x^2}{4} - \dfrac{(y-1)^2}{9} = 1$

5. $4(x-7)^2 - (y+10)^2 = 4$

6. $\dfrac{x^2}{4} + \dfrac{y^2}{9} = 1$

7. $3(x-1)^2 + 12(y+2)^2 = 48$

8. $x^2 - 2y - 2x = -7$

9. $x^2 + y^2 + 8x + 2y = -13$

EQUATIONS OF CONIC SECTIONS

When the equation of a conic section is written in standard form, important features of its graph are apparent.

10. Consider $(x+1)^2 + (y-2)^2 = 16$.

   a. What are the coordinates of the center of the circle?

   b. What is the radius of the circle?

11. Consider $x = \dfrac{1}{2}(y-1)^2 - 2$.

   a. What are the coordinates of the vertex of the parabola?

   b. In which direction does the parabola open?

12. Consider $\dfrac{x^2}{4} + \dfrac{y^2}{16} = 1$.

   a. What are the coordinates of the center of the ellipse?

   b. Is the ellipse horizontal or vertical?

   c. What are the vertices of the ellipse?

13. Consider $\dfrac{(x+2)^2}{9} - \dfrac{(y-1)^2}{4} = 1$.

   a. What are the coordinates of the center of the hyperbola?

   b. In which direction do the branches of the hyperbola open?

   c. What are the dimensions of the central rectangle?

GRAPHING CONIC SECTIONS

Use the results from Exercises 10–13 to graph each conic section.

14. $(x+1)^2 + (y-2)^2 = 16$

15. $x = \dfrac{1}{2}(y-1)^2 - 2$

16. $\dfrac{x^2}{4} + \dfrac{y^2}{16} = 1$

17. $\dfrac{(x+2)^2}{9} - \dfrac{(y-1)^2}{4} = 1$

CHAPTER REVIEW

ELEMENTARY & INTERMEDIATE
Algebra $f(x)$ Now™

| SECTION 13.1 | **The Circle and the Parabola** |

CONCEPTS

Equations of a circle:
$$(x-h)^2 + (y-k)^2 = r^2$$
  center (h, k), radius r

$$x^2 + y^2 = r^2$$
  center $(0, 0)$, radius r

REVIEW EXERCISES

Graph each equation.

1. $x^2 + y^2 = 16$

2. $(x-1)^2 + (y+2)^2 = 4$

3. Write the equation in standard form and graph it.
$$x^2 + y^2 + 4x - 2y = 4$$

4. ART HISTORY Leonardo da Vinci's *Vitruvian Man* (1492) is one of the most famous pen-and-ink drawings of all time. Use the coordinate system that is superimposed on the drawing to write the equation of the circle in standard form.

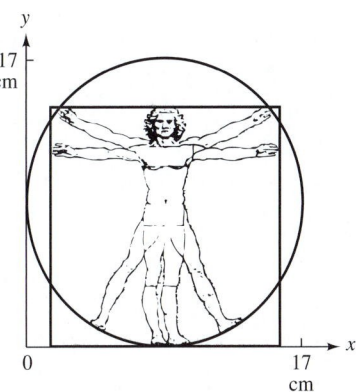

5. Find the center and the radius of the circle whose equation is $(x+6)^2 + y^2 = 24$.

6. Fill in the blanks: A circle is the set of all points in a plane that are a fixed distance from a point called its _____. The fixed distance is called the _____ of the circle.

Equations of parabolas

General forms:

$$y = ax^2 + bx + c$$

$a > 0$: up; $a < 0$: down

$$x = ay^2 + by + c$$

$a > 0$: right; $a < 0$: left

Standard forms:

$$y = a(x - h)^2 + k$$

$a > 0$: up; $a < 0$: down

Vertex: (h, k)

Axis of symmetry: $x = h$

$$x = a(y - k)^2 + h$$

$a > 0$: right; $a < 0$: left

Vertex: (h, k)

Axis of symmetry: $y = k$

Graph each parabola and give the coordinates of the vertex.

7. $x = y^2$ **8.** $x = 2(y + 1)^2 - 2$ **9.** $x = -3y^2 + 12y - 7$

10. Find the axis of symmetry of the graph of each equation.

    a. $x = (y - 4)^2 + 1$ **b.** $y = 2x^2 - 4x + 5$

11. The axis of symmetry, vertex, and two points on the graph of a parabola are shown. Find the coordinates of two other points on the parabola.

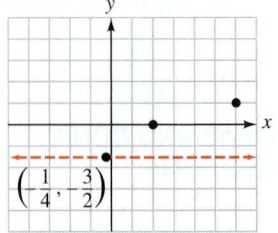

$\left(-\dfrac{1}{4}, \dfrac{3}{2}\right)$

12. LONG JUMP   The equation describing the flight path of the long jumper is $y = -\dfrac{5}{121}(x - 11)^2 + 5$. Show that she will land at a point 22 feet away from the take-off board.

Take-off ← —————— 22 ft —————— → Landing
board

| SECTION 13.2 | **The Ellipse** |

Equations of an ellipse:

Center at $(0, 0)$

$$\frac{x^2}{a^2} + \frac{y^2}{b^2} = 1$$

Center at (h, k)

$$\frac{(x - h)^2}{a^2} + \frac{(y - k)^2}{b^2} = 1$$

Graph each ellipse.

13. $9x^2 + 16y^2 = 144$ **14.** $\dfrac{(x - 2)^2}{4} + \dfrac{(y - 1)^2}{25} = 1$

15. $4(x + 1)^2 + 9(y - 1)^2 = 36$

16. Consider the equation $\dfrac{x^2}{144} + y^2 = 1$. Write each term on the left-hand side with a denominator that is the square of a number.

17. Consider the equation $\dfrac{x^2}{9} + \dfrac{y^2}{4} = 1$. Find two points on the graph of this equation by letting $x = 2$ and finding the corresponding y-coordinates. Express the results as ordered pairs. Give the exact answers and then the approximate answers.

18. SALAMI   When a delicatessen slices a cylindrical salami at an angle, the results are elliptical pieces that are larger than circular pieces. Write the equation of the shape of the slice of salami shown in the illustration if it was centered at the origin of a coordinate system.

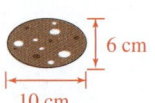

6 cm

10 cm

19. Fill in the blanks: An _____ is the set of all points in a plane for which the sum of the distances from two fixed points is a constant. Each of the fixed points is called a _____.

Ellipses have a *reflective* property such that any light or sound originating at one focus is reflected by the interior of the figure to the other focus.

20. CONSTRUCTION   Sketch the path of the sound when a person, standing at one focus, whispers something in the whispering gallery dome shown below.

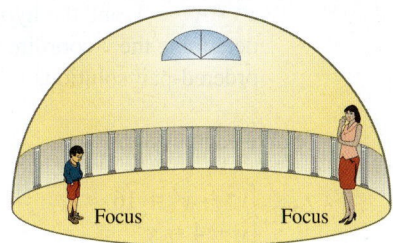

SECTION 13.3

The Hyperbola

Equations of a hyperbola:

Center at $(0, 0)$

$$\frac{x^2}{a^2} - \frac{y^2}{b^2} = 1$$

$$\frac{y^2}{a^2} - \frac{x^2}{b^2} = 1$$

Center at (h, k)

$$\frac{(x - h)^2}{a^2} - \frac{(y - k)^2}{b^2} = 1$$

$$\frac{(y - k)^2}{a^2} - \frac{(x - h)^2}{b^2} = 1$$

Graph each hyperbola.

21. $\dfrac{y^2}{9} - \dfrac{x^2}{1} = 1$

22. $9(x - 1)^2 - 4(y + 1)^2 = 36$

23. $xy = 9$

24. $y^2 - 4y - x^2 - 2x - 22 = 0$

25. ELECTROSTATIC REPULSION
Two similarly charged particles are shot together for an almost head-on collision, as in the illustration. They repel each other and travel the two branches of the hyperbola given by $x^2 - 4y^2 = 4$ on the given coordinate system. How close do they get?

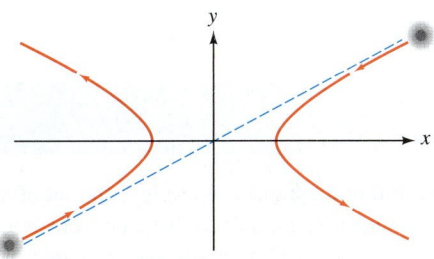

26. Determine whether the equation, when graphed, will be a circle, parabola, ellipse, or hyperbola.

    a. $\dfrac{(x - 4)^2}{16} + \dfrac{y^2}{49} = 1$ **b.** $x^2 + 6x - y^2 + 2y - 16 = 0$

    c. $x = -4y^2 - y + 1$ **d.** $x^2 + 2x + y^2 - 4y = 40$

SECTION 13.4

Solving Nonlinear Systems of Equations

Systems of nonlinear equations are solved by graphing, by substitution, or by elimination.

27. Check to determine whether $\left(-\sqrt{11}, -3\right)$ is a solution of the system $\begin{cases} x^2 + y^2 = 20 \\ x^2 - y^2 = 2 \end{cases}$.

28. The graphs of $y^2 - x^2 = 9$ and $x^2 + y^2 = 9$ are shown. Estimate the solutions of the system

$$\begin{cases} y^2 - x^2 = 9 \\ x^2 + y^2 = 9 \end{cases}$$

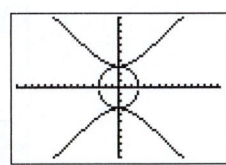

29. Determine the maximum number of solutions there could be for a system of equations consisting of the given curves.

    a. a line and an ellipse   **b.** two hyperbolas

    c. an ellipse and a circle   **d.** a parabola and a circle

30. Suppose the x-coordinate of both points of intersection of the circle, defined by $x^2 + y^2 = 1$, and the hyperbola, defined by $4y^2 - x^2 = 4$, is 0. Without graphing, determine the y-coordinates of both points of intersection. Express the answers as ordered-pair solutions.

Solve each system.

31. $\begin{cases} y^2 - x^2 = 16 \\ y + 4 = x^2 \end{cases}$ **32.** $\begin{cases} y = -x^2 + 2 \\ x^2 - y - 2 = 0 \end{cases}$

33. $\begin{cases} x^2 + 2y^2 = 12 \\ 2x - y = 2 \end{cases}$ **34.** $\begin{cases} 3x^2 + y^2 = 52 \\ x^2 - y^2 = 12 \end{cases}$

35. $\begin{cases} \dfrac{x^2}{16} + \dfrac{y^2}{12} = 1 \\ x^2 - \dfrac{y^2}{3} = 1 \end{cases}$ **36.** $\begin{cases} xy = 4 \\ x^2 + \dfrac{y^2}{2} = 9 \end{cases}$

CHAPTER 13 TEST   ELEMENTARY & INTERMEDIATE Algebra *f(x)* Now™

1. Fill in the blanks: A circle is the set of all points in a plane that are a fixed distance from a point called its _____. The fixed distance is called the _____ of the circle.

2. Find the center and the radius of the circle $x^2 + y^2 = 100$.

3. Find the center and the radius of the circle $x^2 + y^2 + 4x - 6y = 5$

4. TV HISTORY   In the early days of television, stations broadcast a black-and-white test pattern like that shown below during the early morning hours. Use the given coordinate system to write an equation of the large, bold circle in the center of the pattern.

Graph each equation.

5. $(x + 2)^2 + (y - 1)^2 = 9$ **6.** $x = y^2 - 2y + 3$

7. Find the vertex and the axis of symmetry of the graph of $y = -2x^2 - 4x + 5$.

8. SOUND   The equation $x = -\frac{1}{10}y^2$ defines a cross-section view of a parabolic dish. Construct a table of values for the equation, plot the ordered pairs, and connect the points with a smooth curve. Then plot the point $(-2.5, 0)$, which locates the microphone that picks up reflected sound waves. Draw a line parallel to the axis of symmetry coming into the dish and striking the dish at $(-0.9, 3)$ and reflecting into the microphone.

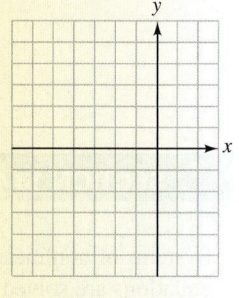

Graph each equation.

9. $9x^2 + 4y^2 = 36$

10. $\dfrac{(x-2)^2}{9} - \dfrac{y^2}{1} = 1$

11. Write the equation in standard form of the ellipse graphed below.

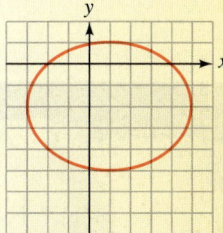

12. Find the center and the vertices of the graph of $25(x+8)^2 + 36(y-10)^2 = 900$.

13. Complete the table of solutions for the equation $\dfrac{x^2}{36} + \dfrac{y^2}{9} = 1$.

x	y
−2	

14. Give an example of the reflective properties of an ellipse. Include a drawing and label it completely.

15. Find the center and the dimensions of the central rectangle of the graph of $x^2 + 2x - y^2 + 2y - 4 = 0$.

16. Graph $xy = -4$.

17. What is the equation in standard form of the hyperbola graphed below?

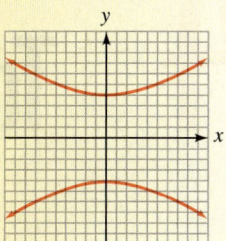

18. Determine whether the equation, when graphed, will be a circle, a parabola, an ellipse, or a hyperbola.

  a. $25x^2 + 100y^2 = 400$

  b. $x^2 - y^2 = 1$

  c. $x^2 + 8x + y^2 - 16y - 1 = 0$

  d. $x = 8y^2 - 9y + 4$

Solve the system graphically.

19. $\begin{cases} x^2 + y^2 = 25 \\ y - x = 1 \end{cases}$

Solve each system.

20. $\begin{cases} 2x - y = -2 \\ x^2 + y^2 = 16 + 4y \end{cases}$

21. $\begin{cases} 5x^2 - y^2 - 3 = 0 \\ x^2 + 2y^2 = 5 \end{cases}$

22. $\begin{cases} xy = -\dfrac{9}{2} \\ 3x + 2y = 6 \end{cases}$

14 Miscellaneous Topics

Getty Images News

On a ballot, the names of the candidates running for a given office can be listed in many different orders. For example, they can be presented alphabetically, by party affiliation, or simply in random order. In this chapter, we will learn how to determine the number of ways in which a set of names can be arranged. This counting concept, known as a *permutation*, has many other important applications in areas such as designing license plates, assigning telephone numbers, and listing combinations for locks.

To learn more about permutations, visit *The Learning Equation* on the Internet at http://tle.brookscole.com. (The log-in instructions are in the Preface.) For Chapter 14, the online lesson is:

• *TLE* Lesson 23: Permutations and Combinations

In this chapter, we introduce several topics with applications in advanced mathematics and in certain occupations. The binomial theorem, permutations, and combinations are used in statistics. Arithmetic and geometric sequences are used in finance.

14.1 The Binomial Theorem

- Raising binomials to powers
- Pascal's triangle
- Factorial notation
- The binomial theorem
- Finding a specific term of an expansion

We have discussed how to raise binomials to positive-integer powers. For example, we have learned that

$$(a + b)^2 = a^2 + 2ab + b^2$$

and that

The Language of Algebra

Recall that two-term polynomial expressions such as $a + b$ and $3u - 2v$ are called *binomials*.

$$
\begin{aligned}
(a + b)^3 &= (a + b)(a + b)^2 \\
&= (a + b)(a^2 + 2ab + b^2) \\
&= a^3 + 2a^2b + ab^2 + a^2b + 2ab^2 + b^3 \\
&= a^3 + 3a^2b + 3ab^2 + b^3
\end{aligned}
$$

In this section, we will learn how to raise binomials to positive-integer powers without performing the multiplications.

■ RAISING BINOMIALS TO POWERS

To see how to raise binomials to positive-integer powers, we consider the following binomial expansions of $a + b$.

The Language of Algebra

To *expand* means to increase in size. When we expand a power of a binomial, the result, called a **binomial expansion.** In general, an expansion has more terms than the original binomial.

$(a + b)^0 =$	1	1 term
$(a + b)^1 =$	$a + b$	2 terms
$(a + b)^2 =$	$a^2 + 2ab + b^2$	3 terms
$(a + b)^3 =$	$a^3 + 3a^2b + 3ab^2 + b^3$	4 terms
$(a + b)^4 =$	$a^4 + 4a^3b + 6a^2b^2 + 4ab^3 + b^4$	5 terms
$(a + b)^5 =$	$a^5 + 5a^4b + 10a^3b^2 + 10a^2b^3 + 5ab^4 + b^5$	6 terms
$(a + b)^6 =$	$a^6 + 6a^5b + 15a^4b^2 + 20a^3b^3 + 15a^2b^4 + 6ab^5 + b^6$	7 terms

Several patterns appear in these expansions:

The Language of Algebra

We can state observation 2 in another way: The *degree* of each term of an expansion is equal to the exponent of the binomial that is being expanded.

1. Each expansion has one more term than the power of the binomial.

2. For each term of an expansion, the sum of the exponents on a and b is equal to the exponent of the binomial being expanded. For example, in the expansion of $(a + b)^5$, the sum of the exponents in each term is 5:

$$
\overset{4 + 1 = 5}{} \quad \overset{3 + 2 = 5}{} \quad \overset{2 + 3 = 5}{} \quad \overset{1 + 4 = 5}{}
$$

$$(a + b)^5 = a^5 + 5a^4b + 10a^3b^2 + 10a^2b^3 + 5ab^4 + b^5$$

3. The first term in each expansion is a, raised to the power of the binomial, and the last term in each expansion is b, raised to the power of the binomial.

4. The exponents on a decrease by one in each successive term, ending with $a^0 = 1$ in the last term. The exponents on b, beginning with $b^0 = 1$ in the first term, increase by one in each successive term. For example, the expansion of $(a + b)^4$ could be written as

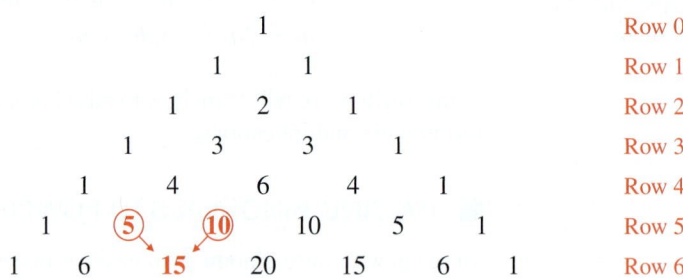

$$a^4 b^0 + 4a^3 b^1 + 6a^2 b^2 + 4a^1 b^3 + a^0 b^4$$

Thus, the variables have the pattern

$$a^n, \quad a^{n-1}b, \quad a^{n-2}b^2, \quad \ldots, \quad ab^{n-1}, \quad b^n$$

5. The coefficients of each expansion begin with 1, increase through some values, and then decrease through those same values, back to 1.

■ PASCAL'S TRIANGLE

To see another pattern, we write the coefficients of each expansion of $a + b$ in a triangular array:

						1							Row 0
					1		1						Row 1
				1		2		1					Row 2
			1		3		3		1				Row 3
		1		4		6		4		1			Row 4
	1		⑤		⑩		10		5		1		Row 5
1		6		**15**		20		15		6		1	Row 6

In this array, called **Pascal's triangle,** each entry between the 1's is the sum of the closest pair of numbers in the line immediately above it. For example, the first 15 in the bottom row is the sum of the 5 and 10 immediately above it. Pascal's triangle continues with the same pattern forever. The next two lines are

1	7	21	35	35	21	7	1		Row 7
1	8	28	56	70	56	28	8	1	Row 8

EXAMPLE 1

Expand: $(x + y)^5$.

Solution   The first term in the expansion is x^5, and the exponents on x decrease by one in each successive term. A y first appears in the second term, and the exponents on y increase by one in each successive term, concluding when the term y^5 is reached. Thus, the variables in the expansion are

$$x^5, \quad x^4 y, \quad x^3 y^2, \quad x^2 y^3, \quad xy^4, \quad y^5$$

Since the exponent of the binomial that is being expanded is 5, the coefficients of these variables are found in row 5 of Pascal's triangle.

$$
\begin{array}{ccccccccc}
 & & & & 1 & & & & \\
 & & & 1 & & 1 & & & \\
 & & 1 & & 2 & & 1 & & \\
 & 1 & & 3 & & 3 & & 1 & \\
1 & & 4 & & 6 & & 4 & & 1 \\
\mathbf{1} & \mathbf{5} & & \mathbf{10} & & \mathbf{10} & & \mathbf{5} & \mathbf{1} \\
1 & 6 & & 15 & 20 & & 15 & 6 & 1 \\
1 & 7 & 21 & 35 & & 35 & 21 & 7 & 1 \\
1 & 8 & 28 & 56 & 70 & 56 & 28 & 8 & 1
\end{array}
$$

$$1 \quad 5 \quad 10 \quad 10 \quad 5 \quad 1$$

Combining this information gives the following expansion:

$$(x + y)^5 = x^5 + 5x^4y + 10x^3y^2 + 10x^2y^3 + 5xy^4 + y^5$$

Self Check 1   Expand: $(x + y)^4$.

....

EXAMPLE 2   Expand: $(u - v)^4$.

ELEMENTARY &
INTERMEDIATE
Algebra *f(x)* **Now**™

Solution   We note that $(u - v)^4$ can be written in the form $[u + (-v)]^4$. The variables in this expansion are

$$u^4, \quad u^3(-v), \quad u^2(-v)^2, \quad u(-v)^3, \quad (-v)^4$$

and the coefficients are given in row 4 of Pascal's triangle:

$$1 \quad 4 \quad 6 \quad 4 \quad 1$$

Thus, the required expansion is

$$(u - v)^4 = u^4 + 4u^3(-v) + 6u^2(-v)^2 + 4u(-v)^3 + (-v)^4$$

The Language of Algebra

To *alternate* means to change back and forth. For example, day alternates with night. In this expansion, the signs + and − alternate.

Now we simplify each term. When $-v$ is raised to an even power, the sign is positive, and when $-v$ is raised to an odd power, the sign is negative. This causes the signs of the expansion to alternate between + and −.

$$(u - v)^4 = u^4 - 4u^3v + 6u^2v^2 - 4uv^3 + v^4$$

Self Check 2   Expand: $(x - y)^5$.

■ **FACTORIAL NOTATION**

Although Pascal's triangle gives the coefficients of the terms in a binomial expansion, it is not the best way to expand a binomial. To develop another way, we introduce **factorial notation.** The symbol $n!$ (read as "n **factorial**") is defined as follows.

Factorial Notation $n!$ is the product of consecutively decreasing natural numbers from n to 1.
For any natural number n,

$$n! = n(n - 1)(n - 2)(n - 3) \cdot \cdots \cdot 3 \cdot 2 \cdot 1$$

Zero factorial is defined as

$$0! = 1$$

EXAMPLE 3

ELEMENTARY &
INTERMEDIATE
Algebra $f(x)$ Now™

Solution

Evaluate each expression: **a.** $4!$, **b.** $6!$, **c.** $3! \cdot 2!$, and **d.** $5! \cdot 0!$.

a. $4! = 4 \cdot 3 \cdot 2 \cdot 1 = 24$ Read as "4 factorial."

b. $6! = 6 \cdot 5 \cdot 4 \cdot 3 \cdot 2 \cdot 1 = 720$

c. $3! \cdot 2! = (3 \cdot 2 \cdot 1) \cdot (2 \cdot 1) = 6 \cdot 2 = 12$ Find each factorial first, then multiply the results.

d. $5! \cdot \mathbf{0!} = (5 \cdot 4 \cdot 3 \cdot 2 \cdot 1) \cdot \mathbf{1} = 120$ $0! = 1$

Self Check 3   Evaluate each expression: **a.** $7!$, **b.** $4! \cdot 3!$, and **c.** $1! \cdot 0!$.

ACCENT ON TECHNOLOGY: FACTORIALS

We can find factorials using a calculator. For example, to find $12!$ with a scientific calculator, we enter

12 $\boxed{x!}$ (You may have to use a $\boxed{\text{2nd}}$ or $\boxed{\text{SHIFT}}$ key first.) $\boxed{479001600}$

To find $12!$ on a TI-83 Plus graphing calculator, we enter

12 $\boxed{\text{MATH}}$ arrow $\boxed{\rightarrow}$ to PRB $\boxed{4}$ $\boxed{\text{ENTER}}$

$\boxed{\begin{array}{l} \text{12!} \\ \qquad \text{479001600} \end{array}}$

The following property follows from the definition of factorial.

Factorial Property

For any natural number n,

$$n(n - 1)! = n!$$

We can use this property to simplify certain expressions involving factorials.

EXAMPLE 4

Simplify each expression: **a.** $\dfrac{6!}{5!}$ and **b.** $\dfrac{10!}{8!(10 - 8)!}$.

Solution   **a.** If we write $6!$ as $6 \cdot 5!$, we can simplify the fraction by removing the common factor $5!$ in the numerator and denominator.

$$\frac{6!}{5!} = \frac{6 \cdot 5!}{5!} = \frac{6 \cdot \overset{1}{\cancel{5!}}}{\underset{1}{\cancel{5!}}} = 6 \quad \text{Simplify: } \frac{5!}{5!} = 1.$$

b. First, we subtract within the parentheses. Then we write $10!$ as $10 \cdot 9 \cdot 8!$ and simplify.

$$\frac{10!}{8! \, \mathbf{(10 - 8)}!} = \frac{10!}{8! \cdot \mathbf{2}!} = \frac{10 \cdot 9 \cdot \overset{1}{\cancel{8!}}}{\underset{1}{\cancel{8!}} \cdot 2!} = \frac{5 \cdot \overset{1}{\cancel{2}} \cdot 9}{\underset{1}{\cancel{2}} \cdot 1} = 45 \quad \begin{array}{l} \text{Simplify: } \frac{8!}{8!} = 1. \text{ Factor 10} \\ \text{as } 5 \cdot 2 \text{ and simplify: } \frac{2}{2} = 1. \end{array}$$

Self Check 4   Simplify: **a.** $\dfrac{4!}{3!}$ and   **b.** $\dfrac{7!}{5!(7-5)!}$.

■ **THE BINOMIAL THEOREM**

The following theorem brings together our observations about binomial expansions and our work with factorials. Known as the *binomial theorem,* it is the most efficient way to expand a binomial.

The Binomial Theorem	For any positive integer n,

$$(a+b)^n = a^n + \frac{n!}{1!(n-1)!}a^{n-1}b + \frac{n!}{2!(n-2)!}a^{n-2}b^2 + \frac{n!}{3!(n-3)!}a^{n-3}b^3$$

$$+ \cdots + \frac{n!}{r!(n-r)!}a^{n-r}b^r + \cdots + b^n$$

In the binomial theorem, the exponents on the variables follow the familiar pattern:

- The sum of the exponents on a and b in each term is n,
- the exponents on a decrease by 1 in each successive term, and
- the exponents on b increase by 1 in each successive term.

The method of finding the coefficients involves factorials. Except for the first and last terms, the numerator of each coefficient is $n!$. If the exponent on b in a particular term is r, the denominator of the coefficient of that term is $r!(n-r)!$.

EXAMPLE 5   Use the binomial theorem to expand $(a+b)^3$.

Solution   We can substitute directly into the binomial theorem and simplify:

$$(a+b)^3 = a^3 + \frac{3!}{1!(3-1)!}a^2b + \frac{3!}{2!(3-2)!}ab^2 + b^3$$

$$= a^3 + \frac{3!}{1!\cdot 2!}a^2b + \frac{3!}{2!\cdot 1!}ab^2 + b^3$$

$$= a^3 + \frac{3\cdot\overset{1}{\cancel{2}}\cdot\overset{1}{\cancel{1}}}{\underset{1}{\cancel{1}}\cdot\underset{1}{\cancel{2}}\cdot 1}a^2b + \frac{3\cdot\overset{1}{\cancel{2}}\cdot\overset{1}{\cancel{1}}}{\underset{1}{\cancel{2}}\cdot\underset{1}{\cancel{1}}\cdot 1}ab^2 + b^3$$

$$= a^3 + 3a^2b + 3ab^2 + b^3$$

Self Check 5   Use the binomial theorem to expand $(a+b)^4$.

EXAMPLE 6

Use the binomial theorem to expand $(x - y)^4$.

Solution   We can write $(x - y)^4$ in the form $[x + (-y)]^4$, substitute directly into the binomial theorem, and simplify:

$$(x - y)^4 = [x + (-y)]^4$$

$$= x^4 + \frac{4!}{1!(4-1)!}x^3(-y) + \frac{4!}{2!(4-2)!}x^2(-y)^2 + \frac{4!}{3!(4-3)!}x(-y)^3 + (-y)^4$$

$$= x^4 - \frac{4 \cdot 3!}{1! \cdot 3!}x^3y + \frac{4 \cdot 3 \cdot 2!}{2! \cdot 2!}x^2y^2 - \frac{4 \cdot 3!}{3! \cdot 1!}xy^3 + y^4$$

$$= x^4 - 4x^3y + 6x^2y^2 - 4xy^3 + y^4 \qquad \text{Note the alternating signs.}$$

Self Check 6   Use the binomial theorem to expand $(x - y)^3$.

EXAMPLE 7

Use the binomial theorem to expand $(3u - 2v)^4$.

Solution   We write $(3u - 2v)^4$ in the form $[3u + (-2v)]^4$ and let $a = 3u$ and $b = -2v$. Then we can use the binomial theorem to expand $(a + b)^4$.

$$(a + b)^4 = a^4 + \frac{4!}{1!(4-1)!}a^3b + \frac{4!}{2!(4-2)!}a^2b^2 + \frac{4!}{3!(4-3)!}ab^3 + b^4$$

$$= a^4 + 4a^3b + 6a^2b^2 + 4ab^3 + b^4$$

Now we can substitute $3u$ for a and $-2v$ for b and simplify:

$$(3u - 2v)^4 = (3u)^4 + 4(3u)^3(-2v) + 6(3u)^2(-2v)^2 + 4(3u)(-2v)^3 + (-2v)^4$$

$$= 81u^4 - 216u^3v + 216u^2v^2 - 96uv^3 + 16v^4$$

Self Check 7   Use the binomial theorem to expand $(4a - 5b)^3$.

■ FINDING A SPECIFIC TERM OF AN EXPANSION

To find a specific term of an expansion, we don't need to write out the entire expansion. A close examination of the binomial theorem and the pattern of the terms suggests the following method for finding a single term of an expansion.

Finding a Specific Term of an Expansion	The $(r + 1)$st term of the expansion of $(a + b)^n$ is $$\frac{n!}{r!(n-r)!}a^{n-r}b^r$$

EXAMPLE 8

Find the 4th term of the expansion of $(a + b)^9$.

ELEMENTARY &
INTERMEDIATE
Algebra $f(x)$ **Now**™

Solution   To use the formula for finding a specific term of $(a + b)^9$, we must determine n and r. If we are to find the fourth term, $r + 1 = 4$, and it follows that $r = 3$.

We also see that $n = 9$. We substitute those values into the formula to find the 4th term of the expansion.

Success Tip

r is always 1 less than the number of the term that you are finding.

$$\frac{n!}{r!(n-r)!}a^{n-r}b^r = \frac{9!}{3!(9-3)!}a^{9-3}b^3$$

$$= \frac{9!}{3!6!}a^6b^3 \qquad \frac{9!}{3!6!} = \frac{9\cdot8\cdot7\cdot\cancel{6!}}{3\cdot2\cdot1\cdot\cancel{6!}} = 84.$$

$$= 84a^6b^3$$

Self Check 8   Find the 3rd term of the expansion of $(a + b)^9$.

EXAMPLE 9

Find the 6th term of the expansion of $\left(x^2 - \dfrac{y}{2}\right)^7$.

ELEMENTARY &
INTERMEDIATE
Algebra $f(x)$ **Now**™

Solution   To use the formula for finding a specific term of $\left(x^2 - \frac{y}{2}\right)^7$, we must determine n, r, a, and b. If we are to find the sixth term, $r + 1 = 6$, and it follows that $r = 5$.

We also see that $a = x^2$, $b = -\frac{y}{2}$, and $n = 7$. We substitute those values into the formula to find the 6th term.

$$\frac{n!}{r!(n-r)!}a^{n-r}b^r = \frac{7!}{5!(7-5)!}(x^2)^{7-5}\left(-\frac{y}{2}\right)^5$$

$$= \frac{7!}{5!2!}(x^2)^2\left(-\frac{y^5}{32}\right) \qquad \frac{7!}{5!2!} = \frac{7\cdot6\cdot\cancel{5!}}{\cancel{5!}\cdot2\cdot1} = 21.$$

$$= -\frac{21}{32}x^4y^5$$

Self Check 9   Find the 5th term of the expansion of $\left(c^2 - \dfrac{d}{3}\right)^7$.

Answers to Self Checks   **1.** $x^4 + 4x^3y + 6x^2y^2 + 4xy^3 + y^4$ **2.** $x^5 - 5x^4y + 10x^3y^2 - 10x^2y^3 + 5xy^4 - y^5$
3. a. 5,040,   **b.** 144,   **c.** 1   **4. a.** 4,   **b.** 21   **5.** $a^4 + 4a^3b + 6a^2b^2 + 4ab^3 + b^4$
6. $x^3 - 3x^2y + 3xy^2 - y^3$ **7.** $64a^3 - 240a^2b + 300ab^2 - 125b^3$ **8.** $36a^7b^2$
9. $\dfrac{35}{81}c^6d^4$

14.1 STUDY SET

ELEMENTARY &
INTERMEDIATE
Algebra *f(x)* Now™

VOCABULARY **Fill in the blanks.**

1. The two-term polynomial expression $a + b$ is called a _____.

2. $a^4 + 4a^3b + 6a^2b^2 + 4ab^3 + b^4$ is the binomial _____ of $(a + b)^4$.

3. We can use the _____ theorem to raise binomials to positive-integer powers without doing the actual multiplication.

4. The array of numbers that gives the coefficients of the terms of a binomial expansion is called _____ triangle.

5. $n!$ (read as "n _____") is the product of consecutively _____ natural numbers from n to 1.

6. In the expansion $a^3 - 3a^2b + 3ab^2 - b^3$, the signs _____ between $+$ and $-$.

CONCEPTS **Fill in the blanks.**

7. Every binomial expansion has _____ more term than the power of the binomial.

8. For each term of the expansion of $(a + b)^8$, the sum of the exponents of a and b is ___.

9. The first term of the expansion of $(r + s)^{20}$ is ___ and the last term is ___.

10. In the expansion of $(m - n)^{15}$, the exponents on m _____ and the exponents on n _____.

11. The coefficients of the terms of the expansion of $(c + d)^{20}$ begin with ___, increase through some values, and then decrease through those same values, back to ___.

12. Complete Pascal's Triangle:

```
                    1
                1       1
            1       2
        1               3       1
    1               6       4       1
1       5       10      10      5       1
1               15              15      6       1
    7       21      35              21      7       1
1   8   28      56      70      56          8
```

13. $n \cdot$ _____ $= n!$

14. $8! = 8 \cdot$ _____

15. $0! =$ _____

16. According to the binomial theorem, the third term of the expansion of $(a + b)^n$ is _____.

17. The coefficient of the fourth term of the expansion of $(a + b)^9$ is $9!$ divided by _____.

18. The exponent on a in the fourth term of the expansion of $(a + b)^6$ is ___ and the exponent on b is ___.

19. The exponent on a in the fifth term of the expansion of $(a + b)^6$ is ___ and the exponent on b is ___.

20. The expansion of $(a - b)^4$ is
$$a^4 \quad 4a^3b \quad 6a^2b^2 \quad 4ab^3 \quad b^4$$

21. $(x + y)^3$
$$= x \quad + \frac{}{1!(3 - 1)!}x^2 + \frac{}{!(3 - 2)!}xy + y$$

22. Fill in the blanks.

  a. The $(r + 1)$st term of the expansion of $(a + b)^n$ is
$$\frac{n!}{r!(n - \quad)!}a^{\quad -r}b^{\quad}.$$

  b. To use this formula to find the 6th term of the expansion of $\left(m + \dfrac{n}{2}\right)^8$, we note that $r =$ ___,
  $n =$ ___, $a =$ ___, and $b =$ ___.

NOTATION **Fill in the blanks.**

23. $n! = n \cdot \left(\right)(n - 2) \cdots 3 \cdot 2 \cdot 1$

24. The symbol $5!$ is read as "_____ _____" and it means $5 \cdot$ _____.

PRACTICE **Evaluate each expression.**

25. $3!$ 26. $7!$

27. $5!$ 28. $6!$

29. $3! + 4!$ 30. $2!(3!)$

31. $3!(4!)$ 32. $4! + 4!$

33. $8(7!)$ 34. $4!(5)$

35. $\dfrac{9!}{11!}$ 36. $\dfrac{13!}{10!}$

37. $\dfrac{49!}{47!}$ 38. $\dfrac{101!}{100!}$

39. $\dfrac{9!}{7!0!}$ 40. $\dfrac{7!}{5!0!}$

41. $\dfrac{5!}{1!(5-1)!}$

42. $\dfrac{15!}{14!(15-14)!}$

43. $\dfrac{5!}{3!(5-3)!}$

44. $\dfrac{6!}{4!(6-4)!}$

45. $\dfrac{7!}{5!(7-5)!}$

46. $\dfrac{8!}{6!(8-6)!}$

47. $\dfrac{5!(8-5)!}{4! \cdot 7!}$

48. $\dfrac{6! \cdot 7!}{(8-3)!(7-4)!}$

Use a calculator to evaluate each expression.

49. $11!$

50. $13!$

51. $20!$

52. $55!$

Expand each expression.

53. $(x+y)^4$

54. $(a-b)^4$

55. $(c-d)^5$

56. $(c+d)^5$

57. $(s+t)^6$

58. $(s-t)^6$

59. $(a-b)^9$

60. $(a+b)^7$

61. $(2x+y)^3$

62. $(x+2y)^3$

63. $(2t-3)^5$

64. $(2b+1)^4$

65. $(5m-2n)^4$

66. $(2m+3n)^5$

67. $\left(\dfrac{x}{3}+\dfrac{y}{2}\right)^3$

68. $\left(\dfrac{x}{2}-\dfrac{y}{3}\right)^3$

69. $\left(\dfrac{x}{3}-\dfrac{y}{2}\right)^4$

70. $\left(\dfrac{x}{2}+\dfrac{y}{3}\right)^4$

71. $(c^2-d^2)^5$

72. $(u^2-v^3)^5$

Find the indicated term of each binomial expansion.

73. $(x-y)^4$; 4th

74. $(x-y)^5$; 2nd

75. $(r+s)^6$; 5th

76. $(r+s)^7$; 5th

77. $(x-y)^8$; 3rd

78. $(x-y)^9$; 7th

79. $(x-3y)^4$; 2nd

80. $(3x-y)^5$; 3rd

81. $(2t-5)^7$; 4th

82. $(2t+3)^6$; 6th

83. $(2x-3y)^5$; 5th

84. $(3x-2y)^4$; 2nd

85. $\left(\dfrac{c}{2}-\dfrac{d}{3}\right)^4$; 2nd

86. $\left(\dfrac{c}{3}+\dfrac{d}{2}\right)^5$; 4th

87. $(a^2-b^2)^6$; 2nd

88. $(a^2+b^2)^7$; 6th

WRITING

89. Describe how to construct Pascal's triangle.

90. Explain why the signs alternate in the expansion of $(x-y)^9$.

91. Explain why the third term of the expansion of $(m+3n)^9$ could not be $324m^7n^3$.

92. Using your own words, write a definition of $n!$.

REVIEW **Assume that $x, y, z,$ and b represent positive numbers. Use the properties of logarithms to write each expression as the logarithm of a single quantity.**

93. $2\log x + \dfrac{1}{2}\log y$

94. $-2\log x - 3\log y + \log z$

95. $\ln(xy+y^2) - \ln(xz+yz) + \ln z$

96. $\log_2(x+1) - \log_2 x$

CHALLENGE PROBLEMS

97. Find the constant term in the expansion of $\left(x+\dfrac{1}{x}\right)^{10}$.

98. Find the coefficient of a^5 in the expansion of $\left(a-\dfrac{1}{a}\right)^9$.

99. a. If we applied the pattern of the coefficients to the coefficient of the first term in a binomial expansion, the coefficient would be $\dfrac{n!}{0!(n-0)!}$. Show that this expression is 1.

b. If we applied the pattern of the coefficients to the coefficient of the last term in a binomial expansion, the coefficient would be $\dfrac{n!}{n!(n-n)!}$. Show that this expression is 1.

100. Expand $(i-1)^7$, where $i = \sqrt{-1}$.

14.2   Arithmetic Sequences and Series

- Sequences   • Arithmetic sequences   • Arithmetic means
- The sum of the first n terms   • Summation notation

The word *sequence* is used in everyday conversation when referring to an ordered list. For example, a history instructor might discuss the sequence of events that led up to the sinking of the *Titanic*. In mathematics, a **sequence** is a list of numbers written in a specific order. When we put a $+$ symbol between the numbers in a sequence, the sum is called a *series*.

■ SEQUENCES

Each number in a sequence is called a **term** of the sequence. **Finite sequences** contain a finite number of terms and **infinite sequences** contain an infinite number of terms. Two examples of sequences are:

Finite sequence: 1, 5, 9, 13, 17, 21, 25

Infinite sequence: 3, 6, 9, 12, 15, . . .   The . . . indicates that the sequence goes on forever.

Sequences are defined formally using the terminology of functions.

Finite and Infinite Sequences	A **finite sequence** is a function whose domain is the set of natural numbers $\{1, 2, 3, 4, \ldots, n\}$, for some natural number n.
	An **infinite sequence** is a function whose domain is the set of natural numbers: $\{1, 2, 3, 4, \ldots\}$.

Instead of using $f(x)$ notation, we use a_n (read as "a sub n") notation to write the value of a sequence at the number n. For the infinite sequence introduced earlier, we have:

1st term	2nd term	3rd term	4th term	5th term
3,	6,	9,	12,	15, . . .
↑	↑	↑	↑	↑
a_1	a_2	a_3	a_4	a_5

To specifically describe *all* the terms of a sequence we can write a formula for a_n, called the **general term** of the sequence. For the sequence 3, 6, 9, 12, 15, . . . , we note that $a_1 = 3 \cdot 1$, $a_2 = 3 \cdot 2$, $a_3 = 3 \cdot 3$, and so on. In general, the nth term of the sequence is found by multiplying n by 3.

$$a_n = 3n \qquad \text{Read } a_n \text{ as "}a \text{ sub } n.\text{"}$$

We can use this formula to find any term of the sequence. For example, to find the 12th term, we substitute 12 for n.

$$a_{12} = 3(12) = 36$$

EXAMPLE 1

ELEMENTARY &
INTERMEDIATE
Algebra $f(x)$ **Now**™

Given an infinite sequence with $a_n = 2n - 3$, find each of the following: **a.** the first four terms   and   **b.** a_{50}.

Solution **a.** To find the first four terms of the sequence, we substitute 1, 2, 3, and 4 for n in $a_n = 2n - 3$ and simplify.

$$a_1 = 2(1) - 3 = -1$$
$$a_2 = 2(2) - 3 = 1$$
$$a_3 = 2(3) - 3 = 3$$
$$a_4 = 2(4) - 3 = 5$$

The first four terms of the sequence are -1, 1, 3, and 5.

b. To find a_{50}, the 50th term of the sequence, we let $n = 50$:

$$a_{50} = 2(50) - 3 = 97$$

Self Check 1 Given an infinite sequence with $a_n = 3n + 5$, find each of the following: **a.** the first three terms   and   **b.** a_{100}.

■ ARITHMETIC SEQUENCES

A sequence where each term is found by adding the same number to the previous term is called an *arithmetic sequence.* Two examples are

The Language of Algebra

We pronounce the adjective *arithmetic* in the term *arithmetic* sequence as: air-rith-met'-ic.

5, 12, 19, 26, 33, 40

Add 7

This is a finite arithmetic sequence where each term is found by adding 7 to the previous term.

3, 1, −1, −3, −5, −7, . . .

Add −2

This is an infinite arithmetic sequence where each term is found by adding −2 to the previous term.

Arithmetic Sequence

An **arithmetic sequence** is a sequence of the form

$$a_1, \quad a_1 + d, \quad a_1 + 2d, \quad a_1 + 3d, \quad \ldots, \quad a_1 + (n-1)d, \ldots$$

where a_1 is the **first term** and d is the **common difference.** The nth term is given by

$$a_n = a_1 + (n-1)d$$

We note that the second term of an arithmetic sequence has an addend of $1d$, the third term has an addend of $2d$, the fourth term has an addend of $3d$, and the nth term has an addend of $(n-1)d$. We also note that the *difference between any two consecutive terms in an arithmetic sequence is d.*

EXAMPLE 2

An arithmetic sequence has a first term 5 and a common difference 4. Find the 25th term of the sequence.

Solution Since the first term is $a_1 = 5$ and the common difference is $d = 4$, the arithmetic sequence is defined by the formula

$$a_n = 5 + (n - 1)4 \qquad \text{In } a_n = a_1 + (n - 1)d \text{, substitute 5 for } a_1 \text{ and 4 for } d.$$

To find the 25th term, we substitute 25 for n and simplify.

$$a_{25} = 5 + (\mathbf{25} - 1)4$$
$$= 5 + (24)4$$
$$= 101$$

The 25th term is 101.

Self Check 2 An arithmetic sequence has a first term 10 and a common difference 8. Find the 30th term of the sequence.

EXAMPLE 3

The first three terms of an arithmetic sequence are 3, 8, and 13. Find the 100th term.

ELEMENTARY &
INTERMEDIATE
Algebra $f(x)$ **Now**™

Solution The common difference d is the difference between any two successive terms. Since $a_1 = 3$ and $a_2 = 8$, we can find d using subtraction.

$$d = a_2 - a_1 = 8 - 3 = 5 \qquad \text{Also note that } a_3 - a_2 = 13 - 8 = 5.$$

Success Tip

The common difference d of an arithmetic sequence is defined to be

$$d = a_{n+1} - a_n$$

To find 100th term, we substitute 3 for a_1, 5 for d and 100 for n in the formula for the nth term.

$$a_n = a_1 + (n - 1)d$$
$$a_{100} = 3 + (100 - 1)5$$
$$= 3 + (99)5$$
$$= 498$$

Self Check 3 The first three terms of an arithmetic sequence are -3, 6, and 15. Find the 99th term.

EXAMPLE 4

The first term of an arithmetic sequence is 12 and the 50th term is 3,099. Write the first six terms of the sequence.

ELEMENTARY &
INTERMEDIATE
Algebra $f(x)$ **Now**™

Solution The key is to find the common difference. Because the 50th term of the sequence is 3,099, we substitute 3,099 for a_{50} and $a_1 = 12$ in the formula for the 50th term and solve for d.

$$a_{50} = a_1 + (50 - 1)d \qquad \text{This gives the 50th term of any arithmetic sequence.}$$
$$3{,}099 = 12 + (50 - 1)d \qquad \text{Substitute 3,099 for } a_{50} \text{ and 12 for } a_1.$$
$$3{,}099 = 12 + 49d \qquad \text{Simplify.}$$
$$3{,}087 = 49d \qquad \text{Subtract 12 from both sides.}$$
$$63 = d \qquad \text{Divide both sides by 49.}$$

Since the first term is 12 and the common difference is 63, the first six terms are

12, 75, 138, 201, 264, 327   Add 63 to a term to get the next term.

Self Check 4   The first term of an arithmetic sequence is 15 and the 12th term is 92. Write the first four terms of the sequence.

■ ARITHMETIC MEANS

If numbers are inserted between two numbers a and b to form an arithmetic sequence, the inserted numbers are called **arithmetic means** between a and b. If a single number is inserted, it is called **the arithmetic mean** between a and b.

EXAMPLE 5   Insert two arithmetic means between 6 and 27.

ELEMENTARY &
INTERMEDIATE
Algebra *f(x)* **Now**™   **Solution**   The first term is $a_1 = 6$ and the fourth term is $a_4 = 27$. We must find the common difference so that the terms

$$6, \quad 6 + d, \quad 6 + 2d, \quad 27$$
$$\uparrow \qquad \uparrow \qquad \uparrow \qquad \uparrow$$
$$a_1 \qquad a_2 \qquad a_3 \qquad a_4$$

form an arithmetic sequence. To find d, we substitute 6 for a_1 and 27 for a_4 in the formula for the 4th term:

$a_4 = a_1 + (4 - 1)d$	This gives the 4th term of any arithmetic sequence.
$27 = 6 + (4 - 1)d$	Substitute.
$27 = 6 + 3d$	Simplify.
$21 = 3d$	Subtract 6 from both sides.
$7 = d$	Divide both sides by 3.

The two arithmetic means between 6 and 27 are

$$6 + d = 6 + 7 \qquad\qquad \text{or} \qquad 6 + 2d = 6 + 2(7)$$
$$= \mathbf{13} \quad \text{This is } a_2. \qquad\qquad = 6 + 14$$
$$= \mathbf{20} \quad \text{This is } a_3.$$

Two arithmetic means between 6 and 27 are 13 and 20.

Self Check 5   Insert two arithmetic means between 8 and 44.

■ THE SUM OF THE FIRST *n* TERMS

When the commas between the terms of a sequence are replaced with + signs, we call the sum a **series.** The sum of the terms of an arithmetic sequence is called an **arithmetic series.** Some examples are

$4 + 8 + 12 + 16 + 20 + 24$ Since this series has a limited number of terms, it is a finite arithmetic series.

$5 + 8 + 11 + 14 + 17 + \cdots$ Since this series has an unlimited number of terms, it is an infinite arithmetic series.

To develop a formula for evaluating the sum of the first n terms of an arithmetic sequence, we let S_n represent the sum of the first n terms of an arithmetic sequence:

$$S_n = \quad a_1 \quad + \quad [a_1 + d] \quad + \quad [a_1 + 2d] \quad + \cdots + [a_1 + (n-1)d]$$

We write the same sum again, but in reverse order:

$$S_n = [a_1 + (n-1)d] + [a_1 + (n-2)d] + [a_1 + (n-3)d] + \cdots + a_1$$

Adding these equations together, term by term, we get

$$2S_n = [2a_1 + (n-1)d] + [2a_1 + (n-1)d] + [2a_1 + (n-1)d] + \cdots + [2a_1 + (n-1)d]$$

Because there are n equal terms on the right-hand side of the preceding equation, we can write

(1)
$$2S_n = n[2a_1 + (n-1)d]$$
$$2S_n = n[a_1 + a_1 + (n-1)d] \qquad \text{Write } 2a_1 \text{ as } a_1 + a_1.$$
$$2S_n = n(a_1 + a_n) \qquad \text{Substitute } a_n \text{ for } a_1 + (n-1)d.$$
$$S_n = \frac{n(a_1 + a_n)}{2} \qquad \text{Divide both sides by 2.}$$

This reasoning establishes the following formula.

Success Tip

An alternate form of the summation formula for arithmetic sequences can be obtained from Equation (1) by combining like terms and dividing both sides by 2.

$$S_n = \frac{n[2a_1 + (n-1)d]}{2}$$

Sum of the First n Terms of an Arithmetic Sequence

The sum of the first n terms of an arithmetic sequence is given by the formula

$$S_n = \frac{n(a_1 + a_n)}{2}$$

where a_1 is the first term, a_n is the nth (or last) term, and n is the number of terms in the sequence.

EXAMPLE 6

Find the sum of the first 40 terms of the arithmetic sequence 4, 10, 16, ...

ELEMENTARY &
INTERMEDIATE
Algebra $f^{(x)}$ **Now**™

Solution   In this example, we let $a_1 = 4$, $n = 40$, $d = 10 - 4 = 6$, and $a_{40} = 4 + (40 - 1)6 = 238$ and substitute these values into the formula for S_n:

$$S_n = \frac{n(a_1 + a_n)}{2}$$

$$S_{40} = \frac{40(4 + 238)}{2} \qquad \text{Substitute } a_1 = 4 \text{ and } a_{40} = 238.$$

$$= 20(242)$$

$$= 4{,}840$$

The sum of the first 40 terms is 4,840.

Self Check 6   Find the sum of the first 50 terms of the arithmetic sequence 3, 8, 13, ...

■ SUMMATION NOTATION

When the general term of a sequence is known, we can use a special notation to write a series. This notation, called **summation notation**, involves the Greek letter Σ (sigma). The expression

$$\sum_{k=1}^{4} 3k \qquad \text{Read as "the summation of } 3k \text{ as } k \text{ runs from 1 to 4."}$$

designates the sum of all terms obtained if we successively substitute the numbers 1, 2, 3, and 4 for k, called the **index of the summation.** Thus, we have

$$\sum_{k=1}^{4} 3k = 3(1) + 3(2) + 3(3) + 3(4)$$

$$\quad\quad = 3 + 6 + 9 + 12$$

$$\quad\quad = 30$$

EXAMPLE 7

ELEMENTARY &
INTERMEDIATE
Algebra $f(x)$ **Now**™

Find each sum: **a.** $\displaystyle\sum_{k=1}^{3} (2k + 1)$ and   **b.** $\displaystyle\sum_{k=2}^{8} k^2$.

Solution **a.** $\displaystyle\sum_{k=1}^{3} (2k + 1) = [2(1) + 1] + [2(2) + 1] + [2(3) + 1]$

$$\quad\quad\quad\quad = 3 + 5 + 7$$

$$\quad\quad\quad\quad = 15$$

b. Here, we substitute the integers from 2 to 8 for k and find the sum.

$$\sum_{k=2}^{8} k^2 = 2^2 + 3^2 + 4^2 + 5^2 + 6^2 + 7^2 + 8^2$$

$$\quad\quad = 4 + 9 + 16 + 25 + 36 + 49 + 64$$

$$\quad\quad = 203$$

Self Check 7 Find the sum: $\displaystyle\sum_{k=1}^{4} (2k^2 - 2)$.

Answers to Self Checks **1. a.** 8, 11, 14, **b.** 305   **2.** 242   **3.** 879   **4.** 15, 22, 29, 36   **5.** 20, 32   **6.** 6,275
7. 52

14.2   STUDY SET   ELEMENTARY & INTERMEDIATE   **Algebra** $f(x)$ **Now**™

VOCABULARY   **Fill in the blanks.**

1. A _____ is a function whose domain is the set of natural numbers.

2. A sequence with an unlimited number of terms is called a(n) _____ sequence.

A sequence with a specific number of terms is called a(n) _____ sequence.

3. Each term of an _____ sequence is found by adding the same number to the previous term.

4. 5, 15, 25, 35, 45, 55, . . . is an example of an _____ sequence. The first _____ is 5 and the common _____ is 10.

5. If a single number is inserted between a and b to form an arithmetic sequence, the number is called the arithmetic _____ between a and b.

6. The sum of the terms of an arithmetic sequence is called an arithmetic _____.

CONCEPTS

7. Write the first three terms of an arithmetic sequence if $a_1 = 1$ and $d = 6$.

8. Given the arithmetic sequence 4, 7, 10, 13, 16, 19, . . . , find a_5 and d.

9. a. Write the formula for a_n, the general term of an arithmetic sequence.

    b. Write the formula for S_n, the sum of the first n terms of an arithmetic sequence.

10. An infinite arithmetic sequence is of the form
$$a_1, a_1 + d, \qquad , a_1 + 3d, \qquad , \ldots$$

NOTATION   Fill in the blanks.

11. The notation a_n represents the _____ term of a sequence.

12. To find the common difference of an arithmetic sequence, we use the formula $d = a\quad - a\quad$.

13. The symbol Σ is the Greek letter _____.

14. In the symbol $\sum\limits_{k=1}^{5} (2k - 5)$, k is called the _____ of summation.

15. We read $\sum\limits_{k=1}^{10} 3k$ as "the _____ of $3k$ as k _____ from 1 to 10."

16. $\sum\limits_{k=1}^{5} k = \quad + \quad + \quad + \quad + \quad$

PRACTICE   Write the first five terms of each sequence.

17. $a_n = 4n - 1$ **18.** $a_n = 5n - 3$

19. $a_n = -3n + 1$ **20.** $a_n = -6n + 2$

Write the first five terms of each arithmetic sequence with the given properties.

21. $a_1 = 3, d = 2$

22. $a_1 = -2, d = 3$

23. $a_1 = -5, d = -3$

24. $a_1 = 8, d = -5$

25. $a_1 = 5$, fifth term is 29

26. $a_1 = 4$, sixth term is 39

27. $a_1 = -4$, sixth term is -39

28. $a_1 = -5$, fifth term is -37

29. $d = 7$, sixth term is -83

30. $d = 3$, seventh term is 12

31. $d = -3$, seventh term is 16

32. $d = -5$, seventh term is -12

33. The 19th term is 131 and the 20th term is 138.

34. The 16th term is 70 and the 18th term is 78.

35. Find the 30th term of the arithmetic sequence with $a_1 = 7$ and $d = 12$.

36. Find the 55th term of the arithmetic sequence with $a_1 = -5$ and $d = 4$.

37. Find the 37th term of the arithmetic sequence with a second term of -4 and a third term of -9.

38. Find the 40th term of the arithmetic sequence with a second term of 6 and a fourth term of 16.

39. Find the first term of the arithmetic sequence with a common difference of 11 if its 27th term is 263.

40. Find the common difference of the arithmetic sequence with a first term of -164 if its 36th term is -24.

41. Find the common difference of the arithmetic sequence with a first term of 40 if its 44th term is 556.

42. Find the first term of the arithmetic sequence with a common difference of -5 if its 23rd term is -625.

43. Insert three arithmetic means between 2 and 11.

44. Insert four arithmetic means between 5 and 25.

45. Insert four arithmetic means between 10 and 20.

46. Insert three arithmetic means between 20 and 30.

47. Find the arithmetic mean between 10 and 19.

48. Find the arithmetic mean between -4.5 and 7.

Write the series associated with each summation.

49. $\sum\limits_{k=1}^{4} (3k)$ **50.** $\sum\limits_{k=1}^{4} (k - 9)$

51. $\sum\limits_{k=2}^{4} k^2$ **52.** $\sum\limits_{k=3}^{5} (-2k)$

Write the summation notation for each sum.

53. $1 + 4 + 9 + 16 + 25$

54. $2 + 4 + 6 + 8$

55. $3 + 4 + 5 + 6$

56. $-1 - 4 - 9 - 16 - 25 - 36$

Find the sum of the first n terms of each arithmetic sequence.

57. $1, 4, 7, \ldots; n = 30$

58. $2, 6, 10, \ldots; n = 28$

59. $-5, -1, 3, \ldots; n = 17$

60. $-7, -1, 5, \ldots; n = 15$

61. Second term is 7, third term is 12; $n = 12$

62. Second term is 5, fourth term is 9; $n = 16$

63. $a_n = 2n + 1$, nth term is 31; n is a natural number

64. $a_n = 4n + 3$, nth term is 23; n is a natural number

65. Find the sum of the first 50 natural numbers.

66. Find the sum of the first 100 natural numbers.

67. Find the sum of the first 50 odd natural numbers.

68. Find the sum of the first 50 even natural numbers.

Find each sum.

69. $\displaystyle\sum_{k=1}^{4} (6k)$

70. $\displaystyle\sum_{k=2}^{5} (3k)$

71. $\displaystyle\sum_{k=3}^{4} k^3$

72. $\displaystyle\sum_{k=2}^{4} (-k^2)$

73. $\displaystyle\sum_{k=3}^{4} (k^2 + 3)$

74. $\displaystyle\sum_{k=2}^{6} (k^2 + 1)$

75. $\displaystyle\sum_{k=4}^{4} (2k + 4)$

76. $\displaystyle\sum_{k=3}^{5} (3k^2 - 7)$

APPLICATIONS

77. SAVING MONEY   Yasmeen puts $60 into a safety deposit box. After each succeeding month, she puts $50 more in the box. Write the first six terms of an arithmetic sequence that gives the monthly amounts in her savings, and find her savings after 10 years.

78. INSTALLMENT LOANS   Maria borrowed $10,000, interest-free, from her mother. She agreed to pay back the loan in monthly installments of $275. Write the first six terms of an arithmetic sequence that shows the balance due after each month, and find the balance due after 17 months.

79. DESIGNING PATIOS   Each row of bricks in the following triangular patio is to have one more brick than the previous row, ending with the longest row of 150 bricks. How many bricks will be needed?

80. FALLING OBJECTS   The equation $s = 16t^2$ represents the distance s in feet that an object will fall in t seconds. After 1 second, the object has fallen 16 feet. After 2 seconds, it has fallen 64 feet, and so on. Find the distance that the object will fall during the second and third seconds.

81. FALLING OBJECTS   Refer to Exercise 80. How far will the object fall during the 12th second?

82. INTERIOR ANGLES   The sums of the angles of several polygons are given in the table. Assuming that the pattern continues, complete the table.

Figure	Number of sides	Sum of angles
Triangle	3	180°
Quadrilateral	4	360°
Pentagon	5	540°
Hexagon	6	720°
Octagon	8	
Dodecagon	12	

WRITING

83. Explain why 1, 4, 8, 13, 19, 26, ... is not an arithmetic sequence.

84. What is the difference between a sequence and a series?

85. What is the difference between a_n and S_n?

86. How is the symbol Σ used in this section?

REVIEW Assume that x, y, z, and b represent positive numbers. Use the properties of logarithms to write each expression in terms of the logarithms of x, y, and z.

87. $\log_2 \dfrac{2x}{y}$

88. $\ln x\sqrt{z}$

89. $\log x^3 y^2$

90. $\log x^3 y^{1/2}$

CHALLENGE PROBLEMS

91. Show that $\displaystyle\sum_{k=1}^{5} 5k = 5 \sum_{k=1}^{5} k$.

92. Show that $\displaystyle\sum_{k=3}^{6} (k^2 + 3k) = \sum_{k=3}^{6} k^2 + \sum_{k=3}^{6} 3k$.

93. Show that $\displaystyle\sum_{k=1}^{n} 3 = 3n$. (*Hint:* Consider 3 to be $3k^0$.)

94. Show that $\displaystyle\sum_{k=1}^{3} \dfrac{k^2}{k} \neq \dfrac{\displaystyle\sum_{k=1}^{3} k^2}{\displaystyle\sum_{k=1}^{3} k}$.

14.3 Geometric Sequences and Series

- Geometric sequences   • Geometric means
- The sum of the first n terms of a geometric sequence   • Infinite geometric series

We have seen that the same number is added to each term of an arithmetic sequence to get the next term. In this section, we will consider another type of sequence where we multiply each term by the same number to get the next term. This type of sequence is called a *geometric sequence*. Two examples are

2, 8, 32, 128, ...   This is an infinite geometric sequence where each term is found by multiplying the previous term by 4.
Multiply by 4

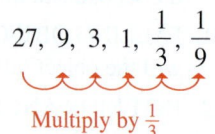
27, 9, 3, 1, $\dfrac{1}{3}$, $\dfrac{1}{9}$ This is a finite geometric sequence where each term is found by multiplying the previous term by $\frac{1}{3}$.
Multiply by $\frac{1}{3}$

■ GEOMETRIC SEQUENCES

Each term of a geometric sequence is found by multiplying the previous term by the same number.

Geometric Sequence	A **geometric sequence** is a sequence of the form

$$a_1, \quad a_1 r, \quad a_1 r^2, \quad a_1 r^3, \quad \ldots, \quad a_1 r^{n-1}, \quad \ldots$$

where a_1 is the **first term** and r is the **common ratio**. The nth term is given by

$$a_n = a_1 r^{n-1}$$

We note that the second term of a geometric sequence has a factor r^1, the third term has a factor r^2, the fourth term has a factor r^3, and the nth term has a factor r^{n-1}. We also note that r *is the quotient obtained when any term is divided by the previous term.*

EXAMPLE 1

ELEMENTARY &
INTERMEDIATE
Algebra $(f(x))$ **Now**™

A geometric sequence has a first term 5 and a common ratio 3. **a.** Write the first five terms of the sequence and **b.** find the ninth term.

Solution **a.** Because the first term is $a_1 = 5$ and the common ratio is $r = 3$, the first five terms are

$$5, \quad 5(3), \quad 5(3^2), \quad 5(3^3), \quad 5(3^4) \qquad \text{Each term is found by multiplying}$$
$$\quad\uparrow \quad\ \uparrow \qquad \uparrow \qquad\ \uparrow \qquad\ \uparrow \qquad\qquad \text{the previous term by 3.}$$
$$\quad a_1 \quad\ a_2 \qquad a_3 \qquad a_4 \qquad a_5$$

or

$$5, 15, 45, 135, 405$$

b. The nth term is $a_1 r^{n-1}$ with $a_1 = 5$ and $r = 3$. Because we want the ninth term, we let $n = 9$:

$$a_n = a_1 r^{n-1}$$
$$a_9 = 5(3)^{9-1}$$
$$= 5(3)^8$$
$$= 5(6,561)$$
$$= 32,805$$

Success Tip

Note the difference: Each term of an *arithmetic* sequence is found by adding the same number to the previous term. Each term of a *geometric* sequence is found by multiplying the previous term by the same number.

Self Check 1 A geometric sequence has a first term 3 and a common ratio 4. **a.** Write the first four terms and **b.** find the eighth term.

EXAMPLE 2

The first three terms of a geometric sequence are 16, 4, and 1. Find the seventh term.

Solution The common ratio r is the ratio between any two successive terms. Since $a_1 = 16$ and $a_2 = 4$, we can find r as follows:

$$r = \frac{a_2}{a_1} = \frac{4}{16} = \frac{1}{4} \qquad \text{Also note that } \frac{a_3}{a_2} = \frac{1}{4}.$$

To find the seventh term, we substitute 16 for a_1, $\frac{1}{4}$ for r, and 7 for n in the formula for the nth term and simplify:

$$a_n = a_1 r^{n-1}$$
$$a_7 = 16\left(\frac{1}{4}\right)^{7-1}$$
$$= 16\left(\frac{1}{4}\right)^6$$
$$= 16\left(\frac{1}{4,096}\right)$$
$$= \frac{1}{256}$$

Success Tip

The common ratio r of a geometric sequence is defined to be

$$r = \frac{a_{n+1}}{a_n}$$

ELEMENTARY &
INTERMEDIATE
Algebra $(f(x))$ **Now**™

Self Check 2 The first three terms of a geometric sequence are 25, 5, and 1. Find the seventh term.

■ GEOMETRIC MEANS

If numbers are inserted between two numbers a and b to form a geometric sequence, the inserted numbers are called **geometric means** between a and b. If a single number is inserted, that number is called the **geometric mean** between a and b.

EXAMPLE 3

ELEMENTARY &
INTERMEDIATE
Algebra $f(x)$ **Now**™

Insert two geometric means between 7 and 1,512.

Solution   In this example, the first term is $a_1 = 7$, and the fourth term is $a_4 = 1,512$. To find the common ratio r so that the terms

$$7, \quad 7r, \quad 7r^2, \quad 1,512$$
$$\uparrow \qquad \uparrow \qquad \uparrow \qquad \uparrow$$
$$a_1 \qquad a_2 \qquad a_3 \qquad a_4$$

form a geometric sequence, we substitute 4 for n and 7 for a_1 in the formula for the nth term of a geometric sequence and solve for r.

$$a_n = a_1 r^{n-1}$$
$$a_4 = 7r^{4-1}$$
$$1,512 = 7r^3$$
$$216 = r^3 \qquad \text{Divide both sides by 7.}$$
$$6 = r \qquad \text{Take the cube root of both sides.}$$

The two geometric means between 7 and 1,512 are

$$7r = 7(6) = \mathbf{42} \quad \text{and} \quad 7r^2 = 7(6)^2 = 7(36) = \mathbf{252}$$

The numbers 7, 42, 252, and 1,512 are the first four terms of a geometric sequence.

Self Check 3   Insert three positive geometric means between 1 and 16.

EXAMPLE 4   Find a geometric mean between 2 and 20.

Solution   We want to find the middle term of the three-termed geometric sequence

$$2, \quad 2r, \quad 20$$
$$\uparrow \qquad \uparrow \qquad \uparrow$$
$$a_1 \qquad a_2 \qquad a_3$$

with $a_1 = 2$, $a_3 = 20$, and $n = 3$. To find r, we substitute these values into the formula for the nth term of a geometric sequence:

$$a_n = a_1 r^{n-1}$$
$$a_3 = 2r^{3-1}$$
$$20 = 2r^2$$
$$10 = r^2 \qquad \text{Divide both sides by 2.}$$
$$\pm\sqrt{10} = r \qquad \text{Use the square root property.}$$

Because r can be either $\sqrt{10}$ or $-\sqrt{10}$, there are two values for a geometric mean. They are

$$2r = 2\sqrt{10} \quad \text{and} \quad 2r = -2\sqrt{10}$$

The sets of numbers 2, $2\sqrt{10}$, 20 and 2, $-2\sqrt{10}$, 20 both form geometric sequences. The common ratio of the first sequence is $\sqrt{10}$, and the common ratio of the second sequence is $-\sqrt{10}$.

Self Check 4 Find the positive geometric mean between 2 and 200.

■ THE SUM OF THE FIRST *n* TERMS OF A GEOMETRIC SEQUENCE

When we add the terms of a geometric sequence, we form a **geometric series.** There is a formula that gives the sum of the first *n* terms of a geometric sequence. To develop this formula, we let S_n represent the sum of the first *n* terms of a geometric sequence.

(1) $S_n = a_1 + a_1 r + a_1 r^2 + a_1 r^3 + \cdots + a_1 r^{n-1}$

We multiply both sides of Equation 1 by *r* to get

(2) $S_n r = \qquad a_1 r + a_1 r^2 + a_1 r^3 + \cdots + a_1 r^{n-1} + a_1 r^n$

We now subtract Equation 2 from Equation 1 and solve for S_n:

$$S_n - S_n r = a_1 - a_1 r^n$$
$$S_n(1 - r) = a_1 - a_1 r^n \qquad \text{Factor out } S_n \text{ from the left-hand side.}$$
$$S_n = \frac{a_1 - a_1 r^n}{1 - r} \qquad \text{Divide both sides by } 1 - r.$$

This reasoning establishes the following formula.

Success Tip

If the common factor of a_1 in the numerator of $\frac{a_1 - a_1 r^n}{1 - r}$ is factored out, the formula can be written in the alternate form:
$S_n = \frac{a_1(1 - r^n)}{1 - r}$.

Sum of the First *n* Terms of a Geometric Sequence	The sum of the first *n* terms of a geometric sequence is given by the formula $$S_n = \frac{a_1 - a_1 r^n}{1 - r} \qquad \text{or} \qquad S_n = \frac{a_1(1 - r^n)}{1 - r} \qquad \text{where } r \neq 1$$ where S_n is the sum, a_1 is the first term, *r* is the common ratio, and *n* is the number of terms.

EXAMPLE 5   Find the sum of the first six terms of the geometric sequence 250, 50, 10, ...

ELEMENTARY & INTERMEDIATE
Algebra *f(x)* **Now**™

Solution   In this sequence, $a_1 = 250$, $r = \frac{1}{5}$, and $n = 6$. We substitute these values into the formula for the sum of the first *n* terms of a geometric sequence and simplify:

$$S_n = \frac{a_1 - a_1 r^n}{1 - r}$$

$$S_6 = \frac{250 - 250\left(\frac{1}{5}\right)^6}{1 - \frac{1}{5}}$$

$$= \frac{250 - 250\left(\frac{1}{15{,}625}\right)}{\frac{4}{5}}$$

$$= \frac{5}{4}\left(250 - \frac{250}{15{,}625}\right)$$

$$= 312.48 \qquad \text{Use a calculator.}$$

The sum of the first six terms is 312.48.

Self Check 5   Find the sum of the first five terms of the geometric sequence 100, 20, 4, ...

EXAMPLE 6   ***Inheritances.*** A father decides to give his son an early inheritance. Each year, on the son's birthday, the father pays the son 15% of what is left of the $100,000 inheritance fund. How much money will be left in the fund after 20 years of payments?

Solution   If 15% of the money in the inheritance fund is given to the son each year, 85% of that amount remains after a payment. To find the amount of money that remains in the fund after a payment is made, we multiply the amount that was in the fund by 0.85. Over the years, the amounts of money that are left in the fund after a payment form a geometric sequence.

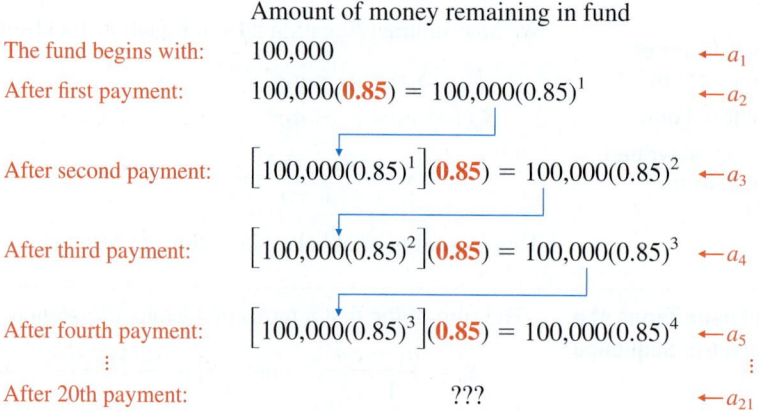

The amount of money remaining in the inheritance fund after 20 years is represented by the 21st term of a geometric sequence, where $a_1 = 100{,}000$, $r = 0.85$, and $n = 21$.

$$a_n = a_1 r^{n-1}$$ The formula for the nth term of a geometric sequence.

$$a_{21} = a_1 r^{21-1}$$ Substitute 21 for n.

$$= \mathbf{100{,}000}(\mathbf{0.85})^{21-1}$$ Substitute 100,000 for a_1 and 0.85 for r.

$$= 100{,}000(0.85)^{20}$$

$$\approx 3{,}876$$ Use a calculator. Round to the nearest dollar.

In 20 years, approximately $3,876 of the inheritance fund will be left.

Self Check 6   How much money will be left in the inheritance fund after 30 years of payments?

■ INFINITE GEOMETRIC SERIES

If we form the sum of the terms of an infinite geometric sequence, we get a series called an **infinite geometric series.** For example, if the common ratio r is 3, we have

Infinite geometric sequence   ***Infinite geometric series***

2, 6, 18, 54, 162, . . . $2 + 6 + 18 + 54 + 162 + \cdots$

As the number of terms of this series gets larger, the value of the series gets larger. We can see that this is true by forming some **partial sums.**

The first partial sum, S_1, of the series is $S_1 = 2$.
The second partial sum, S_2, of the series is $S_2 = 2 + 6 = 8$.
The third partial sum, S_3, of the series is $S_3 = 2 + 6 + 18 = 26$.
The fourth partial sum, S_4, of the series is $S_4 = 2 + 6 + 18 + 54 = 80$.

We can now see that as the number of terms gets infinitely large, the value of the series gets infinitely large. The values of some infinite geometric series get closer and closer to a specific number as the number of terms approaches infinity. One such series is

$$\frac{3}{2} + \frac{3}{4} + \frac{3}{8} + \frac{3}{16} + \frac{3}{32} \cdots \qquad \text{Here, } r = \frac{1}{2}.$$

To see that this is true, we form some partial sums.

The first partial sum is $S_1 = \dfrac{3}{2} = 1.5$.

The second partial sum is $S_2 = \dfrac{3}{2} + \dfrac{3}{4} = \dfrac{9}{4} = 2.25$.

The third partial sum is $S_3 = \dfrac{3}{2} + \dfrac{3}{4} + \dfrac{3}{8} = \dfrac{21}{8} = 2.625$.

The fourth partial sum is $S_4 = \dfrac{3}{2} + \dfrac{3}{4} + \dfrac{3}{8} + \dfrac{3}{16} = \dfrac{45}{16} = 2.8125$.

The fifth partial sum is $S_5 = \dfrac{3}{2} + \dfrac{3}{4} + \dfrac{3}{8} + \dfrac{3}{16} + \dfrac{3}{32} = \dfrac{93}{32} = 2.90625$.

As the number of terms in this series gets larger, the values of the partial sums approach the number 3. We say that 3 is the **limit** of S_n as n approaches infinity, and we say that 3 is the **sum of the infinite geometric series.**

To develop a formula for finding the sum of an infinite geometric series, we consider the formula that gives the sum of the first n terms.

$$S_n = \frac{a_1 - a_1 r^n}{1 - r} \qquad \text{where } r \neq 1$$

If $|r| < 1$ and a_1 is constant, the term $a_1 r^n$ in the above formula approaches 0 as n becomes very large. For example,

$$a_1\left(\frac{1}{2}\right)^1 = \frac{1}{2}a_1, \qquad a_1\left(\frac{1}{2}\right)^2 = \frac{1}{4}a_1, \qquad a_1\left(\frac{1}{2}\right)^3 = \frac{1}{8}a_1$$

and so on. Thus, when n is very large, the value of $a_1 r^n$ is negligible, and the term $a_1 r^n$ in the above formula can be ignored. This reasoning justifies the following formula.

Sum of the Terms of an Infinite Geometric Sequence

If a_1 is the first term and r is the common ratio of an infinite geometric sequence, and if $|r| < 1$, the sum of the terms of the sequence is given by

$$S = \frac{a_1}{1 - r}$$

EXAMPLE 7

Find the sum of the terms of the infinite geometric sequence $125, 25, 5, \ldots$.

Solution   In this geometric sequence, $a_1 = 125$ and $r = \frac{25}{125} = \frac{1}{5}$. Since $|r| = \left|\frac{1}{5}\right| = \frac{1}{5} < 1$, we can find the sum of the terms of the sequence. We do this by substituting 125 for a_1 and $\frac{1}{5}$ for r in the formula $S = \frac{a_1}{1-r}$ and simplifying:

Notation

The sum of the terms of an infinite geometric sequence is also denoted S_∞.

$$S = \frac{a_1}{1-r} = \frac{125}{1 - \dfrac{1}{5}} = \frac{125}{\dfrac{4}{5}} = \frac{5}{4}(125) = \frac{625}{4}$$

The sum of the terms of the sequence $125, 25, 5, \ldots$ is 156.25.

Self Check 7   Find the sum of the terms of the infinite geometric sequence $100, 20, 4, \ldots$.

EXAMPLE 8

Find the sum of the infinite geometric sequence $64, -4, \frac{1}{4}, \ldots$.

Solution   In this geometric sequence, $a_1 = 64$ and $r = \frac{-4}{64} = -\frac{1}{16}$. Since $|r| = \left|-\frac{1}{16}\right| = \frac{1}{16} < 1$, we can find the sum of all the terms of the sequence. We substitute 64 for a_1 and $-\frac{1}{16}$ for r in the formula $S = \frac{a_1}{1-r}$ and simplify:

Caution

If $|r| \geq 1$ for an infinite geometric sequence, the sum of the terms of the sequence, does not exist.

$$S = \frac{a_1}{1-r} = \frac{64}{\dfrac{17}{16}} = 64 \cdot \frac{16}{17} = \frac{1{,}024}{17}$$

The sum of the terms of the geometric sequence $64, -4, \dfrac{1}{4}, \ldots$ is $\dfrac{1{,}024}{17}$.

Self Check 8   Find the sum of the infinite geometric sequence $81, -27, 9, \ldots$.

EXAMPLE 9

Change $0.\overline{8}$ to a common fraction.

Solution   The decimal $0.\overline{8}$ can be written as the infinite series

$$0.\overline{8} = 0.888\ldots = \frac{8}{10} + \frac{8}{100} + \frac{8}{1{,}000} + \cdots$$

where $a_1 = \frac{8}{10}$ and $r = \frac{1}{10}$. Because $|r| = \left|\frac{1}{10}\right| = \frac{1}{10} < 1$, we can find the sum as follows:

$$S = \frac{a_1}{1-r} = \frac{\dfrac{8}{10}}{1 - \dfrac{1}{10}} = \frac{\dfrac{8}{10}}{\dfrac{9}{10}} = \frac{8}{9}$$

Thus, $0.\overline{8} = \dfrac{8}{9}$. Long division will verify that $\dfrac{8}{9} = 0.888\ldots$.

Self Check 9   Change $0.\overline{6}$ to a common fraction.

EXAMPLE 10

Testing steel.   One way to measure the hardness of a steel anvil is to drop a ball bearing onto the face of the anvil. The bearing should rebound at least $\frac{4}{5}$ of the distance from which it was dropped. If a bearing is dropped from a height of 10 inches onto a hard forged steel anvil, and if it could bounce forever, what total distance would the bearing travel?

Solution   The total distance the ball bearing travels is the sum of two motions, falling and rebounding. The bearing falls 10 inches, then rebounds $\frac{4}{5} \cdot 10 = 8$ inches, and falls 8 inches, and rebounds $\frac{4}{5} \cdot 8 = \frac{32}{5}$ inches, and falls $\frac{32}{5}$ inches, and rebounds $\frac{4}{5} \cdot \frac{32}{5} = \frac{128}{25}$ inches, and so on.
The distance the ball falls is given by the sum

$$10 + 8 + \frac{32}{5} + \frac{128}{25} + \cdots \qquad \text{This is an infinite geometric series with } a_1 = 10 \text{ and } r = \frac{4}{5}.$$

The distance the ball rebounds is given by the sum

$$8 + \frac{32}{5} + \frac{128}{25} + \cdots \qquad \text{This is an infinite geometric series with } a_1 = 8 \text{ and } r = \frac{4}{5}.$$

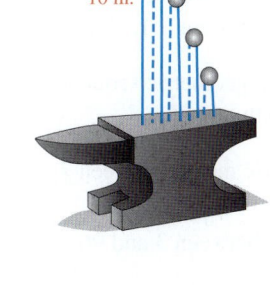

10 in.

Since each of these is an infinite geometric series with $|r| < 1$, we can use the formula $S = \frac{a_1}{1 - r}$ to find each sum.

$$\text{Falling: } \frac{10}{1 - \dfrac{4}{5}} = \frac{10}{\dfrac{1}{5}} = 50 \text{ inches} \qquad \text{Rebounding: } \frac{8}{1 - \dfrac{4}{5}} = \frac{8}{\dfrac{1}{5}} = 40 \text{ inches}$$

The total distance the bearing travels is 50 inches + 40 inches = 90 inches.

Self Check 10   If a bearing was dropped from a height of 15 inches onto a hard forged steel anvil, and if it could bounce forever, what total distance would it travel?

Answers to Self Checks   **1. a.** 3, 12, 48, 192,   **b.** 49,152   **2.** $\dfrac{1}{625}$ **3.** 2, 4, 8   **4.** 20   **5.** 124.96

6. about $763   **7.** 125   **8.** $\dfrac{243}{4}$ **9.** $\dfrac{2}{3}$ **10.** 135 in.

14.3   STUDY SET   ELEMENTARY & INTERMEDIATE   Algebra $f(x)$ Now™

VOCABULARY   Fill in the blanks.

1. Each term of a _____ sequence is found by multiplying the previous term by the same number.

2. 8, 16, 32, 64, 128, . . . is an example of a _____ sequence. The first _____ is 8 and the common _____ is 2.

3. If a single number is inserted between a and b to form a geometric sequence, the number is called the geometric _____ between a and b.

4. The sum of the terms of a geometric sequence is called a geometric _____. The sum of the terms of an infinite geometric sequence is called an _____ geometric series.

CONCEPTS

5. Write the first three terms of a geometric sequence if $a_1 = 16$ and $r = \frac{1}{4}$.

6. Given the geometric sequence $1, 2, 4, 8, 16, 32, \ldots$. Find a_5 and r.

7. Write the formula for a_n, the general term of a geometric sequence.

8. a. Write the formula for S_n, the sum of the first n terms of a geometric sequence.

   b. Write the formula for S, the sum of the terms of an infinite geometric sequence, where $|r| < 1$.

9. Which of the following values of r satisfy $|r| < 1$?

   a. $r = \frac{2}{3}$ **b.** $r = -3$
   c. $r = 6$ **d.** $r = -\frac{1}{5}$

10. Write $0.\overline{7}$ as an infinite geometric series:

$$0.\overline{7} = \frac{7}{} + \frac{7}{} + \frac{7}{}$$

NOTATION   Fill in the blanks.

11. An infinite geometric sequence is of the form

$$a_1, \ a_1r, \ \boxed{}, \ a_1r^3, \ \boxed{}, \ \ldots$$

12. The first four terms of the sequence defined by $a_n = 4(3)^{n-1}$ are $\boxed{}, \boxed{}, \boxed{}, \boxed{}$.

13. To find the common ratio of a geometric sequence, we use the formula $r = \dfrac{a_{\boxed{}}}{a_{\boxed{}}}$.

14. S_8 represents the sum of the first $\boxed{}$ terms of a geometric sequence.

PRACTICE   Write the first five terms of each geometric sequence with the given properties.

15. $a_1 = 3, r = 2$

16. $a_1 = -2, r = 2$

17. $a_1 = -5, r = \frac{1}{5}$

18. $a_1 = 8, r = \frac{1}{2}$

19. $a_1 = 2, r > 0$, third term is 32

20. $a_1 = 3$, fourth term is 24

21. $a_1 = -3$, fourth term is -192

22. $a_1 = 2, r < 0$, third term is 50

23. $a_1 = -64, r < 0$, fifth term is -4

24. $a_1 = -64, r > 0$, fifth term is -4

25. $a_1 = -64$, sixth term is -2

26. $a_1 = -81$, sixth term is $\frac{1}{3}$

27. The second term is 10, and the third term is 50.

28. The third term is -27, and the fourth term is 81.

29. Find the tenth term of the geometric sequence with $a_1 = 7$ and $r = 2$.

30. Find the 12th term of the geometric sequence with $a_1 = 64$ and $r = \frac{1}{2}$.

31. Find the first term of the geometric sequence with a common ratio -3 and an eighth term -81.

32. Find the first term of the geometric sequence with a common ratio 2 and a tenth term 384.

33. Find the common ratio of the geometric sequence with a first term -8 and a sixth term $-1,944$.

34. Find the common ratio of the geometric sequence with a first term 12 and a sixth term $\frac{3}{8}$.

35. Insert three positive geometric means between 2 and 162.

36. Insert four geometric means between 3 and 96.

37. Insert four geometric means between -4 and $-12,500$.

38. Insert three geometric means (two positive and one negative) between -64 and $-1,024$.

39. Find the negative geometric mean between 2 and 128.

40. Find the positive geometric mean between 3 and 243.

41. Find the positive geometric mean between 10 and 20.

42. Find the negative geometric mean between 5 and 15.

43. Find a geometric mean, if possible, between -50 and 10.

44. Find a negative geometric mean, if possible, between -25 and -5.

Find the sum of the first n terms of each geometric sequence.

45. $2, 6, 18, \ldots; n = 6$

46. $2, -6, 18, \ldots; n = 6$

47. $2, -6, 18, \ldots; n = 5$

48. $2, 6, 18, \ldots; n = 5$

49. $3, -6, 12, \ldots; n = 8$

50. $3, 6, 12, \ldots; n = 8$

51. $3, 6, 12, \ldots; n = 7$

52. $3, -6, 12, \ldots; n = 7$

53. The second term is 1 and the third term is $\frac{1}{5}$; $n = 4$.

54. The second term is 1 and the third term is 4; $n = 5$.

55. The third term is -2 and the fourth term is 1; $n = 6$.

56. The third term is -3 and the fourth term is 1; $n = 5$.

Find the sum of each infinite geometric series, if possible.

57. $8 + 4 + 2 + \cdots$

58. $12 + 6 + 3 + \cdots$

59. $54 + 18 + 6 + \cdots$

60. $45 + 15 + 5 + \cdots$

61. $12 - 6 + 3 - \cdots$

62. $8 - 4 + 2 - \cdots$

63. $-45 + 15 - 5 + \cdots$

64. $-54 + 18 - 6 + \cdots$

65. $\frac{9}{2} + 6 + 8 + \cdots$

66. $-112 - 28 - 7 - \cdots$

67. $-\frac{27}{2} - 9 - 6 - \cdots$

68. $\frac{18}{25} + \frac{6}{5} + 2 + \cdots$

Write each decimal in fraction form. Then check the answer by performing a long division.

69. $0.\overline{1}$

70. $0.\overline{2}$

71. $0.\overline{3}$

72. $0.\overline{4}$

73. $0.\overline{12}$

74. $0.\overline{21}$

75. $0.\overline{75}$

76. $0.\overline{57}$

APPLICATIONS   Use a calculator to help solve each problem.

77. DECLINING SAVINGS   John has $10,000 in a safety deposit box. Each year, he spends 12% of what is left in the box. How much will be in the box after 15 years?

78. SAVINGS GROWTH   Sally has $5,000 in a savings account earning 12% annual interest. How much will be in her account 10 years from now? (Assume that Sally makes no deposits or withdrawals.)

79. HOUSE APPRECIATION   A house appreciates by 6% each year. If the house is worth $70,000 today, how much will it be worth 12 years from now?

80. BOAT DEPRECIATION   A boat that cost $5,000 when new depreciates at a rate of 9% per year. How much will the boat be worth in 5 years?

81. INSCRIBED SQUARES   Each inscribed square in the illustration joins the midpoints of the next larger square. The area of the first square, the largest, is 1. Find the area of the 12th square.

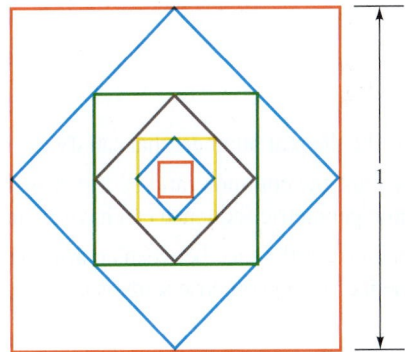

82. GENEALOGY   The following family tree spans 3 generations and lists 7 people. How many names would be listed in a family tree that spans 10 generations?

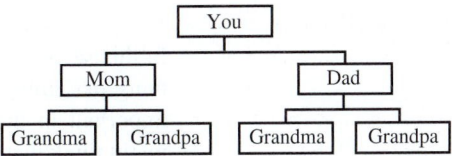

83. BOUNCING BALLS   On each bounce, the rubber ball in the illustration rebounds to a height one-half of that from which it fell. Find the total vertical distance the ball travels.

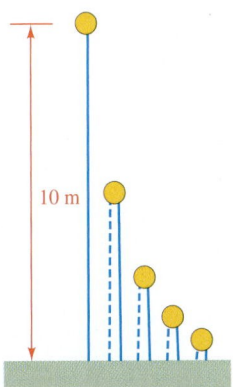

84. BOUNCING BALLS   A golf ball is dropped from a height of 12 feet. On each bounce, it returns to a height that is two-thirds of the distance it fell. Find the total vertical distance the ball travels.

85. PEST CONTROL   To reduce the population of a destructive moth, biologists release 1,000 sterilized male moths each day into the environment. If 80% of these moths alive one day survive until the next, then after a long time the population of sterile males is the sum of the infinite geometric series

$$1,000 + 1,000(0.8) + 1,000(0.8)^2 + 1,000(0.8)^3 + \cdots$$

Find the long-term population.

86. PEST CONTROL   If mild weather increases the day-to-day survival rate of the sterile male moths in Exercise 85 to 90%, find the long-term population.

WRITING

87. Describe the real numbers that satisfy $|r| < 1$.

88. Why must the common ratio be less than 1 before an infinite geometric sequence can have a sum?

89. Explain the difference between an arithmetic sequence and a geometric sequence.

90. Why is $1 - \frac{1}{2} + \frac{1}{4} - \frac{1}{8} + \frac{1}{16} - \frac{1}{32} + \ldots$ called an alternating infinite geometric series?

REVIEW   Solve each inequality. Write the solution set using interval notation.

91. $x^2 - 5x - 6 \leq 0$ **92.** $a^2 - 7a + 12 \geq 0$

93. $\dfrac{x - 4}{x + 3} > 0$ **94.** $\dfrac{t^2 + t - 20}{t + 2} < 0$

CHALLENGE PROBLEMS

95. If $f(x) = 1 + x + x^2 + x^3 + x^4 + \cdots$, find $f\left(\frac{1}{2}\right)$ and $f\left(-\frac{1}{2}\right)$.

96. Find the sum:

$$\frac{1}{\sqrt{3}} + \frac{1}{3} + \frac{1}{3\sqrt{3}} + \frac{1}{9} + \cdots$$

97. If $a > b > 0$, which is larger: the arithmetic mean between a and b or the geometric mean between a and b?

98. Is there a geometric mean between -5 and 5?

14.4   Permutations and Combinations

- The fundamental counting principle   • Permutations   • Combinations
- Alternative form of the binomial theorem

In this section, we will discuss methods of counting the different ways we can do something like lining up in a row or arranging books on a shelf. These kinds of problems are important in the fields of statistics, insurance, and telecommunications. Although one might think that counting problems are easy to solve, they can be deceptively difficult.

■ THE FUNDAMENTAL COUNTING PRINCIPLE

When a student goes to the cafeteria for lunch, he has a choice of three different sandwiches (hamburger, hot dog, or ham and cheese) and four different beverages (cola, root beer, water, or milk). His options are shown in the *tree diagram* on the right.

The tree diagram shows that there are a total of 12 different lunches to choose from. One possibility is a hamburger with a cola, and another is a hot dog with milk.

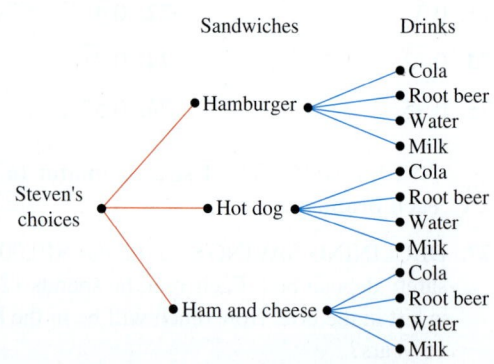

A situation that can have several different outcomes—such as choosing a sandwich—is called an **event.** Choosing a sandwich and choosing a beverage can be thought of as two events. The preceding example illustrates the fundamental counting principle.

Fundamental Counting Principle	If event E_1 can be done in m ways, and if (after E_1 has occurred) event E_2 can be done in n ways, then the event "E_1 followed by E_2" can be done in $m \cdot n$ ways.

EXAMPLE 1

Watching television.   Before studying for an exam, Taylor plans to watch the evening news and then a situation comedy on television. If she has a choice of 4 news broadcasts and 2 comedies, in how many ways can she choose to watch television?

Solution   Let E_1 be the event "watching the news" and E_2 be the event "watching a comedy." Because there are 4 ways to accomplish E_1 and 2 ways to accomplish E_2, the number of choices that Taylor has is $4 \cdot 2 = 8$.

Self Check 1   If Alex has 7 shirts and 5 pairs of pants, how many outfits could he wear?

The fundamental counting principle can be extended to any number of events. In Example 2, we use it to compute the number of ways in which we can arrange objects in a row.

EXAMPLE 2

Arranging books.   In how many ways can we arrange 5 books on a shelf?

Solution   We can fill the first space with any of the 5 books, the second space with any of the remaining 4 books, the third space with any of the remaining 3 books, the fourth space with any of the remaining 2 books, and the fifth space with the remaining 1 (or last) book. By the fundamental counting principle for events, the number of ways that the books can be arranged is

Success Tip

A general guideline is to use the fundamental counting principle to solve problems where consecutive choices are being made.

$$5 \cdot 4 \cdot 3 \cdot 2 \cdot 1 = 120$$

Self Check 2   In how many ways can 4 people line up in a row?

EXAMPLE 3

Signal flags.   If a sailor has 6 flags (each of a different color) to hang on a flagpole, how many different signals can the sailor send by using 4 flags?

Solution   The sailor must find the number of arrangements of 4 flags when there are 6 flags to choose from. The sailor can hang any one of the 6 flags in the top position, any one of the remaining 5 flags in the second position, any one of the remaining 4 flags in the third position, and any one of the remaining 3 flags in the lowest position. By the fundamental counting principle for events, the total number of signals that can be sent is

$$6 \cdot 5 \cdot 4 \cdot 3 = 360$$

Self Check 3   How many different signals can the sailor send if each signal uses 3 flags?

■ PERMUTATIONS

When counting a number of possible arrangements such as books on a shelf or flags on a pole, we are finding the number of **permutations** of those objects. In Example 2, we found that the number of permutations of 5 books, using all 5 of them is 120. In Example 3, we found that the number of permutations of 6 flags, using 4 of them, is 360.

The symbol $P(n, r)$, read as "the number of permutations of n objects taken r at a time," is often used to express permutation problems. In Example 2, we found that $P(5, 5) = 120$. In Example 3, we found that $P(6, 4) = 360$.

EXAMPLE 4

ELEMENTARY &
INTERMEDIATE
Algebra *f(x)* **Now**™

Signal flags. If Sarah has 7 flags (each of a different color) to hang on a flagpole, how many different signals can she send by using 3 flags?

Solution   We must find $P(7, 3)$ (the number of permutations of 7 things 3 at a time). In the top position Sarah can hang any of the 7 flags, in the middle position any one of the remaining 6 flags, and in the bottom position any one of the remaining 5 flags. By the fundamental counting principle for events,

$$P(7, 3) = 7 \cdot 6 \cdot 5 = 210$$

Sarah can send 210 signals using only 3 of the 7 flags.

Self Check 4   How many different signals can Sarah send using 4 flags?

Although it is correct to write $P(7, 3) = 7 \cdot 6 \cdot 5$, there is an advantage in changing the form of this answer to obtain a formula for computing $P(7, 3)$:

$$P(7, 3) = 7 \cdot 6 \cdot 5$$

$$= \frac{7 \cdot 6 \cdot 5}{1} \cdot \frac{4 \cdot 3 \cdot 2 \cdot 1}{4 \cdot 3 \cdot 2 \cdot 1} \qquad \text{Multiply by a form of 1: } \frac{4 \cdot 3 \cdot 2 \cdot 1}{4 \cdot 3 \cdot 2 \cdot 1} = 1.$$

$$= \frac{7!}{4!} \qquad \begin{array}{l}\text{Multiply the numerators and the denominators.}\\ \text{Use factorial notation.}\end{array}$$

$$= \frac{7!}{(7 - 3)!} \qquad \text{Write 4! as } (7 - 3)!.$$

The generalization of this idea gives the following formula.

Finding $P(n, r)$ The number of permutations of n objects taken r at a time is given by the formula

$$P(n, r) = \frac{n!}{(n - r)!}$$

EXAMPLE 5

ELEMENTARY &
INTERMEDIATE
Algebra *f(x)* **Now**™

Compute: **a.** $P(8, 2)$,   and   **b.** $P(n, n)$.

Solution   We use the permutation formula $P(n, r) = \dfrac{n!}{(n - r)!}$.

a. $P(8, 2) = \dfrac{8!}{(8-2)!}$ n = 8 and r = 2.

$= \dfrac{8 \cdot 7 \cdot 6!}{6!}$

$= 8 \cdot 7$

$= 56$

Success Tip

A *permutation* is an ordered arrangement of a given set of objects. To solve counting problems where order is important, use the permutation formula.

b. $P(n, n) = \dfrac{n!}{(n-n)!}$

$= \dfrac{n!}{0!}$

$= \dfrac{n!}{1}$

$= n!$

Self Check 5   Compute: **a.** $P(10, 6)$ and   **b.** $P(10, 10)$.

Part b of Example 5 establishes the following formula.

Finding $P(n, n)$ The number of permutations of *n* objects taken *n* at a time is $n!$.

$$P(n, n) = n!$$

EXAMPLE 6

ELEMENTARY & INTERMEDIATE
Algebra *f(x)* **Now**™

Television schedules.   **a.** In how many ways can a television executive arrange the Saturday night lineup of six programs if there are 15 programs to choose from?   **b.** If there are only six programs to choose from?

Solution   **a.** To find the number of permutations of 15 programs 6 at a time, we use the formula $P(n, r) = \frac{n!}{(n-r)!}$ with $n = 15$ and $r = 6$.

Success Tip

Problems solved with the permutation formula can also be solved by using the fundamental counting principle.

$P(15, 6) = \dfrac{15!}{(15-6)!}$

$= \dfrac{15 \cdot 14 \cdot 13 \cdot 12 \cdot 11 \cdot 10 \cdot 9!}{9!}$

$= 15 \cdot 14 \cdot 13 \cdot 12 \cdot 11 \cdot 10$

$= 3,603,600$

b. To find the number of permutations of 6 programs 6 at a time, we use the formula $P(n, n) = n!$ with $n = 6$.

$P(6, 6) = 6! = 720$

Self Check 6   How many ways are there to arrange the lineup if the executive has 20 programs to choose from?

■ COMBINATIONS

Suppose that Raul must read 4 books from a reading list of 10 books. The order in which he reads them is not important. For the moment, however, let's assume that order is important and find the number of permutations of 10 things 4 at a time:

$$P(10, 4) = \frac{10!}{(10 - 4)!}$$
$$= \frac{10 \cdot 9 \cdot 8 \cdot 7 \cdot 6!}{6!}$$
$$= 10 \cdot 9 \cdot 8 \cdot 7$$
$$= 5{,}040$$

If order is important, there are 5,040 ways of choosing 4 books when there are 10 books to choose from. However, because the order in which Raul reads the books does not matter, the previous result of 5,040 is too big. Since there are 24 (or 4!) ways of ordering the 4 books that are chosen, the result of 5,040 is exactly 24 (or 4!) times too big. Therefore, the number of choices that Raul has is the number of permutations of 10 things 4 at a time, divided by 24:

$$\frac{P(10, 4)}{24} = \frac{5{,}040}{24} = 210$$

Raul has 210 ways of choosing 4 books to read from the list of 10 books.

Success Tip

A *combination* is a distinct group of objects without regard to their arrangement. To solve counting problems where order is not important, use the combination formula.

In situations where order is *not* important, we are interested in **combinations,** not permutations. The symbols $C(n, r)$ and $\binom{n}{r}$ both mean the number of combinations of n objects taken r at a time.

If a selection of r books is chosen from a total of n books, the number of possible selections is $C(n, r)$, and there are $r!$ arrangements of the r books in each selection. If we consider the selected books as an ordered grouping, the number of orderings is $P(n, r)$. Therefore, we have

(1) $r! \cdot C(n, r) = P(n, r)$

We can divide both sides of Equation 1 by $r!$ to get the formula for finding $C(n, r)$:

$$C(n, r) = \binom{n}{r} = \frac{P(n, r)}{r!} = \frac{n!}{r!(n - r)!}$$

Finding $C(n, r)$ The number of combinations of n objects taken r at a time is given by

$$C(n, r) = \frac{n!}{r!(n - r)!}$$

EXAMPLE 7

Compute: **a.** $C(8, 5)$ and   **b.** $\binom{7}{2}$.

Solution   We use the combination formula $C(n, r) = \dfrac{n!}{r!(n - r)!}$.

a. $C(8, 5) = \dfrac{8!}{5!(8 - 5)!}$ $\quad n = 8$
                                            $\quad r = 5$

$\qquad\qquad = \dfrac{8 \cdot 7 \cdot 6 \cdot 5!}{5! \cdot 3!}$

$\qquad\qquad = 8 \cdot 7$

$\qquad\qquad = 56$

b. $\dbinom{7}{2} = \dfrac{7!}{2!(7 - 2)!}$ $\quad n = 7$
                                                $\quad r = 2$

$\qquad\quad = \dfrac{7 \cdot 6 \cdot 5!}{2 \cdot 1 \cdot 5!}$

$\qquad\quad = 21$

Notation

The notation $\dbinom{7}{2}$ means

$C(7, 2)$ and is read as "the number of combinations of 7 objects taken 2 at a time."

Self Check 7   Compute: **a.** $C(9, 6)$ and   **b.** $C(10, 10)$.

EXAMPLE 8

ELEMENTARY &
INTERMEDIATE
Algebra *f(x)* **Now**™

Choosing committees.   If 15 students want to pick a committee of 4 students to plan a party, how many different committees are possible?

Solution   Since the ordering of people on each possible committee is not important, we find the number of combinations of 15 people 4 at a time:

$$C(15, 4) = \frac{15!}{4!(15 - 4)!}$$

$$= \frac{15 \cdot 14 \cdot 13 \cdot 12 \cdot 11!}{4 \cdot 3 \cdot 2 \cdot 1 \cdot 11!}$$

$$= \frac{15 \cdot 14 \cdot 13 \cdot 12}{4 \cdot 3 \cdot 2 \cdot 1}$$

$$= 1{,}365$$

There are 1,365 possible committees.

Success Tip

If, instead, four officers (president, vice president, secretary, and treasurer) were to be selected, order would be important and we would use the permutation formula.

Self Check 8   In how many ways can 20 students pick a committee of 5 students to plan a party?

EXAMPLE 9

ELEMENTARY &
INTERMEDIATE
Algebra *f(x)* **Now**™

Choosing subcommittees.   A committee in Congress consists of 10 Democrats and 8 Republicans. In how many ways can a subcommittee be chosen if it is to contain 5 Democrats and 4 Republicans?

Solution   There are $C(10, 5)$ ways of choosing the 5 Democrats and $C(8, 4)$ ways of choosing the 4 Republicans. By the fundamental counting principle for events, there are $C(10, 5) \cdot C(8, 4)$ ways of choosing the subcommittee:

$$C(10, 5) \cdot C(8, 4) = \frac{10!}{5!(10 - 5)!} \cdot \frac{8!}{4!(8 - 4)!}$$

$$= \frac{10 \cdot 9 \cdot 8 \cdot 7 \cdot 6 \cdot 5!}{120 \cdot 5!} \cdot \frac{8 \cdot 7 \cdot 6 \cdot 5 \cdot 4!}{24 \cdot 4!}$$

$$= \frac{10 \cdot 9 \cdot 8 \cdot 7 \cdot 6}{120} \cdot \frac{8 \cdot 7 \cdot 6 \cdot 5}{24}$$

$$= 17{,}640$$

There are 17,640 possible subcommittees.

Success Tip

To help determine a method of solution, always ask yourself: Does the order in which the objects are chosen matter?

Self Check 9   In how many ways can a subcommittee be chosen if it is to contain 4 members from each party?

■ ALTERNATIVE FORM OF THE BINOMIAL THEOREM

We have seen that the expansion of $(x + y)^3$ is

$$(x + y)^3 = 1x^3 + 3x^2y + 3xy^2 + 1y^3$$

and that the coefficients can be written as

The Language of Algebra

Here, the coefficients of the terms of an expansion are expressed in the form $\binom{n}{r}$.
In this context, we call $\binom{3}{0}, \binom{3}{1}, \binom{3}{2}$, and $\binom{3}{3}$ **binomial coefficients.**

$$\binom{3}{0} = 1, \quad \binom{3}{1} = 3, \quad \binom{3}{2} = 3, \quad \text{and} \quad \binom{3}{3} = 1$$

Combining these facts gives the following way of writing the expansion of $(x + y)^3$:

$$(x + y)^3 = \binom{3}{0}x^3 + \binom{3}{1}x^2y + \binom{3}{2}xy^2 + \binom{3}{3}y^3$$

Likewise, we have

$$(x + y)^4 = \binom{4}{0}x^4 + \binom{4}{1}x^3y + \binom{4}{2}x^2y^2 + \binom{4}{3}xy^3 + \binom{4}{4}y^4$$

The generalization of this idea enables us to write the binomial theorem in an alternative form using combinatorial notation.

The Binomial Theorem   For any positive integer n,

$$(a + b)^n = \binom{n}{0}a^n + \binom{n}{1}a^{n-1}b + \binom{n}{2}a^{n-2}b^2 + \cdots + \binom{n}{r}a^{n-r}b^r + \cdots + \binom{n}{n}b^n$$

EXAMPLE 10   Use the alternative form of the binomial theorem to expand $(x + y)^6$.

Solution $(x + y)^6 = \binom{6}{0}x^6 + \binom{6}{1}x^5y + \binom{6}{2}x^4y^2 + \binom{6}{3}x^3y^3 + \binom{6}{4}x^2y^4 + \binom{6}{5}xy^5 + \binom{6}{6}y^6$

$= x^6 + 6x^5y + 15x^4y^2 + 20x^3y^3 + 15x^2y^4 + 6xy^5 + y^6$

Self Check 10   Use the alternative form of the binomial theorem to expand $(a + b)^5$.

The alternative form for finding a specific term of an expansion is as follows.

Finding a Specific Term   The $(r + 1)$st term of the expansion of $(a + b)^n$ is $\binom{n}{r}a^{n-r}b^r$.

EXAMPLE 11   Find the fifth term of the expansion of $(2x - y)^7$.

ELEMENTARY &
INTERMEDIATE
Algebra $f(x)$ **Now**™   **Solution**   Since $r + 1 = 5$, it follows that $r = 4$. Also, we see that $n = 7$, $a = 2x$, and $b = -y$. The fifth term of the expansion is

Success Tip

Remember that r is always 1 less than the number of the term that you are finding.

$$\binom{n}{r}a^{n-r}b^r = \binom{7}{4}(2x)^{7-4}(-y)^4$$

$$= \frac{7!}{4!(7-4)!}(2x)^3y^4$$

$$= 280x^3y^4 \qquad \frac{7!}{4!(7-4)!} = 35, 2^3 = 8, \text{ and } 35 \cdot 8 = 280.$$

Self Check 11   Find the third term of the expansion of $(2a - 3y)^5$.

Answers to Self Checks   **1.** 35   **2.** 24   **3.** 120   **4.** 840   **5. a.** 151,200,   **b.** 3,628,800   **6.** 27,907,200
7. a. 84,   **b.** 1   **8.** 15,504   **9.** 14,700   **10.** $a^5 + 5a^4b + 10a^3b^2 + 10a^2b^3 + 5ab^4 + b^5$
11. $720a^3y^2$

14.4   **STUDY SET**   ◉   ELEMENTARY &
INTERMEDIATE
Algebra $f(x)$ **Now**™

VOCABULARY   **Fill in the blanks.**

1. A _____ diagram like that shown below can be used to count the number of possible outcomes.

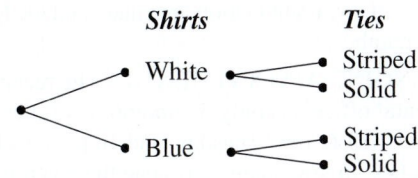

Shirts	**Ties**
White	Striped / Solid
Blue	Striped / Solid

2. Using the fundamental _____ principle, we can determine the number of ways two events can occur.

3. A _____ is an arrangement of objects.

4. When selecting objects when order is not important, we count _____.

CONCEPTS   **Fill in the blanks.**

5. If an event E_1 can be done in p ways and (after it occurs) a second event E_2 can be done in q ways, the event E_1 followed by E_2 can be done in _____ ways.

6. The symbol _____ means the number of permutations of n things taken r at a time.

7. The formula for the number of permutations of n things taken r at a time is _____.

8. $P(n, n) =$ _____

9. The symbol $C(n, r)$ or $\binom{}{}$ means the number of _____ of n things taken r at a time.

10. The formula for the number of combinations of n things taken r at a time is _____.

11. $0! =$ _____   **12.** $1! =$ _____

NOTATION   **Complete each solution.**

13. $P(6, 2) = \dfrac{}{(6-2)!}$

$= \dfrac{6 \cdot 5 \cdot 4!}{}$

$= 6 \cdot $

$= 30$

14. $C(6, 2) = \dfrac{}{(6-2)!}$

$= \dfrac{6 \cdot 5 \cdot 4!}{2 \cdot 1 \cdot }$

$= \cdot 5$

$= 15$

PRACTICE   **Evaluate each permutation or combination.**

15. $P(3, 3)$ **16.** $P(4, 4)$

17. $P(5, 3)$ **18.** $P(3, 2)$

19. $P(2, 2) \cdot P(3, 3)$ **20.** $P(3, 2) \cdot P(3, 3)$

21. $\dfrac{P(5, 3)}{P(4, 2)}$ **22.** $\dfrac{P(6, 2)}{P(5, 4)}$

23. $\dfrac{P(6, 2) \cdot P(7, 3)}{P(5, 1)}$ **24.** $\dfrac{P(8, 3)}{P(5, 3) \cdot P(4, 3)}$

25. $C(5, 3)$ **26.** $C(5, 4)$

27. $\binom{6}{3}$ **28.** $\binom{6}{4}$

29. $\binom{5}{4}\binom{5}{3}$ **30.** $\binom{6}{5}\binom{6}{4}$

31. $\dfrac{C(38, 37)}{C(19, 18)}$ **32.** $\dfrac{C(25, 23)}{C(40, 39)}$

33. $C(12, 0) \cdot C(12, 12)$ **34.** $\dfrac{C(8, 0)}{C(8, 1)}$

35. $C(n, 2)$ **36.** $C(n, 3)$

Use the alternative form of the binomial theorem to expand each expression.

37. $(x + y)^4$

38. $(c - d)^5$

39. $(2x + y)^3$

40. $(2x + y)^4$

41. $(3x - 2)^4$

42. $(3 - x^2)^3$

Find the indicated term of each binomial expansion.

43. $(x - 5y)^5$; fourth term

44. $(2x - y)^5$; third term

45. $(x^2 - y^3)^4$; second term

46. $(x^3 - y^2)^4$; fourth term

APPLICATIONS

47. PLANNING AN EVENING   Kristy plans to go to dinner and see a movie. In how many ways can she arrange her evening if she has a choice of 5 movies and 7 restaurants?

48. TRAVEL CHOICES   Paula has 5 ways to travel from New York to Chicago, 3 ways to travel from Chicago to Denver, and 4 ways to travel from Denver to Los Angeles. How many choices are available if she travels from New York to Los Angeles?

49. LICENSE PLATES   How many 6-digit license plates can be manufactured? Note that there are 10 choices— 0, 1, 2, 3, 4, 5, 6, 7, 8, 9—for each digit.

50. LICENSE PLATES   How many 6-digit license plates can be manufactured if no digit can be repeated?

51. LICENSE PLATES   How many 6-digit license plates can be manufactured if no license can begin with 0 and if no digit can be repeated?

52. LICENSE PLATES   How many license plates can be manufactured with 2 letters followed by 4 digits?

53. PHONE NUMBERS   How many 7-digit phone numbers are available in area code 815 if no phone number can begin with 0 or 1?

54. PHONE NUMBERS   How many 10-digit phone numbers are available if area codes of 000 and 911 cannot be used and if no local number can begin with 0 or 1?

55. LINING UP   In how many ways can 6 people be placed in a line?

56. ARRANGING BOOKS   In how many ways can 7 books be placed on a shelf?

57. ARRANGING BOOKS   In how many ways can 4 novels and 5 biographies be arranged on a shelf if the novels are placed first?

58. BALLOTS   In how many ways can 6 candidates for mayor and 4 candidates for the county board be arranged on a ballot if all of the candidates for mayor must be placed first?

59. LOCKS   How many permutations does a combination lock have if each combination has 3 numbers, no 2 numbers of any combination are equal, and the lock has 25 numbers?

60. LOCKS   How many permutations does a combination lock have if each combination has 3 numbers, no 2 numbers of any combination are equal, and the lock has 50 numbers?

61. ARRANGING APPOINTMENTS   The receptionist at a dental office has only 3 appointment times available before next Tuesday, and 10 patients have toothaches. In how many ways can the receptionist fill those appointments?

62. COMPUTERS   In many computers, a *word* consists of 32 *bits*—a string of thirty-two 1's and 0's. How many different words are possible?

63. PALINDROMES   A palindrome is any word, such as *madam* or *radar,* that reads the same backward and forward. How many 5-digit numerical palindromes (like 13531) are there? (*Hint:* A leading 0 would be dropped.)

64. CALL LETTERS   The call letters of a U.S. commercial radio station have 3 or 4 letters, and the first is either a W or a K. How many radio stations could this system support?

65. PICNICS   A class of 14 students wants to pick a committee of 3 students to plan a picnic. How many committees are possible?

66. CHOOSING BOOKS   Jeffrey must read 3 books from a reading list of 15 books. How many choices does he have?

67. COMMITTEES   The number of 3-person committees that can be formed from a group of persons is 10. How many persons are in the group?

68. COMMITTEES   The number of 3-person committees that can be formed from a group of persons is 20. How many persons are in the group?

69. LOTTERIES   In one state lottery, anyone who picks the correct 6 numbers (in any order) wins. With the numbers 0 through 99 available, how many choices are possible?

70. TAKING TESTS   The instructions on a test read, *Answer any 10 of the following 15 questions. Then choose one of the remaining questions for homework, and turn in its solution tomorrow.* In how many ways can the questions be chosen?

71. COMMITTEES   In how many ways can we select a committee of 2 men and 2 women from a group containing 3 men and 4 women?

72. COMMITTEES   In how many ways can we select a committee of 3 men and 2 women from a group containing 5 men and 3 women?

73. CHOOSING CLOTHES   In how many ways can we select 2 shirts and 3 neckties from a group of 12 shirts and 10 neckties?

74. CHOOSING CLOTHES   In how many ways can we select 5 dresses and 2 coats from a wardrobe containing 9 dresses and 3 coats?

WRITING

75. State the fundamental counting principle.

76. Explain why *permutation lock* would be a better name for a combination lock.

REVIEW   **Solve each equation. Give the solution to four decimal places.**

77. $2^{x+1} = 3^x$

78. $5^{x-3} = 3^{2x}$

79. $e^{3x} = 9$

80. $8^{x^2} = 9^x$

CHALLENGE PROBLEMS

81. How many ways could 5 people stand in line if 2 people insist on standing together?

82. How many ways could 5 people stand in line if 2 people refuse to stand next to each other?

14.5   Probability

- Probability

The probability that an event will occur is a measure of the likelihood of that event. A tossed coin, for example, can land in two ways, either heads or tails. Because one of these two equally likely outcomes is heads, we expect that out of several tosses, about half will be heads. We say that the probability of obtaining heads in a single toss of the coin is $\frac{1}{2}$.

If records show that out of 100 days with weather conditions like today's, 30 have received rain, the weather service will report, "There is a $\frac{30}{100}$ or 30% probability of rain today."

■ PROBABILITY

Activities such as tossing a coin, rolling a die, drawing a card, and predicting rain are called **experiments.** For any experiment, a list of all possible outcomes is called a **sample space.** For example, the sample space S for the experiment of tossing two coins is the set

$S = \{(H, H), (H, T), (T, H), (T, T)\}$ There are four possible outcomes.

where the ordered pair (H, T) represents the outcome "heads on the first coin and tails on the second coin."

An **event** is a subset of the sample space of an experiment. For example, if E is the event "getting at least one heads" in the experiment of tossing two coins, then

$$E = \{(H, H), (H, T), (T, H)\} \qquad \text{There are 3 ways of getting at least one heads.}$$

Because the outcome of getting at least one heads can occur in 3 out of 4 possible ways, we say that the **probability** of E is $\frac{3}{4}$, and we write

$$P(E) = P(\text{at least one heads}) = \frac{3}{4}$$

Probability of an Event   If a sample space of an experiment has n distinct and equally likely outcomes and E is an event that occurs in s of those ways, the **probability of E** is

$$P(E) = \frac{s}{n}$$

Since $0 \le s \le n$, it follows that $0 \le \frac{s}{n} \le 1$. This implies that all probabilities have value from 0 to 1. If an event cannot happen, its probability is 0. If an event is certain to happen, its probability is 1.

EXAMPLE 1

List the sample space of the experiment "rolling two dice a single time."

Solution   We can list ordered pairs and let the first number be the result on the first die and the second number the result on the second die. The sample space S is the following set of ordered pairs:

$$(1, 1)\ (1, 2)\ (1, 3)\ (1, 4)\ (1, 5)\ (1, 6)$$
$$(2, 1)\ (2, 2)\ (2, 3)\ (2, 4)\ (2, 5)\ (2, 6)$$
$$(3, 1)\ (3, 2)\ (3, 3)\ (3, 4)\ (3, 5)\ (3, 6)$$
$$(4, 1)\ (4, 2)\ (4, 3)\ (4, 4)\ (4, 5)\ (4, 6)$$
$$(5, 1)\ (5, 2)\ (5, 3)\ (5, 4)\ (5, 5)\ (5, 6)$$
$$(6, 1)\ (6, 2)\ (6, 3)\ (6, 4)\ (6, 5)\ (6, 6)$$

By counting, we see that the experiment has 36 equally likely possible outcomes.

Self Check 1   How many pairs in the sample space have a sum of 4?

EXAMPLE 2

Find the probability of the event "rolling a sum of 7 on one roll of two dice."

Solution   In the sample space listed in Example 1, the following ordered pairs give a sum of 7:

$$\{(1, 6), (2, 5), (3, 4), (4, 3), (5, 2), (6, 1)\}$$

Since there are 6 ordered pairs whose numbers give a sum of 7 out of a total of 36 equally likely outcomes, we have

$$P(E) = P(\text{rolling a 7}) = \frac{s}{n} = \frac{6}{36} = \frac{1}{6}$$

Self Check 2   Find the probability of rolling a sum of 4.

A standard playing deck of 52 cards has two red suits, hearts and diamonds, and two black suits, clubs and spades. Each suit has 13 cards, including the ace, king, queen, jack, and cards numbered from 2 to 10. We will refer to a standard deck of cards in many examples and exercises.

EXAMPLE 3

ELEMENTARY &
INTERMEDIATE
Algebra $f(x)$ **Now**™

Find the probability of drawing an ace on one draw from a standard card deck.

Solution Since there are 4 aces in the deck, the number of favorable outcomes is $s = 4$. Since there are 52 cards in the deck, the total number of possible outcomes is $n = 52$. The probability of drawing an ace is the ratio of the number of favorable outcomes to the number of possible outcomes.

$$P(\text{an ace}) = \frac{s}{n} = \frac{4}{52} = \frac{1}{13}$$

The probability of drawing an ace is $\frac{1}{13}$.

Self Check 3 Find the probability of drawing a red ace on one draw from a standard card deck.

EXAMPLE 4

ELEMENTARY &
INTERMEDIATE
Algebra $f(x)$ **Now**™

Find the probability of drawing 5 cards, all hearts, from a standard card deck.

Solution The number of ways we can draw 5 hearts from the 13 hearts is $C(13, 5)$, the number of combinations of 13 things taken 5 at a time. The number of ways to draw 5 cards from the deck is $C(52, 5)$, the number of combinations of 52 things taken 5 at a time. The probability of drawing 5 hearts is the ratio of the number of favorable outcomes to the number of possible outcomes.

$$P(5 \text{ hearts}) = \frac{s}{n} = \frac{C(13, 5)}{C(52, 5)}$$

$$P(5 \text{ hearts}) = \frac{\dfrac{13!}{5!8!}}{\dfrac{52!}{5!47!}}$$

$$= \frac{13!}{5! \, 8!} \cdot \frac{5! \, 47!}{52!}$$

$$= \frac{13 \cdot 12 \cdot 11 \cdot 10 \cdot 9 \cdot 8!}{8!} \cdot \frac{47!}{52 \cdot 51 \cdot 50 \cdot 49 \cdot 48 \cdot 47!}$$

$$= \frac{13 \cdot 12 \cdot 11 \cdot 10 \cdot 9}{52 \cdot 51 \cdot 50 \cdot 49 \cdot 48}$$

$$= \frac{33}{66{,}640}$$

The probability of drawing 5 hearts is $\frac{33}{66{,}640}$.

Self Check 4 Find the probability of drawing 6 cards, all diamonds, from a standard card deck.

 **Answers to Self Checks**   **1.** 3   **2.** $\dfrac{1}{12}$ **3.** $\dfrac{1}{26}$ **4.** $\dfrac{33}{391,510}$

14.5   STUDY SET   ELEMENTARY & INTERMEDIATE Algebra *f(x)* Now™

VOCABULARY   Fill in the blanks.

1. An _____ is any activity for which the outcome is uncertain.

2. A list of all possible outcomes for an experiment is called a _____ _____.

CONCEPTS   Fill in the blanks.

3. The probability of an event E is defined as $P(E) = \boxed{}$.

4. If an event is certain to happen, its probability is $\boxed{}$.

5. If an event cannot happen, its probability is $\boxed{}$.

6. All probability values are between $\boxed{}$ and $\boxed{}$, inclusive.

NOTATION   Complete each solution.

7. Find the probability of drawing a black face card from a standard deck.

   a. The number of black face cards is $\boxed{}$.

   b. The number of cards in the deck is $\boxed{}$.

   c. The probability is $\boxed{}$ or $\boxed{}$.

8. Find the probability of drawing 4 aces from a standard card deck.

   a. The number of ways to draw 4 aces from 4 aces is $C(4, 4) = \boxed{}$.

   b. The number of ways to draw 4 cards from 52 cards is $C(52, 4) = \boxed{}$.

   c. The probability is $\boxed{}$.

PRACTICE   List the sample space of each experiment.

9. Rolling one die and tossing one coin

10. Tossing three coins

11. Selecting a letter of the alphabet

12. Picking a one-digit number

An ordinary die is rolled once. Find the probability of each event.

13. Rolling a 2

14. Rolling a number greater than 4

15. Rolling a number larger than 1 but less than 6

16. Rolling an odd number

Balls numbered from 1 to 42 are placed in a container and stirred. If one is drawn at random, find the probability of each result.

17. The number is less than 20.

18. The number is less than 50.

19. The number is a prime number.

20. The number is less than 10 or greater than 40.

Refer to the following spinner. If the spinner is spun, find the probability of each event. Assume that the spinner never stops on a line.

21. The spinner stops on red.

22. The spinner stops on green.

23. The spinner stops on brown.

24. The spinner stops on yellow.

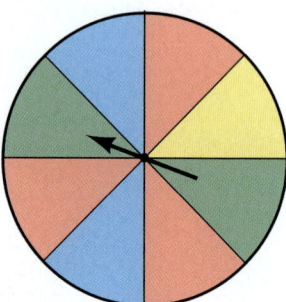

Find the probability of each event.

25. Rolling a sum of 4 on one roll of two dice

26. Drawing a diamond on one draw from a card deck

27. Drawing a red egg from a basket containing 5 red eggs and 7 blue eggs

28. Drawing a yellow egg from a basket containing 5 red eggs and 7 yellow eggs

29. Drawing 6 diamonds from a standard card deck without replacing the cards after each draw

30. Drawing 5 aces from a standard card deck without replacing the cards after each draw

31. Drawing 5 clubs from the black cards in a standard card deck

32. Drawing a face card from a standard card deck

Assume that the probability that an airplane engine will fail a test is $\frac{1}{2}$ and that the aircraft in question has 4 engines. In Exercises 34–38, find each probability.

33. Construct a sample space for the test.

34. All engines will survive the test.

35. Exactly 1 engine will survive.

36. Exactly 2 engines will survive.

37. Exactly 3 engines will survive.

38. No engines will survive.

39. Find the sum of the probabilities in Exercises 34 through 38.

A survey of 282 people is taken to determine the opinions of doctors, teachers, and lawyers on a proposed piece of legislation, with the results shown in the table. A person is chosen at random from those surveyed. Refer to the table to find each probability.

40. The person favors the legislation.

41. A doctor opposes the legislation.

42. A person who opposes the legislation is a lawyer.

	Number that favor	Number that oppose	Number with no opinion	Total
Doctors	70	32	17	119
Teachers	83	24	10	117
Lawyers	23	15	8	46
Total	176	71	35	282

APPLICATIONS

43. QUALITY CONTROL   In a batch of 10 tires, 2 are known to be defective. If 4 tires are chosen at random, find the probability that all 4 tires are good.

44. MEDICINE   Out of a group of 9 patients treated with a new drug, 4 suffered a relapse. If 3 patients are selected at random from the group of 9, find the probability that none of the 3 patients suffered a relapse.

WRITING

45. Explain why all probability values range from 0 to 1.

46. Explain the concept of probability.

REVIEW   **Solve each equation.**

47. $5^{4x} = \dfrac{1}{125}$

48. $8^{-x+1} = \dfrac{1}{64}$

49. $2^{x^2-2x} = 8$

50. $3^{x^2-3x} = 81$

51. $3^{x^2+4x} = \dfrac{1}{81}$

52. $7^{x^2+3x} = \dfrac{1}{49}$

CHALLENGE PROBLEMS   **If $P(A)$ represents the probability of event A, and $P(B \mid A)$ represents the probability that event B will occur after event A, then**

$$P(A \text{ and } B) = P(A) \cdot P(B \mid A)$$

53. In a school, 30% of the students are gifted in math and 10% are gifted in art and math. If a student is gifted in math, find the probability that the student is also gifted in art.

54. The probability that a person owns a luxury car is 0.2, and the probability that the owner of such a car also owns a second car is 0.7. Find the probability that a person chosen at random owns both a luxury car and a second car.

ACCENT ON TEAMWORK

SEATING ARRANGEMENTS

Overview: In this activity, you are to use permutations to determine the number of possible seating arrangements for 4 people.

Instructions: Form groups of 4 students.
a. If possible, arrange your desks in a straight row to help you visualize the following problem: In how many ways can 4 people be seated on a bench?
b. If possible, now arrange your desks in a circle to help you visualize this next situation.

Suppose the same 4 people are seated around a circular table. Some of the possible seating arrangements are really the same, because the relative position of each person to the others is not different. For an example of this, see the illustration below. If we agree not to count any of the duplicate arrangements, in how many different ways can 4 people be seated around a circular table?

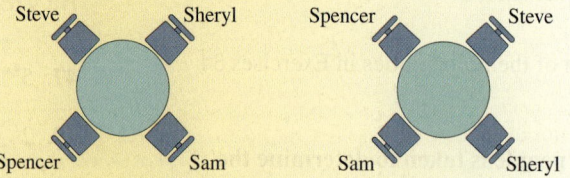

THROWING DICE

Overview: In this activity, you will learn more about probability.

Instructions: Form groups of 2 or 3 students. Your group will need a pair of standard dice.
a. Suppose two dice are rolled, and the sum of the number of dots on the top face is recorded. Complete the table below, which gives the number of ways each sum can occur. For example, there are 3 ways to obtain a sum of 4: a ⚀ on die 1 and a ⚂ on die 2, a ⚂ on die 1 and a ⚀ on die 2, and a ⚁ on die 1 and a ⚁ on die 2.

Sum	2	3	4	5	6	7	8	9	10	11	12
Number of ways			3								

b. Use the table to determine the probability of obtaining a sum of 7 or 11 on one roll of the dice.
c. One-at-a-time, each member of the group should roll the dice until they get a 7 or 11. Keep track of the number of rolls it takes to get one of these outcomes. When finished, compare your results with those of the other members of the class. Which student got 7 or 11 in the least number of rolls? Which student needed the greatest number of rolls to get a 7 or 11?

KEY CONCEPT: THE LANGUAGE OF ALGEBRA

One of the keys to becoming a good algebra student is to know the vocabulary of algebra. Match each instruction in column I with the most appropriate item in column II. Each letter in column II is used only once.

Column I

____ **1.** Use the FOIL method.

____ **2.** Apply a rule for exponents to simplify.

____ **3.** Add the rational expressions.

____ **4.** Rationalize the denominator.

____ **5.** Factor completely.

____ **6.** Evaluate the expression for $a = -1$ and $b = -6$.

____ **7.** Express in lowest terms.

____ **8.** Solve for t.

____ **9.** Combine like terms.

____ **10.** Remove parentheses.

____ **11.** Solve the system by graphing.

____ **12.** Find $f(g(x))$.

____ **13.** Solve using the quadratic formula.

____ **14.** Identify the base and the exponent.

____ **15.** Write without a radical symbol.

____ **16.** Write the equation of the line having the given slope and y-intercept.

____ **17.** Solve the inequality.

____ **18.** Complete the square to make a perfect square trinomial.

____ **19.** Find the slope of the line passing through the given points.

____ **20.** Use a property of logarithms to simplify.

____ **21.** Set each factor equal to zero and solve for x.

____ **22.** State the solution of the compound inequality using interval notation.

____ **23.** Find the inverse function, $h^{-1}(x)$.

____ **24.** Write using scientific notation.

____ **25.** Write the logarithmic statement in exponential form.

____ **26.** Find the sum of the first 6 terms of the sequence.

Column II

a. 2,300,000,000

b. e^3

c. $f(x) = x^2 + 1$ and $g(x) = 5 - 3x$

d. $-2x(3x^2 - 4x + 8)$

e. $4x - 7 > -3x - 7$

f. $\begin{cases} 2x = y - 5 \\ x + y = -1 \end{cases}$

g. $(x^2 - 5)(x^2 + 3)$

h. $(x + 2)(x - 10) = 0$

i. $\dfrac{x - 1}{2x^2} + \dfrac{x + 1}{8x}$

j. $\sqrt{4x^2}$

k. $\ln 6 + \ln x$

l. $\dfrac{10}{\sqrt{6} - \sqrt{2}}$

m. $2x - 8 + 6y - 14$

n. $(3, -2)$ and $(0, -5)$

o. $x^n \cdot x^{3n}$

p. $\dfrac{4x^2 y}{16xy}$

q. $h(x) = 10^x$

r. $\log_2 8 = 3$

s. $m = \dfrac{2}{3}$ and passes through $(0, 2)$

t. 2, 6, 18, . . .

u. $3y^3 - 243b^6$

v. $x + 7 \geq 0$ and $-x < -1$

w. $x^2 - 3x - 4 = 0$

x. $Rt = cd + 2t$

y. $-2\pi a^2 b - 3b^3$

z. $x^2 + 4x$

CHAPTER REVIEW

ELEMENTARY &
INTERMEDIATE
Algebra $f(x)$ Now™

SECTION 14.1	**The Binomial Theorem**

CONCEPTS

Pascal's triangle gives the coefficients of the terms of the expansion of $(a + b)^n$.

The symbol $n!$ (*n factorial*) is defined as

$n! = n(n - 1)(n - 2) \cdots 2 \cdot 1$

$1! = 1$ and $0! = 1$

$n(n - 1)! = n!$

The binomial theorem:

$(a + b)^n =$

$a^n + \dfrac{n!}{1!(n - 1)!} a^{n-1}b$

$+ \dfrac{n!}{2!(n - 2)!} a^{n-2}b^2$

$+ \cdots + b^n$

The $(r + 1)$st term of the expansion of $(a + b)^n$ is

$\dfrac{n!}{r!(n - r)!} a^{n-r}b^r$

Remember that r is always 1 less than the number of the term that you are finding.

REVIEW EXERCISES

1. Complete Pascal's triangle. List the row that gives the coefficients for the expansion of $(a + b)^5$.

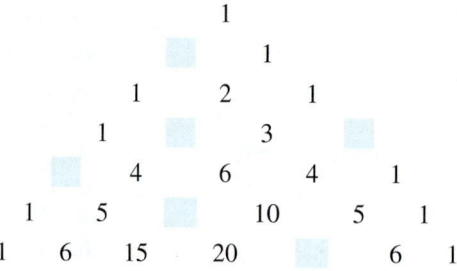

```
                1
              1   1
            1   2   1
          1   3       
        4   6   4   1
      1   5       10   5   1
    1   6   15   20       6   1
```

2. Consider the expansion of $(a + b)^{12}$.

   a. How many terms does the expansion have?

   b. For each term, what is the sum of the exponents on a and b?

   c. What is the first term? What is the last term?

   d. How do the exponents on a and b change from term to term?

Evaluate each expression.

3. $(4!)(3!)$

4. $\dfrac{5!}{3!}$

5. $\dfrac{6!}{2!(6 - 2)!}$

6. $\dfrac{12!}{3!(12 - 3)!}$

7. $(n - n)!$

8. $\dfrac{8!}{7!}$

Use the binomial theorem to find each expansion.

9. $(x + y)^5$

10. $(x - y)^9$

11. $(4x - y)^3$

12. $\left(\dfrac{c}{2} + \dfrac{d}{3} \right)^4$

Find the specified term in each expansion.

13. $(x + y)^4$; third term

14. $(x - y)^6$; fourth term

15. $(3x - 4y)^3$; second term

16. $(u^2 - v^3)^5$; fifth term

| SECTION 14.2 | **Arithmetic Sequences and Series** |

An *arithmetic sequence:*

$$a_1, a_1 + d, a_1 + 2d, \ldots,$$
$$a_1 + (n-1)d, \ldots$$

where a_1 is the first term, d is the common difference, and $a_n = a_1 + (n-1)d$

If numbers are inserted between two given numbers a and b to form an arithmetic sequence, the inserted numbers are *arithmetic means* between a and b.

The sum of the first n terms of an arithmetic sequence is given by

$$S_n = \frac{n(a_1 + a_n)}{2}$$

Summation notation involves the Greek letter sigma Σ. It designates the sum of terms.

17. Find the first four terms of the sequence defined by $a_n = 2n - 4$.

18. Find the eighth term of an arithmetic sequence whose first term is 7 and whose common difference is 5.

19. Write the first five terms of the arithmetic sequence whose ninth term is 242 and whose seventh term is 212.

20. The first three terms of an arithmetic sequence are 6, -6, and -18. Find the 101th term.

21. Find the common difference of an arithmetic sequence if its 1st term is -515 and the 23rd term is -625.

22. Find two arithmetic means between 8 and 25.

23. Find the sum of the first ten terms of the sequence $9, 6\frac{1}{2}, 4, \ldots$.

24. Find the sum of the first 28 terms of an arithmetic sequence if the second term is 6 and the sixth term is 22.

Find each sum.

25. $\displaystyle\sum_{k=4}^{6} \frac{1}{2}k$

26. $\displaystyle\sum_{k=2}^{5} 7k^2$

27. $\displaystyle\sum_{k=1}^{4} (3k - 4)$

28. $\displaystyle\sum_{k=10}^{10} 36k$

29. What is the sum of the first 100 natural numbers?

30. SEATING   The illustration shows the first 2 of a total of 30 rows of seats in an amphitheater. The number of seats in each row forms an arithmetic sequence. Find the total number of seats.

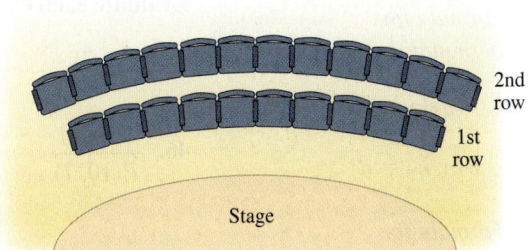

| SECTION 14.3 | **Geometric Sequences and Series** |

A *geometric sequence:*

$$a_1, a_1r, a_1r^2, a_1r^3, \ldots,$$
$$a_1r^{n-1}, \ldots$$

where a_1 is the first term, r is the common ratio, and $a_n = a_1r^{n-1}$.

31. Find the sixth term of a geometric sequence with a first term of $\frac{1}{8}$ and a common ratio of 2.

32. Write the first five terms of the geometric sequence whose fourth term is 3 and whose fifth term is $\frac{3}{2}$.

33. Find the first term of a geometric sequence if it has a common ratio of -3 and the ninth term is 243.

If numbers are inserted between a and b to form a geometric sequence, the inserted numbers are *geometric means* between a and b.

The sum of the first n terms of a geometric sequence:

$$S_n = \frac{a_1 - a_1 r^n}{1 - r} \quad r \neq 1$$

The sum of the terms of an infinite geometric sequence is given by:

$$S = \frac{a_1}{1 - r} \quad |r| < 1$$

34. Find two geometric means between -6 and 384.

35. Find the sum of the first seven terms of the sequence $162, 54, 18, \ldots$.

36. Find the sum of the first eight terms of the sequence $\frac{1}{8}, -\frac{1}{4}, \frac{1}{2}, \ldots$.

37. FEEDING BIRDS   Tom has 50 pounds of birdseed stored in his garage. Each month, he uses 25% of what is left in the bag to feed the birds in his yard. How much birdseed will be left in 12 months?

38. Find the sum of the infinite geometric sequence $25, 20, 16, \ldots$.

39. Change the decimal $0.\overline{05}$ to a common fraction.

40. WHAM-O TOYS   Tests have found that 1998 Superballs© rebound $\frac{9}{10}$ of the distance from which they are dropped. If a Superball© is dropped from a height of 10 feet, and if it could bounce forever, what total distance would it travel?

Permutations and Combinations

The fundamental counting principle for events: If E_1 and E_2 are two events, and if E_1 can be done in m ways and E_2 can be done in n ways, then the event "E_1 followed by E_2" can be done in $m \cdot n$ ways.

Formula for permutations:

$$P(n, r) = \frac{n!}{(n - r)!}$$

$$P(n, n) = n!$$

Formula for combinations:

$$C(n, r) = \binom{n}{r}$$

$$= \frac{n!}{r!(n - r)!}$$

A *permutation* is an ordered arrangement of a given set of objects. To solve counting problems where order is important, use the permutation formula.

41. TRAVEL   If there are 17 flights from New York to Chicago and 8 flights from Chicago to San Francisco, in how many different ways could a passenger plan her trip?

42. LICENSE PLATES   Refer to the illustration. How many different license plates are possible if they are to have 3 letters followed by 3 digits?

Evaluate each expression.

43. $P(7, 7)$

44. $P(7, 0)$

45. $P(8, 6)$

46. $\dfrac{P(9, 6)}{P(10, 7)}$

47. $C(7, 7)$

48. $C(7, 0)$

49. $\dbinom{8}{6}$

50. $C(6, 3) \cdot C(7, 3)$

51. $\dfrac{C(7, 3)}{C(6, 3)}$

52. Use the alternative form of the binomial theorem to expand $(3y - 2z)^4$.

53. LINING UP   In how many ways can 5 persons be arranged in a line?

54. LINING UP   In how many ways can 3 men and 5 women be arranged in a line if the women are placed ahead of the men?

A *combination* is a distinct group of objects without regard to their arrangement. To solve counting problems where order is not important, use the combination formula.

55. CHOOSING PEOPLE   In how many ways can we pick 3 persons from a group of 10 persons?

56. FORMING COMMITTEES   In how many ways can we pick a committee of 2 Democrats and 2 Republicans from a group containing 5 Democrats and 6 Republicans?

| **SECTION 14.5** | **Probability** |

An event that cannot happen has a *probability* of 0. An event that is certain to happen has a probability of 1. All other events have probabilities between 0 and 1.

If S is the *sample space* of an experiment with n distinct and equally likely outcomes, and E is an event that occurs in s of those ways, then the probability of E is

$$P(E) = \frac{s}{n}$$

In Exercises 57–59, assume that a dart is randomly thrown at the colored chart.

57. What is the probability that the dart lands in a blue area?

58. What is the probability that the dart lands in an even-numbered area?

59. What is the probability that the dart lands in an area whose number is greater than 2?

60. Find the probability of rolling an 11 on one roll of two dice.

61. Find the probability of living forever.

62. Find the probability of drawing a 10 from a standard deck of cards.

63. Find the probability of drawing a 5-card poker hand that has exactly 3 aces.

64. Find the probability of drawing 5 cards, all spades, from a standard card deck.

CHAPTER 14 TEST   ELEMENTARY & INTERMEDIATE Algebra *f(x)* Now™

1. Find the first 4 terms of the sequence defined by $a_n = -6n + 8$.

2. Expand: $(a - b)^6$.

3. Find the third term in the expansion of $(x^2 + 2y)^4$.

4. Find the tenth term of an arithmetic sequence whose first 3 terms are 3, 10, and 17.

5. Find the sum of the first 12 terms of the sequence $-2, 3, 8, \ldots$.

6. Find two arithmetic means between 2 and 98.

7. Find the common difference of an arithmetic sequence if the second term is $\frac{5}{4}$ and the 17th term is 5.

8. Find the sum of the first 27 terms of an arithmetic sequence if the 4th term is -11 and the 20th term is -75.

9. PLUMBING   Plastic pipe is stacked so that the bottom row has 25 pipes, the next row has 24 pipes, the next row has 23 pipes, and so on until there is 1 pipe at the top of the stack. If a worker removes the top 15 rows of pipe, how many pieces of pipe will be left in the stack?

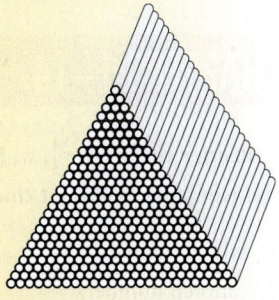

10. Evaluate: $\displaystyle\sum_{k=1}^{3} (2k - 3)$.

11. Find the seventh term of the geometric sequence whose first 3 terms are $-\frac{1}{9}$, $-\frac{1}{3}$, and -1.

12. Find the sum of the first 6 terms of the sequence $\frac{1}{27}$, $\frac{1}{9}$, $\frac{1}{3}$, $\cdots$.

13. Find the first term of a geometric sequence if the common ratio is $-\frac{2}{3}$ and the fourth term is $-\frac{16}{9}$.

14. Find two geometric means between 3 and 648.

15. Find the sum of all of the terms of the infinite geometric sequence 9, 3, 1, ...

16. FALLING OBJECTS   If an object is in free fall, the sequence 16, 48, 80, . . . represents the distance in feet that object falls during the first second, during the second second, during the third second, and so on. How far will the object fall during the first 10 seconds?

Find the value of each expression.

17. $\dfrac{7!}{4!}$

18. 0!

19. $P(5, 4)$

20. $C(6, 4)$

21. $\dbinom{8}{3}$

22. $C(6, 0) \cdot P(3, 3)$

23. PHONE NUMBERS   How many 7-digit phone numbers are available in area code 626 if no phone number can begin with 0, 1, or 2?

24. THE SUPREME COURT   The last names of the members of the U.S. Supreme Court, as of 2004, are shown in one possible seating arrangement below. How many possible seating arrangements in a line are there?

> Bader · Breyer · Kennedy · O'Conner · Rehnquist · Scalia · Souter · Stevens · Thomas

25. CHOOSING PEOPLE   In how many ways can we pick 3 persons from a group of 7 persons?

26. CHOOSING COMMITTEES   From a group of 5 men and 4 women, how many 3-person committees can be chosen that will include 2 women?

Find each probability.

27. Rolling a 5 on one roll of a die

28. Drawing a jack or a queen from a standard card deck

29. Receiving 5 hearts for a 5-card poker hand

30. Tossing 2 heads in 5 tosses of a fair coin

31. Shade an appropriate number of pie-shaped sections of a circle so that the probability of a spinner stopping on an *unshaded* section is $\frac{9}{16}$.

32. Fill in the blanks:

   a. The probability of an event that cannot happen is ☐.

   b. The probability of an event that is guaranteed to happen is ☐.

CHAPTERS 1–14   CUMULATIVE REVIEW EXERCISES

Consider the set $\left\{-\frac{4}{3}, \pi, 5.6, \sqrt{2}, 0, -23, e, 7i\right\}$. List the elements in the set that are

1. whole numbers

2. rational numbers

3. irrational numbers

4. real numbers

5. FINANCIAL PLANNING   Anna has some money to invest. Her financial planner tells her that if she can come up with $3,000 more, she will qualify for an 11% annual interest rate. Otherwise, she will have to invest the money at 7.5% annual interest. The financial planner urges her to invest the larger amount, because the 11% investment would yield twice as much annual income as the 7.5% investment. How much does she originally have on hand to invest?

6. BOATING Use the following graph to determine the average rate of change in the sound level of the engine of a boat in relation to rpm of the engine.

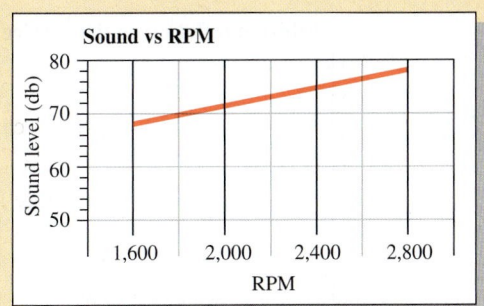

Sound vs RPM

Decide whether the graphs of the equations are parallel or perpendicular.

7. $3x - 4y = 12, y = \dfrac{3}{4}x - 5$

8. $y = 3x + 4, x = -3y + 4$

Write the equation of the line with the given properties.

9. $m = -2$, passing through $(0, 5)$

10. Passing through $(8, -5)$ and $(-5, 4)$

11. Use substitution to solve $\begin{cases} 2x - y = -21 \\ 4x + 5y = 7 \end{cases}$.

12. Use addition to solve $\begin{cases} 4y + 5x - 7 = 0 \\ \dfrac{10}{7}x - \dfrac{4}{9}y = \dfrac{17}{21} \end{cases}$.

13. Use Cramer's rule to solve $\begin{cases} 2(x + y) + 1 = 0 \\ 3x + 4y = 0 \end{cases}$.

14. Solve: $\begin{cases} b + 2c = 7 - a \\ a + c = 8 - 2b \\ 2a + b + c = 9 \end{cases}$

15. The graphs of $y = 4(x - 5) - x - 2$ and $y = -(2x + 6) - 1$ are shown in the illustration. Use the information in the display to solve $4(x - 5) - x - 2 = -(2x + 6) - 1$ graphically.

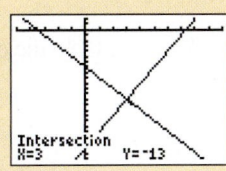

Intersection
X=3   Y=-13

16. MARTIAL ARTS Find the measure of each angle of the triangle shown in the illustration.

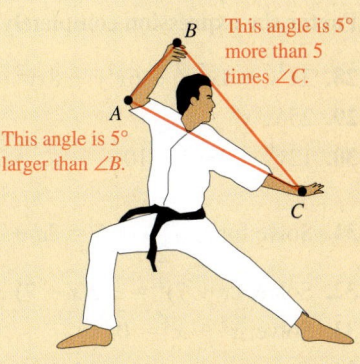

This angle is 5° more than 5 times ∠C.

This angle is 5° larger than ∠B.

17. Explain why the graph does not represent a function.

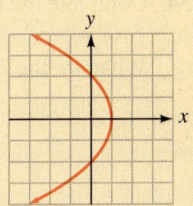

18. If $f(x) = 3x^5 - 2x^2 + 1$, find $f(-1)$ and $f(a)$.

19. Use the graph of function h to find each of the following:

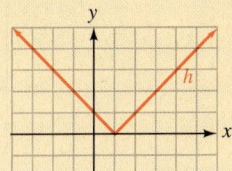

   a. $h(-3)$ **b.** $h(4)$

   c. The value(s) of x for which $h(x) = 1$

   d. The value(s) of x for which $h(x) = 0$

20. Write $173{,}000{,}000{,}000{,}000$ and 0.000000046 in scientific notation.

21. Solve: $\begin{cases} 3x - 2y \le 6 \\ y < -x + 2 \end{cases}$.

Give the solution in interval notation and graph the solution set.

22. Solve: $\left| 5 - 3x \right| - 14 \le 0$.

23. Solve: $4.5x - 1 < -10$ or $6 - 2x \ge 12$.

Perform the operations.

24. $(x - 3y)(x^2 + 3xy + 9y^2)$

25. $(-2x^2y^3 + 6xy + 5y^2) - (-4x^2y^3 - 7xy + 2y^2)$

26. $(9ab^2 - 4)^2$

27. $ab^{-2}c^{-3}(a^{-4}bc^3 + a^{-3}b^4c^3)$

Factor the expression completely.

28. $3x^3y - 4x^2y^2 - 6x^2y + 8xy^2$

29. $256x^4y^4 - z^8$

30. $12y^6 + 23y^3 + 10$

31. Solve for λ: $\dfrac{A\lambda}{2} + 1 = 2d + 3\lambda$.

32. Solve: $(x + 7)^2 = -2(x + 7) - 1$.

33. Solve: $x^3 + x^2 = 0$.

34. Complete the table of function values
$f(x) = -x^3 - x^2 + 6x$ and then graph the function.
What are the x- and y-intercepts of the graph?

x	$f(x)$
-4	
-3	
-2	
-1	
0	
1	
2	
3	

Simplify.

35. $\left(\dfrac{3x^5y^2}{6x^5y^{-2}} \right)^{-4}$

36. $\dfrac{6x^2 + 13x + 6}{6 - 5x - 6x^2}$

37. $\dfrac{p^3 - q^3}{q^2 - p^2} \cdot \dfrac{q^2 + pq}{p^3 + p^2q + pq^2}$

38. $\dfrac{2}{a - 2} + \dfrac{3}{a + 2} - \dfrac{a - 1}{a^2 - 4}$

39. Solve: $\dfrac{x - 4}{x - 3} + \dfrac{x - 2}{x - 3} = x - 3$.

40. Solve: $\dfrac{1}{R} = \dfrac{1}{R_1} + \dfrac{1}{R_2} + \dfrac{1}{R_3}$ for R.

41. TIRE WEAR   See the illustration in the next column.

   a. What type of function does it appear would
model the relationship between the inflation
of a tire and the percent of service it gives?

   b. At what percent(s) of inflation will a tire
offer only 90% of its possible service?

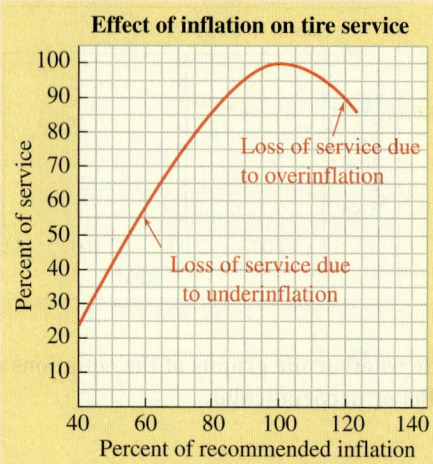

Effect of inflation on tire service

Loss of service due to overinflation

Loss of service due to underinflation

42. CHANGING DIAPERS   The following illustration
shows how to put a diaper on a baby. If the diaper is a
square with sides 16 inches long, what is the largest
waist size that this diaper can wrap around, assuming
an overlap of 1 inch to pin the diaper?

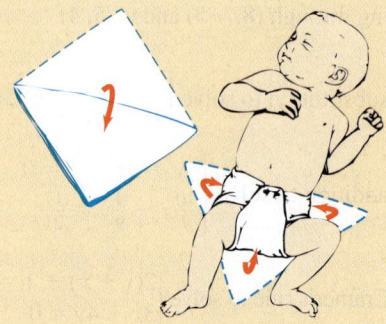

43. Use the long division method to find
$(2x^2 + 4x - x^3 + 3) \div (x - 1)$.

44. LIGHT   The intensity of a light source is inversely
proportional to the square of the distance from the
source. If the intensity is 18 lumens at a distance of
4 feet, what is the intensity when the distance is
12 feet?

45. Graph: $f(x) = \sqrt{x} + 2$. Give the domain and range
of the function.

Simplify each expression.

46. $\sqrt{98} + \sqrt{8} - \sqrt{32}$

47. $3\left(\sqrt{5x} - \sqrt{3}\right)^2$

48. $12\sqrt[3]{648x^4} + 3\sqrt[3]{81x^4}$

49. Evaluate: $\left(\dfrac{25}{49}\right)^{-3/2}$.

Rationalize each denominator.

50. $\dfrac{\sqrt[3]{4a^2}}{\sqrt[3]{2ab}}$

51. $\dfrac{3t - 1}{\sqrt{3t} + 1}$

Write each expression in $a + bi$ form.

52. $\left(-7 + \sqrt{-81}\right) - \left(-2 - \sqrt{-64}\right)$

53. $\dfrac{2 - 5i}{2 + 5i}$

54. Simplify: i^{42}.

Solve each equation.

55. $\sqrt{3a + 1} = a - 1$

56. $\sqrt{x + 3} - \sqrt{3} = \sqrt{x}$

57. $x^4 + 19x^2 + 18 = 0$

58. $4w^2 + 6w + 1 = 0$

59. $2(2x + 1)^2 - 7(2x + 1) + 6 = 0$

60. $3x^2 - 4x = -2$

61. First determine the coordinates of the vertex and the axis of symmetry of the graph of $f(x) = -6x^2 - 12x - 8$ using the vertex formula. Then determine the x- and y-intercepts of the graph. Finally, plot several points and complete the graph.

62. If $f(x) = x^2 - 2$ and $g(x) = 2x + 1$, find $(f \circ g)(x)$.

63. Find the inverse function of $f(x) = 2x^3 - 1$.

64. Graph: $f(x) = \left(\dfrac{1}{2}\right)^x$. Give the domain and range of the function.

65. Graph $y = e^x$ and its inverse on the same coordinate system.

66. Use the properties of logarithms to simplify $\log_6 \dfrac{36}{x^3}$.

67. Write the expression as a single logarithm:

$\dfrac{1}{2} \ln x + \ln y - \ln z$

68. POPULATION GROWTH   As of 2003, the population of Mexico was about 119 million and the annual growth rate was 1.43%. If the growth rate remains the same, estimate the population of Mexico in 25 years.

Find x.

69. $\log_x 25 = 2$

70. $\log 1,000 = x$

71. $\log_3 x = -3$

72. $\ln e = x$

73. Let $\log 7 = 0.8451$ and $\log 14 = 1.1461$. Evaluate $\log 98$ without using a calculator.

74. Find $\log 0$, if possible.

Solve each equation. Round to four decimal places when necessary.

75. $2^{x+2} = 3^x$

76. $\log x + \log (x + 9) = 1$

77. $5^{4x} = \dfrac{1}{125}$

78. $\log_3 x = \log_3 \left(\dfrac{1}{x}\right) + 4$

79. Write the equation of the circle that has its center at $(1, 3)$ and passes through $(-2, -1)$.

80. Write the equation in standard form and graph it: $(x - 2)^2 - 9y^2 = 9$.

81. Graph: $\dfrac{(x-1)^2}{4} + \dfrac{(y-3)^2}{16} = 1$.

82. Complete the square to write the equation $y^2 + 4x - 6y = -1$ in $x = a(y-k)^2 + h$ form. Determine the vertex and the axis of symmetry of the graph. Then plot several points and complete the graph.

83. Use the binomial theorem to expand $(3a - b)^4$.

84. Find the seventh term of the expansion of $(2x - y)^8$.

85. Find the 20th term of an arithmetic sequence with a first term -11 and a common difference 6.

86. Find the sum of the first 20 terms of an arithmetic sequence with a first term 6 and a common difference 3.

87. Evaluate: $\displaystyle\sum_{k=3}^{5} (2k + 1)$.

88. Find the seventh term of a geometric sequence with a first term $\frac{1}{27}$ and a common ratio 3.

89. ▦ BOAT DEPRECIATION   How much will a $9,000 boat be worth after 9 years if it depreciates 12% per year?

90. Find the sum of the first ten terms of the sequence $\frac{1}{64}, \frac{1}{32}, \frac{1}{16}, \ldots$

91. Find the sum of all the terms of the sequence $9, 3, 1, \ldots$

92. LINING UP   In how many ways can 7 people stand in a line?

93. FORMING COMMITTEES   In how many ways can a committee of 3 people be chosen from a group containing 9 people?

94. CARDS   What is the probability of drawing a face card from a standard deck of cards?

Roots and Powers

n	n^2	$\sqrt{n}$	n^3	$\sqrt[3]{n}$	n	n^2	$\sqrt{n}$	n^3	$\sqrt[3]{n}$
1	1	1.000	1	1.000	51	2,601	7.141	132,651	3.708
2	4	1.414	8	1.260	52	2,704	7.211	140,608	3.733
3	9	1.732	27	1.442	53	2,809	7.280	148,877	3.756
4	16	2.000	64	1.587	54	2,916	7.348	157,464	3.780
5	25	2.236	125	1.710	55	3,025	7.416	166,375	3.803
6	36	2.449	216	1.817	56	3,136	7.483	175,616	3.826
7	49	2.646	343	1.913	57	3,249	7.550	185,193	3.849
8	64	2.828	512	2.000	58	3,364	7.616	195,112	3.871
9	81	3.000	729	2.080	59	3,481	7.681	205,379	3.893
10	100	3.162	1,000	2.154	60	3,600	7.746	216,000	3.915
11	121	3.317	1,331	2.224	61	3,721	7.810	226,981	3.936
12	144	3.464	1,728	2.289	62	3,844	7.874	238,328	3.958
13	169	3.606	2,197	2.351	63	3,969	7.937	250,047	3.979
14	196	3.742	2,744	2.410	64	4,096	8.000	262,144	4.000
15	225	3.873	3,375	2.466	65	4,225	8.062	274,625	4.021
16	256	4.000	4,096	2.520	66	4,356	8.124	287,496	4.041
17	289	4.123	4,913	2.571	67	4,489	8.185	300,763	4.062
18	324	4.243	5,832	2.621	68	4,624	8.246	314,432	4.082
19	361	4.359	6,859	2.668	69	4,761	8.307	328,509	4.102
20	400	4.472	8,000	2.714	70	4,900	8.367	343,000	4.121
21	441	4.583	9,261	2.759	71	5,041	8.426	357,911	4.141
22	484	4.690	10,648	2.802	72	5,184	8.485	373,248	4.160
23	529	4.796	12,167	2.844	73	5,329	8.544	389,017	4.179
24	576	4.899	13,824	2.884	74	5,476	8.602	405,224	4.198
25	625	5.000	15,625	2.924	75	5,625	8.660	421,875	4.217
26	676	5.099	17,576	2.962	76	5,776	8.718	438,976	4.236
27	729	5.196	19,683	3.000	77	5,929	8.775	456,533	4.254
28	784	5.292	21,952	3.037	78	6,084	8.832	474,552	4.273
29	841	5.385	24,389	3.072	79	6,241	8.888	493,039	4.291
30	900	5.477	27,000	3.107	80	6,400	8.944	512,000	4.309
31	961	5.568	29,791	3.141	81	6,561	9.000	531,441	4.327
32	1,024	5.657	32,768	3.175	82	6,724	9.055	551,368	4.344
33	1,089	5.745	35,937	3.208	83	6,889	9.110	571,787	4.362
34	1,156	5.831	39,304	3.240	84	7,056	9.165	592,704	4.380
35	1,225	5.916	42,875	3.271	85	7,225	9.220	614,125	4.397
36	1,296	6.000	46,656	3.302	86	7,396	9.274	636,056	4.414
37	1,369	6.083	50,653	3.332	87	7,569	9.327	658,503	4.431
38	1,444	6.164	54,872	3.362	88	7,744	9.381	681,472	4.448
39	1,521	6.245	59,319	3.391	89	7,921	9.434	704,969	4.465
40	1,600	6.325	64,000	3.420	90	8,100	9.487	729,000	4.481
41	1,681	6.403	68,921	3.448	91	8,281	9.539	753,571	4.498
42	1,764	6.481	74,088	3.476	92	8,464	9.592	778,688	4.514
43	1,849	6.557	79,507	3.503	93	8,649	9.644	804,357	4.531
44	1,936	6.633	85,184	3.530	94	8,836	9.695	830,584	4.547
45	2,025	6.708	91,125	3.557	95	9,025	9.747	857,375	4.563
46	2,116	6.782	97,336	3.583	96	9,216	9.798	884,736	4.579
47	2,209	6.856	103,823	3.609	97	9,409	9.849	912,673	4.595
48	2,304	6.928	110,592	3.634	98	9,604	9.899	941,192	4.610
49	2,401	7.000	117,649	3.659	99	9,801	9.950	970,299	4.626
50	2,500	7.071	125,000	3.684	100	10,000	10.000	1,000,000	4.642

Synthetic Division

- Synthetic division • The remainder theorem • The factor theorem

We have discussed how to divide polynomials by polynomials using a long division process. We will now discuss a shortcut method, called **synthetic division,** that we can use to divide a polynomial by a binomial of the form $x - k$.

■ SYNTHETIC DIVISION

To see how synthetic division works, we consider $(4x^3 - 5x^2 - 11x + 20) \div (x - 2)$. On the left below is the long division, and on the right is the same division with the variables and their exponents removed. The various powers of x can be remembered without actually writing them, because the exponents of the terms in the divisor, dividend, and quotient were written in descending order.

$$
\begin{array}{r}
4x^2 + 3x\ -\ 5 \\
x - 2 \overline{)4x^3 - 5x^2 - 11x + 20} \\
\underline{4x^3 - 8x^2} \\
3x^2 - 11x \\
\underline{3x^2 - \ 6x} \\
-5x + 20 \\
\underline{-5x + 10} \\
10 \quad \text{(remainder)}
\end{array}
\qquad
\begin{array}{r}
4 \quad 3 - \ 5 \\
1 - 2\overline{)4 - 5 - 11 \quad 20} \\
\underline{4 - 8} \\
3 - 11 \\
\underline{3 - \ 6} \\
-5 \quad 20 \\
\underline{-5 \quad 10} \\
10 \quad \text{(remainder)}
\end{array}
$$

The numbers printed in color need not be written, because they are duplicates of the numbers above them. Thus, we can write the division in the form shown below on the left. We can shorten the process further by compressing the work vertically and eliminating the 1 (the coefficient of x in the divisor) as shown below on the right.

The Language of Algebra

Synthetic means devised to imitate something natural. You've probably heard of *synthetic* fuels or *synthetic* fibers. *Synthetic* division imitates the long division process.

$$
\begin{array}{r}
4 \quad 3 - 5 \\
1 - 2\overline{)4 - 5 - 11 \quad 20} \\
\underline{- 8} \\
3 \\
\underline{- 6} \\
- 5 \\
\underline{10} \\
10
\end{array}
\qquad
\begin{array}{r}
4 \quad\ 3 \quad -5 \\
-2\overline{)4 \quad -5 \quad -11 \quad 20} \\
\underline{-8 \quad -6 \quad 10} \\
3 \quad -5 \quad 10
\end{array}
$$

If we write the 4 in the quotient on the bottom line, that line gives the coefficients of the quotient and the remainder. If we eliminate the top line, the division appears as follows:

$$\begin{array}{r|rrrr} -2 & 4 & -5 & -11 & 20 \\ & & -8 & -6 & 10 \\ \hline & 4 & 3 & -5 & 10 \end{array}$$

The bottom line was obtained by subtracting the middle line from the top line. If we replace the -2 in the divisor by 2, the division process will reverse the signs of every entry in the middle line, and then the bottom line can be obtained by addition. This gives the final form of the synthetic division.

> **The Language of Algebra**
>
> Synthetic division is used to divide a polynomial by a binomial of the form $x - k$. We call k the **synthetic divisor.** In this example, we are dividing by $x - 2$, so k is 2.

$$\begin{array}{r|rrrr} 2 & 4 & -5 & -11 & 20 \\ & & 8 & 6 & -10 \\ \hline & 4 & 3 & -5 & 10 \end{array}$$

These are the coefficients of the dividend.

These are the coefficients of the quotient and the remainder.

$$4x^2 + 3x - 5 + \dfrac{10}{x-2}$$

Read the result from the bottom row.

Thus,

$$\frac{4x^3 - 5x^2 - 11x + 20}{x - 2} = 4x^2 + 3x - 5 + \frac{10}{x - 2}$$

EXAMPLE 1

Use synthetic division to find $(6x^2 + 5x - 2) \div (x - 5)$.

Solution We write the coefficients in the dividend and the 5 in the divisor in the following form:

Since we are dividing the polynomial by $x - 5$, the synthetic divisor is 5.

$$\begin{array}{r|rrr} 5 & 6 & 5 & -2 \\ \hline \end{array}$$

← This represents the dividend $6x^2 + 5x - 2$.

Then we follow these steps:

> **Success Tip**
>
> In his process, numbers below the line are multiplied by the synthetic divisor and that product is carried above the line to the next column. Numbers above the horizontal line are added.

$$\begin{array}{r|rrr} 5 & 6 & 5 & -2 \\ & \downarrow & & \\ \hline & 6 & & \end{array}$$

Begin by bringing down the 6.

$$\begin{array}{r|rrr} 5 & 6 & 5 & -2 \\ & & 30 & \\ \hline & 6 & & \end{array}$$

Multiply 5 by 6 to get 30.

$$\begin{array}{r|rrr} 5 & 6 & 5 & -2 \\ & & 30 & \\ \hline & 6 & 35 & \end{array}$$

Add 5 and 30 to get 35.

$$\begin{array}{r|rrr} 5 & 6 & 5 & -2 \\ & & 30 & 175 \\ \hline & 6 & 35 & \end{array}$$ Multiply 35 by 5 to get 175.

$$\begin{array}{r|rrr} 5 & 6 & 5 & -2 \\ & & 30 & 175 \\ \hline & 6 & 35 & 173 \end{array}$$ Add -2 and 175 to get 173.

The numbers 6 and 35 represent the quotient $6x + 35$, and 173 is the remainder. Thus,

$$\frac{6x^2 + 5x - 2}{x - 5} = 6x + 35 + \frac{173}{x - 5}$$

Self Check 1   Divide $5x^2 - 4x + 2$ by $x - 3$.

EXAMPLE 2

Use synthetic division to find $\dfrac{x^3 + x^2 - 1}{x - 3}$.

Solution   We begin by writing

$$\begin{array}{r|rrrr} 3 & 1 & 1 & 0 & -1 \\ & & & & \\ \hline \end{array}$$ Write 0 for the coefficient of x, the missing term.

and complete the division as follows.

$$\begin{array}{r|rrrr} 3 & 1 & 1 & 0 & -1 \\ & & 3 & & \\ \hline & 1 & 4 & & \end{array}$$ $$\begin{array}{r|rrrr} 3 & 1 & 1 & 0 & -1 \\ & & 3 & 12 & \\ \hline & 1 & 4 & 12 & \end{array}$$ $$\begin{array}{r|rrrr} 3 & 1 & 1 & 0 & -1 \\ & & 3 & 12 & 36 \\ \hline & 1 & 4 & 12 & 35 \end{array}$$

Multiply, then add.   Multiply, then add.   Multiply, then add.

Thus,

$$\frac{x^3 + x^2 - 1}{x - 3} = x^2 + 4x + 12 + \frac{35}{x - 3}$$

Self Check 2   Use synthetic division to find $\dfrac{x^3 + 3x - 90}{x - 4}$.

EXAMPLE 3

Use synthetic division to divide $5x^2 + 6x^3 + 2 - 4x$ by $x + 2$.

Solution   First, we write the dividend with the exponents in descending order.

$$6x^3 + 5x^2 - 4x + 2$$

Then we write the divisor in $x - k$ form: $x - (-2)$. Thus, $k = -2$. Using synthetic division, we begin by writing

This represents division → $-2\rvert$ $\quad 6 \qquad 5 \quad -4 \qquad 2$
by $x + 2$.

and complete the division.

$$
\begin{array}{r|rrrr}
-2 & 6 & 5 & -4 & 2 \\
   &   & -12 & 14 & -20 \\
\hline
   & 6 & -7 & 10 & -18
\end{array}
$$
The remainder is negative.

Notation

Because the remainder is negative, we can also write the result as

$$6x^2 - 7x + 10 + \dfrac{-18}{x + 2}$$

Thus,

$$\frac{5x^2 + 6x^3 + 2 - 4x}{x + 2} = 6x^2 - 7x + 10 - \frac{18}{x + 2}$$

Self Check 3   Divide $2x - 4x^2 + 3x^3 - 3$ by $x + 1$.

■ THE REMAINDER THEOREM

Synthetic division is important because of the **remainder theorem.**

Remainder Theorem	If a polynomial $P(x)$ is divided by $x - k$, the remainder is $P(k)$.

It follows from the remainder theorem that we can evaluate polynomials using synthetic division. We illustrate this in the following example.

EXAMPLE 4

Let $P(x) = 2x^3 - 3x^2 - 2x + 1$. Find   **a.** $P(3)$ and   **b.** the remainder when $P(x)$ is divided by $x - 3$.

Solution   **a.** To find $P(3)$ we evaluate the function for $x = 3$.

Notation

Naming the function with the letter P, instead of f, stresses that we are working with a polynomial function.

$$
\begin{aligned}
P(3) &= 2(3)^3 - 3(3)^2 - 2(3) + 1 \qquad \text{Substitute 3 for } x. \\
&= 2(27) - 3(9) - 6 + 1 \\
&= 54 - 27 - 6 + 1 \\
&= 22
\end{aligned}
$$

Thus, $P(3) = 22$.

b. We can use synthetic division to find the remainder when $P(x)$ is divided by $x - 3$.

Success Tip

It is often easier to find $P(k)$ by using synthetic division than by substituting k for x in $P(x)$. This is especially true if k is a decimal.

$$
\begin{array}{r|rrrr}
3 & 2 & -3 & -2 & 1 \\
  &   & 6 & 9 & 21 \\
\hline
  & 2 & 3 & 7 & \mathbf{22}
\end{array}
$$
$P(x) = 2x^3 - 3x^2 - 2x + 1$

Thus, the remainder is 22.

The same results in parts a and b show that rather than substituting 3 for x in $P(x) = 2x^3 - 3x^2 - 2x + 1$, we can divide $2x^3 - 3x^2 - 2x + 1$ by $x - 3$ to find $P(3)$.

Self Check 4   Let $P(x) = 5x^3 - 3x^2 + x + 6$. Find   **a.** $P(1)$ and   **b.** use synthetic division to find the remainder when $P(x)$ is divided by $x - 1$.

■ THE FACTOR THEOREM

If two quantities are multiplied, each is called a **factor** of the product. Thus, $x - 2$ is a factor of $6x - 12$, because $6(x - 2) = 6x - 12$. A theorem, called the **factor theorem,** tells us how to find one factor of a polynomial if the remainder of a certain division is 0.

Factor Theorem   If $P(x)$ is a polynomial in x, then

$$P(k) = 0 \quad \text{if and only if} \quad x - k \text{ is a factor of } P(x)$$

If $P(x)$ is a polynomial in x and if $P(k) = 0$, k is called a **zero of the polynomial function.**

EXAMPLE 5

Let $P(x) = 3x^3 - 5x^2 + 3x - 10$. Show that   **a.** $P(2) = 0$ and   **b.** $x - 2$ is a factor of $P(x)$.

Solution   **a.** We can use the remainder theorem to evaluate $P(2)$ by dividing $P(x)$ by $x - 2$.

$$
\begin{array}{r|rrrr}
2 & 3 & -5 & 3 & -10 \\
  &   & 6 & 2 & 10 \\
\hline
  & 3 & 1 & 5 & \mathbf{0}
\end{array}
\qquad P(x) = 3x^2 - 5x^2 + 3x - 10
$$

The remainder in this division is 0. By the remainder theorem, the remainder is $P(2)$. Thus, $P(2) = 0$, and 2 is a zero of the polynomial.

b. Because the remainder is 0, the numbers 3, 1, and 5 in the synthetic division in part a represent the quotient $3x^2 + x + 5$. Thus,

$$\underbrace{(x - 2)}_{\text{Divisor} \cdot} \cdot \underbrace{(3x^2 + x + 5)}_{\text{quotient}} + \underbrace{0}_{+ \text{ remainder} =} = \underbrace{3x^3 - 5x^2 + 3x - 10}_{\text{the dividend, } P(x)}$$

or

$$(x - 2)(3x^2 + x + 5) = 3x^3 - 5x^2 + 3x - 10$$

Thus, $x - 2$ is a factor of $3x^3 - 5x^2 + 3x - 10$.

Self Check 5   Let $P(x) = x^3 - 4x^2 + x + 6$. Show that $x + 1$ is a factor of $P(x)$ using synthetic division.

The result in Example 5 is true, because the remainder, $P(2)$, is 0. If the remainder had not been 0, then $x - 2$ would not have been a factor of $P(x)$.

The Language of Algebra

The phrase *if and only if* in the factor theorem means:

If $P(2) = 0$, then $x - 2$ is a factor of $P(x)$

and

If $x - 2$ is a factor of $P(x)$, then $P(2) = 0$.

Answers to Self Checks **1.** $5x + 11 + \dfrac{35}{x - 3}$ **2.** $x^2 + 4x + 19 - \dfrac{14}{x - 4}$ **3.** $3x^2 - 7x + 9 - \dfrac{12}{x + 1}$

4. 9 **5.** Since $P(-1) = 0$, $x + 1$ is a factor of $P(x)$.

APPENDIX II STUDY SET

VOCABULARY Fill in the blanks.

1. The method of dividing $x^2 + 2x - 9$ by $x - 4$ shown below is called _____ division.

$$\begin{array}{r|rrr} 4 & 1 & 2 & -9 \\ & & 4 & 24 \\ \hline & 1 & 6 & 15 \end{array}$$

2. Synthetic division is used to divide a polynomial by a _____ of the form $x - k$.

3. In Exercise 1, the synthetic _____ is 4.

4. By the _____ theorem, if a polynomial $P(x)$ is divided by $x - k$, the remainder is $P(k)$.

5. The factor _____ tells us how to find one factor of a polynomial if the remainder of a certain division is 0.

6. If $P(x)$ is a polynomial and if $P(k) = 0$, then k is called a _____ of the polynomial.

CONCEPTS

7. a. What division is represented below?

b. What is the answer?

$$\begin{array}{r|rrrr} -2 & 5 & 0 & 1 & -3 \\ & & -10 & 20 & -42 \\ \hline & 5 & -10 & 21 & -45 \end{array}$$

Fill in the blanks.

8. In the synthetic division process, numbers below the line are _____ by the synthetic divisor and that product is carried above the line to the next column. Numbers above the horizontal line are _____.

9. Rather than substituting 8 for x in $P(x) = 6x^3 - x^2 - 17x + 9$, we can divide the polynomial _____ by _____ to find $P(8)$.

10. For $P(x) = x^3 - 4x^2 + x + 6$, suppose we know that $P(3) = 0$. Then _____ is a factor of $x^3 - 4x^2 + x + 6$.

NOTATION Complete each synthetic division.

11. Divide $6x^3 + x^2 - 23x + 2$ by $x - 2$.

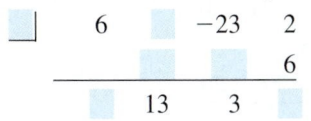

12. Divide $2x^3 - 4x^2 - 25x + 15$ by $x + 3$.

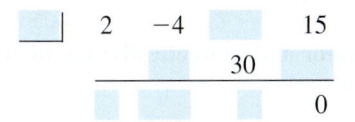

PRACTICE Use synthetic division to perform each division.

13. $\dfrac{x^2 + x - 2}{x - 1}$

14. $\dfrac{x^2 + x - 6}{x - 2}$

15. $\dfrac{x^2 - 7x + 12}{x - 4}$

16. $\dfrac{x^2 - 6x + 5}{x - 5}$

17. $\dfrac{x^2 + 8 + 6x}{x + 4}$

18. $\dfrac{x^2 - 15 - 2x}{x + 3}$

19. $\dfrac{x^2 - 5x + 14}{x + 2}$

20. $\dfrac{x^2 + 13x + 42}{x + 6}$

21. $\dfrac{3x^3 - 10x^2 + 5x - 6}{x - 3}$

22. $\dfrac{2x^3 - 9x^2 + 10x - 3}{x - 3}$

23. $\dfrac{2x^3 - 6 - 5x}{x - 2}$

24. $\dfrac{4x^3 - 1 + 5x^2}{x + 2}$

25. $\dfrac{5x^2 + 4 + 6x^3}{x + 1}$

26. $\dfrac{4 - 3x^2 + x}{x - 4}$

27. $\dfrac{t^3 + t^2 + t + 2}{t + 1}$

28. $\dfrac{m^3 - m^2 - m - 1}{m - 1}$

29. $\dfrac{a^5 - 1}{a - 1}$

30. $\dfrac{b^4 - 81}{b - 3}$

31. $\dfrac{-5x^5 + 4x^4 + 30x^3 + 2x^2 + 20x + 3}{x - 3}$

32. $\dfrac{-6c^5 + 14c^4 + 38c^3 + 4c^2 + 25c - 36}{c - 4}$

33. $\dfrac{8t^3 - 4t^2 + 2t - 1}{t - \dfrac{1}{2}}$ **34.** $\dfrac{9a^3 + 3a^2 - 21a - 7}{a + \dfrac{1}{3}}$

35. $\dfrac{x^4 - x^3 - 56x^2 - 2x + 16}{x - 8}$

36. $\dfrac{x^4 - 9x^3 + x^2 - 7x - 20}{x - 9}$

Use a calculator and synthetic division to perform each division.

37. $\dfrac{7.2x^2 - 2.1x + 0.5}{x - 0.2}$

38. $\dfrac{2.7x^2 + x - 5.2}{x + 1.7}$

39. $\dfrac{9x^3 - 25}{x + 57}$

40. $\dfrac{0.5x^3 + x}{x - 2.3}$

Let $P(x) = 2x^3 - 4x^2 + 2x - 1$. Evaluate $P(x)$ by substituting the given value of x into the polynomial and simplifying. Then evaluate the polynomial by using the remainder theorem and synthetic division.

41. $P(1)$ **42.** $P(2)$

43. $P(-2)$ **44.** $P(-1)$

45. $P(3)$ **46.** $P(-4)$

47. $P(0)$ **48.** $P(4)$

Let $Q(x) = x^4 - 3x^3 + 2x^2 + x - 3$. Evaluate $Q(x)$ by substituting the given value of x into the polynomial and simplifying. Then evaluate the polynomial by using the remainder theorem and synthetic division.

49. $Q(-1)$ **50.** $Q(1)$

51. $Q(2)$ **52.** $Q(-2)$

53. $Q(3)$ **54.** $Q(0)$

55. $Q(-3)$ **56.** $Q(-4)$

Use the remainder theorem and synthetic division to find $P(k)$.

57. $P(x) = x^3 - 4x^2 + x - 2; k = 2$
58. $P(x) = x^3 - 3x^2 + x + 1; k = 1$
59. $P(x) = 2x^3 + x + 2; k = 3$
60. $P(x) = x^3 + x^2 + 1; k = -2$
61. $P(x) = x^4 - 2x^3 + x^2 - 3x + 2; k = -2$
62. $P(x) = x^5 + 3x^4 - x^2 + 1; k = -1$
63. $P(x) = 3x^5 + 1; k = -\dfrac{1}{2}$
64. $P(x) = 5x^7 - 7x^4 + x^2 + 1; k = 2$

Use the factor theorem and determine whether the first expression is a factor of $P(x)$.

65. $x - 3; P(x) = x^3 - 3x^2 + 5x - 15$
66. $x + 1; P(x) = x^3 + 2x^2 - 2x - 3$
    (*Hint:* Write $x + 1$ as $x - (-1)$.)
67. $x + 2; P(x) = 3x^2 - 7x + 4$
    (*Hint:* Write $x + 2$ as $x - (-2)$.)
68. $x; P(x) = 7x^3 - 5x^2 - 8x$
    (*Hint:* $x = x - 0$.)

WRITING

69. When dividing a polynomial by a binomial of the form $x - k$, synthetic division is considered to be faster than long division. Explain why.

70. Let $P(x) = x^3 - 6x^2 - 9x + 4$. You now know two ways to find $P(6)$. What are they? Which method do you prefer?

71. Explain the factor theorem.

72. What is a *zero* of a polynomial function?

REVIEW   Evaluate each expression for $x = -3, y = -5,$ and $z = 0$.

73. $x^2z(y^3 - z)$ **74.** $\left| y^3 - z \right|$

75. $\dfrac{x - y^2}{2y - 1 + x}$ **76.** $\dfrac{2y + 1}{x} - x$

CHALLENGE PROBLEMS   Suppose that $P(x) = x^{100} - x^{99} + x^{98} - x^{97} + \cdots + x^2 - x + 1$.

77. Find the remainder when $P(x)$ is divided by $x - 1$.

78. Find the remainder when $P(x)$ is divided by $x + 1$.

Answers to Selected Exercises

Study Set Section 1.1 (page 6)

1. sum, difference   **3.** Variables   **5.** equation
7. horizontal, vertical   **9.** equation   **11.** algebraic
expression   **13.** algebraic expression   **15.** equation
17. a. multiplication, subtraction   **b.** x **19. a.** addition,
subtraction   **b.** m **21.**

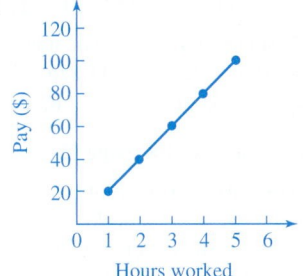

23. They determine that 15-year-old machinery is worth
$35,000.   **25.** $5 \cdot 6$, $5(6)$ **27.** $34 \cdot 75$, $34(75)$ **29.** $4x$
31. $3rt$ **33.** lw **35.** Prt **37.** $2w$ **39.** xy **41.** $\frac{32}{x}$
43. $\frac{90}{30}$ **45.** The product of 8 and 2 is 16.   **47.** The
difference of 11 and 9 is 2.   **49.** The product of 2 and x is 10.
51. The quotient of 66 and 11 is 6.   **53.** $p = 100 - d$
55. $7d = h$ **57.** $s = 3c$ **59.** $w = e + 1{,}200$
61. $p = r - 600$ **63.** $\frac{l}{4} = m$ **65.** 390, 400, 405
67. 1,300, 1,200, 1,100   **69.** $d = \frac{e}{12}$
71. 90, 60, 45, 30, 0

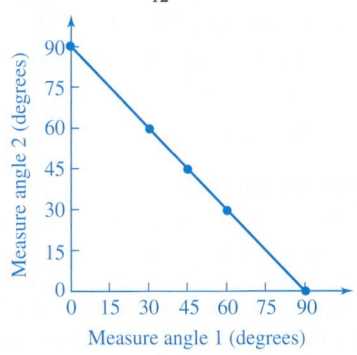

Study Set Section 1.2 (page 18)

1. prime   **3.** numerator, denominator   **5.** equivalent
7. least or lowest   **9.** 60   **11.** $\frac{4}{12} = \frac{1}{3}$ **13.** $\frac{2}{5}$ **15. a.** 1

b. 1   **17. a.** 3 times   **b.** 2 times   **19. a.** $\frac{5}{16}$ **b.** 1
c. $\frac{15}{48}$ **21.** 1, 2, 4, 5, 10, 20   **23.** 1, 2, 4, 7, 14, 28
25. $3 \cdot 5 \cdot 5$ **27.** $2 \cdot 2 \cdot 7$ **29.** $3 \cdot 3 \cdot 13$
31. $2 \cdot 2 \cdot 5 \cdot 11$ **33.** $\frac{3}{9}$ **35.** $\frac{24}{54}$ **37.** $\frac{35}{5}$ **39.** $\frac{1}{2}$ **41.** $\frac{4}{3}$
43. $\frac{3}{4}$ **45.** $\frac{9}{8}$ **47.** lowest terms   **49.** $\frac{4}{25}$ **51.** $\frac{3}{10}$
53. $\frac{8}{5}$ **55.** $\frac{3}{2}$ **57.** 70   **59.** $10\frac{1}{2}$ **61.** $13\frac{3}{4}$ **63.** $\frac{9}{10}$
65. $\frac{5}{8}$ **67.** $\frac{14}{5}$ **69.** 28   **71.** $1\frac{9}{11}$ **73.** $2\frac{1}{2}$ **75.** $\frac{6}{5}$
77. $\frac{5}{24}$ **79.** $\frac{19}{15}$ **81.** $\frac{3}{4}$ **83.** $\frac{17}{12}$ **85.** $\frac{22}{35}$ **87.** $\frac{1}{6}$ **89.** $\frac{9}{4}$
91. $1\frac{1}{4}$ **93.** $\frac{5}{9}$ **95. a.** $\frac{7}{32}$ in.   **b.** $\frac{3}{32}$ in.   **97.** $40\frac{1}{2}$ in.
103. The difference of 7 and 5 is 2.   **105.** The quotient of 30
and 15 is 2.   **107.** 150, 180

Study Set Section 1.3 (page 28)

1. whole   **3.** integers   **5.** negative, positive   **7.** rational
9. irrational   **11.** real   **13.** opposites
15. $\frac{6}{1}, \frac{-9}{1}, \frac{-7}{8}, \frac{7}{2}, \frac{-3}{10}, \frac{283}{100}$ **17.** 13 and -3 **19. a.** $<$
b. $>$ **c.** $>, <$ **d.** $>$ **21.** π in.   **23.** square root
25. is not equal to   **27.** Greek   **29.** $\frac{-4}{5}, \frac{4}{-5}$ **31.** 0.625
33. $0.0\overline{3}$ **35.** 0.42   **37.** $0.\overline{45}$ **39.** $>$ **41.** $>$ **43.** $>$
45. $<$ **47.** $<$ **49.** $=$ **51.** $=$ **53.** $>$ **55.** $>$
57. a. true   **b.** false   **c.** false   **d.** true
59. a. $-5 > -6$ **b.** $-25 < 16$
61.

63. natural, whole, integers: 9; rational: $9, \frac{15}{16}, 3\frac{1}{8}, 1.765$;
irrational: $2\pi, 3\pi, \sqrt{89}$; real: all   **65.** shell 1; $|-6| > |5|$
67. a. '00: $-\$81$ billion; '99: $-\$74$ billion; '02: $-\$70$ billion
b. '90; $-\$40$ billion   **73.** $\frac{4}{9}$ **75.** $\frac{6}{5}$ **77.** $\frac{13}{30}$

Study Set Section 1.4 (page 37)

1. signed   **3.** opposites   **5.** identity
7. $-1 + (-3) = -4$ **9.** -5 **11.** -3 **13. a.** 0   **b.** a
c. a **d.** $(b + c)$ **15.** positive   **17. a.** $1 + (-5)$
b. $-80.5 + 15$ **c.** $(20 + 4)$ **19. a.** 0   **b.** 0   **c.** -6
d. $-\frac{15}{16}$ **e.** 0   **f.** 0   **21.** $x + y = y + x$ **23.** $8 + 9$

25. -2 **27.** 2   **29.** -8 **31.** 8   **33.** 0   **35.** -77
37. 4   **39.** -35 **41.** -8.2 **43.** -20.1 **45.** 0.2
47. $-\frac{1}{8}$ **49.** $\frac{5}{12}$ **51.** $-\frac{9}{10}$ **53.** 16   **55.** -15
57. -21 **59.** -26 **61.** 0   **63.** 4   **65.** -5 **67.** 1
69. -3.2 **71.** 10   **73.** 215   **75.** -112 **77.** 2,150 m
79. $-18, -6, -5, -4$ **81. a.** 3,660.66, 1,408.78
b. $-1,242.86$ **83.** \$89 million   **85.** southward, 132 km
87. \$2.1 million

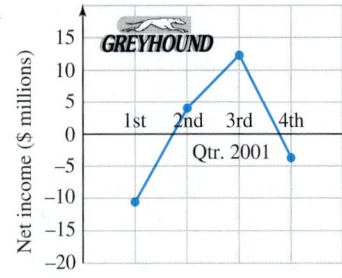

89. true   **91.** -9 and 3

Study Set Section 1.5 (page 44)

1. opposites, inverses   **3.** range   **5. a.** -12 **b.** $\frac{1}{5}$
c. -2.71 **d.** 0   **7. a.** number   **b.** opposite
9. $+ (-9)$ **11.** no   **13.** $-10 + (-8) + (-23) + 5 + 34$
15. a. -500 **b.** y **17.** -3 **19.** -13 **21.** 11
23. -21 **25.** -6 **27.** 1   **29.** 12   **31.** 2   **33.** 40
35. 0   **37.** 5   **39.** -9 **41.** -88 **43.** 12   **45.** 0
47. -4 **49.** -2 **51.** $-\frac{1}{2}$ **53.** $-\frac{5}{16}$ **55.** $-\frac{5}{12}$
57. -1.1 **59.** -3.5 **61.** -2.3 **63.** 4.6   **65.** 22
67. -25 **69.** -11 **71.** -50 **73.** -7 **75.** -1
77. -1 **79.** 0   **81.** 160°F   **83.** 1,030 ft   **85.** $-5.75°$
87. 428 B.C. (-428)   **89.** left; -8 **95.** $2 \cdot 3 \cdot 5$ **97.** $\frac{21}{56}$
99. true

Study Set Section 1.6 (page 54)

1. product, quotient   **3.** commutative   **5.** undefined
7. -5 **9.** positive   **11.** 1   **13. a.** One of the numbers is 0.
b. They are reciprocals (multiplicative inverses).   **15. a.** NEG
b. not possible to tell   **c.** POS   **d.** NEG   **17. a.** a
b. $a(bc)$ **c.** 0   **d.** a **e.** 1   **19. a.** associative property
of multiplication   **b.** multiplicative inverse   **c.** commutative
property of multiplication   **d.** multiplication property of 1
21. a. $-4(-5) = 20$ **b.** $\frac{16}{-8} = -2$ **23.** -16 **25.** 54
27. -60 **29.** -24 **31.** -800 **33.** 36   **35.** 2.4
37. -0.48 **39.** 0.99   **41.** -15.12 **43.** $-\frac{3}{8}$ **45.** $\frac{1}{12}$
47. $-\frac{3}{20}$ **49.** $\frac{15}{16}$ **51.** -520 **53.** 0   **55.** 0   **57.** 60
59. 84   **61.** 120   **63.** $\frac{1}{24}$ **65.** -720 **67.** 10   **69.** 3
71. -2 **73.** -4 **75.** -1 **77.** 1   **79.** -4 **81.** -20
83. -0.005 **85.** 0   **87.** undefined   **89.** $-\frac{5}{12}$ **91.** $\frac{15}{4}$
93. $1\frac{1}{2}$ **95.** -4.7 **97.** 30.3   **99.** -67 **101.** -6
103. $-72°$ **105.** -280 **107.** $-193°$ F   **109.** $-\$43.32$
million   **111.** $-\$614,516$ **113. a.** $5, -10$ **b.** $2.5, -5$
c. $7.5, -15$ **d.** $10, -20$ **119.** -5 **121.** $1.08\overline{3}$
123. $\frac{3}{7}$

Study Set Section 1.7 (page 65)

1. base, exponent   **3.** power   **5.** order   **7. a.** 54, 34
b. 34; multiplication is to be done before addition.
9. innermost: parentheses; outermost: brackets
11. a. subtraction, power, addition, multiplication   **b.** power,
multiplication, subtraction, addition   **13. a.** subtraction
b. division   **15. a.** subtraction   **b.** power   **c.** power
d. power   **17. a.** 3   **b.** x **c.** 1   **d.** 1   **19. a.** -5
b. 5   **21.** 3, 9, 18   **23.** 3^4 **25.** $10^2 k^3$ **27.** $8\pi r^3$
29. $6x^2 y^3$ **31.** 36   **33.** -256 **35.** -125 **37.** $-1,296$
39. 0.16   **41.** $-\frac{8}{125}$ **43.** -17 **45.** 192   **47.** 38
49. 80   **51.** -34 **53.** -28 **55.** 194   **57.** -44
59. 0   **61.** -38 **63.** -8 **65.** 12   **67.** 8
69. undefined   **71.** 201   **73.** 50   **75.** 20   **77.** -396
79. 343   **81.** 360   **83.** 12   **85.** -10 **87.** undefined
89. 59   **91.** 28   **93.** 1,000   **95.** -54 **97.** $\frac{1}{8}$ **99.** -8
101. 31   **103.** -39 **105.** -27 **107.** -8 **109.** 11
111. $-\frac{8}{9}$ **113.** 1   **115.** 10   **117.** 12   **119.** -1
121. 2^2 square units, 3^2 square units, 4^2 square units
123. \$2,106   **125. a.** \$11,875   **b.** \$95   **127.** 81 in.
133. a. ii   **b.** iii   **c.** iv   **d.** i

Study Set Section 1.8 (page 75)

1. expressions   **3.** constant   **5.** evaluate   **7.** expression,
equation   **9. a.** 3   **b.** 11   **c.** -6 **d.** -9
11. a. term   **b.** factor   **c.** factor   **d.** term   **e.** factor
f. term   **13. a.** 7, 14, 21, $7w$ **b.** 1, 2, 3, $\frac{s}{60}$
15. a. $x =$ weight of the car; $2x - 500 =$ weight of the van
b. 3,500 lb   **17.** 5, 30; 10, 10d; 50, 50($x + 5$)
19. 5, 25, 45   **21. a.** $8y$ **b.** $2cd$ **c.** $15sx$ **d.** $-9a^3 b^2$
23. $l + 15$ **25.** $50x$ **27.** $\frac{w}{l}$ **29.** $P + p$
31. $k^2 - 2,005$ **33.** $J - 500$ **35.** $\frac{1,000}{n}$ **37.** $p + 90$
39. $35 + h + 300$ **41.** $p - 680$ **43.** $4d - 15$
45. $2(200 + t)$ **47.** $|a - 2|$ **49.** $300; 60h$ **51. a.** $3y$
b. $\frac{f}{3}$ **53.** $29x$¢   **55.** $\frac{c}{6}$ **57.** $5b$ **59.** $\$5(x + 2)$
61. $-1, -2, -28$ **63.** $41, 11, 2$ **65.** $150, -450$
67. 0, 0, 5   **69.** 20   **71.** -12 **73.** -5 **75.** 156
77. $-\frac{1}{5}$ **79.** 17   **81.** 36   **83.** 230   **85.** 50
87. 48, 64, 48, 0   **89.** $-37°$ C, $-64°$ C   **91.** $1\frac{23}{64}$ in.²
97. 60   **99.** -225

Key Concept (page 81)

1. f   **2.** j   **3.** h   **4.** a   **5.** b   **6.** e   **7.** c   **8.** i
9. d   **10.** g   **11.** $C = p + t$ (Answers may vary depending
on the variables chosen.)   **12.** $b = 2t$ (Answers may vary
depending on the variables chosen.)   **13.** $x + 4 =$ amount of
business (\$ millions) in the year with the celebrity
14. 1, 41, 97

Chapter Review (page 82)

1. 1 hr; 100 cars   **2.** 100   **3.** 7 P.M.   **4.** The difference of
15 and 3 is 12.   **5.** The sum of 15 and 3 is 18.   **6.** The

quotient of 15 and 3 is 5.   **7.** The product of 15 and 3 is 45.
8. $4 \cdot 9$; $4(9)$ **9.** $\frac{9}{3}$ **10.** $8b$ **11.** Prt **12.** equation
13. expression   **14.** 10, 15, 25   **15. a.** $2 \cdot 12, 3 \cdot 8$
(answers may vary)   **b.** $2 \cdot 2 \cdot 6$ (answers may vary)
c. 1, 2, 3, 4, 6, 8, 12, 24   **16.** $3^3 \cdot 2$ **17.** $7^2 \cdot 3$
18. $11 \cdot 7 \cdot 5$ **19.** prime   **20.** 1   **21.** 0   **22.** $\frac{4}{7}$ **23.** $\frac{4}{3}$
24. $\frac{40}{64}$ **25.** $\frac{36}{3}$ **26.** $\frac{7}{64}$ **27.** $\frac{5}{21}$ **28.** $\frac{16}{45}$ **29.** $3\frac{1}{4}$
30. $\frac{2}{5}$ **31.** $\frac{5}{22}$ **32.** $\frac{11}{12}$ **33.** $\frac{5}{18}$ **34.** $\frac{17}{96}$ in.   **35.** 0
36. $-\$65$ billion   **37.** -206 ft   **38.** $<$ **39.** $>$ **40.** $\frac{7}{10}$
41. $\frac{14}{3}$ **42.** 0.004   **43.** $0.7\overline{72}$
44.

45. false   **46.** false   **47.** true   **48.** true   **49.** natural: 8;
whole: 0, 8; integers: 0, -12, 8; rational: $-\frac{4}{5}$, 99.99, 0, -12, $4\frac{1}{2}$,
0.666 . . . , 8; irrational: $\sqrt{2}$; real: all   **50.** $>$ **51.** $<$
52. -82 **53.** 12   **54.** -7 **55.** 0   **56.** -11
57. -12.3 **58.** $-\frac{3}{16}$ **59.** 11   **60.** commutative property
of addition   **61.** associative property of addition
62. addition property of opposites   **63.** addition property of 0
64. -1 **65.** -10 **66.** 3   **67.** $\frac{9}{16}$ **68.** -4 **69.** -19
70. -49 **71.** -15 **72.** 5.7   **73.** -10 **74.** -29
75. 65,233 ft   **76.** -56 **77.** 54   **78.** 12   **79.** 36
80. 6.36   **81.** -2 **82.** $-\frac{2}{15}$ **83.** 0   **84.** associative
property of multiplication   **85.** commutative property of
multiplication   **86.** multiplication property of 1
87. inverse property of multiplication   **88.** 3   **89.** $-\frac{1}{3}$
90. -1 **91.** -4 **92.** 3   **93.** $-\frac{6}{5}$ **94.** undefined
95. -4.5 **96.** high: 2, low: -3 **97.** high: 4, low: -6
98. 8^5 **99.** a^4 **100.** $9\pi r^2$ **101.** x^3y^4 **102.** 81
103. $-\frac{8}{27}$ **104.** 32   **105.** 50   **106.** 4; power,
multiplication, subtraction, addition   **107.** 17   **108.** -48
109. -9 **110.** 44   **111.** -420 **112.** $-\frac{14}{19}$ **113.** 113
114. 0   **115.** \$20   **116.** 3   **117.** 1   **118.** 2, -5
119. 16, -5, 25   **120.** $\frac{1}{2}$, 1   **121.** 9.6, -1 **122.** $h + 25$
123. $s - 15$ **124.** $\frac{1}{2}t$ **125.** $(n + 4)$ in.   **126.** $(b - 4)$ in.
127. $10d$ **128.** $(x - 5)$ years   **129.** 30, $10d$
130. 0, 19, -16 **131.** 40   **132.** -36

Chapter 1 Test (page 90)

1. \$24   **2.** 5 hr   **3.** 3, 20, 70
4. $2 \cdot 2 \cdot 3 \cdot 3 \cdot 5 = 2^2 \cdot 3^2 \cdot 5$ **5.** $\frac{2}{5}$ **6.** $\frac{3}{2} = 1\frac{1}{2}$ **7.** $\frac{25}{36}$
8. $10\frac{1}{15}$ **9.** \$3.57   **10.** $0.8\overline{3}$
11.

12. a. true   **b.** false   **c.** true   **d.** true   **14. a.** $>$
b. $<$ **c.** $<$ **d.** $>$ **15.** a gain of 0.6 of a rating point
16. -2 **17.** $\frac{3}{8}$ **18.** -6 **19.** -30 **20.** 2

21. -2.44 **22.** 0   **23.** -3 **24.** 0   **25.** 50   **26.** $-\frac{27}{125}$
27. 14   **28.** associative property of addition   **29. a.** 9^5
b. $3x^2z^3$ **30.** 170   **31.** 36   **32.** -12 **33.** -100
34. 36   **35.** 4, 17, -59 **36.** $2w + 7$ **37.** $x - 2$
38. $25q\,¢$ **40.** 3; 5

Study Set Section 2.1 (page 100)

1. equation   **3.** check   **5.** equivalent   **7.** isolate
9. a. $x + 5 = 7$ **b.** subtract 5 from both sides
11. a. $x + 6$ **b.** neither   **c.** no   **d.** yes   **13.** 24
15. n **17. a.** c, c **b.** c, c **19. a.** x **b.** y **c.** t
d. h **21.** 15, 15, 30, 30   **23. a.** is possibly equal to
b. $27°$ **25.** yes   **27.** no   **29.** no   **31.** no   **33.** no
35. no   **37.** no   **39.** yes   **41.** yes   **43.** 3   **45.** 71
47. 9   **49.** 0   **51.** -9 **53.** -3 **55.** -2.3 **57.** -36
59. 13   **61.** $\frac{8}{9}$ **63.** $\frac{7}{25}$ **65.** 4   **67.** 41   **69.** 0
71. 1   **73.** -6 **75.** 20   **77.** 0.5   **79.** 45   **81.** 0
83. -105 **85.** 21   **87.** -2.64 **89.** 1,251,989
91. -28 **93.** $65°$ **95.** $38°$ **101.** 0   **103.** $45 - x$

Study Set Section 2.2 (page 110)

1. variable   **3.** variable, equation, solve   **5.** amount,
percent, base   **7.** $x + 371 + 479 = 1,240$
9. $x + 11,000 = 13,500$ **11.** $x + 5 + 8 + 16 = 31$
13. ▨ is ▨ % of ▨ ?   **15.** $12 = 0.40 \cdot x$
17. a. 0.35   **b.** 0.035   **c.** 3.5   **d.** 0.005   **19.** 312
21. 26%   **23.** 300   **25.** 46.2   **27.** 2.5%   **29.** 1,464
31. $6x = 330$ **33.** 63   **35.** \$322.00   **37.** 27 min
39. 16   **41.** 5   **43.** 54   **45.** 975 mi.   **47.** 54 ft
49. $135°$ **51.** 0.48 oz   **53.** \$684 billion   **55.** 78.125%
57. 19%   **59.** 120   **61. a.** 5 g; 25%   **b.** 20 g
63. 1994–1995; about 9.8%   **65.** 12%   **71.** $\frac{12}{5} = 2\frac{2}{5}$
73. no

Study Set Section 2.3 (page 121)

1. simplify   **3.** expressions, equations   **5.** opposite
7. coefficient   **9. a.** 5, 6, 30   **b.** $-8, 2, 4$ **11.** They are
not like terms.   **13.** -1, sign   **15. a.** $3a, 2a$ **b.** 10, 12
c. none   **d.** $9y^2, -8y^2$ **17. a.** $4 + 6, 10$
b. $30 - 50, -20$ **c.** 27   **19. a.** $6(h - 4)$ **b.** $-(z + 16)$
21. no, yes, no, no, yes, yes   **23.** $63m$ **25.** $-35q$
27. $300t$ **29.** $11.2x$ **31.** g **33.** $5x$ **35.** $6y$ **37.** s
39. $-20r$ **41.** $60c$ **43.** $-96m$ **45.** $5x + 15$
47. $36c - 42$ **49.** $24t + 16$ **51.** $0.4x - 1.6$
53. $5t + 5$ **55.** $-12x - 20$ **57.** $-78c + 18$
59. $-2w + 4$ **61.** $9x + 10$ **63.** $9r - 16$ **65.** $-x + 7$
67. $5.6y - 7$ **69.** $40d + 50$ **71.** $-12r - 60$
73. $x + y - 5$ **75.** $6x - 21y - 16z$ **77.** $20x$ **79.** 0
81. 0   **83.** r **85.** $37y$ **87.** $-s^3$ **89.** 5   **91.** $-10r$
93. $3a$ **95.** $-3x$ **97.** x **99.** $\frac{4}{5}t$ **101.** $0.4r$
103. $7z - 15$ **105.** $-2x + 5$ **107.** $20d - 66$
109. $-3c - 1$ **111.** $s - 12$ **113.** $12c + 34$
115. $8x - 9$ **117.** $12x$ **119.** $(4x + 8)$ ft   **123.** 0
125. 2

Study Set Section 2.4 (page 131)

1. equal   **3.** original   **5.** identity   **7.** subtraction, multiplication   **9.** multiplying   **11.** $-\frac{5}{4}$ **13.** 30   **15.** 6
17. 7, 7, 28, 2, 2   **19. a.** -1 **b.** $\frac{3}{5}$ **21.** 6   **23.** 5
25. -7 **27.** -20 **29.** 4   **31.** 2.9   **33.** $\frac{10}{3}$ **35.** -4
37. $-\frac{8}{3}$ **39.** 12   **41.** -48 **43.** -12 **45.** 5
47. $-\frac{17}{4}$ **49.** 0.04   **51.** -6 **53.** 0   **55.** $\frac{1}{7}$ **57.** $\frac{1}{4}$
59. -1 **61.** -41 **63.** 1   **65.** $\frac{9}{2}$ **67.** -7.2
69. -82 **71.** 0   **73.** -20 **75.** 3   **77.** 28   **79.** $-\frac{12}{5}$
81. $\frac{2}{15}$ **83.** $\frac{27}{5}$ **85.** $\frac{52}{9}$ **87.** $\frac{5}{4}$ **89.** -5 **91.** 80
93. 4   **95.** all real numbers   **97.** no solution   **99.** no solution   **101.** all real numbers   **103.** 2,991,980   **109.** 0
111. $\frac{1}{64}$ **113.** $16x$

Study Set Section 2.5 (page 139)

1. formula   **3.** perimeter   **5.** radius   **7.** circumference
9. a. $d = rt$ **b.** $r = c + m$ **c.** $p = r - c$ **d.** $I = Prt$
e. $C = 2\pi r$ **11.** 11,176,920 mi, 65,280 ft   **13. a.** volume
b. circumference   **c.** area   **d.** perimeter
15. a. $(2x + 10)$ cm   **b.** $(2x + 6)$ cm^2
17. Ax, Ax, By, B, B **19. a.** 3.14   **b.** $98 \cdot \pi$ **c.** the radius of the cylinder; the height of the cylinder   **21.** 2.5 mph
23. \$65 million   **25.** 3.5%   **27.** 4,014°F   **29.** \$24.55
31. about 132 in.   **33.** $R = \frac{E}{I}$ **35.** $w = \frac{V}{lh}$ **37.** $r = \frac{C}{2\pi}$
39. $h = \frac{3A}{B}$ **41.** $f = \frac{s}{w}$ **43.** $b = P - a - c$
45. $r = \frac{T - 2t}{2}$ **47.** $x = \frac{C - By}{A}$ **49.** $m = \frac{2K}{v^2}$
51. $c = 3A - a - b$ **53.** $t = T - 18E$ **55.** $r^2 = \frac{s}{4\pi}$
57. $v^2 = \frac{2Kg}{w}$ **59.** $r^3 = \frac{3V}{4\pi}$ **61.** $M = 4.2B + 19.8$
63. $h = \frac{S - 2\pi r^2}{2\pi r}$ **65.** $y = 9 - 3x$ **67.** $y = \frac{1}{3}x + 3$
69. $y = -\frac{3}{4}x - 4$ **71.** $b = \frac{2A}{h} - d$ or $b = \frac{2A - hd}{h}$
73. $c = \frac{72 - 8w}{7}$ **75.** 212°F, 0°C   **77.** 1,174.6, 956.9
79. 36 ft, 48 ft^2 **81.** 50.3 in., 201.1 in.^{2}
83. 56 in., 144 in.^{2} **85.** 2,450 ft^2 **87.** 27.75 in., 47.8125 in.^{2} **89.** 32 ft^2, 128 ft^3 **91.** 348 ft^3
93. 254 in.^{2} **95.** $D = \frac{L - 3.25r - 3.25R}{2}$ **101.** 137.76
103. 15%

Study Set Section 2.6 (page 153)

1. perimeter   **3.** vertex, base   **5. a.** $17, x + 2, 3x$
b. 3 ft, 5 ft, 9 ft   **7.** 180°   **9.** \$30,000, 14%, 1 yr
11. $35t, t, 45t$ **13. a.** $0.06x, 10 - x, 0.03(10 - x), 0.05(10)$
b. $0.50(6), 0.25(x), 6 + x, 0.30(6 + x)$ **15.** To multiply a decimal by 100, move the decimal point two places to the right.
17. Parentheses are needed: $2(2w - 3) + 2w$ **19.** 6,000
21. 4 ft, 8 ft   **23.** 102 mi, 108 mi, 114 mi, 120 mi
25. Australia: 12 wk; Japan: 16 wk; Sweden: 10 wk   **27.** 250 calories in ice cream, 600 calories in pie   **29.** in millions of dollars: \$110, \$229, \$189, \$847   **31.** 7 ft, 7 ft, 11 ft
33. 75 m by 480 m   **35.** 20°   **37.** 22°, 68°   **39.** 17
41. 90   **43.** \$4,900   **45.** \$42,200 at 12%, \$22,800 at 6.2%

47. \$7,500   **49.** 2 hr   **51.** 4 hr into the flights   **53.** 4 hr
55. $1\frac{1}{3}$ liters   **57.** 7.5 oz   **59.** 20 gal   **61.** 50 lb
63. 40 lb lemon drops, 60 lb jelly beans   **69.** $-50x + 125$
71. $3x + 3$ **73.** $16y - 16$

Study Set Section 2.7 (page 167)

1. inequality   **3.** solve   **5.** interval   **7. a.** true
b. false   **c.** true   **d.** false   **9. a.** $-2 \le 17$ **b.** $x > 32$
11. a. same   **b.** positive   **13. a.**
b. $(8, \infty)$ **c.** all real numbers greater than 8   **15.** three
17. a. is less than, is greater than   **b.** is greater than, or equal to   **19.** $[, (, (, \infty, -\infty$ **21.** 5, 5, 12, 4, 4, 3
23. $(-\infty, 5)$
25. $(-3, 1]$
27. $x < -1, (-\infty, -1)$ **29.** $-7 < x \le 2, (-7, 2]$
31. $x > 3, (3, \infty)$
33. $x < -1, (-\infty, -1)$
35. $g \ge 10, [10, \infty)$
37. $x \ge 3, [3, \infty)$
39. $y \le -40, (-\infty, -40]$
41. $x > \frac{6}{7}, \left(\frac{6}{7}, \infty\right)$
43. $x \le 0.4, (-\infty, 0.4]$
45. $y \ge 20, [20, \infty)$
47. $x \ge -24, [-24, \infty)$
49. $n \le 2, (-\infty, 2]$
51. $m < 0, (-\infty, 0)$
53. $x \ge -10, [-10, \infty)$
55. $x < -2, (-\infty, -2)$
57. $x < -\frac{11}{4}, \left(-\infty, -\frac{11}{4}\right)$
59. $x \le -1, (-\infty, -1]$
61. $x \ge -13, [-13, \infty)$
63. $x > 0, (0, \infty)$

65. $x \le 1.5, (-\infty, 1.5]$

1.5

67. $a > 6, (6, \infty)$

6

69. $x \ge \frac{9}{4}, \left[\frac{9}{4}, \infty\right)$

9/4

71. $x \le 20, (-\infty, 20]$

20

73. $n > \frac{5}{4}, \left(\frac{5}{4}, \infty\right)$

5/4

75. $y \le \frac{1}{8}, \left(-\infty, \frac{1}{8}\right]$

1/8

77. $x \le \frac{3}{2}, \left(-\infty, \frac{3}{2}\right]$

3/2

79. $x \ge 3, [3, \infty)$

3

81. $7 < x < 10, (7, 10)$

7   10

83. $-10 \le x \le 0, [-10, 0]$

−10   0

85. $-6 \le c \le 10, [-6, 10]$

−6   10

87. $2 \le x < 3, [2, 3)$

2   3

89. $-5 < x < -2, (-5, -2)$

−5   −2

91. $-1 \le x < 2, [-1, 2)$

−1   2

93. $x \ge 0.03, [0.03, \infty)$

0.03

95. 98% or better   **97.** 27 mpg or better
99. $0 \text{ ft} < s \le 19 \text{ ft}$ **101.** $x \ge 35 \text{ ft}$
103. a. $0° < a \le 18°$ **b.** $18° \le a \le 50°$
c. $30° \le a \le 37°$ **d.** $75° \le a < 90°$
105. a. $26 \text{ lb} \le w \le 31 \text{ lb}$ **b.** $12 \text{ lb} \le w \le 14 \text{ lb}$
c. $18.5 \text{ lb} \le w \le 20.5 \text{ lb}$ **d.** $11 \text{ lb} \le w \le 13 \text{ lb}$
109. -125 **111.** $1, -3, 6$

Key Concept (page 171)

1. a. $2x - 8$ **b.** $x = 6$ **2. a.** $y + 5$ **b.** $y = -5$
3. a. $\frac{2}{3}a$ **b.** $a = \frac{1}{2}$ **4. a.** $-2x - 10$ **b.** $x \le 5$
5. a. 0 **b.** all real numbers   **6.** The mistake is on the third line. The student made an equation out of the answer, which is $x - 6$, by writing $0 =$ on the left. Then the student solved that equation.

Chapter Review (page 172)

1. yes   **2.** no   **3.** no   **4.** no   **5.** yes   **6.** yes
7. variable, true   **8.** 21   **9.** -47 **10.** 13.2   **11.** 107
12. 8   **13.** 1   **14.** -96 **15.** 7.8   **16.** 0   **17.** 0
18. $160°$ **19.** 5   **20.** 429 mi   **21.** $60°$ **22.** $54 billion
23. $26.74   **24.** 192.4   **25.** no   **26.** 1,567%
27. $-28w$ **28.** $15r$ **29.** $24x$ **30.** $2.08f$ **31.** $9a$

32. r **33.** $5x + 15$ **34.** $-4x - 6 + 2y$ **35.** $-a + 4$
36. $3c - 6$ **37.** $20x + 32$ **38.** $-12.6c - 29.4$ **39.** $9p$
40. $-7m$ **41.** $4n$ **42.** $-p - 18$ **43.** $0.1k$ **44.** $-8a^3$
45. w **46.** $4h - 15$ **47.** $(4x + 4) \text{ ft}$ **48.** 2
49. -30.6 **50.** 30   **51.** -19 **52.** 4   **53.** 1   **54.** $\frac{5}{4}$
55. -6 **56.** 6   **57.** $-\frac{22}{75}$ **58.** identity; all real numbers
59. contradiction; no solution   **60.** $176   **61.** $11,800
62. 3.00 hr   **63.** $1,949°F$ **64.** 168 in.   **65.** $1,440 \text{ in.}^2$
66. 76.5 m^2 **67.** 144 in.^2 **68.** 50.27 cm
69. 201.06 cm^2 **70.** $4,320 \text{ in.}^3$ **71.** 9.4 ft^3 **72.** 120 ft^3
73. 381.70 in.^3 **74.** $h = \frac{A}{2\pi r}$ **75.** $G = 3A - 3BC + K$
76. $b^2 = c^2 - a^2$ **77.** $y = \frac{3}{4}x + 4$ **78.** 8 ft   **79.** 12, 4
80. $24.875 \text{ in.} \times 29.875 \text{ in.} \left(24\frac{7}{8} \text{ in.} \times 29\frac{7}{8} \text{ in.}\right)$
81. $76.5°, 76.5°$ **82.** $45x   **83.** $16,000 at 7%, $11,000 at 9%   **84.** 20   **85.** 10 lb of each   **86.** $0.12x$ gal
87. $x < 1, (-\infty, 1)$

1

88. $x \le 12, (-\infty, 12]$

12

89. $d > \frac{5}{4}, \left(\frac{5}{4}, \infty\right)$

5/4

90. $x \ge 3, [3, \infty)$

3

91. $6 < x < 11, (6, 11)$

6   11

92. $-\frac{7}{2} < x \le \frac{3}{2}, \left(-\frac{7}{2}, \frac{3}{2}\right]$

−7/2   3/2

93. $2.40 \text{ g} \le w \le 2.53 \text{ g}$ **94.** The sign length must be 48 inches or less

Chapter 2 Test (page 178)

1. no   **2.** $120°$ **3.** 1,046   **4.** $76,000   **5.** .878, 1.000
6. 3%   **7.** $(4x + 6) \text{ ft}$ **8.** the distributive property
9. $-20x$ **10.** $224t$ **11.** $-4a + 4$ **12.** $-5.9d^2$ **13.** 0
14. -5 **15.** $-\frac{1}{4}$ **16.** 2.5   **17.** $\frac{7}{6}$ **18.** -3
19. $r = \frac{A - P}{Pt}$ **20.** $150   **21.** $-10°\text{ C}$ **22.** 393 in.^3
23. $\frac{3}{5} \text{ hr}$ **24.** 10 liters   **25.** $68°$ **26.** $5,250
27. $x \ge -3, [-3, \infty)$

−3

28. $-3 \le x < 4, [-3, 4)$

−3   4

29. $1.496 \text{ in.} \le w \le 1.498 \text{ in.}$

Cumulative Review Exercises Chapters 1–2 (page 180)

1. a. expression   **b.** equation   **2.** 3, 4, 5
3. $2 \cdot 2 \cdot 2 \cdot 5 \cdot 5 = 2^3 \cdot 5^2$ **4.** $\frac{2}{3}$ **5.** $-\frac{2}{9}$ **6.** 6
7. $\frac{22}{15} = 1\frac{7}{15}$ **8.** $12\frac{11}{24}$ **9.** 0.9375   **10.** 45   **11. a.** 65
b. -12 **12.** the commutative property of multiplication
13. natural number, whole number, integer, rational number, real number   **14.** rational number, real number   **15.** rational number, real number   **16.** irrational number, real number

17. a. 4^3 **b.** $\pi r^2 h$ **18. a.** -10 **b.** -14 **c.** -64
d. 0 **e.** -1 **19. a.** $w + 12$ **b.** $n - 4$ **20.** 4
21. $1, -3, 6$ **22.** $l = \frac{2,000}{d^2}$ (Answers may vary depending on
the variables chosen.)   **23. a.** $6\,\text{ft}^2$ **b.** $1.2\,\text{ft}^2$ **c.** 20%
24. 300   **25.** 0   **26.** -2 **27.** 16   **28.** 0   **29.** $-32d$
30. $10x - 15y + 5$ **31.** $4x$ **32.** $-8a^2$ **33.** $11t - 50$
34. $8t - 20$ **35.** $(x + 3)$ ft   **36.** $3x$ ft   **37.** 9   **38.** 20
39. -0.6 **40.** $\frac{19}{6}$ **41.** -20 **42.** $\frac{5}{4}$ **43.** -2 **44.** no
solution, contradiction   **45.** $65\,\text{m}^2$ **46.** $376.99\,\text{cm}^3$
47. $r^2 = \frac{3V}{\pi h}$ **48.** 37.5 ft-lb   **49.** 9.45 lb   **50.** $55°, 55°$
51. $4,000   **52.** 10 oz
53. $x > -2, (-2, \infty)$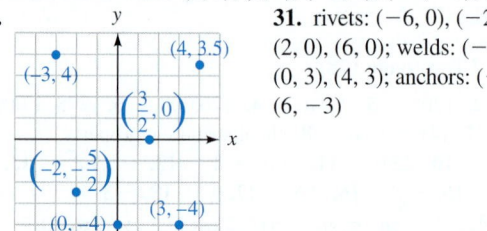
54. $x \le 2, (-\infty, 2]$
55. $x \ge -1, [-1, \infty)$
56. $-1 \le x < 2, [-1, 2)$

Study Set Section 3.1 (page 189)

1. ordered   **3.** x-axis, y-axis, origin   **5.** rectangular
7. a. origin, left, up   **b.** origin, right, down   **9. a.** I and II
b. II and III   **c.** II   **d.** IV   **11.** 60 beats/min   **13.** 140
beats/min   **15.** 5 min and 50 min after starting   **17.** no
difference   **19.** about 6 million vehicles   **21.** $(3, 5)$ is an
ordered pair, $3(5)$ indicates multiplication, and $5(3 + 5)$ is an
expression containing grouping symbols.   **23.** yes
25. horizontal   **27.** $(4, 3), (0, 4), (-5, 0), (-4, -5), (3, -3)$
29.

31. rivets: $(-6, 0), (-2, 0),$
$(2, 0), (6, 0)$; welds: $(-4, 3),$
$(0, 3), (4, 3)$; anchors: $(-6, -3),$
$(6, -3)$

33. $(E, 4), (F, 3), (G, 2)$ **35.** Rockford $(5, B)$, Mount Carroll
$(1, C)$, Harvard $(7, A)$, intersection $(5, E)$ **37.** $(2, 4)$; 12 sq.
units   **39. a.** 35 mi   **b.** 4 gal   **c.** 32.5 mi
41. a. A 3-year-old car is worth $7,000.   **b.** $1,000
c. 6 yr   **43.** 152, 179, 202, 227, 252, 277
49. $h = \frac{3(AC + T)}{2}$ **51.** -1

Study Set Section 3.2 (page 202)

1. two   **3.** table   **5.** linear   **7.** infinitely   **9. a.** 2
b. yes   **c.** no   **d.** infinitely many   **11.** solution, point
13. a. At least one point is in error. The points should lie on a
straight line. Check the computations.   **b.** The line is too
short. Arrowheads are not drawn.   **15. a.** $y = 2x + 1$
b. $y = -\frac{5}{3}x - 2$ **c.** $y = \frac{1}{7}x - 3$ or $y = \frac{x}{7} - 3$
17. $6, -2, 2, 6$ **19.** a, c, a, c **21.** yes   **23.** no   **25.** no

27. yes   **29.** 11   **31.** 4   **33.** 12, $(8, 12)$, 6, $(6, 8)$
35. $-13, (-5, -13), -1, (-1, -1)$
37. 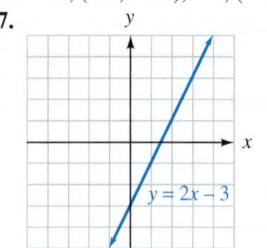   **39.**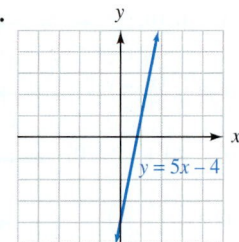

41. 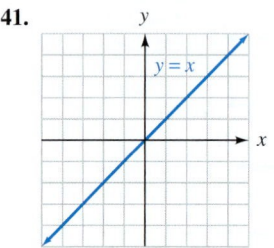   **43.**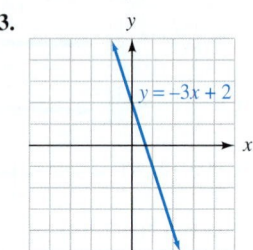

45. 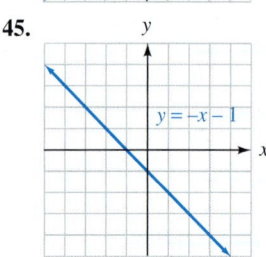   **47.**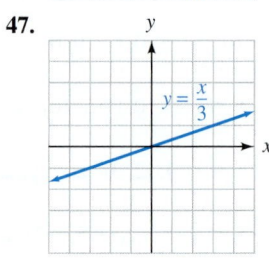

49. 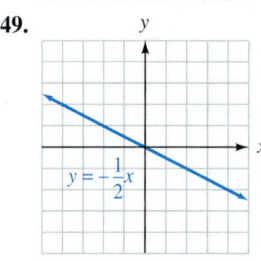   **51.**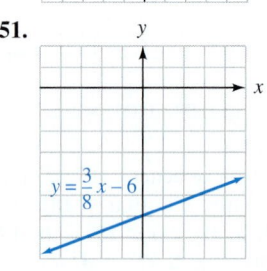

53. 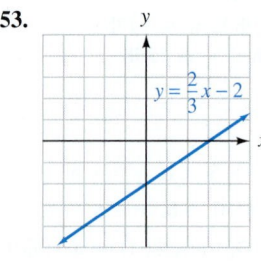   **55.**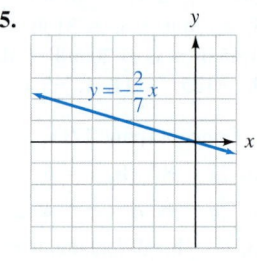

57. 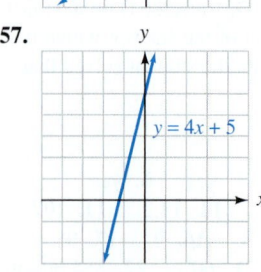   **59.**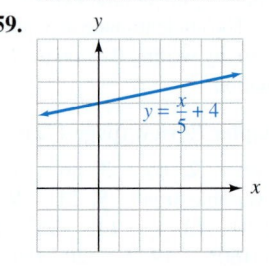

61.

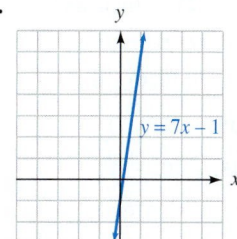

63. -2, $(1, -2)$, 0, $(2, 0)$, 4, $(4, 4)$; 4, $(4, 4)$, 0, $(6, 0)$, -4, $(8, -4)$

65. 3 oz

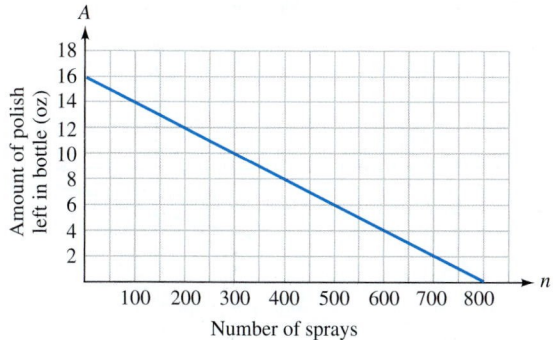

67. about $70   **69.** about 180   **77.** $5 + 4c$ **79.** 0
81. 491

Study Set Section 3.3 (page 214)

1. two   **3.** general/standard   **5.** intersects/crosses
7. x-intercept: $(4, 0)$ y-intercept: $(0, 3)$ **9.** x-intercept:
$(-5, 0)$ y-intercept: $(0, -4)$ **11.** y-intercept: $(0, 2)$
13. y-intercept: $\left(0, \frac{2}{3}\right)$; x-intercept: $\left(-2\frac{1}{2}, 0\right)$ **15. a.** $3x$
b. $(0, 3)$ **17.** 2; 1 **19. a.** ii   **b.** iv   **c.** vi   **d.** i
e. iii   **f.** v   **21. a.** y-axis   **b.** yes   **23.** $y = 0$; $x = 0$
25.

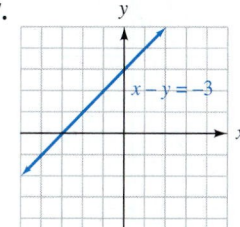

27.

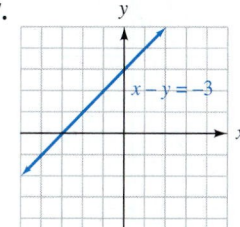

29.

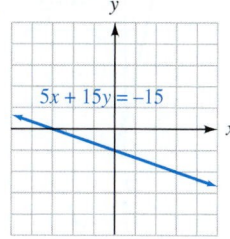

31.

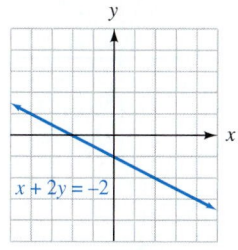

33.

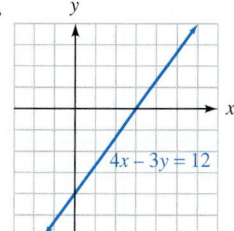

35.

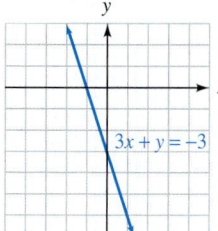

37.

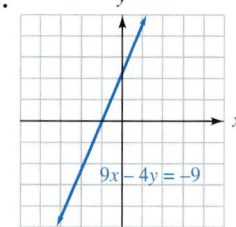

39.

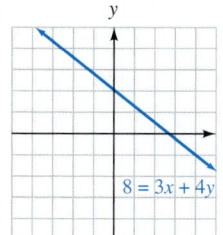

41.

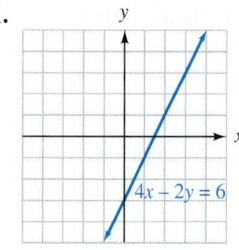

43.

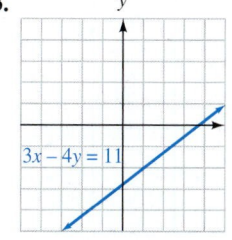

45.

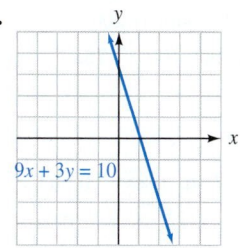

47.

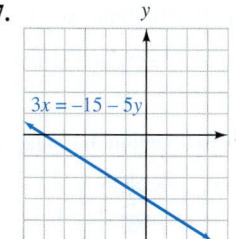

49.

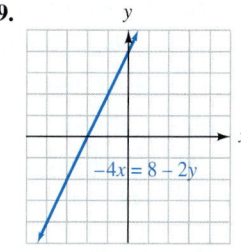

51.

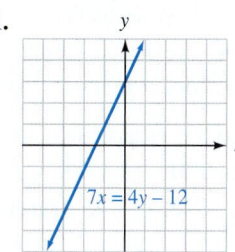

53.

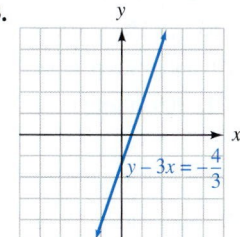

55.

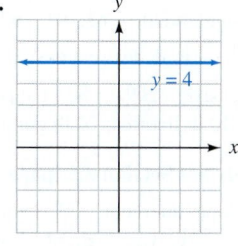

57.

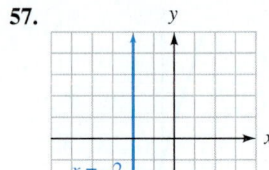

59.

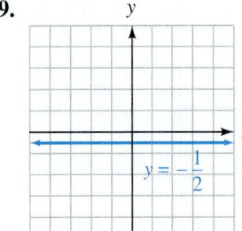

61.

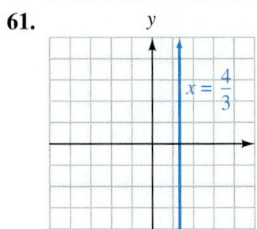

63.

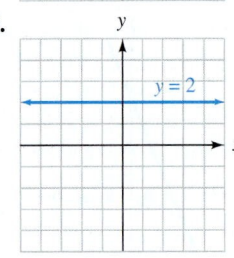

65.

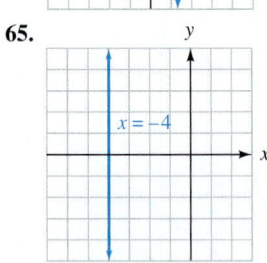

67. a. about $-270°C$
b. 0 milliliters

69. a. If only shrubs are purchased, he can buy 200.
b. If only trees are purchased, he can buy 100.   **75.** $\frac{1}{5}$
77. $2x - 6$

Study Set Section 3.4 (page 224)

1. ratio   **3.** slope   **5.** change   **7. a.** rises   **b.** falls
9. $\frac{2}{15}$ **11. a.** -3 **b.** 8   **c.** $-\frac{3}{8}$ **13. a.** $\frac{1}{2}$ **b.** $\frac{1}{2}$
c. $\frac{1}{2}$ **d.** When finding the slope of a line, any two points on
the line give the same result.   **15. a.** $m = 0$ **b.** undefined
17. a. 40%   **b.** 15%   **19.** $m = \frac{y_2 - y_1}{x_2 - x_1}$ **21.** y^2 means
$y \cdot y$ and y_2 means y sub 2.   **23.** $\frac{2}{3}$ **25.** $\frac{4}{3}$ **27.** -2
29. 0   **31.** $-\frac{1}{5}$ **33.** 1   **35.** -3 **37.** $\frac{5}{4}$ **39.** $-\frac{1}{2}$
41. $\frac{3}{5}$ **43.** 0   **45.** undefined   **47.** $-\frac{2}{3}$ **49.** -4.75
51. $m = \frac{3}{4}$ **53.** $m = 0$ **55.** undefined   **57.** 0
59. undefined   **61.** 0   **63.** $-\frac{2}{5}$ **65.** $\frac{1}{20}$; 5%   **67. a.** $\frac{1}{8}$
b. $\frac{1}{12}$ **c.** 1: less expensive, steeper; 2: not as steep, more
expensive   **69.** -875 gal per hr   **71.** 300 lb per yr
77. 40 lb licorice; 20 lb gumdrops

Study Set Section 3.5 (page 237)

1. slope–intercept   **3.** parallel   **5.** reciprocals   **7. a.** no
b. no   **c.** yes   **d.** no   **e.** yes   **f.** yes   **9. a.** y, $2x$, 4
b. y, $-3x$, +   **11.** $y = -\frac{5}{4}x$ **13. a.** parallel
b. perpendicular   **c.** -1 **15. a.** $-\frac{1}{2}$ **b.** 2   **c.** $-\frac{1}{2}$
d. Line 1 and Line 2   **17.** $2x$, $-2x$, $5y$, 5, 5, 5, 3, $-\frac{2}{5}$, (0, 3)

19. a. $4x$ **b.** $\frac{4x}{3}$ **c.** x **d.** -2 **21.** a right angle
23. 4, (0, 2)   **25.** -5, (0, -8)   **27.** 4, (0, -2)
29. $\frac{1}{4}$, $\left(0, -\frac{1}{2}\right)$ **31.** $\frac{1}{2}$, (0, 6)   **33.** -1, (0, 6)
35. -1, (0, 8)   **37.** $\frac{1}{6}$, (0, -1)   **39.** -2, (0, 7)
41. $-\frac{3}{2}$, (0, 1)   **43.** $-\frac{2}{3}$, (0, 2)   **45.** $\frac{3}{5}$, (0, -3)
47. 1, $\left(0, -\frac{11}{6}\right)$ **49.** 1, (0, 0)   **51.** -5, (0, 0)
53. 0, (0, -2)   **55.** 0, $\left(0, -\frac{2}{5}\right)$
57. $y = 5x - 3$ **59.** $y = \frac{1}{4}x - 2$

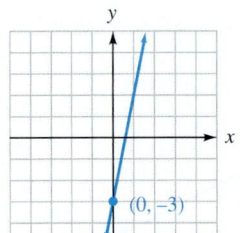

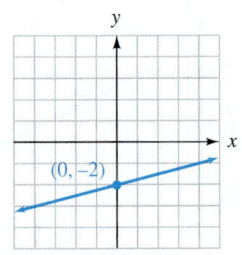

61. $y = -3x + 6$ **63.** $y = -\frac{8}{3}x + 5$

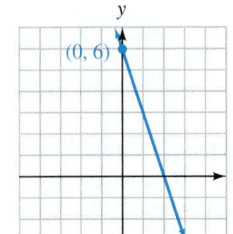

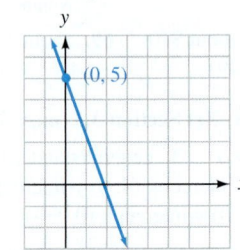

65.

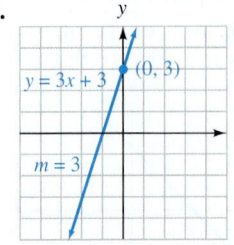

67.

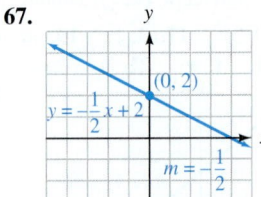

69.

71.

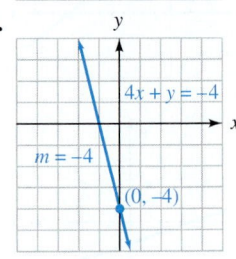

73.

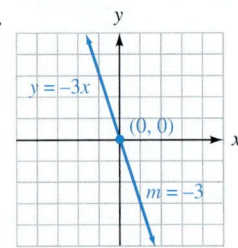

75.

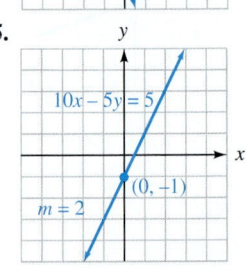

77. parallel   **79.** perpendicular   **81.** neither   **83.** parallel
85. perpendicular   **87. a.** $c = 2{,}000h + 5{,}000$
b. \$21,000   **89.** $F = 5t - 10$ **91.** $c = -20m + 500$
93. a. $c = 5x + 20$
c.

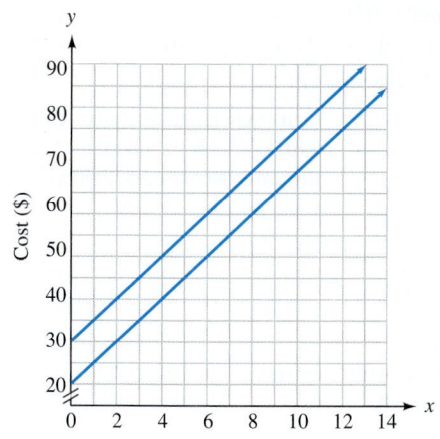

Number of letters

95. $c = 2t + 33.50$ **99.** 42 ft, 45 ft, 48 ft, 51 ft

Study Set Section 3.6 (page 247)

1. point–slope   **3. a.** point–slope   **b.** slope–intercept
5. a. $x + 6$ **b.** $y + 9$ **7.** $-\frac{5}{2}$ **9. a.** yes   **b.** yes
11. a. $(-2, -3)$ **b.** $\frac{5}{6}$ **c.** $(4, 2)$ **13. a.** no   **b.** no
c. yes   **d.** yes   **15.** $(67, 170), (79, 220)$
17. $y - y_1 = m(x - x_1)$ **19.** point–slope, y, slope–intercept
21. $y - 1 = 3(x - 2)$ **23.** $y + 1 = -\frac{4}{5}(x + 5)$
25. $y = \frac{1}{5}x - 1$ **27.** $y = -5x - 37$ **29.** $y = -\frac{4}{3}x + 4$
31. $y = -\frac{11}{6}x - \frac{7}{3}$ **33.** $y = -\frac{2}{3}x + 2$ **35.** $y = 8x + 4$
37. $y = -3x$ **39.** $y = 2x + 5$ **41.** $y = -\frac{1}{2}x + 1$
43. $y = 5$ **45.** $y = \frac{1}{10}x + \frac{1}{2}$ **47.** $x = -8$
49. $y = \frac{1}{4}x - \frac{5}{4}$ **51.** $x = 4$ **53.** $y = 5$
55.

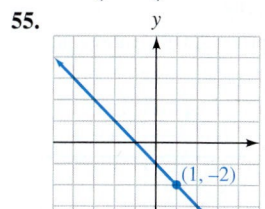

57.

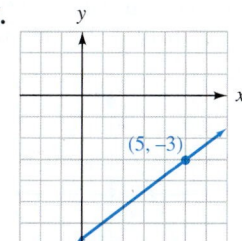

59.

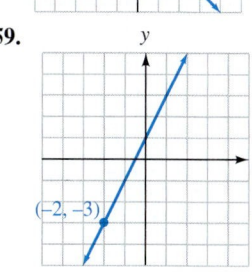

61.

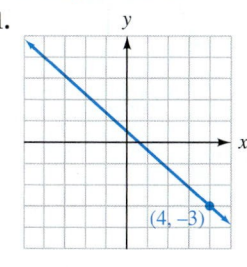

63. $h = 3.9r + 28.9$ **65.** $y = -\frac{2}{5}x + 4, y = -7x + 70,$
$x = 10$ **67. a.** $y = -40m + 920$ **b.** 440 yd^3
69. $l = \frac{25}{4}r + \frac{1}{4}$ **71. a.** $y = -\frac{1}{4}x + \frac{117}{4}$ or
$y = -0.25x + 29.25$ **b.** 19.25 gal   **77.** 17 in. by 39 in.

Study Set Section 3.7 (page 258)

1. inequality   **3.** solution   **5.** boundary   **7.** point
9. a. false   **b.** false   **c.** true   **d.** false   **11. a.** no
b. yes   **c.** no   **d.** no   **13. a.** no   **b.** yes   **15.** the
half-plane opposite that in which the test point lies
17. a. yes   **b.** no   **c.** no   **d.** yes   **19. a.** horizontal
b. vertical   **21. a.** is less than   **b.** is greater than
c. is less than or equal to   **d.** is greater than or equal to
23. $\leq, \geq$ **25.**

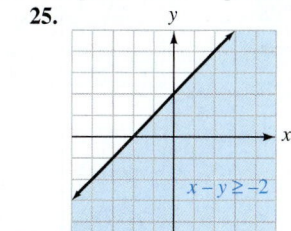

27.

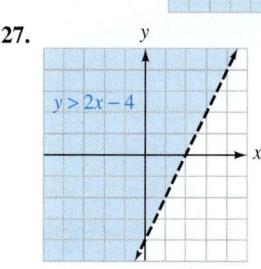

29.

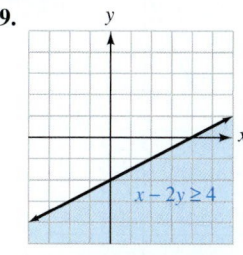

31.

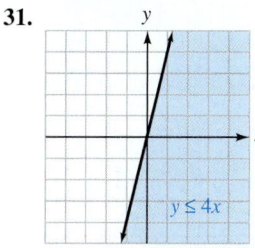

33.

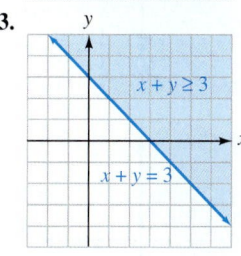

35.

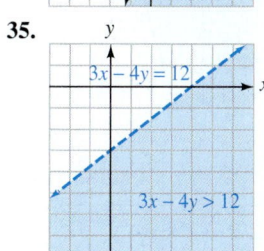

37.

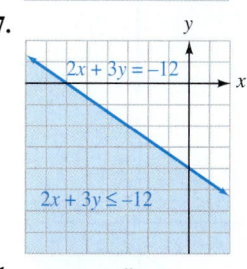

39.

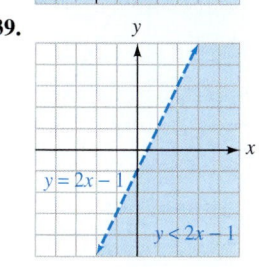

41.

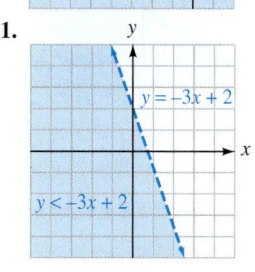

43.
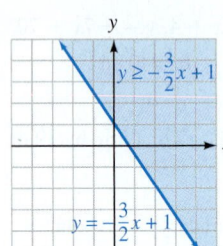

45.
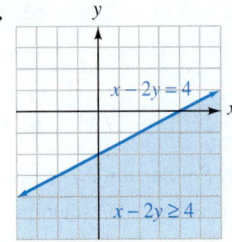

47.
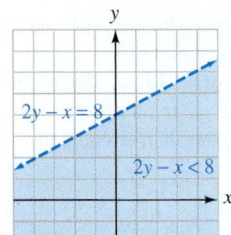

49.
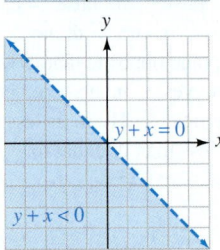

51.
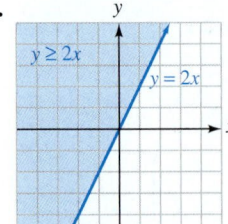

53.
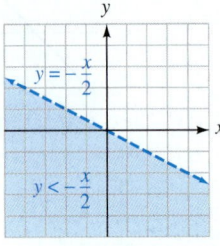

55.
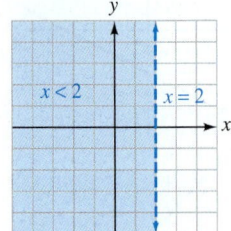

57.
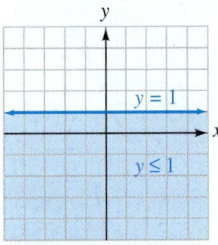

59.
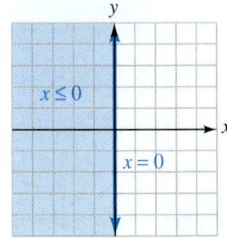

61.
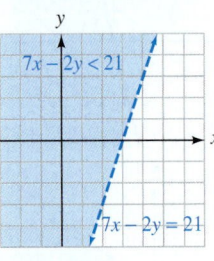

63.
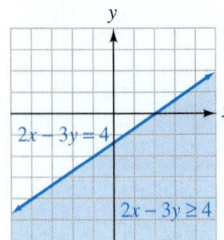

65.
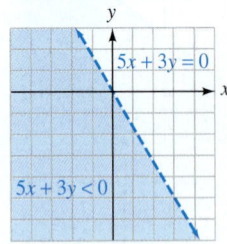

67. ii   **69.** (10, 10), (20, 10), (10, 20); answers may vary
71. (50, 50), (30, 40), (40, 40); answers may vary
77. $t = \frac{A - P}{Pr}$ **79.** $15x + 22$

Key Concept (page 263)

1. $y = -3x - 4$ **2.** $y = \frac{1}{5}x + 1$ **3.** 80°F
4. $y = 209x + 2,660$; $8,930   **5.** $-2, 4, -3$

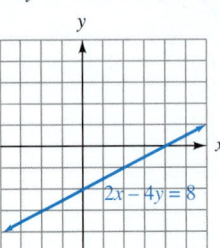

6. a. When new, the press cost $40,000.   **b.** $-5,000$; the value of the press decreased $5,000/yr

7. a. $(-2, -2)$ (answers may vary)   **b.** -2
c. $y = -2x - 6$ **8.** $x = 1$

Chapter Review (page 264)

1.
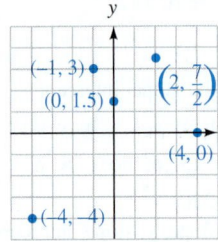

2. (158, 21.5)   **3.** quadrant III
4. (0, 0)   **5.** (1, 4); 36 square units   **6. a.** 2,500; week 2
b. 1,000   **c.** 1st week and 5th week   **7.** yes
8. $-6, (-2, -6), -8, (-8, 3)$
9. $y = x^2 + 1$ and $y - x^3 = 0$

10.
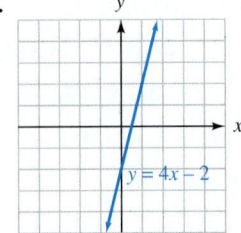

11.
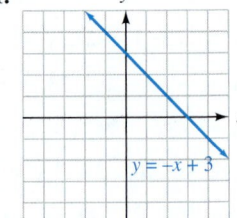

12. a. true   **b.** false   **13.** about $195
14. $(-3, 0), (0, 2.5)$ **15.** x-intercept: $(-2, 0)$; y-intercept: $(0, 4)$

16.
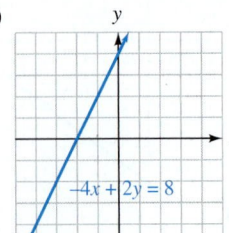

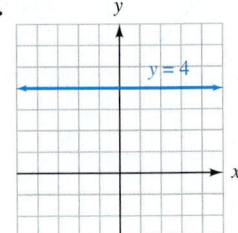

17.

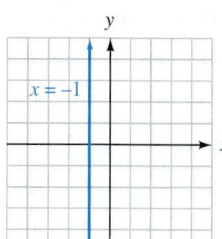

18. (0, 25,000); the equipment was originally valued at $25,000. (10, 0). In 10 years, the sound equipment had no value.

19. $\frac{1}{4}$ **20.** 0   **21.** -7 **22.** $-\frac{3}{2}$ **23.** $\frac{3}{4}$ **24.** 8.3%

25. a. -1.25 million people per yr   **b.** 4.05 million people per yr   **26.** $m = \frac{3}{4}$; y-intercept: $(0, -2)$ **27.** $m = -4$; y-intercept: $(0, 0)$ **28.** $m = \frac{1}{8}$; y-intercept $(0, 10)$

29. $m = -\frac{7}{5}$; y-intercept $\left(0, -\frac{21}{5}\right)$ **30.** $y = -6x + 4$

31. $m = 3$; y-intercept: $(0, -5)$.

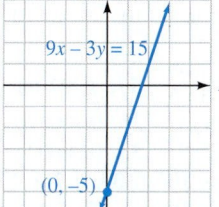

32. a. $c = 300w + 75,000$ **b.** 90,600   **33. a.** parallel   **b.** perpendicular   **34.** $y = 3x + 2$

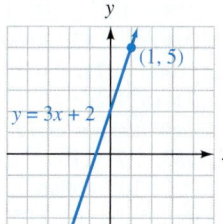

35. $y = -\frac{1}{2}x - 3$

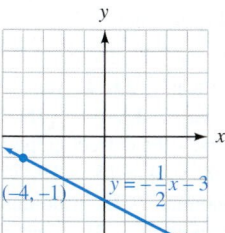

36. $y = \frac{2}{3}x + 5$ **37.** $y = -8$ **38.** $f = -35x + 450$ **39. a.** yes   **b.** yes   **c.** yes   **d.** no

40.

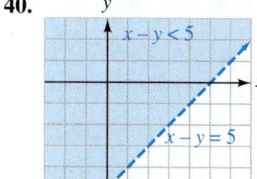

41.

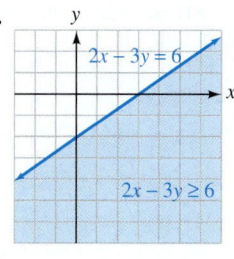

42.

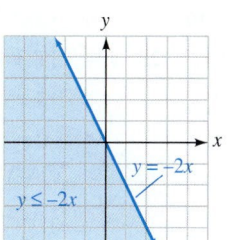

43.

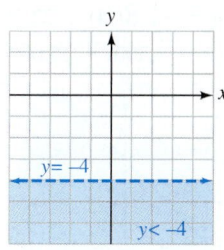

44. a. true   **b.** false   **c.** false

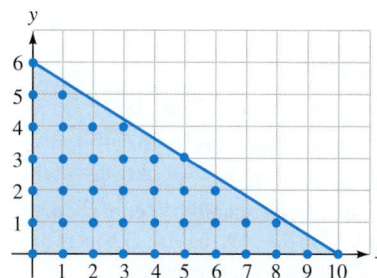

45. $(2, 4)$, $(5, 3)$, $(6, 2)$ answers may vary   **46.** An equation contains an $=$ symbol. An inequality contains one of the symbols $<$, $\le$, $>$, or $\ge$.

Chapter 3 Test (page 269)

1. 10   **2.** 60   **3.** 1 day before and the 3rd day of the holiday   **4.** 50 dogs were in the kennel when the holiday began.

5.

II	I
III	IV

6. 1, $(2, 1)$, -6, $(-6, 3)$ **7.** yes

8. a. false   **b.** true

9.

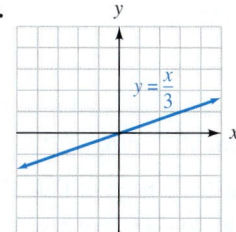

10.

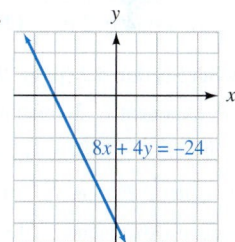

11. x-intercept: $(3, 0)$; y-intercept: $(0, -2)$

12. $\frac{8}{7}$ **13.** -1 **14.** undefined   **15.** $\frac{8}{7}$ **16.** parallel

17. -15 ft per mi   **18.** 25 ft per mi

19.

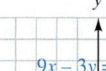

20.

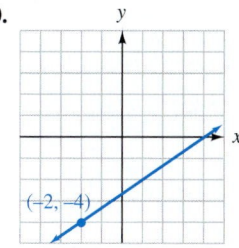

21. $m = -\frac{1}{2}$; $(0, 4)$ **22.** $y = 7x + 19$
23. $v = -1,500x + 15,000$ **24.** yes   **25.** $y = -\frac{1}{5}T + 41$
26.

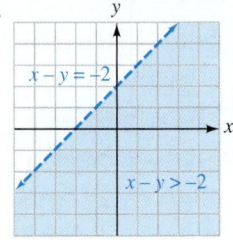

Study Set Section 4.1 (page 279)

1. exponential   **3.** factor   **5.** power   **7. a.** $3x, 3x, 3x, 3x$
b. $(-5y)^3$ **9. a.** $(3x^2)^6$ (answers may vary)   **b.** $\left(\frac{3a^3}{b}\right)^2$
(answers may vary)   **11. a.** $2x^2$ **b.** 0 **c.** x^4 **d.** a^7
13. a. doesn't simplify   **b.** doesn't simplify   **c.** x^5 **d.** x
15. a. 16 **b.** -16 **17.** $x^6, 18$ **19.** base 4, exponent 3
21. base x, exponent 5   **23.** base $-3x$, exponent 2
25. base y, exponent 6   **27.** base $y + 9$, exponent 4
29. base $-3ab$, exponent 7   **31.** $x \cdot x \cdot x \cdot x \cdot x$
33. $\left(\frac{t}{2}\right)\left(\frac{t}{2}\right)\left(\frac{t}{2}\right)$ **35.** $(x - 5)(x - 5)$ **37.** $(4t)^4$ **39.** $-4t^3$
41. $(x - y)^3$ **43.** 12^7 **45.** 2^6 **47.** a^6 **49.** x^7
51. a^9 **53.** $(-7)^5$ **55.** $(8t)^{60}$ **57.** $(n - 1)^3$ **59.** y^9
61. 8^8 **63.** x^{12} **65.** c **67.** $(k - 2)^{14}$ **69.** 3^8
71. y^{15} **73.** m^{500} **75.** a^5b^6 **77.** c^2d^5 **79.** x^3y^3
81. y^4 **83.** c^2d^6 **85.** x^{25} **87.** $243z^{30}$ **89.** $9n^{16}$
91. u^4v^4 **93.** a^9b^6 **95.** $-8r^6s^9$ **97.** $\frac{a^3}{b^3}$ **99.** $\frac{x^{10}}{y^{15}}$
101. $\frac{-32a^5}{b^5}$ **103.** $216k^3$ **105.** $a^{12}b^6$ **107.** a
109. ab^4 **111.** $r^{13}s^3$ **113.** $\frac{y^3}{8}$ **115.** $\frac{27t^{12}}{64}$ **117.** a^{10} mi^2
119. x^9 ft^3 **121. a.** $25x^2$ ft^2 **b.** $9a^2\pi$ ft^2
123. 16 ft, 8 ft, 4 ft, 2 ft   **129.** c   **131.** d

Study Set Section 4.2 (page 288)

1. base, exponent   **3.** reciprocal, 3   **5. a.** $4 - 4, 0, 6, 6, 6,$
$6, 1$ **b.** $1, 1$ **7.** $9, 3, 1, \frac{1}{3}, \frac{1}{9}$ **9. a.** x^{m+n} **b.** x^{m-n}
c. x^{mn} **d.** x^ny^n **e.** $\frac{x^n}{y^n}$ **f.** $\frac{1}{x^n}$ **g.** x^n **h.** $\frac{y^n}{x^m}$ **i.** 1
11. once, no, exponents   **13.** $y^8, -40, 40$ **15.** $4, -2, x, -5,$
$\frac{3}{y}, -8, 7, -1, -2, -3, a, 0$ **17.** 1 **19.** 1 **21.** 2 **23.** 1
25. 1 **27.** $\frac{5}{2}$ **29.** $-15y$ **31.** $\frac{1}{144}$ **33.** $-\frac{1}{4}$ **35.** $\frac{44}{g^6}$

37. $-\frac{1}{1,000}$ **39.** $-\frac{1}{64}$ **41.** $\frac{1}{64}$ **43.** $\frac{1}{x^2}$ **45.** $-\frac{1}{b^5}$
47. 36 **49.** -8 **51.** $\frac{8}{7}$ **53.** 125 **55.** $\frac{3}{16}$ **57.** $\frac{b^2}{a^5}$
59. r^{20} **61.** $-p^{10}$ **63.** $\frac{1}{h^7}$ **65.** $8s$ **67.** h^7 **69.** $\frac{1}{16y^4}$
71. $\frac{1}{a^3b^6}$ **73.** 8 **75.** $\frac{9}{y^8}$ **77.** $\frac{1}{y}$ **79.** $\frac{1}{r^6}$ **81.** $\frac{4t^2}{s^5}$
83. $\frac{125}{d^6}$ **85.** $-2a^4$ **87.** $\frac{1}{x^9}$ **89.** t^{10} **91.** y^5 **93.** 2
95. $\frac{1}{a^2b^4}$ **97.** $\frac{1}{x^6y^3}$ **99.** $\frac{1}{x^3}$ **101.** $-\frac{y^{10}}{32x^{15}}$ **103.** a^{14}
105. $\frac{256x^{28}}{81}$ **107.** $\frac{y^{14}}{9z^{10}}$ **109.** $\frac{9}{4g^2}$ **111.** 2 **113.** $\frac{1}{2}$
115. $\frac{8}{9}$ **117.** $10^2, 10^1, 10^0, 10^{-1}, 10^{-2}, 10^{-3}, 10^{-4}$
119. about \$4,605   **121.** It gives the initial number of
bacteria b.   **125.** 13.5 yr   **127.** $y = \frac{3}{4}x - 5$

Study Set Section 4.3 (page 295)

1. scientific   **3.** notation   **5.** powers   **7. a.** right
b. left   **9. a.** 10^{-7} **b.** 10^9 **11. a.** positive   **b.** negative
13. a. 7.7 **b.** 5.0 **c.** 8 **15. a.** $(5.1 \times 1.5)(10^9 \times 10^{22})$
b. $\frac{8.8}{2.2} \times \frac{10^{30}}{10^{19}}$ **17.** $1, 10,$ integer   **19.** 230 **21.** $812,000$
23. 0.00115 **25.** 0.000976 **27.** $6,001,000$ **29.** 2.718
31. 0.06789 **33.** 0.00002 **35.** $9,000,000,000$
37. 2.3×10^4 **39.** 1.7×10^6 **41.** 6.2×10^{-2}
43. 5.1×10^{-6} **45.** 5.0×10^9 **47.** 3.0×10^{-7}
49. 9.09×10^8 **51.** 3.45×10^{-2} **53.** 9.0×10^0
55. 1.718×10^{18} **57.** 1.23×10^{-14} **59.** $714,000$
61. 0.004032 **63.** $30,000$ **65.** 0.0043 **67.** 0.0308
69. $200,000$ **71.** $9.038030748 \times 10^{15}$
73. $1.734152992 \times 10^{-12}$ **75.** 2.57×10^{13} mi
77. $63,800,000$ mi^2 **79.** 6.22×10^{-3} mi   **81.** g, x, u, v, i,
m, r   **83.** 1.7×10^{-18} g   **85.** 3.09936×10^{16} ft
87. 2.08×10^{11} dollars   **89. a.** 1.7×10^6; $1,700,000$
b. 1986: 2.05×10^6, $2,050,000$ **95.** 5 **97.** $c = 30t + 45$

Study Set Section 4.4 (page 304)

1. polynomial   **3.** term   **5.** degree   **7.** x, decreasing or
descending   **9.** nonlinear   **11. a.** yes   **b.** no   **c.** no
d. yes   **e.** yes   **f.** yes   **13.** no   **15.** Not enough
ordered pairs were found—the correct graph is a parabola.
17. $-2, -2, 4, -2, -8, -14, -15$ **19.** $y = x^2 - 1,$
$y = x^3 - 1$ **21.** binomial   **23.** trinomial   **25.** monomial
27. binomial   **29.** trinomial   **31.** none of these
33. trinomial   **35.** 4th   **37.** 2nd   **39.** 1st   **41.** 4th
43. 12th   **45.** 0th   **47. a.** -44 **b.** 37 **49. a.** 9
b. 8 **51. a.** -22 **b.** -406 **53. a.** 28 **b.** 4
55. a. 2 **b.** 0
57. $9, 4, 1, 0, 1, 4, 9$

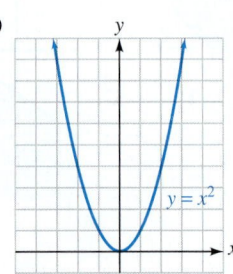

59.

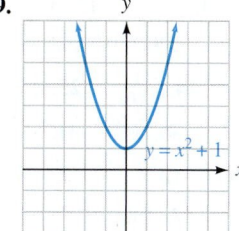

61.

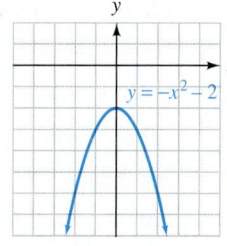

63.

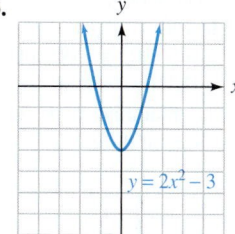

65.

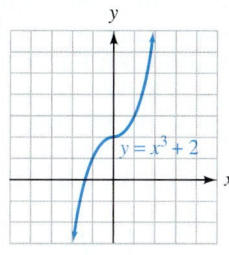

67.

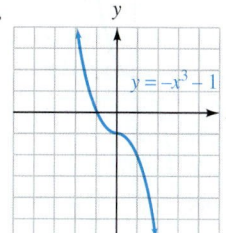

69. 91   **71.** 63 ft

73.

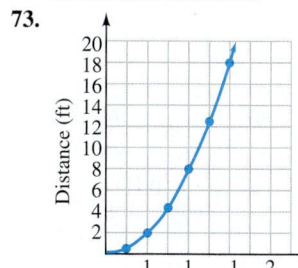

75. a. It costs 8¢ to make a 2-in. bolt.   **b.** 12¢
c. a 4-in. bolt
79. $y \geq -3$; $[-3, \infty)$

81. x^{18} **83.** y^9

Study Set Section 4.5 (page 312)

1. polynomials   **3.** Like   **5.** descending   **7.** combine
9. $-4x^2, +$ **11. a.** $5x^2$ **b.** $14m^3$ **c.** $8a^3b - ab$
d. $6cd + 4c^2d$ **13. a.** $-5x^2 + 8x - 23$
b. $5y^4 - 3y^2 + 7$ **15.** $-, +, -, 2x^2, 2$ **17.** true
19. $12t^2$ **21.** $-48u^3$ **23.** $-0.1x$ **25.** $2st$ **27.** $6r$
29. $-ab$ **31.** $x^2 + x$ **33.** $13x^3$ **35.** $-4x^3y + x^2y + 5$
37. $\frac{7}{12}c^2 - \frac{1}{2}cd + d^2$ **39.** $7x + 4$ **41.** $13a^2 + a$
43. $7x - 7y$ **45.** $3x - 4y$ **47.** $6x^2 + x - 5$
49. $7b + 4$ **51.** $3x + 1$ **53.** $-5h^3 + 5h^2 + 30$
55. $-1.94x^2 + 3.4x + 0.01$ **57.** $\frac{5}{4}r^4 + \frac{7}{3}r^2 - 2$
59. $5x^2 + x + 11$ **61.** $5x^2 + 6x - 8$
63. $-x^3 + 6x^2 + x + 14$ **65.** $-x^3 + 4x^2 + 5x + 6$
67. $7x^3 - 2x^2 - 2x + 15$ **69.** $6x - 2$

71. $-5x^2 - 8x - 4$ **73.** $4y^3 - 12y^2 + 8y + 8$
75. $4s^2 - 5s + 7$ **77.** $3y^5 - 6y^4 + 1.2$
79. $t^3 + 3t^2 + 6t - 5$ **81.** $-3x^2 + 5x - 7$
83. a. $(x^2 - 8x + 12)$ ft   **b.** $(x^2 + 2x - 8)$ ft
85. $(11x - 12)$ ft   **87. a.** $(6x + 5)$ ft   **b.** $(4x^2 + 26)$ ft
89. a. $(22t + 20)$ ft   **b.** 108 ft   **97.** $180°$
99.

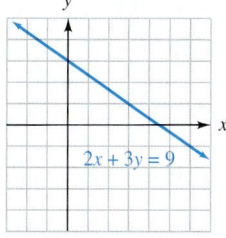

Study Set Section 4.6 (page 321)

1. monomials   **3.** binomials   **5.** like   **7. a.** coefficients,
variable   **b.** each, each   **c.** any, third   **d.** length
9. F, L, I, O, $6x^2, 8x, 15x, 20$ **11. a.** $2x + 4$ **b.** -12
c. $x^2 + 4x - 32$ **13.** $7x, 7x, 7x, 21x^3$ **15.** $15m^2$
17. $12x^5$ **19.** $-24b^6$ **21.** $6x^5y^5$ **23.** $-2a^{11}$ **25.** $6c^6$
27. $\frac{1}{16}a^{10}$ **29.** $-3x^4y^5z^8$ **31.** $3x + 12$ **33.** $-4t^2 - 28$
35. $3x^2 - 6x$ **37.** $-6x^4 + 2x^3$ **39.** $6x^{14} - 72x^{13}$
41. $\frac{5}{8}t^8 + 5t^4$ **43.** $0.12p^9 - 1.8p^7$ **45.** $-12x^4z + 4x^2z^2$
47. $6x^4 + 8x^3 - 14x^2$ **49.** $12a^3 + 9a^2 - 12a$
51. $18a^6 - 12a^5$ **53.** $y^2 + 2y - 15$ **55.** $2t^2 + 5t - 12$
57. $6y^2 - y - 35$ **59.** $4x^2 - 4x - 15$
61. $a^2 + 2ab + b^2$ **63.** $12a^2 - 5ab - 2b^2$
65. $t^4 + t^2 - 12$ **67.** $-6t^2 + 13st - 6s^2$
69. $8a^2 + \frac{17}{9}ar - \frac{5}{12}r^2$ **71.** $8x^2 - 12x - 8$
73. $3a^3 - 3ab^2$ **75.** $x^3 - x + 6$
77. $4t^3 + 11t^2 + 18t + 9$ **79.** $2x^3 + 7x^2 - 16x - 35$
81. $-3x^3 + 25x^2y - 56xy^2 + 16y^3$
83. $r^4 - 5r^3 + 2r^2 - 7r - 15$ **85.** $x^3 - 3x + 2$
87. $12x^3 + 17x^2 - 6x - 8$
89. $6a^4 + 5a^3 + 5a^2 + 10a + 4$ **91.** $(6x^2 + x - 1)$ cm²
93. $(9x^2 + 6x + 1)$ ft²   **95.** $(4x^2 - 6x + 2)$ cm²
97. $(6.28x^2 - 6.28)$ in.²   **99.** $(35x^2 + 43x + 12)$ cm²
101. $(2x^3 - 4x^2 - 6x)$ in.³   **107.** 1   **109.** $-\frac{2}{3}$
111. $(0, 2)$

Study Set Section 4.7 (page 328)

1. term, term   **3.** sum   **5.** difference   **7. a.** $x^{m \cdot n}$
b. $x^n y^n$ **9. a.** 2   **b.** 2   **11.** second, square
13. $x, 4, 4, 8x$ **15.** $s, -, 5, 25$ **17.** $x^2 + 2x + 1$
19. $r^2 + 4r + 4$ **21.** $m^2 - 12m + 36$
23. $f^2 - 16f + 64$ **25.** $d^2 - 49$ **27.** $n^2 - 36$
29. $16x^2 + 40x + 25$ **31.** $49m^2 - 28m + 4$
33. $y^4 + 18y^2 + 81$ **35.** $4v^6 - 32v^3 + 64$
37. $16f^2 - 0.16$ **39.** $9n^2 - 1$ **41.** $1 - 6y + 9y^2$
43. $x^2 - 4xy + 4y^2$ **45.** $4a^2 - 12ab + 9b^2$
47. $s^2 + \frac{3}{2}s + \frac{9}{16}$ **49.** $a^2 + 2ab + b^2$ **51.** $r^2 - 2rs + s^2$
53. $36b^2 - \frac{1}{4}$ **55.** $r^2 + 20rs + 100s^2$

57. $36 - 24d^3 + 4d^6$ **59.** $-64x^2 - 48x - 9$
61. $-25 + 36g^2$ **63.** $12x^3 + 36x^2 + 27x$
65. $-80d^3 + 40d^2 - 5d$ **67.** $4d^5 - 4dg^6$
69. $x^3 + 12x^2 + 48x + 64$ **71.** $n^3 - 18n^2 + 108n - 216$
73. $8g^3 - 36g^2 + 54g - 27$
75. $n^4 - 8n^3 + 24n^2 - 32n + 16$ **77.** $3t^2 + 12t - 9$
79. $2x^2 + xy - y^2$ **81.** $13x^2 - 8x + 5$ **83.** $36m + 36$
85. $\frac{\pi}{4}D^2 - \frac{\pi}{4}d^2$ **87.** $(x^2 + 12x + 36)$ in.^{2}
89. $(36x^2 + 36x + 6)$ ft^2 **95.** $3^3 \cdot 7$ **97.** $\frac{5}{6}$ **99.** $\frac{21}{40}$

Study Set Section 4.8 (page 336)

1. numerator, denominator   **3.** polynomial   **5.** descending
7. placeholder   **9. a.** x^{m-n} **b.** $\frac{1}{x^n}$
11. a. $7x^3 + 5x^2 - 3x - 9$ **b.** $6x^4 - x^3 + 2x^2 + 9x$
13. $-2x$ **15.** It is correct.
17. $7x^2, x^3, 7x^2, 7, 2 - 2, 4x^3, 1$
19. a. $5x^4 + 0x^3 + 2x^2 + 0x - 1$
b. $-3x^5 + 0x^4 - 2x^3 + 0x^2 + 4x - 6$ **21.** $\frac{4}{3}$ **23.** x^3
25. $5m^5$ **27.** $\frac{4h^2}{3}$ **29.** $-\frac{1}{5d^4}$ **31.** $\frac{r^2}{s}$ **33.** $\frac{2x^2}{y}$ **35.** $\frac{4r}{y^2}$
37. $-\frac{13}{3rs}$ **39.** $2x + 3$ **41.** $2x^5 - 8x^2$ **43.** $\frac{h^2}{4} + \frac{2}{h}$
45. $-2w^2 - \frac{1}{w^4}$ **47.** $3s^5 - 6s^2 + 4s$
49. $c^3 + 3c^2 - 2c - \frac{5}{c}$ **51.** $\frac{1}{5y} - \frac{2}{5x}$ **53.** $3a - 2b$
55. $3x^2y - 2x - \frac{1}{y}$ **57.** $5x - 6y + 1$ **59.** $x + 6$
61. $y + 12$ **63.** $3a - 2$ **65.** $b + 3$ **67.** $2x + 1$
69. $x - 7$ **71.** $3x + 2$ **73.** $x^2 + 2x - 1$
75. $2x^2 + 2x + 1$ **77.** $x^2 + x + 1$ **79.** $x + 1$
81. $2x - 3$ **83.** $x^2 - x + 1$ **85.** $a^2 - 3a + 10 + \frac{-30}{a + 3}$
87. $x + 1 + \frac{-1}{2x + 3}$ **89.** $2x + 2 + \frac{-3}{2x + 1}$
91. $x^2 + 2x - 1 + \frac{6}{2x + 3}$ **93.** $2x^2 + x + 1 + \frac{2}{3x - 1}$
95. $(2x^2 - x + 3)$ in.   **97. a.** $t = \frac{d}{r}$ **b.** $3x^2$ **c.** $x + 4$
99. $(3x - 2)$ ft   **101.** $4x^2 + 3x + 7$ **107.** $y = -\frac{11}{6}x - \frac{7}{3}$
109. -80

Key Concept (page 341)

1. a. x, descending, x **b.** 4 **c.** $3, 2, 1, 0$ **d.** 3
e. $1, -2, 6, -8$ **2. a.** binomial   **b.** none of these
c. trinomial   **d.** monomial   **3.** $7x^3$ **4.** m^{10} **5.** $-2a^2b$
6. $-42y^{13}$ **7.** $\frac{2c^2}{d}$ **8.** $25f^6$ **9.** combine   **10.** signs
11. each, each   **12.** term   **13.** long
14. $14x^3 - x^2 - 10x + 4$ **15.** $12s^3t + 10s^2t^2 - 18st^3$
16. $2x^2 - 13x - 24$ **17.** $4x^4 + 12x^2 + 9$
18. $16h^{10} - 64t^2$ **19.** $y^3 + 4y^2 - 3y - 18$
20. $3x^4 + 9x^5 - 6x^3$ **21.** $x^2 + 2x + 3$

Chapter Review (page 342)

1. a. $-3 \cdot x \cdot x \cdot x \cdot x$ **b.** $\left(\frac{1}{2}pq\right)\left(\frac{1}{2}pq\right)\left(\frac{1}{2}pq\right)$ **2. a.** base x,
exponent 6   **b.** base $2x$, exponent 6   **3.** 125 **4.** 64
5. -64 **6.** 4 **7.** 7^{12} **8.** m^2n^2 **9.** y^{21} **10.** $81x^4$

11. b^{12} **12.** $-y^2z^5$ **13.** $256s^3$ **14.** $4x^4y^2$ **15.** x^{15}
16. $\frac{x^2}{y^2}$ **17.** $(m - 25)^{12}$ **18.** $125yz^4$ **19.** $64x^{12}$ in.^{3}
20. $y^4\,m^2$ **21.** 1 **22.** 1 **23.** 9 **24.** $\frac{1}{1,000}$ **25.** $\frac{4}{3}$
26. $-\frac{1}{25}$ **27.** $\frac{1}{x^5}$ **28.** $-\frac{6}{y}$ **29.** $\frac{8}{49}$ **30.** x^{14} **31.** $-\frac{27}{r^9}$
32. $\frac{1}{16z^2}$ **33.** 7.28×10^2 **34.** 9.37×10^{15}
35. 1.36×10^{-2} **36.** 9.42×10^{-3} **37.** 1.8×10^{-4}
38. 7.53×10^5 **39.** $726{,}000$ **40.** 0.0000000391
41. 2.68 **42.** 57.6 **43.** 0.03 **44.** 160
45. $6{,}310{,}000{,}000; 6.31 \times 10^9$ **46.** $1.0 \times 10^5 = 100{,}000$
47. a. 4 **b.** $3x^3$ **c.** $3, -1, 1, 10$ **d.** 10 **48. a.** 7th,
monomial   **b.** 3rd, monomial   **c.** 2nd, binomial   **d.** 5th,
trinomial   **e.** 6th, binomial   **f.** 4th, none of these
49. $3, -13$ **50.** 8 in.
51.   **52.**

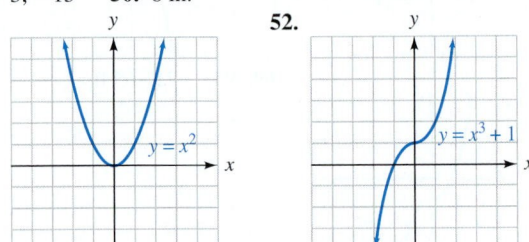

53. $13y^3$ **54.** $-4a^3b + a^2b + 5$ **55.** $\frac{7}{12}x^2 - \frac{3}{4}xy + y^2$
56. $-c^5 - 5c^4 + 12$ **57.** $25r^6 + 9r^3 + 5r$
58. $4a^2 + 4a - 6$ **59.** $4r^3s - 7r^2s^2 - 7rs^3 - 2s^4$
60. $5x^2 + 19x + 3$ **61.** $-z^3 + 2z^2 + 5z - 17$
62. $4x^2 + 2x + 8$ **63.** $8x^3 - 7x^2 + 19x$
64. $(x^2 + x + 3)$ in.   **65.** $10x^3$ **66.** $-6x^{10}z^5$ **67.** $120b^{11}$
68. $2h^{14} + 8h^{11}$ **69.** $x^2y^3 - x^3y^2$ **70.** $9n^4 - 15n^3 + 6n^2$
71. $6x^6 + 12x^5$ **72.** $a^6b^4 - a^5b^5 + a^3b^6 - 7a^3b^2$
73. $x^2 + 5x + 6$ **74.** $2x^2 - x - 1$ **75.** $6a^2 - 6$
76. $6a^2 - 6$ **77.** $2a^2 - ab - b^2$ **78.** $6n^8 - 13n^6 + 5n^4$
79. $8a^3 - 27$ **80.** $56x^4 + 15x^3 - 21x^2 - 3x + 2$
81. $8x^3 + 1$ **82.** $(6x + 10)$ in.; $(2x^2 + 11x - 6)$ in.^{2};
$(6x^3 + 33x^2 - 18x)$ in.^{3} **83.** $x^2 + 6x + 9$ **84.** $4x^2 - 0.81$
85. $a^2 - 6a + 9$ **86.** $x^2 + 8x + 16$ **87.** $4y^2 - 4y + 1$
88. $y^4 - 1$ **89.** $36r^4 + 120r^2s + 100s^2$
90. $-64a^2 + 48a - 9$ **91.** $80r^4s - 80s^5$
92. $36b^3 - 96b^2 + 64b$ **93.** $t^2 - \frac{3}{2}t + \frac{9}{16}$
94. $m^3 + 6m^2 + 12m + 8$ **95.** $24c^2 - 10c + 37$
96. $(x^2 - 4)$ in.^{2} **97.** $2n^3$ **98.** $-\frac{2x}{3y^2}$ **99.** $\frac{a^3}{6} - \frac{4}{a^4}$
100. $3a + 4b - 5$ **101.** $x - 5$ **102.** $2x + 1$
103. $3x^2 + 2x + 1 + \frac{2}{2x - 1}$ **104.** $3x^2 - x - 4$
105. $(y + 3)(3y + 2) = 3y^2 + 11y + 6$ **106.** $x + 5$

Chapter 4 Test (page 347)

1. $2x^3y^4$ **2.** -36 **3.** y^6 **4.** $32x^{21}$ **5.** 3 **6.** $\frac{2}{y^3}$
7. $\frac{1}{125}$ **8.** $(x + 1)^9$ **9.** y^3 **10.** $\frac{64a^3}{b^3}$
11. $1{,}000y^{12}$ in.^{3} **12.** 6.25×10^{18} **13.** 0.000093
14. $9{,}200$ **15.** $x^4, 1, 4, 8x^2, 8, 2, -12, -12, 0, 4$

16. binomial   **17.** 5th degree   **18.**

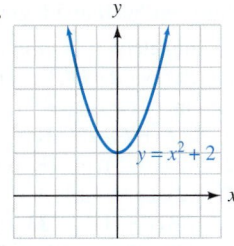

19. 0 ft; The rock hits the canyon floor 18 seconds after being dropped.   **20.** $-4a^3b + a^2b + 5$ **21.** $5a^2 - 5a + 3$
22. $6b^3c - 2bc - 12$ **23.** $-3y^3 + 18y^2 - 28y + 35$
24. $-4x + 8y$ **25.** $-4x^5y$ **26.** $3y^4 - 6y^3 + 9y^2$
27. $6x^2 - 7x - 20$ **28.** $2x^3 - 7x^2 + 14x - 12$
29. $1 - 100c^2$ **30.** $49b^6 - 42b^3 + 9$ **31.** $2x^2 + xy - y^2$
32. $\frac{a}{4b} - \frac{b}{2a}$ **33.** $x - 2$ **34.** $3x^2 + 2x + 1 + \frac{2}{2x-1}$
35. $(x - 5)$ ft

Cumulative Review Exercises Chapters 1–4 (page 349)

1. a. 1993   **b.** 1986   **c.** 1996–1997   **2.** $\frac{5}{8}$ **3.** $\frac{22}{35}$
4. irrational   **5.** $250 - x$ **6.** 18, 16   **7.** $2^2 \cdot 5^2$
8.

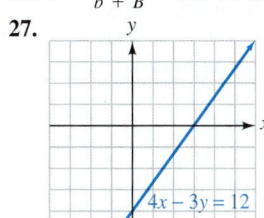

9. $0.\overline{6}$ **10.** associative property of multiplication   **11.** $10d$
cents   **12.** r **13.** $18x$ **14.** $3d - 11$ **15.** $-78c + 18$
16. 0   **17.** 1   **18.** 4   **19.** -2 **20.** 13   **21.** 41
22. $\frac{10}{9}$ **23.** -24
24. $x < -14, (-\infty, -14)$
25. $h = \frac{2A}{b + B}$ **26.** 20 lb of $1.90 candy; 10 lb of $2.20 candy
27.   **28.**

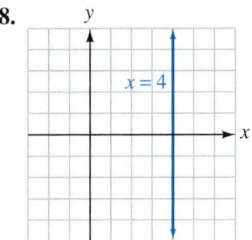

29. $\frac{1}{2}$ **30.** 0   **31.** $\frac{2}{3}$ **32.** $y = \frac{2}{3}x + 5$
33. $3x - 4y = -22$ **34.** $y = 4$ **35.** perpendicular
36. $-6x^4 - 17x^2 - 68x + 11$ **37.** $9x^4y^8$ **38.** $\frac{1}{16y^4}$
39. x^{14} **40.** $a^2b^7c^6$ **41.** a^7 **42.** $\frac{64t^{12}}{27}$ **43.** $7c^2 + 7c$
44. $12x^3 + 36x^2 + 27x$ **45.** $6t^2 + 7st - 3s^2$ **46.** $2x + 1$
47.   **48.** 6.15×10^5
                                 49. 1.3×10^{-6}
                                 50. 1.5 in.

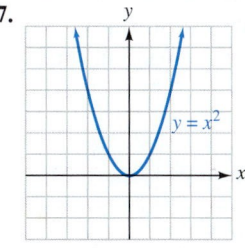

Study Set Section 5.1. (page 358)

1. greatest common factor   **3.** primes   **5.** grouping
7. a. 3   **b.** 2   **c.** 3   **d.** 7   **9.** $3a^2b$ **11. a.** the
distributive property   **b.** $2x$ **13. a.** The GCF is $6a^2$, not $6a$.
b. The 0 in the first line should be 1.   **15.** no   **17.** $8m, 4$
19. $2^2 \cdot 3$ **21.** $2^3 \cdot 5$ **23.** $2 \cdot 7^2$ **25.** $3^2 \cdot 5^2$ **27.** 6
29. m^3 **31.** $4c$ **33.** $3m^3n$ **35.** $2x$ **37.** $3y$ **39.** $2t^2$
41. $8x^2z^2$ **43.** $3(x + 2)$ **45.** $9(2m - 1)$ **47.** $6(3x + 4)$
49. $d(d - 7)$ **51.** $7(2c^3 + 9)$ **53.** $6(2x^2 - x - 4)$
55. $t^2(t + 2)$ **57.** $a(b + c - d)$ **59.** $a^2(a - 1)$
61. $8xy^2(3xy + 1)$ **63.** $6uvw^2(2w - 3v)$
65. $3r(4r - s + 3rs^2)$ **67.** $\pi(R^2 - ab)$
69. $(x + 2)(3 - x)$ **71.** $(14 + r)(h^2 + 1)$ **73.** $-(a + b)$
75. $-(2x - 5y)$ **77.** $-(3r - 2s + 3)$ **79.** $-3x(x + 2)$
81. $-4a^2b^2(b - 3a)$ **83.** $-2ab^2c(2ac - 7a + 5c)$
85. $(x + y)(2 + a)$ **87.** $(p - q)(9 + m)$
89. $(m - n)(p + q)$ **91.** $(2x - 3y)(y + 1)$
93. $(a + b)(x - 1)$ **95.** $x^2(a + b)(x + 2y)$
97. $2z(x - 2)(x^2 + 16)$ **99.** $(y + 9)(x - 11)$
101. $(3p + q)(3m - n)$ **103.** $5x^3y^2z^3(5x^2y^5 - 9z^3)$
105. $4(6m - 3n + 4)$ **107.** $-20P(3P + 4)$
109. a. $6x^3$ cm^2 **b.** $24x^2$ cm^2 **c.** $6x^2(x + 4)$ cm^2
111. $(x^2 + 5)$ ft; $(x + 4)$ ft   **117.** 12%

Study Set Section 5.2 (page 369)

1. trinomial, binomial   **3.** factors   **5.** lead, coefficient, term
7. $4 \cdot 1, 2 \cdot 2, -4(-1), -2(-2)$ **9.** product, sum   **11.** 3, 5
13. a. Last: $5 \cdot 4$ **b.** Outer: $x \cdot 4$ and Inner: $5 \cdot x$ **15. a.** 1
b. 15, 8   **c.** 5, 3   **17. a.** They are both positive or both
negative.   **b.** One will be positive, the other negative.
19. $(x + 1)(x + 5)$ **21.** $-1 + (-8) = -9, -2(-4) = 8$
23. $+3, -2$ **25.** $+, +$ **27.** $-, -$ **29.** $+, -$
31. $(z + 11)(z + 1)$ **33.** $(m - 3)(m - 2)$
35. $(a - 5)(a + 1)$ **37.** $(x + 8)(x - 3)$
39. $(a - 13)(a + 3)$ **41.** prime   **43.** prime
45. $(r - 3)(r - 6)$ **47.** $(x + 2y)(x + 2y)$
49. $(a - 6b)(a + 2b)$ **51.** prime   **53.** $-(x + 5)(x + 2)$
55. $-(t + 17)(t - 2)$ **57.** $-(a + 3b)(a + b)$
59. $-(x - 7y)(x + y)$ **61.** $(x - 4)(x - 1)$
63. $(y + 9)(y + 1)$ **65.** $(r - 2)(r + 1)$
67. $(r + 3x)(r + x)$ **69.** $2(x + 3)(x + 2)$
71. $-5(a - 3)(a - 2)$ **73.** $z(z - 4)(z - 25)$
75. $4y(x + 6)(x - 3)$ **77.** $-(r - 10)(r - 4)$
79. $(y - 14z)(y + z)$ **81.** $(s + 13)(s - 2)$
83. $(a + 9b)(a + b)$ **85.** $-(x - 22)(x + 1)$
87. $d(d - 13)(d + 2)$ **89.** $(x + 9)$ in., x in., $(x + 3)$ in.
97. $\frac{1}{x^2}$ **99.** $\frac{1}{x^{10}}$

Study Set Section 5.3 (page 379)

1. lead, term   **3.** product, sum   **5.** number
7. a. $1 \cdot 9 = 9$ **b.** $-2(-8) = 16$ **c.** $-2(5) = -10$
9. a. $8y$ **b.** $-24y$ **c.** $-16y$ **11.** $10x$ and x, $5x$ and $2x$
13. a. descending, GCF, positive   **b.** $3s^2$
c. $-(2d^2 - 19d + 8)$ **15.** positive, negative, negative
17. negative, different   **19.** $(5x - 2)(3x + 2)$
21. $-1 + (-12) = -13, -2(-6) = 12, -2 + (-6) = -8,$
$-3 + (-4) = -7$ **23. a.** $x^2 + 6x + 1$ **b.** $3x^2 + 6x + 1$

25. a. $12, 20, -9$ **b.** -108 **27.** $+, +$ **29.** $-, -$
31. $-, +$ **33.** $9, 8, 3t, 2, 4t + 3$ **35.** $(3a + 1)(a + 4)$
37. $(5x + 1)(x + 2)$ **39.** $(2x + 3)(2x + 1)$
41. $(6x + 7)(x + 3)$ **43.** $(2x - 1)(x - 1)$
45. $(2t - 1)(2t - 1)$ **47.** $(15t - 4)(t - 2)$
49. $(2x + 1)(x - 2)$ **51.** $(4y + 1)(3y - 1)$
53. $(5y + 1)(2y - 1)$ **55.** $(3y - 2)(4y + 1)$ **57.** prime
59. $(3x + 2)(x - 5)$ **61.** $(3r + 2s)(2r - s)$
63. $(8n + 3)(n + 11)$ **65.** $(4a - 3b)(a - 3b)$
67. $10(13r - 11)(r + 1)$ **69.** $-(5t + 3)(t + 2)$
71. $4(9y - 4)(y - 2)$ **73.** $(2x + 3y)(2x + y)$
75. $(4y - 3)(3y - 4)$ **77.** $(18x - 5)(x + 2)$
79. $-y(y + 12)(y + 1)$ **81.** prime
83. $3r^3(5r - 2)(2r + 5)$ **85.** $(2a + 3b)(a + b)$
87. $(3p - q)(2p + q)$ **89.** $3x(2x + 1)(x - 3)$
91. $(2a - 5)(4a - 3)$ **93.** $2mn(4m + 3n)(2m + n)$
95. $(2x + 11)$ in., $(2x - 1)$ in.; 12 in.   **103.** -49 **105.** 1
107. 49

Study Set Section 5.4 (page 386)

1. perfect   **3. a.** $5x$ **b.** 3   **c.** $5x, 3$ **5. a.** x, y **b.** $-$
c. $+, x, y$ **7.** $1, 4, 9, 16, 25, 36, 49, 64, 81, 100$
9. a. $x^2 - 36, x^2 + 12x + 36, x^2 - 12x + 36$ **b.** no
11. $9x^2 - 16y^2$ **13. a.** $x^2 - 9$ **b.** $(x - 9)^2$
c. $x^2 + 2xy + y^2$ **d.** $x^2 + 25$ **15. a.** $x + 3; 2$
b. $(x - 8)(x - 8)$ **17.** 3   **19.** $+$ **21.** $(x + 3)^2$
23. $(b + 1)^2$ **25.** $(c - 6)^2$ **27.** $(y - 4)^2$ **29.** $(t + 10)^2$
31. $2(u - 9)^2$ **33.** $x(6x + 1)^2$ **35.** $(2x + 3)^2$
37. $(a + b)^2$ **39.** $(5m + 7n)^2$ **41.** $(3xy + 5)^2$
43. $\left(t - \frac{1}{3}\right)^2$ **45.** $(s - 0.6)^2$ **47.** $7, 7$ **49.** t, w
51. $(x + 4)(x - 4)$ **53.** $(2y + 1)(2y - 1)$
55. $(3x + y)(3x - y)$ **57.** $(4a + 5b)(4a - 5b)$
59. $(6 + y)(6 - y)$ **61.** prime   **63.** $(a^2 + 12b)(a^2 - 12b)$
65. $(tz + 8)(tz - 8)$ **67.** prime   **69.** $8(x + 2y)(x - 2y)$
71. $7(a + 1)(a - 1)$ **73.** $(v + 5)(v - 5)$
75. $6x^2(x + y)(x - y)$ **77.** $(x^2 + 9)(x + 3)(x - 3)$
79. $(a^2 + 4)(a + 2)(a - 2)$ **81.** $\left(c + \frac{1}{4}\right)\left(c - \frac{1}{4}\right)$
83. $(p + q)^2$ **85.** $0.5g(t_1 + t_2)(t_1 - t_2)$ **93.** $\frac{x}{y} + \frac{2y}{x} - 3$
95. $3a + 2$

Study Set Section 5.5 (page 392)

1. sum   **3.** binomial   **5. a.** F, L **b.** $-, F^2, L^2$ **7.** $m, 4$
9. $1, 8, 27, 64, 125, 216$ **11.** It is factored completely.
13. a. $x^3 + 8$ **b.** $(x + 8)^3$ **15.** $2a$ **17.** $b + 3$
19. $(y + 1)(y^2 - y + 1)$ **21.** $(a - 3)(a^2 + 3a + 9)$
23. $(2 + x)(4 - 2x + x^2)$ **25.** $(s - t)(s^2 + st + t^2)$
27. $(a + 2b)(a^2 - 2ab + 4b^2)$
29. $(4x - 3)(16x^2 + 12x + 9)$
31. $(a^2 - b)(a^4 + a^2b + b^2)$
33. $2(x + 3)(x^2 - 3x + 9)$ **35.** $-(x - 6)(x^2 + 6x + 36)$
37. $8x(2m - n)(4m^2 + 2mn + n^2)$ **39.** $(1,000 - x^3)$ in.³;
$(10 - x)(100 + 10x + x^2)$ **43.** repeating
45. $\{\ldots, -4, -3, -2, -1, 0, 1, 2, 3, 4, \ldots\}$ **47.** 0

Study Set Section 5.6 (page 397)

1. completely   **3.** prime   **5.** trinomial   **7.** GCF
9. factor out the GCF   **11.** perfect square trinomial
13. sum of two cubes   **15.** trinomial factoring   **17.** $8m^3$
19. $2a(b + 6)(b - 2)$ **21.** $-4p^2q^3(2pq^4 + 1)$
23. $5(2m + 5)^2$ **25.** prime
27. $-2x^2(x - 4)(x^2 + 4x + 16)$ **29.** $(c + d^2)(a^2 + b)$
31. $-(3xy - 1)^2$ **33.** $5x^2y^2z^2(xyz^2 + 5y - 7xz^3)$
35. $(2c + d)(c - 3d)$ **37.** $2xy(2ax + b)(2ax - b)$
39. $(x - a)(a + b)(a - b)$ **41.** $(ab + 12)(ab - 12)$
43. $(c + 2d)(2a + b)$ **45.** $(v - 7)^2$
47. $-(n + 3)(n - 13)$ **49.** $(4y^4 + 9z^2)(2y^2 + 3z)(2y^2 - 3z)$
51. $2(3x - 4)(x - 1)$ **53.** $y^2(2x + 1)^2$
55. $4m^2n(m + 5n)(m^2 - 5mn + 25n^2)$
57. $(x^2 + y^2)(a^2 + b^2)$ **59.** prime
61. $2a^2(2a - 3b)(4a^2 + 6ab + 9b^2)$ **63.** $27(x - y - z)$
65. $(x - t)(y + s)$ **67.** $x^6(7x + 1)(5x - 1)$
69. $5(x - 2)(1 + 2y)$ **71.** $(7p + 2q)^2$ **73.** $4(t^2 + 9)$
75. $(p - 2)(p^2 + 3)$
81.

$$\leftarrow \quad \begin{array}{ccccccccc} & \bullet & & \bullet & & \bullet & & \bullet & \\ -4 & -3 & -2 & -1 & 0 & 1 & 2 & 3 \end{array} \quad \rightarrow$$

83.

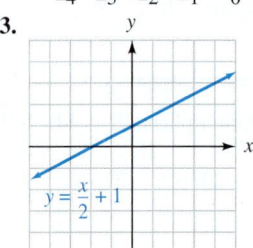

$y = \dfrac{x}{2} + 1$

Study Set Section 5.7 (page 406)

1. quadratic   **3.** consecutive   **5.** hypotenuse, legs
7. a. yes   **b.** no   **c.** yes   **d.** no   **9.** $-\frac{4}{5}$ **11. a.** Add
6 to both sides.   **b.** Distribute the multiplication by x and
subtract 3 from both sides.   **13. a.** $(x - 2)(x + 8)$
b. $2, -8$ **15.** 0   **17.** $a^2 + b^2$ **19.** $7y, y + 2, 0$
21. $2, -3$ **23.** $\frac{5}{2}, -6$ **25.** $7, -8$ **27.** $1, -2, 3$
29. $0, 3$ **31.** $0, \frac{5}{2}$ **33.** $0, 7$ **35.** $0, -\frac{8}{3}$ **37.** $0, 2$
39. $-5, 5$ **41.** $-\frac{1}{2}, \frac{1}{2}$ **43.** $-\frac{2}{3}, \frac{2}{3}$ **45.** $-10, 10$
47. $-\frac{9}{2}, \frac{9}{2}$ **49.** $12, 1$ **51.** $-3, 7$ **53.** $8, 1$ **55.** $-3, -5$
57. $-2, -2$ **59.** $8, 8$ **61.** $\frac{1}{2}, 2$ **63.** $\frac{1}{5}, 1$ **65.** $-\frac{1}{2}, -\frac{1}{2}$
67. $-\frac{2}{3}, -\frac{3}{2}$ **69.** $\frac{2}{3}, -\frac{1}{5}$ **71.** $-\frac{5}{2}, 4$ **73.** $-3, -2$
75. $2, 10$ **77.** $\frac{1}{8}, 1$ **79.** $2, 7, 1$ **81.** $0, -1, -2$
83. $0, 9, -3$ **85.** $0, -1, 2$ **87.** $\frac{3}{2} = 1.5$ sec   **89.** 2 sec
91. 8   **93.** 12   **95.** $(11, 13)$ **97.** 4 m by 9 m
99. $h = 5$ ft, $b = 12$ ft   **101.** 6, 16 ft   **103.** 3 ft, 4 ft, 5 ft
109. 15 min $\leq t <$ 30 min

Key Concept (page 412)

1. Factor $3x + 27$; $3(x + 9)$ **2.** Factor $x^2 + 12x + 27$;
$(x + 3)(x + 9)$ **3.** $-3(a - 3)(a - 4)$
4. $(x + 11y)(x - 11y)$ **5.** $(r + s)(t + 2)$

6. $(v - 2)(v^2 + 2v + 4)$ **7.** $(3t - 5)(2t - 3)$
8. $(5y - 2)^2$ **9.** $2r(r + 5)(r - 5)$ **10.** $2(8w - 1)(w + 3)$

Chapter Review (page 413)

1. $5 \cdot 7$ **2.** $2^5 \cdot 3$ **3.** 7   **4.** $18a^3$ **5.** $3(x + 3y)$
6. $5a(x^2 + 3)$ **7.** $7s^3(s^2 + 2)$ **8.** $\pi a(b - c)$
9. $2x(x^2 + 2x - 4)$ **10.** $xy^2z(x + yz - 1)$
11. $-5ab(b - 2a + 3)$ **12.** $(x - 2)(4 - x)$
13. $-(a + 7)$ **14.** $-(4t^2 - 3t + 1)$ **15.** $(c + d)(2 + a)$
16. $(y + 3)(3x - 2)$ **17.** $(2a^2 - 1)(a + 1)$
18. $4m(n + 3)(m - 2)$ **19.** 1   **20.** $7, 5, -7, -5$
21. $(x + 6)(x - 4)$ **22.** $(x - 6)(x + 2)$
23. $(x - 5)(x - 2)$ **24.** prime   **25.** $-(y - 5)(y - 4)$
26. $(y + 9)(y + 1)$ **27.** $(c + 5d)(c - 2d)$
28. $(m - 2n)(m - n)$ **29.** Multiply to see if
$(x - 4)(x + 5) = x^2 + x - 20$.   **30.** There are no two
integers whose product is 11 and whose sum is 7.
31. $5a^3(a + 10)(a - 1)$ **32.** $-4x(x + 3y)(x - 2y)$
33. $(2x + 1)(x - 3)$ **34.** $(2y + 5)(5y - 2)$
35. $-(3x + 1)(x - 5)$ **36.** $3p(2p + 1)(p - 2)$
37. $(4b - c)(b - 4c)$ **38.** prime   **39.** $(4x + 1)$ in.,
$(3x - 1)$ in.   **40.** The signs of the second terms must be
negative.   **41.** $(x + 5)^2$ **42.** $(3y - 4)^2$ **43.** $-(z - 1)^2$
44. $(5a + 2b)^2$ **45.** $(x + 3)(x - 3)$
46. $(7t + 5y)(7t - 5y)$ **47.** $(xy + 20)(xy - 20)$
48. $8a(t + 2)(t - 2)$ **49.** $(c^2 + 16)(c + 4)(c - 4)$
50. prime   **51.** $(h + 1)(h^2 - h + 1)$
52. $(5p + q)(25p^2 - 5pq + q^2)$ **53.** $(x - 3)(x^2 + 3x + 9)$
54. $2x^2(2x - 3y)(4x^2 + 6xy + 9y^2)$ **55.** $2y^2(3y - 5)(y + 4)$
56. $(t + u^2)(s^2 + v)$ **57.** $(j^2 + 4)(j + 2)(j - 2)$
58. $-3(j + 2k)(j^2 - 2jk + 4k^2)$ **59.** $3(2w - 3)^2$
60. prime   **61.** $2(t^3 + 5)$ **62.** $(20 + m)(20 - m)$
63. $(x + 8y)^2$ **64.** $6c^2d(3cd - 2c - 4)$ **65.** $0, -2$
66. $0, 6$ **67.** $-3, 3$ **68.** $3, 4$ **69.** $-2, -2$ **70.** $6, -4$
71. $\frac{1}{5}, 1$ **72.** $0, -1, 2$ **73.** 15 m   **74.** Streep: 13;
Hepburn: 12   **75.** 5 m   **76.** 10 sec

Chapter 5 Test (page 417)

1. $2^2 \cdot 7^2$ **2.** $3 \cdot 37$ **3.** $15x^3y^6$ **4.** $4(x + 4)$
5. $5ab(6ab^2 - 4a^2b + c)$ **6.** $(q + 9)(q - 9)$ **7.** prime
8. $(4x^2 + 9)(2x + 3)(2x - 3)$ **9.** $(x + 3)(x + 1)$
10. $-(x - 11)(x + 2)$ **11.** $(a - b)(9 + x)$
12. $(2a - 3)(a + 4)$ **13.** $2(3x - 5y)^2$
14. $(x + 2)(x^2 - 2x + 4)$ **15.** $5m^6(4m^2 - 3)$
16. $3(a - 3)(a^2 + 3a + 9)$ **17.** $r^2(4 - \pi)$ **18.** $(5x - 4)$
19. $(x - 9)(x + 6)$; Multiply the binomials:
$(x - 9)(x + 6) = x^2 + 6x - 9x - 54 = x^2 - 3x - 54$.
20. $-3, 2$ **21.** $-5, 5$ **22.** $0, \frac{1}{6}$ **23.** $-3, -3$
24. $\frac{1}{3}, -\frac{1}{2}$ **25.** $9, -2$ **26.** $0, -1, -6$ **27.** 6 ft by 9 ft
28. A quadratic equation is an equation that can be written in
the form $ax^2 + bx + c = 0$; $x^2 - 2x + 1 = 0$. (Answers may
vary.)   **29.** 10   **30.** At least one of them is 0.

Study Set Section 6.1 (page 425)

1. rational   **3.** undefined   **5.** simplify   **7. a.** 0   **b.** 6
c. -6 **9. a.** $x + 1$ **b.** $\frac{1}{1}, 1$ **11. a.** $x - 2, 2 - x$
b. $\frac{-1}{1}, -1$ **13. a.** $\frac{x - 2}{x + 1}$ **b.** $-\frac{y}{9}$ **c.** $m - 5$ **d.** $\frac{1}{x - 30}$
15. yes, yes, yes   **17.** 4   **19.** 0   **21.** $-\frac{2}{11}$ **23.** $-\frac{27}{28}$
25. 0   **27.** 2   **29.** none   **31.** $\frac{1}{2}$ **33.** $-6, 6$
35. $-2, 1$ **37.** $\frac{5}{a}$ **39.** $\frac{3}{2}$ **41.** does not simplify   **43.** $\frac{3x}{y}$
45. $\frac{2x + 1}{y}$ **47.** $\frac{1}{3}$ **49.** $\frac{1}{a}$ **51.** $\frac{4}{5}$ **53.** 4   **55.** -1
57. -6 **59.** $\frac{x + 2}{x + 1}$ **61.** $\frac{x + 1}{x - 1}$ **63.** $\frac{2x}{x - 2}$ **65.** $-\frac{1}{a + 1}$
67. $\frac{1}{b + 1}$ **69.** $\frac{2x - 3}{x + 1}$ **71.** $\frac{10}{3}$ **73.** 9   **75.** $\frac{3(x + 3)}{2x + 1}$
77. $-\frac{x + 11}{x + 3}$ **79.** $\frac{4 - x}{4 + x}$ or $-\frac{x - 4}{x + 4}$ **81.** $\frac{m - n}{2(m + n)}$
83. does not simplify   **85.** $\frac{2u - 3}{u^3}$ **87.** does not simplify
89. $\frac{2}{3}$ **91.** $\frac{3x}{5y}$ **93.** $\frac{x + 2}{x - 2}$ **95.** $85\frac{1}{3}$ **97.** 2, 1.6, and 1.2
milligrams per liter   **103.** $(a + b) + c = a + (b + c)$
105. One of them is zero.

Study Set Section 6.2 (page 434)

1. opposites   **3.** invert   **5.** numerators, denominators, $\frac{AC}{BD}$
7. $\frac{2}{(x + 1)(x - 9)}$ **9.** $\frac{6n}{1}$ **11.** $\frac{1}{18x}$ **13.** $\frac{x^2 + 2x + 1}{x^2 + 1}$ **15.** ft
17. a. m^2 **b.** m^3 **19.** $\frac{3}{2}$ **21.** $\frac{20}{3n}$ **23.** x^2y^2 **25.** $\frac{5r^3t^3}{2s^2}$
27. $\frac{(z + 7)(z + 2)}{7z}$ **29.** $\frac{x}{5}$ **31.** $-x$ **33.** 5   **35.** $x + 1$
37. $2y + 16$ **39.** $x + 5$ **41.** $10h - 30$ **43.** $x + 2$
45. $\frac{3}{2x}$ **47.** $-(x - 2)$ **49.** $\frac{(x - 2)^2}{x}$ **51.** $\frac{(m - 2)(m - 3)}{2(m + 2)}$
53. $\frac{x + 5}{x - 5}$ **55.** $-\frac{2x - 3y}{xy(3y + 2x)}$ **57.** $\frac{x + 1}{2(x - 2)}$ **59.** $\frac{3}{2y}$ **61.** $\frac{6}{y}$
63. $\frac{2}{y}$ **65.** $\frac{4(n - 1)}{3n}$ **67.** $\frac{2}{3x}$ **69.** $\frac{2(z - 2)}{z}$ **71.** $\frac{5(a - 2)}{4a^3}$
73. $-\frac{x + 2}{3}$ **75.** $\frac{1}{3}$ **77.** $\frac{2(x - 7)}{x + 9}$ **79.** $\frac{d(6c - 7d)}{6}$
81. $-(d + 5)$ **83.** $\frac{1}{12(2r - 3s)}$ **85.** 450 ft   **87.** 1,800
meters per min   **89.** $\frac{12x^2 + 12x + 3}{2}$ in.^{2} **91.** 4,380,000
93. 8 yd^2 **95.** $\frac{1}{2}$ mile per minute   **97.** $\frac{1}{4}$ mi^2
105. $w = 6$ in., $l = 10$ in.

Study Set Section 6.3 (page 443)

1. common   **3.** least common denominator   **5.** like
7. numerators, denominator, $A + B, A - B$ **9. a.** $-6x - 9$
b. $2x^2 - 2x - 1$ **c.** $2x + 5$ **d.** $3x^2 - 3x$ **11.** $\frac{x - 7}{x + 7}$
13. a. $2 \cdot 2 \cdot 2 \cdot 5 \cdot x \cdot x$ **b.** $2x(x - 3)$ **c.** $(n + 8)(n - 8)$
15. a. $\frac{5}{21a}$ **b.** $\frac{a + 4}{a + 4}$ **17.** yes   **19.** $\frac{11}{x}$ **21.** $\frac{x + 5}{18}$
23. $\frac{m - 8}{m^3}$ **25.** $\frac{x}{y}$ **27.** $-\frac{2t}{9}$ **29.** $\frac{x}{3}$ **31.** $\frac{r + 50}{r^3 - 25}$
33. $\frac{2a}{a + 2}$ **35.** $-\frac{2}{t + 5}$ **37.** 9   **39.** $\frac{1}{y}$ **41.** $\frac{1}{2}$
43. $\frac{x + 4}{y}$ **45.** $2x - 5$ **47.** 0   **49.** $3x - 2$
51. $\frac{1}{3w(w - 9)}$ **53.** $\frac{1}{a + 2}$ **55.** $\frac{2}{c + d}$ **57.** $\frac{7n - 1}{(n + 4)(n - 2)}$

59. $\frac{t-5}{t^2-2t+1}$ **61.** $6x$ **63.** $3a^2b^3$ **65.** $c(c+2)$

67. $30a^3$ **69.** $12(b+2)$ **71.** $(x+1)(x-1)$

73. $(3x+1)(3x-1)$ **75.** $(x+1)(x+5)(x-5)$

77. $(2n+5)(n+4)^2$ **79.** $\frac{125x}{20x}$ **81.** $\frac{8xy}{x^2y}$ **83.** $\frac{3x^2+3x}{(x+1)^2}$

85. $\frac{2xy+6y}{x(x+3)}$ **87.** $\frac{30}{3(b-1)}$ **89.** $\frac{5t+25}{20(t+2)}$

91. $\frac{4y^2+12y}{4y(y-2)(y-3)}$ **93.** $\frac{36-3h}{3(h+9)(h-9)}$

95. $\frac{6t^2-12t}{(t+1)(t-2)(t+3)}$ **97.** in 90 minutes   **105.** $I=Prt$

107. $A=\frac{1}{2}bh$ **109.** $d=rt$

Study Set Section 6.4 (page 451)

1. unlike   **3. a.** $2\cdot2\cdot5\cdot x\cdot x$ **b.** $(x-2)(x+6)$

5. $(x+6)(x+3)$ **7.** $\frac{5}{5}$ **9. a.** $\frac{5n}{5n}$ **b.** $\frac{3}{3}$ **11.** $8x-24$

13. $7x+10$ **15.** $-x+10=10-x$ **17.** -1 **19.** $\frac{x}{1}$

21. yes, no, yes   **23.** $\frac{13x}{21}$ **25.** $\frac{5y}{9}$ **27.** $\frac{41}{30x}$ **29.** $\frac{7-2m}{m^2}$

31. $\frac{4xy+6x}{3y}$ **33.** $\frac{16c^2+3}{18c^4}$ **35.** $\frac{3y-3}{16}$ **37.** $\frac{2n+2}{15}$

39. $\frac{-9b+28}{42}$ **41.** $\frac{x^3-x^2-9x+18}{9x^3}$ **43.** $\frac{y^2+7y+6}{15y^2}$

45. $\frac{x^2+4x+1}{x^2y}$ **47.** $\frac{7}{24}$ **49.** $\frac{a^2+2a+b^2-2b}{ab}$

51. $\frac{2x^2-1}{x(x+1)}$ **53.** $\frac{35x^2+x+5}{5x(x+5)}$ **55.** $\frac{17t+42}{(t+3)(t+2)}$

57. $\frac{2x^2+11x}{(2x-1)(2x+3)}$ **59.** $\frac{4a+1}{(a+2)^2}$ **61.** $\frac{4m^2+10m+6}{(m-2)(m+5)}$

63. $\frac{14s+58}{(s+3)(s+7)}$ **65.** $\frac{12a+26}{(a+1)(a+3)^2}$ **67.** $-\frac{4c+6}{(2c+1)(c-3)}$

69. $\frac{b-1}{2(b+1)}$ **71.** $\frac{1}{a+1}$ **73.** $\frac{3}{s-3}$ **75.** $\frac{x+2}{x-2}$

77. $-\frac{2}{x-3}$ **79.** $\frac{17h-2}{12(h-2)(h+2)}$ **81.** $\frac{1}{(y+3)(y+4)}$

83. $\frac{s^2+8s+4}{(s+4)(s+1)(s+1)}$ **85.** $\frac{2x+13}{(x-8)(x-1)(x+2)}$

87. $-\frac{1}{2(x-2)}$ **89.** $\frac{6x+8}{x}$ **91.** $\frac{a^2b-3}{a^2}$ **93.** $\frac{x^2-4x+9}{x-4}$

95. $-\frac{4x+3}{x+1}$ **97.** $-\frac{2}{a-4}$ **99.** $\frac{1}{r+2}$ **101.** $\frac{2y+7}{y-1}$

103. $\frac{20x+9}{6x^2}$ cm   **109.** $8;(0,2)$ **111.** 0

Study Set Section 6.5 (page 460)

1. complex, complex   **3.** reciprocal   **5.** single, reciprocal

7. a. $\frac{x-3}{4}$ **b.** yes   **c.** $\frac{1}{12}-\frac{x}{6}$ **d.** no

9. a. $y,3,6,$ and y **b.** $6y$ **c.** $\frac{6y}{6y}$ **11.** $2x$ **13.** $4y^2$

15. $\frac{4x^2}{15}\Big/\frac{16x}{25}$ **17.** $\frac{5x}{12}$ **19.** $\frac{8}{9}$ **21.** $\frac{x^2}{y}$ **23.** $\frac{n^3}{8}$ **25.** $\frac{1-3x}{5+2x}$

27. $\frac{5}{4}$ **29.** $\frac{10}{3}$ **31.** $18x$ **33.** $\frac{2-5y}{6}$ **35.** $\frac{6d+12}{d}$

37. $\frac{1}{2}$ **39.** $\frac{5}{7}$ **41.** $\frac{x^3}{14}$ **43.** $\frac{s^2-s}{2+2s}$ **45.** $-\frac{x^2}{2}$ **47.** $\frac{1+x}{2+x}$

49. $\frac{14x}{9}$ **51.** $-\frac{t}{32}$ **53.** $\frac{8}{4c+5c^2}$ **55.** $\frac{b-5ab}{3ab-7a}$

57. $\frac{3-x}{x-1}$ **59.** $\frac{b+9}{8a}$ **61.** $\frac{32h-1}{96h+6}$ **63.** $\frac{xy}{y+x}$

65. $\frac{1}{x+2}$ **67.** $\frac{1}{x+3}$ **69.** $\frac{5t^2}{27}$ **71.** $\frac{m-2}{6}$ **73.** $\frac{y}{x-2y}$

75. $\frac{m^2+n^2}{m^2-n^2}$ **77.** $\frac{x}{x-2}$ **79.** $\frac{3}{14}$ **81.** $\frac{R_1R_2}{R_2+R_1}$

87. 1 **89.** $\frac{81}{256r^8}$ **91.** $\frac{r^{10}}{9}$

Study Set Section 6.6 (page 469)

1. rational   **3.** clear   **5.** quadratic   **7. a.** yes   **b.** no

9. a. $3,0$ **b.** $3,0$ **c.** $3,0$ **11.** y **13.** $(x+8)(x-8)$

15. $(x+2)(x-2)$ **17.** $12,1,10x$

19. $2a,2a,2a,2a,2a,4,7,3$ **21.** multiplication;

$fpq=f\cdot p\cdot q$ **23.** 1 **25.** $\frac{3}{5}$ **27.** 7 **29.** 3 **31.** no

solution; 5 is extraneous   **33.** $-4,4$ **35.** 1 **37.** no

solution; 0 is extraneous   **39.** $-3,3$ **41.** -12 **43.** -40

45. -48 **47.** $-\frac{5}{9}$ **49.** $-\frac{12}{7}$ **51.** $\frac{11}{4}$ **53.** -1 **55.** 6

57. 1 **59.** 4 **61.** no solution; -2 is extraneous

63. $2,-5$ **65.** $-4,3$ **67.** $-2,1$ **69.** 2 **71.** 0

73. $3,-4$ **75.** $1,-9$ **77.** -5 **79.** $P=nrt$

81. $d=\frac{bc}{a}$ **83.** $A=\frac{h(b+d)}{2}$ **85.** $a=\frac{b}{b-1}$

87. $r=\frac{E-IR}{I}$ **89.** $r=\frac{st}{s-t}$ **91.** $L^2=6dF-3d^2$

93. $R=\frac{HB}{B-H}$ **95.** $r=\frac{r_1r_2}{r_2+r_1}$ **101.** $x(x+4)$

103. $(2x+3)(x-1)$ **105.** $(x^2+9)(x+3)(x-3)$

Study Set Section 6.7 (page 479)

1. distance, rate, time   **3.** interest, principal, rate, time

5. iii   **7.** $t=\frac{d}{r}$ **9. a.** 0.09 **b.** 3.5% **11.** $\frac{1}{3}$ **13.** $\frac{x}{4}$

15. $\frac{50}{r},\frac{75}{r-0.02}$ **17.** $6\frac{1}{9}$ days   **19.** 4 **21.** 2 **23.** 5

25. $\frac{2}{3},\frac{3}{2}$ **27.** 8 **29.** $2\frac{2}{9}$ hr   **31.** $4\frac{4}{9}$ hr   **33.** 8 hr

35. 20 min   **37.** $1\frac{4}{5}$ hr $=1.8$ hr   **39.** 4 mph   **41.** 1st: $1\frac{1}{2}$

ft per sec; 2nd: $\frac{1}{2}$ ft per sec   **43.** $\frac{300}{255+x},\frac{210}{255-x}$, 45 mph

45. Credit union: 4%; bonds: 6%   **47.** 7% and 8%

51. repeating   **53.** $\{\ldots,-4,-3,-2,-1,0,1,2,3,4,\ldots\}$

55. 0

Study Set Section 6.8 (page 489)

1. ratio, rate   **3.** terms   **5.** cross   **7.** unit   **9.** $\frac{6}{5}$

11. a. $60,60$ **b.** $15x,90$ **13.** $25,2,1,000,x$ **15.** 24

twelve-oz bottles   **17.** BC,EF,DE **19.** $x,288,18,18,16$

21. a. $\frac{4}{5}$ **b.** $\frac{9}{7}$ **23.** similar   **25.** $\frac{4}{15}$ **27.** $\frac{5}{4}$ **29.** $\frac{1}{2}$

31. $\frac{13}{8}$ **33.** no   **35.** yes   **37.** no   **39.** 4 **41.** 6

43. 3 **45.** 0 **47.** -8 **49.** $-\frac{3}{2}$ **51.** 2 **53.** $-\frac{1}{5}$

55. -27 **57.** $2,-2$ **59.** $6,-1$ **61.** $-\frac{1}{3},2$ **63.** 15

65. 8 **67. a.** $\frac{3}{2},3{:}2$ **b.** $\frac{2}{3},2{:}3$ **69.** $\$62.50$ **71.** 20

73. 84 **75.** $568,13,14$ **77.** 510 **79.** not exactly, but

close   **81.** 140 **83.** 65 ft, 3 in.   **85.** 10 ft   **87.** 45

minutes for $25   **89.** 150 for $12.99   **91.** 6-pack for $1.50

93. 39 ft   **95.** $46\frac{7}{8}$ ft   **101.** 90%   **103.** 480

Key Concept (page 495)

1. a. $\frac{2x}{x-2}$ **b.** $x-4$ **2. a.** $x+1$ **b.** $x,x-1,x+1$

3. a. $\frac{2x^2-1}{x(x+1)}$ **b.** $\frac{x}{x},\frac{x+1}{x+1}$ **4. a.** $\frac{3(n^2-n-2)}{n^2}$ **b.** $3n$

5. a. 2 **b.** b **6. a.** 5 **b.** $(s+2)(s-1)$ **7. a.** $2,4$

b. $4(y-6)$ **8. a.** $b=\frac{a}{1-a}$ **b.** ab

Chapter Review (page 496)

1. $4, -4$ **2.** $-\frac{3}{7}$ **3.** $\frac{1}{2x}$ **4.** $\frac{5}{2x}$ **5.** $\frac{x}{x+1}$ **6.** $a-2$

7. -1 **8.** $-\frac{1}{x+3}$ **9.** $\frac{x}{x-1}$ **10.** does not simplify

11. $\frac{4}{3}$ **12.** x is not a common factor of the numerator and the denominator; x is a term of the numerator.   **13.** 150 mg

14. $\frac{3x}{y}$ **15.** 96   **16.** $\frac{x-1}{x+2}$ **17.** $\frac{2x}{x+1}$ **18.** $\frac{3y}{2}$

19. $\frac{1}{x(x-1)}$ **20.** $-x-2$ **21. a.** yes   **b.** no   **c.** yes

d. yes   **22.** $\frac{1}{3}$ mile per minute   **23.** $\frac{1}{3d}$ **24.** 1

25. $\frac{2x+2}{x-7}$ **26.** $\frac{1}{a-4}$ **27.** $9x$ **28.** $8x^3$ **29.** $m(m-8)$

30. $(5x+1)(5x-1)$ **31.** $(a+5)(a-5)$

32. $(2t+7)(t+5)^2$ **33.** $\frac{63}{7a}$ **34.** $\frac{2xy+x}{x(x-9)}$ **35.** $\frac{2b+14}{6(b-5)}$

36. $\frac{9r^2-36r}{(r+1)(r-4)(r+5)}$ **37.** $\frac{a-7}{7a}$ **38.** $\frac{x^2+x-1}{x(x-1)}$

39. $\frac{1}{t+1}$ **40.** $\frac{x^2+4x-4}{2x^2}$ **41.** $\frac{b+6}{b-1}$ **42.** $\frac{6c+8}{c}$

43. $\frac{14n+58}{(n+3)(n+7)}$ **44.** $\frac{4t+1}{(t+2)^2}$ **45.** $\frac{1}{(a+3)(a+2)}$

46. $\frac{17y-2}{12(y-2)(y+2)}$ **47.** yes   **48.** $\frac{14x+28}{(x+6)(x-1)}$ units,

$\frac{12}{(x+6)(x-1)}$ square units   **49.** $\frac{n^3}{14}$ **50.** $\frac{r+9}{8s}$ **51.** $\frac{1+y}{1-y}$

52. $\frac{21}{3a+10a^2}$ **53.** x^2+3 **54.** $\frac{y-5xy}{3xy-7x}$ **55.** 3   **56.** no

solution; 5 is extraneous   **57.** 3   **58.** 2, 4   **59.** 0

60. $-4, 3$ **61.** $T_1 = \frac{T_2}{1-E}$ **62.** $y = \frac{xz}{z-x}$ **63.** 3

64. 5 mph   **65.** $\frac{1}{4}$ of the job per hour   **66.** $5\frac{5}{6}$ days

67. 5%   **68.** 40 mph   **69.** no   **70.** yes   **71.** $\frac{9}{2}$ **72.** 0

73. 7   **74.** $4, -\frac{3}{2}$ **75.** 255   **76.** 20 ft   **77.** 5 ft 6 in.

78. 250 for $98

Chapter 6 Test (page 502)

1. 0   **2.** $-3, 2$ **3.** 10 words   **4.** 12,168 yd^2 **5.** $\frac{8x}{9y}$

6. -7 **7.** $\frac{x+1}{2x+3}$ **8.** 1   **9.** $3c^2d^3$

10. $(n+1)(n+5)(n-5)$ **11.** $\frac{5y^2}{4}$ **12.** $\frac{x+1}{3(x-2)}$

13. $-\frac{x^2}{3}$ **14.** $\frac{1}{6}$ **15.** $\frac{13}{2y+3}$ **16.** $\frac{4n-5mn}{10m}$

17. $\frac{2x^2+x+1}{x(x+1)}$ **18.** $\frac{2a+7}{a-1}$ **19.** $\frac{c^2-4c+9}{c-4}$

20. $\frac{1}{(t+3)(t+2)}$ **21.** $\frac{12}{5m}$ **22.** $\frac{a+2s}{2as^2-3a^2}$ **23.** 11

24. no solution, 6 is extraneous   **25.** 1   **26.** 1, 2   **27.** $\frac{2}{3}$

28. $B = \frac{HR}{R-H}$ **29.** yes   **30.** 900   **31.** 171 ft

32. 80 sheets for $3.89   **33.** $3\frac{15}{16}$ hr   **34.** 4 mph   **35.** 5 is

not a common factor of the numerator and denominator.

36. We multiply both sides of the equation by the LCD of the rational expressions appearing in the equation. The resulting equation is easier to solve.

Cumulative Review Exercises Chapters 1–6 (page 504)

1. 36   **2.** about 26%   **3.** $<$ **4.** 77   **5.** 104°F

6. 240 ft^3 **7. a.** false   **b.** false   **c.** true   **d.** true

8. 3   **9.** $6c+62$ **10.** $B = \frac{A-c-r}{2}$

11. $x \geq -1, [-1, \infty)$ **12.** -5

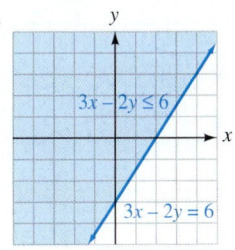

13. 12 lb of the $3.20 tea and 8 lb of the $2 tea   **14.** 500 mph

15.   **16.**

17. $\frac{8}{7}$ **18.** -1 **19.** $y = 3x+2$ **20.** 0.008 mm/m

21. x^7 **22.** x^{25} **23.** $\frac{y^3}{8}$ **24.** $-\frac{32a^5}{b^5}$ **25.** $\frac{a^8}{b^{12}}$ **26.** $3b^2$

27. 2.9×10^5 **28.** 3   **29.** $A = \pi R^2 - \pi r^2$

30.

31. $6x^2 + x - 5$ **32.** $6x^5y^5$

33. $6y^2 - y - 35$

34. $-12x^4z + 4x^2z^2$

35. $2x+3$ **36.** x^2+2x-1

37. $k^2t(k-3)$

38. $(b+c)(2a+3)$

39. $2(a+10b)(a-10b)$

40. $(b+5)(b^2-5b+25)$

41. $(u-9)^2$ **42.** $(2x-9)(3x+7)$ **43.** $-(r-2)(r+1)$

44. prime   **45.** $0, -\frac{1}{5}$ **46.** $\frac{1}{3}, \frac{1}{2}$ **47.** 16 in., 10 in.

48. $5, -5$ **49.** $\frac{x}{3}$ **50.** $-\frac{1}{x-3}$ **51.** $\frac{m-3}{(m+3)(m+1)}$

52. $\frac{x^2}{(x+1)^2}$ **53.** $\frac{17}{25}$ **54.** 2   **55.** $14\frac{2}{5}$ hr   **56.** 34.8 ft

Study Set Section 7.1 (page 512)

1. system   **3.** y-coordinate   **5.** intersection

7. consistent, inconsistent   **9.** $(-4, -1)$ **11.** true

13. false   **15.** $3, 3, (0, -2)$ **17.** no solution; independent

19. is possibly equal to   **21.** yes   **23.** yes   **25.** no

27. no   **29.** yes

31.   **33.**

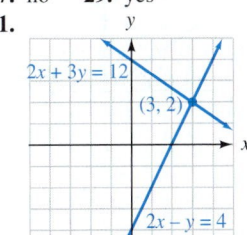   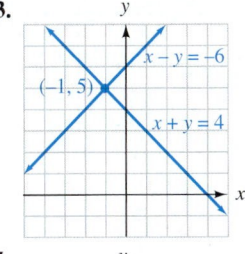

35.   **37.**

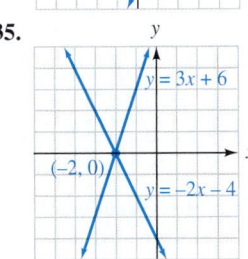   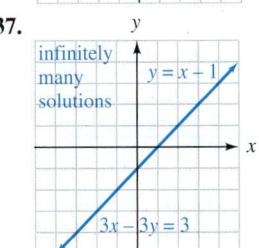

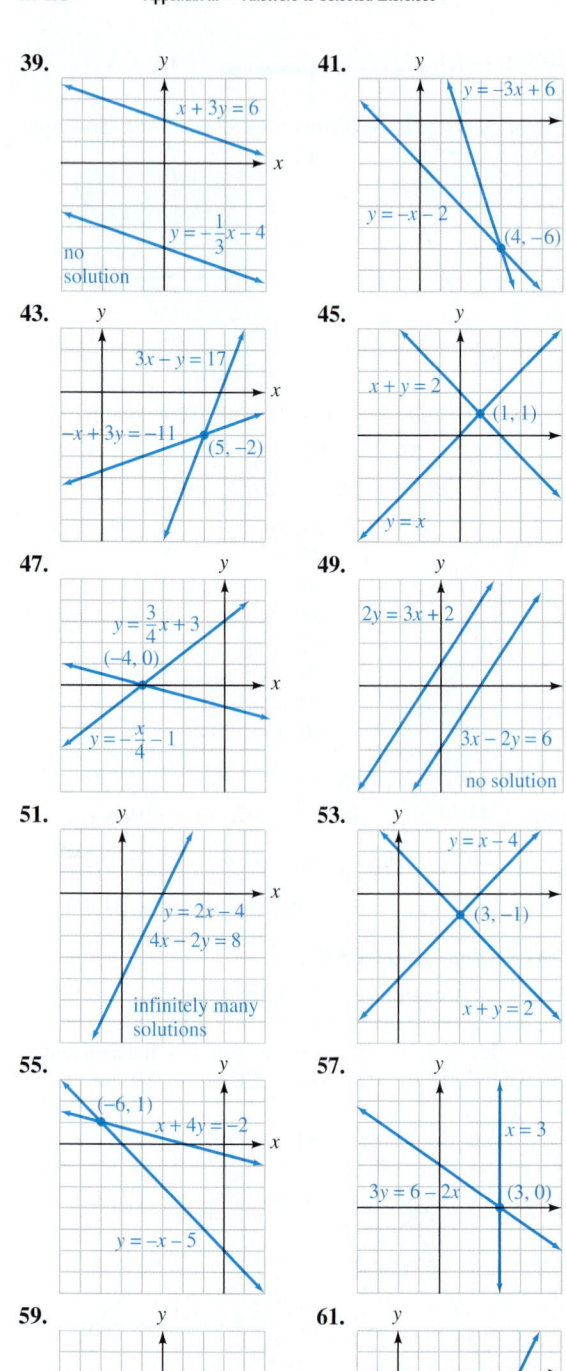

39. $x + 3y = 6$; $y = -\frac{1}{3}x - 4$; no solution

41. $y = -3x + 6$; $y = -x - 2$; $(4, -6)$

43. $3x - y = 17$; $-x + 3y = -11$; $(5, -2)$

45. $x + y = 2$; $(1, 1)$; $y = x$

47. $y = \frac{3}{4}x + 3$; $(-4, 0)$; $y = -\frac{x}{4} - 1$

49. $2y = 3x + 2$; $3x - 2y = 6$; no solution

51. $y = 2x - 4$; $4x - 2y = 8$; infinitely many solutions

53. $y = x - 4$; $(3, -1)$; $x + y = 2$

55. $(-6, 1)$; $x + 4y = -2$; $y = -x - 5$

57. $x = 3$; $3y = 6 - 2x$; $(3, 0)$

59. $-x + 2y = -4$; $(-2, -3)$; $y = -3$

61. $x + 2y = -4$; $(4, -4)$; $x - \frac{1}{2}y = 6$

63. 1 solution   **65.** same line; infinitely many solutions
67. no solution   **69.** 1 solution   **71.** $(1, 3)$ **73.** no
solution   **75.** 1994; 4,100   **77. a.** Houston, New Orleans,
St. Augustine   **b.** St. Louis, Memphis, New Orleans

c. New Orleans   **79. a.** the incumbent; 7%   **b.** November 2
c. the challenger; 3   **81.** 10 mi   **87.** $\frac{x^2 + 7x + 6}{2x^2}$ **89.** z

Study Set Section 7.2 (page 523)

1. system   **3.** original   **5.** substitute   **7.** one
9. infinitely   **11. a.** $y = -3x$ **b.** $x = 2y + 1$
13. a. $x = 2 + 2y$ **b.** $y = -11 + 2x$
15. $x + 3(x - 4) = 8$ **17. a.** 15   **b.** $3x + 10y = 15$
19. a. no   **b.** ii   **21.** $3x, 4, -2, -2, -6, -2, -6$
23. $(2, 4)$ **25.** $(3, 0)$ **27.** $(-3, -1)$ **29.** $(-1, -1)$
31. no solution   **33.** $(-2, 3)$ **35.** $(3, -2)$ **37.** $\left(-4, \frac{5}{4}\right)$
39. $(3, 2)$ **41.** $\left(\frac{2}{3}, -\frac{1}{3}\right)$ **43.** infinitely many solutions
45. $(-2, 0)$ **47.** $(-2, -3)$ **49.** $(3, -2)$ **51.** $(-1, -1)$
53. no solution   **55.** $\left(\frac{1}{2}, \frac{1}{3}\right)$ **57.** $(4, -2)$ **59.** $(-6, 4)$
61. $\left(\frac{1}{5}, 4\right)$ **63.** $\left(10, \frac{15}{2}\right)$ **65.** $(9, 11)$ **67.** infinitely many
solutions   **69.** $(-4, -1)$ **71.** $(4, 2)$ **73.** melon
79. yes   **81.** yes   **83.** no

Study Set Section 7.3 (page 532)

1. opposites   **3.** eliminated   **5.** addition, equal   **7.** $7y$
and $-7y$ **9. a.** $12x + 3y = 6$ **b.** $-2x + 6y = -8$
11. 1   **13.** yes   **15.** $2x, 1, 1, (1, 4)$ **17.** $(3, 2)$
19. $(3, -2)$ **21.** $(-2, -3)$ **23.** $(0, 8)$ **25.** $(-3, 4)$
27. $(0, -2)$ **29.** infinitely many solutions   **31.** $(-12, 1)$
33. $(-5, 10)$ **35.** $(3, 11)$ **37.** no solution   **39.** $\left(\frac{1}{3}, 3\right)$
41. $\left(\frac{1}{3}, 7\right)$ **43.** $(-1, -1)$ **45.** $(1, 1)$ **47.** $\left(1, -\frac{5}{2}\right)$
49. $(-2, 5)$ **51.** $(8, -3)$ **53.** $(3, 0)$ **55.** $\left(\frac{13}{75}, \frac{14}{75}\right)$
57. $(10, 9)$ **59.** $(-4, -1)$ **61.** $(6, 8)$ **63.** no solution
65. $(-1, 2)$ **67.** $(3, 2)$ **69.** $(1, -1)$ **71.** infinitely many
solutions   **73.** $\left(\frac{10}{3}, \frac{10}{3}\right)$ **75.** $(5, -6)$ **77.** 1991
83. $x - 10$ **85.** 0   **87.** 7.5 ft^2

Study Set Section 7.4 (page 545)

1. variable   **3.** system   **5.** complementary
7. $x + y = 20, y = 2x - 1$ **9.** $x + y = 180, y = x - 25$
11. $5x + 2y = 10$ **13. a.** $(x + c)$ mph   **b.** $(x - c)$ mph
15. a. $(x + y)$ mL   **b.** 33%   **17.** 20°, 70°   **19.** 50°, 130°
21. 22 ft, 29 ft   **23.** president: $400,000; vice president:
$192,600   **25.** 65°, 115°   **27.** 96 ft, 70 ft   **29.** 15 m, 10 m
31. $30, $10   **33.** 85, 63   **35. a.** 400 tires
b.

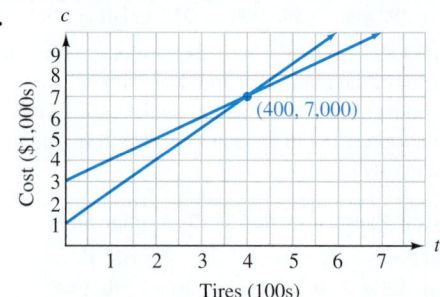

c. the second mold   **37.** 125 min   **39.** 25 mph, 5 mph

41. 180 mph, 20 mph   **43.** nursing: $2,000; business: $3,000
45. 4 L 6% salt water, 12 L 2% salt water   **47.** 52 lb $8.75,
48 lb $3.75   **51.** $(-\infty, 4)$

53. $(-1, 2]$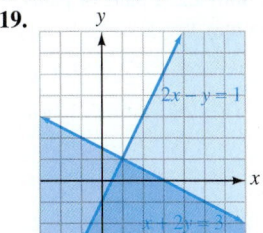

Study Set Section 7.5 (page 556)

1. inequalities   **3.** solve   **5.** intersection
7. a. $3x - y = 5$ **b.** dashed   **9.** slope: $4 = \frac{4}{1}$; y-intercept:
$(0, -3)$ **11. a.** no   **b.** above   **13. a.** yes   **b.** no
c. no   **15. a.** ii   **b.** iii   **c.** iv   **d.** i   **17.** ABC

19.   **21.**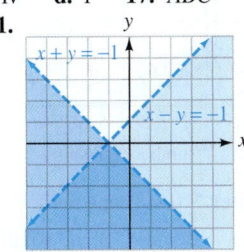

23. 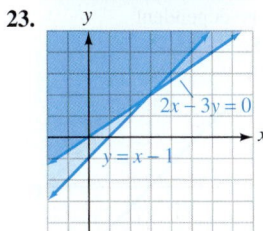   **25.**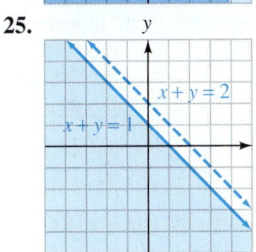

27. 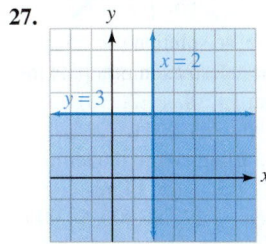   **29.**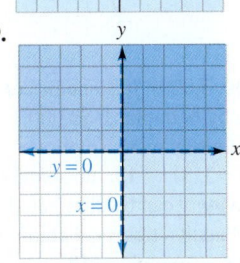

31. 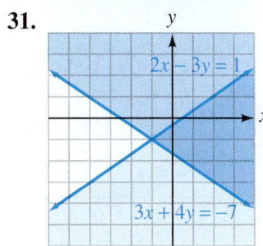   **33.**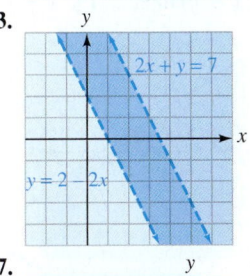

35. 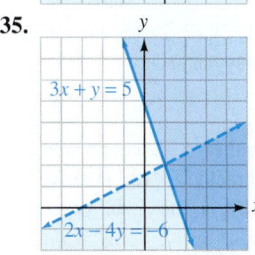   **37.**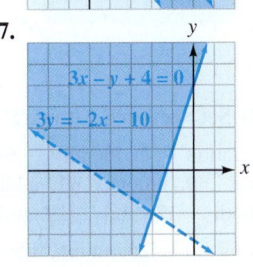

39. 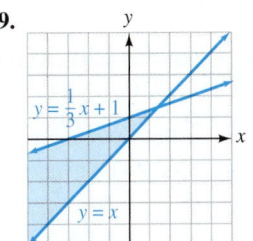   **41.**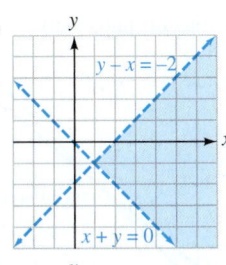

43. 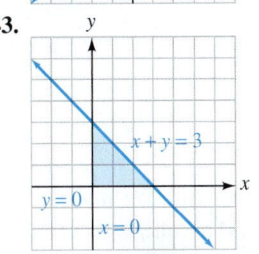   **45.**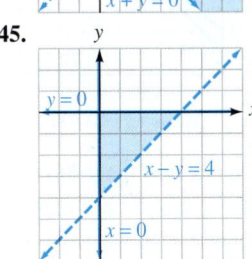

47. (c)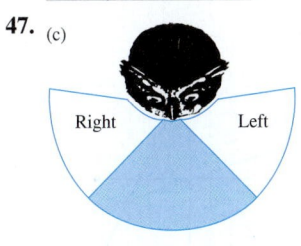

49.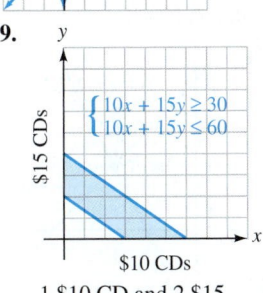
1 $10 CD and 2 $15
CDs; 4 $10 CDs and
1 $15 CD

2 desk chairs and 4 side chairs; 1
desk chair and 5 side chairs

51.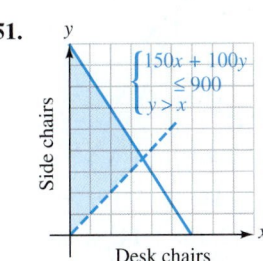

53. 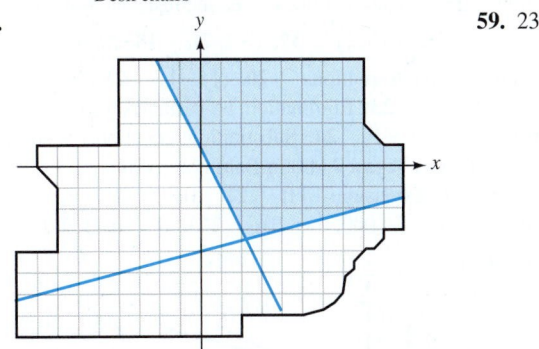   **59.** 230

61. 0.000976   **63.** 2.9×10^5 **65.** 5.1×10^{-6}

Key Concept (page 560)

1.

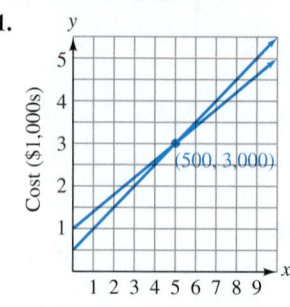

(500, 3,000); 500 meals, $3,000
2. (5, 1)   **3.** (−1, −1)
4. (3, −2)   **5.** (−3, 4)

Chapter Review (page 561)

1. yes   **2.** yes   **3.** (1978, 5,600,000); In 1978, the same number of men as women were enrolled in college, about 5.6 million of each.

4.

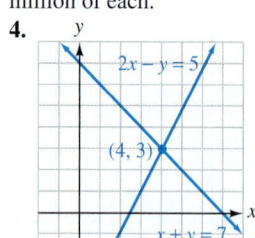

5.

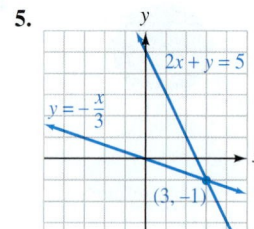

6.

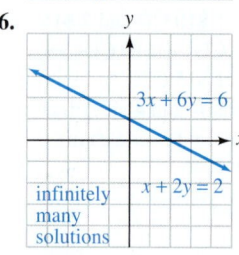

7.

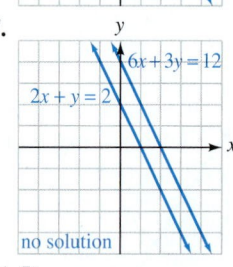

8. (5, 0)   **9.** (3, 3)   **10.** $\left(-\frac{1}{2}, \frac{7}{2}\right)$ **11.** (1, −2)

12. infinitely many solutions   **13.** (12, 10)   **14. a.** no solution   **b.** two parallel lines   **c.** inconsistent system

15. (3, −5)   **16.** $\left(3, \frac{1}{2}\right)$ **17.** (−1, 7)   **18.** (0, 9)

19. infinitely many solutions   **20.** (1, −1)   **21.** (−5, 2)

22. no solution   **23.** Elimination; no variables have a coefficient of 1 or −1   **24.** Substitution; equation 1 is solved for x.   **25.** Las Vegas: 2,000 ft; Baltimore: 100 ft

26. base: 21 ft; extension: 14 ft   **27.** 10,800 yd^2

28. a. $y = 5x + 30,000$, $y = 10x + 20,000$ **b.** 2,000

c. the athlete

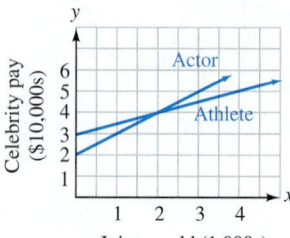

29. 12 lb worms, 18 lb bears   **30.** 3 mph

31. $16.40, $10.20   **32.** $750   **33.** $13\frac{1}{3}$ gal 40%, $6\frac{2}{3}$ gal 70%

34.

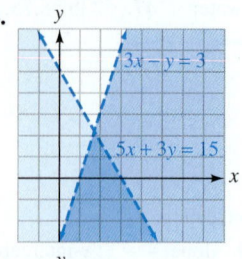

35.

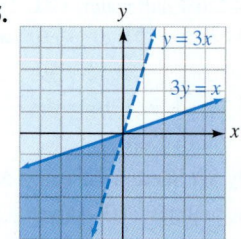

36.

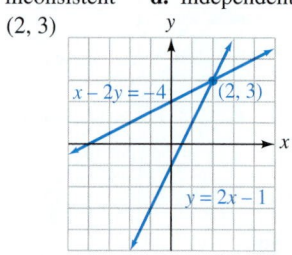

$10x + 20y \geq 40$, $10x + 20y \leq 60$; (3, 1): 3 shirts and 1 pair of pants; (1, 2): 1 shirt and 2 pairs of pants (answers may vary)

Chapter 7 Test (page 565)

1. yes   **2.** no   **3. a.** solution   **b.** consistent
c. inconsistent   **d.** independent   **e.** dependent
4. (2, 3)

5.

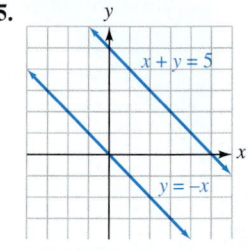

no solution

6. (30, 3,000); if 30 items are sold, the salesperson gets paid the same by both plans, $3,000.   **7.** Plan 1   **8.** (−2, −3)

9. infinitely many solutions   **10.** (−1, −1)   **11.** (2, 4)

12. (−3, 3)   **13.** no solution   **14.** Elimination; the terms involving y can be eliminated easily.   **15.** 8 mi, 14 mi

16. 3 adults' tickets; 4 children's tickets   **17.** $6,000, $4,000

18. $(x - c)$ mph   **19.** $x + y = 90$

20.

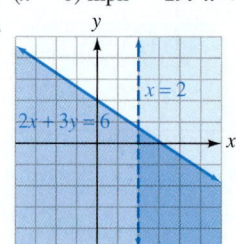

21.

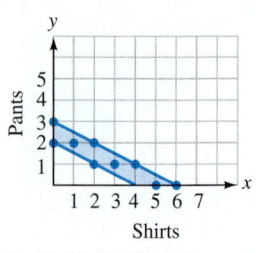

(1, 2), (2, 2), (3, 1) (answers may vary)

Study Set Section 8.1 (page 579)

1. equation   **3.** identity   **5.** inequality   **7.** compound, double   **9.** union   **11.** added, subtracted   **13.** positive

15. both   **17.** and   **19.** union   **21. a.** no   **b.** yes

23. [−2, 1)   **25.** union, intersection   **27.** 6   **29.** $\frac{2}{3}$

31. 4   **33.** 2   **35.** 6   **37.** -4 **39.** -11
41. $\varnothing$; contradiction   **43.** -8 **45.** 2   **47.** -2
49. $\mathbb{R}$; identity   **51.** 24   **53.** 6   **55.** $\frac{21}{19}$
57. $\varnothing$; contradiction   **59.** -5 **61.** $B = \frac{3V}{h}$
63. $w = \frac{p - 2l}{2}$ **65.** $x = z\sigma + \mu$ **67.** $\mu = x - z\sigma$
69. $l = \frac{2S - na}{n}$ or $l = \frac{2S}{n} - a$ **71.** $(2, \infty)$

73. $[-2, \infty)$ **75.** $(-\infty, 1]$

77. $(6, \infty)$ **79.** $\left(-\infty, -\frac{8}{5}\right)$

81. $[-36, \infty)$ **83.** $(-\infty, 6]$

85. $(-2, 5]$ **87.** $(2, 3]$

89. $[5, \infty)$ **91.** $[1, 4]$

93. $[-21, -3)$

95. $(-2, 5)$ **97.** $[-2, 4]$

99. $(-\infty, -2] \cup (6, \infty)$

101. $(-\infty, -1) \cup (2, \infty)$

103. $(-\infty, 2) \cup (7, \infty)$ **105.** 20

107. 20 ft by 45 ft   **109.** $p < \$15$ **111.** 18
113. a. 128, 192   **b.** $32 \leq s \leq 48$ **117.** $\frac{1}{t^{12}}$

Study Set Section 8.2 (page 591)

1. absolute value   **3.** inequality   **5.** isolate
7. solution(s)   **9.** negative   **11.** more than   **13.** 5
15. a. $-2, 2$ **b.** $-1.99, -1, 0, 1, 1.99$
c. $-4, -3, -2.01, 2.01, 3, 4$ **17. a.** $x - 7, 8, x - 7, -8$
b. $x + 10, x - 3, x + 10, -(x - 3)$ **19. a.** $x = 8$ or
$x = -8$ **b.** $x \leq -8$ or $x \geq 8$ **c.** $-8 \leq x \leq 8$
d. $5x - 1 = x + 3$ or $5x - 1 = -(x + 3)$ **21. a.** ii
b. iii   **c.** i   **23.** $(-\infty, -1) \cup (3, \infty)$ **25.** $23, -23$
27. $4, -4$ **29.** $9.1, -2.9$ **31.** $\frac{14}{3}, -6$ **33.** $12, -12$
35. no solution   **37.** $2, -\frac{1}{2}$ **39.** -8 **41.** $-4, -28$
43. $40, -20$ **45.** $\frac{12}{11}, -\frac{12}{11}$ **47.** $-2, -\frac{4}{5}$ **49.** $0, -2$
51. 0   **53.** $\frac{4}{3}$
55. $(-4, 4)$

57. $[-21, 3]$

59. $\left(-\frac{8}{3}, 4\right)$ **61.** no solution

63. $(-\infty, -3) \cup (3, \infty)$

65. $(-\infty, -12) \cup (36, \infty)$

67. $\left(-\infty, -\frac{16}{3}\right) \cup (4, \infty)$

69. $(-\infty, \infty)$

71. $(-\infty, -2] \cup \left[\frac{10}{3}, \infty\right)$

73. $(-\infty, -2) \cup (5, \infty)$

75. $[-10, 14]$

77. $\left(-\frac{5}{3}, 1\right)$

79. $(-\infty, -24) \cup (-18, \infty)$

81. no solution   **83.** $70° \leq t \leq 86°$
85. a. $|c - 0.6°| \leq 0.5°$ **b.** $[0.1°, 1.1°]$
87. a. 26.45%, 24.76%   **b.** It is less than or equal to 1%.
93. $50°, 130°$ **95.** $k < 0$ **97.** x and y must have different
signs.

Study Set Section 8.3 (page 602)

1. factored   **3.** greatest common factor   **5.** completely,
prime   **7.** trinomial   **9.** lead, coefficient   **11.** $6xy^2$
13. $9, 6, -9, -6$ **15.** $-1 + (-12) = -13, -2(-6) = 12,$
$-2 + (-6) = -8, -3 + (-4) = -7$ **17.** $5c^2d^4$
19. $3m, 3$ **21.** $2x(x - 3)$ **23.** $5x^2y(3 - 2y)$
25. $3z(9z^2 + 4z + 1)$ **27.** prime   **29.** $-(5xy - y + 4)$
31. $6s(4s^2 - 2st + t^2)$ **33.** $(x + y)(u + v)$
35. $-6ab(3a + 2b)$ **37.** $\frac{1}{5}x^2(3ax^2 + b - 4ax)$
39. $(a - b)(5 - t)$ **41.** $(m + n + p)(3 + x)$
43. $-7u^2v^3z^2(9uv^3z^7 - 4v^4 + 3uz^2)$ **45.** $2(x^2 + 1)^2(x^2 + 3)$
47. $r_1 = \frac{rr_2}{r_2 - r}$ **49.** $r = \frac{S - a}{S - \ell}$ **51.** $a^2 = \frac{b^2x^2}{b^2 - y^2}$
53. $(x + y)(a + b)$ **55.** $(x + y)(x + 1)$
57. $(a + b)(a - 4)$ **59.** $(x - y)(x - 4)$
61. $(a^2 + b)(x - 1)$ **63.** $(x + y)(x + y + z)$
65. $(1 - m)(1 - n)$ **67.** $(2x^2 + 1)(a - 4)$
69. $x(m + n)(p + q)$ **71.** $y(x + y)(x + y + 2z)$
73. $(x - 3)(x - 2)$ **75.** $(x - 2)(x - 5)$ **77.** prime
79. $(x + 5)(x - 6)$ **81.** $(a - 9)^2$ **83.** $(x + 3y)(x - 7y)$
85. $(s - 2t)(s - 8t)$ **87.** $3(x + 7)(x - 3)$
89. $-(a - 8)(a + 4)$ **91.** $-3a^2(x - 3)(x - 2)$
93. $(y^2 - 10)(y^2 - 3)$ **95.** $x^2(b^2 - 7)(b^2 - 5)$
97. $(3y + 2)(2y + 1)$ **99.** $(4a - 3)(2a + 3)$
101. $(3x - 4y)(2x + y)$ **103.** prime
105. $(2z + 3)(3z + 4)$ **107.** $(2y + 1)^2$
109. $-(3a + 2b)(a - b)$ **111.** $5(4a^2 + 9ab + 12b^2)$
113. $4h^4(8h - 1)(2h + 1)$ **115.** $(m + n)(6a - 5)(a + 3)$
117. $(x + a + 1)^2$ **119.** $(a + b + 4)(a + b - 6)$
121. $(7q - 7r + 2)(2q - 2r - 3)$ **123.** $\pi r^2\left(h_1 + \frac{1}{3}h_2\right)$
125. $x + 3$ **129.** $\$4,900$

Study Set Section 8.4 (page 610)

1. squares   **3.** 1, 4, 9, 16, 25, 36, 49, 64, 81, 100

5. a. $(x + 5)(x + 5) = x^2 + 10x + 25$

b. $(x + 5)(x - 5) = x^2 - 25$ **7.** by the comm. property of mult., $(9t - 4)(9t + 4) = (9t + 4)(9t - 4)$ **9.** $(p - q)$

11. pq **13.** $(p^2 + pq + q^2)$ **15. a.** $x^2 - 4$ **b.** $(x - 4)^2$

c. $x^2 + 4$ **d.** $x^3 + 8$ **e.** $(x + 8)^3$ **17.** $(x + 2)(x - 2)$

19. $(3y + 8)(3y - 8)$ **21.** prime   **23.** $(20 + c)(20 - c)$

25. $(25a + 13b^2)(25a - 13b^2)$ **27.** $(9a^2 + 7b)(9a^2 - 7b)$

29. $(6x^2y + 7z^2)(6x^2y - 7z^2)$ **31.** $(x + y + z)(x + y - z)$

33. $(a - b + c)(a - b - c)$ **35.** $(x^2 + y^2)(x + y)(x - y)$

37. $(16x^2y^2 + z^4)(4xy + z^2)(4xy - z^2)$ **39.** $\left(\frac{1}{6} + y^2\right)\left(\frac{1}{6} - y^2\right)$

41. $2(x + 12)(x - 12)$ **43.** $2x(x + 4)(x - 4)$

45. $5x(x + 5)(x - 5)$ **47.** $t^2(rs + x^2y)(rs - x^2y)$

49. $(a + b)(a - b + 1)$ **51.** $(a - b)(a + b + 2)$

53. $(2x + y)(1 + 2x - y)$ **55.** $(x - 4)(x + y)(x - y)$

57. $(x + 2 + y)(x + 2 - y)$ **59.** $(x + 1 + 3z)(x + 1 - 3z)$

61. $(c + 2a - b)(c - 2a + b)$ **63.** $(r + s)(r^2 - rs + s^2)$

65. $(x - 2y)(x^2 + 2xy + 4y^2)$

67. $(4a - 5b^2)(16a^2 + 20ab^2 + 25b^4)$

69. $(5xy^2 + 6z^3)(25x^2y^4 - 30xy^2z^3 + 37z^6)$

71. $(x^2 + y^2)(x^4 - x^2y^2 + y^4)$ **73.** $5(x + 5)(x^2 - 5x + 25)$

75. $4x^2(x - 4)(x^2 + 4x + 16)$

77. $2u^2(4v - t)(16v^2 + 4tv + t^2)$

79. $(a + b)(x + 3)(x^2 - 3x + 9)$

81. $(x^3 - y^4z^5)(x^6 + x^3y^4z^5 + y^8z^{10})$

83. $(a + b + 3)(a^2 + 2ab + b^2 - 3a - 3b + 9)$

85. $(y + 1)(y - 1)(y - 3)(y^2 + 3y + 9)$

87. $(x + 1)(x^2 - x + 1)(x - 1)(x^2 + x + 1)$

89. $(x^2 + y)(x^4 - x^2y + y^2)(x^2 - y)(x^4 + x^2y + y^2)$

91. $(a + 3)(a - 3)(a + 2)(a - 2)$

93. $\frac{4}{3}\pi(r_1 - r_2)(r_1^2 + r_1r_2 + r_2^2)$ **97.** 45 minutes for \$25

Study Set Section 8.5 (page 622)

1. rational   **3.** opposites   **5.** reciprocal   **7.** build

9. numerators, denominators, reciprocal, $\frac{AC}{BD}, \frac{A}{B} \cdot \frac{D}{C}$

11. factor, greatest   **13. a.** ii   **b.** adding or subtracting rational expressions   **c.** simplifying a rational expression

15. a. twice   **b.** once   **17. a.** c **b.** $(p + 2)(p + 3)$

19. $2x - 1$ **21.** $\frac{4y}{3x}$ **23.** $\frac{3y}{7(y - z)}$ **25.** 1   **27.** $\frac{m - 2n}{n - 2m}$

29. $\frac{x + 1}{x + 3}$ **31.** $\frac{3x - 5}{x + 2}$ **33.** $\frac{x + y}{a - b}$ **35.** $\frac{-3(x + 2)}{x + 1}$

37. $\frac{a^2 - 3a + 9}{4(a - 3)}$ **39.** $\frac{2}{x + 2}$ **41.** $\frac{-p + 2q}{q + 2p}$ **43.** $\frac{3a + b}{y + b}$

45. $\frac{8}{b}$ **47.** $\frac{3m}{2n^2}$ **49.** $2(y + 8)$ **51.** $\frac{x - 4}{x + 5}$ **53.** $\frac{x + 1}{9}$

55. 1   **57.** $x - 5$ **59.** $\frac{n + 2}{n + 1}$ **61.** $-\frac{x + y}{x - y}$ **63.** $-\frac{q}{p}$

65. $\frac{y^2 + xy + x^2}{x + y}$ **67.** $\frac{(x + 1)^2(x + 2)}{x + 2c}$ **69.** $\frac{3c}{2d^2}$ **71.** -2

73. $\frac{x - 7}{x + 7}$ **75.** $\frac{3 - a}{a + b}$ **77.** 2   **79.** 3   **81.** $\frac{1}{2}$

83. $\frac{1}{(x^2 + 9)(x - 3)}$ **85.** $\frac{17}{12x}$ **87.** $\frac{8x - 2}{(x + 2)(x - 4)}$

89. $\frac{7x + 29}{(x + 5)(x + 7)}$ **91.** 2   **93.** $\frac{1}{a - b}$ **95.** $\frac{x^2 + 4x + 1}{x^2y}$

97. $\frac{2x^2 + x}{(x + 3)(x + 2)(x - 2)}$ **99.** $\frac{-x^2 + 11x + 8}{(3x + 2)(x + 1)(x - 3)}$

101. $\frac{-x^3 + 5x^2 - 8}{x(x + 2)(x - 2)}$ **103.** $\frac{x^2 - 5x - 5}{x - 5}$ **105.** $\frac{-x^2 + 3x + 2}{(x - 1)(x + 1)^2}$

107. $\frac{2y}{3z}$ **109.** $-\frac{1}{y}$ **111.** $\frac{1}{c - d}$ **113.** $\frac{b + a}{b}$ **115.** $y - x$

117. $\frac{x + 2}{x - 3}$ **119.** $\frac{a - 1}{a + 1}$ **121.** $\frac{y + x}{y - x}$ **123.** $\frac{x - 9}{3}$

125. $-\frac{h}{4}$ **127.** $\frac{(b + a)(b - a)}{b(b - a - ab)}$ **129.** $k_1(k_1 + 2), k_2 + 6$

131. a. $\frac{3}{r + 5}, \frac{3}{r - 5}$ **b.** $\frac{30}{(r + 5)(r - 5)}$ hr   **137.** 13, 25

Study Set Section 8.6 (page 637)

1. two   **3.** table   **5.** linear   **7.** y-intercept, x-intercept

9. vertical   **11.** $\frac{y_2 - y_1}{x_2 - x_1}$ **13.** x-axis, y-axis   **15.** Parallel

17. $y - y_1 = m(x - x_1)$ **19.** $Ax + By = C$

21. a. $(-3, 0); (0, 4)$ **b.** false   **23.** sub 1

25. 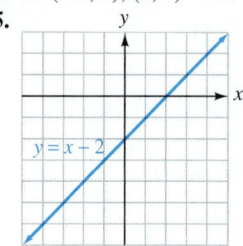   **27.**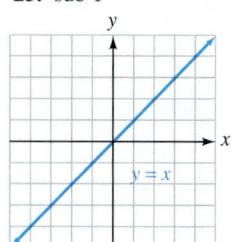

29. 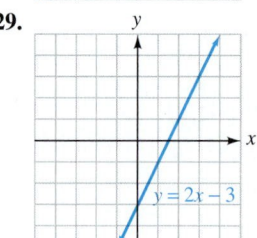   **31.**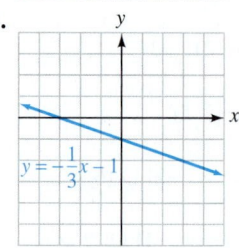

33. 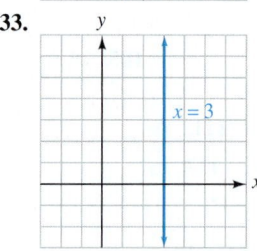   **35.**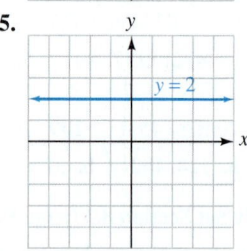

37. 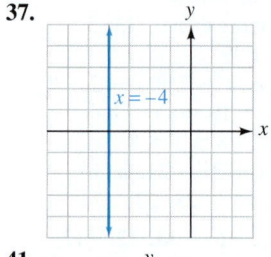   **39.**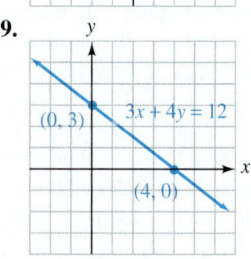

41. 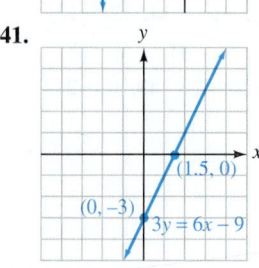   **43.**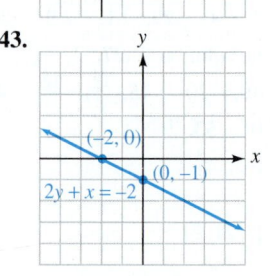

45. $\frac{6}{7}$ **47.** $-\frac{8}{3}$ **49.** 3   **51.** -1 **53.** $-\frac{1}{3}$

55. undefined   **57.** $-\frac{3}{2}$ **59.** $\frac{3}{4}$ **61.** parallel

63. perpendicular   **65.** parallel   **67.** $5x - y = -7$

69. $x - y = 0$ **71.** $7x - 3y = 9$ **73.** $y = 3x + 17$

75. $y = -7x + 54$ **77.** $y = -\frac{1}{2}x + 11$ **79.** $y = 4x$
81. $y = 4x - 3$ **83.** $y = -\frac{1}{4}x$ **85.** $y = -\frac{1}{4}x + \frac{11}{2}$
87. $(3, 4)$ **89.** $(9, 12)$ **91.** $\left(\frac{1}{2}, -2\right)$ **93.** $(4, 1)$
95. $\frac{1}{25} = 4\%$ **97.** 30 students per yr
99. $y = -2{,}298x + 19{,}984$ **101. a.** $B = \frac{1}{100}p - 195$
b. 905   **107.** $29,100

Study Set Section 8.7 (page 649)

1. function, one   **3.** input, output   **5.** value   **9.** $f(8)$
11. 17, $(-3, 17)$, -1, $(0, -1)$, 7, $(2, 7)$
13. a. $(-2, 4), (-2, -4)$ **b.** No; the x-value -2 is assigned
more than one y-value (4 and -4).   **15.** of   **17.** $f(x)$
19. $f(x), y$ **21.** yes   **23.** no; $(4, 2), (4, 4), (4, 6)$ **25.** yes
27. no; $(-1, 0), (-1, 2)$ **29.** no; $(3, 4), (3, -4)$ or
$(4, 3), (4, -3)$ **31.** yes   **33.** yes   **35.** no   **37.** no
39. $9, -3$ **41.** $3, -5$ **43.** $22, 2$ **45.** $3, 11$ **47.** $4, 9$
49. $7, 26$ **51.** $9, 16$ **53.** $6, 15$ **55.** $4, 4$ **57.** $2, 2$
59. $\frac{1}{5}, 1$ **61.** $-2, \frac{2}{5}$ **63.** $3.7, 1.1, 3.4$ **65.** $-\frac{27}{64}, \frac{1}{216}, \frac{125}{8}$
67. $2w, 2w + 2$ **69.** $3w - 5, 3w - 2$ **71.** 0   **73.** 1
75. D: $\{-2, 4, 6\}$, R: $\{3, 5, 7\}$ **77.** D: the set of all real
numbers, R: the set of all real numbers   **79.** D: the set of all
real numbers, R: the set of all nonnegative real numbers
81. D: the set of all real numbers, R: the set of all real numbers
greater than or equal to 0   **83.** D: the set of all real numbers
except 4, R: the set of all real numbers except 0   **85.** not a
function   **87.** a function   **89.** a function   **91.** a function
93.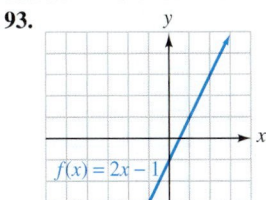   **95.**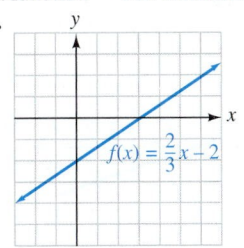

97. between 20°C and 25°C   **99. a.** $I(b) = 1.75b - 50$
b. $142.50   **101. a.** $(200, 25), (200, 90), (200, 105)$
b. It doesn't pass the vertical line test.   **103. a.** $3,400; the
tax on an income of $25,000 is $3,400.
b. $T(a) = 3{,}910 + 0.25(a - 28{,}400)$ **107.** $\frac{-15}{4}$ **109.** $\frac{1}{3}$

Study Set Section 8.8 (page 661)

1. nonlinear   **3.** parabola   **5.** cubing   **7. a.** -4 **b.** 0
c. 2   **d.** -1 **9. a.** 4   **b.** 3   **c.** $0, 2$ **d.** 1
11. a. Range, Domain   **b.** D: all nonnegative real numbers
R: all real numbers greater than or equal to 2   **13.** $4, 0, -2$
15.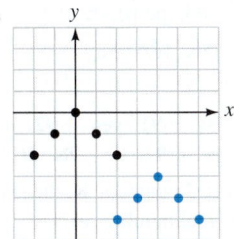   **17.** 4, left   **19.** 5, up

21. D: the set of real numbers,
R: the set of all real
numbers greater than or
equal to -3.

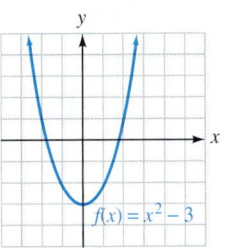

23. D: the set of real
numbers, R: the set of
real numbers.

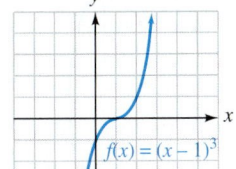

25. D: the set of real numbers,
R: the set of all real
numbers greater than or
equal to -2.

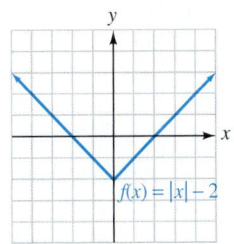

27. D: the set of real
numbers, R: the set of
real numbers greater
than or equal to 0.

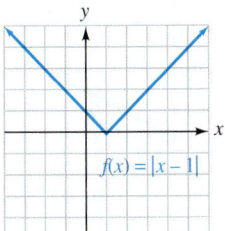

29. D: the set of real numbers,
R: the set of real numbers.

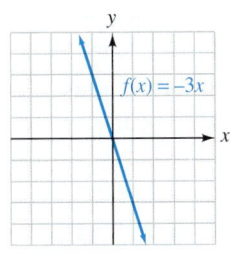

31.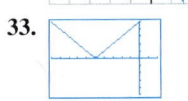

33.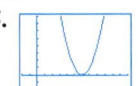   **35.**   **37.**

39.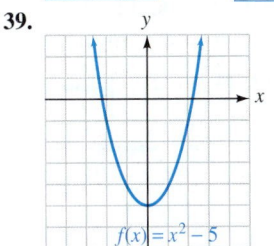   **41.**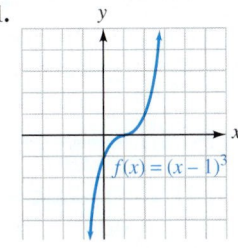

43.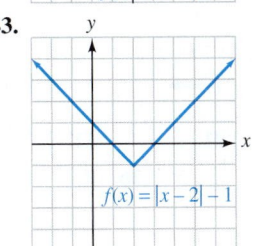   **45.**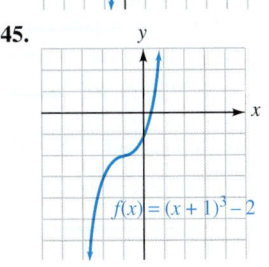

47.
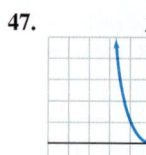

49.
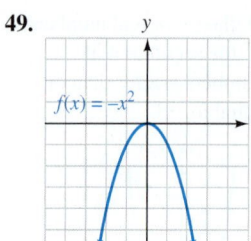

51. $f(x) = |x|$ **53.** a parabola **59.** $W = T - ma$

61. $g = \frac{2(s - vt)}{t^2}$ **63.** $5.4 million

Study Set Section 8.9 (page 670)

1. direct **3.** constant **5.** joint

7.
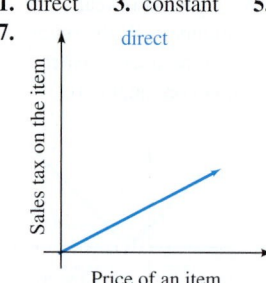

direct

9.
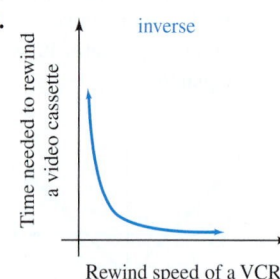

inverse

11. a. yes **b.** no **c.** no **d.** yes **13.** $A = kp^2$

15. $v = k/r^3$ **17.** $B = kmn$ **19.** $P = \frac{ka^2}{j^3}$ **21.** L varies

jointly with m and n. **23.** E varies jointly with a and the
square of b. **25.** X varies directly with x^2 and inversely with
y^2. **27.** R varies directly with L and inversely with d^2.

29. 1,600 ft **31.** 64 **33.** 25 days **35.** 12 in.^{3}

37. $85\frac{1}{3}$ **39.** 12 **41.** 26,437.5 gal **43.** 3 ohms

45. 0.275 in. **47.** 546 Kelvin **53.** 3.5×10^4

55. 0.0025

Key Concept (page 674)

1. u **3.** w **5.** q **7.** v **9.** s **11.** m **13.** z **15.** b
17. d **19.** x **21.** a **23.** t **25.** l

Chapter Review (page 675)

1. $-\frac{12}{5}$ **2.** -9 **3.** 8 **4.** $\frac{11}{7}$ **5.** $\frac{88}{17}$ **6.** 12

7. $\mathbb{R}$; identity **8.** $\varnothing$; contradiction **9.** -8 **10.** 0

11. $h = \frac{V}{\pi r^2}$ **12.** $x = \frac{6v}{ab} - y$ **13.** $[4, \infty)$
![number line at 4]

14. $\left(-\infty, -\frac{51}{11}\right)$
![number line at -51/11]
15. $(-\infty, 20)$
![number line at 20]

16. $\left(-\frac{1}{3}, 2\right)$
![number line -1/3 to 2]

17. $[-10, -4)$
![number line -10 to -4]

18. $(-\infty, -5) \cup (4, \infty)$
![number line -5 0 4]

19. 5 ft from one end **20.** 45 m^2 **21.** 2, -2 **22.** 3, $-\frac{11}{3}$

23. $\frac{26}{3}, -\frac{10}{3}$ **24.** no solution **25.** $\frac{1}{5}, -5$ **26.** $\frac{13}{12}$

27. $[-3, 3]$
![number line -3 to 3]

28. $(-5, -2)$
![number line -5 to -2]

29. $\left[-3, \frac{19}{3}\right]$
![number line -3 to 19/3]

30. no solution

31. $(-\infty, -1) \cup (1, \infty)$
![number line -1 0 1]

32. $\left(-\infty, \frac{4}{3}\right] \cup [4, \infty)$
![number line 4/3 0 4]

33. $(z - 5)(z - 6)$ **34.** $(x^2 + 4)(x^2 + y)$

35. $(4a - 1)(a - 1)$ **36.** $9x^2y^3z^2(3xz + 9x^2y^2 - 10z^5)$

37. $(y + 2)(y + 1 + x)$ **38.** $-(x + 7)(x - 4)$

39. $3(5x + y)(x - 4y)$ **40.** $(w^4 - 10)(w^4 + 9)$

41. $(ry - a + 1)(r - 1)$ **42.** $(7a^3 - 6b^2)^2$ **43.** prime

44. $2a^2(a + 3)(a - 1)$ **45.** $(s + t - 1)^2$

46. $m_1 = \frac{mm_2}{m_2 - m}$ **47.** $(z + 4)(z - 4)$

48. $(xy^2 + 8z^3)(xy^2 - 8z^3)$ **49.** prime

50. $(c + a + b)(c - a - b)$

51. $2c(4a^2 + 9b^2)(2a + 3b)(2a - 3b)$

52. $(k + 1 + 3m)(k + 1 - 3m)$ **53.** $(m + n)(m - n - 1)$

54. $(t + 4)(t^2 - 4t + 16)$

55. $(2a - 5b^3)(4a^2 + 10ab^3 + 25b^6)$

56. $\frac{\pi}{2}h(r_1 + r_2)(r_1 - r_2)$ **57.** $\frac{31x}{72y}$ **58.** -2 **59.** $\frac{cd}{7x^2}$

60. $\frac{2a - 1}{a + 2}$ **61.** $\frac{a + b}{(m + p)(m - 3)}$ **62.** $\frac{3x(x - 1)}{(x - 3)(x + 1)}$

63. $\frac{1}{c - d}$ **64.** $-\frac{2}{t - 3}$ **65.** $\frac{40x + 7y^2z}{112z^2}$ **66.** $\frac{12y + 20}{15y(x - 2)}$

67. $\frac{14y + 58}{(y + 3)(y + 7)}$ **68.** $\frac{5x^2 + 11x}{(x + 1)(x + 2)}$ **69.** $\frac{2bc^3}{7}$

70. $\frac{p - 3}{2(p + 2)}$ **71.** $\frac{x - 2}{x + 3}$ **72.** -1

73.
![graph y = -1/3 x - 1]

74.
![graph x = -2]

75.
![graph 2x + y = 4]

76.
![graph 3x - 4y - 8 = 0]

77. $\frac{2}{3}$ **78.** slope of $l_1 = \frac{4}{5}$; slope of $l_2 = -\frac{8}{5}$ **79.** 1

80. $-\frac{14}{9}$ **81.** 0 **82.** undefined **83.** perpendicular

84. parallel **85.** $3x - y = -29$ **86.** $13x + 8y = 6$

87. $y = \frac{3}{2}x - \frac{1}{2}$ **88.** $y = -\frac{2}{3}x - 7$

89. $y = -1,720x + 8,700$ **90.** $\left(\frac{3}{2}, 8\right)$ **91.** yes **92.** no

93. yes **94.** no **95.** -7 **96.** 18 **97.** 8 **98.** $3t + 2$

99. 3 **100.** $\frac{4}{3}$ **101.** D: the set of real numbers, R: the set of
real numbers **102.** D: the set of all real numbers except 2,
R: the set of all real numbers except 0 **103.** function
104. not a function **105.** $f(t) = 1.37t + 21.2$; about 54%

106.

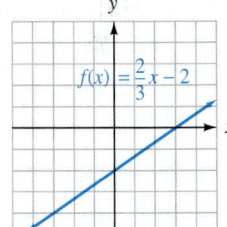

107. a. -4 **b.** 3 **c.** 0

108.

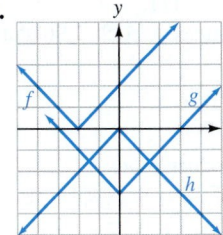

109.

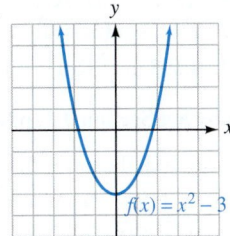

110.

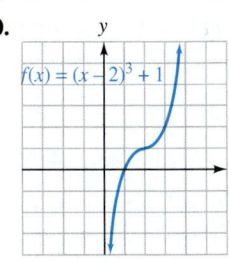

111. D: the set of real numbers, R: the set of real numbers **112.** D: the set of real numbers, R: the set of all real numbers greater than or equal to 1 **113.** $5,460
114. 1.25 amps **115.** $\frac{1}{32}$
116. 126.72 lb **117.** inverse variation **118.** 0.2

Chapter Test (page 682)

1. $\frac{21}{5}$ **2.** -1 **3.** $\varnothing$; contradiction **4.** $a = \frac{180n - 360}{n}$
5. 4 cm by 9 cm **6.** more than 78
7. $(-\infty, -5]$  **8.** $(-2, 16)$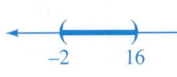

9. $\left[1, \frac{9}{4}\right]$

10. $(-\infty, -3) \cup (8, \infty)$

11. $(-\infty, -9) \cup (13, \infty)$

12. $\left[\frac{4}{3}, \frac{8}{3}\right]$ **13.** $8, -11$ **14.** $4, -4$

15. $3abc(4a^2b - abc + 2c^2)$ **16.** $4(y^2 + 4)(y + 2)(y - 2)$
17. $(b + 5)(b^2 - 5b + 25)$ **18.** $3(u + 2)(2u - 1)$
19. $(a - y)(x + y)$ **20.** $(5m^4 - 6n)^2$ **21.** prime
22. $(x + 3 + y)(x + 3 - y)$
23. $(4a - 5b^2)(16a^2 + 20ab^2 + 25b^4)$
24. $(x - y + 5)(x - y - 2)$ **25.** -3
26. $\frac{2x + y}{4y}$ **27.** $\frac{(x + y)^2}{2}$ **28.** $\frac{13}{x + 1}$ **29.** $\frac{2}{t - 4}$
30. $\frac{6a - 17}{(a + 1)(a - 2)(a - 3)}$ **31.** $\frac{u^2}{2vw}$ **32.** $\frac{3k^2 + 4k + 4}{3k^2 - 9k - 9}$
33. $3x + 2y = -4$ **34.** $m = -\frac{1}{3}, \left(0, -\frac{3}{2}\right)$ **35.** $-\frac{11}{7}$
36. a. $v = -600x + 4,000$ **b.** $(0, 4,000)$; It gives the value of the copier when new: $4,000.

37. $(5, 0), (0, -2)$ **38.**

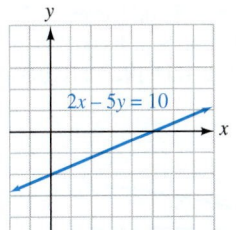

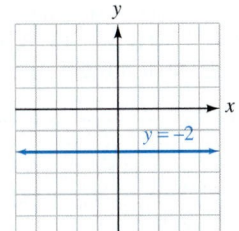

39. yes **40.** no **41.** -1
42. -20 **43.**

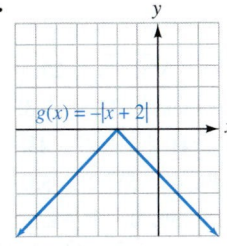

44. D: the set of real numbers, R: the set of all real numbers greater than or equal to -5
45. -2 **46.** 2
47. $\frac{6}{5}$ **48.** 25 decibels

Cumulative Review Exercises, Chapters 1–8 (page 685)

1. $2,016,000,000 **2.** $\{. . . , -3, -2, -1, 0, 1, 2, 3, . . .\}$
3. -37 **4.** 28 **5.** 1.2 ft^3 **6.** rational numbers: terminating and repeating decimals; irrational numbers: nonterminating, nonrepeating decimals **7.** 10 **8.** -6
9. $-26p^2 - 6p$ **10.** $-2a + b - 2$ **11.** 30 **12.** mutual fund: $25,000; bonds: $20,000 **13.** 201 ft^2
14. a. $P = 2l + 2w$ **b.** $A = lw$ **c.** $A = \pi r^2$ **d.** $d = rt$
15. $7x - 13y$ **16.** $6x^4 + 8x^3 - 14x^2$
17. $36p^2 - 60pq + 25q^2$ **18.** $2x^2 + 3x - 9$ **19.** $\frac{1}{8} - \frac{2}{x}$
20. $3x + 1$ **21.** $-3x^3 + 25x^2y - 56xy^2 + 16y^3$
22. $2x^2 + xy - y^2$ **23.** x^{31} **24.** $8a^4$ **25.** $\frac{3}{16}$ **26.** 1
27. $-\frac{u^6}{27v^9}$ **28.** $\frac{16y^{14}}{z^{10}}$ **29.** $-4, 3$ **30.** $16, 8$ **31.** $0, 2$
32. $3, -5$

33. $(-6, 1)$ **34.**

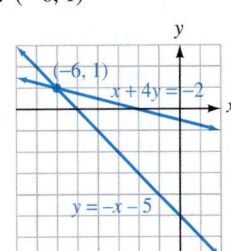

35. $(4, 1)$ **36.** noodles: 2 servings, rice: 3 servings

Study Set Section 9.1 (page 697)

1. square, cube   **3.** index, radicand   **5.** odd, even   **7.** a
9. two, 5   **11.** x **13.** 3, up   **15. a.** 3   **b.** 0   **c.** 6
d. D: $[2, \infty)$, R: $[0, \infty)$ **17.** 0, 1, 2, 3, 4; D: $[0, \infty)$; R: $[0, \infty)$

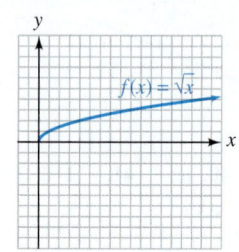

19. $\sqrt{x^2} = |x|$ **21.** $f(x) = \sqrt{x - 5}$ **23.** 11   **25.** -8
27. $\frac{1}{3}$ **29.** 0.5   **31.** not real   **33.** 4   **35.** 3.4641
37. 26.0624   **39.** $2|x|$ **41.** $|t + 5|$ **43.** $5|b|$
45. $|a + 3|$ **47.** 0   **49.** 4   **51.** 2   **53.** -10
55. 4.1231   **57.** 3.3322   **59.** 0, -1, -2, -3, -4;
D: $[0, \infty)$, R: $(-\infty, 0]$ **61.** D: $[-4, \infty)$, R: $[0, \infty)$

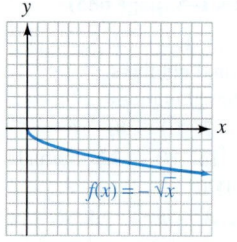 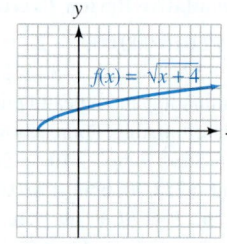

63. D: $(-\infty, \infty)$, R: $(-\infty, \infty)$ **65.** 1   **67.** -5 **69.** $-\frac{2}{3}$
71. 0.4   **73.** $2a$
75. $-10pq$ **77.** $-\frac{1}{2}m^2n$
79. $-0.4s^3t^2$ **81.** 3
83. -3 **85.** not real
87. $\frac{2}{5}$ **89.** $\frac{1}{2}$ **91.** $2a$
93. $2a$ **95.** k^3 **97.** $\frac{1}{2}m$
99. $x + 2$ **101.** 7.0 in.
103. about 61.3 beats/min   **105.** 13.4 ft   **107.** 3.5%
113. 1   **115.** $\frac{3(m^2 + 2m - 1)}{(m + 1)(m - 1)}$

Study Set Section 9.2 (page 709)

1. rational (or fractional)   **3.** index, radicand
5. $25^{1/5}$, $\left(\sqrt[3]{-27}\right)^2$, $16^{-3/4}$, $\left(\sqrt{81}\right)^3$, $-\left(\frac{9}{64}\right)^{1/2}$
7.

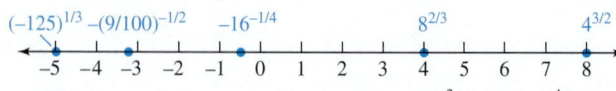

9. $\sqrt[n]{x}$ **11.** $\frac{1}{x^{m/n}}$ **13.** $100a^4$, $10a^2$ **15.** $\sqrt[3]{x}$ **17.** $\sqrt[4]{3x}$
19. $\sqrt[4]{6x^3y}$ **21.** $\sqrt{x^2 + y^2}$ **23.** $m^{1/2}$ **25.** $(3a)^{1/4}$
27. $(8abc)^{1/6}$ **29.** $(a^2 - b^2)^{1/3}$ **31.** 2   **33.** 5   **35.** 3
37. 2   **39.** $\frac{1}{2}$ **41.** -2 **43.** not real   **45.** -6

47. $|x|$ **49.** m^2 **51.** n^3 **53.** $5|y|$ **55.** $2|x|$
57. not real   **59.** $|x + 1|$ **61.** 216   **63.** 27   **65.** 1,728
67. $\frac{1}{4}$ **69.** $125x^6$ **71.** $\frac{4x^2}{9}$ **73.** $\frac{1}{2}$ **75.** $\frac{1}{3,125}$
77. $\frac{1}{64x^3}$ **79.** $\frac{1}{9y^2}$ **81.** $\frac{16}{81}$ **83.** $-\frac{3}{2x}$ **85.** $9^{5/7}$
87. $\frac{1}{36}$ **89.** 3   **91.** a **93.** $a^{2/9}$ **95.** $a^{3/4}b^{1/2}$
97. $\frac{1}{3}x$ **99.** $y + y^2$ **101.** $x^2 - x + x^{3/5}$ **103.** $\sqrt{p}$
105. $\sqrt{5b}$ **107.** $\sqrt[5]{xy}$ **109.** $\sqrt[18]{c}$ **111.** $\sqrt[15]{7m}$
113. 2.47   **115.** 1.01   **117.** 736 ft/sec   **119.** 1.96 units
121. 4,608 in.^{2}, 32 ft^2 **125.** $2\frac{1}{2}$ hr

Study Set Section 9.3 (page 718)

1. like   **3.** factor   **5.** $\sqrt[n]{a}\sqrt[n]{b}$, product, roots
7. a. $\sqrt{4 \cdot 5}$ **b.** $\sqrt{4}\sqrt{5}$ **c.** $\sqrt{4 \cdot 5} = \sqrt{4}\sqrt{5}$
9. a. $\sqrt{5}$, $\sqrt[3]{5}$ (answers may vary); no   **b.** $\sqrt{5}$, $\sqrt{6}$ (answers
may vary); no   **11.** $8k^3$, $8k^3$, $4k$ **13.** $2\sqrt{5}$ **15.** $10\sqrt{2}$
17. $2\sqrt[3]{10}$ **19.** $-3\sqrt[3]{3}$ **21.** $2\sqrt[4]{2}$ **23.** $-2\sqrt[5]{3}$
25. $2\sqrt[6]{5}$ **27.** $\frac{\sqrt{7}}{3}$ **29.** $\frac{\sqrt[3]{7}}{4}$ **31.** $\frac{\sqrt[3]{3}}{10}$ **33.** $\frac{\sqrt[5]{3}}{2}$
35. 10   **37.** 2   **39.** $5x\sqrt{2}$ **41.** $4\sqrt{2b}$ **43.** $-4a\sqrt{7a}$
45. $5ab\sqrt{7b}$ **47.** $-10\sqrt{3xy}$ **49.** $-3x^2\sqrt[3]{2}$
51. $2x^4y\sqrt[3]{2}$ **53.** $2x^3y\sqrt[4]{2}$ **55.** $a\sqrt[5]{a^2}$ **57.** $m\sqrt[6]{m^5}$
59. $2t^2\sqrt[5]{t}$ **61.** $3a$ **63.** $7x$ **65.** $\frac{z}{4x}$ **67.** $\frac{\sqrt[4]{5x}}{2z}$
69. $10\sqrt{2x}$ **71.** $\sqrt[5]{7a^2}$ **73.** $-\sqrt{2}$ **75.** $2\sqrt{2}$
77. $9\sqrt{6}$ **79.** $3\sqrt[3]{3x}$ **81.** $-\sqrt[3]{4}$ **83.** -10
85. $-17\sqrt[4]{2}$ **87.** $16\sqrt[4]{2}$ **89.** $-4\sqrt{2}$ **91.** $3\sqrt{2t} + \sqrt{3t}$
93. $-11\sqrt[3]{2}$ **95.** $y\sqrt{z}$ **97.** $13y\sqrt{x}$ **99.** $12\sqrt[3]{a}$
101. $-7y^2\sqrt{y}$ **103.** $4x\sqrt[5]{xy^2}$ **105.** $8\pi\sqrt{5}$ ft^2; 56.2 ft^2
107. $5\sqrt{3}$ amps; 8.7 amps   **109.** $\left(26\sqrt{5} + 10\sqrt{3}\right)$ in.;
75.5 in.   **115.** $\frac{-15x^5}{y}$ **117.** $3p + 4 - \frac{5}{2p - 5}$

Study Set Section 9.4 (page 729)

1. FOIL   **3.** irrational   **5.** perfect   **7. a.** $6\sqrt{6}$
b. 48   **c.** can't be simplified   **d.** $-6\sqrt{6}$ **9.** Any
number multiplied by 1 is the same number. $\frac{\sqrt{7}}{\sqrt{7}} = 1$.
11. A radical appears in the denominator.   **13.** $\sqrt{6}$, 48, 16, 4
15. 11   **17.** 7   **19.** 4   **21.** $5\sqrt{2}$ **23.** $6\sqrt{2}$ **25.** 5
27. 18   **29.** 8   **31.** 18   **33.** $2\sqrt[3]{3}$ **35.** ab^2
37. $5a\sqrt{b}$ **39.** $-20r\sqrt[3]{10s}$ **41.** $x^2(x + 3)$ **43.** $9b$
45. $a\sqrt[4]{6ab}$ **47.** $2\sqrt[5]{t^2}$ **49.** $12\sqrt{5} - 15$
51. $24\sqrt{3} + 6\sqrt{14}$ **53.** $-8x\sqrt{10} + 6\sqrt{15x}$
55. $-1 - 2\sqrt{2}$ **57.** $\sqrt[3]{25z^2} + 3\sqrt[3]{15z} + 2\sqrt[3]{9}$
59. $3x - 2y$ **61.** $6a + 5\sqrt{3ab} - 3b$ **63.** $18r - 12\sqrt{2r} + 4$
65. $-6x - 12\sqrt{x} - 6$ **67.** $4a\sqrt[3]{2a} - 5\sqrt[3]{4a^2} - 3$
69. $\frac{\sqrt{7}}{7}$ **71.** $\frac{\sqrt{30}}{5}$ **73.** $\frac{\sqrt{10}}{4}$ **75.** $\frac{\sqrt[3]{4}}{2}$ **77.** $\sqrt[3]{3}$

79. $\frac{\sqrt[3]{6}}{3}$ **81.** $\frac{2\sqrt{2xy}}{xy}$ **83.** $\frac{\sqrt{5y}}{y}$ **85.** $\frac{\sqrt[3]{2ab^2}}{b}$

87. $\frac{\sqrt[4]{4}}{2}$ **89.** $\frac{\sqrt[5]{2}}{2}$ **91.** $\frac{\sqrt[4]{27st^2}}{3t}$ **93.** $\frac{t\sqrt[5]{9a^4}}{3a}$

95. $\frac{3\sqrt{2}-\sqrt{10}}{4}$ **97.** $\frac{9-2\sqrt{14}}{5}$ **99.** $\frac{3\sqrt{6}+4}{2}$

101. $\frac{2(\sqrt{x}-1)}{x-1}$ or $\frac{2\sqrt{x}-2}{x-1}$ **103.** $\sqrt{2z}+1$

105. $\frac{x-2\sqrt{xy}+y}{x-y}$ **107.** $\frac{6\sqrt{ab}+a+9b}{a-9b}$ **109.** $\frac{x-9}{x(\sqrt{x}-3)}$

111. $\frac{x-y}{\sqrt{x}(\sqrt{x}-\sqrt{y})}$ **113.** $\frac{\sqrt{2\pi}}{2\pi\sigma}$ **115.** $\frac{\sqrt{2}}{2}$ **119.** $\frac{1}{3}$

Study Set Section 9.5 (page 739)

1. radical **3.** extraneous **5.** n, n **7. a.** Square both sides. **b.** Cube both sides. **c.** Subtract 3 from both sides. **9. a.** x **b.** $x-5$ **c.** $32x$ **d.** $x+3$

11. a. $x-6\sqrt{x}+9$ **b.** $2y+10\sqrt{2y+1}+26$

13. The principal square root of a number, in this case $\sqrt{8x-7}$, is never negative. **15.** 6, 2, 2, $3x$, yes **17.** 2 **19.** 4

21. 0 **23.** 4 **25.** 8 **27.** 1 **29.** -16 **31.** $\cancel{4}$, no solution **33.** $\frac{5}{2}, \frac{1}{2}$ **35.** 1 **37.** 16 **39.** 14, $\cancel{6}$

41. $\cancel{-5}$, 5 **43.** 4, 3 **45.** 2, $\cancel{7}$ **47.** 2, -1

49. $-1, \cancel{X}$ **51.** $\cancel{X}$, no solution **53.** 0, $\cancel{4}$ **55.** $-\frac{1}{2}$

57. $\frac{1}{3}, \frac{4}{5}$ **59.** $-\cancel{3}$, no solution **61.** 1, 9 **63.** 4, $\cancel{8}$

65. 2, $\cancel{142}$ **67.** $\cancel{6}$, no solution **69.** 5 **71.** 1

73. $h=\frac{v^2}{2g}$ **75.** $\ell=\frac{8T^2}{\pi^2}$ **77.** $A=P(r+1)^3$

79. $v^2=c^2\left(1-\frac{L_A^2}{L_B^2}\right)$ **81.** 178 ft **83.** about 488 watts

85. 10 lb **87.** \$5 **95.** 2.5 foot-candles **97.** 0.41511 in.

Study Set Section 9.6 (page 749)

1. hypotenuse **3.** Pythagorean **5.** $a^2+b^2=c^2$

7. $\sqrt{2}$ **9.** $\sqrt{3}$ **11.** $30°, 60°$ **13. a.** 8 **b.** $\sqrt{15}$

c. $2\sqrt{6}$ **15.** 6, 52, 4, 2 **17.** 10 ft **19.** 80 m

21. $h=2\sqrt{2}\approx2.83, x=2$ **23.** $x=5\sqrt{3}\approx8.66, h=10$

25. $x=4.69, y=8.11$ **27.** $x=12.11, y=12.11$

29. $7\sqrt{2}$ cm **31.** 5 **33.** 13 **35.** 10 **37.** $2\sqrt{26}$

41. $(5\sqrt{2}, 0), (0, 5\sqrt{2}), (-5\sqrt{2}, 0), (0, -5\sqrt{2})$; (7.07, 0), (0, 7.07), (−7.07, 0), (0, −7.07) **43.** $10\sqrt{3}$ mm, 17.32 mm

45. $10\sqrt{181}$ ft, 134.54 ft **47.** about 0.13 ft

49. a. 21.21 units **b.** 8.25 units **c.** 13.00 units

51. yes **55.** 7 **57.** 9

Study Set Section 9.7 (page 760)

1. imaginary **3.** real, imaginary **5.** conjugates

7. a. $\sqrt{-1}$ **b.** -1 **c.** $-i$ **d.** 1 **9.** real, imaginary

11. denominator **13. a.** $7i$ **b.** $-\sqrt{6}$ **15.** Complex, Real, Rational, Irrational, Imaginary **17. a.** yes **b.** yes

19. 9, $6i$, 2, 11 **21. a.** true **b.** false **c.** false **d.** false

23. $3i$ **25.** $\sqrt{7}i$ or $i\sqrt{7}$ **27.** $2\sqrt{6}i$ or $2i\sqrt{6}$

29. $-2\sqrt{6}i$ or $-2i\sqrt{6}$ **31.** $45i$ **33.** $\frac{5}{3}i$ **35.** -6

37. $-2\sqrt{3}$ **39.** $\frac{5}{8}$ **41.** -20 **43.** $8-2i$ **45.** $3-5i$

47. $15+2i$ **49.** $15+7i$ **51.** $6-3i$ **53.** $-25-25i$

55. $7+i$ **57.** $12+5i$ **59.** $13-i$ **61.** $14-8i$

63. $8+\sqrt{2}i$ **65.** $3+4i$ **67.** $-9+0i$ **69.** $-6+0i$

71. $0-i$ **73.** $0+\frac{4}{5}i$ **75.** $\frac{1}{8}+0i$ **77.** $0+\frac{3}{5}i$

79. $2+i$ **81.** $-\frac{42}{25}-\frac{6}{25}i$ **83.** $\frac{1}{4}+\frac{3}{4}i$ **85.** $-\frac{4}{13}-\frac{6}{13}i$

87. $\frac{5}{13}-\frac{12}{13}i$ **89.** $\frac{11}{10}+\frac{3}{10}i$ **91.** $\frac{1}{4}-\frac{\sqrt{15}}{4}i$ **93.** i

95. $-i$ **97.** 1 **99.** i **101.** $-1+i$ **105.** 20 mph

Key Concept (page 763)

1. $-3h^2\sqrt[3]{2}$ **2.** $14\sqrt[3]{e}$ **3.** $-4\sqrt{2}$ **4.** $-20r\sqrt[3]{10s}$

5. $3s-2t$ **6.** $-\sqrt{21}+\sqrt{15}$ **7.** $18n-12\sqrt{2n}+4$

8. $\frac{\sqrt[3]{3k^2}}{k}$ **9.** -3 **10.** 5 **11.** 2; 7 is extraneous

12. no solutions **13.** $3^{1/3}$ **14.** $5\sqrt[5]{a^2}$

Chapter Review (page 764)

1. 7 **2.** -11 **3.** $\frac{15}{7}$ **4.** not real **5.** 0.1 **6.** $5|x|$

7. x^4 **8.** $|x+2|$ **9.** -3 **10.** -6 **11.** $4x^2y$ **12.** $\frac{x^3}{5}$

13. 5 **14.** -2 **15.** $4x^2|y|$ **16.** $x+1$ **17.** $-\frac{1}{2}$

18. not real **19.** 0 **20.** 0 **21.** 48 ft **22.** 24 cm^2

23. D: $[-2, \infty)$, R: $[0, \infty)$ **24.** D: $(-\infty, \infty)$, R: $(-\infty, \infty)$

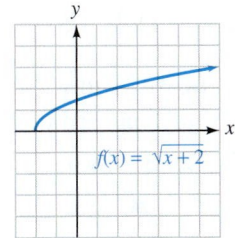

$f(x)=\sqrt{x+2}$

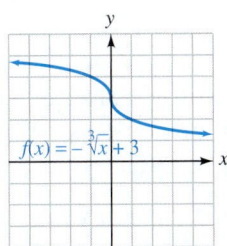

$f(x)=-\sqrt[3]{x}+3$

25. $\sqrt{t}$ **26.** $\sqrt[4]{5xy^3}$ **27.** 5 **28.** -6 **29.** not real

30. 1 **31.** $\frac{3}{x}$ **32.** -2 **33.** 5 **34.** $3cd$ **35.** 27

36. $\frac{1}{4}$ **37.** $-16,807$ **38.** 10 **39.** $\frac{27}{8}$ **40.** $\frac{1}{3,125}$

41. $125x^3y^6$ **42.** $\frac{1}{4u^4v^2}$ **43.** $5^{3/4}$ **44.** $a^{1/7}$ **45.** k^8

46. $3^{2/3}$ **47.** $u-1$ **48.** $v+v^2$ **49.** $\sqrt{a}$ **50.** $\sqrt[6]{c}$

51. 183 mi **52.** Two true statements result: $32=32$.

53. $4\sqrt{15}$ **54.** $3\sqrt[3]{2}$ **55.** $2\sqrt[4]{2}$ **56.** $4\sqrt[5]{3}$

57. $2x^2\sqrt{2x}$ **58.** $r^5\sqrt[3]{r^2}$ **59.** $2xy\sqrt[3]{2x^2y}$ **60.** $9j^2\sqrt[3]{jk}$

61. $4x$ **62.** $\frac{\sqrt{17xy}}{8a^2}$ **63.** $3\sqrt{2}$ **64.** $11\sqrt{5}$ **65.** 0

66. $-8a\sqrt[4]{2a}$ **67.** $29x\sqrt{2}$ **68.** $13x\sqrt[3]{2}$

70. $(6\sqrt{2}+2\sqrt{10})$ in., 14.8 in. **71.** 7 **72.** $6\sqrt{10}$

73. 32 **74.** 72 **75.** $3x$ **76.** $x+1$ **77.** $-2x^3\sqrt[3]{x}$

78. 3 **79.** $42t+9t\sqrt{21t}$ **80.** $-20x^3y^3\sqrt[4]{x^2y^2}$

81. $3b+6\sqrt{b}+3$ **82.** $\sqrt[3]{9p^2}-\sqrt[3]{6p}-2\sqrt[3]{4}$

83. $\frac{10\sqrt{3}}{3}$ **84.** $\frac{\sqrt{15xy}}{5xy}$ **85.** $\frac{\sqrt[3]{u^2}}{u^2v^2}$ **86.** $\frac{\sqrt[4]{27ab^2}}{3b}$

87. $2\left(\sqrt{2}+1\right)$ **88.** $\frac{12\sqrt{xz}-16x-2z}{z-16x}$ **89.** $\frac{a-b}{a+\sqrt{ab}}$

90. $r=\frac{\sqrt[3]{6\pi^2V}}{2\pi}$ **91.** 22   **92.** 16, 9   **93.** $\frac{13}{2}$ **94.** $\frac{9}{16}$

95. 2, −4   **96.** $\varnothing$, no solution   **97.** 1   **98.** $-\frac{3}{2},1$

99. $-\frac{1}{2},4$ **100.** 2   **101.** $P=\frac{A}{(r+1)^2}$ **102.** $I=\frac{h^3b}{12}$

103. 17 ft   **104.** 88 yd   **105.** $7\sqrt{2}$ m   **106.** $6\sqrt{3}$ cm,
18 cm   **107.** 7.07 in.   **108.** 8.66 cm   **109.** 13
110. $2\sqrt{2}$ **111.** $5i$ **112.** $12i\sqrt{2}$ **113.** $-i\sqrt{6}$
114. $\frac{3}{8}i$ **115.** Real, Imaginary   **116. a.** true   **b.** true
c. false   **d.** false   **117.** $3-6i$ **118.** $-1+7i$
119. $0-19i$ **120.** $0+i$ **121.** $8-2i$ **122.** $3-5i$
123. $3+6i$ **124.** $22+29i$ **125.** $-3\sqrt{3}+0i$
126. $-81+0i$ **127.** $0-\frac{3}{4}i$ **128.** $-\frac{5}{13}+\frac{12}{13}i$
129. -1 **130.** i

Chapter Test (page 770)

1. D: $[1,\infty)$, R: $[0,\infty)$

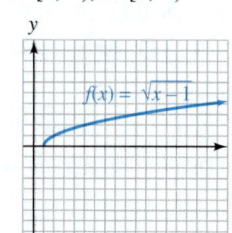

$f(x)=\sqrt{x-1}$

2. 46 feet per second
3. a −1   **b.** 2   **c.** 1
d. D: $(-\infty,\infty)$, R: $(-\infty,\infty)$
4. No real number raised to
the fourth power is −16.
5. $7x^2$ **6.** −9   **7.** $\frac{1}{216}$
8. $\frac{25n^4}{4}$ **9.** $2^{4/3}$ **10.** $a^{1/9}$
11. $|x|$ **12.** $|y+5|$

13. $-4xy^2$ **14.** $\frac{2}{3}a$ **15.** $t+8$ **16.** $5xy^2\sqrt{10xy}$
17. $2x^5y\sqrt[3]{3}$ **18.** $2\sqrt[4]{2}$ **19.** $2y^2\sqrt{3y}$ **20.** $14\sqrt[4]{5}$
21. $5z^3\sqrt[4]{3z}$ **22.** $-6x\sqrt{y}-2xy^2$ **23.** $3-7\sqrt{6}$
24. $\sqrt[3]{4a^2}+18\sqrt[3]{2a}+81$ **25.** $\frac{4\sqrt{10}}{5}$ **26.** $\sqrt{3t}+1$
27. $\frac{\sqrt[3]{18a^2}}{2a}$ **28.** $\frac{1}{\sqrt{2}(\sqrt{5}-3)}$ **29.** $\frac{1}{3}$ **30.** 10
31. 4 is extraneous, no solution   **32.** 3, −3 is extraneous
33. $\sqrt{x-8}$ is the principal square root. It cannot be negative.
34. $G=\frac{4\pi^2r^3}{Mt^2}$ **35.** 9.24 cm   **36.** 8.67 cm   **37.** 25
38. 28 in.   **39.** $i\sqrt{5}$ **40.** −1   **41.** $-1+11i$
42. $4-7i$ **43.** $8+6i$ **44.** $-10-11i$ **45.** $0-\frac{\sqrt{2}}{2}i$
46. $\frac{1}{2}+\frac{1}{2}i$

Study Set Section 10.1 (page 782)

1. quadratic   **3.** perfect   **5.** $\sqrt{c},-\sqrt{c}$ **7.** 5
9. It is.   **11. a.** 36   **b.** $\frac{25}{4}$ **c.** $\frac{1}{16}$ **d.** $\frac{9}{64}$
13. a. Subtract 35 from both sides.   **b.** Add 36 to both sides.
15. $x=-7\pm\sqrt{6}$ **17. a.** 4 is not a factor of the numerator.
Only common factors of the numerator and denominator can be
removed.   **b.** 5 is not a factor of the numerator. Only common
factors of the numerator and denominator can be removed.

19. a. $2;2\sqrt{5},-2\sqrt{5}$ **b.** ±4.47 **21.** $0,-2$ **23.** 5, −5
25. $-2,-4$ **27.** $2,\frac{1}{2}$ **29.** ±6 **31.** $\pm\sqrt{5}$
33. $\pm\frac{4\sqrt{3}}{3}$ **35.** $0,-2$ **37.** 4, 10   **39.** $-5\pm\sqrt{3}$
41. $-2\pm2\sqrt{2}$ **43.** $2,-\frac{4}{3}$ **45.** $\pm4i$ **47.** $\pm\frac{9}{2}i$
49. $3\pm i\sqrt{5}$ **51.** $d=\frac{\sqrt{6h}}{2}$ **53.** $c=\frac{\sqrt{Em}}{m}$ **55.** 2, −4
57. 2, 6   **59.** 1, −6   **61.** $-1,4$ **63.** $-4\pm\sqrt{10}$
65. $1\pm3\sqrt{2}$ **67.** $\frac{7\pm\sqrt{37}}{2}$ **69.** $\frac{1\pm\sqrt{13}}{2}$ **71.** $-\frac{1}{2},1$
73. $\frac{3\pm2\sqrt{3}}{3}$ **75.** $\frac{1\pm2\sqrt{2}}{2}$ **77.** $\frac{-5\pm\sqrt{41}}{4}$
79. $\frac{-7\pm\sqrt{29}}{10}$ **81.** $-1\pm i$ **83.** $-4\pm i\sqrt{2}$
85. $\frac{1}{3}\pm\frac{2\sqrt{2}}{3}i$ **87.** width: $7\frac{1}{4}$ ft; length: $13\frac{3}{4}$ ft
89. 1.6 sec   **91.** 1.70 in.   **93.** 0.92 in.   **95.** 2.9 ft, 6.9 ft
99. $2ab^2\sqrt[3]{5}$ **101.** x^3 **103.** $5ab\sqrt{7b}$

Study Set Section 10.2 (page 792)

1. quadratic   **3. a.** $x^2+2x+5=0$ **b.** $3x^2+2x-1=0$
5. a. true   **b.** true   **7. a.** $2,-4$ **b.** $\frac{1\pm\sqrt{33}}{4}$
9. a. $1\pm2\sqrt{2}$ **b.** $\frac{-3\pm\sqrt{7}}{2}$ **11. a.** The fraction bar
wasn't drawn under both parts of the numerator.   **b.** A $\pm$ sign
wasn't written between b and the radical.   **13.** $-1,-2$
15. $-6,-6$ **17.** $\frac{1}{2},-3$ **19.** $\frac{-5\pm\sqrt{5}}{10}$ **21.** $-\frac{3}{2},-\frac{1}{2}$
23. $\frac{1}{4},-\frac{3}{4}$ **25.** $\frac{4}{3},-\frac{2}{5}$ **27.** $\frac{-5\pm\sqrt{17}}{2}$ **29.** $\frac{3\pm\sqrt{17}}{4}$
31. $5\pm\sqrt{7}$ **33.** $23,-17$ **35.** $\frac{2}{3},-\frac{5}{2}$ **37.** $-1\pm i$
39. $-\frac{1}{4}\pm\frac{\sqrt{7}}{4}i$ **41.** $\frac{2}{3}\pm\frac{\sqrt{2}}{3}i$ **43.** $\frac{1}{3}\pm\frac{2\sqrt{2}}{3}i$
45. $2\pm2i$ **47.** $3\pm i\sqrt{5}$ **49.** $\frac{-3\pm\sqrt{29}}{10}$ **51.** $\frac{9\pm\sqrt{89}}{2}$
53. $\frac{10\pm\sqrt{55}}{30}$ **55.** $-0.68,-7.32$ **57.** $1.22,-0.55$
59. $8.98,-3.98$ **61.** 97 ft by 117 ft   **63.** 0.5 mi by 2.5 mi
65. 34 in.   **67.** \$4.80 or \$5.20   **69.** 4,000   **71.** 9%
73. early 1976   **77.** $n^{1/2}$ **79.** $(3b)^{1/4}$ **81.** $\sqrt[3]{t}$ **83.** $\sqrt[4]{3t}$

Study Set Section 10.3 (page 802)

1. discriminant   **3.** conjugates   **5.** rational, unequal
7. a. x^2 **b.** $\sqrt{x}$ **c.** $x^{1/3}$ **d.** $\frac{1}{x}$ **e.** $x+1$ **9.** $4ac$, 5,
6, 24, rational   **11.** rational, equal   **13.** complex conjugates
15. irrational, unequal   **17.** rational, unequal   **19.** yes
21. $-1,1,-4,4$ **23.** $1,-1,\sqrt{5},-\sqrt{5}$ **25.** $2,-2,i\sqrt{7},$
$-i\sqrt{7}$ **27.** $-i,i,-3i\sqrt{2},3i\sqrt{2}$ **29.** 1   **31.** no solution
33. 16, 4   **35.** $\frac{9}{4}$ **37.** $-8,-27$ **39.** $-1,27$
41. $\frac{243}{32},1$ **43.** $\frac{1}{4},\frac{1}{2}$ **45.** $1\pm2i$ **47.** $-2\sqrt{3},2\sqrt{3},0$
49. $-\frac{3}{2},-\frac{3}{2}$ **51.** 49, 225   **53.** $-4,\frac{2}{3}$ **55.** $-\frac{5}{7},3$
57. $1,1,-1,-1$ **59.** $1\pm i$ **61.** $-1,-4$ **63.** $-1,-\frac{27}{13}$
65. $\frac{3\pm\sqrt{57}}{6}$ **67.** 12.1 in.   **69.** 30 mph   **71.** 14.3 min
75. $x=3$ **77.** $y=\frac{2}{3}x$

Study Set Section 10.4 (page 814)

1. quadratic, parabola   **3.** axis of symmetry   **5. a.** a parabola
b. $(1, 0), (3, 0)$ **c.** $(0, -3)$ **d.** $(2, 1)$ **e.** $x = 2$

7.

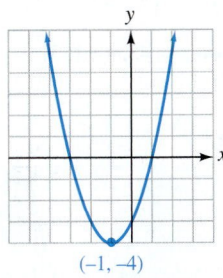

$(-1, -4)$

9. a. 2   **b.** 9, 18
11. $-3, 5$ **13.** $h = -1$;
$f(x) = 2[x - (-1)]^2 + 6$

15.

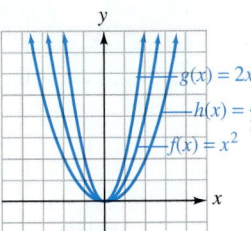

$g(x) = 2x^2$
$h(x) = \frac{1}{2}x^2$
$f(x) = x^2$

17.

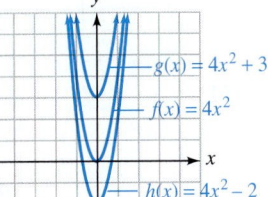

$g(x) = 4x^2 + 3$
$f(x) = 4x^2$
$h(x) = 4x^2 - 2$

19.

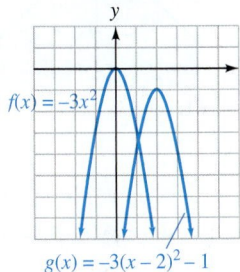

$f(x) = -3x^2$
$g(x) = -3(x - 2)^2 - 1$

21. $(1, 2), x = 1$, upward
23. $(-3, -4), x = -3$,
upward   **25.** $(7.5, 8.5)$,
$x = 7.5$, upward
27. $(1, -2), x = 1$, upward
29. $(2, 21), x = 2$, downward
31. $\left(-\frac{2}{3}, \frac{2}{3}\right), x = -\frac{2}{3}$,
upward

33.

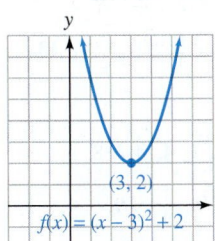

$(3, 2)$
$f(x) = (x - 3)^2 + 2$

35.

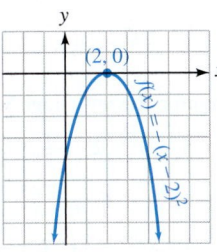

$(2, 0)$
$f(x) = -(x - 2)^2$

37.

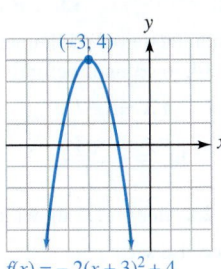

$(-3, 4)$
$f(x) = -2(x + 3)^2 + 4$

39.

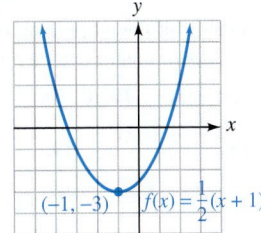

$(-1, -3)$ $f(x) = \frac{1}{2}(x + 1)^2 - 3$

41.

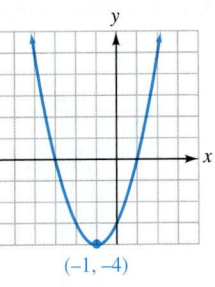

$(-1, -4)$

43.

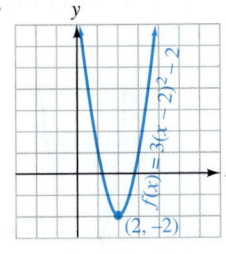

$f(x) = 3(x - 2)^2 - 2$
$(2, -2)$

45.

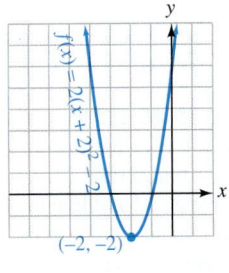

$f(x) = 2(x + 2)^2 - 2$
$(-2, -2)$

47.

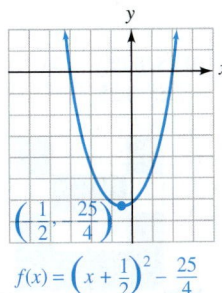

$\left(\frac{1}{2}, \frac{25}{4}\right)$
$f(x) = \left(x + \frac{1}{2}\right)^2 - \frac{25}{4}$

49.

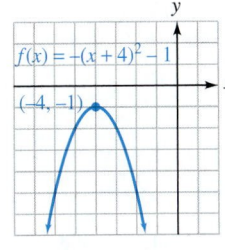

$f(x) = -(x + 4)^2 - 1$
$(-4, -1)$

51.

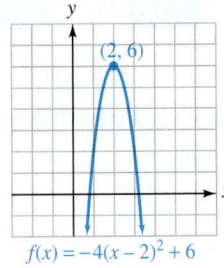

$(2, 6)$
$f(x) = -4(x - 2)^2 + 6$

53. $(1, 0); (0, 1)$
57. $(-2, 0); (-2, 0); (0, 4)$

55. $(-3, 0), (-7, 0); (0, -21)$
59. $(1, 0); (1, 0); (0, -1)$

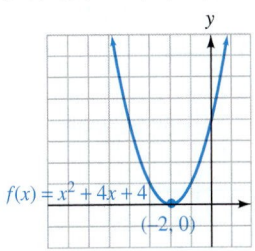

$f(x) = x^2 + 4x + 4$
$(-2, 0)$

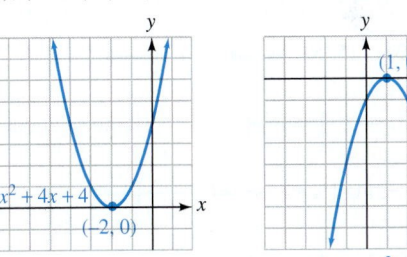

$(1, 0)$
$f(x) = -x^2 + 2x - 1$

61. $(1, -1); (0, 0), (2, 0); (0, 0)$

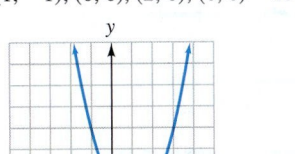

$f(x) = x^2 - 2x$ $(1, -1)$

63. $\left(\frac{3}{2}, 0\right); \left(\frac{3}{2}, 0\right); (0, 9)$

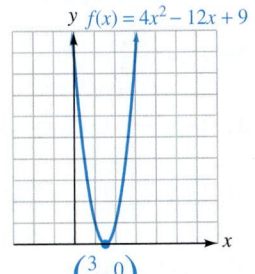

$f(x) = 4x^2 - 12x + 9$
$\left(\frac{3}{2}, 0\right)$

65. $(2, -2)$; $(1, 0)$, $(3, 0)$; $(0, 6)$

67. $(-1, -2)$; no x-intercept; $(0, -8)$

$f(x) = 2x^2 - 8x + 6$

$f(x) = -6x^2 - 12x - 8$

69. $(0.25, 0.88)$ **71.** $(0.50, 7.25)$ **73.** $2, -3$
75. $-1.85, 3.25$

77.

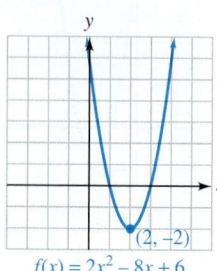

axis of symmetry

79. 3.75 sec, 225 ft
81. 250 ft by 500 ft
83. 15 min, $160
85. 1968, 1.5 million; the U.S. involvement in the war in Vietnam was at its peak
87. 200, $7,000 **95.** $4a^2\sqrt{b}$
97. $\frac{\sqrt{6}}{10}$
99. $15b - 6\sqrt{15b} + 9$

Study Set Section 10.5 (page 827)

1. quadratic **3.** two **5.** $(-\infty, -1), (-1, 4), (4, \infty)$

7. a.   **b.**

9. a. $(-3, 2)$ **b.** $(-\infty, -1] \cup [1, \infty)$ **11. a.** solid
b. yes **13.** $x^2 - 6x - 7 \geq 0$

15. $(1, 4)$

17. $(-\infty, 3) \cup (5, \infty)$

19. $[-4, 3]$ **21.** no solutions

23. $(-\infty, -3] \cup [3, \infty)$

25. $(-5, 5)$

27. $(-\infty, 0) \cup \left(\frac{1}{2}, \infty\right)$

29. $\left(-\infty, -\frac{5}{3}\right) \cup (0, \infty)$

31. $(-\infty, -3) \cup (1, 4)$

33. $\left(-\frac{1}{2}, \frac{1}{3}\right) \cup \left(\frac{1}{2}, \infty\right)$

35. $(0, 2) \cup (8, \infty)$

37. $\left[-\frac{34}{5}, -4\right) \cup (3, \infty)$

39. $(-4, -2] \cup (-1, 2]$

41. $(-\infty, -2) \cup (-2, \infty)$ **43.** $(-1, 3)$

45. $(-\infty, -3) \cup (2, \infty)$

47.

$y = x^2 + 1$

$y < x^2 + 1$

49.

$y = x^2 + 5x + 6$

$y \leq x^2 + 5x + 6$

51.

$y = |x + 4|$

$y < |x + 4|$

53.

$y = -|x| + 2$

$y \leq -|x| + 2$

55. $(-2,100, -900) \cup (900, 2,100)$ **61.** $x = ky$
63. $t = kxy$

Key Concept (page 830)

1. $0, -2$ **2.** $-3, -5$ **3.** $-\frac{3}{2}, -1$ **4.** $\pm 2\sqrt{6}$ **5.** $4, 10$

6. $0 \pm \frac{4\sqrt{3}}{3}i$ **7.** $-5 \pm 4\sqrt{2}$ **8.** $\frac{1 \pm \sqrt{2}}{2}$ **9.** $-1 \pm i$

10. $\frac{3 \pm \sqrt{17}}{4}$ **11.** $23, -17$ **12.** $-\frac{1}{3} \pm \frac{\sqrt{2}}{3}i$ **13.** $1, -2$

Chapter Review (page 831)

1. $-5, -4$ **2.** $-\frac{1}{3}, -\frac{5}{2}$ **3.** $\pm 2\sqrt{7}$ **4.** $4, -8$

5. $0, -\frac{11}{5}$ **6.** $\pm \frac{7\sqrt{5}}{5}$ **7.** $\pm 5i$ **8.** $\frac{1}{4}$ **9.** $-4, -2$

10. $\frac{3 \pm \sqrt{3}}{2}$ **11.** $1 \pm 2i\sqrt{3}$ **12.** $r = \frac{\sqrt{\pi A}}{\pi}$ **13.** 2 is not
a factor of the numerator. Only common factors of the numerator and denominator can be removed. **14.** 6 seconds before midnight **15.** $0, 10$ **16.** $5 \pm \sqrt{7}$ **17.** $\frac{1}{2}, -7$

18. $\frac{13 \pm \sqrt{163}}{3}$ **19.** $-\frac{3}{4} \pm \frac{\sqrt{15}}{4}i$ **20.** $\frac{2}{3} \pm \frac{\sqrt{2}}{3}i$

21. $\frac{-3 \pm \sqrt{29}}{10}$ **22.** $24 or $26 **23.** sides: 1.25 in. wide;
top/bottom: 2.5 in. wide **24.** 0.7 sec, 1.8 sec
25. irrational, unequal **26.** complex conjugates
27. equal rational numbers **28.** rational, unequal

29. $1, 144$ **30.** $8, -27$ **31.** $i, -i, \frac{\sqrt{6}}{3}, -\frac{\sqrt{6}}{3}$
32. $1, -\frac{8}{5}$ **33.** $4 \pm i$ **34.** $1, 1, -1, -1$
35. $-\frac{2}{5}, -\frac{2}{5}$ **36.** about 81 min **37.** $44.3 billion

38.

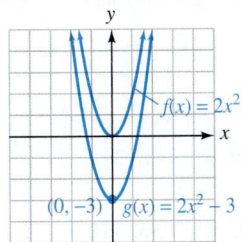

39.

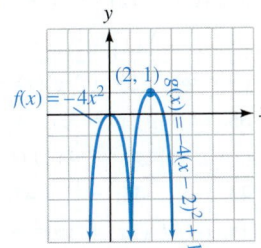

21.

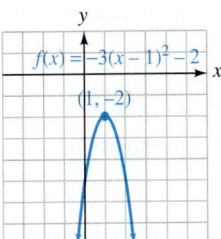

22.

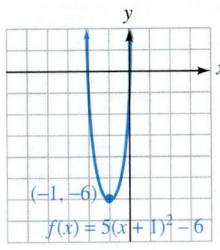

23. $\left(-\frac{1}{4}, -\frac{9}{8}\right), (-1, 0), \left(\frac{1}{2}, 0\right); (0, -1)$ **24.** 211 ft

25. $(-\infty, -2) \cup (4, \infty)$

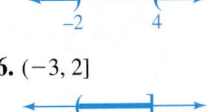

40.

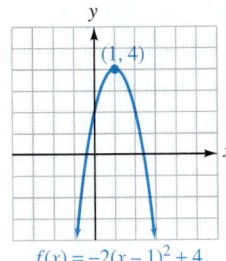

41.

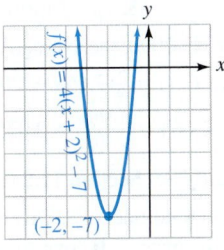

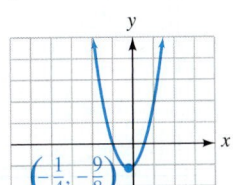

26. $(-3, 2]$

42.

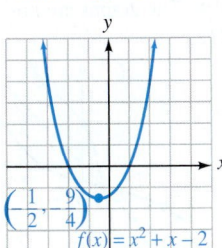

43. 1921; 6,469,326

44. $-2, \frac{1}{3}$

27.

We don't know whether x is positive or negative. When we multiply both sides by x, we don't know whether or not to reverse the inequality symbol.

28.

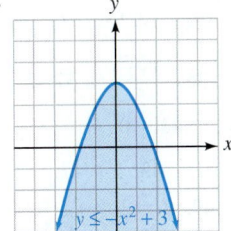

29. $-3, 2$ **30.** $[-3, 2]$

45. $(-\infty, -7) \cup (5, \infty)$

46. $[-9, 9]$

47. $(-\infty, 0) \cup [3/5, \infty)$

48. $(-7/2, 1) \cup (4, \infty)$

49. $\left[-4, \frac{2}{3}\right]$ **50.** $(-\infty, 0) \cup (1, \infty)$

51.

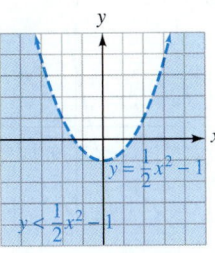

52.

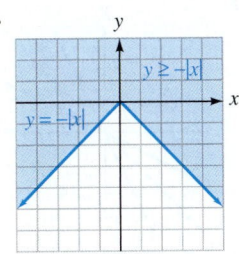

Cumulative Review Exercises, Chapters 1–10 (page 837)

1. $y = 3x + 2$ **2.** $y = -\frac{2}{3}x - 2$ **3.** an increase of about 13,333 a year   **4.**   **5.** $(2, -1, 1)$

Chapter Test (page 835)

1. $0, -6$ **2.** $\pm 2i$ **3.** $-7 \pm 5\sqrt{2}$ **4.** $-\frac{3}{2}, -\frac{5}{3}$ **5.** 144

6. $-3, \frac{8}{3}$ **7.** $\frac{4 \pm \sqrt{6}}{2}$ **8.** $1 \pm \sqrt{5}$ **9.** $2 \pm 3i$

10. $-5, -3$ **11.** $1, \frac{1}{4}$ **12.** $-\frac{1}{2} \pm \frac{\sqrt{3}}{2}i$ **13.** $2, -2,$

$i\sqrt{3}, -i\sqrt{3}$ **14.** $-\frac{4}{5}, \frac{4}{7}$ **15.** $c = \frac{\sqrt{Em}}{m}$

16. a. complex conjugates   **b.** rational, equal   **17.** 4.5 ft by 1,502 ft   **18.** about 46 min   **19.** 20 in.   **20.** iii

6.

7. $\left(-\infty, -\frac{10}{9}\right)$

8. $(-\infty, -10] \cup [15, \infty)$

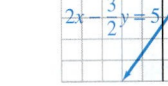

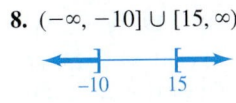

9. $\frac{y^8}{9x^2}$ **10. a.** all real numbers from 0 through 24   **b.** 0.5

c. 1.5   **d.** about -1.4 **e.** The low tide mark was -2.5 m.

f. 0, 2, 9, 17   **11.** $3a^3 + 10a^2 + 6a - 4$ **12.** $\frac{x + y}{x - y}$

13. 0   **14.** $\frac{1}{2r(r+2)}$ **15.** $(x^2 + 4y^2)(x + 2y)(x - 2y)$

16. $2a^2(3a + 2)(5a - 4)$ **17.** $(x - y)(x - 4)$

18. $(2x^2 + 5y)(4x^4 - 10x^2y + 25y^2)$

19. $(x + 5 + y^4)(x + 5 - y^4)$ **20.** $(7s^3 - 6n^2)^2$

21. $2, -\frac{5}{2}$ **22.** $0, \frac{2}{3}, -\frac{1}{2}$ **23.** 5; 3 is extraneous

24. $b = \dfrac{-RTV^2 + aV + PV^3}{PV^2 + a}$

25. D: $(-\infty, \infty)$, R: $(-\infty, \infty)$ **26.** D: $(0, \infty)$, R: $(0, \infty)$

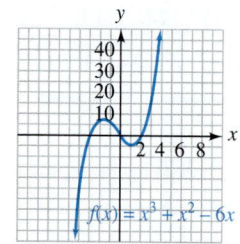

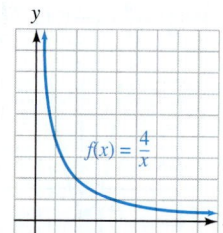

$f(x) = x^3 + x^2 - 6x$ $f(x) = \frac{4}{x}$

27. 9   **28.** D: $[2, \infty)$, R: $[0, \infty)$ **29.** $-3x$

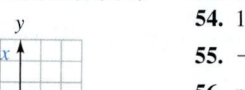

30. $4t\sqrt{3t}$ **31.** $\frac{1}{16}$

$f(x) = \sqrt{x - 2}$

32. $x^{17/12}$

33. $-12\sqrt[4]{2} + 10\sqrt[4]{3}$

34. $-18\sqrt{6}$

35. $\frac{x + 3\sqrt{x} + 2}{x - 1}$

36. $\frac{5\sqrt[3]{x^2}}{x}$ **37.** $2, 7$

38. $\frac{1}{4}$ **39.** $3\sqrt{2}$ in.   **40.** $2\sqrt{3}$ in.   **41.** 10

42. $-i$ **43.** $-5 + 17i$ **44.** $\frac{3}{2} + \frac{1}{2}i$ **45.** $3 + 4i$

46. $0 - \frac{2}{3}i$ **47.** 9   **48.** $1, -\frac{3}{2}$ **49.** $2 \pm 2i$

50. 10 ft by 18 ft   **51.** 50 m and 120 m

52. $(-2, 4)$; $(-4, 0)$, $(0, 0)$; $(0, 0)$ **53.** 9, 16

54. $1, 1, -1, -1$

55. $-\frac{3}{4}$

56. no solution

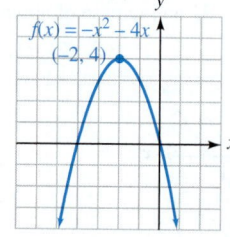

$f(x) = -x^2 - 4x$
$(-2, 4)$

Study Set Section 11.1 (page 846)

1. sum, $f(x) + g(x)$, difference, $f(x) - g(x)$ **3.** domain

5. identity   **7. a.** $g(3)$ **b.** $g(3)$ **9.** $-2, -10, 0$

11. a. 7   **b.** 8   **c.** 12   **d.** 0   **13.** $g(x), (3x - 1), 9x, 2x$

15. $(f + g)(x) = 7x, (-\infty, \infty)$ **17.** $(f \cdot g)(x) = 12x^2$,

$(-\infty, \infty)$ **19.** $(g - f)(x) = x, (-\infty, \infty)$ **21.** $(g/f)(x) = \frac{4}{3}$,

$(-\infty, 0) \cup (0, \infty)$ **23.** $(f + g)(x) = 3x - 2, (-\infty, \infty)$

25. $(f \cdot g)(x) = 2x^2 - 5x - 3, (-\infty, \infty)$

27. $(g - f)(x) = -x - 4, (-\infty, \infty)$ **29.** $(g/f)(x) = \frac{x - 3}{2x + 1}$,

$\left(-\infty, -\frac{1}{2}\right) \cup \left(-\frac{1}{2}, \infty\right)$ **31.** $(f - g)(x) = -2x^2 + 3x - 3$,

$(-\infty, \infty)$ **33.** $(f/g)(x) = \frac{3x - 2}{2x^2 + 1}, (-\infty, \infty)$

35. $(f - g)(x) = 3, (-\infty, \infty)$ **37.** $(g/f)(x) = \frac{x^2 - 4}{x^2 - 1}$,

$(-\infty, -1) \cup (-1, 1) \cup (1, \infty)$ **39.** 7   **41.** 24

43. -1 **45.** $-\frac{1}{2}$ **47.** $(f \circ g)(x) = 2x^2 - 1$

49. $(g \circ f)(2x) = 16x^2 + 8x$ **51.** 58   **53.** 110

55. 2   **57.** $(g \circ f)(x) = 9x^2 - 9x + 2$ **59.** 16

61. $\frac{1}{9}$ **63.** $(g \circ f)(8x) = 64x^2$ **67.** 500, 503, 1,003

69. In 2000, the average combined score on the SAT was 1,019.

71. $C(t) = \frac{5}{9}(2,668 - 200t)$ **73. a.** about \$37.50

b. $C(m) = \frac{1.50m}{20} = 0.075m$ **79.** $\frac{1}{c - d}$

Study Set Section 11.2 (page 857)

1. one-to-one   **3.** inverses   **5.** images, symmetric

7. a. once   **b.** one-to-one   **9.** x **11.** y, interchange

13. yes, no   **15.** yes   **17.** 2   **19.** The graphs are not

symmetric about the line $y = x$.   **21.**

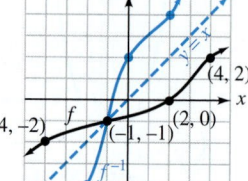

$(4, 2)$
$(-4, -2)$ $(-1, -1)$ $(2, 0)$
f f^{-1} $y = x$

23. $y, 2y, 3, y, f^{-1}(x)$

25. inverse of, inverse

27. yes   **29.** no

31. no   **33.** no   **35.** one-to-one

37. not one-to-one   **39.** not one-to-one   **41.** $f^{-1}(x) = \frac{x - 4}{2}$

43. $f^{-1}(x) = 5x - 4$ **45.** $f^{-1}(x) = 5x + 4$

47. $f^{-1}(x) = \frac{2}{x} + 3$ **49.** $f^{-1}(x) = \frac{4}{x}$

51. $f^{-1}(x) = \sqrt[3]{x - 8}$ **53.** $f^{-1}(x) = x^3$

55. $f^{-1}(x) = \sqrt[3]{x} - 10$ **57.** $f^{-1}(x) = \sqrt[3]{\frac{x + 3}{2}}$

63.

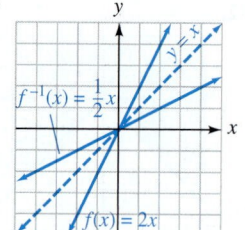

$f^{-1}(x) = \frac{1}{2}x$ $y = x$
$f(x) = 2x$

65.

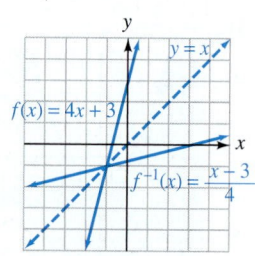

$y = x$
$f(x) = 4x + 3$
$f^{-1}(x) = \frac{x - 3}{4}$

67.

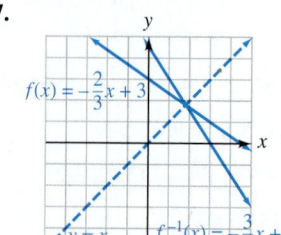

$f(x) = -\frac{2}{3}x + 3$
$y = x$ $f^{-1}(x) = -\frac{3}{2}x + \frac{9}{2}$

69.

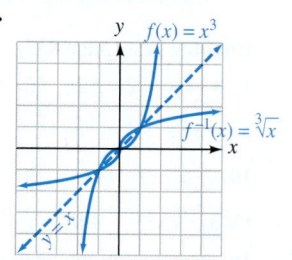

$f(x) = x^3$
$f^{-1}(x) = \sqrt[3]{x}$
$y = x$

71.

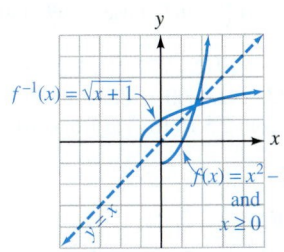

$f^{-1}(x) = \sqrt{x+1}$

$f(x) = x^2 - 1$
and
$x \geq 0$

73. a. yes, no
b. No. Twice during this period, the person's anxiety level was at the maximum threshold value.
79. $3 - 8i$
81. $18 - i$
83. $-28 - 96i$

15. an exponential function

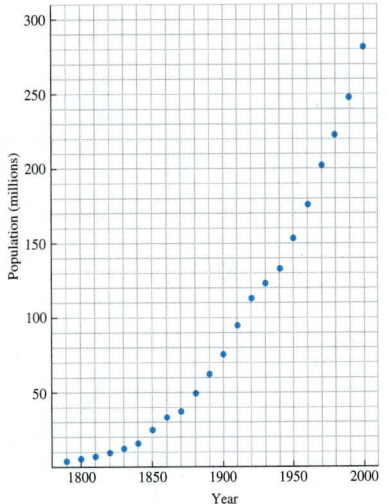

17. $2.7182818\ldots; e$ **19.** $1,000, 10, 0.9, 1,000$

Study Set Section 11.3 (page 870)

1. exponential   **3.** $(0, \infty)$ **5.** yes   **7.** increasing

9.

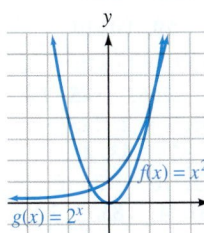

$f(x) = x^2$

$g(x) = 2^x$

11. $A = P\left(1 + \dfrac{r}{k}\right)^{kt}$, $FV = PV(1 + i)^n$
13. $g(x) = 2^x + 3; h(x) = 2^x - 2$
15. $\left(1 + \dfrac{r}{k}\right), kt$
17. 2.6651 **19.** 36.5548
21. 8 **23.** $7^{3\sqrt{3}}$

21.

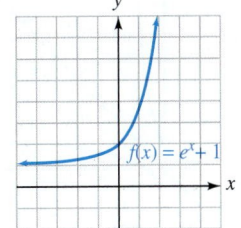

$f(x) = e^x + 1$

23.

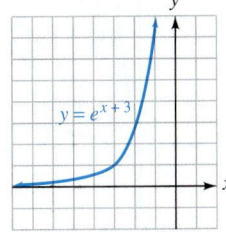

$y = e^{x+3}$

25.

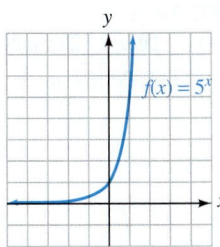

$f(x) = 5^x$

27.

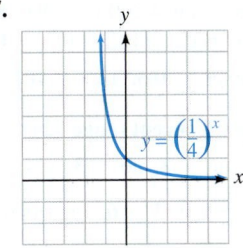

$y = \left(\dfrac{1}{4}\right)^x$

25.

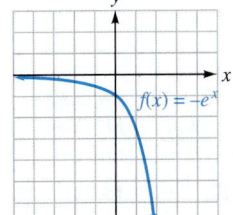

$f(x) = -e^x$

27.

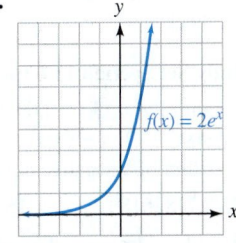

$f(x) = 2e^x$

29.

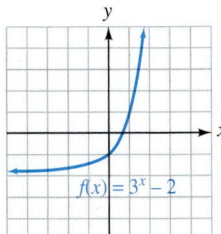

$f(x) = 3^x - 2$

31.

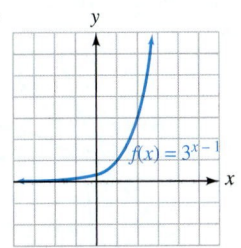

$f(x) = 3^{x-1}$

29. \$13,375.68   **31.** \$7,518.28 from annual compounding, \$7,647.95 from continuous compounding   **33.** \$6,849.16
35. about 9.3 billion   **37.** 315   **39.** 51   **41.** 12 hr
43. 49 mps   **45.** about 72 yr   **51.** $4x^2\sqrt{15x}$ **53.** $10y\sqrt{3y}$

33. increasing

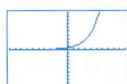

35. decreasing

37. \$22,080.40   **39.** \$32.03   **41.** \$2,273,996.13
43. a. about 1500, about 1825   **b.** 6.5 billion
c. exponential   **45. a.** at the end of the 2nd year
b. at the end of the 4th year   **c.** during the 7th year
47. 1.679046×10^8 **49.** 5.0421×10^{-5} coulombs
51. \$1,115.33   **57.** 40   **59.** $120°$

Study Set Section 11.5 (page 891)

1. logarithmic   **3.** $(-\infty, \infty)$ **5.** yes   **7.** increasing
9. $x = b^y$ **11.** inverse   **13.** $-2, -1, 0, 1, 2$ **15.** none,
none, $-3, \frac{1}{2}, 8$ **17.** $-0.30, 0, 0.30, 0.60, 0.78, 0.90, 1$
19. They decrease.   **21. a.** 10   **b.** x **23.** $3^4 = 81$
25. $10^1 = 10$ **27.** $4^{-3} = \frac{1}{64}$ **29.** $5^{1/2} = \sqrt{5}$
31. $\log_8 64 = 2$ **33.** $\log_4 \frac{1}{16} = -2$ **35.** $\log_{1/2} 32 = -5$
37. $\log_x z = y$ **39.** 3   **41.** 2   **43.** 5   **45.** $\frac{1}{2}$ **47.** -1
49. 6   **51.** 64   **53.** 5   **55.** $\frac{1}{25}$ **57.** $\frac{1}{6}$ **59.** 10
61. $\frac{2}{3}$ **63.** 5   **65.** $\frac{1}{3}$ **67.** 1,000   **69.** 4   **71.** 1
73. 0.5119   **75.** -2.3307 **77.** 25.25   **79.** 0.02

Study Set Section 11.4 (page 879)

1. the natural exponential function   **3.** $(0, \infty)$ **5.** yes
7. increasing   **9.** continuous   **11.** 2.72
13.

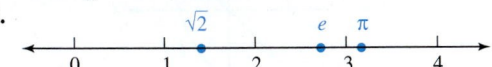

$\sqrt{2}$ e π
0   1   2   3   4

81. increasing

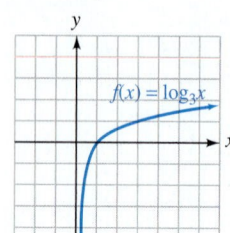

83. decreasing

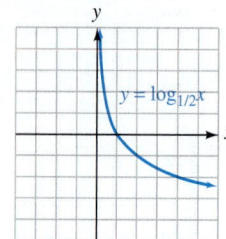

85.

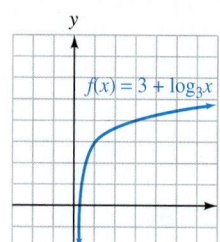

87.

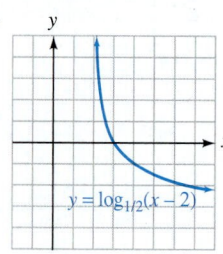

89.

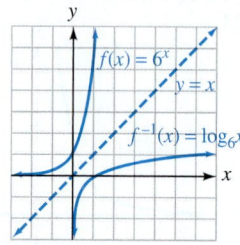

91.

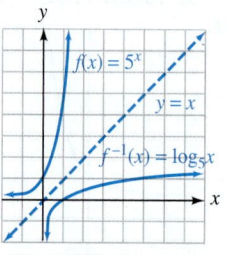

93. 29 db   **95.** 49.5 db   **97.** 4.4   **99.** 2,500 micrometers
101. 3 yr old   **103.** 10.8 yr   **107.** 10   **109.** 0; $-\frac{5}{9}$ does not check

Study Set Section 11.6 (page 898)

1. natural, e **3.** $-0.69, 0, 0.69, 1.39, 1.79, 2.08, 2.30$
5. y-axis   **7.** $(-\infty, \infty)$ **9.** 1, asymptote   **11.** The logarithm of a negative number or zero is not defined.
13. $2.7182818\ldots, e$ **15.** ℓ, n **17.** $\frac{\ln 2}{r}$ **19.** 1   **21.** 6
23. -1 **25.** $\frac{1}{4}$ **27.** 3.5596   **29.** -5.3709 **31.** 0.5423
33. undefined   **35.** 4.0645   **37.** 69.4079   **39.** 0.0245
41. 2.7210   **43.** 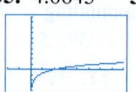   **45.**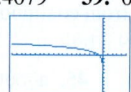

47. 5.8 yr   **49.** 9.2 yr   **51.** about 3.5 hr   **57.** $y = 5x$
59. $y = -\frac{3}{2}x + \frac{13}{2}$ **61.** $x = 2$

Study Set Section 11.7 (page 909)

1. product   **3.** power   **5.** 0   **7.** M, N **9.** $-$ **11.** x
13. $\neq$ **15.** $\log_a b$ **17.** $10^0 = 1, 10^1 = 10, 10^2 = 10^2$
19. 0   **21.** 7   **23.** 10   **25.** 2   **27.** 1   **29.** 7
31. rs, t, r, s **39.** $2 + \log_2 5$ **41.** $\log_6 x - 2$
43. $4 \ln y$ **45.** $\frac{1}{2} \log 5$ **47.** $\log x + \log y + \log z$
49. $1 + \log_2 x - \log_2 y$ **51.** $3 \log x + 2 \log y$
53. $\frac{1}{2}(\log_b x + \log_b y)$ **55.** $\frac{1}{3} \log_a x - \frac{1}{4} \log_a y - \frac{1}{4} \log_a z$
57. $\ln x + \frac{1}{2} \ln z$ **59.** $\log_2 \frac{x+1}{x}$ **61.** $\log x^2 y^{1/2}$

63. $\log_b \frac{z^{1/2}}{x^3 y^2}$ **65.** $\ln \frac{\frac{x}{z} + x}{\frac{y}{z} + y} = \ln \frac{x}{y}$ **67.** false   **69.** false
71. true   **73.** 1.4472   **75.** -1.1972 **77.** 1.1972
79. 1.8063   **81.** 1.7712   **83.** -1.0000 **85.** 1.8928
87. 2.3219   **89.** 4.8   **91.** from 2.5×10^{-8} to 1.6×10^{-7}
95. $-\frac{7}{6}$ **97.** $\left(1, -\frac{1}{2}\right)$

Study Set Section 11.8 (page 920)

1. exponential   **3.** It is a solution.   **5.** yes   **7.** 5
9. a. $10^2 = x + 1$ **b.** $e^2 = x + 1$ **11.** $\log 9$
13. $x \log 7 = 12$ **15. a.** 1.2920   **b.** 2.2619
17. a. $A_0 2^{-t/h}$ **b.** $P_0 e^{kt}$ **19.** log, log 2   **21.** 8
23. $-\frac{3}{4}$ **25.** $3, -1$ **27.** $-2, -2$ **29.** 1.1610
31. 1.2702   **33.** 1.7095   **35.** 0   **37.** ± 1.0878
39. $0, 1.0566$ **41.** 0.7324   **43.** -13.2662 **45.** 1.8
47. 2.1   **49.** 9,998   **51.** -93 **53.** $e \approx 2.7183$
55. 19.0855   **57.** 2   **59.** 3   **61.** -7 **63.** 4
65. $10, -10$ **67.** 50   **69.** 20   **71.** 10   **73.** 10
75. no solution   **77.** 6   **79.** 9   **81.** 4   **83.** 1, 7
85. 20   **87.** 8   **89.** 5.1 yr   **91.** 42.7 days
93. about 4,200 yr   **95.** 5.6 yr   **97.** 5.4 yr
99. because $\ln 2 \approx 0.7$ **101.** 25.3 yr   **103.** 2.828 times
larger   **105.** 13.3   **107.** $\frac{1}{3} \ln 0.75$ **113.** $\sqrt{137}$ in.

Key Concept (page 924)

1. no   **2.** no   **3.** yes   **4.** $f^{-1}(x) = -\frac{x+1}{2}$
5. $-2, 1, 3$ **6.** $\log 1,000 = 3$ **7.** $2^{-3} = \frac{1}{8}$
8. 2   **9.** $\frac{3}{2}$ **10.** 2.71828   **11.** e **12.** 0.7558
13.

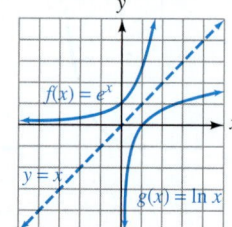

Chapter Review (page 925)

1. $(f + g)(x) = 3x + 1, (-\infty, \infty)$ **2.** $(f - g)(x) = x - 1$,
$(-\infty, \infty)$ **3.** $(f \cdot g)(x) = 2x^2 + 2x, (-\infty, \infty)$
4. $(f/g)(x) = \frac{2x}{x+1}, (-\infty, -1) \cup (-1, \infty)$ **5.** 3   **6.** 5
7. $(f \circ g)(x) = 4x^2 + 4x + 3$ **8.** $(g \circ f)(x) = 2x^2 + 5$
9. a. 12   **b.** 27   **c.** 7   **d.** 0   **10.** $C(m) = \frac{1.85m}{8}$
11. no   **12.** yes   **13.** yes   **14.** no   **15.** yes   **16.** no
17. $-6, -1, 7, 20$ **18.**

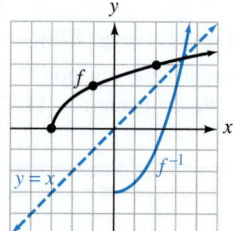

19. $f^{-1}(x) = \dfrac{x+3}{6}$ **20.** $f^{-1}(x) = \dfrac{4}{x} + 1$

21. $f^{-1}(x) = \sqrt[3]{x} - 2$ **22.** $f^{-1}(x) = 6x + 1$

23. 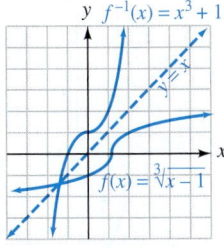   **25.** $5^{4\sqrt{6}}$ **26.** $2^{2\sqrt{7}}$

27. D: $(-\infty, \infty)$, R: $(0, \infty)$ **28.** D: $(-\infty, \infty)$, R: $(0, \infty)$

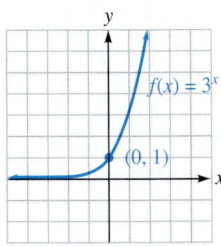

29. D: $(-\infty, \infty)$, R: $(-2, \infty)$ **30.** D: $(-\infty, \infty)$, R: $(0, \infty)$

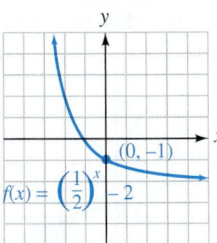 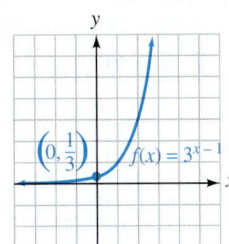

31. the x-axis ($y = 0$)
32. an exponential function

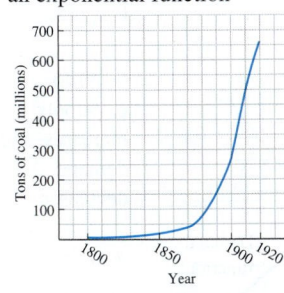

33. $2,189,703.45   **34.** about $2,015
35. 8.22%   **36.** $2,324,767.37
37. D: $(-\infty, \infty)$, R: $(1, \infty)$ **38.** D: $(-\infty, \infty)$, R: $(0, \infty)$

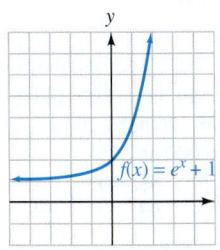 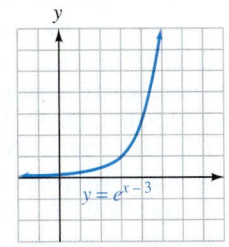

39. about 459,920,041   **40.** The exponent on the base e is negative.   **41.** D: $(0, \infty)$, R: $(-\infty, \infty)$ **42.** Since there is no real number such that $10^? = 0$, $\log 0$ is undefined.   **43.** $4^3 = 64$

44. $\log_7 \frac{1}{7} = -1$ **45.** 2   **46.** -2 **47.** 0   **48.** not possible   **49.** $\frac{1}{2}$ **50.** 3   **51.** 32   **52.** $\frac{1}{81}$ **53.** 4

54. 10   **55.** $\frac{1}{2}$ **56.** 9   **57.** 0.6542   **58.** 26.9153

59. 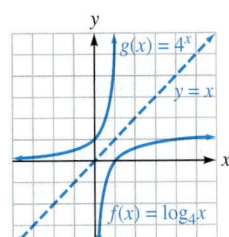   **60.**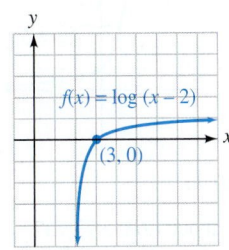

61. 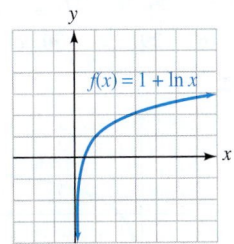   **62.**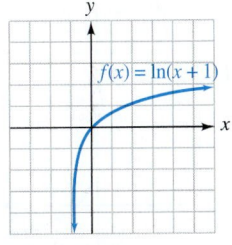

63. about 53   **64.** about 4.4   **65.** 1   **66.** 2   **67.** -5
68. $\frac{1}{2}$ **69.** undefined   **70.** undefined   **71.** 0   **72.** -7
73. 6.1137   **74.** -0.1625 **75.** 10.3398   **76.** 0.0002
77. $\log x = \log_{10} x$ and $\ln x = \log_e x$ **78.** $f^{-1}(x) = e^x$
79. 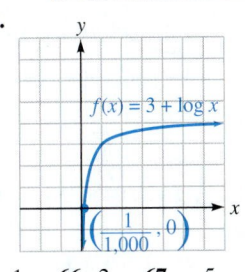   **80.**

81. about $48\frac{1}{2}$ yr   **82.** about 72 in. (6 ft)   **83.** 0
84. 1   **85.** 3   **86.** 4   **87.** $3 + \log_3 x$ **88.** $2 - \log x$
89. $\frac{1}{2}\log_5 27$ **90.** $\log_b 10 + \log_b a + 1$
91. $2\log_b x + 3\log_b y - \log_b z$ **92.** $\frac{1}{2}(\ln x - \ln y - 2\ln z)$
93. $\log_2 \dfrac{x^3 z^7}{y^5}$ **94.** $\log_b \dfrac{\sqrt{x+2}}{y^3 z^7}$ **95.** 2.6609   **96.** 3.0000
97. 1.7604   **98.** about 7.9×10^{-4} gram-ions/liter
99. 2   **100.** $-3, -1$ **101.** 1.7712   **102.** 2.7095
103. 1.9459   **104.** -8.0472 **105.** 104   **106.** 9
107. 25, 4   **108.** 4   **109.** 4, 3   **110.** 2   **111.** 6
112. 31   **113.** $0.76787 \neq -0.27300$ **114.** about 3,300 yr
115. about 91 days   **116.** 2, 5

Chapter Test (page 931)

1. $(f + g)(x) = 4x^2 - 2x + 11, (-\infty, \infty)$
2. $(g/f)(x) = \dfrac{4x^2 - 3x + 2}{x + 9}, (-\infty, -9) \cup (-9, \infty)$ **3.** 76
4. $(f \circ g)(x) = 32x^2 - 128x + 131$ **5.** -16 **6.** 17   **7.** yes

8. no   **9.**

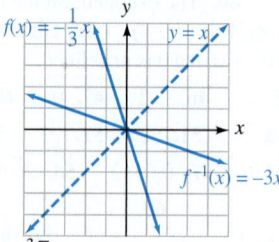

10. $f^{-1}(x) = \sqrt[3]{x} + 15$ **12. a.** yes   **b.** yes
c. 80; when the temperature of the tire tread is 260°,
the vehicle is traveling 80 mph
13. D: $(-\infty, \infty)$, R: $(1, \infty)$ **14.** D: $(-\infty, \infty)$, R: $(0, \infty)$

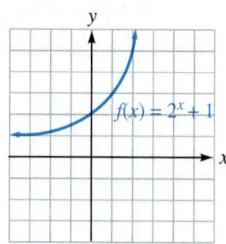   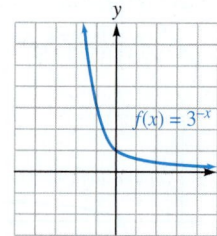

15. $\frac{3}{64}$ g = 0.046875 g   **16.** $1,060.90

17.

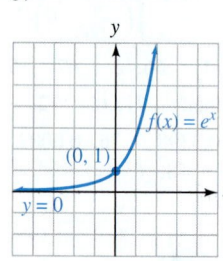

18. about 1,631,973,737
19. $6^{-2} = \frac{1}{36}$ **20. a.** D: $(0, \infty)$,
R: $(-\infty, \infty)$ **b.** $f^{-1}(x) = 10^x$
21. 2   **22.** −2
23. undefined   **24.** −6
25. $\frac{1}{2}$ **26.** 0   **27.** 2
28. 16   **29.** $\frac{1}{27}$ **30.** e

31.   **32.**

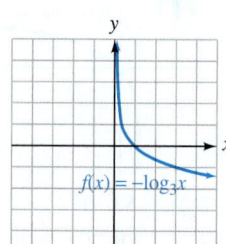   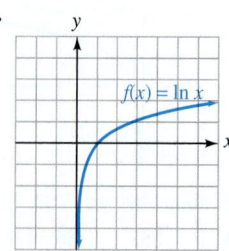

33. 6.4   **34.** about 46   **35.** .0871   **36.** 0.5646

37. $2 \log_b a + 1 + 3 \log_b c$ **38.** $\ln \frac{b\sqrt{a+2}}{c^3}$ **39.** 0.6826
40. 4   **41.** 1   **42.** 10   **43.** 5   **44.** about 20 min

Study Set Section 12.1 (page 945)

1. system   **3.** infinitely   **5. a.** true   **b.** false   **c.** true
d. true   **7. a.** 3; −4 (answers may vary)   **b.** 2; −3
(answers may vary)   **9.** $3x - 7, 5, 4, 12, 3, 2$
11.   **13.**

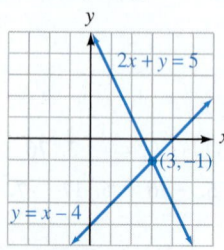   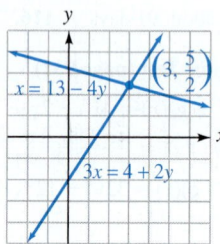

15.

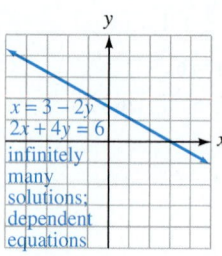

17.

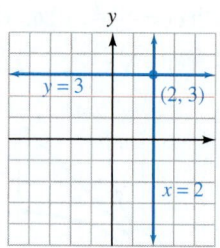

19.

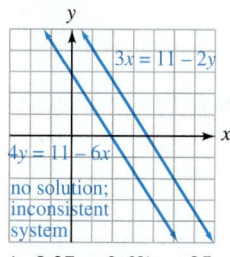

21.

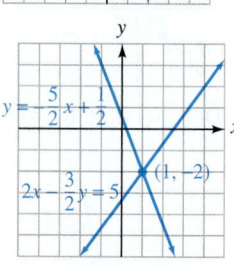

23. $(-0.37, -2.69)$ **25.** $(-7.64, 7.04)$ **27.** $(2, 2)$
29. $(5, 3)$ **31.** $(-2, 4)$ **33.** infinitely many solutions
35. $\left(5, \frac{3}{2}\right)$ **37.** $\left(10, \frac{15}{2}\right)$ **39.** $(5, 2)$ **41.** $(-4, -2)$
43. $(1, 2)$ **45.** $\left(\frac{1}{2}, \frac{2}{3}\right)$ **47.** infinitely many solutions,
dependent equations   **49.** no solution, inconsistent system
51. $(4, 8)$ **53.** $(-6, 16)$ **55.** $(2, -3)$ **57.** $(9, -1)$
59. $(2, 3)$ **61.** $\left(-\frac{1}{3}, 1\right)$

63.   **65.**

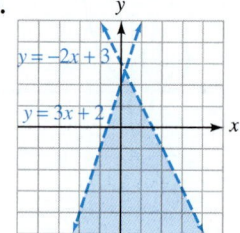   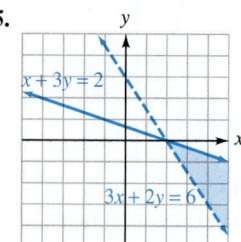

67. $(2,000, 50)$

69. a.   **b.** $140
c. Supply
increases, and
demand
decreases.

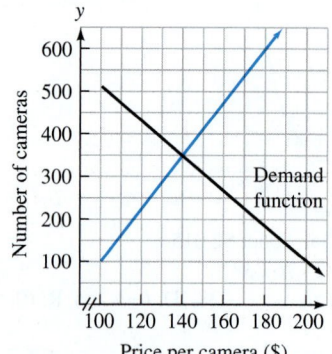

71. $(250, 2,500)$; if the company makes 250 widgets, the cost to
make them and the revenue obtained from their sale will be
equal: $2,500.   **73.** $475, $800   **75.** 16 m by 20 m
77. 150°, 30°   **79.** $6,000 at 6%, $6,000 at 7.5%
81. 55 mph   **83.** 50 g of A, 60 g of B **85.** 103
87. 148 g of the 0.2%, 37 g of the 0.7%   **93.** $r = \frac{A - p}{pt}$
95. $P_1 = \frac{P_2 V_2}{V_1}$

Study Set Section 12.2 (page 959)

1. system   **3.** three   **5.** dependent   **7. a.** no solution
b. no solution   **9.** $x + 2y - 3z = -6$ **11.** yes
13. $(1, 1, 2)$ **15.** $(0, 2, 2)$ **17.** $(3, 2, 1)$ **19.** no solution,
inconsistent system   **21.** $(60, 30, 90)$ **23.** $(2, 4, 8)$
25. infinitely many solutions, dependent equations
27. $(2, 6, 9)$ **29.** 30 expensive, 50 middle-priced, 100
inexpensive   **31.** 2, 3, 1   **33.** 3 poles, 2 bears, 4 deer
35. 78%, 21%, 1%   **37. a.** infinitely many solutions, all lying
on the line running down the binding   **b.** 3 parallel planes
(shelves); no solution   **c.** each pair of planes (cards) intersect;
no solution   **d.** 3 planes (faces of die) intersect at a corner;
1 solution   **39.** $y = \frac{1}{2}x^2 - 2x - 1$
41. $x^2 + y^2 - 2x - 2y - 2 = 0$
43. $A = 40°, B = 60°, C = 80°$ **45.** 12, 15, 21
49.

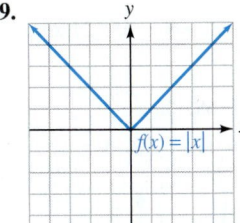

51.

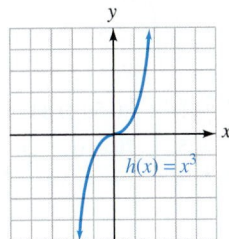

Study Set Section 12.3 (page 971)

1. matrix   **3.** rows, columns   **5.** augmented   **7. a.** 2×3
b. 3×4 **9.** $\begin{cases} x - y = -10 \\ y = 6 \end{cases}$; $(-4, 6)$ **11.** It has no
solution. The system is inconsistent.   **13. a.** multiply row 1
by $\frac{1}{3}$; $\begin{bmatrix} 1 & 2 & -3 & | & 0 \\ 1 & 5 & -2 & | & 1 \\ -2 & 2 & -2 & | & 5 \end{bmatrix}$ **b.** to row 2, add -1 times row 1;
$\begin{bmatrix} 1 & 2 & -3 & | & 0 \\ 0 & 3 & 1 & | & 1 \\ -2 & 2 & -2 & | & 5 \end{bmatrix}$ **15.** $-1, 1, -5, 2, y, 4$ **17.** $(1, 1)$
19. $(2, -3)$ **21.** $(-1, -1)$ **23.** $(0, -3)$ **25.** $(1, 2, 3)$
27. $(4, 5, 4)$ **29.** $(2, 1, 0)$ **31.** $(-1, -1, 2)$
33. $(-3, 2, 1)$ **35.** $\left(\frac{1}{2}, 1, -2\right)$ **37.** no solution,
inconsistent system   **39.** infinitely many solutions, dependent
equations   **41.** $(0, 1, 3)$ **43.** no solution, inconsistent
system   **45.** $(-4, 8, 5)$ **47.** infinitely many solutions,
dependent equations   **49.** 22°, 68°   **51.** 40°, 65°, 75°
53. 262, 144   **55.** 76°, 104°   **57.** founder's circle: 100; box
seats: 300; promenade: 400   **61.** $m = \frac{y_2 - y_1}{x_2 - x_1}$
63. $y - y_1 = m(x - x_1)$

Study Set Section 12.4 (page 982)

1. number   **3.** minor   **5.** rows, columns   **7.** dependent,
inconsistent   **9.** ad, bc **11.** $\begin{vmatrix} 3 & 4 \\ 2 & -3 \end{vmatrix}$ **13.** $\left(\frac{7}{11}, -\frac{5}{11}\right)$
15. 6, 30   **17.** 8   **19.** -2 **21.** 200   **23.** 6   **25.** 1
27. 26   **29.** 0   **31.** -79 **33.** $(4, 2)$ **35.** $\left(-\frac{1}{2}, \frac{1}{3}\right)$
37. no solution, inconsistent system   **39.** $(2, -1)$

41. $(1, 1, 2)$ **43.** $(3, 2, 1)$ **45.** $(3, -2, 1)$
47. $\left(-\frac{1}{2}, -1, -\frac{1}{2}\right)$ **49.** infinitely many solutions, dependent
equations   **51.** no solution, inconsistent system
53. $(-2, 3, 1)$ **55.** 200 of the $67 phones, 160 of the $100
phones   **57.** $5,000 in HiTech, $8,000 in SaveTel, $7,000 in
OilCo   **59.** -23 **61.** 26   **67.** no   **69.** The graph of g
is 2 units below the graph of f.   **71.** y-intercept   **73.** $x; y$

Key Concept (page 987)

1. yes   **2.** no   **3.** $(-1, 2)$ **4.** $\left(\frac{1}{3}, \frac{1}{4}\right)$ **5.** $(3, 2)$
6. $(-1, 0, 2)$ **7.** $(15, 2)$ **8.** $(3, 3, 2)$ **9.** The equations
of the system are dependent. There are infinitely many solutions.
10. The system is inconsistent. There are no solutions.

Chapter Review (page 988)

1. a. $(1, 3), (2, 1), (4, -3)$ (answers may vary)   **b.** $(0, -4)$,
$(2, -2), (4, 0)$ (answers may vary)   **c.** $(3, -1)$ **2.** President
Clinton's job approval and disapproval ratings were the same:
approximately 47% in 5/94 and approximately 48% in 5/95.

3.   **4.**

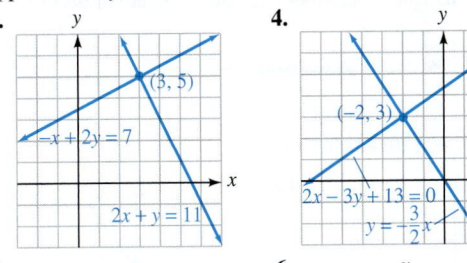

5.   **6.**

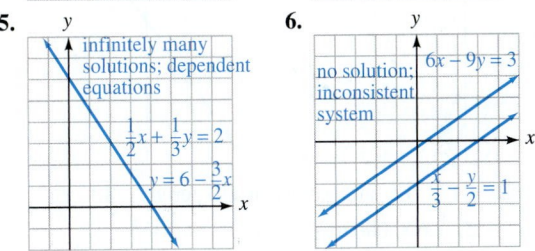

7. 2   **8.** -1 **9.** $(-1, 3)$ **10.** $(-3, -1)$ **11.** $(3, 4)$
12. infinitely many solutions, dependent equations   **13.** $(-3, 1)$
14. no solution, inconsistent system   **15.** $(9, -4)$
16. $\left(4, \frac{1}{2}\right)$ **17.** Using the elimination method, the computations
are easier.   **18.** $(-1, 0.7)$ (answers may vary); $\left(-1, \frac{2}{3}\right)$
19. A-H: 162 mi, A-SA: 83 mi   **20.** 8 mph, 2 mph
21. 17,500 bottles   **22.** no   **23.** $(1, 2, 3)$ **24.** no
solution, inconsistent system   **25.** $(-1, 1, 3)$ **26.** infinitely
many solutions, dependent equations   **27.** yes; infinitely many
solutions   **28.** 25 lb peanuts, 10 lb cashews, 15 lb Brazil nuts

29. $\begin{bmatrix} 5 & 4 & 3 \\ 1 & -1 & -3 \end{bmatrix}$ **30.** $\begin{bmatrix} 1 & 2 & 3 & 6 \\ 1 & -3 & -1 & 4 \\ 6 & 1 & -2 & -1 \end{bmatrix}$
31. $(1, -3)$ **32.** $(5, -3, -2)$ **33.** infinitely many
solutions, dependent equations   **34.** no solution, inconsistent
system   **35.** $4,000 at 6%, $6,000 at 12%   **36.** 18

37. 38 **38.** -3 **39.** 28 **40.** (2, 1) **41.** no solution, inconsistent system **42.** (1, -2, 3) **43.** $(-3, 2, 2)$
44. 2 cups mix A, 1 cup mix B, 1 cup mix C

Chapter Test (page 991)

1. (2, 1) **2.** 3 **3.** (7, 0) **4.** (2, -3) **5.** dependent
6. no **7.** (3, 2, -1) **8.** 55, 70 **9.** 15 gal 40%, 5 gal 80% **10.** 1,375 impressions **11.** (2, 2) **12.** no solution, inconsistent system **13.** 22 **14.** 4 **15. a.** $\begin{vmatrix} -6 & -1 \\ -6 & 1 \end{vmatrix}$
b. $\begin{vmatrix} 1 & -1 \\ 3 & 1 \end{vmatrix}$ **16.** -3 **17.** 3 **18.** -1
19. C: 60, GA: 30, S: 10 **21.** The system has no solution.
22. (2035, 25); in the year 2035, the percent of the U.S. population that is children under age 18 and the percent of the U.S. population that is adults 65 and older will be the same, about 25%.

Cumulative Review Exercises, Chapters 1–12 (page 992)

1. a. true **b.** true **c.** true **2.** 36 **3.** about 98,113
4. 12 **5.** $-\frac{3}{2}$ **6.** -2
7. $x < -2$; $(-\infty, -2)$;

-2

8. $8,000 at 7%, $20,000 at 10%

9. 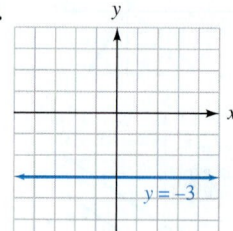 **10.**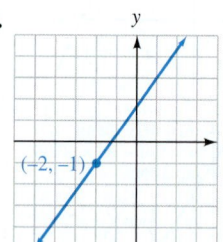

$y = -3$ $(-2, -1)$

11. $y = \frac{1}{4}x - 1$ **12.** an increase of $16 per yr **13.** y^{21}
14. $\frac{x^{10}z^4}{9y^6}$ **15.** 2,600,000 to 1 **16.** 7.3×10^{-4}
17. $6x^3 + 8x^2 + 3x - 72$ **18.** $-6a^5$ **19.** $6b^2 + 5b - 4$
20. $4x^2 + 20xy + 25y^2$ **21.** $6x^3 - 7x^2y + y^3$ **22.** $2x + 1$
23. $6a(a - 2a^2b + 6b)$ **24.** $(x + y)(2 + a)$
25. $(b + 5)(b^2 - 5b + 25)$ **26.** $(t^2 + 4)(t + 2)(t - 2)$
27. $0, -\frac{8}{3}$ **28.** $\frac{2}{3}, -\frac{1}{5}$ **29.** $6x^3 + 4x$ **30.** 5 in.
31. $\frac{x-3}{x-5}$ **32.** $\frac{x^2 + 4x + 1}{x^2y}$ **33.** $\frac{3(x+3)}{x+6}$ **34.** $\frac{9y+5}{4-y}$
35. 1 **36.** $2\frac{6}{11}$ days **37.** 875 days **38.** 39
39. $(-2, 3)$ **40.** $(-3, -1)$ **41.** $(-1, 2)$ **42.** 32 lb of Sweet-n-sour, 16 lb Bit-O-Honey
43. $[-10, 14]$

$-10 \qquad 14$

44. $(-\infty, 2) \cup (7, \infty)$

$2 \qquad 7$

45. $(x + 2 + y)(x + 2 - y)$
46. $(b + 1)(b - 1)(b + 4)(b - 4)$

47.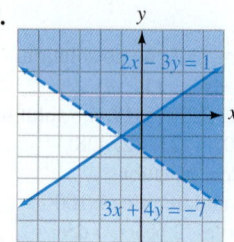

$2x - 3y = 1$

$3x + 4y = -7$

48. -8 **49. a.** 4 **b.** $-1, -5$ **50.** yes
51. 1.2

52. D: $(-\infty, \infty)$; R: $(-\infty, \infty)$

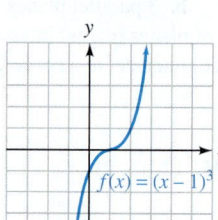

$f(x) = (x-1)^3$

53. $3x^2 - x + 2$ **54.** $5x\sqrt{2}$
55. $10a^3b^2$ **56.** $9\sqrt{6}$
57. $\frac{6x\sqrt{2x}}{y}$ **58.** $4x\sqrt[5]{xy^2}$
59. $\frac{3m}{2n^2}$ **60.** $\frac{1}{16}$ **61.** $\frac{2\sqrt[3]{a^2}}{a}$
62. $\frac{x - 2\sqrt{xy} + y}{x - y}$ **63.** 1, 9
64. $-2, -6$ **65.** 21.2 in.

66. $\frac{1 \pm \sqrt{33}}{8}$; $-0.59, 0.84$ **67.** $7i$ **68.** $3i\sqrt{6}$
69. $1 + 5i$ **70.** $16 - 2i$ **71.** $6 - 17i$ **72.** $1 - i$
73. $0 \pm 4i$ **74.** $2 \pm i$ **75.**

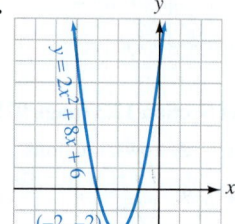

$y = 2x^2 + 8x + 6$

$(-2, -2)$

76. $8, -27$
77. $f^{-1}(x) = -\frac{2}{3}x + 2$
78. 16
79. D: $(-\infty, \infty)$; R: $(0, \infty)$ **80.** D: $(0, \infty)$; R: $(-\infty, \infty)$

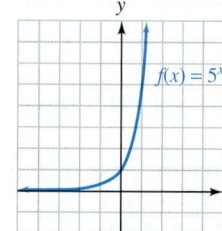

$f(x) = 5^x$

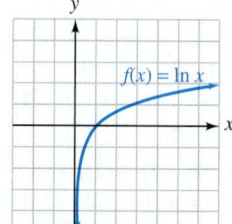

$f(x) = \ln x$

81. 5 **82.** 64 **83.** -2 **84.** $3 \ln y + \frac{1}{2} \ln x - \ln z$
85. -8.2144 **86.** 10 **87.** (2, 0) **88.** $(-2, 0, 2)$

Study Set Section 13.1 (page 1006)

1. conic sections **3.** circle, radius
5. a. $(x - h)^2 + (y - k)^2 = r^2$ **b.** $x^2 + y^2 = r^2$
7. a. $(2, -1); r = 4$ **b.** $(x - 2)^2 + (y + 1)^2 = 16$
9. a. $y = a(x - h)^2 + k$ **b.** $x = a(y - k)^2 + h$
11. a. circle **b.** parabola **c.** parabola **d.** circle
13. $6, -2, 3$

15.
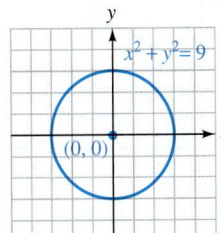

17.
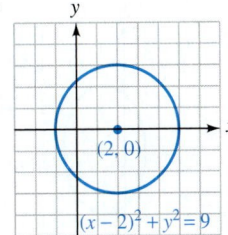

19.
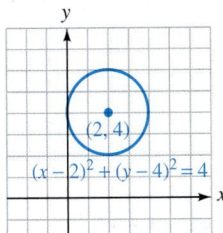

21.
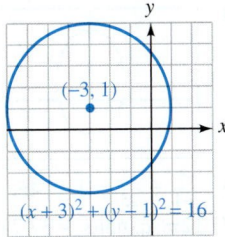

23.
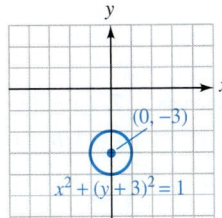

25.
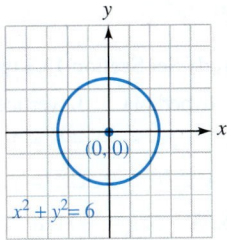

27.
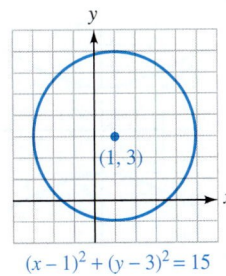

29.
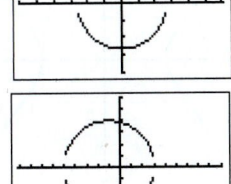

31.
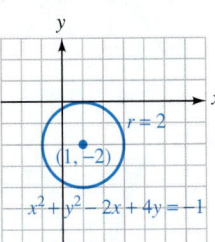

33. $x^2 + y^2 = 1$ **35.** $(x - 6)^2 + (y - 8)^2 = 25$

37. $(x + 2)^2 + (y - 6)^2 = 144$ **39.** $x^2 + y^2 = \frac{1}{16}$

41. $\left(x - \frac{2}{3}\right)^2 + \left(y + \frac{7}{8}\right)^2 = 2$ **43.** $x^2 + y^2 = 8$

45.
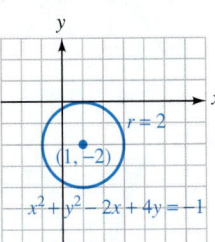

47.
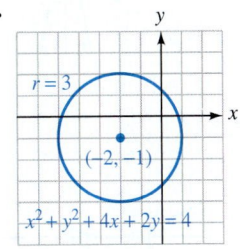

49.
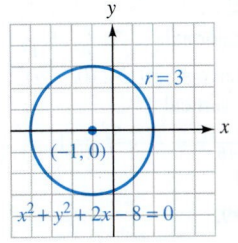

51.
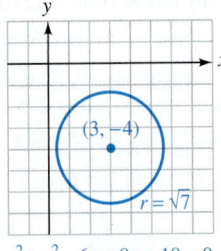

53.
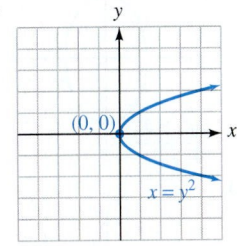

55.
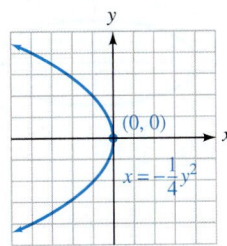

57.
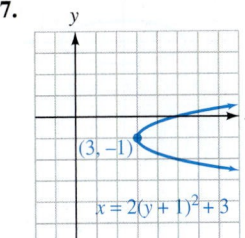

59.
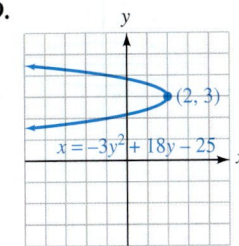

61.
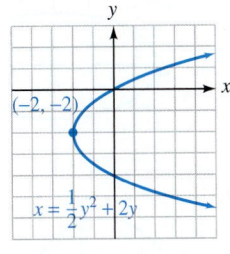

63.
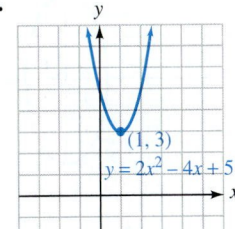

65.
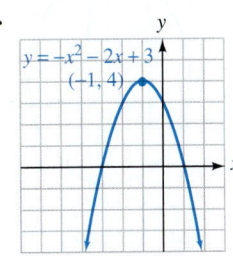

67.
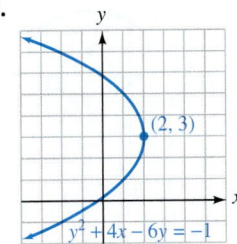

69.
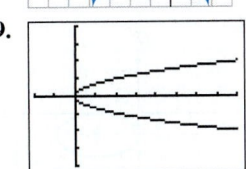

71.
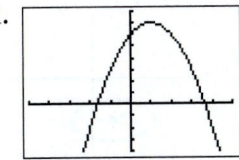

73. $(x - 7)^2 + y^2 = 9$ **75.** no   **77.** 5 ft   **79.** 2 AU

85. $5, -\frac{7}{3}$ **87.** $3, -\frac{1}{4}$

Study Set Section 13.2 (page 1016)

1. ellipse **3.** foci, focus **5.** major axis **7.** $\frac{x^2}{a^2} + \frac{y^2}{b^2} = 1$

9. x-intercepts: $(a, 0)$, $(-a, 0)$; y-intercepts: $(0, b)$, $(0, -b)$.

11. a. $(-2, 1)$; $a = 2$, $b = 5$ **b.** vertical

c. $\frac{(x + 2)^2}{4} + \frac{(y - 1)^2}{25} = 1$ **13.** $\frac{(x - 1)^2}{16} + \frac{(y + 5)^2}{1} = 1$

15. $h = -8$, $k = 6$, $a = 10$, $b = 12$

17.
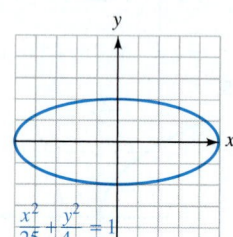
$\frac{x^2}{25} + \frac{y^2}{4} = 1$

19.
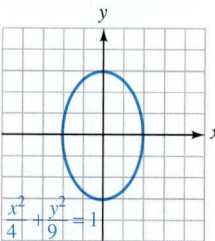
$\frac{x^2}{4} + \frac{y^2}{9} = 1$

21.
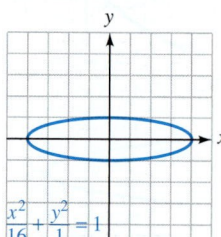
$\frac{x^2}{16} + \frac{y^2}{1} = 1$

23.
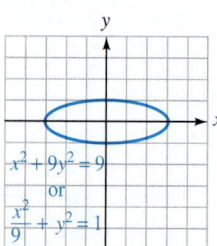
$x^2 + 9y^2 = 9$
or
$\frac{x^2}{9} + y^2 = 1$

25.
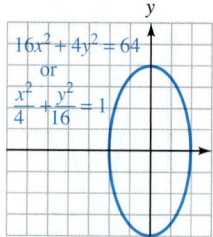
$16x^2 + 4y^2 = 64$
or
$\frac{x^2}{4} + \frac{y^2}{16} = 1$

27.
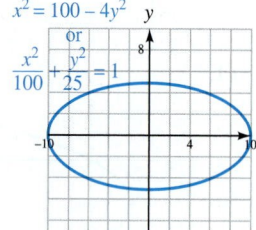
$x^2 = 100 - 4y^2$
or
$\frac{x^2}{100} + \frac{y^2}{25} = 1$

29.
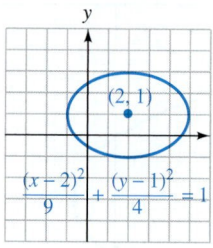
$(2, 1)$
$\frac{(x - 2)^2}{9} + \frac{(y - 1)^2}{4} = 1$

31.
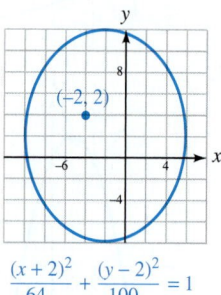
$(-2, 2)$
$\frac{(x + 2)^2}{64} + \frac{(y - 2)^2}{100} = 1$

33.
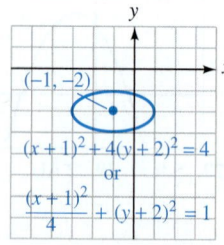
$(-1, -2)$
$(x + 1)^2 + 4(y + 2)^2 = 4$
or
$\frac{(x + 1)^2}{4} + (y + 2)^2 = 1$

35.
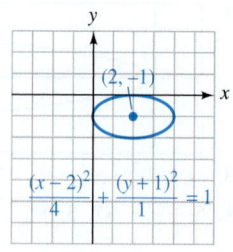
$(2, -1)$
$\frac{(x - 2)^2}{4} + \frac{(y + 1)^2}{1} = 1$

37.
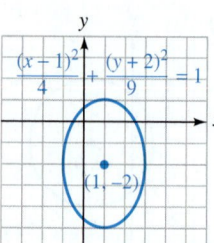
$\frac{(x - 1)^2}{4} + \frac{(y + 2)^2}{9} = 1$
$(1, -2)$

39.
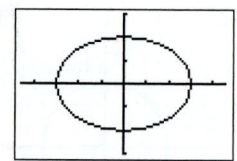

41.
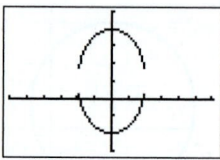

43. $\frac{x^2}{144} + \frac{y^2}{25} = 1$ **45.** $5\sqrt{3}$ ft

47. 12π sq. units ≈ 37.7 sq. units **53.** $12y^2 + \frac{9}{x^2}$

55. $\frac{y^2 + x^2}{y^2 - x^2}$

Study Set Section 13.3 (page 1026)

1. hyperbola **3.** vertices **5.** diagonals **7.** $\frac{x^2}{a^2} - \frac{y^2}{b^2} = 1$

9. $\frac{(x - h)^2}{a^2} - \frac{(y - k)^2}{b^2} = 1$ **11. a.** $(-1, -2)$; $a = 3$, $b = 1$

b. $\frac{(y + 2)^2}{9} - \frac{(x + 1)^2}{1} = 1$ **13.** $\frac{(x + 1)^2}{1} - \frac{(y - 5)^2}{4} = 1$

15. $h = 5$, $k = -11$, $a = 5$, $b = 6$

17.
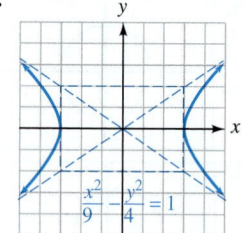
$\frac{x^2}{9} - \frac{y^2}{4} = 1$

19.
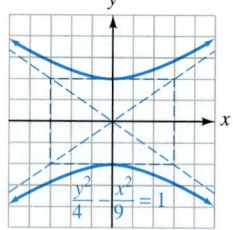
$\frac{y^2}{4} - \frac{x^2}{9} = 1$

21.
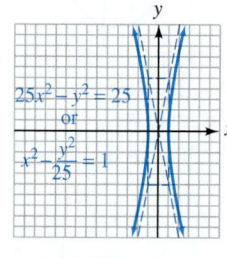
$25x^2 - y^2 = 25$
or
$x^2 - \frac{y^2}{25} = 1$

23.
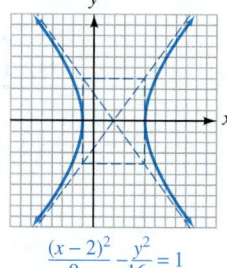
$\frac{(x - 2)^2}{9} - \frac{y^2}{16} = 1$

25.
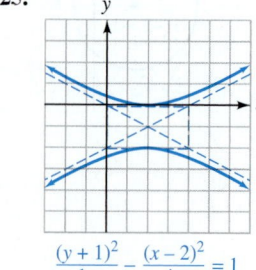
$\frac{(y + 1)^2}{1} - \frac{(x - 2)^2}{4} = 1$

27.
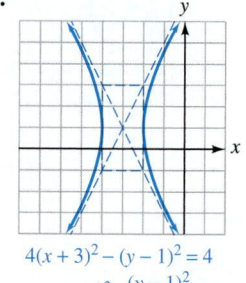
$4(x + 3)^2 - (y - 1)^2 = 4$
or $(x + 3)^2 - \frac{(y - 1)^2}{4} = 1$

29.

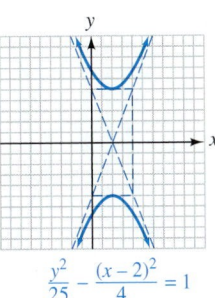

$$\frac{y^2}{25} - \frac{(x-2)^2}{4} = 1$$

31.

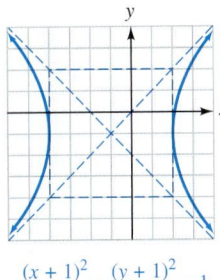

$$\frac{(x+1)^2}{9} - \frac{(y+1)^2}{9} = 1$$

33.

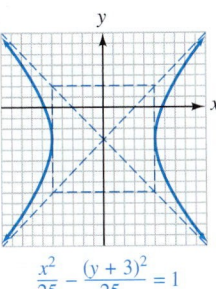

$$\frac{x^2}{25} - \frac{(y+3)^2}{25} = 1$$

35.

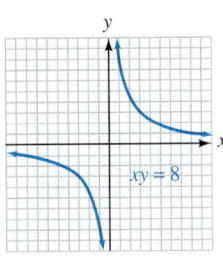

$xy = 8$

37.

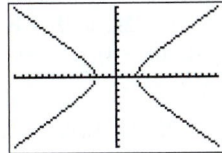

39.

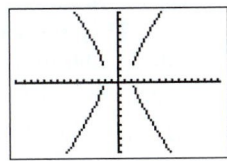

41. 3 units   **43.** $10\sqrt{3}$ miles   **49.** 64   **51.** 3   **53.** $\frac{3}{2}$

Study Set Section 13.4 (page 1032)

1. system   **3.** graphing   **5.** secant   **7. a.** two   **b.** four
c. four   **d.** four   **9.** $(-3, 2)$, $(3, 2)$, $(-3, -2)$, $(3, -2)$
11. a. -4 **b.** -2 **13.** $2x$, 5, 5, 1, 1, -1, -2

15.

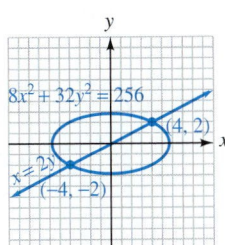

17.

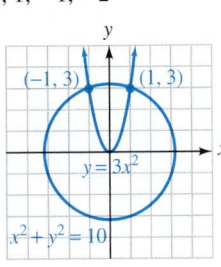

19.

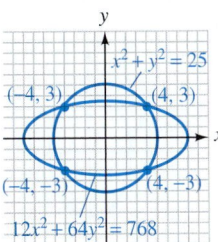

21.

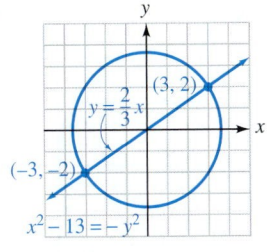

23. $(1, 0)$, $(5, 0)$ **25.** $(3, 0)$, $(0, 5)$ **27.** $(1, 1)$
29. $(1, 2)$, $(2, 1)$
31. $(-2, 3)$, $(2, 3)$
33. $\left(\sqrt{5}, 5\right)$, $\left(-\sqrt{5}, 5\right)$
35. $(3, 2)$, $(3, -2)$, $(-3, 2)$, $(-3, -2)$

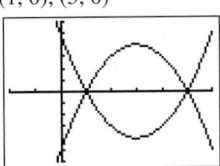

37. $(2, 4)$, $(2, -4)$, $(-2, 4)$, $(-2, -4)$ **39.** $\left(-\sqrt{15}, 5\right)$, $\left(\sqrt{15}, 5\right)$, $(-2, -6)$, $(2, -6)$ **41.** $(0, -4)$, $(-3, 5)$, $(3, 5)$
43. $(-2, 3)$, $(2, 3)$, $(-2, -3)$, $(2, -3)$ **45.** $(3, 3)$
47. $(6, 2)$, $(-6, -2)$, $\left(\sqrt{42}, 0\right)$, $\left(-\sqrt{42}, 0\right)$ **49.** $\left(\frac{1}{2}, \frac{1}{3}\right)$, $\left(\frac{1}{3}, \frac{1}{2}\right)$ **51.** 4, 8   **53.** 7 cm by 9 cm   **55.** either $750 at 9% or $900 at 7.5%   **57.** 68 mph, 4.5 hr   **61.** 2,000   **63.** 7

Key Concept (page 1036)

1. ellipse   **2.** circle   **3.** parabola   **4.** hyperbola
5. hyperbola   **6.** ellipse   **7.** ellipse   **8.** parabola
9. circle   **10. a.** $(-1, 2)$ **b.** 4   **11. a.** $(-2, 1)$
b. right   **12. a.** $(0, 0)$ **b.** vertical   **c.** $(0, 4)$, $(0, -4)$
13. a. $(-2, 1)$ **b.** left and right   **c.** 6 units horizontally, 4 units vertically

14.

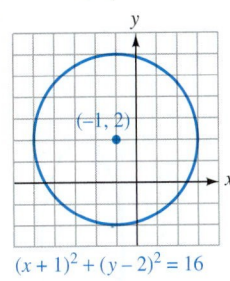

$(x+1)^2 + (y-2)^2 = 16$

15.

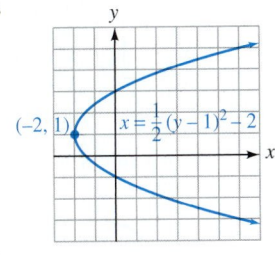

$x = \frac{1}{2}(y-1)^2 - 2$

16.

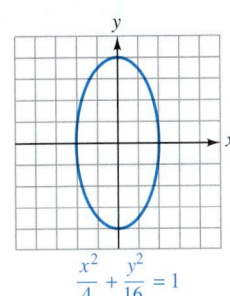

$$\frac{x^2}{4} + \frac{y^2}{16} = 1$$

17.

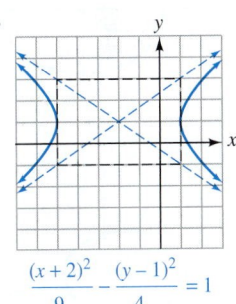

$$\frac{(x+2)^2}{9} - \frac{(y-1)^2}{4} = 1$$

Chapter Review (page 1037)

1.

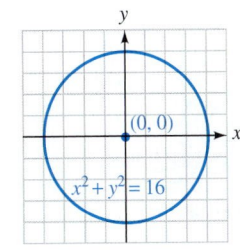

$x^2 + y^2 = 16$

2.

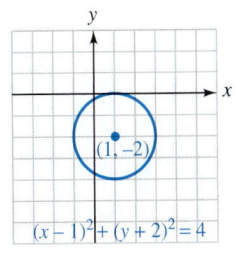

$(x-1)^2 + (y+2)^2 = 4$

3.

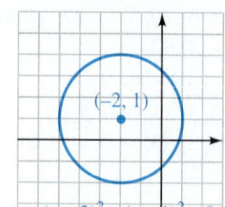

$(x + 2)^2 + (y - 1)^2 = 9$

4. $(x - 8.5)^2 + (y - 8.5)^2 = (8.5)^2$
5. $(-6, 0); r = 2\sqrt{6}$
6. center, radius

7.

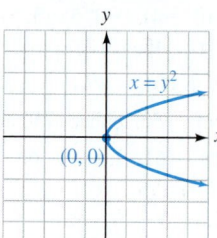

$x = y^2$, $(0, 0)$

8.

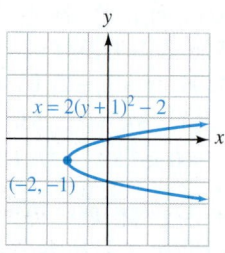

$x = 2(y + 1)^2 - 2$, $(-2, -1)$

9.

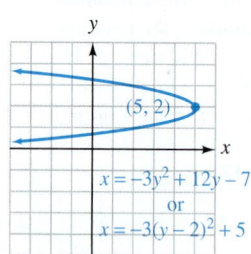

$(5, 2)$
$x = -3y^2 + 12y - 7$
or
$x = -3(y - 2)^2 + 5$

10. a. $y = 4$ **b.** $x = 1$
11. $(2, -3), (6, -4)$
12. When $x = 22$, $y = 0$:
$-\frac{5}{121}(22 - 11)^2 + 5 = 0$

13.

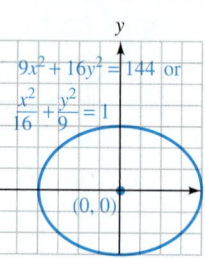

$9x^2 + 16y^2 = 144$ or
$\frac{x^2}{16} + \frac{y^2}{9} = 1$, $(0, 0)$

14.

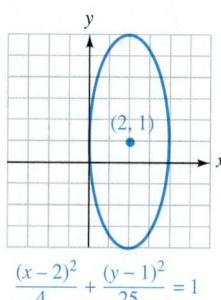

$(2, 1)$
$\frac{(x - 2)^2}{4} + \frac{(y - 1)^2}{25} = 1$

15.

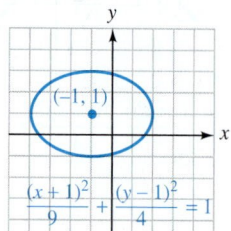

$(-1, 1)$
$\frac{(x + 1)^2}{9} + \frac{(y - 1)^2}{4} = 1$

16. $\frac{x^2}{12^2} + \frac{y^2}{1^2} = 1$
17. $\left(2, \frac{2\sqrt{5}}{3}\right), \left(2, -\frac{2\sqrt{5}}{3}\right)$;
$(2, 1.5), (2, -1.5)$
18. $\frac{x^2}{25} + \frac{y^2}{9} = 1$
19. ellipse, focus

20.

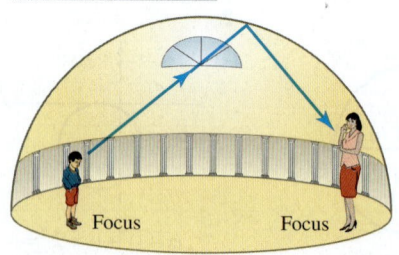

Focus   Focus

21.

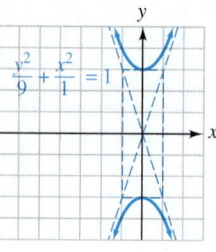

$\frac{y^2}{9} + \frac{x^2}{1} = 1$

22.

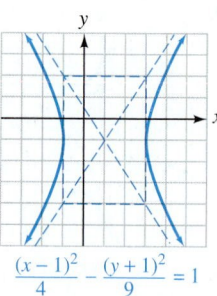

$\frac{(x - 1)^2}{4} - \frac{(y + 1)^2}{9} = 1$

23.

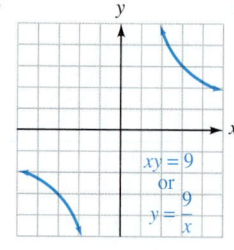

$xy = 9$
or
$y = \frac{9}{x}$

24.

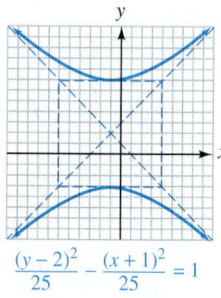

$\frac{(y - 2)^2}{25} - \frac{(x + 1)^2}{25} = 1$

25. 4 units   **26. a.** ellipse   **b.** hyperbola   **c.** parabola
d. circle   **27.** yes   **28.** $(0, 3), (0, -3)$ **29. a.** 2   **b.** 4
c. 4   **d.** 4   **30.** $(0, 1), (0, -1)$ **31.** $(0, -4), (-3, 5),$
$(3, 5)$ **32.** $\left(\sqrt{2}, 0\right), \left(-\sqrt{2}, 0\right)$ **33.** $(2, 2), \left(-\frac{2}{9}, -\frac{22}{9}\right)$
34. $(4, 2), (4, -2), (-4, 2), (-4, -2)$ **35.** $(2, 3), (2, -3),$
$(-2, 3), (-2, -3)$ **36.** $\left(2\sqrt{2}, \sqrt{2}\right), \left(-2\sqrt{2}, -\sqrt{2}\right),$
$(1, 4), (-1, -4)$

Chapter Test (page 1040)

1. center, radius   **2.** $(0, 0); r = 10$
3. $(-2, 3), r = 3\sqrt{2}$ **4.** $(x - 4)^2 + (y - 3)^2 = 9$
5.

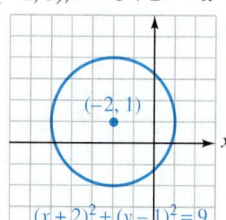

$(-2, 1)$
$(x + 2)^2 + (y - 1)^2 = 9$

6.

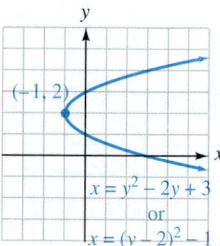

$(-1, 2)$
$x = y^2 - 2y + 3$
or
$x = (y - 2)^2 - 1$

7. $(-1, 7); x = -1$

8.

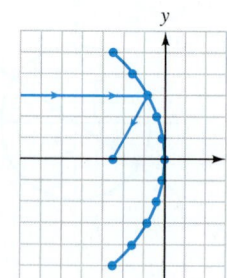

9.

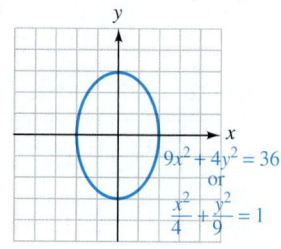

$9x^2 + 4y^2 = 36$
or
$\frac{x^2}{4} + \frac{y^2}{9} = 1$

10.

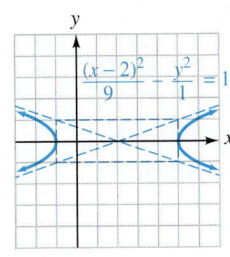

$\dfrac{(x-2)^2}{9} - \dfrac{y^2}{1} = 1$

11. $\dfrac{(x-1)^2}{16} + \dfrac{(y+2)^2}{9} = 1$

12. $(-8, 10)$; $(-2, 10)$, $(-14, 10)$ **13.** $\pm 2\sqrt{2}$

15. $(-1, 1)$; horizontal: 4 units, vertical: 4 units

16.

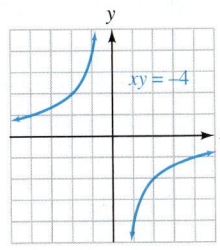

$xy = -4$

17. $\dfrac{y^2}{16} - \dfrac{x^2}{36} = 1$

18. a. ellipse   **b.** hyperbola   **c.** circle   **d.** parabola

19. $(-4, -3)$, $(3, 4)$

20. $(2, 6)$, $(-2, -2)$

21. $\left(1, \sqrt{2}\right), \left(1, -\sqrt{2}\right),$ $\left(-1, \sqrt{2}\right), \left(-1, -\sqrt{2}\right)$

22. $\left(-1, \dfrac{9}{2}\right), \left(3, -\dfrac{3}{2}\right)$

$y + x = 1$ $(3, 4)$ $(-4, -3)$ $x^2 + y^2 = 5$

Study Set Section 14.1 (page 1050)

1. binomial   **3.** binomial   **5.** factorial, decreasing
7. one   **9.** r^{20}, s^{20} **11.** 1, 1   **13.** $(n-1)!$ **15.** 1
17. $3!(9-3)!$ **19.** 2, 4   **21.** 3, 3!, y, 3!, 2, 2, 3
23. $n-1$ **25.** 6   **27.** 120   **29.** 30   **31.** 144
33. 40,320   **35.** $\dfrac{1}{110}$ **37.** 2,352   **39.** 72   **41.** 5
43. 10   **45.** 21   **47.** $\dfrac{1}{168}$ **49.** 39,916,800
51. $2.432902008 \times 10^{18}$ **53.** $x^4 + 4x^3y + 6x^2y^2 + 4xy^3 + y^4$
55. $c^5 - 5c^4d + 10c^3d^2 - 10c^2d^3 + 5cd^4 - d^5$
57. $s^6 + 6s^5t + 15s^4t^2 + 20s^3t^3 + 15s^2t^4 + 6st^5 + t^6$
59. $a^9 - 9a^8b + 36a^7b^2 - 84a^6b^3 + 126a^5b^4$ $- 126a^4b^5 + 84a^3b^6 - 36a^2b^7 + 9ab^8 - b^9$
61. $8x^3 + 12x^2y + 6xy^2 + y^3$
63. $32t^5 - 240t^4 + 720t^3 - 1,080t^2 + 810t - 243$
65. $625m^4 - 1,000m^3n + 600m^2n^2 - 160mn^3 + 16n^4$
67. $\dfrac{x^3}{27} + \dfrac{x^2y}{6} + \dfrac{xy^2}{4} + \dfrac{y^3}{8}$ **69.** $\dfrac{x^4}{81} - \dfrac{2x^3y}{27} + \dfrac{x^2y^2}{6} - \dfrac{xy^3}{6} + \dfrac{y^4}{16}$
71. $c^{10} - 5c^8d^2 + 10c^6d^4 - 10c^4d^6 + 5c^2d^8 - d^{10}$
73. $-4xy^3$ **75.** $15r^2s^4$ **77.** $28x^6y^2$ **79.** $-12x^3y$
81. $-70,000t^4$ **83.** $810xy^4$ **85.** $-\dfrac{1}{6}c^3d$ **87.** $-6a^{10}b^2$
93. $\log x^2y^{1/2}$ **95.** $\ln y$

Study Set Section 14.2 (page 1057)

1. sequence   **3.** arithmetic   **5.** mean   **7.** 1, 7, 13
9. a. $a_n = a_1 + (n-1)d$ **b.** $S_n = \dfrac{n(a_1 + a_n)}{2}$ **11.** nth
13. sigma   **15.** summation, runs   **17.** 3, 7, 11, 15, 19
19. $-2, -5, -8, -11, -14$ **21.** 3, 5, 7, 9, 11
23. $-5, -8, -11, -14, -17$ **25.** 5, 11, 17, 23, 29
27. $-4, -11, -18, -25, -32$ **29.** $-118, -111, -104,$ $-97, -90$ **31.** 34, 31, 28, 25, 22   **33.** 5, 12, 19, 26, 33
35. 355   **37.** -179 **39.** -23 **41.** 12   **43.** $\dfrac{17}{4}, \dfrac{13}{2}, \dfrac{35}{4}$
45. 12, 14, 16, 18   **47.** $\dfrac{29}{2}$ **49.** $3 + 6 + 9 + 12$
51. $4 + 9 + 16$ **53.** $\displaystyle\sum_{k=1}^{5} k^2$ **55.** $\displaystyle\sum_{k=3}^{6} k$ **57.** 1,335
59. 459   **61.** 354   **63.** 255   **65.** 1,275   **67.** 2,500
69. 60   **71.** 91   **73.** 31   **75.** 12   **77.** 60, 110, 160, 210, 260, 310; $6,060   **79.** 11,325   **81.** 368 ft
87. $1 + \log_2 x - \log_2 y$ **89.** $3\log x + 2\log y$

Study Set Section 14.3 (page 1067)

1. geometric   **3.** mean   **5.** 16, 4, 1   **7.** $a_n = a_1 r^{n-1}$
9. a. yes   **b.** no   **c.** no   **d.** yes   **11.** $a_1 r^2, a_1 r^4$
13. $n+1, n$ **15.** 3, 6, 12, 24, 48   **17.** $-5, -1, -\dfrac{1}{5}, -\dfrac{1}{25}, -\dfrac{1}{125}$
19. 2, 8, 32, 128, 512   **21.** $-3, -12, -48, -192, -768$
23. $-64, 32, -16, 8, -4$ **25.** $-64, -32, -16, -8, -4$
27. 2, 10, 50, 250, 1,250   **29.** 3,584   **31.** $\dfrac{1}{27}$ **33.** 3
35. 6, 18, 54   **37.** $-20, -100, -500, -2,500$ **39.** -16
41. $10\sqrt{2}$ **43.** No geometric mean exists.   **45.** 728
47. 122   **49.** -255 **51.** 381   **53.** $\dfrac{156}{25}$ **55.** $-\dfrac{21}{4}$
57. 16   **59.** 81   **61.** 8   **63.** $-\dfrac{135}{4}$ **65.** no sum
67. $-\dfrac{81}{2}$ **69.** $\dfrac{1}{9}$ **71.** $\dfrac{1}{3}$ **73.** $\dfrac{4}{33}$ **75.** $\dfrac{25}{33}$
77. $1,469.74   **79.** $140,853.75   **81.** $\left(\dfrac{1}{2}\right)^{11} \approx 0.0005$
83. 30 m   **85.** 5,000   **91.** $[-1, 6]$ **93.** $(-\infty, -3) \cup (4, \infty)$

Study Set Section 14.4 (page 1077)

1. tree   **3.** permutation   **5.** $p \cdot q$ **7.** $P(n, r) = \dfrac{n!}{(n-r)!}$
9. n, r, combinations   **11.** 1   **13.** $6!, 4!, 5$ **15.** 6
17. 60   **19.** 12   **21.** 5   **23.** 1,260   **25.** 10   **27.** 20
29. 50   **31.** 2   **33.** 1   **35.** $\dfrac{n!}{2!(n-2)!}$
37. $x^4 + 4x^3y + 6x^2y^2 + 4xy^3 + y^4$
39. $8x^3 + 12x^2y + 6xy^2 + y^3$
41. $81x^4 - 216x^3 + 216x^2 - 96x + 16$ **43.** $-1,250x^2y^3$
45. $-4x^6y^3$ **47.** 35   **49.** 1,000,000   **51.** 136,080
53. 8,000,000   **55.** 720   **57.** 2,880   **59.** 13,800
61. 720   **63.** 900   **65.** 364   **67.** 5   **69.** 1,192,052,400
71. 18   **73.** 7,920   **77.** 1.7095   **79.** 0.7324

Study Set Section 14.5 (page 1082)

1. experiment   **3.** $\dfrac{s}{n}$ **5.** 0   **7. a.** 6   **b.** 52
c. $\dfrac{6}{52}, \dfrac{3}{26}$ **9.** {(1, H), (2, H), (3, H), (4, H), (5, H), (6, H), (1, T), (2, T), (3, T), (4, T), (5, T), (6, T)}

11. {a, b, c, d, e, f, g, h, i, j, k, l, m, n, o, p, q, r, s, t, u, v, w, x, y, z}

13. $\frac{1}{6}$ **15.** $\frac{2}{3}$ **17.** $\frac{19}{42}$ **19.** $\frac{13}{42}$ **21.** $\frac{3}{8}$ **23.** 0

25. $\frac{1}{12}$ **27.** $\frac{5}{12}$ **29.** $\frac{33}{391,510}$ **31.** $\frac{9}{460}$ **35.** $\frac{1}{4}$ **37.** $\frac{1}{4}$

39. 1   **41.** $\frac{32}{119}$ **43.** $\frac{1}{3}$ **47.** $-\frac{3}{4}$ **49.** 3, −1

51. −2, −2

Key Concept (page 1085)

1. g   **2.** o   **3.** i   **4.** l   **5.** u   **6.** y   **7.** p   **8.** x

9. m   **10.** d   **11.** f   **12.** c   **13.** w   **14.** b   **15.** j

16. s   **17.** e   **18.** z   **19.** n   **20.** k   **21.** h   **22.** v

23. q   **24.** a   **25.** r   **26.** t

Chapter Review (page 1086)

1. 1, 5, 10, 10, 5, 1   **2. a.** 13   **b.** 12   **c.** a^{12}, b^{12}

d. a: decrease; b: increase   **3.** 144   **4.** 20   **5.** 15   **6.** 220

7. 1   **8.** 8   **9.** $x^5 + 5x^4y + 10x^3y^2 + 10x^2y^3 + 5xy^4 + y^5$

10. $x^9 - 9x^8y + 36x^7y^2 - 84x^6y^3 + 126x^5y^4 - 126x^4y^5 +$ $84x^3y^6 - 36x^2y^7 + 9xy^8 - y^9$ **11.** $64x^3 - 48x^2y + 12xy^2 - y^3$

12. $\frac{c^4}{16} + \frac{c^3d}{6} + \frac{c^2d^2}{6} + \frac{2cd^3}{27} + \frac{d^4}{81}$ **13.** $6x^2y^2$ **14.** $-20x^3y^3$

15. $-108x^2y$ **16.** $5u^2v^{12}$ **17.** −2, 0, 2, 4   **18.** 42

19. 122, 137, 152, 167, 182   **20.** −1,194   **21.** −5

22. $\frac{41}{3}, \frac{58}{3}$ **23.** $-\frac{45}{2}$ **24.** 1,568   **25.** $\frac{15}{2}$ **26.** 378

27. 14   **28.** 360   **29.** 5,050   **30.** 1,170   **31.** 4

32. 24, 12, 6, 3, $\frac{3}{2}$ **33.** $\frac{1}{27}$ **34.** 24, −96   **35.** $\frac{2,186}{9}$

36. $-\frac{85}{8}$ **37.** about 1.6 lb   **38.** 125   **39.** $\frac{5}{99}$ **40.** 190 ft

41. 136   **42.** 17,576,000   **43.** 5,040   **44.** 1   **45.** 20,160

46. $\frac{1}{10}$ **47.** 1   **48.** 1   **49.** 28   **50.** 700   **51.** $\frac{7}{4}$

52. $81y^4 - 216y^3z + 216y^2z^2 - 96yz^3 + 16z^4$ **53.** 120

54. 720   **55.** 120   **56.** 150   **57.** $\frac{3}{8}$ **58.** $\frac{1}{2}$ **59.** $\frac{7}{8}$

60. $\frac{1}{18}$ **61.** 0   **62.** $\frac{1}{13}$ **63.** $\frac{94}{54,145}$ **64.** $\frac{33}{66,640}$

Chapter Test (page 1089)

1. 2, −4, −10, −16   **2.** $a^6 - 6a^5b + 15a^4b^2 - 20a^3b^3 +$ $15a^2b^4 - 6ab^5 + b^6$ **3.** $24x^4y^2$ **4.** 66   **5.** 306

6. 34, 66   **7.** $\frac{1}{4}$ **8.** −1,377   **9.** 205   **10.** 3   **11.** −81

12. $\frac{364}{27}$ **13.** 6   **14.** 18, 108   **15.** $\frac{27}{2}$ **16.** 1,600 ft

17. 210   **18.** 1   **19.** 120   **20.** 15   **21.** 56   **22.** 6

23. 7,000,000   **24.** 362,880   **25.** 35   **26.** 30   **27.** $\frac{1}{6}$

28. $\frac{2}{13}$ **29.** $\frac{33}{66,640}$ **30.** $\frac{5}{16}$ **31.** shade 7 sections

32. a. 0   **b.** 1

Cumulative Review Exercises, Chapters 1–14 (page 1090)

1. 0   **2.** $-\frac{4}{3}, 5.6, 0, -23$ **3.** $\pi, \sqrt{2}, e$ **4.** $-\frac{4}{3}, \pi,$ $5.6, \sqrt{2}, 0, -23, e$ **5.** \$8,250   **6.** $\frac{1}{120}$ db/rpm

7. parallel   **8.** perpendicular   **9.** $y = -2x + 5$

10. $y = -\frac{9}{13}x + \frac{7}{13}$ **11.** (−7, 7)   **12.** $\left(\frac{4}{5}, \frac{3}{4}\right)$

13. $\left(-2, \frac{3}{2}\right)$ **14.** (3, 2, 1)   **15.** 3   **16.** 85°, 80°, 15°

17. It doesn't pass the vertical line test.   **18.** $-4; 3a^5 - 2a^2 + 1$

19. a. 4 **b.** 3   **c.** 0, 2   **d.** 1   **20.** $1.73 \times 10^{14}; 4.6 \times 10^{-8}$

21.

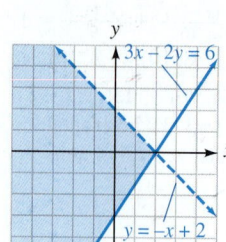

22. $\left[-3, \frac{19}{3}\right]$

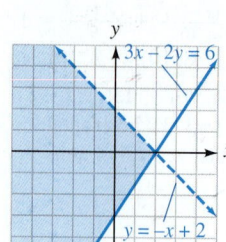

23. $(-\infty, -2)$

24. $x^3 - 27y^3$ **25.** $2x^2y^3 + 13xy + 3y^2$

26. $81a^2b^4 - 72ab^2 + 16$ **27.** $\frac{1}{a^3b} + \frac{b^2}{a^2}$

28. $xy(3x - 4y)(x - 2)$ **29.** $(16x^2y^2 + z^4)(4xy + z^2)(4xy - z^2)$

30. $(3y^3 + 2)(4y^3 + 5)$ **31.** $\lambda = \frac{4d - 2}{A - 6}$ **32.** −8, −8

33. 0, 0, −1   **34.** (−3, 0), (0, 0), (2, 0); (0, 0); 24, 0, −8, −6, 0, 4, 0, −18

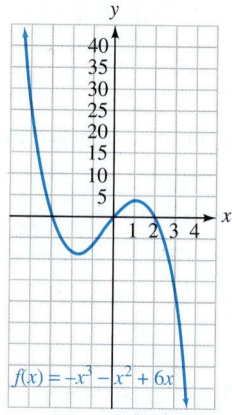

35. $\frac{16}{y^{16}}$ **36.** $-\frac{3x + 2}{3x - 2}$ **37.** $-\frac{q}{p}$ **38.** $\frac{4a - 1}{(a + 2)(a - 2)}$

39. 5; 3 is extraneous   **40.** $R = \frac{R_1R_2R_3}{R_2R_3 + R_1R_3 + R_1R_2}$

41. a. a quadratic function   **b.** at about 85% and 120% of the suggested inflation   **42.** about $21\frac{1}{2}$ in.

43. $-x^2 + x + 5 + \frac{8}{x - 1}$ **44.** 2 lumens

45. D: $[0, \infty)$; R: $[2, \infty)$

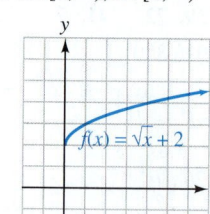

46. $5\sqrt{2}$

47. $15x - 6\sqrt{15x} + 9$

48. $81x\sqrt[3]{3x}$ **49.** $\frac{343}{125}$

50. $\frac{\sqrt[3]{2ab^2}}{b}$ **51.** $\sqrt{3t} - 1$

52. $-5 + 17i$ **53.** $-\frac{21}{29} - \frac{20}{29}i$

54. −1   **55.** 5, 0 does not check   **56.** 0

57. $-i, i, -3i\sqrt{2}, 3i\sqrt{2}$ **58.** $\frac{-3 \pm \sqrt{5}}{4}$ **59.** $\frac{1}{4}, \frac{1}{2}$

60. $\frac{2}{3} \pm \frac{\sqrt{2}}{3}i$

61.

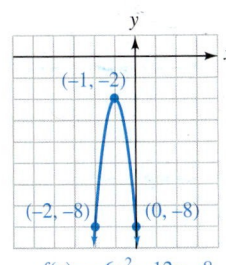

$(-1, -2)$

$(-2, -8)$ $(0, -8)$

$f(x) = -6x^2 - 12x - 8$

62. $4x^2 + 4x - 1$

63. $f^{-1}(x) = \sqrt[3]{\dfrac{x+1}{2}}$

64. D: $(-\infty, \infty)$; R: $(0, \infty)$

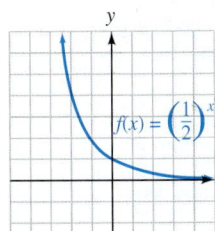

$f(x) = \left(\dfrac{1}{2}\right)^x$

65.

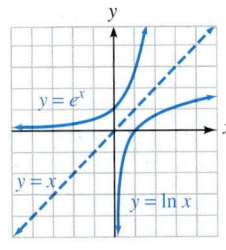

$y = e^x$

$y = x$

$y = \ln x$

66. $2 - 3\log_6 x$ **67.** $\ln \dfrac{y\sqrt{x}}{z}$ **68.** about 170 million

69. 5   **70.** 3   **71.** $\dfrac{1}{27}$ **72.** 1   **73.** 1.9912

74. undefined   **75.** 3.4190   **76.** 1, -10 does not check

77. $-\dfrac{3}{4}$ **78.** 9   **79.** $x^2 + y^2 - 2x - 6y - 15 = 0$

80. $\dfrac{(x-2)^2}{9} - \dfrac{y^2}{1} = 1$

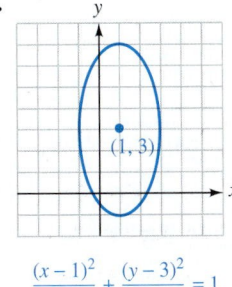

$\dfrac{(x-2)^2}{9} - \dfrac{y^2}{1} = 1$

81.

$(1, 3)$

$\dfrac{(x-1)^2}{4} + \dfrac{(y-3)^2}{16} = 1$

82.

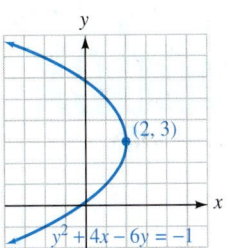

$(2, 3)$

$y^2 + 4x - 6y = -1$

83. $81a^4 - 108a^3b + 54a^2b^2 - 12ab^3 + b^4$ **84.** $112x^2y^6$

85. 103   **86.** 690   **87.** 27   **88.** 27   **89.** \$2,848.31

90. $\dfrac{1,023}{64}$ **91.** $\dfrac{27}{2}$ **92.** 5,040   **93.** 84   **94.** $\dfrac{3}{13}$

Appendix II (page A-7)

1. synthetic   **3.** divisor   **5.** theorem

7. a. $(5x^3 + x - 3) \div (x + 2)$ **b.** $5x^2 - 10x + 21 - \dfrac{45}{x+2}$

9. $6x^3 - x^2 - 17x + 9, x - 8$ **11.** 2, 1, 12, 26, 6, 8

13. $x + 2$ **15.** $x - 3$ **17.** $x + 2$ **19.** $x - 7 + \dfrac{28}{x+2}$

21. $3x^2 - x + 2$ **23.** $2x^2 + 4x + 3$

25. $6x^2 - x + 1 + \dfrac{3}{x+1}$ **27.** $t^2 + 1 + \dfrac{1}{t+1}$

31. $-5x^4 - 11x^3 - 3x^2 - 7x - 1$

33. $8t^2 + 2$ **35.** $x^3 + 7x^2 - 2$ **37.** $7.2x - 0.66 + \dfrac{0.368}{x-0.2}$

39. $9x^2 - 513x + 29,241 - \dfrac{1,666,762}{x+57}$ **41.** -1 **43.** -37

45. 23   **47.** -1 **49.** 2   **51.** -1 **53.** 18   **55.** 174

57. -8 **59.** 59   **61.** 44   **63.** $\dfrac{29}{32}$ **65.** yes   **67.** no

73. 0   **75.** 2

Index